W9-BUC-973
R0125253807
A reader's guide to twentieth-centu

A READER'S GUIDE TO TWENTIETH-CENTURY WRITERS

EDITOR

Peter Parker

CONSULTANT EDITOR

Frank Kermode

New York
OXFORD UNIVERSITY PRESS
1996

First published in Great Britain (as *The Reader's Companion to Twentieth-Century Writers*) in 1995 jointly by Fourth Estate Limited, 289 Westbourne Grove, London W11 and Helicon Publishing Limited, 42 Hythe Bridge Street, Oxford OX1 2EP

Published in the United States of America by Oxford University Press, Inc., 198 Madison Avenue, New York, New York 10016

Library of Congress Cataloging-in-Publication Data available upon request from the Publisher

ISBN 0–19–521215–0

Printing (last digit): 9 8 7 6 5 4 3 2 1

Printed on acid-free paper

Page design by Roger Walker

Typeset by Mendip Communications Ltd
Frome, Somerset

Printed and bound in Great Britain
by The Bath Press Ltd, Bath, Avon

Contents

Contributors

Carol Anderson
Phil Baker
Josie Barnard
Peter Billyard
John R. Bradley
Montague Bream
Niladri R. Chatterjee
Jeremy Clarke
Hazel Coleman
Jo Coriat
Geoffrey Elborn
Graham Grant
Caroline Gonda
Georgina Hammick
Gillian E. Hanscombe
Alison Hennegan
Christie Hickman
C.A.R. Hills
Mark Irving
Dot Lubienska
Lawrence Norfolk
Jeremy Osborne
Lisa Osborne
Georgina Matthews
Kate Parker
Peter Parker
Christopher Potter
Stephen L. Prasher
Jenny Preston
Duncan Priddle
Marylin Purves
Kate Quarry
James Rogers
Vikram Seth
Katharine Slade
Helen Szamuely
Caroline Taverne
Martin Taylor
Ian Warner
Ian Wisniewski
James Wright

Preface

Lady Gregory, included in this volume as a twentieth-century writer, was born in 1852, the year in which, among many other events, the Duke of Wellington died, and Tasmania ceased to be a convict settlement. *Bleak House* was appearing in monthly parts, Herman Melville published the impenetrable *Pierre*, Nathaniel Hawthorne his enigmatic *Blithedale Romance*, and Harrier Beecher Stowe *Uncle Tom's Cabin*. Amitav Ghosh, who also appears here as a twentieth-century writer, was born in 1956, on the whole a year less deserving of congratulation – the year of the Suez disgrace and the crushing of Hungary, though also, as a riposte on behalf of conscience and sanity, the first Aldermaston march. Among the writers who suited the mood of that moment we may remember John Osborne (*Look Back in Anger*) and Angus Wilson (*Anglo-Saxon Attitudes*). Mr Ghosh, though born in a troubled time, has a reasonable expectation of being included, a century hence, in some *Guide to Twenty-First-Century Writers*. Obviously the authors here included, taken together, have lived or will have lived, through much history, public as well as private. This book covers a lot of chronological ground.

Its geographical scope is not less impressive, and is itself the product of history. Now, as never before, the business of continuing the provision of literature in English is multinational, its centres not only in Britain and North America but in Australia, New Zealand, Nigeria, India, the Caribbean. Booker Prizes, even Nobel Prizes, go to South Africans, Australians, Nigerians, Trinidadians, and only rarely to the English or Americans. These changes explain the necessarily broad Anglophone scope of this *Guide*.

It offers brief biographies of the authors of many nations, along with assiduous bibliographies. These, it might be alleged, are for use, while the biographical material is for interest. There are certain purists who might deplore the nature of this interest, on the ground that the books should be the sole object of concern; the lives of the authors might, possibly, by means of very refined arguments, be shown to have a certain restricted relevance to what they write, but how, it is asked, can a simple record of their ordinary lives – their origins, their marriages, their adulteries and other troubles – be thought to have such interest? But less subtle, less tortured readers don't ask such questions; literary biography is a flourishing genre, and those who read it make, as readers have always made, the comfortable assumption that literary lives are interesting in themselves. And users of this book may feel that it justifies that naive confidence.

They may also notice that many of the mini-biographers who contribute to this volume have assumed that in our day their records of fact and opinion may be presented more candidly than earlier custom allowed. A note of asperity, a licence to be critical, even the right, occasionally exercised, to be amusing, enliven these pages. For instance, when an author is reported as having thrown his mother downstairs and broken her leg, the chronicler allows himself to speak of this unusual act as suggesting a 'hint of violence' in the poet's character. A novel by another writer is said to contain a 'particularly graphic description of coprophilia, which doubtless helped to generate some publicity'. Of another celebrated novelist we are informed that her recent works are 'almost universally considered to be in need of a stringent editing which the author will not allow'. The autobiography of a noted but boastful and probably mendacious poet, is here described as 'characteristically boastful and mendacious'. An American writer known to believe himself to be quite simply the greatest in the world is quoted as saying that at Harvard he modestly 'strove to appear of merely normal intelligence'. Nearer home, a famous writer is uncompromisingly described as 'prone to melancholia and intransigence', with a suggestion that the latter condition is also a disease, which experience suggests it may be. Of another lady novelist it is remarked that she 'is likely to have a permanent, although perhaps small, claim on the attention of readers' – the sort of judgment that used, in Downing College, to be described as 'limiting'.

Condemnation can on occasion be much closer to total, as when Jerzy Kosinski's *Painted Bird*, though admitted to be in certain respects a good novel, is anathematised because its 'status as a Holocaust testament . . . has been severely compromised', to the degree that 'the reputation of both the book and its author have been severely damaged'. With only a little less acerbity Fr Knox's translation of the Vulgate Bible, the learned labour of years, is curtly dismissed: 'not enduring'. Even kind verdicts can be very firmly handed down: Somerset Maugham's *The Razor's Edge*, we are instructed, 'will undoubtedly survive'. Such predictions are very assured, but that is surely what we have a right to expect of so authoritative a work of reference.

Merely biographical detail can be recounted with notable verve. One writer, a novelist, is said to be 'married to a manufacturer of widgets', a word I bestirred myself to look up. According to the latest edition of the great *Oxford Dictionary* it is of US origin, first recorded in 1931, and it may have been introduced to England by the Sloane Rangers. It seems, rather disappointingly, to be nothing more than a variant of 'gadget'. For other facts of interest that for all their fascination have only an indirect connection with literature see, for example, the entry on the licentious Anaïs Nin, or the one on the lesser-known Harold Norse (pseud. of Harold George Rosen). Mr Norse, 'born illegitimate in the Bronx', was the lover of Chester Kallman. When W.H. Auden invited them both to his apartment only Kallman turned up: 'the wrong blond', muttered Auden, but the substitute proved so satisfactory that Auden cut Norse out. Thus began that lifelong partnership with Kallman without which we might have been denied, among much else, the libretto of Stravinsky's *The Rake's Progress*. We learn without surprise that Mr Norse didn't come out of it so well: 'had the right blond visited the flat, one can speculate that Norse's career as a poet might have received the boost that distinguished connections can give'. However, he is now almost eighty, having long survived his lover and *his* lover; and 'his raunchy autobiography, *Memoirs of a Bastard Angel*, has further increased his reputation'.

There was no such happy ending for Dorothy Edwards, daughter of a Welsh socialist, who lived in Vienna and Florence, took a degree in Greek and philosophy, and had 'a passion for music, Welsh nationalism and imperfectly understood socialist politics'. Unfortunately, having written some short stories which attracted the attention of Bloomsbury, she went to live as a kind of au pair with David Garnett and his wife, and 'proved herself tactless' in dealing with Virginia Woolf and T.E. Lawrence. Moreover, her clothes displeased Garnett; so she went back to Wales and threw herself under a train. She was thirty-one.

Such revelations are to be expected from a work of this sort, but I was particularly surprised to learn that back in the 1960s my doctor, a genial Irishman, was the long-term lover of a celebrated novelist, a neighbour of us both. In those days the idea of such a liaison never crossed my mind. Perhaps other readers who have been equally imperceptive may be similarly enlightened. But they may more decorously prefer to consider the more general sociological implications of these multitudinous biographies.

They may ask, for instance, whether this array of information about twentieth-century writers permits us to make any general observations about the best class to have been born into, the best sort of school to have attended, if one wants to be worthy of entering the record as a twentieth-century writer. It will be found that the fathers of writers so esteemed can be factory workers, printers, postmen, labourers, drapers, hospital stokers, wages clerks, piano salesmen, cricketers. One father is bluntly described by his gifted daughter as 'a crook'. Of course there are others of much grander lineage, like Elizabeth Bowen, who lived at Bowen's Court; or Sybille Bedford, daughter of a 'south German baron'; or Lord Berners, who was very rich and so far from being a mere writer that he dyed his doves and wrote exquisite joky music. Somewhat in the manner of the celebrated Alec Guinness film, he inherited his title only because a bizarre motor accident at a funeral wiped out all who stood between him and the succession. And then there are such elegant and curious *rentier* writers as Harold Acton, and Maurice Baring, born of a bank. But there seems to be no strong evidence that high class is correlated with high talent; the grammar schools, so improvidently destroyed, seem to have served genius almost as well as Eton and Winchester. And allowing for a few wilful exceptions – Evelyn Waugh comes to mind – the collective Bohemia of the writers seems a pretty democratic place. Certainly the ability to write is not indissolubly associated with birth or money. Of course there must be a downward educational limit, below which talent must remain unsuspected and undeveloped, mute and inglorious; but for all above that limit there is nothing that bars entry into the ranks of the writers, except an inability to write what pleases. Moreover it would seem that those who have shown that they can write have also frequently shown that they can live just as adventurously as their betters, have as many wives or husbands, travel as wantonly, and get into scrapes just as beautiful and entertaining.

It may occur to some readers to ask how the writers included in this work were selected for treatment. They are a large company, and I am willing to admit that there are some I've never so much as heard of. And yet they are, in the nature of the case, selected from a potentially much larger group. One many notice telltale signs: for example, 'his older brother was also a novelist' (but he didn't make it here). There has been a certain deliberation in the choosing. Three Behans are mentioned, but only one achieves fame. Here are three Nicholses, Grace, Peter and Robert; but not Beverley, the fascinatingly juvenile sage of the pre-war years. This is the harshest judgment, to have been considered and yet cast out, to be allowed to be a writer but not a good enough one to be called a Twentieth-Century Writer.

Large as the selection is, it does constitute a canon of sorts, simply because of such exclusions. Canons are nowadays called 'élitist', a foolish charge, since all lists, whether of rock records, comics, westerns, or black women writers, are canonical in their effect. The nearest thing to a truly democratic canon is a list as long as the one this book provides. In fact, so perpetual, so unstoppable is the modern outpouring of books that far more twentieth-century writers must have been left out of this book than are got in.

Of course canons can be called prejudiced, must indeed *be* prejudiced, but any prejudice to be detected in this *Guide* is surely very widely shared. Like everybody else, it takes the word 'writer' to mean somebody who writes fiction or drama or poetry, and although there are a great many such persons there are probably even more who write without, in quite the same way, making things up. Roland Barthes tries to distinguish between the 'scripteur' and the 'écrivain', and, using his distinction without his subtlety, one can say that the writers included in the *Guide* belong to the second category. Useless to deplore it; this is, after all, the sort of writing person we are interested in getting to know more about. Despite all the advice we are nowadays offered to the contrary, we still assume that nothing human is alien to us, even, or especially, the writer, and this large book contains much information concerning these gifted, funny, sad, deplorable and admirable specimens of humanity.

FRANK KERMODE

Editor's Introduction

This book, which contains biographical accounts and bibliographical details of some 1,000 writers in the English language, is intended as a complementary volume to *A Reader's Guide to the Twentieth-Century Novel* (1995). Not only does it frequently amplify material in that volume, and repair some omissions, it is similarly an attempt to reflect a century of literary taste. The two volumes were planned simultaneously and written in tandem. In an aberrant moment we decided to cover poets and playwrights as well as novelists and short-story writers, thus expanding this second volume to what seemed at times unmanageable proportions. Even so, readers are likely to be dismayed, angered or outraged by the exclusion of some writers – and probably by the *inclusion* of others. Although it is fairly easy to pick and choose amongst the great and the dead, the middle-ranking and living call for hard and what must sometimes seem eccentric decisions. On the whole, it seemed preferable that the book be judged wayward rather than be thought cautious. As in the companion volume, the editor's own prejudices may be discerned by noting inclusions rather than exclusions.

While it was a relatively simple matter to select representative writers of fiction and poetry, playwrights provided unforeseen difficulties. The theatrical explosion of the 1960s produced numerous writers who worked, often in collaboration, on intentionally ephemeral projects. Sometimes these plays or shows were published, but often they were not, or have long since slipped out of print, just as plays themselves have vanished from the repertory. The fringe continues to nurture new talent, and plays open and close every week. Unlike fiction and poetry, which live and die by print, plays stand or fall by performance, and this sets them apart. For all these reasons, dramatists are less fully represented than novelists and poets: the leading playwrights of the age are here, along with a selection of those whose work seems promising, of historical importance or worth looking at again, and these make up only about ten per cent of the total entries. Although we have included some television dramatists, we soon realised that once we started to consider writers whose principal work has been in the cinema, we would be straying into a new and alarmingly multitudinous area. Similarly, it could be argued that some of the best and most popular poetry of the century is to be found in the works of such lyricists as Cole Porter, Lorenz Hart and Ira Gershwin. If we started to embrace them, however, we would have needed to consider the singer-songwriters of a later generation: Lennon and McCartney, say, or Bob Dylan – famously championed as a major poet by Christopher Ricks. Tempted though we were to include such writers, we realised that once started it would be hard to stop.

Another category of writers we have excluded comprises essayists, critics and biographers, many of whose works might be judged quite as creative as that of some of those writers we *have* included. Those reluctantly barred include Lytton Strachey, whose little-known play set in Peking during the Boxer uprising, *A Son of Heaven*, and the whimsically erotic novella, *Ermytrude and Esmeralda*, failed to qualify him for inclusion; and A.J.A. Symons, who wrote much original and interesting work, none of which can be counted fiction, poetry or drama – the three principal defining categories. Writers such as Edmund Wilson and Lionel Trilling, although chiefly known for their critical works, were admitted on the grounds that their output of fiction, though tiny, is significant. Other writers and their books simply defy categorisation: while Bruce Chatwin's final book, *Utz*, is undoubtedly a novel, earlier volumes such as *The Songlines* mix fact and fiction; the books of Edward Gorey, a prolific and eccentrically talented cult figure are, I suppose, fiction, but his work is entirely *sui generis*.

Although we are fast approaching the millennium, it is probably too early to judge which authors will survive into the next century. Those who flourished before the mid-century and continue to be read may be judged to have secured a place in the pantheon, but contemporary writers are more difficult to assess. Reputations rise and fall with astonishing swiftness: this entire two-volume project has taken almost a decade to complete, and some of the younger writers who seemed destined for fine things in the 1980s failed to deliver in the 1990s and have thus been ruthlessly discarded. Ruthlessness is an editorial requirement, but I have tried to balance this by championing writers who seemed to me, or to my contributors, undeservedly neglected. A major aim of the book is to rescue those who have fallen by the literary wayside, but for every writer to whom we have extended a hand, there will be dozens of others whose plight we have ignored: eyes averted, we passed by on the other side.

If this seems a grim metaphor for the life of a writer, I can only say that it is borne out by some of the stories related in the pages that follow. For every writer who is rewarded with critical acclaim and substantial sales, there are hundreds who have died without achieving their goal. It does not do to be too sentimental about writers, who need to be reasonably tough to survive at all, and it goes without saying that a lot of people in the latter category get no more than they deserve. But there are a great many who deserve better, and if this book directs readers to authors of whom they have never heard, let alone read, then a large part of our object will have been achieved.

In choosing which authors should be included, it was important to draw the line somewhere: above Beverley Nichols seemed as good a place to start as any. In practice, however, such demarcations prove vague, and disputed territory on either side of the border was fought over long and hard during the years it took to compile the book. Passionate advocacy by a contributor sometimes weakened the editor's resolve, and while authors such as Agatha Christie, H.P. Lovecraft and A.E.W. Mason were admitted, others who had feebly been allowed in were later shown the door – Clemence Dane, Warwick Deeping, Gilbert Frankau, Margaret Kennedy and Michael Sadleir amongst them. As in *A Reader's Guide to the Twentieth-Century Novel*, we have been particularly selective about authors who work in crime, science fiction and other genres, the same rule being applied: the books of those included are enjoyed by the general, not merely the specialist, reader. A very few writers (none of them of great significance) have had to be excluded because, in spite of our best efforts, it proved impossible to find sufficient information to compile a worthwhile entry. In sum, even if readers question the guest list, we hope that they will find this a lively literary gathering.

It was decided from the very beginning that the entries for this book should treat the lives as well as the work of the authors selected, and (where possible) in more detail than is usual in a reference book. Objections will no doubt be raised by those who remain resolutely incurious about human behaviour – even though it is one of the principal subjects of literature. Some objections may come from writers themselves – a handful of those approached to provide facts and figures declined to co-operate. These same writers, however, are quite happy to pillage the lives of family and friends in order to furnish their work and can therefore scarcely complain when they themselves become the object of scrutiny by biographers and journalists – though they do, of course. Those who dislike literary biography frequently enlist the support of W.H. Auden, who famously declared: 'On principle, I object to biographies of artists, since I do not believe that knowledge of their private lives sheds any significant light upon their works.' It is possible to take issue with this much-quoted pronouncement: while a literary text should be able to exist and be appreciated independently of its author, that is not to say that knowledge of a writer's life or the circumstances in which he or she wrote a work cannot add to our enjoyment or understanding of it. In any case, Auden made his statement in the course of a long and admiring review for the *New Yorker* of Robert Gutman's *Richard Wagner: The Man, His Mind and His Music* (1969), and went on to qualify his objection dramatically: 'However, the story of Wagner's life is absolutely fascinating, and it would be so if he had never written a note.'

As with composers, so with writers, some of whom have led lives quite as compelling as anything in their books. Even if the entry on Carl Van Vetchen doesn't send readers scurrying to the libraries, his story seems worth telling; and many people, I suspect, would prefer to read about Laura Riding's grappling with life than read the results of her grappling with the muse. Some of the stories we tell are comic, others less so. Isaac Rosenberg's hapless progress, from a single room in Stepney, shared with his parents and four siblings, through years of ill-health and poverty, to his disappearance at the age of twenty-seven in no-man's land as April Fool's Day dawned, is particularly dispiriting. He, at least, has not been forgotten, and is now regarded as one of the finest of the war poets. Others, fêted in their time, have since sunk almost without trace: Lascelles Abercrombie, one of only two writers to be included in the *Oxford Poets* series during their lifetime, soon fell out of favour and (according to our pessimistic contributor) 'still awaits his audience'. Writers, like farmers, are notorious for grumbling about their lot, but often have good reason to do so. Literary merit has never been any sort of guarantee of recognition or reward. One thinks of William Gaddis, so disheartened by the reception of his enormously long first novel (now regarded as a major work of American fiction), that he spent the next twenty years writing speeches for corporate executives, and was reduced at one point to ghostwriting 'professional journal articles for his dentist in exchange for much-needed root canal work'. Like Rosenberg, he subsequently achieved his just deserts – and is fortunate enough to be alive to enjoy them.

For every writer who has trodden the straight path of school, university, acclaimed first volume, marriage and children, consolidated reputation, major awards, visiting professorships, a tidy death and a sober obituary in *The Times*, there are numerous others who have weaved their luckless way through wretched childhoods, miserable schooldays, expulsions and sendings down, banned or derided books, racketty marriages, inconveniently fecund mistresses or inconveniently sexed lovers, struggles with drink, drugs and publishers, spiralling debts, disappointed promise and a squalid or sensational end. Regardless of what they wrote, those in the first category tend to be less interesting to write and read about than those in the latter. Consequently, the length of each entry often bears little relation to the literary eminence of the author described. Contributors have occasionally been given their heads, and room in which to make their case when the standing of their subject is insecure or in dispute. Furthermore, it often takes more space to introduce the forgotten and the neglected, about whom little is known, than to outline the lives and achievements of the famous. It is a principal contention of this book that while students and academics are obliged to slog their way through the 'major' authors of the century, the general reader has the leisure to explore some of the byways and backwaters of English literature. This *Guide* is intended to help them find their way.

Organisation and Arrangement

Authors are listed alphabetically by the name under which they write. In the headings, names are given in full – e.g.: E(dward) M(organ) FORSTER, Robert (Lee) FROST, (Maurice) Denton WELCH. The real names of pseudonymous authors are noted in brackets – e.g.: George ORWELL (pseud. of Eric Athur Blair); adopted names, shortened forms of original names, or deed-poll changes are differentiated from pseudonyms thus: Stevie (i.e. Florence Margaret) SMITH, Ted (i.e. Edward James) HUGHES, Michael ARLEN (i.e. Dikran Kouyoumdjian). Where writers have published substantial work under more than one name, cross-references direct the reader to the main entry. Within the text, the names of other writers who have their own entries in the book are printed in bold face roman.

Bibliographies are as full as we have been able to make them, although we have limited ourselves to books, passing over sundry items such as uncollected stories, articles and poems and anything that has only appeared in newspapers, magazines or anthologies. We have also tended to ignore privately printed books, unless (as, for example, in the case of W.H. Auden's first volume of *Poems*, hand-printed by Stephen Spender) they are significant in the author's career. Pulp fiction has on the whole been ignored and no attempt has been made to document the full output of writers such as E. Nesbit, who produced reams of childrens stories, often as parts of series. (In Nesbit's case, as in that of many of the writers in this book, a full bibliography may be found in the standard biography.) The title and date of each book listed are those of its first appearance in the country of origin. Alternative titles (for revised or foreign editions), dates of revisions and expansions, and other additional information, such as co-authors (including illustrators and, in the case of musical theatre, composers), appear in square brackets, using the catch-all formula 'with A.N. Other'. Plays and films based on the books of other writers are designated: 'from A.N. Other's work', while 'as A.N. Other' denotes a pseudonym. The titles are listed within categories in chronological order. Where two or more titles appear in the same year, they are ordered alphabetically, unless it is known that one work definitely preceded another (as in the case of first publications). The rare instances where it has been impossible to discover or confirm the date of first publication are noted either with 'nd' or, for example, '1905[?]'.

Bibliographies are divided into the following categories: **Fiction**, **Poetry**, **Non-Fiction**, **Edited**, **Translations**, **For Children**, **Collections and Biography**. Most of these are self-explanatory, but **Collections** lists volumes containing fiction and non-fiction, or poetry and prose, or any mixture of genres. Biographies of the author are listed where they exist: they are either the 'standard' one, the most recent one, or one the contributor or editor can recommend. Some items have been difficult to categorise: the examples of Bruce Chatwin and Edward Gorey have already been mentioned, but a number of avant-garde writers – such as Iain Hamilton Finlay and Richard Kostelanetz – have worked in mixed media, including cards, glass, stone and recording tape, producing work that combines poetry and prose, fiction and non-fiction and generally blurs any boundaries we may have tried to draw. **Collections**, therefore, often contains such anomalies. Authors such as James Thurber and Max Beerbohm have produced books that are principally made up of drawings, and these have been regarded as **Non-Fiction**. We have not listed musical compositions or catalogues of art works by those authors who were also composers or artists. Our main purpose in deciding where such items should be listed has been to keep the bibliographies as clear and as user-friendly as possible. They should be regarded primarily as guides rather than as tools of scholarship.

The categories are arranged in order of importance for each writer. Kingsley Amis, for example, writes poetry, but is principally known as a novelist, and therefore his **Fiction** is listed first, whereas Philip Larkin, who wrote fiction, is principally regarded as a poet, and therefore his **Poetry** is listed first.

Fiction: We do not differentiate between novels and individually published novellas, but an (**s**) following a title denotes a book of short stories. Where individual novels go up to make a trilogy, tetralogy or sequence, they are listed together (separated by commas rather than semi-colons), starting with the first volume. If the books have been collected into a single volume, a date is also given for this.

Plays: This category is used for all dramatic works, including screenplays, television and radio plays and musical theatre. The date given is that of first production or broadcast and is followed by the publication date in brackets only if this is different. Where a play has not been published, the production or broadcast date is followed by [np]. Plays are listed chronologically under the date they first appeared, whether this was on stage or in book form. Collected editions therefore appear under the date on which the earliest play was produced, even if that play was not collected until many years later. Unless marked otherwise, the play was written for the stage: other media are denoted by (**r**) for radio, (**tv**) for television, (**f**) for film and (**l**) for libretto. The abbreviation (**st**) is used when a work originally written for another medium has subsequently been performed on stage. Many writers have worked in cinema principally for financial reasons, their contribution to the finished product often reduced to little more than the occasional line of dialogue. Reflecting this, we have often mentioned such work in the biographical entry, but have only listed films in the bibliographies when they are original work or can be considered a significant part of a writer's oeuvre: Edward Bond's adaptation of *Walkabout* is listed, Aldous Huxley's of *Pride and Prejudice* is not.

Non-Fiction: The abbreviation (**a**) is used to denote an autobiography, and (**c**) is used when the author has made a substantial contribution to a volume edited and contributed to by others. Radio, film and television documentaries are not listed, unless they have subsequently been published in book form.

Translations: The original author's name is noted in roman type, followed by the title or titles, in English, in italics.

Edited: In the case of editions of the works of other writers, the original author's name is noted in roman type, followed by the titles of the works edited, in italics.

While such rules seem easy to apply, we have occasionally run into intractable problems. When *A Reader's Guide to the Twentieth-Century Novel* was published, we were taken to task in the *Times Literary Supplement* for referring to William Burroughs's celebrated novel as *The Naked Lunch* rather than simply *Naked*

Lunch. In fact, the novel was originally published in Paris as *The Naked Lunch*, but the definite article was dropped when an expurgated edition was published in the USA. The first British edition retained the definite article, while subsequent American editions dropped it. To complicate matters further, the current British edition has been issued as *Naked Lunch.* We have decided to stick with *The Naked Lunch*, since this was the form in which the title originally appeared: we have, however, noted the variant in Burroughs's bibliography. This is simply one example of the sort of decisions that had to be made, and while some will be judged arbitrary, each was decided upon individually.

PETER PARKER

Lascelles ABERCROMBIE 1881–1938

Abercrombie was born at Ashton-on-Mersey, Cheshire, the eighth of nine children of a stockbroker (Patrick Abercrombie, the regional planner, was a brother). He attended Malvern College, then read chemistry at Owens College, Manchester (later the University of Manchester): perhaps because his own inclinations were literary, he left without taking a degree. He worked briefly as a quantity surveyor in Birkenhead, but soon gravitated towards the Liverpool daily press, mainly working freelance but also serving for a while on the staff of the *Daily Courier*. His first volume of verse, *Interludes and Poems*, was published in 1908.

In 1909 he married Catherine Gwatkin, with whom he had four children, and with whom he moved in the pre-war years to Herefordshire and then Gloucestershire, where they lived 'economically but happily' and where he blossomed as one of the new Georgian generation of poets. He was one of five poets contributing to the short-lived Georgian magazine *New Numbers* in 1914. He was rejected as unfit for service in the First World War, and spent those years inspecting shells in a Liverpool munitions factory. During the war his financial position was improved by his being one of the three joint beneficiaries of **Rupert Brooke**'s will (**Wilfrid Gibson** and **Walter de la Mare** were the others), and in 1919 a new post was created for him as a lecturer in poetry at Liverpool University. Despite his never having taken a degree by examination, he proved a natural academic, and progressed through various posts before becoming the Goldsmith's Reader in English at Oxford in 1935.

His last years were clouded by chronic diabetes, which brought his considerable literary production to an end by 1932. This had consisted of volumes of verse and verse-drama (he himself considered that his six-act verse-drama *The Sale of St Thomas*, published between 1911 and 1931, was his major work), several volumes on poetics, and an influential study of **Thomas Hardy**, published in 1912. His poetry is abstruse and ostensibly unemotional in style, and although it was highly esteemed in his own era (with **Robert Bridges**, he was one of only two writers to be included in the *Oxford Poets* during their lifetimes), the corresponding fall has been greater. He is one of the comparatively few minor poets of his generation not even to be revived in **Philip Larkin**'s *Oxford Book of Twentieth Century English Verse*, and as that century nears its close he still awaits his audience.

Poetry

Interludes and Poems 1908 [rev. 1928]; *Mary and the Bramble* 1910; *Emblems of Love* 1912; *New Numbers* vol. 1, nos 1–4 [with R. Brooke, J. Drinkwater, W.W. Gibson] 1914; *Twelve Idyls and Other Poems* 1928; *The Poems of Lascelles Abercrombie* 1930; *To Sir Walford Davies at Gregynog, June, 1934* 1934; *Lyrics and Unfinished Poems* 1940; *Vision and Love* 1966

Non-Fiction

Thomas Hardy: a Critical Study 1912; *Speculative Dialogues* 1913; *The Epic* 1914; *Poetry and Contemporary Speech* 1914; *An Essay Towards a Theory of Art* 1922; *Principles of English Prosody* 1923; *Stratford-upon-Avon: a Report on Future Development* [with L.P. Abercrombie] 1923; *The Theory of Poetry* 1924 [rev. 1926]; *The Idea of Great Poetry* 1925; *Romanticism* 1926; *Progress in Literature: the Leslie Stephen Lecture* 1929; *Colloquial Language in Literature* 1931; *An Outline of Modern Knowledge* [ed. W. Rose] (c) 1931; *Revaluations: Studies in Biography* (c) 1931; *Poetry: Its Music and Meaning* 1932; *The Art of Wordsworth* 1952

Plays

Deborah 1913 [rev. 1923]; *The Adder* 1913 [pub. 1922]; *The End of the World* 1914; *The Staircase* 1920; *Four Short Plays* [pub. 1922]; *Phoenix* 1924 [pub. 1923]; *The Sale of Saint Thomas* 1911–1931

Dannie (i.e. Daniel) ABSE 1923–

Born in Cardiff, South Wales, into a talented Jewish-Welsh family, Abse is the brother of the Labour MP Leo Abse and the psychoanalyst Wilfred Abse. He was educated at St Illtyd's College, Cardiff, and at the University College in the city, in an atmosphere of ardent socialism. The milieu of those days is evoked in *Ash on a Young Man's Sleeve*, the first novel that made Abse's reputation. In 1942, he became a medical student in London, enjoying the Bohemian society of wartime Swiss Cottage, where he found lodgings. After qualifying as a doctor at Westminster Hospital in 1949, he entered the Royal Air Force to do his national service, and rose to the rank of squadron leader. Since 1954, he has been a specialist at the Central Medical Establishment, London. He married in 1951 and has three children.

His first volume of poetry, *After Every Green Thing*, was published in 1949, his first mature volume is *Tenants of the House* (1957), and his *Selected Poems* were published in 1970. His poetry is marked by sensitivity and honesty, and in later years has drawn much on his personal experience, for instance as a doctor. Although he is best known as a poet, he has published several novels: *O. Jones, O. Jones*, a humorous novel about a Welsh medical student, is highly regarded. He is also a playwright, many of his plays having been produced by the fringe Questor's Theatre in Ealing, West London: *Gone*, about a man contemplating suicide, is perhaps his best play. He has also been an editor, notably of *Mavericks*, a controversial poetry anthology which reasserted poetic romanticism in the late 1950s.

Poetry

After Every Green Thing 1949; *Walking Under Water* 1952; *Tenants of the House* 1957; *Poems, Golders Green* 1962; *Selected Poems* 1963; *Selected Poems* [with J. Rousselot] 1967; *A Small Desperation* 1968; *Selected Poems* 1970; *Funland and Other Poems* 1973; *Collected Poems* 1977; *Penguin Modern Poets 26* [with D.J. Enright, M. Longley] 1978; *Miscellany One* 1981; *Way Out in the Centre* 1981; *People* 1990; *Remembrance of Crimes Past* 1990

Fiction

Ash on a Young Man's Sleeve 1954; *Some Corner of an English Field* 1956; *O. Jones, O. Jones* 1970; *Ask the Bloody Horse* 1986; *White Coat, Purple Coat: Collected Poems 1948–1988* 1989; *Listening to Voices from Wales* [ed. S. Anstey] 1992

Plays

Fire in Heaven 1956; *Three Questor Plays* [pub. 1967]: *House of Cowards, Gone, In the Cage*; *The Dogs of Pavlov* 1969; *Gone in January* 1978; *Pythagoras* 1979; *The View from Row G: Three Plays* [pub. 1991]

Non-Fiction

Medicine on Trial 1967; *A Poet in the Family* 1974; *A Strong Dose of Myself* 1983; *Journals from the Ant Heap* 1986; *The Music Lover's Literary Companion* [with J. Abse] 1988

Edited

Poetry and Poverty nos 1–7 1949–54; *Mavericks* [with H. Sergeant] 1957; *Modern European Verse* 1964; *Corgi Modern Poets in Focus* 1971 onwards; *Poetry Dimension Annual* nos 2–7 1974–80; *My Medical School* 1978; *Wales in Verse* 1983; *Doctors and Patients* 1984; *Voices in the Gallery* [with J. Abse] 1986; *The Hutchinson Book of Post-War British Poetry* 1989

(Albert) Chinua(lumoga) ACHEBE 1930–

Achebe was born in Ogidi, Eastern Nigeria. His family was Ibo and Christian and his education began at a Church Missionary Society school. He has remarked, however: 'I think I belong to a very fortunate generation ... the old [pre-colonial, pre-Christian] ways hadn't been completely disorganised when I was growing up ... it was easy, especially if you lived in a village, to see ... these old ways of life.' He attended Government College in Umuahia and entered University College, Ibadan, in 1949; awarded a scholarship to study medicine, he changed his course to literature after a year. In 1953 he graduated with a BA, taught for a few months, then began a career with the Nigerian Broadcasting Company, becoming its director of external services in 1961. Three years previously, he had published his first novel, *Things Fall Apart*. In 1966, having published two more novels, he left broadcasting to write full-time. During the Nigerian civil war (1967–70), though, he had little opportunity to pursue his career. Committed to the Biafran cause, he served mainly as a diplomat abroad. In the second month of the war, he, his wife and children narrowly escaped assassination when their flat was bombed.

Since 1971, he has edited *Okike*, the leading journal of Nigerian new writing. Amongst other academic appointments, he has held posts as Professor of English at the University of Nigeria and at the University of Massachusetts, Amherst (where he met **James Baldwin**, also a faculty member), Professor of African Studies at the University of Connecticut, Storrs, and Pro-Chancellor and Chairman of Council, Anambra State University of Technology, Enugu. In 1990 a serious car accident left him paralysed from the waist down.

Things Fall Apart, *No Longer at Ease* and *Arrow of God* comprise *The African Trilogy*, which established his reputation. Its central character, Okonkwo, lives through the early years of the British administration in Nigeria. If the narrative's temporal structure is fragmented, its themes are unified and its purposes clear: it records the distress of a people torn between European and African cultures, absorbing the impact of the former while retaining the traditions of the latter. A highly political drama, it rebukes the apathy of the masses as much as it deplores the arrogance of colonial rule. Achebe has acknowledged a didactic intent: 'The writer cannot expect to be excused from the task of re-education ... I for one would not wish to be excused ... Perhaps what I write is applied art as distinct from pure. But who cares?' Consonant with this aim are his directorships of Heinemann Educational Books (Nigeria) and Nwamife, publishers, for the most part, of children's books.

In addition to his five novels, Achebe has written poetry, essays and children's stories. Diverse and prolific, he has been described as the 'most prominent novelist writing in Africa today'. His importance has been recognised in many awards, including the Margaret Wrong Prize for *Things Fall Apart*, the *New Statesman* Jock Campbell Prize for *Arrow of God* and the Commonwealth Poetry Prize for *Beware, Soul Brother*. *Anthills of the Savannah* was shortlisted for the Booker Prize. When accepting the Scottish Arts Council's Neil Gunn Fellowship, Achebe told his audience: 'I [feel] so completely at home in the world of the peasants and fishermen of Neil Gunn's fiction that I [suspect] he and I must be at heart pre-industrial men.'

Fiction

The African Trilogy 1988: *Things Fall Apart* 1958, *No Longer at Ease* 1960, *Arrow of God* 1964 [rev. 1974]; *The Sacrificial Egg and Other Stories* (s) 1962; *A Man of the People* 1966; *Girls at War and Other Stories* (s) 1972; *Anthills of the Savannah* 1987

Non-Fiction

Morning Yet on Creation Day: Essays 1975; *The Trouble with Nigeria* 1983; *The World of Ogbanje* 1986; *Hopes and Impediments: Selected Essays 1965–1987* 1987; *The University and the Leadership Factor in Nigerian Politics* 1988; *Nigerian Topics* 1989

Poetry

Beware, Soul Brother 1971; *Christmas in Biafra and Other Poems* 1973

For Children

Chike and the River (s) 1966; *How the Leopard Got His Claws* (s) 1972; *The Drum* 1977; *The Flute* 1977

Edited

The Insider: Stories of War and Peace from Nigeria (s) 1971; *Don't Let Him Die: a Tribute to Christopher Okigbo* 1978; *African Short Stories* (s) 1985; *Beyond Hunger in Africa* 1990; *The Heinemann Book of Contemporary African Short Stories* (s) [with C.L. Innes] 1992

Kathy ACKER 1948–

Acker was born and raised in New York, and her literary background is that city's poetry scene of the 1960s, her guru being Jerome Rothenburg of the second generation of Black Mountain School of poets. Much of her work was originally published by American underground presses. Her novels purportedly confront such issues as the suppression and exploitation of women, racial conflict, poverty, education and religion, but she is best known for her handling of the more primal subjects of sex, violence (sexual violence and violent sex), and the quest for self-knowledge. Inspired by **Gertrude Stein** and the 1965 *Paris Review* interview with **William Burroughs**, which gave clear instructions on the methods of cut-up, fold-in, and the confounding of language, Acker writes with determined amorphousness. Her writing ethic is that 'language is more important than words'. A talented self-publicist, she cultivates the appearance of a somewhat elderly punk (*Empire of the Senseless* is dedicated 'to my tattooist'); but there is no doubt that she takes her work seriously, whatever others may make of it.

Opinions as to her merits as a writer are sharply divided. While some see her as a major talent, others find it hard to discern any merit at all in her work. Her determination to shock – peppering her prose with expletives and illustrating her texts with her own (rather poor) drawings of genitalia – has led many (usually male) critics to describe her work as 'hip' and 'visceral', 'direct, fast, sexy, hot, horny, furiously honest'. *Blood and Guts in High School* made a considerable impact in the UK, where Acker was labelled a postmodernist feminist, and later work has been published there by a feminist press. Some commentators, however, feel that she exploits sex too much and language too little. She writes quickly, with an eye to the needs of the moment rather than posterity, sometimes publishing unrevised first drafts. Realistic passages are intercut with hallucinatory visions, poetry with prose. She also draws upon the work of others, most notably Genet and, in *In Memoriam to Identity*, Rimbaud and **William Faulkner**. A less elevated source of inspiration is the popular novelist Harold Robbins, whose prose Acker was accused of copying almost word for word in one of her books. Although this caused quite a stir at the time, Robbins himself seemed not to mind.

Acker's play, *The Birth of the Poet*, is an example of her iconoclastic treatment of plot and structure and was staged both in New York and Rotterdam to rave reviews. Act I depicted the aftermath of a nuclear scourging; Act II sex and violence in Ancient Rome and present-day New York; Act III was a silent piece entitled 'Ali Goes to the Mosque'. She moved to London in 1967, and for some time divided her time between that city and New York. She has taught video and performance art at the San Francisco Institute and writes regularly for *Art Forum.*

Fiction
The Adult Life of Toulouse Lautrec 1978; *The Childlike Life of the Black Tarantula* 1978; *I Dreamt I was a Nymphomaniac* 1980; *Great Expectations* 1982; *Hello, I'm Erika Jong* 1982; *Implosion* 1983; *Blood and Guts in High School* 1984; *Don Quixote* 1986; *Empire of the Senseless* 1988; *The Seven Deadly Sins* **(s)** [with others] 1988; *In Memoriam to Identity* 1990; *The Seven Cardinal Virtues* **(s)** [with others] 1990

Non-Fiction
Kathy Goes to Haiti 1979

Play
The Birth of the Poet 1985 [pub. in *Wordplays 5* 1986]

Collections
Literal Madness 1989; *Young Lust* 1989

J(oe) R(andolph) ACKERLEY 1896–1967

'I was born in 1896 and my parents were married in 1919' is the characteristic opening sentence of Ackerley's 'Family Memoir', *My Father and Myself.* His father (who had a second, secret family of a mistress and three daughters) was one of the founders of Elder & Fyffes, the fruit importers, and was known as 'the Banana King'; his mother was a former actress. His education at Rossall School and Magdalene College, Cambridge, was interrupted by the First World War, in which he served as a captain, was twice wounded, captured in 1917 and eventually exchanged into internment in Switzerland. His experiences there formed the basis of his play, *The Prisoners of War,* one of the first English plays to deal with homosexuality in a contemporary setting. At Cambridge he studied English and wrote many poems, one of which was published in the *London Mercury* in 1922 and caught the attention of **E.M. Forster**, thus becoming 'a not very worthy midwife' to an intimate, lifelong friendship. It was at Forster's suggestion that Ackerley spent five months as secretary-companion to the eccentric Maharajah of Chhatarpur, a jaunt which he recorded in a journal published in 1932, much bowdlerised, as *Hindoo Holiday.*

In 1928 he joined the BBC as an assistant talks producer, and subsequently became literary editor of the *Listener* (1935–59). Much of his time was spent battling with his conservative, philistine superiors ('I am constantly in the dock for everything I do,' he complained to **Leonard Woolf** – 'modernism in art and poetry, leftism in literature…'), but he managed to produce some of the period's liveliest arts pages and is widely considered the most gifted editor of his generation. His energetic, expensive and muddled search for a homosexual partner ('the Ideal Friend') amongst working-class heterosexuals was inconclusive, and was abandoned in 1945 when he acquired a beautiful Alsatian bitch from a criminal boyfriend. Queenie became the emotional focus of his life, and the protagonist of two books: *My Dog Tulip,* a remarkable and unblinking account of her life and habits, and the novel *We Think the World of You,* which won the W.H. Smith Award. To the dismay of many of his friends, Ackerley was to become a vociferous supporter of animal rights. The last years of his life were spent in acrimonious cohabitation with his emotionally unstable sister, Nancy, in a small flat in Putney.

Both as a writer and editor, Ackerley was a perfectionist, and his best books are minor classics. *My Father and Myself,* which he had been working on intermittently for some thirty years, was eventually published the year after his death. A superbly crafted book, it gradually reveals the extraordinary secret lives of both the author and his genial but feckless parent. It is perfectly complemented by *The Secret Orchard of Roger Ackerley* (1975), in which Ackerley's half-sister, Diana Petre, tells the story of their father's two families from her mother's point of view.

Non-Fiction
Hindoo Holiday: an Indian Journal 1932 [rev. 1952]; *My Dog Tulip: Life with an Alsatian* 1965 [rev. 1966]; *My Father and Myself* 1968; *E.M. Forster: a Portrait* 1970; *The Letters of J.R. Ackerley* [ed. N. Braybrooke] 1975; *My Sister and Myself: the Diaries of J.R. Ackerley* [ed. F. King] 1982

Fiction
We Think the World of You 1960

Poetry
Poems by Four Authors [with A.Y. Campbell, E. Davison, F. Kendon] 1923; *Micheldever & Other Poems* 1972

Plays
The Prisoners of War 1925

Edited
Escapers All: the Personal Narratives of Fifteen Escapers from War-Time Prison Camps 1914–1918 1932

Biography
Ackerley by Peter Parker 1989

Rodney ACKLAND 1908–1991

Born in Westcliffe-on-Sea, Essex, Ackland grew up in London, attending Balham Grammar School. His Jewish father was in the rag trade and his mother had danced at the *Folies Bergère*. Ackland was himself stage-struck from an early age, and made his debut, aged fifteen, in the famous Komisarjevsky production of Chekhov's *Three Sisters* at the little Barnes Theatre; Chekhov's work has been the major influence on Ackland's own. As a young man, Ackland worked as a salesman for Swan and Edgar and in an oil company, but he also trained at the Central School of Speech Training and Dramatic Art, and became an actor. He also had his first play, *Improper People*, produced in 1929. During the 1930s he worked widely as an actor (appearing in his own play, *Birthday*), began his career as an author of film scripts, and wrote fourteen plays, all of which were produced. They tended to be seen, however, in small theatre clubs, because Ackland was considered too highbrow for the West End stage: the notices for *Strange Orchestra*, directed by John Gielgud, cost him his job with British International Pictures, who did not wish to employ serious writers. Exploring an atmosphere of genteel seediness, his plays have an atmosphere of piercing melancholy that is not easily forgotten. Hilary Spurling, writing in 1968, called his finest work, *The Dark River*, 'perhaps the one indisputably great play of the past half century in English'. In 1951 the impresario 'Binkie' Beaumont refused to put on *The Pink Room* on the grounds that it was 'a libel on the British people'. An extremely funny, touching and beautifully observed play, it is set during the 1945 General Election in a London drinking club frequented by servicemen, refugees, writers, film people, socialites and elderly eccentrics – a slice of metropolitan Bohemia held together by drink, sex, indigence and natural sympathy. It was astonishingly bold for its time (notably in its frank portrayal of both male and female homosexual characters) and, after a brief and unsuccessful first run, had to wait until 1988 for an unexpurgated version to be produced under the title *Absolute Hell*.

During the 1960s a new theatrical fashion dismissed Ackland, not this time as a highbrow but as a practitioner of the 'well-made play', and a long period began when his plays slipped out of the repertory and no new ones were performed. He was obliged to move from his rooms in Albany to some in Richmond. In the mid-1980s, however, there was a revival of his work, with productions in Windsor and Richmond, and his imaginative adaptation of Ostrovsky's *Too Clever by Half* was presented at the Old Vic in 1988 in an award-winning and highly successful production by Richard Jones. The gratified author was profiled in newspapers, pictured sitting up in bed.

Ackland spent some years living with Arthur Boys, an Australian screenwriter, with whom he had collaborated on a 1948 film of *The Queen of Spades*. (The couple successfully but ruinously sued the film company for giving greater prominence in the credits to Pushkin's name than to their own.) In 1952 he married Mab Lonsdale, daughter of the playwright Frederick Lonsdale, an unexpected turn of events which nevertheless proved enduring, and Ackland was severely shaken by her death twenty years later. He recovered, however, to enjoy a belated but genuine revival in his literary fortunes, and his reputation continues to grow.

Plays

Improper People 1929 [pub. 1930]; *Dance with No Music* 1930 [pub. 1933]; *Marion-Ella* 1930 [np]; *Strange Orchestra* 1931 [pub. 1932]; *Ballerina* [from E. Smith's novel] 1933 [np]; *Birthday* 1934 [pub. 1935]; *The White Guard* [from M. Bulgakov's play] 1934 [np]; *The Old Ladies* [from H. Walpole's novel; US *Night in the House*] 1935; *After October* 1936; *Plot Twenty-One* 1936 [np]; *Yes, My Darling* [from M. Reed] 1937 [np]; *Sixth Floor* [from A. Ghéri's play] 1939 [np]; *The Dark River* 1941 [pub. 1943]; *Blossom Time* 1942 [np]; *The Diary of a Scoundrel* [from A.N. Ostrovsky's play] 1942 [pub. 1948; as *Too Clever By Half* 1988, np]; *Crime and Punishment* [from Dostoevsky's novel] 1946 [pub. 1948]; *Cupid and Mars* [from R.G. Newton's novel; with R.G. Newton] 1947 [np]; *Before the Party* [from Somerset Maugham's story] 1949 [pub. 1950]; *A Multitude of Sins* [with R.G. Newton] 1951 [np]; *The Pink Room* 1952 [np; rev. as *Absolute Hell* 1988, pub. 1990]; *A Dead Secret* 1957 [pub. 1958]; *Farewell, Farewell, Eugene* [from J. Vari] 1959 [pub. 1960]; *The Other Palace* 1964 [np]; *Smithereens* 1985 [np]

Non-Fiction

The Celluloid Mistress (a) [with E. Grant] 1954

Peter ACKROYD 1949–

Ackroyd was born in London. His parents separated when he was young and he was brought up by his mother in a council house in East Acton. He was educated at St Benedict's, a Catholic school in Ealing, and at Clare College,

Cambridge, and was awarded a Mellon Fellowship at Yale University. At the age of twenty-three he became literary editor of the *Spectator*, a post he held for four years; he was also joint managing editor of the magazine, and subsequently its film critic. Since 1986 he has been the chief book reviewer for *The Times* and was for a time that newspaper's television critic. He has been described as one of the last English men-of-letters, supporting himself entirely by his pen. He claims that work consumes his life, once commenting that he had 'sacrificed' himself 'for the sake of English literature'.

His first novel, *The Great Fire of London* (1982), concerns the making of a film of *Little Dorrit*, and reflects his recurring interest in both Dickens and London. His second novel, *The Last Testament of Oscar Wilde*, was an entirely convincing account of the last four months of the dramatist's life as recorded in his own diary, and the first of Ackroyd's books to display his gift for pastiche. Parts of his next novel, *Hawksmoor*, were written in the style of the early eighteenth century. The book was suggested by *Lud Heat*, a poem by **Iain Sinclair**, and the narrative follows two interconnected stories: the building of seven London churches by a Hawksmoor-like architect and the present-day investigation of a series of child murders. His greatest critical and popular success, it won both the Whitbread Novel of the Year Award and the *Guardian* Fiction Award. A similar shuffling of time occurs in *Chatterton*, which centres upon Alfred Wallis's famous painting *The Death of Chatterton*, and looks backwards to the poet's time and forwards to the present. Ackroyd's critics complain that his novels are intellectually stimulating but emotionally arid; that he can tell good stories, but tends to people them with caricatures rather than characters. These reservations were confirmed by *First Light*, an inferior novel set in Dorset with nods to **Thomas Hardy**, which received a generally poor press. *Dan Leno and the Limehouse Golem* was seen by many as a return to form.

Ackroyd's preoccupation with layers of time may have been derived partly from **T.S. Eliot**, whose life and work he investigated in an acclaimed biography. Ackroyd was forbidden by the poet's estate to quote from the work under discussion, but overcame this obstacle and won the Whitbread Award for Biography. He subsequently received an advance of £650,000 to write biographies of Dickens and William Blake. The first of these had a mixed reception, partly because of interpolated passages of imagined conversation between Dickens and his characters and Ackroyd, Wilde and Blake. (These were inserted at a late stage against the advice of his editor and agent.) Ackroyd has also published volumes of poetry and *Dressing Up*, a slim treatise on transvestism.

Fiction

The Great Fire of London 1982; *The Last Testament of Oscar Wilde* 1983; *Hawksmoor* 1985; *Chatterton* 1987; *First Light* 1989; *English Music* 1992; *The House of Doctor Dee* 1993; *Dan Leno and the Limehouse Golem* 1994

Non-Fiction

Notes for a New Culture: an Essay on Modernism 1976; *Dressing Up: Transvestism and Drag: the History of an Obsession* 1979; *Ezra Pound and His World* 1980; *T.S. Eliot* 1984; *Dickens* 1990; *Introduction to Dickens* 1991; *A Biography of William Blake* 1995

Poetry

London Lickpenny 1973; *Country Life* 1978; *The Diversions of Purley and Other Poems* 1987

Edited

PEN New Fiction 1 1984; Oscar Wilde *The Picture of Dorian Gray* 1985; *Dickens' London: an Imaginative Vision* 1987

Harold (Mario Mitchell) ACTON 1904–1994

Acton was born at La Pietra, a fourteenth-century villa near Naples. His father, who was descended from the Actons of Naples, was an art dealer and collector; his mother a Chicago banking heiress, whose money had enabled her husband to purchase the villa and fill it with art treasures. Although educated at Eton and Christ Church, Oxford, Acton had spent most of his early childhood at La Pietra, and it was Italy that had the most profound effect on his aesthetic outlook. While at Eton, he was a founder, with **Henry Green**, of the Eton Society of Arts, and edited with Brian Howard the *Eton Candle*, an avant-garde magazine containing both poetry and prose. Bound in shocking pink and popularly known amongst their schoolfellows as 'The Eton Scandal', it attracted the attention of **Edith Sitwell**, and Acton went up to Oxford with something of a literary reputation. This was consolidated by his editorship of another short-lived magazine, the *Oxford Broom*, so called because of his intention to sweep away the Georgianism to which the university still clung. He affected Victorian side-whiskers and capacious pleated trousers ('Oxford bags'), styled himself 'the scourge of the Philistines', published two volumes of verse, *Aquarium* (1923) and *An Indian Ass*, and purportedly declaimed **T.S. Eliot**'s *The Waste Land* through a megaphone from his window overlooking Christ Church Meadow.

After leaving Oxford, Acton divided his time between London and Paris, grudgingly supported by his father for three years in which he produced three books, the first of which was a lively account in verse of the lives of *Five Saints*, which attracted the attention of **Gertrude Stein**. His first novel, the aptly-titled *Humdrum* (1928), was a misguided attempt to gain a popular audience ('Harold Acton for the masses', as **Evelyn Waugh** put it). It suffered from being reviewed alongside Waugh's infinitely superior *Decline and Fall*, which was published simultaneously. Acton was a remarkable conversationalist, whose vibrato voice enunciated every syllable equally, but his verbal skills did not transfer to the printed page. Aspiring to the heights of **Ronald Firbank**, he tended to land with a dull thud somewhere in the vicinity of **E.F. Benson**. Historical surveys of *The Last Medici* and *The Bourbons of Naples* have their admirers, but were considered more anecdotal than scholarly.

Tired of Europe, Acton spent most of the 1930s in Peking, lecturing and skilfully translating Chinese poetry. *Peonies and Ponies* is a mild satire on the Europeans he encountered there. He was prematurely bald, 'slim and slightly oriental', so that the mandarin guise he adopted wore well. Forced to leave China when Manchuria was invaded by the Japanese in 1939, Acton returned to England, where he served with the Royal Air Force and became a press censor at the Supreme Headquarters Allied Expeditionary Force. After the war, he returned to La Pietra, which had been occupied by the Germans, and found that little was missing, apart from the testicles of the garden statues.

Various books appeared, including novels, histories, collections of short stories, and two volumes of memoirs, which were completely reticent about his private life, but which are perhaps his best work. His insistence on the word 'aesthete' to describe himself fooled no one except Acton. He became one of the 'sights' of Florence and was generous to a fault as a host, particularly towards royalty and towards young men, who were occasionally invited to his bedroom to share an after-dinner siesta. His distinguished involvement with Florence's British Institute, as lecturer, committee member and major benefactor earned him a knighthood, and the library there is named in his honour. He left La Pietra to New York University.

In sum, Acton will be remembered more for what he was than for what he wrote. His influence upon the fruitful 1920s Oxford generation was considerable, and he achieved an unwelcome immortality as the part-model for Anthony Blanche in Waugh's *Brideshead Revisited* (1945).

Poetry

Aquarium 1923; *An Indian Ass* 1925; *Five Saints and an Appendix* 1927; *The Chaos* 1936

Fiction

Humdrum 1928; *Cornelian* 1928; *Peonies and Ponies* 1941; *Prince Isidore* 1950; *Old Lamps for New* 1965; *Tit for Tat* (s) 1972; *The Soul's Gymnasium and Other Stories* (s) 1982

Non-Fiction

The Last Medici 1932; *Memoirs of an Aesthete* (a) 1948; *The Bourbons of Naples (1734–1825)* 1956; *Florence* 1961; *The Last Bourbons of Naples (1825–1861)* 1961; *More Memoirs of an Aesthete* (a) 1970; *Great Houses of Italy* 1973; *Nancy Mitford: a Memoir* 1975; *The Pazzi Conspiracy* 1979; *Three Extraordinary Ambassadors* 1983

Translations

Famous Chinese Poetry 1936; *Modern Chinese Poetry* [with Ch'en Shih-Hsiang] 1936; *Glue and Lacquer: Four Cautionary Tales* 1941; *The Peach-Blossom Fan* 1976

Edited

Florence: a Traveller's Companion 1986

Gilbert ADAIR 1944-

Adair was born in Edinburgh, and does not discuss his early life. He read modern languages at university before moving at the end of the 1960s to Paris where he taught English at the University of Paris (Paris V) for a decade. During this period he worked as a film critic for various magazines including *Sight and Sound* (for whom he interviewed François Truffaut in 1974), and made appearances in two films, Edgardo Cozarinsky's *Les Apprentis sorciers* (1976) and Hugo Santiago's *Ecoute Voir* (1978). In the latter he plays a preacher from a religious cult. His 1981 book *Hollywood's Vietnam* is a seminal study of the cinema's treatment of the Vietnam war. He has also written the screenplay for the Chilean director Raul Ruiz's film *The Territory*, an account of cannibalism in the wake of an air disaster which was described by the director as 'a philosophical exploitation movie'.

Adair's first two works of fiction, *Alice through the Needle's Eye* (1984) and *Peter Pan and the Only Children*, are intellectually dextrous sequels to the children's classics of Lewis Carroll and **J.M. Barrie**, while his third, *The Holy Innocents* (winner of the Author's Club First Novel Award in 1989), is a contemporary reworking of Jean Cocteau's *Les Enfants terribles* (1929). Similarly, *Love and Death on Long Island* echoes and subverts Thomas Mann in a compact, beautifully judged and very funny tale of a distinguished British writer's obsession with a teenage American filmstar. *The Death of the Author* is a satire focusing on a fashionable critical theorist, Léopold Sfax.

Adair has translated a number of books relating to the cinema, as well as Georges Perec's *La Disparition*. This novel is a lipogram which famously avoids the use of the letter 'e', as does Adair in his surprisingly faithful translation, *A Void*. He has written a full-length parody of Alexander Pope, *The Rape of the Cock*, in which the snipped object is *not* a lock of hair, and is a witty cultural commentator who writes for numerous publications. A number of his essays on 'the discourse of contemporary British culture' are collected in *The Postmodernist Always Rings Twice*.

Fiction

Alice through the Needle's Eye 1984; *Peter Pan and the Only Children* 1987; *The Holy Innocents* 1988; *Love and Death on Long Island* 1990; *The Death of the Author* 1992

Non-Fiction

Hollywood's Vietnam 1981; *A Night at the Pictures: Ten Decades of British Films* [with N. Roddick] 1985; *Myths and Memories* 1986; *The Postmodernist Always Rings Twice* 1992

Translations

Andre Bazin *Orson Welles* 1974; Michel Ciment *Kubrick* 1983, *John Boorman* 1986; François Truffaut *Letters* 1990; Georges Perec *A Void* 1994

Poetry

The Rape of the Cock 1990

Richard (George) ADAMS 1920–

Born in Newbury, the son of a country doctor, Adams went to school at Bradfield. He spent five years in the army from 1940, joining first the Royal Army Service Corps then, volunteering for airborne service, becoming a parachutist. It was, he has said, the best time of his life. After the war he went to Worcester College, Oxford, read modern history, but suffered a nervous breakdown before taking his finals. In 1948, the year before his marriage, he joined the civil service, where he worked for twenty years.

Adams remains best known for his first novel, *Watership Down* (1972), an extraordinary story of a group of rabbits searching for a new home after they have been forced to leave their warren by a housing development. (Adams himself worked in both the Ministry of Housing and the Department of the Environment, and the book's title refers to chalkland near Whitchurch, Hampshire, where the author lives.) Although the book is essentially an adventure story with allegorical overtones, Adams drew upon R.M. Lockley's study of *The Private Life of the Rabbit*, so that his characters always behave true to their animal natures. It is this wealth of fascinating detail as much as the gripping narrative which made *Watership Down* a huge bestseller. Not that this happened immediately, for the book was turned down by a number of mainstream publishers before being accepted by Rex Collinge, 'a one-man outfit operating from a top floor in Oxford Street'. Although originally intended for children – Adams wrote the book in his spare time, testing each chapter by reading it aloud to his two daughters, who were aged nine and seven – it appealed as much to adults, becoming a cult book and selling some 30,000,000 copies worldwide.

'I have always thought *Watership Down* was a kind of miracle that couldn't happen twice,' Adams said in the wake of its success. Subsequent novels, aimed more directly at an adult readership though still featuring animals, have had nothing like the same impact. *Shardik* is a somewhat mystical novel about a bear, while *The Plague Dogs* introduces a topical note in a story about animals escaping from an experimental laboratory. *The Girl in the Swing*, which deals frankly with a passionate love-life, marked a change of direction, and was followed by the even more sexually explicit *Maia*.

Adams has also collaborated with the illustrator Nicola Bayley on verse narratives for children, and published an unflattering autobiographical volume covering the first twenty-six years of his life, *The Day Gone By*.

Fiction

Watership Down 1972; *Shardik* 1974; *The Plague Dogs* 1977; *The Girl in the Swing* 1980; *The Iron Wolf and Other Stories* (s) [US *The Unbroken Web*] 1980; *Maia* 1984; *Traveller* 1988

For Children

The Tyger Voyage [with N. Bayley] 1976; *The Ship's Cat* [with A. Aldridge] 1977; *The Bureaucrats* 1985

Non-Fiction

Nature through the Sea [with M. Hooper] 1975; *Nature Day and Night* [with M. Hooper] 1978; *The Watership Down Film Picture Book* 1978; *The Day Gone By* (a) 1990

Edited

Grimm's Fairy Tales 1981; *Richard Adams's Favourite Animal Stories* (s) 1981; *The Best of Ernest Thompson Seton* 1982

(Kareen) Fleur ADCOCK 1934–

Born in Papakura, New Zealand, Adcock was a wartime evacuee for much of her childhood. She was educated in England from 1939 until 1947, when she returned to New Zealand and Wellington Girls' College. She received an MA in classics from Victoria University in Wellington in 1955. She was first married in 1952 to Alistair Campbell, a poet and civil servant, by whom she has two sons. They were divorced in 1958, when Adcock took a post as assistant lecturer in classics at the University of Otago, Dunedin, joining the library staff the next year. She subsequently married the writer Barry Crump, and returned to England in 1963, taking a job at the Foreign and Commonwealth Office Library in London where she worked for the next sixteen years.

Her first book of poems, *The Eye of the Hurricane*, was published in 1964, in New Zealand. Her first British publication, *Tigers*, came in 1967. A less formal collection than its predecessor, reflecting the fashion of the 1960s in the setting of its lines, the book overlaps with her first in its preoccupations: the domestic, the familiar, gardens, children and pets, and the country she had left behind, all observed with a coolly detached eye. Alienated from her homeland and rejecting the call of its landscape in print, she has come to be considered a major poet there. Elsewhere, the classicism and the calm precision of her verse has attracted sometimes stinting praise. *The Incident Book* (1986) marks a move away from the personal to a more narrative form: the section called 'Thatcherland' approaches social commentary.

Adcock has won numerous awards, especially in New Zealand where she was given a National Book Award in 1984. She writes and broadcasts on poetry matters for the BBC, has taken arts fellowships at universities, and contributes to various literary journals between publications of her own. She is the editor of several anthologies, the translator of a collection of medieval verse, and has provided texts and libretti for the composer Gillian Whitehead.

Poetry
The Eye of the Hurricane 1964; *Tigers* 1967; *High Tide in the Garden* 1971; *Modern Poets in Focus 5* [with others] [ed. D. Abse] 1973; *The Scenic Route* 1974; *In Focus* 1977; *A Mordern Tower Reading 5* [with G. Ewart] 1977; *Below Loughrigg* 1979; *The Inner Harbour* 1979; *Selected Poems* 1983; *The Incident Book* 1986; *Meeting the Comet* 1988; *Time Zones* 1991

Translations
The Virgin and The Nightingale: Medieval Latin Poems 1983

Edited
New Poetry 4 [with A. Thwaite] 1978; *The Oxford Book of Contemporary New Zealand Poetry* 1983; *The Faber Book of Twentieth Century Poetry* 1987

AE (pseud. of George William Russell) 1867–1935

An Irish economist, poet, painter, mystic and critic, Russell is best known as a major figure in the Irish intellectual movement, and especially in the Irish literary revival. His beliefs figured strongly in his poetry, which often combined Irish nationalism with mystical imagination, and encompassed Eastern religions.

Born in Lurgan, County Armagh, he studied art in Dublin before making a living as a writer and editor. Active in Irish agrarian and political affairs, he organised the Irish Agricultural Co-operative movement and edited the *Irish Homestead* and the *Irish Statesman*. He also wrote a great many pamphlets on the Irish economy. He was involved in founding the Abbey Theatre in Dublin, and was a friend of **W.B. Yeats**, who encouraged him with *Homeward: Songs by the Way* (1894), his first volume of mystical verse. All his work appeared under the pseudonym 'AE', which evolved from a printer's misprint of the intended 'Æon' on the title-page. He died of cancer in Bournemouth.

Poetry
Homeward 1894; *Deirdre* 1902; *The Divine Vision* 1904; *Gods of War* 1915; *The Interpreters* 1922; *Midsummer Eve* 1928; *The House of the Titans* 1934

Fiction
The Avatars 1933

James (Rufus) AGEE 1909–1955

Although Agee was comparatively unknown during his lifetime, his posthumously published works have earned him a considerable reputation as a talented novelist, film critic and screenwriter. His unsettled life and early death have helped dramatise his image into that of a gifted artist drained of talent and energy by an unappreciative society.

Born into a poor family in Knoxville, Tennessee, he was educated there and in New Hampshire before going to Harvard, from which he graduated in 1932. His principal career was as a journalist, and he served on the staff of *Fortune*, *Time* and the *Nation*, contributing film reviews and features. One of his best-known books, originally intended as an article for *Fortune*, was *Let Us Now Praise Famous Men*, a study of Alabama sharecroppers during the Depression, which was illustrated with photographs by Walker Evans. In 1947 Samuel Barber set a text by Agee for soprano and orchestra. *Knoxville: Summer of 1915* (Op. 24) is an intensely lyrical piece in which Barber's music perfectly complements Agee's nostalgic evocation of American small-town life. It was first performed in 1948.

Apart from his journalism, Agee also wrote poetry, fiction and film scripts, including the one for John Huston's classic *The African Queen* (1951), adapted from the novel by C.S. Forester. His film criticism was collected posthumously along with his screenplays in two volumes as *Agee on Film*. John Huston described Agee as 'the best motion-picture critic this country has ever had'. He is best known, however, for *A Death in the Family*, a semi-autobiographical account of a Tennessee family shattered by the death of the father in a car crash, which was published two years after Agee's own death. The book was in fact left unfinished, with a number of alternative versions of several important passages. The final text had to be collated from these various drafts, a task hampered by Agee's near-illegible handwriting. It is generally agreed that Agee's intentions had been correctly interpreted and the novel was awarded a Pulitzer Prize.

Agee had his first heart attack in 1951, and these became more frequent until he died of one in a taxi in New York. Married three times, he had one son and one daughter.

Fiction

The Morning Watch 1951; *A Death in the Family* 1957; *Four Early Stories* (s) 1964

Non-Fiction

Agee on Film 2 vols 1960; *Letters of James Agee to Father Flye* 1962

Poetry

Permit Me Voyage 1934; *The Collected Poems of James Agee* 1962

Non-Fiction

Let Us Now Praise Famous Men [with W. Evans] 1941

Collections

The Collected Short Prose of James Agee 1962

Robert (Fordyce) AICKMAN 1914–1981

A writer of ghost stories, Aickman was of Scottish and, on his mother's side, partly German-Jewish ancestry (his maternal grandfather, born Bernard Heldmann, was the Victorian mystery writer Richard Marsh, at one time well known for his horror novel *The Beetle*). His father was an eccentric architect, and Aickman grew up in a large house, Langton Lodge, in Stanmore, Middlesex. In his simultaneously revealing and reticent volume of autobiography, *The Attempted Rescue*, he describes the childhood he passed there, pervaded by the memories of Edwardian *luxe* which sometimes feature in his stories, but lonely and unhappy. He attended Highgate School as a weekly boarder, but went to no university, and spent his early manhood living alone at Langton Lodge, as his parents had separated and gone elsewhere. His only literary achievement during these early years of mental torment was to become drama critic for the highbrow magazine *The Nineteenth Century and After*.

He then began to meet women friends in opera and theatre queues, and eventually married Ray Gregorson in 1941, setting up a literary agency with her (he was a conscientious objector, and had won total exemption from war service). In the immediate post-war period he discovered the twin great achievements of his life: the campaign for inland waterways and the ghost story. In 1946, with L.T.C. Rolt (also an author of ghost stories, and of many works on canals and engineering), he set up the Inland Waterways Association, widely considered one of the most successful, as it was one of the most quarrelsome, pressure-groups of recent times. He served as its chairman until 1964; more of the credit for saving Britain's then disused and derelict canals rests with Aickman than with any other single person. His first book of ghost stories was *We Are for the Dark*, published in 1951, to which he and **Elizabeth Jane Howard**, who had also worked for the association, contributed six stories, not identifying who had written which.

After his retirement from active involvement with waterways, he concentrated on his stories, and seven more collections appeared before his death in 1981 (several have also appeared posthumously). His style and subject-matter have been compared, not unjustly, to that of earlier masters such as **M.R. James**, but the shock of the modern world can give his stories a peculiar *frisson*, and there is no doubt that in many tales – 'The Trains', 'Ringing the Changes', 'The Cicerones', 'Meeting Mr Millar', 'The Visiting Star' – he gave new life to what had been, classically, an Edwardian genre. He was also editor of the first eight books of *Fontana Great Ghost Stories* (1964–72), to all but the sixth of which he contributed stimulating introductions.

His other writings include two novels, several books about the waterways, much uncollected criticism and several unperformed plays. His marriage ended in 1957, his wife becoming an Anglican nun, and, as he himself put it, 'loving women, my years became solitary'. He lived for many years in Gower Street, among the last private residents of the area.

Fiction
We Are for the Dark (s) [with E.J. Howard] 1951; *Dark Entries* (s) 1964; *The Late Breakfasters* 1964; *Powers of Darkness* (s) 1966; *Sub Rosa* (s) 1968; *Cold Hand in Mine* (s) 1975; *Tales of Love and Death* (s) 1977; *Painted Devils* (s) 1979; *Intrusions* (s) 1980; *Night Voices* (s) 1985; *The Model* 1987; *The Wine-Dark Sea* (s) 1988; *The Unsettled Dust* (s) 1990

Non-Fiction
Know Your Waterways 1954; *The Story of Our Inland Waterways* 1955 [rev. 1967]; *The Attempted Rescue* 1966; *The River Runs Uphill: a Story of Success and Failure* 1986

Edited
The Fontana Books of Great Ghost Stories (s) vols 1–8 1964–72

Conrad (Potter) AIKEN 1889–1973

Aiken was born in Savannah, Georgia, where he also died. He was brought up in Massachusetts from the age of eleven by a great-great-aunt after his father killed his wife and then himself. Aiken distinguished himself at Middlesex School, Concord, before entering Harvard in 1911, where he shared a class with Walter Lippmann, Robert Benchley and **T.S. Eliot**. The last named was to be a lifelong friend who showed Aiken his early poetry, including those verses later incorporated into *The Waste Land*.

After working as a sports reporter, Aiken, who had a small private income, was able to devote his life to writing. *Earth Triumphant*, his first volume of poems, was published in 1914. As a contributing editor to the *Dial*, Aiken had a long friendship with **Ezra Pound** and other catalystic contemporaries, but was always overshadowed by them. His best-known poetry is contained in his two volumes of *Preludes*, but he was awarded a Pulitzer Prize for his *Selected Poems*, published in 1930, the year he came to live in Rye, England. Aiken was influenced by Freud and William James, but used classical symbolism to celebrate his belief that the human race exists alone in an empty universe. Although Aiken was unfairly dismissed by **Aldous Huxley** as an 'agreeable maker of coloured mists', his philosophy was more approachable through his several volumes of short stories and novels.

After the outbreak of the Second World War, Aiken returned to live in Massachusetts, and published his 'autobiographical narrative' *Ushant* in 1952. Written in the third person with considerable humour, it describes his friendships with **Malcolm Lowry** and Eliot, fictionally called 'Tsetse'. Three times married, Aiken had three children with his first wife Jessie, and a posthumous memoir of himself was written by Clarice, his second wife. The novelist and children's writer Joan Aiken is his daughter.

Poetry
Earth Triumphant and Other Tales in Verse 1914; *The Jig of Forslin* 1916; *Turns and Movies and Other Tales in Verse* 1916; *The Charnel Rose* 1918; *Nocturne of Remembered Spring and Other Poems* 1918; *The House of Dust* 1920; *Punch* 1921; *The Pilgrimage of Festus* 1923; *Priapus and the Pool and Other Poems* 1925; *Poems* [ed. L. Untermeyer] 1927; *Prelude* 1929; *Selected Poems* 1930; *John Deth and Other Poems* 1930; *The Coming Forth by Day of Osiris Jones* 1931; *Preludes for Memnon* 1931; *Landscape West of Eden* 1934; *Time in the Rock* 1936; *And in the Human Heart* 1940; *Brownstone Eclogues and Other Poems* 1942; *The Soldier* 1944; *The Kid* 1947; *The Divine Pilgrim* 1949; *Skylight One* 1949; *Collected Poems* 1953; *A Letter from Li Po and Other Poems* 1955; *The Flute Player* 1956; *Sheepfold Hill* 1958; *Selected Poems* 1961; *The Morning Song of Lord Zero* 1963; *A Seizure of Limericks* 1964; *Preludes* 1966; *Thee* 1967; *The Clerk's Journal* 1971; *Collected Poems 1916–1970* 1970; *A Little Who's Zoo of Mild Animals* 1977

Fiction
Bring! Bring! and Other Stories (s) 1925; *Blue Voyage* 1927; *Costumes by Eros* (s) 1928; *Gehenna* (s) 1930; *Great Circles* 1933; *Among the Lost People* (s) 1934; *King Coffin* 1935; *A Heart for the Gods of Mexico* 1939; *Conversation* 1940; *The Short Stories* (s) 1950; *The Collected Short Stories* (s) 1960

Non-Fiction
Scepticisms: Notes on Contemporary Poetry 1919; *Ushant: an Essay* (a) 1952; *A Reviewer's ABC: Collected Criticism from 1916 to the Present* [ed. R.A. Blanshard] 1958; *Selected Letters* [ed. J. Killorin] 1978

Edited
Modern American Poetry 1922 [rev. 1927; rev. as *Twentieth Century American Poetry* 1945, 1963]; Emily Dickinson *Selected Poems* 1924; *An Anthology of Famous English and American Poetry* [with W.R. Benét] 1945

For Children
Cats and Bats and Things with Wings 1965; *Tom, Sue, and the Clock* 1966

Biography
Conrad Aiken by Edward Butsche 1988

Edward (Franklin) ALBEE 1928–

Born in Washington, DC, Albee was abandoned by his natural parents and adopted a fortnight after his birth by the theatrical impresario Reed Albee. Named after his adoptive grandfather, he grew up in the affluence of Manhattan and Westchester County. He was expelled from Lawrenceville School, New Jersey, and left a military academy before finally graduating from Choate School, Connecticut, in 1946. But after a year at Trinity College, Hartford, Connecticut, 'the college suggested that I not come back, which was fine with me'. In 1948 he moved to Greenwich Village at the start of a decade during which he wrote with no success and held a variety of jobs – office boy in an advertising agency, salesman in Bloomingdale's, continuity writer for a radio station, and so forth. He travelled to Italy with the proceeds of a trust fund left by his grandmother, and served in the US army. A few weeks short of his thirtieth birthday, he left his job as a messenger for Western Union and wrote his first one-act play, *The Zoo Story*, in three weeks.

Rejected by Broadway producers, the play found its way via a friend to a German publisher, and was premiered with **Samuel Beckett**'s *Krapp's Last Tape* in Berlin in 1959. Thus, Albee saw his first play performed in German, a language he does not speak. A steady stream of one-act plays followed in critically praised off-Broadway productions. Most notable is *The American Dream*, a dissection of family life in which the tyrannical Mommy and the submissive Daddy, and the figure of their dead adopted child, echo something of the tensions of Albee's own upbringing. A variation of the theme recurs with the fantasy child 'exorcised' in his first three-act play, *Who's Afraid of Virginia Woolf?*, which opened on Broadway in 1962. The play was subsequently filmed in 1966 by Mike Nicholls, with Richard Burton and Elizabeth Taylor giving memorable performances as the feuding couple, George and Martha.

The success of the play helped Albee to establish in 1963 the Playwrights Unit which assisted the professional production of new and unknown writers such as **Sam Shepard**. In 1971 he founded the William Flanagan Center for Creative Persons on Long Island. Identified with the Theatre of the Absurd, Albee published in 1962 an attack on the complacency of mainstream Broadway entitled 'Which Theatre is the Absurd One?' His plays after *Virginia Woolf* have rarely enjoyed the same success, though *A Delicate Balance* and *Seascape* have both won Pulitzer Prizes, and *Three Tall Women* was hailed by critics as his finest work in thirty years. The play again draws on his family background to present a portrait of his adoptive mother; it was first produced at the English Theatre in Vienna, and then on Broadway and in London, and won a Pulitzer Prize in 1994. Albee has adapted novels for the stage, notably **Vladimir Nabokov**'s *Lolita* and his friend **Carson McCullers**'s *The Ballad of the Sad Café*.

Plays

The Zoo Story 1959 [pub. 1960]; *The Death of Bessie Smith* 1960; *Fam and Yam* 1960 [pub. 1961]; *The Sandbox* 1960; *The American Dream* [with J. Hinton, Jr] 1961; *Bartleby* [with J. Hinton, Jr] 1961 [np]; *Who's Afraid of Virginia Woolf?* 1962; *The Ballad of the Sad Café* [from C. McCullers's story] 1963; *Tiny Alice* 1964 [pub. 1965]; *Malcolm* [from J. Purdy's novel] 1966; *A Delicate Balance* (st) 1966, (r) 1976; *Breakfast at Tiffany's* [from T. Capote's story] 1966 [np]; *Everything in the Garden* [from G. Cooper's play] 1967 [pub. 1968]; *Box-Mao-Box* 1968 [pub. 1969]; *All Over* 1971; *Seascape* 1975; *Counting the Ways* 1976 [pub. 1977]; *Listening* (r) 1976, (st) 1977; *The Lady from Dubuque* 1980; *Lolita* [from V. Nabokov's novel] 1981; *The Man Who Had Three Arms* 1982; *Faustus in Hell* 1985; *Marriage Play* 1986; *Three Tall Women* 1992

Fiction

Straight through the Night 1989

Non-Fiction

Conversations with Edward Albee 1988

Richard (i.e. Edward Godfrey) ALDINGTON 1892–1962

Born in Portsmouth, Aldington grew up in Dover ('Dullborough'), where his father, author of the novel *The Queen's Preferment* (1896), practised law. His mother, whom he was to lampoon as Mrs Winterbourne in *Death of a Hero*, was the author of what Aldington described as 'inconceivable novels', with titles such as *Love Letters to a Soldier*. Aldington adopted the name Richard while a boy, and was educated at Dover College and University College, London. His family had moved to the capital in 1911, and Aldington, who was from an early age a keen admirer of the great English poets (Keats in particular) and the Elizabethan dramatists, was to describe the Charing Cross Road and its bookshops as his true *alma mater*. Literature was a welcome alternative to what he saw as the suffocating drabness of everyday life, and he began to earn his living in journalism, writing sports reports, reviewing and translating.

He became an associate of **Ezra Pound**, Hilda Doolittle (the poet **H.D.**, whom he married in 1913) and T.E. Hulme, and made his literary reputation as a poet, becoming a leading exponent of Imagism. He was associated with several periodicals, such as *New Age*, the *Egoist* and the Chicago-based *Poetry*. He also helped to launch **T.S. Eliot**'s career, and became the friend and champion – and all to often the subsequent enemy and denigrator – of many writers. He was conscripted as a private in 1916 and served as an NCO and, latterly, a second lieutenant in France during the First World War, an experience he was to recall with vividness and savagery in the novel for which he is best remembered, *Death of a Hero*. Although he was gassed in the trenches, the real damage appears to have been psychological. His marriage to H.D., which was undermined by infidelities on both sides, did not survive the war, but, after a necessary estrangement, the two writers remained on friendly terms, Aldington even sending his bisexual former wife art postcards of voluptuous female nudes. Subsequent romantic entanglements eventually led to a second marriage (to Netta Patmore, the daughter-in-law of his long-time mistress, Brigit Patmore) and to almost permanent indigence, since Aldington was obliged to pay out large sums in a divorce settlement. He left England in 1928 to travel around Europe and, harried by lawyers and tax officers, produced book after book in order to keep in funds: he churned out at least one volume, many written in great haste, almost every year of his adult life.

Inevitably, this vast mass of work is variable in quality, but some of the early poetry is exceptionally fine; *Death of a Hero*, although descending far too often to coarse sarcasm, is an energetic and important contribution to the literature of the First World War; and amongst other notable books are characteristically pugnacious biographies of **D.H. Lawrence, Norman Douglas** and, most controversially of all, T.E. Lawrence.

Aldington spent the Second World War in the USA and in 1946 settled permanently in France. He was driven by a curious, self-defeating daemon and, although capable of great generosity, was racked by professional jealousy, believing, like his equally combative friend **Wyndham Lewis**, that he was an outsider, spurned by the British literary establishment. He gave the impression, both in his life and his work, of a punch-drunk boxer up against the ropes, hitting out ineffectually in all directions. Nancy Cunard described his pen as 'an implement for spleen', and although he undoubtedly had considerable gifts, in the end he was unable, through temperament and circumstances, to make the best use of them.

Poetry

Images (1910–1915) 1915 [US *Images Old and New* 1916]; *The Love Poems of the Myrrhine and Konallis* 1917; *Reveries* 1917; *Images* 1919; *Images of Desire* 1919; *Images of War: a Book of Poems* 1919 [rev. 1919]; *War and Love (1915–1918)* 1919; *The Berkshire Kennet* 1923; *Exile and Other Poems* 1923; *A Fool i' the Forest* 1924; *The Love of the Myrrhine and Konallis and Other Prose Poems* 1926; *Hark the Herald* 1928; *Collected Poems* 1929; *The Eaten Heart* 1929; *A Dream in the Luxembourg* 1930; *Love and the Luxembourg* 1930; *Movietones: 1928–1929* 1932; *Collected Poems 1915–1923* 1933; *The Poems of Richard Aldington* 1934; *Life Quest* 1935; *The Crystal World* 1937; *The Complete Poems of Richard Aldington* 1948; *The Poetry of Richard Aldington* [ed. N.T. Gate] 1975

Fiction

Death of a Hero 1929; *At All Costs* (s) 1930; *Roads to Glory* (s) 1930; *Two Stories* (s) 1930; *The Colonel's Daughter* 1931; *Last Straws* (s) 1931; *Stepping Heavenward* (s) 1931; *Soft Answers* 1932; *All Men Are Enemies* 1933; *Women Must Work* 1934; *Very Heaven* 1937; *Rejected Guest* 1939; *The Romance of Casanova* 1946 [UK 1947]

Non-Fiction

Literary Studies and Reviews 1924; *Voltaire* 1925; *French Studies and Reviews* 1926; *D.H. Lawrence: an Indiscretion* 1927; *Rémy de Gourmont: a Modern Man of Letters* 1928; *Balls and Another Book for Suppression* 1930; *The Squire* 1934; *D.H. Lawrence: a Complete List of All His Works together with a Critical Appreciation* [c. 1935?]; *Artifex: Sketches and Ideas* 1935; *W. Somerset Maugham: an Appreciation* 1939; *Life for Life's Sake* (a) 1941; *The Duke: Being an Account of the Life and Achievement of Arthur Wellesley* 1943 [UK *The Duke of Wellington* 1946]; *Four English Portraits 1801–1851* 1948; *Jane Austen* 1948; *The Strange Life of Charles Waterton 1782–1865* 1949; *D.H. Lawrence: an Appreciation* 1950; *D.H. Lawrence: Portrait of a Genius, but ...* 1950; *Ezra Pound and T.S. Eliot* 1954; *Lawrence L'Imposteur: T.E. Lawrence, the Legend and the Man* 1954 [UK *Lawrence of Arabia: a Biographical Inquiry* 1955]; *Pinorman: Personal Recollections of Norman Douglas, Pino Orioli and Charles Prentice* 1954; *A.E. Housman and W.B. Yeats: Two Lectures* 1955; *Introduction to Mistral* 1956; *Frauds* 1957; *Portrait of a Rebel: the Life and Work of Robert Louis Stevenson* 1957; *Rome: a Book of*

Photographs 1960 [UK rev. as *A Tourist's Rome* 1961]; *D.H. Lawrence in Selbstzeugnissen und Bilddokumenten* 1961; *Farewell to Memories* 1963; *Richard Aldington: Selected Critical Writings* [ed. A. Kershaw] 1970; *A Passionate Prodigality: Letters to Alan Bird from Richard Aldington 1949–1962* [ed. M.J. Benkowitz] 1975; *Literary Lifelines: the Richard Aldington–Lawrence Durrell Correspondence* [ed. I.S. McNiven, H.T. Moore] 1981

Edited

A Book of Characters 1924; *Fifty Romance Lyric Poems* 1928; D.H. Lawrence: *Last Poems* [with G. Orioli] 1932, *The Selected Poems* 1934, *The Spirit of Place: an Anthology Compiled from the Prose of D.H. Lawrence* 1935, *Selected Letters* 1950; *The Portable Oscar Wilde* [UK *Oscar Wilde: Selected Works*] 1946; *The Viking Book of Poetry of the English Speaking World* 1947 [UK *Poetry of the English Speaking World* 1961]; Walter Pater *Selected Works* 1948; *The Religion of Beauty: Selections from the Aesthetes* 1950

Translations

The Poems of Anyte of Tegea 1915; Feodor Sologub *The Little Demon* [with J. Cournos] 1916; Folgore Da San Gemignano *The Garland of Months* 1917, *A Wreath for San Gemignano* 1945; *Latin Poems of the Renaissance* 1919; *Greek Songs in the Manner of Anacreon* 1920; *The Poems of Meleager of Gadara* 1920; Carlo Goldoni *The Good-Humoured Ladies* 1922; *French Comedies of the Eighteenth Century* 1923; Cyrano de Bergerac *Voyages to the Moon and Sun* 1923; *A Book of Characters from Theophrastus* 1924; Laclos *Les Liaisons Dangereuses* 1924; Liégeois *The Mystery of the Nativity* 1924; Pierre Custot *Sturly* 1924; Antoine De La Sale *The Fifteen Joys of Marriage* 1926; Voltaire: *Candide and Other Romances* 1927, *Letters of Voltaire and Frederick the Great* 1927, *Letters of Voltaire and Madame du Deffand* 1927; Madame de Sevigné *Letters of Madame de Sevigné to Her Daughter and Her Friends* 1927; Julien Benda *The Great Betrayal* 1928; *Fifty Romance Lyric Poems* 1928; Rémy de Gourmont: *Selections from All His Works* 1928, *Letters to the Amazon* 1931; Euripides *Alcestis* 1930; *The Decameron of Giovanni Boccaccio* 1930; Gérard de Nerval *Aurelia* 1932; *Great French Romances* 1946; Felix Guirard [ed.] *Larousse Encyclopedia of Mythology* [with D. Ames] 1959

Plays

Life of a Lady [with D. Patmore] 1936; *Seven Against Reeves* 1938

Biography

Richard Aldington by Charles Doyle 1989

Brian (Wilson) ALDISS 1925–

Aldiss was born in East Dereham, Norfolk, into a family which had been recorded in the county for 400 years. His father was an outfitter and, when Aldiss was young, the family moved to Norwich and then to Devon. He received a middle-class education at Framlingham College and West Buckland College, and then volunteered for the Second World War, serving in the Royal Signals Corps in the Far East, an experience he had drawn on for his 'Horatio Stubbs' series of novels, the first of which was *The Hand-Reared Boy* (the title is a reference to masturbation). After demobilisation in 1947, Aldiss took work in a bookshop in Oxford and began to write. He contributed a series of articles to the trade journal the *Bookseller* under the pseudonym of Peter Pica, and Faber (in competition with several other publishers) commissioned him to turn these into his first book, which appeared in 1955 as *The Brightfount Diaries*. In the following year he won third prize in a science-fiction competition, and decided to become a full-time writer. Since then, he has been literary editor of the *Oxford Mail* (1958–69), editor of the science-fiction magazine *SF Horizons* from 1964, and the compiler of many science-fiction anthologies, but has concentrated mainly on his own writing, producing roughly a novel a year since his first in the science-fiction field, *Non-Stop* (1958).

For the first fifteen years of his career, he wrote mainly science fiction, originally in a traditional mould but becoming more experimental and complex during the 1960s under the influence of **Michael Moorcock**'s *Two Worlds* magazine. The Science Fiction Association, of which he had been the first president, voted him Britain's most popular science-fiction writer in 1969. Since the early 1970s he has turned increasingly to 'straight' fiction, although *The Malacia Tapestry* was an interesting attempt to mix science fiction with the historical novel, while *The Helliconia Trilogy* in the early 1980s was a distinguished return to his earlier genre. *Forgotten Life* is another novel drawing on his war experience; *Bury My Heart at W.H. Smith's* is an autobiography. He has also written a critical history of science fiction, *Billion Year Spree* (revised as *Trillion Year Spree*), and is known for short stories and travel books. He is notable for his contributions to authors' organisations, having been chairman of the Society of Authors and a Booker Prize judge in 1981. He has married twice and has four children.

Fiction

The Brightfount Diaries 1955; *Space, Time and Nathaniel* (s) 1957; *Non-Stop* 1958 [US *Starship*

1959]; *Canopy of Time* (s) 1959 [US rev. as *Galaxies Like Grains of Sand* 1960]; *Vanguard from Alpha* 1959 [UK *Equator* 1961]; *Bow Down to Nul* 1960 [UK *The Interpreter* 1961]; *The Male Response* 1961; *Primal Urge* 1961; *The Long Afternoon of Earth* 1962 [UK rev. as *Hothouse* 1962]; *The Airs of Earth* (s) 1963; *The Dark Light Years* 1964; *Greybeard* 1964; *Starswarm* (s) 1964; *Earthworks* 1965; *The Saliva Tree and Other Strange Growths* (s) 1966; *An Age* 1967 [US *Cryptozoic!* 1968]; *Report on Probability A* 1968; *Barefoot in the Head* 1969; *Intangibles Inc.* (s) 1969; *The Hand-Reared Boy* 1970; *The Moment of Eclipse* 1970; *Neanderthal Planet* (s) 1970; *A Soldier Erect* 1971; *The Book of Brian Aldiss* 1972 [UK *The Comic Inferno* 1973]; *The Eighty-Minute Hour* 1974; *Frankenstein Unbound* 1974; *Excommunication* (s) 1975; *The Malacia Tapestry* 1976; *Brothers of the Head* 1977; *Last Orders* (s) 1977; *Enemies of the System* 1978; *A Rude Awakening* 1978; *New Arrivals, Old Encounters* (s) 1979; *Life in the West* 1980; *Moreau's Other Island* 1980 [US *An Island Called Moreau* 1981]; *The Helliconia Trilogy*: *Helliconia Spring* 1982, *Helliconia Summer* 1983, *Helliconia Winter* 1985; *Hothouse* 1984; *Seasons in Flight* 1984; *... and the Lurid Glare of the Comet* 1985; *Cracken at Critical* 1987; *The Magic of the Past* 1987; *Non-Stop* 1987; *Ruins* 1987; *Best SF Stories of Brian W. Aldiss* 1988; *Earthworks* 1988; *Forgotten Life* 1988; *Galaxies Like Grains of Sand* 1989; *A Romance of the Equator* 1989; *Science Fiction Blues* 1989; *Starswarm* 1990; *Dracula Unbound* 1991; *Frankenstein Unbound* 1991; *Remembrance Day* 1993; *A Tupolev Too Far* 1993

Edited

Penguin Science Fiction 1961; *Introducing SF* 1964; *Best Fantasy Stories* 1962; *More Penguin Science Fiction* 1962; *Yet More Penguin Science Fiction* 1964; *Nebula Award Stories II* 1967; *The Year's Best Science Fiction* [with H. Harrison] 8 vols 1968–75; *All About Venus* [UK *Farewell Fantastic Venus*] 1968; *The Astounding-Analog Reader* 2 vols 1972, 1973; *Penguin Science Fiction Omnibus* 1973; *Space-Opera* 1974; *Evil Earths: an Anthology of Way-Back-When Futures* 1975; *Hell's Cartographers* 1975; *Science-Fiction Art: the Fantasies of SF* 1975; *Space Odysseys* 1975; *Decade* [with H. Harrison] 3 vols 1975–7; *Galactic Empires* 1976; *Enemies of the System* 1978; *Perilous Planets* 1978; *The Penguin World Omnibus of Science Fiction* [with S.J. Lundwall] 1986; H.G. Wells *The Island of Dr Moreau* 1993

Non-Fiction

Cities and Stones: a Traveller's Yugoslavia 1966; *The Shape of Further Things* 1970; *Billion Year Spree: the History of Science Fiction* 1973 [rev. as *Trillion Year Spree* 1986]; *This World and Nearer Ones* 1979; *Science Fiction Quiz* 1983; *The Penguin Masterquiz* [with others] 1985; *Pale Shadow of Science* 1986; *Bury My Heart at W.H. Smith's* (a) 1990; *Home Life with Cats* [with K. van Heerdin, D. Morins] 1992

Plays

Distant Encounters 1978

Poetry

Pile 1979; *Farewell to a Child* 1982

Nelson (Abraham) ALGREN 1909–1981

Born in Detroit, Algren grew up in Chicago and spent a good part of his youth familiarising himself with the low-life which would feature in his novels. He left the University of Illinois in 1931 with a degree in journalism and moved south, working as a door-to-door salesman. In 1933, stranded outside a derelict petrol station in Texas, he wrote his first short story, 'So Help Me', which was published in *Story*. He subsequently became editor of the *New Anvil*, an experimental magazine. He worked for the Chicago Board of Health, involved in the control of venereal diseases, and during the Second World War served as a private with the Medical Corps.

His first novel, *Somebody in Boots*, was dedicated to 'those innumerable thousands; the Homeless Boys of America'. It was a study of depressed youth in a hostile environment where 'even children drank and smoked', written in the classic documentary style of the 1930s: it sold a mere 750 copies. In 1942 *Never Come Morning* was published: set amongst the Poles of Chicago's West Side, the novel established Algren as a young writer of great promise. Its repellent hero, Bicek, cheats and murders, allowing his girlfriend to be gang-raped, until he is finally betrayed by his clan. Algren was pigeonholed as one of the Chicago realists, a depictor of proletarian depravity along with **James J. Farrell** and **Richard Wright**. He was admired by **Ernest Hemingway**, with whom he spent a Christmas in Cuba, and by the French Existentialists. He was twice married, but his most significant relationship was with Simone de Beauvoir, who, when she arrived in the USA in 1946, headed for Chicago to look him up. At the time Algren was writing *The Man with the Golden Arm*, which would be a bestseller and win a National Book Award in 1950. On the first evening Algren took her to visit a gang of junkie thieves, and they spent the next afternoon touring the bars of the Polish district. De Beauvoir wrote: 'He lived in a hovel – without a bathroom or a refrigerator alongside an alley of steaming trash cans.' They travelled to Latin America together, and in 1949 Algren went to stay with her in Paris where he met Jean-Paul Sartre. In

1950 de Beauvoir returned to the USA and spent a summer with Algren in a cottage on the shore of Lake Michigan. But the affair broke down and they were apart until 1960, when they met in Seville and travelled together to Istanbul, Crete and Athens. De Beauvoir wrote of Algren in her diary, *America Day by Day*, and the semi-autobiographical novel, *The Mandarins* (1956), is dedicated to him. On the publication of her autobiography, *Force of Circumstance*, Algren responded furiously with a review in *Harper's* entitled 'The Question of Simone de Beauvoir'.

In later years, Algren taught creative writing at the universities of Iowa and Florida and wrote a regular column for the *Chicago Free Press*. He left Chicago in 1975 and moved to Long Island, where he lived until his death. *The Man with the Golden Arm* and *A Walk on the Wild Side* were subsequently made into successful films, both of which caused some controversy for their depiction of drug addiction and female homosexuality respectively. *Chicago: City on the Make* and *Who Lost an American?* are books of reportage, the first a portrait of Algren's own city, the latter an account of travels through the less respectable quarters of European and American cities.

Fiction
Somebody in Boots 1935; *Never Come Morning* 1942; *The Neon Wilderness* (s) 1947; *The Man with the Golden Arm* 1949; *A Walk on the Wild Side* 1956; *The Last Carousel* (s) 1973

Non-Fiction
Chicago: City on the Make 1951; *Who Lost an American?* 1963; *Notes from a Sea Diary: Hemingway All the Way* 1965

Edited
Nelson Algren's Own Book of Lonesome Monsters 1962

Biography
Nelson Algren by Bettina Drew 1989

Walter (Ernest) ALLEN 1911–1995

One of the last of the old-fashioned men-of-letters, who lived by writing (with intervals of teaching) ever since graduation in 1932. Allen was born in Birmingham, one of four sons of a silversmith, a working-class man of wide culture. He won a scholarship to King Edward's School, Aston, and then read English at Birmingham University. He began to write for local papers and to broadcast on BBC Midland Region, forming part of a 'Birmingham school' of writers of the mid-1930s which also included **Louis MacNeice** and **John Hampson**. He taught briefly at his old school during 1934, and in 1935 became a lecturer at the State University of Iowa for the summer season, the first of a number of teaching spells at American universities and colleges. In 1937 he went to 'the young writer's traditional bed-sitting-room in Bloomsbury', becoming a reader for Metro-Goldwyn-Mayer and publishing his first novel, *Innocence Is Drowned*, in 1938. During the war he worked in staff jobs in the aircraft industry in Bristol and Birmingham and began to write literary criticism for **John Lehmann**'s influential *Penguin New Writing*. In 1944 he married Peggy Yorke, and they had four children. After the war he became closely associated with the *New Statesman* under Kingsley Martin's editorship, progressing from reviewer to literary editor in 1960–1.

He published eight novels, which mainly deal with traditional working-class life and which, after the first three, published in successive years, appeared at progressively rarer intervals; his first novel after *All in a Lifetime* (1959), a penetrating study of his father and perhaps his best fiction, was *Get Out Early*, published in 1986. He was also well known for two classic studies of the novel, *The English Novel* and *Tradition and Dream*, as well as for his critical biographies of **Arnold Bennett** and George Eliot; his taste was among the most eclectic and informed of its time. He ended his academic career as professor at the New University of Ulster from 1967 to 1973, which was followed by the last of his teaching posts in America until 1975. *As I Walked Down New Grub Street* is an entertaining volume of literary memoirs.

Fiction
Innocence is Drowned 1938; *Blind Man's Ditch* 1939; *Living Space* 1940; *Rogue Elephant* 1946; *Dead Man Over All* 1951; *All in a Lifetime* 1959; *Get Out Early* 1986; *Accosting Profiles* 1989

Non-Fiction
The Black Country 1946; *Arnold Bennett* 1948; *Reading a Novel* 1949; *Joyce Cary* 1953; *The English Novel* 1954; *The Novel Today* 1955; *George Eliot* 1964; *Tradition and Dream* 1964; *The British Isles in Colour* 1965; *The Urgent West: an Introduction to the Idea of the United States* 1969; *Some Aspects of the American Short Story* 1973; *As I Walked Down New Grub Street* (a) 1981; *The Short Story in English* 1981

For Children
The Festive Boiled-Potato Cart and Other Stories 1948; *Six Great Novelists* 1955

Edited
Writers on Writing 1948; *Transatlantic Crossing: American Visitors to Britain and British Visitors to America in the Nineteenth Century* 1971; Wyndham Lewis *The Roaring Queen* 1973

Margery (Louise) ALLINGHAM 1904–1966

Allingham was born at Ealing, into a family of writers: John Till Allingham, the early nineteenth-century author of melodramas, was a forebear; a later John Allingham wrote boys' school stories in the 1890s; and her father, H.J. Allingham, contributed fiction to all the serial magazines of the day, and began to train his daughter as a writer when she was seven (she liked to call herself 'one of the last of the real trained professional writers'). She grew up in Essex, attended the Perse School for Girls, and published her first book, a pirate novel called *Blackkerchief Dick*, when she was nineteen. She wrote many magazine short stories, serials and reviews as well as the detective novels for which she is most famous. In 1927, she married Philip Youngman Carter, the journalist and illustrator, who collaborated with her on much pulp fiction and designed jackets for many of her books. After some years in London, they moved to the Essex village of Tolleshunt D'Arcy (retaining a flat in Bloomsbury) where they lived an English country life until her death. *The Oaken Heart*, her only non-fiction book, is an account of contemporary life in an English village.

She introduced her detective, the aristocratic but self-effacing Albert Campion, in *The Crime at Black Dudley*; the early books about him are slight and much defaced by snobbery (he has hinted connections with royalty, and his Cockney servant, Magersfontein Lugg, is characterised as 'a Vulgarian in the service of Mr Campion'), but from *Flowers for the Judge* onwards, the books rapidly gain distinction in both style and insight, culminating in the remarkable *The Tiger in the Smoke*, a chilling story of a killer adrift in the London fog, and one of the books which, in its excellence, seems to surpass the limits of the detective genre.

Fiction

Blackkerchief Dick 1923; *The Crime at Black Dudley* 1928; *The White Cottage Mystery* 1928; *Mystery Mile* 1930; *Look to the Lady* 1931; *Police at the Funeral* 1931; *Sweet Danger* 1933; *Death of a Ghost* 1934; *Flowers for the Judge* 1936; *Six Against the Yard* [with others] 1936; *The Case of the Late Pig* 1937; *Dancers in Mourning* 1937; *Mr Campion, Criminologist* (s) 1937; *The Fashion in Shrouds* 1938; *Mr Campion and Others* (s) 1939; *Black Plumes* 1940; *Traitor's Purse* 1941; *Dance of the Years* 1943; *Coroner's Pidgin* 1945; *Wanted: Someone Innocent* (s) 1946; *The Case Book of Mr Campion* [ed. E. Queen] 1947; *More Work for the Undertaker* 1948; *Deadly Duo* (s) 1949; *The Tiger in the Smoke* 1952; *No Love Lost* (s) 1954; *The Beckoning Lady* 1955; *Hide My Eyes* 1958; *The China Governess* 1962; *The Mind Readers* 1965; *Cargo of Eagles* [completed by Y. Carter] 1968; *The Allingham Case-Book* 1969; *The Allingham Minibus* 1973

Plays

Dido and Aeneas 1922 [np]; *Water in a Sieve* 1925

Non-Fiction

The Oaken Heart 1941

Biography

Margery Allingham by Julia Thoroughgood 1991

Kenneth ALLOTT 1912–1973

Allott was born in Glamorganshire and spent much of his youth in Cumberland before reading English at Durham University. He continued to study at St Edmund Hall, Oxford, where he co-edited the magazine *Programme* and worked with Tom Harrisson's Mass-Observation. His poems first came to notice in 1935 in **Geoffrey Grigson**'s influential journal, *New Verse*, of which Allott subsequently became assistant editor, from 1936 to 1939. He also made appearances in similar showcases of the period, such as *Contemporary Poetry and Prose* and **Julian Symons**'s *Twentieth Century Verse*.

He published only two collections of verse in his lifetime, *Poems* (1938), followed five years later by *The Ventriloquist's Doll*. Even so, on the strength of his first volume alone, he was included in Francis Scarfe's early survey of the poets of the mid-century, *Auden and After* (1942). His early poetry displayed the customary influences on his generation – chiefly **T.S. Eliot** and **W.H. Auden** – but were also tinged with surrealism, in which he resembles **David Gascoyne**. Like Auden, Allott is witty and stylish, an intellectual poet, who addresses his readers directly. He was nevertheless unimpressed by Auden's Group Theatre collaborations with **Christopher Isherwood**. 'I am sure Auden is a good poet,' he wrote in the 'Auden Double Number' of *New Verse* in November 1937. 'I am equally sure that he is a wrong-headed dramatist.' Grigson printed five of Allott's poems in his *New Verse* anthology of 1939, including his best-known poem, 'Lament for a Cricket Eleven'. Allott's other pre-war work was a comic novel, *The Rhubarb Tree*, written in collaboration with Stephen Tait and notable for the fact that its heroine only ever appears as a voice on the telephone.

After the war, Allott turned his attention to the poetry of others, editing several volumes of such

Victorian writers as Matthew Arnold, Robert Browning and W.M. Praed, *The Penguin Book of Contemporary Verse* (1950), and the five volumes of *The Pelican Book of English Verse*. He wrote studies of the work of Jules Verne and **Graham Greene**, the latter with Miriam Farris, an academic (and editor of *The Longman's Annotated Keats*) who became his wife. He wrote a play, *The Publican's Story*, and his adaptation (with Stephen Tait) of **E.M. Forster**'s *A Room with a View* was produced at the Cambridge Arts Theatre in 1950. In 1964 he became the A.C. Bradley Professor of Modern English Literature at the University of Liverpool, a post he held until his death. A volume of his *Collected Poems* appeared posthumously.

Poetry

Poems 1938; *The Ventriloquist's Doll* 1943; *Collected Poems* 1975

Edited

Poems of William Habington 1948; *The Penguin Book of Contemporary Verse* 1950; Matthew Arnold: *Five Uncollected Essays of Matthew Arnold* 1953, *Matthew Arnold: a Selection of His Poems* 1954, *The Poems of Matthew Arnold* 1965; *Selected Poems of Winthrop Mackworth Praed* 1953; *The Pelican Book of English Verse* 5 vols 1956; *Victorian Prose, 1830–1880* [with M. Allott] 1956; *Selected Poems of Browning* 1967; *Writers and Their Background: Matthew Arnold* 1976

Non-Fiction

Jules Verne 1940; *The Art of Graham Greene* [with M. Farris] 1951.

Fiction

The Rhubarb Tree [with S. Tait] 1937

Plays

A Room with a View [from E.M. Forster's novel; with S. Tait] 1950 [pub. 1951]; *The Publican's Story* 1953 [np]

Lisa ALTHER 1944–

Alther was born in Kingsport, Tennessee, an industrialised town in the region that provides the setting for her major novels. The daughter of a surgeon, she was educated in the American state system and at Wellesley College, Massachusetts, gaining her BA in 1966. In the same year she married Richard Alther, a painter and former advertising copywriter, and they lived in New York City where she worked as an editorial assistant at Atheneum Publishers. After moving to a farm in Vermont, where their daughter Sara was born, Alther got down to some serious writing: she has spoken of writing for as much as fourteen or sixteen hours at a time, 'getting into these highs and just staying there'. She wrote non-fiction for magazines and was for several years a staff writer at Garden Way Publishers, Vermont.

She was slow to break through with fiction and had accumulated some 250 rejection slips before her massively bestselling *Kinflicks* was published in 1976. It is a woman's picaresque, centred on twenty-seven-year-old Ginny Babcock. Ginny's return home to her terminally ill mother in Tennessee frames a collection of flashbacks that present her social and sexual rites of passage, as she moves from being a football cheerleader in the south to a college girl in the North and a participator in the upheavals of the 1960s. The book was widely praised for its freshness of voice and for its picture of a place and a time, as well as for its candid and funny account of female experience.

Alther is often considered to be a woman's writer, but feminists are divided over her work. **Erica Jong** praised *Kinflicks* as a modern feminist classic, while **Doris Lessing** – who also championed it – was more perspicacious: 'No man could have written it but it is very far from being "a woman's book", and it made me wonder what *Tom Jones* would be like, written now.' Germaine Greer disliked it, declaring that Ginny's adventures 'are chiefly remarkable in that they do not raise her consciousness one notch'.

Alther lives in the privacy of rural Vermont with her feet firmly on the ground (one of her less famous works is *Non-Chemical Pest and Disease Control for the Home Orchard*), and she teaches Southern fiction at St Michael's College, Vermont. Subsequent novels have been more serious in tone and have consolidated her reputation. They benefit from her humane shrewdness, her sense of time and place, and the inexhaustible nature of her main subject: American womanhood. Critics have also found them to be solidly crafted, with a sure manipulation of point of view and the different levels of comedy.

Fiction

Kinflicks 1976; *Original Sins* 1981; *Other Women* 1984; *Bedrock* 1990

Non-Fiction

Non-Chemical Pest and Disease Control for the Home Orchard 1973

A(lfred) ALVAREZ 1929–

Born into a family of Sephardic Jews established in Britain since the seventeenth century, Alvarez grew up in London in the comfortable circumstances created by a businessman grandfather.

He attended Oundle School, going on to Corpus Christi, Oxford, where he took a first in English. He then spent several years alternating between further research at Oxford, research fellowships in the USA, and periods as a freelance writer in London. His undergraduate poems had been published by the Fantasy Press, and from the mid-1950s he began to publish in little magazines and to broadcast on the Third Programme (now Radio 3) of the BBC. His first critical work was *The Shaping Spirit* (1958), a study of modern European poetry, which was closely followed by an important study of Donne. While in the USA he had met Frieda Lawrence, and back in England in 1955 he married her granddaughter, Ursula Barr, with whom he had one son. They divorced in 1961, and in the same year Alvarez attempted suicide. He recovered from this to become one of the foremost critics of the 1960s, exerting enormous influence through his Penguin anthology *The New Poetry*, which promoted the credo that the writer should pursue 'his insights to the edge of breakdown and then beyond it', thus exalting the 'confessional' school of **Robert Lowell, John Berryman** and **Sylvia Plath** and deprecating the 'academic-administrative verse' of the Movement. Alvarez was also able to further his views as poetry critic of the *Observer* from 1956 to 1966, and advisory editor for the Penguin Modern European Poets series from 1965 to 1975.

His own creative career as a poet blossomed later; he did not publish a full collection until *Lost* in 1968, and his production since then has been sporadic although quietly distinguished, seeming to owe more in its formal values to the Movement than his earlier critical position would suggest. In 1971 he published *The Savage God*, a study of suicide, and the 1970s also saw two highly regarded novels, *Hers* and *Hunt*. Although not so prominent a cultural figure in the 1980s as in the 1960s, Alvarez continued to appear frequently on television and write prolifically for the Sunday newspapers, publishing books on subjects as diverse as the game of poker, divorce and the world of North Sea oil. He married in 1966 and has two children.

Poetry

Poems 1952; *The End of It* 1958; *Lost* 1968; *Twelve Poems* 1968; *Penguin Modern Poets 18* [with R. Fuller, A. Thwaite] 1970; *Apparition* 1971; *The Legacy* 1972; *Autumn to Autumn and Selected Poems 1953–1976* 1978

Fiction

Hers 1975; *Hunt* 1978; *Day of Atonement* 1991

Non-Fiction

The Shaping Spirit: Studies in Modern English and American Poets 1958; *The School of Donne* 1962; *Under Pressure: the Artist and Society: Eastern Europe and the USA* 1965; *Beyond all this Fiddle: Essays 1955–67* 1968; *The Savage God: a Study of Suicide* 1971; *Beckett* 1973; *Life After Marriage: Scenes from Divorce* 1982; *The Biggest Game in Town* 1983; *Offshore: a North Sea Journey* 1986; *Feeding the Rat: Profile of a Climber – Mo Antoine* 1988

Edited

The New Poetry 1962; *The Faber Book of Modern European Poetry* 1992

Eric AMBLER 1909–

A Londoner by birth, Ambler was educated at Colfe's Grammar School and London University, where he studied engineering. He then took up an apprenticeship in engineering, and an expert technical background is one feature of his accomplished thrillers. He later became a vaudeville comedian, then wrote advertising copy, and by 1937 was the director of a London advertising agency. He was soon able to give this up, however, for between 1937 and 1940 he wrote six classic thriller novels, including *Uncommon Danger, Epitaph for a Spy* and *The Mask of Dimitrios*. These generally portrayed an innocent Englishman wrestling with intrigue against an expertly realised background of European cities, and gave a new direction to the spy and thriller novel with their realism and (in comparison with previous thrillers) their political radicalism.

From 1940 to 1951 Ambler wrote no more thrillers: he entered the British army as a private and emerged a lieutenant-colonel, serving in a combat filming unit and as Assistant Director of Army Kinematography. After the war he took up screenplay-writing for films and scripted several successes for the Rank Organisation, including *The Cruel Sea* (1953). He took up thriller-writing again in 1951 with *Judgement on Deltchev* and has written more than fifteen post-war thrillers which, in contrast to the earlier novels, have worldwide settings, from Africa to the Far East.

Ambler has married twice, the second time (in 1958) to Joan Harrison, who worked as an assistant to the film director Alfred Hitchcock, collaborating on screenplays for some of his best-known films of the 1930s and 1940s (including *Jamaica Inn* and *Rebecca*, both adapted from novels by **Daphne Du Maurier**, and *Suspicion*), then became an independent producer in Hollywood. After she retired, the Amblers lived for a while in Switzerland before returning to England, where Harrison died in 1994. Ambler's unreliable autobiography was published in 1981 under the deliberately ambiguous title *Here Lies Eric Ambler*.

Fiction

The Dark Frontier 1936; *Uncommon Danger* 1937; *Cause for Alarm* 1938; *Epitaph for a Spy* 1938; *The Mask of Dimitrios* 1939; *Journey into Fear* 1940; *Judgement on Deltchev* 1951; *The Night-Comers* 1956; *The Schirmer Inheritance* 1957; *Passage of Arms* 1959; *The Light of Day* 1962; *The Kind of Anger* 1964; *Dirty Story* 1967; *The Intercom Conspiracy* 1969; *The Levanter* 1972; *Doctor Frigo* 1974; *Send No More Roses* 1977 [as *The Siege of the Villa Lipp* 1978]; *The Care of Time* 1981

Non-Fiction

The Ability to Kill and Other Pieces 1963; *Here Lies Eric Ambler* (a) 1985

Edited

To Catch a Spy: an Anthology of Favourite Spy Stories 1964

Kingsley (William) AMIS 1922–

Amis was born in south London and educated at the City of London School and St John's College, Oxford, completing his university education after service in the army with the Royal Corps of Signals. He taught English at the University of Swansea for twelve years and became a fellow of Peterhouse, Cambridge, before abandoning an academic career to devote himself to writing. His suburban London childhood provides a background for *The Riverside Villas Murder* (one of two detective novels), and his time in Wales is reflected in a number of novels, most notably *The Old Devils*, which won the Booker Prize in 1986.

Amis first became known as a poet, with his two collections *Bright November* and *A Frame of Mind*. The former was published by the Fortune Press, whose eccentric and unreliable proprietor, R.A. Caton, is referred to in several of Amis's novels (his initials changed to L.S. for 'Lazy Sod'), until messily killed off in *The Anti-Death League*. Amis belonged to the short lived, loosely bound group of poets known as the Movement, whose members included **Robert Conquest, Elizabeth Jennings** and Amis's lifelong friend and fellow jazz enthusiast, **Philip Larkin**. They adopted an anti-romantic, witty, rational, down-to-earth tone in their poetry. A similar tone informs Amis's first novel, *Lucky Jim*, which placed the author amongst the Angry Young Men when it was published in 1954. Jim Dixon's grimacing rage against almost everyone is a violently comic counterpart to the dour anger of **John Osborne**'s Jimmy Porter. Amis himself has adopted some of Jim's virulent philistinism for his increasingly reactionary public persona, and delights in lampooning the pretentious, the 'arty', liberals, and the cultural establishment. His political allegiance has shifted radically from left to right, although he once claimed that the shift has been in the country and that he has merely stood still. His much-heralded *Memoirs* was a somewhat thin but entertaining compendium of scabrous anecdotes and character assassinations, in which Amis settled a number of old scores but revealed little of himself.

As well as poetry, novels and short stories, Amis has also written on drink (he is a great imbiber of whisky), food (in a restaurant column for *Harper's and Queen*), science fiction, **Rudyard Kipling**, and **Ian Fleming**'s James Bond, compiling *The James Bond Dossier* and writing (under the pseudonym Robert Markham) *Colonel Sun*, a thriller featuring the famous agent. Other genre novels include his detective books and a ghost story about an alcoholic, *The Green Man*. His bluff, anti-cultural, anti-intellectual stance is also reflected in such anthologies as *The New Oxford Book of Light Verse* and *The Popular Reciter*. Later novels, notably *Stanley and the Women*, have been seen by some commentators as misogynistic, although it might be fairer to say that they feature misogynist characters with whom the author appears to identify. The eponymous Stanley may indeed think that all women are mad, but he is a far from admirable character himself. Amis is chiefly admired for his deployment of language, his ability to capture the way people express themselves, particularly when doing so badly.

He had three children from his first marriage to Hilary Bardwell; their elder son is the painter Philip Amis, their younger son is the novelist **Martin Amis**, whose very different books Amis greatly admires. In 1962 Amis met the novelist **Elizabeth Jane Howard**, whom he left his wife to marry in 1965. Complicated by literary rivalry, the marriage was described by onlookers as 'stormy', but when Howard eventually left Amis in 1980, he was devastated. He has said that leaving his first wife was the worst mistake of his life and he subsequently returned to live with her and her third husband, Lord Kilmarnock. Amis was knighted in 1990.

Fiction

Lucky Jim 1954; *That Uncertain Feeling* 1955; *I Like it Here* 1958; *Take a Girl Like You* 1960; *My Enemy's Enemy* (s) 1962; *One Fat Englishman* 1963; *The Egyptologists* [with R. Conquest] 1965; *The Anti-Death League* 1966; *I Want It Now* 1968; *Colonel Sun* [as 'Robert Markham'] 1968; *The Green Man* 1969; *Girl, 20* 1971; *The Riverside Villas Murder* 1973; *Ending Up* 1974; *The Alteration* 1976; *Jake's Thing* 1978; *Collected Short Stories* (s) 1980; *Russian Hide and Seek*

1980; *Stanley and the Women* 1984; *The Old Devils* 1986; *Difficulties with Girls* 1988; *The Folks That Live on the Hill* 1990; *The Russian Girl* 1992; *Mr Barrett's Secret and Other Stories* (s) 1993

Non-Fiction

New Maps of Hell: a Survey of Science Fiction 1960; *The James Bond Dossier* 1965; *What Became of Jane Austen? and Other Questions* 1970; *On Drink* 1972; *Rudyard Kipling and His World* 1975; *Every Day Drinking* 1983; *How's Your Glass? A Quizzical Look at Drinking* 1984; *Memoirs* (a) 1990

Edited

Spectrum: a Science Fiction Anthology [with R. Conquest] 1961; G.K. Chesterton *Selected Stories* 1972; *Tennyson's Poems* 1973; *Harold's Years: Impressions from the New Statesman and the Spectator* 1977; *The Faber Popular Reciter* 1978; *The New Oxford Book of Light Verse* 1978; *The Golden Age of Science Fiction* 1981; *The Great British Songbook* [with J. Cochrane] 1986

Poetry

Bright November 1947; *A Frame of Mind* 1953; *Poems: Fantasy Portraits* 1954; *A Case of Samples: Poems 1946–1956* 1956; *The Evans Country* 1962; *A Look Around the Estate* 1967; *Collected Poems, 1944–1979* 1979

Martin (Louis) AMIS 1949–

Amis *fils* was born in Swansea, where his father, the novelist **Kingsley Amis**, was teaching. He was educated at (and expelled from) a large number of schools in Britain, Spain and America; at several crammers; and at Exeter College, Oxford, where he obtained a first-class degree in English literature. As a teenager he appeared in the film *A High Wind in Jamaica* (1965), adapted from the novel by **Richard Hughes**, but he was to choose literature rather than acting as a career. He has reviewed for a number of magazines and newspapers (notably the *Observer*), spent three years as editorial assistant on the *Times Literary Supplement* and was an inspired literary editor of the *New Statesman* during the late 1970s.

His first novel, *The Rachel Papers*, published when he was twenty-four, established him not only as an *enfant terrible* of British letters, but also as a literary force to be reckoned with. A graphic account of adolescent desire, it won a Somerset Maugham Award and was later made into an ill-conceived film, from which Amis dissociated himself. Amis might have been describing his own literary persona when he wrote that the novel's teenage narrator has 'one of those fashionably reedy voices, the ones with the habitual ironic twang, excellent for the promotion of oldster unease'. Further shocks were delivered in his second novel, provocatively titled *Dead Babies*. It was pusillanimously reissued in paperback as *Dark Secrets*, but has since reappeared under its original title, and recounts a weekend of drink, drugs and debauchery amongst a group of young people. *Other People* was subtitled 'A Mystery Story' and rose to the technical challenge of seeing life through the eyes of someone unable to identify the world in which she moved. The blisteringly comic *Money*, partly set in the film world of America, introduced John Self, one of several memorable grotesques who populate Amis's fiction – notably the dwarfish Keith Whitehead in *Dead Babies* and the darts-obsessed Keith Talent in *London Fields*. Amis has been a champion of, and been influenced by, American fiction, notably **Vladimir Nabokov** and **Saul Bellow**, whose notion of 'The Moronic Inferno' he borrowed for the title of a collection of pieces about America.

He married in 1984 and said that fatherhood has focused his concern for the future of the world. The theme of the nuclear threat, he has written, 'is a background which then insidiously foregrounds itself', and it informs both *London Fields* and his collection of stories, *Einstein's Monsters*. Like his father, Amis has been criticised by feminists, and it was rumoured that *London Fields*, which features a woman in search of her own murderer (a theme treated some nineteen years earlier, and more succinctly, in **Muriel Spark**'s *The Driver's Seat*), was vetoed for the Booker Prize shortlist by women on the judging panel. *Time's Arrow* did make the Booker shortlist, but only after one of the judges, **Nicholas Mosley**, resigned. Like *Other People*, it is constructed upon a literary device; in this case, time runs backwards. Set partly in the Nazi concentration camps, the book was strongly criticised in some quarters for using the Holocaust for what the *Spectator* described as 'an elegant and trivial fiction'. (The magazine's cover bore the legend 'Designer Gas Ovens', and a furious correspondence ensued.) The novel contained a postscript thanking a number of people who had discussed the Holocaust with the author, but this caused as much controversy as the book itself, in particular Amis's theory about the death of the Italian novelist Primo Levi. Some people greatly admired the novel, while others concurred with **Tom Paulin**'s savage attack, delivered on television, in which he described it as 'bone-headed', 'stupid' and 'morally obtuse'.

In 1994, Amis left his wife for the American writer Isobel Fonseca, a domestic matter which became headline news, partly perhaps because of the author's earlier pronouncements about fatherhood and family. This was followed by another furore when negotiations began for the publication rights of his novel, *The Information*. Amis was originally rumoured to be seeking an advance of some £460,000 for a two-book deal, and subsequently sacked his agent, Pat Kavanagh (wife of his erstwhile close friend **Julian Barnes**, who was also popularly but erroneously believed to be the model for one of the novel's principal characters), in pursuit of a rumoured £500,000. The exact amount Amis was paid was never made public, but the entire affair became a literary *cause célèbre*, which Amis's publishers astutely turned to their advantage in an enormous publicity campaign, with profiles of the author in innumerable newspapers and magazines. What the majority of readers wanted to know, however, was not whether the novel was worth £500,000 of a publishing empire's money, but whether it was worth £15.99 of theirs. Most critics thought that the book deserved to be read, but they were far from unanimous in their praise of this story of rivalry between two bad novelists, one successful the other a failure. The 'information' of the title is the realisation when a man reaches forty that he will die and, although Amis strenuously denied that the book was in any way autobiographical, some felt that this theme of mid-life crisis had more personal than universal relevance. While Amis was as usual praised for his comic scenes and his extravagant use of language, there was a feeling that the sententiousness which had marred *London Fields* was in danger of undermining a talent which nevertheless remains one of the most controversial and admired of the late twentieth century.

Fiction

The Rachel Papers 1973; *Dead Babies* 1975; *Success* 1978; *Other People* 1981; *Money* 1984; *Einstein's Monsters* (s) 1987; *London Fields* 1989; *Time's Arrow* 1991; *Two Stories* (s) 1994; *The Information* 1995

Non-Fiction

My Oxford [ed. A. Thwaite] (c) 1977; *Invasion of the Space Invaders* 1982; *The Moronic Inferno* 1986; *Visiting Mrs Nabokov and Other Excursions* 1993

A(rchie) R(andolph) AMMONS 1926–

Ammons was born on a North Carolina farm, and his family endured considerable hardship during the years of the Depression. He graduated from Whiteville High School in 1943 and worked at a shipyard in Wilmington installing fuel pumps in freighters. In 1944 he joined the navy and saw action in the South Pacific. Under the terms of the GI Bill he studied science at Wake Forest College, graduating in 1949, and then taught for a year at the tiny elementary school in Hatteras, North Carolina.

In 1950 he went to Berkeley to study English and there met the poet Josephine Miles who encouraged him to submit his poetry to magazines. Leaving Berkeley in 1952 without taking a degree, he moved to New Jersey, where he worked for a decade for a manufacturer of biological and medical glassware, rising to the post of vice-president. In a 1974 interview, Ammons recalled: 'I never dreamed of being a Poet poet. When you have something else to do to earn money, you can do just what the hell you want to with your poems.' In this sentiment, Ammons echoes **Wallace Stevens**, who famously worked in the insurance business throughout his life. Ammons, like Stevens, writes in the American Romantic tradition of Emerson and Whitman. His primary theme is man's relationship to nature, and echoes of his rural upbringing can be seen throughout his work.

Ammons published his first poems in the *Hudson Review* in 1953, and two years later, at his own expense, printed 100 copies of his first collection, *Ommateum, with Doxology*. In five years it sold sixteen copies and received one review. The turning-point in his career came in 1964. A reading at Cornell led to his becoming an assistant professor teaching English there, and his second collection, *Expressions of Sea Level*, was more widely noticed than his first. It contains poems such as 'Raft' and 'Hymn', which strive towards realising the Emersonian transcendental vision. Ammons has continued to teach at Cornell, rising to the post of Goldwin Smith Professor of Poetry in 1973. Of his latter collections, *Corsons Inlet*, *Uplands* and *Briefings* have been critically praised, and are discussed in an essay in Harold Bloom's influential study of the Romantic tradition, *The Ringers in the Tower*. *Tape for the Turn of the Year* is one of Ammons's more unusual experiments, a 200-page single poem which was typed on a roll of continuous adding matching tape. It is not included in his *Collected Poems*, which won a National Book Award in 1973. *Garbage* won the same prize in 1995, and his other awards include the Levinson Prize in 1970, the Bollingen Prize in 1974, a MacArthur Fellowship in 1981, and the National Book Critics Circle award in 1982. Ammons was married in 1949 and has one son.

Poetry

Ommateum, with Doxology 1955; *Expressions of Sea Level* 1964; *Corsons Inlet* 1965; *Tape for the*

Turn of the Year 1965; *Northfield Poems* 1966; *Selected Poems* 1968; *Uplands* 1970; *Briefings* 1971; *Collected Poems 1951–1971* 1972; *Sphere* 1974; *Diversifications* 1975; *For Doyle Fosco* 1977; *Highgate Road* 1977; *Poem* 1977; *The Selected Poems 1951–1977* 1977; *The Snow Poems* 1977; *Six-Piece Suite* 1979; *Selected Longer Poems* 1980; *A Coast of Trees* 1981; *Worldly Hopes* 1982; *Lake Effect Country* 1983; *Easter Morning* 1986; *Sumerian Vistas* 1987; *Garbage* 1994

Mulk Raj ANAND 1905–

Anand was born in Peshawar, India, the son of a coppersmith and soldier. He was educated at Khalsa College in Amritsar, the University of Punjab, University College London (where he did his PhD) and Cambridge. He was also a student – and later a lecturer – at the League of Nations School of Intellectual Co-operation at Geneva. During the Second World War he worked as a broadcaster and scriptwriter in the films division of the BBC in London.

Anand was driven to literature at an early age, when he wrote free verse in Punjabi and Urdu railing against God as a means of dealing with the death of a nine-year-old cousin. He wrote his first prose in reaction to the trauma of the suicide of an aunt who had been excommunicated for dining with a Muslim woman. Later, his calf-love for a Muslim girl, who was married, produced more poetry. Even late in life, whenever he picked up his pen, it would usually be to give vent to social or personal anguish. His beliefs were moulded by his reading of the Urdu poet Md. Iqbal, Nietzsche (which confirmed his atheism) and his association in London with **T.S. Eliot, Herbert Read**, John Strachey, Harold Laski and the trade unionist Alan Hutt. Among his early literary influences may be counted Dante, **James Joyce** and Rimbaud. However, the most important influence upon him was Gandhi, who shaped his social conscience and ideas of duty and ethics.

An obviously Gandhian imperative marks his first novel, *Untouchable*, published in 1935 with a preface by **E.M. Forster**, after being turned down by nineteen London publishers. In the central character of eighteen-year-old Bakha, Anand gives a moving portrait of the dignity of labour. The protagonist of his second novel, *The Coolie*, is Munoo, a child-labourer who dies of tuberculosis aged fifteen, while *Two Leaves and a Bud* takes as its subject the exploitation of workers at an Assam tea-plantation. The publication in 1939 of *The Village*, the first volume of the Lai Singh trilogy, marked a new turn in Anand's career. The grimness of earlier books is less in evidence in the trilogy, as the reader is taken through the story of Lai Singh's life, from his acts of adolescent rebellion, through his experiences in the trenches of the First World War, to his return home, where he elopes with the village landlord's daughter and plunges into revolutionary activities. 'My knowledge of Indian life at various levels had always convinced me that I should do a *comédie humaine*,' Anand once said. 'In this the poor were only one kind of outcasts. The middle sections and the rajas were also to be included as a species of untouchables.' Hence the novel that is widely regarded as his best: *Private Life of an Indian Prince*. Although the novel has its origins in the betrayal of a hill-woman with whom the author was romantically involved while married to his first wife (the actress Kathleen van Gelder), it is rightly seen today as a true picture of the state of Indian princely life in the face of gradually diminishing powers.

Anand is credited with being the first Indian writer in English to dispel the myths accumulated around his country by imperfect Western perception. But he does not do so by resorting to mere polemics or adopting a chauvinist stance. Instead he gives us the gritty realism of the Indian condition, suffusing it with gentle humour. Nowhere is this more evident than in his many volumes of short stories. He has also written a play, *India Speaks*, and edited numerous volumes on subjects as diverse as Marx and Engels in India, the Kama Sutra and Aesop's Fables. His non-fiction output is also huge, treating topics ranging from Persian and Indian painting, curries, bridal make-up and Indian theatre to erotic sculpture, Nehru and Tagore.

Since the 1950s, Anand has been intermittently working on a projected seven-volume autobiography, four volumes of which have been published to date. He has been a director of Kutub Publishing, Bombay, and editor of the arts journal *Maro* since 1946. The man who fought with the Republicans in the Spanish Civil War leads a quiet life today at an elegant Cuffe Parade address in Bombay. His high forehead, heavy jowls and Hapsburg lip make his an unmistakable face at the meetings of the Indian Academy of Letters. In his nineties, his profound humanism remains unshaken.

Fiction

The Lost Child and Other Stories (s) 1934; *Untouchable* 1935; *The Coolie* 1936 [as *Coolie* 1945]; *Two Leaves and a Bud* 1937; *Lament on the Death of a Master of Arts* 1939; *The Village* 1939; *Across the Black Waters* 1940; *The Sword and the Sickle* 1942; *The Barber's Trade Union and Other Stories* (s) 1944; *The Big Heart* 1945 [rev. 1980]; *Reflections on the Golden Bed* (s) 1947; *The Tractor and the Corn Goddess and Other Stories* (s) 1947; *Seven Summers* 1951; *Private Life of an Indian Prince* 1953; *The Power of Darkness and Other Stories* (s) 1958; *The Old Woman and the Cow* 1960 [as *Gauri* 1976]; *The Road* 1961; *Death of a Hero* 1963; *Lajwanti and Other Stories* (s) 1966; *Morning Face* 1968; *Between Tears and Laughter* (s) 1973; *Selected Short Stories* (s) 1973; *Confession of a Lover* 1976; *Selected Short Stories* (s) 1977; *The Bubble* 1984

Plays

India Speaks 1943; *Little Plays of Mahatma Gandhi* 1991

Non-Fiction

Persian Painting 1930; *Curries and Other Indian Dishes* 1932; *The Golden Breath: Studies in Five Poets of the New India* 1933; *The Hindu View of Art* 1933 [rev. 1957]; *Letters on India* 1942; *Apology for Heroism: an Essay in Search of Faith* 1946; *Homage to Tagore* 1946; *The Bride's Book of Beauty* [with K. Hutheesing] 1947 [rev. as *The Book of Indian Beauty* 1981]; *On Education* 1947; *The King-Emperor's English* 1948; *Letters Written to an Indian Air: Essays* 1949; *The Indian Theatre* 1950; *The Dancing Foot* 1957; *The Hindu View of Art* 1957 [rev. 1988]; *Indian Colour* 1959; *Homage to Khajuraho* 1960; *Is There a Contemporary Indian Civilization?* 1963; *Kama Kala: Some Notes on the Philosophical Basis of Indian Erotic Sculpture* 1963; *The Third Eye: a Lecture on the Appreciation of Art* 1963; *Bombay* 1965; *The Humanism of M.K. Gandhi: Three Lectures* 1967; *Delhi, Agra, Sikri* 1968; *Konorak* [with others] 1968; *The Volcano: Some Comments on the Development of Rabindranath Tagore's Aesthetic Theories* 1968; *Indian Ivories* 1970; *Ajanta* 1971; *Album of Indian Paintings* 1973; *Author to Critic: the Letters of Mulk Raj Anand* 1973; *Apology for Heroism: a Brief Autobiography of Ideas* 1975; *Roots and Flowers: Two Lectures on the Metamorphosis of Technique and Content in the Indian-English Novel* 1975; *Tantra Magic* [with A. Mookerjee] 1978; *The Humanism of Jawaharlal Nehru* 1978; *Only Connect: Letters to Indian Friends – E.M. Forster, a Profile* [ed. E.M. Forster, S. Hamid Husain, E. Morgan] 1979; *Seven Little-Known Birds of the Inner Eye* 1978; *The Humanism of Rabindranath Tagore* 1979; *Conversations in Bloomsbury* (a) 1981; *Ellora: a Guide to Ellora Caves* 1984; *Madhubani Painting* 1984; *Pilpali Sahib: the Story of a Childhood under the Raj* (a) 1985; *Poet-Painter: Paintings by Rabindranath Tagore* 1985; *Ajanta: a Guide to Ajanta Caves* 1988; *Amrita Sher-Gil* 1989; *Three Eminent Personalities on the Ram Janambhoomi* 1989; *Pilpali Sahib: the Story of a Big Ego in a Small Boy* (a) 1990; *Caliban and Gandhi: Letters to 'Bapu' from Bombay* 1991; *Old Myth and New Myth: Letters from Mulk Raj Anand to K.V.S. Murti Anand* 1991

For Children

Indian Fairy Tales Retold 1946; *The Story of India* 1948; *The Story of Man* 1952; *More Indian Fairy Tales* 1961; *The Story of Chacha Nehru* (s) 1965; *Maya of Mohenjo-Daro* 1980

Edited

Marx and Engels on India 1933; *Indian Short Stories* (s) [with I. Singh] 1947; A.K. Coomaraswamy *Introduction to Indian Art* 1956; *Annals of Childhood* 1968; *Contemporary World Sculpture* 1968; *Experiments: Contemporary Indian Short Stories* (s) 1968; *Grassroots* 1968; *Folk Tales of Punjab* 1974; *Homage to Amritsar* 1977; *Homage to Jaipur* 1977; *Alampur* 1978; *Tales from Tolstoy* (s) 1978; *Homage to Kalamkari* 1979; *Golden Goa* 1980; *Splendours of Tamil Nadu* 1980; *Splendours of the Vijayanagara* 1980; *Maharaga Ranjit Singh as Patron of the Arts* 1981; *Treasures of Everyday Art* 1981; *Kama Sutra of Vatsyayana* [with L. Dane] 1982; *Panorama: an Anthology of Modern Indian Short Stories* (s) [with S.B. Rao] 1986; Edward Thompson *The Other Side of the Medal* [with S. Sarkar] 1989

Maxwell ANDERSON 1888–1959

Anderson was born in Atlantic, Pennsylvania, the son of a Baptist minister. He was educated in various places where his father held a living before attending the University of North Dakota from which he graduated in 1911. In 1914 he received an MA from Stanford University, having previously taught at schools in North Dakota and California; while at Stanford he was an instructor in English. His first love was poetry and in 1920 he joined with Padraic Colum, Genevieve Taggard and others in founding the magazine *Measure*. On leaving Stanford he entered journalism, working on newspapers in Grand Forks, North Dakota, and in San Francisco before joining the *New Republic* in New York in 1918, where he became drama critic. His first play, *White Desert*, a tragedy in blank verse, flopped when it was first produced in Stamford, Connecticut, but the following year *What Price Glory?*, a play about two marines and their women in wartime, written in collaboration with Lawrence Stallings, was a Broadway hit. Another collaborator was Kurt Weill, and whom Anderson wrote the musical dramas *Knickerbocker Holiday* and *Lost in the Stars*, a version of **Alan Paton's** celebrated novel *Cry, the Beloved Country*. *Both Your Houses*, attacking political corruption, won him a Pulitzer Prize. In 1938 he became, with **Elmer Rice** and Robert Sherwood, a founder of the Playwrights Company, which backed his later plays; these included several costume dramas – *Elizabeth the Queen, Mary of Scotland, Joan of Lorraine* and *Anne of the Thousand Days*. The last two of these made successful films, though by then his best work was behind him. The verse drama *Winterset* is concerned with miscarriage of justice. He married twice and had four children.

Plays
White Desert 1923; *What Price Glory?* [with L. Stallings] 1924; *The Buccaneer* [with L. Stallings] 1925; *Outside Looking In* 1925; *Saturday's Children* 1927; *Gypsy* 1928; *Gods of the Lightning* [with H. Hickerson] 1928; *Elizabeth the Queen* 1930; *Night Over Taos* 1932; *Both Your Houses* 1933; *Mary of Scotland* 1933; *Valley Forge* 1934; *Winterset* 1935; *The Masque of Kings* 1936; *The Wingless Victory* 1936; *High Tor* 1937; *The Star-Wagon* 1937; *Knickerbocker Holiday* (l) [with K. Weill] 1938; *Key Largo* 1939; *Journey to Jerusalem* 1940; *Eleven Verse Plays* 1940; *Candle in the Wind* 1941; *The Miracle of the Danube* 1941; *The Eve of St Mark* 1942; *Storm Operation* 1944; *Joan of Lorraine* 1946; *Truckline Café* 1946; *Anne of the Thousand Days* 1948; *Lost in the Stars* (l) [from A. Paton's *Cry, the Beloved Country*; with K. Weill] 1950; *Barefoot in Athens* 1951; *The Bad Seed* 1955

Poetry
You Who Have Dreams 1925

Non-Fiction
The Essence of Tragedy 1939; *The Bases of Artistic Creation* 1942; *Broadway* 1947

Biography
The Life of Maxwell Anderson by Alfred E. Shivers 1983

Sherwood ANDERSON 1876–1941

Born in Clyde, Ohio, Anderson did not write a word of fiction until 1910. Largely self-educated, he held more jobs than he was later able to remember, including labouring, military service in the Spanish-American War, managing a paint factory, and copywriting for advertising, supporting himself by this last until 1923. The strains of working and writing caused a nervous breakdown in 1912, from which he emerged more determined than ever to become a successful artist. He gave up his job, left his wife, and went to Chicago, becoming involved with the many writers who were at that time making the city something of a cultural capital.

His first novel, *Windy McPherson's Son*, published in 1916, sold moderately well, and he published a volume of poety, *Mid-American Chants*, in 1918; but it was *Winesburg, Ohio*, his collection of twenty-three interrelated short stories, which made his name the following year. A small Midwestern town and its inhabitants are observed by the central character, a young man named George Willard, and the book's innovative form, which bears some similarities to contemporary Post-Impressionist and Cubist art, was seen as an influence upon **Ernest Hemingway** and **Thomas Wolfe**. By the time of the book's publication, Anderson was living in New York and involving himself in the literary scene there. During visits to Paris in 1919 and 1926, he met **James Joyce** and **Gertrude Stein**, but he remained resolute in his desire to work within America, treating American themes. Amongst numerous books written during the 1920s were two autobiographical volumes, *A Story Teller's Story* and *Tar: A Midwest Childhood.*

The structure of *Winesburg, Ohio* was the one most suited to Anderson's talents, and later novels repeatedly demonstrated that sustained narrative was not one of his strengths. His short stories are generally considered to be his finest achievement. He remained a prolific writer of novels and short stories, poetry, plays, and non-fiction about America, but during the last twenty years of his life he became discouraged by the slow sales of his books, and in 1927 he went to Virginia, where he became editor of two local newspapers. He flirted with Communism in the 1930s and wrote articles on the Depression for *Today* magazine. In his final years, Anderson and his wife travelled extensively around America.

Since his sudden death, he has been acknowledged as an important influence by a number of writers – notably **William Faulkner** and **John Steinbeck** – and his work has received a general, if belated, recognition.

Fiction
Windy McPherson's Son 1916; *Marching Men* 1917; *Winesburg, Ohio* (s) 1919; *Poor White* 1920; *The Triumph of the Egg* (s) 1921; *Horses and Men* (s) 1923; *Many Marriages* 1923; *Dark Laughter* 1925; *Alice and the Lost Novel* (s) 1929; *Beyond Desire* 1932; *Death in the Woods and Other Stories* (s) 1933; *Kit Brandon* 1936; *Sherwood Anderson: Short Stories* (s) [ed. M. Geismar] 1962

Poetry
Mid-American Chants 1918; *A New Testament* 1927; *Five Poems* 1939

Plays
Plays: Winesburg and Others 1937

Non-Fiction
A Story Teller's Story (a) 1924; *The Modern Writer* 1925; *Sherwood Anderson's Notebook* 1926; *Tar: A Midwest Childhood* (a) 1926; *Hello Towns!* 1929;

Nearer the Grass Roots and Elizabethan 1929; *The American County Fair* 1930; *Perhaps Women* 1931; *No Swank* 1934; *Puzzled America* 1935; *A Writer's Conception of Realism* 1939; *Home Town* 1940; *Sherwood Anderson's Memoirs* (a) 1942; *Letters of Sherwood Anderson* [ed. H.M. Jones, W.B. Rideout] 1953; *Return to Winesburg* [ed. R.L. White] 1967; *The Buck Fever Papers* [ed. W.D. Taylor] 1971; *Sherwood Anderson/Gertrude Stein: Correspondence and Personal Essays* [ed. R.L. White] 1972; *France and Sherwood Anderson* [ed. M. Fanning] 1976

Biography
Sherwood Anderson by Kim Townsend 1987

Maya ANGELOU (pseud. of Marguerite Ann Johnson) 1928–

Angelou's adopted surname derives from that of a husband of Greek origin, while the forename was given to her as a child by her brother Bailey. She was born black in St Louis, into an era of severe prejudice, the daughter of a doorman, later naval hospital worker. Her parents divorced during her infancy, and when she was three she was sent to Stamps, Arkansas, a town practising strict racial segregation, to be brought up by her grandmother, Annie ('Momma') Henderson, who kept a general store. When she was eight, during a brief visit to St Louis, she was raped by her mother's lover (he was later murdered by her uncles), and for five years was mute; she was later knifed by her father's mistress. When she was a teenager, her mother – now a professional gambler who ran a rooming-house – moved her children to San Francisco. Angelou graduated from the Mission High School aged sixteen, and shortly afterwards gave birth to her son Guy. She had a variety of occupations, including that of a street-car conductor, night-club waitress, prostitute, and madame for a pair of lesbian prostitutes. At twenty-two she married a sailor, Tosh Angelos, but was divorced within three years; in 1973 she was married to, and later divorced from, Paul de Feu (she is reticent about what further marriages she may have contracted).

Angelou entered upon a happier period when she became a singer at the prestigious Purple Onion nightclub; she studied dance with Martha Graham, and became part of the company in the twenty-two-nation tour of *Porgy and Bess* in 1954–5. By her early thirties she had developed twin commitments to becoming a writer and to the civil rights movement, and this took her to New York, where she became northern co-ordinator for the organisation run by Martin Luther King. Then, travelling with the African freedom fighter Vusumze Make, she went to Egypt, where she worked as associate editor of the English-language *Arab Observer*, and to Ghana, where she stayed between 1963 and 1966, teaching and lecturing.

On her return to America, she obtained her first teaching post there, at the University of California in Los Angeles, and began her television documentary work in 1968. In 1970, the first volume of her autobiography, *I Know Why the Caged Bird Sings*, was published to great acclaim, and has been followed by several further volumes, the whole, despite its occasional *longueurs*, quickly achieving recognition as a modern classic. She has also published several volumes of poetry, which is more simplistic than her prose, and written and directed several plays and works for television. She has become an extremely celebrated lecturer and personality, with activities ranging from directing at the Almeida Theatre in London to acting in the television series *Roots*. Since 1982 she has held a lifetime appointment as Professor of American Studies at the University of North Carolina, but continues to travel and lecture widely.

Non-Fiction
I Know Why the Caged Bird Sings (a) 1969; *Gather Together in My Name* (a) 1974; *Singin' and Swingin' and Getting Merry Like Christmas* (a) 1976; *The Heart of a Woman* (a) 1981; *All God's Children Need Traveling Shoes* (a) 1986; *Conversations with Maya Angelou* [ed. J.M. Eliot] 1989; *Wouldn't Take Nothing for My Journey Now* (a) 1993

Plays
Cabaret For Freedom [with G. Cambridge] 1960 [np]; *The Least of These* 1966 [np]; *Blacks, Blues, Black* (tv) 1968; *Ajax* [from Sophocles' play] 1974 [np]; *Assignment America* (tv) 1975; *And Still I Rise* 1976 [np]; *The Inheritors* (tv) 1976; *The Legacy* (tv) 1976; *Sister Sister* (tv) 1982

Poetry
Just Give Me a Cool Drink of Water 'Fore I Diiie 1971; *Oh Pray My Wings Are Gonna Fit Me Well* 1975; *And Still I Rise* 1978; *Shaker, Why Don't You Sing?* 1983; *Poems* 4 vols 1986; *Now Sheba Sings the Song* 1987; *I Shall Not Be Moved* 1990; *On the Pulse of the Morning* 1993; *The Complete Collected Poems of Maya Angelou* 1994

For Children
Life Doesn't Frighten Me [ed. S.J. Boyers] 1993; *My Painted House, My Friendly Chicken & Me* 1994

John ARDEN 1930–

The son of a glassworks manager, Arden grew up in Barnsley, where he was born. After leaving Sedbergh School, he did national service as a lance-corporal in the Intelligence Corps of the British Army from 1949 to 1950. He went to King's College, Cambridge, in 1950, graduating in architecture there in 1953, and to the Edinburgh College of Art from 1953 to 1955 where he took a diploma.

He had already been writing plays at university, and *All Fall Down* (1955) was produced in Edinburgh by some of his fellow students. He worked in London as an architectural assistant until 1957, when he married the Irish actress and playwright, **Margaretta D'Arcy**. After the success of *The Waters of Babylon* in 1957, he decided to become a full-time writer. His work gained attention with *Live Like Pigs*, an examination of class conflict. A family called Sawney are evicted from their broken-down, filthy tramcar home to live on a housing estate. They feud with their neighbours, who eventually drive them out because of their social habits. The play was interspersed with songs and ballads, and Arden was mistakenly compared to Brecht. In fact Arden regarded this technique as an extension of language, and not an interruption for dramatic emphasis. The other label of Angry Young Man was equally inappropriate, for he has confirmed rather than rejected his Marxist stance with successive work. *Serjeant Musgrave's Dance* was regarded by **Sean O'Casey** as 'far and away the finest play of the present day' and remains Arden's best-known work. An exploration of attitudes to violence in society, it centres on the maniacal Musgrave, a nineteenth-century British officer who kills many time over, regardless of moral considerations, to exact revenge for a soldier, shot by a sniper, and another soldier who dies accidentally.

Arden's ideas of presentation, and improvisation in a 'non-theatre' environment, have much in common with **John McGrath**, with whom he has worked. He had done much to break down theatrical conventions, in the hope of creating a relationship between the audience and the actors. Arden and D'Arcy have written many plays together. Where their name appears on the title-page, it is important to stress, their collaboration is equal, and not as is sometimes suggested, with an emphasis on Arden. One such work was *The Non-Stop Connolly Show*, a marathon six-part dramatic cycle which was performed continuously, and lasted nearly twenty-four hours. Based on the life of James Connolly, the Irish patriot who was shot in the Easter Rising, it refers to dozens of characters involved in the traditions of violence and pacifism in Ireland. Premiered in Dublin, it attracted an audience of varying ages, those who, the writer Desmond Hogan wrote, 'had long tired of trying to make something out of their history'. Less consciously 'literary' than some of his contemporaries, Arden has nevertheless held academic posts in universities abroad, including that of writer-in-residence at the University of New England, Australia. Of his own position in the theatre, he feels that: 'modern dramatists must abandon their solitary status and learn to combine together to secure conditions-of-work and artistic control over the products of their imagination ... they must be prepared to combine not only with their fellows, but also with *actors* ... playwrights should be members of theatrical troupes'. He has written two novels: *Silence among the Weapons*, which was shortlisted for the 1982 Booker Prize, and *Books of Bale: a Fiction of History*. He jointly wrote with D'Arcy a volume called *Awkward Corners: Essays, Papers, Fragments*. A volume of short stories, *Cogs Tyrannic*, was published in 1991. Arden has lived for many years in County Galway, Ireland.

Plays

All Fall Down 1955; *Three Plays* [pub. 1964]: *The Waters of Babylon* 1957, *Live Like Pigs* 1958, *The Happy Haven* [with M. D'Arcy] 1960; *Soldier, Soldier and Other Plays* [pub. 1967]: *When Is a Door Not a Door?* 1958, *Soldier, Soldier* (**tv**) 1960, *Wet Fish* (**tv**) 1961, *Friday's Hiding* 1966 [with M. D'Arcy]; *Serjeant Musgrave's Dance* 1959 [pub. 1960]; *Ironhand* [from Goethe's play *Götz von Berlichingen*] 1963 [pub. 1965]; *The Workhouse Donkey* 1963 [pub. 1964]; *Armstrong's Last Goodnight* 1964 [pub. 1965]; *Left-Handed Liberty* 1965; *Two Autobiographical Plays* [pub. 1971]: *Squire Jonathan and His Unfortunate Treasure* 1968, *The Bagman* (**r**) 1970; *Pearl* (**r**) 1978 [pub. 1979]; *The Old Man Sleeps Alone* (**r**) 1982 [pub. in *Best Radio Plays of 1982* 1983]

With Margaretta D'Arcy: *The Business of Good Government: a Christmas Play* 1960 [pub. 1963]; *The Happy Haven* 1960 [pub. in *New English Dramatists 4* ed. T. Maschler 1962]; *Ars Longa, Vita Brevis* 1964 [pub. 1965]; *Friday's Hiding* 1966 [pub. in *Soldier, Soldier and Other Plays* 1967]; *The Royal Pardon* 1966 [pub. 1967]; *Harold Muggins Is a Martyr* 1968 [np]; *The Hero Rises Up* 1968 [pub. 1969]; *Granny Welfare and the Wolf* [with R. Smith] 1971 [np]; *Rudy Dutschke Must Stay* 1971 [np]; *Two Hundred Years of Labour History* 1971 [np]; *The Ballygombeen Bequest* 1972 [np; rev. as *The Little Gray Home in the West* 1978, pub. 1982]; *The Island of the Mighty* 1972 [pub. 1974]; *Keep These People Moving* (**r**) 1972;

Portrait of a Rebel: Sean O'Casey (r), (tv) 1973; *The Crown Strike Play* 1974 [np]; *The Devil and the Parish Pump* 1974 [np]; *The Non-Stop Connolly Show* 1975 [pub. 1977–8]; *The Mongrel Fox* 1976 [np]; *No Room at the Inn* 1976 [np]; *Sean O'Scrudu* 1976 [np]; *Blow-In Chorus for Liam Cosgrave* 1977 [np]; *Mary's Name* 1977 [np]; *A Pinprick of History* 1977 [pub. in *Awkward Corners* 1988]; *Silence* 1977 [np]; *Vandaleur's Folly* 1978 [pub. 1981]; *The Making of Muswell Hill* 1984 [np]; *Manchester Enthusiasts* (r) 1984; *The Mother* [from B. Brecht] 1984 [np]; *Whose Is the Kingdom?* (r) 1988

Fiction

Silence among the Weapons 1982; *Books of Bale* 1988; *Cogs Tyrannic* (s) 1991

Non-Fiction

To Present the Pretence: Essays on the Theatre and Its Public 1977; *Awkward Corners* [with M. D'Arcy] 1988

Michael ARLEN (pseud. of Dikran Kouyoumdjian) 1895–1956

Kouyoumdjian adopted the name Michael Arlen when he became a naturalised British citizen in 1922. Born at Rustchuk, Bulgaria, of Armenian parentage, he emigrated to England with his parents when a child, and was educated at Malvern College and at the University of Edinburgh, where he briefly studied medicine. After the First World War, he began to publish clever, cynical stories reflecting the changed milieu of fashionable life, while *Piracy* was his first major success with a novel. But the book on which his reputation is still precariously based was *The Green Hat*, which perfectly caught the mood of the 1920s with its veiled references to sex and decadence, couched in an urbane, mannerist prose. (In those days, its wildly melodramatic plot and superficiality were less obvious defects.) Arlen quickly became an essential part of the Mayfair society he had chronicled. 'His ties and socks are a gracefully subdued symphony. His barber is the best in town,' a friend recorded enthusiastically, and he was seen everywhere, getting in and out of his enormous yellow Rolls Royce. In 1928, he married the Countess Atalanta Mercati, and retired to live in the South of France; they had two children, one of whom Michael J. Arlen. became a well-known writer and his father's biographer.

During the Second World War, Arlen returned to help his adopted country: he became Civil Defence Public Relations Officer for the West Midlands, was in Coventry when it was bombed, and suffered injury. He then went to Hollywood as a scriptwriter, and lived in New York during the latter part of his life, dying there after a long illness. His last novel, *The Flying Dutchman*, had been published in 1939 and, of the others, *Hell! Said the Duchess* may be remembered, if only for its striking title. He also wrote a few plays including an adaptation of *The Green Hat*.

Fiction

The London Venture 1920; *The Romantic Lady and Other Stories* (s) 1921; *Piracy* 1922; *These Charming People* (s) 1923 [rev. as *The Man with the Broken Nose and Other Stories* 1927]; *The Green Hat* 1924; *May Fair* (s) [rev. as *The Ace of Cads and Other Stories*] 1925; *Ghost Stories* (s) 1927; *Lily Christine* 1928; *Babes in the Wood* (s) 1929; *Men Dislike Women* 1931; *A Young Man Comes to London* 1932; *Man's Mortality* 1933; *The Short Stories of Michael Arlen* (s) 1933; *Hell! Said the Duchess* 1934; *The Crooked Coronet* (s) 1937; *The Flying Dutchman* 1939

Plays

Dear Father 1924; *The Green Hat* [from his novel] 1925; *The Zoo* [with W. Smith] 1927; *Good Loser* [with W. Hackett] 1931 [pub. 1933]

Biography

Exiles by Michael J. Arlen 1971

Simon ARMITAGE 1963–

Armitage was born in 1963 in Huddersfield, West Yorkshire. Not a particularly bookish child (he liked to browse through catalogues, instruction manuals and 'lists' of all sorts), he has said that until the upper sixth form he was a very poor student, and even then only read the poets all schoolboys of his generation have taken to, namely **Philip Larkin** and **Ted Hughes**. He went on to read geography at Portsmouth Polytechnic before returning to north-west England to train in social work. He did write a number of poems while at Portsmouth, but has described these as 'uninformed teenage nonsense'. He also wrote a thesis on the psychology of television violence.

In the late 1980s, Armitage published three pamphlets, and subsequently won an Eric Gregory Award in 1988. This seems to have been a turning-point in his life, as it gave him a little financial freedom and also, more importantly, formal recognition as a poet. 'I'd been given the okay', he later reflected, 'by a group of people I'd just about hero-worshipped: it gave me the guts to go on writing poetry.' The publishing company Bloodaxe signed him shortly afterwards, and his first collection, *ZOOM!*, appeared as a Poetry Book Society

Choice. One of the most celebrated first books of poetry published in the 1980s, its distinctive imagery of West Riding urban decay and wilder moor landscapes reflects his surroundings, and he has further commented that in his work 'the back street meets the sheep dip'.

Armitage works as a Probation officer in Manchester. Though his poetry deals with popular culture and television, incorporates images of teenage angst and urban decay, and is not averse to raising party political issues (one poem was written as a protest against the poll tax), he refuses to be categorised as a political poet, though he does call himself 'vaguely leftist'. After the success of *ZOOM!*, two subsequent collections appeared: *Kid*, which won a Forward Poetry Prize, and *Book of Matches*, which includes two sequences of short poems, 'Book of Matches' and 'Reading the Banns' (about his own marriage), as well as fourteen individual poems. He gained a wider readership after writing a poem for television, which was published as *Xanadu: a Poem Film for Television*. He is now recognised as one of Britain's most promising young poets, sits on the panels that award major poetry prizes, and frequently appears at major (and minor) poetry readings.

Poetry

ZOOM! 1989; *Kid* 1992; *Xanadu* 1992; *Book of Matches* 1993

Elizabeth (pseud. of Mary Annette) von ARNIM 1866–1941

Von Arnim was born Mary Annette Beauchamp in Sydney, Australia, in 1866. Her father was a successful trade and shipping merchant. In 1869 she and her family moved to London where she was educated at Miss Summerhayes' School in Ealing and the Royal College of Music. Travelling in Europe with her father in 1889 she met the widowed Count Henning August von Arnim, whom she married in London the following year. After five years of married life in Berlin, they moved to the family estate, Nassenheide in Pomerania, and in 1898 *Elizabeth and Her German Garden* – which celebrates the freedom of the countryside after the formality of the city – was published. It quickly became a bestseller, running to over twenty editions in little more than a year. The success of the novel lies in its combination of pastoral idyll, social comedy and feminist protest, done with a lightness of touch worthy of Jane Austen. It was published anonymously to protect Count von Arnim, a respectable figure in *Junker* society, and thereafter all her novels appeared with the statement 'By the author of *Elizabeth and Her German Garden*'. Count von Arnim, who appears in the early novels as 'The Man of Wrath' and in *The Caravaners* (1909) as Baron von Ottringe, died in 1910, and Elizabeth subsequently based herself at the Château Soleil in Switzerland. There she entertained such friends as **E.M. Forster** and **Hugh Walpole** (both of whom had been tutors to her children at Nassenheide), her cousin **Katherine Mansfield, H.G. Wells** and Francis, Earl Russell, whom she married in 1916.

Her early books were increasingly sophisticated versions of her first success; notably *Fräulein Schmidt and Mr Anstruther*, a poised revival of the epistolary novel, and *The Pastor's Wife*, a vivid portrait of life in Germany and of her later affair with H.G. Wells. Her most acclaimed novel was *Vera* – more Charlotte Brontë than Jane Austen – based on her disastrous and short-lived second marriage. Katherine Mansfield thought it 'amazingly good' and **Rebecca West** wrote perceptively of von Arnim's entire *oeuvre*: 'The author has so little heart that when she writes of sentiment she often writes like a humbug, but she has a clear and brilliant head that enables her to write a particular kind of witty and well-constructed fiction, a sort of sparkling Euclid, which nobody else can touch.' *The Enchanted April* was a confirmation of West's judgement; but *Love*, based on her affair with the younger A.S. Frere, which she found difficult to write, is painfully moving. The last novels were tired exercises, although done with her usual wit and charm. Von Arnim spent her later years on the French Rivera until the outbreak of the Second World War, when she moved to America. She died in Charleston.

Fiction

Elizabeth and Her German Garden 1898; *The Benefactress* 1901; *The Solitary Summer* 1901; *The Adventures of Elizabeth in Rügen* 1904; *Fräulein Schmidt and Mr Anstruther* 1907; *The Caravaners* 1909; *The Pastor's Wife* 1914; *Christopher and Columbus* 1919; *In the Mountains* 1920; *Vera* 1921; *The Enchanted April* 1922; *Love* 1925; *Introduction to Sally* 1926; *Expiation* 1929; *Father* 1931; *The Jasmine Farm* 1934; *Mr Skeffington* 1940

For Children

The April Baby's Book of Tunes [with K. Greenaway] 1900; *The Princess Priscilla's Fortnight* 1905

Non-Fiction

All the Dogs of My Life (a) 1936

Biography

Elizabeth of the German Garden by Leslie De Charms 1956; *'Elizabeth'* by Karen Usborne 1986

John (Lawrence) ASHBERY 1927–

Born in Rochester, New York, Ashbery studied at Deerfield Academy, Harvard and Columbia University. He wrote his honours thesis on **W.H. Auden** who, together with **Wallace Stevens**, is the most important poetic influence in his work. He was awarded a Fulbright Scholarship, and just before the publication of his first widely available collection in 1956 he travelled to Paris where he was to live for the next ten years. The uncompromising avant-gardism of his second volume can be traced to his interests during this decade, which included the paintings of Willem de Kooning and Jackson Pollock, the music of John Cage and Anton Webern, and the writings of Raymond Roussel. He was to write art criticism, becoming editor of both *ARTnews* (1966–72) and *Partisan Review* (1976–80). On his return from Paris, he found his friends **Kenneth Koch** and **Frank O'Hara** in the poetic ascendant in New York. Together they formed the nucleus of a group of artists later dubbed the 'New York School', which lasted into the early 1970s.

Ashbery's poetry after *The Tennis Court Oath* gradually regained the appearance of normality, although bafflement remains, even now, an initial reaction of many readers. The success in 1975 of *Self-Portrait in a Convex Mirror*, which won the National Book Critics Circle Prize, the National Book Award for Poetry and the Pulitzer Prize for Poetry, expanded Ashbery's audience considerably and since then he has been seen as one of America's foremost poets. He retains the ability to shock; 'Litany', a seventy-page 'dialogue' printed in *As We Know*, outraged as many critics as *The Tennis Court Oath* had done seventeen years before. Generally though, his poetry is oblique, fastidious, aesthetic and formally innovative. It is perhaps a measure of his stature that, despite very rarely being polemical himself, he has polarised America into two highly argumentative camps: those who love him and those who hate him. His only novel, *A Nest of Ninnies*, written in collaboration with James Schuyler, is a satire on literary life.

Poetry
Turandot and Other Poems 1953; *Some Trees* 1956; *The Poems* 1961; *The Tennis Court Oath* 1962; *Rivers and Mountains* 1966; *Selected Poems* 1967; *Three Madrigals* 1968; *The Double Dream of Spring* 1970; *Three Poems* 1972; *Self-Portrait in a Convex Mirror* 1975; *The Vermont Notebook* 1975; *Houseboat Days* 1977; *As We Know* 1979; *Kitaj* [with R.B. Kitaj] 1981; *Shadow Train* 1981; *A Wave* 1984; *Selected Poems* 1985; *April Galleons* 1987; *The Ice Storm* 1987; *Selected Poems* 1987; *Flow Chart* 1991; *Hotel Lautréamont* 1992; *Three Books* 1993; *And the Stars Were Shining* 1994

Fiction
A Nest of Ninnies [with J. Schuyler] 1969

Plays
Three Plays [pub. 1978]

Non-Fiction
Fairfield Porter: Realist Painter in an Age of Abstraction [with others] 1982; *Reported Sightings, Art Chronicles 1957–1987* [ed. D. Bergman] 1989

Translations
Selected Poems: Pierre Reverdy [with M.A. Caws, P. Terry] 1991

Edited
The American Literary Anthology 1968; *Penguin Modern Poets 24* 1974

Daisy (i.e. Margaret Mary Julia) ASHFORD 1881–1972

Ashford was born in Petersham, Surrey, and educated by governesses and at the Priory, Hayward's Heath. (The name by which she is known arose from a daisy-chain woven for her to wear at her christening by a half-brother.) Her mother came from a wealthy mine-owning family in Nottinghamshire, but had broken off an engagement to a peer to run away to Ireland with a penniless Hussar, whom she married after being received into the Catholic church. They had five children before he died, after which she returned to England and met and married William Ashford, a founder of the Catholic Truth Society who worked in the War Office. (A distinct Catholic thread runs through Daisy's books.) They had three daughters: Daisy; Vera, who painted; and Angela, who also wrote. Daisy's first story, since lost, was 'Mr Chapmer's Bride', which, like its successor, 'The Life of Father McSwiney' (purportedly composed when the author was four), was dictated by her to her father. Further stories followed, but it was *The Young Visiters* which made Ashford's literary reputation. Written when she was nine, it was rediscovered in 1919 when the author, by now retired from literary composition and married to a Mr Devlin, was going through her late mother's papers. Amused by the manuscript, Ashford showed it to a friend who passed it on to **Frank Swinnerton**. **J.M. Barrie** agreed to write a preface on condition he could meet the author and authenticate the manuscript, and the book was published (original spelling and punctuation intact) on 22 May 1919; by July it was in its eleventh impression. Emboldened by this success, Ashford hunted out further manuscripts and in 1920 published *Daisy Ashford: Her Book*, which included four of her own stories and 'The Jealous Governes', written by her sister Angela at the age of eight. (Angela's later story, 'Treacherous Mr Campbell', had been lost.) One of these stories, 'The Hangman's Daughter', was Daisy's last completed work, written at the advanced age of thirteen. She worked very hard at this manuscript and was dismayed when her family laughed at its melodramatic style. An autobiography, begun in maturity, was never finished.

After leaving school, which she greatly enjoyed, she worked briefly as a secretary, sharing a flat with Vera, who was at the Slade School of Art. She was a shy woman, quite different from the character who emerges from her childhood books; their publication apart, her life was entirely uneventful.

Fiction

The Young Visiters 1919; *Daisy Ashford: Her Book* (s) [with A. Ashford] 1920; *Love and Marriage* (s) [with A. Ashford] 1965; *The Hangman's Daughter and Other Stories* (s) 1983

Sylvia ASHTON-WARNER 1908–1984

Ashton-Warner (who sometimes contributed short stories and poetry to magazines under her married name, Sylvia Henderson, or simply as 'Sylvia') was born the middle child of nine at Stratford, Taranaki, on the North Island of New Zealand. Her father was a permanent invalid and her mother a schoolteacher, and she was educated at country schools wherever her mother taught, but mainly at Wairarapa College. She trained as a teacher in Auckland in 1928–9 and in 1932 she married Keith Henderson, a headmaster. She taught with him until 1958 in the sort of remote country schools in which she herself had been educated, often taking mixed classes of white and Maori children. In response to this, she developed her controversial ideas, which included throwing out the *Janet and John* readers as culturally unsuitable, developing a 'key vocabulary' reading scheme, and encouraging the children, in the interests of racial harmony, in free self-expression, even if this was of a sexual or scatalogical nature. These ideas brought her into constant conflict with the New Zealand Education Department, who persisted in classifying her as a 'low-ability teacher', but her philosophy nevertheless proved widely influential, and she continued to propagate it after her retirement from teaching in the late 1950s.

In 1958, when she was almost fifty, she published her first – and probably her best – novel, *Spinster*, which had great success, both for its uninhibited picture of an unconventional teacher (similar, but not identical, to Ashton-Warner herself) and for its daring stream-of-consciousness style. She published four further novels, which develop the theme of the creative individual in conflict with a narrow and hide-bound society. It is difficult to know whether *Myself* should be classified as fiction or non-fiction, but it gives an account, based on her own experience, of a passionate wartime relationship with a young doctor. *I Passed This Way* is more easily classifiable as an autobiography. After her husband died in 1969 (they had three children), she began a phase of her life during which she travelled widely, teaching at an experimental school in Colorado in 1970–1 (an experience out of which she wrote *Spearpoint*, largely an attack on the influence of television in America) and becoming Professor of Education at the Simon Fraser University, Vancouver. She returned to live at Tauranga, New Zealand, before her death. Her reputation has suffered an eclipse in recent years.

Fiction

Spinster 1958; *Incense to Idols* 1960; *Bell Call* 1964; *Greenstone* 1966; *Three* 1970

Non-Fiction

Teacher 1963; *Myself* (a) 1967; *Spearpoint: 'Teacher' in America* 1972; *I Passed This Way* (a) 1979

For Children

O Children of the World 1974

Biography

Sylvia! by Lynley Hood 1988

Isaac ASIMOV 1920–1992

Born in Petrovichi in the Soviet Union, Asimov emigrated with his family to the USA when he was three and settled in New York, where his parents opened a sweet-shop. After leaving the boys' high school in Brooklyn, he studied chemistry at Columbia University, New York, where he graduated in 1939 and received his MA in 1941. When the USA entered the Second World War, he went to the Naval Air Experimental Station in Philadelphia as a chemist and, towards the end of the war, served in the army as a corporal. He returned to Columbia for his PhD and, in 1949, joined the teaching staff of Boston University School of Medicine where he became an associate professor in 1955.

Four months after finishing his first story in 1938, he sold it to a science-fiction magazine and soon became a regular contributor to the many journals in this field; he published his first science-fiction novel, *Pebble in the Sky*, in 1950. He soon became very popular and prolific and in 1958 quit full-time teaching for writing; he remained an associate professor until 1979, and subsequently held the title of professor, but his teaching duties in later years were limited to giving one lecture a year.

Asimov was among the most prolific writers of our time: by 1969 he had published his one-hundredth book, in 1973 the total exceeded 120, and by early 1989 it was around 370. His output spans not only science fiction but crime fiction (he pioneered the science-fiction detective story) as well as extensive science non-fiction – books such as the *New Intelligent Man's Guide to Science* – and books on historical and literary topics (not surprisingly, in one reference book, in the section devoted to leisure interests, Asimov said that he had 'no leisure'). Of his science-fiction novels, the early *Foundation* trilogy is still the best known and, in its theme of empire building on an intergalactic scale, it is typical of his work. Asimov himself admitted that he was much stronger on plot than on human character or emotion, dealing with which he scrupulously avoided. Asimov married twice, and had two children.

Fiction

I, Robot (s) 1950; *Pebble in the Sky* 1950; *Foundation* 1951; *The Stars Like Dust* 1951; *The Currents of Space* 1952; *Foundation and Empire* 1952; *Second Foundation* 1953; *The Caves of Steel* 1954; *The End of Eternity* 1955; *The Martian Way and Other Stories* (s) 1955; *Earth is Room Enough* (s) 1957; *The Naked Sun* 1957; *The Death Dealers* 1958 [UK *A Whiff of Death* 1968]; *Nine Tomorrows* (s) 1959; *The Rest of the Robots* (s) 1964; *Fantastic Voyage* 1966; *Through a Glass, Clearly* (s) 1967; *Asimov's Mysteries* (s) 1968; *Nightfall and Other Stories* (s) 1969; *The Early Asimov; or Eleven Years of Trying* (s) 1972; *The Gods Themselves* 1972; *The Best of Isaac Asimov (1939–1972)* (s) 1973; *Have You Seen These?* (s) 1974; *Tales of the Black Widowers* (s) 1974; *Buy Jupiter and Other Stories* (s) 1975; *The Bicentennial Man and Other Stories* (s) 1976; *More Tales of the Black Widowers* (s) 1976; *Murder at the ABA* [UK *Authorized Murder*] 1976; *Three Short Stories* (s) 1976; *Good Taste* (s) 1977; *Casebook of the Black Widowers* (s) 1980; *3 by Asimov* (s) 1981; *The Complete Robot* (s) 1982; *Foundation's Edge* 1982; *The Robots of Dawn* 1983; *The Union Club Mysteries* (s) 1983; *The Winds of Change and Other Stories* (s) 1983; *Banquets of the Black Widowers* (s) 1984; *The Edge of Tomorrow* (s) 1985; *Robots and Empire* 1985; *Alternative Asimovs* (s) 1986; *Foundation and Earth* 1986; *Robot Dreams* (s) 1986; *Fantastic Voyage II* 1987; *Prelude to Foundation* 1988; *Azazel* 1989; *Nemesis* 1989; *Nightfall* 1990; *Puzzles of the Black Widowers* (s) 1990; *Robot Visions* 1990

For Children

As 'Paul French': *David Starr, Space Ranger* 1952; *Lucky Starr and the Pirates of the Asteroids* 1953; *Lucky Starr and the Oceans of Venus* 1954 [UK *The Oceans of Venus* as Isaac Asimov 1973]; *Lucky Starr and the Big Sun of Mercury* 1956 [UK *The Big Sun of Mercury* as Isaac Asimov 1974]; *Lucky Starr and the Moons of Jupiter* 1957 [UK *The Moons of Jupiter* as Isaac Asimov 1974]; *Lucky Starr and the Rings of Saturn* 1958 [UK *The Rings of Saturn* as Isaac Asimov 1974]; *Limericks for Children* 1981

Poetry

Lecherous Limericks 1975; *More Lecherous Limericks* 1976; *Still More Lecherous Limericks* 1977; *Asimov's Sherlockian Limericks* 1978; *Limericks: Too Gross* [with J. Ciardi] 1978; *A Grossery of Limericks* [with J. Ciardi] 1981

Margaret (Eleanor) ATWOOD 1939–

Atwood was born in Ottawa and spent her childhood summers in the Quebec wilderness where her father, a forest entomologist, undertook research. The subject of the wilderness recurs in both her poetry and her fiction, notably in *Surfacing* and *Wilderness Tips*. She was eleven before she attended school full-time, but as a young child she had started writing 'morality plays, poems, comic books, and an unfinished novel about an ant'. Having graduated from Leaside High School in 1957, she attended Victoria College at the University of Toronto where her English teachers included the distinguished Canadian critic Northrop Frye. In 1961 she graduated and published her first volume of poetry, *Double Persephone*. She took her MA at Radcliffe College, Cambridge, Massachusetts, the following year, and studied Victorian literature at Harvard. Her time at Harvard was interrupted while she wrote and worked in various jobs and she left in 1967 having failed to complete her dissertation on 'The English Metaphysical Romance'. Since then she has held a wide variety of academic posts and has been writer-in-residence at numerous American and Canadian universities.

Atwood first rose to prominence with her 1966 volume of poetry, *The Circle Game* (a revised edition of the shorter 1964 volume), which won her the Governor-General's Award for poetry. In the introduction to the 1979 English edition of her first novel, *The Edible Woman* (1969), she tells the story of how an embarrassed Canadian publisher agreed to publish the novel without having read it because he had lost the manuscript some eighteen months earlier. While working as an editor at the Toronto publishing house Anansi in the early 1970s Atwood published *Survival*, a critical study which enraged many scholars by asserting that Canadian literature remained blighted by a subservient, colonial mentality. The controversy resulted in the book selling a remarkable 85,000 copies over the next ten years.

Atwood has gained a wider international readership with feminist novels such as *Lady Oracle*, *The Robber Bride* and *The Handmaid's Tale*, the third of which won the Governor-General's Award in 1986, was filmed in 1990 by Volker Schlondorff from a screenplay by **Harold Pinter**, and sold over 1,000,000 copies in the USA. Atwood is a prominent member of Amnesty International and has been involved in a wide range of political campaigns. She has said that she started out 'as a profoundly apolitical writer, but then I began to do what all novelists and some poets do: I began to describe the world around me.' She has two daughters and lives with the Canadian novelist Graeme Gibson.

Fiction

The Edible Woman 1969; *Surfacing* 1972; *Lady Oracle* 1976; *Dancing Girls* **(s)** 1977; *Life Before Man* 1979; *Bodily Harm* 1982; *Encounters with the Element Man* **(s)** 1982; *Bluebeard's Egg* **(s)** 1983; *Murder in the Dark* **(s)** 1983; *Unearthing Suite* **(s)** 1983; *The Handmaid's Tale* 1986; *Cat's Eye* 1988; *Wilderness Tips and Other Stories* **(s)** 1991; *Good Bones* **(s)** 1992; *The Robber Bride* 1993

Poetry

Double Persephone 1961; *The Circle Game* 1964 [rev. 1966]; *Kaleidoscopes* 1965; *Talismans for Children* 1965; *Expeditions* 1966; *Speeches for Doctor Frankenstein* 1966; *Who Was in the Garden?* 1969; *The Animals in That Country* 1969; *The Journals of Susanna Moodie* 1970; *Procedures for Underground* 1970; *Power Politics* 1971; *You Are Happy* 1974; *Selected Poems* 1976; *Mash, Mawk* 1977; *Two-Headed Poems* 1978; *True Stories* 1981; *Notes Towards a Poem That Can Never Be Written* 1981; *Snake Poems* 1983; *Interlunar* 1984; *Poems Selected and New, 1976–1986* 1986

Non-Fiction

Survival 1972; *Days of the Rebels 1815–1840* 1977; *Second Words: Selected Critical Prose* 1982; *The Mission of the University* 1985; *Margaret Atwood: Conversations* [ed. E. E. Ingersoll] 1990

Plays

Heaven on Earth [with P. Pearson] **(tv)** 1986 [np]

For Children

Up in the Tree 1978; *Anna's Pet* 1980

Edited

The Oxford Book of Canadian Short Stories in English **(s)** [with R. Weaver] 1986; *The Canlit Food Book* 1987; *The Best American Short Stories 1989* **(s)** [with S. Ravenel] 1989

Louis (Stanton) AUCHINCLOSS 1917–

Born in Lawrence, NY, Auchincloss was educated at the Bovee School for Boys, Gorton and Yale. Apart from service in the navy during the war (1941–5), he has spent his entire life in New York City, dividing his time between writing and practising the law (he was admitted to the New York Bar in 1941). He has been associated with the same law firm – Partner, Hawkins, Delafield and Wood – since 1954. Among other positions he has since 1966 been president of the Museum of the City of New York and a member of the National Institute of Arts and Letters.

Steeped in **Henry James** (he once called himself a 'Jacobite') and **Edith Wharton**, he has depicted the New York 'aristocracy' in nearly thirty novels and several collections of stories, starting with *The Indifferent Children*, a book set during the war, which he published under the pseudonym Andrew Lee; unlike James and Wharton, however, he has remained part of this social group. He is quite clear about his method: 'I am neither a satirist nor a cheerleader. I am strictly an observer', he wrote in his book of essays, *Life, Law and Letters*. In the course of his life he has had much to observe, particularly the slackening of ethical standards, and the manipulation that is rife on Wall Street and elsewhere. The central figure of *The Dark Lady*, Elesina Dart, is a failed actress who amasses a large fortune through the designs of Ivy Trask (names that could have come from one of Wharton's novels); the *New Statesman* talked of its 'sharp moral observation'. More recently *Diary of a Yuppie*, as the title suggests, is concerned with greed and ambition. However, Auchincloss is careful not to judge his characters; he is particularly skilful in depicting women. The British reviewers have in general been particularly kind to him, talking of 'his talent for compression'; the *Sunday Times* remarked of *The Winthrop Cove-*

nant: 'Jamesian worldly wisdom is hard to come by, and Mr Auchincloss looks like one of its last professional exponents.' Besides *Life, Law and Letters*, he has written on his two mentors, as well as on Shakespeare, Henry Adams, Queen Victoria, and an autobiography, *A Writer's Capital*. He married in 1957, and has three sons.

Fiction

The Indifferent Children [as 'Andrew Lee'] 1947; *The Injustice Collectors* (s) 1950; *Sybil* 1952; *A Law for the Lion* 1953; *The Romantic Egoists* (s) 1954; *The Great World and Timothy Colt* 1956; *Venus in Sparta* 1958; *The House of Five Talents* 1960; *Pursuit of the Prodigal* 1960; *Portrait in Brownstone* 1962; *Powers of Attorney* (s) 1963; *The Rector of Justin* 1965; *The Embezzler* 1966; *Tales of Manhattan* (s) 1967; *A World of Profit* 1969; *Second Chance* (s) 1971; *Come as a Thief* 1973; *The Partners* 1974; *The Winthrop Covenant* 1976; *The Dark Lady* 1977; *The Country Cousin* 1978; *The House of the Prophet* 1980; *The Cat and the King* 1981; *Watchfires* 1982; *Narcissa and Other Fables* (s) 1983; *The Book Class* 1984; *Exit Lady Masham* 1984; *Honorable Men* 1985; *Diary of a Yuppie* 1986; *Skinny Island* 1987; *The Golden Calves* 1988; *Fellow Passengers* 1989; *The Lady of Situations* 1990; *False Gods* 1993; *Three Lives* 1993

Plays

The Club Bedroom 1964 [np]

Non-Fiction

Edith Wharton 1961; *Reflections of a Jacobite* 1961; *Ellen Glasgow* 1964; *Pioneers and Caretakers: a Study of Nine American Women Novelists* 1965; *Motiveless Malignity* 1969; *Edith Wharton: a Woman in Her Time* 1971; *Henry Adams* 1971; *Richelieu* 1972; *A Writer's Capital* (a) 1974; *Reading Henry James* 1975; *Life, Law and Letters: Essays and Sketches* 1979; *Persons of Consequence: Queen Victoria and her Circle* 1979; *Three 'Perfect Novelists' and What They Have in Common* 1981; *Unseen Versailles* 1981; *False Dawn: Women in the Age of the Sun King* 1984

Edited

An Edith Wharton Reader 1965; Trollope: *The Warden, Barchester Towers* 1966; *Fables of Wit and Elegance* 1975; Florence Adele Sloane *Maverick in Mauve* 1983; *The Hone & Strong Diaries of Old Manhattan* 1989

W(ystan) H(ugh) AUDEN 1907–1973

The son of a doctor, Auden was born in York and brought up in Solihull in the West Midlands, an industrial landscape which was to remain important to him as a poet. He was educated at Gresham's School, Holt (which he later described as 'a Fascist state'), and Christ Church, Oxford. He was to become the voice of his generation, his name forever associated with his friends and fellow writers **Christopher Isherwood**, with whom he collaborated on several projects, and **Stephen Spender**, who printed Auden's first volume of poems on a hand-press in 1928.

After leaving university, he spent a year in Berlin, where he took advantage of the city's relaxed attitude to homosexuality and, through the anthropologist John Layard, was introduced to the liberating theories of the American psychologist Homer Lane. During the 1930s he taught in prep schools, where he was both popular and successful as a master, then joined John Grierson's GPO film unit, making documentaries such as the celebrated *Night Mail* (1935). Music for this film was provided by the young composer Benjamin Britten, whom Auden took under his wing, and who was to write music for the Group Theatre productions of the Auden–Isherwood plays. Auden and Britten also collaborated on the song-cycle *Our Hunting Fathers*, several songs and the unsuccessful American folk-opera *Paul Bunyan*. The poems he wrote before the war, like those of Spender, **Louis MacNeice** and **C. Day Lewis**, reflected the *Zeitgeist* of the 1930s. At the same time, his highly intellectual poetry is personally allusive, and the private jokes it contains (notably in *Letters from Iceland*, written with MacNeice, and the plays written with Isherwood) emphasise a group loyalty. He was influenced by Norse sagas (he believed himself to be of Icelandic descent) and found parallels in them with prep schools, often drawing upon both for imagery in his poetry.

In 1935, in order to provide her with a British passport, Auden married Thomas Mann's daughter, Erika, a predominantly lesbian actress, journalist and rally-driver who was the proprietor of a satirical anti-Nazi cabaret club. He went to Spain as an ambulance driver during the Civil War, but ended up making a few broadcasts and filing some copy; it was an experience he never discussed. In 1938 he and Isherwood travelled to China in order to write a book about the Sino-Japanese war, and in January 1939 they emigrated to America, a decision which caused a great deal of controversy. Auden remained in New York and endured a difficult 'marriage' (his own term) with Chester Kallman, who was to be his collaborator in a number of projects, notably the libretto for Stravinsky's *The Rake's Progress*. He taught at various universities and in late 1940 returned to the Anglican faith of his childhood.

In 1945 he went to Germany as part of the Morale Division of the US Strategic Bombing Survey, and the following year he became a US citizen. This prevented him from being considered for the British Laureateship but not for the post of Professor of Poetry at his old university, to which he was appointed in 1956. Through Kallman he became a devotee of the opera, and he continued to write poetry, winning numerous awards (including a Pulitzer Prize for *The Age of Anxiety*), and to produce a great deal of criticism.

Auden's verse is at once personal and public, private and universal, commemorating the romantic difficulties of his friends, and confronting the psychological and political confusions of the century. He believed that poetry was important but had little real influence upon events: 'All the verse I wrote,' he later acknowledged, 'all the positions I took in the Thirties didn't save a single Jew.' His range was enormous: he wrote short love lyrics, long philosophical poems, deeply felt celebrations of landscape, playful tributes to other writers, ballads, sonnets and squibs (he edited an acclaimed *Oxford Book of Light Verse*). He was a dedicated craftsman and his poetic voice is perhaps the most distinctive of the century, instantly recognisable and yet inimitable, colloquial yet magisterial.

In character, Auden was autocratic, his bossiness somehow enhanced rather than diminished by a distinctly camp manner and by sartorial and domestic slovenliness. Smoking heavily, he shuffled around his disordered New York apartment in carpet slippers, his clothes as crumpled as his famously lined face, and insisted upon going to bed at about nine p.m., even when there were guests for dinner. Kallman's persistent infidelity caused Auden a great deal of anguish, and may have been responsible for his heavy drinking. In later years they spent much of their time apart, although they had a house in Kirchstetten, Austria, and in 1972 Auden left New York and returned to Oxford, living in a cottage provided by Christ Church. Arguably the century's greatest poet, Auden worked right until the end, dying in his sleep of a heart-attack after giving a poetry reading in Vienna.

Poetry

Poems 1928; *Poems* 1930 [rev. 1933]; *The Orators* 1932; *Look, Stranger!* 1936 [US *On This Island* 1937]; *Spain* 1937; *Another Time* 1940; *The Double Man* [UK *New Year Letter*] 1941; *The Collected Poetry of W.H. Auden* 1945; *The Age of Anxiety* 1947; *For the Time Being* 1947; *Collected Shorter Poems, 1930 to 1944* 1950; *Nones* 1951; *The Shield of Achilles* 1955; *The Old Man's Road* 1956; *Homage to Clio* 1960; *About the House* 1965; *Collected Shorter Poems, 1927 to 1957* 1966; *Collected Longer Poems* 1968; *City Without Walls* 1969; *Academic Graffiti* 1971; *Epistle to a Godson and Other Poems* 1972; *Thank You, Fog* 1974; *Collected Poems* [ed. E. Mendelson] 1976; *Juvenilia: Poems 1922–28* [ed. K. Bucknell] 1994

Plays

The Dance of Death 1934 [pub. 1933]; *The Dog Beneath the Skin* [with C. Isherwood] 1936 [pub. 1935]; *The Ascent of F6* [with C. Isherwood] 1937 [pub. 1936]; *On the Frontier* [with C. Isherwood] 1938; *For the Time Being* (l) [with M.D. Levy] 1944; *The Rake's Progress* (l) [with C. Kallman, I. Stravinsky] 1951; *Elegy for Young Lovers* (l) [with H.W. Henze] 1961; *The Bassarids* (l) [with H.W. Henze] 1966; *Paul Bunyan* (l) [with B. Britten] 1940 [pub. 1958]; *Plays and Other Dramatic Writings* [with C. Isherwood; ed. E. Mendelson; pub. 1988]; *Libretti and Other Dramatic Writings* [with C. Kallman; ed. E. Mendelson; pub. 1993]

Non-Fiction

Letters from Iceland [with L. MacNeice] 1937; *Education Today and Tomorrow* [with T.C. Worsley] 1939; *Journey to a War* [with C. Isherwood] 1939; *The Enchafèd Flood: the Romantic Iconography of the Sea* 1950; *The Dyer's Hand and Other Essays* 1962; *Selected Essays* 1964; *Secondary Worlds* 1968; *A Certain World: a Commonplace Book* 1970; *Forewords and Afterwords* [ed. E. Mendelson] 1973; *The Prolific and the Devourer* 1981

Translations

E. Schikaneder *The Magic Flute* (l) 1956; Jean Cocteau *The Knights of the Round Table* 1957; *Collected Poems of St John Perse* 1971; Goethe: *Italian Journey* 1962, *The Sorrows of Young Werther* 1971; Dag Hammarskjöld *Markings* 1964; *The Elder Edda: a Selection* 1969

Edited

Oxford Poetry [with C. Plumb] 1926; *Oxford Poetry* [with C. Day Lewis] 1927; *The Poet's Tongue* [with J. Garrett] 1935; *The Oxford Book of Light Verse* 1938; *Tennyson* 1946; John Betjeman *Slick But Not Streamlined* 1947; *Selected Prose and Poetry of Poe* 1950; *Poets of the English Language* [with N.H. Pearson] 5 vols 1952; *Kierkegaard* 1955; *The Faber Book of American Verse* 1956; *The Selected Writings of Sidney Smith* 1957; *An Elizabethan Song Book* [with N. Greenberg, C. Kallman] 1955; *The Faber Book of Aphorisms* [with L. Kronenberger] 1962; *Poet to Poet: Herbert* 1973

Collections

The English Auden: Poems, Essays and Dramatic Writings [ed. E. Mendelson] 1977

Biography
W.H. Auden by Humphrey Carpenter 1981; *Auden* by Richard Davenport-Hines 1995

Paul AUSTER 1947–

Auster was born in New Jersey to parents of East European origin. He graduated from Columbia University with an MA in literature and began writing poetry in 1970, while working as a merchant seaman on an oil tanker. He spent four years in Paris earning a living as a translator, before returning to New York, where he supported himself by writing reviews and articles.

City of Glass, the first novella in *The New York Trilogy*, was rejected by seventeen publishers before an independent press in Los Angeles picked it up. His first published novel, however, was *The Invention of Solitude*, a meditation on the process of writing autobiography and on the break-up of his first marriage and his separation from his son, Daniel. *The New York Trilogy* consists of three obsessively unsolvable mysteries – three stories of great dramatic suspense. Auster describes his protagonists as 'seekers', prepared to drop everything to find what they want most, an idea which, for him, is linked with the act of writing. He is concerned with the nature of identity, with language and with the way driving obsession creates its own internal logic. On the publication of *The New York Trilogy*, critics grasped for new categories to define Auster's work: 'Post-post-modernist' was one of them; the *Times Literary Supplement* came up with 'seductive metaphysical thrillers'. Everyone agreed that this was a new departure for contemporary fiction, that Auster's literary fame promised to endure and that he was a phenomenon who was not comparable with the American 'Brat Pack' writers, who seemed to burn out almost as quickly as they shot to celebrity status.

With *In the Country of Last Things*, Auster evokes a shattered, broken-down world. But it is still the people who matter most. He comments: 'What I try to do is leave enough room in the prose for the reader to enter it fully. All the books I've most enjoyed, the writers I most admire, have given me the space in which to imagine the details for myself. Identity is the main question I deal with.'

Fiction
The Invention of Solitude 1982; *The New York Trilogy* 1986: *City of Glass* 1985, *Ghosts* 1986, *The Locked Room* 1986; *In the Country of Last Things* 1987; *Moon Palace* 1989; *The Music of Chance* 1990; *Leviathan* 1992; *Mr Vertigo* 1994

Translations
A Little Anthology of Surrealist Poems 1972; Jacques Dupin: *Fits and Starts* 1974; *Jacques Dupin: Selected Poems* [with S. Romer, D. Shapiro] 1992; *Arabs and Israelis: a Dialogue by Saul Friedlander and Mahmoud Hussein* 1975; *The Uninhabited: Selected Poems of André de Bouchet* 1976; Jean-Paul Sartre *Life Situations* 1978; Jean Chesneaux *China: the People's Republic* 1979, *China from the 1911 Revolution to Liberation* [with others] 1979; *The Notebooks of Joseph Joubert: a Selection* 1983: Mallarmé *A Tomb for Anatole* 1983; Maurice Blanchot *Various Circles* 1985; Philippe Petit *On the High Wire* 1985

Poetry
Unearth: Poems 1970–72 1974; *Wall-Writing: Poems 1971–75* 1976

Edited
The Random House Book of Twentieth Century French Poetry 1982

Collections
Groundwork: Selected Poems and Essays 1990

Alan AYCKBOURN 1939–

Ayckbourn was born in Hampstead, where his father was the first violinist in the London Symphony Orchestra and his mother was a journalist and magazine editor whose parents had both been music-hall entertainers. He was five when the marriage ended, and much of the next ten years was spent moving from town to town in Sussex following his stepfather, a bank manager, as he moved from job to job. Middle-class gentility and the sacrifices it demands of those who affect it has since been the principal target of his comedy. Ayckbourn was educated at Haileybury School and the Imperial Service College in Hertfordshire, leaving at the first opportunity with the encouragement of a teacher who had organised school drama.

Determined to be an actor, Ayckbourn learnt stage management and technique with a succession of small repertory companies before joining the Stephen Joseph Theatre in the Round Company in 1957, the first of Britain's fringe venues. The company encouraged Ayckbourn to write, which he did, he says, in order to supply himself with leading roles – the rock singer in *The Square Cat*, for instance. His first four plays, Feydeau-like farces, written under the pseudonym Roland Allen, were all premiered in Scarborough and written to satisfy the demands of a holiday audience and the slim budget of a small independent company. The unsuccessful London transfer of *Mr Whatnot* persuaded him to leave the company temporarily and he worked as a drama producer for the BBC in Yorkshire from 1965 to 1970, whilst continuing to write for the Theatre in the Round, to which he returned as artistic director upon the death of Stephen Joseph in 1970.

Despite the international success Ayckbourn has enjoyed since the West End transfer of *Relatively Speaking* in 1967, he still prefers to open his plays either in Scarborough or at the Stephen Joseph's sister theatre in Stoke-on-Trent. He remains a prolific writer: *Jeeves*, a musical collaboration with the composer Andrew Lloyd Webber in 1975, stands out as a rare failure, whilst stints as visiting playwright for, and director to, the National Theatre in London in 1977, 1980 and 1986–8 have elevated his critical standing. His collection of London *Evening Standard* Best Play Awards – three of them, won for *The Norman Conquests*, *Just between Ourselves* and *A Small Family Business* – confirm his popularity with West End audiences and critics alike. Wider praise followed the opening of *Woman in Mind*, a serious examination of mental illness further exploring the dark areas which had begun to appear in earlier plays such as *Way Upstream*. Ayckbourn was awarded a CBE in 1987. He married in 1959, and has two sons.

Plays

As 'Roland Allen': *Love After All* 1959 [np]; *The Square Cat* 1959 [np]; *Standing Room Only* 1961 [np]

As Alan Ayckbourn: *Mr Whatnot* 1963 [pub. 1992]; *Xmas v. Mastermind* 1963 [np]; *Meet My Father* 1965 [as *Relatively Speaking* 1967, pub. 1968]; *The Sparrow* 1967 [np]; *How the Other Half Loves* 1969 [pub. 1972]; *Mixed Doubles* **(c)** 1969 [pub. 1970]; *The Story So Far* 1970 [as *Me Times Me Times Me* 1972, *Family Circles* 1987] [np]; *Time and Time Again* 1971 [pub. 1973]; *Absurd Person Singular* 1972 [pub. 1974]; *The Norman Conquests* 1973 [pub. 1975]; *Absent Friends* 1974 [pub. 1975]; *Confusions* 1974 [pub. 1977]; *Service Not Included* **(tv)** 1974; *Bedroom Farce* 1975 [pub. 1977]; *Jeeves* [from P.G. Wodehouse's novels; with A. Lloyd Webber] 1975 (np); *Just Between Ourselves* 1976 [pub. 1978]; *Ten Times Table* 1977 [pub. 1979]; *Joking Apart* 1978 [pub. 1979]; *Men on Women on Men* [with P. Todd] 1978 [np]; *Sisterly Feelings* 1979 [pub. 1981]; *Taking Steps* 1979 [pub. 1981]; *First Course* [with P. Todd] 1980 [np]; *Season's Greetings* 1980 [pub. 1982]; *Second Helping* [with P. Todd] 1980 [np]; *Suburban Strains* [with P. Todd] 1980 [pub. 1982]; *Making Tracks* [with P. Todd] 1981 [np]; *Me, Myself and I* [with P. Todd] 1981 [pub. 1989]; *Way Upstream* 1981 [pub. 1983]; *Intimate Exchanges* 1982 [pub. 1985]; *A Trip to Scarborough* [from Sheridan's play] 1982 [np]; *Incidental Music* [with P. Todd] 1983 [np]; *It Could Be Any One Of Us* 1983 [np]; *A Chorus of Disapproval* 1984 [pub. 1986]; *A Cut in the Rates* **(tv)** 1984 [pub. 1991]; *A Game of Golf* 1984 [np]; *The Seven Deadly Virtues* [with P. Todd] 1984 [np]; *The Westwoods* [with P. Todd] 1984 [np]; *Boy Meets Girl, Girl Meets Boy* [with P. Todd] 1985 [np]; *Woman in Mind* 1985 [pub. 1986]; *Mere Soup Songs* [with P. Todd] 1986 [np]; *Henceforward* 1987 [pub. 1988]; *A Small Family Business* 1987 [pub. 1988]; *Man of the Moment* 1988 [pub. 1990]; *The Inside Outside Slide Show* 1989 [np]; *The Revengers' Comedies* 1989 [pub. 1991]; *Wolf at the Door* [from H. Becque's play; with D. Walker] 1989 [pub. 1993]; *Body Language* 1990 [np]; *Wildest Dreams* 1991 [pub. 1994]; *Dreams from a Summer House* [with J. Pattison] 1992 [np]; *Time of My Life* 1992 [pub. 1993]; *Communicating Doors* 1994 [np]

For Children

Dad's Tale [as 'Roland Allen'] 1960 [np]; *Ernie's Incredible Illucinations* 1971 [pub. in *Playbill One* ed. A. Durbrand 1969]; *Mr A's Amazing Maze Plays* 1988 [pub. 1989]; *Invisible Friends* 1989 [pub. 1991]; *Callisto 5* 1990 [np]; *This Is Where We Came In* 1990 [np]; *My Very Own Story* 1991 [np]

Non-Fiction

Conversations with Ayckbourn [with I. Watson] 1981

B

Enid (Algerine) BAGNOLD 1889–1981

Bagnold (who published her earliest novels under the pseudonym 'A Lady of Quality') was the daughter of a colonel in the Royal Engineers and spent part of her childhood in Jamaica before attending Prior's Field, Godalming, a school run by the mother of **Aldous Huxley**. After going to finishing schools in Switzerland, Paris and Marburg, she attended Walter Sickert's school of drawing and painting, simultaneously plunging into London smart and Bohemian society. She met **Katherine Mansfield** and Henri Gaudier-Brzeska (who sculpted her head), lost her virginity to **Frank Harris** in an upstairs room at the Café Royal, avoided being seduced by **H.G. Wells** and had an affair with Prince Antoine Bibesco. (Bagnold's connections with this distinguished Romanian family also included translating Princess Marthe Bibesco's *Alexandre Asiatique* and writing a play, *The Last Joke*, based on the suicide of Prince Emmanuel Bibesco.)

During the First World War, Bagnold was a nurse in Woolwich until her frank book about the job, *A Diary without Dates*, caused her dismissal; she then served as an ambulance driver with FANY (the First Aid Nursing Yeomanry) in France. In 1920 she married the head of Reuters, Sir Roderick Jones, and was thenceforth often known as Lady Jones. Despite frequent flirtations on both sides, the marriage proved a lasting one until his death in 1962. They lived largely at Rottingdean, were well known as grand hosts, and had four children.

Bagnold's early career was as a novelist, and her best-known work in this field was *National Velvet*, the story of a young girl who rides her horse to victory in the Grand National, a book whose huge popularity was enhanced by the film (1945) starring the young Elizabeth Taylor. The best of the novels, however, may be *The Loved and Envied*, the only novel from her later years, a perceptive study of an ageing society beauty. From the early 1940s Bagnold had turned to writing plays, of which she produced eight: the best is undoubtedly *The Chalk Garden*, rich with scintillating dialogue and human insight. In old age she wrote a frank autobiography, both sharp and sympathetic; she also published poetry (her second book was of poems). Her long last years were passed in morphia-induced decline until her death at the age of ninety-one.

Fiction
The Happy Foreigner 1920; *Serena Blandish* 1924; *Alice and Thomas and Jane* 1930; *National Velvet* 1935; *The Squire* 1938; *The Loved and the Envied* 1950

Plays
Two Plays: Lottie Dundas, Poor Judas [pub. 1951]; *Gertie* 1952 [np]; *The Chalk Garden* 1956; *National Velvet* 1961; *Four Plays: The Chalk Garden, The Last Joke, The Chinese Prime Minister, Call Me Jacky* [pub. 1970]; *A Matter of Gravity* 1978

Non-Fiction
A Diary without Dates 1918; *Enid Bagnold's Autobiography* (a) 1969; *Letters to Frank Harris and Others* 1980

Poetry
The Sailing Ships and Other Poems 1918; *Poems* 1978

Paul (i.e. Peter Harry) BAILEY 1937–

As he explains in *An Immaculate Mistake*, a frank and funny volume of 'Scenes from Childhood and Beyond', Bailey was the third (and unplanned) son of a fifty-four-year-old road sweeper and his wife, who worked in domestic service. He grew up in a traditional London working-class environment, but was already an intellectual and outsider at Walter St John's Grammar School in Battersea, where he became interested in acting. He trained at the Central School of Speech and Drama and subsequently worked as an actor, taking a leading role in **Ann Jellicoe**'s *The Sport of My Mad Mother* (1958) and appearing with the Royal Shakespeare Company in Stratford. He changed his first name to avoid confusion with another actor, and drew upon his experiences for his novel *Peter Smart's Confessions*, which was shortlisted for the Booker Prize in 1977.

He left the theatre and was working as a salesman in the book department of Harrods when his first novel, *At the Jerusalem*, was published in 1967, winning him a Somerset Maugham Award and an Arts Council Award for the best first novel published between 1963 and 1967. (The novelist **Elizabeth Taylor** was so intrigued by the idea that a young man should write a novel set in a home for elderly women that she went to observe him at work, gaining inspiration for her own novel, *Mrs Palfrey at the Claremont*, 1971.) Shortly afterwards, Bailey became a freelance writer and is now a well-known reviewer and broadcaster, contributing regularly to national newspapers and writing television criticism for the *Guardian*. He became a notable interviewer of other writers on BBC Radio 3's comparatively short-lived and much-missed *Third Ear*. Bailey has received numerous prizes (he was the first recipient of the E.M. Forster Award and won a George Orwell Prize for his essay 'The Limitations of Despair') and was a Literary Fellow of the Universities of Newcastle and Durham. A Bicentennial Fellowship took him to the American Midwest; he was a visiting lecturer at North Dakota; and is a Fellow of the Royal Society of Literature.

Bailey usually sets his tragi-comic novels in London and is an adept chronicler of the city's ostracised and the dispossessed: the social misfits encountered by the protagonist of *Trespasses*, for example, or the First World War veterans haunted by their past experiences in *Old Soldiers*. He has also edited an anthology, *The Oxford Book of London*. His most highly regarded novel is the long, complex *Gabriel's Lament*, a haunting study of grief which was shortlisted for the Booker Prize in 1986. (It is dedicated to his partner of twenty years, the costumier David Healy, who died that year.) The protagonist of this novel reappears in *Sugar Cane* as the lover of a female doctor who becomes involved with a rent-boy. Bailey has also written a television documentary about **J.R. Ackerley**, started but abandoned a biography of **Henry Green**, and is the author of *An English Madam*, a sympathetic account of the well-known procuress Cynthia Payne, which formed the basis of two successful films, *Personal Services* and *Wish You Were Here* (both 1987).

Fiction

At the Jerusalem 1967; *Trespasses* 1970; *A Distant Likeness* 1973; *Peter Smart's Confessions* 1977; *Old Soldiers* 1980; *Gabriel's Lament* 1986; *Sugar Cane* 1993

Non-Fiction

An English Madam: the Life and Works of Cynthia Payne 1982; *An Immaculate Mistake* (a) 1990 [rev. 1991]

Edited

The Oxford Book of London 1995

Beryl (Margaret) BAINBRIDGE 1934–

The daughter of a socialist businessman, who claimed to have imported the first safety matches to Berlin and subsequently worked in shipping, cotton and property and as a commercial traveller, Bainbridge was born in Liverpool and educated locally and at the Cone-Ripman Ballet School in Tring, Hertfordshire. While still at school, she appeared regularly on the BBC's enormously popular *Children's Hour* radio programme, and at the age of fifteen she joined the Liverpool Playhouse: her novel *An Awfully Big Adventure* (inspired by **J.M. Barrie**'s *Peter Pan*, shortlisted for the Booker Prize in 1989, and filmed by Mike Newell in 1995 from a screenplay by **Charles Wood**) draws upon her experiences there as an assistant stage manager and 'character juvenile'. She ran away to London a year later. Perhaps her most famous acting role was as 'some sort of ban-the-bomb marcher' who was briefly a girlfriend of the then radical Ken Barlow in Granada Television's long-running soap opera, *Coronation Street*.

At the age of nineteen, while appearing in repertory in Dundee, she became a Roman Catholic, but lapsed before her marriage in 1954 to the artist Austin Davies, with whom she had three children, one of whom, Rudi Davies, also became an actress; they divorced in 1959. Bainbridge later worked as a clerk for Duckworth's, who were subsequently her publishers; the company's director and his wife, Colin and Anna Haycraft, became close friends and 'Beryl' made regular appearances in 'Home Life', the column Anna Haycraft wrote for the *Spectator* under her *nom de plume* **Alice Thomas Ellis**. Bainbridge also worked, briefly, in a bottle factory, an experience that she put to good use in one of her best-known novels, *The Bottle Factory Outing*, which (like many of her novels) is a black comedy in which someone is murdered, and was awarded the *Guardian* Fiction Prize. It was also shortlisted for the Booker Prize, as was *The Dressmaker* (the title refers to a trade pursued by the author's Auntie Margo), while *Injury Time*, in which a catastrophic dinner-party is interrupted by armed criminals, won the Whitbread Prize.

Although many of her novels draw upon her own experiences, others are taken from history, notably *Watson's Apology*, based on a famous Victorian murder case, and *The Birthday Boys*, a remarkable fictional recreation of Captain Scott's doomed expedition to the South Pole. Equally extraordinary is *Young Adolf*, in which the adolescent Adolf Hitler spends a traumatic and hallucinatory period in 1920s' Liverpool with his Irish sister-in-law Bridget. *Filthy Lucre* is a work of juvenilia. Her novels tend to be short, sharp and very funny, a laconic narrative voice frequently detailing the most unexpected – and frequently unpleasant – events. Although they range widely in subject-matter and setting, they have a remarkable consistency of both tone and quality and are quite unlike the books of anyone else, except perhaps **Barbara Comyns**, whose characters stand at a similarly odd angle to the world. For each novel, Bainbridge paints a series of pictures related to the characters and story, but these have never been exhibited.

For some time she wrote a diary for London's *Evening Standard* newspaper, providing readers with a deadpan account of her eccentric life and circle. Rather more rewarding are her *English Journey*, in which she follows the footsteps of **J.B. Priestley** fifty years on, and *Forever England*, which investigates Britain's 'two nations' of the 1980s: the North and the South. Both books grew out of BBC television programmes and mix travel, reportage and autobiography. Bainbridge lives in north London, where, amongst her other accomplishments, she decorates furniture.

Fiction

A Weekend with Claude 1967 [rev. 1981]; *Another Part of the Wood* 1968; *Harriet Said* 1972; *The Dressmaker* 1973; *The Bottle Factory Outing* 1974; *Sweet William* 1975; *A Quiet Life* 1976; *Injury Time* 1977; *Young Adolf* 1978; *Winter Garden* 1980; *Watson's Apology* 1984; *Mum and Mr Armitage* (s) 1985; *Filthy Lucre* 1986; *An Awfully Big Adventure* 1989; *The Birthday Boys* 1991

Non-Fiction

English Journey 1984; *Forever England* 1987

Dorothy BAKER 1907–1968

When Jane Rule wrote about Baker in her 1975 book *Lesbian Images*, she confessed to knowing 'only the barest facts' about her life: 'she was married; she had children; she loved music; she wrote four novels, two of which are of interest for this study.' Ironically (but perhaps not surprisingly) for someone who chose to write about flamboyant or dramatic central characters, information about her own life, particularly her later years, is hard to come by.

Baker was born Dorothy Dodds in Missoula, Montana, where her father was 'chief division dispatcher for the Northern Pacific Railroad'. Her mother was 'very good at dry fly casting. So I went around with my mother and caught a lot of trout.' When she was still at school, the family moved to California, where her father went into the oil business, her mother 'took up fancywork', and Baker went to grammar school, high school and 'parts of three colleges'. She graduated with a BA from Whittier College, UCLA, in 1928, and then went to France, supposedly to be a writer. While there, she married the poet and critic Howard Baker. Despite taking out a ninety-nine-year lease on an apartment in Paris, the Bakers soon returned to California, where Howard taught English at Berkeley and Dorothy did an MA in the French department.

Her experience as a teacher of Latin in a private school provided the basis for her first short story; its acceptance by a new Californian literary magazine encouraged her to continue writing short stories, and she soon gave up her job to write full-time. Many of Baker's central characters are involved with the arts, literature and particularly music. Her first novel, *Young Man with a Horn* (1938 – and filmed in 1950, although Baker didn't like the film), was inspired by the music (and life) of Bix Beiderbecke; the photographer hero of *Trio* is passionate about jazz and classical music; *Our Gifted Son* is a classical musician; and Judith, in *Cassandra at the Wedding*, is a talented pianist whose marriage to an unmusical young doctor will mean the end of her career.

Undeterred by her unsuccessful attempts to adapt *Young Man with a Horn* for the stage (a project suggested by two Broadway producers), the Bakers collaborated on a stage adaptation of *Trio*, whose first run on Broadway in 1944 was curtailed by a manufactured moral protest (apparently engineered by the theatre manager, who wanted to replace the play with a commercial success). It was later produced in Los Angeles, San Francisco, and at the Arts Theatre, London. The Bakers also co-wrote *The Ninth Day*, a television play 'about the end of the world'.

On the fringes of academic life (Howard Baker lectured at Harvard and Berkeley), Baker remained ambivalent towards it in her fiction. Pauline Maury, in *Trio*, is described by her devoted lover, Janet Logan, as 'the first woman in the department that ever got up to a full professorship'; but Ray MacKenzie, who rescues Janet for heterosexuality, sees Pauline as a fraud and parasite: 'She'd never write about poetry if she could write poetry, don't you know that? Nobody would.' Pauline is revealed not only as a sado-masochistic lesbian vampire – a type familiar from Clemence Dane's *Regiment of Women* (1917) onwards – aping the lives of the French decadent poets she writes about, but as a plagiarist even of her academic work, stolen from another, now dead, woman lover. In Baker's last – and best – novel, *Cassandra at the Wedding*, the reader senses that Cassandra (a far more sympathetic portrait of a lesbian intellectual) is outgrowing academic writing and that she will eventually become a true writer.

In her own writing, the young Baker had found herself 'seriously hampered by an abject admiration for **Ernest Hemingway** ... I didn't allow myself any quotation marks until I felt confident that I could write, for good or ill, my own stuff.' Her later work excels precisely in that confidence of a distinctive, individual voice, ironic and sharp, mocking and vulnerable by turns. She achieved what she said she admired 'above all else' in fiction: 'simplicity and clarity in both phrase and story'.

Baker's career was cut short by terminal cancer, but her writing won her several awards in her lifetime: a Houghton Mifflin Fellowship for creative writing in 1935; a Guggenheim Fellowship in 1942; the Gold Medal for Literature of the Commonwealth Club of California for *Trio*; and a National Institute of Arts and Letters Fellowship in 1964. Despite these successes, her work has yet to receive sustained critical attention – although given her opinion of academics as a breed, it is doubtful whether she would have minded.

Fiction

Young Man with a Horn 1938; *Trio* 1943; *Our Gifted Son* 1948; *Cassandra at the Wedding* 1962

Plays

Trio [from her novel; with H. Baker] 1944 [np]; *The Ninth Day* [with H. Baker] (tv) 1957; (st) 1962 [pub. 1967]

Nicholson BAKER 1957–

Baker was brought up in Rochester, New York State. His father ran his own advertising agency, which is where Baker claims to have acquired his obsessional interest in brands and design: 'I'd read issues of *Advertising Age* and get very excited about the latest Rice Krispies box.' After high school he studied music and originally wanted to be a composer. He had short stories published in the *New Yorker* and the *Atlantic Monthly*, and supported himself with a series of jobs including Wall Street analyst, technical writer and stockbroker (the latter for two weeks: he sold one stock, to his former bassoon teacher).

Panicked by the approach of his thirtieth birthday, he left work to write a novel with the original title of 'Desperation'. The novel was published in 1988 as *The Mezzanine*, a plotless and microscopic exploration of one man's lunch-break which was remarkable for its fastidious yet candid attention to the minutiae of modern life, and the baroque descriptive precision – complete with long footnotes – with which it celebrated the manufactured world. Baker lives in upstate New York with his wife and two children, and in his second book *Room Temperature* (in which, he confessed, 'even less happens') a man meditates on the nature of marital intimacy while feeding his baby daughter, who is known – in the book and in life – as 'the Bug'.

While working on Wall Street, Baker was shot at by a mugger at point-blank range, but the cartridge failed to fire: he denies, however, that this reprieve has given him his extraordinary pleasure in the small details of life. Critics have suggested recherché forerunners such as Francis Ponge, the French poet, but Baker cites **Vladimir Nabokov** and **John Updike** as the writers he likes. His third book, *U and I*, was a meditation on his hopeless admiration for Updike: Updike read it and sent Baker one of his own books inscribed 'From U to you'.

At first, Baker could do little wrong where critics were concerned; his books, however, have made a steadily diminishing impact. *Vox*, his genuinely witty and erotic novel about telephone sex, was unjustly dismissed by some as merely pornographic, but their complaints seemed to be justified when Baker followed this book with *The Fermata*, a novel largely devoted to female masturbation, in which the narrator finds that he is able to stop time and uses this for the purposes of sexual voyeurism. A small minority of critics have found even his earlier work to be smug and trivial, or a bland paean to consumer-capitalism: his attention to brand names, for example, is entirely devoid of the satirical thrust that it carries in some other American writers. Despite the manifest idiosyncrasy of his books, Baker stresses his realism and denies any avant-garde intentions: 'When I was trying to be a composer I wanted to resurrect the tonal tradition, and now I really want to be a normal, conservative novelist.'

Fiction

The Mezzanine 1988; *Room Temperature* 1990; *Vox* 1992; *The Fermata* 1994

Non-Fiction

U and I: a True Story 1991

Nigel (Marlin) BALCHIN 1908–1970

Balchin was born in Potterne, Wiltshire, and educated at Dauntsey School and Peterhouse College, Cambridge, where he took a degree in natural science. On leaving university, he successfully combined two careers: those of a research scientist in industry and a novelist. His books frequently drew upon his experiences at work, and the first two, *How to Run a Bassoon Factory* and *Business for Pleasure*, took the form of spoofs on industrial efficiency; they were written under the pseudonym Mark Spade and had originally appeared in *Punch*. At the same time, he was writing novels, but it was not until the war that he had real success with his fiction. His first wartime job was as a psychologist in the personnel section of the War Office, and *Darkness Falls from the Air* (a title adapted from Thomas Nashe's celebrated 'Adieu!' to his play *Summer's Last Will and Testament*, 1600) is an evocative account of the London blitz. Balchin's laconic narrative suggestively juxtaposes the crippling bureaucracy of a ministry department and the *laissez-faire* attitude of the protagonist to his wife's adultery. Satirical and tragic by turns, it is one of his best books and is ranked by some critics alongside contemporary novels by **Elizabeth Bowen** and **Henry Green**. He subsequently worked as deputy scientific adviser to the Army Council, and *The Small Back Room* is the story of a bomb disposal expert whose intricate work is threatened by his alcoholism. It was made into a highly successful film (1949) by the maverick team of Michael Powell and Emeric Pressburger. Also filmed (in 1947, from his own screenplay) was Balchin's other well-known novel, *Mine Own Executioner*, a melodrama about a psychoanalyst and his patient, and *A Sort of Traitors*, a spy story which he adapted as *Suspect* (1960), and which was also made into a radio play.

His adaptation of the works of others was less successful. He worked in Hollywood without a great deal of distinction (scripting, for example, an ill-advised, updated remake of Von Sternberg and Dietrich's *The Blue Angel* in 1959), and wrote a novel about his experiences there, *In the Absence of Mrs Petersen*. Other novels include *A Way through the Wood*, which was adapted for the stage by Ronald Miller as *Waiting for Gillian* and won the Play of the Year Award in 1954, and *Kings of Infinite Space* (his last), about the first British astronaut. He once claimed that 'there is practically nothing in which I am not or cannot be intensely interested', a remark that his prolific and uneven output amply supports. He was married twice, had three children and lived largely in Sussex.

Fiction

No Sky 1934; *Simple Life* 1935; *Lightbody on Liberty* 1936; *Darkness Falls from the Air* 1942; *The Small Back Room* 1943; *Mine Own Executioner* 1945; *Lord, I Was Afraid* 1947; *The Borgia Testament* 1948; *A Sort of Traitors* [US *Who Is My Neighbour?*] 1949; *The Anatomy of Villainy* 1950; *A Way through the Wood* 1951; *Private Interest* 1953; *Sundry Creditors* 1953; *Last Recollections of my Uncle Charles* (s) 1954; *The Fall of the Sparrow* 1955; *Seen Dimly Before Dawn* 1962; *In the Absence of Mrs Petersen* 1966; *Kings of Infinite Space* 1967

Non-Fiction

How to Run a Bassoon Factory [as 'Mark Spade'] 1934; *Business for Pleasure* [as 'Mark Spade'] 1935; *Income and Outcome: a Study of Personal Finance* 1936

James (Arthur) BALDWIN 1924–1987

Baldwin's adopted surname was that of his stepfather, a factory worker and storefront preacher in Harlem. Illegitimate, he never knew his own father and was brought up in great poverty: his mother had eight further children and worked as a cleaner, being paid daily so that the family could eat. His stepfather, a brutal man who eventually died in a mental hospital, overshadowed Baldwin's childhood and had a profound influence on his subsequent life.

Baldwin's gifts were recognised at primary school, where he was encouraged to write (on grocery bags, since he could not afford writing paper) by a sympathetic teacher. He continued to develop his craft in stories and articles at the De Witt Clinton High School in the Bronx, where a fellow student introduced him to the painter Beauford Delaney, who taught him about the black culture of the Harlem Renaissance. His style, particularly in his polemical writing, was influenced by his experiences as a teenager preacher at the Fireside Pentecostal Assembly. Recognising the young man's talent, **Richard Wright** secured him a grant, and Baldwin embarked upon an autobiographical book, which would draw upon his religious family background and his homosexuality. Considerably revised, it was published to great critical acclaim in 1953 as *Go Tell It on the Mountain*. White liberals hailed Baldwin as the voice of Black America, a reputation consolidated by his first collection of essays published the same year as *Notes of a Native Son* (the title of which was derived from Wright's famous novel). He divided his time between Greenwich Village and Paris, which formed the background to *Giovanni's Room*, a frank novel of bisexual entanglements.

Although fêted by the literary establishment, Baldwin continued to be a passionate 'witness' (as he put it) to the endemic racism of American society; the comparative failure of his play *Blues for Mister Charlie* suggested that his bleak vision of race relations had begun to alienate former champions of his work. He was involved in the civil rights movement in the 1960s, aligning himself with such integrationists as Martin

Luther King, whose assassination (coming shortly after that of Malcolm X) deeply affected him. Never in good health, now drinking and smoking self-destructively, he seemed burned out by the late 1960s. He suffered a writer's block, and endured much violent criticism from younger black activists, such as Eldridge Cleaver. Commissioned work was often delivered late or not at all. Although he continued to be an eloquent campaigner for the black cause, his literary reputation declined, and his later work, in particular his account of the Atlanta murders, *Evidence of Things Not Seen*, disappointed the critics. He became Five College Professor in the Afro-American Studies department of the University of Massachusetts at Amherst in 1983, and spent his latter years in France at St Paul-de-Vence, where he died of cancer.

Fiction
Go Tell It on the Mountain 1953; *Giovanni's Room* 1956; *Another Country* 1962; *Going to Meet the Man* (s) 1965; *Tell Me How Long the Train's Been Gone* 1968; *If Beale Street Could Talk* 1974; *The Devil Finds Work* 1976; *Little Man, Little Man* [with Y. Cazac] 1976; *Just Above My Head* 1979

Non-Fiction
Notes of a Native Son 1953; *Nobody Knows My Name: More Notes of a Native Son* 1961; *The Fire Next Time* 1963; *Nothing Personal* 1964; *A Rap on Race* [with M. Mead] 1971; *No Name in the Street* 1972; *A Dialogue* [with N. Giovanni] 1973; *Evidence of Things Not Seen* 1983; *The Price of the Ticket* 1985

Plays
The Amen Corner 1955; *Blues for Mister Charlie* 1964; *The Woman at the Well* 1972; *One Day I Was Lost* 1973

Biography
James Baldwin: Artist on Fire by W.J. Weatherby 1990; *Talking at the Gates* by James Campbell 1991

J(ames) G(raham) BALLARD 1930–

Ballard was born in the International Settlement in Shanghai where his father was an executive with a textile manufacturer. In his autobiographical novel *The Kindness of Women* he describes his childhood, cycling around the cosmopolitan city, and eagerly awaiting the outbreak of war. Following the Japanese occupation of Shanghai, Ballard and his parents were interned in the camp at Lunghua for over three years. This experience is the foundation of *Empire of the Sun* (which was shortlisted for the Booker Prize in 1984), though in the novel the boy Jim is without his parents, a state Ballard considers more psychologically accurate. Coming to England – 'a foreign country' – in 1946, Ballard was educated at the Leys School in Cambridge, and spent two years studying medicine at King's College, Cambridge, much of the time engaged in the dissection of corpses. He left in 1951, having won a short story competition, and wrote while working as an advertising copywriter and Covent Garden porter before joining the RAF in 1955, a year after his marriage. After his discharge from the RAF in 1957 he worked on a science journal for six years before the income from his fiction enabled him to write full-time.

Ballard had discovered science fiction while stationed in Canada with the RAF and immediately felt that it was 'the only form of fiction which was trying to make head or tail of what was going on in our world'. Like many of his early short stories, his first novel, *The Drowned World* (1962), describes environmental catastrophe, this one wrought by the melting of the polar ice-caps. The sudden and traumatic death of his wife in 1964 may have influenced the change to a more mystical tone in *The Crystal World*. In the late 1960s Ballard was the leading light at the literary magazine *Ambit*, and provoked controversy by running a competition for the best story written under the influence of drugs. By way of researching the theories about the relationship between sex and technology outlined in *Crash*, he organised an exhibition of wrecked cars with a topless model at the Arts Laboratory in Camden. *High-Rise* and *Hello America* continue his exploration of the contemporary technological 'media landscape', while *The Unlimited Dream Company* is a fantasy set in suburban Shepperton, where Ballard lives. The success of *Empire of the Sun*, his most conventional narrative to date, was enhanced by Steven Spielberg's film version (1987, with a screenplay by **Tom Stoppard**), in which Ballard briefly appears dressed as John Bull.

Fiction
Billenium (s) 1962; *The Drowned World* 1962; *The Voices of Time* (s) 1962; *The Wind from Nowhere* 1962; *The Four-Dimensional Nightmare* (s) 1963 [rev. as *The Voices of Time* 1974]; *Passport to Eternity* (s) 1963; *The Drought* 1964; *The Terminal Beach* (s) 1964; *The Crystal World* 1966; *The Impossible Man* (s) 1966; *The Day of Forever* (s) 1967; *The Disaster Area* (s) 1967; *The Overloaded Man* (s) 1967; *The Atrocity Exhibition* (s) 1970; *Chronopolis* (s) 1971; *Vermilion Sands* (s) 1971; *Crash* 1973; *Concrete Island* 1974; *High-Rise*

1975; Low-Flying Aircraft and Other Stories (s) 1976; *The Unlimited Dream Company* 1979; *The Venus Hunters* (s) 1980; *Hello America* 1981; *Myths of the Near Future* (s) 1982; *Empire of the Sun* 1984; *The Day of Creation* [US *The Act of Creation*] 1987; *Running Wild* 1988; *War Fever* (s) 1990; *The Kindness of Women* 1991; *Rushing to Paradise* 1994

Sara BANERJI 1932–

The English wife of a Brahmin tea-planter, Banerji runs a garden design company in Oxford and writes novels about incest, child abuse and murder. She was born in Buckinghamshire, in a house adjoining Stoke Poges church, the daughter of a baronet and the novelist Mary Anne Fielding (descendant of Henry). She was sent to a number of different schools, none of them good, she says, because her mother did not believe in education for girls. After the war her father decided that the family should leave their moated Oxfordshire manor for Rhodesia, where he started a tobacco plantation. He went bankrupt almost immediately, and the family moved into two mud huts in a remote area where the bush fires were so bad that wildebeest and zebra had to cram into their compound for safety. Banerji got tick-fever and was taken out of school completely.

Her parents' marriage did not survive long, and the four children returned to England with their divorced mother who insisted that Sara be 'brought out' at eighteen by an aunt who was married to Prince Blücher. A reluctant debutante, she left England to hitch-hike round Europe, and spent some time in Austria trying to track down Lippizaner dressage horses which had been hidden from the Russians after the Second World War. She failed to find them, returned to Oxford and started work at La Roma coffee bar where she met Ranjit Banerji, an Oxford undergraduate, whom she married.

They moved to India and stayed there for seventeen years, first on a tea-planting estate in the south. They had three children. In addition, Sara rode as a jockey on the flat, held exhibitions of her paintings in Madras and Delhi, and began to experiment with writing. It all went wrong when they moved to Bengal to start a dairy farm – 'a dream that turned into a nightmare. It was the time of the Naxalites and open murder.' The family abandoned everything and came back to England with only five pounds each.

Sara got a job in a market garden and later set up her own business. Her first published novel, *Cobwebwalking* (1986), established her as a novelist who interweaves fantasy and mysticism with the comic banalities of life. Her rich and varied experience informs her work. Her characters include a gawky deb and a tea-planter's daughter who hears tongueless yogis speak; her settings include a crumbling aristocratic home in rural England and a Calcutta society wedding. But it is transcendental meditation that has focused the theme she identifies as common to all her books. In *Absolute Hush* she crystallises the consciousness itself into a baby, Lump, who is the product of incest. Critics were particularly shocked by this novel, but Banerji replied that 'I didn't think of it as shocking when I wrote it ... What I'm trying to say is that there's something very optimistic about life. I think evolution is always going for something better. That is a Hindu concept. The human race is heading towards something wonderful.'

Fiction

Cobwebwalking 1986; *The Wedding of Jayanthi Mandel* 1987; *The Teaplanter's Daughter* 1988; *Shining Agnes* 1990; *Absolute Hush* 1991; *Writing on Skin* 1993

Iain BANKS 1954–

Banks was born in Dunfermline, Scotland, the only child of an admiralty officer and a former professional ice skater. He was educated at schools in North Queensferry, Gourock and Greenock, and then at Stirling University. While at university he decided he wanted to be a writer, rather than a teacher as his mother had hoped. After graduating he took several different jobs to support himself, including that of a testing technician for British Steel and a legal costs draughtsman in his spare time. He continued to write, and decided that when he reached thirty it would be make or break for his literary career. And so it proved to be, with *The Wasp Factory* published on his thirtieth birthday, after which he became a full-time writer. 'I like obsessive characters,' he has said. 'I find them good fun to write, because they make sense to people, because people are more weird than they let on a lot of the time.' Consequently, he has become as well known for his innovative, macabre, violent and controversial subject-matter as for his literary merits. His work reflects his philosophy that it is good insurance to regard the world as not only incompetent but absolutely malevolent. Critical opinion of his work varies widely. Some hail him as the best new writer to emerge in the 1980s, while others dole out savage criticism.

Science fiction is also a significant interest, with his science-fiction novels published under the differentiating name of Iain M. Banks: *Consider Phlebas* and *The Player of Games*. 'SF is full of dystopias,' he says. 'There are no utopias ... I wanted to explore the possibilities and make a convincing case for this sort of utopian society or future.'

His novels, apart from science fiction, have all been very different, which the author says is deliberate, as he wants to keep facing new challenges and to stop himself from 'getting bored'.

Fiction

The Wasp Factory 1984; *Walking on Glass* 1985; *The Bridge* 1986; *Consider Phlebas* 1987; *Espedair Street* 1987; *The Player of Games* 1988; *Canal Dreams* 1989; *The Use of Weapons* 1990; *The Crow Road* 1992; *Against a Dark Background* 1993; *Complicity* 1993

Non-Fiction

The State of the Art 1991

Lynne Reid BANKS 1929-

Banks was born in London, England; her father was a doctor and her mother, Muriel Alexander, an actress. She attended Queens' Secretarial College, London, from 1945 to 1946 when she decided to follow her mother on to the stage. She studied at the Italia Conti Stage School for a year, worked at various theatre-related jobs, and then gained a place at the Royal Academy of Dramatic Art, where she stayed until 1949. Although she gave up the idea of acting after working in repertory for five years, the experience had taught her how to write and in 1954 she had her first play, *It Never Rains*, produced by the BBC. She also worked as a freelance journalist before joining the ITV network as a news reporter in 1955. In 1958 she moved to news scriptwriting and then, in 1962, after the publication and great success of her first novel, *The L-Shaped Room* (1960), she moved to Israel where she taught English to Hebrew-speaking children. Whilst there she met and married Chaim Stephenson, a sculptor. They have three children. After giving up teaching in 1971 to work as a full-time writer, she began a successful second career writing children's fiction: fantasies of escape, usually involving magic of some kind. *The Indian in the Cupboard*, concerning the adventures of a plastic Indian and Omri, the Indian's owner, was much awarded, as was its sequel, *The Return of the Indian*.

The L-Shaped Room remains her most famous novel for adults, examining as it does the still-controversial topic of single parenthood. It was adapted for the screen by the author and filmed in 1962, and a revised version was published in 1977. *The Backward Shadow* and *Two Is Lonely* complete the story of Jane Graham, the novel's heroine. Banks has also written radio and television plays, *Dark Quartet*, an award-winning study of the Brontë sisters, and a series of books addressing the complex history of Israel and its place in Middle Eastern politics. She now lives in Dorset, England, and describes herself as 'a practising atheist'.

Fiction

The L-Shaped Room 1960; *An End to Running* [US *House of Hope*] 1962; *Children at the Gate* 1968; *The Backward Shadow* 1970; *Two Is Lonely* 1974; *Defy the Wilderness* 1981; *The Warning Bell* 1984; *Casualties* 1986; *The Broken Bridge* 1994

Plays

It Never Rains (tv) 1954; *Miss Pringle Plays Portia* 1955; *All in a Row* 1956; *The Killer Dies Twice* 1956; *Already It's Tomorrow* (tv) 1962; *The L-Shaped Room* (f) [from her novel] 1962 [rev. 1977]; *The Unborn* 1962 [np]; *The Wednesday Caller* (tv) 1963; *Last Word on Julie* (tv) 1964; *The Gift* 1965 [np]; *The Stowaway* (r) 1967; *The Eye of the Beholder* (tv) 1977; *Lame Duck* (r) 1978; *Purely from Principle* (r) 1984

For Children

One More River 1973; *Sarah and After* 1975; *The Adventures of King Midas* 1976; *The Farthest-Away Mountain* 1976; *My Darling Villain* 1977; *I, Houdini* 1978; *The Indian in the Cupboard* 1980; *The Writing on the Wall* 1981; *Maura's Angel* 1984; *The Fairy Rebel* 1985; *The Return of the Indian* 1986; *Melusine* 1988; *The Secret of the Indian* 1989; *The Magic Hare* 1992; *The Indian Trilogy* 1993; *The Mystery of the Cupboard* 1993

Non-Fiction

The Kibbutz 1972; *Dark Quartet* 1976; *Path to the Silent Country* 1977; *Letters to My Israeli Sons* 1979; *Torn Country: an Oral History of the Israeli War of Independence* 1982

John BANVILLE 1945-

Banville was born in Wexford, Ireland, where his father managed a garage. He was educated at the Christian Brothers School in Wexford and at St Peter's College. Banville's career started in journalism including a spell as sub-editor at the *Irish Press*, and night copy editor of the *Irish Times* (while writing novels during the day). Currently he is literary editor of the *Irish Times*, and lives in Dublin with his wife and two sons.

His early work was published in literary magazines, and his first book was a collection of short stories, *Long Lankin*, published in 1970. Since then his award-winning career includes the James Tait Black Memorial Prize in 1976 for *Doctor Copernicus*; the *Guardian* Fiction Prize in 1981 for *Kepler*; and the Guinness Peat Aviation Prize for *The Book of Evidence* in 1989. *The Newton Letter* has been filmed for Channel 4 television.

Science is an important theme in many of his novels, which are concerned with the paradoxical relationship between the systems we create in order to live in the world, and the chaos that is everywhere outside these systems. *Kepler*, for example, features a hero who has discovered the elliptical movement of the planets, and is struggling to sustain his vision of cosmic harmony amid degrees of personal and social chaos. *The Newton Letter* continues this theme, focusing on a historian who is writing about Newton and the nervous breakdown he suffers. The bleakness of a later book, *Mefisto*, Banville attributes to the deaths of his parents in the late 1970s and early 1980s. In *The Book of Evidence*, a man who gives up being a scientist has to face the fact that there is no morality, and that all the things he lived by he had invented himself. 'All my books are about people who cannot live fully,' he has said.

Hallmarks of Banville's work are his literary intelligence and craftsmanship, for which he has received consistently high critical praise. 'I've always likened writing a novel to having a very powerful dream that is going to haunt you for days,' he has commented. 'If you sit down ... and try to explain the dream to someone ... they can't understand what you're on about. But if you said I'm going to sit down for three years and I'm going to write something and at the end when you read it, then you'll have that dream. Then you're close to the impulse of my novels.'

Fiction

Long Lankin (s) 1970 [rev. 1984]; *Nightspawn* 1971; *Birchwood* 1973; *Doctor Copernicus* 1976; *Kepler* 1981; *The Newton Letter* 1982; *Mefisto* 1987; *The Book of Evidence* 1989; *Ghosts* 1993

Amiri (Imamu) BARAKA 1934–

Baraka was born Everett LeRoi Jones in Newark, New Jersey, where his father was a postman and lift operator. He was educated at Rutgers University and Howard University, Washington, and served for three years in the US airforce. He has taught at a number of universities and is currently professor of African studies at the State University of New York. He married Hettie Cohen, who was white, in 1958 (they divorced in 1965), with whom he had two children, and subsequently married Sylvia Robinson (now Amina Baraka) in 1967. He originally wrote as LeRoi Jones, but changed his name in 1968, when he converted to Islam.

His first play, *A Good Girl is Hard to Find*, was produced in 1958, and he has been prolific as a playwright, poet, editor and polemicist. His major success was *Dutchman* in 1964, which was awarded an Obie by the *Village Voice* as the best play of the season. In it a young black is hounded by a vampire and realises that the only cure for his neuroses is the killing of a white, but 'white middle-class values' inhibit him and he dies himself as a result. By contrast, in *The Slave* the black protagonist enters the house where his former white wife and two children are living with a white liberal professor and kills the professor. After his ex-wife is killed by a falling beam he leaves the house and children to the artillery fire of the black army. In the following year *Experimental Death Unit #1* featured two white homosexuals and a black whore. After they are caught by black militants and executed as degenerates, their severed heads are paraded on poles.

He was the literary lion of black militancy and his writing is inseparable from his career as an activist. He has founded several organisations including the Black Arts Repertory Theatre, which was disbanded in 1966 after he was found to be financing his war on white America through it, and he has had several brushes with the law, including a conviction (overturned on appeal) for carrying a gun in the 1967 Newark riots. In 1968 he stood for a council seat in Newark, but was not elected, and in 1969 he founded the Committee for Unified Newark, advocating Islam and the speaking of Swahili. He is also founder of the black community theatre Spirit House (also known as the Heckalu Community Centre) and has been Secretary-General of the National Black Political Assembly and Chairman of the Congress of Afrikan People.

Liberals soon found Baraka's agitprop calls for the unmetaphorical extermination of 'Whitey' or 'Charley' increasingly hard to defer to: 'We must eliminate the white man before we can draw a free breath on this planet.' He has called for a new black mathematics with no trace of whiteness in it, and for the end of the Christian God ('the dead Jew'). He is a self-avowed opponent of the middle classes, whether white or black, and has said that the likes of **James Baldwin** are too 'hip' to be truly black. His detractors have suggested that his work is crude, racist and bigoted, and pointed out that he has done well out of the liberal society he hates so bitterly (he has been the recipient of numerous awards and fellowships). He is less highly regarded now than he was in the 1960s, but his champions emphasise the political significance of his work, its dramatic power and verbal poetry, and its links with magic and ritual, such as the shedding of sacrificial blood on the audience. He says simply: The black people will judge me. History will absolve me.'

Plays

A Good Girl Is Hard to Find 1958 [pub. 1965]; *Dante* 1961 [pub. 1965]; *The Baptism* 1964 [pub. 1967]; *Dutchman* **(st)** 1964, **(f)** 1967; *The Slave* 1964; *The Toilet* 1964 [pub. 1967]; *Four Black Revolutionary Plays* [pub. 1969]: *Experimental Death Unit #1* 1965, *A Black Mass* 1966, *Great Goodness of Life* 1967, *Madheart* 1967; *Jello* 1965 [pub. 1970]; *Arm Yrself or Harm Yrself* 1967 [pub. 1969]; *Black Spring* **(f)** 1967; *Slave Ship* 1967 [pub. 1969]; *Home on the Range* 1968; *Police* 1968; *The Death of Malcolm X* 1969; *Insurrection* 1969; *Bloodrites* and *Junkies Are Full of (SHHH ...)* 1970 [pub. in *Black Drama Anthology* ed. W. King, R. Milner 1971]; *A Fable* **(f)** 1971; *BA-RA-KA* [pub. in *Spontaneous Combustion* ed. R. Owens 1972]; *Columbia, the Gem of the Ocean* 1973 [np]; *A Recent Killing* 1973 [np]; *The New Ark's a Moverin* 1974 [np]; *The Sidnee Poet Heroical* 1975 [pub. 1979]; *S–I* 1976; *The Motion of History* 1977 [pub. in *The Motion of History and Other Plays* 1978]; *What Was the Relationship of the Lone Ranger to the Means of Production?* 1979 [np]; *At the Dim'crackr Convention* 1980 [np]; *Boy and Tarzan Appear in a Clearing* 1981 [np]; *Weimar 2* 1981 [np]; *Money* **(l)** [with G. Gruntz] 1982 [np]; *Primitive World* **(l)** [with D. Murray] 1984 [np]

Fiction

The System of Dante's Hell 1965; *Tales* **(s)** 1967

Poetry

Spring and Soforth 1960; *Preface to a Twenty Volume Suicide Note* 1961; *The Dead Lecturer* 1964; *Black Art* 1966; *A Poem for Black Hearts* 1967; *Black Magic* 1969; *In Our Terribleness* 1970; *It's Nation Time* 1970; *Spirit Reach* 1972; *Afrikan Revolution* 1973; *Hard Facts* 1976; *AM/TRAK* 1979; *Selected Poetry* 1979; *Reggae or Not!* 1982; *Thoughts for You!* 1984

Non-Fiction

Cuba Libre 1961; *Blues People* 1963; *Home: Social Essays* 1966; *Black Music* 1968; *Trippin': a Need for Change* [with L. Neal, A.B. Spellman] 1969; *A Black Value System* 1970; *Raise Race Rays Raze* 1971; *Strategy and Tactics of a Pan African Nationalist Party* 1971; *Beginning of National Movement* 1972; *Kawaida Studies: the New Nationalism* 1972; *Afrikan Free School* 1974; *Crisis in Boston!!!!* 1974; *National Liberation and Politics* 1974; *Toward Ideological Clarity* 1974; *The Creation of the New Ark* 1975; *Spring Song* 1979; *Daggers and Javelins* 1984; *The Autobiography of LeRoi Jones/Amiri Baraka* **(a)** 1984; *The Artist and Social Responsibility* 1986; *The Music: Reflections on Jazz and Blues* 1987; *Gary and Miami: Before and After* [nd]

Edited

Four Young Lady Poets 1962; *The Moderns: New Fiction in America* 1963; *Black Fire: an Anthology of Afro-American Writing* [with L. Neal] 1968; *African Congress* 1972; *The Floating Bear* [with D. di Prima] 1974; *Confirmation: an Anthology of African American Women* [with A. Baraka] 1983

Collections

Selected Plays and Prose 1979

Maurice BARING 1874–1945

Baring was born in Mayfair, into the powerful banking family, the English branch of which had migrated from north Germany in the early eighteenth century. His father became the first Lord Revelstoke, his mother was the granddaughter of the second Earl Grey, and the first Earl of Cromer was his uncle. After Eton, he attended Trinity College, Cambridge, for a year without taking a degree, but spent more time on the Continent learning languages, for which he had a marked gift (according to Sir Edward Marsh, Russians appealed to him for elucidation of their own grammar), and which formed the basis of his vast and cosmopolitan culture. In 1898 he passed the Diplomatic Service examination at the third attempt (he had been let down previously by arithmetic), and was posted to Paris, Copenhagen and Rome – in the latter two cities he was third secretary – before quitting the service in 1904.

While in Paris, he published *Hildesheim*, four parodies of French authors written in French; other early books include volumes of verse, verse drama, fairy-tales and volumes of reminiscences. He took up journalism in 1904, going out to Russia as correspondent of the *Morning Post*; he spent much of the next eight years there, also reporting from Constantinople and on the Balkan War of 1912 for *The Times*. During the First World War, he joined the Royal Flying Corps, being personal secretary to Lord Trenchard from 1915 to 1918, and eventually attaining the rank of wing commander.

After the war, he turned to the novel, publishing his first venture in this field in 1921; his novels, such as *C*, *Cat's Cradle* and *Daphne Adeane*, refined, charming and subtle portraits of his own milieu, are not much read today. His total publications number around eighty volumes, covering almost all genres; his poetic elegies for friends killed in the First World War (he had a genius for friendship) were among his strongest work. In 1936 he published his last major book, *Have You Anything to Declare?*, an anthology in eight languages of passages he could quote from memory, but he was already suffering from *paralysis agitans*, which forced him to move from his house in Rottingdean in 1940 to be cared for during his last years by his friends the Lord Lovats at Beaufort Castle, Inverness-shire. He was a lifelong bachelor.

Fiction
Damozel Blanche and Other Faery Tales (s) 1891; *Hildesheim* 1899; *The Story of Forget-Me-Not and Lily of the Valley* 1905; *Orpheus in Mayfair and Other Stories and Sketches* (s) 1909; *Dead Letters* 1910; *The Glass Mender and Other Stories* (s) 1910 [as *The Blue Rose Fairy Book* 1911]; *Lost Diaries* 1913; *The Brass Ring* 1917; *Passing By* 1921; *Overlooked* 1922; *A Triangle* 1923; *C* 1924; *Cat's Cradle* 1925; *Half a Minute's Silence and Other Stories* (s) 1925; *Daphne Adeane* 1926; *Tinker's Leave* 1927; *Comfortless Memory* 1928; *When They Love* 1928; *The Coat Without Seam* 1929; *Robert Peckham* 1930; Friday's Business 1932; *The Lonely Lady of Dulwich* 1934; *Darby and Joan* 1935

Plays
Gaston de Foix and Other Plays 1903; *Mahasena* 1905; *Desiderio* 1906; *Diminutive Dramas* (pub. 1912; rev. as *Ten Diminutive Dramas* 1951]; *The Grey Stocking and Other Plays* [pub. 1912]; *Palamon and Arcite* 1913; *Manfroy* 1920; *His Majesty's Embassy and Other Plays* [pub. 1923]; *Fantasio* 1927

Poetry
Pastels and Other Rhymes 1891; *Northcourt Nonsense* 1893; *Triolets* 1893; *A Litany for Those in the Train* [with E. Marsh; *c.* 1895]; *Poems* 1897; *Poems* 1899; *The Black Prince and Other Poems* 1903; *Poems* 2 vols 1905; *Sonnets and Short Poems* 1906; *The Collected Poems of Maurice Baring* 1911; *Sonnets* 1914; *Fifty Sonnets* 1915; *In Memoriam Auberon Herbert* 1917; *Poems 1914–17* 1918 [rev. as *Poems 1914–1919* 1920]; *I.M.H.* [as 'C'] 1923; *Ivo Grenfell* 1927; *Cecil Spencer* 1928; *Poems 1892–1929* 1929 [rev. as *Selected Poems of Maurice Baring* 1930]

Non-Fiction
With the Russians in Machuria 1905; *A Year in Russia* 1907 [rev. 1917]; *Russian Essays and Stories* 1908; *Landmarks in Russian Literature* 1910; *The Russian People* 1911; *Letters from the Near East, 1909 and 1912* 1913; *What I Saw in Russia* 1913 [rev. 1927]; *The Mainsprings of Russia* 1914; *An Outline of Russian Literature* 1914; *Round the World in Any Number of Days* 1914; *The R.F.C. Alphabet* 1915; *A Place of Peace, 'Somewhere in London'* 1922; *The Puppet Show of Memory* (a) 1922; *Punch and Judy and Other Essays* 1923; *French Literature* 1927; *Per Ardua 1914–1918* 1928; *In My End Is My Beginning* 1931; *The Last Cruise of H.M.S. Tiger* 1931; *Lost Lectures* 1932; *Sarah Bernhardt* 1933; *Have You Anything to Declare?* 1936

Translations
Leonardo da Vinci *Thoughts on Art and Life* 1906; *Song of the Nameless* 1907; Count Paul Benckendorff *Last Days at Tsarskoe Selo* 1927; *Poems from Pushkin* 1931; *Russian Lyrics* 1943

Edited
English Landscape: an Anthology 1916; *The Oxford Book of Russian Verse* 1924; *Algae: an Anthology of Phrases* 1928; *Translations Found in a Commonplace Book* 1916 [rev. as *Translations Ancient and Modern* 1918]

Collections
Unreliable History 1934; *Baring Restored: Selections from His Work* [ed. P. Horgan] 1970

Biography
Maurice Baring by Emma Letley 1991

A(udrey) L(illian) BARKER 1918–

Barker was born in St Paul's Cray, Kent; her father was an engineer, and her mother took work as a cleaning lady to help support the family. She was educated at primary and secondary county schools, first in Kent and then at Wallington in Surrey, but her father, who disapproved of her schooling, escorted her to her first job, with a firm of City clockmakers, when she was sixteen. She later found work with a literary agency, and then wrote and edited stories for girls' annuals for the Amalgamated Press. During the Second World War she worked first in the Land Army and then with the National Fire Service. She then worked in a car showroom in Croydon; she also married during this time, although the marriage, to a naval man, was short-lived.

Her early stories began to appear in magazines soon after the war, and in 1946 she was offered an Atlantic Award in Literature of £200 to devote herself solely to writing for a year; she declined to accept it, however. Her first volume of short stories, *Innocents* (1947), won the first Somerset Maugham Award. After the war she worked for a few years as a publisher's reader, but in 1949 entered the BBC as a secretary, where she was to remain until her retirement in 1978, for many years a 'general dogsbody' and latterly sub-editor on the *Listener*. Her first novel, *Apology for a Hero*, appeared in 1950, and she did not publish another until 1965; since then they have appeared at regular intervals, interspersed with volumes of shorter fiction. Her highly individual works take place within a conventional (even old-fashioned) English setting, but concern themselves with the surrealistic worlds of the odd and the lonely. They are by no means always successful, but when they are – as in such strange and utterly compelling stories as 'The Whip Hand', 'The Iconoclasts', 'Watch It' or 'Romney' – the effect is unusually memorable and haunting. Many of her stories have been broadcast on the radio and she has always been highly praised by critics and reviewers, but it was her eighth novel, *The Gooseboy*, that brought her a wider audience.

Fiction

Innocents (s) 1947; *Apology for a Hero* 1950; *Novelette with Other Stories* (s) 1951; *The Joy-Ride and After* (s) 1963; *Lost Upon the Roundabouts* (s) 1964; *A Case Examined* 1965; *The Middling* 1967; *John Brown's Body* 1969; *Femina Real* (s) 1971; *A Source of Embarrassment* 1974; *A Heavy Feather* 1978; *Life Stories* (s) 1981; *Relative Successes* 1984; *No Word of Love* (s) 1985; *The Gooseboy* 1987; *The Woman who Talked to Herself* 1989; *Any Excuse for a Party* 1991; *Element of Doubt* 1992

Plays

Pringle (tv) [from her story] 1958

George (Granville) BARKER 1913–1991

Barker was born to Catholic parents in Loughton, Essex, but moved as a young child to a Chelsea tenement. His father had served in the Coldstream Guards and spent the 1920s in and out of work. Barker was educated at Marlborough Road Council School and Regent's Park Polytechnic, and at sixteen had already decided he was a poet. 'I affected a great long blue cloak down to my ankles and a huge Spanish-style hat so that I looked exactly like the advertisement for Sandeman's port,' he recalled. He worked briefly as a garage mechanic and as a wallpaper designer, was sacked by a firm of accountants because of his extravagant dress, and wrote copy for **John Betjeman**'s *Shell Guides*.

The encouragement of John Middleton Murry, to whom Barker had written, led to his first collection, *Thirty Preliminary Poems*, being published by David Archer's Parton Press in 1933. His second collection was published by **T.S. Eliot** at Faber, and in 1936 he was youngest poet to appear in **W.B. Yeats**'s *Oxford Book of Modern Verse*. Despite his lack of formal education, he was Professor of English Literature at the Imperial Tohoku University in Sendai, Japan, from 1939 to 1940, and subsequently did short spells as visiting professor at various American universities. Aside from this, he lived – sometimes very precariously – by his writing alone.

Barker was married to Jessica Woodward in 1933, and in 1940 the couple were flown to the USA by **Elizabeth Smart**, who had fallen in love with Barker through reading a volume of his poetry, never having met him. The ensuing affair of Barker and Smart is described in the latter's novel *By Grand Central Station I Sat Down and Wept* (1945). Barker subsequently lived with and had children with other women, and from 1964 he lived in Norfolk with Elspeth Langlands, whom he married in 1989 and who writes under her married name (her first novel *O Caledonia* was published to great acclaim in 1991). Estimates of the number of his children vary, but one daughter guesses at 'about seventeen'.

In the 1940s and 1950s he was an *habitué* of the Fitzrovia pubs also frequented by **Dylan Thomas**, and by the artists John Minton, Robert Colquhoun and Robert MacBryde. Barker attracted critical attention with his collections *Lament and Triumph* and *Eros in Dogma*. His autobiographical long poem *The True Confession of George Barker* was published in 1950 (though not by Faber, Eliot having considered the poem too sexually candid for public taste), and provoked outraged accusations of pornography in Parliament when it was broadcast by BBC radio in 1958. Barker said of it: 'I wanted to be thoroughly vulgar because I was sick of chi-chi verse.' The publication of his *Collected Poems* in 1987 led to a widespread critical reappraisal. He died of emphysema in 1991.

Poetry

Third Preliminary Poems 1933; *Poems* 1935; *Calamiterror* 1937; *Lament and Triumph* 1940; *Eros in Dogma* 1944; *Love Poems* 1947; *News of the World* 1950; *The True Confession of George Barker* 1950; *A Vision of Beasts and Gods* 1954; *The View from a Blind I* 1962; *Dreams of a Summer Night* 1966; *The Golden Chains* 1968; *Runes and Rhymes and Tunes and Chimes* 1968; *At Thurgarton Church* 1969; *Poems of Places and People* 1971; *In Memory of David Archer* 1973; *Dialogues* 1976; *Villa Stellar* 1978; *Anno Domini* 1983; *Collected Poems* [ed. R. Fraser] 1987; *Seventeen* 1988; *Three Poems* 1988

Fiction

Alanna Autumnal 1933; *Janus* (s) 1935; *The Dead Seagull* 1950

Plays
Two Plays [pub. 1958]

Non-Fiction
Essays 1970

For Children
The Alphabetical Zoo 1970; *To Aylsham Fair* 1970

Howard BARKER 1946–

Barker is often compared with other obviously political playwrights such as **Edward Bond**, **Howard Brenton**, **David Edgar** and **David Hare**, all of whom came to prominence in the early 1970s. Barker's theatrical initiation came after seeing Bond's *Saved* (1965). Although disappointed by its 'artificial imitation of working-class speech', he conceived that it was possible 'to write naturally about one's own background, using authentic dialogue'.

Barker's father worked in a printing factory and his mother was a part-time cashier. He was born in Dulwich and educated at Battersea Grammar School and the University of Sussex. After receiving his MA, Barker worked at a succession of jobs, such as labouring and van driving, whilst developing an association with London's burgeoning fringe theatre; his first stage plays were produced in small theatre clubs such as the Open Space and the Theatre Upstairs at the Royal Court, where his first play, *Cheek*, was produced in 1970. Stage work was augmented with commissions from the BBC, writing for both television and radio; the radio play *Scenes from an Execution* won a Prix Italia in 1984, and has since been rewritten as a stage play for Glenda Jackson. He sees himself as 'a playwright first and a socialist second': as an avid lampoonist, whose stage victims have included Sir Francis Chichester, Edward Heath, the Duke of Windsor, Enoch Powell and Lord Shinwell, he could hardly have expected to endear himself to the establishment in either incarnation. Although still very much a state-subsidised writer, his most frequent patrons remain the Royal Court and the Almeida. More recently the Wrestling School has been established from the Leicester Haymarket and the Sheffield Crucible to produce his work.

Perhaps his greatest success, partly one of scandal, came in 1986 with *Women Beware Women*, a rewriting of Thomas Middleton's seventeenth-century masterpiece which uses the historical setting as a launchpad for an attack on the present-day state of England in vivid, frequently obscene language. It follows the original – Middleton's first act is left unchanged – in its obsessive linkage of money, power and sex and adds to its source in suggesting the regenerative possibilities of desire. The use of 'imaginary history' – he does not research – and of aggressive imagery is not unusual in Barker's plays: one memorable scene in *Victory* has Nell Gwynn attempting to French kiss the severed, maggoty head of the regicide Bradshaw; the plot is built around the widow's attempts to reunite and bury her disinterred husband's scattered remains.

His essays have been collected in *Arguments for a Theatre*. He also writes dramatic poetry intended for performance and has exhibited as a graphic artist; his drawings adorn many of his book-jackets.

Plays
Cheek 1970 [pub. 1972]; *One Afternoon in the North Face of the 63rd Level of the Pyramid of Cheops the Great* (r) 1970; *Edward: the Final Days* 1971 [np]; *Faceache* 1971; *Henry V in Two Parts* (r) 1971; *No One Was Saved* 1971 [np]; *Alpha Alpha* 1972 [np]; *Cows* (tv) 1972; *Herman, with Millie and Mick* (r) 1972; *Private Parts* 1972 [np]; *Bank* 1973 [np]; *Rule Britannia* 1973 [np]; *Skipper, and My Sister and I* 1973 [np]; *Mutinies* (tv) 1974; *Claw* 1975 [pub. 1977]; *Stripwell* 1975 [pub. 1977]; *Wax* 1976 [np]; *Fair Slaughter* 1977 [pub. 1978]; *That Good Between Us* 1977 [pub. 1980]; *The Hang of the Gaol* 1978 [pub. 1982]; *The Love of a Good Man* 1978 [pub. 1980]; *Birth on a Hard Shoulder* 1980 [pub. 1982]; *Credentials of a Sympathizer* 1980; *The Loud Boy's Life* 1980 [pub. 1982]; *No End of Blame* 1981; *The Poor Man's Friend* 1981; *Heaven* 1982; *A Passion in Six Days* 1983 [pub. 1985]; *Victory* 1983; *The Power of the Dog* 1984 [pub. 1985]; *Scenes from an Execution* (r), (st) 1984 [pub. 1985]; *The Castle* 1985; *Crimes in Hot Countries* 1985; *Downchild* 1985; *Pity in History* (tv) 1985, (st) 1986 [pub. 1987]; *Women Beware Women* [from T. Middleton's play] 1986 [pub. 1987]; *The Possibilities* 1987; *The Bite of the Night* 1988; *The Last Supper* 1988; *Collected Plays* 2 vols [pub. 1990, 1993]; *The Europeans* 1990; *Golgo: Sermons on Pain and Privilege* 1990; *Judith: a Parting from the Body* 1990; *Seven Lears: the Pursuit of the Good* 1990; *The Early Hours of a Reviled Man* (r) 1992; *A Hard Heart* 1992; *Rome* 1992; *All He Fears* 1993; *Hated Nightfall; and Wounds to the Face* 1994

Poetry
Don't Exaggerate 1985; *The Breath of the Crowd* 1986; *Gary the Thief: Gary Upright* 1987; *Lullabies for the Impatient* 1988; *The Ascent of Monte Grappa* 1990

Non-Fiction
Arguments for a Theatre 1989

Pat BARKER 1943–

Born at Thornaby-on-Tees, to working-class parents, Barker was brought up by her grandmother. She was educated at the local grammar school, and then at the London School of Economics, and she has been a teacher of history and politics. She married in 1968, and left work to have her son in 1970; her daughter was born four years later. She wrote several unpublished novels which were middle-class comedies of manners but, on a writing course, the novelist **Angela Carter** advised her to write out of the working-class experience of her childhood. Two years later Barker completed *Union Street*, seven linked stories rather than a novel, which graphically illustrated women's working-class experience in the north-east; it was published by Virago in 1982, and attracted great acclaim, winning the Fawcett Prize, being runner-up for the *Guardian* Fiction Prize and getting Barker named among twenty 'Best of Young British Novelists'. It was filmed as *Stanley and Iris*, with the scene transferred to Boston, USA, and the principals played by Robert De Niro and Jane Fonda.

Perhaps Barker's most ambitious novel is *The Century's Daughter*, an attempt to recount one woman's experience throughout the whole of the century, using the spare, economical, humorous style which has won its author a considerable reputation, particularly in the USA. *Regeneration* and *The Eye in the Door* (which won the *Guardian* Fiction Prize in 1993) are the first two volumes of a projected trilogy set during the First World War. The two central characters are Dr W.H. Rivers, a real person who worked at the Craiglockhart War Hospital outside Edinburgh treating soliders suffering from war trauma (including **Wilfred Owen** and **Siegfried Sassoon**), and Billy Prior, a (fictional) working-class officer, who is one of his patients. Skilfully mixed fictional and historical characters (Sassoon, Owen and **Robert Graves** all appear), Barker performs a remarkable feat of imaginative sympathy by looking at the war from the perspective of the disaffected and the (sexually and politically) dissident.

Fiction
Union Street 1982; *Blow Your House Down* 1984; *The Century's Daughter* 1986; *The Man Who Wasn't There* 1989; *Regeneration* 1991; *The Eye in the Door* 1993

Djuna (Chappell) BARNES 1892–1982

Born in Cornwall-on-the-Hudson, New York, Barnes was educated at home by her parents, a painter and a writer, then studied art at the Pratt Institute in New York City, where she became part of a Bohemian set in Greenwich Village. She made her reputation initially as a journalist: she contributed an illustrated column to the *Brooklyn Daily Eagle* from 1913, and a collection of her creative interviews with assorted sporting and artistic celebrities – amongst them, Diamond Jim Brady, Florenz Ziegfeld, Alfred Stieglitz, Gaby Delys, Frank Harris, the Lunts and D.W. Griffiths – was published posthumously as *Interviews*. She also began writing for the theatre, and in the early 1920s had three experimental plays performed in New York by the influential Provincetown Players. Her first publication book was a volume of poems, *The Book of Repulsive Women* (1915). Marriage to the editor Courtenay Lemon proved a mistake and lasted only briefly.

Armed with letters of introduction to **James Joyce** and **Ezra Pound**, she went to Paris in 1915, and spent much of the next quarter of a century there, associated with other American exiles of the 1920s and 1930s, such as **F. Scott Fitzgerald** and **Gertrude Stein**, who – to Barnes's disgust – much admired her legs. Her work was published in Margaret Anderson's *Little Review*, which had relocated in Paris from New York in 1922, and by Robert McAlmon's Contact Publishing Company, which had been set up with money from his wife-of-convenience, the novelist Bryher (Winifred Ellerman). A volume collecting Barnes's stories, play and poems was published in 1923 as *A Book*.

Barnes lived for some ten years with the artist Thelma Wood in a relationship undermined by drink and drugs, and in 1928 wrote the *Ladies' Almanack*, a lesbian pastiche of eighteenth-century erotica. Published anonymously the following year, and banned by the US customs, the volume was sold on the Paris streets by Barnes and her friends, and achieved a considerable *sub rosa* reputation. Her first novel, the satirical and experimental *Ryder* was published the same year.

Her relationship with Wood finally floundered in 1931, and its splendours and miseries were commemorated in Barnes's most famous novel, *Nightwood*, which was published in 1936. It had an introduction by **T.S. Eliot**, who judged that it would 'appeal primarily to readers of poetry', and discerned qualities in its style, wit, characterisation and pervasive sense of doom which recalled Elizabethan tragedy. Like all Barnes's work, *Nightwood* is experimental and challenging – a complex, dreamlike series of interrelated narratives concerning the shifting relationships between four people in Paris and America. Even Eliot was obliged to admit: 'it took me, with this book, some time to come to an appreciation of its meaning as a whole'. Barnes had found a patron in Peggy Guggenheim, one of the novel's dedicatees, but after the book was published she became something of a recluse, returning to live in Greenwich Village in 1940. She produced only one further substantial work, a surrealist verse play, *The Antiphon*, although several collected volumes appeared, including one of stories, *Spillway*.

In spite of Eliot's imprimatur, and a revival of interest in Barnes's work amongst feminists in the 1970s and 1980s, which resulted in such volumes as *Smoke*, a collection of juvenilia, her literary reputation remains insecure. This is partly because her work is both diverse and difficult, but she was undoubtedly an influential figure, and it seems that *Nightwood* at least will continue to be read.

Fiction

Ryder 1928; *Nightwood* 1936; *Spillway* (s) 1962

Plays

The Antiphon 1958; *Three from the Earth* 1919; *An Irish Triangle* 1921; *Kurzy from the Sea* 1920

Non-Fiction

Ladies' Almanack 1928; *I Could Never Be Lonely without a Husband: Interviews by Djuna Barnes* [ed. A. Barry] 1987; *Djuna Barnes's New York* 1989

Poetry

The Book of Repulsive Women 1915; *Creatures in a Wood* 1982

Collections

A Book 1923 [rev. as *Night among the Horses* 1929]; *Selected Works* 1962; *Smoke and Other Early Stories* 1982

Biography

The Formidable Miss Barnes by Andrew Field 1983

Julian (Patrick) BARNES 1946–

The son of two French-language teachers, Barnes was born in Leicester, but the family moved to west London when he won a scholarship to the City of London School for Boys in 1957. He felt isolated from his schoolfellows and disliked the social life at Magdalen College, Oxford, where he matriculated in 1964. Changing courses twice, he failed to graduate with the expected first and, rejected by the Foreign Office, he became an editor of the supplement to the *Oxford English Dictionary*. He wrote a literary guide to Oxford, which was never published, and studied for the Bar, but abandoned law the day he passed his exams.

By this time, Barnes had already been reviewing for the *Times Literary Supplement*, selling his review copies in order to stay financially afloat, and he wrote a column for **Ian Hamilton**'s *New Review* under the pseudonym of Edward Pygge. In 1977, he became deputy to **Martin Amis**, who was then literary editor of the *New Statesman*, and who was for many years a close friend. Barnes had a strained relationship with Amis's successor, **David Caute**, and eventually resigned. He took his revenge by writing a letter critical of Caute to the magazine (which printed it) purportedly from the Agnes Varda Women's Collective. He has also been deputy literary editor of the *Sunday Times* and the television critic of the *Observer*.

His first novel, *Metroland*, was autobiographical, drawing upon memories of his own suburban upbringing in the story of two youths who rebel against their stuffy middle-class upbringing. Published in 1980, it won a Somerset Maugham Award, and was followed by another conventional novel, *Before She Met Me*, a tragicomic tale of obsessive retrospective jealousy. In 1983 Barnes was chosen as amongst the 'Best of Young British Novelists' in an influential Arts Council promotion, and in *Granta 7*, a sampler of those chosen, he published part of a novel-in-progress, *Flaubert's Parrot*. This unconventional book combines elements of literary criticism, biography and fiction, and concerns a melancholy doctor who is obsessed by Flaubert (a fascination shared by the author). Rather than tell his own story, the doctor pursues his obsession, sharing with the reader assorted arcane facts and speculations, until finally he is ready to describe his own experiences with a wife who was unfaithful and committed suicide – a story that has parallels with that of the great French novelist. 'The novel is an example of displacement activity,' Barnes has commented, linking his art with the narrative of this novel. The book proved to be a great critical and (less expectedly) popular success, making Barnes's reputation as one of the leading novelists of his generation and winning him the Geoffrey Faber Memorial Prize. He also won an E.M. Forster Award, and became the first British writer to be awarded the French *Prix Medicis Etrangère*.

Meanwhile, under another pseudonym, Dan Kavanagh, Barnes was writing detective fiction, featuring a bisexual detective, Duffy. In spite of jacket copy, changing with each book, which claimed that Kavanagh had been (amongst other things) 'an assistant marshal at Romford Greyhound Stadium', the author's real identity was widely known. Indeed, Barnes borrowed his *alter ego*'s surname from his wife, the literary agent, Pat Kavanagh. Other novels published under Barnes's own name include the philosophical *A History of the World in 10½ Chapters*, in which fact and fiction are once again mingled; *Talking It Over*, a story of adultery narrated in turn by each of the three protagonists, two best friends and the woman they both love; and *The Porcupine*, a blackly comic account of Bulgaria after the collapse of Communism. Barnes has said of his work that with each new book 'you have to convince yourself that it's not only a new departure for you but for the history of the novel'.

Although Barnes is personally retiring, he has been the subject of much public speculation concerning his role as the model for characters in novels by other writers, and has been caught up in a great deal of unwelcome publicity. In spite of the promptings of persistent journalists, he has maintained a dignified silence. He was for several years the London correspondent of the *New Yorker*, and collected his pieces in *Letters from London*.

Fiction

As Julian Barnes: *Metroland* 1980; *Before She Met Me* 1982; *Flaubert's Parrot* 1984; *Staring at the Sun* 1986; *A History of the World in 10½ Chapters* 1989; *Talking It Over* 1991; *The Porcupine* 1992

As 'Dan Kavanagh': *Duffy* 1980; *Fiddle City* 1981; *Putting the Boot In* 1985; *Going to the Dogs* 1987

Non-Fiction

Letters from London 1995

Translations

Volker Kriegel *The Truth About Dogs* 1988

Peter BARNES 1931–

Described by Malcolm Hay in *Plays and Players* in 1990 as 'one of the great unrecognised (or, at best, only grudgingly acknowledged) geniuses of the English theatre of the past couple of decades', Barnes has a reputation that has fluctuated over the years.

He was born in London and educated at Stroud Grammar School, which he left to join the London County Council, where he remained – with a year off to do national service in the RAF – until 1953. The following year he became a freelance contributor to *Films and Filming* and a story editor for Warwick Films. He began writing screenplays in the late 1950s and writing for the theatre in the early 1960s, with his first play, *The Time of the Barracudas*, produced in 1963. His first big success came in 1968 with his 'baroque comedy' *The Ruling Class*, an anarchic and extremely funny satire upon the aristocracy, in which the deranged 14th Earl of Gurney quite literally gets away with murder. Drawing upon the stories of Jack the Ripper and *La Dame aux Camélias* (as well as the Gospels), it is a comedy of the blackest hue, and was joint winner of the John Whiting Award. It was made into a successful film from Barnes's own screenplay in 1972. Harold Hobson recalled that in twenty years of reviewing plays the excitement at 'being suddenly and unexpectedly faced with the explosive blaze of an entirely new talent of a very high order' had only happened four times: at the opening nights of *Waiting for Godot*, *Look Back in Anger*, *The Birthday Party* and *The Ruling Class*. Barnes's next two plays, *Leonardo's Last Supper* and *Noonday Demons*, appeared to fulfil this initial promise, while *The Bewitched* was acclaimed by Martin Esslin as 'a masterpiece ... a feast for intellectuals as well as a rollicking example of folk theatre'. Irving Wardle, reviewing *Laughter!*, placed Barnes with **Samuel Beckett**, **John Osborne** and **Peter Nichols** as 'among those who have given audiences this sense of the earth moving under their feet'.

In the 1970s and 1980s, Barnes spent much of his time adapting the works of other dramatists, including Jonson, Feydeau, Wedekind and Brecht. His most important original work during this period was *Red Noses*, a historical black comedy which attracted mixed notices. He has also done much work for radio, including further adaptations of Jacobean and other plays and a celebrated series of monologues, *Barnes' People*. *Revolutionary Witness*, written to mark the bicentenary of the storming of the Bastille, is another set of radio monologues, delivered by characters whose lives were fundamentally changed by the French Revolution. For television, he has written *Nobody Here But Us Chickens*, three short plays about disabled people.

Plays

A Man with a Feather in His Hat (tv) 1960; *The Time of the Barracudas* 1963 [np]; *Sclerosis* 1965 [np]; *The Ruling Class* (st) 1968, (f) 1972 [pub. 1969]; *Leonardo's Last Supper, and Noonday Demons* 1969 [pub. 1970]; *The Alchemist* [from Jonson's play; with T. Nunn] 1970 [rev. 1977] [np]; *Lulu* [from F. Wedekind; with C. Beck] 1970 [pub. 1971]; *The Devil Is an Ass* [from Jonson's play] 1973 [rev. 1976; np]; *Eastward Ho!* [from G. Chapman, Jonson and J. Marston's play] 1973; *My Ben Jonson* (r) 1973; *The Bewitched* 1974; *The Frontiers of Farce* [from G. Feydeau and F. Wedekind] 1976 [pub. 1977]; *For Those Who Get Despondent* [from B. Brecht and F. Wedekind]

1976 [rev. as *The Two Hangmen: Brecht and Wedekind* (r) 1978] [np]; *Antonio* [from plays *Antonio* and *Antonio's Revenge* (r) 1977, (st) 1978 [np]; *Laughter* 1978; *A Chaste Maid in Cheapside* [from T. Middleton's play] (r) 1979; *The Devil Himself* [from F. Wedekind's play; with C. Davis, S. Deutsch] 1980 [np]; *Eulogy on Baldness* (r) [from Synesius of Cyrene] 1980; *The Atheist* (r) 1981; *Barnes' People* (r) 1981; *Collected Plays* [pub. 1981]; *For the Conveyance of Oysters* (r) [from Gorky] 1981; *The Singer* (r) [from F. Wedekind] 1981; *The Soldier's Fortune* (r) [from T. Otway's play] 1981; *Somersaults* 1981 [np]; *The Dutch Courtesan* (r) [from J. Marston's play] 1982; *The Magician* (r) [from Gorky] 1982; *A Mad World, My Masters* (r) [from T. Middleton's play] 1983; *Barnes' People II* (r) 1984 [pub. 1984]; *The Primrose Path* (r) [from G. Feydeau's play] 1984; *Red Noses* 1985; *A Trick to Catch the Old One* (r) [from T. Middleton's play] 1985; *Barnes' People III* (r) 1986 [pub. in *The Real Long John Silver and Other Plays* 1986]; *The Old Law* (r) [from T. Middleton and W. Rowley's play] 1986; *Scenes from a Marriage* [from G. Feydeau's play] 1986 [np]; *Woman of Paris* (r) [from H. Becque] 1986; *No End to Dreaming* (r) 1987; *More Barnes' People* (r) 1989; *Nobody Here But Us Chickens* (tv) 1989; *Plays* 2 vols [pub. 1989, 1993]; *The Spirit of Man* (tv) 1989; *Revolutionary Witness* (r) 1989; *Sunsets and Glories* 1990; *Tango at the End of Winter* [from K. Shimizu's play] 1991

Alan BARNSLEY

see Gabriel FIELDING

J(ames) M(atthew) BARRIE 1860–1937

The son of a weaver, Barrie was born in humble circumstances in Kirriemuir, a small town in north-east Scotland, and was educated at the Dumfries Academy and Edinburgh University. Before he became one of the theatre's most popular and successful writers, he worked as a journalist, and wrote short stories based upon his Scottish background, and a number of novels.

Although primarily a dramatist for adults, Barrie made his most enduring contribution to the twentieth century with a play ostensibly written for children. His own brother had died in a skating accident on the eve of his fourteenth birthday, a tragedy which deeply affected their mother. She took comfort, however, from the notion that by dying young, her son would remain a boy forever. This was the germ of *Peter Pan; or The Boy Who Wouldn't Grow Up*, the play which was to make Barrie's fortune and which became part of the theatrical calendar for decades after its first performance at Christmas 1904. Barrie's story of the ethereal Pan, who takes the children of the Darling family to the Never Land, where they become involved with 'the Lost Boys', a tribe of noble redskins and a band of ferocious but incompetent pirates led by the dastardly Captain Hook, underwent yearly revisions and in 1911 was turned by the author into a novel, *Peter and Wendy*. Many of the elements of the story, including the eponymous hero, had first appeared in Barrie's 1902 novel *The Little White Bird*. 'Nothing that happens after we are twelve matters very much,' Barrie said, and his personal obsession with boyhood was one shared by his Edwardian audience, and echoes the stories of **Saki**, the poems of **A.E. Housman** and the plethora of school stories published in the years before the First World War. Over the years the play has been reduced to pantomine, set to undistinguished music, illustrated by Mabel Lucie Atwell, made into a Disney cartoon, abridged, bowdlerised, trivialised and traduced; it is, however, a great deal darker and more psychologically acute (and uncomfortable) than its reputation as whimsy would suggest, as became clear when it was restored and revived by the Royal Shakespeare Company in 1982.

Among Barrie's other plays, the most interesting are *The Admirable Crichton*, a comedy about class in which the eponymous butler takes charge when shipwrecked on a desert island with his employers (but returns to servitude when they are rescued); *Dear Brutus*, in which characters who are given a second chance of life by the agency of magic, but make the same mistakes; the proto-feminist *What Every Woman Knows*, in which the heroine is shown to be the real power and brains behind a male politician; and *Mary Rose*, in which a young mother mysteriously vanishes from an island to return, unchanged, years later when her son has grown up (a theme developed from *The Little White Bird*).

Barrie married the actress Mary Ansell in 1894, but the marriage was never consummated and ended in divorce after Mary had been found by a gardener in bed with the novelist Gilbert Cannan, whom she subsequently married. Apart from his mother, about whom he wrote in *Margaret Ogilvy*, at the centre of Barrie's emotional life were five boys, the children of Arthur and Sylvia Llewelyn Davies, whom he befriended and eventually adopted after their parents' premature deaths. He described *Peter Pan* as 'streaky with' these boys, and it was a hideous irony that his two favourites, like his most famous creation, never grew up: George was killed in the war and Michael drowned while an undergraduate, and Barrie never really recovered from this double blow. Having turned down a knighthood, he accepted a baronetcy in 1913. He was elected Rector of St Andrew's University in 1919 and was awarded the OM in 1922.

Plays

Ibsen's Ghost 1891; *Richard Savage* 1891; *The Professor's Love Story* 1892; *Walker, London* 1892 [pub. 1907]; *Becky Sharp* 1893; *Jane Annie* (l) [with A. Conan Doyle] 1893; *The Little Minister* 1897 [pub. 1898]; *A Platonic Friendship* 1898; *The Wedding Guest* 1900; *The Admirable Crichton* 1902 [pub. 1914]; *Quality Street* 1902 [pub. 1913]; *Little Mary* 1903; *Peter Pan* 1904; *Alice Sit-by-the-fire* 1905; *Josephine* 1906; *Punch* 1906; *What Every Woman Knows* 1908 [pub. 1918]; *When Wendy Grew Up* 1908; *Old Friends* 1910; *A Slice of Life* 1910; *The Dramatists Get What They Want* 1912; *The Adored One* 1913; *Half-an-Hour* 1913; *The Will* 1913; *Der Tag* 1914; *Half Hours* [pub. 1914]; *The Fatal Typist* 1915; *Rosy Rapture* 1915; *Irene Vanbrugh's Pantomime* 1916; *The Real Thing at Last* 1916; *A Kiss for Cinderella* 1916 [pub. 1920]; *Shakespeare's Legacy* 1916; *Dear Brutus* 1917; *Reconstructing the Crime* 1917; *Echoes of the War* [pub. 1918]; *Mary Rose* 1920 [pub. 1924]; *Shall We Join The Ladies?* 1921 [pub. 1927]; *Neil and Tintinnabulum* 1925; *The Boy David* 1936; *The Two Shepherds* 1936

Fiction

Better Dead 1887; *Auld Licht Idylls* (s) 1888; *When a Man's Single* 1888; *A Window in Thrums* (s) 1889; *The Little Minister* 1891; *A Holiday in Bed and Other Sketches* (s) 1892; *A Powerful Drug* (s) 1893; *Sentimental Tommy* 1896; *Tommy and Grizel* 1900; *The Little White Bird* 1902; *Peter Pan in Kensington Gardens* 1906; *Peter and Wendy* 1911

Non-Fiction

An Edinburgh Eleven 1889; *My Lady Nicotine* 1890; *Margaret Ogilvy: a Memoir by Her Son* 1896; *George Meredith: an Appreciation* 1909; *Charles Frohman: a Tribute* 1915; *Courage* 1922; *The Greenwood Hat* [as 'James Anon'] 1930; *Letters of J.M. Barrie* [ed. V. Meynell] 1934; *M'Connachie and J.M.B.: Speeches* 1938

Poetry

Scotland's Lament 1895

Collections

Plays and Stories [ed. R.L. Green] 1962

Biography

The Story of J.M.B. by Denis MacKail 1941; *J.M. Barrie and the Lost Boys* by Andrew Birkin 1970

Stan(ley) BARSTOW 1928-

Barstow was born the son of a miner in the Yorkshire village of Horbury, in an environment where there were few cultural amenities and the prospect of a literary career seemed remote. He attended the high school in the nearby town of Ossett, and in 1944 left to work as a draughtsman for an engineering firm in the same town. In 1951 he married and subsequently had two children. He began writing, at first without much encouragement, in his spare time, and by 1960 was able to publish his first novel, *A Kind of Loving*. Set, like many of Barstow's novels, in the fictional Yorkshire town of Cressley, it is the humorous but perhaps ultimately pessimistic story of the troubled love-affair between Vic Brown and Ingrid Rothwell, and caught public notice at a time when first novels by such fellow northerners as **John Braine**, **Alan Sillitoe**, **Keith Waterhouse** and **David Storey** were also being published with great success. Like several of the others, it was quickly turned into a popular film (1962, directed by John Schlesinger from a screenplay by Waterhouse and Willis Hall).

Barstow contrasted with writers such as Braine and Sillitoe, however, in that he resisted the temptation to move south. He continued to live in Yorkshire (latterly in the town of Ossett, near his birthplace), and while, like them, he has continued writing prolifically, and has never again quite achieved the success of his first book, his abiding subject-matter has remained fairly exclusively the northern working-class life he knows well. He left the engineering firm in 1961 (he was by then a sales executive), and has lived since as a professional writer.

His novels include two trilogies: one about Vic Brown, begun in *A Kind of Loving*; the other, dating from the later 1980s onward, about the Palmer family in wartime. He has also published five collections of short stories. His work has attracted consistent praise from the critics, who prize its brilliant creation of scene and warm but unsentimental sympathy with character. Comparisons have been made with **D.H. Lawrence**, but Barstow entirely lacks the fierce criticism Lawrence turns on the world. Like Lawrence, however, he is overwhelmingly concerned with love and sex, although less metaphysically. From the 1970s onward Barstow has increasingly diversified his work to include many dramatisations for theatre, radio and television, adapting both his own work and that of others (he has adapted *An Enemy of the People* by Ibsen, for instance); in this field, he attracted especial notice for his popular 1970s television dramatisation of **Winifred Holtby**'s novel *South Riding* (1936).

Fiction

Vic Brown Trilogy: A Kind of Loving 1960, *The Watchers on the Shore* 1966, *The Right True End* 1976; *The Desperadoes and Other Stories* (s) 1961; *Ask Me Tomorrow* 1962; *Joby* 1964; *A Raging Calm* 1968; *The Human Element and Other Stories* (s) 1969; *A Season with Eros* (s) 1971; *A Casual Acquaintance and Other Stories* (s) 1976; *The Glad Eye and Other Stories* (s) 1978; *A Brother's Tale* 1980; *Just You Wait and See* 1986; *B-Movie* 1987; *Give Us This Day* 1989; *A Season with Eros* 1990; *Next of Kin* 1991

Plays

Ask Me Tomorrow [from his novel; with A. Bailey] 1966; *A Kind of Loving* [from his novel] 1970; *Listen for the Trains, Love* 1970; *Springer's Last Stand* 1972; *South Riding* (tv) [from W. Holtby's novel] 1974; *Joby* [from his novel; with M. Davies] 1977; *An Enemy of the People* [from Ibsen's play] 1978; *The Human Element, and Albert's Part* 1984

John (Simmons) BARTH 1930-

Barth has lived on the east coast of America for most of his life. A native of Maryland, he has pursued his dual career of novelist and academic through the universities around that region. He was educated at the Juilliard School of Music (he has been a professional drummer) and Johns Hopkins University where his work as a writer was given its first impetus. To pay for his tuition, the undergraduate Barth shelved books at the university library, where he necessarily came across a wide and arcane selection of works normally outside the university syllabus: *The Thousand and One Nights* was a particular favourite.

His own works have been similarly unconventional in form, and include *Lost in the Funhouse*, a volume of stories 'for Print, Tape, Live Voice'; *The Sot-Weed Factor*, an immensely long pastiche of an eighteenth-century novel; and a truly labyrinthine epistolatory novel, *LETTERS*. As an academic, he has championed the claims of postmodernism as vigorously as he has practised them in his novels. His essays are theoretically and polemically formidable; however, their obvious commitment is at variance with Barth's plea to be spared from 'Social-Historical responsibility, and in the last analysis every other kind as well'.

Barth is best seen as a counter-realist, constructing worlds outside the everyday. That he has lived most of his life protected within the university system fits with the insularity of his novels, but his works are more than hugely elaborate in-jokes or intellectual monuments. Barth is at once one of America's bestselling novelists and at the forefront of the literary avant-garde; his popularity attests to the relevance of even his most experimental fiction outside the privileged enclave of the university.

Fiction

The Floating Opera 1956 [rev. 1967]; *The End of the Road* 1958 [rev. 1967]; *The Sot-Weed Factor* 1960; *Giles Goat-Boy* 1966; *Lost in the Funhouse* (s) 1968; *Chimera* 1972; *LETTERS* 1979; *Sabbatical* 1982; *The Tidewater Tales* 1988; *The Last Voyage of Somebody the Sailor* 1991

Donald BARTHELME 1931–1989

Born in Philadelphia, Pennsylvania, Barthelme grew up largely in Houston, Texas, where his father, an architect, became a university professor of architectural design. After attending the University of Houston, Barthelme became a reporter on the Houston *Post* until he was drafted into the US army at the time of the Korean War. He arrived in Korea on the day the truce was signed, but was employed to edit an army newspaper. On his return he took various jobs in his home town, including working in the public relations department of the university and as the curator of a gallery of modern art.

In the early 1960s he moved to New York as the managing editor of a short-lived review, *Locations*, and at about the same time his early stories and non-fictional pieces began to appear in the *New Yorker*: ten of them were published there between 1963 and 1964. The magazine really made his name in 1967 by devoting a whole issue to the novella *Snow White*, a reworked fairy-tale set in New York's Greenwich Village. Barthelme had the ideal characteristics to become fashionable in the 1960s, declaring that 'the principle of collage is the central principle of all art in the twentieth century in all media', and seeming to offer a literary equivalent of pop art in stories written in one hundred numbered sentences, or a single sentence, or packed with word games and random allusions. Beyond this, however, he was a considerable artist who had found a method of mirroring the confusions and dislocation of the century in which he found himself. He wrote eight short-story collections, of which *Come Back, Dr Caligari* was the first, as well as four novels and children's fiction; he illustrated a number of his own books. In person, he was quiet and conservatively dressed, often to be seen in the afternoons around Greenwich Village, where he lived, prospecting for additions to his collection of prints. He was married four times and had two daughters; he died in Houston, Texas.

His younger brother, Frederick Barthelme, is also a novelist and the editor of the *Mississippi Review*.

Fiction

Come Back, Dr Caligari (s) 1964; *Snow White* 1967; *Unspeakable Practices, Unspeakable Acts* (s) 1968; *City Life* (s) 1970; *Sadness* (s) 1972; *Guilty Pleasures* (s) 1974; *Amateurs* (s) 1976; *The Dead Father* 1976; *Great Days* (s) 1979; *The Emerald* (s) 1980; *Presents* (s) 1980; *Sixty Stories* (s) 1981; *Overnight to Many Distant Cities* (s) 1983; *Paradise* 1986; *Forty Stories* (s) 1989; *Sixty Stories* (s) 1989; *The King* 1991

For Children

The Slightly Irregular Fire Engine 1971

H(erbert) E(rnest) BATES 1905–1974

Bates was born in Rushton, Northamptonshire, and educated at Kettering Grammar School, which – as he explained in an essay contributed to Graham Greene's celebrated symposium, *The Old School* (1934) – failed to live up to expectations raised by public-school stories. However, the influence of one inspirational master altered the course of Bates's career. He had intended becoming a farmer or a footballer but was now inspired to become a writer. Although he was academically accomplished, 'gaining illustrious commendation in subjects that have since been of not the slightest use to me', his family was poor and he was obliged to leave school in order to work. He became a clerk in a leather warehouse, then a journalist on a local paper, and started writing fiction. When he was just twenty, he attracted the notice of Edward Garnett ('another genius – Miss Bates'), who recommended his first novel, *The Two Sisters*, for publication in 1926. Bates went on to become one of the most widely read writers of his time, producing a large number of novels and short stories. His subject was England, its landscape and characters, which he described in lyrical prose, but any suggestion that he was parochial or insular must be set against the fact that his greatest influences were foreign (his idol being Turgenev) and that his books have been widely translated.

In 1931 Bates married and moved to Kent; with his wife, Margaret, he had four children. During the Second World War, he was recuited to the RAF to write morale-boosting short stories under the pseudonym 'Flying Officer X'. In this capacity, he was to immortalise the Battle of Britain and the pilots who took part in it, rising above propagandist requirements to produce distinguished fiction. *The Greatest People in the World* depicts life at a bomber station, while *Fair Stood the Wind for France* (published under his own name and considered to be amongst his finest achievements) is a novel about a British air crew brought down in occupied France. In 1945 he was sent to Burma in order to write stories which would make the Americans aware of the military campaign there. This resulted in two of his best novels, *The Purple Plain* and *The Jacaranda Tree*.

Shortly after the war, Bates glimpsed a family which inspired his sequence of comic stories (beginning with *The Darling Buds of May*) about the irrepressible Larkins, a junk-dealing family 'gargantuan of appetite, unenslaved by conventions, blissfully happy', whose philosophy is 'all *carpe diem* and the very antithesis of the Welfare State'. Popular in their times, these stories reached a new audience when made into a hugely successful television series in 1991. Pop Larkin's famous term of approbation, 'Perfick!' was adopted by the tabloid press and appeared with monotonous regularity in headlines for several months.

Enormously prolific, Bates wrote over forty volumes of fiction, as well as plays, books for children and numerous volumes of non-fiction including a critical survey of *The Modern Short Story*, a memoir of his early mentor, Edward Garnett, a three-volume autobiography, and assorted books on gardening and the English countryside. He was awarded the CBE in the year before his death.

Fiction

The Two Sisters 1926; *Day's End* (s) 1928; *Catherine Foster* 1929; *Seven Tales and Alexander* (s) 1929; *The Hessian Prisoner* (s) 1930; *Charlotte's Row* 1931; *The Black Boxer* (s) 1932; *The Fallow Land* 1932; *A Threshing Day* (s) 1932; *The House with the Apricot* (s) 1933; *The Woman Who Had Imagination* (s) 1934; *Cut and Come Again* (s) 1935; *The Poacher* 1935; *A House of Women* 1936; *Something Short and Sweet* (s) 1937; *Country Tales* (s) 1938; *Spella Ho* 1938; *The Flying Goat* (s) 1939; *The Beauty of the Dead* (s) 1940; *The Greatest People in the World* (s) [as 'Flying Officer X'] 1942; *How Sleep the Brave* (s) [as 'Flying Officer X'] 1943; *Fair Stood the Wind for France* 1944; *The Cruise of the Breadwinner* 1946; *The Purple Plain* 1947; *The Bride Comes to Evensford* (s) 1949; *Dear Life* 1949; *The Jacaranda Tree* 1949; *The Scarlet Sword* 1950; *Colonel Julian*

(s) 1951; *Selected Short Stories* (s) 1951; *Love for Lydia* 1952; *The Face of England* (s) 1953; *The Nature of Love* 1953; *The Feast of July* 1954; *The Daffodil Sky* (s) 1955; *The Sleepless Moon* 1956; *Sugar for the Horse* (s) 1957; *The Darling Buds of May* 1958; *A Breath of French Air* 1959; *The Watercress Girl* (s) 1959; *When the Green Woods Laugh* 1960; *The Day of the Tortoise* 1961; *Now Sleeps the Crimson Petal* (s) 1961; *A Crown of Wild Myrtle* 1962; *The Golden Oriole* 1962; *Oh! To Be In England* 1963; *A Moment in Time* 1964; *The Wedding Party* (s) 1965; *The Distant Horns of Summer* 1967; *The Four Beauties* (s) 1968; *The Wild Cherry Tree* (s) 1968; *A Little of What You Fancy* 1970; *Triple Echo* 1971; *The Song of the Wren* 1972

Non-Fiction

Flowers and Faces 1935; *Through the Woods* 1936; *Down the River* 1937; *The Modern Short Story: a Critical Survey* 1941; *In the Heart of the Country* 1942; *Country Life* 1943; *O! More Than Happy Countryman* 1943; *There's Freedom in the Air* 1944; *The Tinkers of Elstow* 1947; *Edward Garnett: a Memoir* 1950; *The Country of White Clover* 1952; *The Face of England* 1952; *The Vanished World* (a) 1969; *The Blossoming World* (a) 1971; *A Love of Flowers* 1971; *The World in Ripeness* (a) 1972; *A Fountain of Flowers* 1974

Plays

The Last Bread 1926; *The Day of Glory* 1945

For Children

The Seasons and the Gardener 1940; *Achilles and the Donkey* 1962; *Achilles and Diana* 1963; *Achilles and the Twins* 1964

Nina (Mary) BAWDEN 1925–

Bawden was born in London and educated at Ilford High School, where she remained until the war separated her from her parents, Charles and Ellalaine Mabey. She was evacuated to Wales and completed her schooling, after which she won a scholarship to Somerville College, Oxford. She was awarded her BA in 1946 and went on to study for an MA in 1951. She did additional graduate study at Salzburg University in 1960. She worked as an assistant for the Town and Country Planning Associates from 1946 to 1947, and served as a Justice of the Peace in Surrey, England, from 1968 to 1976. She has been married twice; first in 1946 to Henry Bawden, with whom she had two sons (the elder of whom was schizophrenic and died in his twenties); and then in 1954 to Austen Kark, a BBC executive who went on to become head of the World Service: they have a daughter.

Although Bawden is now best known for her children's writing, much of which has been dramatised for television, and for the quiet dissection of middle-class life of her later adult fiction, her first novels were experiments in genre: murder mysteries such as her first novel, *Who Calls the Tune* (1953), *The Odd Flamingo*, and *Change Here for Babylon*; and Gothic romance, such as *The Solitary Child*. With the publication of *Just Like a Lady* in 1960, she found the themes which have occupied her ever since: the relationships, infidelities and domestic unease of the educated middle class. 'Social comedies with modern themes', the book concentrates on the minutely observed ordinary lives of her north London neighbours.

Her first children's novel, *The Secret Passage*, was published in 1963. *Carrie's War*, a children's story drawing on the author's wartime evacuation to Wales (a theme which she returned to in *Keeping Henry*), was commended for the Carnegie Medal in 1973. *The Peppermint Pig* won the *Guardian* Award for Children's Fiction in 1975. Of her many adult novels, *Afternoon of a Good Woman* won the *Yorkshire Post* Novel of the Year for 1976, and *Circles of Deceit* was nominated for the Booker Prize in 1987. A concern for social justice and passion for travel combine in the setting of *Rebel on a Rock*, a condemnation of the Greek military junta. She has also written about child abuse in the novel *Anna Apparent* and the children's novel *Squib* (1971). She reviews for the *Daily Telegraph* and *Evening Standard*, is a member of PEN and a continually prolific writer with over thirty novels to her name. She gave her own account of her life and career in *In My Own Time*, which is subtitled 'Almost an Autobiography'.

Fiction

Who Calls the Tune 1953; *The Old Flamingo* 1954; *Change Here for Babylon* 1955; *The Solitary Child* 1956; *Devil by the Sea* 1957; *Just Like a Lady* 1960; *In Honour Bound* 1961; *Tortoise by Candlelight* 1963; *Under the Skin* 1964; *A Little Love, a Little Learning* 1965; *A Woman of My Age* 1967; *The Grain of Truth* 1969; *The Birds on the Trees* 1970; *Anna Apparent* 1972; *On the Edge* 1973; *George Beneath a Paper Moon* 1974; *Afternoon of a Good Woman* 1976; *Rebel on a Rock* 1978; *Familiar Passions* 1979; *Walking Naked* 1981; *The Ice House* 1983; *Circles of Deceit* 1988; *Family Money* 1988; *Real Plate Jones* 1994

For Children

The Secret Passage 1963; *On the Run* 1964; *The White Horse Gang* 1966; *The Witch's Daughter* 1966; *A Handful of Thieves* 1967; *The Runaway Summer* 1969; *Squib* 1971; *Carrie's War* 1973; *The Peppermint Pig* 1975; *The Robbers* 1979; *Robert on a Rock* 1979; *Kept in the Dark* 1982; *Princess Alice* 1985; *The Finding* 1985; *Keeping Henry* 1988; *The Outside Child* 1991; *Humbug* 1992

Non-Fiction
In My Own Time (a) 1994

Walter BAXTER 1915–1994

The son of a farmer and sausage maker, Baxter was born in Margate, Kent, and educated at Ramsgate Junior School. After graduating from Trinity Hall, Cambridge, in 1936, he was articled as a clerk to a London firm of solicitors for three years. At the outbreak of the Second World War, he served first in Burma with the King's Own Yorkshire Light Infantry and later in India. He was ADC to Viscount Slim, and on the staff of a corps HQ during the reconquest of Burma. War stress, and especially the difficulties of coping with the bigoted British attitude towards homosexuality, compelled Baxter to write *Look Down in Mercy*, a novel on this theme. After it was completed he briefly returned to India to work on a mission. *Look Down in Mercy* was a commercial success and was, surprisingly, unmolested by the press. A fan-letter from **E.M. Forster**, whose identity was at first unclear owing to his illegible signature, led to friendship with him and with **J.R. Ackerley**.

When Baxter published *The Image and the Search* it was attacked by the Beaverbrook press, resulting in prosecution for obscene libel in 1954. Baxter had attempted to depict the degradation and despair felt by a woman after her husband is lost during the war. The jury twice failed to reach a verdict, and a second jury told a new judge that they would never reach agreement. Finally, at a third trial, the jury were directed to reach a verdict of 'not guilty'. This undoubtedly prepared the way for the liberal attitudes which freed **D.H. Lawrence**'s *Lady Chatterley's Lover* from legal bondage, but the effect on Baxter was devastating. Because he felt he could never express his feelings about sex in the way he wished, he stopped writing. Despite the crippling costs of the trial, Baxter opened a London restaurant in 1955, which he ran successfully, and which he sold in 1978. He subsequently lived mainly in Wiltshire with his dogs, cultivating orchids and vegetables, cooking and 'much else'. Tall, well built and handsome, he had a noticeable drawling voice which some considered an affectation, but which he regarded as a 'minor affliction'.

Fiction
Look Down in Mercy 1951; *The Image and the Search* 1953

Non-Fiction
People or Penguins 1974; *The Political Economy of Antitrust* 1980

Samuel (Barclay) BECKETT 1906–1989

Beckett was born into a prosperous Protestant family in the Dublin suburb of Foxrock, where his father was a surveyor, and read modern languages at Trinity College, Dublin. He travelled widely in Europe during the 1930s and settled in Paris, where he was a close disciple of **James Joyce**. The decade was dominated for Beckett by his dislike of Ireland, his difficult relationship with his mother, and the recurrent mental distress that led him into psychoanalysis during 1935–6. Fortunately this failed to 'cure' him of anything, and an unfeigned unhappiness and pessimism formed the lifelong core of his work.

His first separately published work was the poem *Whoroscope* (1930), which won the Nancy Cunard £10 Competition, followed in 1931 by an important study of Proust in which much of Beckett's own subsequent programme is already discernible. His major work of the decade was *Murphy*, a novel about a young Irishman irresistibly drawn towards interiority, madness and padded cells, which **Dylan Thomas** reviewed as 'Freudian blarney: Sodom and Begorrah'.

Beckett opted to remain in France during the war and was active in the French Resistance, for which he was subsequently decorated. His major works date from an extraordinary creative burst just after the war, and his great novel *Molloy* explores the mysterious Jekyll-and-Hyde relation between petit-bourgeois, punctilious Maron, who is hunting Molloy, and bear-like, shambling Molloy, who in turn is hunting for his mother. The trilogy of novels comprising *Molloy*, *Malone Dies* and *The Unnamable* were originally written in French, as was much of his work.

Beckett considered himself to be primarily a novelist, but he became a celebrity in 1954 with the play *Waiting for Godot*. Its fame has skewed Beckett's critical reception: he was misleadingly associated with existentialism and later corralled into the so-called Theatre of the Absurd, but these have both waned and he is now regarded as a postmodernist. The fatal fascination he has held for critics is bound up with his fertile avoidance of finite meaning: it has been said that his work is a kind of Rorschach blot in which people see whatever they want, and Beckett commented that the most important word in his work is 'perhaps'.

Notable later works include *Krapp's Last Tape*, originally written for the actor Patrick McGee, his bleakest play *Endgame* (his own favourite), and his 'second trilogy' of female drama, associated with the actress Billie Whitelaw: *Not I*, *Footfalls* and *Rockaby*. Much of the prose from the 1950s and 1960s is almost unreadable, but it underwent a late renaissance with slim works such as *Company*, *Ill Seen Ill Said* and *Stirrings Still*. Beckett lived since the 1930s with Suzanne Deschevaux-Dumesnil, communicating (according to his biographer) by telephone. They had no children. She seems to have shared his outlook, and her comment when he won the Nobel Prize in 1969 was: 'This is a catastrophe.'

Fiction

More Pricks Than Kicks **(s)** 1934; *Murphy* 1936; *Malone meurt* 1951; *Molloy* 1951; *L'Innommable* 1953; *Watt* 1953; *Trilogy* 1959; *Molly* [Eng.] 1955, *Malone Dies* 1956, *The Unnameable* 1958; *Nouvelles et textes pour rien* 1955 [as *Stories and Texts for Nothing* 1967]; *From an Abandoned Work* 1958; *Comment c'est* 1961 [as *How It Is* 1968]; *Imagination morte imaginez* [and as *Imagination Dead Imagine*] 1965; *Assez* 1966 [as *Enough* 1970]; *Bing* 1966; *No Knife* **(s)** 1967; *L'Issue* 1968; *Sans* 1969 [as *Lessness* 1971]; *Mercier et Camier* 1970 [as *Mercier and Camier* 1974]; *Premier Amour* 1970 [as *First Love* 1973]; *Séjour* 1970; *Le Depeupleur* 1971 [as *The Lost Ones* 1972]; *Abandonne* 1972; *The North* 1972; *Still* 1974; *All Strange Away* 1976; *For to End Yet Again and Other Fizzles* **(s)** 1976; *Four Novellas* 1977; *Six Residua* **(s)** 1978; *Nohow On* 1981: *Company* 1980, *Ill Seen Ill Said* 1982 [as *Mal vu mal dit* 1981], *Worstward Ho* 1983; *Stirrings Still* 1988 [as *Soubresauts* 1989]; *Dream of Fair to Middling Women* 1992

Plays

Le Kid 1931 [np]; *En attendant Godot* 1953 [pub. 1952; as *Waiting for Godot* 1955, pub. 1954]; *Acte sans paroles* 1957 [as *Act without Words* 1957, pub. 1958]; *All That Fall* **(r)** 1957; *Fin de partie* 1957 [as *Endgame* pub. 1958]; *Acte sans paroles II* [pub. 1963; as *Act without Words II* 1960, pub. 1959]; *Krapp's Last Tape* 1958 [pub. 1959]; *Embers* **(r)** 1959; *Happy Days* 1961; *Words and Music* **(r)** 1962; *Cascando* **(r)** 1963; *Play* 1963 [pub. 1964]; *Film* **(f)** 1965 [pub. 1967]; *Eh Joe* **(tv)** 1966 [pub. 1967]; *Va et vient* 1966 [as *Come and Go* 1968, pub. 1967]; *Breath* 1969 [pub. 1970]; *Not I* 1972 [pub. 1973]; *... but the clouds ...* 1976 [pub. 1977]; *Ghost Trio* **(tv)** 1977 [pub. 1976]; *Rough for Radio I* **(r)** [pub. 1976]; *Rough for Radio II* **(r)** 1976; *Rough for Theatre I* [pub. 1976]; *Rough for Theatre II* [pub. 1976]; *That Time* 1976; *Ends and Odds* [pub. 1977]; *A Piece of Monologue* 1980 [pub. 1982]; *Ohio Impromptu* 1981 [pub. 1982]; *Rockaby* 1981 [pub. 1982]; *Catastrophe* 1982 [pub. 1984]; *Quad* **(tv)** 1982 [pub. 1984]; *Nacht und Träume* **(r)** 1983; *What Where* 1983 [pub. 1984]; *Complete Dramatic Works* [pub. 1986]; *Teleplays* **(tv)** [pub. 1988]

Non-Fiction

Our Exagmination Round His Factification for Incamination of Work in Progress **(c)** 1929; *Proust* 1931; *Bram Van Velde* **(c)** 1958; *Proust and Three Dialogues* [with G. Duthuit] 1965; *Hommage à Jack B. Yeats* 1988; *Le Monde et le Pantalon* 1989

Poetry

Whoroscope 1930; *Echo's Bones and Other Precipitates* 1935; *Poems in English* 1961; *Collected Poems in English and French* 1977; *Poèmes suivi de mirlitonnades* 1978; *Collected Poems 1930–1978* 1984; *Comment dire* 1989

Translations

Negro [ed. N. Cunard] 1934; *Anthology of Mexican Poetry* [ed. O. Paz] 1958; Robert Pinget *The Old Tune* 1960; Guillaume Apollinaire *Zone* 1972; Arthur Rimbaud *The Drunken Boat* 1976; René Crevel *The Negress in the Brothel* 1989; Paul Éluard *Seconde Nature* 1990

Collections

Disjecta: Miscellaneous Writings and a Dramatic Fragment [ed. R. Cohn] 1983; *Collected Shorter Prose 1945–80* 1984; *As the Story Was Told* 1987; *As the Story Was Told: Uncollected and Late Prose* 1990

Biography

Samuel Beckett by Deirdre Bair 1978

Sybille BEDFORD 1911–

Born in Charlottenburg, Berlin, the daughter of a south German baron, Maximilian von Schoenebeck, and his part-English, part-Jewish wife, Bedford passed her early childhood in a manor house at Feldkirch, Baden, amid the formality and violence (an uncle was killed by a brother officer) of the Wilhelminian Germany evoked in her first and most famous novel, *A Legacy* (1956). Her parents separated and, after her father's death, she lived with her mother and art-dealer stepfather in Italy. The milieu was cultured and cosmopolitan, and for her girlhood Bedford spent much time in England, soon adopting English as her literary language. In the mid-1920s the family moved to Sanary in the south of France, where she soon became a fledgling member of the artistic colony gathered there, in particular developing a close friendship with **Aldous Huxley** and his first wife, Maria Nys. Bedford was later chosen as Huxley's official biographer, and her two-volume life was published to considerable acclaim. To the Sanary period also belong her mother's growing morphine addiction, movingly chronicled in her fourth and most autobiographical novel, *Jigsaw*, and her short-lived marriage to Walter Bedford in 1935. In 1940, a minor anti-fascist record made it expedient for her to leave France, and she joined the Huxleys in the USA, living there for some years and, in the 1950s, in Rome. Several early novels were rejected, and she supported herself by ghost-writing and translating, publishing her first book, *The Sudden View* (about travelling in Mexico, and later republished as *A Visit to Don Otavio*) in 1953. Three novels followed in the 1950s and 1960s, all Jamesian evocations of the vanished European society of her youth, winning high praise for beauty of style and insight into human motives, although dismissed by a few as snobbish and insubstantial. From the later 1950s she developed a new career as a reporter of trials, publishing *The Best We Can Do*, a classic account of the Bodkin Adams poisoning trial, and a comparative study of European systems of justice. Her accounts of several more trials, commissioned by newspapers and magazines, have reappeared in 1990 in the volume of essays *As It Was*. After periods spent living in East Anglia, and between Britain and the south of France, Bedford has now been settled for some years in Chelsea, London. Her shortlisting for the Booker Prize in 1989 with *Jigsaw* coincided with a revival of interest in her work and distinguished literary career, which resulted in her being awarded the Golden Pen Award for long service to literature by the PEN Literary Foundation.

Fiction

A Legacy 1956; *A Favourite of the Gods* 1962; *A Compass Error* 1968; *Jigsaw* 1989

Non-Fiction

The Sudden View 1953 [as *A Visit to Don Otavio* 1960]; *The Best We Can Do* 1958; *The Faces of Justice* 1961; *Aldous Huxley* 2 vols 1973, 1974; *As It Was* 1990

Patricia BEER 1924–

Beer spent her early childhood in Withycombe Raleigh, a village near Exmouth, where she was born. Her father, Andrew, was a railway clerk, and her mother, Harriet, a schoolteacher. In fact neither of her parents called each other by their given name, and were known as John and Queen. Socially superior to her husband, Queen was a member of the Plymouth Brethren and dominated the family with her religious foibles. She taught her daughter at home, considering the local infant school to be too rough for her delicate children. With some misgivings, Queen later sent her to be taught in a kindergarten, run by a Roman Catholic. When she was tough enough to walk the two miles to the Council School, Beer found a conflict between what she had learned at home and what she was now taught. She was kept apart from other children to an extent, and frowned upon for taking part in school plays complete with make up. Her initial introduction to literature was through hymns. Later, after winning a scholarship to Exmouth Grammar School, she received every school prize, and after going up to Oxford, graduated in English literature.

Beer lived in Italy from 1946 to 1953, where she lectured in English, later lecturing at Goldsmith's College, London. Life in Devon and the unusual childhood Beer spent there became something of an obsession with her, providing subject-matter for much of her poetry, and also an autobiography, *Mrs Beer's House*. Family legends, some of dubious origin, find their way into her first volume of poem, *Loss of the Magyar* (1959), which showed an original talent that had not been influenced by any school of poetry. Beer has said that after publishing her autobiography, her verse became much freer in style, and the work she produced after 1968 depends less on hymn or ballad form than her early work, which she felt suffered from 'verbal debauchery'. A novel, *Moon's Ottery*, was preceded by a book of literary criticism, *Reader, I Married Him*. Beer herself married an architect in 1964.

Poetry

Loss of the Magyar and Other Poems 1959; *The Survivors* 1963; *Just Like the Resurrection* 1967; *The Estuary* 1971; *Spanish Balcony* 1973; *Driving West* 1975; *Selected Poems* 1979; *The Lie of the Land* 1983; *Collected Poems* 1988; *Friend of Heraclitus* 1993

Fiction

Moon's Ottery 1978

Non-Fiction

Mrs Beer's House (a) 1968; *An Introduction to the Metaphysical Poets* 1972; *Reader, I Married Him* 1974; *Patricia Beer's Devon* 1984; *Wessex* 1985

Edited

New Poems [with T. Hughes, V. Scannell] 1962; *New Poems 1975* 1975; *New Poetry 2* [with K. Crossley-Holland] 1976; *Poetry Supplement* 1978

Plays

Pride, Prejudice, and the Woman Question (r) 1975; *The Enterprise of England* 1979 [np]

(Henry) Max(imilian) BEERBOHM 1872–1956

The youngest member of an Anglo-German family of merchants, Beerbohm was given a traditional English education at Charterhouse and Merton College, Oxford. He emerged into the London of the 1890s, a dandified figure who became an intimate (and an affectionate critic) of the Wilde circle. Equally talented as a writer and draughtsman, he contributed elegant essays and caricatures to the *Yellow Book* and other periodicals. **George Bernard Shaw**, whom he succeeded as drama critic of the *Saturday Review*, dubbed him 'the incomparable Max', a sobriquet which was both appropriate and enduring. He published numerous stories, essays, parodies and a single, popular novel of Oxford life, *Zuleika Dobson*. He is best remembered for his delicately coloured line-and-wash cartoons which depicted recent and contemporary representatives of English cultural life with feline wit, and which fulfilled his own definition of the perfect caricature: 'that which, on a small surface, with the simplest means, most accurately exaggerates, to the highest point, the peculiarities of a human being, at his most characteristic moment, in the most beautiful manner.' He haunted the fashionable Café Royal, where he often drew his unsuspecting subjects, reflected in the open back of his pocket watch which rested on a table.

Uninterested in psychology or philosophy, Beerbohm hardly allowed himself to be tempted out of the nineteenth century into modernity, but this brought him young friends such as **Siegfried Sassoon**, and an admiring **Henry James**. Beerbohm married an American actress, Florence Kahn, in 1910, and despite settling near Rapallo in Italy where he retired for life, he never attempted to learn Italian. He exhibited his caricatures frequently in London, which he visited to make a series of radio broadcasts in 1936. Because of the threat of war, the Beerbohms returned to England in 1938, and the following year to his great satisfaction, he was knighted. Few traces remained, however, of the *fin-de-siècle* world which he had inhabited; and he returned to Italy in 1947, never to visit England again.

Beerbohm remained to the end of his life, as he appeared to **Virginia Woolf** in 1938, '... like a Cheshire cat. Orbicular. Jowld. Blue eyed. Eyes grow vague ... all curves'. In 1951, Florence died, and he was cared for by Elizabeth Jungmann whom he had known since 1927; they married less than a month before Beerbohm's death. His ashes were buried in St Paul's Cathedral, an honour which would have amused but pleased him.

Fiction

The Happy Hypocrite (s) 1897; *Zuleika Dobson* 1911; *The Dragon of Hay Hill* (s) 1928

Non-Fiction

Caricatures of Twenty-Five Gentlemen 1896; *More* 1899; *The Poet's Corner* 1904 [rev. 1943]; *A Book of Caricatures* 1907; *Yet Again* 1909; *Cartoon: 'The Second Childhood of John Bull'* 1911; *Fifty Caricatures* 1913; *And Even Now* 1920; *A Survey* 1921; *A Defence of Cosmetics* 1922; *Rossetti and His Circle* 1922; *A Peep into the Past* 1923; *Things Old and New* 1923; *Around Theatres* 1924; *The Gueron* 1925; *Observations* 1925; *Leaves from the Garland* 1926; *Heroes and Heroines of Bitter Sweet* 1931; *Mainly on the Air* 1946 [rev. 1957]; *Sherlockiana: a Reminiscence of Sherlock Holmes* 1948; *Caricatures by Max* 1958; *Max's Nineties* 1958; *Letters to Reggie Turner* [ed. R. Hart-Davis] 1964; *More Theatres: 1898–1903* 1969; *Last Theatres* 1970; *A Catalogue of the Caricatures of Max Beerbohm* 1972; *Max and Will: Max Beerbohm and William Rothenstein, Their Friendship and Letters 1893–1945* [ed. M.M. Lago, K. Beckson] 1975; *Beerbohm's Literary Caricatures: from Homer to Huxley* [ed. J.G. Riewald] 1977; Siegfried Sassoon *Letters to Max Beerbohm: with a Few Answers* [ed. R. Hart-Davis] 1986; *The Letters of Max Beerbohm* [ed. R. Hart-Davis] 1988

Poetry

A Luncheon 1946; *Max in Verse* [ed. J.G. Riewald] 1963

Edited

Herbert Beerbohm Tree: Some Memories of Him and His Art 1920

Collections

A Christmas Garland 1912; *Seven Men* 1919 [rev. as *Seven Men and Two Others* 1950]; *A Variety of Things* 1928

Biography

Max by Lord David Cecil 1964

Brendan (Francis) BEHAN 1923–1964

Behan was the most famous of the three sons, all of whom became writers, of a Dublin house-painter and IRA man, and his wife, the powerful and passionate Kathleen Behan. His father first saw the infant Brendan through the bars of Kilmainham Jail, and it was perhaps inevitable that Brendan, after being schooled by the French Sisters of Charity in Dublin, should follow his father into the IRA. Aged sixteen, he was arrested with a suitcase full of homemade explosives in Liverpool and, being too young for prison, sentenced to three years of Borstal training. When his sentence was completed, he was expelled from England and returned to Dublin, where, having shot at two policemen, he spent a further period of captivity in an Irish jail. Released in 1945, be became a house-painter, a seaman and, increasingly, a freelance journalist. His frank and vivid autobiography *Borstal Boy* won him success and reputation in Britain, although it was banned in Ireland and Australia.

Simultaneously, he was gaining fame as a playwright. *The Quare Fellow*, the story of a man condemned to execution, had its Dublin premier in 1954 (it was originally written in Gaelic) and was followed four years later by *The Hostage*, about an English soldier imprisoned in a Dublin brothel. Both plays gained further currency when produced by Joan Littlewood at the Theatre Royal, Stratford East, during 1958 and 1959. Behan also wrote a thriller, *The Scarperer*, several compilations of shorter pieces, and translated Marlowe's verse into Gaelic. A boisterous, sometimes obstreperous yet sensitive man, he was also, as his obituary in *The Times* put it, 'an inebriate of some standing', and died, aged only forty-one, of diabetes, drink and jaundice. He married the painter Beatrice Salkeld in 1955 and they had one daughter.

Plays

The Quare Fellow **(st)** 1954, **(f)** 1961 [pub. 1956]; *The Big House* **(r)** 1957 [pub. 1958]; *The Hostage* 1958; *Moving Out* and *A Garden Party* [ed. R. Hogan; pub. 1967]; *Richard's Cork Leg* 1972 [pub. 1973]

Fiction

The Scarperer 1964

Non-Fiction

Borstal Boy **(a)** 1958; *Brendan Behan's Island: an Irish Sketchbook* 1962; *Hold Your Hour and Have Another* 1963; *Brendan Behan's New York* 1964; *Confessions of an Irish Rebel* 1965

Collections

After the Wake 1981

Biography

Brendan Behan by Ulick O'Connor 1970

Adrian (Hanbury) BELL 1901–1980

Born in London, Bell was the son of a Scottish journalist who wrote for the *Observer* newspaper. Early inclinations towards a scholarly career vanished when Bell found that he was miserable at Uppingham School. At home he had studied classics and music, and had a vague desire to be a poet. After a year of illness when he was nineteen, Bell decided that he would be a writer, but in 1920 he became apprenticed to a Suffolk farmer for a year. He put these experiences into his first novel, *Corduroy* (1930), and when the book was an unexpected bestseller, he borrowed money from his father, and bought a fifty-acre farm of his own. His life there is described in *Silver Ley*, in which he recalls a neighbour remarking of the difficulties he encountered: 'What a pity when you've got a brain for other things.' These included the setting of the first crossword published in *The Times*, which appeared on 1 February 1930. It was a remarkable achievement since Bell had neither compiled nor even solved one previously. Wearing out several dictionaries, he was a frequent compiler until his death. *Corduroy* brought Bell many admirers, including Marjorie Gibson, whom he married in 1931. In fifteen novels and essays Bell charted the changing ways of agricultural practice and country folk-lore; his writing was popular, perhaps because his appeal to nostalgia was never marred by sentimentality. Always preferring independence to office hours, he moved to Beccles in Suffolk after the Second World War, where he wrote the 'Countryman's Notebook' in a weekly column for the local newspaper; these pieces were later collected in a book of the same name. A lightweight but well-crafted autobiography was published in 1961.

Fiction

Corduroy 1930; *Silver Ley* 1931; *The Cherry Tree* 1932; *Folly Field* 1933; *Balcony* 1934; *By-Road* 1937; *Shepherd's Farm* 1939; *Apple Acre* 1942; *Sunrise to Sunset* 1944; *The Budding Morrow* 1947; *The Black Donkey* 1949; *The Path by the Window* 1952; *Music in the Morning* 1954; *A Young Man's Fancy* 1955; *The Mill House* 1958

Non-Fiction

Men and Fields 1939; *A Suffolk Harvest* 1956; *A Street in Suffolk* 1964; *My Own Master* **(a)** 1961

Poetry

Poems 1935

Edited

The Open Air 1936

(Mary Eirene) Frances BELLERBY 1899–1975

Bellerby was born Frances Parker in Bristol, the second child of an Anglo-Catholic clergyman. In her unpublished autobiography, over which she laboured painfully between 1957 and 1968, finally abandoning it because she had 'some fear of insanity', she wrote that her father was 'possessed by a powerful desire to take ... the gospel to the poor and by it to comfort the sick and dying. He had no wish whatever to bother about the rich'. She and her brother had an isolated childhood. She was taught by her mother until the age of nine, when she went to Mortimer House, Clifton, where she stayed until she was nineteen. In 1915 her brother Jack was killed in France – the poem 'Anniversary', as finally revised in 1968, opens 'Never mourn the deathless dead' – and by the end of the war her mother had suffered a mental breakdown; she was to commit suicide in 1932. Both these events and the fact that she felt her father had neglected her mother were to haunt Bellerby for the rest of her life; after her father's death she came to feel that, as the last of the family, she was the ghost and they were alive.

In 1929 she married John Bellerby, then a Fellow of Caius College, Cambridge; six months later she suffered a spinal injury that was to leave her in permanent pain. Though she published a novel and a collection of stories, she wrote no poetry during her marriage; then suddenly, living in Cornwall apart from her husband, reliving the scenes of childhood holidays, it 'returned, unpremeditated, after a total of "12 lost years"'. She began publishing in magazines, especially the *Listener*, where her poems attracted attention, as did her first collection, *Plash Mill* (named after her Cornish cottage). In 1950 she contracted breast cancer. After an operation in June 1951 she regarded herself as a Lazarus. The most complete expression of this appeared some twenty years later in 'The Sunset of that Day', in her last collection, *The First-Known*, which opens 'Lazarus dare not raise his eyes / above the hooded valley'. The poems she wrote in the 1950s were largely based on her hospital notebooks; *The Stone Angel and the Stone Man* is the collection in which she began to find her true voice.

In 1955 she moved to Goveton in Devon, where she lived for the rest of her life. She became increasingly self-critical, but her last poems suggest that she had arrived at a calm acceptance of her agonised life. She died a month before her seventy-sixth birthday.

Her friend **Charles Causley** described her as 'one of the true originals of our time': 'Her memory is long, particularized, sensual ... Her infant self could have been described as sacramentalist; and, this, to a large extent, she has remained.'

Poetry

Plash Mill 1946; *The Brightening Cloud and Other Poems* 1949; *The Stone Angel and the Stone Man* 1958; *The Sheltering Water and Other Poems* 1970; *Selected Poems* [ed. C. Causley] 1971; *The First-Known and Other Poems* 1975; *Selected Poems* [ed. A. Stevenson] 1986

Fiction

The Unspoilt [as Frances Parker] 1928; *Shadowy Bricks* 1932; *Come to an End and Other Stories* (s) 1939; *Hath the Rain a Father?* 1946; *The Acorn and the Cup with Other Stories* (s) 1948; *A Breathless Child and Other Stories* (s) 1952; *Selected Short Stories* (s) [ed. J. Hooker] 1986

Non-Fiction

Perhaps? [as Frances Parker] 1927; *The Neighbours* 1931

(Joseph) Hilaire (Pierre René) BELLOC 1870–1953

Of mixed French, English and Irish ancestry, Belloc was born at St Cloud, near Paris. His father's family had a French military tradition going back to Napoleon's time but, when Belloc was left fatherless at the age of two, his English mother brought him and his sister to England, and from 1878 he grew up at Slindon in Sussex. Thus were formed the dual loyalties of his life, to an idealised rural England (with an especial attachment to the 'South Country'), and to the traditions of Gallic and Latin Europe, particularly as focused by the Roman Catholic church. After leaving the Oratory School, Birmingham, Belloc drifted for some years: he spent one period training for the French navy, and another serving in the French artillery. His sister (later the popular novelist Mrs Belloc Lowndes) persuaded W.T. Stead of the *Pall Mall Gazette* to advance a small sum for him to travel around France on a bicycle and record his impressions. He also travelled to America in quest of his future wife, Elodie Hogan, an Irish Californian, with whom he had three sons and two daughters.

He went up to Balliol College, Oxford, in 1893, where he had great social success and became president of the Union. After failing to win a Fellowship, he quickly settled to the career of man-of-letters, eventually writing more than a hundred books. He achieved Fleet Street and public prominence, becoming literary editor of the *Morning Post* from 1906 to 1910, while also a Liberal MP campaigning against 'the modern Anglo-Judaic plutocracy under which we live', and pouring out a stream of books with such titles as *On Something*, *On Everything* and *On Nothing*. He was closely associated with fellow Catholic journalist **G.K. Chesterton** and the two conducted many campaigns in support of the roistering English peasantry and against humanist writers such as **H.G. Wells** and **George Bernard Shaw**. The death of his wife in 1914 was a shattering blow, and in later life he was prone to melancholia and intransigence, pouring out a stream of polemical Catholic works. Personal decline after the age of seventy put a final stop to his voluminous *oeuvre*. Of his many works, perhaps his cautionary verses for children, a handful of haunting poems (such as 'Ha'nacker Mill' and 'Tarantella'), his early travel book *The Path to Rome* and his volume of political analysis, *The Servile State*, best entitle him to a permanent readership.

For Children

The Bad Child's Book of Beasts: Verses 1896; *More Beasts for Worse Children: Verses* 1897; *A Moral Alphabet* 1899; *Cautionary Tales for Children* 1907; *Economics for Helen* 1924; *New Cautionary Tales: Verses* 1930

Fiction

Emmanuel Burden 1904; *Mr Clutterbuck's Election* 1908; *A Change in the Cabinet* 1909; *Pongo and the Bull* 1910; *The Girondin* 1911; *The Green Overcoat* 1912; *The Mercy of Allah* 1922; *Mr Petre* 1925; *The Emerald of Catherine the Great* 1926; *The Haunted House* 1927; *Belinda* 1928; *But Soft – We Are Observed* [USA *Shadowed!*] 1928; *The Missing Masterpiece* 1929; *The Man Who Made Gold* 1931; *The Postmaster-General* 1932; *The Hedge and the Horse* 1936; *Bona Mors* (s) 1953

Poetry

Verse and Sonnets 1896; *The Modern Traveller* 1898; *Verses* 1910; *More Peers* 1911; *Sonnets and Verse* 1923; *The Chanty of the Nona* 1928; *The Praise of Wine* 1931; *Ladies and Gentlemen* 1932; *Songs of the South Country* 1951

Non-Fiction

Danton a Study 1899; *Lambkin's Remains* 1900; *Paris* 1901; *Robespierre a Study* 1901; *The Path to Rome* 1902; *The Aftermath; or, Gleanings from a Busy Life* 1903; *The Great Inquiry (Only Authorised Version) Faithfully Reported* 1903; *Avril, Being Essays on the Poetry of the French Renaissance* 1904; *The Old Road* 1904; *Esto Perpetua: Algerian Studies and Impressions* 1906; *Hills and the Sea* 1906; *An Open Letter on the Decay of Faith* 1906; *Sussex, Painted by Wilfred Ball* 1906; *The Historic Thames* 1907; *The Catholic Church and Historical Truth* 1908; *An Examination of Socialism* 1908; *The Eye-Witness* 1908; *On Nothing and Kindred Subjects* 1908; *Marie Antoinette* 1909; *On Everything* 1909; *The Pyrenees* 1909; *The Church and Socialism* 1910; *The Ferrer Case* 1910; *The International* 1910; *On Anything* 1910; *On Something* 1910; *First and Last* 1911; *The French Revolution* 1911; *The Party System* [with C. Chesterton] 1911; *Socialism and the Servile State* 1911; *Six British Battles* 6 vols 1911–13; *The Four Men: a Farrago* 1912; *The River of London* 1912; *The Servile State* 1912; *This and That and the Other* 1912; *Warfare in England* 1912; *The Stane Street: a Monograph* 1913; *Anti-Catholic History: How It Is Written* 1914; *The Book of the Bayeux Tapestry* 1914; *Three Essays* 1914; *A General Sketch of the European War: the First and Second Phase* 2 vols 1915, 1916; *High Lights of the French Revolution* 1915; *The History of England from the First Invasion by the Romans to the Accession of King George the Fifth* 1915; *'Land and Water' Map of the War and How to Use It* 1915; *Two Maps of Europe and Some Other Aspects of the Great War* 1915; *At the Sign of the Lion and Other Essays* 1916; *The Last Days of the French Monarchy* 1916; *The Second Year of the War* 1916; *The Free Press* 1918; *Religion and Civil Liberty* 1918; *The Catholic Church and the Principle of Private Property* 1920; *Europe and Faith* 1920; *The House of Commons and Monarchy* 1920; *Pascal's Provincial Letters* 1921; *Catholic Social Reform Versus Socialism* 1922; *The Jews* 1922; *The Contrast* 1923; *On* 1923; *The Road* 1923; *The Campaign of 1812 and the Retreat from Moscow* 1924; *The Political Effort* 1924; *The Cruise of the 'Nona'* 1925; *England and the Faith* 1925; *A History of England* 2 vols 1925, 1931; *Miniatures of French History* 1925; *The Catholic Church and History* 1926; *A Companion to Mr Wells's 'Outline of History'* 1926; *The Highway and Its Vehicles* [ed. G. Holme] 1926; *Mr Belloc Still Objects to Mr Wells's 'Outline of History'* 1926; *Mrs Markham's New History of England* 1926; *Short Talks with the Dead and Others* 1926; *Oliver Cromwell* 1927; *Towns of Destiny* [UK *Many Cities*] 1927; *How the Reformation Happened* 1928; *James the Second* 1928; *A Conversation with an Angel and Other Essays* 1929; *Joan of Arc* 1929; *Richelieu* 1929; *Survivals and New Arrivals* 1929; *Wolsey* 1930; *A Conversation with a Cat and Others* 1931; *Cramner* 1931; *Essays of a Catholic Layman in England* 1931; *On Translation* 1931; *How We Got the Bible* 1932; *Napoleon* 1932; *The Question and the Answer* 1932; *Becket* 1933; *Charles the First, King of England* 1933; *The Tactics and Strategy of the Great Duke of Marlborough* 1933; *William the Conqueror* 1933; *Cromwell* 1934; *A Shorter History of England* 1934; *Milton* 1935; *The Battle Ground* 1936; *Characters of the Reformation* 1936; *An Essay on the Restoration of Property* 1936; *The Crusade: The World's Debate* 1937; *The Crisis of Our Civilization* 1937; *An Essay on the Nature of Contemporary England* 1937; *The Issue* 1937; *The Case of Doctor Coulton* 1938; *The Great Heresies* 1938; *Monarchy: a Study of Louis*

XIV 1938; *Return to the Baltic* 1938; *Charles II: the Last Rally* 1939; *The Test Is Poland* 1939; *The Catholic and the War* 1940; *On the Place of Gilbert Chesterton in English Letters* 1940; *The Silence of the Sea and Other Essays* 1940; *Places* 1941; *Elizabethan Commentary* 1942; *One Thing and Another: a Miscellany from His Uncollected Essays* [ed. P. Cahill] 1955; *Letters from Hilaire Belloc* [ed. R. Speaight] 1958; *Advice* 1960; *Hilaire Belloc's Prefaces, Written for Fellow Authors* [ed. J.A. De Chantigny] 1971

Edited

Extracts from the Diaries and Letters of Hubert Howard 1899; *The Footpath Way: an Anthology for Walkers* 1911; James Murray Allison *Travel Notes on a Holiday Tour in France* 1931; *The Calvert Series* 13 vols 1926–30

Biography

The Life of Hilaire Belloc by Robert Speaight 1957; *Hilaire Belloc* by A.N. Wilson 1989

Saul BELLOW 1915–

Bellow was born in Montreal, the youngest child of a Russian-Jewish family who had left St Petersburg in 1913. His father was a businessman who enjoyed mixed fortunes, and in 1924 the family moved to Chicago, where Bellow was educated at Tuley High School and the University of Chicago, before transferring to Northwestern University, where he graduated in 1937 with honours in anthropology and sociology. After a brief spell at the University of Wisconsin, Madison, he gave up his thesis on anthropology to concentrate on writing. He taught at Pestalozzi-Froebel Teachers' College, Chicago, from 1938 to 1942, and worked for the editorial department of the *Encyclopaedia Britannica* from 1943 to 1944. While serving with the Merchant Marine, he published his first novel, *Dangling Man*, which describes the intellectual and spiritual vacillations of a young Chicago man waiting to be drafted. After the war Bellow returned to teaching, holding various posts at the Universities of Minnesota, New York, Princeton and Puerto Rico. From 1960 to 1962 he was a co-editor of the literary magazine *The Noble Savage*, and since 1962 he has been a professor on the Committee of Social Thought at the University of Chicago.

Travelling in Paris and Rome in 1948 with the proceeds of a Guggenheim Fellowship, Bellow began work on *The Adventures of Augie March*, a long comic narrative which departed from the introspective and pessimistic tone of his earlier novels. The best-selling *Herzog* is perhaps the quintessential Bellow novel, a comic examination of 'the mire of post-Renaissance, post-humanistic, post-Cartesian dissolution, next door to the void'. Many of Bellow's novels are partially autobiographical, none more so than *Humboldt's Gift*, which draws on the young writer's friendship with the poet **Delmore Schwartz**.

In 1975 Bellow visited Israel and recorded his impressions of a people 'actively, individually involved in universal history' in his only substantial work of non-fiction, *To Jerusalem and Back*. He has written a number of plays, including *The Last Analysis*, which enjoyed a short run in New York in 1964, and his novella *Seize the Day* was adapted for television (PBS-TV, 1987) with Bellow making a cameo appearance alongside Robin Williams. Bellow has won many awards for his work, including three National Book Awards (for *The Adventures of Angie March*, *Herzog* and *Mr Sammler's Planet*), a Pulitzer Prize for *Humbolt's Gift*, and the Nobel Prize for Literature in 1976. In his Nobel acceptance lecture he spoke of the need for literature to encourage the belief 'that the good we hang onto so tenaciously – in the face of evil, so obstinately – is no illusion'. He has three sons from his first four marriages, all of which ended in divorce, and has been married since 1989 to Janis Freedman.

Fiction

Dangling Man 1944; *The Victim* 1947; *The Adventures of Augie March* 1953; *Seize the Day* 1956; *Henderson the Rain King* 1959; *Herzog* 1964; *Mosby's Memoirs, and Other Stories* (s) 1968; *Mr Sammler's Planet* 1970; *Humboldt's Gift* 1975; *The Dean's December* 1982; *Him with His Foot in His Mouth, and Other Stories* (s) 1984; *More Die of Heartbreak* 1987; *The Bellarosa Connection* 1989; *A Theft* 1989; *Something to Remember Me By* (s) 1992

Plays

The Wrecker (tv) 1964 [pub. in *Seize the Day* 1956]; *The Last Analysis* 1964 [pub. 1965]; *Out from Under* 1966 [np]; *A Wen* 1966 [pub. in *Traverse Plays* ed. J. Haynes 1966]; *Orange Soufflé* 1966 [pub. in *Traverse Plays* ed. J. Haynes 1966]

Non-Fiction

Dessins, by Jesse Reichek [with C. Zervos] 1960; *Recent American Fiction: a Lecture* 1963; *Technology and the Frontiers of Knowledge: the Frank Nelson Doubleday Lectures 1972–3* (c) 1973; *To Jerusalem and Back: a Personal Account* 1976; *Nobel Lecture* 1979; *Modernity and Its Discontents* (c) [ed. B. Bourne, U. Eichler, D. Herman] 1987; *It All Adds Up: From the Dim Past to the Uncertain Future* 1994

Edited
Chicago in Fiction 1963 onwards; *Great Jewish Short Stories* 1963

Collections
Like You're Nobody: the Letters of Louis Gallo to Saul Below 1961–62, Plus Oedipus-Schmoedipus, the Story That Started It All 1966

Stephen Vincent BENÉT 1898–1943

Benét was born in Bethlehem, Pennsylvania, into an army family. His father was an avid reader and so Benét grew up with a love of books. He was educated at Yale, where his contemporaries included **Thornton Wilder** and **Archibald MacLeish** and the Sorbonne, where he met his wife, the writer Rosemary Carr. He had started to write verse at school and his first collection, *Five Men and Pompey*, appeared when he was still in his teens. He was a great collector of prizes: 'King David' won the poetry prize in the *Nation* magazine and, *John Brown's Body*, a long poem about the Civil War reeking of patriotism, a Pulitzer Prize. *The Times* said of it: 'the poem breathed the spirit of freedom under which America grew up', and suggested that thereafter Benét become the equivalent to a poet laureate. He also won the 1932 Shelley Award and the 1933 Roosevelt Medal.

He was a strong advocate of America's entry into the Second World War: in the United Nations Day speech President Roosevelt made on 14 July 1942, he read a prayer specially written by Benét. In addition to poetry he wrote five novels and five volumes of short stories. His best-known story is 'The Devil and Daniel Webster', a Faustian tale set in nineteenth-century rural America, which was made into a celebrated film by William Dieterle under the title *All That Money Can Buy* (1941). He also made a number of radio broadcasts.

After his early death in New York the first part of another narrative poem, *Western Star*, was published. It is nevertheless complete in itself, and also won a Pulitzer Prize.

His brother, William Rose Benét, was also a poet and edited *The Reader's Encyclopedia*, the standard American guide to world literature.

Poetry
Five Men and Pompey 1915; *Young Adventure* 1918; *Heaven and Earth* 1920; *Tiger Joy* 1925; *John Brown's Body* 1928; *Ballads and Poems, 1915–1930* 1931; *A Book of Americans* [with R. Carr Benét] 1933; *Burning City* 1936; *Western Star* 1943

Fiction
The Beginning of Wisdom 1921; *Young People's Pride* 1922; *Jean Huguenot* 1923; *Spanish Bayonet* 1926; *The Barefoot Saint* (s) 1929; *James Shore's Daughter* 1934; *The Devil and Daniel Webster* 1937; *Thirteen o'Clock* (s) 1937; *Johnny Pye and the Fool Killer* (s) 1938; *Tales Before Midnight* (s) 1943

Non-Fiction
We Stand United 1945

Alan BENNETT 1934–

The son of a butcher, Bennett was born in Leeds and educated at the Leeds Modern School and Exeter College, Oxford. He did postgraduate work on 'Richard II's retinue from 1388 to 1399' and, after national service in the Joint Services School for Linguists, he was a junior lecturer in modern history at Magdalen College, Oxford, from 1960 to 1962. He spent much of this latter period performing in the celebrated revue *Beyond the Fringe*, which he created with Jonathan Miller, Peter Cook and Dudley Moore, and which is generally held to have inaugurated the 'satire boom' of the early 1960s.

His first stage play, *Forty Years On* (1968), is clearly influenced by revue: 'an elegy for the passing of a traditional England ... constructed round a series of literary parodies'. Set in a boys' public school and taking the form of a play within a play, it set the agenda and tone for much of his subsequent work, which at once celebrates and deprecates English manners and mores. 'A writer's normal situation is to be in two minds,' he has said, 'because that is where the poetry or whatever comes from.' Much of this has to do with Bennett's sense of what it is to be English: 'An ironic attitude towards one's country and a scepticism about one's heritage are a part of that heritage,' he has said. Such equivocal feelings have led to his writing plays on the subject of espionage: *The Old Country* and *Single Spies*, a double bill consisting of *An Englishman Abroad* and *A Question of Attribution*, about Guy Burgess and Anthony Blunt respectively. His association with the National Theatre has also led to a sophisticated and idiosyncratic adaptation of **Kenneth Grahame**'s *The Wind in the Willows* and – perhaps his most impressive play to date – *The Madness of George III*, which he adapted for the cinema in 1995. Most of his plays have been published with illuminating (and occasionally rueful) prefaces, in which he explores the roots and themes of his work, and describes how each piece evolved during rehearsal or filming.

His extensive work for television is often set in his native Yorkshire, featuring formidable Northern women, and showing a highly acute ear for the way people speak. Bennett, who is occasionally prevailed upon to act, sometimes appears in these plays, as he does on stage. He has turned the national trait of self-deprecation into an art form, and much of his work for television takes the form of tragi-comedies about ordinary people, although *The Old Crowd* (1979) was a notable work of experimentation, made in collaboration with the director Lindsay Anderson. Amongst his most acclaimed work for television is the subtle series of monologues, *Talking Heads*, in which it is apparent that there are other, equally valid (but perhaps unrecognised), versions of the story each narrator tells. He has also made a number of television documentaries and has been a notable and eloquent advocate of television as a medium, particularly during a period when it has been seen to be under attack by the government.

Bennett has also written plays and films about other writers, notably Kafka (*The Insurance Man* and *Kafka's Dick*), Proust (*102 Boulevard Haussmann*) and **Joe Orton** (*Prick Up Your Ears*). He has had a long association with the *London Review of Books*, for which he has written reviews, diaries and features, including *The Lady in the Van*, an account of the elderly vagrant who lived for several years in a broken-down van in his Camden garden. Many of these pieces, along with prefaces to the plays, were collected to form a characteristically oblique, funny and fragmentary autobiography, *Writing Home*, which was chosen by many critics as their Book of the Year in 1994. The book's large sales confirmed Bennett's reputation as one of the country's most popular as well as considerable playwrights.

Plays

Beyond the Fringe [with P. Cook, D. Moore, J. Miller; pub. 1963; rev. as *The Complete Beyond the Fringe* 1987]; *Forty Years On* 1968 [pub. 1969]; *Getting On* 1971 [pub. 1972]; *Habeas Corpus* 1973; *The Old Country* 1977 [pub. 1978]; *Enjoy* 1980; *Office Suite* **(tv)** [pub. 1981]: *Green Forms* 1978, *A Visit from Miss Prothero* 1978; *Objects of Affection and Other Plays for TV* **(tv)** [pub. 1982; rev. 1984]: *A Day Out* 1972, *An Englishman Abroad* 1982, *Intensive Care* 1982, *Marks* 1982, *Our Winnie* 1982, *Rolling Home* 1982, *Say Something Happened* 1982, *A Woman of No Importance* 1982; *A Private Function* **(f)** 1984; *The Writer in Disguise* **(tv)** [pub. 1985]: *Me, I'm Afraid of Virginia Woolf* 1978, *Afternoon Off* 1979, *All Day on the Sands* 1979, *The Old Crowd* 1979, *One Fine Day* 1979; *Prick Up Your Ears* 1987; *Two Kafka Plays* [pub. 1987]: *The Insurance Man* **(tv)** 1986, *Kafka's Dick* **(st)** 1986; *Talking Heads* **(tv)** 1987 [pub. 1988]; *Single Spies* [pub. 1989]: *An Englishman Abroad and A Question of Attribution* 1988; *102 Boulevard Haussmann* 1990; *The Wind in the Willows* 1990 [from K. Grahame's novel; pub. 1991]; *The Madness of George III* **(st)** 1991 [pub. 1992; rev. as *The Madness of King George* **(f)** 1995]

Non-Fiction

The Lady in the Van 1990; *Writing Home* 1994

(Enoch) Arnold BENNETT 1867–1931

Born in Hanley in the Staffordshire Potteries, the son of a self-made solicitor, Bennett was brought up in a cultured but strictly Wesleyan Methodist home. On leaving school, he went to work in his father's office as a clerk. In 1899, at the age of twenty-one, he secured a job in a London lawyer's office and left for the metropolis intending to be a writer. Although he never returned to his native area of the Potteries (the 'Five Towns' of his novels), it is the background to almost all his best work. In London, he contributed stories to *Tit-Bits* and the *Yellow Book*, and soon became a professional writer and journalist; he was assistant editor and later editor of the magazine *Woman*, and his first novel, *A Man from the North*, was published in 1898.

In 1902, already deeply imbued with French culture and an admirer of the French realistic novel, he went to live in Paris, where he remained for several years and married Marie Marguerite Soulié in 1907. He published his best novel, *The Old Wives' Tale*, in 1908; soon acclaimed as a masterpiece, it placed him in the forefront of contemporary realistic novelists with **H.G. Wells** and **John Galsworthy**.

Bennett wrote more than eighty books and an enormous amount of journalism: in 1908 he noted methodically that his output that year had totalled 423,500 words. This brought the provincial lad with the stammer the country-house, yacht and acquaintance with fashionable people which he had always craved and Bennett eventually died in the type of luxury hotel that he had celebrated in one of his more light-hearted novels. Many of his works were inevitably potboilers, but some are excellent, for example *Clayhanger* and *Riceyman Steps*, the story of a London clerk. Bennett was separated from his wife in 1921, but in his last years found happiness with the actress Dorothy Chesterton. He was acknowledged as one of the leading literary men in England during this time, writing an exceptionally influential weekly literary *causerie* for the London *Evening Standard* from 1926.

Although his realism was deprecated by the modernist **Virginia Woolf**, he was one of the first Englishmen to praise Proust. Although admitted to be not quite a writer of the first rank, Bennett has never lacked admirers since his death, and his influence on the inter-war and post-war English regional novel can hardly be over-estimated.

Fiction

A Man from the North 1898; *Anna of the Five Towns* 1902; *The Grand Babylon Hotel* 1902; *The Gates of Wrath* 1903; *A Great Man* 1904; *Leonora* 1904; *The Loot of Cities* (s) 1904; *Sacred and Profane Love* 1905; *Tales of the Five Towns* (s) 1905; *Teresa of Watling Street* 1905; *Hugo* 1906; *Whom God Hath Joined* 1906; *The Sinews of War* 1906; *Buried Alive* 1907; *The City of Pleasure* 1907; *The Ghost* 1907; *The Grim Smile of the Five Towns* (s) 1907; *The Old Wives' Tale* 1908; *The Statue* 1908; *The Glimpse* 1909; *Helen with Her High Hand* 1910; *Clayhanger* 1910; *The Card* 1911; *Hilda Lessways* 1911; *The Matador of the Five Towns and Other Stories* (s) 1912; *The Regent* 1913; *The Price of Love* 1914; *These Twain* 1916; *The Lion's Share* 1916; *The Pretty Lady* 1918; *The Roll Call* 1918; *Lilian* 1922; *Mr Prohack* 1922; *Riceyman Steps* 1923; *Elsie and the Child and Other Stories* (s) 1924; *Lord Raingo* 1926; *The Woman Who Stole Everything and Other Stories* (s) 1927; *Accident* 1928; *The Strange Vanguard* 1928; *Imperial Palace* 1930; *The Night Visitor and Other Stories* (s) 1931; *Dreams of Destiny* [unfinished] 1932

Plays

Polite Farces for the Drawing-Room [pub. 1900]; *Cupid and Commonsense* 1909; *What the Public Wants* 1909; *The Honeymoon* 1911; *Milestones* [with E. Knoblock] 1912; *The Great Adventure* 1913; *The Title* 1918; *Judith* 1919; *Sacred and Profane Love* [from his novel] 1919; *Body and Soul* 1921; *The Love-Match* 1922; *Don Juan de Marana* (st) 1923, (l)1935; *The Bright Island* 1924; *London Life* 1924; *Mr Prohack* [with E. Knoblock; from his novel] 1927; *Judith* (l) 1929; *Piccadilly* 1929; *Flora* [pub. in *Five 3-Act Plays* 1933]; *The Snake-Charmer* [pub. in *Eight One-Act Plays* 1933]; *The Ides of March* [with F. Alcock; pub. in *One-Act Plays for Stage and Study* 1934]

Non-Fiction

Academy 1898; *Journalism for Women* 1898; *Fame and Fiction* 1901; *How to Become an Author* 1903; *How to Live on 24 Hours a Day* 1907; *The Reasonable Life* 1907; *The Human Machine* 1908; *Literary Taste: How to Form It* 1909; *The Feast of St Friend* 1911; *Mental Efficiency and Other Hints to Men and Women* 1912; *Those Untitled States* 1912; *Paris Nights* 1913; *The Plain Man and His Wife* 1913 [as *Married Life* 1916]; *The Author's Craft* 1914; *Friendship and Happiness* 1914; *Liberty: a Statement of the British Case* 1914; *Over There: Warscenes on the Western Front* 1915; *Books and Persons: Being Comments on a Past Epoch 1908–1911* 1917; *Self and Self-Management* 1918; *Our Women* 1920; *Things That Have Interested Me* 1921–6; *How to Make the Best of Life* 1923; *Essays of Today and Yesterday* 1926; *Mediterranean Scenes* 1928; *The Savour of Life: Essays in Gusto* 1928; *Journal, 1929* 1930; *Journals* [ed. N. Flower] 3 vols 1932–3; *Arnold Bennett's Letters to His Nephew* 1935; *Arnold Bennett: a Portrait Done at Home* 1935; *Arnold Bennett and H.G. Wells: a Record of a Personal and Literary Friendship* [ed. H. Wilson] 1960; *The Letters of Arnold Bennett* [ed. J.G. Hepburn] 3 vols 1966–70; *The Author's Craft and Other Critical Writings* [ed. S. Hynes] 1968; *Arnold Bennett in Love* 1972; *Sketches for Autobiography* (a) [ed. J. Hepburn] 1979

Biography

Arnold Bennett by Margaret Drabble 1974

E(dward) F(rederic) BENSON 1867–1940

Benson was born at Wellington College, where his father, Edward White Benson, later Archbishop of Canterbury, was headmaster. 'Fred' Benson was the fifth of six children, all of whom were to become writers. The eldest, Martin White Benson (1860–78), showed perhaps the most promise of all at the time of his sudden death from meningitis aged eighteen. At first, intellectually he had performed less than averagely well, but he was put under great pressure by his phenomenally industrious father and suddenly flowered after his twelfth birthday. His father, an important figure in ecclesiastical history, never really recovered from the shock of Martin's death, for which he blamed himself. *The Martin*, a moving poem, was discovered amongst his papers seven years after his death.

Arthur Christopher Benson (1862–1925), the eldest surviving son, was a notable writer in his day, and, like Fred, extraordinarily prolific. Though professing attachment to no church he wrote many books of a devotional nature. *From a College Window* and *Thread of Gold*, both published in 1906, are typical. He wrote a malicious and phenomenally long diary; at five million words, it is perhaps the longest diary ever kept. He was made master of Magdalene College, Cambridge, from 1915. Magdalene became one of the wealthiest colleges at this time almost entirely as a result of a penfriendship struck between Arthur and an American heiress who became a patron of the college. He wrote the verses 'Land of Hope and Glory'.

Remorse for Martin's death meant a reversal in their father's educational methods. Robert Hugh (1871–1914), the youngest son, was not actively encouraged to mature. The result was also fairly disastrous, and he retained certain child-like qualities all his life. He wrote many novels, though they are not much read today. He converted to Rome and came under the influence of the self-styled 'Baron' Corvo (**Frederick Rolfe**). Corvo, at one period, would send Robert obscene postcards, to the delight of the Benson family who displayed them on the mantelpiece. When Robert died his will revealed an obsessional fear of being buried alive. His elaborate precautions (including the installation of a bell) were ignored.

It seems that none of the Benson family was particularly enamoured of the opposite sex. When Arthur was asked if he had ever kissed a woman he is said to have replied: 'Yes. Once. On the brow'. Fred almost certainly found any physical contact distasteful and is sure to have concurred with the description 'that nasty thing Freud calls sex', as exclaimed by his most enduring creation, Emmeline ('Lucia') Lucas. Of his mother, Mary, Fred wrote: 'No man, except my father, had ever counted in her life.' After the death of her husband, Mary (known as 'Minnie', then as 'Ben') lived for the rest of her life with Lucy Tait, the daughter of a former Archbishop of Canterbury.

The elder of Fred's sisters, also called Mary (1863–90) but known as Nellie, wrote one novel during her short life: *At Sundry Times and in Diverse Manners*. The younger, Margaret (1865–1916), known as Maggie, also wrote but is best known as an archaeologist. Notably, she excavated the Temple of Mut in Karnak. She lived with Nettie Gourlay: 'one of the most silent of human beings'. The whole Benson family was susceptible to black depressions, but Maggie suffered a complete breakdown for the last nine years of her life.

E.F. Benson was educated at Marlborough, where he edited for a time *The Marlburian*. After King's College, Cambridge, and a brief period as a gentleman scholar at the British School of Archaeology in Athens (1892–5), he settled to the career of man-of-letters that he was to pursue until his death. In 1893 his first novel, *Dodo*, was an immediate success, and inspired the clerihew by **E.C. Bentley**: 'Archbishop Odo / Was just in the middle of *Dodo* / When he realised it was Sunday, / Sic transit gloria mundi.'

He wrote from the 1890s what he called 'an incredible amount of light fiction', including over seventy novels. Altogether he wrote over one hundred books, perhaps the most obscure of which is *Winter Sports in Switzerland*. (He had represented England in figure skating at the Winter Olympics.) The sports of his youth – tennis, rugby, squash and curling – gave way in later years to golf, walking and birdwatching. For the last twenty years of his life he suffered from acute arthritis.

His manservant, Charlie Tomlin, with whom he lived in his final years, died in 1980. They lived at Lamb House, once the home of **Henry James**, in Rye, Sussex, which was the model for the town of Tilling, where his most popular fictional creations, Miss Mapp and Lucia, first met. Benson most valued his biographies – subjects ranging from Charlotte Brontë to Alcibiades – but it is for the 'Lucia' novels, novels written in a similar vein such as *Paying Guests* and *Secret Lives*, and his ghost stories that he is remembered. He is buried in Playden, near Rye, grave No. 6223.

Fiction

Dodo 1893; *Six Common Things* (s) 1893; *A Double Overture* (s) 1894; *The Rubicon* 1894; *The Judgement Books* 1895; *Limitations* 1896; *The Babe, B.A.* 1897; *The Money Market* 1898; *The Vintage* 1898; *The Caspina* 1899; *Mammon and Co.* 1899; *The Princess Sophia* 1900; *The Luck of the Vails* 1901; *Scarlet and Hyssop* 1902; *An Act in a Backwater* 1903; *The Book of Months* 1903; *The Relentless City* 1903; *The Valkyries* [from Wagner's opera] 1903; *The Challoners* 1904; *The Image in the Sand* 1905; *The Angel of Pain* 1906; *The House of Defence* 1906; *Paul* 1906; *Shearers* 1907; *The Blotting Book* 1908; *The Climber* 1908; *A Reaping* 1909; *The Osbornes* 1910; *Daisy's Aunt* 1910; *A Sweeping* [from his novel *A Reaping*; as 'Edward of the Golden Heart'] 1911; *Juggernaut* 1911; *Account Rendered* 1911; *Mrs Ames* 1912; *The Room in the Tower* (s) 1912; *Thorley Weir* 1913; *The Weaker Vessel* 1913; *Dodo the Second* 1914; *Arundel* 1914; *The Oakleyites* 1915; *David Blaize* 1916; *Mike/Michael* 1916; *The Freaks of Mayfair* (s) 1916; *An Autumn Sowing* 1917; *David Blaize and the Green Door* 1918; *Up and Down* 1918; *Across the Stream* 1919; *Robin Linnet* 1919; *The Countess of Lowndes Square* (s) 1920; *Queen Lucia* 1920; *Dodo Wonders* 1921; *Lovers and Friends* 1921; *Miss Mapp* 1922; *Peter* 1922; *Colin* 1923; *Alan* 1924; *David of King's* 1924; *Colin II* 1925; *Rex* 1925; *Mezzanine* 1926; *Pharisees and Publicans* 1926; *Lucia in London* 1927; *The Male Impersonator* (s) 1928; *Spook Stories* (s) 1928; *Visible and Invisible* (s) 1923; *Paying Guests* 1929; *The Inheritor* 1930; *The Step* (s) 1930; *Mapp and Lucia* 1931; *Secret Lives* 1932; *Travail of Gold* 1933; *More Spook Stories* (s) 1934; *Ravens' Blood* 1934; *Lucia's Progress* 1935; *Trouble for Lucia* 1939

Non-Fiction
Sketches from Marlborough [as 'Anon'] 1888; *Daily Training* 1902; *The Mad Annual* [with E.H. Miles] 1903; *Diversions Day by Day* [with E. Miles] 1905; *English Figure Skating* 1908; *Winter Sports in Switzerland* 1913; *Bensoniana* 1912; *Thoughts from E.F. Benson* [ed. E.E. Morton] 1913; *Thoughts from E.F. Benson* [ed. H.B. Elliott] 1917; *Crescent and Iron Cross* 1918; *Mother* 1918; *The White Eagle of Poland* 1918; *Our Family Affairs* 1920; *Sir Francis Drake* 1927; *The Life of Alcibiades* 1928; *Ferdinand Magellan* 1929; *As We Were* (a) 1930; *As We Are* (a) 1932; *Charlotte Brontë* 1932; *King Edward VII* 1933; *The Outbreak of War, 1914* 1933; *Queen Victoria* 1935; *The Kaiser and English Relations* 1936; *Old London* 4 vols 1937; *Queen Victoria's Daughters* 1938; *Final Edition* (a) 1940

Plays
Aunt Jeannie 1902; *Dinner for Eight* 1915

Biography
The Life of E.F. Benson by Brian Masters 1991

Stella BENSON 1892–1933

Born at Lutwych Hall, Much Wenlock, Shropshire, her father's country home, Benson was a delicate child, educated privately and abroad, an experience which gave her a lifelong taste for travel. In 1912 she sailed to the West Indies, which gave her material for her first novel, *I Pose* (1915). Returning to England before the First World War, she took up women's suffrage and social reform, going into partnership with a local woman in Hoxton, east London, to sell paper bags. After a period as a land worker, she sailed to California for her health in 1918, and occupied herself there in a wide variety of professions ranging from lady's maid to lecturer at the University of California. A return trip to England by way of China found her experiencing a revolution and working in the X-ray department of a Peking hospital, as well as meeting her future husband, James Anderson, of the Chinese customs service. Once married to him, according to her obituary in *The Times*, she 'settled down to a quiet life in Yunnan, not too far from the brigands'. She lived in remote parts of China until her death from pneumonia at the age of forty.

Her most famous novel, *Tobit Transplanted*, which won the Femina Vie Heureuse Prize, was about a colony of White Russian exiles in China, while that country also provided the setting for several of her uneven but sometimes powerful short stories. Benson is not much read today, although her reputation has survived to an extent that many others have not; her interesting combination of fantasy and realism make her an apt candidate for revival. Besides novels and stories, she also wrote poetry, travel books and collections of articles.

Fiction
I Pose 1915; *This is the End* 1917; *Living Alone* 1919; *The Poor Man* 1922; *Pipers and a Dancer* 1924; *The Awakening* 1925; *Goodbye, Stranger* 1926; *The Man Who Missed the 'Bus* 1928; *The Far-Away Bride* 1930 [UK *Tobit Transplanted* 1931]; *Hope Against Hope and Other Stories* (s) 1931; *Christmas Formula and Other Stories* (s) 1932; *Mundos* [unfinished] 1935; *Collected Short Stories* (s) 1936

Non-Fiction
The Little World 1925 [rev. 1925]; *Worlds within Worlds* 1928

Poetry
Twenty 1918; *Poems* 1935

Plays
Kwan-Yin 1922

Edited
Come to Eleuthera 1925; Nicolas de Toulouse Lautrec de Savine *Pull Devil – Pull Baker* 1933

Biography
Stella Benson by Joy Grant 1987

E(dmund) C(lerihew) BENTLEY 1875–1956

Author of a classic detective novel and inventor of a verse form, Bentley was born in Shepherd's Bush, London, the son of an English civil servant and a Scottish mother, whose maiden name, Clerihew, has now entered the dictionary. He was educated at St Paul's, where he met **G.K. Chesterton**, who became a lifelong friend. It was while he was at school that he invented the short biographical poem consisting of four lines of uneven length, rhyming aa, bb, that he named the 'clerihew'. The earliest of the many he wrote that survive is the famous 'Sir Humphry Davy / Detested gravy / He died in the odium / Of having discovered sodium'. These verses, which he continued to write throughout his life, were later collected at long intervals in *Biography for Beginners* (under the partial pseudonym E. Clerihew) and *More Biography* and *Baseless Biography* (under his own name, the latter illustrated by his younger son, the cartoonist Nicolas Bentley).

In 1894 Bentley won a history scholarship to Merton College, Oxford, where he was president of the Union and captain of the University Boat Club (although he failed to get a blue). He was called to the Bar in 1901, but never practised, becoming instead a journalist. He joined the radical *Daily News* in 1902 (the same year that he married Violet Boileau), then became leader writer for the Tory *Daily Telegraph* in 1912, remaining there until his first retirement in 1934 (he returned for a spell as literary editor in 1939, replacing Harold Nicolson, who was on war service).

His famous novel, *Trent's Last Case*, was written in the small hours after the newspaper had been put to bed and published in 1912. One of the most original of detective novels, it broke the formula in many ways, not least in the detective's failure to solve the case. As its title suggested, the book was not intended to be the beginning of a series, and its success proved unrepeatable. Persuaded by friends, he wrote two further novels in the 1930s: *Trent's Own Case* (in collaboration with H. Warner Allen) and *Trent Intervenes*. Neither really succeeded, nor did *Elephant's Work*, which he published towards the end of his life. Far more successful were his volumes of clerihews, a 'definitive edition' of which ('with some suppressions and some textual changes') was published in the last decade of his life as *Clerihews Complete*. They have been much imitated but rarely surpassed, although one distinguished practitioner was **W.H. Auden**, who published an entire volume of them in 1971 under the title *Academic Graffiti*.

Fiction

Trent's Last Case 1912; *Trent's Own Case* [with H. Warner Allen] 1936; *Trent Intervenes* 1938; *Elephant's Work* 1950

Poetry

Biography for Beginners [as 'E. Clerihew'] 1905; *More Biography* 1929; *Baseless Biography* [with N. Bentley] 1939; *Clerihews Complete* 1951

Non-Fiction

Those Days (a) 1940

Edited

A Second Century of Detective Stories 1938

John (Peter) BERGER 1926–

Primarily a Marxist art critic, Berger has diversified with considerable success into the novel and other forms such as screenplays. Born in London, he studied at the Central School of Art and Chelsea School of Art, and became a drawing teacher and professional painter. He was an influential art critic for the *New Statesman* for ten years and was also a prolific broadcaster. Critical works include *The Success and Failure of Picasso*, which found Picasso guilty of 'a failure of revolutionary nerve', *The Moment of Cubism and Other Essays* and *Ways of Seeing*, a 1972 BBC Television series and book which presented Marxist art criticism at its most accessible. Designed as a riposte to Kenneth Clark's *Civilisation* (1969), the book established Berger as a Marxist populariser, who does for the visual arts what Terry Eagleton has done for literary criticism. Berger himself has said: 'I never considered *Ways of Seeing* an important work.' Collections of essays include *Permanent Red*. He was a major influence on the young Peter Fuller, who later recorded his disenchantment in *Seeing through Berger* (1988).

Berger's first novel was *A Painter of Our Time* (1958), but he achieved wide recognition in 1972 with *G*. It is the story of a Don Juan-like figure, written in a manner that seems to owe more than a little to the French *nouveau roman*, and it won the Booker Prize. In his speech of acceptance Berger attacked Booker–McConnell Ltd as imperialists, and announced his intention of giving half the prize money to the Black Panthers.

Berger has been married twice and has three children. For some years he has lived in France, which has given rise to *Pig Earth* (1979), a realist novel (with poems and a political essay) about the French peasantry. At once analytic, compassionate and authentic, this may be Berger's major literary work, although some commentators found it to be too self-consciously earthy and affectedly Lawrentian in its search for primal realities. The reception of Berger's fiction has been very mixed: Marxist academic Arnold Kettle saw *G* as 'one of the few serious attempts of our time to do for the novel what Brecht did for drama', while Duncan Fallowell felt that 'acne bursts from every page'.

Berger has also written non-fiction, including *A Fortunate Man*, the story of a country doctor, and *A Seventh Man: a Book of Images and Words about the Experience of Migrant Workers in Europe* (1975), and screenplays including *Jonas qui aura 25 ans en l'an 2000*. Berger continues to write, but his most lasting and influential work to date is likely to be his art criticism of the 1960s and 1970s. He lives in France in the Haute Savoie.

Non-Fiction

Marcel Frishman [with G. Besson] 1958; *Permanent Red: Essays in Seeing* [US *Towards Reality*] 1960; *The Success and Failure of Picasso* 1965; *A Fortunate Man* [with J. Mohr] 1967; *Art and Revolution: Ernst Neigvestny and the Role of the Artist in the USSR* 1969; *The Moment of Cubism and Other Essays* 1969; *Ways of Seeing* 1972; *A Seventh Man* [with J. Mohr] 1975; *Alinari – Photographs of Florence* 1978; *About Looking*

1980; *Another Way of Telling* [with J. Mohr] 1982; *And Our Faces, My Heart, Brief as Photos* 1984; *The White Bird* 1985 [US *The Sense of Sight* 1986]; *Keeping a Rendezvous* 1992

Fiction

A Painter of Our Time 1958; *The Foot of Clive* 1962; *Corker's Freedom* 1964; *G* 1972; *Into Their Labours Pig Earth* (s) 1979; *Once in Europe* (s) 1987; *Lilac and Flag* 1990

Translations

Bertolt Brecht; *Helene Weigel, Actress* [with A. Bostock] 1961, *Poems on the Theatre* [with A. Bostock] 1972; Nella Bielski: *Oranges for the Son of Asher Levy* [with L. Appignanesi] 1982, *After Arkadia* [with J. Steffen] 1991; Aimé Cesaire *Return to My Native Land* [with A. Bostock] 1969

Plays

A Question of Geography [with N. Bielski] 1984 [pub. 1987]; *Les Trois Chaleurs* 1985 [np]; *Boris* 1985 [np]; *Goya's Last Portrait* [with N. Bielski] 1989

Thomas BERGER 1924–

Berger was born in Cincinatti, Ohio, the only child of an artist and the business manager of a small local town. His mother encouraged him to read from a very early age. As a teenager he wrote for the Lockland High School yearbook and worked as a hotel desk clerk and a cinema usher. After graduating from high school, he enrolled at Miami University at Oxford, Ohio, and then joined the Army Medical Corps, and was stationed in England and West Germany.

After the war he resumed his studies, graduating with honours from the University of Cincinatti in 1948, after which he enrolled in a writing workshop at the new School for Social Research. A year of postgraduate work at Columbia University followed, though he abandoned his master's thesis on **George Orwell** when he married in 1950. Over the next few years he worked as a librarian at the Rand School of Social Science, summarised the Korean War for the *New York Times Index* and 'translated' technical language for *Popular Science Monthly.*

In 1954, encouraged by his wife, he based himself at their New York home to devote six days a week to writing his first novel, *Crazy in Berlin,* while freelancing as a copy-editor for publishing houses. A prolific writer, Berger has completed several novels, a play and many short stories since 1958. 'I need some rest between novels', he has said. 'But I can never take much, because real life is unbearable to me unless I can escape from it into fiction.' Since writing is a very intense experience for Berger, he never writes for more than four hours a day, working without plot or plans, starting with nothing to say and proceeding by a series of revelations.

His work has been described as aggressively intelligent, and it resists sentimentalising idealism and despair. His fiction reveals a love of language and a dark comic sensibility with a wide-ranging subject- matter, embracing literature, history and popular culture. He refuses to 'subscribe to a codified philosophy, whether romantic, existential or absurd', which may help to explain why mainstream success has been denied him, despite considerable praise from critics.

Fiction

Crazy in Berlin 1958; *Reinhart in Love* 1962; *Little Big Man* 1964; *Killing Time* 1967; *Vital Parts* 1970; *Regiment of Women* 1973; *Sneaky People* 1975; *Who Is Teddy Villanova?* 1977; *Arthur Rex* 1978; *Neighbours* 1980; *Reinhart's Women* 1981; *The Feud* 1983; *Granted Wishes* (s) 1984; *Nowhere* 1985; *Being Invisible* 1987; *The Houseguest* 1988; *Changing the Past* 1989; *Orrie's Story* 1990

Plays

Other People 1970 [np]

Steven BERKOFF 1937–

Berkoff was born in Stepney, east London, and insists that as a schoolboy he was 'delicate, sensitive, shy, introverted', adding 'perhaps I had to fight my way out of it'. He was brought up in a family that was ethnically mixed: 'all of the grand-parents came from Russia or the Eastern bloc'. In his mid-teens he left Hackney Grammar School and endured a series of menial jobs which he describes as 'hell, utter degradation and torture'. He was in and out of the employment exchange, joined the merchant navy and polished silver at Claridges – a job which may have turned out to be good source material for *Decadence,* his abrasive caricature of the upper classes, but which gave him ulcers.

At nineteen he joined the City Literary Institute and started acting with the zeal of a convert. Within a year he was at the Webber Douglas School of Drama, and he then worked extensively in repertory theatre in England and Scotland.

Two things set Berkoff off on his own path: his failed audition for a part in a Peter Brook stage production and his discovery of the work of Antonin Artaud. Overwhelmed by Artaud's vision of theatre, which Berkoff saw as 'a platform for most exotic rites, profanities, rituals, confessions ... a theatre of passions, sins and vices, to see the mechanism of the human spirit and purge it', he knew what he wanted to do. In 1968 he formed his own company, the London Theatre Group. Through mime, gymnastics and voice the group liberated itself from the conventions of mainstream theatre and started to evolve an innovative, more integrated theatrical language. On the fringes of a theatrical scene dominated by directors, Berkoff reinvented the actor-manager. The first production was an adaptation of Kafka's *In the Penal Colony*, which was followed by *Metamorphosis*.

Berkoff also began to write – in particular plays in verse. In his writing he maintains a romantic loyalty to the working classes, which he regards as: 'the most articulate, the most poetic, the most imagistic, most creative, the most interesting, the most fascinating, the most earthy, the most powerful, the most daring'. Middle-class people are for him 'the most dull-witted, the most boring, the most stupid, the most atrophied, the most impotent'. In the mid-1970s his original works – for example, *East*, *Decadence*, *Agememnon* – attracted cult audiences and radical criticism.

Outraged by, and contemptuous of, the critics, Berkoff always gives as good as he gets. He dismisses charges of misogyny in his work by suggesting that critics are confusing 'the vigour with which I express the male energy with some kind of macho attitude', and that, on the contrary, he is in fact a feminist writer.

Berkoff's play *Greek*, which transfers the Oedipus Rex story to 1980s' Tufnell Park, was a huge success in Los Angeles in 1983, winning several awards, and he achieved the kind of status in the USA that had so far eluded him at home. Mark-Anthony Turnage's opera, *Greek*, inspired by the play and with a libretto by Berkoff, was first performed in 1993. In 1989, Berkoff's production of Wilde's *Salome* transferred from the Edinburgh Festival to the National Theatre, which is perhaps an indication that he has finally been absorbed into mainstream theatre.

In addition to writing and directing, he has acted in several films, mainly taking the role of screen villain. However, he is ambitious to play more female parts, including the lead role of a female prostitute in one of his own works, *Massage*. He has also published a volume of short stories, which in style and subject-matter complement his work in the theatre.

Plays

In the Penal Colony 1968, *Metamorphosis* 1969 and *The Trial* 1971 [from Kafka; pub. 1988]; *Knock at the Manor Gate* [from Kafka's story] 1972 [np]; *Agamemnon* [from Aeschylus's play] 1973 [pub. 1977]; *Miss Julie versus Expressionism* [from Strindberg's play] 1973 [np]; *The Fall of the House of Usher* [from E.A. Poe's story] 1974 [pub. 1977]; *East* 1975 [pub. 1977]; *Massage* [pub. 1977]; *Greek* 1979, *Decadence* 1981 [pub. 1982]; *The Tell Tale Heart* [from E.A. Poe's story] 1981 [np]; *Lunch* 1983, *West* 1983 and *Harry's Christmas* 1985 [pub. 1985]; *Kvetch* 1986 and *Acapulco* 1990 [pub. 1986]; *Sink the Belgrano!* 1986 [pub. 1987]; *Actor* **(tv)** 1989; *Collected Plays* 2 vols [pub. 1994]

Fiction

Gross Intrusion and Other Stories **(s)** 1979

Non-Fiction

Steven Berkoff's America [with G. Dean] 1988; *I am Hamlet* 1989; *A Prisoner in Rio* 1989; *The Theatre of Steven Berkoff* [with R. Morton and others] 1992

Lord BERNERS 1883–1950

Sir Gerald Hugh Tyrwhitt-Wilson, 5th baronet and 14th baron Berners, generally known as Lord Berners, was of the ancient nobility: Sir John Bourchier de Berners had been created 1st baron in the middle of the fifteenth century. The son of a naval commodore, Berners was born at Arley Park in Shropshire and brought up against a conventional fox-hunting background. His love of and competence in the arts were developed at late nineteenth-century Eton. He received a musical education largely abroad, studying briefly with Stravinsky, then entered the diplomatic service as an honorary attaché, and served in the embassies at Constantinople and Rome between 1909 and 1919.

On succeeding to the barony in 1918 (after a motor accident during a funeral cortège had removed several of those in line to the title), he sold most of the family property as old-fashioned and inconvenient, and lived for the rest of his life at Faringdon House in Berkshire. He also owned a series of *pieds-à-terre* on the Continent, including a house that overlooked the Roman forum. He devoted his later years to his talents as artist, impresario and *farceur*: he was as famous for having dyed his pigeons at Faringdon pink and blue and for having a harpsichord installed in his motor-car, as he was for having written the music for the Diaghilev ballet *The Triumph of Neptune* to a scenario by **Sacheverell Sitwell**. His neighbour, **Nancy Mitford**, portrayed him in her novels as Lord Merlin, and he appeared in his own novel *Far from the Madding War* as Lord FitzCricket. Of this *alter ego*, Berners wrote: 'He was astute enough to realise that, in Anglo-Saxon countries, art is more highly appreciated if accompanied by a certain measure of eccentric publicity. This fitted in well with his natural inclinations.'

Of the six volumes of fiction he wrote – light satires on the social scene – four were published in the same year, 1941. Perhaps more enduring were his two sensitive memoirs of his early years, *First Childhood* and *A Distant Prospect*. His musical compositions included an opera, *La Carrosse du Saint Sacrement* (after Merimée, 1924); several ballets, including *The Wedding Bouquet*, a collaboration with **Gertrude Stein**, for which he also designed the sets; orchestral and chamber works; and songs. He also painted – in the manner, he said, of early Corot.

Berners remained unmarried – 'Handsomeness he greatly admired in both sexes,' as **Peter Quennell** delicately put it – and when he died after several years of suffering from a terrifying brain tumour, the baronetcy became extinct.

Fiction
The Camel 1936; *The Girls of Radcliff Hall* [as 'Adela Quebec'] 1937; *Count Omega* 1941; *Far From the Madding War* 1941; *Percy Wallingford and Mr Pidger* (s) 1941; *The Romance of a Nose* 1941

Non-Fiction
First Childhood 1934; *A Distant Prospect* 1945

Elizabeth BERRIDGE 1921–

Berridge was born in Wandsworth, south London, the daughter of an agent who, from his firm's offices in Grosvenor Gardens, managed some of the great London estates before they were broken up. She was educated at Clapham County School, where she was lucky enough to have an inspiring English teacher. After leaving school she spent a year in Paris and Geneva, then took a commercial course and started work in the Bank of England. It was to be the briefest of flirtations with the financial world: 'I crept out of the Bank secretly,' she recalled, 'because of parental disapproval of the change from a "steady" job to the insecure and rather "shady" one of journalism.'

Appropriately enough, she met her husband, the editor, novelist and short-story writer Reginald Moore, in a bookshop: they remained together for half a century until his death in 1990 and had a son and a daughter. It was Moore – who with Maurice Fridberg started the magazine *Modern Reading* in an attempt to keep the short story alive during the war – who taught Berridge to love the form: she has had many stories broadcast and published in magazines both in England and abroad, and has collected them in two volumes.

In 1942 the Moores moved to a primitive cottage in Wales, without electricity or mains water. It was here that she began writing novels by lamplight, and her life at the time is reflected in *Upon Several Occasions*, a delightful novel of village life, which was serialised by the BBC. They returned to London in the 1950s and Berridge worked as an editor for the publisher Peter Owen and reviewed fiction on the radio and in assorted literary journals. For twenty years she was a regular contributor to the *Daily Telegraph*.

Her reputation was established in the 1960s with three witty, beautifully written and highly praised novels. Set against a background of 1950s' Bohemian London, *Rose Under Glass* is the study of a widow coming to terms with being a single person once again. 'What a very good thing prose is when you use it as well as you do,' **John Betjeman** wrote to her of this book. *Across the Common* won the *Yorkshire Post* Novel of the Year Award in 1964. It is a sinister comedy in which the Wandsworth of Berridge's childhood is evoked when a woman returns to the house where she was born, now inhabited by three eccentric aunts. *Sing Me Who You Are* (the title of which is taken from **Louis MacNeice**'s 'Ecologue by a Five-Barred Gate') also deals with the influence of past events upon the present, when a woman travels to the country in order to take possession of a converted bus left to her by an aunt.

In spite of the praise she received for these novels, she did not publish another for fifteen years. During that time, however, she put together a second collection of short stories, published two children's books and compiled *The Barretts at Hope End* from Elizabeth Barrett Browning's early diaries, which she edited with a long introduction.

Fiction
House of Defence 1945; *The Story of Stanley Brent* 1945; *Selected Stories* (s) 1947; *Be Clean Be Tidy* 1949; *Upon Several Occasions* 1953; *Rose Under Glass* 1961; *Across the Common* 1964; *Sing Me Who You Are* 1967; *Family Matters* (s) 1980; *People at Play* 1982

For Children
That Surprising Summer 1972; *Run for Home* 1981

Edited
The Barretts at Hope End 1974

James BERRY 1924–

Born in Jamaica, Berry was in the first wave of immigrants from the Caribbean to Britain, arriving in London at the age of twenty-four. He worked for many years in a telegraph office, not publishing his first volume of poetry, *Fractured Circles* (1979), until two years after he left his job. He had, however, been writing poetry before this, making a name for himself in the West Indian community, and editing *Bluefoot Traveller*, a pioneering anthology of black British poetry. He revised the anthology in 1981 and produced a second one, *News from Babylon*, in 1984. He also edited an anthology of poems from the 1983 Brixton Poetry Festival in south London. He has worked in education and become a leader of the West Indian community.

In 1981 his poem 'Fantasy of an African Boy' won the National Poetry Competition, and that same year he published his second collection, *Cut-Way Feelins, Loving, Lucy's Letters*. His poetry is often – though not exclusively – colloquial, making use of the rich West Indian patois, and is very much a product of his own background and culture. *Chain of Days* is a volume about colonisation, its effect upon both the coloniser and the colonised, drawing upon black history, in particular slavery, in both Britain and America. Although some of his poetry understandably gives vent to anger and frustration, his vision is always a humane one.

Berry is also a critic and has published short stories and books for children. It is as a poet that he remains best known, however, the leading figure amongst the black poets whose work has enriched British literature.

Poetry
Fractured Circles 1979; *Cut-Way Feelins, Loving, Lucy's Letters* 1981 [rev. as *Lucy's Letters and Loving* 1982]; *Chain of Days* 1985; *Hot Earth Cold Earth* 1995

Fiction
The Girls and Yanga Marshall (s) 1987; *The Future-Telling Lady* 1991

For Children
A Thief in the Village (s) 1987; *Anancy-Spiderman* 1988; *When I Dance* 1988; *Isn't My Name Magical?* 1991

Edited
Bluefoot Traveller: an Anthology of West Indian Poets in Britain 1976 [rev. 1981]; *Dance to a Different Drum: Brixton Poetry Festival 1983* 1983; *News from Babylon: the Chatto Book of West Indian-British Poetry* 1984

Wendell BERRY 1934–

Born in Henry County, Kentucky, where his family had farmed for five generations, Berry graduated from the University of Kentucky in 1957 and in the same year he married. The following year a Wallace Stegner Writing Fellowship took him to Stanford University, where he remained until 1960. A brief period teaching at New York University came to an end with Berry's conscious decision to turn his back on the literary circles in that city. He returned to Kentucky in 1964, accepting the post of professor of English at the university there.

His life in Kentucky since then has revolved about his small farm near Port Royal. Berry works the land organically and sees his writing as a complementary activity, insisting that there is no conflict between the two. The North Kentucky farmland provides a setting for his four novels, and assumes as much importance in his fiction as Yoknapatawpha County does in the work of **William Faulkner**. Berry's principal theme is the relationship between man and the land and much of his work is overtly didactic. He rejects as sentimental any notion of farming as a quaint rural idyll, notably in *The Memory of Old Jack*, his best-known novel, in which two generations attempt to achieve the right sort of relationship with the land they farm.

His increasing interest in ecological and environmental issues led Berry to resign his professorship in 1977 and take up a post as contributing editor to the Rodale Press. The press's titles include *The New Farm* and *Organic Gardening and Farming*: Berry has published several of his poems in these unlikely showcases. Since the mid-1970s, Berry's increasing environmental concerns have led him to publish his work in essay form, although he concedes that 'a good argument cannot make you happy'. Berry still writes poetry at the farm where he works and lives with his wife and two children.

Fiction
Nathan Coulter 1960; *A Place on Earth* 1967; *The Memory of Old Jack* 1974; *Remembering* 1988

Poetry
The Broken Ground 1964; *November Twenty-Six Nineteen Hundred Sixty-Three* 1964; *Openings* 1968; *Findings* 1969; *Farming* 1970; *The Country of Marriage* 1973; *Horses* 1975; *Sayings and Doings* 1975; *To What Listens* 1975; *The Kentucky River* 1976; *There is Singing Around Me* 1976; *Clearing* 1977; *Three Memorial Poems* 1977; *A Part* 1980; *The Wheel* 1982; *The Landscape of Harmony* 1987; *Sabbaths* 1987; *Traveling at Home* [with J. DePol] 1989; *The Discovery of Kentucky* 1991; *Sabbaths 1987–90* 1992

Non-Fiction
The Long-Legged House 1969; *The Hidden Wound* 1970; *The Unforeseen Wilderness* [with E. Meatyard] 1971; *A Continuous Harmony: Essays Cultural and Agricultural* 1972; *The Unsettling of*

America: Culture and Agriculture 1977; *The Gift of Good Land: Further Essays Cultural and Agricultural* 1981; *Recollected Essays: 1965–1980* 1981; *Standing by Words: Essays* 1983; *Home Economics: Fourteen Essays* 1987; *Harlan Hubbard, Life and Work* 1990; *What are People For?* 1990; *Standing on Earth: Selected Essays* 1991

John BERRYMAN 1914–1972

Born John Smith in Oklahoma, Berryman took his stepfather's name. His own father had committed suicide, as Berryman himself was to do.

His first collection of poetry, entitled simply *Poems*, came out in 1942. Of the subsequent volumes the most important ones are *Homage to Mistress Bradstreet*, a series supposedly addressed to Anne Bradstreet, the seventeenth-century colonial poet, and *77 Dream Songs*, which received a Pulitzer Prize. These centred on Henry, an imaginary protagonist – clearly the poet's *alter ego* – and were followed by a further volume of 'Dream Songs', *His Toy, His Dream, His Rest*, four years later. They were published in a single volume in 1969.

For many years Berryman was professor of humanities at the University of Minnesota. He became involved in one or two political controversies: in 1948–9 he was one of those who supported the award of the Bollingen Prize for Poetry to **Ezra Pound**; in 1965 he protested against the bombing of North Vietnam.

Berryman's poetry is anguished, confessional and, according to some – notably his great friend **Robert Lowell** – disorderly. He often uses American diction, such as the rhythms of black speech. His work is nevertheless complex and technically polished as well as learned.

Shortly before his death, Berryman returned to the Catholic church. He also published his own selection of his work, *Selected Poems 1938–1968*. He committed suicide by jumping off a bridge into the Mississippi. He has always felt that he belonged to a doomed generation of poets, but a more direct explanation of his despair and death may be found in an incomplete autobiographical novel, *Recovery*, which deals with an intellectual's addiction to alcohol. This affliction was no secret and Berryman's appearances on television, much the worse for drink but still intellectually impressive, became notorious.

Poetry

Poems 1942; *The Dispossessed* 1948; *Homage to Mistress Bradstreet* 1956; *His Thought Made Pockets and The Plane Buckt* 1958; *77 Dream Songs* 1964; *Berryman's Sonnets* 1967; *His Toy, His Dream, His Rest* 1968; *The Dream Songs* 1969; *Love and Fame* 1970; *Delusions, etc.* 1972; *Selected Poems 1938–1968* 1972; *Henry's Fate and Other Poems, 1967–1972* [ed. J. Haffenden] 1977; *Collected Poems 1937–1971* [ed. C. Thornbury] 1990

Fiction

Recovery 1973

Non-Fiction

Stephen Crane 1950; *The Freedom of the Poet* 1976; *One Answer to a Question* 1981; *We Dream of Honour: John Berryman's Letters to His Mother* [ed. R.J. Kelly] 1988

Edited

The Arts of Reading [with R. Ross, A. Tate] 1960

Biography

Poets in Their Youth by Eileen Simpson 1982

John BETJEMAN 1906–1984

The son of a furniture manufacturer (who invented the tantalus, a lockable rack for decanters intended to prevent servants from pilfering their employers' spirits), Betjeman was born in Highgate, north London. His surname, of Dutch origin, was officially 'Betjemann', but he dropped the final 'n' in his teens, having been taunted during the First World War for being 'a German spy'. He was educated at Marlborough and at Magdalen College, Oxford. While at school, he associated in literary enterprises with left-wing aesthetes such as **Louis MacNeice** and Anthony Blunt, both of whom were his contemporaries there. At Oxford he studied Divinity with **C.S. Lewis**, whom he found deeply antipathetic, wrote poetry, learned a great deal about architecture and failed to take a degree. He is also rumoured to have been discovered by a college servant in bed with **W.H. Auden**, who was obliged to pay the scout £5 not to report the matter (and later remarked that the experience had not been worth the money). Betjeman's childhood and education were later recalled in his long narrative poem *Summoned by Bells*.

After a brief spell as a schoolmaster, he became in 1930 an assistant editor of the *Architectural Review* (or 'Archie Rev', as he dubbed it), where he met his future wife, the writer and traveller Penelope Chetwode, whose father, a Field Marshal who was Commander-in-Chief in India, thoroughly disapproved of the match. Betjeman's first book of poems, *Mount Zion*, bound in firework paper, was published in 1931 by an Oxford friend, the wealthy and eccentric patron of the surrealists, Edward James, and this was followed by an architectural pamphlet, *Ghastly Good Taste*. He subsequently worked in the publicity department of the Shell petroleum company, where he was responsible for the *Shell Guides* to the counties of England, edited and illustrated by such people as Robert Byron, **Peter Quennell** and Paul Nash. (Betjeman himself edited the volume on his beloved Cornwall.) He also contributed prose and verse to a number of newspapers and magazines, including the *Evening Standard* (where he was also film critic) and **Graham Greene**'s short-lived but influential *Night and Day*.

His poetry is traditional in style and celebrates English themes, often casting an acutely satirical eye on national foibles and revealing a great love of architecture, brand-names, railways and churches. A troubled Anglican, he edited the excellent *Collins Guide to English Parish Churches* and wrote a characteristic book about the religious travails of his teddy bear, *Archie and the Strict Baptists*. Although Betjeman's work is affectionate and celebratory, an underlying melancholia – in part prompted by very real and terrifying intimations of mortality (which he shared with his friend **Philip Larkin**) – prevents it from ever being merely cosy.

During the Second World War, Betjeman went to Dublin as a press attaché. Back in England, from the 1950s onwards, his work reached a wide audience through television, for which he made a number of documentaries, usually on architectural themes and often accompanied by narratives in verse. He was to become one of the country's most popular broadcasters. A fervent conservationist, he founded the Victorian Society in 1958, at a time when nineteenth-century architecture was held in very low esteem. Two years later he received a CBE, and he was knighted in 1969. His poetry, although it always had its distinguished champions, had fallen out of fashion in the 1960s, but underwent a revaluation when he was appointed Poet Laureate in 1972. During his final years, he was increasingly incapacitated by Parkinson's disease. At his memorial service in Westminster Abbey, during which a selection of 'favourite school songs' was played and thanks were given 'for his delight in trains and railways and the Underground', the Dean of Westminster rightly said that Betjeman's death had 'eclipsed the gaiety of nations'. The architectural writer Candida Lycett-Green, who edited a volume of his letters in 1994, is his daughter.

Poetry

Mount Zion 1931; *Continual Dew* 1937; *Old Lights for New Chancels* 1940; *Selected Poems* 1940; *New Bats in Old Belfries* 1945; *Slick But Not Streamlined* 1947; *A Few Late Chrysanthemums* 1954; *Poems in the Porch* 1954; *John Betjeman: a Selection of Poems* 1958; *John Betjeman's Selected Poems* [ed. Earl of Birkenhead] 1958 [rev. 1962, 1970, 1980]; *Summoned by Bells* 1960; *A Ring of Bells* [ed. I. Slade] 1962; *High and Low* 1966; *Collected Poems* 1971; *A Nip in the Air* 1974; *Metro-land* **(tv)** 1973 [as *A Souvenir of Metroland* 1977]; *Church Poems* [with J. Piper] 1981; *Uncollected Poems* 1982; *John Betjeman: Selected Poems* [ed. H. Blamires] 1992

Non-Fiction

Ghastly Good Taste 1933; *An Oxford University Chest* 1938; *Vintage London* 1942; *John Piper* 1944; *First and Last Loves* 1952; *The English Town in the Last Hundred Years* 1956; *Sir John Betjeman's Guide to English Parish Churches* 1958 [rev. 1993]; *Ground Plan to Skyline* [as 'Richard M. Farran'] 1960; *English Churches* [with B. Clarke] 1964; *The City of London Churches* 1965; *Victorian and Edwardian London from Old Photographs* 1969; *Victorian and Edwardian Oxford from Old Photographs* [with D.G. Vaisey] 1971; *London's Historic Railway Stations* 1972; *Victorian and Edwardian Brighton from Old Photographs* 1972; *A Pictorial History of English Architecture* 1973; *Victorian and Edwardian Cornwall from Old Photographs* [with A.L. Rowse] 1974; *Betjeman's Cornwall* 1984; *Betjeman's London* [ed. P. Denton] 1988; *Collected Letters* [ed. C. Lycett-Green] 1994

Edited

Cornwall Illustrated 1934; *English, Scottish and Welsh Landscape 1700–c. 1860* [with G. Taylor] 1944; *English Cities and Small Towns* 1945; *Murray's Buckinghamshire Architectural Guide* 1948; *Murray's Berkshire Architectural Guide* 1949; *English Love Poems* [with G. Taylor] 1957; *Collins Guide to English Parish Churches* 1968; *Altar and Pew: Church of England Verses* 1959

Fiction

Archie and the Strict Baptists 1977

Collections

The Best of Betjeman [ed. J. Guest] 1978; *My Favourite Betjeman* [ed. A. Kilminster, D. Lenox] 1985

Biography

Young Betjeman by Bevis Hiller 1988

George A. BIRMINGHAM (pseud. of James Owen Hannay) 1865–1950

The son of a Belfast clergyman and a clergyman's daughter, Hannay was educated at Haileybury and Trinity College, Dublin, where he graduated in modern literature in 1887. The following year he was ordained deacon and took up a curacy in Wicklow, and with that security married in 1889 Ada, the daughter of a third cousin, Richard Wynne, later Bishop of Killaloe. Short stories and scholarly works such as *The Spirit and Origin of Christian Monasticism* appeared under his own name, as well as novels such as *Hyacinth*, which were often politically controversial. Hannay moved to Mayo in 1902 where he was appointed rector of Westport. He adopted a pseudonym (the surname Birmingham was a common one in the area) under which he wrote several comic novels, of which the first, *Spanish Gold*, had a red-headed, wild and talkative vicar as its main character.

Novels led to a stage play, *General John Regan*, which, featuring Charles Hawtrey, was a London West End success in 1913. The play's loquacity, wildness and humour were not appreciated in Westport, where its performance led to a riot; when Hannay's parishioners penetrated his pseudonym, they boycotted his church. Their turbulent priest immediately quit Mayo to go on a lecture tour in America, which resulted in the book *From Connaught to Chicago*. *A Padre in France* described Hannay's experiences as chaplain on the Western Front between 1916 and 1917. Through the politically stormy years of 1918 to 1922, Hannay was chaplain to the Lord-Lieutenant of Ireland, while holding the rectory of the uncomplaining parish of Carnlway in Kildare. Restlessness took him to Budapest as chaplain to the British Legation for two years, before which he relinquished his appointment of the canonry of St Patrick's Cathedral, which he had held for ten years.

Despite Hannay's declaration that he was 'more interested in Ireland than in anything else', he settled as rector in Mells, Somerset, from 1924 to 1933, and remained in England until his death. His interest in Irish affairs was recognised by the conferment of an honorary doctorate by Trinity College, Dublin, in 1946. Personal tragedy, including the destruction of his rectory by fire in 1929 and the death of his wife in 1933, led to a move to London, where he became rector of Holy Trinity Church, Kensington, an appointment he held for the rest of his life. Hannay regarded himself as a priest first and a novelist second, but nevertheless wrote an enormous number of books as George A. Birmingham, completing the last just before his death. Robust and tall, with curly grey hair he was an amusing conversationalist, much appreciated in the Garrick and the Athenaeum, where he entertained his friends.

Fiction

The Seething Pot 1905; *Hyacinth* 1906; *Benedict Kavanagh* 1907; *The Northern Iron* 1913; *The Bad Times* 1908; *Spanish Gold* 1908; *The Search Party* 1909; *Lalage's Lovers* 1911; *The Major's Niece* 1911; *The Simpkins Plot* 1911; *Dr Whitty's Patient and Mrs Chalmer's Public Meeting* 1912; *The Inviolable Sanctuary* 1912; *The Red Hand of Ulster* 1912; *The Adventures of Dr Whitty* 1913; *The Lost Tribes* 1914; *Minnie's Bishop and Other Stories* (s) 1915; *The Island Mystery* 1918; *Our Casualty, and Other Stories* (s) 1919; *Up the Rebels* 1919; *Adventures of the Night* 1920; *Good Conduct* 1920; *Inisheeny* 1920; *Lady Bountiful* (s) 1921; *The Lost Lawyer* 1921; *The Great Grandmother* 1922; *A Public Scandal* (s) 1922; *Found Money* 1923; *King Tommy* 1923; *Send for Dr O'Grady* 1923; *The Grand Duchess* 1924; *Bindon Parva* 1925; *The Gun-Runners* 1925; *Goodly Pearls* 1926; *The Lady of the Abbey* 1926; *The Smuggler's Cave* 1926; *Fidgets* 1927; *Gold Gore and Gehenna* 1927; *The Runaways* 1928; *The Major's Candlesticks* 1929; *The Hymn Tune Mystery* 1930; *Wild Justice* 1930; *Fed Up* 1931; *Elizabeth and the Archdeacon* 1932; *Angel's Adventures* 1933; *The Silver-Gilt Standard* 1933; *Two Fools* 1934; *Love or Money and Other Stories* (s) 1935; *Millicent's Corner* 1935; *Daphne's Fishing* 1937; *Mrs Miller's Aunt* 1937; *Magilligan Strand* 1938; *Appeasement* 1939; *Miss Maitland's Spy* 1940; *The Search for Susie* 1941; *Over the Border* 1942; *Lieutenant Commander* 1943; *Poor Sir Edward* 1943; *Good Intentions* 1945; *The Piccadilly Lady* 1946; *Golden Apple* 1947; *A Sea Battle* 1948; *Laura's Bishop* 1949; *Two Scamps* 1950

Non-Fiction

As the Revd James Owen Hannay: *The Spirit and Origin of Christian Monasticism* 1903; *The Wisdom of the Desert* 1904; *From Connaught to Chicago* 1914; *A Padre in France* 1918; *Can I Be a Christian?* 1923; *A Wayfarer in Hungary* 1925; *Murder Most Foul* 1929; *Isaiah* 1937; *God's Iron: a Life of the Prophet Jeremiah* 1939

As 'George A. Birmingham': *The Lighter Side of Irish Life* 1911; *An Irishman Looks at His World* 1919; *Spillikins: a Book of Essays* 1926; *Now You Tell Me One: Stories of Irish Wit and Humour* 1927; *Ships and Sealing Wax* 1927; *Murder Most Foul! A Gallery of Famous Criminals* 1929; *Pleasant Places* (a) 1934

Plays

General John Regan 1913; *Send for Dr O'Grady* 1923

Edited

Irish Short Stories 1932

Elizabeth BISHOP 1911–1979

One of the finest and most original poets of the twentieth century, Bishop was born in

Worcester, Massachusetts, the child of William Thomas Bishop, who died in the year of her birth, and Gertrude Bulmer Bishop. On William's death, from Bright's disease, Gertrude took her infant daughter to the Bulmer family home in Great Village, Nova Scotia. Following a series of breakdowns, Gertrude was committed to a mental institution and Elizabeth was cared for by her maternal aunts until, aged seven, she was uprooted from rural and coastal Great Village (a life and place she would celebrate in the poems, stories and prose pieces she wrote in Brazil), and returned to Worcester by her better-off paternal grandparents. Bishop never saw her mother again, and the trauma of the double loss of parent and home manifested itself in a number of chronic ailments, including asthma, that interrupted her schooling and kept her a semi-invalid until her mid-teens. Asthma would plague her for the rest of her life, its effects compounded, as time went on, by depression, by anti-depressant drugs and by bouts of alcoholic bingeing – a destructive pattern that often landed her in hospital. In her long and anguished battle against alcoholism, Bishop was supported by Anny Baumann, her New York doctor and friend.

At sixteen, she was sent to Walnut Hill boarding-school in Natick and thence to Vassar. It was while still a student that she met **Marianne Moore**, her first mentor and lifelong friend. Moore's encouragement of the shy and modest younger poet (it was on Moore's recommendation that she received a Lucy Donnelly Fellowship for 1951–2), was crucial and her influence on Bishop's work considerable. However, Bishop, whose deeper influences included George Herbert and Gerard Manley Hopkins, would eventually free herself from Moore's quirky prosodic strictures.

Between graduation and the publication of her first book of poems, Bishop travelled extensively, settling eventually, in 1939, in Key West, where she spent much of the next decade. The title of her first book, *North & South*, published in 1946, reflects her constant travels, which can be seen, in part, as a quest for a place she could call home. *North & South* which contains many of the much-anthologised early poems, including 'The Man-Moth', 'Roosters' and 'The Fish', established her voice and her reputation, and, when published in a combined edition with *A Cold Spring* in 1955, won a Pulitzer Prize. Among the poems' admirers was **Robert Lowell**, who wrote of her 'famous eye' for meticulously observed description and 'humorous commanding genius for picking up the unnoticed'. Bishop, who was homosexual, told friends she would have liked to have had a child by the charismatic, brilliant, but unstable Lowell. Their friendship, one of the most important of her life and which lasted until Lowell's death in 1977, had its rough patches. Bishop's own work, love poems included, is notable for its subtlety and restraint, and she mistrusted the 'confessional' poetry, originated by Lowell, practised by **Anne Sexton** and others. 'You just wish they'd keep some of these things to themselves,' she famously said in a *Time* magazine interview in 1967. Five years later she was deeply upset when, in his collection *The Dolphin*, Lowell made poetic capital out of the personal letters of his estranged wife, **Elizabeth Hardwick**. 'But art just isn't worth that much,' she wrote to him, underlining the credo.

In 1951, on a voyage round South America, Bishop stopped off at Rio, suffered an allergic illness, and stayed on. While convalescing, she fell in love with Maria Carlota Costellat de Macedo Soares (Lota), and they set up house together in Petropolis, Brazil. Fine poems and stories, as well as translations from the Spanish and Portuguese (notably *The Diary of 'Helena Morley'*), were written during the next, happy and productive, decade. *Questions of Travel*, the 1965 collection that succeeded *A Cold Spring*, was dedicated to Lota. In the early 1960s, however, the relationship began to founder, and in 1967, following a severe mental breakdown brought on by overwork, Lota committed suicide.

Vacations apart, Bishop's remaining twelve years were spent in Cambridge, Massachusetts, and Boston, teaching at Harvard, giving readings and sharing her life with Alice Methfessel, a young university administrator whom she made her literary executor. It was to Methfessel that she dedicated her last, acclaimed 1976 volume, *Geography III*, which contains the masterpieces 'Crusoe in England' and 'The Moose', together with the celebrated villanelle 'One Art'. That same year, she won the Books Abroad/Neustadt International Prize for Literature. Bishop's awards – she collected seven honorary degrees – are too numerous to list: worth mention are the National Book Award won by the *Collected Poems* in 1970, and the Order of the Rio Branco, awarded her by the Brazilian government in 1971.

Since her death (of a cerebral aneurysm) and the publication, in 1983, of *The Complete Poems 1927–1979* (and, in 1984, of *The Collected Prose*), Bishop's fame and critical reputation have grown. Her voluminous selected letters, aptly entitled *One Art*, chosen and edited by her friend and editor Robert Giroux, received universal praise for their artful naturalness, their wit and their painterly topographical description (Bishop was a gifted water colourist). The insights they give into her poetry and her life make them essential reading. Brett C. Millier's biography deepens and darkens the picture of a woman who, for all her sense of humour and ability to inspire strong and lasting friendships, once described herself as 'the loneliest person whoever lived'.

Poetry
North & South 1946; *Poems: North & South, and A Cold Spring* 1955; *Questions of Travel* 1965; *The Ballad of the Burglar of Babylon* 1968; *The Complete Poems* 1969; *Geography III* 1976; *The Complete Poems 1927–1979* 1983

Non-Fiction
Brazil 1962; *The Collected Prose* 1984; *One Art* [ed. R. Giroux] 1994

Translations
Alice Bradt *The Diary of 'Helena Morley'* 1957

Edited
An Anthology of Twentieth Century Brazilian Poetry [with E. Brasil] 1972

Biography
Becoming a Poet by David Kalstone 1989; *Elizabeth Bishop, Life and the Memory of It* by Brett C. Millier 1993

Neil BISSOONDATH 1955–

Bissoondath was born at Arima, Trinidad. West Indian by birth and Indian by descent, he is the nephew of the writer **V.S. Naipaul**. In 1973 he emigrated to Canada to study French at York University, Toronto; after graduating, he taught French and English at the Inlingua Institute of Languages. He became assistant director of The Language Workshop, Toronto, in 1982 and the following year attended the Advanced Writing Workshop at the Banff School of Fine Arts in Alberta. He had been writing full-time since 1984 and his first book, *Digging Up the Mountains*, a collection of short stories set in the Caribbean and dealing with the themes of exile, immigration and alienation, was published in 1985.

The action in his first novel, the much praised *A Casual Brutality*, takes place on the fictional Caribbean island of Casaquemada and in Canada. It is the story of the disintegration of a banana republic that has overdosed on sudden wealth and gone into decline, and the viewpoint is that of a Hindu doctor, Raj Ramsingh, who has returned from Toronto to his birthplace with wife and small son, only to find that the oil-rich island has metamorphosed into a place where murders and bombings are part of the daily routine. The novel was shortlisted for the *Guardian* Fiction Prize.

Although *A Casual Brutality* is clearly questioning whether independence has been a good thing for the Caribbean colonies, ill-fated Casaquemada has been seen to have wider implications, as Bissoondath hoped. He is anxious to avoid being labelled as an 'immigrant writer' and declares that he has little left to say about the Caribbean. The stories in his third book, a collection entitled *On the Eve of Uncertain Tomorrows*, are set in Canada and Spain.

Fiction
Digging Up the Mountains (s) 1985; *A Casual Brutality* 1988; *On the Eve of Uncertain Tomorrows* (s) 1990

Algernon (Henry) BLACKWOOD 1869–1951

Born at Shooters Hill, Kent, Blackwood was the second son of Sir Arthur Blackwood, financial secretary to the Post Office, and Sydney, widow of the 6th Duke of Manchester. He was educated within a fanatically evangelical atmosphere by his parents and sent for a while to a Moravian school in the Black Forest, Germany, before attending the more conventional Wellington and Edinburgh University. Aged twenty, he was packed off by his parents, with a small sum, to Canada, where he first worked on a Methodist magazine in Toronto, then lost all his money in ill-advised ventures into dairy farming and hotel-keeping. Penniless, he drifted to New York where, after doing very varied jobs and suffering periods of near-starvation, he eventually became a newspaperman on the New York *Sun* and *Times*, and later private secretary to James Speyer, the youthful millionaire banker. He returned to England aged thirty, and his early adventures are engagingly recorded in his autobiographical volume *Episodes before Thirty*. In England, he worked at first in the dried milk business, but began to write seriously when aged thirty-six; his first volume of stories, *The Empty House*, enthusiastically championed by **Hilaire Belloc**, launched his career as a full-time writer. In later life he lived largely on the Continent, particularly Switzerland; he never married.

His novels and stories centre overwhelmingly on the themes of the supernatural and mystical communion with nature. Although some of his novels, such as *The Centaur*, are sensitive studies of the child's mind, they tend towards diffuseness; on the other hand, his finest supernatural stories, such as 'The Wendigo', are masterpieces of that supremely difficult genre. His writing became steadily less prolific from the 1920s onwards, but in his last years he developed a substantial second reputation as a teller of his own ghost stories on television, a role in which, according to *The Times*, he was 'terrifyingly effective'. His collected *Tales of the Uncanny and Supernatural* were published in 1949.

Fiction

The Empty House and Other Ghost Stories (s) 1906; *The Listener and Other Stories* (s) 1907; *John Silence, Physician Extraordinary* (s) 1908; *The Education of Uncle Paul* 1909; *Jumbo* 1909; *The Human Chord* 1910; *The Lost Valley and Other Stories* (s) 1910; *The Centaur* 1911; *Pan's Garden* (s) 1912; *A Prisoner in Fairyland* 1913; *Incredible Adventures* (s) 1914; *Ten Minute Stories* (s) 1914; *The Extra Day* 1915; *Julius le Vallon* 1916; *The Wave* 1916; *Day and Night Stories* (s) 1917; *The Garden of Survival* 1918; *The Promise of Air* 1918; *The Bright Messenger* 1921; *The Wolves of God and Other Stories* (s) [with W. Wilson] 1921; *Tongues of Fire and Other Stories* (s) 1924 [rev. 1925]; *The Dance of Death and Other Tales* (s) 1927; *Sambo and Smitch* 1927; *The Willows and Other Queer Tales* (s) 1927; *Mr Cupboard* 1928; *Dudley and Gilderoy* 1929; *Full Circle* 1929; *Strange Stories* (s) 1929; *By Underground* (s) 1930; *Stories of Today and Yesterday* (s) 1930; *The Italian Conjuror* 1932; *Maria – of England – in the Rain* 1933; *The Fruit Stoners* 1934; *Sergeant Poppett and Policeman James* 1934; *Shocks* (s) 1935; *How the Circus Came to Tea* 1936; *The Doll and One Other* (s) 1946; *Tales of the Uncanny and Supernatural* (s) 1949; *In the Realm of Terror* (s) 1957

Plays

Karma 1918; *Through the Crack* 1925

Non-Fiction

Episodes Before Thirty 1923 [as *Adventures Before Thirty* 1974]

Caroline (Hamilton Temple) BLACKWOOD 1931–

Blackwood was born in Northern Ireland, the daughter of the Marquis and Marchioness of Dufferin and Ava, and a descendant of Richard Brinsley Sheridan. She was brought up on the dilapidated family estate at Clandeboye, County Down (the model for the crumbling stately home in her short comic novel *Great Granny Webster*), and educated at a number of private schools in Ulster and England. In London in the early 1950s she met and married the painter Lucian Freud, becoming one of his most important models. Following their divorce she married the American composer Israel Citkovitz with whom she lived in New York for many years whilst raising their three daughters. Divorced again, she met the poet **Robert Lowell** (then married to **Elizabeth Hardwick**) in 1970, and a year later bore his son before they married in 1972. Lowell, who died in 1977, wrote about their relationship in *The Dolphin*.

Blackwood did not begin writing until the 1970s and published her first collection of stories and journalism, *For All That I Found There*, in 1973. The book is divided into three sections, 'Fact', 'Fiction' and 'Ulster', the last of which deals largely with memories of her childhood. Both the stories and the essays, which deal with such subjects as a free school in Harlem and a Surrey hospital for burns patients, display the sardonic and icy wit which characterises most of Blackwood's work. Her first novel, *The Stepdaughter*, which won the David Higham Prize in 1976, takes the form of an epistolary novel in which an abandoned wife living in a luxurious Manhattan apartment writes imaginary letters about her fat and loathsome stepdaughter. Subsequent novels include the macabre *The Fate of Mary Rose*, another account of the disintegration of fragile lives.

Blackwood has written several volumes of non-fiction, including a book about the Greenham Common peace protestors, and *In the Pink*, a savagely funny, courageously even-handed, but inevitably controversial study of foxhunting. Equally contentious was her book about the last years of the Duchess of Windsor, which grew out of a commission by the *Sunday Times* magazine. Blackwood describes the book, which required numerous excisions on the grounds of libel, as 'an entertainment, an examination of the fatal effects of myth, a dark fairy-tale', and it contains a merciless portrait of the Duchess's lawyer and protectress, Maître Suzanne Blum, as well as piquant sketches of others in the Windsor circle, notably Lady Moseley, who unsurprisingly gave the book a less than favourable review. Blackwood also collaborated with Anna Haycraft (better known as the novelist **Alice Thomas Ellis**) on a facetious cookery book, full of time-saving hints for celebrities, entitled *Darling, You Shouldn't Have Gone to So Much Trouble*. In an interview Blackwood recalled a friend suggesting a companion volume on the subject of sex.

Fiction

The Stepdaughter 1976; *Great Granny Webster* 1977; *The Fate of Mary Rose* 1981; *Good Night Sweet Ladies* (s) 1983; *Corrigan* 1984

Non-Fiction

Darling, You Shouldn't Have Gone to So Much Trouble [with A. Haycraft] 1980; *On the Perimeter* 1984; *In the Pink* 1987; *The Last of the Duchess* 1995

Collections
For All That I Found There 1973

Nicholas BLAKE

see C. DAY LEWIS

Alan BLEASDALE 1946–

Bleasdale is of the generation of British playwrights who turned to the medium of television to create a new writing form. Following the success and critical acclaim of his BBC series *The Boys from the Black Stuff* in 1982, which won him a BAFTA award, he has written further television series, as well as several stage plays.

Bleasdale was born in Liverpool and, apart from three years teaching in the Gilbert and Ellice Islands, has always lived on Merseyside. He attended the Wade Deacon Grammar School and left Padgate Teacher's Training College in 1967. He gave up teaching in 1975 and has since held the posts of playwright-in-residence at the Liverpool Playhouse and the Contact Theatre in Manchester. He later became associate director at the Liverpool Playhouse. He was married in 1970 and has three children.

Essentially a 'people's' writer (he once said: 'I should have been a social worker'), he deals with struggling, despairing lives in his native territory of north-west England. His representation of near-tragedy is usually mixed up with farce and surrealism. In 1985 he temporarily veered from ironic dramatic misery to flagrant sentimentality with the West End hit *Are You Lonesome Tonight?*, the musical story of Elvis Presley's last hours, which won the *Evening Standard* Award for Best Musical. *Having a Ball* is a succinctly punchy view of the professional classes; three men await vasectomies in a private clinic and, amidst much high comedy, are exposed as terrified victims. Bleasdale's first screenplay was *No Surrender*, which had as its protagonists two parties of old age pensioners, one Catholic, the other Protestant, who have double-booked a Liverpool nightclub and uproariously fight it out.

Scully was a drama (which grew out of a novel) about a Liverpudlian adolescent, while *G.B.H.* was a complex and highly praised series about the grassroots antagonism between a Militant-style councillor and an old-fashioned socialist headmaster, in which the audience's sympathies were skilfully manipulated and altered. Although *G.B.H.*'s contemporary setting and its suggestion of destabilising right-wing conspiracies might have seemed controversial, the series did not attract nearly as much hostility as *The Monocled Mutineer*, the alternately hilarious and harrowing account of a conscienceless rogue's adventures during the First World War. Based on the true story of Percy Toplis, who led a major mutiny of British troops at Etaples in September 1917 and was dubbed 'the most wanted man in Britain', it was criticised in some quarters for historical inaccuracy. It soon became clear, however, that what really enraged people was the play's unflinching depiction of war's brutality (there is a peculiarly horrifying sequence in which a soldier is inexpertly executed for cowardice) and scenes of rape and pillage during the mutiny which struck some viewers as a libel upon the British Army.

Plays
Early to Bed **(tv)** 1975; *Fat Harold and the Last 26* 1975 [np]; *The Party's Over* 1975 [np]; *Scully* [from his novel] **(st)** 1975, **(tv)** 1984 [pub. 1984]; *Dangerous Ambition* **(tv)** 1976; *Down the Dock Road* 1976 [np]; *Franny Scully's Christmas Stories* 1976 [np]; *It's a Madhouse* 1976 [pub. 1986]; *Should Auld Acquaintance* 1976 [np]; *No More Sitting on the Old School Bench* 1977 [pub. 1979]; *Crackers* 1978 [np]; *Pimples* 1978 [np]; *Scully's New Year's Eve* **(tv)** 1978; *Love is a Many Splendoured Thing* 1986 [pub. in *Act I* ed. D. Self, R. Speakman 1979]; *The Black Stuff* **(tv)** 1980; *Having a Ball* 1981 [pub. 1986]; *The Muscle Market* **(tv)** 1981; *The Boys from the Black Stuff* **(tv)** 1982 [pub. 1985]; *Are You Lonesome Tonight?* 1985; *The Monocled Mutineer* **(tv)** 1986; *No Surrender* **(f)** 1987 [pub. 1986]; *GBH* **(tv)** 1991; *On the Ledge* 1993

Fiction
Scully 1975; *Who's Been Sleeping in My Bed?* 1977; *The Boys from the Black Stuff* [from his tv series] 1985

Edward (William) BLISHEN 1920–

Born at Whetstone, near Barnet on the fringes of north London (the area where he has lived all his life), Blishen was the son of a civil servant. On leaving school, he started work as a journalist on a local paper in Muswell Hill, and then, being a conscientious objector, worked on the land from 1941 to 1946, an experience he has described in his autobiographical volume *A Cackhanded War*). He subsequently taught in a prep school from 1946–1949, following this with a decade as a teacher in a secondary modern school in Islington, north London. He began writing pieces describing his experiences from the *Manchester Guardian*, and as a result several publishers encouraged him to turn them into a book: the result was *Roaring Boys*, published in 1955, which he calls the first of his volumes of autobiography, although it uses some of the techniques of fiction. It was a great success and established him overnight as an educational pundit. After leaving teaching in 1963, he spent two years as a part-time lecturer in the Department of Education at the University of York, and since then has been a professional writer and broadcaster: he worked for the Overseas Services for many years, and he currently presents *A Good Read* on Radio 4.

He has published thirteen autobiographical volumes: honest and humorous records of disaster narrowly averted, they are deservedly popular. Blishen has also edited the *Junior Pears Encyclopaedia* for many years, and in the 1960s edited the *Oxford Miscellanies* for children. His imaginative retelling of Greek myth, *The God Beneath the Sea*, written with Leon Garfield, was awarded the Carnegie Medal in 1971. He married Nancy Smith in 1948; they have two sons.

Non-Fiction

Roaring Boys (a) 1955; *Town Story* [with Ionicus] 1964; *Hugh Lofting* 1968; *Encyclopaedia of Education* 1969; *This Right Soft Lot* (a) 1969; *A Cackhanded War* (a) 1972; *Uncommon Entrance* (a) 1974; *The Writer's Approach to the Novel* 1976; *Sorry, Dad* (a) 1978; *Pears Guide to Today's World* [with C. Cook] 1979; *A Nest of Teachers* (a) 1980; *Shaky Relations* (a) 1981; *Lizzie Pye* (a) 1982; *Donkey Work* (a) 1983; *A Second Skin* (a) 1984; *The Outside Contributor* (a) 1986; *The Disturbance Fee* (a) 1988; *The Penny World* (a) 1990

For Children

Robin Hood 1969; *The God Beneath the Sea* [with L. Garfield] 1970; *The Golden Shadow*, [with L. Garfield] 1973; *A Treasury of Stories for Seven Year Olds* [with N. Blishen] 1988; *A Treasury of Stories for Six Year Olds* [with N. Blishen] 1988; *A Treasury of Stories for Five Year Olds* [with N. Blishen] 1989; *A Treasury of Stories for Under Fives* [with N. Blishen] 1994

Edited

Junior Pears Encyclopaedia 1961–75; *Education Today* 1963; *Oxford Book of Poetry for Children* 1963; *Junior Pears Leisure Book* 1965; *Oxford Miscellany* 6 vols 1964–9; *Come Reading* 1968; *The School That I'd Like* 1969; *The Thorny Paradise* 1975; *Science Fiction Stories* (s) 1988

Karen BLIXEN

see Isak DINESEN

Edmund (Charles) BLUNDEN 1896–1974

The son of a financially unstable London schoolmaster, Blunden was brought up in Yalding, Kent, and won scholarships to Christ's Hospital and Queen's College, Oxford. He read classics and began writing pastoral verse which, unlike much Georgian poetry, displayed a genuine involvement in and understanding of the countryside. In 1915 he enlisted in the Royal Sussex Regiment, with whom he served ('a harmless young shepherd in a soldier's coat') as a second lieutenant, winning the Military Cross. In the spring of 1918 he was invalided home after being gassed.

'My experiences in the First World War have haunted me all my life,' he confessed in 1973, echoing his friend **Siegfried Sassoon**, 'and for many days I have, it seemed, lived in that world rather than this.' His understated but powerful memoir, characteristically entitled *Undertones of War*, contained 'A Supplement of Poetical Interpretations and Variations' which further contrast the gentle countryside and the man-made violence which destroyed it. The pastoral impetus behind his life and work is reflected in the titles of his books: *The Waggoner*, *Cricket Country*, *English Poems* and *The Shepherd*, which was awarded the Hawthornden Prize.

After the war he set about becoming a man of letters, working for some time on the *Athenaeum*, writing reviews, a literary gossip column and other short articles. He had married Mary Daines, a Suffolk barmaid with little formal education, in 1918 – largely to rescue her from the sort of 'unsavoury associations' traditionally associated with her job. This misguided act of chivalry did not make for a happy marriage, and Mary did not accompany her husband when in 1921 he went to Tokyo University as professor of English Literature. In Japan Blunden had an affair with Aki Hayashi, a plump, plain but devoted woman whom he brought back to England as his secretary. By this time Mary had found another man and the marriage was dissolved. The unfortunate Hayashi was installed in a London flat, from which she set forth each day to the British Museum to do research for her neglectful former lover.

In 1931 Blunden became a Fellow of Merton College, Oxford, where a war poet of a later generation, **Keith Douglas**, was one of his pupils. In 1933 he embarked upon a second unsuitable marriage, to Sylva Norman (*née* Nahabedian), an Armenian novelist and literary journalist. His third wife was Claire Poynting, an Oxford undergraduate twenty-two years his junior whom he married in 1945. He pursued an academic career, lecturing both in England and abroad, and in 1945 joined the staff of the *Times Literary Supplement*, for whom he was to write over 1,000 reviews. He remained enormously prolific, producing reams of poetry on all manner of subjects (including one 'written on the occasion of the Eighth Congress of the International Society of Blood Transfusion in Tokyo'), essays, biography and criticism. He was also a renowned editor of other poets, notably John Clare, whom he rescued from oblivion, the Romantics, and his own contemporaries, **Wilfred Owen** and **Ivor Gurney**. In 1966 he was elected Professor of Poetry at Oxford.

Blunden remained a Georgian, a haunter of pubs, cricket pavilions and secondhand bookshops. He was a much loved figure, his physical slightness bringing out the protective qualities of his friends. Although the best of his poetry is extremely fine, he published far too much that was inferior. In his later years he suffered from high blood pressure, drank heavily and was overcome by violent rages, eventually withdrawing from the world as he slipped towards death.

Poetry
Poems 1914; *The Barn* 1916; *Pastorals* 1916; *Three Poems* 1916; *The Waggoner* 1920; *The Shepherd* 1922; *To Nature* 1923; *Edmund Blunden: Twenty-One Poems* 1925; *Masks of Time* 1925; *English Poems* 1926 [rev. 1929]; *Japanese Garland* 1928; *Retreat* 1928; *Near and Far* 1929; *Poems 1914–1930* 1930; *A Summer's Fancy* 1930; *To Themis* 1931; *Halfway House* 1932; *Choice or Chance* 1934; *An Elegy and Other Poems* 1937; *On Several Occasions* 1939; *Poems 1930–1940* 1941; *Shells by a Stream* 1944; *After the Bombing* 1949; *Eastward* 1950; *Poems of Many Years* 1957; *A Hong Kong House: Poems 1951–1961* 1962; *Eleven Poems* 1966; *Poems on Japan* 1967

Non-Fiction
The Bonadventure: a Random Journal of an Atlantic Holiday 1922; *Christ's Hospital: a Retrospect* 1923; *On the Poems of Henry Vaughan* 1927; *Undertones of War* 1928 [rev. 1956, 1964]; *Nature in English Literature* 1929; *De Bello Germanico: a Fragment of Trench History* 1930; *Leigh Hunt: a Biography* 1930; *Votive Tablets: Studies Chiefly Appreciative of English Authors and Books* 1931; *The Face of England: in a Series of Occasional Sketches* 1932; *Charles Lamb and His Contemporaries* 1933; *The Mind's Eye* 1934; *Edward Gibbon and His Age* 1935; *Keats's Publisher: a Memoir of John Taylor* 1936; *English Villages* 1941; *Thomas Hardy* 1942; *Cricket Country* 1944; *Shelley: a Life-Story* 1946; *Shakespeare to Hardy: Short Studies of Characteristic English Authors* 1948; *Two Lectures on English Literature* 1948; *Addresses on General Subjects* 1949; *Poetry and Science* 1949; *Sons of Light* 1949; *Chaucer to 'B.V.'* 1950; *Favourite Studies in English Literature* 1950; *Hamlet, and Other Studies* 1950; *Influential Books* 1950; *John Keats* 1950; *Records of Friendship* 1950; *A Wanderer in Japan* 1950; *Sketches and Reflections* 1951; *Essayists of the Romantic Period* 1952; *Charles Lamb* 1954; *War Poets 1914–1918* 1958; *Three Young Poets: Critical Sketches of Byron, Shelley and Keats* 1959

Fiction
We'll Shift Our Ground [with S. Norman] 1933

Edited
John Clare: Poems, Chiefly from Manuscript [with A. Porter] 1920; *Madrigals and Chronicles* 1924; *Shelley and Keats as They Struck their Contemporaries* 1925; *A Hundred English Poems from the Elizabethan to the Victorian Age* 1927; *The Autobiography of Leigh Hunt* 1928; Charles Lamb *Last Essays of Elia* 1929; *The Poems of William Collins* 1929; *Coleridge: Studies by Several Hands on the Hundredth Anniversary of his Death* [with E.L. Griggs] 1934; *Return to Husbandry* 1943; *Hymns for the Amusement of Children* 1947; Shelley: *Shelley's Defence of Poetry* 1948, *Shelley: Selected Poems* 1955; *Christ's Hospital Book* 1953; *Keats: Selected Poems* 1955; *Tennyson: Selected Poems* 1960; *Wayside Poems of the Seventeenth Century* 1963

Biography
Edmund Blunden by Barry Webb 1990

Robert (Elwood) BLY 1926–

Bly was born in Madison, Minnesota, the son of a second generation Norwegian farmer who was an active member of the Lutheran church and encouraged his son's early reading. After attending Madison High School, Bly served in the US Navy from 1944 to 1946, during which time (he recalls in his autobiographical essay 'In Search of the American Muse') he first thought of writing poetry. He studied literature at St Olaf's College, Northfield, and at Harvard where his peers included **John Ashbery** and **John Hawkes**. Inspired by **W.B. Yeats**'s declaration that a poet must find his own 'inner space', Bly lived in solitude in a log cabin for some time before moving to New York, and then to Cambridge,

Massachusetts, where he directed the first American production of Yeats's *Player Queen*. He taught at the University of Iowa from 1954 to 1955 and was awarded his MA in 1956 for a collection of verse, 'Steps Toward Poverty and Death'. Having travelled in Norway on the proceeds of a Fulbright grant, Bly returned in 1957 to rural Minnesota (where he has been based ever since) and published the first issue of his magazine *The Fifties* (subsequently titled *The Sixties*, *The Seventies*, and so on).

His first collection of poems, *Silence in the Snowy Fields* (1962), reflects an intense concern with the natural world and is written in a simple, often repetitive language. The award of an **Amy Lowell** Fellowship allowed Bly to travel in Europe, and on his return to the USA he founded the organisation American Writers against the Vietnam War. The poems in *The Light around the Body*, such as 'Counting Small-Boned Bodies', reflect his new political engagement. With other poets, including **Lawrence Ferlinghetti**, he performed readings in colleges and public halls, and in 1968 he gave the money from his National Book Award to an anti-draft organisation. Further collections sought to synthesise Bly's political and pastoral concerns and increased his reputation as one of America's leading poets.

In the 1970s his growing concern with environmentalism and reading of Carl Jung led him to conclude that 'what we call masculine consciousness is a very recent creation'; and in the 1980s he began to organise seminars in which men studied ancient myths about male initiation. This led to the publication of *Iron John: A Book about Men*, which became a bestseller and added fuel to the debate about sexual identity. Bly has translated prose and poetry from Norwegian, Spanish, German, Chinese and Hindi. He has been married twice, and has four children from his first marriage.

Poetry

The Lion's Tail and Eyes [with J. Wright, W. Duffy] 1962; *Silence in the Snowy Fields* 1962; *Chrysanthemums* 1967; *The Light around the Body* 1967; *Ducks* 1968; *The Morning Glory* 1969 [rev. 1970, 1975]; *The Teeth Mother Naked at Last* 1970; *Poems for Tennessee* [with W. Matthews, W. Stafford] 1971; *Christmas Eve Service at Midnight at St Michael's* 1972; *Water under the Earth* 1972; *The Dead Seal Near McClure's Beach* 1973; *Jumping Out of Bed* 1973; *Sleepers Joining Hands* 1973; *The Hockey Poem* 1974; *Point Reyes Poems* 1974; *Grass from Two Years, Let's Leave* 1975; *Old Man Rubbing His Eyes* 1975; *The Loon* 1977; *This Body is Made of Camphor and Gopherwood* 1977; *This Tree Will Be Here for a Thousand Years* 1979; *Visiting Emily Dickinson's Grave and Other Poems* 1979; *Finding an Old Ant Mansion* 1981; *The Man in the Black Coat Turns* 1981; *The Eight Stages of Translation* 1983; *Four Ramages* 1983; *The Whole Moisty Night* 1983; *Mirabai Versions* 1984; *Out of the Rolling Ocean and Other Love Poems* 1984; *In the Month of May* 1985; *A Love of Minute Particulars* 1985; *Loving a Woman in Two Worlds* 1985; *Selected Poems* 1986; *This Tree Will Be Here for a Thousand Years* 1992; *What Have I Ever Lost by Dying?* 1993

Non-Fiction

A Broadcast against the New York Times Book Review 1961; *Talking All Morning: Collected Conversations and Interviews* 1979; *A Little Book on the Human Shadow* [ed. W. Booth] 1986; *The Pillow and the Key Commentary on the Fairy Tale Iron John* 1987; *American Poetry: Wildness and Domesticity* 1990; *Iron John: a Book about Men* 1991; *The Spirit Boy and the Insatiable Soul* 1994

Edited

A Poetry Reading against the Vietnam War [with D. Ray] 1966; *The Sea and the Honeycomb: A Book of Poems* 1966; *Forty Poems Touching on Recent American History* 1970; *Leaping Poetry: an Idea with Poems and Translations* 1975; *Selected Poems by David Ignatow* 1975; *News of the Universe: Poems of Twofold Consciousness* 1980; *Ten Love Poems* 1981; *The Winged Life: Selected Poems and Prose of Thoreau* 1986

Translations

Hans Hvass *The Illustrated Book about Reptiles and Amphibians of the World* 1960; *Twenty Poems of Georg Trakl* [with J. Wright] 1961; Selma Lagerlöf *The Story of Gösta Berling* 1962; *Twenty Poems of César Vallejo* [with J. Knoepfle, J. Wright] 1962; Tomas Tranströmer: *Three Poems of Tomas Tranströmer* [with T. Buckman, E. Sellin] 1966, *Night Vision* 1971, *Twenty Poems of Tomas Tranströmer* 1971, *Elegy, Some October Notes* 1973, *Truth Barriers* 1980, *Selected Poems 1954–1986* [ed. R. Hass] 1987; Knut Hamsun *Hunger* 1967; Gunnar Ekelöf: *I Do Best Alone at Night* [with C. Paulston] 1967, *Late Arrival on Earth: Selected Poems of Gunnar Ekelöf* [with C. Paulston] 1967; Pablo Neruda: *Twenty Poems of Pablo Neruda* [with J. Wright] 1967, *Letter to Miguel Otero Silva, in Caracas (1948)* 1982; *Selected Poems by Yvan Goll* [with others] 1968; *Forty Poems of Jan Ramón Jiménez* 1969; *Ten Poems by Issa Kobayashi* 1969; *Neruda and Vallejo: Selected Poems* [with J. Knoepfle, J. Wright] 1971; Kabir: *The Fish in the Sea Is Not Thirsty* 1971, *Grass from Two Years* 1975, *Twenty-Eight Poems* 1975, *Try to Live to See This!* 1976, *The Kabir Book* 1977; Rainer Maria Rilke: *Ten Sonnets to Orpheus* 1972, *The Voices* 1977, *I Am Too Alone in the World* 1980, *Selected Poems*

1981; *Lorca and Jiménez: Selected Poems* 1973; *Basho* 1974; *Friends, You Drank Some Darkness: Three Swedish Poets, Henry Martinson, Gunnar Ekelöf, Tomas Tranströmer* 1975; *Twenty Poems of Rolf Jacobson* 1977; *Twenty Poems of Vicente Aleixandre* [with L. Hyde] 1977; *Mirabai Versions* 1980; Antonio Machado: *Canciones* 1980, *Selected Poems and Prose* [with W. Kirkland] 1983, *Times Alone: Selected Poems of Antonio Machado* 1983; Rumi *Night and Sleep* [with C. Barks] 1981; Olav H. Hauge *Trusting Your Life to Water and Eternity* 1987

Louise BOGAN 1897–1970

Bogan was born in Livermore Falls, Maine, to working-class parents of Irish descent. She was educated at the Girls' Latin School and spent a year at Boston University, leaving in 1916 to marry Curt Alexander, an army officer. Alexander died in 1920 shortly after the death of their daughter, and Bogan married the poet Raymond Holden in 1925. Her two mental breakdowns led to conflict and the couple were divorced in 1937, and Bogan had a relationship with the poet **Theodore Roethke**.

Her first collection of poetry, *Body of This Death*, demonstrates the quiet control of language for which she is noted and suggests the influence of the metaphysical poets. From 1931 to 1968 she was a reviewer and poetry editor of the *New Yorker*, where her influence on taste was considerable, and she published three volumes of criticism, notably *Achievement in American Poetry*. Between 1948 and 1965 she spent various short periods as visiting professor at four American universities. She won the Bollingen Prize in Poetry in 1955 for her *Collected Poems*. *The Blue Estuaries* collects the 105 poems by which she wished to be remembered. Her literary executor Ruth Limmer has edited a volume of her letters and compiled an autobiographical 'mosaic' from various of Bogan's writings.

Poetry
Body of This Death 1923; *Dark Summer* 1929; *The Sleeping Fury* 1937; *Poems and New Poems* 1941; *Collected Poems, 1923–1953* 1954; *The Blue Estuaries: Poems, 1923–1968* 1968

Non-Fiction
Women 1929; *Achievement in American Poetry, 1900–1950* 1951; *Selected Criticism: Prose, Poetry* 1955; *Emily Dickinson: Three Views* [with A. MacLeish, R. Wilbur] 1960; *A Poet's Alphabet: Reflections on the Literary Art and Vocation* **(c)** [ed. R. Phelps, R. Limmer] 1970; *What the Woman Lived: Selected Letters of Louise Bogan, 1920–1970* [ed. R. Limmer] 1973; *Journey Around My Room* **(a)** [ed. R. Limmer] 1980

Translations
Yvan Goll: *Elegy of Ihpetonga* 1954, *Masks of Ashes* 1954, *The Myth of the Pierced Rock* 1962; Ernest Juenger *The Glass Bees* [with E. Mayer] 1961; Goethe: *Elective Affinities* [with E. Mayer] 1963, *The Sorrows of Young Werther, and Novella* [with E. Mayer] 1971; Jules Renard *Journal* [with E. Roget] 1964

Edited
The Golden Journey: Poems for Young People [with W.J. Smith] 1965

Biography
Louise Bogan: a Portrait by Elizabeth Frank 1985

Robert (Oxton) BOLT 1924–1995

Bolt was born in Sale, near Manchester, into what he described as 'an atmosphere of Northern Nonconformity'. His father kept a small furniture shop and his mother taught at primary school. Bolt's own school career was undistinguished. He left Manchester Grammar School in 1940 and worked in an insurance office until a special wartime entrance scheme enabled him to study history at Manchester University. He was drafted into the RAF in 1943 and taken from London to South Africa where, suffering from 'air-sickness', he failed his flying exams. Moved to Sandhurst and the Army Training College, he went on to serve in Ghana, leaving the army in 1946 having attained the rank of lieutenant. Resuming his studies, he joined the Communist Party, became chairman of the Socialist Society, and graduated in 1950. He took a teaching diploma at Exeter University and taught in a village school in Devon – he wrote his first play for the children there – before moving to Millfield School in Somerset in 1952.

Bolt stayed at Millfield until 1958, by which time the BBC had produced eight of his radio plays for adults – including an early version of *A Man for All Seasons*, which treated the life and death of Sir Thomas More – and seven of his plays for children. His first theatrical success came in 1957 with the London production of *Flowering Cherry* which won an *Evening Standard* Award for the most promising play. A greater success came in 1960 with the London opening of the reworked *A Man for All Seasons*, a success which was repeated in America that same year when the play won a New York Drama Critics Circle Award. The conflict between the individual conscience and the demands of authority was also the theme of his next play, *The Tiger and the Horse.* In 1961, Bolt had been arrested and imprisoned after a CND march had been stopped by police. The play dramatises and expands upon that same situation.

Bolt's career as a screenwriter began, after the brief interruption of imprisonment when he had to be persuaded to keep the peace so that production could continue, with the much awarded and Oscar-nominated *Lawrence of Arabia*, the first of his collaborations with David Lean. His next two screenplays, *Doctor Zhivago* and *A Man for All Seasons*, were both awarded Oscars. It was whilst working with David Lean on a planned version of *The Bounty* in 1979 that Bolt needed open heart surgery after suffering a stroke that left him paralysed and speechless. The film, which was not in the end directed by Lean, was eventually released in 1984. More successful was his screenplay for *The Mission*, which won the Palme d'Or at Cannes in 1986.

Bolt was first married in 1949 to the painter Celia Ann Roberts. He had four children and subsequently married Sarah Miles, the actress.

Plays

Fifty Pigs (r) 1953; *The Master* (r) 1953; *Ladies and Gentlemen* (r) 1954; *A Man for All Seasons* (r) 1954, (st) 1960, (f) 1966; *The Last of the Wine* (r) 1955 [pub. 1957]; *Mr Sampson's Sundays* (r) 1955; *The Critic and the Heart* (r) 1957; *Flowering Cherry* 1957 [pub. 1958]; *The Drunken Sailor* (r) 1958; *The Tiger and the Horse* 1960 [pub. 1961]; *The Banana Tree* (r) 1961; *Lawrence of Arabia* (f) [with D. Lean] 1962; *Gentle Jack* 1963 [pub. 1964]; *Doctor Zhivago* (f) [from B. Pasternak's novel; with D. Lean] 1965; *Ryan's Daughter* (f) [with D. Lean] 1970; *Vivat! Vivat Regina!* 1970 [pub. 1971]; *Lady Caroline Lamb* (f) 1972; *State of Revolution* 1977; *The Bounty* (f) 1984; *The Mission* (f) 1986

For Children

The Thwarting of Baron Bolligrew 1965 [pub. 1966]

Edward BOND 1934–

Bond was born in Holloway, north London, the son of working-class parents who had moved there from East Anglia during the Depression. Evacuated during the blitz in 1940, he spent four years in Cornwall and with his grandparents in East Anglia, an area which provides the setting for *The Pope's Wedding*, *The Sea* and *The Fool*, his play about the poet John Clare. He left Crouch End Secondary Modern School when he was fifteen, worked in factories and offices, and did two years' national service from 1953 to 1955. He considers his lack of formal education to be an advantage as a playwright, and has stated: 'I think universal education is one of the worst disasters that has hit Western society since the Black Death.' In 1949 he saw Donald Wolfit in *Macbeth* at the Bedford Theatre in Camden Town and recalls the experience as 'the first thing that made sense of my life for me; the first time I'd found something beautiful and exciting and alive'.

During the 1950s he wrote some fifteen unproduced plays, two of which he submitted to the English Stage Company at the Royal Court in 1958. He joined their Writers' Group, which included **John Arden**, **Arnold Wesker** and **Ann Jellicoe**, and read plays for the theatre. *The Pope's Wedding* was given a single performance for the Royal Court's private theatre club in 1962 and the artistic director George Devine commissioned a new play. *Saved* was premiered in 1965 – again in the private club because Bond had refused to make the cuts demanded by the Lord Chamberlain – and was almost universally damned by the critics for its depiction of violence, notably the gratuitous stoning to death of a baby. When the Lord Chamberlain banned the play after three months as unsuitable even for a private club it soon became the subject of a minor *cause célèbre*, with Kenneth Tynan, Laurence Olivier and **Penelope Gilliatt** among those who sprang to its defence. A similar ban on Bond's third party, *Early Morning*, after one performance in April 1968 – it contained scenes of cannibalism, and when Florence Nightingale dies after being raped by Queen Victoria, she ascends to heaven and opens a brothel there which is patronised by Gladstone and Disraeli – shortly preceded the abolition of the Lord Chamberlain's powers of censorship, and three of Bond's plays were promptly produced at the Royal Court. His subsequent plays include *Lear*, an audacious, politically motivated rewriting of Shakespeare, *Bingo*, in which Shakespeare is the principal character, and *Restoration*, which takes a critical look at the cruelties and social hierarchies which underpin Restoration comedy. Bond has written numerous screenplays, including those for Nicolas Roeg's *Walkabout* and, in collaboration, Antonioni's *Blow-Up*. He was married in 1971 to the Austrian theatre critic, Elisabeth Pablé.

Plays

The Pope's Wedding 1962 [pub. 1969]; *Saved* 1965 [pub. 1966]; *A Chaste Maid in Cheapside* [from T. Middleton's play] 1966; *Blow-up* (f) [with M. Antonioni, T. Guerra] 1967; *Early Morning* 1968; *Laughter in the Dark* (f) 1968; *Michael Kohlhaas* (f) 1968; *Narrow Road to the Deep North* 1968; *Laughter in the Dark* [from V. Nabokov's novel] 1969; *Black Mass* 1970; *The Lady of Monza* 1970 [np]; *Lear* 1971 [pub. 1972]; *Nicholas and Alexandra* (f) [with J. Goldman] 1971; *Passion* 1971; *Walkabout* (f) [from J.V. Marshall's novel] 1971; *Bingo* 1973 [pub. 1974]; *Fury* (f) [with A. Calenda, U. Pirro] 1973; *The Sea* 1973; *The Fool* 1975 [pub. 1976]; *The Master Builder* (tv) [from Ibsen's play] 1975; *A-A-America!* 1976; *Stone* 1976; *We Come to the River* (l) [with H.W. Henze] 1976; *The White Devil* [from J. Webster's play] 1976 [np]; *The Bundle* 1978, *The Woman* 1978 [pub. 1979]; *Orpheus* (l) [with H.W. Henze] 1979; *The Worlds* 1979 [pub. 1980]; *Restoration* 1981; *Derek* 1982 [pub. 1983]; *Summer* 1982; *After the Assassinations* 1983 [np]; *The English Cat* (l) [with H.W. Henze] 1983 [pub. 1982]; *The War Plays* [pub. 1985]: *Red, Black and Ignorant* 1984, *The Tin Can People* 1984, *Great Peace* 1985; *Human Cannon* 1986; *Jackets* 1989 [pub. 1990]; *September* 1989 [pub. 1990]; *In the Company of Men* 1990; *Olly's Prison* 1993; *Tuesday* 1993

Translations

Chekhov *Three Sisters* 1967; Frank Wedekind *Spring Awakening* 1974 [pub. 1980]

Fiction

Fables (s) 1982

Poetry

The Swing Poems 1976; *Choruses from After the Assassinations* 1983; *Poems 1978–1985* 1987

Non-Fiction

The Activist Papers 1980; *Notes on Post Modernism* 1990

For Children

Burns 1986

Collections

Poem, Stories and Essays for The Woman 1979

H(erman) C(harles) BOSMAN 1905–1951

An Afrikaner who wrote almost exclusively in English, and the most outstanding South African fictional author of the earlier part of the century, Bosman was born at Kuils River, near Cape Town, but was brought up in the Witwatersrand, where his father was killed in a mining accident while Bosman was a boy. He attended English junior and high schools, the University of the Witwatersrand in Johannesburg, and the Normal College. He later became a demonstrator for left-wing youth organisations and a teacher. He had married while still a student, but this marriage soon broke up when his wife refused to accompany him on his first teaching post in the Groot Marico, a famine-ridden area of the New Transvaal near the Bechuanaland border. Bosman's period in this remote community was to become a potent source of inspiration for his sharply realistic stories of Afrikaner life.

On holiday in the Rand, still only twenty-one, Bosman killed his brother-in-law with a hunting-rifle during a family brawl, and was condemned to death: the sentence was commuted to four years' hard labour for 'culpable homicide'. While in prison, Bosman wrote many of the brilliant, laconic stories later collected in *Mafeking Road*, and he was later to write up his prison experience in his classic work, *Cold Stone Jug*. On his release, Bosman helped edit a series of literary journals, and married a young pianist, Ella Manson, with whom he spent nine years living in London, Paris and Brussels. She attempted to set herself up as a concert artiste, he as a writer, but neither met with much success. In 1939, Bosman returned, alone, to South Africa, working first as a journalist and an advertising space salesman, then after 1943 as editor of the local paper in Pietersburg. Here he married his third wife, a teacher. In his last years he became better known as a writer and was the editor of the journal *South Africa Opinion*. In 1951, soon after moving from Cape Town to Johannesburg, he and his wife gave a housewarming party: during the night after the party, he suffered a fatal stroke.

The strange and difficult first novel, *Jacaranda in the Night* published in 1947, was not a success. His fame spread slowly, and publication of his major work, the novel *Willemsdorp*, which records the rise to power of an underprivileged Afrikaner after the 1948 elections, had to wait until 1977. His *Collected Works* were issued in 1981. Bosman is a cold-eyed writer who makes few judgments on the cruelty he sometimes describes: his clarity of vision clothed in outstanding prose ensures the quality of his work. He also wrote poetry, published three volumes in the early 1930s under the pseudonym of Herman Malan.

Fiction

Jacaranda in the Night 1947; *Mafeking Road* (s) 1947; *Cold Stone Jug* 1949; *Unto Dust* (s) [ed. L. Abrahams] 1963; *A Bekkersdal Marathon* (s) [ed. L. Abrahams] 1971; *Jurie Steyn's Post Office* (s) [ed. L. Abrahams] 1971; *Willemsdorp* 1977; *Almost Forgotten Stories* (s) [ed. V. Rosenberg] 1979; *Selected Stories* (s) [ed. S. Gray] 1980 [rev. 1982]; *Makapan's Cave and Other Stories* (s) 1987; *Romoutsa Road and Other Recollected Stories* (s) 1990

Poetry
As 'Herman Malan': *The Blue Princess* 1931; *Rust* 1932; *Jesus* 1933
As H.C. Bosman: *The Earth is Waiting* [ed. L. Abrahams] 1974; *Death Hath Eloquence* [ed. A.J. Blignant] 1981

Non-Fiction
Uncollected Essays [ed. V. Rosenberg] 1981

Plays
The Urge of the Primordial 1925; *Street-woman* [pub. in *Theatre One: New South African Drama* ed. S. Gray] 1978

Edited
Veld-trails and Pavements: an Anthology of South African Short Stories (s) [with C. Bredell] 1949

Collections
Mara [as 'Herman Malan'] 1932; *A Cask of Jerepigo: Sketches and Essays* [ed. L. Abrahams] 1957; *Bosman at His Best: a Choice of Stories and Sketches* [ed. L. Abrahams] 1965; *The Bosman I Like* [ed. P. Mynhardt] 1981; *Collected Works of Herman Charles Bosman* [ed. L. Abrahams] 1981; *Bosman's Johannesburg* [ed. S. Gray] 1986; *The Illustrated Bosman* [with P. Badcock] 1988

Elizabeth (Dorothea Cole) BOWEN 1899–1973

The daughter of a barrister, Bowen was born in Dublin into a family of the Anglo-Irish gentry and spent much of her childhood at Bowen's Court, the family's Georgian house in County Cork. Her childhood and youth were overshadowed by her mother's death, her father's insanity, and the circumstances leading to her own breakdown; yet childhood provided an emotional landscape to which she was continually to return in her best novels and stories. She was educated at Harpenden Hall School, Hertfordshire, and Downe House, Kent, and in 1923 she married Alan Cameron, an educational administrator, with whom she enjoyed a childless but happy marriage, diversified by a number of affairs, including one unhappy one with the man-of-letters Goronwy Rees, which she utilised in one of her best novels, *The Death of the Heart*.

Bowen lived with her husband in Oxford and later in Regent's Park, London, and she became well known for her novels, the first of which, *The Hotel*, was published in 1927. Maurice Bowra said that if she had not been a creative artist, she would have had the right presence and authority to be head of an Oxford or Cambridge women's college. Instead, with **Virginia Woolf** and **I. Compton-Burnett**, she became one of the three formidable doyennes of the inter-war literary scene. She was distinguished by her aristocratic manners, horselike features, slight stammer and incessant smoker's cough. During the war, she remained in London, and some of her best stories (such as 'Mysterious Kôr') and her novel *The Heat of the Day* unforgettably evoke London under fire.

Her early novels, such as *Friends and Relations*, tend to be finely observed comedies of manners; later ones chart delicately graded and shifting emotions in relationships complicated by guilt, uncertainty and misapprehension. She is a mistress of nuance and is especially adept at creating the physical and psychic atmosphere of houses (particularly large houses in Ireland and London), and at entering the world of children, notably in *The House in Paris* and in her stories. Her complex and subtle style, which owes something to **Henry James**, is at its most opaque and convoluted in *The Heat of the Day* and *A World of Love*, while in later novels, such as *The Little Girls* and *Eva Trout*, it is the narratives themselves that are strikingly bizarre. Her short stories were published throughout her life in numerous magazines and collected in several volumes. Her *Collected Stories* were published posthumously, with an admiring introduction by **Angus Wilson**, who described her as 'a writer of instinctive formal vision'. Her journalism and essays have also been collected, and she wrote several volumes of non-fiction, including histories of Dublin's famous hotel, the Shelbourne, and her own home, Bowen's Court.

She had inherited Bowen's Court in 1930 (Alan Cameron died there in 1952), but by 1959, despite earnings from much literary journalism and lecture tours in the USA, she was obliged to sell it. Its new owner promptly pulled it down. In later years, she lived at Hythe in Kent. She was awarded the CBE in 1948 and was made a Companion of Literature by the Royal Society of Literature in 1973.

Fiction
Encounters (s) 1923; *Ann Lees and Other Stories* (s) 1926; *The Hotel* 1927; *Joining Charles* (s) 1929; *The Last September* 1929; *Friends and Relations* 1931; *To the North* 1932; *The Cat Jumps* (s) 1934; *The House in Paris* 1935; *The Death of the Heart* 1938; *Look at All Those Roses* (s) 1941; *The Demon Lover* (s) 1945 [as *Ivy Gripped the Steps* 1946]; *Selected Stories* (s) 1946; *The Heat of the Day* 1949; *Early Stories* (s) 1951; *A World of Love* 1955; *Stories by Elizabeth Bowen* (s) 1959; *The Little Girls* 1964; *A Day in the Dark* (s) 1965; *Eva Trout* 1965; *Collected Stories* (s) 1981

For Children
The Good Tiger (s) 1965

Plays

Anthony Trollope: a New Judgement 1946

Non-Fiction

Bowen's Court 1942 [rev. 1964]; *English Novelists* 1942; *Seven Winters: Memoirs of a Dublin Childhood* 1942; *Why Do I Write?* [with G. Greene, V. Pritchett] 1948; *The Shelbourne: a Centre in Dublin Life for More Than a Century* 1951; *A Time in Rome* 1960; *Afterthought: Pieces About Writing* 1962; *The Mulberry Tree* 1986

Collections

Collected Impressions 1950; *Pictures and Conversations* 1975

Biography

Elizabeth Bowen by Victoria Glendinning 1977

John (Griffith) BOWEN 1924–

Born on Guy Fawkes' Day in India, Bowen was sent 'home' at the age of six to live with an unwelcoming uncle and aunt in Whitehaven, Cumberland. During his early years, he was shunted between assorted relatives and boarding-schools, seeing little of his parents, except when they came back to England on leave. He spent much of his childhood alone and became a voracious reader. In 1940 he went back to India, where he joined the army as a private two years later. He was immediately sent to an officers' training school and became an officer in the 5th Mahratha Light Infantry Regiment.

In 1948 he returned to England to study history at Pembroke College, Oxford, and became involved in undergraduate magazines and plays. After a postgraduate year at St Antony's College, he was awarded a history fellowship at the Ohio State University and spent a year in America. Back in London, he worked as an assistant editor on the fortnightly *Sketch* magazine, an undemanding job which left him time to write his first novel, *The Truth Will Not Help Us* (1956). Subtitled 'Embroidery on an Historical Theme', it was concerned with HUAC and the execution of an early eighteenth-century ship's captain. This was followed by *Pegasus* (a picture book for children) and a second novel with a futuristic setting, *After the Rain*. Neither of these, nor reviewing ballet on the radio, brought in much money, and he earned his living in advertising, working first at the J. Walter Thompson Company and then, as copy chief, at S.T. Garland Advertising. During this period, he published a third novel, *The Centre of the Green*, which was an Alternative Book Society Choice, and he left advertising to become a freelance writer shortly afterwards. The world of advertising forms the background of his novel *Storyboard*.

He began writing drama for television and was given a contract by ATV to work as a script consultant and provide the company with two plays a year. His next novel, *The Birdcage*, drew upon these experiences. Between 1960 and 1981 he produced an average of one television play every nine months, as well as numerous plays for the stage. The first of these, *I Love You, Mrs Patterson*, was written during a two-month sabbatical in France, and although well reviewed, did not enjoy a long run. *A World Elsewhere* was the only novel he published during this period.

By this time he was living with **David Cook**, who had started out as an actor but whom Bowen encouraged to write. They acquired some pigeons, which feature in Cook's paired novels *Walter* (1978) and *Winter Doves* (1979) and Bowen's extraordinary *Squeak*. Described by the author as a 'biography', but generally thought of as a novel, *Squeak* recounts the life of a pigeon he and Cook had hand-reared. The pigeon 'became imprinted, wished to mate with us, wished to be with us always, was only with great difficulty persuaded to accept a partner'. Delighted to rediscover writing that did not involve the inevitable collaborative compromises of drama, Bowen returned to the novel, producing *The McGuffin* about the world of a Hitchcock film (the title refers to the device whereby the action of a film is precipitated), *The Girls*, which deals with female homosexuality, and *Fighting Back*, which draws upon his childhood.

Plays

Digby **(r)** [as 'Justin Blake'; with J. Bullmore] 1959; *The Essay Prize* **(tv)** 1960; *A Holiday Abroad* **(tv)** 1960; *The Candidate* **(tv)** 1961; *The Jackpot Question* **(tv)** 1961; *Nuncle* [from J. Wain's story] **(tv)** 1962; *The Truth about Alan* **(tv)** 1963; *A Case of Character* **(tv)** 1964; *I Love You, Mrs Patterson* 1964; *The Corsican Brothers* [from D. Boucicault] **(tv)** 1965 [rev. 1970]; *Mr Fowlds* **(tv)** 1965; *After the Rain* [from his novel] **(st)** 1966 [pub. 1967; rev. 1972]; *Fall and Redemption* 1967 [pub. as *The Fall and Redemption of Man* 1968]; *Finders Keepers* **(tv)** 1967; *Silver Wedding* **(tv)** 1967 [rev. as *Mixed Doubles* 1969, pub. 1970]; *The Whole Truth* **(tv)** 1967; *Little Boxes* 1968; *A Most Unfortunate Accident* **(tv)** 1968; *Varieties of Love* **(r)** 1968; *The Disorderly Woman* [from Euripides] 1969; *Flotsam and Jetsam* **(tv)** [from Somerset Maugham's story] 1970; *Robin Redbreast* **(tv)** 1970, **(st)** 1974 [pub. in *The Television Dramatist* ed. R. Muller 1973]; *The Waiting Room* 1970; *A Woman Sobbing* **(tv)** 1972; *Diversions* 1973 [rev. in *Play* 9 ed. R. Cook 1981]; *The Emergency*

Channel (tv) 1973; *Roger* 1973 [pub. 1976]; *Young Guy Seeks Part-Time Work* (tv) 1973, (st) 1978 [np]; *Miss Nightingale* (tv) 1974 [as *Florence Nightingale* 1975, 1976]; *Heil Caesar!* [from Shakespeare's play *Julius Caesar*] (st), (tv) 1974; *The Snow Queen* (tv) 1974; *the Treasure of Abbot Thomas* (tv) [from M.R. James's story] 1974; *A Juicy Case* (tv) 1975; *Brief Encounter* (tv) [from N. Coward's screenplay] 1976; *Which Way Are You Facing* 1976 [rev. in *Play 9* ed. R. Cook 1981]; *A Photograph* (tv) 1977; *Singles* 1977 [np]; *Bondage* 1978 [np]; *A Dog's Ransom* (tv) [from P. Highsmith's play] 1978; *Games* (tv) 1978; *The Inconstant Couple* [from Marivaux] 1978 [pub. 1988]; *Rachel in Danger* (tv) 1978; *Spot the Child* [rev. as *Uncle Jeremy*] 1978 [np]; *The Letter of the Law* (tv) 1979; *Dying Day* (tv) 1980; *The Specialist* (tv) 1980; *Dark Secret* (tv) 1981; *Honeymoon* (tv) 1985; *The Geordie Gentleman* [from Molière] 1988 [np]; *The False Diaghilev* (r) 1989; *The Oak Tree Room Siege* 1990 [np]

Fiction

The Truth Will Not Help Us 1956; *After the Rain* [US *After the Flood*] 1958; *The Centre of the Green* 1959; *Storyboard* 1960; *The Birdcage* 1962; *A World Elsewhere* 1965; *Squeak* 1983; *The McGuffin* 1984; *The Girls* 1986; *Fighting Back* 1989; *The Precious Gift* 1992

For Children

Pegasus 1957; *The Mermaid and the Boy* 1958

Jane (Stajer Auer) BOWLES 1917–1973

The daughter of second-generation American Jews, whose parents had migrated from Austria-Hungary, Bowles grew up, in affluent circumstances, largely in Manhattan. She attended Stoneleigh School for Girls but, soon after going there, fell from a horse and broke her leg, which led to tuberculosis of the knee; the ultimate result was a limp and a legacy of bad health which was to be lifelong. She spent two years in a sanatorium in Switzerland between the ages of fifteen and seventeen, mastered French, and wrote her first novel (subsequently lost) in French on her return. Back in New York, she attended drama school and did a typing course, but early entered both the lesbian and intellectual life of the city. She re-encountered the writer and musician **Paul Bowles** at the house of **e e cummings** and immediately decided to depart with him on a tour of Mexico; they were married in February 1938, and their partnership until her death was a devoted one although they were both largely homosexual. The couple travelled widely, joined the Communist Party, and for a time formed part of a Bohemian household in New York which included Benjamin Britten, Peter Pears, **Carson McCullers**, the striptease artist Gypsy Rose Lee and **W.H. Auden**, who persuaded Jane to get up regularly soon after 6 a.m. to do his typing.

Her major novel, *Two Serious Ladies*, appeared in 1943 and, although it soon became a rare book, also acquired a considerable underground reputation as an experimental work of great seriousness and importance. The 1940s were a wandering decade for the Bowleses and they were often apart, but in 1952 Jane joined her husband more or less permanently in Tangier, where she found it difficult to write (the place was 'good for Paul, but not for me'), and conducted a passionate relationship with a Moroccan woman, Cherifa, who worked in the local market.

Jane Bowles was afflicted by severe alcoholism, and in 1957 she suffered a cerebral haemorrhage (there was rumours that Cherifa had poisoned her), which made it impossible for her to write again and left her progressively crippled. From the late 1960s she spent much of her time in clinics in Malaga, Spain, where she died. Her whole published work consists of the novel, a play, *In the Summer House*, a short puppet-play, some stories and a little poetry, but its individuality ensures her still-growing reputation. Her complex style unites naturalism and surrealism, and is almost entirely in dialogue; all major characters are women. Alfred Kazin wrote: 'The world in her altogether dry, vaguely jeering pages becomes even more heartless than one could have imagined it.'

Fiction

Two Serious Ladies 1943; *Plain Pleasures* (s) 1966

Plays

In the Summer House 1954; *A Quarreling Pair* 1966

Collections

The Collected Works of Jane Bowles [ed. T. Capote] 1966 [rev. 1984]; *Feminine Wiles* [ed. T. Williams] 1974; *Everything is Nice* 1989

Biography

A Little Original Sin by Millicent Dillon1981

Paul (Frederick) BOWLES 1910–

The son of a dentist, Bowles was born in Jamaica, New York City, and from an early age failed to meet the standards of his father. He was educated at the University of Virginia, but twice ran away to Paris, where he was taken up by

Gertrude Stein. He is an artist whose career resists, or ignores, lines of classification. A pupil of Virgil Thomson, Aaron Copland and (later) Nadia Boulanger, he made his reputation first of all as a composer, and in the 1930s he was known primarily for his music. Of his operas, two are adapted from Lorca, while the surrealist *The Wind Remains* was put on at the Museum of Modern Art in New York when he was only thirty-four. His work has been influenced by the folk music he has collected during his travels, and he has composed ballets, chamber and orchestral music, songs, film scores, and music for plays by **Tennessee Williams**, **Lillian Hellman** and others.

In 1931, he went to Europe to study music with Copland, and in Berlin met **Christopher Isherwood**, who borrowed his new friend's surname for his most famous Berlin character, Sally Bowles. He spent much of the 1930s travelling, principally in Morocco, returning to America at intervals, during one of which in 1938, he married Jane Auer, who, as **Jane Bowles**, became a well-known writer. Bowles has acknowledged his wife's importance, both as inspiration and critic, and it was the success of her novel *Two Serious Ladies* in 1943 that prompted him to embark upon his own literary career. His early stories attracted some controversy owing to their subject-matter, which although coolly observed was nevertheless shocking: in the classic 'Pages from Cold Point' a father recounts his equivocal relationship with his homosexual adolescent son, while 'The Delicate Prey' features the emasculation and mutilation of a youth, whose murderer is subsequently buried up to his neck in the sand. Bowles's reputation was swiftly established with such stories, but he was also during the 1940s working as a journalist, contributing a music column to the *New York Herald Tribune*, and as a translator: *No Exit*, his version of Jean-Paul Sartre's *Huis Clos*, enjoyed a Broadway run in 1946. That same year the *Partisan Review* published 'A Distant Episode', a story about an academic who is captured and tortured by nomadic tribesmen. On the strength of this, Bowles persuaded a publisher to pay him an advance to return to Tangier in order to write the book which became his first and most highly regarded novel, *The Sheltering Sky*, which was later made into an epic film by Bernardo Bertolucci (1990). The principal characters, Port and Kit Moresby, are partly modelled on Bowles and his wife.

Predominantly homosexual, Paul and Jane Bowles were nevertheless devoted to one another, locked into an almost symbiotic relationship. In spite of estrangements, and although they lived apart – Paul with the painter Ahmed Ben Driss el-Yacoubi, Jane with the sinister Cherifa, whose influence over her horrified many of her friends – they remained close, and Jane's gradual decline through illness towards insanity and death profoundly affected her husband, who found it difficult to work. He produced no further novels, although he continued to publish poetry and short stories and to translate the works of Moroccan and other writers. Several of his stories – notably in the collection *Midnight Mass* – take the form of modern folk-tales, directly told, and written from the inside. He has also produced an unrevealing autobiography, *Without Stopping*, and an impressionistic book about Morocco, *Points in Time*.

Jane drifted in and out of various hospitals, until in 1968 her husband insisted that she left Tangier altogether. She spent the last five years of her life in Spanish psychiatric hospitals, where Paul visited her regularly until her death in 1973. He continues to live in Tangier, a doyen of the expatriate community.

Fiction

The Sheltering Sky 1949; *The Delicate Prey and Other Stories* (s) 1950; *A Little Stone* (s) 1950; *Let it Come Down* 1952; *The Spider's House* 1955; *The Hours After Noon* (s) 1959; *A Hundred Camels in the Courtyard* (s) 1962; *Up Above the World* 1966; *The Time of Friendship* (s) 1967; *Pages from Cold Point and Other Stories* (s) 1968; *Three Tales* (s) 1975; *Things Gone and Things Still Here* (s) 1977; *Collected Stories 1939–1976* (s) 1979; *Midnight Mass* (s) 1981; *A Thousand Days for Mokhtar* 1989; *Too Far from Home* 1994

Poetry

Scenes 1968; *The Thicket of Spring: Poems 1926–1969* 1972; *Next to Nothing* 1976; *Next to Nothing: Collected Poems 1926–1977* 1981

Non-Fiction

Yallah 1956; *Their Heads Are Green* 1963; *Without Stopping* (a) 1972; *Points in Time* 1982; *Two Years Beside the Strait: Tangier Journal 1987–1989* 1990

Translations

Jean-Paul Sartre *No Exit* 1946; R. Firson-Roche *Lost Trail of the Sahara* 1956; Driss ben Hamed Charhadi *A Life Full of Holes* 1964; Mohammed Mrabet: *Love with a Few Hairs* 1967, *The Lemon* 1969, *Mhashish* 1969, *Hadidan Aharam* 1975, *Harmless Poisons, Blameless Sins* 1976, *Look and Move On* 1976, *The Bid Mirror* 1977, *The Beach Café and The Voice* 1980, *The Chest* 1983; Mohamed Chouki: *For Bread Alone* 1974, *Jean Genet in Tangier* 1974, *Tennessee Williams in Tangier* 1979; *The Boy Who Set the Fire and Other Stories* (s) 1974; Isabelle Eberhardt *The*

Oblivion Seekers 1975; *Five Eyes: Short Stories by Five Moroccans* (s) 1979; Rodrigo Rey Rosa: *The Beggar's Knife* 1985, *She Woke Me Up So I Killed Her* 1985

Opera

Denmark Vesey 1937; *The Wind Remains* 1941

Biography

The Invisible Spectator by Christopher Sawyer-Lauçanno 1989

William (Andrew Murray) BOYD 1952–

The son of an expatriate medical officer of Scots descent, Boyd was born at Accra, Ghana, and brought up there and in Nigeria, but was sent to a boarding-school in Scotland at the age of nine. Of Gordonstoun, he commented that 'nine years in a males-only British public-school took its toll', but it should also be said that they provided him with a great deal of material for his writing, notably the two television plays, *Good and Bad at Games* and *Dutch Girls*, which recreate the atmosphere of a public school of the period with unflinching accuracy. Boyd himself admitted that Gordonstoun provided the atmosphere of the First World War public-school battalion in which the protagonist of his novel *The New Confessions* served. He went on to the universities of Nice, Glasgow and Oxford, and was a lecturer in English literature at St Hilda's College, Oxford, from 1981 to 1983. He is married to the journalist and editor Susan Boyd.

Like several of his contemporaries, he first began publishing short stories in the *London Magazine*, although his first book was a novel. *A Good Man in Africa* (1981) is set in the imaginary city of Nkognsamba in West Africa and its eponymous anti-hero is the overweight, dissolute and accident-prone diplomat Morgan Leafy. A brilliantly funny novel in the tradition of **Evelyn Waugh**, **Kingsley Amis** and **Tom Sharpe**, it won the Whitbread Award for Best First Novel and a Somerset Maugham Award. The success of the novel led to the publication of a volume of short stories, *On the Yankee Station*, several of which had been published in such magazines as *Mayfair* and *Punch* or broadcast on the radio; it also included a school story and further misadventures of Morgan Leafy. His second and arguably his best novel was *An Ice-Cream War*, which was shortlisted for the Booker Prize and won the John Llewelyn Rhys Prize. A sophisticated adventure story, it is set in East Africa, one of the least remembered theatres of the First World War. Hilarious and shocking by turn, it is also a serious book about the unpredictable and arbitrary in life. Boyd himself witnessed the Biafran War and has noted that: 'In war is to be found the randomness, chaos and real character of the world.'

Stars and Bars, set in America, was less highly regarded and was made into an unmemorable film (1988) from Boyd's own screenplay. *The New Confessions*, however, is an epic novel covering the first half of the twentieth century, taking in the trenches of the Western Front and the early Hollywood movie industry. (Like most of Boyd's books, it is meticulously researched.) As the novel's title suggests, its protagonist, John James Todd, is obsessed by Jean-Jacques Rousseau, and it draws parallels between the dilemmas of the eighteenth-century philosopher and twentieth-century man. It won the Scottish Arts Council Book Award and conquered the French, being sold to a publisher there for some £150,000. *Brazzaville Beach* was similarly ambitious, and like its predecessor tipped by many to win the Booker Prize, although in the event neither novel reached the shortlist. The heroine of *Brazzaville Breach* is an anthropologist studying chimpanzees, but there is also a sub-plot about hedge-dating in England and some rather voguish references to mathematics. While the book was widely praised, a few dissenting voices suggested that it was rather hollow and written in a prose that lacked any real distinction. He returned to form with *The Blue Afternoon*, a bestselling novel in which a female architect in 1930s' Los Angeles is approached by a man who claims to be her father. (Once again, Boyd pursues the theme of order being disrupted by the unexpected in life.) The bulk of the novel is a love story and murder mystery, set in late nineteenth-century Manila and containing a fascinating wealth of detail about pioneer surgery and aviation.

Boyd's work for film and television includes an overlong adaptation of Evelyn Waugh's *Scoop* and a disappointing, coarsened movie of *A Good Man in Africa* (1994), directed by Bruce Beresford.

Fiction

A Good Man in Africa 1981; *On the Yankee Station and Other Stories* (s) 1981; *An Ice-Cream War* 1982; *Stars and Bars* 1984; *The New Confessions* 1987;*Brazzaville Beach* 1990; *The Blue Afternoon* 1993; *The Destiny of Nathalie X and Other Stories* (s) 1995

Plays

School Ties (tv) [pub. 1985]: *Good and Bad at Games* 1983, *Dutch Girls* 1985; *Scoop* (tv) [from E. Waugh's novel] 1987; *Stars and Bars* (f) [from his novel] 1988; *A Good Man in Africa* (f) [from his novel] 1994

Clare BOYLAN 1948–

Boylan was born in Dublin, and had 'a totally urban middle-class childhood'. Her mother wrote essays and short stories, and would often keep her daughters home from school for several days at a stretch and write while the children undertook 'curious and elaborate, if often useless, pieces of housework'. At the age of fourteen, Boylan formed a pop group with her two sisters, touring the Dublin variety shows under the name 'The Girl Friends' and earning extra pocket money. She left school at sixteen, and her first job was in a bookshop. She was next employed by the *Irish Press* as a reporter, after which her career as a journalist in print and on radio and television escalated. From 1973 to 1978 she wrote for the *Dublin Evening Press*; she went on to edit the Dublin-based magazines *Young Woman* and *Image*, and has written for *Cosmopolitan*, *Good Housekeeping* and New York *Vogue*. In 1983 she won the Dublin Journalist of the Year award. She gave up journalism to write fiction full-time, but subsequently returned to it, interviewing other writers in the *Guardian*.

Her first novel, *Holy Pictures*, was published in 1983 to extremely favourable reviews. In order to get started on it she interviewed her favourite writers (including **Molly Keane**, **William Trevor** and **Jennifer Johnston**), asking them: 'What's the trick of it?'. *Holy Pictures* is the story of two young girls growing up in Dublin in the 1920s, closeted by dream-thwarting elders. The 'holy pictures' are the photographs of film stars who represent escape, beauty and fulfilment. The novel was praised for its skilful handling of the teenage viewpoint. *Last Resorts*, published the following year, transposes the vision of *Holy Pictures*: the children are seen as selfish and restricting as they dominate the single-parent protagonist. In *Black Baby*, a Dublin fairy-tale, an elderly woman is visited by a black woman whom she had sponsored when a covent-school girl.

Boylan has also published short stories in a wide variety of magazines in Ireland, England, Europe and America and has produced two collections. *A Nail in the Head* confirmed her reputation, while *Concerning Virgins*, takes as its theme the lawlessness of innocence.

She lives in County Wicklow with her journalist husband.

Fiction
Holy Pictures 1983; *A Nail in the Head* (s) 1983; *Last Resorts* 1984; *Black Baby* 1988; *Concerning Virgins* (s) 1990; *Home Rule* 1992

Edited
The Agony and the Ego: the Art and Strategy of Fiction-Writing Explored 1993

Kay BOYLE 1903–1992

From an old and well-off American family (her grandfather was editor-in-chief of a legal publishing firm and another forebear had been a major-general under George Washington), Boyle received little formal schooling, but spent much of her childhood moving around America and Europe with her feckless father and exceptionally cultivated mother. She studied music and architecture in Cincinnati, but graduated in neither, getting a job in 1922 as assistant to the Australian poet Lola Ridge who was New York editor of *Broom* magazine, which printed Boyle's earliest poetry. In the same year she married Richard Brault, a Frenchman who had studied engineering in Cincinnati, and in 1923 the couple went to live in France, initially settling in Brault's home region of Brittany. Boyle soon made contact with expatriate Paris, having an affair with Ernest Walsh, editor of *This Quarter*, with whom she had a daughter, and who published her poetry. Soon after this she became a regular contributor to *transition*.

Her first collection of short stories was published in 1930, and her first novel, *Plagued by the Nightingale*, in 1931. Separated from Brault, and with Walsh dead, she spent some time ghost-writing the memoirs of the sister-in-law of the white rajah of Sarawak (Gladys Palmer, the biscuit heiress), and then went to live in Raymond Duncan's artistic colony at Neuilly, from which she had to escape when Duncan demanded retention of her daughter. In 1931 she married the writer Laurence Vail, a man of substantial means, with whom she had three children (one of them was called Apple-Joan, while his son by Peggy Guggenheim was called Sindbad) and with whom she eventually settled in a château in the Mégève. In 1941 they returned to the USA, where she divorced Vail and married in 1941 an Austrian baron, Joseph von Franckenstein, who had been tutor to her children. After the war he did official work in Germany, where she was a correspondent for the *New Yorker*, but her left-wing convictions cost him his job in the McCarthyite era. They had two further children and he died in 1963. In the same year, Boyle, having worked in various schools and colleges, became a teacher of creative writing at San Francisco State College, where she played a leading role in the student agitation of 1967–8, going to prison briefly after demonstrating against the Vietnam draft and taunting the principal of her college as 'Hayakawa-Eichmann'.

She retired in 1979, but continued to live in the San Francisco Bay area. Her many publications include, besides fiction and poetry, books for children, translations of Radiguet and Crevel, and the well-known *Being Geniuses Together*, her amplification of the memoirs of Robert McAlmon. Her novels and short stories are directly autobiographical and probably more interesting for that than as fiction *per se*. One of her novels, *Gentlemen, I Address You Privately* (1933), is an early treatment of male homosexuality.

Fiction

Wedding Day and Other Stories (s) 1930; *Plagued by the Nightingale* 1931; *Year Before Last* 1932 [rev. 1986]; *The First Lover and Other Stories* (s) 1933; *Gentlemen, I Address You Privately* 1933; *My Next Bride* 1934; *Death of a Man* 1936; *The White Horses of Vienna and Other Stories* (s) 1936; *Monday Night* 1938; *The Crazy Hunter* 1940; *Primer for Combat* 1942; *Avalanche* 1944; *A Frenchman Must Die* 1946; *Thirty Stories* (s) 1946; *1939* 1948; *His Human Majesty* 1949; *Smoking Mountain* (s) 1951; *The Seagull on the Step* 1955; *Three Short Novels* (s) 1958; *Generation Without Farewell* 1960; *Nothing Ever Breaks Except the Heart* (s) 1966; *The Underground Woman* 1974; *Fifty Stories* (s) 1980; *The Crazy Hunter* 1993

Non-Fiction

Breaking the Silence: Why a Mother Tells Her Son about the Nazi Era 1962; *Being Geniuses Together* [with R. McAlmon] 1969; *Long Walk at San Francisco State* 1970; *Words That Must Somehow Be Said: Selected Essays 1927–1984* [ed. E.S. Bell] 1985

Poetry

A Statement 1932; *A Glad Day* 1938; *American Citizen Naturalized in Leadville, Colorado* 1944; *Collected Poems* 1962; *Testament for My Students and Other Poems* 1970; *This Is Not a Letter and Other Poems* 1985

Edited

Enough of Dying! An Anthology of Peace Writings [with J. Van Gundy] 1972

Translations

Joseph Delteil *Don Juan* 1931; René Crevel: *Mr Knife, Miss Fork* 1931, *Babylon* 1985; Raymond Radiguet *Devil in the Flesh* 1932

For Children

The Youngest Camel 1939; *Pinky, the Cat who Liked to Sleep* 1966; *Pinky in Persia* 1968

T(homas) Coraghessan BOYLE 1948–

Born in Peekskill, New York, the son of Irish immigrant parents, Boyle claims not to have read a book until he was eighteen. He majored in music at college and became a high-school teacher, chiefly 'to stay out of Vietnam'. His main interests at this period – again by his own account – were just 'hanging out [and] taking a lot of drugs'. When he did start reading he was impressed by **Thomas Pynchon**, **John Barth** and Gabriel Garcia Marquez, as well as the plays of Ionesco and Genet.

In 1972 Boyle joined the Iowa Writers' Workshop, where he studied creative writing under Vance Bourjaily and **John Irving**. He gained a PhD in 1977 with his short stories, and in 1979 he published his first collection, *Descent of Man*, which won the St Lawrence Award for Short Fiction. Boyle had already been publishing work since the early 1970s, and was a fiction editor of the *Iowa Review*. He now teaches creative writing at the University of Southern California, lives in Los Angeles with his wife and two children, and plays the saxophone in a band.

Boyle's pyrotechnical writing has attracted notice from the very beginning of his career. He won the Coordinating Council of Literary Magazines Award for Fiction and a Creative Writing Fellowship from the National Endowment for the Arts before his first collection was published, and extracts in the *Paris Review* from his first novel won the Aga Khan Award. He writes with extreme virtuosity and ingenuity on an extraordinary range of subjects. His first collection included stories about a man whose wife leaves him for a chimpanzee who is translating Nietzsche; an inventor who begins his career by inventing the non-defecating cat, and ends it by photographing the death of God; a grotesque eating competition; a man ruinously obsessed by his exclusion from a woman-only restaurant; and the Dadaist gesture of bringing Idi Amin Dada to a New York Dada art show.

Consistent themes and tendencies include excess and cultural collapse, violence, the otherness of women, the blurring of the animal–human boundary, politics and popular culture at their most overblown, bodily functions, and bizarre juxtapositions and 'what if' situations – a later collection of stories includes a romance between Nina Khrushchev and Eisenhower.

Critics have for the most part been dazzled, with some dissenting voices complaining about satirical obviousness, unpleasantness, flashy cleverness, and an ultimate bleakness and 'punitive grimness' in the black humour. 'Glitterature' was the word used by one critic, who also compared Boyle to the film director Mel Brooks and noted his 'mad scientist skew on character and situation'. In his shining virtuosity, fine-tuned grossness, and his twisted, off-the-wall take on contemporary life, Boyle resembles Frank Zappa. His satirical novel about the founder of Kelloggs and the American health obsession, *The Road to Wellville*, was filmed in 1995 by Alan Parker.

Fiction

Descent of Man (s) 1979; *Water Music* 1982; *Building Prospects* 1984; *Greasy Lake and Other Stories* (s) 1985; *World's End* 1988; *If the River Was Whiskey* (s) 1989; *East Is East* 1990; *The Road to Wellville* 1993; *Without a Hero* 1994

Malcolm (Stanley) BRADBURY 1932–

Born in Sheffield, Bradbury was educated at West Bridgeford Grammar School, and Queen Mary's College, London, where he obtained a first in English. It was in the course of a year at Indiana University that he 'fell in love with America and American writers'. In 1959, he took a PhD in American studies at Manchester University and, after teaching at the Universities of Hull and Birmingham, joined the new University of East Anglia where, since 1970, he has held the Chair of American Studies while writing in tandem.

His first novel, *Eating People Is Wrong* (1959), centred on academic life led in the Midlands of the 1950s, and took eight years to complete: 'I felt the book darken as it went on, and I think that in this darkening I actually found a reason for writing.' University life features in his second novel, *Stepping Westward*, which is characterised by a blend of exaggeration, comedy and self-consciously artful writing. Indeed, Bradbury has stated his commitment to comedy, believing it to be 'the essential tradition of the English novel'. In *The History Man*, which was adapted for BBC Television in 1981 by **Christopher Hampton**, the radicalism prevalent in universities of the 1970s is sharply satirised, and his anti-hero Howard Kirk displays an intriguing mix of thrusting professional and sexual amibition. Bradbury has said how his own writing for television has influenced him considerably: 'I think the imagery and grammar of film and television has brought new concepts of presentation and perception to the novel.' With *Rates of Exchange*, shortlisted for the Booker Prize in 1983, he breaks away from the campus scenario to follow the adventures of the linguist Dr Patworth, a lecturer on a British Council tour of Slaka, a fictitious Eastern bloc state. Linguistics dominate this encounter of Thatcherite Britain with Eastern Europe, at a time when the latter's disintegration was imminent. Its key word, Bradbury notes, is 'exchange ... political exchanges, sexual exchanges, economic exchanges, exchanges of every sort' – including linguistic ones. Whereas in *The History Man* there was, he says, an 'ironic distance', in *Rates of Exchange* he tries 'to get the source of pain into the experience of the characters'. This pain affected the writer himself, who suffered a breakdown on its completion: 'It explored areas of great pain to me, and I found it very painful to write.' *Cuts* is a short novel on academic life in an era of cutbacks in spending.

In *Who's Who*, Bradbury lists his recreations as 'none'. His energy goes into his work which is almost entirely concerned with the modern novel. It was with this concern in mind that he established, with **Angus Wilson**, an MA degree course in creative writing at the University of East Anglia in 1969, and from which he retired in 1995. Its alumni include writers such as **Kazuo Ishiguro**, **Rose Tremain**, **Clive Sinclair** and **Ian McEwan**. He has produced books of criticism, edited others, written radio and television series (*Anything More Would Be Greedy* and *The Gravy Train*) and adaptations of **Stella Gibbons**'s *Cold Comfort Farm* and two novels by **Tom Sharpe**: he was awarded an International Emmy Award for his work on *Porterhouse Blue* (1987). His favourite among his own books is *My Strange Quest for Mensonge*, which he sees as 'a satire on Structuralism and Deconstruction'. He lives in Norwich with his wife, whom he married in 1959, and has two sons. He is a CBE.

Fiction

Eating People Is Wrong 1959; *Stepping Westward* 1965; *The History Man* 1975; *Who Do You Think You Are?* (s) 1976 [rev. 1986]; *Rates of Exchange* 1983; *Cuts* 1987; *Dr Criminale* 1992

Non-Fiction

Phogey! or, How to Have Class in a Classless Society 1960; *Evelyn Waugh* 1964; *What Is a Novel?* 1969; *The Social Context of Modern English Literature* 1971; *Possibilities: Essays on the State of the Novel* 1973; *The Outland Dart: American Writers and European Modernism* 1977; *All Dressed Up and Nowhere to Go: the Poor Man's Guide to Affluent Society* 1982; *The Expatriate Tradition in American Literature* 1982; *Saul Bellow* 1982; *The Modern American Novel* 1983 [rev. 1992]; *My Strange Quest for Mensonge* 1987 [as *Mensonge* 1993]; *No, Not Bloomsbury* 1987; *Why Come to Slaka?* 1987; *The Modern World: Ten Great Writers* 1988; *Unsent Letters* 1988; *From Puritanism to Postmodernism* [with R. Ruland] 1993; *The Modern British Novel* 1993

Plays
Between These Four Walls [with D. Lodge, J. Duckett] 1963 [np]; *The After Dinner Game* **(tv)** 1975 [pub. 1982; rev. 1987]; *Blott on the Landscape* **(tv)** [from T. Sharpe's novel] 1981; *Porterhouse Blue* **(tv)** [from T. Sharpe's novel] 1987; *Anything More Would Be Greedy* **(tv)** 1989; *The Gravy Train* **(tv)** 1990; *Cold Comfort Farm* **(tv)** [from S. Gibbons's novel] 1995

Edited
E.M. Forster: *A Collection of Critical Studies* 1966, *A Passage to India—a Casebook* 1970; *The Penguin Companion to Literature* vol. 3 [with E. Mottram] 1971; *Modernism* [with J.W. McFarlane] 1976; *The Novel Today* 1977; *An Introduction to American Studies* [with H. Temperley] 1981; *Contemporary American Fiction* [with S. Ro] 1987; *The Penguin Book of Modern British Short Stories* **(s)** 1987; *New Writing* [with J. Cooke] 1992; Stephen Crane *The Red Badge of Courage* 1993; Washington Irving *The Sketch Book of Geoffrey Crayon, Gent* 1993

Ray(mond Douglas) BRADBURY 1920–

America's best-known writer of science fiction does not fit the mould of most of that genre's practitioners. Bradbury's roots lie firmly in Waukegan, Illinois, where he was born and whose values and landscape (that of the American Midwest) permeate even his most outlandish stories. His fiction first appeared during the 1940s in *Weird Tales* and other pulp science-fiction magazines, but he quickly made his reputation and on the eve of his marriage in 1947 felt sufficiently confident to burn some one million words of writing which did not come up to scratch.

Since his marriage, Bradbury has lived in Los Angeles, and it was there in 1950, in a bookshop, that he met **Christopher Isherwood**, upon whom he pressed a copy of his volume of short stories, *The Martian Chronicles*. Isherwood was no admirer of science fiction, but nevertheless read the book and was so impressed that he offered to review it in *Tomorrow* magazine. Bradbury recalled that this long, praising review marked a turning-point in his career, since his book had hitherto been all but ignored by the mainstream press.

Bradbury has consistently defied the expectations aroused by the label 'science-fiction writer'. His stories remain resolutely unfascinated by technology, even suspicious of it, and Bradbury himself at one stage in his life refused to use cars or aeroplanes. Furthermore, he is a stylist in a genre which values invention over expression. In the late 1950s he made the logical step implied by this and began to write realistic stories alongside his works of fantasy. The themes of alienation and separation which recur in his earlier work were revealed at the roots in his tales of Midwestern American life.

From the 1960s onwards, Bradbury has been producing plays and poetry to augment his novels and huge volume of short stories. His travels in Mexico and Ireland inform many of them; others reflect childhood experiences. He has also written a large number of television plays and several screenplays, most notably that for John Huston's celebrated 1956 film adaptation of Herman Melville's *Moby Dick*. Several of his own stories have been made into films, the most distinguished of which was François Truffaut's *Fahrenheit 451* (1966). Bradbury's original novel, which is generally recognised to be his best book beside *The Martian Chronicles*, is a characteristically humane allegory which envisages a future dominated by the state media, where books are banned (the title refers to the degrees of heat needed to burn paper).

Fiction
Dark Carnival **(s)** [UK *The Small Assassin*] 1947; *The Martian Chronicles* **(s)** [UK *The Silver Locusts*] 1950; *The Illustrated Man* **(s)** 1951; *Fahrenheit 451* 1953; *The Golden Apples of the Sun* **(s)** 1953; *The October Country* **(s)** 1956; *Dandelion Wine* 1957; *The Day It Rained Forever* **(s)** 1959; *A Medicine for Melancholy* **(s)** 1959; *Something Wicked This Way Comes* 1963; *The Machineries of Joy* **(s)** 1964; *The Autumn People* **(s)** 1965; *The Vintage Bradbury* **(s)** 1965; *Tomorrow Midnight* **(s)** 1966; *3 to the Highest Power* **(c)** [ed. W.F. Nolan] 1968; *I Sing the Body Electric!* **(s)** 1969; *Bloch and Bradbury* **(s)** [with R. Bloch; UK *Fever Dreams and Other Fantasies*] 1970; *Long After Midnight* **(s)** 1976; *To Sing Strange Songs* **(s)** 1979; *The Last Circus and The Electrocution* **(s)** 1980; *Dinosaur Tales* **(s)** 1983; *A Memory for Murder* **(s)** 1984; *Death Is a Lonely Business* 1985; *The Toynbee Convector* **(s)** 1989; *A Graveyard for Lunatics* 1990; *Green Shadows, White Whale* 1992

Plays
The Meadow 1948 [pub. in *Best One-Act Plays of 1947–48* ed. M. Mayorga]; *It Came from Outer Space* **(f)** [with D. Schwartz] 1952; *Moby Dick* **(f)** [from H. Melville's novel; with J. Huston] 1956; *Shopping for Death* **(tv)** 1956; *Design For Loving* **(tv)** 1958; *The Gift* **(tv)** 1958; *Special Delivery* **(tv)** 1959; *The Tunnel to Yesterday* **(tv)** 1960; *Icarus Mongolfier Wright* **(f)** [with G.C. Johnston] 1961;

The Faith of Aaron Menefee (**tv**) 1962; *The Jail* (**tv**) 1962; *The Anthem Sprinters and Other Antics* 1968 [pub. 1963]; *The Life and Work of Juan Díaz* (**tv**) 1963; *The World of Ray Bradbury* 1964 [np]; *The Wonderful Ice-Cream Suit and Other Plays* [pub. 1972]: *The Wonderful Ice-Cream Suit* 1965, *The Veldt* 1965, *To the Chicago Abyss* 1965; *The Day It Rained Forever* 1966; *The Pedestrian* 1966; *Christus Apollo* [with J. Goldsmith] 1969 [np]; *The Groom* (**tv**) 1971; *Leviathan 99* 1972 [np]; *Picasso Summer* (**f**) [as 'Douglas Spaulding'; with E. Booth] 1972; *The Foghorn, Kaleidoscope, Pillar of Fire* [pub. in *Pillar of Fire and Other Plays* 1975]; *The Ghost, That Bride of Time: Excerpts from a Play-in-Progress* [pub. 1976]; *Dandelion Wine* [from his novel] 1980; *Forever and the Earth* (**r**) 1984

Poetry

Old Ahab's Friend, and Friend to Noah, Speaks His Piece 1971; *When Elephants Last in the Dooryard Bloomed* 1975; *That Son of Richard III* 1974; *Where Robot Mice and Robot Men Run Round in Robot Towns* 1977; *Twin Hieroglyphs that Swim the River Dust* 1978; *The Bike Repairman* 1978; *The Author Considers His Resources* 1979; *The Aqueduct* 1979; *The Attic Where Meadow Greens* 1980; *The Haunted Computer and the Android Pope* 1981; *Imagine* 1981; *The Complete Poems of Ray Bradbury* 1982; *Two Poems* 1982; *The Love Affair* 1983

For Children

Switch on the Night 1955; *R is for Rocket* 1962; *S is for Space* 1968; *The Halloween Tree* 1972; *The Ghost of Forever* 1981; *The Million Year Picnic and Other Stories* (**s**) [ed. K. Jones] 1986

Non-Fiction

Teacher's Guide: Science Fiction [with L. Olfson] 1968; *Mars and the Mind of Man* 1973; *Zen and the Art of Writing and The Joy of Writing* 1973; *The Mummies of Guanajuato* [with A. Lieberman] 1978; *About Norman Corwin* 1979; *Beyond 1984: Remembrance of Things Future* 1979; *Los Angeles* 1984; *Orange Country* [with others] 1985; *In the Stream of Stars* [with others] 1990; *Yestermorrow: Obvious Answers to Impossible Futures* 1991

Edited

Timeless Stories for Today and Tomorrow 1952; *The Circus of Dr Lao and Other Improbable Stories* 1956

Joan BRADY 1939–

Brady was born in San Francisco, California; her father was Professor of Economics at the University of California at Berkeley, her mother a journalist specialising in consumer economics. Both her brother, Michael, and sister, Judith, are writers. She attended state schools until she was thirteen, and then went to a private school. Her first love was ballet, and she trained at the San Francisco Ballet School, making her first public appearance in the opera *Aida* when she was fourteen. She danced with the San Francisco Ballet Company for three years before moving to New York to study with George Balanchine at his School of American Ballet. She danced with the New York City Ballet in 1960, and then, partly because of her disastrous relationship with her mother, lost her confidence and gave it all up to attend Columbia University from 1961 to 1965, majoring in philosophy.

When she was eighteen, she moved in with fifty-year old Dexter Masters, an old family friend she had adored since childhood. Later, she discovered that not only had he been her mother's lover, but that after her father's death when she was seventeen, her mother had earmarked Masters to be her next husband. They married and had a son, Alexander, in 1965. Her mother died the same year, and the family moved to England, partly for Alexander's education.

She started writing in the 1970s, and published two short stories before her first novel *The Impostor* was published in 1979. Her second book, *The Unmaking of a Dancer*, is an autobiography of the first half of her life. Perhaps its most poignant pages are those where she revisits her obsession with dancing and tries to take it up again professionally at the age of thirty-nine with her husband's support. (It was published in the UK in 1994 as *Prologue: An Unconventional Life*.) Soon after its publication, she began work on *Theory of War*, a novel which took her ten painful years to research and write. It is based on the story of her own grandfather, Nathaniel Brady, who was sold for $15 at the age of four to a white farmer who treated him with such cruelty that the scars have reverberated through the lives of his seven children and beyond. Four of them (including Brady's father) committed suicide, one was mentally handicapped and one drank himself to death.

Dexter Masters died in 1989 after a long illness, and two years later, Brady finally finished *Theory of War*, which was published in 1993 and won the Whitbread Book of the Year Award. She admitted that she found it hard to write in her true voice until after Masters died. 'He liked me to write in a "feminine" way,' she has said, 'and many of what I regard as the best passages in the book are ones he took out originally.' *Theory of War* is a classic of its kind. Stark and uncompromising, it probes the darkest corners of human nature, and proves that severe harm is handed on through generations. It is too soon to tell whether Brady will ever move away totally from her life and family as subject matter for her books, because both have yielded such rich material.

Brady has lived quietly in Totnes, Devon, since the mid-1960s and is now a British citizen.

Fiction

The Impostor 1979; *Theory of War* 1993

Non-Fiction

The Unmasking of a Dancer (a) 1982 [rev. as *Prologue* 1994]

Melvyn BRAGG 1939–

Born in Carlisle, Bragg was brought up in nearby Wigton where his father was a fitter at an aerodrome and his mother worked in various part-time jobs. His father served in the RAF during the war, and was later a bookmaker and publican. At eleven Bragg won a scholarship to Nelson-Tomlinson Grammar School, and he went on to read modern history at Wadham College, Oxford, graduating in 1961. It was while at Oxford – 'in the gentle depression of adolescent academic life' – that he first decided to write. He joined BBC Radio as a general trainee in 1961, and first worked in television on the arts programme *Monitor*, writing scripts about Claude Debussy and the painter Henri Rousseau among others. Bragg's first novel, *For Want of a Nail* (1965), chronicles the adolescent years of a Cumberland boy, and describes the disruption caused by his family's move from country to city, a theme also taken up in *The Hired Man*, the first novel in *The Cumbrian Trilogy*. *The Hired Man* was adapted as a musical by Bragg and composer Howard Goodall in 1984.

With the publication in 1966 of his second novel, *The Second Inheritance*, Bragg left the BBC to become a freelance writer, and wrote scripts for Ken Russell and collaborated with Norman Jewison on the screenplay of Andrew Lloyd Webber's *Jesus Christ Superstar*. He returned to the BBC as a presenter of *Second House* (1973–7) and *Read All About It* (1976–7), and since 1978 has been presenter and editor of the influential arts programme *The South Bank Show*, and head of arts at London Weekend Television since 1982. In 1993 he received a reported £2.9 million in consequence of a controversial management share scheme.

With his 1971 novel *The Nerve*, Bragg moved away from his familiar Cumbrian terrain to describe a London teacher's nervous breakdown, though both the teacher and his friend – a successful novelist and television producer – hail from Cumberland. In the 1980s Bragg wrote successful theatrical biographies of Laurence Olivier and Richard Burton, the latter being authorised by Burton's widow, Sally, who granted him full access to Burton's papers. His novel *A Time to Dance* chronicles the affair between a middle-aged man and a young girl, and created something of a furore when it was adpated for television in 1992, largely because of the explicit sex scenes. Bragg's first wife, Marie-Elisabeth Roche, died in 1971, and he married the writer Cate Haste in 1973. He has one daughter from his first marriage, and a son and daughter from his second.

Fiction

For Want of a Nail 1965; *The Second Inheritance* 1966; *Without a City Wall* 1968; *The Cumbrian Trilogy* 1984: *The Hired Man* 1969, *A Place in England* 1970, *Kingdom Come* 1980; *The Nerve* 1971; *Josh Lawton* 1972; *The Silken Net* 1974; *A Christmas Child* 1976; *Autumn Manoeuvres* 1978; *Love and Glory* 1983; *The Maid of Buttermere* 1987; *A Time to Dance* 1990; *Crystal Rooms* 1992; *The Nerve* 1992

Non-Fiction

Speak for England: an Essay on England, 1900–1975 1976; *Land of the Lakes* 1983; *Laurence Olivier* 1984; *Abuses of Literacy?* [with M. Holroyd] 1985; *Rich: the Life of Richard Burton* 1988; *The Seventh Seal* 1993

Edited

My Favourite Stories of Lakeland 1981

Plays

The Hired Man [from his novel; with H. Goodall] 1984; *A Time to Dance* (tv) [from his novel] 1992

John (Gerard) BRAINE 1922–1986

Braine was born in Bradford, Yorkshire. His father was a works superintendent for Bradford Council and of Nonconformist tradition; his mother came from an Irish Catholic background. After St Bede's Grammar School, Bradford, Braine did a succession of dead-end jobs – selling furniture and books, working in a factory – and was invalided out of wartime service in the navy. He later became a librarian in Bingley, but in 1951 gave up his secure position to become a literary man in London, although having little experience. His mother's death and an attack of tuberculosis brought him back to Yorkshire, and during his period in a sanatorium he started writing his bestselling novel, *Room at the Top*. He remained in Yorkshire and became a librarian again for some years, but when *Room at the Top* was published in 1957 its combination of tough northern aggressiveness, political radicalism and rebellion and (for the time) explicit sex made it an overnight bestseller, and Braine was enabled to give up his job for good and become a full-time writer. The novel, and its sequel *Life at the Top*, were both filmed: the first very successfully in 1958 by Jack Clayton from a screenplay by Neil Patterson (which won an Academy Award); the second far less so by Ted Kotcheff from a screenplay by **Mordecai Richler**.

During the 1960s he moved south and lived at Woking with his wife, Helen Wood, and their four children. He had considerable popular success and wrote a dozen novels; nevertheless, none of them made quite the impact that *Room at the Top* had and, while some, such as *The Jealous God*, received good reviews, others were judged to be merely potboilers. His conversion, like many of his generation, to a vociferous rightism may also have damaged his reputation. In the early 1980s, because of an affair he was having, Braine went to live in a small bedsitter in Hampstead. His last years saw financial difficulty, and a much-publicised auction of his literary manuscripts in late 1985 to raise money to buy a flat in Hampstead failed to find a bidder. The events of these last years of his life were chronicled by Braine in two frankly autobiographical novels, *One and Last Love* and *These Golden Days*. Braine died in London of a burst stomach ulcer.

Fiction

Room at the Top 1957; *The Vodi* 1959 [as *From the Hand of the Hunter* 1960]; *Life at the Top* 1962; *The Jealous God* 1964; *The Crying Game* 1968; *Stay With Me Till Morning* 1970 [as *The View from Tower Hill* 1971]; *The Queen of a Distant Country* 1972; *The Pious Agent* 1975; *Waiting for Sheila* 1976; *Finger of Fire* 1977; *One and Last Love* 1981; *The Two of Us* 1984; *These Golden Days* 1985

Non-Fiction

Writing a Novel 1974; *J.B. Priestley* 1978

E(dward) K(amau) BRATHWAITE 1930–

The son of a warehouse clerk, Brathwaite was born in Bridgetown, Barbados. Encouraged by his mother, he stayed on at school and won a Barbados scholarship to Pembroke College, Cambridge, to read history; **Ted Hughes, Thom Gunn** and **Peter Redgrove** were among his contemporaries. On going down in 1954, he went to Ghana as an education officer, and founded a children's theatre there. His years in Ghana were an important experience that put him in touch with his pre-colonial African roots.

He returned to the Caribbean in 1962 as a tutor for the University of the West Indies Extra Mural Department in St Lucia. He subsequently became a lecturer and reader at the university and published his first major work, a trilogy of long poems, during the late 1960s. *The Arrivants* consists of *Rights of Passage*, *Masks* and *Islands* and is concerned with the rebirth and recovery of the West Indian's identity through an examination of his African roots. He insists that the acceptance of these roots will begin to heal the traumas of middle passage, plantation and colonial life. To accomplish this, he considers it essential to introduce the denigrated Creole language spoken by most West Indians into the education system, an idea that has met with some resistance from the establishment, his own university not excepted. During this same period he also wrote a thesis on the Jamaican slave and Creole society in the eighteenth century.

In his later work – especially in the trilogy of *Mother Poem*, *Sun Poem* and *X/Self* – he has taken this process further, often abandoning the notations of the written word in an attempt to reproduce the rhythms of everyday speech. His study *The History of the Voice* demonstrates the vital link between language and cultural freedom in the development of Caribbean poetry, and without him the dub poetry movement could hardly have made progress.

His standing, alongside **Derek Walcott**, as a leading West Indian poet is reflected in several international awards, including the Cholmondely Award in 1970 and Guggenheim and Fulbright fellowships in 1983. He married Doris Welcome in 1960, has one son, and since 1983 has been professor of social and cultural history at the University of the West Indies in Kingston, Jamaica. He should not be confused with the older, Guyanese-born writer E.R. Braithwaite, author of the popular novel *To Sir, with Love* (1959).

Poetry

The Arrivants 1973: *Rights of Passage* 1967, *Masks* 1968, *Islands* 1969; *Panda No. 349* 1969; *Penguin Modern Poets 15* [with A. Bold, E. Morgan] 1969; *Days and Nights* 1975; *Other Exiles* 1975; *Black + Blues* 1976; *Mother Poem* 1977; *Soweto* 1979; *Wordmaking Man* 1979; *Sun Poem* 1982; *Third World Poems* 1983; *Jah Music* 1986; *X/Self* 1987; *Middle Passages* 1992; *DreamStories* 1994

Plays

Four Plays for Schools 1961–2 [pub. 1964]; *Odale's Choice* 1962 [pub. 1967]

Non-Fiction

The People Who Came vols 1–3 1968–72; *The Folk Culture of the Slaves in Jamaica* 1970; *The Development of Creole Society in Jamaica 1770–1820* 1971; *Caribbean Man in Space and Time* 1974; *Contradictory Omens: Cultural Diversity and Integration in the Caribbean* 1974; *Our Ancestral Heritage: a Bibliography of the Roots of Culture in the English-speaking Caribbean* 1976; *Wars of Respect: Nanny, Sam Sharpe and the Struggle for People's Liberation* 1977; *Barbados Poetry: a Checklist, Slavery to the Present* 1979; *Jamaica Poetry: a Checklist 1686–1978* 1979; *Gods of the Middle Passage* 1982; *Kumina* 1982; *The Colonial Encounter: Language* 1984; *The History of the Voice* 1984; *Roots* 1986

Edited

Iouanaloa: Recent Writing from St Lucia 1963; *Dream Rock* 1987

Richard (Gary) BRAUTIGAN 1935–1984

By the late 1960s Brautigan had achieved controversial recognition as *the* hippy spokesman of that era. He was vigorously celebrated or damned, according to varying opinion on the social phenomenon he was supposed to be expressing: the bittersweetness of the American Dream as seen through the eyes of a disillusioned but reasonably good-humoured San Francisco 'flower child'. He was born in Washington, and described his father as a 'casual labourer'. As a boy, he spent much time alone, fishing. It was in San Francisco, where he performed public readings and handed out photocopied poems on the street, that he became a mascot of the flower children, and his first novel, *A Confederate General from Big Sur*, was published in 1964. *Trout Fishing in America* (1967) brought him critical and commercial success.

His influences were the Imagists, Japanese aesthetics and the French Symbolists, who inspired his use of synesthesia. From the Romantic poets he took his major theme of the transforming power of the imagination, particularly that of the artist. His plots are bizarre, often devoid of dramatic action, while his analogies and images come from unrelated ordinary objects which when linked become astonishing. The surreal, language-distorting, snapshot-like elements of his writing led some to consider him the literary equivalent of the drug experience, while many of his older readers hoped he would provide an insight into the youth culture. *Life* magazine called his work 'a gentle carnival'. His obituary in the Los Angeles *Times* was headlined: 'Brautigan, literary guru of the 60s ...'

Never at any point was he forthcoming about his private life or his attitude as a writer. He married in 1957, divorced in 1970 and had a daughter. He travelled throughout the USA and in later years spent much time in Tokyo. In one of his rare interviews he explained that he wrote poetry only as a means of finding the perfect sentence. In his later work he revealed a blacker humour and a brooding, melancholic narrator, depressed at growing old. In *Sombrero Fallout*, the narrator describes his worries following him round 'like millions of trained white mice'.

Brautigan committed suicide by shooting himself in the head at his home in Bolinas, California. His body was not found until four weeks after his death, for his friends had believed him to be writing and therefore living in self-imposed seclusion, as was his habit. Shortly before his death he had been drinking heavily and was extremely depressed. Certainly his popularity as a writer had faded in the USA, although it lasted somewhat longer in the UK, where he was a campus favourite. His friend and fellow novelist Tom McGuane commented: 'when the sixties ended, he was the baby thrown out with the bathwater'. It is said of Brautigan that he never cared about the critics but was broken-hearted at the loss of his readers, and it is still argued as to whether or not he ranks as a 'minor novelist', or was just a fad of the 1960s.

Fiction

A Confederate General from Big Sur 1964; *The Abortion* 1966; *Trout Fishing in America* 1967; *In Watermelon Sugar* 1968; *Revenge of the Lawn: Stories 1962–1970* (s) 1971; *The Hawkline Monster* 1974; *Willard and His Bowling Trophies* 1974; *Dreaming of Babylon* 1976; *Sombrero Fallout* 1976; *The Tokyo–Montana Express* 1980; *So the Wind Won't Blow It All Away* 1982

Poetry

The Galilee Hitch-Hiker 1958; *Lay the Marble Tea* 1959; *The Octopus Frontier* 1960; *All Watched Over by Machines of Loving Grace* 1967; *The Pill versus the Springhill Mine Disaster* 1968; *Please Plant This Book* 1968; *Rommel Drives on Deep into Egypt* 1970; *Loading Mercury with a Pitchfork* 1976; *June 30th – June 30th* 1978

Howard BRENTON 1942–

Brenton was born in Portsmouth and was educated at Chichester High School and St Catherine's College, Cambridge, from which he graduated with a BA in English in 1965. His first play, *Ladder of Fools*, which he had written while an undergraduate, was produced at Cambridge that same year.

Starting out as a stage manager and actor, he was involved in repertory and fringe theatre, and came to public attention in 1969 with *Revenge*, produced at the Theatre Upstairs at the Royal Court (where, three years later, he became resident dramatist for a year). Since then he has become one of Britain's most prolific, political and controversial playwrights. An early play, provocatively titled *Christie in Love*, took as its subject the necrophiliac murderer John Reginald Christie (winning the John Whiting Award in 1970), but much of his work – while using experimental theatrical techniques – has been more or less straightforward left-wing agitprop, attacking the class system, right-wing politics, and other elements of the establishment. He has adapted the works of Shakespeare (a version of *Measure for Measure*), Gorki and Rabelais as well as major European dramatists – Brecht (a considerable influence), Genet, Büchner – and he has successfully collaborated with other playwrights, notably **David Edgar** and **David Hare**. With the latter he wrote *Pravda*, a savage comedy about the press and its barons, which was a popular success.

His plays have been attacked for their politics, sex and violence, and these criticisms came to a head with *The Romans in Britain*, which caused an outcry when produced at the National Theatre in 1980. Voices were raised not so much against the theme, which linked the Roman occupation of Britain with the presence of British troops in Northern Ireland, but against a scene in which a male Druid is raped by a Roman soldier. Since the National Theatre is funded by the Arts Council, preposterous objections were made that ratepayers were funding homosexual acts, a protest which prefigured the Conservative government's later introduction of the notorious Clause 28 which attempted to legislate against the 'promotion' of homosexuality by local government authorities. Those who were outraged by the play (few of whom, it transpired, had actually been to see it) found a figurehead in Mary Whitehouse, a veteran campaigner against 'permissiveness'. Whitehouse brought a private prosecution under the Sexual Offences Act 1967, claiming that the play's director had attempted to procure a homosexual act in public. Not only was this a misuse of the law, but the act in question was of course simulated rather than actual, and after much publicity the case collapsed. The result of the controversy is that a not very interesting play remains the one for which Brenton is best known.

In 1989, Brenton responded to the *fatwa* imposed upon **Salman Rushdie** by writing, with Tariq Ali, *Iranian Nights*, which was produced at the Royal Court Theatre (amid some protests from the British Muslim community) and broadcast on Channel 4 television.

Brenton has also published a novel. *Diving for Pearls*, which continued to explore themes he had confronted in his plays. He was married in 1970 to Jane Fry and has two children.

Plays

Ladder of Fools 1965 [np]; *It's My Criminal* 1966 [np]; *A Sky Blue Life* [from Gorky] 1966 [rev. 1971, pub. 1989]; *Winter* 1966 [np]; *Christie in Love* 1969 [pub. 1970]; *The Education of Skinny Spew* 1969 [pub. 1970]; *Gargantua* [from Rabelais's novel] 1969 [np]; *Gum and Goo* 1969 [pub. 1972]; *Heads* 1969 [pub. 1970]; *Revenge* 1969 [pub. 1970]; *Fruit* 1970 [np]; *Wesley* 1970 [pub. 1972]; *Lay By* [with B. Clark, T. Griffiths, D. Hare, S. Poliakoff, H. Stoddart, S. Wilson] 1971; *Scott of the Antarctic* 1971 [pub. 1972]; *England's Ireland* [with T. Bicat, B. Clark, D. Edgar, F. Fuchs, D. Hare, S. Wilson] 1972 [np]; *Hitler Dances* 1972 [pub. 1982]; *How Beautiful with Badges* 1972 [pub. 1989]; *Lushly* **(tv)** 1972; *Measure for Measure* [from Shakespeare's play] 1972 [pub. 1989]; *Brassneck* [with D. Hare] 1973; *A Fart for Europe* [with D. Edgar] 1973 [np]; *Magnificence* 1973; *Mug* 1973 [np]; *The Screens* [from J. Genet] 1973 [np]; *The Churchill Play* 1974; *Jedefrau* [from H. von Hofmannsthal's play *Jedermann*] 1974 [np]; *A Saliva Milkshake* **(tv)** 1975 [pub. 1977]; *Paradise Run* **(tv)** 1976; *Weapons of Happiness* 1976; *Epsom Downs* 1977; *Deeds* [with K. Campbell, T. Griffiths, D. Hare] 1978; *Sore Throats* 1979; *The Romans in Britain* 1980; *A Short Sharp Shock* [with T. Howard] 1980 [pub. 1981]; *Thirteenth Night* 1981; *Danton's Death* [from G. Büchner's play] 1982; *Conversations in Exile* [from B. Brecht] 1983 [pub. 1986]; *The Genius* 1983; *Sleeping Policemen* [with T. Ikoli] 1983 [pub. 1984]; *Blood Poetry* 1984 [pub. 1985]; *Desert of Lies* **(tv)** 1984; *Pravda* [with D. Hare] 1985; *Dead Head* **(tv)** 1986 [pub. 1987]; *Plays: One* [pub. 1986]; *Greenland* 1988; *Iranian Nights* [with T. Ali] 1989; *H.I.D.: Hess is Dead* 1989; *Plays: Two* [pub. 1989]; *Three Plays* [ed. J. Bull; pub. 1989]; *Moscow Gold* [with T. Ali] 1990; *Berlin Bertie* 1992

For Children

The Thing 1982

Translations

Bertold Brecht *The Life of Galileo* 1980

Fiction

Diving for Pearls 1989

Poetry

Notes from a Psychotic Journal and Other Poems 1989; *Sore Throats; & Sonnets of Love and*

Opposition 1979; *Nail Poems* 1981; *Bloody Poetry* 1988

Robert (Seymour) BRIDGES 1844–1930

One of nine children of a prosperous Kentish landowner and the daughter of a baronet, Bridges was educated conventionally at Eton and Corpus Christi College, Oxford, after which he spent two years travelling on the Continent. He trained as a physician and practised in London for some years, but never had any real interest in medicine. In 1881 he became ill, gave up his medical practice, and retired to the manor house at Yattendon in Berkshire, devoting the rest of his long life to poetry. In 1884 he married Monica Waterhouse, daughter of a well-known architect, and they had a son and a daughter. His first book, *Poems*, was published in 1873, and he continued to publish over many years, but sparsely, continually refining his style: one typical decade saw only the publication of a single poem in Latin elegiacs. Verse dramas, now forgotten, studies of writers such as Milton and Keats, words for musical works by Hubert Parry and editions of hymns made up the rest of his fastidious *oeuvre*.

In 1913 he was appointed Poet Laureate in succession to Alfred Austin, although the general public would have preferred **Rudyard Kipling**. In this office, he became known as 'the silent Laureate' because he wrote so few publicly inspired works. He devoted himself to the Society for Pure English and, in his last years, to his poetical masterpiece, the long poem *The Testament of Beauty*, which was published in 1929, when he was eighty-four. In latter years he lived at Boar's Hill near Oxford, an intimate of many of the leading scholars of the university, and an impressive aristocratic figure with his great height, thick white hair and white beard. A major achievement was his edition of the poems of his friend Gerard Manley Hopkins (1918), who had been neglected until then: the great modern reputation of Hopkins owes much to Bridges, who had met him at Oxford and encouraged him continually during his lifetime. Bridges' own reputation as a poet now stands much lower than that of his friend, and he is known mainly for several shorter pieces, but as Joseph Bain has put it, 'to a small and not particularly influential readership he seems a fine, even a great, poet'.

Poetry

Poems 1873; *Carmen Elegiacum* 1876; *The Growth of Love* 1876 [rev. 1889]; *Poems* 1879; *Poems* 1880; *Prometheus the Firegiver* 1883; *Poems* 1884; *Eros and Psyche* 1885 [rev. 1894]; *The Shorter Poems of Robert Bridges* 1890; *November Drear* 1892; *Founders Day* 1893; *Shorter Poems* 1893; *Invocation to Music* 1895; *Ode for Henry Purcell and Other Poems* 1896; *Poetical Works of Robert Bridges* vols. 1–6 1898; *A Song of Darkness and Light* 1898; *Hymns* 1899; *Now in Wintry Delights* 1903; *Peace Ode* 1903; *Poems Written in the Year MCMXIII* 1914; *Ibant Obscuri* 1916; *Lord Kitchener* 1916; *Ode on the Tercentenary Commemoration of Shakespeare* 1916; *Britannia Victrix* 1918; *October and Other Poems* 1920; *Poor Poll* 1923; *New Verse* 1925; *The Tapestry* 1925; *The Testament of Beauty* 1929; *Verse Written for Mrs Daniel* 1932

Plays

Nero 2 parts 1885, 1894; *The Feast of Bacchus* 1889; *Palicio* 1890; *Achilles in Skyros* 1890; *The Christian Captives* 1890; *The Return of Ulysses* 1890; *Eden* (l) [with C.V. Stanford] 1891; *The Humours of the Court* 1893; *Demeter* 1905

Non-Fiction

On the Elements of Milton's Blank Verse 1887; *On the Prosody of Paradise Regained and Samson Agonistes* 1889; *John Keats: a Criticial Essay* 1896; *On English Homophones* 1919; *On the Present State of English Pronounciation* 1913; *Original Prospectus of the Society for Pure English* 1913; *The Necessity of Poetry* 1918; *The Dialectial Words in Blunden's Poems* 1921; *What is Pure French?* [as 'M. Barnes'] 1922; *Classical* 1923; *Grotesque* 1923; *Pictorial* 1923; *Picturesque* 1923; *Romantic* 1923; *Poetry in Schools* 1924; *Henry Bradley: a Memoir* 1926; *The Influence of the Audience* 1926; *Collected Essays, Papers &c.* 10 vols 1927–36; *Poetry* 1929; *Three Friends* 1932; *Correspondence of Robert Bridges and Henry Bradley* 1940; *XXI Letters between R. Bridges and R.C. Trevelyan* 1955

Translations

Sonnet XLIV of Buonarotti 1912

Edited

Gerard Manley Hopkins *Poems* 1918

Biography

Robert Bridges by Catherine Phillips 1992

James BRIDIE (pseud. of Osborne Henry Mavor) 1888–1951

The son of an electrical engineer, Bridie took his pseudonym from family names. Educated at Glasgow Academy, he studied medicine at Glasgow University, where his eccentric, dandified dress and conversation hinted at his love of theatre. During the First World War, he served in the Royal Army Medical Corps in France and later in Mesopotamia, where he wrote plays to entertain his unit. Discontented as a doctor, he dutifully married the daughter of a lawyer in 1923. Although artistic and clever, she was – inappropriately for a doctor's wife – a lifelong hypochondriac. Constantly 'dying', she confounded her own diagnosis by living to the age of eighty-nine. Her neuroses, and the anxiety they created domestically, goaded Bridie into writing drama. (His wife generally managed to wheeze through her husband's first nights.) Following the success of *The Anatomist* (1930), directed by Tyrone Guthrie, Bridie hesitated about disposing of his medical practice to become a full-time writer. The doctor's dilemma became the theme of his plays, for he could not reconcile the responsibility of exploiting his artistic vision of society with the risk of impoverishing his family. Despite rich box-office returns from *A Sleeping Clergyman*, presented at the Malvern Festival in 1933, and *Storm in a Teacup*, three years later, he continued to combine his two careers.

In 1936, unsettled in accommodation and occupation, he bought a Scots baronial pile he nicknamed 'Tinned Salmon Castle', one of ten different houses he lived in during his marriage. Despite (or because of) the opulence of country living, the quality of Bridie's plays declined. Finally relinquishing his civilian medical duties, he moved in 1939 to a more modest house in Glasgow, and was accepted as a lieutenant in the RAMC when the Second World War broke out. Returning early from Alexandria in 1942, Bridie had written nothing of significance while in the army. Living now in a hideous suburban Glasgow mansion, he wrote screenplays for Alexander Korda and Alfred Hitchcock, but his career only revived when he collaborated with the actor Alastair Sim in *Mr Bolfry* (1943). This play earned the praise of **John Betjeman**, who saw it with Miss Joan Hunter Dunn.

Bridie was a founder of the Glasgow Citizen's Theatre, which opened in 1945 with his *Holy Isle*. By now he had retired, but he was nevertheless unable to completely detach himself from the Glasgow bourgeois temperament. He exhibited a certain pitiless quality of personality, typified by absence of emotion when he received the news that his favourite younger son had been killed during the war. Nevertheless the Jekyll-and-Hyde duality triumphed in a Laurence Olivier production of *Daphne Laureola* in 1949, in which Peter Finch made his London debut. Bridie declared of life that, 'this Sinbadism of the Soul was no more lively a matter than collecting passports and tickets and insuring the luggage'.

Plays

The Pardoner's Tale [pub. 1930]; *The Sunlight Sonata* [pub. 1930]; *The Switchback* [pub. 1930; rev. 1932]; *The Amazed Evangelist* 1931; *The Anatomist* 1931; *Tobias and the Angel* 1931; *Jonah and the Whale* 1932; *A Sleeping Clergyman* 1933; *A Sleeping Clergyman and Other Plays* [pub. 1934]; *Marriage is No Joke* 1934; *Moral Plays* [with C. Gurney; pub. 1936]: *Mary Read* 1935; *The Black Eye* 1935; *Mrs Waterbury's Millennium* 1935; *Storm in a Teacup* 1936; *The King of Nowhere and Other Plays* 1938; *The Letter Box Rattles* 1938; *What Say They?* 1939; *Susannah and the Elders and Other Plays* [pub. 1940]; *Mr Bolfry* 1943 [with A. Sim]; *Plays for Plain People* [pub. 1944]; *Holy Isle* 1945; *It Depends What You Mean* 1948; *Daphne Laureola* 1949; *John Knox and Other Plays* [pub. 1949]; *Mr Gillie* 1950; *The Queen's Comedy* 1952; *The Baikie Charivari* 1953; *Meeting at Night* 1956

Non-Fiction

Some Talk of Alexander (a) 1926; *Mr Bridie's Alphabet for Little Glasgow Highbrows* 1934; *One Way of Living* (a) 1939; *The British Drama* 1945; *A Small Stir: Letters on the English* [with M. McLaren] 1949

Collections

Tedious and Brief 1944

Walter BRIERLEY 1900–1972

The son of an engine-winder at Denby Hall pit, Brierley was born at Waingroves, Derbyshire. He left school to work in the local pit at the age of thirteen, but continued his education on a part-time basis, attending classes in French, English and mathematics. Between 1927 and 1931 he studied two days a week at Nottingham University College, but failed to gain a full-time place and shortly afterwards became unemployed. He remained out of work until 1935 and in August 1933 contributed an article about his experiences, 'Frustration and Bitterness', to *The Listener*. This led to further pieces in the *Spectator* and the *London Mercury* and won him the admiration of **Walter Allen**.

Unemployment also provided the background for his first novel, *Means Test Man* (1935), which described a week in the life of a Derbyshire miner, Jack Cook. Having spent two years on the dole, Jack is awaiting the arrival of a means-test inspector to whom he must reveal every detail of his personal finances. Unable to hold up his head in the community, he is deprived of all dignity and zest for life. The novel brought Brierley brief fame, although the reviewer in the *Daily Worker* regretted that it failed to reflect 'the fighting spirit of the unemployed'. Brierley's second novel, *Sandwich Man*, was also based on personal experience of attempting to pursue further education in the face of the hostility of family and workmates. After allegedly causing an accident in the pit, the protagonist loses his job and fails his examinations and ends up carrying a sandwich-board. The novel covers some of the same ground as *Sons and Lovers* (1913), but Brierley's conclusions are more pessimistic than those of **D.H. Lawrence**.

Brierley's other two novels were less successful and he published nothing after 1940. At the beginning of the Second World War he left the pit to become a child welfare officer and remained in this job until 1965.

Fiction
Means Test Man 1935; *Sandwich Man* 1937; *Dalby Green* 1938; *Danny* 1940

Poetry
Poems 1926

André (Philippus) BRINK 1935–

Born of Afrikaner stock in the village of Vrede in the Orange Free State, Brink matriculated in the Transvaal at Lydenbergh High School and Potchefstroom University, where he took degrees in history, English, Afrikaans, Dutch and French. He spent a further two years doing postgraduate work at the Sorbonne, since when, with sporadic trips to Europe, he has taught modern literature and drama at Rhodes University in Cape Town, where he has held the chair of Afrikaans and Dutch literature from 1980. He was appointed Professor of English at the University of Cape Town in 1991.

Between 1963 and 1965 he was editor of the avant-garde literary magazine *Sestiger* and in 1986 he became editor of *Standpunte* for a year. His work has earned him numerous prizes, both in his own country and abroad, including South Africa's prestigious CNA Award in 1965, 1979 and 1983, the Martin Luther King Memorial Prize and the Prix Medicis Etrangers. He was appointed Chevalier de Légion d'Honneur in 1982 and Officier de l'Ordre des Arts et des Lettres in 1987. He has also been shortlisted for the Booker Prize and nominated for a Nobel Prize.

A leading spokesman of the 'Sestigers' (i.e. 1960s) generation of South African writers, he has confronted in his work many of his country's taboos about sex, religion and race, and inevitably many of his books were banned under the former political regimes there, starting with *Looking on Darkness* (1974). This novel, which deals with miscegenation, was hailed as a milestone by **Alan Paton**. Brink's greatest success is undoubtedly *A Dry White Season*, which won him international acclaim when it was published in 1979. The story of a man who considers himself apolitical until his friend, a teacher, is killed in mysterious circumstances, it bears out Brink's belief that simply to exist in South Africa at that time was to be political, and that no one could avoid it, however much they might try. It has been suggested that the novel is rather simplistic, but, as a stirring story of the birth of political consciousness, it proved extremely popular and was made into a feature film in 1990.

Brink has always seen himself 'not as a member of a small white enclave, but as a writer of a country belonging more to Africa than to Europe'. Afrikaans is his native language, but he also writes in English, translating his own work. He sometimes uses both languages in his initial compositions, but the majority of his books are published in the Afrikaans version first, and some remain untranslated. *States of Emergency* (1988) is the first of his novels to be written entirely in English. Brink also translates the works of others from English, French, German and Spanish into Afrikaans, and has won several translation prizes.

Fiction
Dowry for Life 1962; *The Ambassador* 1964; *Orgy* 1965; *Looking on Darkness* 1974; *An Instant Wind* 1976; *Rumours of Rain* 1978; *A Dry White Season* 1979; *A Chain of Voices* 1982; *The Wall of the Plague* 1984; *States of Emergency* 1988

For Children
The Gang 1961; *Broke* 1962

Plays
The Bond Around Our Hearts [pub. 1959]; *Caesar* 1965 [pub. 1961]; *The Guardian Angel and Other One Act Plays* [pub. 1962]; *Baggage* 1965; *Elsewhere Fair and Warm* 1969 [pub. 1965]; *The Rebels* [pub. 1970]; *The Trial* 1975 [pub. 1970]; *Knots in the Cable* [from Shakespeare's play *Much Ado About Nothing*] [pub. 1971]; *Afrikaners Make Merry* [pub. 1973]; *Pavane* 1980 [pub. 1974]; *The Hammer of the Witches* [pub. 1976]; *Toings on the Long Road* [pub. 1979]

Non-Fiction
Order and Chaos: Study of Germanicus and the Tragedies of Shakespeare 1962; *Sempre diritto: Italian Travel Journal* 1963; *Olé: a Travel Book of Spain* 1965; *Aspects of the New Fiction* 1967; *Midi: Travelling through the South of France* 1969; *Paris–Paris Return* 1969; *Fado: a Journey through Northern Portugal* 1970; *The Poetry of*

Breyten Breytenbach 1971; *Portrait of a Woman as a Young Girl* 1973; *Aspects of the New Drama* 1974; *Desert Wine in South Africa* 1974; *I've Been There: Conversations in South Africa* [with others] 1974; *A Stroke from the Mill* 1974; *The Wine from Up There* 1974; *Preliminary Report: Views on Afrikaans Literature* 1976; *Second Preliminary Report* 1980; *Why Literature?* 1980; *Mapmakers: Writing in a State of Siege* [US *Writing in a State of Siege*] 1983; *Literature in the Arena* 1985

Edited

Land Apart: a South African Reader [with J.M. Coetzee] 1986

John BRODERICK 1927–1989

Broderick was born in Athlone, Northern Ireland, and his early education was as unsettled as his later career. He attended six different schools, but never completed any formal studies in any of them. His family owned a prosperous bakery company, which he eventually inherited and ran successfully for many years. A comfortable income allowed him the freedom to travel widely abroad, and his first novel, *The Pilgrimage*, was first published in France, before appearing in London in 1961. In it, he stated themes which he would develop and expand in his future work. Placing his characters in the Irish midlands, his own area, Broderick deals with those who have a 'small-town' mentality, and are 'narrow minded, money-grubbing, hypocritical, furtive'. His satirical acounts of what he termed the 'little grocer's republic', characterised by a gift for crisp dialogue, and featuring sexual frustration, local gossip and the hypocrisy of the Roman Catholic church, did not make him popular. His novels, he felt, were all written from a 'negative point of view' attacking 'things I dislike most in Irish life'.

The novels also revealed his gift for psychological insight, perfected in *The Waking of Willie Ryan*. A homosexual drinker, who is also an atheist, Ryan refuses to comply with the rules of the Roman Catholic church or society, and is incarcerated in a lunatic asylum at the instigation of a local priest. When he is released twenty-five years later, Ryan realises he can be in a more powerful position than he was before, if he externally appears to adapt to society. His beliefs are unaltered, and with them he triumphs over the shallow hypocrisy of 'the system'. 'When you don't believe in hell it's easy to conform to local ritual, if you're forced to', Ryan tells the priest who had him locked up.

Broderick's hopes of producing a novel 'written with love' were defeated by an increasing bitterness with society and his own alcoholism, which he readily admitted. Shy and solitary, he was a perceptive critic, who had a wide knowledge of literature. Although he had an acerbic wit, he was kind and generous to young writers. Disgusted with Ireland, he made his home in England towards the end of his life, preferring to be anonymous there. He died after a long illness, from complications as a result of his alcoholism.

Fiction

The Pilgrimage 1961 [as *The Chameleons* 1965]; *The Fugitives* 1962; *Don Juaneen* 1963; *The Waking of Willie Ryan* 1965; *An Apology for Roses* 1973; *The Pride of Summer* 1976; *London Irish* 1979; *The Trial of Father Dillingham* 1982; *A Prayer for Fair Weather* 1984; *The Rose Tree* 1985; *The Flood* 1987; *The Irish Magdalen* 1991

Plays

The Enemies of Rome (r) 1981; *A Share of the Light* (r) 1981

Harold BRODKEY 1930–

Brodkey was born in Illinois, and originally named Aaron Roy Weintraub. His mother died after a six-month illness when he was two, and he became withdrawn and mute for over two years following her death. He was adopted by his father's second cousin, when he was given his new name, but his adoptive parents both died while he was an adolescent. His parents figure prominently in his writing, along with the excavation of his trauma. He emerged from his silent cocoon as a prodigy: 'I was so abominably bright as a child there was no limit to my social acceptability. If I played with a child ... their mother was intensely honoured.' Of the effect of this, he says: 'It fucks you up. Ha! What do you think? You can see it in the prose.' He attended Harvard, where he strove to appear of merely normal intelligence. He married, moved to New York in 1953, and had a daughter. Divorce followed in 1960 and he later married Ellen Schwamm, also a novelist.

Brodkey's short stories for the *New Yorker* were collected in his 1957 *First Love and Other Sorrows*. This was his only significant book until a second collection of stories in 1988, and his long-awaited novel *The Runaway Soul* three years later. Meanwhile Brodkey developed one of the most extraordinary reputations in modern letters, and was compared by serious critics to Milton, Wordsworth, Freud and Proust. This reputation has only added to Brodkey's burdens, because his work contains an oppressive sense of himself as an intellectual and sexual champion whose perfection has become the cross he has to bear. Brodkey was very attractive from adolescence on, and his extensive bisexual career is rumoured to include a brief dalliance with Marilyn Monroe.

Brodkey's major subject is himself, but his narcissism is redeemed by his candour and sincerity, as well as his extreme articulacy. His interest in consciousness involves a microscopic attention to 'ordinary' experience rather than fictive invention, and he is equally attentive to intersubjectivity and family dynamics. His prose has been compared to that of Proust in its rambling precision, and he has justified his style theoretically by factoring time into the equation and suggesting that conventional western syntax denies the reality of the passing moment.

Throughout Brodkey's life and work runs an obsession with greatness and standing – largely his own – in every sphere from intellectual to genital. Fortunately he is not humourless, and has related how he once lay in Central Park gazing up at the sky and thinking: 'If *only* I'd been tall'. After getting up to walk home he remembered that in fact he was. The critical reception of *The Runaway Soul* was a major blow to Brodkey, although it had its champions, including **Salman Rushdie**. Brodkey's bizarre reputation as the greatest living writer has given way to a more sober and qualified evaluation of his genius. He now has Aids, and his *New Yorker* essays about this attracted considerable affection and respect.

Fiction

First Love and Other Sorrows (s) 1957; *Women and Angels* 1985; *Stories in an Almost Classical Mode* (s) [UK *The Abundant Dreamer*] 1988; *The Runaway Soul* 1991; *Profane Friendship* 1994

Poetry

A Poem about Testimony and Argument 1986

Non-Fiction

Avedon: Photographs 1947–1977 1978

Louis (Brucker) BROMFIELD 1896–1956

Bromfield was born on a farm in Mansfield, Ohio, and retained throughout his life a passionate concern with farming and the plight of the farmer. He left school at sixteen and worked as a reporter before studying agriculture at Cornell for a year, after which he studied at the Columbia University School of Journalism. His education was cut short by the First World War, during which he was an ambulance driver in the French Army. Awarded the Légion d'Honneur and the Croix de Guerre for his distinguished service, he returned to the USA where Columbia granted him an honorary degree because of his war record. He held various posts as journalist, critic and advertising manager for the publishers G.P. Putnam's Sons, and was an original staff member of *Time* magazine. In 1923 he married Mary Appleton Wood – with whom he had three daughters – and moved to a village outside Paris to write. He did not return to the USA until 1939.

The Green Bay Tree was the first of a succession of novels set in his native Midwest and written in a simple, direct style which prompted one critic to remark: 'He greatly prefers the writer who has a good deal to say and says it clumsily, to the artist in words who has little to say.' Throughout the 1920s and early 1930s he enjoyed considerable critical acclaim, winning a Pulitzer Prize for *Early Autumn* in 1926. Perhaps his most accomplished novel was *The Farm*, a semi-autobiographical chronicle spanning four generations of Ohio farmers. After a couple of years in India (which provided the material for his two novels *The Rains Come* and *Night in Bombay*), he returned to the USA in 1939 and established the co-operative Malabar Farm in Ohio. During the war he was an active critic of Roosevelt's agriculture policy, prophesying famine and economic disaster. His later fiction continued to explore the threat of industrialism and the decline of American individualism, themes also addressed in an increasing volume of essays and journalism, but his reputation had declined enormously by the time of his death.

Fiction

The Green Bay Tree 1924; *Possession* 1925 [as *Lilli Barr* 1926]; *Early Autumn* 1926; *A Good Woman* 1927; *The Strange Case of Miss Annie Spragg* 1928; *Awake and Rehearse* (s) 1929; *Tabloid News* (s) 1930; *Twenty-Four Hours* 1930; *A Modern Hero* 1932; *The Farm* 1933; *Here Today and Gone Tomorrow* (s) 1934; *The Man Who Had Everything* 1935; *It Had to Happen* 1936; *The Rains Come* 1937; *Night in Bombay* 1940; *Wild is the River* 1941; *Until the Day Break* 1942; *Mrs Parkington* 1943; *Bitter Lotus* 1944; *What Became of Anna Bolton* 1944; *The World We Live In* (s) 1944; *Colorado* 1947; *Kenny* 1947; *McLeod's Folly* 1948; *The Wild Country* 1948; *Mr Smith* 1951

Non-Fiction

The Work of Robert Nathan 1927; *England, a Dying Oligarchy* 1939; *Pleasant Valley* 1945; *A Few Brass Tacks* 1946; *Malabar Farm* 1948; *Out of the Earth* 1950; *The Wealth of the Soil* 1952; *A New Pattern for a Tired World* 1954; *Animals and Other People* 1955; *From My Experience: the Pleasures and Miseries of Life on a Farm* 1955; *Walt Disney's Vanishing Prairie* 1956

Plays
The House of Women [from his novel *The Green Bay Tree*] 1927 [np]; *DeLuxe* [with J. Gearnen] 1934 [np]; *Times Have Changed* 1935 [np]

(Bernard) Jocelyn BROOKE 1908–1966

Ten years separated Brooke, the delicate and precocious son of a wine-merchant, from his older siblings, so that he grew up almost as an only child, spending much of his time in solitary botanising, overseen and cosseted by 'Ninnie', his adoring nanny. The family spent the winter in Sandgate and the summer in Kent's Elham Valley, places that were to become the backdrops of his private mythology.

After a false start at King's Canterbury, from which he was removed after two weeks, Brooke was educated at Bedales and Worcester College, Oxford. Numerous jobs – including a brief spell in the family firm – proved unsatisfactory and Brooke suffered a breakdown. He joined the Royal Army Medical Corps during the war and served as a 'pox wallah' in Italy and North Africa, treating troops for venereal disease. Demobilisation left Brooke feeling insecure about his future and he decided to re-enlist in the ranks, referring to his military career as a 'liaison with the armed forces (and I use the word in its erotic rather than military sense)'. He served until 1948 when the success of his novel *The Military Orchid* enabled him to buy himself out and devote his life to writing.

He returned to the Kent of his childhood, the land of lost content which he explored exhaustively in his writing, and lived there with his mother and Ninnie. He contributed articles and short stories to a number of periodicals (*Penguin New Writing, New Statesman*, The Listener, etc.), and wrote critical studies of **Aldous Huxley, Elizabeth Bowen, John Betjeman** and **Ronald Firbank**. He was a champion both of Firbank and **Denton Welch**, and edited a rather fainthearted edition of the latter's journals.

As well as books on botany, Brooke published several volumes of poetry and fiction, including two further volumes of *The Orchid Trilogy*, the work for which he is best known. An autobiography treated as fiction, it has some affinities with the early novels of **Christopher Isherwood**, not least in its veiled allusiveness to the author's homosexuality. Of his other books, perhaps the best are *The Image of a Drawn Sword*, an eerily atmospheric novel with echoes of Kafka and Lewis Carroll; the snobbish but hilarious *Conventional Weapons*; and *The Dog at Clambercrown* an evocative meditation upon the author's childhood, family history (his great-grandfather was a 'polyphiloprogenitive' and 'singularly unsuccessful' clergyman who wrote '"humorous" novels ... deservedly forgotten' under the name 'Peter Priggins'), botany and literature. Brooke was a minor master whose highly literary and distinctive work fell into neglect after his death, but has recently undergone a deserved revival.

Fiction
The Orchid Trilogy: The Military Orchid 1948, *A Mine of Serpents* 1949, *The Goose Cathedral* 1950; *The Scapegoat* 1948; *The Wonderful Summer* 1949; *The Image of a Drawn Sword* 1950; *The Passing of a Hero* 1953; *Private View* (s) 1954; *Conventional Weapons* 1961

Non-Fiction
The Wild Orchids of Great Britain 1950; *Ronald Firbank* 1951; *Elizabeth Bowen* 1952; *The Flower in Season* 1952; *Aldous Huxley* 1953 [rev. 1958]; *The Dog at Clambercrown* 1955; *The Crisis in Bulgaria* 1956; *John Betjeman and Ronald Firbank* 1962

Poetry
Six Poems 1928; *December Spring* 1946; *The Elements of Death* 1952

Edited
The Denton Welch Journals 1952; *Denton Welch: a Selection from His Published Works* 1963

Rupert (Chawner) BROOKE 1887–1915

Brooke was born at Rugby, Warwickshire, where his father taught classics and was a housemaster at the famous public school. After a successful school career, where he had posed as a Wildean decadent, Brooke accurately prophesied that he 'could not ... hope for or even imagine such happiness elsewhere'. He won a scholarship to read classics at King's College, Cambridge, where he was elected an Apostle, acted, wrote poetry and developed an interest in politics, becoming President of the Fabian Society in 1909. His enthusiasm for the works of the Jacobean playwright John Webster (upon whom he wrote a fellowship dissertation, published posthumously in 1916) and John Donne lent a strain of macabre, metaphysical wit to his best verse. His only job was that of housemaster at Rugby, a temporary post he held after his father had died in 1910; thereafter he lived on an allowance from his domineering mother of £150 a year.

His personal loyalties were divided between Bloomsbury (swimming naked with **Virginia Woolf**, being loved hopelessly by James Strachey) and a group of youthful idealists known as the Neo-Pagans. Amongst the latter group was a young girl called Noel Olivier, with whom Brooke had a protracted, idealistic and ultimately frustrating affair. (Their letters were published in 1991.) He subsequently fell in love with Ka Cox, who almost certainly bore his stillborn child, and the actress Cathleen Nesbitt. His flagrant good looks certainly attracted the attention of as many men as women; amongst his admirers were **Henry James** (who wrote a glowing preface to Brooke's *Letters from America*) and Edward Marsh, whose doting patronage led to Brooke's poems appearing in the *Georgian Poetry* volumes and also to a commission in the Royal Naval Division when war was declared (Marsh was Winston Churchill's private secretary). His sexuality appears to have been somewhat uncertain, and he once wrote a long and detailed letter about seducing a young man – although some people have claimed that his account was in fact made up.

In 1913 he had gone to America, sponsored by the *Westminster Gazette*, for whom he wrote articles, eventually collected as *Letters from America*. He subsequently travelled to the South Seas, spending three months on Tahiti where he wrote some of his finest poems and had an affair with a woman called Taatamata, commemorated in 'Tiare Tahiti'. He returned to England two months before the outbreak of war.

He is best, and unfairly, remembered for his volume, *1914 and Other Poems*, which included his sequence of patriotic sonnets ('The Soldier', 'The Dead' and others) and caught the mood of the time, becoming an enormous bestseller. It bore a frontispiece photograph of the bare-shouldered poet – the very icon of self-sacrificing Youth (but referred to by Brooke's friends when first circulated amongst them in 1913 as 'Your Favourite Actress'). The cause of Brooke's death (septicaemia as a result of a mosquito bite) was less impressive than the timing and circumstances: on St George's Day, on his way to Gallipoli, shortly after Dean Inge had quoted 'The Soldier' in his Easter Day service at St Paul's. The legend was established, and enhanced when Winston Churchill joined those writing fulsome (and fanciful) obituaries. 'One fears his memory being brought to the poster-grade,' **Harold Monro** justly complained. ' "He did his duty. Will you do yours?" is hardly the moral to be drawn.'

Brooke was far more complex than this and wrote far better poems than the '1914' sequence, notably the South Seas poems, 'The Fish', 'Heaven' and 'The Old Vicarage, Grantchester', a delightful and slyly witty celebration of his Cambridge home. His immediate posthumous reputation far outweighed his achievements, but shortly fell into decline. More recently he has been re-established as a gifted, if minor, lyrical poet.

Poetry

The Pyramids 1904; *The Bastille* 1905; *Poems 1911* 1911; *1914 and Other Poems* 1915; *The Collected Poems of Rupert Brooke* 1915 [rev. 1918, 1928]; *The Poetical Works of Rupert Brooke* [ed. G. Keynes] 1946 [rev. 1969, 1974]; *Poems of Rupert Brooke* [ed. G. Keynes] 1952; *The Poems of Rupert Brooke: a Centenary Edition* [ed. T. Rogers] 1987

Non-Fiction

The Authorship of the Latter Apius and Virginia 1913; *John Webster and the Elizabethan Drama* 1916; *Letters from America* 1916; *A Letter to the Editor of the Poetry Review* 1929; *Democracy and the Arts* 1946; *The Letters of Rupert Brooke* [ed. G. Keynes] 1968; *Letters from Rupert Brooke to His Publisher 1911–1914* 1975; *Rupert Brooke in Canada* 1978; *Song of Love: the Letters of Rupert Brooke and Noel Olivier* [ed. P. Harris] 1991

Plays

Lithuania 1915

Collections

The Prose of Rupert Brooke 1956; *Rupert Brooke: a Reappraisal and Selection* [ed. T. Rogers] 1971

Biography

Rupert Brooke by Christopher Hassall 1964; *The Neo-Pagans* by Paul Delaney 1987

Christine BROOKE-ROSE 1926–

Brooke-Rose was born in Geneva, and spent most of her childhood in Brussels after her Swiss-American mother's separation from her English father. Following her father's death, she lived in England for the duration of the war, attending St Stephen's College, Folkestone, before going up to Somerville College, Oxford, in 1946. There she studied English philology and medieval literature, and was married in 1948 to the novelist Jerzy Peterkiewicz, from whom she was divorced in 1975. Having graduated in 1949 she spent four years at the University of London working on her PhD thesis, a grammatical analysis of metaphor in Old French and Middle English poetry. Her book *A Grammar of Metaphor* applies the same methods to the study of fifteen poets from Chaucer to **Dylan Thomas**. She was a freelance writer, translator and literary journalist until taking up the post of lecturer in English and American literature at the University of Paris in 1969. From 1975 until her retirement in 1988 she was a professor there.

In 1956, when her husband was seriously ill, Brooke-Rose wrote her first novel, *The Languages of Love*. A comic satire about philologists, it set the tone for her next three novels, of which she has written: 'I was somehow dissatisfied with them. Wit and satire seemed to me too easy, not just technically but morally.' Her reading of the French novelists Nathalie Sarraute and Alain Robbe-Grillet, and of scientific books, suggested a new direction, first pursued in *Out*, an almost plotless description of global catastrophe which is rich in technical and scientific language. Her next novel, *Such*, a joint winner of the James Tait Black Memorial Prize, describes the rebirth of an astronomer who comes to view people the way he once viewed the motions of the stars and planets. *Xorandor* is a fable about the nuclear threat involving a stone which consumes radiation, and its sequel *Verbivore* addresses contemporary society's information overload. An alien force destroys broadcast messages, with the result that the world reverts to the condition of the nineteenth century. The more playful *Textermination* was described by one reviewer as 'not six characters in search of an author, but several hundred characters in search of readers'.

Fiction

The Languages of Love 1957; *The Sycamore Tree* 1958; *The Dear Deceit* 1960; *The Middlemen* 1961; *Out* 1964; *Such* 1966; *Between* 1968; *Go When You See the Green Man Walking* (s) 1970; *Thru* 1975; *Amalgamemnon* 1984; *Xorandor* 1986; *Verbivore* 1990; *Textermination* 1991

Non-Fiction

A Grammar of Metaphor 1958; *A ZBC of Ezra Pound* 1971; *A Structural Analysis of Pound's 'Usura Canto': Jakobson's Method Extended and Applied to Free Verse* 1976; *A Rhetoric of the Unreal: Studies in Narrative and Structure, Especially of the Fantastic* 1981; *Stories, Theories and Things* 1991

Translations

Juan Goytisilo *Children of Chaos* 1958; Alfred Sauvy *Fertility and Survival: Population Problems from Malthus to Mao Tse Tung* 1960; Alain Robbe-Grillet *In the Labyrinth* 1968

Poetry

Gold 1955

Anita BROOKNER 1928–

Brookner was born in London into a family of European Jewish origins and was educated at James Allen's Girls' School and King's College, London. She made her name initially as an art historian, specialising in late eighteenth- and early nineteenth-century French painters, publishing studies of Watteau, Greuze and David. She was a Reader at the Courtauld Institute of Art and in 1967 was appointed the first woman Slade Professor of Art at Cambridge. Her work at the Courtauld brought her into contact with Anthony Blunt, and after he had been unmasked as a spy, there were newspaper reports that he had used the unwitting Brookner to obtain information from Phoebe Pool, another colleague who had acted as a courier for Blunt in the 1930s. In fact, M15 had arranged for Brookner to interview Pool for them.

Her first novel, *A Start in Life*, was published to very good reviews in 1981, and thereafter she has continued to publish one novel a year. Typically, the early novels centre on a single woman who is intelligent, even intellectual, but who lacks self-confidence and has too many scruples to get what she wants in life. What she wants is usually the love of a patently inadequate but handsome man. The lesson derived from these books is that the good and the vulnerable do not win in life, but are exploited and deceived by those more ruthless and less moral than themselves. Brookner was rightly praised for her elegant but unostentatious prose, her wit and the precision of her observation of human behaviour. *Look at Me*, in which a young woman is taken up, then brutally dropped, by a glamorous couple, is perhaps her best novel from this period, although she made her name with *Hotel du Lac*, which was an unexpected winner of the Brooker Prize in 1984 and was subsequently adapted for television by **Christopher Hampton**. More ambiguous in tone than earlier novels, it is concerned with a longing for moral rectitude, nobility of purpose and the perfect union of two souls – all of which prove to be unattainable.

Family and Friends in 1985 marked a turning-point in Brookner's career. Drawing upon her own background, it opens with the contemplation of a wedding photograph, and describes the subsequent, interwoven lives of the eponymous family and their circle. If her early novels owe something to **Barbara Pym** (not a comparison Brookner welcomes), *Family and Friends* is more like the work of **Edith Templeton**, three of whose novels from the early 1950s were reissued in the same year with admiring prefaces by Brookner. Subsequent novels have disappointed some of her early admirers. The sharp humour that enlivened her early books is less in evidence, and their tone of plangent melancholy has given way to mere gloom. *A Friend from England*, with its outdated and unconvincing revelation of the homosexuality of a male character, was a particularly poor effort.

While Brookner retains a loyal readership, and the admiration of several notable critics, others have commented that the novels have become almost indistinguishable. Thinking about his agoraphobic wife, the protagonist of *Lewis Percy* seems to echo some of Brookner's critics: 'It now seemed to Lewis that his wife would remain for ever within the tight circle of her limitations, and even that it was policy on her part to reject anything or anyone outside that circle.'

Brookner has also translated works on art and has been a regular reviewer of fiction for the *Spectator*. She was awarded a CBE in 1990. She is unmarried.

Fiction

A Start in Life 1981; *Providence* 1982; *Look at Me* 1983; *Hôtel du Lac* 1984; *Family and Friends* 1985; *A Friend from England* 1988; *Lewis Percy* 1989; *Brief Lives* 1990; *Fraud* 1992; *Family Romance* 1993

Non-Fiction

The Genius of the Future: Studies in French Art Criticism 1971; *Watteau* 1971; *Greuze: the Rise and Fall of an Eighteenth Century Phenomenon* 1972; *Jacques-Louis David* 1980

Edited

Edith Wharton: The Stories of Edith Wharton (s) 2 vols 1988, 1989, *Stories* (s) 2 vols 1993, *Reef* 1994

Translations

Gauguin 1962; Jean P. Crespelle *Les Fauves* 1962; Alfred Werner *Maurice Utrillo 1883–1955* 1960

Brigid (Antonia Susan) BROPHY 1929–

Brophy was born in London, the daughter of the writer and editor John Brophy. 'Any ability I have as an adult to write English prose I learned from him and from reading the masters he directed me to, **Bernard Shaw** and **Evelyn Waugh**,' she has said. Her mother, who also published a novel, was born in Chicago of Scottish parents, but Brophy has always been more aware of her father's Irish heritage. She was educated at a number of state and private schools, including a boys' preparatory school, where her mother was headmistress, St Paul's School in Hammersmith, and the Abbey School, Reading. She also attended Camberwell School of Art and a secretarial college, where she learned shorthand and typing, before winning a senior scholarship to St Hugh's College, Oxford, where she read classics until sent down at the end of her fourth term.

Her first book was a volume of short stories, *The Crown Princess* (1953), which she has since disowned. It was published the same year as her first novel, *Hackenfeller's Ape*, which concerns plans to send an ape into space and signals one of the principal themes of her work: the rights of animals. It was inspired by living within howling distance of the Regent's Park Zoo and won the Cheltenham Festival Prize for a first novel. She has published six further novels, ranging from the relatively straightforward *The King of a Rainy Country* ('my only work of fiction to contain sizeable chunks of autobiography') to the highly experimental *In Transit*, 'a transexual comedy' set in an airport. 'My designs are, invariably, baroque,' she has written. 'They all proceed by contraposition; and in a reductive analysis the elements contraposited are always Eros and Thanatos.' The most overtly baroque, and one of the best, of her novels is *The Snow Ball*, in which the influence of Mozart is dominant; she published simultaneously with this novel a study of his operas, *Mozart the Dramatist*. Another important figure in her intellectual life has been Freud, most noticeably in *Black Ship to Hell*, a work of non-fiction which she had described as 'an exploitation of my discovery that, whereas religion is, art is not in opposition to reason', and in the novel *Flesh*. She has championed the work of **Ronald Firbank** in the dazzling and vast *Prancing Novelist* (the subtitle of which is 'A defence of fiction in the form of a critical biography in praise of Ronald Firbank'), and his influence may be seen in her own fiction, not only the pastiche novel, *The Finishing Touch*, which is set in a girls' school, but also in *Palace without Chairs*, 'a baroque novel' set in a modern, Kafkaesque Ruritania. Firbank described his books as 'aggressive, witty, & unrelenting', a description that equally applies to Brophy's. Her seven novels are amongst the most stylish, intelligent and underrated fictions of the century.

In the 1960s, Brophy was an eloquent and controversial broadcaster, critic and essayist, whose work was published initially in the *London Magazine* and the *New Statesman*, and thereafter in many newspapers and journals. Many of these pieces were collected in three volumes: *Don't Never Forget*, *Baroque-'n'-Roll* and *Reads*. What she describes as her 'Irishly rational tone of voice' was mistaken by some for pugnaciousness, but **Edward Blishen** summed up her method more perceptively: 'She has always the gift of a most stirring sort of firmness. It is not the tone of a know-all, it is not remotely bossy; it is, I suppose, basically, the sound logic makes.'

She married the art historian and novelist Michael Levey in 1954, on her twenty-fifth birthday, the day she also became a vegetarian. They have one daughter. As well as championing the rights of animals and homosexuals (she is herself bisexual), she has also fought for the proper remuneration of authors, carrying on her father's campaign to introduce Public Lending Right. With this aim in mind, in 1972 she founded the Writers' Action Group with the novelist **Maureen Duffy**; the group was disbanded in 1982 after successfully lobbying Parliament. Two years later, Brophy developed multiple sclerosis, which has progressively incapacitated her and caused her husband to take early retirement from the directorship of the National Gallery in order to care for her.

Her other books include two studies of Aubrey Beardsley; a volume of fables titled *The Adventures of God in His Search for the Black Girl*; the script of her unorthodox 'bedroom farce', *The Burglar* (produced unsuccessfully in the West End in 1967), which is published with a long, illuminating and discursive preface; and *The Prince and the Wild Geese*, an imaginative reconstruction of the relationship between a Russian prince and a young Irish woman, based upon the latter's paintings.

Fiction

The Crown Princess and Other Stories (s) 1953; *Hackenfeller's Ape* 1953; *The King of a Rainy Country* 1956; *Flesh* 1962; *The Finishing Touch* 1963; *The Snow Ball* 1964; *In Transit* 1969; *The Adventures of God in His Search for the Black Girl and Other Fables* (s) 1973; *Palace without Chairs* 1978

Non-Fiction

Mozart the Dramatist 1964; *Don't Never Forget* 1966; *Black Ship to Hell* 1962; *Fifty Works of English and American Literature We Could Do Without* [with M. Levey, C. Osborne] 1967; *Black and White: a Portrait of Aubrey Beardsley* 1968; *Pop Art, Exhibition, London* [with M. Duffy] 1969; *Prancing Novelist* 1973; *Beardsley and His World* 1976; *A Guide to Public Lending Right* 1983; *The Prince and the Wild Geese* 1983; *Baroque-'n'-Roll* 1987; *Reads* 1989

Plays

The Burglar 1968; *The Waste Disposal Unit* 1968

For Children

Pussy Owl 1976

George Mackay BROWN 1921-

The son of a postman and part-time tailor, Brown was born in Stromness, Orkney, and has lived there most of his life. Educated at Stromness Academy, he felt he was only good at composition. Intermittently dogged by tuberculosis, he worked for the local newspaper, where his reports owed more to imagination than fact. In 1951, he met fellow Orcadian **Edwin Muir**, who encouraged him to attend Newcastle Abbey, an adult college of education near Edinburgh, where the older writer was warden. After an initial year there, Brown returned later for two further years, then read English at Edinburgh University, privately publishing a collection of poetry, *The Storm*, in 1954. After graduating in 1960, he stayed on to do postgraduate work on Gerard Manley Hopkins. Meanwhile, Muir sent a selection of Brown's poems to the Hogarth Press, which was published in 1959 as *Loaves and Fishes*.

During the six years Brown lived in Edinburgh, he wrote almost nothing, but became friends with **Norman MacCaig** and **Hugh MacDiarmid**, who met regularly in a public house called Milne's Bar, famous as a centre for literary debate. Brown began training as a teacher but after a short and horrendous spell of teaching practice, he abandoned schools for ever. He became a Roman Catholic in 1961, and although he never 'preaches' religion, the use of symbols and ritual is important in his work.

Three years later, Brown returned to Orkney. The powerful *Orkneyinga Saga* and local fables fascinated him, while the treeless island, rich in Stone Age and Viking remains, nourished his imagination. In 1967, he published his first short stories, *A Calendar of Love*. This established what would become a familar pattern, where he wrote of the lives of farmers, fishermen – 'ordinary' people, all affected by the eternal battle with the seasons and elements. His second collection of short stories, *A Time to Keep*, was awarded the Katherine Mansfield Prize. All of Brown's *oeuvre* is concerned with Orkney life, whether of the past or the present. He has rarely travelled beyond his own island, feeling he has more than enough material to draw on. Although not anti-progress, he reflects the importance of tradition and a desire for a certain simplicity, embodying a natural wisdom, rather than the diversions of much of twentieth-century 'culture'.

Seamus Heaney has written that Brown's work 'transforms everything by passing it through the eye of the needle of Orkney. His sense of the world and his way with words are powerfully at one with each other.' Slight in appearance, Brown would not be recognised on a beach of fishermen. Quiet in manner, he lives alone in a book-cluttered but comfortable flat. Even in discussing the banal, Brown has a sense of innocent wonder, and is a patient listener. He begins writing every day at 8.30 in the morning, on his kitchen table, after sweeping away the toast crumbs. He was awarded a James Tait Black Memorial Prize in 1988, and his novel *Beside the Ocean of Time* was nominated for the 1994 Booker Prize.

Poetry

The Storm 1954; *Loaves and Fishes* 1959; *The Year of the Whale* 1965; *The Five Voyages of Arnor* 1966; *Twelve Poems* 1968; *Fishermen with Ploughs* 1971; *Lifeboat and Other Poems* 1971; *Poems New and Selected* 1971; *Penguin Modern Poets 21* [with I. Crichton Smith, N. MacCaig] 1972; *Winterfold* 1976; *Selected Poems* 1977; *Voyages* 1983; *Christmas Poems* 1984; *The Scottish Bestiary* 1986; *Stone* 1987; *Songs for St Magnus Day* 1988; *Two Poems for Kenna* 1988; *Tryst on Egilsay* 1989; *The Wreck of the Archangel* 1989; *Selected Poems 1954–1983* 1991; *Brodgar Poems* 1992; *The Lost Village* 1992

Fiction

A Calendar of Love (s) 1967; *A Time to Keep* (s) 1969; *Greenvoe* 1972; *Magnus* 1973; *Hawkfall* (s) 1974; *Andrina and Other Stories* (s) 1983; *Time in a Red Coat* 1984; *Christmas Stories* (s) 1985; *The Hooded Fisherman* (s) 1985; *The Golden Bird* (s) 1987; *The Keepers of the House* (s) 1987; *The Masked Fisherman* (s) 1989; *The Sea King's Daughter* (s) 1991; *Vinland* (s) 1992; *Beside the Ocean of Time* 1994

Non-Fiction

Let's See the Orkney Islands 1948; *Stromness Official Guide* 1956; *An Orkney Tapestry* 1969; *Letters from Hamnavoe* 1975; *Edwin Muir: a Brief Memoir* 1975; *Under Brinkie's Brae* 1979; *Portrait of Orkney* 1981; *Letters to Gypsy* 1990; *In the Margins of Shakespeare* 1991

Plays

A Spell for Green Corn 1970; *Three Plays* [pub. 1984]; *The Loom of Light* 1986; *A Celebration for Magnus* 1987

For Children

The Two Fiddlers (s) 1974; *Pictures in the Cave* 1979; *Witch and Other Stories* (s) 1977; *Six Lives of Fankle the Cat* 1980

Edited

Edwin Muir: Selected Prose 1987

Alan (Charles) BROWNJOHN 1931–

Brownjohn was born in Catford, south London, where his father was a printer. He attended Brockley County Grammar School before going to Merton College, Oxford, where he read modern history, graduating in 1953. He was a schoolteacher at Beckenham and Penge Boys' Grammar School from 1957 to 1965, and senior lecturer in English at Battersea College of Education from 1965 to 1979.

Brownjohn emerged out of the loosely associated collection of poets known as the Group, which met during the late 1950s and early 1960s and included **Edward Lucie-Smith, Peter Redgrove, Peter Porter** and **George MacBeth**. In contrast to the Movement poets, they were more concerned with addressing social and political issues, and more perhaps than any of their number Brownjohn has continued to do this, often with a barbed satirical edge. He has described himself as being concerned with 'watching the shifts in the social and cultural fabric of England in my time, interpreting the interrelation of private living and public events'. Nowhere is this more evident than in the ironic vignettes of modern life in *A Song of Good Life*, which prompted one reviewer to wonder whether Brownjohn's gifts were put to best use in 'executive-bashing'. His later collections, notably *A Night in the Gazebo* and *The Old Flea-Pit*, show an increasing interest in the narrative possibilities of verse, and in 1990 he published his first adult novel, *The Way You Tell Them*, a scabrous vision of life in 1999 under a fifth successive Conservative administration. Brownjohn was a borough councillor in Wandsworth in the early 1960s and an unsuccessful Labour parliamentary candidate in Richmond, Surrey, in 1964. He was poetry critic for the *New Statesman* from 1968 to 1976, a member of the Arts Council Literature Panel from 1973 to 1977, and Chair of the Poetry Society from 1982 to 1988. He received a Cholmondeley Award in 1979. He has been married twice, in 1960 to the writer Shirley Toulson with whom he has a son, and in 1972 to Sandy Willingham, with whom he has collaborated on various books and a radio adaptation of Goethe's play *Torquato Tasso*.

Poetry

Travellers Alone 1954; *The Railings* 1961; *The Lions' Mouths* 1967; *Being a Garoon* 1969; *A Day by Indirections* 1969; *Penguin Modern Poets 14* [with M. Hamburger, C. Tomlinson] 1969; *Sandgrains on a Tray* 1969; *Woman Reading Aloud* 1969; *Frateretto Calling* 1970; *Synopsis* 1970; *An Equivalent* 1971; *Transformation Scene* 1971; *Warrior's Career* 1972; *She Made of It* 1974; *A Song of Good Life* 1975; *A Night in the Gazebo* 1980; *Nineteen Poems* 1980; *Collected Poems 1952–1983* 1983; *The Old Flea-Pit* 1987; *Collected Poems 1952–1988* 1988; *The Observation Car* 1990; *In the Cruel Arcade* 1994

Edited

Six Women Poets [with J. Adlard] 1953; *First I Say This: a Selection of Poems for Reading Aloud* 1969; *New Poems 1970–1971* [with S. Heaney, J. Stallworthy] 1971; *New Poetry 3* [with M. Duffy] 1977; *Meet and Write* [with S. Brownjohn] 3 vols 1985–7; *The Gregory Anthology* [with K.W. Gransden] 1990

Non-Fiction
The Little Red Bus Book 1972; *A Tribute to W.H. Auden* [with others] 1973; *Philip Larkin* 1975

Fiction
The Way You Tell Them 1990

Translations
Goethe *Torquato Tasso* (r) [with S. Brownjohn] 1982 [pub. 1985]

For Children
To Clear the River [as 'John Berrington'] 1964; *Oswin's Word* (l) 1967; *Brownjohn's Beasts* 1970

John BUCHAN 1875–1940

Born in Perth, the son of a minister of the Free Church, Buchan was educated at Glasgow and Brasenose College, Oxford. Intensely ambitious, he published a considerable amount of journalism and his first book (an edition of Bacon) while still at Oxford, where he gained a first, was President of the Union, and won the Newdigate Poetry Prize. His career in public service began when he went to South Africa as Lord Milner's secretary (1901–2).

Back in London he continued to work as a prolific journalist and became a barrister and an MP. In 1906 he became a partner in Nelson's and later compiled that publisher's history of the First World War. During the war he also worked for the Foreign Office, was a correspondent for *The Times*, joined the Intelligence Corps and in 1915 published the first of his adventure novels, *The Thirty-Nine Steps*, which featured his gentlemanly hero Richard Hannay. Hannay's further adventures, foiling the dastardly plots of assorted foreign powers, were described in a number of subsequent 'shockers', as Buchan described them. By the end of the war he had become Director of Intelligence.

He continued to write biographies, essays, works of history and criticism, poetry and short stories, but it is for the 'shockers' that he is best remembered. He was a complex and contradictory man who, for example, supported Zionism but peppered his novels with the casual anti-Semitism and xenophobia of their period. His work fell into disfavour in the 1960s, when it was inaccurately dismissed as racist and reactionary; but it has subsequently enjoyed something of a revival. His public career culminated in a peerage, when he became 1st Baron Tweedsmuir in 1935 and was appointed Governor-General of Canada.

Fiction
Sir Quixote of the Moors 1895; *Grey Weather* (s) 1899; *John Burnet of Barns* 1898; *A Lost Lady of Old Years* 1899; *The Half-Hearted* 1900; *The Watcher by the Threshold* (s) 1902; *A Lodge in the Wilderness* 1906; *Prester John* [US *The Great Diamond Pipe*] 1910; *The Moon Endureth* (s) 1912; *Salute to Adventurers* 1915; *The Thirty-Nine Steps* 1915; *Greenmantle* 1916; *The Power House* 1916; *Mr Standfast* 1919; *The Path of the King* 1921; *Huntingtower* 1922; *Midwinter* 1923; *The Three Hostages* 1924; *John Macnab* 1925; *The Dancing Floor* 1926; *Witch Wood* 1927; *The Runagates Club* (s) 1928; *The Courts of the Morning* 1929; *Castle Gay* 1930; *The Blanket of the Dark* 1931; *The Gap in the Curtain* 1932; *The Magic Walking-Stick* 1932; *A Prince of the Captivity* 1933; *The Free Fishers* 1934; *The House of the Four Winds* 1935; *Sick Heart River* [US *Mountain Meadow*] 1941

Non-Fiction
Scholar Gypsies 1896; *Sir Walter Ralegh* 1897; *A History of Brasenose College* 1898; *The African Colony* 1903; *The Law Relating to the Taxation of Foreign Income* 1905; *Eighteenth Century Byways* 1908; *Sir Walter Raleigh* 1911; *Andrew Jameson, Lord Ardwall* 1913; *The Marquis of Montrose* 1913; *The Achievement of France* 1915; *Nelson's History of the War* 24 vols 1915–19 [rev. as *A History of the Great War* 4 vols 1921–2; *The Battle of the Somme, First Phase* 1916; *The Battle of the Somme, Second Phase* 1917; *Poems Scots and English* 1917; *The Battle-Honours of Scotland* 1919; *The Island of Sheep* [with S. Buchan; as 'Cadmus and Harmonia'] 1919; *These for Remembrance* 1919; *Francis and Riversdale Grenfell* 1920; *The History of the South African Forces in France* 1920; *A Book of Escapes and Hurried Journeys* 1922; *Days to Remember: the British Empire in the Great War* 1923; *The Last Secrets: the Final Mysteries of Exploration* 1923; *Lord Minto* 1924; *The Man and the Book: Sir Walter Scott* 1925; *Homilies and Recreations* 1926; *The History of the Royal Scots Fusiliers (1678–1918)* 1925; *The Fifteenth (Scottish) Division* 1926; *The Causal and the Casual in History* 1929; *The Kirk in Scotland 1560–1929* [with G.A. Smith] 1930; *Montrose and Leadership* 1930; *An Oxfordshire Theatrical Pageant* [with others] 1931; *Julius Caesar* 1932; *Sir Walter Scott* 1932; *The Massacre of Glencoe* 1933; *Gordon at Khartoum* 1934; *Oliver Cromwell* 1934; *The King's Grace 1910–1935* 1935; *Men and Deeds* 1935; *Augustus* 1937; *Canadian Occasions* 1940; *Comments and Characters* 1940; *Memory Hold-the-Door* (a) 1940

Edited
Essays and Apothegms of Francis Lord Bacon 1894; *Selected Essays of Augustine Bell* 1909; *The Long Road to Victory* 1920; *Great Hours in Sport* 1921; *A History of English Literature* 1923; *The Northern Muse: an Anthology of Scots Vernacular Poetry* 1924; *Essays and Studies by*

Members of the English Association 1926; *South Africa* 1927; *A General Survey of British History* 1928; *The Poetry of Neil Munro* 1931

Collections

The Long Traverse [US *Lake of Gold*] 1941

Biography

John Buchan by Janet Adam Smith 1965

Pearl (Comfort) S(ydenstricker) BUCK 1892–1973

Pearl Sydenstricker was born in Hillsboro, West Virginia, the daughter of Presbyterian missionaries who returned to their mission in China when their daughter was three months old. Her father was a scholar who translated the New Testament into vernacular Chinese. She grew up in Chinkiang on the Yangtse River and learnt to speak Chinese before she could speak English. She was educated at a boarding school in Shanghai from 1907 to 1909, and then attended Randolph-Macon College, Lynchburg, Virginia, where she graduated in 1914 before returning to China where her mother was ill. In 1917 she met and married Dr John Lossing Buck, an agricultural missionary, and from 1922 to 1930 taught English at a number of Chinese schools and universities. She returned to the USA in 1924 to seek medical care for her first daughter who was mentally retarded, an experience recounted in *The Child Who Never Grew*. She received her MA from Cornell in 1926. Her marriage was unhappy; she was divorced in 1935, and married in the same year to Richard J. Walsh, president of the John Day Company, her publishers. They lived in Pennsylvania, had one child, and adopted a further nine.

Buck had written since childhood and published magazine articles in the mid-1920s. Her first novel, *East Wind: West Wind*, was published in 1930, but it was with her second story of Chinese peasant life, *The Good Earth*, that she achieved enormous popular and critical success. The novel sold 1,800,000 copies in its first year, was filmed in 1937, and won a 1932 Pulitzer Prize. It was followed by *Sons* and *A House Divided*, the three novels constituting the *House of Earth* trilogy. Biographies of her father and her mother, *Fighting Angel* and *The Exile*, were cited by the Nobel committee when she won the prize in 1938 'for rich and genuine epic portrayals of Chinese peasant life and masterpieces of biography'. She remains the only American woman to have won the prize. From this time on she devoted much of her considerable energy to public services, working for the mentally handicapped and establishing the Pearl S. Buck Foundation to assist the plight of Asian children. Increasingly her novels were set in America, five of them published under the pseudonym of John Sedges because, she said, she was tired of being classed as 'a writer of Chinese stories'.

Fiction

As Pearl S. Buck: *East Wind: West Wind* 1930; *House of Earth* 1936: *The Good Earth* 1931, *Sons* 1932, *A House Divided* 1935; *The Young Revolutionist* 1932; *The First Wife and Other Stories* (s) 1933; *The Mother* 1934; *This Proud Heart* 1938; *The Patriot* 1939; *Other Gods* 1940; *Today and Forever* (s) 1941; *China Sky* 1942; *Dragon Seed* 1942; *The Promise* 1943; *China Flight* 1945; *Portrait of a Marriage* 1945; *Pavilion of Women* 1946; *Far and Near* (s) 1948; *Peony* 1948; *Kinfolk* 1949; *God's Men* 1950; *The Hidden Flower* 1952; *Satan Never Sleeps* 1952; *Come, My Beloved* 1953; *Imperial Woman* 1956; *Letter from Peking* 1957; *Command the Morning* 1959; *Fourteen Stories* (s) 1961; *Hearts Come Home and Other Stories* (s) 1962; *The Living Reed* 1963; *Stories of China* (s) 1964; *Death in the Castle* 1965; *The Time is Noon* 1967; *The New Year* 1968; *The Good Deed and Other Stories of Asia, Past and Present* (s) 1969; *The Three Daughters of Madame Liang* 1969; *Mandala* 1970

As 'John Sedges': *The Townsman* 1945; *The Angry Wife* 1947; *The Long Love* 1949; *Bright Procession* 1952; *Voices in the House* 1953

Non-Fiction

East and West and the Novel: Sources of the Early Chinese Novel 1932; *Is There a Case for Foreign Missions?* 1932; *The Exile* 1936; *Fighting Angel* 1936; *The Chinese Novel* 1939; *Of Men and Women* 1941; *American Unity and Asia* 1942; *What America Means to Me* 1943; *Talk about Russia* [with M. Scott] 1945; *Tell the People: Talks with James Yen about the Mass Education Movement* 1945; *How It Happens: Talk about the German People, 1914–1933* [with E. von Pustau] 1947; *American Argument* [with E.G. Robeson] 1949; *The Child Who Never Grew* 1950; *The Man Who Changed China: the Story of Sun Yat Sen* 1953; *My Several Worlds* (a) 1954; *Friend to Friend* [with C.P. Romulo] 1958; *The Delights of Learning* 1960; *A Bridge for Passing* (a) 1962; *The Joy of Children* 1964; *Welcome Child* 1964; *Children for Adoption* 1965; *The Gifts They Bring: Our Debt to the Mentally Retarded* [with G.T. Zarfoss] 1965; *For Spacious Skies: Journey in Dialogue* [with T.F. Harris] 1966; *My Mother's House* (c) 1966; *The People of Japan* 1966; *To My Daughters, with Love* 1967; *The People of China* 1968; *China as I See It* [ed. T.F. Harris] 1970; *The Kennedy Women* 1970; *Pearl Buck's America* 1971; *The Story Bible* 1971; *China Past and*

Present 1972; *A Community Success Story: the Founding of the Pearl Buck Center* 1972; *Pearl S. Buck's Oriental Cookbook* 1972

For Children

Stories for Little Children 1940; *When Fun Begins* 1941; *The Chinese Children Next Door* 1942; *The Water Buffalo Children* 1943; *The Dragon Fish* 1944; *Yu Lan: Flying Boy of China* 1945; *The Big Wave* 1947; *One Bright Day* 1950; *The Beech Tree* 1954; *Johnny Jack and His Beginnings* 1954; *Christmas Miniature* 1957; *The Christmas Ghost* 1960; *The Big Fight* 1965; *The Little Fox in the Middle* 1966; *Matthew, Mark, Luke and John* 1967

Plays

Flight into China 1939; *Sun Yat Sen* 1944; *The First Wife* 1945; *A Desert Incident* 1959; *Christine* 1960; *The Guide* [from N.K. Narayan's novel] 1965

Edited

China in Black and White: an Album of Woodcuts by Contemporary Chinese Artists 1945; *Fairy Tales of the Orient* 1965; *Pearl Buck's Book of Christmas* 1974

Translations

All Men Are Brothers [by S.H. Chan] 1933

Biography

Pearl S. Buck by Theodore F. Harris, 2 vols 1969, 1971

Charles (Henry) BUKOWSKI 1920–1994

Bukowski was born in Andernach, Germany, the son of a US soldier and a German woman, and the family moved to Los Angeles in 1923. His father, in and out of work during the Depression years, regularly beat the boy for supposed misdemeanours, an experience painfully recounted in Bukowski's autobiographical novel, *Ham on Rye*. After graduating from Los Angeles High School in 1939, Bukowski spent a year at Los Angeles City College taking courses in journalism and literature when he wasn't hanging out in bars and engaging in whiskey drinking competitions, which he usually won.

He left home in 1941 after his father had read his stories and reacted by throwing his possessions on to the lawn. Hank, as he was known to friends, began a peripatetic life travelling across America while writing stories. One, aptly named 'Aftermath of a Lengthy Rejection Slip', was published by *Story* in 1944, the same year that he returned to Los Angeles. In 1946 he met Jane Cooney Baker, with whom he lived on and off for the next decade, a period of casual labour, drinking, gambling and fighting during which he wrote little. His friend and biographer, Neeli Cherkovski, quotes from a typical argument: Hank: 'I'm a genius and nobody knows it but me!' Jane: 'Shit! You're a fucking asshole.'

After a brief marriage to Barbara Frye, Bukowski began in 1958 twelve years as a Post Office clerk, and his growing reputation as a poet saw the publication of his first volume, *Flower, Fist and Bestial Wail*, two years later. His first volume of prose, *All the Assholes in the World and Mine*, is 'a humorous account of one man's hemorrhoid operation', and *Notes of a Dirty Old Man* a collection of pieces from his column of the same name in the underground newspaper *Open City*. The autobiographical *Confessions of a Man Insane Enough to Live with Beasts* introduced his *alter ego* Henry Chinaski, who is the hero of all his subsequent novels, including *Post Office* and *Factotum*. He left his job in 1970 after John Martin of the Black Sparrow Press had offered him $100 a month for life to write full-time.

His reputation spread with an award-winning television documentary by Taylor Hackford shown in 1973, and in 1977 he was approached by the Swiss-German director Barbet Schroeder who wanted to make a film based on his work. Bukowski's reaction was characteristic: 'Fuck off, you French frog!' Schroeder persisted, and Bukowski eventually wrote the screenplay for the successful film *Barfly* (1987) which starred Mickey Rourke and Faye Dunaway. The cult success which Bukowski enjoyed in his last years prompted the *American Book Review* to comment that: 'Those who despised him as a drunken bum, now despise him as a rich drunken bum.' He had one daughter, Marina Louise, by one-time girlfriend Frances Smith, and was married in 1985 to Linda Lee Beighle, a health-food store proprietor twenty-five years his junior who exerted something of a stabilising influence on his life.

Poetry

Flower, Fist and Bestial Wail 1960; *Longshot Poems for Broke Players* 1962; *Poems and Drawings* 1962; *Run with the Hunted* 1962; *It Catches My Heart in Its Hands* 1963; *Cold Dogs in the Courtyard* 1965; *Crucifix in a Deathhand* 1965; *The Genius of the Crowd* 1966; *At Terror Street and Agony Way* 1968; *Poems Written Before Jumping Out of an 8 Story Window* 1968; *The Days Run Away Like Wild Horses Over the Hills* 1969; *Me and Your Sometimes Love Poems* 1972; *Mockingbird Wish Me Luck* 1972; *Burning in Water Drowning in Flame* 1974; *Love is a Dog from Hell* 1977; *Play the Piano Drunk/Like a Percussion Instrument/Until the Fingers Begin to Bleed a Bit* 1979; *Dangling in the Tournefortia* 1981; *Under the Influence* 1984; *War All the Time*

1984; *Alone in a Time of Armies* 1985; *The Day it Snowed in L.A.* 1986; *Gold in Your Eye* 1986; *You Get So Alone at Times That It Just Makes Sense* 1986; *Luck* 1987; *The Movie Critics* 1988; *We Ain't Got No Money, Honey, But We Got Rain* 1990; *In the Shadow of the Rose* 1991; *The Last Night of the Earth Poems* 1992; *Supposedly Famous* 1992; *Run with the Hunted: a Charles Bukowski Reader* [ed. J. Martin] 1993

Fiction
All the Assholes in the World and Mine 1966; *Notes of a Dirty Old Man* 1969; *Fire Station* 1970; *Post Office* 1971; *Erections, Ejaculations, Exhibitions and General Tales of Ordinary Madness* (s) 1972; *Life and Death in the Charity Ward* (s) 1973; *South of No North* (s) 1973; *Factotum* 1975; *Women* 1978; *Ham on Rye* 1982; *Bring Me Your Love* (s) [with R. Crumb] 1983; *Hot Water Music* (s) 1983; *There's No Business* 1984; *The Most Beautiful Woman in Town* (s) 1986; *Hollywood* 1989

Non-Fiction
Confessions of a Man Insane Enough to Live with Beasts (a) 1965; *Shakespeare Never Did This* (a) 1979; *The Bukowski–Purdy Letters, 1964–74* [ed. S. Cooney] 1983

Plays
The Movie: 'Barfly' (f) 1987

Collections
Septuagenarian Stew: Stories and Poems [nd]

Biography
Hank: the Life of Charles Bukowski by Neeli Cherkovski 1991

Basil BUNTING 1900–1985

Bunting regarded himself as a 'minor poet'; others regarded him as the successor to **Ezra Pound** and one of the leaders of the new British avant-garde of the 1960s. Born in Scotswood-on-Tyne, Northumberland, he wrote musical and visual poetry often rooted in the countryside of his birth. Imprisoned during the First World War as a conscientious objector, he later moved to Paris where, with **Ford Madox Ford**, he edited the *Transatlantic Review*. In Italy he subsequently befriended **W.B. Yeats** and Pound, and his association with the latter affected the reception of his poetry: none of his work was published in Great Britain until 1964. His journalistic work included several years as the Persian correspondent for *The Times*, and in 1953 he returned to Northumberland to work for the *Newcastle Evening Chronicle*. After his acceptance and increasing recognition in the 1960s, he lectured in poetry at several universities in the USA, Canada and the UK, and, from 1972 to 1976, was president of the Poetry Society in London. He married twice and died in Hexham, Northumberland.

Poetry
Redimiculum Metellarum 1930; *Poems* 1950; *The Spoils* 1965; *Loquitur* 1965; *Briggflatts* 1966; *First Book of Odes* 1966; *Collected Poems* 1968 [rev. 1977]; *Uncollected Poems* [ed. R. Caddel] 1991

Non-Fiction
A Note on Briggflatts 1989

Edited
Selected Poems of Joseph Skipsey 1976

Translations
Nizam al-Din 'Ubayd Zakani' *The Pious Cat* 1986

(John) Anthony BURGESS (Wilson) 1917–1993

Burgess was born in Manchester. His ancestors were Catholic, 'obstinate in their adherence to Rome'. (He adopted the name Anthony at his confirmation.) His father was a cashier and pub pianist; his mother died in the flu pandemic of 1919. Consequently, Burgess was brought up by a maternal aunt and, later, by a stepmother. Aged eleven, he gained a scholarship to the Xaverian College. There he began seriously to question the Catholic faith and by 1937, when he entered Manchester University to read English language and literature, he was an apostate. He graduated in 1940 and joined the Royal Army Medical Corps; 1942 saw his marriage to Llewela Isherwood Jones and his transfer to the Army Educational Corps. Demobilised in 1946, he spent the next eight years teaching English. His first novel, *A Vision of Battlements*, published in 1965, was written in 1953. From 1954 to 1959 Burgess taught in Malaya. The experience was used in *The Malayan Trilogy* (republished in 1982 as *The Long Day Wanes*), the modest success of which Burgess attributed to its authenticity.

After collapsing in a classroom, he was sent back to England. Diagnosed as having a cerebral tumour, and given twelve months to live, he wrote rapidly: reviews, literary and pedagogical studies as well as novels. The threat of death receded but financial pressures dictated continuing haste; and, while Burgess has often been criticised for writing too much too fast, some of his better fiction was produced in the period up to 1968. *A Clockwork Orange* – famously filmed by Stanley Kubrick (who subsequently withdrew the film from distribution) – and *The Wanting Seed* both offer violent, dystopian prospectives; *Tremor of Intent* parodies the espionage genre yet manages simultaneously to examine the nature of good and evil; *Inside Mr Enderby* and *Enderby Outside* present, on the one hand, a poet's creative (and digestive) processes and, on the other, a wide-sprayed satire of 1960s' culture.

In 1968 Llewela died of alcoholic cirrhosis and Burgess remarried: his second wife, Liana Macellari, was already the mother of his four-year-old son. The new family moved immediately to Malta, then to Italy in 1970 and to Monaco in 1975, where he remained until his death from cancer eighteen years later. Visiting Fellow at Princeton University (1970–1), Distinguished Professor at the City College of New York (1972–3) and writer-in-residence at the University of New York at Buffalo (1976), Burgess was also appointed literary adviser to the Guthrie Theatre, Minneapolis, in 1972. An accomplished composer, largely self-taught, he heard his Third Symphony performed at the University of Iowa in 1975 and his musical version of *Ulysses, Blooms of Dublin*, performed on British and Irish radio on the centenary of Joyce's birth. He translated Rostand's *Cyrano de Bergerac* for stage and screen and created an ur-argot for Jean-Jacques Annaud's film *Quest for Fire*.

The 1970s and 1980s also witnessed the publication of about thirty books: the Enderby sequence was completed, as was *Earthly Powers*, generally regarded as Burgess's finest novel. It failed narrowly to win the 1980 Booker Prize but, in France, was honoured as 'le meilleur livre étranger de l'année'. Narrated by a figure based loosely on **Somerset Maugham**, it strives – with a very un-Miltonic humour – to 'justify the ways of God to Men'. In spite of its author's avowed apostasy, the novel's theological co-ordinates are Catholic. Burgess recognised that he could not wholly break free from that faith. Accordingly, he rejected England's Protestant culture; as a lower middle class provincial, he felt excluded from London's literary coteries; as a man of letters, he was repelled by the technocratic philistinism of – particularly – the Wilson era. Thus, triply alienated, he became a voluntary exile and the variety of his textual concerns exceeds even his actual wanderings. Equally confident whether predicting the future or reconstructing the past, whether lambasting **Geoffrey Grigson** or lauding **James Joyce**, his true homeland was language. Polyglot and wildly inventive, he affirmed in his work that words possessed sufficient resources to portray humanity in all its flaws and glory. His two volumes of autobiography, *Little Wilson and Big God* and *You've Had Your Time*, reveal a persona more vulnerable and less bombastic than the one that frequently appeared to the public.

Fiction

The Malayan Trilogy 1964 [as The Long Day Wanes 1982]; *Time for a Tiger* 1956, *The Enemy in the Blanket* 1958, *Beds in the East* 1959; *The Doctor is Sick* 1960; *The Right to an Answer* 1960; *Devil of a State* 1961; *One Hand Clapping* [as 'Joseph Kell'] 1961 [as Anthony Burgess 1974]; *The Worm and the Ring* 1961; *A Clockwork Orange* 1962; *The Wanting Seed* 1962; *Honey for the Bears* 1963; *Enderby* 1982: *Inside Mr Enderby* [as 'Joseph Kell'] 1963 [as Anthony Burgess 1966], *Enderby Outside* 1968, *The Clockwork Testament* 1974; *The Eve of St Venus* 1964; *Nothing Like the Sun* 1964; *A Vision of Battlements* 1965; *Tremor of Intent* 1966; *MF* 1971; *Napoleon Symphony* 1974; *Beard's Roman Women* 1976; *Moses* 1976; *Abba Abba* 1977; *1985* 1978; *Man of Nazareth* 1979; *Earthly Powers* 1980; *The End of the World News* 1982; *Enderby's Dark Lady* 1984; *The Kingdom of the Wicked* 1985; *The Piano Players* 1986; *Any Old Iron* 1989; *The Devil's Mode* 1989; *A Dead Man in Deptford* 1993

Non-Fiction

English Literature: a Survey for Students 1958; *The Novel Today* 1963; *Language Made Plain* 1964 [rev. 1975]; *Here Comes Everybody: an Introduction to James Joyce* 1966; *The Novel Now: a Student's Guide to Contemporary Fiction* 1967; *Urgent Copy: Literary Criticism* 1968; *Shakespeare: a Biography* 1970; *Joysprick: an Introduction to the Language of James Joyce* 1973; *New York* 1976; *Ernest Hemingway and His World* 1978; *On Going to Bed: Essays on Beds and Sleeping Customs* 1982; *This Man and Music* 1982; *Flame into Being: the Life and Work of D.H. Lawrence* 1985; *Oberon – Old and New* 1985; *Homage to QWERT YUIOP: Selected Journalism 1978–1985* 1986; *Little Wilson and Big God* (a) 1987; *You've Had Your Time* (a) 1990; *A Mouthful of Air: Language and Languages, Especially English* 1992

Translations

The New Aristocrats [with L. Burgess] 1962; *The Olive Trees of Justice* 1962; *The Man Who Robbed Poor Boxes* 1965; Edward Rostand *Cyrano de Bergerac* 1971; Sophocles *Oedipus the King* 1972; Ludovic Halévy and Henri Meilhac *Carmen* (l) 1986

Edited

The Age of the Grand Tour (1720–1820) 1966; Daniel Defoe *Journal of the Plague Year* [with C. Bristow] 1966; James Joyce
A Shorter Finnegan's Wake 1966; *The Coaching-Days of England (1750–1850)* 1966

Plays
Jesus of Nazareth (f) 1977; *Blooms of Dublin* [from J. Joyce's *Ulysses*] 1986; *Mozart and the Wolf Gang* 1991

For Children
The Land Where the Ice-Cream Grows 1979

Frances (Eliza) Hodgson BURNETT 1849–1924

Born Frances Burnett Hodgson in Manchester, the daughter of a hardware wholesaler, she emigrated to Knoxville, Tennessee, with her mother in 1865, after her father died, and the business collapsed. She wrote for magazines when she was seventeen, after an unsuccessful attempt to run a private school, and married Dr Swan Burnett in 1877. In the same year she published her first novel, *That Lass o' Lowrie's*, probably her best book, but not the most widely known. The couple moved to Washington where two sons were born, but her writing career strained relations with her husband, and they divorced in 1898. A second marriage also ended in divorce three years later.

Burnett's childhood experience of poverty was reflected in both her most famous children's books. *Little Lord Fauntleroy*, based on her younger son, Vivian, ruined his life, and was cursed by many small boys who were made to dress in his lordship's flamboyant style. *The Secret Garden* brought Burnett the adoration of children, whom she genuinely loved, and to whom she was generous. After living in Kent for some years, she settled in Long Island, where she maintained a famous garden. Very much the image of the popular Victorian lady novelist, 'Fluffy' (as she was nicknamed) dressed in frilly clothing and a series of wigs, and was the Barbara Cartland of her day. Nevertheless, her candy-floss appearance belied a self-centred, snobbish and despotic nature. Gushing and over-blown, she remained a schoolgirl at heart all her life. Her refusal to confront everyday reality was the recipe for her popular success, and the escapist pleasure she gave to millions.

For Children
Little Lord Fauntleroy 1886; *Sara Crewe* 1887; *Editha's Burglar* 1888; *Little Saint Elizabeth and Other Stories* (s) 1890; *Children I Have Known* (s) 1892; *The One I Knew Best of All* 1893; *Piccino and Other Child Stories* (s) 1894; *Two Little Pilgrims' Progress* 1895; *A Little Princess* 1905; *Queen Silver-Bell* 1906 [as *The Troubles of Queen Silver-Bell* 1907]; *Racketty-Packetty House* 1906; *The Cozy Lion, as Told by Queen Crosspatch* 1907; *The Good Wolf* 1908; *The Spring Cleaning, as Told by Queen Crosspatch* 1908; *Barty Crusoe and His Man Saturday* 1909; *The Land of the Blue Flower* 1909; *The Secret Garden* 1911; *The Lost Prince* 1915; *Little Hunchback Zia* 1916; *The Way to the House of Santa Claus* 1916

Fiction
That Lass o'Lowrie's 1877; *Dolly* 1877; *Pretty Polly Pemberton* (s) 1877 [as *Vagabondia* 1883]; *Surly Tim and Other Stories* (s) 1877; *Theo* (s) 1877; *Kathleen* (s) 1878; *Earlier Stories* (s) [2 vols] 1878; *Miss Crespigny* (s) 1878; *Our Neighbour Opposite* (s) 1878; *A Quiet Life and The Tide on the Moaning Bar* (s) 1878; *Haworth's* 1879; *Jarl's Daughter and Other Stories* (s) 1879; *Louisiana* 1879; *Natalie and Other Stories* (s) 1879; *A Fair Barbarian* 1881; *Through One Administration* 1883; *A Woman's Will* 1887; *The Fortunes of Philippa Fairfax* 1888; *The Pretty Sister of José* 1889; *A Lady of Quality* 1896; *His Grace of Osmonde* 1897; *In Connection with the De Willoughby Claim* 1899; *The Making of a Marchioness* 1901; *The Methods of Lady Walderhurst* 1901; *In the Closed Room* 1904; *The Dawn of a To-morrow* 1906; *The Shuttle* 1907; *T. Tembaron* 1913; *The White People* 1917; *The Head of the House of Coombe* 1922; *Robin* 1922

Non-Fiction
The Drury Lane Boys' Club 1892; *The One I Knew Best of All* (a) 1893; *In the Garden* 1925

Plays
That Lass o'Lowrie's [with J. Magnus] 1878; *Esmeralda* [with W. Gillette] 1881 [as *Young Folk's Ways* 1883]; *Editha's Burglar* [with S. Townesend] 1888 [also as *Nixie* 1890]; *The Real Little Lord Fauntleroy* 1888; *Phyllis* [from her novel *The Fortunes of Philippa Fairfax*] 1889; *The First Gentleman of Europe* [with C. Fletcher] 1897; *A Lady of Quality* [with S. Townesend] 1897; *The Showman's Daughter* [with S. Townesend] 1897; *That Man and I* 1903; *The Pretty Sister of José* 1903; *The Dawn of a To-morrow* 1910; *The Little Princess* 1911 [from her novel *Sara Crewe*; as *A Little Unfairy Princess* 1902]

Biography
Waiting for the Party by Ann Thwaite 1974

John Horne BURNS 1916–1953

The first of seven children, Burns was born in Andover, Massachusetts, educated at Saint Augustine's School and Phillips Academy in Boston, and Harvard. He graduated Phi Beta Kappa in English literature in 1937, after which he taught English at Loomis School in Windsor, Connecticut until he was drafted in 1942. After a year in the infantry Burns was sent to the Adjutant General's School in Washington, where he was commissioned as a second lieutenant and then sent overseas. He served in North Africa and Italy as a military intelligence officer, principally engaged in censoring the mail of prisoners of war. Released from the army in 1946, he went back to teaching – the subject of his second novel – until the enormous success of his first novel, *The Gallery*, enabled him to become a full-time writer.

Loosely based on his experiences in liberated Naples, *The Gallery* is considered by many, including **Gore Vidal**, to be the finest American novel of the Second World War, or at least that conflict's equivalent to *A Farewell to Arms* (1929). Like **Ernest Hemingway**, Burns contrasts a romantic vision of Italy with the moral bankruptcy of his own culture, and creates for his post-war generation a comparably powerful literary statement of the loss of innocence. In a series of 'Portraits' the novel describes the lives of the habitués of the Galleria Umberto in Naples in 1944, which represent correlatively the diverse behaviour of a war-torn city, and, by implication, of war-torn Europe. These 'Portraits' are interspersed with 'Promenades', where the author-narrator reflects upon his experiences and observations. The 'Promenades' are less successful than the 'Portraits' which, in their vividness, undermine the purpose of the novel: an overt affirmation of the power of human dignity and love to triumph even in the midst of war. The 'Portraits' nearly all end in defeat, and their accumulated power tends to swamp the humanistic ruminations of the author-narrator of the 'Promenades'.

This unresolved tension between emotional instinct and intellectual conviction becomes more obvious and less productive in Burns's subsequent novels. *Lucifer with a Book* (1949) is a bitter and angry work, but its subject is not weighty enough to justify its emotions. To an insubstantial satire on an American private school Burns inappropriately brings a technique developed to accommodate the immense range of material presented by the Second World War. *A Cry of Children* (1952), in reaction to *Lucifer with a Book*, marks a return to his concern for the inner life of isolated individuals and describes a complex love affair between two ill-matched former Catholics. Operatic in scale, it is as overblown in its emotionalism as *Lucifer with a Book* is in its sarcasm. As with **Richard Aldington** and *Death of a Hero* (1929), Burns was unable to find a subject to equal that of his first novel, and his subsequent books became increasingly shrill and maudlin.

After the critical drubbing accorded *Lucifer with a Book*, he left America for Italy, where he settled outside Florence. His disappointment at the reception of his second novel led to alcoholism, a condition worsened by the silence that greeted *A Cry of Children*. He appears also to have been pursued by demons other than drink and critical neglect, and his homosexuality, judging by his treatment of the subject in his novels, may have contributed to his unhappiness. At the time of his death at Leghorn from a cerebral haemorrhage he was working on 'The Stranger's Guise', a novel about Francis of Assisi in modern life; perhaps Hal, one of the 'Portraits' in *The Gallery*, who came to believe that he was Jesus Christ, had convinced Burns that he was at least a saint. The novel remains unpublished; but its subject and the circumstances under which it was written do not suggest that it would have revived either his critical or personal fortunes.

Fiction

The Gallery 1947; *Lucifer with a Book* 1949; *A Cry of Children* 1952

William (Seward) BURROUGHS (II) 1914–

Burroughs was born into the faded patrician atmosphere of St Louis, a scion of the 'Burroughs adding machine' family. He was educated at Los Alamos Ranch School (then the most expensive school in America) and Harvard, but from boyhood he was haunted by a sense of disaffection and deviance that led him into the criminal low-life of New York and the heroin addiction dispassionately recorded in *Junky*.

During the 1940s and 1950s he was an older and wiser mentor to his younger 'Beat' friends **Allen Ginsberg** and **Jack Kerouac**, and he appears as 'Bull Lee' in Kerouac's novel *On the Road* (1957). The search for homosexual and pharmaceutical freedom took him across the world, exploring the South American jungle for the drug *yage* and settling for long periods in Mexico and Tangier, as well as Paris and London. He was married twice and accidentally shot his second wife Joan Vollmer dead in Mexico in a bizarre 'William Tell' stunt. After ballistics experts had been bribed, he was acquitted and subsequently bought an ounce of heroin from his lawyer. Their son William Burroughs III died at thirty-three from drink and drug abuse.

Burroughs achieved notoriety and cult status in 1959 with the publication of *The Naked Lunch*, a collection of twenty-one outrageous satirical pieces that purport to lay bare the horrors of reality: hence the title, coined by Kerouac, and

Burroughs's comment: 'Let them see what they eat.' It was admired for its black humour and maniacal insight but it became more famous for its ground-breaking obscenity and the scatological and sadistic elements that led to the so-called 'Ugh' debate in the *Times Literary Supplement*. **Norman Mailer** championed Burroughs as 'the only living American writer who may conceivably be possessed of genius'.

Burroughs has been centrally concerned with control and freedom, using the metaphors of addiction and cure. Despite its orgiastic elements his work has a strong ascetic streak, advocating freedom from the body and from verbal consciousness as well as from all legal and national authority. Burroughs experimented extensively with the 'cut-up' method, a form of verbal collage advocated by his associate Brion Gysin, which was intended both to break the addictive authority of words and to generate unexpected new texts.

Among Burroughs's less persuasive opinions is his suggestion that until male cloning is perfected women should have their heads cut off and be kept alive artificially for breeding purposes, and his work has a vein of paranoid cracker-barrel philosophising which blurs the boundary between fiction and non-fiction. His towering position as the Wicked Uncle of post-modernism seems assured and he now lives in Kansas, devoted to his cats and his firearms.

Fiction

Junkie [as 'William Lee'] 1953 [rev. *Junky* 1977]; *The Naked Lunch* 1959 [US as *Naked Lunch]*; *Exterminator* [with B. Gysin] 1960; *The Soft Machine* 1961 [rev. 1968; rev. as *The Streets of Chance* 1981]; *The Ticket That Exploded* 1962 [rev. 1967]; *Dead Fingers Talk* 1963; *Roosevelt After Inauguration* [as 'Willy Lee'] 1964; *Nova Express* 1964; *White Subway* 1965; *Exterminator!* 1967; *So Who Owns Death TV* [with C. Pelieu, C. Weissner] 1967; *They Do Not Always Remember* 1968; *Ali's Smile* 1969; *The Dead Star* 1969; *The Last Words of Dutch Schulz* 1970 [rev. 1975]; *Electronic Revolution 1970–71* 1971; *The Wild Boys* 1971; *Brion Gysin Let the Mice In* (c) 1973; *Exterminator!* 1973; *Port of Saints* 1973 [rev. 1980]; *White Subway* 1973; *The Book of Breeething* 1974; *Cobble Stone Gardens* 1976; *Ah, Pook Is Here and Other Texts* 1979; *Blade Runner: a Movie* 1979; *Early Routines* 1981; *Cities of the Red Night* 1981; *The Place of Dead Roads* 1983; *Four Horsemen of the Apocalypse* 1984; *Queer* 1985; *Routine* 1987; *The Western Lands* 1987; *Apocalypse* [with K. Haring] 1988; *The Cat Inside* [with B. Gysin] 1988; *Interzone* 1989; *Tornado Alley* 1989

Non-Fiction

The Yage Letters [with A. Ginsberg] 1963; *Health Bulletin: Apo 33: a Metabolic Regulator* 1965; *The Job* [with D. Odier] 1969 [rev. 1974]; *Snack* [with E. Mottram] 1975; *Colloque de Tanger* [with B. Gysin; ed. G.-G. Lemaire] 1976; *The Retreat Diaries* 1976; *The Third Mind* [with B. Gysin] 1978; *Naked Scientology* 1978; *Colloque de Tanger II* [with B. Gysin; ed. G.-G. Lemaire] 1979; *Letters to Allen Ginsberg 1953 to 1957* 1981; *Sidetripping* [with C.S. Gatewood] 1982; *The Adding Machine: Collected Essays* 1985; *Paintings and Drawings* 1988; *William Burroughs: Painting* [with J. Grauerholz] 1988; *Selected Letters 1945–59* [ed. O. Harris] 1992

Poetry

Minutes to Go 1960 [with S. Beiles, G. Corso, B. Gysin]; *Time* 1965

Collections

The Burroughs File 1984

Biography

Literary Outlaw: the Life and Times of William S. Burroughs by Ted Morgan 1988

Dorothy (i.e. Dorothea) BUSSY 1865–1960

Bussy was the sister of Lytton Strachey and the third of the ten children of Lieutenant-General Sir Richard Strachey, of a family long prominent in the service of the British Raj, and his wife Jane Grant (of the family of the painter Duncan Grant) of Rothiemurchus. She grew up first in Clapham and later at Lancaster Gate, in the gloomy and eccentric family atmosphere there. In 1882 she followed her elder sister Elinor to the famous and fashionable school Les Ruches, kept at Fontainebleau by Marie Souvestre, a charismatic Frenchwoman who had a great effect on the whole Strachey family (she was instrumental in directing Lytton Strachey's attention to all things French) and who is fictionalised as Mlle Julie in Bussy's one novel, *Olivia*. Marie Souvestre later ran a school in London called Allenswood, where Bussy taught as a young adult, years during which she also coached Lytton and accompanied him on several foreign trips, including one to Egypt.

In 1903 she married the impecunious French painter Simon Bussy, who had come to the house to paint portraits of the Strachey family. The match was originally opposed by her parents (they were particularly shocked to see Bussy mopping up his plate with bread), but when Dorothy proved adamant, Sir Richard gave the couple La Souco, a beautiful house in Roquebrune in the Alpes Maritimes which had originally been built for his own use by the village carpenter. Here they lived until Bussy's death in 1954, having one daughter, Jane, and enjoying a happy union little disturbed by Dorothy's passionate attachment to André Gide, whose chief English translator she became. (Their letters have been published.) She also translated other works, painted (as did her daughter), travelled widely and enjoyed the friendship of Matisse as well as many other distinguished cultural figures.

The Bussys remained in France during the Second World War, when Jane played a role in the French Resistance. Bussy's two original books came out in successive years when she was in her mid-eighties; *Olivia*, which was published in 1949 under the pseudonym 'Olivia' and proved very popular, with its passionate picture of schoolgirl friendships; and *Fifty Nursery Rhymes, with a Commentary on English Usage for French Students*, a book more interesting than it may sound, as it contains asides on such matters as what Gide thought when he came across the sentence in *Robinson Crusoe* 'I was but a young sailor'. Jane Bussy continued to live with her mother into middle age, and looked after her when she became senile. In April 1960 Jane Bussy was killed in an accident involving a water-heater in her London flat, and her mother, who never knew her daughter was dead, died in a nursing home only weeks later. A complicated inheritance wrangle ensued because Dorothy Bussy had died intestate.

Translations

A. Bréal *Velazquez* 1905; Camille Mauclair *Antoine Watteau* 1906; André Gide: *Strait is the Gate* 1924, *The Counterfeiters* 1927 [as *The Coiners* 1950], *The Vatican Cellars* 1927, *The Immoralist* 1930, *The School for Wives; Robert; Genevieve* 1930, *Travels in the Congo* 1930, *Two Symphonies: Isabelle, The Pastoral Symphony* 1931, *Back from the USSR* 1937, *Afterthoughts: a Sequel to Back from the USSR* 1938, *Fruits of the Earth* 1949, *If It Die* 1950, *The Return of the Prodigal, and, Saul* 1953; Jean Schlumberger *The Seventh Age* 1933; Paul Valéry *Dance and the Soul* 1951

Fiction

Olivia 1949

Non-Fiction

Eugène Delacroix 1907; *Selected Letters of André Gide and Dorothy Bussy* [ed. R. Tedeschi] 1983

Edited

Fifty Nursery Rhymes 1950

Guy (Frederick) BUTLER 1918–

The most distinguished South African poet to have remained in his native county in the latter part of the century, Butler was born in Cradock, Cape Province, educated at the local high school, and then at Rhodes University, Grahamstown, where he has also taught for much of his adult life. After receiving his MA in 1939, he served in the South African Army during the Second World War in Egypt, the Lebanon, Italy and England, and was then at Brasenose College, Oxford, for two years, from 1945 to 1947, before returning to South Africa. The confrontation between Africa and Europe in his own life has always formed a prime topic of his poetry. He married in 1940, and has three sons and one daughter. From 1948 to 1950 he was lecturer in English at the University of the Witwatersrand in Johannesburg, and from 1952 and 1986 he was professor of English at his *alma mater*, Rhodes University, becoming a leading figure in South African letters and aiming to encourage creative work by writers of all races. He became an Honorary Research Fellow of the university in 1987.

He is the former editor of the magazine *New Coin*, and among many honours awarded to him was his becoming first president of the Shakespeare Society of South Africa in 1985. He is a poet of great subtlety and lyricism who has experimented in a wide variety of genres – ballad, elegy, narrative verse, poetry of metaphysical debate – and rarely fallen below the high standard set in his first collection, *Stranger to Europe* (1952). Among his finest poems is the long, metaphysical 'On First Seeing Florence' (Italy is a potent source of inspiration in his writing). His other works include plays, criticism, two volumes of autobiography, and his editorship of several poetry anthologies.

Poetry

Stranger to Europe: Poems 1939–1949 1952; *South of the Zambezi* 1966; *On First Seeing Florence* 1968; *Selected Poems* 1975; *Songs and Ballads* 1978; *Pilgrimage to Diaz Cross* 1987; *Out of the Africa Dark* 1988

Non-Fiction

An Aspect of Tragedy 1953; *The Republic of the Arts* 1964; *Karoo Morning: an Autobiography 1918–1935* (a) 1977; *Bursting World: an Autobiography 1936–1945* (a) 1983; *South Africa: Landshapes, Landscapes, Manshapes* 1990

Plays

The Dam 1953; *The Dove Returns* 1954 [pub. 1956]; *Take Root or Die* 1966 [pub. 1970]; *Cape*

Charade 1967 [pub. 1970]; *Richard Gush of Salem* 1970 [pub. 1982]; *Demea* 1990

Edited

A Book of South African Verse 1959; *When Boys Were Men* 1969; *The 1820 Settlers: an Illustrated Commentary* 1974; *A New Book of South African Verse in English* [with C. Main] 1980; S.C. Cronwright Schreiner [with N. Visser] *My Diary 7–15 June 1921 and 8th to 29 August 1921* 1985; *The Magic Tree* [with J. Opland] 1989

Fiction

A Rackety Colt 1989; *Tales of the Old Karoo* (s) 1989

A(ntonia) S(usan) BYATT 1936–

Byatt was born Antonia Drabble in Sheffield, the eldest of four children of a judge and his wife, an English teacher. Both parents, although originally of working-class origins, were Cambridge-educated, as were to be all their children; several of the family became writers, including the judge, who was himself a novelist. Like her sister, the novelist **Margaret Drabble**, Byatt attended the Mount School, York, a Quaker boarding-school, and Newnham College, Cambridge, where she took a first in English in 1957. This was followed by a year at Bryn Mawr College, Pennsylvania, and then a year's study towards a doctorate at Somerville College, Oxford. In 1959, she married Ian Byatt, an economist, and had two children with him.

In 1964 she became a part-time teacher of literature at the University of London and at the Central School of Arts and Crafts. The same year, her autobiographical first novel, *Shadow of a Sun*, was published, and a pattern was inaugurated in her writing in which fiction tended to alternate with critical works, the first of the latter being a study of **Iris Murdoch**, *Degrees of Freedom*, published in 1965. Her second novel, *The Game* (1967), depicting the rivalry of two sisters, was widely thought to be based on her relations with Margaret Drabble, who had herself treated a similar theme in *A Summer Bird-Cage* four years earlier.

A gap of more than a decade followed before her third novel, attributable to a series of crises and catastrophes in her domestic life. In 1969 she divorced Ian Byatt and married Peter Duffy, a property consultant with whom she had two more children. Her son by her first marriage, however, was killed in a road accident in 1972, which robbed her of her ability to write for several years (she eventually exorcised the terrible event in her frequently anthologised short story 'The July Ghost'). Also in 1972, she accepted a full-time lectureship at University College London, and became a senior lecturer before retiring to write in 1983.

In 1978, her third novel, *The Virgin in the Garden*, the first volume of a projected tetralogy dealing with the 'new Elizabethan age', was published; it has been followed so far only by the second book, *Still Life*. These dense, complex and intensely serious novels, following the fortunes of the three Potter children, are her major fictional achievement, but have been eclipsed in esteem by *Possession*, the first of her books to win wide popularity, as well as the Booker Prize and the *Irish Times*-Aer Lingus International Fiction Prize. It enmeshes a modern romantic detective story with many learned pastiches of Victorian writing, most notably Browning's dramatic monologues (which she has edited), and only a minority felt that she had here fallen far below the standard of seriousness set by her earlier books.

Byatt's other most recent fictions have been mainly short stories and novellas. Other works include several volumes edited and a volume of essays, *Passions of the Mind*. She is well known as a cultural polymath and commentator.

Fiction

Shadow of a Sun 1964; *The Game* 1967; *The Virgin in the Garden* 1978; *Still Life* 1985; *Possession* 1990

Non-Fiction

Degrees of Freedom – The Novels of Iris Murdoch 1965; *Wordsworth and Coleridge in Their Time* 1970 [rev. as *Unruly Times: Wordsworth and Coleridge in Their Time* 1989]; *Iris Murdoch* 1976; *Passions of the Mind: Selected Writings* 1991

Edited

George Eliot: The Mill on the Floss 1979; *George Eliot: Selected Passages, Poems and Other Writings* [with N. Warren] 1990

C

James M(allahan) CAIN 1892–1977

Cain was born in Annapolis, Maryland. Educated at Washington College, Chesterton, Maryland, he obtained his BA in 1910 and his MA in 1917. Between degrees he was a coal-miner, a teacher of mathematics and a reporter. From 1917 to 1919 he served as a private in France, editing (in spite of his rank) the *Lorraine Cross*, the newspaper of the US 79th Division and filing stories for the *Baltimore American*. Subsequently he returned to journalism, working for the *Baltimore Sun*, then as professor of journalism at St John's College, Annapolis, and as an editorial writer for the *New York World* under the supervision of Walter Lippmann. For ten months in 1931 he was managing editor of the *New Yorker*. Next he moved to Hollywood, where he remained for seventeen years. The co-author of three screenplays, he was to admit, '[There] was one kind of writing I was no good at: I couldn't write pictures.' He supported himself instead through syndicated articles and columns. However, the success of *The Postman Always Rings Twice* relieved him of financial worries and, after 1948, he earned his income solely through his fiction, receiving the Mystery Writers of America Grand Master Award in 1970. He was married four times – once to the actress Aileen Pringle – and once described himself thus: 'I ... weigh 220 pounds, and look like the chief inspector of a long-distance hauling concern. I am a registered Democrat. I drink.'

Alongside **Dashiell Hammett** and **Raymond Chandler**, he is often called a founder of the 'hard-boiled' school. Certainly, his style is terse and explicit and his subject the melodramas of 'ordinary' lives. Nevertheless, he resented the label. He claimed to have read fewer than twenty pages of Hammett's, and disliked them. Moreover, he wrote his crime stories not from the detective's viewpoint ('the least interesting angle') but from the criminal's. *Double Indemnity* typifies his technique. Walter Huff – an insurance salesman and Cain's first-person narrator – conspires with Phyllis Nirdlinger to murder her husband and to defraud Huff's employers. No sooner is the murder committed than the conspirators' relationship is poisoned. Mutual suspicions multiply: nemesis in many shapes pursues. Flawlessly plotted, the tale derives its power in part from its attention to detail (Cain was a meticulous researcher) and in part from the plausibility of its characters' corruption. Like suburban Macbeths, they cannot extricate themselves from the consequences of their deed and the traps of their fatal desires.

Cain's reputation rests on his crime novels. There are few US crime writers who do not bear traces of his influence. Elements of his technique may be discerned in 'dirty realism', and Camus declared that had he not read *The Postman Always Rings Twice*, he would never have written *L'Etranger*. Cain's range, though, encompassed other genres. *Past All Dishonour* and *Mignon* are historical novels, set in the 1861–5 Civil War; *The Butterfly* is a Faulknerian exercise in incest, misconception and despair; *Mildred Pierce*, his most ambitious book, dissects a mother's obsessive, masochistic love for her child.

Cain himself may not have been able to write successful screenplays, but several of his novels were made into classic films. Analysing the appeal of his stories, he wrote: '[They] have some quality of the opening of a forbidden box, and it is this, rather than violence or sex ... that gives them [their] drive.'

Fiction

The Postman Always Rings Twice 1934; *Serenade* 1937; *Mildred Pierce* 1941; *Love's Lovely Counterfeit* 1942; *Career in C Major* 1943; *Double Indemnity* 1944; *The Embezzler* 1944; *Past All Dishonour* 1946; *The Butterly* 1947; *Sinful Woman* 1947; *The Moth* 1948; *Jealous Woman* 1950; *The Root of His Evil* 1952 [as *Shameless* 1958]; *Galatea* 1953; *Mignon* 1962; *The Magician's Wife* 1965; *Rainbow's End* 1975; *The Institute* 1976; *The Baby in the Ice Box* (s) 1981; *Cloud Nine* 1984; *The Enchanted Isle* 1985

Plays

The Postman Always Rings Twice 1936

Non-Fiction

Our Government 1930; *Sixty Years of Journalism* [ed. R. Hoopes] 1986

Edited

79th Division Headquarters Troop: a Record [with M. Gilbert] 1919; *For Men Only: a Collection of Short Stories* (s) 1944

Biography

Cain by Roy Hoopes 1982

Arthur CALDER-MARSHALL 1908–1992

Calder-Marshall was born at Wallington, Surrey, the younger son of a consulting engineer and merchant. In *The Magic of My Youth* he describes his adolescent revolt against his parents; also his time in Oxford in the 1920s where he was a contemporary of **Evelyn Waugh, Harold Acton** and **Peter Quennell**; and his quest for Aleister Crowley, self-styled Knight Elect of the Sangrail, Master of Thelema, the Beast of the Apocalypse. On going down he spent some time in Bloomsbury, then bursting with artists, writers and their hangers-on, taught for a while at Denstone College, Staffordshire and, in 1937, went to Hollywood as a scriptwriter. Between 1942 and 1945 he served in BP's Wartime Department.

He published his first novel, *Two of a Kind*, when he was twenty-five. A love story, it is said to contain what was for its time a hidden obscenity, 'they came, together', but if anybody noticed the inserted comma it went unremarked. It was followed by *About Levy*, and altogether in the years up to 1940 he published six novels as well as many short stories, collected in three volumes. As a writer of fiction he was never to be so productive again but his last novel, *The Scarlet Boy*, is generally regarded as his best. It is the moving story of the effect on a group of people of a house haunted by two boys who committed suicide in different eras.

After the war he turned to biography. A key book in his development was *No Earthly Command*, which is described as 'being an enquiry into the life of Vice-Admiral the Reverend Alexander Riall Wadham Wood, DSO and bar, of whom it was said that during the Battle of Jutland while Signals Officer to the Admiral of the Grand Fleet, Sir John Jellicoe, he received an "interposed message" telling him to serve God'. In 1931, aged fifty-one, Wood entered theological college; he died aged seventy-four, 'a poor parson in London's East End'. Calder-Marshall also published three children's books and edited editions of Smollett, Jane Austen and **Jack London**. In 1934 he married Violet Nancy Sale; they had two daughters of whom Anna has a distinguished career as an actress.

Fiction

Two of a Kind 1933; *About Levy* 1933; *At Sea* 1934; *Crime Against Cania* **(s)** 1934; *Dead Centre* 1935; *A Pink Doll* **(s)** 1935; *A Date with a Duchess* **(s)** 1937; *Pie in the Sky* 1937; *The Way to Santiago* 1940; *A Man Reprieved* 1949; *Occasion of Glory* 1955; *The Scarlet Boy* 1961 [rev. 1962]

Non-Fiction

Glory Dead 1939; *The Watershed* 1947; *Challenge to Schools: Public School Education* 1935; *The Changing Scene* 1937; *The Book Front* **(c)** [ed. J. Lindsay] 1947; *The Magic of My Youth* **(a)** 1951; *No Earthly Command* 1957; *Havelock Ellis* 1959; *The Enthusiast* 1962; *The Innocent Eye* 1963; *Wish You Were Here: the Art of Donald McGill* 1966; *Prepare to Shed Them Now ...: the Biography and Ballads of George R. Sims* 1968; *Lewd, Blasphemous and Obscene* 1972; *The Two Duchesses* 1976; *The Grand Century of the Lady* 1976

Edited

Tobias Smollett *Selected Writings* 1950; Jack London: *The Bodley Head Jack London* vols 1–4 1963–6, *The Call of the Wild and Other Stories* 1969; Charles Dickens: *David Copperfield* 1967, *Nicholas Nickleby* 1968, *Oliver Twist* 1970, *Bleak House* 1976; *The Life of Benvenuto Cellini* 1968; Jane Austen *Emma* 1970; Thomas Paine *Common Sense and The Rights of Man* 1970

For Children

The Man from Devil's Island 1958; *Fair to Middling* 1959; *Lone Wolf: the Story of Jack London* 1961

Erskine (Preston) CALDWELL 1903–1987

Born in a remote part of rural Georgia, from birth until the age of twenty Caldwell rarely lived for more than six months in the same place, because his father (lovingly portrayed in the biographical volume *Deep South*) was a travelling Presbyterian minister whose work took him all over the American South. There was little time to attend school, and Caldwell was mainly taught by his mother. He briefly attended Erskine College in South Carolina, but left aged eighteen to go on a gun-running boat to South America. He also went for some semesters to the universities of Virginia and Pennsylvania, but his true education, as for many American writers, was in a wide variety of casual jobs, ranging in his case from playing professional football to working in a poolroom, from arranging a lecture tour for a British soldier of fortune to acting as a bodyguard to a Chinese man. He was also for a time a cub reporter on the *Atlanta Journal*. From 1930 to 1934 he did his first stint as a screenwriter in Hollywood.

An early success with his fiction was in 1933 when his story 'Country Full of Swedes' won the *Yale Review* $1,000 award for fiction, but by this time he had already written his two most famous novels, *Tobacco Road* and *God's Little Acre*, which used his knowledge of the poor whites and blacks of the South to earthy, realistic, devastating effect. Both novels were prosecuted for obscenity – Caldwell appeared in court more often than any American writer of his time – but, helped by the unprecedented Broadway run in the mid-1930s of the dramatisation of *Tobacco Road* by Jack Kirkland, they also helped to make him outstandingly successful. From 1938 to 1941 he travelled widely as a foreign correspondent, and he and his second wife, the photographer Margaret Bourke-White (who illustrated several of his travel books), were in the Soviet Union during the German invasion of 1941: he wrote an excellent book describing his experiences. He continued publishing novels into the 1970s, but they were mostly uninspired and often semi-pornographic. An occasional title, such as *Georgia Boy* and *Miss Mamma Aimee*, returned to his original high standard, but his work, which had been extremely popular, fell into obscurity. He also wrote hundreds of short stories, many very fine, as well as two volumes of autobiography, several of travel essays, and several topographical coffee-table books. He was general editor from 1941 to 1955 of the series of regional books *American Folkways*.

After a wandering life he settled down during his last years at Scottsville, Arizona. He married four times and had four children; his first three marriages ended in divorce while the fourth proved lasting. A heavy smoker since the age of fifteen, he eventually died of inoperable cancer.

Fiction

The Bastard 1930; *Poor Fool* 1930; *American Earth* (s) 1931 [UK *A Swell-Looking Girl* 1935]; *Mama's Little Girl* (s) 1932; *Tobacco Road* 1932; *God's Little Acre* 1933; *A Message for Genevieve* (s) 1933; *We Are the Living* (s) 1933; *Journeyman* 1935 [rev. 1938]; *Kneel to the Rising Sun and Other Stories* (s) 1935; *The Sacrilege of Alan Kent* (s) 1936; *Southways* (s) 1938; *Jackpot: the Short Stories of Erskine Caldwell* (s) 1940 [rev. as *Midsummer Passion* 1948]; *Trouble in July* 1940; *All Night Long* 1942; *Georgia Boy* (s) 1943; *A Day's Wooing and Other Stories* (s) 1944; *Tragic Ground* 1944; *Stories by Erskine Caldwell* (s) [ed. H. Seidel Canby] 1944 [US *The Pocket Book of Erskine Caldwell Stories* 1947]; *A House in the Uplands* 1946; *The Sure Hand of God* 1947; *This Very Earth* 1948; *Where the Girls Were Different and Other Stories* (s) [ed. D.A. Wollheim] 1948; *Place Called Estherville* 1949; *A Woman in the House* (s) 1949; *Episode in Palmetto* 1951; *The Humorous Side of Erskine Caldwell* (s) 1951; *The Courting of Susie Brown* (s) 1952; *A Lamp for Nightfall* 1952; *The Complete Stories* (s) 1953; *Love and Money* 1954; *Gretta* 1955; *Gulf Coast Stories* (s) 1956; *Certain Women* (s) 1957; *Claudelle Inglish* 1959 [UK *Claudelle* 1959]; *When You Think of Me* (s) 1959; *Jenny by Nature* 1961; *Men and Women* (s) 1961; *Close to Home* 1962; *The Bastard and Poor Fool* 1963; *The Last Night of Summer* 1963; *Miss Mamma Aimee* 1967; *Summertime Island* 1968; *The Weather Shelter* 1969; *The Earnshaw Neighbourhood* 1971; *Annette* 1973; *Stories* (s) 1980; *Stories of Life* (s) 1983; *The Black and White Stories of Erskine Caldwell* (s) 1984

Non-Fiction

In Defence of Myself 1930; *Some American People* 1935; *Tenant Farmer* 1935; *You Have Seen Their Faces* [with M. Bourke-White] 1937; *North of the Danube* [with M. Bourke-White] 1939; *Say! Is This the USA?* [with M. Bourke-White] 1941; *All-Out on the Road to Smolensk* 1942 [UK *Under Fire: A Wartime Diary, 1941–1942*]; *Russia at War* [with M. Bourke-White] 1942; *Call It Experience: the Years of Learning How to Write* (a) 1951; *Around about America* 1964; *In Search of Bisco* 1965; *In the Shadow of the Steeple* 1966; *Writing in America* 1967; *Deep South: Memory and Observation* 1968; *Afternoons in Mid-America: Observations and Impressions* 1976; *With All My Might* (a) 1987

For Children

Molly Cottontail 1958; *The Deer at Our House* 1966

Collections

The Caldwell Caravan: Novels and Stories 1946

Biography

Erskine Caldwell by James E. Devlin 1984

Hortense CALISHER 1911–

Born in New York City, of Jewish parents, the daughter of a prosperous manufacturer (one of her brothers was formerly a vice-president of Chanel), Calisher grew up as part of the tight-knit community of German-Jewish Yorkville. Middle-class Manhattan apartment-dwellers form the characteristic personnel of her fiction, and her complex Jewish family heritage one of its characteristic subjects. She attended Hunter College High School in New York, and attained her BA from Barnard College, New York in 1932. After college, she worked for a while as a sales clerk, model and social worker, but in 1935 married Heaton Bennet Heffelfinger, an engineer, and they travelled widely around the USA, eventually settling in 1946 in Nyack, a town in the Hudson River valley: they had a son and a daughter.

Calisher had written since childhood, but was not published until she was thirty-six: she enrolled in a writing course while a housewife in Nyack, and her first story was published in the *New Yorker* (which subsequently published many of her stories) in 1948. Her first published volume, of stories, was *In the Absence of Angels* (1951). During the 1950s, with the aid of Guggenheim Fellowships, she was able to devote herself to writing and also started, at Barnard College in 1956, the career she has followed in later life of visiting professor at many American colleges and universities. In 1958, she divorced her first husband and in 1959 married Curtis Harnack, the novelist and director of the arts foundation Yaddo: they have lived for many years in New York.

Calisher is probably most distinguished as a short-story writer, with an enormous range of subject-matter and a dense, highly wrought style reminiscent of **Henry James** and Proust: a novella, *The Railway Police* about a bald woman facing the world, is also one of her best works. Her novels, from *False Entry* to *The Bobby-Soxer*, are generally regarded as less successful, containing brilliant passages, but lapsing frequently into obscurity and sentimentality. *Herself* is a volume of memoirs containing idiosyncratic views about artistic creation.

Fiction

In the Absence of Angels (s) 1951; *False Entry* 1961; *Tale for the Mirror* (s) 1962; *Textures of Life* 1963; *Extreme Magic* (s) 1964; *Journal from Ellipsia* 1965; *The Railway Police and The Last Trolley* (s) 1966; *The Last Trolley Ride* 1966; *The New Yorkers* 1969; *Queenie* 1971; *Standard Dreaming* 1972; *Eagle Eye* 1973; *The Collected Stories of Hortense Calisher* (s) 1975; *On Keeping Women* 1977; *Mysteries of Motion* 1983; *Saratoga, Hot* (s) 1985; *The Bobby-Soxer* 1986; *Age* 1987

Non-Fiction

Herself (a) 1972; *What Novels Are* 1969

Edited

Best American Short Stories (s) [with S. Ravevel] 1981

Morley (Edward) CALLAGHAN 1903–1990

Born in Toronto, Canada, of Irish descent on both sides, Callaghan was educated at the University of Toronto and Osgoode Hall Law School, and was admitted to the Ontario Bar in 1928. Between college and law school, he worked as a reporter on the Toronto *Daily Star*, and there met **Ernest Hemingway**, who liked him (at school Callaghan had distinguished himself as an athlete and boxer) and encouraged his work. Callaghan lived in Paris from 1928 to 1929 and was introduced by Hemingway to fellow expatriate writers such as **F. Scott Fitzgerald** and **James Joyce**. Early stories were published in *This Quarter* and *transition*, and his memoir of this period, *That Summer in Paris*, is one of the most evocative accounts of the Montparnasse *quartier* in the late 1920s.

Callaghan returned to North America and, apart from brief periods spent in New York and Pennsylvania, made his home for the rest of his life in his native Toronto. His energies were increasingly devoted solely to his writing, which is almost all fiction, but he practised law to an extent for some years, worked as a journalist and, more widely, as a radio broadcaster. During the Second World War, he worked for the Canadian Navy on assignment for the National Film Board and toured Canada as chairman of the radio forum 'Of Things to Come'. He married in 1929 and had two children.

In his early years, he was best known for his masterly short stories and tense moralistic novellas: his first considerable work was *Such Is My Beloved* (1934), about a priest who attempts to redeem two prostitutes (Callaghan was a devout Roman Catholic and much of his work approaches the world from that perspective). From the early 1950s on, he turned to much more extended works in the full-blown realist manner, but opinion is divided over the merits of novels such as *The Loved and the Lost*, which treats of black–white relations in Montreal. With *Close to the Sun Again*, a late and masterly novella about a dying tycoon, he returned to the spare and austerely humane style of earlier works. Callaghan has been called the most unjustly neglected of modern masters: perhaps only Canada's slight provincialism on the world cultural scene has prevented full recognition of his merits.

Fiction

Strange Fugitive 1928; *A Native Argosy* (s) 1929; *It's Never Over* 1930; *No Man's Meat* (s) 1931; *A Broken Journey* 1932; *Such Is My Beloved* 1934; *They Shall Inherit the Earth* 1935; *Now That April's Here and Other Stories* (s) 1936; *More Joy in Heaven* 1937; *The Varsity Story* 1948; *The Loved and the Lost* 1951; *Morley Callaghan's Stories* (s) 1959; *The Many Colored Coat* 1960; *A Passion in Rome* 1961; *A Fine and Private Place* 1975; *Close to the Sun Again* 1977; *No Man's Meat and The Enchanted Pimp* (s) 1978; *A Time for Judas* 1983; *The Lost and Found Stories of Morley Callaghan* (s) 1985; *Our Lady of the Snows* 1986; *A Wild Old Man on the Road* 1988

Plays

Turn Again Home [from his novel *They Shall Inherit the Earth*] 1940 [np]; *To Tell the Truth* 1949 [np]; *And Then Mr Jones* (tv) 1974; *Season of the Witch* 1976

For Children

Luke Baldwin's Vow 1948

Non-Fiction

That Summer in Paris: Memories of Tangled Friendships with Hemingway, Fitzgerald and Some Others 1963; *Winter* 1974

(Ignatius) Roy(ston Dunnachie) CAMPBELL 1901–1957

The fourth child of a South African doctor, Campbell was born in Durban and educated at the city's Boys' High School, where he began writing poetry. He came to England after the Armistice and entered Merton College, Oxford, where he was introduced by a wide literary circle by William Walton and by the future critic T.W. Earp, with whom he had an affair. He left Oxford without taking a degree, travelled for eighteen months in Europe, doing casual work and writing, and in 1922 married Mary Garman, a beautiful young woman who lived on the fringes of Bohemian London and whose mother was the illegitimate daughter of Lord Grey of Falloden and his Irish housekeeper. They had two daughters.

Campbell's first publication was *The Flaming Terrapin*, a verse allegory which won him considerable acclaim in 1924. He returned to South Africa that same year and, with **William Plomer**, founded *Voorslag* ('Whiplash'), a controversial satirical magazine, and wrote *The Wayzgoose*, a verse satire. In 1927 he returned to England, where he produced further volumes of poetry and became a combative member of the literary scene. **Edith Sitwell**, who later became a friend, described him at this period as 'noisy, frothing little Mr Roy Campbell (that typhoon in a beer-bottle)', and **Wyndham Lewis** portrayed him in *The Apes of God* (1930) as 'Zulu Blades', a dashing, colonial womaniser. His wife's affair with **Vita Sackville-West** prompted a coarse assault in verse upon the whole of Bloomsbury entitled *The Georgiad*, a poem coloured and compromised by his own bisexuality. His posturing 1934 autobiography, *Broken Record*, was characteristically boastful and mendacious.

From the 1930s, he spent much of his life abroad, and his love of Spain led him to embrace bull-fighting, *machismo*, Roman Catholicism and fascism. He fought for, and wrote poems in praise of, Franco, and despised the left-wing poets of the 1930s, whom he bracketed together into a four-headed hydra, 'MacSpaunday' (i.e. **Louis MacNeice**, **Stephen Spender**, **W.H. Auden**, **C. Day Lewis**). Turned down by the navy and the Intelligence section of the RAF, he served as a sergeant in East Africa during the Second World War, but suffered from malaria and lamed himself permanently after injuring his hip during a training exercise.

Back in London, he worked as a temporary assistant to the War Damage Commission before joining the BBC in 1946 as a radio producer for the Third Programme. He resigned in 1949 to become editor of *The Catacomb*, a right-wing Catholic magazine with a peak circulation of 200 copies, which survived less than eighteen months. His violent nature exacerbated by alcohol, ill health and paranoia, he was involved in a number of celebrated literary and physical scuffles. On one occasion he punched Stephen Spender on the nose during a poetry reading; on another he threatened **Geoffrey Grigson** with his walking-stick. He spent his last years in Portugal, where he was killed in a car crash on his way back from Holy Week celebrations in Seville.

Poetry

The Flaming Terrapin 1924; *The Wayzgoose* 1928; *Adamastor* 1930; *The Gum Trees* 1930; *Poems* 1930; *Choosing a Mast* 1931; *The Georgiad* 1931; *Nineteen Poems* 1931; *Pomegranates* 1932; *Flowering Reeds* 1933; *Mithraic Emblems* 1936; *Flowering Rifle* 1939; *Sons of the Mistral* 1941; *Talking Bronco* 1946; *Collected Poems* vol. 1 1949; *Nativity* 1954; *Collected Poems* vol. 2 1957

Non-Fiction

Taurine Provence 1932; *Broken Record* (a) 1934; *Light on a Dark Horse: an Autobiography 1901–1935* (a) 1951; *Lorca: an Appreciation of His Poetry* 1952; *The Mamba's Precipice* 1953; *Portugal* 1957

Translations

Helge Krog *Three Plays* 1934; *The Poems of St John of the Cross* 1951; Baudelaire *Poems; a Translation of Les Fleurs du mal* 1952; Eca de Queiros: *The City and the Mountains* 1953, *Cousin Basilio* 1953; *The Classic Theatre: Six Spanish Plays* vol. 3 [ed. E. Bentley] 1959; Calderon de la Barca *The Surgeon of His Honour* 1960; *Collected Poems* vol. 3 1960; Paco d'Arcos *Nostalgia: a Collection of Poems* 1960

Biography

Roy Campell by Peter Alexander 1982

Truman CAPOTE (pseud. of Truman Streckfus Persons) 1924–1984

The only child of a clerk for a steamboat company and his much younger wife, Capote was born in New Orleans and brought up in Monroeville, Alabama. His father never stuck at any job for long and was always leaving home in pursuit of ambitious and usually fruitless schemes, and this led to the gradual disintegration of the marriage. A precocious child, Capote started writing fiction at the age of eight, and was portrayed by his lifelong friend **Harper Lee** as the wildly imaginative Dill in her novel *To Kill a Mockingbird* (1960). He left school early and at the age of seventeen secured a job on the *New Yorker*. He later claimed that one of his duties was to escort the purblind **James Thurber** to an assignation with one of the magazine's secretaries, helping him in and out of his clothes.

'My life – as an artist, at least – can be charted as precisely as a fever,' Capote once wrote: 'the highs and lows, the very definite cycles.' It certainly had its peaks and troughs, but started brilliantly in 1948 with his first novel, *Other Voices, Other Rooms*, a shimmering and ambiguous exercise in Southern Gothic about a boy growing up in the Deep South. Its success was immediate, and enhanced by both the controversy surrounding its treatment of homosexuality and the photograph of the author on the jacket, in which he appeared both very young and very louche. *A Tree of Night*, published the following year, gathered together short stories on Southern themes which he had published in *Harper's Bazaar*, *Mademoiselle* and elsewhere. His second novel, *The Grass Harp*, also drew upon childhood memories, but was more straightforward and whimsical than his first. His own stage adaptation of the story had a brief and unsuccessful run on Broadway in 1951.

The novella *Breakfast at Tiffany's* signalled a change of direction: a camp metropolitan comedy, it owes something in both style and character to **Christopher Isherwood**'s *Sally Bowles* (1937). There followed a succession of journalistic assignments and several works of reportage, including *The Muses Are Heard*, a dazzlingly funny account of Gershwin's *Porgy and Bess* on tour in the USSR. *Observations* is a book of portraits in collaboration with the photographer Richard Avedon. He also wrote two film scripts: John Huston's *Beat the Devil* (1953) and Jack Clayton's *The Innocents* (1961), an extremely fine adaptation of **Henry James**'s *The Turn of the Screw* (1898).

In 1959, he read about the murder of a wealthy family in Holcomb, Kansas, and immediately saw a subject for a long piece of non-fiction. Sponsored by the *New Yorker*, and accompanied by Harper Lee, he set off for the small community and began interviewing local people. Two drifters were arrested, and Capote realised that this would alter the whole thrust of his project and that the story had potential as a full-length 'nonfiction novel' (a genre he claimed to have invented), in which he would minutely recreate the lives of both the murderers and their victims. This raised a moral dilemma, since he could not finish the book until the case was resolved, either by pardon or execution, and he had got to know the killers well. They were eventually convicted and executed, and Capote published *In Cold Blood*, which is probably his masterpiece.

'It took me five years to write *In Cold Blood*,' he said, 'and a year to recover – *if* recovery is the word.' It made him a social celebrity, and he planned to write a Proustian novel to be called *Answered Prayers*, in which he would dissect New York's high society. Four chapters published in advance in *Esquire* caused an outcry since the characters he had depicted were recognisably based upon well-known socialites. Problems with drink and drugs sapped Capote's creative energies, as did bitter wrangles with fellow writers such as **Gore Vidal**. Although he produced further work – including a musical based on one of his stories, *House of Flowers*; a fine collection of stories and reportage, *The Dogs Bark*; and another nonfiction novel about a bizarre series of murders, *Music for Chameleons* – he never completed his intended masterwork, *Answered Prayers*, which at his death was discovered to be little more than a few jottings and fragments, but was published in 1986.

Fiction

Other Voices, Other Rooms 1948; *A Tree of Night and Other Stories* (s) 1949; *The Grass Harp* 1951; *Breakfast at Tiffany's* (s) 1958; *A Christmas Memory* 1966; *In Cold Blood* 1966; *The Thanksgiving Visitor* 1968; *Music for Chameleons* 1980; *One Christmas* 1982; *Answered Prayers* (s) 1986

Plays

The Grass Harp [from his novel] 1951; *House of Flowers* [with H. Arlen] 1968

Non-Fiction

Local Color 1950; *The Muses Are Heard* 1956; *Observations* [with R. Avedon] 1959; *Trilogy: an Experiment in Multimedia* [with E. and F. Perry] 1969

Collections

Selected Writings of Truman Capote 1963; *The Dogs Bark: Public People and Private Places* 1973; *A Capote Reader* 1987

Biography
Capote by Gerald Clarke 1988

Peter CAREY 1943–

Carey was born and raised in Bacchus Marsh, Victoria, Australia. His parents, who had a General Motors dealership, sent him at the age of ten to Geelong Grammar School, 'where the children of Australia's Best Families all spoke with English accents', and then Timbertop. He went from there to Monash University where he failed a science degree. He entertained briefly the idea of going on to study architecture, but suddenly could not bear the thought of his parents paying for anything more, so in 1962 he got a job in an advertising agency. He had no moral qualms about this – 'I came from a long line of people who'd sold motor cars and done it honestly and I wasn't ashamed of anything to do with selling' – and became regarded as one of Australia's best copywriters.

It was at the advertising agency that he met the Australian writer Barry Oakley, who passed on his cast-off review copies of books and, for the first time in his life, Carey read voraciously. Then in 1964 he began to write. He married at twenty-one, and would go home from the job to write every night and at weekends. By the time he left the country in 1968, to go travelling through Greece and Italy, he had written three unpublished novels. He lived in London for a couple of years, before returning to Australia, where he divided his time between Sydney and an alternative community in the rainforest of Queensland, near Yandina.

Carey's collection of stories, *The Fat Man in History* (first published in Australia in 1974 and in England in 1980), won immediate recognition as the emergence of a remarkable talent. This was followed by *Bliss*, which won two major prizes (the Miles Franklin Award and the NSW Premier Award) and was made into a feature film, for which Carey wrote the screenplay.

Carey's writing is, he says, not really autobiographical, but it often draws from stories that he grew up around. For example, his next novel, *Illywhacker*, drew on his grandfather's experiences delivering the first airmail around South Australia in a Blériot plane, while *The Tax Inspector* is set amongst a family of car dealers. He felt that *Illywhacker* represented 'me at last trying to come to grips with what it means to be an Australian, and what Australia is'. For the literary world, it established him as a major writer – it was shortlisted for the Booker Prize. After *Illywhacker* (some 600 pages), Carey intended to write a much shorter book – three years later, *Oscar and Lucinda* weighed in at 511 pages. Set in mid-nineteenth-century England and Australia (and drawing in particular on Edmund Gosse's *Father and Son*, 1907), it is Victorian in size and scope, though modern in sensibility. It tells the interlocking stories of two compulsive gamblers and a wager over the construction and transportation through the Australian bush of a glass church. At once naturalistic and fantastical, the novel was a popular and critical success, and won Carey the 1988 Booker Prize.

Subsequent books have made rather less impact, although *The Tax Inspector* attracted a certain notoriety for one scene of shocking violence. Carey, who lives in Sydney and continues to copywrite part-time, remains one of the most exciting and impressive talents to emerge from Australia this century.

Fiction
The Fat Man in History (s) 1974; *Bliss* 1981; *Illywhacker* 1985; *Oscar and Lucinda* 1988; *The Tax Inspector* 1991; *The Unusual Life of Tristan Smith* 1994

Plays
Bliss (f) [from his novel] 1986

Edward CARPENTER 1844–1929

Born in Brighton, the son of a naval officer who had retired early from the service, Carpenter attended Brighton College and Trinity Hall, Cambridge, becoming a fellow of his college and taking orders. For a while he was the curate of the well-known Cambridge liberal clergyman F.D. Maurice, but he soon became discontented with the priesthood and, after publishing his early poem *Narcissus*, resigned both his fellowship and his orders in 1874.

Carpenter was a bisexual with a romantic (and soon to be socialist) devotion to the working-class, and he furthered his ideals by working from 1874 to 1881 on the staff of the university extension movement, mainly in Sheffield. He made the first of two formative visits to the USA in 1877, meeting the poet Walt Whitman, who became the leading cultural influence on him. Carpenter's best-known poem, *Towards Democracy*, with its gospel of the march of humanity towards socialist brotherhood, while perhaps not wholly successful as poetry, can be seen as the most important attempt to introduce the manner and ethos of Whitmanesque verse into English poetry. The death of his father in 1883 and a small legacy enabled him to buy a farm at Millthorpe, near Chesterfield, where he lived until 1922, in the earlier years with a working-class lover, Albert Fernehough, and his family. There, he adopted market-gardening and sandal-making, becoming meanwhile a well-known speaker for the nascent socialist movement and publishing many pamphlets and books advocating a multitude of progressive causes including vegetarianism, anti-vivisection, prison reform, women's rights and interest in Eastern religion. Homosexuality was his most controversial cause: he published a pamphlet on it at the height of the Oscar Wilde trials, and later a book, *The Intermediate Sex*. Perhaps surprisingly, he was never persecuted by the authorities. Other well-known books were *Civilisation: Its Cause and Cure* and his autobiography *My Days and Dreams*. On his eightieth birthday he achieved the honour of an address from the Trades Union Congress.

From 1898, his lover was George Merrill, a former coal-miner, and in 1922, the couple moved to Guildford, where Merrill predeceased him by a year. Carpenter was perhaps more important as an influence than for his own creative work: that influence, first on intellectuals such as Havelock Ellis and **E.M. Forster** and, beyond that, on the libertarian ideals that came to the fore in the 1960s, was of a profundity that it is difficult to overestimate.

Non-Fiction

The Religious Influence of Art 1870; *Co-operative Production with Reference to the Experiment of Leclaire* 1883; *Modern Money-Lending and the Meaning of Dividends* 1883; *Progress* 1883; *Modern Science: a Criticism* 1885; *A Letter to the Employees of the Midland and Other Railway Companies* 1886; *The Sheffield Socialists* 1886; *England's Ideal and Other Papers on Social Subjects* 1887; *Civilisation: Its Cause and Cure* 1889; *Our Parish and Our Duke: a Letter to the Parishioners of Holmesfield* 1889; *The Smoke Nuisance and Smoke-Preventing Appliances* 1889; *Manifesto of the Sheffield Socialists: an Appeal to the Workers* 1890; *From Adam's Peak to Elephanta* 1892; *Homogenic Love and Its Place in Free Society* 1894; *Woman, and Her Place in a Free Society* 1894; *Marriage in a Free Society* 1894; *Sex-Love, and Its Place in a Free Society* 1894; *Love's Coming of Age* 1896; *The Need of a Rational and Humane Science* 1896; *An Unknown People* 1897; *Angels' Wings: a Series of Essays on Art and Its Relation to Life* 1898; *Boer and Briton* 1990; *Empire: In India and Elsewhere* 1900; *Sketches from Life in Town and Country* 1903; *The Art of Creation* 1904; *Vivisection* 1904; *Prisons, Police and Punishment: an Enquiry into the Causes and Treatment of Crime and Criminals* 1905; *Days with Walt Whitman* 1906; *British Aristocracy and the House of Lords* 1908; *The Intermediate Sex* 1908; *The Wreck of Modern Industry and Its Reorganisation* 1910; *The Drama of Love and Death: a Study of Human Evolution and Transfiguration* 1912; *The Inner Self* 1912; *Letter to Prince Kropotkin on his 70th Birthday* 1912; *Intermediate Types among Primitive Folk* 1914; *The Healing of Nations* 1915; *My Days and Dreams* (a) 1916; *Never Again!* 1916; *Towards Industrial Freedom* 1917; *Pagan and Christian Creeds: Their Origin and Meaning* 1920; *Some Friends of Walt Whitman: a Study in Sex-Psychology* 1924; *The Psychology of the Poet Shelley* [with G.C. Barnard] 1925

Poetry

Narcissus and Other Poems 1873; *Towards Democracy* 1883–1902 [rev. in 1 vol. 1905]; *England Arise!* 1886; *Three Ballads* 1917

Plays

Moses 1875; *The Promised Land* 1909

Edited

Chants of Labour: a Song Book of the People 1888; *Forecasts of the Coming Century* 1897; Ponnambalam Arunáchalam *Light from the East: Being Letters on Gñánam, the Divine Knowledge* 1927

Translations

The Story of Eros and Psyche from Apuleius, and the First Book of the Iliad of Homer 1900

For Children

St George and the Dragon 1895

Collections

Ioläus: an Anthology of Friendship 1902; *Edward Carpenter: Selected Writings* 1985

J(ames Joseph) L(loyd) CARR 1912–1994

Born at Thirsk and brought up at Carlton Miniott in the North Riding of Yorkshire, Carr completed his education at Castleton Grammar School. He became a schoolteacher, initially in Hampshire and then in Birmingham, with a break between 1938 and 1939 when he taught in the high school in Hurron, South Dakota, USA. He was in the Royal Air Force from 1940 to 1946, in which he became a flight lieutenant and served mainly in Africa in coastal command squadrons. During the war he married Sally Sexton, a Red Cross nurse; they had one son and their marriage ended with her death in 1981. After the war Carr returned to teaching, which included another spell in South Dakota and a period in Northamptonshire.

He wrote his first novel when he was fifty; this was *The Battle of Pollocks Crossing*, based on his South Dakotan experiences, and eventually published in 1985, when it was shortlisted for the Booker Prize. His first published novel, however, was *A Day in Summer* (1963) which, after being turned down by many publishers, was eventually brought out by the now defunct house, Barrie and Rockcliff. Six novels followed this, of which the most celebrated is *A Month in the Country*, a delicate and evocative story of rural Yorkshire just after the First World War which was widely translated, was shortlisted for the Booker Prize, won the *Guardian* Fiction Prize, and was turned into a feature film (1987), directed by Pat O'Connor from a screenplay by **Simon Gray**. Carr's other novels fall into a comparable pattern in that they are mostly short, based on personal experience and much concerned with traditional England, which is described in a plain, non-experimental but finely honed style, with much humour and an abiding obsession with women.

In the early 1960s he ventured into publishing, producing a series of inexpensive booklets from his home in Kettering, Northamptonshire. Designed to fit into an 'envelope 4.5″ by 6.3″, a common size', these books include selections from the major poets as well as arcane dictionaries of (amongst other things): *Extra-ordinary Cricketers*; *Eponymists*; *Brief Lives of the Frontier*; *English Kings, Consorts, Pretenders, Usurpers, Unnatural Claimants & Royal Athelings*; *Prelates, Parsons, Vergers, Wardens, Sidesmen & Preachers, Sunday-School Teachers, Hermits, Ecclesiastical Flower-arrangers, Fifth Monarchy Men and False Prophets*. The success of these books, some of which have sold over 20,000 copies, and the county maps he also published allowed Carr to give up teaching in 1977. Ten years later, tired of the ways of mainstream publishers, he decided to publish *What Hetty Did* himself under the imprint of the Quince Tree Press. His son continues to run the company.

Apart from publishing and writing, Carr was a painter and sculptor, specialising in providing free of charge limestone effigies for church niches, replacing those destroyed during the Reformation.

Fiction

A Day in Summer 1963; *A Season in Sinji* 1967; *The Harpole Report* 1972; *How Steeple Sinderby Won the FA Cup* 1975; *A Month in the Country* 1980; *The Battle of Pollocks Crossing* 1985; *What Hetty Did* 1987

For Children

The Dustman 1969; *The Red Wind Cheater* 1971; *The Old Cart* 1973; *Red Foal's Coat* 1973; *Ear-ring for Anna Beer* 1976; *The Green Children of the Woods* 1977; *Gone with the Whirlwind* 1980

Non-Fiction

The Old Timers of Beadle County 1957; *Dictionary of English Kings* 1979; *Dictionary of English Queens* 1979; *Dictionary of Extra-ordinary English Cricketers* 1977; *Sydney Smith: the Smith of Smiths* 1980

Angela CARTER 1940–1992

Britain's fairy godmother of magic realism was born in Eastbourne, but was removed by her grandmother (an indomitable woman, to whom Carter credited her own determination) from the wartime coast to the relative safety of a Yorkshire coalfield. She was a chubby child nicknamed 'Tubs', but when she went back south and rejoined her mother she became anorexic. Her father was a journalist, and when Carter was in the process of flunking her A-levels – something she later interpreted as a rebellion against her over-possessive mother – he found her a job on the Croydon *Advertiser* instead. She was far from being a born news journalist, but she found a role writing features and record reviews. In 1960 she married, and when her industrial chemist husband took a teaching job at Bristol Technical College, she moved there with him. She explored Bristol's provincial Bohemia, but became bored with her housewifely existence. A breakthrough came when, at the suggestion of an uncle, she went to Bristol University to read English. She began writing as an undergraduate and wrote her first novel, *Shadow Dance* (1966), in the summer vacation of her second year. Her third book, *Several Perceptions*, won a Somerset Maugham Award. She used the money to go to Japan, which was to be another formative experience. She worked for the English language branch of a broadcasting company and had a number of other jobs including bar hostess on the Ginza where, as she later told a friend, 'I could hardly call my breasts my own'.

'Pleasure has always had a bad press in Britain,' she writes in her book of critical essays, *Expletives Deleted*; 'I'm all for pleasure'. Carter's pleasure, though, was often of a dangerous and ambivalent kind, and a strong theme of female masochism is explored throughout her work. As an opening book for Virago she wrote *The Sadeian Woman* (1979), a study of women, sexuality and power. Some feminist critics found Carter's taste for the pornographic and macabre to be at odds with what they saw as the correct interests of the progressive woman, and her *Times* obituary declared approvingly – perhaps with an element of reactionary relish – that: 'She squarely faced the possibility that sex is ultimately a violent business and that women can acquiesce in that.'

Carter's work is distinguished by its lush writing, its eroticism and its Gothic elements.

The Bloody Chamber is characteristic in its distinctively adult, Freudian-inflected retellings of fairy-tales. She regarded the fantasy element of her work as 'social satire' or 'social realism of the unconscious'. Others have suggested that her vein of fantasy, particularly evident in novels such as *Nights at the Circus*, appeals to the submerged adolescent in all of us. Carter's status as the godmother of British women's writing had become virtually official in her lifetime, and her influence can be seen in such writers as **Jeannette Winterson**. Her personal manner had become very grand by the time of her death, by which time she had also acquired a second, younger, husband, and a much-loved young son.

Her work has enjoyed immense popularity with young women: by 1994 more than 80 per cent of the demand-oriented 'new' universities were teaching it, and she had become the most popular modern British author studied on university English courses. In 1992–3 there were forty proposals for doctorates on her work, which was more than for the entire eighteenth century. Carter herself was a wonderfully readable critic, and a feisty one: her judgement on Shakespeare was that he was 'a lovable man, but I'm afraid not very clever'.

Fiction

Shadow Dance 1966; *The Magic Toyshop* 1967; *Several Perceptions* 1968; *Heroes and Villains* 1969; *Love* 1971 [rev. 1987]; *The Infernal Desire Machines of Doctor Hoffman* 1972; *Fireworks* (s) 1974 [rev. 1987]; *The Passion of New Eve* 1977; *The Bloody Chamber and Other Stories* (s) 1979; *Martin Leman's Comic and Curious Cats* 1979; *Moonshadow* [with J. Todd] 1982; *Nights at the Circus* 1984; *Black Venus* (s) 1985; *Come Unto These Yellow Sands* (s) 1985; *Wise Children* 1991

Non-Fiction

The Sadeian Women 1979; *Nothing Sacred* 1982; *Expletives Deleted: Selected Writings* 1992; *American Ghosts and Old World Wonders* 1993

Translations

The Fairy Tales of Charles Perrault 1977; *Sleeping Beauty and Other Favourite Fairy Tales* 1991

Edited

Wayward Girls and Wicked Women (s) 1986; *The Virago Book of Fairy Tales* 2 vols 1990, 1992

Jim CARTWRIGHT 1959–

Cartwright was born and brought up in Lancashire where he still lives today. The son of a factory worker, he left school aged sixteen and hung around his hometown of Farmworth, which would be the inspiration for the setting of *Road*, his first play. After several years' unemployment, he decided to become an actor and moved to London to attend drama school. He left after a few months and set up a three-man theatre company. 'It was like dial-a-pizza,' he explained, 'only this was dial-a-theatre. You'd call up and we'd come and perform plays in your home.' Eventually he returned to Lancashire and began to write seriously.

Road opened in 1986 at the Royal Court Theatre Upstairs in London and was soon transferred to the Royal Court's main auditorium. It won *Drama* magazine's Best Play Award, the Samuel Beckett Award, the *Plays and Players* Best New Play Award, and was joint winner of the George Devine Award. *Road* is quite literally a trip through a grim Northern village where unemployment has instilled anger and spiritual aridity. The simple pleasures of pub-crawling and casual sex take on a desperate, almost sinister edge. Under the guidance of player/narrator Scullery, the audience found themselves face to face with the short snatches of life that make up the play. The auditorium had been covered with plywood, forming a 'promenade' theatre and the performances took place all over the open, seatless main floor space. The BBC film version of *Road* (1987) won the Golden Nymph Award for Best Film at the Monte Carlo Television and Film Festival.

His second play, *Two*, has two actors playing the bickering landlord and landlady of a Northern pub as well as all the pub's customers. It opened at the Bolton Octagon, then transferred to the Young Vic in London, and won the Manchester *Evening News* Award for the best new play. Indeed, the number of awards Cartwright has won in a relatively short career testifies to the outstanding and quirky originality of his work. His plays, which tend to have monosyllabic titles, are characterised by rambling, apparently inconsequential, bitterly humorous scripts. Both *Bed* and *The Rise and Fall of Little Voice* were put on at the National Theatre. The latter concerned a withdrawn young woman who spends her time listening to torch singers, notably Judy Garland, and developing a large voice of her own. It won Olivier and *Evening Standard* Awards for comedy and subsequently played on Broadway, in Australia and in many European countries. *Vroom* was a modern 'road movie', first shown at the London Film Festival and then on Channel 4 television.

Plays

Road (st) 1986, (tv) 1987 [pub. 1986]; *Vroom* (f) 1988 [np]; *Bed* 1989 [pub. 1991]; *Two* 1989 [pub. 1991]; *Baths* **(r)** 1990 **(st)** 1990 [np]; *Wedded* **(tv)** 1990 [np]; *The Rise and Fall of Little Voice* 1992; *Stone Free* 1994 [np]

Justin CARTWRIGHT 1943-

Cartwright was born in Cape Town, the son of A.P. Cartwright, editor of the *Rand Daily Mail* and an authority on the history of South Africa's mining industry. He was educated at the Diocesan College, Cape Town, the University of the Witwatersrand and Trinity College, Oxford, where he played polo for the university. There followed three years as an advertising copywriter before he graduated to making advertising films. He has lived in London since 1973, has made a number of documentaries, including one on the ANC, as well as wildlife films for the BBC and ITV, and was media adviser to the Liberal SDP Alliance, directing three television campaigns for them in 1979, 1983 and 1987.

'When young I was very keen on movies', he said in 1990. 'My whole life has been a process of avoiding writing – until about ten years ago. But I think I was always cut out to write rather than make movies.' His first two novels, *Deep Six* and *The Horse of Darius* were thrillers, the third, *Freedom for the Wolves*, was set in South Africa. *Interior*, also set in Africa, was shortlisted for the Whitbread, *Sunday Express* and CNA awards, and received an outstanding press. A coolly ironic, intelligently reflective, extremely funny and beautifully written book, it knowingly combines, and subverts, Rider Haggard, **Joseph Conrad** and **Evelyn Waugh** in its account of a journey into the dark heart of an imaginary African country. The narrator, Curtiz, is following in the footsteps of his father who had disappeared during an earlier expedition, and Cartwright piquantly contrasts the relentless 1950s' *National Geographical* optimism of Curtiz Snr with the 1980s' cynicism of his son, who sees all too clearly the unpicturesque and alarming savagery of Banguniland. *Look At It This Way* is set in London, the 'unreal city' of the 1980s, where a lion has escaped from the Regent's Park Zoo. It presents a devastating picture of the contemporary scene – 'the capital is sick to its soul', he once wrote – and some of the less savoury elements of the media. It was also shortlisted for the Whitbread and CNA awards, as well as the James Tait Black Memorial Prize; it was subsequently made into a television series. With *Masai Dreaming*, an ambitious, multi-layered novel, Cartwright returns to two of his principal themes: Africa and the cinema. Tim Curtiz, protagonist of *Interior*, is sent to Africa to write the screenplay for a film about a Jewish anthropologist, Claudia Cohn-Casson, who studied the Masai and died in Auschwitz.

Fiction

Deep Six 1978; *The Horse of Darius* 1980; *Freedom for the Wolves* 1983; *Interior* 1988; *Look At It This Way* 1990; *Masai Dreaming* 1993

Raymond CARVER 1938-1988

Carver came from the sort of background he wrote about in his short stories. He was born in Clatskanie, Oregon, where his father worked in a sawmill, spending much of his pay on drink, and his mother worked as a waitress. His parents eventually separated and his father ended up living in a trailer, after which he suffered a nervous breakdown. Carver's father did, however, interest his son in narrative, reading to him and telling him stories about his own hunting and fishing exploits. In 1957, shortly after leaving high school, Carver married sixteen-year-old Maryann Burk and for the next decade, in order to support his wife and two children, took a number of jobs, including those of janitor, sawmill worker and salesman. He wrote in his spare time and in 1958 took a course in creative writing under **John Gardner** at Chico State College, California. He subsequently attended Humbolt State University, California, graduating in 1963, and the Writers' Workshop at the University of Iowa. Many of his stories were first published in relatively obscure literary magazines, but his reputation took off in 1967 when 'Will You Please Be Quiet, Please?' appeared in an anthology of *Best American Short Stories*. That year he also began work as a textbook editor at Science Research Associates at Palo Alto, remaining there until he was fired in 1970, by which time his reputation had grown sufficiently for him to become a visiting lecturer, then a visiting professor, at various universities.

His first collection, *Put Yourself in My Shoes*, was published in 1974, and 'Will You Please Be Quiet, Please?' became the title story of his second collection, which was nominated for a National Book Award two years later. Subsequent volumes were also nominated for major literary awards. He had become a heavy drinker and in 1977 he and his wife separated, divorcing five years later. He managed to stop drinking after joining Alcoholics Anonymous, and in 1987, when the lung cancer for which he had been treated recurred, he married the poet Tess Gallagher, with whom he had lived for several years. From 1980 to 1984 he was professor of English at Syracuse University, New York.

'I began as a poet,' Carver said; 'so I suppose on my tombstone I'd be very pleased if they put "Poet and short story writer – and occasional essayist". In that order.' Although he produced several volumes of poetry, it is for his short stories that Carver will be remembered. He depicts ordinary American lives in a spare, unadorned prose: *Newsweek* described his usually inarticulate blue-collar characters as 'not only real but representative'. Much of what he wrote about came from his own experience and struggles, and he once said that he did not 'think there should be any barriers, artificial or otherwise, between life as it's lived and life as it's written about'. His tough background, hard drinking and heavy smoking, and his terse prose led to his being considered a 'man's writer', inheritor of the mantle of **Ernest Hemingway**. Before publishing stories in volume form, he frequently revised them, usually paring them down 'to the marrow', as he put it – although in later years he occasionally expanded them. His reputation continued to grow after his death at the age of fifty, fuelled by compilations, a biographical symposium, *Remembering Ray*, and even a photographic album entitled *Carver Country*. His popularity was further enhanced by Robert Altman's much praised film *Short Cuts* (1993), based on several of Carver's stories.

Fiction

Put Yourself in My Shoes (s) 1974; *Will You Please Be Quiet, Please?* (s) 1976; *Furious Seasons and Other Stories* (s) 1977; *What We Talk about When We Talk about Love* (s) 1981; *The Pheasant* (s) 1982; *Cathedral* (s) 1983; *If It Please You* (s) 1984; *The Stories of Raymond Carver* (s) 1985; *Elephant and Other Stories* (s) 1988; *Where I'm Coming From* (s) 1988; *New Path to the Waterfall* (s) 1989; *Short Cuts* (s) 1993

Poetry

Near Klamath 1968; *Winter Insomnia* 1970; *At Night the Salmon Move* 1976; *Two Poems* 1982; *Where Water Comes Together With Other Water* 1985; *Ultramarine* 1986; *In a Marine Light* 1988; *A New Path to the Waterfall* 1989

Plays

Carnations 1962

Non-Fiction

Conversations with Raymond Carver [ed. M.B. Gentry, W.L. Stull] 1990

Collections

Fires: Essays, Poems, Stories 1984; *We Are Not in This Together* [with W. Kittredge] 1987; *Carver Country* [with B. Adelman] 1991; *No Heroics, Please: Uncollected Writings* [ed. W.L. Stull] 1991

(Arthur) Joyce (Lunel) CARY 1888–1957

Cary was born in Londonderry into an 'intensely Loyalist' family whose fortunes had risen with the land grants and shrunk in the 1880s land war. His childhood was divided between stays with his grandparents in Ireland – idyllic, rural interludes – and the stricter regime of life with his parents in England. His education, however, at Clifton College in Bristol, was 'wholly English'. Cary was both shy and frequently unwell, suffering from asthma, rheumatism, pleurisy and near-blindness in one eye, despite which, and under the influence of the French Impressionists, he had decided to become a painter. He left school to study at the Edinburgh College of Art, staying for two years until, despairing of greatness, he switched his attention to literature. At his father's request and hoping for financial support, he studied law at Trinity College, Oxford, graduating without distinction in 1912 and immediately abandoning the legal profession.

With the outbreak of hostilities in the Balkans, Cary enlisted in a Montenegran battalion – he was decorated by that country – before joining the British Red Cross. After the war, he joined the Nigerian Service for whom he spent seven years collecting taxes and acting as judge, census taker, road builder and, with the onset of war, troop leader. He fought in the Cameroons Campaign of 1915–16, was wounded at Mora Mountain, and awarded a second decoration, after which he served as a magistrate in Borgu and the remoter regions of West Africa. In 1916 he married Gertrude Ogilvie, with whom he was to have four sons.

Cary's first success as a writer came with the sale of short stories to the *Saturday Evening Post*, a lucrative market at the time. Hoping to subsidise more serious work with short, commercial fiction, he retired from the service in 1920, citing poor health, and gave himself up to his writing and 're-education'. 'After a study of the masters', he destroyed most of his early efforts, and promptly lost the magazine market which had begun to find his writing too difficult. He had to wait until 1930 for his first novel, *Aissa Saved*, to be published. In 1939 *Mister Johnson*, his fifth novel, brought him critical success. *A House of Children* won him the James Tait Black Memorial Prize in 1941 and *The Horse's Mouth*, the last volume of the trilogy including *Herself Surprised* and *To Be a Pilgrim*, became a bestseller.

Cary also published two volumes of poetry. His liberal convictions were not confined to his novels; as a political theorist and pamphleteer his publications included *The Case for African Freedom*, and *The Process of Real Freedom*. A second trilogy, exploring the 'conflict between creative freedom and established order' – *Prisoner of Grace*, *Except the Lord* and *Not Honour More* – was completed in 1955 at which time Cary was diagnosed as suffering from the gradual paralysis which killed him two years later. Plans for a third trilogy were abandoned as he worked to complete a volume of essays on the creative process, posthumously published as *Art and Reality*.

Fiction
Aissa Saved 1932; *An American Visitor* 1933; *The African Witch* 1936; *Castle Corner* 1938; *Mister Johnson* 1939; *Charley Is My Darling* 1940; *A House of Children* 1941; *Herself Surprised* 1941, *To Be a Pilgrim* 1942, *The Horse's Mouth* 1944; *The Moonlight* 1946; *A Fearful Joy* 1949; *Prisoner of Grace* 1952, *Except the Lord* 1953, *Not Honour More* 1955; *The Captive and the Free* 1959; *Spring Song and Other Stories* (s) 1963

Poetry
Marching Soldier 1945; *The Drunken Sailor* 1947

Non-Fiction
Power in Men 1939; *The Case for African Freedom* 1941; *The Process of Real Freedom* 1943; *Britain and West Africa* 1946; *Art and Reality: Ways of the Creative Process* 1958; *The Case for African Freedom and Other Writings on Africa* 1962

Biography
Gentleman Rider by Alan Bishop 1988

Will(el)a (Sibert) CATHER 1873–1947

Cather spent her earliest years on a hill farm in the Shenandoah Valley of Virginia, but, aged nine, was taken by her family to the Nebraska settler country which forms the background for half her novels and many short stories. Her father first tried to farm the prairie in Webster County, but soon moved into the insurance business in the nearby town of Red Cloud, whose European immigrants, adjusting developed cultures to the realities of a frontier, form the basis for Cather's characters. She attended the University of Nebraska, but then moved east, living in Pittsburgh between 1896 and 1906, working first as a journalist and later as a teacher. Her early work consisted largely of poetry and short stories, and her first volume was of verse, *April Twilights* (1903), followed by the stories of *The Troll Garden*. In 1906 she was hired to help edit the influential *McClure's Magazine*, and moved to New York, where she lived for the rest of her life. Her fastidious writerly spirit was not much at ease with the muckraking ethics of the magazine, but she remained there until 1911, publishing her first novel, *Alexander's Bridge*, the following year, when she was almost forty. Thereafter she devoted herself to writing, averaging a novel every three years, the cornerstone of an output which also included much journalism and comment, several volumes of short stories and a certain amount of poetry. She lived in her Park Avenue apartment with her lifelong companion, Edith Lewis, and this quiet existence, which perhaps parallels that simultaneously being led by **I. Compton-Burnett** in England, was enriched by summers in the country and travel to Europe, particularly France.

The twelve novels (and short stories) fall into three groups: Jamesian tales, often early in her career, of sophisticated Easterners (such as *A Lost Lady*); works dealing with immigrant life in the West, often Nebraska but later sometimes the Southwest (such as *O Pioneers!*); and a late group of historical novels (such as *Death Comes for the Archbishop*), which demonstrate her growing commitment to Christianity and distaste for the modern world. This distaste, and a corresponding classicism and lack of immediacy in style, perhaps account for the comparative neglect into which a writer universally considered major by all who have studied her has fallen. Cather never fully recovered from a major operation performed in 1942, and published little in her last years, which were solaced by the close friendship of Yehudi Menuhin and his sisters.

Fiction
The Troll Garden (s) 1905; *Alexander's Bridge* 1912; *O Pioneers!* 1913; *The Song of the Lark* 1915; *My Antonía* 1918; *Youth and the Bright Medusa* (s) 1920; *One of Ours* 1922; *A Lost Lady* 1923; *The Professor's House* 1925; *My Mortal Enemy* 1926; *Death Comes for the Archbishop* 1927; *Shadows on the Rock* 1931; *Obscure Destinies* (s) 1932; *Lucy Gayheart* 1935; *Not under Forty* (s) 1936; *Sapphira and the Slave Girl* 1940; *The Old Beauty and Others* (s) 1948; *Collected Short Fiction, 1892–1912* (s) [ed. V. Faulkner] 1970; *Uncle Valentine and Other Stories: Willa Cather's Uncollected Short Fiction* (s) 1973

Poetry
April Twilights and Other Poems 1903

Non-Fiction
Willa Cather on Writing 1949; *Willa Cather in Europe: Her Own Story of the First Journey* 1956; *The World and the Parish: Willa Cather's Articles and Reviews* [ed. W.M. Curtin] 1968

Edited
Georgine Milimine *The Life of Mary Baker G. Eddy* 1909

Collections
Writings from Willa Cather's Campus Years 1950

Biography
Willa Cather by Sharon O'Brien 1987

Christopher CAUDWELL (pseud. of Christopher St John Sprigg) 1907–1937

Born at Putney, south London, Caudwell grew up mainly at East Hendred in the Berkshire Downs and was sent from there to Ealing Priory School, a Benedictine establishment, where he showed a precocious aptitude for both poetry and science. His father and grandfather were newspaper journalists and, aged only fifteen, he left school to become a cub reporter on the *Yorkshire Observer*, where his father was then literary editor. Two years later he became editor of *British Malaya*, and two years after that founded an aeronautical publishing company with his brother, publishing designs in *Automobile Engineer*, launching his own journal, *Aircraft Engineering*, and inventing an infinitely variable gear that impressed experts. At the same time he was writing a large amount of poetry, but did not seek to publish it (except for one early piece which appeared in *The Dial*). He remained single and lived mainly with his brother, even after the latter's marriage.

In 1934 Caudwell became a freelance writer. For some years previously he had been helping to support himself in London by writing, in very quick succession, a series of aeronautical books, beginning with *The Airship* (1931), and also a series of seven detective novels, which he published under his own name and which, rather in the manner of the 'Nicholas Blake' novels of **C. Day Lewis**, feature an up-to-date detective, Charles Venable, and advanced social attitudes. He was rapidly converted to Marxism in 1934, having previously had few political interests, and in 1935 he joined the Poplar branch of the Communist Party, simultaneously going to live in the East End of London. During 1936 he spent some time in Paris to study the Popular Front movements, but in December of that year he joined the British Battalion of the International Brigade to fight in Spain. He was killed on his first day of action, in February 1937, as he covered with machine-gun fire the retreat of his comrades from the 'suicide hill' in the notoriously bloody battle of the Jarama River.

A telegram had just been dispatched from a high Party official in London recommending his recall to England as a valuable intellectual. The official had read the proofs of his first major critical book, *Illusion and Reality*, which, like all the books of Marxist literary criticism for which he is nowadays best remembered, was published posthumously. In such works as *Studies in a Dying Culture*, Caudwell showed himself a libertarian Marxist, in many ways anticipating the New Left criticism of the 1960s in his emphasis on the youthful Marx's theories of alienation, but being the first to apply Marxism to literary criticism. Also posthumously published were his *Poems*, which show him as a sincere, anti-romantic, but still developing craftsman, whose dogmatism is tempered by precise attention to imagery and the flow of verse.

Fiction [as Christopher St John Sprigg]
Crime in Kensington [US *Pass the Body*] 1933; *Fatality in Fleet Street* 1933; *Death of an Airman* 1934; *The Perfect Alibi* 1934; *The Corpse with the Sunburned Face* 1935; *Death of a Queen* 1935; *This My Hand* [as 'Christopher Caudwell'] 1936; *The Six Queer Things* 1937

Poetry
Poems 1939; *Collected Poems* [ed. A. Young] 1986

Non-Fiction
The Airship: Its Design, History, Operation and Future 1931; *Fly with Me: an Elementary Textbook on the Art of Piloting* [with H.D. Davis] 1932; *British Airways* 1934; *Great Flights* 1935; *Illusion and Reality: a Study of the Sources of History* 1937; *Let's Learn to Fly* 1937; *Studies in a Dying Culture* 1938; *The Crisis in Physics* [ed. H. Levy] 1939; *Further Studies in a Dying Culture* [ed. E. Rickword] 1949; *The Concept of Freedom* [ed. G. Thomson] 1965; *Romance and Realism: a Study in English Bourgeois Literature* 1970; *The Breath of Discontent: a Study in Bourgeois Religion* [nd]; *Consciousness: a Study in Bourgeois Psychology* [nd]; *Liberty: a Study in Bourgeois Illusion* [nd]; *Men and Nature: a Study in Bourgeois History* [nd]; *Reality: a Study in Bourgeois Philosophy* [nd]

Edited
Uncanny Stories (s) 1936

Collections
Scenes and Actions: Unpublished Manuscripts [ed. J. Duparc, D. Margolis] 1987

Charles (Stanley) CAUSLEY 1917–

Causley was born in Launceston in Cornwall where he has lived for his entire life. His Canadian-born father, who had returned to his native Devon and served in the Royal Army Medical Corps in France, died in 1924 having been invalided out of the war. Causley was educated at Horwell Grammar School and Launceston College, and left school in 1933 to work in a number of office jobs while writing stories, poems and plays, one of which was broadcast in 1939. 'I found the work stiflingly boring, and felt I was doomed for life,' he has written of this period. He served in the navy from 1940 to 1946, rising to the rank of petty officer, but has described himself as 'a poor sailor, often seasick, and almost as apprehensive of the sea ... as of the enemy mine, bomb and torpedo'. After a year at the Peterborough Training College he returned to Launceston in 1947 where he lived with his mother and was a schoolteacher until 1976. He has been literary editor of various radio programmes in the West Country, and a Visiting Fellow in Poetry at Exeter University from 1973 to 1974. He received a Cholmondeley Award in 1971, and was awarded a CBE in 1986.

Causley attributes the impetus behind his poetry to his experience of the war, particularly to the memory of a friend who joined the navy on the same day and was killed in an Archangel convoy. His early collections, such as *Farewell, Aggie Weston* and *Survivor's Leave*, demonstrate the use of traditional ballad forms for which he is well-known, but also display the influence of **W.H. Auden**, **A.E. Housman** and particularly **Siegfried Sassoon** who, he has said, showed him that poetry could be written in 'a simple, terse, and entirely personal manner of one's own'. *Union Street* consists of the best of these two volumes plus nineteen new poems.

Much of his later verse springs from his experience as a teacher, and poems such as the ballad 'Timothy Winters' from the collection *Johnny Alleluia*, show a strong concern with social issues. Causley has never married, and in 'Trusham', from *Underneath the Water*, he describes returning to the village where his grandfather was born and being reprimanded by an old family acquaintance for his failure to continue the family line. He has published numerous volumes of verse for children, but intends that a collection such as *Figure of 8* should also be read by adults.

Poetry

Farewell, Aggie Weston 1951; *Hands to Dance* 1951 [as *Hands to Dance and Skylark* 1979]; *Survivor's Leave* 1953; *Union Street* 1957; *Johnny Alleluia* 1961; *Penguin Modern Poets No. 3* [with G. Barker, M. Bell] 1962; *Underneath the Water* 1968; *Figure of 8* 1969; *Collected Poems 1951–1975* 1975; *Here We Go Round the Round House* 1976; *Early in the Morning* 1986; *Twenty-One Poems* 1986; *A Field of Vision* 1988; *Secret Destinations: Selected Poems 1977–1988* 1989

For Children

Figgie Hobbin 1971; *The Tail of the Trinosaur* 1972; *The Hill of the Fairy Calf* 1976; *The Animals' Carol* 1978; *The Gift of a Lamb* 1978; *The Last King of Cornwall* 1978; *Three Heads Made of Gold* 1978; *The Ballad of Aucassin and Nicolette* [from the 13th-century story] 1981; *Secret Destinations* 1984; *Quack! Said the Billy-Goat* 1986; *Jack the Treacle Eater and Other Poems* 1987; *The Young Man of Cury* 1991; *Bring in the Holly* [with L. Kopper] 1992

Translations

Hamdija Demirovic *Twenty-Five Poems* 1980; *Schondilie* 1982; *King's Children and Other German Ballads in English Versions* 1986

Edited

Peninsula 1957; *Dawn and Dusk* 1962; *Rising Early* 1964; *Modern Folk Ballads* 1966; *The Puffin Book of Magic Verse* 1974; *The Puffin Book of Salt-Sea Verse* 1978; *Batsford Book of Stories in Verse for Children* (s) 1979; *The Sun, Dancing* 1982

Plays

Jonah (l) [with W. Mathias] 1990

(John) David CAUTE 1936–

Caute was born in Alexandria where his father was serving as an officer in the army. He was educated at the Edinburgh Academy and Wellington College. After doing national service in Africa he went up to Wadham College, Oxford, in 1956 and read modern history, graduating in 1959. After a week at St Antony's College he was made a Fellow of All Souls and was awarded his D.Phil. in 1962, his thesis subsequently being published as *Communism and the French Intellectuals 1914–1960*. In 1965, along with Tariq Ali and others, he organised the 'Oxford Teach-In', a protest against the Vietnam War modelled on similar protests at Berkeley and the London School of Economics, and in the same year resigned from All Souls. He was a visiting professor at New York University and Columbia, and a reader in social and political theory at Brunel. Since 1970 he has dedicated most of his time to writing, briefly holding posts at the University of California in 1974 and the University of Bristol in 1985, and as literary editor of the *New Statesman* from 1979 to 1980. He has been married twice, has four children and lives in London.

Caute is best known as a journalist and historian of left-wing politics, and his major works include *The Great Fear*, a study of the McCarthy era, *Sixty-Eight: The Year of the Barricades*, and the first biography of the once blacklisted film director Joseph Losey. His first novel, *At Fever Pitch*, which won the John Llewelyn Rhys Memorial Prize in 1960, is set in a thinly disguised Ghana in the run-up to independence in 1957 and draws upon his experience of national service in Africa. His subsequent novels include *News from Nowhere*, which concerns an LSE academic in the 1960s who becomes embroiled in the turmoil of the Rhodesian civil war, and two thrillers written under the *nom de plume* of John Salisbury. His novel *Dr Orwell and Mr Blair* retells the story of **George Orwell**'s *Animal Farm* (1945) from the point-of-view of Mr Jones's son Alex, and includes a witty portrait of the 'very tall stranger in a baggy tweed jacket' who intrudes into the life of Manor Farm. Caute's plays include *The Fourth World*, produced at the Royal Court in 1973, and he has written a television documentary, *Brecht and Company*, which was first broadcast in 1979.

Fiction

At Fever Pitch 1959; *Comrade Jacob* 1961; *The Decline of the West* 1966; *The Occupation* 1971; *The Baby Sitters* (as 'John Salisbury') 1978; *Moscow Gold* (as 'John Salisbury') 1980; *The K-Factor* 1983; *News from Nowhere* 1986; *Veronica* 1989; *The Women's Hour* 1991; *Dr Orwell and Mr Blair* 1994

Non-Fiction

Communism and the French Intellectuals 1914–1960 1964; *The Left in Europe Since 1789* 1966; *Fanon* 1970; *The Illusion* 1971; *The Fellow-Travellers* 1973; *Collisions: Essays and Reviews* 1974; *Cuba, Yes?* 1974; *The Great Fear: the Anti-Communist Purge under Truman and Eisenhower* 1978; *Under the Skin: the Death of White Rhodesia* 1983; *The Espionage of the Saints: Two Essays on Silence and the State* 1986; *Sixty-Eight: The Year of the Barricades; a Journey through 1968* 1988; *Joseph Losey: a Revenge on Life* 1994

Plays

Songs for an Autumn Rifle 1961 [np]; *The Demonstration* 1969 [pub. 1970]; *Fallout* (r) 1972; *The Fourth World* 1973 [np]; *The Zimbabwe Tapes* (r) 1983; *Henry and the Dogs* (r) 1986; *Sanctions* (r) 1988

Edited

Karl Marx *Essential Writings* 1967

Raymond (Thornton) CHANDLER 1888–1959

Chandler was born in Chicago, where his father was a railway engineer. His parents were divorced when he was seven, and with his mother he returned to London, where her Irish-born Protestant family lived. He was educated at Dulwich College (where the headmaster was the novelist A.H. Gilkes, who had also taught **P.G. Wodehouse**), and spent a year in Paris and Germany preparing for the civil service entrance examination.

After six months working as a clerk at the Admiralty, Chandler angered his family by resigning to pursue his chosen career as a writer. He published poems in magazines such as *Chambers' Journal*, of which his biographer Frank MacShane is moved to say: 'The less said of Chandler's early poems the better.' Of his brief stint as a reporter on the *Daily Express*, Chandler himself later commented: 'I was a complete flop, the worst man they ever had.' After a spell with the *Westminster Gazette* he borrowed £500 from an uncle and travelled to the USA where he worked in various mundane jobs before enlisting with the Canadian army in 1917 and seeing action in France. After the war he enjoyed some success working for the South Basin Oil Company in Los Angeles, and in 1924 married Cissy Pascal, who was eighteen years older than him. Despite his periodic bouts of womanising, they remained married until her death in 1954.

The decisive break in Chandler's life occurred in 1932 when, after a period of depression and heavy drinking, he was fired by the oil company. He began writing detective stories for pulp magazines such as *Black Mask* and *Dime Detective*, publishing his first story, 'Blackmailers Don't Shoot', in 1933. Taking **Dashiell Hammett** as his model and incorporating elements from various of his stories, he wrote *The Big Sleep*, the first of his novels to feature the private detective Philip Marlowe. Chandler was dismissive of a good deal of detective fiction – particularly that of **Agatha Christie** – but equally contemptuous of the supposed distinction between highbrow literature and lowbrow entertainment. He commented that: 'The mystery writer is looked down upon as sub-literary mainly because he is a mystery writer rather than, for instance, a writer of social-significance twaddle.' The subsequent Marlowe novels, particularly *Farewell, My Lovely* and *The Lady in the Lake*, are notable for their sophisticated characterisation and eloquent dialogue, and paint a powerful portrait of the seedy, corrupt underbelly of Los Angeles society, though Chandler was the first to deny their realism, stating that a detective like Marlowe 'does not and could not exist'.

Many of the Marlowe novels were filmed, none more successfully than the 1946 Howard Hawks version of *The Big Sleep*, starring Humphrey Bogart and scripted by, among others, **William Faulkner**. In the 1940s Chandler worked as a Hollywood scriptwriter, notably on

Billy Wilder's film of **James M. Cain**'s *Double Indemnity* and Alfred Hitchcock's film of **Patricia Highsmith**'s *Strangers on a Train*, the plot of which he altered considerably to lend it psychological credibility. His letters to his London publisher Hamish Hamilton, some of which can be found in *Raymond Chandler Speaking*, provide a witty picture of Hollywood. In his last years Chandler was a somewhat reclusive figure, though after the death of his wife he spent some time in London where his social circle included **Stephen Spender** and **Ian Fleming**.

Fiction
The Big Sleep 1939; *Farewell, My Lovely* 1940; *The High Window* 1942; *The Lady in the Lake* 1943; *Five Murderers* (s) 1944; *Five Sinister Characters* (s) 1945; *Finger Man and Other Stories* (s) 1946; *The Little Sister* 1949 [as *Marlowe* 1969]; *The Long Good-Bye* 1953; *Playback* 1958; *Killer in the Rain* (s) [ed. P. Durham] 1964; *The Smell of Fear* (s) 1965

Plays
And Now Tomorrow (f) [with F. Partos] 1944; *Double Indemnity* (f) [from J.M. Cain's book; with B. Wilder] 1944 [pub. in *Best Film Plays 1945* ed. J. Gassner, D. Nichols 1946]; *The Unseen* (f) [with K. Englund, H. Wilde] 1945; *The Blue Dahlia* (f) 1946 [pub. 1976]; *Strangers on a Train* (f) [from P. Highsmith's book; with W. Cook, C. Ormonde] 1951; *Raymond Chandler's Unknown Thriller: the Screenplay of Playback* [pub. 1985]

Non-Fiction
Raymond Chandler Speaking [ed. D. Gardiner, K.S. Walker] 1962; *The Notebooks of Raymond Chandler, and English Summer: a Gothic Romance* [ed. F. MacShane] 1976; *Raymond Chandler and James M. Fox: Letters* 1979; *Selected Letters of Raymond Chandler* [ed. F. MacShane] 1981

Biography
The Life of Raymond Chandler by Frank MacShane 1976

Sid(ney) CHAPLIN 1916–1986

Chaplin was born in a colliery row at Shildon, County Durham. His early life was dominated by Shildon Lodge Colliery, where his grandfathers had been an enginewright and a winding engineer, and his father had worked as an electrician. As a child, he became a voracious reader, inheriting from his grandfather 'a passion for good bad books': Rider Haggard, **A.E.W. Mason**, **John Buchan**. Although his education was peripatetic – he attended six elementary schools in various mining villages – he was lucky in always finding a teacher who encouraged him to write. He left school at fourteen, worked in a bakery for a year, went to the pit as a screener, then became apprenticed to a blacksmith. He also became involved in Methodism, politics and Workers' Educational Association classes, which led to his winning a scholarship to Fircroft, where he read English literature and started writing seriously between shifts at the Dean and Chapter Colliery.

He remained at the colliery for eighteen years, apart from a period of eighteen months spent writing full-time on a £300 Atlantic Award in Literature, which he had won for his first volume of short stories, *The Leaping Lad* (1946). He was branch secretary of the Miner's Federation of Great Britain between 1943 and 1945, and in 1950 became public relations officer for the northern division of the Coal Board, a job he held until his retirement in 1972.

His early stories were published in such journals as *Penguin New Writing*, *English Story* and *Modern Reading*, and drew upon his early experiences and the tradition he had inherited. A prologue to *The Leaping Lad* gives a brief account of how the stories came to be written and dedicates them to family and friends, concluding: 'They belong to all those who believe in the human heart, and to these I return my stories with all my love and thanks.' This essential humanity also characterises his first two novels, of which *The Thin Seam*, along with several stories, was used by **Alan Plater** as a basis for his play *Close the Coalhouse Door* (1968). *The Thin Seam* was followed by a silence of ten years, which gave rise in the early 1960s to his most creative period, in which he published four novels and a stream of short stories. *The Smell of Sunday Dinner* and *A Tree with Rosy Apples* collect the weekly column he wrote for the *Guardian*.

Chaplin was often urged by such friends as **Stan Barstow** to write a 'big novel about a people caught in the industrial decline of several decades', but he was a modest man, who was content to leave a less ambitious but faithful record of the mining industry in his fiction, particularly in his short stories, which are perhaps his finest achievement. He was awarded an OBE for services to the arts in the north-east in 1977. He had married Rene Rutherford on New Year's Day 1941, and after his death at Grasmere, she and their son, Michael, celebrated Chaplin's life with a miscellany of his work, *In Blackberry Time*.

Fiction
The Leaping Lad (s) 1946; *The Fate Cries Out* 1949; *The Thin Seam* 1950; *The Big Room* 1960; *The Day of the Sardine* 1961; *The Watchers and*

the Watched 1962; *Sam in the Morning* 1965; *The Mines of Alabaster* 1971; *On Christmas Day in the Morning* (s) 1978; *The Bachelor Uncle and Other Stories* (s) 1980

Non-Fiction

The Lakes to Tyneside 1951; *The Smell of Sunday Dinner* 1971; *A Tree with Rosy Apples* 1972

Collections

In Blackberry Time [ed. M. and R. Chaplin] 1987

(Charles) Bruce CHATWIN 1940–1989

Chatwin was born in Sheffield; his father, later a solicitor, was then a naval officer and the writer's first decade was spent in a variety of sea ports or housed with eccentric relatives. From 1951 to 1958 he attended Marlborough College, finding its classical bias uncongenial but developing a fascination for French literature and painting. On leaving school, he joined Sotheby's, initially as a porter. His rise was rapid, and he was soon, in his own words, 'an instant expert, flying here and there to pronounce, with unbelievable arrogance, on the value or authority of works of art'. In 1965 – the year in which he married – he was appointed director of the Impressionist Art Department. Increasingly, however, his interests were reaching beyond the 'civilised' West and, at the age of twenty-six, he abandoned his career to read archaeology at Edinburgh University. Field study permitted him to pursue his fascination with nomadic tribes; he never, though, completed his degree. Instead, in 1968, he began the explorations which were to occupy most of the rest of his life.

His first extended period of travelling – lasting until 1972 – took him from Mauretania to Afghanistan. For the next three years he was employed as a peripatetic correspondent for the *Sunday Times*. Among the subjects on whom he filed copy were Indira Gandhi and André Malraux. (Examples of his journalism are collected in *What Am I Doing Here*.) When he severed his association with the paper, it was to trek through the deserts of Argentina and Chile. That expedition, undertaken alone, was to provide the material for his first book, *In Patagonia* (1977). Notable for its anecdotal concision and eclectic scholarship, it won the Hawthornden Prize, the Somerset Maugham Award and the E.M. Forster Award.

Chatwin's first fiction, *The Viceroy of Ouidah*, was a historically based account of the life of a monstrous Brazilian slave trader; its style owes less to English than to Latin American precedents – and more to Borges's terseness than to Marquez's exuberance. His second, *On the Black Hill*, earned him comparisons with **D.H. Lawrence** and **Thomas Hardy** and won the Whitbread Award. Strikingly unexotic, set in a few square miles of farming Wales, it acknowledges the tenacious hold exerted on people by place, while examining the sorts of 'backward' culture that persist in the interstices of modern society. Similar themes are extrapolated in *The Songlines*, where Chatwin links elements of autobiography into a narrative of Australian Aboriginal myth and despair. His final novel, *Utz*, returns him to a European scene, to the ambiguities of porcelain collecting. Its eponymous hero is an internal exile, a baron in communist Prague, estranged from his country yet bound to it by his hoard of Meissen china. Terrified at the prospect of breakage or owning a fake, he perceives how oppressed he is by his pieces and destroys them. Obliquely, the story elucidates Chatwin's decision to quit Sotheby's: the acquisition and consumption of the aesthetic degenerates to disabling gluttony. Utz's ultimate tendency is to renounce: so his creator, in his wanderings, continually sought astringency, occasionally danger – an alternative to the surfeits of the gravy train upon which he might readily have ridden.

Outside his immediate circle, Chatwin's homosexuality was a closely guarded secret, so that when he contracted the Aids virus, it was put about that he was dying from a rare and wasting bone-marrow disease supposedly picked up in Tibet. This was the official version of events presented in obituaries and tributes when he died in Nice, but the true cause of his death emerged shortly afterwards.

Fiction

The Viceroy of Ouidah 1980; *On the Black Hill* 1982; *The Songlines* 1987; *Utz* 1988

Non-Fiction

In Patagonia 1977; *Lady: Lisa Lyon* [with R. Mapplethorpe] 1983; *What Am I Doing Here* 1989; *Photographs and Notebooks* [ed. D. King, F. Wyndham] 1993

John CHEEVER 1912–1982

Cheever was born in Quincy, Massachusetts, where his father was a prosperous shoe manufacturer who lost his fortune in the stock-market collapse of 1929. Cheever's formal education ended when he was seventeen and was expelled from the Thayer Academy. He moved to New York and made immediate use of the incident for his first published story, 'Expelled', which Malcolm Cowley bought for *New Republic* in 1930. In her biography of her father, *Home Before Dark*,

Susan Cheever warns that 'there are many different versions of these events' and suggests that Cheever's natural storytelling talent may sometimes have obscured the literal truth. For the next few years Cheever supported himself by writing synopses for MGM and by selling stories to various magazines including *Collier's* and *Atlantic Monthly*, and in 1935 he began a lifelong association with the *New Yorker*. More perhaps than any other writer, he created the genre which came to be known as 'the *New Yorker* story'. In 1934 he spent some time at the recently established writers' colony at Yaddo in Saratoga Springs, to which he made periodic returns throughout his life.

Cheever was married in 1941 and served for four years as an infantry gunner and member of the Signal Corps with the US Army. His first collection of stories, *The Way Some People Live*, was published during this period, and established his reputation as an ironic chronicler of suburban mores, 'the Chekhov of the suburbs' as John Leonard memorably described him. His long- gestated first novel, *The Wapshot Chronicle*, which won a National Book Award in 1958, is a collage of stories concerning different characters, and draws on aspects of his own family's genteel decline. When Leander Wapshot's wife converts his storm-damaged boat into a 'Floating Gift Shoppe' there is a distinct echo of the gift shop which Cheever's mother ran in the 1920s in an effort to restore the family's dwindling resources. A sequel, *The Wapshot Scandal*, describes some of the characters a few years later struggling in a world now dominated by computers and a new missile base. Cheever's mundane suburbia is often disrupted by outbursts of irrationality and violence, nowhere more so than in *Bullet Park*. Critical hostility to this seems to have contributed to a period of depression during which he was treated for his long-standing alcoholism. His fourth novel, *Falconer*, although not explicitly autobiographical, reflects something of this experience, and in describing the hero Farragut's discovery of religion while in prison draws on Cheever's time teaching prisoners in Sing-Sing prison. The posthumous publication of his letters and journals revealed his guilt-ridden promiscuous bisexuality, though his work always retained an old-fashioned decorum regarding sexual matters. In a 1968 letter, referring to friends such as **John Updike** and **Philip Roth**, he commented: 'While all my friends are describing orgasms I still dwell on the beauty of the evening star.'

In the 1960s Cheever worked briefly as a Hollywood scriptwriter on a film version of **D.H. Lawrence**'s *The Lost Girl*, and his story 'The Swimmer' was made into a rather curious 1968 film starring Burt Lancaster. He taught English at Barnard College, New York, from 1956 to 1957, and was a visiting professor at Boston University from 1974 to 1975. His numerous awards included the Howells Medal for *The Wapshot Scandal* in 1965, and the Pulitzer Prize for *The Stories of John Cheever* in 1979. He had three children, two of whom, Susan and Benjamin, have become novelists. Benjamin Cheever's *The Plagiarist* (1992) contains a portrait of an ageing, alcoholic grand old man of American letters.

Fiction

The Way Some People Live (s) 1943; *The Enormous Radio and Other Stories* (s) 1953; *The Day the Pig fell into the Well* (s) 1954; *Stories* (s) 1956; *The Wapshot Chronicle* 1957; *The Housebreaker of Shady Hill and Other Stories* (s) 1958; *Some People, Places and Things That Will Not Appear in My Next Novel* (s) 1961; *The Brigadier and the Golf Widow* (s) 1964; *The Wapshot Scandal* 1964; *Homage to Shakespeare* (s) 1965; *Bullet Park* 1969; *The World of Apples* (s) 1973; *Falconer* 1977; *The Stories of John Cheever* (s) 1978; *The Leaves, the Lionfish and the Bear* 1980; *Oh What A Paradise It Seems* 1982; *The Uncollected Stories of John Cheever* (s) 1990

Non-Fiction

Homage to Shakespeare 1965; *The Letters of John Cheever* [ed. B. Cheever] 1989; *John Cheever: the Journals* 1991

Biography

Home Before Dark by Susan Cheever 1984; *John Cheever* by Scott Donaldson 1988

G(ilbert) K(eith) CHESTERTON 1874–1936

Born in Campden Hill, London, the elder son of a retired estate agent, Chesterton went to St Paul's School, London, and then studied desultorily: first art at the Slade, then English literature at London University, but without taking a degree. Simultaneously he began to review books in the *Bookman*, Hodder and Stoughton's monthly magazine, and by the age of twenty-one he was well-launched as a reviewer and journalist. He was to describe journalism as 'the easiest of all professions', and he continued in it until his death, writing, for instance, a regular weekly essay for the *Illustrated London News* from 1905 to 1930, missing only two issues in those twenty-five years. He published his essays – a now largely defunct form which he made peculiarly his own – in many volumes with titles such as

Tremendous Trifles. He also produced prolifically poetry, fiction, general works and criticism. Visits to Ireland, America, Rome and Palestine each infallibly brought forth a book, and in later years he delivered a weekly *causerie* on BBC radio.

Although Chesterton was not officially received into the Roman Catholic church until 1922, he had taken a strong Catholic line long before that, and became closely associated with fellow Catholic writer and journalist, **Hilaire Belloc**, so much so that **George Bernard Shaw** dubbed them 'the Chesterbelloc'. Both men combined vast output with equally vast bulk, but, in contrast to Belloc's square head and aggressive jaw, Chesterton cultivated a more lovable bespectacled unkemptness, and became famous for his larger-than-life eccentricities ('Am in Market Harborough. Where ought I to be?' he once cabled his wife when travelling to a speaking engagement by train). He rejected socialism, evolving an alternative theory, 'Distributism', and founded the journal *G.K.'s Weekly* as its official organ. He married in 1901 and lived for many years in Beaconsfield; he died there at the age of sixty-two.

Chesterton's work has already been winnowed by time, but he still collects influential admirers: **W.H. Auden** and **Kingsley Amis** have both edited selections. Among other works, his surrealistic novels, of which *The Man Who Was Thursday* is outstanding, his short stories about the Catholic priest and detective Father Brown, perceptive critical books on such authors as Browning and Dickens, and some fine individual poems continue to be read and admired.

Non-Fiction

The Defendant 1901; *Thomas Carlyle* [with J.E.H. Williams] 1902; *Twelve Types* 1902; *Charles Dickens* [with F.G. Kitton] 1903; *Leo Tolstoy* [with G.H. Perris] 1903; *Robert Browning* 1903; *Thackeray* [with L. Melville] 1903; *Tennyson* [with R. Garnett] 1903; *G.F. Watts* 1904; *Heretics* 1905; *Charles Dickens* 1906; *All Things Considered* 1908; *Varied Types* 1908; *George Bernard Shaw* 1909; *Tremendous Trifles* 1909; *Alarms and Discussions* 1910; *What's Wrong with the World* 1910; *William Blake* 1910; *Appreciations and Criticisms of the Works of Charles Dickens* 1911; *A Miscellany of Men* 1912; *The Victorian Age in Literature* 1913; *The Barbarism of Berlin* 1914; *The Crimes of England* 1915; *Utopia of Usurers* 1917; *Irish Impressions* 1919; *The Superstition of Divorce* 1920; *The Uses of Diversity* 1920; *The New Jerusalem* 1920; *Eugenics and Other Evils* 1922; *What I Saw in America* 1922; *Fancies versus Fads* 1923; *St Francis of Assisi* 1923; *The End of the Roman Road* 1924; *William Cobbett* 1925; *The Outline of Society* 1926; *The Catholic Church and Conversion* 1927; *Robert Louis Stevenson* 1927; *Generally Speaking* 1928; *G.K.C. as M.C. – a Collection of 37 Introductions by G.K. Chesterton* [ed. J.P. de Fonseka] 1929; *The Thing* 1929; *Come to Think of It* [ed. J.P. de Fonseka] 1930; *The Resurrection of Rome* 1930; *All Is Grist* 1931; *Chaucer* 1932; *Christendom in Dublin* 1932; *Sidelights on New London and Newer York* 1932; *All I Survey* 1933; *St Thomas Aquinas* 1933; *Avowals and Denials* 1934; *The Well and the Shallows* 1935; *As I Was Saying* 1936; *The Paradoxes of Mr Pond* 1937; *Essays* [ed. J. Guest] 1939; *The End of the Armistice* [ed. F.J. Sheed] 1940; *Selected Essays of G.K. Chesterton* [ed. D. Collins] 1949; *The Common Man* 1950; *Chesterton's Essays* [ed. K.E. Whitehorn] 1953; *A Handful of Authors: Essays on Books and Authors* [ed. D. Collins] 1953; *The Glass Walking-Stick and other Essays from the Illustrated London News 1905–1936* [ed. D. Collins] 1955; *Lunacy and Letters: Essays from the Daily News 1901–1911* [ed. D. Collins] 1958; *The Catholic Church and Conversion* 1960; *The Man Who Was Orthodox: a Selection from the Uncollected Writings of G.K. Chesterton* [ed. A.L. Maycock] 1963; *The Spice of Life and Other Essays* [ed. D. Collins] 1964; *G.K. Chesterton: a Selection from His Non-Fictional Prose* [ed. W.H. Auden] 1970; *Chesterton on Shakespeare* [ed. D. Collins] 1972

Fiction

The Napoleon of Notting Hill 1904; *The Club of Queer Trades* (s) 1905; *The Man Who Was Thursday* 1908; *Orthodoxy* 1908; *The Ball and the Cross* 1910; *The Innocence of Father Brown* (s) 1911; *Manalive* 1912; *The Flying Inn* 1914; *The Wisdom of Father Brown* (s) 1914; *The Man Who Knew Too Much and Other Stories* (s) 1922; *The Everlasting Man* 1925; *Tales of the Long Bow* (s) 1925; *The Incredulity of Father Brown* (s) 1926; *The Secret of Father Brown* (s) 1926; *The Return of Don Quixote* 1927; *The Poet and the Lunatics* (s) 1929; *Four Faultless Felons* (s) 1930; *The Scandal of Father Brown* (s) 1935

Poetry

Greybeards at Play 1900; *The Wild Knight and Other Poems* 1900; *The Ballad of the White Horse* 1911; *Poems* 1915; *Wine, Water and Song* 1915; *The Ballad of St Barbara* 1922; *The Queen of Seven Swords* 1926; *Collected Poems of G.K. Chesterton* 1927

Plays

Magic 1913; *The Judgement of Dr Johnson* 1927; *The Surprise* 1952

Edited

The Readers' Classics [with H. Jackson, R.B. Johnson] 1922; *Royal Society of Literature of the United Kingdom Transactions* 1929 onwards

Collections

A Chesterton Calendar 1911; *The Wit and Wisdom of G.K. Chesterton* 1911; *Thoughts from G.K. Chesterton* [ed. E.E. Morton] 1913; *The G.K. Chesterton Calendar* [ed. H.C. Palmer] 1916; *A Shilling for My Thoughts* [ed. E.V. Lucas] 1916; *A Gleaming Cohort* [ed. E.V. Lucas] 1926; *A Chesterton Catholic Anthology* [ed. P. Braybrooke] 1928; *Essays and Poems* [ed. W. Sheed] 1958; *The Spirit of Christmas* [ed. M. Smith] 1984

Biography

G.K. Chesterton by Dudley Barker 1973

(Robert) Erskine CHILDERS 1870–1922

Although Childers was born in London, he lived until he married in Glendalough, County Wicklow, Ireland. Named after his father, who was a famous Pali scholar, he was educated at Haileybury. He went to Trinity College, Cambridge, where he took the Law Tripos and a BA in 1893. He developed a taste for sailing when, shortly after leaving university, he spent long holidays with friends navigating a small boat off the Danish and Baltic coasts. This was to provide practical material for his novel *The Riddle of the Sands*, but writing was far from Childers' mind at the time, for he became a clerk to the House of Commons in 1895. He interrupted this in 1899 to serve in the Boer War, in the Civil Imperial Volunteers of Honourable Artillery Company. His account of his experience was published in 1900 as *In the Ranks of the C.I.V.*, and three years later followed by *The H.A.C. in South Africa*, in collaboration with others. He also wrote the fifth volume of *The Times History of the War in South Africa*.

It is for *The Riddle of the Sands* (1903), however, that Childers is remembered as a writer. Strangely prophetic in its account of the discovery of a planned invasion of England by the German Navy, the novel was reprinted in 1914 (and – emphatically so – by Faber and Faber in 1940). It was extremely influential in drawing attention to the rapid programme of German naval re-armament. In spite of later events, Childers' career in the First World War never wavered from strong British patriotism. He did reconnaissance work on the sea-plane carrier HMS *Engadine*, and was involved in the Cuxhaven Raid in 1914. Promoted to major, he spent the remainder of the war as an intelligence and training officer for the Royal Naval Air Service, for which he received the Distinguished Service Cross.

When he was demobbed in 1919, Childers, who was a Protestant, promoted the cause of Irish independence. He had resigned as House of Commons clerk in 1910, to write *The Framework of Home Rule*, and had married an American, Mary Osgood, who supported his views. Shortly after Asquith passed the Home Rule for Ireland Bill in 1914, Childers had brought arms in his yacht from Europe to Ireland. His efforts on behalf of Irish independence led to his being elected to the self-constituted Irish Parliament, the Dail Eireann, in 1921. As propaganda minister, he demanded republican status for all Ireland. When the 1922 civil war broke out, Childers took the Republican side against the Irish Free State Government. After he took part in military action in the South, his house in Glendalough was surrounded by Free State soldiers, and he was arrested. He refused to recognise the authority of the court-martial in which he was tried for treason the following week. Exceptionally courteous, and considered even by his enemies to have great charm, he shook hands with each member of the firing squad who executed him. His son, Erskine Hamilton Childers, was President of Ireland from 1973 to 1974.

Fiction

The Riddle of the Sands 1903

Non-Fiction

In the Ranks of the C.I.V. 1900; *The H.A.C. in South Africa* [with A.F.B. Williams] 1903; *War and the Arme Blanche* 1910; *The Framework of Home Rule* 1911; *German Influence on British Cavalry* 1911; *The Form and Purpose of Home Rule* 1912; *Military Rule in Ireland* 1920

Edited

The Times History of the War in South Africa vol. 5 1907; *Who Burnt Cork City?* [with A. O'Rahilly] 1921

Agatha (Mary Clarissa) CHRISTIE 1890–1976

Born Agatha Miller in Torquay, Christie had a conventional late Victorian upbringing, educated at home by her mother until the age of sixteen and then sent to Paris to study music. During the First World War, she served as a VAD (volunteer nurse) in Torquay, and this medical background was to prove useful when she started to write her murder mysteries. Her first marriage, to Col. Archibald Christie in 1914, was not a success, and, in considerable personal distress, Christie disappeared for ten days in 1926. This event initiated a huge manhunt and much contemporary speculation; she herself never offered an explanation of the event and, for once, the mystery remains unsolved. (The solution put forward in Michael Apted's 1978 film *Agatha* is ingenious but entirely fictional and fairly implausible.)

Christie had one daughter from this first marriage, then, following her divorce in 1928, she met the archaeologist Max Mallowan, whom she married in 1930. Her experiences with him at various excavations in the Middle East provide backgrounds to such novels as *Death on the Nile*. Her first detective story, however, was of the 'country house' sort: *The Mysterious Affair at Styles* was published in 1920 and introduced the first of her two most popular sleuths, Hercule Poirot. It was followed by over sixty further novels in this genre, as well as a number of plays, several volumes of short stories and six romantic novels published under the pseudonym Mary Westmacott.

Although Christie dubbed herself 'the Duchess of Death', her work exhibits realistic rather than Gothic features: plot construction and moral values are more important to her than sensational effects. Her two principal detectives are Hercule Poirot and Miss Marple. Poirot, retired from the Belgian police force, is a fastidious and vain man, proud of his elaborate moustache. His command of English idiom is shaky, and he claims to solve mysteries by the use of 'the little grey cells'. By way of contrast, Jane Marple is an inspired – though equally successful – amateur. An innocent-seeming, elderly English spinster, she lives in the emblematic village of St Mary Mead, and draws upon her close observation of local people to provide parallels of motive and behaviour in criminal cases. Both detectives are practitioners of common sense and unflagging upholders of tradition.

The novels have been made into numerous films and television plays, some more successful than others. Poirot has been impersonated most accurately by David Suchet in a long-running television series, but has also found notable interpreters in Albert Finney and – an inspired and rewarding bit of miscasting – Peter Ustinov. In the cinema Miss Marple was memorably played by Margaret Rutherford, but it was on television that Joan Hickson provided a definitive performance.

With one extraordinary exception, Christie's own forays into drama are now all forgotten. *The Mousetrap* (the title of which is derived from *Hamlet*) opened in the West End in 1952, and was still playing there forty years later. It is for her novels, however, that Christie will be remembered, and for which she was dubbed, with some justification, 'the Queen of Crime'. While many of her minor characters are stock figures and her prose is undistinguished, her plotting is always ingenious and it is for this rather than any strictly literary merits that she remains the most popular crime writer of the century.

Fiction

As Agatha Christie: *The Mysterious Affair at Styles* 1920; *The Secret Adversary* 1922; *The Murder on the Links* 1923; *The Man in the Brown Suit* 1924; *Poirot Investigates* (s) 1924; *The Secret of Chimneys* 1925; *The Murder of Roger Ackroyd* 1926; *The Big Four* 1927; *The Mystery of the Blue Train* 1928; *Partners in Crime* (s) 1929; *The Seven Dials Mystery* 1929; *The Under Dog* (s) 1929; *The Murder at the Vicarage* 1930; *The Mysterious Mr Quin* (s) 1930; *The Sittaford Mystery* 1931; *Peril at End House* 1932; *The Thirteen Problems* (s) 1932; *The Hound of Death and Other Stories* (s) 1933; *Lord Edgware Dies* 1933; *The Listerdale Mystery and Other Stories* (s) 1934; *Murder in Three Acts* 1934; *Murder on the Orient Express* 1934; *Parker Pyne Investigates* (s) 1934; *Why Didn't They Ask Evans* 1934; *Death in the Clouds* 1935; *The A.B.C. Murders* 1936; *Cards on the Table* 1936; *Murder in Mesopotamia* 1936; *Dumb Witness* 1937; *Murder in the Mews and Three Other Poirot Cases* (s) 1937; *Appointment with Death* 1938; *Death on the Nile* 1938; *Hercule Poirot's Christmas* 1938; *Murder Is Easy* 1939; *The Regatta Mystery and Other Stories* (s) 1939; *Ten Little Niggers* 1939 [as *And Then There Were None* 1991]; *One, Two, Buckle My Shoe* 1940; *Sad Cypress* 1940; *Evil Under the Son* 1941; *N or M?* 1941; *The Body in the Library* 1942; *Five Little Pigs* 1942; *The Moving Finger* 1943; *The Mystery of the Baghdad Chest* (s) 1943; *The Mystery of the Crime in Cabin 66* (s) 1943; *Poirot and the Regatta Mystery* (s) 1943; *Poirot on Holiday* (s) 1943; *Problem at Pollensa Bay and Christmas Adventure* (s) 1943; *Death Comes as the End* 1944; *Towards Zero* 1944; *Sparkling Cyanide* 1945; *The Hollow* 1946; *Poirot Knows the Murderer* (s) 1946; *Poirot Lends a Hand* (s) 1946; *The Labours of Hercules* (s) 1947; *Taken at the Flood* 1948; *Witness for the Prosecution and Other Stories* (s) 1948; *Crooked House* 1949; *The Mousetrap and Other Stories* (s) 1949; *A Murder Is Announced* 1950; *They Came to Baghdad* 1951; *Mrs McGinty's Dead* 1952; *They Do It with Mirrors* 1952; *After the Funeral* 1953; *A Pocket Full of Rye* 1953; *Destination Unknown* 1954; *Hickory Dickory Dock* 1955; *Dead Man's Folly* 1956; *4.50 from Paddington* 1957; *Ordeal by Innocence* 1958; *Cat Among the Pigeons* 1959; *Double Sin and Other Stories* (s) 1961; *The Pale Horse* 1961; *The Mirror Crack'd from Side to Side* 1962; *The Clocks* 1963; *A Caribbean Mystery* 1964; *At Bertram's Hotel* 1965;*Third Girl* 1966; *Endless Night* 1967; *By the Pricking of My Thumbs* 1968; *Hallowe'en Party* 1969; *Passenger to Frankfurt* 1970; *The Golden Ball and Other Stories* (s) 1971; *Nemesis* 1971; *Elephants Can Remember* 1972;

Postern of Fate 1973; *Poirot's Early Cases* (s) 1974; *Curtain: Hercule Poirot's Last Case* 1975; *Sleeping Murder* 1976; *Miss Marple's Final Cases and Two Other Stories* (s) 1979

As 'Mary Westmacott': *Giant's Bread* 1930; *Unfinished Portrait* 1934; *Absent in the Spring* 1944; *The Rose and the Yew Tree* 1948; *A Daughter's a Daughter* 1952; *The Burden* 1956

Plays

Alibi [with M. Morton] 1928 [pub. 1929]; *Black Coffee* 1930 [pub. 1934]; *Ten Little Niggers* 1943 [pub. 1944; as *Ten Little Indians* 1946]; *Appointment with Death* 1945 [pub. 1956]; *Death on the Nile* 1945 [pub. 1946]; *The Hollow* 1951 [pub. 1952]; *The Mousetrap* 1952 [pub. 1954]; *Witness for the Prosecution* 1953 [pub. 1954]; *Spider's Web* 1954 [pub. 1957]; *Towards Zero* 1956 [pub. 1958]; *The Unexpected Guest* 1958 [pub. 1978]; *Verdict* 1958; *Go Back for Murder* 1960; *Afternoon at the Seaside* 1962 [pub. 1963]; *The Patient* 1974 [pub. 1963]; *The Rats* 1978 [pub. 1963]; *Fiddlers Three* 1971; *Akhnaton* 1979 [pub. 1973]

Poetry

The Road of Dreams 1925; *Star Over Bethlehem* 1965; *Poems* 1973

Non-Fiction

Come Tell Me How You Live 1946; *An Autobiography* (a) 1977

Richard (Thomas) CHURCH 1893–1972

Born and brought up in London, the son of a post office worker, Church went to Dulwich Hamlet School and entered the external customs and excise branch of the civil service at the age of sixteen. Soon afterwards, he began to write, and his first book of poems, *The Flood of Life*, was published in 1917. He experienced conflict between the demands of his job and his desire to lead a literary life, and in 1933, at the age of forty, retired from the civil service, where he was a customs clerical officer. After that, he was for many years an editorial adviser to J.M. Dent and Sons, supplementing his income with much reviewing and general writing.

Until the 1930s he wrote mainly poetry; later, he turned to the novel: his novels are now largely forgotten, but in his Georgian-style verse he often achieved, as his obituary in *The Times* put it, a 'lyrical contemplation of telling quietness and candour'. From the 1940s onwards he wrote many volumes of essays, topographical writing and works for children, and also the three volumes of autobiography – *Over the Bridge, The Golden Sovereign* and *The Voyage Home* – that are generally regarded as his major work.

He was married three times and had four children. He lived for many years in Kent, but his tall, spare figure and studious, fastidious manner were familiar in London literary society. He had been a co-founder of the *Criterion* in 1921, was a president of PEN, vice-president of the Royal Society of Literature, and recipient of many honours.

Poetry

The Flood of Life and Other Poems 1917; *Hurricane and Other Poems* 1919; *Philip and Other Poems* 1923; *The Portrait of the Abbot* 1926; *The Dream and Other Poems* 1927; *Mood without Measure* 1927; *Theme with Variations* 1928; *The Glance Backward* 1930; *News from the Mountain* 1932; *Twelve Noon* 1936; *The Solitary Man and Other Poems* 1941; *The Lamp* 1946; *Poems for Spring* 1950; *The Inheritors: Poems 1948–1955* 1957; *Forty-Seven Poems* 1959; *The Little Kingdom* 1964; *The Burning Bush: Poems 1958–1966* 1967; *The French Lieutenant* 1971

Non-Fiction

Mary Shelley 1928; *Calling for a Spade* 1939; *Eight for Immortality* 1941; *Plato's Mistake* 1941; *British Authors: a Twentieth Century Gallery* 1943; *Green Tide* 1945; *Kent* 1948; *The Growth of the English Novel* 1951; *A Portrait of Canterbury* 1953; *Over the Bridge* (a) 1955; *The Royal Parks of London* 1956; *The Golden Sovereign* (a) 1957; *Small Moments* 1957; *A Country Window 1958; Calm October* 1961; *The Voyage Home* (a) 1964; *A Stroll Before Dark* 1965; *London, Flower of Cities All* 1966; *Speaking Aloud* 1968; *A Harvest of Mushrooms* 1970; *The Wonder of Words* 1970; *London in Colour* 1971; *Kent's Contribution* 1972

Fiction

Oliver's Daughter 1930; *High Summer* 1931; *The Prodigal Father* 1933; *The Apple of Concord* 1935; *The Porch* 1937; *The Stronghold* 1939; *The Room Within* 1940; *The Sampler* 1942; *The Dangerous Years* 1956; *The Crab-Apple Tree* 1959; *Prince Albert* 1963; *Little Miss Moffat* 1969

For Children

A Squirrel Called Rufus 1941; *The Cave* 1950; *Dog Toby* 1953; *Down River* 1957; *The Bells of Rye* 1960; *The White Doe* 1968

Edited

Poems and Prose by A.C. Swinburne 1940; *Poems by P.B. Shelley* 1949; *The Spoken Word: a Selection from Twenty-Five Years of the Listener* 1955

Plays

The Prodigal 1953

Caryl CHURCHILL 1938–

Churchill was born in London; her father was a political cartoonist and her mother a model, actress and secretary. In 1948 the family moved to Montreal, where she attended the Trafalgar School. Nine years later, she returned to England to read English language and literature at Lady Margaret Hall, Oxford. Her first play, *Downstairs*, was performed by Oriel College Dramatic Society in 1958. She graduated in 1960 and married in 1961.

Throughout the following decade – in which she had three sons – she mainly restricted herself to radio drama. However, in 1972, she was commissioned by Michael Codron to write for the Royal Court Theatre. The result was *Owners*, a fast-moving discussion of the connections between financial, sexual and physical power. If the play looks back to her radio work, where the accent tends to be on personal crisis, it also anticipates the broader range of what was to come. Her two subsequent plays for the Royal Court (she was resident dramatist there for a year from autumn 1974) were not, though, successes: both were criticised for verbosity. It was not until *Light Shining in Buckinghamshire* that she seemed to have settled how best to balance speech and action. Partly the result of improvisation workshops conducted with the Joint Stock Theatre Group, *Light Shining* marked the beginning of a long professional relationship with the director Max Stafford-Clark. It describes the attempts of puritan sectarians in the England of the 1640s to build the New Jerusalem on earth. It has no individual hero or heroine and no plot in the conventional sense. Rather, its subject is the progress of uptopianism from unbound hope to conflict, defeat and – finally – a renewal of hope. An emotive piece, it debates enduring and complex political questions: reform versus revolution; the role of women in revolutionary organisations; the status and purpose of the law; the radicalism of religious dissent.

Cloud Nine which again emerged from Joint Stock workshops, is a provocative sexual comedy which contrasts the supposedly strait-laced Victorian family (in fact rife with sexual variety) with the flexible (but not limitlessly so) structure of relationships in late twentieth-century Britain. Churchill's next major play, however, *Top Girls*, proposed that for most women economic independence remains unattained: the success of the few has not ameliorated the plight of the many. Obliquely it commented on the social impact of the UK's first woman prime minister, whereas *Serious Money* observed directly the dizziest peak of the 1980s' boom. Set among the City's brokers and dealers, composed in (very) mock-heroic couplets, it satirises the frenzied habits of its protagonists and, simultaneously, displays the webs of deceit spun by their greed. Her 1990 play, *Mad Forest*, is a sombre reflection on the fall of Ceaucescu. It was written after she and a group of students from the London Central School of Speech and Drama had visited and researched in Romania.

On her innovative approach, Churchill has said: 'Playwrights don't give answers, they ask questions. We need to find new questions, which may help us to answer the old ones or make them unimportant, and this means new subjects and new form.'

Plays

Downstairs 1958 [pub. 1990]; *Having a Wonderful Time* 1960 [pub. 1990]; *You've No Need to Be Frightened* **(r)** 1961 [pub. 1990]; *The Ants* **(r)** 1962 [pub. 1968]; *Easy Death* 1962 [pub. 1990]; *Identical Twins* **(r)** 1967 [pub. 1990]; *Lovesick* **(r)** 1967 [pub. 1990]; *Abortive* **(r)** 1971 [pub. 1990]; *Not ... Not ... Not ... Not ... Not ... Enough Oxygen* **(r)** 1971 [pub. 1990]; *Henry's Past* **(r)** 1972 [pub. 1990]; *The Judge's Wife* **(tv)** 1972; *Owners* 1972 [pub. 1973]; *Schreber's Nervous Illness* [from D.P. Schreber's book *Memoirs of My Nervous Illness*] **(r)**, **(st)** 1972 [pub. 1990]; *Perfect Happiness* **(r)** 1973, **(st)** 1975 [pub. 1990]; *Turkish Delight* **(tv)** 1974 [pub. 1990]; *Moving Clocks Go Slow* 1975 [np]; *Objections to Sex and Violence* 1975 [pub. 1985]; *Save It for the Minister* [with M. O'Malley, C. Potter] **(tv)** 1975; *Light Shining in Buckinghamshire* 1976 [pub. 1978]; *Vinegar Tom* 1976 [pub. 1978]; *Traps* 1977 [pub. 1978]; *The After Dinner Joke* **(tv)** 1978; *Floorshow* [with D. Bradford, B. Lavery, M. Wandor] 1978 [np]; *The Legion Hall Bombing* **(tv)** 1978; *Cloud Nine* 1979; *Three More Sleepless Nights* 1980 [pub. 1990]; *Crimes* **(tv)** 1982 [pub. 1990]; *Top Girls* 1982 [rev. 1984]; *Fen* 1983; *Midday Sun* 1984 [np]; *Softcops* 1984; *Plays: One* 1985; *A Mouthful of Birds* [from Euripides' play *The Bacchae*; with D. Lan] 1986 [np]; *Serious Money* 1987; *Fugue* **(tv)** 1988; *Hot Fudge* 1989 [pub. 1990]; *Ice Cream* 1989 [pub. 1990]; *Churchill: Shorts* 1990; *Mad Forest* 1990 [pub. 1992]; *Plays: Two* 1990; *The Lives of the World's Great Poisoners* 1991 [np]

Non-Fiction

Contemporary Women's Drama 1992

Amy CLAMPITT 1920–1994

The daughter of a farmer, Clampitt grew up on a 125-acre farm near the small town of New Providence, Iowa. She was educated at Grinnell

College, where she took her BA in 1941, then went to Columbia University, New York, on a graduate fellowship, but dropped out before the end of the academic year. She worked at various jobs in New York: first as a secretary and promotion director of the Oxford University Press there; then as a reference librarian at the National Audubon Society; as a freelance editor and researcher; and, finally, as an editor at the publisher E.P. Dutton from 1977 to 1982.

She won a trip to England by entering an essay competition and said that: 'England changed my perspective on life.' Her career as a poet began late: her first collection appeared in 1975, published at her own expense, but it was not until 1978, by which time she was in her mid-fifties, that her poems started to appear in the *New Yorker*. Her first commercially published collection, *The Kingfisher*, did not appear until 1983, but was immensely successful, establishing her reputation so quickly that she is now regarded as the most significant of those poets who emerged in America in the 1970s and 1980s. *The Kingfisher* takes its title poem and epigraph from Gerard Manley Hopkins, and Clampitt shares his sense of delighted wonder in the natural world; her packed, alliterative lines may also owe something to the Victorian poets. Her next collection, *What the Light Was Like*, equally displays an exceptional sense of place (her native Iowa is memorably evoked), while *Archaic Figure* reveals a new emphasis on the struggle and experience of women.

Clampitt's work appeared in many influential magazines, and after her retirement she undertook writerships-in-residence at the College of William and Mary, Williamburg, and at Amherst College. She lived mainly in a Greenwich Village apartment, but enjoyed travelling, and criss-crossed the USA many times by Greyhound bus. Late in life she married Harold Korn, who had been her companion for many years.

Poetry

Multitudes, Multitudes 1974; *The Isthmus* 1982; *The Kingfisher* 1983; *The Summer Solstice* 1983; *A Homage to John Keats* 1984; *What the Light Was Like* 1985; *Archaic Figure* 1987; *Westward* 1990; *Predecessors Et Cetera* 1991; *Silence Opens* 1994

Walter van Tilberg CLARK 1909–1971

Clark was born in East Orland, Maine, and spent his early childhood in West Nyack, New York State. In 1917 the family moved to Reno, Nevada, when his father, Dr Walter Ernest Clark, became president of the University of Nevada. He was educated in Reno and at the University of Nevada from which he graduated in 1932. He took an MA from the University of Vermont, and spent ten years teaching in Cazenovia, New York, before giving up his job and returning to the West with Barbara Morse, whom he had married in 1933, and their two children.

He had begun by writing verse and published his first collection *Ten Women in Gale's House*, while still at university. His first novel, *The Ox-Bow Incident*, made his name and remains a minor classic. It is the story of the lynching of three innocent men while the sheriff is out of town. Seen from the viewpoint of an outsider who becomes involved, if only as a bystander, the characters are drawn with accuracy and in depth; the terrifying nature of mob rule, the stifling of reason in the face of blind passion, is what makes the novel remarkable. The moment when the sheriff returns after the hanging, riding with the man who is supposed to have been murdered is one of the most chilling in all fiction. It was made into a moving film (1943), starring Henry Fonda, directed by William Wellman, but without the impressive cumulative detail of the novel. Clark's other novels are *The City of Trembling Leaves* about a boy growing up in Reno and *The Track of the Cat*, the story of the pursuit of a panther. He also wrote short stories; 'The Wind and the Snow' won the 1945 O. Henry Prize.

Fiction

The Ox-Bow Incident 1940; *The City of Trembling Leaves* 1945 [as *Tim Hazard* 1951]; *The Track of the Cat* 1949; *The Watchful Gods and Other Stories* (s) 1950

Poetry

Christmas Comes to Hjalsen, Reno 1930; *Ten Women in Gale's House and Shorter Poems* 1932

Edited

Alfred Doten *The Journals, 1849–1903* 3 vols 1974

Arthur C(harles) CLARKE 1917–

Clarke was born in the year his great-grandfather, approaching one hundred, died. Like other officers returning from the trenches, his father took (unsuccessfully) to farming, and died when Arthur was fourteen. His mother, left with four children, gave riding lessons to augment the family income. He was educated at Huish Grammar School, Taunton, where, aged thirteen, he began writing 'fantastic' stories for the school magazine. In his 'science-fiction autobiography'

Astounding Days, he describes his addiction to the magazine *Astounding Stories*, and how as a boy he rummaged through the boxes of pulp magazines at Woolworth's in Taunton. On leaving school he worked in the Exchequer and Audit Department in London, sharing a flat at 88 Gray's Inn Road with two other enthusiasts. This became the headquarters of the British Interplanetary Society, which held meetings in pubs, and was to be the inspiration for the stories in *Tales from the White Hart*. He joined the Royal Air Force in 1941 and worked on the first trials of ground control approach radar; his only non-science-fiction novel, *Glide Path*, is based on this experience. In 1945 he wrote a technical paper that was the forerunner of communication satellites. After the war he went up to King's College, London, where he took a first in mathematics and physics. He has written that: 'An sf writer is allowed to invent not yet existing technologies as long as they are plausible, but he must not state as a scientific fact something that is wholly untrue.' Strict adherence to this principle has been the key to his success.

His first non-fiction book was *Interplanetary Space*, followed a year later by his first novel *The Sands of Mars*. Since the early 1950s he has published more than twenty novels, seven volumes of stories, and over thirty non-fiction books; they have sold millions of copies worldwide. In the early 1950s he acquired a passion for skin-diving and with his friend Mike Wilson has explored and filmed the Great Barrier Reef, from which his novel *The Deep Range* (one of the few science-fiction novels to be reviewed in the *Wall Street Journal*) derives. This has also resulted in several volumes of popular science in collaboration with Wilson. Of his novels *2001: a Space Odyssey*, derived from an early story, 'The Sentinel', with elements from 'Encounter with the Dawn', was a huge success both as a film (in collaboration with Stanley Kubrick) and as a novel. He was guest of honour at the 1956 World Science Fiction Convention when he won a Hugo for his story 'The Star'. *Rendezvous with Rama* won the Nebula and Hugo Awards, and the John W. Campbell Memorial Award. Clarke has also won the Franklin Gold Medal, and in 1962 the UNESCO-Kalinga Prize for popularising science. He has been married and divorced, and has lived in Sri Lanka for many years, now only occasionally visiting Britain.

Fiction

The Sands of Mars 1951; *Islands in the Sky* 1952; *Against the Fall of Night* 1953; *Prelude to Space* 1953; *Childhood's End* 1954; *Expedition to Earth* 1954; *Earthlight* 1955; *The City and the Stars* 1956; *The Deep Range* 1957; *Tales from the White Hart* (s) 1957; *Across the Sea of Stars* (s) 1959; *A Fall of Moondust* 1961; *The Other Side of the Sky* (s) 1961; *From the Oceans, from the Stars* (s) 1962; *Reach for Tomorrow* (s) 1962; *Tales from Ten Worlds* (s) 1962; *Dolphin Island* 1963; *Prelude to Mars* (s) 1965; *The Nine Billion Names of God* (s) 1967; *2001: a Space Odyssey* 1968; *Glide Path* 1969; *The Lion of Comarre and Against the Fall of Night* 1970; *Of Time and Stars* (s) 1972; *The Wind from the Sun* 1972; *Rendezvous with Rama* 1973; *Imperial Earth* 1975; *The Fountains of Paradise* 1979; *2010: Odyssey Two* 1982; *The Sentinel* (s) 1983; *The Songs of Distant Earth* 1986; *Cradle* 1988; *2061: Odyssey Three* 1988; *Rama II* [with G. Lee] 1989; *Tales from the Planet Earth* (s) 1989; *The Garden of Rama* 1991; *The Ghost from the Grand Banks* 1991; *The Fantastic Muse* 1992; *By Space Possessed* 1993; *The Hammer of God* 1993; *Rama Revealed* [with G. Lee] 1993

Plays

2001: a Space Odyssey (f) [with S. Kubrick] 1968

Non-Fiction

Interplanetary Space 1950 [rev. 1960]; *The Exploration of Space* 1951 [rev. 1959]; *The Exploration of the Moon* [with R.A. Smith] 1954; *The Young Traveller in Space* 1954; *The Coast of Coral* 1956; *The Making of the Moon: the Story of the Earth Satellite Programme* 1957 [rev. 1958]; *The Reefs Of Taprobane: Underwater Adventures Around Ceylon* 1957; *Boy Beneath the Sea* [with M. Wilson] 1958; *Voice Across the Sea* 1958; *The Challenge of the Sea* 1960; *The Challenge of the Spaceship: Previews of Tomorrow's World* 1960; *The First Five Fathoms: a Guide to Underwater* [with M. Wilson] 1960; *Indian Ocean Adventure* 1962; *Profiles of the Future: an Inquiry into the Limits of the Possible* 1962; *Man and Space* 1964; *The Treasure of the Great Reef* 1964; *Voices from the Sky: Previews of the Coming Space Age* 1966; *The Promise of Space* 1968; *First on the Moon: with the Astronauts* 1970; *Beyond Jupiter* [with C. Bonestell] 1972; *Indian Ocean Treasure* [with M. Wilson] 1972; *Into Space* [with R. Silverberg] 1972; *The Lost Worlds of 2001* 1972; *Report on Planet Three* 1972; *Ascent To Orbit: a Scientific Autobiography* (a) 1984; *The Odyssey File: the Making of 2010* [with P. Hyams] 1984; *Arthur C. Clarke's Chronicles of the Strange and Mysterious* [with J. Fairley, S. Welfare] 1987; *Astounding Days: a Science Fiction Autobiography* (a) 1989; *How the World Was One* 1992

Edited

The Coming of the Space Age: Famous Accounts of Man's Probing of the Universe 1967; *Time Probe: the Science in Science Fiction* 1967; *Arthur C. Clarke's July 20, 2019* 1986; H.G. Wells: *First Men in the Moon* 1993, *War of the Worlds* 1993

J(ohn) M(ichael) COETZEE 1940–

Born in Cape Town, Coetzee (pronounced Coot'zeer) took a first degree at the university there and obtained his PhD from the University of Texas. Trained as a computer scientist and linguist, he was in England from 1962 to 1965, first as an applications programmer for IBM in London, then as systems programmer for International Computers at Bracknell. He was an assistant professor of English at the State University of New York, Buffalo, from 1968 to 1971 and in 1972 returned to Cape Town, since when he has taught at the university and held the chair of general literature since 1984. He had been a visiting professor at various American universities, which, together with the wide acclaim that his novels have received, has helped to build him a formidable international reputation, the first requirement for a dissident in a repressive regime.

His first book, *Dusklands*, which consists of two novellas, was originally published in South Africa in 1974 and reissued in the UK in 1982; they have widely different settings but an interlocking theme. Both 'The Vietnam Project' and 'Jacobus Coetzee' are concerned with the death of the spirit, although the two protagonists – an expert in psychological warfare and a megalomaniac Boer respectively – achieve this by different means. **Nadine Gordimer** wrote of it, 'J.M. Coetzee's vision goes to the very nerve-centre of being. What he finds there is more than most people will ever know about themselves.' His first full-length novel, *In the Heart of the Country*, which won the CNA (South Africa), is the impassioned diary of a sexually deprived woman living on a remote South African farm, a powerful portrait of loneliness, frustration and anger. In *Waiting for the Barbarians*, which again won him the CNA, as well as the Geoffrey Faber and James Tait Black Memorial Prizes, the arrival of interrogation experts is what drives a magistrate into what the state sees as an act of treason. Bernard Levin wrote of it: 'I have known few authors who can evoke such a wildness in the heart of man ... Coetzee knows the elusive terror of Kafka.' In the haunting *Life and Times of Michael K*, winner of the 1983 Booker Prize and the 1985 Prix Etrangers Fémina, civil war has been raging for years. It is the story of a gardener who takes his ailing mother on a long trek to find the farm where she was born. After she dies, he lives in isolation (is it the right farm?) but even here the war catches up with him. In the 'rehabilitation' camp to which he is taken his long fast angers, baffles and finally awes his captors. (Significantly it is only towards the end of the book that we learn what the war is about – 'so that minorities will have a say in their destinies'.) The short novel *Foe* is a reworking of *Robinson Crusoe*. In *Age of Iron* Coetzee has felt himself able to attack apartheid directly for the first time. An elderly white woman who has just been diagnosed as having a terminal cancer returns home to find an evil-smelling black tramp at the bottom of her garden. From here on we are in the world of *The Divine Comedy*. Acting as Virgil to her Dante, he conducts her through the 'Inferno' of the Cape Town shanties. On the strength of five novels and two novellas, Coetzee, who also won the Jerusalem Prize (1987), is one of the most impressive novelists writing in English today. It is a measure of his subtlety that none of his books have been banned in South Africa.

Fiction

Dusklands (s) 1974; *In the Heart of the Country* 1977; *Waiting for the Barbarians* 1980; *Life and Times of Michael K* 1983; *Foe* 1986; *Age of Iron* 1990; *The Master of Petersburg* 1994

Non-Fiction

White Writing: On the Culture of Letters in South Africa 1988; *Doubling the Point: Essays and Interviews* 1992

Isabel COLEGATE 1931–

Born in Lincolnshire, Colegate was educated in boarding-schools in Norfolk and Shropshire. She married Michael Briggs, a manufacturer of widgets, in 1953. In the early 1950s, Colegate worked for Anthony Blond, then a literary agent, who subsequently became a publisher. In 1958, Blond published her first novel, *The Blackmailer*, which, like most of her work examined the British attitude to an aristocracy as it slipped into decline. The first three novels are set in a post Second World War welfare state. The frayed aristocrats of *The Blackmailer* feel threatened by the new opportunities given to the 'lower orders', while the *nouveau-riche* protagonist of her second novel, *A Man of Power*, is able to enter upper-class society because of his money. *The Orlando Trilogy* is the most complex and wide-ranging of Colegate's fiction, drawing upon the Oedipus myth, and politically covering the Spanish Civil War, the England of Suez, and the Burgess–Maclean affair. The problem of 'new' capitalism challenging the old order recurs, but both the aristocracy and the radicals they oppose are seen to be equally futile in purpose.

The First World War provided the background for *The Shooting Party*, Colegate's best-known novel. Her meticulous attention to period detail in the book was repeated in a successful film (1984, directed by Alan Bridges from a screenplay by Julian Bond), in which Sir Randolph Nettleby, a crusty landowner, was played by James Mason, making his last film appearance. It is set in 1913, the passing of an epoch seen clearly by Nettleby, who states: 'An age, perhaps a civilisation, is coming to an end.' Violence, insistence on blind tradition, but also a regard for 'standards' and a certain courtesy to servants, were factors which contributed to make the novel more powerful than the social satire she had previously written. Colegate herself has remarked that she is interested in the concept of 'the gentleman', believing that 'there was a sense that certain responsibilities went with the privilege, and now that's entirely gone. I'm ambivalent about it, whether it was altogether a good thing.'

With a house in London and a farmhouse in Tuscany, Colegate prefers English country living, spending most of the time in an eighteenth-century Gothic castle near Bath. Having no regard whatsoever for 'fashions in fiction, authors' launch parties, the whole London literary life', she has said she writes because she enjoys it, and for no other reason.

Fiction

The Blackmailer 1958; *A Man of Power* 1960; *The Great Occasion* 1962; *Statues in a Garden* 1964; *The Orlando Trilogy* 1984: *Orlando King* 1968, *Orlando at the Brazen Threshold* 1971, *Agatha* 1973; *News from the City of the Sun* 1979; *The Shooting Party* 1980; *A Glimpse of Sion's Glory* (s) 1985; *Deceits of Time* 1988; *The Summer of the Royal Visit* 1991

John (Henry Noyes) COLLIER 1901–1980

Born and educated in London, Collier began his writing career as a poet. Supported by his father, he led the life of a *flâneur* during the 1920s, visiting art galleries, sitting in cafés and gambling, occasionally writing poetry. He was struggling, as he put it, 'to reconcile ... the intensely visual experience opened to him by the **Sitwells** and the modern painters, with the austerer preoccupations of those classical authors who were [then] fashionable'. This resulted in a single, short volume of poems, *Gemini* (1931) He was nevertheless deemed qualified to be poetry editor of *Time and Tide* during this period, when he was living on a farm in Hampshire. He also edited Aubrey's *Brief Lives* under the title *The Scandal and Credulities of John Aubrey*, and wrote *Just the Other Day*, 'a resumée of post-war events and tendencies', and a book on writing, *Please Excuse Me, Comrade.*

He came to notice with his first novel, *His Monkey Wife*, a charming fantasy about a chimpanzee called Emily, who falls in love with and eventually marries a colonial schoolmaster, Mr Fatigay. Various claims have been made for the novel as an allegory, but it remains *sui generis*. Subsequent novels include *Tom's A-Cold* (published in the USA as *Full Circle*), in which Collier depicts Britain in 1955, after it has been reduced to barbarism by warfare, and *Defy the Foul Fiend*, a rumbustious account of the life and adventures of the by-blow of a dissolute aristocrat, detailing in particular his disastrous relations with a young woman (it is subtitled 'The Misadventures of a Heart'). He also published several volumes of short stories, often on macabre or fantastic themes. **Anthony Burgess** described him as 'chiefly a creator of very wayward miniatures'.

He moved to America in the 1930s and married there in 1936. He returned to America in 1942 and spent some time at Warner Bros, where he became a friend of **Christopher Isherwood**. Collier only really liked writing at night, and so spent much of his time in the studio office distracting Isherwood with conversation and literary games. In spite of this, he wrote, or collaborated on, several screenplays, none of them particularly distinguished. The best known of these is his version of *I Am a Camera* (1955), based on **John van Druten**'s stage adaptation of Isherwood's Berlin stories; it was regretfully dismissed by his old friend and colleague as 'disgusting ooh-la-la near pornographic trash'. Collier was divorced in 1945, but subsequently married for a second time.

He continued to publish stories in magazines, such as the *New Yorker*, *Esquire*, *Atlantic Monthly* and *Harper's Bazaar*, winning the Edgar Allen Poe Award in 1951 and the International Fantasy Award in 1952. He later lived in France for twenty years, though he returned to California to die. A remarkably modest man, he rarely talked about his work, which he undervalued, merely marvelling that 'a third-rate writer like me has been able to palm himself off as a second-rate writer'.

Fiction

His Monkey Wife 1930; *Green Thoughts* 1932; *Tom's A-Cold* [US *Full Circle*] 1933; *Defy the Foul Fiend* 1934; *The Devil and All* (s) 1935; *Variations on a Theme* 1935; *Witch's Money* 1940; *Presenting Moonshine* (s) 1941; *Fancies and Goodnights* (s) 1951; *Pictures in the Fire* (s) 1958; *Of Demons and Darkness* (s) 1965

Non-Fiction

Just the Other Day: an Informal History of Great Britain Since the War [with I. Lang] 1932; *Please Excuse me, Comrade* 1933

Plays

I Am a Camera (f) 1955; *Milton's Paradise Lost* (f) 1973

Poetry

Gemini 1931

Edited

The Scandal and Credulities of John Aubrey 1931

I(vy) COMPTON-BURNETT 1884–1969

Compton-Burnett was born at Pinner, Middlesex, the daughter of James Compton Burnett, a crusading homeopathic doctor ('Homeopathy is the winning horse in the Medical Derby of the world', he wrote), and his second wife, daughter of a mayor of Dover, who added the hyphen to the family surname. Her life divides sharply into two parts at 1919: during the first part, she played her role, both as victim and aggressor, in the sort of Victorian family tyranny that appears in her work; in the second, living more uneventfully, she transmuted this experience into fiction. Dr Compton Burnett had five children by his first marriage, seven by his second, and the two families did not get on; three of the children were to commit suicide and all eight daugthers were to remain unmarried. Within the family, Ivy formed a group with her brothers Guy and Noel, both of whom died young, to her great grief.

After attending Addiscombe College in Hove, a town to which the family had moved, she read classics at Royal Holloway College, London. On coming down, her father having died, she returned to the unhappy home ruled by her mother, occupying much of her time in writing the uncharacteristic early novel *Dolores* (1911). After her mother's death, Compton-Burnett replaced her as family ruler, and tyrannised to such effect that when two of her sisters set up house in London with Dame Myra Hess (their piano teacher), they refused to have her living with them. This was only one of the psychological blows she suffered during the First World War: her brother Noel's death in action; the mysterious death by veronal poisoning (presumed suicide) of her two youngest sisters; her own near death in the influenza epidemic.

From 1919, however, she shared a flat in Braemar Gardens, Kensington, with Margaret Jourdain, an expert on furniture, and their gruff but loving companionship lasted until Jourdain's death in 1951. The minutiae of their shared lives – the gloomy flat with its black furniture and blind mirrors, the formidable teatime hospitality to fellow writers, Compton-Burnett's hairnets and curious *coiffure* – have become part of the iconography of the century. After her friend died, Compton-Burnett lived on at Braemar Mansions, pestered by rising rents, until her own death.

After the false start of *Dolores*, she did not produce another book for fourteen years, but in 1925 published *Pastors and Masters*, in which her mature style was immediately apparent, although the novel has an academic setting. Thereafter, eighteen further novels (she wrote nothing else) appeared at intervals as regular as their literary devices: the formulaic titles – *A House and Its Head*, *A Family and a Fortune*, *Two Worlds and Their Ways*; the settings in a country house of the fecund Victorian gentry; the pages of stylised, frequently aphoristic, dialogue and little else. Her principal subject, drawing upon her own experiences of family life, is domestic tyranny, with battle-lines drawn up between a bullying head of a household and those under his or her dominion. Compton-Burnett is particularly good at portraying children – notably three-year-old Nevill in *Parents and Children*, who always refers to himself in the third person, and the younger Lambs in *Manservant and Maidservant*, who stick pins in a wax effigy of their father. Her humour is both sophisticated and savage, with bad behaviour usually prevailing, while her plots are often melodramatic and perfunctory, turning upon the revelation of sexual or matrimonial irregularities of one sort or another.

Her work gradually collected a following, and although she once remarked that it was difficult not to put down one of her books once you have picked it up, she has many admirers amongst her fellow novelists (including her friend **Elizabeth Taylor**) and is considered by some critics to be one of the finest novelists of the century. She is undoubtedly one of the most original, and her reputation seems secure.

Fiction

Dolores 1911; *Pastors and Masters* 1925; *Brothers and Sisters* 1929; *Men and Wives* 1931; *More Women Than Men* 1933; *A House and Its Head* 1935; *Daughters and Sons* 1937; *A Family and a Fortune* 1939; *Parents and Children* 1941; *Elders and Betters* 1944; *Manservant and Maidservant* 1947 [US *Bullivant and the Lambs* 1948]; *Two Worlds and Their Ways* 1949; *Darkness and Day* 1951; *The Present and the Past* 1953; *Mother and Son* 1955; *A Father and His Fate* 1957; *A Heritage and Its History* 1959; *The Mighty and Their Fall* 1961; *A God and His Gifts* 1963

Biography
Ivy When Young by Hilary Spurling 1974 [rev. 1983]; *Secrets of a Woman's Heart* by Hilary Spurling 1984

Barbara COMYNS 1909–

Comyns was born Barbara Bailey; her authorial name is derived from her second husband, Comyns Carr, whom she married in 1945. Her father was the semi-retired managing director of a Midland chemical firm, and she grew up as one of six children in his large house in Warwickshire. Her mother became deaf at the age of about twenty-five, and the family grew up largely unsupervised or under the care of governesses, spending a lot of time on the River Avon, the world evoked in Comyns's first novel, *Sisters by a River* (1947). Comyns's father died when she was seventeen, and she used the money he left her to attend art schools, first in Stratford-upon-Avon and then Heatherley's in London: she has remained a visual artist as well as a writer throughout her life and exhibited at the London Group, which showed the Vorticists. She married a fellow artist in 1931 and had two children, but parted from him before the Second World War. She worked at a wide variety of jobs in the 1930s and 1940s including modelling for artists, breeding poodles, converting flats, running a garage and selling antique furniture.

The unconventional evocation of childhood in *Sisters by a River* led to its being serialised in *Lilliput* as 'The Novel Nobody will Publish'; Eyre and Spottiswoode eventually brought it out with the unorthodox spelling and punctuation deliberately left uncorrected. It was the first of ten novels, seven of them written in the first person, 'plaintive, strange and robust all at once' as Patricia Craig has called them, making a special feature of a calculated *naïvety*: 'This book does not seem to be growing very large although I have got to Chapter Nine. I think this is partly because there isn't any conversation.'

Comyns's second husband was a Foreign Office official, and when he retired, the couple moved to Spain for eighteen years, living near Barcelona. This period coincided with a long break in the publication of her novels (from 1967 to 1985), when she acted as amanuensis to her husband in his own literary career. The publication of her eighth novel, *The Juniper Tree* sparked a revival of interest in her work, and two further novels have since appeared (both written earlier in her career). Her most famous novel remains, however, *The Vet's Daughter*, which was turned into a musical, *The Clapham Wonder*, by Sandy Wilson. Besides her novels, Comyns has also written a book based on her experiences in Spain. She survives her second husband, and lives in London.

Fiction
Sisters by a River 1947; *Our Spoons Came from Woolworths* 1950; *Who Was Changed and Who Was Dead* 1954; *The Vet's Daughter* 1959; *The Skin Chairs* 1962; *Birds in Tiny Cages* 1964; *A Touch of Mistletoe* 1967; *The Juniper Tree* 1985; *Mr Fox* 1987; *The House of Dolls* 1989

Non-Fiction
Out of the Red into the Blue 1960

Richard (Thomas) CONDON 1915–

Condon was born in New York and on leaving school he served in the United States Merchant Navy. His long career as a film publicist began in 1936, and he did not write his first novel, *The Oldest Confession*, until he was forty-two, by which time he was a master of plot, pace and suspense. For twenty-one years he had studied over eight feature films a week. The production companies he worked for include Walt Disney, Twentieth Century Fox and Paramount Pictures.

The *New York Times* has described his novels as 'cynical, hip political thrillers that contain an extravagance of invention'. He is best known for his second novel, *The Manchurian Candidate*, which American critics decided was one of the best 'bad books' ever to have been written. It tells the story of GI Raymond Shaw, captured in Korea and innocently brainwashed into committing crimes against US Intelligence on behalf of his former captives. Naturally, Condon novels translate well into screenplays: *The Manchurian Candidate* was directed by John Frankenheimer in 1962 and is a cinema classic to this day.

For the next ten years, although Condon published prolifically, the critics were less enthusiastic. Then came the acclaimed *Winter Kills* in 1974, a novel paralleling the lives of members of the Kennedy family with a prevalent theme that murdering presidents is a good idea for world leaders who wish to better themselves. The subsequent film opened in New York in 1979 and closed after a short run. One producer was murdered, the other sent down for forty years on a drugs charge. It reopened in 1983 and again disappeared mysteriously.

More recently, Condon has achieved fame for the 'Prizzi' trilogy, which began with *Prizzi's Honour*, a tale of Mafia highlife and the romance of two psychopaths (played by Jack Nicholson and Kathleen Turner in the 1985 film directed by John Huston from Condon's own screenplay), one obsessed with domestic cleanliness, the other a freelance assassin and tax consultant.

Condon, his wife and two daughters have lived in France, Spain, Ireland and Switzerland. On mythologised fact and political satire, Condon comments: 'Politics is a form of high entertainment and low comedy. It has everything: it's melodramatic, it's sinister and it has wonderful villains.'

Fiction

The Oldest Confession 1958; *The Manchurian Candidate* 1959; *Some Angry Angel* 1960; *A Talent for Loving* 1961; *Any God Will Do* 1964; *An Infinity of Mirrors* 1964; *The Ecstasy Business* 1967; *Mile High* 1969; *The Vertical Smile* 1971; *Arigato* 1972; *The Star-Spangled Crunch* 1974; *Winter Kills* 1974; *Money Is Love* 1975; *The Whisper of the Axe* 1976; *The Abandoned Woman* 1977; *Bandicoot* 1978; *Death of a Politician* 1978; *The Entwining* 1980; *Prizzi's Honour* 1982; *A Trembling Upon Rome* 1983; *Prizzi's Family* 1986; *Prizzi's Glory* 1988; *Emperor of America* 1990; *The Final Addiction* 1991; *The Venerable Bead* 1993

Plays

Men of Distinction 1953 [np]; *Prizzi's Honour* (f) 1985

Non-Fiction

And Then We Moved to Rossenarra, or, The Art of Emigrating 1973; *The Mexican Stove: a History of Mexican Food* 1973

Evan S(helby) CONNELL Jr 1924–

Connell was born in Kansas City, Missouri, where his father was a surgeon, and educated at Southwest High School, Kansas, and Dartmouth College, Hanover, New Hampshire, before joining the US Navy as an aviation cadet in 1943. He graduated on VE-Day and worked for a year as a flight instructor before reading English at the University of Kansas, Lawrence, on the GI Bill. Having graduated in 1947 he studied at Stanford University, California, and Columbia University, New York. He published his first short story, 'A Cross to Bear', in 1947, and travelled in Europe in the early 1950s before moving to San Francisco where he has lived ever since.

His first collection of short stories, *The Anatomy Lesson* (1957), includes several stories about the Bridges, the Kansas couple who are the subject of his first novel, *Mrs Bridge*, and the subsequent *Mr Bridge*. The two novels, generally regarded as Connell's best work, are montages of short sketches describing the affluent torpor of Midwest upper-middle-class life. (A highly regarded film based on the two novels and directed by James Ivory was released in 1990 as *Mr and Mrs Bridge*.) *The Patriot*, published in 1960, is far more conventional in form, a *Bildungsroman* concerning a young naval cadet during the war, which was probably written before *Mrs Bridge*. Connell's subsequent works include *The Diary of a Rapist*, the story of a San Francisco civil servant who keeps 'a scrapbook of monstrous events' which he is driven to imitate before betraying himself in a bid for attention; two long poems based on the author's study of ancient cultures; and a detailed study of General Custer and the battle of Little Bighorn, *Son of the Morning Star*. Of writing Connell has said: 'I enjoy doing it, and don't mind the long hours. I am irritated by the effusions of certain authors who claim that writing is some sort of agony.'

Fiction

The Anatomy Lesson and Other Stories (s) 1957; *Mrs Bridge* 1959; *The Patriot* 1960; *At the Crossroads* (s) 1965; *The Diary of a Rapist* 1966; *Mr Bridge* 1969; *The Connoisseur* 1974; *Double Honeymoon* 1976; *St Augustine's Pigeon: the Selected Stories* (s) [ed. G. Blaisdell] 1980; *The Alchymist's Journal* 1991

Non-Fiction

A Long Desire 1979; *The White Lantern* 1980; *Son of the Morning Star* 1984

Poetry

Notes from a Bottle Found on the Beach at Carmel 1963; *Points for a Compass Rose* 1973

Edited

Jerry Stoll *I Am a Lover* 1961; Joanne Leonard *Woman by Three* 1969

Cyril (Vernon) CONNOLLY 1903–1974

Born in Coventry, the son of an army officer and well-known conchologist, and connected to the Anglo-Irish aristocracy on his mother's side, Connolly was educated at Eton (which, despite his early unhappiness there, always remained the Eden of his life) and at Balliol College, Oxford, where he cultivated his personality and took a third in history. After a brief spell of tutoring in Jamaica (the only regular job he ever did), he became a salaried protégé of the American man-of-letters Logan Pearsall Smith. Through him, and through Desmond MacCarthy, he obtained a start in literary London. Literary journalism was to remain his chief support throughout life, eked out by a good deal of private patronage and one rich wife (the American Jean Bakewell, whom he married in 1930).

In the 1930s, the Connollys lived first on the Riviera and then in Chelsea, and he contributed to the *New Statesman*. From 1939 to 1950 – Jean Bakewell having left him and returned to America – he was founder and editor of the celebrated literary magazine *Horizon*; from 1951 until his death in 1974, he was one of the two chief book reviewers of the *Sunday Times*.

Connolly cultivated a myth of himself (particularly in the sparkling critical book-cum-early autobiography *Enemies of Promise*) as a creative writer *manqué*, frustrated by circumstances and temperament; and indeed the one novel he completed, *The Rock Pool*, seems to indicate (although some would disagree) that he had the potential to write fiction as well as *belles-lettres*. Sloth, hedonism, melancholy and perfectionism were causes endlessly diagnosed by himself and others for his failure to create more. In fact, his final *oeuvre* became quite large and includes, besides the books already mentioned, *The Unquiet Grave*, a book of melancholy wartime *pensées* which is sometimes considered his major achievement. His journalism, collected in four volumes, contains much that is enjoyable.

From 1950 to 1954 he was married to Barbara Skelton, who has recounted their life together in her racy memoirs, and in 1959 – both his earlier marriages having ended in divorce – he married Deirdre Craig, with whom he lived during his later years in Eastbourne and with whom he had a son and a daughter. He was a celebrated (not universally loved) figure of literary London, about whom anecdotes abounded, ranging from his having allegedly stolen prize avocados from **W. Somerset Maugham** to his failure with a mistress during an air-raid because, as he said, 'perfect fear casteth out all love'. Since his death there have been published *A Romantic Friendship*, his letters to Noel Blakiston, a fellow Etonian to whom he was attached in his youth, and *Shade Those Laurels*, a novel which he almost completed and which was finished by his widow's subsequent husband, **Peter Levi**.

Non-Fiction

Enemies of Promise (a) 1938; *The Unquiet Grave* [as 'Palinurus'] 1944 [rev. 1967]; *The Condemned Playground: Essays 1927–1944* 1945; *The Missing Diplomats* 1952; *Ideas and Places: a Selection of Cyril Connolly's Editorial Comments from 'Horizon'* 1953; *Les Pavillons: French Pavilions of the Eighteenth Century* [with J. Zerbe] 1962; *Previous Convictions* 1963; *The Modern Movement: One Hundred Key Books from England, France and America, 1880–1950* 1966; *The Evening Colonnade* 1973; *A Romantic Friendship* [ed. N. Blakiston] 1975; *Cyril Connolly, Journal and Memoir* [ed. D. Pryce-Jones] 1983

Fiction

The Rock Pool 1936; *Shade Those Laurels* [with P. Levi] 1990

Translations

Jean Bruller [as 'Vercors'] *Put Out the Light* 1944; Alfred Jarry *The Ubu Plays* 1968

Edited

Horizon Stories (s) 1943; *The Golden Horizon* 1953; *Great English Short Novels* 1953

Biography

Cyril Connolly: a Nostalgic Life by Clive Fisher 1995

(George) Robert (Acworth) CONQUEST 1917–

Conquest was born in Great Malvern, the son of an American father of slight but independent means and an Anglo-Indian mother, and spent much of his childhood in France. After Winchester he spent a year at the University of Grenoble before going up to Magdalen College, Oxford, in 1936 to read politics, philosophy and economics. After graduating in 1939 he joined the Oxfordshire and Buckinghamshire Light Infantry and served in Europe and the Balkans until 1946 when he joined the Foreign Service. Until 1956 he worked as a press attaché and second secretary in Sofia (during which time he knew **Lawrence Durrell**, then a press attaché in Belgrade), at the Foreign Office in London, and as first secretary at the United Nations.

Conquest's reputation was quickly established in 1955 with the publication of *Poems* and *A World of Difference*, a jokey science-fiction novel which uses the names of literary friends for spaceships. In the 1960s he was to edit five volumes of the science-fiction anthology *Spectrum* with his friend **Kingsley Amis**. A section in Amis's *Memoirs* (1991) recounts Conquest's talents for elaborate practical jokes and obscene limericks, and recalls 'the so-called Fascist lunches' in Charlotte Street which both men attended during the 1960s. Conquest's editorship in 1956 of the influential poetry anthology *New Lines* brought the work of the Movement poets – including Amis, **Philip Larkin** and **D.J. Enright** – to the attention of a wider public. In the same year he became a research fellow at the London School of Economics, thereby beginning a distinguished career as a scholar of Soviet history. He has subsequently held academic posts at the University of Buffalo, New York, Columbia University, the Smithsonian Institute and the Hoover Institution in Stanford, where he has been curator of the Russian and East European collection since 1981. He has been a research associate at Harvard since 1983, and was literary editor of the *Spectator* from 1962 to 1963.

The Great Terror, published in 1968, was a ground-breaking account of Stalin's purges of the 1930s based on government documents and the accounts of Soviet defectors. The book was substantially rewritten in the light of new material and published as *The Great Terror: A Reassessment* in 1990, and much of it has been published in Soviet magazines in the wake of *glasnost*. *The Harvest of Sorrow* is a similarly detailed account of the collectivisation of Soviet agriculture and Stalin's murderous purge of the *kulaks*. Conquest has been married four times and has two sons by his first marriage.

Poetry

Poems 1955; *Between Mars and Venus* 1962; *Arias from a Love Opera* 1969; *Forays* 1979; *Collected Poems* 1986; *New and Collected Poems* 1988

Non-Fiction

Common Sense about Russia 1960; *Courage of Genius: the Pasternak Affair* 1961; *Power and Policy in the USSR* 1961; *The Last Empire: On Soviet Rule in the Minority Republics* 1962; *Marxism Today* 1964; *Russia After Khrushchev* 1965; *The Great Terror* 1968 [rev. as *The Great Terror: a Reassessment* 1990]; *The Nation Killers: the Soviet Deportation of Nationalities* 1970; *Lenin* 1972; *Kolyma: the Arctic Death Camps* 1978; *The Abomination of Moab* 1979; *Present Danger: Towards a Foreign Policy* 1979; *We and They* 1980; *What to Do When the Russians Come* [with J. Manchip White] 1984; *Inside Stalin's Secret Police: NKVD Politics 1936–1939* 1985; *The Harvest of Sorrow* 1986; *Tyrants and Typewriters* 1988; *Stalin and the Kirov Murder* 1989; *Stalin: Breaker of Nations* 1991

Fiction

A World of Difference 1955; *The Egyptologists* [with K. Amis] 1965

Translations

Alexander Solzhenitsyn *Prussian Nights* 1974

Edited

New Lines 1956; *Back to Life: Poems from Behind the Iron Curtain* 1958; *New Lines II* 1963; *Spectrum* [with K. Amis] 1963; Petr Ionovich Iakir *A Childhood in Prison* 1972; *The Robert Shechley Omnibus* 1973; Tibor Szamuely *The Russian Tradition* 1974; *The Last Empire* 1986

Joseph CONRAD (pseud. of Josef Teodore Konrad Nalecz Korzeniowski) 1857–1926

Conrad was born in the Ukraine, the only son of Apollo and Ewa Korzeniowski, both of whom would die before Conrad reached puberty. By 1861 the family had moved to Warsaw, where Apollo edited a political magazine and became embroiled in political activities to such a degree that he and his wife were arrested and imprisoned for seven months. In the following year, on a journey to Vologda (about 300 miles north-east of Moscow) following the family's deportation, Conrad developed pneumonia and shortly after their arrival his mother fell seriously ill. She died of this illness in 1865, after the family were allowed to move south to Kiev. Four years later, Conrad's father died and Conrad was consequently sent to Switzerland to be tutored by his uncle, though his education would be largely self-directed through an almost constant reading of literary classics.

Conrad was supported by a patrimony sent by his uncle, and the 1870s witness Conrad's first movements by sea, mainly between France and St Pierre, Martinique, on the *Mont Blanc*, but also – this time as a steward – on more extensive voyages on the *Saint-Antoine*, during which Conrad developed an anal abscess. The same year he lost a great deal of money gambling and, in the following year, gambled away an even larger fortune, events which seem to have determined a self-inflicted gunshot wound in the chest. Though Conrad was not seriously injured, the wound left lifelong scars; the experiences would be drawn on in *The Arrow of Gold*. Immediately after the event, Conrad's uncle arrived in Marseilles to bail out his nephew.

In 1878 Conrad heard his first words of spoken English on the *Mavis*, which he had boarded when it was bound for Malta and Constantinople. When the ship eventually arrived in Lowestoft, Conrad left it and set foot on English soil for the first time. After finding his way to London, he was employed as an ordinary seaman and made three trips on the *Skimmer of the Sea* between Lowestoft and Newcastle. His knowledge of the English language increased, but he later recalled that 'no explorer could have been more lonely' during this time.

Following visits to Australia and Greece. Conrad moved back to his lodging in Holloway and studied for the examination which would qualify him as a second mate, an examination he passed in 1884. The time between had been spent travelling to such places as Bangkok, Singapore, Madras and Dunkirk, and in 1886 he had been given British citizenship. He first captained his own ship, the *Otago*, in 1888, after spending six weeks in the European Hospital at Samarang, Java, because of a serious back injury. In 1990 he sailed to the Congo, 'the most traumatic journey of his life', during which he would suffer from repeated attacks of dysentery and malaria. The journey provided much of the material for his great novel *Heart of Darkness*.

Conrad had been working on his first novel, *Almayer's Folly*, for five years before it was finally finished in 1895. Edward Garnett accepted it on behalf of the publishing firm Fisher Unwin, and in the same year Conrad received the first part of his inheritance (about £120). This was the first time he published under the pseudonym Joseph Conrad; the novel was reviewed widely and in a favourable light, but sold very badly. *An Outcast of the Islands*, his second book, was published a year later, the same year that he married.

From then on Conrad wrote almost continuously, and entered what might be called his mature phase as an author with *The Nigger of the 'Narcissus'*, which was serialised in the *New Review* between August and September 1897, a very eventful year for him in other respects. His stories appeared in such publications as *Blackwood's Magazine*, *Cornhill Magazine* and *Cosmopolis* and he met **Henry James**, **Stephen Crane** and R.B. Cunningham Graham. In 1898 his son Borys Conrad was born.

Conrad's reputation as a deeply serious and very modern literary novelist now rests with *Heart of Darkness* (serialised in *Blackwood's Magazine* between February and April 1899), *Nostromo* and *Under Western Eyes*, but these works were read at the time as simple sea stories, and Conrad did not gain a wide readership until the publication of *Chance* in serial form in 1912. By 1919 he had become so acclaimed that he could sell his film rights for the then enormous sum of £3,080, and in 1924 he declined the offer of a knighthood (as he had earlier declined honorary degrees from five universities). He died of a heart-attack and was buried in Canterbury. He is now considered to be one of the first, and one of the finest, modern novelists writing in English.

Fiction

Almayer's Folly 1895; *An Outcast of the Islands* 1896; *The Nigger of the 'Narcissus'* 1897; *Tales of Unrest* (s) 1898; *Heart of Darkness* 1899; *Lord Jim* 1900; *The Inheritors* [with F. Madox Ford] 1901; *Typhoon* 1902; *Youth* (s) 1902; *Romance* [with F. Madox Ford] 1903; *Typhoon and Other Stories* (s) 1903; *Nostromo* 1904; *The Secret Agent* 1907; *A Set of Six* (s) 1908; *The Secret Sharer* 1910; *Under Western Eyes* 1911; *Twixt Land and Sea* (s) 1912; *Chance* 1913; *Victory* 1915; *Within the Tides* (s) 1915; *The Shadow-Line* 1917; *The Arrow of Gold* 1919; *The Rescue* 1920; *The Rover* 1923; *The Nature of a Crime* [with F. Madox Ford] 1924; *Suspense* [unfinished] 1925; *Tales of Hearsay* (s) 1925; *The Sisters* [unfinished] 1928

Non-Fiction

The Mirror of the Sea (a) 1906; *Some Reminiscences* (a) 1912; *Notes on Life and Letters* 1921; *Last Essays* 1926; *Joseph Conrad's Letters to His Wife* 1927; *The Life and Letters of Joseph Conrad* [ed. G. Jean-Aubry] 2 vols 1927; *Letters from Conrad 1895–1924* 1928; *Conrad to a Friend: 150 Letters from Joseph Conrad to Richard Curle* 1928; *Letters to William Blackwood and David S. Meldrum* 1958; *Conrad's Polish Background: Letters to and from Polish Friends* [ed. Z. Najder] 1964; *The Collected Letters of Joseph Conrad* [ed. L. Davies, F.R. Karl] 3 vols 1983–8

Plays

The Secret Agent 1923; *Laughing Anne and One Day More* [pub. 1924]

Collections

Congo Diary and Other Uncollected Pieces [ed. Z. Najder] 1978

Biography

Joseph Conrad by Jeffrey Meyers 1991

David (John) CONSTANTINE 1944–

Constantine was born in Salford, Lancashire, and went to Manchester Grammar School before going to Wadham College, Oxford, where he gained a BA in modern languages in 1966 and a PhD in 1971. He married in 1966 and has two children. He and his wife, Helen, have also produced a book of collaborative translations of Henri Michaux. He was a lecturer at Durham University from 1969 to 1981, and returned to Oxford in 1981 as a fellow in German at Queen's College.

Constantine's first book was an academic work, *The Significance of Locality in the Poetry of Friedrich Hölderlin*. He waited until the age of thirty-six to publish his first collection of his own poetry, *A Brightness to Cast Shadows*. It was praised for its direct address and showed his subsequently enduring concern with love of different kinds. Both sensuality and social concern are to the forefront in his work and the critic Bernard O'Donoghue has commended his freedom from 'the elegant, self-constraining ironies characteristic of English poetry since the Movement', and his distance from what **A. Alvarez** has called 'the gentility principle'. The spirit of place and a pleasure in tactile immediacy are prominent in his work, sometimes refracted more obliquely through mythic frameworks. Although Constantine would see the liberal humanity and essential seriousness of his work as timeless features of English poetry, he may have escaped certain English traps because his influences are principally German and classical writers (he taught himself Greek while a student). He has continued to publish aademic works on Greek and German topics.

One poem in *A Brightness to Cast Shadows* dominates the volume: 'In Memoriam 8571 Private J.W. Gleave', which concerns the death of Constantine's grandfather in the First World War. Later books have had a more integrated wholeness and complexity, and *Watching for Dolphins* marked a significant advance in unity of themes and voice. *Madder* is widely regarded as his best collection to date. It develops the engagement with social injustice prominent in his earlier poetry, and which was also strongly a part of his novel *Davies*, about a turn-of-the-century criminal. Constantine has said that he sees his work as something of a revolt against the kind of ideology that prevailed in the Thatcher years, and that for him 'poetry is intrinsically, in its rhythms, a gesture in favour of a generous and passionate life'.

Poetry

A Brightness to Cast Shadows 1980; *Watching for Dolphins* 1983; *Talitha Cumi* [with R. Pybus] 1983; *Mappa Mundi* 1984; *Madder* 1987; *Selected Poems* 1991; *Caspar Hauser* 1994

Fiction

Davies 1985; *Back at the Spike* 1994

Non-Fiction

The Significance of Locality in the Poetry of Friedrich Hölderlin 1976; *Early Greek Travellers and the Hellenic Ideal* 1984; *Hölderlin* 1988

Edited

German Short Stories 2 (s) 1976; *New Worlds: Prize Winning Poems from the 1992 Berkshire Literature Festival* [with M. Lomax] 1992

Translations

Selected Poems of Hölderlin 1990; Henri Michaux *Deplacements, Degagements* [with H. Constantine] 1990; Philippe Jaccottet *Under Clouded Skies/Beauregard* [with M. Treharne] 1994; Goethe *Elective Affinities* 1994

David COOK 1940–

Cook was born in Preston, where he attended Rishton Secondary Modern School. He left at fifteen and worked variously in the storeroom of a factory that made bedroom slippers, as a weaver in a cotton factory, unloading lorries at Woolworth's, as a scaffolder and a plumber's mate. In 1959 he went to the Royal Academy of Dramatic Art: 'I'd read no Shakespeare,' he recalled, 'I didn't know what Restoration Comedy was ... it took me three weeks to read a play.' He nevertheless emerged from the academy as a professional actor and had what he describes as a 'fairly successful' career on stage and television.

He began writing during a period of unemployment when he was thirty, publishing his first novel, *Albert's Memorial*, in 1972. It is a quirky, touching story of the dislocated lives of, and eventual alliance between, an old female tramp and a young homosexual man. It was inspired by a bag-lady he had seen in doorways near South Kensington underground station, and in constructing a life for his fictional character, Cook 'found [himself] talking to doctors and long-term patients at mental hospitals'. His next novel, *Happy Endings*, is another study in isolation, in which a twelve-year-old boy is taken into care after 'interfering with' a child and twenty years later seduces one of his girl pupils: it won the **E.M. Forster** Award. *Walter*, which won the Hawthornden Prize in 1978, and its sequel, *Winter Doves*, record the life and loves of a mentally handicapped man with a passion for racing pigeons. Cook and his partner, the novelist **John Bowen**, had themselves kept pigeons, and Walter was inspired by two people he had known twenty years apart: a shelf-stacker at Woolworth's and a stagehand at Ludlow. Cook creates his characters by research and empathy: 'I gave him my own birthday,' Cook said of Walter, 'as the first step to becoming him.' He has likened writing to acting, claiming that he has 'brought an actor's concern with character to the task of writing fiction'. He adapted *Walter* for a television play, which caused considerable controversy when broadcast on the opening night of Channel 4; it received a Special Award from the jury at the Monte Carlo festival and was nominated for an International Emmy. *Winter Doves* was subsequently adapted for television as *Walter and June*.

The only departure Cook has made in his novels from the shadow world of the contemporary underprivileged is *Sunrising*, which is set in the nineteenth century. 'I probably never shall, as a writer, get away from the walking wounded,' he has said; 'they press in so closely.' Besides adaptations of his own novels (notably *Second Best*, 1994, directed by Chris Menges), he has written a number of television plays, including a series for BBC Schools, and *Closing Numbers*, a drama about a woman coming to terms with the implications of her husband's affair with another man, which was broadcast by Channel 4 on World Aids Day in 1993.

Fiction

Albert's Memorial 1972; *Happy Endings* 1974; *Walter* 1978; *Winter Doves* 1979; *Sunrising* 1984; *Missing Persons* 1986; *Crying Out Loud* 1988; *Soldiers of Ice* 1993

Plays

Square Dance 1968 [np]; *Willy* **(tv)** 1973; *Jenny Can't Work Any Faster* **(tv)** 1975; *Couples* **(tv)** 1976; *A Place Like Home* **(tv)** 1976; *Why Here?* **(tv)** 1976; *Repent at Leisure* **(tv)** 1978; *Mary's Wife* **(tv)** 1980; *Walter* **(tv)** [from his novel] 1982; *Walter and June* **(tv)** [from his novel *Winter Doves*] 1984; *If Only* [pub. in *Scene Scripts 3* ed. Roy Blatchford] 1984; *Singles Weekend* **(tv)** 1984; *Love Match* **(tv)** 1986; *Missing Persons* **(tv)** 1990; *Pity* **(r)** 1990; *Closing Numbers* **(tv)** 1993; *Second Best* **(f)** 1994

Robin (i.e. Robert) COOK 1931–1994

Cook was born in 1931 in London, son of a company director, and educated at Eton, which he hated. He left at sixteen but his Etonian 'front' stood him in good stead in his later criminal career. He had numerous jobs, including waiter, grape picker and taxi driver, but his most significant work experience was his involvement with crime and dodgy businesses. These include smuggling, Soho pornography and fraud: he was the seemingly respectable member of a gang working a 'long firming' racket. He was also married several times and lived on the Continent for long periods.

Cook's first book, *The Crust on Its Uppers* (1962), is a louche and semi-autobiographical novel of Chelsea life, where Bohemia meets the underworld. It is remarkable for its idiosyncratic use of a slang lexicon partly invented by Cook, with terms such as 'morries' (people one respects) and 'slag' (the rest: the public and the punters and the losers). Eric Partridge is said to have recognised it as the greatest source of slang for a quarter of a century.

Cook's novels of upper-middle-class moral decay are only half of his career; the other was his writing of much darker crime novels under the name of Derek Raymond. Whereas wide-boy crime has a certain sleazy glamour in some of the Cook titles, the Raymond books are concerned with deep moral rottenness and psychopathic evil, and they mine a rather more mad and melancholy seam in Cook's psyche. His 'factory' novels (factory being slang for a police station) have become infamous for sheer nastiness; *I Was Dora Suarez* is perhaps the most extreme of them (a publishing editor is said to have been made physically ill by reading the manuscript). The factory novels are highly regarded in France, where Cook lived for some years, and the first of them, *He Died with His Eyes Open*, has been filmed with Charlotte Rampling. Cook achieved media celebrity in the UK during 1992, the year which also saw the publication of his autobiography, *The Hidden Files*. This is sparing in its details of his criminal career, but contains a great deal of lucubration about life and the rationale of 'the black novel' (a formulation evidently owing something to the French *roman noir*). It is, said Cook, the 'novel in mourning' that seeks to alert people to the 'vile psychic weather outside their front doors'. He wrote that he was 'often terrified by the work I do on the lost. I do not mean the dead, so much as those whose misery, as formless as it is profound, is so great they wish they were dead.' Cook is an absolutely distinctive writer and both sides of his output have minor classic status; not great, but not replaceable either.

Fiction

As Robin Cook: *The Crust on Its Uppers* 1962; *Bombe Surprise* 1963; *The Legacy of the Stiff Upper Lip* 1965; *Public Parts and Private Places* 1967; *A State of Denmark* 1970; *The Tenants of Dirt Street* 1971

As 'Derek Raymond': *Le Soleil qui s'éteint* 1980; *He Died with His Eyes Open* 1983; *The Devil's Home on Leave* 1984; *How the Dead Live* 1986; *Cauchemar dans la rue* 1988; *I Was Dora Suarez* 1990; *Dead Man Upright* 1993; *Not Till the Red Fog Rises* 1994

Plays

On ne meurt que deux fois **(f)** 1985; *Les Mois d'avril sont meurtriers* **(f)** 1987; *I Was Dora Suarez* **(f)** [from his novel] 1994

Non-Fiction

The Hidden Files 1992

Lettice (Ulpha) COOPER 1897–1994

Cooper was born in Eccles, the eldest of three children of middle-class parents, but was brought up and spent the first half of her life in

Leeds. Her father, a structural engineer who reviewed occasionally for the *Manchester Guardian* and believed in equal education for girls, guided her towards an early love of books. She won an exhibition to Lady Margaret Hall, Oxford, where she read classics; she loved the clarity and objectivity of Greek and this influence was to be crucial in the attainment of objectivity in her work. On going down she worked for a few years in her father's business, but writing was what she had always wanted to do – she had begun aged seven – and she published *The Lighted Room*, the first of her twenty novels, when she was twenty-eight.

She was a devout Freudian and socialist – converted to socialism during the Depression – and her involvement with working-class people in Leeds was crucial to her development. Her pre-war novels culminated in *The New House* and *National Provincial*. The former is an intricate study of the psychological effects of a removal on members of an extended family. **Maureen Duffy** writes of it: 'the fundamental problems which the book embodies of conservatism versus progress, of the influence of upbringing and the conflicts between parent and child and the whole nature and future of English society are still very much present'. *National Provincial*, another novel of the North, was written as war was approaching and mirrors the times; it is seen by many as her best book.

There was a break during the war years when she was first editorial assistant and drama critic of Lady Rhondda's *Time and Tide*, then public relations officer in the Ministry of Food. It was after the war that she began her love affair with Tuscany which became her second spiritual home. The result was *Fenny*, with its splendid Florentine settings. By the 1960s she had begun writing most successfully for children, but eight more novels were to come, notably *Three Lives*, set in an adult education college, and *Late in the Afternoon* in which she somewhat daringly penetrates the hippy world.

Cooper never married. For many years she lived in London with her sister Barbara (who had been an assistant to **John Lehmann**) and their much-loved cat, 'Patch'. When Barbara died she stayed on there until her eyesight began to fail, and then moved to Norfolk to live with her brother, the writer Leonard Cooper. She was awarded the OBE in 1980. An active member of PEN for many years, she was president from 1979 to 1981; in September 1987 the society celebrated her ninetieth birthday with a huge dinner-party to which many distinguished authors went. From 1958 to 1978 she was president – in the last three years vice-chairman – of the Robert Louis Stevenson Society. She wrote an excellent short biography of him, as well as a monograph on George Eliot, her other great literary enthusiasm. *National Provincial*, *The New House* and *Fenny* have been reissued by Virago, thus reviving interest in her work for a new generation.

Fiction

The Lighted Room 1925; *The Old Fox* 1927; *Good Venture* 1928; *Likewise the Lyon* 1928; *The Ship of Truth* 1930; *Private Enterprise* 1931; *Hark to Rover!* 1933; *We Have Come to a Country* 1935; *The New House* 1936; *National Provincial* 1938; *Black Bethlehem* 1947; *Fenny* 1953; *Three Lives* 1957; *A Certain Compass* 1960; *The Double Heart* 1962; *Late in the Afternoon* 1971; *Tea on Sunday* 1973; *Snow and Roses* 1976; *Desirable Residence* 1980; *Unusual Behaviour* 1986

For Children

Great Men of Yorkshire (West Riding) 1955; *Blackberry Kitten* 1961; *The Young Victoria* 1961; *The Bear Who Was Too Big* 1963; *Bob-a-Job* 1963; *James Watt* 1963; *Conradino* 1964; *Garibaldi* 1964; *The Young Edgar Allan Poe* 1964; *The Fugitive King* 1965; *The Twig of Cypress* 1966; *We Shall Have Snow* 1966; *A Hand upon the Time: a Life of Charles Dickens* 1968; *Robert Louis Stevenson* 1969; *Gunpowder: Treason and Plot* 1970; *Robert the Spy Hunter* 1973; *Parkin* 1977

Non-Fiction

Robert Louis Stevenson 1947; *Yorkshire: West Riding* 1950; *George Eliot* 1951 [rev. 1960, 1964]

William COOPER (pseud. of Harry Summerfield Hoff) 1910–

Cooper was born in Crewe, where both his parents were teachers in public elementary schools. By his own account, he was 'sprung from generations of Dissent', and this inheritance is evident in his novels. He broke with his background by passing from secondary school to Cambridge, where he became a physicist. From 1933 to 1940 he taught physics at Alderman Newton's School in Leicester. He served in the Royal Air Force during the Second World War, emerging as a squadron leader, after which he became an assistant commissioner in the civil service, specialising in matching research scientists with jobs. He did similar work while holding part-time personnel consultancies with the Atomic Energy Authority and the Central Electricity Generating Board: over the years he claims to have interviewed some 50,000 scientists.

He was a close friend and colleague of **C.P. Snow** (about whom he wrote a pamphlet for the

British Council's 'Writers and Their Work' series), and their fiction is often said to bear similarities, though Cooper's is a good deal livelier. He initially published four novels under the name H.S. Hoff in the 1930s, but he made his reputation in 1950 with *Scenes from Provincial Life.* An irreverent and funny story about a rebellious and rather caddish young provincial teacher, the novel was a seminal influence upon the 'Angry Young Men' of the 1950s, but he did not think a senior civil servant could admit to writing it, and so adopted the *nom de plume* under which all his subsequent books have been published. Cooper has taken the book's anti-hero, Joe Lunn, through three further novels: *Scenes from Metropolitan Life* (which was not published for thirty years because of a libel suit), *Scenes from Married Life* and *Scenes from Later Life.*

Scenes from Provincial Life was one of the first novels in which a homosexual character was treated in exactly the same way as the heterosexual ones, without sensationalism or special pleading, and *The Ever-Interesting Topic* is an equally level-headed and amusing account of sexual matters at a boys' public school. Cooper's other novels include *Disquiet and Peace*, which is about a severely depressed upper-class woman in Edwardian times and is greatly admired by discerning critics, though not well known. Cooper has also adapted for the stage Lady Murasaki's eleventh-century Japanese novel *The Tale of Genji*; published *Shall We Ever Know?*, an account of the trial of the Hussein brothers for the murder of Muriel McKay, the wife of a newspaper editor; and written a volume of memoirs, *From Early Life*, which was published on his eightieth birthday.

He married Joyce Harris in 1951 and has two daughters. Joyce Hoff's death from cancer in 1988 is movingly described in 'A History', published in *Granta* 27 (Summer 1989).

Fiction

As: H.S. Hoff: *Three Girls*: *Trina* 1934, *Rhéa* 1935, *Lisa* 1936; *Three Marriages* 1946

As 'William Cooper': *Scenes from Life* series: *Scenes from Provincial Life* 1950, *Scenes from Married Life* 1961, *Scenes from Metropolitan Life* 1982, *Scenes from Later Life* 1983; *The Ever-Interesting Topic* 1953; *The Struggles of Albert Woods* 1953; *Disquiet and Peace* 1956; *Young People* 1958; *Memoirs of a New Man* 1966; *You Want the Right Frame of Reference* 1971; *Love on the Coast* 1973; *You're Not Alone* 1976

Plays

Prince Genji [from Lady Murasaki's novel *The Tale of Genji*] 1960

Non-Fiction

C.P. Snow 1959 [rev. 1971]; *Shall We Ever Know?* 1971; *From Early Life* (a) 1990

Robert (Lowell) COOVER 1932-

Coover was born in Charles City, Iowa, a place he has described as a 'small, rural, mid-American classic town – the kind that Reagan and Eisenhower were nostalgic for'. The family subsequently moved to Indiana, and to Herrin, Illinois, a mining town where Coover's father was the editor of a small newspaper. Cover was educated at the Southern Illinois University, Carbondale, and Indiana University, Bloomington, where he graduated in 1953. He was drafted into the navy and served for four years rising to the rank of lieutenant, after which he holed up in a cabin in Canada to make his first serious effort at writing: 'I found my mentor in [**Samuel**] **Beckett**, in the way he erased the slate, allowed everything to start over again.' He returned to the USA to study history and philosophy at the University of Chicago, and in 1959 married a Spanish woman, Maria del Pilar Sans-Mallafré, with whom he has three children. Since receiving his MA in 1965, Coover has had a succession of short-term academic posts at American universities, and has lived in Spain and elsewhere, believing that a sense of expatriation is necessary to his writing.

Published in *Noble Savage* in 1961, his first story 'Blackdump' was inspired by a mining disaster in Herrin which he had covered for his father's newspaper. The same incident provided the seed for his first novel, *The Origin of the Brunists*, in which Giovanni Bruno, the sole survivor of a mine collapse, believes himself to have been saved by divine intervention and establishes a religious cult. The novel won the Faulkner Award in 1966. In common with much of his later work, it explores the human need to impose order on the chaotic experience of life, and in this sense reflects Coover's view of fiction: 'We need story; otherwise the tremendous randomness of experience overwhelms us.' Coover achieved considerable notoriety with his third novel, *The Public Burning*, a political satire concerning the trial and execution of the Rosenbergs, narrated by the then Vice-President Richard Nixon. He had great difficulty getting the novel published because of fear of libel proceedings, and many reviewers found it overblown and repetitive, but its notoriety ensured good sales figures.

Coover claims literary kinship with other anti-realist American novelists such as **John Barth, John Hawkes** and **William Gass**, but denies that their work is experimental. 'Most of what we call experimental actually has been precisely traditional in the sense that it's gone back to old forms to find its new form,' he has said. In *Gerald's Party* he subverts the classic English detective novel into a frantic tale laced with sex and violence, and in *Pinocchio in Venice* he returns to the most traditional of forms, and wildly transforms the fable of the animated puppet into a farrago of elaborate wordplay and fantastic grotesquerie. Coover has been the recipient of Rockefeller and Guggenheim Fellowships, and won the American Academy of Arts and Letters Award in 1976.

Fiction

The Origin of the Brunists 1966; *The Universal Baseball Association, J. Henry Waugh, Prop.* 1968; *Pricksongs and Descants* (s) 1969; *The Water Pourer* (s) 1972; *The Public Burning* 1977; *Hair o' the Chine* (s) 1979; *After Lazarus* (s) 1980; *Charlie in the House of Rue* (s) 1980; *A Political Fable* 1980; *Spanking the Maid* 1981; *In Bed One Night and Other Brief Encounters* (s) 1983; *Gerald's Party* 1985; *A Night at the Movies* (s) 1987; *Whatever Happened to Gloomy Gus of the Chicago Bears?* 1987; *Pinocchio in Venice* 1991

Plays

A Theological Position [pub. 1972]: *The Kid* 1972, *Love Scene* 1973, *Rip Awake* 1975, *A Theological Position* 1977

Edited

The Stone Wall Book of Short Fiction (s) [with K. Dixon] 1973; *Minute Stories* (s) [with E. Anderson] 1976

Wendy COPE 1945–

Cope was born in Erith, Kent, and was educated at Farringtons School. She graduated from St Hilda's College, Oxford, in 1966 with a degree in history and went on to train as a teacher. From 1967 to 1981 she worked as a London primary-school teacher, eventually becoming deputy headteacher of a school in the Old Kent Road. In 1982 she was seconded as arts and review editor to *Contacts*, a magazine for teachers, and until 1986 taught music one day a week at a south London primary school. 'The best thing about being a poet,' she commented, 'was that it enabled me to give up being a teacher.'

In this she was luckier than many poets, but her work has been popular from the first. She started writing in 1973 and attended several Arvon courses, publishing poems in a number of magazines. Her work appeared in Faber's *Poetry Introduction* series, and that firm became her publisher. Her first collection, the cheekily titled *Making Cocoa for Kingsley Amis*, was published to great critical acclaim in 1986 and was a Poetry Book Society recommendation. It immediately became a bestseller, achieving an astonishing sale of 33,000 copies. Witty, clever and accessible, the poems fell into two categories: brilliant parodies of the daunting likes of Wordsworth and **T.S. Eliot** (*The Waste Land* in every sense reduced to five limericks); and humorous, anguished and unquestionably personal love poems. The second section of the book was made up of work by Jason Strugnell, a bathetic Tulse Hill poet Cope had invented for a BBC Radio 3 programme, *Shall I Call Thee Bard?*, and consists of a series of very funny and accurate parodies of, amongst others, **Philip Larkin, Ted Hughes** and Omar Khayyám. The love poetry is largely colloquial and rueful ('There are so many kinds of awful men – One can't avoid them all'), but also includes 'My Lover', a touching catalogue based upon Christopher Smart's consideration of his cat, Jeoffry. The brief title-poem apparently arose from a dream; asked to comment, **Kingsley Amis** praised Cope's precise regard for metre and added: 'My only real complaint is that one wants more.'

More came in two limited-edition pamphlets (*Men and Their Boring Arguments* and *Does She Like Word-Games?*) and in *Serious Concerns*, the cover of which depicted a teddy bear slumped with a copy of Eliot's *Notes towards a Definition of Culture*. Once again, there were poems about the inadequacy of men ('Bloody Men' and 'Faint Praise' are characteristic titles), parodies, and further verses by Strugnell (now working on a song-cycle), as well as several verses on the unfortunate lot of poets. While Cope's verse is undoubtedly light (some have complained that her work is a substitute for real poetry), it is mostly light in a similar way to that of **John Betjeman**'s verse. The poems display a genuine craftsmanship, and beneath the glittering, seductive surface of her lines there are indeed serious concerns. She won a Cholmondeley Award for poetry in 1987, and her poems have appeared in numerous journals and anthologies, and been broadcast on radio.

Famed for her addiction to nicotine-flavoured chewing-gum and her elegant wit, Cope has worked as a freelance writer and journalist since 1986. As television critic for the *Spectator* from 1986 to 1990, she was notorious for having a faulty set. She has also written a book for children and edited *Is That the New Moon?*, an anthology of poems by women. She is unmarried and describes herself as a 'non-joining, non-lentil-worshipping feminist'.

Poetry

Across the City 1980; *Poetry Introduction* [with others] 1982; *Making Cocoa for Kingsley Amis*

1986; *Men and Their Boring Arguments* 1988; *Does She Like Word-Games?* 1988; *The River Girl* 1991; *Serious Concerns* 1992

For Children

Twiddling Your Thumbs 1988

Edited

Is That the New Moon? 1989; *The Orchard Book of Funny Poems* 1993

A(lfred) E(dgar) COPPARD 1878–1957

Coppard was born in Folkestone, the eldest of four children; his father was a journeyman tailor and his mother a housemaid. The family was poor but, as he recalls in the first volume of his autobiography, *It's Me, Oh Lord!* (he had just finished reading the proofs before he died), the two rooms they lived in were snug enough. He attended Lewes Road Board School, Brighton, but his father's death made it necessary to go out and earn a living at the age of nine. He was employed at one time by a trouser manufacturer in Whitechapel, at another by a paraffin seller; by the age of twenty he had had innumerable jobs. He then became a clerk and accountant, in his spare time competing in athletic meetings and using his prize money to buy books. In 1907 he landed a job with the Eagle Ironworks in Oxford, a turning point in his life; for the first time he met people who stimulated him intellectually.

In 1919 he gave up his job to become a full-time writer, collecting mostly rejection slips but surviving somehow. His first book *Adam and Eve and Pinch Me* appeared two years later – it was the first title to come from the Golden Cockerel Press – and was followed by *Hips and Haws*, a collection of verse, and a second book of stories, *Clorinda Walks in Heaven*, heavily influenced by Chekhov. He found his own voice in his next collection, *The Black Dog*. At their best his stories are informed with delicacy and poetic truth. His people are farm workers, publicans, craftsmen, poachers; he caught their essence and rhythms of speech to perfection. In his later work, however, he seemed to lose the trick. He told a friend that he had learned all there was to know about his craft and there was no fun in it anymore; by then he was earning a decent living, which may also have had something to do with it. Besides *Hips and Haws* he published several volumes of verse, but his best poetry probably went into his stories. He finally settled near Dunmow, Essex, with his second wife, Winifred de Kok from South Africa (his first wife, whom he married in 1905, having died). She became a county medical officer and broadcast on family matters. They had a son and a daughter. A keen footballer, he wrote delightedly to a friend telling him that his son had given him a new pair of football boots for his seventieth birthday.

Fiction

Adam and Eve and Pinch Me (s) 1921; *Clorinda Walks in Heaven* (s) 1922; *The Black Dog* (s) 1923; *Fishmonger's Fiddle* (s) 1925; *The Field of Mustard* (s) 1926; *Count Stephan* (s) 1928; *Silver Circus* 1928; *Pink Furniture* (s) 1930; *Easter Day* (s) 1931; *My Hundredth Story* (s) 1931; *Nixey's Harlequin* (s) 1931; *Crotty Shinkwin* (s) 1932; *Rummy* (s) 1932; *Dunky Fitlow* (s) 1933; *Emergency Exit* (s) 1934; *Ring the Bells of Heaven* (s) 1934; *Polly Oliver* (s) 1935; *Ninepenny Flute* (s) 1937; *Tapster's Tapestry* (s) 1938; *You Never Know, Do You?* (s) 1939; *Collected Stories* (s) 1946; *Fearful Pleasures* (s) 1946; *Selected Tales* (s) 1946; *The Dark-Eyed Lady* (s) 1947; *Lucy in Her Pink Jacket* (s) 1954

Poetry

Hips and Haws 1922; *Pelegea and Other Poems* 1926; *Yokohama Garland* 1926; *Collected Poems* 1928; *Cherry Ripe* 1935

Non-Fiction

It's Me, Oh Lord (a) 1957

Frances (Crofts) CORNFORD 1886–1960

The granddaughter of Charles Darwin and only child of Francis and his second wife, Ellen (*née* Crofts) Darwin, Frances was born into one of Victorian England's most redoubtable intellectual dynasties. The Cambridge in which she spent most of her life, and the exhilarating, if exhausting, family tribe of Darwins, Wedgwoods and Jebbs who peopled her childhood, have been enchantingly described and illustrated by her cousin, the artist and writer Gwen Raverat, in *Period Piece* (1952). Her father was a distinguished botanist, and her mother (a great-niece of the poet Wordsworth) was a Newnham College lecturer.

In 1908 a judicious piece of matchmaking by Jane Harrison, Newnham's formidable pioneering classical anthropologist, resulted in marriage between one of her favourite young graduates, Francis Cornford, and Frances. This confusing proliferation of homonymous forenames inspired a little-known dramatic work, *The Importance of Being Frank*. Gleefully penned by young Darwin cousins, *all* characters were called either Francis or Frances. The youthful playwrights' recommended ploy for preventing further confusion – that Francis the Hero should hitherto be known as Frank – was not adopted by the young Cornford couple.

Frances, though quiet and gentle in manner, proved an irresistible hostess and her considerable powers as an amusing but thoughtful conversationalist drew a remarkably varied group of visitors to her Cambridge house, including Eric Gill, William Rothenstein, Bertrand Russell and Rabindranath Tagore.

Her first book, *Poems*, was published in 1910, and seven more were to follow, the last appearing very shortly after her death in 1960. Her publishers included **Harold Monro** and **Leonard** and **Virginia Woolf**; *Different Days* appeared under the Hogarth Press imprint. Her seventh book, *Collected Poems*, was a Poetry Book Society Choice, and in 1959 she was awarded the Queen's Medal for Poetry.

Rigorously self-critical, she was revising work to the last and versions printed in the 1954 collected edition show that she was constantly refining and pruning what she came to consider excesses in earlier diction and imagery. In the years preceding the First World War she was a valued member of the group known as the Neo-Pagans. She was particularly close to **Rupert Brooke** who gloomily predicted that when their work was jointly reviewed, they would inevitably be referred to as Mr Cornford and Miss Brooke (author of *Dead Pansy-Leaves: and Other Flowerets*). On Brooke's death his executor, Edward Marsh, pleaded with Cornford to act as a mediator between himself and Brooke's fiercely hostile mother – a demanding task which she successfully fulfilled.

As was customary for a young woman of her class and period, she had been educated at home (no hardship in a family whose intellectual standards were very much higher than most schools could hope to provide). Linguistically gifted, she later produced a number of translations from French and Russian, and **Stephen Spender** declared her to be 'one of the best translators living'. Keenly alive to Cambridge's history and its seductively austere physical beauty, she was also painfully aware that it and the literary tradition which Cambridge-educated poets had done so much to create through the centuries often excluded women with barely concealed hostility. From the earliest, her work (wrongly dismissed and so often condescended to as 'sweetly pretty', feminine and domestic) confronted pain and loss, loneliness and fear. Throughout her life she suffered periods of intense depression, often following upon bereavement. Raised in a household which firmly ignored religion, she was nevertheless beset from an early age by religious doubt. A quiet but strong strain of mysticism runs through her work, and she died a convinced Anglican.

The terrors and rewards of motherhood are amongst her constant themes. Poetry and parenthood combined painfully in her dealings with her son, **John Cornford**, born in 1915 and the oldest of her five children. Himself a poet, and today romantically enshrined as the archetypal young Briton who died for Spain (fighting for the Republican side in 1936), his increasingly modernist loyalties made him loudly contemptuous of his mother's verse and values. Mercifully, his attitudes changed before his death, and he was able to tell her so.

Today she receives little serious critical attention, although there have been recent signs of fresh and more sympathetic interest from feminist critics, who note that her preference for 'old-fashioned' forms did not diminish her poetry's consistent integrity or its undoubted capacity to dissect and convey thoroughly 'modern' pains and perplexities.

Poetry

Poems 1910; *Spring Morning* 1915; *Autumn Midnight* 1923; *Different Days* 1928; *Mountains and Molehills* 1934; *Travelling Home and Other Poems* 1948; *Collected Poems* 1954; *On a Calm Shore* 1960

Translations

Poems from the Russian [with E.P. Salaman] 1943; Paul Éluard *Le dur désir de durer*, [with S. Spender] 1950; *Fifteen Poems from the French* 1976

Plays

Death and the Princess 1912

(Rupert) John CORNFORD 1915–1936

Cornford was the son of F.M. Cornford, a Cambridge professor, and the poet **Frances Cornford**. Despite winning a scholarship to Stowe when he was thirteen, he rebelled against the public-school system, and after refusing to join the Officers' Training Corps, became interested in socialism, fiercely opposing family Liberalism. Friendship with Sydney Schiff, the translator of Proust, encouraged Cornford to write poetry when he was fifteen, but he abandoned this for a time to study Marxism, an interest he actively promoted when he became an undergraduate at Cambridge in 1933. Untidy by habit, he tramped about 'wrapped in portentous gloom and an antique raincoat', but his political speeches and writing were penetrating and methodically prepared; they included *Communism in the Universities*, published in the year of his death. The widespread development of socialist awareness in British universities was almost entirely due to him.

After graduating with first class honours, he volunteered as a soldier in the Spanish civil war, serving for a month in Aragon, before returning to England to form a company which later developed into the British section of the International Brigade. Popularity elected Cornford as the reluctant leader of a machine-gun unit which fought in Madrid, and in University City, where he was wounded in the head by shrapnel. After a day in hospital, he was back at the front, but was later forced to retreat from the bombardment of German troops in Boadilla del Monte. He was killed on 28 December, the day after his twenty-first birthday, while leading his men into action in Cordova. Cornford was anything but sentimental, but his early death made him something of a legendary figure – he had, after all, been named after that First World War icon of youthful self-sacrifice, **Rupert Brooke**, who had been a close friend of his mother. The few poems that survive, however – including 'Full Moon at Tierz: Before the Storming of Huesca', written in Spain – are evidence of a substantial and genuine talent.

Non-Fiction

Communism in the Universities 1936

Collections

John Cornford: a Memoir [ed. P. Sloan] 1938; *Understand the Weapon, Understand the Wound: Selected Writings of John Cornford* [ed. J. Galassi] 1976

Biography

Journey to the Frontier: Julian Bell and John Cornford by Peter Stansky and William Abrahams 1966

Gregory (Nunzio) CORSO 1930–

The youngest and cutest of the Beat poets, the Italian-American Corso got through five sets of foster parents and three years in jail before he was twenty. In Clinton, New York (charged with attempted armed robbery), he read extensively and on his release worked as a labourer, then as a reporter on the *LA Examiner*, before heading for Harvard where he gate-crashed lectures and published his first book of poems, *The Vestal Lady on Brattle* (1955). At this point, he was claimed by **Jack Kerouac, Allen Ginsberg** and **Lawrence Ferlinghetti**. Ginsberg, in the introduction to the City Lights Pocket Poet edition of *Gasoline* (1958) observed: 'he is a great word-slinger, first naked sign of a poet, a scientific master of mad mouthfuls of language'; and Kerouac wrote: 'he is a tough young kid from the Lower East Side who rose like an angel over the rooftops and sang Italian songs ... Amazing and beautiful Gregory. Read slowly and see'.

Corso travelled throughout Europe in the early 1960s (he speaks of having '50 poems to one pair of underwear') and in 1965, back in New York, he appeared in Andy Warhol's film *Couch*, and moved in with Ginsberg for a while.

True to the Beat style, Corso's poetry is deeply personal, full of inner experience and observations on day-to-day living. His 'I' is deliberately naïve and constantly bewildered by a passionless, secretive and destructive world in which he finds beauty and searches for truth. He aspires to transcendental freedom and pure creativity, states of consciousness which he feels are held back by religion and American society. In 'Bomb', the notorious poem written in the shape of a mushroom cloud, he juxtaposes unrelated images to illustrate his feeling of need for complete revolution: 'Turtles exploding over Istanbul ... / The top of the Empire State / Arrowed in a broccoli field in Sicily'. Corso speaks of automatic writing as 'an entranced moment in which the mind accelerates'. He was greatly inspired by Kerouac's stream-of-consciousness method, and while his experiments with sound were not as cataclysmic as Ginsberg's, most of his work is clearly written to the moods and patterns of 'bop' and for reading aloud.

As the Beats grew older, less sensational and less wild, so they wrote less. Then, in 1975, Ginsberg and Corso appeared on their old podium at Columbia University before an anxious audience. Corso had married three times and had three children. Kerouac had been dead for fifteen years. 'Sixteen years ago we were put down for being filthy beatnik sex commie dope fiends,' said Corso in his prologue that evening. Beat writing had already greatly influenced song-writing and journalism and demonstrated to the new generation of writers that immediacy and the reduction of retrospective makes literature closer to life.

Poetry

The Vestal Lady on Brattle and Other Poems 1955; *Bomb* 1958; *Gasoline* 1958; *A Pulp Magazine for the Dead Generations* [with H. Marsman] 1959; *The Happy Birthday of Death* 1960; *Minutes to Go* [with others] 1960; *Long Live Man* 1962; *Selected Poems* 1962; *Penguin Modern Poets 5* [with A. Ginsberg, L. Ferlinghetti] 1963; *The Mutation of the Spirit* 1964; *There is*

Yet Time to Run through Life and Expiate All That's Been Sadly Done 1965; *The Geometric Poem* 1966; *10 Times a Poem* 1967; *Elegiac Feelings American* 1970; *Ankh* 1971; *Egyptian Cross* 1971; *(Poems)* 1971; *The Night Last Night Was at Its Nightest* 1972; *Earthegg* 1974; *Herald of the Autochthonic Spirit* 1981; *Mind Field* 1989

Plays

In This Hung-Up Age 1955 [pub. 1964]; *Standing on a Streetcorner* [pub. 1962]; *That Little Black Door on the Left* [pub. in *Pardon Me Sir, But Is My Eye Hurting Your Elbow?* 1968]; *Way Out* 1974

Fiction

The American Express 1961

Non-Fiction

The Minicab War [with A. Hollo, T. Raworth] 1961; *The Japanese Notebook* 1974; *Writings from Oxford* 1981

Noël (Pierce) COWARD 1899–1973

An exemplar of the glamorous and sophisticated inter-war period, Coward was born prosaically enough in Teddington, Middlesex, the son of a piano salesman. Encouraged by his mother to enter the theatre, he became a child actor, appearing as Slightly in the 1913 revival of *Peter Pan*. In 1924 his melodrama *The Vortex*, which concerned a young man's addiction to drugs and his mother, had a *succès de scandale*, but it was the brittle comedy *Hay Fever* (1925) which made his reputation as a leading playwright.

He was an extremely accomplished songwriter whose haunting melodies and witty, bittersweet lyrics captured the world-weary cynicism of the period. He acted in his own plays, notably *Private Lives*, in which he was partnered for the first time by Gertrude Lawrence, a frequent and popular interpreter of his work. *Design for Living*, his great success of the 1930s, concerned a three-way relationship between two men and a woman, and combined the 'daring' qualities of *The Vortex* with the aphoristic style for which he had become famous.

During the Second World War he wrote and appeared in plays and films which patriotically charted the fortunes of England and the English, '*This Happy Breed*', as he sentimentalised and simplified them. His excursions into prose, notably some long short stories and a comic novel, were less successful. Out of fashion in the post-war theatre (and writing distinctly below par), he turned to cabaret, in which he triumphed, particularly in America. These performances and a number of bizarre appearances in films in the 1960s emphasised a high-camp element of his life, the source of which was left unrevealed in two volumes of autobiography. By now he was once more a leading socialite and, so his posthumously published diaries maintained, an intimate of royalty. He was knighted in 1970, retired to Jamaica and took up painting, his reputation firmly re-established.

Plays

I'll Leave It to You 1920; *Sirocco* 1921 [pub. 1927]; *The Young Idea* 1922 [pub. 1924]; *The Rat Trap* 1926 [pub. 1924]; *The Vortex* 1924 [pub. 1925]; *Chelsea Buns* [as 'Hernia Whittlebot'] 1925; *Easy Virtue* 1925 [pub. 1936]; *Fallen Angels* 1925; *Hay Fever* 1925; *The Queen Was in the Parlour* 1926; *The Marquise* 1927; *Home Chat/Sirocco/This Was a Man* [pub. 1928]; *Bitter Sweet* 1929; *Private Lives* 1930; *Cavalcade* 1931 [pub. 1932]; *Post-Mortem* 1968 [pub. 1931]; *Design for Living* 1933; *Conversation Piece* 1934; *Play Parade: the Collected Plays of Noël Coward* 1934; *Point Valaine* 1934; *Tonight at 8.30* 3 vols [pub. 1936]; *Operette* 1938; *Blithe Spirit* 1941; *Present Laughter* 1943; *This Happy Breed* 1943; *Brief Encounter* [from his play *Still Life*] (f) 1945 [pub. in *Three British Screenplays* ed. R. Manvell 1950]; *Peace in Our Time* 1947; *Relative Values* 1951 [pub. 1952]; *Quadrille* 1952; *After the Ball* [from O. Wilde's play *Lady Windermere's Fan*] 1954; *Nude with Violin* 1956 [pub. 1957]; *South Sea Bubble* 1956; *Look After Lulu!* [from G. Feydeau's play *Occupe-toi d'Amélie*] 1959; *Waiting in the Wings* 1960; *Suite in Three Keys* 1966; *The Plays of Noël Coward* 5 vols [pub. 1979–83]

Fiction

A Withered Nosegay (s) 1922; *To Step Aside* (s) 1939; *Star Quality* (s) 1951; *Pomp and Circumstance* 1960; *The Collected Short Stories* (s) 1962; *Pretty Polly Barlow and Other Stories* (s) 1964; *Bon Voyage and Other Studies* (s) 1967; *Collected Short Stories of Noël Coward* (s) 1969; *The Complete Short Stories of Noël Coward* (s) 1985

Non-Fiction

Present Indicative (a) 1937; *Middle East Diary: July–October 1943* 1944; *Future Indicative* (a) 1954; *The Noël Coward Diaries* 1982

Poetry

Poems by Hernia Whittlebot [as 'Hernia Whittlebot'] 1923; *The Lyrics of Noël Coward* 1965; *Collected Verse* [ed. G. Payn, M. Tickner] 1984

Edited

Spangled Unicorn 1932

Collections

Collected Sketches and Lyrics 1931

Biography
The Life of Noël Coward by Cole Lesley 1976; *Noël Coward* by Clive Fisher 1992

James Gould COZZENS 1903–1978

Cozzens was born in Chicago and brought up on Rhode Island where the family had been established since the seventeenth century (his great-grandfather had been the Mayor of Newport and Governor of Rhode Island during the civil war). In 1916 he was sent to Kent School, Connecticut, an Episcopal prep school, and in 1920 published an essay, 'A Democratic School', in the *Atlantic Monthly*. In his diary the next year he wrote: 'If I cannot write I shall have lived my life worth nothing, leaving the world no better no wiser for me.'

When he entered Harvard in 1922 he had already started his first novel, *Confusion*, the story of a young woman's quest for enduring spiritual and intellectual values. It was published in April 1924, whereupon Cozzens left Harvard, considering himself 'an author of far too much promise to waste time any longer at schoolboy work'. His biographer, Matthew J. Bruccoli, however, also cites debt and ill health as factors. Moving to New Brunswick he wrote *Michael Scarlett* an ambitious depiction of Elizabethan England. He worked as a schoolteacher in Cuba and travelled in Europe with his mother before returning to the USA in 1927 where he married his literary agent, Bernice Baumgarten, at the end of the year.

In Cozzens's experimental novel *Castaway*, the protagonist, Mr Lecky, is unaccountably alone in a vast department store; although the text offers no explanation of his plight, the blurb of the English edition stated that he is the sole survivor of a catastrophe which has destroyed New York. The novel contains some suggestion of the self-confessed 'more or less illiberal' political views which were later to rankle with his detractors. He worked for a year on *Fortune* magazine in 1938 and, having completed his most highly regarded novel to date, *The Just and the Unjust*, volunteered for the Army Air Forces in 1942. Assigned to writing manuals and special reports in Washington, he compiled a voluminous diary which furnished him with material for *Guard of Honor*, which won a Pulitzer Prize in 1949.

Cozzens produced little fiction in the last twenty-five years of his life and became in the public mind an arrogant and eccentric recluse. A *Time* cover story in 1957 entitled 'The Hermit of Lambertville' assisted sales of *By Love Possessed*, but added to this impression. Cozzens died in 1978, a few months after his wife.

Fiction
Confusion 1924; *Michael Scarlett* 1925; *Cock Pit* 1928; *The Son of Perdition* 1929; *SS San Pedro* 1931; *The Last Adam* 1933 [UK *A Cure of Flesh* 1934]; *Castaway* 1934; *Men and Brethren* 1936; *Ask Me Tomorrow* 1940; *The Just and the Unjust* 1942; *Guard of Honor* 1948; *By Love Possessed* 1957; *Children and Others* (s) 1964; *Morning, Noon and Night* 1968

Non-Fiction
A Flower in Her Hair 1974; *A Rope for Dr Webster* 1976; *Just Representations: a James Gould Cozzens Reader* [ed. M.J. Bruccoli] 1978; *Selected Notebooks, 1960–1967* [ed. M.J. Bruccoli] 1984; *A Time of War: Air Force Diaries and Pentagon Memos, 1943–45* [ed. M.J. Bruccoli] 1984

Biography
James Gould Cozzens: a Life Apart by Matthew J. Bruccoli 1983

Jim CRACE 1946–

Crace's background is working class; he was born in 1946 and brought up in north London, where his father was a Co-op insurance agent. After school, between 1965 and 1968, Crace went to the Birmingham College of Commerce and took an English literature honours degree externally at the University of London. He then decided to 'do the travel thing'. Through VSO he got posted to Khartoum as an assistant in Sudanese Educational Television. From there he went on to explore the Middle East and the Indian Ocean islands and, after a spell teaching geography in Botswana, he returned to England in 1970. The following year he started writing scripts for BBC Schools Radio and gradually moved into writing adult plays for Radio 4.

In 1974 he moved back to Birmingham and began to write fiction. He found success early on: everything he has ever written, he says with pride, has been published. In 1975 a story appeared in Faber's *Introductions 6*, which led to approaches from publishers and offers of advances should he be writing a novel. He wasn't, but quickly said that he was and got an advance of £1,500 from Heinemann. Nevertheless, between 1975 and 1986, Crace calculates that ninety per cent of his time was dedicated to earning a living as a journalist, doing literary criticism and features for magazines and newspapers including the *Radio Times* and the *Sunday Times*. During this period, he also worked as the writer-in-residence at the Midland Arts Centre and in 1983 he was the initiator and director of the first Birmingham Festival of Readers and Writers.

His publisher was meanwhile pressuring him to finish the long-awaited first novel. The manuscript was sent to **John Fowles** and **David Lodge**, who both responded enthusiastically, comparing his work to that of Jorge Luis Borges. Heinemann adopted a bold marketing strategy, halving the hardback price and quadrupling the print run, and before *Continent* had even reached the bookshops Crace had been offered an advance for his next novel of £20,000. Crace could clearly make money from his fiction: he gave up journalism completely, immediately. The novel won the David Higham, *Guardian* Fiction and Italian Chianti Ruffino/Antico Fattore Prizes and was winner of the Whitbread Award for a first novel. His second novel, *The Gift of Stones*, won the American 1990 GAP International Prize for Literature. His stories have appeared in the *New Review* and the *London Review of Books*.

He is a politically active person. A member of the Labour Party since the age of fourteen, he has also long been involved in CND and third world politics. For his fiction he creates extraordinary 'mythical lands' which serve as settings through which he can explore ideas. That he is happiest with fantasy as his structural base does not surprise him. As a child, he 'spent hours drawing maps of places that didn't exist. They were so detailed they were convincing.' He lives in Birmingham with his wife and two children.

Fiction
Continent 1986; *The Gift of Stones* 1988; *Arcadia* 1992; *Signals of Distress* 1994

Plays
The Bird Has Flown (r) 1977; *A Coat of Many Colours* (r) 1979

(Harold) Hart CRANE 1899–1932

The son of the owner of a large sweet-manufacturing company, Crane grew up in Cleveland, Ohio. His father early grew to distrust what he saw as Crane's effeminacy, and the parents' marriage ended in divorce: in this childhood were sowed the traumas which were to dominate Crane's adult life. As an adolescent, he spent time, with his mother or alone, in Cuba, Paris and New York, backgrounds which gave him his twin symbols of the sea and the city. During the First World War, he worked as a labourer in munition yards and shipyards. His father hoped to knock the 'poetic nonsense' out of him, and took him into the family business, but Crane rebelled against this several times, and in the early 1920s was in New York, producing advertising copy for a living, but beginning to write poetry seriously. Alcoholism was growing on him, however, which he combined with aggressive, loveless homosexuality and a penchant for brawls: while in Paris, for instance, a scrap with a waiter led to his spending a night cooling off in the Santé.

White Buildings (1926) was an outstandingly achieved first volume of poetry. Shortly before it was published, the philanthropy of the banker Otto Kahn, who donated $2,000, enabled Crane to devote his time not only to his wild lifestyle but also to composing an American poetic epic, *The Bridge*, which, if successful only in parts, is still acknowledged to contain some of the century's most beautiful poetry. Towards the end of his life, Crane travelled to Mexico, hoping to write a Mexican epic to complement his North American one. He took ship to the USA with this still unachieved, but accompanied by a woman with whom for the first time he had achieved a long-term heterosexual relationship. However, after a night on board spent gambling with some sailors, who eventually beat him up, he threw himself overboard and was drowned. Crane's work is difficult and modernistic, shot through with strong strains of lyricism and personal melancholy: he is an uneven, imperfectly educated but emphatically major poet.

Poetry
White Buildings 1926; *The Bridge* 1930; *Collected Poems of Hart Crane* [ed. W. Frank] 1933; *Seven Lyrics* 1966; *Ten Unpublished Poems* 1972; *The Poems of the Hart Crane* [ed. M. Simon] 1986

Non-Fiction
The Letters of Hart Crane, 1916–1932 [ed. B. Weber] 1952; *Robber Rocks: Letters and Memories of Hart Crane 1923–1932* [ed. S. Jenkins Brown] 1969; *Letters of Hart Crane and His Family* [ed. T.S.W. Lewis] 1974; *Hart Crane and Yvor Winters: Their Literary Correspondence* [ed. T. Parkinson] 1978

Collections
The Complete Poems and Selected Letters and Prose of Hart Crane [ed. B. Weber] 1966 [rev. 1984]

Biography
Voyager: a Life of Hart Crane by John Unterecker 1970

Robert CREELEY 1926–

Creeley was born in Arlington, Massachusetts, the son of a physician who died when his son was four; Creeley and his sister were raised by their mother on a small farm. He was admitted to

Harvard in 1943 but spent the last two years of the war as a field ambulance driver in India and Burma, before returning to his studies. He married and left Harvard without graduating, becoming a subsistence farmer and attempting to launch a small magazine. This attempt failed, but it was the start of a long correspondence with **Charles Olson**, who was to be a major influence.

The Creeleys moved to France and then to Majorca, where they stayed until their divorce in 1955. Creeley wrote a novel there (*The Island*, not published until 1963) and set up his own small press, the Divers Press. His first book of poems, *Le Fou*, was published in 1952, followed by *The Kind of Act of* in 1953. Olson invited Creeley to teach at the Black Mountain College in 1954–5, and he was to become a major member of the Black Mountain School of poets, founding and editing the *Black Mountain Review*. He moved to Albuquerque and married again, schoolteaching and taking an MA at the University of New Mexico in 1960. Although he had published numerous small-press books, his career really took off in 1962 with the nationally noticed collection *For Love: Poems 1950–1960*.

Creeley's free verse was strongly influenced by Olson, and the two men developed the concept of 'projective verse', which would reject traditional form in favour of an open-ended construction that was to be discovered in the process of composition. Olson credited Creeley with formulating one of the basic tenets of this poetry, that 'form is never more than an extension of content'. Creeley was influenced not only by Olson, **William Carlos Williams**, and **Allen Ginsberg**, but also by jazz musicians, with their communication of feeling in free extemporaneous forms. He was at one time associated with the Beat movement and with the San Franciso Poetry Renaissance, and his work appeared in the influential Beat anthology edited by Donald Allen, *The New American Poetry: 1945–1960* (1960).

Creeley's most characteristic work rejects traditional poetic elements such as irony, lyricism, metaphor, symbolism and closure. It is also centrally concerned with relationships and with his marriages: *For Love* is particularly concerned with the break-up of one relationship and the beginning of another. This second marriage lasted nearly twenty years, with a third marriage in 1977. His academic career has consolidated into a professorship at the State University of New York, Buffalo. His later work has mellowed to some extent, becoming retrospective and concerned with ageing, but no less concerned with the realm of immediate feeling and 'visible truth' as 'the apprehension of the absolute condition of present things' – a definition from Herman Melville that Creeley quotes with approval in his introduction to *The New Writing in the USA*. His reputations is now assured, and he has become one of the leading figures in American poetry.

Poetry

Le Fou 1952; *The Kind of Act of* 1953; *The Immoral Proposition* 1953; *A Snarling Garland of Xmas Verses* 1954; *All That is Lovely in Men* 1955; *Ferrini and Others* [with others] 1955; *If You* 1956; *The Whip* 1957; *A Form of Woman* 1959; *For Love: Poems 1950–1960* 1962; *Distance* 1964; *Two Poems* 1964; *Hi There!* 1965; *Words* 1965; *About Women* 1966; *For Joel* 1966; *Poems 1950–1965* 1966; *A Sight* 1967; *Words* 1967; *The Boy* 1968; *The Charm* 1968; *Divisions and Other Early Poems* 1968; *The Finger* 1968; *5 Numbers* 1968; *Numbers* 1968; *Pieces* 1968; *Hero* 1969; *Mazatlan: Sea* 1969; *A Wall* 1969; *America* 1970; *As Now It Would Be Snow* 1970; *For Benny and Sabina* 1970; *For Betsy and Tom* 1970; *Christmas: May 10* 1970; *The Finger: Poems 1966–1969* 1970; *In London* 1970; *Mary's Fancy* 1970; *For the Graduation* 1971; *1.2.3.4.5.6.7.8.9.0.* 1971; *St Martin's* 1971; *Sea* 1971; *Change* 1972; *A Day Book* 1972; *One Day After Another* 1972; *For My Mother* 1973; *His Idea* 1973; *Kitchen* 1973; *Sitting Here* 1974; *Thirty Things* 1974; *Backwards* 1975; *Away* 1976; *Myself* 1977; *Thanks* 1977; *The Children* 1978; *Desultory Days* 1978; *Later* 1978; *Later: New Poems* 1979; *Corn Close* 1980; *Mother's Voice* 1981; *Echoes* 1982; *A Calendar* 1983; *Mirrors* 1983; *Four Poems* 1984; *Memories* 1984; *Memory Garden* 1986; *The Company* 1988; *Windows* 1990

Plays

Listen 1972

Fiction

The Gold Diggers (s) 1954; *The Island* 1963; *Mister Blue* 1964; *Mabel* (s) 1976

Non-Fiction

An American Sense 1965; *Contexts of Poetry* 1968; *A Day Book* 1970; *A Quick Graph: Collected Notes and Essays* 1970; *Contexts of Poetry: Interviews 1961–1971* [ed. D. Allen] 1973; *The Creative* 1973; *Inside Out: Notes on the Autobiographical Mode* 1973; *A Sense of Measure* 1973; *Presences: a Text for Marisol* 1976; *Was That a Real Poem or Did You Just Make It Up Yourself* 1976; *Was That a Real Poem and Other Essays* [ed. D. Allen] 1979; *Charles Olson and Robert Creeley: the Complete Correspondence* [ed. G.F. Butterick] 8 vols 1980–87; *The Collected Prose of Robert Creeley* 1984; *The Collected Essays of Robert Creeley* 1989; *Irving Layton and Robert Creeley: the Complete Correspondence 1953–1978* [ed. E. Faas, S. Reed 1990

Edited
Charles Olson: *Mayan Letters* 1953, *Selected Writings* 1966, *Selected Poems* 1993; *New American Story* [with D. Allen] 1965; *The New Writing in the USA* [with D. Allen] 1967; *Whitman* 1973; Joanne Kyger *Going On: Selected Poems 1958–1980* 1983; *The Essential Burns* 1989

Richmal CROMPTON (Lamburn) 1890–1969

The daughter of a curate-schoolmaster, Crompton was born in Bury, Lancashire, and educated at St Elphin's School for the daughters of the clergy, Warrington, and at Royal Holloway College in Surrey. She obtained a BA in classics in 1914 and returned to her old school as a teacher. After three years, she took the post of classics mistress at Bromley High School in Kent, but in 1923 contracted polio, which left her right leg paralysed, and she was obliged to retire from teaching. Fortunately, by this time she was able to support herself by her writing, as she was to do for the remainder of her life. By 1927, she could afford to have a substantial house built for herself in Bromley, where she lived with her mother.

Her first published story appeared in the *Girl's Own Paper* in 1918 and featured a small boy who is clearly the forerunner of her most famous creation, William Brown, who made his debut in a story entitled 'Rice-Mould' in the February 1919 edition of *Home Magazine*. The first of the thirty-eight William books, *Just – William*, appeared in 1922; the last, *William – the Lawless*, in 1970. Crompton also wrote a play, *William and the Artist's Model*, and a book about the 1939 film *Just William* (one of several adaptations of the stories for cinema and television). Apart from the William books, she published three volumes of stories about a younger boy called Jimmy (these had first appeared in the *Star*, a London evening newspaper), several volumes of short stories about female juveniles, and forty-one novels for adults, none of which is read or remembered today. In fact, the William stories were also originally intended for an adult audience, but Crompton's habit of always taking the side of her anarchic protagonist rather than that of the adult characters ensured that children eventually made up her principal readership. Her skill as a writer is such, however, that these sophisticated stories, characterised by a crisp style and an ironic wit, continue to appeal as much to adults as they do to children.

William lives in Kent, is middle class and (forever) eleven years old. Time passes (in later books William is involved with evacuees, the Brains Trust, television, pop singers and moon rockets), but William remains the cheerful, scruffy, indomitable boy he always was, his image brilliantly captured in Thomas Henry's vigorous illustrations, which became an integral part of the books until he died in his eighties in 1962. (The last five books were illustrated rather less vividly by Henry Ford.) William has a mongrel named Jumble and a gang called the Outlaws: Henry, Ginger and Douglas. The bane of his life is soppy, lisping Violet Elizabeth Bott, daughter of a 'sauce magnate' who has bought the local big house. Crompton rejoices in Plato's assertion that 'a boy is, of all wild beasts, the most difficult to tame', and William's exploits cause endless anxiety to his parents and annoyance to his older brother and sister, Robert and Ethel. Time may have stood still in William's quintessentially English village, but the books have not dated at all, and remain as funny today as they did when first published. Attempts to bring William to the screen have been largely unsuccessful, although popular. Robbed of Crompton's authorial voice, the stories seem mundane, and William has never come alive in film as he does on the page.

Crompton's life was dedicated to writing and remained uneventful. Undaunted by her disability, she continued to cycle, with her 'dead' leg (as she put it) sticking out at an odd angle, and travelled abroad. She developed cancer in her forties but underwent a mastectomy, surviving into an active old age. She served with the Auxilliary Fire Service and at a Toc H canteen during the Second World War. She did not marry, but remained close to her family, finding in nephews new inspiration for the William stories.

For Children
Just – William (s) 1922; *More William* (s) 1923; *William Again ...* (s) 1923; *William – the Fourth* (s) 1924; *Still – William* (s) 1925; *Kathleen and I, and of course, Veronica* (s) 1926; *William – the Conqueror* (s) 1926; *William – in Trouble* (s) 1927; *William – the Outlaw* (s) 1927; *William – the Good* (s) 1928; *William ...* (s) 1929; *William – the Bad* (s) 1930; *William's Happy Days* (s) 1930; *William's Crowded Hours* (s) 1931; *William – the Pirate* (s) 1932; *William – the Rebel* (s) 1933; *William – the Gangster* (s) 1934; *William – the Detective* (s) 1935; *Sweet William* (s) 1936; *William – the Showman* (s) 1937; *William – the Dictator* (s) 1938; *William and ARP* (s) 1939 [as *William's Bad Resolution* 1956]; *William and the Evacuees* (s) 1940 [US *William the Film Star* 1956]; *William Does His Bit* (s) 1941; *William Carries On ...* (s) 1942; *William and the Brains Trust* (s) 1945; *Just William's Luck* 1948; *Jimmy* (s) 1949; *William*

– *the Bold* (s) 1950; *Jimmy Again* (s) 1951; *William and the Tramp* (s) 1952; *William and the Moon Rocket* (s) 1954; *William and the Space Animal* (s) 1956; *William's Television Show* (s) 1958; *William – the Explorer* (s) 1960; *William's Treasure Trove* (s) 1962; *William and the Witch* (s) 1964; *Jimmy the Third* (s) 1965; *William – the Ancient Briton* (s) 1965; *William and the Cannibal* (s) 1965; *William and the Monster* (s) 1965; *William and the Pop Singers* (s) 1965; *William – the Globetrotter* (s) 1965; *William and the Masked Ranger* (s) 1966; *William – the Superman* (s) 1968; *William – the Lawless* (s) 1970

Fiction

The Innermost Room 1923; *The Hidden Light* 1924; *Anne Morrison* 1925; *The Wildings* 1925; *David Wilding* 1926; *The House* 1926 [US *Dread Dwelling* 1926]; *Enter – Patricia* (s) 1927; *Leaden Hill* 1927; *Millicent Dorrington* 1927; *A Monstrous Regiment* (s) 1927; *Felecity – Stands By* (s) 1928; *The Middle Things* (s) 1928; *Mist and Other Stories* (s) 1928; *Roofs Off* 1928; *The Thorn Bush* 1928; *Abbots' End* 1929; *The Four Graces* 1929; *Ladies First* (s) 1929; *Sugar and Spice* (s) 1929; *Blue Flames* 1930; *Naomi Godstone* 1930; *Portrait of a Family* 1931; *The Silver Birch and Other Stories* (s) 1931; *Marriage of Hermione* 1932; *The Odyssey of Euphemia Tracey* 1932; *The Holiday* 1933; *Chedsy Place* 1934; *The Old Man's Birthday* 1934; *Quartet* 1935; *Caroline* 1936; *The First Morning ... Second Impression* (s) 1936; *There Are Four Seasons* 1937; *Journeying Wave* 1938; *Merlin Bay* 1939; *Steffan Green* 1940; *Narcissa* 1941; *Mrs Frensham Describes a Circle* 1942; *Weatherley Parade* 1944; *Westover* 1946; *The Ridleys* 1947; *Family Roundabout* 1948; *Frost at Morning* 1950; *Linden Rise* 1952; *Gypsy's Baby* 1954; *Four in Exile* 1955; *Matty and the Dearingroydes* 1956; *Blind Man's Buff* 1957; *Wiseman's Folly* 1959; *The Inheritor* 1960; *The Ridleys* 1970

Plays

William and the Artist's Model 1956

Non-Fiction

Just William: the Story of the Film 1939

Biography

Just Richmal: the Life and Work of Richmal Crompton Lamburn by Kay Williams 1986; *Richmal Crompton* by Patricia Craig 1986 *Richmal Crompton: the Woman Behind William* by Mary Cadogan 1986

Countee CULLEN 1903–1946

One of the two towering figures of the Harlem Renaissance, Cullen, was born Countee LeRoy Porter in Louisville, Kentucky. He was adopted by the Revd Frederick A. Cullen, an important figure in Harlem who later became head of the Harlem chapter of the NAACP (National Association for the Advancement of Colored People) and was a shaping force on the young man who took his name.

Cullen was educated at New York University and Harvard, obtaining his MA in 1926. His meteoric career was launched in this period, and praise and prizes were coming thick and fast when he published his first collection of poems, entitled *Color*, in 1925. Over the next few years he was a major celebrity, with four volumes of poetry in print by 1929. His 1928 marriage to Nina Yolande DuBois – daughter of W.E.B. DuBois, the leading black intellectual – was a colossal celebration uniting two leading Harlem families, but they divorced in 1930. One reason the marriage failed was that he was also involved in an intense relationship with the handsome Harold Jackman, a teacher whom **Carl Van Vechten** had used as a model for the protagonist of his 1926 novel *Nigger Heaven*. Cullen and Jackman were known as 'David and Jonathan', and their friendship is commemorated by the Countee Cullen-Harold Jackman Memorial Collection of material relating to black Americans, which is housed at Atlanta University. Cullen subsequently married Ida Mae Robertson in 1940.

He did not ignore racial themes in his poetry ('The Black Christ', for example, is about a lynching), but he was against the development of a specifically black writing and insisted poetry was raceless. This has not in general been the way that black writing has gone, and it is this more than anything else which has caused the decline in his reputation. His fine Romantic verse is formally conservative and it is modelled on white and largely nineteenth-century poets: his favourite was Keats.

As the two great poets of the Harlem Renaissance, Cullen and **Langston Hughes** represent two tendencies. Cullen urged Hughes to avoid black jazz rhythms in his poetry, while Hughes's 1926 essay 'The Negro Artist and the Racial Mountain' took Cullen devastatingly to task for having said: 'I want to be a poet – not a Negro poet.' These racial contradictions largely undid Cullen – as in his notorious lines on the mysterious trick God had played with him: 'Yet do I marvel at this curious thing / To make a poet black, and bid him sing' – and his reputation had already declined in his lifetime. Cullen has not been favoured by posterity, but his work is central to the Harlem Renaissance and his aesthetic Romanticism is best seen as a groundbreaking contrast to the largely primitive or comical black stereotypes in pre-war America.

Poetry
Color 1925; *Copper Sun* 1927; *The Black Christ and Other Poems* 1929; *The Medea and Some Poems* 1935; *The Lost Zoo* 1940; *On These I Stand* 1947

Fiction
One Way to Heaven 1932; *My Lives and How I Lost Them* 1942

Non-Fiction
The Ballad of the Brown Girl 1927

Plays
St Louis Woman [from A. Bontemps's novel *God Sends Sunday*; with A. Bontemps] 1946 [pub. 1971]; *The Third Fourth of July* [with O. Dodson; pub. 1946]

e(dward) e(stlin) CUMMINGS 1894–1962

Cummings was born in Cambridge, Massachusetts. His father was a Harvard teacher and later a Unitarian minister who was introduced to his future wife by William James. After the Cambridge High and Latin School, cummings attended Harvard from 1911 to 1916 where he met **John Dos Passos**. With Dos Passos and others he published *Eight Harvard Poets* in 1917, and dated 'the beginning of my style' to his use of the lower case in these poems. (His name is usually rendered e e cummings though he was less insistent upon this than is sometimes supposed.) In the same year he volunteered for the ambulance service in France and was falsely charged with treasonable correspondence and imprisoned. His experience of a French detention camp furnished the material for his only novel, *The Enormous Room*. Throughout the 1920s cummings spent much of his time in Paris – where he met and was encouraged by **Ezra Pound** – and he supported himself by painting portraits and writing humorous articles for *Vanity Fair* (often under pseudonyms such as Helen Whiffletree and Professor Dunkell of Colgate University). His poems appeared in *The Dial*, and his first volume, *Tulips and Chimneys*, includes the sensuous love poem 'Puella Mea' which is mysteriously absent from the ill-titled *Collected Poems* of 1938. Later volumes such as *&* and *is 5* introduced many of the innovative typographical techniques with which he is identified. The quirky gimmickry of this seems to have distracted many readers from the subtle ironies and lyrical power of cummings's best verse.

In 1931 cummings spent a month in the Soviet Union and recorded his caustic impressions in a travel diary, *Eimi*, thereby earning the disapprobation of many of his erstwhile friends from the American intellectual left. 'Here lies one whose name was writ in lower case' was the Keatsian epitaph suggested by one of the disgruntled 'kumrads' in the *Daily Worker*. His 1935 volume *no thanks* was self-published and dedicated to a lengthy list of the firms who refused to publish him. His reputation increased in the 1950s and he spent a year as Charles Eliot Norton Professor of Poetry at Harvard from 1952 to 1953 (his charactistically idiosyncratic lectures from this year are collected in *i: six nonlectures*).

Cummings was married three times, and had a daughter from his first marriage. The *Complete Poems 1904–1962* (1994) finally gathers all of cummings's verse into one volume.

Poetry
Puella Mea 1923; *Tulips and Chimneys* 1924 [rev. 1976]; *& [And]* 1925; *XLI Poems* 1925; *is 5* 1926; *CIOPW* 1931; *W [Vi Va]* 1931; *no thanks* 1935; *One Over Twenty* 1937; *Collected Poems* 1938; *New Poems* 1938; *50 Poems* 1940; *1 x 1* 1944; *XAIPE* 1950; *Poems 1923–1954* 1954; *95 Poems* 1958; *73 Poems* 1964; *Complete Poems 1910–1962* [ed. G.J. Firmage] 2 vols 1968 [rev. 1981]; *Poems 1905–1962* [ed. G.J. Firmage] 1973; *Uncollected Poems (1910–1962)* 1981; *Etcetera: the Unpublished Poems of e e cummings* [ed. G.J. Firmage, R.S. Kennedy] 1983; *Complete Poems 1904–1962* [ed. G.J. Firmage] 1994

Fiction
The Enormous Room 1922

Plays
Him 1927; *Tom* [a ballet from H.E.B. Stowe's novel *Uncle Tom's Cabin*] 1935; *Santa Claus* 1946; *Three Plays and a Ballet* [ed. G.J. Firmage] 1968

Non-Fiction
'by e e cummings' [no title] 1930; Eimi 1948; *i, six nonlectures* (a) 1953; *Selected Letters of e e cummings* [ed. F.W. Dupree, G. Stade] 1969

For Children
Hist Whist and Other Poems for Children 1983

Collections
A Miscellany 1958 [rev. as *A Miscellany Revised* ed. G.J. Firmage 1966]

Biography
The Magic-Maker by Charles Norman 1958; *Dreams in a Mirror* by Richard S. Kennedy 1980

D

H.D. (i.e. Hilda Doolittle) 1886–1961

Doolittle was born and brought up in a conventional patriarchal household within the Moravian community in Pennsylvania. Through her father's employment at the university there, she came into contact with the young **Ezra Pound** and **William Carlos Williams**. These were the first of many literary friendships, which later included **W.B. Yeats**, **T.S. Eliot**, **Ford Madox Ford**, **Wyndham Lewis** and **D.H. Lawrence**. Doolittle's relationship with Pound, to whom she was at one time engaged, is described in her novel *HERmione*, posthumously published in 1981, and was only one of several painful affairs with both men and women which were to complicate her life. However, it was Pound who first encouraged and bullied her to publish her work, and who introduced her to literary circles in London when she moved there in 1911.

The following year, Proust introduced Doolittle to **Richard Aldington**, and in a Kensington 'bun shop' the three writers founded the Imagist movement in poetry, setting out a brief manifesto. Pound later said the Imagism had been 'invented to launch Doolittle and [Aldington] before either had enough stuff for a volume', and their work appeared in his anthology, *Des Imagistes* (1914). Her first collection of poems, *Sea-Garden*, was published in 1916, and was followed by several further volumes and a *Collected Poems* in 1925. She also published two novels during the 1920s, both set in the ancient world.

In October 1913, Doolittle and Aldington had married, but their relationship foundered during the First World War, while Aldington was serving at the Front. (Doolittle is the principal model for the faithless Elizabeth in Aldington's *Death of a Hero*, 1929.) She had given birth to a stillborn child in 1915, and in 1918 discovered that she was pregnant again, the result of a liaison with the composer Cecil Gray (with whose wife Aldington had enjoyed an affair). Her daughter was born in 1919, by which time she was living with Winifred Ellerman, who wrote novels under the pen-name of Bryher and was to remain Doolittle's companion for the remainder of her life. Although officially separated, after an initial stand-off Aldington and Doolittle continued to keep in touch, Aldington going so far as to send his estranged wife numerous postcards of female nudes. They were finally divorced in 1938, after a bitter court case, but afterwards remained on friendly terms, keeping up a voluminous correspondence.

In 1933 Doolittle underwent a course of psychoanalysis with Sigmund Freud, which eventually resulted in her *Tribute to Freud* as well as several more or less autobiographical works. She also returned to writing poetry, publishing several volumes during the 1940s, including a long sequence, *The Walls Do Not Fall*. What is probably her finest, and certainly most ambitious, work, *Helen in Egypt*, was begun in 1953, completed three years later, and published in the year of her death. She received the Brandeis Gold Medal for poetry in 1960, the year her best-known novel, *Bid Me to Live*, was published. Drafted in 1927, and frequently revised, it is another *roman à clef*, set in Bloomsbury towards the end of the First World War, detailing the collapse of her marriage and containing portraits of Aldington, Pound, Gray, the Lawrences, and other friends and lovers.

In spite of encomiums from Pound, Eliot and Lawrence, Doolittle's work was little read during her lifetime, and much of it was published posthumously. She has more recently been championed by the feminist movement and has gradually emerged from the shadow of her better-known (and male) friends and associates to be considered not merely an important woman writer, but a major figure of modernism.

Poetry

Sea Garden 1916; *The Tribute and Circe* 1917; *Heliodora and Other Poems* 1921; *Hymen and Other Poems* 1921; *Collected Poems of H.D.* 1925; *Red Roses for Bronze* 1929 [rev. 1931]; *Collected Poems of H.D.* 1940; *The Walls Do Not Fall* 1944; *What Do I Love?* 1944; *Tribute to the Angels* 1945; *The Flowering of the Rod* 1946; *By Avon River* 1949; *Helen in Egypt* 1961; *Collected Poems* 1981; *Selected Poems* [ed. L.L. Martz] 1988

Fiction

Palimpsest 1926; *Heydlus* 1928; *Kora and Ka* 1934; *The Usual Star* 1934; *Nights* [as 'John Helforth'] 1935; *The Hedgehog* 1936; *Bid Me to Live* 1960; *Asphodel* [ed. R. Spoo] 1992

Translations

Euripides *Ion* 1937

Non-Fiction

Tribute to Freud 1956; *Hermetic Definition* 1972; *End to Torment: a Memoir of Ezra Pound* 1979;

Notes on Thought and Vision and The Wise Sappho 1982

Collections

HERmione and *The Gift, a Childhood Memoir* 1981

Roald DAHL 1916–1990

Dahl's outspoken right-wing attitudes, and the allegations of misogyny, anti-Semitism and racism, vigorously denied by his friends, make him a slightly alarming successor to Enid Blyton in the affection of his readership. 'Writing,' he once claimed, 'is all propaganda in a sense.' In the macabre and sometimes unsettling books he wrote for children the targets of his propaganda tend to be guilty, in the general way, merely of being adult or foreign. We read the books pruned of the excess and sometimes brutal racial stereotyping with which his long-suffering American editor was presented, and graced by Quentin Blake's witty illustrations. Children love them.

Always the outsider, Dahl was born in Wales of Norwegian parents. His father had been a farmer near Oslo before emigrating to Llandaff, where he settled as a shipbroker. He died when Dahl was three. Dahl's elder sister died three weeks later. The family were still comfortably off, although jewellery had to be sold to pay for Dahl's upkeep at Repton School in Derbyshire. He was not happy there and, declining the chance of a place at university, left school at eighteen to join a Public School's Exploring Society expedition to Newfoundland. Returning to England, he took a job with Shell and, after training in London, he was posted to Dar-es-Salaam. At the outbreak of war he drove to Nairobi and enlisted in the Royal Air Force. He was shot down in Libya and hospitalised before rejoining his squadron in Greece. Flying against Vichy planes in Syria he was wounded again, invalided and posted to Washington as an assistant air attaché to British Security.

It was while he was posted in America that he began to write. Challenged by C.S. Forester to produce an account of his most dramatic wartime experience, Dahl produced the short story 'A Piece of Cake'. The result so impressed Forester that he handed it on to the *Saturday Evening Post*, which paid Dahl $1,000 for publication. Dahl was a keen gambler and womaniser, and legend has it that he promptly lost the money playing poker. The same story was later included in *Over to You: Ten Stories of Flyers and Flying* (1946), all of which had previously been published by the *Post*. *The Gremlins* (1943), Dahl's first children's novel, originally conceived as a script for his friend Walt Disney, was also based on his wartime flying experience. Despite the failure in 1948 of his first novel for adults, *Sometime Never: A Fable for Supermen*, Dahl was able to support himself writing the macabre and often violent stories for the *New Yorker* that preceded his success as a children's author, many of which went into the collection *Someone Like You*. The book was a bestseller, as was the follow-up collection, *Kiss Kiss* and the two books were serialised for television, both in America, as *Way Out*, and later in the UK, as *Tales of the Unexpected*.

A rare professional failure had come in 1955 with the New York premiere of his first and only stage play, *The Honeys*. The financial independence Dahl longed for did not arrive until the 1960s, when he began to write the children's novels for which he is most famous, *James and the Giant Peach* and *Charlie and the Chocolate Factory*, which he said were written 'from the nursery'. Dahl had married Patricia Neal, a successful and wealthy actress, in 1953, but his family life was run through with tragedy. His eldest son was permanently disabled in a road accident in New York, after which the family settled in Buckinghamshire, where his seven-year-old daughter, Olivia, died of measles in 1962. His wife suffered the first of three catastrophic strokes whilst pregnant with their fifth child in 1965. Her recovery was slow and, instigated by Dahl's dominating presence, it placed strains on the marriage which led to a divorce in 1983. Dahl then married Felicity Crossland with whom he had been having an affair for some years, keeping the house his first wife had paid for, and omitting her name from the family tree which decorates the cookery book named after the family home. In wealthy later life – his estate was eventually valued at over three million pounds – the singularity of his character became yet more marked, culminating in a notorious letter to *The Times* over the **Salman Rushdie** affair. He died of a viral infection in 1990. The recipient of three Edgar Allan Poe Awards – in 1954, 1959 and 1980 – Dahl also adapted many of his novels for the screen.

Fiction

Over to You (s) 1946; *Sometime Never* 1948; *Someone Like You* (s) 1954 [rev. 1961]; *Kiss Kiss* (s) 1959; *Twenty-Nine Kisses from Roald Dahl* (s) 1969; *Switch Bitch* (s) 1974; *My Uncle Oswald* 1979; *Tales of the Unexpected* (s) 1979; *More Tales of the Unexpected* (s) 1980; *Taste and Other Tales* (s) [ed. M. Caldon] 1979; *The Way Up to Heaven and Other Stories* (s) 1980; *Roald Dahl's Book of Ghost Stories* (s) 1983; *Two Fables* 1983; *Boy* 1984; *Going Solo* (s) 1986; *Ah, Sweet Mystery of Life* 1989; *The Collected Short Stories of Roald Dahl* (s) 1991

For Children

The Gremlins 1943; *James and the Giant Peach* 1961; *Charlie and the Chocolate Factory* 1964; *The Magic Finger* 1966; *Fantastic Mr Fox* 1970; *Danny, the Champion of the World* 1975; *The Wonderful World of Henry Sugar, and Six More* (s) 1977; *Charlie and the Great Glass Elevator* 1978; *The Complete Adventures of Charlie and Mr Willy Wonka* (s) 1978; *The Enormous Crocodile* 1978; *The Twits* 1980; *George's Marvellous Medicine* 1981; *The BFG* 1982; *Roald Dahl's Revolting Rhymes* 1982; *Dirty Beasts* 1983; *Rhyme Stew* 1983; *The Witches* 1983; *The Giraffe and the Pelly and Me* 1985; *Matilda* 1988; *Esio Trot* 1990; *The Muipuis* 1991; *The Vicar of Nibbleswick* 1991; *My Year* 1993

Plays

The Honeys 1955

Non-Fiction

Memories with Food at Gipsy House [with F. Dahl] 1991

Edited

The Dahl Collection of Nursery Verse 1992

Biography

Roald Dahl by Jeremy Treglown 1994

Margaretta (Ruth) D'ARCY 1934–

D'Arcy, who began her career as an actress, was born in Ireland. She had read much contemporary drama, and introduced the work of Brecht, Strindberg and others to **John Arden**, who hitherto was unfamiliar with them. They married in 1957. Arden has stated his debt to her, feeling that his attitude to the stage was academic, and even 'pompous'. Their first joint venture was *The Happy Haven* (1960), and all Arden's subsequent work for the stage has been written in collaboration with D'Arcy, their names on the title-pages devoting an equal share in the plays. D'Arcy also worked with other writers in New York on a thirteen-hour marathon, *Vietnam War Games*.

Entirely motivated by what has been termed 'socio-political' issues, D'Arcy does not separate her personal beliefs from her work, wholly and emotionally dedicating herself to the causes she strenuously espouses. She has extraordinary energy in making in her views known through theatre workshops and numerous theatrical productions. She has said that her 'creative spark is largely derived from my experience of British imperialism in Ireland, and the suffering and bitterness that it causes'. She was imprisoned twice in Armagh for vociferously protesting against injustices in Northern Ireland, and turned her experience of this into the semi-autobiographical *Tell Them Everything*. Generally, D'Arcy expresses her feelings 'through comedy and satire', but an exception to this was her mammoth radio play *Whose Is the Kingdom?*, commissioned by the BBC, and six years in the making. First broadcast in 1988, this series of nine linked plays covers thirty years, leading to the death of Constantine the Great, but with reference to the time of Christ. D'Arcy was conscious of the precedence of **Dorothy L. Sayers**' *The Man Born to Be King*, a work written for the BBC in 1941, which contained hidden references to the threat of Hitler. *Whose Is the Kingdom?* is not a retelling of the life of Christ, but more an examination of the way in which Christianity has been interpreted. Although the play is narrated by an impartial Greek to an audience of Irish Druids, the characters throughout the sequence debate the events of which they are a part. D'Arcy, who has set up an autonomous women's group, the Galway Women's Entertainment, found the historical questions of the period an ideal means to investigate feminist parallels between historical and modern times, just one issue that is prominent in the play. Through her examination of the hundreds of historical accounts, and research in ten countries, D'Arcy found that the suppression of women in the church, and the abolition of female bishops, was a result of the Council of Nicaea under the reign of Constantine.

A member of the Dublin Asodana, a council of Irish artists, D'Arcy was outraged by what she saw as the violations of its democratic principles, where communal decisions transferred 'into the hands of a clutch of misogynist malevolent goblins hunkered over their crock of knowledge'. She received with Arden an award from the Arts Council for *The Ballygombeen Bequest* and *The Island of the Mighty*.

Plays

Vietnam War Games 1967; *My Old Man's a Tory* 1971 [np]

With John Arden: *The Happy Haven* 1960 [pub. in *New English Dramatists 4* ed. T. Maschler 1962]; *The Business of Good Management* 1960 [pub. 1963]; *Ars Longa, Vita Brevis* 1964 [pub. 1965]; *Friday's Hiding* 1966 [pub. in *Soldier, Soldier and Other Plays* 1967]; *The Royal Pardon* 1966 [pub. 1967]; *Harold Muggins Is a Martyr* 1968 [np]; *The Hero Rises Up* 1968 [pub. 1969]; *Granny Welfare and the Wolf* [with R. Smith] 1971 [np]; *Rudy Dutschke Must Stay* 1971 [np]; *Two Hundred Years of Labour History* 1971 [np]; *The Ballygombeen Bequest* 1972 [np; rev. as *The*

Little Gray Home in the West 1978, pub. 1982]; *The Island of the Mighty* 1972 [pub. 1974]; *Keep These People Moving* (r) 1972; *Portrait of a Rebel: Sean O'Casey* (r), (tv) 1973; *The Crown Strike Play* 1974 [np]; *The Devil and the Parish Pump* 1974 [np]; *The Non-Stop Connolly Show* 1975 [pub. 1977–8]; *The Mongrel Fox* 1976 [np]; *No Room at the Inn* 1976 [np]; *Sean O'Scrudu* 1976 [np]; *Blow-In Chorus for Liam Cosgrave* 1977 [np]; *Mary's Name* 1977 [np]; *A Pinprick of History* 1977 [pub. in *Awkward Corners* 1988]; *Silence* 1977 [np]; *Vandaleur's Folly* 1978 [pub. 1981]; *The Making of Muswell Hill* 1984 [np]; *Manchester Enthusiasts* (r) 1984; *The Mother* [from B. Brecht] 1984 [np]; *Whose Is the Kingdom?* (r) 1988

Non-Fiction

Tell Them Everything (a) 1981; *Awkward Corners* [with J. Arden] 1988

Eleanor DARK 1901–1985

Dark was born in Burwood, Sydney, daughter of Dowell O'Reilly, who wrote poetry and short stories, and during whose brief spell as a Labour MP (1894–8), introduced into the New South Wales Legislative Assembly the first women's suffrage bill. She was educated at private schools, including Redlands, Cremorne, and worked briefly as a stenographer until 1922, when she married Eric Payten Dark, a medical practitioner and writer on physiotherapy and sociological medicine. In 1923 the Darks settled in Katoomba, in the Blue Mountains west of Sydney, where they remained until her death. Their son Mick subsequently donated the house, 'Varuna', to the Eleanor Dark Foundation for use as a writers' centre and retreat.

Dark's first published work appeared under the pseudonym Patricia O'Rane or P.O'R. Her first published poem appeared in *Triad* in 1921 and first short story, 'Take Your Choice', in 1923. Further work – short stories, articles and verse – appeared in a range of periodicals, including *Bulletin*, *Home* and *Australian Women's Mirror*. Her first novel, *Slow Dawning*, written in 1923, found no publisher until 1932. Initial earnings from her early work were two guineas for articles of 1,000 words, and 7*s* 6*d* each for sonnets, 'the same price as a bag of manure for the garden', she later told an interviewer. Between 1935 and 1940 her average weekly income from four novels was half that of a junior journalist. She received consistent and unfailing support from her husband, who guarded the work periods she spent in her studio, even telling their son that she was never to be disturbed 'except on matters of life and death'.

Her second novel, *Prelude to Christopher*, won the Australian Literature Society's Gold Medal for the best novel by an Australian published in 1934, and two years later, she won the award again for her third novel, *Return to Coolami*.

Like her contemporaries **Miles Franklin**, Marjorie Barnard and Nettie Palmer, Dark became strongly committed to social justice issues. In the Cold War period of the late 1940s, the Darks were vilified for their undisguised leftist sympathies, receiving death threats and poison-pen letters, and being generally ostracised by the local community. These experiences, together with despair at the materialism and Americanisation of the 1950s, caused Dark to become reclusive and despondent; and from having enjoyed an international reputation in the 1930s, her work fell into obscurity until the 1970s. In 1977 she was made an Officer of the Order of Australia for services to Australian literature, and in 1978 received the first Alice Award from the Society of Women Writers of Australia.

The 1930s' novels are psychological studies in modern settings, in which characters in crisis resolve inner dilemmas through social interactions. The pressure of the Depression, of rising fascism and of war forced innovations, so that the fiction could be opened to political rhetoric, Utopian images and historical 'facts'. These expanded fictional modes produced novels of large scope, where ideological struggles between optimism and despair could be explored.

In her historical trilogy, written during the war period, consisting of *The Timeless Land*, *Storm of Time* and *No Barrier*, Dark made the first protracted attempt to understand the consciousness of the Aborigine and to internalise it. Mixing fact and fiction, and juxtaposing public and private elements of the narrative, the novels cover the period from the beginnings of white colonisation to the departure of Governor Phillip in 1814. The writing involved extensive research into official records, accounts of Aboriginal life, manuscripts, letters and diaries. Her trilogy was adapted for television under the title *The Timeless Land* in 1980. In a 1945 interview, she explained: 'Writing is not only entertainment, but an *essential* contribution. The products of the writer's creative impulse – books – are an interpretation to the people of what they are doing and why, and how what they are doing is related and reacts upon what other people are doing.'

Fiction

Slow Dawning 1932; *Prelude to Christopher* 1934; *Return to Coolami* 1936; *Sun Across the Sky* 1937; *Waterway* 1938; *The Little Company* 1938; *Tales by Australians* (c) [ed. E.M. Fry] 1939; *The Timeless Land* 1941; *Storm of Time* 1949; *No Barrier* 1954; *Lantana Lane* 1959

Biography

Eleanor Dark by A. Grove Day 1976

Donald (Grady) DAVIDSON 1893–1968

Davidson was born in Campbellsville, Tennessee, the son of a teacher, and educated at Branham and Hughes School before going to Vanderbilt University in 1909. His education was cut short and he spent various spells as a teacher before graduating in 1917, a year before his marriage to Theresa Sherrer, with whom he had one child. He served in France with the US Army from 1917 to 1919 before returning to Vanderbilt where he remained for the rest of his life, becoming professor emeritus upon his retirement from teaching in 1964.

As editor of *Fugitive* magazine from 1922 to 1925, Davidson was a leading member of the Fugitive Group of poets, which also included **John Crowe Ransom**, **Robert Penn Warren** and **Allen Tate**. The group, according to the *New York Times*, was dedicated to the proposition that 'the South must achieve its own destiny out of its own traditions and resources and not be an imitative stepchild of the nation'. Davidson was, perhaps, more dedicated to this cause than any of his colleagues, and remained so throughout his life. His first publication was in 1923, with *Avalon*, published with work by Ransom and William Percy (the uncle and adoptive father of the novelist **Walker Percy**), and his first collection, *An Outland Piper*, appeared a year later. In his poetry, as in his journalism (he wrote a weekly column, 'Spyglass', for the *Nashville Tennessean*, of which he was literary editor from 1924 to 1930), he continually addressed the declining rural tradition of the South and argued that 'the present struggle must be to retain spiritual values "against the fiery gnawing of industrialism".' This quest is perhaps best represented in his long narrative poem *The Tall Men* and in the much-anthologised 'Lee in the Mountains'. The diction of his poetry is often rather archaic, and his tone is frequently nostalgic as he evokes the antebellum South, so much so that he was sometimes accused of racism by critics who saw him as hankering for the days of slavery. In the introduction to *Poems 1922–1961* he argued in favour of racial segregation.

Poetry

Avalon [with others] 1923; *An Outland Piper* 1924; *The Tall Men* 1927; *Lee in the Mountains and Other Poems* 1938; *The Long Street* 1961; *Poems 1922–1961* 1966

Non-Fiction

The Attack on Leviathan: Essays on Regionalism and Nationalism in the United States 1938; *American Composition and Rhetoric* 1939 [rev. as *Concise American Composition and Rhetoric* 1964]; *The Tennessee* 1946; *Twenty Lessons in Reading and Writing* 1955; *Still Rebels, Still Yankees and Other Essays* 1957; *Southern Writers in the Modern World* 1958; *The Spyglass: Views and Reviews, 1924–1930* [ed. J.T. Fain] 1963; *It Happened to Them: Character Studies of New Testament Men and Women* 1965; *The Literary Correspondence of Donald Davidson and Allen Tate* [with A. Tate; ed. J.J. Fain, T.D. Young] 1974

Edited

British Poets of the Eighteen-Nineties 1937; *Readings for Composition* [with S.E. Glenn] 1942; John Donald Wade *Selected Essays and Other Writings* 1966

Plays

Singin' Billy (l) [with C.F. Bryan] 1952 [np]

Donald (Alfred) DAVIE 1922–

Davie was born in Yorkshire, was educated at the Barnsley Holgate Grammar School and St Catharine's College, Cambridge, and served in the navy between 1941 and 1946. Since then, he has combined the careers of poet, critic, editor, translator and academic. He is the author of several studies on the technical aspects of poetry, and books on Sir Walter Scott, **Thomas Hardy**, **Ezra Pound**, hymnology and dissent. He was a lecturer in English, then Fellow of Trinity College, Cambridge, before moving to the University of Dublin in 1958. He subsequently held the chair of English at the University of Essex, acting for three years as Pro-Vice- Chancellor. In 1968 he left for America, becoming first Professor of English and Palmer Professor in Humanities at Stanford, then Andrew Mellon Professor of Humanities at Vanderbilt University, Nashville. He married in 1945 and has two sons and a daughter.

He was associated with the Movement poets during the 1950s, and has written an important history of British poetry of the following three decades, *Under Briggflats*. His own poetry has been described as academic: it is certainly objective, urbane and meticulous in its prosody, while its subject-matter is often historical or biographical. One influence has been the poets of the eighteenth century, whose works he has edited. He is also a poet of place: *The Shires* consists of poems about the English counties, arranged in alphabetical order from Bedfordshire to Yorkshire, and written after he had left Britain for America. His conversion to Christianity in later life has produced some impressive if discomfiting verse. His work has often been praised for its integrity, and Andrew Swarbrick, reviewing Davie's *Collected Poems* of 1990, concluded that: 'No contemporary poet has been so demanding of himself as Davie and there is none whose achievement is so rigorously humane.'

Poetry

Poems 1954; *Brides of Reason* 1955; *A Winter Talent and Other Poems* 1957; *The Forests of Lithuania* 1959; *New and Selected Poems* 1961; *A Sequence for Francis Parkman* 1961; *Events and Wisdoms: Poems 1957–1963* 1964; *Poems* 1969; *Essex Poems 1963–1970* 1972; *Orpheus* 1974; *The Shires* 1975; *In the Stopping Train* 1977; *Trying to Explain* 1980; *Three for Water-Music* 1981; *Collected Poems 1970–1983* 1983; *Selected Poems* 1985; *To Scorch or Freeze* 1988; *Collected Poems* 1990; *Ghosts in the Corridor* [with others] 1992

Non-Fiction

Purity of Diction in English Verse 1952; *Articulate Energy: an Enquiry into the Syntax of English Poetry* 1955; *The Heyday of Sir Walter Scott* 1961; *The Language of Science and the Language of Literature 1700–1740* 1963; *Ezra Pound: Poet as Sculptor* 1964; *Thomas Hardy and British Poetry* 1972; *Poetry in Translation* 1975; *The Poet in the Imaginary Museum: Essays of Two Decades* [ed. B. Alpert] 1977; *A Gathered Church: the Literature of the English Dissenting Interest 1700–1930* 1978; *Trying to Explain* 1979; *Dissentient Voice: the Ward-Phillips Lecture* 1980; *English Hymnology in the Eighteenth Century* 1980; *Kenneth Allott* and the *Thirties* 1980; *These the Companions: Recollections* (a) 1982; *Under Briggflats: a History of Poetry in Great Britain 1960–1988* 1989; *Polish Literature* 1990; *Older Masters: Essays and Reflections on English and American Literature* 1992

Edited

The Late Augustans: Longer Poems of the Later Eighteenth Century 1958; *Poems: Poetry Supplement* 1960; *Poetics Poetyka* 1961; William Wordsworth *Selected Poems* 1962; *Russian Literature and Modern English Fiction* 1965; *Pasternak* [with A. Livingstone] 1969; *Agenda: Thomas Hardy Issue* 1972; *Augustan Lyric* 1974; Elizabeth Daryush *The Collected Poems* 1975; Yvor Winters *Collected Poems* 1978; *The New Oxford Book of Christian Verse* 1981; *Studies in Ezra Pound* 1991

Translations

Boris Pasternak *The Poems of Doctor Zhivago* 1965

Andrew (Wynford) DAVIES 1936–

One of Britain's leading writers of television drama, Davies was born in Cardiff and read English at University College, London. He subsequently taught at a London grammar school until 1963, when he became a lecturer in English at the University of Warwick, a post he held for over twenty years.

His years in the academic world provided material for one of his early successes, the BBC television series *A Very Peculiar Practice*, a richly inventive comedy set amongst the doctors in a university health centre. He had in fact been writing television plays for some considerable time before this: his first, *Is That Your Body, Boy?*, was broadcast in 1971. As well as writing his own original plays, he has become well known for adapting the works of others, starting in 1979 with a very distinguished series based on R.F. Delderfield's not very distinguished public-school novel, *To Serve Them All My Days*. He has adapted both the classics (George Eliot's *Middlemarch* and, amidst a welter of tabloid outrage about reported nude scenes, Jane Austen's *Pride and Prejudice*) and more popular works, such as Liza Cody's Anna Lee novels and Michael Dobbs's political thrillers. He has also adapted novels by **Mary Wesley**, **Molly Keane** and **Angus Wilson**, while his series based on **Kingsley Amis**'s *The Old Devils*, was thought by some to be superior to the novel. He won the BAFTA Writer's Award in 1989 and has tried his hand at situation comedy, though *Game On*, about three young, sex-obsessed flatmates, had mixed reviews.

Although Davies has concentrated upon television work, he has also written for the stage. *Rose* was his most successful play, produced at the Duke of York's Theatre in London, with Glenda Jackson, and later transferred to Broadway. Of his other stage plays, *Prin*, which centred on a schoolmistress and starred Sheila Hancock, is perhaps the most highly regarded.

Davies also writes fiction for both adults and children. His best book for children is *Conrad's War*, which won the *Guardian* Award for Children's Fiction in 1979, although he is probably better known for the books about Marmalade Atkins. His novel *Getting Hurt* is the story of a once debauched and successful lawyer who is undone by his hopeless passion for a young Polish student he meets in an after-hours drinking club. Davies had taken the traditional story of a woman getting hurt by a man and turned it on its head, and the novel was widely praised for its intense depiction of male vulnerability and the loss of clarity which can result from prolonged jealousy and suffering. Davies has also written a volume of short stories, appropriately titled *Dirty Faxes*.

Plays

Is That Your Body, Boy? **(tv)** 1971; *Filthy Fryer* 1974; *Grace* **(tv)** 1974; *The Water Maiden* **(tv)** 1974; *The Imp of the Perverse* **(tv)** 1975; *Randy Robinson* 1976; *The Signalman* **(tv)** 1976; *The Eleanor Marx Trilogy* 1977; *Happy in War* **(tv)** 1977; *The Legend of King Arthur* **(tv)** 1979; *Brainstorming With the Boys* 1977; *Going Bust* 1977; *To Serve Them All My Days* **(tv)** [from R.F. Delderfield's novel] 1979; *Diana* **(tv)** [from R.F. Delderfield's novel] 1980; *Rose* 1981; *Heart Attack Hotel* 1983; *Time After Time* **(tv)** [from M. Keane's novel] 1986; *A Very Peculiar Practice* **(tv)** 1986; *Inappropriate Behaviour* 1987; *Ball-Trap on the Côte Sauvage* 1989; *Mother Love* **(tv)** [from D. Taylor's novel] 1989; *House of Cards* **(tv)** [from M. Dobbs's novel] 1990; *Prin* 1990; *A Private Life* 1990; *A Very Polish Practice* 1992; *Anglo-Saxon Attitudes* **(tv)** [from A. Wilson's novel] 1992; *Filipina Dreamers* 1991; *The Old Devils* **(tv)** [from K. Amis's novel] 1992; *Anna Lee* **(tv)** [from L. Cody's novels] 1993; *Harnessing Peacocks* **(tv)** [from M. Wesley's novel] 1993; *To Play the King* **(tv)** [from M. Dobbs's novel] 1993; *A Few Short Journeys of the Heart* 1994; *Middlemarch* **(tv)** [from G. Eliot's novel] 1994; *The Final Cut* **(tv)** [from M. Dobbs's novel] 1995; *Game On* **(tv)** [with B. Davis] 1995; *Pride and Prejudice* **(tv)** [from J. Austen's novel] 1995

Fiction

A Very Peculiar Practice [from his TV series] 1986; *The New Frontier* 1987; *Getting Hurt* 1989; *Dirty Faxes* **(s)** 1990; *B. Monkey* 1992

For Children

The Fantastic Feats of Doctor Boox 1972; *Conrad's War* 1978; *Marmalade and Rufus* 1980; *Marmalade Atkins in Space* 1981; *Educating Marmalade* 1982; *Danger – Marmalade at Work* 1983; *Marmalade Hits the Big Time* 1984; *Alfonso Bonzo* 1986; *Poonam's Pets* 1990 [with D. Davies]; *Raj in Charge* [with D. Davies] 1994; *Marmalade on the Ball* 1995

Rhys (i.e. Rees Vivian) DAVIES 1901–1978

The son of a grocer and a teacher, Davies was born in Tonypandy in the Rhondda Valley. His partly autobiographical novel, *Tomorrow to Fresh Woods*, gives a picture of his early life. He attended local schools, then went to London at the age of eighteen. When three of his stories were published in Charles Lahr's quarterly *New Coterie*, they attracted the attention of a publisher. With the advance for his first novel, *The Withered Root*, he went to France for six months. While there he received an invitation from **D.H. Lawrence** to come as his guest to a hotel in Bandol. His account of Lawrence, first published in *Horizon* and subsequently included in his autobiography, *Print of a Hare's Foot*, was described by Goronwy Rees as 'one of the most sympathetic and understanding portraits of a man of genius ever written'.

Davies's subsequent career was extremely prolific: eighteen novels and as many volumes of short stories. Most of his early work has a Welsh setting, notably *Jubilee Blues* in which a servant is left some money by her JP employer, with which she marries and buys a pub near a colliery. *The Dark Daughters* is a powerful story of the hatred nurtured by three spinsters for their father. In later books he broke away from Wales: in *The Painted King* to explore the musical comedy scene in London; and in *Nobody Answered the Bell* to an English seaport, the setting of a tense psychological drama about a lesbian affair. His final novel, *Honeysuckle Girl* concerns heroin addiction. Of his short stories, **C. Day Lewis** wrote in 1946 that Davies 'possesses the gift, uncommon today, of story telling; his stories have plot, we wonder what is going to happen next, we are startled or amused by the dénouement ... the Welsh turbulence of spirit, the exaggeration, the darts into fantasy, the slyness and obliqueness ... all these help to create a personal style not to be imitated by the mass fiction factories'. He was awarded the OBE in 1968 and died in London.

Fiction

Anon **(s)** 1927; *The Song of Songs and Other Stories* **(s)** 1927; *The Withered Root* 1927; *A Bed of Feathers* **(s)** 1929; *Rings on Her Fingers* 1930; *The Stars, the World, and the Women* **(s)** 1930; *Tale* **(s)** 1930; *Arfon* **(s)** 1931; *A Pig in a Poke* **(s)** 1931; *A Woman* **(s)** 1931; *Count Your Blessings* 1932; *Daisy Matthews and Three Other Tales* **(s)** 1932; *The Red Hills* 1932; *Love Provoked* **(s)** 1933; *Honey and Bread* 1935; *One of Norah's Early Days* **(s)** 1935; *The Skull* 1936; *The Things Men Do* **(s)** 1936; *A Time to Laugh* 1937; *Jubilee Blues* 1938; *Under the Rose* 1940; *Tomorrow to Fresh Woods* 1941; *A Finger in Every Pie* **(s)** 1942; *The Black Venus* 1944; *Selected Stories* **(s)** 1945; *The Trip to London* **(s)** 1946; *The Dark Daughters* 1947; *Boy with a Trumpet* **(s)** [US *Boy with a Trumpet and Other Selected Short Stories*] 1949; *Marianne* 1951; *The Painted King* 1954; *Collected Stories* **(s)** 1955; *The Perishable Quality* 1957; *Girl Waiting in the Shade* 1960; *The Chosen One and Other*

Stories (s) 1967; *The Darling of Her Heart and Other Stories* (s) 1967; *Nobody Answered the Bell* 1971; *Honeysuckle Girl* 1975; *The Best of Rhys Davies* (s) 1979

Plays

No Escape [with A. Batty] 1954 [pub. 1955]

Non-Fiction

My Wales 1938; *Sea Urchin: Adventures of Jorgen Jorgensen* 1940; *The Story of Wales* 1943; *Print of a Hare's Foot* (a) 1969

Biography

Rhys Davies by David Rees 1975

(William) Robertson DAVIES 1913–

Davies was born in a 'humble dwelling' in Thamesville, Ontario, the son of a poor but enterprising printer, William Davies, who eventually rose to become a senator. William gradually prospered and founded the *Peterborough Journal*, and his son earned his first wage (of twenty-five cents) at the age of eleven by reporting a local event. He was educated at Queen's University, Toronto, and at Balliol College, Oxford.

Davies's creative life has been divided between three careers: actor, writer and academic. While at Oxford, he became interested in the theatre and was for two years from 1938 a teacher and actor, working as Tyrone Guthrie's assistant at the Old Vic Theatre in London, a job he was given because of his encyclopaedic knowledge of the works of Shakespeare. In 1940 he returned to Toronto, accompanied by Brenda Mathews, an Australian stage manager at the Old Vic, whom he married in that year. He was appointed literary editor of the *Saturday Review*, then in 1942 became editor of his father's newspaper, for which he also wrote a daily humorous column under the name of Samuel Marchbanks, later publishing two collections, *The Diary of Samuel Marchbanks* and *The Table Talk of Samuel Marchbanks*. He also wrote a large number of plays, but became bored with having to insert wisecracks into the text on the insistence of actors; his last play to date was produced in 1977. None of his plays were produced outside Canada, English agents telling him firmly that: 'Nobody, but *nobody*, is interested in Canada.'

Davies was to prove them wrong by becoming a novelist, although international recognition did not come immediately. He published his first novel, *Tempest-Tost*, in 1951, and it became the first volume of *The Salterton Trilogy*. In spite of winning several major Canadian awards (including the Leacock Award in 1954 for *Leaven of Malice* and the Governor-General's Award in 1973 for *The Manticore*), he remained a cult author for some years, much praised by aficionados, but denied the wide readership his work deserved. *The Deptford Trilogy* made more of a mark than its predecessor, but it was not until *What's Bred in the Bone*, the central volume of *The Cornish Trilogy*, was shortlisted for the 1986 Booker Prize that Davies earned universal praise and an international readership.

Davies is an expansive and generous novelist, whose books tend to form energetic trilogies, firmly rooted in Canadian life, notably the theatre and academia. *The Cornish Trilogy*, which is perhaps his finest achievement, combines both elements: a deliriously Gothic campus novel in which forgery, murder, magic and opera feature, as he investigates disguise and deception, art and artifice, instinct and creativity. He has criticised certain aspects of Canada (which he once described as 'the Home of Modified Rapture'), and claims that the national prayer runs: 'Oh God, grant me mediocrity and comfort; protect me from the radiance of Thy light.' His work, which combines elements of fantasy, religion, astrology, alchemy, Jungian theory and much arcane lore, effectively combats this dismal plea and exemplifies his own impassioned cry: 'I don't believe in the minimum. I *hate* the minimum!' Modest about his work, he describes himself as 'an entertainer who just wants to make people think'.

Davies's work has earned him a reputation as a sort of mystical Victorian eccentric, something his appearance has done nothing to dispel. With his long and luxuriant white beard falling over an antique jacket, Davies relishes his image. During his period as the Master of the University of Toronto's Massey College (1963–81), he was unable to suppress his roguish sense of humour, and a typical examination question set by him read: 'An Irish critic has dismissed **Eugene O'Neill**'s *Long Day's Journey into Night* as a "long, snot-ridden whine". Discuss.' He nevertheless holds honorary doctorates from a number of Canadian universities. His journalism and criticism have been collected in several volumes, with appropriate titles such as *The Enthusiasms of Robertson Davies* and *The Well-Tempered Critic*.

Fiction

The Salterton Trilogy 1986: *Tempest-Tost* 1951, *Leaven of Malice* 1954, *A Mixture of Frailities* 1958; *The Deptford Trilogy* 1983: *Fifth Business* 1970, *The Manticore* 1972, *World Of Wonders* 1975; *The Cornish Trilogy* 1991: *The Rebel Angels* 1981, *What's Bred in the Bone* 1985, *The Lyre of Orpheus* 1988; *High Spirits* (s) 1982; *Murther & Walking Spirits* 1991; *The Cunning Man* 1995

Non-Fiction
Shakespeare's Boy Actors 1939; *Shakespeare for Young Players* 1942; *The Diary of Samuel Marchbanks* 1947; *The Table Talk of Samuel Marchbanks* 1949; *Renown at Stratford* [with T. Guthrie, G. MacDonald] 1953; *Twice Have the Trumpets Sounded* [with T. Guthrie] 1954; *Thrice the Brindled Cat Hath Mew'd* [with T. Guthrie, B. Neal, T. Moiseiwitsch] 1955; *A Voice From the Attic* 1960 [rev. 1990]; *Le Jeu de centaire* 1967; *Samuel Marchbanks' Almanack* 1967; *The Heart of a Merry Christmas* 1970; *Stephen Leacock* 1970; *One Half of Robertson Davies* 1977; *The Enthusiasms of Robertson Davies* [ed. J.S. Grant] 1979; *The Mirror of Nature* 1983; *Conversations with Robertson Davies* [ed. J. Madison] 1989; *Reading and Writing* 1993; *Robertson Davies, the Well-Tempered Critic* 1981

Plays
A Play of Our Lord's Nativity 1946; *Overlaid* 1947 [pub. 1949]; *At the Gates of the Righteous* 1948 [pub. 1949]; *Eros at Breakfast* 1948 [pub. 1949]; *Fortune My Foe* 1948 [pub. 1949]; *Hope Deferred* 1948 [pub. 1949]; *The Voice of the People* 1948 [pub. 1949]; *At My Heart's Core* 1950 [pub. 1972]; *King Phoenix* 1950 [pub. 1972]; *A Masque of Aesop* 1952; *A Jig for Gypsy* 1954; *Hunting Stuart* 1955 [pub. 1972]; *Leaven of Malice* 1960 [rev. 1973; pub. 1981]; *A Masque of Mr Punch* 1962 [pub. 1963]; *Centennial Play* [with others] 1967; *Hunting Stuart and Other Plays* [pub. 1972]; *Brothers in the Black Art* 1974 [pub. 1981]; *Question Time* 1975; *Pontiac and the Green Man* 1977; *Dr Damon's Cure* (l) [with D. Holman] 1982

Edited
Feast of Stephen: an Anthology of Some of the Less Familiar Writings of Stephen Leacock 1970

Collections
The Papers of Samuel Marchbanks 1985

W(illiam) H(enry) Davies 1871–1940

Born in Newport, Monmouthshire, Davies was the son of a moulder, who died of tuberculosis at thirty-one. His paternal grandparents were scandalised at his mother's remarriage within a year of her husband's death, and adopted the three-year-old William. Although his grandmother was strict and forbade books, Davies enjoyed a happy childhood. Caught shoplifting, he left school at thirteen to work for an ironmonger, a job he held until his grandfather died in 1885.

After apprenticeship to a framemaker, Davies went to London in 1891, drinking heavily in the slums where he lived. Two years later, his grandmother's death rescued him from squalor with a small allowance, and he took off to America. Physically tough, he worked casually for five years, returning to Newport to squander his accumulated allowance, In 1899, Davies sailed for the gold-mines of Klondike but shortly after arriving in Canada, fell from a train, and severed a foot. His leg was amputated, and he returned to England, living in doss-houses in London and elsewhere, writing poetry he could not publish. He later tramped the Midlands on his wooden leg.

In 1905, he gathered his verse and scraped money to finance *The Soul's Destroyer and Other Poems* which he sent to influential writers, including **George Bernard Shaw** who in 1908 wrote the preface to *The Autobiography of a Super-Tramp*, Davies's account of his wanderings in England and America. Simplicity of expression, a Blake-like innocence, and a natural lyric gift attracted the praise and help of **Edward Thomas** who found Davies accommodation in Kent. Inclusion in Edward Marsh's *Georgian Poetry* series ensured Davies some comfort and enabled him to live in London in 1914.

Arrogant and conceited, modest and shy, Davies enjoyed the friendship of the **Sitwells** and **D.H. Lawrence**, and cohabitation with prostitutes. He resembled his cousin Henry Irving, with high cheekbones and a large nose, and he was sculpted by Epstein. In 1924, Davies virtually abandoned his literary friends, after his marriage to the twenty-two-year-old Helen Payne, whom he declined to mention in his *Who's Who* entry. *Young Emma*, a remarkably honest account of their early relationship, and his life with prostitutes, was withdrawn after Davies had second thoughts, but was eventually published in 1980. He died from a stroke, near Stroud. Davies wrote too much, but poems such as 'Fancy's Home' remain engaging because his work was never contrived, and survives the stigma of 'Georgianism'.

Poetry
The Soul's Destroyer and Other Poems 1905; *New Poems* 1907 [rev. 1922]; *Nature Poems and Others* 1908; *Farewell to Poesy and Other Pieces* 1910; *Songs of Joy and Others* 1911; *Foliage* 1913 [rev. 1922]; *The Bird of Paradise and Other Poems* 1914; *Child Lovers and Other Poems* 1916; *Forty New Poems* 1918; *Raptures* 1918; *The Song of Life and Other Poems* 1920; *The Captive Lion and Other Poems* 1921; *The Hour of Magic and Other Poems* 1922; *Secret* 1924; *A Poet's Alphabet* 1925; *The Song of Love* 1936; *A Poet's Calendar* 1927; *Moss and Feather* 1928; *Ambition and Other Poems* 1929; *In Winter* 1932; *Poems 1930–31* 1932; *The Lovers' Song-Book* 1933; *Love Poems* 1935; *The Birth of Song: Poems 1935–36* 1936;

The Loneliest Mountain and Other Poems 1939; *Common Joys and Other Poems* 1941

Non-Fiction

The Autobiography of a Super-Tramp (a) 1908 [rev. 1920]; *Beggars* 1909; *The True Traveller* 1912; *Nature* 1914; *A Poet's Pilgrimage* 1918; *Later Days* 1925; *The Adventures of Johnny Walker, Tramp* 1926; *My Birds* 1933; *My Garden* 1933; *Young Emma* (a) 1980

Fiction

A Weak Woman 1911; *Dancing Mad* 1927; *True Travellers* 1923

Edited

Shorter Lyrics of the Twentieth Century 1900–1922 1922

Dan(iel Marcus) DAVIN 1913–1990

Davin was born in Invercargill, New Zealand, into a family of Irish extraction, and was educated at the Marist Brothers' School, the Sacred Heart College in Auckland, and Otago University, Dunedin, from which he graduated in 1935. He completed his education in England, having won a Rhodes Scholarship to Balliol College, Oxford, where he took a first in Greats in 1939. His distinguished war career, which was to have a profound effect upon his work as a writer, began in the same year, when he joined the Royal Warwickshire Regiment, transferring in 1940 to the New Zealand Expeditionary Force, with whom he served in Greece and Crete. He was wounded in 1941 and was later to write the official history of the Crete campaign. Recovered from his injuries, he spent the remainder of the war as an Intelligence Officer attached to GHQ in the Middle East. He was three times mentioned in dispatches and left the army with the rank of major and an MBE.

He returned to Oxford in 1946 and became junior assistant at the Clarendon Press, where he worked until his retirement in 1978, eventually becoming director of the publisher's academic division. He was elected Fellow of Balliol in 1965, was awarded a DLitt by Otago University in 1985, and was made CBE in 1987.

Davin's fiction largely reflects the stages of his life: for example, his evocative early short stories, collected in *The Gorse Blooms Pale*, and his novel about Catholic life, *Roads from Home*, are based on his upbringing in New Zealand; *For the Rest of Our Lives* draws upon his war service with the New Zealand divison; and his final novel, *Brides of Price*, is a comedy set in the academic world. Although he lived most of his life in England, he remained a New Zealander, who returned to his native land in both fiction and non-fiction. He was the co-author of a standard two-volume guide to the New Zealand novel and wrote on **Katherine Mansfield**. He was also the editor of several volumes of short stories, notably *Short Stories from the Second World War*, which is perhaps the best of its kind.

A robust raconteur – he gave 'talking' as one of his recreations in *Who's Who* – he was a colourful figure in Oxford, and his best-known work of non-fiction is a lively volume of memoirs, *Closing Time*, which contains portraits of **Louis MacNeice**, **Joyce Cary**, **Dylan Thomas**, Enid Starkie and others. He married in 1939 and had three daughters.

Fiction

Cliffs of Fall 1945; *For the Rest of Our Lives* 1947; *The Gorse Blooms Pale* (s) 1947; *Roads from Home* 1949; *The Sullen Bell* 1956; *No Remittance* 1959; *Not Here, Not Now* 1970; *Brides of Price* 1972; *Breathing Space* (s) 1975; *Selected Stories* (s) 1981

Non-Fiction

An Introduction to English Literature [with J. Mulgan] 1947; *Crete* 1953; *Writing in New Zealand: the New Zealand Novel* [with W.K. Davin] 2 vols 1956; *Katherine Mansfield in Her Letters* 1959; *Closing Time* 1975

Edited

New Zealand Short Stories (s) 1953; *Selected Stories by Katherine Mansfield* (s) 1953; *English Short Stories of Today* (s) 2nd series 1982; *From Oasis to Italy* 1983; *Short Stories from the Second World War* (s) 1982; *The Salamander and the Fire* (s) 1987

C(ecil) DAY(-)LEWIS 1904–1972

Day Lewis was born at Ballintubbert in Ireland, the son of a Church of Ireland minister. His mother, who died when he was four, was of Anglo-Irish stock, collaterally descended from Oliver Goldsmith. He grew up in England, and was educated at Sherborne School, where he was a games player and Company Sergeant-Major of the Officer Training Corps, and at Wadham College, Oxford, where he became part of the circle gathered around **W.H. Auden**, whom he helped to edit *Oxford Poetry 1927*. Day Lewis's early verse, of which the first important volume is *Transitional Poem* (1929), is often indistinguishable in manner from that of Auden. After Oxford, he taught for several years at various prep and public schools, and married the daughter of a Sherborne master, with whom he had two children.

From the mid-1930s he was able to make his living by writing. The detective stories he began to publish from 1935 under the pseudonym of Nicholas Blake, of which *The Beast Must Die* and *A Tangled Web* are perhaps the best, were instrumental in this. (His amateur sleuth, Nigel Strangeways, is partly based on Auden.) In the 1930s Day Lewis was an avowedly Communist poet (it was at this period that he dropped the supposedly upper-class hyphen from his surname), yet he was also a traditionalist, and perhaps always something of an establishment man. After working at the Ministry of Information during the war, he became a well-known public man-of-letters, serving as Professor of Poetry at Oxford from 1951 (Auden's predecessor in the post and the first reasonably distinguished poet to hold it since Matthew Arnold), chairman of the Arts Council Literature Panel, vice-president of the Royal Society of Literature, director of the publishers Chatto and Windus and so forth. But he also found time to write, and later poetry, such as *The Whispering Roots* (1970), demonstrates continuing growth in feeling and technique. He also wrote juvenile fiction, of which *The Otterbury Incident* is particularly good, a few non-detective novels, varied non-fiction, and was a distinguished verse translator, especially of Virgil. His many extra-marital affairs included one with the novelist **Rosamond Lehmann**. He was divorced from his first wife in 1951 and married the actress Jill Balcon, with whom he lived in a large Georgian house in Greenwich and with whom he also had two children, one of whom is the actor Daniel Day-Lewis. His last years were marred by ill-health, but his public career was crowned by his appointment to the Laureateship in 1968 in succession to **John Masefield**. He died, of cancer, in the Hertfordshire home of **Kingsley Amis** and **Elizabeth Jane Howard**, where he and his wife were staying. A great admirer of **Thomas Hardy**, he had arranged that he should be buried as close as possible to the poet's grave in Stinsford churchyard.

Poetry

Beechen Vigil and Other Poems 1925; *Country Comets* 1928; *Transitional Poem* 1929; *From Feathers to Iron* 1931; *The Magnetic Mountain* 1933; *A Time to Dance and Other Poems* 1935; *Noah and the Waters* 1936; *Overtures to Death and Other Poems* 1938; *Poems in Wartime* 1940; *Word Over All* 1943; *Poems 1943–1947* 1948; *An Italian Visit* 1953; *Pegasus and Other Poems* 1957; *The Gate and Other Poems* 1962; *Requiem for the Living* 1964; *The Room and Other Poems* 1965; *The Abbey That Refused to Die* 1967; *The Whispering Roots* 1970; *Posthumous Poems* 1979

Non-Fiction

A Hope for Poetry 1934; *Revolution in Writing* 1935; *We're Not Going To Do Nothing* 1936; *The Poetic Image* 1946; *The Colloquial Element in English Poetry* 1947; *Enjoying Poetry* 1947; *The Poet's Task* 1951; *The Grand Manner* 1952; *The Lyrical Poetry of Thomas Hardy* 1953; *Notable Images of Virtue: Emily Brontë; George Meredith; W.B. Yeats* 1954; *The Poet's Way of Knowledge* 1957; *The Buried Day* (a) 1960; *The Lyric Impulse* 1965; *A Need for Poetry?* 1968

Translations

Virgil: *The Georgics* 1940, *The Aeneid* 1952, *The Eclogues* 1963; Paul Valéry *The Graveyard by the Sea* 1946

Fiction

As 'Nicholas Blake': *A Question of Proof* 1935; *Thou Shell of Death* 1936; *There's Trouble Brewing* 1937; *The Beast Must Die* 1938; *The Smiler with the Knife* 1939; *The Case of the Abominable Snowman* 1940; *Malice in Wonderland* 1940; *Minute for Murder* 1947; *Head of a Traveller* 1949; *The Dreadful Hollow* 1953; *The Whisper in the Gloom* 1954; *A Tangled Web* 1956; *End of Chapter* 1957; *A Penknife in my Heart* 1958; *The Widow's Cruise* 1959; *The Worm of Death* 1961; *The Deadly Joker* 1963; *The Sad Variety* 1964; *The Morning After Death* 1966; *The Private Wound* 1968

As C. Day Lewis: *The Friendly Tree* 1936; *Starting Point* 1937; *Child of Misfortune* 1939

For Children

Dick Willoughby 1933; *Poetry for You* 1944; *The Otterbury Incident* 1948

Edited

Oxford Poetry [with W.H. Auden] 1927; *The Mind in Chains: Socialism and the Cultural Revolution* 1937; *Ralph Fox: a Writer in Arms* [with J. Lehmann, T. Jackson] 1937; *The Chatto Book of Modern Poetry: 1915–1955* [with J. Lehmann] 1956; *A Book of English Lyrics* 1961; Wilfred Owen *Collected Poems* 1963

Biography

C. Day-Lewis by Sean Day-Lewis 1980

E.M. DELAFIELD (pseud. of Edmée Elizabeth Monica de la Pasture) 1890–1943

Delafield (whose *nom de plume* anglicises her real surname) was born in Monmouthshire, the daughter of Count Henry de la Pasture (whose family had fled to Wales in the wake of the French Revolution) and the popular novelist who wrote under the name Mrs Henry de la Pasture. During the First World War, Delafield worked as a VAD (voluntary nurse) in England and wrote her first novel, *Zella Sees Herself* (1917), during snatched moments of free time. Her war service forms the background of her second novel, *The War Workers*, in which the action transfers from Devon to the Midlands. She married Major Arthur Dashwood, the younger son of a baronet, in 1919 and spent two years in the Far East before returning to Devon, where she was to remain for the rest of her life.

Delafield was a prolific and popular writer of witty and gently satirical novels of middle-class life. Her greatest and most enduring success was *Diary of a Provincial Lady* (1930) and its three sequels, in which she guyed the sort of life she herself led, revolving around husband and children (she had two, who were, she admitted, models for her characters), good works for the Women's Institute and as a magistrate, domestic crises and social embarrassments of one sort or another. It is a comedy in the very English tradition of George and Weedon Grossmith's *The Diary of a Nobody* (1892), and like that masterpiece began life as a magazine serial. Lady Rhondda had recruited Delafield to the feminist weekly *Time and Tide*, where she contributed sketches and articles and eventually became a director. The first volume of diaries was followed by three others in which the unnamed heroine *Goes Further* (to London), travels to America (this volume, serialised in *Punch*, is based on Delafield's own American lecture tour), and meets the exigencies of the Second World War. The Provincial Lady is a parody of a recognisable type of between-wars Englishwoman who appears in countless novels and films of the period. Well-meaning, self-deprecating, equal to every occasion and muddling through, she is a comic archetype whom Delafield observes with affectionate mockery. It is appropriate that she wrote an introduction for *The British Character* (1938), a volume of drawings by the English cartoonist of *Punch*, 'Pont', whose work covers similar territory.

Delafield also wrote three plays, several volumes of non-fiction, including reportage and literary criticism, and an enormous amount of journalism. Apparently indefatigable, she died before her time, collapsing during a lecture tour at the age of fifty-three.

Fiction

Zella Sees Herself 1917; *The Pelicans* 1918; *The War Workers* 1918; *Consequences* 1919; *Tension* 1920; *The Heel of Achilles* 1921; *Humbug* 1921; *The Optimist* 1922; *A Reversion to Type* 1923; *Miss Harter* 1924; *The Chip and the Block* 1925; *Jill* 1926; *The Entertainment* (s) 1927; *The Way Things Are* 1927; *The Suburban Young Man* 1928; *What is Love?* 1928; *Women Are Like That* (s) 1929; *Diary of a Provincial Lady* 1930; *Turn Back the Leaves* 1930; *Challenge to Clarissa* 1931; *The Provincial Lady Goes Further* 1932; *Thank Heaven Fasting* 1932; *Gay Life* 1933; *The Provincial Lady in America* 1934; *The Bazalgettes* 1935; *'Faster! Faster!'* 1936; *As Others Hear Us* (s) 1937; *Nothing is Safe* 1937; *When Women Love* (s) 1938; *Love has No Resurrection and Other Stories* (s) 1939; *Three Marriages* 1939; *The Provincial Lady in Wartime* 1940; *No One Will Know How* 1941; *Late and Soon* 1943

Non-Fiction

General Impressions 1933; *A Note by the Way* 1933; *Ladies and Gentlemen in Victorian Fiction* 1937; *Straw Without Bricks: I Visit Soviet Russia* 1937

Plays

Messalina of the Suburbs 1924; *The Glass Wall* 1933

Edited

The Time and Tide Album (s) 1932

Biography

The Life of a Provincial Lady by Violet Powell 1988

Walter (John) DE LA MARE 1873–1956

Born in the Kent village of Charlton, the son of a Bank of England official and churchwarden of Huguenot stock, de la Mare was related on his mother's side to Robert Browning. He attended St Paul's School in London, then went into the city offices of the Anglo-American (Standard) Oil Company, where he remained for almost twenty years. Meanwhile, he affected a velvet coat in the fashion of the 1890s and began to contribute to magazines. His first book of poetry, *Songs of Childhood*, was published in 1902 under the pseudonym of Walter Ramal, but attracted little attention. In 1908, however, the Asquith government, on the recommendation of Sir Henry Newbolt, recognised de la Mare's gifts with a small grant and a civil list pension of £100 per annum.

De la Mare retired to Taplow in Buckinghamshire, where he lived with his wife and four children and, apart from some book reviewing, devoted the remainder of his long life to his own work as a poet, novelist, short-story writer and anthologist. Further funds were provided when he became one of the legatees of his fellow poet **Rupert Brooke**, who died in 1915. His reputation as a poet was consolidated in 1912 by his volume *The Listeners*, the eerie title poem of which, much anthologised subsequently, is characteristic of de la Mare's style and subject-matter. Traditional in form, much of his poetry explores, in haunting, beautifully crafted verse, his chosen territory, where the numinous borders on the everyday. Many of his poems are about the English countryside (his final volume, written in his eighties, was *O Lovely England*), but there is also a distinctly macabre quality to his work, notably in the outstanding volume of poems for children, *Peacock Pie*, with its tales of mad princes, persecuted and solitary children, voracious giants and mocking fairies. (It was reissued in 1958 with exquisitely atmospheric line drawings by Edward Ardizzone.) **W.H. Auden** described de la Mare's poetry for children as 'unrivalled', and it was eventually gathered as *Collected Rhymes and Verses*, a complementary volume to his *Collected Poems* for adults (although the two categories richly overlap). The supernatural and other-worldly is also an important element of his short stories, which were published in numerous journals and collected in several volumes, and of his best novel, *Memoirs of a Midget*, an extraordinary tale of passion and obsession which won a James Tait Black Memorial Prize in 1921.

De la Mare was one of the century's best and most inventive anthologists, casting his net wide and providing lengthy introductions and commentaries. Amongst several fine volumes are *Love* (with decorations by Barnett Freedman), *Early One Morning* (on childhood), *Desert Islands*, and *Behold, This Dreamer!* ('Of Reverie, Night, Sleep, Dream, Love-Dreams, Nightmare, Death, the Unconscious, the Imagination, Divination, the Artist, and Kindred Subjects'). *Come Hither: a Collection of Rhymes and Poems for the Young of All Ages* remains unsurpassed as an anthology for juveniles.

Belonging to no school, influencing other writers only slightly, de la Mare is one of the notable originals of English literature. A man of great personal fascination, with a 'dark Roman profile rather like a bronze eagle', he received honorary degrees from many universities, was made a Companion of Honour and awarded the Order of Merit, and is buried in St Paul's Cathedral.

Poetry

Poems 1906; *The Listeners and Other Poems* 1912; *The Sunken Garden and Other Poems* 1917 [rev. as *The Sunken Garden and Other Verse* 1931]; *Motley and Other Poems* 1918; *Flora* 1919; *Poems 1901–1918* 1920; *The Veil and Other Poems* 1921; *A Ballad of Christmas* 1924; *Twenty-Nine Poems* 1926; *Stuff and Nonsense and So On* 1927 [rev. 1946]; *The Captive and Other Poems* 1928; *The Fleeting and Other Poems* 1933; *Poems 1919–1934* 1935; *Memory and Other Poems* 1938; *Collected Poems* 1942 [rev. 1979]; *The Burning-Glass and Other Poems* 1945; *The Traveller* 1945; *Inward Companion* 1950; *Winged Chariot* 1951; *O Lovely England and Other Poems* 1953; *The Winnowing Dream* 1954; *Complete Poems* 1969

For Children

Songs of Childhood [as 'Walter Ramal'] 1902; *The Three Mulla-Mulgars* 1910 [rev. as *The Three Royal Monkeys* 1935]; *A Child's Day: a Book of Rhymes* 1912; *Peacock Pie* 1913 [rev. 1924, 1958]; *Crossings: a Fairy Play* 1921; *Down-Adown-Derry: a Book of Fairy Poems* 1922; *Broomsticks and Other Tales* (s) 1925; *Miss Jemima* 1925; *Told Again* (s) 1927; *Old Joe* 1927; *Stories from the Bible* (s) 1929; *Poems for Children* 1930; *The Dutch Cheese* (s) 1931; *Old Rhymes and New* 1932; *The Lord Fish and Other Stories* (s) 1933; *Letters from Mr Walter de la Mare to Form Three* 1936; *This Year: Next Year* 1937; *Bells and Grass* 1941; *The Magic Jacket and Other Stories* (s) 1943; *Collected Rhymes and Verses* 1944 [rev. 1947, 1970]; *The Scarecrow and Other Stories* (s) 1945; *Jack and the Beanstalk* 1951; *Dick Whittington* 1951; *Snow-White* 1952; *Cinderella* 1952

Fiction

Henry Brocken 1904; *The Return* 1910; *Memoirs of a Midget* 1921; *Lispet, Lispett and Vaine* 1923; *The Riddle and Other Stories* (s) 1923; *Ding Dong Bell* (s) 1924; *Two Tales* (s) 1925; *The Connoisseur and Other Stories* (s) 1926; *At First Sight* 1928; *On the Edge* (s) 1930; *A Froward Child* (s) 1934; *The Wind Blows Over* (s) 1936; *The Picnic and Other Stories* (s) 1941; *A Beginning and Other Stories* (s) 1955; *Ghost Stories* (s) 1956

Non-Fiction

M.E. Coleridge: an Appreciation 1907; *Rupert Brooke and the Intellectual Imagination: a Lecture* 1919; *The Printing of Poetry* 1931; *Poetry in Prose: a Lecture* 1936; *Pleasures and Speculations* 1940; *Private View: Selected Reviews* 1953

Edited

Come Hither: a Collection of Rhymes and Poems for the Young of All Ages 1923; *Christina Rossetti* 1930; *Desert Islands and Robinson Crusoe* 1930; *The Eighteen-Eighties: Essays by Fellows of the Royal Society of Literature* (c) 1930; *Tom Tiddler's Ground* 1932; *Early One Morning* 1935; *Animal Stories* (s) 1939; *Behold, This Dreamer!* 1939; *Love* 1943

Biography

Imagination of the Heart by Theresa Whistler 1993

Shelagh DELANEY 1939–

Of Irish heritage, Delaney is the daughter of a bus inspector and was born in Salford, Lancashire. After an undistinguished academic career at Broughton Secondary School, she left aged seventeen for a series of undemanding jobs. Two years later, however, she achieved immediate recognition with her first play, *A Taste of Honey*, a funny and touching tragi-comedy which **Graham Greene** hailed as superior to **John Osborne**'s *Look Back in Anger*. The impetus for writing the play (which had originally been conceived as a novel) was provided by **Terence Rattigan**'s *Variations on a Theme* (1958), which seemed to the nineteen-year-old author to have little to do with real life. Her own play, written in two weeks, opens with the arrival of a Salford schoolgirl and her mother ('a semi-whore', as the stage directions put it with characteristic bluntness) at their latest accommodation, 'a comfortless flat in Manchester'. The girl becomes pregnant after an affair with a black naval rating, who subsequently disappears, and her only friend is a homosexual shop assistant. This was certainly a world away from Rattigan (at least, the public perception of him), and the play provoked extreme reactions when put on by Joan Littlewood at the Theatre Royal, Stratford East, but it transferred to the West End and was produced in New York two years later. It won the Charles Henry Foyle New Play Award and the New York Drama Critics Circle Award and was subsequently made into an award-winning film (1961), from a screenplay by Delaney and the director, Tony Richardson.

Although Delaney is far from being a one-play-wonder, she has not to date surpassed her astonishingly assured and precocious debut. Her second play, *The Lion in Love*, was produced on both sides of the Atlantic, but was considered less successful than its predecessor. She has subsequently written plays for radio and television and produced a volume of short stories. Her greatest success, however, has been in the cinema, where her screenplays have won several awards. She has collaborated with distinguished directors on unusual and quirky projects: Lindsay Anderson on *The White Bus*, adapted from her own short story; Albert Finney on *Charlie Bubbles*, in which a novelist returns to his northern, working-class roots; and – after a long absence – Mike Newell on the atmospheric *Dance with a Stranger*, a film about Ruth Ellis, the last woman to be hanged in Britain.

Plays

A Taste of Honey **(st)** 1958, **(f)** 1961 [pub. 1959]; *The Lion in Love* 1960 [pub. 1961]; *The White Bus* **(f)** 1966; *Charlie Bubbles* **(f)** 1968; *Did Your Nanny Come from Bergen?* **(tv)** 1970; *St Martin's Summer* **(tv)** 1974; *The House That Jack Built* **(tv)** 1977, **(st)** 1979; *Find Me First* **(tv)** 1981; *So Does the Nightingale* **(r)** 1981; *Don't Worry about Matilda* **(r)** 1983; *Dance with a Stranger* **(f)** 1985; *The Railway Man* [from J. Johnston's novel] **(f)** 1993

Fiction

Sweetly Sings the Donkey **(s)** 1963

Don DeLILLO 1936–

DeLillo was born in New York City, and brought up by Italian immigrant parents in the Bronx. He rarely gives interviews and is reluctant to discuss his personal background, but did reveal after the publication of *Libra* that his interest in the novel's protagonist, Lee Harvey Oswald, was initially sparked by the discovery that he and Oswald 'had lived about six or seven blocks from each other in the Bronx, when he was 13 and I was 16'. DeLillo's reticence is evidently well planned: in 1979 an interviewer tracked him down in Athens and, the moment the conversation strayed from the subject of DeLillo's fiction, was handed a business card engraved with the words: 'Don DeLillo. I don't want to talk about it.'

DeLillo studied history, philosophy and theology at the Jesuit-run Fordham University from 1954 to 1958. After working for three years as a copywriter at the Ogilvy & Mather advertising agency in New York, in 1966 he started his first novel, *Americana*, the writing of which was interrupted by a series of temporary jobs. Published in 1971, the novel details the life of a young television executive who travels across the USA attempting to impose order on his chaotic experience by filming everything with a movie camera. DeLillo's subsequent novels are mordant black satires of contemporary America comparable with the work of **Thomas Pynchon** and **William Gaddis** (a friend of DeLillo's). *End Zone* concerns a college football captain who becomes obsessed with the parallels between his sport and strategic nuclear thinking. *Great Jones Street* explores the peculiarly American notion of celebrity through the figure of a rock star attempting to avoid the premature death his fans expect of him. *Ratner's Star* is a long and exhaustive satire of science revolving around a cartoon-like child prodigy. *Running Dog* brings together many of DeLillo's central preoccupations, focusing on the efforts of the security services, art dealers and the Mafia to lay hands on a pornographic film reputedly shot in Hitler's bunker. Despite his repeated treatment of the themes of paranoia and violence, DeLillo stresses that his primary concern is with the exploration of language: 'What writing means to me is trying to make interesting, clear, beautiful language.'

The award of a Guggenheim Fellowship in 1979 enabled DeLillo to travel for three years in India, the Middle East and Greece, the last of which provides the setting for *The Names*. *White Noise*, which won the American Book Award in 1985, details the efforts of Jack Gladney, a Professor of Hitler Studies, to protect his family from a cloud of industrial waste, an 'airborne toxic event' as it is called in DeLillo's jargon-filled world. His most successful novel, *Libra*, recreates events leading up to the assassination of John F. Kennedy, and won the first annual International Fiction Prize in 1989. *Mao II* describes a reclusive writer re-emerging into public view and becoming embroiled with terrorists. DeLillo has published short stories in the *New Yorker*, *Esquire*, *Atlantic* and other magazines, and lives in the New York suburb of Westchester with his wife.

Fiction

Americana 1971; *End Zone* 1972; *Great Jones Street* 1973; *Ratner's Star* 1976; *Players* 1977; *Running Dog* 1978; *Amazons* [as 'Cleo Birdwell'] 1980; *The Names* 1982; *White Noise* 1985; *Libra* 1988; *Mao II* 1991

Plays

The Engineer of Moonlight 1979; *The Day Room* 1986

William (Frend) DE MORGAN 1839–1917

De Morgan (whose family, of Huguenot origin, had taken to writing the *particule* with a capital letter) is an unusual figure in the arts, in that he first achieved fame as a potter, and only after his retirement from this, in old age, took up a separate career as a novelist. He was born in London, the son of Augustus De Morgan, who was professor of mathematics at University College London, which he had helped to found. De Morgan himself attended the college after University College School, but as his primary interest was in art he went on to attend the Royal Academy School, where he made the acquaintance of the chief Pre-Raphaelites, who were to remain his lifelong friends.

As an artist, he began by designing stained glass and tiles, but from the early 1870s found his true *métier* as a potter, rediscovering the process of making coloured lustres, and becoming famous for his thickly glazed blue and green pots. He was also particularly successful in the construction of decorative panels; a house in Addison Road, Holland Park, London, is a well-known example of his work. From 1872 onwards he lived in Chelsea, with one short interval, and in 1887 married another artist, Evelyn Pickering, the daughter of the recorder of Pontefract, who was eighteen years his junior, but with whom his marriage, although childless, was happy. He entered several partnerships in his artistic work, including one with William Morris, and had workshops in Chelsea, Wimbledon and Fulham, but was never financially successful, and in 1905 was forced to bring the last of these ventures to an end.

Then began, at age sixty-seven, his second career, as a novelist. In 1906, he published *Joseph Vance*, and this rambling but immensely lively story of a boy from a working-class home who goes to Oxford and becomes a prosperous inventor was an immediate success with the public. De Morgan followed it up with six more novels until the outbreak of the First World War in 1914, but then became increasingly preoccupied with aircraft and submarine defence (he was also a notable inventor) and largely ceased to write. He died in 1917, it is said because a soldier and admirer of his works, who visited him, gave him 'trench fever', and two more novels were expertly completed for publication by his wife, who had always closely advised him, before her own death in 1919.

De Morgan – tall and thin with a straggly beard, his voice swooping from bass to falsetto, wintering in Florence for his health – was an archetypal Victorian figure, and his novels, which owe much to Dickens, are really robust mid-Victorian works which, by a quirk of fate, were produced in the early twentieth century. The contemporary *Times Literary Supplement* said: 'One would as soon think of criticising the Bank of England as of criticising one of his novels'; and while we may consider today that the charms of their humour and easy knowledge of all social classes are outweighed by excessive prolixity, it is also true that books such as *Alice-for-Short* attract their modern admirers. De Morgan published two treatises on the craft of pottery and was a minor poet. It is not likely that he will be forgotten, for his story is among the strangest in the history of the arts.

Fiction

Joseph Vance 1906; *Alice-for-Short* 1907; *Somehow Good* 1908; *It Never Can Happen Again* 1909; *An Affair of Dishonour* 1910; *A Likely Story* 1911; *When Ghost Meets Ghost* 1914;

The Old Madhouse (s) [unfinished; with E. De Morgan] 1919; *The Old Man's Youth* (s) [unfinished; with E. De Morgan] 1921

Biography

William De Morgan by M.D.E. Clayton-Stamm and William Gaunt 1971

Nigel (Forbes) DENNIS 1912–1989

Dennis was born at Bletchingley, Surrey, but when his father died in 1918 his mother married again and the family moved to what was then Northern Rhodesia. He was educated partly there, partly in Austria and Germany. There followed 'the usual Depression jobs for the next few years, selling hosiery door-to-door and tweed suits'. His literary career started with reviews and short stories for *Time and Tide*. After helping to translate Alfred Adler's *Social Interest*, he went to live in New York, where he worked on the National Board of Reviews of Motion Pictures and for the *New Republic* as review editor. Finally he joined the staff of *Time* magazine, leaving New York in 1949.

In London he became the staff reviewer of the *Sunday Telegraph* and the dramatic critic of *Encounter*. His first novel, *Boys and Girls Come Out to Play*, which won the Anglo-American Novel Contest of 1949, satirises American 'liberalism' and the American attitude to Europe on the eve of the Second World War. As editor of the wealthy Mrs Morgan's progressive paper *Forward*, the American journalist Max Divver goes to investigate the Polish crisis in the spring of 1939; there, without really knowing what is happening to him, he is carried virtually into the enemy camp. The *New Statesman* called it 'extremely intelligent, always entertaining and sometimes very funny indeed'. *Cards of Identity*, with its country-house setting, is another satire concerned with something called the 'Identity Club'; it was the book that made his name. **W.H. Auden** praised it in *Encounter*; John Davenport compared it to Stevenson's 'Suicide Club'; **Walter Allen** talked of its 'exuberance of comic invention'. Dennis adapted it for the stage and it was published with one of his other plays, *The Making of Moo*, in *Two Plays and a Preface*. Of it **Philip Toynbee** wrote: 'The brilliantly funny preface ... is in the tradition of **George Bernard Shaw**'s best prefaces – clever, perceptive and talking a great deal of sense ... Mr Dennis is a genuine satirist, richly equipped for his enjoyable task.' It can be no accident that he published a life of Swift. He left all too little, but nothing that is not enjoyable. He married twice, and had two daughters.

Fiction

Boys and Girls Come Out to Play 1949; *Cards of Identity* 1955; *A House in Order* 1966

Plays

Two Plays and a Preface [pub. 1958]: *Cards of Identity* [from his novel] 1956, *The Making of Moo* 1957; *August for the People* 1961 [pub. 1962]; *Swansong for 7 Voices* (r) 1985

Poetry

Exotics 1970

Non-Fiction

Dramatic Essays 1962; *Jonathan Swift: a Short Character* 1964; *An Essay on Malta* 1972

Anita DESAI 1937–

Desai was born Anita Mazumdar in Mussoorie, a hill station north of Delhi, the daughter of a Bengali businessman and his German wife. She grew up in Delhi speaking the English language and began to write in it at the age of seven, publishing her first story at the age of nine. She was educated at Queen Mary Higher Secondary School, Miranda House (a women's college) and the University of Delhi, where she took her BA degree in English literature in 1957. In the following year, she married a businessman; they have four children.

Desai had begun to publish short stories regularly before her marriage, and her first novel came out in 1963, when she was in her mid-twenties. She comments on her work: 'My novels are no reflection of Indian society, politics or character. They are part of my private attempt to seize upon the raw material of life.' The books are mostly set in North India (one, *Bye Bye, Blackbird*, is set in Britain) and deal mainly with the spiritual stresses affecting educated Indians, particularly women. There is little agreement as to which are the best, although the earlier four are often considered melodramatic, and *Clear Light of Day*, which concerns one sister's visit to another in a decaying mansion outside Delhi, is generally thought the outstanding exemplar of its author's cool style, potent imagery and human insight. It was shortlisted for the Booker Prize, as was *In Custody*; the latter was also filmed by Ismail Merchant (1994), with a scenario by Desai herself. She has published a volume of short stories and, unusually for an Indian author, several books for children.

Desai has long divided her life between India, the USA (where she has taught creative writing at Smith College and Mount Holyoke University, and has been professor of writing at Massachusetts Institute of Technology since 1993) and Britain, where she is based in Cambridge (she has had an association with Girton College for many years, and has been an honorary fellow there since 1988). She is a member of the advisory board of the National Academy of Letters in Delhi and a Fellow of the Royal Society of Literature in London.

There is in Desai's work a clear sense of the growth of a distinguished *oeuvre* over time. *Baumgartner's Bombay*, the story of German expatriate adrift in India, saw a darkening in tone, and was followed by a gap of seven years before *Journey to Ithaca*, set in India and Italy.

Fiction

Cry, the Peacock 1963; *Voices in the City* 1965; *Bye Bye, Blackbird* 1971; *Where Shall We Go This Summer?* 1975; *Fire on the Mountain* 1977; *Games at Twilight* (s) 1978; *Clear Light of Day* 1980; *The Village by the Sea* 1982; *In Custody* 1984; *Baumgartner's Bombay* 1988; *Journey to Ithaca* 1995

For Children

The Peacock Garden 1974; *Cat on the Houseboat* 1976

Plays

In Custody (f) [from her novel] 1994

Peter DE VRIES 1910–1993

De Vries was once widely regarded, especially in the metropolitan circles of his native America, as a purveyor of gags. His finest work, however, was *The Blood of the Lamb*, a harrowing story which made explicit the serious purpose behind even his wildest creations: 'I don't think that the comic and the serious can be separated in talking about human reality, any more than you can separate hydrogen and oxygen and still be talking about water.'

Born in Chicago to immigrants of the strict Calvinist Dutch Reformed persuasion, he duly showed a flair for basketball and an interest in politics – neither of which might be suspected by his readers, nor that, as a child, he was forbidden movies, dancing, cards, indeed anything secular. 'We were the product of schism and we produced schism,' he commented.

On graduating from Calvin College in Great Rapids, Michigan, he worked on a community newspaper, and was sacked for revealing that a businessman was a Freemason. Laid low for much of 1932 with tuberculosis, he then turned to Democratic reform politics, all the while nurturing a desire to write, and fending off parental pressure to join the ministry. In the Depression, he became a wireless actor, candy salesman, women's-club lecturer, furniture-removals man, ice-cream salesman and coal distributor. His father, spiritual succour notwithstanding, had a severe breakdown. How much this influenced De Vries, he did not care to ponder: 'Who knows? Who cares? I don't. One just sits in the corner and secretes the stuff. Don't ask a cow to analyse milk.' Such bluff belies a man sensitive to language. By 1938 he became assistant editor, then editor, of *Poetry*, a magazine which even at the time seemed venerable. He wrote some verse himself, but later chose to leave it in obscurity. Also now impossible to find are three early novels, none of which was published in the UK.

Marriage to Katinka Loesser in 1943 was as much a new direction in his life as a move to the *New Yorker* that year, which came about after inviting **James Thurber** to lecture. Much of the magazine's post-war tone and style was set by his part-time work as cartoon editor, reflected in his return to fiction of a different kind from the apprentice efforts a decade earlier. Time and again, suburban man is vexed by the weighty questions of life, especially when middle age makes itself felt. Like many of the magazine's writers, De Vries chose not to live in the city.

If not as nifty as those of **P.G. Wodehouse**, his plots (which are impossible to summarise) provide numerous opportunities for puns and bizarre, fitting metaphor. *Reuben, Reuben* has three sections each told by a different narrator (including a **Dylan Thomas**-like poet), while *The Tents of Wickedness*, has chapters consisting of parodies of authors from **James Joyce** to **James Jones**. De Vries, however, insisted that 'a one-liner should be inevitable'. Wrenched from context, these include a man who, hurling rocks at birds, declares: 'I'm leaving no tern unstoned'; 'Stop looking at my legs,' says a woman to a parson, who replies: 'Don't worry madam, my thoughts were on higher things'; another philosophises: 'We're all like the cleaning woman. Come to dust.' The novels came in swift succession during the 1950s, peaking with *The Blood of the Lamb* in 1961, one year after the death of his young daughter from leukaemia. Pared of puns, but not lacking in wit, the novel is distinctly autobiographical (he returned to autobiography in the twin novellas of *The Cat's Pajamas* and *Witch's Milk*). The world changes about the narrator, and for De Vries himself the 1960s obliged him to take stock, aware that he was less trenchant than the likes of **Joseph Heller**. 'Satirists shoot to kill,' he remarked 'but the humorist brings his prey back alive.'

Novels continued at regular intervals, and perhaps the best of them is 1983's splendidly titled *Slouching towards Kalamazoo*, which echoes an earlier remark that: 'the standards for immorality are getting progressively steeper ... a hundred years ago Hester Prynne of *The Scarlet Letter* was given an A for adultery. Today she would rate no better than a C-plus.' Only towards the end – he never stopped – did De Vries go below B-plus, the only wisecracker to find repeated inspiration in Sir Thomas Browne.

Fiction
But Who Wakes the Burglar? 1940; *The Handsome Heart* 1943; *Angels Can't Do Better* 1944; *No, But I Saw the Movie* (s) 1952; *The Tunnel of Love* 1954; *Comfort Me with Apples* 1956; *The Mackerel Plaza* 1958; *The Tents of Wickedness* 1959; *Through the Fields of Clover* 1961; *The Blood of the Lamb* 1961; *Reuben, Reuben* 1964; *Let Me Count the Ways* 1965; *The Vale of Laughter* 1967; *The Cat's Pajamas and Witch's Milk* 1968; *Mrs Wallop* 1970; *Into Your Tent I'll Creep* 1971; *Without a Stitch in Time* (s) 1972; *Forever Panting* 1973; *The Glory of the Hummingbird* 1974; *I Hear America Swinging* 1976; *Madder Music* 1977; *Consenting Adults* 1981; *Sauce for the Goose* 1982; *Slouching Towards Kalamazoo* 1983; *Peckham's Marbles* 1986; *The Prick of Noon* 1986

Play
The Tunnel of Love 1957

Biography
Peter De Vries by J.H. Bowden 1983

Pete DEXTER 1943–

Dexter was born in Pontiac, Michigan, and brought up in Georgia and South Dakota. He finally graduated from the University of South Dakota in 1970 after eight years of intermittent attendance during which he worked as a truck driver and in various labouring jobs. The next year he became a reporter on the *West Palm Beach Post* but left after two years. 'I wasn't the best writer there,' he has explained in an interview. He worked briefly in a gas station and left: 'I wasn't even the best writer in the gas station.' From 1972 to 1984 he was a columnist on the *Philadelphia Daily News*. In 1982 he was severely beaten in a bar by a group of baseball-bat-wielding citizens offended at his coverage of a local murder. According to Dexter this was the impetus behind his decision to write fiction: with an altered sense of taste from a head injury he was no longer able to engage in his favourite occupation of drinking, so turned to writing instead.

Dexter's first novel, *God's Pocket*, is a comic romp set in motion when a man is murdered on a construction site and includes a sub-plot in which an alcoholic Philadelphia newspaper columnist is beaten to death by an irate mob of God's Pocket citizens. *Deadwood* is a similarly violent and bawdy comedy set in the Dakota gold rush town of Deadwood in 1876. In Dexter's third novel, *Paris Trout*, which won a National Book Award in 1988, the tone is darker, though a comic streak persists. An account of racial bigotry and class antagonism set in Georgia in the 1950s, the novel was inevitably compared with **Williams Faulkner** and **Flannery O'Connor**. Dexter insisted, however, that the events of the novel could take place anywhere and commented, from bitter experience: 'In fact, South *Philadelphia* is more violent than the South.' As if to confirm the point, his fourth novel, *Brotherly Love*, concerns a violent power struggle between union leaders and mobsters in 1960s' Philadelphia. A film version of *Paris Trout*, directed by Stephen Gyllenhaal, was released in 1991. Since 1985 Dexter has been a columnist on the *Sacramento Bee*, and he is married with one daughter.

Fiction
God's Pocket 1984; *Deadwood* 1986; *Paris Trout* 1988; *Brotherly Love* 1992

Michael DIBDIN 1947–

Dibdin was born in Wolverhampton. The early years of his childhood were peripatetic, as his father, a folk-dancing instructor and later a physics lecturer, and his mother, a nurse, travelled throughout England and Scotland. They settled in County Antrim in the mid-1950s and Dibdin attended the Friends' School, Lisburn. He studied English at the University of Sussex, and moved to Canada where he obtained his MA from the University of Alberta, Edmonton, in 1970. Remaining in Canada for another five years, he took various casual jobs and at one point ran an ill-fated painting and decorating business. His first marriage, from which he has one daughter, lasted from 1971 to 1979. He returned to England, where he wrote a play, *Rough Music*, which was never performed, and his first novel, *The Last Sherlock Holmes Story* (1978). The novel is a Victorian pastiche, narrated by Dr Watson, which brings together Sherlock Holmes and Jack the Ripper.

Dibdin moved to Italy where in the early 1980s he taught English at International House in Perugia and worked as a language assistant at the University of Perugia. In Italy he met his second wife, with whom he has one daughter; they live in Oxford. *A Rich Full Death* is another exercise in historical pastiche, set in Victorian Florence and featuring Robert and Elizabeth Barrett Browning.

With his third novel, *Ratking*, which won the Crime Writers' Association Gold Dagger Award in 1988, Dibdin had his first substantial success. It is the first in a series of sophisticated and stylish crime novels about the Italian police detective Aurelio Zen. The series continues with *Vendetta*, *Cabal* and *Dead Lagoon*. *The Tryst* and *Dirty Tricks* are ironic and frequently comic novels set in contemporary England which fit only loosely into the category of crime fiction. Both involve dense and interwoven plots, and the crime element is, in Dibdin's words, 'emblematic of large social and historical forces'. *The Dying of the Light* is an ironic subversion of the traditional **Agatha Christie** 'country house' murder investigation. Dibdin contributes to the *Independent on Sunday* and has edited an anthology of crime writing.

Fiction

The Last Sherlock Holmes Story 1978; *A Rich Full Death* 1986; *Ratking* 1988; *The Tryst* 1989; *Vendetta* 1990; *Dirty Tricks* 1991; *Cabal* 1992; *The Dying of the Light* 1993; *Dead Lagoon* 1994

Edited

The Picador Book of Crime Writing 1993

Kay DICK 1915–

The daughter of an unmarried mother, Dick was born in London and taken at once to the Café Royal in Regent Street, where she received her only baptism, toasted by the restaurant's Bohemian *habitués*. Mother and child were then taken to 'some outer London suburb' to live with an artists' model. For many years, she was encouraged by her mother to believe that her father had strayed from the pages of *Burke's Peerage* or *Debrett*, or was a dead war hero, but his identity remains unknown. When she was four, her mother met a rich Swiss man and lived with him for three years, travelling abroad; but when they married, Dick had to endure two miserable years, from the age of seven, in a boarding-school in Bognor Regis. Her stepfather then placed her with a Swiss-French family in Geneva, to be educated at a progressive day-school until the age of sixteen, years during which she enjoyed the temporary security of family life. She subsequently returned to England, where her education was completed at the Lycée Français de Londres and an English tutorial college.

Dick had decided to be a writer at the age of ten, but her stepfather, who had lost his money, forced her at the age of twenty to earn her living. She entered the world of publishing and bookselling, working for a number of companies, including Foyle's, where she ran the mail publicity department. During the war, a period she describes in her second novel, *Young Man*, she worked for the *New Statesman* and then managed the publisher P.S. King & Son, becoming the first woman director in English publishing. She subsequently became assistant editor to John Brophy on *John O'London's Weekly*. It was while working here that she met several of the people she wrote about in her volume of 'Conversations and Reflections', *Friends and Friendship*, including **Brigid Brophy** and **Pamela Hansford Johnson**. She became a tireless campaigner for Public Lending Right, each of her books carrying an appeal for its support. Dick subsequently edited *The Windmill*, a literary quarterly, using the pseudonym Edward Lane, and publishing (amongst others) **Olivia Manning**, Malcolm Muggeridge and **George Orwell**.

She lived in a circle of artists and writers in Hampstead's Flask Walk, sharing her home with the novelist Kathleen Farrell, but later moved to live alone in Brighton. An elegant, striking figure, with softly cropped hair, she invariably wears men's shirts and trousers and sports a monocle. She has suffered prolonged periods of writer's block and her output has been small. In novels such as *Sunday*, based loosely upon her parents, the emphasis is on the necessity of sincere relationships in a humanist world, while *They* (published after a fifteen-year silence) investigated the subject of grief and was inspired by a recent psychiatric treatment in which emotions were 'burned out ... to expel grief'. Dick's novel condemns this method, because she felt it destroyed creative work and encouraged a *Nineteen Eighty-Four* atmosphere of fear, and the novel ends with the triumph of hope and a faith in human love. Her most recent novel to date is *The Shelf* (1984).

Perhaps her best book, however, is *Ivy and Stevie*, another volume of conversations with two of her friends, **I. Compton-Burnett** and **Stevie Smith**, which provides a remarkable record of these two writers.

Fiction

By the Lake 1949; *Young Man* 1951; *An Affair of Love* 1953; *Solitaire* 1958; *Sunday* 1962; *They* 1977; *The Shelf* 1984

Non-Fiction
Pierrot: an Investigation into the Commedia Dell'Arte 1960; *Ivy and Stevie* 1971; *Friends and Friendship* 1974

Edited
The Mandrake Root (s) [as 'Jeremy Scott'] 1946; *At Close of Eve* (s) [as 'Jeremy Scott'] 1947; *The Uncertain Element: an Anthology of Fantastic Conceptions* 1950; *Writers at Work: the Paris Review Interviews* 1972

James (Lafayette) DICKEY 1923–

Dickey was born in Atlanta, Georgia, and graduated from Vanderbilt University, Tennessee in 1950. His career began with six years as an advertising copywriter for McCann-Erickson in New York, but with his first publication in 1960, *Into the Stone and Other Poems*, Dickey decided to earn a living as poet and teacher, though he expressed horror at the drastic reduction in salary. In 1948 he married, but his wife died in 1976. He has since remarried and has two sons – one of whom is the journalist and writer Christopher Dickey – and one daughter.

As teacher and writer-in-residence, he held posts at Rice University, Reed College, San Fernando State College, the University of Wisconsin and as Professor of English at the University of South Carolina. He was a fighter-bomber pilot in the US Air Force during the Second World War and a training officer in Korea. In 1961 he received a Guggenheim Fellowship and visited Europe. Between 1966 and 1968 he was poetry consultant to the Library of Congress. *Life* magazine once dubbed him 'the most unlikely poet', but in 1977 it was Dickey who read his latest work, *The Strength of Fields*, live on television at the inauguration ceremony of President Jimmy Carter, and was generally considered to be the finest public reader of his own poetry since **Dylan Thomas**.

He is as well known for being an athlete and a hunter with bow and arrow as he is for being a poet obsessed with technique, but he hates the idea of a poem been dissected 'like a cat in a biology lab'. *Buckdancer's Choice* won a National Book Award in 1965 and was cited by some critics as a good example of Dickey's fascination with violence. Dickey himself explains that his fascination is with 'survivorship' and 'intensified man'.

In 1971 he reached celebrity status with his appearance as the sheriff in the film adaptation of his bestselling novel *Deliverance* (directed by John Boorman from Dickey's own screenplay). He published little in the early 1970s and his latest work has been increasingly experimental. He speaks of wanting to write 'the conclusionless poem ... the unwell made poem', and of playing 'metaphysical scrabble' with words and phrases when writing. One of his most well-known announcements was that he had discovered 'the creative possibilities of the lie', and this has led to even more use of dialogue and multiple roles, and poetry that was essentially fiction.

Dickey has always strived for epic quality. His themes are turbulent and strange with intense, sustained images – like the air-hostess hurtling through the sky to her death in a field in Kansas in *Falling*. He writes of violence, power, fertility and the primeval instincts of nature and says of himself: 'The Poet is not trying to tell the truth. He's trying to make it.'

Poetry
Into the Stone and Other Poems 1960; *Drowning with Others* 1962 [rev. as *The Owl King* 1977]; *Helmets* 1964; *Two Poems of the Air* 1964; *Buckdancer's Choice* 1967; *The Eye-Beaters, Blood, Victory, Madness, Buckhead and Mercy* 1977; *The Zodiac* 1976; *The Strength of Fields* 1977; *Head-Deep in Strange Sounds* 1979; *Veteran Birth: the Gadfly Poems 1947–1949* 1979; *Falling, May Day Sermon and Other Poems* 1981; *The Early Motion* 1981; *Puella* 1982; *The Central Motion: Poems 1968–1979* 1983; *Summons* 1988

Fiction
Deliverance 1970; *Alnilam* 1987

Non-Fiction
The Suspect in Poetry 1964; *A Private Brinkmanship* 1965; *Spinning the Crystal Ball: some guesses at the future of American Poetry* 1967; *Metaphor as Pure Adventure* 1968; *Babel to Byzantium: Poets and Poetry Now* 1968; *Self-Interviews* (a) 1970; *Sorties* 1971; *Exchanges ... Being in the Form of a Dialogue with Joseph Trumbull Stickney* 1971; *Jericho: the South Beheld* [with H. Shuptrine] 1974; *God's Images: the Bible: a New Vision* [with M. Hayes] 1977; *Tucky the Hunter* 1978; *The Enemy from Eden* 1978; *In Pursuit of the Gray Soul* 1981; *The Water-Bug's Mittens: Ezra Pound, What We Can Use* 1979; *The Starry Place Between the Antlers: Why I Live in South Carolina* 1981; *The Eagle's Mile* 1981; *False Youth: Four Seasons* 1982; *The Poet Turns on Himself* 1982; *For a Time and Place* 1983; *Bronwen, the Traw and the Shape-Shifter* 1986; *Wayfarer* 1988; *To the White Sea* 1994

Collections
Night Hurdling: Poems, Essays, Conversations, Commencements and Afterwards 1983

Patric (Thomas) DICKINSON 1914–1994

Dickinson was born in India, but never saw his father, an officer in the Indian Army, who was killed in Mesopotamia in 1915. Brought to Petersfield by his mother, Dickinson was educated at Grenham House, Birchington, and at Plox House in Bruton. Despite the low academic standards of the latter, in 1932 Dickinson won an exhibition in classics at St Catharine's College, Cambridge. Glad to be free of organised team games, he was from an early age a passionate golfer, a sport which earned him a Blue in 1935.

After graduating in 1936, he became a schoolmaster, and when the Second World War began, he enlisted in the Artists' Rifles, until typhoid forced him out after one year. During a long convalesence, Dickinson wrote poems, with the encouragement of **Walter de la Mare**. His first volume, *The Seven Days of Jericho*, was published in 1944. Early work such as the play *Stones in the Midst* was influenced by his classical background, and received mixed reviews. Later work became looser in form, and reflected his interest in geology and nature. He also translated Aristophanes, and Virgil.

In 1946, he married Sheil Shannon, who edited with W.J. Turner the Britain in Pictures series, and they bought a house in Rye, Sussex. An Atlantic Award obliged Dickinson to leave the BBC, where he worked in the radio features and drama department, and later as poetry editor, where he produced the long-running and popular programme, *Time for Verse*, but he continued to work for them as a freelance broadcaster. He did much to promote the acceptance of poets reading their own work.

Particularly drawn to **Wilfred Owen**, he persuaded Oxford University Press to publish Harold Owen's *Journey from Obscurity* (1963–5). As well as publishing numerous volumes of poetry, Dickinson also edited the works of Byron, **C. Day Lewis** and, unfashionably, Henry Newbolt. *The Good Minute* is a lively and evocative 'Autobiography of a Poet-Golfer'.

Poetry

The Seven Days of Jericho 1944; *The Sailing Race* 1952; *The Scale of Things* 1955; *The World I See* 1960; *This Cold Universe* 1964; *Selected Poems* 1968; *More Than Time* 1970; *A Wintering Tree* 1973; *The Bearing Beast* 1976; *Chapter and Verse* 1979; *Our Living John and Other Poems* 1979; *Poems from Rye* 1979; *Winter Hostages* 1980; *A Rift in Time* 1982; *To Go Hidden* 1984

Translations

Aristophanes against War 1957; *The Aeneid of Virgil* 1963; *Aristophanes* 2 vols 1970, 1976

Plays

Stones in the Midst 1948; *A Durable Fire* 1962

Non-Fiction

A Round of Golf Courses 1951; *The Good Minute* (a) 1965

Edited

Soldier's Verse 1945; *Byron* 1949; *Poems to Remember* [with S. Shannon] 1958; *C. Day Lewis: Selections from His Poetry* 1967; *Poet's Choice: an Anthology of English Poetry from Spenser to the Present Day* 1967; *Selected Poems of Henry Newbolt* 1981

Collections

Theseus and the Minotaur, and Poems 1946; *Stone in the Midst, and Poems* 1948

Joan DIDION 1934–

Didion was born in the Sacramento Valley, California, where the family had lived for five generations and her father was in the real-estate business. She graduated from the University of California at Berkeley in 1956 having majored in English literature, and in the same year won the first prize in a *Vogue* competition for an article on William Wilson Wurster, the originator of the San Francisco style of architecture. She moved to New York where she worked at *Vogue*, slowly working her way up to the post of associate features editor. She also wrote freelance for *Mademoiselle*, the *National Review*, the *New Yorker* and other magazines. Some of her journalism from the 1960s is collected in *Slouching Towards Bethlehem*, the title piece of which is an account of Haight-Ashbury in 1967. Much of her best work concerns the extremes of life in her native California. She took leave from *Vogue* to write her first novel, *Run River*, which depicts the gradual self-destruction of an old Sacramento family. She has said of the area that it is 'a place in which a boom mentality and a sense of Chekhovian loss exist in uneasy suspension'.

In 1964 Didion married the writer John Gregory Dunne, then working at *Time*, and they moved to Southern California where they adopted a daughter. The marriage was evidently somewhat rocky in its early years, as is attested to by a 1969 article in *Life* in which Didion describes herself as sitting in a Honolulu hotel room in lieu of filing for divorce. Like *Run River*, her second novel, *Play It As It Lays*, examines the cultural disintegration of contemporary America, though the novel is notable for a taut prose style which contrasts markedly with the elaborate descriptiveness of the first. It was a critical and commercial success and won a National Book Award in 1971. In collaboration with Dunne, she adapted the novel for a film version, and together they have also written scripts for *Panic in Needle Park, True Confessions* (which is based on Dunne's novel) and the remake of *A Star Is Born*. None of their work on this last survives in the final script, but they still receive royalties for their early development of the idea.

Didion's subsequent journalism includes a book-length study of the plight of El Salvador, and *After Henry*, a collection of essays on culture and politics which emphasises the gulf between the rhetoric and reality of American life. Since 1976 she has been a visiting lecturer in English literature at Berkeley.

Fiction

Run, River 1963; *Play It As It Lays* 1970; *A Book of Common Prayer* 1977; *Democracy* 1984; *Miami* 1987

Non-Fiction

Slouching Towards Bethlehem 1968; *The White Album* 1979; *Salvador* 1983; *After Henry* 1992 [UK *Sentimental Journeys* 1993]

Isak DINESEN (pseud. of Karen Christentze Blixen-Finecke) 1885–1962

Born into the Danish landed gentry in Rungstedlund, northern Denmark, Dinesen was the daughter of Wilhelm Dinesen, who had left the army after the Franco-Prussian war and gone to America to live amongst the Plains Indians as a fur trader; he was the author of several books, including the hunting classic *A Sportsman's Letters*. Dinesen studied art in Paris, Rome and Copenhagen before marrying her Swedish second cousin, Baron Bror Blixen-Finecke in 1914. He gave her the title for which she yearned and the syphilis from which she was to die. They emigrated to British East Africa, where they started a coffee farm near Nairobi. The marriage was short-lived and the couple divorced in 1921, but Dinesen stayed on to run the plantation and enjoyed a long affair with the English hunter-explorer Denys Finch-Hatton. The story of these years is related in *Out of Africa*, which reached a new audience almost of quarter of a century after the author's death when it was made into an award-winning film (1985).

When the coffee market collapsed and Finch-Hatton died, the near-bankrupt Dinesen returned to the family home in Denmark and, financed by her brother, began to write in earnest, making her name in 1934 with *Seven Gothic Tales*. These fantastic stories, some of which she had begun as early as 1909, were initially rejected by a British publisher, and the book was first published in America. Her narrative gifts, allied with an intricate prose style (she wrote in both Danish and English), won her a considerable following, and in 1954 she was nominated for a Nobel Prize. (It was in fact won by **Ernest Hemingway**, who upon hearing the news generously praised Dinesen, describing her as 'a damn sight better writer than any Swede they ever gave it to'.) She published novels, short stories and non-fiction during her lifetime under several names: Isak Dinesen, Karen Blixen and – for *The Angelic Avengers*, a novel dictated to a stenographer and composed largely to stave off the boredom of life in occupied Denmark, she claimed – Pierre Andrézel. The manuscript of *Winter's Tales* had to be smuggled out of Denmark and was sent to the USA via Stockholm in the British embassy's diplomatic bag. Dinesen had no idea of its fate until after the war.

Dinesen's health was poor, the result of her untreated venereal infections, and as the years passed she endured long spells in bed, both at her home (which had once belonged to the Danish poet Johannes Ewald) and in hospitals. Her appearance, which she emphasised by her outlandish costumes, became as gothic as her stories: gaunt, sharp-nosed, hollow-eyed, with parchment skin. She once remarked: 'I really am three thousand years old and have dined with Socrates.'

Fiction

As 'Isak Dinesen': *Seven Gothic Tales* (s) 1934; *Winter's Tales* (s) 1942; *The Angelic Avengers* [as 'Pierre Andrézel'] 1946; *Last Tales* (s) 1957; *Anecdotes of Destiny* (s) 1958; *Shadows on the Grass* 1960; *Ehrengard* (s) 1963; *Carnival* (s) 1977

Non-Fiction

As Karen Blixen: *Out of Africa* (a) 1937; *Essays* 1965; *Daguerrotypes and Other Essays* 1979; *Letters from Africa, 1914–1931* 1981

Biography

Isak Dinesen by Judith Thurman 1982

Jenny DISKI 1947–

Diski was born Jenny Simmonds in the East End of London and is descended from Russian Jews on one side of the family and Polish Jews on the other. She describes her father's occupation as 'crook': his work involved various jobs 'of an entrepreneurial nature' and a spell in prison. Diski was sent to a number of different schools, including a primary school in Camden and a boarding-school in Letchworth, Hertfordshire, from which she was expelled 'for unruliness'. She spent a year working in Banbury at Cullens food store and at the Freeman, Hardy and Willis shoe shop next door, but was sacked from both jobs and went back to school, this time to a co-ed day-school in London, to take her O-levels. She went on to study for A-levels but left without sitting the exams in 1966, the same year her father died.

She then decided to 'hang out' for three years, during which she did 'a bit of this and a bit of that and nothing at all in the way people did then'. In 1970 she took a course in teacher training, and from 1970 to 1972 ran Freightliners, a free-school for 'hopeless truant people'. In 1973 she started teaching history then English at a girls' comprehensive in Hackney.

In 1977 Diski had a daughter and left full-time teaching in favour of home-tuition, work that she did for the Inner London Education Authority up until 1982 when she went to University College London as an undergraduate studying for a degree in anthropology. Instead of completing the course she wrote her first novel, *Nothing Natural*, which begins with the central character opening a newspaper to discover a photofit picture of a rapist who may or may not be the man with whom she has been having a sado-masochistic relationship. The novel established Diski as a 'ground-breaking' writer, a label which surprised her: 'It was received as much more sexually outrageous than it was. What's desperate in the book is the psychological relationship; the sex was a kind of pointing up of that, and very mild too.'

She found equally unsatisfactory the label 'intelligent', which the critics used on publication of both her second and fourth novels, *Rainforest* and *Then Again*. She freely confesses to being more interested in science and ideas than stories (she reads populist textbooks avidly). She thinks the English are not 'terribly keen on ideas' and believes that 'a polite way of saying that is to call people intelligent – it has the quality of "She's too clever for her own good." '

'Feminist' is another label Diski dislikes. Her third novel, *Like Mother*, which is narrated by an anhydranencephalic baby, was reviewed largely in terms of feminism, motherhood and gender, 'which for me were things only in passing. … It does bother me that I should be perceived as a "woman writer", because I think it limits people's expectations.' In insisting that she does not want her work to be categorised, she volunteers her own definition of her novels as 'long things I make up'. Diski is a regular contributor of reviews, stories and diary-pieces to the *London Review of Books*, and lives with her husband and her daughter in north London.

Fiction

Nothing Natural 1986; *Rainforest* 1987; *Like Mother* 1988; *Then Again* 1990; *Happily Ever After* 1991; *Monkey's Uncle* 1994

Plays

Seduction **(tv)** 1991; *Fair and Easy Passage* **(tv)** 1990

E(dgar) L(awrence) DOCTOROW 1931–

Doctorow was born in the Bronx, New York, where his father owned a record store. He attended the Bronx High School of Science and graduated from Kenyon College, Ohio, in 1952 with a BA in philosophy. He met his future wife while studying for a year at Columbia University, and they married in 1954. After serving in the US Army between 1953 and 1955), he worked as a reservations clerk at LaGuardia airport and as a reader for a film company. He was an editor at New American Library from 1959 until 1964, when he was appointed editor-in-chief of Dial Press. Among his authors were **James Baldwin** and **Norman Mailer**. He left publishing in 1971 and has been a full-time writer since, although he has held a number of university posts, including creative writing fellow, Yale School of Drama and visiting senior fellow, Princeton University.

His first novel, *Welcome to Hard Times*, was a western exploring the foundations and morality of American progress. It announced a theme that was to recur in later work: the brutalisation of commerce by crime. *Loon Lake*, set in the Depression, further examined the interpenetration of business and gangsterism, while *Billy Bathgate* charted the decline of a Jewish gangster, Dutch Schultz, crushed between the WASP establishment and the Mafia: the individual has no resources to match those deployed against him by corporate interests. An earlier version of the same theme is to be found in *The Book of Daniel*, a fictionalised account of the Rosenberg case. If the improbable spies of that book are victims of government, their plight is only a dramatisation of a contemporary phenomenon. As Doctorow has said: 'in the twentieth century one of the most personal relationships to have developed is that of the person and the state … most always to [the person's] detriment.'

Much of Doctorow's work is characterised by the fusion of historical and fictional events and figures (notably in *Ragtime*, where Freud, Houdini and Henry Ford appear). Given an authentic political and intellectual *locus*, the authority of the narratives is enhanced at the expense of subjectivity. In this respect, Doctorow is a social realist rather than a modernist: 'Modernism made us think of writing as an act of ultimate individualism ... in fact, every writer speaks for a community ... Writers are witnesses ... to this terrifying century.'

No post-war writer of the USA has borne more eloquent witness than Doctorow, and none has more consistently received both popular and critical acclaim. His standing has been formally acknowledged in the award of the National Book Critics Circle Award for *Ragtime*, the American Book Award for Fiction for *World's Fair* and the Howells Medal and the PEN/Faulkner Award for *Billy Bathgate*. He and his wife live in New Rochelle and Sag Harbor, New York. They have two daughters and a son. Doctorow has disclosed: 'I have few vices, but one of them is moderation.'

Fiction

Welcome to Hard Times 1960 [UK *Bad Man from Bodie* 1961]; *Big as Life* 1966; *The Book of Daniel* 1971; *Ragtime* 1975; *Loon Lake* 1980; *Lives of the Poets* (s) 1984; *World's Fair* 1985; *Billy Bathgate* 1989; *The Waterworks* 1994

Plays

Drinks Before Dinner 1978 [pub. 1979]

Non-Fiction

Poets and Presidents 1993

J(ames) P(atrick) DONLEAVY 1926–

Donleavy was born in Brooklyn, New York, to Irish-born Catholic parents. His father was a building inspector with the New York City Fire Department. After a chequered education at various local schools, Donleavy served with the US Naval Reserve in the last year of the war, and under the GI Bill studied biology at Trinity College, Dublin, from 1946 to 1949. During this period he met his 'close enemy' **Brendan Behan** and a fellow student, Gainor Crist, who was to provide the model for the character of Sebastian Dangerfield in *The Ginger Man*. Having left Trinity, he lived with his first wife in County Wicklow and made a little money as a painter 'notorious for my nerve producing risqué female nudes'. In his autobiography, *The History of the Ginger Man*, he relates how he was spurred to begin writing seriously by the rejection of his (rather indifferent) paintings by Cork Street galleries. In 1952 he returned for a year to New York where numerous publishers rejected *The Ginger Man* because of 'the frank depiction of certain episodes and the very heavy use of four letter words, etc.' Disenchanted with America, he moved to London and published humorous sketches in the *Manchester Guardian* and *Punch*, and, at Behan's suggestion, sent his novel to Maurice Girodias's Olympia Press in Paris. His first intimation that his publisher did not regard the novel as a serious work of literature came when he was paid his advance in used fivers in a Soho basement. The novel was published in 1955 in the pornographic Traveller's Companion series, which boasted such titles as *The Whip Angels* and *Rogue Women*. Donleavy went to Paris where the unflappable Girodias told him that 'the book does a brisk trade in the Arab quarter of Jerusalem'. Twenty-odd years of litigation over rights to the novel ensued, ending only after the Olympia Press was bankrupted and Donleavy bought it at auction.

His play *Fairy Tales of New York* recounts the disastrous return of an expatriate to America and won an *Evening Standard* drama award in 1961. *The Ginger Man* was published in an unexpurgated English edition in 1963 and successfully adapted for the Broadway stage in the same year. Donleavy's subsequent novels are similarly eccentric fantasies which rarely achieve the same level of raw comic energy. His autobiography is a fine evocation of the drunken, promiscuous and pugilistic times which inspired *The Ginger Man*, and contains a wealth of anecdotes concerning Behan and Gainor Crist. Donleavy became an Irish citizen in 1967 and has four children from his two marriages.

Fiction

The Ginger Man 1955; *A Singular Man* 1963; *Meet My Maker the Mad Molecule* (s) 1964; *The Saddest Summer of Samuel S.* 1966; *The Beastly Beatitudes of Balthasar B.* 1969; *The Onion Eaters* 1971; *A Fairy Tale of New York* 1973; *The Destinies of Darcy Dancer, Gentleman* 1977; *Schultz* 1979; *Leila* 1983; *Are You Listening, Rabbi Löw* 1987; *That Darcy, That Dancer, That Gentleman* 1990

Plays

Helen (r) 1956; *The Ginger Man* [from his novel] 1959 [pub. 1961]; *Fairy Tales of New York* 1960 [pub. 1961]; *A Singular Man* [from his novel] 1964 [pub. 1965]; *J.P. Donleavy: the Plays* [pub. 1972]

Non-Fiction

The Unexpurgated Code: a Complete Manual of Survival and Manners 1975; *De Alfonce Tennis* 1984; *Donleavy's Ireland* 1986; *A Singular Country* [with P. Prendergast] 1989; *The History of the Ginger Man* (a) 1994

Ed(ward) DORN 1929–

Dorn has described himself as 'a poet of the [American] West – not by nativity but by orientation'. He was born and raised in Villa Grove, Illinois, on the banks of the Wabash river tributary, the Embarass. He never knew his father and was raised by his mother. His grandfather was a railroad worker, and Dorn attended a one-room schoolhouse before going to the University of Illinois. He did not complete a degree there, receiving his major education instead at the Black Mountain College, where he studied from 1951 to 1954. He was taught by **Charles Olson**, a central figure in the Black Mountain school of poets, who was to be a major influence. Dorn's verse bears comparison with the breath-governed free verse of Olson – 'projective verse' – and at the start of his published career he wrote *What I See in the Maximus Poems* (1960), about Olson's work. Dorn has held a series of teaching posts at universities including Idaho, Kansas, California and Essex in the UK, where he taught in the mid-1960s. He is now a professor at the University of Boulder, Colorado.

The mythical American West has been Dorn's special subject, and his particular take on it is that of a wry 1960s' radicalism. He has produced a considerable breadth of writing, but his major work is unquestionably *Gunslinger* (four volumes, collected as *Slinger*). Characters include the eponymous Gunslinger, a kind of mythical cowboy demigod; a brothel madam named Liz; Kool Everything, a hipster; the academic Dr Flamboyant; and a dope-smoking talking horse called Claude Lévi-Strauss. They collage together different idioms and jargon, and address anti-captialist, anti-industrial themes within an ambivalent nostalgia for the idea of America, and a sense of the great American myth.

Dorn has always been a strongly political poet, taking up the cause of the Native Americans and writing against privilege in all forms (in 'Oxford', a poem written in England, he sees an upper-class woman on a train and declares: 'The woman opposite / by no act other than Murder / is permitted existence'). He has had an unlikely champion in **Peter Ackroyd**, who has called Dorn 'the only plausible political poet in America'. Other critics have complained about the chopped-prose preachiness of Dorn's polemical ruminations (one felt he 'might do better to publish his fulminations against America ... as prose'), but his place in an alternative canon seems assured: **Iain Sinclair** has written that: 'Howard Hughes was finished when he became a character in Ed Dorn's *Gunslinger*.'

Poetry

Paterson Society 1960; *The Newly Fallen* 1961; *Hands Up!* 1964; *From Gloucester Out* 1964; *Idaho Out* 1965; *Geography* 1965; *The North Atlantic Turbine* 1967; *Song* 1968; *Gunslinger* 1968 *Gunslinger Book II* 1969 *The Midwest Is That Space Between the Buffalo Statler and the Lawrence Eldridge* 1969; *The Cosmology of Finding Your Spot* 1969; *Ed Dorn Sportscasts Colonialism* 1969; *Twenty-Four Love Songs* 1969; *Gunslinger I & II* 1970; *Songs* 1970; *The Cycle* 1971; *The Kultchural Exchange* 1971; *A Poem Called Alexander Hamilton* 1971; *Spectrum Breakdown* 1971; *Gunslinger, Book III: the Winterbook, Prologue to the Great Book IV Kornerstone* 1972; *The Hamadryas Baboon at the Lincoln Park Zoo* 1972; *Old New Yorkers Really Get My Head* 1972; *Recollections of Gran Apacheria* 1974; *Collected Poems: 1956–1974* 1975; *Manchester Square* [with J. Dunbar] 1975; *Slinger* 1975; *Hello, La Jolla* 1978; *Selected Poems* [ed. D. Allen] 1978; *Yellow Lola* 1981; *Captain Jack's Chaps* 1983; *Abhorrences* 1984

Non-Fiction

What I See in the Maximus Poems 1960; *Prose I* [with M. Rumaker, W. Tallman] 1964; *The Rites of Passage: a Brief History* 1965 [rev. as *By the Sound* 1971]; *The Shoshoneans: the People of the Basin Plateau* 1966; *Bean News* 1972; *The Poet, the People, the Spirit* [ed. B. Rose] 1976; *Roadtesting the Language: an Interview* [with S. Fredman] 1978; *Views, Interviews* [ed. D. Allen] 2 vols 1980

Fiction

Some Business Recently Transacted in the White World (s) 1971

Translations

With Gordon Brotherston: *Our Word: Guerrilla Poems from Latin America* 1968; José Emilio Pacheco *Tree Between Two Walls* 1969; *Selected Poems of César Vallejo* 1976; *Image of the New World: the American Continent Portrayed in Native Texts* 1979

Collections

Way West: Stories, Essays and Verse Accounts 1963–1993 1993

John (Roderigo) DOS PASSOS 1896–1970

The son of a lawyer, Dos Passos was born in Chicago and spent his early life travelling with his family in Mexico and Europe. He was educated on the east coast of America and in England, then attended Harvard from the age of fifteen. It was here that he came into contact with a literary milieu that included **e e cummings**. He subsequently went to Spain with the intention of studying architecture, but when the First World War broke out, he served with the French as an ambulance driver, seeing action at Verdun. He later joined the US Medical Corps, in which he served as a private. The war informed both his early literary work and his nascent political consciousness. His first novel, *One Man's Initiation – 1917* (1920), was based upon his experiences as an ambulance driver, and he also treated the war in *Three Soldiers*, which follows the lives of three men who, taken together, represent a cross-section of American society, and his great first trilogy consisting of *The 42nd Parallel, 1919* and *Big Money*. To Dos Passos, the war was a betrayal of American men and women by self-serving capitalist power-mongers, and his rage found forceful expression in his greatest and most innovative works: *Manhattan Transfer*, an impressionistic novel about 1920s' New York, which made use of techniques borrowed from the cinema, and the *USA* trilogy.

Despite scattered protestations of liberalism and individualism, Dos Passos was widely seen as a champion of the left in the 1920s and 1930s, when he made his living as a journalist. His social realist novels made him something of a hero in post-war, radical Greenwich Village, and **Edmund Wilson** provides a portrait of him at this period as Hugo Bamman in his novel *I Thought of Daisy* (1929). He also wrote plays during this period and published several volumes of travel writing, reportage and essays on political and social themes. His unease (apparent in the novels) concerning the totalitarian aspects of socialism surfaced dramatically in his work of the 1940s and 1950s, which anatomised the failure of collective government and its subjugation of the individual. His work took a more personal turn, becoming self-analytical and introspective, after the death of his wife in a car accident in 1947.

Dos Passos's last books turned back towards the collage-like interweaving of idioms and narratives of the best work of the 1930s. In his final years he began to give interviews and readings, many of which threw light on his complex career and clarified work which had been read more for its political stance than its considerable artistic merits.

Fiction

One Man's Initiation – 1917 1920; *Three Soldiers* 1921; *Streets of Night* 1923; *Manhattan Transfer* 1925; *USA* 1938: *The 42nd Parallel* 1930, *1919* 1932, *Big Money* 1936; *District of Columbia* 1952: *Adventures of a Young Man* 1939, *Number One* 1943, *The Grand Design* 1949; *Chosen Country* 1951; *Most Likely to Succeed* 1954; *The Great Days* 1958; *Midcentury* 1961; *Century's Ebb* 1975

Non-Fiction

Rosinante to the Road Again 1922; *Facing the Chair: Story of the Americanization of Two Foreign-Born Workmen* 1927; *Orient Express* 1927; *In All Countries* 1934; *Journeys Between Wars* 1938; *The Villages Are the Heart of Spain* 1938; *The Ground We Stand On: Some Examples from the History of a Political Creed* 1941; *State of the Nation* 1944; *Tour of Duty* 1946; *The Prospect Before Us* 1950; *The Head and Heart of Thomas Jefferson* 1954; *The Theme Is Freedom* 1956; *The Men Who Made the Nation* 1957; *Prospects of a Golden Age* 1959; *Brazil on the Move* 1963; *A Dramatic Review* 1963; *Occasions and Protests* 1964; *The Portugal Story: Three Centuries of Exploration and Discovery* 1966; *The Shackles of Power: Three Jefferson Decades* 1966; *World in a Glass: a View of Our Century* 1966; *Easter Island: Island of Enigmas* 1971; *The Fourteenth Chronicle: Letters and Diaries of John Dos Passos* 1975

Plays

The Garbage Man 1926; *Airways Inc.* 1928; *Fortune Heights* 1934

Poetry

A Pushcart at the Curb 1927

For Children

Thomas Jefferson: the Making of a President 1964

Biography

John Dos Passos by Townsend Ludington 1980

Keith (Castellain) DOUGLAS 1920–1944

The son of a First World War veteran intermittently in employment and an artistic mother of uncertain health who did freelance work for publishers, Douglas had a financially precarious childhood. He was born in Tunbridge Wells and spent his childhood in Surrey. In 1928, his father left home; Douglas never saw him again. He won a scholarship to Christ's Hospital and achieved distinction there in art and literature and upon the games-field. At the age of sixteen he put together his first collection of poetry. In 1938 he won an Open Exhibition in history to Merton College, Oxford, with an option to study English literature, which he took up on arrival, becoming a tutee of **Edmund Blunden**. He continued to play rugger and write poetry, publishing verse in *The Cherwell*, which he edited, and other university periodicals. He assembled a second collection of his work, was co-editor of *Augury*, a literary miscellany, and was represented in Michael Meyer and **Sidney Keyes**'s *Eight Oxford Poets*.

A keen horseman, he joined the mounted section of the Officers' Training Corps, and in

July 1940 he enlisted as a cavalry trooper; entrusting his poems to Blunden, he began training at the Army Equitation School. In November of that year he went to Sandhurst, which he left as a second lieutenant in 1941 to train with the 2nd Derbyshire Yeomanry. His ironic-romantic notions of chivalry and horsemanship found little scope in the war, since cavalry regiments such as his own now trained in tanks, a circumstance which was to colour his best poems. Posted to Egypt, he was transferred to the Nottinghamshire (Sherwood Rangers) Yeomanry. In 1942 his poetry began to be published in *Poetry (London)* and in the Anglo-Egyptian magazine *Citadel*. He took part in the battle of El Alamein and was wounded by a mine at Wadi Zem Zem, hence the title of his vivid prose account of desert welfare, *Alamein to Zem Zem* which he began while in hospital at El Ballah, Palestine, where he also wrote many of his war poems.

Highly conscious of the legacy of the First World War poets ('Rosenberg I only repeat what you were saying'), Douglas wrote some of the finest poetry of his own war, indeed of any war, poetry which reflects his anti-authoritarian and romantic character: 'Aristocrats', 'Cairo Jag', 'How to Kill', 'Gallantry', 'Vergissmeinnicht', 'Behaviour of Fish in an Egyptian Tea Garden'. In February 1943 his poems appeared alongside those of his friend J.C. Hall and **Norman Nicholson** in a volume of *Selected Poems*. He was promoted to captain that summer and in November returned to England. He was sure that he would not survive the war and unsentimentally made arrangements for the posthumous publication of his work, giving the projected volume the title 'Bête Noire'. He was killed in June 1944 during the Normandy invasion.

Poetry

Eight Oxford Poets [with others; ed. S. Keyes, M. Meyer] 1941; *Selected Poems* [with J.C. Hall, N. Nicholson] 1943; *The Collected Poems of Keith Douglas* [ed. G.S. Fraser, J. Waller] 1951; *Selected Poems* [ed. T. Hughes] 1964; *Collected Poems* [ed. G.S. Fraser, J.C. Hall, J. Waller] 1966; *The Complete Poems of Keith Douglas* [ed. D. Graham] 1978 [rev. 1987]

Non-Fiction

Alamein to Zem Zem 1946 [rev. 1969]; *Keith Douglas: a Prose Miscellany* [ed. D. Graham] 1985

Edited

Augury: an Oxford Miscellany of Verse and Prose (c) [with A.M. Hardie] 1940

Biography

Keith Douglas 1920–1944 by Desmond Graham 1974; *Keith Douglas, a Study* by William Scammell 1988

(George) Norman DOUGLAS 1868–1952

Of mixed Danish and titled Scottish parentage, Douglas was born at Thüringen, near the Swiss border in Austria, where his father (who spelled his surname 'Douglass') was running a family cotton mill. His father died in 1874, and when his mother remarried, Douglas was sent to relatives in Britain. He was educated at Uppingham and at Karlsruhe Gymnasium in Germany. Although he spoke fluent English, German had been his first language, and he soon mastered Italian and Russian. This linguistic proficiency led to his entering the Foreign Office in 1893, and he served in the embassy at St Petersburg, but resigned after three years in order to devote himself to writing.

His principal interest was zoology, and he published several articles on the subject while living at Posilipo, Naples. He married a cousin, Elsa Fitzgibbon, in 1896 and collaborated with her, under the joint pseudonym of 'Normyx', on a hugely unsuccessful volume of *Unprofessional Tales*. (He later claimed that the book sold only eight copies.) Douglas was voraciously bisexual, with a particular taste for the under-aged, and his marriage ended in divorce in 1904, after which he moved to Capri and dropped the final 's' from his surname. Short of funds, he sold his house in Capri and returned to London in 1910, becoming between 1912 and 1915 assistant editor of the *English Review*, which had been founded under the editorship of **Ford Madox Ford** in 1908, but was now edited by Austin Harris. In 1911, he published *Siren Land*, the first of his many travel books, most of which were about Italy.

His interest in children found expression in an exhaustive account of *London Street Games*, published in 1916, the same year that an encounter at the Natural History Museum in South Kensington led to his being charged with 'committing an act of gross indecency' with a sixteen-year-old youth. Before the case came to trial, Douglas slipped bail and went into permanent exile in a more tolerant Europe. He later quipped (somewhat inaccurately) that he had left England under a cloud 'no bigger than a small boy's hand'. He endured a year of poverty in France, but in 1917 published his first and best-known novel, *South Wind*. Set on the island of Nepenthe (clearly drawn from Capri), it is a hedonistic novel of character and atmosphere rather than of plot, for Douglas was not by temperament a fiction writer. 'Superbly aloof from the catastrophes of the time,' as **Graham Greene** put it, the book became something of a cult amongst the younger generation and was cited (with three other Douglas titles) in **Cyril Connolly**'s list of 'Books in the Modern Movement' in *Enemies of Promise* (1938).

In the 1920s Douglas settled in Florence and entered a long association with the publisher and writer Guiseppe 'Pino' Orioli. He continued to publish travel books, fiction, works of scholarship such as his *Birds and Beasts of the Greek Anthology*, a notoriously gamey anthology of limericks, and an autobiography, *Looking Back*. In 1937 he went to live in France, then moved to Portugal and England. After the Second World War, he returned to Capri, where he spent the remaining six years of his life. A notable cook, he was befriended by Elizabeth David, who often cites him in her *Italian Food* (1954), and was himself the 'editor' (i.e. author) of the splendid *Venus in the Kitchen*, a book of extravagant and purportedly aphrodisiac recipes.

Fiction

Unprofessional Tales (s) [as 'Normyx'; with E. Fitzgibbon] 1901; *South Wind* 1917; *They Went* 1920; *In the Beginning* 1927

Non-Fiction

Siren Land 1911; *Fountains in the Sand: Rambles among the Oases of Tunisia* 1912; *Old Calabria* 1915; *London Street Games* 1916 [rev. 1931]; *Alone* 1921; *Together* 1923; *Experiments* 1925; *Birds and Beasts of the Greek Anthology* 1927; *How about Europe? Some Footnotes on East and West* 1929; *Capri: Materials for a Description of the Island* 1930; *Paneros: Some Words on Aphrodisiacs and the Like* 1930; *Summer Islands: Ischia and Ponza* 1931; *Looking Back* (a) 1933; *Late Harvest* 1946; *Footnote on Capri* 1952; *Venus in the Kitchen* [as 'Pilaff Bey'] 1952

Poetry

Some Limericks 1928

Collections

Three of Them 1930; *An Almanac* 1941

Biography

Norman Douglas by Mark Holloway

A(rthur) Conan DOYLE 1859–1930

Doyle – keen sportsman, failed politician and doctor who came to believe in fairies – is chiefly famous for creating Sherlock Holmes, the hero of stories which make up only a small percentage of his prodigious literary output, and a character he came to hate. He was born in Edinburgh to a mother he adored and a father who had squandered artistic abilities to become a civil servant. They were devout Roman Catholics, so Doyle was sent to Jesuit institutions: Hodder, Stonyhurst and Feldkirch in Austria for a final year, after which he entered the University of Edinburgh to study medicine. One of his professors there, Dr Joseph Bell, was the model for Sherlock Holmes.

In 1882, after a brief and acrimonious partnership with an old friend, Doyle opened his own practice in Southsea, a suburb of Portsmouth. It was here that his writing really took off, because business was so bad: he had few patients to see, lots of time on his hands, and he needed to supplement his income. His first book featuring Sherlock Holmes, *A Study in Scarlet*, was published in 1888, and the character achieved great popularity when stories about him started appearing in the newly founded *Strand* magazine in 1891. Further commissions followed, and although Doyle intended to write six at most (he considered himself a serious writer and said that Holmes 'takes my mind from better things'), demand forced his hand.

After writing two dozen stories, Doyle killed off Holmes by sending him over the Reichenbach Falls with Dr Moriarty, his most formidable adversary. He wrote to a friend of the relief he felt: 'I have had such an overdose of him that I feel towards him as I do towards *pâté de foie gras*, of which I once ate too much, so that the name of it gives me a sickly feeling to this day.' Doyle created another hero, the Brigadier Gerard, in an attempt to displace his violin-playing detective, but eight years later cries of public protest persuaded Doyle to revive Holmes, which he did with *The Hound of the Baskervilles*.

In 1893, his wife became seriously ill with consumption and for her health they moved with their two children to Switzerland. She died in 1900. In the same year, Doyle sailed for South Africa as both a doctor and an unofficial attaché, eager to witness the Boer War. Two publications came from this experience: *The Great Boer War* and a small pamphlet, *The War in South Africa: Its Cause and Conduct*. In 1900 and 1906 Doyle unsuccessfully ran for Parliament, as a Liberal Unionist and tariff reformer respectively. Always a campaigner for social justice, he then championed the cause of a man called Oscar Slater, who was sentenced to death in 1909 for murder and robbery. Doyle fought hard, believing it was a cause of mistaken identity, and in 1929 the sentence was quashed. Even though Slater received £6,000 in compensation, he refused to pay Doyle's legal fees which amounted to several hundred pounds.

Doyle had remarried in 1907, and he had three children with his second wife. When his first son, Kingsley, was badly wounded on the Somme and later died in London of pneumonia, the loss completed Doyle's conversion to spiritualism, to which he devoted the remainder of his life.

Fiction

A Study in Scarlet 1888; *Micah Clarke* 1889; *Mysteries and Adventures* (s) 1889; *The Mystery of Cloomber* 1889; *The Captain of the Pole-Star* (s) 1890; *The Firm of Girdlestone* 1890; *The Sign of Four* 1890; *The White Company* 1891; *The Adventures of Sherlock Holmes* (s) 1892; *The Doings of Raffles Haw* 1892; *The Refugees* 1893; *The Memoirs of Sherlock Holmes* (s) 1894; *The Parasite* 1894; *Round the Red Lamp* (s) 1894; *The Stark Munro Letters* 1895; *The Exploits of Brigadier Gerard* (s) 1896; *Rodney Stone* 1896; *Uncle Bernac* 1897; *The Tragedy of the Korosko* 1898; *A Duet with an Occasional Chorus* 1899; *The Green Flag* (s) 1900; *The Hound of the Baskervilles* 1902; *The Adventures of Gerard* (s) 1903; *The Return of Sherlock Holmes* (s) 1905; *Sir Nigel* 1906; *Round the Fire Stories* (s) 1908; *The Last Galley* (s) 1911; *The Lost World* 1912; *The Poison Belt* 1913; *The Valley of Fear* 1915; *His Last Bow* (s) 1917; *Danger* (s) 1918; *Tales and Adventures of Medical Life* (s) 1922; *The Land of Mist* 1926; *The Case-book of Sherlock Holmes* (s) 1927; *The Maracot Deep* 1929

Non-Fiction

The Great Boer War 1900; *The War in South Africa* 1902; *Through the Magic Door* 1907; *The British Campaign in France and Flanders* 1916–19; *The New Revelation* 1918; *Vital Message* 1919; *The Wanderings of a Spiritualist* 1921; *The Coming of the Fairies* 1922; *Our American Adventure* 1923; *Three of Them* (a) 1923; *Memories and Adventures* (a) 1924; *Our Second American Adventure* 1923; *The History of Spiritualism* 2 vols 1926; *Pheneas Speaks* 1927; *Our African Writer* 1929; *The Edge of the Unknown* 1930

Poetry

Songs of Action 1898; *Songs of the Road* 1911; *The Guards Came Through* 1919

Biography

The Quest for Sherlock Holmes by Owen Dudley Edwards 1983

Roddy DOYLE 1958–

Doyle was born in Kilbarrack, Dublin, and his childhood shares most of its incidental details with the world of his novels: urban, working-class, estate-bound (he still lives there). Only the adult career that he has built from it seems so remarkable. Having graduated from University College, Dublin, with a degree in English and geography, he drifted into teaching in 1980 at the Greendale Community School, Kilbarrack, a post from which he retired thirteen years later as a highly successful author.

A £5,000 loan from the bank enabled him to publish his first book, *The Commitments* (1987), under the imprint King Farouk (Dublin rhyming slang for 'book') when an earlier effort had been rejected by all the London publishers. The book attracted favourable reviews but little in the way of sales before being taken up by Heinemann in 1988, by which time Doyle had completed a play, *Brownbread*, to help pay off the bank loan. Set in the fictional Dublin suburb of Barrytown, the novel is a comic and dialogue-rich account of local boy Jimmy Rabbite's efforts to become a soul singer after the example of Otis Redding and James Brown. The book was successfully filmed by Alan Parker in 1991, by which time Doyle had netted some £70,000 on his original investment. In the following novel, *The Snapper*, later filmed by Stephen Frears, Jimmy's sister has a baby. The third in the series, *The Van*, has Jimmy's father setting himself up in the burger business: a darker book than its predecessors, it increased Doyle's standing with the critics and was nominated for the Booker Prize in 1991.

A second play, *War*, was premiered in 1989. Its author was still teaching geography and English in Kilbarrack. A father, Doyle married Belinda Moller, his PR on the King Farouk enterprise. *Paddy Clarke Ha Ha Ha*, his fourth novel and a bestselling Booker Prize winner, goes back to the 1960s to recreate the Dublin childhood of its hero. Doyle then retired from teaching to take up writing full-time. He wrote a four-part television script, *The Family*, broadcast by the BBC in 1994, and contributed a recollection of Eire's 1990 World Cup campaign – the 1988 European Championship victory over England having already been memorialised in *The Van* – to an anthology of football writing. He has two sons, and is a lifelong Chelsea fan.

Fiction

The Commitments 1987; *The Snapper* 1990; *The Van* 1991; *Paddy Clarke Ha Ha Ha* 1993

Plays

Brownbread 1987 [pub. 1992]; *War* 1989; *The Family* (tv) 1994

Margaret DRABBLE 1939–

Drabble was born in Sheffield, into a talented family of Cambridge-educated lawyers, writers and teachers, but the grand-daughter of working-class Methodists (she has said that in her childhood she always regarded their small cottage as her home). Like her sister, the novelist

A.S. Byatt, she was educated at the Mount School (a Quaker boarding establishment) and Newnham College, Cambridge, where, like Byatt, she took a first in English. She married the actor Clive Swift the week after she left Cambridge in 1960, and initially worked with him, as an actress, at the Royal Shakespeare Company.

Her first three novels coincided with the three pregnancies of her marriage, and the first two were written while she was still an actress at Stratford-upon-Avon. From the time the first, *A Summer Bird-Cage*, was published in 1963, she attracted consistent critical and popular attention, and her themes – the struggles of educated, independent women; the conflicts between motherhood and such women's lives – appealed strongly to the tastes of the 1960s. She was soon fully established as a writer and media personality in Hampstead, north London, while her husband continued to tour; they were divorced in 1975.

Drabble has continued to be a popular and prolific novelist, and from the time of her first critical book, on Wordsworth, in 1966, she has also written a considerable body of non-fiction (her biographies of **Arnold Bennett** and **Angus Wilson** are her major works in this field). As a novelist, she has moved from the feminine themes of her earlier books to an attempt to encompass a large, omniscient vision of society; *The Ice Age* in 1977 was the first of her novels to feature a male protagonist, and in the later 1980s she embarked on a trilogy intended to give a panoramic and critical vision of Margaret Thatcher's Britain. But the three novels – *The Radiant Way*, *A Natural Curiosity* and *The Gates of Ivory* – were disappointingly received, and Drabble was caught in a strange reversal of fortunes whereby her sister, A.S. Byatt, whose career as a novelist had taken off much more slowly in the 1960s, had by the early 1990s decisively overtaken her in both critical esteem and popular audience. The critic Martin Seymour-Smith has unkindly described Drabble as 'pure middlebrow', and what is certainly true is that her fiction, which she often derives from such sources as newspaper articles and interviews, has the character of illustrating experience rather than emerging from it. Her work, however, is so varied – encompassing such fields as journalism, broadcasting, lecturing, film and television scripts, a book for children, a play, and her editorship of the revised *Oxford Companion to English Literature* – that it would be difficult not to find something of interest within it. She married the biographer Michael Holroyd in 1982, and the couple became famous for living in separate houses and consulting once a month about when they could fit each other into their schedules. Drabble was appointed CBE in 1980.

Fiction

A Summer Bird-Cage 1963; *The Garrick Year* 1965; *The Millstone* 1966; *Jerusalem the Golden* 1967; *The Waterfall* 1969; *The Needle's Eye* 1972; *The Realms of Gold* 1975; *The Ice Age* 1977; *The Middle Ground* 1980; *The Radiant Way* 1987; *A Natural Curiosity* 1989; *The Gates of Ivory* 1991

Non-Fiction

Wordsworth 1966; *Arnold Bennett* 1974; *A Writer's Britain: Landscape in Literature* 1979; *Angus Wilson* 1995

Edited

London Consequences: a Novel [with B.S. Johnson] 1972; Jane Austen: *Lady Susan / The Watsons / Sanditon* 1974; *The Genius of Thomas Hardy* [with C. Osborne] 1976; *New Stories I* 1976; *The Oxford Companion to English Literature* 1985

For Children

For Queen and Country: Britain in the Victorian Age 1978

Theodore (Herman Albert) DREISER 1871–1945

Dreiser was born into a large and chaotic immigrant family in Terre Haute, Indiana, and his early life was characterised by frequent upheavals, removals and struggles against poverty. His education was necessarily fragmentary – he spent only one year at Indiana University – but he later redressed this with a rigorous programme of self-education. In his teens, alone in Chicago, he worked at various menial jobs. supporting a hand-to-mouth existence. He returned to the city after university and embarked upon a career as a journalist there and in St Louis. His view of America as an arena in which only the fittest survive was arrived at through his reports on the changing face of the industrialised USA.

In 1894, he went to New York as editor of *Every Week*, a music journal. (His brother was a well-known writer of popular songs under the Anglicised name of Paul Dresser, and Dreiser himself provided lyrics for the barber-shop song 'On the Banks of the Wabash'.) Four years later he embarked upon a career as a freelance journalist, and it was during this period that he started writing novels, publishing his first, *Sister Carrie*, in 1900. An unflattering portrait of the late nineteenth-century era of expansionism and *laissez-faire* economics, the book proved controversial. Its eponymous heroine takes a wealthy lover rather than do demeaning work, and rises from poverty to fame on the New York stage.

Her rise is both ruthless and spectacular, but leaves her dissatisfied, and Dreiser offers no moral judgment, content merely to depict her life naturalistically. The book was widely condemned for its amorality, but opened the way for a new school of American naturalism, the most notable practitioner of which was **Sinclair Lewis**.

Later novels – such as *An American Tragedy*, the story of a man who murders his pregnant girlfriend – share a similar moral landscape. Dreiser typically pits an individual against the hostile environment of urban America in a struggle with heroic, even tragic, overtones. That he often chooses weak, ineffectual or disadvantaged central characters naturally gives his work a pessimistic tone, and there is a certain clumsiness about much of his prose; but his importance in the development of the twentieth-century American novel can hardly be underestimated.

Dreiser also wrote plays, poetry, essays and short stories, and is the author of such non-fiction, socialist works as *Tragic America* and *Dreiser Looks at Russia*. Latterly, his reputation consolidated, he has been seen as the documenter and anatomist of a hitherto invisible side of American life. Outward changes in American society notwithstanding, his work remains as disturbingly relevant today as it was when first published.

Fiction

Sister Carrie 1900; *Jennie Gerhardt* 1911; *The Financier* 1912 [rev. 1927]; *The Titan* 1914; *The 'Genius'* 1915; *Free and Other Stories* (s) 1918; *Twelve Men* (s) 1919; *An American Tragedy* 1925; *Chains* (s) 1927; *The Bulwark* 1946; *The Stoic* 1947

Poetry

Moods Cadenced and Declaimed 1926; *The Aspirant* 1929; *Epitaph* 1930; *Moods Philosophic and Emotional* 1935

Plays

Plays of the Natural and the Supernatural [pub. 1916]; *The Hand of the Potter* 1921 [pub. 1919]

Non-Fiction

A Traveller at Forty (a) 1913; *A Hoosier Holiday* 1916; *Life, Art and America* 1917; *Hey Rub-A-Dub-Dub* 1920; *A Book about Myself* (a) 1922; *The Color of a Great City* 1923; *Dreiser Looks at Russia* 1928; *Tragic America* 1931; *America Is Worth Saving* 1941; *Letters of Theodore Dreiser* [ed. R.H. Elias] 1959; *Letters to Louise Campbell from Theodore Dreiser* [ed. L. Campbell] 1959; *Notes on Life* [ed. J.A. MacAleer, M. Tjader] 1974; *An Amateur Labourer* [ed. R.W. Dowell] 1983; *American Diaries, 1902–1926* (a) [ed. P. Riggio] 1983

Biography

Theodore Dreiser by Ellen Moers 1989

John DRINKWATER 1882–1937

The son of a schoolmaster turned actor who was determined that his son should not follow him on to the stage, Drinkwater left Oxford High School in 1897, and was put to work as an office boy in insurance. He pursued this career, rising through various grades and in various provincial cities, until 1909. His interest in literature and the theatre developed meanwhile: in 1903 he paid a bookseller to print his first volume of poems, and, while in Birmingham, he met Sir Barry Jackson, with whom he founded the Pilgrim Players in 1907, a company that was to develop, from 1909 onwards, into the successful Birmingham Repertory Theatre. After giving up insurance, and having worked briefly in the advertising department of a trade paper, Drinkwater became general manager of the theatre, for which he produced, acted and wrote: Of many historical plays, *Abraham Lincoln* (1918) enjoyed a notable popular success.

He developed a parallel career as a busy man-of-letters, doing editorial work for Routledge's Muses' Library, reviewing widely, and appearing in all five volumes of Edward Marsh's *Georgian Poetry*. He also published many volumes of verse with such titles as *Cromwell and Other Poems, Swords and Ploughshares, Cotswold Characters, All about Me* and *More about Me*, as well as many critical and general books. His first marriage ended in divorce; his second, to the violinist Daisy Kennedy, the former wife of Russian pianist Benno Moiseiwitsch, produced a daughter. Although always taken with limited seriousness by intellectuals and now largely a name, Drinkwater enjoyed considerable public success in his day. He is buried in the churchyard at Piddington, the Oxfordshire village where he spent his holidays as a boy.

Poetry

Poems 1903; *The Death of Leander and Other Poems* 1906; *Lyrical and Other Poems* 1908; *Poems of Men and Hours* 1911; *Poems of Love and Earth* 1912; *Cromwell and Other Poems* 1913; *Lines for the Opening of the Birmingham Repertory Theatre* 1913; *Swords and Ploughshares* 1915; *Olton Pools* 1916; *Poems 1908–1914* 1917; *Tides* 1917; *Loyalties* 1918; *Cotswold Characters* 1921; Persuasion 1921; *Seeds of Time* 1921; *Preludes 1921–1922* 1922; *From an Unknown Isle* 1924; *From the German* 1924; *At Pisa* 1925; *New Poems* 1925; *A Graduation Song for the University of London* 1926; *Persephone* 1926; *All about Me* 1928; *Thomas Hardy, June 2nd 1925: his 85th Birthday* 1928; *More about Me* 1930; *Christmas Poems* 1931; *American Vignettes, 1860–1865* 1933; *Summer Harvest: Poems 1924–1933* 1933

Plays

Cophetua 1911; *An English Medley* 1911; *The Only Legend* 1913; *The Pied Piper* 1913; *Puss in Boots* 1913; *Rebellion* 1914; *Robin Hood and the Pedlar* 1914; *The Storm* 1915; *The God of Quiet* 1916; *Pawns* [pub. 1917]; *Abraham Lincoln* 1918; *Mary Stuart* 1921; *Oliver Cromwell* 1921; *Robert E. Lee* 1923; *Robert Burns* 1925; *Bird in Hand* 1927; *John Bull Calling* 1928; *Midsummer Eve* **(r)** 1932; *Napoleon: The Hundred Days* [from G. Forzano and B. Mussolini] (An Adaptation) 1932; *Laying the Devil* 1933; *A Man's House* 1934; *Garibaldi* 1936

Non-Fiction

William Morris: a Critical Study 1912; *Swinburne: an Estimate* 1913; *The Lyric* 1915; *Rupert Brooke* 1916; *Politics and Life* 1917; *Prose Papers* 1917; *Lincoln: the World Emancipator* 1920; *Claud Lovat Fraser: a Memoir* [with A. Rutherston] 1923; *The Poet and Communication* 1923; *Victorian Poetry* 1923; *Patriotism in Literature* 1924; *The Muse in Council* 1925; *The Pilgrim of Eternity: Byron – a Conflict* 1925; *A Book for Bookmen* 1926; *Mr Charles, King of England* 1926; *Cromwell: a Character Study* 1927; *The Gentle Art of Theatre Going* 1927; *Charles James Fox* 1928; *'The Other Point of View ...' 1928; The World's Lincoln* 1928; *Art and the State* 1930; *Pepys: His Life and Character* 1930; *Poetry and Dogma* 1931; *Inheritance* **(a)** 1931; *Discovery* **(a)** 1932; *John Hampden's England* 1933; *Shakespeare* 1933; *This Troubled World: Four Essays* 1933; *Speeches in Commemoration of William Morris* [with H. Jackson, H. Laski] 1934; *The King's Reign: On the Reign of George V* 1935; *English Poetry: an Unfinished History* 1938

Fiction

Robinson of England 1937

Edited

The Poems of Sir Philip Sidney 1910; *The Way of Poetry* 1919; *The Outline of Literature* 1923; *An Anthology of English Verse* 1924; *The John Drinkwater Series for Schools* 1924; *Selected Poems of Lord de Tabley* 1924; *The Way of Prose* 1924; *Royal Society of Literature Transactions* 1925; *Interlude* 1927; *Collins Modern School Shakespeare* 1927–31; *Collins New Stage Shakespeare* 1927; *The 1860's: Essays by Fellows of the RSL* 1932; *A Pageant of England's Life* 1934; Henry Kirke White *Poems, Letters and Prose Fragments* [nd]

Carol Ann DUFFY 1955–

Duffy was born in Glasgow, the daughter of an engineer, but the family moved to Stafford when she was seven. She was educated at St Joseph's Convent School, Stafford (run by Irish nuns), and Stafford Girls' High School. She took a degree in philosophy at the University of Liverpool in 1977, but had already published with a small press her first poetry collection, *Fleshweathercock* (1973), while still at school.

She has not followed any profession other than writer, and made her reputation first in Liverpool in the early 1980s as a playwright, with two successful plays – *Take My Husband* and *Cavern of Dreams* – produced at the Liverpool Playhouse. The influence of her work as a dramatist can be seen in her first mainstream poetry collection, *Standing Female Nude*, which of all her books makes the greatest speciality of the dramatic monologue – an English tradition going back to Robert Browning but to which she brings a modernist twist: portraits of 1980s' outsiders such as sexually ambiguous vicars, murderous daughters, rapists and psychopaths.

Her reputation has grown rapidly through three further collections, with optimism breaking in from time to time, some urgent lesbian love poetry, and a colloquial, satirical style which suits the taste of the age. Her 1993 collection, *Mean Time*, won the Whitbread Poetry Award, and a *Selected Poems* appeared the following year. Duffy has held many of the residencies, fellowships and workshops which are the lifeblood of the modern poet, and has lived for some years in London.

Poetry

Fleshweathercock 1973; *Beauty and the Beast* [with A. Henri] 1977; *Fifth Last Song* 1982; *Standing Female Nude* 1985; *Thrown Voices* 1986; *Selling Manhattan* 1987; *The Other Country* 1990; *William and the Prime Minister* 1990; *Mean Time* 1993; *Selected Poems* 1994

Plays

Take My Husband 1982 [np]; *Cavern of Dreams* 1985 [np]; *Loss* **(r)** 1986; *Little Women, Big Boy Boys* 1987

Edited

Home and Away: New Poetry from the South 1989; *I Wouldn't Thank You for a Valentine: Anthology of Women's Poetry* 1992

Maureen (Patricia) DUFFY 1933–

Duffy was born illegitimate in Worthing, Sussex, into a working-class background, and her early life became the subject of her evocative first novel, appropriately titled *That's How It Was* (1962). She never met her father, who was Irish, a sometime builder and a member of the IRA. Her mother suffered from tuberculosis, and this led to long separations and spells with foster parents or in homes; but she was also very well-read and encouraged her daughter, who won a scholarship to Trowbridge High School in Wiltshire, where she sang tenor in the school choir and developed a passion for music. She moved to London, and read English at King's College, where she also started writing poetry and drama. After graduating, she taught in various state schools, and subsequently became a teacher in adult education, both in Italy and in evening classes in London.

Highly prolific, she has written novels, poetry, biography, plays and works of non-fiction, and has served on numerous committees, most of which are dedicated to improving the lot of her fellow writers. She was co-founder with **Brigid Brophy** of the Writers' Action Group, which successfully fought for the introduction of Public Lending Right, was both Joint-Chair and President of the Writers' Guild of Great Britain, and has been a long-serving member of the British Copyright Council, frequently travelling abroad to conferences and lectures. She is also a well-known reader of her own work and has appeared on many discussion panels.

Duffy's fiction is both bold in its experimentation and wide ranging in style and subject-matter. She has written in, and subverted, many genres, including those of science fiction (*Gor Saga*, rendered unrecognisable in *First Born*, a 1988 television adaptation) and crime (*Scarborough Fear*, written under the pseudonym D.M. Cayer). One of her best-known novels is *The Microcosm* (1966), a pioneering and stylistically inventive novel set in London's then largely hidden lesbian community. Amongst her other novels are her London trilogy, consisting of *Wounds*, *Capital* and *Londoners; Change*, which also takes London as its subject; *Love Child*, an astonishing technical *tour de force* in which the gender of two characters (the child narrator and his/her mother's lover) is unrevealed; and *Illuminations*, a novel on the theme of Europe, which counterpoints the stories of an eighth-century nun and a present-day academic.

A passionate advocate of animal-rights – a theme that has surfaced in her poetry and such novels as *Gor Saga* and *I Want to Go to Moscow* – Duffy is the author of *Men and Beasts*, a book of essays and resources she describes as 'a blueprint for the way forward and a programme for action'. Other non-fiction books include *The Erotic World of Faery*, a popularising study of the fey in art, literature and folklore; *Inherit the Earth*, a social history of an ordinary Essex family (that of her great-uncle); and biographies of Aphra Behn and Henry Purcell.

Her first volume of poetry, *Lyrics for the Dog Hour*, appeared in 1968; subsequent volumes include *The Venus Touch*, a collection of erotic verse with illustrations by Gustav Klimt and Egon Schiele, and her *Collected Poems 1949–1984*. She won the City of London Festival Playwright's Prize for her first stage play, *The Lay-Off*, in 1962, and was awarded an Arts Council Bursary the following year in order to write a play about the balladeer François Villon. Amongst her other plays are *Rites*, which was performed at the National Theatre in 1969, and is one of several based on Greek mythology but with contemporary settings, and *A Nightingale in Bloomsbury Square* about **Virgina Woolf**.

Fiction
That's How It Was 1962; *The Single Eye* 1964; *The Microcosm* 1966; *The Paradox Players* 1967; *Wounds* 1969; *Love Child* 1971; *I Want to Go to Moscow* 1973; *Capital* 1975; *Housespy* 1978; *Gor Saga* 1981; *Scarborough Fear* [as 'D.M. Cayer'] 1982; *Londoners* 1983; *Change* 1987; *Illuminations* 1991; *Occam's Razor* 1993

Poetry
Lyrics for the Dog Hour 1968; *The Venus Touch* 1971; *Evesong* 1975; *Memorials of the Quick and the Dead* 1979; *Collected Poems* 1949–1984

Plays
Josie (tv) 1961; *The Lay-Off* 1962 [np]; *The Silk Room* 1966 [np]; *Rites* [pub. in *New Short Plays 2*] 1969; *Old Tyme* 1970 [np]; *Solo* 1970 [np]; *A Nightingale in Bloomsbury Square* 1973 [pub. in *Factions* ed. G. Gordon, A. Hamilton 1974]; *Only Goodnight* (r) 1981

Non-Fiction
The Erotic World of Faery 1972; *The Passionate Shepherdess: Aphra Behn, 1640–1689* 1977; *Inherit the Earth* 1980; *Men and Beasts* 1984; *A Thousand Capricious Chances* 1989; *Henry Purcell* 1994

Edited
New Poetry 3 1977; *Oroonoko and Other Stories by Aphra Behn* [with A. Brownjohn] 1986

Translations
Domenico Rea *A Blush of Shame* 1968

Alfred (Leo) DUGGAN 1903–1964

Duggan was born in Buenos Aires, Argentina; his father was a rich Argentine settler of largely Irish descent, while his mother was the daughter of an American diplomat who had become Consul-General in Rio de Janeiro. He was taken to England when he was two. In 1917 his mother (who had been widowed two years earlier) married Lord Curzon, one of Britain's premier statesmen; Duggan was thus brought up within the English aristocracy and went from Eton to Balliol College, Oxford, where he is described by **Evelyn Waugh** as having been 'a full-blooded rake of the Restoration'. He was 'almost always' drunk, kept a string of hunters, and travelled up to the '43' Club in London in the evening, 'having a woman' there before returning to climb into his Oxford rooms.

After Oxford, he travelled under sail to the Galapagos Islands to collect specimens on behalf of the British Museum, but was to spend much of the next fifteen years becoming a spectacular alcoholic, dissipating his large fortune, and also travelling widely, particularly among the Crusader castles of the Levant, laying the foundations of the eclectic knowledge of ancient and medieval history that he was to turn to advantage in his novels. He enlisted as a private soldier in 1939 and was posted to Norway, where he endured the bloody campaign of 1940 before being invalided out of the army; he spent the rest of the war working at the bench of an aeroplane factory.

After the war, his alcoholism cured, he turned to writing historical novels, of which the first, *Knight with Armour* (written in 1946, published in 1950), deals with a subject to which he often returned, the Crusades. It was followed by fourteen others, which appeared annually, the last posthumously. He received considerable encouragement from Evelyn Waugh, who had arranged for Duggan's younger brother, Hugh, a Conservative MP to receive unction on his deathbed, a scene he subsequently recreated for Lord Marchmain in *Brideshead Revisited* (1945). Duggan's novels cover no period later than 1300 and display an extraordinary familiarity and identification with the remote past, describing its cruelties and treacheries in a deadpan yet curiously powerful style. The best of them, such as *Family Favourites*, which deals with the reign of Elagabalus, are also very funny, and even moving in an impersonal fashion. Duggan wrote three historical biographies and seven historical studies for younger readers, including the splendid *The Story of the Crusades*; he also wrote many articles on medieval history for the children's magazine *Look and Learn*. He married, in 1953, Laura, daughter of St Quintin Hill, and they had one son; they lived in Ross-on-Wye, Herefordshire. He was a devout Roman Catholic.

Fiction

Knight with Armour 1950; *Conscience of the King* 1951; *The Little Emperors* 1951; *The Lady for Ransom* 1953; *Leopards and Lilies* 1954; *God and My Right* 1955; *Winter Quarters* 1956; *Three's Company* 1958; *Founding Fathers* 1959; *The Cunning of the Dove* 1960; *Family Favourites* 1960; *The King of Athelney* 1961; *Lord Geoffrey's Fancy* 1962; *Elephants and Castles* 1963; *Count Bohemond* 1964

Non-Fiction

Thomas Becket of Canterbury 1952; *Devil's Brood: the Angevin Family* 1957; *He Died Old: Mithradates Eupator, King of Pontus* 1958

For Children

Julius Caesar 1955; *Look at Castles* 1960; *Look at Churches* 1961; *Growing Up in the Thirteenth Century* 1962; *The Story of The Crusades, 1097–1291* 1963; *Growing Up with the Norman Conquest* 1965; *The Romans* 1965

Daphne DU MAURIER 1907–1989

Du Maurier was born in London, daughter of the successful actor-manager Gerald Du Maurier, and granddaughter of the artist and novelist George Du Maurier. She was educated by governesses and at a finishing school near Paris, and was encouraged in her writing from a young age by family friends such as **J.M. Barrie**, Edgar Wallace and Sir Arthur Quiller-Couch ('Q'). She had two sisters, one of whom, Angela, also became a novelist. Her early 'dark and dismal' short stories explore problematic human relationships (the power of men and the weakness of women), a theme which recurs throughout her writing. Her first published story, 'And Now to God the Father', appeared in *The Bystander* in 1929, the year in which she had an affair with the film director Carol Reed. Her parents had acquired a house in Cornwall, and her first novel, *The Loving Spirit*, drew on stories she heard from a neighbour to describe four generations of a Cornish family. In 1932 she met and married Major Frederick Browning (usually known as 'Boy' or 'Tommy'). Browning rose to the rank of major-general, was knighted in 1946, and became treasurer to the Duke of Edinburgh. The couple had two daughters and a son, and Browning died in 1965.

Following the death of her father in 1933, Du Maurier was encouraged to write his biography

(*Gerald: A Portrait*) by her energetic new publisher, Victor Gollancz. She achieved great success with her two most famous romantic novels, *Jamaica Inn* ('a tale of adventure ... set in Cornwall, full of smugglers and steeped in atmosphere') and *Rebecca*, both of which were filmed by Alfred Hitchcock (1939 and 1940). Other films from her work include *The Birds* (Hitchcock, 1963) and *Don't Look Now* (Nicolas Roeg, 1973), both adapted from short stories. From 1943 until her death she lived mainly at Par in Cornwall (originally in Menabilly, the house which inspired *Rebecca* and her Civil War novel *The King's General*). According to her biographer, **Margaret Forster**, Du Maurier had long felt a sense of sexual ambiguity (describing herself in a letter as a 'half-breed'), and had an affair with the actress Gertrude Lawrence, who starred in the first production of her play *September Tide*. Others question that there was any physical side to the relationship. Although successful as a 'romantic' novelist, Du Maurier is a more profound writer than most practitioners of the genre, and her best novels (Forster cites *Rebecca*, *The Scapegoat* and *The House on the Strand*) are complex psychological dramas reminiscent of the Brontës. She died in 1989 after a period of depression and failing memory.

Fiction

The Loving Spirit 1931; *I'll Never Be Young Again* 1932; *The Progress of Julius* 1933; *Jamaica Inn* 1936; *Rebecca* 1938; *Come Wind, Come Weather* 1940; *Happy Christmas* (s) 1940; *Frenchman's Creek* 1941; *Consider the Lilies* (s) 1943; *Hungry Hill* 1943; *Nothing Hurts for Long* (s) 1943; *Spring Picture* (s) 1944; *Leading Lady* 1945; *London and Paris* (s) 1945; *The King's General* 1946; *The Parasites* 1949; *My Cousin Rachel* 1951; *The Apple Tree* (s) [US *Kiss Me Again Stranger*] 1953; *Early Stories* 1954; *Mary Anne* 1954; *The Scapegoat* 1957; *The Breaking Point* (s) 1959; *The Lover* (s) 1961; *The Glass-Blowers* 1963; *The Flight of the Falcon* 1965; *The House on the Strand* 1969; *Not After Midnight* [US *Don't Look Now*] 1971; *Rule Britannia* 1972; *Echoes from the Macabre* (s) 1976; *The Rendezvous* (s) 1980; *Classics from the Macabre* (s) 1987

Non-Fiction

Gerald: a Portrait 1934; *The Du Mauriers* 1937; *The Infernal World of Branwell Brontë* 1960; *Vanishing Cornwall* 1967; *Golden Lads: Sir Francis Bacon, His Rise and Fall* 1976; *Growing Pains: the Shaping of a Writer* (a) [US *Myself When Young*] 1977; *The Rebecca Notebook and Other Memories* 1981; *Letters from Menabilly: Portrait of a Friendship* [ed. O. Malet] 1993

Plays

Rebecca 1940; *September Tide* 1948

Edited

The Young George Du Maurier; a Selection of His Letters 1860–1867 1951; Phyllis Bottome *Best Stories* (s) 1963

Biography

Daphne: a Life of Daphne Du Maurier by Judith Cook 1991; *Daphne Du Maurier* by Margaret Forster 1993

Andrea DUNBAR 1961–1990

Born in Bradford, one of eight children of an unemployed father, Dunbar grew up amid the deprivation of Buttershaw, a district of Bradford containing a large council estate. Her teachers at Buttershaw Comprehensive School encouraged her to write, and her first play, *The Arbor* (Brafferton Arbor is part of Buttershaw), was written when she was fifteen. She left school aged sixteen with three CSEs and expecting a baby, but a car crash led to the child being stillborn. She later had three children, all by different fathers, as well as several miscarriages, and, as her first child was born while she was still in her teens, she did not have a job but lived on social security in her native Buttershaw, apart from one period which she spent in a battered women's home in Keighley.

She sent *The Arbor* to the 1980 Young Writers' Festival, an annual competition run by the Royal Court Theatre in London, and it was produced with great success, the theatre encouraging her to write an amplified version. The Royal Court also produced her second play, *Rita, Sue and Bob Too*, in 1982. Her work was directly autobiographical, uninhibitedly sexual, both funny and sad. The London literary and media world regarded this genuinely working-class writer as a piece of exotica (the *Mail on Sunday* described her, inaccurately, as 'a genius displaying black teeth'), but Dunbar remained faithful to Buttershaw, where she suffered many domestic crises and spent much time in her local pub, the Beacon, where she did her writing. Over several years *Rita, Sue and Bob Too* (enriched by material from *The Arbor*) was turned into a film. It was released with great success in 1987 and was directed by Alan Clarke (who, like Dunbar, died young in 1990). The highly explicit film caused some protests in Buttershaw, but, more seriously for the author, her failure to disclose her earnings to either the tax or social security authorities led to her being prosecuted and forced to pay back much of the money she had earned, itself not a large sum in film terms. In her later years she lived with a steady boyfriend, Jim Wheeler. Her last completed work was her third play, *Shirley*, produced at the Royal Court in

1986, and she was working on a fourth script at her death. This occurred when she suffered a brain haemorrhage at the Beacon in December 1990; she died later in hospital. She was twenty-nine years old.

Plays

The Arbor 1980 [pub. 1980]; *Rita, Sue and Bob Too* 1982 [pub. 1982]; *Shirley* 1986 [pub. 1988]; *Rita, Sue and Bob Too* **(f)** [from her play and *The Arbor*] 1987

Robert (Edward) DUNCAN 1919–1988

Duncan was at first named Edward Howard Duncan by his father, a labourer of that name. His mother had died giving birth, and he was adopted in early infancy by an architect and his wife, Edwin and Minnehaha Symmes. The Symmes adopted him on the basis of his astrological chart, being keen enthusiasts of the occult and also of European philosophy; things that were to be a lifelong influence on Duncan. His adopted name was Robert Edward Symmes, but he changed it back to Duncan in 1941, the same year that his brief brush with military service ended in a discharge on psychological grounds. He was educated at the University of California, Berkeley, from 1936 to 1938, and went back in 1948–50, when he studied medieval civilisation. He taught at the Black Mountain College in the mid-1950s.

Duncan was a leading figure among the Black Mountain school of poets, and he shares the open form of **Charles Olson** and **Robert Creeley**. He published his best-received work in the 1960s, when the Black Mountain school was at the height of its influence. His work was a delicate musical quality which shows the lasting influence of **Ezra Pound**, although his first volume, *Heavenly City, Earthly City* (1947), was more strongly influenced by **George Barker**. He played a pivotal role in establishing the San Francisco Bay area as a major centre of poetry in North America, and was affiliated with the Beat movement, the Black Mountain school and the Romantic current in general.

Duncan's work was bound up with his interest in the occult, and he saw poetry as a ritual emerging from a deep and pagan divinity. He said that he saw his poems as 'intentions' towards one great poem, and in his autobiographical essay 'Towards an Open Universe' he wrote: 'Our consciousness and the poem as a supreme effort of consciousness comes in a dancing organization between personal and cosmic identity.' Free unbounded dance is an important concept for him. His poetry is a musical and mystical art of process rather than product, producing evolving magical evocations rather than finished statements, and involving an extreme romanticism that privileges intuition and feeling over rationality and knowledge.

In addition to traditionally Romantic themes of self and environment, Duncan's homosexuality has been central to his work, and he was the partner of Jess (Jess Collins), the San Francisco collage artist. After *Bending the Bow* (1968) Duncan announced that he would not publish another major collection for fifteen years, hoping to concentrate on 'process' work rather than the 'overcomposed' (as he felt) poetry of books. This probably contributed to his relative obscurity compared with poets such as Olson and Creeley. His admirers were so distressed when his 1984 *Ground Work: Before the War* failed to achieve the notice they felt it deserved that they founded the National Poetry Award and made Duncan its first recipient.

Poetry

Heavenly City, Earthly City 1947; *Medieval Scenes* 1947; *Poems 1948–9* 1950; *The Song of the Border Guard* 1951; *The Artist's View* 1952; *Fragments of a Disordered Devotion* 1952; *Caesar's Gate* 1956; *Letters* 1958; *Selected Poems 1942–50* 1959; *O!* 1960; *The Opening of the Field* 1960; *Elegies and Celebrations* 1962; *Roots and Branches* 1964; *Writing Writing* 1964; *Uprisings* 1965; *A Book of Resemblances* 1966; *Boob* 1966; *Christmas Present, Christmas Presence* 1966; *Of the War* 1966; *The Years as Catches* 1966; *Audit: Robert Duncan* 1967; *Epilogos* 1967; *Bending the Bow* 1968; *Derivations* 1968; *The First Decade* 1968; *My Mother Would Be a Falconress* 1968; *Names of People* 1968; *Achilles' Song* 1969; *1942, a Story* 1969; *Playtime, Pseudo Stein* 1969; *Bring It Up from the Dark* 1970; *Poetic Disturbances* 1970; *A Selection of 65 Drawings From One Drawing Book* 1970; *Tribunals* 1970; *The Massacre of Glencoe* 1972; *Poems from the Margins of Thom Gunn's Moly* 1972; *A Seventeenth Century Suite* 1973; *An Ode and Arcadia* [with J. Spicer] 1974; *The Venice Poem* 1978; *Veil, Turbine, Cord and Bird* 1979; *The Five Songs* 1981; *Towards an Open Universe* 1982; *Ground Work: Before the War* 1984; *The Regulators* 1985; *A Paris Visit* 1985

Plays

Faust Foutu 1955 [pub. 1958]; *O-B-A-F-G etc!* 1964; *Medea at Kolchis* 1965; *Adam's Way* 1966

Non-Fiction

As Testimony: the Poet and the Scene 1964; *Wine* 1964; *Poets on Poetry* **(c)** [ed. H. Nemerov] 1966; *Six Prose Pieces* 1966; *The Truth and Life of Myth* 1968; *The Artistic Legacy of Walt Whitman* **(c)** [ed. E. Haviland Miller] 1970; *Notes on Grossinger's 'Solar Journal'* 1970; *A Prospectus for the Prepublication of Ground Work to Certain Friends of the Poet, R.D.* 1971; *Structure of Rime XXVIII* 1972; *Dante* 1974; *Fictive Certainties* 1979

For Children

The Cat and the Blackbird 1967

Douglas (Eaglesham) DUNN 1942–

Born at Inchinnan in Renfrewshire, the son of a factory worker, Dunn qualified at the Scottish School of Librarianship before going on to read English at the University of Hull, graduating with a first-class BA in 1969. He spent ten years as a librarian, and the path from library assistant to assistant librarian to librarian (and back to assistant librarian again) took him to Renfrew County Library in Paisley; Strathclyde University Library; Andersonian Library, Glasgow; Akron Library, Ohio; the Chemistry Department Library of the University of Glasgow; and the Brynmor Jones Library at the University of Hull, where he was assistant to **Philip Larkin**. Meanwhile, Larkin-like, Dunn was pursuing another career as a poet, which was to lead to writer's residences and creative writing fellowships at the University of Hull; the University of New England, Australia; Duncan of Jordanstone College of Art; and the University of Dundee. He is now a professor at the University of St Andrews.

Dunn's first collection, *Terry Street* (1969), was immediately acclaimed for its atmospherically down-at-heel evocation of life in working-class Hull. Many of the poems had already been published in late 1960s' issues of the *New Statesman*, *Times Literary Supplement* and *London Magazine*. Inventories of detail were a prominent feature, and the poems seemed to inhabit the desolate spiritual landscape of Larkin. But Dunn differs sharply from Larkin (whom he has discussed in his 1987 book *Under the Influence*) in a number of ways – not least politically – and already in *Terry Street* there were positive inflections and quietly ambitious effects that were deviating from the dour and blokeish plainness of Larkin and the Movement tradition.

Terry Street won Dunn a partisan following, some of whom found that his subsequent books had moved disappointingly on from simple loyalties to place and class, and into an unsatisfying complexity. But the same note of celebrating proletarian life comes through in more recent work such as 'Poor People's Cafés', which first appeared in *Poll Tax: the Fiscal Fake*, a Chatto and Windus 'Counterblast' pamphlet. One of the more successful of his ambitious works occurs in his 1993 collection *Dante's Drum Kit*: readers will look in vain for the title-poem, since it is named after the 'drumming' *terza rima* used in 'Disenchantments', a longish poem on the prospect of mortality. Reviewers have sometimes found his work patchy, and he is sometimes capable of rendering ordinariness with a polysyllabic awkwardness.

Dunn's first wife, Lesley, an artist and gallery curator, died of cancer at the age of thirty-five in 1981. Her death became the subject of his most moving and well-received book, *Elegies*. He married again in 1985 to another Lesley, also an artist, and they have two children. Notable recent works include his televised poem *Dressed to Kill*, about Scottish participation in imperial wars, and his place in the Scottish cultural establishment seems secure.

Poetry

Terry Street 1969; *The Happier Life* 1972; *Love or Nothing* 1974; *Barbarians* 1979; *St Kilda's Parliament* 1981; *Europa's Lover* 1982; *Elegies* 1985; *Selected Poems: 1964–1983* 1986; *Northlight* 1988; *Dante's Drum Kit* 1993; *Dressed to Kill* 1994

Fiction

Secret Villages 1985

Non-Fiction

Under the Influence: Douglas Dunn on Philip Larkin 1987; *Poll Tax: the Fiscal Fake* 1990; *Australian Dream* 1993; *Garden Hints* 1993

Edited

New Poems, 1972–1973 1973; *A Choice of Byron's Verse* 1974; *Two Decades of Irish Writing: a Critical Survey* 1975; *The Poetry of Scotland* 1979; *A Rumoured City: New Poets from Hull* 1982; *To Build a Bridge* 1982

Nell (Mary) DUNN 1936–

Born in London, the daughter of Sir Philip Gordon Dunn, second baronet Bathurst – the family is of New Brunswick, Canada, origin – Dunn grew up on an aristocratic estate in Wiltshire. She attended convent school, but came to London at the age of seventeen to study art history and experience Bohemian life. She married the writer Jeremy Sandford in 1957, and they have three sons. At the beginning of the 1960s she left her large house in Chelsea's Cheyne Row to live across the Thames in nearby Battersea among the factory workers. Out of this experience came her two highly successful 1960s' novels, the episodic *Up the Junction* and *Poor Cow*, about how a young working-class woman is progressively destroyed by her environment. These novels were admired for their crisp realism, remarkable feel for dialogue and uninhibited exploration of sexuality. Both novels were made into films in 1967, the latter (more successfully) from a screenplay by Dunn and the director Ken Loach, a noted practitioner in the field of cinematic social realism.

Later novels such as *The Incurable* and *The Only Child* move back to the middle class and the rich, but explore similar themes of the pressures on female experience. Dunn has also written books for children and, in recent years, has moved from the novel to the theatre: her best-known stage play is probably *Steaming*, a comedy with an all-female cast set in a Turkish bath. *The Little Heroine*, about drug problems, was performed in 1988, as was her television play, *Every Breath You Take*, based on her own experience of being the mother of a diabetic child.

Fiction

Up the Junction 1963; *Poor Cow* 1967; *The Incurable* 1971; *I Want* [with A. Henri] 1972; *Tear His Head Off His Shoulders* 1974; *The Only Child* 1978

Plays

Poor Cow **(f)** [from her novel; with K. Loach] 1967; *Steaming* 1981; *Every Breath You Take* **(tv)** 1988; *The Little Heroine* 1988

For Children

Freddy Gets Married 1968

Non-Fiction

Talking to Women 1965; *Grandmothers Talking to Nell Dunn* 1991

Edited

Living Like I Do [rev. as *Different Drummers*] 1977

Paul DURCAN 1944–

Durcan was born in Dublin, the son of a barrister (later judge) from County Mayo, and with a mother who was also a lawyer. He was educated by the Jesuits (at the splendidly named Gonzaga College, Dublin), but did not immediately go on to university, spending his earliest adult years attempting to establish himself as a poet in London. His several jobs there include a period working as a stellar manipulator at the London Planetarium, but he also founded, with the critic Martin Green, the literary quarterly *Two Rivers* in 1969, made recordings of his work for Harvard University and the British Council, and published, with Brian Lynch, an early volume of poems, *Endsville* (1967).

In the early 1970s, Durcan read archaeology and modern history at University College, Cork, taking a first-class degree in 1974. Recognition for his work came relatively slowly, although he won the Patrick Kavanagh Award and received creative writing bursaries from the Arts Council of Ireland in 1976 and 1980. After publishing several collections, his breakthrough came in 1982 with *The Selected Paul Durcan*, edited by Edna Longley, one of the critics who has been most influential in promoting the Irish poetic renaissance. With his 1985 volume, *The Berlin Wall Café*, he was firmly established among the most celebrated poets of his country, and the one, above all, who makes a feature of public performance of his work all over the world. In common with other performance poets, he has been accused of superficiality, and of treating poetry as if it were newspaper headlines, but his admirers point to his energy, warmth and sensitivity to the often bizarre nuances of Irish life. He is also a trenchant critic of that life, and of his country's politics, handing out even-handed brickbats to the British government and the IRA, the Catholic church and other institutions.

The political is not at the expense of the personal, however, as he has shown in his volume about the death of his father, *Daddy, Daddy*, which won the Whitbread Poetry Award in 1990, or in his moving poetry about his separation, in 1983, from his wife Nessa O'Neill, whom he had married in 1969, and with whom he had two daughters. His love of visual art is another strand, celebrated in two unusual poetry books of recent years, *Crazy about Women* and *Give Me Your Hand*, responses to paintings in the National Gallery of Ireland and the National Gallery, London, respectively. The publication of a substantial poetic retrospective, mixed with new work, in *A Snail in My Prime*, is another tribute to Durcan's ready accessibility.

Poetry

Endsville [with B. Lynch] 1967; *O Westport in the Light of Asia Minor* 1975; *Poetry Now – 1* [with others] 1975; *Teresa's Bar* 1976; *Sam's Cross* 1978; *Ark of the North* 1982; *The Selected Paul Durcan* [ed. E. Longley] 1982; *Jesus, Break His Fall* 1982; *The Berlin Wall Café* 1985; *Going Home to Russia* 1987; *In the Land of the Punt* [with G. Lambert] 1988; *Jesus and Angela* 1988; *Daddy, Daddy* 1990; *Crazy about Women* 1991; *A Snail in My Prime* 1993; *Give Me Your Hand* 1994

Lawrence (George) DURRELL 1912–1990

Durrell was born in India, in the foothills of the Himalayas, where his father was a prosperous civil engineer of Irish descent. At the age of twelve he was sent to England where he was educated at St Edmund's School, Canterbury, and repeatedly failed the qualifying exams for Oxbridge entry. 'Intellectually I was brilliant, but like all Irishmen I was dreadfully lazy. And above all my subconscious was set on proving to my father ... that it was unfair to send me back to prison,' he said in a 'dialogue' with Marc Alyn published in 1973 as *The Big Supposer*. He moved to London where he worked as a jazz pianist at the Blue Peter night club and published his first volume of poems, *Quaint Fragment*, in 1931. In 1935 he published his first novel, *Pied Piper of Lovers*, a portrait of Bohemian life in Bloomsbury, married Nancy Myers, and persuaded his widowed mother to move the entire family from Bournemouth to Corfu. This period is amusingly – and somewhat fancifully – described by Durrell's younger brother, the naturalist Gerald Durrell, in *My Family and Other Animals* (1956). In Corfu he wrote *The Black Book*, the novel in which he felt that 'I first heard the sound of my own voice'. It was admired by **T.S. Eliot** at Faber and Faber, but not published for fear of obscenity charges, and Durrell's new friend **Henry Miller** arranged publication in Paris. In the late 1930s in Paris, Durrell and Miller edited *The Booster* with Alfred Perlès, publishing the work of **Anaïs Nin**, **David Gascoyne**, **William Saroyan** and others.

Durrell was a press officer at the British Information Office in Cairo from 1941 to 1944, moving for the last year of the war to Alexandria. For another decade he worked for various government agencies in Argentina, Belgrade and Cyprus before settling in the South of France in 1957 with his third wife, to make his living from writing. In that year he published a children's adventure story, *White Eagles over Serbia*, a travel book *Bitter Lemons*, and *Justine*, the first novel of *The Alexandria Quartet*. Intended as 'an investigation of modern love', the *Quartet* draws on his year in Alexandria to paint an exotic and lavish picture of the cosmopolitan city and its people. It was praised by critics such as **Cyril Connolly** and **George Steiner**, though some were put off by its opulent, almost Paterian, prose.

Over the next twenty-five years he completed two further novel sequences, *The Revolt of Aphrodite* and *The Avignon Quintet*, which explore an eclectic range of ideas drawn from Zen Buddhism, psychoanalysis, science and European history. Like the *Quartet*, Durrell's travel books about Corfu, Rhodes, Cyprus, Sicily and Provence powerfully evoke what he called 'the spirit of place'. His three volumes of comic stories about diplomatic life are gathered in *Antrobus Complete* and display a fine Wodehousian wit. Durrell was married four times and had two daughters from his first two marriages. His second daughter, Sappho, committed suicide in 1985, and it was later alleged on the basis of some of her writings that she had had an incestuous relationship with Durrell. This seems exceedingly unlikely; in all probability the worst Durrell could be charged with was encouraging his mentally fragile daughter to read too much Freud. Durrell died of a stroke at his home in Sommières following a lengthy struggle with emphysema.

Fiction

Pied Piper of Lovers 1935; *Panic Spring* [as 'Charles Norden'] 1937; *The Revolt of Aphrodite*: *The Black Book* 1938, *Cefalû* 1947 [as *The Dark Labyrinth* 1958], *Tunc* 1968, *Numquam* 1970; *The Alexandria Quartet* 1962: *Justine* 1957, *Balthazar* 1958, *Mountolive* 1958, *Clea* 1960; *Esprit de Corps* (s) 1957; *Stiff Upper Lip* (s) 1959; *Sauve qui peut* (s) 1969; *The Best of Antrobus* (s) 1974; *The Avignon Quintet* 1992: *Monsieur* 1977, *Livia* 1978, *Constance* 1982, *Sebastian* 1983, *Quinx* 1985; *Antrobus Complete* (s) 1985

Poetry

Quaint Fragment 1931; *Ballade of Slow Decay* 1932; *Ten Poems* 1932; *Transition* 1934; *Mass for the Old Year* 1935; *A Private Country* 1943; *Cities, Plains and People* 1946; *On Seeming to Presume* 1948; *Deus Loci* 1950; *Private Drafts* 1955; *Collected Poems* 1960 [rev. 1968]; *Penguin Modern Poets 1* [with E. Jennings, R.S. Thomas] 1962; *Selected Poems 1935–1963* 1964; *The Tree of Idleness and Other Poems* 1966; *Vega and Other Poems* 1970; *The Red Limbo Lingo* 1971; *Lifelines* 1974; *Selected Poems of Lawrence Durrell* [ed. A. Ross] 1977; *Sicilian Carousel* 1977; *Collected Poems 1931–1974* [ed. J.A. Brigham] 1985

Non-Fiction

Zero, and Asylum in the Snow: Two Excursions into Reality 1946; *Key to Modern Poetry* 1952; *Prospero's Cell: a Guide to the Landscape and Manners of the Island of Corcyra* 1953; *Reflections of a Marine Venus: a Companion to the Landscape of Rhodes* 1953; *Bitter Lemons: an Account of Life in Cyprus, 1953–1956* 1957; *Art and Outrage: a Correspondence about Henry Miller* 1959; *Lawrence Durrell – Henry Miller: a*

Private Correspondence [ed. G. Wicks] 1963; *Spirit of Place: Letters and Essays on Travel* [ed. A.G. Thomas] 1969; *The Big Supposer* [with M. Alyn] 1973; *The Greek Islands* 1978; *A Smile in the Mind's Eye* 1980; *Literary Lifelines: The Richard Aldington–Lawrence Durrell Correspondence* [ed. I.S. MacNiven, H.T. Moore] 1981; *The Durrell–Miller Letters 1935–1980* [ed. I.S. MacNiven] 1988; *Letters to Jean Fanchette 1958–1963* 1988; *Caesar's Vast Ghost: Aspects of Provence* 1990

Plays

Sappho 1950; *An Irish Faustus* 1963 [rev. 1987]; *Acte* 1965

Edited

The Henry Miller Reader 1959; *The Best of Henry Miller* 1960

For Children

White Eagles over Serbia 1957

E

David EDGAR 1948–

Edgar was born in Birmingham; his background, he has said, was 'fairly conventional, more or less upper middle class'. From 1961 to 1965 he attended Oundle (where he acted the part of Miss Prism in *The Importance of Being Earnest*) and – after three terms teaching at a prep school – he read drama at Manchester University. He graduated in 1969, and until 1972 worked as a journalist on the Bradford *Telegraph and Argus* (he was assistant reporter on the first exposé of the Poulson scandal). A two-year association with Bradford University began in 1970, during which Edgar wrote eighteen plays. The first of these, *Two Kinds of Angel*, borrows the lives of Rosa Luxemburg and Marilyn Monroe to lend substance to the arguments of a pair of female students. Its resolution proposed that neither political nor sexual power is enough of itself: to each must be added disinterested love.

Since 1972, Edgar has been a writer full-time. He has held residential writing posts at Leeds Polytechnic and at Birmingham Repertory Theatre and has taught a course in play-writing at Birmingham University. In 1989 he was appointed chair of the same university's play-writing course. The recipient of a Bicentennial Fellowship, he spent 1978–9 in the USA.

His early work is characterised by its relevance to its historical context: his 1970s' subjects included Charles Manson, the Meriden motorcycle factory, Edward Heath's administration, the *Oz* trials and the Vietnam War. To review his range is to be reminded of the legion causes espoused by the left. The forms employed were diverse too – pantomine, agitprop, naturalism and tragedy – and, if the results have been ephemeral, Edgar applauds that very quality: in opposition to 'old', 'middle-class' theatre, 'the new Theatre ... must be temporary, immediate, specific, functional'. Towards the end of the decade, however, Edgar's prolificness was diminishing while his work assumed a less disposable air. *Destiny* (1976, and revised a decade later) addresses not merely the issue of contemporary racism but also its historical foundation; and thus, by acknowledging lasting dilemmas while recording their transient mutations, it signals Edgar's readiness to examine what it is that links present to past. In *Mary Barnes* he followed the treatment of a schizophrenic under 'alternative' psychiatry. Protracted and harrowing, the play asserts the therapeutic potential of communal relations and foreshadows the displacement of radicalism from the arena of politics proper into the marginal spaces of consciousness-raising and ideological struggle. Indeed, *Maydays* – a three-and-a-half-hour survey of post-war European protest and dissent – is an attempt to diagnose the repeated failures of liberal democratic socialism. Although it embodies Edgar's many strengths – commitment, energy, breadth, sardonic wit – its very urgency disables it and renders it dramatic journalism, as powerful as it is impermanent.

Edgar entered a more popular domain when he scripted the stage version of *Nicholas Nickleby* for the Royal Shakespeare Company. An enormous success on both sides of the Atlantic, it was, according to its author, 'an early ... shot in the war ... [over] the uses of history and heritage ... and the nature of Victorian values'.

Plays

Two Kinds of Angel 1970 [pub. 1975]; *A Truer Shade of Blue* 1970 [np]; *Acid* 1971 [np]; *Bloody Rosa* 1971 [np]; *Conversation in Paradise* 1971 [np]; *The National Interest* 1971 [np]; *Still Life* 1971 [np]; *Tedderella* 1971 [np]; *Death Story* 1972 [np]; *The End* 1972 [np]; *England's Ireland* 1972 [np]; *Excuses, Excuses* 1972 [np]; *Not with a Bang but with a Whimper* 1972 [np]; *Rent* 1972 [np]; *Road to Hanoi* 1972 [np]; *The Rupert Show* 1972 [np]; *Baby Love* 1973 [pub. 1989]; *The Case of the Workers' Plane* 1973 [np]; *A Fart for Europe* 1973 [np]; *The Eagle Has Landed* **(tv)** 1973; *Gangsters* 1973 [np]; *Liberated Zone* 1973 [np]; *Operation Iskra* 1973 [np]; *Up Spaghetti Junction* 1973 [np]; *The All-Singing, All-Talking Golden Oldie Rock Revival Ho Chi Minh Peace Love and Revolution Show* 1974 [np]; *Dick Deterred* 1964; *The Dunkirk Spirit* 1774 [np]; *I Know What I Meant* **(tv)** 1974; *Man Only Dines* 1974 [np]; *The Owners* **(r)** 1974; *Bad Guy* **(r)** 1975; *Censors* **(tv)** 1975; *Hero or Villain* **(r)** 1975; *The Midas Connection* **(r)** 1975; *The National Theatre* 1975 [pub. 1989]; *O Fair Jerusalem* **(r)** 1975; *Summer Sports* 1975 [as *Blood Sports* 1976]; *Destiny* 1976 [rev. 1985, pub. 1986]; *Events Following the Closure of a Motorcycle Factory* 1976 [np]; *The Perils of Bardfrod* 1976 [np]; *Our Own People* 1977 [pub. 1987]; *Saigon Rose* **(st)** 1976, **(r)** 1979 [pub. 1987]; *Do Something – Somebody* **(r)** 1977; *Ecclesiastes* **(r)** 1977 [pub. 1990]; *Wreckers 1977; The Jail Dairy of Albee Sachs* [from A. Sachs's memoirs] **(st)** 1978, **(tv)** 1981 [pub. 1982]; *Mary Barnes* [from M.

Barnes's and J. Berke's *Two Accounts of a Journey through Madness*] 1978 [pub. 1979]; *Teendreams* 1979; *Nicholas Nickleby* [from Dickens's novel] **(st)** 1980, **(tv)** 1982 [pub. 1982]; *Maydays* 1983; *Entertaining Strangers* 1985 [pub. 1986]; *Lady Jane* **(f)** 1986; *Plays: One* [pub. 1987]; *That Summer* 1987; *Vote for Them* **(tv)** 1989; *Heartlanders* 1989; *Plays: Two* [pub. 1989]; *Shorts* [pub. 1989]; *The Shape of the Table* 1990; *The Mill on the Floss* **(f)** [from G. Eliot's novel] 1991; *A Movie Starring Me* **(r)** 1991; *Plays: Three* [pub. 1991]; *The Strange Case of Dr Jekyll and Mr Hyde* [from R.L. Stevenson's novel] 1991; *Pentecost* 1994

Non-Fiction

The Second Time as Farce: Reflections on the Drama of Mean Times 1988

Dorothy EDWARDS 1903–1934

Born at Ogmore Vale, Glamorgan, Edwards was the daughter of an ardent Welsh socialist and vegetarian (he was a friend of Keir Hardie), who brought his daughter up to share his beliefs and to attend the boys' primary school of which he was headmaster. She was later educated more conventionally at Howell's School, Cardiff, and the University College in that city, where she read a degree in Greek and philosophy. After university, she rejected the conventional career of teaching, and spent six months living with the family of a Viennese socialist bookseller, and then nine months in Florence and other parts of Italy. Back in Wales she lived with her widowed mother in the village of Rhiwbina outside Cardiff, writing short stories, attempting to help the miners, and indulging her passions for music, Welsh nationalism and imperfectly understood socialist politics (she joined a series of evening classes on the Gold Standard because 'gold is such a brilliant, beautiful colour').

From the mid-1920s her unexpected, *faux naïf* short stories began to appear in magazines, and in 1927, a volume of them was published, *Rhapsody*. This was followed the next year by her only other book, the novel *Winter Sonata*, a Chekovian study of aimless people living in an English village. Her books led to her being taken up by the London intelligentsia. Raymond Mortimer had recommended her books to **David Garnett**, who around 1930 wrote to her asking whether he could adopt her as a sister. She visited Garnett and his wife, Ray, in London several times, and in 1933 went to live in their house, partly to look after their small son. But familiarity bred contempt; Edwards proved herself tactless in dealing with distinguished friends such as **Virginia Woolf** and T.E. Lawrence, while Garnett became increasingly irritated by her 'clumsy homemade dresses and her lack of any form of corset'. In despair, she returned to Wales and threw herself under a train.

Her writing remained fashionable during the 1930s (when she was a favourite writer of **Cyril Connolly**), but after that she was almost entirely forgotten until reprints of her two books by Virago in the 1980s won her a modest revival. Hers is minor writing, but it has a flavour of its own, and is unlikely now to vanish completely.

Fiction

Rhapsody **(s)** 1927; *Winter Sonata* 1928

Cyprian EKWENSI 1921–

A member of the Ibo tribe, Ekwensi was born in Minna, Nigeria, and educated at Government College, Ibadan; Yaba Higher College; Achimota College in Ghana; the School of Forestry, Ibadan; and the Chelsea School of Pharmacy, London. He worked initially as a forestry officer in northern Nigeria (the background to his novel *Burning Grass*), but his main career has been in broadcasting and journalism. He has held a variety of posts including head of features at the Nigerian Broadcasting Corporation, and, after the Nigerian civil war (in which he was an active campaigner on behalf of the Biafran cause), chairman of the East Central State Library Board and managing director of the state-owned Star Printing and Publishing Company. Since 1979 he has concentrated largely on journalism and literature.

His first book, *When Love Whispers* (1948), which grew out of his experiences as a teacher in Lagos, is characteristic of much of his later work. A chronicler of city life, he has been called an African Defoe, although several of his novels are set in rural parts of Nigeria and others include volumes of native folk-tales and children's fiction. Ekwensi sees himself rather as a 'national novelist', 'because I know Nigeria backwards'. He also sees himself as a *popular* novelist, writing for 'the ordinary working man ... the masses'. This deliberate accessibility is a strategic inducement to the serious matter of his novels, which are preoccupied with the problems facing many African countries in the twenteith century: the loss of traditional values; the new materialism; the legacy of empire; and the effects of political instability and civil war. A social purpose is inherent in all his writings: for Ekwensi the modern African writer 'must be committed to the exposure of the ills of society. And he must be committed to pointing the way to the future, as he understands it.' Like many modern African writers he has found himself in the anomalous position of writing about African issues in a non-African tongue. But Ekwensi defines his English as African English, which conveys the essence of Nigerian culture without limiting its appeal to a specifically Nigerian audience. This has ensured his popularity in English-speaking countries. His works are also much translated. He is one of the most famous of modern African writers, and in 1968 was awarded the Dag Hammarskjöld International Prize in Literature.

Ekwensi's most popular novel, *Jagua Nana*, features the best of his many sympathetic female characters: an authentically Nigerian Moll Flanders. *Iska* has another compelling heroine: a Nigerian Grace Darling. Notable amongst his other works are *Survive the Peace* and *Divided We Stand*, which deal with his experiences in the Nigerian civil war. At the heart of all his work is the figure of a wanderer, caught between the unknown future and the abandoned past; the dilemma of modern Africa.

Fiction

When Love Whispers 1948; *People of the City* 1954; *Jagua Nana* 1961; *Burning Grass* 1962; *Beautiful Feathers* 1963; *Iska* 1966; *Lokotown and Other Stories* (s) 1966; *Restless City and Christmas Gold and Other Stories* (s) 1975; *Survive the Peace* 1976; *Divided We Stand* 1980; *Jagua Nana's Daughter* 1986; *For a Roll of Parchment* 1987; *Gone to Mecca* 1991; *King Forever!* 1992

For Children

Ikolo the Wrestler and Other Ibo Tales (s) 1947; *The Leopard's Claw* 1950; *The Drummer Boy* 1960; *The Passport of Malam Ilia* 1960; *An African Night's Entertainment* 1962; *Yaba Roundabout Murder* 1962; *The Great Elephant-Bird* 1965; *The Rainmaker and Other Stories* (s) 1965; *The Boa Suitor* 1966; *Juju Rock* 1966; *Trouble in Form Six* 1966; *Coal Camp Boy* 1973; *Samankwe in the Strange Forest* 1973; *The Rainbow-Tinted Scarf and Other Stories* (s) 1975; *Samankwe and the Highway Robbers* 1975; *Motherless Baby* 1980; *Masquerade Time* 1991

Edited

Festac Anthology of Nigerian New Writing 1977

T(homas) S(tearns) ELIOT 1888–1965

Born at St Louis, Missouri, Eliot was the seventh and youngest child of local businessman Henry Ware Eliot. When the poet was eleven, his father built a house in Massachusetts and the whole family subsequently spent their summers in New England. Eliot's education was thorough and traditional and led to the completion of a thesis on the nineteenth-century philosopher F.H. Bradley. He was officially a student at Harvard between 1906 and 1916, but this period was punctuated by visits to the Sorbonne (1910–11), the University of Marburg in Germany (1914) as well as Merton College, Oxford, in the same year. The years 1912–14 were intensely creative and allowed Eliot to arrive in London in 1914 with a number of what would become his major early poems in manuscript form, including 'Preludes', 'Prufrock', 'Portrait of a Lady', 'Rhapsody on a Windy Night' and 'La Figlia che Piange'. The publication of these poems in book form in 1917 as *Prufrock and Other Observations* marks a pivotal moment in the history of English poetry: fragmented, bleak and intense, and made up of a vocabulary and imagery drawn from the everyday and popular as well as from a broad range of literature and history, the collection was championed by **Ezra Pound**, a figure who would be instrumental in guiding Eliot (and countless other writers of the same generation) through his literary career.

In July 1915 Eliot married Vivien Haigh-Wood and began earning a living by teaching and reviewing. Both his marriage and his finances would cause him suffocating difficulties throughout the next decade. His wife proved to be possessive, jealous and socially embarrassing and was eventually revealed to be suffering from a nervous illness. Eliot worked at Lloyds Bank as of 1917 and became assistant editor of the *Egoist* during the same year. In 1920 he published *The Sacred Wood*, the first in a series of critical works which would, collectively, be more influential in the development of literary taste in England than any other writing produced this century. He also began work on *The Waste Land*, generally thought to be the greatest poem of the modernist period in the same way that *Ulysses* (1922) by **James Joyce** (whom Eliot had earlier visited in Paris) is the monumental work of prose. *The Waste Land* was brilliantly edited by Pound and was published over two issues in the *Criterion* in 1922, along with notes. It is now fashionable to argue that Eliot's disastrous marriage and nervous disposition were manifested in easily traceable ways in the poem, though it is wise to be cautious when forging these links as a great many are based on pure speculation, while the poet famously believed the writing of poetry to be a way of escaping the limitations of the personal and the relative.

The largest and most prestigious publishing house in England, Faber and Faber, had offered Eliot a post in 1925, which he accepted. In 1927 he converted, publicly and controversially, to Anglo-Catholicism. The following year he wrote the first version of his best-known conversion poem, *Ash Wednesday*. He returned to America in 1932 for the first time since leaving, and while at Harvard delivered some of his most famous lectures. During the same year Eliot finally managed to instigate the break-up of his unhappy marriage. Over the next three decades he would publish a series of plays, including *Murder in the Cathedral*, *The Family Reunion* and *The Cocktail Party*, and also *Four Quartets*, which deals with the theme of time and redemption. This series of successes culminated in his being awarded the Order of Merit and the Nobel Prize for Literature in 1948. By this time Eliot was a celebrity: crowds of people gathered whenever he made a public appearance and *Time* magazine ran a cover story on him in 1950. His personal circumstances also improved and he married Valerie Fletcher in 1957. He died in the knowledge that he was amongst the most important poets and critics of his century, though he claimed that his creativity had not been worth the immense personal suffering it had caused him.

Poetry

Prufrock and Other Observations 1917; *Poems* 1919; *Ara Vos Prec* 1920; *The Waste Land* 1922; *Poems 1909–1925* 1925; *Journey of the Magi* 1927; *A Song for Simeon* 1928; *Animula* 1929; *Ash Wednesday* 1930; *Marina* 1930; *Triumphal March* 1931; *Sweeney Agonistes* 1932; *Collected Poems 1909–1935* 1936; *Old Possum's Book of Practical Cats* 1939; *East Coker* 1940; *Burnt Norton* 1941; *The Dry Salvages* 1941; *Little Gidding* 1942; *Four Quartets* 1944; *The Undergraduate Poems* 1949; *Poems Written in Early Youth* 1950; *The Cultivation of Christmas Trees* 1954; *Collected Poems 1909–1962* 1963

Plays

The Rock 1934; *Murder in the Cathedral* **(st)** 1935, **(f)** [with G. Hoellering] 1952; *The Family Reunion* 1939; *The Cocktail Party* 1949 [pub. 1950]; *The Confidential Clerk* 1953 [pub. 1954]; *The Elder Statesman* 1958 [pub. 1959]

Non-Fiction

The Sacred Wood 1920; *For Lancelot Andrewes* 1928; *Tradition and Experiment in Present Day Literature* **(c)** 1929; *Humanism and America* **(c)** [ed. N. Foerster] 1930; *A Garland for John Donne* **(c)** [ed. T. Spencer] 1931; *John Dryden: the Poet, the Dramatist, the Critic* 1932; *Selected Essays 1917–1932* 1932; *After Strange Gods* 1933; *English Critical Essays: Twentieth Century* **(c)** [ed. P.M. Jones] 1933; *The Use of Poetry and the Use of Criticism* 1933; *A Companion to Shakespeare Studies* **(c)** [ed. H. Granville-Baker, G.B. Harrison] 1934; *Elizabethan Essays* 1934; *Faith that Illuminates* **(c)** [ed. V.A. Delmant] 1935; *Essays Ancient and Modern* 1936; *Revelation* **(c)** [ed. J. Baillie] 1937; *Seventeenth Century Studies Presented to Sir Herbert Grierson* **(c)** [ed. J. Purvis] 1938; *The Idea of a Christian Society* 1939; *The Church Looks Ahead* **(c)** [ed. E.L. Mascall] 1941; *Malvern 1941: the life of the Church and the Order of Society* **(c)** 1942; *The Shock of Recognition: The Development of Literature in the United States* **(c)** [ed. E. Wilson] 1943; *James Joyce: Two Decades of Criticism* **(c)** [ed. S. Givens] 1948: *Notes Towards the Defence of Culture* 1948; *Ezra Pound* **(c)** [ed. P. Russell] 1950; *The Value and Use of Cathedrals in England Today* 1952; *Selected Prose* 1953; *Religious Drama: Medieval and Modern* 1954; *On Poetry and Poets* 1957; *George Herbert* 1962; *Knowledge and Experience in the Philosophy of F.H. Bradley* 1964; *To Criticize the Critic and Other Writings* 1965

Edited

Ezra Pound: Selected Poems 1928; *Selected Poems by Marianne Moore* 1935; *A Choice of Kipling's Verse* 1941; *Introducing James Joyce* 1942; *Literary Essays of Ezra Pound* 1954; *Selected Poems of Edwin Muir* 1965; *The Criterion 1922–1939* 18 vols 1967

Translations

St John Perse *Anabasis* 1930

Biography

T.S. Eliot by Peter Ackroyd 1984

Stanley (Lawrence) ELKIN 1930–

Elkin was born into a Jewish family in Brooklyn, where his father was a salesman. They moved to Chicago when Elkin was three, and he was educated at the South Shore High School there before entering the University of Illinois, Urbana, where he initially studied journalism – he thought newspapers were 'where stories were written' – before switching to English. He graduated in 1952 and one of his instructors during his MA year was the poet **Randall Jarrell**.

He served with the US Army from 1955 to 1957 at Fort Lee, Virginia, during which time his literary endeavours included writing a training manual on the use of forklift trucks. Upon returning to Illinois, he completed his doctoral dissertation, 'Religious Themes and Symbolism in the Novels of **William Faulkner**', and published short stories in *Accent* magazine, for which he also read manuscripts. He became an instructor in English at Washington University in St Louis in 1960 and, with periodic breaks, has remained there since, becoming Merle Kling Professor of Modern Letters in 1983. A friend and colleague is **William Gass**, another writer whose primary concerns are with language and metaphor. Elkin has said of his own writing: 'What I like best about it, I suppose, are the sentences.'

Elkin's first novel, *Boswell: a Modern Comedy*, was written in 1962 while he spent a year in Rome and London. The hero of this parodic comedy is a contemporary figure, loosely based on Samuel Johnson's biographer, who on the advice of his psychiatrist seeks some kind of immortality by pursuing and trying to ingratiate himself with various celebrities, among them a Nobel Prize-winning anthropologist and the world's wealthiest man. One of Elkin's most highly regarded novels is his satire *The Dick Gibson Show*, which concerns a disc jockey and talk-show host who celebrates the wondrous nature of quotidian existence and seeks to give voice to the ordinary American, which proves to be a rather hazardous enterprise.

Elkin's increasing lack of interest in realism is evident in *George Mills*, winner of the National Book Critics Circle Award in 1983, a narrative spanning 1,000 years and featuring forty generations of characters all called George Mills. Much of his work focuses on contemporary popular culture, notably the darkly comic *The Magic Kingdom* which describes a party of terminally ill children touring Disney World. His novella 'The Bailbondsman' was made into the film *Alex and the Gypsy* (1976), which starred Jack Lemmon. He has been a visiting professor at numerous American universities, and won the American Academy of Arts and Letters award in 1974. He married in 1953 and has three children.

Fiction

Boswell 1964; *Criers and Kibitzers, Kibitzers and Criers* (s) 1966; *A Bad Man* 1967; *The Dick Gibson Show* 1971; *The Making of Ashenden* (s) 1972; *Searches and Seizures* (s) 1973; *The Franchiser* 1976; *The Living End* (s) 1979; *Stanley Elkin's Greatest Hits* 1980; *George Mills* 1982; *Early Elkin* (s) 1985; *The Magic Kingdom* 1985; *The Rabbi of Lud* 1987; *The Macguffin* 1991

Plays

The Coffee Room (r) 1987; *The Six-Year-Old Man* (f) [pub. 1987]

Non-Fiction

Why I Live Where I Live 1983

Edited

Stories from the Sixties (s) 1971; *The Best American Short Stories 1980* (s) [with S. Ravenel] 1980

Janice ELLIOTT 1931–

Elliott was born in Derbyshire and brought up in wartime Nottingham, the background of her bestseller, *Secret Places*, a touching and funny novel about a German refugee girl in an English boarding-school, which won the Southern Arts Award for Literature and was made into a prize-winning film (1984), directed by Zelda Barron. Between leaving school and going up to St Anne's College, Oxford, she worked in an East End settlement, an experience that found its way into *A State of Peace*, the first of her *England Trilogy*. This sequence took up again an interest in modern history and politics that was foreshadowed in *Angels Falling*, a saga of English life from 1901 to 1968, and has informed, tangentially or centrally, much of her work.

Elliott read English at Oxford; on going down she went into glossy journalism before joining the *Sunday Times*. By now she was married to an oil executive, and when her first novel, *Cave with Echoes*, was accepted, she moved in 1962 to Sussex. Here she had one son and began reviewing, which she continued for twenty years, seventeen of them for the *Sunday Telegraph*. She now lives in a house overlooking the River Fowey in Cornwall.

She has many novels, a collection of stories, *The Noise from the Zoo*, and five children's books to her credit. Throughout her many years of publishing she has received consistent critical acclaim. She has been described as 'one of the best novelists writing in England' and as 'one of the most resourceful and imaginative living English novelists'. In 1989 she was elected a Fellow of the Royal Society of Literature.

Although her novel *Necessary Rites*, a story of buried mourning set in a Thatcherite England, is centred upon a family circle, she shuns writing any novel that might be described as purely domestic. In *The Italian Lesson* and *The Country of Her Dreams* individuals abroad find themselves caught up alarming in the kind of situation from which they would normally be insulated. *Life on the Nile* tells of one woman's search for a great-aunt murdered in Egypt in 1924 and in so doing confronts universal issues of life, death and history.

The global dimension to her novels has been remarked on, and a sense of comedy has made her books increasingly accessible. In *Dr Gruber's Daughter* high farce carried through the proposition that Hitler was alive in Coronation year, hidden in a north Oxford attic. This novel exemplifies Elliott's determination never to write the same book twice, also her reluctance to admit to reiterated themes. Her versatility works both against and for her. She cannot be pigeonholed. At the same time she retains the right to spring surprises.

Fiction

Cave with Echoes 1962; *The Somnambulists* 1964; *The Godmother* 1966; *The Buttercup Chain* 1967; *The Singing Head* 1968; *Angels Falling* 1969; *The*

Kindling 1970; *England Trilogy: A State of Peace* 1971, *Private Life* 1972, *Heaven on Earth* 1975; *A Loving Eye* 1977; *The Honey Tree* 1978; *Summer People* 1980; *Secret Places* 1981; *The Country of Her Dreams* 1982; *Magic* 1983; *The Italian Lesson* 1985; *Dr Gruber's Daughter* 1986; *The Sadness of Witches* 1987; *Life on the Nile* 1989; *Necessary Rites* 1990; *The Noise from the Zoo* (s) 1991

For Children

The Birthday Unicorn 1970; *Alexander in the Land of Mog* 1973; *The Incompetent Dragon* 1982; *The King Awakes* 1987; *The Empty Throne* 1988

Alice Thomas ELLIS (pseud. of Anna Haycraft) 1932–

Ellis was born Anna Margaret Lindholm; under her married name, Anna Haycraft, she has published a cookery book in collaboration with **Caroline Blackwood**. Her father was of Russo-Finnish, her mother Welsh ancestry. She was born in Liverpool, but grew up largely at Penmaenmawr on the North Wales coast (this period was covered in her volume of autobiography, *A Welsh Childhood*). She attended Bangor Grammar School and then Liverpool College of Art. Her parents belonged to the Church of Humanity, which is based on the rationalist ideas of Auguste Comte, but she rebelled against this at the age of nineteen, converting to Roman Catholicism, which she continues to practise, being conservative in matters of doctrine and ritual, but combining this with strong feminism.

For a while after leaving college, she was a postulant in a Liverpool convent, but then went to work in Chelsea, London, becoming part of the Bohemian society gathered around the King's Road in the 1950s, and in 1956 meeting and marrying the publisher Colin Haycraft, who was then a director of Weidenfeld and Nicolson. They had seven children, of whom five survive (one daughter died in infancy, while a young adult son died by falling on to a railway line). Haycraft died in 1994, but for more than twenty years before that had been chairman and managing director of the publisher Duckworth; Ellis remains a director and fiction editor of the company. Duckworth also published her earlier novels, and it was suggested that she was the leading member of the 'Duckworth gang', a group of women writers bringing a touch of the macabre to their brief novels of domestic life. The group was also said to consist of **Beryl Bainbridge** (long a close personal friend) and Caroline Blackwood.

Ellis's own first novel, appearing when she was in her mid-forties, was *The Sin Eater* (1977), the grim if brief story of the patriarch of a Welsh family. Later novels, such as *The 27th Kingdom* (shortlisted for the Booker Prize in 1982), exploit more the witty, eccentric side of her talent introducing elements of magic and miracle. The settings of her novels alternate between London and the country (usually Wales), and her Catholicism and feminism provide constant themes. *The Clothes in the Wardrobe*, *The Skeleton in the Cupboard* and *The Fly in the Ointment* form a trilogy. She has also become well-known for her magazine columns about her 'Home Life' in the large house in Gloucester Terrace, Camden Town, which she shared with her family and husband (featuring in the columns as 'Someone'); this household also became familiar through innumerable colour supplement and feature articles.

Besides the works already mentioned, she has published two books about juvenile delinquency in collaboration with the psychiatrist Tom Pitt-Aikens, several volumes of her collected articles, an anthology about Wales, and a study of the current position of the Catholic church, *Serpent on the Rock*. She also edited for publication *Mrs Donald*, a novel written by her mother under the pseudonym Mary Keene, which appeared posthumously in 1983.

Fiction

The Sin Eater 1977; *The Birds of the Air* 1980; *The 27th Kingdom* 1982; *The Other Side of the Fire* 1983; *Unexplained Laughter* 1985; *The Summerhouse Trilogy* 1991: *The Clothes in the Wardrobe* 1987, *The Skeleton in the Cupboard* 1988, *The Fly in the Ointment* 1989; *The Inn at the Edge of the World* 1990; *Pillars of Gold* 1992; *This Rock* 1993

Non-Fiction

Darling, You Shouldn't Have Gone to So Much Trouble [as Anna Haycraft; with C. Blackwood] 1980; *Home Life* 4 vols 1986–9; *Secrets of Strangers* [with T. Pitt-Aikens] 1986; *Loss of the Good Authority* [with T. Pitt-Aikens] 1989; *A Welsh Childhood* (a) 1990; *Serpent on the Rock* 1994

Edited

Mary Keene *Mrs Donald* 1983; *Wales: an Anthology* [with K. Williams] 1989

Ralph (Waldo) ELLISON 1914–1994

The grandson of slaves, Ellison was born into the black community of Oklahama City, in a state which practised segregation but where racism was not at its most intense. His father, who was a construction worker, died when Ellison was three, and his mother went out to work as a domestic servant; it was the discarded magazines and records that she brought home from the white houses which fired his love for the arts. He attended the Tuskegee Institute, Alabama, studying music, between 1933 and 1936, but failed to complete the course through lack of funds. Instead he went north to New York City, where he met fellow black writers such as **Richard Wright** and **Langston Hughes**, and worked from 1938 to 1942 on the Federal Writers' Project, an initiative which was part of the New Deal. In the latter part of the Second World War, he served in the US Merchant Marine.

After the war, a Rosenwald Fellowship enabled him to spend several years writing the long and complex novel which was published as *Invisible Man* (1952). The book caused a sensation such as few other post-war fictions could match, won a prestigious National Book Award, and in 1965, more than a decade after it was published, was voted by a poll of *Book Week* critics the best novel to have been published since the conflict ended. It is the semi-autobiographical story of a deliberately unnamed hero who wins through to the discovery of his own identity and, while it is acknowledged as a central text relating to the black American experience, its dense network of symbolic patterns and allusions means that it is also a universal novel in the grand tradition of Dostoevsky and Kafka. It made its author's name, and led to him obtaining a series of academic appointments, culminating in a professorship at New York University between 1970 and 1979, but it was not followed up in writerly terms.

The black radicals of the 1960s increasingly rejected Ellison's integrationist approach, and it was rumoured that his reluctance to protest more led to him failing to win the Nobel Prize. He preferred to live quietly with the wife he had married in 1946 and who survived him, in their Manhattan flat; here he amused himself with a series of complex hobbies, including playing the jazz trumpet, photography, furniture-making and hi-fi.

His only other published books were two volumes of essays, which were generally felt to be disappointing; he reviewed extensively, and published around fifteen short stories, so far uncollected. For forty years, in a pattern reminiscent of several other American writers, he was rumoured to be working on a vast second novel, from which several extracts were published. At one stage, a fire at his second home in Plainsfield, Massachusetts, destroyed large portions of the manuscript, and the book was still uncompleted at his death.

Fiction

Invisible Man 1952

Non-Fiction

Shadow and Act 1964; *Going to the Territory* 1986

Buchi (Florence Onye) EMECHETA 1944–

Emecheta was born of Ibuza parentage in Lagos, Nigeria. She won a scholarship to the local Girls' High School and in 1962 she arrived in Britain with her student husband. By the time she was twenty-two her marriage had broken up and she had five children. She took a BSc degree in sociology at London University, found part-time employment at the British Library and as a social worker, fought for proper housing and embarked on a writing career.

Her first novel, *In the Ditch*, appeared as a series of articles in the *New Statesman*. Largely autobiographical, it is the story of a young Nigerian woman, a single parent, living in London in the 1960s. Since then, Emecheta has written further novels as well as radio and television plays, and has won several literary awards, including the *New Statesman*'s Jock Campbell Award for *The Slave Girl* in 1978.

Her principal theme is female survival. Her protagonists are underprivileged black immigrants who must endure poverty, racism and cultural dislocation in present-day Britain. Emecheta is praised for her rancourless realism, her vivid, direct narrative and her skilful irony. When she wrote her autobiography in 1986, she gave it the title *Head Above Water*. *Gwendolen* is a characteristic work, the story of a young girl of Afro-Caribbean origin who is sexually abused by her 'Uncle Johnny', when her parents immigrate to the 'moder kontry' and leave her behind in Jamaica with her grandmother. On her arrival in Britain, Gwendolen's father becomes her sexual oppressor and eventually the father of her child.

Emecheta has lectured at numerous American universities and has been a fellow at the Universities of London and Calabar, Nigeria. She now lives in London but often returns to her birthplace, Lagos. She comments: 'If I were to make a statement about myself I would prefer to say that I campaign for world peace and for people to respect each other for what they are, whatever their race.'

Fiction

In the Ditch 1972; *Second-Class Citizen* 1974; *The Bride Price* 1976; *The Slave Girl* 1977; *The Joys of*

Motherhood 1979; *Destination Biafra* 1982; *Naira Power* 1982; *Adah's Story* 1983; *Double Yoke* 1983; *The Rape of Shavi* 1985; *Gwendolen* 1989; *Kehinde* 1994

Plays

The Ju Ju Landlord (tv) 1976; *A Kind of Marriage* (tv) 1976

For Children

Tich the Cat 1979; *The Moonlight Bride* 1980; *Nowhere to Play* 1980; *The Wrestling Match* 1981

Non-Fiction

Our Own Freedom 1981; *Head Above Water* (a) 1986

William EMPSON 1906–1984

Born into the Yorkshire squirearchy at Yokefleet Hall, Empson was educated at Winchester and Magdalene College, Cambridge, where he first graduated in mathematics and then took a first in English during a fourth year, studying under the influential critic I.A. Richards. Richards was credited with having introduced the American New Criticism, centring on the close reading of literary texts, to England, and Empson's own, immensely influential, first critical book, *Seven Types of Ambiguity* (1930), was seen as a first English monument of the school.

Empson spent the years from 1931 to 1934 as a professor of English literature in Tokyo, and from 1937 to 1939 was a professor at the National University of Peking, which was in exile in various parts of the East during that period. The reason for this period abroad, when Empson had made such a brilliant start in England, long remained a mystery. It was eventually revealed that in mid-1929 condoms had been found in his college room by a 'bedder', who reported Empson to the Master, A.B. Ramsay. Shocked by the discovery of what he termed 'engines of love', and fearing that other young men might be morally contaminated, Ramsay ordered Empson to leave Magdalene immediately, and expunged his name from the college records. Empson spent some time in the country finishing *Seven Types of Ambiguity*, before finding work abroad. Returning to wartime England, he joined the monitoring department of the BBC and later became Chinese editor in the Far Eastern section. In 1941 he married a South African, and they had two sons, each of whom was given an Afrikaans name, an English name, and the name of a town captured by the Allies on the day he was born (one of them was thus called Jacobus Arthur Calais). In 1947, Empson returned with his family to Peking University, came home in 1952, and was professor of English literature at Sheffield University from 1953 to 1971, retiring as emeritus. He was knighted in 1979.

Empson had a dual reputation, as critic and poet: his poetry, the great bulk of which is contained in two early volumes, *Poems* and *The Gathering Storm*, is formally very difficult, but its combination of intellectuality and suppressed passion makes it major. In later life, he confined himself to criticism: influential books such as *Some Versions of Pastoral*, *The Structure of Complex Words* and *Milton's God* ranged broadly from semantic analysis to socialist and anti-Christian polemic. He was noted as one of the great literary characters of his time: assailed by a vast range of illnesses in his last years, he was often a difficult and tormented, yet formidably upright, man, equally well known for extreme absent-mindedness as for legendary bluntness and caustic wit.

Poetry

Poems 1935; *The Gathering Storm* 1940; *Collected Poems* 1955

Non-Fiction

Seven Types of Ambiguity 1930; *Some Versions of Pastoral* 1935; *The Structure of Complex Words* 1951; *Milton's God* 1961; *Using Biography* 1984; *Essays on Shakespeare* 1986

Edited

Coleridge's Verse: a Selection [with D. Pirie] 1972

Collections

The Royal Beasts (A Fable) and Other Works 1986

Isobel ENGLISH (pseud. of June Guesdon Braybrooke) 1920–1994

English was born June Jolliffe. Her father was a Welsh businessman, working largely in the import and export of tinned fruit, who had gone out to Australia at the turn of the century and married her part-Huguenot mother there. English was born late in her parents' marriage, when they had returned to England and were mainly living apart. As a young child she suffered from tuberculosis of the spine, the beginning of a history of debility which was to continue through her life, and her mother took her to stay in Brittany. From the age of six to sixteen she attended a convent school in Somerset, leaving to learn typing and be presented at court. However, she had only one job as a secretary, in the early days of the war, with MI5. After only a couple of days she wrote a letter to occupied Belgium describing her new job, and it was decided she was not suited to the secret service.

In 1941, she married a civil servant, then at Scotland Yard, Ronald Orr-Ewing, with whom she had a daughter, and she did not work again, except to do some reading for a publisher. She was divorced from Orr-Ewing in the early 1950s, and in 1953 married the writer Neville Braybrooke.

From the early days of her marriage also dated her conversion to Catholicism and the beginning of her writing, which bore its main fruit in three novels published between 1954 and 1961: *The Key That Rusts*, *Every Eye* and *Four Voices*. English was one of the group of novelists who were interviewed by **Kay Dick** in her book *Friends and Friendship* (1973), who were all personally acquainted and who explored a similar territory of middle-class personal relationships. As fine and initially highly admired as any of this circle, English became much more obscure than **Francis King**, **Olivia Manning** or **Pamela Hansford Johnson** because her perfectionism and recurrent writer's block meant that she could not be so prolific. Besides the three novels, one volume of short stories, *Life After All* (often considered her finest work), a play which appeared in *The New Review*, a book of jokes about gifts, and some uncollected short stories, all published under the name Isobel English, constitute the *oeuvre* that appeared during her lifetime. Many other projects were started only to be abandoned: one children's book is as yet unplaced, and, in the last seven years of her life, she and her husband were at work on a biography of Olivia Manning which it is planned eventually to publish under the names of Neville and June Braybrooke. It remains to be seen whether the appearance of this work, and the process of re-evaluation that can follow a writer's death, will lead to her small but distinctive body of work achieving a continued readership.

Fiction

The Key That Rusts 1954; *Every Eye* 1956; *Four Voices* 1961; *Life After All and Other Stories* (s)

Plays

Meeting Point [pub. 1976]

Non-Fiction

The Gift Book [with B. Jones] 1964

D(enis) J(oseph) ENRIGHT 1920–

Born in Leamington, the son of a postman, Enright found education to be a ticket that would take him not only from his class but his country. He read English at Downing College, Cambridge, under F.R. Leavis, and contributed to Leavis's journal *Scrutiny*. He took a teaching post in 1947 at the University of Alexandria, beginning a university teaching career which (apart from a three-year stint in Birmingham at the start of the 1950s) gave him twenty years in Egypt, Japan, Germany, Thailand and Singapore. He returned to England in 1970, becoming co-editor of *Encounter* magazine, a director of Chatto and Windus, and an Honorary Professor of Warwick University. He married in 1949 and has one daughter.

The signature of Enright's literary criticism has been a hearty, no-nonsense humanism. He has praised Leavis and **T.S. Eliot** for the way in which their innovative critical work still retained a sense of 'literature's *human* meaningfulness', whereas with newer criticism, 'where reading had been too, too easy, now it became impossible'. It is entirely characteristic of Enright to praise the 'loving wholeheartedness' of Goethe, since the same quality animates his own best work. Admirers of his criticism have been disappointed that he has never attempted a really major critical work, contenting himself with book reviews and essays. Some of these have been gathered together in his best-known collection, *Man Is an Onion*. He has also written on irony in *The Alluring Problem*, an untheoretical, engaging and anecdotal book.

Enright's poetry has an unpretentious, almost casually conversational quality which has caused some critics to see it as a species of light verse. He published his first volume, *Season Ticket*, in Alexandria in 1948, and much of his work has been written from the compassionate and witty perspective of a not-so-innocent abroad, often addressing himself to social injustice and squalor rather than the tourist exoticism of countries such as Japan. As a poet he has been associated with the anti-modernism of the Movement.

Enright has also written four novels, all concerned with British academics in foreign countries; the first of them, *Academic Year*, was received as 'an Alexandrian *Lucky Jim*'. His amiable, pipe-smoking persona and his wide-ranging erudition combine with a dash of the entertainer to make him a perfect anthologist, and he has edited *The Oxford Book of Death*, *The Oxford Book of Friendship*, and *The Faber Book of Fevers and Frets*. He has also been involved in a new translation of Proust, paying characteristically down-to-earth special attention to colloquialism and swearing in *A la recherche du temps perdu*.

Poetry

Seaon Ticket 1948; *The Laughing Hyena* 1953; *Bread Rather Than Blossoms* 1956; *Some Men Are*

Brothers 1960; *Addictions* 1962; *The Old Adam* 1965; *Unlawful Assembly* 1968; *Selected Poems* 1969; *Daughters of Earth* 1972; *Foreign Devils* 1972; *The Terrible Shears* 1973; *Sad Ires* 1975; *Penguin Modern Poets 26* [with D. Abse, M. Longley] 1978; *Paradise Illustrated* 1978; *A Faust Book* 1979; *Collected Poems* 1981 [rev. 1987]; *Instant Chronicles* 1985; *Selected Poems* 1990; *The Way of the Cat* 1992; *Old Men and Comets* 1993

Fiction

Academic Year 1955; *Heaven Knows Where* 1957; *Insufficient Poppy* 1960; *Figures of Speech* 1965

For Children

Rhyme Times Rhyme 1974; *The Joke Shop* 1976; *Wild Ghost Chase* 1978; *Beyond Land's End* 1979

Non-Fiction

The World of Dew: Aspects of Living Japan 1955; *The Apothecary's Shop* 1957; *Conspirators and Poets* 1966; *Memoirs of a Mendicant Professor* (a) 1969; *Shakespeare and the Students* 1970; *Man Is an Onion* 1972; *The Alluring Problem* 1986; *Fields of Vision: Essays on Literature, Language and Television* 1988

Translations

Colette Portal *Nature Alive* [with M. Enright] 1979; Proust: *In Search of Lost Time* [with C.K. Scott-Moncrieff, T. Kilmartin] 6 vols 1992

Edited

The Poetry of Living Japan 1952; *English Critical Texts: From the Sixteenth to the Twentieth Century* [with E. de Chickera] 1962; *A Choice of Milton's Verse* 1975; Samuel Johnson *The History of Rasselas, Prince of Abyssinia* 1976; *The Oxford Book of Contemporary Verse, 1945–1980* 1980; *A Mania for Sentences* 1983; *The Oxford Book of Death* 1983; *Fair of Speech: the Uses of Euphemism* 1985; *The Faber Book of Fevers and Frets* 1989; *The Oxford Book of Friendship* 1991; Proust *In Search of Lost Time* 1992; *The Oxford Book of the Supernatural* 1994

Collections

Under the Circumstances 1991

Louise (Karen) ERDRICH 1954–

Erdrich was born in Little Falls, Minnesota, the eldest of seven children, and the daughter of two teachers in the Bureau of Indian Affairs. Her mother was largely of Chippewa Indian descent, while her father was German-American. She grew up largely near the Turtle Mountain Chippewa Reservation, North Dakota, where her maternal grandparents lived, and attended boarding-school in Wapheton, North Dakota. It was not until she entered Dartmouth College, and enrolled on the Native American Studies Programme (run by her future husband, Michael Dorris, himself largely of Modoc Indian ancestry) that she began to think about the indigenous heritage which has later become one of the chief inspirations for her writing. After graduating in 1976, she worked at a wide variety of jobs (waitress, teacher, sugar-beet hoer, lifeguard, flag-signaller on a construction site) before being awarded her MA in creative writing from Johns Hopkins in 1979.

She had already begun to publish short stories and poems in little magazines, and first came to public notice by winning several important awards in 1982. She had married Dorris, himself a writer (the author of a novel, *A Yellow Raft on Blue Water*), in 1981, and became adoptive mother of the three children he had adopted as a single parent, one of them suffering from foetal alcohol syndrome (Dorris has written about this in the book *A Broken Cord*). Dorris and Erdrich also have three children of their own. In 1984 Erdrich became widely successful with publication of her first novel, *Love Medicine*, the first of a Northern Plains tetralogy, dealing with a group of linked Indian families throughout the century. The series, of which the second volume, *The Beet Queen*, is perhaps the most successful, has received high praise for its complex grappling with issues, although some censure it for insufficiently realised characters and a tendency towards rhetorical over-inflation in the writing. The other novels in the tetralogy are *Tracks* and *The Bingo Palace*. Erdrich has also published two volumes of poetry. In 1991 the first full collaboration between Dorris and Erdrich, *Crown of Columbus*, a novel about an American campus couple who discover a lost diary of Christopher Columbus, was published in good time for the quincentenary. Dorris and Erdrich live near Dartmouth, where he continues his teaching activities, while she has also taught in various colleges.

Fiction

Love Medicine 1984; *The Beet Queen* 1986; *Tracks* 1988; *Crown of Columbus* [with M. Dorris] 1991; *The Bingo Palace* 1994

Poetry

Jacklight 1984

Caradoc EVANS 1878–1945

The son of an auctioneer, Evans was brought up in the Cardiganshire village of Rhydlewis, leaving it – and school – at fourteen for the draper's Jones Brothers in Carmarthen, which was owned by his uncle; his sister Mary was already in charge of the millinery department. After some years he moved to James Howell in Cardiff. By 1899 he had reached London, working first in Kentish Town, then at Wallis of Holborn and Whiteley of Bayswater. It was at the Working Man's College in Camden Town that he acquired his insight into literature which enabled him to get into journalism. He sub-edited for the *Daily Mirror* and joined *T.P.'s Weekly* under T.P. O'Connor; by 1927 he was effectively the editor. He became one of the most colourful Fleet Street personalities of his time. Thomas Burke wrote of him: 'He goes about with spurts and dashes, bursting into places and bursting out before you knew he had been there. He talks in cascades, words tumbling over each, precisely the opposite to the caustic manner of his work'. It all ended in 1930 when he left his wife for Marguerite Barcińsky, who wrote romantic novels under the pseudonyms Countess Barcyńska and Oliver Sandys.

Evans had burst upon the literary scene during the First World War with his first collection of stories, *My People*, attacking the establishment in rural Wales. His fellow countrymen hated him for it – later, jeering crowds of Welshmen were to wreck productions of his play *Taffy* – but Mrs Asquith invited him to lunch and the *Sunday Express* offered him a column. The Welsh hated him even more for his famous novel *Nothing to Pay*. Notable for its biblical echoes, it is the story of a miserly Welsh draper clawing his way eastward, as its author had, from Nonconformist Cardiganshire to Kentish Town. Again the English press was full of praise, but the *Western Mail* accused him of not depicting 'Wales as it is. His method is to eliminate space and time and to give, regardless of history and justice, a synthesis of everything he has heard or seen or imagined detrimental to his country ... he will receive his pieces of silver' – from the English. The Welsh reviewer went on to describe it as 'filth masquerading as truth', and filled with 'lewd imagery and frank obscenity'. Comparison with *Kipps* (1905) is inevitable, but Evans's satire is the more savage. Nevertheless, **H.G. Wells** praised the book: 'Caradoc Evans, like myself, has been a draper, and the scene he draws of a draper's existence in the meaner shops of London ... is, I know, true in all substantial particulars.' True to life or not, everything about Evans seems to have been larger than life. Returning to Wales, he spent his last years at Aberystwyth.

Fiction

My People (s) 1915; *Capel Sion* (s) 1916; *My Neighbours* (s) 1920; *The Earth Gives All and Takes All* (s) 1942; *Pilgrim in a Foreign Land* (s) 1942; *Nothing to Pay* 1930; *This Way to Heaven* 1934; *Wasps* 1935; *Mother's Marvel* 1949

Plays

Taffy 1923 [np]

Gavin (Buchanan) EWART 1916–

The son of a surgeon, Ewart was born in London and educated at Wellington, along with Giles and Esmond Romilly, to whose anti-public-school magazine, *Out of Bounds*, he contributed his schoolboy verse. One poem, 'The Fourth of May', was considered such a libel upon the system that the headmaster informed Ewart he should not think of visiting his old school for some time. (Ewart was later to make some amends for this early rebelliousness by editing a fine anthology of public-school songs.) He continued to write verse while an undergraduate at Christ's College, Cambridge, publishing *Poems and Songs* with the celebrated Fortune Press in 1939. Ewart saw active war service in North Africa and Italy, during which he got out of the habit of writing poetry. He nevertheless wrote some very fine war poems, including one classic, 'When a Beau Goes In', but he was to give up publishing poetry for two decades until **Alan Ross**, editor of the *London Magazine*, encouraged him to take it up again in 1959, leading to his acclaimed reappearance in 1964 with the volume *Londoners*. In the meantime he worked as an assistant in the book review department of the British Council and spent twenty years as an advertising copywriter; a suitable job for someone with such an extreme and ingenious facility with language.

T.S. Eliot, **Ezra Pound** and **W.H. Auden** were all influences on the early Ewart, and he produced a characteristically sharp but loving parody of Auden's style in his poem 'Audenesque for an Initiation', published in *New Verse* in 1933. His poetic rebirth since the early 1960s has been associated with the writing of light and irreverent verse that shows a total mastery of form; he seems able to mimic almost any style of writing, and other writers, including Byron, W.S. Gilbert, **John Betjeman** and Auden. There is an element of keen social comment too, as in his 1974 poem 'The Gentle Sex', about a real-life incident in which a group of Belfast women beat another woman to death for political reasons: it gains much of its ironic effect from being written in the metrical and stanzaic pattern of Gerard Manley Hopkins's poem about the death of five nuns, 'The Wreck of the Deutschland' (1875).

Ewart keeps a worldly eye on the eternal verities, chief among which is sex. He writes in

'Office Friendships' that: 'Sex suppressed will go berserk, / But it keeps us all alive. / It's a wonderful change from wives and work / And it ends at half past five.' His poem 'Eight Awful Animals' presents an erotic bestiary with creatures such as 'The Panteebra', 'The Stuffalo' and 'The Spirokeet'. His work is often scabrous, like his poetical, calypso-styled wish that Mrs Thatcher should be stung on her clitoris by a bee.

Ewart has been modest about his work, as in the epitaph he composed for himself: 'He wrote some silly poems, and some of them are funny', or the cheeky opening of *Pleasures of the Flesh*: 'A small talent, like a small penis, / Should not be hidden lightly under a bushel, / But shine in use, or exhibitionism. / Otherwise how should one know it was there?' None the less he was a close runner-up for the Poet Laureateship in 1984 (he was beaten by **Ted Hughes**) and has many admirers. **Anthony Thwaite** wrote in a 1978 *Times Literary Supplement* that 'one of the few bright features' about recent poetry was that 'Gavin Ewart is growing old disgracefully'.

Poetry

Poems and Songs 1939; *Londoners* 1964; *Throwaway Lines* 1964; *Pleasures of the Flesh* 1966; *Two Children* 1966; *The Deceptive Grin of the Gravel Porters* 1968; *Twelve Apostles* 1970; *Folio* [with others] 1971; *The Gavin Ewart Show* 1971; *Alphabet Soup* 1972; *The Select Party* 1972; *Venus* 1972; *An Imaginary Love Affair* 1974; *Penguin Modern Poets 25* [with B.S. Johnson, Z. Ghose] 1974; *Be My Guest!* 1975; *No Fool Like an Old Fool* 1976; *A Question Partly Answered* 1976; *The First Eleven* 1977; *A Mordern Tower Reading 5* [with F. Adcock] 1977; *All My Little Ones* 1978; *Or Where a Young Penguin Lies Screaming* 1978; *The Collected Ewart 1933–1980* 1980; *More Little Ones* 1982; *The New Ewart Poems 1980–1982* 1982; *Capital Letters* 1983; *The Ewart Quarto* 1984; *Festival Nights* 1984; *A Cluster of Clerihews* 1985; *The Young Pobble's Guide to His Toes* 1985; *The Complete Little Ones* 1986; *Late Pickings* 1987; *Penultimate Poems* 1989; *Selections* 1989; *Collected Poems 1980–1991* 1991; *Like It or Not* 1992; *85 Poems* 1993

For Children

The Learned Hippopotamus 1986; *Caterpillar Stew* 1990

Plays

Tobermory [from Saki's story; with J. Gardner] 1977 [np]

Edited

Forty Years On: an Anthology of School Songs 1964; *The Batsford Book of Children's Verse* 1976; *New Poems 1977–78* 1977; *The Batsford Book of Light Verse for Children* 1978; *The Penguin Book of Light Verse* 1980; *Other People's Clerihews* 1983

Frederick (Earl) EXLEY 1929–

Exley grew up in Watertown, New York, the son of a father whose memory has dominated his life and writing. Early Exley was an American football player who was much loved in their small town but whose seemingly charmed life was cut short by cancer while his son was still young. Exley was educated at Hobart College and the University of Southern California, beginning a career that he was to characterise in terms of abject failure to live up to his father's model. Although his occupations – editor, public-relations executive, college English teacher – could only be characterised as the mildest of underachievement by most people, to Exley they represented unbearable anonymity and 'life's hard fact of famelessness'. He drank heavily and was repeatedly hospitalised and subjected to electroconvulsive therapy and insulin coma therapy. The only compensation in his life was an almost religious devotion to the New York Giants football team, and his vicarious involvement with the career of running-back Frank Gifford.

He literally turned his failure into success by writing an autobiographical novel about it, *A Fan's Notes* (1968). The title comes from the revelation 'that it was my destiny – unlike that of my father, whose fate it was to hear the roar of the crowd – to sit in the stands with most men and acclaim others. It was my fate … to be a fan.' The book made Exley's reputation and it won the William Faulkner Award and the Rosenthal Award from the National Institute of Arts and Letters, as well as being nominated for a National Book Award. It also won him the real-life friendship of Frank Gifford, who contacted him after reading it.

Exley became a full-time writer and has taught creative writing at the University of Iowa. *A Fan's Notes* has become the keystone of an autobiographical trilogy, being followed by *Pages from a Cold Island*, which charts Exley's experiences in a literary world which, in the wake of the death of **Edmund Wilson**, is depicted as a moral vacuum, and *Last Notes from Home*, an account of visiting Hawaii to see his dying brother. The latter (which was partly serialised in *Rolling Stone*) is much concerned with booze, buddies and broads and was felt by most critics to be a bit of a mess. He has had two broken marriages and his first wife is portrayed in *A Fan's Notes* under the name of 'Patience'.

Critical opinion is divided over Exley. His detractors find his subject-matter (himself) rather limited and his writing overblown, while his admirers praise his style, his self-lacerating honesty and above all his humour: his indictment of American life often produces bitterly funny writing. After the cult status of his first book the others have been less well received, with some critics complaining of boorish machismo and a lack of form and theme. Perhaps as a reaction to this, and perhaps to protect those he writes about, he has increasingly insisted that the 'Exley' persona of the books is fictional.

Fiction

A Fan's Notes 1968; *Pages From a Cold Island* 1975; *Last Notes From Home* 1988

Ruth (Esther) FAINLIGHT 1931–

Fainlight was born in New York City to a complex heritage: her parents were both Jewish; her father British, born in London; her mother born in what was then the Austro-Hungarian empire and later became the Soviet Union. Fainlight herself grew up partly in Britain and partly in the USA, strongly attached to her younger brother, Harry, who also became a poet, but lived a life tormented by drug addiction, breakdowns and the repression of his homosexuality, and who died in a lonely Welsh cottage in 1982. (Four years later, she edited a volume of his selected poems.)

Her own passage through adult life has been a more tranquil one, but began with a break for freedom. After school, she attended Birmingham and Brighton Colleges of Arts and Crafts for two years, and made an early marriage, but in 1951 met fellow writer **Alan Sillitoe**, with whom she decided to live on the Continent. They spent the greater part of the 1950s developing their writing in France, Majorca and mainland Spain, and were first published at around the same time, Fainlight's first volume of poems, *A Forecast, a Fable*, appearing in 1958. She and Sillitoe married in 1959; they have a son and an adopted daughter. In the wake of Sillitoe's great success as a novelist, the couple lived in London during the early 1960s, and during this period Fainlight became a close friend of the poet **Sylvia Plath**. However, during the last, isolated, period of Plath's life, before her suicide in 1963, the Sillitoes were in Morocco, on the first of several foreign sojourns punctuating their married life, and thus could not offer her the support that might have saved her.

Fainlight's individuality as a poet developed fairly slowly, and it is perhaps with the volume *Cages* that she finds her mature voice. Since then, she has established herself as a direct and honest writer, whose work, metrically accomplished and strongly rooted in personal experience and her own unsentimental version of feminism, is perhaps slightly unglamorous but always commands respect. *This Time of Year* (1993) is the latest of eight collections. She has also published two volumes of short stories, has broadened her work to the writing of libretti, and with her husband has translated Lope de Vega's play *Fuenteovejuna* under the title *All Citizens Are Soldiers*. Her Iberian interests extend also to Portugal, and she is the chief English translator of the Portuguese poet and socialist activist Sofia de Melo Breyner Andresen.

Poetry

A Forecast, A Fable 1958; *Cages* 1967; *18 Poems for 1966* 1967; *To See the Matter Clearly and Other Poems* 1968; *Poems* [with T. Hughes, A. Sillitoe] 1971; *The Region's Violence* 1973; *21 Poems* 1973; *Another Full Moon* 1976; *Two Fire Poems* 1977; *The Function of Tears* 1979; *Sibyls and Others* 1980; *Two Wind Poems* 1980; *Climates* 1983; *Fifteen to Infinity* 1983; *Selected Poems* 1987; *Three Poems* 1988; *The Knot* 1990; *On the Coast Road* 1991; *This Time of Year* 1993

Plays

All Citizens Are Soldiers [from L.F. de Vega's play; with A. Sillitoe] 1967 [pub. 1969]

Fiction

Daylife and Nightlife (s) 1971; *Penguin Modern Stories 9* (s) [with others] 1971; *Dr Clock's Last Care* 1994

Translations

Sofia de Melo Breyner Andresen: *Coral* 1982, *Navigacions* 1983

Edited

Harry Fainlight *Selected Poems* 1986

John Meade FALKNER 1858–1932

The son of a curate, Falkner was born at Manningford Bruce, Wiltshire, but the family soon moved to Dorchester. His ten years there, including time at a private school run by the Revd Moule (inventor of a celebrated earth-closet) and at Dorchester Grammar School, made an indelible impression upon him; equally important was the tutoring in Greek and Latin he received, aged five, from Edward Stone, a master at Eton.

The family moved to Weymouth, where his well-meaning but improvident father took up a curacy under the tyrannical evangelist Talbot Greaves. An outbreak of typhoid fever due to drinking contaminated water – from a jug which appeared to have in it a coil of black string, later found to be a rat's tail – laid the children low. It subsequently killed their mother who, mindful of a son earlier lost to croup, exhausted herself in nursing them.

Penury struck (her money had kept them afloat), but Falkner was never resentful. As a clergyman's son, he won a place at Marlborough, where his enthusiasm for architecture, crucial to his fiction and poetry, led to the lifelong habit of touring churches. An adept scholar, winning poetry prizes and given to practical jokes, he was popular, but none the less was expelled early for allegedly making a homosexual advance (most likely, he was merely welcoming back a friend from the sanatorium).

He read history at Hertford College, Oxford, but left without achieving academic distinction, preoccupied as he was in gaining a wider education. After a few stints at schoolmastering, chance took him to Jesmond, outside Newcastle, and the congenial home of Sir Andrew Noble, a partner in the shipbuilding and armaments firm of Armstrong's. His son John was at Eton but required tutoring.

Falkner had landed on his feet. He stayed with the family, later joining first the firm and then the board. Such was the erudite teasing charm of this gangling man that he travelled the world in this capacity, a most unlikely salesman for bellicose products (sultans and emperors awarded him medals). His interest in matters ecclesiastical had grown in tandem with his business career, and he was able to peruse the world's libraries to satisfy it.

Fittingly, his first book was an anonymous *Handbook for Travellers in Oxfordshire* (1894) in Murray's well-known series, followed two years later by the first of three masterpieces. Each is markedly different from the others but shares a nerve-edge control of narrative and a preoccupation with human beings at their mental and physical limits, shot through with topographical exactitude, humour and effortlessly wide-ranging allusion. Just as *The Lost Stradivarius* can be enjoyed as an Oxford ghost-story, so *Moonfleet* – a staple of every schoolroom – is a popular tale of smuggling on the Dorset coast; but it is in his last novel, *The Nebuly Coat*, that all his interests come together. This novel, an account of all that seethes – past and present – beneath the sleepy surface of a small coastal town during the renovation of its minster, was successful in its year of publication; one of the century's greatest novels, it has since had an underground reputation, much admired by the diverse likes of **E.M. Forster**, **Philip Larkin** and **Dorothy L. Sayers**.

Falkner said that the manuscript of a fourth novel was stolen on a train, perhaps mistaken for secrets by an enemy agent, but one might suspect that this was of a piece with a man who called himself 'the eternal romantic', both of the world and not, cultivating a 'medieval little atmosphere' (his letters, often written on trains, are in an ornate script), indeed – like his friend **Thomas Hardy** – thriving on mystery: many did not realise that he was married, and perhaps neither did he. Riches (he left £250,000) meant that he could cultivate fine tastes while loyalty to the Nobles (and alarm over possible business revelations) kept him at Armstrong's when part of him yearned to break free. Depression and ailments dogged him (he described an operation for piles in close, almost Joycean, detail), a situation worsened by the First World War and his rise to chairman of what was then the largest firm in the world. His relief on resigning in 1921 is palpable. By then he was too exhausted for more than slight work, although his letters, as yet unpublished, comprise a fourth masterpiece.

Fiction

The Lost Stradivarius 1896; *Moonfleet* 1898; *The Nebuly Coat* 1903

Non-Fiction

A Handbook for Travellers in Oxfordshire 1894; *A History of Oxfordshire* 1899; *A Handbook for Travellers in Berkshire* 1902; *Bath in History and Social Tradition* 1918

Poetry

Poems c. 1930/1933

U(rsula) A(skham) FANTHORPE 1929–

Fanthorpe was born in Kent 'of middle-class but honest parents and correctly educated at Oxford' (St Anne's College). She went on to take a Diploma in Education at the University of London, and between 1954 and 1970 taught English at Cheltenham Ladies' College, spending the last eight years as head of department. She says that she 'became a middle-aged drop-out in order to write', and had a number of temporary jobs. Between 1974 and 1983 she worked full time as a clerk in a hospital in Bristol, an experience recorded in many of the poems in her first collection, the aptly-titled *Side Effects*, which was published in 1978. In these poems, many of which she says were written during her lunch-hours, she describes unblinkingly the effects of illness and accident. Other poems are more humorous, taking sideswipes at the egregious poetaster Patience Strong: 'No doubt such rubbish sells / She must be feathering her inglenook'. 'Not My Best Side' was inspired by Uccello's painting of St George and the dragon, which hangs in the National Gallery: each of the three figures – dragon, maiden and saint – are given their say, and it is St George (who has 'diplomas in Dragon Management and Virgin Reclamation') who comes off worst. **Charles**

Causley praised *Side Effects*, describing Fanthorpe as 'a new and welcome voice in English poetry; clear, distinctive and remarkably assured'. She is also witty and accessible, and the book was – in terms of a first volume of poetry – a commercial success, reprinting several times.

Causley was one of the judges who awarded Fanthorpe third prize in the 1980 *Observer*/Arvon/*South Bank Show* poetry competition. The others were **Seamus Heaney**, **Ted Hughes** and **Philip Larkin**, and she has always attracted the praise of her fellow poets. Of *A Watching Brief*, **Roy Fuller** noted: 'Some of these poems are not only the fruit of observation but also about the value of observation – observation so accurate that it becomes an intellectual and emotional exercise.' Other notable volumes include *Standing To*, *Voices Off* and a book of *Selected Poems*, which was published by Penguin in 1986.

Other awards include two Society of Authors travelling scholarships in 1983 and a Hawthornden Scholarship in 1987, after which she became the Northern Arts Literary Fellow, based at the universities of Durham and Newcastle-upon-Tyne. She has also held an Arts Council Writer's fellowship at St Martin's College, Lancaster. Fanthorpe is one of the three contributors to *The Crystal Zoo*, a volume of poems for children.

Poetry

Side Effects 1978; *Four Dogs* 1980; *Four Poems* 1981; *Standing To* 1982; *Voices Off* 1984; *Selected Poems* 1986; *A Watching Brief* 1987; *Neck-Verse* 1992

For Children

The Crystal Zoo [with L.J. Anderson, J. Cotton] 1985

J(ames) G(ordon) FARRELL 1935–1979

Born in Liverpool, the son of an accountant, and of Anglo-Irish stock on both sides of his family, Farrell attended Rossall School and Brasenose College, Oxford, where he took a third in French and Spanish. During his first term, after playing a rugby game for his college, he contracted polio, a disease which almost killed him and which left one arm partially paralysed. He was to claim that it was this experience which turned him from a typical public-school hearty into the sensitive and thoughtful man who became a novelist. (Friends thought this an over-simplication.)

After graduating, he taught for a while, unhappily, in France, while working on his first novel, *A Man from Elsewhere* (1963), which is set in that country. Returning to London, he determined on the life of a writer and, doing little reviewing or literary hackwork, lived at first in a greenhouse, later in a single room in Notting Hill, and for the last nine years of his life in a tiny two-roomed flat in Knightsbridge, where he entertained the many writers and intellectuals who were his friends.

His first three novels are brief and deal with contemporary life: his literary reputation rests rather on the three later and longer novels: *Troubles*, *The Siege of Krishnapur*, which won the Booker Prize in 1973, and *The Singapore Grip*. Sometimes referred to as the 'Empire Trilogy', they are all historical novels illuminating various aspects of the British empire, and marked by an absurdist, ironic, oblique standpoint and technique which made many critics consider Farrell one of the most important novelists of the 1970s. A further, unfinished, novel, *The Hill Station*, was published two years after his death. Farrell remained single, and spent much of his life travelling widely, in America (where he held a Harkness Fellowship during the 1960s), and in India and the Far East, where he researched his imperial novels. Four months before his death he moved from London to a cottage beside Bantry Bay in Ireland: here he was drowned while fishing off a rock.

Fiction

A Man from Elsewhere 1963; *The Lung* 1965; *A Girl in the Head* 1967; *Troubles* 1970; *The Siege of Krishnapur* 1973; *The Singapore Grip* 1978; *The Hill Station, and An Indian Diary* [ed. J. Spurling] 1981

James T(homas) FARRELL 1904–1979

Farrell was born in Chicago, Illinois. Because of his parents' poverty he was sent, when he was three, to live with his grandparents in the middle-class South Side of the city – a neighbourhood that was the setting for much of his fiction. He attended a series of parochial schools and graduated in 1923 from St Cyril High. Subsequently, he studied for six months at De Paul University and for three years at the University of Chicago; but his formal education was hampered by his need to help support his family. He held jobs wrapping shoes in a department store, clerking for a cigar company, selling advertising and assisting in an undertaker's parlour.

His first publication – a short story, 'Slob' – appeared in 1929, and the following year he turned to writing full-time. In 1931 he married.

(He was to divorce his wife in 1935 and to remarry her in 1955: a second marriage, to someone else, intervened.) The couple spent ten months in Paris, where they were briefly friends of **Ezra Pound**. On his return to the USA, Farrell moved to New York. He was awarded a Guggenheim Fellowship in 1936 and a Book of the Month Club prize (worth $2,500) in 1937 for the *Studs Lonigan* trilogy. In 1941 he was made a member of the American Academy. From 1964 to 1965 he was Adjunct Professor, St Peter's College, New Jersey. Other appointments included being writer-in-residence at Richmond College, Virginia, and at Glassboro College, New Jersey. He received doctorates from the Universities of Miami and Oxford, Ohio, and from Columbia College, Chicago. Shortly before his death, he was honoured with the Emerson-Thoreau Medal.

Brought up a Catholic, Farrell seems to have grown disillusioned with the faith around 1922. Although a supporter of the Communist Party from 1932 to 1935, he opposed it thereafter, but remained a socialist all his adult life. His *oeuvre* was monumental in design, conceived of as a single scheme, 'loosely integrated through the associations of the characters depicted, and the scenes and the environments recreated. The purpose [is] ... to recreate a sense of American life.' Simultaneously, Farrell set out to record his own 'psychological life cycle'. There is an autobiographical strain in most of his books, from Studs Lonigan's nigh-illiterate miseries to Bernard Carr's religious and political revisions. If this conveys a powerful sense of place and personality, it scarcely permits an escape from its author's experience. Indeed, in spite of his prolixity, Farrell's range is limited. Geographically, he confines himself to the urban East: racially, to the Irish. Motifs recur without development – notably, sickness, a lost sweetheart and the volatile bonding of father and son. The accumulated effect suggests little of the dynamism of US society: rather, it presents a *lumpenproletariat*'s grinding decline. Let down by the Catholic church, industrial capitalism and the Communist Party, Farrell's protagonists in the main go under. Those who survive – like Danny O'Neill, a dedicated writer and his author's closest *doppelgänger* – pay the price in ethical deracination and emotional isolation.

That Farrell's reputation is currently at a low ebb may be attributed to his style. In an era of experiment, he adhered to an often brutal naturalism, while the range of his subject-matter generates the impression that much might gainfully have been omitted. Nevertheless, the immensity and detail of his output has secured him a place in the tradition of North American sociological novelists.

A prodigious worker, Farrell once said: 'Just about my only hobby is baseball. I like to follow baseball from year to year, to see games, and to keep track of records such as batting and pitching averages.'

Fiction

Studs Lonigan 1936: *Young Lonigan* 1932, *The Young Manhood of Studs Lonigan* 1934, *Judgement Day* 1935; *Gas-House McGinty* 1933; *Calico Shoes and Other Stories* (s) 1934; *Guillotine Party and Other Stories* (s) 1935; *Danny O'Neill* series: *A World I Never Made* 1936, *No Star Is Lost* 1938, *Father and Son* 1940 [as a Father and His Son 1943], *My Days of Anger* 1943, *Can All This Grandeur Perish? and Other Stories* (s) 1937; *Tommy Gallagher's Crusade* (s) 1939; *Ellen Rogers* 1941; *$1,000 a Week and Other Stories* (s) 1942; *To Whom It May Concern and Other Stories* (s) 1944; *The Face of Time* 1953; *Bernard Carr* 1960: *Bernard Clare* 1946, *The Road Between* 1949, *Yet Other Waters* 1952; *When Boyhood Dreams Come True* (s) 1946; *The Life Adventurous and Other Stories* (s) 1947; *A Misunderstanding* (s) 1949; *An American Dream Girl* (s) 1950; *This Man and This Woman* 1951; *French Girls Are Vicious and Other Stories* (s) 1955; *A Dangerous Woman and Other Stories* (s) 1957; *Saturday Night and Other Stories* (s) 1958; *The Girls at the Sphinx* (s) 1959; *Looking 'em Over* (s) 1960; *Boarding House Blues* 1961; *Side Street and Other Stories* (s) 1961; *A Universe of Time* [unfinished]: *The Silence of History* 1963, *What Time Collects* 1964, *When Time Was Born* 1966, *Lonely for the Future* 1966, *A Brand New Life* 1968, *Judith* 1969, *Invisible Swords* 1971, *The Dunne Family* 1976, *The Death of Nora Ryan* 1978; *Sound of a City* (s) 1962; *New Year's Eve/1929* 1967; *Childhood Is Not Forever and Other Stories* (s) 1969; *Sam Holman* 1983

Poetry

The Collected Poems of James T. Farrell 1965

Non-Fiction

A Note on Literary Criticism 1936; *The League of Frightened Philistines and Other Papers* 1945; *The Fate of Writing in America* 1946; *Literature and Morality* 1947; *The Name Is Fogarty: Private Papers on Public Matters* [as 'Jonathan Titlescu Fogarty, Esq.'] 1950; *Poet of the People: an Evaluation of James Whitcomb Riley* [with H. Gregory, J.C. Nolan] 1951; *Reflections at Fifty and Other Essays* 1954; *My Baseball Diary* 1957; *It Has Come to Pass* 1958; *Dialogue on John Dewey* [with others] 1959; *Literary Essays 1954–1974* [ed. J.A. Robbins] 1976; *On Irish Themes* [ed. D. Flynn] 1982; *Hearing Out Farrell* [ed. D. Phelps] 1985

Edited

H.L. Mencken *Prejudices: a Selection* 1958; *A Dreiser Reader* 1962

M.J. FARRELL

see Molly KEANE

William (Cuthbert) FAULKNER 1897–1962

Faulkner was born in New Albany, Mississippi, into a long-established Southern family of Scottish extraction, and raised in nearby Oxford, where his father owned a livery stable and was business manager for the University of Mississippi. Faulkner was named after his great-grandfather, Colonel William Clark Falkner [sic], who fought in the Civil War and wrote a popular novel, *The White Rose of Memphis* (1880). After a desultory education during which he frequently played truant, he worked briefly in his grandfather's bank and discovered the poetry of Swinburne and **A.E. Housman**, which he sought to imitate. In 1918 he trained as a cadet with the Canadian Royal Air Force and was discharged (though in his cups in later years he was not above inventing a rather more glorious military career). In the same year he lost his fiancée, Estelle Oldham, to a rival; they eventually married in 1929 after her divorce and had two daughters, one of whom died in infancy. In 1919 he attended the University of Mississippi and began to publish poems in local magazines. Over the next few years he worked variously as a postmaster, bookstore clerk and scoutmaster (a post from which he was dismissed for drinking).

His first volume of verse, *The Marble Faun*, was published in 1924, the year in which he met the novelist **Sherwood Anderson**, who suggested he try his hand at fiction. The resulting novel, *Soldiers' Pay*, written in six weeks, tells of a wounded pilot's return to his Southern family. His third novel, *Sartoris*, dealt with a similar theme, and was the first to introduce the fictional Yoknapatawpha County, which provides the setting for all of his subsequent fiction.

In the late 1920s and early 1930s Faulkner produced his most technically innovative and perhaps enduring works, *The Sound and the Fury*, *As I Lay Dying* and *Light in August*. *Sanctuary* is a violent thriller apparently written with the sole purpose of making money, though it transcends the genre and further extends Faulkner's dark vision of human depravity. His subsequent novels and short stories deal with the troubled history of the South (especially with two families, the Compsons and the Sartorises) and explore themes of race, violent sexuality and religion.

In 1932 he travelled to Hollywood for the start of a lengthy career as a scriptwriter. In his compendious biography of Faulkner Joseph Blotner narrates the story that on Faulkner's first day as an MGM contract writer he announced that he had a good idea for a Mickey Mouse cartoon, and had to be informed of the distinction between the studios. Having gained his pilot's licence in 1933 he often flew his own plane back and forth between Hollywood and his farm in Mississippi, and enjoyed twenty years working at Universal, 20th Century Fox and Warner Bros on films such as *To Have and Have Not* (1945), *The Big Sleep* (1946) and Howard Hawks's 1955 epic *Land of the Pharaohs* (for which he went to Cairo in 1954). In 1950 Faulkner travelled to Stockholm to receive the belatedly announced 1949 Nobel Prize for Literature and spoke in his acceptance speech of the imperative for the writer to 'create out of the materials of the human spirit something which did not exist before'. His legendary rivalry with another hard-drinking Nobel Prize-winner, **Ernest Hemingway**, is sometimes exaggerated, though on hearing of Hemingway's suicide he is reported to have quipped: 'I don't like a man who takes the short way home.' In his last years many felt that Faulkner was taking the long way home with the assistance of the bottle. He eventually died of a heart attack after falling from his horse.

Fiction

Soldiers' Pay 1926; *Mosquitoes* 1927; *Sartoris* 1929; *The Sound and the Fury* 1929; *As I Lay Dying* 1930; *Idyll in the Desert* 1931; *Sanctuary* 1931; *These 13* (s) 1931; *Light in August* 1932; *Miss Zilphia Gant* 1932; *Doctor Martino and Other Stories* (s) 1934; *Pylon* 1935; *Absalom, Absalom!* 1936 [rev. 1964]; *The Unvanquished* 1938; *The Wild Palms* 1939; *The Hamlet* 1940; *Go Down, Moses* 1942; *Intruder in the Dust* 1948; *Knight's Gambit* 1949; *The Collected Stories of William Faulkner* (s) 1950; *Notes on a Horsethief* (s) 1951; *The Penguin Collected Stories of William Faulkner* (s) 1951; *Requiem for a Nun* 1951; *Mirrors of Chartres Street* (s) 1953; *A Fable* 1954; *Big Woods* (s) 1955; *The Town* 1957; *Uncle Willy and Other Stories* (s) 1958; *The Mansion* 1960; *The Reivers* 1962; *Barn Burning and Other Stories* (s) 1971; *Uncollected Stories of William Faulkner* (s) [ed. J. Blotner] 1979

Poetry

The Marble Faun 1924; *Salmagundi* 1932; *A Green Bough* 1933; *Mississippi Poems* 1979; *Vision in Spring* [ed. J.L. Sensibar] 1984

Non-Fiction

New Orleans Sketches 1959; *William Faulkner: Early Prose and Poetry* [ed. C. Collins] 1962; *The Faulkner–Cowley File: Letters and Memories, 1944–1962* [ed. M. Cowley] 1966; *Essays, Speeches and Public Letters* [ed. J.B. Meriwether] 1967;

Selected Letters of William Faulkner [ed. J. Blotner] 1977

Plays

Requiem for a Nun 1959

Collections

The Portable Faulkner [ed. M. Cowley] 1946 [rev. 1966, 1976]; *The Faulkner Reader* 1954

Biography

William Faulkner by Joseph Blotner 2 vols 1974 [rev. 1984]; *William Faulkner: American Writer* by Frederick R. Karl 1989

Elaine FEINSTEIN 1930–

Feinstein originated from a Russian-Jewish family who emigrated to England at the beginning of the twentieth century. Born in Bootle, she went to Wyggeston Grammar School in Leicester, then studied English at Cambridge in the 1950s, and afterwards read law, living permanently in Cambridge since her marriage in 1956.

She began writing poetry in the 1960s which from the first has shown a strong awareness of her family's Russian origins. She was briefly a member of the Cambridge group of poets, which included **J.H. Prynne**, but left, feeling that their concern with Englishness in poetry clashed with her own, more international, cluster of influences. These included Marina Tsvetayeva, whose work she translated in the early 1970s and whose biography she wrote in 1987, and the Black Mountain poets in America, particularly **Charles Olson**, with whom she maintained an important correspondence.

Her own poetry exhibits many of the features of Objectivist poetry: the economy of line and the mixing of narrative and lyric modes are important techniques in her work. Her translations of other poets led her to explore more traditional verse-forms in the 1970s, while *The Feast of Eurydice* reworks one of the most traditional stories of all, the Orpheus myth.

As well as her novels, which often concentrate upon the experience of women, she has written a life of the great blues singer Bessie Smith, and a play based upon the life of Marie Stopes. Feinstein's talents are dispersed between prose, poetry and translation while her poetic influences are international in scope. These may be the reasons why she has yet to reach a wide cross-section of an English readership sometimes parochial in its tastes.

Fiction

The Circle 1970; *The Glass Alembic* 1973 [as *The Crystal Garden* 1974]; *The Amberstone Exit* 1972; *Matters of Chance* (s) 1972; *Children of the Rose* 1975; *The Ecstasy of Dr Miriam Garner* 1976; *The Shadow Master* 1978; *The Silent Areas* (s) 1980; *The Survivors* 1982; *The Border* 1984; *Mother's Girl* 1988; *All You Need* 1989; *Dreamers* 1994

Poetry

In a Green Eye 1966; *The Magic Apple Tree* 1971; *At the Edge* 1972; *The Celebrants and Other Poems* 1973; *Some Unease and Angels* 1977; *The Feast of Eurydice* 1980; *Badlands* 1986; *City Music* 1990; *Selected Poems* 1994

Translations

Marina Tsvetayeva *Selected Poems* [with S. Franklin, A. Livingstone and others] 1971 [rev. 1981]; *Three Russian Poets: Margarita Aliger, Yunna Moritz, Bella Akhmadulina* 1979; Nika Turbina *First Draft* [with A.W. Bouis] 1988

Plays

Breath (tv) 1975; *Echoes* (r) 1980; *Lunch* (tv) 1981; *A Late Spring* (r) 1982; *A Day Off* (r) [from S. Jameson] 1983; *Country Diary of an Edwardian Lady* [from E. Holden] 1984; *A Brave Face* (tv) 1985; *A Captive Lion* (r) 1985; *If I Ever Get on My Seat Again* (r) 1987; *The Brecht Project* (tv) 1990; *The Man in Her Life* (r) 1990; *A Passionate Woman: Marie Stopes* (r)1990

Non-Fiction

Bessie Smith 1985; *A Captive Lion: the Life of Marina Tsvetayeva* 1987; *Marina Tsvetayeva* 1989; *Loving Brecht* 1992; *Lawrence's Women: the Intimate Life of D.H.* 1993

Edited

Selected Poems of John Clare 1968; *New Stories 4* [with F. Weldon] 1979; *P.E.N. New Poetry II* 1988

James (Martin) FENTON 1949–

The son of a vicar, who eventually rose to become Honorary Canon Emeritus at Christ Church, Oxford, Fenton was born in Lincoln and educated at the Choristers' School, Durham, and at Repton. Between leaving school and going up to Magdalen College, Oxford, to read psychology and physiology, he spent six months studying in Florence. He began writing and publishing poetry while an undergraduate, winning the Newdigate Prize in 1968 for 'Our Western Furniture', a sequence of twenty-one sonnets about the opening up of Japan to Western trade. The sequence was broadcast on the BBC's Third Programme, and was published in book form by the Sycamore Press, started by his friend and contemporary, the poet **John Fuller**, as were the two poems, *Put Thou My Tears Into Thy Bottle*.

His first collection of poems, *Terminal Moraine*, was published in 1972, and won the Eric Gregory Award, the proceeds of which Fenton used to fund a trip to Cambodia and

Vietnam, where he reported on the Vietnam War for the *New Statesman*, the assistant literary editor of which he had been for a brief spell after leaving university. As a correspondent, he achieved fame when, as Saigon fell, he rode to the presidential palace on a North Vietnamese tank. His war reportage was later published along with other political writings in the collection *All the Wrong Places*. He returned to England, where he continued to write for the *New Statesman*, then spent some time as German correspondent for the *Guardian*. He started writing poetry again, drawing upon his experiences in Cambodia to produce the volume that made his reputation, *The Memory of War: Poems 1968–1982*.

He succeeded Bernard Levin as theatre critic of the *Sunday Times*, collecting his reviews in *You Were Marvellous*. Not everyone thought well of Fenton's tenure: 'Mr Fenton's subsequent abandonment of dramatic criticism to become the *Independent*'s correspondent in the Philippines was one of the more cheering developments in the theatre in the eighties,' commented **Alan Bennett**. Before this, in 1983, Fenton accompanied the naturalist and travel writer Redmond O'Hanlon through the jungles of Borneo to Mt Batu Tiban. O'Hanlon's hilarious and hair-raising account of their journey, *Into the Heart of Borneo* (1984), provides an affectionate and amusing portrait of Fenton. While there, Fenton read Hugo's *Les Misérables*, which he had been asked to adapt for a musical. Although he did not eventually provide a script, he was rewarded for his work by a tiny royalty, which – thanks to the subsequent worldwide success of the musical – made him a wealthy man.

Meanwhile, he was writing more travel pieces for *Granta*, spending much of his time abroad. He also served as lead reviewer on the books pages of *The Times*, translated Verdi's *Rigoletto* for Jonathan Miller's celebrated English National Opera production, and produced further volumes of poetry (including *Partingtime Hall*, a collaboration with John Fuller). By the late 1980s he was considered one of Britain's leading poets, frequently compared with **W.H. Auden**, whose wit, dexterity, playfulness and intellectual rigour he shares. In 1994, like Auden before him, he was elected Professor of Poetry at Oxford, and he now lives in a house with extensive grounds just outside the city, where he has created a fine garden. He continues to write for the *Independent* and has produced *Out of Danger*, a volume of poetry which is notable for a new lyricism in the love poems it includes along with poems on his more usual subjects of war and politics.

Poetry

Our Western Furniture 1968; *Put Thou My Tears into Thy Bottle* 1969; *Terminal Moraine* 1972; *A Vacant Possession* 1978; *Dead Soldiers* 1981; *A German Requiem* 1981; *The Memory of War: Poems 1968–1982* 1982; *Memory of War and Children in Exile: Poems 1968–83* 1984; *Partingtime Hall* [with J. Fuller] 1987; *Manila Envelope* 1989; *Out of Danger* 1993

Non-Fiction

You Were Marvellous 1983; *All the Wrong Places* 1988

Translations

Verdi: *Rigoletto* 1982, *Simon Boccanegra* 1985

Edited

The Original Michael Frayn 1983; *Cambodian Witness: the Autobiography of Someth May* 1986

Lawrence FERLINGHETTI 1919–

A popular American poet, influential avant-garde editor and political activist, Ferlinghetti was born in Yonkers, New York, the last of the three sons of Clemence Mendes-Monsanto, a French-Sephardic Jew, and her Italian immigrant husband Charles Ferling (as the name was adapted in America). His father died before Ferlinghetti was born and his mother suffered a nervous breakdown soon after the birth and was institutionalised. He was adopted first by a maternal aunt and taken to Strasbourg, France, then spent six months in an orphanage before being adopted, finally, by his 'parents emeritus', Presley and Anna Bisland, an elderly couple of New Yorkers. On graduating from the University of North Carolina with a degree in journalism, he enlisted and served in the US Navy throughout the war, notably at Normandy and at Nagasaki six weeks after the atom bomb had been dropped. After the war he continued his studies at Columbia University and the Sorbonne in Paris, where his dissertation (written in French) was entitled 'The City as a Symbol in Modern Poetry'. Here he also wrote a first version of the novel *Her* (eventually published in 1960) and a series of (unpublished) poems influenced by **Ezra Pound**'s *Cantos*.

In 1950, Ferlinghetti returned to the USA and married an English literature student, Selden Kirby-Smith, with whom he bought a waterfront studio in San Francisco, attracted by the city's cosmopolitan atmosphere and the flourishing artistic community. He painted, wrote poetry and contributed arts reviews to the *San Francisco Chronicle* and the *Arts Review*, and in 1953 he co-launched the City Lights Bookstore and publishing house. The enterprise became a focus for writers and dissident intellectuals and the launch of its Pocket Poets series in a cheap paperback edition heralded the so-called San Francisco Poetry Renaissance. First in the series was a collection of Ferlinghetti's own poems, *Pictures of the Gone World* (1955), and at this point he adopted his father's full Italian name, perhaps marking a stage in his search for the identity that had eluded him in his insecure childhood.

When the poet **Allen Ginsberg** and the writer **Jack Kerouac** arrived on the scene in 1955, the live poetry readings organised by Ferlinghetti became popular, public affairs and Ginsberg's reading of his new poem *Howl* became a seminal event in the mythology of the growing Beat movement, which Kerouac went on to fictionalise in his novel *On the Road* (1957). In 1957, *Howl* was published in the Pocket Poets series and Ferlinghetti was arrested and charged with obscenity. The legal action was the most celebrated case of censorship in the USA since the 1933 ban on **James Joyce**'s *Ulysses*. After a political furore, Ferlinghetti was cleared and he and Ginsberg found themselves national figures as spokesmen for a revolution in thinking as well as writing. After the trial he continued to co-ordinate the new wave of experimental writing and poetry from the City Lights Bookstore and to evolve his own recognisable poetic style: unpunctuated, spatially unjustified, but direct and accessible. His *A Coney Island of the Mind* was an immediate and popular success and remains his bestselling work to date.

Ferlinghetti travelled extensively in Latin America in the early 1960s. His growing belief in the benefits of humanitarian socialism and his outspoken attacks on US foreign policy in Cuba and later in Vietnam earned him the opprobrium of the political and literary establishments and the scrutiny of the FBI. Undaunted, he continued to champion the avant-garde and the radical in the *City Lights Journal* and the *Journal for the Protection of All Beings*. As the Beats gave way to the hippy movement of a younger generation, Ferlinghetti continued to push back the limits. He took LSD in 1967 and described his experiences in the prose 'document' *After the Cries of Birds*; published *Eagle Brief* (1970), the prison writings of Timothy Leary, the self-styled high priest of psychedelic culture, and *The First Third* (1978), a posthumous collection of the writing of the Beat legend, Neal Cassady. In 1967 he was arrested with the folk singer Joan Baez and imprisoned for nineteen days for demonstrating at the Oakland Army Induction Center against the military draft for the Vietnam War. Despite the strain of an emotional divorce and a custody battle over the two children, his energy and commitment to producing radical art continued unabated throughout the 1970s. He produced his largest collection of poems, *Open Eye, Open Heart* in 1973, launched the Poet's Theatre as a venue for mass poetry readings and worked in support of such diverse causes as the United Farmworkers of America, Greenpeace and the anti-nuclear campaign.

Ferlinghetti remains a prodigious and vital spokesman for activist and radical writing in America. As well as his own output, he has published work by **Paul Bowles**, **Malcolm Lowry**, **Norman Mailer**, **William Carlos Williams** and Kenneth Patchen, and had the temerity to reject an early version of Kerouac's *On the Road* and **William Burroughs**' *The Naked Lunch* (1959). Once marginalised by critics owing to the experimental nature of his work, his leftist political stance and perhaps his popular appeal, he is now being re-evaluated as a major American poet.

Poetry

Pictures of the Gone World 1955; *A Coney Island of the Mind* 1958; *Tentative Description of a Dinner Given to Promote the Impeachment of President Eisenhower* 1958; *Berlin* 1961; *One Thousand Fearful Words for Fidel Castro* 1961; *Starting from San Francisco* 1961 [rev. 1967]; *Penguin Modern Poets 5* [with G. Corso, A. Ginsberg] 1963; *Christ Climbed Down* 1965; *To Fuck is to Love Again* 1965; *Where is Vietnam?* 1965; *After the Cries of the Birds* 1967; *An Eye on the World* 1967; *Moscow in the Wilderness, Segovia in the Snow* 1967; *Repeat After Me* 1967; *Fuclock* 1968; *Reverie Smoking Grass* 1968; *Tyrannus Nix?* 1969 [rev. 1973]; *Back Roads to Far Towns After Basho* 1970; *Sometime During Eternity* 1970; *The World is a Beautiful Place* 1970; *Back Roads to Far Places* 1971; *The Illustrated Wilfred Funk* 1971; *Love is No Stone on the Moon* 1971; *A World Awash with Fascism and Fear* 1971; *Open Eye* [with *Open Head* by Allen Ginsberg] 1972; *Assassination Raga* **(tv)** 1973; *Constantly Risking Absurdity* 1973; *Open Eye, Open Heart* 1973; *Director of Alienation* 1975; *The Jack of Hearts* 1975; *Populist Manifesto* 1975; *Soon It Will be Night* 1975; *The Old Italians Dying* 1976; *Who Are We Now* 1976; *White on White* 1977; *Adieu à Charlot* 1978; *Northwest Ecolog* 1978; *The Sea and Ourselves at Cape Ann* 1979; *Landscapes of Living and Dying* 1979; *The Love Nut* 1979; *Mule Mountain Dreams* 1980; *A Trip to Italy and France* 1981; *The Populist Manifestos/An Interview with Jean-Jacques Lebel* 1981; *Endless Life* 1981; *Over All the Obscene Boundaries* 1984; *Seven Days in Nicaragua Libre* 1984; *Since Man Began to Eat Himself* 1986; *Love in the Days of Rage* 1988; *Wild Dreams of a New Beginning* 1988; *Cool Eye* [with A. Lykiard] 1993

Plays

The Alligation 1962 [pub. 1963]; *The Customs Collector in Baggy Pants* 1964 [pub. 1963]; *Motherlode* 1963; *The Nose of Sisyphus* 1965 [pub. 1963]; *Soldiers of No Country* 1965 [pub. 1963]; *Three Thousand Red Ants* 1963; *Victims of Amnesia* 1965 [pub. 1963]; *Routines* 1964

Fiction
Her 1960

Non-Fiction
Dear Ferlinghetti/Dear Jack: the Spicer–Ferlinghetti Correspondence 1964; *After the Cries of Birds* 1967; *The Mexican Night: Travel Journal* 1970; *A Political Pamphlet* 1975; *The Sea and Ourselves at Cape Ann* [with J. Morgan] 1979; *Literary San Francisco: a Pictorial History from Its Beginnings to the Present Day* [with N.J. Peters] 1980; *Leaves of Life: Fifty Drawings from the Model* 1984; *When I Look at Pictures* 1990

Edited
Beatitude Anthology 1960; *Journal for the Protection of All Beings* [with M. McClure, D. Meltzer] 2 vols 1961, 1969; *City Lights Journal* 4 vols 1963–78; *City Lights Anthology* 1974

Translations
Selections from Paroles by Jacques Prévert 1958; Yevegeny Yevtushenko *Flowers and Bullets* and *Freedom to Kill* [with A. Kahn] 1970; Karl Marx *Love Poems* [with R. Lettan] 1977

Biography
Ferlinghetti by Neeli Cherkovski 1979

Gabriel FIELDING (pseud. of Alan Gabriel Barnsley) 1916–1986

The pseudonym under which Barnsley wrote all his books except the first, was chosen because his mother, an unsuccessful playwright who dominated his childhood, was a descendant of the novelist Henry Fielding. His father was a 'sporting parson' at Hexham in the north of England who put Fielding through the conventional (in his case bruising) middle-class education at prep school, St Edward's, Oxford, and Trinity College, Dublin. He then studied medicine on his parents' advice at St George's Hospital, London. He married in 1943 and had five children (including a girl called Gabriel) as well as a ward. During the war he served in the Royal Army Medical Corps, and after it he set up in private practice in Maidstone (always being known as Dr Alan Barnsley), a practice which included attendance at Maidstone Prison.

His first volume, *The Frog Prince and Other Poems* (1952), was of verse, and his first novel, *Brotherly Love* (the first of an autobiographical trilogy concerning the character John Blaydon), was published in 1954. He really attracted notice, however, with his fifth novel, *The Birthday King*, a chilling and penetrating study of a rich Jewish family in Nazi Germany, and it is for this that he is known today, although high claims are also often made for his other novels and his poetry. He had steadily run down his medical practice to write, and in 1966 he accepted an offer to go out as writer-in-residence to Washington State University, Pullman. He proved so popular that this was converted to a full professorship in 1967 and he remained in America for the rest of his life, retiring at Pullman as Professor Emeritus in 1981. In 1968 he suffered a nervous breakdown, and after 1966 he published only two more novels (in 1979 and 1986) to add to his existing six, a fact which, with his exile, may help to account for his reputation being less than his talent deserved.

Fiction
Brotherly Love 1954; *In the Time of Greenbloom* 1956; *Eight Days* 1958; *Through Streets Broad and Narrow* 1960; *The Birthday King* 1962; *Gentlemen in Their Season* 1966; *New Queens for Old* (s) 1972; *Pretty Doll-Houses* 1979; *The Women of Guinea Lane* 1986

Poetry
The Frog Prince and Other Poems [as Alan Barnsley] 1952; *XXVIII Poems* 1955; *Songs without Music* 1979

Eva FIGES 1932–

From a well-to-do Jewish family, Figes was born Eva Unger in Berlin, but emigrated to England shortly before the war. In her autobiography, *Little Eden*, she describes how her family arrived in England penniless, and how she was sent at the age of eight to an eccentric boarding-school in Gloucestershire to escape the bombing. She was subsequently educated at Kingsbury Grammar School and Queen Mary College, London, where she read English. In spite of having lived most of her life, and been educated, in England, she says that her 'literary influences and affinities have always been European rather than English'. She sees herself as outside and in conflict with the literary establishment, which she describes as 'not so much hostile as irrelevant in its response to new writing'.

She married shortly after leaving university, but divorced nine years later and brought up her two children on her own. Her first novel, *Equinox* (1966), the study of a woman's mind through the year in which her marriage is coming painfully to an end, reflects this experience. Her second novel, *Winter Journey*, describes the experiences of an old man dying in a council house and won the *Guardian* Fiction Prize in 1967. She moved away from British settings with *Konek Landing*, a novel about post-war Europe, but returned to it in *The Seven Ages*, which presents English history from the Saxon period onwards from the viewpoint of women, a theme which recurs in *The Tree of Knowledge*, in which Milton's youngest daughter looks back in old age to the troubled England of her youth. Of her other novels, perhaps the best known is *Light*, which features another historical character, the painter Monet, and describes his life and that of his household at Giverny in what amounts to a ninety-one-page prose poem.

Her works of non-fiction include *Patriarchal Attitudes*, a study of women in society which has been described as 'a seminal text for the Women's Liberation Movement', and *Sex and Subterfuge*, a study of early female novelists, covering the period up to 1850. In the 1960s and early 1970s she translated numerous novels from the German as well as George Sand's *La Petite Fadette* from the French. She has also written books for children and several radio plays.

Fiction

Equinox 1966; *Winter Journey* 1967; *Konek Landing* 1969; *B* 1972; *Days* 1974; *Nelly's Version* 1977; *Light* 1981; *Waking* 1981; *The Seven Ages* 1986; *Ghosts* 1988; *The Tree of Knowledge* 1990; *The Tenancy* 1993

Plays

Time Regained (r) 1980; *Days* (r) [from her novel] 1981; *Dialogue Between Friends* (r) 1982; *Punch-Flame and Pigeon-Breast* (r) 1983; *The True Tale of Margery Kempe* (s) 1985

Non-Fiction

Patriachal Attitudes 1970; *Tragedy and Social Evolution* 1976; *Little Eden* (a) 1978; *Sex And Subterfuge* 1982

For Children

The Musician of Bremen 1967; *The Banger* 1968; *Scribble Sam* 1971

Translations

Martin Walser *The Gadarene Club* 1960; Elisabeth Borchers *The Old Car* 1967; Bernhard Grzimek *He and I and the Elephant* 1967; Renate Rasp *A Family Failure* 1970; George Sand *Little Fadette* 1970; Manfred von Conta *The Deathbringer* 1971

Edited

Classic Choice 2 vols 1965, 1966; *Women Their World* 1980

Timothy FINDLEY 1930–

Findley has taken his mother's surname; his father's name was Gilmore. His grandfather had been president of Massey-Harris, the forerunner of Massey-Ferguson, a large Canadian agricultural machinery business, but the family had lost money and Findley grew up in Toronto, where his father was a stockbroker, in fairly modest circumstances. When he was three, his infant brother died, and he himself became so seriously ill it was thought he would not survive; the pattern was set for a succession of childhood illnesses which finally led to his formal education, in public and private schools, being cut short at the age of sixteen when he had glandular fever. Illness was one factor contributing to a lonely childhood, as was his father's absence at the war, and he spent much time speaking to no one else but the maid.

After school, he determined to act, a field in which he was soon successful, being a professional actor between 1950 and 1962 in Canada, Britain and the USA. Alec Guinness noticed him during the first Shakespeare season at Stratford, Ontario, and invited him to study at the Central School in London. While in London, in 1956, he wrote his first short story, prompted by the actress Ruth Gordon and by **Thornton Wilder**. For many years, however, he was to be known first as an actor, then as an author of television and radio scripts, rather than as a novelist. He went briefly to Hollywood, where he wrote scripts for CBS, and then back to Toronto, where he married the actress Janet Reid (they parted, amicably, after less than two years together). He subsequently met the CBC science reporter William Whitehead, with whom he has lived for more than thirty years, and with whom he has collaborated on many television scripts and in the company Pebble Productions.

His first novel, *The Last of the Crazy People* (1967), was the story of a lonely boy in southern Ontario who eventually guns down his entire family, but he did not win lasting success until the Penguin edition of his third novel, *The Wars* (1977), dealing with treason and heroism in the First World War. Since then, his reputation in modern Canadian fiction has become second only to **Margaret Atwood**'s, and there is no doubt that his fiction – studded with violence, returning obsessively to images of fire, boldly traversing the history of the century, written in a style full of sound and spectacle – is among the more distinctive of the age. Perhaps his major novel so far is *Famous Last Words*, whose central character is **Ezra Pound**'s Hugh Selwyn Mauberley, and which features a large gallery of twentieth-century 'fascists' as well as Sir Harry Oakes (whose murder is fictionally solved). Findley has also published volumes of short stories and a little non-fiction, while he is a prolific playwright, with perhaps rather less success (his best-known work in this genre is *Can You See Me Yet?*, set in an asylum). He has a strong public presence as one of Canada's leading writers, having been president of the English-language chapter of PEN, Canada, in 1986–7. He lives in the Ontario countryside near Toronto.

Fiction
The Last of the Crazy People 1967; *The Butterfly Plague* 1969; *The Wars* 1977; *Island/The Long Hard Walk* [pub. in *The Newcomers* (s) [ed. C.E. Israel] 1979; *Famous Last Words* 1981; *Dinner Along the Amazon* (s) 1984; *Not Wanted on the Voyage* 1984; *The Telling of Lies* 1986; *Stones* (s) 1988

Plays
The Paper People (tv) 1967 [pub. in *Canadian Drama* vol. 9, no. 1, 1983]; *Can You See Me Yet?* 1976 [pub. 1977]; *John A., Himself* 1979; *Daybreak at Pisa: 1945* 1982; *Strangers at the Door* (r) 1982; *The Wars* (f) 1983

Non-Fiction
Imaginings [with H. Cooper, J. Rapaport] 1982

Ian Hamilton FINLAY 1925–

Finlay was born in Nassau, in the Bahamas, where his father worked as a bootlegger. At about the age of three, he was sent to Larchfield Academy, a preparatory school at Helensburgh, Scotland, where **W.H. Auden** was a master, and later, Dollar Academy. He had to leave the latter school at the age of thirteen, when his father, who was now farming in Florida, lost his money and had to return to Scotland to take labouring jobs. Finlay attended various other schools, then took a variety of uncongenial unskilled jobs, before spending three years in the British Army from 1944.

After attending Glasgow College of Art, he settled in Perthshire. His first writings were plays and short stories, which were broadcast by the BBC. His art training drew him to the visual side of poetry, and with *The Dancers Inherit the Party* (1960) he became established as Britain's leading 'concrete poet'. Six years later, he moved to Stonypath, a small farm in Dunsyre, Lanarkshire, where he has since published several hundred books and booklets of his work, as well as cards and 'poem/prints'. Finlay resents the label 'concrete poet'. He has said he is interested in poetry which for him is 'the best words, in the best possible order, *in the best materials*': this may include glass or stone. His work has evolved with the constant aim of clarity, and has used many different forms, from traditional rhymed verse to poems designed as complete gardens. He has said that he does not expect poems to solve his problems, but considers 'the seasons, nature, inland waterways and oceans' as proper subjects for poetry.

In the 1980s, he produced bibliographies of what would appear to some as graphic work, in order to 'categorise' it as 'poetry', or, as he has said, 'in part to alert the viewer (reader?) to the subject beyond the object'. Stonypath, renamed 'Little Sparta', was created by Finlay with the help of Susan, his partner of twenty-five years. Described as a 'philosophy park', it was in part inspired by the concept of Roman gardens and the ideas of Jean-Jacques Rousseau. The mixture of wild flowers, inscriptions lettered in stone, sundials and broken columns present multiple visual puns.

Finlay was appreciated in France, where his work was exhibited in different places between 1985 and 1987. The French Government commissioned a work for Versailles in 1987, to be called 'A Revolutionary Garden', and intended to celebrate the Declaration of the Rights of Man by the Estates General in 1789. This was abruptly cancelled in early 1988 following publicity arising from unfounded and false charges in the French press that Finlay was a Nazi sympathiser, because one of his works, *Osso* (Italian for bone), used a Nazi stylised design of 'SS'. A lengthy and often angry public debate, conducted by letters and broadcasts, ensued, and Finlay was further charged with anti-Semitism. He was supported by Alan Bowness, then director of the Tate Gallery, among others, but the battle raged until 1989, when a Civil Tribunal in Paris agreed that he had been libelled. He was awarded only a symbolic one franc in compensation. The resultant strain of the long litigation caused Finlay and his partner to separate.

Finlay has never lacked the energy to staunchly defend his beliefs, although he is extremely mild-mannered. In 1961, he was attacked in the pamphlet *Ugly Birds without Wings* by his near neighbour, **Hugh MacDiarmid**, who objected to Finlay's use of Glaswegian in his book *Glasgow Beasts, an' a Burd.* MacDiarmid sent his friend, the Cuban ambassador, to rap Finlay on the knuckles, who responded by announcing an 'Anti-MacDiarmid March', safe in the knowledge that it would be banned by the police. Literary disagreements apart, the two men remained friends.

Poetry
The Dancers Inherit the Party 1960; *Glasgow Beasts, an' a Burd* 1961; *Concertina* 1962; *Rapel* 1963; *Canal Stripe Series 3, 4* 1964; *Telegrams from My Windmill* 1964; *Cythera* 1965; *Ocean Stripe 2, 3, 4, 5* 1965–7; *Autumn Poem* 1966; *4 Sails* 1966; *6 Small Pears for Eugen Gomringer* 1966; *6 Small Songs in 3's* 1966; *Tea-Leaves and*

Fishes 1966; *Canal Game* 1967; *Headlines Eavelines* 1967; *Stonechats* 1967; *Air Letters* 1968; *The Blue and the Brown Poems* 1968; *The Collected Coaltown of Callange Tri-kai* 1968; *After the Russian* 1969; *3/3's* 1969; *'Fishing News' News* 1970; *Rhymes for Lemons* 1970; *Evening/ Sail 2* 1971; *From 'An Inland Garden'* 1971; *A Memory of Summer* 1971; *The Olsen Excerpts* 1971; *Poems to Hear and See* 1971; *A Sailor's Calendar* 1971; *30 Signatures to Silver Catches* 1971; *The Weed Boat Master's Ticket, Preliminary Text (Part Two)* 1971; *Jibs* 1972; *Sail/Sundial* 1972; *Butterflies* 1973; *A Family* 1973; *Honey by the Water* 1973; *Straiks* 1973; *Exercises X* 1974; *Homage to Robert Lax* 1974; *A Pretty Kettle of Fish* 1974; *Silhouettes* 1974; *Airs Waters Graces* 1975; *The Axis* 1975; *Homage to Watteau* 1975; *A Mast of Hankies* 1975; *So You Want to Be a Panzer Leader* 1975; *Three Sundials* 1975; *Trombone Carrier* 1975; *The Wild Hawthorn Wonder Book of Boats* 1975; *Imitations, Variations, Reflections, Copies* 1976; *The Boy's Alphabet Book* 1977; *Heroic Emblems* 1977; *The Wartime Garden* 1977; *The Wild Hawthorn Art Test* 1977; *Homage to Poussin* 1978; *Trailblazers* 1978; *Dzaezl* 1979; *Peterhead Fragments* 1979; *'SS'* 1979; *Two Billows* 1979; *Woods and Seas* 1979; *The Anaximander Fragment* 1981; *A Litany, A Requiem* 1981; *Romances, Emblems, Enigmas* 1981; *Anticipations* 1982; *3 Developments* 1982; *Little Sermon Series: Cherries* 1982; *Little Sermon Series: Volume Makes Beauty* 1982; *The Mailed Pinkie* 1982; *Midway 3* 1982; *The Errata of Ovid* 1983; *Interpolations in Hegel* 1984; *Talismans and Signifiers* [with R. Grasby, N. Sloan] 1984; *Heraclitean Variations* 1986; *Thoughts on Waldemar* 1986; *Loaves* 1987; *A Concise Classical Dictionary* 1988; *Swastika, n.* 1988; *After* 1989; *Bicentenary Texts* 1989; *Blades* 1989; *3 Texts* 1989; *Thermidor* 1989; *4 Baskets* 1990; *Flakes* 1990; *Woodpaths* 1990; *Detached Sentences on Friendship* 1991; *Four Monostichs* 1991; *Jacobin Definitions* 1991; *Myths* 1991; *The Old Stonypath Hoy* 1991; *Picturesque* 1991; *Scud* 1991; *3 Stitches* 1991; *Calligrammes and Sonnets* 1992; *The Happy Catastrophe* 1992; *Images from the Arcadian Dream Garden* 1992; *More Proverbs for Jacobins* 1992; *Prosaic Proverbs* 1992; *4 Corners* 1993; *Glider Echoes Series* 1993; *Spring Verses* 1993; *Four Trees* 1994; *Lights* 1994; *Redoubt* 1994

Fiction

The Sea-Bed and Other Stories (s) 1958

(Arthur Annesley) Ronald FIRBANK 1886–1926

Firbank was born, appropriately, in Mayfair, London, the second son of Thomas Firbank, a recent member of the landed gentry and later the Unionist MP for East Hull, who was knighted in 1902. Firbank's grandfather, Joseph, had started his working life at the age of seven as a 'putter' in a Durham colliery, but had educated himself at evening classes and secured labouring contracts in the expanding railway system, eventually making a substantial fortune. The greatest influence upon Firbank's life, however, was his mother, 'Baba', to whom he was devoted. Within eighteen months of Firbank's birth, the family moved to Chislehurst in Kent, where two more children were born, the younger of them a sister, Heather, who remained close to Firbank throughout his life. Destined for a career in the diplomatic service, Firbank was educated at Uppingham (which he left after one and a half terms), in France, and at Trinity Hall, Cambridge (which he left after five terms). He became a Roman Catholic in 1907, but failed in his ambition to become a Papal Guard.

His literary career, which had been flourishing since the completion of a novel at the age of ten, now grew apace, but his boldly innovative books were considered hopelessly uncommercial, and he was obliged to pay the publication costs of all but two of them. His first publication, in 1905, was a volume containing the pallid religious fairy-tale 'Odette D'Antrevernes' and the rather more promising comic story 'A Study in Temperament'. Both tales showed – for worse and for better, respectively – the strong influence of Oscar Wilde, who was a crucial figure in Firbank's personal and literary development. Firbank soon outgrew his distinguished but uneven master, however. Although not fully achieved, his next work, *The Artificial Princess* (started in 1906, revised 1909–12, completed 1925, but not published until 1934) has all the hallmarks of his mature novels: *Inclinations*, *Caprice*, *Valmouth* (which became a cult musical in 1958, written by Sandy Wilson), *The Flower beneath the Foot*, *Prancing Nigger* and his final masterpiece, *Concerning the Eccentricities of Cardinal Pirelli*. These are unmistakably modernist works, which in turn influenced the younger writers of the 1920s, most notably **Evelyn Waugh**. The novels' thematic links with the 'naughty nineties' (Catholicism, high society and sexual heterodoxy), and the author's distinctly camp personal demeanour (as much to do with shyness as homosexuality), led to charges of 'frivolity', which were not properly rebuffed even by such champions of his work as **Jocelyn Brooke** and **Anthony Powell**. In 1973, however, **Brigid Brophy** published a large and persuasive work of revaluation, *Prancing Novelist: a Defence of Fiction in the Form of a Critical Biography in Praise of Ronald Firbank*. Dazzlingly speculative, properly combative and enormously entertaining, Brophy's book was amongst other things a brilliant piece of literary detection (many of Firbank's papers, including manuscripts of early work, were unavailable to her), and a work of intense imaginative sympathy, which unapologetically and convincingly presented Firbank as one of the century's most important, indeed great, writers.

For each of his books, Firbank kept extensive notebooks in which he jotted down ideas, images and exchanges, usually in violet ink. The resulting tragi-comic novels are painstakingly constructed upon passages of witty but often oblique dialogue, interspersed with vividly impressionistic scene-painting. (*Inclinations*, which details the thwarted love of a female biographer for a heartless young woman, is written almost entirely in dialogue.) Form and content are perfectly matched, and the use of compression and ellipsis means that Firbank's brief narratives suggest far more than they ever state. Firbank himself accurately described his style as 'aggressive, witty, & unrelenting'.

'Research' and persistent ill-health kept Firbank abroad for much of his life, and many of his novels have exotic settings: for example, *The Flower beneath the Foot*, an unorthodox fictional hagiography, is set in 'some imaginary Vienna' apparently relocated in North Africa. His last completed novel, *Concerning the Eccentricities of Cardinal Pirelli*, was written in the knowledge that he was dying of tuberculosis. The book is suffused with intimations of mortality, and it is the century's finest example of high camp literature: highly wrought, unashamedly artificial, brilliantly funny, but altogether serious.

Firbank died in Rome and was buried briefly and mistakenly in the Protestant cemetery there, the muddled funeral arrangements overseen by **Lord Berners**. Four months later, his body was removed to the Catholic cemetery of Campo Verano. Plans to turn *Prancing Nigger* into a 'jazz fantasy', with music by George Gershwin, were abandoned, and Firbank's one play, an elegant bisexual comedy titled *The Princess Zoubaroff*, remained unperformed until 1952. His novel set in New York, *The New Rythum* (Firbank was an incorrigibly bad speller), was unfinished at his death, but was later published along with several other fragments. *The Early Firbank* is a fascinating and impressive volume of juvenilia.

Fiction

Odette D'Antrevernes and A Study in Temperament (s) 1905 [rev. as *Odette: a Fairy Tale for Weary People* 1916]; *Vainglory* 1915; *Inclinations* 1916; *Caprice* 1917; *Valmouth* 1919; *Santal* 1921; *The Flower beneath the Foot* 1923; *Prancing Nigger* [UK *Sorrow in Sunlight*] 1924; *Concerning the Eccentricities of Cardinal Pirelli* 1926; *The Artificial Princess* 1934; *The New Rythum and Other Pieces* 1962; *When Widows Love and A Tragedy in Green* 1980; *The Early Firbank* [ed. S. Moore] 1991

Plays

The Princess Zoubaroff 1952 [pub. 1920]

Biography

Prancing Novelist Brigid Brophy 1973

Roy FISHER 1930–

Fisher was born in Birmingham, where all his family had lived for generations. He was the son of a jewellery-maker there, and attended Wattville Road Elementary School, Handsworth Grammar School and Birmingham University. He has recorded that he was thirteen before he spent a night out of Birmingham and that he lived in the same house until he was twenty-three; naturally, a sense of his own city and native habitat is a constant in all his work. After graduating, he taught at various schools, then at Dudley College of Education from 1953 to 1963, and Bordesley College of Education from 1963 to 1971, where he rose to become head of the department of English and drama. From 1972 to 1982 he was a member of the Department of American Studies at Keele University, first as lecturer and then senior lecturer; in the latter year, he left to become a freelance writer and musician, his profession ever since.

He had begun writing when he was nineteen or twenty, and his first book of poems, *City* (perhaps more exactly a collage of poetry and prose), was published in 1961. The city in question is partly Birmingham and partly a quintessential city, and he here explores with characteristic power and subtlety the destruction and *anomie* that such an environment causes. The book was heavily edited for original publication, and he later published a revised edition in his own Migrant Press, the form in which the poem is usually enjoyed today. Most of Fisher's early works were published in literary magazines and by small presses, almost twenty collections (some of them prose-poems) appearing in this form. Not surprisingly, he was almost entirely ignored by the metropolitan press, and not until the Oxford University Press published his *Poems 1955–1980* in 1980 (the first of two substantial collections they issued) did his largely underground reputation emerge into the full light of day.

He is now recognised as a leading poet, one who perhaps resembles **Philip Larkin** in his determined independence and provinciality (like Larkin, he had an early and continuing love of jazz, and is moreover an accomplished jazz pianist), but who differs from him in that his verse is complex and modernistic, making direct personal statement only with difficulty. But the coldness is illusory, and, as with Larkin, the moments of hard-won pathos are the finest. In 1986, he returned to the themes of *City* in his fine long poem *A Furnace*, where the influence of **John Cowper Powys** transforms the verse and gives it greater universality. An occasional satiric tone, particularly about the humbug of the poetry scene, and a love of the visual arts, are other themes emerging in his work, but he has written little love or metaphysical poetry. His non-poetic works are few, but he has compiled a film, *Birmingham Is What I Think With*. He has been married twice, and has two sons. For some years, he has lived near Buxton in Derbyshire.

Poetry

City 1961; *Ten Interiors with Various Figures* 1966; *The Ship's Orchestra* 1966; *The Memorial Fountain* 1967; *Collected Poems 1968* 1969; *Correspondence* 1970; *The Cut Pages* 1971; *Matrix* 1971; *Metamorphoses* 1971; *Three Early Pieces* 1971; *Also There* 1972; *Bluebeard's Castle* 1972; *Cultures* 1975; *Neighbours!* 1976; *Nineteen Poems and an Interview* 1976; *Four Poems* 1976; *Widening Circles: Five Black Country Poets* [with others; ed. E. Lowbury] 1976; *Barnardine's Reply* 1977; *Scenes from the Alphabet* 1978; *The Thing about Joe Sullivan: Poems 1971–1977* 1978; *Comedies* 1979; *Wonders of Obligation* 1979; *Poems 1955–1980* 1980; *Talks for Words* 1980; *Consolidated Comedies* 1981; *From Diversions* 2nd series 1981; *Running Changes* 1983; *Turning the Prism* 1985; *A Furnace* 1986; *Near Garmsley Camp* 1988; *Poems 1955–1987* 1988; *Top Down Bottom Up* 1990; *Birds United* 1992; *Birmingham River* 1994

F(rancis) Scott (Key) FITZGERALD 1896–1940

Like his most famous character, Jay Gatsby, Fitzgerald, who came to epitomise the glamour of the American Jazz Age, was in fact born in the American Midwest, at St Paul's, Minnesota. Of mixed Southern and Irish descent, he was given his three names after the author of 'The Star Spangled Banner', to whom he was distantly related. While an undergraduate at Princeton, Fitzgerald – an artist, but one who lived by the world's values – aimed for sporting, social and literary success: only the first was to elude him. Having left university in 1917 without taking a degree, he joined the US Army, but never saw action, or even got to France. Demobilised in 1919, he worked as an advertising copywriter in New York while writing his first novel, *This Side of Paradise*, which was accepted by Scribner's after two weeks' consideration and published in 1920 to enormous popular success. Autobiographical in background and experimental in construction (part of the book is written in the form of drama), the novel described the life of a young man from the Midwest at Princeton and in the First World War, and instantly established Fitzgerald as the leading chronicler of his own generation. He followed this with two volumes of short stories and a second novel, *The Beautiful and Damned*, the title of which came to seem increasingly prophetic.

In 1920 Fitzgerald married the beautiful, talented but unstable Zelda Sayre. Newly rich and fashionable, the couple embarked upon a decade of drinking, socialising and high-living, first in the USA, and later in Europe, particularly on the Riviera. This was to turn them in the popular imagination into almost mythical representatives of their era. Fitzgerald's novels and short stories drew on this myth to become flawed but lastingly resonant works of art. Of these, the best is undoubtedly *The Great Gatsby*, which, despite its brevity, has some claim to being *the* Great American Novel.

The late 1920s and early 1930s were overshadowed by money troubles, the alcoholism of both Fitzgeralds, his frequent illnesses and her successive breakdowns, the last of which, in 1934, left her a permanent invalid. (She was to die in a hospital fire in 1948.) Their unhappy circumstances inspired *Tender Is the Night*, which charts the involvement of a brilliant but alcoholic psychoanalyst with a beautiful young schizophrenic. Dissatisfied with the book, and its unfavourable reception, Fitzgerald revised it repeatedly. From 1937 he attempted, with limited success, to become a Hollywood scriptwriter: his final, unfinished, novel *The Last Tycoon*, draws on this experience. *The Pat Hobby Stories*, which he wrote between 1939 and his death, and were published in *Esquire*, feature a 'scenario hack' and represented, his editor said, his 'last word from his last home, for much of what he felt about Hollywood and about himself permeated these stories'. A heart-attack killed Fitzgerald before he was forty-five, and at the time of his death he was considered a forgotten figure of the 1920s. Posterity has proved the obituarists wrong, however, and Fitzgerald is now regarded as one of America's greatest twentieth-century writers.

Fiction

This Side of Paradise 1920; *The Beautiful and Damned* 1922; *Flappers and Philosophers* (s)

1922; *Tales of the Jazz Age* (s) 1922; *All the Sad Young Men* (s) 1926; *The Great Gatsby* 1925; *Tender Is the Night* 1934; *Taps at Reveille* (s) 1935; *The Crack-Up with Other Pieces and Tales* (s) 1945; (s) *The Last Tycoon* [unfinished; ed. E. Wilson] 1949; *The Pat Hobby Stories* (s) 1967; *Bernice Bobs Her Hair* (s) 1968; *Bits of Paradise* (s) 1973; *The Price Was High* (s) [ed. M. Bruccoli] 1979

Plays

The Vegetable 1923

Non-Fiction

Letters of F. Scott Fitzgerald [ed. A. Turnbull] 1964; *F.Scott Fitzgerald's Letters to His Daughter* 1965 [ed. A. Turnbull] 1965; *As Ever, Scott Fitz* [ed. J. Atkinson, M. Bruccoli] 1972; *The Correspondence of F. Scott Fitzgerald* [ed. M. Bruccoli, M. Duggan] 1980

Collections

The Portable F. Scott Fitzgerald [ed. D. Parker] 1945; *Afternoon of an Author: a Selection of Uncollected Essays and Stories* [ed. A Mizener] 1957

Biography

Some Sort of Epic Grandeur by Matthew J. Bruccoli 1981

Penelope FITZGERALD 1916–

Born in Lincoln, Fitzgerald is the daughter of Edmund George Valpy Knox, one of a remarkable quarter of brothers about whom she wrote a group biography in 1977. Her father was editor of *Punch* and contributed to the magazine under the name 'Evoe'; one uncle was the Catholic novelist and theologian **Ronald Knox**. She was educated at Somerville College, Oxford, where she took a first in English in 1939. She began her career in early wartime as a Recorded Programmes Assistant at the BBC, the background of her novel *Human Voices*. She married Desmond Fitzgerald in 1941; they had three children and he died in 1976.

Fitzgerald came to writing late, publishing her first novel when she was sixty-one, and her varied earlier career, some of which she has utilised in her writing, has included working in a bookshop, running an all-night coffee stall, working in journalism and, in latter years, teaching at a London tutorial agency, Westminster Tutors. Her first book was a study of the Victorian painter Edward Burne-Jones, published in 1975, and her first novel, *The Golden Child*, followed two years later, since when she has been prolific. Almost from the first, her novels have attracted high critical estimation, both in Britain and in America. The *Los Angeles Times* has proclaimed that she 'may be the best English novelist who is at present at the prime of her power', and in her own country she had been a regular contender for the Booker Prize. Her second novel, *The Bookshop*, was the first of four novels (to date) to be shortlisted. *Offshore*, set on a houseboat on the Thames (and drawn from Fitzgerald's own experiences as a river-dweller), was an unexpected winner in 1979 – it is arguably her slightest book, and certainly less deserving of the prize than *The Beginning of Spring*, which is set in pre-revolutionary Russia and won enormous acclaim, or *The Gate of Angels*, a characteristically elliptical story set in an imaginary Cambridge college in 1912, both of which have reached the shortlist. Fitzgerald describes her books as 'microchip novels', as they are written with extreme economy of style and are almost always very short; their complex allusiveness, sensitivity to material experience and their humour give them their great literary distinction.

The houseboat on which Fitzgerald and her family lived eventually sank, and she is now settled between London and Somerset.

Fiction

The Golden Child 1977; *The Bookshop* 1978; *Offshore* 1979; *Human Voices* 1980; *At Freddie's* 1982; *Innocence* 1986; *The Beginning of Spring* 1988; *The Gate of Angels* 1990

Non-Fiction

Edward Burne-Jones 1975; *The Knox Brothers* 1977; *Charlotte Mew and Her Friends* 1984

Edited

William Morris *The Novel on Blue Paper* 1982

Mary FLANAGAN 1943–

Flanagan was born in Rochester, New Hampshire, the only child of first-generation Irish immigrant parents, one Catholic and one Protestant: her father was a housing director and her mother worked as a secretary. She was educated in a convent school by the Sisters of Mercy, and then won a place studying history of art at Brandeis University in Boston, from which she graduated, after a year out in New York following the death of a boyfriend, in 1965. Moving back to New York, she worked as an actress in various avant-garde short films and science-fiction spoofs and then in various jobs in and around publishing. She spent a year travelling in Morocco, from 1968 to 1969, before resettling in England. She now lives in Hackney, London.

She did not begin to write until she was thirty-six. Her literary debut came in 1984 with

Bad Girls, a much-praised collection of short stories, which was published after a mutual friend had shown the manuscript to the publisher Liz Calder. The book has now been translated into several languages. Written to exorcise the author's own 'fantastic anger', Flanagan's first novel, *Trust* is an examination into the nature of love which concentrates on the many small betrayals which undermine it. It was also the first book to be published by Calder's newly founded house, Bloomsbury. *Rose Reason*, Flanagan's second novel, plunders the author's own past, from the convent to Bohemian New York, to present the reader with a confession from the eponymous victim to her lover. It took three and a half years to write, and was eventually published in 1991. *Blue Woman*, a collection of short stories, was published in 1994.

Flanagan is a keen gardener and piano player and she also writes screenplays, and articles and reviews for the *New York Times*, and *Financial Times* and the *Evening Standard*. She is an active supporter of the Campaign for Nuclear Disarmament and various animal rights and environmental groups.

Fiction

Bad Girls (s) 1984; *Trust* 1987; *Rose Reason* 1991; *Blue Woman and Other Stories* (s) 1993

Thomas (James Bonner) FLANAGAN 1923–

Flanagan has taken the name of his mother; his father was Owen de Salus. All four of his grandparents had emigrated to America in the later nineteenth century from County Fermanagh in the north of Ireland, where his two grandfathers had been members of the Irish Republican Brotherhood. He learned about Ireland during his childhood from one of his grandmothers and developed a lifelong identification with his ancestral country. He was born in Greenwich, Connecticut, and was educated at Amherst College, Massachusetts, where he took his BA in 1945. His education had been interrupted by war service in the Pacific as part of the US Naval Reserve between 1942 and 1944. He married in 1949, and has two daughters.

Flanagan made his early academic career at Columbia University in New York, where he took his MA and PhD and where he became an assistant professor between 1952 and 1959. He specialised in the field of Irish writing in English, particularly in the eighteenth and nineteenth centuries, and his major academic work is *The Irish Novelists: 1800–1850*, in which he surveys five writers ranging from Maria Edgeworth to neglected figures such as John Banion and Gerald Griffin. The second phase of his academic career was spent at the University of California at Berkeley, where he became a professor and eventually chairman of the Department of English between 1973 and 1976. Since 1978 he has been a professor at the State University of New York at Stony Brook (he lives in Long Island).

He has spent most summers of his adult life in Ireland and has chosen his closest friends among Irish writers, an association that led him, in his fifties, to write his first novel. He had wanted to be a writer as a young man, and had published stories, but gave up for thirty years, until he began *The Year of the French* one day in his office in Berkeley, 'keeping going almost as if in a trance ... I wrote the whole first draft with incredible happiness'. The novel is a complex multi-narrational structure centring on the Irish rebellion of 1798 in County Mayo, when a contingent of the French helped the insurgents; when both groups surrendered, the French were evacuated but the Irish driven into a bog and massacred. When published in 1979, *The Year of the French* was hailed by many as a masterpiece, tragic but far from narrowly partisan in its tone, rich in character and vivid description.

He followed it up with *The Tenants of Time*, another multi-narrational novel, this time dealing with the complex politics of late nineteenth-century Ireland, but failing, initially at least, to achieve the same impact as its predecessor. Nevertheless, few writers have achieved a reputation comparable to Flanagan's on such a small body of work.

Fiction

The Year of the French 1979; *The Tenants of Time* 1988; *The End of the Hunt* 1994

Non-Fiction

Louis 'David' Riel: Prophet of the New World 1979; *The Irish Novelists, 1800–1850* 1959; *Riel and the Rebellion 1885* 1983

Edited

The Diaries of Louis Riel 1976; *Theories of Property: Aristotle to the Present* [with A. Parell] 1979

James (Herman) Elroy FLECKER 1884–1915

Flecker was christened Herman but adopted the name James at Oxford. The eldest of four children, he was born at Lewisham, his father, Revd W.H. Flecker, being then headmaster of the City of London School. When Flecker was two, his father was appointed headmaster of Dean Close and the family moved to Cheltenham. He had a strict evangelical upbringing and remained emotionally and financially dependent on his parents all his life. He went first to Dean Close until his parents came to see that it would not do to have his father for headmaster; they sent him to Uppingham at which he won a scholarship to Trinity College, Oxford, where he wrote much verse – he had begun aged thirteen. A fellow undergraduate described him as 'extraordinarily undeveloped, even for an English Public Schoolboy, when he first went up', but goes on: 'I had … a growing feeling that, in spite of his immaturity and occasional bad taste, he was the most important of any of us … By 1906 he had developed greatly – largely thanks to the companionship of an Oxford friend whom … he loved deeply till the end of his life.' This friend was J.D. Beazley, who went on to become the foremost scholar of Greek vases in Europe, and the friendship, which was undoubtedly homosexual in feeling (if not in act), was commemorated by Flecker in such poems as 'The Ballad of the Student in the South'.

After a spell of teaching in a private school in Hampstead, he went up to Caius College, Cambridge, in 1908 to read Persian and Arabic in preparation for entry into the consular service. He was posted to Constantinople in 1910 but contracted tuberculosis. Invalided home later that year, he returned in 1911 and was transferred to Smyrna; later he became vice-consul in Beirut. In 1911 he married the Greek poet Hellé Skiadaressi. By then he had published several volumes of verse, the first, *The Best Man*, was privately printed. He was working on his first play, *Don Juan*, and had also written *The Golden Journey to Samarkand* from which his most famous work, the verse play *Hassan*, germinated. By 1913, the year he published his bizarre fantasy novel, *The King of Alsander*, his health had broken down and he was sent to sanatoria in Switzerland. There he was able to complete *Hassan*, though it was not finally produced – with music by Delius and choreographed by Fokine – until after the First World War. Indeed all his mature work appeared posthumously. He died, aged thirty, at Davos.

The unsigned obituary in *The Times* of 6 January 1915 is by his friend **Rupert Brooke**, with whom he shared the intense patriotism of the pre-war years. Brooke wrote: 'He sought beauty everywhere, but preferred, for the most of his life, to find her decoratively clad. He loved England, Greece, and the East, always with more passion than affection … His work gained in strength and clearness as he went on, and his craftsmanship became singularly accurate.' His best poems like 'The Golden Journey' itself, 'The Old Ships', and the much anthologised 'To a Poet a Thousand Years Hence' combine colour and simplicity; for all his fascination with the East, Flecker's essential Englishness remains. One knows all too well what would have happened to this romanticism had he survived a few more years. Reprinting steadily between September 1916 and October 1924, sales of his *Collected Poems* had by then reached 10,500.

Poetry

The Best Man 1908; *The Bridge of Fire* 1908; *The Last Generation* 1908; *Thirty-Six Poems* 1911; *The Golden Journey to Samarkand* 1913; *Burial in England* 1915; *The Old Ships* 1915; *Collected Poems* [ed. J.C. Squire] 1916; *God Save the King* 1918

Plays

Hassan 1922; *Don Juan* 1925

Fiction

The King of Alsander 1913

Non-Fiction

The Grecians: a Dialogue on Education 1910; *The Scholar's Italian Book* 1911

Collections

Collected Prose 1920

Ian (Lancaster) FLEMING 1908–1964

Fleming was born in London, grandson of the Scottish merchant banker Robert Fleming, and son of Valentine Fleming, a Tory MP who was killed in action in 1917. He was educated at Eton and Sandhurst, and studied privately in Austria, Germany and Switzerland before becoming a reporter with Reuters, working in the Soviet Union and elsewhere. He worked briefly as a stockbroker and banker, and in 1939 became personal assistant to the Director of Naval Intelligence. One of his more outlandish wartime schemes was an attempt to recruit the self-styled 'Great Beast' Aleister Crowley to interrogate Rudolf Hess in 1941. Fleming travelled widely during the war, and in 1945 built his house, Goldeneye, on Jamaica, and became foreign manager for Kemsley Newspapers (the group including the *Sunday Times*). In 1952 he married Ann Charteris (who had previously been married to Lord O'Neill and Viscount Rothermere) and the couple had two children. *The Letters of Ann Fleming* (1985), edited by Mark Amory, provides a valuable picture of their life and times, and includes some amusing stories about their neighbour in Jamaica, **Noël Coward**.

Fleming had long spoken of writing, but seems to have felt that he lived in the shadow of his older brother, Peter, whose travel books had enjoyed great success since the 1930s. Drawing on his experience in Naval Intelligence – albeit with considerable fantastical embellishment – Fleming wrote *Casino Royale*, which he sent to his friend **William Plomer**, who was then editorial advisor to Jonathan Cape. Published in 1953, the novel is the first of the secret service thrillers in which James Bond almost invariably saves the world from a megalomaniac. The novels were widely praised for their style and wit – in Washington in 1960 Fleming met an admiring John F. Kennedy – though dissenting voices included that of Paul Johnson who coined the phrase 'sex, sadism and snobbery' in a dismissive 1958 review of *Dr No*. Harry Saltzman and Albert R. Broccoli's film versions (starting with *Dr No* in 1962) furthered Bond's popularity whilst laying greater emphasis on the technical gadgetry, and since Fleming's death there have been further Bond adventures by John Gardner. The Bond in **Kingsley Amis**'s 1968 *Colonel Sun* (written under the pseudonym Robert Markham) was described by Ann Fleming as 'a petit bourgeois redbrick Bond'. Fleming also wrote three volumes for children about the car Chitty-Chitty-Bang-Bang.

Fiction

Casino Royale 1954; *Live and Let Die* 1954; *Moonraker* 1955; *Diamonds Are Forever* 1956; *From Russia with Love* 1957; *The Diamond Smugglers* 1958; *Dr No* 1958; *Goldfinger* 1959; *For Your Eyes Only* (s) 1960; *Thunderball* 1961; *The Spy Who Loved Me* 1962; *On Her Majesty's Secret Service* 1963; *You Only Live Twice* 1964; *The Man with the Golden Gun* 1965; *Octopussy and The Living Daylights* (s) 1966

Non-Fiction

Thrilling Cities 1963; *Ian Fleming Introduces Jamaica* 1965

For Children

Chitty-Chitty-Bang-Bang 3 vols 1964

Biography

Ian Fleming by Andrew Lycett 1995

Helen FLINT 1952–

Flint was born in Cambridge, Massachusetts, to a peripatetic academic father, as a result of whose postings to various universities around the world, she was brought up in America, Canada, Nigeria and England. She was educated at the universities of Dalhousie and Oxford, and has had a variety of jobs including a spell as a waitress on the ferry that crosses the legendary Bay of Fundy. Her career in England has revolved around education: she has taught English, drama, classical studies and Russian in English schools, been a schools examiner, and has taught writing on Arvon Foundation residential courses.

Her outstanding first novel, *Return Journey*, published in 1987, drew on something of the author's own experience to tell the story of Jane Wells, newly married and an Oxford graduate, called upon to deliver her disturbed and disabled mother from Canada to a nursing home in England. Confined for hours in the darkened rear of the plane, Jane discovers with horror and revulsion the full extent of her mother's condition: 'terminally ill without hope of a terminus'. Throughout the long flight she tries to replay the good times of their shared past – events from her mother's previous life as a nurse and wife of a travelling academic – but what the story reveals is the collapse of a family under intolerable stress, and the mother's madness and cruelty which come with her progressive loss of control. *Return Journey* won a Betty Trask Award in 1987.

The black comedy and exact deployment of language that marked out *Return Journey* also characterise Flint's other books. Set in Sheffield, *In Full Possession* (the punning title of which is an example of Flint's attention to the ambiguities and nuances of words) is a dark tale of revenge. Benedict Ashe conceives a long and elaborate plot against the female lodger whose refusal to move out of his house caused the death of his mother. He befriends and seduces the lodger in order to beget a child, who will become an instrument of retribution. *Making the Angels Weep* is a study in domestic tyranny, exerted by a man who plays host to a group of friends during a bank holiday. This contemporary narrative is interwoven with the story of a thirteenth-century monk.

Flint, who lives in Dorset with her husband and two children, has also published a children's novel, *Not Just Dancing*.

Fiction

Return Journey 1987; *In Full Possession* 1989; *Making the Angels Weep* 1992

For Children

Not Just Dancing 1993

'Flying Officer X'

see H.E. BATES

Ford Madox FORD (formerly Ford Hermann Hueffer) 1873–1939

Ford was born in Merton, Surrey. His father, who had emigrated from Germany four years previously, was then music editor of *The Times*. He began his education at the Praetorius School, Folkestone, but in 1889 his father died and the family moved into the London house of his maternal grandfather, the Pre-Raphaelite painter Ford Madox Brown, from where Ford continued his education at University College School, and amongst the cosmopolitan intelligentsia that frequented London before the First World War.

In 1891, he showed his fairy story 'The Brown Owl' to his revered grandfather, who liked it, illustrated it, and arranged for it to be published. The following year he had a novel, *The Shifting of the Fire*, published, and was received into the Roman Catholic church 'as a relief from the gospel of perfect indifference to everything'. Two years later, he eloped with his fiancée and settled in Kent, where they led a bucolic, impoverished life. Ford continued to entertain local eminent writers, including **Henry James**, **Rudyard Kipling**, **Stephen Crane**, **H.G. Wells**, **W.H. Hudson** and **Joseph Conrad**. He established a close relationship with Conrad and went on to collaborate with him on *The Inheritors* (1901) and *Romance* (1903). These novels were unsuccessful, but the exchange of ideas enabled both men to develop theories of fiction which would prove useful for their later work. Ford did all that he could for Conrad: lent him money, managed his affairs, took dictation and even wrote sections of *Nostromo* (1904) in perfect imitation of Conrad's style when the author was in a state of collapse over his book. Meanwhile, ill-health, financial anxiety and social ostracism (as a result of an affair with his wife's sister) led to a nervous breakdown in 1904.

In 1908 'the itch of trying to meddle in English literary affairs' led to his founding editorship of the *English Review*; he had a flair for every aspect of the business, except that he could not make it pay. Under his colourful and perceptive editorship, the review attracted most of the leading writers of the day. Before it went under, he had 'discovered' **D.H. Lawrence**, **Ezra Pound** and **Wyndham Lewis**.

Ford was a great talker and liked women: it has been said that he only seduced women in order to carry on a conversation with them. While editing the *English Review* he fell in love with the novelist Violet Hunt, but his wife refused his pleas for divorce and, after a messy public court case, he eventually was sent to Brixton prison for eight days because he refused to pay a maintenance order.

In 1915 Ford published what is regarded as his finest achievement, *The Good Soldier*, a novel about adultery and deceit that has come to be regarded by some as one of the greatest of twentieth-century novels. Although over-age, he enlisted in the army and served in France until he was invalided home in 1917. His war experiences inspired some poetry and propaganda pieces, and also, more significantly, his other major work of fiction, the *Parade's End* tetralogy, which was published between 1924 and 1928.

He took his new partner, Stella Bowen, to France to 1922, where he founded the *Transatlantic Review*, which published work by **James Joyce**, **Ezra Pound**, **e e cummings**, **Gertrude Stein** and **Jean Rhys**. The magazine folded after a year and, following a brief affair with Rhys, Ford went to America in order to try to resolve his financial problems. The last decade of his life was spent travelling between France and America with another partner. In spite of a heart-attack in 1930, he continued to write poetry, criticism, novels and reminiscences. By the time he died at Deauville, France, he had published over eighty books. He is remembered as one of English literature's most devoted servants as well as a major novelist.

Fiction

The Shifting of the Fire 1892; *The Inheritors* [with J. Conrad] 1901; *Romance* [with J. Conrad] 1903; *The Benefactor* 1905; *The Fifth Queen* 1982: *The Fifth Queen* 1906, *Privy Seal* 1907, *The Fifth Queen Crowned* 1908; *An English Girl* 1907; *Mr Apollo* 1908; *The 'Half Moon'* 1909; *A Call* 1910; *The Portrait* 1910; *The Simple Life Limited* [as 'Daniel Chaucer'] 1911; *Ladies Whose Bright Eyes* 1911 [rev. 1935]; *The New Humpty-Dumpty* [as 'Daniel Chaucer'] 1912; *The Panel* 1912; *Mr Fleight* 1913; *The Young Lovell* 1913; *The Good Soldier* 1915; *The Marsden Case* 1923; *The Nature of a Crime* [with J. Conrad] 1924; *Parade's End* 1950: *Some Do Not . . .* 1924, *No More Parades* 1925, *A Man Could Stand Up* 1925, *Last Post* 1928; *A Little Less Than Gods* 1928; *When the Wicked Men* 1931; *The Rash Act* 1933; *Henry for Hugh* 1934; *Vive Le Roy* 1936

Poetry

The Questions at the Well [as 'Fenil Haig'] 1983; *Poems for Pictures* 1900; *The Face of the Night* 1904; *From Inland* 1907; *Songs from London* 1910; *High Germany* 1912; *Collected Poems* 1916; *On Heaven* 1918; *Mr Bosphorus and the Muses* 1923; *New Poems* 1927; *Buckshee* 1966

Non-Fiction

Ford Madox Brown 1896; *The Cinque Ports* 1900; *Rossetti* 1902; *Hans Holbein* 1905; *The Soul of London* 1905; *The Heart of the Country* 1906; *The Pre-Raphaelite Brotherhood* 1907; *The Spirit of the People* 1907; *The Critical Attitude* 1911; *The Desirable Alien: Impressions of Germany* [with V. Hunt] 1913; *Between St Dennis and St George* 1915; *When Blood Is Their Argument* 1915; *Zeppelin Nights* [with V. Hunt] 1915; *Thus to Revisit* 1923; *Women and Men* 1923; *Joseph*

Conrad: a Personal Remembrance 1924; *A Mirror to France* 1926; *New York Essays* 1927; *New York Is Not America* 1927; *No Enemy* (a) 1929; *The English Novel: From the Earliest Days to the Death of Joseph Conrad* 1930; *Return to Yesterday* (a) 1931; *It Was the Nightingale* (a) 1934; *Great Trade Route* 1937; *The March of Literature: From Confucius' Day to Our Own* 1938; *Mightier Than the Sword* 1938; *Provence* 1938; *Critical Writings* [ed. F. MacShane] 1964; *The Letters of Ford Madox Ford* [ed. R.M. Ludwig] 1965

For Children

The Brown Owl 1891; *The Feather* 1892; *The Queen Who Flew* 1894; *Christina's Fairy Book* 1906

Biography

Ford Madox Ford by Alan Judd 1990

Richard FORD 1944–

Ford was born in Jackson, Mississippi, the son of a travelling salesman of laundry starch. 'Probably it was arbitrary where I was born,' he has said. 'But my literary instincts were provoked at an early stage on account of living in the South.' His father died in his arms when Ford was sixteen, and after college he volunteered for the Marines. He was invalided out when he contracted hepatitis, and in 1968 worked very briefly for the CIA. ('I was not particularly enlightened,' he has commented.) He attended law school for a brief period before giving it up to write fiction. 'Being a very conventional boy I went to school for it,' he has said. He had met Kristina Hensley when he was nineteen, and they were married in 1968; it was her encouragement and financial support which enabled him to dedicate himself full-time to writing. Her work as an economic researcher necessitated a great deal of travel, and Ford has accompanied her to different parts of the USA. The idea of continual travel, which he sees as an essential component of the American psyche, is a theme which recurs throughout his work. He has taught creative writing at Michigan and Princeton, and was the recipient of a Guggenheim Fellowship in 1977.

His first novel, *A Piece of My Heart* (1976) is a melodramatic, archetypically Southern story set in Mississippi which invited inevitable comparison with **William Faulkner** and **Flannery O'Connor**. It is not a terrain to which he has subsequently returned. In the early 1980s he began to publish short stories in *Granta* and *Esquire* and, along with his friends **Tobias Wolff** and **Raymond Carver**, was identified as one of the practitioners of a new strain in American literature dubbed 'dirty realism'.

With the successful publication of *The Sportswriter* in 1986 Ford shook off the limitations inherent in that definition and found his own distinctive voice. 'In that book,' he has said, 'I tried to resurrect a mystery for facts. To me, life is a matter of human beings taking pleasure in what's available.' In contrast to the sprawling loquaciousness of *The Sportswriter*, his fourth novel, *Wildlife*, is a slight and taut story narrated by a sixteen-year-old boy who witnesses his mother's adultery.

Fiction

A Piece of My Heart 1976; *The Ultimate Good Luck* 1981; *The Sportswriter* 1986; *Rock Springs* (s) 1988; *Wildlife* 1990; *Independence Day* 1995

Edited

The Best American Short Stories 1990 (s) 1991; *The Granta Book of the American Short Story* 1992

E(dward) M(organ) FORSTER 1879–1970

Forster was born in London, the son of an architect who died before his only child was two years old. Consequently, his childhood – and indeed much of his adult life – was overseen by his formidable but doting mother. He was educated at Tonbridge School and King's College, Cambridge. The influence of King's, and in particular its 'secret society', the Apostles, was profound, instilling in him concepts of individualism, personal loyalty and what Keats designated 'the holiness of the heart's affections'. His Cambridge connections also led to his association with the Bloomsbury Group and to the development of his creed of liberal humanism.

Extended holidays with his mother in Italy provided material for his first novel, *Where Angels Fear to Tread* (1905), and for *A Room with a View*. In both books the restrictions, repressions and hypocrisies of English society are set against and exposed by the emotional openness of life in southern Europe. The latter, a social comedy that nevertheless treats serious themes, remains his most popular book, although his own favourite was *The Longest Journey*, in which, he said: 'I have managed to get nearer than elsewhere to what was on my mind'. The mystical sense of landscape in this book is a characteristic of Forster's novels, and of the short stories collected in *The Celestial Omnibus* and *The Eternal Moment*.

After the publication of *Howard's End* in 1910, Forster felt 'weariness of the only subject that I both can and may treat – the love of men for women & vice versa'. He embarked upon a new novel with a homosexual theme, but felt that it could not be published 'until my death and England's'. Shown to friends such as **J.R. Ackerley**, T.E. Lawrence, **Siegfried Sassoon** and **Christopher Isherwood**, *Maurice* was revised several times during his life, and eventually published posthumously. Many critics thought the book unworthy of the author, accusing it of proselytism, melodrama and sentimentality, but whatever its weaknesses (which have often been exaggerated), it remains crucial to an understanding and assessment of his work.

A trip to India in 1912–13 resulted in Forster's final and most highly regarded novel, *A Passage to India*, as well as a non-fiction book, *The Hill of Devi*. He also wrote two books about Egypt, where he had spent part of the First World War working as a searcher for the 'Wounded and Missing' bureau of the Red Cross: *Alexandria* and *Pharos and Pharillon*. He published no novels after 1924, but spent the rest of his long life contributing distinguished reviews and essays to numerous journals (most notably *The Listener*), writing biographies of his friend Goldsworthy Lowes Dickinson and his great-aunt Marianne Thornton, tinkering with 'unpunishable' (because homosexual) short stories (eventually collected in *The Life to Come*), and campaigning for assorted liberal and literary causes. The Clark Lectures he gave in Cambridge in 1927 were published the same year as *Aspects of the Novel*, and resulted in his being offered a three-year fellowship at King's. He made many broadcasts for the BBC and collaborated with Eric Crozier on the libretto for Benjamin Britten's opera *Billy Budd*. Something of a national monument, he turned down a knighthood, but was made a Companion of Honour in 1953 and accepted an Order of Merit in 1969. He was widely regarded as keeper of the liberal conscience, his philosophy expounded movingly and eloquently in his famous essay 'What I Believe', printed originally in the New York *Nation* in 1938 as (a characteristic title) 'Two Cheers for Democracy', and republished in England on the eve of the Second World War as the first of the Hogarth Press Sixpenny Pamphlets. His selected journalism appeared in two volumes, *Abinger Harvest* and *Two Cheers for Democracy*. 'My defence at the Last Judgement would be: "I was trying to connect up and use all the fragments I was born with",' he wrote in 1915, and this was something he succeeded in doing conspicuously well.

Forster's emotional life was initially unhappy, but a series of more-of-less unsatisfactory liaisons with working-class men culminated in 1930 in a relationship with Bob Buckingham, a London policeman. He was plunged into despair when Buckingham married shortly after they had met, but equilibrium was soon restored, thanks largely to Buckingham's wife, May, and the friendship endured for the rest of Forster's life.

Evicted after her death from the home he shared with his mother at Abinger Hammer in Surrey, Forster returned in 1946 to his old college as an Honorary Fellow and divided his time between rooms there and the Buckinghams' house in Coventry, where he died. An unobtrusive figure, customarily dressed in a shabby raincoat and flat cap, Forster was in fact wealthy. He lived frugally, writing letters on envelopes, pages torn from diaries, old proofs or any other scrap of paper that came to hand, but he was generous to others, often sending them gifts of money, which sometimes came with notes of advice or admonishment. His posthumous reputation remains high, and was enhanced by a number of very successful film adaptations of his novels in the 1980s and 1990s.

Fiction

Where Angels Fear to Tread 1905; *The Longest Journey* 1907; *A Room with a View* 1908; *Howard's End* 1910; *The Celestial Omnibus and Other Stories* (s) 1911; *A Passage to India* 1924; *The Eternal Moment and Other Stories* (s) 1928; *Maurice* 1971; *The Life to Come and Other Stories* (s) 1977; *Arctic Summer and Other Fiction* [ed. E. Heine, O. Stallybrass] 1980

Non-Fiction

Alexandria: a History and a Guide 1922; *Pharos and Pharillon* 1923; *Aspects of the Novel* 1927; *Goldsworthy Lowes Dickinson* 1934; *Abinger Harvest* 1936; *Two Cheers for Democracy* 1951; *The Hill of Devi* 1953; *Marianne Thornton, 1797–1887* 1956; *Commonplace Book* [ed. P. Garner] 1979; *Selected Letters* [ed. P.N. Furbank, M. Lago] 2 vols 1983, 1985

Plays

Billy Budd (l) [from H. Melville's story; with B. Britten, E. Crozier] 1951

Biography

E.M. Forster: a Life by P.N. Furbank, 2 vols 1977, 1978

Margaret FORSTER 1938–

'The most important thing about me,' Forster has written, 'is that I am a Cumbrian.' She was born in Carlisle, educated at the County High School, and won an open scholarship to Somerville College, Oxford. She gained an honours degree in history in 1960 and in that same year married the writer and broadcaster Hunter Davies, with whom she has three children. The couple have lived in London for over thirty years, but spend part of each year in the Lake District;

Forster complains that the return south becomes harder and harder. After leaving university, she worked as a teacher at Barnsbury Girls' School in Islington, before becoming a full-time writer in 1963.

Forster's first novel, *Dame's Delight*, was published the following year. Set in Oxford, it is not in the customary mould of nostalgic recollections of golden days, attacking as it does the entire Oxbridge mystique. Her second novel, *Georgy Girl*, though not her best, remains her most popular, largely because it was made into a popular film in 1966, directed by Silvio Narizzano from a screenplay by Forster and **Peter Nichols**, and with a hit theme-song by The Seekers. Set in a London that was famously 'swinging', it captured a particular period and its mores. The film starred two icons of the age, Lynn Redgrave as the plain but good-hearted Georgy and Charlotte Rampling as her beautiful but selfish flatmate. More than a dozen novels have followed. Forster says that she writes compulsively and quickly, each book taking on average six to nine weeks to complete. Her only requirement for this work is absolute silence. The best known of her novels are *Have the Men Had Enough?*, which drew upon personal experience in its depiction of the effects of Alzheimer's disease upon a family, and which was shortlisted for the *Sunday Express* Book of the Year Award in 1989, and *Lady's Maid*, in which the elopement of Robert and Elizabeth Barrett Browning is described from the viewpoint of the latter's maid.

Forster's earlier biography of E.B. Browning won the Royal Society of Literature Award in 1988, and she had also edited a selection of her poems. Other non-fiction works include biographies of Thackeray and the Young Pretender, Charles Edward Stuart, and an important account of 100 years of feminism, from 1839 to the outbreak of the Second World War. Her biography of **Daphne Du Maurier** was much admired, but proved controversial. The relevation of Du Maurier's homosexuality – and in particular an alleged affair with the actress Gertrude Lawrence – caused consternation amongst Du Maurier's fans and 'the ladies of Cornwall', and complaints from some commentators who felt so contentious a biography should have had source-notes.

Fiction

Dame's Delight 1964; *The Bogeyman* 1965; *Georgy Girl* 1965; *The Travels of Maudie Tipstaff* 1967; *The Park* 1968; *Miss Owen-Owen is at Home* 1969; *Fenella Phizackerley* 1970; *Mr Bone's Retreat* 1971; *The Seduction of Mrs Pendlebury* 1974; *Mother Can You Hear Me?* 1979; *The Bride of Lowther Fell* 1980; *Marital Rites* 1981; *Private Papers* 1986; *Have the Men Had Enough?* 1989; *Lady's Maid* 1990; *The Battle for Christabel* 1991; *Mother's Boys* 1994

Plays

Georgy Girl (f) [from her novel; with P. Nichols] 1966

Non-fiction

The Rash Adventurer: the Rise and Fall of Charles Edward Stuart 1973; *William Makepeace Thackeray: Memoirs of a Victorian Gentleman* 1978; *Significant Sisters: Grassroots of Active Feminism 1839–1939* 1984; *Elizabeth Barrett Browning: a Biography* 1988; *Daphne Du Maurier* 1993

Edited

Elizabeth Barrett Browning *Selected Poems* 1988

John (Robert) FOWLES 1926–

Fowles was born at Leigh-on-Sea in Essex, the son of the owner of a small cigar-making and tobacconist's business. He has reacted in his fiction against the lack of freedom in his middle-class upbringing, and a subsequent distaste for authority was inculcated when, as captain of prefects at Bedford School, he allowed himself to exercise a tyranny over the younger boys. He did two years' national service with the Royal Marines before studying French and German and, less formally, the philosophy of Camus and Sartre at New College, Oxford. After graduating, Fowles taught English at the University of Poitiers and later on the Greek island of Spetsai, which was to feature as the backdrop for *The Magus*. It was there that Fowles began to write and also where he met his future wife.

In 1952, he left Greece to spend a decade teaching in schools around London; he was still writing, and in 1963 his first novel, *The Collector*, was published. This was considered promising, but it was really the success of *The Magus* three years later which enabled him to live as a writer.

In 1966, he moved to Lyme Regis, where he has since resided in a state of voluntary 'exile', self-confessedly remaining remote both from literary London and his *haut bourgeois* neighbours. Although Fowles has published novellas, poetry and general non-fiction, his reputation rests largely on his first six novels. These are characteristically long, but otherwise very varied, ranging from the magic realism of *The Magus* to the alternative endings and nineteenth-century pastiche of *The French Lieutenant's Woman* (set in Lyme Regis and made into a very successful film in 1981) to the more conventional *Bildungs-*

roman in *Daniel Martin*. More than any other writer of his period, he seemed to bridge the gap between serious experimental fiction and the bestseller list; his sales have been large in Britain, and even larger in the USA.

He has inevitably had his detractors – Martin Seymour-Smith comments that he 'has wrecked his novels by trying to be a Dostoevsky or a Hesse'; perhaps this is a way of saying that he is a writer who particularly appealed to a certain period but will be less likely to please a succeeding one. He has not published a novel since *A Maggot* in 1985. His notorious discontent with work in progress (he published a revised version of *The Magus* a decade after the original one) may be one reason for this; others are his grief over the death of his wife from cancer in 1990 and his own recurrent bouts of illness in recent years.

Fiction

The Collector 1963; *The Aristos* 1965 [rev. 1980]; *The Magus* 1966 [rev. 1977]; *The French Lieutenant's Woman* 1969; *The Ebony Tower* 1974; *Daniel Martin* 1977; *Mantissa* 1982; *A Maggot* 1985

Non-Fiction

Shipwreck 1974; *Islands* 1978; *The Tree* [with F. Horvat] 1979; *The Enigma of Stonehenge* [with B. Brukoff] 1980

Poetry

Poems 1973

Edited

Streep Holm: a Case History in the Study of Evolution [with the Kenneth Allsop Trust] 1978; *Monumenta Britannica: a Miscellany of British Antiquities* 1980; *Thomas Hardy's England* 1984

Translations

C. Perrault *Cendrillon* 1974

Janet (Paterson) FRAME (Clutha) 1924–

Widely regarded as the leading living New Zealand writer, Frame was born, a fourth-generation New Zealander, at Dunedin and grew up largely at Oamaru (which became 'Waimaru' in her fiction). Her father was a railway worker and her youth was spent in railway houses – 'a background of poverty, drunkenness, attempted murder and near-madness'. She attended Waitaki Girls' High School, Otago University and Dunedin Teacher Training College, but taught for less than a year. She took work looking after four elderly women in a boarding-house and wrote in a linen cupboard, but she was deeply affected by the drowning of two of her sisters in separate accidents in 1937 and 1947, made a suicide attempt and had several nervous breakdowns.

From 1947 to 1955 she spent most of the time in mental hospitals, emerging sometimes to take menial jobs in hotels and attempt to 'mix and conform': another hospitalisation was the invariable result. She was rescued from this life by New Zealand's senior writer, **Frank Sargeson**, who offered her an army hut in his garden, where she wrote his first novel, *Owls Do Cry*, an account of her childhood which was immediately recognised as an important work when published in 1957. It had followed her first book of short stories, *The Lagoon*, which had been written when she was twenty-one, and won the Hubert Church Award, thus saving the author from undergoing a leucotomy. Before the novel was published, Sargeson had been instrumental in getting her a Literary Fund grant, and she used the money to go to Europe, living for most of 1957 in Ibiza and Andorra, then spending seven years in England, publishing three more novels. On her return to New Zealand, she lived in a number of small towns, was for a time Burns Fellow at Otago University, then settled in a cottage in Dunedin, and has since moved to Auckland; she also spent time at a writers' colony in the USA in the late 1960s.

Her novels, of which the fourth, *Scented Gardens for the Blind*, perhaps still remains the most remarkable, are dense and difficult, much preoccupied with madness, death and man's capacity for self-destruction. She is also a poet, although her work here is more conventional, drawing inspiration from everyday scenes and objects, often affirmatively. Besides novels and poetry, she has published writing for children, several collections of stories and sketches, and three volumes of autobiography (which have been used as the basis for the film *An Angel at My Table* made by Jane Campion in 1990). Frame has legally adopted the surname of Clutha, taken from a river which featured in her childhood, but does not use it in her writing.

Fiction

The Lagoon (s) 1952 [rev. as *The Lagoon and Other Stories* 1961]; *Owls Do Cry* 1957; *Faces in the Water* 1961; *The Edge of the Alphabet* 1962; *Scented Gardens for the Blind* 1963; *The Reservoir: Stories and Sketches* (s) 1963 [rev. as *The Reservoir and Other Stories* 1966]; *Snowman, Snowman* (s) 1963; *The Adaptable Man* 1965; *A State of Siege* 1966; *The Rainbirds* 1968 [US *Yellow Flowers in the Antipodean Room* 1969]; *Intensive Care* 1970; *Daughter Buffalo* 1972;

Living in the Maniototo 1981; *You Are Now Entering the Human Heart* (s) 1983; *The Carpathians* 1988

Poetry
The Pocket Mirror 1967

For Children
Mona Minim and the Smell of the Sun 1969

Non-Fiction
An Autobiography (a) 1989: *To the Is-land* (a) 1982, *An Angel at My Table* (a) 1984, *The Envoy from Mirror City* (a) 1985

Plays
An Angel at My Table (f) [from her book] 1990

Ronald FRAME 1953–

Frame was brought up in the genteel western Glasgow suburb of Bearsden, and was educated in Glasgow and at Oxford. His first short stories appeared in literary magazines when he was seventeen, and he has been a prolific writer ever since.

His first novel, *Winter Journey*, was joint winner in 1984 of the first ever Betty Trask Award, a much disputed prize ostensibly for traditional fiction, though a number of positively experimental novels have subsequently won it. Frame's novel is narrated by a young girl who is travelling through Central Europe with her mother to visit her father in Prague. He adapted the book for the radio, as he was to do with several of his short stories, a volume of which, *Watching Mrs Gordon*, was published in 1985. The volume was aptly titled, for one of Frame's principal themes is voyeurism – sometimes, though not always, in the sexual sense. The excellent title novella of *A Woman of Judah* draws upon the biblical story of Susannah and the Elders in the story of a young lawyer's involvement with a doctor and his wife in a Wessex village in the 1930s. 'The Corner Table', one of the fifteen stories collected with it, describes a writer who sits in a café observing the other customers in a mirror. Other stories in the collection display a wide range of style and subject, from 'Sur la Plage à Trouville', based upon a famous painting, through 'Fludde's Ark', about a cabinet of curiosities, to 'Begun at Midnight', a surprising variation on a theme by Dickens. Less successful (and, astonishingly, published the same year) is the long novel *Sandmouth People*, in which the inhabitants of a south coast resort in the 1950s are observed going about their – often illicit – business. In title, setting and mood, it unwisely invited comparisons with **William Trevor**'s infinitely superior *The Children of Dynmouth* (1976).

In later novels, Frame has typically explored the lives of women: *Penelope's Hat* concerns a female novelist with a passion for disguising millinery (women's clothes, minutely and sensuously described, are something of a fetish in Frame's books); *Bluette* details the numerous roles played in her life by a woman, from clergyman's housekeeper to night-club hostess. In spite of the insistence of publishers that only big novels sell, Frame's career suggests that talented writers of short fiction should be allowed to keep doing what they do best. His novels are on the whole far less distinguished than his short stories, and perhaps his best volumes are *A Woman of Judah* and *A Long Weekend with Marcel Proust*, both of which consist of a novella and several short stories.

Frame's stories have appeared in numerous publications, including the *London Magazine*, *Fiction Magazine* and the *New Edinburgh Review*, and he was one of the authors represented in Faber's *Introduction 8*. He has written a number of stories for radio and adapted others as plays on both radio and television. His first original television play, *Paris*, won both the Samuel Beckett Prize and Pye's Most Promising New Writer to Television Award.

Fiction
Introduction 8 (s) [with others] 1983; *Winter Journey* 1984; *Watching Mrs Gordon* (s) 1985; *A Long Weekend with Marcel Proust* (s) 1986; *Sandmouth People* 1988; *A Woman of Judah* (s) 1988; *Penelope's Hat* 1989; *Bluette* 1990; *Underwood and After* 1991; *Walking with My Mistress in Deauville* 1992

Plays
Paris (tv) 1985 [pub. 1987]; *Winter Journey* [from his novel] (r) 1985; *Twister* (r) 1986; *Out of Time* (tv) 1987; *Rendezvous* (r) 1987; *Cara* (r) 1989; *Marina Bray* (r) 1989

Pamela FRANKAU 1908–1967

Frankau was a member of an extensive literary dynasty. She was the daughter of the novelist and poet Gilbert Frankau; her grandmother, Julia Frankau, wrote novels under the pseudonym of Frank Danby; her uncle Ronald Frankau was the author and performer of numerous music-hall songs; her cousin was Ernest Raymond, author of the bestselling First World War novel, *Tell England* (1922), and other popular works. The family was of Jewish origin, but she, like her father, was brought up in the Church of England. Her parents separated when she was a child, and thereafter she and her sister Ursula, who also wrote novels, were brought up by their mother, with their father making spasmodic, but on the whole constructive, appearances in her life. (Gilbert's novel *The Love Story of Aliette Brunton*, 1922, argues for more liberal divorce laws.) At the age of seven, she was sent to boarding-school, where, though never happy, she won the swimming prize, was an energetic Girl Guide, and by last-minute cramming achieved first-class honours in her School Certificate.

To her later regret, she turned down the offer of a place at Girton College, Cambridge, and joined the Amalgamated Press, where she met Edgar Wallace's daughter, Pat, who was to become a close friend. Her first novel, *Marriage of Harlequin*, was written on the commuter train between her home and work, and was published in 1927, while she was still in her teens. From then on she produced a steady stream of novels and short stories, as well as a somewhat premature autobiography, *I Find Four People* (1935).

Although predominantly lesbian, in the early 1930s she fell in love with the poet Humbert Wolfe. His death in 1940 proved devastating, but she found consolation in the Roman Catholic church, with a special devotion for St Antony of Padua, patron saint of lost things. Wolfe is first portrayed in her novels in *Tassell-Gentle* (1934), the book which also marked a turning-point in her development. Her memories of him were to give her most famous novel, *The Willow Cabin*, much of its power. The story of a young actress who abandons a promising career when she becomes involved with a middle-aged surgeon, it drew upon Frankau's knowledge of the theatre, gained through friendships with actors and dramatists such as **Noël Coward** and **John van Druten**.

Shaken in the Wind is the first novel she wrote after being received into the church, and it loudly proclaims her new faith. She was also, rather less orthodoxly, psychic – a holiday in Elsa Maxwell's villa at Cannes was disturbed by 'a nightly visitation of a pair of clasped, white, female, beringed hands and an effluvium similar to drains in the last stages of dilapidation' – and second sight is a theme of *Slaves of the Lamp*, for which she was taught correct Tarot procedure by her nephew, the author Timothy d'Arch Smith.

In 1945 she married an American academic, Marshall Dill Jr, and went to live in California. The marriage was not a success, however, and her only child died in infancy. She returned to England in 1952, and continued to publish novels, amongst them *The Winged Horse* (dedicated to her last great love, the theatrical producer and daughter of Dame May Whitty, Margaret Webster, with whom she lived during the final ten years of her life), *A Wreath for the Enemy* and (her own favourite) *The Bridge*, in which the gamine Linda Platt echoes her creator by observing that: 'All intelligent people are bisexual.' She also wrote short stories, became a broadcaster on BBC radio and spent some of her time composing poetic squibs about her pet aversions, such as **Elizabeth Bowen** ('She was writin' a story so tender and passionate/She scarce had the strength left to put one more dash in it').

Frankau died of cancer, working right up to the end of her life. Long out of fashion, she reached a new audience when several of her novels were reissued by the feminist publisher Virago in the 1980s.

Fiction

Marriage of Harlequin 1927; *The Fig Tree* 1928; *The Black Minute and Other Stories* (s) 1929; *Three* 1929; *She and I* 1930; *Born at Sea* 1931; *Letters From a Modern Daughter to Her Mother* 1931; *'I was the Man'* 1932; *Women Are So Serious and Other Stories* (s) 1932; *The Foolish Apprentices* 1933; *Tassell-Gentle* 1934; *Fifty-Five and Other Stories* (s) 1936; *Villa Anodyne* 1936; *Some New Planet* 1937; *No News* 1938; *A Democrat Dies* 1939; *The Devil We Know* 1939; *Shaken in the Wind* 1948; *The Willow Cabin* 1949; *The Offshore Light* 1952; *The Winged Horse* 1953; *A Wreath for the Enemy* 1954; *The Bridge* 1957; *Ask Me No More* 1958; *Road Through the Woods* 1960; *Sing for Your Supper* 1963; *Slaves of the Lamp* 1965; *Over the Mountains* 1967; *Colonel Blessington* [ed. D. Raymond] 1968

Non-Fiction

A Manual of Modern Manners 1933; *I Find Four People* (a) 1935; *Jezebel* 1937; *Pen to Paper: a Novelist's Notebook* 1961; *A Letter From R*b*cc* W*st* 1986

(Stella Maria Sarah) Miles FRANKLIN 1879–1954

Franklin was born into a pioneering family (of Irish descent on her father's side, of German and English on her mother's), which had settled in the mountain valleys of the Australian Alps in New South Wales. She lived on her father's cattle station, Brindabella, growing up with her six younger brothers and sisters and sharing with them and neighbouring cousins a Scottish tutor who liked his drink too well but nevertheless gave her a good grounding in literature: 'Shakespeare, the Bible, Dickens, Aesop's Fables – that's what I was brung (or drugged) up on.'

When she was ten, reduced circumstances forced her family to move down from the mountains, first into dairying, then to a small property, Talbingo. Conditions were harsh, and she clashed with her austere mother and was jealous of her pretty younger sister. These were years of growing rebellion, which brought her when she was sixteen to write *My Brilliant Career*, 'conceived and tossed off in a matter of weeks', using her grandfather's name, Miles. She was writing at a time of intense Australian nationalism, and this – as well as life in the bush – affected her work.

My Brilliant Career was finally published in 1901, and brought her instant fame and a notoriety so unwelcome that she withdrew the book and forbade its republication until ten years after her death. In the local community it was taken for fact rather than fiction, and Franklin's distress caused her to leave home, moving to Sydney in 1904, where she worked as a domestic servant for a year. She wrote for the *Bulletin* under a new pseudonym (Mary Anne), and continued to work on the novel she had begun in the intervening years: *My Career Goes Bung*, intended as a corrective sequel to *My Brilliant Career*. Rejected by publishers as too outspoken, this book was not published until 1946.

Franklin then went to America where she stayed for the next ten years, disillusioned with her own country. She worked for the Women's Trade Union League, managing its national office and editing its magazine, *Life and Labour*. During that time she wrote only one book, *Some Everyday Folk and Dawn*, which she dedicated to 'English *men* who believe in votes for *women*'. With the outbreak of the First World War, she went to London, working in slum nurseries before going to Salonika as part of the Scottish Women's Hospital Unit. She returned to London in 1918, and until 1933 worked as a political secretary for the National Housing Council. During this period she published three novels, two of them part of a sequence of six novels forming an interconnected saga of Australian pioneering life. Issued under the pseudonym Brent of Bin Bin, they are considered her best work. Because they made use of childhood memories, Franklin refused to admit that she was Brent of Bin Bin, unable to forget the reaction to her first novel that she had found so creatively stifling.

In 1933 she returned to Australia, settled in Carlton (a suburb of Sydney), and published three novels under her own name. She lived the rest of her life in the house she inherited from her mother, putting her energies into encouraging young Australian writers and into campaigning – for rights for women, for greater opportunities for fellow writers, for the underprivileged, and against war. She never married – (her views on that were made clear in *My Brilliant Career*: a 'most horribly tied-down and unfair-to-women existence') – but had a great many friends. She was described by Eleanor Witcombe (who wrote the screenplay for Gillian Armstrong's acclaimed 1979 film of *My Brilliant Career*) as 'a very gutsy woman, my God she was. A tough old biddy, but she has a marvellous sense of humour. All those 1890s types were tough ladies.'

Fiction

As Miles Franklin: *My Brilliant Career* 1901; *Some Everyday Folk and Dawn* 1909; *Old Blastus of Bandicoot* 1931; *Bring the Monkey* 1933; *All That Swagger* 1936; *Pioneers on Parade* [with D. Cusack] 1939; *My Career Goes Bung* 1946; *On Durban Street* 1981

As 'Brent of Bin Bin': *Up the Country* 1928; *Ten Creeks Run* 1930; *Back to Bool Bool* 1931; *Cockatoos* 1945; *Prelude to Waking* 1950; *Gentlemen at Gyang Gyang* 1956

Non-Fiction

Joseph Furphy: the Legend of a Man and His Book [with K. Baker] 1944; *Laughter, Not for a Cage: notes on Australian Writing* 1956; *Childhood at Brindabella* (a) 1963

Michael FRAYN 1933–

Frayn was born in the London suburb of Mill Hill, the son of a sales rep for an asbestos firm, and he has had a lifelong loyalty to the idea of the suburbs. He was educated at Kingston Grammar School and Emmanuel College, Cambridge, where he read philosophy and was influenced by the work of Wittgenstein. Before university he spent his national service learning to be a Russian interpreter, which has borne late fruit in his acclaimed translations of Chekhov.

He began his working life in 1957 as a *Guardian* journalist, and in 1960 he married his first wife, a social worker. His journalism turned increasingly away from news and towards the writing of humorous columns, something he had already done for the university publications *Varsity* and *Granta*. He went on to write a column for the *Observer*, and his newspaper pieces were collected in *The Day of the Dog*, *The Book of Fub* and *At Bay in Gear Steeet*. He has traced his humour back to his grammar school days, and an early talent for mockery.

Frayn's first public work was a musical comedy, *Zounds!*, written in collaboration with John Edwards and produced in Cambridge in 1957. He has been prolific in several fields, writing acclaimed novels such as *The Russian Interpreter* and *The Trick of It*, numerous plays, a film screen play, *Clockwise*, and even a book of philosophy, *Constructions*. His plays include *Wild Honey*, a highly acclaimed adaptation from Chekhov. The seriousness of *Benefactors*, a play about failing marriage and the dilemmas of being a neighbour, was noted by some critics as a departure from such works as *Noises Off*, which is perhaps his most unashamedly comic play and deals with the unintentionally comic aspects of a disaster-prone production of a second-rate sex farce. Frayn had his own experience of unintentional farce during an isolated stint of acting, when he jammed a door and was unable to leave the set. The audience built up a slow hand clap while technicians on the other side of the door struggled to open it.

Serious undercurrents of social and ethical concern have come increasingly to the fore in Frayn's work. He has been immensely successful as a distinctive satirist, distinguished from the start by his humanity, erudition and intellectual agility. The exceedingly cerebral and civilised quality of his work, together with his left-liberal conscience and a tendency to write about middle-class London social mores, perhaps suggests everything that is good and bad about the adjective 'Hampstead'. Frayn's first marriage broke up in 1989, and he now lives with the biographer and critic Claire Tomalin.

Plays

'Zounds!' [with J. Edwards]; *Jamie on a Flying Visit* (**tv**) 1968 [pub. 1990]; *Birthday* (**tv**) 1969 [pub. 1990]; *The Two of Us* 1970; *The Sandboy* 1971; *Alphabetical Order* 1975 [pub. 1976]; *Clouds* 1976 [pub. 1977]; *Donkeys' Years* 1976 [pub. 1977]; *Liberty Hall* 1979 [rev. 1987; pub. as *Balmoral* 1987]; *Make and Break* 1980; *Noises Off* 1982; *Benefactors* 1984; *Clockwise* (**f**) 1986; *First and Last* (**tv**) 1989; *Listen to This* 1990; *Look Look* 1990; *Audience* 1991; *Here* 1993

Translations

Chekhov: *The Cherry Orchard* 1978, *Wild Honey* 1984, *Three Sisters* 1985 [pub. 1983], *The Seagull* 1986, *Uncle Vanya* 1988 [pub. 1987], *The Sneeze* 1989; Leo Tolstoy *The Fruits of Enlightenment* 1979; Jean Anouilh *Number One* 1984 [pub. 1985]; Yuri Trifonov *Exchange* 1990

Fiction

The Tin Men 1965; *The Russian Interpreter* 1966; *Towards the End of the Morning* 1967; *A Very Private Life* 1968; *Sweet Dreams* 1973; *The Trick of It* 1989; *A Landing on the Sun* 1991; *Now You Know* 1992

Non-Fiction

The Day of the Dog 1962; *The Book of Fub* 1963; *On the Outskirts* 1964; *At Bay in Gear Street* 1967; *Constructions* 1974; *Great Railway Journeys of the World* [with others] 1981; *The Original Michael Frayn: Satirical Essays* 1983

Edited

The Best of Beachcomber 1963; *Timothy: the Drawings and Cartoons of Timothy Birdsall* [with B. Gascoigne] 1964

Brian (i.e. Bernard Patrick) FRIEL 1929–

Friel was born into the Catholic community of Northern Ireland just outside the town of Omagh in County Tyrone, the son of a headmaster. In 1939 the family moved to Derry, where his father took over a larger school, and where Friel passed much of the earlier part of his life. He was educated at St Columb's College in Derry between 1941 and 1946, and then attended St Patrick's College, Maynooth, a seminary for the priesthood. Giving up his plan to become a priest, he trained as a teacher at St Mary Training College (now St Joseph's College of Education) in Belfast between 1949 and 1950, and for ten years worked as a teacher in Derry. He began writing short stories while a teacher, which were quickly accepted by the *New Yorker*. Although he is now best known as a playwright, Friel has remained faithful to writing stories, which have also appeared in volume form.

By the end of the decade, his first radio plays, *A Sort of Freedom* and *To This Hard House*, had been produced by BBC Belfast, and his first stage play, *This Doubtful Paradise*, was also showing in the city. Encouraged by this, and by his regular acceptances from the *New Yorker*, he gave up teaching in 1960, and has since then been a full-time writer.

In 1963, he spent five months with Tyrone Guthrie at the New Guthrie Theater, Minneapolis, where he wrote *Philadelphia, Here I Come!*, in which a young Irishman bids farewell to an inarticulate father, a few friends and an old teacher on the eve of his departure for America. This was his first major success, opening at the Gaiety Theatre, Dublin, but later enjoying a long and successful run in New York, and being filmed. Friel quickly showed himself a consistent master of stagecraft and of dramatic dialogue in the Irish idiom, bringing great humanity to the protrayal of his often rural characters. With **Hugh Leonard** and **Thomas Murphy** he forms the prolific trinity of the leading Irish playwrights.

In 1980, he was co-founder with Stephen Ray of the Field Day Theatre Company, an influential Irish institution. Its first production, in 1980, was *Translations*, which is often considered to be Friel's finest play. It is set in 1833, and deals with the suppression by England of traditional Gaelic culture in Ireland, material handled in a concerned but not narrowly partisan spirit. It enjoyed an extremely successful run at the National Theatre in London. Since then, Friel has devoted a fair amount of his energy to adaptations of Russian classics – there are three: *Three Sisters, A Month in the Country* and *Fathers and Sons* – but he has also continued to write original plays, including the enormously successful *Dancing at Lughnasa* and *Molly Sweeney*, the story of a partially sighted woman who regains her sight with unexpected consequences.

Friel is the recipient of many honours. He became a member of the Irish Academy of Letters in 1972, of Aosdána, the cultural affiliation which is important in Ireland, in 1983, and of the Irish Senate in 1987. In 1954 he married; he and his wife have four daughters and a son, and have lived for some years in County Donegal.

Plays

A Sort of Freedom (r) 1958; *To This Hard House* (r) 1958; *The Francophile* 1960 [rev. as *This Doubtful Paradise*, 1960]; *The Enemy Within* 1962 [pub. 1979]; *The Blind Mice* 1963; *Philadelphia, Here I Come!* 1964 [pub. 1965]; *Loves of Cass McGuire* 1966 [pub. 1967]; *Lovers* 1967 [pub. 1968]; *Crystal and Fox* 1968 [pub. 1970]; *The Mundy Scheme* 1969 [pub. 1970]; *The Gentle Island* 1971 [pub. 1973]; *The Freedom of the City* 1973 [pub. 1974]; *Volunteers* 1975 [pub. 1979]; *Living Quarters* 1977 [pub. 1978]; *Aristocrats* 1979 [pub. 1980]; *Faith Healer* 1979 [pub. 1980]; *American Welcome* 1980 [pub. 1981]; *Translations* 1980 [pub. 1981]; *The Communication Cord* 1982 [pub. 1983]; *Selected Plays of Brian Friel* [pub. 1986]; *Making History* 1988 [pub. 1989]; *Dancing at Lughnasa* 1990; *The London Vertigo* 1990; *Living Quarters* [from Hippolytus] 1992; *The Freedom of the City* 1993; *The Gentle Island* 1993; *Wonderful Tennessee* 1993; *Molly Sweeney* 1994

Translations

Chekhov *Three Sisters* 1981; Turgenev: *Fathers and Sons* 1987, *A Month in the Country* 1992

Fiction

A Saucer of Larks (s) 1962; *The Gold in the Sea* (s) 1966; *Selected Stories* (s) 1979; *The Diviner* (s) 1983

Edited

Charles McGlinchey *The Last of the Name* 1987

Robert (Lee) FROST 1874–1963

Frost was born in San Francisco, the son of a Harvard-educated newspaperman, of Scottish and New England ancestry. His father died when he was ten, and his mother took him to Lawrence, Massachusetts, in the New England country where he was to live for almost the whole of his long life, and which was the major inspiration for his poetry. He received higher education briefly at Dartmouth and Harvard, but took no degree, and filled his early manhood with a variety of jobs: the humblest, as a bobbin-boy in a Massachusetts mill; others as a schoolteacher, newspaper editor and cobbler; the longest, as a farmer in Derry, New Hampshire. Aged twenty-one, he married Eleanor White, and their long life together ended with her death in 1938; they had six children.

He had a book of poems, *Twilight*, privately printed in 1894, but editors were slow to respond to his work, and it was his move to England, in 1912, where he and his wife settled at a farm in Beaconsfield, which was to prove the vital stimulus to gaining recognition as a poet. He became friendly with the Georgian poets and attracted the sponsorship of **Ezra Pound**; his first two volumes of verse were published in England, and appeared in the USA through the representations of Pound to **Harriet Monroe**, editor of *Poetry*. Even more significant was Frost's close friendship with **Edward Thomas**. It was he who turned Thomas into a poet by the simple expedient of suggesting he rewrite his prose in lines of verse, while the relationship was almost equally significant in the broadening of Frost's own talent.

Frost returned to America in 1915, and from then on was famous. The rest of his life was divided between many teaching appointments (he was the first of the major American campus writers), notably at Amherst, Harvard and Yale, and a succession of farms in New Hampshire, Vermont and, finally, Florida. He continued writing poetry prolifically into old age, and in later years was adopted as a sort of unofficial laureate of America, the winner of four Pulitzer Prizes, a representative, as J.F. Kennedy put it, of 'the spirit which informs and controls our strength'. But beneath his crackerbarrel New England image, Frost was less simple than he seemed: his private and professional life contained much grief (a difficult marriage, his own depression, four children who died young or tragically, another daughter who almost broke with him, the continual self-doubt which led him desperately to desire a Nobel Prize that never came), and the country idyll of his verse conceals emptiness and terror, a spirit uncertainly poised between the nineteenth century and modernism. He wrote much indifferent verse, but all his volumes, up to the last, *In the Clearing*, contain fine works, and the best of his poems – 'Mending Wall', 'Birches', 'The Death of the Hired Man',

'The Road Not Taken', 'A Servant to Servants' – are among the most significant of the century.

Poetry
Twilight 1894; *A Boy's Will* 1913; *North of Boston* 1915; *Mountain Interval* 1916; *New Hampshire* 1923; *Selected Poems* 1923; *Collected Poems* 1930; *A Further Range* 1936; *A Witness Tree* 1942; *Come In and Other Poems* 1943; *A Masque of Reason* 1945; *A Masque of Mercy* 1947; *Steeple Bush* 1947; *Complete Poems of Robert Frost* 1949; *Aforesaid* 1954; *You Come Too* 1959; *In the Clearing* 1962; *The Poetry of Robert Frost* 1969; *The Poetry of Robert Frost* [ed. E.C. Lathem] 1972

Plays
The Cow's in the Corn 1929; *A Way Out* 1929; *The Guardeen* 1943

Fiction
Stories for Lesley (s) [ed. R.D. Snell] 1984

Non-Fiction
Robert Frost and John Bartlett: the Record of a Friendship [ed. M.B. Anderson] 1963; *The Letters of Robert Frost to Louis B. Untermeyer* [ed. L.B. Untermeyer] 1963; *Selected Letters of Robert Frost* [ed. L. Thompson] 1964; *Interviews with Robert Frost* [ed. E.C. Lathem] 1966; *Family Letters of Robert and Elinor Frost* [ed. A. Grade] 1972; *Robert Frost on Writing* [ed. E. Barry] 1973; *Robert Frost: a Time to Talk* [ed. R. Francis] 1973; *Robert Frost and Sidney Cox: Forty Years of Friendship* [ed. W.R. Evans] 1981

Collections
Selected Prose of Robert Frost [ed. H. Cox, E.C. Lathem] 1966

Christopher FRY 1907–

Largely as a matter of euphony, Fry took his surname from his mother, who was a member of the famous Quaker family. His father was Charles Hammond, an architect and lay missionary in Bristol, where his son was born. Fry was educated at Bedford Modern School where he began writing verse drama and poetry, but had no particular success at his English class studies. When he left school at the age of eighteen, he taught briefly, and acted for a year in 1927 with the Bath Repertory Company under his newly acquired name. A second acting career, lasting for eight years, followed a brief spell of teaching at Hazelwood Preparatory School (1928–31).

Fry has declared that his life is 'too complicated for tabulation' from 1935 to 1939, but he founded the Tunbridge Wells Repertory Players in 1932, worked for the Dr Barnado's organisation as a magazine editor between 1934 and 1939, and in 1936 married the journalist Phyllis Hart. In 1938, the couple were rescued from penury by a legacy, and Fry wrote a pageant, *The Boy with a Cart*, which was produced in the same year. He temporarily abandoned directorship of the Oxford Playhouse owing to the war, and, as a conscientious objector, travelled England to clear bomb damage, and fought blitz fires at Liverpool Docks. When the war ended, he staged *A Phoenix Too Frequent* at the London Arts Theatre Club in 1946, after which it successfully transferred to the West End.

Fry often shares equal honours with **T.S. Eliot** for the revival of verse drama, which **Terence Rattigan** considered had 'rescued the theatre from the thraldom of middle-class vernacular'. Unlike Eliot, Fry is often exuberant, with lines crammed full of imagery. His greatest success was *The Lady's Not for Burning*, which was awarded the Shaw Prize Fund for the best play of 1948. Of considerable importance are his translations of Anouilh, including *The Lark* and *Ring Round the Moon*. Fry lives in West Sussex, and invariably writes late at night through to the early morning, straight on to a typewriter.

Plays
She Shall Have Music (l) [from F. Eyton; with M. Crick, R. Frankau] 1934; *Youth and the Peregrines* 1934; *To Sea in a Sieve* 1935; *Open Door* 1936; *The Boy with a Cart* 1938 [pub. 1939]; *Thursday's Child* 1939; *The Tower* 1939; *The Firstborn* (r) 1947 [pub. 1946]; *A Phoenix Too Frequent* 1946; *The Lady's Not for Burning* 1948 [pub. 1949]; *Rhineland Journey* (r) 1948; *Thor, with Angels* 1948; *The Canary* (tv) 1950; *Venus Observed* 1950; *A Sleep of Prisoners* 1951; *The Beggar's Opera* (f) [from J. Gay's opera; with D. Cannon] 1953; *The Dark Is Light Enough* 1954; *Ben Hur* (f) [from Lewis Wallace's novel] 1959; *Curtmantle* 1961; *Barabbas* (f) 1962; *The Bible: In the Beginning* [with J. Griffin] (f) 1966; *The Tenant of Wildfell Hall* (tv) [from A. Brontë's novel] 1968; *A Yard of Sun* 1970; *The Brontës of Haworth* (tv) 1973; *The Best of Enemies* (tv) 1976; *Sister Dora* (tv) [from J. Manton] 1977; *Paradise Lost* (l) [from Milton's poem; with K. Penderecki] 1978; *One Thing More* 1986

Translations
Jean Anouilh: *Ring Round the Moon* 1950 [pub. 1948], *The Lark* 1955; Jean Giraudoux: *Tiger At the Gate* 1955 [*The Trojan War Will Not Take Place* 1983], *Duel of Angels* 1958, *Judith* 1962 [pub. 1963]; Colette *The Boy and the Magic* 1964; Ibsen *Peer Gynt* [with J. Fillinger] 1970; Edmond Rostand *Cyrano de Bergerac* 1975

Non-Fiction
An Experience of Critics 1952; *Can You Find Me: a Family History* 1978; *Death Is a Kind of Love* 1979; *Genius, Talent and Failure* 1987

For Children

The Boat That Mooed 1966

Athol (Harold) FUGARD (Lannigan) 1932–

Fugard was born in Cape Province, the semi-desert Karroo region of South Africa. In 1935 the family moved to Port Elizabeth and opened a small boarding-house. On graduating from the University of Cape Town, where he studied French literature and social anthropology, he travelled to the Far East as a crew member on a steamer. When he returned to South Africa he worked as a journalist on the Port Elizabeth *Evening Post*. In 1956 he met and married the actress Sheila Meirling. They moved to Johannesburg in 1957 and Fugard worked as a clerk at the Fordsburg Native Commissioners' Court, where pass-law offenders were tried and where he says he found his 'basic pessimism'.

In the multi-racial ghetto of Sophiatown, Fugard and his wife set up a theatre workshop. His first play, *No-Good Friday*, a study of deprived lives in South African townships, was performed there in 1958. Although in the following years his plays commanded audiences all over the world, their staging has retained the theatrical limitations imposed by township poverty. In 1959 the Fugards moved to London, working as office cleaners so that they could afford to buy theatre tickets. However, in March 1960, seventy-two people were massacred in Johannesburg for contesting the pass-laws, and the couple immediately returned to South Africa, where they have lived ever since, in Port Elizabeth, leaving only for Fugard to hold short-term posts in the literature departments at the Universities of Yale and Georgetown in the USA. They have one daughter.

The action of Fugard's plays takes place in the landscapes and ghettos of South Africa and is instigated by South African politics. His characters tend to be victims – whatever their race – and the intensely compelling human situations become metaphors that embody the tensions of South African society and illuminate the politics. Most of his work directly confronts apartheid and is based on his own experiences and encounters. Some plays testify to the effects of specific laws. *Sizwe Bansi Is Dead*, for example (written in collaboration with the African actors Winston Nshota and John Kani), deals with the effect on individuality of dividing people into three separate categories; the play begins with a death and ends with the symbolic exchanging of identity cards with the dead man. *'Master Harold' and the Boys* is a very personal drama that refers to a single incident when, as a boy, Fugard aggressively asserted what he believed to be his racial superiority over Sam Semela, who worked as a waiter at his mother's boarding-house.

The Blood Knot was the first of his plays to be performed in the USA and the *New York Times* voted it the best play of the year. In 1967 it was shown on British television. As a playwright, Fugard is renowned for increasing the input of his actors by working on improvisations and then writing minimum dialogue. He has acted in many of his plays himself and in 1972 founded The Space experimental theatre in Cape Town. His awards have been numerous and include an Obie in 1971 and the Commonwealth Award in 1984. He is as esteemed as a theatrical director as he is as a political author and humanist and ranks amongst the most significant playwrights of the twentieth century.

Plays

Dimetos and Two Early Plays [pub. 1977]: *Nongogo* 1959, *No-Good Friday* 1958, *Dimetos* 1975; *Three Port Elizabeth Plays* [pub. 1974]: *The Blood Knot* 1961, *Hello and Goodbye* 1965, *Boesman and Lena* **(st)** 1969, **(f)** 1973 [first pub. 1969]; *The Coat* 1966 [pub. 1971]: *People Are Living There* 1968 [pub. 1970]: *Orestes* [from Aeschylus' plays] 1971 [pub. 1978]: *Statements* [pub. 1974]: *Statements After an Arrest under the Mortality Act* 1972, *Sizwe Bansi Is Dead* [with W. Nshota, J. Kani] 1972, *The Island* 1973; *The Guest* 1977; *Boesman and Lena and Other Plays* [pub. 1978]; *A Lesson from Aloes* 1978 [pub. 1981]: *The Drummer* 1980 [np]; *Marigolds in August* **(f)** 1980 [pub. 1982]; *'Master Harold' and the Boys* 1982; *The Road to Mecca* 1984 [pub. 1985]; *A Place with Pigs* 1987 [pub. 1988]; *Selected Plays* [pub. 1987]; *My Children! My Africa!* 1990; *Playland* 1992; *The Township Plays* [ed. D. Walder] 1993

Fiction

Tsotsi 1980

Non-Fiction

Notebooks 1960–1977 [ed. M. Benson] 1983

John (Leopold) FULLER 1937–

Fuller was born in Ashford, Kent, the only child of the poet and solicitor **Roy Fuller**. His early childhood was mostly spent in Lancashire, during the war, but then he grew up in London, attending St Paul's School and seeing his first poem published in *The Listener* when he was sixteen. After national service in the Royal Air

Force, he read English at New College, Oxford, graduating in 1960 and winning the Newdigate Prize for Poetry in that year. His first collection, *Fairground Music*, was published in 1961, immediately winning a reputation.

After Oxford, Fuller spent a brief period as a freelance writer, but was then a lecturer at the State University of New York at Buffalo between 1962 and 1963, an assistant lecturer at the University of Manchester from 1963 to 1966, and from that date a fellow and tutor in English at Magdalen College, Oxford, where he remains. He married in 1960, and has three daughters, one of whom is Sophie Fuller, the translator.

Fuller has published many volumes of verse, several novels, books for children, numerous musical plays in collaboration with the composer Bryan Kelly and a variety of non-fiction. He has gone on record as saying that he has 'never felt that there was any inherent virtue in powerful feelings and simplicity', and his work is what one would expect from such a statement, quintessentially donnish in its combination of the virtuosic and the tight-lipped. His books are often fantastical, especially in writing to do with Wales, where he lives for part of each year. His first novel, *Flying to Nowhere* (1983), which was shortlisted for the Booker Prize and won a Whitbread Award, was a dazzlingly inventive fable set in a sixteenth-century monastery on a Welsh island. Fuller's verse is full of technical challenges surmounted, such as the long poem *The Illusionists*, where he utilises the stanza form of Pushkin's *Eugene Onegin* (1831), very difficult to bring off in English. He treats the unlikely subject of nineteenth-century chess in another well-known poem, 'The Most Difficult Position'. He has won both the Geoffrey Faber Memorial Prize for poetry in 1974 and a Cholmondeley Award in 1983 for *The Beautiful Inventions*.

Among his later novels most attention was gained by *The Burning Boys*, where the interesting subject of the relationship between a boy and an airman during the Second World War was somewhat vitiated by Fuller's distance. He founded the Sycamore Press in 1968 and this, combined with his teaching at Oxford, has enabled him powerfully to influence a generation of younger poets. **James Fenton** was first published by him, is a close friend, has collaborated with him on the work *Partingtime Hall*, and has described him as a 'secret guru'. **Alan Hollinghurst**, whose career began with poetry, and who was Fuller's student at Magdalen, is another to be influenced (he now edits the literary competition *Nemo's Almanac*, once run by Fuller). Fuller's other works include the invaluable *A Reader's Guide to W.H. Auden*, *The Worm and the Star*, a book of very brief stories, and editorship of *The Chatto Book of Love Poetry*.

Poetry
Fairground Music 1961; *The Tree That Walked* 1967; *The Labours of Hercules* 1969; *The Wreck* 1970; *Boys in a Pie* 1972; *Cannibals and Missionaries* 1972; *Epistles to Several Persons* 1973; *Penguin Modern Poets 22* [with A. Mitchell, P. Levi] 1973; *The Mountain in the Sea* 1975; *Selected Poems 1954–1982* 1985; *Bel and the Dragon* 1977; *Lies and Secrets* 1979; *The Illusionists* 1980; *The January Divan* 1980; *Waiting for the Music* 1982; *The Beautiful Inventions* 1983; *Partingtime Hall* [with J. Fenton] 1987; *The Grey Among the Green* 1988; *Tell It Me Again* 1988; *A Floribundum* [with others] 1990; *The Mechanical Body* 1991

Fiction
Flying to Nowhere 1983; *The Adventures of Speedfall* (s) 1985; *The Burning Boys* 1989; *Look Twice* 1992; *The Worm and the Star* (s) 1993

For Children
A Bestiary 1974; *Squeaking Crust* 1974; *The Last Bid* 1975; *The Extraordinary Wool Mill and other Stories* (s) 1980; *Come Aboard and Sail Away* 1983

Plays
With B. Kelly: *Herod, Do your Worst* 1967 [pub. 1968]; *Half a Fortnight* 1970 [pub. 1973]; *Fox-Trot* 1971; *The Spider Monkey Uncle King* 1971 [pub. 1975]; *The Queen in the Golden Tree* 1974 [np]; *Adam's Apple* 1975 [np]; *Linda* 1975 [np]; *The Ship of Sounds* 1975 [pub. 1981]; *St Francis of Assissi* 1981 [np]

Non-Fiction
A Reader's Guide to W.H. Auden 1970; *The Sonnet* 1975

Edited
New Poems 1967 [with H. Pinter, P. Redgrove] 1968; *The Dramatic Works of John Gay* 1982; *New Poetry 8* 1982; *The Gregory Poets 1983–1984* [with H. Sergeant] 1984; *The Chatto Book of Love Poetry* 1990

Roy (Broadbent) FULLER 1912–1991

Fuller was born at Failsworth, Lancashire, the son of a businessman, and of Portuguese Jewish ancestry. He attended Blackpool High School, and became a solicitor. He prospered in both his business and his literary career, in which respect he resembles the American poet **Wallace Stevens**. He first came to attention as a poet during the Second World War, when he was serving in the navy, and two volumes were published by the Hogarth Press. His legal career was with the Woolwich Building Society in

London, first as assistant solicitor between 1938 and 1958, then solicitor from 1958 to 1969, then as a director after his retirement. What is often considered his best novel, *Image of a Society* (1956), is based on his experience at the Woolwich. In later life, he also functioned as a cultural bureaucrat, serving a controversial term as literature chairman of the Arts Council (he disliked the funding of community arts projects and other avant-garde ideas) as well as being a governor of the BBC. He carried his cultural conservatism into his tenure of the Professorship of Poetry at Oxford between 1968 and 1973, although, curiously in this context, he was until almost the end of his life a Marxist. He published two volumes of his Oxford lectures.

In the 1930s Fuller was influenced by **W.H. Auden** both artistically and politically. This shows in his first collection, *Poems* (1939). His work was first published in *Penguin New Writing* and the *Listener*; he became a close friend of **J.R. Ackerley** and a regular contributor of reviews to the BBC magazine, of which Ackerley was literary editor. After the war, he became connected with the Movement, although he was older than its other members. His early poetry was sensuous, his later work more ironic and detached, and his reputation can be gauged from the fact that his *Collected Poems* were published as early as 1962.

Fuller used a break from poetry after 1945 to establish himself as a novelist. He wrote first a novel for children, *With My Little Eye* (1948), perhaps the only really good detective story about a boy detective. This was followed by three excellent crime novels for adults and then six non-genre novels. With the exception of one late work, *Stares*, he wrote no more novels after *The Carnal Island* in 1970. The posthumous publication of his *Last Poems*, however, was the final fruit of a remarkable late flowering as a poet, a phase in which he once again became more personal, and more concerned with the individual's approach to problems such as age.

In his last decade, he also wrote several guarded autobiographies, which have a complex relationship to each other. He first published three volumes – *Souvenirs, Vamp Till Ready* and *Home and Dry* – which appeared in a subsequent redaction, *The Strange and the Good*. In the year of his death, he added a further, separate, volume, *Spanner and Pen*.

There is something rather dry about Fuller as an artistic personality which, however honourable, will always limit his appeal (it was perhaps symptomatic that, as a young man, he grew a moustache to hide his good looks). He married in the late 1930s and his only son is the poet, novelist and Oxford don **John Fuller**.

Poetry

Poems 1939; *Epitaphs and Occasions* 1940; *The Middle of a War* 1942; *A Lost Season* 1944; *Counterparts* 1954; *Brutus's Orchard* 1957; *Collected Poems 1936–1961* 1962; *Buff* 1965; *New Poems* 1968; *Tiny Tears* 1973; *From the Joke Shop* 1975; *The Joke Shop Annexe* 1975; *An Ill-Governed Coast* 1976; *The Reign of Sparrows* 1980; *More about Tompkins and Other Light Verse* 1981; *As from the Thirties* 1983; *Upright Downfall* [with B. Giles, A. Rumble] 1983; *Mianserin Sonnets* 1984; *New and Collected Poems 1934–1984* 1985; *Twelfth Night: a Personal View* 1985; *Subsequent to Summer* 1985; *Outside the Canon* 1986; *Consolations* 1987; *Last Poems* 1993

Fiction

The Second Curtain 1953; *Fantasy and Fugue* 1954; *Image of a Society* 1956; *The Ruined Boys* 1959; *The Father's Comedy* 1961; *The Perfect Fool* 1963; *My Child, My Sister* 1965; *The Carnal Island* 1970; *Stares* 1990

For Children

Savage Gold 1946; *With My Little Eye* 1948; *Catspaw* 1966; *Seen Grandpa Lately?* 1972; *Poor Roy* 1977; *The Other Planet* 1979; *Upright Downfall* 1983

Non-Fiction

Owls and Artificers 1971; *Professors and Goods* 1973; *Souvenirs* (a) 1980, *Vamp Till Ready* (a) 1982, *Home and Dry* (a) 1984 (3 vols [rev. in 1 vol. as *The Strange and the Good* 1989]; *Spanner and Pen* 1991; Rudyard Kipling *Proofs of Holy Writ* 1981

Edited

The Penguin New Writing, 1940–1950 [with J. Lehmann] 1985

G

William GADDIS 1922–

Gaddis was born in Manhatten and brought up by his Quaker mother on Long Island where the family owned property. After attending a boarding-school at Berlin, Connecticut, and the Farmingdale High School, Long Island, he went in 1941 to Harvard where he edited *Lampoon* magazine. He was disappointed when a kidney ailment disqualified him from serving in the Second World War, and was 'asked to leave' Harvard in 1945 following a run-in with the local police. He soon found work as a fact-checker at the *New Yorker*, and while living in New York became acquainted with the as yet unknown group of writers which included **Allen Ginsberg**, **Jack Kerouac** and **William Burroughs**. From 1947 he travelled widely in Central America, North Africa and Europe while working on his first novel, *The Recognitions* (1955). The novel takes its title from a third-century religious romance, and is a huge and ambitious work filled with arcane scholarship. Through the central figure of Wyatt Gwyon, an art forger, Gaddis explores themes of imitation and hypocrisy, and repeatedly highlights the inauthenticity of American experience.

The almost universal critical hostility which greeted the novel upon its publication forced Gaddis into a twenty-year silence during which he supported himself by writing speeches for corporate executives and scripts for industrial and military films. At one particularly low point he was obliged to ghostwrite professional journal articles for his dentist in exchange for much-needed root canal work. Despite its initial failure, *The Recognitions* achieved a minor cult status on college campuses (where obscure and obsessive journals devoted to the work circulated) and clearly influenced **Thomas Pynchon**'s novel *V* (1963). Critical reappraisal, notably in Tony Tanner's *City of Words* (1971), and two grants awarded in 1963 and 1967 encouraged Gaddis to work on his second novel, *JR*, in which little of his experience of corporate and military jargon is wasted. The novel is a mordant satire on American life revolving around the figure of a precocious eleven-year-old boy who comes to control a vast business empire. In stark contrast to the reception which greeted *The Recognitions, JR* was lauded, one reviewer being moved to describe, it as 'the long-awaited great American novel', and it won a 1976 National Book Award.

In 1982 Gaddis received a MacArthur Foundation award and was given a five-year stipend of $280,000 which enabled him to write *Carpenter's Gothic*, a shorter and more accessible novel which nevertheless addresses itself to the familiar Gaddis themes of deception, falsehood, media saturation, usury, and what he has called 'the amoral – which is to say simple, naked, cheerful greed'. Similar matters are addressed in *A Frolic of His Own*, which anatomises the American legal system.

Gaddis has two children from an early marriage, but in common with his friend **Don DeLillo** is secretive about his personal life. With the publication of *Carpenter's Gothic* he agreed to give interviews for the first time and explained his reticence about personal publicity thus: 'I don't want to be seen in *People* magazine romping with my dog.'

Fiction

The Recognitions 1955; *JR* 1975; *Carpenter's Gothic* 1985; *A Frolic of His Own* 1994

Mary GAITSKILL 1954–

Gaitskill was born in Lexington, Kentucky, and brought up near Detroit. Her father is a teacher, and her mother a social worker. After a troubled adolescence, including a brief spell in psychiatric counselling, she moved to New York at the age of sixteen and became a stripper. From there she travelled around North America 'selling flowers and jewelry or begging on the streets', before deciding to qualify herself to return to full-time education. She graduated from the University of Michigan in 1981, having won a Jule and Avery Hopwood Award for a college-published selection of stories, *The Women Who Knew Judo and Other Stories*, and then returned to New York, the setting for all of the stories in *Bad Behaviour*. Originally published by Poseidon Press, the collection was glowingly reviewed and quickly taken up by a larger publishing house.

Gaitskill has complained of the attention many reviewers have paid to her own briefly outlined biography which she considers to be 'of limited relevance' to her writing, explaining that her interest in the book was to describe the inner worlds and safe havens created by her characters – prostitutes, drug addicts, bored clerics – in the mistaken search for intimacies they have never enjoyed.

Her first novel, *Two Girls, Fat and Thin*, was published in 1991. The novel is less concerned with misfits than with victims: the two girls of the title are brought together, after variously traumatic upbringings, by a shared interest in the philosophy of the 1960s seer Anna Granite, a barely fictionalised portrait of writer Ayn Rand. The novel was praised as a bitter and perceptive critique of the 'Greed Is Good' ethos and the doublethink of American schooling. Gaitskill now lives in California and New York.

Fiction

Bad Behaviour (s) 1989; *Two Girls, Fat and Thin* 1992

Mavis GALLANT 1922–

Gallant was born Mavis de Trafford Young in Montreal and was sent to a Jansenist convent at the age of four. 'They were extremely strict,' she has said. 'Toys were forbidden, we dressed in black, and our sole recreation consisted of walking up and down a wooden platform in the garden.' After the premature death of her father, and her mother's remarriage, she was sent to a series of seventeen different schools in Canada and the USA. It was a solitary and often lonely childhood which informs much of her subsequent fiction. Aged eighteen she returned from New York to Montreal and was briefly married to Johnny Gallant, a pianist. Following her divorce she began work in 1944 as a reporter and critic on the Montreal *Standard*. She published short stories in a number of Canadian magazines and in 1950 had a story, 'Madeline's Birthday', accepted by the *New Yorker*, where she has continued to publish ever since. She used the $600 she earned from the story to finance her move to Paris where she has continued to live whilst travelling extensively in Europe.

The stories in her first collection, *The Other Paris*, explore themes which recur throughout her work: the contrast between European and American values, and the unhappiness of those whose dreams are never matched by their experience of real life. Her first novel, *Green Water, Green Sky*, describes an American mother and daughter drifting through Europe and explores the self-destructive nature of their mutual love. Like all of her work, it is notable for its elegant language and poignant descriptions of landscape. In 1981 Gallant received the Governor-General's Award for her collection *Home Truths*, and in 1985 the Canada–Australia Literature Prize.

In addition to her fiction, she is a noted essayist and political commentator, and her only play, *What Is to Be Done?*, is a satire on the political position of women in wartime Canada.

Fiction

The Other Paris (s) 1956; *Green Water, Green Sky* 1959; *My Heart Is Broken* (s) 1964 [UK *An Unmarried Man's Summer* 1965]; *A Fairly Good Time* 1970; *The Pegnitz Junction* (s) 1973; *The End of the World and Other Stories* (s) 1974; *From the Fifteenth District* (s) 1979; *Home Truths* (s) 1981; *Overhead in a Balloon* (s) 1985; *In Transit* (s) 1988; *Across the Bridge* (s) 1994

Non-Fiction

Paris Notebooks: Essays and Reviews 1986

Plays

What Is to Be Done? 1982 [pub. 1983]

John GALSWORTHY 1867–1933

Born in Kingston-upon-Thames, Surrey, Galsworthy was the son of a successful Devonshire lawyer with substantial City interests. He went from Harrow to New College, Oxford – where his main interests were horse-racing and gambling – and then began desultory legal studies, interrupting them with an extended tour of Australasia and the South Seas. Sailing to Cape Town on the clipper *Torrens*, he was encouraged to write by the first mate, who happened to be **Joseph Conrad** ('The first mate is a Pole called Conrad and is a capital chap, though queer to look at,' he wrote in a letter home). Returning to England, he practised briefly as a barrister and, collecting rents for his father in London slum districts, was fired by an abiding passion for social justice, but soon turned more and more to a writing career.

His earliest books, the first of which appeared in 1897, were novels and short stories, and published under the pseudonym of John Sinjohn: he began using his own name for his first substantial novel, *The Island Pharisees* (1904). From the mid-1890s he had begun a liaison with Ada Nemesis Pearson Cooper, who was married to his cousin, an army major; after the couple were divorced, Galsworthy married her in 1905, and these circumstances contributed greatly to the plot of *The Man of Property*, which was to become (although not planned as such at the time) the first volume of his Edwardian family chronicle, *The Forsyte Saga*. The same year saw Galsworthy's first play, *The Silver Box*; he was to become one of the most popular of Edwardian and Georgian playwrights, and one who specialised in confronting thorny moral and social issues. *Justice*, an attack on the practice of solitary confinement in prisons, contributed powerfully to reform.

Galsworthy's wife was both a hypochondriac and addicted to ceaseless travel, while he lent himself to innumerable good causes, and this may account for the fact that his *oeuvre*, although quite large, increasingly failed to fulfil his early promise. The second two volumes of *The Forsyte Saga*, published in 1920 and 1921, and the three of a further saga, showed a growing sentimental love of what he had once satirised. He was showered with public honours, and awarded the Nobel Prize in 1932, but had already fallen into the premature senile dementia which killed him, and could not collect the prize in person. His reputation did not long survive his death, although the television dramatisation of *The Forsyte Saga* in the 1960s attained a popularity which has never been equalled in British television, and his plays are still often revived. Besides novels and plays, he also published poetry, essays, literary criticism and letters.

Fiction

As 'John Sinjohn': *From the Four Winds* (s) 1897; *Jocelyn* 1898; *A Man of Devon* (s) 1901 [rev. as *Villa Rubein and Other Stories* 1909]

As John Galsworthy: *The Island Pharisees* 1904; *The Forsyte Saga* 1922: *The Man of Property* 1906, *In Chancery* 1920, To Let 1921; *The Country House* 1907; *A Commentary* (s) 1908; *Fraternity* 1909; *A Motley* (s) 1910; *The Patrician* 1911; *The Inn of Tranquillity* (s) 1912; *The Dark Flower* 1913; *The Freelands* 1915; *The Little Man* 1915; *Beyond* 1917; *Five Tales* (s) 1918; *The Burning Spear* 1919; *Saint's Progress* 1919; *Captures* (s) 1923; *A Modern Comedy* 1929: *The White Monkey* 1924, *The Silver Spoon* 1926, *Swan Song* 1928; *Two Forsyte Interludes* 1927; Caravan (s) 1925; *Four Forsyte Stories* (s) 1929; *On Forsyte Change* (s) 1930; *Maid in Waiting* 1931; *Flowering Wilderness* 1932; *Over the River* 1933; *Forsytes, Pendyces and Others* 1935

Plays

The Silver Box 1906 [pub. 1910]; *Joy* 1907 [pub. 1910]; *Strife* 1909 [pub. 1910]; *Justice* 1910; *The Little Dream* 1911; *The Eldest Son* 1912; *The Pigeon* 1912; *The Fugitive* 1913; *The Mob* 1914; *A Bit o' Love* 1915; *The Foundations* 1917 [pub. 1920]; *The Skin Game* 1920; *A Family Man* 1921 [pub. 1922]; *Six Short Plays* [pub. 1921]; *Loyalties* 1922; *Widows* 1922; *The Forest* 1924; *Old English* 1924; *The Show* 1925; *Escape* 1926; *Exiled* 1929; *The Roof* 1929; *The Winter Garden* [pub. 1935]

Poetry

Moods, Songs and Doggerels 1912; *The Bells of Peace* 1921; *Verses New and Old* 1926; *The Collected Poems of John Galsworthy* 1934

Non-Fiction

A Justification for the Censorship of Plays 1909; *The Spirit of Punishment* 1910; *For Love of Beasts* 1912; *The Slaughter of Animals for Food* 1913; *Treatment of Animals* 1913; *A Sheaf* 1916; *Your Christmas Dinner Is Served!* 1916; *The Christmas Jewel Fund* 1918; *The Land* 1918; *Addresses in America* 1919; *Another Sheaf* 1919; *International Thought* 1923; *Memorable Days* 1924; *On Expression* 1924; *Is England Done?* 1925; *Castles in Spain* 1927; *Horses in Mines* 1927; *The Way to Prepare Peace* 1927; *Gentles, Let Us Rest* 1928; *The Creation of Character in Literature* 1931; *Letters from John Galsworthy 1900–1932* [ed. E. Garnett] 1934; *Glimpses and Reflections* 1937; *Letters to Margaret Morris* 1967; *John Galsworthy's Letters to Leon Lion* [ed. A.B. Wilson] 1968

Jane (Mary) GARDAM 1928–

Gardam was born Jane Pearson, and grew up in Coatham, Yorkshire, where her father was a housemaster at a school for boys for forty-two years; she herself attended Saltburn High School for Girls. Although the Yorkshire and Northumberland scenes of her girlhood recur frequently in her work, she came south for good in 1946, when she began reading English at Bedford College, London. After graduation, and a period of postgraduate research, she spent two years as a librarian with the Red Cross hospital libraries. She then became a journalist, first as an editorial assistant on *Weldon's Ladies Journal* from 1952 to 1953, and then, for two years, assistant literary editor for *Time and Tide*. In 1952, she had married David Gardam, a barrister and later QC, with whom she has had three children; she gave up working in 1956 when the first of these (Tim Gardam, later well known for his television career) was born.

For a long period she was primarily a housewife and mother, and her first novel, *A Few Fair Days*, was not published until 1971. It appeared, as did its two successors, in Hamish Hamilton's children's list, although she had not set out initially to write for children, and several books which appeared first in this form, such as *A Long Way from Verona* and *Bilgewater*, have since been republished for adults. Her first unequivocally adult novel, *God on the Rocks*, was shortlisted for the Booker Prize in 1978, while *The Queen of the Tambourine* won the Whitbread Award for a novel in 1991. Two years previously, she had become involved in controversy as a judge for the same category of the Whitbread, when she successfully blocked the giving of the award to Alexander Stuart's *The War Zone* because of its treatment of incest (this was in spite of the fact that Stuart had already been informed he had won the prize).

She has published several collections of short stories, of which the first, *Black Faces, White Faces* won both the David Higham and Winifred Holtby Prizes. Some of these stories were inspired by a trip to Jamaica; her husband's practice has been international, and their travels, particularly to the Far East, are another background for her work. Of her other collections, *The Pangs of Love* won the Katherine Mansfield Award, and *Going Into a Dark House* won PEN's Macmillan Silver Pen Award. Gardam is much admired for her spare and humorous style and the unsentimental but affecting tone of her work. She has published one non-fictional work, *The Iron Coast*, a highly illustrated portrait of the East Yorkshire coast she knew in her girlhood.

Fiction

A Few Fair Days 1971; *A Long Way from Verona* 1971; *The Summer After the Funeral* 1973; *Black Faces, White Faces* (s) 1975; *The Sidmouth Letters* (s) 1975; *Bilgewater* 1977; *God on the Rocks* 1978; *The Pangs of Love and Other Stories* (s) 1983; *Crusoe's Daughter* 1985; *Showing the Flag* (s) 1989; *The Queen of the Tambourine* 1991; *Trio* (s) 1993; *Going into a Dark House and Other Stories* (s) 1994

Non-Fiction

The Iron Coast [with P. Burton, H. Walshaw] 1994

John (Champlin) GARDNER 1933–1982

Gardner was born in Batavia, a small town near Rochester, New York, which provides the setting for two of his novels, *The Resurrection* and *The Sunlight Dialogues*. His father was a farmer and his mother an English teacher, and both were immersed in literature. 'When I was a little kid,' Gardner once recalled, 'I was writing poetry and, in fact, novels.' He studied at DePauw University from 1951, graduated from Washington University, St Louis, in 1955, and received his PhD in classical and medieval literature from Iowa State University in 1958. From 1958 he taught at Oberlin College, Chico State College, California, and San Fransisco State College, before settling in 1965 at Southern Illinois University, Carbondale, where he remained until 1976. From 1978 Gardner was head of the creative writing course at the State University of New York, Binghampton. He was married in 1958 to Joan Louise Patterson, a pianist and composer, who collaborated closely on his work. In addition to fiction, he published modern English versions of medieval poetry and a number of scholarly introductions to medieval literature.

His first novel, *The Resurrection*, describes a young philosopher dying of leukaemia who returns to his native town. It is, in Gardner's words, 'the story of modern man's "absurd" search for the answer to what he knows beforehand is a meaningless question: "What is the meaning of life?" ' Philosophical concerns recur in his subsequent novels *The Wreckage of Agathon*, which is set in ancient Sparta, and *Grendel*, which retells the story of *Beowulf* from the point of view of the monster. Despite being compared with other academic novelists such as **John Barth** and **William Gass**, Gardner was in many ways a fierce traditionalist, and in *On Moral Fiction* he decried the absence of moral thought in many of his contemporaries, declaring that art is good 'only when it has a clear positive moral effect, presenting valid models for imitation, eternal verities worth keeping in mind, and a benevolent vision of the possible which can incite human beings towards virtue.'

Gardner was a notable philanderer, around whom many legends and stories grew. On one occasion, his wife flew over the grounds of the college where he was teaching and dropped leaflets listing his infidelities. Even his death remains a mystery: he was killed when his motorcycle crashed near his home in Susquehanna, Pennsylvania, but it has never been fully established whether this was an accident or suicide.

Fiction

The Resurrection 1966; *The Wreckage of Agathon* 1970; *Grendel* 1971; *The Sunlight Dialogues* 1972; *Nickel Mountain* 1973; *The King's Indian* (s) 1974; *October Light* 1976; *Freddy's Book* 1980; *The Art of Living* (s) 1981; *Mickelsson's Ghosts* 1982

Non-Fiction

The Construction of the Wakefield Cycle 1974; *The Construction of Christian Poetry in Old English* 1975; *The Life and Times of Chaucer* 1977; *The Poetry of Chaucer* 1977; *On Moral Fiction* 1978; *The Art of Fiction: Notes on Craft for Young Writers* 1984; *On Writers and Writing* [ed. S. O'Nan] 1994

Poetry

Jason and Medeia 1973

For Children

Dragon, Dragon 1975; *Gudekin the Thistle Girl* 1976; *A Child's Bestiary* 1977; *In the Suicide Mountains* 1977

Plays
Three Libretti (l) [with J. Baber] 1979: *Frankenstein, Rumpelstiltskin, William Wilson*

Translations
The Complete Works of the Gawain-Poet 1965; *The Alliterative Morte Arthure, The Owl and the Nightingale, and Five Other Middle English Poems* 1971

Edited
The Forms of Fiction [with L. Dunlap] 1961; *Papers on the Art and Age of Geoffrey Chaucer* [with N. Joost] 1967

Alan GARNER 1934–

Born at Congleton, Cheshire, Garner grew up in the Cheshire village of Alderley Edge, speaking a local dialect. He has lived in his native county for almost the whole of his life, and its history, archaeology and folklore are the mainsprings of his work. After an early childhood of severe illness, and attending the village school, he won a scholarship to Manchester Grammar School (the first of his family to attain such education) and became an outstanding young athlete; at eighteen he was considered by many to be the best young sprinter in Britain. He did his national service in the Royal Artillery, attaining the rank of second lieutenant, and then read classics at Magdalen College, Oxford, coming down however, without taking a degree. He returned to near Alderley Edge in 1956 (where he still lives) and began to write. In the same year, he married his first wife, with whom he had three children, and from whom he was divorced. He married again in 1972, and has two children from his second marriage.

Garner's first book for children, *The Weirdstone of Brisingamen*, was published in 1960 and immediately hailed as an outstanding debut. He was quickly to become the most fashionable writer for children of the 1960s and early 1970s, although this first book and its successor, *The Moon of Gomrath*, were in fact relatively conventional fantasy stories for children. He ventured into more complex waters with *The Owl Service*, which won him both the Carnegie Medal and the *Guardian* Award. Here a mythical structure based on the medieval Welsh epic *The Mabinogion* is wedded to the clipped, elliptical dialogue of three teenagers. He went even further with *Red Shift*, which the *Times Literary Supplement* called 'probably the most difficult book ever to be published on a children's list': the story is told almost entirely in dialogue, and takes place at three different historical periods, which are not clearly differentiated. In 1976, Garner published the first of his *Stone Book Quartet*, four much shorter and simpler books for younger children, which were originally intended for reluctant readers; they are based on his own family history, and are considered by many critics to be his best work. Garner was never prolific, and his production slowed in the 1980s: he has written some plays, and his recent children's fiction includes *A Bag of Moonshine*. An earlier work often considered among his best is the nativity play, *Holly from the Bongs*, which he wrote for his local school. He has also written collections of stories about goblins, tricksters and fairies, poetry for children and an opera libretto.

For Children
The Weirdstone of Brisingamen 1960; *The Moon of Gomrath* 1963; *Elidor* 1965; *Holly from the Bongs* 1966; *The Old Man of Mow* 1967; *The Owl Service* 1967; *The Hamish Hamilton Book of Goblins* 1969; *Potter Thompson* 1972; *Red Shift* 1973; *The Guizer* 1975; *The Stone Book Quartet* 1983: *The Stone Book* 1976, *Granny Reardun* 1977, *Tom Fobble's Day* 1977, *The Aimer Gate* 1978; *Alan Garner's Fairytales of Gold* 1979; *The Lad of the Gad* 1980; *Alan Garner's Book of British Fairytales* (s) 1984; *A Bag of Moonshine* 1986; *Jack and the Beanstalk* 1992; *Once Upon a Time* 1993

David GARNETT 1892–1981

While awaiting the birth of her only child, Constance Garnett learned Russian and later translated Dostoevsky. David Garnett's father, Edward, a publishers' reader, warned him to have nothing to do with books, and after private education, he studied botany at the Royal College of Science in London, where he discovered a new species of mushroom. During part of the First World War, 'Bunny' (as he was known) helped in France with the Friends' War Victims' Relief Mission; the remaining war years he spent working on the land at Charleston with some members of the Bloomsbury Group. There, he had a brief affair with Duncan Grant whom he shared with Vanessa Bell, and after the war ended, he opened a bookshop in Soho.

Predominantly heterosexual, in 1921 Garnett married Ray Marshall, who illustrated the first book he wrote, *Lady into Fox*, about a man whose wife turns into a vixen; it won both the Hawthornden and the James Tait Black Memorial Prizes in 1923. The similarly fantastic *A Man in the Zoo* was also acclaimed, but the more conventional novels that Garnett later wrote were never so popular. (*Aspects of Love*, however, reached a new audience when it was turned into a musical by Andrew Lloyd Webber after Garnett's death.)

Briefly involved in publishing in the 1920s with The Nonesuch Press and Rupert Hart-Davis, Garnett was as much drawn to outdoor activities, and enjoyed fishing and running a cattle farm. His blond and blue-eyed appearance attracted many lovers, and he was reckless and adventurous. In his literary output, this found an outlet in his sympathetic editing of the letters of T.E. Lawrence. Domestically, he was domineering, and began an affair with Angelica Bell, the daughter of Vanessa and his former lover Duncan Grant, which he conducted in his home, where his wife was dying of cancer. When Angelica was born in 1918, Garnett had declared he would marry her, an intention he fulfilled in 1942, two years after his wife's death. This was partly designed to hurt Duncan Grant and Vanessa Bell, and permanently estranged mother and daughter. He had managed to persuade an extremely doubtful Angelica that she loved him, although she found his assertive personality difficult to cope with.

During the Second World War, Garnett was a flight lieutenant in the RAFVR, then worked for Air Intelligence. His second marriage gradually collapsed, and the couple stayed together only for the sake of their four daughters. Unable to compete with Garnett's rages often directed at his children, Angelica finally left him in 1971, after which he abandoned England to spend his last years in France.

Fiction

Lady into Fox 1922; *A Man in the Zoo* 1924; *The Sailor's Return* 1925; *Go She Must!* 1927; *To the Dovecote and Other Stories* (s) 1928; *No Love* 1929; *The Grasshoppers Come* 1931; *A Terrible Day* 1932; *Beany-Eye* 1935; *Aspects of Love* 1955; *A Shot in the Dark* 1958; *A Net for Venus* 1959; *Two by Two* 1964; *Ulterior Motives* 1966

Non-Fiction

A Rabbit in the Air: Notes from a Diary Kept While Learning to Handle an Aeroplane 1932; *Pocahontas; or, the Nonpareil of Virginia* 1933; *War in the Air: September 1919–May 1941* 1941; *The Golden Echo* (a) 1953; *The Flowers of the Forest* (a) 1955; *The Familiar Faces* (a) 1962; *Great Friends: Portraits of Seventeen Writers* 1979

Edited

The Letters of T.E. Lawrence 1938; *The Novels of Thomas Love Peacock* 1963; *Carrington: Letters and Extracts from Her Diaries* 1979

David (Emery) GASCOYNE 1916–

Gascoyne was born in Harrow and brought up in Hampshire where his father was a bank manager. He was educated at Salisbury Cathedral Choir School, which he has described as 'a lasting influence on my life', and Regent Street Polytechnic. He found his voice as a poet at a precociously young age and, with the proceeds of a small legacy, financed the publication of his first volume, *Roman Balcony*, while still at school in 1932. These poems were strongly influenced by the Imagists and anticipate Gascoyne's subsequent interest in Surrealism. He wrote a short introduction to Surrealism and translated works by Salvador Dali and André Breton, and the poems in his 1936 collection, *Man's Life Is This Meat*, make explicit use of Surrealist techniques.

In 1936 he joined the Communist Party and with Roland Penrose flew to Spain. In his autobiography, *World within World* (1951), **Stephen Spender** recalls his surprise at hearing Gascoyne broadcasting from Barcelona for the Catalonian Government Propaganda Bureau. On his return from Spain, as is related in his *Journal 1936–1937*, Gascoyne visited Picasso in Paris to bring the artist news of his family's safety.

From 1937 to 1939 he lived in Paris where he was encouraged by **Lawrence Durrell** and **Anaïs Nin** to publish excerpts from the journal in their magazine *The Booster*. In a characteristically self-critical letter to Durrell he lamented its adolescent tone: 'I'm afraid a lot of this stuff must give the impression of a dank and feeble creature with great fish-like masturbatory eyes.'

Gascoyne achieved widespread recognition with his 1943 collection, *Poems 1937–1942*, in which his mystical concerns find a specifically Christian context. Similar themes are explored in his solitary verse play, *Night Thoughts*, which was originally commissioned and broadcast by the BBC in 1956. Since this time he has written little new verse and has suffered severe bouts of depression and three prolonged mental breakdowns. During one such period he met Judy Tyler Lewis, whom he married in 1975. His two volumes of youthful journals were published in 1978 and 1980 and are a remarkably frank and mature self-portrait as well as a revealing account of his milieu in the late 1930s. His only novel, *Opening Day*, is an accomplished work of juvenilia, published when he was seventeen.

Poetry

Roman Balcony and Other Poems 1932; *Man's Life Is This Meat* 1936; *Hölderlin's Madness* 1938; *Poems 1937–1942* 1943; *A Vagrant and Other Poems* 1950; *Night Thoughts* [with H. Searle]

1956; *Requiem* [with P. Rainier] 1956; *The Sun at Midnight* 1970; *Three Poems* 1976; *Early Poems* 1980; *Collected Poems* 1988

Fiction

Opening Day 1933

Non-Fiction

A Short Survey of Surrealism 1936; *Thomas Carlyle* 1952; *Paris Journal 1937–1939* 1978; *Journal 1936–1937: Death of an Explorer* 1980; *Journal de Paris et d'ailleurs (1936–1942)* 1984; *Recontres avec Benjamin Fondane* 1984; *Roland Penrose: Recent Collages Catalogue* (c) 1984; *David Gascoyne: Collected Journals, 1936–42* 1991

Translations

Salvador Dali *Conquest of the Irrational* 1935; Benjamin Péret *A Bunch of Carrots/Remove Your Hat* 1936; Paul Éluard *Thorns of Thunder* [with S. Beckett, D. Devlin, E. Jolas, M. Ray, G. Reavey, R. Todd] 1936; André Breton *What Is Surrealism?* 1936; with Philippe Soupault *The Magnetic Fields* 1985

William H(oward) GASS 1924–

Gass was born in Fargo, North Dakota, the son of a teacher of engineering drawing, whom he has described as 'crippled by arthritis and his own character', and an alcoholic mother. At the age of eight he announced his intention to become a writer and read voraciously: 'I read whatever came to hand, and what came to hand were a lot of naughty French novels, some by Zola, detective stories and medical adventures.' His education at Kenyon College, Ohio, and the Ohio Wesleyan University was interrupted by a call-up from the navy, and he was stationed for three years in China and Japan. In 1947 he went to Cornell and wrote his dissertation, 'A Philosophical Investigation of Metaphor': 'I love metaphor the way some people love junk food,' he has said. While at Cornell he also attended a series of seminars given by Wittgenstein, whose linguistic theories have exerted a considerable influence on Gass's work.

A full-time academic career teaching philosophy at Washington University, St Louis, where he has been a professor since 1969, has saved Gass from the rude financial constraints which hamper many writers, and his work has never reached a public beyond academic and literary circles. His first novel, *Omensetter's Luck*, was not published until 1966, twelve years after its inception, and after a string of rejections. This highly stylised novel embodies the ideas which run through Gass's work: the treatment of words as what he calls 'musical notes'; a belief in the self-governing, self-referential nature of fiction; and a stalwart rejection of realism. His subsequent work defies easy categorisation, the boundary between fiction and philosophical enquiry being blurred in works such as *Willie Masters' Lonesome Wife*. For many years Gass had spoken in interviews about an epic work-in-progress which, with characteristic perversity, he hoped would be so good no publisher would accept it. In the event, *The Tunnel* was published in 1994. The novel is narrated by an elderly historian, William Frederick Kohler, author of *Guilt and Innocence in Hitler's Germany*, and is a melange of philosophical and historical musings, limericks, cartoons and all manner of textual trickery designed to undermine any suggestion of realism.

Gass has been married twice, in 1952 and 1969, and has five children from his two marriages. He was the recipient in 1975 of the National Institute for Arts and Letters Prize for Literature.

Fiction

Omensetter's Luck 1966; *In the Heart of the Heart of the Country and Other Stories* (s) 1968; *Willie Master's Lonesome Wife* 1971; *The First Winter of My Married Life* (s) 1979; *The Tunnel* 1994

Non-Fiction

Fiction and the Figures of Life 1970; *On Being Blue: a Philosophical Inquiry* 1976; *The World within the Word* 1979; *The House VI Book* [with P. Eisenman] 1980; *Habitations of the Word* 1985

Carlo GÉBLER 1954–

The son of writers Ernest Gébler and **Edna O'Brien**, Gébler was born in Dublin. He has deliberately not read his parents' work, as he was anxious to find his own voice, and not draw on what might have been shared experiences. After graduating from York University, where he took a BA in English, Gébler took a postgraduate diploma at the National Film and Television School at Beaconsfield in 1979. He directs and writes feature films, notably *Croagh Patrick* (1977), a documentary about the annual pilgrimage, and *Rating Notman* (1981), a drama about prisoners-of-war, which was commended by the Cork Film Festival and nominated for a BAFTA award. His documentary *Two Lives*, a portrait of **Francis Stuart**, was broadcast on Channel 4 in 1985.

In the same year, his first novel, *The Eleventh Summer*, was published, and established a major theme of his work: the presentation of the pain of everyday life through 'ordinary' people. The narrator, Paul Weismann, recalls as an adult a summer when he was eleven, and sent to live with his grandparents in the country after the sudden death of his mother. Far from finding tranquillity, Paul witnesses his grandmother's difficult relationship with her drunken husband, and loses his childhood innocence during an awkward sexual encounter with a cousin. The difficulties of family relationships were further explored in *Malachy and His Family*. In each of his novels, the *Times Literary Supplement* noted, 'the fundamentals of the narrative ... are themselves so intrinsically absorbing ... that our recognition of external similarities is subordinated to a fascination for the present as it unfolds'. As a result of a rather idiosyncratic tour of Cuba, with his children and his partner (whom he married in 1990), Gébler published *Driving through Cuba*. Deliberately personal and selective, its view of Cuba and Fidel Castro was generally admiring in tone. Since Gébler moved to Enniskillen in Northern Ireland in 1990, he has continued to write both books and films. *The Glass Curtain*, published the following year, was a non-fiction book about life in Enniskillen, and in 1993 he made a six-part documentary for the BBC entitled *Plain Tales from Northern Ireland*. Influenced by Chekhov, Kafka and Camus, amongst others, Gébler has found in them an ability to 'arrange words very simply to tell stories which yet deeply affect the feeling faculty of the reader'. Committed politically to the Labour party, Gébler teaches creative writing at the Maze prison in Lisburn, County Antrim.

Fiction

The Eleventh Summer 1985; *August in July* 1986; *Work and Play* 1987; *Malachy and His Family* 1990; *Life of a Drum* 1991; *The Cure* 1994

Non-Fiction

Driving through Cuba 1988; *The Glass Curtain* 1991

For Children

The TV Genie 1989; *The Witch That Wasn't* 1991

Plays

The Beneficiary (f) 1979; *Over Here* (f) 1980; *Rating Notman* (f) 1981; *Country & Irish* (f) 1982; *Life After Death* (f) 1994; *The Joint's Not Jumpin'* (tv) 1995; *No Other Purpose* (tv) 1995

Maggie (Mary) GEE 1948–

The daughter of teachers, Gee was born in Poole, Dorset, and attended Hersham High School for Girls. She began writing poetry at the age of six, and wrote a novel while on holiday at the age of nineteen. She studied English literature at Somerville College, Oxford, and edited the magazine *Elsevier International Press*. She gained an MLitt in 1972, and a PhD in English from Wolverhampton Polytechnic. In 1980 she moved to London in answer to an advertisement and became companion to an elderly woman in Camden Town. She worked as a night receptionist at a hotel (an experience recalled in *Grace*), and spent her days in the British Library. Her first novel, *Dying, in Other Words*, was published in 1981. The following year she was awarded a writing fellowship at the University of Norwich. In 1983 she married Nicholas Rankin, an arts producer for the BBC's World Service, and was named one of the Book Marketing Council's Best of Young British Novelists. She has one daughter.

Her second novel, *The Burning Book*, was concerned with the nuclear threat, a subject that recurs in her work, while *Light Years* was a touching and witty account of essentially selfish people in love, set against a cosmic background suggested by the work of Carl Sagan. Her themes are often sinister, with a surreal edge, while her writing tends to be experimental; she has been described as a 'literary scientist'. In some of Gee's work 'issues' intrude too obviously, as in *Grace*, a dyspeptic novel of Thatcher's Britain, inspired by the unsolved murder in 1984 of the rose-grower and anti-nuclear campaigner, Hilda Murrell, and by the birth of Gee's own child (to whom it is dedicated). She has described the novel as 'carefully patterned', adding: 'I hope a fan of thrillers could read *Grace* and enjoy it as a fast-moving, exciting story, but I also hope it leaves behind a stronger after-taste than the average thriller.' It might be argued that the patterning is overdone (with characters named Grace Stirling and Faith, and hotels called the Empire and the Albion), whilst the thriller element, which involves a psychopathic transvestite, requires considerable suspension of disbelief. Whereas *Light Years* ended on a note of genuine optimism, *Grace* owes its happy ending to the novel's scheme, and it thus seems rather glib (babies solve everything). At her best, however, Gee is a powerful and challenging writer, 'attracted', as she acknowledges, 'to extremes, to what stirs my imagination'.

Fiction

Dying, in Other Words 1981; *The Burning Book* 1983; *Light Years* 1985; *Grace* 1988; *Where Are the Snows?* 1991; *The Lost Children* 1994

Plays

Over and Out 1984

Maurice (Gough) GEE 1931–

Gee was born at Whakatane in the North Island of New Zealand, educated at Avondale College and the University of Auckland Teachers College. Between 1955 and 1957 he was a schoolteacher. Leaving the profession, he worked at various jobs before holding a number of librarian's posts at Wellington, Napier and Auckland between 1967 and 1976. Since the publication of his fourth novel, *Games of Choice*, he has been a full-time writer. *Plumb*, the first in a trilogy of novels covering over a century of New Zealand history, won the James Tait Black Memorial Prize and the Sir James Wattie Award, establishing him with **Janet Frame** among New Zealand's leading novelists. The *Sunday Times* said of him; 'It's as though New Zealand's uncluttered shoreline and sinuous forests have influenced his prose, making it wild, free and unselfconscious.' In *The Burning Boy*, his small-town saga, fire in one form or another plays a leading part, as it does in *Prowlers*. Of this novel's genesis he told his publisher, 'In 1978 I wrote a history of the school my children attended, and unearthed two characters, a brilliant unorthodox headmaster and a fireraiser, who burned down the school among other things'. He has also produced distinguished short stories, and books for children, as well as television plays including the *Mortimer's Patch* series. Among many other awards and honours, he has won the New Zealand Book Award for 1983, and he was Robert Burns Fellow at the University of Otago in 1964. He is married and has a stepson and two daughters.

Fiction
The Big Season 1962; *A Special Flower* 1965; *In My Father's Den* 1972; *A Glorious Morning, Comrade* 1975; *Games of Choice* 1976; *Plumb* 1978; *Sole Survivor* 1983; *Collected Stories* **(s)** 1986; *Prowlers* 1987; *The Burning Boy* 1990; *Going West* 1992

For Children
Under the Mountain 1979; *The World Around the Corner* 1981; *The Halfman of O* 1982; *The Priest of Ferris* 1984; *Motherstone* 1985; *The Fire-Raiser* 1986; *The Champion* 1989; *Mortimer's Patch* **(tv)** 1980; *The Fire-Raiser* **(tv)** [from his novel] 1986; *The Champion* **(tv)** [from his novel] 1989

Non-Fiction
Nelson Central School: a History 1978

Martha (Ellis) GELLHORN 1908–

Gellhorn was born in St Louis, Missouri, daughter of one of the city's most eminent gynaecologists and from its wealthy and distinguished families on both sides. After attending the John Burroughs School there, she went to Bryn Mawr College, Pennsylvania, but left after one year, determined to enter journalism and see the world. After two brief jobs during 1929 on the *New Republic* and the *Hearst Times Union*, she went to Paris, where she worked for various papers and contracted her first, short-lived, marriage to Bertrand de Juvenel, a French journalist and marquis. Returning to America in 1934 (the year her highly autobiographical first novel, *What Mad Pursuit*, was published), she was hired by Harry Hopkins to report on the Federal Emergency Relief programme, work which led to her meeting the President and First Lady and to the publication of a book of four didactic novellas, *The Trouble I've Seen.*

At the end of 1936 she travelled with her family to Key West in Florida, where she met **Ernest Hemingway**, and in 1937, the Spanish Civil War in progress, she joined him in Madrid. Here she became a war correspondent for *Collier's Weekly*, work that continued until 1946 and led to her covering five major wars: in Spain, Finland, China and Java, and the Second World War. In 1940 she married Hemingway, after the three-year struggle of his second wife, Pauline Pfeiffer, to avoid a divorce; but although Hemingway and Gellhorn travelled to China together (riding on horses so small Hemingway mainly carried his), they grew increasingly distant during the war and divorced in 1946.

Gellhorn published six novels, between 1934 and 1967, as well as several books of short stories and novellas and, although her early fiction was largely undigested autobiography, her writing matured enough in novels such as *Liana* and *His Own Man*, to become effective. She is perhaps better known for her several books of collected journalism and memoirs, among which is the outstandingly vivid *Travels with Myself and Another* (the Other in question is Hemingway, who features as 'U.C.' or 'Unwilling Companion'). In 1949 Gellhorn adopted a son from an Italian orphanage, and the reluctance of the US authorities to accept the entry of this child led to her living several years in Cuernavaca, Mexico, while later wanderings took her to Kenya. She married Thomas Matthews in 1954 and they were divorced in 1963. She returned in the 1960s to cover more wars in Vietnam and Israel for the *Guardian*, and has continued as a writer, publishing a volume of collected non-fiction in 1988.

Fiction

What Mad Pursuit 1934; *The Trouble I've Seen* (s) 1936; *A Stricken Field* 1940; *The Heart of Another* (s) 1941; *Liana* 1944; *The Wine of Astonishment* 1948; *The Honeyed Peace* (s) 1953; *Two by Two* (s) 1953; *His Own Man* 1961; *Pretty Tales for Tired People* (s) 1965; *The Lowest Trees Have Tops* 1967; *The Weather in Africa* 1978; *The Novellas of Martha Gellhorn* 1993

Non-Fiction

The Face of War 1959 [rev. 1986); *Vietnam: a New Kind of War* 1966; *Travels with Myself and Another* 1978; *A View from the Ground* 1988

(Iris) Pam(ela) GEMS 1925–

Born Pam Price in Bransgore in the New Forest, Gems was brought up by her mother and two widowed grandmothers after the death of her father in 1929. She won a scholarship to Brockenhurst County Grammar School in 1936 and, having left school in 1941, served in the Women's Royal Naval Service, a period she has described as 'all sex and high adrenalin'. Her war service enabled her to read psychology at Manchester University, and she graduated in 1949, the same year that she married Keith Gems, a model manufacturer. The couple lived in Paris and the Isle of Wight before moving to London, where she worked briefly as a research assistant for the BBC, and had one of her television plays, *A Builder by Trade*, produced in 1961. They have four children, including the writer Jonathan Gems.

Gems' first staged production was the children's play, *Betty's Wonderful Christmas*, which was produced in London in 1972. Working with Al Berman's Almost Free Theatre she wrote a number of plays, including the monologues *My Warren* and *After Birthday* which concern, respectively, an elderly woman who receives a vibrator in the post as a practical joke, and a girl who puts her miscarried baby down the lavatory. Originally entitled *Dead Fish* when it was produced at the Edinburgh Festival in 1976, *Dusa, Fish, Stas and Vi* was Gem's first play to reach the West End, and was produced in Los Angeles in 1978, and New York in 1980. The play is a coarse comedy in which a group of radical feminists search for Dusa's children who have been kidnapped by her divorced husband. Turning to historical subjects in later plays such as *Queen Christina* (about the seventeenth-century Swedish ruler), *Piaf*, Dumas *fils' Camille* and *The Danton Affair*, Gems continued to examine the position of women in a patriarchal society. She has translated and adapted Marianne Auricoste's play about Rosa Luxembourg, and has adapted work by Marguerite Duras, Chekhov and Ibsen.

Plays

A Builder by Trade (tv) 1961; *Betty's Wonderful Christmas* 1972 [pub. 1973]; *After Birthday* 1973 [as *Sandra* 1979] [np]; *The Amiable Courtship of Miz Venus and Wild Bill* 1973 [np]; *My Warren* 1973 [np]; *Sarah B. Divine* [with T. Eyen, J. Kramer] 1973 [np]; *Go West Young Woman* 1974 [np]; *Up in Sweden* 1975 [np]; *Dead Fish* 1976 [pub. as *Dusa, Fish, Stas and Vi* 1977]; *Guinevere* 1976 [np]; *My Name is Rosa Luxembourg* [from M. Auricoste's play] 1976 [np]; *The Project* 1976 [np] [rev. as *Loving Women* 1984, pub. 1985]; *The Rivers and Forests* [from M. Duras] 1976 [np]; *Franz into April* 1977 [np]; *Queen Christina* 1977 [pub. 1982]; *Piaf* 1978 [pub. 1979]; *Ladybird, Ladybird* 1979 [np]; *Uncle Vanya* [from Chekhov's play] 1979; *We Never Do What They Want* (tv) [from her play *The Project*] 1979; *A Doll's House* [from Ibsen's play] 1980 [np]; *Aunt Mary* 1982 [pub. in *Plays by Women 3* ed. M. Wandor 1984]; *The Treat* 1982 [np]; *Camille* [from A. Dumas *fils*] 1984 [pub. 1985]; *The Cherry Orchard* [from Chekhov's play] 1984 [np]; *Pasionaria* [with P. Sand] 1985 [np]; *The Danton Affair* [from S. Przybyszewska] 1986 [np]; *The Blue Angel* [from H. Mann's novel *Professor Unraat*] 1991 [pub. 1992]; *Uncle Vanya* [from Chekhov's play] 1992; *Ghosts* [from Ibsen's play] 1993; *Yerma* [from Lorca's play] 1993 [np]; *Deborah's Daughter* 1994 [np]; *The Seagull* [from Chekhov's play] 1994 [np]

Fiction

Mrs Frampton 1989; *Bon Voyage, Mrs Frampton* 1990

William (Alexander) GERHARDIE 1895–1977

Gerhardie was born in St Petersburg, the son of a cotton-spinning manufacturer of English origin. His original surname was Gerhardi, but he added a final 'e' towards the end of his life, either, according to different versions, because it was an ancestral spelling or because other great writers, such as Shakespeare and Blake, had a final 'e'. He attended schools in St Petersburg and grew up in a polyglot household, speaking English, French, German and Russian. According to his own account, his father was tied up in a sack by revolutionaries in 1905 and only escaped death through being mistaken for the socialist Keir Hardie. As a teenager, Gerhardie was sent to a commercial school in London, but enlisted at the beginning of the First World War and, after a while, was commissioned and posted to the staff of the British embassy in what was now Petrograd.

Having served with the British military mission in Siberia from 1918 to 1920 and been awarded the OBE, he read Russian at Worcester College, Oxford, and published the first English study of Chekhov as well as his first novel, *Futility* in 1922. (After trying it on thirteen publishers, he sent it to **Katherine Mansfield**, who found a publisher for it within a fortnight.) This, and his second novel, *The Polyglots*, both of which concern the English in Russia, are generally considered his best fiction. He published a number of other novels until the late 1930s, and high claims are sometimes made for *Of Mortal Love*, but it is generally thought that his essentially autobiographical talent became exhausted. During the 1920s he had travelled widely, but in 1931 he moved into a flat near Broadcasting House in central London, where he lived for the rest of his life. During the Second World War he worked for the European service of the BBC, but after that he became a recluse, publishing nothing of any importance after a study of the Romanov dynasty in 1940. He was understood to be working on a vast novel tetralogy, but after his death only a large card-index to this work was found. A highly eccentric portrait of the years from 1890 to 1940 was eventually assembled and published as *God's Fifth Column*. He was unmarried.

Fiction

Futility 1922; *The Polyglots* 1925; *Pretty Creatures* (s) 1927; *Jazz and Jasper* 1928; *Pending Heaven* 1930; *The Casanova Fable* [with H. Kingsmill] 1934; *The Memoirs of Satan* [with B. Lunn] 1932; *Resurrection 1934; Of Mortal Love* 1936; *My Wife's the Least of It* 1938

Plays

Donna Quixote 1929

Non-Fiction

Anton Chekhov 1923; *Memoirs of a Polyglot* (a) 1931; *Meet Yourself as You Really Are* [with Prince L. zu Loewenstein] 1936; *The Romanovs* 1940; *God's Fifth Column* 1981

Biography

William Gerhardie by Dido Davies 1990

Zulfikar GHOSE 1935–

Ghose was born in Sialkot, a Punjab city then in India and now in Pakistan. In 1942 his extended Muslim family moved to Bombay, where Partition was soon to fuel rising religious conflict with the Hindu community. In his autobiography, *Confessions of a Native-Alien*, he recalls: 'When we left Bombay in 1952 for England, we were leaving two countries, for in some ways we were alien to both and our emigration to a country to which we were not native only emphasised our alienation from the country in which we had been born.' His subsequent marriage to a Brazilian, and his move to the USA in 1969, increased the sense of deracination and alienation which he has explored in much of his work.

In England Ghose attended the Sloane School in Chelsea before going to the University of Keele, where he read English and philosophy, graduating in 1959. He became a teacher and wrote book reviews for the *Guardian* and the *Times Literary Supplement*, and from 1960 to 1965 was a cricket correspondent for the *Observer*. Having married Helena de la Fontaine, a Brazilian artist, in 1964, he took a teaching post at the University of Texas, Austin, in 1969, and has remained there since, rising to the post of professor of English.

Ghose's early poetry and short stories deal largely with pre-independence India, exploring the themes of rural displacement and cultural conflict, as does his first novel, *The Contradictions* (1966). His trilogy *The Incredible Brazilian* spans 400 years of Brazilian history in a traditional, linear fashion, but also anticipates his later, more experimental, fiction. His novels of the 1980s seem intent upon undermining their apparent mimetic function. *Hulme's Investigations into the Bogart Script*, for example, is an assembly of nine imaginary Humphrey Bogart scripts which poses questions about the nature of language and artifice. *The Triple Mirror of the Self* once again draws on elements of autobiography, and is the story of a man variously known as 'Roshan', 'Urion', 'the scattered one' and 'Shimmers' who is born in India, and exiled to London and Arizona. The novel is part spiritual journey, and part allegory of twentieth-century history.

Fiction

Statement against Corpses (s) [with B.S. Johnson] 1964; *The Contradictions* 1966; *The Murder of Aziz Khan* 1967; *The Incredible Brazilian* trilogy: *The Native* 1972, *The Beautiful Empire* 1975, *A Different World* 1978; *Crump's Terms* 1975; *The Texas Inheritance* [as 'William Strang'] 1980; *Hulme's Investigations into the Bogart Script* 1981; *A New History of Torments* 1982; *Don Bueno* 1983; *Figures of Enchantment* 1986; *The Triple Mirror of the Self* 1992

Poetry

The Loss of India 1964; *Jets from Orange* 1967; *The Violent West* 1972; *Penguin Modern Poets 25* [with G. Ewart, B.S. Johnson] 1974; *A Memory of Asia* 1984; *Selected Poems* 1991

Non-Fiction

Confessions of a Native-Alien (a) 1965; *Hamlet, Prufrock, and Language* 1978; *The Fiction of Reality* 1983; *The Art of Creating Fiction* 1991; *Shakespeare's Mortal Knowledge* 1993

Amitav GHOSH 1956–

Ghosh was born in Calcutta, into a family whose origins lay in what is now Bangladesh, but which had been established in India for some generations. His father, after seeing service in the British army during the Second World War, became a civil servant in the Indian Foreign Ministry, and was much posted abroad, so that Ghosh's childhood was largely spent either outside India or at an Indian boarding-school, Doon School in Dehra Dun (where one of his teachers was **Vikram Seth**). He spent most of his earliest years in Dhaka (then the capital of East Pakistan, now of Bangladesh), then lived, from 1965, in Colombo, while his parents later went to Tehran. He studied history at Delhi University, taking his BA in 1976, and then worked briefly on the *Indian Express*, the only press opposition to Indira Gandhi's then State of Emergency.

Although these were exciting times, Ghosh quickly became convinced that he wanted to be a writer and not a journalist, and studied towards a postgraduate degree in social anthropology. In 1978, he was able to begin a doctorate in this subject at Oxford, but spent most of his period of doctoral study outside England: first travelling slowly through North Africa; then doing a period of fieldwork in an Egyptian village; and finally spending a year at Oxford itself in 1981–2. After receiving his doctorate, he returned to India, where he first held a fellowship for a year in Trivandrum, Kerala, and then became a research associate at Delhi University. His salary was very small, and he lived for a period in a small rented room on a rooftop, but there completed his first novel, *The Circle of Reason*, which was published in India and England in 1985, and in the USA in 1986. He was marked out as a writer of promise, and, after briefly holding a lectureship in sociology at Delhi University in 1987, he departed for the USA, where he taught literature and sociology for a year at the University of Virginia in 1988. Since then he has held various academic appointments in both America and India, and briefly in Cairo, and since 1993 has taught at Columbia University, New York, where he now lives for much of the year.

His first novel was perhaps a little too directly a contribution to the fashionable school of magic realism, but his second, *The Shadow Lines*, described by its reviewer in the *Times Literary Supplement* as 'polished and profound', brought a more individual art to its complex story linking India and England in the 1930s with the anti-Sikh riots in India following the assassination of Indira Gandhi in 1984. *In an Antique Land* is a work of non-fiction partly based on the Egyptian fieldwork he had carried out a decade earlier but linking this with a reconstruction of the life of a twelfth-century Indian slave of a Jewish trader. Ghosh's other works include many contributions to influential journals such as *Granta*, and his several honours include the prestigious French prize, the Prix Medicis Etrangère, in 1990 (he was the first Indian to win the award). In 1990 he married the American biographer Deborah Baker, best known for her life of **Laura Riding**; they have two children.

Fiction

The Circle of Reason 1985; *The Shadow Lines* 1988

Non-Fiction

The Slave of Ms. H.6 1990; *In an Antique Land* 1992; *Infidel in Egypt* 1992

Lewis Grassic GIBBON (pseud. of James Leslie Mitchell) 1901–1935

Gibbon was born on his father's croft in Auchterless, Aberdeenshire, and lived there for seven years. After a brief period in Aberdeen, the family settled to farm in the rugged countryside of Kincardineshire. Gibbon's youthful scholarly interests, including archaeology, infuriated his dour father, who expected his son to work for him. Miserably unhappy, he ran away from his secondary school and unsuccessfully tried to join the army. Instead he became a junior reporter for an Aberdeen newspaper, but his journalistic career ended in 1918 when he was sacked in Glasgow for falsifying his expense account when with another paper. He enlisted as a clerk in the Royal Army Service Corps for six years, and took the same job in the Royal Air Force in Surrey.

Still a thwarted writer, he published a short story in 1924, the same year he married Ray Middelton, a childhood friend. Despite help from **H.G. Wells**, it was five years before he published further short stories. The *Cornhill Magazine* accepted a series of twelve, on the day that happily coincided with his release from the RAF. In the meantime *Hanno*, a book of exploration, had appeared, but novels, short stories and an archaeological study of South America (all published under the name J. Leslie Mitchell) were overshadowed by the reception of *Sunset Song*, the first volume of a trilogy. Published under his *nom de plume*, the novel attracted international praise. A lyrical, realistic account of a Kincardineshire community, it aimed, as he wrote, to 'mould the English language into the rhythms and cadences of Scots spoken speech'. His mother, from whom he inherited his gifts, nevertheless considered his books 'muck' and kept them wrapped in brown paper. Like C.M. Grieve and Ronald Mavor, who also used pseudonyms, the novelist evolved a dual personality. Gibbon was generous and kind and in conversation tolerated views he disagreed with, but he could also be touchy and defensive, often replying to hostile reviews from his home in Welwyn Garden City. The last of the trilogy was regarded as uneven by those who disliked the emphasis on communism, but as a whole *A Scots Quair* was considered a work of genius. He died after an operation for a perforated ulcer, one week before his thirty-fourth birthday. H.G. Wells wrote that the novelist would have been 'a very outstanding writer. Of his quality there can be no doubt.'

Fiction

As James Leslie Mitchell: *Stained Radiance 1930; The Calends of Cairo* (s) 1931; *The Thirteenth Disciple* 1931; *The Lost Trumpet* 1932; *Persian Dawns, Egyptian Nights* (s) 1932; *Three Go Back* 1932; *Image and Superscription* 1933; *Spartacus* 1933; *Gay Hunter* 1934 *Nine against the Unknown* [with 'Lewis Grassic Gibbon'] 1934

As 'Lewis Grassic Gibbon': *A Scots Quair* 1946: *Sunset Song* 1932, *Cloud Howe* 1933, *Grey Granite* 1934

Non-Fiction

As James Leslie Mitchell: *Hanno: or the Future of Exploration* 1928; *The Conquest of the Maya* 1934

As 'Lewis Grassic Gibbon': *Scottish Scene: or the Intelligent Man's Guide to Albyn* [with H. MacDiarmid] 1934; *Niger: the Life of Mungo Park* 1934

Stella (Dorothea) GIBBONS 1902–1989

The daughter of a doctor, Gibbons was born in Kentish Town, north London, into a household rent by sexual intrigue and domestic violence. She was initially educated by a series of governesses, then attended the North London Collegiate School for Girls, after which she took a two-year course in journalism at University College, London. She worked in Fleet Street for ten years, first as a cable decoder for the British United Press, then as a feature writer for the *Evening Standard*, a job from which she was fired 'for coming in late and crying about a love affair'. Between 1928 and 1931 she was a drama and literary critic on *The Lady*. She published her first volume of poetry, *The Mountain Beast and Other Poems*, in 1930.

Instant success came with her first novel, *Cold Comfort Farm*, for which she was paid £32; published in 1932 it has remained in print ever since. The novel is an immensely funny parody of the school of rural novelists represented by writers such as **Mary Webb** and **Sheila Kaye-Smith**, and of the earthy passions of **D.H. Lawrence**. It may also reflect the passions which raged in her family: 'I realized they all revelled in it. They enjoyed it.' She described herself as 'a committed pantheist' and much of her poetry is about the natural world. Indeed, she considered herself primarily a poet and published her *Collected Poems* in 1950. She was also a prolific writer of fiction, producing more than twenty novels and several volumes of short stories. Apart from *Cold Comfort Farm*, her novels are little read or regarded now. The majority of them have domestic settings, usually taking place in Hampstead or Highgate, where she lived in later life. Her own favourite, however, was *Ticky*, a satire on army life. She was married for twenty-six years to the actor and singer Alan Bourne Webb, who died in 1959; they had one daughter. Gibbons was made a Fellow of the Royal Society of Literature in 1950.

Fiction

Cold Comfort Farm 1932; *Bassett* 1934; *Enbury Heath* 1935; *Miss Linsey and Pa* 1936; *Roaring Tower* (s) 1937; *Nightingale Wood* 1938; *My American* 1939; *Christmas at Cold Comfort Farm* (s) 1940; *The Rich House* 1941; *Ticky* 1943; *The Bachelor* 1944; *Westwood* 1946; *Conference at Cold Comfort Farm* 1949; *The Matchmaker* 1949; *The Swiss Summer* 1951; *Fort of the Bear* 1953; *Behind the Pearly Water* (s) 1954; *The Shadow of a Sorcerer* 1955; *Here Be Dragons* 1956; *White Sand and Grey Sand* 1958; *A Pink Front Door* 1959; *The Weather at Tregulla* 1962; *The Wolves Were in the Sledge* 1964; *The Charmers* 1965; *Starlight* 1967; *The Snow Woman* 1969; *The Woods in Winter* 1970

Poetry

The Mountain Beast and Other Poems 1930; *The Priestess* 1934; *The Lowland Venus* 1938; *Collected Poems* 1950

For Children

The Untidy Gnome 1935

Wilfrid (Wilson) GIBSON 1878–1962

Gibson was born at Hexham, Northumberland, one of many children of a chemist who was also an amateur archaeologist and photographer. He was educated at private schools and partly at home by his half-sister, and began his lifelong dedication to poetry in childhood. He was never to attend a university or follow a profession, but his first collection of pastoral verse-plays was published in the *Spectator* before he was twenty in 1897, and his first volume of verse, *Urlyn the Harper*, in 1902. Early books of poetry dealt with the mythological past and romantic heroes, but from *Daily Bread* in 1910 he found his voice as the spokesman in verse for the inarticulate poor, particularly in the north of England: he was a poet who tried to write, in his own phrase, of 'the heart-break in the heart of things'.

The turning-point in Gibson's life was his move to London in 1912, where he found rooms in Bloomsbury above **Harold Monro**'s Poetry Bookshop, was introduced to Edward Marsh, and was included in his anthologies of *Georgian Poetry*. Gibson's wife, whom he married in 1915, had been Monro's assistant at the bookshop. This period saw the height of Gibson's considerable production of verse and verse-drama; in the ten years after 1907 his volumes were annual and had one-word titles (a feature dropped later) such as *Fires*, *Livelihood*, *Friends* and *Battle*. During the First World War, Gibson tried four times to enlist, and was rejected on grounds of poor eyesight, but eventually accepted for the Service Corps in 1917. He had earlier benefited as one of three Georgian writers named as heirs to **Rupert Brooke**'s estate, and any financial difficulties in later years were met by taking in paying guests.

Gibson's *Collected Poems* were published in 1931, and from this date his reputation began to decline, never substantially to recover, although he continued to publish until 1950, the year of his wife's death. Perhaps his poetry's interesting combination of colloquialism and symbolism will some day occasion a revival. His native North Country remained his inspiration, although he lived with his wife and three children in Gloucestershire, Pembrokeshire and around London, dying at Weybridge in Surrey. His daughter, Audrey, was tragically killed in a landslide in Italy in 1939, and his son, Michael, published several books on the history of roses, among other subjects.

Poetry

Urlyn the Harper and Other Song 1902; *Mountain Lovers* 1902; *The Queen's Vigil and Other Song* 1902; *Song* 1902; *The Golden Helm and Other Verse* 1903; *The Nets of Love* 1905; *On the Threshold* 1907; *The Stonefolds* 1907; *The Web of Life* 1908; *Akra the Slave* 1910; *Daily Bread* 1910 [rev. 1923]; *Fires* 1912; *Borderlands* 1914; *Thoroughfares* 1914; *Battle* 1915; *Battle and Other Poems* 1916; *Friends* 1916; *Livelihood* 1917; *Whin* 1918; *Twenty-Three Selected Poems* 1919; *Home* 1920; *Neighbours* 1920; *Krindlesyke* 1922; *I Heard a Sailor* 1925; *Sixty-Three Poems* 1926; *The Early Whistler* 1927; *The Golden Room and Other Poems* 1928; *Hazards* 1930; *Collected Poems 1905–1925* 1931; *Highland Dawn* 1932; *Islands* 1932; *Fuel* 1934; *A Leaping Flame – a Sail!* 1935; *Coming and Going* 1938; *The Alert* 1941; *Challenge* 1942; *The Searchlights* 1943; *The Outpost* 1944; *Solway Ford and Other Poems* 1945; *Coldknuckles* 1947; *Within Four Walls* 1950

Plays

Womenkind 1912 [rev. 1922]; *Kestrel Edge and Other Plays* [pub. 1924]

Fiction

Between Fairs 1928

Ellen GILCHRIST 1935–

Gilchrist was born and brought up in the Mississippi Delta, the locale which provides the background for much of her writing. The second child of a professional baseball player turned engineer and the 'Most Popular Girl' of 1927 in Ole Miss, she spent her childhood in Issaquena County before moving to Harrisburg, Illinois. The region and the relationships fostered there, particularly the fierce sibling rivalry between her and brother Dooley, are recalled in the 'Rhoda' stories, which have been included in all Gilchrist's volumes of short stories. It was the accuracy and intensity of her portrayal of the idealised hopes and rampant acquisitiveness of youth which initially attracted the attention of critics. Her one-time tutor and fellow Southerner, **Eudora Welty**, once described the art of short-story writing as requiring 'not the creating of an illusion, but the restoring of one', a belief which is particularly appropriate to the imaginative understanding of childhood as Gilchrist manages it in these stories.

She did not begin to write seriously until, at the age of forty, after three marriages, three divorces and with three children, she sent some of her poetry to Jim Whitehead who asked her to join his creative writing class run through the University of Arkansas. To do so she left New Orleans, her home of fourteen years, and her job as contributing editor to the *New Orleans Courier*, and moved to Fayetteville, a small town in the

Ozarks where academicians live in close community with local workers. It was from here that she published a volume of poetry, *The Land Surveyor's Daughter*, in 1979. In the same year she received a National Endowment Grant to write a book of short stories. *In the Land of Dreamy Dreams* was published in 1981 under the University of Arkansas imprint and sold over 10,000 copies within a matter of months in the south-west alone, with none of the advertising or promotional back-up which a major publishing house might have provided. That this was evidence of more than the interest which Southerners take in their own home ground was proven when the book was taken up by Little, Brown, and the book's success was repeated on a national level. Its popularity is not hard to fathom: the volume's fourteen stories return the form to an earlier era, that of **Ring Lardner**, **Dorothy Parker** and before them Mark Twain; no problems of interpretation are presented to the reader, the stories are not self-consciously highbrow; Gilchrist's voice is rooted in the everyday experience of a readership about whom it can be as impolite as she wishes it to be. Her third book, *Victory Over Japan*, won the National Book Award for Fiction in 1984 and returns to many of the characters first introduced in *In the Land of Dreamy Dreams*. Applause has been more muted since. Her novels in particular have had a mixed reception. Some critics have detected a slackening of tone from the sardonic to the sentimental in her most recent work.

Fiction

In the Land of Dreamy Dreams (s) 1981; *The Annunciation* 1983; *Victory Over Japan* (s) 1984; *Drunk with Love* (s) 1986; *The Anna Papers* 1988; *Light Can Be Both Wave and Particle* (s) 1989; *The Blue-Eyed Buddhist and Other Stories* (s) 1990; *I Cannot Get You Close Enough* (s) 1990; *Net of Jewels* 1993; *Star Carbon* 1994

Poetry

The Land Surveyor's Daughter 1979; *Riding Out the Tropical Depression: Selected Poems 1975–1985* 1986

Non-Fiction

Falling Through Space: the Author's Journals 1987; *Anabasis: a Journey to the Interior* 1994

Penelope (Ann Douglass) GILLIATT 1932–1993

Gilliatt was born Penelope Conner in London, the daughter of a barrister who became head of Commonwealth and foreign relations at the BBC. She grew up partly in Northumberland and, after her parents' separation, with her father in London, where she attended Queen's College in Harley Street between 1942 and 1947. She then spent a year at Bennington College, Vermont, and entered journalism, joining British *Vogue*, becoming its features editor, and establishing a early reputation as a reviewer. In 1954 she married the neurologist Roger Gilliatt, who became professor of clinical neurology at the University of London. This marriage ended in divorce, and she married the playwright **John Osborne**, with whom she had one daughter, Nolan, and from whom she was also divorced, in 1968. She became well known from 1961 onwards as film critic for the *Observer*; from 1967 to 1979 she was film critic for six months a year for the *New Yorker* (which also published many of her short stories), and she retained her links with the magazine. Since the mid-1960s she divided her time between the USA and London, and throughout her adult life travelled widely.

She is most admired perhaps for her many volumes of short stories, where her terse dialogue and spare but distinguished narrative are used to best effect. Her novels, of which the first was *One by One*, adumbrating her common theme of young couples attempting to hold their marriages together under pressure, are generally considered more uneven. She was also a film scenarist (her most famous work in this area is the distinguished script for John Schlesinger's film, *Sunday, Bloody Sunday*) and a television scriptwriter and playwright; her writings on the subject of film include a study of Jacques Tati and a collection of reviews (also including theatre reviews), *Unholy Fools*.

Fiction

One By One 1965; *A State of Change* 1967; *What's It Like Out? and Other Stories* (s) 1968; *Penguin Modern Stories* (s) [with others] 1970; *Nobody's Business* (s) 1972; *Splendid Lives* (s) 1977; *The Cutting Edge* 1978; *Quotations from Other Lives* (s) 1982; *Mortal Matters* 1983; *They Sleep without Dreaming* (s) 1985; *Twenty-Two Stories* (s) 1986; *A Woman of Singular Occupation* 1988; *Luigo* (s) 1990

Plays

Sunday, Bloody Sunday (f) 1972 [rev. as *Making Sunday Bloody Sunday* 1986]; *In Trust* [pub. 1980]; *Property, and Nobody's Business* 1980; *Beach of Aurora* 1981 [np]; *But When All's Said and Done* 1981 [np]

Non-Fiction

Unholy Fools 1975; *Jacques Tati* 1977; *Jean Renoir: Essays, Conversations, Reviews* 1977; *Three-Quarter Face: Profiles and Reflections* 1980; *To Wit: In Celebration of Comedy* 1990

Charlotte (Anna) Perkins (Stetson) GILMAN 1860–1935

Although Gilman is now remembered mostly as a writer of feminist and Utopian fiction, in her own time she was widely known as a lecturer, polemicist, journalist, activist and economic theorist. Born in Hartford, Connecticut, into what one biographer calls 'an upper-crust New England family', Charlotte Perkins was the great-granddaughter of political and religious leader Lyman Beecher, and her great-aunts include the suffragist Isabella Beecher Hooker and the writer Harriet Beecher Stowe. Her father had dabbled in law, schoolteaching and journalism before choosing a career in librarianship. He never really settled into marriage and fatherhood, abandoning his wife and their two surviving children to the care of relatives for increasing periods of time until the final separation in 1869. Gilman later recalled that her mother had been 'forced to move nineteen times in eighteen years, fourteen of them from one city to another'.

She had little or no formal education; her one year at Rhode Island School of Design was disrupted by family problems. Largely self-taught, she read extensively, sometimes following her father's suggestions for reading programmes in history and science. Her self-consciousness as a writer took early and mocking form in the 'Literary and Artistic Vurks of Princess Charlotte' (written when she was ten), and in the 'Autobiography of C.A. Perkins' (begun when she was nineteen). Her first publication, the poem 'In Duty Bound', appeared in the *Woman's Journal* in December 1883.

In May 1884, after more than two years of vacillation, she married Charles Walter Stetson, an artist who had proposed to her seventeen days after they first met. After the birth of their daughter in 1885, Gilman suffered nearly three years of severe depression, for which she was treated by the nerve specialist Dr S. Weir Mitchell, whose 'cure' – forbidding her all intellectual activity and enforcing passive domesticity – brought her, she later recalled, 'so near the borderline of utter mental ruin that I could see over'. The experience prompted one of her best-known works, *The Yellow Wall-paper*, in which an intelligent woman suffering from post-natal depression is slowly driven mad by her doctor husband's treatment: 'I read the thing to three women here ... and I never saw such squirms!' she wrote to a friend: 'Daylight too. It's a simple tale, but highly unpleasant.'

Gilman had rebelled not only against Mitchell's treatment but against domesticity, leaving her husband and moving to California. She supported herself by teaching, periodical journalism and lecturing, becoming a prominent writer and speaker in the feminist and Nationalist (that is, Utopian socialist) causes. During this period she returned to the pattern of life before her marriage, one of intense relationships with women. Despite a much happier second marriage in 1900, to her cousin, George Houghton Gilman, she remained deeply hostile to conventional family values and sex roles: 'The essential indecency of the dependence of one sex upon the other for a living is in itself sexual immorality.'

Women and Economics, an influential text of feminist theory attacking the 'artificial diversity between the sexes', was translated into six languages. In her fiction, much of it published in her own magazine, *The Forerunner* (1909–16), she envisaged alternative ways of life through role reversal, role redefinition or social reorganisation, co-operative living and communities of women. The most extreme version of this, *Herland* (written in 1915), is a feminist Utopia in which parthenogenesis has solved the problem of men; the problem of motherhood, of fit and unfit mothers and their place in the social hierarchy, remains unsolved. Throughout her writing career, Gilman the 'unnatural mother', who had let her ex-husband and his second wife bring up her daughter, would wrestle with the problem of motherhood and its meaning, motherhood and its relation to women's work. 'The one predominant duty is to find one's work and do it, and I have striven mightily at that', she wrote in her final attempt at autobiography, *The Living of Charlotte Perkins Gilman*. It was to be published posthumously; knowing she had inoperable cancer, she committed suicide in 1935.

Gilman's reputation, which stood so high in the late 1890s and early 1900s, suffered a diminution in the 1920s, when her mixture of socialism and feminism proved unpalatable to many Americans. With the resurgence of feminist energy in the late 1960s, a new generation of readers and critics has discovered her work, whose reputation continues to grow steadily.

Non-Fiction

Gems of Art for the Home & Fireside 1888; *The Labor Movement* 1893; *Women and Economics* 1898; *Concerning Children* 1900; *The Home: Its Work and Influence* 1903; *Human Work* 1904; *The Punishment That Educates* 1907; *Women and Social Service* 1907; *The Crux* 1911; *The Man-Made World* 1911; *Moving the Mountain* 1911; *His Religion and Hers: a Study of the Faith of Our Fathers and the Work of Our Mothers* 1923; *Sex in Civilization* (c) [ed. V.F. Calverton, S.D. Schmalhausen] 1929; *Our Changing Morality* (c) [ed. F. Kirchway] 1930; *Women's Coming of Age* (c) [ed. S.D. Schmalhausen] 1932; *The Living of Charlotte Perkins Gilman* (a) 1935

Fiction

The Yellow Wall-Paper 1891; *What Diantha Did* 1910; *The Crux* 1911; *Moving the Mountain* 1911; *Herland* 1979; *The Charlotte Perkins Gilman Reader* [ed. A. Lane] 1980

Poetry

In This Our World 1893; *Suffrage Songs and Verses* 1911

Allen GINSBERG 1926–

Ginsberg was born in Newark, New Jersey. His father, Louis, was a poet and teacher, his Russian-born mother was an ardent, lifelong Marxist whose political commitment began to degenerate into paranoia during Ginsberg's childhood. She was institutionalised, lobotomised, and died in an asylum in 1956, her blighted life becoming the subject of the long memorial poem, 'Kaddish', which many critics consider to be Ginsberg's greatest poetic achievement. The family lived in Depression-era Paterson, where Ginsberg discovered the poetry of Whitman, attended public high school and, upon graduation in 1943, determined to study law. Having won a scholarship to Columbia University, he then changed his major to English literature, becoming the star student in a liberal-minded department headed by **Lionel Trilling**.

Ginsberg had also attached himself to an off-campus underground that constituted the core of the soon-to-be christened Beat movement: **Jack Kerouac**, an ex-Columbia student, Neal Cassady, the movement's patron saint, and **William Burroughs** provided an alternative syllabus for Ginsberg. It was Burroughs, a heroin addict at the time, who introduced them to the New York underworld (not to mention the works of Rimbaud and **W.B. Yeats**, Proust and Céline): Ginsberg was suspended from Columbia for a year and ordered to undergo psychiatric counselling in the aftermath of a murder investigation in which both Kerouac and Burroughs were arrested as accomplices and Lucien Carr, Ginsberg's friend and fellow student, was convicted. Ginsberg then went to work as a welder in the Brooklyn naval yards, as a dishwasher in Times Square, then night porter, and copy boy, before returning to college. He eventually graduated in 1948.

A series of mystical visitations began when Ginsberg was haunted by a disembodied voice reciting Blake. A second brush with the law, when his flatmate, the writer and thief Herbert Huncke, began using the house as a repository for stolen goods, resulted in his arrest. He pleaded insanity – having first claimed to be gathering material for a 'realistic' story (the two had been arrested after a car chase) – and was released on the condition that he attend the Columbia Psychiatric Institute where he met Carl Solomon, a disciple of Artaud, and the dedicatee of 'Howl'. Committed for stealing a sandwich and then owning up to the theft, Solomon was treated with a coma-inducing course of insulin shocks. Eight months later, at liberty again, Ginsberg returned to Paterson where he met and became friendly with **William Carlos Williams**. A brief flirtation with respectability – reviewing for *Newsweek*, working as a market research consultant in New York (1951–3) and then in San Francisco – came to an end on the advice of his psychotherapist, and Ginsberg resolved to devote himself to 'writing and contemplation, to Blake and smoking pot and doing whatever I want'.

Surrounded by like-minded friends and artists, Ginsberg made his literary debut in 1955 at a poetry reading hosted by **Kenneth Rexroth**. He read the manifesto-like, incantatory 'Howl', and immediately became the leader of a movement. Rexroth offered to publish the poem in the City Lights Pocket Poets series, Williams provided the book with an introduction, and the police seized the entire edition on the grounds of obscenity: Ginsberg's loudly declared homosexuality was explicitly expressed in parts of the poem. **Lawrence Ferlinghetti**, joint owner of City Lights, was arrested, and the matter went to trial. The resultant national publicity set the seal on Ginsberg's fame, a success which Ginsberg then began to use to his friends' advantage, helping to find a publisher for Kerouac's *On the Road* (1957) and assisting Burroughs with the editing and publication in Paris of *The Naked Lunch* (1959). Ginsberg based himself in New York and embarked on a hectic round of poetry readings and 'art happenings', travelling widely both at home and abroad, experimenting with hallucinogens and campaigning for the liberalisation of American anti-drug laws. A personal demon was exorcised in *Kaddish and Other Poems* (1961).

Trips to the Far East and India with his lover, Peter Orlovsky, persuaded him to renounce some of the visionary excess of his youth in favour of the meditation-based discipline of 'direct vision' and 'sense perception', the subject of *The Change*. Armed with his mantras and anti-establishment credentials, Ginsberg had become something of a father figure to the 1960s counter-culture, coining the term 'flower-power', lecturing in the universities, demonstrating against the Vietnam War and getting arrested in the riots that surrounded the 1968 Democratic Convention in Chicago. He testified before a Senate committee on narcotics that the use of LSD had 'made it possible for him to stop hating

President Lyndon B. Johnson as a criminal and to pray for him'. A similar spirit presided over the San Francisco Human Be-Ins he inaugurated in 1967. Cuba deported him after he protested at the regime's treatment of homosexuals. The students of Prague elected him the 'King of May', at which point the Czech authorities also deported him: the title was restored in 1990 when Ginsberg visited Harvel's post-communist democracy.

Ginsberg continued to be a focus for and leader of dissent, and the 1970s saw him jailed for his part in an anti-Nixon protest, touring with Bob Dylan, and campaigning on ecological issues – he wrote 'Plutonium Ode' to be read aloud at a public demonstration in Colorado and was arrested yet again. The 1980s found him decrying the growth of the 'surveillance state' and Reagan's covert policies in Nicaragua, still seeking to reform the drug laws, with greater resignation, perhaps, and turning increasingly to Buddhism in the light of lost faith. Having stayed with the small presses throughout his long writing career, he then signed his first major contract, a $160,000 six-book deal with Harper and Row. An 800-page *Collected Poems 1947–1980* was published in 1984. He was visiting professor at Columbia in 1986–7 and took a teaching job at Brooklyn College in 1987.

Poetry

Howl and Other Poems 1956; *Siesta in Xbalba and Return to the States* 1956; *Empty Mirror* 1961; *Kaddish and Other Poems, 1958–1960* 1961; *The Change* 1963; *Reality Sandwiches* 1963; *A Strange New Cottage in Berkeley* 1963; *Kral Majales* 1965; *Wichita Vortex Sutra* 1966; *TV Baby Poems* 1967; *Ankor Wat* 1968; *Message II* 1968; *Planet News, 1961–1967* 1968; *Scrap Leaves, Hasty Scribbles* 1968; *Wales – a Visitation* 1968; *The Moments Return* 1970; *Notes after an Evening with William Carlos Williams* 1970; *The Fall of America: Poems of These States, 1965–1971* 1972; *The Gates of Wrath: Rhymed Poems, 1948–1952* 1972; *Iron Horse* 1972; *New Year Blues* 1972; *Open Head* 1972; *First Blues: Rags, Ballads & Harmonium Songs, 1971–74* 1975; *Sad Dust Glories* 1975; *Mind Breaths: Poems 1972–1977* 1978; *Mostly Sitting Haiku* 1978; *Poems All Over the Place* 1978 *Plutonium Ode: Poems 1977–1980* 1982; *Collected Poems, 1947–1980* 1984; *Many Loves* 1984; *Old Love Story* 1986; *White Shroud: Poems 1980–1985* 1986; *Cosmopolitan Greetings* 1994

Non-Fiction

Mystery in the Universe: Notes on an Interview with Allen Ginsberg [ed. E. Lucie-Smith] 1965; *The Yage Letters* [with W.S. Burroughs] 1963; *Airplane Dreams: Compositions from Journals* 1968; *Indian Journals, March 1962–May 1963* 1970; *Declaration of Independence for Dr Timothy Leary* 1971; *Improvised Poetics* [ed. M. Robinson] 1972; *Allen Verbatim: Lectures on Poetry, Politics, Consciousness* [ed. G. Ball] 1974; *The Visions of the Great Rememberer* 1974; *Chicago Trial Testimony of Allen Ginsberg* 1975; *To Eberhart from Ginsberg* 1976; *As Ever: the Collected Correspondence of Allen Ginsberg and Neal Cassady* [ed. B. Gifford] 1977; *Journals: Early Fifties, Early Sixties* [ed. G. Ball] 1977; *Composed on the Tongue: Literary Conversations, 1967–77* [ed. D. Allen] 1980

Plays

Don't Go Away Mad [pub. in *Pardon Me, Sir, But Is My Eye Hurting Your Elbow?* ed. B. Booker, G. Foster 1968]; *Kaddish* 1972 [np]; *The Hydrogen Jukebox* [with P. Glass] 1990 [np]

Collections

Straight Hearts Delight: Love Poems and Selected Letters 1947–1980 [ed. W. Leyland] 1980

Biography

Allen Ginsberg by Barry Miles 1989

Nikki GIOVANNI 1943–

Variously hailed as 'the Princess of Black Poetry', 'the poet laureate of young Black women', *and* voted Woman of the Year by the readers of magazines as diverse as *Mademoiselle*, the *Ladies' Home Journal* and *Ebony*, Giovanni has proved one of the most prolific and widely read black American poets to emerge during the 1960s. She was born in 1943 in Knoxville, Tennessee, to middle-class parents, but was brought up in Cincinnati, Ohio. The literary and political were pleasingly combined in her maternal grandparents: a scholarly grandfather, a Latinist and steeped in mythology; and a grandmother whose spirited resistance to white racism had necessitated a speedy exit from Georgia. Giovanni's own academic career was ultimately successful (although Fisk University, from which she eventually graduated in 1967 after studying history, literature and art, found that 'her attitudes did not suit'). Graduate work at the universities of Pennsylvania and Columbia was succeeded by academic appointments in various institutions, including the City University of New York, and Rutgers. She also holds honorary doctorates from, amongst others, the University of Maryland and Smith College.

Politically active from an early age (and on a picket-line by 1956), she was involved with the Student Non-Violent Co-ordinating Committee (SNCC) in the late 1960s. Her first (and initially self-published) work, *Black Feeling, Black Talk* (1968), reflected the passion and anger of the emerging 'Black Revolution'. In later works political concerns have often been explored in the context of personal relationships, especially the one she has with her son, very consciously borne outside marriage in 1969. Giovanni fiercely repudiates, however, any suggestion that her work has become less political: 'The assumption inherent [there] ... is that the self is not a part of the body politic. There's no separation'.

'God wrote one book. The rest of us are forced to do a little better', she said in an interview with Claudia Tate, for *Black Women Writers At Work* (1983). Giovanni's efforts at that point already included a dozen or so remarkably wide-ranging books: of poetry, for adults and for children; an 'autobiographical essay', *Gemini*; literary dialogues with **James Baldwin** and Margaret Walker; critical discussions of key figures in black American cultural history, from Phillis Wheatley, the eighteenth-century slave poet, to Lena Horne, one of the many black musicians whom Giovanni reveres and celebrates throughout her work; and an anthology of black women poets, *Night Comes Softly*. Keenly alive to the importance of the African oral tradition, and the part women have played in creating and transmitting it, she has also made a number of successful recordings of her own work. including the album, *Truth Is on Its Way*.

No stranger to controversy, she has often aroused strong and hostile reactions from other black activists, fellow writers and academic critics. *Sacred Cows and Other Edibles* is an appropriate title for a woman who has at various times audaciously challenged some of her assumptions most commonly cherished by recent black, feminist, and black feminist writers and thinkers. She is impatient with the quest for a 'black aesthetic' and rejects easy idealisations of a golden pre-slavery Africa: 'People get upset because [Wheatley] talks about Africa in terms of how delighted she was to discover Christianity. Well, from what little I know she might have been damned delighted. Life for an African woman can be very difficult even today, and she was writing in the eighteenth century.' Unfashionably and early, she denied, as a writer, one of the enshrined 'truths' of the period's emerging 'identity politics', that 'truthful' and 'authentic' writing can spring only from the author's 'lived experience': 'Writers don't write from experience, though they are hesitant to admit that they don't ... Writers write from empathy. We cheapen anything written when we consider it as experience. Because if it's someone else's experience we don't have to take it seriously.' Revolutionary, indeed, in the context of 1970s and 1980s cultural politics!

Poetry

Black Feeling, Black Talk 1968; *Black Judgement* 1968; *Poem of Angela Yvonne Davis* 1970; *Re: Creation* 1970; *My House* 1972; *The Women and the Men* 1975; *Cotton Candy on a Rainy Day* 1978; *Those Who Ride the Night Winds* 1983

For Children

Spin a Soft Black Song 1971 [rev. 1985]; *Ego Tripping and Other Poems for Young Readers* 1973; *Vacation Time* 1980

Non-Fiction

Gemini 1971; *A Dialogue: James Baldwin and Nikki Giovanni* 1973; *A Poetic Equation* 1974; *Sacred Cows and Other Edibles* 1988; *Racism* 1994

Edited

Night Comes Softly 1970

Ellen (Anderson Gholson) GLASGOW 1873–1945

Born in Richmond, Virginia, where she lived most of her life, Glasgow was one of ten children of the South's largest foundry and munitions supplier (he had been associated with the Tredegar Iron Works during the Civil War), while her mother was from an aristocratic Tidewater family. Her father was a strict Calvinist, and family conflict led to his wife suffering a nervous breakdown. Glasgow, who only intermittently attended school, sought solace in extensive reading in the family library, where she early became influenced by Darwin and Nietzsche.

Her first novel, *The Descendant*, appeared anonymously in 1897, and she also published an early volume of verse. From 1900 she published novels regularly and was eventually to write more than twenty works of fiction. From 1911 and 1916 she lived in New York City, travelling abroad frequently, and during her trip to England in 1914 met **Henry James**, **John Galsworthy**, **Joseph Conrad**, **Arnold Bennett** and **Thomas Hardy**. In 1916, her father died, and she returned to the family home, where she lived until her death. She had had several unhappy relationships with men, and soon after her return to Richmond became engaged to a local politician, Henry Watkins Anderson, but his relationship with Queen Marie of Roumania (he was officer in charge of the Red Cross Commission in the Balkans) led Glasgow to break off the engagement and to attempt suicide. She lived for the rest of her life with a devoted female companion, Anne Virginia Bennett.

Many of her novels deal with the tragedy of women who remain single, but she also emphasised women who struggle and achieve. Her gallery of female characters, indeed, is outstanding in American fiction, and she was the first modern novelist of the South, presenting its social problems and wide range of types trenchantly and realistically. Critics are agreed that her best work was done from the 1920s on; her own favourite was *Barren Ground*, which was followed by a group of ironic social comedies such as *The Sheltered Life*, after which her last novels became more didactic. She also wrote short stories and essays, while *A Certain Measure* was a collection of prefaces to the Virginia edition of her novels rather in the manner of Henry James's prefaces to his New York edition. Glasgow had suffered from partial deafness since early adulthood and, although her work collected many honours, her last years were marred by a series of heart attacks which led to repeated hospitalisations. She eventually died at home after suffering a heart attack in her sleep. Her autobiography, *The Woman Within*, was published posthumously, as was one further novel, *Beyond Defeat*. Her reputation, in decline for some years, has been significantly revived by feminism.

Fiction

The Descendant 1897; *Phases of an Inferior Planet* 1898; *The Voice of the People* 1900; *The Battle-Ground* 1902; *The Deliverance* 1904; *The Wheel of Life* 1906; *The Ancient Law* 1908; *The Romance of a Plain Man* 1909; *The Miller of Old Church* 1911; *Virginia* 1913; *Life and Gabriella* 1916; *The Builders* 1919; *One Man in His Time* 1922; *The Shadowy Third and Other Stories* (s) 1923 [As *Dare's Gift and Other Stories* 1924]; *Barren Ground* 1925; *The Romantic Comedians* 1926; *They Stooped to Folly* 1929; *The Sheltered Life* 1932; *Vein of Iron* 1935; *In This Our Life* 1941; *Collected Stories* (s) 1963; *Beyond Defeat* 1966

Non-Fiction

A Certain Measure: an Interpretation of Prose Fiction 1943; *The Woman Within* (a) 1954; *Letters* [ed. B. Rouse] 1958

Poetry

The Freeman and Other Poems 1902

(Margaret) Rumer GODDEN 1907–

In her novel *The River* (which Jean Renoir made into a film in 1951) Godden paints a portrait of her upbringing with her three sisters on a remote river station in Bengal, where her father ran the local steamship company. The girls led an isolated life and developed the dreaded sing-song Eurasian accent, which made them the subjects of ridicule when she and her older sister, Jon, were sent to a convent boarding-school in England. Godden was not happy there and swore to pay out her tormentors when she grew up: 'Yet, when ... I did write a book about nuns, *Black Narcissus*, I mysteriously could not take that revenge.' This book was to be her first bestseller, and the material for the classic Powell-and-Pressburger film of the same title (1946), which she hated: 'Everything about it was phoney.'

In 1925, she returned to India to open a dancing-school in Calcutta. Once again she was a social misfit, because 'nice girls' did not earn a living there. She began to write because it offered an escape from her compatriots' snobbery, philistinism and restrictive practices. She also married a stockbroker, Laurence Foster, 'one of the most charming people you could hope to meet, but completely amoral'. They had two children, but the marriage was doomed. He began gambling on the stock exchange, embezzled his firm's funds, lost his job and, without a word of warning or farewell, went into the army in 1941, leaving Godden to face the bailiffs. She felt honour-bound to use the proceeds from *Black Narcissus* to pay off her debts, then removed herself to a life of extreme and primitive frugality in the Himalayan foothills. The diary written at this time was later published as *Rungli-Rungliot*.

Godden's determination to support herself through writing resulted in hardship. When she travelled back to England with her children to begin a new life, she refused the divorce court's order of maintenance and several offers of help, and lived in a damp house in Sussex.

Things changed for the better when she met James Haynes-Dixon, who had a job in one of the ministries and, in post-war times of rationing, was able to woo her with chocolates, peat and logs, pâté, wine and whisky. In the winter of 1947 he became part of her life.

In the second part of her autobiography, *A House with Four Rooms*, Godden talks briefly of the decline in health of her sister Jon (who was also a writer, and upon whom she modelled the elder sister in *The Greengage Summer*), of Haynes-Dixon's illness, and of her conversion to Roman Catholicism. The book ends with her seventieth birthday in Rye, Sussex, in the house that once belonged to **Henry James**. She subsequently moved to Dumfriesshire in Scotland, to be near her family, where she leads a busy life with her Pekinese dogs.

Godden has always considered herself a storyteller. She and her sister were never, she says, Rolls-Royce writing engines like **E.M. Forster** or **Paul Scott**, more motor scooters, 'what the Indians call "putputtis" because of their uncertain stroke, but we knew the authentic hum.'

Fiction
Chinese Puzzle 1936; *The Lady and the Unicorn* 1938; *Black Narcissus* 1939; *Gypsy, Gypsy* 1940; *Breakfast with the Nikolides* 1942; *Take Three Tenses* 1945; *The River* 1946; *A Candle for St Jude* 1948; *A Breath of Air* 1950; *Kingfishers Catch Fire* 1953; *An Episode of Sparrows* 1955; *The Greengage Summer* 1958; *China Court* 1961; *The Battle of the Villa Fiorita* 1963; *Swans and Turtles* (s) 1968; *In This House of Brede* 1969; *The Peacock Spring* 1975; *Five for Sorrow, Ten for Joy* 1979; *The Dark Horse* 1981; *Thursday's Children* 1984

Non-Fiction
Rungli-Rungliot 1943; *Hans Christian Andersen: a Great Life in Brief* 1955; *Two Under the Indian Sun* [with J. Godden] 1966; *The Raphael Bible* 1969; *Shiva's Pigeons: an Experience of India* [with J. Godden] 1971; *The Tale of Tales: the Beatrix Potter Ballet* 1971; *The Butterfly Lions: the Story of the Pekinese in History, Legend and Art* 1977; *Gulbadan: Portrait of a Rose Princess at the Mughal Court* 1980; *A Time to Dance, No Time to Weep* (a) 1989; *Indian Dust* [with J. Godden] 1989; *A House with Four Rooms* (a) 1989; *Coromandel Sea Change* 1991

For Children
The Doll's House 1947; *In Noah's Ark* 1949; *The Mousewife* 1951; *Impunity Jane* 1955; *The Fairy Doll* 1956; *Mouse House* 1958; *The Story of Holly and Ivy* 1958; *Candy Floss* 1960; *Miss Happiness and Miss Flower* 1961; *St Jerome and the Lion* 1961; *Home is the Sailor* 1963; *Little Plum* 1963; *The Kitchen Madonna* 1968; *A Letter to the World: Poems for Young People* 1968; *Operation Sippacik* 1969; *The Diddakoi* 1972; *The Old Woman Who Lived in a Vinegar Bottle* 1972; *Mr McFadden's Hallowe'en* 1975; *The Rocking Horse Secret* 1977; *A Kindle of Kittens* 1978; *The Dragon of Og* 1981; *The Valiant Chatti-Maker* 1983; *Fu-Dog* 1989; *Great Grandfather's House* 1992; *Listen to the Nightingale* 1992

Poetry
The Beasts' Choir 1965

Translations
Carmen Bernos de Gasztold: *The Creatures' Choir* 1963, *Prayers from the Ark* 1963

Collections
Mooltiki: Stories and Poems from India 1957

Oliver St John GOGARTY 1878–1957

The son of a doctor, Gogarty was born in Dublin, and educated at Trinity College, Oxford, and Vienna, later qualifying as a surgeon. His student days in Dublin were distinguished by his athleticism, clowning wit and skill in winning prizes for his poetry, then less memorable than his limericks. His Rabelaisian nature made him a natural ally of **James Joyce** whom he befriended; both were united in their sceptical view of the mysticism of **W.B. Yeats** and **AE**. Gogarty felt betrayed when he discovered he was the source for the 'plump and stately' Buck Mulligan of *Ulysses*. By 1922 he was one of Ireland's most distinguished surgeons, and his respectable appointment to the Irish Senate, the Upper House of the Free State Parliament, in that year clashed with his fictional *alter ego*.

Appalled by the Dublin slums, Gogarty focused on the problem in 1917 with a play, *Blight*, and, as a politician, he fought for adequate housing schemes. Like all senators, he was under the death threat from the old IRA, and escaped an assassination attempt by swimming the Liffey early in 1923. After his country house was burned down, he was for safety obliged to live in London for several months, where he delighted fashionable society with his conversation and charm. Returning to Dublin in 1924, he presented two swans to the Liffey as a symbolic gesture for his life, a ceremony which inspired the title poem of his book *An Offering of Swans* (1924). Yeats praised his classical style of verse, perhaps unaware that his friend's version of 'Leda' was in the same metre as 'Goosey Goosey Gander'.

Gogarty published three volumes of autobiography, the most notable of which was *As I Was Going Down Sackville Street*. With the help of a youthful **Samuel Beckett** appearing for the prosecution, this book was successfully sued for libel. The action was brought by a man who claimed that both he and his grandfather were referred to in verses Gogarty quoted, which identified them as having a predilection for small girls. Nevertheless, this was Gogarty's best book, confirming both Asquith's impression that he was 'the wittiest man in London', and Yeats's opinion that he was 'one of the great lyric poets of the age'. His last years were spent in America, reading detective fiction. He died in New York, and was buried in Connemara.

Poetry
Hyperthuleana 1916; *Requiem and other Poems* [as 'S. O'Sullivan'] 1917; *The Ship and Other Poems* 1918; *An Offering of Swans* 1923; *Wild Apples* 1928; *Others to Adorn* 1938; *Elbow Room* 1939; *Perennial* 1946; *Collected Poems* 1951; *Unselected Poems* 1954

Fiction
Mad Grandeur 1943; *Mr Petunia* 1946; *Mourning Became Mrs Spendlove and Other Portraits, Grave and Gay* (s) 1948

Non-Fiction
As I Was Going Down Sackville Street (a) 1937; *I Follow St Patrick: On the Journeys of St Patrick* 1938; *Tumbling in the Hay* (a) 1939; *Going Native* 1941; *Intimations* 1950; *Rolling Down the Lea* 1950; *It Isn't This Time of Year At All!* (a) 1954; *Start from Somewhere Else: an Exposition of Wit and Humour, Polite and Perilous* 1955; *A Weekend in the Middle of the Week and Other Essays on the Bias* 1958; *William Butler Yeats: a Memoir* 1963

Plays
Blight [with J. O'Connor] 1917; *The Enchanted Trousers* [as 'Gideon Ouseley'] 1919; *A Serious Thing* [as 'Gideon Ouseley'] 1919

Biography
Oliver St John Gogarty – a Poet and His Times by Ulick O'Connor 1964

William (Gerald) GOLDING 1911–1993

Golding was born near Newquay in Cornwall, one of a family which had provided schoolmasters for generations. He grew up in Marlborough in Wiltshire, the county in which he lived for much of his life, and attended the local grammar school, at which his father taught. His intellectual and writerly interests began early: aged seven, he taught himself hieroglyphics in order to write a play set in ancient Egypt. He went up to Brasenose College, Oxford, where, at his parents' wish, he first read natural science, and then switched to English.

After coming down, he became a social worker, and also wrote, acted in and produced plays for small theatres. In 1934, he published a volume of poems, but did not attempt poetry in later life. In 1939, he married, and in the same year took a job teaching English at Bishop Wordsworth's School in Salisbury. In 1940, he joined the Royal Navy, and ended the war as a lieutenant in charge of a rocket ship. He then returned to teach at the school, and began writing seriously, completing several novels which were not accepted. Eventually, *Lord of the Flies* was published in 1954 (although one publisher's reader had dismissed it as merely a story about small boys on an island). In fact, it is the first of Golding's allegorical dystopias, and as such achieved enormous success, finding a response among young people which in the 1950s was only matched by **J.D. Salinger**'s *The Catcher in the Rye*. In 1961, Golding retired from teaching and, after a year as writer-in-residence at an American college, continued as a full-time writer. He had two children.

Golding is one of the few post-war English authors for whom greatness is regularly claimed, an estimate reflected by the award of the Nobel Prize to him in 1983: the estimate will be accepted by those who prefer novels to be moral and symbolic dramas exploring a great theme – in Golding's case, the sinfulness of humanity – and will arouse doubts in those who value the creation of characters and engagement with the society in which an author lives. There is a critical consensus that Golding's finest novel is *The Spire*, which deals with the building of Salisbury Cathedral in the thirteenth century, and also that his later work is a falling off from the earlier. Besides his novels, he wrote three plays, a volume of short stories, an account of his travels in Egypt, and two illuminating volumes of essays. His last major fictional work was a trilogy about life on an eighteenth-century ship travelling to Australia, the first volume of which, *Rites of Passage*, won the Booker Prize in 1980, and which was completed by *Fire Down Below* in 1989.

Fiction
Lord of the Flies 1954; *The Inheritors* 1955; *Pincher Martin* 1956 [US The Two Deaths of Christopher Martin 1957]; *Free Fall* 1959; *The Spire* 1964; *The Pyramid* 1967; *The Scorpion God* (s) 1971; *Darkness Visible* 1979; *To the Ends of the Earth* 1991: *Rites of Passage* 1980, *Close Quarters* 1987, *Fire Down Below* 1989; *The Paper Men* 1984; *The Double Tongue* [unfinished] 1995

Plays
The Brass Butterfly 1958; *Miss Pulkinhorn* (r) 1960; *Break My Heart* (r) 1962

Poetry
Poems 1934

Non-Fiction
The Hot Gates and Other Occasional Pieces 1965; *A Moving Target* 1982; *An Egyptian Journal* 1985

Paul GOODMAN 1911–1972

Goodman was born in New York City, 'an orphan who had had a home'. His father left home when Paul was still a baby, his elder brother as soon as he could. He described his mother as 'a bourgeois gipsy'. His sister Alice, ten years older, saw him through college. He was educated at the City College of New York, and the University of Chicago, taught at colleges and universities in New York, North Carolina, Wisconsin and California. He was at different times editor of *Complex* magazine, film editor of *Partisan Review*, television critic of *New Republic* and, from 1960, editor of *Liberation* magazine. He practised as a Gestalt therapist, and in addition wrote poems, stories, plays, novels, literary criticism and educational theory. A 'utopian thinker' – 'my experience of radical community is that it does not tolerate my freedom' – he describes himself as 'a man of letters in the old sense'. It wasn't until his forty-ninth year, with the publication of *Growing Up Absurd*, his study of youth in the 'Organized System', that he became generally known, at least in America. The book had been turned down by the publisher who had commissioned it and several others, as was his major novel *The Empire City*, much praised when it was eventually published. His poems, often on themes of love and death, have been described as 'distinguished by their immediacy of feeling'. They and his stories were collected and published posthumously. **Denise Levertov** in the *Nation* said of the latter, 'Paul Goodman has written, besides his better known novels, short stories that I believe stand with the greatest American stories.' In a 'serious attempt to achieve for Paul Goodman the recognition so long denied him in Britain', Wildwood House reissued a number of his books in the 1960s and 1970s. They had never appeared previously in the UK. He has been described as 'ostentatiously homosexual and ostentatiously heterosexual at the same time'. He married twice, had three children, one of whom, a son, died aged twenty, saddening his later years. He died in New Hampshire, about whose countryside some of his last lyrics were written.

Poetry

Stop-light 1941; *Hawkweed* 1942 [rev. 1967]; *Pieces of Three* [with M. Liben, E. Roditi] 1942; *Five Young American Poets* [with others] 1945; *Red Jacket* 1956; *The Well of Bethlehem* 1957; *The Lordly Hudson* 1962; *Day and Other Poems* 1965; *North Percy* 1968; *Homespun of Oatmeal Grey* 1970; *Collected Poems* 1973; *Little Prayers and Finite Experience* 1973

Plays

Three Plays: The Young Disciple, Faustina and Jonah [pub. 1965]; *Tragedy & Comedy* [pub. 1970]

Fiction

The Grand Piano 1942; *The Facts of Life* (s) 1945; *The State of Nature* 1946; *The Break-up of Our Camp and Other Stories* (s) 1949; *The Dead of Spring* 1950; *Parents' Day* 1951; *The Empire City* 1959; *Our Visit to Niagara* (s) 1960; *Making Do* 1963; *Adam and His Works* (s) 1968; *Collected Stories* (s) 1978

Non-Fiction

Art and Social Nature 1946; *The Copernican Revolution* 1946; *Kafka's Prayer* 1947; *Communitas: Means of Livelihood and Ways of Life* [with P. Goodman] 1947 [rev. 1960]; *Gestalt Therapy* [with R. Hefferline, F.S. Perls] 1951; *The Structure of Literature* 1954; *Censorship and Pornography on the Stage and Are Writers Shirking Their Political Duty?* 1959; *Growing Up Absurd: Problems of Youth in the Organized System* 1960; *The Community of Scholars* 1962; *Drawing the Line* 1962; *Utopian Essays and Practical Proposals* 1962; *The Society I Live in Is Mine* 1963; *Compulsory Mis-education* 1964; *People or Personnel: Decentralizing and the Mixed System* 1965; *Five Years* (a) 1967; *Like a Conquered Province: the Moral Ambiguity of America* 1967; *A Message to the Military Industrial Complex* 1969; *The Open Look* 1969; *New Reformation: Notes of a Neolithic Conservative* 1971; *Speaking and Language: Defence of Poetry* 1971

Edited

Seeds of Liberation 1965

Nadine GORDIMER 1923–

Gordimer was born in the small mining town of Springs, Transvaal, the daughter of a Lithuanian-Jewish watchmaker and an English-born mother. She was educated at a convent, and spent a year at the University of Witwatersrand, Johannesburg, without taking a degree. She began to write in childhood, and her first published story, 'Come Again Tomorrow', appeared in the Johannesburg magazine *Forum* in 1939 when she was only fifteen. Another early story, 'The Amateurs', satirises the futility of liberal 'good works' by describing a white amateur dramatic society presenting *The Importance of Being Earnest* to an uncomprehending black township audience. Gordimer has written of how her writing emerged out of the realisation that: 'I was not merely part of a suburban white life aping Europe, I lived with and among a variety of colours and kinds of people.' This discovery hardened into an increasingly powerful antipathy to apartheid, a process evident in the development of her work.

His first novel, *The Lying Days* (1953), is partially autobiographical and describes the young life of narrator Helen Shaw who, mirroring Gordimer's literary ambition, reflects: 'I had never read a book in which I myself was recognizable; in which there was a "girl" like Anna who did the housework and the cooking and called the mother and father Missus and Baas.' Helen rebels against the bigotry of her parents' world only to watch powerless from a car as a man is killed by the police during a black township riot. The impotence of middle-class liberalism in the face of apartheid is a consistent theme in her work, not least in the short, powerful novel *The Late Bourgeois World*, which is widely regarded as one of her finest. Its explicit condemnation of South African society ensured that it was banned there. Though she has never joined a political party, Gordimer was active in the campaign against censorship, taking the opportunity of her acceptance speech when she won the CNA Award (one of South Africa's most prestigious literary prizes) in 1980 to castigate censorship, describing it as a weapon of apartheid as potent 'as the hippo cars that went through the streets of Soweto in '76'. In an interview with *The Times* in 1974 she declared: 'I am a white South African radical. Please don't call me a liberal.' This remark led to something of a furore, with **Alan Paton**, a former leader of the Liberal Party, furiously berating Gordimer, and other writers such as **André Brink** weighing in. Everyone calmed down when it was recognised that her remark was intended with a measure of irony.

The Conservationist, joint winner of the 1974 Booker Prize, describes an industrialist's struggle to reclaim a tract of wilderness, and then unsuccessfully trying to defend it from the black squatters who gradually take over. *July's People* explores a similar theme, and is the first of Gordimer's novels to look towards the possibility of a post-apartheid future. *My Son's Story* is seen through the eyes of an adolescent boy, the son of a black activist who is in love with a white woman. Though the race laws have now been abolished the natural laws governing relationship within families still apply, and the affair is inextricably bound up with the political struggle that has led to the change in South African society.

Gordimer has written several books of non-fiction on South African subjects, and made television documentaries, notably collaborating with her son Hugo Cassirer on the television film *Choosing for Justice: Allan Boesak*. She has won numerous prizes, including the W.H. Smith Award in 1961, the Thomas Pringle Award in 1969, a James Tait Black Memorial Award in 1972, and the Nobel Prize for Literature in 1991. In the 1960s and 1970s she taught at various American universities, though in common with Brink and **J.M. Coetzee** she never considered leaving South Africa, even at the height of the apartheid regime. She has been married twice, and has two children.

Fiction

Face to Face (s) 1949; *The Lying Days* 1953; *Six Feet of the Country* (s) 1956; *A World of Strangers* 1958; *Friday's Footprint and Other Stories* (s) 1960; *Occasion for Loving* 1963; *Not for Publication and Other Stories* (s) 1965; *The Late Bourgeois World* 1966; *A Guest of Honour* 1970; *Penguin Modern Stories 4* (s) [with others] 1970; *Livingstone's Companions* (s) 1971; *The Conservationist* 1974; *Selected Stories* (s) 1975 [as *No Place Like* 1978]; *Some Monday for Sure* (s) 1976; *Burger's Daughter* 1979; *A Soldier's Embrace* (s) 1980; *Town and Country Lovers* (s) 1980; *July's People* 1981; *Something Out There* (s) 1984; *A Sport of Nature* 1987; *My Son's Story* 1990; *Jump and Other Stories* (s) 1991; *Why Haven't You Written?* 1992

Non-Fiction

Lifetimes: Under Apartheid 1986; *Crimes of Conscience* 1991

Edited

South African Writing Today [with L. Abrahams] 1967

Edward (St John) GOREY 1925–

Gorey was born in Chicago and educated at Harvard, where he spent much of his time reading the novels of **I. Compton-Burnett**, but nevertheless took his BA in 1950. He worked as a book illustrator for Doubleday from 1953 to 1960, but his first published book of his own was the illustrated narrative *The Unstrung Harp: or, Mr Earbrass Writes a Novel* (1953). This details the horrors of literary life as experienced by Mr Clavius Frederick Earbrass, author of such works as *A Moral Dustbin* and *The Hipdeep Trilogy*. It already shows the distinctive style and sensibility of subsequent work, with its atmosphere of *fin-de-siècle* morbidity and wistful desuetude. Gorey's work is not quite Gothic or camp, though it may appeal to people who like either. **Edmund Wilson** describes it as 'a whole little personal world, equally amusing and sombre, nostalgic ... claustrophobic ... poetic and poisoned'. Gorey has been prevailed upon to admit some similarity between the period atmosphere of his work and that of **Henry James**, a writer he professes to loathe.

The death of children is a favourite subject. Sometimes they are particularly good (like little Henry Clump in *The Pious Infant*), sometimes they are particularly bad (like the hideously indestructible *Beastly Baby*), and sometimes they are just small, like the catalogue of fatalities that makes up *The Gashlycrumb Tinies*: 'A is for Amy who fell down the stairs, B is for Basil assaulted by bears ...' and so on. Gorey plays down the macabre aspect and claims that much of his work is really intended for children.

A good deal of Gorey's output is claimed by expressive, anagrammatic pseudonyms like Ogdred Weary, Mrs Regera Dowdy, Dreary Wodge and Roy Grewdead. The man behind them is tall and bearded and habitually wore a fur coat (now discarded for ecological reasons), deer stalker and tennis shoes. He lives alone and is devoted to cats and ballet, both of which figure in the work. He attended the New York City Ballet almost daily between 1956 and 1976.

Among his prolific illustrations for other writers are those for works by **Samuel Beckett**, including the first edition of *All Strange Away*, which have been signed by both men. Underneath its funereal sprightliness Gorey's world shares something of Beckett's hatred of procreation and his feel for diminution and abjection, as shown in works like *The Abandoned Sock* and *The Iron Tonic* ('The people in the grey hotel / Are either aged or unwell'). Gorey has won major prizes for illustration and designed a Broadway set for *Dracula*. For some time now his unique work has looked set to break into the commercial mainstream, a prospect that fills longtime enthusiasts with gloom.

Fiction

The Unstrung Harp 1953; *The Listing Attic* 1954; *The Doubtful Guest* 1957; *The Object-Lesson* 1958; *The Bug Book* 1960; *The Fatal Lozenge* 1960; *The Hapless Child* 1961; *The Curious Sofa* [as 'Ogdred Weary'] 1961; *The Beastly Baby* [as 'Ogred Weary'] 1962; *The Willowdale Handcar* 1962; *The Vinegar Works* 1963; *The Wuggly Ump* 1963; *15 Two* 1964; *The Remembered Visit* 1965; *The Evil Garden* [as 'Edward Blutig'] 1966; *The Gilded Bat* 1966; *The Inanimate Tragedy* 1966; *The Pious Infant* [as 'Mrs Regera Dowdy'] 1966; *Fletcher and Zenobia* [with V. Chess] 1967; *The Utter Zoo Alphabet* 1967; *The Blue Aspic* 1968; *The Epileptic Bicycle* 1968; *The Other Statue* 1968; *The Secrets* 1968; *The Deranged Cousins* 1969; *The Eleventh Episode* [as 'Raddory Gewe'] 1969; *The Iron Tonic* 1969; *The Chinese Obelisks* 1970; *Donald Has a Difficulty* [with P.F. Neumeyer] 1970; *The Osbick Bird* 1970; *The Sopping Thursday* 1970; *Why We Have Day and Night* [with P.F. Neumeyer] 1970; *The Disrespectful Summons* 1971; *Fletcher and Zenobia Save the Circus* [with V. Chess] 1971; *The Untitled Book* 1971; *Amphigorey* 1972; *The Abandoned Sock* 1972; *The Awdrey-Gore Legacy* 1972; *Leaves From a Mislaid Album* 1972; *The Vinegar Works: Seven Volumes of Moral Instruction* 1972; *The Lavender Leotard: or, Going a Lot to the New York City Ballet* 1973; *A Limerick* 1973; *The Lost Lions* 1973; *The Glorious Nosebleed* 1974; *Amphigorey Too* 1975; *L'Heure bleue* 1975; *The Ominous Gathering* 1975; *The Broken Spoke* 1976; *The Loathsome Couple* 1977; *The Fantod Words* 1978; *The Green Beads* 1978; *The Catafalque Works* 1979; *Dancing Cats and Neglected Murderesses* 1980; *The Gashlycrumb Tinies* 1981; *Mélange funeste* 1981; *The Dwindling Party* 1982; *The Water Flowers* 1982; *Amphigorey Also* 1983; *Donald and the —:* [with P. Neumeyer] 1983; *The Eclectic Abecedarium* 1983; *The Prune People* 1983; *The Raging Tide* 1987; *The Dripping Faucet* 1989; *The Helpless Doorknob* 1989

Non-Fiction

Categorey: Fifty Drawings 1973; *Gorey Endings: a Calendar for 1978* 1978; *Dracula: a Toy Theatre* 1978; *Gorey Games* [with L. Evans] 1979; *Gorey Posters* 1979; *The Betrayed Confidence: Seven Series of Dogear Wryde Postcards* 1992

Edited

The Haunted Looking-Glass: Ghost Stories (s) 1959

Translations

Alphonse Allais *Story for Sara* [as 'Edward Pig'] 1971

W(illiam) S(ydney) GRAHAM 1918–1986

Born in Greenock, Scotland, Graham was educated at the high school there before winning a scholarship to Newbattle Abbey College, near Edinburgh. He originally intended to take up his father's profession, and served an apprenticeship as a structural engineer while studying literature and philosophy at night classes. Before he started publishing poetry in the early 1940s he worked as an engineer in Scotland and as a casual labourer, crofter and fisherman in Cornwall. He made Cornwall his home, and it proved the landscape for many of his later poems. The shipyards of his native Clydeside also influenced his work.

A poet concerned with isolation and the difficulties in communication, he claimed that writing poetry 'makes a life for me to live in and, at the same time, I hope, communicates with other people'. His earlier poems, with their twisted syntax, are reminiscent of **Dylan Thomas**. While praised for experimenting with the relationship between poet and reader, he was criticised for being obsessed with the theme of language. He married in 1954 and had one daughter.

Poetry

Cage without Grievance 1942; *The Seven Journeys* 1944; *Second Poems* 1945; *The White Threshold* 1949; *The Nightfishing* 1955; *That Ye Inherit* 1968; *Malcolm Mooney's Land* 1970; *Penguin Modern Poets 17* 1970; *Implements in Their Places* 1977; *Collected Poems 1942–1977* 1979

Kenneth GRAHAME 1859–1932

Born in Edinburgh, Grahame was the son of a lawyer, from an old Scottish family which traced its descent from Robert the Bruce. In his early years, he lived with his family in the Western Highlands, but his mother died when he was five, and his father could not look after the children (he eventually died of drink in Le Havre), so they were sent to live with their maternal grandmother in the Thames-side village of Cookham Dene, where Grahame also lived for a time in later life, and which is the chief setting for *The Wind in the Willows*. He attended St Edward's School, Oxford, but his relations decided they were too poor to let him attend the university, and procured him a clerkship in the Bank of England, which he entered in 1879. While working there, he began to write: essays of his began appearing in London magazines from 1888, and soon after he became a frequent contributor to W.E. Henley's *National Observer* and the *Yellow Book*. His essays appeared in a volume, *Pagan Papers*, in 1893, and two years later a number of fictional sketches from this, together with additional ones, about a family of orphaned children living in a large house, were published as *The Golden Age*. This, with its successor, *Dream Days*, enjoyed a vast success, despite the fact that these are really books *about* rather than *for* children. In the meantime, in 1898, he was promoted to secretary at the Bank, and in 1899 he married Elspeth Thomson. They had one son, Alistair, who was known as 'Mouse', and the stories told to him as a child about the animals of the river bank eventually found publication as *The Wind in the Willows*. The book became an established classic only slowly: it was rejected at first, and Grahame's public was not ready for his change of style. In the same year he retired from the Bank, either because of ill-health or under pressure from his employers (his brother-in-law Lord Courtauld-Thomson remarked: 'his duties as Secretary may have suggested the titles *Golden Age* and *Dream Days*').

He and his wife spent the rest of their lives in increasingly eccentric idleness. Their son Alistair committed suicide while an undergraduate at Oxford by lying down at night on a railway track. *The Wind in the Willows* went through 100 editions in its first sixty years of publication, proof if any were needed of its status in hearts and minds. Part Edwardian rural dream-idyll, part apprehension of the menace of the world, this subtle, complex and beautiful work appeals as much to adults as it has been loved by generations of children.

Fiction
The Golden Age (s) 1895; *Dream Days* (s) 1898; *The Headswoman* 1898

For Children
The Wind in the Willows 1908

Edited
The Cambridge Book of Poetry for Children 1916

Collections
Pagan Papers 1893; *Kenneth Grahame: Life, Letters and Unpublished Work* [ed. P.R. Chalmers] 1933

Biography
The First Whisper of 'The Wind in the Willows' by Elspeth Grahame 1944; *Kenneth Grahame: An Innocent in the Wild Wood* by Alison Prince 1994

Harley GRANVILLE-BARKER 1877–1946

Regarded as an exponent of what was in the early twentieth century called 'New Drama', Granville-Barker was educated privately in London, where he was born. He left school at thirteen because, from an early age, he was fascinated by the theatre. After coaching in acting by his mother, Mary Bozzi-Granville, who taught elocution, he was admitted to a drama school at Margate; his father was content to leave Harley's career to his wife. The following year the boy appeared under the direction of Charles Hawtrey in a London play.

Although he had an amateur play put on stage when he was sixteen, his first professional production was in 1899, with *Weather Hen*, written in collaboration with Herbert Thomas. In 1900, he was involved with the Stage Society both as an actor and producer; the following year it put on his play *The Marrying of Ann Leete* to critical success. It was with a move to the Court Theatre in 1905 that Granville-Barker's real talent as both actor and manager were given expression. He was the first to introduce plays by **George Bernard Shaw** and Ibsen, who had been unable to find a commercial stage for their work. With great imagination, he chose Gordon Craig, the son of Ellen Terry, to be responsible for scenery and lighting, both of which were regarded as innovative. He married an actress in his company, Lillah McCarthy, but they later divorced. Following his marriage in 1918, to Helen Huntingdon, an American writer, he played the part of a country squire in Devon, and it was at this point that he hyphenated his name. Preferring seclusion, the couple moved to France, to collaborate in translating many foreign plays including those of Jules Romain.

Granville-Barker had given up acting in 1910, and directing in 1914, preferring to concentrate on writing plays and books about the theatre. The plays, which were concerned with social issues in the manner of Shaw, often aroused the wrath of the censor. The best, *The Voysey Inheritance*, dealt with problems of class, avarice and self-deception, and his knowledge of the theatre was demonstrated by his management of twelve speaking characters on the stage at the same time. His most important and lasting contribution to the theatre are his *Prefaces to Shakespeare*; both the director Peter Hall and the actor Ian McKellen have written of Granville-Barker's critical perception. A bust of the writer by Kathleen Scott, which matched a contemporary description of him looking like a cross between 'a bishop and a butler', was presented to the National Theatre in 1911, but mysteriously disappeared in the same year.

Plays

Weather Hen [with H. Thomas] 1899; *The Marrying of Ann Leete* 1901 [pub. 1909]; *Prunella* 1906; *The Voysey Inheritance* 1909; *Waste* 1909; *The Madras House* 1911; *Farewell to the Theatre* 1917; *Rococo* 1917; *Harlequinade* [with D. Calthorp] 1918; *The Secret Life* 1923; *His Majesty* 1928; *Plays: One* 1993

Translations

Arthur Schnitzler *Anatol* 1911; Sacha Guitry *Deburau* 1921; Gregorio Martinez Sierra *Four Plays* [with H. Granville-Barker] 1923; Jules Romains: *Doctor Knock* 1925, *Six Gentlemen in a Row* 1927; Serafin and Joaquin Alvarez Quintero: *Four Plays* [with H. Granville-Barker] 1927, *Four Comedies* [with H. Granville-Barker] 1932

Non-Fiction

A National Theatre [with William Archer] 1907; *The Exemplary Theatre* 1922; *Prefaces to Shakespeare* 5 vols 1927–47; *A National Theatre* 1930; *On Dramatic Method* 1931; *The Study of Drama* 1934; *On Poetry in Drama* 1937; *The Use of Drama* 1945

Fiction

Souls on Fifth (s) 1917

Biography

Granville-Barker: a Secret Life by Eric Salmon 1983

Robert (von Ranke) GRAVES 1895–1985

Born in Wimbledon, south London, Graves was the son of A.P. Graves, a school inspector and man-of-letters (author of the popular song 'Father O'Flynn') of Irish Scots descent; his mother was a great-niece of Germany's greatest historian, Leopold von Ranke (1795–1866). Graves passed a normal upper-middle-class childhood (one highlight was being patted on the head, while a baby in his pram, by Swinburne), and was educated at Charterhouse, where he became a keen boxer and devoted himself to poetry.

When the First World War broke out, he enlisted in the Royal Welch Fusiliers, serving alongside **Siegfried Sassoon** (who portrayed him as David Cromlech in *Memoirs of an Infantry Officer*, 1929), and eventually becoming a captain. He was badly wounded on the Somme in 1916: mistakenly reported dead, he had the odd experience of reading his own obituary in *The Times*. His first volume of poems, *Over the Brazier*, was published in 1916, and he became, along with Sassoon and **Wilfred Owen**, one of the most distinguished of the war poets. He was responsible for 'saving' Sassoon from a court martial when the latter published his celebrated 'Soldier's Declaration'.

Although he had been (emotionally at least) homosexual during the war – causing Sassoon some anguish – he subsequently 'recovered by a shock' (that of his Charterhouse love being arrested for procuring) and in 1918 married the painter and feminist Nancy Nicholson, whose father, William, and brother, Ben, were also well-known artists. They lived in Oxford, where Nancy ran a shop on Boar's Hill for the community of writers and artists there, while Graves published further volumes of poetry and took his BLitt in 1926. Always short of money, he relied upon hand-outs from the ever-generous Sassoon, but that same year he took the only salaried job of his career and was briefly Professor of English Literature at Cairo University. He was accompanied by his wife and four children and by **Laura Riding**, the combative American poet, with whom he embarked upon a long, complicated and painful relationship.

In 1929 he signalled his break with England by abandoning his family and going with Riding to Deya, a small fishing village in Mallorca (where – apart from the war years – he was to live for the rest of his life), and by publishing *Goodbye to All That*, a controversial autobiography which is also one of the liveliest (and most inventive) memoirs of the First World War. It became a bestseller, but alienated several of his friends, notably Sassoon and **Edmund Blunden**, who furiously annotated a copy of the book, marking its many inaccuracies and exaggerations. The domineering Riding was an important influence upon Graves's development as a poet, and they collaborated on a number of literary projects. Their personal partnership was less beneficial, however, and Riding's jealousy of the success of Graves's two novels, *I, Claudius* and *Claudius the God*, was indicative of their basic incompatibility. There were infidelities on both sides, and the relationship had disintegrated by the end of the 1930s, by which time they had returned to England, where Graves fell in love with the wife of his friend and literary collaborator, Alan Hodge. Graves and Beryl Hodge returned to Mallorca in 1946, had three children and eventually married in 1950.

Graves's reputation grew apace after the Second World War. He was enormously prolific, producing some 140 volumes in all: novels, short stories, drama, history, biography, anthropology, mythology, essays, translations, works for children, and above all poetry. He is widely regarded as one of the major English-language poets of the century, albeit one standing slightly outside the main currents of poetic development. His poems are traditional in form and – the war poems apart (he was inclined to omit them from later volumes of collected work) – largely concerned with love. He was elected Professor of Poetry at Oxford in 1961.

Graves, who outlived most of his generation, was a man of commanding physical presence and aristocratic hauteur. He was a formidable literary controversialist, and his most important prose works, apart from his novels, are his vast and highly contentious account of early Christianity, *The Nazarene Gospel Restored*, his two volumes of *The Greek Myths*, and his wide-ranging cultural study, *The White Goddess*. This last was partly an exploration of the source of his own poetry, which was inspired by a series of women whom he regarded as the earthly personifications of his poetic muse, the White Moon Goddess.

Poetry

Goliath and David 1916; *Over the Brazier* 1916; *Fairies and Fusiliers* 1917; *Treasure Box* 1919; *Country Sentiment* 1920; *The Pier-Glass* 1921; *The Feather Bed* 1923; *Whipperginny* 1923; *Mock Beggar Hall* 1924; *The Marmosite Miscellany* [as 'John Doyle'] 1925; *Welshman's Hose* 1925; *Poems, 1914–1927* 1927; *Poems 1929* 1929; *Ten Poems More* 1930; *Poems, 1926–1930* 1931; *To Whom Else* 1931; *Poems, 1930–1933* 1933; *Collected Poems* 1938; *No More Ghosts* 1940; *Poems* 1945; *Collected Poems, 1914–1947* 1948; *Poems and Satires* 1951; *Poems* 1953; *Collected Poems* 1959; *More Poems* 1961; *The More Deserving Cases* 1962; *New Poems* 1962; *Man Does, Woman Is* 1964; *Love Respelt* 1965; *Seventeen Poems Missing from 'Love Respelt'* 1966; *Colophon to 'Love Respelt'* 1967; *The Crane Bag* 1969; *Advice from a Mother* 1970; *Queen-Mother to New Queen* 1970; *The Green-Sailed Vessel* 1971; *Poems: Abridged for Dolls and Princes* 1971; *Poems, 1968–1970* 1971; *At the Gate* 1974; *Collected Poems* 1975; *New Collected Poems* 1977; *Eleven Songs* 1983

Fiction

My Head! My Head! 1925; *The Shout* (s) 1929; *No Decency Left* [as 'Barbara Rich'; with L. Riding] 1932; *Claudius the God* 1934; *I, Claudius* 1934; *Antigua, Penny, Puce* 1936; *Count Belisarius* 1938; *Sergeant Lamb of the Ninth* 1940; *Proceed, Sergeant Lamb* 1941; *The Story of Marie Powell, Wife to Mr Milton* 1943; *The Golden Fleece* 1944; *King Jesus* 1946; *The Islands of Unwisdom* 1949; *Seven Days in New Crete* 1949; *Homer's Daughter* 1955; *Catacrok!* (s) 1956; *They Hanged My Saintly Billy* 1957; *Collected Short Stories* (s) 1964

For Children

The Penny Fiddle 1960; *The Big Green Book* 1962; *Ann at Highwood Hall* 1964; *Two Wise Children* 1966; *The Poor Boy Who Followed His Star* 1968; *An Ancient Castle* 1980

Non-Fiction

On English Poetry 1922; *The Meaning of Dreams* 1924; *Contemporary Techniques of Poetry: a Political Analogy* 1925; *Poetic Unreason and Other Studies* 1925; *Impenetrability; or, The Proper Habit of English* 1926; *Another Future of Poetry* 1926; *The English Ballad* 1927; *Lars Porsena, or The Future of Swearing* 1927; *Lawrence and the Arab* 1927; *A Survey of Modernist Poetry* [with L. Riding] 1927; *Mrs Fisher, or The Future of Humour* 1928; *A Pamphlet against Anthologies* [with L. Riding] 1928; *Goodbye to All That* (a) 1929; *But It Still Goes On* (a) 1930; *T.E. Lawrence to His Biographer* 1938; *The Long Week-End: a Social History of Great Britain 1918–1939* [with A. Hodge] 1940; *The Reader Over Your Shoulder* [with A. Hodge] 1943; *The White Goddess* 1948; *The Common Asphodel – Collected Essays on Poetry* 1949; *Occupation: Writer* 1950; *The Nazarene Gospel Restored* [with J. Podro] 1953; *Adam's Rib* 1955; *The Crowning Privilege* 1955; *The Greek Myths* 1955; *Jesus in Rome – an Historical Conjecture* [with J. Podro] 1957; *Five Pens in Hand* 1958; *Greek Gods and Heroes* 1960; *Nine Hundred Iron Chariots* 1963; *Mammon* 1964; *The Hebrew Myths: The Book of Genesis* [with R. Patai] 1964; *Majorca Observed* 1965; *Mammon and the Black Goddess* 1965; *Poetic Craft and Principle* 1967; *The Crane Bag, and Other Disputed Subjects* 1969; *On Poetry: Collected Talks and Essays* 1969; *Difficult Questions, Easy Answers* 1973; *In Broken Images: Selected Letters of Robert Graves 1914–1946* [ed. P. O'Prey] 1982; *Between Moon and Moon: Selected Letters of Robert Graves 1946–1972* [ed. P. O'Prey] 1984; *Dear Robert, Dear Spike: the Graves–Milligan Correspondence* [ed. P. Scudamore 1991]

Plays

John Kemp's Wager 1925

Collections

Steps: Stories, Talks, Essays, Poems 1958; *Food for Centaurs: Stories, Talks, Critical Studies, Poems* 1960

Biography

Robert Graves by Martin Seymour-Smith 1982 [rev. 1995]; *Robert Graves* by Richard Perceval Graves, 2 vols 1986, 1990; *Robert Graves* by Miranda Seymour 1995

Alasdair (James) GRAY 1934–

Gray was born in Glasgow where his father was a blacksmith, and educated at Whitehill Senior Secondary School and the Glasgow Art School, where he studied painting and design, graduating in 1957. As a child he was frequently bedridden with asthma and eczema, and recalls: 'I had this vision of thousands and thousands of men in dirty work clothes, and streaming in to a hard, grinding, rather painful and frequently boring job, and I thought, "This is what keeps the world going, although don't let it be me." ' He worked variously as a part-time art teacher and scene painter (including a spell at Glasgow's Citizens' Theatre) before becoming in 1963 a freelance 'artist in words and pictures'. Aside from brief periods as artist recorder at the People's Palace Local History Museum (1976–7) and writer-in-residence at Glasgow University (1977–9), he has since made his living as a painter and writer.

Gray wrote a number of radio and television plays in the 1960s and 1970s, and worked for ten years on his novel *Lanark*, a big, fantastical exploration of Glasgow and of Unthank, a nightmarish parallel city to which the narrator is condemned after his death. The novel is digressive, ironic and self-consciously in the tradition of Rabelais and Sterne, and even contains 'an index of diffuse and imbedded Plagiarisms'. It won wide critical praise, and Gray's characteristically jokey response was to advertise his next novel, *1982, Janine*, as containing 'every stylistic excess and moral defect which critics conspired to ignore in the author's first books'. Narrated by an alcoholic, insomniac supervisor of security installations, the novel abounds in typographical trickery, lurid 'sadomasochistic fetishistic fantasy', and ferocious political diatribe, much of it targeted at his native country: 'The truth is that we are a nation of arselickers, though we disguise it with surfaces.' His strongly nationalist views are propounded in *Why Scots Should Rule Scotland*. His later novels include *The Fall of Kelvin Walker*, a tale about 'a monstrously pushy young Scot getting rich quickly in London', which is based on his television play broadcast by the BBC in 1968, and *Poor Things*, a satire loosely modelled on the memoirs of a nineteenth-century Scottish public health officer, and is extravagantly illustrated by the author.

Gray has published a number of short stories, notably the collection *Lean Tales* which also includes work by fellow Scots writers **James Kelman** and Agnes Owens. He is widely seen as the leading figure in the renaissance of Scottish literature in the 1980s, and has established his own small publishing house, the Dog & Bone Press. In a short autobiographical pamphlet commissioned by the Saltire Society in 1988, Gray reveals little autobiographical detail, and asks instead: 'What does this ordinary-looking, eccentric-sounding, obviously past-his-best person exist to do apart from eat, drink, publicize himself, get fatter, older and die?' Gray was married in 1962 and has one son.

Fiction

The Comedy of the White Dog **(s)** 1979; *Lanark* 1981; *Unlikely Stories, Mostly* **(s)** 1983; *1982, Janine* 1984; *The Fall of Kelvin Walker* [from his TV play] 1985; *Lean Tales* **(s)** [with J. Kelman, A. Owens] 1985; *Something Leather* 1990; *Poor Things* 1992; *Ten Tales Tall and True* **(s)** 1993

Poetry

Old Negatives 1989

Plays

Jonah 1956 [np]; *Under the Helmet* **(tv)** 1965; *The Fall of Kelvin Walker* **(tv)** 1968, **(st)** 1972 [pub. 1985]; *Quiet People* **(r)** 1968; *The Night Off* **(r)** 1969; *Dialogue* 1971 [np]; *Triangles* **(tv)** 1972; *Homeward Bound* 1973 [np]; *The Loss of the Golden Silence* 1973 [np]; *The Man Who Knew about Electricity* **(tv)** 1973; *Honesty* **(tv)** 1974; *The Loss of Golden Silence* **(r)** 1974; *The Harbinger Report* **(r)** 1975; *Today and Yesterday* **(tv)** 1975; *Beloved* **(tv)** 1976; *McGrotty and Ludmilla* **(r)** 1976; *The Gadfly* **(tv)** 1977; *Tickly Mince* [with T. Leonard, L. Lochhead] 1982 [np]; *The Pie of Damocles* [with others] 1983 [np]; *The Story of a Recluse* **(tv)** 1986

Non-Fiction

Thomas Muir of Huntershill **(r)** 1970; *Alasdair Gray* **(a)** 1988; *Why Scots Should Rule Scotland* 1992

Simon (James Holliday) GRAY 1936–

Gray was born at Hayling Island, Hampshire, the son of a pathologist who was a first-generation Canadian. During the Second World War he was evacuated to his Scottish grandparents in Montreal. After an education at Westminster School, he returned with his parents to Canada where he took a BA in English at Dalhousie

University. After a year spent lecturing at the university in Clermont-Ferrand, France, he took a second English BA at Cambridge, England, and then did postgraduate work, spending more time on his first novel, *Colmain*, published in 1963.

During his early career, Gray published four novels, one of them an espionage thriller written under the name of Hamish Reade; it is generally agreed that his dialogue is the best thing about these books. In the early 1960s he held several brief fellowships and lectureships, including one in British Columbia, where commissions for radio scripts from the Canadian Broadcasting Corporation gave him his first experience of playwriting. He married in 1964, and has two children.

In 1965 he became a lecturer in English at Queen Mary College, London, and he taught there for twenty years, never attaining promotion because he published no works of scholarship; his field was largely Charles Dickens and the Victorian novel, and after his retirement he was made an honorary fellow. His first stage play, originally intended for television, was *Wise Child*, in 1967, and his first West End success was *Butley* (1971), about an acerbic and frustrated university lecturer. A second big success was *Otherwise Engaged* in 1975, but Gray has enjoyed a troubled relationship with the critics. His plays, which often explore a milieu of intellectual and media folk in London, are much concerned with themes of dominance and submission, personal and political guilt, and suppressed homosexuality. Of these, probably the best are *The Rear Column*, which deals with an ill-fated expedition in colonial Africa, and *Quartermaine's Terms*, an engrossing and moving study of a failed and self-deceived teacher in a down-at-heel tutorial college. *The Common Pursuit* traced the lives of a group of Oxbridge friends and was performed on the London stage by several well-known actor-comedians, one of whom, Stephen Fry, subsequently walked out of the first production of Gray's *Cell Mates* (in which he was playing the role of the traitor George Blake) and temporarily disappeared. The play was hastily recast but, closed after a short run. Gray gives his own version of these much publicised events in *Fat Chance*.

Since retiring from academe, Gray has directed many of his own plays in the USA (an experience entertainingly chronicled in his book *How's That for Telling 'Em, Fat Lady?*) and in London. In the theatre he has often collaborated with **Harold Pinter** as his director and with the actor Alan Bates. He is also well known for television and film work, notably two fine black comedies for the BBC, *After Pilkington* and *They Never Slept*, and a sensitive adaptation of **J.L. Carr**'s *A Month in the Country* for the cinema.

Plays

Up in Pigeon Lake [from his novel *Colmain*] **(r)** 1963; *The Caramel Crisis* **(tv)** 1966; *Death of a Teddy Bear* **(tv)** 1967 [rev. as *Molly* **(st)** 1977, pub. 1978]; *Sleeping Dog* **(tv)** 1967 [pub. 1968]; *A Way with the Ladies* **(tv)** 1967; *Wise Child* 1967 [pub. 1972]; *Spoiled* **(tv)** 1968, **(st)** 1970 [pub. 1971]; *The Dirt on Lucy Lane* **(tv)** 1969; *Dutch Uncle* 1969; *Pig in a Poke* **(tv)** 1969 [pub. 1980]; *The Idiot* [from Dostoevsky's novel] 1970 [pub. 1971]; *The Princess* **(tv)** 1970; *Style of the Countess* **(tv)** 1970; *Butley* 1971, **(f)** 1976; *Man in a Side-Car* **(tv)** 1971 [pub. 1978]; *Otherwise Engaged* 1975; *Plaintiffs and Defendants* **(tv)** 1975; *Two Sundays* **(tv)** 1975; *Dog Days* 1976; *The Rear Column* 1978; *Close of Play* 1979 [pub. 1980]; *Stage Struck* 1979; *Quartermaine's Terms* 1981 [rev. 1983], **(tv)** 1983 **(r)** 1988; *Chapter 17* 1982 [np]; *Tartuffe* [from Molière's play] 1982 [np]; *The Common Pursuit* 1984; *After Pilkington* **(tv)** 1987; *Melon* 1987 [rev. as *The Holy Terror* **(r)** 1989]; *A Month in the Country* [from J.L. Carr's novel] **(f)** 1987; *Flames* **(tv)** 1990; *Hidden Laughter* 1990; *They Never Slept* **(tv)** 1991; *The Rector's Daughter* **(r)** [from F.M. Mayor's novel] 1992; *Suffer the Little Children* **(r)** 1993; *With a Nod and a Bow* **(r)** 1993; *Cell Mates* 1995

Fiction

Colmain 1963; *Simple People* 1965; *Little Portia* 1967; *A Comeback for Stark* [as 'Hamish Reade'] 1968

Non-Fiction

An Unnatural Pursuit and Other Pieces: a Playwright's Journal **(a)** 1985; *How's That for Telling 'Em, Fat Lady?* **(a)** 1988; *Fat Chance*1995

Edited

Selected English Prose [with K. Walker] 1967

F(rederick) L(awrence) GREEN 1902–1953

Green was born in Portsmouth, Hampshire; his father was Irish and his mother of Huguenot descent. He was brought up a Roman Catholic and sent to Salesian College, Farnborough, but his schooling was much broken up by periods of illness and extended convalescence, during which he spent his time swimming and wandering around shipyards. He worked in an accountant's office for a while, but gave this up in his mid-twenties to lead an itinerant life, picking up odd jobs in hotels, breweries, factories and theatres. He married in 1929, and in 1933 he went to Ireland, initially on a holiday, but ended up staying eighteen years, living mainly in Belfast, and devoting himself entirely to writing.

He published fourteen novels, of which the first, *Julius Penton*, appeared in 1934 and the rest one a year from 1939 to 1952, with the exception of 1949. Popular in their day, but now largely forgotten, they are mainly psychological thrillers of chase and pursuit with a great deal of theological and mysical speculation added. His best novel, *Odd Man Out*, is a tense story about an IRA man on the run; a celebrated film was made from it by Carol Reed in 1947, for which Green himself wrote the script. Green was a noted expert on the Irish contribution to modern English literature, but did not publish in this field. Towards the end of his life he moved to Somerset, and he died in Bristol.

Fiction

Julius Penton 1934; *On the Night of the Fire* 1939; *The Sound of Winter* 1940; *Give Us the World* 1941; *Music in the Park* 1942; *A Song for the Angels* 1943; *On the Edge of the Sea* 1944; *Odd Man Out* 1945; *A Flask for the Journey* 1946; *A Fragment of Glass* 1947; *Mist on the Waters* 1948; *Clouds in the Wind* 1950; *The Magician* 1951; *Ambush for the Hunter* 1952

Plays

Odd Man Out [from his novel; with R.C. Sherriff] 1947 [pub. in *Three British Screenplays* ed. R. Manvell 1950]

G(eorge) F(rederick) GREEN 1910–1977

'Dick' Green was born in Derbyshire, the setting for many of his short stories, and educated at Repton and Magdalene College, Cambridge. His father was an ironfounder, but Green always wanted to be a writer. His only other job was as 'a sort of amateur tutor' to the son of a publisher during the 1930s, and he was frequently penniless, relying upon the generosity of friends and a wealthy uncle. His short stories of working-class life have been much admired by **Alan Sillitoe**, **Christopher Isherwood**, **Elizabeth Bowen** and **John Lehmann**, and were published in *Penguin New Writing*, *The Listener* and elsewhere.

He was called up in 1940 and served first in the 3rd Suffolk Infantry Regiment, then as a second lieutenant in the Border Regiment. He was posted to Ceylon and, after serving as an aide-de-camp, transferred to the public relations department of the Commander-in-Chief in Colombo, where he edited *Veera Lanka*, a magazine for the Ceylonese forces which appeared in two different editions, Sinhalese and Tamil. This job was scarcely more onerous than his tutoring, and he divided most of his time between trips into the countryside to gather material for contributions to *Veera Lanka* and his own stories, and 'verandahism', a term he coined for passing the time on his private verandah with smoking, drinking, benzedrine and sex. It was this last pursuit which landed him in trouble in 1944, when he was caught *in flagrante* with a Ceylonese youth. He was court-martialled, cashiered and sentenced to two years' imprisonment. An account of his experiences as a prisoner in Ceylon was later published pseudonymously in *Penguin New Writing* (No. 31, 1947) as 'Military Detention' by 'Lieutenant Z'.

He served the larger part of his sentence in Wakefield prison, and after he was released in 1946, he became a patient of Dr Charlotte Wolff, the celebrated psychiatrist, who helped him to rediscover his identity. Disorientated by his experiences, he wandered from place to place, living in Chiswick, Wiltshire and Devon. He was still drinking heavily, but nevertheless published a collection of short stories, *Land without Heroes* (1948); an anthology of stories on the theme of childhood, *First View*, dedicated to the memory of **Denton Welch**; and *In the Making*, a remarkable novel which drew upon his memories of prep school (Wells House, Malvern) and was highly praised by **E.M. Forster**. His principal themes were childhood and relationships between youths and men from different social or cultural backgrounds, and his work has some affinity with the poems of **A.E. Housman**, which he much admired. In 1953 he gave up drink and in 1957 bought a house in Somerset with a legacy from his uncle. He divided his time between gardening and writing until he became terminally ill with lung cancer. After his death, his sister-in-law and his publisher compiled a collection of his writings, interspersed with memoirs and criticism by several people, including Sillitoe, Lehmann, **Frank Tuohy** and Michael Redgrave. It was published in 1980 as *A Skilled Hand*.

Fiction

Land without Heroes (s) 1948; *In the Making* 1952; *The Power of Sergeant Streater* (s) 1972

Edited

First View [rev. as *Tales of Innocence*] 1950

Collections

A Skilled Hand [ed. C. Green, A.D. Maclean] 1980

Henry GREEN (pseud. of Henry Vincent Yorke) 1905–1973

The youngest son of a wealthy industrialist, Green was born ('a mouthbreather with a silver spoon') near Tewkesbury and educated at Eton and Magdalen College, Oxford. He published his first novel, *Blindness* (1926), while he was still at university. The story of a boy who is accidentally blinded, it is an astonishingly assured work based upon a story he had written at Eton; it contains an account of the school's Society of Arts (of which he had been secretary) and an amusing portrait of his forthright mother. Like **John Betjeman**, Green hated his tutor, **C.S. Lewis**, and left Oxford without taking a degree. He worked on the shop floor of his father's Birmingham engineering works, H. Pontifex & Co., living in digs and writing a novel of factory life, *Living*. He evolved an impressionistic style of writing which dispensed with traditional grammar, often omitting articles, nouns or verbs, and which relied upon vividly realised colloquial dialogue. Although he claimed a debt to Charles Montagu Doughty's *Travels in Arabia Deserta*, his prose remained *sui generis*. In 1929 he married the Hon. Adelaide ('Dig') Biddulph; they had one son, who later married the novelist **Emma Tennant**. *Party Going*, which took most of the 1930s to write, pursued his interest in class in the form of a comedy of manners set in a fog-bound railway station. This was followed by a premature autobiography, *Pack My Bag*, written in conviction of his impending annihilation in the war. It is largely concerned with his childhood and the nature of memory, writing and autobiography. He served throughout the war in London with the Auxiliary Fire Service, which provided the background of *Caught*, eventually published in 1943 after he had been obliged to revise it in order to eliminate an adulterous relationship.

After the war Green returned to Pontifex & Co. as managing director, working in the shadow of his father, who continued to come to the office in extreme old age, sometimes on the arm of a nurse. He produced five further novels particularised by their idiosyncratic titles and style. *Loving* is set amongst servants at an Irish castle during the war; *Back* concerns a one-legged soldier returning from the war in search of his lost love; and *Concluding*, the most Firbankian of his books, is set in a girls' college. His last two novels, *Nothing* and *Doting* are attempts to write almost entirely in dialogue, as explained in a radio talk (published in *The Listener*, 9 November 1950). Brilliant comedies of manners, they disappointed some ('I think nothing of *Nothing*', complained his friend, **Evelyn Waugh**), but are as experimental in their way as earlier novels. For the last twenty-one years of his life, he completed no book, although he was working on a history of the Auxiliary Fire Service, and a satirical play (variously titled *The Only Man* and *All On His Lonesome*) set in the future, where men have become virtually extinct, and the all-female cabinet attempt to persuade a bank clerk to repopulate the country. Green became increasingly deaf and reclusive, and drank very heavily. Quintessentially English – he disliked reading any book set abroad – he had the greatest potential of the writers of his generation, and his achievement is surpassed only by Waugh. His grandson, Matthew Yorke, who is also a novelist, edited a posthumous volume of 'uncollected writings' entitled *Surviving* in 1992.

Fiction

Blindness 1926; *Living* 1929; *Party Going* 1939; *Caught* 1943; *Loving* 1945; *Back* 1946; *Concluding* 1948; *Nothing* 1950; *Doting* 1952

Non-Fiction

Pack My Bag (a) 1940

Collections

Surviving: the Uncollected Writings of Henry Green [ed. M. Yorke] 1992

(Henry) Graham GREENE 1904–1991

Greene was born in Hertfordshire and educated at Berkhamsted School where his father became headmaster in 1910. In his first volume of autobiography, *A Sort of Life*, Greene recalls the green baize door separating the family home from the school and evokes the early sense of divided loyalties which permeates much of his fiction. During a severely disturbed adolescence he poisoned himself, underwent psychoanalysis, and, 'when boredom had reached an intolerable depth', played Russian roulette with his elder brother's revolver. He read modern history at Balliol College, Oxford, from 1922–1925, and published a book of verse, *Babbling April*, in 1925.

After a year as an apprentice on the *Nottingham Journal*, he worked as a sub-editor on *The Times* from 1926 to 1930. He converted to Roman Catholicism in order to marry Vivien Dayrell-Browning in 1927; the couple had a son and a daughter and were later separated. Of his conversion, Greene wrote: 'Vivien was a Roman Catholic, but to me religion went no deeper than the sentimental hymns in the school chapel.' Despite this, he remained a doubtful Catholic throughout his life and his novels, notably *Brighton Rock* and *The Heart of the Matter*, frequently exploit theological dilemmas.

Greene enjoyed some success with his first novel, *The Man Within* (1929), a tale of nineteenth-century smuggling, and on the strength of this was contracted to write three more by his publisher. Only the third of these, *Stamboul Train*, was successful, and this after parts of it had been reprinted following the threat of a libel action by **J.B. Priestley**. Four years later

Greene again provoked a libel action with a review of the 1937 film *Wee Willie Winkie*, which appeared in the magazine *Night and Day*, of which he was co-editor. He had described the 'dimpled depravity' of Shirley Temple and its libidinous effect upon 'middle-aged men and clergymen', and was obliged to pay damages of £3,500. Throughout his life Greene courted similar controversy: as late as 1982 his exposé of police and local government corruption, *J'Accuse: The Dark Side of Nice*, was banned in France.

Beginning in the 1930s Greene travelled widely in dangerous and politically unstable parts of the world. A trip to Liberia resulted in his first travel book, *Journey without Maps*, and a commission to report on the persecution of the Catholic church in revolutionary Mexico resulted in *The Lawless Roads* and one of his greatest novels, *The Power and the Glory*, which won the Hawthornden Prize. During the Second World War he was an MI6 agent in Sierra Leone (which provides the setting for *The Heart of the Matter*) and in the 1950s and 1960s he travelled in the Far East, Africa and Latin America. Greene became particularly hostile to the USA and his concern with left-wing Latin American politics was reflected in his friendships with political leaders such as Fidel Castro, Daniel Ortega and the Panamanian General Omar Torrijos Herrera. As if further to irritate his detractors, he also remained loyal to the spy Kim Philby, with whom he had worked in Sierra Leone during the war.

Nearly all of Greene's novels have been filmed and he enjoyed a long and fruitful involvement with the cinema, notably collaborating with Carol Reed on the screenplay of *The Third Man*, and with Terence Rattigan on *Brighton Rock*. He also had a cameo role as an insurance agent in François Truffaut's *La Nuit Américaine* (1973).

The legendary but short-lived *Night and Day* (July–December 1937) was perhaps Greene's most celebrated excursion into journalism. Greene and his co-editor, John Marks, partly modelled this weekly magazine on the *New Yorker* and attracted one of the most distinguished gallery of contributers ever assembled for a periodical. **Evelyn Waugh** reviewed books, **Elizabeth Bowen** was theatre critic, Osbert Lancaster wrote about art and **John Betjeman** about architecture. Others who contributed to the magazine included **Adrian Bell**, **Cyril Connolly**, **William Empson**, Peter Fleming, **Chrisopher Isherwood**, **Hugh Kingsmill**, Constant Lambert, **Rose Macaulay**, **Louis MacNeice**, **William Plomer**, **Anthony Powell**, **V.S. Pritchett**, **Herbert Read**, **Stevie Smith**, A.J.A. Symons, and **Antonia White** – 'a whole literary generation', as Greene later put it. Greene was also literary editor of the *Spectator* from 1937 to 1940 and a director of The Bodley Head from 1958 to 1968. He wrote a number of plays, and in 1931 a biography of John Wilmot, Second Earl of Rochester, which was not published until 1974 for fear of obscenity charges. Under a false name, he once won a *New Statesman* literary competition for a parody of his own style, and in later years his footsteps were apparently dogged by an impostor calling himself Graham Greene. Greene moved to Antibes in France in 1966 and for many years had a relationship with a married woman, Yvonne Cloetta. He was made a Companion of Honour in 1966, and awarded the Order of Merit in 1986.

Fiction

The Man Within 1929; *The Name of Action* 1930; *Rumour at Nightfall* 1931; *Stamboul Train* 1932 [US *Orient Express* 1933]; *It's a Battlefield* 1932; *The Basement Room and Other Stories* (s) 1935; *The Bear Fell Free* (s) 1935; *England Made Me* 1935 [US *The Shipwrecked* 1953]; *A Gun for Sale* [US *This Gun for Hire*] 1936; *Brighton Rock* 1938; *The Confidential Agent* 1939; *Twenty-Four Stories* (s) [with H. Laver, S. Townsend Warner] 1939; *The Power and the Glory* [US *The Labyrinthine Ways*] 1940; *The Ministry of Fear* 1943; *Nineteen Stories* (s) 1947 [rev. *Twenty-One Stories* 1954]; *The Heart of the Matter* 1948; *The Third Man* 1950; *The End of the Affair* 1951; *Loser Takes All* 1955; *The Quiet American* 1955; *Our Man in Havana* 1955; *A Visit to Morin* (s) 1959; *A Burnt-Out Case* 1961; *A Sense of Reality* (s) 1963; *The Comedians* 1966; *May We Borrow Your Husband?* (s) 1967; *Travels with My Aunt* 1969; *The Honorary Consul* 1973; *Shades of Greene* (s) 1975; *The Human Factor* 1978; *Doctor Fisher of Geneva* 1980; *Monsignor Quixote* 1982; *The Tenth Man* 1985; *The Captain and the Enemy* 1988; *The Last Word* (s) 1990

Plays

The First and Last (f) 1937; *The New Britain* (f) 1940; *Brighton Rock* (f) [from his novel; with T. Rattigan] 1947; *The Third Man* (f) 1949; *The Living Room* 1953; *The Potting Shed* 1957; *Saint Joan* (f) [from G.B. Shaw's play] 1957; *The Complaisant Lover* 1959; *Our Man in Havana* (f) [from his novel] 1960; *Three Plays* [pub. 1961]; *Carving a Statue* 1964; *The Comedians* (f) [from his novel] 1967; *Alas, Poor Maling* (tv) [from his story] 1975; *The Return of A.J. Raffles* 1975

Non-Fiction

Journey without Maps: the Narrative of a Journey in Liberia 1936; *The Lawless Roads: a Mexican Journey* [US *Another Mexico*] 1939; *British Dramatists* 1942; *Why Do I Write? An Exchange of Views between Elizabeth Bowen,*

Graham Greene and V.S. Pritchett 1948; *After Two Years* 1949; *The Lost Childhood and Other Essays* 1951; *Essais catholiques* 1953; *In Search of a Character: Two African Journals* 1961; *The Revenge* (a) 1963; *Collected Essays* 1969; *A Sort of Life* (a) 1971; *The Pleasure-Dome: the Collected Film Criticism, 1935–40* [ed. J.R. Taylor] 1972; *Lord Rochester's Monkey* 1974; *Ways of Escape* (a) 1980; *J'Accuse: the Dark Side of Nice* 1982; *Why the Epigram?* 1989; *Yours Etc.: Letters to the Press 1945–89* [ed. C. Hawtree] 1989; *Reflections* [ed. J. Adamson] 1990; *A Weed among the Flowers* 1990; *A World of My Own: a Dream Diary* 1992; *Mornings in the Dark: the Graham Greene Film Reader* [ed. D. Parkinson] 1993

Edited

The Old School: Essays by Divers Hands 1934; *The Best of Saki* 1950; *The Spy's Bedside Book: an Anthology* [with H. Greene] 1957; *The Bodley Head Ford Madox Ford* 1962; *An Impossible Woman: the Memoirs of Dottoressa Moor of Capri* 1975; *Victorian Villainies* [with H. Greene] 1984; *Night and Day* [with C. Hawtree, J. Marks] 1985

Poetry

Babbling April 1925; *For Christmas* 1951

For Children

The Little Train 1946; *The Little Fire Engine* 1950; *The Little Horse Bus* 1952; *The Little Steamroller* 1953

Biography

The Life of Graham Greene by Norman Sherry, 2 vols 1989, 1994; *Graham Greene* by Michael Shelden 1994

Walter GREENWOOD 1903–1974

For a year before he left his council school at the age of thirteen, Greenwood had worked as a pawnbroker's assistant and part-time milk boy, an experience that typified the dead-end quality of all his jobs. Born in Salford, Lancashire, he was the son of a master hairdresser, who died when his son was nine. For eighteen years, Greenwood had nearly as many jobs. These included office boy, stable lad, sign-writer, chauffeur and warehouse man, only once earning more than thirty-five shillings (£1.75) a week (from a car factory). In an effort to keep sane, he read and listened to music when he could, and took advantage of cheap excursions to the Peak District, which he loved.

Frequently unemployed, Greenwood set down his experiences of destitution and of his locality in his first novel, *Love on the Dole* (1933). It found immediate success, bringing its author a living income, and, by causing questions to be asked in Parliament, helped to instigate social reform. Greenwood collaborated with Ronald Gow, the future husband of the actress Wendy Hiller, on a dramatisation of the novel in 1934. Hiller played the leading female role, in a play that had a long run in England and in New York in 1936.

Greenwood became a councillor for Salford, and it was widely expected that he would marry his fiancée, upon whom he had modelled Sally Hardcastle in his novel. He changed his mind, and was sued for breach of promise, settling financially out of court. He married instead Pearl Osgood, in 1937, and they took off to London. Later novels and short stories followed, but none had the popularity of *Love on the Dole*. All his work, from plays to film scripts and later television scripts such as *The Secret Kingdom*, was notable for the robust evocation of working-class life; it never aimed to be 'literature'. Greenwood lived for many years in the Isle of Man where he died of a heart-attack.

Fiction

Love on the Dole 1933; *His Worship the Mayor* 1934; *Standing Room Only* 1936; *The Cleft Stick* (s) 1937; *Only Mugs Work* 1938; *The Secret Kingdom* 1938; *Something in My Heart* 1944; *So Brief the Spring* 1952; *What Everybody Wants* 1954; *Down by the Sea* 1956; *Saturday Night at the Crown* 1959

Plays

Love on the Dole [from his novel; with R. Gow] 1934, (f) 1940 [pub. 1935]; *My Son's My Son* [from D.H. Lawrence's play] 1936 [np]; *Give Us This Day* [from his novel *His Worship the Mayor*] 1940 [np]; *The Cure for Love* (st) 1945, (f) 1949 [pub. 1947]; *So Brief the Spring* 1946 [np]; *Chance of a Lifetime* (f) 1947; *Never a Dull Moment* 1950 [rev. as *Too Clever For Love* 1951, pub. 1952]; *Saturday Night at the Crown* 1954 [pub. 1958]; *Happy Days* 1958 [np]; *The Secret Kingdom* (f), (tv) [from his novel] 1960; *Fun and Games* 1963 [np]; *This Is Your Wife* 1964 [np]; *There Was a Time* [from his autobiography] 1967 [np]

Non-Fiction

How the Other Man Lives 1939; *Lancashire* 1951; *There Was a Time* (a) 1967

Lady (Isabella Augusta) GREGORY 1852–1932

The daughter of a rich Irish landowner, Augusta Persse married her neighbour Sir William Gregory of Coole Park, Galway, in 1880. Gregory, who was sixty-three, later fathered their only child, Robert, but the marriage was otherwise unsatisfactory. Lady Gregory was for a short time the mistress of Wilfrid Scawen Blunt, when she discovered sex could be a pleasure rather than a duty. Nevertheless, she did not share Blunt's support for Home Rule, for she then feared the Protestant ruling class had too much to lose.

When Gregory died in 1892, his widow donned black for life in an effort to convince herself that she was in mourning, and all but fell in love with **W.B. Yeats** when he visited her at Coole in 1896. Quick to realise that he was at the centre of an Irish literary renaissance, she at first collected Irish folklore to please him, but in doing so she became genuinely interested in Irish literature. The following year, she planned with Yeats and others the formation of the Irish Literary Theatre with the purpose of performing Irish drama in Dublin. Collaboration with Yeats resulted in *Cathleen Ni Houlihan* (1902), a play which confirmed her Irish patriotism. Despite her larger contribution, Yeats claimed it for his own, speculating in Easter 1916: 'Did that play of mine send out / Certain men the English shot?'

Lady Gregory by now made Irish literature and the care of the poet her mission in life. Yeats was hopelessly in love with Maud Gonne, but it was rumoured that he spent chaste nights in bed with Lady Gregory, using her broad back as a table for his notebook when roused by nocturnal inspiration.

Between entertaining literary figures at Coole, Lady Gregory translated two volumes of Gaelic epic sagas published in 1904 and 1906. It was as a playwright, however, that she felt most comfortable. Her first drama, *Twenty Five*, was unsuccessful, but a natural comic talent was exploited in many plays such as *The Bogie Men*, when she evoked the speech of country people. With Yeats and **J.M. Synge**, she founded the Abbey Theatre, and defended the performance of Synge's *The Playboy of the Western World* when it was staged there in 1907 amidst public outcry. Lady Gregory translated Goldoni, and wrote a history of the Abbey Theatre in 1913. A love affair with John Quinn, the champion of modernism, was unresolved, and she successfully began a long fight to reclaim for Dublin the collection of paintings owned by her nephew Sir Hugh Lane, originally bequeathed to London. Her son was killed during the First World War, and in 1924, despite failing health, she championed the young **Sean O'Casey**. She had already sold most of her land to her tenants, and the house itself, immortalised by Yeats, was demolished in 1941. Her gossipy journals illuminate the period of the renaissance of Irish literature which she energetically helped to bring about.

Plays

Cathleen Ni Houlihan [with W.B. Yeats] 1902; *Seven Short Plays* [pub. 1909]; *The Kiltartan Molière* [pub. 1910]; *Irish Folk History Plays* [pub. 1912]; *New Comedies* [pub. 1913]; *The Dragon* 1916; *The Golden Apple* 1916; *The Image and Other Plays* [pub. 1922]; *Three Wonder Plays* [pub. 1923]; *Mirandolina* 1924; *The Story Brought by Brigit* 1924; *On the Racecourse* 1926; *Three Last Plays* [pub. 1928]; *My First Play* 1930; *Selected Plays* [ed. M. Fitzgerald; pub. 1983]

Non-Fiction

Arabi and His Household 1882; *Mr Gregory's Letter Box* 1898; *Poets and Dreamers* 1903; *A Book of Saints and Wonders* 1908; *The Kiltartan History Book* 1910; *On Irish Theatre* 1913; *Visions and Beliefs in the West of Ireland* 1920; *Hugh Lane's Life and Achievement* 1921; *Coole* 1931; *Journals 1916–1930* [ed. L. Robinson] 1946; *Lady Gregory's Journals* [ed. D.J. Murphy] 2 vols 1978, 1987

Edited

Sir William Gregory, an Autobiography 1984; *Ideals in Ireland* 1901; *The Kiltartan Poetry Book* 1918

Translations

Cuchulain of Muirthemne 1902; *Gods and Fighting Men* 1904

Collections

The Coole Edition [ed. C. Smythe] 1970–82

Biography

Lady Gregory – a Literary Portrait by Elizabeth Coxhead 1966; *Lady Gregory – the Woman Behind the Irish Renaissance* by Mary Lou Kohfeldt 1985

Trevor GRIFFITHS 1935–

Griffiths was born in Ancoats, East Manchester. Sent to live with his maternal grandmother near Bradford when he was two – unemployment had forced his father to surrender his home – he was educated at a Catholic primary school until, in 1945, he returned to Manchester, having won a scholarship to St Bede's College. Subsequently, he read English at Manchester University, graduating in 1955. National service followed and, next, four years teaching at a private school. He spent a further period at Manchester University – starting but not completing an MA – before going back to teaching, this time at Stockport Technical College. From 1965 he was for seven years an education officer for the BBC, assessing the effectiveness of the organisation's educational broadcasts. Since 1972 he has supported himself wholly by writing. He married in 1961.

The first of his plays to be performed was *The Wages of Thin* (1969), a black comedy set in a public lavatory, depicting the incrimination of a homosexual by a pair of (apparent) policemen. It resulted from a bet with **Stan Barstow** and Griffiths has admitted: 'I hadn't really worked out what it was about or for.' In that respect, it is untypical, for Griffiths's writing is characterised by its clarity of intention. Unswervingly committed to radical socialism, it addresses moments of struggle within political history. Hence *Occupations*, the play which brought its author to prominence in 1970, is both a dramatisation of the Turin workers' strikes of 1920 and a discussion of the student leftism of 1968. Ruthless, pragmatic Leninism debates against a humane (and thus enfeebled) Trotskyism. If the consequence is sermonising, by the time of *Comedians* Griffiths had refined his strategy. Five would-be stand-ups rehearse, perform and criticise their turns. Their conflicts – internal and external – unfold subtely: two capitulate to the bigotries of the northern night-club circuit; two strive to keep some shred of dignity; the fifth performs a sinister routine – a mixture of class hatred, symbolic rape and self-disgust. As in *Oi for England*, Griffiths is concerned with the degradation of *lumpen* culture: what he cannot suggest is a means of escape from its trap. Indeed, increasingly his attention has turned toward the obverse territory: the tactics and crises of the ruling class. His version of Chekhov's *The Cherry Orchard* and *Piano* examine the frailty of Russia's *haute-bourgeoisie* on the precipice of revolution. *Country* portrays an aristocratic dynasty which, as Attlee's victory in the 1945 election is declared, must choose an heir to manage its affairs. Griffiths points up the parallels between the Carlions (brewers and landowners) and Coppola's Corleones, identifying a licit mafia. More controversial was his assault on Captain R.F. Scott's reputation in the six-part Central Television series *The Last Place on Earth*. Transmitted not long after the Falklands War, it was interpreted as – and was intended to be – an indictment of the myths underpinning British nationalism. Its accuracy is still in dispute: its anti-orthodox potency is not.

Almost all of Griffiths' work has been televised and much of it was written for the medium. He has adapted *Sons and Lovers* for the BBC and, in an uncomfortable partnership with Warren Beatty, wrote the screenplay for *Reds*. His writing for theatre, he has said, is 'largely … the necessary means of sustaining a reputation which would enhance [my] bargaining power when dealing with television'. This proved to be of no avail when in 1991 *God's Armchair*, a two-part play commissioned by the BBC for its Schools Television, was cancelled before it went into production on the grounds that it 'was not helpful for the emotional development of sixteen-year-olds'. Griffiths' own explanation is that the BBC had been 'frightened of the Government because of what the play says about the conditions for young people'.

Plays

The Big House **(r)** 1969 [pub. 1972]; *The Wages of Thin* 1969 [np]; *Occupations* **(st)** 1970, **(tv)** 1974, [pub. 1972, rev. 1980]; *Apricots* 1971 [pub. 1979]; *Jake's Brigade* **(r)** 1971; *Lay By* [with H. Brenton, B. Clark, D. Hare, S. Poliakoff, H. Stoddart, S. Wilson] 1971 [pub. 1972]; *Thermidor* 1971 [pub. 1979]; *Sam, Sam* 1972; *Adam Smith* **(tv)** [as 'Ben Rae'] 1972–3; *Gun* 1973 [np]; *The Party* 1973, **(tv)** 1988 [pub. 1974]; *The Silver Mask* **(tv)** 1973; *Absolute Beginners* **(tv)** 1974 [pub. 1977]; *All Good Men* **(tv)** 1974 [pub. 1977]; *Comedians* **(st)** 1975, **(tv)** 1979, [pub. 1976]; *Don't Make Waves* [with K. Coyne, S. Wilson] **(tv)** 1975; *Through the Night* **(tv)** 1975 [pub. 1977]; *Bill Brand* **(tv)** 1976; *The Cherry Orchard* [from Chekhov's play; with H. Rappaport] **(st)** 1977, **(tv)** 1981 [pub. 1978]; *Deeds* [with H. Brenton, K. Campbell, D. Hare] 1978; *Country* **(tv)** 1981; *Reds* **(f)** 1981; *Sons and Lovers* **(tv)** [from D.H. Lawrence's novel] 1981 [pub. 1982]; *Oi for England* **(st)** 1982, **(tv)** 1982 [pub. 1983]; *The Last Place on Earth* **(tv)** 1985 [pub. as *Judgement Over the Dead* 1986]; [pub. 1986]; *Real Dreams* 1984, **(tv)** 1987 [pub. 1988]; *The Last Place on Earth* **(tv)** 1985 (pub. as *Judgment over the Dead* 1985]; *Fatherland* **(f)** 1987 [pub. 1988]; *Collected Plays for Television* **(tv)** 1988; *Piano* 1990; *The Gulf Between Us* 1992; *Thatcher's Children* 1993; *Hope in the Year Two* **(tv)** 1994

For Children

Tip's Lot 1972

Geoffrey GRIGSON 1905–1985

Born the seventh son of a Church of England clergyman in rural Cornwall, Grigson came from an old Norfolk family which had produced parsons for generations. His father was over sixty when Grigson was born, and his grandfather had been a noted scholar of the reign of George III. Of his six brothers, three were killed in the First World War, two in the Second, and the remaining shortly afterwards in an air crash in Pakistan. Grigson attended St John's School, Leatherhead, and St Edmund's Hall, Oxford. On leaving, he considered working in a mustard factory but instead joined the Fleet Street staff of the *Yorkshire Post*.

Simultaneously he began establishing himself in literary London as a propagandist for the **W.H. Auden** generation of poets; he was lodged in a powerful position to further them in the mid-1930s as the founding editor of the influential poetry magazine *New Verse* (1933–9) and the literary editor of the *Morning Post*. He married, in 1929, Frances Galt of St Louis; they had one daughter before his wife's early death in 1937. In the following year he married an Austrian, Berte Kunert; there were two children of this marriage, which ended in dissolution. During the war Grigson was employed in various capacities by the BBC, and after it he lived as a freelance man-of-letters, becoming very well known as a critic and anthologist. He had published his own first volume of poems, *Several Observations*, in 1939, and produced poetry, much of it celebrating his native Cornwall and the delights of flora and fauna, throughout his life; he was generally less regarded as poet than critic, although there is consensus that his poetry matured greatly with age, and that almost all the best of it postdates his *Collected Poems* of 1963. In the literary world he was known as a ferocious controversialist, although he was more kindly in person; **Edwin Muir** was only one of the writers who, savaged by Grigson in print, became his friend. More positive results of his taste were the many anthologies of verse from Faber and editions of individual poets in which he brought neglected work to public attention. Of many miscellaneous volumes, perhaps outstanding was the autobiographical *The Crest on the Silver*; the later *Recollections* found him more uncharitable. In later years he lived at a farmhouse in Swindon, Wiltshire, with his third wife, the cookery writer Jane Grigson, with whom he had one daughter, Sophie, also a cookery writer.

Poetry

Several Observations 1939; *Under the Cliff and Other Poems* 1943; *The Isles of Scilly and Other Poems* 1946; *Legenda Suecana* 1953; *Collected Poems of Geoffrey Grigson, 1924–1962* 1963; *A Skull in Salop and Other Poems* 1967; *Ingestion of Ice-Cream and Other Poems* 1969; *Discoveries of Bones and Stones* 1971; *Penguin Modern Poets 23* [with E. Muir, A. Stokes] 1973; *Sad Grave of an Imperial Mongoose* 1973; *Angels and Circles* 1974; *The Fiesta and Other Poems* 1978; *History of Him* 1980; *Collected Poems 1963–1980* 1982; *The Cornish Dancer and Other Poems* 1982; *Montaigne's Tower and Other Poems* 1984

Non-Fiction

Wild Flowers in Britain 1944; *The Harp of Aeolus and Other Essays on Art, Life and Nature* 1947; *Samuel Palmer: the Visionary Years* 1947; *An English Farmhouse and Its Neighbourhood* 1948; *The Scilly Isles* 1948; *Places of the Mind* 1949; *Flowers of the Meadow* 1950; *The Crest on the Silver* (a) 1950; *Essays from the Air* 1951; *A Master of Our Time: a Study of Wyndham Lewis* 1951; *Thornton's Temple of Flora* 1951; *Gardenage; or, The Plants of Ninhursaga* 1952; *The Female Form in Painting* [with J. Casson] 1953; *Freedom of the Parish: On Pelynt, East Cornwall* 1954; *The Shell Guide to Flowers of the Countryside* 1955; *England* 1957; *The Living Rocks* 1957; *Old Stone Age* 1957; *Painted Caves: On the Cave Art of France and Spain* 1957; *The Wiltshire Book* 1957; *English Villages in Colour* 1958; *The Shell Guide to Trees and Shrubs* 1958; *English Excursions* 1960; *Poets in Their Pride* 1962; *The Shell Country Book* 1962; *The Shell Book of Roads* 1964; *The Shell Nature Book* 4 vols 1955–9; *The Shell Country Alphabet* 1966; *Poems and Poets* 1969; *Notes from an Odd Country* 1970; *The Contrary View* 1974; *A Dictionary of English Plant-Names* 1974; Britain Observed: the Landscape through Artists' Eyes 1975; *The Englishman's Flora* 1975; *Blessings, Kicks and Curses: a Critical Collection* 1982; *The Goddess of Love: the Birth, Triumph, Death and Return of Aphrodite* 1982; *The Private Art: a Poetry Notebook* 1982; *Country Writings* 1984; *Recollections: Mainly of Writers and Artists* 1984

For Children

Looking and Finding and Collecting and Reading and Investigating and Much Else 1958; *Shapes and Stories: a Book about Pictures* 1964; *Shapes and Adventures* [with J. Grigson] 1967

Edited

The Arts Today 1935; *New Verse: an Anthology* 1939; *The Journals of George Sturt* 1941; *The Mint: a Miscellany of Literature, Art and Criticism* 1946–8; John Clare: *Poems of John Clare's Madness* 1949, *Selected Poems* 1950; *Selected Poems of John Dryden* 1950; *Selected Poems of William Barnes, 1800–1866* 1950; *People, Places and Things* 4 vols 1954; *Fossils, Insects and Reptiles* 1957; *The Concise Encyclopedia of Modern World Literature* 1963; W.S. Landor *Poems* 1964; *The Faber Book of Popular Verse* 1971; *The Penguin Book of Unrespectable Verse* 1971; *Rainbows, Fleas and Flowers: a Nature Anthology for Children* 1971; *The Faber Book of Love Poems* 1973; *The Faber Book of Epigrams and Epitaphs* 1977; *The Faber Book of Poems and Places* 1980; *The Faber Book of Reflective Verse* 1984

Philip GROSS 1952–

Gross was born near the slate quarries of Delabole in Cornwall, the son of a wartime refugee from Estonia and a Cornish schoolteacher's daughter. He was brought up in Plymouth and went up to Sussex University in 1970 to read English. He remarks that he stopped writing poetry there ('a coincidence?') not to return to it until 1978, shortly after the birth of his first child ('another coincidence?'). He worked for a while 'half-time as a librarian in Croydon, and half-time as a "housewife" caring for a daughter and a son. In the third half, I wrote.' The family now live in Bristol.

He began to acquire a reputation in the early 1980s, winning the Drake 400 Poetry Competition in 1980, an Eric Gregory Award in 1981, and a first prize of £1,000 in the 1982 National Poetry Competition for his poem 'The Ice Factory', which became the title for his second book, his first to be published by Faber. The opening poem, 'Ben's Shed', closes with the richly suggestive lines: 'In his hand he weighed / a clutch of dried corms – hoarded, papery / shrunk hearts – the past / and future sealed in them; held them to me.' *Manifold Manor* was shortlisted for the Signal Poetry Award. Reviewing it in *The Listener*, Peter Forbes wrote: 'Gross is one of the most gifted of younger poets, and his inventiveness makes him an ideal children's poet'. The poems in *The Son of the Duke of Nowhere* emerge from the shadow of his Estonian family background. He has also published an adventure novel for children entitled *The Song of Gail and Fludd.*

Poetry

Familiars 1983; *The Ice Factory* 1984; *Cat's Whisker* 1987; *The Air Mine of Mistila* [with S. Kantaris, K. Lewis] 1988; *The Son of the Duke of Nowhere* 1991; The All-Nite Café 1993

For Children

Manifold Manor [with C. Riddell] 1989; *The Song of Gail and Fludd* 1991

Thom(son William) GUNN 1929–

Gunn was born in Gravesend, England. His mother, he tells us, read Gibbon while he was in the womb, while his father was respected as a journalist, and became editor of the tabloid *Daily Sketch.* Gunn's parents divorced when he was eight and his mother died when he was fifteen – an apparently traumatic event for Gunn, though it has not found its way into the subject-matter of his poetry. He was brought up in Hampstead and educated there and in Hampshire as an evacuee from the blitz. He read with great passion the poetry of Keats and the plays of Marlowe during his adolescence, and towards the end of his teens he began to admire **W.H. Auden** – later a seminal influence on his own poetry – and also discovered Baudelaire, who was subsequently to remain one of his favourite poets.

After spending two years in the British army (where he read Proust), Gunn went up to Trinity College, Cambridge, the most intellectually sound of the Cambridge colleges in the 1950s. Here, during his first year, he and his fellow undergraduates discovered **W.B. Yeats**, then (like all other poets) suffering from being in the shadow of the domineering **T.S. Eliot**, though widely seen as a great modernist poet. His first collection was written while he studied at Trinity, and brought him into contact with those who would later form the Movement – **Kingsley Amis**, **Philip Larkin** and **John Wain** being the most important. Gunn has commented that he never did see himself as being part of the Movement, and indeed has shunned all such attempts to locate him in a particular school of poetry; this labelling may have been part of the reason why he decided to move to California in 1954, where he has lived ever since.

It is a measure of the seriousness of Gunn's poetic career that it is not defined by the knowledge we have of his personal life or his public persona, but through his poetry alone. Two aspects of that poetry, though superficially unrelated, run parallel, and by looking at them together it is possible to trace his maturation over the past thirty-five years.

The first area to trace is the form which his poetry has taken. He began with rhyme, then added a form of half-rhymed syllabic verse in his third book, *My Sad Captains and Other Poems*, and finally managed to present his own characteristic free verse. This trajectory is in no small way linked to his move to America. He has complained that the British cannot hear free verse, and it is true that no British poet of significance apart from **D.H. Lawrence** has used it well. Alternatively, from Whitman onwards, American poetry has naturally found its home in free verse. But this 'mid-Atlantic' position in which Gunn has found himself has inevitably had an effect on his reputation in both countries: British readers do not on the whole read what they consider to be American poets, and vice versa, and it is likely that the readership for his poetry has consequently not been as great as it would have been if he had stayed in England.

This may have been alleviated somewhat by the fact that, since he is the most famous openly homosexual poet since Auden, he has commanded a loyal homosexual readership, both in Britain and abroad, and it is in this context that a second poetic emergence can be traced over the years. When he was writing metrical verse, the subject of his poetry was, as Neil Powell has argued, on the whole the need for sexual containment and order; the second phase of poems – up to *Talbot Road* – which are more formally experimental, often feature what might be called liberation poems; while the final phase, incorporating *The Passages of Joy* and *The Man with Night Sweats* and their largely free verse poetics, are completely sexually open. This is especially the case with the last book, since it deals with the subject of Aids. Living in the gay community of Los Angeles has meant that Gunn has continually been in close proximity to both the political aspects and the personal consequences of the disease. *The Man with Night Sweats* is probably the most eloquent, beautiful and informed literary manifestation of the crisis yet produced in either Britain or America, and with its publication he has consolidated his position amongst the two or three most important poets now writing in English.

Poetry

Poems 1953; *Fighting Terms* 1954; *The Sense of Movement* 1957; *My Sad Captains and Other Poems* 1961; *Selected Poems* [with T. Hughes] 1962; *Positives* [with A. Gunn] 1966; *Touch* 1967; *Poems 1950–1966* 1969; *Sunlight* 1969; *Moly* 1971; *Mandrakes* 1973; *Songbook* 1973; *To the Air* 1973; *Jack Straw's Castle and Other Poems* 1976; *The Missed Beat* 1976; *Games of Chance* 1979; *Selected Poems 1950–1975* 1979; *Talbot Road* 1981; *The Passages of Joy* 1982; *Sidewalks* 1985; *Undesirables* 1988; *The Man with Night Sweats* 1992; *Old Stories* 1992; *Collected Poems* 1993

Non-Fiction

The Occasions of Poetry 1982; *Shelf Life* 1994

Edited

Poetry from Cambridge 1952; *Five American Poets* [with T. Hughes] 1963; *Selected Poems of Fulke Greville* 1968

Ivor (Bertie) GURNEY 1890–1937

The son of a Gloucester tailor, Gurney never lost his sense of place, seeing himself as the county's equivalent of **A.E. Housman**'s Shropshire lad. Brought up in a cathedral city rich in musical activity, and encouraged by a local clergyman and two spinster sisters called Hunt, Gurney decided early on to become a professional musician. He won a scholarship to the Royal College of Music where he met Marion Scott, another spinster who was to become his confidante and champion. He was to write chamber works and many songs, setting his own poems and those of Housman, **Rupert Brooke** and the Elizabethans.

In spite of poor health, he enlisted as a private in his local regiment, the 5th Gloucesters, and fought on the Somme and at Passchendaele, where he was gassed. Unable to write music during the war he concentrated upon his poetry and published a volume of verse, *Severn and Somme*, in 1917. Popularly believed to have been driven mad by his experience in the trenches, Gurney had in fact displayed symptoms of mental disturbance before the war, earning himself the nickname 'Batty Gurney'. However, a breakdown in 1918 was ascribed to 'deferred shell-shock', and after an erratic but productive period, his illness was diagnosed in 1922 as 'Delusional Insanity (Systematized)', or paranoid schizophrenia, and he was sent to a private mental hospital near Gloucester. He had claimed that voices were urging him to kill himself, and had made at least one suicide bid before being hospitalised. He was later transferred to the City of London Mental Hospital at Dartford in Kent, where his condition showed no signs of improvement. He believed that he was being persecuted by electric waves coming from the wireless, and claimed that he had written all Shakespeare's plays and composed the entire works of both Beethoven and Haydn. He spent the rest of his life in Dartford, eventually dying of pulmonary tuberculosis. At first he protested about his incarceration in letters to bewildered friends, acquaintances and occasionally complete strangers, as well as to the Metropolitan Police Force, who were also addressed in long poems. His moments of lucidity became less frequent, although he often played the piano and produced reams of letters and poetry which became less coherent as the years passed. Eccentricities of syntax and punctuation led to the mistaken dismissal of much of his work as 'mad', but recent editing has re-established a large corpus of impressive poetry, while his music has grown in popularity, and been published in many volumes.

Poetry

Severn and Somme 1917; *War's Embers* 1919; *Poems by Ivor Gurney* [ed. E. Blunden] 1954; *Poems of Ivor Gurney, 1890–1973* [ed. L. Clark] 1973; *Collected Poems of Ivor Gurney* [ed. P.J. Kavanagh] 1982

Non-Fiction

War Letters: a Selection [ed. R.K.R. Thornton] 1983; *Stops in the Dark Night: the Letters of Ivor Gurney to the Chapman Family* 1986

Biography

The Ordeal of Ivor Gurney by Michael Hurd

(Marguerite) Radclyffe(-)HALL 1880–1943

Born in Bournemouth, Hall was the unwanted daughter of Mary and Radclyffe Radclyffe-Hall. Her parents divorced, and Mary taunted her daughter with only possessing Radclyffe family characteristics, forcing an identity with her father. At twenty-one, Hall inherited her grandfather's large fortune, and moved into a house of her own. She described herself as a 'congenital invert', but the love poetry she wrote, much of which was set to music, was genderless. When she was twenty-three, she fell in love with Mabel Batten, a plump married amateur singer, who, at fifty, was past her musical peak. Known as 'Ladye', she invented the name 'John' for Hall, but the relationship did not develop fully until Ladye was widowed, after which they shared a home.

In 1915, Hall had an affair with Una Troubridge, her lover's cousin, but was devastated when Ladye died a year later, and kept in touch with her through a spiritualist for many years. Hall and Troubridge travelled extensively, collected oak furniture and bred dogs, and it was not until 1924 that Hall wrote *The Forge*, the first of seven novels. *Adam's Breed* followed two years later, winning the Femina Vie Heureuse and the James Tait Black Memorial Prizes. After consulting Ladye, who reluctantly gave her consent from the other side, Hall cut her hair short and adopted a severely masculine appearance. In 1928, after some hesitation, Cape published *The Well of Loneliness*, which pleaded for tolerance for lesbians. The *Sunday Express* led an abusive press attack with the headline 'A Book That Must Be Suppressed'. The novel was withdrawn on the instructions of the Home Secretary who prosecuted, and found *The Well of Loneliness* guilty of obscene libel. Hall was supported by numerous distinguished writers – **H.G. Wells**, **Arnold Bennett**, **George Bernard Shaw**, **E.M. Forster** – on the grounds of literary freedom, but they were reluctant to express an opinion on the novel's literary merit. Hall was furious but undaunted. None of her later work achieved the popularity of the banned book.

Hall and Troubridge took a house in Rye in 1930 and, untroubled by the ghost of **Henry James**, were befriended by **E.F. Benson** and other homosexuals, and were drawn into a local artistic community. Hall's final affair was with an ungrateful nurse; to Troubridge's relief, this ended after five years. Their remaining years were spent in tranquil fidelity, until Hall died of inoperable cancer. She was interred in the Highgate vault she had built for Ladye, and shelved next to her. Troubridge wrote an artless and inaccurate life of her lover, and by dying in Rome in 1963, forfeited the place she had reserved for herself in Highgate.

Fiction

The Forge 1924; *The Unlit Lamp* 1924; *A Saturday Life* 1925; *Adam's Breed* 1926; *The Well of Loneliness* 1928; *The Master of the House* 1932; *Miss Ogilvy Finds Herself* 1934; *The Sixth Beatitude* 1936

Poetry

'Twixt Earth and Stars 1906; *A Sheaf of Verses* 1908; *Poems of the Past and Present* 1910; *Songs of Three Counties and Other Poems* 1913; *The Forgotten Island* 1915

Biography

Our Three Selves by Michael Baker 1985

Michael (Peter Leopold) HAMBURGER 1924–

A poet, translator and critic, Hamburger was born in Berlin into a prosperous German-Jewish family. His father was a professor of medicine whose work included a pioneering study of the dietary cause of rickets. On Hitler's rise to power, the family moved to London in 1933, and Hamburger was educated at Westminster School, where he began to write poetry in English and undertook his first translations of Hölderlin. He went up to Christ Church, Oxford, to read modern languages and, as he relates in his volume of memoirs, *A Mug's Game*, came close to being sent down during his first term when a drunken **Dylan Thomas** started a fight in his rooms after a poetry reading. His friends in Oxford included **Philip Larkin**, **David Wright** and **John Mortimer**. His education was disrupted by service as an infantryman with the Royal Army Educational Corps, most of it spent being shunted around Britain, but he believes the experience cured him of his 'monomaniacal literariness'.

Having gained his MA in 1948, he travelled in Europe and worked as a freelance writer before marrying the poet Anne Beresford in 1951 and starting an academic career, lecturing in German at University College, London (1952–5), and the University of Reading (1955–64). He has subsequently been a visiting professor at various American and English universities. Hamburger and Beresford were divorced in 1970, remarried in 1974, and have three children.

During the war Hamburger's verse was published by Tambimutti, and he received encouragement from **Stephen Spender** and **Herbert Read**. His first collection, *Flowering Cactus*, gathers his early verse in which, he has said, he used 'traditional forms to protect myself from the pressure and intensity of my feelings'. He turned to freer forms in his 1963 volume *Weather and Season* with a consequent intensification of emotion. His verse is profoundly concerned with ideas and addresses the violence of twentieth-century history and the dehumanising influence of technology whilst also seeking solace in the natural world. As a translator he has been praised for establishing the reputation of Hölderlin with English readers and introducing the work of other German poets such as Hans Magnus Enzensberger and Nelly Sachs, and his critical work includes the ground-breaking study of poetry since Baudelaire, *The Truth of Poetry*. He has won a number of awards including the Goethe Medal in 1986.

Poetry

Later Hogarth 1945; *Flowering Cactus: Poems 1942–1949* 1950; *Poems 1950–1951* 1952; *The Dual Site* 1957; *Weather and Season* 1963; *In Flashlight* 1965; *In Massachusetts* 1967; *Feeding the Chickadees* 1968; *Home* 1969; *Penguin Modern Poets 14* [with A. Brownjohn, C. Tomlinson] 1969; *Travelling: Poems 1963–1968* 1969; *In Memoriam: Friedrich Hölderlin* 1970; *Travelling I–IV* 1972; *Conversations with Charwomen* 1973; *Ownerless Earth: New and Selected Poems 1950–1972* 1973; *Babes in the Wood* 1974; *Travelling VI* 1975; *Travelling VII* 1976; *Moralities* 1977; *Palinode* 1977; *Variations in Suffolk IV* 1980; *Variations* 1981; *In Suffolk* 1982; *Collected Poems 1941–1983* 1984

Plays

Struck by Apollo (r) [with A. Beresford] 1965; *The Tower* [from P. Weiss's play] 1974

Non-Fiction

Reason and Energy: Studies in German Literature 1957 [rev. 1971]; *Hugo von Hofmannsthal: Zwei Studien* 1964 [as *Hofmannsthal: Three Essays* 1970]; *From Prophecy to Exorcism: the Premisses of Modern German Literature* 1965; *The Truth of Poetry* 1969; *A Mug's Game: Intermittent Memoirs 1924–1964* (a) 1973; *Art as Second Nature: Occasional Pieces 1950–1974* 1975; *A Proliferation of Prophets: German Literature from Nietzsche to the Second World War* 1983; *After the Second Flood: Essays on Post-War German Literature* 1986

Translations/Edited

Beethoven: Letters, Journals and Conversations 1951 [rev. 1966, 1984]; Hugo von Hofmannsthal: *Poems and Verse Plays* [with others] 1961, *Selected Plays and Libretti* [with others] 1963; Friedrich Hölderlin *Selected Verse* 1961; *Modern German Poetry, 1910–1960: an Anthology with Verse Translation* [with C. Middleton] 1962; *East German Poetry: an Anthology in German and English* 1972; *German Poetry, 1970–1975: an Anthology* 1977

Translations

Friedrich Hölderlin: *Poems* 1943 [rev. 1952, 1961, 1966, 1980], *Poems and Fragments* 1966; *Twenty Prose Poems of Baudelaire* 1946 [rev. 1968]; *Decline: Twelve Poems by Georg Trakl* 1952; Albrecht Goes *The Burnt Offering* 1956; Goethe: *Egmont* 1959, *Urworte Orphisch: Five Poems by Goethe* 1982, *Poems and Epigrams* 1983; Bertolt Brecht *Tales from the Calendar* (s) [with Y. Kapp] 1961; Günther Grass: *Selected Poems* [with C. Middleton] 1966, *Poems* [with C. Middleton] 1969, *In the Egg and Other Poems* [with C. Middleton] 1977; Hans Magnus Enzensberger: *Poems* 1966, *Poems for People Who Don't Read Poems* [with H.M. Enzensberger, J. Rothenberg] 1968; Georg Büchner: *Lenz* 1966, *Leonce and Lena, Lenz, Woyzeck* 1972; Nelly Sachs: *O the Chimneys* [with others] 1967, *The Seeker and Other Poems* [with M. Mead] 1970; Peter Bichsel *And Really Frau Blum Would Very Much Like to Meet the Milkman* 1968, *Stories for Children* (s) 1971; *Journeys: Two Radio Plays by Günter Eich* (r) [pub. 1968]; Paul Celan: *Selected Poems* [with C. Middleton] 1972; *Poems* 1980, *32 Poems* 1985; Peter Huchel *Selected Poems* 1974, *The Garden of Theophrastus and Other Poems* 1983; *German Poetry 1910–1975* 1977; Helmut Heissenbittel *Texts* 1977; P. Jaccottet *Seedtime: Extracts from the Notebooks 1954–1967* [with A. Lefèvre] 1977; Franco Fortini *Poems* 1978; *An Unofficial Rilke: Poems 1912–1926* 1981; Martin Soresen *Selected Poems* 1982

Edited

Jesse Thoor *Das Werk: Sonette, Lieder Erzählungen* 1965; Thomas Good *Selected Poems* 1973

Collections

Zwischen den Sprachen: Essays und Gedichte 1966

Cicely (Mary) HAMILTON (pseud. of Hammill) 1872–1952

An actress, suffragist, feminist playwright, dystopian novelist, cautiously optimistic pacifist, disillusioned socialist and unexpected Christian, Hamilton was born in Paddington, west London. She was the product of a soldier father of Scots descent and an Irish mother from the landed gentry. Although she was a voracious, self-taught reader by three years of age, her early formal education was largely neglected, partly because of her military father's many moves, partly because a succession of bank failures and embezzling solicitors had depleted the family coffers. Her learning reflected a passionate but patchy autodidacticism (the facts of life she 'acquired fairly early from the Bible'). Feminism came early when she discovered, at six, that she could not be a railway guard. It was later reinforced when she first understood, via Shakespeare's *Lucrece*, that a woman's 'honour' could be lost to any man strong enough to take it by force.

Her 1935 autobiography, *Life Errant*, though vigorous and attractive, has many well-concealed gaps and silences (much of the 1920s, for example, simply disappears) and she allowed few of her personal papers to survive. We can only speculate on why the young Hamilton was left for several years with a much hated stepmother, nor have biographers been able to discover why the parting with her much loved mother was one 'whose finality was not recognised at the time': death, divorce, insanity or suicide?

During adolescence she entered a girls' boarding-school at Malvern where her intense love of acting found expression. In her late teens she was sent to the German resort of Bad Homburg, her first acquaintance with a country she was to revisit and shrewdly observe, both under occupation after the First World War and in the very early years of the Nazis' rise to power. Of the latter visit she wrote particularly well in a 1931 book, *Modern Germanies, as seen by an Englishwoman*, the first in her impressive series of 'nation' books which included France, Russia, Italy, Austria and Scotland.

With schooldays over, and after a miserable period as a teacher in the Midlands, she turned to acting, though her 'unconventional' looks for years confined her to the gruelling regimen of the Edwardian touring company. The romantic comedies and melodramas in which she played served her well, however, when she began to turn their comfortably familiar conventions to feminist advantage in a series of her own highly successful plays. Chief amongst them were the constantly performed one-acter *How the Vote Was Won* and *Diana of Dobson's*. Her plays explore issues further developed in her still startling 1909 book, *Marriage as a Trade*, in which she insisted on the economic constraints, often akin to prostitution, which forced many women into essentially dishonest and immoral matches.

Hating all her life the 'herd-instinct', she quickly broke with the Women's Social and Political Union (WSPU), joining instead the veteran campaigner Charlotte Despard in the Women's Freedom League. Later she compared WSPU members' adulation of their leader, Emmeline Pankhurst, to fascist and Nazi worship of Mussolini and Hitler. (Such *lèse-majesté* did not, surprisingly, lose her the friendship of Dame Ethel Smyth for whose suffragette battle hymn, 'The March of the Women', she had provided the words.)

For Hamilton the key to women's advancement was 'voluntary motherhood' (contraception), and during the First World War she did much praised administrative work for the Scottish Women's Hospitals in France. She later returned to acting, organising theatrical performances for the troops.

Loving England but loathing jingoism, Hamilton spent most of the rest of her life grappling, via journalism, fiction and political action, with the intensifying evils caused by 'herd-instinct' – especially tyranny and totalitariansim, in both their fascist and communist forms. Her novel *William, an Englishman*, which won the Femina Vie Heureuse Prize in 1919, the first year it was offered, explored and linked different types of war: sexual, political and 'bullets and blood and high explosives'. During the 1920s she was a constant contributor to the feminist weekly, *Time and Tide*. Her bleakly dystopian novel *Theodore Savage* reflected her increasingly deeply held conviction that 'when our civilization comes to ruin, the destructive agent will be Science'. Her early beliefs – pacifism, socialism, internationalism – gave way to a scepticism which was never allowed to degenerate into cynicism or inertia. Her feminism remained undiminished, and she saw in the 1930s alarming signs that the century's earlier gains were fast being eroded.

The love and support of female friends provided the emotional mainstay of her life, a fact she happily acknowledged, although the names of several of the women who mattered most are entirely absent from her autobiography. In her late sixties the war which she had so long predicted began. Characteristically she took her part in firefighting during the London blitz. Although her health began to fail she continued to work for anti-fascist causes during and after the war, drinking champagne to the end (thoughtfully provided, for medicinal purposes, by more affluent friends).

Plays

The Sixth Commandment 1906 [np]; *Mrs Vance* 1907 [np]; *The Serjeant of Hussars* 1907 [np]; *Diana of Dobson's* 1908 [pub. 1925]; *How the Vote Was Won* [with C. St John] 1909; *The Pot and the Kettle* 1909 [np]; *The Homecoming* 1910 [np]; *Just to Get Married* 1910 [pub. 1911]; *A Pageant of Great Women* 1910; *The Cutting of the Knot* 1911; *Jack and Jill and a Friend* 1911; *The Constant Husband* 1912 [np]; *Lady Noggs* 1913; *Phyl* 1913 [np]; *The Lady Killer* 1914 [np]; *The Child in Flanders* 1919 [pub. 1922]; *The Brave and the Fair* 1920 [np]; *The Human Factor* 1924; *Mrs Armstrong's Advisor* 1920 [np]; *The Beggar Prince* 1929 [pub. 1936]; *Caravan* 1932; *Mr Pompous and the Pussy-Cat* 1948

Non-Fiction

Marriage as a Trade 1909; *Senlis* 1917; *Modern Germanies, as seen by an Englishwoman* 1931 [rev. 1933]; *Modern Italy, as seen by an Englishwoman* 1932; *Modern France, as seen by an Englishwoman* 1933; *Modern Russia, as seen by an Englishwoman* 1934; *Life Errant* (a) 1935; *Modern Austria, as seen by an Englishwoman* 1935; *Modern Ireland, as seen by an Englishwoman* 1936; *Modern Scotland, as seen by an Englishwoman* 1937; *Modern England, as seen by an Englishwoman* 1938; *Modern Sweden, as seen by an Englishwoman* 1939; *The Englishwoman* 1940; *Lament for Democracy* 1940; *Holland Today* 1950

Fiction

Diana of Dobson's 1908; *William, an Englishman* 1919; *Theodore Savage* 1922 [rev. as *Lest We Forget* 1928] *The Old Vic* [with L. Baylis] 1926; *Full Stop* 1931; *Little Arthur's History of the Twentieth Century* 1933

Biography

The Life and Rebellious Times of Cicely Hamilton by Lis Whitelaw 1990

(Robert) Ian HAMILTON 1938–

Hamilton was born in King's Lynn and educated at Darlington Grammar School and Keble College, Oxford, where he was editor from 1959 to 1960 of *Tomorrow*. This marked the start of a distinguished career as an editor, which he has combined with that of a poet and biographer. After leaving Oxford, he founded and edited the *Review*, a poetry magazine which ran from 1962 to 1972. He was also a regular poetry reviewer for the *London Magazine* and the *Observer*, a fiction reviewer for the *Times Literary Supplement*, and spent a year (1972–3) as a lecturer in poetry at the University of Hull. His reviews and articles have been collected in two volumes: *A Poetry Chronicle* and *Walking Possession*. He has also been a presenter of books programmes on BBC television.

Hamilton's principal achievement as an editor was the short-lived but influential *New Review*, which he founded in 1974 and which closed after fifty issues in 1979. Heavily subsidised by the Arts Council (and criticised in some quarters for this dependency), the magazine achieved a reputation far beyond its small circulation. It published articles, reviews, interviews, poems, spoofs (most famously by **Julian Barnes** under the pseudonym 'Edward Pygge') and the complete texts of plays by such authors as **Simon Gray**, **Harold Pinter** and **Dennis Potter**. *The New Review Anthology* of pieces from the magazine testifies to its wide range and the quality of its contributors: stories by **Ian McEwan**, **John McGahern**, **Shena Mackay**, **Jean Rhys** and **William Trevor**; poems by **Douglas Dunn**, **Gavin Ewart**, **John Fuller**, **Seamus Heaney**, **Andrew Motion**, **Peter Porter** and **Craig Raine**; autobiography, profiles, reportage and polemic by **Beryl Bainbridge**, **Robert Lowell**, **Hugo Williams** and John Carey.

Hamilton's output as a poet has, he acknowledges, been small: 'Fifty poems in twenty-five years,' he commented in the preface to his *Fifty Poems*; 'not much to show for half a lifetime, you might think.' Of these, thirty had already been published in his first collection, *The Visit* (1970) – brief, laconic verses addressed to 'you', the reader. His later lyrics spring from emotional turmoil, 'a kind of anguished incredulity, a refusal to believe that fathers die, that wives go mad, that love – however certain of itself – is not enough, not always'. He has received the Eric Gregory Award and the Poetry Society of America's Melville Cane Award. Hamilton is also the editor of *The Oxford Companion to Twentieth-Century Poetry*.

He is the author of several volumes of non-fiction, including *The Little Magazines: a Study of Six Editors*, a biography of Robert Lowell, a far from exhaustive account of *Writers in Hollywood*, and a surprisingly disappointing book about literary estates and their executors, *Keepers of the Flame*. His own difficulties as a biographer of the famously reclusive **J.D. Salinger** resulted not in the book he had planned, which got to proof stage but had to be withdrawn, but a fascinating compromise: *In Search of J.D. Salinger*. Perhaps wishing to avoid further legal battles, Hamilton spent much of *Keepers of the Flame* treating controversies surrounding the estates of writers long dead (such as John Donne) at the expense of more recent cases, although the book includes a wary chapter on **Sylvia Plath**.

Hamilton has been married twice and has a son from each marriage.

Poetry
Pretending Not to Sleep 1964; *The Visit* 1970; *Anniversary and Vigil* 1971; *Fifty Poems* 1990

Non-Fiction
A Poetry Chronicle 1973; *The Little Magazines* 1976; *Robert Lowell: a Biography* 1981; *In Search of J.D. Salinger* 1988; *Keepers of the Flame* 1991; *Writers in Hollywood, 1915–1951* 1990; *Walking Possession: Essays and Reviews 1968–93* 1994

Edited
The Poetry of War, 1939–45 1965; *Selected Poetry and Prose by Alun Lewis* 1966; *Eight Poets* 1968; *The Modern Poet: Essays from* The Review 1968; *Selected Poems by Robert Frost* 1973; *Poems since 1900: an Anthology of British and American Verse in the Twentieth Century* [with C. Falck] 1975; *Yorkshire in Verse* 1984; *The New Review Anthology* 1985; *The Faber Book of Soccer* 1992; *The Oxford Companion to Twentieth-Century Poetry* 1994

(Anthony Walter) Patrick HAMILTON 1904–1962

The younger brother of the detective novelist Bruce Hamilton (his closest ally throughout his life), Hamilton grew up in upper-middle-class circumstances in Hove, the son of Bernard Hamilton, a barrister turned popular novelist; but the pattern of alcoholic decline in the father's life was to be tragically echoed in that of the son. He was educated at Westminster School and at the age of seventeen went on the stage, playing small parts and acting as assistant stage manager for the repertory company of Andrew Melville, which was the last to put the old-fashioned Lyceum melodramas on the English stage. The influence of this can be seen in Hamilton's own two hugely successful plays, *Rope* (based on the famous Leopold and Loeb 'Killing for Kicks' murder) and *Gaslight*, which use the conventions of the Victorian melodrama to new and frightening effect.

He learned typing and shorthand by correspondence course, hoping to better his financial position, but was soon overtaken by his success as a writer: his second novel, *Craven House* (1926), made almost as much mark as his plays were to do. However, in 1932, at the height of his early success, he suffered a serious road accident, which left him permanently disfigured, a fact which probably contributed to the severe alcoholism which began from about this period. His best novels, however, such as *Hangover Square* and *The Slaves of Solitude*, date from his mid-life, when he also achieved success in the relatively new field of radio drama, and produced another distinguished play, *The Duke in Darkness*.

His relationship with women tended to be obsessive and unhappy. In 1927 he fell disastrously in love with Lily Connolly, a prostitute, and the circumstances of this relationship are recorded in *The Midnight Bell*, the first novel in his London trilogy, *Twenty Thousand Streets Under the Sky*. A similarly hopeless, and unrequited, passion for the actress Geraldine Fitzgerald in the mid-1930s inspired George's murderous obsession with Netta in *Hangover Square*. By this time, Hamilton had a wife, having married in 1930 Lois Martin, with whom he was unable to sustain sexual relations. While still married to her, he began a liaison with Lady Ursula Stewart, who became his second wife after his divorce in 1954; during his last years, he moved back and forth between the two women (whose mutual antagonism was clearly evident at his funeral). Hamilton's work declined seriously from the 1950s, and in his last year his alcoholism had rendered him immobile. His permanent achievement will be his novels, which even now are too little read; dealing in a masterful way with the banalities and loneliness of modern urban life, both tender and disturbing, they are major monuments of the fiction of their time.

Fiction
Monday Morning 1925; *Craven House* 1926 [rev. 1943]; *Twopence Coloured* 1928; *Twenty Thousand Streets Under the Sky* 1935: *The Midnight Bell* 1929, *The Siege of Pleasure* 1932, *The Plains of Cement* 1934; *Impromptu in Moribundia* 1939; *Hangover Square* 1941; *The Slaves of Solitude* [US *Riverside*] 1947; *The West Pier* 1951; *Mr Stimpson and Mr Gorse* 1953; *Unknown Assailant* 1955

Plays
Rope 1929 [pub. 1930]; *John Brown's Body* 1930 [np]; *The Procurator of Judea* [from A. France's story] 1930 [np]; *Gaslight* 1938 [pub. 1941]; *Money with Menaces* (r) 1937 [pub. 1939]; *To the Public Danger* (r) 1939; *This Is Impossible* (r) 1941 [pub. 1942]; *The Duke in Darkness* 1942 [pub. 1943]; *The Governess* 1946 [np]; *Caller Anonymous* (r) 1952; *The Man Upstairs* 1953 [pub. 1954]; *Miss Roach* (r) [from his novel *The Slaves of Solitude*] 1958; *Hangover Square* (r) [from his novel] 1965

Biography
Through a Glass Darkly by Nigel Jones 1991; *Patrick Hamilton* by Sean French 1993

(Samuel) Dashiell HAMMETT 1894–1961

Hammett was born in Saint Mary's County, Maryland; the family moved to Baltimore when he was seven. There he attended Public School

No. 72 and, for a single term in 1908, the city's Polytechnic Institute. From 1908 to 1915 he was, in his own words, 'the unsatisfactory and unsatisfied employee of various railroads, stock brokers, machine manufacturers, canners, and the like. Usually I was fired'. Next he became an operative for Pinkerton's National Detective Service, where an old hand, James Wright, taught him his craft. Sadly, Hammett's case reports were destroyed and no records survive of his agency work. After a spell in the US Army Motor Ambulance Company (1918–19) he returned to Pinkerton's. Hospitalised in 1920 with pulmonary tuberculosis, he met Josephine Annas Dolan, a nurse, and married her in 1921. That December, he quit Pinkerton's for an eighteen-month course at Munson's Business College, San Francisco, and around this time he started to write, churning out short stories, reviews and advertising copy. His first piece, a 100-word anecdote, 'The Parthian Shot', was published in *The Smart Set* in October 1922, five years before his first novel, *Red Harvest*, was serialised in *Black Mask*. By 1934 he had written his fifth and final novel, *The Thin Man*.

He moved to New York in 1929, but, under contract to write screenplays to Warner Brothers and later MGM, he spent much of the 1930s in Hollywood. In 1930 he began a long relationship with **Lillian Hellman**. His wife obtained a divorce in a Mexican court in 1937; apparently it had no legal standing in the USA. During the same decade he became an active supporter – and probably a member – of the Communist Party: the involvement was to lead to his being gaoled for five months in 1951. Meanwhile, in 1942, he had rejoined the army. Ill-health notwithstanding (he continued to suffer tubercular attacks) he was posted to the Aleutian Islands, served in the Signals Corps and edited *The Adakian*, a forces newspaper. In 1945 he was discharged with the rank of sergeant and in 1946 was appointed to teach courses in mystery writing at the Jefferson School of Social Science, New York. His own writing, though, had effectively ceased. He smoked and drank heavily, suffered from emphysema, and eventually died of lung cancer. Since 1941 he had produced nothing of note, relying on reissues, collections and film and radio adaptations for what was still a significant income. But, in 1951, the Internal Revenue Service obtained a judgement against him for $111,008.60 back taxes. Now on the studios' blacklist, Hammett had few markets for his talents and at his death he was over $200,000 in debt.

Although Hammett's best-known character is Sam Spade (of *The Maltese Falcon*), the anonymous Continental Detective Agency operative, or 'op' (based on James Wright) figures most often in Hammett's *oeuvre*. According to his author, the op 'is a little man going forward day after day through mud and blood and death and deceit – as callous and brutal and cynical as necessary – towards a dim goal, with nothing to push or pull him towards it except he's been hired to reach it'. Less introspective and scrupulous than **Raymond Chandler**'s Marlowe, the op is distinct from the criminals he pursues chiefly because he does not seek personal gain. Indifferent even to himself, he represents an extra-legal justice – and so foreshadows the sociopathic enforcers whose corpses litter the pages of pulp *noir*.

The father of the 'hard-boiled' detective story, Hammett soon came to believe he had exhausted the genre. His last prose sentence resignedly reads: 'If you are tired you ought to rest, I think, and not try to fool yourself or your customers with colored bubbles.'

Fiction

The Dain Curse 1929; *Red Harvest* 1929; *The Maltese Falcon* 1930; *The Glass Key* 1931; *Secret Agent X-9* [with A. Raymond] 1934; *The Thin Man* 1934; *Blood Money* (s) 1943; *The Adventures of Sam Spade and Other Stories* (s) 1944; *The Continental Op* (s) 1945; *The Return of the Continental Op* (s) 1945; *Hammett Homicides* (s) 1946; *Dead Yellow Women* (s) 1947; *Nightmare Town* (s) 1948; *The Creeping Siamese* (s) 1950; *Woman in the Dark* (s) 1951; *A Man Named Thin and Other Stories* (s) 1962; *The Big Knockover* (s) 1966

Plays

City Streets (f) [with O.H.P. Garrett, M. Marcin] 1931; *Mister Dynamite* (f) [with H. Clork, D. Malloy] 1935; *Satan Met a Lady* (f) [with B. Holmes] 1936; *After the Thin Man* (f) [with F. Goodrich, A. Hackett] 1939; *Another Thin Man* (f) [with F. Goodrich, A. Hackett] 1939; *Watch on the Rhine* (f) [from L. Hellman's play] 1943

Non-Fiction

The Battle of the Aleutians [with R. Colodny] 1944

Edited

Creeps by Night (s) [UK *Modern Tales of Horror*] 1931

Biography

Shadow Man: the Life of Dashiell Hammett by Richard Layman 1981; *Hammett: a Life on the Edge* by William F. Nolan 1983; *The Life of Dashiell Hammett* by Diane Johnson 1984

Georgina HAMMICK 1939–

The daughter of a professional soldier, Hammick was born Georgina Heyman in Aldershot,

Hampshire. She is one of a pair of identical twins: her sister, Amanda Vesey, is a well-known children's writer and illustrator. Her early life was somewhat peripatetic, and she lived, and was 'semi-educated', in England, America and Kenya. She began training as a painter, attending the Academie Julian in Paris and Salisbury Art School, but eventually chose to make her career as a writer. (All three of her sisters are accomplished artists.) She taught in various schools, both state and private, before marrying Charles Hammick, a soldier, in 1961. Together they had three children (one of whom is the painter Tom Hammick), and also brought up his two children from a previous marriage. Her husband left the army in 1964 and became a company secretary, but after suffering a near-fatal heart-attack in 1967 he was obliged to find other employment. The following year the Hammicks set up a bookshop in Farnham. Hammick's was the first bookshop to use a computerised ordering system, and soon expanded to become a small independent chain.

Hammick began publishing poetry in magazines and anthologies in the 1970s, and was one of five poets whose work appeared in *A Poetry Quintet* (1976). Her poems were singled out by one reviewer as 'clean, unpretending, to the point, and unusually assured in their restraint'. For many years she took part in the Poetry Society's 'Poet in Schools' scheme, and she contributed to **Gavin Ewart**'s anthology of *Other People's Clerihews* (1983).

Once her children had 'more or less' grown up, she began writing short stories. 'People for Lunch' won the *Stand* Magazine Short Story Competition in 1985, and became the title-story of her first volume, which was published to enormously enthusiastic reviews in 1987, described by the critic Peter Lewis as 'surely one of the best collections by an English writer published during the 1980s'. She has a faultless ear for dialogue and idiom, particularly when people are behaving impossibly (as they frequently do in her work), but the high comedy of her stories is often shot through with menace and melancholy, as in 'Tales from the Spare Room', about cold-eyed children taken to the home of the family servants, and discovering a different, frightening world running parallel to their own. The volume contains stories which vary in mood and style: from the exact observation of the title-story, in which a woman, on the second anniversary of her husband's death, attempts to prepare lunch and organise her recalcitrant adolescent children, to the unorthodox 'A Few Problems in the Day Case Unit', about the humiliations of a gynaecological examination.

Hammick's standing ('as good as anyone now writing', as **Susan Hill** put it) was consolidated by *Spoilt*, a second volume which extended her range, while maintaining the consistently high standard of *People for Lunch*. Her stories are distinguished by unobtrusive craftsmanship, and the quest for the exact word, which is a hallmark of her work, is reflected in 'The Dying Room', in which a family wrangle over the use of language, its social and cultural shading. Other outstanding stories in this volume are two which investigate truth and fiction, what happens and how it is described or remembered: 'Lying Doggo', in which an indulgent owner is described and exposed by her pet, and 'The American Dream', in which a woman appropriates a childhood experience from her twin brother – or so he believes.

Her considerable reputation rests upon these two volumes, the stories of which have appeared in numerous magazines and anthologies (including *The Penguin Book of Modern Women's Short Stories* and several volumes of the annual *Best Short Stories*), and have frequently been broadcast on BBC radio. Something of what she attempts in her own work and looks for in the work of others may be seen in *The Virago Book of Love and Loss*, which she edited in 1992. The anthology contains stories by, amongst others, **Alice Munro**, **Elizabeth Bowen** and **Elizabeth Taylor**. She has also been fiction editor of *Critical Quarterly* magazine.

Fiction

People For Lunch (s) 1987; *Spoilt* (s) 1992

Edited

The Virago Book of Love and Loss 1992

Poetry

A Poetry Quintet [with others] 1976

John HAMPSON (Simpson) 1901–1955

After the collapse of the family brewing firm, Simpson's family moved from Birmingham to Leicester, where he was brought up in poverty and educated at home, since ill-health prevented him from attending school. He ran away from home and, after a succession of casual jobs and a spell in Wormwood Scrubs for stealing books, eventually found employment with a wealthy family near Birmingham, looking after a mentally handicapped boy.

Simpson was devoted to his charge, but the job was time-consuming, and he was obliged to do his writing before the day's work started. His first and best-known novel was *Saturday Night at the Greyhound* (1931), which started out as a play based upon his experiences helping his sister run a pub in Derbyshire, and was dedicated to

Forrest Reid. A grim story, which has some of the qualities of Greek tragedy, it is written in an unadorned prose and became an unexpected best seller for the Hogarth Press. Encouraged by **Leonard** and **Virginia Woolf**, Reid, **E.M. Forster** and **William Plomer**, Simpson overcame the disadvantages of his haphazard education (although he never mastered spelling), producing a number of novels and stories of social realism, and becoming a leading figure in the 'Birmingham Group' of writers. He had met **Walter Allen** in 1933 and through him was introduced to other leading figures of the 1930s, including **W.H. Auden** and **Louis MacNeice**.

Although homosexual, in 1936 Simpson married the Austrian actress Therese Giehse in order to give her a British passport. This event was instigated and overseen by Auden (who had married in similar circumstances), and he afterwards led the wedding party to a pub, where he was prevented from celebrating the match by playing hymns on the piano when he discovered the landlord's corpse stretched out on the billiard table.

Simpson's later novels included *Care of 'The Grand'*, which was set in a hotel and told the same story from different perspectives, as had his earlier novel, *Family Curse*. None of these was as successful as *Saturday Night at the Greyhound*, and attempts to turn that novel into a play and film were frustrated. Simpson wrote documentaries for the BBC during the Second World War and later went to India, having fallen under the influence of an educational psychologist, James Ford Thomson.

An eccentric figure, small and bespectacled, with a Hapsburg chin, and invariably dressed in brown, Simpson was plagued throughout his life by poor health. His death came about while he was in the process of discharging himself from hospital: climbing out of bed in order to dress, he immediately dropped dead.

Fiction

Saturday Night at the Greyhound 1931; *Two Stories* (s) 1931; *O Providence* 1932; *Strip Jack Naked* 1934; *Man about the House* 1935; *Family Curse* 1936; *The Larches* [with L.A. Pavey] 1938; *Care of 'The Grand'* 1939; *A Bag of Stones* (s) 1951

Non-Fiction

The English at Table 1944

Christopher (James) HAMPTON 1946–

Hampton was born on Fayal Island, one of the Azores, where his father was working as a marine telecommunications engineer. His play *White Chameleon* is based upon his childhood. He attended schools in Aden and Alexandria before coming to England with his mother in 1956 when the Suez crisis erupted. He has described the shock of 'this very strange country in which everybody ate cabbage and it was cold and they didn't turn the heat on', and attributes his interest in *émigrés* and outsiders to this experience. At Lancing College from 1959 to 1963, he was an exact contemporary of the playwright **David Hare**. In 1964 he entered New College, Oxford, to read modern languages.

Before going up he wrote his first play, *When Did You Last See My Mother?*, which was performed at the Oxford Playhouse in 1966, and for one night at the Royal Court in the same year before transferring to the West End. The play is a bleak tale of youthful despair and tangled bisexual relationships set in a London bedsit, where two young friends are waiting to go to university. During a year off from Oxford Hampton travelled in Germany and France where he worked as a translator while writing *Total Eclipse*, a play about the relationship between Rimbaud and Verlaine which was produced at the Royal Court in 1968, the year he graduated from Oxford. He achieved considerable success in 1971 with his next original play, *The Philanthropist*, which was conceived as 'a riposte' to Molière's *Le Misanthrope*.

In addition to his own plays, Hampton has adapted work by Molière, Ibsen, Chekhov and Odön von Horvath. His play *Tales from Hollywood* imagines von Horvath (who died in 1936) in Hollywood during the war. Hampton's screenplays include *The Honorary Consul* from **Graham Greene**'s novel, and he has adapted **Malcolm Bradbury**'s *The History Man* and **Anita Brookner**'s *Hotel du Lac* for television. His hugely successful 1985 play *Les Liaisons Dangereuses*, adapted from Laclos's novel, was subsequently filmed by Stephen Frears, and won an Academy Award for best screenplay in 1988. He has collaborated with Andrew Lloyd Webber in adapting the classic film *Sunset Boulevard* as a musical, and wrote and directed the film *Carrington*, which is based on Michael Holroyd's biography of Lytton Strachey.

Hampton was married in 1971 to Laura de Holesch, has two daughters, and lives in London. Much to his chagrin, he is sometimes confused with another writer called Christopher Hampton. On one occasion the other Hampton wrote a letter to the *Guardian* denouncing **Harold Pinter**, which a surprised Pinter took to be the work of his friend. The entry in the book *Authors Take Sides on the Falklands* is also the work of the other Hampton, whom the editors thought was the better-known playwright. It may not be coincidental that Hampton has adapted Molière's play *Tartuffe; or, The Impostor.*

Plays

When Did You Last See My Mother? 1966 [pub. 1967]; *Marya* [from I. Babel] 1967 [pub. 1969]; *Total Eclipse* 1968 [pub. 1969]; *Hedda Gabler* [from Ibsen's play] 1970 [pub. 1972]; *The Philanthropist* 1970; *Uncle Vanya* [from Chekhov's play] 1970 [pub. 1971]; *A Doll's House* [from Ibsen's play] **(st)** 1971 **(f)** 1973 [pub. 1972]; *Savages* 1973 [pub. 1974]; *Signed and Sealed* [from M. Desvallières, G. Feydeau] 1976 [np]; *Treats* 1976; *Able's Will* **(tv)** 1977 [pub. 1979]; *Ghosts* [from Ibsen's play] 1978 [pub. 1983]; *Geschichten aus dem Wiener Wald* **(f)** [from Odön von Horváth with M. Schell] 1979; *The Wild Duck* [from Ibsen's play] 1979 [pub. 1980]; *After Mercer* [from D. Mercer] 1980 [np]; *The Prague Trial* **(r)** [from P. Chereau, A. Mnouchkine] 1980; *The History Man* **(tv)** [from M. Bradbury's novel] 1981; *A Night of the Day of the Imprisoned Writer* [with R. Harwood] 1981 [np]; *The Portage to San Cristobal of A.H.* [from G. Steiner] 1982 [pub. 1983]; *The Honorary Consul* **(f)** [from G. Greene's novel] 1983; *Tales from Hollywood* 1983; *Les Liaisons Dangereuses* [from Laclos's novel] 1985 [pub. 1986]; *The Good Father* [from P. Prince] **(f)** 1986; *Hotel du Lac* **(tv)** [from A. Brookner's novel] 1986; *Wolf at the Door* **(f)** 1986; *Dangerous Liaisons* **(f)** [from Laclos's novel] 1989; *The Ginger Tree* **(tv)** [from O. Wynd's novel] 1989; *White Chameleon* 1991; *Sunset Boulevard* [from C. Brackett, D.M. Marsham Jr; with D. Black, A. Lloyd Webber] 1993 [np]; *Alice's Adventure's Under Ground* 1994 [pub. 1995]; *Carrington* **(f)** 1994

Translations

Odön von Horváth: *Tales from the Vienna Woods* 1977, *Don Juan Comes Back from the War* 1978, *Faith, Hope and Charity* 1989; Molière: *Don Juan* **(r)**, **(st)** 1972 [pub. 1974], *Tartuffe* 1983 [pub. 1984]

Non-Fiction

Two Children Free to Wander **(r)** 1969

HAN Suyin (pseud. of Zhou Guanghu) 1917–

Han's relationship to the China whose intricate and bloodstained twentieth-century history she has done so much to illuminate for Western readers has always been painfully ambivalent. Her father was a member of the Zhou family, part of China's feudal landowning class, her mother was a Belgian woman whom he met when studying engineering in Belgium. Cultural tensions between husband and wife left their permanent mark upon their Chinese-born daughter. Her mixed race brought confusion of identity and considerable social hostility, in both her native China and her mother's harshly colonial Belgium, and, sad to say, sometimes embarrassed liberal Britons during the Second World War, including the otherwise admirable Lady Cripps, wife of Sir Stafford and President of the Aid to China Fund. Most painfully of all, her vicious and physically violent first husband constantly flung the word 'Eurasian' at her during his terrifying tirades.

Her grasp of the Chinese language itself was shaky enough during adolescence and early womanhood to necessitate private lessons, paid for from the salary she earned as a 'native' typist. The later decision to write in English was, however, dictated by reason of political urgency, not linguistic incapacity: Western readers must be able to hear the stories she had to tell of a China grievously beleaguered by poverty and faction within, menaced and brutally invaded by enemies beyond its borders. Childhood was full of the sights and sounds of horror: public whippings, unburied corpses, casual violence meted out to the blind and helpless. In *The Crippled Tree* she describes 'the enormous indifference of the enormous, strange land around her, an indifference more suffocating than any enmity ... where she turned on herself ... imprisoned among crowds, imprisoned in indifference, in dirt, in squalor, an enormous rat-cage without beginning or end'. The practice of medicine seemed to offer the only salvation and in 1933 she entered Peking's Yenjing University, transferring later, on a scholarship, to Belgium. There she was repelled by Belgian racism and the blithely pro-Nazi, anti- Semitic attitudes of her mother's friends. When Japan invaded China in 1938 she returned, and met on the boat home the man she would so disastrously marry, Tan Paohuang, an officer in the Chinese army, whose already right-wing politics would eventually become full-blown fascism.

Her work as a midwife in Sichuan Province revealed to her the stoicism of the Chinese peasantry confronting ceaseless grinding poverty, and their courage in the face of Japanese invasion. Her first book, *Destination Chungking* (1943), is in large measure her tribute to them and, as such, anathema to her increasingly extreme husband. Accordingly she published it under the pseudonym by which she is now solely known: Han Suyin, a name which might fittingly be translated as 'Little Voice of China'.

Almost ten years separate her first and second book, *A Many-Splendoured Thing*, the highly autobiographical and enormously successful novel for which she is still best known. In that decade she lived first in London, where her husband had been posted as cultural attaché, and where she completed her medical studies and stored away the memories and experience which, reworked, would surface late in what many fellow authors regard as her finest fictional work, the novella *Winter Love*. From London she went as a surgeon, in 1949, to Hong Kong, the back door to a China in which she still hoped to live and work. Her welcome was uncertain; her husband had been killed in 1947, fighting on the wrong (anti-Communist) side.

Not until 1956 was she permitted to return to Peking. Emotionally moved and intellectually persuaded by Zhou Enlai's vision of Chinese Socialism, she threw herself into serving the new China, taking on the role of a widely travelling, highly visible cultural ambassador. She also became China's unofficial historian, chronicling, explaining, excusing, celebrating, in a remarkable series of books which include biography, fiction, and five volumes cast in so individual a form that 'autobiography' seems an inadequate classification. From *The Crippled Tree* to *The Wind in My Sleeve* she has perfected her own unique synthesis of history, politics, legend, folklore, oral testimony, family chronicle and personal memoir.

Such epic endeavours forced her, at last, to give up the medical practice which hitherto she has so fruitfully combined with writing, to the benefit of both. Han now lives in Switzerland, with Colonel Vincent Ruthnaswamy, her second husband for many years. In recent, post-Tiananmen Square days her unabashed support of China's 'old men' has aroused furious hostility against her in many quarters, and editorial decisions to exclude her from most contemporary literary companions and guides to women's writing must surely have been made on political rather than aesthetic grounds. So remarkable a body of work deserves better.

Non-Fiction

See Singapore 1955; *The Crippled Tree* (a) 1965; *A Mortal Flower* (a) 1966; *China in the Year 2000* 1967; *Birdless Summer* (a) 1968; *Asia Today: Two Outlooks* 1969; *The Morning Deluge: Mao Tsetung and the Chinese Revolution 1893–1953* 1972; *Lhasa, Étoile Fleur* [with M. Olivier-Lacamp] 1976; *Wind in the Tower: Mao Tsetung and the Chinese Revolution 1949–1975* 1976 [rev. 1978]; *Lhasa, the Open City: a Journey to Tibet* 1977; *My House Has Two Doors* (a) 1980; *Han Suyin's China* 1987; *A Share of Loving* (a) 1987; *Fleur du Soleil* (a) 1988; *Tigers and Butterflies* [ed. A. Hussein] 1990; *The Wind in My Sleeve* (a) 1992; *Eldest Son: Zhon Enlai and the Making of Modern China, 1898–1976* 1994

Fiction

Destination Chungking 1943; *A Many-Spendoured Thing* 1952; *... and the Rain My Drink* 1956; *The Mountain is Young* 1958; *Cast But One Shadow and Winter Love* 1962; *The Four Faces* 1963; *Till Morning Comes* 1982; *The Enchantress* 1985

Gerald HANLEY 1916–

Hanley was born in Ireland, into a working-class seafaring family. He is the brother of **James Hanley**, though, being fifteen years younger, he is virtually of another generation. On leaving school he went out to farm in Kenya, aged sixteen. He served in the Royal Irish Fusiliers for seven years, during the Second World War in Africa and Burma; the latter part of his service yielded his first novel, *Monsoon Victory*. At the end of the war he worked on films in India and with the BBC in London. In 1950, he settled for a while in the Punjab. He finally returned to his native Ireland where he is a member of the Irish Academy of Letters.

Most productive in the 1950s and 1960s, Hanley published only nine novels in thirty-six years and none since 1982. He writes out of his varied youthful experience; *The Consul at Sunset*, perhaps his best-known novel, *The Year of the Lion* and *Drinkers of Darkness* are all set in Africa and deal in one way or another with racial tension. In *The Journey Homeward* and *Noble Descents*, published after a silence of thirteen years, he tackles the political problems of post-Partition India. His travel book *Warriors and Strangers* throws retrospective light on his African experience.

Fiction

Monsoon Victory 1946; *The Consul at Sunset* 1951; *The Year of the Lion* 1954; *Drinkers of Darkness* 1955; *Without Love* 1957; *The Journey Homeward* 1961; *Gilligan's Last Elephant* 1962; *See You in Yasukuni* 1969; *Noble Descents* 1982

Plays

A Voice from the Top (r) 1962

Non-Fiction

Warriors and Strangers 1971

James HANLEY 1901–1985

Born in Dublin, Hanley was the son and grandson of working-class Irish seamen; the name was originally Hanly. He grew up in Liverpool, where he received only a board-school education before going to sea aged thirteen as a stoker. He sailed around the world, deserted ship in New Brunswick, joined the Canadian Expeditionary Force, saw service with them on the battlefields in France, then returned to sea as a deckhand. In 1924 he became a landsman, originally finding work as a railway porter, and doing other menial jobs, but progressing to become a journalist (he undertook a programme of self-education and became skilled in, among other areas, music). Hanley's first published story was *The German Prisoner* (1930), an astonishingly violent and explicit account of the sexual torture and eventual murder of a German soldier by two British privates, who are subsequently killed by a shell. It was published privately, with an introduction by **Richard Aldington**.

His first novel, *Drift*, was also published in 1930, after being offered to nineteen publishers: he was paid a fee of £5 for all rights. His second novel, *Boy*, the grim story of a cabin boy who is abused and finally murdered aboard ship, was seized by police in Britain because it was allegedly obscene: it was eventually available in an expurgated edition in England, and was published in full in Paris. Soon after this, Hanley became a full-time writer, entering a prolific period between 1930 and 1941 when he published eleven novels and ten collections of stories. His final tally was to be twenty-nine novels and thirteen volumes of stories and he also wrote many plays, as well as numerous general works, working through many publishers. He was always highly praised by critics and fellow writers – his list of admirers ranges from **E.M. Forster** to **William Faulkner**, from **Henry Green**, who encouraged him early in his career, to T.E. Lawrence, and he was regularly compared to **Joseph Conrad** and Dostoevsky – but his work did not make a breakthrough with the general public; he was never published in paperback, and through much of the 1970s and 1980s most of his novels were unavailable. His consistent themes of life among the working class or outcasts or seamen, as well as the grim, often despairing, tone of his work, were perhaps responsible for his failure to win popularity. Among his most admired novels was *The Closed Harbour*, and he also wrote a five-volume saga (1935–58) of a Dublin working-class family, the Furys.

During the 1960s, on a retainer from the BBC, he turned for ten years to writing television and radio plays: one of his novels, *Say Nothing*, set in a boarding-house, first appeared in a play version. He returned to the novel for four final works in the 1970s which showed no flagging of power. After living in London during the 1930s, he spent thirty years in a cottage in Wales, but spent his final years in a council flat off the Highgate Road, London. He was long married to the painter and illustrator Liam Hanley, who died in 1980. His younger brother, **Gerald Hanley**, is also a novelist.

Fiction

Drift 1930; *The German Prisoner* (s) 1930; *A Passion Before Death* 1930; *Boy* 1931 [rev. 1932]; *The Last Voyage* (s) 1931; *Men in Darkness* (s) 1931; *Aria and Finale* (s) 1932; *Ebb and Flood* 1932; *Stoker Haslett* (s) 1932; *Captain Bottell* 1933; *Quartermaster Clausen* (s) 1934; *Resurrexit Dominus* 1934; *At Bay* (s) 1935 [rev. 1944]; *The Furys Chronicle: The Furys* 1935, *The Secret Journey* 1936, *Our Time is Gone* 1940 [rev. 1949], *Winter Song* 1950, *An End and a Beginning* 1958; *Stoker Bush* 1935; *Half an Eye* (s) 1937; *Hollow Sea* 1938; *People Are Curious* (s) 1938; *Between the Tides* 1939; *The Ocean* 1941; *New Directions* 1943; *Sailor's Song* 1943; *Crilley and Other Stories* (s) 1945; *What Farrar Saw* 1946; *Selected Stories* (s) 1947; *Emily* 1948; *A Walk in the Wilderness* (s) 1950; *The House in the Valley* [as 'Patrick Shone'] 1951 [as *Against the Stream* as James Hanley 1981]; *The Closed Harbour* 1952; *Collected Stories* (s) 1953; *Don Quixote Drowned* (s) 1953; *The Welsh Sonata 1954; Levine* 1956; *Lost* (s) [as 'Patrick Shone'] 1956 [as James Hanley 1979]; *Say Nothing* 1962; *The Face of Winter* 1969; *Another World* 1972; *The Darkness* (s) [ed. R. Hayman] 1973; *A Woman in the Sky* 1973; *Dream Journey* 1976; *A Kingdom* 1978

Plays

Don Quixote Drowned (r) [from his story] 1953; *Man in the Mirror* (r) 1954; *Sailor's Song* (r) [from his novel] 1954; *The Welsh Sonata* (r) [from his novel] 1956; *A Letter in the Desert* (r) 1958 *The Ocean* (r) [from his novel] 1958; *Gobbet* (r) 1959; *Levine* (r) [from his novel] 1959; *Miss Williams* (r) 1960; *A Moment for Reflection* (r) 1961; *Say Nothing* [from his novel] (r) 1961, (tv) 1964; *The Furys* (r) [from his novel] 1961; *The Inner Journey* 1962 [pub. in *Plays One* 1968] *A Dream* (r) 1963; *What Farrar Saw* (r) [from his novel] 1963; *Another Port, Another Town* (tv) 1964; *The Inner World of Miss Vaughn* (tv) 1964; *Day Out for Lucy* (tv) 1965; *Mr Ponge* (tv) 1965; *A Walk in the Sea* (tv) 1966; *One Way Only* (r) 1967; *That Woman* (tv) 1967; Nothing Will Be the Same Again (tv) 1968; *The Silence* (r) 1968; *A Stone Flower* 1968; *It Wasn't Me* (tv) 1969; *The House in the Valley* (tv) [from his novel] 1971; *A Terrible Day* (r) 1973; *A Dream Journey* (r) [from his novel *New Directions*] 1976

Non-Fiction

Broken Water (a) 1937; *Grey Children* 1937; *John Cowper Powys: a Man in the Corner* 1969; *Herman Melville: a Man in the Customs House* 1971

Edited

Chaliapin: an Autobiography as Told to Maxim Gorky [with N. Froud] 1968

For Children

A Pillar of Fire (r) 1962

James Owen HANNAY

see George A. BIRMINGHAM

Barbara HANRAHAN 1939–1991

Hanrahan was a novelist, short-story printmaker and artist, who was born in the Adelaide suburb of Thebarton. Her father died just after her first birthday and she was brought up by her mother, a commercial artist, her grandmother, and her great-aunt, who had Down's syndrome. Educated at the South Australian School of Art and Adelaide Teachers' College, she taught for three years, then, in 1963, set out for the Central School of Art, London. Dividing her time between Adelaide and London, she established an international reputation as a painter and printmaker, with exhibitions in Tokyo, Florence, London and throughout Australia. Eighteen of her prints and three drawings are held in the permanent collection of the Australian National Gallery, Canberra. She never married, but from 1966 onwards lived with the Australian sculptor Jo Steele. They had no children.

Although she had written extensive diaries since she was a young woman, it was the death of her grandmother in 1968 that prompted *The Scent of Eucalyptus*, an autobiographical novel based on childhood memories of 1950s Adelaide. Encouraged by **D.J. Enright** at Chatto and Windus, she revised the manuscript for publication and embarked on a second memoir, *Sea-Green*, which she later described as a failure. Most of her subsequent novels have Victorian or Edwardian settings, periods which had held tremendous fascination for her since early childhood. She recounted, for example:

> At the bottom of the yard we had a corrugated iron shed covered in lavatory creeper and bridesmaid's fern. In summer, I'd be in the shed where there was a big box of my great-grandmother's and my grandmother's books. I'd sit there and read the *Girl's Own Annual* of 1895 and look at pictures of pompadour hair and leg-o'-mutton sleeves and read about snow storms in old England – while in Adelaide it was a heatwave day ... I'm attracted to the Victorian period. It suits me exactly. There's the weird combination of the spiritual existing side by side with the material. To me, this juxtaposition of the concrete and the immaterial is poetic ... I like my angels weighed down in marble.

Admiring of **D.H. Lawrence**, **Katherine Mansfield**, **Flannery O'Connor** and **Janet Frame**, she developed a suggestive prose style capable of counterpointing the sinister and the everyday, the inner torments of emotional crisis and the outer frameworks of convention and respectability. Her most constant theme is the divided self, torn between the need for conformity and the dread of its consequences.

Hanrahan wrote painstakingly, first in longhand, then on a battered manual typewriter, making many drafts, and always reading her work aloud 'to get the sound'. In addition, she did a great deal of newspaper research and listened to many people's remembrances and recollections, intent on accuracy both of everyday details and of atmosphere. Though favourably reviewed, her work won no major prizes, about which she remarked: 'It doesn't bother me personally, though I feel sorry for the writing sometimes.' Since her comparatively early death in 1991, her work has enjoyed new critical attention among commentators interested in contemporary Australian writing.

Fiction

The Scent of Eucalyptus (a) 1973; *Sea-Green* 1974; *The Albatross Muff* 1977; *Where the Queens All Strayed* 1978; *The Peach Groves* 1979; *The Frangipani Gardens* 1980; *Dove* 1982; *Kewpie Doll* 1984; *Annie Magdalene* 1985; *Dream People* (s) 1987; *A Chelsea Girl* 1988

Edited

Some Poems of Shaw Neilson 1984

Lorraine (Vivian) HANSBERRY 1930–1965

Hansberry was born in Chicago, the daughter of black middle-class parents. Her father Robert, a real-estate broker and banker, attempted to move his family into a restricted white neighbourhood in the late 1930s, and subsequently fought a famous civil rights case which, in Hansberry's words, 'he fought all the way up to the Supreme Court, and which he won after the expenditure of a great deal of money and emotional strength'. The family moved there, despite intimidation from white neighbours, but Robert Hansberry died in 1945 while preparing to move to Mexico.

Hansberry first became interested in the theatre while at Englewood High School, and did a course in stage design at the University of Wisconsin before studying painting at the Art

Institute of Chicago, Roosevelt College and in Guadalajara in Mexico. In 1950 she moved to New York, having decided to abandon painting in favour of writing. She worked at various jobs while writing a number of unfinished plays, short stories and articles for Paul Robeson's *Freedom* magazine. In 1953 she married Robert Nemiroff, a Jewish music publisher and songwriter, whom she had met during a protest against discrimination at New York University.

With her husband's encouragement Hansberry worked on her first play, *A Raisin in the Sun*. Drawing upon her own background, the play depicts a black family's attempt to move into a white neighbourhood with the fruits of an insurance policy left by their dead father. After trial performances in various cities, the play opened on Broadway in 1959. It starred her old friend Sidney Poitier, and ran for 530 performances. Hansberry was the first black woman to have a play produced on Broadway, and became the youngest writer to win the New York Drama Critics Circle Award. The success of the production led to a film version in 1961 which won a special award at the Cannes Film Festival. Hansberry's second play, *The Sign in Sidney Brustein's Window*, concerns the political vacillations of a Jewish intellectual. It opened in 1964, a few months before her early death from cancer. Despite their divorce in 1964, Robert Nemiroff has remained a champion of her work and has edited much of her writing for posthumous production and publication.

Plays

A Raisin in the Sun 1959; *The Sign in Sidney Brustein's Window* 1964 [pub. 1965]; *Les Blancs* 1970 [pub. 1966; as *Lorraine Hansberry's 'Les Blancs'* with R.B. Nemiroff 1972]; *Les Blancs: the Collected Last Plays of Lorraine Hansberry* [ed. R.B. Nemiroff pub. 1972]

Non-Fiction

The Movement: Documentary of a Struggle for Equality 1964 [UK *A Matter of Colour: Documentary of the Struggle for Racial Equality in the USA* 1965]; *American Playwrights on Drama* (c) [ed. H. Frenz] 1965; *To Be Young, Gifted and Black: Lorraine Hansberry in Her Own Words* [ed. R.B. Nemiroff] 1969

Elizabeth HARDWICK 1916–

The author of three novels, published at wide intervals during her career, and of numerous short stories, Hardwick is perhaps best known as an essayist and representative of the left-liberal intelligentsia of America. She was born and grew up in Lexington, Kentucky, one of eleven children of the owner of a plumbing and heating (later oil-furnaces) business. She attended local schools and then the University of Kentucky, gaining her MA in 1939. In the same year she moved to New York to study at Columbia, and has lived in the North since then, never becoming identified as a specifically Southern writer. She studied at Columbia until 1941, and then began a career as a freelance writer and journalist, soon beginning to contribute to *Partisan Review* and becoming well known for her trenchant notices there.

Her first novel, *The Ghostly Lover*, the seemingly autobiographical story of the daughter of an emotionally unresponsive Kentucky family who moves to New York, was published in 1945. In 1949 Hardwick married the poet **Robert Lowell**, already well established on his career of manic–depressive illness. He was admitted to a clinic soon after their honeymoon and, although Hardwick did not suffer the physical attacks which had literally disfigured the life of Lowell's first wife, **Jean Stafford**, her time with him was difficult enough. During their residence in Europe in the early 1950s he once went missing on the German–Austrian frontier; she had to arrange for his admission to clinics several times; and his affairs with other women were frequent. They had a daughter, Harriet, born when her mother was forty, and lived in various Midwestern cities and Boston before returning to New York in 1960.

Hardwick's second novel, *The Simple Truth*, the story of an Iowa murder trial, was published in 1955. In 1972 Lowell divorced Hardwick to marry the English writer **Caroline Blackwood** (events which he immortalised in two books of poetry, *For Lizzie and Harriet* and *The Dolphin*), but for much of 1977, the last year of his life, he returned to Hardwick, having met her again at a writer's conference in Moscow (an echo of their first meeting, which also took place at a writer's conference). He eventually died, on his way from Blackwood to Hardwick, in a New York taxi. In 1979 Hardwick published her third, and most highly praised, novel, *Sleepless Nights*, a lonely old woman's meditations on various people who helped form her life. She has also published three collections of essays, largely consisting of contributions to *Partisan Review* and the *New York Review of Books*, which she and Lowell helped to found in 1963, as well as an edition of the letters of William James and a multi-volume anthology of American women's fiction. She has taught widely, particularly at Barnard College. She has continued to live in the house on New York's Upper West Side which she once shared with Lowell.

Fiction

The Ghostly Lover 1945; *The Simple Truth* 1955; *Sleepless Nights* 1979

Non-Fiction

A View of My Own: Essays in Literature and Society 1962; *Seduction and Betrayal: Women and Literature* 1974; *Bartleby in Manhattan and Other Essays* 1983

Edited

William James *Selected Letters* 1961

Thomas HARDY 1840–1928

Born in a cottage at Higher Bockhampton, Dorset, Hardy was the son of a master mason and a cook. Many of his relations were bricklayers and labourers, and from them he heard folk-tales and traditional verse. Equally influential in his development as a writer was a love of music, which he inherited from his father, and he memorised folk-tunes, which he played for hours at country dances. His mother taught him to read at the age of three, and when he left Dorchester Grammar School at sixteen he knew classics, French and German. He made friends with William Barnes, whose dialect poetry motivated him to write verse, and his reading list was overseen by Horace Moule, a clergyman whose eventual suicide contributed to Hardy's bleak view of life.

He became apprenticed to a Dorchester architect in 1856, but despite winning awards while he worked as an architect in London, he had determined to be a writer. He returned to Dorset in 1867, but early efforts in both poetry and fiction proved fruitless until 1871, when his first novel, *Desperate Remedies*, was published. The previous year, he had visited Cornwall in order to draw up plans for the restoration of a church and had fallen in love with the rector's sister-in-law, Emma Gifford. Hitherto, Hardy had suffered a number of unrequited love-affairs, and so was grateful for Emma's attention, as well as being impressed by her vitality and social position. When in 1874 his fourth novel, *Far from the Madding Crowd*, was serialised to great acclaim, the couple married.

Hardy's novels are largely set in Wessex, an area in south-west England, centred upon Dorset, but also taking in parts of Hampshire, Devon, Cornwall, Somerset and Wiltshire. They detail the often hard lives of country people and are suffused with a sense of *genius loci*, particularly in such novels as *The Return of the Native*, set around Egdon Heath. Most of Hardy's characters are pitted against a malign fate which rules and destroys their lives, and some of the novels – such as *Jude the Obscure*, in which the protagonist's children commit suicide – have plots which trip over into melodrama. The books seem modern, however, in their subject-matter and attitudes. He became one of the leading and most controversial novelists of the late nineteenth century, and his books have always remained popular.

Essentially shy and reclusive, Hardy was nevertheless appreciative of his own celebrity. Before *Tess of the d'Urbervilles* firmly established his reputation in 1891, he concealed his origins from his wife and cut himself off from his family, even though their way of life had inspired his best work: *Under the Greenwood Tree, The Return of the Native, The Mayor of Casterbridge* and *Wessex Tales*. Emma was a religious bigot, and her disgust at her atheist husband's later novels – the last of which she attempted to suppress – was indicative that their marriage, which had been insecure from the start, was a failure. In 1885, the couple moved near Dorchester to Max Gate, an ugly house of Hardy's own design which proved cramped, but large enough for the unhappy couple to remain largely apart. As Emma lost her looks, Hardy ignored her presence and formed close attachments with other women.

Severe public criticism of *Jude the Obscure*, largely on moral grounds, caused Hardy to renounce fiction for poetry, which he published from 1898, and he had a second career as a highly esteemed and extremely prolific poet. He also wrote a long, three-part verse drama about the Napoleonic War, *The Dynasts*. Emma was too ill to accompany Hardy when he accepted the Order of Merit in 1910, and two years later she died in his arms. Hardy pretended that her death had been unexpected, and his hypocrisy was rewarded when he discovered a vicious account of him written by this undevoted wife. In 1914 he married the 35-year-old Florence Dougdale, whom he had installed as his secretary within a month of Emma's death, and whose talentless scribblings he had helped into print. Overcome with remorse about Emma, he venerated her memory with brilliant poetry and dismal visits to her grave. Dismayed, bored and neglected, Florence quickly tired of Hardy and of living with a ghostly rival. Desperate to preserve the public self-image he had carefully manufactured, Hardy dictated a fabricated version of his life to Florence, which was published under her name after his death.

In spite of an unedifying life, Hardy remains one of the great figures in English literature. His poetry, which was traditional in form, often colloquial, occasionally clumsy but more often beautifully crafted and deeply felt, found little favour with the modernists, but occupied a central position in **Philip Larkin**'s controversial and revisionist *Oxford Book of Twentieth Century English Verse* (1973), in which Hardy was represented by twenty-seven poems (as compared with nine by **T.S. Eliot** and sixteen by **W.H. Auden**). Like **John Betjeman**, he remains a poet as much loved as admired, and amongst those who have felt his influence are Auden, **C. Day Lewis**, **Dylan Thomas** and of course Larkin himself.

Fiction
Desperate Remedies 1871; *Under the Greenwood Tree* 1872; *A Pair of Blue Eyes* 1873; *Far from the Madding Crowd* 1874; *The Hand of Ethelberta* 1876; *The Return of the Native* 1878; *The Trumpet Major* 1880; *The Laodicean* 1881; *Two on a Tower* 1882; *The Mayor of Casterbridge* 1886; *Wessex Tales* 1886; *The Woodlanders* 1887; *A Group of Noble Dames* (s) 1891; *Tess of the d'Urbervilles* 1891; *Life's Little Ironies* (s) 1894; *Jude the Obscure* 1896; *The Well-Beloved* 1897; *A Changed Man and Other Tales* (s) 1913; *The Excluded and Collaborative Stories* (s) [ed. P. Dalziel], 1992

Poetry
Wessex Poems 1898; *Poems of the Past and Present* 1902; *Time's Laughing Stock* 1909; *Satires of Circumstances* 1914; *Selected Poems* 1916; *Moments of Vision* 1917; *Late Lyrics and Earlier* 1922; *Human Shows* 1925; *Winter Words* 1928; *Complete Poems* [ed. J. Gibson] 1976

Plays
The Three Wayfarers 1893 [np]; *The Dynasts* 3 parts 1903–8; *The Famous Tragedy of the Queen of Cornwall* 1923

For Children
Our Exploits at West Poley 1883

Non-Fiction
The Early Life of Thomas Hardy 1840–1891, *The Later Life of Thomas Hardy 1892–1928* (a) [as 'Florence Hardy'] 1928, 1930; *The Collected Letters of Thomas Hardy* [ed. M. Millgate, R. Purdy] 7 vols. 1978–88

Biography
The Younger Thomas Hardy by Robert Gittings 1975; *The Older Thomas Hardy* by Robert Gittings 1978.

David HARE 1947–

Hare was born at St Leonard's-on-Sea, Sussex. When he was five his family moved to Bexhill and he was educated at Lancing College. He graduated from Jesus College, Cambridge, in 1968 with a BA in English. For a few weeks he worked at A.B. Pathé, locating material for sex-education films, then, with Tony Bicât, he founded Portable Theatre, a touring company.

His first play, *Inside Out*, a collaboration with Bicât, was staged by Portable in September 1968. Like the two that followed, it was largely unnoticed; but Hare's fourth, *Slag*, brought him to prominence. Set in an all-female boarding-school, it shows three teachers' futile and wild enthusiasms. Fashionably, it conflated political and sexual revolution; but – while it won an *Evening Standard* award – some critics took its fantasy for caricature and its comedy for cruelty. Responding, Hare has described it as 'really a play about institutions'. Whatever his intentions, he had signalled that, even if he was a left-wing playwright, he wasn't interested in the regurgitation of socialist pieties. Indeed, much of his work is a lament for Labour's failure to revitalise post-1945 Britain: victory in war produced no peacetime fruits. Hence, typically, Hare's protagonists suffer from a deprivation more spiritual than economic. When Curly, the arms dealer 'hero' of *Knuckle*, searches for his revolutionary sister, his obsession suggests the pursuit of a vanished ideal. Amoral as he is, he despises the smugness of his stockbroker father and, by extension, what it represents. A related theme is explored in *Plenty*. Susan Traherne, having served with the Special Operations Executive in occupied France, returns to an England that can offer her no comparable experience. Frustration shades into neurosis, neurosis into full-blown self-destruction. Traherne's crises are linked to moments of national decline. Yet, if the structure is schematic, the language disturbs this level of representation. 'The theatre is the best way of showing the gap between what is said and what is seen to be done,' Hare has argued. Accordingly, his characters speak with unidiomatic exactness – a heightened clarity reminiscent of dream. Language is at odds with the true situation: sometimes a gloss, sometimes an icy distortion, it permits his people to act in bad faith; syntactical elegance covers shabby behaviour and truth becomes a piercing violence. By the same token, violence accompanies moments of truth, notably in *Licking Hitler*, Hare's much praised play about black propaganda, which he directed for the BBC.

In 1985 Hare's first feature film, *Wetherby*, won the Golden Bear Award in Berlin. In the same year, he was made a Fellow of the Royal Society of Literature. *Pravda*, co-authored with **Howard Brenton**, was the first of a series of plays which investigated British institutions, in this case the press, in particular the role of newspaper tycoons. *Racing Demon* would take the Church of England as its subject, *Murmuring Judges* the judiciary. *Pravda* was widely interpreted as an attack on what Hare considered the ethics of Thatcherism. Two further projects, *The Secret Rapture* and *Paris by Night*, sustained that direction. However, *Heading Home* and *Murmuring Judges*, indicated a return to the regrets, enigmas and ambivalences of his work of the late 1970s.

In 1970 Hare married Margaret Matheson, with whom he had two sons and one daughter. The couple were divorced in 1980. Since the end of his long relationship with Kate Nelligan, who appeared in many of his plays, Hare has been associated with, successively, the actresses Penny Downie and Blair Brown.

Plays

Inside Out [from Kafka; with T. Bicât], 1968 [np]; *How Brophy Made Good* 1969 [pub. 1971]; *Slag* 1970 [pub. 1971]; *What Happened to Blake* 1970; *Deathsheads* 1971 [np]; *Lay By* [with H. Brenton, B. Clark, T. Griffiths, S. Poliakoff, H. Stoddart, S. Wilson] 1971 [pub. 1972]; *The Rules of the Game* [from Pirandello's play *La Règle du jeu* 1971; *England's Ireland* [with T. Bicât, H. Brenton, B. Clark, D. Edgar, F. Fuchs, S. Wilson] 1972 [np]; *The Great Exhibition* 1972; *Brassneck* [with H. Brenton] 1973 [pub. 1974]; *Man Above Men* **(tv)** 1973; *Knuckle* 1974; *Fanshen* [from W. Hinton] 1975 [pub. 1976]; *Teeth 'n' Smiles* [with N. and T. Bicât] 1975 [pub. 1976]; *Deeds* [with H. Brenton, K. Campbell, T. Griffiths] 1978; *Licking Hitler* **(tv)** 1978; *Plenty* 1978; *Dreams of Leaving* **(tv)** 1980 [pub. 1979]; *A Map of the World* 1982; *Saigon* **(tv)** 1983; *The Madman Theory of Deterrence* **(st)** 1983 [np]; *Plenty* **(st)** 1978, **(f)** 1985; *Pravda* [with H. Brenton] 1985; *Wetherby* **(f)** 1985; *The Bay at Nice* 1986; *Wrecked Eggs* 1986; *The Knife* **(l)** [with N. Bicât, T.R. Price] 1987; *The Secret Rapture* 1988; *Paris By Night* **(f)** 1989 [pub. 1988]; *Strapless* **(f)** 1989; *Racing Demon* 1990; *Heading Home* **(tv)** 1991; *Murmuring Judges* 1991; *Damage* **(f)** 1992; *The Absence of War* 1993; *Skylight* 1995

Non-Fiction

Writing Left-Handed 1991; *Asking Around* 1993

(Theodore) Wilson HARRIS 1921–

Harris was born in New Amsterdam, British Guyana, and was educated at Queen's College, Georgetown. He subsequently studied land surveying and practised as a surveyor until 1958, having been appointed Senior Surveyor, Projects, for the Government of British Guyana in 1955. In 1959 he moved to London, where he still lives, and began a career as a full-time writer. Since then he has held a number of residencies and lectureships at various universities, including those of the West Indies, Toronto, Leeds, Texas, Newcastle (Australia), California and Queensland. He is married and has a son and a daughter.

His first published work was a volume of poetry, *Fetish* (1951), and, although now better known as a novelist, Harris's writing is still markedly poetic in its compression and opacity. Eschewing straightforward narrative, he prefers to present an often perfunctory story through the medium of the consciousness on which events register. That consciousness, in turn, articulates itself through a highly wrought symbolism, translating its perceptions into a shared stock of images, metaphors and fables. One effect of this is that all Harris's characters think and feel in similar vocabularies, prone to statements such as: 'Fire is the womb of brain and mind'; but it is also a persistent challenge to the Western European doctrine of the individual. It emphasises what human beings have in common at the expense of what divides and differentiates them. Jungian thought, Latin American myth and the mystical Christianity of **T.S. Eliot's** *Four Quartets* are ever-present influences.

In the early *Guyana Quarter*, Harris was concerned also with 'the uncertain economic fate of [Guyanese] society and the majority of his protagonists continue to survive on the margins of society, whether as fishermen, holy men or unemployed. In recent work, however, material conditions have diminished in importance – an attenuated reality serving merely as a pretext for the communication of ghosts, legends and dreams. Complex and experimental, Harris's fiction has never enjoyed a wide audience – nor is it likely to do so. Nevertheless, it is admired by academics, critics and fellow writers of the *avant garde*.

Harris was awarded the 1987 Guyana Prize for Literature (Fiction Category) for *The Infinite Rehearsal*. He holds an honorary doctorate from the University of the West Indies.

Fiction

The Guyana Quartet 1985: *Palace of the Peacock* 1960, *The Far Journey of Oudin* 1961, *The Whole Armour* 1962, *The Secret Ladder* 1963; *Heartland* 1964; *The Eye of the Scarecrow* 1965; *The Waiting Room* 1967; *Tumatumari* 1968; *Ascent to Omai* 1970; *The Sleepers of Roraima* **(s)** 1970; *The Age of the Rainmakers* **(s)** 1971; *Companions of the Day and Night* 1975; *De Silva da Silva's Cultivated Wilderness; and Genesis of the Clowns* 1977; *The Tree of the Sun* 1978; *The Angel at the Gate* 1982; *The Carnival Trilogy* 1993: *Carnival* 1985, *The Infinite Rehearsal* 1987, *The Four Banks of the River of Space* 1990

Poetry

Fetish 1951; *Eternity to Season* 1954 [rev. 1978]

Non-Fiction

Tradition and the West Indian Novel 1965; *Tradition, the Writer and Society: Critical Essays* 1967; *Explorations: a Selection of Talks and Articles* 1981; *The Womb of Space: the Cross-Cultural Imagination* 1983

Tony HARRISON 1937–

Harrison was born in Leeds, the only son of a bakery worker, and attended Cross Flats County Primary, the local school. Aged eleven, however, he won one of only a few scholarships available at Leeds Grammar School, and thus began the transition from his original background to the world of intellectual aspiration, a tension which has remained his principal subject. He went on to read classics at Leeds University, publishing his poems in the university magazine, and enjoying the presence of such contemporaries as **Geoffrey Hill, Jon Silkin** and **Wole Soyinka**. He then did a postgraduate diploma in linguistics before beginning to teach in Dewsbury, Yorkshire.

In 1960 he married his first wife, with whom he had two children, as well as one who was stillborn (the subject of his only short story, published by Silkin, together with several early poems, in his magazine, *Stand*). In 1962 he went to Nigeria as lecturer in English at Anmadu Bello University in Zaria, northern Nigeria, and it was there that he began the artistic process which has proved most fruitful for him: the adaptation of classic plays to modern verse versions. His first attempt in this line was *Aikin Mata*, written in collaboration with James Simmons, which was an adaptation for a student performance in 1965 of Aristophanes' *Lysistrata*. The previous year he had published his first volume of poetry, *Earthworks*, and his careers as playwright and poet have proceeded in tandem.

From 1966 he taught for a year at the Charles University in Prague, was then briefly a fellow in poetry in Newcastle, while his growing renown allowed him to spend much of 1969–70 visiting Cuba, Brazil, Senegal and Gambia on a UNESCO fellowship. His first major breakthrough as a dramatist came when his version of Molière's *Le Misanthrope* was performed at the National Theatre in 1973; he soon established a continuing connection with the house, and was its resident dramatist between 1977 and 1979. Since then, he has been able largely to earn a living as a verse dramatist, although for several years he continued to hold various fellowships, often in foreign places. Perhaps the most famous of his adaptations are of *The Oresteia* of Aeschylus and of the York Mystery Plays. It is always said of them that they are supremely actable and that he is a real man of the theatre.

His poetry usually appeared from small presses until his *Selected Poems* in 1984 augmented his reputation in this field; while offering much political comment, his verse will not appeal to those who enjoy euphony. From the later 1980s he became a well-known figure on television, but caused considerable controversy when he read his poem *V* on the medium in 1987, because its response to the miners' strike of three years earlier (it had been written then) contained much scatological language. Harrison has also translated the fifth-century AD Greek poet Palladas. He married the soprano Teresa Stratas in 1984, was the President of the Classical Association in 1987–8, and has for some years divided his time between his long-time home of Newcastle upon Tyne and America.

Poetry

Earthworks 1964; *Newcastle is Peru* 1969; *The Loiners* 1970; *Corgi Modern Poets in Focus Four* [with others] 1971; *Ten Poems from the School of Eloquence* 1976; *From the School of Eloquence and Other Poems* 1978; *Looking Up* [with P. Sharpe] 1979; *Continuous* 1981; *A Kumquat for John Keats* 1981; *U.S. Martial* 1981; *Selected Poems* 1984; *The Fire-Gap* 1985; *V* 1985; *Anno Forty Two* 1987; *Ten Sonnets from the School of Eloquence* 1987; *The Mother of the Muses* 1989; *Losing Touch* 1990; *A Cold Coming* 1991; *The Gaze of the Gorgon* 1992; *Square Rounds* 1992; *Black Daisies for the Bride* 1993

Plays

Aikin Mata [from Aristophanes's play *Lysistrata*; with J. Simmons] 1965 [pub. 1966]; *The Misanthrope* [from Molière's play] 1973 [pub. 1974]; *Phaedra Britannica* [from Racine's play] 1975; *Bow Down* [with H. Birtwistle] 1977; *The Passion* [from the York mystery plays] 1977 [pub. 1978]; *The Bartered Bride* [from B. Smetana's opera] 1978 [pub. 1985]; *The Oresteia* [from Aeschylus's plays; with H. Birtwistle] 1981; *The Big H* (**tv**) [with D. Muldowney] 1984 [pub. 1985]; *The Mysteries* [from the York mystery plays] 1985; *The Trackers of Oxyrhynchus* (**l**) 1990; *The Common Chorus* [from Aristophanes's play *Lysistrata*] 1992; *Poetry or Bust* 1993

Translations

Palladas *Poems* 1975

L(eslie) P(oles) HARTLEY 1895–1972

Born in Whittlesea, Cambridgeshire, the son of a solicitor who retired early and made a fortune as director of a brickworks, Hartley was brought up at the family home, Fletton Tower, near Peterborough. After attending Harrow, where he was a conformist but intellectual schoolboy, he went up to Balliol College, Oxford, where his course was interrupted by service in France during the First World War. After being invalided out of the army in 1918, he returned to Oxford, where he edited the *Oxford Outlook* and, as a guest of figures such as Lady Ottoline Morrell and Margot Asquith, first experienced the fashionable society which was always to remain a natural milieu for him.

When he came down, his private income allowed him not to work, and he spent the inter-war period living between England and Venice, meanwhile becoming known as a frequent reviewer for many London papers, but particularly the *Saturday Review* and *Sketch*. He published two books of short stories, mainly ghost stories, during this period, but the fact that they were rather little noticed inhibited the growth of his creative talent. During the Second World War, however, when he was living in Wiltshire and Dorset, he found his voice, and wrote most of the three novels (really one long novel) which were published separately (1944–47) and then as the *Eustace and Hilda* trilogy (1958). This account of the intense relationship from childhood to young adulthood of a brother and sister is a masterpiece, and is the fullest expression of Hartley's distinctive voice as a novelist: a voice which combines pensive charm and humour with a stern morality, an other-worldly detachment with a strong sense of the social and particular.

After the success of this work Hartley became very prolific, writing fourteen other novels; he recaptured the success of *Eustace and Hilda* with one other complex and beautiful work about the loss of innocence, *The Go-Between*, but his last novels, increasingly hinting at his carefully veiled homosexuality, were also increasingly trivial, class-ridden and unconvincing. His short stories, too, only intermittently suggest his real quality.

In later life he lived between a London flat and a large house near Bristol, overlooking the Avon. He was a big man (in his last years very fat), a president of English PEN, a man known for his charm in the literary world, although one of extreme reactionary views (he proposed that known criminals be branded, and liked to refer to working-class people as 'WCs'). His best work, however, made few judgements, especially not crude ones; its full stature has yet to be appreciated.

Fiction

Night Fears and Other Stories (s) 1924; *Simonetta Perkins* 1925; *The Killing Bottle* (s) 1932; *Eustace and Hilda* 1958: *The Shrimp and the Anemone* 1944, *The Sixth Heaven* 1946, *Eustace and Hilda* 1947; *The Boat* 1949; *My Fellow Devils* 1951; *The Travelling Grave and Other Stories* (s) 1951; *The Go-Between* 1953; *The White Wand and Other Stories* (s) 1954; *A Perfect Woman* 1955; *The Hireling* 1957; *Facial Justice* 1960; *Two for the River* (s) 1961; *The Brickfield*, 1964; *The Betrayal* 1966; *The Collected Short Stories of L.P. Hartley* (s) 1968; *Poor Clare* 1968; *The Love-Adept* 1969; *My Sister's Keeper* 1970; *The Harness Room* 1971; *Mrs Cartaret Receives and Other Stories* (s) 1971; *The Collections* 1972; *The Will and the Way* 1973

Non-Fiction

The Novelist's Responsibility: Lectures and Essays 1967

Ronald HARWOOD 1934–

Born Ronald Horwitz in Cape Town, Harwood was educated at Sea Point Boys' High School and, coming to London where he has remained ever since, the Royal Academy of Dramatic Art. He joined Donald Wolfit's Shakespeare Company in 1953, first as Wolfit's dresser, an experience from which his most successful play derives, before becoming an actor. He has many credits to his name. He was the presenter of the radio arts programme *Kaleidoscope* in 1973, the television series *Read All About It* (1978–9) and scripted *All the World's a Stage* (1984). He has been artistic director of the Cheltenham Festival (1975), visitor in theatre at Balliol College, Oxford (1986), and President of International PEN. He married in 1956 and has a son and two daughters.

In the early 1960s Harwood published his first novel, *All the Same Shadows*, television plays and screenplays, including an adaptation of **Richard Hughes**'s *A High Wind in Jamaica*. Much of his early work for the stage was adaptation. His first big success was with *The Dresser* in which the much put upon young dresser panders to the needs and moods of a tyrannical old actor. A powerful piece of theatre and Harwood's most significant achievement to date, it won the 1980 *Evening Standard* Award. Of his other plays *Tramway Road* discusses the racial laws in South Africa; *The Deliberate Death of a Polish Priest* is a courtroom drama based on the case of Father Jerzy Popielusko; *Interpreters* is a comedy and love story. His novel *Articles of Faith* won the 1974 Winifred Holtby Prize. With thirteen stage plays, six novels, a volume of stories, innumerable television and screenplays to his credit, besides non-fiction books on the theatre (including a biography of Wolfit), Harwood is a prolific writer. A useful club cricketer, in his younger days he appeared whenever he could for the Jesters and other actors' and writers' teams.

Plays

The Barber of Stamford Hill **(tv)** 1960, **(f)** 1962; *Private Potter* [with C. Wrede] **(tv)** 1961, **(f)** 1962; *Take a Fellow Like Me* **(tv)** 1961; *The Lads* **(tv)** 1963; *Convalescence* **(tv)** 1964; *Guests of Honour* **(tv)** 1965; *A High Wind in Jamaica* [from R. Hughes's novel] **(f)** [with D. Cannon, S. Mann] 1965; *Drop Dead, Darling* **(f)** [with K. Hughes] 1966; *The Paris Trip* **(tv)** 1966; *The New Assistant* **(tv)** 1967; *Diamonds for Breakfast* **(f)** 1968; *Country Matters* 1969 [np]; *Cromwell* **(f)** [with K. Hughes] 1970; *Eyewitness* **(f)** 1970; *One Day in the Life of Ivan Denisovich* **(f)** [from A. Solzhenitsyn's novel] 1970; *All the Same shadows* **(r)** [from his novel] 1971; *The Guests* **(tv)** 1972; *Long Lease of Summer* **(tv)** 1972; *The Good Companions* [from J.B. Priestley's novel; with J. Mercer, A. Previn] 1974; *Operation Daybreak* **(f)** 1975; *The Ordeal of Gilbert Pinfold* [from E. Waugh's novel] 1977 [pub. 1983]; *A Family* 1978; *The Way Up to Heaven* **(tv)** 1979; *The Dresser* **(st)** 1980, **(f)** 1984 [pub. 1980]; *Evita Péron* **(f)**, **(tv)** 1981; *A Night of the Day of the Imprisoned Writer* [with C. Hampton] 1981 [np]; *After the Lions* 1982 [pub. 1983]; *Tramway Road* 1984; *The Deliberate Death of a Polish Priest* 1985; *Interpreters* 1985 [pub. 1986]; *The Doctor and the Devils* **(f)** 1986; *Parson's Pleasure* **(tv)** 1986; *The Umbrella Man* **(tv)** 1986; *Breakdown at Reykjavik* **(tv)** 1987; *Countdown to War* **(tv)** 1987; *J.J. Farr* 1987 [pub. 1988]; *Mandela* **(tv)** 1987; *Another Time* 1989; *Ivanov* [from Chekhov's play] 1989 [np]; *Reflected Glory* 1992; *Poison Pen* 1994

Fiction

All the Same Shadows [US *George Washington, September, Sir!*] 1961; *The Guilt Merchants* 1963; *The Girl in Melanie Klein* 1969; *Articles of Faith* 1973; *The Genoa Ferry* 1976; *César and Augusta* 1978; *One. Interior Day* **(s)** 1978; *Home* 1993

Non-Fiction

Sir Donald Wolfit, CBE: His Life and Work in the Unfashionable Theatre 1971; *A Sense of Loss* **(tv)** [with J. Selwyn] 1978; *All the World's a Stage* 1984; *Mandela* 1987

Edited

New Stories 3 [with F. King] 1978; *A Night at the Theatre* 1982; *The Ages of Gielgud: an Actor at Eighty* 1984; *Dear Alec Guinness at Seventy-Five* 1989; *The Faber Book of the Theatre* 1993

Christopher HASSALL 1912–1963

Hassall, who was born in London, grew up in a congenial artistic environment. His father was John Hassall, a successful artist and illustrator, and his elder sister, Joan, became a well-known wood engraver. Educated at the famous choristers' school, St Michael's College, Tenbury Wells, Hassall was influenced by the economy of the choral settings, and developed a strong interest in writing verse. He read English literature and music at Wadham College, Oxford, in 1931, but a family financial crisis ended his undergraduate studies the following year. He had gained useful experience in the Oxford University Dramatic Society, for he became an actor, and toured abroad with an English company in 1933. On his return, he met Ivor Novello, and was cast as his understudy in Novello's *Proscenium*. The friendship which developed changed Hassall's career. A collection of verse, *Poems of Two Years* (1935), impressed Novello, who asked the young actor to write the libretto for the musical *Glamorous Night*, the first of six such collaborations which included *The Dancing Years*. Thick-set and heavily built, Hassall was nevertheless enthusiastically befriended by the sponsor of the Georgian poets, Edward Marsh, after an introduction by Novello. Marsh, whose social elegance belonged to an earlier age, appreciated the courtesy and geniality which was Hassall's hallmark, and encouraged him to work with several composers, including William Walton, for whose *Troilus and Cressida* he wrote the libretto. Marsh's guiding influence ensured that Hassall's narrative poem *Penthesperon* received both the A.C. Benson Medal and the Hawthornden Prize in 1939. The poem was considered by some to be dull and old-fashioned, but at its best contained memorable images.

During the war, Hassall served in the Education Corps, and after, took up freelance lecturing, becoming poetry editor for the BBC Third Programme in 1950. The plays and poems which followed are largely forgotten, but his biography of Marsh, long and verbose, provided a useful portrait of its age. Less successful, but no less ambitious, was his life of **Rupert Brooke**, where he was undoubtedly hampered by the restrictions of the Brooke Estate. It was posthumously published in 1964. Hassall, who married in 1938 and had two children, had died suddenly on a train at Chatham, Kent, the previous year.

Plays

Glamorous Night **(l)** [with I. Novello] 1935; *Careless Rapture* **(l)** [with I. Novello] 1936; *Devil's Dyke* 1937; *Christ's Comet* 1938; *The Dancing Years* **(l)** [with I. Novello] 1939; *King's Rhapsody* **(l)** [with I. Novello] 1949; *The Rainbow* **(l)** [with T. Wood] 1951; *The Player King* 1952; *Voices of Night* **(l)** [with F. Reizenstein] 1952; *Out of the Whirlwind* 1953; *Salutation* **(l)** [with E. Rubbra]

1953; *Troilus and Cressida* (l) [with W. Walton] 1954; *Genesis* (l) [with F. Reizenstein] 1958; *Tobias and the Angel* (l) [with A. Bliss] 1962; *Mary Magdala* (l) [with A. Bliss] 1963; *Valley of Song* (l) [with I. Novello] 1964

Poetry

Poems of Two Years 1935; *Penthesperon* 1938; *Crisis* 1939; *S.O.S. ... 'Ludlow'* 1940; *The Slow Night and Other Poems* 1948; *The Red Leaf* 1957; *Bell Harry and Other Poems* 1963; *Poems for Children* 1963

Non-Fiction

Notes on the Verse Drama 1948; *This Timeless Quest: Stephen Haggard* 1948; *Eddie Marsh: Sketches for a Composite Literary Portrait* [with D. Mathews] 1953; *Edward Marsh, Patron of the Arts* 1959; *Ambrosia and Small Beer* 1964; *Rupert Brooke* 1964

Edited

The Unpublished Poems of Stephen Haggard 1945; *P.E.N.: New Poems* [with L. Lee, R. Warner] 1954; *The Prose of Rupert Brooke* 1956

Collections

Words by Request 1952

John (Clendennin Burne) HAWKES 1925–

Born in Stamford, Connecticut, Hawkes spent the early years of his adolescence in Alaska where he has said he 'acquired a permanent preoccupation with the alien nightmare landscape of darkness, rain, high wind, mountains, fragments of glacier, distant bears, wild strawberries, one-legged Indians, and the terrifying ruins of abandoned mining towns'. He returned to Connecticut in 1940 and attended Trinity School and Pawling High School before enrolling at Harvard in 1943. Before attending Harvard he spent two years as an ambulance driver for the American Field Service in Italy and Germany, an experience which would furnish something of the surreal, war-ravaged landscape of his first novel.

In 1947 Hawkes married (he has four children), and graduated two years later, in the same year that his first novel was published. *The Cannibal*, like much of his subsequent work, is a bleakly entropic vision of a world dominated by violence and terror. Cannibalism provides the central metaphor of negation. The novel was originally written in the third person, but Hawkes stumbled upon the first-person narrative of the neo-Nazi leader Zizendorf 'by a series of semi-conscious impulses and sheer accidents'. While continuing to write fiction he worked as assistant to the production manager at Harvard University Press until 1955, after which he held a succession of academic posts at Harvard culminating in a full professorship. Since 1973 he has been a professor at Brown University, Providence, Rhode Island.

For many years Hawkes had no more than a small cult readership, but the publication of *Second Skin* in 1964 brought him to the attention of a wider audience. The novel lost that year's National Book Award to **Saul Bellow's** *Herzog* by one vote. Again dominated by a bizarre imaginary landscape, the novel's more conventional narrative structure may none the less have helped ensure its greater popularity. Hawkes has been the recipient of a Guggenheim Fellowship, in 1962, and a Rockefeller Fellowship in 1968.

Fiction

The Cannibal 1949; *The Beetle Leg* 1951; *The Goose on the Grave/The Owl* 1954; *The Lime Twig* 1961; *Second Skin* 1964; *Lunar Landscapes: Stories and Short Novels 1949–1963* (s) 1969; *The Blood Oranges* 1971; *Death, Sleep and the Traveller* 1974; *Travesty* 1976; *The Universal Fears* (s) 1978; *The Passion Artist* 1979; *Virginie, Her Two Lives* 1982; *Humors of Blood and Skin: a John Hawkes Reader* 1984; *Adventures in the Alaskan Skin Trade* 1985; *Innocence in Extremis* 1985; *Whistlejacket* 1988; *Hawkes Scrapbook* 1991

Plays

The Innocent Party: Four Short Plays [pub. 1966]: *The Questions* 1966; *The Wax Museum* 1966; *The Undertaker* 1967; *The Innocent Party* 1968

Edited

The Personal Voice: a Contemporary Prose Reader [with others] 1964; *The American Literature Anthology 1* 1968

Poetry

Fiasco Hall 1943

Shirley HAZZARD 1931–

Hazzard was born in Sydney, Australia, where her father was a government official, and educated at Queenswood School there. In 1946 she went with her family to Hong Kong where her father was on a diplomatic mission, and worked for two years in the office of the British Combined Services Intelligence. She followed her father to New Zealand in 1948 and worked for the United Kingdom Commissioner's Office in Wellington, and again in 1951 to New York where she joined the staff of the United Nations Secretariat. 'As many people did in the early years of the UN, I joined in a spirit of idealism and from a desire to do something useful,' she has said. 'I don't think I did one useful thing in my entire time there.' The disillusionment wrought by a decade with the UN is described in *Defeat of an Ideal*, a scrupulously researched polemic against the bureaucracy and American control of the organisation.

Hazzard had written copiously during her youth, and returned to writing in the late 1950s. She published a story, 'Woollahra Road', in the *New Yorker* in 1961, and left the UN the next year. In 1963 she married Francis Steegmuller, the writer and translator, and in the same year published her first collection of stories, *Cliffs of Fall.* With a close attention to the nuances of language, the stories anatomise the tensions and misunderstandings of relationships. Since 1956 she has been a frequent visitor to Italy, and these stories and her first novel, *The Evening of the Holiday* (1966), are set there. The novel recounts a short and passionate affair between a separated Italian architect and a single English woman. Eight satirical sketches of a vast UN-like bureaucracy which originally appeared in the *New Yorker* were reworked as *People in Glass Houses*, while her second novel, *The Bay of Noon*, combines the romantic Italian setting with a scathing portrait of a NATO establishment in Naples. Hazzard's undoubted masterpiece, however, is her third novel, *The Transit of Venus*, winner of a US National Book Critics' Award in 1980. The subject of deracination, which recurs in her work, is in the forefront of this complex narrative spanning the lives of two Australian-born sisters, Grace and Caro Bell, who move to England in the 1920s. She had more recently returned to the subject of the UN in *Countenance of Truth*, a critical account of the Waldheim affair.

Fiction

Cliffs of Fall (s) 1963; *The Evening of the Holiday* 1966; *People in Glass Houses* (s) 1967; *The Bay of Noon* 1970; *The Transit of Venus* 1980

Non-Fiction

Defeat of an Ideal 1973; *Coming of Age in Australia* 1985; *Countenance of Truth: the United Nations and the Waldheim Case* 1990

Seamus (Justin) HEANEY 1939-

Heaney was born near Castledawson, County Derry, and grew up on his father's farm. His early life has a great deal in common with that of the playwright **Brian Friel**, his senior in Derry by ten years; the two men attended St Columb's College, Derry (Heaney between 1951 and 1957), and were later to train as teachers at the same teacher training college (by Heaney's time called St Joseph's College of Education), but, in Heaney's case, after taking a first in English at Queen's University, Belfast. Heaney's period as a secondary school teacher was briefer – only one year – and, after returning to St Joseph's to lecture for three years, he became a lecturer at Queen's in 1966. The previous year he had married; the couple subsequently had three children.

Also in 1965 came his first poetic publication, a pamphlet of *Eleven Poems*, but the following year his first collection, *Death of a Naturalist*, containing powerful poems relating to his farming background, marked him out as a poet of remarkable promise. He received the Eric Gregory Award in 1966, the first of what were to be many honours, and his second volume, *Door into the Dark*, consolidated his reputation. *Wintering Out* contained his first response, perhaps a rather hesitant one, to the Northern Irish Troubles, but with *North* he was firmly established as the major and representative voice among a remarkable generation of Irish poets and other writers. This reputation has been fully maintained, and he is widely considered the finest Irish poet since **W.B. Yeats** and, with **Ted Hughes**, a writer with whom he has much in common, among the leading poets in English. He is the possessor of what one critic has called 'a trustworthy, declarative voice', and his thoughtful but always concerned response to his country's history and recent political difficulties is undoubtedly praiseworthy. Only a minority consider him overrated, detecting a falling-off in quality from the 1970s, although others believe he has moved from clotted and 'difficult' work towards greater (and welcome) clarity.

His non-poetic career has kept pace with his great fame. In 1972, he gave up his job at Queen's and moved from Northern Ireland to County Wicklow, where for three years he was a freelance writer; he then taught at Carysfort College of Education until 1981. He had begun to spend frequent periods as a guest professor at American universities, and in 1982 he became a visiting professor at Harvard. Since 1985 he has been there as Boylston Professor of Rhetoric and Oratory. He also held a successful tenure of the Oxford Professorship of Poetry between 1989 and 1994.

His works are overwhelmingly volumes of poetry, but he has also written a play, *The Cure at Troy* (a version of the *Philoctetes* of Sophocles), which was produced by the influential Field Day Theatre Company in 1990, and published two volumes of critical prose. For some years he has divided his life between Dublin and Harvard.

Poetry

Eleven Poems 1965; *Death of a Naturalist* 1966; *The Island People* 1968; *Door into the Dark* 1969;

The Last Mummer 1969; *A Lough Neagh Sequence* 1969; *A Boy Driving His Father to Confession* 1970; *Catherine's Poem* 1970; *Night Drive* 1970; *Chaplet* 1971; *Land* 1971; *Servant Boy* 1971; *January God* 1972; *Wintering Out* 1972; *Bog Poems* 1975; *North* 1975; *Stations* 1975; *Four Poems* 1976; *Glanmore Sonnets* 1977; *After Summer* 1978; *Christmas Eve* 1978; *A Family Album* 1978; *Field Work* 1979; *Gravities* 1979; *Ugolimo* 1979; *Changes* 1980; *Selected Poems 1965–1975* 1980; *Toome* 1980; *Holly* 1981; *Sweeney Praises the Trees* 1981; *The Names of the Hare* 1982; *Remembering Malibu* 1982; *Sweeney and the Saint* 1982; *Sweeney Astray* 1982; *Verses for a Fordham Commencement* 1982; *A Hazel Stick for Catherine Ann* 1983; *Hailstones* 1984; *Station Island* 1984; *From the Republic of Conscience* 1985; *Clearances* 1986; *The Haw Lantern* 1987; *An Upstairs Outlook* [with M. Longley] 1989; *The Earth House* 1990; *New Selected Poems 1966–1987* 1990; *The Tree Clock* 1990; *Seeing Things* 1991; *Sweeney's Flight* 1992; *Joy or Night* 1993

Non-Fiction

The Fire i' The Flint: Reflections on the Poetry of Gerard Manley Hopkins 1975; *Two Decades of Irish Writing* (c) [ed. D. Dunn] 1975; *Richard Murphy, Poet of Two Traditions* (c) [ed. M. Harman] 1978; *Robert Lowell: a Memorial Address* 1978; *The Making of a Music* 1980; *Preoccupations: Selected Prose 1966–1987* 1980; *Chekhov on Sakhalin* 1982; *Contemporary Irish Art* (c) [ed. R. Knowles] 1982; *James Joyce and Modern Literature* (c) [ed. W.J. McCormack, A. Stead] 1982; *Among Schoolchildren* 1983; *An Open Letter* 1983; *Place and Displacement: Recent Poetry of Northern Ireland* 1985; *Towards a Collaboration* 1986; *The Government of the Tongue* 1988; *The Redress of Poetry* 1990

Edited

1980 Anthology: Arvon Foundation Poetry Competition [with T. Hughes] 1982; *The Rattle-Bag* [with T. Hughes] 1982; *The May Anthology of Oxford and Cambridge Poetry* 1993

Translations

The Midnight Verdict [from B. Merriman and Ovid] 1993

Plays

The Cure at Troy [from Sophocles' play *Philoctetes*] 1991

Roy (Aubrey Kelvin) HEATH 1926–

Born in Guyana, then British Guiana, the only English-speaking country of South America, Heath is the son of a headmaster who died before he was two. He was brought up by his mother, a teacher, and their extended family in the village of Agricola and in the middle-class black society of the capital, Georgetown. He has given a vivid picture of his early milieu in his autobiographical volume, *Shadows Round the Moon*. He was educated at the Central High School in Georgetown, one of the few schools from which candidates were accepted for the civil service. He began work at sixteen, initially as a commissary clerk, issuing petrol coupons to vehicle owners; then, from 1944 to 1951, he was a clerk in the Treasury. In the later year he migrated to Britain, and was a clerical worker in London until 1958, but also found time to gain his BA in French from London University in 1956, and later to take up legal studies, being called to the Bar at Lincoln's Inn in 1963, but not practising. From 1959 to 1968 he was a primary school teacher for the Inner London Education Authority, and since 1968 he has been a teacher of French and German for Barnet Borough Council, London.

His first novel, *A Man Come Home*, was published in 1974, and he is best known for *The Georgetown Trilogy* about the Armstrong family. Heath has said: 'I see myself as a chronicler of Guyanese life in this century', and his novels, although predominantly set in Georgetown, also range broadly over the rural hinterlands and over all classes and races in a complex society. He has been compared to **V.S. Naipaul** in the way he uses traditional life and myths to illuminate essentially modern concerns, both social and political. Among the novels, his second, *The Murderer*, is perhaps especially notable as a chilling study of a psychotic temperament. Heath is married and has three children.

Fiction

A Man Come Home 1974; *The Murderer* 1978; *The Georgetown Trilogy: From the Heat of the Day* 1979, *One Generation* 1981, *Genetha* 1981; *Kwaku* 1982; *Orealla* 1984; *The Shadow Bride* 1988

Non-Fiction

Ministry of Education, Guyana 1984; *Shadows Round the Moon* (a) 1990

Plays

Inez Combray 1972 [np]

John (Francis Alexander) HEATH-STUBBS 1918–

Heath-Stubbs comes from a family descended from West country gentry; his mother was a professional pianist. Born in London, grew up near Bournemouth, attending local schools and then, because he had defective eyesight, a progressive school, Bembridge, and Worcester College for the Blind. He went up to Queen's College, Oxford, in 1939, where his early verse appeared in *Eight Oxford Poets* (1941). He became established there as a brilliant undergraduate poet, leader of poets, and classic university figure, wandering around with his stick and a too-small blue trilby on his abundant hair. His first volume, *Wounded Thammuz* (1942), was issued by a London publisher in the year he took first class honours.

After Oxford, however, he found it difficult to get work, but he eventually became an English master at the Hall School in Hampstead (1944–5) and then an editorial assistant on *Hutchinson's Twentieth Century Encyclopaedia* (1945–6), for which he wrote almost all the articles on zoology, music, cookery and theology. At the same time, he became an enthusiastic member of Fitzrovia and, although his eventually total blindness has made him in many ways a solitary figure, he has also been among the most clubbable of English poets.

During the late 1940s and early 1950s he lived on what he could make from freelance writing and a small legacy, before embarking upon an academic life. He was Eric Gregory Fellow in Poetry at Leeds University (1952–3), and has held posts at the University of Alexandria, the University of Michigan at Ann Arbor and the College of St John and St Mark in Chelsea, London. Subsequently, he has done some part-time tutorial work. Volumes of verse have appeared from him at regular intervals and his *Collected Poems* were published to great acclaim in 1988, soon after which he was awarded the OBE. He has not, however, always been a fashionable poet, being traditional in manner, and deriving much of his material from mythology and the classics. Despite his consistently literary inspiration, he is often simple, lyrical and touching. Among his most impressive poems is the ironic epic *Artorius*, based on the King Arthur legend, while the exquisite lyrics in *A Parliament of Birds* represent the other extreme of his talent. He has also published criticism, numerous translations and verse drama and an evocative autobiography.

Poetry

Wounded Thammuz 1942; *Beauty and the Beast* 1943; *The Divided Ways* 1946; *The Charity of the Stars* 1949; *The Swarming of the Bees* 1950; *A Charm against the Toothache* 1954; *The Triumph of the Muse and Other Poems* 1958; *The Blue-Fly in His Head* 1962; *Selected Poems* 1965; *Satires and Epigrams* 1968; *Artorius, Book I* 1970; *Four Poems in Measure* 1973; *The Twelve Labours of Hercules* 1974; *A Parliament of Birds* 1975; *The Watchman's Flute: New Poems* 1978; *The Mouse, the Bird, and the Sausage* 1978; *Birds Reconvened* 1980; *Buzz Buzz* 1981; *This Is Your Poem* 1981; *Naming the Beasts* 1982; *The Immolation of Aleph* 1985; *Cats' Parnassus* 1987; *Collected Poems* 1988; *A Partridge in a Pear Tree* 1988

Non-Fiction

The Darkling Plain: a Study of the Later Fortunes of Romanticism in English Poetry from George Darkley to W.B. Yeats 1950; *Charles Williams* 1955; *The Ode* 1969; *The Pastoral* 1969; *The Verse Satire* 1969; *Hindsights* (a) 1993

Edited

Selected Poems of Shelley 1947; *Selected Poems of Swift* 1947; *Selected Poems of Tennyson* 1947; *The Forsaken Garden: an Anthology of Poetry 1824–1909* [with D. Wright] 1950; *Mountains Beneath the Horizon: Selected Poems by William Bell* 1950; *The Faber Book of Twentieth Century Verse* [with D. Wright] 1953 [rev. 1965, 1975]; *Images of Tomorrow: an Anthology of Recent Poetry* 1953; *Selected Poems of Alexander Pope* 1964; *Homage to George Barker on His Sixtieth Birthday* [with M. Green] 1973; *Selected Poems by Thomas Gray* 1981; *Poems of Science* [with P. Salman] 1984

Translations

Poems from Giacomo Leopardi 1946; *Aphrodite's Garland* 1952; *Thirty Poems of Hafiz of Shiraz* [with P. Avery] 1952; *Selected Poetry and Prose by Giacomo Leopardi* [with I. Origo] 1966; Alfred de Vigny *The Horn* 1969; *Dust and Carnations: Traditional Funeral Chants and Wedding Songs from Egypt* [with S. Megully] 1977; *Anyte* [with C. Whitside] 1979; *The Rubáiyát of Omar Khayyám* [with P. Avery] 1979

Plays

The Talking Ass 1953 [pub. 1958]; *The Harrowing of Hell* 1958; *Helen in Egypt* 1958;

Anthony (Evan) HECHT 1923–

Hecht was born in New York City where his father was a businessman. He was apparently deemed so talentless while at school that his anxious Jewish parents sent him for tests at the Pratt Institute which proved that he lacked any 'aptitudes whatsoever' – though this story should be taken with a pinch of salt, as Hecht is not above teasing interviewers who pry into his biography. He was educated at Bard College, Annandale, where he 'fell in love with poetry'. When he announced his intention of becoming a poet, his distressed parents consulted their friend Theodore Geisel, who is better known as the children's author Dr Seuss.

Having graduated in 1944, Hecht served with the US Army in Japan and Europe where he witnessed the sight of the liberated concentration camps, an experience evident in recurrent images of the Holocaust in his poetry. Under the GI Bill, he went to study at Kenyon College, Ohio, with the poet **John Crowe Ransom**, whose other poet protégés have included **Robert Lowell** and **Allen Tate**. He taught at Kenyon from 1947, and has since held a succession of academic posts at the University of Iowa, New York University, Smith College, Northampton, Bard College, Annandale, Rochester University, New York, and lastly at Georgetown University, Washington, where he became a professor in 1985.

On the strength of early poems published in magazines such as the *Kenyon Review* and the *New Yorker*, Hecht won the Prix de Rome awarded by the American Academy of Arts and Letters, and thereby travelled to Rome where he translated some of Rainer Maria Rilke's verse; these translations were later set to music by Lukas Foss for the cantata *A Parable of Death*. His first collection, *A Summoning of Stones* (1954), was widely praised for its technical virtuosity, though many reviewers noted a shortage of worthy subject-matter. *The Hard Hours*, which won a Pulitzer Prize in 1968, contains poems such as 'More Light! More Light!', a moving study of human suffering and betrayal. Without losing sight of his often tragic vision, Hecht discovered a gentler, sometimes comic, tone in the two widely praised collections *Millions of Strange Shadows* and *The Venetian Vespers*. *The Seven Deadly Sins* and *Struwwelpeter* are illustrated 'cautionary tales' rooted in children's literature.

Hecht is also credited with having invented a new light verse form called the 'double dactyl' which became something of a craze on American campuses. With **John Hollander** he has edited an anthology of these, *Jiggery-Pokery*. From 1982 to 1984 he was Consultant in Poetry to the Library of Congress, and he has won various awards, including Guggenheim Fellowships and the Bollingen Prize in 1983. He has been married twice, and has three children.

Poetry

A Summoning of Stones 1954; *The Seven Deadly Sins* 1958; *Struwwelpeter* 1958; *A Bestiary* 1962; *Aesopic* 1967; *The Hard Hours* 1967; *Millions of Strange Shadows* 1977; *The Venetian Vespers* 1979; *A Love for Four Voices* 1983; *Collected Earlier Poems* 1990; *The Transparent Men* 1990

Non-Fiction

Robert Lowell 1983; *The Pathetic Fallacy* 1985; *Obbligati: Essays in Criticism* 1986

Edited

Jiggery-Pokery [with J. Hollander] 1967; Jonathan Aaron *Second Sight: Poems* 1982; Susan Donnelly *Eve Names the Animals* 1985; *The Essential Herbert* 1987

Translations

Aeschylus *Seven against Thebes* [with H. Bacon] 1973; Voltaire *Poem upon the Lisbon Disaster* 1977

Ben HECHT 1893–1964

Of the various careers open to him, Hecht chose the least sensational when he became a writer. The son of Russian immigrants, he was born in New York, but moved to Racine in Wisconsin. Offered a glittering future after giving a violin concert at ten, Hecht preferred circus life, and toured as an acrobat during summer vacations. At sixteen he ran away from home to Chicago, where he managed a theatre, and then, in 1910, was given jobs first on the Chicago *Journal*, then, four years later, on the Chicago *News*. He also contributed to literary magazines including the *Little Review*, interrupting this for a year to become Berlin correspondent to the Chicago *News* in 1918.

A daily column, *1001 Afternoons in Chicago* (later collected in a book), brought him fame, and he left the Chicago *News* in 1923 to found his own newspaper, the Chicago *Literary Times*, which he edited for two years. Hecht has been described as a swashbuckler, 'tough, short and stubby with sharp blue eyes and stringy black hair', who somehow managed to find time to study botany, sail a small boat and play baseball. His unlimited energy typifies his fiction, which is often carelessly written but driven by verve and dash.

His first novel, *Erik Dorn* (based upon his time in Berlin), was received with acclaim in 1921, and suggested a promising future as a novelist, which he did not fulfil. He attempted everything but poetry, but is mainly remembered for his collaboration with Charles MacArthur in plays and screenplays, notably *The Front Page*, set in a newsroom, a milieu he knew inside-out.

He married twice, first in 1915 and again, after a divorce, in 1925. Jenny, a daughter by his first marriage, achieved prodigious success as an actress from the age of eight. A passionate believer in an independent Jewish state, Hecht advocated violent and swift action to attain this, views he explained in *A Guide for the Bedevilled*, and which gained him many enemies. His first autobiography, *A Child of the Century*, provoked the comment from **Saul Bellow** that: 'Among the pussy cats who write of social issues ... he roars like an old-fashioned lion.'

Plays

The Wonder Hat [with K.S. Goodman] 1916 [pub. 1920]; *The Hero of Santa Maria* [with K.S. Goodman] 1917 [pub. 1920]; *The Master Poisoner* [with M. Bodenheim] 1918; *The Egoist* 1920 [pub. 1925]; *The Hand of Siva* [with K.S. Goodman] 1920; *The Stork* [from L. Foder] 1925; *The Wonder Hat and Other One-Act Plays* [with K.S. Goodman; pub. 1925]; *Underworld* [with others] **(f)** 1927; *Man Eating Tiger* 1927 [np]; *Christmas Eve* 1939 [pub. 1928]; *The Big Noise* **(f)** [with T. Geraghty, G. Marion Jr] 1928; [np]; *The Front Page* [with C. MacArthur] 1928; *The Great Gabbo* **(f)** [with H. Herbert] 1929; *River Inn* **(f)** [with G. Fort] 1930; *The Unholy Garden* **(f)** [with C. MacArthur] 1931; *The Great Magoo* [with G. Fowler] 1932 [pub. 1933]; *Scarface, the Shame of the Nation* **(f)** [with others] 1932; *Design for Living* **(f)** [from N. Coward's play] 1933; *Hallelujah, I'm a Bum* **(f)** [with S.N. Behrman] 1933; *Turn Back the Clock* **(f)** [with E. Selwyn] 1933; *Crime without Passion* **(f)** [with C. MacArthur] 1934; *Twentieth Century* **(f)** [with C. MacArthur] 1934; *Upperworld* **(f)** [with others] 1934; *Viva Villa!* **(f)** [with H. Hawks] 1934; *Barbary Coast* **(f)** [with C. MacArthur] 1935; *Jumbo* [with L. Hart, C. MacArthur, R. Rodgers] 1935; *Once in a Blue Moon* **(f)** [with C. MacArthur] 1935; *The Scoundrel* **(f)** [with C. MacArthur] 1935; *Soak the Rich* **(f)** [with C. MacArthur] 1936; *Nothing Sacred* **(f)** 1937 [pub. as *Hazell Flagg* 1953]; *To Quito and Back* 1937; *It's A Wonderful World* **(f)** [with H. Mankiewitz] 1939; *Ladies and Gentlemen* [from L. Bus-Fekete; with C. MacArthur] 1939 [pub. 1941]; *Lady of The Tropics* **(f)** 1939; *Let Freedom Ring* **(f)** 1939; *Wuthering Heights* **(f)** [with C. MacArthur] 1939 [pub. 1943]; *Angels Over Broadway* **(f)** 1940; *Fun to Be Free* [with C. MacArthur] [pub. 1941]; *The Black Swan* **(f)** [with S.I. Miller] 1942; *China Girl* **(f)** [with M. Crossman] 1942; *Lily of the Valley* 1942 [np]; *Tales of Manhattan* **(f)** [with others] 1942; *A Tribute to Gallantry* 1943; *The Common Man* 1944 [np]; *Spellbound* **(f)** [with A. MacPhail] 1945 [pub. 1946]; *Watchtower Over Tomorrow* **(f)** 1945; *A Flag Is Born* [with K. Weill] 1946; *Notorious* **(f)** 1946; *Spectre of the Rose* **(f)** 1946; *Swan Song from H. Hinsdale and R. Romero;* with C. MacArthur] 1946 [pub. 1974]; *Her Husband's Affairs* **(f)** [with C. Lederer] 1947; *Ride the Pink Horse* **(f)** [with C. Lederer] 1947; *The Miracle of the Bells* **(f)** [with Q. Reynolds] 1948; *Whirlpool* **(f)** [with A. Solt] 1950; *Actors and Sin* **(f)** 1951; *The Iron Petticoat* **(f)** 1956; *Miracle in the Rain* **(f)** 1956; *A Farewell to Arms* **(f)** 1957; *Legend of the Lost* **(f)** [with R. Presnell Jr] 1957; *Queen of Outer Space* **(f)** [with C. Beaumont] 1958; *Winkelberg* 1958; *Simon* [from Brecht and L. Feuchtwanger] 1962; *The Magnificent Showman* **(f)** 1964

Fiction

Erik Dorn 1921; *Fantazius Mallare* 1922; *Gargoyles* 1922; *A Thousand and One Afternoons in Chicago* **(s)** 1922; *The Florentine Dagger* 1923; *Broken Necks and Other Stories* **(s)** 1924; *Cutie, a Warm Mamma* [with M. Bodenheim] 1924; *Humpty Dumpty* 1924; *The Kingdom of Evil* 1924; *Tales of Chicago Streets* **(s)** 1924; *Broken Necks, Containing More 1001 Afternoons* **(s)** 1926; *Count Bruga* 1926; *Infatuation and Other Stories of Love's Misfits* **(s)** 1927; *Jazz and Other Stories of Young Love* **(s)** 1927; *The Policewoman's Love-Hungry Daughter and Other Stories of Chicago Life* **(s)** 1927; *The Sinister Sex and Other Stories of Marriage* **(s)** 1927; *The Unlovely Sin and Other Stories of Desire's Pawns* **(s)** 1927; *The Champion Far Away* **(s)** 1931; *A Jew in Love* 1931; *A Book of Miracles* **(s)** 1939; *1001 Afternoons in New York* **(s)** 1941; *Miracle in the Rain* 1943; *I Hate Actors!* 1944 [as *Hollywood Mystery* 1946]; *The Collected Stories* **(s)** 1945; *Concerning a Woman of Sin and Other Stories* **(s)** 1947; *The Sensualists* 1959; *In the Midst of Death* 1964

Non-Fiction

A Guide for the Bedevilled 1944; *A Child of the Century* **(a)** 1954; *Charlie: the Improbable Life and Times of Charles MacArthur* 1957; *Perfidy* 1961; *Gaily, Gaily* **(a)** 1963; *Letters from Bohemia* 1964

For Children

The Cat That Jumped Out of the Story 1947

Collections

A Treasury of Hecht 1959

Biography

The Five Lives of Ben Hecht by Doug Fetherling 1977; *Ben Hecht* by William MacAdams 1990

Joseph HELLER 1923–

Heller was born in Brooklyn, New York, the son of poor Jewish parents. His Russian-born father was a bakery truck driver who died when Heller was four. Having graduated from Abraham Lincoln High School in 1941, he attended officer cadet school before serving with the US Air Force in Corsica. He flew sixty missions as a B-25 bombardier, the experience which was to furnish him with the material for his phenomenally successful first novel, *Catch-22*. Under the terms of the GI Bill he studied English at the University of Southern California, New York University and Columbia University, and in

1949–50 spent a year at Oxford University as a Fulbright scholar. He taught for two years at Pennsylvania State University before returning to New York where he worked as a magazine advertising writer on *Time* (1952–6) and *Look* (1956–8) and promotion manager at *McCall's* magazine (1958–61).

Heller began writing after the war and published short stories in *Atlantic Monthly*, *Esquire* and other magazines. The product of seven years' work, his burlesque satire *Catch-22* was published in 1961 but did not receive much attention until 1962 when its English publication brought lavish critical praise from reviewers such as **Philip Toynbee**. In subsequent interviews Heller has denied that the novel reflects his own experience of the war: 'I was so brainwashed by Hollywood's image of heroism that I was disappointed when nobody shot back at us.' A film version directed by Mike Nichols (1970) consolidated its success. In the 1960s Heller was involved with the anti-Vietnam war protest movement, and made a brief foray into the theatre with *We Bombed in New Haven*. His second novel, *Something Happened*, concerns the non-eventful, angst-ridden existence of a corporate executive, while *Good as Gold* satirises the workings of the American government (reserving a special bile for Henry Kissinger). In 1981 Heller went down with Guillain-Barré syndrome, a paralysing disease of the nervous system. *No Laughing Matter* is a surprisingly cheerful account of his illness (it was co-written with Speed Vogel, the friend who took care of his affairs during this period), and describes his recuperation and visits from friends such as Mario Puzo, Dustin Hoffman and the manically hypochondriac Mel Brooks. Heller has two children by his first marriage (to Shirley Held in 1945), and his divorce is recounted in *No Laughing Matter*, his wife's lawyer at one point describing *Good as Gold* as 'the *Mein Kampf* of matrimonial warfare'. In 1989 he married Valerie Humphries, a nurse he met while ill.

Fiction

Catch-22 1961; *Something Happened* 1974; *Good as Gold* 1979; *God Knows* 1984; *Picture This* 1988; *Closing Time* 1994

Plays

We Bombed in New Haven 1967 [pub. 1968]; *Catch-22* [from his novel; pub. 1973]; *Clevinger's Trial* [from his novel *Catch-22*] 1974 [pub. 1973]; *Casino Royale* (f) [from I. Fleming's novel] 1967; *Dirty Dingus Magee* [from D. Markson's novel; with F. and T. Waldman] 1970

Non-Fiction

No Laughing Matter (a) [with S. Vogel] 1986

Lillian (Florence) HELLMAN 1905–1984

Hellman came of a Jewish family which had emigrated in the 1840s from Germany to New Orleans, where she herself was born. Her mother came from a plutocratic Alabama background, but unwise speculation destroyed most of this money, and her father was forced to become a travelling salesman based in New York, where she largely grew up, with frequent visits to the South. She attended university for two years in New York but, deciding not to complete the course, quickly found work as a reader for the publisher Horace Liveright. She gave this up to marry the minor writer Arthur Kober and, after a brief sojourn in Paris, the couple went to Hollywood in 1930, she to read manuscripts, he to turn them into scenarios.

Here her life changed fatefully when one evening she met the crime novelist **Dashiell Hammett**, who was 'getting over a five-day drunk', and quickly returned with him to New York. Divorce from her husband followed in 1932, and for the rest of Hammett's life he lived with Hellman in a 'marriage' which lacked only the legal sanction. Under Hammett's tutelage, Hellman quickly became one of the most well-known playwrights of the 1930s, enjoying a string of successes from *The Children's Hour* (1934), with its daring approach to the subject of lesbianism, to *The Little Foxes*, a compelling study of Southern decadence, and *Watch on the Rhine*, the first American anti-Nazi play. Hellman's left-wing politics were apparent in these works, which were usually made into successful films with the didacticism left out (particularly well-known was *The Loudest Whisper*, 1961, one of two films made from *The Children's Hour*).

During these years, Hellman also travelled widely as a left-wing emissary, visiting Republican Spain during the civil war and meeting Stalin in 1944; although she was always to deny it, there is evidence that she was a member of the Communist Party from 1938 to 1940. In 1952, she was called to testify before the House of Representatives Committee on Un-American Activities, but refused to appear; once again it seems that her own, heroic, account of these events was not entirely accurate.

From the later 1950s, she turned from original plays to adaptations, and was also to edit Hammett's stories and Chekhov's letters, but her greatest glory came with the three principal volumes of her autobiography, *An Unfinished Woman*, which ended with a sentence consisting of one word, 'However'; *Pentimento*, her memoirs of other people, including her supposedly close friend, the anti-Fascist fighter 'Julia'; and *Scoundrel Time*, her account of her ordeal in the McCarthy years. They were published in one volume, *Three*, in 1979, but in the same year their reputation began to disintegrate, when **Mary McCarthy** accused Hellman on television of lying, and the latter sued. In the course of the investigations, embarrassing inaccuracies emerged, and matters were made worse when the original of 'Julia', Muriel Gardiner (also the 'Margaret' of **Stephen Spender**'s *World within World*, 1951), published a book in 1983 making it clear that she had never known Hellman. Now only the plays, sharp and well-made studies of auto-destruction, remain.

Plays
The Children's Hour 1934; *Days to Come* 1936; *The Little Foxes* 1939; *Watch on the Rhine* 1941; *The Searching Wind* 1944; *Another Part of the Forest* 1946 [pub. 1947]; *The Autumn Garden* 1951; *Candide* [from Voltaire's story; with L. Bernstein and others] 1957; *Toys in the Attic* 1960; *My Mother, My Father and Me* [from B. Blechman's novel *How Much*] 1963

Translations
Emmanuel Roblès *Montserrat* 1950; Jean Anouilh *The Lark* 1956; Voltaire *Candide* 1957

Non-Fiction
Three (a) 1979: *An Unfinished Woman* 1969, *Pentimento* 1973, *Scoundrel Time* 1976

Edited
The Selected Letters of Anton Chekhov 1955; *The Dashiell Hammett Story Omnibus* (s) 1966

Ernest (Miller) HEMINGWAY 1899–1961

Born at Oak Park, a middle-class suburb of Chicago, Hemingway was the son of a doctor who was also a keen sportsman. Hemingway Sr introduced his son to the masculine pleasures of hunting, shooting and fishing which, with writing, were to dominate his life. The lakeside hunting lodge in Michigan, near Indian settlements, where father and son went together was to remain one of the most important imaginative landscapes of his life, figuring most directly in the autobiographical stories where Hemingway appears as the youthful Nick Adams.

After joining the Kansas City *Star* as a cub reporter in 1917, Hemingway volunteered as an ambulance driver on the Italian front and was badly wounded. While in war hospital, he conceived the passion for an American nurse, Agnes von Kurowsky, which is celebrated in one of his best novels, *A Farewell to Arms*. It was perhaps to remain the major emotional experience of his life, despite four subsequent marriages. In the 1920s he took up the profession of American foreign correspondent in Europe and, while in Paris, began to attract attention as a writer: he came into contact with **Gertrude Stein**, **Erza Pound** and **James Joyce**. The novel *The Sun Also Rises* (1926, UK title *Fiesta*) established an international reputation for him, and from then on his life became that of an internationally celebrated writer. He was devoted to bull-fighting, boxing, big-game hunting and deep-sea fishing, and his bullying, drunken, mock-heroic style won him great notoriety and many enemies, beginning the process whereby response to his brilliant, laconic writing became confused in many minds by distaste for his personality and image.

He contrived to become involved in the Spanish Civil War (where he met his third wife, **Martha Gellhorn**) and the Second World War; in the latter, although officially a war correspondent, he took part in the liberation of Paris at the head of a small irregular force. Despite his winning the Nobel Prize in 1954, his last years, lived largely in Cuba, were marked by literary and personal decline, and his eventual suicide by shooting was widely seen as a response to contradictions which had become too much for him to bear. Hemingway was, despite the flawed nature of his achievement, one of the major English-language writers of the century, and his simple yet immensely evocative prose style has exerted an influence second to none on the development of modern writing.

Fiction
In Our Time (s) 1925; *The Torrents of Spring* 1926; *The Sun Also Rises* 1926 [UK *Fiesta* 1927]; *Men without Women* (s) 1927; *A Farewell to Arms* 1929; *Death in the Afternoon* 1932; *Winner Takes Nothing* (s) 1933; *Green Hills of Africa* 1935; *To Have and to Have Not* 1937; *For Whom the Bell Tolls* 1941; *Across the River and into the Trees* 1950; *The Old Man and the Sea* 1952; *The Short Happy Life of Francis Macomber and Other Stories* (s) 1963; *The Snows of Kilimanjaro and Other Stories* (s) 1963; *Islands in the Stream* 1970; *The Nick Adams Stories* (s) 1972; *Collected Stories* [ed. J. Fenton] 1995

Poems
88 Poems [ed. N. Gerogiannis] 1979

Plays
The Spanish Earth (f) 1938

Non-Fiction
The Spanish War 1938; *The Wild Years: Articles from the Toronto Star* (ed. G.Z. Hanrahan) 1962; *A Moveable Feast* (a) 1964; *By-Line: Selected Articles and Dispatches of Four Decades* [ed. W. White] 1967; *Selected Letters, 1917–1961* [ed. C.

Baker] 1981; *Ernest Hemingway on Writing* [ed. W. Phillips] 1984

Edited

Men at War: the Best War Stories of All Time (s) 1942

Collections

Three Stories and Ten Poems 1923; *The Fifth Column (a Play) and the First Forty-Nine Stories* 1938; *Hemingway Selections* [ed. M. Cowley] 1944

Biography

Hemingway by Kenneth S. Lynn 1988

Adrian (Maurice) HENRI 1932–

Henri was born in Birkenhead, the son of a clerk in what was later to become the Ministry of Defence; the unusual surname derives from Henri's paternal grandfather, a French colonial sailor from Mauritius, who settled down with a Birkenhead woman. In 1938 the family moved to Rhyl in North Wales, where Henri's childhood was mainly passed. He attended the St Asaph Grammar School, and from 1951 to 1955 King's College, Durham, where he studied fine art. His earliest jobs were as a teacher in various schools in the north-west, and he initially varied this with summers spent working at the Rhyl fairground (an activity he had begun while at school).

He also worked during the late 1950s as a scenic artist at the Liverpool Playhouse, and in 1958 had his first exhibition as a painter, an activity in which he matured almost a decade earlier than as a poet. From 1961 to 1968 he was a lecturer at Manchester then Liverpool colleges of art, and became fully part of the remarkable art-school-led Bohemia of Liverpool in the 1960s. A series of performances with other poets, such as **Roger McGough**, led finally to the publication of *Penguin Modern Poets 10: The Mersey Sound* (1967), the immensely popular anthology which established the names of three 'Liverpool poets': McGough, Henri and **Brian Patten**. Henri's poetic performances had usually been with music, and from 1968 to 1970 he led the poetry/rock group 'The Liverpool Scene'. In the latter year, however, the group broke up, and a series of other traumas affected his life: both his parents and his two surviving grandparents died within a period of five weeks, and he himself was diagnosed with serious heart trouble, which forced a major retrenchment of his activities. Since 1970, in fact, he has worked as a freelance writer, performer and painter, although he has had a couple of residencies – at the Tattenhall Centre, Cheshire, from 1980 to 1982, and Liverpool University in 1989 – as well as some regular activities, such as teaching writing courses at the Arvon Foundation since 1972. He had married in 1957 but was separated from his wife in the late 1960s and later divorced (she died in 1987). He has had a number of relationships with other women, who have inspired his verse; these include the poet **Carol Ann Duffy**, while his present partner is Catherine Marcangeli, a Frenchwoman who lives largely in Paris. As might be expected, he specialises in love poetry. In the 1960s, his works tended to approximate to pop lyrics and, as with McGough and Patten, there have always been those who have refused to take him seriously as a poet; this is perhaps to underestimate the increasing complexity of his work, which has included several long poems, including what is probably his masterpiece, *Autobiography*, an account of his life to the age of thirty. He has now published almost twenty volumes of verse, and has proved prolific in other fields. From the late 1980s, for instance, he became active as an author for children and as a dramatist; his best-known work in the latter field in his modern-language version of *The Wakefield Mysteries*. He has also written a novel in collaboration with **Nell Dunn**, *I Want* (1972).

Poetry

Penguin Modern Poets 10 [with R. McGough, B. Patten] 1967; *Twenty-One Poems* 1967 [rev. 1974, 1983]; *Tonight at Noon* 1968; *Autobiography* 1971; *The Best of Henri* 1975; *One Year* 1976; *Beauty and the Beast* [with C.A. Duffy] 1977; *City Hedges: Poems 1970–1976* 1977; *Words without a Story* 1978; *From the Loveless Matel: Poems 1970–1979* 1980; *Penny Arcade: Poems 1978–1982* 1983; *Holiday Snaps* 1985; *Collected Poems* 1986; *Box and Other Poems* 1990; *Wish You Were Here* 1990; *The Cerise Swimsuit* 1992; *Dinner with Spratts* 1993; *Not Fade Away, Poems 1989–1994* 1994; *One of Your Legs is Both the Same* [with others] 1994

Plays

I Wonder [with M. Kustow] 1968; *Yesterday's Girl* 1973; *The Husband, the Wife and the Stranger* 1986; *The Wakefield Mysteries* 1991

Fiction

I Want [with N. Dunn] 1972

Non-Fiction

Environments and Happenings 1974

For Children

Eric the Punk Cat [with R.W. Walker] 1982; *Eric and Frankie in Las Vegas* [with R.W. Walker] 1986; *The Phantom Lollipop Lady and Other Poems* [with T. Ross] 1986; *Eric and Frankie in Paris* [with R.W. Walker] 1987; *Rhinestone Rhino and Other Poems* [with T. Ross] 1989

O. HENRY (pseud. of William Sydney Porter) 1862–1910

Henry was born in North Carolina, the son of a doctor obsessed by the idea of 'perpetual motion'. His mother died when he was three. He left school at fifteen, was employed in a drug store, worked on a Texas ranch, lived in Houston for ten years where he had a number of jobs including that of bank clerk. He eloped with his first wife in 1887. In 1894 he bought a weekly magazine, which he ran briefly, and sold a few stories. The story of how he was wanted for embezzlement from the First National Bank, escaped to New Orleans, then Honduras, where he met up with one Al Jennings, lived it up in South America and Mexico on the proceeds of Jennings's robbery, only returning to stand trial when news reached him of his wife's illness, is colourful stuff. Sentenced to five years in the Ohio State Penitentiary, he did just over three, wrote a number of stories, including 'A Retrieved Reformation' which later became a successful play, *Alias Jimmy Valentine*, and was filmed. He is said to have acquired his pseudonym from a warder called Orrin Henry.

On leaving prison he wandered again before being invited to New York by Gilham Hall, an editor of *Ainslee's Magazine.* In the last six years of his life he wrote something like 600 stories and drank an amazing quantity of whisky. He married three years before he died from tuberculosis. He had the knack of writing a magazine story and was master of the final twist. Unlike others who led rackety lives and wrote about them, he never seemed to run out of material and the quality of his best work is such that it has survived when so much that was popular in its day has not. In 1918 the first O. Henry Awards were given for magazine stories, and these continue to be awarded each year.

Fiction

Cabbages and Kings (s) 1905; The Four Million (s) 1906; Waifs and Strays (s) 1906; Hearts of the West (s) 1907; The Trimmed Lamp and Other Stories (s) 1907; The Gentle Grafter (s) 1908; The Voice of the City (s) 1908; Options (s) 1909; Roads of Destiny (s) 1909; Strictly Business (s) 1910; Whirligigs (s) 1910; Sixes and Sevens (s) 1911; Rolling Stones (s) 1913

(John) Rayner HEPPENSTALL 1911–1981

The son of a draper and a domestic servant, Heppenstall grew up in the industrial Midlands where he was born. An unexpected visit to France, paid for by his otherwise impoverished parents when he was fifteen, made him a lifelong Francophile. He graduated in modern languages at Leeds University in 1933, but abandoned teaching after one year, to become a freelance writer on the strength of an unusually sympathetic study of Middleton Murry. In a ceremony witnessed by **Herbert Read**, he married Margaret Edwards in 1937; she alone was to receive his loyal devotion.

Described by **Evelyn Waugh** as a 'disappointed novelist', Heppenstall began his literary life as a poet, and published a fourth collection of verse, *Blind Men's Flowers Are Green*, while serving in the Royal Artillery during the Second World War. He remains best known, however, for his first novel, *The Blaze of Noon*, which caused a scandal when published in 1939 because of what was then thought of as sexual explicitness. It is the story of a blind man, and Heppenstall researched the book by walking round blindfolded. His later fiction, especially *The Woodshed*, has something in common with the work of Alain Robbe-Grillet, of whom he was an early champion. From 1946 to 1965, Heppenstall worked with distinction for the BBC, first as feature-writer and producer, then for the two years before his retirement as drama producer. His early socialism, which had brought him temporary friendship with **George Orwell** (one of the *Four Absentees* recalled in a lively memoir), deteriorated into quirky reactionary attitudes: for example, he accepted homosexuality because it might become 'rife among the "working" class and coloured immigrants', and would therefore limit their breeding. His obsessive interest in crime publicly resulted in excellent studies in criminology. In private, Heppenstall's plots at various times against neighbours (including **Alan Sillitoe**) revealed a lonely man, not however above sensibly and professionally fictionalising these foibles in his novels. His posthumously published novel, *The Pier*, appeared to be a wishful fantasy of revenge, as can be seen by comparing it with *The Master Eccentric*, a volume of journals covering the last twelve years of his life. He was obsessive, waspish and difficult, but oddly likeable. He considered taking his own life and kept a lethal potion ready for the purpose. In the event, a stroke after a short illness rendered this unnecessary.

Fiction

The Blaze of Noon 1939; *Saturnine* 1943 [rev. as *The Greater Infortune* 1969]; *The Lesser Infortune* 1953; *The Connecting Door* 1962; *The Woodshed* 1962; *The Shearers* 1969; *Two Moons* 1977; *The Pier* 1986

Non-Fiction
Middleton Murry 1934; *Apology for Dancing* 1936; *The Double Image Seekers* 1947; *Leon Bloy* 1954; *My Bit of Dylan Thomas* 1957; *Four Absentees* 1960; *The Intellectual Part* (a) 1963; *Raymond Roussel* 1966; *A Little Pattern of French Crime* 1969; *Portrait of the Artist as a Professional Man* (a) 1969; *French Crime in the Romantic Age* 1970; *Bluebeard and After* 1972; *The Sex War and Others* 1973; *Reflections on the Newgate Calendar* 1975; *Tales from the Newgate Calendar* 1981; *The Master Eccentric* 1986

Poetry
Patina 1932; *First Poems* 1935; *Sebastian* 1937; *Blind Men's Flowers Are Green* 1940; *Poems 1933–1945* 1946

Translations
Chateaubriand *Atala and René 1963;* R. Roussel *Impressions of Africa* 1969; Balzac *A Harlot High and Low* 1970; René Floriot *When Justice Falters* 1972

A(lan) P(atrick) HERBERT 1890–1971

The son of an Irish civil servant in the Indian Office and maternal grandson of a lord justice of appeal, Herbert was educated at Winchester and at New College, Oxford, where he took a first in jurisprudence. While at Winchester he published the first of his many volumes of light verse, and he became a freelance contributor to *Punch* in the year he went up to Oxford. He joined the Royal Naval Volunteer Reserve on the outbreak of war, saw active service as a sub-lieutenant at Gallipoli and on the Western Front, and was invalided home in 1917. While convalescing, he wrote his first novel, *The Secret Battle*, a stark novel about the Gallipoli Campaign and an unjust court martial.

He had married the artist Gwendolen Quilter, a cousin of the composer Roger Quilter, in 1914; they had four children (one of whom married the poet **John Pudney**) and lived for fifty-four years on the river at Hammersmith Terrace, west London. This stretch of the Thames was something of an artists' colony, centred on The Dove pub. **Naomi Mitchison** recalled 'neighbours with similar tastes and standards all along the river right up to Chiswick', and the Herberts became famous for their parties, where guests refreshed themselves by diving into the Thames.

Herbert was called to the Bar in 1918, but worked first as a private secretary to the man-of-affairs Sir Leslie Scott; he joined the staff of *Punch* in 1924, where he made the initials 'A.P.H.' equally well known as entertainer and public campaigner, functions he also performed through an endless succession of letters to *The Times*. His 'Misleading Cases', first published in *Punch*, were popular in book form, while another well-loved serial concerned the featherbrained Topsy; *The Water Gipsies* was a popular novel of canal life, and he wrote lyrics for revues and operettas such as *La Vie Parisienne* and *Tantivy Town*.

He entered Parliament as a junior burgess for Oxford in 1935 (he was to lose his seat when the university franchise was abolished in 1950), and within a year had got his Matrimonial Causes Act 1936, easing the divorce laws, on to the statute book. During the Second World War, he worked with the River Emergency Service, once personally engaging with a German plane, and defeated a purchase tax on books; after it, he wrote more words for musical plays and took up many new causes, including public lending right for authors and a Thames barrage. He was knighted in 1945 and appointed CH in 1970.

Fiction
The Secret Battle 1919; *The House by the River* 1920; *The Old Flame* 1925; *The Trials of Topsy* 1928; *Topsy, MP* 1929; *The Water Gipsies* 1930; *Holy Deadlock* 1934; *Topsy Turvy* 1947; *Number Nine* 1951; *Why Waterloo?* 1952; *Made for Man* 1958; *The Singing Swan* 1968

Poetry
Poor Poems and Rotten Rhymes 1910; *Play Hours with Pegasus* 1912; *Half Hours at Helles* 1916; *The Bomber Gipsy and Other Poems* 1918; *The Wherefore and the Why* 1921; *'Tinker Tailor'* 1922; *Laughing Ann and Other Poems* 1925; *She-Shanties* 1926; *Plain Jane* 1927; *Ballads for Broadbrows* 1930; *A Book of Ballads* 1931; *Siren Song* 1940; *Let Us Be Glum* 1941; *Bring Back the Bells* 1943; *A.T.I.: There Is No Cause for Alarm* 1944; *'Less Nonsense'* 1944; *Light the Lights* 1945; *'Full Enjoyment' and Other Verses* 1952; *Silver Stream* 1962

Plays
Double Demon 1923 [pub. in *Four One-Act Plays* 1924]; *The Blue Peter* [pub. 1925]; *King of the Castle* 1924 [np]; *At the Same Time* 1925 [np]; *The White Witch* 1926 [np]; *Riverside Nights* [with N. Playfair] 1926; *The Policeman's Serenade* 1926; *Fat King Melon and Princess Caraway* 1927; *Two Gentlemen of Soho* 1927 [pub. 1928]; *Plain Jane* 1927 [pub. 1929]; *La Vie Parisienne* (l) [from Offenbach's operetta] 1929; *Derby Day* (l) 1932 [pub. 1931]; *Tantivy Towers* (l) 1931; *Helen* (l) [from H. Meilhac and L. Halévy's operetta] 1932; *Mother of Pearl* [from A. Grunwald's play] 1933; *Perseverance* 1934; *Streamline* [with R. Jeans] 1934 [np]; *Home and Beauty* 1937 [np]; *Paganini* [from B. Jenbach and P. Knepler; with R. Arkell] 1937 [np]; *Big Ben* 1946; *Bless the Bride* 1947;

Tough at the Top 1949 [np]; *Come to the Ball* (l) [from J. Strauss; with R. Arkell] 1951; *The Water-Gipsies* [from his novel] 1951 [pub. 1957]; *Better Dead* 1962 [np]

Non-Fiction

Light Articles Only 1921; *The Man about Town* 1923; *Misleading Cases in the Common Law* 1927; *Honeybubble and Co.* 1928; *More Misleading Cases* 1930; *'No Boats on the River'* 1932; *A.P. Herbert* [ed. E.V. Knox] 1933; *Still More Misleading Cases* 1933; *Mr Pewter* 1934; *Uncommon Law* 1935; *What a Word!* 1935; *Mild and Bitter* 1936; *The Ayes Have It: the Story of the Marriage Bill* 1937; *Sip! Swallow!* 1937; *General Cargo* 1939; *Let There Be Liberty* 1940; *'Well, Anyhow'* 1942; *A Better Sky* 1944; *The War Story of Southend Pier* 1945; *The Point of Parliament* 1946; *Leave My Old Morale Alone* 1948; *Mr Gay's London* 1948; *Independent Member* 1950; *Cold's Last Case and Other Misleading Cases* 1952; *Pools Pilot* 1953; *The Right to Marry* 1954; *'No Fine on Fun': The Comical History of the Entertainments Duty* 1957; *Anything But Action: a Study of the Uses and Abuses of Committees of Inquiry* 1960; *Look Back and Laugh* 1960; *'Public Lending Rights': Authors, Publishers and Libraries* 1960; *Libraries: Free for All?* 1962; *Bardot, MP? and Other Modern Misleading Cases* 1964; *The Thames* 1966; *Wigs at Work: Selected Cases* 1966; *Sundials Old and New* 1967; *A.P.H.: His Life and Times* (a) 1970; *In the Dark* 1970; *More Uncommon Law* 1982

Edited

Watch This Space 1964

Biography

A.P. Herbert by Reginald Pound 1976

James Leo HERLIHY 1927–1993

Of German–Irish descent, with two grandfathers who worked with horses (one a blacksmith, the other a harness-maker), Herlihy was himself the son of a construction engineer and inspector for Detroit. He was born in Detroit, and grew up both there and in Chillicothe, Ohio, attending the John J. Pershing High School in Detroit. In 1945–6 he served in the US Navy, joining as an ordinary seaman and rising to petty officer, third class. After the war, he studied for one year at Black Mountain College, North Carolina, where a creative writing instructor, Isaac Rosenfield, advised him to follow a career other than that of a writer.

Believing that acting might be more his vocation, he went to work at the Pasadena Playhouse, where he acted in some fifty roles between 1948 and 1952 and where his first play, *Streetlight Sonata*, was staged in 1950. Throughout his career as a writer and in middle life, Herlihy continued to work as an actor at intervals, sometimes appearing in his own plays. A fantasy play, *Moon in Capricorn*, reached off-Broadway in 1953, and a period followed when he wrote mainly television scripts and studied playwriting under John Gassner for a year at Yale University.

His early *curriculum vitae* also includes the American writer's normal progression through a variety of colourful jobs, in his case including foil for a carnival medical man, snake exhibitor and inspector of guided missile instruments. In 1958 he had his great success as a playwright when *Blue Denim*, a study of adolescent sexuality which he wrote in collaboration with W.A. 'Bill' Noble, began its run on Broadway. Later plays were never so successful. His first prose fiction consisted of the seven bizarre stories in *The Sleep of Baby Filbertson* (1959). The theme of crippled emotions among society's rejects continued in his first novel, *All Fall Down*, the study of a tortured family, and in his subsequent work. His second novel, *Midnight Cowboy*, was made into a hugely successful film by John Schlesinger in 1969, while his third, *The Season of the Witch*, was a study of the youth culture of the period, perhaps inspired by his own frequent hitch-hiking around the USA. His publishing career stopped in the early 1970s, after which he spent his time gardening, painting portraits and writing letters. Estimates of his work vary greatly: highly praised by some for his insight into a phantasmagoric America of lonely drive-ins and chance encounters, he was described by John Simon as 'quite a good raconteur, a sort of after-orgy speaker'. After many years living mainly in Key West, Florida, he moved in the 1980s to Los Angeles, where he eventually committed suicide.

Fiction

The Sleep of Baby Filbertson and Other Stories (s) 1959; *All Fall Down* 1960; *Midnight Cowboy* 1965; *A Story That Ends with a Scream and Eight Others* (s) 1967; *The Season of the Witch* 1971

Plays

Streetlight Sonata 1950; *Moon in Capricorn* 1953 [np]; *Blue Denim* [with W.A. Noble] 1958 [pub. 1959]; *Crazy October* 1958 [pub. 1959]; *Stop, You're Killing Me* 1968 [pub. 1970]: *Terrible Jim Fitch* 1965, *Bad Bad Jo Jo* 1970, *Laughs, Etc.* 1973

Robert HERRICK 1868–1938

Herrick was born in Cambridge, Massachusetts, where his father was a lawyer 'descended from an unbroken line of New England farmers dating back to 1636'. He attended the Cambridge Latin School and then Harvard, where he began to publish stories in the *Harvard Advocate* and the *Harvard Monthly*, of which he later became editor. Having graduated in 1890, he was instructor in English at the Massachusetts Institute of Technology from 1890 to 1893, after which he began his long academic career at the University of Chicago, where he was professor of English at the time of his retirement in 1923. He was married in 1894 to Harriet Emery (the exact namesake of his mother, though no relation) and had one son. During the First World War he travelled in Europe as a writer for the Chicago *Tribune*. Having retired in 1923 to dedicate more time to his writing and travelling, Herrick took the post of secretary to the Governor of the Virgin Islands in 1935 and died three years later from a heart-attack.

The title story of Herrick's first collection, *The Man Who Wins* (1895), concerns a scientist whose belief in man's free will is undermined by the seemingly deterministic influence of his family and professional pressures. The themes of free will and individual responsibility recur in his first novel, *The Gospel of Freedom*, the first in a succession of novels which made Herrick, in the words of Alfred Kazin, 'a pioneer realist' in American fiction. 'I belong to the realistic school of novelists,' Herrick once commented, 'nevertheless I recognise the futility of critical labels.' Novels such as *The Common Lot*, about the corruption of a young Chicago architect, and *The Master of the Inn*, about the breakdown of health under the pressures of modern city life, continue his examination of contemporary American society. In his later work there emerges an increasingly satirical streak, notably in *Chimes*, a novel – published three years after his retirement – about the dubious business methods of a university not dissimilar to the University of Chicago, and in the Utopian story *Sometime*.

Fiction

The Man Who Wins (s) 1895; *Literary Love Letters and Other Short Stories* (s) 1897; *The Gospel of Freedom* 1898; *Love's Dilemmas* (s) 1898; *The Web of Life* 1900; *The Real World* 1901; *Their Child* (s) 1903; *The Common Lot* 1904; *The Memoirs of an American Citizen* 1905; *The Master of the Inn* 1908; *Together* 1908; *A Life for a Life* 1910; *The Healer* 1911; *His Great Adventure* 1913; *One's Woman's Life* 1913; *Clark's Field* 1914; *The Conscript Mother* 1916; *Homely Lilla* 1923; *Waste* 1924; *Wanderings* (s) 1925; *Chimes* 1926; *The End of Desire* 1932; *Sometime* 1933

Non-Fiction

Composition and Rhetoric for Schools [with L.T. Damon] 1899 [rev. 1902, 1911, 1922]; *The World Decision* 1916

(Edwin) DuBose HEYWARD 1885–1940

Mistakenly acclaimed as a 'member of Harlem's intellectual community ... a Southern Negro of the old tradition', Heyward was in fact a direct descendent of Thomas Heyward, a signatory of the Declaration of Independence. Born in Charleston, Southern Carolina, he was given his mother's maiden name of DuBose. Family fortunes had been depleted by the Civil War, and poverty followed his father's death in 1887. He became a clerk in a hardware shop after he left school at fourteen, but infantile paralysis kept him an invalid for three years, after which he worked as a warehouse clerk, and then in a family insurance company, until he had a breakdown due to overwork.

He was a founder of the influential Poetry Society of South Carolina, and wrote verse in collaboration with Hervey Allen. Meanwhile, he had married Dorothy Kuhns in 1923, and she persuaded him to abandon his job in 1924. Thin, pale and fragile, with soulful brown eyes, the Heywards were said to resemble each other. Penniless, they moved to the Great Smokies, where Heyward wrote his first novel, *Porgy*, which was a best seller in 1925. He collaborated with his wife to produce the play of the same name, and in 1935 collaborated with George and Ira Gershwin to turn the play into the celebrated opera, *Porgy and Bess*. Heyward understood the blacks of South Carolina and, unusually for his time, depicted them without sentimentality, hence the contemporary assumption that he must himself be black. Other novels and poetry followed but, although perfectly competent, they were overshadowed by the fame of *Porgy and Bess*. A disciplined writer who regarded his output as small, Heyward died of a heart-attack in Tryon, North Carolina, after a long illness.

Fiction

Porgy 1925; *Angel* 1926; *The Half Pint Flask* (s) 1929; *Mamba's Daughters* 1929; *Peter Ashley* 1932; *Lost Morning* 1936; *Star Spangled Virgin* 1939

Plays

Porgy [from his novel; with D.K. Heyward] 1927; *Brass Ankle* 1931; *Porgy and Bess* (l) [with G. and I. Gershwin] 1935; *Mamba's Daughters* [with D.K. Heyward] 1939

Poetry

Carolina Chansons: Legends of the Low Country [with H. Allen] 1922; *Skylines and Horizons* 1924; *Jasbo Brown and Selected Poems* 1931

Non-Fiction

Fort Sumter [with H.R. Sass] 1938

For Children
The Country Bunny and the Little Gold Shoes 1939

Edited
Year Book of the Poetry Society of South Carolina [with others] 1921–4

F(rederick) R(obert) HIGGINS 1896–1941

Higgins was born in Foxford, County Mayo, in Ireland, the son of an engineer who was a Protestant Unionist. He moved to Dublin where he started a Clerical Workers' Union, and attempted to establish the first women's magazine in Ireland, which collapsed after two issues. His real interest was poetry, and he was responsible for publishing **Francis Stuart**'s first book of verse, *We Have Kept the Faith*, in 1923. With the encouragement of **W.B. Yeats**, he was a co-founder of a literary magazine edited by Stuart called *To-morrow*. This was extremely controversial and, denounced from Roman Catholic pulpits as indecent and blasphemous, it folded abruptly.

Higgins contributed rather florid poetry to a pamphlet, *Salt Air*, and then published *Island Blood*, in which his interest in the folk tradition was obvious. His best volume of poems, *The Dark Breed*, confirms Higgins's own belief that Irish poets 'must work more and more out of that realistic beauty found only in the folk, fusing nature with a personal emotion that incidentally reveals the all important quality of a racial memory'.

Marrying Beatrice Moore in 1921, Higgins took up 'gallivanting in the country' as a leisure pursuit. He became a founder member of the Irish Academy of Letters at the invitation of Yeats and **George Bernard Shaw**.

Poetry
Salt Air (c) 1923; *Island Blood* 1925; *The Dark Breed* 1927; *Arable Holdings* 1933; *The Gap of Brightness* 1940

Edited
Cuala Press Broadsides, New Series nos. 1–12 [with W.B. Yeats] 1935; *Progress in Irish Printing* 1936

Patricia HIGHSMITH 1921–1995

Highsmith was born in Fort Worth, Texas. Her mother was a commercial artist, her father a man of German descent called Plangman, whom she did not meet until she was twelve; she took her surname from her stepfather. She was educated at the Julia Richmond High School in New York, then at Columbia, where she studied English, Latin and Greek. As a teenager she wrote short stories of a 'bizarre or psychopathic nature' influenced by Karl Menninger's *The Human Mind* (1930), a study of psychological aberration she had discovered at the age of nine. She also painted, and after leaving college worked on comic books, supplying the writers with plots.

Although she remained a talented (if self-deprecating) sculptor, she was determined to be a writer, and took a number of jobs to earn money, including that of a saleswoman at a New York store. It was here, in 1948, that she experienced a *coup de foudre* when a female customer purchased a doll from her. When she got home, she decided that her experience may have been the first symptoms of flu, but she nevertheless sketched out the plot of a novel which was to become *The Price of Salt*. Her first published novel was *Strangers on a Train* (1950), the story of two men who 'swap' murders; it was made into a classic film the following year by Alfred Hitchcock, from a screenplay by **Raymond Chandler**. That same year she finished *The Price of Salt*, but her publishers turned it down; it was eventually published under a pseudonym in 1953 and sold almost a million copies. Unusually for the period, it was a homosexual love story with a happy ending, and she received a great many letters from grateful readers; it was subsequently reissued with an afterword as *Carol*.

Highsmith is best known, however, for her psychological thrillers, notably those featuring Tom Ripley, a bisexual murderer and fraudster, first introduced in 1956 as *The Talented Mr Ripley*. **Graham Greene** captured some of the flavour of Highsmith's novels when he described her as 'the poet of apprehension'. The books are disturbingly objective, often blackly funny, written in a cool, clear prose. *The People Who Knock on the Door* is an uncharacteristic late novel about the effect upon a family when the father becomes a born-again Christian. She also published numerous volumes of short stories, and much of her work has been adapted for the cinema and television. An insight into her methods, and the means by which she came to write several of her books, can be found in her manual *Plotting and Writing Suspense Fiction*.

A reclusive, private person, Highsmith spent most of her life alone, a circumstance she found necessary for her work. She appeared to be fonder of animals than she was of people, with a particular devotion to cats and snails. (One of her best volumes of short stories is *The Animal Lover's Book of Beastly Murder*.) She lived in East Anglia and France, but spent her final years in

Switzerland in an isolated house she had built for herself near Locarno on the Swiss–Italian border.

Her final novel, *Small g: a Summer Idyll*, was published within weeks of her death from leukemia. The 'Small g' of the title is a bar in Zurich, marked thus in a guide book to mean gay but not exclusively so. Its clientele include a number of homosexual, heterosexual and bisexual characters, most of them in love with the wrong people, but as the subtitle suggests, this is a warm-hearted novel. There may be a murder on the first page, and the ending may leave doubts as to the future happiness of the bar's denizens, but the novel proved to be a gentle farewell from a fierce talent.

Fiction

Strangers on a Train 1950; *The Price of Salt* [as 'Claire Morgan'] 1952 [rev. as *Carol* as Patricia Highsmith 1991]; *The Blunderer* 1954 [US *Lament for a Lover* 1956]; *Deep Water* 1957; *The Talented Mr Ripley* 1956; *A Game for the Living* 1958; *This Sweet Sickness* 1960; *The Cry of the Owl* 1962; *The Glass Cell* 1964; *The Two Faces of January* 1964; *The Story Teller* 1965 [UK *A Suspension of Mercy*]; *Those Who Walk Away* 1967; *The Tremor of Forgery* 1969; *Eleven* (s) 1970 [UK *The Snail-Watcher and Other Stories*]; *Ripley Under Ground* 1970; *A Dog's Ransom* 1972; *Ripley's Game* 1974; *The Animal Lover's Book of Beastly Murder* (s) 1975; *Edith's Diary* 1977; *Kleine Geschichte für Weiberfeinde* (s) 1974 [UK *Little Tales of Misogyny* 1977]; *Slowly, Slowly in the Wind* (s) 1979; *The Boy Who Followed Ripley* 1980; *The Black House and Other Stories* (s) 1981; *The People Who Knock on the Door* 1983; *Mermaids on the Golf Course and Other Stories* (s) 1985; *Found in the Street* 1986; *Tales of Natural and Unnatural Catastrophes* (s) 1987; *Ripley Under Water* 1991; *Small g: a Summer Idyll* 1995

Non-Fiction

Plotting and Writing Suspense Fiction 1966

For Children

Miranda the Panda is on the Veranda [with D. Sanders] 1958

Geoffrey (William) HILL 1932–

Hill was born in Bromsgrove, Worcestershire, in the historic 'Mercian' region he often celebrates in his poetry, and was educated at Bromsgrove County High School and Keble College, Oxford, where he took a first in English in 1953. His earliest poems appeared in 1952 in a Fantasy Press pamphlet (the medium in which many undergraduate talents of that time were first published), immediately establishing him as an original voice and one with a powerful poetic technique. He was appointed a lecturer in English at Leeds University in 1954, beginning a career which has been made entirely within academe. In 1976 he became a professor of the university, and was at Leeds until 1980.

His first book of poetry was *For the Unfallen* (1959), in which the crabbed difficulty of much of his work, its 'formality under duress', to use his own phrase, was already evident. This volume went far to make Hill's reputation as one of the most important and individual of the younger generation of British poets (he was the latest to be born of the poets represented by **Kenneth Allott** in 1962 in *The Penguin Book of Contemporary Verse*). His work has consistently remained difficult, and has used a concern with history, myth and the remaining substance of British tradition to produce a poetry of both relentless questioning and affirmation. Perhaps his major achievement is *Mercian Hymns*, a series of thirty prose poems celebrating Offa, the eighth-century king of Mercia, while a later work of importance is *The Mystery of the Charity of Charles Péguy*, a long, meditative poem which uses the life, beliefs and death in action of the French poet to investigate the debate between poetry and politics. A *Collected Poems* appeared in 1985, and has been followed by little new work.

Hill exemplifies perhaps more than any other British poet the paradox of having the highest of reputations among fellow professionals but being all but unknown to the general public. His poetry was, however, the original subject of a celebrated controversy, developing in the *London Review of Books* from 1986, waged by **Tom Paulin**, **Craig Raine** and the academic Martin Dodsworth, and centring on the question of whether metrics can of their nature be political.

Hill has been married twice and has five children. From 1981 to 1988 he was a lecturer at Emmanuel College, Cambridge, and from the latter date professor of literature and religion at Boston University, Massachusetts. He has also been a visiting lecturer at Michigan and Ibadan, Nigeria. Other works beside his own poetry are an English version of Ibsen's *Brand* and a book of academic essays, *The Lords of Limit*.

Poetry

For the Unfallen 1959; *Penguin Modern Poets 8* [with E. Brock, S. Smith] 1966; *King Log* 1968; *Mercian Hymns* 1971; *Somewhere Is Such a Kingdom; Poems 1952–1971* 1975; *Tenebrae* 1978; *The Mystery of the Charity of Charles Péguy* 1983; *Collected Poems* 1985

Plays
Brand [from Ibsen's play] 1978

Non-Fiction
The Lords of Limit: Essays on Literature and Ideas 1984; *Enemy's Country: Words, Contexture and Other Circumstances of Language* 1991

Susan (Elizabeth) HILL 1942–

Born in Scarborough, Yorkshire, Hill was educated at grammar schools there and in Coventry and at King's College, London, from which she graduated with a degree in English literature. She published her first novel, *The Enclosure*, at the age of nineteen, while still an undergraduate, and was able to earn her living as a writer from then on.

Her books invariably have an English setting, such as the shabby tearooms of a seaside resort in *A Change for the Better* or the university city of *Air and Angels*. Her fictional world is far from cosy, however. *I'm the King of the Castle* (which won a Somerset Maugham Award) is a sinister and tragic story about two small boys, unwillingly thrown together by circumstances. The calculated cruelties of children are unflinchingly recorded, and build to a truly horrifying climax. *Strange Meeting* is a remarkable feat of imaginative sympathy in which Hill convincingly recreates the physical and psychological world of the trenches of the First World War. Essentially a love story, the book details the relationship between two officers; it takes its title from **Wilfred Owen**'s celebrated poem, and contains echoes of the work of **Siegfried Sassoon**. Perhaps her most ambitious novel is *The Bird of Night*, which won the Whitbread Award in 1972. Exploring the relationship between insanity and genius, it takes the form of the recollections of Harvey Lawson, who as a young man had become the companion and protector of a famous poet, Francis Croft, who suffers from bouts of insanity and eventually commits suicide. She has also published several volumes of short stories, of which *The Albatross and Other Stories* won the John Llewellyn Rhys Memorial Prize in 1971.

Hill's long relationship with a much older man, and the effect upon her of his death, lies behind *In the Springtime of the Year*, a novel about a young widow gradually emerging from grief; it proved to be her last work of fiction for nine years. She announced that she had retired as a novelist and throughout the 1970s and early 1980s, she wrote many radio plays and a trilogy of books based on her life in the Oxfordshire countryside: *The Magic Apple Tree*, *Through the Kitchen Window* and *Through the Garden Gate*. She also wrote scripts for BBC radio's long-established and hugely popular rural saga, *The Archers*. Her return to the novel came with *The Woman in Black*, which she firmly (but accurately) described as 'a ghost story'. It proved very successful and was subsequently adapted for a long-running stage play. She followed it with another ghost story, *The Mist in the Mirror*. Her next novel, *Air and Angels*, was more like the books upon which she had made her reputation, but although it had its champions, it was savaged by several reviewers. Hill has also written books for children and edited several collections, including a volume of stories by **Thomas Hardy**. She is married to the Shakespearean scholar Stanley Wells, with whom she had three daughters. The premature birth and brief survival of her second daughter are harrowingly but movingly described in *Family*.

Fiction
The Enclosure 1961; *Do Me a Favour* 1963; *Gentleman and Ladies* 1968; *A Change for the Better* 1969; *I'm the King of the Castle* 1970; *The Albatross and Other Stories* (s) 1971; *Strange Meeting* 1971; *The Bird of Night* 1972; *The Custodian* (s) 1972; *A Bit of Singing and Dancing* (s) 1973; *In the Springtime of the Year* 1974; *The Woman in Black* 1983; *The Mist in the Mirror* 1992; *Air and Angels* 1994

Plays
Chances (r) 1971, (st) 1983; *Taking Leave* (r) 1971; *Winter Elegy* (r) 1973; *The Cold Country and Other Plays for Radio* (r) [pub. 1975); *The Summer of the Giant Sunflower* (r) 1977; *The Badness within Him* [from her story] (tv) 1980; *Here Comes the Bride* (r) 1980; *The Sound That Time Makes* (r) 1980; *Out in the Cold* (r) 1982; *Autumn* (r) 1985; *Winter* (r) 1985

Non-Fiction
The Magic Apple Tree 1982; *Through the Kitchen Window* 1984; *Through the Garden Gate* 1986; *The Lighting of the Lamps: Essays and Preview* 1987; *Shakespeare Country* 1987; *Lanterns Across the Snow* 1987; *The Spirit of the Cotswolds* 1988; *Family* 1989

For Children
The Ramshackle Company (r) 1981; *One Night at a Time* 1984; *Mother's Magic* 1986; *Can It Be True?* 1988; *Suzy's Shoe* 1989; *I've Forgotten Edward* 1990; *I Won't Go There Again* 1990; *Pirate Roll* 1990; *Septimus Honeydew* 1990; *Stories from Codling Village* 1990

Edited
New Stories 5 [with I. Quigly] 1980; *Ghost Stories* (s) 1983; *People: Essays and Poems* 1983; Thomas

Hardy *The Distracted Preacher and Other Tales* (s) 1984; *The Walker Book of Ghost Stories* (s) 1990

James HILTON 1900–1954

The creator of Mr Chips was himself the son of a schoolmaster, and was born at Leigh in Lancashire. He was educated at the Leys School and Christ's College, Cambridge, where he read English and history, but had already embarked upon a writing career by the age of seventeen, when he became a contributor to the Manchester *Guardian*. He subsequently wrote a column for the Irish *Independent* and reviewed fiction for the *Daily Telegraph*. He was as precocious a novelist as he was a journalist, publishing his first novel, *Catherine Herself* (1920), while still an undergraduate.

This early start notwithstanding, he did not have much commercial success until 1931, when he published his sixth novel, *And Now Good-bye*. Like most of his books, this one is now forgotten, and Hilton's literary reputation rests upon *Lost Horizon* (which won the Hawthornden Prize for 1934), *Good-bye, Mr Chips* and *Random Harvest*. Of these, *Lost Horizon*, in which survivors of a plane crash in Tibet discover an earthly paradise (and which introduced the term 'Shangri-la' into the language), and *Random Harvest*, a melodrama about a shell-shocked aristocrat and a music-hall artiste, are chiefly remembered because well-known films were based upon them. *Good-bye, Mr Chips* was also filmed twice – extremely well in 1931, by Sam Wood from a screenplay by **R.C. Sheriff**, and as an ill-advised musical in 1969 by Herbert Ross from a screenplay by **Terence Rattigan** – but achieved a great success in its original form, serialised in both the *British Weekly* and *Atlantic Monthly*. A sentimental paean to the extraordinary devotion of an ordinary schoolmaster, and to the place of the English public schools in English history, this brief, telegraphic novel was a reassuring text for the unsettled 1930s, about as far removed from the prevailing intellectual *Zeitgeist* as it is possible to imagine. Although Hilton reputedly wrote it in four days, he was a genuine craftsman and the book still reads surprisingly well.

In 1939, he went to Hollywood as a screenwriter, achieving his greatest success with his screenplay of Jan Struther's *Mrs Miniver*. Hilton's own qualities of modesty and optimism were ideally suited to adapting this propagandist tale of a British housewife enduring the dangers and privations of wartime London. He continued to write novels, few of which are still read, reckoning on an average of 3,000 words a day. He married twice, and died at Long Beach, California.

Fiction

Catherine Herself 1920; *The Passionate Year* 1923; *The Meadows of the Moon* 1926; *Terry* 1927; *The Silver Flame* 1928; *And Now Good-Bye* 1931; *Murder at School* [as 'Trevor Glen'] 1931; *Contango* 1932; *Rage in Heaven* 1932; *Knight without Armour* 1933; *Lost Horizon* 1933; *Good-bye, Mr Chips* 1934; *Dawn of Reckoning* 1937; *We Are Not Alone* 1937; *To You, Mr Chips* 1938; *Random Harvest* 1941; *The Story of Dr Wassall* 1944; *So Well Remembered* 1947; *Nothing So Strange* 1948; *Twilight of the Wise* 1949; *Morning Journey* 1951; *Time and Time Again* 1953

Thomas HINDE (pseud. of Thomas Willes Chitty) 1926–

Hinde was born in Felixstowe, Suffolk, and from the age of six to twelve attended a boarding-school there run by his father, Sir Thomas Chitty, about whom he later wrote a lively memoir, *Sir Henry and Sons*. Hinde was not particularly interested in literature as a young man, despite (or perhaps because of) the fact that Sir Henry Newbolt was a great-uncle. At Winchester he excelled as a gymnast, but at little else, and was forced to join the Royal Navy by his parents. He eventually got out after three miserable years, and in 1944 he went to University College, Oxford, to read modern history. He has said he spent a largely idle time, resulting in a third-class degree, and published some short stories which he later considered poor.

It was at this time that Hinde decided he wanted to be a writer, but the nine years he spent as a rating officer left little time for anything else. He managed to write *Mr Nicholas* while tutoring children on a farm, and it was published in 1952. The novel, about a greedy, retired businessman who exploits a family, was widely praised for the quality of its writing. **Walter Allen** noted that its strength lay in 'the exactness of the observation of one section of middle-class life in transition'. No further novels were published for five years, as Hinde worked for Shell, spending two years in Nairobi where he became ill with glandular fever. Although a baronet, Hinde has had to earn an income, and is not ashamed of having to produce journalism and other factual books simply for the money. The dilemma of social change and its effect on individuals, permeates all of his fiction. *The Day the Call Came* is a powerful allegory about the nature of perception, in which a rather ordinary fruit-farmer is convinced he is a secret agent. *The Village*, in which Hinde attempted to present a detailed analysis of all the inhabitants in an English village which is due to be demolished, was considered a brilliant failure. A sense of personal detachment reminiscent of **Henry James** has led the *Times Literary Supplement* to suggest that the 'intensity of his artistic conviction has made him the one *pure* novelist of today'.

Hinde was taught writing and English literature at the University of Illinois from 1965 to 1967, and was visiting professor of English literature at Boston in 1969 for one year. He married the biographer Susan Chitty (the daughter of **Antonia White**) in 1951. They have collaborated on several books, writing in the morning, to leave the afternoon free for their four children. Hinde adopted his pen-name to avoid the use of his title, which he finds an embarrassment.

Fiction

Mr Nicholas 1952; *Happy as Larry* 1957; *For the Good of the Company* 1961; *A Place Like Home* 1962; *The Cage* 1962; *Ninety Double Martinis* 1963; *The Day the Call Came* 1964; *Games of Chance* 1965; *The Village* 1966; *High* 1968; *Bird* 1970; *Generally a Virgin* 1972; *Agent 1974; Our Father* 1975; *Daymare* 1980

Non-Fiction

On Next to Nothing: a Guide to Survival Today [with S. Hinde] 1976; *The Great Donkey Walk* [with S. Chitty] 1977; *The Cottage Book* 1979; *Sir Henry and Sons: a Memoir* 1980; *Stately Gardens of Britain* 1983; *A Field Guide to the English Country Parson* 1984; *Forests of Britain* 1985; *Just Chicken* [with S. Chitty] 1985; *Capability Brown* 1986; *Courtiers* 1986; *Tales from the Pump Room* 1988; *Imps of Promise* 1990

Edited

Spain: a Personal Anthology 1963; *The Domesday Book: England's Heritage, Then and Now* 1985; *Lewis Carroll: Looking-Glass Letters* 1991

(Melvin) Barry HINES 1939–

Born in Barnsley into a family of miners, Hines was educated at Ecclesfield Grammar School. As a boy he excelled at football, and several clubs were keen to sign him, but he realised that he was not up to professional standard and went instead to Loughborough College to train as a physical-education teacher. He taught PE at schools in London, Barnsley and South Yorkshire during the 1960s and early 1970s.

His first novel, *The Blinder* (1966), was about football, and met with limited success, although comparisons were drawn with such established chroniclers of northern working-class life as **Alan Sillitoe** and **David Storey**. He made his name, however, two years later with his second novel, *A Kestrel for a Knave*, the funny and moving story of a solitary boy who comes out of his shell when he adopts and trains a kestrel. A striking parallel is drawn between the boy and his hawk, both of them wild and isolated creatures, and the novel shows a genuinely reciprocal relationship between them: just as the boy trains the bird, so the bird educates the boy. It was made into a very fine film by the realist director Ken Loach from a screenplay by Hines, Loach and Tony Garnett which won the Writers' Guild Screenplay Award. The film was released as *Kes*, the title under which the novel was subsequently reissued. It remains his best-known book.

Amongst Hines's other novels are *The Gamekeeper*, which describes a year in the life of its urban protagonist on a large country estate, and demonstrates the author's skill at evoking landscape and his concern with class boundaries and loyalties, and *Looks and Smiles*, which is about a school-leaver looking for work. He adapted both books for the screen, the former as a television play, the latter for the cinema, with Loach once again directing. His original television play about the miners' plight during the 1970s, *The Price of Coal*, was broadcast in 1979, and he subsequently rewrote it as a novel. His work remains firmly rooted in his own background, and he is a leader of the second wave of social realist writers from the north of England. Although his subject-matter is almost invariably economic or educational deprivation, his novels have always been leavened by humour and optimism. He has also had a career teaching creative writing at various universities and colleges in both England and Australia.

Fiction

The Blinder 1966; *A Kestrel for a Knave* 1968 [as *Kes* 1974]; *First Signs* 1972; *The Gamekeeper* 1975; *The Price of Coal* 1979; *Looks and Smiles* 1981; *Unfinished Business* 1983

Plays

Billy's Last Stand **(tv)** 1970, **(st)** 1985; *Kes* **(f)** [from his novel; with T. Garnett, K. Loach] 1970 [with A. Stronach; pub. 1976]; *Speech Day* 1973; *Two Men from Derby* **(tv)** 1976 [pub. in *Act Two* ed. D. Self, R. Speakman 1979]; *The Price of Coal* **(tv)** 1977 [pub. 1979]; *The Gamekeeper* [from his novel] 1979; *Looks and Smiles* **(f)** 1981; *Threads* 1984

Edward HIRSCH 1950–

Hirsch was born in Chicago and graduated from Grinnell College in 1972. He attained his PhD at the University of Pennsylvania in 1979, and since then has taught English at Wayne State University, Detroit, and at the University of Houston since 1985.

He began publishing poems in various magazines in the 1970s and won immediate critical praise for his first collection, *For the Sleepwalkers* (1981). Many of the poems in this and his second collection, *Wild Gratitude*, which won the National Book Critics Circle Award in 1987, are tight vignettes of urban life, while other poems explore his other primary concern which is with the work of writers and artists such as Rilke, Matisse and Rimbaud. In *The Night Parade* his interests broaden to embrace the history of his family and of Chicago, an exercise which did not meet with universal critical approval. In the *New York Review of Books* a critic wrote: 'When he is not being historically stagy, he is being familially prosaic, as he recalls stories told by his parents.' Hirsch has described the poems in *Earthly Measures* as 'a soul journey' which seeks to explore the redemptive power of art, and the collection ends with a homage to the seventeenth-century Dutch painters.

He has received numerous awards, notably a Guggenheim Poetry Fellowship for 1985–6, and the American Academy and Institute of Arts and Letters Rome Prize in 1988. He married in 1977 and has one son.

Poetry

For the Sleepwalkers 1981; *Wild Gratitude* 1986; *The Night Parade* 1989; *Earthly Measures* 1994

Russell (Conwell) HOBAN 1925–

Hoban grew up in Pennsylvania, the son of a Russian Jewish immigrant who had become advertising manager of the city's *Jewish Daily Forward*. He attended Temple University for five weeks before going to the Philadelphia Museum School of Industrial Art from 1941 to 1943. He then served in the army for two years, mainly in Italy, winning the Bronze Star. On his return to America he did a wide variety of jobs ranging from messenger to freight handler. Moving to New York with his first wife, Lillian Aberman, whom he had married in 1944, he became a commercial artist for magazines and advertising agencies, and eventually television art director for an advertising agency. From 1956 he was a freelance illustrator.

His writing career began late, in 1959, and for more than a decade he wrote only for children. His first book was *What Does It Do and How Does It Work?* about heavy machines, and he followed this up with a series of picture books, many of them featuring the well-loved character Frances the Badger and illustrated by his wife, herself a professional artist. His first novel for children was *The Mouse and His Child* (1967), about two clockwork mice let loose into a hostile world, a book that has quickly established itself as a classic. He became a freelance writer in 1967. He moved from Connecticut to London in 1969, and at about the same time his marriage broke up, his wife remaining in America with their four children.

Soon afterwards he began work on the first of his novels for adults, although by no means giving up writing for children. *Turtle Diary*, about attempts to liberate giant sea turtles at London Zoo, was considered the most successful of his early novels, but he won a very substantial reputation in 1980 with *Riddley Walker*, a novel set in a post-nuclear-holocaust future and narrated in fragmented phonetic English by a twelve-year-old boy; it was hailed in Britain as a masterpiece. Also well received was *Pilgermann*, about a Jewish pilgrim on the First Crusade, but *The Medusa Frequency* was considered slighter.

His writing method is unusual in that he reads much of his work daily to a psychoanalyst, which has helped him to be 'good friends' with his head. He still produces about four children's books a year, and has written more than sixty books altogether. He follows a 'strictly regimented life' to produce them at his house in London, where he lives with his second wife, Gundula Ahl, a German bookseller, and their three children. His sister, Tana Hoban, is also a writer and illustrator for children.

Fiction

The Lion of Boaz-Jachin and Jachin-Boaz 1973; *Kleinzeit* 1974; *Turtle Diary* 1975; *Riddley Walker* 1980; *Pilgermann* 1983; *The Medusa Frequency* 1987

For Children

What Does It Do and How Does It Work? 1959; *The Atomic Submarine* 1960; *Bedtime for Frances* 1960; *Herman the Loser* 1961; *London Men and English Men* 1962; *The Song in My Drum* 1962; *Some Snow Said Hello* 1963; *The Sorely Trying Day* 1964; *A Baby Sister for Frances* 1964; *Bread and Jam for Frances* 1964; *Nothing to Do* 1964; *The Story of Hester Mouse Who Became a Writer* 1965; *Tom and the Two Handles* 1965; *What Happened when Jack and Daisy Tried to Fool the Tooth Fairies* 1965; *Goodnight* 1966; *Henry and the Monstrous Din* 1966; *The Little Brute Family* 1966; *Charlie the Tramp* 1967; *Save My Place* 1967; *The Mouse and His Child* 1967; *A Birthday for Frances* 1968; *The Pedalling Man and Other Poems* 1968; *The Stone Doll of Sister Brute* 1968;

Best Friends for Frances 1969; *Harvey's Hideout* 1969; *The Mole Family's Christmas* 1969; *Ugly Bird* 1969; *A Bargain for Frances* 1970; *Emmet Otter's Jug-Band Christmas* 1971; *Egg Thoughts and Other Frances Songs* 1972; *The Sea-Thing Child* 1972; *Letitia Rabbit's String Song* 1973; *How Tom Beat Captain Najork and His Hired Sportsmen* 1974; *Ten What?* 1974; *Crocodile and Pierrot* [with S. Selig] 1975; *Dinner at Alberta's* 1975; *A Near Thing for Captain Najork* 1975; *Arthur's New Power* 1978; *The Twenty-Elephant Restaurant* 1978; *The Dancing Tigers* 1979; *La Corona and the Tin Frog* 1979; *Ace Dragon Ltd* 1980; *Flat Cat* 1980; *The Great Fruit Gum Robbery* 1981; *The Serpent Tower* 1981; *They Came From Aargh!* 1981; *The Battle of Zormla* 1982; *The Flight of Bembel Rudzuk* 1982; *Big John Turkle* 1983; *Jim Frog* 1983; *Charlie Meadows* 1984; *Lavinia Bat* 1984; *The Marzipan Pig* 1986; *The Rain Door* 1986; *Jim Hedgehog's Supernatural Christmas* 1989; *Monsters* 1989; *Jim Hedgehog and the Lonesome Tower* 1990

Janet (Konradin) HOBHOUSE 1948–1991

Born in New York City, Hobhouse was the daughter of a beautiful but unstable American mother and a dour English father whose messy transatlantic marriage and acrimonious divorce culminated in a tug-of-war over two-year-old Janet in the departure lounge of Heathrow airport. She lived with her mother in New York until she was sixteen, when she returned to England to try to establish a relationship with her father. She read English at Oxford, and after graduation worked for a while in New York, copy-editing for various publishers and writing about art.

Love brought her back to London, and at a party where she was introduced to Lord Weidenfeld, she charmingly sold him the idea for an illustrated biography of Gertrude Stein. *Everybody Who Was Anybody* was published in 1975, when Hobhouse was only twenty-four, and nominated for a Pulitzer Prize.

That same year, she married Nick Fraser, a wealthy old Etonian whom she had met at Oxford, and they lived a peripatetic lifestyle on both sides of the Atlantic, interspersed with devastating periods of tragedy. Her first novel, *Nellie without Hugo* (1982), is the least autobiographical of her books, but was written mostly in 1977 just before the suicide of her mother in 1978. Hobhouse and her mother shared the closest of relationships, and she never fully recovered from the shock and guilt that the suicide engendered. Six months after this, the Frasers' accountant absconded with a vast amount of their money, and the marriage began to disintegrate. They were divorced in 1983.

Two subsequent novels, *Dancing in the Dark* and *November*, are much bleaker books, and although they lack neither humour nor wit, they expose a vulnerability which enabled her to write about extreme emotional states with great conviction. Her perceptions about love, marriage and friendship are both accurate and unsentimental. *Dancing in the Dark* deals with a time when her marriage was foundering, while being shored up with the time and attention of a good friend who found himself subsequently dumped when things got better. The friend never forgave them the double betrayal, both in life and on the page. In 1988, she returned to art criticism and published a well-received study of interpretations of the female nude in the twentieth century, *The Bride Stripped Bare.*

The Furies, published posthumously, is really Hobhouse's autobiography, and is a lucid, impassioned account of coming to terms with a life of alternating deprivation and privilege, devastating talent, and an emotional world that seesawed between blind devotion and cool detachment. Most importantly, it is a love letter to her mother, and an attempt to reconcile that sad and futile death. It tells us as much as we will probably ever know about Hobhouse's tortured inner spirit. She was popular, if a little formidable socially, and her friends talk generously of her beauty and intellect, her love of luxury, parties and the company of celebrities (she had affairs with both a bestselling American author and a famous English actor). But she could be demanding, and while she had a gift for empathy and for making people feel special, her friend Elisa Segrave said she had 'the egotism of the only child and the vulnerability of the abandoned'.

She was first diagnosed as having cancer at the age of thirty-six, but with characteristic strength and determination, recovered. When it recurred in 1990, she was forty-two and hard at work on *The Furies*, now generally acknowledged to be her best and most powerful novel. The short career which produced four exceptional works of fiction, a biography and a book of art criticism pointed to even greater things to come. She was buried next to her mother on a Connecticut hillside. She left no will and there were no last wishes.

Fiction

Nellie without Hugo 1982; *Dancing in the Dark* 1983; *November* 1987; *The Furies* [unfinished] 1992

Non-Fiction

Everybody Who Was Anybody: a Biography of Gertrude Stein 1975; *The Bride Stripped Bare* 1988

Philip (Dennis) HOBSBAUM 1932–

Hobsbaum was born in London to a Russian mother and a Jewish father, who was an engineer. Educated at Downing College, Cambridge, under F.R. Leavis in the 1950s, he founded there the Group, an association of poets including **Peter Redgrove** (and later **George MacBeth** and **Edward Lucie-Smith**). They reacted against certain aspects of the Movement (another grouping of poets, centred on **Philip Larkin**) and the Black Mountain school in the USA. He was editor, with Lucie-Smith, of *A Group Anthology*. Hobsbaum published his first collection, *The Place's Fault* (1964), while he was a lecturer at the Queen's University, Belfast, where he galvanised **Michael Longley**, **Derek Mahon** and **Seamus Heaney** into poetic action. Longley and Heaney have both written appreciatively of Hobsbaum's role in their careers.

His own poetry has not been as successful as that of many poets he has encouraged, and since the 1970s he has concentrated his energies largely on critical works on poetry, literary theory and the novels of Dickens and **D.H. Lawrence**. In his tireless efforts to get others into print, he might be seen as a latter-day **Ezra Pound**. Since 1972 he has published no poetry, but has been composing a large work entitled 'Poems for Several Voices': he intimates that his best work is yet to come.

Since 1966 Hobsbaum has been an academic at the University of Glasgow, living in that city. Initially lecturer, then senior lecturer and reader, he has been a titular professor since 1985. He has married twice. Besides his Cambridge degree, he also studied music, being a licentiate of both the Royal Academy of Music and the Guildhall School.

Poetry

The Place's Fault and Other Poems 1964; *Snapshots* 1965; *In Retreat and Other Poems* 1966; *Coming Out Fighting* 1969; *Some Lovely Glorious Nothing* 1969; *Women and Animals* 1972

Plays

Children in the Woods (r) 1974; *Round the Square* (r) 1976

Non-Fiction

A Theory of Communication: a Study of Value in Literature 1970; *A Reader's Guide to Charles Dickens* 1972; *Tradition and Experiment in English Poetry* 1979; *A Reader's Guide to D.H. Lawrence* 1981; *Essentials of Literary Criticism* 1983; *Theory of Criticism* 1970; *A Reader's Guide to Robert Lowell* 1988; *Channels of Communication* 1992

Edited

A Group Anthology [with E. Lucie-Smith] 1963; *Ten Elizabethan Poets* 1969

Michael HOFMANN 1957–

Hofmann was born in Freiburg, West Germany, the son of the writer Gert Hofmann, and came to England at the age of four. He attended schools in Bristol, Edinburgh and Winchester before going up to Magdalene College, Cambridge, where he graduated in English in 1979. He undertook graduate study at Regensburg University and at Trinity College, Cambridge, from 1980 to 1983, and since then has been a freelance writer and translator, and a teacher of creative writing at the University of Florida, Gainesville, since 1990.

Hofmann published poetry in a number of small magazines before his first collection, *Nights in the Iron Hotel*, was published in 1983. With his second collection, *Acrimony*, he achieved widespread critical praise. The first half of the collection gathers a group of more or less political poems which describe Britain in the 1980s with a close attention to the seedy, grubby aspects of contemporary life, while the second half comprises nineteen poems about the personal and domestic in which Hofmann addresses his troubled relationship with his father. With **Christopher Middleton** he has translated a collection of his father's stories, *Balzac's Horse*, and in 1990 he published in the USA *K.S. in Lakeland*, which brings together poems from his earlier British collections with new poetry. He won a Cholmondeley Award in 1984, and the Geoffrey Faber Memorial Prize in 1988.

Poetry

Nights in the Iron Hotel 1983; *Acrimony* 1986; *K.S. in Lakeland: New and Selected Poems* 1990; *Corona, Corona* 1993

Plays

The Double-Bass [from P. Süskind's play] 1989 [pub. 1987]; *The Good Person of Sichuan* [from B. Brecht's play] 1989

Translations

Kurt Tucholsky *Castle Gripsholm* 1985; Gert Hofmann *Balzac's Horse and Other Stories* (s) [with C. Middleton] 1988; Beat Sterchi *Blösch* 1988 [US *Cow* 1990]; Joseph Roth *The Legend of the Holy Drinker* 1989; Wim Wenders *Emotion Pictures: Reflections on the Cinema* [with S. Whiteside] 1990; Wolfgang Koeppen *Death in Rome* 1992; Patrick Süskind *The Story of Mr Sommer* 1992

John HOLLANDER 1929–

Hollander was born in New York City into a Jewish family. His father was a research physiologist and his mother a schoolteacher, and he has described himself as having grown up 'in a home that buzzed with ideals and enlightenment'. After attending the Bronx High School of Science he entered Columbia University at the age of sixteen to study English and art history. His teachers included **Lionel Trilling** and the art critic Jacques Barzun, and an influential contemporary was the older **Allen Ginsberg**. After graduating in 1950, he travelled in Europe before returning to Columbia to study the history of music. He spent two years at Indiana University, Bloomington, where he began to publish poems in *New Directions* and *Discovery*, and in 1954 was elected to the Society of Fellows at Harvard where he furthered his interest in medieval, Renaissance and baroque music. After stints as an English lecturer at Connecticut College and Yale he became Professor of English at Hunter College, New York, and in 1977 he returned to Yale, becoming A. Bartlett Giametti Professor of English in 1986.

According to Hollander, the influence of **W.H. Auden** was so naked in his first collection of poems, *A Crackling of Thorns* (1958), that he 'had to rewrite in proof many lines ... to expunge embarrassingly blatant echoes of his voice'. Curiously, the collection was published in the Yale Series of Younger Poets which was edited that year by Auden. The poems are remarkable for their technical virtuosity and a distinctively metaphysical turn of mind. Hollander is dismissive of much of his early verse as mere rhetoric, and considers the poems in *The Night Mirror* to be his first mature work. They are densely allusive and syntactically complex, so much so that many critics found them impenetrable, though Harold Bloom has praised the emotional power with which Hollander addresses the theme of mortality.

Since then Hollander has consistently broadened his range, often mining a surprising vein of wit, notably in the book-length narrative poem *Reflections on Espionage* in which a master spy called Cupcake transmits coded messages to agents with names such as Gland, Thumbtack and Steampump. Evidently it is an allegory on the state of contemporary poetry. Hollander relishes a formal challenge, nowhere more so than in *Powers of Thirteen*, a sequence of 169 poems of thirteen lines each in which each line contains thirteen syllables. One reviewer described this as an 'infuriating example of academic jokiness'.

Hollander was an editorial assistant on *Partisan Review* from 1959 to 1965 and a contributing editor on *Harper's* from 1969 to 1971. His award include the Levinson Prize in 1974, a Guggenheim fellowship in 1979 and the Bollingen Prize in 1983. He has been married twice, and has two daughters.

Poetry

A Crackling of Thorns 1958; *A Beach Vision* 1962; *Movie-Going and Other Poems* 1962; *Visions from the Ramble* 1965; *Philomel* 1968; *Types of Shape* 1969; *The Night Mirror* 1971; *Selected Poems* 1972; *Town and Country Matters* 1972; *The Head of the Bed* 1974; *Tales Told of the Fathers* 1975; *Reflections on Espionage* 1976; *In Place* 1978; *Spectral Emanations* 1978; *Blue Wine and Other Poems* 1979; *Looking Ahead* 1982; *Powers of Thirteen* 1983; *A Hollander Garland* 1985; *In Time and Place* 1986; *Harp Lake* 1988; *Some Fugitives Take Cover* 1988

Non-fiction

The Untuning of the Sky: Ideas of Music in English Poetry 1500–1700 1961; *Images of Voice: Music and Sound in Romantic Poetry* 1970; *Vision and Resonance: Two Senses of Poetic Form* 1975; *The Figure of Echo: a Mode of Allusion in Milton and After* 1981; *Rhyme's Reason: a Guide to English Verse* 1981; *Dal Vero* [with S. Steinberg] 1983; *Melodious Guide: Fictive Pattern in Poetic Language* 1988

Edited

Jonson *Selected Poems* 1961; *The Wind and the Rain: an Anthology of Poems for Young People* [with H. Bloom] 1961; *Jiggery-Pokery: a Compendium of Double Dactyls* [with A. Hecht] 1967; *American Short Stories Since 1945* (s) 1968; *Modern Poetry: Essays in Criticism* 1968; *Poems of Our Moment* 1968; *I.A. Richards: Essays in His Honor* [with R. Brower, H. Vendler] 1973; *The Oxford Anthology of English Literature* [with others] 1973; *Literature as Experience* [with D. Bromwich, I. Howe] 1979; *Poetics of Influence* [with H. Bloom] 1988; *The Essential Rossetti* 1990; *Animal Poetry* 1994

For Children

A Book of Various Owls 1963; *The Quest for the Gole* 1966; *The Immense Parade on Supererogation Day* 1972

Plays

An Entertainment for Elizabeth 1969 [pub. 1972]

Alan HOLLINGHURST 1954-

Hollinghurst was born at Stroud, Gloucestershire. His father was a bank manager, who moved frequently with the job, so he was later brought up at Faringdon, then in Berkshire, and Cirencester, Gloucestershire. He was educated at a public school, Canford in Dorset, and Magdalen College, Oxford, where he read English and took a first in 1975. While an undergraduate, he won the prestigious Newdigate Prize for poetry, and his earliest published works were poems: a selection in Faber's *Poetry Introduction Four* in 1979 and the sequence *Confidential Chats with Boys*, published by **John Fuller**'s Sycamore Press in 1982.

He stayed on at Oxford to do a postgraduate BLitt (subsequently MLitt) thesis on the effect of their homosexuality on the works of **E.M. Forster**, **L.P. Hartley** and **Ronald Firbank**, and meanwhile held junior college lectureships at Magdalen and Somerville Colleges. He then received a year's busary from the Southern Arts Association, during which time he wrote the greater part of a subsequently abandoned novel. He moved to London in September 1981, and taught for one term at University College London before getting the job of assistant editor at the *Times Literary Supplement* in March 1982. He became the paper's deputy editor in 1985.

Hollinghurst achieved sudden fame in 1988, when his extremely accomplished first novel, *The Swimming-Pool Library*, was published to immediate *réclame*. This exuberant story of a handsome and arrogant young aristocrat enjoying the gay clubs, porno cinemas and outdoor cruising spots of pre-Aids London both gave a remarkably vivid idea of the period in which it was set and caught the public mood of a later time. Its cultural references were balanced by the many passages of explicit sex, and only the attempt to provide a gay history of the twentieth century by flashbacks to the life of an older lord fell a little flat.

Hollinghurst's second novel, *The Folding Star*, came six years later, and featured another gay man, more mature and introspective, but also achieving frequent conquests, this time in a Belgian city based on Bruges. Naturally, the explosive public impact of the first novel could not be matched, but there was general agreement that Hollinghurst had passed a test many of his contemporaries failed: he had written a second novel to match his first. The novel was shortlisted for the Booker Prize.

Hollinghurst's other work includes a blank-verse translation of Racine's tragedy *Bajazet*, a short story (so far the only one he has written), published in *Firebird* in 1983, and an entertaining, discursive body of literary journalism. Since 1988, he has been editor of the long-established (founded in 1882) literary competition *Nemo's Almanac*. He gave up full-time work on the *TLS* in 1990, but continues as a part-time member of staff and the poetry editor.

Fiction

The Swimming-Pool Library 1988; *The Folding Star* 1994

Poetry

Confidential Chats with Boys 1982

Translations

Jean Racine *Bajazet* 1991

(Christopher) John HOLLOWAY 1920-

Holloway was born in Hackney and brought up in various parts of south-east London where his father was a hospital stoker and his mother a nurse. His memoir of his early years, *A London Childhood*, is a vivid portrait of growing up in the Depression. He won a scholarship to New College, Oxford, where he graduated in modern greats in 1941. He worked briefly as a social scientist before being seconded to Army Intelligence where he 'had a safe, dull war'. He was a fellow of All Souls from 1946 but a desire to escape the dry study of philosophy and switch to literature prompted him to take up an English lectureship at Aberdeen University in 1949. While there he published *The Victorian Sage*, a major study of Victorian prose writers. In 1954 he moved to Queen's College, Cambridge, where he was Professor of Modern English from 1972 to 1982, and is now a Life Fellow and Emeritus Professor. He has also travelled widely to teach in Athens, Chicago, the Middle East and the Indian sub-continent.

Holloway's critical works have certainly been more widely noticed than his poetry, and have been influenced by his study of philosophy and history. A colleague, Frank Kermode, once defined his relationship to the critical establishment as resembling 'one who enters to an unruly company and sternly asks what all this is about'. His most important critical works include *The Story of the Night*, a penetrating analysis of Shakespeare's tragedies, and *The Charted Mir-*

ror, which questions the assumptions behind the New Criticism.

Holloway contributed poems to **Robert Conquest**'s Movement anthology *New Lines* in 1956, the same year in which he published his first collection, *The Minute*. His verse shares with the Movement poets a concern with the use of traditional forms, but he has described himself as having been 'right on the edge of the group'. *The Landfallers* is a lengthy narrative poem, praised for its craftsmanship, which seeks to explore philosophical ideas and which many readers found obscure. For all of their intellectual enquiries, his poems continually tap what he has called 'the inexhaustible soil of common speech'. Holloway has been married twice, and has two children. With his second wife, Joan Black, he has edited two volumes of broadside ballads.

Poetry

The Minute, and Longer Poems 1956; *The Fugue and Shorter Pieces* 1960; *The Landfallers* 1962; *Wood and Windfall* 1965; *New Poems* 1970; *Planet of Winds* 1977

Non-Fiction

Language and Intelligence 1951; *The Victorian Sage* 1953; *The Charted Mirror* 1960; *The Story of the Night* 1961; *The Colours of Clarity: Essays on Contemporary Literature and Education* 1964; *The Lion Hunt: a Pursuit of Poetry and Reality* 1964; *A London Childhood* (a) 1966; *Widening Horizons in English Verse* 1966; *Blake: the Lyric Poetry* 1968; *The Establishment of English* 1972; *The Proud Knowledge: Poetry and Self, 1620–1920* 1977; *Narrative and Structure: Explanatory Essays* 1979; *The Slumber of Apollo: Reflections on Recent Art, Literature, Language and the Individual Consciousness* 1983

Edited

Poems of the Mid-Century 1957; Shelley *Selected Poems* 1959; Dickens *Little Dorrit* 1967; *Later English Broadside Ballads* [with J. Black] 2 vols 1975, 1979; *The Oxford Book of Local Verses* 1987

John Clellon HOLMES 1926–1988

Holmes was born in Holyoke, Massachusetts, the son of a sales representative. He attended Columbia University and the New School for Social Research in New York, his education being interrupted by war service in the US Navy Hospital Corps. Despite the outcast Bohemianism which forms the subject of his major work, he pursued a respectable academic career: he began as a lecturer in England at Yale and ended up as professor of English at the University of Arkansas.

Holmes has been credited as the inventor of 'the Beat Generation', a phrase which he introduced into print in his first and most important book *Go* (1952; published in the UK as *The Beat Boys* in 1959). He took the word 'Beat' from his friend **Jack Kerouac**, who seems in turn to have taken it from Herbert Huncke (a.k.a. 'Huncke the Junky'). Holmes may not have invented the phrase, but he was an important chronicler of the lifestyle, values and milieu. The 1940s' New York scene in *Go* contains pictures of his friends Kerouac ('Gene Pasternak') and **Allen Ginsberg** ('David Stofsky'), and his 1952 essay 'This is the Beat Generation' is also an important account of its subject.

He followed *Go* with *The Horn*. This told of the rise and fall of black saxophonist Edgar Pool, who seems to be loosely based on Charlie Parker. All Beat writers dreamed of catching the soul of jazz, and *The Horn* is one of the most notable attempts. It recently drew a preface from Archie Shepp, the great tenor saxophonist.

Holmes said that: 'The rebel, the outcast, the artist, the young – all those whose extremes of consciousness match their extremes of experience – are my subjects.' Contemporary blurbs talk of 'a searing story of youth in search of "kicks" ' and a 'strange, savage world of night people, hot music and cool jazz'. For all that, Holmes was a writer of rather traditional seriousness who saw himself as following **D.H. Lawrence**, and he was slightly less 'hip' than the group he chronicled, 'living an instant behind everyone else', as he confessed in *Go*. His fame has dimmed since his heyday as the first major Beat spokesman, when Kerouac was so jealous of him that it ruined their friendship. His critical reputation remains relatively intact and his books genuinely merit the over-used term 'minor classics'. Their place seems assured, not least as period documents.

Fiction

Go 1952 [UK *The Beat Boys* 1959]; *The Horn* 1958; *Get Home Free* 1964

Poetry

The Bowling Green Poems 1977; *Death Drag* 1979

Non-Fiction

Nothing More to Declare 1967; *Visitor: Jack Kerouac in Old Saybrook* 1980; *Displaced Person* 1987

Winifred HOLTBY 1898–1935

The daughter of farming stock on both sides, Holtby spent her early childhood at Rudston in the East Riding of Yorkshire, where she was born. Due to her father's ill health, the household was dominated by women, and in particular her mother Alice, who later became the first woman alderman on the East Riding Council. As a child, she took part in village life, and loved the Yorkshire countryside, experiences and memories she assimilated into her best-known novel, *South Riding*. Educated at Queen Margaret's School, Scarborough, Holtby was astonished to find a book of her own poems in a nearby bookshop when she was thirteen. Publication had been arranged by her mother, who 'prodded' her into authorship.

During the First World War, she worked for a year in a London nursing home, where Beerbohm Tree died in her arms. Her university career at Somerville College, Oxford, which began in 1917, was interrupted until 1919, while she served with the Women's Auxiliary Army Corps in the Signals Unit. It was at Oxford that she met Vera Brittain, whose name has become inseparably linked with hers. Both held the same political values, had a passionate concern for feminism, and wanted to be writers. Spurning academic careers, after they went down from Somerville in 1921 Holtby and Brittain shared a Bloomsbury flat with a tortoise. To supplement a small allowance, Holtby lectured and wrote journalism, meanwhile publishing her first novel, *Anderby Wold* (1923). One of the magazines she contributed to was the feminist *Time and Tide*, of which she later became a director. Her journalistic success led to a lecture tour in South Africa under the auspices of the League of Nations Union in 1926. Her observations of racism and imperialism supplied her with incidents which inspired *The Land of Green Ginger* and *Mandoa, Mandoa!*, and socialism and her work for peace were to be constant themes in her writing. Brittain did not let her semi-detached marriage to Professor George Catlin impinge on her relationship with Holtby with whom she continued to share a home. Holtby and Brittain so often analysed marriage and projected their own friendship through fictional characters that it was assumed they were lovers. Brittain vehemently denied having had sexual relations with any woman, and when she edited a selection of Holtby's correspondence, removed all terms of endearment to her, such as 'sweetieheart'. Nevertheless Brittain's *Testament of Friendship* implicitly states their mutual emotional dependency.

Holtby was tall, golden-haired and gregarious, but her liveliness and driving energy were gradually diminished by Bright's disease. She completed *South Riding* a month before she died, unaware that her mother tried to suppress the novel because it incorporated a portrait of her and her work as a county councillor. Posthumously published, *South Riding* was awarded the James Tait Black Memorial Prize in 1936.

Fiction

Anderby Wold 1923; *The Crowded Street* 1924; *The Land of Green Ginger* 1927; *Poor Caroline* 1931; *Mandoa, Mandoa!* 1933; *Truth Is Not Sober and Other Stories* (s) 1934; *South Riding* 1936; *Pavements at Anderby* (s) [ed. V. Brittain, H.S. Reid] 1937

Non-Fiction

Eutychus, or, The Future of the Pulpit 1928; *A New Voter's Guide to Party Programmes: Political Dialogues* 1929; *Virginia Woolf: a Portrait* 1932; *The Astonishing Island* 1933; *Women, and a Changing Civilisation* 1934; *Letters to a Friend* [ed. A. Holtby, J. McWilliam] 1937; *Selected Letters of Winifred Holtby and Vera Brittain 1920–1930* [ed. V. Brittain, G. Handley-Taylor] 1960; *Testament of a Generation: the Journalism of Vera Brittain and Winifred Holtby* [ed. P. Berry, A. Bishop] 1985

Plays

Take Back Your Freedom [with N. Ginsburg; pub. 1939]

Poetry

My Garden and Other Poems 1911; *The Frozen Earth and Other Poems* 1935

Biography

Testament of Friendship: the Story of Winifred Holtby by Vera Brittain 1940

A(lec) D(erwent) HOPE 1907–

Hope was born in Cooma, New South Wales, the son of a Presbyterian minister. He spent his childhood in Tasmania and Bathurst, NSW, where he went to a school run by the Society of Friends. He studied English and philosophy at Sydney University and won a travelling scholarship to Oxford, where he took the Language Schools in Old and Middle English, receiving a second BA in 1931. Returning to Australia he was briefly unemployed during the Depression years and claims to have lived in a tent while writing verse. In 1933 he started as an English teacher with the NSW Department of Education and as a guidance counsellor for the Youth Employment Bureau. From that time until 1968, when he retired from his post as professor of English at the Australian National University, Canberra, he held a succession of academic posts in Sydney, Melbourne and Canberra. He is now Professor Emeritus at ANU, and the recipient of numerous awards, including an OBE in 1972 and a Companion, Order of Australia, in 1981. He was married in 1938 and has three children.

Despite writing verse since his youth (at eight he wrote a fifty-two-stanza poem for his mother on the virtues of Christianity), Hope did not publish his first collection, *The Wandering Islands*, until he was nearly fifty. Writing firmly in the English tradition according to strict metrical rules, he consciously echoes Byron, Donne and Pope, and has maintained an implacable war against modernism. 'I have consistently attacked the influence of men like **T.S. Eliot**, while admiring the man, and shown irreverent contempt for New Criticism and Old Balderdash alike,' he once remarked. His reputation as Australia's most vituperative literary iconoclast was ensured in 1954 when he reviewed **Patrick White**'s *The Tree of Man*, describing it as 'pretentious and illiterate verbal sludge'.

In his satirical vein, Hope has produced poems such as the mock-heroic *Dunciad Minor*, a defence of Pope against modern detractors which develops into an assault on 'envious Leavis' and other critics. But in his sonnet sequences and in long poems such as 'Vivaldi, Bird and Angel' there is ample evidence of a powerful visionary imagination. His erotic verse (he was the first Australian to have a poem – the bawdy yet scholarly 'The Ballad of Dan Homer' – published in *Playboy*) equally demonstrates his rich lyrical gift.

Poetry

The Wandering Islands 1955; *Poems* 1960; *Collected Poems, 1930–1965* 1966; *New Poems: 1965–1969* 1969; *Dunciad Minor* 1970; *Collected Poems 1930–1970* 1972; *The Damnation of Byron* 1973; *A Late Picking: Poems 1965–1974* 1975; *A Book of Answers* 1978; *A Drifting Continent and Other Poems* 1979; *Antechinus: Poems, 1975–1980* 1981; *The Age of Reason* 1985; *Orpheus* 1991

Non-Fiction

The Structure of Verse and Prose 1943; *The Study of English* 1952; *Australian Literature, 1950–1962* 1963; *The Cave and the Spring: Essays on Poetry* 1965; *The Literary Influence on Academics* 1970; *A Midsummer Eve's Dream: Variations on a Theme by William Dunbar* 1970; *Harry Kendall: a Dialogue with the Past* 1971; *Native Companions: Essays and Comments on Australian Literature, 1936–1966* 1974; *Judith Wright* 1975; *The Pack of Autolycus* 1978; *The New Cratylus: Notes on the Craft of Poetry* 1979; *Poetry and the Art of Archery* 1980; *Directions in Australian Poetry* 1984; *Chance Encounters* (a) 1992

Plays

The Tragical History of Doctor Faustus [from Marlowe's play] 1982; *Ladies from the Sea* [from Homer's *The Odyssey*] 1987

Edited

Australian Poetry, 1960 1960; Harry Kendall *Poems* [with L. Kramer] 1976

Christopher (David Tully) HOPE 1944–

Born in Johannesburg of English-speaking parents, Hope grew up in Pretoria, and was educated at the Christian Brothers' College there, the University of Witwatersand, Johannesburg, and the University of Natal. He served briefly in the South African Navy, was an underwriter for the South British Insurance Company in Johannesburg in 1966, an editor at the publishers Nasionale Pers, and a copywriter at Lintas and at Lindsay Smithers, Durban. He was editor of *Bolt*, Durban, from 1972 to 1973. In 1967 he married and in 1975 he and his wife moved with their two sons to London. He has since been writer-in-residence at Gordonstoun, and has written on African affairs for *London Magazine*, for which he also reviews poetry, and the *Guardian*.

He started by publishing verse, winning the English Academy of Southern Africa Pringle Award for 1974. Of the poems in *Cape Drives*, he wrote: 'My work is an attempt to gauge the stress under which the English-speaking white South African exists ... As I see it, my poems reflect the struggle of one English-speaking white South African to become part of the world again.' *Englishmen* is a long poem, divided into fourteen parts, giving the facts and fictions of South African life in general and its English settlers in particular. He won a Cholmondeley Award for Poetry in 1978.

His novels have been an extension of this exploration. The starting point of *Kruger's Alp*, which won the Whitbread Award, is *The Pilgrim's Progress*. Like Bunyan's Christian, the protagonist, Blanchaille ('white-bait') embarks on an allegorical journey – through South Africa *The Hottentot Room* is concerned with an exclusive club for exiles in Earls Court where Frau Katie, a refugee from Nazi Germany, is queen. She lies on her deathbed while, all around her, her subjects yearn for their past and their African homeland. The novella *Black Swan* is the personal odyssey of a teenage boy from his township to a guerilla training-camp in East Germany. In *My Chocolate Redeemer*, Bella, a teenage chocolate addict, lives by an idyllic lake in France with her aristocratic grandmother who adores the charismatic leader of a right-wing political group dedicated to ridding France of its blacks.

Hope has also published a collection of stories on South African subjects; *White Boy Running*, his first non-fiction book, which was the result of his first real visit to South Africa in twelve years; and another travel book, *Moscow! Moscow!* This body of work has put him in the forefront of the generation of writers to emerge in the 1980s.

Fiction

A Separate Development 1980; *Private Parts and Other Tales* (s) [also as *Learning to Fly*] 1982; *Kruger's Alp* 1984; *The Hottentot Room* 1986; *Black Swan* 1987; *My Chocolate Redeemer* 1989; *Serenity House* 1992;

Poetry

Whitewashes [with M. Kirkwood] 1971; *Cape Drives* 1974; *In the Country of the Black Pig and Other Poems* 1981; *Englishmen* 1985; *The Love Songs of Nathan J. Swirsky* 1993

Plays

Bye-Bye Booyens (tv) 1976; *Ducktails* (tv) 1976; *An Entirely New Concept in Packaging* (tv) 1983

Non-Fiction

White Boy Running 1988; *Moscow! Moscow!* 1990

For Children

The King, the Cat and the Fiddle [with Y. Menuhin] 1983; *The Dragon Wore Pink* [with A. Barrett] 1985

Frances (Margaret) HOROVITZ 1938–1983

Horovitz was born Frances Hooker in Walthamstow, where her father was a 'suppressed and not very articulate' wages clerk. She describes in 'Elegy' her struggle to come to terms with their relationship at the end of his life. In 1942 he was given charge of a small munitions factory and the family moved into his mother-in-law's house in Nottingham. Horovitz has recalled how she used to lie in 'a huge spreading bed of lilies' in her grandmother's garden, 'immersing herself in the scent'; later, when the family returned to Walthamstow after the war, she had visions of fairies dancing on the flowers of a favourite bush in her parents' garden.

The first member of her family to go to grammar school, she attended Walthamstow High School for girls, then read English and drama at Bristol University before going to the Royal Academy of Dramatic Art. On graduating she became the principal reader for the poetry programmes **George MacBeth** edited for the BBC until his retirement in 1976. Her first stage performance was in MacBeth's *The Doomsday Show* at the Establishment Club, Soho. She also taught at schools three or four days a week, giving poetry readings in the evenings. In 1964 she married the poet **Michael Horovitz**, but by 1968 the marriage was in trouble. She yearned for a child and to live in the country, which their circumstances would not at first allow. Her son was born in 1971, and she found a cottage near Slad in the Cotswolds, within sight of the cottage where **Laurie Lee** had grown up. In 1980, she left Horovitz to live with Roger Garfitt, previously editor of the *Poetry Review*. They moved to Sunderland, which she found 'worse than Walthamstow', but the northern landscape inspired some of her last poems, such as 'Rain Birdoswald'. In 1982 skin cancer in her left ear was diagnosed. Her divorce came through on 28 July 1983; she married Garfitt from her sick-bed on 6 September, and died on 2 October.

Horovitz was essentially a poet of place: the Cotswolds, the Black Mountains, Cornwall and finally the North provided her settings, though her last, personal, poems, such as 'Cancer Ward', are heartrending. Her voice was unique and she was a fine reader. As a person she was much loved. When Garfitt's earning power had been restricted while he was looking after her, leading figures in the world of poetry collaborated in a poetry reading to raise money for her family. Her friend and fellow poet **Anne Stevenson** described her both physically and intellectually as 'without exception beautiful'.

Poetry

Poems 1967; *The High Tower* 1970; *Letters to Be Sent by Air* 1974; *Dream* 1976; *Elegy* 1976; *Water Over Stone* 1976; *Wall* [with others] 1980; *Rowlstone Haiken* [with R. Garfitt] 1982; *Snow Light, Water Light* 1982; *Tenfold* 1983; *A Celebration of and for Frances Horovitz* [with others; ed. M. Horovitz] 1984; *Collected Poems* [ed. R. Garfitt] 1985

Michael HOROVITZ 1935–

Horovitz was born in Frankfurt, the youngest of ten children 'descended from cantors, rabbis, vintners and Chasidim of Hungary, Czechoslovakia and Bohemia'. He was brought to England aged two, and went to William Ellis School in London and Brasenose College, Oxford. In 1959 he abandoned postgraduate work on William Blake, one of his idols, to found *New Departures*, called by the *Times Literary Supplement* 'the most substantial avant-garde magazine in Great Britain'; in May 1983 he defended it in an open letter to **John Wain** in the *Times Educational Supplement*, attacking the literary Establishment.

In 1964 he married the poet **Frances Horovitz**, who died in 1983. His first book of verse, *Declaration*, 'in twelve takes', was described by *Manifold* magazine as giving 'a new slant on pacifism ... Here war appears to be the coward's way out, and peace the challenge to heroes'. In 1965 he filled the Albert Hall with an event including seventeen poets from nine countries, and there have been innumerable *New Departure* 'road shows' up and down the country; at these he plays a kazoo and does mime, as well as reading poetry.

It is impossible to separate the poet from the performer. Essentially of the 'Beat Generation', he is a close friend of **Allen Ginsberg**; his other supporters have included **Samuel Beckett**, **William Burroughs**, **Gregory Corso** and **Adrian Mitchell**. Mitchell has said of him: 'His poems are written to be read aloud, chanted, sung, even danced ... Sometimes he fails because he takes large risks in his poems. Risk taking is a rare quality in poetry today ... Few poets can match him for sheer joy.' His most ambitious book is the sequence *The Wolverhampton Wanderer*. *Nude Lines for Larking* is a satire on the neon travesty of the female form as presented in night-time Soho. *Midsummer Morning Jog Log*, a romp around the Cotswolds, has been variously described as a 'slushy poem of praise to flowers, fruits and insects', and 'a song of innocence'; it is dedicated to the memory of his wife. *Growing Up* is his selected poems between 1951 and 1979. He has written a monograph on the painter Alan Davie, translated (with Stefan Themerson) Anatol Stern's *Europa* – described as the first Polish Futurist poem— and edited *Children of Albion*, a famous anthology of 'Underground' poetry in Britain.

Poetry

Declaration 1963; *Nude Lines for Larking in Present Night Soho* 1965; *Strangers* 1965; *High Notes from When I Was Rolling in Moss* 1966; *Poetry for the People* 1966; *Bank Holiday* 1967; *Love Poems* 1971; *The Wolverhampton Wanderer* 1971; *A Contemplation* 1978; *Growing Up* 1979; *Midsummer Morning Jog Log* 1986

Non-Fiction

Alan Davie 1963; *Big League Poets* 1978

Translated

Anatol Stern *Europa* [with S. Thermerson] 1962; Arno Schmidt *The Egghead Republic:* [ed. M. Boyars, E. Kraswlh] 1979

Edited

Children of Albion 1969; *'Big Huge' Reunion Anthology* 1975; *Poetry Olympics* 3 vols 1980–3; *A Celebration of and for Frances Horovitz* 1984; *Grandchildren of Albion* 1992

Janette Turner HOSPITAL 1942–

Hospital was born in Melbourne, Australia, the daughter of Adrian Turner, a painter. Hers was a strict Evangelical Fundamentalist family and she has unhappy memories of school, where she was regarded as a 'religious freak'.

She took a BA from Queensland in 1966 and went to the USA to be a librarian at Harvard from 1967 to 1971, then to Canada where she received an MA and a doctoral fellowship. She married a Methodist minister and university professor Clifford Hospital in 1965, and they have two children with the rather Chaucerian names of Geoffrey and Cressida. She taught English at various institutions while building up a reputation as a writer of short stories for magazines and anthologies in America, Canada, Australia and England. Her short story 'Waiting' won a citation from *Atlantic Monthly* in 1978, and in 1982 she published her first novel, *The Ivory Swing*, which won the Canadian Seal First Novel Award.

Her first collection of short stories was appropriately called *Dislocations*, and she comments: 'I have lived for extended periods in Australia, the USA, Canada, England, and India, and I am very conscious of being at ease in many countries but belonging nowhere. All my writing reflects this. My characters are always caught between worlds or between cultures or between subcultures.' Her sense of mutually contradictory worlds extends to epistemology: 'I feel that I write for global nomads who are integrating slippery and contradictory acts of recording and trying to find some meaning and coherence in that.'

Her novels characteristically have a simple basis – a family reunion in *The Tiger in the Tiger Pit*, or Charade Ryan's search for her father in *Charades* – but they become shifting and uncertain in the telling. Feelings of disorientation are integral to her work and she is concerned to bring out the problematic aspects of both memory and truth with multiple perspectives, although she does believe that 'somewhere absolute truth resides'. In *Charades*, where one of the characters is a physics professor, the old Newtonian certainties have been replaced by Chaos Theory and the Uncertainty Principle.

She divides her year between Canada and Australia, and teaches at universities in both countries. Critics have admired her writing for its style, woman's sensibility and sensuous intelligence, as well as for its global postmodernity.

Fiction

The Ivory Swing 1982; *The Tiger in the Tiger Pit* 1983; *Borderline* 1985; *Dislocations* (s) 1986; *Charades* 1989; *Isobars* (s) 1990; *The Last Magician* 1992

Geoffrey (Edward West) HOUSEHOLD 1900–1988

The son of a barrister who became secretary of education for Gloucester, Household received a conventional education at Clifton and Magdalen College, Oxford, where he took a first in English in 1922. Chance then made him a globetrotter for many years: a friend's father was managing director of the Bank of Romania and Household became secretary there. Four years later he went to Spain with Elders and Fyffes to market bananas and, from there, went to New York in 1929 where, among other things, he wrote articles for a children's encyclopaedia and historical playlets for American educational broadcasting. He married his first wife in 1930. In 1933 he returned to England, but soon became an overseas representative in Europe, the Middle East and South America, then living for a period in Spain.

The outbreak of the Second World War found him back in Romania with a group of commandos waiting to blow up the oil wells, and his subsequent career as an intelligence officer took him all over the Middle East.

He had been making sporadic attempts to become a professional writer since the 1920s, and his first novel, *The Third Hour*, was published in 1938; the following year the novel that was always to remain his most famous, *Rogue Male*, was serialised in *Atlantic Monthly*. This story of a lone English gentleman stalking to kill an unnamed dictator inaugurated Household's successful formula of pursuit, tension and revenge. He wrote more than twenty other novels in this style (one of them, *Face to the Sun*, was published in the year of his death), some frankly neo-Buchanesque but others making imaginative and more updated use of a variety of foreign locations. He also published several collections of short stories, which he himself considered his best work. His writing enabled Household to spend his last forty years as a hard-shooting country gentleman, first in Dorset and later in Buckinghamshire; the main excitement of his later years was when the threat of a third London airport in his area in 1970 turned him into a key member of the Cublington Underground Resistance Movement. He had married again, in 1942, and his second wife survived him with their three children.

Fiction

The Salvation of Pisco Gabar (s) 1938; *The Third Hour* 1938; *Rogue Male* 1939; *Arabesque* 1948; *The High Place* 1950; *A Rough Shoot* 1951; *A Time to Kill* 1951; *Tales of Adventurers* (s) 1952; *Fellow Passenger* 1955; *The Brides of Solomon* (s) 1958; *Watcher in the Shadows* 1960; *Thing to Love* 1963; *Olura* 1965; *Sabres on the Sand* (s) 1966; *The Courtesy of Death* 1967; *Dance of the Dwarfs* 1969; *Doom's Caravan* 1971; *The Three Sentinels* 1972; *The Lives and Times of Bernardo Brown* 1973; *The Cats to Come* 1975; *Red Anger* 1975; *Hostage: London* 1977; *The Last Two Weeks of Georges Rivac* 1978; *The Europe That Was* (s) 1979; *Summon the Bright Water* 1981; *The Sending* 1980; *Capricorn and Cancer* (s) 1981; *Rogue Justice* 1983; *Arrows of Desire* 1985; *Face to the Sun* 1988

For Children

The Terror of Villadonga 1936 [as *The Spanish Cave* 1940]; 1936; *Xenophon's Adventure* 1955; *Prisoner of the Indies* 1967; *Escape into Daylight* 1976

Non-Fiction

Against the Wind (a) 1958

A(lfred) E(dward) HOUSMAN 1859–1936

Housman was the eldest of seven children (including the popular historical playwright, **Laurence Housman**) of a prosperous solicitor practising in Bromsgrove, Worcestershire. The idealised rural Shropshire of Housman's poetry was thus the 'western horizon' of his childhood, eternally near and just out of reach. His mother died when he was twelve, but he subsumed his grief in a growing attachment to the classics, developed at Bromsgrove High School, and he went up to St John's College, Oxford, in 1877 already marked down as a brilliant classical scholar. While at Oxford, he fell deeply in love with his contemporary Moses Jackson, and it is believed that it was the unhappiness connected with Jackson's failure to respond to him which led him to plough his Finals and take only a Pass degree.

He took the civil service examination, and became a clerk in the Patent Office in London from 1882 to 1892, continuing to read the classics in the British Museum after work, and sharing bachelor lodgings with Jackson until the latter left to work in India in 1887. Housman began to publish esteemed scholarly papers in the 1880s, and in 1892 seventeen eminent scholars supported his application, which was successful, to become professor of Latin at University College, London.

A great upsurge of poetic inspiration in the early months of 1895 led to the poems of *A Shropshire Lad*, published at the author's expense in 1896. The great love of Housman's life, while contributing to the ferocious personal reserve he maintained in later years, had also inspired a totally individual poetry, combining stoic bitterness with passionate lyricism, inhabiting a world of beautiful young soldiers doomed to die and rustics doomed to hang. The book was not an immediate success, but its romantic pessimism gradually gained admirers, particularly those like **E.M. Forster** who discerned a homosexual subtext. 'To my generation,' wrote **W.H. Auden**, 'no other English poet seemed so perfectly to express the sensibility of the male adolescent.' Reflecting the mood of the times, sales boomed in the years before the First World War: in 1905 the book sold 886 copies, but by 1911 the average yearly sale was 13,500 copies. The popularity of the work was increased when composers began to set the poems to music. Housman did not approve of this, but never refused permission (or indeed accepted any fee). Amongst the better-known settings of poems, either individually or as song cycles, were those by Somervell, **Ivor Gurney**, Vaughan Williams, Butterworth and Ireland. During the war, copies of the book were to be found 'in every pocket', according to **Robert Nichols**, and some of the poems were published as a broadsheet for the troops by *The Times*. In 1918, in spite of the fact that the book had doubled in price from sixpence to one shilling, some 16,000 copies were sold.

Housman had become professor of Latin at Trinity College, Cambridge, in 1911 and remained there until his death in 1936, discreetly indulging his homosexual tastes during frequent visits to the Continent, particularly Venice. His *magnum opus* as a scholar was his five-volume edition (1903–30) of the minor Augustan poet M. Manilius, and his scathing comments on other men's work made him the terror of English and German classicists. In 1922, news of Jackson's impending death inspired many of the verses in *Last Poems*. A volume of *More Poems* followed posthumously. Housman as a poet is one of the great minors: although his range is limited and his pessimism sometimes veers on self-parody, his best lines have a doomed music which is at once unmistakable and unforgettable.

Poetry

A Shropshire Lad 1896; *Last Poems* 1922; *Three Poems* 1935; *More Poems* [ed. L. Housman] 1936; *The Manuscript Poems of A.E. Housman* [ed. T.B. Haber] 1955; *Complete Poems* [ed. B. Davenport, T. Burns Haber] 1959

Edited

M. Manilii Astronomica 5 vols 1903–30; *D. Iunii Invenalis Saturae* 1905; *Lucani Belli Civile* 1926

Non-Fiction

Introductory Lecture 1892; *Nine Essays* (c) [ed. A. Platt] 1927; *The Name and Nature of Poetry* 1933; *Thirty Housman Letters to Witter Bynner* [ed. T.B. Haber] 1957; *A.E. Housman: Selected Prose* [ed. J. Carter] 1961

Collections

A.E. Housman: Some Poems, Some Letters and a Personal Memoir by His Brother Laurence Housman 1937

Biography

A.E. Housman: the Scholar-Poet by Richard Perceval Graves 1979

Laurence HOUSMAN 1865–1959

Housman was born in Bromsgrove in the West Midlands, the sixth of seven children of a solicitor, of whom the eldest was the poet **A.E. Housman**; more important to Laurence was a sister, Clemence, also a writer, with whom he was to live for the greater part of his life. Despite the early death of the children's mother, and their father's remarriage, family life was happy, if increasingly impoverished, which probably accounted for Housman not attending university after leaving Bromsgrove School. Instead, he went with Clemence to London to study art, and the earliest six years of his career were passed as a book designer and illustrator, working for Kegan Paul and John Lane. His first book was a collection of fairy-tales, *A Farm in Fairyland* (1894), and he went on to publish several volumes of poetry in the 1890s, but his first substantial success was with the anonymously published novel *An Englishwoman's Love-Letters*, in 1900.

Shortly afterwards, encouraged by **Harley Granville-Barker**, he launched into the drama, the field in which he was to become best known. His first play was biblical, *Bethlehem*, the first of many dramatic works to involve him in battles with the Lord Chamberlain. Censorship was imposed not for reasons of sexual explicitness but because Housman chose scriptural subjects and those involving the royal family. At that time there was a ban on impersonating biblical characters and members of the royal family on stage. (Even a play about George IV's rejected wife, Caroline of Brunswick, was initially censored.) Therefore many of his plays were first produced in private theatre clubs, at Rutland Boughton's Glastonbury Festival, and by the Dramatic Society of London University, before achieving commercial productions when the censorship was lifted, as it usually was. Housman's most original contribution to the stage was to group large numbers of short plays about a single character into longer works, and two series in particular became famous: *The Little Plays of St Francis*, eventually sixth brief dramas, and the series of plays about Queen Victoria, the most substantial part of which was consolidated as *Victoria Regina*. Edward VIII himself intervened to lift the censorship from the latter, in which the young Vincent Price played Prince Albert. Housman also wrote propaganda plays about the many radical causes (including woman's suffrage, Irish Home Rule and pacifism) to which he was committed, as well as 'dialogues', not intended for stage performance, often on legendary and literary subjects (his *Echo de Paris*, for instance, was based on the encounter he had had with Oscar Wilde in Paris in 1899). Several novels, many miscellaneous essays and much poetry completed an *oeuvre* which ran to almost 100 separate works. But, despite the skill he showed, he seems mainly interesting now as an essentially 1890s' figure who adopted himself into a prolific twentieth-century writer; the plangent, deeply individual manner in which his brother gave voice to a repressed homosexuality which they shared was lacking. His production slowed from the early 1940s and his last skirmish with the censor was in 1950. From 1924 he had lived with Clemence in the village of Street in Somerset, where she died in 1955. He followed her four years later, aged ninety-three, the last of the seven Housmans.

Plays

Bethlehem 1902; *Prunella* [with H. Granville Barker] 1904 [pub. 1906]; *The Vicar of Wakefield* [from Goldsmith's novel] 1906; *Oxford Historical Pageant* [with L. Binyon, R. Bridges and others] 1906 [pub. 1907]; *A Chinese Lantern* 1908; *A Likely Story* 1910; *The Lord of the Harvest* 1910 [pub. 1916]; *Lysistrata* [from Aristophanes' play] 1910 [pub. 1911]; *Alice in Ganderland* 1911; *Pains and Penalties* 1911; *Bird in Hand* 1918 [pub. 1916]; *Angels and Ministers* 1921; *The Fairy* 1921 [pub. 1916]; *The Little Plays of St Francis* 1921 [pub. 1922]; *Possessions* 1921; *St Martin's Pageant* 1921; *False Premises* 1922; *A Fool and His Money* 1922; *The Torch of Time* 1922; *The Followers of St Francis: Four Plays of the Early Franciscan Legend* 1923; *The Death of Socrates* 1925; *The Comments of Jupiter* 1926; *Ways and Means* 1926; *His Favourite Flower* 1927; *Cornered Poets* 1929; *The Queen: God Bless Her!* 1929; *Mr Gladstone's Comforter* 1929; *The New Hangman* 1930; *Palace Plays* 1930; *The Queen's Progress* 1932; *Nunc Dimittis* 1933; *Victoria and Albert* 1933; *Four Plays of St Clare* 1934; *Victoria Regina* 1935; *Palace Scenes* 1937; *Twelve One-Act Plays* [with R. Latcham] 1938; *Gracious Majesty* 1941; *Jacob* 1942; *Palestine Plays* 1942; *Samuel the King-Maker* 1944; *The Family Honour* 1950; *Consider Your Verdict* 1950

Fiction

All-Fellows (s) 1896; *Gods and Their Makers* (s) 1897 [rev. 1920]; *An Englishwoman's Love-Letters* 1900; *The Missing Answers to An Englishwoman's Love-Letters* 1901; *A Modern Antaeus* 1901; *Tale of a Nun* [with L. Simons] 1901; *Sabrina Warham* 1904; *The Cloak of Friendship* 1905; *The Wheel* 1910; *John of Jingalo* 1912; *The Bawling Brotherhood* 1913; *The Royal Runaway and Jingalo in Revolution* 1914; *Good as Gold* 1916; *Nazareth* 1916; *Return of Alcestis* 1916; *The Snow Man* 1916; *Dethronements* (s) 1922; *Trimblerigg* 1924; *Odd Pairs* (s) 1925; *The Open Door* 1925; *Ironical Tales* 1926; *The Life of H.R.H. the Duke of Flamborough* 1928; *Ye Fearful Saints!* 1932; *What Next?* 1938; *Big Powers and Little Powers* 1944; *Strange Ends and Discoveries* (s) 1948; *The Kind and the Foolish* (s) 1952

Non-Fiction

Articles of Faith in the Freedom of Women 1910; *The Immoral Effects of Ignorance in Sex Relations* 1911; *The 'Physical Force' Fallacy* 1913; *Be Law Abiding!* 1914; *The Moving Spirit in Womanhood* 1914; *What is 'Womanly'?* 1914; *Great Possessions* 1915; *Christianity a Danger to the State* 1916; *Ploughshare and Pruninghook* 1919; *Echo de Paris: a Study from Life* 1923; *The New Humanism* 1923; *A Thing to be Explained* 1926; *A Substitute for Capital Punishment* 1929; *Histories: Introductory to Marten and Carter's Histories* [with C.H.K. Marten] 4 vols 1931–2; *The Long Journey: the Tale of Our Past* [with C.H.K. Marten] 1933; *The Unexpected Years* (a) 1936; *What Can We Believe?* [with H.R.L. Sheppard] 1939; *The Preparation of Peace* 1940; *Military Necessity in the Middle Ages* 1941; *What Price Salvation Now?* 1949

For Children

A Farm in Fairyland (s) 1894; *The House of Joy* 1895; *The Field of Clover* (s) 1898; *Seven Young Goslings* 1899; *The Blue Moon* (s) 1904; *The Sheepfold* 1918; *Uncle Tom Pudd* 1927; *Etheldrinda's Fairy* (s) 1928; *Wish to Goodness!* (s) 1928; *The Boiled Owl and The Ass's Mouth* (s) 1930; *Busybody Land . . . and The Cuckoo Whose Nest Was Made for Her* (s) 1930; *Cotton-Woolleena* (s) 1930; *A Gander and His Geese and The Promise of Beauty* (s) 1930; *Little and Good and The Giant and the Pigmy* (s) 1930; *Turn Again Tales* 1930; *A Clean Sweep* (s) 1931; *What-o'Clock Tales* 1932

Poetry

Green Arras 1896; *Spikenard* 1898; *The Little Land* 1899; *Rue* 1899; *Mendicant Rhymes* 1906; *The New Child's Guide to Knowledge* 1911; *The Winners* 1915; *The Heart of Peace* 1918; *The Death of Orpheus* 1921; *Puss-in-Boots* 1926; *The Love Concealed* 1928; *Hop-o'-me-Heart* 1938

Edited
The Venture 1903; *War Letters of Fallen Englishmen* 1930

Translated
Of Aucassin and Nicolette and Amabel and Amoris 1902

Elizabeth Jane HOWARD 1923–

Howard was born in London, the daughter of David Liddon and Katherine Howard, a dancer with the Diaghilev ballet. She was privately educated before training as an actress at the London Mask Theatre School. During the war she acted with the Scott Thorndyke Student Repertory, had roles in radio and television, and worked as a model. She also served as an air-raid warden. Married in 1942 to the naturalist Peter Scott, with whom she had one daughter, she was secretary of the Inland Waterways Association from 1947 to 1950, after which she worked as a publisher's reader, and later as an editor at Chatto and Windus, and Weidenfeld and Nicolson. She was book critic of *Queen* magazine from 1957 to 1960, honorary artistic director of the Cheltenham Festival in 1962, and has since contributed to magazines such as the *New Yorker* and *Vogue*. Having divorced Scott, she was married to James Douglas-Henry in 1959, but her most famous husband was **Kingsley Amis**, to whom she was married from 1965 to 1983. Howard and Amis were a glamorous and formidable literary pair who were much in the public eye. The marriage has been described as 'stormy', but Amis declined to discuss it in his otherwise score-settling *Memoirs* (1991).

Howard has said she started writing plays when she was fourteen, and 'before that I wrote 400 immensely dull pages (since destroyed) about a horse'. Her first novel, *The Beautiful Visit*, which won the John Llewelyn Rhys Memorial Prize in 1950, is a delicate and introspective account of a young girl's passage to maturity in which, as in all of her subsequent novels, the subtleties of relationships are closely examined. Howard has said of her work: 'I do not write about "social issues or values" – I write simply about people, by themselves and in relation to others.' An element of the comedy of manners enters her work with *After Julius* (some passages of which Amis is alleged to have co-operated on) and *Something in Disguise*, both of which Howard later adapted for television series. *The Light Years, Marking Time* and *Confusion* are the first three novels in *The Cazalet Chronicle*, a projected quartet which traces the lives of members of a prosperous middle-class family in the years around the Second World War. In addition to fiction Howard has written a biography of Bettina von Arnim, two film scripts, radio plays and television scripts (including an episode of the popular period soap-opera *Upstairs, Downstairs*).

Fiction
The Beautiful Visit 1950; *We Are for the Dark* [with R. Aickman] 1951; *The Long View* 1956; *The Sea Change* 1959; *After Julius* 1965; *Something in Disguise* 1967; *Odd Girl Out* 1972; *Mr Wrong* (s) 1975; *Getting It Right* 1982; *The Cazalet Chronicle*: *The Light Years* 1990; *Marking Time* 1991; *Confusion* 1993

Non-Fiction
Bettina: a Portrait [with A. Helps] 1957

Edited
The Lover's Companion 1978; *Howard and Maschler on Food* [with F. Maschler] 1987; *Green Shades: an Anthology of Plants, Gardens and Gardeners* 1991

W(illiam) H(enry) HUDSON 1841–1922

Hudson was born at Quilmes, near Buenos Aires, the son of working-class Anglo-American parents who, after an accident in the Massachusetts brewery where the father worked, had come to Argentina to farm. Their son grew up wild on the ranches of the Río de la Plata, but rheumatic fever at fifteen made an active life difficult and concentrated his interest on the observation of animals, particularly birds. He spent years roving on horseback round Brazil and Uruguay, achieving renown as an ornithologist, and finally making his famous journey to Patagonia, where he discovered a new species, the *Enipolegus hudsoni*.

He came to England (authorities differ as to whether this happened in 1869 or 1874), and for some years lived in poverty in London, working as secretary to the archaeologist Chester Walters for a time. In 1876 he married Emily Wingrave, who was fifteen years his senior. She earned a small living giving music lessons at their home in Westbourne Park, and they also managed a series of boarding-houses; the marriage was reputedly unhappy, and yielded no children.

Hudson's first book, published in 1885, was *The Purple Land That England Lost*, a collection of stories set in South America, and this was followed by a Utopian novel, *A Crystal Age*, in 1887; but for many years he wrote chiefly as a naturalist, drawing on his South American experiences in a simple, pleasant style, with titles such as *The Naturalist in La Plata* and *Idle Days in Patagonia*. His reputation grew only slowly, but in 1901 he gained the support of the famous editor Richard Garnett and was also awarded a civil list pension, which enabled him to spend his latter years in ornithological observations around Britain and write a series of books about the experience. He also made a number of (not particularly successful) experiments in verse.

Towards the end of his life his fame was very great, and he crowned his literary production in 1918 with his autobiography of boyhood, *Far Away and Long Ago*. This probably remains his most famous book, but he is also remembered for several novels, particularly *Green Mansions* (1904). His fiction almost always has South American settings, varies little in plot, and typically concerns the search for nature and the perfect woman; it is simple but strong. Probably much of Hudson's lasting appeal comes from his remote and outdoor background, so unusual in a British literary man, but there is no doubt that he is still read for pleasure.

Fiction

The Purple Land That England Lost (s) 1885; *A Crystal Age* 1887; *Fan* [as 'Henry Harford'] 1892; *El Ombu* 1902; *Green Mansions* 1904; *A Little Boy Lost* 1905; *Dead Man's Plak* (s) 1920

Non-Fiction

The Naturalist in La Plata 1892; *Birds in a Village* 1893; *Idle Days in Patagonia* 1893; *British Birds* 1895; *Birds in London* 1898; *Nature in Downland* 1900; *Birds and Man* 1901; *Hampshire Days* 1903; *The Land's End* 1908; *Afoot in England* 1909; *A Shepherd's Life: Impressions of the South Wiltshire Downs* 1910; *Adventures among Birds* 1913; *Far Away and Long Ago* (a) 1918; *Birds in Town and Village* 1919; *The Book of a Naturalist* 1919; *Birds of La Plata* 1920; *A Traveller in Little Things* 1921; *A Hind in Richmond Park* 1922

Biography

W.H. Hudson by Ruth Tomalin 1984

Ford Madox HUEFFER

see Ford Madox FORD

David (John) HUGHES 1930–

Hughes was born in Alton, Hampshire, and, owing to moves dictated by wartime, grew up partly there and partly in Wimbledon, south London; established at the latter after 1945, he attended a minor public school, King's College School, Wimbledon. After military service in the Royal Air Force between 1949 and 1951 (he rose to pilot officer), he attended Christ Church, Oxford, where he read English, and signalled his literary interests by editing the student magazine *Isis*. After graduating, he worked for two years, from 1953 to 1955, as an assistant on the recently founded *London Magazine*; then, between 1956 and 1960, he was a reader for Rupert Hart-Davis among other publishers. Since his editorship of the magazine *Town*, between 1960 and 1961, he has been a full-time writer.

He had published his first novel, *A Feeling in the Air* (US title, *Man Off Beat*), in 1957, and in the following year he married the Swedish actress and director Mai Zetterling, then near the end of the early part of her career in which she filled undemanding roles in minor films. Her real ambition was to direct, and from 1961 to 1968 the couple lived in Sweden, where Hughes developed a second career scripting the feature films and documentaries that his wife made both in Sweden and for the BBC in London. From 1970 to 1974, they retired to France to live off the land, but they returned to London in the latter year, and were divorced in 1976 (she died in 1994). Since then, Hughes has largely divided his time between London and South Wales, and in 1980 he remarried and subsequently had two children.

Hughes's first three novels (the first really a novella) saw him exploring his talent, and he found form with his fourth, *The Major* (1964), the powerful story of an army officer whose harshness eventually leads him to commit suicide. Succeeding novels explored a broad range of setting, subject-matter and characters, with perhaps the legacy of the century's wars providing a unifying theme. This was the background of the most successful of them, *The Pork Butcher* (1984), where a German officer finds redemption in returning to the scene of his wartime crimes in occupied France; the novel won the W.H. Smith Award and was filmed as *Souvenir* in 1989. However, while he often found critical favour, the public reception for his novels was disappointing, and he has not published one since *But for Bunter* (1985). Hughes's witty set of variations upon a theme by Frank Richards is in fact one of his best novels, ostensibly the story of a man who claims to be the model for Bunter, but also an exuberant investigation of fact and fiction, truth and falsehood, history and literature and romance and reality.

His non-fictional output is not extensive but varied. An early study of **J.B. Priestley** (1958) was judicious but warm in its tone. *The Road to Stockholm and Lapland* was an excellent travel book written for the BP touring service, while *The Seven Ages of England* was the by-product of a series Hughes wrote for Swedish television. *The Rosewater Revolution* is unclassifiable: neither fact nor fiction, partly writer's reflections, partly recipe for personal growth. Since 1986, Hughes has edited the annual volumes of *Best Short Stories* with the literary agent Giles Gordon. He has himself written a number of short stories, taught writing at several American universities, and held various journalistic posts, most durably that of fiction critic of the *Mail on Sunday* since 1982.

Fiction

A Feeling in the Air [US *Man Off Beat*] 1957; *Sealed with a Loving Kiss* 1959; *The Horsehair Sofa* 1961; *The Major* 1964; *The Man Who Invented Tomorrow* 1968; *Memories of Dying* 1976; *A Genoese Fancy* 1979; *The Imperial German Dinner Service* 1983; *The Pork Butcher* 1984; *But for Bunter* 1985

Non-Fiction

J.B. Priestley: an Informal Study 1958; *The Road to Stockholm and Lapland* 1964; *The Seven Ages of England* 1968; *The Rosewater Revolution – Notes on a Change of Attitude* 1971

Edited

Winter's Tales: New Series I (s) 1985; *Best Short Stories* (s) [with G. Gordon] 1986–9; *The Best of Best Short Stories* [with G. Gordon] 1995

Glyn HUGHES 1935–

Hughes was born and brought up in rural Cheshire. His father was a ticket inspector on the buses, a lover of nature, **H.G. Wells** and socialism; his mother had gone into service at the age of fifteen. Hughes gives an affectionate picture of their agonised marriage in an autobiographical piece, 'My Father's Roses'. He was educated at Altrincham Grammar School and the Regional College of Art, Manchester. 'My childhood', he says, 'was, by choice, fairly solitary.' For a time he earned a living as an art teacher, hating it; he now lives by his writing and, his wife being Greek, he has since 1974 divided his time between Yorkshire and Greece.

In the late 1960s and throughout the 1970s he published many pamphlets and books of verse, his work springing directly from the harsh northern landscape and its people. *The Stanedge Bull* was the first; *Neighbours* was a Poetry Book Society choice, *Presence* a Poet-of-the-Month choice; *Best of Neighbours* contains the new and selected poems he wished to keep.

In the 1980s he started to publish novels. *Where I Used to Play on the Green*, a study in religious fanaticism, is set at Haworth in the eighteenth century. The central figure is the Revd William Grimshaw, the leader of northern Methodism and predecessor of the Revd Patrick Brontë. It won the 1982 *Guardian* Fiction Prize and the David Higham Award. *The Hawthorn Goddess*, also set in West Yorkshire, is the first of a trilogy. Its heroine Anne Wilde is a symbolic figure, the personification of Man's exploitation of Nature, her wildness tamed as the commons are to be enclosed and the weavers confined in the mills. *The Rape of the Rose* sees her in the nineteenth century. The protagonist of *The Antique Collector* is a drag artiste.

Hughes has been called 'the northern **Thomas Hardy**' by reviewers keen on labels, presumably because they are both poets who turned to writing fiction from their native countryside. There the comparison ends; Hughes has his own voice, which has been described as 'eerie and rhapsodic'.

Poetry

The Stanedge Bull and Other Poems 1966; *Almost-Love Poems* 1968; *Love on the Moor: Poems 1965–8* 1968; *Neighbours: Poems 1965–9* 1970; *Presence* 1971; *Towards the Sun* 1971; *Rest the Poor Struggler: Poems 1969–71* 1972; *The Beast* 1973; *Alibis and Convictions* 1978; *Best of Neighbours* 1979

Fiction

Where I Used to Play on the Green 1982; *The Hawthorn Goddess* 1984; *The Rape of the Rose* 1987; *The Antique Collector* 1990

Plays

Mary Hepton's Heaven (r) 1984; *The Yorkshirewomen* 1978; *Dreamers* 1979

Non-Fiction

Millstone Grit 1975; *Fair Prospects: Journey in Greece* 1976

Edited

Selected Poems of Samuel Laycock 1981

(James Mercer) Langston HUGHES 1902–1967

A leading figure of the Harlem Renaissance, which flourished in New York in the 1920s and 1930s, Hughes was born in Joplin, Missouri. His parents separated and his mother had to move from city to city in search of work: before he was twelve Hughes had lived in Mexico, Topeka, Kansas, Colorado, Indiana and Buffalo. In 1923 he enrolled at Columbia University, New York, but was soon distracted by Broadway, jazz and the blues artists of Harlem. He boarded a freighter as cabin-boy and, dispatching a box of books into the sea off New Jersey, set sail for West Africa, in search of his roots. In 1924 he was in Paris, a doorman at a Montmartre night-club, then he travelled through Italy. His work was brought to public attention by **Vachel Lindsay**, and his first book of poetry, *The Weary Blues*, was published in New York in 1926.

Hughes declared that his poetry was about 'workers, singers and job hunters in NY, Washington and Chicago. People up today and down tomorrow, working this week and fired the next ... hoping to get a new suit for Easter and pawning that suit before the 1st July.' He further said that his poems were 'to be read aloud, crooned, shouted and sung. None with a faraway voice.' He is Afro-Americana in the blues mood, and many of his poems have been set to music. Throughout his career, Hughes sought to depict the lives of ordinary black people in the USA and said that he wrote 'only for the hearts of my people'. To this end he created the character of Jesse B. Semple, a plain-speaking, ironic, street-wise Harlem dweller whose comments upon life first appeared in the Chicago *Whip*, and were later collected into four volumes beginning with *Simple Speaks His Mind.* Contemporary black academics considered Hughes's presentation of black people to be unattractive and adverse to racial integration in literature. He defied them further by publishing the satirical shorts entitled *The Ways of White Folks* in 1934.

His first novel, *Not without Laughter*, is set in Kansas and recounts the story of a mother and her three daughters. It came out in 1930 and the following year Hughes bought a Ford and toured the colleges of southern America as teacher and poet. In 1932, he travelled again, this time to Russia, returning to San Francisco in 1933 via the Orient, where his interest in the backstreet life of Tokyo caused the local police to detain and expel him. Two years later he won a Guggenheim Fellowship and his first play, *The Mulatto*, opened on Broadway. He went on to found black theatre groups in Harlem, Chicago and Los Angeles. During the Spanish Civil War, as Madrid correspondent for the *Baltimore Afro-American*, Hughes stayed at the Alianza de Intelectuales, looking out on the gunfire from the room above **Ernest Hemingway**, with whom he became friends and with whom he attended several bullfights. He recognised in the vital, undefeated music of the Spanish something akin to the blues.

There has been considerable, though inconclusive, speculation about Hughes's sexuality. Several of his friends and patrons among the leading figures of the Harlem Renaissance were homosexual – including Lindsay, Alain Locke and **Carl Van Vechten**, for whose controversial novel *Nigger Heaven* (1926) Hughes provided snatches of blues lyrics – and he never married. In later years he held posts at the Universities of Chicago and Atlanta. His work has been translated into many languages and has won a secure place in American literature. He restored the rhythmical language of Africa into black writing and consistently demonstrated inextinguishable hope to his fellow African-Americans, remaining perhaps the most significant black writer of the century.

Poetry

The Weary Blues 1926; *Fine Clothes to the Jew* 1927; *Dear Lovely Death* 1931; *The Negro Mother and Other Dramatic Recitations* 1931; *The Dream Keeper and Other Poems* 1932; *Scottsboro Limited* 1932; *A New Song* 1938; *Shakespeare in Harlem* 1942; *Freedoms' Plow* 1943; *Jim Crow's Last Stand* 1943; *Lament for Dark Peoples and Other Poems* 1944; *Fields of Wonder* 1947; *One-Way Ticket* 1949; *Montage of a Dream Deferred* 1951; *Ask Your Mama* 1961; *The Panther and the Lash* 1967

Fiction

Not without Laughter 1930; *The Ways of White Folks* (s) 1934; *Simple Speaks His Mind* (s) 1950; *Laughing to Keep from Crying* (s) 1952; *Simple Takes a Wife* (s) 1953; *Simple Stakes a Claim* (s) 1957; *Tambourines to Glory* 1958; *Something in Common and Other Stories* (s) 1963; *Simple's Uncle Sam* (s) 1965

Plays

Little Ham 1935; *The Mulatto* 1935; *Emperor of Haiti* 1936; *Troubled Island* 1936; *When the Jack Hollers* 1936; *Front Porch* 1937; *Joy to My Soul* 1937; *Soul Gone Home* 1937; *Don't You Want to Be Free?* 1938; *The Em-Fuehrer Jones* 1938; *Limitations of Life* 1938; *Little Eva's End* 1938; *The Organizer* 1939; *The Sun Do Move* 1942; *Way Down South* (f) 1942; *For This We Fight* 1943; *The Barrier* 1950; *Simply Heaven* 1957; *Black Nativity* 1961; *Gospel Glory* 1962; *Five Plays by Langston Hughes* [ed. W. Smalley] 1963; *Jericho Jim Crow* 1963; *The Prodigal* 1965; *Soul Yesterday and Today* 1965; *Angelo Herndon-Jones* 1966; *Mother and Child* 1966; *Outshines the Sun* 1966; *Trouble with the Angels* 1966

Non-Fiction

The Big Sea (a) 1940; *The Sweet Flypaper of Life* [with R. de Cavara] 1955; *A Pictorial History of the Negro in America* [with M. Meltzer] 1956; *I Wonder as I Wonder* (a) 1956; *Fight for Freedom: the Story of the NAACP* 1962; *Black Magic: a Pictorial History of the Negro in American Entertainment* [with M. Meltzer] 1967; *Black Misery* 1969; *Good Morning Revolution: the Uncollected Social Protest Writing of Langston Hughes* [ed. F. Berry] 1973

For Children
Popo and Fifina: Children of Haiti [with A. Bontemps] 1932; *The First Book of Negroes* 1952; *Famous American Negroes* 1954; *The First Book of Rhythms* 1954; *Famous Negro Music Makers* 1955; *The First Book of Jazz* 1955; *The First Book of the West Indies* 1956; *Famous Negro Heroes of America* 1958; *The First Book of Africa* 1960

Translations
Jacques Roumain *Masters of Dew* [with M. Cook] 1947; Nicholas Guillen *Cuba Libre* [with F. Carruthers] 1948; *Selected Poems of Gabriel Mistral* 1957

Biography
The Life of Langston Hughes: I, Too, Sing America by Arnold Rampersad 1986

Richard (Arthur Warren) HUGHES 1900–1976

Of Welsh descent, Hughes was born in Surrey and educated at Charterhouse and Oriel College, Oxford. His first appearance in public print was in 1917 when a school essay he wrote attacking *The Loom of Youth* by **Alec Waugh** was sent by his form master to the *Spectator*, where it was published as a youthful counterblast to Waugh's attack on the public-school system.

While still at Oxford Hughes varied a busy vacation career of begging, pavement artistry and political intrigue in central Europe with publishing his first volume of poems and having a play produced in the West End. **George Bernard Shaw** described it as 'the finest one-act play ever written'.

His later career, while distinguished, seems almost an anticlimax after this brilliant start, being characterised by extreme slowness of production and, consequently, a reputation not quite commensurate with his gifts. The inter-war years were marked by poor health and extensive foreign travel: he published two highly praised novels, *A High Wind in Jamaica* (1929), a classic story of childhood, and *In Hazard* (1938). In 1932, he married the painter Frances Bazley: they had five children and lived largely in Wales. During the war he served in the Admiralty. In 1961, he published *The Fox in the Attic*, a novel intended as the first part of 'The Human Predicament', a projected sequence of possibly four parts. He knew that he was never likely to finish it, and in spite of financial encouragement by his publisher, he completed only one further volume – *The Wooden Shepherdess* – before his death.

Hughes filled his life outside writing with a devotion to theatre and a penchant for adventurous travel, and was a Merlin-like man with a sharp red beard, of gnomic impressiveness. **Rebecca West** tells the story of a servant who feared to show him in to her mistress because 'he looked so much like Our Lord'. The beauty of his prose style and originality of his conceptions will ensure him the status of a writer of minor classics.

Fiction
A Moment of Time (s) 1926; *A High Wind in Jamaica* 1929; *In Hazard* 1938; *The Fox in the Attic* 1961; *The Wooden Shepherdess* 1973; *Into the Lap of Atlas* (s) 1979

Poetry
Gipsy-Night and Other Poems 1922; *Confessio Juvenis: Collected Poems* 1926

Plays
The Sisters' Tragedy 1922; *The Sisters' Tragedy and Three Other Plays* [pub. 1924]

Non-Fiction
The Administration of War Production [with J.D. Scott] 1955; *Fiction as Truth: Selected Literary Writings* [ed. R. Poole] 1983

For Children
The Spider's Palace and Other Stories (s) 1932; *Don't Blame Me! and Other Stories* (s) 1940; *Gertrude's Child* (s) 1967; *The Wonder Dog* (s) 1977

Edited
John Skelton: Poems 1924

Biography
Richard Hughes by Richard Perceval Graves 1994

Ted (i.e. Edward James) HUGHES 1930–

Hughes was born in the remote Yorkshire town of Mytholmroyd. His father was a carpenter, and later a shopkeeper, and one of only seventeen from his regiment to have survived at Gallipoli during the First World War. Both the bleak moorland landscape of his childhood and his father's recollections of the suffering and violence of the war were to exert a powerful influence on the development of Hughes's poetic sensibility.

After leaving Mexborough Grammar School, Hughes did two years' national service with the Royal Air Force, much of it spent on an isolated radio station in Yorkshire with 'nothing much to do but read and reread Shakespeare and watch the grass grow.' At Pembroke College, Cambridge, he switched his studies from English to archaeology and anthropology, and immersed himself in the poetry of **W.B. Yeats** and – an important influence – **Robert Graves**'s *The White Goddess* (1948). After graduating in 1954, he moved to London where he worked variously as a zoo attendant, gardener and script reader for J. Arthur Rank.

In Cambridge in 1956, he and some friends launched a literary magazine, *St Botolph's Review*, with lasted for only one issue, and at the inaugural party Hughes met the as yet unknown poet **Sylvia Plath**. They married within a few months, and went to America the following year where Hughes taught English and creative writing at the University of Massachusetts, Amherst. Of this period Hughes has recalled: 'We would write poetry every day. It was all we were interested in, all we ever did.' Plath helped with the preparation of Hughes's first collection, *The Hawk in the Rain*, (1957) which was published after having won a competition for new poets judged by **W.H. Auden**, **Stephen Spender** and **Marianne Moore**. Together with *Lupercal*, these poems established his reputation as one of the most important poets of the post-war period. Poems such as 'Hawk Roosting' and 'Pike' describe a resplendent and violent natural world, and draw on Hughes's preoccupation with folk-tales and myths. The rhythm and vigour of their language owes something to Gerard Manley Hopkins and **Dylan Thomas**. *Lupercal* won a Somerset Maugham Award in 1960 and the 1961 Hawthornden Prize.

The circumstances surrounding Hughes's marriage, the suicide of Plath, and the omissions from Plath's journals and letters, which were edited by Hughes, have aroused a considerable amount of biographical speculation and dispute, little of which is relevant to the work of either poet. Janet Malcolm's 1993 study *The Silent Woman* provides an overview of the different biographical accounts and the acrimony they have generated. The bare facts are as follows. Having returned to England in 1959, Hughes and Plath had two children, and moved in 1961 to Devon. Plath committed suicide in London in 1963. Hughes was at the time having an affair with Assia Wevill, who in 1969 also killed herself and their child. In 1970 he married Carol Orchard. In the aftermath of these personal tragedies, Hughes has sought to protect his and Plath's children, and in a letter to the *Times Literary Supplement* in 1992 he berated one of Plath's biographers, Jacqueline Rose, for treating Plath as 'only a dead, peculiar person in a book' and having no regard to the way in which her speculations about Plath's 'sexual identity' might hurt their children. In the same year he is reported to have taken legal action to prevent a proposed film about his and Plath's life together.

Hughes's first major volume of verse after Plath's suicide was *Wodwo*, which borrows its title from a character in the medieval romance *Sir Gawain and the Green Knight* and shows an increasing concern with mythology. The tone of this and the subsequent *Crow* is dark and turbulent, but there is little to suggest any direct biographical influence. In 1971 Hughes travelled to Iran where he wrote the verse drama *Orghast* for the director Peter Brook. This work, much of it written in an invented language, and the poems in *Prometheus on His Crag*, are explorations of the Prometheus myth which point towards the possibility of a reconciliation between man and his suffering. The influence of *The White Goddess* is most clearly seen in *Cave Birds* and *Gaudete*, where Hughes explores the idea that modern man has lost touch with the primordial aspect of his nature. A gentler, more humane tone is evident in some of his later verse.

Hughes has written an adaptation of Seneca's *Oedipus* which was produced at the National Theatre in 1969, two librettos for the composer Gordon Crosse, and numerous books of children's verse, notably *The Iron Man*. Together with **Seamus Heaney** he edited the influential anthology *The Rattle Bag*. In 1970 he was one of the founders of the Arvon Foundation, and in 1977 was awarded an OBE. In 1984 he was appointed Poet Laureate, a marked contrast to the previous Laureate, **John Betjeman**. His Laureate poems are gathered in *Rain-Charm for the Duchy* and show a passionate engagement with the mythic notion of Englishness, a theme which is also central to his critical work *Shakespeare and the Goddess of Complete Being*, in which he describes Shakespeare as 'modern England's creation story, our sacred book, closer to us than the Bible'.

Poetry

The Hawk in the Rain 1957; *Pike* 1959; *Lupercal* 1960; *Selected Poems* [with T. Gunn] 1962; *The Burning of the Brothel* 1966; *Animal Poems* 1967; *Gravestones* 1967; *Recklings* 1967; *Scapegoats and Rabies* 1967; *Wodwo* 1967; *I Said Goodbye to Earth* 1969; *Amulet* 1970; *Crow* 1970 [rev. 1972]; *A Crow Hymn* 1970; *A Few Crows* 1970; *Fighting for Jerusalem* 1970; *Four Crow Poems* 1970; *The Martyrdom of Bishop Farrar* 1970; *Crow Wakes* 1971; *Poems* [with R. Fainlight, A. Sillitoe] 1971; *In the Little Girl's Angel Gaze* 1972; *Selected Poems 1957–1967* 1972; *Prometheus on His Crag* 1973; *Cave Birds* 1975 [rev. *Cave Birds: an Alchemical Cave Drama* 1978]; *The Interrogator: a Titled Vulturess* 1975; *The New World* 1975; *Eclipse* 1976; *Chiasmadon* 1977; *Gaudete* 1977; *Orts* 1978; *Moortown Elegies* 1978; *A Solstice* 1978; *Adam and the Sacred Nine* 1979; *Broadsides* 1979–83; *Four Tales Told by an Idiot*

1979; *In the Black Chapel* 1979; *Moortown* 1979 [as *Moortown Diary* 1989]; *Remains of Elmet* 1979 [rev. as *Elmet* 1994]; *A Primer of Birds* 1981; *Sky-Furnace* 1981; *Selected Poems 1957–1981* [US *New Selected Poems*] 1982 [rev. as *New Selected Poems 1957–1994* 1995]; *River* 1984; *Flowers and Insects* 1986; *The Cat and the Cuckoo* 1988; *Wolfwatching* 1989; *Rain-Charm for the Duchy* 1992

Plays

The House of Aries 1960 [pub. 1961]; *The Calm* 1961; *A Houseful of Women* 1961; *The Wound* 1962 [pub. 1967]; *Difficulties of a Bridegroom* 1963; *Epithalamium* 1963; *Dogs* 1964; *The House of Donkeys* 1965 [pub. 1974]; *Orghast* 1968 [pub. 1971] *Seneca's Oedipus* 1968 [pub. 1969]; *Eat Crow* 1971; *The Story of Vasco* 1974

Fiction

The Threshold (s) 1979

Non-Fiction

Shakespeare's Poem 1971; *Henry Williamson* 1979 *T.S. Eliot: a Tribute* 1987; *Shakespeare and the Goddess of Complete Being* 1992; *Winter Pollen: Occasional Prose* [ed. W. Scammell] 1994

Edited

New Poems 1962 [with P. Beer, V. Scannell] 1962; *Five American Poets* [with T. Gunn] 1963; *Here Today* 1963; Keith Douglas *Selected Poems* 1964; Sylvia Plath: *Ariel* [with O. Hughes] 1966, *Poetry in the Making: an Anthology of Poems and Programmes from 'Listening and Writing'* 1967, *Selected Poems* [with J. Csokits] 1976, *Johnny Panic and the Bible of Dreams, and Other Prose Writings* 1977, *The Collected Poems of Sylvia Plath* 1981, *The Journals of Sylvia Plath* [with F. McCullough], *Sylvia Plath's Selected Poems* 1985; *A Choice of Emily Dickinson's Verse* 1968; *A Choice of Shakespeare's Verse* 1971; János Pilinszky Yehuda Amichai *Amen* [with Y. Amichai] 1977; *New Poetry 6* 1980; *1980 Anthology: Arvon Foundation Poetry Competition* [with S. Heaney] 1982; *The Rattle Bag* [with S. Heaney] 1982; *A Dancer to God: Tributes to T.S. Eliot* 1992

Translations

Yehuda Amichai: *Selected Poems* [with A. Gutmann] 1968, *Poems* 1969, *Time* [with Y. Amichai] 1979; János Pilinszky *The Desert of Love* [with J. Csokits] 1989

For Children

Meet my Folks! 1961 [rev. 1987]; *The Earth-Owl and Other Moon People* 1963; *How the Whale Became* 1963; *The Coming of the Kings* 1964 [pub. 1970]; *Nessie the Mannerless Monster* 1964 [US *Nessie the Monster* 1974]; *Beauty and the Beast* 1965 [pub. 1970]; *The Tiger's Bones* 1965 [pub. 1970]; *The Price of a Bride* 1966 [pub. 1968]; *The Demon of Adachigahara* 1968 [pub. 1969]; *The Iron Man* 1968 [US *The Iron Giant* 1968]; *Sea, the Fool, the Devil and the Cats* 1968 [pub. 1970]; *Five Autumn Songs for Children's Voices* 1969; *Autumn Song* 1971; *Orpheus* 1971 [pub. 1973]; *Spring, Summer, Autumn, Winter* 1974 [rev. *Season Songs* 1975, UK 1976, 1985]; *Earth-Moon* 1976; *Moon-Whales and Other Poems* 1976 [rev. UK as *Moon-Whales* 1988]; *Moon-Bells and Other Poems* 1978; *The Pig Organ* 1980; *Under the North Star* 1981; *What is the Truth?* 1984; *Ffangs the Vampire Bat and the Kiss of Truth* 1986; *Tales of the Early World* 1988; *The Iron Woman* 1993

Keri HULME 1947–

Hulme was born in Christchurch, New Zealand, of mixed Maori, Scottish and English parentage, but she identifies strongly with her Kai Tahu Maori origins. At the age of eighteen she left home to work on a tobacco plantation at Montueka before attending Canterbury University, Christchurch. It was not until she was twenty-five, after working as a fish-and-chip cook and a postwoman, that she began to concentrate on writing.

Since 1973 she has won many New Zealand literary grants and awards – including the Maori Trust Fund Prize in 1978 – and her work has appeared in numerous New Zealand publications. One short story, 'Hooks and Feelers', was filmed for television in 1984. In 1985 she was writer-in-residence at her old university. As a founder member of the Wellington Women's Gallery, she has exhibited regularly, and her paintings were included in 'Mothers', which toured major public galleries in New Zealand and Australia.

The Bone People, her best-known work, began as a short story. It was twelve years in the making and was eventually published in 1984 by Spiral, a Wellington-based feminist collective. The first printing sold out immediately, as did the second several weeks later. It won the 1984 New Zealand Book Award and the Pegasus Award for Maori literature, before taking the Booker Prize in 1985. Hulme did not attend the award ceremony and the prize was accepted on her behalf by a small group of chanting women in men's evening wear and Maori feather cloaks. It is the story of an isolated woman, Kerewin Holmes, who lives in a tower and befriends a young, mute boy. The novel was praised for the intrinsic New Zealandness of its language which successfully invokes the rhythms and accents of the Maori idiom, for its accounts of portentous dreams and powerful images personifying aspects of the characters' inner life, for its emphasis on myth, and for its susceptibility to Jungian interpretation.

Hulme lives in Westland, New Zealand. She comments: 'What I write is fantasy solidly based in reality. I have a grave suspicion that Life is a vast joke and we are unwitting elements of the joke.'

Fiction

The Bone People 1984; *The Windeater / Te Kaihau* (s) 1982; *Lost Possessions* 1985

Poetry

The Silences Between 1982

Zora Neale HURSTON 1903–1960

America's pioneering black folklorist and novelist was born in Eatonville, Florida, the first incorporated all-black town in the USA. Her father was a carpenter and lay preacher, and her mother a seamstress. **Langston Hughes** remembered her as 'full of sidesplitting anecdotes, humorous tales, and tragicomic stories remembered out of her life in the South as the daughter of a travelling minister of God'.

She was taken out of school at thirteen to look after her brother's children, but she became a maid to a white woman who arranged for her to attend high school in Baltimore. She went on to study anthropology at Barnard College and Columbia University with Franz Boas. It was anthropological training ('the spyglass of anthropology' as she called it) that provided the necessary distance from the cultural heritage of Brer Rabbit and the rest that she had grown up with; a heritage of which she said: 'I couldn't see it for wearing it'. She collected stories in her native Florida and in Jamaica, Haiti, Bermuda and Honduras, and they fed into her four novels as well as her two books of folklore, *Mules and Men* and *Tell My Horse*.

Her first separately published work was a musical play, *Fast and Furious* (1931), and her first novel, published three years later, was *Jonah's Gourd Vine*. Her best novel is generally held to be *Their Eyes Were Watching God*, a positive and loving picture of blackness. It has been seen as the prototypical black novel of affirmation, but the angrier black writer **Richard Wright** called it 'counter-revolutionary'.

In her heyday Hurston was closely associated with the Harlem Renaissance, and her work served to remind some of the more cosmopolitan members of their down-home heritage. At the same time the elements of folksiness and exotic primitivism in her work were something of a double-edged gift. Some of her anthropological work sounds dated today: Hughes remembered her walking the streets of Harlem and taking cranial measurements with 'a strange looking anthropological device'. No one else, he said, could have tried this on strangers without getting bawled out for the attempt 'except Zora, who used to stop anyone whose head looked interesting and measure it'. The story says as much about her spunk and sass as anything else.

She was an assertive woman, without self-pity. She wrote in 1928: 'I am not tragically coloured. There is no great sorrow dammed up in my soul, nor lurking behind my eyes. I do not mind at all. I do not belong to the sobbing school of Negrohood who hold that nature has somehow given them a lowdown dirt deal and whose feelings are all hurt about it.' She remains a woman of contradictions – a black nationalist and a Republican conservative – but her self-acceptance was a contribution to black culture in the USA, although it may have involved less progressive acceptances as well. She was out of fashion for some decades before a revival of interest in the 1970s, and she has been an influence on writers such as **Ralph Ellison** and **Toni Morrison**.

Fiction

Jonah's Gourd Vine 1934; *Their Eyes Were Watching God* 1937; *Seraph on the Sewanee* (s) 1948; *Spunk: the Selected Short Stories of Zora Neale Hurston* (s) 1985

Non-Fiction

Mules and Men 1935; *Dust Tracks on a Road* (a) 1939; *Tell My Horse* 1938 [rev. as *Voodoo Gods: an Inquiry into Native Myths and Magic in Jamaica and Haiti* 1939]; *Moses, Man of the Mountain* 1939

Plays

The First One [pub. in *Ebony and Topaz: a Collectanea* ed. C.S. Johnson 1927]; *Fast and Furious* [with C. Fletcher, T. Moore] 1931; *Mule Bone* [with L. Hughes] 1931; *Stephen Kelen-d'-Oxylion Presents Polk County* 1944

Collections

I Love Myself When I Am Laughing . . . and Then Again When I am Looking Mean and Impressive: a Zora Neale Hurston Reader [ed. A. Walker] 1979

Biography

Zora Neale Hurston: a Literary Biography by Robert E. Hemenway 1977

R(ay) C(oryton) HUTCHINSON 1907–1975

Hutchinson was born in Watford and educated at Monkton Coombe School and Oriel College, Oxford. His first job was in the advertising department of Colman, of mustard fame. In the evening he wrote short stories, some of which appeared in the *English Review*. As a boy, he had completed a 20,000-word novel, but his first book, *Thou Hast a Devil*, was published in 1930, the year after he married Margaret Jones, with whom he had two sons and two daughters.

Testament, a large-scale novel about the Russian Revolution was published in 1938, and won him an appreciative public, and the *Sunday Times* Gold Medal for fiction. During the war, he was put in command of a company of a newly formed battalion of the Buffs, and later returned to London to the War Office. In 1944 he wrote the famous speech delivered by George VI at the Home Guard standing-down ceremony, and in 1945, was sent to Egypt, Palestine, Persia and Iraq, to record his experiences of 'Paiforce', the then Persian and Iraq command. *A Child Possessed*, about a Russian émigré's love for his retarded daughter, won the W.H. Smith Literary Annual Award in 1966, two years after its publication. A volume of memoirs, *Interim*, appeared at the end of the war, but his final complete novel was the comic *Origins of Cathleen*, an instant bestseller. *Rising*, although unfinished, was published posthumously.

Shock-haired, husky-voiced and with sunken grey eyes, Hutchinson once remarked that while fiction was his full-time job, he 'varied it with amateurish attention to church work and to an entrancing family'. He regarded his own life as uneventful, and 'all negatives'. He disliked literary society, preferring to research the background details for his novels, which were generally set abroad. Foreign railway stations were described with meticulous accuracy, and characters with an often surprising use of outrageous imagery ('Like a sausage fried too fast, her wooden face was split transversely by a monstrous smile.'). He was personally modest about his own achievements, but his admirers included **Edwin Muir**, Julian Huxley and **I. Compton-Burnett**. Despite his recent neglect, Hutchinson was declared to be a genius by Rupert Hart-Davis, who wrote that of the novels, 'at least seven are major works, fit to stand beside the finest of this century in any language'.

Fiction

Thou Hast a Devil 1930; *The Answering Glory* 1932; *The Unforgotten Prisoner* 1933; *One Light Burning* 1935; *Shining Scabbards* 1936; *Testament* 1938; *The Fire and the Wood* 1940; *Interim* 1945; *Elephant and Castle* 1949; *Recollection of a Journey* 1952; *The Stepmother* 1955; *March the Ninth* 1957; *A Child Possessed* 1964; *Image of my Father* 1964; *Johanna at Daybreak* 1969; *Origins of Cathleen* 1971; *Rising* 1976; *The Quixotes* (s) 1984

Non-Fiction

Paiforce: the Official Story of the Persia and Iraq Command, 1941–1946 [as 'anon.'] 1949

Aldous (Leonard) HUXLEY 1894–1963

Born near Godalming in Surrey, Huxley was the third son of Leonard Huxley, later editor of the *Cornhill Magazine*. He united the blood of two of the most distinguished intellectual families of nineteenth-century England, the Huxleys and the Arnolds: the biologist and champion of Darwin T.H. Huxley was his grandfather, Dr Arnold of Rugby was his great-grandfather, and the novelist Mrs Humphrey Ward was his aunt. He was educated at Eton, but while there contracted a serious eye disease which left him almost blind and forced him to leave the school. The death of his mother and the suicide of his brother Trevenen were other traumas of early years, but he fought these and his near-blindness with courage, and attended Balliol College, Oxford.

Having graduated with a first in English, he taught briefly at Eton and Repton while he embarked on a career as a writer, chiefly as a poet, and then lived in financial stringency as a literary journalist with his Belgian wife, Maria Nys (whom he had married in 1919), and their son. He worked for the *Athenaeum* and was drama critic for the *Westminster Gazette*. By 1920 he had published three volumes of poetry and one of short stories, *Limbo*. It was, however, his first novel, *Crome Yellow* (1921), that began a process by which he became one of the most fashionable literary figures of the decade. He published a dozen books in eight years, and made enough money to live abroad during the 1930s, mainly in France at Sanary, near Toulon. His novels were largely satires of contemporary social and literary life, often containing recognisable portraits of his contemporaries. It has been argued that he was more of an essayist than he was a novelist, and that his novels are undoubtedly clever, but have little plot and are not deeply felt. This is certainly untrue of *Point Counter Point*, with its harrowing account of the death of the protagonist's child and of various thwarted love-affairs. Later novels became increasingly philosophical, and included *Brave New World*, a dystopian view of the future, which remains his most popular and best-known book.

In 1937, Huxley emigrated to America along with the guru-figure Gerald Heard, and became interested in mysticism. He settled in California, where he and Heard were sought out two years later by **Christopher Isherwood**, also in search of a spiritual life. He was a regular contributor to *Vedanta and the West*, the magazine Isherwood edited while a disciple of Swami Prabhavananda, and collaborated with the younger novelist on several film projects, none of which reached the screen. His own work for Hollywood included a screenplay for MGM's *Pride and Prejudice* (1940). Huxley's eyesight was improved by his use of the Bates Method of exercises, and he began experimenting with consciousness-expanding drugs such as mescalin and LSD, writing about his experiences in *The Doors of Perception* and *Heaven and Hell*.

Maria Huxley died in 1955, and the following year Huxley married the violinist and psychotherapist Laura Archera. He continued to write fiction, although his reputation as a novelist had begun to decline, and he spent much of his time lecturing, becoming a visiting professor at numerous American universities. Amongst his later books are *The Devils of Loudon*, an account of mass hysteria and exorcism in seventeenth-century France; *Island*, an overly optimistic (and, some argue, ill thought-out) return to the territory of *Brave New World*; and *Brave New World Revisited*, in which he compares the predictions he had made in his novel with subsequent developments in science and society. He suffered a severe loss in 1961 when his house and all its contents, including his papers and his library, were destroyed in a bush-fire. By this time he was suffering from cancer of the throat, although he continued to work right up until the day he died, 22 November 1963. News of his death was overshadowed in the media by that of the assassination of President J.F. Kennedy, but his reputation subsequently underwent a revaluation and he is now considered one of the most interesting and influential writers of his time.

Fiction

Limbo (s) 1920; *Crome Yellow* 1921; *Mortal Coils* (s) 1922; *Antic Hay* 1923; *Little Mexican and Other Stories* (s) 1924; *Those Barren Leaves* 1925; *Two or Three Graces and other Stories* (s) 1926; *Point Counter Point* 1928; *Brief Candles* (s) 1930; *Brave New World* 1932; *Eyeless in Gaza* 1936; *After Many a Summer* 1939; *Time Must Have a Stop* 1945; *Ape and Essence* 1949; *The Devils of Loudon* 1952; *The Genius and the Goddess* 1955; *Collected Short Stories* (s) 1957; *Island* 1962

Non-Fiction

On the Margin 1923; *Along the Road* 1925; *Essays New and Old* 1926; *Jesting Pilate* 1926; *Proper Studies* 1927; *Do What You Will* 1929; *Holy Face and Other Essays* 1929; *Music at Night* 1931; *Beyond the Mexique Bay* 1934; *The Olive Tree* 1936; *Ends and Means* 1937; *Vulgarity in Literature* 1939; *Grey Eminence: a Study in Religion and Politics* 1941; *The Art of Seeing* 1943; *Science, Liberty and Peace* 1947; *Themes and Variations* 1950; *Joyce the Artificer* [with S. Gilbert] 1952; *The Doors of Perception* 1954; *Adonis and the Elephant* 1956; *Heaven and Hell* 1956; *Collected Essays* 1958; *Brave New World Revisited* 1959; *Literature and Science* 1963; *Letters of Aldous Huxley* [ed. G. Smith] 1969

Poetry

The Burning Wheel 1916; *The Defeat of Youth* 1918; *Leda* 1920; *Arabia Infelix* 1929; *The Cicadas* 1931; *The Collected Poetry of Aldous Huxley* [ed. D. Watt] 1971

Plays

The World of Light 1931; *The Gioconda Smile* 1948

For Children

The Crows of Pearblossom 1967

Edited

The Letters of D.H. Lawrence 1932; *Texts and Pretexts* 1932; *An Encyclopaedia of Pacifism* 1937; *The Perennial Philosophy: an Anthology of Mysticism* 1946

Collections

Rotunda: a Selection from the works of Aldous Huxley 1935; *Stories, Essays and Poems* 1937; *Verses and Comedy* 1946; *The World of Aldous Huxley* [ed. C.J. Rolo] 1957; *The Collected Works of Aldous Huxley* 1970

Biography

Aldous Huxley by Sybille Bedford, 2 vols 1973, 1974

I

Gary INDIANA 1950–

'Indiana', who remains discreet about his real name, was born in New Hampshire, where his father ran a lumber company and his mother was a town clerk; he left the family home at an early age. He read philosophy at Berkeley, but dropped out after his second year, and worked in a succession of mundane jobs, including in an insurance firm, a plastic surgery clinic and a psychiatric hospital. During the 1970s he sold popcorn in a theatre at night, while working for Legal Aid in the Watts ghetto during the day. He has said of his Watts years that his world view was formed by commuting every day between America's worst ghetto and its richest suburb (the suburb being Westwood, Los Angeles).

Indiana got involved with theatre in the late 1970s as an actor and playwright, working in New York and Europe. His first play, *Alligator Girls Go to College*, was performed at New York's Mudd Club in 1979, and he followed it with *The Roman Polanski Story* in 1982. He has also had a career as an art critic, which began in 1983 after someone who knew of Indiana's interest in the artist John Chamberlain asked him to write a piece on him for *Art in America*. He has subsequently written for *Details*, *Interview* and *Artforum* as well, and is contracted as art critic on the *Village Voice*. He travelled for several years around Europe, North Africa, South America and South-East Asia, and now lives in New York.

Indiana followed two volumes of short stories, *Scar Tissue* and *White Trash Boulevard*, with his first novel, the controversial *Horse Crazy*, in 1988. The novel's protagonist, a young homosexual man with an aesthetic sensibility, finds himself curtailed and terrified by Aids, and the book follows his awkwardly celibate relationship with his lover. It is a document of the Aids era, but it differs sharply from the more 'considered' and relatively genial writing of, say, **Armistead Maupin** or **Edmund White** in being a product of the postmodern panic it depicts, with a texture which is nihilistic, excessive, dandified and junk-cultured. It is this particular brand of postmodernity which has caused Indiana to be numbered among the so-called 'blank generation' of obliquely traumatised American writers whose spiritual home is the shopping mall and the television set. To borrow **Samuel Beckett**'s comment on **James Joyce**, if the subject is the death-ridden and chaotic spectacle of late twentieth-century American life, then Indiana's work 'is not *about* something; *it is that something itself*'.

As critics have noted, it is not that Indiana uses Aids as a metaphor, but rather that his depiction of Aids-haunted homosexual life, often at its most extreme fringes, has a wider relevance to mortality and relationships at the postmodern *fin-de-siècle*. This is particularly the case with *Rent Boy*, which takes 'Life is very precious, even right now' as one of its epigraphs. It is a tale of murder which features (among other things) grotesque sex, a criminal doctor who runs a homicidal organ-transplant ring, and satirical creations such as 'Sandy Miller' (a memorably awful female avant-gardist who is clearly not a million miles removed from **Kathy Acker**). The narrator is an architecture student and male prostitute, with a wit that is not only waspish but so pragmatic and worldly-wise that at least one critic has compared his voice to that of the early **William Burroughs**. Like all of Indiana's work it is savagely intelligent, matter-of-factly gross, and often very funny, doing full justice to Indiana's observation that 'These are peculiar times'.

Fiction

Scar Tissue (s) 1987; *White Trash Boulevard* (s) 1988; *Horse Crazy* 1988; *Gone Tomorrow* 1993; *Rent Boy* 1994

Plays

Alligator Girls Go to College 1979; *The Roman Polanski Story* 1982

Non-Fiction

Roberto Juarez 1986; *Hybrid Neutral* (c) 1988

Edited

Living with the Animals 1994

William (Motter) INGE 1913–1973

Inge (not to be confused with his British contemporary William Ralph Inge, better known as Dean Inge) was born in Independence, Kansas, the son of a travelling salesman who was often absent and a devoted mother. He was educated at the local high school, and in 1930 went to the University of Kansas, taking part in much student drama there and intending to become an actor. But, on graduation, believing that the life of a teacher would be safer, he enrolled at George Peabody College in Nashville, Tennessee.

His unhappiness at college led to him deserting the course, and spending a year working variously as a labourer on the roads, a scriptwriter and announcer for a radio station and a high-school teacher, but he eventually completed his studies, and in 1938 was able to begin an instructorship in English and drama at Stephens College in Columbia, Missouri. He was only a moderately successful teacher, however, began to drink heavily and had a nervous breakdown before escaping in 1943 to become drama critic for the St Louis *Star-Times*. While there, he was impressed by the plays of **Tennessee Williams**, met the playwright and was encouraged by him to write, an ambition that had been latent previously.

In 1946, the colleague for whom he had been covering as drama critic returned from the war. Inge therefore had to return to teaching, this time at Washington University in St Louis, but his first play, *Farther Off from Heaven*, was performed in Dallas in 1947, and more was to come. He gave up his post when *Come Back, Little Sheba* was accepted for Broadway in 1949, and its rapturous reception the following year heralded a decade of extraordinary success for Inge, when this and three subsequent plays – *Picnic*, *Bus Stop* and *The Dark at the Top of the Stairs* – were all made into films and grossed their author well over a million dollars.

Inge made few innovations in stagecraft or language, but he was original in that his subject-matter was the repressed and intensely conventional life of small Midwestern towns, which had already figured widely in the novel but had up to then not penetrated Broadway. Inge's greatest strength was the touching honesty with which he handled themes such as alcoholism, sexual frustration and the quest for happiness (although homosexual he avoided homosexuality as a subject more than Tennessee Williams did).

He was not himself made much happier by his success, and began regular psychoanalysis before losing much of the money he had made in backing his first Broadway failure, *A Loss of Roses*, in 1959. Two further flops in the early and mid-1960s put an end to his Broadway career, but he moved to Hollywood, where he lived with his widowed sister Helene Inge Connell, wrote filmscripts, and advised on theatre workshops at the University of California at Los Angeles. Very late in his life he published two novels, which used his familiar Midwestern material but made limited impact. He committed suicide.

Plays

Farther Off from Heaven 1947 [rev. as *The Dark at the Top of the Stairs* 1957]; *Come Back, Little Sheba* 1950 [pub. 1951]; *Picnic* 1953 [rev. as *Summer Brave* 1962]; *Bus Stop* 1955; *Glory in the Flower* 1959 [pub. in *24 Favorite One-Act Plays* ed. V.H. Cartmell, B.Cerf 1958]; *A Loss of Roses* 1959 [pub. 1960]; *Splendor in the Grass* (f) 1961; *All Fall Down* (f) 1962; *Summer Brave and Eleven Short Plays* [pub. 1962]; *Out on the Outskirts of Town* (tv) 1964; *Bus Riley's Back in Town* (f) 1965; *Family Things, Etc.* 1965 [rev. as *Where's Daddy?* 1966]; *Don't Go Gentle* 1968 [rev. as *The Last Pad* 1972, as *The Disposal* 1973]; *Two Short Plays* [pub. 1968]; *Love Death Plays* [pub. 1975]

Fiction

Good Luck, Miss Wyckoff 1971; *My Son Is a Splendid Driver* 1972

Michael INNES

see J.I.M. STEWART

John (Winslow) IRVING 1942–

Irving was born in Exeter, New Hampshire, where his stepfather was a teacher of Russian history and his mother a social worker. His grandfather, Frederick Irving, was a professor of obstetrics and author of *The Expectant Mother's Handbook*. Irving was educated at the Phillips Exeter Academy and, between 1961 and 1967, the Universities of Pittsburgh, Vienna (where he looked after a bear), New Hampshire and Iowa. His first novel, *Setting Free the Bears* (1969), was a freewheeling comic adventure set in Vienna which culminates in a plan to liberate the bears from the Heitzinger Zoo. He worked briefly with the director Irvin Kershner on a proposed film adaptation of the novel which came to nothing, after which he taught at Mount Holyoake College, Massachusetts, and subsequently at Iowa University, in order to finance his writing and support his family. He had married the photographer Shyla Leary in 1964 and they had two sons. He was a successful amateur wrestler and also coached wrestling, an activity he has said he found 'more copacetic with writing than teaching literature because it didn't exercise the same muscles'.

Irving's fourth novel, *The World According to Garp*, spent six months on the US bestseller lists and enabled him to give up his teaching job. A manic comedy laced with liberal doses of sex, violence and death, the novel contains many of the subjects which have become familiar Irving trademarks: Vienna, wrestling, teaching, medicine and, of course, bears. Like much of his work it draws heavily on elements of his own life but is too fantastical to be autobiographical. He has described it as 'artfully disguised soap opera ... I mean to make you laugh, to make you cry.' *Garp* and *The Hotel New Hampshire* were both filmed.

The concern with moral choices implicit in Irving's earlier work becomes more explicit in his later novels, which consider issues such as abortion and the effects of the Vietnam War. He has addressed often hostile student audiences on what he regards as the disturbing rightward drift in contemporary American politics, and was somewhat reluctantly dragged into giving speeches for pro-abortion groups following the publication of *The Cider House Rules*. 'I'm still angry about it because abortion is such a personal and private decision,' he has said. Irving was divorced in 1981 and in 1987 married the literary agent Janet Turnbull. They live in Vermont and Toronto.

Fiction

Setting Free the Bears 1969; *The Water-Method Man* 1972; *The 158-Pound Marriage* 1974; *The World According to Garp* 1978; *The Hotel New Hampshire* 1981; *The Cider House Rules* 1985; *A Prayer for Owen Meany* 1989; *Trying to Save Piggy Sneed* (s) 1993; *A Son of the Circus* 1994

Christopher (William Bradshaw-) ISHERWOOD 1904–1986

The son of a professional soldier, but heir to a large house and estate, Isherwood was born near High Lane in Cheshire. The family was descended from Judge Bradshawe, who signed the death warrant of Charles I, and through his mother's family Isherwood was a cousin of **Graham Greene**. His father's career meant that he had a peripatetic childhood – in Cheshire, Yorkshire, Surrey and Ireland – and this restlessness continued to be a feature of his life. He was educated at St Edmund's preparatory school in Surrey, where he first met **W.H. Auden**, at Repton, where he met **Edward Upward**, and at Corpus Christi College, Cambridge, where he read history.

His father had been killed in action in 1915, and subsequent relations with his mother, who lived on into her nineties, were strained. He spent much of his time at Cambridge with Upward creating and writing about a fantastical village called Mortmere, 'the mad nursery' where the two young writers developed their craft. Although little of the saga was published until 1994, the world they created earned a considerable reputation and influenced the early poetry of Auden, who sent almost all his work to Isherwood for criticism during the 1920s. After deliberately failing his tripos, Isherwood worked as secretary to André Mangeot's string quartet, was a private tutor, and briefly attended King's College, London, as a medical student. By this time, he had met **Stephen Spender**, and with Auden they formed a triumvirate that came to dominate and exemplify the progressive, left-wing literature of the 1930s in the popular imagination.

Isherwood's first novel, *All the Conspirators* (1928), was an accomplished study of 'the great war between the old and young', influenced by **Virginia Woolf**, **E.M. Forster** and **James Joyce**. The following year, he went to Berlin, where he plunged into the homosexual subculture and wrote his second novel, *The Memorial*, which continues to analyse the battle between the generations and draws upon his own family background. He was to capture the essence of late Weimar Berlin, and establish his characteristic tone of sardonic detachment, in *Mr Norris Changes Trains*, the comic story of a shameless con-man and masochist, and *Goodbye to Berlin*. The latter book was a collection of interrelated stories, one of which featured the irrepressible would-be actress Sally Bowles, a character who achieved a life independent of her creator when the stories were adapted for the stage by **John van Druten** as *I Am a Camera* (1951, filmed 1955) and as a musical, *Cabaret* (1966, filmed 1972). Both books were to have formed part of a large novel called 'The Lost', which was never completed.

He left Berlin in 1933 and spent several years wandering around Europe with his lover, Heinz Neddermeyer, trying to evade the latter's military call-up. The couple parted in 1937, after Neddermeyer was arrested and returned to Germany. During this period, Isherwood had been collaborating with Auden on three experimental plays for the Group Theatre: *The Dog Beneath the Skin*, *The Ascent of F.6* and *On the Frontier*. In 1938 the two men set off for China to write a book about the Japanese invasion, *Journey to a War*. On their way back to England they stopped in New York, and over the next few months decided to emigrate to America. Isherwood said farewell to England in a lively autobiography of his youth, *Lions and Shadows*.

He and Auden travelled to the USA in January 1939, and were later accused of 'running away' from the war in Europe. Isherwood had become a pacifist, and settled in California, where he associated with the guru-figure Gerald Heard, and through him met Swami Prabhavananda, whose disciple he became. He worked as a screenwriter for MGM and other Hollywood companies (collaborating with, amongst others, **Aldous Huxley**), but his film work was largely undistinguished. He lived for a while in a Vedantan monastery, where he assisted Prabhavananda in the translation of Hindu texts, edited the magazine *Vedanta and the West*, and wrote

Prater Violet, a novel about his experiences during the 1930s working on a film with the Austrian director Berthold Viertel.

He spent part of the war working with Quakers in Pennsylvania helping to settle German refugees, and this became the background for *The World in the Evening*, a novel which he found difficult to write, and which proved wholly unworthy of his effort. By the time this book was published in 1954, he had met Don Bachardy, a student thirty years his junior, with whom he was to spend the rest of his life. (Bachardy later became a portraitist and painted hundreds of pictures of Isherwood and other writers.) Fears that Hollywood had ruined Isherwood's talent were calmed by *Down There on a Visit*, which was made up of four episodes from his past and shocked contemporary reviewers with its frank treatment of homosexuality. What is perhaps his best novel, *A Single Man*, is the account of a day in the life of a bereaved, angry, homosexual, Anglo-American academic, and brought together most of the themes he had been pursuing in his largely autobiographical fiction. It was followed by a less successful last novel, *A Meeting by the River*, about two brothers, one a worldly writer, the other a Hindu monk: it was later adapted by Isherwood and Bachardy for the stage, but lasted for one performance only in a disastrous Broadway production.

In his last years, taking advantage of more enlightened attitudes towards homosexuality, Isherwood turned to undisguised autobiography: *Kathleen and Frank* was a moving and reconciliatory account of his parents and their times; *Christopher and His Kind* returned to the 1930s; and *My Guru and His Disciple* described his relationship with his swami. He lived in a house overlooking Santa Monica Canyon, and was taken up by the emergent gay liberation movement as a figurehead. He remained extraordinarily youthful until his last years, when he was suffering from the prostate cancer which eventually killed him.

Fiction

All the Conspirators 1928; *The Memorial* 1932; *Mr Norris Changes Trains* [US *The Last of Mr Norris*] 1935; *Sally Bowles* 1937; *Goodbye to Berlin* (s) 1939; *Prater Violet* 1946; *The World in the Evening 1954*; *Down There on a Visit* 1962; *A Single Man* 1964; *A Meeting by the River* 1967; *The Mortmere Stories* (s) [with E. Upward] 1994

Non-Fiction

Lions and Shadows (a) 1938; *Journey to a War* [with W.H. Auden] 1939; *The Condor and the Cows* 1949; *An Approach to Vedanta* 1963; *Ramakrishna and His Disciples* 1965; *Essentials of Vedanta* 1969; *Kathleen and Frank* (a) 1971; *Christopher and His Kind* (a) 1977; *My Guru and His Disciple* (a) 1980; *October* (a) [with D. Bachardy] 1983; *The Wishing Tree: Christopher Isherwood on Mystical Religion* [ed. R. Adjemain] 1987

Plays

The Dog Beneath the Skin [with W.H. Auden] 1935; *The Ascent of F.6* [with W.H. Auden] 1936; *On the Frontier* [with W.H. Auden] 1938; *The Adventures of the Black Girl in Her Search for God* [from G.B. Shaw] 1969 [np]; *Frankenstein: the True Story* (tv) [from M. Shelley's novel; with D. Bachardy] 1973; *A Meeting by the River* [with D. Bachardy] 1979 [np]

Translations

Baudelaire *Intimate Journals* 1930; Bertolt Brecht *A Penny for the Poor* [with D. Vesey] 1938; *The Song of God: Bhagavad-Gita* [with S. Prabhavananda] 1944; *Shankara's Crest-Jewel of Discrimination* [with S. Prabhavananda] 1947; *How to Know God: the Yoga Aphorisms of Patanjali* [with S. Prabhavananda] 1953

Edited

Vedanta for the Western World 1945; *Vedanta for Modern Man* 1951; *Great English Short Stories* (s) 1957

Poetry

People One Ought to Know [with S. Mangeot] 1982

Collections

Exhumations: Stories, Articles and Verses 1966

Biography

Isherwood by Jonathan Fryer 1977; *Christopher Isherwood* by Brian Finney 1979

Kazuo ISHIGURO 1954–

Ishiguro was born in Nagasaki, and in 1960 his family moved from Japan to England where his father worked as an oceanographer assisting with the development of the North Sea oil fields. He enjoyed what he has called a typical 'southern English upbringing' in Guildford, Surrey, though he spoke Japanese at home and learned English through mimicry. He was educated at Woking County Grammar School, during which time his ambition was to pursue a career as a musician. He has described doing the rounds of the London record companies with a demo tape as 'all very dispiriting'. In 1973 he worked as a beater at Balmoral Castle in Scotland before hitchhiking across the USA and Canada. A spell as a social worker in Renfrew preceded his return to education at the University of Kent where he read English and philosophy. Having graduated in 1978 he returned briefly to social work, running a hostel for the homeless in London, and in 1979, on the strength of a radio play he had written, he was accepted for the creative writing course at the University of East Anglia where his tutors included **Malcolm Bradbury** and **Angela Carter**. In 1990 he recalled: 'I suddenly panicked because I'd never really written anything before and I was afraid I'd be exposed as a fraud.' In the event, his time at UEA was so successful that he gained a contract for his first novel before finishing the course.

A Pale View of Hills, which won the Winifred Holtby Award in 1983, is narrated by a Japanese widow living in England and reflecting on her life in Japan in the aftermath of the war and the bombing of Hiroshima and Nagasaki. The novel was widely praised as a sensitive study of emotional dislocation and the conflict between traditional and modern views of the world. *An Artist of the Floating World* is also set in post-war Japan and explores similar themes; it won the Whitbread Award in 1986.

Before 1989 Ishiguro had never returned to Japan and he has emphasised his relative ignorance about the country and the degree to which these two novels emerged out of distant memory and imagination. With *The Remains of the Day* he turned to English subject-matter, describing in precise and telling detail an elderly butler's recollection of his service in the 1930s and the painful realisation that his revered employer was a Nazi sympathiser. The novel won the Booker Prize in 1989 and was made into a successful film (1993) directed by James Ivory from a screenplay by **Ruth Prawer Jhabvala**. *The Unconsoled*, Ishiguro's longest and perhaps most ambitious novel to date, is a comedy set in contemporary Europe, and was published to extremely mixed reviews. Ishiguro has also written two television plays. He was married in 1986.

Fiction

Introduction 7 **(s)** [with others] 1981; *A Pale View of Hills* 1982; *An Artist of the Floating World* 1986; *The Remains of the Day* 1989; *The Unconsoled* 1995

Plays

A Profile of Arthur J. Mason **(tv)** 1985; *The Gourmet* **(tv)** 1987

J

Shirley JACKSON 1919–1965

Jackson was born in San Francisco and educated at Rochester University (which she had to leave because of her parents' financial difficulties during the Depression) and Syracuse University, New York, where she took her BA in 1940. Not a great deal is known about her life, and she once responded to a request for information by stating: 'I very much dislike writing about myself or my work, and when pressed for autobiographical material can give only a bare chronological outline which contains, naturally, no pertinent facts.' In the year of her graduation she married Stanley Edgar Hyman, a writer and numismatist with whom she had founded and edited a short-lived magazine, *The Spectre*, the previous year. She brought up four children in Vermont, 'in a quiet rural community with fine scenery and comfortably far away from city life'.

Jackson published her first novel, *The Road through the Wall*, in 1948 (it was retitled *The Other Side of the Street* when reissued in 1956), and the following year a collection of stories, *The Lottery*. Originally published in the *New Yorker* in 1948, and much anthologised since, the title story is her best-known work, a macabre, Gothic tale set in a small town where an annual lottery dating back to time immemorial decides which one of the town's citizens is going to be stoned to death. She adapted the story as a television play in 1950. Her subsequent novels similarly explore the violent and fantastic that lurk beneath the surface of everyday life. *Hangsaman* deals with a mentally disturbed girl who has four different voices, and *The Sundial* describes the bizarre Halloran family who await the end of the world in the expectation that they alone will survive.

Despite demonstrating an imagination which has been compared to that of Jorge Luis Borges, Jackson's primary concern in all of her work is with storytelling, a point reinforced by the lectures collected in the posthumous volume, edited by her husband, *Come Along with Me*. She also wrote two humorous works, *Life Among the Savages* and *Raising Demons*, which she described as a 'disrespectful memoir' of her offspring.

Fiction

The Road through the Wall 1948 [as *The Other Side of the Street* 1956]; *The Lottery* (s) 1949; *Hangsaman* 1951; *The Bird's Nest* 1954; *The Sundial* 1958; *The Haunting of Hill House* 1959; *We Have Always Lived in the Castle* 1962

Non-Fiction

Life Among the Savages (a) 1953; *Raising Demons* (a) 1957

For Children

The Witchcraft of Salem Village 1956

Plays

The Lottery (tv) [from her novel] 1950

Collections

The Magic of Shirley Jackson [ed. S.E. Hyman] 1966; *Come Along with Me* [ed. S.E. Hyman] 1968

Dan JACOBSON 1929–

Jacobson was born in Johannesburg at a time when more Jews were emigrating to South Africa than to any country apart from the USA. He then moved on with his family, at the age of four, to Kimberley, where his father owned a butter factory. At this period a diamond 'pipe' and the subsequent Big Hole inaugurated a mineral revolution that transformed the country. It was in Kimberley, as well, that John Cecil Rhodes made his initial fortune. The influence on Jacobson of the ideologies that have come and gone in South Africa, as well as the extraordinary landscape of the Northern Cape veld, can be traced throughout his writing career, and is evident in *The Electronic Elephant: a Southern African Journey* (1994). Part autobiography, part travelogue, but also part meditation on South African history, it is an intensely personal inquiry into remembered states of being.

Jacobson matriculated from the University of Witwatersrand in 1949 and, amongst other jobs during the next few years, worked on a newspaper in Johannesburg and did some translation work from Afrikaans into English. But he had already defined himself as a writer of novels and stories, a vocation he says was always associated in his mind with leaving South Africa. This he did in 1954, when he arrived in London. The literary motivation behind the journey can be gauged from the fact that, during the flight, Jacobson read some of **Virginia Woolf**'s letters, posted from an address in Tavistock Square; after leaving the airport, he and his brother, before doing anything else, visited the Square, to find that the house for which they were looking had been blown up during the war, leaving only a crater which had yet to be filled.

Having already published a number of short stories, he set about writing works of greater length and substance: *The Trap* (1955) and *A Dance in the Sun* (1956), Jacobson's first two novels, were published at the same time as he took up a fellowship in creative writing at Stanford University. Both novels are set in South Africa and dramatise the relationship between blacks and whites, their strictly hierarchical and segregated environment, and the way that people are unwillingly brought into conflict and unison as a result of impulsive human acts like falling in love or committing a crime.

All of Jacobson's fiction and essays up to *The Beginners* (1966) had been set in, or had heavily involved the theme of, South Africa. This novel, however, proved to be transitional. It is the first not to be set in that country, and breaks with the previous fiction in terms of narrative (it is unclear with whom the authority – if there is any – lies in the novel) and structure (Jacobson has said that he wanted it 'to break away sharply from the conventions of naturalism'). This radically new approach to the role of the writer in composing his work and to the role of the reader in constructing its meaning led shortly afterwards to *The Rape of Tamar*, probably Jacobson's most praised novel, and also one of his most popular. It tells the biblical story of the rape of King David's daughter by her brother Amnon and the subsequent murder of Amnon by his brother Absalom. The narrator, like narrators in the novels of **Vladimir Nabokov** (a writer Jacobson greatly admires), is cynical, manipulative and devious, and is very much not to be trusted. The novel provided the source for **Peter Shaffer**'s play *Yonadab* (1986), which ran at the National Theatre, London. A later novel, *The Confessions of Josef Baisz*, in which the authority and sincerity of the narrator is also highly questionable, won the 1978 *Jewish Chronicle*/H.H. Wingate Literary Award.

Though Jacobson has taught at numerous universities in America and elsewhere, his appointment as lecturer in English at University College London in 1976 is by far the most significant. Many of his lectures have been collected in *Adult Pleasures*, while his lengthy award-winning poem-sequence 'A Month in the Country' was published in Britain in the *UCL Book Review*. Meanwhile, his creative output continued to be prolific and was greeted with much praise: in 1986 he was awarded the J.R. Ackerley Prize for autobiography for *Time and Time Again*, while his novel *The God-Fearer* was shortlisted for the Whitbread Award. Jacobson's devotion to teaching at UCL was rewarded in 1988, when he was made professor. He officially retired in 1994, but still teaches part-time at UCL. Although he earned his living by writing freelance for twenty or so years before arriving at UCL, the eighteen years he has taught there have inevitably made him more institutionalised than he could have imagined when he began. He has nevertheless produced works of fiction that have secured his reputation as one of Britain's leading literary novelists.

Fiction

The Trap 1955; *A Dance in the Sun* 1956; *The Price of Diamonds* 1957; *A Long Way from London* (s) 1958; *The Zulu and the Zeide* (s) 1959; *The Evidence of Love* 1960; *Beggar My Neighbour* (s) 1964; *The Beginners* 1966; *Through the Wilderness* (s) 1968; *Penguin Modern Stories 6* (s) [with others] 1970; *The Rape of Tamar* 1970; *A Way of Life and Other Stories* (s) [ed. A. Pirani] 1971; *Inklings* (s) 1973; *The Wonder-Worker* 1973; *The Confessions of Josef Baisz* 1977; *Her Story* 1987; *Hidden in the Heart* 1991; *The God-Fearer* 1992

Non-Fiction

No Further West 1961; *Time of Arrival and Other Essays* 1963; *The Story of the Stories: The Chosen People and Its God* 1982; *Time and Time Again* (a) 1985; *Adult Pleasures* 1988; *Biblical Narratives and Novelists' Narratives* 1988; *The Electronic Elephant: a Southern African Journey* 1994

Plays

The Caves of Adullan (r) 1972

Howard (Eric) JACOBSON 1942–

Jacobson was born in Manchester; according to his mother: 'He weighed 10½lb and was a lovely peach colour, but very restless.' His linguistic precocity was noted when he was still in infant school, and when at the age of fourteen he won the Stand Grammar School prize for literature, he chose *Ulysses*. He was also a champion table-tennis player. From 1961 to 1964 he read English (under F.R. Leavis) at Downing College, Cambridge, and, over the next fifteen years, taught at Selwyn College, Cambridge, Sydney University and Wolverhampton Polytechnic. During the early 1970s he spent a period in Australia working as a plasterer's labourer and a publisher's representative; his second wife is a native of Perth. The couple's 1986 return to Australia has been the subject of a book, *In the Land of Oz*, and a Channel 4 television documentary. Formerly dividing their time between Cornwall – where they used to run a bookshop – and London, they are presently settled in the latter.

Each of Jacobson's novels contains some element of autobiography – albeit sardonically distorted. His first protagonist, Sefton Goldberg in *Coming from Behind*, is a polytechnic lecturer in Wrottesley, a dismal Midlands town. Mired in this intellectual pit, Goldberg clings to visions of escape. He applies for a post at Cambridge, writes a book about failure and cheers himself up with a less than heady cocktail of sexual adventure and academic intrigue. A downmarket rendition of the university satires of **David Lodge** and **Malcolm Bradbury**, the tale achieves its most fluent effects through the evocation and demolition of stereotypes. Goldberg, though, is incongruously complex: convinced of his superiority and at the same time demoralised by the terms of his existence. The format is not big enough to hold him, but in *Peeping Tom*, Jacobson found a subject of suitable dimensions for a similarly anguished character: Barney Fugelman. Fugelman's queasy, comic examination of the passage that leads from writing to voyeurism via fantasy reveals a destructive addiction to women and a profound terror at their sexual power. If he is drawn to necrophilia, that is because a corpse offers the ultimately undemanding object and the temptation of the last taboo.

The climate in *Redback* is healthier and the laughter easier. The story sends its hero, Leon Forelock, Down Under. He is to teach the natives how to live. Possessing a degree in Moral Decencies, he has no other qualifications for the task. Indeed, his *locus standi* is a riposte to Leavis's opinions on the civilising force of English letters. The picaresque takes Forelock through a vivid assemblage of antipodeans, while his creator's regard for Australia is too high to allow a lapse into cliché. The appeal of the country becomes plain: like Goldberg, it is sure of its position as 'God's own' and gnawed at by perceived inadequacies. Forelock is taught a salutory lesson by a feminist and her metaphorical agent, the titular spider. The spider, however, could equally stand as an image of Jacobson's fiction – painful and moral in the poison it bears.

Co-author, with Wilbur Sanders, of a critical study, *Shakespeare's Magnanimity*, Jacobson started writing fiction in 1981, explaining that he did so 'out of blind panic, because I thought I would die soon and I am not a good man or a good father'. For two years he was an urbane and acerbic television reviewer for the *Independent on Sunday*.

Fiction
Coming from Behind 1983; *Peeping Tom* 1984; *Redback* 1986; *The Very Model of a Man* 1992

Non-Fiction
Shakespeare's Magnanimity [with W. Sanders] 1978; *In the Land of Oz* 1987; *Roots Schmoots, Journey among Jews* 1993

Henry JAMES 1843–1916

James was born in Washington Square in New York, but spent his childhood and youth travelling back and forth between Europe and America, circumstances initiated by his father, Henry James Sr, whose decisions to travel were typically determined by spiritual crises or emotional impulses. James Sr was one of the best-known intellectuals in mid-nineteenth-century America, and his friends included Thoreau, Emerson and Hawthorne. A passionate advocate of Swedenborg's theological system, he published several books, including *Moralism and Christianity* (1850) and *Society, the Redeemed Form of Man* (1879), none of which found a readership.

The immediate, personal consequence of his beliefs was the liberal and peripatetic education of his children, of whom Henry was the second to be born. Though the two youngest had little involvement with literary or any other creative media, James's sister Alice was an important diarist, and his elder brother William (with whom Henry would have an intense, competitive relationship for the remainder of his life) was a founder of modern psychology and one of the most influential philosophers of his day. The James family was the most creative in nineteenth-century America, and perhaps the most remarkable family America has ever known.

At the age of six months James was taken to live near Windsor Castle and in the following year to Paris. Between 1855 and 1860 he lived at various times in Geneva, London, Paris, Boulogne, Bonn and Newport, Rhode Island, where he spent a term at Harvard Law School in 1862–3. By the time he was twenty-one, James had spent approximately one-third of his life in foreign countries and almost all of his time had been spent learning languages, including Latin. From an early age he knew all the classics of English, American and French literature (the latter, of course, he read in the original); the most important German texts, also in the original; as well as the Russian classics in translation.

He arrived in England a relatively independent man in 1869, having published his first short story, 'A Tragedy of Error', in New York's *Continental Monthly* in 1864, and soon introduced himself to Darwin, George Eliot, G.H. Lewes, Frederic Harrison, Ruskin, Rossetti, Morris and others. Soon afterwards he was again touring Europe. It is unsurprising that one of James's central, recurring themes is the contrast between New World innocence and Old World sophistication, and the comedy and tragedy that can result when an individual from one environment is immersed in the other.

In 1870 the unexpected death of his cousin Minny Temple, for whom he had an idealistic adoration, shocked him greatly. James's first novel, *Watch and Ward* (published in serial form from 1871), tells the story of a bachelor who adopts a twelve-year-old girl and quickly puts into action a long-term plan to marry her. The novel is not very successful, either in terms of characterisation or structure, and James would later with some justification call *Roderick Hudson* (1875) his 'first real novel'. The latter also presents the symbolic 'adoption' by an older man of a vulnerable person – though in the second novel the beloved is a young lad. This difference in sexual emphasis is in no small way responsible for the difference in climax in the two novels: in the first the relationship between the man and the girl is successful because socially sanctioned by marriage, whereas in *Roderick Hudson* the young man dies, and there is a sense of hopelessness whenever the social context of the relationship between the two protagonists is referred to.

James wrote the novel while travelling through Venice and Paris, when he also came across Walter Pater's *The Renaissance* (1873). It can be argued that his explicit exploration of English aestheticism culminates in his novel *The Tragic Muse* (1890), which deals with a life lived for art alone. Associated with aestheticism in the 1870s and 1880s, of course, was homosexuality. It is far from clear whether James was homosexually inclined: he suffered what he would later refer to as an 'obscure hurt' during an accident on the eve of the Civil War while serving as a volunteer fireman, which he claimed resulted in his asexuality. However, he explores the subject of homosexuality extensively in his fiction. The protagonist of 'The Author of Beltraffio', first published in the *English Illustrated Magazine* in 1884, is based on John Addington Symonds, with whom James was acquainted and who was the leading defender of homosexuality of his time.

In the 1890s James tried for a while to write plays, but these were badly received. Particularly upsetting was the reception given to *Guy Domville* in 1895, when the crowd hissed and booed him when he took the stage. Though he was intensely depressed for a while, in the long term his experience of the theatre proved fruitful, because it helped him to approach the novel from a new perspective, one which initiated the movement towards modernism in fiction. In works such as *The Spoils of Poynton* (first published as 'The Old Things'), *What Maisie Knew* and *The Sacred Fount*, characters are presented as though they are in a play: little background information is offered, and there is no objective narrative voice; instead, characters develop through dialogue, and plot is generated by the suspicions and guess-work of characters about one another. *The Turn of the Screw*, a ghost story in which the question of childhood corruption obsesses a governess and in which absolutely nothing is certain, is the most elusive novella ever written, and is James's best-known work.

In 1898 James had taken a long lease at Lamb House, Rye, and between 1902 and 1904 wrote there the three great novels of his 'late' period: *The Wings of the Dove*, *The Ambassadors* and *The Golden Bowl*. His literary acquaintance was vast, and he developed friendships with, among others, **Joseph Conrad** and **H.G. Wells**; as well as, in a more romantic context: Henderik Anderson, Jocelyn Persse and **Hugh Walpole**. In 1905 he visited America for the first time in twenty-five years, was shocked at the architectural and cultural developments in New York especially, and wrote 'The Jolly Corner', one of his best short stories, based, on one level, on his observations of New York; but also – in its nightmarish presentation of a man who is haunted by a *doppelgänger* with a horribly disfigured face and mutilated fingers – a vision of what he himself might have become if he had stayed in America.

Between 1906 and 1910, James revised many of his tales and novels for the New York Edition of his complete works. Comparing these revisions with the original versions has proved a useful way of unearthing the meaning in many of his works. In 1913 John Singer Sargent painted his portrait as a seventieth-birthday present, and in the same year James wrote his autobiographies *A Small Boy and Others* and *Notes of a Son and Brother*. A third volume of autobiography, *The Middle Years*, was published posthumously in 1917. He died in Chelsea, one year after finally becoming a British subject.

Though James wrote a great deal on painting, produced his famous *Notebooks*, has had his letters and much of his travel writing collected, it is as a novelist that he is now remembered. To say that he is the best novelist of the twentieth century is contentious, but not absurd. The collected prefaces to the New York Edition – published as *The Art of Fiction* – remain the fullest and most profound exploration and defence of the novelist's craft ever likely to be produced.

Fiction

A Passionate Pilgrim and Other Tales (s) 1875; *Roderick Hudson* 1875; *The American* 1877; *Daisy Miller* 1878; *The Europeans* 1878; *Watch and Ward* 1878; *Confidence* 1879; *An International Episode* (s) 1879; *The Madonna of the Future* (s) 1879; *The Diary of a Man of Fifty* (s) 1880; *Washington Square* 1880; *The Portrait of a Lady* 1881; *The Siege of London* (s) 1883; *Tales of Three Cities* (s) 1884; *The Aspern Papers* (s) 1885;

The Author of Beltraffio (s) 1885; *Stories Revived* (s) 3 vols 1885; *The Bostonians* 1886; *The Princess Casamassima* 1886; *The Reverberator* 1888; *A London Life* (s) 1889; *The Tragic Muse* 1890; *The Lesson of the Master* (s) 1892; *The Private Life* (s) 1893; *The Real Thing* (s) 1893; *Terminations* (s) 1895; *Embarrassments* (s) 1896; *The Other House* 1896; *The Spoils of Poynton* 1897; *What Maisie Knew* 1897; *In the Cage* (s) 1898; *The Turn of the Screw* 1898; *The Two Magics* (s) 1898; *The Awkward Age* 1899; *The Soft Side* (s) 1900; *The Sacred Fount* 1901; *The Wings of the Dove* 1902; *The Ambassadors* 1903; *The Better Sort* (s) 1903; *The Golden Bowl* 1904; *The Finer Grain* (s) 1909; *The Outcry* 1911; *The Ivory Tower* 1917; *The Sense of the Past* 1917; *Gabrielle de Bergerac* (s) 1918; *Travelling Companion* (s) 1919; *The Ghostly Tales of Henry James* (s) 1949; *Eight Uncollected Tales* (s) 1950; *The Complete Tales of Henry James* (s) 12 vols 1962–4

Non-Fiction

Transatlantic Sketches 1875; *French Poets and Novelists* 1878; *Hawthorne* 1880; *Portraits of Places* 1883; *The Art of Fiction* [with W. Besant] 1885; *Partial Portraits* 1888; *Essays in London and Elsewhere* 1893; *William Wetmore Story and His Friends* 1903; *English Hours* 1905; *The Question of Our Speech* 1905; *The American Scene* 1907; *View and Reviews* 1908; *Italian Hours* 1909; *A Small Boy and Others* (a) 1913; *Notes of a Son and Brother* (a) 1914; *Notes on Novelists* 1914; *The Question of the Mind* 1915; *The Middle Years* (a) 1917; *Within the Rim* 1919; *The Notebooks of Henry James* 1947; *Letters of Henry James* [ed. L. Edel] 4 vols 1974–84

Plays

Pyramus and Thisbe [pub. 1869]; *Still Waters* [pub. 1871]; *A Change of Heart* [pub. 1872]; *Daisy Miller* [pub. 1883]; *The Album* [pub. 1894]; *Disengaged* [pub. 1894]; *The Reprobate* [pub. 1894]; *Tenants* [pub. 1894]; *Monologue Written for Ruth Draper* [pub. 1922]; *The American* 1891 [pub. 1949]; *Guy Domville* 1895 [pub. 1949]; *The High Bid* 1907 [pub. 1949]; *Note for The Chaperon* [pub. 1949]; *The Other House* [pub. 1949]; *The Outcry* [pub. 1949]; *Rough Statement for Three Acts Founded on The Chaperon* [pub. 1949]; *The Saloon* [pub. 1949]; *Summersoft* [pub. 1949]

Collections

Complete Works 1907–9

Biography

Henry James by Leon Edel, 5 vols 1953–72

M(ontague) R(hodes) JAMES 1862–1936

Born in Goodnestone in Kent, James was brought up in Suffolk, where his clergyman father had a living. A precocious child, who at the age of six was discovered studying a Dutch Bible and three years later was composing sermons, he was educated at Eton and King's College, Cambridge, two places which were to shape his entire life. He took a double first, but spent much time at the Fitzwilliam Museum, studying illuminated manuscripts, and was appointed assistant director there in 1887, the year he was elected a fellow of King's. He lectured in divinity, made several catalogues of manuscripts, and in 1889 was elected junior dean. In 1893 he became director of the Fitzwilliam and in 1895 he published the first of his ghost stories, 'Canon Alberic's Scrapbook', which set the tone of his fiction: stories told in gentlemen's clubs or university common rooms, frequently featuring academics and presented in a spirit of scientific, scholarly inquiry. Many of them were first read to colleagues as part of the Christmas Eve festivities at King's. The first volume, *Ghost Stories of an Antiquary* was published in 1904; several of the stories had already appeared in periodicals.

James continued to pursue a life of distinguished scholarship, producing bibliographical and paleographical works, and became provost of King's in 1905 and vice-chancellor of the university in 1913. He spent holidays bicycling (sometimes tricycling) abroad, studying ecclesiastical architecture and gaining inspiration for his stories. In 1918 he became provost of Eton, where he remained until his death; he was awarded the Order of Merit in 1930.

A donnish, bespectacled figure, James remained unmarried, enjoying intense but asexual relationships with a number of young men, devoted to his school and university. 'It's odd that the Provost of Eton should still be aged sixteen,' Lytton Strachey remarked. 'A life without a jolt.' Although chiefly remembered for his four volumes of ghost stories ('I never cared to try any other kind'), which include several classics of the genre, James also translated and edited *The Apocryphal New Testament*, wrote works of theology, topography, memoirs, guidebooks and an unsuccessful children's book.

Fiction

Ghost Stories of an Antiquary (s) 1904; *More Ghost Stories of an Antiquary* (s) 1911; *A Thin Ghost* (s) 1919; *A Warning to the Curious* (s) 1925; *The Collected Ghost Stories of M.R. James* (s) 1931

For Children

Old Testament Legends 1913; *The Five Jars* 1922

Non-Fiction

The Life and Miracles of St William of Norwich [with A. Jessop] 1896; *Guide to the Windows of King's College Chapel* 1899; *The Sculptured*

Bosses in the Roof of the Bauchun Chapel 1908; *Abbeys* 1925; *Eton and King's: Recollections, Mostly Trivial* (a) 1926; *Suffolk and Norfolk* 1930

Translations

The Gospel According to Peter and the Revelation of Peter [with J. Armitage Robinson] 1892; *The Biblical Antiquities of Philo* 1917; Walter Map *De Nugis Curialium* 1923; *The Apocryphal New Testament* 1924

Biography

M.R. James: an Informal Portrait by Michael Cox 1983

P(hyllis) D(orothy) JAMES 1920–

James, who has inherited the mantle of **Agatha Christie** as the *doyenne* of British crime novelists, was born on the Suffolk coast at Southwold, and was educated at Cambridge Girls' High School. During the Second World War, she served as a nurse for the Red Cross, then worked in the Ministry of Food. She married a doctor in 1941 (he died after a long illness in 1964), and has two daughters. After the war, she embarked upon a career as a civil servant, first as administrator in the National Health Service, then as a principal at the Home Office, where she worked in the Police Department from 1968 to 1972, and then in the Criminal Policy Department until her retirement in 1979.

This career proved a useful background for her novels, the first of which, *Cover Her Face*, was published in 1962. The novel introduced Detective Chief-Inspector Adam Dalgliesh, who is also a well-known if minor poet. Dalgliesh's successful career at Scotland Yard (he rises to become a Commander) and his private anguish have developed through subsequent novels, and his somewhat lugubrious face has been fixed in the public imagination by the actor Roy Marsden, who has portrayed him in a number of successful television adaptations. Dalgliesh is a conscientious policeman, but is also a sensitive man who often finds sympathetic links with the victim or the suspects. James has wisely refrained from quoting any of his supposedly excellent poetry (although one of the television adaptations did so). Several of the novels are set in the Suffolk of James's childhood.

In *An Unsuitable Job for a Woman*, which is set in Cambridge, she introduced a second series detective, Cordelia Gray, a young woman who runs her own detective agency in London. James's novels are more ambitious than the average detective story, and although she constructs complicated, gripping and largely satisfying plots, her concern is as much with psychological motivation as with actual detection. She describes the day-to-day details of police and forensic work and also deals with the disruptive, sordid, personal aspects of a crime and its subsequent investigation. Large claims have been made for her as a novelist who transcends the genre in which she works, and it has even been suggested that some of her books should be considered for the Booker Prize (the jury of which she chaired in 1987). The literary quality of her work does not in fact bear very close inspection, and later novels have attracted adverse criticism in some quarters, for hers is essentially (and in every sense) a conservative talent. *Shroud for a Nightingale* and *The Black Tower* both won the Silver Dagger, presented by the Crime Writers' Association.

As well as the books featuring Dalgliesh and Gray, James has written non-genre novels, such as *Innocent Blood* (one of her best), in which a young woman attempts to trace her estranged mother, who had been imprisoned for the murder of a child, and *The Children of Men*, which is set in the future, where the population is dwindling. She has also written numerous short stories and a non-fiction account of a famous nineteenth-century crime.

James, who has served as a Justice of the Peace in London, is very much an establishment figure. She has been Chairman of the Society of Authors and the Arts Council's Literary Advisory Panel, has been on the board of the British Council, and has received several honorary degrees, including an Honorary DLitt from the 'private' University of Buckingham. Her involvement with the BBC has been less happy: as a governor, she oversaw the closure of the corporation's long-established magazine, *The Listener*, and in 1990 she seemed uncomfortable as the presenter of *Speaking Volumes*, one of television's least successful books programmes. She was created an OBE in 1983 and a life peer in 1991, and now styles herself Baroness James of Holland Park, taking this title from the area in London where she has lived for many years.

Fiction

Cover Her Face 1962; *A Mind to Murder* 1963; *Unnatural Causes* 1967; *Shroud for a Nightingale* 1971; *An Unsuitable Job for a Woman* 1972; *The Black Tower* 1975; *Death of an Expert Witness* 1977; *Innocent Blood* 1980; *The Skull Beneath the Skin* 1982; *A Taste for Death* 1986; *Devices and Desires* 1989; *The Children of Men* 1992; *Murder in the Dark* (s) [with others] 1994; *Original Sin* 1994

Non-Fiction

The Maul and the Pear Tree: the Ratcliffe Highway Murders 1811 [with T.A. Critchley] 1971

(Margaret) Storm JAMESON 1891–1986

Jameson was born and grew up in Whitby, Yorkshire, into a family of ship-owners and builders of Viking descent. Her formidable mother (a model for the Victorian matriarchs who feature in her family chronicles) was a devout Congregationalist who frequently thrashed her daughter for reading before breakfast, and Jameson's Nonconformist background is uncompromisingly reflected in the severe, sometimes didactic, honesty of her work. She was educated largely at home, and attended Leeds University, where she took a first and was the first woman to graduate in English. She did postgraduate work at King's College, London, then became a copywriter for the Carlton advertising agency, the editor of a magazine, *New Commonwealth*, and, for several years, English representative and later London co-manager for the New York publisher Knopf.

Her first novel, *The Pot Boils*, was published in 1919. She had one son from her first marriage, which ended in divorce in 1925: the following year she married Guy Chapman, who became a professor of history at Leeds University and who is remembered for his account of trench warfare in France, *A Passionate Prodigality* (1933). A determined and hot-tempered Yorkshirewoman, of strong liberal views, Jameson made an outstanding contribution to literary life when she was president of the English branch of International PEN from 1938 to 1944 and became noted for her work with refugees and imprisoned writers. She wrote well over forty novels, and was herself the first to admit that not all were of high quality; but the best of them are excellent, and still undeservedly neglected. *Cousin Honoré* is perhaps outstanding, set in Alsace between the wars, and showing an ability to paint Frenchmen and Frenchwomen unusual in an English writer. Fine later novels from the 1960s were *The Early Life of Stephen Hind* and *The White Crow*. She also wrote an outstanding two-volume autobiography, *Journey from the North*, and much distinguished non-fiction, largely on literary subjects: her last book, published when she was eighty-eight, was a study of Stendhal. For many years she and her husband lived between Whitby and London; when her husband retired in 1952 she moved with him to Cambridge and, despite her northern roots, remained there after his death in 1972, dying herself at the age of ninety-five.

Fiction

The Pot Boils 1919; *Happy Highways* 1920; *The Clash* 1922; *Lady Susan and Life* 1923; *The Pitiful Wife* 1923; *The Lovely Ship* 1927; *Farewell to Youth* 1928; *The Voyage Home* 1930; *A Richer Dust* 1931; *The Single Heart* 1932; *That Was Yesterday* 1932; *A Day Off* 1933; *The Mirror in Darkness Trilogy* 1994: *Company Parade* 1934, *Love in Winter* 1935, *None Turn Back* 1936; *In the Second Year* 1936; *Delicate Monster* 1937; *The Moon is Making* 1937; *Here Comes a Candle* 1938; *Farewell Night, Welcome Day* 1939; *Cousin Honoré* 1940; *Europe to Let* 1940; *The Fort* 1941; *Then We Shall Hear Singing* 1942; *Cloudless May* 1943; *The Journal of Mary Hervey Russell* 1945; *The Other Side* 1946; *Before the Crossing* 1947; *The Black Laurel* 1948; *The Moment of Truth* 1949; *The Green Man* 1952; *The Hidden River* 1955; *The Intruder* 1956; *A Cup of Tea for Mr Thorgill* 1957; *A Ulysses Too Many* 1958; *A Day Off and Other Stories* (s) 1959; *Last Score* 1961; *A Month Soon Goes* 1962; *The Road from the Monument* 1962; *The Aristide Case* 1964; *The Early Life of Stephen Hind* 1966; *The White Crow* 1968; *There Will Be a Short Interval* 1972

Non-Fiction

Modern Drama in Europe 1920; *The Georgian Novel and Mr Robinson* 1929; *The Decline of Merry England: an Essay on Puritanism* 1930; *No Time Like the Present* 1933; *Women against Men* 1933; *The Novel in Contemporary Life* 1938; *Civil Journey* 1939; *The End of This War* 1941; *The Writer's Situation and Other Essays* 1950; *Morley Roberts: the Last Eminent Victorian* 1961; *Journey from the North* (a) 2 vols 1969, 1970; *Parthian Words* 1970; *Speaking of Stendhal* 1979

Plays

Full Circle 1928

Randall JARRELL 1914–1965

Though he was born in Nashville, Tennessee, Jarrell spent much of his childhood in California and throughout his life remained romantically attached to the state, as well as to his memories of the childhood abandon and freedom he experienced there. He studied psychology at Vanderbilt University, where he met his lifelong friend **Robert Lowell**. After a brief spell of teaching at Kenyon College, he entered the US Army Air Corps in 1942 and served with them until 1946, spending most of the time training in Texas and Illinois but finally moving on to operate a celestial-navigational tower at Davis-Monthon Field in Tucson, Arizona. He had already published one collection of poems, *Blood for a Stranger*, in 1942, but now drew on his wartime experiences to produce his better-known (and better-written) poems published in *Little Friend, Little Friend* and *Losses*, which include 'Lines', 'Absent with Official Leave' and, perhaps his most famous poem, 'The Death of the Ball-Turret Gunner'. But it would be wrong to see

Jarrell's poetry of this period as providing simple documentary evidence of the horrors of war. Innocence, and the loss of innocence, was always an obsession for him, and he saw the global conflict as a symbol of how innocence is lost through suffering and experience. In a more general way, his insistence on pursuing the mythological, dream-like and archetypal amidst the very real world of mass destruction and annihilation pointed to the direction American poetry would take in the post-war period.

Jarrell was of the generation of American poets who grew up after a first wave of modernists, most notably **Hart Crane** and **Wallace Stevens**, had emphasised a particularly American type of poetics, as opposed to the European classical tradition favoured by those who had left America for Europe, such as **T.S. Eliot** and **Ezra Pound**. By exploring the work of these poets and others, as well as the relationship his own generation had forged with them, Jarrell quickly established himself as the leading critic of his age. His essays have been collected in numerous volumes, the most important of which are *Poetry and the Age* and *Kipling, Auden & Co.* He was not known only as the most incisive and serious of critics, however: the frequent expression of wit, insult and bitter aside in his reviews caused considerable amusement and gained him a reputation for controversy.

Jarrell went on to publish three more collections of poetry, *The Seven-League Crutches*, *The Woman at the Washington Zoo* and *The Lost World*. The title-poem of the last collection is a modern American classic. He had earlier published a highly accomplished novel, *Pictures from an Institution*, which is about a progressive American college for women, and since the end of the Second World War he had been teaching creative writing at various colleges and universities in the USA, including the University of Texas, Sarah Lawrence College, Princeton, and the Women's College of the University of North Carolina. Despite the fact that his second marriage in 1952 proved to be more successful than the first, Jarrell seems to have been permanently unhappy and was eventually diagnosed as being clinically depressed. He suffered a nervous breakdown in 1965, for which he was hospitalised, and later that year he was killed in a car accident. Whether or not he committed suicide is still a matter of considerable debate. The motorist driving the car testified that the poet lunged towards the vehicle deliberately and for no obvious reason.

Poetry

Five Young American Poets [with others] 1940; *Blood for a Stranger* 1942; *Little Friend, Little Friend* 1945; *Losses* 1948; *The Seven-League Crutches* 1951; *The Woman at the Washington Zoo* 1960; *The Lost World* 1965 [rev. 1966]; *The Complete Poems* 1968

Fiction

Pictures from an Institution 1954

Non-Fiction

Poetry and the Age 1953; *A Sad Heart at the Supermarket* 1962; *The Third Book of Criticism* 1969; *Kipling, Auden & Co.: Essays and Reviews 1935–64* 1979; *Randall Jarrell's Letters* [ed. M. Jarrell, S. Wright] 1984

Translations

Grimm: *The Golden Bird and Other Fairy Tales of the Brothers Grimm* 1962, *Snow-White and the Seven Dwarfs* 1972, *The Juniper Tree and Other Tales from Grimm* 1973; *The Rabbit Catcher and Other Tales by Ludwig Beckstein* 1962; Chekhov *The Three Sisters* 1969; Goethe *Faust: Part I* 1976

For Children

The Bat-Poet 1964; *The Gingerbread Letters* 1964; *The Animal Family* 1965; *Fly by Night* 1976; *A Bat Is Born* 1977

Biography

Randall Jarrell by William H. Pritchard 1990

(John) Robinson JEFFERS 1887–1962

Jeffers was born in Pittsburgh, Pennsylvania. His father was a Presbyterian minister and classical scholar who travelled widely in Europe where much of Jeffers's early education took place. The family settled in California and Jeffers studied a wide range of subjects (including forestry and medicine) at Occidental College, the University of Zurich and other universities. He attracted little attention with his first two collections, *Flagons and Apples* and *Californians*, but *Tamar and Other Poems* was praised by **T.S. Eliot** and established his reputation. The narrative title poem draws loosely on the biblical story of King David's daughter to tell a contemporary tale of incest and death. Incest figures frequently in Jeffers's verse as a symbol of man's destructive self-obsession.

In 1913 Jeffers caused something of a furore by marrying a divorcee, Una Call Kuster, with whom he had twin sons. The following year a legacy from an uncle enabled him to build (with his own hands) a stone house and observation tower on the coast in Carmel, California. Here he retreated to the contemplation of the dramatic landscape of sea and mountains which dominates much of his verse. After a visit to London in 1928, during which he saw his English publisher, **Leonard Woolf**, Jeffers lived an increasingly isolated life and developed a reputation as a recluse. This did not, however, prevent him from being the first poet to appear on the cover of *Time* magazine (in 1932).

Borrowing from Schopenhauer and Nietzsche, he developed a personal philosophy of 'inhumanism', which aimed at 'a shifting of emphasis from man to not man, the rejection of human solipsism and recognition of the transhuman magnificence'. Jeffers's misanthropic thinking is manifest in a poem such as 'Roan Stallion', wherein a woman allows a stallion to trample her husband to death but then shoots it 'out of some obscure human fidelity'. Alfred Kazin (in *On Native Grounds*, 1942) was not alone in feeling uneasy with Jeffers's 'disgust with the human species in toto', though the posthumous publication of *The Beginning and the End* revived some interest in the poet. Two of his shorter poems, 'Hawk Hurts' and 'November Surf', are still widely anthologised.

Poetry

Flagons and Apples 1912; *Californians* 1916; *Tamar and Other Poems* 1924; *Roan Stallion, Tamar and Other Poems* 1925; *The Women at Point Sur* 1927 [rev. 1977]; *Poems* 1928; *Cawdor and Other Poems* 1929; *Dear Judas and Other Poems* 1929; *Stars* 1930; *Descent to the Dead* 1931; *Thurso's Landing and Other Poems* 1932; *Give Your Heart to the Hawks and Other Poems* 1933; *Solstice and Other Poems* 1935; *The Beaks of Eagles* 1936; *Such Counsels You Gave Me and Other Poems* 1937; *Selected Poetry of Robinson Jeffers* 1938; *Two Consolations* 1940; *Be Angry at the Sun* 1941; *The Double Axe and Other Poems* 1948 [rev. 1977]; *Hungerfield and Other Poems* 1954; *The Loving Shepherdess* 1956; *The Beginning and the End and Other Poems* 1963; *The Alpine Christ and Other Poems* [ed. W. Everson] 1973; *Brides of the South Wind: Poems 1917–1922* [ed. W. Everson] 1974; *The Double Axe and Other Poems* 1977; *What Odd Expedients and Other Poems* 1981; *The Collected Poetry of Robinson Jeffers* [ed. T. Hunt] 2 vols 1988, 1989; *Songs and Heroes* 1988

Non-Fiction

Poetry, Gongorism and a Thousand Years 1949; *Themes in My Poems* 1956; *The Selected Letters of Robinson Jeffers* [ed. A.N. Ridgeway] 1968; *Tragedy Has Obligations* 1973

Plays

Medea [from Euripides' play] 1946

(Patricia) Ann JELLICOE 1927–

Jellicoe was born in Middlesbrough, Yorkshire, and educated at a boarding-school in Castle Howard. At the age of four she played the title role in a kindergarten production of *Sleeping Beauty* after which, she has said, 'I had no doubt, and always said theatre was what I wanted to do.' She studied at the Central School of Speech and Drama, London, from 1944 to 1947, and worked variously as an actress, stage manager and director before founding the Cockpit Theatre Club in 1950. This was a Sunday theatre club which, she believes, used 'the first open stage in London for about four hundred years'. She returned to teach at Central in 1953 and wrote her first play, *The Sport of My Mad Mother*, in 1956 in response to an *Observer* playwriting competition. It was staged in 1958 at the Royal Court where, in the words of Philip Locke, it was something of a 'flop *d'estime*'. An apparently shapeless examination of arbitrary violence which draws on Artaudian ritual, the play was praised by Kenneth Tynan, but proved rather too radical for audiences. Jellicoe's fortunes were reversed by her second play, the fast-moving comedy *The Knack*, of which she has said: 'A comedy with one set, four actors, about sex: it's absolutely cast iron.' The play was first produced in Cambridge in 1961 and was subsequently produced at the Royal Court Theatre and in New York by Mike Nichols. Richard Lester's startling black-and-white film version was scripted by **Charles Wood** and is substantially different from the play.

Following divorce from her first husband, Jellicoe remarried, and devoted herself to bringing up the couple's son and daughter. During this period she wrote two further full-length plays, *Shelley* and *The Giveaway*. She was literary manager at the Royal Court from 1973 to 1975 before moving to Lyme Regis where she produced and directed *The Reckoning*, the first of her 'community plays'. In her book *Community Plays: How to Put Them On* she gives a comprehensive account of the history and practice of staging these plays. In 1979 she founded the Colway Theatre Trust which, in addition to her own work, has funded and organised community plays written by **David Edgar**, **Howard Barker**, Sheila Yeger and Charles Wood. She was awarded the OBE in 1984.

Plays

Rosmersholm [from Ibsen's play] 1952 [pub. 1960]; *The Sport of My Mad Mother* 1958 [pub. in *The Observer Plays* 1958]; *The Knack* 1961 [pub. 1962]; *The Lady from the Sea* [from Ibsen's play] 1961 [np]; *The Seagull* [from Chekhov's play; with A. Nicolaeff] 1964 [np]; *Shelley* 1965 [pub. 1966]; *The Rising Generation* 1967 [pub. in *Playbill 2* ed. A. Durbrand 1969]; *The Giveaway* 1968 [pub. 1970]; *The Reckoning* 1978 [np]; *The Bargain* 1979 [np]; *The Tide* 1980 [np]; *The Western Women* [from F. Weldon's story; with N. Brace]

1984 [np]; *Mark Og Mont* 1988 [np]; *Under the God* 1989 [np]; *Changing Places* 1992 [np]

For Children

3 Jelliplays [pub. 1975]: *You'll Never Guess* 1973, *Clever Elsie, Smiling John, Silent Peter* and *A Good Thing or a Bad Thing* 1974; *Flora and the Bandits* 1975 [np]

Translations

Der Freischütz [from F. Kind's libretto; with C. von Weber] 1964 [np]

Non-Fiction

Some Unconscious Influences in the Theatre 1967; *Devon: a Shell Guide* [with R. Mayne] 1975; *Community Plays: How to Put Them On* 1987

Alan (Fitzgerald) JENKINS 1955–

Jenkins was born in London, growing up in East Sheen. He was educated at the London Nautical School, and developed an interest in verse as a teenager, writing his first poems on a typewriter given him by the father of an early girlfriend. He attended the University of Sussex, where he read English and French, taking a first in 1978, and staying on to do an MA in modern English literature. He then obtained work briefly with IPC Magazines, working on *Woman's Own* and even at one point writing the agony aunt column. In 1980 he joined the staff of the *Times Literary Supplement*, initially having responsibility for poetry, later becoming fiction and poetry editor, and in 1990 deputy editor, a position he continues to hold.

As a poet, he early won a prestigious Eric Gregory Award in 1981, but his first book-length appearance did not come until 1986, when he was one of a number of writers represented in *New Chatto Poets*. His collection *In the Hot-House* followed in 1988, the first of three so far published. Jenkins is a fluent, technically adroit poet, whose broad range of subject-matter nevertheless allots a special place to sex. The second collection, *Greenheart*, saw a move towards political and ecological themes, but *Harm* was something of a return to the more personal mood of his first book, with several poems relating to the death of his father. Foreign places (Jenkins has travelled widely, and has particularly spent time in France) and homage to other writers are other themes running through all three collections. Technically, the poet's startling use of rhyme and (in early works) an exploration of the possibilities of enjambment are notable features. Jenkins has also gained notice as a critic, being from 1987 to 1990 the regular poetry reviewer for the *Observer*, and has judged literary competitions, taught writing in various venues, contributed to reference works and broadcast frequently. The 1992 Quartet Reprint of Gilbert Cannan's edition of Valery Larbaud's *A.O. Barnabooth: His Diary* contains an introduction by Jenkins which incorporates translations of several 'Barnabooth' poems.

Poetry

New Chatto Poets [with others] 1986; *In the Hot-House* 1988; *Greenheart* 1990; *Harm* 1994

Edited

Contemporary Poetry 1994

(Margaret) Elizabeth (Heald) JENKINS 1905–

The daughter of the headmaster of a boys' preparatory school, Jenkins grew up in Hertfordshire, and was educated at St Christopher's School, Letchworth, and Newnham College, Cambridge, where she read English and history and where she was, according to her own account, 'gauche, prim, covered with ink, and wrapped up in work'. She became a teacher, and from 1924 to 1939 was senior English mistress at King Alfred School, Hampstead, while writing in her spare time.

Virginia Water (1930) was her first novel, but she found her form four years later with her fourth novel, *Harriet*, an engrossing account of a Victorian murder case. During the war she worked in the civil service and, at its conclusion, she withdrew to her Hampstead home, where she has devoted the rest of her long life to reading and writing. Her best novel is *The Tortoise and the Hare*, a penetrating and underrated study of the limitations involved in traditional femininity.

Jenkins has become as well known as a biographer as she is as a novelist, and has published distinguished lives of Jane Austen, Elizabeth I and Lady Caroline Lamb. In latter years she has devoted herself more to non-fiction than to the novel. *Dr Gully* is a novel based on the life of the doctor in the celebrated Balham murder case: Jenkins claims to have had the help of Dr Gully himself in this book through automatic writing. *The Shadow and the Light*, a defence of the Victorian medium Daniel Douglas Home, was followed by a late novel (her first for twenty years), *A Silent Joy*. To some extent, this reprises the theme of *The Tortoise and the Hare* in its depiction of the developing friendship between an adult and a neglected child. Jenkins received an OBE in 1981.

Fiction

Virginia Water 1930; *The Winters* 1931; *Portrait of an Actor* 1933; *Harriet* 1934 [as *Murder by*

Neglect 1960]; *Doubtful Joy* 1935; *The Phoenix' Nest* 1936; *Robert and Helen* 1944; *Young Enthusiasts* 1946; *The Tortoise and the Hare* 1954; *Brightness* 1963; *Honey* 1968; *Dr Gully* 1972; *A Silent Joy* 1992

Non-Fiction
Lady Caroline Lamb 1932; *Jane Austen* 1938; *Henry Fielding* 1947; *Six Criminal Women* 1949; *Ten Fascinating Women* 1955; *Elizabeth the Great* 1958; *Elizabeth and Leicester* 1961; *The Mystery of King Arthur* 1975; *The Princes in the Tower* 1978; *The Shadow and the Light* 1982

For Children
Joseph Lister 1960

Elizabeth (Joan Cecil) JENNINGS 1926–

Jennings is the daughter of a doctor, and was born in Boston, Lincolnshire. Her early memories of this place inform some of her poetry, but when she was six the family moved to Oxford, the scene of most of the rest of her life. She was educated at Oxford High School before reading English at St Anne's College, Oxford, graduating in 1949. She had begun to write poetry early and, as an undergraduate, became part of the lively poetry scene associated with the university; in the year of her graduation, her first poems were chosen by **Kingsley Amis** and James Michie for *Oxford Poetry*. From Oxford, she obtained work briefly as a copywriter for an advertising agency in London, but then worked as an assistant at Oxford City Library from 1950.

Her poetry began to appear widely in magazines in the early 1950s, and in 1953 her volume *Poems* was brought out by the Fantasy Press, which first published in book form many of the younger Oxford poets of the time. It was well received, and her reputation was further consolidated in 1955 with *A Way of Looking*, which won a Somerset Maugham Award. She gained leave of absence from the library, and used the award to travel for a few months to Rome, an immensely fructifying experience for her, laying the foundation for a love of Italy which she has retained. In 1956 she was included with eight other poets (all male, and including **Philip Larkin** and Amis) by **Robert Conquest** in *New Lines*, the anthology of the Movement, a poetic tendency emphasising clarity, sobriety and order. At first sight, Jennings might seem the odd woman out, her lifelong Catholicism contrasting with the Movement's general agnosticism, her love of Italy with their insularity. However, it could be argued that she of all these poets has retained most faithfully the original ideals, continuing to write a formal (often rhymed) verse which is as unpretentious as it is clearly and lyrically felt. Fashion has perhaps neglected her, but the publication of almost thirty poetic volumes, including several substantial collections, attests to the respect in which she is held. Childhood and age, art and religion are among the themes she treats, and her work has been complemented by several general non-fiction books on poetry and anthologies edited.

In 1958, she left the library, and became a reader for Chatto & Windus, living in a convent in London, and in 1961 was briefly a lecturer at Columbia University in New York. But around this period she suffered a breakdown, later spending time in a psychiatric hospital and suffering illness for two decades. She has therefore held no further appointments (although she lectured at Barnard College, Columbia University, in the 1970s), but has lived as a freelance writer in Oxford. In later years she has become mildly celebrated for her unconventional lifestyle there: writing in bed at night in her small rented room in North Oxford; doing more writing in a café during the day; carrying many plastic bags and paying little attention to her dress. One could say it was the lifestyle of one who has chosen poetry as her vocation and knows no other.

Poetry
Poems 1953; *A Way of Looking* 1955; *The Child and the Seashell* 1957; *A Sense of the World* 1958; *Song for a Birth or a Death and Other Poems* 1961; *Penguin Modern Poets 1* [with L. Durrell, R.S. Thomas] 1962; *Recoveries* 1964; *The Mind Has Mountains* 1966; *Collected Poems* 1967; *The Animals' Arrival* 1969; *Hurt* 1970; *Lucidities* 1970; *Folio* [with others] 1971; *Relationships* 1972; *Growing Points* 1975; *Consequently I Rejoice* 1977; *Moments of Grace* 1979; *Selected Poems* 1979; *Winter Wind* 1979; *A Dream of Spring* 1980; *Celebrations and Elegies* 1982; *Extending the Territory* 1985; *In Shakespeare's Company* 1985; *Collected Poems 1953–1985* 1986; *Tributes* 1989; *Times and Seasons* 1992; *Familiar Spirits* 1994

Non-Fiction
Let's Have Some Poetry 1960; *Every Changing Shape: On the Relationship Between Poetry and Mystical Experience* 1961; *Poetry Today 1957–1960* 1961; *Robert Frost* 1964; *Christianity and Poetry* 1965; *Seven Men of Vision* 1976

For Children
The Secret Brother 1966; *After the Ark* 1978; *A Quintet* [with others] 1985

Edited
The Batsford Book of Children's Verse 1958; *A Choice of Christina Rossetti's Verse* 1970; *The Batsford Book of Religious Verse* 1981; *In Praise of Our Lady* 1982

Translations

A Translation of Michelangelo's Sonnets 1961

F(ryniwyd) (pseud. of Wynifried Margaret) Tennyson JESSE 1888–1958

Jesse (the final 'e' is silent) was born in Chislehurst, Kent, the second of three daughters of a High Anglican clergyman; her paternal grandmother was a sister of Alfred, Lord Tennyson, and had once been engaged to A.H. Hallam. The family emigrated to South Africa, where Jesse's younger sister died of meningitis; her mother subsequently entrusted her elder sister to wealthy relations, became an invalid and shared 'passionate friendships' with a number of women. (*The Alabaster Cup* contains a chilling portrait of Mrs Jesse.) Jesse studied art under Stanhope and Elizabeth Forbes at the Newlyn School in Cornwall (where she acquired her new name of Fryniwyd), but at the age of twenty-three left to become a writer in London.

A short story, 'The Mask', was accepted by the *English Review* and on the strength of this her first novel, *The Milky Way*, was commissioned by William Heinemann and published in 1913. During a joyride in an aeroplane over Lake Windermere, she caught her right hand in a propeller while waving; a series of operations failed to save her fingers and resulted in an addiction to morphia. She was fitted with a mechanical contraption, which she called her 'pandy', and taught herself to write again. Undaunted by her experiences, she travelled to America and the West Indies, working as a correspondent for the *Daily Mail*. During the First World War she was one of the few women reporters to penetrate the war zone, reporting on the siege of Antwerp and subsequently working for the Ministry of Information.

In 1918, she married the playwright H.M. ('Tottie') Harwood, who had written to ask her if he could dramatise 'The Mask'. The marriage had to be kept secret because Harwood wanted to retain access to his son by his married mistress. The Harwoods collaborated on a number of plays, including *Billeted*, a wartime success in 1917, but the marriage was unhappy; Harwood was still involved with his mistress and son, while Jesse had several miscarriages, became involved with a succession of doting female secretaries, and repeatedly attempted suicide.

Through Alfred Mond, owner of the *English Review*, she had met the Attorney-General, Sir Rufus Isaacs, who instructed her in law, and criminology became one of her chief interests. Her best known novel, *A Pin to See the Peepshow*, is based upon the Thompson–Bywaters case of 1922–3, a notorious miscarriage of justice in which a woman was hanged after her young lover had killed her husband. Jesse also edited six volumes of the *Notable British Trials* series, and wrote the acclaimed *Murder and Its Motives*, as well as producing volumes of poetry and *belles-lettres*, *The Baffle Book* (an English version of an American volume of detective puzzles), a history of Burma and several novels, the most highly-regarded of which is *The Lacquer Lady*, an adventure story (based on fact) concerning a schoolgirl who becomes favourite of the Queen of Burma in the 1880s.

Fiction

The Milky Way 1913; *Beggars on Horseback* (s) 1915; *Secret Bread* 1917; *Tom Fool* 1926; *Moonraker* 1927; *Many Latitudes* (s) 1928; *The Lacquer Lady* 1929; *The Solange Stories* 1931; *A Pin to See the Peepshow* 1934; *Act of God* 1937; *The Alabaster Cup* 1950; *Dragons in the Heart* 1956

Plays

With H.M. Harwood: *The Mask* 1912; *Billeted* 1917; *The Hotel Mouse* 1921; *The Pelican* 1924; *How to be Healthy though Married* 1930; *A Pin to See the Peepshow* [from her novel] 1948; *The Thin Line* [with E. Atiyah] 1953

Quarantine 1922; *Anyhouse* 1925; *Birdcage* 1949

Non-Fiction

The Sword of Deborah 1919; *The White Riband* 1921; *Murder and Its Motives* 1924; *Notable British Trials: Madeleine Smith* 1927, *Samuel Herbert Dougal* 1928, *Sidney Harry Fox* 1934, *Rattenbury and Stoner* 1935, *Ley and Smith* 1947, *Evans and Christie* 1957; *Sabi Pas* 1935; *London Front* [with H.M. Harwood] 1939; *While London Burns* [with H.M. Harwood] 1940; *The Saga of San Demetrio* 1942; *The Story of Burma* 1946; *Comments on Cain* 1948

Poetry

The Happy Bride 1920; *The Compass* 1951

Edited

Randle McKay and Lassiter Wren *The Baffle Book* 1930

Biography

A Portrait of Fryn by Joanna Colenbrander 1984

Ruth Prawer JHABVALA 1927–

Jhabvala describes herself as 'practically born a displaced person', most of her novels, short stories and film scripts bearing a characteristic trademark of cultural interaction, conflict and exile. She was born in Cologne, Germany, her father, a lawyer, of Polish-Jewish, and her mother of German-Jewish, origin. The beginning of her schooling coincided with the Nazi era in Germany, and she attended Jewish segregated schools before emigrating with her family to Britain in 1939. Here they lived in Hendon, north-west London, and Jhabvala attended Hendon County School between 1940 and 1945 before going up to Queen Mary College, University of London, where she eventually took an MA in 1951. Her elder brother S.S. Prawer was to take an academic course after university, and became a professor of German literature at Oxford (he is known for many publications, including his editorship of *The Penguin Book of Lieder*, 1964). She, however, married Cyrus Jhabvala, an Indian architect, in the year of her graduation, and went to live with him in India. They were to have three daughters.

As a housewife in Delhi, she began to devote much of her time to writing, and was soon successful, publishing her first novel, *To Whom She Will*, in 1955, and also becoming a regular contributor of stories to the *New Yorker* (four volumes of her stories, plus a selection, have now appeared). Her earliest novels dealt with a purely Indian scene, but from her third, the popular *Esmond in India* (1958), she began to make a speciality of describing the confrontation between Indians and westerners, thus placing herself in a tradition deriving from **E.M. Forster**. Comparisons with the work of Jane Austen have also been made, chiefly arising from Jhabvala's delicate social satire, her concern with manners, and her classic, traditional style.

The same respect for the past is evident in her long collaboration with the film-makers Ismail Merchant and James Ivory, which dates from 1963 when she scripted the film for them of her own novel, *The Householder*. She has since become their principal scenarist, responsible, among others, for several highly successful adaptations of the works of **Henry James** and Forster. Her career as novelist has continued, although recently at a slower rate, and perhaps reached its apogee when *Heat and Dust*, a novel contrasting British attitudes to India during and after the Raj, won the Booker Prize in 1975.

In the same year, a major change took place in her life: she had become increasingly disenchanted with India, and had suffered serious jaundice there; therefore she moved her principal home to New York, while her husband continued with his architectural practice in Delhi. The marriage has been maintained with frequent visits between the two countries, as well as much time spent in London. More recent novels, such as *Poet and Dancer*, increasingly have purely American settings, marking a new direction for this always fertile writer.

Fiction

To Whom She Will 1955 [US *Amrita* 1956]; *The Nature of Passion* 1956; *Esmond in India* 1958; *The Householder* 1960; *Get Ready for Battle* 1962; *Like Birds, Like Fishes* (s) 1962; *A Backward Place* 1965; *An Experience of India* (s) 1966; *A Stronger Climate* (s) 1968; *A New Dominion* 1972 [US *Travelers* 1973]; *Penguin Modern Stories 11* [with others] 1972; *Heat and Dust* 1975; *How I Became a Holy Mother and Other Stories* (s) 1976; *In Search of Love and Beauty* 1983; *Out of India* (s) 1986; *Three Continents* 1987; *Poet and Dancer* 1993; *Shards of Memory* 1995

Plays

The Householder (f) [from her novel] 1963; *Shakespeare Wallah* (f) [with J. Ivory] 1965 [pub. 1973]; *The Guru* (f) 1969; *Bombay Talkie* (f) 1970; *Autobiography of a Princess* (f) [with J. Ivory, J. Swope] 1975; *The Place of Peace* (tv) 1975; *Roseland* (f) 1977; *Hullabaloo Over Georgie and Bonnie's Pictures* (f) 1978; *The Europeans* (f) [from H. James's novel] 1979; *Jane Austen in Manhattan* (f) 1980; *A Call from the East* 1981 [np]; *Quartet* (f) [from J. Rhys's novel] 1981; *Heat and Dust* (f) [from her novel] 1983; *The Bostonians* (f) [from H. James's novel] 1984; *A Room with a View* (f) [from E.M. Forster's novel] 1986; *Maurice* (f) [from E.M. Forster's novel] 1971; *Madame Sousatska* (f) [from B. Rubens's novel, with J. Schlesinger] 1988; *Slaves of New York* (f) 1989; *Mr and Mrs Bridge* (f) [from Evan S. Connell Jr's novels] 1990; *Howards End* (f) [from E.M. Forster's novel] 1992; *The Remains of the Day* [from K. Ishiguro's novel] 1993; *Jefferson in Paris* (f) 1995

Non-Fiction

Meet Yourself at the Doctor 1949

B(ryan) S(tanley William) JOHNSON 1933–1973

Johnson was the most experimental British novelist of his generation, although his experimental status is something he played down: 'Certainly I make experiments,' he said, 'but the unsuccessful ones are quietly hidden away.' Formal experimentation was used to incorporate real life into the books, usually Johnson's own. The son of a bookseller's stock-keeper, Johnson was born in Hammersmith and lived in London for most of his life. After working as a bank clerk he read English at King's College, London, and was later a teacher and sports journalist before becoming a novelist, poet, television producer, playwright and editor. During the 1960s he was a poetry editor of *Transatlantic Review*, which ceased publication at his death. In 1964 he married Virginia Ann Kimpton, who appears as Ginnie in *Trawl*. They had two children.

Johnson's first novel, *Travelling People* (1963), already shows his characteristic subversion of the conventions of fiction through extreme self-reflexivity. At one point he writes of his protagonist, Henry Henry, that: 'Henry went into the back of his mind, a private world where he could live untouched and where no one else could ever penetrate: always excepting myself, of course.' It is a characteristic line: in his most popular novel, *Christie Malry's Own Double Entry*, featuring a bank clerk with a spectacular grudge against society, he writes: 'For the following passage it seems to me necessary to attempt transcursion into Christie's mind; an illusion of transcursion, of course, since you know only too well in whose mind it all really takes place.' Johnson was influenced by **James Joyce** and by **Samuel Beckett** (with whom he corresponded, and who admired his work). He complained that the majority of novelists still try to tell stories 'as though *Ulysses* (let alone *The Unnamable*) had never happened'.

Johnson set himself rigorously against what he saw as the element of 'lying' in fiction, or illusionistic storytelling. In *Albert Angelo*, which features a hard-up architect whose architecture is a metaphor for Johnson's writing, the author suddenly bursts through: 'OH, FUCK ALL THIS LYING! … what i'm really trying to write about is writing not all this stuff about architecture …'. He wrote that: 'The two terms *novel* and *fiction* are not, incidentally, synonymous … The novel is a form in the same way that the sonnet is a form; within that form, one may write truth or fiction. I choose to write truth in the form of a novel.' He backed up this honesty with a range of devices and tropes in the tradition of (among others) Sterne, such as metafictional direct addresses to the reader, a page of black to represent the death by heart-attack of a character in *Travelling People*, and blank pages to represent sleep or mental confusion in *House Mother Normal*.

The Unfortunates has become one of the best-known avant-garde novels, consisting of twenty-seven loose-bound sections unbound in a box: apart from the first and last, they may be read in any order that the reader chooses. The novel's narrator is called B.S. Johnson and its concerns are painfully realistic, since it is about the death of Johnson's real-life friend Tony Tillinghurst from cancer. Johnson's last work, *See the Old Lady Decently*, was written after his own mother died of cancer in 1971. It was to be the first volume in 'The Matrix Trilogy', the next two of which were to be titled 'Buried Although' and 'Amongst Those Left Are You', so that the full title could be read continuously across their spines. Johnson's mother had been a domestic servant in Belgravia, and he made the book into a combined work on his mother, on archetypal motherhood in general, and on the decline of the mother country. He committed suicide in a fit of despair shortly after completing it.

Fiction

Travelling People 1963; *Albert Angelo* 1964; *Statement against Corpses* [with Z. Ghose] **(s)** 1964; *Trawl* 1966; *The Unfortunates* 1969; *House Mother Normal* 1971; *London Consequences* [with M. Drabble] 1972; *Christie Malry's Own Double Entry* 1973; *A Dublin Unicorn* 1973; *Everybody Knows Somebody Who's Dead* 1973; *See the Old Lady Decently* 1975

Plays

A Quiet Night **(tv)** 1963 [pub. in *Z Cars: Four Scripts* ed. M. Marland 1968]; *See the Pretty Lights* **(tv)** 1963, **(st)** 1970 [pub. in *Theatre Choice: a Collection of Modern Short Plays* ed. M. Marland 1972]; *The Seventh Day of Arthur* **(r)** 1963; *Entry* **(tv)** 1965; *You're Human Like the Rest of Them* **(f)** 1967 [pub. in *New English Dramatists 14* ed. E. Morgan 1970]; *Up Yours Too, Guillaume Appollinaire* **(f)** 1968; *Bath* **(tv)** 1969; *Charlie Whildon Talking, Singing, Playing* **(tv)** 1969; *Paradigm* **(f)** 1969; *The Unfortunates* **(tv)** [from his book] 1969; *The Smithsons on Housing* **(tv)** 1970; *B.S. Johnson versus God* 1971 [np]; *On Reflection: Alexander Herzen* **(tv)** 1971; *On Reflection: Sam Johnson* **(tv)** 1971; *Whose Dog Are You?* 1971 [np]; *Hafod a Henref* **(tv)** 1972; *Not Counting the Savages* **(tv)** 1972

Poems

Poems 1964; *Poems Two* 1972; *Penguin Modern Poets 25* [with G. Ewart, Z. Ghose] 1975

Non-Fiction

Street Children [with J.T. Oman] 1964; *Aren't You Rather Young to be Writing Your Memoirs?* **(a)** 1973

Edited

The Evacuees 1968; *All Bull: the National Servicemen* 1973; *You Always Remember the First Time* 1975

Pamela Hansford JOHNSON 1912–1981

The daughter of a railway official on the Gold Coast, and from a theatrical family on her mother's side (her grandfather was treasurer to Sir Henry Irving), Johnson grew up in Clapham, south London, an area evoked in many of her novels, including the first, *This Bed Thy Centre*

(1935), and the last, *A Bonfire* (1981). On leaving Clapham County Secondary School (her family having suffered a reduction in income after her father's death), she went to work as a junior in a bank.

Her first published volume was of verse, and through this interest she met the young **Dylan Thomas**, with whom she had an important early relationship. Her first novel had a *succès de scandale*, and soon afterwards she became a professional writer, marrying an Australian in 1936, with whom she had two children and from whom she was subsequently divorced. She married the novelist and politician **C.P. Snow** in 1950, and they had one son.

She quickly became a well-known mid-century woman-of-letters, appearing on BBC radio's *The Critics*, reviewing widely, and (particularly after her second marriage) becoming known as a literary hostess in Cromwell Road and Eaton Terrace, London. Known for her generosity to fellow writers, she did not entirely escape the acerbities of literary life (she was commissioned, for instance, to write a British Council pamphlet on the works of **I. Compton-Burnett**, but encountered little but the scorn of that formidable figure).

Johnson wrote almost thirty novels, often hurriedly and, from the late 1960s on, under the shadow of ill health. Her production was inevitably uneven and sometimes frankly middlebrow, but the richly comic *The Unspeakable Skipton*, the central character of which is based on **Fr. Rolfe**, and the psychologically penetrating *An Error of Judgement* were among many excellent novels. *The Unspeakable Skipton*, *Night and Silence! Who is Here?* and *Cork Street, Next to the Hatter's* all feature the poet Dorothy Merlin as well as several other recurrent characters and are sometimes referred to as the 'Dorothy Merlin trilogy', although their stories are not in fact closely linked.

Among other writings were *Six Proust Reconstructions*; *On Iniquity*, a report on the Moors' murders in which she argued that they were linked to a general climate of permissiveness (a theme of her husband's novel, *The Sleep of Reason*, published the following year); and an engaging volume of personalia, *Important to Me*.

Fiction

This Bed Thy Centre 1935; *Blessed Above Women* 1936; *Here Today* 1937; *World's End* 1937; *The Monument* 1938; *Girdle of Venus* 1939; *Too Dear for My Possessing* 1940; *The Family Pattern* 1942; *Winter Quarters* 1943; *The Trojan Brothers* 1944; *An Avenue of Stone* 1947; *A Summer to Decide* 1948; *The Philistines* 1949; *Catherine Carter* 1952; *An Impossible Marriage* 1954; *The Last Resort* 1956; *The Humbler Creation* 1959; *The Unspeakable Skipton* 1959; *An Error of Judgement* 1962; *Night and Silence! Who is Here?* 1963; *Cork Street, Next to the Hatter's* 1965; *The Survival of the Fittest* 1968; *The Honours Board* 1970; *The Holiday Friend* 1972; *The Good Listener* 1975; *The Good Husband* 1978; *A Bonfire* 1981

Plays

Corinth House 1950; *Six Proust Reconstructions* (r) 1958

With C.P. Snow: *Family Party* 1951; *Her Best Foot Forward* 1951; *The Pigeon with the Silver Foot* 1951; *Spare the Rod* 1951; *The Supper Dance* 1951; *To Murder Mrs Mortimer* 1951

Non-Fiction

Thomas Wolfe. A Critical Study 1947; *I. Compton-Burnett. A Critical Study* 1951; *On Iniquity* 1967; *Important to Me* 1974

Poetry

Symphony for a Full Orchestra 1934

Translations

Jean Anouilh *The Rehearsal* [with K. Black] 1961 [pub. 1962]

Biography

C.P. Snow: an Oral Biography by John Halperin 1983

Jennifer (Prudence) JOHNSTON 1930–

The eldest daughter of the playwright Dennis Johnston and the actress and theatre director Shelagh Richardson, Johnston was born in Dublin and educated there at Park House School and Trinity College, where she read English and French. Her parents separated when she was eight, and she has commented that her father's absence was partly responsible for a recurring theme in her fiction: 'searching for a father or substitute'. Typically, a central character is rejected by someone who is near their own age and forms a relationship with an older person. She left university without a degree in 1951 in order to marry, and began writing when she already had two small children, with a third on the way. Following her divorce in 1976, she married a farmer and solicitor and now lives in Derry.

Her first novel, *The Captains and the Kings* (1972), is a sensitive account of a friendship between a young boy and an old man, a relationship which is misunderstood and ends in tragedy. Although the main action of the novel is set in the present, the book is chiefly concerned with the past, and the majority of novels that follow have historical settings, usually the 1920s with the 'Big House' still dominant. This repetition is deliberate, for Johnston believes that she may 'go on writing the same book over and over again and exploring those themes' in order to become a better writer. Her short, compact and beautifully written novels frequently deal with Ireland's troubled history, but they are rarely overtly political. 'I find it impossible to write about here and now,' she has said; 'using the past you can take yourself outside it all.' One of her few novels with a contemporary setting (Northern Ireland) is *Shadows on Our Skin*, but although it won widespread critical acclaim and was shortlisted for the Booker Prize, it did not entirely satisfy Johnston. 'I seem to be able to disentangle emotions by setting things in the past,' she commented, and her next novel, *The Old Jest*, is set in the 1920s. It won the Whitbread Award for fiction in 1979, confirming Johnston as one of Ireland's leading novelists.

Perhaps her best book is *How Many Miles to Babylon?*, a deeply moving account of the friendship between two men which transcends the barriers of class and rank. Set against the backdrop of the First World War, and suffused by the poetry of **W.B. Yeats**, it is a spare, elliptical novel which subtly draws parallels between the war and Ireland's relations with England. It also treats another of the principal themes of her fiction: 'at certain moments we can create relationships that are nothing to do with sex, but where for brief moments you can reach another person and that way you yourself become illuminated'.

A staunch Republican, Johnston is deeply opposed to political violence, believing that 'the only sane and sensible thing that can happen in Ireland is that it is united'. She has also written a number of plays, including *O An Anais Azarias and Misael*, which won the Giles Cooper Award when broadcast on BBC radio.

Fiction

The Captains and the Kings 1972; *The Gates* 1973; *How Many Miles to Babylon?* 1974; *Shadows on Our Skin* 1977; *The Old Jest* 1979; *The Christmas Tree* 1981; *The Railway Station Man* 1984; *Fool's Sanctuary* 1987; *The Invisible Worm* 1991

Plays

The Nightingale and Not the Lark 1980 [pub. 1981]; *Andante un poco mosso* [pub. in *The Best Short Plays 1983* ed. R. Delgado 1983]; *Billy* **(r)**, **(st)** 1984; *Indian Summer* **(r)**, **(st)** 1984 [pub. 1988]; *The Porch* 1986 [pub. 1988]; *The Invisible Man* 1987 [pub. 1988]; *O An Anais Azarias and Misael* **(st)** 1989, **(r)** 1990; *Triptych* 1989

(Monica) Elizabeth (Knight) JOLLEY 1923–

Born in Birmingham, Jolley is of mixed English–Austrian parentage: her father had met her mother, the daughter of an Austrian general, while engaged on famine relief work in Vienna in 1919. Both her parents were teachers. She was educated at home until the age of eleven, in a household where German was spoken, by a series of French, Swiss and German governesses and, perhaps because of this background, the theme of the displaced person, often of central European origin, recurs frequently in her work. Her later schooling was at a Quaker boarding school, and she then trained as a nurse in London and Birmingham. She married Leonard Jolley, a university librarian, with whom she has had three children, and in 1959, when her husband was appointed as librarian at the University of Western Australia, the family emigrated there. She became a naturalised Australian citizen in 1978 and has said that she regards herself as an Australian writer.

During the 1960s and early 1970s she did a great variety of jobs, including nurse, door-to-door salesperson and flying domestic. She had been writing since childhood but, although her stories began appearing in anthologies and magazines in the 1960s, the breakthrough into book publication was slow. In 1974 she began attending writing classes at Fremantle Arts Centre, and its energetic small press published her first two volumes of short stories. Her first novel, *Palomino*, a study of lesbianism, followed in 1980, also from a small press, and attracted little notice. It was with her third novel, *Mr Scobie's Riddle*, set in an old people's home, and with her fourth, the highly comic *Miss Peabody's Inheritance*, that her reputation began to be a substantial one, both in Australia and Britain.

Her novels are quirky, comical treatments of life's outsiders and misfits, generally from a feminine viewpoint, and handle such themes as surrogate motherhood and incest. In her habit of reintroducing characters from one novel into subsequent ones she echoes **Barbara Pym**. She has been a writer-in-residence at Curtis University of Technology from 1978, and at the advanced education college at Needlands since 1983. She has also conducted frequent writing workshops at prisons and community centres. She lives in a suburb of Perth, Western Australia, where she keeps a farm for fruit and geese. Her tenth novel, *Cabin Fever*, was published in 1990, and she is also a writer of poetry and, prolifically, of radio plays, some of them adaptations of her own stories.

Fiction

Five Acre Virgin (s) 1976; *The Travelling Entertainer* (s) 1979; *Palomino* 1980; *The Newspaper of Claremont Street* 1981; *Miss Peabody's Inheritance* 1983; *Mr Scobie's Riddle* 1983; *Woman in a Lampshade* (s) 1983; *Foxybaby* 1984; *Milk and Honey* 1984; *The Well* 1986; *The Sugar Mother* 1988; *My Father's Moon* 1989; *Cabin Fever* 1990; *George's Wife* 1994

Plays

Night Report (r) 1976; *The Performance* (r) 1976; *The Shepherd on the Roof* (r) 1977; *The Well-Bred Thief* (r) 1977; *Woman in a Lampshade* (r) 1979; *Two Men Running* (r) 1981; *Paper Children* (r) 1989; *Little Lewis Has Had a Lovely Sleep* (r) 1990; *The Newspaper of Claremont Street* [with A. Belcher, D. Britton] 1991

Non-Fiction

Travelling Notebook: Literature Notes 1978; *Central Mischief* [ed. C. Lurie] 1992; *Diary of a Weekend Farmer* [with E. Kotai] 1993

(Walter) David (Michael) JONES 1895–1974

One of the select company of British writers who were equally distinguished as visual artists, Jones (who was christened Walter, but soon discarded the name) was born at Bromley, Kent, the son of a printer and of a former governess. He was of Welsh ancestry on his father's side, and his sense of Welsh identity was very important to him, although he only lived in the country briefly, in the 1920s. He had little formal education, but drew brilliantly before he could read, and attended Camberwell School of Art from the age of fourteen. In 1915 he enlisted in the Royal Welch Fusiliers and served throughout the First World War as a private, being wounded in the leg in July 1916 during the attack on Mametz Wood on the Somme.

The decency of the serving soldier remained the linchpin of his personal morality, and his first major literary work, *In Parenthesis* (1937) – part-novel, part-memoir, part-epic poem – portrays his fellow soldiers as representatives of a type whose long line of descent stretches back through Shakespeare, Malory, *The Mabinogion* and the *Chanson de Roland* to *Y Gododdin*, a Welsh epic poem of the sixth century. An enormous range of reference to these works, and to others in between, as well as to Catholic rites and the primitive myths and rituals of Fraser's *The Golden Bough*, necessitated some thirty-five pages of notes.

Jones taught at Westminster School of Art after the war, became a Roman Catholic in 1921, and went to work for a while under the famous Catholic artist Eric Gill. He was engaged to Gill's daughter, Petra, but she broke off the engagement: Jones never married, although he had several relationships with women. He lived alone, largely in poverty, in various parts of the British Isles, receiving some patronage, particularly from Helen Sutherland.

He began to write only in the late 1920s, and was famous during the inter-war period as an engraver and water-colourist. His second major literary work, *The Anathemata*, was a heroic retelling of the history of Britain, also in mixed prose and verse. It considerably increased the select but influential public that had admired *In Parenthesis*, which included – as **T.S. Eliot** noted in his introduction to the latter book's 1961 reissue – 'a number of writers whose opinions usually command attention'. Jones suffered a second nervous breakdown in 1947 and, after that, lived largely in Harrow, in a private hotel for part of the time, finally dying in a convent nursing home. Fragments of a third major work, *The Sleeping Lord*, were then published. Jones was a dedicated and difficult artist, whose complexity has denied him a wide audience, but his profundity and importance to modern poetry are indisputable.

Poetry

In Parenthesis 1937; *The Anathemata* 1952; *The Fatigue* 1965; *The Tribune's Visitation* 1969; *The Sleeping Lord and Other Fragments* 1974; *The Kensington Mass and Other Fragments* 1975

Non-Fiction

Epoch and Artist: Selected Writings [ed. H. Grisewood] 1959; *An Introduction to the Rime of the Ancient Mariner* 1972; *David Jones: Letters to Vernon Watkins* [ed. R. Pryor] 1976

Gwyn JONES 1907–

Jones was born at Blackwood, Monmouthshire, educated at Tredegar Grammar School and University College, Cardiff. On going down he became a schoolmaster until returning as lecturer to his old university in 1935. He held the chair of English language and literature at the University College of Wales, Aberystwyth, from 1940 to 1965, when he returned to Cardiff where he remained until his retirement in 1975. He was the Ida Beam Visiting Professor at the University of Iowa in 1982. From 1939 to 1960 he was editor of *Welsh Review* in Cardiff, and he was president of the Viking Society for Northern Research (1951–2) and chairman of the Welsh Arts Council (1957–67). He received a Welsh Arts Council Award in 1973 and the Christian

Gauss Award for non-fiction in the same year. He has been awarded the DLitt by the University of Wales, Cardiff, and the Universities of Nottingham and Southampton. He was appointed Knight of the Order of the Falcon, Iceland, in 1963 and CBE in 1965.

Jones has employed his breadth of learning to create novels with a broad historical sweep. *Richard Savage* has as its centre the one-time swaggering eighteenth-century poet whose extravagant life ended in ruin. Having killed a man in a tavern brawl, though pardoned, Savage died in a debtor's prison and Jones gives full weight to his hero's tragic flaws. *Times Like These* has as its theme the sufferings of the Welsh miners and their families in the 1926 strike as they held out for months after the General Strike collapsed. At the centre of *Garland of Bays* is the dissolute sixteenth-century poet, playwright and pamphleteer Robert Greene; like Savage, he died in poverty. *The Flowers Beneath the Scythe* embraces the terrible years 1914–45 for the men, women and children of a Welsh valley, while *The Walk Home* takes a look at nineteenth-century Wales. His short stories are also impressive. As a scholar he is a leading authority on the Norse voyages of discovery and the work that he published in the mid-1960s has led to further fundamental research, resulting in a new edition of *The Norse Atlantic Saga* more than twenty years later. His translations of the Icelandic sagas are also standard works. He lives with his second wife in Aberystwyth, and in his mid-eighties published a book of essays, *Background to Dylan Thomas and Other Explorations*.

Fiction

Richard Savage 1935; *Times Like These* 1936; *The Nine-Days' Wonder* 1937; *Garland of Bays* 1938; *The Buttercup Field and Other Stories* (s) 1945; *The Green Island* 1946; *The Still Waters and Other Stories* (s) 1948; *The Flowers Beneath the Scythe* 1952; *Shepherd's Hey and Other Stories* (s) 1953; *The Walk Home* 1962; *Selected Short Stories* (s) 1974

Non-Fiction

A Prospect of Wales [with K. Rowntree] 1948; *The First Forty Years: Some Notes on Anglo-Welsh Literature* 1957; *The Norse Atlantic Saga* 1964 [rev. 1986]; *A History of the Vikings* 1968 [rev. 1984]; *The Legendary History of Olaf Tryggvason* 1968; *Kings, Beasts, and Heroes* 1972; *Being and Belonging: Some Notes on Language, Literature and the Welsh* 1977; *Babel and the Dragon's Tongue* 1981; *The Novel and Society* 1981; *Three Poetical Prayer-Makers of the Island of Britain* 1981; *Background to Dylan Thomas and Other Explorations* 1991

Translations

Four Icelandic Sagas 1935; *The Vatnsdalers' Saga* 1944; *The Mabinogion* [with T. Jones] 1948; *Sir Gawain and the Green Knight* 1952; *Two Folk-Tales* 1952; *Welsh Legends and Folk-Tales* 1955; *Scandinavian Legends and Folk-Tales* 1956; *Egil's Saga* 1960; *Eirik the Red and Other Icelandic Sagas* 1961

Edited

Narrative Poems for Schools [with E.M. Silvanus] 3 vols 1935; *Prose of Six Centuries* 2 vols 1935; *Poems of Six Centuries* 1936; *Welsh Short Stories* (s) 1940; Alun Lewis *Letters from India* [with G. Lewis] 1946; *Salmacus and Hermaphroditus* 1951; William Browne *Circe and Ulysses* 1954; *Songs and Poems of John Dryden* 1957; *The Metamorphosis of Publius Ovidius Naso* 1958; *The Poems and Sonnets of Shakespeare* 1960; *Twenty-Five Welsh Short Stories* (s) [with I.F. Elis] 1971; *The Oxford Book of Welsh Verse in English* 1977; *Fountains of Praise: University College, Cardiff 1883–1983* 1983

Biography

Gwyn Jones by Cecil Price 1976

James JONES 1921–1977

Jones was born in Robinson, Illinois, into a bookish, middle-class family. His father was a dentist who was unable to find work during the Depression when the family suffered considerable hardship. Jones joined the peace-time army in 1939, rose to the rank of sergeant, and won a Purple Heart and a Bronze Star. Stationed in Hawaii he studied briefly at the University of Hawaii in 1942, read **Thomas Wolfe** for the first time and decided he had been a writer all along without knowing it. He was discharged from the army in 1944 following an unfavourable psychiatric report and, after further studies at the University of New York, submitted a manuscript to Maxwell Perkins at Scribner. Perkins turned it down but gave Jones an advance for an unwritten second novel. Living with Lonney Handy, he began work on *From Here to Eternity* whilst being subjected to her unorthodox teaching methods. Believing there to be no substitute for the close study of the American classics, she had Jones copy out lengthy passages of **F. Scott Fitzgerald** and other writers.

From Here to Eternity won a National Book Award in 1951 and became an instant bestseller. A brutal picture of the army based in Hawaii, it catapulted Jones into the New York literary scene and he was to be seen enjoying the high life with his new friends, **Norman Mailer** and **William Styron**. In 1957, Jones told the novelist Budd

Schulberg he was looking for a woman, and Schulberg – according to Jones's biographer Frank McShane – asked what kind. 'Well, she has to be interested in writers and writing, but don't give me one of those New York intellectual highbrows. I'd like her to look like Marilyn Monroe.' Schulberg obliged and introduced him to the actress Gloria Mosolino. They were married after a short romance and settled in Sagapouack, New York, where they had two children.

Jones's subsequent novels did not achieve the same critical acclaim as his first, though they were commercially successful and two – *Some Came Running* and *The Thin Red Line* – were also made into films. *The Thin Red Line* and *Whistle* were sequels to *Eternity*. The last chapters of *Whistle* were uncompleted at the time of Jones's death and the novel was prepared for posthumous publication by his friend Willie Morris. Jones rejected elegant prose and said of himself: 'I'm the common man's novelist. I'm not writing for PhDs at Harvard ... I'm the last of the proletarian novelists.'

Fiction

From Here to Eternity 1951; *Some Came Running* 1957; *The Pistol* 1959; *The Thin Red Line* 1962; *Go to the Widow Maker* 1967; *The Ice-Cream Headache and Other Stories* (s) 1968; *The Merry Month of May* 1971; *A Touch of Danger* 1973; *Whistle* 1978

Non-Fiction

Viet Journal 1974; *World War II: a Chronicle of Soldiering* [with A. Wetthas] 1975

Biography

James Jones by Frank McShane 1987

LeRoi JONES

see Amiri BARAKA

Erica JONG 1942–

Jong was born Erica Mann in Manhattan where her father – a sometime vaudeville act – ran an import business. She was educated at the High School of Music and Art in New York and at Barnard College, Columbia. After graduating in 1963 she taught English at the City College, New York, whilst working at Columbia on her MA thesis on 'Women in the Poems of Alexander Pope'. She was married to Michael Werthman in 1963, and a fictionalised account of their marriage and his mental breakdown can be found in her novel *Fear of Flying*. In 1966 she married Allan Jong, a Chinese-American Freudian psychoanalyst, and the couple lived from 1966 to 1969 in Heidelberg, Germany, where he had been posted by the army. Jong has described her time in Germany as when she first became fully aware of her Jewishness. She taught at Columbia upon their return to the USA and published poems in *Mademoiselle* and *Poetry Journal* before publishing her first collection, *Fruits and Vegetables*, in 1971.

Various early efforts at writing fiction – notably a Nabokovian tale of revenge entitled 'The Man Who Murdered Poets' – had all proved abortive, but when Jong turned to her own experience for fictional material the result was the hugely successful *Fear of Flying*. Reactions to this sexually frank, picaresque novel varied: the original typesetter refused to finish setting it, and the *New Statesman* described it as a 'crappy novel' whose heroine resembled 'a mammoth pudenda'. Others such as **John Updike** were kinder, and Jong was credited with giving a voice to female sexual liberation. An enthusiastic letter from **Henry Miller** in 1974 led to a friendship between the two which is recounted in Jong's book about Miller, *The Devil at Large*.

Predictably the notoriety had its downside, not least with her family: in her autobiographical memoir, *Fear of Fifty*, she records her father asking: 'Why do they call you a pornographer, darling?' The further adventures of heroine Isadora Wing are recounted in two sequels, *How to Save Your Own Life* and *Parachutes and Kisses*, neither of which achieves quite the same impact, though both have their comic moments. Jong's study of eighteenth-century literature is evident in the comic romp *Fanny*, which was inspired by the question: 'What if Tom Jones had been a woman?' The heroine's adventures include spells in a brothel and a witches' coven, and dalliances with Swift and Pope. A second, even more fantastical, venture into historical fiction, *Serenissima*, is set in Venice where a middle-aged film star travels back in time to meet Shakespeare and – rather predictably – witness 'Will's stiff staff' as he performs various sexual acrobatics.

Jong was married in 1977 to the writer Jonathan Fast, with whom she has a daughter, and in 1989 to Kenneth Burrows, a divorce lawyer. She won the International Sigmund Freud Prize in 1979, and became President of the Author's Guild in 1991.

Fiction

Fear of Flying 1973; *How to Save Your Own Life* 1977; *Fanny* 1980; *Parachutes and Kisses* 1984; *Serenissima* 1987; *Any Woman's Blues* 1990

Poetry

Fruits and Vegetables 1971; *Half-Lives* 1973; *Here Comes and Other Poems* 1975; *Loveroot* 1975; *The Poetry of Erica Jong* 1976; *Selected Poems 2* vols 1977, 1980; *At the Edge of the Body* 1979; *Ordinary Miracles* 1983; *Becoming Light* 1991

Non-Fiction

Four Visions of America [with others] 1977; *Megan's Book of Divorce* 1984; *The Devil at Large: Erica Jong on Henry Miller* 1993; *Fear of Fifty: a Mid-Life Memoir* (a) 1994

Collections

Witches 1981

Neil JORDAN 1950–

The son of a university lecturer, Jordan was born in Sligo and raised in Dublin. Both his mother and sister are painters. He studied history and English at University College, Dublin, but instead of graduating toured the USA with a Dublin fringe theatre company, and started writing plays. In the early 1970s he moved to London, where he lived in a squat and played saxophone in a rock band. He earned money working as a labourer on a demolition site, as a porter in Fortnum and Mason's wine cellar, and as a bath-house attendant.

He returned to Dublin in 1974 and was co-founder of the Irish Writers' Publishing Cooperative. His play *Miracles and Miss Langan* was performed on RTE radio in 1977, and the following year he sold a screenplay, *Mr Solomon Wept*, to BBC television. His first book was *Night in Tunisia*, a collection of short stories published by the Cooperative; it won the *Guardian* Fiction Prize in 1979.

Jordan became involved in film-making when the director John Boorman hired him as creative consultant on his Arthurian epic, *Excalibur* (1981). He made a documentary film about the shooting of *Excalibur*, and subsequently persuaded Channel 4 Television and the Irish Film Board to fund his first feature film, *Angel* (1982). A thriller set against the Irish troubles, it featured a rock saxophonist bent on revenge for the killing of a mute girl he has befriended. It won Jordan the *London Standard* Award for Most Promising Newcomer, and the following year he collaborated with **Angela Carter** on *The Company of Wolves*, a sensuous horror fairy-tale based on the darker aspects of the 'Little Red Riding Hood' story. He returned to thriller territory with *Mona Lisa*, a much admired love story about a black prostitute and her driver in the criminal underworld of London.

By now Jordan was firmly established as a film writer-director, and went to Hollywood to direct a commercial comedy-fantasy about 'ghost weekends', *High Spirits*. The film was an undoubted failure, but *We're No Angels*, a period comedy-thriller about two escaped convicts who disguise themselves as priests, is much underrated. *The Crying Game* was seen as a return to form, however: another thriller about Irish terrorism, it was also a surprising and unconventional love story, and pleased both critics and audiences. Its success led to Jordan being hired to direct a film version of **Anne Rice**'s cult novel, *Interview with the Vampire*. Rice took out advertisements complaining about the casting of the film (Tom Cruise seemed an unlikely choice as the vampire hero), but eventually relented, giving the finished product her blessing. The critics were less impressed, although Jordan's extraordinary visual flair was much in evidence.

He has continued to write fiction, and foremost in this work is a determination to define Ireland and its political history. He described his novel *The Dream of the Beast* as 'revenge on Dublin', as well as 'a novel of pure sensation'. He says that he wrote 'filmically' long before he made films, and in his fiction, as in his films, he recreates the physical world and infiltrates it with a dream-like illogicality. His work on the page and the screen is closely related. He writes of both inner and territorial netherlands: memories, mythic cities, the power of the imagination and inevitable metamorphosis.

Fiction

Night in Tunisia (s) 1976; *The Past* 1980; *The Dream of the Beast* 1983; *A Neil Jordan Reader* (s) 1993; *Sunrise with Sea Monster* 1995

Plays

Miracles and Miss Langan (r) 1977; *Mr Solomon Wept* (tv) 1978; *Angel* (f) 1982 [pub. 1989]; *The Company of Wolves* (f) [with A. Carter] 1983; *Mona Lisa* (f) [with D. Leland] 1986; *High Spirits* (f) 1988 [pub. 1989]; *The Miracle* (f) 1991; *The Crying Game* (f) 1992; *Interview with the Vampire* (f) [from A. Rice's novel] 1994

James (Augustine Aloysius) JOYCE 1882–1941

Joyce was born in Dublin, where his father was a rates collector. He was educated at Jesuit schools and University College, Dublin, where he studied philosophy and language. When he was still an undergraduate, in 1900, his long review of Ibsen's last play was published in the *Fortnightly Review*. At this time he also began writing lyric poems which were later collected in *Chamber Music*, published in 1907.

In 1902 Joyce left Dublin for Paris, but returned the following year as his mother was dying. From 1904 he lived with Nora Barnacle, whom he married in 1931 (the year his father died); a son was born in 1905, and a daughter in 1908. Their home from 1905 to 1915 was Trieste, where Joyce taught English at the Berlitz school. In 1909 and 1912 he made his final trips to Ireland, attempting to arrange the publication of his first book, *Dubliners*, which finally appeared in England in 1914. It was during this time that he was contacted by **Ezra Pound**, a leading champion of modernist writers who helped organise financial payments to keep Joyce writing during his most poverty-stricken periods. *Dubliners* is a series of short, interrelated stories which deal with the lives of ordinary people, whose actions are invested with a symbolic profundity. Joyce explores what would become central themes in his work: youth, adolescence, adulthood and maturity, and how identity is affected by these different stages in life.

The following year, Joyce wrote *Exiles*, his only play, and went into permanent exile himself. He is taken, in fact, as the quintessential exiled writer of the twentieth century, who obsessively relates to his past by distancing himself from it. The year 1914 was an intensely productive one for Joyce; he had two books in print and began work on his greatest achievement, *Ulysses*. In 1916 *A Portrait of the Artist as a Young Man* appeared (it had been published in serial form in the *Egoist* from 1914 to 1915), and established Joyce's reputation as a writer of genius. The fullest and most accomplished product to have emerged from the modernist movement in European literature, it presented the world of Dublin solely through the consciousness of the narrator, and charted his growth from Catholic boyhood to an early adulthood defined by a yearning to be an artist.

It was in this year that Joyce and his family moved to Zurich, where they lived in great poverty while he worked on *Ulysses*, despite undergoing surgery on his eyes. It began to appear in serial form in the *Little Review* in 1918, but was suspended in 1920 following prosecution. It eventually appeared in book form in 1922 in Paris, where Joyce and his family had settled, in a limited edition of 1,000 copies, and was followed by an English edition of 2,000 copies, also printed in Paris. The first unlimited edition followed in 1924, again in Paris, but there was no American edition until ten years later, and no British edition until 1937. The novel traces the experiences of Mr Leopold Bloom, his wife Molly (whose erotic reverie towards the book's close is what caused most of the legal difficulties) and the poet Stephen Dedalus from *A Portrait of the Artist* during a single day in Dublin in 1904. As its title suggests, however, the book is an epic, loosely analogous to Homer's *Odyssey*, which is echoed in several episodes. Enormously long and complex, using a variety of styles – notably the 'stream-of-consciousness' method – *Ulysses* is one of the great literary achievements of the century, and has been described as the greatest novel ever written.

Joyce's other major novel, *Finnegans Wake*, is even more uncompromising than *Ulysses*, written in a language of his own devising, a salmagundi of linguistic fragments and borrowings. It was published in 1939, the year after the Joyces returned to Switzerland from France. Joyce died the following year. His reputation has grown immeasurably since his death, partly because of the growth in academia. He is the one novelist in whom we can be sure to place our absolute trust, the single figure we can also be sure will be remembered, if any are, in 1,000 years' time. As one critic famously wrote: 'James Joyce was and remains almost unique among novelists in that he published nothing but masterpieces.'

Fiction

Dubliners (s) 1914; *A Portrait of the Artist as a Young Man* 1916; *Ulysses* 1922 [rev. 1961]; *Finnegans Wake* 1939; *Stephen Hero* 1944; *'Anna Livia Plurabella' – the Making of a Chapter* 1960; *A First Draft Version of 'Finnegans Wake'* 1963

Plays

Exiles 1918

Poetry

Chamber Music 1907; *Pomes Penyeach* 1927; *Collected Poems* 1936

Non-Fiction

The Critical Writings of James Joyce 1959; *The Letters of James Joyce* [ed. R. Ellmann, S. Gilbert] 3 vols 1957–66; *Giacomo Joyce* 1968; *Selected Letters of James Joyce* 1975

Collections

The James Joyce Archive 63 vols 1977–80

Biography

James Joyce by Richard Ellmann 1959

Alan JUDD (pseud.) 1946–

The invisible correction of computer records obscures the hiccups of literary history. For a while, the British Library catalogue's entry for Judd's hard-hitting novel of military service in Northern Ireland, *A Breed of Heroes* (1981), recorded: 'pseudonym; real name, **A.N. Wilson**.' This pleasingly incongrous notion is explained by the literal-mindedness of the library trade. It had merely been reported that, during a book-promotion bash, the pseudonymous Judd and the very real Wilson had exchanged name-tags.

Born in Kent, Judd attended a secondary modern school which he left at seventeen for a local technical and then teacher-training college. Intermittent work teaching physical education and English was duly exchanged for the calmer world of the army (he had been in the Territorials). This he left in 1972 to take a degree in philosophy and theology at Oxford.

Such experience fitted him for 'Government work' and the writing of fiction. It is this Intelligence work that has led to Judd keeping a low profile, adopting a pseudonym and appearing as a silhouette in the photographic line-up of those writers selected as representing 'The Best of Young British Novelists' in an influential 1982 promotion. *A Breed of Heroes*, which was based upon his own tours of duty in the province as a lieutenant, was begun in 1972, set aside and taken up again at the end of the decade. It won the Winifred Holtby Award and a Royal Society of Literature Award. Three years later came *Short of Glory*, which is perhaps his best novel to date. This tale of diplomatic life in 'Lower Africa' features the customary novice to the trade and, amid the bantering incompetence, there runs a distinct strand of compassion and fine ear for dialogue. Less successful was the Oxford novella *The Noonday Devil*, which has an air of dust being blown off, while *Tango* is a spirited romp in South America which owes something to **Graham Greene**'s *Our Man in Havana* (1958).

Judd then quixotically took two years' unpaid leave to write an excellent biography of the writer's writer **Ford Madox Ford**. He had chanced upon *The Good Soldier* (1915) and was so taken by it that he wanted to explore further. The work included a visit to **Graham Greene** and, with it, an idea grew. This was galvanised by a recommendation that he read another enigmatic figure, **John Meade Falkner**. Judd's novella *The Devil's Own Work*, inspired by Falkner's *The Lost Stradivarius* (1895), concerns the curse of literature passed on by one O.M. Tyrrell; a Waughian cigar little disguises the inspiration: 'It was easy to forget how old he was. But the face was alive and beneath his monstrously sprouting white eyebrows he had very blue eyes.' Deceptively short, not quite as chilling as Falkner's tale, this book certainly heralds a new direction in Judd's work.

Fiction

A Breed of Heroes 1981; *Short of Glory* 1984; *The Noonday Devil* 1987; *Tango* 1989; *The Devil's Own Work* 1991

Non-Fiction

Ford Madox Ford 1990

Anna KAVAN (pseud. of Helen Emily Woods) 1901–1968

Kavan, who adopted her pseudonym by deed poll, was born at Cannes on the Riviera, where her rich and idle parents were temporarily living. They played little role in her life (except, in the case of the mother, to grant an allowance) after early childhood, which was passed wandering through England, Europe and the USA. Later she attended Malvern College, and at eighteen was married off to Donald Ferguson, a junior member of the family of the Earls of Mar and Kelly, who took her to live on his plantation in Burma. Her later novel *Who Are You?* gives an account of the brutalisation to which he subjected her and his hobby of playing tennis with a live rat as the ball. A son was born (killed in the Second World War), but the marriage soon broke up, and she returned to Europe, where her lifelong drug-addiction began, she took up first painting and later writing, made her first suicide attempt and had her first abortion.

She married Stuart Edmonds, another wealthy man and a fellow amateur painter, and they spent some time travelling before settling down to live an *haut bourgeois* life in Buckinghamshire. During this period, beginning in 1929, she began to publish a series of six early novels under the name Helen Ferguson, fairly conventional in manner although sombre in tone. But during the 1930s her marriage collapsed, breakdowns and more suicide attempts began to punctuate her life, and she spent several periods in clinics. She emerged from one of these, in Surrey, in 1938, dyed her auburn hair white-blonde and adopted the name Anna Kavan (taken from one of the heroines of her earlier novels). The first work published under this name was a volume of short stories, *Asylum Piece* (1940), the first of the strange and experimental books that deliberately break down the barriers between conventional reality and the phantasmagoric on which her reputation now rests.

She spent the war years once again mainly wandering the globe, although the considerable vogue for her writing also secured her a job for a period as one of the assistant editors on **Cyril Connolly**'s magazine, *Horizon*. Subsequent detoxifications were to take her as far afield as South Africa, but after the war she lived mainly in Kensington in London, having a modernistic house built for herself, where she became known for her often erratic entertaining, and renovating several older ones on Campden Hill. Her work entered a period of neglect during the realistic 1950s, and the novel *A Scarcity of Love* was brought out by a vanity publisher, who then promptly went bankrupt and pulped the edition. The 1960s proved more sympathetic, however, and her last novel, *Ice*, an impressive foray into the world of science fiction, received excellent reviews when published in 1967. But by then she was sunk in physical decline and taking increasing amounts of heroin, eventually dying of a heart attack but, according to some accounts, with a loaded syringe in her hand. Two posthumously published volumes of short stories were her first works to tackle the subject of addiction, completing an *oeuvre* of poignant and macabre interest.

Fiction

As 'Helen Ferguson': *A Charmed Circle* 1929; *The Dark Sisters* 1930; *Let Me Alone* 1930; *A Stranger Still* 1935; *Goose Cross* 1936; *Rich Get Rich* 1937

As Anna Kavan: *Asylum Piece and Other Stories* **(s)** 1940; *Change the Name* 1941; *I Am Lazarus* **(s)** 1945; *The House of Sleep* [UK *Sleep Has His House*] 1947; *The Horse's Tale* [with K.T. Bluth] 1949; *A Scarcity of Love* 1956; *Eagles' Nest* 1957; *A Bright Green Field and Other Stories* **(s)** 1958; *Who Are You?* 1963; *Ice* 1967; *Julia and the Bazooka and Other Stories* **(s)** 1970; *My Soul in China* **(s)** 1975

Dan KAVANAGH

see Julian BARNES

Patrick (Joseph) KAVANAGH 1904–1967

Kavanagh was the eldest son of a County Monaghan shoemaker who eked out a living with a smallholding in Inniskeen. Hardly troubling to attend the local school, Kavanagh left it at thirteen to be apprenticed to his father, but learned as little as possible about cobbling or farming. Inniskeen seemed embedded in the Middle Ages, and Kavanagh, who started writing verse at the age of twelve, was regarded with suspicion. In 1929, he astonished the community by earning a guinea for a poem published in **AE**'s *Irish New Statesman*, but his decision to become a writer was thwarted when that same year his father died and he inherited family responsibilities. His eventual solution was characteristically reckless: he sold the farm after the death of his mother and subsequently lost all his financial assets.

He spent some time in Dublin, contributing poems to assorted magazines, before moving to London, where a volume of his work, *Ploughman and Other Poems*, was published in 1936. His experiences as a poet struggling for recognition inform his 1938 volume of autobiography, *The Green Fool*, which was suppressed after **Oliver St John Gogarty** brought a libel suit. (Kavanagh had mistaken Gogarty's maid for his mistress.) Disillusioned, Kavanagh returned to Dublin and thereafter shuttled between temporary accommodation in the Irish capital, London and Inniskeen. Further recognition came in 1942 when **Cyril Connolly** printed his poem 'The Old Peasant' in *Horizon*. (It was subsequently published as *The Great Hunger*.) The Irish literary renaissance centred on **W.B. Yeats**, AE and **Lady Gregory** had left a large gap in its evocation of Ireland, and it was this gap that Kavanagh filled. He rejected idealising mysticism and uncritical sentimentality for a poetry rooted in the rude particularity of his background. Although he was regarded by some as the Robert Burns of contemporary Ireland, others regarded him with suspicion because of his aggressive social behaviour, which was fomented by alcohol. Many editors refused his work, especially after the appearance of the controversial *Kavanagh's Weekly*, the magazine he founded with his brother Peter in 1952.

After surviving an operation for lung cancer in 1955, he deteriorated in squalor for the last decade of his life, and eventually died of pneumonia. He nevertheless lectured at University College, Dublin, during 1957 and 1958, and produced two collections and a volume of *Collected Poems*. Six months before he died, he married Katherine Barry, with whom he had lived in London for over ten years. He had stated in 1964 that his brother was 'the sacred keeper of his sacred conscience', but, assuming proprietorial rights after Kavanagh's death, Peter Kavanagh wrote a biography of the poet which sneered at this late marriage and omitted Katherine's name. This enmity was carried quite literally to the grave, for at Katherine's funeral in 1989, Peter Kavanagh halted the proceedings, announcing that she was not to be buried alongside her husband, and that another grave had been dug for her. After an unseemly argument with Katherine's family, he was overruled and the mourners were obliged to wait in a community hall while Kavanagh's grave was re-opened. Peter Kavanagh subsequently removed stones and a cross from the grave (which he had designed himself) to a field one mile away, posting a notice to explain this desecration. A dispute between him and Katherine Kavanagh's estate over copyright ownership remains unresolved, but Patrick Kavanagh's literary legacy is of undoubted importance, influencing a later generation of poets that includes **Seamus Heaney**, **Derek Mahon** and **Michael Longley**.

Poetry

Ploughman and Other Poems 1936; *The Great Hunger* 1942; *A Soul for Sale* 1947; *Recent Poems* 1958; *Come Dance with Kitty Stobling and Other Poems* 1960; *Collected Poems* 1964; *The Complete Poems* [ed. P. Kavanagh] 1972

Fiction

Tarry Flynn 1948; *By Night Unstarred* 1977

Non-Fiction

The Green Fool (a) 1938; *Self-Portrait* (a) 1964; *Collected Prose* 1967; *Lapped Furrows: Correspondence 1933–1967 Between Patrick and Peter Kavanagh* [ed. P. Kavanagh] 1969; *Love's Tortured Headland* 1978

Collections

Collected Prose 1967; *November Haggard: Uncollected Prose and Verse* [ed. P. Kavanagh] 1971

Biography

Sacred Keeper by Peter Kavanagh 1979

P(atrick) J(oseph Gregory) KAVANAGH 1931–

Kavanagh is of Irish ancestry on both sides, but his father, who became a scriptwriter for the wartime comedy series *ITMA*, grew up in New Zealand, and he himself was born in Worthing and has lived his life predominantly in England. (He has told the story of his search for his ancestral roots in *Finding Connections*.) He went to Douai, the Catholic public school, and the Lycée Jaccard in Lausanne in Switzerland; he then attended drama school in Paris and did his national service, taking part in the Korean War and suffering a slight wound. He went up to Merton College, Oxford, and afterwards spent a year in Barcelona as an employee of the British Institute there. Periods as an assistant floor manager for BBC television and a production trainee in a publishing company followed.

He had met his future wife, Sally Philipps, daughter of the novelist **Rosamond Lehmann**, when they were both at Oxford (they had a spell doing Christmas postal deliveries together), and in 1957 they were married. In the same year they sailed out to Djakarta, Indonesia, where Kavanagh (employed by the British Council) had a post as a lecturer in English literature at the university. Sally died in Java, aged twenty-four, in 1958, and Kavanagh's autobiography of his early life, *The Perfect Stranger*, contains a moving account of their relationship.

His first book of verse, *One and One*, was published in 1959. His poetry, while taking some time to mature, has developed a quiet distinction, being particularly strong on the evocation of both nature and feeling. During the 1960s, Kavanagh spent some years working as an actor, but from 1970 he has been a full-time writer. He has published four novels for adults and two for older children, and written television plays and documentaries, and edited the works of **Ivor Gurney** and **G.K. Chesterton**. He is also well known as a journalist (perhaps particularly for his *Spectator* column) and as a television personality. He married again in 1965; the couple have two sons. Kavanagh (who should not be confused with **Patrick Kavanagh**, the famous Irish poet of the mid-century) lives in Cheltenham, Gloucestershire.

Poetry
One and One 1959; *On the Way to the Depot* 1967; *About Time* 1970; *Edward Thomas in Heaven* 1974; *Life Before Death* 1979; *Real Sky* 1980; *Selected Poems* 1982; *Presences* 1987; *An Enchantment* 1991; *Collected Poems* 1992

Fiction
A Song and Dance 1968; *A Happy Man* 1972; *People and Weather* 1978; *Only by Mistake* 1986

For Children
Scarf Jack 1978 [US *The Irish Captain* 1979]; *Rebel for Good* 1980

Non-Fiction
The Perfect Stranger **(a)** 1966; *People and Places: a Selection 1975–1987* 1988; *Finding Connections* **(a)** 1990; *Voices in Ireland* 1994

Edited
Ivor Gurney: *Collected Poems of Ivor Gurney* 1982, *Selected Poems of Ivor Gurney* 1990; *The Oxford Book of Short Poems* [with J. Michie] 1985; G.K. Chesterton: *The Bodley Head G.K. Chesterton* 1985, *The Essential G.K. Chesterton* 1987; *A Book of Consolations* 1992

Plays
Scarf Jack **(tv)** [from his novel] 1981

Sheila KAYE-SMITH 1887–1956

Born at St Leonards-on-Sea near Hastings, the daughter of a prominent Hastings doctor and surgeon, Kaye-Smith spent much of her youth at farms in Sussex and Scotland, and was educated privately. She wrote no fewer than thirteen novels during her last two years at school, and her first published novel, *The Tramping Methodist*, appeared when she was only twenty-one and inaugurated her many books about the past and present of a Sussex family, the Alards.

She was always a very prolific writer, publishing a book a year on average, usually fiction; among her best-known titles are *Tamarisk Town* and *Joanna Godden*, and she returned again and again to the theme of innocence besieged by passion, her stories played out against an earthy Sussex setting. With **Mary Webb** and some others, she formed a 'rural' school of novelists who were successfully parodied by **Stella Gibbons** in her *Cold Comfort Farm* (1932), dealing with the Starkadder family of Sussex and now much more widely read than Kaye-Smith (although the latter has been republished by Virago).

Kaye-Smith remained at St Leonards-on-Sea until her mid thirties, when she married Theodore Penrose Fry, who had been the local High Anglican rector, and was later a priest at St Stephen's, Gloucester Road, London and at Norland. In 1929 the couple became Roman Catholics and Fry resigned his ordination; they moved to Northiam, fifteen miles from Hastings, where they farmed their own land and built a chapel dedicated to Sainte Thérèse de Lisieux, and where Kaye-Smith wrote until her death. Among her many books are short stories; poetry; an autobiography, *Three Ways Home*; a book of mingled cookery and reminiscences, *Kitchen Fugue*; a well-known study of Jane Austen written with **G.B. Stern**, *Talking of Jane Austen*; a book on the Weald of Kent and Sussex contributed to the 'Regional Books' series; and *Quartet in Heaven*, a study of four women who either had already been raised to sainthood by the Roman Catholic church or who might be in the future.

Fiction
The Tramping Methodist 1908; *Starbrace* 1909; *Spell Lane* 1910; *Isle of Thorns* 1913; *Three against the World* 1914; *Sussex Gorse* 1916; *The Challenge to Sirius* 1917; *Little England* 1918; *Tamarisk Town* 1919; *Green Apple Harvest* 1920; *Joanna Godden* 1921; *The End of the House of Alard* 1923; *Saints in Sussex* 1923; *The George and the Crown* 1925; *Joanna Godden Married and Other Stories* **(s)** 1926; *Iron and Smoke* 1928; *A Wedding Morn* **(s)** 1928; *The Village Doctor* 1929; *Shepherds in Sackcloth* 1930; *The Children's Summer* 1932; *The Ploughman's Progress* 1933; *Gallybird* 1934; *Superstition Corner* 1934; *Selina Is Older* 1935; *Rose Deeprose* 1936; *Faithful Stranger and Other Stories* **(s)** 1938; *The Valiant*

Woman 1938; *Ember Lane* 1940; *The Hidden Son* 1941; *Tambourine, Trumpet and Drum* 1943; *The Lardners and the Laurelwoods* 1947; *The Happy Tree* 1949; *Mrs Gailey* 1951; *The View from the Parsonage* 1954

Poetry

Willow's Forge and Other Poems 1914; *Songs Late and Early* 1931

Non-Fiction

John Galsworthy 1916; *Anglo-Catholicism* 1925; *The Mirror of the Months* 1925; *Sin* 1929; *The History of Susan Spray, the Female Preacher* 1931; *Three Ways Home* (a) 1937; *Dropping the Hyphen: a Story of a Conversion* 1938; *Talking of Jane Austen* [with G.B. Stern] 1943; *Kitchen Fugue* 1945; *More about Jane Austen* [with G.B. Stern] 1949; *Quartet in Heaven* 1952; *Weald of Kent and Sussex* 1953; *All the Books of My Life* 1956

Plays

Mrs Adis, with The Mock-Beggar [with J. Hampsen; pub. 1929]

Molly (i.e. Mary Nesta) KEANE 1905–

Keane was born Mary Skrine in County Kildare, Ireland. Her mother was a successful poet who wrote under the pseudonym Moira O'Neill, and the family was part of the Anglo-Irish landowning classes about whom Keane has written throughout her career. She was educated by governesses until the age of fourteen, when she was sent away to a 'prim, suburban school', which she found restrictive after the comparative freedom of her childhood. She was also very lonely. 'I might never have become a writer had it not been for the isolation in which I suffered as an unpopular schoolgirl,' she has written. She wrote her first novel, *The Knight of the Cheerful Countenance*, at the age of seventeen. A romance set in the world she knew best, it featured as a heroine 'the girl I most wished to be myself'. She sold it for £75, and it was published by Mills and Boon in 1926.

This, and the ten novels which followed, were published under the pseudonym M.J. Farrell (borrowed from a pub sign) because literary aspirations were regarded as highly suspect among the fox-hunting gentry Keane knew socially. (Far more suitable was her volume of hunting reminiscences, *Red Letter Days*, illustrated by the famous equestrian artist 'Snaffles'.) Later novels set amongst the Anglo-Irish gentry in their beautiful country houses are as much a critique as a celebration of this society and the often miserable and insular life that passed in such elegant surroundings. She captures the aimlessness of this milieu: husbands and fathers shoot and ride to hounds or pursue such mild interests as genealogy; mothers tend to bully or exert a malign influence, often, it seems, out of sheer boredom. In general, people in Keane's novels are more devoted to their dogs and horses than they are to each other. Love affairs are usually ill-judged, one-sided or otherwise doomed. One of the best and most characteristic of these novels is *The Rising Tide*, which traces four decades in the life of a 'Big House', Garonlea, haunted by its unhappy past. *Devoted Ladies*, as its title suggests, is a gently mocking melodrama about a lesbian relationship, set partly in London, and was considered daring in its time.

In 1938 she married Robert Keane, whose father ran a bacon factory, and lived with him and their two daughters in a large country house. His sudden death eight years later at the age of only thirty-six was devastating, and led to a long creative silence. The family had to move to a smaller house, and Keane's morale was further dented by the failure of her play *Dazzling Prospect*. She had enjoyed great success with three earlier comedies, all written with John Perry, produced on the London stage, and directed by John Gielgud. The first of these, *Spring Meeting*, was particularly notable for the triumphant performance of Margaret Rutherford as Bijou Furze, an elderly spinster who is a secret gambler. The play made the reputations of both actress and playwright.

During the late 1970s, Keane began work on a black comedy, which took the material of her early novels but treated it with a great deal less respect. The completed book was rejected by a publisher as too dark, but some time later the actress Peggy Ashcroft read the manuscript while staying with Keane in her County Waterford home and encouraged her to show it to other publishers. It was thus, at the age of seventy-seven, that Keane embarked upon a second career, this time under her own name. *Good Behaviour*, narrated by the hapless Aroon, 'a fool who doesn't see what's happening', was an enormous and deserved success, shortlisted for the Booker Prize and serialised on television in 1983. It was followed by another black Anglo-Irish comedy, *Time After Time*, which was also made into a television film (1986), starring her old associate John Gielgud. Her last novel to date, *Loving and Giving*, came five years later, ill health preventing her from producing further new work. A more sombre work than its predecessors, it nevertheless shared the theme of those who are deceived by others or themselves in its story of Nicandra (named after her father's prize racehorse), who squanders love upon people who can never return it in full measure – from her mother, who runs away with another man, to her feckless husband, Andrew, who deserts her for her best friend. These late novels rekindled interest in her earlier books, all of which have since been reissued in new editions, including *The Knight of the Cheerful Countenance*, which had been forgotten, but was rediscovered in the British Library in 1992.

Fiction

As M.J. Farrell: *The Knight of the Cheerful Countenance* 1926; *Young Entry* 1928; *Taking Chances* 1929; *Mad Puppetstown* 1931; *Conversation Piece* 1932; *Devoted Ladies* 1934; *Full House* 1935; *The Rising Tide* 1937; *Two Days in Aragon* 1941; *Loving without Tears* 1951; *Treasure Hunt* [from her play] 1952

As Molly Keane: *Good Behaviour* 1981; *Time After Time* 1983; *Loving and Giving* 1988

Plays

With J. Perry: *Spring Meeting* 1938; *Ducks and Drakes* 1942; *Treasure Hunt* 1950; *Dazzling Prospect* 1961

Non-Fiction

Red Letter Days [as M.J. Farrell; with 'Snaffles'] 1933; *Molly Keane's Nursery Cooking* 1985

Garrison KEILLOR 1942–

Keillor, 'Minnesota's best-known shy person' and also, at six foot four inches, 'America's tallest radio humorist', was the third of six children to be born to the wife of a railroad clerk in Anoka, a small town just outside Minneapolis. His family was descended from Scots on both sides and belonged to the Plymouth Brethren. Radio and television were discouraged and befriending heathen neighbours was strictly forbidden, so Keillor learned the art of storytelling, largely from his Uncle Lew who kept everyone entertained through long evenings with tales of his time as a salesman.

In 1966 he graduated from the University of Minnesota where he had earned his tuition working at the campus radio station. Nurturing an ambition to write, he continued working in radio until, three years later, he sold a story to the *New Yorker* and became a journalist. In 1974 he returned to radio after a trip to Nashville for a *New Yorker* piece on the Grand Ole Opry inspired the conception of the live evening show that was to become *A Prairie Home Companion*. When it was first broadcast it had fewer listeners in the audience than performers, but by 1987 it had become a phenomenal success, able to boast a committed listenership of more than two million.

A Prairie Home Companion is an eclectic mixture of music and humour dominated by Keillor's half-hourly improvised monologue about the inhabitants of a mythical Midwestern town, Lake Wobegon. 'Most men wear their belts low here, there being so many outstanding bellies, some big enough to have names of their own and be formally introduced.'

Lake Wobegon Days and *Leaving Home* (both of which were drawn from the radio show), were very successful in the USA, and in Britain *Lake Wobegon Days* remained in the *Sunday Times* bestseller list for over twenty weeks, a unique phenomenon for a then little-known American author. In reviews, critics conjure the names of **James Thurber** and E.B. White, and comparisons are being drawn with Mark Twain for, like that earlier illustrious yarn spinner and lecture-hall tiger, Keillor likes sometimes to wear a white suit on stage.

A very private man, Keillor is good at delivering versions of his life and emotions in witty one-liners. But beneath the humour there is bitterness in some of his semi-autobiographical recollections of Lake Wobegon. An angry son returns to the town intending to nail ninety-five complaints about his repressive parents to the door of the Lutheran church. The thirty-fourth of these accusations is that the parents made it impossible for him to accept a compliment. Certainly, any comments like 'good speech' are dismissed by Keillor because, he says, 'good' is not good enough. He goes on: 'Under this thin veneer of modesty lies a monster of greed. I drive away faint praise, beating my little chest, waiting to be named Sun God, King of America, Idol of Millions. I don't want to say "Thanks, glad you liked it." I want to say "Rise my people." '

Divorced in 1976 after eleven years of marriage (he has one son), Keillor lives in St Paul where he spends most of his time working.

Fiction

Happy to Be Here (s) 1981; *Lake Wobegon Days* 1985; *Leaving Home* (s) 1987; *We Are Still Married* (s) 1989; *Radio Romance* 1991; *The Book of Guys* (s) 1993

Joseph KELL

see Anthony BURGESS

James KELMAN 1946–

Kelman was born and grew up in the Glasgow of his stories, where he still lives. He drinks and smokes hard, frequents bars, likes his tea strong, often has a beard and is reputed to be a fierce interviewee. He spent time on the dole and did a variety of casual jobs, including those of apprentice compositor, bus driver and solvents salesman. He finds it easy to remain close to the lives and experiences of his subjects: 'I don't earn much money,' he explains, 'so I'm involved in the culture I write about. Glasgow's less broken up than London. There's more mixing in pubs. It's harder to become divorced from people.'

As a child, he read voraciously at the local library, but most important to his writing, he thinks, is that he left school at fifteen, 'so I had no *a priori* opinions about books'. He began writing at twenty-two, while working as a builder, but it was not until 1972 that an American press collected thirteen of his stories, sending him 200 books as payment. After that he remained unpublished until the volume of stories *Not Not While the Giro* in 1983. In 1987 his first stage play, *The Busker*, was performed at Edinburgh's Traverse Theatre and, in the same year, *Into the Night* was produced as part of BP's Young Director Festival at Battersea Arts Centre.

He writes about the world he inhabits and lives for his books, which are, he says, political. He describes them as 'anti-imperialist': they reject a set of accepted values which are primarily London-based. He is seeking to give authentic literary voice to the West of Scotland working class. He rejects the received pronunciation type of English (as spoken by the ruling classes), writes in the vernacular, and has been victim of the censors for his generous use of expletives. He is anti-parliamentarian and outspoken about it, considering it the intellectual's responsibility to stimulate dialogue. All this, he claims, makes him in every way at odds with Anglo-American literary society, which 'invents all these wee categories in its field', and because he cannot be fitted conveniently into one of them, he is 'seen as a fluke'. Even when he did get the establishment seal of approval – *A Disaffection* was shortlisted for the Booker Prize – he was referred to disparagingly during a television broadcast as 'Billy Connolly with philosophy'. Elsewhere, he has been compared to **James Joyce**, **Samuel Beckett**, Kierkegaard and Zola. In 1987 his volume of stories, *Greyhound for Breakfast*, won the Cheltenham Prize and a Scottish Arts Council Book Award, and in 1994 *How Late It Was, How Late* was a controversial Booker Prize-winner.

Kelman has been described as Scotland's Kafka, 'one of Britain's most important writers'. 'It matters to me that my work should be received favourably,' he has said, 'because it might improve the level of criticism, open critical routes for other writers committed to different, anti-imperialist ways of writing.' He lives in a tenement with his wife and two daughters.

Fiction

Not Not While the Giro (s) 1983; *The Busconductor Hines* 1984; *A Chancer* 1985; *Lean Tales* (s) [with A. Gray, A. Owen] 1985; *Greyhound for Breakfast* (s) 1987; *A Disaffection* 1989; *The Burn* (s) 1991; *How Late It Was, How Late* 1994

Plays

Hardie and Baird (r) 1978, (st) 1990 [pub. in *Hardie and Baird and Other Plays* 1991]; *The Busker* 1985 [np]; *Le Rodeur* [from E. Cormann's play] 1987 [np]; *In the Night* 1988 [np]; *The Return* (f) 1990

Non-Fiction

Some Recent Attacks: Essays Cultural and Political 1992

Edited

An East End Anthology 1988

Thomas (Michael) KENEALLY 1935–

The descendant on both sides of Irish Roman Catholics, Keneally was born, and spent his earliest years, in Kempsey, a dairy town of northern New South Wales. His family moved to Sydney when his father joined the Australian Air Force during the Second World War, and he was educated at St Patrick's College, Strathfield. On leaving school, he trained for the diocesan priesthood, but left two weeks before he was due to be ordained. He then studied law, without practising it, and later worked as a high-school teacher in Sydney.

His first short story had been published in *The Bulletin*, a magazine which regularly served as a nursery for Australian writers, and his first novel, *The Place at Whitton*, was published in 1964. He married in 1965, and has two daughters. After giving up teaching, he took up part-time work to finance his writing, but received a grant from the Australian government, a Miles Franklin Award (repeated twice), to write his third novel, *Bring Larks and Heroes*. A story of the penal colony at Sydney in the 1790s, the novel made his name. He has since become a prolific and popular novelist with around twenty titles to his credit. He has also been a lecturer in drama at the University of New England, New South Wales, from 1968 to 1970; lived in England and the USA for periods during the 1970s; was a fellow in creative writing at California University in 1985; and Berg Professor of English at New York University in 1988.

He won the Booker Prize for *Schindler's Ark* in 1982, a choice which aroused controversy because many considered this story of heroism against Nazism a work of non-fiction. The book was subsequently made into an epic and hugely acclaimed, monochrome film under its American title, *Schindler's List*. Directed by Steven Spielberg, the film was thought by many to be the best (non-documentary) cinematic evocation of the Holocaust ever made. That Keneally had been twice previously shortlisted for the Booker Prize – for *The Chant of Jimmie Blacksmith* in 1972 and *Gossip from the Forest* in 1975 – attests to his stature, although even his admirers tend to admit that he is not a stylist. Raymond Sokolov has called him an 'honest workman', but Keneally brings vigour to his often historical subjects for novels, which include Joan of Arc, the armistice of 1918 and the American Civil War in *Confederates*. War is also a recurring theme, and *Towards Asmara* is a committed account of the Eritrean conflict.

Keneally is a well-known figure in Australian life, having been connected with the rise to power of the Labour politician Gough Whitlam in the 1970s and chairman of the Australian Society of Authors in 1987.

Fiction

The Place at Whitton 1964; *The Fear* 1965; *Bring Larks and Heroes* 1967; *Three Cheers for the Paraclete* 1969; *The Survivor* 1970; *A Dutiful Daughter* 1971; *The Chant of Jimmie Blacksmith* 1972; *Blood Red, Sister Rose* 1974; *Gossip from the Forest* 1975; *Season in Purgatory* 1976; *A Victim of the Aurora* 1977; *Confederates* 1979; *Passenger* 1979; *The Cut-Rate Kingdom* 1980; *Schindler's Ark* [US *Schindler's List*] 1982; *A Family Madness* 1985; *The Playmaker* 1987; *By the Line* 1989; *Towards Asmara* [US *To Asmara*] 1989; *Flying Hero Class* 1991; *The Woman of the Inner Sea* 1992; *Jacko* 1994; *A River Town* 1995

Plays

Halloran's Little Boat [from his novel *Bring Larks and Heroes*] 1966 [pub. in *Penguin Australian Drama 2* 1975]; *Childermass* 1968 [np]; *An Awful Rose* 1972 [np]; *Essington* **(tv)** 1974; *Bullie's House* 1980 [pub. 1981]; *Gossip from the Forest* [from his novel] 1983 [np]; *Silver City* **(f)** [with S. Turkiewicz] 1985; *Child of Australia* **(l)** [with P. Sculthorpe] 1987; *Our Country's Good* [from his novel *The Playmaker*; pub. 1988]

For Children

Moses the Lawgiver 1975; *Ned Kelly and the City of the Bees* 1978

Non-Fiction

Outback 1983; *Australia: Beyond the Dreamtime* [with P. Adam-Smith, R. Davidson] 1987; *Now and in Time to Be* 1991; *Ireland* 1992; *The Place Where Souls are Born* 1992; *With Yellow Shoes* 1992; *Memoirs from a Young Republic* 1993

James (Peeble Ewing) KENNAWAY 1928–1968

Kennaway was born at Kenwood Park, Auchterarder, Perthshire, the ancient capital of Strathearn. His father, who died when he was twelve, was a leading local solicitor and factor, his mother a doctor. His father's early death affected the boy deeply; he always believed that he too would die young, and the belief was later reinforced when a great friend was killed by a booby trap while doing his national service in Jerusalem, and an old schoolfriend died in a road accident. He was educated at Cargilfield Preparatory School, Edinburgh, and Glenalmond. On leaving school in 1946, he did his national service largely with the Gordon Highlanders in Essen; from this experience came his first and best-known novel, *Tunes of Glory* (1956). On leaving the army he went up to Trinity College, Oxford, to read philosophy, politics and economics (his tutor was Anthony Crosland). He married in 1951, and had two sons and two daughters. In the same year he joined Longmans Green, who later published him. He left in 1957 to become a full-time writer.

In *Tunes of Glory*, which examines the tensions in an officers' mess when two senior officers from different military and social backgrounds vie for the leadership of a regiment, the characters are drawn from life though the setting is removed to Scotland. **Peter Quennell**, **John Betjeman**, **Pamela Hansford Johnson** and **Compton Mackenzie** all praised it, and it became an Oscar-winning film starring Alec Guinness and John Mills. *Household Ghosts*, which caused him much pain to write, also has a Scottish setting – though critical of the Scots and insisting that he was 'a writer from Scotland' not 'a Scottish writer', he frequently returned there for his books. The story revolves round a triangular situation – a favourite theme – in a crumbling country house in Strathearn, the characters out of touch with the world outside. *The Mind Benders*, originally a screenplay, is concerned with sensory deprivation and subsequent brain-washing. Essentially a figure of the 'swinging sixties', earning vast fees from films, some of which were never made, Kennaway had a number of casual affairs. He saw himself as a dual personality: James the family man and Jim the deceiver. In 1955 the recurring theme of the triangular relationship invaded his life when his greatest friend David Cornwell (**John le Carré**) fell in love with his wife. The last novel to be published in his lifetime, *Some Gorgeous Accident*, uses this traumatic event, which is also described in the posthumously published *The Kennaway Papers*. Three years later he died on the M4 motorway when his car went out of control after he had suffered a massive heart-attack. His prediction that he would die young had been realised. The other posthumous publications were *The Cost of Living Like This*, his last completed novel, and *Silence*, put together from four fragments.

Fiction

Tunes of Glory 1956; *Household Ghosts* 1961; *The Bells of Shoreditch* 1963; *The Mind Benders* 1963; *Some Gorgeous Accident* 1967; *The Cost of Living Like This* 1969; *Silence* [ed. L. Hughes] 1972; *The Dollar Bottom and Taylor's Finest Hour* (s) [ed. T. Royle] 1981

Plays

Violent Playground (f) 1958; *Tunes of Glory* (f) [from his novel] 1960; *The Mind Benders* (f) 1963; *Country Dance* [from his novel *Household Ghosts*] (st) 1967, (f) 1969; *The Shoes of the Fisherman* (f) [from M. West's novel; with J. Patrick] 1968; *Battle of Britain* (f) [with W. Greatorex] 1969

Non-Fiction

The Kennaway Papers [with S. Kennaway] 1981

Biography

James and Jim by Trevor Royle 1983

William KENNEDY 1928–

Kennedy was, unusually, the only child of working-class Irish-American parents in the Albany district of New York. His father was a barber, foundry worker and deputy sheriff whose involvement with the Democratic political machine of Daniel P. O'Connell afforded his son an early glimpse of the political clubs and gaming rooms which are such an important feature of his fiction. After attending the Christian Brothers Academy in Albany, he took his BA in English in 1949 from the Franciscan Siena College, Loudonville, New York, and commenced work the same year as a sports writer on the Glen Fall *Post-Star*. After serving in the army, he worked for three years as a reporter on the Albany *Times-Union*, and thereafter as assistant managing editor of a short-lived Puerto Rican newspaper. While living in Puerto Rico, Kennedy started to write fiction and studied creative writing with **Saul Bellow** at the university there. In 1957 he was married to Ana Daisy Sosa, a former actress and dancer, with whom he has three children.

Returning to Albany in 1961 to look after his sick father, Kennedy was drawn towards the area and its history as a possible source for his fiction. While working as a part-time feature writer on the Albany *Times-Union*, he wrote his first novel, *The Ink Truck* (1969), a sardonic and comic account of a newspaper strike. At Dial Press the novel was edited by **E.L. Doctorow** who, together with Bellow, has proved an important mentor to Kennedy. In 1974 he became a lecturer at the State University, Albany, and in 1982 a professor. During this period he wrote the five novels which thus far constitute his Albany Cycle. The novels tell the story of three generations of the Irish-American Phelan family, and bear out Kennedy's contention that the area provides him with a microcosm of life 'as abundant in mythical qualities as it is in political ambition, remarkably consequential greed, the genuine fear of the Lord'. *Legs* and *Billy Phelan's Greatest Game* were praised by critics but enjoyed little commercial success, with the result that Kennedy's publisher, Viking, felt unable to publish the third novel in the sequence, *Ironweed*. The novel was unsuccessfully submitted to a further twelve publishers before Bellow intervened and persuaded Viking to reconsider. *Ironweed* was published in 1983, sold over 100,000 copies in hardback, won a 1984 Pulitzer Prize, was made into a film (1987) starring Jack Nicholson and Meryl Streep, and secured for Kennedy an award of $264,000 from the MacArthur Foundation. Viking proved only too happy to publish the next two novels in the cycle, *Quinn's Book* and *Very Old Bones*. Kennedy has also written film versions of *Legs* and *Billy Phelan's Greatest Game*, and collaborated with Francis Ford Coppola on the screenplay for *The Cotton Club*.

Fiction

The Ink Truck 1969; *Legs* 1975; *Billy Phelan's Greatest Game* 1978; *Ironweed* 1983; *Quinn's Book* 1988; *Very Old Bones* 1992

Non-Fiction

O Albany! Improbable City of Political Wizards, Fearless Ethnics, Spectacular Aristocrats, Splendid Nobodies, and Underrated Scoundrels 1984; *Riding the Yellow Trolley Car* 1993

For Children

Charlie Malarkey and the Belly Button Machine [with B. Kennedy] 1986

Maxwell KENTON

see Terry SOUTHERN

Jack (i.e. Jean Louis Libris) KEROUAC 1922–1969

Kerouac was born in Lowell, Massachusetts, the third child of working-class French-Canadian *émigrés*. His first language was the French-Canadian dialect *joual*, until he went to parochial school where he was educated by Jesuits. In 1939 he matriculated to Columbia University on a football scholarship, but broke his leg in the first season, dropped out, and signed on as a merchant seaman. Shortly afterwards he joined the navy but was discharged as being of indifferent character. While hanging around the Columbia campus in 1944, he began to mix with a group of New York based intellectuals including **William Burroughs** and **Allen Ginsberg**, whose Bohemian lifestyle and search for a new philosophy profoundly influenced him. When he was arrested in connection with the homicide of one of the group, his girlfriend, Edie Parker, offered to bail him out on condition that he married her – which he did, but they separated almost immediately. Around this time, Kerouac was hospitalised after excessive use of his favourite drug, Benzedrine – a drug he continued to use for most of his life.

His father had been critical of his son's shiftless lifestyle and, when he died of cancer in 1946, Kerouac began the 'huge novel' he hoped would redeem him. *The Town and the City* (1950) received tepid reviews and he was disappointed with his inability to break away from imitation of his early literary heroes, **Jack London** and **Thomas Wolfe**, and develop a voice and style of his own. However, his meeting with the man whose legend he would create, Neal Cassady, and the drug-fuelled cross-country car rides that they made together, inspired a three-week stint at the typewriter which produced the 120-foot-long, single-spaced paragraph that was eventually published in 1957 as *On the Road*. It was an immediate bestseller, and its articulation of what Kerouac called the 'best generation' catapulted him to fame not only as a writer but as a television personality. A blend of fiction and autobiography, *On the Road* is an exhilarating 'road' book written in what he described as 'spontaneous prose', in which he manages to convey the intoxicating intensity of the Benzedrine 'rush' and the virtuoso jazz solo. **Truman Capote** condemned it as 'just typing', but the book caught the mood of middle-class American youth uneasy with the material culture of the 1950s, and quickly assumed cult status. The central character, Dean Moriarty, based on Neal Cassady, became the archetype of a rebellious alternative lifestyle and the novel's marathon, anarchic car journey became a modern manifestation of the pioneer spirit. In spite of having found fame, financial security and a voice, Kerouac was disillusioned by the way the book was misunderstood by the literary establishment and idealistic enthusiasts alike. Although characterised by the media as 'King of the Beats' and the original 'beatnik', he was at heart a conservative, living quietly at home and being looked after by his mother, only occasionally going on the wild, drunken cross-country adventures for which he became famous.

He went on to write a series of autobiographical novels comprising what he called 'the legend of Duluoz', his fictitious French-Canadian name for himself. He continued to write his spontaneous prose – *The Subterraneans* was written on Benzedrine in three days – and he flirted with Buddhism for a while. *The Dharma Bums*, *Doctor Sax* and *Maggie Cassidy* are all autobiographical novels in which the characters of *On the Road* reappear under different names. *Mexico City Blues* is a collection of 'sensory meditations' written in Mexico City under the influence of morphine and marijuana and which attempts to capture the tempo and rhythm of jazz (it was dedicated to the memory of the saxophonist Charlie Parker). Virtually a recluse in his final years, Kerouac suffered an abdominal haemorrhage whilst vomiting in his lavatory and died at home aged only forty-seven.

Fiction

The Town and the City 1950; *On the Road* 1957; *The Dharma Bums* 1958; *The Subterraneans* 1958; *Doctor Sax* 1959; *Excerpts from Visions of Cody* 1959; *Maggie Cassidy* 1959; *Mexico City Blues* 1959; *Book of Dreams* 1960; *Lonesome Traveller* 1960; *Tristessa* 1960; *Big Sur* 1962; *Visions of Gerard* 1963; *Desolation Angels* 1965; *Sartori in Paris* 1966; *Vanity of Duluoz* 1968; *Pic* 1971; *Visions of Cody* 1973

Poetry

The Scripture of the Golden Eternity 1960; *Scattered Poems* 1971; *Heaven and Other Poems* 1977

Plays

Pull My Daisy (f) 1961

Biography

Jack's Book by Barry Gifford and Lawrence Lee 1978; *Memory Babe* by Gerald Nicosia 1983

Ken (Elton) KESEY 1935–

Kesey was born in La Junta, Colorado, and brought up in Eugene, Oregon, where his father built up a creamery business after the war. At the University of Oregon he developed his interests in creative writing, drama and all-in wrestling and married his girlfriend from high school. On graduating he won a scholarship to Stanford University and enrolled in the creative writing course, but dropped out as he became involved with a Bohemian community flourishing at Stanford which modelled itself on San Francisco's Beat culture.

Kesey wholeheartedly embraced the alternative lifestyle, began experimenting with drugs and wrote an unpublished novel, 'Zoo'. During this time he volunteered to take part in drug experiments being conducted at the nearby Veterans Hospital in which he was paid to take a number of hallucinatory drugs including LSD, which the US Army was then considering for use in interrogation procedures as a 'truth' drug. The sessions were a revelation to him. His experiences as an orderly at a psychiatric hospital and the insights into the nature of psychosis offered by the LSD sessions were the inspiration for *One Flew Over the Cuckoo's Nest* (1962), a novel set in a psychiatric ward narrated from the viewpoint of a mute, schizophrenic Native American, Chief Bromden, who Kesey claims appeared to him in a drug-induced hallucination. The book is a complex but humorous examination of the dehumanising effects of the social conformity demanded by American society in the 1950s and was a huge critical and commercial success. Its comic-book quality, with unambiguous hero and arch-villain, lent itself well to dramatic representation and the 1975 film, directed by Milos Forman and starring Jack Nicholson, was also immensely popular.

Kesey followed this up with *Sometimes a Great Notion*, a long novel about a feud between two brothers in a logging community that could be described as an *Absalom, Absalom!* set in Oregon, especially given Kesey's acknowledged debt to **William Faulkner**. Like its predecessor, it is a muscular, masculine book but more ambitious technically and thematically and remains undeservedly overshadowed by the cult status of *Cuckoo's Nest*.

After completing it, Kesey felt the urge to transcend the novel as an art form, gave up writing and turned his attention to experimenting with LSD. He formed the 'Merrie Pranksters', bought an old bus, put 'Further' on the destination sign, and toured America dispensing LSD at communal 'acid tests', recording the 'happenings' in an uncompleted film called 'The Movie'. Kesey recruited the Beat hero Neal Cassady as the driver. The bus trip and later escapades of the Pranksters are reported in **Tom Wolfe**'s *The Electric Kool-Aid Acid Test* (1973), a book written in the New Journalism style which soon established itself in the hippy canon and increased Kesey's reputation as a pioneer of the new drug culture. His practical 'farm boy' self-image has always sat uneasily with the intellectual radicalism of the hippy counterculture however, and many of the pilgrims who visited his Oregon farm in search of a guru were disappointed. The short story 'The Day After Superman Died' (included in the autobiographical collection *Demon Box*, a personal assessment of his life and art described by himself as 'form in transit') describes one such occasion. In 1965 Kesey was arrested for possession of marijuana and fled to Mexico. Nine months later, he gave himself up and served a five-month prison sentence at the San Mateo County Jail.

His interest in writing returned in the early 1970s. *Kesey's Garage Sale* is a miscellaneous collection of writings by and about him with a foreword by **Arthur Miller**, and includes Kesey's screenplay *Over the Border*, about his time on the run. Other items of his short fiction, some of which were previously published in Kesey's magazine *A Spit in the Ocean*, are included in *Demon Box*. His novel *Sailor Song* is a futuristic fable about the efforts of an Alaskan fishing village to resist the corrupting effects of a Hollywood film crew.

Fiction

One Flew Over the Cuckoo's Nest 1962; *Sometimes a Great Notion* 1964; *Seven Prayers by Grandma Whittier* 1974–9; *Demon Box* (s) 1986; *The Further Inquiry* [with R. Bevirt] 1990; *Sailor Song* 1992; *Last Go Round* 1994

Collections

Kesey's Garage Sale 1973

Daniel KEYES 1927–

Keyes was born in Brooklyn, New York, and took a degree in psychology at Brooklyn College, because he was 'fascinated by the complexities of the human mind'. It was not until the mid-1960s, however, that he wrote his first novel, *Flowers for Algernon*. From 1945 to 1947 he was senior assistant purser in the US Maritime Service, and later was fiction editor for two years in a New York publishing company. He married the fashion stylist and photographer Aurea Vazquez in 1952, and became co-owner of Fenko and Keyes, Photography Inc. the following year. For some years he held various lecturing posts, teaching English in Brooklyn in the mid-1950s, and lecturing from 1966 in Ohio University until he was appointed professor of English there from 1972, and also director of creative writing until 1976.

Flowers for Algernon, which was an instant success, was unusual because it began life as a short story in the *Magazine of Fantasy and Science Fiction* and won an award from the World Science Fiction Convention in that form, and the Nebula Award from the Science Fiction Writers of America when it was expanded to a novel. It was regarded as remarkable in its hypothesis of examining moral and social issues through the story of Charlie, a mentally retarded man seen as a moron by 'intelligent' society. He is changed from a moron into a genius by psychosurgery, only to gradually regress, giving him the awareness of his condition and the knowledge that he will end in a home for the feeble-minded. Keyes's sensitive handling of his subject-matter and astute examination of complex issues ensure that the novel is more than a work of pure science fiction. *Flowers for Algernon* was adapted as a stage play which has been performed throughout the world.

His interest in disturbed minds continued in his third novel, *The Fifth Sally*, which was a fictional treatment of an actual case concerning the four contrasting multiple personalities of one woman, Sally Porter, and her doctor's efforts to merge them into one. Keyes's most successful book is a work of non-fiction, *The Minds of Billy Milligan*, in which he examines the case of a man who was tried for rape in Ohio in 1977, and was the first person in American legal history to be acquitted for a major crime for reasons of possessing multiple personalities – in his case, twenty-four. Several of Milligan's personalities had read *Flowers for Algernon*, and they agreed to co-operate with Keyes in the writing of the book. All twenty-four personalities are 'interviewed' by Keyes, who reveals how Milligan had been forced to invent the various inhabitants of his mind to protect himself from the physical and mental abuse he suffered from his stepfather.

Fiction

Flowers for Algernon 1966; *The Touch* 1968 [UK *The Contaminated Man* 1973]; *The Fifth Sally* 1980

Non-Fiction

The Minds of Billy Milligan 1981; *Unveiling Claudia: the True Story of a Serial Murder* 1986

Plays

Flowers for Algernon [from his novel; with D. Rogers] 1969

Sidney (Arthur Kilworth) KEYES 1922–1943

The son of an army officer and his second wife, Keyes was born in Dartford, Kent. His mother died of peritonitis shortly after his birth and his father was largely absent during his childhood; he was brought up by his paternal grandparents. A sickly, solitary and precocious child, he soon learned to read, rejecting Arthur Mee's famous *Children's Encyclopaedia* as 'inaccurate' at the age of five. He was educated at Dartford Grammar School and Tonbridge before going up to Queen's College, Oxford, as a history scholar in 1940. Encouraged by a Tonbridge schoolmaster, and inspired by a visit to France, he had already begun to write poetry, much of it influenced by the Romantics and visionaries. He had an affinity with nature, and a death-wish haunts much of his work; he claimed Wordsworth, Rilke and Jung amongst his influences.

At Oxford he became particularly friendly with fellow poets **John Heath-Stubbs** and Michael Meyer; he formed a dramatic society, wrote poems and short stories, and became editor of the *The Cherwell*. He also edited (with Meyer) *Eight Oxford Poets*, proclaiming: 'We are all, with the possible exception of [J.A.] Shaw, *Romantic* writers ... we have, on the whole, little sympathy with the Audenian school'. His work appeared in numerous magazines, such as *Poetry (London)*, the *New Statesman, Horizon* and *The Listener* (where his poem 'Remember Your Lovers' earned the literary editor, **J.R. Ackerley**, a rebuke from the Director-General of the BBC because it contained the word 'lust'). His first volume of poetry, *The Iron Laurel*, was published in 1942.

He gained a commission in the Queen's Own West Kent Regiment, and was posted to Tunisia in March 1943, but saw only two weeks' active service before being killed a month short of his twenty-first birthday. It was long believed that he had been captured and died 'of unknown causes' in enemy hands, but it is now known he was killed outright while covering his platoon's retreat during a counter-attack. Before going to Tunisia, he had assembled another volume of poetry, *The Cruel Solstice*, which was published posthumously and followed by his *Collected Poems*, edited by Meyer.

Keyes was undoubtedly one of the finest poets of the Second World War, although the bulk of his work did not actually deal with that conflict (poems he was working on in Tunisia were never recovered). His range was large: poems of nature and love, tributes to fellow writers (including an elegy for **Virginia Woolf**), and blues lyrics. He was awarded the Hawthornden Prize posthumously in 1944.

Poetry
The Iron Laurel 1942; *The Cruel Solstice* 1944; *Collected Poems* [ed. M. Meyer] 1945 [rev. 1988]

Edited
Eight Oxford Poets (c) [with M. Meyer] 1941

Francis (Henry) KING 1923–

King was born in Adelboden, Switzerland, where his tubercular father, a member of the Indian Police who became Deputy Director of the Intelligence Bureau, had come in search of a cure. King spent his childhood in India, before being educated at Shrewsbury School in England. His father died during King's first term at school, and he spent school holidays 'shuttling between relatives, often in boarding-houses'.

A conscientious objector, he spent the war working on the land, before going up to Balliol College, Oxford, to study classics. While still an undergraduate, he published three novels, the first of them, *To the Dark Tower*, in 1946 when he was twenty-three. In spite of a career in the British Council, long spells as a theatre and book reviewer, and service on numerous committees, King has kept up the high rate of literary production he inaugurated as a student. He is, by his own admission, a compulsive writer, and he has published well over thirty volumes of fiction, several volumes of non-fiction, travel guides, plays (including adaptations of the work of his friend **C.H.B. Kitchin**), an early volume of poetry and an autobiography, as well as editing and translating the works of other writers.

King's work for the British Council informs many of his novels and short stories. He was posted first to Florence in 1949, and this provides the milieu for both his fourth novel, *The Dividing Stream* (which won a Somerset Maugham Award), and the much later *The Ant Colony*, a comedy set amongst the city's English community (the character of Ivor Luce is based on **Harold Acton**). He then went to Salonika, before spending some time in Athens, the setting of *The Firewalkers*. A lightly fictionalised account of sexual unorthodoxy in Greece, it is perhaps King's most genial novel, but was disapproved of by the Council, to whom all employees were obliged to submit manuscripts, and it therefore had to be published under the pseudonym of Frank Cauldwell (a name borrowed from that of the young writer in his first novel). The Greek years also provided material for the background of the central character of *The Man on the Rock*, a bleak novel of bisexual manipulation and betrayal. After a spell as assistant representative in Helsinki, King became the Council's regional director on Kyoto, Japan. This period resulted in several books, including *The Custom House, The Waves Behind the Boat* and a volume of short stories, *The Japanese Umbrella* (which won the Katherine Mansfield Prize), and King later edited the Japanese writings of Lafcadio Hearn.

In 1970, King published *A Domestic Animal*, an autobiographical novel detailing the narrator's erotic obsession with an Italian man. The book was threatened with an injunction by a Brighton eccentric and former Labour MP called Tom Skeffington-Lodge, who had decided (quite correctly) that one of the novel's female characters was based upon him. The publishers were obliged to withdraw the book at the last moment and King had to rewrite it, incurring considerable costs in the process. Characteristically, he used this imbroglio as the basis for his novel *The Action*, in which a female character imagines that she has been portrayed as a man in a novel (based on King's own novel *The Last of the Pleasure Gardens*). Amongst other novels are *The Widow*, which is in part a portrait of King's mother (who lived into her early hundreds) and *Act of Darkness*, perhaps his best novel (and winner of the *Yorkshire Post* Book of the Year Award), which is based upon the Constance Kent murder case, but is set in India in the 1930s.

Although his novels have attracted widespread praise, and have immense narrative drive, many critics feel that his finest achievements have been in the short-story form. His first volume announced a frequent theme of his work in its title, *So Hurt and Humiliated*, for he often deals with small treacheries and the great anguish they cause. Indeed, most of his fiction, although lightened by comic scenes, explores the darker recesses of the human psyche, and the picture of the world that emerges is pessimistic, characterised by physical and emotional squalor, with characters making a mess of their lives both literally and metaphorically. Many of his stories contain a supernatural element, a theme that also emerges in the novel *Voices in an Empty Room*.

King has written studies of **E.M. Forster** and **Christopher Isherwood** and edited the diaries of **J.R. Ackerley**, to whose pages in *The Listener* he was a frequent contributor. He has known most of the major writers of the century and many of them are entertainingly recalled in *Yesterday Came Suddenly*, a volume of memoirs which ends with an unblinking account of the death from Aids of his partner, David Atkin. He was for several years the theatre critic of the *Sunday Telegraph* and continues to review fiction for the *Spectator*. He has been president of the English Centre of PEN, as well as the organisation's international president, and is a familiar figure on conference platforms in many countries. (His comic novel *Visiting Cards* is based upon his experiences in this role.) He remains an inveterate traveller, a fact reflected in the title he chose for his own selection of his best short stories, *One is a Wanderer*.

Fiction

To the Dark Tower 1946; *An Air That Kills* 1948; *Never Again* 1948; *The Dividing Stream* 1951; *The Dark Glasses* 1954; *The Firewalkers* [as 'Frank Cauldwell'] 1956; *The Man on the Rock* 1957; *The Widow* 1957; *So Hurt and Humiliated* (s) 1959; *The Custom House* 1961; *The Japanese Umbrella* (s) 1964; *The Last of the Pleasure Gardens* 1965; *The Waves Behind the Boat* 1967; *The Brighton Belle* (s) 1968; *A Domestic Animal* 1970; *Flights* 1973; *A Game of Patience* 1974; *The Needle* 1975; *Hard Feelings* (s) 1976; *Danny Hill* 1977; *The Action* 1978; *Indirect Method* (s) 1980; *Act of Darkness* 1982; *Voices in an Empty Room* 1984; *One is a Wanderer* (s) 1985; *Frozen Music* 1987; *The Woman Who Was God* 1988; *Punishments* 1989; *Visiting Cards* 1990; *The Ant Colony* 1991; *Hidden Lives* [with P. Gale, T. Wakefield] 1991; *A Several World* 1991; *The One and Only* 1994

Plays

The Prisoner (r) 1967; *Corner of a Foreign Field* (r) 1969; *A Short Walk in Williams Park* [from C.H.B. Kitchin] (r) 1972; *Death of My Aunt* [from C.H.B. Kitchin's novel] (r) 1973; *Desperate Cases* (r) 1975; *Far East* 1980 [np]

Non-Fiction

Japan 1970; *Christopher Isherwood* 1976; *E.M. Forster and His World* 1978; *Florence* 1982; *Florence: a Literary Companion* 1991; *Yesterday Came Suddenly* (a) 1994

Edited

Introducing Greece 1956 [rev. 1968]; Osbert Sitwell *Collected Short Stories* (s) 1974; *New Stories 3* (s) [with R. Harwood] 1978; *Prokofiev by Prokofiev: a Composer's Memoirs* 1979; *My Sister and Myself: the Diaries of J.R. Ackerley* 1982; Lafcadio Hearn *Writings from Japan* 1984; *Twenty Stories* (s) 1985

Poetry

Rod of Incantation 1952

Hugh KINGSMILL (Lunn) 1889–1949

Kingsmill was the son of Sir Henry Lunn – first a Methodist missionary and later an immensely successful pioneer of organised tourism – and the brother of Arnold Lunn, the well-known Catholic apologist and author of the school story *The Harrovians*. Because of his father's business (largely Alpine tours), Kingsmill grew up partly in Switzerland, but also attended Harrow as a day boy and then New College, Oxford, where he failed to take a degree; the omission was repaired at Trinity College, Dublin. He worked for his father's firm until he was thirty-eight, a period interrupted by the First World War, when he served briefly in France but spent a longer period as a prisoner-of-war in Germany. In his youth he was a protégé of the 1890s' man-of-letters and journalist Frank Harris, who helped launch his literary career, which was originally as a novelist and author of short stories. His first novel, *A Will to Love* (1919), was published under his full name.

Kingsmill had married Eileen Turpin in 1914, with whom he had one child and with whom his relations were consistently unhappy. When he left her in 1928, it became impossible for him to remain with his father's Free Church Touring Guild, and he spent the rest of his life mainly as an impecunious man-of-letters (he once touched **George Bernard Shaw** for £10). He wrote a life of Matthew Arnold in 1928, and became best known as a biographer and anthologist; *Samuel Johnson* is perhaps the most durable of his biographies. During the Second World War he became a master at Marlborough and Merchant Taylors schools, and he was later literary editor of *Punch*.(He is the author of the best parody of **A.E. Housman**'s *A Shropshire Lad*, with a verse which begins: 'What, still alive at twenty-two, /A clean upstanding chap like you?'.)

Although little read today, Kingsmill had ideas of some influence and interest, and was passionately opposed to what he called 'Dawnism', political prescriptions for a perfect society; in many ways he continued the trend in English letters slightly earlier represented by **G.K. Chesterton** and **Hilaire Belloc**. He was a close friend of two other men-of-letters, Hesketh Pearson and Malcolm Muggeridge, and the trio are the subject of a book by Richard Ingrams, *God's Apology* (1977). Kingsmill collaborated on books with both Pearson and Muggeridge, and they jointly published *About Kingsmill* (1951), based on their letters. Kingsmill's second, more durable, marriage was to Dorothy Vernon; they had three children.

Fiction

A Will to Love [as Hugh Kingsmill Lunn] 1919; *The Dawn's Delay* (s) 1924; *Blondel* 1927; *The Return of William Shakespeare* 1929; *The Fall* 1948

Non-Fiction

Matthew Arnold 1928; *After Puritanism 1850–1900* 1929; *Behind Both Lines* (a) 1930; *Frank Harris* 1932; *Samuel Johnson* 1933; *The Table of Truth* 1933; *The Casanova Fable* [with W. Gerhardie] 1934; *The Sentimental Journey: a Life of Charles Dickens* 1934; *Brave Old World* [with M. Muggeridge] 1936; *Next Year's News* [with M. Muggeridge] 1937; *Skye High* [with H. Pearson] 1937; *D.H. Lawrence* 1938; *The Blessed Plot* [with H. Pearson] 1942; *The Poisoned Crown* 1944; *Talking of Dick Whittington* [with H. Pearson] 1947; *The Progress of a Biographer* 1949; *About Kingsmill* [with M. Muggeridge, H. Pearson] 1951

Edited

An Anthology of Invective and Abuse 1929; *More Abuse* 1930; *The Worst of Love* 1931; *What They Said at the Time* 1935; *Parents and Children* 1936; *Made on Earth: a Panorama of Marriage* 1937; *The English Genius* 1938; *Courage* 1939; *Johnson without Boswell* 1940; Leonard Dobbs *Shakespeare Revealed* 1948; *The High Hill of the Muses* 1955

Collections

The Best of Hugh Kingsmill [ed. M. Holroyd] 1970

Biography

Kingsmill by Michael Holroyd 1964

Thomas KINSELLA 1928–

Kinsella was born in Dublin and educated at O'Connels and University College, Dublin. He joined the Irish civil service in 1946, retiring as assistant principal officer in the Department of Finance to go to America in 1965. He was at the South Illinois University, Carbondale, first as writer-in-residence then as professor of English. Since then he has been professor of English at Temple University, Philadelphia. He is a director of the Dolmen Press in Dublin and founder in 1972 of its associate Peppercanister, which have between them been responsible for the first publication of nearly all his work. He has used Peppercanister for first publication of his poems of protest, like 'Butcher's Dozen' – about Bloody Sunday 1972 when thirteen civil-rights demonstrators were shot by British soldiers – and his reflections on the assassination of John F. Kennedy, ten years after the event. He is also artistic director of the Lyric Players, Belfast. He married in 1955 and has two daughters and a son.

His early poems were mainly lyrical, the collection *Another September* being a Book Society Choice. With *Downstream*, Kinsella broadened his canvas, tackling political themes, but there are also poignant poems like 'Cover Her Face', about a girl who died suddenly, aged twenty-nine. In *Nightwalker*, a Poetry Book Society Choice, he describes his mother's deathbed in another moving poem. He has written of 'Phoenix Park', a love poem in four parts and a farewell to Dublin: 'Love ... it seems, will continue until we fail: in the sensing of the wider scope, in the growth towards it, in the swallowing and absorption of bitterness, in the resumed innocence ... The positive aspect of these ideas is dealt with in "Phoenix Park" '. His translations from the Irish into English verse include the early epic *The Táin*.

Kinsella received the Guinness Award in 1958, the Dennis Devlin Memorial Award in 1967, Guggenheim Fellowships in 1968 and 1971, the Before Columbus Foundation in 1983, and a DLitt from the National University of Ireland in 1984. He now divides his time between Dublin and Philadelphia.

Poetry

The Starlit Eye 1952; *Three Legendary Sonnets* 1952; *The Death of a Queen* 1956; *Poems* 1956; *Another September* 1958 [rev. 1962]; *Moralities* 1960; *Poems and Translations* 1961; *Downstream* 1962; *Six Irish Poets* [with others; ed. R. Skelton] 1962; *Wormwood* 1966; *Nightwalker* 1967; *Nightwalker and Other Poems* 1968; *Poems* [with D. Livingstone, A. Sexton] 1968; *Tear* 1969; *Butcher's Dozen* 1972; *Finistère* 1972; *Notes from the Land of the Dead* 1972; *The Good Fight* 1973; *New Poems* 1973; *Notes from the Land of the Dead and Other Poems* 1973; *Selected Poems 1956–1968* 1973; *Vertical Man* 1973; *One* 1974; *Poems 1956–1973* 1974; *A Technical Supplement* 1976; *The Messenger* 1978; *Song of the Night and Other Poems* 1978; *Fifteen Dead* 1979; *One and Other Poems* 1979; *Peppercanister Poems 1972–1978* 1979; *Poems 1956–1973* 1980; *One Fond Embrace* 1981; *Her Vertical Smile* 1985; *Songs of the Psyche* 1985; *Out of Ireland* 1987; *St Catherine's Clock* 1987; *Blood and Family* 1988; *Personal Places* 1990; *Poems from Centre City* 1990; *Madonna and Other Poems* 1991; *Open Court* 1991; *From Centre City* 1994

Non-Fiction

Davis, Manan, Ferguson: Tradition and the Irish Writer 1970; *Dual Tradition* 1995

Translations

Longes mac n-Usnig, Being the Exile and Death of the Sons of Usnech 1954; *Thirty-Three Triads* 1955; *Faeth Fiaddh: the Breastplate of St Patrick* 1957; Cúalnge *The Táin* 1969; *An Duanaire: Gaelic Poetry 1600–1900* [with S. Ó Tuama] 1981

Edited

Selected Poems of Austin Clarke 1976; S. Ó Riada *Our Musical Heritage: Lectures on Irish Traditional Music* 1982; *The New Oxford Book of Irish Verse* 1986

(Joseph) Rudyard KIPLING 1865–1936

Kipling was born in Bombay, where his father, an arts and crafts teacher, taught applied pottery and architectural sculpture at the Jeejeebhoy School of Art. His mother was a sister-in-law of the painter Edward Burne-Jones, and Kipling's earliest writings at the age of thirteen were influenced by the Pre-Raphaelites, a group he later regarded as degenerate. At the age of six, he was taken to England and spent several unhappy years in the care of foster parents in what he called 'The House of Desolation' in Southsea. His stepmother regarded him as spoilt and backward and treated him accordingly, and Kipling later recalled the experience in his story 'Baa, Baa, Black Sheep'. He attended a local school before going to United Services College, a new and minor public school, where he was protected by the headmaster, a friend of his father, who tolerated his arrogance and excused him from games, at which he was hopelessly inept, partly because his eyesight was poor. He was to recall his schoooldays in one of his most popular books, the episodic *Stalky & Co.*

Kipling was a mediocre student and did not attempt university entrance. From 1882, he spent seven years as a journalist in India, and produced a volume of verse, *Departmental Ditties*, and one of short stories, *Plain Tales from the Hills*. These contained shrewd observations of society on various levels, but were characterised by an inflexible belief in British imperialism. He found that his fame had preceded him when he returned to England (via America) in 1890, but that same year his novel *The Light That Failed* (published in different versions in England and the USA) proved a failure. It revealed a profound misogyny (it was Kipling who famously observed in a 1919 poem that: 'the female of the species is more deadly than the male') and contained a veiled attack upon his mother, whom he had never forgiven for neglecting him as a child.

In 1892, Kipling married Carrie Balestier, the sister of his friend (and literary collaborator on the novel *The Naulahka*) Wolcott Balestier, an American literary agent based in London, who had died a month previously. They had two daughters (the eldest of whom died at the age of seven) and a son, and these became the first audience for his work for children. Although assertive in masculine company, he was dominated by his bullying wife who exhausted him with her hysteria. She disliked the vulgar aspects of her husband's character and encouraged him to disguise details of his own life. Kipling invented a persona acceptable to his public, and in his work developed his ideal man of action, whose character was often moulded by violence.

Volumes of colloquial and vigorous poetry, such as *Barrack-Room Ballads*, show the influence of the music-halls he attended in his youth and celebrate the life of the ranker in the British army. They have served as popular recital pieces ('Gunga Din', 'Mandalay'), although his most anthologised poem is the sententious 'If', which recommends the ideal qualities of manhood. By the beginning of the twentieth century, Kipling was a famous poet, known as 'the soldier's friend' and judged by one enthusiastic journalist to have 'contributed more than anyone perhaps towards the consolidation of the British Empire'. He spent part of the Boer War in South Africa as a journalist for *The Times*, and wrote several stories about the campaign (collected in *Traffics and Discoveries*), all strongly anti-Boer in tone. He returned to England in 1902, settling at Bateman's, a seventeenth-century house near Burwash, Sussex, but lost touch with the national political mood (casting doubt, for example, on the true intentions of an electorate which had voted the Liberals to power in a landslide victory in 1906), and this alienated many of his readers. In 1907, however, the year he made what amounted to a state visit to Canada, he became the first Englishman to win the Nobel Prize for Literature. Deeply lonely and isolated following the death of his son in the First World War, Kipling ranted against both the Germans and the Russians, and regarded his cousin, the Conservative Prime Minister Stanley Baldwin, as left-wing. News of his death was eclipsed by the demise of the King-Emperor, George V, but although his reputation was in decline, he was buried in Westminster Abbey.

His work was gathered shortly after his death in a thirty-five-volume Sussex Edition (1937–9), which included previously uncollected poetry and prose. A posthumous autobiography was selectively and appropriately titled *Something of Myself*. Later efforts to produce an objective biography were thwarted by his wife and daughter, the latter heavily censoring Lord Birkenhead's 'official' study. In fact, Kipling was a much more complicated man than his popular image as the imperial laureate suggests, and a major work of sympathetic (though far from uncritical) revaluation, **Angus Wilson**'s *The Strange Ride of Rudyard Kipling*, appeared in 1977.

Perhaps his most popular works remain *The Jungle Book* and *Just So Stories*, both of which were written for children. *Puck of Pook's Hill*, a collection of stories in which two children are taken back through history by Shakespeare's sprite, has lasted less well. Of his work for adults, the novel about a boy spy, *Kim*, is the most highly regarded, with its vivid evocation of Indian life and landscape. It is judged by many Indians as one of the best books ever written about their country. Some of his finest work was done in short stories, particularly the ones set in India, such as *Plain Tales from the Hills*. Of the later stories, perhaps the best known is 'Mary Postgate' (collected in *A Diversity of Creatures*), in which a lady's companion during the First World War refuses to give succour to a dying German airman. This shares with *Stalky & Co.* a worrying sadism, and many of Kipling's attitudes and opinions are deemed 'unacceptable' today. He remains a controversial figure, but one whose reputation has survived considerable denigration and seems likely to endure.

Fiction

The Phantom Rickshaw (s) 1888; *Plain Tales from the Hills* (s) 1888; *In Black and White* (s) 1890; *The Light That Failed* 1890 [rev. 1890, 1898]; *Soldiers Three* (s) 1890; *The Story of the Gadsbys* (s) 1890; *Under the Deodars* (s) 1890; *Wee Willie Winkie* (s) 1890; *Life's Handicap* (s) 1891; *The Naulahka* [with W. Balestier] 1892; *Many Inventions* (s) 1893; *Captains Courageous* 1897; *The Day's Work* (s) 1898; *Kim* 1901; *Traffics and Discoveries* (s) 1904; *Actions and Reactions* (s) 1909; *A Diversity of Creatures* (s) 1917; *Debits and Credits* (s) 1926; *Limits and Renewals* (s) 1932

For Children

The Jungle Book 1894; *The Second Jungle Book* 1895; *Stalky & Co.* 1899; *Just So Stories* (s) 1902; *Puck of Pook's Hill* (s) 1906; *Rewards and Fairies* 1910

Poetry

Schoolboy Lyrics 1881; *Echoes* [with A. Kipling] 1884; *Departmental Ditties and Other Verses* 1886; *Barrack-Room Ballads and Other Verse* 1892; *The Seven Seas* 1896; *The Five Nations* 1903; *Collected Verse* 1912; *Songs from Books* 1912; *Inclusive Edition* 1919; *The Years Between* 1919; *Definitive Edition* 1940; *A Choice of Kipling's Verse* [ed. T.S. Eliot] 1941; *Complete Barrack Room Ballads of Rudyard Kipling* [ed. C. Carrington] 1973

Non-Fiction

A Fleet in Being 1898; *From Sea to Sea* 1900; *Letters of Travel* 1920; *The Irish Guards in the Great War* 2 vols 1923; *Something of Myself* (a) 1937; *O Beloved Kids: Rudyard Kipling's Letters to His Children* [ed. E.L. Gilbert] 1983

Collections

Quartette [with A., A. and J. Kipling] 1885

Biography

Rudyard Kipling: His Life and Work by Charles Carrington 1955 [rev. 1970]; *The Strange Ride of Rudyard Kipling* by Angus Wilson 1977

James (Falconer) KIRKUP 1918–

The son of a carpenter, Kirkup grew up at odds with his working-class environment in South Shields, County Durham. He wanted to be a dancer and then a poet, and was a solitary figure at South Shields Secondary School. He went on to Durham University, then, refusing military service, spent the Second World War as a lumberjack and farm labourer. His poems first appeared in Tambimuttu's *Poetry (London)* in 1940 and caught the eye of **William Plomer**, who suggested to his friend **J.R. Ackerley**, who was then literary editor of *The Listener*, that he ask Kirkup to send him some poems. After refusing several, Ackerley eventually accepted 'Mortally', which was published in the BBC magazine in February 1943. Thereafter, Kirkup became a regular – and frequently controversial – contributor to *The Listener*. (His fine poem describing open-heart surgery, 'A Correct Compassion', was considered 'not nice' by the BBC hierarchy, but was published anyway; however, 'The Convenience', a poem about a men's public lavatory, had to be spiked after protests by female members of the magazine's staff.) His first volume of poems, *The Drowned Sailor*, appeared in 1947.

Kirkup has spent his life largely as a visiting professor in many countries. He held an Atlantic Award for literature in 1950 and was the first Eric Gregory Poetry Fellow at Leeds University (1950–2) before becoming a lecturer for various periods in Sweden, Spain and Malaya. For many years he lived mainly in Japan as a lecturer at several universities, most notably being professor of English literature at Kyoto University. In 1961 he was appointed literary editor of *Orient/West*, a bi-monthly magazine devoted to Far Eastern topics. He retired to Andorra, where he still lives.

He remains best known as a poet, having published many volumes in which his fluent and pleasing gift is put to use, but his total output as a writer is enormous and includes a novel, *The Love of Others*, several volumes of autobiography, travel books (most notably *Tropic Temper*, about Malaya), numerous translations, several plays and operas, and works with such titles as *The Bad Boy's Bedside Book of Do-It-Yourself Sex*. A crowned ollave of the Order of Bards, Ovates and Druids, who gave his recreation in *Who's Who* first as 'standing in shafts of sunlight' and then as 'standing in shafts of moonlight', Kirkup is an inspired fantasist and a professed teaser of the serious-minded. His later volumes of autobiography contain passages of sustained silliness and are entirely and frankly without modesty. His most famous poem – though very far from his best – is 'The Love That Dares to Speak Its Name', which was published in *Gay News* in 1976 and was the subject of the first prosecution for blasphemous libel in fifty years. This had been brought by the veteran 'moral crusader',

Mary Whitehouse, who objected to a poem in which a centurion enjoyed sexual fantasies about the crucified Christ. Whitehouse won the case, which was tried at the Old Bailey in 1977, and the editor of the magazine was fined and given a suspended prison sentence. The poem cannot be reprinted in the UK and is represented in *The Penguin Book of Homosexual Verse* (1983) by a note to this effect.

Poetry

Indications [with J. Bayliss, J.O. Thomas] 1942; *The Cosmic Shape* [with R. Nichols] 1946; *The Drowned Sailor and Other Poems* 1947; *The Submerged Village and Other Poems* 1951; *The Creation* 1951; *A Correct Compassion and Other Poems* 1952; *A Spring Journey and Other Poems of 1952–1953* 1954; *The Descent into the Cave and Other Poems* 1957; *The Prodigal Son: Poems 1956–1959* 1959; *The Refusal to Conform* 1963; *Japan Marine* 1965; *Paper Windows* 1968; *Japan Physical* [ed. F. Miura] 1969; *White Shadows, Black Shadows* 1970; *The Body Servant* 1971; *Broad Daylight* 1971; *A Bewick Bestiary* 1971; *Transmental Vibrations* 1971; *Zen Garden* 1973; *Many-Lined Poem* 1973; *Enlightenment* 1978; *Scenes from Sesshu* 1978; *Prick Prints* 1978; *Steps to the Temple* 1979; *Zen Contemplations* 1979; *The Tao of Water* [with B. Skiold] 1980; *Cold Mountain Poems* 1980; *Dengonban Messages* 1981; *Ecce Homo – My Pasolini* 1981; *No More Hiroshimas* 1982; *To the Ancestral North* 1983; *The Sense of the Visit* 1984; *So Long Desired* [with J. McRae] 1986; *Three Poems* 1988; *First Fireworks* 1992; *Shooting Stars* 1992; *Throwback* 1992; *Words for Contemplation* 1993; *Blue Bamboo* 1994

Plays

The Peach Garden (**tv**) 1954; *Two Pigeons Flying High* (**tv**) 1955; *Upon This Rock* 1955; *Masque* 1955 [np]; *Peer Gynt* [from Ibsen's play] 1973; *The Magic Drum* 1974 [np]; *An Actor's Revenge* (**l**) [with M. Miki] 1978 [pub. 1979]; *Ghost Mother* (**r**) 1978; *Friends in Arms* 1980 [np]; *The Damask Drum* [with P. Heininen] 1984

Translations

Paul Christian *The History and Practice of Magic* [with J. Shaw] 2 vols 1952; Todja Tartschoff *The Vision and Other Poems* [with L. Sirombo] 1953; Camara Laye: *The Dark Child* 1955, *The Radiance of the King* 1956, *A Dream of Africa* 1968, *The Guardian of the Word* 1980; Pierre Boileau and Thomas Narcejac *The Sleeping Beauty* 1959; Doan-Vinh-Thal *Ancestral Voices* 1956; Hertha Von Gebhardt *The Girl from Nowhere* 1956; Simone de Beauvoir *Memories of a Dutiful Daughter* 1959; Kleist *The Prince of Homburg* [pub. in *Classic Theatre 2* ed. E. Bentley 1959]; Henry Klier *A Summer Gone* 1959; Jeanne Loisy *Don Tiburcio's Secret* 1959; Schiller *Don Carlos* [pub. in *Classic Theatre 2* ed. E. Bentley 1959]; Arnoul and Simon Grelan: *The True Mistery of the Nativity* (**tv**) 1960 [pub. 1956], *The True Mistery of the Passion* (**r**), (**tv**) 1961 [pub. 1962]; Margot Benary *Dangerous Spring* 1961; S. Martin-Chauffier *The Other One* 1961; Ernst Von Salomon *The Captive* 1961; Herbert Wendt *It Began in Babel* 1961; Jerzy Andrzejewski *The Gates of Paradise* 1962; Fritz Brunner *Trouble in Brusada* 1962; Christian Giessler *Sins of the Fathers* 1962; Gine Victor Leclercq: *Fast as the Wind* 1962, *My Friend Carlo* 1963; L.N. Navolle *Nuno* 1962; Friedrich Dürrenmatt: *The Physicists* 1963 [pub. 1964], *The Meteor* 1966 [pub. 1974], *Play Strindberg* 1971 [pub. nd], *The Anabaptists, Period of Grace* 1976 [np], *Frank the Fifth* 1976 [np]; *Modern Malay Verse* [with A. Majid, O. Rice] 1963; Jacques Heurgon *Daily Life of the Etruscans* 1964; James Kruss *My Great-Grandfather and I* 1964; Jean Robiquet *Daily Life in the French Revolution* 1964; Theodor Storm *Immensee* 1965; May D'Alençon *Red Renard* 1966; Erich Kästner *The Little Man* 1966; *Tales of Hoffmann* 1966; Heinrich von Kleist *Michael Kohlhaas: From an Old Chronicle* 1967; Paul Valéry *The Eternal Virgin* 1970; Ibsen *Brand* [pub. in *The Oxford Ibsen* ed. J.W. McFarlane] 1972; *Selected Poems of Takagi Kyozo* [with M. Nakano] 1973; Rostand *Cyrano de Bergerac* 1975 [np]; *Modern Japanese Poetry* 1978; Tete-Michel Kpomassie *An African in Greenland* 1983; *Miniature Masterpieces of Kawabata Yasunari* 1983; Jean-Noël Pancrazi *Vagabond Winter* 1992; Pascal Quignard *All the World's Mornings* 1992; Patrick Drevet *My Micheline* 1993; Hervé Guibert *Compassion Protocol* 1993, *Man in the Red Hat* 1993; Georges Arthur Goldschmidt *Worlds of Difference* 1993; Tahar Ben Jelloun *State of Absence* 1994; Marcelle Lagesse *Isabelle* 1995

Fiction

The Love of Others 1962; *The Bad Boy's Bedside Book of Do-It-Yourself Sex* 1978; *Gaijin on the Ginza* 1991; *Queens Have Died Young and Fair* 1993

Non-Fiction

The Only Child: an Autobiography of Infancy (**a**) 1957; *Sorrows, Passions and Alarms: an Autobiography of Childhood* (**a**) 1959; *These Horned Islands: a Journal of Japan* 1962; *Tropic Temper* 1963; *England Now* 1964; *Japan Now* 1966; *Frankly Speaking* 1966; *Tokyo* 1966; *Filipinescas: Travels through the Philippine Islands* 1966; *Bangkok* 1968; *One Man's Russia* 1968; *Aspects of the Short Story* (**s**) 1969; *Streets of Asia* 1969; *Hong Kong and Macao* 1970; *Japan Behind the Fan* 1970; *Nihon Bungaku Eiyaku no Yuga na Gijutsu* 1973; *Heaven, Hell, and Hara-Kiri: the Rise and Fall of the Japanese Superstate* 1974; *America Yesterday and Today* 1977;

Mother Goose's Britain 1977; *The Britishness of the British* 1978; *Eibungaku Saiken* 1980; *Scenes from Sutcliffe: Twelve Meditations upon Photographs by Frank Meadow Sutcliffe* 1981; *Folktales Japanesque* 1982; *I Am Count Dracula* 1982; *I Am Frankenstein's Monster* 1983; *When I Was a Child: a Study of Nursery Rhymes* 1983; *I, of All People: an Autobiography of Youth* (a) 1988; *A Poet Could Not But Be Gay* (a) 1991; *Me All Over* (a) 1993

For Children

Insect Summer: an Introduction to Haiku Poetry 1971; *Look at It This Way! Poems for Young People* 1994

Lincoln KIRSTEIN 1907–

Kirstein was born in Rochester, New York, the son of a spectacles merchant who became extremely wealthy after going into partnership with the owner of a Boston department store. 'Socially, my parents sprang from liberal, un-Orthodox Jews of northeastern Teutonic origin, intellectually and morally, from sources in Protestant Reformation, the Enlightenment, and the revolutions of 1848,' Kirstein has written. He was brought up in Boston in an atmosphere of cultured anglophilia, educated at Phillips Exeter Academy and the Berkshire School, and spent his holidays in the galleries and museums of Europe. He was briefly apprenticed to a stained-glass manufacturer before, from 1926 to 1930, continuing his education at Harvard, where he founded and edited the quarterly 'Miscellany', *Hound & Horn*, decorated the Liberal Club with a modernist mural, and mounted exhibitions of the works of leading European painters such as Matisse and Picasso.

His chief passion, however, was the ballet, and he studied dance with Fokine, although he never performed professionally. Instead, he promoted ballet as an art, founding with George Balanchine the School of American Ballet in 1934. He went on to found several other companies, which culminated in 1947 with the New York City Ballet, from which he retired at the age of eighty-two, taking the title of General Director Emeritus. He founded the magazine *Dance Index* in 1942 and wrote several books on ballet – including a collaboration with Romola Nijinsky on a biography of her husband – and on other arts, notably photography. He was also closely involved with New York's Museum of Modern Art.

He married Fidelma Cadmus, the sister of his friend the painter Paul Cadmus, in 1941 (they are the dedicatees of **W.H. Auden**'s *The Shield of Achilles*, 1955) and was drafted into the army as a private first class in 1943. He spent part of the war in Europe cataloguing works of art lost in action or by theft. His volume of poems, *Rhymes of a Pfc.*, was one of the most distinguished American books to emerge from the war. 'I cannot believe', wrote Auden, 'that any poet, no matter how accomplished, will read these poems without admiration and envy.' Often cast in the form of dramatic monologues, the poems are vital and colloquial. 'I never aimed "to be a poet",' Kirstein has commented. 'I liked to write verse; this was always play without pretension.' His acknowledged masters and models range from Gerard Manley Hopkins and Heinrich Heine, through **Rudyard Kipling** and Cavafy, to **John Betjeman** and **Gavin Ewart**. Two further volumes and a collected edition followed.

Kirstein has also written two novels, the second of which, *For My Brother*, was published by **John Lehmann** on the recommendation of **Christopher Isherwood**, whom Kirstein had befriended when the British novelist came to New York. Almost the entire stock of the book was destroyed in the London Blitz. *By With To & From* is a useful sampler of Kirstein's work and demonstrates his considerable range, while *Mosaic* is an autobiography of his youth. He now lives in New York and Connecticut.

Non-Fiction

Fokine 1934; *Dance: a Short History of Classical Theatrical Dancing* 1935; *Blast at Ballet: a Corrective for the American Audience* 1938; *Ballet Alphabet: a Primer for Laymen* 1939; *The Classic Ballet: Basic Techniques and Terminology* [with M. Stuart] 1952; *What Is Ballet All About?* 1959; *For Fania: December 23 1964* 1964; *Movement and Metaphor: Four Centuries of Ballet* 1970; *For John Martin: Entries from an Early Diary* 1973; *The New York City Ballet* 1973 [rev. 1978]; *Coppelia: The New York City Ballet* [with N. Goldner] 1974; *Nijinsky Dancing* 1975; *Union Jack: the New York City Ballet* 1977; *Ballet: Bias and Belief* 1982; *Paul Cadmus* 1984; *Quarry: a Collection in Lieu of Memoirs* 1986; *Memorial to a Marriage* 1989; *Mosaic* (a) 1994

Fiction

Flesh Is Heir 1932; *For My Brother* [with J.M. Berlanga] 1943; *White House Happening* 1966

Poetry

Notre Dame des cadres 1933; *Low Ceiling* 1935; *Rhymes of a Pfc.* 1964 [rev. 1966]; *The Poems of Lincoln Kirstein* 1987

Edited

Pavel Tchelitchew 1947

Collections

By With To & From 1991

C(lifford) H(enry) B(enn) KITCHIN 1895–1967

Born in Harrogate, Kitchin was educated at Clifton and Exeter College, Oxford, to which he won a classical scholarship. He served in France with the Durham Light Infantry during the First World War, was called to the Bar in 1924, and was for some years a member of the London stock exchange until, released from having to earn a living as a result of shrewdly investing inherited money, he became a full-time writer.

His first two novels, *Streamers Waving* and *Mr Balcony*, published for him by **Leonard** and **Virginia Woolf** at the Hogarth Press, were experimental, the first remarkable for its ice-cold wit, the second described as 'the most enigmatic of his novels'. He seemed to be anticipating **Patrick White**, whom he came to admire greatly. In *The Auction Sale* he put to use his considerable knowledge as a collector of china and furniture. If *The Secret River* is the most accessible of his novels, *The Book of Life* is generally considered the best. It is set in 1909–10, largely in Whitgate-on-Sea, Kent. The first-person narrator is a nine-year-old boy, and 'The Book of Life' a ledger in which his paternal grandfather records those parts of his estate which are to be allocated to different members of the family. It seems that a good deal of money is coming to the boy and that he is likely to find himself a baronet which, being a great snob, he anticipates with relish; but of course fate takes a hand. *The Book of Life* was a critical success, but sales were poor and he never wrote another novel. His humour and lyricism resembled those of his lifelong friend **L.P. Hartley**, but the sardonic quality in his writing probably prevented any of his books from meeting with the success of *The Shrimp and the Anemone* (1944) and *The Go-Between* (1953). Kitchin also wrote detective stories, of which the most successful was *Death of My Aunt* – he always referred to it as 'that wretched book' – and some short stories, collected as *Jumping Joan*.

Hartley described Kitchin as 'the most talented man I have ever known'; as well as a novelist and investor, he was a first-class bridge and chess player and an excellent pianist. **Dorothy L. Sayers** praised 'his shrewd psychology and excellence of style'. He never married, and died in Brighton where he had lived for many years.

Fiction

Streamers Waving 1925; *Mr Balcony* 1927; *Death of My Aunt* 1929; *The Sensitive One* 1931; *Crime at Christmas* 1934; *Olive E.* 1937; *Birthday Party* 1938; *Death of His Uncle* 1939; *The Auction Sale* 1949; *The Cornish Fox* 1949; *Jumping Joan* (s) 1954; *The Secret River* 1956; *Ten Pollit Place* 1957; *The Book of Life* 1960

John KNOWLES 1926–

Knowles was born in Fairmont, West Virginia, and educated at Yale, where he gained a BA in 1949. He worked as a reporter, editor and travel writer until the reception of his first novel enabled him to write fiction full-time.

A Separate Peace was published in 1959, and won a Rosenthal Award from the National Institute of Arts and Letters and a William Faulkner Award. It is about life in a New England prep school (in the American rather than English sense) and it centres on the intense and ambivalent relationship between two adolescents, Gene Forrester and his friend Finney. Gene's love–hate attitude to Finney causes Finney to have a serious accident and is later linked to his death, but by then the two are reconciled. The emotional focus is on the experience of Gene and his transition from adolescence to manhood, which is tempered by the period of impending war (1942) in which the book is set. The novel perfectly captures the pseudo-sophistication and private languages of adolescence, as well as the almost romantic quality of adolescent friendship. It has proved to have considerable appeal for teenage readers, and it also won an award from the National Association of Independent Schools in 1961.

It seems clear that the peculiar strengths of *A Separate Peace* are related to the weaknesses of Knowles's subsequent novels. None of them have had such popular success and critics have generally found them unsatisfying. The characters have been judged hollow and implausible and there are – perhaps significantly – no memorable women characters, while the theme of male friendship becomes mechanical. Above all it seems that Knowles's heart is not in his other settings (such as Gatsbyesque Yale in *The Paragon*, homosexual France in *Spreading Fires*, or coal-boom West Virginia in *Vein of Riches*) as it is in prep-school life. As if in acknowledgement of this, his seventh novel, *Peace Breaks Out*, returns to the same school of *A Separate Peace* and centres on a death from bullying. Prep-school life is clearly Knowles's forte and one can only speculate at an autobiographical element: the protagonist of *A Separate Peace* is of Knowles's generation and the name of the fictional school – Devon School, New Hampshire – is linked to Knowles's own school: Phillips Exeter Academy,

Exeter, New Hampshire. This ludic nod at English regional geography may have been less obvious to Knowles's American readership.

Knowles's declared interest in rootlessness underlies his prowess as a travel writer and, more obliquely, his empathy with adolescence. He is a fine writer and a consummate stylist, but he seems set to be remembered chiefly for his first book.

Fiction

A Separate Peace 1959; *Morning in Antibes* 1962; *Indian Summer* 1966; *Phineas* (s) 1968; *The Paragon* 1971; *Spreading Fires* 1974; *Vein of Riches* 1978; *Peace Breaks Out* 1981; *A Stolen Past* 1983; *The Private Life of Axie Reed* 1986

Non-Fiction

Double Vision: American Thoughts Abroad 1964

Ronald A(rbuthnott) KNOX 1888–1957

One of four sons of an Anglican bishop of Manchester, and grandson on his mother's side of a bishop of Lahore, Knox was educated at Eton, becoming well known while still at school for the publication of *Signa Severa* (1906), a collection of English, Greek and Latin verses. He was regarded as the most brilliant of Edwardian Etonians, and was the founder of *The Outsider*, an Etonian magazine of which six numbers appeared during the summer of 1906. He was largely responsible for its contents, and its other editors included the best-known Etonians of the day, most of whom died in the First World War: Patrick Shaw-Stewart, Edward Horner, Julian Grenfell and others. Knox also edited the less ephemeral publication *The Eton Chronicle*. His poem 'The Wilderness', a comical plea for the planting of a garden in School Yard, was included in the *Eton Poetry Book*, and learned for reciting by many generations of Etonians.

He took a first in *literae humaniores* at Balliol College, Oxford, in 1910, and was immediately made a fellow of Trinity, becoming its chaplain in 1912 on taking Anglican orders. During the First World War he taught at Shrewsbury School and worked at the War Office. He was a leading controversialist on behalf of the Anglo-Catholic party in the Church of England, but in 1917 was converted to Roman Catholicism, resigning his fellowship of Trinity and teaching for some years, on the instructions of Cardinal Bourne, at St Edmund's School, Ware. In 1926 he became chaplain to the Catholic undergraduates at Oxford, a post in which he blossomed as a priest with a special mission to the well-connected, known to successive generations as 'Ronnie' and always seen pulling on his pipe.

He poured forth an immense stream of books and articles, including many works of Catholic apologetics and, in the manner of an earlier day, some volumes of collected sermons. His chief talents, however, were probably as a *farceur* and parodist, expressed in his six detective stories (featuring the sleuth Miles Bredon), his skits on Trollope and the German higher criticism, and perhaps most enduringly in *Let Dons Delight*, a clever work in which six imaginary conversations in an Oxford common room over the centuries are used to demonstrate the breakdown of a common culture.

In 1939 he resigned the Catholic chaplaincy and spent the rest of his life as private chaplain in aristocratic houses, first to Lord and Lady Acton at Aldenham and then to Katharine Asquith at Mells. He devoted his later life mainly to his translation of the Bible, which immense labour has not proved enduring. He was a lifelong celibate: in youth he cherished a number of romantic friendships, including one for Harold Macmillan (who is identified simply as 'C' in his book *A Spiritual Aeneid*), and in later years he was devoted to Lady Acton. His many distinctions ranged from a gamut of Oxfordian prizes to being created Monsignor in 1936 and a protonotary apostolic in 1951. His biography was published by his friend and literary executor **Evelyn Waugh**, and there is also *The Knox Brothers* (1977) by the novelist **Penelope Fitzgerald**, daughter of his brother E.V. Knox, who was editor of *Punch*.

Fiction

A Still More Sporting Adventure! [from E.F. Jourdain and C.A.E. Moberley's *An Adventure*; with C.R.L. Fletcher] 1911; *Studies in the Literature of Sherlock Holmes* 1912; *Reunion All Round* 1914; *Memories of the Future* 1923; *Sanctions* 1924; *The Viaduct Murder* 1925; *Other Eyes Than Ours* 1926; *The Three Taps* 1927; *Essays in Satire* 1928; *The Footsteps at the Lock* 1928; *The Rich Young Man* 1928; *Caliban in Grub Street* 1930; *The Floating Admiral* [with others] 1931; *The Body in the Silo* [US *Settled Out of Court*] 1934; *Still Dead* 1934; *Barchester Pilgrimage* 1935; *Six against the Yard* [with others; US *Six against Scotland Yard*] 1936; *Double Cross Purposes* 1937; *Let Dons Delight* 1939; *The Scoop and Behind the Scenes* [with others] 1983

Non-Fiction

Naboth's Vineyard in Pawn 1913; *Some Loose Stones: Being Consideration of Certain Tendencies in Modern Theology* 1913; *Authority in the Church* 1914; *The Church in Bondage* 1914; *An Hour at the Front* 1914 [rev. as *Ten Minutes at the Front* 1916]; *Bread or Stone: Four Conferences on Imperative Prayer* 1915; *The Essentials of Spiritual Unity* 1918; *A Spiritual*

Aeneid (a) 1918; *Meditations on the Psalms* 1919; *Patrick Shaw-Stewart* 1920; *A Book of Acrostics* 1924; *An Open-Air Pulpit* 1926; *The Belief of Catholics* 1927; *Anglican Cobwebs* 1928; *Miracles* 1928; *The Mystery of the Kingdom and Other Sermons* 1928; *The Church on Earth* 1929; *On Getting There* 1929; *Broadcast Minds* 1932; *Difficulties: Being a Correspondence about the Catholic Religion between Ronald Knox and Arnold Lunn* 1932; *Heaven and Charing Cross: Sermons on the Holy Eucharist* 1935; *Captive Flames: a Collection of Panegyrics* 1940; *Nazi and Nazarene* 1940; *In Soft Garments: a Collection of Oxford Conferences* 1942; *I Believe: the Religion of the Apostles' Creed* 1944; *God and the Atom* 1945; *A Retreat for Priests* 1946; *The Mass in Slow Motion* 1948; *On Englishing the Bible* 1949; *A Selection from the Occasional Sermons* [ed. E. Waugh] 1949; *Enthusiasm: a Chapter in the History of Religion* 1950; *The Gospel in Slow Motion* 1950; *St Paul's Gospel* 1950; *Stimuli* 1951; *A New Testament Commentary for English Readers* 3 vols 1952–6; *The Hidden Stream: a Further Collection of Oxford Conferences* 1953; *Off the Record* 1954; *A Retreat for Lay People* 1955; *The Window in the Wall and Other Sermons on the Holy Eucharist* 1956; *Bridegroom and Bride* 1957; *On English Translation* 1957; *Literary Distractions* 1958; *The Priestly Life: a Retreat* 1958; *Lightning Meditations* 1959; *Proving God: a New Apologetic* 1959; *Occasional Sermons, The Pastoral Sermons, University and Anglican Sermons* [ed. P. Caraman] 3 vols 1960–3; *Retreat for Beginners* 1960 [UK *Retreat in Slow Motion* 1961]; *The Layman and His Conscience: a Retreat* 1961

Edited

The Miracles of King Henry VI [with S. Leslie] 1923; *The Best Detective Stories of the Year 1928* (s) [with H. Harrington; US *The Best English Detective Stories of 1928*] 1929; *The Best Detective Stories of the Year 1929* (s) [with H. Harrington; US *The Best English Detective Stories of the Year*] 1930; *The Holy Bible: an Abridgement and Rearrangement* 1936; *The Westminster Hymnal* [with others] 1940; *The Manual of Prayers* [with others] 1942; G.K. Chesterton *Father Brown: Selected Stories* (s) 1955

Translations

Virgil *Aeneid* 1924; *The Holy Gospel of Jesus Christ According to Matthew* 1941; *The New Testament of Lord and Saviour Jesus Christ* 1944; *The Epistles and Gospels for Sundays and Holidays* 1946; *The Book of Psalms in Latin and English, with the Canticles Used in the Divine Office* 1947; *The Old Testament* 2 vols 1948–50; *The Missal in English* [with H.P.R. Finberg, J. O'Connell] 1949; *Encyclical Letter – Humani Genesis – of His Holiness Pius XII* 1950; *Holy Week* 1951; *The Holy Bible* 1955; St Thérèse of Lisieux *Autobiography of a Saint* 1958; Thomas à Kempis *The Imitation of Christ* [with M. Oakley] 1959

Poetry

Signa Severa 1906; *Remigium Alarum* 1910; *Q. Horati Flacci Carminum Liber Quintus a Rudyardo Kipling et Carolo Graves Anglice Redditus* [from Horace's *Odes*; with others; ed. A.D. Godley] 1920; *In Three Tongues* [ed. L. Eyres] 1959

Collections

Juxta Salices 1910

Biography

The Life of the Right Reverend Ronald Knox by Evelyn Waugh 1959; *The Knox Brothers* by Penelope Fitzgerald 1977

Kenneth (Jay) KOCH 1925–

Koch was born in Cincinnati, Ohio, and after high school served from 1943 to 1946 as a rifleman with the US Army in the Pacific. At Harvard he studied writing under the tuition of **Delmore Schwartz** and met **John Ashbery** who has been a lifelong friend. He graduated in 1948 and undertook graduate studies at Columbia University where he, Ashbery and **Frank O'Hara** formed what came to be known as the New York School of poets. The group were distinguished by a close kinship with the theories and methods of the abstract expressionist painters then flourishing in the city. Koch taught at Rutgers University College, New Jersey, until 1959 when he received his doctorate for a thesis on 'The Reception and Influence of Modern American Poetry in France, 1918–1950'. Since then he has taught at Columbia, becoming professor of English in 1971, and was director of the Poetry Workshop at the New School for Social Research in New York until 1966. He has been actively concerned with pioneering educational work, and has taught poetry writing both to schoolchildren and to the members of a Manhattan old people's nursing home. Accounts of both ventures can be found in *Wishes, Lies and Dreams* and *I Never Told Anybody*.

Koch's early plays consisted largely of short surrealistic farces and parodies, some of which were staged off-off-Broadway. His first full poetry collection, *Ko*, is a parodistic comic epic with a multiplicity of plots and a range of characters which includes a Japanese baseball player called Ko. Subsequent major collections such as *The Art of Love* and *The Duplications* are also characterised by their verbal inventiveness and the witty celebration of the power of the poetic imagination, and Koch considers himself to have been less successful when attempting directly to address serious 'issues', as he does the

Vietnam War in the title-poem of *The Pleasures of Peace*.

His collections of anarchic short plays include *A Change of Hearts*, which parodies Noh drama, morality plays, opera libretti and other forms, and *One Thousand Avant-Garde Plays*, which in fact contains only 112 verse plays. He was involved with the magazine *Locus Solus* in the early 1960s, and his awards include the American Academy award in 1976. He was married in 1952 (his wife died in 1981) and he has one daughter.

Poetry

Poems 1953; *Ko* 1960; *Permanently* 1961; *Thank You and Other Poems* 1962; *Poems from 1952 and 1953* 1968; *The Pleasures of Peace and Other Poems* 1969; *Sleeping with Women* 1969; *When the Sun Tries to Go On* 1969; *Penguin Modern Poets 24* [with K. Elmslie, J. Schuyler] 1973; *The Art of Love* 1975; *The Duplications* 1977; *The Burning Mystery of Anna in 1951* 1979; *From the Air* 1979; *Days and Night* 1982; *Selected Poems 1950–1982* 1985; *On the Edge* 1986; *Seasons on Earth* 1987; *Selected Poems 1950–1987* 1991; *One Train* 1994; *On the Great Atlantic Railway* 1994

Plays

Bertha 1959 [pub. 1966]; *The Election* 1960 [pub. 1973]; *Pericles* 1960 [pub. 1966]; *The Construction of Boston* 1962 [pub. 1966]; *George Washington Crossing the Delaware* 1962 [pub. 1966]; *Guinevere* 1964 [pub. 1966]; *The Tinguely Machine Mystery* 1965 [pub. 1973]; *Bertha and Other Plays* [pub. 1966]; *The Scotty Dog* **(f)** 1967 [pub. 1973]; *The Apple* **(f)** 1968 [pub. 1973]; *The Gold Standard* 1969 [pub. 1966]; *The Moon Balloon* 1969 [pub. 1973]; *The Artist* [from his poem; with P. Reif] 1972 [np]; *A Little Light* 1972 [np]; *A Change of Hearts* [with D. Hollister] 1985 [pub. 1973]; *A Change of Hearts: Plays, Films, and Other Dramatic Works 1951–1971* [pub. 1973]; *Rooster Redivivius* 1975 [np]; *The Art of Love* [from his poem] 1976 [np]; *The Red Robins* [from his novel] 1978 [pub. 1979]; *The New Diana* 1984 [np]; *Popeye Among the Polar Bears* 1986 [np]; *One Thousand Avant-Garde Plays* [pub. 1988]

Fiction

Interlocking Lives **(s)** [with A. Katz] 1970; *The Red Robins* 1975; *Hotel Lambosa and Other Stories* **(s)** 1993

Non-Fiction

John Ashbery and Kenneth Koch: a Conversation 1965; *Wishes, Lies and Dreams: Teaching Children to Write Poetry* 1970; *Rose, Where Did You Get That Red? Teaching Great Poetry to Children* 1973; *I Never Told Anybody: Teaching Poetry Writing in a Nursing Home* 1977; *Collaborations with Artists* 1993

Edited

Sleeping on the Wing: an Anthology of Modern Poetry, with Essays on Reading and Writing [with K. Farrell] 1981; *Talking to the Sun: an Illustrated Anthology of Poems for Young People* [with K. Farrell] 1985

Arthur KOESTLER 1905–1983

Born in Budapest, Hungary, the son of a promoter and would-be inventor, Koestler spent his childhood between Hungary, Austria and Germany: he is one of the very few writers to have changed his language of writing twice, from Hungarian to German, and then to English. He attended the Polytechnic High School and University in Vienna, but left before completing his degree, in 1925, to work on a kibbutz in Israel. Later, he became a journalist and foreign correspondent, continued this career back in Europe, and became foreign editor and assistant editor-in-chief of the influential Berlin paper *BZ am Mittag* at the early age of twenty-five.

In 1931 he joined the Communist Party, and travelled in the Soviet Union and Central Asia, and lived in Paris, before going to Spain in 1936 at the outbreak of the Civil War as a correspondent and communist propagandist. He managed to get out of Francoist Spain one hour before a warrant was issued for his arrest, but he was later captured by the Nationalists and spent three months in jail under sentence of execution before being exchanged for another prisoner. His early book, *Spanish Testament*, published in 1937, describes these experiences. At the beginning of the Second World War, he suffered another period of imprisonment, this time at Le Vernet in France, but was then able to join the French Foreign Legion and, after reaching England, the British Army. Koestler's later life was spent mainly in England, although he frequently travelled.

His first novel, *The Gladiators* (1939), about the Spartacus slave revolt in ancient Rome, investigated the mechanism of the revolutionary process from the point of view of one who had become disillusioned with communism; but real fame came to Koestler with his second novel, *Darkness at Noon*, about an old Bolshevik being forced to confess to crimes he did not commit during the Moscow show trials. This book is recognised as among the greatest political novels of the century. Koestler wrote six novels, and more than twenty books of other types; his work exists mainly in the border country between literature and journalism. Among his other notable works, which investigate a wide range of scientific and literary topics, were *The Sleep-*

walkers, a history of Renaissance science; *The Yogi and the Commissar*, an influential volume of essays; and his contribution to *The God That Failed*, in which prominent intellectuals described their loss of faith in communism.

A notable philanderer, Koestler was married three times and pursued numerous extra-marital affairs: among other escapades, he was the co-respondent named in Bertrand Russell's divorce from his third wife. None of his wives was the mother of his only child. His first wife, Dorothy Asher, whom he had married in 1935, died in 1950, the year he married Mamaine Paget, from whom he was divorced in 1953. His third wife was Cynthia Paterson, the daughter of a South African surgeon and twenty-two years his junior. She became his secretary in 1948, and he seduced her (as he did all his secretaries). In an attempt to break away from him she married another man, becoming Cynthia Jefferies. The marriage failed, and in 1965 she and Koestler finally married.

Koestler was an outspoken supporter of euthanasia and both he and Cynthia were members of the society EXIT. When Koestler became terminally ill with leukaemia, the couple committed suicide at their London home. Cynthia's death was a matter of some controversy, since she was not ill and some people felt that she had been coerced by Koestler to die with him. Koestler left his fortune to found a university chair in parapsychology, which has been set up in Edinburgh. He is also commemorated by the 'Arthur Koestler Annual Awards for Arts from United Kingdom Prisons and Special Hospitals'.

Fiction

The Gladiators 1939; *Darkness at Noon* 1940; *Arrival and Departure* 1943; *Thieves in the Night* 1946; *The Age of Longing* 1951; *The Call-Girls* 1972

Non-Fiction

Spanish Testament (a) [rev. as *Dialogue with Death*] 1937; *Scum of the Earth* (a) 1941 [rev. 1955]; *The Yogi and the Commissar and Other Essays* 1945; *The Challenge of Our Time* [with others] 1948; *The God That Failed* (c) [ed. R. Crossman] 1949; *Insight and Outlook: an Inquiry into the Common Foundations of Science, Art, and Social Ethics* 1949; *Promise of Fulfillment: Palestine 1917–1949* 1949; *Arrow in the Blue* (a) 1952; *The Invisible Writing* (a) 1954; *The Trial of the Dinosaur and Other Essays* 1955; *Reflections on Hanging* 1956; *The Sleepwalkers* 1959; *The Lotus and the Robot* 1960; *Hanged by the Neck: an Exposure of Capital Punishment in England* [with C.H. Rolph] 1961; *The Act of Creation* 1964; *The Ghost in the Machine* 1967; *Drinkers of Infinity: Essays 1955–1967* 1968; *The Case of the Midwife Toad* 1971; *The Roots of Coincidence* 1972; *Challenge of Chance* [with A. Hardy, R. Harvie] 1973; *The Heel of Achilles: Essays 1968–1973* 1974; *Life After Death* [with others] 1976; *The Thirteenth Tribe: the Khazar Empire and Its Heritage* 1976; *Janus: a Summing Up* 1978; *Strangers on the Square* (a) [with C. Koestler] 1984

Plays

Twilight Bar 1945

Edited

Suicide of a Nation? An Enquiry into the State of Britain Today 1963; *Beyond Reductionism: the Alpbach Symposium* [with C.H. Rolph] 1969

Collections

Bricks to Babel: Selected Writings 1980

Arthur (Lee) KOPIT 1937–

Kopit was born in the prosperous New York suburb of Lawrence, Long Island. His father worked in the jewellery trade before becoming a real-estate consultant. His mother has worked for her son, as his business manager. Kopit graduated from Lawrence High School in 1955 and won a scholarship to study electrical engineering at Harvard where he began writing one-act plays, the first of which, *The Questioning of Nick*, was televised in 1959 whilst Kopit was still a student. He graduated with honours shortly afterwards, having already attracted the notice of off-Broadway producers on the strength of his college productions, and won a travelling fellowship.

He left for Europe and wrote the grandly titled *Oh Dad, Poor Dad, Mamma's Hung You in the Closet and I'm Feelin' So Sad: a Pseudoclassical Tragifarce in a Bastard French Tradition*. The play was given short runs in Massachusetts and London before successfully opening off-Broadway in 1962. It was filmed in 1967. Kopit's next opening was not so successful. Awarded a $70,000 Rockefeller Grant, he had two one-act plays due to be produced in Minneapolis when he withdrew co-operation, accusing the university authorities of censorship. The plays eventually opened off-Broadway in 1965, but to mixed reviews.

Kopit married, withdrew to Vermont, and wrote *Indians*. A Brechtian essay on the violence of American history, the play uses the life story of Buffalo Bill Cody, as mythologised in the famous travelling Wild West Show, to invite a comparison with Vietnam. Because of the large cast needed, Kopit opened the play in London, where it was well received, moving to Washington and then to New York in October 1969. The play closed in January 1970, a victim of its own production costs, before reappearing in 1976 as the inspiration behind Robert Altman's film *Buffalo Bill and the Indians, or Sitting Bull's History Lesson.*

His next theatrical success came in 1978, with the stage production of *Wings*. Originally conceived for and first broadcast on radio, the play describes the slow emergence from aphasia of an aviatrix confined to a hospital bed by stroke-induced paralysis. Harking back to the institution-based background of *Chamber Music*, the play owes much to the intensity of Kopit's personal commitment. His own father had suffered a stroke in 1976 and, as he explains in his preface, his intention in writing had been to find out 'to what extent he was aware of what had befallen him? What was it like inside?'. *Nine*, a musical inflation of the Fellini film *8½*, with dialogue by Kopit, was premiered in Waterford, Connecticut before opening in New York in 1982. The play was then published with added text. *The End of the World: with Symposium to Follow*, a play about nuclear proliferation, was premiered in 1984.

Plays

Don Juan in Texas [with W. Lawrence] 1957 [np]; *Gemini* 1957 [np]; *On the Runway of Life* 1957 [np]; *The Day the Whores Came Out to Play Tennis and Other Plays* [pub. 1965]: *The Questioning of Nick* 1957, *Sing to Me through Open Windows* 1959 [rev. 1965], *Asylum* [rev. as *Chamber Music* 1971] 1963, *The Conquest of Everest* 1964, *The Day the Whores Came Out to Play Tennis* 1964, *The Hero* 1964; *Across the River and into the Jungle* 1958 [np]; *Aubade* 1958 [np]; *Oh Dad, Poor Dad, Mamma's Hung You in the Closet and I'm Feelin' So Sad* 1960; *Mhil'daim* 1963 [np]; *The Conquest of Television* **(tv)** 1966; *Indians* 1968 [pub. 1969]; *An Incident in the Park* [pub. in *Pardon Me, Sir, But Is My Eye Hurting Your Elbow?* ed. B. Booker, G. Foster 1968]; *Promontory Point Revisited* **(tv)** 1969; *What's Happened to the Thornes' House* 1972 [np]; *Louisiana Territory* 1975 [np]; *Secrets of the Rich* 1976 [pub. 1978]; *Wings* **(r)**, **(tv)** 1977, **(st)** 1978 [pub. 1978]; *Starstruck* **(tv)** 1979; *Nine* [from F. Fellini's film *8½*; with M. Fratti, M. Yeston] 1981 [pub. 1983]; *Good Help Is Hard to Find* 1981 [pub. 1982]; *Ghosts* [from Ibsen's play] 1982 [pub. 1984]; *The End of the World* 1984 [UK *The Assignment* 1985]; *Hands* **(tv)** 1987; *Bone-the-Fish* 1989 [rev. as *The Road to Nirvana* pub. 1991]; *Phantom* [from G. Leroux's *The Phantom of the Opera*; with M. Yeston] 1991 [pub. 1992]

Bernard KOPS 1926–

Kops was born in Stepney Green of Dutch Jewish working-class parents. 'I was born into a Jewish world', he has said, 'and until the age of eight I didn't know anything else. As a kid "up West" was a magical place ... Beyond the Aldgate Pump it was a case of "Here be Dragons".' He was educated at elementary schools to the age of thirteen and afterwards worked as a docker, trainee chef, door-to-door salesman, waiter, lift man, barrow boy, leather worker, bookseller and, for a time in the first years after the war, as an actor with various fit-up theatre companies. 'If I hadn't been a writer I would have been walking around mogadonned under the chemical cosh, probably shouting "Get cultured before the cataclysm!" ' Instead he found **T.S. Eliot**, Sophie Tucker – the subject of one of his plays – and, in the late 1940s, Soho. 'My parents warned me that it was full of thieves and they were right. It was marvellous, I could experiment with drugs and fall madly in love four times a week.' In the second half of the 1950s he published a book of poems, a first novel, *Awake for Mourning*, and a play, *The Hamlet of Stepney Green*, first produced in Oxford, in which Sam, having wasted his own life, returns to haunt his son and persuade him not to waste his. Again and again in his plays Kops is saying: 'Life is all confusion but we must embrace it all'. In *Sophie – The Last of the Red Hot Mamas*, he does so as it were at source, since Tucker was 'childhood icon and sacred monster'. Perhaps the best of his novels, *On Margate Sands*, is about the violence of the mentally disturbed. His poems are informed with the same affirmation of life and its confusions, as is his splendid autobiography, *The World is a Wedding*. He has also written much for radio and television. He married in 1956, and has four children.

Plays

The Hamlet of Stepney Green 1958 [pub. 1959]; *Goodbye World* 1959 [np]; *Change for the Angel* 1960 [np]; *The Dream of Peter Mann* 1960; *Enter Solly Gold* [with S. Myers] 1962 [pub. in *Satan, Socialites, and Solly Gold* 1961]; *Stray Cats and Empty Bottles* **(st)** 1961, **(tv)** 1964 [pub. 1967]; *Home Sweet Honeycomb* **(r)** 1962 and *The Lemmings* **(r)** 1963 [pub. in *Four Plays* 1964]; *Born in Israel* **(r)** 1963; *I Want to Go Home* **(tv)** 1963; *The Dark Ages* **(r)** 1964; *Israel* **(r)** 1964; *The Boy Who Wouldn't Play Jesus* [pub. in *Eight Plays: Book I* ed. M.S. Fellows] 1965; *The Lost Years of Brian Hooper* **(tv)** 1967; *David, It's Getting Dark* 1970; *Alexander the Greatest* **(tv)** 1971; *Just One Kid* **(tv)** 1974; *The Boy Philosopher* and *Why the Geese Shrieked* **(tv)** [from I.B. Singer's stories]

1974; *It's a Lovely Day Tomorrow* **(tv)** [with J. Goldschmidt] **(tv)** 1975, **(st)** 1976; *Moss* **(tv)** 1975, **(st)** 1991; *Rocky Marciano Is Dead* **(tv)** 1976; *Bournemouth Nights* **(r)** 1979; *I Grow Old, I Grow Old* **(r)** 1979; *More Out Than In* 1980 [np]; *Over the Rainbow* **(r)** 1980; *Ezra* 1981 [np]; *Simon at Midnight* **(r)** 1982, **(st)** 1985 [np]; *Night Kids* **(tv)** 1983; *Trotsky Was My Father* **(r)** 1984; *Kathy Kropotkin* **(r)** 1988; *Colourblind* **(r)** 1989; *Congress in Manchester* **(r)** 1990; *Sophie – The Last of the Red Hot Mamas* 1990 [np]; *Playing Sinatra* 1991 [pub. 1992]; *Soho Nights* **(r)** 1991; *Dreams of Anne Frank* [pub. 1993]

Fiction

Awake for Mourning 1958; *Motorbike* 1962; *Yes from No-Man's Land* 1965; *The Dissent of Dominick Shapiro* 1966; *By the Waters of Whitechapel* 1969; *The Passionate Past of Gloria Gaye* 1971; *Settle Down Simon Katz* 1973; *Partners* 1975; *On Margate Sands* 1978

Poetry

Poems 1955; *Poems and Songs* 1958; *An Anemone for Antigone* 1959; *Erica, I Want to Read You Something* 1967; *For the Record* 1971; *Barricades in West Hampstead* 1988

Non-Fiction

The World Is a Wedding **(a)** 1963; *Neither Your Money Nor Your Sting: an Offbeat History of the Jews* 1985

Edited

Poetry Hounslow 1981

Jerzy (Nikodem) KOSINSKI 1933–1991

Kosinski was six years old when Hitler invaded Poland, where he was born. According to his own account, his Jewish parents went into hiding and he was placed in care, although he spent most of his childhood wandering the ravaged countryside of Eastern Europe, alone. When he was reunited with his parents, he was traumatised to the point of being rendered mute, but recovered his voice in a skiing accident in 1942. These experiences, which were used as the basis of his first novel, *The Painted Bird*, have since been disputed by those who claim to have known Kosinski during this period.

He graduated from the University of Lodz, Poland, in 1955 with MAs in history and political science. Studying at the Academy of Sciences, he engineered his escape to the West: four fictitious maestros of academia from diverse Polish institutions came up with a sponsorship for him to conduct a research project in the USA, and in 1957 he arrived in New York, a penniless immigrant.

Within a year, however, he could speak the language and was pursuing his studies in social science at Columbia University. By 1962 he had pseudonymously written two books on collective behaviour under communism, *The Future Is Ours, Comrade* and *No Third Path*, both of which were best-sellers. He had also married Mary Wier, the vastly wealthy young widow of a steel baron, whose legacy included an apartment on Park Avenue, a villa in Florence, permanent suites at the Paris Ritz and the Connaught, and a stable full of polo ponies.

On its publication in 1965 *The Painted Bird* was considered by some critics to be the finest piece of literature to come out of the Second World War. It is an intense and disturbing account of an abandoned boy's five-year odyssey through war-torn Eastern Europe, and was published at the risk of being considered too brutal and too horrific by the public. The communist press demanded the book be banned, denounced it as salacious Western propaganda subsidised by the US government, and declared Kosinski a renegade. The repercussions of this went on for years and the American press, as well as encouraging rumour regarding his authenticity as a writer, discussed at length his position as 'a man under ideological attack'. However, the most intriguing aspect of the sensation caused by *The Painted Bird* was the question of whether or not it was entirely autobiographical. Kosinski did not commit himself to an answer either way, but the new edition in 1975 showed a photograph of Kosinski as a child on the front cover and in the foreword he wrote: 'The remembered event becomes a fiction, a structure made to accommodate certain feelings.' He nevertheless insisted that the basic facts about his childhood were true. In the wake of the book's publication, Kosinski's claims were frequently and virulently challenged in the Polish press, which is hardly surprising since Western reviewers had seen the book as confirming evidence of Polish anti-Semitism during the war. No Polish translation of the book appeared until 1989, but one of Kosinski's fiercest and most tenacious critics has been the Polish journalist Joanna Siedlecka, whose claims have been attacked in their turn, both in Poland and in the USA. Perhaps more damaging was an article in the *Village Voice* in 1982, which also disputed Kosinski's version of events. What is or is not true is still being disputed, and Kosinski is now dead, having committed suicide in 1991. As far as *The Painted Bird* is concerned, it has been argued that the facts do not much matter: it is, after all, a novel, and an undeniably powerful one. Its status as a Holocaust testament, however, relied on its supposed authenticity. Now that this has been severely compromised, the reputation of both the book and its author have been irreparably damaged.

In subsequent novels, Kosinski never strayed far from his own extraordinary life to invent his protagonists. He takes them to the frontiers of human endurance where all things cultural and social cease to exist and the survival of the opportunist is the only thing that matters; thus the essential self emerges. 'The whole didactic point of my novels', he wrote, 'is how you redeem yourself if you are threatened by the chances of daily life'. Much of his work was condemned as overtly shocking and violent. He retaliated: 'I have a vision that I will not sacrifice for sentimental critics brought up on *The Fiddler on the Roof*'. That said, his best-known novel after *The Painted Bird* is *Being There*, which concerns a gardener whose banal utterances are taken for the most profound philosophy and who consequently achieves national celebrity. Kosinski wrote the screenplay for a film of the novel (1979), which was thought by many to be deeply sentimental.

Fiction

The Painted Bird 1965; *Steps* 1968; *Being There* 1970 [rev. 1981]; *The Devil Tree* 1973; *Cockpit* 1975; *Blind Date* 1977; *Passion Play* 1979; *Pinball* 1982; *The Hermit* 1986; *The Hermit of 69th Street* 1987

Plays

Being There [from his novel] **(st)** 1973, **(f)** 1979; *Passion Play* [from his novel] 1982

Non-Fiction

Documents of the Struggle of Man 1955; *Jakob Jaworski's Program of People's Revolution* 1955; *The Future Is Ours, Comrade* [as 'Joseph Novak'] 1960; *No Third Path* [as 'Joseph Novak'] 1962; *Notes of the Author on 'The Painted Bird'* 1965; *The Art of the Self* 1968; *The Literature of Violation* 1992

Richard (Cory) KOSTELANETZ 1940–

Kostelanetz was born in New York City, the son of a lawyer, and gained a BA in American civilisation at Brown University in 1962. He went on to take an MA at Columbia, where he was a Woodrow Wilson Fellow, and to do graduate work at King's College, London, where he was a Fulbright Fellow from 1964 to 1965. While in England he was producer–interviewer of the BBC programme *New Release* (1965–6), and since returning to the USA he has held a number of visiting professorships, artist's residencies, and grants, including a Guggenheim Fellowship, a Pulitzer Fellowship and a number of National Endowment for the Arts grants.

Kostelanetz's chief claim on our interest is that he is probably the world's most experimental writer, or at least that he represents the farthest extreme of the formalist approach within the broader field of 'experimental writing'. He goes much farther along the route more popularly associated with Georges Perec, who wrote a novel without the letter 'e'. Kostelanetz's work includes a novella with no more than two words to a paragraph, a story with only single word paragraphs, a 'novel' of 1,000 blank pages, stories composed exclusively of cut-up photographs, 'narratives' – one of book length – composed entirely of numerals, and a good deal more, often of some complexity and including film, video and audio-tape pieces. His output in 'visual poetry', a medium between poetry and painting which differs from most concrete poetry by being non-linear and non-syntactic, is among his most significant work.

He considers his work to be 'the purest oeuvre of fiction, as fiction, uncompromised by vulgar considerations, that anyone has ever done' and is pleased to have produced 'no conventional fiction – absolutely none'. His radically alternative formalist approach is backed up by an extensive theorising which has the appearance of rigour, and assumes a historicity that sees 'experimental literature [as] a history of formal innovation'. His creative work is bound up with his prolific career as a critic and editor: he has edited over thirty books including *The Yale Gertrude Stein, The Avant-Garde Tradition in Literature, John Cage, Moholy-Nagy* and *On Contemporary Literature* (1964), which was his first publication.

His uncompromising extremeness has left him as perhaps the foremost critic of his own work, and certainly one of the most appreciative. In complaining that his work has not been widely reviewed, he asked: 'Does anyone care? Should anyone care? ... (Should I care? If so, how? Should I have written this?) What should be made of the fact that no one else – absolutely no one else visible to us – is making fiction in these ways?'

Fiction

In the Beginning 1971; *Accounting* **(s)** 1972; *Ad Infinitum* **(s)** 1973; *Obliterate* 1974; *Come Here* **(s)** 1975; *Constructs* **(s)** 1975; *Extrapolate* **(s)** 1975; *Modulations* **(s)** 1975; *Openings & Closings* **(s)** 1975; *One Night Stood* 1977; *Three Places in New Inkland* **(s)** [with others] 1977; *Constructs Two* **(s)** 1978; *Foreshortenings and Other Stories* **(s)** 1978; *Inexistences* **(s)** 1978; *Tabula Rasa* **(s)** 1978; *And So Forth* **(s)** 1979; *Exhaustive Parallel Intervals* 1979; *More Short Fictions* **(s)** 1980; *Reincarnations* 1981; *Epiphanies* **(s)** 1983; *Constructs Three – Six* 4 vols 1991; *March* **(s)** 1991; *Fifty Constructivist Fictions* **(s)** 1991; *Flipping* 1991; *Intermix* 1991; *Minimal Fictions* **(s)** 1994; *More Openings & Closings* **(s)** 1995

Plays
Openings & Closings **(tv)** [with B. Weiss] 1975; *Three Prose Pieces* **(tv)** 1975; *Constructivist Fictions* **(f)** [with P. Longauer] 1976; *Audio Art* **(r)** 1978; *Declaration of Independence* **(tv)** 1979; *Epiphanies* **(f)**, **(st)**, **(tv)** 1980; *Text-Sound in North America* **(r)** 1981; *Seductions* **(st)** 1982, **(tv)** 1987; *Hörspiel USA* **(r)** 1983; *Audio Writing* **(r)** 1984; *Nach Weissensee* **(r)** [with M. Koerber, M. Maassen] 1984; *A Berlin Lost* **(f)** [with M. Koerber] 1985; *Partitions* **(tv)** 1986; *Home Movies Reconsidered: My First Twenty-Seven Years* **(tv)** 1987; *Relationships* **(tv)** 1987; *Video Writing* **(tv)** 1987; *New York City Radio* **(r)** 1988; *Kinetic Writings* **(tv)** 1989; *Turfs/Grounds/Lawns* **(f)** 1989; *Video Strings* **(f)** 1989; *Collected Performance Texts* 1991; *Lovings* 1991 [np]; *More Complete Audio Writing* 5 vols 1992; *More Partitions* 1994; *Tranimations* **(f)** 1995; *Video Fictions* **(tv)** 1995; *Video Poems* **(tv)** 1995

Non-Fiction
The Theatre of Mixed Means 1968; *Master Minds: Portraits of Contemporary American Artists and Intellectuals* 1969; *The End of Intelligent Writing: Literary Politics in America* 1974; *Recyclings: a Literary Autobiography* **(a)** 1974 [rev. 1984]; *Grants and the Future of Literature* 1978; *Prunings/Accruings* 1978; *Wordsand* 1978; *'The End' Essentials/'The End' Appendix* 1979; *Twenties in the Sixties: Previously Uncollected Critical Essays* 1979; *Metamorphosis in the Arts* 1980; *Autobiographies* **(a)** 1981; *The Old Poetries and the New* 1981; *Reincarnations* 1981; *American Imaginations* 1983; *Autobiographien New York Berlin* **(a)** 1986; *The Grants-Fix: Publicly Funded Literary Granting in America* 1987; *The Old Fictions and the New* 1987; *Prose Pieces/Aftertexts* 1987; *Conversing with Cage* 1988; *On Innovative Music(ian)s* 1989; *The New Poetries and Some Old* 1990; *Unfinished Business: an Intellectual Non-History 1963–1989* 1990; *On Innovative Art(ist)s* 1991; *Politics in the African-American Novel* [with others] 1991; *References* 1991; *The Dictionary of the Avant Gardes* 1993; *On Innovative Performance* 1994; *One Million Words of Book Notes 1958–1993* **(a)** 1995; *Twenty-Five Years After: Recollections of Rock Theatre* 1995

Poetry
Visual Language 1970; *I Articulations* 1974; *Portraits from Memory* 1975; *Word Prints* 1975; *Rain Rains Rain* 1976; *Illuminations* 1977; *Number Two* 1977; *Richard Kostelanetz* 1980; *Turfs Arenas Fields Pitches* 1980; *Arenas Fields Pitches Turfs* 1982; *Solos, Duets, Trios, and Choruses* 1991; *Wordworks* 1993

Edited
On Contemporary Literature 1964; *The New American Arts* 1965; *Twelve from the Sixties* 1967; *The Young American Writers* 1967; *Beyond Left & Right: Radical Thought for Our Time* 1968; *Assembling, and Second through Eleventh Assembling* 12 vols 1970–81; *Imaged Words & Worded Images* 1970; *John Cage* 1970; *Moholy-Nagy* 1970; *Possibilities of Poetry: an Anthology of American Contemporaries* 1970; *Future's Fictions* 1971; *Human Alternatives: Visions for Our Time* 1971; *Social Speculations: Visions for Us Now* 1971; *In Youth* 1972; *Seeing through Shuck* 1972; *Breakthrough Fictioneers: an Anthology* 1973; *The Edge of Adaptation: Man and the Emerging Society* 1973; *Essaying Essays* 1975; *Language and Structure* 1975; *Esthetics Contemporary* 1977; *Younger Critics of North America: Essays on Literature and the Arts* 1977; *Assembling Assembling* 1978; *Visual Literature Criticism* 1979; *Scenarios* 1980; *Text-Sound Texts* 1980; *The Yale Gertrude Stein* 1980; *Aural Literature Criticism* 1981; *American Writing Today* 2 vols 1981 [rev. in 1 vol. 1991]; *The Avant-Garde Tradition in Literature* 1982; *The Literature of SoHo* 1983; *The Poetics of the New Poetries* [with B. Hrushovski] 1983; *Precisely Complete* [with S. Scobie] 6 vols 1985; *Gertrude Stein Advanced: an Anthology of Criticism* 1991; *Merce Cunningham: Dancing in Time and Space* 1992; John Cage: *John Cage, Writer: Previously Uncollected Pieces* 1993, *Writings about John Cage* 1993; *Nicolas Slominsky: the First Hundred Years* [with J. Darby] 1994

Collections
Numbers: Poems and Stories 1976

Larry KRAMER 1935–

Kramer was born in Bridgeport, Connecticut, but grew up in Washington DC before attending Yale University. He lived in England for ten years after graduating, and took up a job as an executive with Columbia (British) Productions, Ltd, with whom he made his first film, *Here We Go Round the Mulberry Bush* (1967), a 'swinging sixties' sex comedy based on a novel by Hunter Davies. Kramer co-wrote the script, and subsequently combined the roles of producer and writer on Ken Russell's celebrated 1969 screen version of **D.H. Lawrence**'s *Women in Love* (1920). According to the maverick director, this collaboration was not altogether happy, but it remains one of his finest films (achieving a certain notoriety for the nude wrestling scene) and both he and Kramer were nominated for Academy Awards.

In terms of the Aids crisis in the 1980s and 1990s, Kramer has been one of the most influential figures to emerge from the gay community. A play by David Duke, called *The Night Larry*

Kramer Kissed Me, which is about growing up gay in the 1970s and dealing with the impact of Aids in the 1980s, indicates, in terms of the title, how much of an impression Kramer has made on young men of that generation. The gay cultural critic David Bergman has written that 'no one is more responsible for the rhetoric of Aids than Larry Kramer, whose pronouncements, ultimatums, vilifications, lampoons, and dramatizations seemed ubiquitous in the early years of the epidemic'. Kramer is best known for his play *The Normal Heart* (1985), which almost single-handedly put Aids on the political agenda in the USA and afterwards in Britain (where it opened in 1986 with Martin Sheen in the lead role). The play proved to be a huge commercial success and was awarded the George and Elisabeth Marton Award by the American Dramatists Guild. Despite this, it is possible to argue that, from a strictly artistic point of view, the play is not a great accomplishment – the death-bed marriage at its climax being particularly melodramatic – but Kramer would probably settle for being remembered more as a political activist than a writer of great literature. One would have to balance this comment, however, with the suggestion that his painfully autobiographical play *The Destiny of Me* deserves serious attention on artistic terms alone. But it is not contentious to say that he has seen the advent of Aids as a call to arms and his weapon, as an intellectual, has been primarily the written word.

More generally, he has secured his place in the history of Aids by founding the Gay Men's Health Crisis (GMHC), the largest private service organisation dedicated to helping people with Aids, and also by founding the protest groups Aids Coalition to Unleash Power (ACT UP). But despite his reputation as a radical proponent of confrontational politics, Kramer has produced a series of what might be seen as reactionary – even hysterical – comments on Aids and how best to avoid it, most notably the suggestion that total abstinence is the only way to stop its spread. This attitude might be related to a slightly paranoid psyche; as he writes in his collection *Reports from the Holocaust: the Making of an Aids Activist*: 'It is very difficult for me to make love, even "safely", when the very act is now so inextricably bound up with death.' This underlying distaste for casual sex can be traced back to his 1978 novel *Faggots*, in some ways a parody of the nature of the gay community that had emerged in America in the 1970s, but also a deeper, more critical reflection on the lovelessness and intellectual bankruptcy that he believes often characterises the sexual encounters and general lifestyle of individuals within that community.

Plays

Here We Go Round the Mulberry Bush **(f)** [from H. Davies's novel; with H. Davies] 1967; *Women in Love* **(f)** [from D.H. Lawrence's novel] 1969; *The Sissies' Scrapbook* 1972 [np]; *The Normal Heart* 1985; *Just Say No* 1988 [np]; *The Furniture of Home* 1989 [np]; *The Destiny of Me* 1991 [pub. 1993]

Fiction

Faggots 1978

Non-Fiction

Reports from the Holocaust: the Making of an Aids Activist 1989

Hanif KUREISHI 1954–

Kureishi was born in Bromley, Kent, to an English mother and a Pakistani father who was a clerk and an occasional journalist. In an autobiographical essay, 'The Rainbow Sign', Kureishi has written of the conflict of identity he suffered as a result of his mixed parentage: 'From the start I tried to deny my Pakistani self. It was a curse and I wanted to be rid of it.' He attended Ravenswood School for Boys where he began to write as a means of confronting these conflicts and the casual racism he witnessed in friends. While reading philosophy at King's College, London, he became involved with the London fringe theatre scene, and recalls in the introduction of *Outskirts*, a collection of his early plays, how he was encouraged by Donald Howarth, literary manager at the Royal Court Theatre. During rehearsals for **Samuel Beckett**'s *Footfalls* Kureishi 'sometimes sat with Beckett in the pub next to the Court. Once he gave me £50 when I needed money for a course I wanted to do.' After leaving university he worked in various menial jobs in London theatres and supported himself at one point by writing for pornographic magazines under the pseudonym Antonia French.

In 1980 two of his plays were produced: *The Mother Country* (which won the Thames Television Playwrights Award) at the Riverside Studios, and *The King and Me*, a short study of a couple's obsession with Elvis Presley, at the Soho Poly. Kureishi's subsequent plays include *Borderline*, a study of the difficulties faced by Asians living in London, written in collaboration with the Joint Stock Company while he was writer-in-residence at the Royal Court in 1981, and *Birds of Passage*, which describes the problems of a lower-middle-class family.

He came to the attention of a wider audience when Channel 4 Television commissioned the film *My Beautiful Laundrette* which he wrote while visiting his family in Karachi in 1985. The film was criticised by some Asians for painting a negative portrait of their lives, and many were undoubtedly upset by its celebratory depiction of an inter-racial homosexual relationship. It won an *Evening Standard* Award for best film, and the screenplay was nominated for an Oscar. Kureishi again collaborated with the director Stephen Frears on *Sammy and Rosie Get Laid*, an equally contentious exploration of racial violence and sexual discovery set in contemporary London. Kureishi himself directed *London Kills Me*, a study of drugs and homelessness among young people which lacks the structure of the earlier films.

Kureishi's first novel, *The Buddha of Suburbia*, which won the Whitbread Award for a first novel in 1990, is a comic *Bildungsroman* set in the 1970s which takes its title from the protagonist's father's ruthless exploitation of supposed Eastern mysticism among the suburban housewives of south-east London. Kureishi adapted the novel for a fine, if controversial, BBC television series in 1993. *The Black Album* contains a familiar cocktail of sexual exploration, drugs, music and youth culture in a fast-moving urban thriller. Kureishi has also published a number of short stories, including 'My Son, the Fanatic', which he had adapted for a film, and has edited with Jon Savage *The Faber Book of Pop*. He is married and has twin sons.

Plays

Soaking Up the Heat 1976 [np]; *The King and Me* 1980 [pub. 1983]; *The Mother Country* 1980 [np]; *You Can't Go Home* (r) 1980; *Cinders* [from J. Glowacki's play] 1981 [np]; *Borderline* 1981 [pub. in *Outskirts and Other Plays* 1992]; *Janusz Glowacki* [with others] 1981; *Outskirts* 1981 [pub. 1983]; *Tomorrow – Today!* 1981 [pub. 1983]; *Artists and Admirers* [from A. Ostrovsky's play; with D. Leveaux] 1982 [np]; *The Trial* (r) [from Kafka's novel] 1982; *Birds of Passage* 1983; *Mother Courage* [from B. Brecht's play] 1984; *My Beautiful Laundrette* (f) [pub. with *The Rainbow Sign* (a)] 1986; *Sammy and Rosie Get Laid* (f) 1988; *London Kills Me: Screenplays* (f) 1991; *The Buddha of Suburbia* (tv) [from his novel] 1993

Fiction

The Buddha of Suburbia 1990; *The Black Album* 1995

Edited

The Faber Book of Pop [with J. Savage] 1995

Tony KUSHNER 1957–

Although a New Yorker by birth, Kushner was brought up in Louisiana, the middle child of three in a literate and political household. His parents were both musicians. He attended Lake Charles High School and excelled in the school's debating team. Being white, Jewish and homosexual, and growing up in the turmoil of the American South in the 1960s, it is perhaps not surprising that Kushner should later choose the theme of community and collectivity for his writing. His best-known play, *Angels in America*, is designed to 'test the limits of tolerance', limits which Kushner himself felt unsure of when, as an undergraduate at Columbia University, New York, he attempted to psychoanalyse himself into the ranks of the heterosexual majority. The experiment failed, as his psychoanalyst had assured him it would, and he continued his studies as an openly homosexual postgraduate student at New York University. His first major play, *A Bright Room Called Day*, had an unsuccessful run at the New York Theater Workshop in 1985 and was subsequently presented at the Bush Theatre in London in 1988, a production which the writer recalls as 'a catastrophe'. Set in Weimar Germany, the play is an indictment of Reaganite trickle-down economics. Some British reviewers found the comparisons between Hitler's Germany and Thatcher's Britain hard to take, a resentment which found its way into later reviews of *Angels in America*. The play was subsequently picked up by Oskar Eustis, artistic director of the Eureka Theatre in San Francisco, who gave it its first successful showing, and commissioned Kushner to begin work on *Angels in America*.

The two successfully applied for a National Endowment for the Arts grant, which saw the play through a lengthy workshop stage at the Mark Taper Forum in Los Angeles and the Sundance Institute before the triumphant opening of the first part, *Millennium Approaches*, in 1991 at the Eureka. The play's second half, *Perestroika*, was given a staged reading at the same time. *Millennium Approaches* went on to be seen in hugely acclaimed productions both on Broadway and at London's National Theatre. The two plays were eventually performed together at the Mark Taper Forum in November 1992, earning $32,804 on the first day, more than doubling the previous house record. Subtitled 'A Gay Fantasia on National Themes', running in total for well over six hours and dealing with Aids, Mormonism and the deeply unsympathetic Roy Cohn (a homosexual but homophobic, violently right-wing lawyer, involved in the McCarthy hearings and the prosecution and execution of the Rosenbergs), the play did not seem a natural Broadway hit, but it was both a popular and a critical success, winning numerous awards, including two Tonys, a Pulitzer Prize, and the Drama Critics'

Circle Awards in San Francisco, Los Angeles, New York and London. It is considered by many to be one of the century's most important plays and immediately placed Kushner in the front rank of twentieth-century dramatists.

Kushner's other work includes adaptations of Goethe, Corneille and Ariel Dorfman, his play *Slavs!*, and a collection of plays, essays and poetry, *Thinking about the Long-Standing Problems of Virtue and Happiness*.

Plays

A Bright Room Called Day 1985 [pub. 1995]; *Yes, Yes, No, No* 1985 [pub. 1987]; *Hydriotaphia* 1987 [np]; *Stella* [from Goethe's play] 1987 [np]; *The Illusion* [from Corneille's play] 1988 [pub. 1991]; *Angels in America*: *Millennium Approaches* 1991 [pub. 1992], *Perestroika* 1992 [pub. 1993]; *Widows* [from A. Dorfman's novel] 1991; *Slavs!* 1994 [pub. in *Thinking about the Long-Standing Problems of Virtue and Happiness* 1995]

Collections

Thinking about the Long-Standing Problems of Virtue and Happiness 1995

L

George (William) LAMMING 1927–

Lamming was born in Carrington Village, a former sugar estate close to Bridgetown, the capital of Barbados, the son of an unmarried mother, and of mixed African and English blood. Later his mother married, and he grew up partly in his native village and partly in St David's Village, where his stepfather worked. He began to move away from this village world when he won a scholarship to attend Combermere High School, where a teacher, Frank Collymore, encouraged his interest in writing. In 1946, when the course ended, Collymore helped him get a job at El Colegio de Venezuela, a boarding-school for boys of South American origin in Port of Spain, Trinidad. The teacher also published Lamming's early poems and other pieces in a magazine he edited, *Bim*.

In 1950, Lamming left, with many other West Indian immigrants, for London, where he initially worked briefly in a factory but soon found work on a programme of the BBC's Overseas Service, *Caribbean Voices*. His autobiographical first novel, *In the Castle of My Skin*, appeared in 1953, with an enthusiastic preface by **Richard Wright**, and was immediately recognised as being an outstanding account of growing up in a colonial society undergoing rapid change. Its success opened a successful literary career for Lamming, and he followed it up with three more novels, as well as an influential volume of essays looking at the Caribbean heritage, *The Pleasures of Exile*. His second novel, *The Emigrants*, is again autobiographical, following a group of young men of his generation to London, while the next two are set on the fictitious island of San Cristobal, examining problems of personal identity and politics in the Caribbean context.

The 1950s also saw Lamming on extended tours of North Africa, the Caribbean and North America, and the pattern was set for the next three decades, when he was based largely in London but got lectureships and grants which took him to many countries, including Australia, India, Tanzania, Kenya, Denmark, Jamaica and (often) the USA. His writing career came to a halt during the 1960s, but in 1971 and 1972 he published two ambitious further novels, *Water with Berries*, which shows West Indians turning towards violence in London, and *Natives of My Person*, a dense, allegorical account of a sixteenth-century voyage of plunder. Some critics considered the latter his major work, although others complained of an increasing tendency to explore tendentious themes at the expense of credible life. Since then, Lamming has published no more fiction, but in 1974 he edited an anthology dealing with the black response to white racism, while a volume of *Conversations* with him was published by a small London press in 1992. He has returned from London to Barbados, continuing to travel widely and teaching each summer at the University of Miami's Institute for Caribbean Creative Writing.

Fiction

In the Castle of My Skin 1953; *The Emigrants* 1954; *Of Age and Innocence* 1958; *Season of Adventure* 1960; *Water with Berries* 1971; *Natives of My Person* 1972; *The Most Important People* [with K. Drayton] 1981

Non-Fiction

The Pleasures of Exile 1960; *Conversations: Essays, Addresses and Interviews, 1953–90* [ed. Andaiye, R. Drayton] 1992

Edited

New World: Guyana Independence Issue 1966; *Cannon Shot and Glass Beads: Modern Black Writing* 1974

Biography

The Novels of George Lamming by Sandra Pouchet Paquet 1982

Ring(gold Wilmer) LARDNER 1885–1933

Lardner's fluency in the redneck language of the baseball circuit was not that of a native. His own background was comfortably middle class: his father was a prosperous agricultural businessman and his mother wrote poetry and organised theatrical evenings when not caring for her nine children, of whom Ring was the youngest. In the early 1900s the business hit a slump and, at his father's wish, Lardner went to study engineering at the Armour Institute of Technology in Chicago, but the experiment was not successful. He returned home within a year and, after brief spells as a freight agent and book-keeper, found work on the South Bend *Times* with whom he served a two-year apprenticeship before moving back to Chicago.

Between 1907 and 1919 he worked on the Chicago *Inter-Ocean*, the *Examiner* and the *Tribune*. In 1911 he married; he had four children. Hired as a sportswriter, Lardner had come to specialise in baseball reports, travelling to games with the Chicago White Sox, reporting from the dug-out and the practice field and, from 1913, writing the daily column 'In the Wake of the News'. Baseball was still considered a slightly disreputable sport at the time; Lardner made its idioms and characters familiar and almost respectable before himself becoming disillusioned with the game after the Black Sox Scandal of 1919 in which it emerged that the Chicago team had contrived to lose the World Series. The legendary 'fix' is mentioned in awed tones in *The Great Gatsby* (1925); **F. Scott Fitzgerald** and Lardner had become friends and drinking partners whilst living as neighbours in Great Neck, Long Island, in 1923.

By this time, Lardner was already a well-known writer; the first of his short stories about a rookie baseball player called Jack O'Keefe had appeared in the *Saturday Evening Post* in 1914 and the first collection of them, *You Know Me, Al*, had been published in 1916. His magazine work had made him a household name. It was at Fitzgerald's insistence that he collected the best of his work for a volume entitled, with characteristic self-deflation, *How to Write Short Stories (with Samples)*. The book introduced him to a serious audience and attracted commendations from H.L. Mencken, **Virginia Woolf** and, more guardedly, **Edmund Wilson**. 'Readers,' Lardner wrote, 'might think I was having an affair with some of the critics', and he remained wary of critical acclaim and deaf to the advice of those who were urging him to stretch his talents beyond the boundaries set for it in the newspapers and magazines in which he had trained himself. Instead, and in addition to the short stories, he wrote numerous song lyrics, a cartoon strip, a series of short, Dadaesque plays and, in collaboration with George S. Kaufman, *June Moon*, a play satirising the songwriters of Tin Pan Alley. It had a successful Broadway run in 1929 but Lardner was too ill to enjoy the play's success; he spent much of his last seven years in various hospitals, dying in New York of tuberculosis and heart trouble aggravated by heavy drinking.

His literary standing remained high; **Ernest Hemingway** cited him as an influence whilst simultaneously distancing himself from the more exaggerated tributes elicited by Lardner's premature death. He argued that Lardner had considered himself the superior of his subject-matter in all but one or two of the stories, that they lack compassion, the satire is cruel and, by implication, that of a snob. More generous appreciation is found in recent criticism, much of which emphasises Lardner's importance as a link between the vernacular of *Huckleberry Finn* and the modernised but equally homegrown voices of twentieth-century American fiction.

Fiction
You Know Me, Al (s) 1916; *Gullible's Travels* (s) 1917; *Treat 'Em Rough* (s) 1918; *The Real Dope* (s) 1919; *Own Your Own Home* (s) 1919; *Regular Fellows I Have Met* (s) 1919; *The Big Town* (s) 1921; *How to Write Short Stories (with Samples)* 1924; *What of It?* (s) 1925; *The Love Nest and Other Stories* (s) 1926; *Round-Up* (s) 1929; *Lose with a Smile* (s) 1933; *First and Last* (s) 1934

Non-Fiction
My Four Weeks in France 1918; *The Young Immigrunts* 1920; *Symptoms of Being Thirty-Five* 1921; *Say It with Oil: a Few Remarks About Wives* 1923; *The Story of a Wonder Man* (a) 1927; *Ring Around Max: the Correspondence Between Ring Lardner and Maxwell Perkins* [ed. C.M. Caruthers] 1973; *Letters from Ring* [ed. C.M. Caruthers] 1979

Plays
Elmer the Great 1928; *June Moon* [with G.S. Kaufman] 1929 [pub. 1930]

Poetry
Bib Ballads 1915

Collections
The Portable Lardner 1946; *Shup Up, He Explained* 1962; *The Ring Lardner Reader* 1963; *Some Champions* 1976

Biography
Ring Lardner by Donald Elder 1956

Philip (Arthur) LARKIN 1922–1985

The son of a local government official, Larkin grew up in Coventry, and belongs to the fairly large body of writers for whom their childhood is not an inspiration ('only where my childhood was unspent', he remarks deprecatingly of Coventry in his poem 'I Remember, I Remember'). After King Henry VIII School, Coventry, he went to St John's College, Oxford, as a member of a wartime generation that also included **Kingsley Amis** and **John Wain**. These writers were to remain a group in later life, being associated with the Movement school of poetry in the 1950s, and shifting politically distinctly to the right in later years.

On leaving Oxford, Larkin became a librarian: first in the library of an urban district council in Shropshire, later in university libraries in Leicester and Belfast, and from 1955 as librarian of the Brynmor Jones library at the University of

Hull, which he built up from a staff of eleven to one of over 100. Despite increasing fame as a writer, Larkin retained his job until his death within a few years of retirement age; in his poems 'Toads' and 'Toads Revisited' he explored the idea that having a job, however dreary, is preferable to an unemployed or freelance existence.

Although he had a number of affairs, Larkin never married, and lived for much of his life in university rooms and flats before buying a house in Hull in 1974, which he shared with his companion Monica Jones. In his poetry he cultivated a provincialism and bareness of outlook: 'Deprivation is for me what daffodils were for Wordsworth', he once remarked. Nevertheless, his lyrical poetry of ordinary lives quickly became recognised as a major, perhaps *the* major, contribution to post-war English poetry. His three mature collections, *The Less Deceived*, *The Whitsun Weddings* and *High Windows*, are generally acknowledged as constituting, in small compass, an English poetic *oeuvre* of the first rank. He also wrote in the 1940s two novels, *Jill* and *A Girl in Winter*, while his lifelong love of jazz was celebrated in the collection *All What Jazz*, and his critical essays and reviews were collected in *Required Writing*. His edition of *The Oxford Book of Twentieth-Century English Verse* was revisionist, giving what some considered excessive space to **Thomas Hardy** at the expense of the more revered gods of modernism.

Despite Larkin's provincialism, well-publicised dislike of modernism and the fact that from the mid-1970s he had largely given up writing poetry, his stature was such that when his friend **John Betjeman** died in 1984, there was widespread agreement that he ought to be offered the Laureateship. However, when this *was* offered he declined it, even though this meant (or so he imagined) disappointing the prime minister, Margaret Thatcher, whom he held in high regard. The terrifying approach of death, which he had contemplated in several poems in *High Windows*, was realised when shortly after refusing the Laureateship, he underwent surgery for cancer of the oesophagus, and died within the year.

Although he had frequently said that he wanted all his papers burned, his will was judged by Queen's Counsel to be 'repugnant' (i.e. contradictory). His voluminous diaries were destroyed, but other manuscripts were saved and, amid considerable controversy, one of his literary executors, **Anthony Thwaite**, edited volumes of *Collected Poems* and *Selected Letters*, while another, **Andrew Motion**, wrote a long biography. The revelations in both the letters and the biography of Larkin's casual racism, his taste for pornography, and other politically incorrect attitudes and habits caused much comment, although few felt that any of this in any way diminished his standing as one of the century's leading poets.

Poetry

The North Ship 1945; *XX Poems* 1951; *Poems* 1954; *The Less Deceived* 1955; *The Whitsun Weddings* 1964; *High Windows* 1974; *Femmes Damnées* 1978; *Aubade* 1980; *Collected Poems* [ed. A. Thwaite] 1988

Fiction

Jill 1946; *A Girl in Winter* 1947

Non-Fiction

All What Jazz: a Record Diary, 1961–1968 1970 [rev. as *All What Jazz: a Record Diary, 1961–1971* 1985]; *Required Writing: Miscellaneous Pieces, 1955–1982* 1983; *Selected Letters* [ed. A. Thwaite] 1992

Edited

New Poems 1958 [with B. Dobrée, L. MacNeice] 1958; *The Oxford Book of Twentieth-Century English Verse* 1973; *Poetry Supplement* 1974

Biography

Philip Larkin by Andrew Motion 1982

Maura LAVERTY 1907–1966

Laverty was born Maura Kelly in Rathangan, County Kildare, in central Ireland, one of nine surviving children of a family of thirteen. Her father was a prosperous gentleman farmer who became poor through gambling, and her mother had to take up dressmaking to support the family. Laverty attended school at the Brigidine Convent in Tullow, County Carlow, and, on leaving, went to Spain as a governess (here her life parallels that of another writer, **Kate O'Brien**, and indeed the pattern was a fairly common one for girls in Ireland at that time). Soon tiring of her subordinate position, she became secretary to Prince Bibesco, later working for the Banco Calamarte and for the Madrid newspaper *El Debate*. Returning to Ireland in 1928, she very soon married the journalist James Laverty, with whom she lived in Dublin and with whom she had three children.

Laverty wrote from an early age, but when her husband lost money through an agency he ran for the Irish Sweepstake, this gave her the impetus to write her first novel, *Never No More* (1942), a sentimental but richly realistic picture of an Irish childhood in the 1920s which had a huge success on publication. It was followed by three more novels for adults, all published during the mid-1940s, one of which, *No More Than*

Human, is a sequel to her first success (it takes her young heroine to Spain). Laverty's novels are interspersed with recipes and references to cooking, and she became equally famous in Ireland as an author of cookery books, with such titles as *Kind Cooking*, *Flour Economy* and *Feasting Galore*. She also wrote two novels for children during the 1940s.

From the 1950s she developed a new career as a playwright, mainly for radio, and in the 1960s she wrote a highly successful television soap opera, *Tolka Row*. Throughout her adult life she wrote much journalism, editing a women's magazine at one time, and in her last years she was a well-known radio agony aunt. She died of a heart-attack at the relatively early age of fifty-eight.

Fiction

Never No More 1942; *Alone We Embark* [US *Touched by the Thorn*] 1943; *No More Than Human* 1944; *Gold of Glanaree* 1945; *The Cottage in the Bog* 1946

Non-Fiction

Flour Economy 1941; *Maura Laverty's Cookbook* 1946; *Kind Cooking* 1955; *Full and Plenty* 1960; *Feasting Galore* 1961

For Children

Lift Up Your Gates 1946 [US *Liffey Lane* 1947]; *The Green Orchard* 1949

Mary LAVIN 1912–

Lavin was born of Irish parents in East Walpole, Massachusetts, where she went to school. At fourteen, she came to live in Ireland when her father left America to manage the Bective House estate in County Meath. After education at a Dublin convent, Lavin completed a thesis on Jane Austen at the National University of Ireland and, after gaining first class honours in 1937, intended to continue with a PhD on **Virginia Woolf**, but instead went to America. When she returned to Ireland, she had a short story published in a Dublin magazine, and renounced her academic career after encouragement by Lord Dunsany, who wrote a preface to her first collection of stories, *Tales from Bective Bridge*. This won the James Tait Black Memorial Prize in 1942.

Apparently simple and low key in tone, and entirely apolitical, Lavin's fiction set in Ireland often examines the experience of the impossibility of love for lonely characters, and stresses the validity of communal values in terms of individual circumstances. Although she has written novels, the best of which is *Mary O'Grady*, she is not completely at ease with the form. As she has written: 'it is in the short story that a writer distils the essence of his thought.'

Married in 1942 to William Walsh, a Dublin lawyer who died in 1954, Lavin wrote a series of short stories concerned with the struggles of widowhood. In 1969, she married Michael MacDonald Scott, a former Jesuit priest. She divides her time between Dublin and a farm on the Bective estate.

Fiction

Tales from Bective Bridge (s) 1942; *The Long Ago* (s) 1944; *The House in Clewe Street* (s) 1945; *The Becker Wives* (s) 1946; *Mary O'Grady* 1950; *A Single Lady* (s) 1951; *The Patriot Son* (s) 1956; *Selected Stories* (s) 1959; *The Stories of Mary Lavin* (s) 3 vols 1964–85; *The Great Wave* (s) 1967; *Happiness* 1969; *Collected Stories* (s) 1971; *A Memory* (s) 1971; *A Family Likeness* (s) 1985

For Children

A Likely Story 1957; *The Second Best Children in the World* 1972

D(avid) H(erbert Richards) LAWRENCE 1885–1930

Lawrence was born in the mining village of Eastwood in the Nottinghamshire coalfield, the son of a foreman of a mine and his more educated wife. His childhood was dominated by poverty and friction between his parents, as well as his intense love for his mother, who figures as Mrs Morel in his first fictional masterpiece, *Sons and Lovers*. After leaving Nottingham High School, to which he had won a scholarship, Lawrence worked first as a clerk in a surgical appliance factory and then for four years as a pupil-teacher (he was later to speak of 'the savage teaching of collier lads'). A course of study at University College, Nottingham, enabled him to move to a school in Croydon in south London, where he taught between 1908 and 1911 while developing his literary career.

In 1909, his childhood sweetheart, Jessie Chambers, sent some of his poems to **Ford Madox Ford** at the *English Review*, and they were immediately accepted; Ford was also influential in arranging the publication of his first novel, *The White Peacock*, in 1911, the year he decided to live entirely as a writer. In 1912, he called on Professor Ernest Weekley, who had been his teacher of modern languages at the university college, and immediately fell in love with Weekley's wife, the German aristocrat Frieda von Richthofen. The two eloped to Bavaria, and from there travelled through Austria, Germany and Italy before arriving back in England to marry in 1914.

Lawrence already had a substantial reputation, but the literary and establishment world that had been so quick to take him up proved as little able to assimilate him as he was able to tolerate its hypocrisy. During the First World War, despite exemption from military service on grounds of the ill health which had plagued him from childhood, he and his wife, unable to obtain passports, were subject to constant harassment from the authorities; most of his later novels, from *The Rainbow* (1915) onwards, were either banned or not published on grounds of obscenity; his paintings were confiscated from an art gallery.

The war over, the couple began the world travels which were to take them to Florence, Capri, Sicily, Ceylon, Australia, the USA and Mexico. From 1925, his tuberculosis became life-threatening, and the travels were confined to Europe. Lawrence eventually died, aged only forty-four, at Vence, in the hills above Nice.

Despite his wandering and difficult life, his production was enormous, and he contributed with distinction to genres as diverse as the essay, the drama and poetry. Four novels – *Sons and Lovers*, *The Rainbow*, *Women in Love* and (on a lower level of achievement) *Lady Chatterley's Lover* – are among the major fictions of their time. Three others – *Aaron's Rod*, *Kangaroo* and *The Plumed Serpent* – utilise his travels in Italy, Australia and Mexico respectively, and are lesser works, although containing much of interest. Lawrence's short stories, where he was free from the structural problems that sometimes weaken his novels, are, by common consent, superb. Despite his controversial ideas of 'blood-knowledge' and 'the dark gods', which some see as foreshadowing fascism, his status as a great and always supremely vital writer is unassailable.

Fiction

The White Peacock 1911; *The Trespasser* 1912; *Sons and Lovers* 1913; *The Prussian Officer and Other Stories* (s) 1914; *The Rainbow* 1915; *The Lost Girl* 1920; *Women in Love* 1920; *Aaron's Rod* 1922; *England, My England and Other Stories* (s) 1922; *Kangaroo* 1923; *The Ladybird* (s) 1923; *The Boy in the Bush* [with M.L. Skinner] 1924; *St Mawr with The Princess* (s) 1925; *The Plumed Serpent* 1926; *Lady Chatterley's Lover* 1928; *The Woman who Rode Away and Other Stories* (s) 1928; *The Escaped Cock* (s) [rev. as *The Man Who Died*] 1929; *Love Among the Haystacks and Other Pieces* (s) 1930; *The Virgin and the Gipsy* (s) 1930; *The Lovely Lady* (s) 1933; *A Modern Lover* (s) 1934; *The Man Who Was through with the World* [unfinished; ed. J.R. Elliot Jr] 1959; *Mr Noon* 1984

Poetry

Love Poems and Others 1913; *Amores* 1916; *Look! We Have Come Through!* 1917; *New Poems* 1918; *Bay* 1919; *Tortoises* 1921; *Birds, Beasts and Flowers* 1923; *Pansies* 1929; *Nettles* 1930; *Last Poems* 1932; *Complete Poems* [ed. V. de Sola Pinto, W. Roberts] 1964

Plays

The Widowing of Mrs Holroyd 1914; *Touch and Go* 1920; *David* 1926; *A Collier's Friday Night* 1934

Non-Fiction

Twilight in Italy 1916; *Movements in European History* [as 'Lawrence H. Davison'] 1921; *Psychoanalysis and the Unconscious* 1921; *Sea and Sardinia* 1921; *Fantasia of the Unconscious* 1922; *Studies in Classic American Literature* 1923; *Reflections on the Death of a Porcupine and Other Essays* 1925; *Mornings in Mexico* 1927; *My Skirmish with Jolly Roger* 1929 [rev. as *A Propos of Lady Chatterley's Lover* 1930]; *Pornography and Obscenity* 1929; *Assorted Articles* 1930; *Apocalypse* 1931; *We Need Another* 1933; *The Letters of D.H. Lawrence* [ed. R. Aldington] 1950; *Letters to Thomas and Adele Seltzer* [ed. G.M. Lacy] 1976; *The Letters of D.H. Lawrence* [ed. J.T. Boulton] 3 vols 1979–84

Translations

Leo Shestov *All Things Are Possible* [with S.S. Koteliansky] 1920; L.A. Bunin *The Gentleman from San Francisco and Other Stories* (s) [with S.S. Koteliansky, L. Woolf] 1922; Giovanni Varga: *Mastro-Don Gesualdo* 1923, *Little Novels of Sicily* 1925, *Cavalleria Rusticana and Other Stories* (s) 1928; *The Story of Dr Manente* 1929; Dostoevsky *The Grand Inquisitor* [with S.S. Koteliansky] 1930

Collections

The Spirit of Place [ed. R. Aldington] 1924; *Phoenix: the Posthumous Papers of D.H. Lawrence* [ed. E.D. MacDonald] 1936; *Phoenix II: Uncollected, Unpublished and Other Prose Works of D.H. Lawrence* 1968

Biography

D.H. Lawrence: a Literary Life by John Worthen 1989; *D.H. Lawrence: the Early Years, 1885–1912* by John Worthen 1991; *The Married Man* by Brenda Maddox 1994

David (Adam) LEAVITT 1961–

Leavitt was born in Pittsburgh, Pennsylvania, where his father was teaching at the time (he was later appointed to the faculty of the Stanford Business School). The family moved to Northern California in 1965, and Leavitt grew up in this region, the scene of some of his fiction. He moved to the East Coast to study English at Yale, and developed an interest in the literature of the Renaissance (he completed a thesis on Spenser's *The Faerie Queene*). His first story was published in the *New Yorker* when he was still an undergraduate, laying the foundations for a literary career which has burgeoned remarkably early and brought him worldwide success. He has been able to live entirely as a writer since graduation, with the exception of a brief period as an editorial assistant at Viking-Penguin (later his publisher) and some teaching appointments, the latest, in 1992, at Princeton.

His first volume of short stories, *Family Dancing*, was published in 1984, when he was only twenty-three, and announced the characteristic territory of his fiction with remarkable maturity: the tortured and closely observed relationships of the middle-class American family, and the lives of homosexual men and women. This narrow but nevertheless very real world has since been explored in two novels and a further volume of stories. His work has been published in the *New Yorker*, *Harpers* and other magazines. *Family Dancing* includes 'Counting Months', which won an O. Henry Award, and this volume was nominated for both the National Book Critics Circle Award and the PEN/Faulkner Prize.

His first novel, *The Lost Language of Cranes*, concerns a father and son who each discover the other to be homosexual. The action transferred to England, it was made into a television film (1992). *Equal Affections*, which takes its title from **W.H. Auden**'s poem 'The More Loving One', is the study of tensions within a family as the mother dies of cancer. Leavitt's second volume of stories, *A Place I've Never Been* is, if anything, even better than his first. As in most recent gay fiction, several of the stories, including the fine title one, are concerned with Aids. Leavitt himself had been involved in Aids work in New York (and has a sister who is a psychiatric social worker at an Aids outpatient clinic in San Francisco).

Far less successful was *While England Sleeps*, a novel set in England and Spain during the period of the Spanish Civil War and based upon events in the life of **Stephen Spender**. It had been partly inspired by Leavitt's term as writer-in-residence at the Institute of Catalan Letters in Barcelona during the winter of 1990–1, and seemed less secure in tone and ambience than his earlier books. The novel became the subject of legal proceedings, when Spender objected that it plagiarised his autobiography, *World within World* (1951). Before the case came to the courts in London, Leavitt's publishers agreed to withdraw the book and pay all legal costs. The novel was also withdrawn in the USA, and both Leavitt and Spender have written about this unhappy episode. He eventually rewrote the novel, publishing it in a revised edition.

Leavitt has written extensively for various magazines as a book-reviewer and essayist, and with Mark Mitchell, with whom he lives in Italy, has edited an imaginative anthology of gay fiction, characterised by the high literary quality of the pieces selected.

Fiction

Family Dancing (s) 1984; *The Lost Language of Cranes* 1986; *Equal Affections* 1989; *A Place I've Never Been* (s) 1990; *While England Sleeps* 1993 [rev. 1995]

Edited

The Penguin Book of Gay Short Stories (s) [with M. Mitchell] 1994

John le CARRÉ (pseud. of David Moore John Cornwell) 1931–

Le Carré was born in Poole, Dorset, the son of Ronnie Cornwell, a shady businessman who engaged in several million-pound swindles and was imprisoned for fraud. The author has said that his fascination for secrecy is the result of this background, which also inspired his novel, *The Honourable Schoolboy*. He attended Sherborne School, but, dissatisfied there, persuaded his father to send him to school in Switzerland, after which he spent a year at Berne University, from 1948 to 1949. He did military service in Austria, and then read modern languages at Lincoln College, Oxford, an experience he has credited with redeeming him from his background. After graduating in 1956, he worked for two years as a master at Eton, and then joined the Foreign Service in 1960, serving in Germany, latterly as consul in Hamburg. It is now known that during his early career le Carré worked for the secret service, although he has denied that his work in Germany also had an espionage character.

His first novel, *Call for the Dead* (1961), in some ways set the tone for all his spy thrillers, although it is the most conventional; it introduced his most successful character, George Smiley, the quiet, conscientious, fiercely patriotic but morally weary spymaster with an improbably glamorous and promiscuous wife. It states le Carré's themes: the futility of the espionage game; the erosion of morality it entails; the poisonous destruction of human relations by state requirements. Le Carré's second novel, the excellent detective novel set in a boys' school, *A Murder of Quality*, seemed to move away from the spy theme, but that quickly returned with his first international success, *The Spy Who Came in from the Cold* (1963), which established the new, unglamorous character of the spy story, and profited from huge public interest in the unmasking of the Cambridge-educated traitors.

Le Carré married in 1954; after his literary success enabled him to give up his career in 1964, the couple went to live on various Greek islands, but soon returned to England, and were divorced in 1971. He married again in 1972, and has four children, three from his first marriage. Le Carré's success turned into a cult in the 1970s with the three novels *Tinker, Tailor, Soldier, Spy*, *The Honourable Schoolboy* and *Smiley's People*, sometimes known as the 'Search for Karla trilogy', because they concern the struggle between Smiley and the Soviet spymaster Karla. Like many of the novels, the first two were turned into hugely successful television dramatisations, and they exemplify the fact that his Byzantine plots often have a simple unravelling mechanism. His novel *The Little Drummer Girl* initiates a tendency towards greater length but shows him at his most ingenious, while *The Secret Pilgrim* in 1991 was billed as Smiley's last bow. Le Carré's one non-genre novel, *The Naive and Sentimental Lover*, was considered less successful. He now divides his time between Britain, especially Cornwall, and the Continent. Controversy has attended his biography: Graham Lord was obliged to give up, and Robert Harris is now working on the project. The actress Charlotte Cornwell is le Carré's sister.

Fiction

Call for the Dead 1961; *A Murder of Quality* 1962; *The Spy Who Came in from the Cold* 1963; *The Looking-Glass War* 1965; *A Small Town in Germany* 1968; *The Naive and Sentimental Lover* 1971; *Tinker, Tailor, Soldier, Spy* 1974; *The Honourable Schoolboy* 1977; *Smiley's People* 1979; *The Little Drummer Girl* 1983; *A Perfect Spy* 1986; *The Russia House* 1989; *The Secret Pilgrim* 1991; *Our Game* 1995

(Nelle) Harper LEE 1926–

A descendant of the great Confederate general Robert E. Lee, Lee was born in Monroeville, Alabama, where her father – a former newspaper editor and proprietor, who had served as a state senator – practised as a lawyer. As a child she twice had to be saved by an older sister from being drowned by her mother, who was mentally unstable. Lee grew up alongside **Truman Capote**, whom she depicted in *To Kill a Mockingbird* as the precocious Dill; the character of Idabel Tompkins in Capote's *Other Voices, Other Rooms* (1948) is drawn from Lee. A curious pair of children – she far more masculine than he – they spent much of their childhood in a tree-house, planning their literary futures. Lee was educated locally and at the University of Alabama, spending a year as an exchange student at Oxford, but leaving college before taking the law degree for which she had been studying.

Intending to become a writer, she moved to New York and supported herself by working as an airline reservations clerk for BOAC. Friends clubbed together to finance her for a year out of work in which she could write, and the result was her only novel, *To Kill a Mockingbird* (1960), which won a Pulitzer Prize in 1961, was made into an award-winning film the following year, and has sold some 30,000,000 copies worldwide. Based upon her own childhood in Monroeville ('Maycomb'), it is narrated by a young girl, Scout Finch, whose lawyer father is called upon to defend a black man falsely accused of raping a white woman. The setting and several of the characters are drawn from life (Finch was the maiden name of Lee's mother), but the case is entirely fictional. Since Lee has never completed another novel (although she apparently started one during the 1960s) it has been suggested that this one was in fact written largely by Capote. There is no evidence to support this theory, and Lee, who refuses all attempts to interview her, has never commented publicly on the matter.

In 1959, she accompanied Capote to Holcombe, Kansas, as research assistant for an article he wanted to write about the murder of a farmer and his family. Capote thought that the locals would respond better to her, since she was still rooted in small-town life and had not acquired his off-putting metropolitan gloss. This project eventually resulted in Capote's classic 'non-fiction novel', *In Cold Blood* (1966). Lee conducted and wrote up many of the interviews, and her contribution to the book can hardly be underestimated.

Lee eventually left her Upper East Side apartment in New York and returned to Monroeville, where the Chamber of Commerce now has a mockingbird as its symbol. She continues to live there, apparently alone, a complete recluse.

Fiction

To Kill a Mockingbird 1960

Laurie LEE 1914–

The youngest but one of a family of eight, Lee was born in Stroud, Gloucestershire. His father deserted the family when Lee was very young, so they moved to the Gloucestershire village of Slad, where the pattern of life was still semi-feudal, and they lived in an overcrowded cottage, cooking on wood-fires and going to bed by candlelight. As Lee has said, he caught the very end of a way of life that had been going on for a thousand years, and he commemorated it in his most famous book, the autobiographical *Cider with Rosie*. He went to the village school and to Stroud Central School, leaving at the age of fifteen to become an errand-boy, and also giving lessons on the violin.

He had begun to write poetry in his early teens, but at twenty, he left for London, travelling on foot, and worked for a year as a builder's labourer. He then spent four years travelling in Spain (on the brink of the Civil War) and the eastern Mediterranean. He returned to England at the outbreak of the Second World War, and spent some years writing film scripts for, amongst others, the Crown Film Unit, and working as publications editor at the Ministry of Information. His first poem had been accepted in 1940 by **Cyril Connolly**'s magazine *Horizon*, and his first collection, *The Sun My Monument*, was published in 1944. During the earlier part of his writing career he published several more slim collections of verse, containing poems noted for their strong lyric and personal impulse but perhaps rather slight in content.

He married, in 1950, Catherine Polge, a Provençal woman, daughter of a Martigues fisherman, with whom he has one daughter. From 1950 to 1951 Lee was caption-writer-in-chief for the Festival of Britain (for which service he was awarded the MBE) and since then has lived as a freelance writer.

Immense success came as a result of *Cider with Rosie*, which won the W.H. Smith Award and which has been highly praised for its lyrical and joyful account of childhood, although some critics found it over-lush. Its sequels, *As I Walked Out One Midsummer Morning* and *A Moment of War*, describe his early years of travelling. Lee has written other memoirs, screenplays and travel books, but has published sparsely in recent years. For some time he has divided his life between his homes in London and Gloucestershire.

Non-Fiction

Land at War 1945; *We Made a Film in Cyprus* [with R. Keene] 1947; *An Obstinate Exile* 1951; *A Rose for Winter: Travels in Andalusia* 1955; *Cider with Rosie* (a) 1959; *Man Must Move: the Story of Transport* [with D. Lambert] 1960 [rev. as *The Wonderful World of Transport* 1969]; *The Firstborn* 1964 [rev. as *Two Women: a Book of Words and Pictures* 1983]; *As I Walked Out One Midsummer Morning* (a) 1969; *I Can't Stay Long* (a) 1975; *Innocence in the Mirror* 1978; *A Moment of War* (a) 1991

Poetry

The Sun My Monument 1944; *The Bloom of Candles* 1947; *My Many-Coated Man* 1955; *Poems* 1960; *Pergamon Poets 10* [with C. Causley; ed. E. Owen] 1970; *Selected Poems* 1983

Plays

Cyprus Is an Island (f) 1946; *The Voyage of Magellan* (r) 1946 [pub. 1948]; *Peasant's Priest* 1947; *A Tale in a Teacup* (f) 1947

Edited

New Poems 1954 [with C. Hassall, R. Warner] 1954

Translations

Avigdor Dagan *The Dead Village* 1943

Collections

I Can't Stay Long 1975

Vernon LEE (pseud. of Violet Paget) 1856–1935

A mystery surrounds the extraction of Lee's father, who at some point assumed the name of Henry Ferguson Paget: he had grown up in Poland, and had been forced to leave there after taking part in the failed insurrection of 1848, but whether he was of Polish, French aristocratic or English ancestry seems uncertain. Once in England, he became tutor to a boy called Eugene Lee-Hamilton, and eventually married his pupil's widowed mother; Vernon Lee was the only child of this union. Mrs Paget came from a Welsh landowning, originally slave-owning, family, which had been called Adams but which adopted the name of Abadam. Her considerable wealth enabled the family to live a wandering life around Europe, mainly in Germany until 1866, when they began wintering in Rome, before settling in Florence in 1873, thenceforward the centre of Lee's life.

Lee was trained as a precocious author by her mother, and published her first essay, written in French, when only thirteen, but grew up in an atmosphere lacking affection. Aged fifteen, she discovered the deserted meeting-place of the eighteenth-century Arcadian Academy of Rome, and this started the train of thought which led to the publication of *Studies of the Eighteenth Century in Italy* (1880), a brilliantly erudite collection of essays on art and music which immediately made her famous and gained her the entrée, in 1881, on the first of her many annual visits to England, to the cultivated society of the day. It was a signal for a flood of books, but her novel *Miss Brown*, an attempted satire on London aesthetic circles, led to a social setback ('as it is her first attempt at a novel, so it is to be hoped that it may be her last,' wrote **Henry James** in a private letter). The breakdown of her passionate relationship with Mary Robinson, who left her to get married, led to a nervous collapse in 1887, but she found temporary consolation with Clementina 'Kit' Anstruther-Thomson, and wrote several works on aesthetics with her, a course which led to another controversy when Bernard Berenson accused the authors of plagiarism. A third major scandal was to follow during the First World War when, although living in England, she refused to support the Allies and published an anti-war play, *The Ballet of the Nations*.

Lee's more than forty works fall mainly into four classes: fiction, including historical novels and short (often ghost) stories; travel books such as *Genius Loci*; volumes of essays on history and art; and works on aesthetics, such as *The Beautiful*; other works include a play, a puppet-show and a biography of the Young Pretender's wife. A ghost story such as 'Oke of Okehurst' still has genuine power, but it is as a great Florentine hostess that Lee is most remembered: operating first with her parents and her half-brother, a failed poet and diplomat who developed a psychosomatic illness which left him paralysed from the waist down, and later, increasingly deaf and ill, alone.

Opinions about her differed, from **Maurice Baring**, who considered her the best talker on earth, to those who thought her a self-centred bore; but her place in cosmopolitan literary history and gossip seems assured.

Fiction

Tuscan Fairy Tales (s) 1880; *Ottilie* 1883; *Miss Brown* 1884; *A Phantom Lover* 1886; *Hauntings* (s) 1892; *Vanitas* (s) 1892; *Penelope Brandling* 1903; *Pope Jacynth and Other Fantastic Tales* (s) 1904; *Sister Benevenuta and the Christ Child* (s) 1906; *Louis Norbert* 1914; *For Maurice* (s) 1927; *The Snake Lady and Other Stories* (s) 1954; *Supernatural Tales* (s) 1955 [rev. 1987]; *Pope Jacynth and More Supernatural Tales* (s) 1956

Non-Fiction

Studies of the Eighteenth Century in Italy 1880; *Belcaro, Being Essays on Sundry Aesthetical Questions* 1881; *The Countess of Albany* 1884; *Euphorion: Being Studies of the Antique and the Medieval in the Renaissance* 1884; *Juvenilia: Being a Second Series of Essays on Sundry Aesthetical Questions* 1887; *Renaissance Fancies and Studies: Being a Sequel to Euphorion* 1895; *Limbo and Other Essays* 1897; *Genius Loci* 1899; *Le Rôle de l'élément moteur dans la perception ésthétique visuelle* [with C. Anstruther-Thomson] 1901; *Hortus Vitae: Essays on the Gardening of Life* 1904; *The Enchanted Woods and Other Essays* 1905; *The Spirit of Rome: Leaves from a Diary* 1906; *Gospel of Anarchy and Other Contemporary Studies* 1908; *The Sentimental Traveller* 1908; *Laurus Nobilis: Chapters on Art and Life* 1909; *Vital Lies: Beauty and Ugliness and Other Studies in Psychological Aesthetics* [with C. Anstruther-Thomson] 1912; *Studies of Some Varieties of Recent Obscurantism* 1912; *The Beautiful* 1913; *The Tower of Mirrors: and Other Essays on the Spirit of Places* 1914; *The Handling of Words: and Other Studies in Literary Psychology* 1923; *The Golden Keys and Other Essays on the Genius Loci* 1925; *Proteus: or the Future of Intelligence* 1925; *The Poet's Eye* 1926; *Music and Its Lovers* 1932; *Letters* 1937

Plays

The Prince of the Hundred Soups 1883; *Baldwin* 1886; *Althea* 1894; *Ariadne in Mantua* 1916 [pub. 1903]; *The Ballet of the Nations* 1915 [rev. as *Satan the Waster* 1920]

Edited

Clementina Anstruther-Thomson *Art and Man* 1924

Collections

A Vernon Lee Anthology [ed. I. Cooper Willis] 1929

Biography

Vernon Lee by Peter Gunn 1964

(Rudolph) John (Frederick) LEHMANN 1907–1987

Lehmann was the son of Rudolph Chambers Lehmann, a Liberal Member of Parliament of German Jewish origin and (as 'R.C.L.') famous contributor to *Punch* and his American wife. His sisters included the novelist **Rosamond Lehmann** and the actress Beatrix Lehmann. He was educated at Eton, where he was a disciplinarian Captain of Chamber, and Trinity College, Cambridge, where he made the friendship of Julian Bell, **Virginia Woolf**'s nephew. When he came down, the Woolfs' Hogarth Press both published his first volume of poems, *A Garden Revisited* (1931), and offered him a job. Thus was inaugurated a career during which he was to write in almost every medium but was always most regarded as an editor and inspirer of others.

During the 1930s Lehmann lived in Vienna for long periods; published various volumes, such as an early novel, *Evil Was Abroad*, and a study of Danubian problems, *Down River*;

became general manager and a partner in the Hogarth Press in 1938; and founded his own literary magazine, *New Writing*, in 1936. The magazine was particularly well known for its promotion of the **[W.H.] Auden** group of poets and the proletarian writing fashionable at the time. During the Second World War, it became the enormously influential *Penguin New Writing*, and Lehmann was, with **Cyril Connolly**, one of the two leading editors of the day. In 1946 he tried to win financial control of the Hogarth Press but failed, and so started his own publishing company, John Lehmann Ltd, which produced fine-quality books but lost the support of its backers in 1953. The same year, Lehmann became founding editor of the *London Magazine*, holding this position until 1961, when he relinquished it to **Alan Ross**. This ended his period as a leading editor; in later life, although much less influential, he continued to review and to publish his works, held numerous visiting professorships at American universities, and was president of the Royal Literary Fund.

Later volumes included accomplished if minor verse, literary appreciations of friends such as the **Sitwells** and **Christopher Isherwood** (the latter, his last book, was published posthumously), a flimsy study of the poetry of the First World War, a short biography of **Rupert Brooke**, and a rather colourless autobiography in three volumes. He was unmarried and a lifelong practising homosexual; during the war he had had a relationship with the Greek poet Demetrios Capetanakis who died in 1944, and throughout his life he took a strong interest in young writers of talent and good looks. His late, homosexual novel, *In the Purely Pagan Sense*, demonstrated once again that he had little talent for fiction.

Poetry

A Garden Revisited and Other Poems 1931; *The Noise of History* 1934; *Forty Poems* 1942; *The Sphere of Glass and Other Poems* 1944; *The Age of the Dragon: Poems 1930–1951* 1951; *Collected Poems 1930–1963* 1963; *Christ the Hunter* 1965; *Holborn* 1970; *The Reader at Night and Other Poems* 1974; *New and Selected Poems* 1985

Non-Fiction

Prometheus and the Bolsheviks 1937; *Down River* 1939; *Demetrios Capetanakis: a Greek Poet in London* 1947; *The Open Night: Essays on Literature* 1952; *In My Own Time: Memoirs of a Literary Life* (a) 1969: *The Whispering Gallery* 1955, *I Am My Brother* 1960, *The Ample Proposition* 1966; *Ancestors and Friends* 1962; *A Nest of Tigers: Edith, Osbert and Sacheverell Sitwell in Their Times* 1968; *Virginia Woolf and Her World* 1975; *Edward Lear and His World* 1977; *Thrown to the Woolfs* 1978; *Rupert Brooke: His Life and His Legend* 1980; *The English Poets of the First World War* 1981; *Three Literary Friendships: Byron and Shelley, Rimbaud and Verlaine, Robert Frost and Edward Thomas* 1983; *Christopher Isherwood: a Personal Memoir* 1987

Edited

Ralph Fox: a Writer in Arms [with T.A. Jackson, C. Day Lewis] 1937; *Poems for Spain* [with S. Spender] 1939; *New Writing in Europe* 1940; *Poems from New Writing 1936–1946* 1946; *Shelley in Italy: an Anthology* 1947; *French Stories from New Writing* (s) 1947; *The Year's Work in Literature 1949–1950* 1950; *English Stories from New Writing* (s) 1951; *Pleasures of New Writing* 1952; *The Chatto Book of Modern Poetry, 1915–1955* [with C. Day Lewis] 1956; *The Craft of Letters in England: a Symposium* 1956; *Modern French Stories* (s) 1956; *Coming to London: Reminiscences by Various Writers* 1957; *Italian Stories of Today* (s) 1959; *Selected Poems of Edith Sitwell* [with D. Parker] 1965; *Edith Sitwell: Selected Letters* 1970; *The Penguin New Writing, 1940–1950* [with R. Fuller] 1985

Fiction

Evil Was Abroad 1938; *In the Purely Pagan Sense* 1976

Rosamond (Nina) LEHMANN 1901–1990

Lehmann was one of four children of the Liberal Member of Parliament, founder of *Granta* and editor of *Punch*, Rudolph Chambers Lehmann. She was half-American and sister to the poet and editor **John Lehmann** and the actress Beatrix Lehmann (both of whom also wrote novels). The family was brought up at Fieldhead, a riverside house at Bourne End, Buckinghamshire, and she was educated privately before becoming a scholar of Girton College, Cambridge. She met her first husband at Cambridge, and married him in 1922, after which the couple lived in Newcastle.

Her first novel, the autobiographical *Dusty Answer* (1927), had an immense popular success and established her as a front-rank novelist. She followed it up with several others, all noted for their sensuous and passionate exploration of women's experience and often touching on themes generally taboo in the inter-war period: *Dusty Answer* and *A Note in Music* both dealt with aspects of homosexuality, while *The Weather in the Streets* was one of the first novels in which the heroine has a backstreet abortion.

After divorcing her first husband, she married in 1928 the painter and communist Wogan Philipps (later Lord Milford), with whom she had a son and a daughter; this marriage also ended in divorce. She was a full member of the high-society literary Bohemia which dominated the cultural scene in the inter-war period, and was described as the 'youngest and prettiest of British novelists'. In 1941 she embarked upon a nine-year relationship with **C. Day Lewis**, with whom she lived in London, where he was working for the Ministry of Information. Although Day Lewis had a wife and two children in Devon, this affair was as important to him as it was to Lehmann, whose marriage had broken up before the war. Lehmann was the dedicatee of Day Lewis's *Word over All* (1943) and the subject of several of the volume's poems, while Lehmann's *The Echoing Grove* draws upon the affair. Day Lewis shuttled between wife and lover until in 1950 he fell in love with the actress Jill Balcon, whom he married in 1953. Lehmann was devastated – her original title for *The Echoing Grove*, published the same year, was 'Buried Day', the significance of which had to be pointed out to her. (Day Lewis himself used it as the title of his 1960 autobiography.)

After the tragically early death of her daughter Sally (wife of the poet **P.J. Kavanagh**) in 1958, she entered a period of silence and withdrawal from the literary world, broken by *The Swan in the Evening*, a meditation on her daughter's death which had turned her towards spiritualism. *A Sea-Grape Tree*, generally considered an unsatisfactory work, was a late return to the novel in 1976.

Her many honours included the CBE and the Companion of Literature, and the republication of her novels by Virago in the 1980s led to a major revival of interest in her work. In these last years she divided her time between a house near Aldeburgh, Suffolk, and one in South Kensington, London.

Fiction
Dusty Answer 1927; *A Note in Music* 1930; *Invitation to the Waltz* 1932; *The Weather in the Streets* 1936; *The Gypsy's Baby* (s) 1939; *The Ballad and the Source* 1944; *The Echoing Grove* 1953; *A Man See Afar* [with W. Tudor Pole] 1965; *A Sea-Grape Tree* 1976

Plays
No More Music 1939

Non-Fiction
Letters from Our Daughters [with Cynthia, Baroness Sandys] 2 vols 1971; *The Awakening Letters* [with Cynthia, Baroness Sandys] 1978; *My Dear Alexias: Letters from Wellesley Tudor Pole to Rosamond Lehmann* [ed. E. Gaythorpe] 1979; *The Swan in the Evening* 1967; *Rosamond Lehmann's Album* 1985

Edited
Orion: a Miscellany 1945

Translations
Jacques Lemarchand *Geneviève* 1947; Jean Cocteau *Les Enfants Terribles* 1955

Mike LEIGH 1943–

A doctor's son, Leigh was born in Salford, 'a middle-class kid right in the middle of a working-class area'. He went to the local grammar school and then took a scholarship to the Royal Academy of Dramatic Art at the age of seventeen. His two years at drama school were a disappointment, and it was not until his time at Camberwell School of Arts and Crafts in 1963 that he realised why: 'We never made an organic or truthful statement about what we were experiencing – everything was secondhand or borrowed or learnt. Nobody ever confronted themselves with in-the-street experience, or tried to distil or express it. Now you'll find all the importance of that in Stanislavsky, but to me it was a revelation: suddenly you are in an area of creative investigation instead of mere reproduction.'

After Camberwell, Leigh attended the Central School of Art in 1964 and the London Film School in 1965, before becoming associate director of the Midlands Arts Centre. In 1967, he landed a job as assistant director in Peter Hall's last Stratford season with the Royal Shakespeare Company, but his individual working method – relying on collaborative scripts achieved with the actors and only completed on the first night of performance – precluded him from working on many of the company's schedules. 'So I went off into the wilderness for a while.' In 1968, he directed the E15 Acting School, and then spent a year teacher training in Manchester before returning to directing plays.

His first major success was *Abigail's Party* (1977), set in a garish suburban sitting-room. Although Leigh maintains firm directorial control, his working method is perilously individual, requiring that actors work independently to develop their character roles before reassembling to shape the play. He insists that 'there's no committee involved'. None the less, there tends to be little actual plot structure to his films or plays, rather the dramatic realisation of snatched conversation, observed mannerisms and seemingly minute, but genuinely terrible, tragic moments. In his *Greek Tragedy* Leigh translates his watchful eye on the farcical nature of human values to Australian culture. It is a *tour de force* of dramatic realism, and examines the vulnerability of human hopes and the monstrous effects of financial success.

Much of his best work has been done for film and television. He adapted his aptly titled stage play *Bleak Moments* for a feature film debut in 1972 and has since taught regularly at the London Film School and the National Film School. He had a notable success with the television play *Nuts in May*, a comedy set amongst campers, but perhaps more impressive are films set amongst the hopeless and dispossessed, which manage to combine comic scenes with near-tragedy. He has been accused of patronising and caricaturing his characters, but films such as *High Hopes*, in which grotesque Thatcherite aspirations are set against the last vestiges of family and community, are deeply felt and genuinely moving. *Life Is Sweet* once again included broad comedy (a hopelessly over-ambitious and incompetent restaurateur) and genuine anguish (a disaffected daughter suffering from bulimia). Leigh offers no solutions, but there is an underlying faith in humane values and sheer resilience.

Naked confirmed his talent for mixing social comment with disturbing and profound characterisation. David Thewlis's portrayal of an unemployed misogynist, who rapes a girl and then leads a life of increasing destruction in the alleyways of London, is monumentally effective and won him an award for best actor at Cannes that year, with Leigh taking the award for best director. Some critics have felt that the central character's enmity towards women is too forced, and claim to see a heavy-handedness in his treatment of this issue.

In 1973, he married the actress Alison Steadman, with whom he has two sons, and who has acted in many of his plays and films.

Plays

The Box Play 1965; *The Last Crusade of the Five Little Nuns* 1966; *My Parents Have Gone to Carlisle* 1966; *Waste Paper Guards* 1966; *NENAA Stratford on Avon* 1967; *Down Here and Up There* 1968; *Individual Fruit Pies* 1968; *Big Basil* 1969; *Epilogue* 1969; *Glum Victoria and the Lad with the Specs* 1969; *Bleak Moments* **(st)** 1970, **(f)** 1972; *A Rancid Pong* 1971; *Dick Whittington and His Cat* 1973; *Hard Labour* **(tv)** 1973; *A Mug's Game* **(tv)** 1973; *Wholesome Glory* 1973; *Babies Grow Old* 1974; *The Silent Majority* 1974; *Five Minute Films* **(tv)** 1975; *The Permissive Society* **(tv)** 1975; *Knock for Knock* **(tv)** 1976; *Nuts in May* **(tv)** 1976; *Abigail's Party* 1977; *The Kiss of Death* **(tv)** 1977; *The Jaws of Death* 1978; *Ecstasy* 1979; *Too Much of a Good Thing* **(r)** 1979; *Who's Who* **(tv)** 1979; *Grown Ups* **(tv)** 1980; *Goose-Pimples* 1981; *Home Sweet Home* **(tv)** 1982; *Meantime* **(tv)** 1983; *Four Days in July* **(tv)** 1985; *The Short and Curlies* **(f)** 1987; *High Hopes* **(f)** 1988; *Greek Tragedy* 1989; *Smelling a Rat* 1989; *Life Is Sweet* **(f)** 1991; *Naked* **(f)** 1993

Hugh LEONARD (pseud. of John Byrne) 1926–

Leonard was born illegitimate in Dublin, and grew up in a small cottage at the nearby seaside resort of Dalkey, where his adoptive father worked from childhood to old age as a gardener in the service of the Jacob family of manufacturers. He later added the surname of his adopted father as his own middle name and called himself John Keyes Byrne; he is still commonly known in Ireland as Jack Byrne. In 1941, he won a scholarship to Presentation College, Dun Laoghaire, a school run by a religious order, and he remained there until 1945, when, having worked briefly in a film rental office, he entered the Irish Land Commission as a clerk. In 1955, he married Paule Jacquet, Belgian by birth; they have one daughter.

While at the Land Commission, he became involved with the organisation's amateur dramatic society, which put on his earliest plays. In 1956, one of these, *The Big Birthday*, was accepted by the Abbey Theatre, and launched him on the professional Irish stage. (At this point he adopted his pseudonym, which was the name of a character in an earlier play rejected by the Abbey.) After two more plays had been produced, he decided to leave the civil service, in 1959, and initially supported himself by writing serials for sponsored radio. In 1961, he went to Manchester to work for Granada Television as a script editor, and from 1963 to 1970 was a freelance writer in London. From this period date the majority of his many contributions to British television, which have taken the form of adaptations of classic authors, contributions to popular series and whole series.

He was also during the same period developing a profitable association with the Dublin Theatre Festival, where his earliest big success, *Stephen D*, an adaptation of **James Joyce**, was put on in 1962. In 1970, he took the decision to return to Ireland, where he lives at his childhood home of Dalkey, and maintains a phenomenal rate of productivity, which has led to the production of more than twenty original plays as well as many adaptations and screenplays. He is, with **Brian Friel**, acknowledged as the leading Irish playwright, and, while he takes Ireland as his subject, this is in no parochial sense; however, he has never really established himself in the London West End, and in the USA mainly enjoyed a close relationship with the theatre group at Olney, Maryland, before conquering Broadway in the late 1970s with *Da*, his touching play about his adoptive father. In his mastery of witty dialogue and stagecraft which does not exclude deep feeling, and his relative freedom from innovation, Leonard is perhaps reminiscent of earlier veterans of the 'well-made play' such as **Terence Rattigan**.

He was the director of the Abbey Theatre for a brief, stormy period in the mid-1970s, and from 1978 has been programme director of the Dublin Theatre Festival. Since the age of sixty, he has published *Parnell and the Englishwoman*, relating the fall of Charles Stewart Parnell, which also exists as a television series. His other works include two volumes of memoirs and several collections of trenchant columns contributed to Irish newspapers.

Plays

The Italian Road [as John Byrne] 1954 [np]; *The Big Birthday* [as John Byrne] 1956 [np]; *A Leap in the Dark* **(st)** 1957, **(tv)** 1960 [np]; *Madigan's Lock* 1958 [pub. 1987]; *A Walk on the Water* 1960 [np]; *The Passion of Peter Ginty* [from Ibsen's play *Peer Gynt*] 1961 [np]; *The Irish Boys* **(tv)** 1962 [np]; *Saki* **(tv)** 1962; *Stephen D.* [from J. Joyce] **(st)**, **(tv)** 1962 [pub. 1965]; *Dublin One* [from J. Joyce's novel *Dubliners*) 1963; *Family Solicitor* **(tv)** 1963; *Jezebel* **(tv)** 1963; *A Kind of Kingdom* **(tv)** 1963; *The Poker Session* **(st)** 1963, **(f)** 1967; *Do You Play Requests?* **(tv)** 1964; *The Family Way* [from E. Labiche] 1964 [np]; *The Hidden Truth* **(tv)** 1964; *The Late Arrival of the Incoming Aircraft* **(tv)** 1964 [pub. 1968]; *My One True Love* **(tv)** 1964; *Realm of Error* **(tv)** 1964; *The Second Wall* **(tv)** 1964; *A Triple Irish* **(tv)** 1964; *Undermind* **(tv)** 1964; *A View from the Obelisk* **(tv)** 1964, **(st)** [in *Scorpions*] 1983; *Blackmail* **(tv)** 1965; *Great Big Blond* **(tv)** 1965; *I Loved You Last Summer* **(tv)** 1965; *Public Eye* **(tv)** 1965; *The Saints Go Cycling In* [from F. O'Brien's novel *The Dalkey Archive*] 1965 [np]; *The Informer* **(tv)** 1966; *Insurrection* **(tv)** 1966; *The Liars* **(tv)** 1966; *The Lodger and the Judge* **(tv)** [from G. Simenon] 1966; *Mick and Mick* 1966; *The Retreat* **(tv)** 1966; *Second Childhood* **(tv)** 1966; *Silent Song* [from F. O'Connor] **(tv)** 1966; *Out of the Unknown* **(tv)** 1966–7; *Great Catherine* **(f)** 1967; *Great Expectations* **(tv)** [from Dickens's novel] 1967; *Interlude* **(f)** [with L. Langley] 1967; *Love Life* **(tv)** 1967; *The Quick, and the Dead* 1967 [np]; *A Time of Wolves and Tigers* **(tv)** 1967, **(st)** [in *Irishmen*] 1975 [pub. 1983]; *Wuthering Heights* **(tv)** [from E. Brontë's novel] 1967; *Assassin* **(tv)** 1968; *The Au Pair Man* 1968 [pub. 1974]; *The Hound of the Baskervilles* and *A Study in Scarlet* **(tv)** [from A. Conan Doyle] 1968; *The Corpse Can't Play* **(tv)** 1968; *Lord Dismiss Us* **(f)** 1968; *A Man and His Mother-in-Law* **(tv)** 1968; *Nicholas Nickleby* **(tv)** [from Dickens's novel] 1968; *No Such Thing as a Vampire* **(tv)** 1968; *The Barracks* [from J. McGahern's novel] 1969; *Dombey and Son* **(tv)** [from Dickens's novel] 1969; *Hunt the Peacock* **(tv)** [from H.R.F. Keating's novel] 1969; *P&O* **(tv)** [from Somerset Maugham] 1969; *The Possessed* **(tv)** [from Dostoevsky] 1969; *The Scarperers* **(tv)** 1969; *Talk of Angels* **(tv)** 1969; *Jane* **(tv)** [from Somerset Maugham] 1970; *Percy* **(f)** [with T. Feely] 1970; *A Sentimental Education* **(tv)** [from Flaubert's novel] 1970; *Whirligig* **(f)** 1970; *Me Mammy* **(tv)** 1970–71; *The Sinners* **(tv)** 1970–71; *Pandora* **(tv)** 1971; *The Patrick Pearse Motel* 1971 [pub. 1972]; *The Removal Person* **(tv)** 1971; *White Walls and Olive Green Carpets* **(tv)** 1971; *The Ghost of Christmas Present* **(tv)** 1972; *The Moonstone* **(tv)** [from W. Collins's novel] 1972; *Our Miss Fred* **(f)** 1972; *The Sullen Sisters* **(tv)** 1972; *Tales from the Lazy Acres* **(tv)** 1972; *The Truth Game* **(tv)** 1972; *The Virgins* **(tv)** 1972; *The Watercress Girl* **(tv)** [from H.E. Bates's story] 1972; *Another Fine Mess* **(tv)** 1973; *The Bitter Pill* **(tv)** 1973; *Da* **(st)** 1973, **(f)** 1986 [pub. 1976]; *The Higgler* **(tv)** 1973; *High Kampf* **(tv)** 1973; *Judgement Day* **(tv)** 1973; *Milo O'Shea* **(tv)** 1973; *Stone Cold Sober* **(tv)** 1973; *The Travelling Woman* **(tv)** 1973; *The Actor and the Alibi*, *The Eye of Apollo*, *The Forbidden Garden*, *The Hammer of God*, *The Quick One*, *The Three Tools of Death* **(tv)** [from G.K. Chesterton] 1974; *Rake's Progress* **(f)** 1974; *Summer* 1974 [pub. 1979]; *Irishmen* 1975 [pub. as *Suburb of Babylon* 1983]; *Some of My Best Friends Are Husbands* [from E. Labiche] 1976 [np]; *Liam Liar* [from W. Hall and K. Waterhouse's play *Billy Liar*] 1976 [np]; *Time Was* 1976 [pub. 1981]; *Bitter Suite* **(tv)** 1977; *The Fur Coat*, *Teresa* and *Two of a Kind* **(tv)** [from S. O'Faolain] 1977; *Herself Surprised* **(f)** 1977; *London Belongs to Me* **(tv)** [from N. Collins's novel] 1977; *The Last Campaign* **(tv)** [from J. Johnston] 1978; *The Ring and the Rose* **(tv)** 1978; *A Life* 1979 [pub. 1980]; *Strumpet City* **(tv)** [from J. Plunkett's novel] 1980; *The Little World of Don Camillo* **(tv)** [from G. Guareschi] 1981; *Kill* 1982 [np]; *Good Behaviour* **(tv)** [from M. Keane] 1983; *O'Neill* **(tv)** 1983; *Pizzaz* [in *Scorpions*] 1983 [pub. 1986]; *Scorpions* 1983 [np]; *Hunted Down* **(tv)** [from Dickens] 1985; *The Irish R.M.* **(tv)** 1985; *The Mask of Moriarty* [from A. Conan Doyle] 1985 [pub. 1987]; *Widows' Peak* **(f)** 1986; *Troubles* **(tv)** 1987; *Parnell and the Englishwoman* **(tv)** [from his novel] 1991; *Moving* 1992 [pub. 1994]; *The Lilly Lally Show* 1994 [np]; *Senna for Sonny* 1994 [np]

Fiction

Parnell and the Englishwoman 1990

Non-Fiction

Leonard's Last Book 1978; *Home Before Night: Memoirs of an Irish Time and Place* **(a)** 1979; *A Peculiar People and Other Foibles* 1979; *Leonard's Year* 1985; *Out After Dark* **(a)** 1989; *Rover and Other Cats* 1992

Doris (May) LESSING 1919–

Lessing was born Doris Tayler in Kermanshah, Persia (now Iran), where her English-born father was a bank clerk with the Imperial Bank of Persia. After a visit to London in 1925 her father bought a maize farm in Southern Rhodesia where Lessing spent the remainder of her 'hellishly lonely' childhood. She was educated at a Roman Catholic convent in Salisbury, and left the Girls' High School there when she was fourteen to immerse herself in reading and writing. In her first, unforgiving, volume of autobiography, *Under My Skin*, she describes herself as 'a drop-out, long before the term had been invented'. She wrote two 'bad novels' and worked in Salisbury as a nursemaid, telephone operator and clerk. She was married in 1939, had a son and a daughter, and divorced in 1943.

She joined the Communist Party, and this period of her life is reflected in *A Ripple from the Storm*, 'the most directly autobiographical' novel in the five-volume *Children of Violence* sequence. She married the German political activist Gottfried Lessing in 1943 'because in those days people could not have affairs, let alone live together, without attracting unpleasant comment'. (Gottfried later became the German ambassador to Uganda and was accidentally killed in the 1979 revolt against Idi Amin.) The marriage was as ill-fated as her first, and in 1949, with the son from her second marriage and the manuscript of *The Grass Is Singing*, she travelled to London. Published in 1950 to immediate acclaim, the novel describes the breakdown of a poor Rhodesian farming couple's marriage and the wife's obsessive relationship with a black houseboy who eventually murders her.

There are elements of similarity between the events of Lessing's life and those of Martha Quest, heroine of the *Children of Violence* sequence, but it is best seen as a psychological autobiography, and has been described by the author as 'a study of the individual conscience in its relation with the collective'. Lessing's best-known novel, *The Golden Notebook*, was praised by **Joyce Carol Oates** as 'predating and superseding even the most sophisticated of all "women's liberation" works'. It is also a complex formal experiment which explores political assumptions and the theme of mental breakdown. Various forms of madness and mental instability recur in Lessing's work, notably in the frequently bizarre *Briefing for a Descent into Hell* and in the rather more comic world of squatters-turned-terrorists in *The Good Terrorist*. The range of Lessing's interests is evident in the ambitious *Canopus in Argos* sequence which employs the techniques and machinery of traditional science fiction.

In the 1980s Lessing provoked a minor storm in the world of publishing when she pseudonymously submitted the first of her two 'Jane Somers' novels to her usual publisher, and used its rejection to highlight the difficulties faced by unestablished writers. Her other books include two accounts of return visits to Zimbabwe (formerly Rhodesia) and a study of Afghanistan which reflects her interest in Sufism. In 1981 David Gladwell adapted *The Memoirs of a Survivor* as a film, and Lessing has collaborated with the composer Philip Glass on an opera based on her novel *The Making of the Representative for Planet 8*. She has won numerous prizes, including a Somerset Maugham Award in 1956 and the W.H. Smith Award in 1986.

Fiction

The Grass Is Singing 1950; *This Was the Old Chief's Country* (s) 1951; *Children of Violence*: *Martha Quest* 1952, *A Proper Marriage* 1954, *A Ripple from the Storm* 1965, *Landlocked* 1965, *The Four-Gated City* 1969; *No Witchcraft* (s) 1956; *Retreat to Innocence* 1956; *Five* (s) 1953; *The Habit of Loving* (s) 1957; *The Golden Notebook* 1962; *A Man and Two Women* (s) 1963; *African Stories* (s) 1964; *The Black Madonna* (s) 1966; *Nine African Stories* (s) [ed. M. Marland] 1968; *Briefing for a Descent into Hell* 1971; *The Story of a Non-Marrying Man* (s) [US *The Temptation of Jack Orkney*] 1972; *The Summer Before the Dark* 1973; *The Memoirs of a Survivor* 1975; *Collected Stories* (s) 1978; *Collected African Stories* (s) 2 vols 1973; *The Canopus in Argos Archives*: *Re: Colonised Planet 5, Shikasta* 1979, *The Marriages Between Zones Three, Four and Five* 1980, *The Sirian Experiment* 1980, *The Making of the Representative for Planet 8* 1982, *Documents Relating to the Sentimental Agents in the Volyen Empire* 1983; *The Diaries of Jane Somers* [as 'Jane Somers'] 1984: *Diary of a Good Neighbour* 1983, *If the Old Could ...* 1984; *The Good Terrorist* 1985; *The Fifth Child* 1988; *London Observed* (s) 1990

Plays

Before the Deluge 1953 [np]; *Each His Own Wilderness* 1958 [pub. in *New English Dramatists* 1959]; *Mr Dollinger* 1958 [np]; *The Truth about Billy Newton* 1960 [np]; *Play with a Tiger* 1962; *The Grass Is Singing* (tv) [from her novel] 1962; *Care and Protection* (tv) 1966; *Do Not Disturb* (tv) 1966; *The Storm* (tv) [from A. Ostrovsky's play] 1966 [np]; *Between Men* (tv) 1967; *The Singing Door* [pub. in *Second Playbill 2* ed. A. Durband 1973]; *The Making of the Representative for Planet 8* (l) [from her novel; with P. Glass] 1988 [np]

Poetry

14 Poems 1959

Non-Fiction

Particularly Cats 1967; *Going Home: Personal Reminiscences* (a) 1957 [rev. 1968]; *In Pursuit of the English: a Documentary* 1960; *A Small Personal Voice* [ed. P. Schlueter] 1974; *Prisons We Choose to Live Inside* 1986; *The Wind Blows Away Our Words and Other Documents Relating to Afghanistan* 1987; *African Laughter* 1992; *Under My Skin: Volume One of My Autobiography* (a) 1994; *Playing the Game* 1995

Collections

The Doris Lessing Reader 1989

Ada (Esther) LEVERSON 1862–1933

Dubbed by Grant Richards 'the Egeria of the whole Nineties movement', Leverson survived the 1890s to become a considerable novelist of the Edwardian era. She was one of eight children; her father was Samuel Beddington, whose wealth came from wool-trading and property, while her mother came from a distinguished Jewish family and was the daughter of the Liberal Member of Parliament John Simon. Her youngest sister, Violet, married Sydney Schiff, who wrote novels under the name Stephen Hudson. She was educated privately, and at nineteen, against her family's wishes, married Ernest Leverson, the son of a diamond merchant and twelve years her senior. The Beddingtons' concern was justified, since it was shortly revealed that Leverson had an illegitimate daughter in France. He worked in the City, but his principal recreation was gambling. His cousin, Marguerite, married Brandon Thomas, and the two young couples became very close, Thomas borrowing the name of one of Ada's brothers for his play *Charley's Aunt* (1892). A rather more distinguished playwright friend was Oscar Wilde, whom Ada met in 1892, and whose works she parodied in *Punch*. (Her own Wildean play, *The Triflers*, which she spent some twenty years adapting from the French, was never staged.) Wilde called her 'the Sphinx', a nickname she retained throughout her life. The Leversons proved Wilde's staunchest friends, offering him refuge while he was on bail, and greeting him upon his release from prison: 'Sphinx, how marvellous of you to know exactly the right hat to wear at seven o'clock in the morning to meet a friend who has been away!' the playwright commented. Her loyalty was all the more remarkable since she had a horror of scandal; another test of it came in 1897, when her husband was cited as co-respondent in a divorce suit.

The Leversons had two children, a son who died young and a daughter, the writer Violet Wyndham; the marriage was not happy and she had many admirers, including William, 4th Earl of Desart, the novelist **George Moore**, Prince Henri d'Orléans and her publisher Grant Richards, any of whom might have proved a better match. In the event, her husband lost his money in a business enterprise and was sent by his father to live in Canada, where he was looked after by his illegitimate daughter until his death in 1922. Leverson remained in London where her literary career took off. She had contributed stories to the *Yellow Book*, a weekly women's column entitled 'White and Gold' to the *Referee* (under the name 'Elaine'), and interviewed **Max Beerbohm** for the *Sketch*; now she became a novelist, publishing between 1907 and 1916 six witty comedies of manners, three of them featuring 'the little Ottleys', a young couple much like the Leversons. She wrote in bed 'in a confusion of foolscap, newspapers, cigarettes and oranges', assisted by a 'tall, gaunt stenographer'. Most of her friends were dead, but she became involved with a younger generation, in particular the **Sitwells**, whose work she championed. She wrote a short story, 'The Blow', for the *English Review*, contributed articles on Wilde and other writers to **T.S. Eliot**'s *Criterion*, edited a volume of *Letters to the Sphinx from Oscar Wilde*, and appeared as 'the Sib' in **Wyndham Lewis**'s *The Apes of God* (1930). She became plump and rather deaf in old age, and died after catching pneumonia in Italy where she often spent the winter months. The writer **Francis Wyndham** is her grandson.

Fiction

The Twelfth Hour 1907; *The Little Ottleys* 1962: *Love's Shadow* 1908, *Tenterhooks* 1912, *Love at Second Sight* 1916; *The Limit* 1911; *Bird of Paradise* 1914

Edited

Letters to the Sphinx from Oscar Wilde 1930

Biography

The Sphinx and Her Circle by Violet Wyndham 1963

Denise LEVERTOV 1923–

The daughter of a Russian Jew who had converted to Christianity at university in Germany, then came to England and became an Anglican clergyman, Levertov was born in London. Her mother was Welsh, and this rich international background informed both her upbringing and her subsequent career as a writer. She was educated at home, apart from a period at a ballet school, and had her first poem published in

Poetry Quarterly when she was just sixteen. She sent her work to **Herbert Read**, and his encouragement and criticism led to her deciding to pursue a career as a poet.

She spent the war working as a nurse in London, and had her first volume of poems, *The Double Image*, published in 1946. During a walking holiday in France shortly after the war, she met the American writer Mitchell Goodman, whom she married and with whom she lived in France and Italy before settling in America in 1948. They had one son, and have since divorced. Levertov's second volume of poems, *Here and Now*, was published by **Lawrence Ferlinghetti** in the 'Pocket Poets' series and was followed two years later by her third, *Overland to the Islands*. Since then, she has published over thirty volumes of poetry, as well as two volumes of essays and one of short stories. Her work is notable for its clarity and immediacy, and the New York *Times* described her as 'the most subtly skilful poet of her generation, the most profound, the most modest, the most moving'. She has received several prizes and been awarded numerous honours by American universities.

She has also had a distinguished career as an academic, notably as professor of English at Stanford University in the 1980s. In 1961 and again between 1963 and 1965, she was editor of the *Nation*, and was subsequently editor of the San Francisco magazine *Mother* for three years from 1975. In the mid-1960s she became a leading activist in pacifist and anti-nuclear movements, and this has resulted in a political dimension to her work, notably her poems on the Vietnam War.

Poetry

The Double Image 1946; *Here and Now* 1956; *5 Poems* 1958; *Overland to the Islands* 1958; *With Eyes at the Back of Our Heads* 1960; *The Jacob's Ladder* 1961; *City Psalm* 1964; *O Taste and See* 1964; *Psalm Concerning the Castle* 1966; *Penguin Modern Poets 9* [with K. Rexroth, W.C. Williams] 1967; *The Sorrow Dance* 1967; *A Marigold from North Viet Nam* 1968; *Three Poems* 1968; *A Tree Telling of Orpheus* 1968; *The Cold Spring and Other Poems* 1969; *Embroideries* 1969; *Relearning the Alphabet* 1970; *A New Year's Garland for My Students, MIT 1969–70* 1970; *Summer Poems 1969* 1970; *To Stay Alive* 1971; *Footprints* 1972; *The Freeing of the Dust* 1975; *Chekhov on the West Heath* 1977; *Modulations for Solo Voice* 1977; *Life in the Forest* 1978; *Collected Earlier Poems 1940–1960* 1979; *Pig Dreams* 1981; *Wanderer's Daysong* 1981; *Candles in Babylon* 1982; *Poems 1960–1967* 1983; *El Salvador* 1984; *The Menaced World* 1984; *Olique Prayers* 1984; *Selected Poems* 1986; *Poems 1968–1972* 1987; *Breathing Water* 1988; *A Door in the Hive* 1993

Fiction

In the Night (s) 1968

Non-Fiction

Conversation in Moscow 1973; *The Poet in the World* 1973; *Denise Levertov: an Interview* [with J.K. Atchity] 1980; *Light Up the Cave* 1981; *New and Selected Essays* 1992

Translations

Selected Poems by Guillevic 1969

Edited

Out of the War Shadow: an Anthology of Current Poetry 1967; *In Praise of Krishna: Songs from the Bengali* [ed. and tr. with E.C. Dimock Jr] 1967

Ira LEVIN 1929–

Levin was born in New York City, educated at Drake University, Des Moines, Iowa, and at New York University. Between 1953 and 1956 he served in the US Signal Corps, by which time he had published his first novel, *A Kiss Before Dying*, in which a young man murders his pregnant girlfriend in order to marry her sister. It reaches its unforgettable climax in a copper-smelter, 'a giant heart of American industry'. **Julian Symons** described the novel as 'a masterpiece of the genre' and it won the Mystery Writers of America Edgar Allan Poe Award for 1954.

Levin's subsequent novels might all be described as superior thrillers, although they often include elements of science fiction and the supernatural. His most famous novel remains *Rosemary's Baby*, in which a young woman finds herself pregnant with the Devil's child. The story cleverly plays upon the anxieties which inevitably accompany pregnancy and is told in a wholly convincing, matter-of-fact manner. **Truman Capote** called it 'a darkly brilliant tale of modern devilry that, like [**Henry**] **James**'s *Turn of the Screw*, induces the reader to believe the unbelievable'. *The Stepford Wives* is another novel of justified paranoia, in which the women of a small community are gradually replaced by slave-like replicants. In *The Boys from Brazil*, Levin plays a variation upon the popular theme of Hitler having survived the war: in a South American hideout a doctor who worked in Auschwitz plans with his fellows to rekindle Nazism using clones of the Führer. All these novels have been made into films, with two versions (the latter execrable) of *A Kiss Before Dying*. The most distinguished of these are Roman Polanski's classic version of *Rosemary's*

Baby (1968) – which earned Levin a fortune – and Bryan Forbes's quirkily feminist adaptation of *The Stepford Wives* (1974). His other novels are less well known, although *Sliver* was praised by Stephen King as 'the ultimate *fin-de-siècle* horror novel, a fiendish goodbye wave to trendy urban living in the last decade of the twentieth century'.

Levin has also written a number of plays, of which the complex thriller *Deathtrap*, in which a prominent playwright attempts to steal the plot of a younger rival, was a great success both on stage and as a film.

He has been married twice, and has three sons from his first marriage.

Fiction

A Kiss Before Dying 1953; *Rosemary's Baby* 1967; *This Perfect Day* 1970; *The Stepford Wives* 1972; *The Boys from Brazil* 1976; *Sliver* 1991

Plays

No Time for Sergeants [from M. Hyman's novel] 1955 [pub. 1956]; *Interlock* 1958; *Critic's Choice* 1960 [pub. 1961]; *General Seeger* 1962; *Drat! The Cat!* [with M. Schafer] 1965; *Dr Cook's Garden* 1967 [pub. 1968]; *Veronica's Room* 1973 [pub. 1974]; *Break a Leg* prod 1974 [1981]; *Deathtrap* 1978 [pub. 1979]; *Cantorial* 1984 [pub. 1990]

Alun LEWIS 1915–1944

Lewis was born in Cwmaman, a mining village in South Wales; his father, a schoolteacher, was the only one of four brothers never to work in the pits. He was educated at Cowbridge Grammar School and the University College of Wales at Aberystwyth, taking his BA in 1935; he then did postgraduate work at the University of Manchester, and trained as a teacher. He hoped to find work as a journalist, but instead, in 1938, became a supply teacher at Pengam School near Aberdare. When war came he had pacifist leanings, but nevertheless enlisted in the Royal Engineers in 1940, finding himself at Longmoor Depot in Hampshire, inspecting railway engines.

The period between the winter of 1939 and the autumn of 1940 was his maturing-time as a poet, and in 1942 his *Raiders' Dawn and Other Poems* was published, immediately establishing him as one of the leading war poets. In the same year he published a book of observant and passionate short stories, *The Last Inspection*, and many authorities consider Lewis's fiction at least as promising as his poetry. **Julian Maclaren-Ross**, who met him at this time, wrote an evocative memoir of him, 'Second Lieutenant Lewis', which is collected in his *Memoirs of the Forties* (1965). In 1941, Lewis married Gweno Ellis, a fellow schoolteacher; his fine love poetry is addressed to her. He obtained a lieutenantship in the infantry, and in November 1942 sailed for India, but for some time did not see action. In early 1944 he had been moved to the Burmese front and it seemed that battle experience was imminent, but on his way to it, at Arakan in Lower Burma, he was killed in a mysterious incident involving his own pistol, a death which it is thought may have been self-inflicted. A second book of poems, *Ha! Ha! Among the Trumpets*, which he had worked on with the help of criticism from **Robert Graves**, was published posthumously.

Although his early verse was romantic and conventional, Lewis quickly became a poet who described the loneliness of military life, the stimulus offered by places, and the experience of love, with extraordinary honesty and accuracy; he is undoubtedly, with **Keith Douglas**, one of the two finest English poets of the Second World War.

Poetry

Raiders' Dawn and Other Poems 1942; *Ha! Ha! Among the Trumpets* 1945

Fiction

The Last Inspection (s) 1942

Non-Fiction

Letters from India 1946

Collections

In the Green Tree 1948; *Alun Lewis: Selected Poetry and Prose* [ed. I. Hamilton] 1966

C(live) S(taples) LEWIS 1898–1963

Born in Belfast, the son of a solicitor, Lewis attended a private school in Hertfordshire (which he afterwards called 'Belsen') and Malvern College (where bullying horrified him). He won a triple first at University College, Oxford, and from 1925 to 1954 held a fellowship at Magdalen College, thereafter becoming Professor of Medieval and Renaissance English at Cambridge. His most influential academic works were *The Allegory of Love*, a study of courtly love in early romance literature, and *English Literature in the Sixteenth Century*.

He was brought up a Protestant but became an atheist in adolescence. However, after a trip to Whipsnade Zoo in the side-car of his brother's motor bike in 1931, he converted to Anglicanism. At Oxford he became a close friend of **J.R.R. Tolkien** and **Charles Williams**. Along with other friends, they formed a group known as 'the Inklings', like-minded Christian writers who met at Magdalen and in Woodstock Road public houses to discuss religion and literature, to read work in progress, and to smoke pipes and drink beer.

Lewis preferred the company of men, often idealising the male form and male friendships in his writings. He was bellicose, not to say bullying, and was said to have fought a duel with a pupil over a point in Matthew Arnold. **John Betjeman** and **Henry Green** were among his tutees, and both detested him. Lewis was also a misogynist, of the opinion that women's minds are intrinsically inferior to men's. Not surprisingly, his relationships with women were mostly fraught. He had been very close to his mother, whose death caused him acute anguish. For many years, he lived with a much older woman, Mrs Jamie Moore, whom he referred to as his 'mother'. Mrs Moore was the mother of Edward 'Paddy' Moore, with whom Lewis had shared rooms for just a few weeks at Keble in the summer of 1917. Paddy Moore's death in the First World War also affected Lewis deeply. Much surprise was caused by Lewis's marriage in his late fifties to a Jewish American divorcee, Joy Davidson, who died a few years later of cancer. Their relationship is the subject of William Nicholson's acclaimed play *Shadowlands* (1989), later made into an equally successful film.

During the Second World War Lewis's broadcasts about Christianity made him popular, as did *The Screwtape Letters*, a series of letters from a devil to his nephew which presents Christianity from their point of view. *Out of the Silent Planet* is the first of a trilogy of science-fiction novels which take Christianity into outer space. His children's stories, *The Chronicles of Narnia*, are also Christian allegories; they did not at first find much of a readership, but later became extremely popular. Lewis and Tolkien, despite their friendship, despised each other's writings for children. *Surprised by Joy* is an autobiographical account of religious faith.

For Children

The Chronicles of Narnia: *The Lion, the Witch and the Wardrobe* 1950, *Prince Caspian* 1951, *The Voyage of the 'Dawn Treader'* 1952, *The Silver Chair* 1953, *The Horse and His Boy* 1954, *The Magician's Nephew* 1955, *The Last Battle* 1956; *Letters to Children* [ed. L.W. Dorsett, M. Lamp Mead] 1985

Non-Fiction

The Pilgrim's Progress 1933; *The Allegory of Love* 1936; *The Personal Heresy* [with E.M.W. Tillyard] 1939; *Rehabilitations* 1939; *The Problem of Pain* 1940; *Broadcast Talks: Right and Wrong* 1942 [US *The Case for Christianity* 1943]; *Hamlet* 1942; *Preface to 'Paradise Lost'* 1942 [rev. 1960]; *The Weight of Glory* 1942; *The Abolition of Man* 1943; *A Christian Behaviour: a Further Series of Broadcast Talks* 1943; *The Screwtape Letters* 1943 [rev. *The Screwtape Letters and Screwtape Proposes a Toast* 1961]; *Beyond Personality* 1944; *The Great Divorce* 1946; *Miracles: a Preliminary Study* 1947; *Vivisection* 1947; *Transposition and Other Addresses* 1949; *The Literary Impact of the Authorized Version* 1950; *Hero and Leander* 1952; *Mere Christianity* 1952; *English Literature in the Sixteenth Century, Excluding Drama* 1954; *De Descriptione Temporum* 1955; *Surprised by Joy: the Shape of My Early Life* (a) 1956; *Reflections on the Psalms* 1958; *Shall We Lose God in Outer Space?* 1959; *The Four Loves* 1960; *Studies in Words* 1960; *The World's Last Night* 1960; *An Experiment in Criticism* 1961; *A Grief Observed* (a) [as 'N.W. Clerk'] 1961; *They Asked for a Paper* 1962; *Beyond the Bright Blur* 1963; *The Discarded Image: an Introduction to Medieval and Renaissance Literature* [ed. W. Hooper] 1964; *Letters to Malcolm, Chiefly on Prayer* 1964; *Letters* [ed. W.H. Lewis] 1966; *Studies in Medieval and Renaissance Literature* [ed. W. Hooper] 1966; *Christian Reflections* [ed. W. Hooper] 1967; *Letters to an American Lady* [ed. C.S. Kilby] 1967; *Mark vs. Tristram: Correspondence Between C.S. Lewis and Owen Barfield* [ed. W. Hooper] 1967; *Spenser's Images of Life* [ed. A. Fowler] 1967; *Selected Literary Essays* [ed. W. Hooper] 1969; *God in the Dock: Essays on Theology and Ethics* [ed. W. Hooper] 1970 [UK *Undeceptions* 1971]; *The Humanitarian Theory of Punishment* 1972; *Fern-Seed and Elephant and Other Essays on Christianity* [ed. W. Hooper] 1975; *The Joyful Christian: 127 Readings* [ed. W. Griffin] 1977; *They Stand Together: the Letters of C.S. Lewis to Arthur Greeves 1914–1963* [ed. W. Hooper] 1979; *C.S. Lewis at the Breakfast Table and Other Reminiscences* [ed. J.T. Como] 1979; *The Visionary Christian: 131 Readings* [ed. C. Walsh] 1981; *On Stories and Other Essays on Literature* [ed. W. Hooper] 1982; *Of This and Other Worlds* [ed. W. Hooper] 1982; *The Cretaceous Perambulator* [ed. W. Hooper] 1983; *The Business of Heaven: Daily Readings from C.S. Lewis* [ed. W. Hooper] 1984; *Boxen: the Imaginary World of the Young C.S. Lewis* [ed. W. Hooper] 1985; *Present Concerns* [ed. W. Hooper] 1986; *Timeless at Heart* [ed. W. Hooper] 1987; *Letters: C.S. Lewis and D.G. Calabria* [ed. M. Moynihan] 1989

Edited

George MacDonald: an Anthology 1946; Charles Williams *Arthurian Torso, Containing the Posthumous Fragment of 'The Figure of Arthur'* 1948

Fiction
Out of the Silent Planet 1943; *Perelandra* 1943; *That Hideous Strength* 1945; *Till We Have Faces* 1956; *Of Other Worlds* (s) [ed. W. Hooper] 1966; *The Dark Tower* (s) [ed. W. Hooper] 1969

Poetry
Spirits in Bondage [as 'Clive Hamilton'] 1919; *Dymer* [as 'Clive Hamilton'] 1926; *Poems* [ed. W. Hooper] 1964; *Narrative Poems* [ed. W. Hooper] 1969

Collections
Of Other Worlds: Essays and Stories [ed. W. Hooper] 1967; *A Mind Awake: an Anthology of C.S. Lewis* [ed. C.S. Kilby] 1968; *The Essential C.S. Lewis* [ed. L.W. Dorsett] 1988

Biography
C.S. Lewis by A.N. Wilson 1990

Norman LEWIS 1908–

The son of a retail chemist who also became a spiritualist medium, Lewis was brought up partly at his parents' home in Enfield, north of London (where he participated in his father's seances), and partly by three eccentric aunts in Carmarthen, Wales. Both environments are vividly described in his autobiographical volume, *Jackdaw Cake*. On leaving Enfield Grammar School, he worked at a number of jobs including that of a motor racer. In 1935 he opened a camera shop in Holborn and soon built up a chain of ten shops.

From the mid-1930s he began his lifelong habit of travelling widely, particularly after he married his first wife, Ernestina Corvaja, who came from a Sicilian Mafia family which was domiciled in Bloomsbury. His early travels in Arabia gave rise to his first book, *Sand and Sea in Arabia* (1938), consisting mainly of photographs. (Lewis's photographs have continued to feature as powerful illustrations to his later work.) The outbreak of the Second World War found him and his wife in Cuba; she remained in the region, but he returned to join up, was drafted into the Field Security Police and then its overseas service, and worked as an Intelligence officer in Algeria, Tunisia and Naples. His diary of his time in Italy was the basis for the celebrated *Naples '44*.

His long separation from his wife led to their divorce. Lewis lived for a time in Tenby in Wales and Farol in Spain (his classic account of traditional Spain is *Voices of the Old Sea*), published his first novel, *Samara*, in 1949, and from the early 1950s began a series of travels in the Far East, which led to two books, *A Dragon Apparent* and *Golden Earth*, which made his reputation as a travel writer. A highly successful novel, set in Guatemala, was *The Volcanoes Above Us*, and for a time Lewis concentrated on fiction, but he has spent much of his later life travelling extensively, often on behalf of Sunday newspapers, and predominantly in Latin America. As an author/traveller he can in some ways be compared to **Graham Greene** – their paths have sometimes crossed – although he does not have quite Greene's celebrity.

Lewis has developed a special concern for threatened environments and peoples, and one major achievement was 'Genocide in Brazil', an article in the *Sunday Times* in 1968 about the fate of the Indians, which led to the founding of organisations such as Survival International. Lewis has been married three times (his present wife is an Australian), has several children, and lives with his family 'in introspective, almost monastic calm' in Essex, where he has been based for many years. Among his other notable books is his account of the Sicilian Mafia, *The Honoured Society*.

Fiction
Samara 1949; *Within the Labyrinth* 1950; *A Single Pilgrim* 1953; *The Day of the Fox* 1955; *The Volcanoes Above Us* 1957; *Darkness Visible* 1960; *The Tenth Year of the Ship* 1962 [as *Dragon Tree Island* 1964]; *A Small War Made to Order* 1966; *Every Man's Brother* 1967; *Flight from a Dark Equator* 1972; *The Sicilian Specialist* 1974; *The German Company* 1979; *The Man in the Middle* 1984; *A Suitable Case for Corruption* 1984

Non-Fiction
Sand and Sea in Arabia 1938; *A Dragon Apparent* 1951; *Golden Earth* 1952; *The Changing Sky* 1959; *The Honoured Society* 1964; *Naples '44* 1978; *Cuban Passage* 1982; *Voices of the Old Sea* 1984; *Jackdaw Cake* (a) 1985 [as *I Came, I Saw* 1994]; *The Missionaries* 1988; *To Run Across the Sea: Travels* 1989; *Goddess in the Stories: Travels in India* 1991; *An Empire of the East: Travels in Indonesia* 1993

(Henry) Sinclair LEWIS 1885–1951

The son of a country doctor, Lewis was born in Sauk Center, Minnesota, a small Midwestern town which is usually thought to be the model for Gopher Prairie in his novel *Main Street*. He went from the local high school to Yale, where he began writing and from which he graduated in 1908. He had several jobs in publishing houses and on newspapers, from a number of which he was fired, and worked for a while as a ghost-writer to **Jack London**.

In 1916, after the publication of two undistinguished novels, he became a freelance writer and achieved massive success in 1920 with *Main Street*, which exposed the hypocrisy of American small-town life. Similar themes marked the other novels written in the 1920s – *Babbitt*, *Arrowsmith*, *Elmer Gantry* – which established him as perhaps the most famous American writer of his time. In 1930 he became the first American to win the Nobel Prize. In later life, he became a celebrity, but the ten novels published after 1930 are generally agreed to be inferior to his earlier work. He married twice, and had a son by each marriage. His second wife was the celebrated columnist Dorothy Thompson; both marriages ended in divorce. Lewis spent his last years travelling around Europe, and died in Rome, of paralysis of the heart.

He was a notoriously difficult man, frequently drunk, restlessly moving from place to place and haunted by a fear of loneliness, but he impressed many of those who knew him as a person of integrity. Martin Seymour-Smith described him as one of the worst writers to win the Nobel Prize, but **E.M. Forster** said of him that he 'lodged a piece of a continent in the world's imagination'. A considered judgement might be that he was a journalistic type of novelist, but possessed vigour within that limitation.

Fiction

Hike and the Aeroplane 1912; *Our Mr Wrenn* 1914; *The Trail of the Hawk* 1915; *The Innocents* 1917; *The Job* 1917; *Free Air* 1919; *Main Street* 1920; *Babbitt* 1922; *Arrowsmith* 1925; *Mantrap* 1926; *Elmer Gantry* 1927; *The Man Who Knew Coolidge* 1928; *Dodsworth* 1929; *Ann Vickers* 1933; *Work of Art* 1934; *Selected Short Stories* (s) 1935; *It Can't Happen Here* 1935; *The Prodigal Parents* 1938; *Bethel Merriday* 1940; *Gideon Planish* 1943; *Cass Timberlane* 1945; *Kingsblood Royal* 1947; *The God-Seeker* 1949; *World So Wide* 1951; *I'm a Stranger Here Myself and Other Stories* (s) 1962

Plays

Dodsworth [from his novel] 1934; *Jayhawker* [with L. Lewis] 1934; *It Can't Happen Here* [from his novel] 1938; *Storm in the West* (f) [with D. Shary] 1963

Non-Fiction

John Dos Passos' Manhattan Transfer 1926; *Cheap and Contented Labor* 1929; *The American Fear of Literature* 1930; *From Main Street to Stockholm: Letters of Sinclair Lewis 1919–1930* [ed. H. Smith] 1952; *The Man from Main Street* [ed. M.H. Cane, H.E. Maule] 1952

(Percy) Wyndham LEWIS 1882–1957

Lewis was born on his father's yacht moored off the coast of Nova Scotia, and retained Canadian nationality for his entire life. The family moved to the Isle of Wight in 1888 and five years later Lewis's American father – sometime soldier, lawyer, sportsman and dilettante gentleman – ran off with a housemaid. Raised by his mother in genteel poverty in a series of London suburbs, Lewis briefly attended Rugby School before going in 1898 to the Slade School of Art where he first met Augustus John, four years his senior and already something of a legend.

In 1902 Lewis moved to Paris to continue his study of painting and travelled in Spain, Holland and Germany throughout the decade. He returned to London in 1908 after fathering the first of five illegitimate children – none of whose welfare he ever greatly cared about – and published his first short stories in the *English Review* the following year. He exhibited with the Camden Town Group of artists, began work on his novel *Tarr*, conducted an acrimonious dispute with Roger Fry (which would result in the brief establishment of the Rebel Art Centre in 1914) and launched *Blast*, the first of his three magazines (the others were *The Tyro* and *Enemy*). His acquaintance with **Ezra Pound**, **T.S. Eliot**, **Ford Madox Ford** and other writers stems from this period. In 1916, recovered from one of many debilitating bouts of venereal disease, he volunteered for the Royal Artillery, and later worked as an artist for Beaverbrook's Canadian War Memorial scheme.

Tarr, which was influenced by Nietzsche and based upon his experiences as a painter in Paris, was originally serialised in the *Egoist* between 1916 and 1917 and appeared in book form in 1918. As with subsequent books, Lewis himself designed its typography. It established him as a polemical satirist, and it was followed over the next two decades by a relentless stream of publications expressing his disgust for modern civilisation – a 'moronic inferno of insipidity and decay'. The best-known of his novels is *The Apes of God*, an unwieldy satire following the progress of a young aspirant writer (based on **Stephen Spender**) through the talentless 'Ape-world' of London's literary Bohemia. It exemplifies both Lewis's paranoia and his method of delineating characters from their outward appearance and actions, a system related to his theory of painting. Written in an extremely elaborate and ornamented prose, it contains caricature portraits of many of Lewis's contemporaries, notably the **Sitwells**, Clive and Vanessa Bell, Eliot, and his friend and fellow right-winger **Roy Campbell**. It also, with characteristic ingratitude and a dose of anti-Semitism, lampoons his patrons, Sydney and Violet Schiff. (Another of Lewis's many benefactors who failed to send a cheque on time received a terse postcard demanding: 'Where's the fucking stipend?') *The Roaring Queen*, which covered similar territory (caricaturing, amongst others, **Arnold Bennett**, **Virginia Woolf** and Nancy Cunard), was considered so libellous that it was withdrawn by the publishers at proof stage in 1936, and remained unpublished during Lewis's lifetime. Lewis himself was memorably satirised in the character of Casimir Lypiatt in **Aldous Huxley**'s *Antic Hay* (1923), which draws upon the two men's rivalry for the affections of Cunard.

Lewis's advocacy of fascism in the early 1930s did him little good in literary circles, but was recanted after a trip to Berlin in 1937. With the outbreak of war, he moved first to Canada and then the USA with his wife Anne – 'Froanna', as she was known – whom he had married in 1930. He returned to London in 1946 and continued the struggle to make a living. He was an extremely distinguished art critic for *The Listener*, a job he was obliged to give up when he started to go blind in 1951. His farewell piece, 'The Sea-Mists of Winter', which was published in the magazine in May of that year, is one of the best and most moving things he wrote. His eyesight was affected by the brain tumour which eventually killed him, and in his last years he was completely blind. He nevertheless completed his novels *Monstre Gai* and *Malign Fiesta*, which with the much earlier *The Childermass* form a trilogy, *The Human Age*.

There has been a revival of interest in Lewis's writing in recent years, with many of his books being reissued in facsimile editions by the Black Sparrow Press in America. His reputation as a painter – in particular the mechanistic Vorticist work he did in the First World War and his portraits of such figures as Eliot, Pound, Spender and Edith Sitwell – continues to grow.

Fiction

Tarr 1918; *The Wild Body* (s) 1927; *The Human Age* 1928; *The Childermass* 1928; *Monstre Gai* 1955; *Malign Fiesta* 1955; *The Apes of God* 1930; *Snooty Baronet* 1932; *The Revenge for Love* 1937; *The Vulgar Streak* 1941; *Rotting Hill* (s) 1951; *Self Condemned* 1954; *The Red Priest* 1956; *The Roaring Queen* 1973; *Unlucky for Pringle* (s) 1973; *Imaginary Letters* 1977; *Mrs Dukes' Million* 1977

Non-Fiction

The Caliph's Design 1919; *Harold Gilman* 1919; *The Art of Being Ruled* 1926; *The Lion and the Fox: the Role of the Hero in the Plays of Shakespeare* 1927; *Time and Western Man* 1928; *Paleface* 1929; *Satire and Fiction* 1930; *Hitler* 1931; *The Diabolical Principle and The Dithyrambic Spectator* 1931; *The Doom of Youth* 1932; *Filibusters in Barbary* 1932; *The Old Gang and the New Gang* 1933; *Men without Art* 1934; *Left Wings Over Europe* 1936; *Blasting and Bombardiering* (a) 1937; *Count Your Dead: They Are Alive!* 1937; *The Mysterious Mr Bull* 1938; *The Jews – Are They Human?* 1939; *The Hitler Cult, and How It Will End* 1939; *Wyndham Lewis the Artist: From 'Blast' to Burlington House* 1939; *America, I Presume* 1940; *Anglosaxony* 1941; *The Role of Line in Art* 1941; *America and Cosmic Man* 1948; *Rude Assignment* (a) 1950; *The Writer and the Absolute* 1952; *The Demon of Progress in the Arts* 1954; *The Letters of Wyndham Lewis* [ed. W.K. Rose] 1963; *Wyndham Lewis on Art* [ed. C.J. Fox, W. Michael] 1969; *Enemy Salvoes* 1976; *Journey into Barbary: Morocco Writings and Drawings* [ed. C.J. Fox] 1983; *Correspondence* [ed. T. Materer] 1985

Plays

Enemy of the Stars 1932

Poetry

One-Way Song 1933

Collections

Collected Poems and Plays [ed. A. Munton] 1979

Biography

The Enemy by Jeffrey Meyers 1980

(John) Robert LIDDELL 1908–1993

Liddell was born in Tunbridge Wells. Because his father, an army major, was Inspector-General of the Egyptian State Telegraphs, his earliest years were largely spent in that country. His childhood was dominated by the death of his mother and the impact on his and his brother's life of a hated stepmother (who appears as Elsa in his painful autobiographical novel, *Stepsons*). He attended Haileybury and then Corpus Christi College, Oxford, where he formed part of the circle which included the young **Barbara Pym**: he is referred to as 'Jock' in her letters and diaries, published as *A Very Private Eye*. From 1933 to 1938 he worked as senior assistant in the Department of Western Manuscripts of the Bodleian Library, Oxford. From this period dates the beginning of his important friendship with **I. Compton-Burnett**. He was to publish, in 1955, the standard critical study of her novels and, despite the fact that they met less than a dozen times in all, he was among the six of her closest friends to each of whom, as residuary legatees, she left a looking-glass.

Liddell's own first novel, *The Almond Tree*, was published in 1938. At about the same time he gained employment with the British Council and for many years lectured for them in various foreign countries. He initially went to Helsingfors (as Helsinki was then generally known), but during wartime left Finland for first Greece and later Egypt, where he became part of the literary society anatomised by **Olivia Manning** in her two trilogies. He was a lecturer and assistant professor at the universities in Cairo and Alexandria from 1940 to 1951, and then moved to Athens, where he became head of the English department of the university between 1963 and 1968. He did not visit England again after 1947.

Liddell published ten novels, of which perhaps *Unreal City*, based on his Egyptian experiences and featuring the poet Cavafy as a central character, is outstanding; his fiction, however, is vitiated by snobbery and, although it was well known in the 1940s and 1950s, it later became difficult to publish and generally went out of print. Another novel, *The Aunts*, set in the England of his youth, was finally published in 1987, and from the same general period date several brief memoirs and literary studies, such as *Elizabeth and Ivy*, the account of his relations with the novelists **Elizabeth Taylor** and I. Compton-Burnett. His other books include *A Treatise on the Novel* (long regarded, with Percy Lubbock's *The Craft of Fiction*, as a classic study of the subject), and a translation of *L'Abbé Tigrane* by Ferdinand Fabre, the nineteenth-century novelist of the Cevennes.

Fiction

The Almond Tree 1938; *Kind Relations* [US *Take This Child*] 1939; *The Gantillons* 1940; *Watering Place* (s) 1945; *The Last Enchantments* 1948; *Unreal City* 1952; *The Rivers of Babylon* 1959; *An Object for a Walk* 1966; *The Deep End* 1968; *Stepsons* 1969; *The Aunts* 1987

Non-Fiction

A Treatise on the Novel 1947; *Some Principles of Fiction* 1953; *Aegean Greece* 1954; *The Novels of I. Compton-Burnett* 1955; *Byzantium and Istanbul* 1956; *The Morea* 1958; *The Novels of Jane Austen* 1963; *Mainland Greece* 1965; *Cavafy: a Critical Biography* 1974; *The Novels of George Eliot* 1977; *Elizabeth and Ivy* 1986; *A Mind at Ease: Barbara Pym and Her Novels* 1989; *Twin Spirits: the Novels of Emily and Anne Brontë* 1990

Translations

Demetrios Sicilianos *Old and New Athens* 1960; Linos Politos *A History of Modern Greek Literature* 1973; *Neoklasika Ereipia* [ed. S.B. Skopelitis] 1977; Ferdinand Fabre *The Abbé Tigrane* 1988

Jack LINDSAY 1900–1990

Lindsay was born in Melbourne, Australia, into a family of visual artists. Most well known among them was his father, Norman Lindsay, primarily an illustrator in a style reminiscent of Aubrey Beardsley, and widely considered the leading Australian artist of his time. He was also a writer, and had three sons who followed him in this (of the others, Philip Lindsay achieved some fame as a historical novelist). Lindsay spent his teens largely in Brisbane, where he attended the boys' grammar school and the University of Queensland, graduating with a first-class degree in classics, and gaining the interest in ancient Greece and Rome which runs like a connecting thread through his writings. In his own words, he then made 'some largely misdirected efforts to start a literary movement in Sydney', a modest way of referring to the magazine *Vision*, founded with other members of his family in 1924 and playing a seminal role in the development of Australian writing. His own work began with poetry – *Fauns and Ladies*, published in 1923 – and was followed by several verse dramas and translations from Greek and Roman classics during the 1920s.

Meanwhile, in 1926, he had migrated to England, where he founded the Fanfrolico Press, devoted to fine printing of his translations, as well as another magazine, the *London Aphrodite*, one of a number of magazines he was to be associated with during his long life. However, he lost most of his money in the Depression, and moved to Cornwall, where he lived in poverty, devised a personal philosophy which synthesised Marxism with the work of Romantic and idealist thinkers, and turned to the historical novel, at first largely set in ancient Rome but later in many periods of British history. He was to publish almost forty works of fiction until the 1960s, the latter ones mostly contemporary novels of the 'British way', and continued production of non-fiction, mostly cultural, biographical and historical studies, into old age in the 1980s.

He wrote, edited and translated more than 170 books and plays, which puts him among the most prolific of authors with any claim to seriousness, and if no single one of them made a lasting impact, most were marked by learning, skill and personal commitment. His lifelong Marxism, which led to vast sales and honours in the Soviet Union, may have accounted for comparative neglect in his adopted country. Perhaps his three volumes of early autobiography, collected in one volume, *Life Rarely Tells*, in 1982, are the best candidates for continued life. During the war, he served first in the Signals Corps and then as a scriptwriter for the theatre produced by the

Army Bureau of Current Affairs. After it, he lived first at an old farmhouse at Halstead in Essex, Castle Hedingham, and latterly at Cambridge. He married three times, and with his third wife, Meta Waterdrinker, a potter, he had two children. One minor claim to fame was that in the later 1930s he was a pioneer of 'mass-declamation poems', once orating to a crowd of thousands in Trafalgar Square, an audience larger than that usually achieved by the performance poets of a later age.

Fiction

Cressida's First Lover 1931; *Caesar Is Dead* 1934; *Rome for Sale* 1934; *Despoiling Venus* 1935; *Last Days with Cleopatra* 1935; *Storm at Sea* 1935; *Adam of a New World* 1936; *Come Home At Last* (s) 1936; *Shadow and Flame* (as 'Richard Preston') 1936; *Wanderings of Wenamen* 1936; *End of Cornwall* (as 'Richard Preston') 1937; *Sue Verney* 1937; *1649* 1938; *Brief Light* 1939; *Lost Birthright* 1939; *Hannibal Takes a Hand* 1941; *Light in Italy* 1941; *The Stormy Violence* 1941; *We Shall Return* 1942; *Beyond Terror* 1943; *The Barriers Are Down* 1945; *Hullo Stranger* 1945; *Time to Live* 1946; *The Subtle Knot* 1947; *Men of Forty-Eight* 1948; *Fires in Smithfield* 1950; *The Passionate Pastoral* 1951; *Betrayed Spring* 1953; *Rising Tide* 1953; *Moment of Choice* 1955; *The Romans Were Here* 1956; *The Great Oak* 1957; *A Local Habitation* 1957; *1764* 1959; *Revolt of the Sons* 1960; *The Writing on the Wall* 1960; *All on the Never Never* 1961; *The Way the Ball Bounces* 1962; *Choice of Times* 1964; *The Blood Vote* 1985

Non-Fiction

William Blake: Creative Will and the Poetic Image 1927; *Dionysos; or, Nietzsche contra Nietzsche* 1928; *The Romans* 1935; *Marc Anthony* 1936; *John Bunyan, Maker of Myths* 1937; *England, My England* 1939; *A Short History of Culture* 1939; *Perspective in Poetry* 1944; *British Achievement in Art and Music* 1945; *Song of a Falling World: Culture During the Break-Up of the Roman Empire* 1948; *Marxism and Contemporary Science* 1949; *Charles Dickens* 1950; *Byzantium into Europe* 1952; *Rumanian Summer* [with M. Cornforth] 1953; *Civil War in England* 1954; *After the Thirties: the Novel in Britain and Its Future* 1956; *George Meredith: His Life and Work* 1956; *Arthur and His Times* 1958; *The Discovery of Britain* 1958; *Life Rarely Tells* (a) 1982: *Life Rarely Tells* 1958, *The Roaring Twenties* 1960, *Fanfrolico and After* 1962; *The Death of the Hero: Painting from David to Delacroix* 1960; *William Morris, Writer* 1961; *Celtic Heritage* 1962; *Daily Life in Roman Egypt* 1963; *The Paintings of D.H. Lawrence* (c) [ed. M. Levy] 1964; *The Clashing Rocks: a Study of Early Greek Religion and Culture and the Origins of Drama* 1965; *Leisure and Pleasure in Roman Egypt* 1965; *J.M.W. Turner: His Life and Work* 1966; *The Ancient World: Manners and Morals* 1968; *Meetings with Poets* 1968; *Cézanne: His Life and Art* 1969; *Cleopatra* 1971; *Gustav Courbet: His Life and Work* 1972; *Blast Power and Ballistics* 1974; *Faces and Places* 1974; *Helen of Troy: Woman and Goddess* 1974; *William Morris: His Life and Work* 1975; *Decay and Renewal: Critical Essays on Twentieth Century Writing* 1976; *Hogarth, His Art and His World* 1977; *The Monster City: Defoe's London 1699–1730* 1978; *William Blake: His Life and Work* 1978; *The Crisis in Marxism* 1981; *Thomas Gainsborough: His Life and Art* 1981; *William Morris, Dreamer of Dreams* [ed. D. Gerard] 1991

Translations

Aristophanes: *Lysistrata* 1925, *Women in Parliament* 1929; *Gaius Petronius: Complete Works* 1927; *Propertius in Love* 1927; *A Homage to Sappho* [with N. Lindsay] 1928; Catullus: *The Complete Poetry of Gaius Catullus* 1929, *The Complete Poems* 1948; *The Complete Poems of Theocritos* 1929; *The Mimiambs of Herondas* 1929; *Homer's Hymns to Aphrodite* 1930; *Patchwork Quilt: Poems by Decimus Magnus Ausonius* 1930; *I Am a Roman* 1934; *Medieval Latin Poets* 1934; Longus *Daphnis and Chloe* 1948; Vitezslav Nezval *Song of Peace* [with S. Jolly] 1951; *Russian Poetry 1917–1955* 1956; *Poems of Adam Mickiewicz* 1957; Apuleius *The Golden Ass* 1960; *Asklepiades in Love* 1960; *Modern Russian Poetry* 1960; Petronius Arbiter *The Satyricon and Poems* 1960; Giordano Bruno *Cause, Principle and Unity* 1962; *Ribaldry of Ancient Greece* 1965; *Ribaldry of Ancient Rome* 1965; Eleonore Bille de Mot *The Age of Akhenaten* 1966; Andreas Paulou *Greece, I Keep My Vigil for You* 1968; *The Troubadours and Their World of the Twelfth and Thirteenth Centuries* 1976; Alexander Blok *The Twelve and The Scythians* 1982

Edited

Poetry in Australia [with K. Slessor] 1923; John Harington *The Metamorphosis of Aiax* [with P. Warlock] 1927; *Loving Mad Tom: Bedlamite Verses of the XVI and XVII Centuries* 1927; Robert Herrick: *Delighted Earth* 1928, *Poems by Robert Herrick* 1948; *Inspiration* 1928; John Eliot *The Parlement of Pratlers* 1928; *The London Aphrodite* [with P.R. Stephenson] 1929; *The Letters of Philip Stanhope, Second Earl of Chesterfield* 1930; *Handbook of Freedom* [with E. Rickword] 1939; *Giuliano the Magnificent* [from D. Johnson] 1940; *New Lyrical Ballads* [with H. Arundel, M. Carpenter] 1945; *Anvil: Life and the Arts* 1947; *William Morris: a Selection* 1948; *Key Poets* [with R. Swingler] 1950; *Paintings and Drawings by Leslie Hurry* 1950; Zaharia Stancu *Barefoot* 1951; *The Sunset Ship: the Poems of J.M.W. Turner* 1966

For Children
Runaway 1935; *Rebels of the Goldfields* 1936; *To Arms* 1938; *The Dons Sight Devon* 1941

Poetry
Fauns and Ladies 1923; *The Passionate Neatherd* 1926; *We Are the English* 1936; *Into Action: the Battle of Dieppe* 1942; *Second Front* 1944; *Clue of Darkness* 1949; *Peace Is Our Aim* 1950; *Three Letters to Nikolai Tikhonov* 1951; *Three Elegies* 1956; *Collected Poems* 1981

Plays
Helen Comes of Age [pub. 1927]; *Marino Faliero* [pub. 1927]; *Hereward* (l) [with J. Gough; pub. 1930]

(Nicholas) Vachel LINDSAY 1879–1931

Lindsay was born fourteen years after the end of the American Civil War in Springfield, Illinois, in a house once owned by a sister-in-law of Abraham Lincoln. The town of Springfield was split between Union and Confederate sympathies, and racial tension persisted. It remained close to his heart and work, emblematic to him of America, and he was to die in the room immediately above the one in which he was born. His parents were members of an evangelical sect called 'The Disciples of Christ', and his fundamentalist upbringing had a considerable influence upon his life and work. His father was a doctor (and protagonist of the poem 'Doctor Mohawk'), and intended to hand on his medical practice to his son, but Lindsay's mother was determined right from the start that he should be an artist of some sort. 'I am practically the person she made me when I was eight,' he later acknowledged.

He was educated at Hirman College, Ohio, which he had entered in 1897 ostensibly to read medicine, but where he read widely. He subsequently attended the Chicago Art Institute and, from 1905, the Chase School in New York. Failing to earn his living as a pen-and-ink artist, he remained financially dependent upon his father and on a number of menial jobs, none of which he held for long. He became a hobo, peddling his poems on walking tours, and finally setting out from Springfield in 1912 along the Santa Fe Trail, through Missouri into Kansas and New Mexico, sending letters home which were later published as *Adventures While Preaching the Gospel of Beauty*. The poems which resulted from these travels were published in his first two volumes, *The Tramp's Excuse* (1909) and *Rhymes to be Traded for Bread* (1912), but he made his reputation with his next two volumes, *General William Booth Enters into Heaven* and *The Congo*, the title-poems of which have been much anthologised. These poems showed the principal influences on his work – Salvation Army hymns and the rhythms of Afro-American dance and song – and are natural recital pieces. Lindsay confirmed his reputation with a number of tours in which he declaimed his poems and preached the gospel of art. 'We are planning not an economic, but Art Revolution,' he declared. 'Once the Poets and Artists are in power, goodbye to the business men and tariff senators and such forevermore. We must make this a *Republic of Letters*.' Although his work had a considerable vogue during the years before and during the First World War, it fell out of favour in the 1920s, and his Utopian ideal remained unrealised. He continued to lecture, however, and began promoting the work of black writers, such as **Langston Hughes**, whom he had found working as a busboy in Washington in 1925.

Lindsay's sexuality has been much debated, and he is supposed to have been infatuated for a while with the Australian pianist and composer Percy Grainger. Teetotal, he lectured on the evils of drink for the YMCA and the Anti-Saloon League, and was almost certainly a virgin until his late marriage in 1925 to a woman twenty years his junior. He continued to publish volumes of poetry, including *The Daniel Jazz*, which owes a great deal to black spirituals (and was later set to music), and books of social criticism, such as *The Art of the Moving Picture* and *The Golden Book of Springfield*, which were largely ignored. His reputation failed to recover, and, impoverished, exhausted and despairing, he eventually committed suicide by drinking Lysol.

Because of its strong rhythms, simple subject-matter and sheer exuberance, Lindsay's poetry has often been used in classrooms, although teachers today might hesitate to promote such poems as 'The Congo', which is subtitled 'A Study of the Negro Race' and opens with an impressionistic account of 'Their Basic Savagery'. His espousal of black art, like that of **Carl Van Vechten**, was entirely sincere, but very much of its period and seems today distinctly equivocal. Lindsay remains, however, one of the most important folk poets of the century, the creator of a genuinely popular but nevertheless sophisticated art form.

Poetry
The Tramp's Excuse and Other Poems 1909; *Rhymes to be Traded for Bread* 1912; *General William Booth Enters into Heaven and Other Poems* 1913; *The Congo and Other Poems* 1914; *The Chinese Nightingale and Other Poems* 1917; *The Daniel Jazz and Other Poems* 1920; *The Golden Whales of California and Other Rhymes*

in the American Language 1920; *Going-to-the-Sun* 1923; *Collected Poems* 1923 [rev. 1925]; *The Candle in the Cabin* 1926; *Going-to-the-Stars* 1926; *Johnny Appleseed and Other Poems* 1928; *Every Soul is a Circus* 1929; *Selected Poems* [ed. H. Spencer] 1931; *Selected Poems* [ed. M. Harris] 1963; *The Poetry* [ed. D. Camp] 2 vols 1984, 1985

Non-Fiction

The Village Magazine 1910; *Adventures While Preaching the Gospel of Beauty* 1914; *The Art of the Moving Picture* 1915 [rev. 1922]; *A Handy Guide for Beggars, Especially Those of the Poetic Fraternity* 1916; *The Golden Book of Springfield* 1920; *The Litany of Washington Square* 1929; *Letters to A. Joseph Armstrong* [ed. A. Joseph Armstrong] 1940; *Letters* [ed. M. Chenetier] 1979

Biography

Lindsay: a Poet in America by Edgar Lee Masters 1935; *The West-Going Heart* by Eleanor Ruggles 1959; *Lindsay: Fieldworker for the American Dream* by Ann Massa 1970

Eric (Robert Russell) LINKLATER 1899–1974

It was not until near the end of his life, in his third volume of autobiography, *Fanfare for a Tin Hat*, that Linklater publicly admitted that he had been born in Penarth, Glamorgan: 'I have never said that I was born in Orkney, but my close connections with the islands prompted that assumption.' His father, the son of an Orcadian crofter, was a master mariner working out of Cardiff, and the family moved back to Orkney when Linklater was very young. He attended the Intermediate School for Boys, before going to Aberdeen Grammar School when the family moved to Scotland; but from 1906 they made annual trips to Orkney and he regarded the islands as his spiritual home. In 1916 he went to Aberdeen University to read medicine before joining the Black Watch the following year. As a sniper, he records with characteristic irony, 'I earned, very honestly, a private's pay by killing several Germans.' He was himself wounded in the head – the helmet that saved his life is reproduced on the jacket of *Fanfare for a Tin Hat*. The consequent nightmares from which he suffered form a recurring theme in his writing, notably in his last novel, *A Terrible Freedom*. On getting out of the army he resumed his medical studies for a time before switching to English literature.

He began by writing plays and verse. His first book of poetry, *Poobie*, was published locally in 1925 by Porpoise Press Broadsheets, Aberdeen, and Linklater hankered after writing poetry all his life. As his biographer Michael Parnell points out, a number of the central figures in his novels – Saturday Keith in *Poet's Pub*, the eponymous Magnus Merriman, Stephen Sorely in *Ripeness Is All*, Albyn in *A Spell for Old Bones* and Hector MacRae in *The Merry Muse* – are poets. His first, frankly autobiographical, novel was *White-Maa's Saga* in which a young Orcadian, Peter Fleet, attends medical school in 'Inverdoon'. Linklater returned to Orcadian themes in *The Men of Ness* and *Magnus Merriman*, but in 1928 he won a Commonwealth Fellowship to travel to the USA, which resulted in the picaresque *Juan in America*, the forerunner of a whole genre of novels featuring innocents abroad in that country. Of his other novels written between the wars, *Ripeness Is All* was described by Alan Bold as 'a black comedy, dedicated to the proposition that the human race is hardly worth running'.

He was also a prolific writer of non-fiction of various kinds. An ardent Scottish nationalist, he stood unsuccessfully as a parliamentary candidate in the East Fife by-election of February 1933. *The Lion and the Unicorn*, about Scotland's historical relations with England, comes down in favour of Scottish autonomy, and in the early 1930s he wrote several short biographies of key figures in Scotland's past. During the Second World War, in which he served first in Orkney, then at the War Office, finally in Italy, he wrote several propaganda pamphlets, and his time in Italy was responsible for what many consider his finest novel, *Private Angelo*, an anti-war book that has been compared to Hašek's *The Good Soldier Schweik* (1921–3). Besides three volumes of autobiography, he wrote plays, many short stories and three children's books.

An inveterate traveller and a man of many interests, he was a marvellous companion. He always claimed to be one of nature's idlers, which is odd when one considers what a prolific, if somewhat uneven, writer he was. He was married for over forty years, died in Aberdeen and, appropriately, is buried in Orkney.

Fiction

White-Maa's Saga 1929; *Poet's Pub* 1929; *Juan in America* 1931; *The Men of Ness* 1932; *The Crusader's Key* (s) 1933; *The Fairies Return* (s) 1934; *Magnus Merriman* 1934; *The Revolution* (s) 1934; *God Likes Them Plain* (s) 1935; *Ripeness Is All* 1935; *Juan in China* 1937; *The Sailor's Holiday* 1937; *The Impregnable Woman* 1938; *Judas* 1939; *Private Angelo* 1946; *Sealskin Trousers and Other Stories* (s) 1947; *A Spell for Old Bones* 1949; *Mr Byculla* 1950; *Laxdale Hall* 1951; *The House of Gair* 1953; *The Faithful Ally* 1954; *The Dark of Summer* 1956; *A Sociable Plover and Other Stories and Conceits* (s) 1957; *Position at Noon* 1958; *The Merry Muse* 1959; *Roll of Honour* 1961;

Husband of Delilah 1962; *A Man over Forty* 1963; *A Terrible Freedom* 1966; *The Stories of Eric Linklater* (s) 1968

Non-Fiction

Ben Johnson and King James 1931; *Mary, Queen of Scots* 1933; *Robert the Bruce* 1934; *The Lion and the Unicorn* 1935; *The Defence of Calais* 1941; *The Northern Garrisons* 1941; *The Man on My Back* (a) 1941; *The Highland Division* 1942; *The Art of Adventure* 1946; *The Campaign in Italy* 1951; *Our Men in Korea* 1952; *A Year of Space* (a) 1953; *The Ultimate Viking* 1955; *Edinburgh* 1960; *Guller's Sweden* 1964; *Memorandum on the Conditions in the North of Scotland* 1964; *The Prince in the Heather* 1965; *Orkney and Shetland* 1965 [rev. 1971, 1980]; *The Conquest of England* 1966; *Notes for a Scottish Pantheon* 1967; *Scotland* [with E. Smith] 1968; *The Survival of Scotland* 1968; *The Secret Larder or How the Salmon Lives and Why It Dies* 1969; *Fanfare for a Tin Hat* (a) 1970; *The Music of the North* 1970; *The Royal House of Scotland* 1970; *The Corpse on Clapham Common* 1971; *The Voyage of the Challenger* 1972; *The Black Watch* 1977

Poetry

Poobie 1925; *A Dragon Laughed and Other Poems* 1930

Plays

Rosemount Nights 1923; *The Prince Appears* 1924; *The Devil's in the News* 1934; *The Cornerstones* 1941; *The Raft & Socrates Asks Why* 1942; *Crisis in Heaven* 1944; *The Great Ship & Rabelais Replies* 1944; *Two Comedies* [pub. 1950]; *The Mortimer Touch* 1952; *Breakspear in Gascony* 1958

For Children

The Wind on the Moon 1944; *The Pirates in the Deep Green Sea* 1949; *Karina with Love* 1958

Edited

The Thistle and the Pen: an anthology of Scottish Writers 1950; *John Moore's England* 1971

Biography

Eric Linklater: a Critical Biography by Michael Parnell 1984

Penelope (Margaret) LIVELY 1933–

The daughter of a bank manager, Lively was born Penelope Low in Cairo. Her memoir *Oleander, Jacaranda* recalls a childhood spent absorbing the sights and smells of Egypt before being sent to England at the age of twelve to begin her formal education at a boarding-school. She was an undergraduate at St Anne's College, Oxford, and has for many years lived near the city, frequently setting her fiction there and in Oxfordshire. She married Jack Lively, an academic at the University of Warwick, in 1957, the year after she graduated, and has followed no career other than that of a freelance writer. She has two children, one of whom is the novelist Adam Lively.

Like **Jane Gardam** and **Ann Schlee**, Lively started as a children's novelist. Her first book, *Astercote* (1970), is the story of a medieval village, the original inhabitants of which died during the Black Death, but continue to influence their modern successors. Of her subsequent novels for children, the best is probably *The House in Norham Gardens*, while *The Ghost of Thomas Kempe* won the Carnegie Medal and *A Stitch in Time* won the Whitbread Award for children's fiction.

Her first novel for adults was *The Road to Lichfield*, published in 1977. This and subsequent novels share similar themes with her work for children: the influence on people of memory as it survives in landscape, houses or pictures (one of her works of non-fiction is a book about the history of landscape), and the organic connection between past and present. These themes are apparent in her second novel, *Treasures of Time*, a sharp comedy with a background of archaeology, which won the Arts Council National Book Award. Memory in Lively's novels may be the collective historical one of the bleak *Judgement Day*, the grief of a bereaved family in *Perfect Happiness*, the life of a minor writer as reasearched by his would-be biographer in *According to Mark*, or the attempts of one woman to understand her life in *Moon Tiger*. This last novel, which makes use of Lively's Egyptian background and is perhaps her most ambitious and serious to date, won the Booker Prize in 1987. Lively has also distinguished herself as a writer of short stories, which have been collected in three volumes.

Fiction

The Road to Lichfield 1977; *Nothing Missing but the Samovar* (s) 1978; *Treasures of Time* 1979; *Judgement Day* 1980; *Next to Nature, Art* 1982; *Perfect Happiness* 1983; *According to Mark* 1984; *Corruption* (s) 1984; *Pack of Cards: Stories 1978–86* (s) 1986; *Moon Tiger* 1987; *Passing On* 1989; *City of the Mind* 1991; *Cleopatra's Sister* 1993

For Children

Astercote 1970; *The Wild Hunt of Hagworthy* 1971 [US *The Wild Hunt of the Ghost Hounds* 1972]; *The Driftway* 1972; *The Ghost of Thomas Kempe* 1973; *The Whispering Knights* 1973; *The House in Norham Gardens* 1974; *Boy without a*

Name 1975; *Going Back* 1975; *The Stained Glass Window* 1976; *A Stitch in Time* 1976; *The Voyage of QV 66* 1978; *Fanny and the Battle of Potter's Piece* 1980; *The Revenge of Samuel Stokes* 1981; *Fanny and the Monsters and Other Stories* (s) 1982; *Uninvited Ghosts and Other Stories* (s) 1984; *A House Inside Out* 1987

Non-Fiction

The Presence of the Past: an Introduction to Landscape History 1976; *Oleander, Jacaranda: a Childhood Perceived* (a) 1994

Plays

Boy Dominic (tv) 1974; *Time Out of Mind* (tv) 1976

Richard (Daffyd Vivian) LLEWELLYN (Lloyd) 1907–1983

Llewellyn was born in St David's, Pembrokeshire, and educated locally and in London. After school, he spent some time in Italy, supposedly learning hotel management, but also spending time studying art and working in the film industry. In 1926 he joined the British army and served in the Far East, returning home in 1932 in the depths of the Depression. He was repeatedly down and out in London and often slept rough. Eventually he found work in film journalism, then in production, scriptwriting and acting.

In 1938 he established himself in the theatre with his play *Poison Pen*, a psychological thriller (subsequently filmed), but it was his first novel, *How Green Was My Valley*, which really made his name the following year and is the work by which he will be remembered. Set in a South Wales mining community, it was an instant bestseller both in Britain and America and in 1941 was made into a film, directed by John Ford, which won the Academy Award for best picture. Llewellyn had another stage success in 1947 with *Noose*, which concerned a black-market racket in Soho, and which he adapted himself almost immediately for a film.

Llewellyn spent the early part of the war as a chief transport officer for the Entertainments National Services Association, then in 1940 was commissioned in the Welsh Guards. His second novel, *None But the Lonely Heart*, was published during the war and thought by some to have more literary merit than his first, but neither this, nor his third novel, *A Few Flowers for Shiner*, met with the same success. In other novels Llewellyn attempted to move away from what was clearly his principal subject – either abroad, as in *A Flame for Doubting Thomas*, which has an American setting, or back in time, as in his fiction for children: *The Flame of Hercules* is set in the Roman Empire in the first century AD, while *Warden of the Smoke and Bells* is about Marco Polo's return from Cathay to encounter further dangers in Assisi. *Mr Hamish Gleave* was inspired by the Burgess and Maclean affair, and in the late 1960s and early 1970s he wrote a series of espionage novels featuring the same central character, Edmund Trothe.

Llewellyn married twice. Although he spent much of his later life abroad, principally in America, he returned to his Welsh background in his fiction, eventually writing three sequels to *How Green Was My Valley*: *Up, Into the Singing Mountain*, *Down Where the Moon is Small* and *Green, Green My Valley Now*.

Fiction

How Green Was My Valley 1939; *None But the Lonely Heart* 1943; *A Few Flowers for Shiner* 1950; *A Flame for Doubting Thomas* 1953; *Mr Hamish Gleave* 1956; *Chez Pavan* 1958; *Up, Into the Singing Mountain* 1963; *A Man in a Mirror* 1964; *Sweet Morn of Judas' Day* 1965; *Down Where the Moon Is Small* 1966; *The End of the Rug* 1968; *But We Didn't Get the Fox* 1969; *The Night is a Child* 1972; *White Horse to Banbury Cross* 1972; *Bride of Israel, My Love* 1973; *A Hill of Many Dreams* 1974; *Green, Green My Valley Now* 1975; *At Sunrise, the Rough Music* 1976; *Tell Me Now and Again* 1977; *A Night of Bright Stars* 1979; *I Stand on a Quiet Shore* 1982

For Children

Sweet Witch 1955; *Warden of the Smoke and Bells* 1956; *The Flame of Hercules* 1957

Plays

Poison Pen 1938; *Noose* (st) 1947, (f) 1948; *The Scarlet Suit* 1962; *Ecce!* 1974; *Hat!* 1974; *Oranges and Lemons* (tv) 1980

Liz LOCHHEAD 1947–

Lochhead was born in Motherwell, Scotland, 'the daughter of two white-collar hopes from blue-collar backgrounds', and educated at Dalziel High School, Motherwell, and Glasgow School of Art, where she trained as a painter but began writing poems. In 1972, the year in which her first collection, *Memo for Spring*, was published, she won a BBC radio competition and the Scottish Arts Council New Writing Award. For eight years she taught art at Bishopbriggs High School and other schools in Glasgow and Bristol, until in 1978 selection as the first holder of a Scottish exchange fellowship enabled her both to visit Canada and to become a full-time writer. In the same year her revue *Sugar and Spice* was performed and she published another collection of poems, *Islands*.

Lochhead aims above all at accessibility; the *Scotsman* described *Dreaming Frankenstein and Collected Poems* as 'a rare thing; a book of poems which sparkle'. In *True Confessions and New Clichés* she selected the best of the raps, songs, sketches and monologues from her plays and revues. In characters such as Mrs Abernethy with her shortbread, and Sharon who has a crush on the English teacher, she pokes fun at the seriousness with which women can take themselves. 'I don't really think a man could ever be as irreverent about women or take the piss out of them in the same way', she has said. She is an effective satirist, and her musical, *Same Difference*, contains touching songs and sharp comments on the tension between the sexes. Lochhead has also adapted works of Molière for the stage, translating *Tartuffe* into vernacular Scots. She lives in Glasgow but spends much of her time travelling the country to give readings and performances of her work.

Poetry

Memo for Spring 1972; *Islands* 1978; *The Grimm Sisters* 1981; *Dreaming Frankenstein and Collected Poems* 1984; *True Confessions and New Clichés* 1985; *Bagpipe Muzak* 1991; *Three Scottish Poets* [with N. MacCaig, E. Morgan; ed. R. Watson] 1993; *Penguin Modern Poets 4* [with R. McGough, S. Olds] 1995

Plays

Now and Then **(f)** 1972 [pub. 1982]; *Sugar and Spice* 1978 [np]; *Blood and Ice* **(st)** 1982, **(r)** 1990; *Disgusting Objects* 1982; *True Confessions* 1982; *A Bunch of Fives* [with S. Hardie, T. Leonard] 1983; *Rosaleen's Baby Shanghai'd* 1983; *Same Difference* 1984; *Silver Service* 1984 [pub. 1985]; *Sweet Nothings* **(tv)** 1984; *Dracula* [from B. Stoker's story] 1985 [pub. 1989]; *Tartuffe* [from Molière's play] 1985; *Fancy You Minding That* **(r)** 1986; *Mary Queen of Scots Got Her Head Chopped Off* 1987 [pub. with *Dracula* 1989]; *The Big Picture* 1988; *Pattermerchants* [from Molière] 1989

Ross LOCKRIDGE 1914–1948

Lockridge was Indiana's state champion in shorthand and typing, skills he learned at high school. He was born and educated in Bloomington, the son of parents who were graduates of Indiana University. His father, also called Ross, wrote two noted histories of Indiana, and it seemed natural that his son would also become a writer. Lockridge had a distinguished academic career: after a year studying in Paris, he graduated from Indiana State University in 1935 with a BA in English, and taught there while completing his master's degree which he received in 1939. The following year, he was awarded a fellowship at Harvard, and moved to Boston with his wife, whom he had married in 1937.

A long epic poem was rejected by publishers, and Lockridge began a seven-year task of research and writing his historical novel, *Raintree County*. Despite the accumulation of four children in a small apartment, he revised, and his wife retyped, the book, which was first offered to a publisher in 1946; after further revision, it was eventually published in January 1948.

Raintree County won the Metro-Goldwyn-Mayer award for fiction, and became a sensational bestseller. Over 1,000 pages long, the novel is set on one day in 1892, and brilliantly reconstructs the intertwined lives of a community, with flashbacks to detail the background and the past. The sweep and breadth of action provide both the appeal and the limitations of the novel. Some critics complained that Lockridge's research material was too obvious, but it was agreed that he had ended a slump in American fiction.

Described by a friend as smouldering with 'a boyish eagerness which never relaxes', Lockridge suffered a nervous breakdown after the effort of concentrated work. Rich and famous, he moved back to Bloomington to a house he had built, but committed suicide in March, two months after his only book was published.

Fiction

Raintree County 1948

David (John) LODGE 1935–

Lodge was born in Dulwich, south London, the son of a dance-band musician. His mother was an Irish-Belgian Catholic, and he was educated at a Catholic school, St Joseph's Academy in Blackheath, before reading English at University College, London, from 1952 to 1955. His experience of national service with the Royal Armoured Corps is reflected in his novel *Ginger, You're Barmy*. In 1957 he returned to UCL and wrote a 700-page MA thesis on 'Catholic Fiction Since the Oxford Movement', its bulk evidently causing the authorities to impose a limit on the length of future theses. He married in 1959, and, after a year teaching English at the British Council's overseas students' centre, he became an assistant lecturer at Birmingham University. It was in Birmingham that he met **Malcolm Bradbury** with whom he co-operated on two comic revues. In 1964 he won a Harkness

Travelling Scholarship which enabled him to travel in America whilst completing his first comic novel, *The British Museum Is Falling Down*. He was made professor of modern English literature at Birmingham in 1976, and in 1987 retired from full-time teaching to concentrate on writing.

Lodge's first novel, *The Picturegoers* (1960), is a realistic portrait of a lower-middle-class adolescence in pre-war south London which he started while doing national service. The more ambitious *Out of the Shelter* similarly draws on his childhood in describing the experience of wartime evacuation. *How Far Can You Go?*, which was the 1980 Whitbread Book of the Year, addresses the moral dilemmas faced by Catholics in the wake of the second Vatican Council, and again draws on personal experience in describing a child with Down's syndrome (as does the youngest of Lodge's three children).

Lodge has described himself as 'a domesticator of more extreme types of continental criticism', and these ideas are reflected in his novels of academic life, *Changing Places* and *Small World*, which make playful use of the theories of critics such as Roman Jakobson and Jacques Derrida. Lodge's first solo stage play, *The Writing Game*, was performed at the Birmingham Repertory Theatre in 1990, and he has adapted for television his own novel *Nice Work* (1989) and Dickens's *Martin Chuzzlewit* (1994).

Fiction

The Picturegoers 1960; *Ginger, You're Barmy* 1962; *The British Museum Is Falling Down* 1965; *Out of the Shelter* 1970 [rev. 1985]; *Changing Places* 1975; *How Far Can You Go?* [US *Souls and Bodies*] 1980; *Small World* 1984; *Nice Work* 1988; *Paradise News* 1991

Plays

Between These Four Walls [with M. Bradbury, J. Duckett] 1963 [np]; *Slap in the Middle* [with others] 1965 [np]; *Big Words ... Small Worlds* **(tv)** 1987; *Nice Work* [from his novel] **(tv)** 1989; *The Writing Game* 1990 [pub. 1991]; *Martin Chuzzlewit* [from Dickens's novel] **(tv)** 1994

Non-Fiction

Language and Fiction 1966 [rev. 1984]; *Graham Greene* 1966; *Evelyn Waugh* 1971; *The Novelist at the Crossroads and Other Essays on Fiction and Criticism* 1971; *Modernism, Antimodernism, and Postmodernism* 1977; *The Modes of Modern Writing* 1977; *Working with Structuralism* 1981; *Write On: Occasional Essays 1965–1985* 1986; *After Bakhtin: Essays on Fiction and Criticism* 1990; *The Art of Fiction* 1992

Edited

Jane Austen's 'Emma': a Casebook 1968; Jane Austen *Emma* [with J. Kinsley] 1971; *Twentieth Century Literary Criticism: a Reader* 1972; George Eliot *Scenes of Clerical Life* 1973; Thomas Hardy *The Woodlanders* 1974; *The Best of Ring Lardner* 1984; Henry James *The Spoils of Poynton* 1987; *Modern Criticism and Theory: a Reader* 1988

For Children

About Catholic Authors 1958

Christopher LOGUE 1926–

Logue was born in Portsmouth and educated at the local grammar school. He served in the army from 1944 to 1948, spending sixteen months in military detention, and in the early 1950s lived in France for five years. He is primarily known as a poet and playwright, although he has also had a parallel career as an actor, appearing in a number of films and on the stage. His association with Ken Russell led to his playing Swinburne in *Dante's Inferno* (1967) and Cardinal Richelieu in *The Devils* (1971), and he wrote the screenplay (adapted from the book by H.S. Ede) for Russell's film about Henri Gaudier-Brzeska, *Savage Messiah* (1972). In 1985 he married the critic Rosemary Hill. He was until 1994 a regular contributor to *Private Eye*, where he edited the 'True Stories' column, which garnered newspaper accounts of bizarre events from around the world.

His first volume of poetry, *Wand and Quadrant*, was published in 1953 and inaugurated a prolific career. An advocate of 'making poetry pay', he led with **Adrian Mitchell** the radical poetry-reading movement of the late 1950s and 1960s, protesting against the Vietnam War and attacking imperialism, capitalism and nuclear weapons. His contributions included poems such as 'The Song of a Dead Soldier', with its provocative subtitle 'One Killed in the Interests of Certain Tory Senators in Cyprus'. Some critics felt that Logue's public readings affected the poems themselves. Of his volume *Songs* (containing such verses as 'Song of the Road' about the attempts of Third World peasants to build a road in defiance of authority), **A. Alvarez** commented: '[Logue] has replaced rhythm by a histrionic flow, like platform oratory'. Later volumes, in which he turned his attention to the permissive society, have been attacked for glibness; *New Numbers* was described by his detractors as 'journalistic', but Logue would argue that he is describing a world that trivialises itself. A collected volume of twenty-five years' worth of poems (1953–78) was published in 1981 as *Ode to the Dodo*.

His most successful work for the stage has been his adaptation of Homer's *Iliad*, starting in 1981 with *War Music*, a graphic version of Books

XVI to XIX, which went beyond the original in the violence of its language. This was followed by *Kings* and *The Husbands*. *Patrocleia* and *Pax* also form part of this long work, originally published separately, but later incorporated in *War Music*. Logue has also written several books for children, edited several collections from his 'True Stories' column, and written a confessedly pornographic novel, *Lust*, which was published by the Olympia Press under the pseudonym of Count Palmiro Vicarion. He has also used this pseudonym for two collections of bawdy verse.

Poetry

Wand and Quadrant 1953; *First Testament* 1955; *The Weekdream Sonnets* 1955; *Devil, Maggot and Son* 1956; *She Sings, He Sings* 1957; *A Song for Kathleen* 1958; *Songs* 1959; *Memoranda for Marchers* 1960; *Songs from 'The Lily-White'* 1960; *I Shall Vote Labour* 1966; *Logue's ABC* 1966; *Selections from a Correspondence Between an Irishman and a Rat* 1966; *The Words of Christopher Logue's Establishment Songs, Etcetera* 1966; *Poem* 1967; *Gone Ladies* [with W. Southam] 1968; *Hermes Flew to Olympus* 1968; *Rat, Oh Rat* 1968; *The Girls* 1969; *SL* 1969; *New Numbers* 1970; *For Talitha* 1971; *The Isles of Jessamy* 1971; *Duet for a Mole and a Worm* 1972; *Twelve Cards* 1972; *What* 1972; *Singles* 1973; *Mixed Rushes* 1974; *Urbanal* 1975; *Abecedary* 1977; *Red Bird* 1979; *Ode to the Dodo: Poems from 1953 to 1978* 1981; *Lucky Dust* 1985

Plays

The Trial of Cob and Leach 1959 [np]; *The Lily-White Boys* **(l)** [with others] 1960 [np]; *Trials by Logue* 1960 [np]; *The Arrival of the Poet* [pub. 1963; rev. with G. Nicholson 1985, pub. 1983]; *The End of Arthur's Marriage* **(tv)** [with S. Myers] 1965; *Friday* [from H. Klaus] 1971 [pub. 1972]; *Savage Messiah* **(f)** 1972; *War Music* [from Homer's *Iliad*] [with D. Fraser] 1977 [pub. 1981]; *Baal* [from B. Brecht's play] 1986 [np]; *The Seven Deadly Sins* [from B. Brecht's play] 1986 [np]; *Crusoe* **(f)** [with W. Green] 1989; *Strings* **(r)** [with J. Osborn] 1989; King's [from Homer's *Iliad*] 1991; *The Husbands* 1994

Fiction

Lust [as 'Count Palmiro Vicarion'] 1959; *Christopher Logue's Bumper Book of True Stories* **(s)** 1980

For Children

The Crocodile 1976; *Puss-in-Boots Pop-Up* 1976; *Ratsmagic* 1976; *The Magic Circus* 1979

Translations

Pablo Neruda *The Man Who Told His Love* 1958; Homer: *Patrocleia* 1962, *Pax* 1967; H. Claus *Friday* 1971

Edited

Count Palmiro Vicarion's Book of Limericks [as 'Count Palmiro Vicarion'] 1959; *Count Palmiro Vicarion's Book of Bawdy Ballads* [as 'Count Palmiro Vicarion'] 1962; *True Stories* 1966; *True Stories from Private Eye* 1973; *The Children's Book of Comic Verse* 1979; *London in Verse* 1982; *The Oxford Book of Pseuds* 1983; *Sweet and Sour: an Anthology of Comic Verse* 1983; *The Children's Book of Children's Rhymes* 1986

Jack (i.e. John Griffith) LONDON 1876–1916

Reflecting on his childhood, London once observed: 'my body and soul were starved'. He was born 'a bastard' (as he insisted on putting it) in San Francisco, the son of a 'professor' of astrology and a medium. He took his surname from his stepfather, a farmer, and was brought up in Oakland in a family continually beset by financial problems. His independence of mind and his determination were necessities before they were virtues. His childhood was spent on the waterfront, and he started work, as a newspaper vendor, at the age of ten. He subsequently worked in a canning factory before becoming 'King of the Oyster Raiders' in the bay. His early life reads like a compendium of American adventure: sailor, hobo, seal-hunter, political activist and Klondike gold prospector. In 1895 he began a furious process of self-education (he spent a single semester at Berkeley University before poverty forced him to leave), and three years later was beginning to turn his experiences of life into books.

His first book was a volume of short stories, *The Son of the Wolf* (1900), and led to his being dubbed 'The Kipling of the Klondike'. His first novel, *A Daughter of the Snows* (1902), also described the harsh life of the prospectors in dramatic, economical prose, but it was with *The Call of the Wild* the following year that he established his reputation. The vivid and unsentimental story of a sledge-dog in the Yukon, it is a fable in which the relationship between man and beast is used as a metaphor for the initiation of children into the world of adults. Its celebration of the pioneer spirit proved extremely popular and London soon became a bestselling author, earning an astonishing one million dollars by the time he was thirty. Success did not diminish his capacity for sustaining vast workloads, however, and he remained an enormously prolific writer to the end of his brief life. He also continued to widen his experience, most famously in his foolhardy project to sail round the world in a schooner, *The Snark*, which he had constructed himself at ruinous expense.

He was also a campaigning journalist, who reported on the Russo-Japanese war of 1904–5, and, in the guise of a sailor, investigated London's East End in 1902, recording the lives of the slum-dwellers in *The People of the Abyss* the following year. His view of the world was contradictory. **George Orwell** noted that: 'London could foresee fascism [in his political novel, *The Iron Heel*] because he had a fascist streak in himself: or at any rate a marked strain of brutality and an almost unconquerable preference for the strong man as against the weak man.' He saw himself as a socialist revolutionary (he once stood as socialist candidate for the mayorship of Oakland), but – influenced by Nietzsche and Darwin's theory of the survival of the fittest – he espoused white, Anglo-Saxon supremacy. A believer in social justice, but also in what Orwell called 'a "natural aristocracy" of strength, beauty and talent', he was at once sympathetic towards and repelled by the lives of those at the bottom of society's teeming pile. 'They are out of place,' he wrote of the English cockneys on their annual outing to work in the hopfields of Kent. 'As they drag their squat misshapen bodies along the highways and byways, they resemble some vile spawn from the underground.' Some of this ambivalence may be explained by his own degradation, as he squandered his huge earnings – he never completely reconciled his political beliefs with his financial success – and battled with his addiction to whisky, recording his experiences in his volume of 'Alcoholic Memoirs', *John Barleycorn*.

London was not treated seriously by contemporary critics, and claimed that he wrote only for money. Already suffering from uraemia, he hastened his death at the age of forty with an overdose of morphine. He remains best known for his animal fables, but *The Sea Wolf* (a novel about the ruthless captain of a sealing vessel), *The Iron Heel* and the autobiographical *Martin Eden* seem likely to survive, and writers as diverse as Orwell and **Tom Wolfe** have testified to the importance and seriousness of his work.

Fiction

The Son of the Wolf (s) 1900; *The God of His Fathers* (s) 1901; *Children of the Frost* (s) 1902; *A Daughter of the Snows* 1902; *The Cruise of the 'Dazzler'* 1902; *The Call of the Wild* 1903; *The Faith of Men* (s) 1904; *The Sea Wolf* 1904; *The Game* 1905; *Tales of the Fish Patrol* (s) 1905; *Moon-Face and Other Stories* (s) 1906; *White Fang* 1906; *Before Adam* 1907; *Love of Life and Other Stories* (s) 1907; *The Road* (s) 1907; *The Iron Heel* 1908; *Martin Eden* 1909; *Burning Daylight* 1910; *Lost Face* (s) 1910; *Adventure* 1911; *When God Laughs* (s) 1911; *South Sea Tales* (s) 1911; *The House of Pride and Other Tales of Hawaii* (s) 1912; *Smoke Bellew Tales* (s) 1912; *A Son of the Sun* (s) 1912; *The Abysmal Brute* 1913; *The Night Born* (s) 1913; *The Valley of the Moon* 1913; *The Strength of the Strong* (s) 1914; *The Mutiny of the Elsinore* 1914; *The Jacket* 1915; *The Scarlet Plague* 1915; *The Little Lady of the Big House* 1916; *Turtles of Tasman* (s) 1916; *Jerry of the Islands* 1917; *Michael, Brother of Jerry* 1917; *Hearts of Three* 1920; *Dutch Courage* (s) 1922; *The Assassination Bureau Ltd* [with R.L. Fish] 1963; *Curious Fragments* (s) [ed. D.L. Walker] 1975

Non-Fiction

The People of the Abyss 1903; *War of the Classes* 1905; *Revolution* 1910; *The Cruise of the 'Snark'* 1911; *John Barleycorn* (a) 1913; *The Human Drift* 1917; *Jack London, American Rebel* [ed. P.S. Foner] 1947; *Letters from Jack London* [ed. K. Hendricks, I. Shepard] 1966; *No Mentor but Myself* [ed. D.L. Walker] 1979

Plays

Scorn of Women 1906; *Theft* 1910; *The Acorn Planter* 1916

Collections

The Kempton-Wace Letters [with A. Strunsky] 1904

Biography

Sailor on Horseback by Irving Stone 1938; *Jack London* by Richard O'Connor 1965; *Jack* by Andrew Sinclair 1978

Michael (George) LONGLEY 1939–

Longley was born in Belfast, into the Protestant community of the city, but his parents were both originally from Clapham, south London, and came to live in Northern Ireland in 1927. His father, who had been shell-shocked in the First World War (an event that Longley sometimes links with the Troubles in his poems), was a commercial traveller. Longley grew up in the Belfast district of Malone, where he has lived for much of his life, and attended primary school there before going to the Royal Belfast Academical Institution between 1951 and 1958. He then read classics at Trinity College, Dublin, where he met **Derek Mahon**, one of the three other Ulster poets (James Simmons and **Seamus Heaney** are the remaining two) with whom his name is often associated and with whom he can be said to form part of a recognisable school.

From 1963, he worked as a teacher for seven years, first briefly in Dublin and (for one summer term) in Erith, south-east London, and then more extensively in Belfast, where he taught mainly at his old school, the Royal Belfast Academical Institution. He married Edna Broderick in 1964, with whom he has three children; as Edna Longley, she is a well-known critic, specialising particularly in modern Ulster poetry, and, since 1963 lecturing at Queen's University, Belfast, where from 1991 she has held a chair.

Longley himself began gradually to be known as a poet during his teaching years; he had won the prestigious Eric Gregory Award as early as 1965, but his first works consisted of two pamphlets and two collaborations, before his first full collection, *No Continuing City*, appeared in 1969, the first of four substantial collections he was to publish during the first decade of the Troubles. Largely absent from the first collection, the Troubles are movingly and straightforwardly to the fore by the second volume, *An Exploded View*, but do not preclude the exploration of a wide range of civilised themes, among which animals (particularly birds), classics, jazz and North American poetry are prominent, all treated in a formal, skilful but never lifeless verse which stands somewhere between the reticence of Mahon and the exuberance of Heaney.

From 1970, Longley was assistant director (later combined arts director) at the Arts Council of Northern Ireland, but a period of increasing disputes there during the latter part of his tenure was largely responsible for the twelve-year poetic silence before his fifth substantially new collection, *Gorse Fires*, appeared in 1991. In that year, he took early retirement from the Arts Council to devote himself to his writing. His major work is poetry, but he has also published an important revisionist edition of the poems of **Louis MacNeice** in 1988, while among other works edited is *Under the Moon, Over the Stars*, a collection of young people's writing from Ulster. He has been the author of numerous scripts for the BBC Schools Department, and is well known as a broadcaster and reader of his own verse.

Poetry

Ten Poems 1965; *Room to Rhyme* [with D. Hammond, S. Heaney] 1968; *Secret Marriages* 1968; *Three Regional Voices* [with I. Crichton Smith, B. Tebb] 1968; *No Continuing City: Poems 1963–1968* 1969; *Lares* 1972; *An Exploded View: Poems 1968–1972* 1973; *Fishing in the Sky* 1975; *Penguin Modern Poets 26* [with D. Abse, D.J. Enright] 1975; *Man Lying on a Wall: Poems 1972–1975* 1976; *The Echo Gate: Poems 1975–1978* 1979; *Selected Poems 1963–1980* 1980; *Patchwork* 1981; *Poems 1963–1983* 1985; *Poems* 1986; *Gorse Fires* 1991; *Baucis and Philemon: After Ovid* 1993; *Ghost Orchid* 1995

Edited

Causeway: the Arts in Ulster 1971; *Under the Moon, Over the Stars* 1971; *Selected Poems of Louis MacNeice* 1988; W.R. Rodgers *Poems* 1993

Anita LOOS 1888–1981

Best known as the author of the classic comic novel *Gentlemen Prefer Blondes*, but also one of the most prolific Hollywood film scenarists of the century, Loos was born in Sisson (now Mount Shasta), California, where her father edited the local newspaper. He later became a theatrical producer in San Francisco, and his young daughter acted in his company, playing Little Lord Fauntleroy among other roles (she claimed at the age of five, but was accustomed to give her date of birth, incorrectly, as 1893).

Later she began sending film scenarios to the influential film director D.W. Griffith; one of the first to be accepted was for *The New York Hat* (1912), which starred Mary Pickford and Lionel Barrymore. Between 1912 and 1915 Loos is said to have written more than 100 scenarios for the nascent Hollywood film industry, most of them brief slapstick comedies, with names (according to her own account: there are few records) such as *A Hicksville Romance*, *The Making of a Masher*, *Nellie, the Female Villain* and *A No Bull Spy*. A little later she adapted *Macbeth* and wrote the subtitles for Griffith's famous *Intolerance*, among other films.

In 1915 she married Frank Pallma Jr, but left him after a few weeks. In 1920 she married John Emerson, the director who made Douglas Fairbanks Sr famous. They collaborated on many films starring Fairbanks or Constance Talmadge, but were less happy personally, usually living apart and with Emerson confined to a home as an invalid many years before his death in 1956. In the early 1920s Loos wrote several Broadway plays, but became a household name in 1925 with *Gentlemen Prefer Blondes*, the story of the little gold-digger Lorelei Lee, which collected praise from **Aldous Huxley** and **Edith Wharton** and gave its author the entrée to **H.G. Wells** and Mussolini as the archetypal flapper. In the course of an interview on British television in the early 1960s, she was asked: 'If you were to write such a book today, what would be your theme?' 'Gentlemen Prefer Gentlemen', she replied. *But Gentlemen Marry Brunettes* is a sequel.

In the early 1930s she returned to Hollywood as a scenarist for Metro-Goldwyn-Mayer, writing many of the Jean Harlow films. From the mid-1940s she turned her attention to Broadway, scoring a hit in 1946 with *Happy Birthday* and adapting Colette for the stage in *Gigi*. The last phase of her writing included two satirical novels about Hollywood as well as a series of memoirs with titles such as *A Girl Like I* and *Kiss Hollywood Goodbye*. Her later life was passed in an apartment opposite Carnegie Hall in New

York, where she was cared for by her devoted black companion Gladys Tipton, rose at 4 a.m., and followed a busy social schedule before retiring at supper-time.

Fiction

Gentlemen Prefer Blondes 1925; *But Gentlemen Marry Brunettes* 1928; *A Mouse Is Born* 1951; *No Mother to Guide Her* 1961

Non-Fiction

How to Write Photoplays [with J. Emerson] 1920; *Breaking into the Movies* [with J. Emerson] 1921; *A Girl Like I* **(a)** 1966; *Twice Over Lightly: New York Then and Now* [with H. Hayes] 1972; *Kiss Hollywood Goodbye* **(a)** 1974; *Cast of Thousands* **(a)** 1977; *The Talmadge Girls: a Memoir* 1978; *Fate Keeps on Happening* [ed. R.P. Corsini] 1984

Plays

The New York Hat **(f)** 1912; *The Whole Town's Talking* [from F. Arnold and E. Bache; with J. Emerson] 1922 [pub. 1924]; *Gentlemen Prefer Blondes* [from her novel] **(st)** 1926, **(f)** 1928 [np]; *The Struggle* [with J. Emerson] **(f)**, **(st)** 1931; *San Francisco* **(f)** [with R. Hopkins] 1936 [pub. 1979, ed. M.J. Bruccoli]; *The Women* **(f)** [with J. Murfin] 1939 [pub. in *20 Best Film Plays* ed. J. Gassner, D. Nichols 1943]; *Happy Birthday* 1946 [pub. 1947]; *Gigi* [from Colette's novel] 1951 [pub. 1952; rev. 1956]; *The Amazing Adele* **(l)** [from P. Barillet and J.-P. Gredy's play] 1955 [np]; *Chéri* [from Colette] 1959 [np]; *Gogo Loves You* **(l)** 1964 [np]; *Something about Anne* **(l)** [from J. Canolle's *The King's Mare*] 1966 [pub. 1967]

Translations

Jean Canolle *The King's Mare* 1966

Biography

Anita Loos by Gary Carey 1988

H(oward) P(hillips) LOVECRAFT 1890–1937

Lovecraft was born in Providence, Rhode Island, where he lived for most of his life, and where his sensitivity and sickliness as a child and young man prevented him from attending college. His father was confined in a lunatic asylum when Lovecraft was a child, and the same thing happened to his mother when he was twenty-nine; thenceforward he lived mainly with a couple of aunts.

He had begun writing when young, and in his twenties became involved with the United Amateur Press Association, which first published his stories. Although his work appeared fairly widely in literary magazines, it was the publication of his story 'Dagon' by the magazine *Weird Tales* in October 1923 which launched him with the horror-reading public. Most of his subsequent stories appeared in that magazine, as well as in *Amazing Stories* and *Astounding Stories*, while he also began to write novellas. In 1924 he married a New York writer, Sonia Greene, and lived with her for a while in Brooklyn, but the partnership lasted less than two years and they were divorced in 1929. He returned to Providence, where he lived the life of an extreme recluse, keeping the curtains closed all day and only going out at night, and writing letters, of which he is believed to have produced more than 100,000.

He worked mainly as a ghost writer and reviser of manuscripts for others, only publishing four books, including the novella *The Shadow over Innsmouth* (1936), during his lifetime. Two years after his death, the horror writers August Derleth and Donald Wandrei formed the specialist press Arkham House to perpetuate his work, and Derleth in particular was successful in creating a widespread cult of Lovecraft from the late 1940s on. Many collections have now been published, which include, besides stories and novellas, the poetry Lovecraft wrote in a pastiche of eighteenth-century style and such essays as 'Supernatural Horror in Fiction'. Lovecraft has passionate admirers, but most reputable critics consider his work sub-literary. Most of his later writings concerned the Cthulhu Mythos, a world dominated by evil Ancient Ones who have been expelled from Earth but still meddle in its affairs. Perhaps Lovecraft's chief interest can be said to be that he was a pioneer of the area where crude horror and science fiction merge, and certainly the Cthulhu Mythos has been carried on by many later writers.

Fiction

The Shunned House **(s)** 1928; *The Battle That Ended the Century* **(s)** 1934; *The Cats of Ulthar* **(s)** 1935; *The Shadow over Innsmouth* 1936; *The Outsider and Others* **(s)** [ed. A. Derleth, D. Wandrei] 1939; *The Weird Shadow over Innsmouth and Other Stories of the Supernatural* **(s)** 1944; *The Best Supernatural Stories of H.P. Lovecraft* **(s)** [ed. A. Derleth] 1945; *The Dunwich Horror* **(s)** 1945; *The Dunwich Horror and Other Weird Tales* **(s)** 1945; *The Lurker at the Threshold* [with A. Derleth] 1945; *The Lurking Fear and Other Stories* **(s)** 1947; *The Haunter of the Dark and Other Tales of Horror* **(s)** 1951; *The Case of Charles Dexter Ward* **(s)** 1952; *The Curse of Yig* **(s)** 1953; *The Dream Quest of Unknown Kadath* **(s)** 1955; *The Survivor and Others* **(s)** 1957; *At the Mountains of Madness and Other Novels* **(s)** 1964; *The Colour Out of Space* **(s)** 1964;

The Lurking Fear and Other Stories (s) 1964; *Dagon and Other Macabre Tales* (s) [ed. A. Derleth] 1965; *The Dark Brotherhood and Other Pieces* (s) [ed. A. Derleth] 1966; *Three Tales of Horror* (s) 1967; *The Shadow Out of Time and Other Tales of Horror* (s) [with A. Derleth] 1968 [rev. as *The Shuttered Room and Other Tales of Horror* 1970]; *Ex Oblivione* (s) 1969; *The Tomb and Other Tales* (s) 1969; *The Horror in the Museum and Other Revisions* (s) [ed. A. Derleth] 1970; *Nyarlathotep* (s) 1970; *What the Moon Brings* (s) 1970; *The Dream-Quest of Unknown Kaddath* (s) [ed. L. Carter] 1970; *Memory* (s) 1970; *The Doom That Came to Sarnath* (s) [ed. L. Carter] 1971; *The Lurking Fear and Other Stories* (s) 1971; *The Shadow over Innsmouth and Other Tales of Horror* (s) 1971; *The Shuttered Room and Other Tales of Terror* (s) [with A. Derleth] 1971; *The Watchers Out of Time and Others* (s) [with A. Derleth] 1974; *The Horror in the Burying Ground and Other Tales* (s) 1975; *Collapsing Cosmoses* (s) 1977; *Herbert West Reanimator* (s) 1977; *Bloodcurdling Tales of Horror and the Macabre: the Best of H.P. Lovecraft* (s) 1982; *The Dunwich Horror and Others* (s) [ed. S.T. Joshi] 1985

Poetry
The Crime of Crimes 1915; *A Sonnet* 1936; *H.P.L.* 1937; *Fungi from Yuggoth* 1941; *Collected Poems* 1963; *A Winter Wish* [ed. T. Collins] 1977; *Saturnalia and Other Poems* [ed. S.T. Joshi] 1984; *Medusa and Other Poems* [ed. S.T. Joshi] 1986

Non-Fiction
Looking Backward 1920[?]; *The Materialist Today* 1926; *Further Criticism of Poetry* 1932; *Charleston* 1936; *Some Current Motives and Practices* 1936[?]; *A History of the Necronomicon* 1938; *The Notes and Commonplace Book* [ed. R.H. Barlow] 1938; *Beyond the Wall of Sleep* [ed. A. Derleth, D. Wandrei] 1943; *Marginalia* [ed. A. Derleth] 1944; *Supernatural Horror in Literature* 1945 [rev. 1975]; *Something about Cats and Other Pieces* [ed. A. Derleth] 1949; *The Lovecraft Collector's Library* [ed. G.T. Wetzel] 5 vols 1952–5; *Dreams and Fancies* 1962; *Some Notes on Nonentity* (a) 1963; *Selected Letters 1911–1937* [ed. A. Derleth, D. Wandrei] 5 vols 1965–76; *Hail, Karkash-Ton!* 1971; *Ec'h-Pi-El Speaks: an Autobiographical Sketch* 1972; *Medusa: a Portrait* 1975; *The Occult Lovecraft* 1975; *Lovecraft at Last* 1975; *First Writings: Pawtuxet Valley Gleaner 1906* [ed. M.A. Michaud] 1976; *To Quebec and the Stars* [ed. L. Sprague de Camp] 1976; *Writings in The United Amateur 1915–1925* [ed. M.A. Michaud] 1976; *The Californian 1934–1938* 1977; *The Conservative: Complete 1915–1923* [ed. M.A. Michaud] 1977; *Memoirs of an Inconsequential Scribbler* (a) 1977; *Writings in the Tryout* [ed. M.A. Michaud] 1977; *H.P. Lovecraft in 'The Eyrie'* [ed. S.T. Joshi, M.A. Michaud] 1979; *Science Versus Charlatanry: Essays on Astrology* [with J.F. Hartmann; ed. S. Connors, S.T. Joshi] 1979

Edited
The Poetical Works of Jonathan E. Hoag 1923; John Ravenor Bullen *White Fire* 1927; Eugene B. Kuntz *Thoughts and Pictures* 1932

Collections
Uncollected Prose and Poetry [ed. S.T. Joshi, M.A. Michaud] 1978

Biography
Lovecraft by L. Sprague de Camp 1976

Amy (Lawrence) LOWELL 1874–1925

Born at 'Sevenels', Brookline, Massachusetts, the house in which she also died, Lowell was a member of one of the most distinguished New England families: one of her brothers was president of Harvard, another a noted astronomer, and two other well-known American poets were also Lowells. She was educated at private schools and spent much of her early womanhood travelling abroad (in 1897 she went up the Nile in a *dahabiyeh*), but from the age of thirty, after both her parents had died, she lived an independent life at the family estate.

In 1902 she discovered poetry (previously her main interest had been civic affairs, as a diehard Republican), and she spent ten years in silent preparation before publishing her first book of verse, *A Dome of Many-Coloured Glass* (1912). In 1913, during a visit to England, she met **Ezra Pound** and was immediately converted to Imagism; she grew to consider herself the movement's leader, while Pound retorted by nicknaming it 'Amygism'. She lectured widely on poetry throughout the USA and became a notorious figure, at least as much for her appearance and habits as her views: a glandular disorder had caused her to become immensely fat and she smoked black cigars; her many eccentricities included demanding that twelve pitchers of iced water be on hand wherever she went, and guarding her house with seven savage sheepdogs, which she killed, however, when they became a nuisance. Her companion and amanuensis during her later life was Ada Russell.

Lowell wrote more than 650 poems, which are not much read today – the two best known are 'Patterns' and 'Lilacs' – and made use of a free-verse form which she called 'unrhymed cadence' (she also experimented with 'polyphonic prose'). Her last four years were mainly occupied with her enormous biography of John

Keats; highly praised on its first appearance in 1925, it has not maintained its reputation.

Poetry
A Dome of Many-Coloured Glass 1912; *Sword Blades and Poppy Seed* 1914; *Men, Women and Ghosts* 1916; *Can Grande's Castle* 1918; *Pictures of the Floating World* 1919; *Legends* 1921; *A Critical Fable* 1922; *What's O'Clock* [ed. A.D. Russell] 1925; *East Wind* [ed. A.D. Russell] 1926; *Ballads for Sale* [ed. A.D. Russell] 1927; *The Madonna of Carthagena* 1927; *Selected Poems* 1928; *The Complete Poetical Works* 1955; *A Shard of Silence: Selected Poems* [ed. G.R. Ruilhey] 1957

Non-Fiction
Six French Poets: Studies in Contemporary Literature 1915; *Tendencies in Modern American Poetry* 1917; *John Keats* 2 vols 1925; *Poetry and Poets: Essays* [ed. F. Greenslet] 1930; *Florence Ayscough and Lowell: Correspondence of a Friendship* [ed. H.F. MacNair] 1946

Plays
Weeping Pierrot and Laughing Pierrot [from E. Rostand] 1914

Fiction
Dream Drops (s) [with E. and K.B. Lowell] 1887

Translations
Fir-Flower Tablets 1921

Robert (Traill Spence) LOWELL 1917–1977

Lowell was born into a renowned family of 'Boston Brahmins' which included the poet **Amy Lowell**. His father was a naval officer and executive with Lever Brothers, his mother a leading socialite. He was educated at the Episcopalian St Mark's School, Southborough, Massachusetts, and spent an unhappy year at Harvard where **Robert Frost** dismissed an epic he was writing on the Crusades with the words: 'It goes on a bit.' As a child he was nicknamed 'Caligula' on account of his unruly temper, and throughout a life punctuated with episodes of mental breakdown and psychiatric treatment, violent drunkenness, affairs and turbulent marriages, he was known as 'Cal'.

In 1937 he travelled south to Kenyon College, Ohio, where he was taught by **John Crowe Ransom** and **Allen Tate** in what he called 'the heyday of the New Criticism', and met **Randall Jarrell** and **Peter Taylor**, who were to be lifelong friends. In 1940 he graduated and married the future novelist **Jean Stafford**, and his intellectual development was influenced by a year at Louisiana State University studying under Cleanth Brooks and **Robert Penn Warren**. For a while he converted to Catholicism, partially in rebellion against what he saw as the puritan, capitalist ethic of his parents' world, and in 1943, a year before the publication of his first collection of poems, *Land of Unlikeness*, he achieved notoriety by being imprisoned after having written a letter to President Roosevelt explaining his moral objections to being drafted into the US Army.

With *Lord Weary's Castle* he achieved widespread acclaim, winning a Pulitzer Prize in 1947 and featuring in *Life* magazine. The volume contains Lowell's first fully mature work, notably the Miltonic lament 'The Quaker Graveyard at Nantucket'. Divorced from his first wife, Lowell married the writer **Elizabeth Hardwick** in 1949, this despite Allen Tate having warned her that 'there are definite homicidal implications in his world'. Hardwick's novel *Sleepless Nights* (1979) provides some rather oblique glimpses – 'How we fight after too much gin' – of their sometimes strained marriage. The couple travelled widely in Europe and, after the death of Lowell's mother and his inheritance in 1953, settled in Boston. Their daughter was born in 1957, an event Lowell anticipated thus: 'We're so excited we can hardly speak, and expect a prodigy whose first words will be "Partisan Review".'

Recovering from another breakdown and electric-shock treatment in 1954, Lowell began work on '91 Revere Street', the prose remembrance which eventually formed part of the autobiographical sequence of poems *Life Studies*. This ground-breaking work, which won a National Book Award in 1960, is written in a fluent free verse, and inaugurated the so-called 'confessional' school of poetry. It contains 'Waking in the Blue', a vivid evocation of the poet's incarceration in a mental hospital, and the powerful, ambiguous 'Skunk Hour'. In his later verse, beginning with *For the Union Dead*, Lowell adopted an increasingly public and political voice, seeming to see himself as a sort of latter-day Henry Adams, and many were disappointed by the 'seedy grandiloquence' of his rather desultory late collections.

In the 1960s Lowell adapted a number of classical plays and wrote the trilogy of plays *The Old Glory*, based on stories by Herman Melville and Nathaniel Hawthorne, two of which were directed by Jonathan Miller in New York in 1964. With other writers such as **Norman Mailer** and Noam Chomsky he was active in the campaign against the Vietnam War. In 1966 he was nominated for election as Professor of Poetry at Oxford and lost out to **Edmund Blunden** in one of the more hotly contested elections of recent years, but he did take up a professorship at All Souls College in 1970, thereby meeting **Caroline Blackwood**. They had a son and married in 1972. Lowell died of a heart-attack in a New York taxi on his way to see Hardwick, carrying a portrait of Blackwood wrapped in brown paper.

Poetry
Land of Unlikeness 1944; *Lord Weary's Castle* 1946; *Poems 1939–1949* 1950; *The Mills of the Kavanaughs* 1951; *Life Studies* 1959; *For the Union Dead* 1964; *Selected Poems* 1965; *Near the Ocean* 1967; *Notebook 1967–1968* 1969 [rev. 1970]; *The Dolphin* 1973; *For Lizzie and Harriet* 1973; *History* 1973; *Selected Poems* 1976; *Day by Day* 1977

Plays
Phaedra [from Racine's play] 1961; *The Old Glory* [from N. Hawthorne, H. Melville] 1964 [rev. 1968]; *Prometheus Bound* [from Aeschylus's play] 1969; *The Oresteia* [from Aeschylus's plays] 1978

Translations
Baudelaire *Imitations* 1962 [rev. as *The Voyage and Other Versions of Poems by Baudelaire* 1968]

Biography
Robert Lowell by Ian Hamilton 1982

(Clarence) Malcolm (Boden) LOWRY 1909–1957

Born in Liscard, Cheshire, Lowry was the fourth son of a wealthy cotton broker. After a period during his childhood when he was almost blind through ulceration of the cornea of both eyes, he attended the Leys School, Cambridge. On leaving school, he persuaded his father to let him take ship to the China seas, a 'before the mast' adventure which provided material for his first novel, *Ultramarine* (1933). He then spent a year studying in Bonn and went to Massachusetts to meet his hero, the American poet **Conrad Aiken**, before attending St Catherine's College, Cambridge, from 1929 to 1932, where the pattern of drinking which was to mark his whole adult life was already established. Recognising his son's alcohol problem, Lowry's father made him an allowance so that he need not work, and he was to spend much of his life wandering about various parts of the globe, drinking and intermittently writing.

In 1933, while in Paris, he married a young American writer, Jan Gabrial, and went with her to New York, Los Angeles and, in 1936, Mexico. He was to be jailed in and deported from the latter country, but it gripped his imagination, and he began work on his masterpiece, the story of the last day in the life of an alcoholic British consul in Mexico, which was published as *Under the Volcano* in 1947, almost immediately winning acclaim as one of the most extraordinary visionary novels of the century. Lowry's wife had deserted him because of his drinking and the marriage was dissolved in 1940; the same year, he married the glamorous ex-starlet and aspirant writer Margerie Bonner, and the couple moved to a seaside shack at Dollarton, British Columbia (the setting of the idyllic and haunting novella 'The Forest Path to the Spring', published in the collection *Hear Us O Lord From Heaven Thy Dwelling Place*). The shack burnt down in 1944 and Lowry and his wife attempted to rebuild it themselves. Exhausted by this effort, they went back to Mexico in late 1945, where Lowry's residence was again cut short by the authorities. They then returned to Canada until 1954, when they moved to England, to the small village of Ripe in Sussex. Lowry had made several suicide attempts in later years, and in 1957 the last of these was successful. (Although a verdict of misadventure was recorded, rumours persisted that he had in fact been murdered.)

Lowry had intended that all his work should form part of a vast integrated corpus called 'The Voyage That Never Ends', but in fact only published two novels during his lifetime. After his death, his mass of unfinished writing was worked on by his widow and several more works were published, such as *Dark as the Grave Wherein My Friend Is Laid* and *Lunar Caustic*. Intermittently remarkable, these books are not completely successful, and how far they represent Lowry's final intentions is uncertain.

Fiction
Ultramarine 1933; *Under the Volcano* 1947; *Hear Us O Lord from Heaven Thy Dwelling Place* (s) 1961; *Dark as the Grave Wherein My Friend Is Laid* [ed. D. Day, M. Lowry] 1968; *Lunar Caustic* [ed. E. Birney, M. Lowry] 1968; *October Ferry to Gabriola* 1970

Poetry
Selected Poems [ed. E. Birney, M. Lowry] 1962; *Malcolm Lowry: Psalms and Songs* [ed. M. Lowry] 1975

Non-Fiction
Selected Letters [ed. H. Breit, M. Lowry] 1965

Biography
Malcolm Lowry by Douglas Day 1973; *Pursued by Furies* by Gordon Bowker 1993

(John) Edward (McKenzie) LUCIE-SMITH 1933–

Lucie-Smith was born in Kingston, Jamaica: his father was an official in the British colonial service, who came of a family that had arrived in

Barbados from Britain in the 1650s. An ancestor on the paternal side was Bonnie Prince Charlie's physician during the '45 rebellion, while on his mother's side he is descended from 'Lush', the drunken chaplain mentioned in *Brief Lives*, the seventeenth-century book of anecdotes and gossip collected by John Aubrey. Lucie-Smith's early childhood was spent in Jamaica, but he returned to Britain aged fourteen to be educated at King's School, Canterbury, and from there went to Merton College, Oxford, where he formed part of a notable undergraduate generation of poets. He had his first pamphlet of verse, as did many others, published by the Fantasy Press. He then spent two years as an education officer in the Royal Air Force, and after that joined a London advertising agency, a profession in which he 'strove not to rise'; in 1966 he resigned to become a full-time writer. (Among his colleagues at the agency had been **Peter Redgrove** and **William Trevor**.)

In 1956 **Philip Hobsbaum** invited him to join the Group, an association of poets that met regularly in reaction to the controlled restraint of the Movement. From 1959 to 1965 (when it was disbanded) the Group met at Lucie-Smith's Chelsea flat and he was its organiser; with Hobsbaum he edited *A Group Anthology* in 1963. His own first book of verse was *A Tropical Childhood* in 1961, and he has developed as a poet from traditionalism and emotional reticence towards experimentation; his book *The Well-Wishers* makes use of a complex prosody where an attempt is made to reproduce the effect of classical Greek and Latin poetry, and is notable for its erotic poems and many poems about visual art (Lucie-Smith is also a noted and prolific art critic). Since the mid-1970s he has produced poetry at a reduced rate, and has diversified into a novel (*The Dark Pageant*), an autobiography (*The Burnt Child*), as well as a great deal of history and criticism; he is also well known as an anthologist, particularly for *British Poetry Since 1945*, an influential and unbiased overview. He founded the small Turret Press in 1965. Among his recreations he lists 'malice'.

Poetry

A Tropical Childhood and Other Poems 1961; *Confessions and Histories* 1964; *Penguin Modern Poets 6* [with J. Clemo, G. MacBeth] 1964; *Borrowed Emblems* 1967; *Towards Silence* 1968; *The Well-Wishers* 1974; *Beasts with Bad Morals* 1984

Non-Fiction

What Is a Painting? 1966; *Thinking about Art* 1968; *Movements in Art Since 1945* 1969; *Art in Britain 1969–1970* [with P. White] 1970; *A Concise History of French Painting* 1971; *Criticism in Western Art* 1972; *Symbolist Art* 1972; *The First London Catalogue* 1974; *The Burnt Child* (a) 1975; *The Invented Eye* 1975; *World of the Makers* 1975; *How the Rich Lived: the Painter as Witness, 1870–1914* [with C. Dars] 1976; *A Life of Joan of Arc* 1976; *Art Today: From Abstract Expressionism to Superrealism* 1977; *Fantin-Latour* 1977; *Toulouse-Lautrec* 1977 [rev. 1983]; *Work and Struggle: The Painter as Witness, 1870–1914* [with C. Dars] 1977; *Outcasts of the Sea: Pirates and Piracy* 1978; *Cultural Calendar of the Twentieth Century* 1979; *Furniture: a Concise History* 1979; *Art in the Seventies* 1980; *Superrealism* 1980; *The Art of Caricature* 1981; *The Body: Images of the Nude* 1981; *The Story of Craft: the Craftsman's Role in Society* 1981; *Jan Vanriet* 1982; *Masterpieces from the Pompidou Centre* 1982; *The Sculpture of Helaine Blumenfeld* 1982; *A History of Industrial Design* 1983; *The Thames and Hudson Dictionary of Art Terms* 1984; *American Art Now* 1985; *Art in the Thirties* 1985; *Lives of the Great Twentieth Century Artists* 1986; *Sculpture Since 1945* 1987; *Impressionist Women* 1989; *Art Deco Painting* 1990; *Art in the Eighties* 1990; *Harry Holland: the Painter and Reality* 1991; *Ruskin* 1991; *Sexuality in Western Art* 1991; *Art & Civilization* 1992; *Elizabeth Fritsch: Vessels from Another World* 1993; *Latin American Art of the Twentieth Century* 1993

Fiction

The Dark Pageant 1977

Translations

Robert Day *Manet* 1962; Paul Claudel *Five Great Odes* 1967; Jean Paul de Dadelsen *Jonah* 1967

For Children

Bertie and the Big Red Ball [with B. Cook] 1982

Edited

A Group Anthology [with P. Hobsbaum] 1963; *The Penguin Book of Elizabethan Verse* 1965; *A Choice of Browning's Verse* 1967; *The Liverpool Scene* 1967; *Penguin Book of Satirical Verse* 1967; *British Poetry Since 1945* 1969; *Holding Your Eight Hands: an Anthology of Science-Fiction Verse* 1970; *A Primer of Experimental Verse* 1971; *French Poetry: the Last Fifteen Years* [with S.W. Taylor] 1971; *The Real British: an Anthology of the New Realism of British Painting, October–November 1981* 1981

Alison LURIE 1926–

Lurie was born in Chicago, the daughter of a Latvian-born teacher, scholar and socialist who became the founder of the Council of Jewish Federations and Welfare Funds. The family moved to New York City in 1930, and shortly thereafter to the suburb of West Plains. She began to write at an early age and has recalled: 'I was encouraged to be creative past the usual age because I didn't have much else going for me. I was a skinny, plain, odd-looking little girl, deaf in one badly damaged ear from a birth injury, and with a resultant atrophy of the facial muscles that pulled my mouth sideways whenever I opened it to speak and turned my smile into a sort of sneer.'

Having graduated in 1943 from the progressive Cherry Lawn School in Darien, Connecticut, she studied at Radcliffe College, Cambridge, Massachusetts, receiving her BA in 1947. She worked as an editorial assistant for the Oxford University Press in New York, and was married in 1948. Continuing to write while bringing up her family of three sons, she saw only a handful of short stories published. Her first book was a memoir of her friend, the poet and actress V.R. Lang, and was privately printed in 1959.

Lurie's first novel, *Love and Friendship* (1962), is a comedy about adultery and the relationship between the sexes, matters to which she has repeatedly returned. The title derives from a piece of juvenilia by Jane Austen, with whom she has frequently been compared for her ironic, detached wit, and precision of expression. *The Nowhere City* deals with similar material and is set in Los Angeles where Lurie and her husband lived from 1957 to 1961. *The War between the Tates* is set at the end of the 1960s in Corinth University (based on Cornell where she lived since 1961) and describes in corrosive terms the domestic breakdown of the Tate family against the background of the Vietnam War, Eastern mysticism and experiments with drugs.

Lurie separated from her husband in 1975 and was divorced in 1985. Since 1968 she has taught at Cornell, and has been a full professor since 1976, though she teaches only part-time to allow time for writing and travel. In addition to her novels, she has written a semiotic study of clothes and *Don't Tell the Grown-Ups*, a study of children's literature, and has edited volumes of children's literature. Perhaps her most accomplished novel, *Foreign Affairs* won a Pulitzer Prize in 1985.

Fiction

Love and Friendship 1962; *The Nowhere City* 1965; *Imaginary Friends* 1967; *The War between the Tates* 1974; *Real People* 1978; *Only Children* 1979; *Foreign Affairs* 1985; *The Truth about Lorin Jones* 1988; *Women and Children* (s) 1994

For Children

The Heavenly Zoo 1979; *Clever Gretchen* 1981; *Fabulous Beasts* 1981

Non-Fiction

The Language of Clothes 1982; *Don't Tell the Grown-Ups* 1990

Edited

The Oxford Book of Modern Fairy Tales (s) 1993

M

(Emilie) Rose MACAULAY 1881–1958

Born in Rugby, the daughter of a master at Rugby School who later became a university lecturer at Cambridge, Macaulay not only numbered the historian Lord Macaulay among her antecedents but was descended through her mother from the equally intellectually distinguished family of the Conybeares. After spending much of her childhood living near Genoa in Italy, an experience which helped form her love of Mediterranean countries, she returned to England with her family in her early teens and was educated at Oxford High School and Somerville College, Oxford, where she was awarded an *aegrotat* in 1903.

She returned home to live with her family for some years, and published her first novel, *Abbots Verney*, in 1906, but then gravitated to London where she launched herself into pre-war literary society, an inveterate talker who entertained both hosts and guests with her torrent of words. She had published nine novels by 1918, largely Victorian in theme and execution, but it was with the brilliantly satirical *Potterism* (1920) that she found her form, and similarly successful satirical novels followed during the 1920s, after which her fiction, published at longer intervals, had a more serious message. An exception amongst her early novels is *Non-Combatants and Others*, a dissident novel set, and published, during the First World War. She was a member of the mandarin literary society of the inter-war period, connected with the Bloomsbury Group (she published a study of **E.M. Forster**), and a well-known literary journalist for the *Spectator* and many other papers.

Her life took a darker turn during the Second World War when her flat was bombed, destroying many of her possessions, and when Gerald O'Donovan, the former priest and married man with whom she had had a long relationship, died after a long illness. This troubled affair, and the reconciliation to her long abandoned Anglican faith which followed, later became part of the subject-matter of her last and best novel, *The Towers of Trebizond*, a work of fiction which in its blend of fantastic comedy and deep religious seriousness is quite unique. Her two other outstanding novels are the penultimate one, *The World My Wilderness*, and *They Were Defeated*, a wonderfully sympathetic vision of the seventeenth century. Other works included several books of essays, literary studies, and travel books such as *Fabled Shore*. In person, she was a lanky English gentlewoman who enjoyed swimming in the Serpentine in winter and possessed an irrepressible, sometimes alarming, sense of humour. ('I think I'm going to die in a fortnight,' she said to a friend shortly before her death. 'When are *you* pushing off?') She was appointed Dame Commander of the British Empire in 1958. She never married.

Fiction
Abbots Verney 1906; *The Furnace* 1907; *The Secret River* 1909; *The Valley Captives* 1911; *The Lee Shore* 1912; *Views and Vagabonds* 1912; *The Making of a Bigot* 1914; *Non-Combatants and Others* 1916; *What Not* 1918; *Potterism* 1920; *Dangerous Ages* 1921; *Mystery at Geneva* 1922; *Told by an Idiot* 1923; *Orphan Island* 1924; *Crewe Train* 1926; *Keeping Up Appearances* 1928; *Staying with Relations* 1930; *They Were Defeated* 1932; *Going Abroad* 1934; *I Would Be Private* 1937; *And No Man's Wit* 1940; *The World My Wilderness* 1950; *The Towers of Trebizond* 1956

Poetry
The Two Blind Countries 1914; *Twenty-Two Poems* 1927; *Three Days* 1932

Non-Fiction
A Casual Commentary 1925; *Catchwords and Claptrap* 1926; *Some Religious Elements in English Literature* 1931; *Milton* 1934; *Personal Pleasures* 1935; *The Writings of E.M. Forster* 1938; *Life Among the English* 1942; *They Went to Portugal* 1946; *Fabled Shore: From the Pyrenees to Portugal* 1949; *Pleasure of Ruins* 1953; *Letters to a Friend, 1950–52* [ed. C. Babington Smith] 1961; *Last Letters to a Friend, 1952–58* [ed. C. Babington Smith] 1962; *Letters to a Sister* [ed. C. Babington Smith] 1964; *They Went to Portugal Too* 1990

Edited
The Minor Pleasures of Life 1934

Biography
Rose Macaulay by Jane Emery 1991

George (Mann) MacBETH 1932–1992

MacBeth was born, the son of a miner, in the village of Shotts in Lanarkshire, Scotland, but when he was four the family moved to Sheffield. Here his childhood was punctuated by a series of disasters and losses – his father was killed by an anti-aircraft shell during the war; the family were bombed out of their home; he spent a year of his early teens in hospital with rheumatic fever; his mother died before he was twenty – and it is perhaps these events which account for the climate of nightmare which ran through his poetry. Otherwise, he attended King Edward VII School, Sheffield, took a first in greats at New College, Oxford, and immediately after coming down in 1955 was recruited into the BBC.

While still at Oxford he had become an influential judge of poetry as an editor for the Fantasy Press, by which many of the most important talents of his generation were first published, and his career as poetic impresario gained its full substance between 1958 and 1976 when, in various posts as BBC editor and producer, he was the person chiefly responsible for deciding which poems were broadcast. Even the quarrelsome world of poetry was in broad agreement that he fulfilled this task with sympathy and judgement, and he also found time to be an extremely prolific writer in verse and, more latterly, prose.

His first collection, *A Form of Words* (1954), was in the dryly judicious style typical of the early 1950s, but his second, *The Broken Places*, which followed nine years later, showed the full influence of the Group (the association of poets who met first at the home of **Philip Hobsbaum** and later **Edward Lucie-Smith**) in its macabre fantasies, a mood enhanced by his editorship of *The Penguin Book of Sick Verse*, published in the same year. After that, for some time, he produced a volume a year, giving his opinion that 'the important thing is to thrash out huge quantities of fairly well-written poetry'. Inevitably, he attracted the same accusations of lack of seriousness which, two decades later, were to pursue the rather similar figure of **Craig Raine**, but MacBeth's poetry was as often memorable as it was occasionally careless.

His life changed its pattern from the mid-1970s. In 1955, he had married the distinguished biologist Elizabeth Robson, and they lived in Richmond, Surrey, for many years until their divorce in 1975. His first novel, *The Transformation*, was published in that year, and his resignation from the BBC was partly intended to further his career as a novelist, which never fully gathered steam. He also moved into a series of vast country homes – in Norfolk, Yorkshire and Ireland – and remarried twice, having a child by either marriage. He married first the exotic novelist **Lisa St Aubin de Teran**, and their divorce in 1989 was the subject of extensive literature on both sides. His third marriage, to Penelope Ronchetti-Church, was quickly followed by his death from motor neurone disease. His *oeuvre* also included several books for children, autobiography, and many anthologies, among which *Poetry 1900–1965* introduced a generation of children to modern verse.

Poetry

A Form of Words 1954; *The Broken Places* 1963; *Penguin Modern Poets 6* [with J. Clemo, E. Lucie-Smith] 1964; *A Doomsday Book* 1965; *The Colour of Blood* 1967; *The Night of Stones* 1968; *A War Quartet* 1969; *The Burning Cone* 1970; *Collected Poems 1958–1970* 1971; *The Orlando Poems* 1971; *My Scotland* 1973; *A Poet's Year* 1973; *Shrapnel* 1973; *In the Hours Waiting for the Blood to Come* 1975; *The Journey to the Island* 1975; *Buying a Heart* 1977; *Poems of Love and Death* 1980; *Poems from Oby* 1982; *The Long Darkness* 1983; *The Cleaver Garden* 1986; *The Anatomy of Divorce* 1988; *Collected Poems, 1958–1982* 1989; *Trespassing* 1991; *The Patient* 1992

Fiction

The Transformation 1975; *The Samurai* 1976; *The Survivor* 1977; *The Seven Witches* 1978; *The Born Losers* 1981; *A Kind of Treason* [from J. Beeby] 1981; *Anna's Book* 1983; *The Lion of Pescara* (s) 1984; *Dizzy's Women* 1986; *Another Love Story* 1991; *The Testament of Spencer* 1992

For Children

Jonah and the Lord 1969; *The Rectory Mice* 1982; *The Story of Daniel* 1986

Edited

The Penguin Book of Sick Verse 1963; *Penguin Book of Animal Verse* 1965; *Poetry 1900–1965* 1967 [rev. as *Poetry 1900–1975* 1979]; *The Penguin Book of Victorian Verse* 1969; *The Falling Splendour* 1970; *The Book of Cats* [with M. Booth] 1976; *Poetry for Today* 1984

Non-Fiction

A Child of War 1987

Norman (Alexander) MacCAIG 1910–

MacCaig was born into a poor family in Edinburgh, his father, who was of west Highland stock, working in and, much later, owning a chemist's shop. His mother was a native of Scalpay in the Outer Hebrides, and had come to Edinburgh at the age of sixteen speaking only Gaelic. MacCaig first visited his mother's country when he was twelve and, although he has lived in Edinburgh all his life, he has always felt drawn to the traditional Scottish inheritance, although content with English as a native and poetic language. Similarly, although he was friendly with **Hugh MacDiarmid** and other Scottish nationalist writers, the political dimension is largely absent from his work. He attended the Royal High School in Edinburgh and then read classics at Edinburgh University. By his own confession he has lived in 'a happy and uneventful way', being a primary school teacher from 1934 to 1967, except for some years during the war when he suffered penalties as a conscientious objector. In 1940 he married, and subsequently had two children; his wife died in 1990.

MacCaig first came to notice as a poet as a member of the New Apocalypse movement, and his first two volumes (1943 and 1946) display this tendency at its most extreme. More than any other poet associated with the movement, however, he has demonstrated the ability to develop, and has gone through at least two other major phases. *Riding Lights* and the following four collections see him as a strictly metrical and formal poet who uses precise observation of native scenes (particularly Sutherland in the western Highlands, to which he has returned most years of his adult life) to illumine cosmic, abstract concerns in a way which led critics to label him as 'metaphysical' and speak of the influence of John Donne and **Wallace Stevens**. In 1964 he received a Society of Authors Travelling Scholarship, which enabled him to visit Italy and America, and the intellectual stimulus this provided coincided, from *Surroundings* onward, with a major move towards free verse and a deepening of subject-matter which suggests that MacCaig has reached his apogee as a poet. Like **Thomas Hardy**'s, however, his *oeuvre* suffers for want of a single, central poem which would focus attention on the whole, and this fact, coupled with his reluctance to court an English metropolitan audience, may account for a relative but still puzzling neglect.

His *Collected Poems* appeared in 1985; he has written little but poetry, although he has edited two anthologies. From 1967 to 1969 he was a writing fellow of Edinburgh University, and then returned to teaching for one year as a headmaster. He ended his career in academe, as first lecturer in English studies (1970–2) and then reader in poetry (1972–7) at Stirling University. He has been the recipient of many Scottish Arts Council awards, and other honours include his being awarded an OBE in 1979.

Poetry

Far Cry 1943; *The Inward Eye* 1946; *Riding Lights* 1955; *The Sinai Sort* 1957; *A Common Grace* 1960; *A Round of Applause* 1962; *Measures* 1965; *Surroundings* 1966; *Rings on a Tree* 1968; *A Man in My Position* 1969; *Selected Poems* 1971; *Penguin Modern Poets 21* [with G.M. Brown and I. Crichton-Smith] 1972; *The White Bird* 1973; *The World's Room* 1974; *Three Poems* 1975; *Nothing Too Much and Other Poems* 1976; *Tree of Strings* 1977; *Inchnadamph and Other Poems* 1978; *Old Maps and New: Selected Poems* 1978; *The Equal Skies* 1980; *A World of Difference* 1983; *Collected Poems* 1985; *Voice-Over* 1988; *Three Scottish Poets* [with L. Lochhead, E. Morgan] 1992

Edited

Honour'd Shade 1959; *Contemporary Scottish Verse 1959–1969* [with A. Scott] 1970

(Charles) Cormac McCARTHY (Jr) 1933–

Since McCarthy has stayed out of the public gaze for so long – he rarely agrees to be interviewed and will not lecture – one might plausibly expect him to have gained in public profile, especially in America where the reclusive writer has become something of a media obsession. As it is, despite the success of *All the Pretty Horses* in 1992, he has remained largely a cult figure. His virtually unpunctuated prose, with its lack of commas and speech marks, label him 'difficult'; he is well reviewed but stigmatised as a writer's writer, popular in England and the Southern states. Comparisons have often been drawn between his writing and that of **William Faulkner, Flannery O'Connor** and **Carson McCullers** and, in his writing as in theirs, the universality of theme transcends regionalism. He also shares the dark humour and the taste for violent epiphanies to be found in much Southern fiction: *Blood Meridian* was described by one reviewer as 'perhaps the bloodiest book ever penned by an American author'.

McCarthy was born in Providence, Rhode Island, but his upbringing was in the South. The family moved to Knoxville, Tennessee, when he was four. He graduated from high school in 1951 and moved to the University of Tennessee, but stayed for only one year. He then enlisted in the US Air Force. After his discharge in 1956 he returned to college and completed his degree. In 1960 he was awarded an Ingram–Merrill Foundation grant in creative writing. He has married twice, and has a son by his first marriage. His first novel, *The Orchard Keeper*, won the William Faulkner Foundation Award in 1965. The second and third, *Outer Dark* (1968) and *Child of God* (1973) were equally well received and he was cited as 'the best undiscovered novelist of his generation'. Each book has a Southern background, and the theme of justice and retribution runs through them with a biblical insistence. The fourth, *Suttree*, may explain some of the lost time between publications. Described by Stanley Booth as taking 'the closest approach to autobiography' to be found in McCarthy's fiction, the book is set in a grim underworld of outcasts and alcoholics in which the central character has become immersed after walking out on his home. Sustained by grants and fellowships (he had also won a travelling fellowship to Europe from 1965 to 1966), McCarthy moved from Tennessee to El Paso and relocated his fiction after his own footsteps. *Blood Meridian* follows a group of bounty hunters on a scalping expedition across the territories of the Old West. *All the Pretty Horses*, the first of a planned Border trilogy, won both a National Book Award and the National Book Critics Circle Award.

Fiction

The Orchard Keeper 1965; *Outer Dark* 1968; *Child of God* 1973; *Suttree* 1979; *Blood Meridian* 1985; *All the Pretty Horses* 1992; *The Crossing* 1994

Plays

The Gardener's Son (tv) 1977; *The Stonemasons* [pub. 1994]

Mary (Therese) McCARTHY 1912–1989

McCarthy was born in Seattle, Washington, to a complex heritage. On her father's side, the family were Catholic Irish farm-settlers who became rich through grain elevators in Minnesota. Her maternal grandfather, on the other hand, was a WASP senator who married a Jew. The background became important when both McCarthy's parents died in the influenza epidemic of 1918, and with her three brothers she went to live with a severe great-aunt and her sadistic husband, an experience later recounted in *Memories of a Catholic Girlhood*, which many consider her best book. Aged eleven, she was rescued from this life by her patrician grandfather, who gave her an expensive convent education, from which she went to Vassar. She graduated in 1933, and three weeks later married her first husband, Harold Johnsrud, an actor.

Almost immediately, she became part of intellectual society in New York, writing book reviews for the *New Republic*, the *Nation* and the *Partisan Review*, mixing extensively with writers, and becoming converted to Trotskyism (moderated to libertarian socialism in later years). All her life she was the prototype of an American intellectual, intervening frequently in public life in a way which was sometimes shrill, occasionally had a touch of the ridiculous, but was always honest and humane. Her first marriage collapsed after three years, and she married the critic **Edmund Wilson** in 1938, with whom she had one son. It was another temporary arrangement in the lives of two much-married writers; its importance was that Wilson encouraged her to write fiction. Her first novel, *The Company She Keeps*, appeared in 1942, and she was to write novels intermittently throughout the rest of her life, the most famous of which is *The Group*, which charts the life of eight Vassar students of her own generation in the eight years after their graduation. All McCarthy's novels tended to satirise intellectuals, and perhaps their chief originality lay in a treatment of sex both intimate and detached, and from the woman's point of view, which was to set a trend for later writers. Her copious non-fiction included collected journalism, drama criticism, autobiography and evocations of Venice and Florence.

Her third marriage was to a Manhattan school official, Bowden Broadwater, with whom she lived happily for several years despite his homosexuality. In the late 1940s she had taught at various colleges, but from the 1950s onwards tended to spend increasing time in Europe, first Italy, and from 1962 the Left Bank in Paris, where she established a second home beside that in Castine, Maine. She had made her fourth marriage in 1961, to the diplomat James West, and it endured to her death.

Her later years were enlivened by a legal battle with **Lillian Hellman**, whom she had long disliked on political and personal grounds, and whose honesty she questioned; Hellman died in 1984 with the suit still undecided and her reputation in tatters, while McCarthy's triumph was signalled in the same year by the award of the National Medal for Literature. The few years that remained were marked by ill health and frequent operations, but saw her intellectual energy undimmed.

Fiction

The Company She Keeps 1942; *The Oasis* 1949; *Cast a Cold Eye* (s) 1952; *The Groves of Academe* 1952; *A Charmed Life* 1955; *The Group* 1963; *Birds of America* 1971; *Cannibals and Missionaries* 1979; *The Hounds of Summer and Other Stories* (s) 1981

Non-Fiction

Venice Observed 1956; *Memories of a Catholic Girlhood* (a) 1957; *Sights and Spectacles: Theater Chronicles 1937–1956* 1957 [rev. as *Mary McCarthy's Theater Chronicles 1937–1962* 1963]; *The Stones of Florence* 1959; *On the Contrary: Articles of Belief 1946–61* 1961; *Vietnam* 1967; *Hanoi* 1968; *The Writing on the Wall and Other Literary Essays* 1970; *Letter from Portugal* 1971;

Medina. On the Court-Martial of Ernest L. Medina 1972; *The Mask of State: Watergate Portraits* 1974; *The Seventeenth Degree* 1974; *Can There Be a Gothic Literature?* 1975; *Ideas and the Novel* 1980; *How I Grew* (a) 1987; *Intellectual Memoirs* (a) 1992

Translations
Simone Weil *The Iliad, or, The Poem of Force* 1973; Henri Bosco *Culotte the Donkey* 1978

Plays
La Traviata (l) [from Verdi's opera; with F.M. Piave] 1983

Biography
The Company She Kept by Doris Grumbach 1976; *Mary McCarthy* by Carol Geldman 1988; *Writing Dangerously* by Carol Brightman 1992

(Lula) Carson McCULLERS 1917–1967

Born Lula Carson Smith in the small town of Columbus, Georgia, McCullers was the daughter of a well-to-do watchmaker and jeweller of French Huguenot extraction. At the urging of her doting mother she took piano lessons from the age of five and was considered so promising that she was sent when seventeen to New York to study at the Juilliard School of Music. She never attended Juilliard, having lost her tuition fees in rather murky circumstances, and took a succession of menial jobs while studying creative writing at Columbia and New York Universities. Her precocious teenage efforts had included a verse dialogue between Christ and Nietzsche, but she now turned to autobiographical material for 'Wunderkind', a story about a musical prodigy's failure which was published in *Story* magazine in 1936. The story's acute depiction of adolescent insecurity anticipates much of her later fiction.

In 1937 she married Reeves McCullers and they lived for two years in North Carolina where she wrote *The Heart Is a Lonely Hunter*. Theirs was a troubled and sometimes violent relationship which inevitably suggests parallels with the violent and grotesque world of her fiction. Both were involved in homosexual liaisons and they divorced in 1940 when she moved to New York to live with George Davis, the editor of *Harper's Bazaar*. During the 1940s, Davis's large Brooklyn house was occupied at various times by a prestigious and Bohemian collection of writers and artists which included **W.H. Auden**, **Paul** and **Jane Bowles**, Benjamin Britten and Peter Pears, **Louis MacNeice** and – rather more improbably – the striptease artiste Gypsy Rose Lee. McCullers and Reeves remarried in 1945 and in 1948, depressed by her continued ill health, she attempted suicide. Reeves later suggested a double suicide, and in a Paris hotel in 1953 after a lengthy period of alcoholism he killed himself with an overdose of sleeping pills. Despite all these traumas, it was during this time that McCullers produced the small body of work for which she is remembered.

The Heart Is a Lonely Hunter describes a deaf-mute who is the largely uncomprehending confidante of four residents of a small Georgia town who talk to him about their loneliness and misery. Whilst fitting within the bracket of the school of Southern Gothic, the novel is also a compassionate and psychologically convincing portrait of human isolation and thwarted love, themes to which her work repeatedly returns. Her second novel, *Reflections in a Golden Eye*, which has a military setting, is a morbid tale of murder, voyeurism, infidelity, self-mutilation, sado-masochism and bisexuality which might almost be a primer in Southern grotesquerie and prompted one reviewer to comment that 'not even the horse is normal'. John Huston's film version starring Marlon Brando and Elizabeth Taylor was released in 1967. *The Ballad of the Sad Café* is a study of love and isolation which was originally published serially by George Davis in *Harper's Bazaar* in 1943. **Edward Albee** adapted it for the stage in 1963, and Simon Callow directed a 1990 film version which starred Vanessa Redgrave. McCullers adapted what is perhaps her best novel, *The Member of the Wedding*, for the Broadway stage in 1950 and it won the New York Drama Critics Circle Award and two Donaldson Awards that year. A further play, *The Square Root of Wonderful*, and her last novel were both commercial and critical failures.

Throughout her life McCullers was plagued by a variety of ailments including rheumatic fever, pneumonia, breast cancer and a succession of strokes which partially paralysed her. She died in New York after a stroke and a resultant brain haemorrhage.

Fiction
The Heart Is a Lonely Hunter 1940; *Reflections in a Golden Eye* 1942; *The Member of the Wedding* 1946; *The Ballad of the Sad Café* 1951; *Clocks without Hands* 1961

Plays
The Member of the Wedding [from her novel] 1950; *The Square Root of Wonderful* 1958

Poetry
Sweet as a Pickle and Clean as a Pig 1964

Collections
The Mortgaged Heart [ed. M.G. Smith] 1971

Biography
The Lonely Hunter by Virginia Spencer Carr 1975

Hugh MacDIARMID (pseud. of Christopher Murray Grieve) 1892–1978

The son of a rural postman in Langholm, Dumfriesshire, MacDiarmid was educated at Langholm Academy, and at Broughton Pupil Teaching Centre, from which he was expelled for stealing books. He also lost his first job, as a journalist on an Edinburgh newspaper, for operating a scheme with review copies in order to pay for his own work to be typed. Several more sackings from papers followed, until he enlisted in 1915, and was posted to Salonika, 'the forgotten front', where he saw little action, but much misery, disease and death.

In 1918 he married, and had a son and a daughter. A brief job followed as caretaker on the Highlands estate of Charles Perrins, who had made his fortune from Worcester Sauce, and this was followed by eight years as a journalist at Montrose, a period which coincided with the flowering of both his writing and political activity. His first book, consisting largely of poems written at Salonika, *Annals of the Five Senses*, was published in 1923. Shortly afterwards, he adopted the pseudonym, and started to write in a literary version of Scots called by MacDiarmid 'synthetic Scots' and later known as Lallans. In this, he composed the exquisite lyrics of *Sangschaw* and *Penny Wheep,* as well as his major work, *A Drunk Man Looks at the Thistle*, a long poem-sequence exploiting a wide variety of verse forms, allusions and subject-matter, often considered to form, with **T.S. Eliot**'s *The Waste Land* (1922) and **James Joyce**'s *Ulysses* (1922), the vital troika of English-language modernism in the 1920s.

MacDiarmid's political activity was more controversial: a founder member of the Scottish National Party in 1928, he flirted for a time with a 'species of Scottish fascism', was expelled from the SNP and joined the Communist Party, only to be ejected from that in turn in 1937 for nationalist deviation. He was to rejoin in 1957 after the Hungarian uprising, and to defend the invasion of Czechoslovakia in 1968; anti-British feeling was his main political consistency. Meanwhile, in the 1930s and 1940s, he had suffered much hardship. A brief spell in London, from 1929 to 1932, proved disastrous: the magazine on which he was working, *Vox*, folded; he fell from a bus and injured his head; and his marriage broke up (he was denied access to his children until they were adults). In 1934, he remarried, had another son, and lived for some years in poverty in the Shetlands. In the Second World War, he was called up under the industrial conscription scheme, and put to work in a heavy engineering factory in Glasgow, where he damaged his leg in an industrial accident, before being transferred to lighter duties. In 1945, he signed on the newly established dole, but later received a civil list pension.

His later poetry, often in English, and including such works as *In Memoriam James Joyce*, shows a loss of power, but from the appearance of his *Collected Poems* in 1962 he was recognised as the major Scottish poet of the century. He lived from 1951 until his death at a cottage near Biggar. His non-poetic works include two volumes of autobiography as well as several of scholarship and polemic.

Poetry
Annals of the Five Senses [as Christopher Murray Grieve] 1923; *Sangschaw* 1925; *A Drunk Man Looks at the Thistle* 1926; *Penny Wheep* 1926; *The Lucky Bag* 1927; *To Circumjack Cencratus* 1930; *First Hymn to Lenin and Other Poems* 1931; *Scots Unbound and Other Poems* 1932; *Tarras* 1932; *Scottish Scene* 1934; *Stony Limits and Other Poems* 1934; *The Birlinn of Clauranald* 1935; *Cornish Heroic Song for Valda Trevlyn* 1943; *Poems of the East–West Synthesis* 1946; *A Kist of Whistles* 1947; *In Memoriam James Joyce* 1955; *The Battle Continues* 1957; *Three Hymns to Lenin* 1957; *The Kind of Poetry I Want* 1961; *Bracken Hills in Autumn* 1962; *Collected Poems* 1962; *Poems to Paintings by William Johnstone, 1933* 1963; *Poet at Play and Other Poems* 1965; *Whuchulls* 1966; *A Lap of Honour* 1967; *Early Lyrics* 1968; *A Clyack* 1969; *More Collected Poems* 1970; *Direadh* 1974; *The Complete Poems of Hugh MacDiarmid* [ed. W.R. Aitken, M. Grieve] 1978; *The Socialist Poems of Hugh MacDiarmid* [ed. T. Berwick, T.S. Law] 1978

Non-Fiction
Contemporary Scottish Studies 1925; *Albyn; or, Scotland and the Future* 1927; *The Present Condition of Scottish Arts and Affairs* 1927; *The Present Position of Scottish Music* 1927; *The Scottish National Association of April Fools* 1928; *Fidelity in Small Things* 1929; *At the Sign of the Thistle* 1934; *Scotland in 1980* 1935; *Charles Doughty and the Need for Heroic Poetry* 1936; *The Islands of Scotland* 1939; *Lucky Poet: a Self-Study in Literary and Political Ideas* (a) 2 vols 1943; *Cunningham Graham: a Centenary Study* 1952; *Francis George Scott* 1955; *Burns Today and Tomorrow* 1959; *David Hume, Scotland's Greatest Son* 1962; *The Man of (Almost) Independent Mind* 1962; *When the Rat Race Is Over* 1962; *Sidney Goodsir Smith* 1963; *Tribute to Harry Miller* 1963; *The Company I've Kept* 1966; *The*

MacDiarmids. A Conversation with Duncan Glen 1970; *Metaphysics and Poetry* 1975; *The Letters of Hugh MacDiarmid* [ed. A. Bold] 1984

Collections
The Uncanny Scot: a Selection of Prose [ed. K. Buthlay] 1968; *The Hugh MacDiarmid Anthology* 1972; *The Thistle Rises: an Anthology of Poetry and Prose* [ed. A. Bold] 1984

Biography
MacDiarmid by Alan Bold 1988

A(rchibald) G(ordon) MacDONELL 1895–1941

The son of a Scottish lawyer, Macdonell was born in Poona, India, but brought up in Scotland until sent to Winchester. After serving in the Royal Field Artillery of the 51st Highland Division during the First World War, he joined the staff of the League of Nations, where he remained from 1922 to 1927, during which period he twice unsuccessfully attempted to become the Liberal parliamentary candidate for Lincoln. He married in 1926, but the marriage was dissolved in 1937.

His literary career had begun in 1919, when he worked as a drama critic for the *London Mercury*. He achieved success in both Britain and the USA with a series of crime novels written under the name of Neil Gordon, featuring the detective Peter Corrigan. Notable amongst these is *The Shakespeare Murders*, which involves a cypher drawn from *Hamlet*. As secretary of the London-based Sherlock Holmes Society, he took this unpaid post seriously enough to attend an annual dinner of the American Baker Street Irregulars in New York. He recounted his experiences there in *A Visit to America*. In 1940 he married a member of the Warburg banking family.

He also made frequent broadcasts for BBC radio on such topics as 'A Scotsman Looks at England'. This particular script was rejected by the Talks Department in 1930, but it was a similar project that really made Macdonell's name three years later. *England, Their England* is the story of a Scottish soldier, Donald Cameron, who returns from the Western Front with the intention of writing an anthropological study of English manners and mores. It was recognised at once as a classic book of humour, and it was said that Macdonell 'awoke and found himself famous'. The most celebrated scene in the book describes a cricket match between a village team and one made up of beery journalists, and is based upon those played by a team captained by J.C. Squire, editor of the *Mercury*, and dedicatee of the book.

Macdonell's other books include a work of history, *Napoleon and His Marshals*, which demonstrates his grasp of strategy. Tall and energetic, an enthusiastic sportsman with a military moustache, Macdonell was a characteristic example of the 'Squirarchy', as Squire's friends were called. Shortly after his forty-fifth birthday, he was killed during an air raid while visiting Oxford.

Fiction
As 'Neil Gordon': *The Factory on the Cliff* 1928; *The New Gun Runners* 1928; *The Professor's Poison* 1928; *The Silent Murders* 1929; *The Big Ben Alibi* 1930; *Murder in Earl's Court* 1931; *The Shakespeare Murders* 1933

As 'John Cameron': *The Seven Stabs* 1929; *Body Found Stabbed* 1932

As A.G. Macdonnell: *England, Their England* 1933; *How Like an Angel* 1934; *Lords and Masters* 1936; *Autobiography of a Cad* 1938; *The Spanish Pistol* (s) 1939; *The Crew of the Anaconda* 1940

Non-Fiction
Napoleon and His Marshals 1934; *A Visit to America* 1935; *My Scotland* 1937; *Flight from a Lady* 1939

Plays
What Shall I Do Next, Baby? 1939; *The Fur Coat* 1943

Ian (Russell) McEWAN 1948–

McEwan was born in Aldershot, Hampshire, the army town where his father was a sergeant-major. His childhood followed the chances of army life: at the age of three he was taken to Singapore, at five to Tripoli, and summer holidays from Woolverstone Hall School near London were spent at camps in north Germany (he has described himself as psychologically an only child). He read English at Sussex University and then took an MA in creative writing at the University of East Anglia, where he was the first (and, in his year, the only) student in what has since become the highly successful writing course run until 1995 by **Malcolm Bradbury**. After leaving, he drifted for several years, first claiming the dole in Cambridge, later buying a secondhand bus in which he travelled with American students to Afghanistan, and even at one point working as a dustman in north London.

His early writing was financed by freelance work for the *Radio Times*. Early stories appeared in such diverse publications as the *Transatlantic Review* and the Hungarian *Nagy Világ*. His first published book, the short stories in *First Love,*

Last Rites, won a Somerset Maugham Award in 1976 and was followed by another successful collection of stories, *In Between the Sheets*; McEwan is among the very few British writers of recent times to build a successful career on the back of the short story. His stories dealt in a matter-of-fact way with such themes as incest, child abuse and infanticide, and one of his early plays for television, *Solid Geometry*, was banned by the BBC in 1979 because of its 'grotesque and bizarre sexual elements'. Nevertheless, McEwan was now firmly established as the most fashionable writer of his generation, and the reputation of his early novel *The Cement Garden*, another cold-eyed stare into the face of horror, has survived accusations that it was largely plagiarised from an earlier novel, *Our Mother's House* (1963) by Julian Gloag. (McEwan has strenuously denied ever having read this book.) A chilly novella set in Venice, *The Comfort of Strangers*, followed. In 1982 he married the astrologer and healer Penny Allen, with whom he moved from London to North Oxford; they have four children, two adopted from her earlier marriage.

McEwan once again won great prominence in 1987 when his new novel, *The Child in Time*, won a Whitbread Award. He attempted to develop a more humane style, with a genuinely moving story of a child who is abducted, but it was felt by some reviewers that the book was vitiated by its many improbabilities. The feminism, first espoused in *The Comfort of Strangers* and developed here, was unsympathetically but acutely analysed in **Adam Mars-Jones**'s 'Counterblast' pamphlet, *Venus Envy* (1990). Two later novels, *The Innocent* and *Black Dogs*, although routinely successful, did not make the same impact. His other works include several screenplays (some of his own work, the highly regarded *The Ploughman's Lunch*, contributions to a screenplay of **Timothy Mo**'s *Sour Sweet*), television plays, an oratorio on the anti-nuclear theme and a children's book. He has become well known as a campaigner for a bill of rights and other radical causes.

Fiction

First Love, Last Rites (s) 1975; *In Between the Sheets* (s) 1978; *The Cement Garden* 1978; *The Comfort of Strangers* 1981; *The Child in Time* 1987; *The Innocent* 1990; *Black Dogs* 1992; *The Daydreamer* 1994

Plays

Conversation with a Cupboardman (r) 1975; *The Imitation Game* (tv) [pub. 1981]: *Jack Flea's Birthday Celebration* 1976, *The Imitation Game* 1980, *Solid Geometry* [nd]; *The Last Day of Summer* (tv) 1983; *The Ploughman's Lunch* (f) 1983 [pub. 1985]; *Soursweet* (f) [from T. Mo's novel] 1989 [pub. 1988]; *Or Shall We Die?* (l) 1982

For Children

Rose Blanche [from C. Gallaz] 1985

John McGAHERN 1934–

McGahern was born in Dublin and brought up in the west of Ireland. When he was nine, his schoolteacher mother died, and he went with his five sisters to live with their father, a police sergeant, in the local barracks. All McGahern's novels are rooted in his childhood, although he denies fiercely that they are autobiographical: 'There is a dilemma at the heart of writing – that self expression is not expression. You can't just pour out what happened to you. It has to be made different and to respect the shapes and forms of language.' The primary images of his past are oppressive ones: a dying mother; barracks; a violent and unpredictable father; the domineering and guilt-inducing Catholic church. Dark sexuality and passionate tension, presented in a terse and dispassionate narrative, characterise his work. Critics draw comparisons with **James Joyce**, and McGahern is often described as the 'Irish Chekhov'. In France he is revered, yet he has mainly been treated with wary respect on the British literary scene, and is still notorious in his native Ireland as a banned writer.

His first book, *The Barracks*, was respectfully reviewed in 1963, and earned him an A.E. Memorial Award and an Arts Council Fellowship. In order to write, he had become a primary teacher in Dublin, attracted by the short hours and long holidays. This church-controlled job ended in a blaze of publicity when his second novel, *The Dark*, was banned by the Irish State Censorship Board as 'indecent or obscene'. McGahern was fired and fled to London.

He believes that all a writer needs is a room to work in and a good, dull life, and so it is not surprising that he found it difficult to write for four or five years after the banning. He met, married and broke up with a Finnish theatre director (neither could adopt the other's country); he spent time in France, Spain, the USA; and travelled to university fellowships at Reading and Newcastle. After seven years, the ban was lifted. In the early 1970s McGahern returned with his second wife, Madeleine Green, an American magazine photographer, to the drumlins of Leitrim where they still live.

With *Amongst Women*, which was shortlisted for the 1990 Booker Prize, McGahern received the widespread recognition in Britain that had been previously denied him: McGahern himself thinks it his best book. Normally a reluctant interviewee, on its publication he spoke about his work, himself and the process of writing: 'It often takes me two hours with several pots of tea to lead into the two hours of writing ... It's the most intense form of life that I know, which is part of its attractiveness. While you hate having to do it, you'd never experience such intensity otherwise.'

Fiction

The Barracks 1963; *The Dark* 1965; *Nightlines* **(s)** 1970; *The Leavetaking* 1974 [rev. 1984]; *Getting Through* **(s)** 1978; *The Pornographer* 1979; *High Ground* **(s)** 1985; *Amongst Women* 1990; *The Power of Darkness* 1991; *The Collected Stories of John McGahern* **(s)** 1992

Plays

Sinclair **(r)** 1971, **(st)** 1972; *Swallows* **(tv)** 1975; *The Rockingham Shoot* **(tv)** 1987; *A Search for Happiness* **(tv)** 1989

Roger McGOUGH 1937–

McGough was born in Liverpool, of Irish ancestry on both sides, and the son of a docker who eventually rose to the position of foreman. He was educated at St Mary's College in Crosby, a school of the Irish Christian Brothers, and went to Hull University, where he took a BA in French and geography. He then trained as a teacher, and from 1959 to 1961 was an assistant schoolmaster at St Kevin's Comprehensive School in Kirkby. From 1961 to 1963 he was an assistant lecturer at a technical college, the Mabel Fletcher College in Liverpool.

He became increasingly involved in the popular artistic movements originating in Liverpool in the 1960s, and from 1963 until 1973, with intervals, was a member of the pop group the Scaffold, much of whose material he wrote. The group eventually had great success with 'Thank You Very Much' and 'Lily the Pink', but McGough was also developing during this period as a poet, often performing his work in public, and a playwright. His *annus mirabilis* as a writer came in 1967: in this year, Michael Joseph published his book *Frinck, A Day in the Life Of, and Summer with Monika* (consisting of the novella *Frinck* and the poem-sequence *Summer with Monika*); he was included in the influential anthology *The Liverpool Scene*; and, with **Adrian Henri** and **Brian Patten**, he was one of three poets included in *Penguin Modern Poets 10: The Mersey Sound*. This last proved an extremely successful volume, and McGough was immediately established as one of the 'Liverpool poets', a classification which has proved difficult to dispel, despite the extremely prolific nature of his talent. Although he knew Henri and Patten well, and although the three poets were once again linked in *New Volume*, McGough denies that they ever in fact formed a cohesive group.

He has gone on to work in the genres of poetry, books for children (both stories and poetry) and plays, for the stage, radio and television; despite his early novella, he has not returned to prose fiction *per se*. His considerable success as a writer has enabled him to live as one, since his last period as a teacher, between 1973 and 1975, when he was a fellow of poetry at the University of Loughborough. Some critics, such as **Ian Hamilton**, are inclined to dismiss the work of McGough and the other two Liverpool poets as part of the history of performance rather than of poetry, but this is perhaps to ignore McGough's lyric gift and empathy with the sad and lonely of the urban scene. It would perhaps be more accurate to say that, in overtly populist form, he is part of a recognisable tradition in English poetry running through **John Betjeman** and **Philip Larkin**.

McGough has married twice and has two children. He has lived in London since the later 1970s, continues to perform his works, and travels abroad frequently as a performer and lecturer for the British Council.

Poetry

Penguin Modern Poets 10 [with A. Henri, B. Patten] 1967 [rev. 1974, 1983]; *Frinck, A Day in the Life Of, and Summer with Monika* 1967 [rev. as *Summer with Monika* 1978]; *Watchwords* 1969; *After the Merrymaking* 1971; *Out of Sequence* 1972; *Gig* 1973; *Sporting Relations* 1974; *In the Glassroom* 1976; *Holiday on Death Row* 1979; *Unlucky for Some* 1980; *Waving at Trains* 1982; *New Volume* [with A. Henri, B. Patten] 1983; *Crocodile Puddles* 1984; *Melting into the Foreground* 1986; *Blazing Fruit: Selected Poems 1967–1987* 1989; *Defying Gravity* 1992; *Pen Pals* 1994; *Penguin Modern Poets 4* [with L. Lochhead, S. Olds] 1995

Plays

Birds, Marriages and Deaths 1964 [np]; *The Chauffeur-Driven Rolls* 1966 [np]; *The Commission* 1967 [np]; *The Puny Little Life Show* 1969 [pub. in *Open Space Plays* ed. C. Marowitz 1974]; *Zones* 1969 [np]; *Stuff* 1970 [np]; *PC Plod* 1971 [np]; *Plod* **(f)** 1972; *The Lifeswappers* **(tv)** 1976, **(st)** 1980; *Gruff: a TV Commercial* **(r)** 1977; *Summer with Monika* 1978; *World Play* 1978; *Watchwords* 1979; *Golden Nights and Golden Days* 1979 [np]; *All the Trimmings* 1980 [np]; *Lifeswappers* 1980; *Like Father, Like Son, Like* 1980 [np]; *Walking the Dog* **(r)** 1981; *Behind the Lines* 1982 [np]; *The Mouthtrap* [with B. Patten] 1982 [np]; *Kurt, BP, Mungo and Me* **(tv)** 1983; *The Narrator and Scenes from a Poet's Life* **(r)** 1983; *Fast Forward* **(tv)** 1986; *A Matter of Chance* [from V. Nabokov's story] 1988; *FX* **(r)**

1989; *Waving at the Tide* 1989; *Mistaken Identity* (tv) 1990

For Children

Mr Noselighter 1976; *You Tell Me* [with M. Rosen] 1979; *The Great Smile Robbery* 1982; *Sky in the Pie* 1983; *Wind in the Willows* (l) [from K. Grahame's novel; with J. Iredale, W. Perry] 1984 [np]; *The Stowaways* 1985; *Noah's Ark* 1986; *Nailing the Shadow* 1987; *An Imaginary Menagerie* 1988; *Counting by Numbers* 1989; *Helen Highwater* 1989; *Pillow Talk* 1990; *The Lighthouse That Ran Away* 1990; *The Oxford ARC Picture Dictionary* 1990; *Pillow Talk* 1990; *You at the Back: Selected Poems 1967–1987* 1991; *My Dad's a Fire-Eater* 1992; *The Oxford 123 Book of Number Rhymes* 1992; *Another Custard Pie* 1993; *Lucky* 1993; *The Magic Fountain* 1995; *Stinkers Ahoy!* 1995

Edited

Strictly Private: an Anthology of Poetry 1981; *The Kingfisher Book of Comic Verse* 1986

John (Peter) McGRATH 1935–

McGrath was born in Birkenhead, Cheshire, and although educated at the Alun Grammar School, Mold in Wales, did not immediately go up to university after he left school. After working on a Cheshire farm in 1951, he began his period of national service in the army in 1953, which he completed in 1955. This experience inspired his 1966 play *Events While Guarding the Bofors Gun*, filmed two years later as *The Bofors Gun*, directed by Jack Gold from McGrath's own screenplay. He was admitted as an open exhibitioner to St John's College, Oxford, where, in 1958, he directed his own first original play, *A Man Has Two Fathers*.

After graduating in 1959, he was a play-reader for the Royal Court Theatre, London, before joining the BBC television department. Jointly responsible for creating the popular *Z-Cars* police series, for which he wrote the first eight episodes, McGrath laid the foundations for what became known as 'television naturalism'. This was a concept which he later rejected because he felt it to be politically restricting. His strongly held socialism, which motivates the themes of all his work, was unsympathetically received by London West End theatre audiences, but it was not until he witnessed the student unrest in Paris in 1968 that he completely changed his theatrical purpose.

In 1971, McGrath founded the 7:84 Theatre Company in Scotland, so named because 7 per cent of the population own 84 per cent of the country's land. His 7:84 productions often take place in non-theatre environments, such as public houses, to bring plays to working-class people rather than 'a metropolitan middle-class audience'. Determined to remove the 'obstacle' of the theatre itself, he often invites audience response and dispenses with much theatrical convention. Very largely concerned with the historical exploitation of Scotland, as a parable of modern society, he nevetheless believes in entertainment as well as enlightenment. These features were strongly apparent in his most important play, *The Cheviot, The Stag, and the Black, Black Oil*, which toured in Scotland in 1973. This took the form of a presentation via a *ceilidh* (a traditional Highland evening entertainment) of Scottish history, with reference to English domination, up to the oil boom of the 1970s. This was during a period of strong Scottish nationalism, when 'It's Scotland's Oil' was a popular political slogan. McGrath's extraordinary energy has resulted in over fifty published plays as well as screenplays and television adaptations. He married in 1962.

Plays

The Invasion [from A. Adamov; with B. Cannings] 1958 [np]; *A Man Has Two Fathers* 1958 [np]; *The Tent* 1958 [np]; *Why the Chicken* 1959 [np]; *Take It* 1960 [np]; *Tell Me Tell Me* 1960 [pub. in *New Departures* 1960]; *The Seagull* [from Chekhov's play] 1961 [np]; *People's Property* (tv) 1962; *Basement in Bangkok* [with D. Moore] 1963 [np]; *Diary of a Young Man* (tv) [with T.K. Martin] 1964; *The Day of Ragnarok* (tv) 1965; *Events While Guarding the Bofors Gun* (st) 1966 [rev. as *The Bofors Gun* (f) 1968]; *Diary of a Nobody* (tv) [from G. and W. Grossmith's novel; with K. Russell] 1966; *Shotgun* (tv) [with C. Williams] 1966; *Billion Dollar Brain* (f) 1967; *Bakke's Night of Fame* [from W. Butler's *A Danish Gambit*] 1968 [pub. 1973]; *Comrade Jacob* [from D. Caute's novel] 1969 [np]; *The Virgin Soldiers* (f) [with J. Hopkins, I. La Fresnais] 1969; *Random Happenings in the Hebrides* 1970 [pub. 1972]; *The Reckoning* (f) 1970; *Sharpeville Crackers* 1970 [np]; *Orkney* (tv) [from G. Mackay Brown] 1971; *Soft of a Girl* 1971 [np]; *Trees in the Wind* 1971 [np]; *Unruly Elements* 1971 [np]; *Bouncing Boy* (tv) 1972; *The Caucasian Chalk Circle* [from B. Brecht's play] 1972 [np]; *Fish in the Sea* [with M. Brown] 1972 [pub. 1977]; *Prisoners of the War* [from P. Terson's play] 1972 [np]; *Sergeant Musgrave Dances On* [from J. Arden's play] 1972 [np]; *Underneath* 1972 [np]; *The Cheviot, the Stag, and the Black, Black Oil* 1973; *Boom* 1974 [np]; *The Game's a Bogey* 1974 [pub. 1975]; *Pulse* 1974 [np]; *Lay Off* 1975 [np]; *Little Red Hen* 1975 [pub. 1977]; *Oranges and Lemons* 1975 [np]; *Yobbo Nowt* [with M. Brown] 1975 [pub. 1978]; *Out of Our Heads* [with M. Brown] 1976 [np]; *The*

Rat Trap [with M. Brown] 1976 [np]; *Trembling Giant* 1977 [np]; *The Life and Times of Joe of England* 1977 [rev. as *The Adventures of Frank* **(tv)** 1979] [np]; *Once Upon a Union* **(tv)** 1978; *Big Square Fields* [with M. Brown] 1979; *Bitter Apples* [with M. Brown] 1979 [np]; *If You Want to Know the Time* 1979 [np]; *Joe's Drum* 1979; *Two Plays for the Eighties* [pub. 1981]: *Swings and Roundabouts* 1980, *Blood Red Roses* **(st)** 1980, **(tv)** 1986; *The Catch* [with M. Brown] 1981 [np]; *Nightclass* [with R. Lloyd] 1981 [np]; *Rejoice!* [with M. Brown] 1982 [np]; *On the Pig's Back* [with D. MacLennan] 1983 [np]; *The Women of the Dunes* 1983 [np]; *Women in Power* [from Aristophanes; with T. Mikroutsikos] 1983 [np]; *The Baby and the Bathwater* 1984 [np]; *Six Men of Dorset* [from H. Brooks and M. Malleson; with J. Tams] 1984 [np]; *The Albannach* [from F. MacColla; with E. McGuire] 1985 [np]; *Behold the Sun* **(l)** [with A. Goehr] 1985 [np]; *All the Fun of the Fair* [with others] 1986 [np]; *The Dressmaker* **(f)** 1987; *There Is a Happy Land* **(st)**, **(tv)** 1987 [np]; *Mairi Mhor* 1988 [np]; *Border Warfare* **(st)**, **(tv)** 1989 [np]; *John Brown's Body* **(st)**, **(tv)** 1990 [np]; *Watching for Dolphins* 1991 [np]; *The Wicked Old Man* 1992 [np]; *The Long Roads* **(tv)** 1993; *Half the Picture* [with R. Norton Taylor] 1994 [np]; *Reading Rigoberta* 1994 [np]; *The Silver Darlings* [from N. Gunn's novel] 1994

Non-Fiction

A Good Night Out: Popular Theatre, Audience, Class and Form 1981; *The Bone Won't Break: On Theatre and Hope in Hard Times* 1990

Translations

Jean Renoir *The Rules of the Game* [with M. Teitelbaum] 1970

Arthur (Llewellyn Jones) MACHEN 1863–1947

Born at Carleon-on-Usk, the only child of a Welsh Anglican clergyman, Machen was a lonely boy and from an early age a voracious reader. He attributed everything to his upbringing, 'finding happiness in books and exploring the wooded countryside of Gwent'. All his writing, he said, was an attempt to recreate 'those vague impressions of wonder and awe and mystery that I myself had received from the form and shape of the land of my boyhood and youth'. He attended Hereford Cathedral School. In 1880 he went to London, getting a job in a publishing house, then teaching; he also translated Casanova's memoirs. A period of employment as a cataloguer of diabolistic and occult books introduced him to the world of the secret society. He later joined the Order of the Golden Dawn, a group of theosophists involved in cabalistic magic, of which **W.B. Yeats** and Aleister Crowley were also to become members.

He was miserably poor for most of his life, but in the 1890s a small legacy enabled him to concentrate on writing. Always searching for spiritual truth – he researched extensively the legend of the Holy Grail – his best work is contained in *The Great God Pan, The Inmost Light* and *The Three Impostors*. In *Hieroglyphics* he expounds his literary philosophy. He turned to acting in his late thirties, touring with Sir Frank Benson's Shakespeare Repertory Company, and (his first wife having died in 1899) married a member of the company; they had a son and a daughter. At the age of fifty he became a regular contributor to the *Evening News* and the First World War brought him his only commercial success, the wholly fictional account of the 'Angel of Mons'. Two volumes of autobiography, *Far Off Things* and *Things Near and Far*, written after the war, have been extravagantly praised. In the 1920s his principal works to that time were collected in the nine-volume Carleon Edition, and an indication of continued interest is that his literary remains, some critical work, some letters and his pieces for the *Evening News*, were published in the 1950s. In a somewhat gushing obituary *The Times* concludes: 'He fished perhaps in a small pool, but his line went deep.' He turned to Roman Catholicism in later life.

Fiction

The Chronicle of Clemendy 1888; *The Great God Pan and The Inmost Light* **(s)** 1894; *The Three Impostors* **(s)** 1895; *The House of Souls* **(s)** 1906; *The Hill of Dreams* 1907; *The Angel of Mons* **(s)** 1915; *The Great Return* **(s)** 1915; *The Terror* 1917; *The Secret Glory* 1922; *The Shining Pyramid* **(s)** 1923; *Precious Balms* **(s)** 1924; *The Glitter of the Brook* **(s)** 1932; *The Cosy Room and Other Stories* **(s)** 1936; *The Children of the Pool and Other Stories* **(s)** 1936

Non-Fiction

The Anatomy of Tobacco [as 'Leolinus Siluriensis'] 1884; *Don Quijote de la Mancha* 1887; *Thesaurus Incantus* 1888; *Hieroglyphics* 1902; *The House of the Hidden Light* [with others] 1904; *Dr Stiggins: His Views and Principles* 1906; *Parsifal: the Story of the Holy Grail* 1913; *War and the Christian Faith* 1918; *Far Off Things* **(a)** 1922; *Things Near and Far* **(a)** 1923; *The Collector's Craft* 1923; *The Grand Trouville: a Legend of Pentonville* 1923; *Strange Roads* 1923; *Dog and Duck* 1924; *The London Adventure* 1924; *The Glorious Mystery* [ed. V. Starret] 1924; *The Canning Wonder: the Case of Elizabeth Canning* 1925; *Dread and Drolls* 1926; *Notes and Queries* 1926; *A Souvenir of*

Cadby Hall 1927; *Parish of Amersham* 1930; *Tam O'Bedlam and His Song* 1930; *Beneath the Barley: a Note on the Origins of Eleusinia* 1931; *In the 'Eighties* 1931; *Letters of Arthur Machen* 1931; *A Few Letters* 1932; *A Critical Essay* 1953; *A.L.S.: Letters between Arthur Machen and J.H. Stewart, jr* 1956; *A Receipt for Fine Prose* 1956; *A Note on Poetry* 1959; *From the London* Evening News 1959; *Essays* [ed. F.B. Sewell] 1960; *Selected Letters* [ed. G. Brangham, R. Dobson, R.A. Gilbert] 1988

Translations
The Heptameron; or, Tales and Novels of Marguerite, Queen of Navarre 1886; Casanova: *The Memoirs of Casanova* 12 vols 1894, *Casanova's Escape from the Leads* 1925; Lady Hester Stanhope *Remarks upon Hermodactylus* 1933

Poetry
Eleusinia 1881

Biography
Arthur Machen by William Charlton and Aidan Reynolds 1963

William McILVANNEY 1936–

Born in Kilmarnock, Scotland, McIlvanney is the son of a miner who left the pits in the 1930s and became a general labourer. Although uneducated, his father was highly intelligent, and with his wife, who read constantly, he stimulated serious discussion in the family. (McIlvanney's elder brother, Hugh McIlvanney, is a well-known sports writer.) McIlvanney felt compelled to write when he was fourteen, and during his secondary education at Kilmarnock Academy, dropped Greek because he felt it 'disturbed his creativity'. Despite this, Greek myth and tragedy, along with Elizabethan drama, has been the greatest influence on his work.

After graduating in English from Glasgow University in 1959, he became a teacher, a career he finally abandoned in 1971. *Docherty*, which won a Whitbread Award in 1975, established McIlvanney as a major novelist, who realistically examined aspects of Scottish life without sentimentality. The destruction of a mining village and the sense of displacement of those powerless to defeat the machinations of capitalism was portrayed starkly and without political compromise in *The Big Man*, which was filmed in 1990 from McIlvanney's own screenplay. A series of novels centred on a policeman started with *Laidlaw*, and continued with *Strange Loyalties*, in which McIlvanney reintroduced characters from previous books, although the impact of the novel does not depend on that knowledge.

McIlvanney, who is a committed socialist, considers that: 'Scottish Literature has a much stronger democratic tradition than English Literature ... has tended to take ideas much more seriously than in English Literature where they mustn't be allowed to interfere with the social plumbing, and perhaps less solemnly than in French Literature, say, where the person is sometimes expected to inhabit the idea rather than the other way around.' Gregarious by nature, he nevertheless believes that being such an intense writer is a handicap to maintaining a serious relationship. A marriage, which produced two children, ended after seventeen years. Enjoying literary company as eagerly as 'Dracula in search of a crucifix', he prefers sport, music and the art of conversation.

Fiction
Remedy Is None 1966; *A Gift from Nessus* 1968; *Docherty* 1975; *Laidlaw* 1977; *Papers of Tony Veitch* 1983; *The Big Man* 1985; *Walking Wounded* (s) 1989; *Strange Loyalties* 1991

Poetry
Landscapes and Figures 1973; *Weldings and After* 1984; *In Through the Head* 1988

Non-Fiction
Shades of Grey: Glasgow 1956–1987 1987

Plays
The Big Man (f) [from his novel] 1990; *Dreaming* (tv) 1991

Colin MacINNES 1914–1976

MacInnes (who himself slightly altered the spelling of the family name) was born in London, and was the son of the singer James McInnes and the novelist Angela Thirkell (this being the surname of the Australian captain she married after divorcing her first husband). She had been born Mackail; her father was a distinguished classical scholar, and they came of a Victorian intellectual dynasty which included Burne-Jones, **Rudyard Kipling** and Stanley Baldwin among its connections. Thirkell took her children to Australia with their stepfather, and MacInnes grew up largely in suburban Melbourne, a shared experience described by his brother Graham McInnes, also a novelist, in four outstanding volumes of autobiography.

Graham McInnes was the conformist of the brothers, and ended up as the Canadian ambassador to UNESCO, but Colin MacInnes, in revolt against the middle-class ideals expressed by his mother in her increasingly successful novels, returned to Europe to drift, first spending five years working for a gas company in Brussels and then attempting to become a painter. War service interrupted this, and after it MacInnes moved towards writing, having a first, autobiographical novel, *To the Victors the Spoils*, published in 1950, and soon afterwards establishing himself as a noted journalist, specialising in sympathetic coverage of the new popular and youth culture of the time and of black immigrants. These were also the themes he handled with marked originality in his *London Trilogy* of novels – *City of Spades*, *Absolute Beginners* and *Mr Love and Justice* – which were published between 1957 and 1962 and which, although only loosely linked in plot, constitute his major fictional achievement.

Claims have been made that his journalism, which appeared in almost every major intellectual journal during the 1960s, and which was later collected in several volumes, is of equal importance, and that he is the successor to **George Orwell** as a chronicler of the condition of England. The claim may be exaggerated, but he was certainly one of the last examples of the English man-of-letters in his unbuttoned form: unmarried, bisexual and of shifting abode; well known on the London streets and an *habitué* of the Colony Room Club when not banned; smoking, drinking and engaging in argument until stomach cancer killed him in his early sixties.

His later novels, which rather unexpectedly included two historical romances, did not achieve the impact of the trilogy, and general works on such subjects as music hall, London and bisexuality were of only moderate interest. After his death, it seemed likely that he would be quickly forgotten, but a much heralded film in 1986 of *Absolute Beginners* coincided with the reissue of the trilogy. The film was unsuccessful, however, and in the mid-1990s MacInnes appeared to stand among a group of minor writers whose prospects for survival were good but not certain.

Fiction

To the Victors the Spoils 1950; *June in her Spring* 1952; *London Trilogy*: *City of Spades* 1957, *Absolute Beginners* 1959, *Mr Love and Justice* 1960; *All Day Saturday* 1966; *Sweet Saturday Night* 1967; *Westward to Laughter* 1969; *Three Years to Play* 1970; *Out of the Garden* 1974

Non-Fiction

England, Half English 1961; *London, City of Any Dream* 1962; *Out of the Way: Later Essays* 1979

Biography

Inside Outsider by Tony Gould 1983

Shena MACKAY 1944–

Mackay was born in Edinburgh, brought up in Kent, and was educated at Tonbridge Girl's Grammar School and Kidbrooke Comprehensive School. She grew up in a literary household – her mother was a schoolteacher – and started writing poetry at the age of thirteen. At the age of seventeen, she left home to live with a friend in Earl's Court, London, where she drifted from job to job, eventually ending up in an antiques shop in Chancery Lane. The manager of the silver department was Frank Marcus, author of the famous lesbian play, *The Killing of Sister George* (1967), and Mackay showed him a novella she had written at the age of fifteen. This resulted in *Toddler on the Run* being published in the same volume as another novella, *Dust Falls on Eugene Schlumburger*, in 1964. This book established her as a quirky new talent, and *Toddler* was adapted for the BBC's 'Wednesday Play' series.

The following year she married, and subsequently had three daughters; although she published three novels from 1965 to 1971, this period of early creativity was temporarily halted while she brought up her family in Surrey, the setting of many of her short stories. Of these early novels, *Music Upstairs* is a bisexual rites-of-passage story set in Earl's Court, and *Old Crow* the bleak tale of an old woman; they firmly established themes which would be familiar in her work: loneliness, hysteria, revenge and the menace of everyday life.

She continued to write short stories, which appeared in assorted magazines during the 1970s, but after *An Advent Calendar* (1971), she did not produce another book until 1983. Her sense of herself as a professional writer was reaffirmed in 1980 when her story 'A Stained-Glass Door' won a competition organised by BBC Radio 3 and *The Listener*. Her first volume of short stories, *Babies in Rhinestones*, was published to great acclaim in 1983 and resulted in an Arts Council bursary. A new novel, *A Bowl of Cherries*, was published the following year, and in 1987 her black comedy *Redhill Rococo* won the Fawcett Prize, which is awarded to a book 'which has made a substantial contribution to the understanding of women's position in society today'. The novel's title suggests in its curious juxtaposition of concepts Mackay's world, which is largely that of suburban domestic life, described in ornate, inventive and at times almost surreal prose. She admits no influences, but **Brigid Brophy** (the dedicatee of *A Bowl of Cherries*) has compared her style to that of **Ronald Firbank**, while the way in which Mackay transforms the quotidian by her vivid powers of description recalls the work of **Henry Green**.

Subsequent books have confirmed Mackay's reputation as a leading chronicler of apparently ordinary (but in fact extraordinary) everyday life in the late twentieth century. Even *Dunedin*, a novel which was structurally flawed, seemed more inventive, alive and rewarding than the work of many of her contemporaries. Her outstanding short stories, which have been widely broadcast and published in numerous journals and anthologies, were the subject of a BBC *Bookmark* programme in 1994 and were collected in a single volume the same year.

Mackay has also edited a volume of stories by women on the theme of sisterhood, and has reviewed fiction in various national newspapers. Divorced in the early 1980s, she moved back to London, and the city has become once more the setting for much of her fiction.

Fiction

Dust Falls on Eugene Schlumburger/Toddler on the Run 1964; *Music Upstairs* 1965; *Old Crow* 1967; *An Advent Calendar* 1971; *Babies in Rhinestones* (s) 1983; *A Bowl of Cherries* 1984; *Redhill Rococo* 1986; *Dreams of Dead Women's Handbags* (s) 1987; *Dunedin* 1992; *The Laughing Academy* (s) 1993; *Collected Short Stories* (s) 1994

Edited

Such Devoted Sisters 1993

(Edward Montague) Compton MACKENZIE 1883–1972

Born at West Hartlepool, Mackenzie was the son of the founder actor-manager of the Compton Comedy Company and his American wife. (Many members of this theatrical family, including Mackenzie's sister, the actress Fay Compton, used the family name of Compton, although the surname was more correctly Mackenzie.) He was related to many leading figures in the various arts, and grew up in a cosmopolitan atmosphere. He was also an infant prodigy, who taught himself to read at twenty-two months and remembered meeting the Empress of Austria when he was two, as well as even earlier events. Having attended St Paul's and Magdalen College, Oxford, Mackenzie soon established himself as a writer, publishing his first book of poems in 1907 and his first novel, *The Passionate Elopement*, in 1911.

He achieved enormous fame with *Sinister Street* (1913–14), a two-volume *Bildungsroman* which was considered very daring at that date (the American publishers felt themselves obliged to excise such words as 'tart' and 'bitch'); it remained one of his best novels. He became a lieutenant in the Royal Marines in 1915, served at the Dardanelles, and then became director of the Aegean Intelligence Service; he was later to stand trial at the Old Bailey under the Official Secrets Act for his book about this work.

After the war he became part of the Bohemian society gathered at Capri, and wrote two comic novels about the island, *Vestal Fire* and *Extraordinary Women*, which were notable for their very frank treatment for the time of both male and female homosexuality. After leaving Capri, he acquired the small Channel islands of Herm and Jethou, setting the pattern for a wandering life. During the Second World War he commanded the Home Guard on the Hebridean island of Barra (gaining the material for his comic novel *Whisky Galore*), and in his last years divided his time between Edinburgh and his house at Lot in France.

Mackenzie was a man of many enthusiasms and attachments: he was the founder-editor of *Gramophone* magazine from 1923 to 1961; president of the Siamese Cat Club and the Croquet Association; and in old age he visited all the battlefields of the Indian Army in the Second World War. Perhaps he was more a phenomenon than a literary figure, and it is doubtful which of his 100 books, which include plays, many biographical studies and a ten-volume autobiography, will survive: his best novels, perhaps. He was married from 1905 to 1960 to Faith Stone; in 1962 he married Christina MacSween, who died the following year, after which he married her younger sister Lilian. He was knighted in 1952.

Fiction

The Passionate Elopement 1911; *Carnival* 1912; *Sinister Street* 2 vols 1913, 1914; *Guy and Pauline* 1915; *The Early Life and Adventures of Sylvia Scarlett* 1918; *Sylvia and Michael* 1919; *Poor Relations* 1919; *The Vanity Girl* 1920; *Rich Relatives* 1921; *The Altar Steps* 1922; *The Parson's Progress* 1923; *The Seven Ages of Woman* 1923; *The Heavenly Ladder* 1924; *The Old Men of the Sea* 1924 [UK *Paradise for Sale* 1963]; *Coral* 1925; *Fairy Gold* 1926; *Rogues and Vagabonds* 1927; *Vestal Fire* 1927; *Extraordinary Women* 1928; *Extremes Meet* 1928; *The Three Couriers* 1929; *April Fools* 1930; *Told* (s) 1930; *Buttercups and Daisies* 1931; *Our Street* 1931; *Water on the Brain* 1933; *The Darkening Green* 1934; *Figure of Eight* 1936; *The Four Winds of Love: The East Wind of Love* 1937, *The South Wind of Love* 1937, *The West Wind of Love* 1940, *West to North* 1940, *The North Wind of Love* 2 vols 1944; *The Monarch of the Glen* 1941; *The Redtape Worm* 1941; *Keep the Home Guard Turning* 1943; *Whisky Galore* 1947; *Hunting the Fairies* 1949;

The Rival Monster 1952; *Ben Nevis Goes East* 1954; *Thin Ice* 1956; *Rockets Galore* 1957; *The Lunatic Republic* 1959; *Mezzotint* 1961; *The Stolen Soprano* 1965; *Paper Lives* 1966

Non-Fiction

New Humour and Old Comedy 1910; *Gramophone Nights* [with A. Marshall] 1923; *First Athenian Memories* (a) 1931; *Gallipoli Memories* (a) 1931; *Greek Memories* (a) 1932; *Prince Charlie* 1932; *Marathon and Salamis* 1934; *Prince Charlie and his Ladies* 1934; *Literature in My Time* 1935; *Catholicism and Scotland* 1936; *Pericles* 1937; *The Windsor Tapestry, Being a Study of the Life, Heritage and Abdication of HRH The Duke of Windsor* 1938; *Calvary* [with F. Compton Mackenzie] 1942; *Mr Roosevelt* 1943; *Wind of Freedom: the History of the Invasion of Greece by the Axis Powers* 1943; *Brockhouse: a Study in Industrial Evolution* 1945; *Dr Benes* 1946; *The Vital Flame* 1947; *All Over the Place: Fifty Thousand Miles by Sea, Air and Rail* 1949; *Eastern Epic* 1951; *House of Coalport (1750–1950)* 1951; *Took a Journey: a Tour of National Trust Properties* 1951; *The Queen's House: a History of Buckingham Palace* 1953; *The Savoy of London* 1953; *Realms of Silver: One Hundred Years of Banking in the East* 1954; *Sublime Tobacco* 1957; *Cat's Company* 1960; *Greece in My Life* 1960; *On Moral Courage* 1962; *My Life and Times* (a) 10 vols 1963–71

Edited

A Dictionary of Modern Music 1924; *Corbett's Cyclopaedic Survey of Music* 1929

For Children

Santa Clause in Summer 1924; *Mabel in Queer Street* 1927; *Unpleasant Visitors* 1928; *Adventures of Two Chairs* 1929; *The Enchanted Blanket* 1930; *The Conceited Doll* 1931; *Fairy in the Window Box* 1932; *The Dining Room Battle* 1933; *Enchanted Island* 1934 [UK *Secret Island* 1969]; *Naughtymobile* 1936; *The Stairs That Kept on Going Down* 1937; *The Strongest Man on Earth* 1968; *Butterfly Hill* 1970; *Achilles* 1972; *Jason* 1972; *Perseus* 1972; *Theseus* 1972

Plays

Columbine [from his novel *Carnival*] (st) 1922, (r) 1929; *The Lost Cause* (r) 1931 [pub. 1933]

Poetry

Poems 1907; *Kensington Rhymes* 1912; *A Posy of Sweet Months* 1955

Collections

Unconsidered Trifles 1932; *Reaped and Bound* 1933; *The Musical Chair* 1939; *Echoes* 1954; *My Record of Music* 1955; *Catmint* 1961

Biography

Compton Mackenzie by Andro Linklater 1987

Jay McKINERNEY 1955–

The young American writers known as 'the Brat Pack' tell tales of yuppie helplessness, self-destruction and remorse in a frenetic and decadent world. Their style is sparse, fluid and full of slang. McKinerney, born in Hertford, Connecticut, the son of a marketing executive for a multinational paper company, is often cited as their pacemaker.

As a result of his father's travels, McKinerney attended eighteen schools and has lived all over the USA, in Switzerland and in Oakshott, Surrey, in England. He studied Asian languages, English literature and philosophy at Princeton before moving to Tokyo, where he lived for two years, translating advertising copy and teaching English. On his return to New York, he worked as a fact-checker on the *New Yorker* and fell into a lifestyle of missing weeks and squandered talent, which he later chronicled in the autobiographical *Bright Lights, Big City*. He was rescued from what he now calls 'The Brotherhood of Unfulfilled Early Promise' by the novelist **Raymond Carver**, who whisked him away to Syracuse University to study creative writing. Here, McKinerney began drafting *Ransom*, but it was a piece reminiscing about a night in New York which grabbed the attention of the *Paris Review* and which McKinerney expanded into *Bright Lights, Big City*, the phenomenally successful first novel which was written in six weeks. It went on to become a film starring the (strangely cast) teen-idol Michael J. Fox, was translated into twelve languages, and tentatively compared to **J.D. Salinger**'s *The Catcher in the Rye* (1951). *Ransom*, based on the years in Tokyo where McKinerney learned karate (his other hobbies include skiing and white-water rafting), is a vehicle for a study of an *au fait* American living abroad and at one with the natives. *Story of My Life* – debauched glamour in Manhattan – established McKinerney as a style guru. When visiting London McKinerney frequents modish clubs and is often chased by magazine editors to write racy articles on London nightlife. So far he has not conceded. His acknowledged 'turf' is New York (he lives on the East Side with his wife), and he is genuinely fascinated by the moral wasteland of America.

Fiction

Bright Lights, Big City 1984; *Ransom* 1985; *Story of My Life* 1988; *Brightness Falls* 1992

Edited

Cowboys, Indians and Commuters 1994; *The Penguin Book of New American Voices* 1995

J(ulian) MACLAREN-ROSS 1912–1964

Famous as the model for the mysterious novelist X. Trapnel in **Anthony Powell**'s *A Dance to the Music of Time* (1951–75), Maclaren-Ross was himself a considerable and much underrated writer. His father had been born in Havana and his mother in Calcutta, but he was born in South Norwood, London. A Roman Catholic, he was christened James, but subsequently exchanged this name for the more elegant Julian. The family's background was military, and his father, who had been an officer in the Boer War, had very little money. Maclaren-Ross spent his early childhood on the south coast, with a spell in the south of France, the setting of his first novel, *Bitten by the Tarantula*.

In the 1930s, he spent some time in Bognor Regis as a vacuum-cleaner salesman, during which period he was married for six months. He was subsequently sacked for selling off second-hand cleaners, and embarked with his landlord on a short career as an incompetent jobbing gardener. In spite of his perpetual indigence, Maclaren-Ross was always elegant, and a friend recalled him 'behind a mower in his cream suit, the cigarette-holder tilted skywards'. He then spent a period on National Assistance, while unsuccessfully attempting to sell his short stories, adapt the works of others (amongst them **Graham Greene**) for the radio and find a more secure job as a screenwriter. One problem with his stories was their style and subject-matter. The first story he ever wrote, 'Five Finger Exercises', detailed the seduction of a man by a sixteen-year-old girl, while the original opening of 'A Bit of a Smash in Madras' ran: 'Absolute fact, I knew fuck-all about it ...' He was attempting, he said 'to create a completely English equivalent to the American vernacular used by such writers as [**Ernest**] **Hemingway**, [**James M.**] **Cain** and [**John**] **O'Hara**, concentrating in my case on the middle and lower-middle classes, an area cornered so far by **V.S. Pritchett** and **Patrick Hamilton**'. The milieu of his fiction is similar to that of Hamilton, but his prose is far less mannered: the language is unadorned, the tone dry, laconic and cynical.

He was called up in June 1940, by which time 'A Bit of a Smash in Madras' had been sufficiently expurgated to be accepted for publication in *Horizon* (**Stephen Spender** provided acceptable alternatives for the original expletives). He served as a private, stationed at Blandford in Essex and then at Ipswich, always carrying his portable typewriter with him. In 1942 he met **Alun Lewis**, about whom he wrote a characteristically evocative memoir-essay. He was regarded as insubordinate and his war ended in a court martial, a spell in the glasshouse and discharge in 1943. This inglorious military career resulted in some of the finest short stories to emerge from the Second World War, collected in *The Stuff to Give the Troops* and *Better Than a Kick in the Pants*. His work had started to appear in *Lilliput*, *English Story* and other magazines, and he got a job with Strand Films working on a documentary about the Home Guard with **Dylan Thomas**. The two men spent much of their time in pubs or planning feature films that never got made.

Bitten by the Tarantula was written in 1942 'as a relaxation' from the army stories which occupied most of his time then, and published in 1945. Just over 100 pages long, it describes a fraught holiday the narrator spends with a group of ill-matched people in the remote French house of a morphine-addict. The ending is abrupt and melodramatic, but this is in keeping with the book's terse style. In spite of a disclaimer, his second novel, *Of Love and Hunger*, was based very closely on his work as a vacuum-cleaner salesman, and is one of the best British novels of the Depression, depicting conditions from the less usual viewpoint of the middle (rather than the working) classes.

During the 1940s, Maclaren-Ross – wearing his teddy-bear coat and dark glasses, a carnation in his button-hole and a gold-topped cane in his hand – was a familiar figure in Fitzrovia, the Bohemian quarter of London where many writers and artists congregated. Always short of funds, he described himself as being 'a professional writer as opposed to being a professional literary man', and his work appeared in numerous magazines and anthologies, such as *Penguin New Writing* and the *London Magazine*. He wrote at night and often submitted his work in longhand, his immaculate manuscript bearing no signs of emendations or erasures. 'He wrote everything out twice,' **Alan Ross** recalled, 'fast and fluently, in the way he talked.'

He had numerous affairs, but never again married, dying unexpectedly of a heart-attack at the age of fifty-two. His *Memoirs of the Forties*, an unsurpassed account of Fitzrovia containing superb portraits of Greene, Thomas, **Cyril Connolly**, Tambimuttu and assorted writers, painters and publishers, was only half complete, and only the first of four projected volumes of autobiography was ever published.

Fiction

The Stuff to Give the Troops (s) 1944; *Better Than a Kick in the Pants* (s) 1945; *Bitten by the Tarantula* 1945; *The Nine Men of Soho* (s) 1946; *Of Love and Hunger* 1947; *The Funny Bone* 1956;

Until the Day She Dies [from his radio play] 1960; *The Doomsday Book* [from his radio play *The Girl in the Spotlight*] 1961

Non-Fiction

The Weeping and the Laughter (a) 1953; *Memoirs of the Forties* (a) 1965

Plays

Until the Day She Dies (r) 1958; *The Girl in the Spotlight* (r) 1961

Bernard MacLAVERTY 1942-

MacLaverty was born in Belfast, the son of a commercial artist, and worked for ten years as a laboratory technician before deciding to go to university. He initially thought he might study biology in order to improve his qualifications, but then resolved 'to hell with it, I'll do what I like, and I went into the degree in English'. After a BA at Queen's University, Belfast, he took a teacher training course, and taught first in Edinburgh and then on the isolated island of Islay, where he was head of an English department. He is married with four children, for whom he has written children's stories.

MacLaverty found his authentic voice as a writer slowly, at least by his own account: 'over those ten years I worked in the lab, I wrote very badly, mostly copies of the stuff I'd been reading: I would write a Kafka story, or I'd write a **D.H. Lawrence** story, you know, full of things like "the teats of the cows pulsed in the hands of the men". And this was from a wee boy in the middle of the town. It was an attempt to see what writing was about.' He won immediate notice with his first book of short fictions, *Secrets and Other Stories* (1977), which was praised in the *Times Literary Supplement* for its 'surprising tenderness' and its vivid sense of 'the squalor of loneliness'. He followed it with his first novel, *Lamb*, which tells the story of a reformatory teacher's affection for an epileptic pupil. It has since been made into a very distinguished film (1986), directed by Colin Gregg from MacLaverty's own screenplay, as has his second novel, *Cal*, which was directed by Pat O'Connor in 1984. Set in the Belfast troubles, *Cal* deals with the love that the title-character feels for the widow of a murdered policeman, despite his own complicity in the policeman's death.

There is a certain bleakness in MacLaverty's work: in *A Time to Dance* a boy discovers that his mother is a stripper, for example, and another boy finds that the priest he admires is a drunk, while the title-story of *The Great Profundo* is about the death of a sword swallower who is a secret transvestite. This bleakness has earned him comparison with the **James Joyce** of *Dubliners* (1914), but he denies any attempt to follow in Joyce's footsteps. Above all MacLaverty's work is notable for the restraint with which he handles extremely moving material. He has said: 'If I've learned anything, it's to underwrite as much as possible and to rewrite.' There are few dissenting voices to the universally high regard in which his work is held: a *TLS* reviewer has acclaimed it for showing 'what the great Profundo claimed for his act: genuineness and lack of trickery'.

Fiction

Secrets and Other Stories (s) 1977; *Lamb* 1980; *A Time to Dance and Other Stories* (s) 1982; *Cal* 1983; *The Great Profundo and Other Stories* (s) 1983; *Walking the Dog and Other Stories* (s) 1994

For Children

A Man in Search of a Pet 1978; *Storyline Scotland Book 4* [ed. M. Miller] 1985; *Andrew McAndrew* 1988

Plays

Cal (f) [from his novel] 1984; *Lamb* (f) [from his novel] 1986

Sorley MACLEAN 1911-

Maclean (whose name in the Gaelic form is Somhairle MacGhill-Eain) was born in Osgaig on the island of Raasay off the east coast of Skye, where his parents struggled to keep a croft and a small tailoring business. After attending a local school, he left to continue his education at Portree Secondary School (now Portree High School), staying at Portree during the week, and going home to Raasay on Saturday. During the 1930s, Maclean's family moved to a larger house, abandoning the croft, and with only the tailoring business for income, suffered financially through the Depression. Maclean's political views were affected by this, and by the experience of others he knew who tried to eke a living out of infertile crofts. He was educated at Edinburgh University from 1929 to 1933, and graduated with first-class honours in English, returning the following year to Skye to his first teaching post. Profoundly affected by the events in Spain, Maclean began *Poems to Eimir* (*Dain do Eimhir*), which describes a love affair set against the background of the Spanish Civil War. He married in 1943, the year this volume was published, and the book would establish him as the leading writer in Gaelic of the twentieth century.

Maclean served with the Signals Corps in North Africa, where he was badly wounded at the battle of El Alamein in 1943. After recovering, he resumed the Edinburgh teaching job he had left, finally moving to Plockton in Wester Ross in 1956, where he was appointed headmaster of a school, holding this post until his retirement in 1972.

Maclean was friendly with **Hugh MacDiarmid**, another political rebel, and like him is gentle and courteous, with a shy reticence when reading in public that conceals a romantic nature. His radical communism, tempered by Scottish nationalism, does not dominate his poetry, and he has also written sensitive lyrics. He writes in both Gaelic and English, and his poems are often published in parallel editions. In 1986, a cairn was built at Hallaig, Raasay, to celebrate the 100th anniversary of the Crofters Act and Maclean's seventy-fifth birthday. He was awarded the Queen's Medal for Poetry in 1991.

Poetry

17 Poems for Sixpence 1940; *Dain do Eimhir agus Dain Eile* 1943 [as *Poems to Eimir and Other Poems* 1971]; *Four Points of a Saltire* 1970; *Barran agus Asbhuin* 1973; *Spring Tide and Neap Tide: Selected Poems 1932–72* 1977; *From Wood to Ridge: Collected Poems in Gaelic and English* 1989

Non-Fiction

Ris a'bhruthaich: Criticism and Prose Writings [ed. W. Gillies] 1985

Archibald MacLEISH 1892–1982

One of the few twentieth-century English-language writers to combine a literary reputation with a high-level political career, MacLeish was born in Glencoe, Illinois, the son of a Glaswegian who had migrated to Chicago before the Civil War and made money in business there. He attended the Hotchkiss School, and then Yale, where he took his BA in 1915. His subsequent studies at Harvard Law School were interrupted by the First World War, where he served in the ambulance service and the field artillery and was discharged with the rank of captain (his brother was killed). While still a student he married Ada Hitchcock, a singer; they had four children. After graduating from Harvard, he taught there for a year, then practised law in Boston from 1920 to 1923.

He had published a juvenile book of poems in 1917, and six years later he decided to join the contingent of American *émigré* writers in Paris; 'I date the beginning of my life from 1923,' he said later, and now he began to produce a yearly volume of verse, heavily influenced by **T.S. Eliot** and **Ezra Pound**. He achieved fame in 1928 with *The Hamlet of A. MacLeish*, which **Edmund Wilson** ridiculed in a splendid parody, *The Omelet of A. MacLeish*. Returning to America, he became the editor of *Fortune* from 1929 to 1938, winning the first of his three Pulitzer Prizes for his long poem *Conquistador*, a paraphrase of the prose work *True History of the Conquest of New Spain* written in 1568 by Bernal Diaz del Castillo, a companion of Cortez. MacLeish was once described as 'a phenomenon of our time and culture': in the 1930s his work reflected the politics of the New Deal, and he combined verse with a ballet (*Union Pacific*), stage drama and radio. His growing closeness to President Roosevelt led to him being appointed the Librarian of Congress in 1939. His earnest demands that American writers should produce positive inspiriting works about warfare and reject the literature of the First World War (which, he claimed, had 'done more to disarm democracy in the face of fascism than any other single influence') led to another attack upon him by his old enemy, Edmund Wilson. MacLeish subsequently became the director of the Office of Facts and Figures (an appointment he used for frankly propagandist purposes) in 1941, and Assistant Secretary of State from 1944 to 1945.

The change of administration effectively ended his political career, but he was chairman of the American delegation to the first general conference of UNESCO in 1946 and Boylston Professor of Rhetoric and Oratory at Harvard from 1949 to 1962. He continued to publish poetry and verse into old age; his last notable success was with *J.B.*, a verse drama reworking of the biblical Book of Job. *A Continuing Journey* is an autobiography.

Poetry

Songs for a Summer's Day 1915; *Tower of Ivory* 1917; *The Happy Marriage and Other Poems* 1924; *The Pot of Earth* 1925; *Streets in the Moon* 1926; *The Hamlet of A. MacLeish* 1928; *Einstein* 1929; *New Found Land* 1930; *Conquistador* 1932; *Before March* 1932; *Frescoes for Mr Rockefeller's City* 1933; *Poems 1924–1933* 1933; *Poems* 1935; *Public Speech* 1936; *Land of the Free* 1938; *America Was Promises* 1939; *Active and Other Poems* 1948; *Collected Poems 1917–52* 1952; *Songs for Eve* 1954; *The Collected Poems of Archibald MacLeish* 1963; *The Wild Old Wicked Man and Other Poems* 1968; *New and Collected Poems 1917–1976* 1976

Non-Fiction

Housing America 1932; *Jews in America* 1936; *Libraries in the Contemporary Crisis* 1939; *Deposit of the Magna Carta in the Library of Congress on November 28, 1939* 1939; *The American*

Experience 1939; *The Irresponsibles* 1940; *The American Cause* 1941; *The Duty of Freedom* 1941; *The Free Company Presents: the States Talking* 1941; *The Next Harvard* 1941; *Prophets of Doom* 1941; *A Time to Speak* 1941; *American Opinion and the War* 1942; *A Free Man's Books* 1942; *In Honour of a Man and an Ideal: Three Talks on Freedom* [with E.R. Murrow, W.S. Paley] 1942; *Report to the Nation* 1942; *A Time to Act* 1943; *The American Story: Ten Broadcasts* 1944; *The Son of Man* 1947; *Martha Hillard MacLeish, 1836–1947* 1949; *Poetry and Opinion: the Pisan Cantos of Ezra Pound* 1950; *Freedom Is the Right to Choose: an Inquiry into the Battle for the American Future* 1951; *Poetry and Journalism* 1958; *Poetry and Experience* 1960; *The Dialogues of Archibald MacLeish and Mark Van Doren* 1964; *A Continuing Journey* **(a)** 1968; *Champion of a Cause: Essays and Addresses on Librarianship* [ed. E.M. Goldschmidt] 1971; *Riders on the Earth: Letters and Recollections* 1978; *Letters of Archibald MacLeish 1970–1982* [ed. R.H. Winnick] 1983; *Reflections* [ed. B. A. Drabeck, H.E. Ellis] 1986

Edited

Gerald Fitzgerald *The Wordless Flesh* 1960; Edwin Muir *The Estate of Poetry* 1962; Felix Frankfurter *Law and Politics: Occasional Papers, 1913–38* [with E.F. Prichard Jr] 1963; Leonard Baskin *Figures of Dead Men* 1968

Plays

Nobodaddy 1926; *Union Pacific* **(l)** [with N. Nabakoff] 1934 [np]; *Panic* 1935; *The Fall of the City* **(r)** 1937, **(tv)** 1962 [pub. 1937]; *Air Raid* **(r)** 1938 and *The Secret of Freedom* **(tv)** 1960 [pub. in *Three Short Plays* 1961]; *The Trojan Horse* **(r)** 1950 [pub. 1952]; *This Music Crept by Me upon the Waters* 1953: *J.B.* 1958; *The Eleanor Roosevelt Story* **(f)** 1965; *An Evening's Journey to Conway, Massachusetts* 1967; *Herakles* 1965 [pub. 1967]; *Magic Prison* **(l)** 1967 [np]; *Scratch* [from S.V. Benét's story *The Devil and Daniel Webster*] 1971; *The Great American Fourth of July Parade* **(r)** 1975

Larry (Jeff) McMURTRY 1936–

Texas is both McMurtry's birthplace and the special subject of his fiction. In the Pulitzer Prize-winning *Lonesome Dove* (1985) he re-created the great cattle drives of frontier legend. His first novels tend to focus more on the uneasy tension that exists between the Old West of myth and the uneasily urbanised, small-town life of his own upbringing.

He was born in Wichita Falls and brought up on a cattle ranch, the third generation of his family to have worked there, and although he has since bought a ranch of his own, he has written of the relief he felt upon leaving rural Archer County for the comparative freedom of university. He was awarded a BA from North West State College in 1958, and an MA from Rice University in 1960. An early draft of his first novel, *Horseman, Pass By* (1961), written whilst still at college, won him a grant to enter a creative writing course at Stanford University. He gained experience and extra money by trading in rare books, scouting for dealers in Texas and California to subsidise his own buying: he has owned a bookshop in Washington since 1970. Although since disowned by its author, *Horseman, Pass By*, was critically successful, winning the Jesse H. Jones Award in 1962. The following year it was filmed as *Hud*, a classic of the genre, directed by Martin Ritt and the winner of three Academy Awards. A vividly compressed, semi-autobiographical novel, it is centred on ranch life, like its successor, *Leaving Cheyenne*, filmed by Sidney Lumet in 1973 as *Lovin' Molly*. Both books had been written by the time McMurtry was twenty-three. His next novel, *The Last Picture Show*, was set in the small town of Thalia and describes the frustrations and boredom that he had felt in Wichita. The tone of the book was much softened in an Oscar-winning screenplay which he wrote in collaboration with director Peter Bogdanovich. Shot on location in Archer City, the film went on to win the Academy Award for best picture in 1972.

McMurtry had been teaching to support his wife and child since leaving college – he had married in 1959 and was divorced in 1966 – but felt able to abandon academia in 1969 and move to Washington, where he wrote a trilogy of urban novels, beginning in 1970 with *Moving On*, centred on Houston. Some critics found these novels too accurate a reflection of the aimless lives that they described: characters circulate within the trilogy, travel to California and back, and generally seem to lose themselves in the endless distractions of motion in between. *Terms of Endearment*, the last of the three, was the winner of five Academy Awards and four Golden Globes when filmed by James L. Brooks in 1983. McMurtry's own passion for cinema, already reflected in numerous articles and reviews, was given a new platform with the editorship of *American Film* which he took up in 1975: a collection of his essays on Hollywood were published as *Film Flam* in 1987.

The move from Texas also introduced a new backdrop to his fiction. Hollywood, Las Vegas and the shifting life of a cross-country dealer were all explored, before he returned both in time and place to the Texas of legend to write the bestselling *Lonesome Dove. In a Narrow Grave:*

Essays on Texas collects his non-fiction on the subject. Since then he has written similarly anti-mythic retellings of the lives of Calamity Jane and Billy the Kid, alternating these with contemporary novels. In *Texasville* McMurtry finally returned to Thalia, Texas, and the characters of *The Last Picture Show*. Bogdanovich was unable to repeat the success of his earlier screen translation when the book was adapted for release in 1990.

Fiction

Horseman, Pass By 1961 [UK *Hud* 1963]; *Leaving Cheyenne* 1963; *The Last Picture Show* 1966; *Moving On* 1970; *All My Friends Are Going to Be Strangers* 1972; *Terms of Endearment* 1977; *Somebody's Darling* 1978; *Cadillac Jack* 1982; *The Desert Rose* 1983; *Lonesome Dove* 1985; *Texasville* 1987; *Anything for Billy* 1988; *Some Can Whistle* 1989; *Buffalo Girls* 1990; *The Evening Star* 1992; *Streets of Laredo* 1993

Plays

The Last Picture Show (f) [with P. Bogdanovich] 1971

Non-Fiction

In a Narrow Grave: Essays on Texas 1968; *It's Always We Rambled: an Essay on Rodeo* 1974; *Larry McMurtry: Unredeemed Dreams* [ed. D. Schmidt] 1980; *Film Flam: Essays on Hollywood* 1987

(Frederick) Louis MacNEICE 1907–1963

The son of a rector who rose to become Bishop of Down and Connor and Dromore, MacNeice was born in Belfast. His mother died when he was young, and his father was remote; the melancholy and sense of impermanence which impregnates some of MacNeice's best verse perhaps owes something to this early background. He was educated at Marlborough and at Merton College, Oxford, where he took a first in *literae humaniores* in 1930, having published his first book of poems, *Blind Fireworks*, in the previous year. He then became a lecturer in classics for the next decade, first in Birmingham and later at Bedford College, London. A chief fruit of this was his acclaimed verse translation of Aeschylus' *Agamemnon*.

In the 1930s, MacNeice was generally associated with the group of poets including **W.H. Auden**, **Stephen Spender** and **C. Day Lewis**, but alone of the group MacNeice was never a Marxist, and from the first his poetry achieved a nice balance between the political and personal, ranging from moving lyric in 'Prayer Before Birth' to almost slapstick comedy in the famous 'Bagpipe Music'. Among his best earlier work was the long poem *Autumn Journal*, and the unorthodox travel book, *Letters from Iceland*, written in collaboration with Auden, which combines verse and prose and includes the two poets' celebrated 'Last Will and Testament', an extended (and largely private) joke in verse. MacNeice continued to grow as a poet throughout his life, and some of his most intense works are in the posthumously published *The Burning Perch*, where the melancholy note predominates.

MacNeice had a child by each of his marriages, the first of which ended in divorce, the second of which – to the singer and actress Hedli Anderson, who was associated with the Group Theatre and was a notable performer of works by Auden and Benjamin Britten – in separation. In the first year of the Second World War he was in America, lecturing at Cornell University, but he returned to England and, in early 1941, joined the features department of the BBC; he continued as a full-time radio writer and producer until 1961, a job diversified by several overseas assignments and a year-long appointment as director of the British Institute in Athens. But increasing unhappiness and alcoholism marked his later years. He continued as a freelance radio producer after resigning from his full-time job; going down a mine as part of his work led to him catching a chill, from which he died. Besides his poetry, MacNeice wrote plays for stage and radio including the powerful *The Dark Tower*, an early novel, some critical works, and a guarded autobiography, *The Strings Are False*.

Poetry

Blind Fireworks 1929; *Roundabout Way* 1932; *Poems* 1937; *The Earth Compels* 1938; *Autumn Journal* 1939; *The Last Ditch* 1940; *Plant and Phantom* 1941; *Poems 1925–1940* 1941; *Springboard: Poems 1941–1944* 1944; *Holes in the Sky: Poems 1944–1947* 1948; *Collected Poems 1925–1948* 1949; *Ten Burnt Offerings* 1952; *Autumn Sequel* 1954; *Visitations* 1957; *Solstices* 1961; *The Burning Perch* 1963; *The Collected Poems of Louis MacNeice* [ed. E.R. Dodds] 1966; *Selected Poems of Louis MacNeice* [ed. M. Longley] 1988

Plays

Out of the Picture (r) 1937; *Christopher Columbus* (r) 1944; *The Dark Tower and Other Radio Scripts* (r) [pub. 1947]: *The Nosebag* 1944, *Sunbeams in His Hat* 1944, *The March Hare Reigns* 1945, *The Dark Tower* 1946, *Salute to All Fools* 1946; *The Administrator* 1961 [pub. 1964]; *The Mad Islands* 1962 [pub. 1964]; *One for the Grave* 1966 [pub. 1968]; *Persons from Porlock and Other Plays for the Radio* (r) [pub. 1969]: *Enter Caesar* 1946, *East of the Sun and West of the Moon* 1959, *They Met on Good Friday* 1959, *Persons from Porlock* 1963; *Selected Plays* [ed. A. Hewer, P. McDonald] 1993

Non-Fiction

I Crossed the Minch 1938; *Modern Poetry: a Personal Essay* 1938; *Zoo* 1938; *The Poetry of W.B. Yeats* 1941; *Astrology* 1964; *The Strings Are False* (a) 1965; *Varieties of Parable* 1965

Fiction

Roundabout Way [as 'Louis Malone'] 1932

For Children

The Sixpence That Rolled Away [US *The Penny That Rolled Away*] 1954

Translations

The Agamemnon of Aeschylus 1936; *Goethe's Faust* 1951

Collections

Letters from Iceland [with W.H. Auden] 1937

Biography

Louis MacNeice by John Stallworthy 1995

Candia McWILLIAM 1955–

McWilliam was born in Edinburgh, the daughter of the architectural historian Colin McWilliam, and was educated there until the age of thirteen, when she went to school in England. She has written of her childhood, and of her mother's early death, in 'The Many Colours of Blood', the piece by which she was represented in the 1993 issue of *Granta* (No. 48) devoted to the 'Best of Young British Novelists'. She read English at Girton College, Cambridge, where she took a first, then went on to work for *Vogue* (she had won the magazine's talent contest as a teenager), and as an advertising copywriter. She married Quentin Wallop, Earl of Portsmouth, in 1981, with whom she had two children. They divorced in 1985 and the following year she married Fram Dinshaw, a biographer and bursar of an Oxford college; they have one son.

McWilliam became an instant celebrity with her first novel, *A Case of Knives*, which was published in 1988, shortlisted for a Whitbread Award and won a Betty Trask Award. It is a complex story of emotional entanglements concerning a homosexual heart surgeon, the young estate agent with whom he is in love, and the estate agent's fiancée, who is herself fascinated by the surgeon. Its element of moral enquiry recalled the novels of **Iris Murdoch**, while its stylish savagery suggested that McWilliam is a natural heir to **Muriel Spark**, whose preoccupation with a Scottish heritage she also shares. The book was praised for the extreme acuity of its writing, and although it drew mockery from some quarters for its occasionally arcane vocabulary, it was widely enjoyed for its shrewdness and wit, as well as for McWilliam's sophisticated and worldly-wise way of writing about the lives of the privileged.

Her second novel, *A Little Stranger*, made less impact, but was equally impressive: short, sharp and shocking, it describes the relationship between a wilfully unobservant woman and her apparently perfect, though not very likeable, nanny. McWilliam's skill lies in the way she manipulates the narrative, bringing it to a wholly unexpected conclusion, clues to which have been artfully placed throughout. Once again, the book was distinguished by its wit – the narrator is a compulsive joker – and by the dazzling inventiveness of its prose. Her third novel, *Debatable Land*, won the *Guardian* Fiction Prize in 1994, and is her most densely written and complex work to date. Set aboard a yacht sailing the South Seas from Tahiti to New Zealand, it is partly concerned with memory, identity and nationality, and owes much to McWilliam's Scottish roots (the epigraph is taken from Robert Louis Stevenson's *Songs of Travel*). As with her earlier novels, this one was replete with extraordinary images, notably when the principal character, a painter, recalls his Edinburgh childhood. McWilliam has also written short stories, which have appeared in various magazines and anthologies, and is noted for the seriousness and generosity of the book reviews she writes, principally for the *Independent on Sunday*.

Fiction

A Case of Knives 1988; *A Little Stranger* 1989; *Debatable Land* 1994

Charles (Henry) MADGE 1912–

Madge was born in Johannesburg, South Africa. His father – one of 'Milner's young men' in the period of reconstruction after the Boer War – was killed in the First World War, and Madge was brought up in England, attending Winchester and Magdalene College, Cambridge, where he read English under I.A. Richards. He began to write poetry early, and **T.S. Eliot** helped him to get his first poems published as well as to get a job as a reporter on the *Daily Mirror* in 1935. With Humphrey Jennings, Madge founded the well-known sociological organisation Mass Observation; several books arose from this experience, co-written with Tom Harrisson. During the 1930s he married fellow poet **Kathleen Raine**, whom he had met at Cambridge; there were two children of this marriage, which was dissolved.

Madge's earlier poetry was collected in two volumes, *The Disappearing Castle* and *The Father Found*, after which he continued to write poetry at a reduced rate of production: it took almost fifty years for a third volume to appear, by which time Madge was in his late seventies.

He was among the most influential Marxist poets of the 1930s, complex and indirect but resonant in expression. During the Second World War he directed a survey of wartime saving and spending in industrial cities under the auspices of J. Maynard Keynes. He subsequently joined Pilot Press to edit a series of volumes on post-war reconstruction, while from 1947 to 1950 he was social development officer for the new town of Stevenage. From 1950 to 1970 he was professor of sociology at Birmingham University, and most of his later volumes are sociological works. After retiring, he lived for six years in a village in the Languedoc region of southern France; since then he has lived in London.

Madge's second wife was Inez Pearn, whom he detached from her then husband, **Stephen Spender**, in 1939. They had two children, and she wrote novels under the name Elizabeth Lake. Both she and a third wife, Evelyn Brown, have predeceased him.

Poetry

The Disappearing Castle 1937; *The Father Found* 1941; *Of Love, Time and Places* 1990

Non-Fiction

Britain by Mass Observation [with T. Harrisson] 1939; *Industry After the War* [with D. Tyerman] 1943; *War Time Patterns of Saving and Spending* 1943; *To Start You Talking* (c) 1945; *Village Communities in North East Thailand* 1955; *Survey Before Development in Thai Villages* 1957; *Village Meeting Places: a Pilot Inquiry* 1958; *Evaluation and the Technical Assistance Expert: an Operational Analysis* 1961; *Society in the Mind: Elements of Social Eidos* 1964; *Art Students Observed* [with B. Weinberger] 1973; *Inner City Poverty in Paris and London* [with P. Willmott] 1981

Edited

May the Twelfth. Mass Observation Day-Surveys 1937 [with H. Jennings] 1937; *First Year's Work 1937–1938, by Mass Observation* [with T. Harrisson] 1938: *War Begins at Home, by Mass Observation* [with T. Harrisson] 1940; *Pilot Guide to the General Election* 1945; *Pilot Papers: Social Essays and Documents 1945–1947* 1947; Humphrey Jennings *Pandaemonium 1660–1886: the Coming of the Machine as Seen by Contemporary Observers* [with M.L. Jennings] 1986

Derek MAHON 1941–

Mahon was born in Belfast, where his father – as his grandfather had – worked in the shipyards. There are several references to the *Titanic* in his work: 'My grandfather worked on it before it was launched,' he has said. 'He was a boilermaker – and somehow it loomed large in the background.' Mahon was educated at the Royal Belfast Academical Institution and Trinity College, Dublin, graduating in 1965. He spent a term at the Sorbonne, as an *auditeur libre*, and taught at Belfast High School and the Language Centre of Ireland in Dublin. He was co-editor of *Atlantis* in Dublin from 1970 to 1974, and has been drama critic of the *Listener* and features editor of *Vogue*. He has been writer-in-residence at the University of East Anglia, Emerson College, Boston, and the New University of Ulster, Coleraine.

While still an undergraduate Mahon won an Eric Gregory Award for his first book, *Twelve Poems* (1965), which was to form the nucleus of his first collection, *Night-Crossing*, in its turn a Poetry Book Society Choice. The *Times Literary Supplement* called it 'an attractive book, lively and engaging ... there are poems of unusual exactitude, the feeling beautifully adjusted, nothing too much, nothing too little'. In the Introduction to *Ecclesiastes* he writes: 'the poems constitute, in some measure, the poetic record of an attempt by an uprooted Ulsterman to come to terms with his background and with something, glimpsed through that background, that may contain some kind of final answer'. In *Lives* the poems move away from the romanticism of his early work towards a personal and historical view of Ulster and the world at large. Many of the poems are sombre but Mahon's wit and lyricism remain. He wrote of *Poems 1962–78* as 'the work of a young man who thought he knew quite a lot about one thing and another ... and discovered that he knew, in fact, nothing of importance'. Such self-criticism, it seems, can be salutary. **Seamus Heaney** wrote of his next collection, *The Hunt by Night*, 'Some creative tremor has given him deepening access to his sources of power; it is as if the very modernity of his intelligence has goaded a primitive stamina in his imagination. There is a copiousness and excitement about these poems that is to be found only in work of the highest order.' Mahon has also translated Molière, Gerard de Nerval and Philippe Jacottet, as well as adapting a number of novels for television. He is married and has two children.

Poetry

Twelve Poems 1965; *Design for a Grecian Urn* 1967; *Night-Crossing*1968; *Beyond Howth Head*

1970; *Ecclesiastes* 1970; *Lives* 1972; *The Man Who Built His City in Snow* 1975; *Light Music* 1977; *In Their Element* [with S. Heaney] 1977; *The Sea in Winter* 1979; *Poems 1962–78* 1979; *Courtyards in Delft* 1981; *The Hunt by Night* 1982; *A Kensington Notebook* 1984; *Antarctica* 1985; *Selected Poems* 1991; *The Yaddo Letter* 1994

Plays

Shadows on Our Skin **(tv)** [from J. Johnston's novel] 1980; *How Many Miles to Babylon?* **(tv)** [from J. Johnston's novel] 1981; *First Love* **(tv)** [from Turgenev] 1982; *The Demon Lover* **(tv)** [from E. Bowen's story] 1983; *The Cry* **(tv)** [from J. Montague; with C. Menaul] 1984; *A Moment of Love* **(tv)** [from B. Moore's novel] 1984; *The School for Wives* [from Molière's play] 1986; *Bacchae* [pub. 1991]; *High Time* [from Molière] 1984 [pub. 1985]

Edited

The Sphere Book of Modern Irish Poetry 1972; *The Penguin Book of Contemporary Irish Poetry* [with P. Fallon] 1990

Translations

Gérard de Nerval *The Chimeras* 1982; *Selected Poems of Philippe Jaccottet* 1988

Norman (Kingsley) MAILER 1923–

Mailer was born Nachum Malech Mailer in New Jersey, the son of an accountant and a devoted and famously loyal mother. The family moved to Brooklyn, 'the most secure Jewish community in the world', when Mailer was a child. He attended Brooklyn schools, graduated at the age of sixteen, and then went to Harvard to study aeronautical engineering. Aged eighteen, he won *Story Magazine*'s annual college short-story writing contest, by which time his ambition as a writer had been formed. He wrote two apprentice novels, married, was drafted in 1944 and set off for war with the intention of finding his book.

Trained as an artillery surveyor, Mailer requested transfer to a front-line Intelligence and Reconnaissance infantry unit, in which he proved himself both brave and foolhardy and learned the common language of his first novel, *The Naked and the Dead.* Discharged in 1946, he took fifteen months off to write, and then to find a publisher willing to take the book in full, profanities and all. Mailer and his first wife then set off for the Sorbonne, taking advantage of the GI Bill. Telegrams of congratulation interrupted the journey. Published in 1948, the book was a bestseller and Mailer bade 'farewell to an average man's experience'. He had, he has written, 'lobotomised the past' in one book, and so seemed to set about reinventing himself to find the material for more.

Formerly a self-described anarchist, he was briefly persuaded in Paris of the necessity of collective action. His second novel, *Barbary Shore*, shows Mailer struggling more with message than with pacing or plot. Jean Malaquais, Mailer's friend and teacher in France, dismissed the book as a 'political tract'. American reviews were more or less hysterical: 'It is relatively rare to discover a novel whose obvious intention is to debauch as many readers as possible, mentally, morally, physically and politically,' wrote one critic. Far from curbing the author's ambition, the controversy confirmed his scorn of 'the totalitarian plague' in America. For Mailer the war had begun. He separated from his wife and moved to Greenwich Village in 1951, helped found the *Village Voice*, began a turbulent affair with Adele Morales, the Peruvian painter who was to be his second wife, and after four years of dispiriting work on *The Deer Park*, he provoked another scandal. Originally conceived as the prologue to an eight-novel cycle, as announced in his collection *Advertisements for Myself*, the book drew upon Mailer's brief, unhappy experience as a Hollywood screenwriter. Rinehart, his regular publishers, refused it. The novel was then hawked around eight houses before finding a home at Putnam's. Upon publication, after much critical vilification, Mailer underlined future intentions by taking out a full-page advertisement in the *Village Voice* composed entirely from the worst of his reviews.

He now began a public life of near unrelieved controversy: as bar brawler and drinker, arm-wrestling specialist, film-maker, actor, would-be politician and polemicist on behalf of the home-grown American existentialist in articles like 'The White Negro' and 'Reflections on Hipsterism'. He first announced his candidacy as Democratic nominee for the mayoralty of New York in 1960, advocating that the city secede from the state. The campaign was abandoned soon after the pre-campaign party at which Mailer was arrested for stabbing his wife. He was released when she refused to press charges and the two were divorced in 1962. In need of funds, he wrote his next novel in eight serial instalments for *Esquire.* Although the critics were unconvinced, *An American Dream* was a bestseller. Mailer then abandoned the novel in favour of lengthy, journalistic narratives. Film projects occupied him for a while, the first of which, *Wild 90* (written, produced and directed by Mailer himself, with friends as actors and himself in the lead role), is another adaptation of *The Deer Park.* (Six wives on and with eight children to support, such profligacy is past history, although the experience helped him to get Zoetrope backing for a self-directed screen version of *Tough Guys Don't Dance.*) A second, more successful, campaign to become the mayor of New York was fought alongside the writer Jimmy Breslin in 1969. The two proposed the reintroduction of the death penalty by gladiatorial combat. Cited by Kate Millett in 1970 as the perfect chauvinist pig, the author went on stage to defend his ground in a filmed public debate.

Mailer won his second Pulitzer Prize – the first, for non-fiction, had been given to *The Armies of the Night* in 1969 – when *The Executioner's Song* was published to great acclaim in 1979. A massive, documentary-style account of the life and execution of Gary Gilmore in 1977, the book was also the beginning of his petitioning to secure the release of long-term convict Jack Henry Abbott. Taken up by the New York literary establishment, Abbott was released only to commit a second murder. Abbott's autobiography, *In the Belly of the Beast* (1981), with its introduction by Mailer, will forever stand as the reply to all Mailer's wild attitudinising, even though his influence in securing Abbott's release was much exaggerated. A plagiarism suit, settled out of court, was another low point, coming at a time when crippling alimony payments seemed to be dictating his need to write. Typically, when he returned to publishing fiction with *Harlot's Ghost* in 1991 it was on the grand scale, with a 1,200-page book. The Great Bitch, as he describes the American novel in *Cannibals and Christians*, had not entirely deserted him.

Fiction

The Naked and the Dead 1948; *Barbary Shore* 1951; *The Deer Park* 1955; *An American Dream* 1965; *Why Are We in Vietnam?* 1967; *The Short Fiction of Norman Mailer* (s) 1967; *A Transit to Narcissus* [ed. H. Fertig] 1978; *The Short Fiction of Norman Mailer* (s) 1981; *Ancient Evenings* 1983; *The Last Night* 1984; *Tough Guys Don't Dance* 1984; *Harlot's Ghost* 1991

Non-Fiction

The White Negro 1957; *The Presidential Papers* 1963; *Cannibals and Christians* 1966; *The Bullfight* 1967; *The Armies of the Night: the Novel as History, History as a Novel* 1968; *The Idol and the Octopus: Political Writings on the Kennedy and Johnson Administrations* 1968; *Miami and the Siege of Chicago: an Informal History of the Republican and Democratic Conventions of 1968* 1968; *King of the Hill: On the Fight of the Century* 1971; *Of a Fire on the Moon* [UK *A Fire on the Moon*] 1971; *The Prisoner of Sex* 1971; *Existential Errands* 1972; *St George and the Godfather* 1972; *Marilyn: a Novel Biography* 1973; *The Faith of Graffiti* [with M. Kurlansky, J. Naar] 1974 [UK *Watching My Name Go By*] 1975; *The Fight* 1975; *Genius and Lust: a Journey through the Major Writings of Henry Miller* 1976; *Some Honorable Men: Political Conventions 1960–1972* 1976; *The Executioner's Song: a True Life Novel* 1979; *Of Women and Their Elegance* [with M.H. Greene] 1980; *The Essential Mailer* 1982; *Pieces and Pontifications* 1982; *Huckleberry Finn: Alive at 100* 1985; *Conversations with Norman Mailer* [ed. J.M. Lennon] 1988; *Pablo and Fernande* 1994

Plays

The Deer Park [from his novel] 1960 [pub. 1967]; *A Fragment from Vietnam* 1967 [pub. in *Existential Errands* 1972]; *Beyond the Law* (f) 1968; *Wild 90* (f) 1968; *Maidstone* (f) 1971; *The Executioner's Song* (f) 1982; *Tough Guys Don't Dance* (f) [from his novel] 1987

Poetry

Deaths for the Ladies and Other Disasters 1962

Collections

Advertisements for Myself 1959; *The Long Patrol: 25 Years of Writing from the Works of Norman Mailer* 1971

Sara MAITLAND 1950–

Maitland was born in London and educated there and in Scotland. Shortly after leaving Oxford she married an Anglican priest and settled to a combination of family life and freelance academic research and writing which caused her to describe herself as a feminist-socialist-Christian-wife-mother.

Her writing occupies a unique position in post-war British letters in that it represents an attempt to reconcile what many regard as the irreconcilable: Christianity and feminism. While remaining within the body of the church, she challenges the orthodoxy that offers an absolutely patriarchal religious world view. Some of her earliest published writing appeared in an anthology of prose entitled *Tales I Tell My Mother* which also included work by Zoe Fairbairns, **Michèle Roberts** and Michelene Wandor. This was published in 1978, the year in which she published her first, highly acclaimed, novel, *Daughter of Jerusalem*. The novel explores the dilemma of feminism as its central character struggles to reconcile her principles with her obsessive desire to have a child. When Elizabeth asks for help to conceive she is told that her problems are all in the mind, and that it is her reluctance to submit to a traditional feminine role that is causing her to suppress ovulation. She is forced to reassess her life to date and to face the possibility that her mind is suppressing her body in this way.

Maitland's second novel, *Virgin Territory*, focuses on rape and its aftermath – the rape of a nun which opens the novel and the metaphorical rape of native peoples by the Spanish Conquistadors. In it the protagonist, Sister Anna, searches for an identity through the opposing women's cultures of the Catholic Sisterhood and the lesbian sisterhood to which she is introduced through her attraction to Karen. *Three Times Table* vividly illustrates Maitland's versatility as she bends the saga form to her own ends, using it to tell the story of three generations of women in one family, each facing a particular crisis of womanhood: a young woman tries to shake off her childhood; her mother to hang on to her life through the trial of cancer; and her grandmother to acknowledge the possibility that her life has been based on a lie. The short stories collected in *Telling Tales*, *Weddings and Funerals* and *Women Fly When Men Aren't Watching* have given her an opportunity to explore a rich variety of literary forms and to range freely over subject-matter that encompasses folk-stories and myth, historical detail and contemporary ideas. Her contribution to **Adam Mars-Jones**'s collection of lesbian and gay short stories, *Mae West Is Dead* (1983), provides a fresh perspective on issues of gender and sexuality.

Maitland's first work of non-fiction continues many of the themes of her fiction. *A Map of the New Country*, published in 1983, offers a timely overview of women's role in the church through history and into the present. She has also written a book about the 1960s and a biography of the Victorian music-hall star and male impersonator Vesta Tilley, whose extraordinary life is done full justice by a storyteller fascinated by concepts of gender and its effect on one's position in society.

Maitland lived for a number of years in the East London vicarage of her husband's parish and brought up their two children there. She now lives in Kettering, Northamptonshire.

Fiction

Daughter of Jerusalem 1978; *Tales I Tell My Mother* (s) [with others] 1978; *Mae West Is Dead* (c), (s) [ed. A. Mars-Jones] 1983; *Telling Tales* (s) 1983; *Virgin Territory* 1984; *Weddings and Funerals* (s) 1984; *Arky Types* [with M. Wandor] 1987; *A Book of Spells* 1988; *Three Times Table* 1990; *Home Truths* 1993; *Women Fly When Men Aren't Watching* (s) 1993

Non-Fiction

Happy Unicorns (c) [ed. S. Purcell, L. Purves] 1970; *A Map of the New Country: Women and Christianity* 1983; *Fathers* (c) [ed. U. Owen] 1983; *Gender and Writing* (c) [ed. M. Wandor] 1983; *Vesta Tilley* 1986; *Very Heaven: Looking Back at the 1960s* 1988

Edited

Walking on Water: Women Talk About Spirituality [with J. Garcia] 1983

Adewale MAJA-PEARCE 1953-

Maja-Pearce was born in London, the son of an English mother and a Nigerian father then studying medicine. He was educated at St Gregory's College in Lagos where, despite the recent independence of Nigeria, he received a largely European education which he has described as turning him into 'a proper little colonial with a faultless English accent'. Returning to England, he received his BA from the University College of Swansea, and an MA in 1986 from the School of Oriental and African Studies at the University of London. Since then he has worked as the African researcher in the London office of Index on Censorship.

His single collection of stories, *Loyalties*, explores themes of cultural and racial identity against a Nigerian setting. Most of the stories are short, almost fable-like vignettes: the title-story, notably, describes the effects of the civil war on a group of villagers who learn that they are now Biafrans, and, a few days later, Nigerians once more.

In his subsequent work Maja-Pearce has favoured the extended essay and the travel book as forms with which to explore 'the nature of my double inheritance, Nigeria and Great Britain'. *In My Father's Country* describes a return trip to Nigeria which was commissioned as a result of an article he had published in the *London Magazine*, and paints a sad picture of a country riddled with political corruption and a resultant apathy. He has also written on the Nigerian novelist **Wole Soyinka**.

Fiction

Loyalties and Other Stories (s) 1986

Non-fiction

In My Father's Country 1987; *How Many Miles to Babylon?* 1990; *Who's Afraid of Wole Soyinka* 1991; *Wole Soyinka: an Appraisal* 1994

Edited

The Heinemann Book of African Poetry in English 1990

Bernard MALAMUD 1914–1986

Born in Brooklyn, New York, Malamud was the son of Jewish immigrants from Tsarist Russia. His father was a grocer, and Malamud worked in the store after his mother died during his teenage years. His home milieu, poor but close-knit, is reflected in his best novel, *The Assistant*. He was educated at Erasmus Hall, New York, and New York City College, and also attended Columbia University. After graduating, he worked in a factory, in stores and as a clerk in the Census Bureau. During the 1940s he taught in various New York night schools and began to write seriously during the day. He married in 1945, and had two children. Between 1949 and 1961, he taught at Oregon State College, and published his first novels. He began with *The Natural* (1952), one of his few novels not to feature Jewish characters, and about the American love of baseball. *The Assistant* was more characteristic in its warm Jewish humour and insistence that graceless human beings can find redemption through love and suffering. Malamud became generally estimated as the second Jewish-American novelist, after **Saul Bellow**, and was always somewhat conscious of standing in the other novelist's shadow. *The Fixer*, set in Tsarist Russia, reads almost like a conscious attempt to write the Great Jewish Novel, but it met with a lukewarm critical reception. Later, and more successful, novels include *Dubin's Lives* and *God's Grace*.

Some of Malamud's best work was in his short stories: in such a collection as *Rembrandt's Hat* he is able to distil whole human lives into a few spare but lyrical pages. In 1961, he moved back east, and in later years taught at Bennington State College, Vermont.

Fiction

The Natural 1952; *The Assistant* 1957; *The Magic Barrel* 1958; *A New Life* 1961; *Idiots First* 1963; *The Fixer* 1966; *Pictures of Fidelman* 1969; *The Tenants* 1971; *Rembrandt's Hat* (s) 1973; *The Jewbird, Two Fables* (s) 1978; *Dubin's Lives* 1979; *God's Grace* 1982; *The Stories of Bernard Malamud* (s) 1983

Biography

Bernard Malamud by Sheldon Hershinow 1980

Herman MALAN

see H.C. BOSMAN

David (George) MALOUF 1934–

Malouf was born in Brisbane of Lebanese and English parents, and was educated at Brisbane Grammar School and the University of Queensland. In 1958 he sailed to Britain, where he was to live for ten years, working as a teacher at St Anselm's College in Cheshire. On his return to Australia in 1968 he was appointed senior tutor in English at the University of Sydney.

His work first appeared with that of other new Brisbane-based poets in *Four Poets*. His first collection, *Bicycle and Other Poems*, was published in 1970 to critical acclaim and established him as one of Australia's leading poets. Malouf is an imagistic writer; his themes are man's search for self-knowledge, the relationship of past and present, and the mysteries of the natural world. His writing is effortlessly visual with a surrealistic edge, and he is praised for his lavish, elegant language. In *First Things Last*, one poem is about wild lemon trees, another about a man surveying a crab in a restaurant before he eats it. Throughout the collection the image of the Garden of Eden is sustained.

It is Malouf's novels, however, which have aroused the most critical attention. *An Imaginary Life* is a fictional account of the last years of Ovid, following his exile to the Black Sea on the orders of the Emperor Augustus. The novel is in the form of a letter from the cosmopolitan poet, which he will bury in rock and cast upon the centuries. In this lyrical novel, Malouf creates a barbaric village of mud huts, threatened by the savages of the surrounding grasslands, where the elegant Roman poet must live, cruelly humbled, isolated from all he has known, and unable to communicate with his fellow villagers. Eventually Ovid meets a child raised by wolves and attempts to educate him, but instead finds himself transformed and finally at one with nature. Thus Ovid's own theme of metamorphosis is celebrated.

Malouf's preoccupation with the juxtaposition and transformation of opposing modes of existence and perception can be seen again in *Fly Away Peter*, where the calm of a Queensland bird sanctuary is set against the horror of the First World War, and in *Remembering Babylon*. This dense and poetic novel, shortlisted for the 1993 Booker Prize and the *Irish Times* International Fiction Prize, is set for the most part in nineteenth-century Queensland, and describes the arrival of Gemmy Fairley in a colony of farmers. Though English-born, Fairley has spent the last sixteen years as part of an Aboriginal tribe, and his presence provokes suspicion and eventually violence. The novel's contrasting perspectives suggest the many ways in which being, consciousness and environment may infuse each other.

Malouf has received many literary awards, among them the Australian Literature Society Gold Medal in 1974, the New South Wales Premier's Prize for Fiction in 1979 and the Melbourne *Age* Book of the Year Award in 1982. He lives in Italy.

Fiction

Johnno 1975; *An Imaginary Life* 1978; *Child's Play with Eustace and the Prowler* 1982; [US *Child's Play, The Bread of Time to Come*]; *Fly Away Peter* 1982; *Harland's Half Acre* 1984; *Antipodes* (s) 1985; *12 Edmondstone Street* 1985; *The Great World* 1990; *Remembering Babylon* 1993

Poetry

Four Poets [with others] 1962; *Bicycle and Other Poems* 1970 [US *The Year of the Foxes and Other Poems* 1979]; *Neighbours in a Thicket* 1974; *Poems 1975–76* 1976; *First Things Last* 1980; *Selected Poems* 1980; *Wild Lemons* 1980; *Poems 1959–89* 1992

Plays

Voss (l) [from P. White's novel; with R. Meale] 1986; *Blood Relations* 1987; *Baa Baa Black Sheep* (l) [with M. Berkeley] 1993

Non-Fiction

New Currents in Australian Writing [with K. Brisbane, R.F. Brissenden] 1978

Edited

We Took Their Orders and Are Dead: an Anti-War Anthology [with others] 1971; *Gesture of a Hand* 1975

Collections

Johnno, Short Stories, Poems, Essays, and Interview [ed. J. Tulip] 1990

David (Alan) MAMET 1947–

Mamet claims that he owes most of his early inspiration to growing up in Chicago. His first job was bus-boy at the comedy cabaret club Second City, and from the nightly improvisations of future cabaret and acting stars he came to understand the rhythms and patterns of speech. He began writing dialogues as a teenager, encouraged by his father, a lawyer obsessed with semantics. Mamet studied literature and drama at Goddard College (where he returned in 1971 as drama teacher), taking a year off to attend the Neighbourhood Playhouse School in New York, where he learned the Stanislavsky acting method and the dramatic principles of continuous action and 'moment-to-moment'. His early jobs were many and diverse: cab driver, real-estate agent, fast-food cook, window washer, factory worker and office cleaner. He was eventually employed as a drama teacher at Marlboro College, Vermont, and with his best students formed the St Nicholas Theater Company, who performed his first play, *Lakeboat*, in 1970.

Chicago was seething with small theatres and undiscovered talent and Mamet soon opened his joint-production of *Sexual Perversity in Chicago* and *Duck Variations*. In 1975 this double-bill made it to the off-Broadway Cherry Lane Theater and was included in *Time* Magazine's list of the top ten plays of that year. Meanwhile, *American Buffalo* had opened at the Goodman Theater, Chicago, to great critical acclaim. It opened on Broadway in 1977, won first an Obie, then the New York Drama Critics Circle Award, and established Mamet as one of the most important young dramatists in America. He has written prolifically since and in 1984 he won the Pulitzer Drama Prize for *Glengarry Glen Ross*.

Mamet is known as a writer of character and master of American idiom and situation rather than plot. His dialogue is thoroughly naturalistic, his characters unerringly authentic whether they are Chicago gangsters, estate agents or office secretaries. His vision of America is one of a place that is rootless, restless and overwhelmed by capitalist exploitation. In the worlds he creates, the bonds of friendship, love and family are paramount. *Newsweek* described Mamet as 'a cosmic eavesdropper' and commented: 'Mamet has heard the ultimate Muzak ... People yammering at one another and not connecting.' Mamet himself says: 'All my plays attempt to bring out the poetry in the plain, everyday language people use.'

His most controversial play is *Oleanna*, an equivocal piece about sexual politics. A distressed female student comes to see her male lecturer, who attempts to comfort her and is then accused by her of sexual harassment. There are only two characters – although a telephone which interrupts the main action of the play almost counts as a third – and Mamet constantly challenges the assumptions the audience is making as the play goes along, so that sympathy shifts from one character to the other and back again. Some theatregoers, however, failed to register the play's subtlety: regarding it as a welcome backlash against feminism, they stood to applaud the lecturer. Reports of these scenes naturally caught the public's attention, and the play attracted large audiences in both the USA and in the UK, where it was directed by **Harold Pinter**.

In 1978 Mamet became associate artistic director at the Goodman Theater, Chicago, and has since written the screenplays of several successful films, most notably *The Postman Always Rings Twice*, *The Verdict* and *The Untouchables*. He has also directed two films, *House of Games* and *Things Change*, and says of his experiences in Hollywood: 'you find yourself dealing with characters you would not normally want to be associated with'. He does not intend to give up playwriting for screenwriting and plans to write 'more classically structured' plays in the future. He has started a theatre company in Boston, USA, with his wife, the actress Lindsay Crouse.

Plays

Lakeboat 1970 [pub. 1981]; *Duck Variations* 1972 [with *Sexual Perversity in Chicago* 1974; pub. 1978]; *The Poet and the Rent* 1974 [pub. 1986]; *Squirrels* 1974 [pub. 1982]; *American Buffalo* 1975 [pub. 1977]; *Reunion* 1976 [with *Dark Pony* 1977; pub. 1979]; *A Life in the Theatre* 1977 [pub. 1978]; *The Water Engine* 1977 [with *Mr Happiness* 1978; pub. 1978]; *The Woods* 1977 [pub. 1979]; *Lone Canoe* 1979; *The Sanctity of Marriage* 1979; *The Postman Always Rings Twice* (f) 1981; *Short Plays and Monologues* [pub. 1981]; *Esmond* 1982 [pub. 1983]; *The Verdict* (f) 1982; *The Disappearance of the Jews* 1983; *Glengarry Glen Ross* 1983 [pub. 1984]; *The Cherry Orchard* [from Chekhov's play] 1985 [pub. 1987]; *Dramatic Sketches and Monologues* [pub. 1985]; *The Shawl* [with *Prairie du chien*] 1985; *The Spanish Prisoner* 1985; *Vint* 1985 [pub. 1986]; *House of Games* (f) 1987; *The Untouchables* (f) 1987; *Speed-the-Plow* 1988; *Things Change* (f) 1988; *Oleanna* 1993; *Vanya on 42nd Street* (f) [from Chekhov's play *Uncle Vanya*] 1994

For Children

Three Children's Plays [pub. 1986]; *The Owl* 1987

Non-Fiction

Writing in Restaurants 1986; *Some Freaks* 1989

Biography

David Mamet by Dennis Carroll 1987

Frederic MANNING 1882–1935

Manning was born in Sydney, the son of a solicitor of Irish Catholic stock who served four terms as the city's mayor. The chronic asthma which was to circumscribe his life meant that he was largely educated at home. In 1898 he came to England with one of his tutors and settled in the Lincolnshire village of Edenham with the intention of entering literary life, supported by his parents. He began writing poetry and was introduced into literary and artistic circles, meeting in 1909 **Ezra Pound**, with whom he embarked upon a troubled but enduring friendship. There was some suggestion that the two men should found a magazine, publishing the works of the Imagists and others, but this came to nothing.

Manning was something of an aesthete and in 1909 he published *Scenes and Portraits*, a collection of prose pieces largely on historical themes, which was well received, and was followed in 1910 by a volume of *Poems*. He began work on 'The Golden Coach', a 'romance' set in France during the reign of Louis XIV, and became a reviewer for the *Spectator*.

In October 1915 Manning enlisted as a private in the King's Shropshire Light Infantry and served on the Somme. The fruits of his experience first appeared in 1917 in *Eidola*, a volume of verse which included several highly wrought and impressive war poems. That same year he gained a commission and transferred to the Royal Irish Regiment. but while training in Ireland he was diagnosed as suffering from shellshock and did not return to the war zone. After the war, Manning continued to pursue the life of a man of letters, contributing articles and reviews to the *Chapbook*, the *Little Review* and the *Criterion*. His reputation was made, however, with his war novel, *The Middle Parts of Fortune*, which was published in a limited edition of 520 copies for subscribers only in 1929, and subsequently (with the soldiers' language considerably modified) as *Her Privates We*. The author was given as 'Private 19022', Manning's regimental number, but most people penetrated this disguise. The unexpurgated version of the novel was not republished until 1977.

He spent the early 1930s travelling in Europe in an attempt to find relief for his deteriorating health. Surrounded by oxygen cylinders, he struggled to complete 'The Golden Coach', but died of pneumonia before completing more than 100 pages.

Fiction

The Middle Parts of Fortune 1929 [rev. as *Her Privates We* as 'Private 19022' 1930]

Poetry

Poems 1910; *Eidola* 1917

Non-Fiction

Scenes and Portraits 1909 [rev. 1930]

Olivia (Mary) MANNING 1908–1980

Confusion has surrounded Manning's exact date of birth – 1911, 1915 and 1917 were also contenders – but the earliest date is now established as correct. She was born in Portsmouth, the daughter of a retired naval officer and his part-American, part-Irish wife. The family struggled to keep up middle-class appearances on a small income, and the parents frequently quarrelled, leaving Manning to derive most solace from her relationship with her younger brother, Oliver, who was later killed in the Second World War. (Their parents were also called Oliver and Olivia, leading to some confusion). Several of her novels and stories derive from this family background. After attending

Portsmouth Grammar School, she went to the town's technical college to study art, but she could not afford to remain there, and so embarked upon a career as a writer, submitting romantic fiction to various magazines under the pseudonym Jacob Morrow. She also worked for a while as a typist in an architect's office.

At the age of twenty-three, she moved to London, where she got a job typing delivery lists for the van drivers at Peter Jones department store. She lived almost at subsistence level, and was once taken to hospital suffering from malnutrition; she would suffer from poor health for the remainder of her life. After being mistaken for someone with a similar name who had had an exhibition in Bond Street, she was promoted to painting furniture for the store, a period in her life recalled in her novel *The Doves of Venus* (which contains in the character of Nancy Claypole a portrait of **Stevie Smith**, who was to be bridesmaid at Manning's wedding). Various other short-term jobs followed, such as working at the Medici Society and reading scripts for Metro-Goldwyn-Mayer.

Her first novel, *The Wind Changes* (1937), was set in Ireland, where she had spent much of her youth, but her career as a novelist was interrupted by the Second World War. Shortly before war broke out, she signed up as an ambulance driver, despite the fact that she had never learned to drive. Before this was discovered, however, she married R.D. ('Reggie') Smith, and within days was travelling with him to Bucharest, where he had a job as a British Council lecturer. They arrived in Romania as war was declared, and there followed six years of wartime peregrinations with her husband, which took them to Athens, Egypt and, finally, Jerusalem, usually one step ahead of the advancing Nazi forces. This wartime experience, and an unsparing but curiously moving and very funny analysis of her marriage, formed the basis of her novel sequence *Fortunes of War*, made up of two trilogies, *The Balkan Trilogy* and *The Levant Trilogy*. The sequence is unusual in that it views the war largely from the viewpoint of expatriate civilians, although in the second trilogy she performs a remarkable feat of empathy in showing the war in the desert through the eyes of a very young officer. These novels are undoubtedly Manning's finest achievements and are generally regarded as monuments of post-war English fiction. Under the sequence title, they were adapted by **Alan Plater** for a fine, if somewhat compressed, BBC television series in 1987.

After the war, Manning and Smith returned to London, where he became a BBC drama producer, and later a professor at the New University of Ulster. She settled to 'the necessarily unadventurous life of a creative writer', although her marriage was enlivened by infidelities on both sides. Manning had affairs with both **Henry Green** and **William Gerhardie**, and claimed that **Anthony Burgess** had proposed to her shortly after the death of his first wife, but her principal lover was a married doctor called Jerry Slattery, who predeceased her. Smith was also a noted philanderer, but despite this, and the couple's widely divergent political views (Smith was a communist, Manning a Tory), they remained true to each other in their fashion, living in St John's Wood with the Siamese and Burmese cats to which she was devoted. She was known as a generous but haphazard party-giver and was famous for her sometimes acerbic interest in other writers. Although inclined to slip away to the pub during his wife's parties, Smith was in other ways supportive, reading and commenting on everything she wrote. Manning wrote a series of pieces about him for *Punch*, which were subsequently collected as *My Husband Cartwright*.

She wrote seven other novels beside the trilogies, but often complained, quite rightly, that she did not receive her due from the reviewers. She was not above telephoning literary editors to harangue them for publishing critical notices of her books. Amongst the best of these is the comic and touching *School for Love*, which is about a young orphaned boy marooned in Jerusalem in 1945. Her excellent short stories were collected in two volumes, and she also wrote several works of non-fiction about cats, Ireland and the explorer Stanley. She wrote *The Levant Trilogy* when she was mortally ill, and the final volume was published posthumously. She died at Ryde on the Isle of Wight and left a large part of her estate to the Wood Green Animal Sanctuary.

Fiction

The Wind Changes 1937; *Growing Up* (s) 1948; *Artist Among the Missing* 1949; *School for Love* 1951; *A Different Face* 1953; *The Doves of Venus* 1955; *The Balkan Trilogy* 1981: *The Great Fortune* 1960, *The Spoilt City* 1962, *Friends and Heroes* 1965; *A Romantic Hero* (s) 1967; *The Play Room* 1969; *The Rain Forest* 1974; *The Levant Trilogy* 1982: *The Danger Tree* 1977, *The Battle Lost and Won* 1978, *The Sum of Things* 1980

Non-Fiction

The Remarkable Expedition 1947; *The Dreaming Shore* 1950; *My Husband Cartwright* 1956; *Extraordinary Cats* 1967

Katherine (i.e. Kathleen) MANSFIELD (Beauchamp) 1888–1923

Born in Wellington, New Zealand, the daughter of Harold Beauchamp, a banker and industrialist, Mansfield spent her early years in the nearby village of Karori, and an idyllically rendered rural New Zealand forms the background to many of her best short stories. In 1903 she was sent to England and was a music student at Queen's College, Harley Street. She returned to New Zealand briefly but, chafing at provincialism, she persuaded her father to send her back to England in 1908 with an allowance of £100 a year.

She began to write short stories, but at first experienced the proverbial poor fortune of the literary aspirant. In 1909 she married a music teacher, but left him after a few days. She toured for a while as an extra in opera, became pregnant by another man, and, to avoid scandal, went to a pension in Bavaria, where she had a stillborn baby. Her German experience was utilised in her first collection of short stories, *In A German Pension* (1911), several of the pieces in which had previously been published in A.R. Orage's influential magazine the *New Age*.

In 1911 she also met the rising literary man John Middleton Murry, who became first a tenant in her flat, then her lover ('Why don't you make me your mistress?' she challenged him after a long evening of literary discussion), and from 1918 her husband. Murry helped her by publishing her stories in his magazines *Rhythm* and the *Blue Review* and later giving her reviewing work for the *Athenaeum*, although it was not until 1920, with her second collection, *Bliss and Other Stories*, that her reputation was fully established.

From early wartime years she was diagnosed as consumptive and after that spent much time in the south of France and Switzerland for her health. The death in action of her beloved brother, 'Chummie', shattered her, but also released a flood of her finest writing about New Zealand. During the war she and Murry became closely associated with **D.H. Lawrence** and his wife Frieda, until Lawrence turned against them. Her health became inexorably worse and in October 1922 she entered the Gurdjieff Institute at Fontainebleau, where controversial faith-healing was practised. She died there of a pulmonary haemorrhage at the age of thirty-four.

Mansfield is among the select band of writers to have made their reputation almost entirely through the short story (although several volumes of letters and her journals have been published). In this field, she influenced the English short story strongly against the well-made tale of Maupassant towards the more impressionistic treatment of Chekhov. She is usually accounted a virtuoso of the genre, although some critics – notably **Virginia Woolf** and **Frank O'Connor** – have considered her overpraised.

Fiction

In a German Pension (s) 1911; *Bliss and Other Stories* (s) 1920; *The Garden-Party and Other Stories* (s) 1922; *The Dove's Nest and Other Stories* (s) 1923; *Something Childish and Other Stories* (s) 1924; *The Short Stories of Katherine Mansfield* (s) 1937; *Collected Stories* (s) 1945 [rev. 1981); *Complete Short Stories of Katherine Mansfield* (s) 1974

Non-Fiction

The Journal of Katherine Mansfield [ed. J. Middleton Murray] 1927; *The Letters of Katherine Mansfield* [ed. J. Middleton Murry] 1928; *Novels and Novelists. Reviews Reprinted from 'The Athenaeum'* [ed. J. Middleton Murry] 1930; *Katherine Mansfield's Letters to John Middleton Murry, 1913–1922* [ed. J. Middleton Murry] 1951; *Undiscovered Country. The New Zealand Letters of Katherine Mansfield* [ed. I.A. Gordon] 1974; *Passionate Pilgrimage. Katherine Mansfield's Letters to John Middleton Murry from the South of France, 1915–1920* 1976; *Collected Letters of Katherine Mansfield* [ed. M. Scott, V. O'Sullivan] 3 vols 1984–93

Poetry

Poems 1923

Translations

Gorky *Reminiscences of Leonid Andreyev* [with S.S. Koteliansky] 1928

Collections

Publications in Australia, 1907–1909 [ed J.E. Stone] 1977

Biography

Katherine Mansfield: A Secret Life by Claire Tomalin 1987

Hilary (Mary) MANTEL 1952–

Mantel was born of Irish Catholic parents in Hadfield, Derbyshire, where her father was an engineer. When she was eleven, her family moved to Cheshire; an amusing account of Mantel's attempts to lose her Derbyshire accent is given in 'Learning to Talk', published in the *London Magazine* (Vol. 27, 1 and 2). She was educated at the Convent of the Nativity, Romiley, where she became head girl and joined the Young Communist League. She subsequently attended the London School of Economics, where she studied law, resigned from the YCL and worked in student politics for the Labour Party. In 1972 she married Gerald McEwen, a geologist, and transferred to Sheffield University, where she graduated as a Bachelor of Jurisprudence.

She spent ten years as a social-work assistant in a geriatric hospital, an experience which inspired her first novel, *Every Day Is Mother's Day* (1985), a black comedy about a mentally deranged young woman, Muriel Axon, and the social services. Her second novel, *Vacant Possession*, is an equally dark and funny sequel, in which Muriel has grown up and become even more disruptive. Between 1974 and 1977 Mantel worked as a saleswoman while researching a novel about the French Revolution, which eventually ran to 220,000 words. By this time she had 'given up active for contemplative politics'.

In 1977 she went to Botswana, where her husband was scientific editor of *Geological Survey*. She taught English there, but became seriously ill and returned to England in 1979, where she failed to find a publisher for her French Revolution novel. In 1982, her husband got a job in Saudi Arabia at the Ministry of Petroleum and Mineral Resources. While in Jeddah, Mantel wrote *Every Day Is Mother's Day*, and the city is the setting for her third novel, *Eight Months on Ghazzah Street*, a witty thriller in which a young woman goes to Jeddah and attempts to fight against the role Islam forces upon women. For years later, she returned to England to live in Berkshire, and in 1987 she won the first Shiva Naipaul Prize, awarded by the *Spectator* for travel writing. She subsequently became the magazine's film critic, and she has also written widely for other magazines and newspapers in both England and America.

Fludd, which is perhaps her best novel to date, draws upon childhood memories. Set in a Northern mill village in 1956, it is a considerable departure from earlier works, a remarkable account of a Catholic community disrupted by the arrival of the eponymous Fludd, who may be a curate, an episcopal spy, an alchemist, a saint or the Devil. Like all her books, it is beautifully written, and it won the Winifred Holtby, Southern Arts and Cheltenham Festival of Literature Prizes, placing her in the front rank of contemporary British novelists. It was followed by a revised version of her novel about the French Revolution, published as *A Place of Greater Safety*. Once again, this was quite unlike anything else she had written: a long, highly detailed historical novel based upon contemporary accounts, with an enormous cast of characters, but centred principally upon Danton, Robespierre and Desmoulins. It won the *Sunday Express* Book of the Year Award. The question of good and evil, which has often been a concern of this writer, came to the fore in *A Change of Climate*, in which a family's life in present-day Norfolk is affected by an act of almost literally unspeakable horror, which took place in the Kalahari in the 1950s and has been buried ever since. *An Experiment in Love* returns to the world of Mantel's childhood, tracing the relationship between two girls (one the child of refugee parents, whose past experiences once again leave a dark stain upon the present) from their early friendship at a school in Lancashire, through university in London during the 1960s, to a terrible nemesis. The influence of **Muriel Spark**, with whom she shares several concerns and qualities, is more apparent (and acknowledged) here than ever before. Mantel has also written a number of fine short stories which have yet to be collected.

Fiction

Every Day Is Mother's Day 1985; *Vacant Possession* 1986; *Eight Months on Ghazzah Street* 1988; *Fludd* 1989; *A Place of Greater Safety* 1992; *A Change of Climate* 1994; *An Experiment in Love* 1995

Kamala (Purnaiya) MARKANDAYA (Taylor) 1924–

Markandaya was born into a Brahmin family of Mysore in south India; her maternal grandfather had been a judge, while her father's family could trace its ancestry back to the seventeenth century. Her father was a rail transport officer whose work took him abroad, so she travelled as a child both in India and Europe. She attended various schools, and entered Madras University when aged sixteen, but left after three years without a degree to work on a small weekly newspaper. This soon folded, and she then spent time living in an Indian village, experience that has borne fruit in her novels which often explore such a life.

She did some freelance writing in Madras and Bombay in the later 1940s, but since her marriage to an Englishman in 1948 she has lived in England; she has one daughter. She tried and failed to become a Fleet Street journalist, and worked in various 'dull but amiable jobs', such as proof-reader and secretary, before the third novel she wrote, *Nectar in a Sieve*, became her first published novel in 1954.

Her novels, which cover a wide social range of rich and poor, are for the most part set in India (although *The Nowhere Man* deals with an elderly Brahmin suffering racism in a London suburb), and excel in the depiction of individual consciousness, the minutiae of social life and relations between Indians and English. Her critical reputation is high, both in India (where she has featured on an Indira Gandhi Open University course) and in other English-language countries, but she has not made the breakthrough into a wide audience. She published novels regularly into the 1970s, but has not to date published one since *Pleasure City* (1982). She has appeared occasionally on British television and undertaken adjudication for the Arts Council. She lives in London, but has continued to make frequent visits to her native country.

Fiction
Nectar in a Sieve 1954; *Some Inner Fury* 1955; *A Silence of Desire* 1960; *Possession* 1963; *A Handful of Rice* 1966; *The Coffer Dams* 1969; *The Nowhere Man* 1972; *Two Virgins* 1973; *The Golden Honeycomb* 1977; *Pleasure City* [US *Shalimar*] 1982

Robert MARKHAM

see Kingsley AMIS

Don(ald Robert Perry) MARQUIS 1878–1937

Of Scottish-Irish ancestors who had emigrated to Virginia in the eighteenth century, Marquis was born in Walnut, Illinois, and grew up there, the son of a country doctor. He attended a local high school and, very briefly, Knox College, before undertaking a variety of occupations such as schoolteaching, clerking and hay baling. Soon, however, he gravitated towards the (largely unrecorded) world of American country newspapers, and in 1900 went to Washington DC, where for a period he was simultaneously a student at the Corcoran School of Art, a clerk in the Census Bureau and a reporter for the *Washington Times*. (He was also briefly on the stage.)

His first real break as a journalist came after he had moved to Atlanta, when he was taken up by Joel Chandler Harris – in those days very well known for his 'Uncle Remus' stories – who made him associate editor of *Uncle Remus's Home Magazine*, which had been founded in 1907: now Marquis's writing expanded to book reviews and parodies. He married in 1909, and had two children. Also in 1909 he moved to New York, hoping to become a columnist there and, after some freelance years, was able to begin the 'Sun Dial' column of the *New York Evening Sun* in 1913. He had published his first novel, *Danny's Own Story*, a book heavily influenced by Mark Twain's *Huckleberry Finn*, in 1912, and was to follow it with several other novels as well as short stories, plays and poetry. His well-known newspaper column – which from 1922 was called 'The Lantern' and appeared in the *Herald Tribune* – incorporated material which would nowadays be out of place, such as poetry and parody. It served as the basis for books of essays, and led to the creation of characters such as Hermione and the Old Soak (who themselves expanded into books and plays), very well known in the 1920s, but now forgotten. Two characters have, however, triumphantly survived: archy the cockroach, who wrote philosophical free-verse poems on Marquis's typewriter, using only lower-case letters because he could not work the shift-key, and the subject of much of archy's poetry, mehitabel the alley cat. Their adventures were collected into three books, *archy and mehitabel*, *archys life of mehitabel* and *archy does his part*, which are acknowledged as classics both comic and thoughtful. (The titles were originally published with upper-case letters, subsequently not.)

Despite his success, Marquis's later life was plagued by tragedy: both his children died young, as did his first wife, and even his second wife, the actress Marjorie Vonnegut, whom he married in 1926, predeceased him. He himself retired from newspaper columns in 1925, but was forced to spend unhappy periods as a scriptwriter for Hollywood and in his last years was incapacitated by a series of strokes and a final cerebral haemorrhage. 'It would be one on me if I should be remembered longest for creating a cockroach character,' he once said: despite a powerful, posthumous, autobiographical novel, *Sons of the Puritans*, and *The Dark Hours*, his play about the Crucifixion, this gloomy prophecy has come true.

Fiction
Danny's Own Story 1912; *The Cruise of Jasper B* 1916; *Hermione and Her Little Group of Serious Thinkers* 1916; *Prefaces* 1919; *The Old Soak, and Hail and Farewell* 1921; *Carter and Other People* 1921; *The Revolt of the Oyster* 1922; *The Old Soak's History of the World* 1924; *Pandora Lifts the Lid* [with C.D. Morley] 1924; *The Almost Perfect State* 1927; *When Turtles Sing* 1928; *A Variety of People* 1929; *Off the Arm* 1930; *Chapters for the Orthodox* 1934; *Sun Dial Time* 1936; *Sons of the Puritans* [unfinished] 1939

Plays
The Old Soak [from his story] 1922 [pub. 1926]; *The Dark Hours* 1932 [pub. 1924]; *Words and Thoughts* [pub. 1924]; *Out of the Sea* 1927; *Everything's Jake* 1930 [np]; *Master of the Revels* [pub. 1934]

Poetry
Dreams and Dust 1915; *Noah an' Jonah an' Cap'n John Smith* 1921; *Poems and Portraits* 1922; *Sonnets to a Red-Haired Lady (by a Gentleman with a Blue Beard) and Famous Love Affairs* 1922; *The Awakening and Other Poems* 1924; *archy and mehitabel* 1927; *Love Sonnets of a Caveman*

and Other Verses 1928; *archys life of mehitabel* 1933; *archy does his part* 1935

Biography

O Rare Don Marquis by Edward Anthony 1962

(Edith) Ngaio MARSH 1899–1982

Born in Christchurch, New Zealand, the daughter of an official of the Bank of New Zealand and granddaughter of one of the first English settlers (an aristocrat), Marsh was educated at St Margaret's College, Christchurch, and Christchurch University College School of Art. She then became an actress for two years and a writer and producer for New Zealand repertory companies for five. In 1928 she came to London, where she ran an interior decorating business in Knightsbridge in partnership with the Honourable Mrs Tahu Rhodes, one of an eccentric aristocratic family with whom she had stayed on her arrival in England, and who served as a model for the family in one of her best thrillers, *Surfeit of Lampreys*.

In 1932 her mother's sudden mortal illness forced her to return to New Zealand; as she departed, she left her detective novel, *A Man Lay Dead*, with a literary agent, and it was published in 1934, the first of some thirty witty and well-written detective stories featuring her gentleman detective, Superintendent Roderick Alleyn. Using characteristic settings of the theatre, art and high society, they helped to bring a new literary status to the detective genre, and much later, in the 1990s, were successfully adapted for television.

During the 1930s and 1940s she kept house for her widowed father and during the Second World War drove a hospital bus for the Red Cross. From 1944 she was a producer with D.D. O'Connor's Theatre Management, and became a key figure in keeping theatre alive in New Zealand and developing its links with Britain. In her own country, she was more famous for this work than for her writing, and it was largely for theatre work that she was created a Dame Commander of the Order of the British Empire in 1966. In later life she visited Britain often (and many more of her novels are set there than in New Zealand), but continued to live in her father's house overlooking Christchurch. She was unmarried.

Fiction

A Man Lay Dead 1934; *Enter a Murderer* 1935; *Nursing-Home Murder* [with H. Jellett] 1936; *Death in Ecstasy* 1937; *Vintage Murder* 1937; *Artists in Crime* 1938; *Death in a White Tie* 1938; *Overture to Death* 1939; *Death at the Bar* 1940; *Surfeit of Lampreys* 1941; *Death and the Dancing Footman* 1942; *Colour Scheme* 1943; *Died in the Wool* 1945; *Final Curtain* 1947; *Swing, Brother, Swing* 1948; *Opening Night* 1951; *Spinsters in Jeopardy* 1953; *Scales of Justice* 1954; *Off with His Head* 1957; *Singing in the Shrouds* 1959; *False Scent* 1960; *Hand in Glove* 1962; *Dead Water* 1964; *Death at the Dolphin* 1967; *Clutch of Constables* 1968; *When in Rome* 1970; *Tied Up in Tinsel* 1972; *Black as He's Painted* 1974; *Last Ditch* 1977; *Grave Mistake* 1978; *Photo Finish* 1980; *Light Thickens* 1982

Plays

A Unicorn for Christmas 1962; *Murder Sails at Midnight* [from her novel *Singing in the Shrouds*] 1973

Non-Fiction

Black Beech and Honeydew 1965

Biography

Ngaio Marsh by Margaret Lewis 1991

Adam MARS-JONES 1954–

The son of a well-known high-court judge, Mars-Jones was born in London and educated at Westminster School, Cambridge, where he read English, and the University of Virginia, where he was awarded the Benjamin C. Moomaw Prize for Oratory. A story inspired by his time in America, 'Trout Day (by pumpkin light)', was published in an anthology of 'stories by gay men', *Cracks in the Image* (1981). Its title echoed **J.D. Salinger**, but its style and tone were entirely Mars-Jones's own: witty, cool and sophisticated. Although the story was not particularly substantial, it cut an elegant dash amidst the rest of the anthology's down-at-heel reportage.

The previous year, he had published a rather more interesting story in *Quarto*, a distinguished but short-lived British literary periodical. 'Lantern Lecture' was based on the life of Philip Yorke, the eccentric master of Erddig Hall, near Wrexham in Wales, and lent its title to Mars-Jones's first volume. Published in 1981, this collection of three stories was one of the most remarkable debuts of recent years and earned its author a place amongst the 'Best of Young British Novelists' in an influential Book Marketing Council promotion in 1983. He was represented in the accompanying *Granta* anthology by 'Trout Day by Pumpkin Light' (as his story was now titled), and some commentators thought his inclusion was based on an insufficiently large body of work. The other stories in

Lantern Lecture were 'Hoosh-Mi', a fantasy in which Queen Elizabeth II develops rabies after being bitten by a corgi, and 'Bathpool Park', a chilling account of the crimes and trial in the 1970s of Donald Neilson ('the Black Panther'), a famous case tried by Mars-Jones's father. All three stories were based on fact, but developed with the freedom of fiction, in which device their author resembles **Bruce Chatwin**. The volume won him a Somerset Maugham Award.

In 1983, he edited *Mae West Is Dead*, an anthology of 'recent lesbian and gay fiction' distinguished by the quality of the stories included and by the fact that it was published by a mainstream publisher, Faber. It contained stories by, amongst others, **James Purdy**, **David Cook**, **John Bowen**, **Sara Maitland**, Simon Burt, Ann Leaton and **Jane Rule**, and Mars-Jones contributed a long introductory essay, 'Gay Fiction and the Reading Public', which analysed the genre of gay popular fiction. Four years later, he collaborated with **Edmund White** on a volume of 'Stories from a Crisis', *The Darker Proof*. The crisis in question was the Aids epidemic, which had personally affected both writers (each had lovers who later died as a result of complications arising from the syndrome). Mars-Jones contributed four stories, White two, although the book was later revised to include additional stories from both writers. The word 'Aids' was not mentioned in any of Mars-Jones's stories: 'The problems attaching to the subject turned out overwhelmingly to be attached to the name,' he later commented. 'By suppressing that, I was suddenly able to write about the epidemic.' He subsequently produced an entire volume of short stories concerned with HIV and Aids, *Monopolies of Loss*. Some critics had previously admired his undoubted cleverness as a writer but had questioned whether he had any heart. *Monopolies of Loss* answered this: a grim, funny, touching, eloquent and brave collection of stories that constituted, with the American Allen Barnett's *The Body and Its Dangers* (1990), the most outstanding literary response to Aids thus far. In particular, the story 'Baby Clutch' (which, like much of his work, originally appeared in *Granta*) is a small masterpiece, demonstrating that his concern is as much with language as with the ostensible subject-matter.

Books about Aids now form an identifiable sub-genre of gay fiction, and Mars-Jones has been hailed by the *Guardian* as 'the **Wilfred Owen** of this new long-drawn and deadly trench warfare'. His long-awaited first novel, *The Waters of Thirst*, still featured a homosexual protagonist, but one suffering from a kidney disease rather than the expected virus. Some commentators thought that this constituted a metaphor rather than a new departure, but the novel (which is in the form of a monologue) was highly praised, not least for its black humour – notably a sustained aria about the dangers of motor-cycling, which the narrator hopes will result in a kidney donor being found for him.

Mars-Jones has also worked as a journalist, reviewing books for newspapers and the *Times Literary Supplement*. His 'CounterBlast' pamphlet, *Venus Envy* ('On the WOMB and the BOMB'), is a characteristically stylish and salutary dissection of **Martin Amis**'s *Einstein's Monsters* (1987) and **Ian McEwan**'s *The Child in Time* (1987). He has for some years been the film critic of the *Independent*.

Fiction

Lantern Lecture (s) [US *Fabrications*] 1981; *The Darker Proof* (s) [with E. White] 1987 [rev. 1988]; *Monopolies of Loss* (s) 1992; *The Waters of Thirst* 1993

Non-Fiction

Venus Envy 1990

Edited

Mae West Is Dead: Recent Lesbian and Gay Fiction (s) 1983

John (Edward) MASEFIELD 1878–1967

Poet Laureate from 1930 until his death, Masefield was also a prolific author of novels, plays and numerous other volumes. One of the six children of a solicitor, he was born in Ledbury, Herefordshire, but the idyllic childhood which he described as 'living in Paradise' (and recalled in his last book, the fragmentary *Grace Before Ploughing*) was shattered when his mother died in 1884 and his father suffered financial difficulties and became insane. He was subsequently brought up by unsympathetic relatives and left school at thirteen to join the Merchant Navy. His career at sea did not last long, owing to poor health, but it inspired much of his best-known work. He deserted ship in New York in 1895, lived as a vagrant, then had a series of jobs including that of a 'mistake-catcher' in a carpet factory. He returned to London in 1897 and became a bank clerk, subsequently working as a journalist for the Manchester *Guardian*.

He achieved early prominence in 1902 with his first volume of poems, *Salt Water Ballads*. The following year he married, and he gained wide recognition with his 1905 volume of short stories, *A Mainsail Haul*. Inspired by his friendship with **W.B. Yeats**, he also began writing plays, the first of which, *The Tragedy of Nan* produced in 1908, remains his best-known work for the theatre. Other poems, plays and works of non-fiction followed, including a testament of *My Faith in Women's Suffrage* in 1910. The following year his long narrative poem about the salvation of the depraved Saul Kane, *The Everlasting Mercy*, caused a sensation with its supposed blasphemy, its attacks upon the clergy and its use of colloquialisms. **Siegfried Sassoon** borrowed the protagonist's name when he pseudonymously published his parodic *The Daffodil Murderer* in 1913, its title derived from Masefield's *The Daffodil Fields* of the same year – a volume in which he depicted the harsh realities of country life.

His reputation stood high when the First World War came in 1914, and his work was particularly admired by the soldier-poets, not only Sassoon, but also **Robert Graves** and **Charles Sorley**, who felt that Masefield's poetry 'express[ed] the spirit of the age ... the upheaval of the masses of the population'. Although thirty-six, he volunteered for active service, but was rejected for health reasons. Nevertheless, in 1915 he went to France to serve with the Red Cross, an experience which was to have a profound effect upon him. He subsequently went to Gallipoli with the same organisation, and during the first half of 1916 embarked on a lecture tour in the USA, literary in essence but propagandist in intent. Heckled about the failure of the Dardanelles campaign, he returned to England and wrote *Gallipoli*, an almost mystical account of the engagement, which did not underplay the suffering of the troops, but was suffused with chivalric echoes and ideals. (Later, and privately, Masefield was to refer to 'that insane move on Gallipoli'.) He subsequently returned to France as an observer for British Military Intelligence and wrote *The Old Front Line* and *The Battle of the Somme*.

His long narrative poem *Reynard the Fox*, which was influenced by Chaucer, was published in 1919 and remains the work by which he is best known. His achievement as a poet was to use popular, folkloric language which broke the somewhat etiolated and upper-class traditions of Victorian poetry, reinvigorating an interest in verse throughout the English-speaking world. Such poems as 'Sea Fever', 'Up on the Downs' and 'Cargoes' have been frequently anthologised, but he was considered old-fashioned to the rising generation of poets and was omitted, for example, from **Michael Roberts**'s *Faber Book of Modern Verse* (1936), although he was the laureate and had been given the Order of Merit the previous year. He continued to produce poetry, novels, plays, criticism, books for children and non-fiction works throughout his long life. Notable amongst his post-First-World-War works are the two excellent children's novels *The Midnight Folk* and *The Box of Delights*, which have survived far better than once popular novels for adults such as *Sard Harker*. At his memorial service in Westminster Abbey, Graves said that 'the fierce flame of true poetry had truly burned' in Masefield, and that 'he never lost what is the supreme poetic quality: an unselfish love for his fellow men'. He remains a much loved poet of a particularly English kind, who celebrated rural and seafaring life without sentimentality and, as the century ends, his reputation seems likely to endure.

Poetry

Salt Water Ballads 1902; *Ballads* 1903; *Ballads and Poems* 1910; *The Everlasting Mercy and The Widow in the Bye Street* 1911; *The Story of a Roundhouse and Other Poems* 1912 [rev. 1913]; *The Daffodil Fields* 1913; *Dauber* 1913; *Good Friday and Other Poems* 1916; *Philip the King and Other Poems* 1916; *Salt-Water Poems and Ballads* 1916; *Sonnets and Poems* 1916; *The Cold Cotswolds* 1917; *Lollingdon Downs and Other Poems* 1917; *Rosas* 1918; *Reynard the Fox* 1919 [rev. 1946]; *Animula* 1920; *Enslaved and Other Poems* 1920; *Right Royal* 1920; *King Cole* 1921; *The Dream* 1922; *Collected Poems* 1923; *King Cole and Other Poems* 1923; *Midsummer Night and Other Tales in Verse* 1928; *Oxford Recitations* 1928; *The Wanderer of Liverpool* 1930; *Minnie Maylow's Story and Other Tales and Scenes* 1931; *A Tale of Troy* 1932; *Poems* 1935; *A Letter from Pontus and Other Verse* 1936; *The Country Scene* 1937; *Tribute to Ballet* 1938; *Some Verses of Some Germans* 1939; *Gautama the Enlightened and Other Verse* 1941; *Land Workers* 1942; *Natalie Masie and Pavilastukay* 1942; *A Generation Risen* 1943; *Wonderings* 1943; *On the Hill* 1949; *In Praise of Nurses* 1950; *Poems* 1953; *Bluebells and Other Verse* 1961; *The Western Hudson Shore* 1962; *Old Rainger and Other Verse* 1965

Plays

The Tragedy of Nan and Other Plays [pub. 1909]; *The Tragedy of Pompey the Great* 1910; *The Faithful* 1915; *The Locked Chest and The Sweeps of Ninety-Eight* [pub. 1916]; *Melloney Holtspur* 1922; *A King's Daughter* 1923; *The Trial of Jesus* 1925; *Tristan and Isolt* 1927; *The Coming of Christ* 1928; *Easter* 1929; *End and Beginning* 1933; *A Play of St George* 1948

Fiction

A Mainsail Haul (s) 1905 [rev. 1913, 1954]; *A Tarpaulin Muster* (s) 1907; *Captain Margaret* 1908; *Multitude and Solitude* 1909; *The Street of Today* 1911; *The Taking of Helen* 1923; *Sard Harker* 1924; *Odtaa* 1926; *The Hawbucks* 1929; *The Bird of Dawning* 1933; *The Taking of the Gry* 1934; *Victorious Troy* 1935; *Eggs and Baker* 1936; *The Square Peg* 1937; *Dead Ned* 1938; *Live and*

Kicking Ned 1939; *Basilissa* 1940; *Conquer* 1941; *Badon Parchments* 1947

Non-Fiction

Sea Life in Nelson's Time 1905; *On the Spanish Main* 1906; *My Faith in Women's Suffrage* 1910; *William Shakespeare* 1911; *Chaucer* 1913; *John Synge* 1915; *Gallipoli* 1916; *The Old Front Line* 1917; *The Battle of the Somme* 1919; *Shakespeare and Spiritual Life* 1924; *Poetry* 1931; *The Conway from Her Foundation to the Present Day* 1933; *Some Memories of W.B. Yeats* 1940; *The Nine Days' Wonder* 1941; *In the Mill* 1941; *I Want! I Want!* 1944; *New Chum. Experiences on the Training Ship 'Conway'* 1944; *A Macbeth Production* 1945; *Thanks Before Going: Notes on Some of the Original Poems of D.G. Rossetti* 1946; *St Katherine of Ledbury and Other Ledbury Papers* 1951; *So Long to Learn* **(a)** 1952; *Words on the Anniversary of the Birthday of William Blake* 1957; *Grace Before Ploughing* **(a)** 1966; *Letters to Reyna* [ed. W. Buchan] 1983; *John Masefield's Letters from the Front 1915–1917* [ed. P. Vansittart] 1984; *Letters to Margaret Bridges 1915–1919* [ed. D. Stanford] 1984

Translations

H. Wiers-Jenssen *Anne Pedersdotter* 1917; Racine *Esther* 1922

For Children

A Book of Discoveries 1910; *Lost Endeavour* 1910; *Martin Hyde, the Duke's Messenger* 1910; *Jim Davis* 1911; *The Midnight Folk* 1927; *The Box of Delights* 1935

Edited

Lyricists of the Restoration from Sir Edward Sherburne to William Congreve 1905; *Dampier's Voyages* 1906; *Essays Moral and Polite, 1660–1714* [with C. Masefield] 1906; *Lyrics of Ben Jonson, Beaumont and Fletcher* 1906; *The Poems of Robert Herrick* 1906; *A Sailor's Garland* 1906; *An English Prose Miscellany* 1907; *Hakluyt's Voyages* 1907; *Defoe. Selections* 1909

Biography

John Masefield by Constance Babbington Smith 1978

A(lfred) E(dward) W(oodley) MASON 1865–1948

Mason was born in Dulwich, south London, and educated at Dulwich College and at Trinity College, Oxford, where he became involved in the University Dramatic Society (OUDS). While taking part in a production of Euripides' *Alcestis*, he became friendly with Oscar Wilde, who was later to advise him about his first novel. (Novels by Mason were amongst the books Wilde asked to be sent to him while he languished in Reading Gaol.) After leaving university, he embarked upon a career as an actor, touring the provinces. Although he gave up acting when he became a writer, he later wrote several plays, as well as a biography of the actor-manager George Alexander, who was responsible for putting on Wilde's later plays at the St James's Theatre, London.

In 1906 Mason was elected Member of Parliament for Coventry, a job he held for three years. During the First World War, he served as a captain in the Manchester Regiment, later becoming a major in the secret service, the activities of which furnished him with a number of ideas for his novels. The first of these, *A Romance of Wastdale*, was published in 1895, and with *The Courtship of Morrice Buckler*, published the following year, he established himself as a bestselling author. His work falls into two categories: historical and contemporary romances and detective stories. His best-known novel is *The Four Feathers*, a stirring tale of honour and heroism set in the Sudan campaign of the 1890s. Enormously popular, the book has been filmed four times (first in 1921, but most memorably in the Korda brothers' version of 1939 from a script by **R.C. Sheriff**), and adapted for television. Amongst his other historical novels, the best-known are *Fire over England*, which describes the defeat of the Spanish Armada, and *The Drum*, in which the British Army comes to the aid of a young Indian prince, whose uncle wants to take power. These too were made into popular films in the 1930s. Lord Curzon praised Mason for his faithful and atmospheric portrayals of India.

Rather more distinguished, and certainly more original, are Mason's detective stories, the first of which, *At the Villa Rose*, was published in 1910. Partly based on a real case (Mason often attended murder trials at the Old Bailey), it is a classic of the genre and introduced Inspector Hanaud of the Paris Sûreté and his friend Mr Ricardo. Julius Ricardo is a somewhat stuffy retired tea-broker from Mincing Lane and a man of considerable means. He spends much of his time in France, either at the casinos or at the great wine châteaux, and has a chauffeur-driven Rolls Royce, which M. Hanaud greatly envies. Hanaud is a vain Frenchman with a passion for, but shaky grasp of, English idiom ('It is not the sort of news you would write to the house'), and a somewhat obstreperous sense of humour, both of which pain his fastidious friend, from whose point of view the action of the novels is described. Hanaud is also a brilliant detective, forever one step ahead of Ricardo, who always thinks that he may have noticed some vital clue overlooked by the cocksure inspector. Apart from the fine plotting and the foreign locale, one of the chief pleasures of these novels is the relationship between the two men: the teasing but usually triumphant Hanaud, who has the appearance of 'a prosperous comedian', and the enthusiastic but distinctly amateur Ricardo. Indeed, Mason is far more interested in character than many lesser practitioners in the genre. Mason himself adapted *At the Villa Rose* for the stage, and it was filmed three times, as was *The House of the Arrow*.

Another striking element of the books is their savagery. In *The Prisoner in the Opal*, for example, a young woman is found not only murdered, but with her hand hacked off, having met her death during a satanic mass. Mason's villains are usually true embodiments of evil, believable fiends who glory in their work. Of his later historical novels, the most bizarre is undoubtedly *Musk and Amber*, written when he was in his late seventies, and featuring an aristocratic hero who, castrated as a boy, becomes a great opera star, then wreaks his revenge upon his mutilators.

Fiction

A Romance of Wastdale 1895; *The Courtship of Morrice Buckler* 1896; *Lawrence Clavering* 1897; *The Philanderers* 1897; *Miranda of the Balcony* 1899; *The Watchers* 1899; *Parson Kelly* [with A. Lang] 1900; *Clementina* 1901; *Ensign Knightley and Other Stories* (s) 1901; *The Four Feathers* 1902; *The Truants* 1904; *The Broken Road* 1907; *Running Water* 1907; *At the Villa Rose* 1910; *The Turnstile* 1912; *The Witness for the Defence* 1913; *The Four Corners of the World* (s) 1917; *The Summons* 1920; *The Winding Stair* 1923; *The House of the Arrow* 1924; *No Other Tiger* 1927; *The Prisoner in the Opal* 1929; *The Dean's Elbow* 1930; *The Three Gentlemen* 1932; *The Sapphire* 1933; *Dilemmas* (s) 1934; *They Wouldn't Be Chessmen* 1935; *Fire Over England* 1936; *The Drum* 1937; *Königsmark* 1938; *Musk and Amber* 1942; *The House in Lordship Lane* 1946

Plays

Blanche de Malétroit [from R.L. Stevenson's story] 1894; *Green Stockings* 1910; *The Witness for the Defence* [from his novel] 1913; *At the Villa Rose* [from his novel] 1928; *A Present from Margate* [with I. Hay] 1934

Non-Fiction

Sir George Alexander and the St James's Theatre 1935; *The Life of Francis Drake* 1941

Anita (Frances) MASON 1942–

An only child, Mason was born in Bristol; her father was an aeroplane engineer at Filton. She has said that her parents were surprised when she won a place to read English at St Hilda's College, Oxford, but is otherwise reticent about her childhood. On leaving university in 1963, she went to London where she got a job editing medical textbooks and worked as a freelance journalist. In 1968, she went to live in Cornwall with a friend, and stayed there until 1984.

At first she wrote features for the *Cornish Times*, but after attending an Arvon Foundation course in neighbouring Devon, she began to write her first novel, *Bethany*, which is set in Cornwall during the 1960s and was published to wide acclaim in 1981. Her second novel, *The Illusionist*, was shortlisted for the Booker Prize. Set in biblical Judea, it is the story of the magician Simon of Gitta, better known as Simon Magus, and combines, as Bernard Levin observed when choosing it as one of his books of the year in the *Observer*, 'a remarkable imagination with a keen sense of history'. It is a characteristically intelligent and finely written book, and remains the one by which she is best known.

Mason is not a novelist who repeats herself, and each new book represents a new departure. *The War against Chaos* was a futuristic novel, which she found unsatisfactory. It did, however, lead unexpectedly to her next novel, *The Racket*, which was inspired by a publicity tour she had undertaken in São Paulo. Set in Brazil, the novel concerns a history teacher whose regard for truth prevents her from toeing the political and religious line, in spite of intimidation.

In 1984, Mason was awarded a writer's fellowship at Leeds University, and she has lived there ever since.

Fiction

Bethany 1981; *The Illusionist* 1983; *The War against Chaos* 1988; *The Racket* 1990; *Angel* 1994 [US *Reich Angel* 1995]

Bobbie Ann MASON 1940–

Mason was the daughter of a Kentucky dairy farmer. The world and issues she grew up with – the small towns of America's Southern states, the steady incursion of the modern world into rural communities, the strong ties people hold to their land – form the core of her writing. As soon as she could, however, she went to college. The ambivalence in her attitude towards her roots is reflected in her work: she is proud of characters who are prepared to risk all for a new life, but she is equally drawn to (or, in her words, 'haunted by') those who don't escape.

After graduate school at the University of Connecticut, Mason settled in the Pennsylvania countryside with her husband, also a writer, whom she married in 1970.

Mason's first collection, *Shiloh and Other Stories*, won her the PEN/Faulkner Award in 1982, and three years later her first novel, *In Country*, received the PEN/Hemingway Award for best fiction. The literary world needed a label for her, and Bill Buford (then editor of *Granta*) coined the phrase 'dirty realism', bracketing her with **Raymond Carver** and **Tobias Wolff**. She feels comfortable with the phrase: 'It gives a very interesting and different perspective, because it focuses on content, and on the ways people are seeing the world.' In the USA, the term 'minimalism' is used, which she dislikes 'because it's derogatory. It implies that there's not much substance.'

One of Mason's concerns is popular culture. She complains that other writers mostly insert references to particular television programmes and pop groups 'to make a comment on the culture and the comment is always negative: popular culture is trash. I don't do that. I think I use the references more organically to say this is the world of the characters and this is what they like ... It's not interesting or fair for me or the reader to condemn them for drinking beer and watching the superbowl. I prefer to ask why their only pleasure is booze and TV.'

Mason herself has a special fondness for pop music, and when she is writing she listens to rock-and-roll, played very loud.

In Country has been filmed, and Mason feels none of the outrage about Hollywood ignorance and insensitivity that writers usually express. 'I love it, I love it, it's just wonderful', she has said. 'Seeing it was just overwhelming.'

Fiction

Shiloh and Other Stories (s) 1982; *In Country* 1985; *Spence & Lila* 1988; *Love Life* (s) 1989; *Feather Crowns* 1993

Non-Fiction

Nabokov's Garden: a Nature Guide to Ada 1974; *The Girl Sleuth: a Feminist Guide to the Bobbsey Twins, Nancy Drew, and Their Sisters* 1975

Allan (Johnstone) MASSIE 1938–

Of Scottish ancestry, Massie was born in Singapore, as his father was a rubber planter in Malaya. The war intervened shortly after his birth, and he grew up in his father's native Aberdeenshire. He attended a preparatory school, Drumtochty Castle School, the public school Trinity College, Glenalmond, and Trinity College, Cambridge, where he studied history. After graduating, he taught for more than a decade, from 1960 to 1971, at his old prep school, then taught English to foreign students in Rome between 1972 and 1975. He married in 1973 and has three children.

On his return from Rome to Scotland, he became the principal fiction reviewer for the *Scotsman*, a post which laid the foundations for his successful journalistic career. In the 1980s, he reviewed for various Scottish newspapers and also held creative writing fellowships at Scottish universities. From 1991, he has been at the *Daily Telegraph*, writing mainly but not exclusively on literary subjects, and is among the paper's most well-known contributors. From 1982 to 1984 he was editor of the *New Edinburgh Review*.

He published his first novel, *Change and Decay in All Around I See*, in 1978. A comedy of manners rather in the style of **Evelyn Waugh**, it did not indicate the direction in which he was to develop. His books have in fact been marked by their extreme variety. Although his time-span is mainly modern, there is a group of historical novels (for example, the three books about Roman emperors, *Augustus, Tiberius* and *Caesar*, and *King David*, about the biblical hero); novels set in Scotland compete with such books as *The Death of Men*, set in Rome and loosely based on the kidnapping and murder of Italian premier Aldo Moro, or *A Question of Loyalties*, which deals with Vichy France in 1940. To this broad range of sympathies he has added an unceasing experimentation, and it is a measure of the esteem in which he is held that **Nicholas Mosley** resigned from the Booker Prize panel in 1991 largely because Massie's *The Sins of the Father* had not been shortlisted. It can be argued, however, as with **David Hughes**, that the very variety and unpredictability of Massie's work has militated against his receiving quite the audience he deserves.

Massie's first non-fiction work was a study of **Muriel Spark** in 1979, and since then he has written many works in this field, mainly elegant *belles-lettres* such as *The Caesars* (about whom he also wrote and presented a radio programme in 1994) and *Byron's Travels*. A work edited is *PEN New Fiction II* (1987), one of two anthologies of new writing published under the auspices of the writers' organisation PEN in the mid-1980s, after which the series was discontinued.

Fiction

Change and Decay in All Around I See 1978; *The Last Peacock* 1980; *The Death of Men* 1981; *One Night in Winter* 1984; *Augustus* 1986; *A Question of Loyalties* 1989; *The Hanging Tree* 1990; *Tiberius* 1990; *The Sins of the Father* 1991; *Caesar* 1993; *These Enchanted Woods* 1993; *The Ragged Lion* 1994; *King David* 1995

Non-Fiction
Muriel Spark 1979; *Ill-Met by Gaslight: Five Edinburgh Murders* 1980; *The Caesars* 1983; *Portrait of Scottish Rugby* 1984; *Colette* 1986; *101 Great Scots* 1987; *Byron's Travels* 1988; *Glasgow: Portraits of a City* 1989; *The Novel Today* 1990; *Edinburgh* 1994

Edited
Edinburgh and the Borders: In Verse 1983; *PEN New Fiction II* 1987

Plays
Quintet in October (r) [nd]; *The Minstrel and the Shirra* 1989; *First-Class Passengers* 1995

John MASTERS 1914–1973

A novelist who explored all aspects of the relationship between Britain and India from 1600 to independence in 1947, and of the fifth generation of his family to grow up in India, Masters spent the first half of his life as soldier, the second as writer. Sent home by his father, a captain in the 16th Rajputs, to Wellington and the Royal Military College, Sandhurst, he returned to join the Indian army in 1934 as a second lieutenant. During his fourteen-year army career he served in many parts of the sub-continent and the Middle East, commanded a brigade of Orde Wingate's Chindits in Burma, and was a member of the 19th Indian Division at the taking of Mandalay. He retired as a lieutenant-colonel in 1948, having spent the last years of his career teaching jungle warfare at the Camberley staff college in England. He married in 1945, and had three children, one of whom died young, as well as two step-children. The terms of Indian independence had left him unable to carry on his career in the Indian army, and he decided to move his family to the USA.

On arriving in America, his original plan was to run treks through the Himalayas, but a *New Yorker* writer, Rex Lardner, had suggested he write up India; he wrote ten pages, which Lardner sent to an agent, and soon he was planning thirty-five novels about a family called the Savages who lived in India during the colonial centuries. He completed only about half this scheme, but adventure stories such as *Nightrunners of Bengal* and *Bhowani Junction* became popular middlebrow reading. He also wrote autobiographical books, a life of Casanova, and *Loss of Eden*, a late trilogy about the First World War, in which real characters and events were introduced into a fictional narrative. Towards the end of his life his fiction became very salacious, and had to be toned down by his editors. In America, he eventually settled in New Mexico, becoming an American citizen in 1954. He died in Albuquerque of complications following a heart-bypass operation.

Fiction
Nightrunners of Bengal 1951; *The Deceivers* 1952; *The Lotus and the Wind* 1953; *Bhowani Junction* 1954; *Coromandel!* 1955; *Far, Far the Mountain Peak* 1957; *Fandango Rock* 1959; *The Venus of Konpara* 1960; *To the Coral Strand* 1962; *Trial at Monomoy* 1964; *Fourteen Eighteen* 1965; *The Breaking Strain* 1967; *The Rock* 1970; *The Ravi Lancers* 1972; *Thunder at Sunset* 1974; *The Field Marshal's Memoirs* 1975; *The Himalayan Concerto* 1976; *Bedlam* 1977; *The Indian Trilogy* 1978; *Loss of Eden*: *Now God Be Thanked* 1979, *Heart of War* 1980, *By the Green of Spring* 1981; *Man of War* 1983

Non-Fiction
Bugles and a Tiger (a) 1956; *The Road Past Mandalay* (a) 1961; *Casanova* 1969; *Pilgrim Son* (a) 1971; *The Glory of India* [with W. MacQuitty] 1982

Harry (Burchell) MATHEWS 1930–

Mathews was born to well-to-do parents in Manhattan, and briefly attended Princeton and served in the US Navy before graduating in music from Harvard in 1952. He moved to Paris to continue his study of music at the École Normale de Musique, and there met the American poet **John Ashbery** in 1956. He spent some time in Nice and Majorca where he was associated with **Robert Graves**'s artists' colony at Deya. With Ashbery, **Kenneth Koch** and James Schuyler he founded and edited the magazine *Locus Solus* which lasted from 1960 to 1962.

Mathews's first novel, *The Conversations* (1962), is a typically perverse exercise in playful postmodernist caprice, filled with the games, riddles, futile quests, absurd puzzles and black humour which characterise his early fiction. His work has been described by his friend Georges Perec as operating in 'a narrative world determined by rules from another planet'. It was not until 1973 that Mathews was elected a member of the Paris-based group known as OuLiPo, the first American to be so honoured, but from the start his work has epitomised the aims of the group. OuLiPo is short for *Ouvroir de Littérature Potentielle* (the Workshop of Potential Literature) and was founded in 1960 by the novelist Raymond Queneau and the mathematician François Le Lionnais. Other members have included Italo Calvino and Perec (who is remembered in Mathews's memoir *The Orchard*).

Referring to the elaborate word games in which OuLiPo specialises, Mathews has said that 'the abstract procedures allow me to express "myself" more easily than if I set out with that goal'. His 1983 'short fiction' (much of his work defies easy categorisation) *Plaisirs singuliers* describes sixty-one masturbatory techniques and involves an organisation which 'encourages its members to invent obstacles to overcome while masturbating'. This is conceivably a reference to the literary philosophy of OuLiPo.

Mathews's 1987 novel *Cigarettes* deals in more conventional narrative matter but is none the less built around a complex formal structure. He leads an affluent, cosmopolitan life divided between Paris and New York which is clearly not funded by his writing, most of which is published by small presses. He has two children from his first marriage, and is now married to the writer Marie Chaix whose novel *The Laurels of Lake Constance* he has translated.

Fiction

The Conversions 1962; *Tlooth* 1966; *The Sinking of the Odradek Stadium and Other Novels* 1975; *Selected Declarations of Dependence* (s) [with A. Katz] 1977; *Country Cooking and Other Stories* (s) 1980; *Plaisirs singuliers* [with M. Chaix] 1983 [as *Singular Pleasures* 1988]; *Cigarettes* 1987; *The American Experience* (s) 1991; *The Journalist* 1994

Poetry

The Ring: Poems 1956–1969 1970; *The Planisphere* 1974; *Le Savoir des rois* 1976; *Trial Impressions* 1977; *Armenian Papers: Poems 1954–1984* 1987; *Out of Bounds* 1989; *A Mid-Season Sky: Poems 1954–1989* 1991

Non-Fiction

Niki at Nassau [with D. Bourdon, J. Cage] 1987; *The Orchard* 1988; *20 Lines a Day* 1988; *Immeasurable Distances* 1991

Translated

Marie Chaix *The Laurels of Lake Constance* 1977; Georges Bataille *Blue of Noon* 1978; Jeanne Cordelier *The Life: Memoirs of a French Hooker* 1978

Edited

Georges Perec *'53 Days'* [with J. Roubaud] 1992

Collections

The Way Home: Collected Longer Prose 1988

Mustapha (pseud. of Noel Cuthbert) MATURA 1939–

One of the leading playwrights dealing with black experience in Britain who came to prominence in the 1970s, Matura was born in Trinidad. His father was a car salesman, his mother a shop assistant, and he is of mixed African, Indian and European descent. He had little formal education, attending a Roman Catholic intermediate school for some years. He worked as an office boy, stock clerk and insurance salesman in Trinidad before coming to Britain in 1961. He then worked as a hospital porter, cosmetics display assistant and stockroom assistant in England until 1968, when he became a full-time writer.

He quickly became well known as a playwright with such works as *Black Pieces*, a collection of short plays, and *As Time Goes By, Rum an' Coca Cola, Nice* and *Welcome Home Jacko.* He received a John Whiting Award from the Arts Council in 1970 and has won several other awards. In the 1980s he has turned increasingly to television work, co-writing the series *Black Silk* and *No Problem.* Other theatre work includes *Trinidad Sisters*, an adaptation of Chekhov.

He married in 1961, and had two children from that marriage; he has also had two children with his present partner. He lives in London and was the co-founder in 1973 of the Black Theatre Co-operative, one of the best known of black theatre groups. He has written one children's book, *Moonjump*, but his reputation rests on his plays, which mix poignancy and humour in their exploration of the confrontation between the Caribbean black and white society both in Britain and the West Indies.

Plays

Black Pieces 1970 [pub. 1972]; *As Time Goes By* 1971 [pub. 1972]; *Bakerloo Line* 1972 [np]; *Murders of Boysie Singh* (f) 1972; *Nice* 1973 [pub. 1980]; *Play Mas* 1974 [pub. 1976]; *Black Slaves, White Chains* 1975 [np]; *Bread* 1976 [np]; *Rum an' Coca Cola* 1976 [pub. 1980]; *More, More* 1978 [np]; *Another Tuesday* 1978 [np]; *Independence* 1979 [pub. 1982]; *Welcome Home Jacko* 1979 [pub. 1980]; *A Dying Business* 1980 [np]; *Meetings* 1981 [pub. 1982]; *One Rule* (l) [with J. Laddis, V. Romero] 1981 [np]; *Trinidad Sisters* [from Chekhov's play *The Three Sisters*] 1988 [np]; *The Coup* [pub. 1991]; *Six Plays* [pub. 1992]

For Children

Moonjump 1988

Robin (i.e. Robert Cecil Romer) MAUGHAM 1916–1981

The 2nd Viscount Maugham, son of Frederic Herbert Maugham who was briefly Lord Chancellor in 1938–9, nephew of the novelist **W. Somerset Maugham**, and from a distinguished legal family on both sides, Maugham was educated at Eton, where he was unhappy, and at Trinity College, Cambridge. His father, who intended him for the law, procured him a job as a judge's marshal. He joined up in 1939 as a trooper in the Inns of Court Regiment, was commissioned in 1940, fought in the Western Desert where he was severely wounded in the arm and head, and was invalided out in 1944 with the honorary rank of captain.

He was called to the Bar but, against strong paternal opposition, determined to live as a writer. For some years after the war he carried out unofficial Intelligence work in the Middle East while publishing a large number of early novels and travel reminiscences. In 1948 his powerful novel *The Servant*, successfully filmed by Joseph Losey in 1963 from a screenplay by **Harold Pinter**, did much to establish his reputation. From the 1950s onwards he lived on a combination of a diminishing private income and his writing, which included plays, film scenarios, short stories, journalism, family history and autobiography, as well as novels. He spent long periods abroad (living, for instance, on Ibiza and Gozo); his British home was Brighton, while he also maintained a *pied-à-terre* in Charing Cross Road, London. He was predominantly homosexual and achieved a series of sustained relationships with men usually much younger than himself and sometimes of a different social class. The same theme of sexual relations across age (and sometimes class or race) barriers recurs obsessively in his many novels, which are marked by strong plotting and exotic locales but marred by a portentous, as well as undistinguished, style. He inevitably suffered comparisons with his famous uncle, himself not considered in the front rank among novelists.

Maugham succeeded to his father's viscounty in 1958 and made his maiden speech in the House of Lords on the slave trade, an evil against which he campaigned. His last years were darkened by illness, particularly diabetes, which was exacerbated by his heavy drinking. Some years after his death the journalist Peter Burton, a former secretary, claimed that he himself wrote much of Maugham's last novel, *The Deserters*, and helped him compose *Conversations with Willie*, a book Maugham published about his uncle, despite the latter having paid him money not to write his biography. The extent of Burton's contributions was subsequently disputed by some close to Maugham.

Fiction

The 1946 MS 1943; *The Servant* 1948; *Line on Ginger* 1949 [as *The Intruder* 1960]; *The Rough and the Smooth* 1951; *Behind the Mirror* 1955; *The Man with Two Shadows* 1958; *November Reef* 1962; *The Green Shade* 1966; *The Wrong People* 1967; *The Second Window* 1968; *The Link* 1969; *The Black Tent and Other Stories* **(s)** [ed. P. Burton] 1972; *The Last Encounter* 1972; *Testament: Cairo 1898* **(s)** 1972; *The Barrier* 1973; *The Sign* 1974; *Knock on Teak* 1976; *Lovers in Exile* **(s)** 1977; *The Dividing Line* 1979; *The Corridor* 1980; *The Deserters* 1981; *The Boy from Beirut and Other Stories* **(s)** [ed. P. Burton) 1982

Plays

Thirteen for Dinner 1935; *The Walking Stick* 1935; *He Must Return* 1944; *The Rising Heifer* 1952; *The Intruder* **(f)** [from his novel *Line on Ginger*; with J. Hunter] 1953; *The Leopard* 1955; *The Black Tent* **(f)** [from his novella; with B. Forbes] 1956; *Mister Lear* 1956 [pub. 1963]; *The Last Hero* 1957; *Odd Man In* [from C. Magnier's *Monsieur Masure*] 1957 [pub. 1958]; *The Lonesome Road* [with P. King] 1957 [pub. 1959]; *The Servant* [from his novel] 1958 [pub. 1972]; *The Hermit* [with P. King] 1959; *It's in the Bag* [from C. Magnier's *Oscar*] 1960; *The Two Wise Virgins of Hove* **(tv)** 1960; *Azouk* [from A. Rivemale; with W. Hall] 1962; *The Claimant* 1962; *Winter in Ischia* 1964; *Enemy!* [pub. in *Plays of the Year* vol. 39, ed. J.C. Trewin 1971]; *A Question of Retreat* **(r)**, **(st)** 1981

Non-Fiction

Come to Dust 1945; *Approach to Palestine* 1947; *Nomad* 1947; *North African Notebook* 1948; *Journey to Siwa* [with D. Papadimou] 1950; *The Slaves of Timbuktu* 1961; *The Joyita Mystery* 1962; *Somerset and all the Maughams* 1966; *Escape from the Shadows* **(a)** 1972; *Search for Nirvana* **(a)** 1975; *Conversations with Willie: Recollections of W. Somerset Maugham* 1978

W(illiam Henry) Somerset MAUGHAM 1874–1965

Born in the English embassy in Paris, the son of the legal adviser there, Maugham came from a distinguished legal family (one of his three brothers, father of the novelist **Robin Maugham**, was to become Lord Chancellor). He lived in France until the age of ten, and French was his first language. His mother died when he was eight, a psychological blow from which he never wholly recovered. Two years later, his father died, and he was sent to live with an uncle who was a clergyman in Whitstable, Kent. The displaced Maugham hated both his uncle and the King's School, Canterbury, and developed a lifelong stammer.

He left the King's School at the age of sixteen, and studied for a year in Heidelberg, where he experienced his first homosexual affair, with the Cambridge aesthete John Ellingham Brooks (later the husband of the lesbian painter Romaine Brooks). Maugham was to portray Brooks as Hayward in what is perhaps his finest and certainly his most personal novel, *Of Human Bondage*, in which his own stammer (and perhaps his homosexuality) is represented by the protagonist's club foot. For five years from 1892, he studied medicine at St Thomas's Hospital, London, and by the time he had graduated, he had already published his first novel, *Liza of Lambeth* (1897), which drew upon his experiences as a doctor in the slums of south London. Having come into two legacies, he decided to abandon medicine and devote himself entirely to writing.

Early novels and plays won him only a limited reputation, but in 1911 *Lady Frederick* was a success on the West End stage, and in the following year he achieved the unprecedented triumph of having four plays running there simultaneously. He had arrived as a fashionable writer and, although not primarily a playwright, he was to have many West End successes before giving up the theatre in 1933.

During the First World War, he was first an ambulance driver in France, then a secret agent, sent to Russia with a mission to prevent the Revolution, an experience which resulted in his stories about the spy Ashenden. Some time during the first six months of the war, purportedly on a battlefield, he met Gerald Haxton, a debonair American brought up in England, and eighteen years his junior. The two men embarked upon a relationship, scarcely disturbed by Maugham's marriage in 1916 to Syrie Wellcome (the divorced daughter of the philanthropist Dr Barnardo), who was bearing his child. As Maugham's 'secretary', the engaging but somewhat dissolute Haxton accompanied him on his extensive travels, while Syrie Maugham took up a highly successful career as an interior decorator, famed for her all-white schemes. The couple had one daughter, named after the heroine of Maugham's first novel, but spent most of their time apart: the marriage was dissolved in 1927. Maugham's travels resulted in the short stories, the best of which – 'Rain', 'The Alien Corn', 'The Force of Circumstance' – are among his most permanent literary legacy. 'There must be thousands of readers who have wished themselves, as I often have, at Maugham's side on board some small freighter, as it steams up a tropical river or into the harbour of a Pacific island,' wrote **Christopher Isherwood**, who was to become one of the author's many literary friends. Maugham's exotic locales, his emblematic characters and his cynical, world-weary tone earned him an enormous readership, but one which could be discriminating. He was a thoroughly professional writer, working to a strict routine, which sometimes resulted in his stopping in mid-sentence in order not to miss the cocktail hour at Villa Mauresque, the home he had made in 1928 at Cap Ferrat on the French Riviera. He plotted his career in fine detail and rarely deviated from his plan or allowed his imagination to lead him from a chosen path. This resulted in a distinctly mechanical quality in some of his work, and a 1940 volume of stories was ominously titled *The Mixture As Before*. He nevertheless persisted in regarding himself as one of the century's greatest writers, and although he disliked theorising about writing, he outlined his methods in *A Writer's Notebook*.

Maugham lived at Cap Ferrat for the remainder of his life, apart from a period in America during the Second World War, but he continued to travel widely. Haxton died of complications arising from his alcoholism in 1944 and was succeeded by the rather more reliable Alan Searle. In spite of a career which won him both popular and critical acclaim – *Of Human Bondage* alone was said to have sold 10,000,000 copies – Maugham did not enjoy a relaxed old age. In 1962 his memoirs, in which he vilified his former wife (who had died in 1955) and failed to take any responsibility for the failure of his marriage (Haxton was described merely as a secretary and travelling companion), were serialised first in the American glossy magazine *Show*, then in the *Sunday Express* under the title 'Looking Back'. They lost him a number of old friends and were never published in book form, which was probably as well. At the same time he became involved in an unedifying and very public legal battle to disinherit Liza, who he claimed was not his daughter, in favour of Alan Searle, whom he had recently adopted. Some of these events have been ascribed to the course of 'cellular therapy' he underwent in order to stave off old age (it involved his being injected with material derived from lamb foetuses); this resulted in his body far outliving his mind and he ended his days in a welter of senile confusion.

Maugham's status as a writer is essentially middlebrow, but time will winnow his massive output, which also included travel books, essays and an early autobiography. His finest things – novels such as *The Moon and Sixpence*, which is based on the life of Paul Gaugin, *Cakes and Ale*, which mercilessly satirises **Hugh Walpole** and **Thomas Hardy**, and *The Razor's Edge*, a novel of mysticism which people mistakenly thought was based on the experiences of Isherwood – will undoubtedly survive.

Fiction
Liza of Lambeth 1897; *The Making of a Saint* 1898; *Orientations* (s) 1899; *The Hero* 1901; *Mrs Craddock* 1902; *The Merry-Go-Round* 1904; *The Bishop's Apron* 1906; *The Explorer* 1907; *The Magician* 1908; *Of Human Bondage* 1915; *The Moon and Sixpence* 1919; *The Trembling of a Leaf* (s) 1921; *The Painted Veil* 1925; *The Casuarina Tree* (s) 1926; *Ashenden* 1928; *Cakes and Ale* 1930; *First Person Singular* (s) 1931; *The Narrow Corner* 1932; *Ah King* (s) 1933; *Altogether* (s) 1934; *Cosmopolitans* (s) 1936; *Theatre* 1937; *Christmas Holiday* 1939; *The Round Dozen* (s) 1939; *The Mixture as Before* (s) 1940; *Up at the Villa* 1941; *The Hour Before the Dawn* 1942; *The Razor's Edge* 1944; *Then and Now* 1946; *Creatures of Circumstance* (s) 1947; *Catalina* 1948; *The Complete Short Stories of W. Somerset Maugham* 3 vols 1951 [rev. in 4 vols 1976–7]

Plays
Jack Straw 1911; *Lady Frederick* 1911; *A Man of Honour* 1911; *The Explorer* 1912; *Mrs Dot* 1912; *Penelope* 1912; *Landed Gentry* 1913; *The Land of Promise* 1913; *Smith* 1913; *The Tenth Man* 1913; *The Unknown* 1920; *The Circle* 1921; *Caesar's Wife* 1922; *East of Suez* 1922; *Home and Beauty* 1923; *Our Betters* 1923; *The Unattainable* 1923; *Loaves and Fishes* 1924; *The Constant Wife* 1927; *The Sacred Flame* 1928; *The Breadwinner* 1930; *For Services Rendered* 1932; *Sheppey* 1933; *The Collected Plays* 3 vols 1952; *The Noble Spaniard* 1953

Non-Fiction
The Land of the Blessed Virgin. Sketches and Impressions in Andalusia 1905; *On a Chinese Screen* 1922; *The Gentleman in the Parlour. A Record of a Journey from Rangoon to Haiphong* 1930; *Don Fernando: or Variations on Some Spanish Themes* 1935; *The Summing Up* (a) 1938; *Books and You* 1940; *France at War* 1940; *Strictly Personal* (a) 1941; *Great Novelists and Their Novels* 1948; *A Writer's Notebook* 1949; *The Vagrant Mood* 1952; *Points of View* 1958; *Purely for My Pleasure* 1962; *Selected Prefaces and Introductions of W. Somerset Maugham* 1964; *Essays on Literature* 1967; *The Letters of W. Somerset Maugham to Lady Juliet Duff* [ed. L. Rothschild] 1982

Edited
C.H. Hawtrey *The Truth at Last* 1924; *Tellers of Tales* 1939; *A Choice of Kipling's Prose* 1952

Collections
The Wit and Wisdom of Somerset Maugham [ed. C. Hewetson] 1966; *A Traveller in Romance. Uncollected Writings 1901–1964* [ed. J. Whitehead] 1984

Biography
Willie by Robert Calder 1989

Armistead MAUPIN 1944–

The son of a lawyer, Maupin was born in Washington DC, but brought up in Raleigh, North Carolina. He was educated at the University of North Carolina, from which he graduated in 1966. He attended law school but left after failing his first-year exams. In 1967 he began his national service in the US Navy, serving as a communications officer in the Mediterranean and then as a lieutenant during the Vietnam War, earning the Navy Commendation Medal. After the war, he organised a small group of American veterans to return to Vietnam to take part in a community project building houses for disabled Vietnamese veterans at Cat Lai. This earned him the Freedom Leadership Award in 1972, and a Presidential Commendation.

By this time, Maupin had moved from Charleston, South Carolina, where he had worked as a newspaper journalist, to San Francisco, the city he was to celebrate in his fiction. He worked as a reporter for the Associated Press, then took a number of short-term jobs, including that of an accounts executive for a public relations firm, a publicist for San Francisco Opera (for whom he wrote an English libretto for the 1976 production of Offenbach's *La Perichole*), and a commentator for KRON-TV. His first job as a columnist was with the *Pacific Sun* in 1974, and two years later he started writing a daily fiction serial in the *San Francisco Chronicle*. This acutely observed, deliriously inventive, and consistently entertaining story of San Francisco life, centred on a ramshackle rooming-house in Barbary Lane, soon became a cult. Developing his story day by day, Maupin was able to incorporate current news stories, fads and fashions, so that the saga was both up to the minute and gradually building to become a historical record of a certain time and place. Although the story featured several homosexual characters, it was never in any sense 'ghetto' fiction aimed at a purely gay audience. Characters such as Michael 'Mouse' Tolliver, Mary Ann Singleton, Anna Madrigal, Mona Ramsey and Brian Hawkins represented a rich cross-section of San Francisco life, and their adventures soon attracted a large following.

Maupin stopped writing the serial in 1977, and the following year revised it to make a novel, *Tales of the City*. Maupin did not find a substantial book audience until the remainder of the saga was published two years later as *More Tales of the City*, after which the books attracted a wide and devoted readership in both America and in the UK. A second series of instalments started appearing in the *Chronicle* between 1981 and 1983, by which time the newspaper was running a large number of stories about Aids, which was cutting a swathe through the city's large homosexual population. As the epidemic took hold, so Maupin's story darkened, with several of the characters becoming infected with HIV and some dying from Aids-related illnesses. Even so, the general trajectory of the series was optimistic, and later volumes proved as popular as ever. The serial subsequently transferred to the *San Francisco Examiner*, and the final volume, *Sure of You*, was published in book form in 1990. In all there were six books, which have been reissued in various omnibus editions in the USA and the UK. An attempt to turn the saga into a television series met with limited success. No American television company would handle a story in which homosexuality played so prominent and approved a role, and it was left to the UK's Channel 4 to make and broadcast an adaptation of the first volume. This was subsequently sold to America, where it attracted the largest ever audience for a drama series on public service broadcasting. In spite of this, and the promise of considerable funding, plans to adapt the remainder of the saga have thus far come to nothing, a state of affairs generally ascribed to anti-homosexual prejudice among television executives.

Despite pleas that he should continue the series, Maupin has so far resisted, writing instead *Maybe the Moon*, which takes the form of a journal kept by a Hollywood actress, who is also 'the world's shortest woman'. Some critics felt that the novel was marred by a sentimentality that had been kept firmly at bay in the *Tales*, and it inevitably proved less popular. Maupin has also written journalism for such magazines as *Interview*, the *Advocate, Village Voice*, the *New York Times* and the *Los Angeles Times*. He wrote San Francisco's longest-running theatrical show, the legendary *Beach Blanket Babylon*, and adapted his short story 'Suddenly Home' (published in **Edmund White**'s *The Faber Book of Short Gay Fiction*) for a musical, *Heart's Desire*, which had its world premiere at the Cleveland Playhouse in 1990. He lives in San Francisco with Terry Anderson, a gay rights activist.

Fiction

Tales of the City 1978; *More Tales of the City* 1980; *Further Tales of the City* 1982; *Babycakes* 1984; *Significant Others* 1987; *Sure of You* 1990; *Maybe the Moon* 1992

Plays

Beach Blanket Babylon 1975; *La Perichole* (l) 1976; *Heart's Desire* (l) [from his story 'Suddenly Home'] 1990 [np]

F(lora) M(acdonald) MAYOR 1872–1932

Mayor did not model the formidable Canon Jocelyn of her finest novel, *The Rector's Daughter*, on her father, the Revd Joseph Mayor, professor of classics and later moral philosophy at King's College, London. Not only was her father a brilliant academic, but her mother, Jessie, was an outstanding linguist who rendered a Zulu grammar in Danish into English. Mayor was born in Kingston Hill, and grew up with a twin sister and two brothers. During her education at Surbiton High School, she discovered her talent for acting, but winning the Sixth Form Latin Prize at seventeen destined her for Cambridge. After a year at a finishing school in Switzerland, she went up to Newnham College in 1882, where she enjoyed herself for four years. The result of cycling, dancing and acting was a depressing third, which was to chain her to Kingston and the routine of parish life.

In an effort to break away, Mayor travelled with a theatrical touring company; this proved to be a disastrous failure, and she turned to writing instead. Much of her tedious upper-middle-class existence was transmuted into her first novel, *Mrs Hammond's Children*, which had few sales when published under the pseudonym Mary Stafford in 1901. A second attempt at acting also failed, and the only means of rescue was marriage, which was proposed by Ernest Shepherd, an old friend. At first, Mayor hardly loved him, although they grew closer through correspondence while he was in India, but during the finalisation of the wedding plans Ernest died suddenly of malaria in 1903. Mayor was devastated, and suffered a physical and nervous breakdown. In *The Third Miss Symons*, she projected her fears about her own future in a portrait of a woman who longs for love, but is incapable of achieving it. *The Rector's Daughter*, published when Mayor was fifty-two, relates the self-sacrificing life of a spinster with an unfulfilled passion for a man; it was compared to Jane Austen in its treatment of a small section of society and manners. Praise from **E.M. Forster** and **John Masefield** encouraged her to write *The Squire's Daughter*, but although distinguished, this novel could not match the individuality of the earlier book. A collection of stories, *The Room Opposite*, was published after Mayor died of pneumonia. Neglected until recent years, Mayor's work has now been reassessed.

For **Susan Hill**, Mayor 'achieved total mastery of her style ... most fluently expressing every shade of human emotion'.

Fiction

Mrs Hammond's Children [as 'Mary Stafford'] 1901; *The Third Miss Symons* 1913; *The Rector's Daughter* 1924; *The Squire's Daughter* 1924; *The Room Opposite* (s) 1935

Biography

Spinsters of this Parish by Sybil Oldfield 1984

David MERCER 1928–1980

Mercer grew up in Wakefield, Yorkshire, the son of an engine-driver who was ambitious for his sons to escape from the working class. Mercer, however, unlike his elder brother (who went on to become a nuclear scientist), failed the 11-plus twice, and left school aged fourteen to become a laboratory technician, performing post-mortems. According to his own account, his early years were spent as 'a kind of dourly functioning animal', possessed of 'huge violence'. From 1945 to 1948, he was in the Royal Navy, performing the work of a pathological laboratory technician. He then went to Durham University to read chemistry but soon transferred to study art at King's College, Newcastle, graduating in 1953. He then spent a period of two years in Paris as a Bohemian, copying old masters in the Louvre and selling them to tourists, writing novels 'because that's what everybody wrote', and marrying a wealthy *émigrée* Czech. This marriage soon failed, however, and his return to London was followed by a breakdown, a period when he lived in a colourful household which also included novelist **Bernice Rubens**, poet **Jon Silkin** and 'a hundred other people'. Rescue from his psychological problems came with a second marriage to a northern civil servant.

He spent the later 1950s working as a teacher, first in a private language school and then as a supply teacher and in a technical college, but he was also discovering his *métier* as a playwright for the developing medium of television. He attracted the interest of the producer Don Taylor, working with him during 1959 and 1960 on *Where the Difference Begins*, the first of the television trilogy *The Generations*. He received an unprecedented contract from the BBC where they promised to produce anything he wrote for them, and he was able to give up teaching in 1961. Thenceforth he lived as a writer, following this career prolifically for twenty years, branching out into the theatre a few years after beginning his work for television, and achieving a success which enabled him to live in St John's Wood, a prosperous district of London.

His work, which often centres on the affairs, socialist struggles and nervous breakdowns of the educated children of working-class parents, is clearly autobiographical in inspiration, and heavily influenced by the work of fashionable thinkers such as R.D. Laing; if it now seems rather too much of its time, its foreshadowing of the breakdown of Communism was certainly prescient. Besides the plays for television (*A Suitable Case for Treatment* is perhaps the best) and the generally less successful ones for theatre, there were also film scenarios, including the Alain Resnais film *Providence* (1977), and a little fiction.

Mercer's personal life was stormy, undermined by severe alcohol and nicotine addiction. His second marriage broke down, and he had many affairs, including one with the actress Kika Markham, and another with **Penelope Mortimer**, described in her autobiographical volume, *About Time Too* (1993). Late in his life, he married the Israeli Dafna Mercer-Hadari, with whom he had a daughter; he had another daughter with a French actress. In his last years he lived largely in Israel, and died in Haifa of a massive heart-attack.

Plays

Where the Difference Begins (tv) 1961 [pub. 1964]; *The Buried Man* (st) 1962, (tv) 1963; *A Climate of Fear* (tv) 1962 [pub. 1964]; *A Suitable Case for Treatment* (tv) 1962, (f) [as *Morgan! A Suitable Case for Treatment*] 1966; *Birth of a Private Man* (tv) 1963 [pub. 1964]; *For Tea on Sunday* (tv) 1963 [pub. 1966]; *A Way of Living* (tv) 1963; *And Did Those Feet?* (tv) 1965 [pub. 1966]; *The Governor's Lady* 1965 [pub. 1968]; *Ride a Cock Horse* 1965 [pub. 1966]; *Blecher's Luck* 1966 [pub. 1967]; *In Two Minds* (tv) 1967, (f) 1971; *Let's Murder Vivaldi* (tv) 1968 [pub. 1967]; *The Parachute* (tv) 1968 [pub. 1967]; *On the Eve of Publication* (tv) 1968 [pub. 1970]; *After Haggerty* 1970; *The Cellar and the Almond Tree* (tv) 1970; *Emma's Time* (tv) 1970; *Flint* (st) 1970, (tv) 1978; *White Poem* 1970; *Blood on the Table* 1971; *The Bankrupt* (tv) 1972 [pub. 1974]; *A Doll's House* (f) [from Ibsen's play] 1972; *You and Me and Him* (tv) 1973 [pub. 1974]; *An Afternoon at the Festival* (tv) 1973 [pub. 1974]; *Barbara of the House of Grebe* [from Hardy] (tv) 1973; *The Arcata Promise* (tv) 1974 [pub. 1977]; *Duck Song* 1974; *Find Me* (tv) 1974; *Folie à Deux* (r) 1974; *Huggy Bear* (tv) 1976 [pub. 1977]; *Providence* (f) 1977; *A Superstition* (tv) 1977; *Shooting the Chandelier* (tv) 1977 [pub. 1978]; *Cousin Vladimir* 1978; *For Tea on Sunday* (tv) 1978; *The Ragazza*

(tv) 1978; *The Monster of Karlovy Vary* 1979; *Then and Now* 1979; *No Limits to Love* 1980 [pub. 1981]; *A Rod of Iron* **(tv)** 1980

Fiction

Ninety Degrees in the Shade 1965

James (Ingram) MERRILL 1926–1995

Merrill was born to well-to-do parents in Greenwich Village, New York. His father was a founder of the Merrill Lynch brokerage firm, and his mother published her own small newspaper. As a boy he spent his winters in Florida and his summers on Long Island, but poems such as 'The Broken Home' (from *Nights and Days*) reflect the unhappiness engendered by his parents' divorce. Merrill was educated at Lawrenceville School, New Jersey, where he first met the future novelist Frederick Buechner, and saw *Jim's Book*, a collection of his poems and short stories, privately printed by his father. His study at Amherst College, Massachusetts, was interrupted in 1944 by a year as an infantry private in the US Army, and his first collection, *The Black Swan*, was published by the Athens publisher Icarus a year before he graduated in 1947. Dissatisfied with life in New York, and unable to write there, Merrill travelled widely in Europe and the Orient before moving in 1954 to Stonington, Connecticut, where he lived with his partner David Jackson.

His first collection to be published in the USA, *First Poems*, demonstrates his technical gifts but was judged by many critics, notably **Louise Bogan** of the *New Yorker*, 'frigid and dry as diagrams'. Merrill subsequently turned his attention to other media. His play *The Bait*, about a brother and sister who attempt to avoid emotional involvement, was presented off-Broadway in 1953, and *The Immortal Husband*, a retelling of the Greek myth of Tithonus, in Greenwich Village in 1955. His first novel, *The Seraglio*, is a comedy about an ageing businessman surrounded by predatory women. With *The Country of a Thousand Years of Peace* (which contains many poems from *Short Stories*) Merrill perhaps found his voice as a poet for the first time, although the tone is still highly formal and stylised.

From 1959 Merrill spent half of each year in Greece, and his subsequent work demonstrated a warmth and personal intimacy previously lacking. His long narrative poem 'The Book of Ephraim' (from *Divine Comedies*, winner of a Pulitzer Prize in 1977) describes the poet and his lover communicating with the spirit of Ephraim through a Ouija board, and persuaded even so stern a critic as Harold Bloom to re-evaluate Merrill. *Mirabell: Books of Number* and *Scripts for the Pageant* continue the narrative of that poem, and the three poems are published together – an epic running to some 560 pages – in *The Changing Light at Sandover. A Different Person* is an outstanding autobiography.

Poetry

The Black Swan and Other Poems 1946; *First Poems* 1951; *Short Stories* 1954; *The Country of a Thousand Years of Peace and Other Poems* 1959 [rev. 1970]; *Selected Poems* 1961; *Water Street* 1962; *The Thousand and Second Night* 1963; *Violent Pastoral* 1965; *Nights and Days* 1966; *The Fire Screen* 1969; *Braving the Elements* 1972; *Two Poems* 1972; *Yannina* 1973; *The Yellow Pages* 1974; *Divine Comedies* 1976; *Metamorphosis of 741* 1977; *Mirabell: Books of Number and Scripts for the Pageant* 1978; *Ideas, Etc.* 1980; *Scripts for the Pageant* 1980; *The Changing Light at Sandover* 1982; *From the First Nine: Poems 1947–1976* 1982; *Marbled Paper* 1982; *Peter* 1982; *Santorini* 1982; *Bronze* 1984; *Occasions and Inscriptions* 1984; *Plays of Light* 1984; *Rendezvous* 1984; *Souvenirs* 1984; *Late Settings* 1985; *The Inner Room* 1988; *Three Poems* 1988

Plays

The Bait 1953 [pub. in *Artists' Theatre: Four Plays* ed. H. Machiz 1960]; *The Immortal Husband* 1955 [pub. in *Playbook: Plays for a New Theatre* 1956]; *The Image Maker* 1986

Fiction

The Seraglio 1957; *The (Diblos) Notebook* 1965

Non-Fiction

Heroes and Other Enlisted Men 1983; *Recitative* [ed. J.D. McClatchy] 1986; *A Different Person* **(a)** 1993

Edited

Jeffrey Harrison *The Singing Underneath* 1988

Collections

Jim's Book: a Collection of Poems and Short Stories 1942

W(illiam) S(tanley) MERWIN 1927–

Merwin was born in New York City, but his father was a Presbyterian minister, whose calling took the family first to Union City, New Jersey, and then to their native region of Scranton, Pennsylvania, where Merwin was educated. He went on a scholarship to Princeton in 1944 (his career there was interrupted by a year in the US Navy Air Corps), but, according to his own account, he was not a good student, spending much of his time horse-riding, and only being saved from expulsion by the intervention of his teacher, the respected critic R.P. Blackmur. Nevertheless, he spent a year in the Princeton graduate school, studying romance languages, before being advised that an academic life was not for him. He got work overseas, as a tutor, first in France and Portugal, and then tutoring the son of **Robert Graves** in Majorca. Here he met an English woman with whom he made the first of his several marriages, in 1954.

By this time, he had returned with her to England, where he supported himself doing translations from French and Spanish for what was then the Third Programme of the BBC. His first book of poetry, *A Mask for Janus*, was published in 1952 with the *imprimatur* of a foreword by **W.H. Auden** and made an immediate impact. In 1956, he returned to America to write plays for the Poets' Theatre, Cambridge, Massachusetts, and completed four, one in collaboration with his second wife, Dido Milroy, before losing interest in drama except as a vehicle for translation. Merwin is unusual among modern American poets in that he has made his career outside the campus, and has written, therefore, little criticism; his work, always highly regarded, has been sustained by a long series of prizes and awards, enabling him to live as a writer (and performance poet) from the first. He has spent long periods abroad, first at a farmhouse in Lot, France, in the 1960s and early 1970s, and later in Haiku, Hawaii, among other places.

Like many poets whose careers began in the early 1950s, Merwin started with technically controlled and formal verse, moving in the 1960s to more open utterance. His work is characteristically spare and difficult, but remains fundamentally lyrical, developing a strain of love poetry from the late 1970s, and gaining its fundamental unity from a continuing protest against the dehumanisation and destruction which he associates with his country's cultural imperialism. A minority of critics dismiss him as a poet, but he is generally accepted as major. He is almost equally celebrated as a translator, particularly from French and Spanish, but also other languages, and has published several prose collections, including a rather unrevealing volume of childhood autobiography, *Unframed Originals*.

Poetry

A Mask for Janus 1952; *The Dancing Bears* 1954; *Green with Beasts* 1956; *The Drunk in the Furnace* 1960; *The Moving Target* 1963; *The Lice* 1967; *Three Poems* 1968; *Animae* 1969; *The Carrier of Ladders* 1970; *Signs* 1971; *Asian Figures* 1973; *Writings to an Unfinished Accompaniment* 1973; *The First Four Books of Poems* 1975; *Three Poems* 1975; *The Compass Flower* 1977; *Feathers from the Hill* 1978; *Finding the Islands* 1982; *Opening the Hand* 1983; *Regions of Memory* [ed. E. Folsom, C. Nelson] 1987; *The Rain in the Trees* 1988; *Selected Poems* 1988; *The Lost Upland* 1992; *Travels* 1993

Plays

Darkling Child 1956 [np]; *Favor Island* 1957 [np]; *Eufemia* [from L. de Rueda's play; pub. 1956]; *The False Confession* [from Marivaux's play] 1963 [pub. in *The Classic Theatre* vol. 4, ed. E. Bentley 1961]; *The Gilded West* 1961 [np]; *Turcaret* [from A. Lesage's play; pub. in *The Classic Theatre* vol. 4, ed. E. Bentley 1961]; *Yerma* [from Lorca's play] 1966 [np]; *Iphigenia at Aulis* [from Euripides' play; with G.E. Dimock Jr] 1982

Non-Fiction

The Miner's Pale Children 1970; *Houses and Travellers* 1977; *Unframed Originals* (a) 1982; *Regions of Memory: Uncollected Prose 1949–1982* [ed. E. Folsom, C. Nelson] 1987; *A New Right Arm* [nd]

Edited

West Wind: Supplement of American Poetry 1961; *The Essential Wyatt* 1989

Translations

The Poems of the Cid 1959; *The Satires of Persius* 1961; *Some Spanish Ballads* 1961; *The Life of Lazarillo de Tormes* 1962; *The Song of Roland* 1963; *Selected Translations 1948–1968* 1968; *Transparence of the World: Poems of Jean Follain* 1969; *Products of the Perfected Civilization: Selected Writing by Sebastian Chamfort* 1969; *Voices: Selected Writings of Antonio Porchia* 1969 [rev. 1988]; Pablo Neruda: *Twenty Love Poems and A Song of Despair* 1969, *Selected Poems: a Bilingual Edition* (c) [ed. N. Tarn] 1969; *Chinese Figures: Second Series* 1971; *Japanese Figures* 1971; *Asian Figures* 1973; *Selected Poems of Osip Mandelstam* [with C. Brown] 1973; Roberto Juarroz *Vertical Poems* 1977 [rev. 1988]; *Sanskrit Love Poetry* [with J. Moussaieff Mason] 1977 [as *The Peacock Egg: Love Poems from Ancient India* 1981]; *Selected Translations 1968–1978* 1979; *Robert the Devil* 1981; *Four French Plays* 1985; *From the Spanish Morning* 1985; Muso Soseki *Sun at Midnight* [with S. Shigemetsu] 1989; Lorca *Blood Wedding; and, Yerma* [with L. Hughes] 1994

Charlotte (Mary) MEW 1869–1928

Mew's work, though small in quantity, is striking in its emotional sincerity and as a record of a life filled with a sense of loss, isolation and fear. She was born in London. Her father was an architect who had married the daughter of his partner; her mother, who felt she had married beneath her, was concerned with keeping up appearances at all costs and was seen as a silly, superficial woman. A wilful girl, Mew attended Lucy Harrison's School for Girls in Gower Street, but insisted on learning only what appealed to her, namely English literature and a little art and music. After leaving school she went to lectures at the 'Female School of Art' in Queen's Square. Three of her brothers died during childhood, and the presence of mental instability in both her surviving older brother and younger sister (they were confined to institutions in later life) remained a constant shadow. Her recognition of the dislocation of the mentally afflicted informs some of her most interesting poetry, notably 'Ken' and 'On the Asylum Road', the latter providing the memorable lines: 'Theirs is a house whose windows – every pane – / Are darkly stained or made of coloured glass.' Her family history made her and her remaining sister decide never to marry or bear children, although her own decision was also influenced by her lesbianism.

Her poetry and short stories were published in a variety of literary magazines, including Henry Harland's *Yellow Book*, the *Englishwoman*, the *Nation*, and the *Egoist*. Harland's encouragement brought her into the circles of **Max Beerbohm**, **Kenneth Grahame** and **Henry James**, and she struck a singular figure, with her bobbed hair, short stature and habit of swearing and smoking. A friendship developed between the relatively unknown Mew and the novelist **May Sinclair**, then approaching the height of her career. Sinclair brought Mew's work to the attention of influential writers and publishers such as Austin Harrison of the *English Review* and **Ezra Pound** of the *Egoist*. Another champion of her work was **Harold Monro**, who in 1916 published a collection of seventeen of her poems under the title *The Farmer's Bride*. It was highly regarded by many in the literary establishment, but it sold slowly, even though it was republished in 1921 in England and the USA. **Edith Sitwell** praised Mew for her lack 'of the self-conscious or the self-protective weakness of emotional poems by women', and **Virginia Woolf** regarded her as 'the greatest living poetess'. Other critics felt that Mew's work was often obscure. Her poetry adopts conventional stanzaic and rhyming patterns, but its effectiveness lies in the way that the emotional immediacy of those of whom she writes breaks through and shapes these patterns and cadences.

Poverty continued to characterise Mew's daily life, particularly after her father's death, which left the family in very reduced circumstances. The successful efforts of **Thomas Hardy**, **Siegfried Sassoon** and others to secure her a government pension of £75 in 1923 were a reflection of the high regard in which her work was held: her friends' petition stated that 'her work stands alone in power, quality and suggestion'. During this period, Mew looked after her ailing mother, and later her sister who died of cancer in 1927. She lived in increasingly shabby and depressing accommodation, and visitors noted her gradual withdrawal from life and company. Her suicide, by taking poison, in 1928 was the act of a despairing woman, burdened with the fate of her unstable family. Her work continued to be published and, although not widely known, it has retained its original expressive power.

Poetry

The Farmer's Bride 1916 [rev. 1921; US *Saturday Market* 1921]; *The Rambling Sailor* 1929; *Collected Poems* [ed. A. Monro] 1953

Collections

Collected Poems and Prose [ed. V. Warner] 1981

Biography

Charlotte Mew and Her Friends by Penelope Fitzgerald 1984

(John) Christopher MIDDLETON 1926–

Middleton was born in Truro, Cornwall, and brought up in Cambridge. He was educated at Felsted School and Merton College, Oxford, his education interrupted by service with the Royal Air Force between 1944 and 1948. After graduating, he embarked upon an academic career, becoming a lecturer in English at Zürich University, then, between 1955 and 1966, a lecturer (later senior lecturer) in German at King's College, London. He then moved to America to become professor of Germanic languages and literature at the University of Texas at Austin.

His first two volumes of poetry were published in 1944 and 1945, but he has subsequently dismissed them as 'quite irrelevant'. He therefore dates his work as a poet from 1962, when *Torse 3* was published. A collection of poems from the period 1949–61, it was hailed by **A. Alvarez** as 'a work of genuine distinction'. Alvarez noted that Middleton had 'an aesthete's concern with hardness of definition and pure, unencumbered language. But unlike most modern aesthetes, he also tackles the big, perennial themes of poetry.' Middleton, a leading translator of German poetry, has criticised the parochialism of many of his British contemporaries, who have, he maintains, 'cut themselves off from the tradition of European modernism'. This tradition, he says, has linked 'a strong sense of social revolution, a catastrophic view of history' with an 'interest in the radical remaking of techniques'. In volumes such as *The Lonely Suppers of W.V. Balloon*, reprinting poems from earlier pamphlets, he explores the nature of time, but human distress as witnessed in the century's killing fields – Buchenwald. Treblinka, Vietnam, Biafra – keeps intruding. Other volumes, such as *Pataxanadu* (which owes much to Aesop's fables), contain prose poems. As befits a translator, he is an innovative poet above all distinguished by his use of language, and the range of his work is exemplified by his collection of *111 Poems* and his volume of *Selected Writings*.

As a translator, he has worked on English editions of the poems of Günter Grass with fellow poet **Michael Hamburger**, with whom he has also edited and translated a volume of *Modern German Poetry*, covering half a century. *Andalusian Poems* is a volume of translations from Spanish versions of medieval Arabic poems. As well as poetry, he has also translated plays, fiction and letters, in particular a number of books by the German novelist Gert Hofmann, father of the poet **Michael Hofmann**. His own poems have been translated by other hands into Dutch and Swedish.

Poetry

Poems 1944; *Nocterne in Eden* 1945; *Torse 3: Poems 1949–1961* 1962; *Penguin Modern Poets 4* [with D. Holbrook, D. Wevill] 1963; *Nonsequences* 1966; *Die Taschenelefant* 1969; *Our Flowers and Nice Bones* 1969; *Fractions for Another Telemachus* 1974; *The Lonely Suppers of W.V. Balloon* 1975; *Wildhorse* 1975; *Razzmatazz* 1976; *Eight Elementary Inventions* 1977; *Anasphere le torse antique* 1978; *Carminalenia* 1980; *Wooden Dogs* 1982; *111 Poems* 1983; *Serpentine* 1984; *Two Horse Wagon Going By* 1986; *Andalusian Poems* [with L. Garza-Falcón] 1991; *The Balcony Tree* 1993; *Ballad of the Putrefaction* 1994; *Fishing Boats at Assos* 1994; *Some Dogs* 1994

Plays

The Metropolitans (l) [with H. Vogt] 1964

Non-Fiction

Bolshevism in Art and Other Expository Writings 1978; *The Troubled Sleep of America: 40 Collages* 1982; *The Pursuit of the Kingfisher: Essays* 1983; *Munich* 1993

Translations

Robert Walser *The Walk and Other Stories* (s) 1957, *Jakob von Gunten* 1969; Gottfried Benn *Primal Vision* [with others] 1960; Hugo von Hofmannsthal *Poems and Verse Plays* 1961; Günter Grass *Selected Poems* [with M. Hamburger] 1966 [rev. as *In the Egg and Other Poems* 1977], *Inmarypraise* 1974; Nietzsche *Selected Letters* 1969; Christa Wolf *The Quest for Christa T.* 1971; Paul Celan *Selected Poems* [with M. Hamburger] 1972; Friedrich Hölderlin and Eduard Mörike *Selected Poems* 1972; Elias Canetti *Kafka's Other Trial: the Letters to Felice* 1974; *Selected Stories* by Robert Walser [with others] 1982; Gert Hofmann: *Our Conquest* 1985, *The Spectacle at the Tower* 1985, *The Parable of the Blind* 1986, *Balzac's Horse and Other Stories* (s) 1988

Translations/Edited

Modern German Poetry 1910–1960: an Anthology with Verse Translations [with M. Hamburger] 1962; *The Poet's Vocation: Selections from the Letters of Hölderlin, Rimbaud, and Hart Crane* [tr. with W. Burford] 1967; Goethe *Selected Poems* [tr. with others] 1983; *The Figure on the Boundary Line: Selected Prose by Christoph Meckel* [tr. with others] 1983

Edited

German Writing Today 1967; Georg Trakl *Selected Poems* 1968; Lars Gustafsson *The Stillness of the World Before Bach: New Selected Poems* 1988

Collections

Pataxanadu and Other Prose 1977; *Selected Writings* 1989

Stanley MIDDLETON 1919–

Born in Bullwell, Nottingham, the son of a guard on the railway, Middleton was educated at High Pavement School, Nottingham, and at the then University College there. His education was interrupted by the war; he served in the army, which is the only time he has been out of Nottingham for a lengthy period. He married in 1951, has two daughters and has lived in the same Edwardian house in Sherwood for over thirty years.

Since 1958, when he published his first novel, *A Short Answer* – said by the *Sunday Times* to 'attempt a little more than the [**John**] **Wain** / [**Kingsley**] **Amis** approach' – hardly a year has gone by without a new novel, all from the same publisher. There is a regularity about Middleton's life so that, simply looking at the facts, one wonders where the variety comes from. He maintains that a writer does not have to travel if he is prepared to dig deep; in brief, he limits himself to what he knows. When he was a teacher he wrote three nights a week in term time and every day in the holidays; his only ambition, he said in 1974, was 'to write seriously the whole time'. Since he retired he has done exactly that.

A very private person, he hates being interviewed because it takes him from his work. *Holiday*, which shared the 1974 Booker Prize with **Nadine Gordimer**'s *The Conservationist*, is the story of a university teacher who goes to an east coast seaside resort resembling Skegness to try to come to terms with his separation from his wife; it is probably his best-known novel. Of the others, *A Man Made of Smoke* is about an ex-sergeant who returns to the army to run a small factory. 'It delighted me when the man next door, who is a businessman, read it and said, "Yes, that's right. That's just what it's like," ' he said.

Fiction

A Short Answer 1958; *Harris's Requiem* 1960; *A Serious Woman* 1961; *The Just Exchange* 1962; *Two's Company* 1963; *Him They Compelled* 1964; *Terms of Reference* 1966; *The Golden Evening* 1968; *Wages of Virtue* 1969; *Apple of the Eye* 1970; *Brazen Prison* 1971; *Cold Gradations* 1972; *A Man Made of Smoke* 1973; *Holiday* 1974; *Distractions* 1975; *Still Waters* 1976; *Ends and Means* 1977; *Two Brothers* 1978; *In a Strange Land* 1979; *The Other Side* 1980; *Blind Understanding* 1982; *Entry into Jerusalem* 1983; *The Daysman* 1984; *Valley of Decision* 1985; *An After Dinner's Sleep* 1986; *After a Fashion* 1987; *Recovery* 1988; *Vacant Places* 1989; *Changes and Chances* 1990; *Beginning to End* 1991; *A Place to Stand* 1992; *Married Past Redemption* 1993; *Catalysts* 1994

Plays

The Captain from Nottingham (r) 1972; *Harris's Requiem* (r) [from his novel] 1972; *A Little Music at Night* (r) 1972; *Cold Gradations* (r) [from his novel] 1973

Edna St Vincent MILLAY 1892–1950

Millay (whose middle name derives from the French priest St Vincent de Paul) was born at Rockland, Maine. Her father was a teacher with a weakness for poker playing, and in 1900 his wife divorced him, taking her three daughters to live in Camden, Maine, and supporting them by nursing. Millay attended Camden High School, but her most important stimulus came from her mother, who encouraged her musical and poetic ambitions. After leaving school, she remained at home, but in 1912 submitted a long poem, 'Renascence', to an anthology called *The Lyric Year*; although the poem was judged only the fourth-best submitted, it caused a sensation, and, on the strength of it, Caroline B. Dow of the National Training School of the YWCA arranged for Millay to go as an undergraduate to Vassar. She graduated in 1917, the year her first volume, *Renascence and Other Poems*, was published.

She migrated to New York, where she soon became part of the Bohemian society of Greenwich Village, and initially tried to support herself as an actress. **Edmund Wilson**, then editor of *Vanity Fair*, printed many of the early journalistic pieces, written under the pseudonym Nancy Boyd and collected in *Distressing Dialogues*; he later proposed marriage to her, and portrayed her as the heroine of his novel, *I Thought of Daisy* (1929). From the early 1920s her poetry became very popular, partly because she seemed to epitomise the liberated 'new woman' of the period, and in later life, although critics became hostile to her, she retained a very wide audience.

From 1921 to 1923 she travelled in Europe, writing for *Vanity Fair*, and, on her return, married Eugen Jan Boissevain, a much older Dutch businessman and widower; with him, in the mid-1920s, she bought the farm at Austerlitz, New York State, that was their main home for the rest of their lives.

Millay's poetry is largely forgotten today, and certainly it abounds in ridiculous lines such as 'O world, I cannot hold thee close enough!', but some of the more child-like poems based on fairy-tales and some of her verse-plays have charm, and she is probably at her most accomplished as a sonneteer, in such poems as 'Euclid alone has looked on beauty bare'. From the late 1920s she became interested in social protest, and during the war in Allied propaganda verse, but such poems as her elegy about the massacre of Lidice are not notable. She also translated Baudelaire's *Les fleurs du mal*, not very competently, and wrote an opera libretto. Her later life was passed in ill health, partly due to a serious car accident in 1936 and a nervous breakdown in 1944; she survived her husband by one year, dying alone at their farm.

Poetry

Renascence and Other Poems 1917; *A Few Figs From Thistles* 1921; *Second April* 1921; *Ballad of the Harp-Weaver* 1922; *Poems* 1923; *The Buck in the Snow and Other Poems* 1928; *Fatal Interview* 1931; *Wine From These Grapes* 1934; *Conversation at Midnight* 1937; *Huntsman, What Quarry?* 1939; *There Are No Islands Anymore* 1940; *The Murder of Lidice* 1942; *Poem and a Prayer for an Invading Army* 1944; *Mine the Harvest* [ed. N. Millay] 1954

Plays

Aria Da Capo 1920; *The Lamp and the Bell* 1921; *Two Slatterns and a King* 1924; *The King's*

Henchman 1927; *Three Plays* [pub. 1927]; *The Princess Marries the Page* 1932; *The Maid of Orleans* (l) [with M.S. McLain] 1942

Fiction

Distressing Dialogues [as 'Nancy Boyd'] 1924

Non-Fiction

Make Bright the Arms 1940; *Letters of Edna St Vincent Millay* [ed. A.R. Macdougall] 1952

Translations

Baudelaire *Flowers of Evil* [with G. Dillon] 1936

Biography

The Poet and Her Book by Jean Gould 1969

Arthur (Asher) MILLER 1915–

Miller was born in Harlem where his father, a Jewish immigrant of Austro-Hungarian extraction, ran a clothing manufacturing business. When Miller graduated from Abraham Lincoln High School in 1932 he had a poor academic record and little interest in reading, but, having read *The Brothers Karamazov*, resolved to write and spent two years working – first in his father's factory and later as a shipping clerk in a Manhattan warehouse – to save the tuition fees for college. He enrolled at the University of Michigan, Ann Arbor, in 1934 and studied drama under Kenneth Rowe.

His precarious finances were helped by university awards for his plays *No Villain* and *Honors at Dawn*, and in 1937 by the $1,250 he received from the Theater Guild's Bureau of New Plays Prize. 'One of the other winners,' Miller recalls in his autobiography, *Timebends*, 'was a fellow from St Louis with the improbable name **Tennessee Williams**, whom I envisioned in buckskins, carrying a rifle.' Having graduated in English in 1938 he returned to New York where he worked for the Federal Theater and wrote radio plays for CBS and NBC. He was exempt from the draft because of a football injury, and in 1940 married a Catholic girl, Mary Slattery, with whom he has two children.

Miller's first professionally produced play, *The Man Who Had All the Luck*, opened on Broadway in November 1944 and closed after four performances. The play depicts a young man whose success and happiness are undermined by his sense of impending disaster. Looking back at the play, Miller sees in its exploration of the father–son relationship the roots of a theme he was to explore more successfully in his subsequent two plays, *All My Sons* and *Death of a Salesman*. One of the great American plays of the century, *Salesman* shows Willy Loman, a pathetic and disillusioned travelling salesman, whose sons turn against him and accuse him of failing as a father. It was an immediate success, won a Pulitzer Prize in 1949, and was filmed by Elia Kazan in 1952. Miller's next original play, *The Crucible*, uses the seventeenth-century Salem witch hunts as an allegory for the McCarthy era, which did not help his cause when he was indicted for contempt by the House Un-American Activities Committee in 1956.

His fame increased in the same year when he divorced his first wife and married Marilyn Monroe, who was to star in John Huston's film from Miller's screenplay, *The Misfits*. Their marriage ended five years later and Miller married Inge Morath, a photographer with whom he has co-operated on two books about China and Russia. During the period of his marriage to Monroe, Miller wrote nothing for the theatre, and many critics felt that Maggie, the self-destructive central character in *After the Fall*, was modelled on Monroe, though Miller predictably denied this. Many of his later plays, such as *The Ride Down Mount Morgan* and *Broken Glass* have had their first productions in London. Miller has been politically active throughout his life and was a delegate at the 1968 Democratic Convention.

Plays

Honors at Dawn 1936 [np]; *No Villain (They Too Arise)* 1937 [np]; *The Pussycat and the Expert Plumber Who Was a Man* (r) 1941 [np]; *William Ireland's Confession* (r) 1941; *The Man Who Had All the Luck* 1944; *That They May Win* 1944 [pub. 1945]; *Grandpa and the Statue* (r) 1945; *The Story of G.I. Joe* (f) 1945; *The Story of Gus* (r) 1947; *The Guardsman* (r) [from F. Molnar] 1947; *Three Men on a Horse* (r) [from G. Abbott and J.C. Holm] 1947; *All My Sons* (st) 1947, (f) 1948; *Death of a Salesman* (st) 1949 (f) 1952; *An Enemy of the People* [from Ibsen's play] 1950 [pub. 1951]; *The Crucible* (st) 1953, (f) [as *The Witches of Salem*] 1957, (l) [with R. Ward] 1961; *A View from the Bridge* (st) 1955 [rev. 1965], (f) 1961; *A Memory of Two Mondays* 1955; *The Misfits* (f) 1961; *After the Fall* 1964; *Incident at Vichy* 1964 [pub. 1965]; *The Price* 1968; *Fame* and *The Reason Why* 1970; *The Creation of the World and Other Business* 1972 [pub. 1973; rev. as *Up from Paradise* 1974, pub. 1984]; *The Archbishop's Ceiling* 1977 [rev. 1984]; *The American Clock* 1980 [pub. 1983]; *Playing for Time* [from F. Fenelou] (tv) 1980, (st) 1985 [pub. 1981]; *Elegy for a Lady* 1982 [pub. 1984]; *Some Kind of Love Story* (st) 1982, (f) [as *Everybody Wins*] 1990 [pub. 1984]; *Danger! Memory!* 1987 [pub. 1986]: *Clara* 1987, *I Can't Remember Anything* 1987; *The Golden Years* (r) 1987 [pub. 1989]; *The Last Yankee* 1990; *The Ride Down Mount Morgan* 1991; *Gillbury* 1993; *Broken Glass* 1994

Fiction

Focus 1945; *The Misfits* [from his screenplay] 1961; *I Don't Need You Any More* (s) 1967; *'The Misfits' and Other Stories* (s) 1987

Non-Fiction

Situation Normal 1944; *The Theatre Essays of Arthur Miller* [ed. R.A. Martin] 1978; *Timebends: a Life* (a) 1987

With Inge Morath: *In Russia* 1969; *In the Country* 1977; *Chinese Encounters* 1979; *Salesman in Beijing* 1984

For Children

Jane's Blanket (s) 1963

Henry (Valentine) MILLER 1891–1980

Miller was born in New York, the son of a tailor of German extraction, and raised in Brooklyn. After high school he entered the City College in 1909 and left after two months, apparently in response to Spenser's *The Faerie Queene*: 'If I have to read stuff like that, I give up.' He worked briefly for a cement company before fleeing west to escape a difficult love affair. He later claimed to have been greatly changed by meeting the radical Emma Goldman at a rally in San Diego in 1913, though the historical veracity of this is doubtful. In his novels, letters and conversation, Miller often tended to make myths out of his own life, with scant regard to the literal truth.

From 1914 he assisted in his ailing father's tailor's shop, and from 1920 was employment manager for Western Union, an experience later transmuted into his description of the 'Cosmodemonic Telegraphic Company' in *Tropic of Capricorn*. His first marriage ended in 1924 when he married June Edith Smith, a Broadway taxi dancer who lent passionate encouragement to his literary aspirations. Living on the money June obtained from a succession of wealthy lovers, Miller wrote stories and two novels, *Crazy Cock* and *Moloch*, which remained unpublished until 1991 and 1992. Their tempestuous marriage is described in the trilogy *The Rosy Crucifixion*, and was characterised by Miller's friend Alfred Perles thus: 'She put him through the tortures of hell, and he was masochistic enough to enjoy it.'

The decisive moment in Miller's development came when he moved to Paris in 1930. Saved from destitution and near-starvation by Perles, Miller met **Anaïs Nin**, who subsequently underwrote the publication of *Tropic of Cancer* with 5,000 francs borrowed from another of her lovers, her psychiatrist Otto Rank. The torrid triangular relationship between Miller, June and Nin is described in Nin's unexpurgated volume of journals, *Henry and June*, which later formed the basis of a successful film. *Tropic of Cancer* is a riotous, sexually frank celebration of the author's 'heroic descent to the very bowels of the earth', which, predictably, remained unpublished in the USA until 1961, by which time its notoriety ensured huge sales. Another important friendship was with **Lawrence Durrell**, whom Miller met in 1937. Together with Nin they produced the magazine *The Booster*, and in 1939 Miller stayed with Durrell on Corfu and wrote *The Colossus of Maroussi*, arguably his best travel book.

With the outbreak of war Miller returned to the USA and his mordant observations on his native land are recorded in *The Air-Conditioned Nightmare*, a fierce diatribe against a soulless, materialistic country where 'a corn-fed hog enjoys a better life than a creative writer, painter or musician'. He settled at Big Sur on the Californian coast where he wrote *The Rosy Crucifixion*, a less successful venture in fictionalised autobiography than the *Tropic* novels. In a frank letter regarding the first volume, *Sexus*, Durrell asked: 'What on earth possessed you to leave so much twaddle in?' Having survived puritan censorship in his early writing life, in the 1970s Miller was castigated for sexism by feminists, notably Kate Millett in *Sexual Politics* (1969). These charges are forcefully rebutted by Erica Jong in her appreciation of Miller, *The Devil at Large* (1993). Miller was married five times, lastly in 1967 to a young Japanese cabaret singer, Hoki Tokuda, who subsequently ran a Tokyo night-club called 'Tropic of Cancer'.

Fiction

Tropic of Cancer 1934; *Tropic of Capricorn* 1939; *The Smile at the Foot of the Ladder* 1948; *The Rosy Crucifixion*: *Sexus* 1949, *Plexus* 1952, *Nexus* 1960; *Quiet Days in Clichy* 1956; *Under the Roofs of Paris* 1983; *Crazy Cock* 1991; *Moloch* 1992

Non-Fiction

Aller Retour New York 1935; *What Are You Going to Do about Alf?* 1935; *Black Spring* 1936; *Max and the White Phagocytes* 1938; *The Cosmological Eye* 1939; *Hamlet* 2 vols 1939, 1941; *The World of Sex* 1940; *The Colossus of Maroussi* 1941; *Wisdom of the Heart* 1941; *Plight of the Creative Artist in the United States of America* 1944; *Semblance of a Devoted Past* 1944; *Sunday After the War* 1944; *The Air-Conditioned Nightmare* 1945; *Henry Miller Miscellanea* [ed. B. Porter] 1945; *Maurizius Forever* 1946; *Of, By and About Henry Miller* 1947; *Remember to Remember* 1947; *The Waters Reglitterized* 1950; *The Books in My Life* 1952; *Rimbaud* 1952; *Nights of Love and Laughter* [ed. K. Rexroth] 1955; *The Time of the Assassins* 1956; *Big Sur and the Oranges of Hieronymous Bosch* 1957; *The Red Notebook* 1958; *Art and Outrage* 1959; *The Henry Miller Reader* [ed. L. Durrell] 1959; *The*

Intimate Henry Miller 1959; *To Paint is to Love Again* 1960; *Joseph Delteil: Essays in Tribute* 1962; *The Michael Fraenkel–Henry Miller Correspondence Called Hamlet* 1962; *Stand Still Like a Humming Bird* 1962; *A Private Correspondence: Lawrence Durrell and Henry Miller* 1963; *Greece* 1964; *Journey to an Antique Land* 1965; *Letters to Anaïs Nin* [ed. G. Stuhlmann] 1965; *Order and Chaos Chez Hans Reichel* 1966; *Collectors Quest: the Correspondence of Henry Miller and J. Rives Childs 1947–1965* [ed. R.C. Wood] 1968; *Writer and Critic: a Correspondence with Henry Miller* [W.A. Gordon] 1968; *Entretiens de Paris avec Georges Belmont* 1970; *On Turning Eighty* 1972; *Reflections on the Death of Mishima* 1972; *First Impressions of Greece* 1973; *Letters of Henry Miller and Wallace Fowlie 1943–1972* 1974; *My Life and Times* 1975; *The Nightmare Notebook* 1975; *Book of Friends*: *Book of Friends* 1976, *My Bike and Other Friends* 1978, *Joey* 1979; *Flashback: Entretiens de Pacific Palissandes avec Christian Bartillat* 1976; *Gliding into the Everglades and Other Essays* 1976; *The Ineffable Frances Steloff* [with A. Nin] 1976; *J'Suis pas plus con qu'un autre* 1976; *Our America* [with A. Rattner] 1976; *Four Visions of America* (c) 1977; *Sextet* 1977; *Henry Miller: Years of Trial and Triumph: the Correspondence of Henry Miller and Elmer Gertz* 1978; *Love Between the Sexes* 1978; *The Theatre and Other Pieces* 1979; *Correspondence privée de Henry Miller et Joseph Delteil 1935–78* [ed. F.J. Temple] 1980; *Notes on Aaron's Rod and Other Notes on Lawrence from the Paris Notebooks* [ed. S. Cooney] 1980; *The World of Lawrence: a Passionate Appreciation* [ed. E.J. Hinz, J.J. Teunissen] 1980; *From Your Capricorn Friend* 1984; *Dear Brenda* [ed. G.S. Sindell] 1986; *Letters from Henry Miller to Hoki Tokuda Miller* [ed. J. Howard] 1986; *A Literate Passion: Letters of Anaïs Nin and Henry Miller* [ed. G. Stuhlmann] 1987; *The Durrell–Miller Letters 1935–1980* [ed. S. MacNiven] 1988; *Henry Miller's Hamlet Letters* [ed. M. Hargraves] 1988; *Letters to Emil* [ed. G. Wickes] 1989; *Henry Miller – The Paintings* [ed. D. Johansen] 1991; *Nothing but the Marvelous* 1991; *Octet* 1991; *A Devil in Paradise* 1993; *Henry Miller and James Laughlin* 1995

Plays

Just Wild about Harry 1963

Collections

A Henry Miller Reader [ed. J. Calder] 1983; *Henry Miller* 1992

Biography

Henry Miller by Robert Ferguson 1991; *The Devil at Large* by Erica Jong 1993

A(lan) A(lexander) MILNE 1882–1956

The son of a preparatory school headmaster of Scots origin, Milne was born in London, and educated at Westminster School and Trinity College, Cambridge, where he edited *Granta*. On coming down in 1903 he set about establishing himself as a journalist in London, and by 1906 was assistant editor of *Punch*. During the First World War he was first a signalling officer in England, was posted to France briefly in 1916 and then wrote propaganda for the Intelligence service. He used wartime to write his early fantasy plays and, by 1918, had enough success with this not to return to *Punch* but to become a freelance writer.

He was to write around twenty plays for the West End stage until 1938 when the vogue for his work in this form faded; with titles such as *Mr Pim Passes By* and *The Dover Road*, they were mainly pleasant, undemanding comedies and are now forgotten. Milne essayed almost every literary form – he wrote a detective novel, *The Red House Mystery*, non-genre novels, many volumes of verse and essays, short stories, two volumes of autobiography – and performed gracefully in most, but he is remembered today solely for the four books he wrote for children between 1924 and 1928 and for his stage adaptation of **Kenneth Grahame**'s *The Wind in the Willows*, *Toad of Toad Hall* (1929); he wrote nothing for children after this date.

He married Dorothy de Selincourt in 1931, and they had one son, Christopher Robin, whom Milne portrayed in the children's books for which he is best known. Milne was a shy, pipe-smoking Englishman, who seems not to have been able to communicate well with his son, but the two books of verse, *When We Were Very Young* and *Now We Are Six*, and the two volumes of stories about a teddy-bear, Winnie-the-Pooh, show a remarkable ability to understand the world of children. Although occasionally mawkish, the poems are often very funny and have good, strong rhythms, which make them perfect for reading aloud. The stories featuring Pooh, a bear of very little brain but with a 'hum' or poem to suit every occasion, show a real gift for characterisation, particularly when Milne depicts the supporting cast: the supposedly wise Owl and the splendidly lugubrious Eeyore (a toy donkey apparently drawn from Milne's boss at *Punch*, Owen Seaman). All the books benefited enormously from extremely fine 'decorations' by Ernest H. Shepard. They have been translated into Latin and (less successfully) into Disney cartoon films.

Milne has had his critics, notably **Dorothy Parker**, whose notorious review of *The House at*

Pooh Corner written for the *New Yorker*'s 'Constant Reader' column was headed 'Far from Well' and concluded: 'Tonstant Weader Fwowed up'. The lost world of 1920s' middle-class London, with its uniformed nannies and their besmocked charges, is seen by some as outmoded and 'irrelevant' to the late twentieth-century child, but these books have the enduring qualities of classics and will ensure that Milne is remembered as more than simply a cheerful professional writer.

He lived for much of his life between Chelsea in London and a country house in Sussex on the edge of Ashdown Forest, where the Pooh books are set. An operation on his brain in 1952 left him an invalid during the last four years of his life and ended his writing career. His son, who survived being immortalised to run an independent bookshop in Dartmouth, recalled his childhood in *The Enchanted Places* (1974).

For Children

Once Upon a Time 1917; *When We Were Very Young* 1924; *A Gallery of Children* **(s)** 1925; *The King's Breakfast* 1925; *Winnie-the-Pooh* **(s)** 1926; *Now We Are Six* 1927; *The House at Pooh Corner* **(s)** 1928; *Prince Rabbit and The Princess Who Could Not Laugh* **(s)** 1966

Plays

Belinda 1918; *First Plays* [pub. 1919]; *Mr Pim Passes By* 1921; *The Dover Road* 1921; *Second Plays* [pub. 1921]; *The Stepmother* 1921; *Three Plays* [pub. 1922]; *The Artist* 1923; *Success* 1923; *Ariadne* 1925; *To Have the Honour* 1925; *The Man in the Bowler Hat* 1925; *Four Plays* [pub. 1926]; *Miss Marlowe at Play* 1926; *The Ivory Door* 1928; *The Fourth Wall* 1929; *Toad of Toad Hall* [from K. Grahame's novel *The Wind in the Willows*] 1929; *Other People's Lives* 1935; *Miss Elizabeth Bennett* 1936; *The General Takes Off His Helmet* 1939; *The Ugly Duckling* 1941; *Before the Flood* 1951

Fiction

Lovers in London 1905; *The Holiday Round* 1912; *Once a Week* 1914; *Happy Days* 1915; *Mr Pim* 1921; *The Red House Mystery* 1922; *The Secret and Other Stories* **(s)** 1929; *Two People* 1931; *Four Days' Wonder* 1933; *Chloe Marr* 1946; *Birthday Party and Other Stories* **(s)** 1948; *A Table Near the Band* **(s)** 1950

Poetry

For the Luncheon Interval 1925; *Behind the Lines* 1940; *The Norman Church* 1948

Non-Fiction

Not That It Matters 1919; *If I May* 1920; *The Ascent of Man* 1928; *By Way of Introduction* 1929; *When I Was Very Young* **(a)** 1930; *Peace with Honour* 1934; *It's Too Late Now* **(a)** 1939; *War with Honour* 1940; *War Aims Unlimited* 1941; *Year In, Year Out* 1952

Collections

The Day's Play 1910; *The Sunny Side* 1921; *A.A. Milne: a Collection of Shorter Works* 1933; *The Pocket Milne: a Selection of Shorter Works* 1941

Biography

A.A. Milne by Anne Thwaite 1990

Anthony MINGHELLA 1954–

Minghella was born of Anglo-Italian parentage in Ryde, Isle of Wight, and educated at the University of Hull, graduating in 1975; between 1976 and 1981 he was a lecturer in drama there. Of his early plays, in *Whale Music* a student faces up to an unwanted pregnancy; *A Little Like Drowning* discusses the divided loyalties of an Italian Catholic family split by adultery; and *Two Planks and a Passion* follows the machinations of the citizens of York while rehearsing the Mystery Cycle to be performed before Richard II. Michael Coveney has said of him that he is 'an unclassifiable dramatist. He is not knowingly, or overtly, witty, or fashionable, or blatantly left wing. He is, on the other hand, intriguing, genuinely experimental, unpredictable and dedicated.' In *Made in Bangkok*, a bittersweet comedy with a moral question at its heart, we see a group of tourists grasping the sexual opportunities offered by that city.

Minghella has written widely for television and radio, notably several episodes of the *Inspector Morse* television series, based on the popular crime novels of Colin Dexter. *What If It's Raining?*, a play for television, was described by Mark Lawson as 'wonderfully depicting the difficulties of grown-ups learning to be adults'. His radio play *Cigarettes and Chocolate* won the Best Drama, Sony and Giles Cooper awards; another radio play, *Hang-Up*, won the Prix Italia for Radio Fiction in 1988. His screenplay for the film *Truly, Madly, Deeply*, which he directed, was also much praised.

Plays

Mobius the Stripper [from G. Josipovici's story] 1975; *Child's Play* 1978; *Whale Music and Other Plays* [pub. 1987]: *Whale Music* 1980, *A Little Like Drowning* 1982, *Two Planks and a Passion* 1983; *Love Bites* 1984; *Made in Bangkok* 1986; *Interior: Room: Exterior: City* [pub. 1989]: *What If It's Raining?* **(tv)** 1986, *Hang-Up* **(r)** 1987, *Cigarettes and Chocolate* **(r)** 1988; *The Dead of Jericho* **(tv)** 1987; *Last Seen Wearing* **(tv)** [as 'Thomas Ellis']

1988; *Deceived by Flight* **(tv)** 1989; *Living with Dinosaurs* **(tv)** 1989 [pub. 1991]; *Second Thoughts* **(tv)** 1989; *Driven to Distraction* **(tv)** 1990; *Truly, Madly, Deeply* **(f)** 1991

For Children

Jim Henson's Storyteller 1988

Adrian MITCHELL 1932–

Poet, novelist, playwright and performance-artist, especially well known for his contribution to the pop-poetry movement of the late 1960s and early 1970s, Mitchell was born in London. His father was a scientist and his mother a teacher, and he was educated at Dauntsey's School in Wiltshire and then, after national service, at Christ Church, Oxford. In 1955 a pamphlet of his poems was published in the Fantasy Press series, based in Oxford. After university, Mitchell worked as a reporter on the *Oxford Mail* for two years and then, until 1959, on the London *Evening Standard.*

Since that time, he has lived as a professional writer, largely by various temporary fellowships and writer's residenceships, during which he has often encouraged people to group-write songs and shows. To take one period of his professional life, for instance: from 1978 to 1980 Mitchell was a visiting writer at Billericay Comprehensive School; in 1980 and 1981 he was Judith E. Wilson Fellow at Cambridge University, and in 1982 and 1983 he was writer-in-residence at the Unicorn Theatre for children in London. He has published several volumes of verse, in which the themes of political and social protest are prominent; he is on record as saying that 'the lowest common denominator is the only audience worth bothering about', and his poetry, which is often intended for performance, entirely lacks the density and complexity associated with most contemporary verse. It is generally agreed that his work is more interesting for exemplifying a particular historical movement than for any intrinsic merit.

He has published three novels, of which the first was *If You See Me Comin'* in 1962, and has written many plays and stage adaptations; in recent years a concentration on theatre work, and writing for children, has supplanted poetry and fiction.

Poetry

(Poems) 1954; *Poems* 1964; *Peace Is Milk* 1966; *Out Loud* 1968 [rev. *The Annotated Out Loud* 1976]; *Cease-fire* 1973; *Penguin Modern Poets 22* [with J. Fuller, P. Levi] 1973; *The Apeman Cometh* 1975; *For Beauty Douglas: Collected Poems 1953–1979* 1982; *On the Beach at Cambridge* 1984; *Adrian Mitchell's Greatest Hits* 1991

Plays

Animals Can't Laugh **(tv)** 1961; *The Ledge* **(l)** [with R.R. Bennett] 1961 [np]; *The Island* **(l)** [with W. Russo], **(r)** 1963; *The Persecution and Assassination of Jean-Paul Marat* [from P. Weiss's play] 1964, **(f)** 1966 [pub. 1965]; *The Magic Flute* **(l)** [from Mozart's opera] 1966 [np]; *US* [with others] 1966 [pub. 1968; in US as *Tell Me Lies*]; *Alive and Kicking* **(tv)** 1971; *Tamburlane the Mad Hen* 1971; *Tyger* [with M. Westbrook] 1971; *Man Friday* [from his novel; with M. Westbrook] **(tv)** 1972, **(st)** 1973, **(f)** 1976 [pub. 1974]; *Mind Your Head* [with A. Roberts] 1973 [pub. 1974]; *The Inspector General* [from Gogol's play] 1974 [rev. as *The Government Inspector* 1985]; *Someone Down There Is Crying* **(tv)** 1974; *Daft as a Brush* **(tv)** 1975; *The Fine Art of Bubble Blowing* **(tv)** 1975; *Silver Giant, Wooden Dwarf* **(tv)** 1975; *Total Disaster* **(tv)** [with B. Clark] 1975; *A Seventh Man* [from J. Berger and J. Mohr with D. Brown] 1976 [np]; *Houdini* **(l)** [with P. Schat] 1977; *Round the World in Eighty Days* **(l)** [from J. Verne's novel; with B. Scott] 1977; *White Suit Blues* [from M. Twain; with M. Westbrook] 1977 [np]; *Uppendown Mooney* 1978; *The White Deer* [from J. Thurber] 1978 [np]; *Glad Day* [with M. Westbrook] **(tv)** 1979; *Hoagy, Bix and Wolfgang Beethoven Bunkhaus* 1979 [np]; *In the Unlikely Event of an Emergency* 1979 [np]; *Peer Gynt* [from Ibsen's play] 1980 [np]; *Juno and Avos* **(tv)** [from A. Rybnikov and A. Voznesensky] 1983; *The Tragedy of King Real* **(f)** 1983 [pub. in *Peace Plays 1* ed. S. Lowe 1985]; *C'mon Everybody* 1984 [np]; *The Travels of Lancelot Quail* **(l)** [pub. 1984]; *Satie Day/Night* 1986 [np]; *The Last Wild Wood in Sector 88* 1987; *Love Songs of World War Three* 1987 [pub. 1988]; *Mirandolina* [from Goldoni] 1987 [np]; *Anna on Anna* 1988; *The Patchwork Girl of Oz* 1988 [np]; *Woman Overboard* 1988 [np]; *Triple Threat* 1989 [np]; *Vasilisa the Fair* [from S. Prokofieva and I. Tokmatova] 1991 [np]

For Children

Mowgli's Jungle [from Kipling] 1981 [pub. 1993]; *You Must Believe All This* **(tv)** [from Dickens's 'Holiday Romance'; with N. Bicat, A. Dickson] 1981; *The Wild Animal Song Contest* 1982 [pub. 1993]; *A Child's Christmas in Wales* [from D. Thomas; with J. Brooks] 1983 [pub. 1986]; *Animal Farm* **(l)** [from G. Orwell's novel; with P. Hall, R. Peaslee] 1984 [pub. 1985]; *Nothingmas Day* 1984; *The Adventures of Baron Munchausen* 1985; *The Baron Rides Out* 1985; *The Baron on the Island of Cheese* 1986; *Leonardo, the Lion from Nowhere* 1986; *The Pied Piper* [from R. Browning's poem; with D. Muldowney] 1986 [pub. 1988]; *The Baron All at Sea* 1987; *Our Mammoth* 1987; *Our*

Mammoth Goes to School 1987; *Our Mammoth in the Snow* 1989; *Strawberry Drums* 1989; *All My Own Stuff* 1991; *The Ugly Duckling* 1994

Fiction

If You See Me Comin' 1962; *The Bodyguard* 1970; *Wartime* 1973; *Man Friday* 1975

Non-Fiction

Naked in Cheltenham 1978; *Tourist Snapshots of Chile* 1985

Edited

Oxford Poetry [with R. Selig] 1955; Tim Daley *Jump, My Brothers, Jump: Poems from Prison* 1970; *The Oxford Book of Australian Literature* [with L. Kramer] 1985; *The Orchard Book of Poems* 1993

Translations

Jose Triana *The Criminals* 1967 [np]; *Victor Jara: His Life and Songs* [with J. Jara] 1976; Calderón: *The Mayor of Zalamea* 1981, *Life's a Dream* [with J. Barton] 1983, *The Great Theatre of the World* 1984 [pub. in *Three Plays by Calderón* 1990]; Lope de Vega Carpio: *Fuente Ovejuna* and *Lost in a Mirror* [pub. 1989]

Collections

Ride the Nightmare: Verse and Prose 1971

Gladys (Maude Winifred) MITCHELL 1901–1983

One of the most original of the detective-story practitioners of the 'Golden Age' of the 1920s and 1930s, and maintaining her productivity until her death in the early 1980s, Mitchell was born in Cowley, then a little outside Oxford, into a family of partly Scottish origin. In 1909, the family moved to the Brentford and Isleworth area on the western outskirts of London, where she attended the Rothschild School, Brentford, and the Green School, Isleworth. The area, and memories of her childhood, feature in one of the best of all her detective novels, *The Rising of the Moon*.

She attended Goldsmith's College, London, between 1919 and 1921, and then became a secondary school teacher in the early 1920s, also continuing to study, at University College London, where she took a diploma in European history in 1926. She taught until reaching the retirement age of sixty, except for a brief period between 1950 and 1953, after which boredom with living simply as a writer made her take up teaching again. Her various schools were in the western London outskirts, at Brentford, Hanwell and Staines; her subjects English, history and games (with, at one period, the unlikely addition of Spanish). In her work, her rationalist attitude mixed with eccentric enthusiasms, unsentimental sympathy for young people, and frequent school settings are all an obvious reflection of her occupation. In retirement she lived in the village of Corfe Mullen in Dorset. She never married.

Mitchell's first four novels, written in early adulthood, were rejected, but she published *Speedy Death* in 1929, which launched her series detective, Mrs (later Dame) Beatrice Adela Lestrange Bradley, the sybilline Home Office psychiatrist who is sometimes known, because of her saurian appearance, as 'Mrs Croc', and who is unusual amongst fictional detectives in having committed a murder herself. Dame Beatrice features, with other recurring characters, in all of the more than sixty novels which Mitchell published under her own name at a rate of one (sometimes two) a year until her death, a few more appearing posthumously.

She wrote two other series, under pseudonyms: a group of 'straight' novels in the 1930s, published under the name Stephen Hockaby, including the excellent historical novel, *Grand Master*, which centres on the Knights of Malta; and a series of detective novels in the late 1960s and early 1970s, as Malcolm Torrie, and featuring the detective Timothy Herring. She also wrote detective novels for children, and her books in total number eighty-six.

Her no-nonsense style, elaborate plotting and early membership of the Detection Club make her sound a typical literate detective author of her period who went on writing into old age, perhaps another Elizabeth Ferrars, but she is not. **Philip Larkin**, who appreciated her strangeness, called her 'The Great Gladys', and she almost seems to parody the detective genre, certainly to play games with it. Her proper place, therefore, is as a literary writer, and she is likely to have a permanent, although perhaps small, claim on the attention of readers.

Fiction

As Gladys Mitchell: *Speedy Death* 1929; *Mystery of a Butcher's Shop* 1929; *The Longer Bodies* 1930; *The Saltmarsh Murders* 1932; *Death at the Opera* 1934; *The Devil at Saxon Wall* 1935; *Dead Man's Morris* 1936; *Come Away Death* 1937; *St Peter's Finger* 1938; *Printer's Error* 1939; *Brazen Tongue* 1940; *Hangman's Curfew* 1941; *When Last I Died* 1941; *Laurels Are Poison* 1942; *Sunset Over Soho* 1943; *The Worsted Viper* 1943; *My Father Sleeps* 1944; *The Rising of the Moon* 1945; *Here Comes a Chopper* 1946; *Death and the Maiden* 1947; *The Dancing Druids* 1948; *Tom Brown's Body* 1949; *Groaning Spinney* 1950; *The Devil's Elbow* 1951; *The Echoing Strangers* 1952; *Merlin's Furlong* 1953; *Faintley Speaking* 1954;

Watson's Choice 1955; *Twelve Horses and the Hangman's Noose* 1956; *The Twenty-Third Man* 1957; *Spotted Hemlock* 1958; *The Man Who Grew Tomatoes* 1959; *Say It with Flowers* 1960; *The Nodding Canaries* 1961; *My Bones Will Keep* 1962; *Adders on the Heath* 1963; *Death of a Delft Blue* 1964; *Pageant of Murder* 1965; *The Croaking Raven* 1966; *Skeleton Island* 1967; *Three Quick and Five Dead* 1968; *Dance to Your Daddy* 1969; *Gory Dew* 1970; *Lament for Leto* 1971; *A Hearse on May Day* 1972; *The Murder of Busy Lizzie* 1973; *A Javelin for Jonah* 1974; *Winking at the Brim* 1974; *Convent of Styx* 1975; *Late, Late in the Evening* 1976; *Fault in the Structure* 1977; *Noonday and Night* 1977; *Mingled with Venom* 1978; *Wraiths and Changelings* 1978; *The Mudflats of the Dead* 1979; *Nest of Vipers* 1979; *Uncoffin'd Clay* 1980; *The Whispering Knights* 1980; *The Death-Cap Dancers* 1981; *My Father Sleeps* 1981; *Lovers, Make Moan* 1981; *Death of a Burrowing Mole* 1982; *Here Lies Gloria Mundy* 1982; *Cold, Lone and Still* 1983; *The Greenstone Griffins* 1983; *The Crozier Pharoes* 1984; *No Winding-Sheet* 1984

As 'Stephen Hockaby': *Marsh Hay* 1934; *Seven Stars and Orion* 1935; *Shallow Brown* 1936; *Grand Master* 1939

As 'Malcolm Torrie': *Heavy as Lead* 1966; *Late and Cold* 1967; *Your Secret Friend* 1968; *Churchyard Salad* 1969; *Shades of Darkness* 1970; *Bismarck Herrings* 1971

For Children

The Three Fingerprints 1940; *Holiday River* 1948; *Outlaws of the Border* 1948; *The Seven Stones Mystery* 1949; *The Malory Secret* 1950; *Pam at Storne Castle* 1951; *Caravan Creek* 1954; *On Your Marks* 1954; *The Light-Blue Hills* 1959

J. Leslie MITCHELL

see Lewis Grassic GIBBON

(Charles) Julian MITCHELL 1935–

Born in Epping, Essex, the son of a London solicitor, Mitchell was educated at Winchester and at Wadham and St Antony's Colleges, Oxford. He has followed no career other than that of a writer, publishing four novels before he was thirty, and later becoming a well-known playwright. He was first published in the *Faber Introductions* series in 1960. His first novel, *Imaginary Toys* (1961), was conventionally set during an Oxford summer term, but contained essayistic touches reminiscent of the early **Aldous Huxley**. *The White Father*, partly set in a remote African territory, is among the most interesting of Mitchell's novels; *The Undiscovered Country* was by far the most experimental.

Mitchell subsequently turned to drama, beginning with stage adaptations of the novels of **I. Compton-Burnett** in the late 1960s, and he has written more than thirty broadcast plays, both original stories and adaptations. He has also achieved West End success, with plays such as *Half-Life* and, particularly, *Another Country*, his best-known work, which is set in an English public school in the 1930s and loosely based on the imagined schooldays of the spy Guy Burgess. The play was notable for its stylishness and wit, its touching and funny depiction of adolescent sexuality, and for launching the careers of several young actors. Mitchell himself adapted the play for a vastly inferior film (1984), which preserved Rupert Everett's remarkable performance in the central role, but dispensed with one of the principal characters and in almost every other way coarsened the original. Other plays since then include *Francis*, recounting the life of the saint, and *After Aida*, about how Verdi came to write his last opera, *Falstaff*.

Mitchell has had a successful career as a television dramatist, notably for his adaptation of **Paul Scott**'s 1977 book *Staying On*, which reunited Trevor Howard and Celia Johnson over thirty years after their appearance in the film *Brief Encounter* (1945), and his scripts for episodes of the hugely popular *Inspector Morse* series, based on the crime novels of Colin Dexter. He has also been a frequent reviewer and broadcaster, becoming well known as a contributor to BBC radio's long-running *Critics' Forum*. He was a member of the Arts Council's literature panel between 1966 and 1969.

Fiction

Faber Introductions [with others] 1960; *Imaginary Toys* 1961; *A Disturbing Influence* 1962; *As Far as You Can Go* 1963; *The White Father* 1964; *A Circle of Friends* 1966; *The Undiscovered Country* 1968

Plays

A Heritage and Its History [from I. Compton-Burnett's novel] 1965; *A Family and a Fortune* [from I. Compton-Burnett's novel] 1975; *Half-Life* 1977; *Staying On* **(tv)** [from P. Scott's novel] 1979; *The Enemy Within* 1980; *Another Country* **(st)** 1981, **(f)** 1984; *Francis* 1983 [pub. 1984]; *After Aida* [pub. 1986]

Non-Fiction

Jennie, Lady Randolph Churchill: a Portrait with Letters [with P. Churchill] 1974

Translations

Pirandello *Henry IV* 1979

Naomi (Mary Margaret) MITCHISON 1897–

Born in Edinburgh into the famous Haldane family, Mitchison was educated at the Dragon School in Oxford, where her father, the physiologist and philosopher J.B.S. Haldane, was a fellow of New College. During the First World War she served as a VAD (volunteer nurse) in London, and in 1916 married Richard (Dick) Mitchison, who became a well-known barrister. The couple's first house was in Cheyne Walk, and it was here that Mitchison began her prolific writing career, publishing the first of her historical novels, *The Conquered*, in 1923.

After the birth of their first two children, the Mitchisons moved to Hammersmith, which in the 1920s was something of an artists' and intellectuals' colony. Although socialist in outlook, they had nannies and servants, a paradox she later explored in her evocative memoir, *You May Well Ask*. Whilst pursuing her writing, Mitchison was becoming involved with the Fabians and the Labour Party, and campaigned for birth control, a subject which led to Jonathan Cape refusing to publish her contemporary novel *We Have Been Warned*: 'I don't suppose any reputable writer before my time had mentioned the unpleasantness of the touch of rubber,' she later commented. Mitchison also campaigned for her husband who was standing as a Labour candidate.

During the Second World War she worked for Mass Observation, making notes about the lives of the people of Carradale in Scotland, where she had a house; extracts from the diaries she kept during this period were later published as *Among You Taking Notes*. Her husband was elected Member of Parliament for Kettering in the 1945 election and she continued to write and publish books in many genres: plays, children's books, autobiographies, travel books, social studies, and novels and stories set in the ancient world, Scotland and Africa. *The Corn King and the Spring Queen*, set in ancient Greece and Sparta, is her best-known novel. In 1963 she was adopted as 'tribal mother' to the Ba Kgatla of Botswana and spent some time with them. She now divides her time between Carradale and Kensington. The journalist Val Arnold-Foster is her daughter.

Fiction

The Conquered 1923; *When the Bough Breaks* (s) 1924; *Cloud Cuckoo Land* 1925; *Black Sparta: Greek Stories* (s) 1928; *Barbarian Stories* (s) 1929; *The Corn King and the Spring Queen* 1931; *We Have Been Warned* 1935; *The Blood of the Martyrs* 1939; *The Bull Calves* 1947; *Lobsters on the Agenda* 1952; *Travel Light* 1952; *To the Chapel Perilous* 1955; *Behold Your King* 1957; *Memoirs of a Spacewoman* 1962; *When We Become Men* 1965; *Cleopatra's People* 1972; *Solution Three* 1975; *Images of Africa* (s) 1981; *What Do You Think Yourself?* (s) 1982; *Not by Bread Alone* 1983; *Beyond This Limit* [ed. I. Murray] (s) 1986; *Early in Orcadia* (s) 1987; *A Girl Must Live* (s) 1990; *The Oath Takers* 1991; *Sea-Green Ribbons* 1991

For Children

Nix-Nought-Nothing 1928; *The Hostages* 1930; *Boys and Girls and Gods* 1931; *Kate Crackernuts: a Fairy Play* [pub. 1931]; *The Powers of Light* 1932; *Historical Plays for Schools* 1939; *Nix-Nought-Nothing and Elfin Hall* 1948; *The Big House* 1950; *Graeme and the Dragon* 1954; *The Swan's Road* 1954; *The Land the Ravens Found* 1955; *Little Boxes* 1956; *The Far Harbour* 1957; *Judy and Lakshmi* 1959; *The Rib of the Green Umbrella* 1960; *The Young Alexander the Great* 1960; *Karensgaard: the Story of a Danish Farm* 1961; *The Fairy Who Couldn't Tell a Lie* 1963; *The Young Alfred the Great* 1963; *Alexander the Great* 1964; *Ketse and the Chief* 1965; *Friends and Enemies* 1966; *The Big Surprise* 1967; *Highland Holiday* 1967; *African Heroes* 1968; *Don't Look Back* 1969; *The Family at Ditlabeng* 1969; *Sun and Moon* 1970; *Sunrise Tomorrow* 1973; *The Danish Teapot* 1973; *The Little Sister* [with I. Kirby, K. Masogo] 1976; *Snake!* 1976; *The Brave Nurse and Other Stories* (s) 1977; *The Wild Dogs* [with M. Biesele] 1977; *The Two Magicians* [with D. Mitchison] 1978; *The Vegetable War* 1980

Non-Fiction

Anna Comnena 1928; *The Home and a Changing Civilisation* 1934; *Naomi Mitchison's Vienna Diary* 1934; *Socrates* [with R.H.S. Crossman] 1937; *The Moral Basis of Politics* 1938; *The Kingdom of Heaven* 1939; *Men and Herring: a Documentary* [with D. Macintosh] 1949; *Other People's Worlds* 1958; *A Fishing Village on the Clyde* [with G.W.L. Paterson] 1961; *Presenting Other People's Children* 1961; *Return to the Fairy Hill* 1966; *The Africans* 1970; *Small Talk of an Edwardian Childhood* (a) 1973; *Sunrise Tomorrow. A Story of Botswana* 1973; *All Change Here: Girlhood and Marriage* (a) 1975; *You May Well Ask: a Memoir, 1920–1940* (a) 1979; *Mucking Around: Five Continents over Fifty Years* 1981; *Margaret Cole 1893–1980* 1982; *Among You Taking Notes* [ed. D. Sheridan] 1985; *Naomi Mitchison* (a) 1986; *As It Was* (a) 1988

Plays

With L.E. Gielgud: *The Price of Freedom* 1931 [pub. 1949]; *Full Fathom Five* 1932 [np]; *As It Was in the Beginning* 1939 [np]

An End and a Beginning and Other Plays [pub. 1937]; *The Corn King* [with B. Easdale] 1951; *Spindthrift* [with D. Macintosh] 1951

Collections
The Delicate Fire: Short Stories and Poems 1933; *The Fourth Pig: Stories and Verses* 1936; *Five Men and A Swan: Stories and Verse* 1957

Biography
Naomi Mitchison by Jill Benton 1990

Nancy (Freeman) MITFORD 1904–1973

The eldest daughter of the second Baron Redesdale, Mitford enjoyed an erratic upbringing, which has been described by her sister Jessica Mitford in *Hons and Rebels* (1960) and, lightly fictionalised, in her own novels. The eccentric child of eccentric parents she was educated largely at home. With her brother Tom, and her five sisters, she formed a close-knit group which spent much of its time antagonising her parents, 'Muv' and 'Farve'. She was presented at Court and briefly enrolled at the Slade School of Art.

Through her sister Diana she met **Evelyn Waugh** who guided her literary career, often correcting or making suggestions for her novels (she never mastered spelling or punctuation). Her first novel, *Highland Fling*, was published in 1931, and was followed by several other light comedies. *The Pursuit of Love* introduced the Radlett family, a rich gallery of comic characters, and was a perfect antidote to the privations being suffered by a war-weary public in 1945. With *Love in a Cold Climate*, which featured many of the same characters, it remains her most popular book.

After a brief and unsuccessful marriage to Peter Rodd, the model for Waugh's anti-hero Basil Seal, Mitford went to live in France, largely to be near Colonel Gaston Palewski, the original of Fabrice in her novels, with whom she had a long-standing but unsatisfactory affair. She wrote popular and lively biographies of such figures as Voltaire, Madame de Pompadour and Louis XIV. This last was the first 'coffee-table' book – 'one of those boring books millionaires give each other for Christmas', she observed, unfairly. She introduced the terms 'U' and 'Non-U' to a wider audience in her hugely successful sociological 'tease', *Noblesse Oblige* (a collaboration with such friends as Waugh, **John Betjeman** and Peter Fleming). She was awarded the Légion d'Honneur and the CBE in 1972, but her last years were spent in considerable pain, since she was suffering from cancer and Palewski had married someone else. She wrote a great deal of journalism, which has been collected under the fitting title *A Talent to Annoy*, and her splendidly idiosyncratic and lively letters have also been published.

Fiction
Highland Fling 1931; *Christmas Pudding* 1932; *Wigs on the Green* 1935; *Pigeon Pie* 1940; *The Pursuit of Love* 1945; *Love in a Cold Climate* 1949; *The Blessing* 1951; *Don't Tell Alfred* 1960; *The Water Beetle* 1962

Non-Fiction
Madame de Pompadour 1954; *Voltaire in Love* 1957; *The Sun King* 1966; *Frederick the Great* 1970; *A Talent to Annoy: Essays, Journalism and Reviews* [ed. C. Mosley] 1986; *Letters* [ed. C. Mosley] 1993

Edited
The Ladies of Alderley 1938; *The Stanleys of Alderley: Their Letters from 1851–1865* 1939; *Noblesse Oblige: an Enquiry into the Identifiable Characteristics of the English Aristocracy* [with A.S.C. Ross] 1956

Translations
Madame de La Fayette *The Princesse de Clèves* 1950; Alfred Roussin *The Little Hut* 1951

Biography
Nancy Mitford by Selina Hastings 1985

Timothy MO 1950–

Mo was born in Kowloon, Hong Kong, the son of a Cantonese barrister and an English mother. He was educated at the Convent of the Precious Blood and his first language was the Cantonese he learned from his parents' house servants. In 1960 he moved to London with his mother (who had remarried) and attended Mill Hill School before reading history at St John's College, Oxford. He worked as a journalist on the *Times Education Supplement*, the *New Statesman* and *Boxing News*. Boxing, along with scuba diving and weight training, remains one of his enthusiasms.

Mo's first novel, *The Monkey King*, which won the Geoffrey Faber Memorial Award in 1978, is a dark comedy reminiscent of early **Evelyn Waugh**, concerning a young man from the Portuguese colony of Macao who marries into a large Hong Kong Chinese family. *Sour Sweet*, which won the Hawthornden Prize in 1982, is set in London and describes an industrious family of Chinese immigrants who open a restaurant but live under threat from the Triads. The novel was made into a film by Mike Newell (1988) with a screenplay by **Ian McEwan**. Mo expanded his range of concerns with the ambitious *An Insular Possession*, a panoramic account of the Opium Wars and the founding of Hong Kong, and again in *The Redundancy of Courage*, which describes guerillas in East Timor, and won the E.M. Forster Award from the American Academy in 1991.

With these novels – the last three of which were all shortlisted for the Booker Prize – Mo enjoyed considerable critical acclaim and a fair measure of commercial success, which made it all the more surprising when he decided to publish *Brownout on Breadfruit Boulevard* himself, under his own imprint, Paddleless Press. He explained this as a response to the poor paperback sales figures for his earlier novels and to his general disillusionment with conglomerate publishing where 'you're playing office politics, you're pinning your flag to the correct superior, you're intimidating your subordinates and you're bull-shitting the authors, in that order'. Beginning with a modest print-run, Mo oversaw every stage of the publishing process from typesetting to promotion, and seemed to relish doing so. The novel is a political comedy set in the Philippines, and contains a particularly graphic description of coprophilia, which doubtless helped to generate some publicity.

Fiction

The Monkey King 1978; *Sour Sweet* 1982; *An Insular Possession* 1986; *The Redundancy of Courage* 1991; *Brownout on Breadfruit Boulevard* 1995

Deborah MOGGACH 1948–

Moggach was born Deborah Hough, one of the four daughters of the historian Richard Hough and his wife, Charlotte, a well-known writer and illustrator of children's books. She was educated at the Camden School for Girls in north London and subsequently spent a year in America teaching art and horseback riding in the Rockies. She attended a college in Massachusetts, where she was the only female student, then returned to England to read English at Bristol University, graduating in 1971. After leaving college she worked at the Oxford University Press before spending two years in Pakistan, where she became a journalist, winning the Young Journalist of the Year Award in 1975. During this period she married a publisher, with whom she has a son and a daughter.

She started writing her first novel, *You Must Be Sisters*, while she was in Pakistan, but it was not published until 1978. Her second, *Close to Home*, was a witty, faintly sinister comedy of family life in north London, which reflected her admission that her favourite hobby is 'walking around cities at dusk, looking into people's windows'. She drew upon her experiences in Pakistan to write *Hot Water Man*, a characteristically intelligent comedy of manners, which is set in the expatriate community of Karachi. It was with *Porky*, a disturbing novel about incest between a father and his daughter, that Moggach made her name as a novelist unafraid to deal with contemporary and controversial social issues. *To Have and to Hold*, for example, is about surrogate mothers (in this case a woman who decides to have a baby for her own sister), *Driving in the Dark* is about a coach driver searching England for his long-lost illegitimate son, and *Stolen* deals with a custody battle when a mixed marriage breaks down and two children are abducted and taken to Pakistan. Her novels are never merely 'issue' books, however, and are usually leavened by her acute wit.

The success of these novels led to a second career as a screenwriter. Moggach wrote such books as both *To Have and to Hold* and *Stolen* in tandem with highly successful television series, and has written several other screenplays. Her experiences in the film world resulted in *The Stand-In*, a thriller about an American film star and the Englishwoman who is employed by a studio as her double. 'I wanted to frighten myself,' Moggach said of the novel, 'and get near the edge of things I don't want to think about.' She has also written numerous short stories, which have been widely broadcast, published in various anthologies (including the annual *Best Short Stories* series) and collected in two volumes.

She was divorced in 1986 and was subsequently the partner, until his sudden death in 1994, of the well-known cartoonist Mel Calman.

Fiction

You Must Be Sisters 1978; *Close to Home* 1979; *A Quiet Drink* 1980; *Hot Water Man* 1982; *Porky* 1983; *To Have and to Hold* 1986; *Smile* (s) 1987; *Driving in the Dark* 1988; *Stolen* 1990; *The Stand-In* 1991; *The Ex-Wives* 1993; *Changing Babies* 1995

Plays

Crown Court (tv) 1983; *To Have and to Hold* (tv) 1986; *Doub' -Take* 1990; *Stolen* (tv) 1990; *Goggle Eyes* (tv) 1993

Harold (Edward) MONRO 1879–1932

Born in Belgium, Monro came to England at the age of seven and later went to Radley College, from which he was expelled for possessing wine.

Disgrace was a boon, however, for it led to him discovering poetry at Cannes, where his mother was taking a cure. He studied modern and medieval languages at Gonville and Caius College, Cambridge, where, on the face of it, he was more interested in horse-racing; he kept his poetry a secret for some while. On a walking tour in Germany with Maurice Browne, Monro fell in love with his companion's sister, Dorothy – a preoccupation which, allied with his poetry, led to his abandoning the Bar. In an odd move, he became a land-agent for a farmer in Ireland, a country which held no allure for Dorothy, who was, moreover, as much taken with hockey as she was with Monro.

They married in 1907 and moved from Ireland to Haslemere in Surrey. His first volume, *Poems*, which contained a small proportion of his output, appeared in 1906. Meanwhile, Monro had been in close correspondence with Browne, discussing poetry, and they now set up the Samurai Press. The first publication was a collaborative effort, the grimly Wellsian *Proposals for a Voluntary Nobility*. There cannot have been much of a sale among poets, for it outlawed promiscuity, alcohol, smoking, gambling and beards. Monro's own first publication with the press was hardly more appetising: a long blank-verse poem titled *Judas*. However, a number of poets had contacted the press, amongst them **John Drinkwater** and **Wilfrid Gibson**, whose work it published. In low spirits, Monro went abroad for a couple of years, partly and lumpenly recording his experiences in *The Chronicle of a Pilgrimage: Paris to Milan on Foot* and in the verse of *Before Dawn*. Travel brought contact with Maurice Hewlett and **Edward Carpenter**, and he also encouraged a young student called Arundel del Re, who accompanied him back to London. His marriage had failed, and he was eventually divorced in 1916. Browne, meanwhile, had gone to Chicago, where he established the Little Theater.

Monro was thirty-two and felt it, but was galvanised when in 1912 the Poetry Society asked him to edit a magazine, the *Poetry Review*. Demanding independence, he had no truck with woolly committeedom. He found an uneasy ally in **Ezra Pound**, and both men were keen to cut through the puffery of contemporary reviewing. Although weighed down by its titular, critical nature, the magazine also printed poetry, ranging from works by Pound to **Rupert Brooke**'s 'The Old Vicarage, Grantchester'. Monro eventually fell out with the Society, but in 1913, buoyed by family funds from a private lunatic asylum, he had established the Poetry Bookshop in a dank corner of Holborn: 35 Devonshire Street. The shop held readings and became a famous meeting place for poets, particularly during the First World War, when those on leave (including **Wilfred Owen**) could stay in bed-sitting rooms on the premises. The Poetry Bookshop was also a publishing house: one of its earliest successes was the bestselling *Georgian Poetry 1911–1912*, and Monro also started *Poetry and Drama*, a journal which was halted by the war, during which Monro was bizarrely sent to an anti-aircraft station in Manchester, before returning to work in Intelligence and run the bookshop.

Apart from the *Georgian Poetry* series (edited by Brooke's patron, Edward Marsh) and its obverse, Pound's *Des Imagistes*, the bookshop's most notable publications were **Robert Graves**'s first book of poetry, *Over the Brazier* (1916), and volumes by **Charlotte Mew** and **Anna Wickham**. Unfortunately, Monro turned down **Edward Thomas** and **T.S. Eliot** (who bore no grudge and later wrote an introduction for Monro's *Collected Poems*), and in their stead indulged his penchant for the voluminous, collaborative work of an aunt and niece who wrote under the joint pseudonym of Michael Field. Much of the shop routine fell to Alida Klemantaski, who had hoped variously to be a doctor, go on the stage or rescue prostitutes, but ended up as Monro's housekeeper. She was twenty years his junior and they married in 1920.

The next decade saw the continuing but less frequent appearance of a serviceable journal, *The Chapbook*. After this closed in 1925, Monro was beset by the painful ailments which resulted in his death seven years later. Despite near-blindness, alcoholism and despair, however, he worked on. His posthumous volume of *Collected Poems* is some tribute to a man split between the mystical, the whimsical and the effective. This last category embraces both the popular 'Milk for the Cat' and many of those poems – notably 'Midnight Lamentation' and the nightmarish 'Bitter Sanctuary' – written in that late, terrible period, which was made no easier by increasing awareness of his sexual ambivalence. As Eliot said, no one poem encapsulates the man. A single line, however, might be said to do so: 'I think too much of death.'

Poetry

Poems 1906; *Judas* 1907; *Before Dawn* 1911; *Children of Love* 1914; *Strange Meetings* 1917; *Real Property* 1922; *The Earth for Sale* 1928; *Collected Poems* 1933

Edited

Some Contemporary Poets 1920; *Twentieth-Century Poetry: an Anthology* 1929

Non-Fiction

The Chronicle of a Pilgrimage: Paris to Milan on Foot 1909–12

Harriet MONROE 1860–1936

Best known as one of the great animators of the modern movement through her magazine *Poetry*, but also a poet in her own right, Monroe was born in Chicago, the city with which she was to be identified all her life. She was the daughter of one of its pioneer attorneys and of Scottish stock on both sides, and her early background was affluent. Although her father lost a lot of money in the Chicago fire of 1871, he was still able to send her to school at the Visitation Convent, Georgetown.

From 1879 until 1912 when, aged over fifty, she founded her magazine, she undertook a wide variety of activities: she was an art critic for the Chicago *Tribune*, helped to found the Chicago Art Institute, reported in various papers, lectured, and carried on a long-distance correspondence with Robert Louis Stevenson. In 1891 she paid for the publication of her first volume of verse, *Valeria and Other Poems*. In 1892 she scored a poetic *coup* when her *Columbian Ode* was chanted by five thousand voices in the new auditorium built when Chicago was chosen as the site for the World Columbia Exposition. When the New York *World* printed her ode without permission, she collected $5,000 in damages with which she indulged her lifelong penchant for travel.

The next two decades were spent publishing a biography of her brother-in-law John Wellborn Root, one of the pioneers of the skyscraper, in writing several verse dramas and a long poem, *The Dance of Seasons*; but in 1911, on returning from a trip to Siberia and Peking, she conceived the idea of a magazine devoted simply to verse. She secured wealthy sponsors and mailed likely poets, including **Ezra Pound**, who became European correspondent for the first six years. *Poetry* quickly became, together with the *Little Review* and the *Dial*, one of the most influential American literary magazines of its time, a leading player in the Chicago literary renaissance, broadly modernist in tendency but offering a home to most schools of poetry, printing all the leading poets of its time, and securing perhaps its greatest triumph in being first to publish **T.S. Eliot**'s 'The Love-Song of J. Alfred Prufrock' in 1915. Monroe was fully occupied in editing the magazine until her death, but found time to publish two more volumes of her own clear and lyrical poetry, much of it occasional in inspiration, as well as a volume of essays on poetry and an anthology.

In 1936 she attended a PEN conference in Buenos Aires and, on the way back, climbing to inspect the ruins of Cuzco, the Inca city in the Peruvian Andes, she suffered a fatal cerebral haemorrhage. She is buried in the Pantheon of the Peruvian city of Arequipa at the foot of Mount Misti. Her autobiography, *A Poet's Life: Seventy Years in a Changing World*, which is very informative about the literary history of her time, and which she had almost finished at her death, was completed and published by her assistant, Morton Dauven Zabel, who succeeded her as editor of *Poetry*.

Poetry

Valeria and Other Poems 1891; *Commemoration Ode* 1892 [as *The Columbian Ode* 1893]; *The Dance of Seasons* 1911; *You and I* 1914; *The Difference and Other Poems* 1924; *Chosen Poems* 1935

Plays

The Passing Show: Five Modern Plays in Verse [pub. 1903]

Non-Fiction

John Wellborn Root: a Study of His Life and Work 1896; *Poets and Their Art* 1926 [rev. 1932]; *A Poet's Life: Seventy Years in a Changing World* (a) 1938

Edited

The New Poetry: an Anthology [with A.C. Henderson] 1917 [rev. as *The New Poetry: an Anthology of Twentieth-Century Verse in English* 1923, 1932]

C(harles) E(dward) MONTAGUE 1867–1928

Montague was born at Ealing, the son of a former priest from Tyrone who had lost his faith. He was educated at the City of London School and Balliol College, Oxford, where he read greats. He played rugby and rowed for the college. He also rescued a man from drowning; for this he was awarded the Royal Humane Society's bronze medal. On going down he joined the *Manchester Guardian*, in 1898 becoming chief assistant to its great editor C.P. Scott. In the same year he married Scott's daughter Madeline, with whom he had five sons and two daughters. His first publication was a collection of dramatic criticism with W.T. Arnold, Oliver Elton and Allan Monkhouse; his first novel, *A Hind Let Loose*, a satire on Fleet Street, did not appear until ten years later. Another collection of his dramatic criticism appeared a year after that. Though well into his forty-eighth year when war broke out he volunteered for the army. He went to France in 1915, was soon invalided home but returned in 1916 as an Intelligence officer. Much of his best writing came out of the war: the stories in *Fiery Particles*, the novels *Disenchantment* and *Rough*

Justice. He retired from the *Guardian* in 1925 and died of pneumonia in Manchester three years later. He is regarded as a good journalist and a workmanlike rather than electrifying writer. His enthusiasm for rock climbing appears in many of his books.

Fiction

A Hind Let Loose 1910; *The Morning's War* 1913; *Disenchantment* 1922; *Fiery Particles* (s) 1923; *Rough Justice* 1926; *Right Off the Map* 1927; *Action* (s) 1928

Non-Fiction

The Manchester Stage [with W.T. Arnold, O. Elton, A. Monkhouse] 1900; *Dramatic Values* 1911; *The Right Place* 1924; *A Writer's Notes on His Trade* 1930

Biography

C.E. Montague by Oliver Elton 1929

Michael (John) MOORCOCK 1939-

Moorcock was born in Mitcham, Surrey. His parents separated when he was young and his schooling was desultory. Brought up among women, he has acknowledged the blitz and mystical Christianity as formative influences. By the age of twelve he had produced his first magazine, *Outlaw's Own*. After leaving school he worked as an editor for *Tarzan Adventures* and the Sexton Blake Library. His first novel, *Caribbean Crisis*, a thriller, was published by the latter. For a brief period he earned his living as a night-club blues singer then, in 1962, he began to contribute science-fiction stories to *SF Adventures* and *Science Fantasy*. He edited *New Worlds* from 1964 until its demise as a magazine in 1971. Subsequently he edited a series of short-story collections under *New Worlds*' imprint: among his writers were **J.G. Ballard** and **Brian Aldiss**. In so far as Moorcock promoted fiction which broke with the mechanistic, futuristic strictures of the form, he is generally acknowledged to have extended the boundaries of British science fiction.

Although his most popular books are his 'sword and sorcery' epics, he has said that he dislikes science fiction and prefers his output to be labelled 'fantasy, if anything'. His fiction is characterised by its promiscuous allusiveness – from Billy Bunter to Captain Ahab – and formal unpredictability. This tactic (at its best in *The Cornelius Chronicles*) enables Moorcock to engage in a broad-ranging satire and functions as a corollary of his avowed anarchism. Literary conventions, cultural icons, social pieties – all are treated with equal disrespect.

Since the appearance of his complex mock-historical novel *Gloriana*, he has gained acceptance as an important – but eccentric – mainstream novelist. This reputation has been secured by *Byzantium Endures* and its sequel, *The Laughter of Carthage*, the partial memoirs of Pyatnitski, Tsarist colonel and Munchausen-like self-mythologiser. *Mother London*, however, is his finest achievement to date. Its central characters, traumatised by the blitz, record the moods of their city, itself a casualty of war. From the late 1940s to the 1980s, London is tormented by its past while neurotically responding to its present. If Moorcock suggests that people and place are damaged and reciprocally damaging, he nevertheless manages a Utopian conclusion in defiance of the symptoms he has observed. *Behold the Man* – a bleak revision of the Crucifixion story – won the Nebula Award for best novella in 1967. The *Guardian* Fiction Prize was awarded to *The Condition of Muzak* in 1977, and the World Fantasy Award to *Gloriana* in 1979.

Moorcock has been married three times (once to the novelist Hilary Bailey); he has a son and two daughters. Throughout the early 1970s he frequently performed with the rock group Hawkwind and recorded an album, *The New Worlds Fair*, with his own band, the Deep Fix, in 1975. Now active in the Campaign against Pornography and Censorship, Moorcock believes that 'the Women's Liberation Movement .. is the most important political force of modern times'. He has added: 'Andrea Dworkin's work remains a great influence.'

Fiction

Caribbean Crisis [as 'Desmond Reid'; with J. Cawthorn] 1962; *The Stealer of Souls and Other Stories* (s) 1963; *The Fireclown* 1965 [US *The Winds of Limbo* 1969]; *Stormbringer* 1965; *The Sundered Worlds* 1965 [as *The Blood Red Game* 1970]; *The Deep Fix* (s) [as 'James Colvin'] 1966; *Printer's Devil* [as 'Bill Barclay'] 1966; *Somewhere in the Night* [as 'Bill Barclay'] 1966; *The Twilight Man* 1966 [as *The Shores of Death* 1970]; *Hawkmoon: the History of the Runestaff* 1984: *The Jewel in the Skull* 1967, *Sorcerer's Amulet* 1968 [UK *The Mad God's Amulet* 1969], *The Sword of the Dawn* 1968, *The Runestaff* [US *The Secret of the Runestaff*] 1969; *The Wrecks of Time* 1967; *The Cornelius Chronicles* 1977: *The Final Programme* 1968, *A Cure for Cancer* 1971, *The English Assassin* 1972, *The Condition of Muzak* 1977; *Behold the Man* 1969; *The Black Corridor* 1969; *The Ice Schooner* 1969; *The Time Dweller* (s) 1969; *The Chinese Agent* 1970; *The Eternal Champion* 1970 [rev. 1978]; *Phoenix in Obsidian* 1970 [US *Silver Warriors* 1971]; *The Singing Citadel* (s) 1970; *The King of the Swords* 1971; *The Knight of the Swords* 1971; *The Queen *of the Swords* 1971; *The Rituals of Infinity* 1971; *The Nomad of Time* 1984: *The Warlord of the Air* 1971, *The Land Leviathan* 1974, *The Steel Tsar* 1980; *The Dancers at the End of Time* 1981: *An Alien Heat* 1972, *The Hollow Lands* 1974, *The End of All Songs* 1976; *Breakfast in the Ruins* 1972; *The Dreaming City* 1972; *Elric of Melnibone* 1972; *The Sleeping Sorceress* 1972 [as *The Vanishing Tower* 1977]; *The Chronicles of Corum* 1986: *The Bull and the Spear* 1973, *The Oak and the Ram* 1973, *The Sword and the Stallion* 1974; *Hawkmoon: the Chronicles of Castle Brass* 1986: *Count Brass* 1973, *The Champion of Garathorm* 1973, *The Quest for Tanelorn* 1975; *The Jade Man's Eyes* (s) 1973; *The Distant Suns* [with P. James] 1975; *The Adventures of Una Persson and Catherine Cornelius in the Twentieth Century* 1976; *Legends from the End of Time* (s) 1976; *The Lives and Times of Jerry Cornelius* (s) 1976; *Moorcock's Book of Martyrs* (s) [with M. Butterworth] 1976; *The Sailor on the Seas of Fate* 1976; *The Time of the Hawklords* [with M. Butterworth] 1976; *The Bane of the Black Sword* 1977; *Queens of Deliria* [with M. Butterworth] 1977; *Sojan the Swordsman* 1977; *The Transformation of Miss Mavis Ming* 1977 [US *Messiah at the End of Time* 1978]; *The Weird of the White Wolf* 1977; *Dying for Tomorrow* (s) 1978; *Gloriana* 1978 [rev. 1993]; *Byzantium Endures* 1980; *The Golden Barge* 1980; *The Great Rock and Roll Swindle* 1980; *My Experiences in the Third World War* (s) 1980; *The Russian Intelligence* 1980; *The Entropy Tango* 1981; *The Warhound and the World's Pain* 1981; *The Brothel in Rosenstrasse* 1982; *Elric at the End of Time* (s) 1984; *The Laughter of Carthage* 1984; *The Opium General* (s) 1984; *The City in the Autumn Stars* 1986; *The Crystal and the Amulet* 1986; *The Dragon in the Sword* 1986; *Mother London* 1988; *Casablanca and Other Stories* (s) 1989; *The Fortress of the Pearl* 1989; *The Revenge of the Rose* 1991; *The Pleasure Gardens of Philippe Sagittarius* 1992; *Earl Aubec and Other Stories* (s) 1993; *Blood* 1994

As 'E.P. Bradbury': *Barbarians of Mars* 1965 [as *Masters of the Pit* as Michael Moorcock] 1970]; *Blades of Mars* 1965; [as *Lord of the Spiders* as Michael Moorcock] 1970; *Warriors of Mars* 1965 [as *The City of the Beast* as Michael Moorcock 1970]

Plays

The Land That Time Forgot (f) [with J. Cawthorn] 1974

For Children

Sojan 1977

Non-Fiction

Epic Pooh 1978; *Letters from Hollywood* 1986; *Wizardry and Wild Romance: a Study in Epic Fantasy* 1987

Edited

The Best of New Worlds (s) 1965; *The Best SF Stories from New Worlds 1–8* (s) 8 vols 1967–74; *The Traps of Time* 1968; *The Inner Landscape* [as 'anon.'] 1969; *New Worlds Quarterly 1–5* (s) 5 vols 1971–3; *The Nature of the Catastrophe* [with L. Jones] 1971; *New Worlds 6* (s) [with C. Platt] 1973; *Before Armageddon: an Anthology of Victorian and Edwardian Imaginative Fiction Published Before 1914* 1975; *England Invaded: a Collection of Fantasy Fiction* 1977; *The Retreat from Liberty* 1983; *Fantasy: The 100 Best Books* [with J. Cawthorn] 1988; H.G. Wells: *The Island of Doctor Moreau* 1993, *The Time Machine* 1993

Brian MOORE 1921–

Moore (whose first name is pronounced Bree-ann) was born into the middle-class Belfast Catholic world that he has depicted in several of his books, and from which he has spent a lifetime escaping. Son of a surgeon, he was educated at St Malachy's College, Belfast. During the Second World War he served in the Belfast Fire Service and then with the British Ministry of War Transport in North Africa, Italy and France. After the war he worked with the United Nations Relief and Rehabilitation Administration (UNRRA) Mission to Poland, until 1947. He then went to Canada, becoming a journalist on the Montreal *Gazette*, and a full-time writer in 1952. He took Canadian citizenship but moved to America in 1959. Married twice, with one son, he has been a professor at the University of California, Los Angeles, since 1974, and lives in Malibu.

Moore had already written pseudonymous thrillers when he began publishing serious novels: his first book was *The Executioners*, under the name of Michael Bryan. His first serious novel was *Judith Hearne* (1955; later reissued under its US title, *The Lonely Passion of Judith Hearne*). It has never been out of print and concerns a lonely and pious middle-aged spinster with a weakness for the bottle who attempts to win the love of a man returned from New York (where he was, unknown to her, only a doorman). This precipitates her into a violent alcoholic and spiritual crisis, and we leave her not knowing whether she will regain her faith or whether she has only emotional destitution to look forward to. Moore handles his material with a certain bleak comedy, and it is the first of several books to deal with failure and unlived lives.

In some respects Moore's world bears comparison with that of **James Joyce**'s *Dubliners* (1914) – Moore has been described as 'the

Laureate of Irish drabness' – and the short fictions of **William Trevor**. A subsequent book, *The Luck of Ginger Coffey*, deals with the failed dreams of an Irish emigrant to Canada. Instead of getting rich in business, as he had hoped, Coffey works as a proof-reader and the driver of a van full of nappies. Finally he gets into trouble with the police for drunkenness and indecent exposure, but he sacrifices himself to protect his family; throughout the book, his dreams of success have been for the sake of his wife and daughter, and in a moving conclusion his wife comes to see that it is Ginger's well-meaning concern for others that is the motor of his self-deceptions.

The respect with which Moore treats his characters is a hallmark of his writing, as is his ability to portray female characters in books such as *I Am Mary Dunne*. Moore has rejected the Catholicism of his past, but he remains fascinated by its effects in books such as *Catholics*, *Black Robe* and *Cold Heaven*, the last of which is an eerie story which reflects his statement that 'always at the back of my mind, I've wondered what if all this stuff was true and you didn't want it to be true and it was happening in the worst possible way?' He has been equally esteemed by literary and popular readers, and critics have commended the combination of simple style with complexity, authenticity and depth in his treatment of character and relationships. He has been nominated three times for the Booker Prize, and his critical standing is suggested by a piece by Christopher Ricks in the *New Statesman*: 'The Simple Excellence of Brian Moore'.

Fiction

As 'Michael Bryan': *The Executioners* 1951; *Wreath for a Redhead* 1951; *Intent to Kill* 1956; *Murder in Majorca* 1957

As Brian Moore: *Judith Hearne* 1955 [US *The Lonely Passion of Judith Hearne* 1956]; *The Feast of Lupercal* 1958 [as *A Moment of Love* 1965]; *The Luck of Ginger Coffey* 1960; *An Answer from Limbo* 1963; *The Emperor of Ice-Cream* 1968; *I Am Mary Dunne* 1968; *Fergus* 1970; *The Revolution Script* 1971; *Catholics* 1972; *The Great Victorian Collection* 1975; *The Doctor's Wife* 1976; *Two Stories* (s) 1978; *The Mangan Inheritance* 1979; *The Temptation of Eileen Hughes* 1981; *Cold Heaven* 1983; *Black Robe* 1985; *The Colour of Blood* 1987; *Lies of Silence* 1990; *No Other Life* 1993

Plays

The Luck of Ginger Coffey (f) 1964; *Torn Curtain* 1966; *The Slave* (f) 1967; *Catholics* (f), (tv) 1973, (st) 1980; *The Blood of Others* (f) 1984; *The Sight* (f) 1985; *The Black Robe* (f) 1991

Non-Fiction

Canada 1963 [rev. 1968]

George (Augustus) MOORE 1852–1933

The eldest son of a Roman Catholic Irish landowner, Moore was born at Moore Hall, the family's seat in County Mayo. He was educated privately and at Oscott College, Birmingham, from which he was expelled for making advances to a maid, thus inaugurating a lifetime of philandering. The plan to send him to a military school was happily thwarted by his father's death when Moore was eighteen. Instead, he left the administration of his inherited property in the hands of an agent and went to live in Paris for seven years, ostensibly to become a painter. He indulged in café society, published some poetry, was drawn in a Bohemian pose by Manet, but eventually abandoned any hope of becoming an artist himself, though he later became an art critic, publishing a study of *Modern Painting* in 1893.

In 1883 he came to live in London, publishing his first novel, *A Modern Lover*, that same year. Influenced by Zola and French realism, it was based upon his own Parisian experiences and was banned by the circulating libraries. This very naturally drew a great deal of attention to the book, and started the first of several battles Moore fought against censorship. After reading Walter Pater's *Marius the Epicure* (1885), Moore changed his literary allegiance and wrote several novels which presented various approaches to the dilemma of paganism versus religion. *Esther Waters* was a sympathetic account of the suffering endured by a religious young woman who bears an illegitimate child, and – remarkably for the time – had a happy ending. Praised by Gladstone, but attacked by **Katherine Mansfield**, the book brought Moore the fame he wanted, but he declared that it bored him. The results of his efforts to produce fiction that expressed universal themes were poorly received, partly because he kept changing his style. He also published several distinguished volumes of short stories. The best-known novels of his later years are *The Brook Kerith*, about the life of Christ, and *Héloïse and Abélard*, about the twelfth-century French lovers.

Serious about reforming English drama, he became an influential theatre critic, and from 1891 to 1897 was instrumental in the Independent Theatre, which aimed to produce original, national work. Returning to Ireland in 1900, he took part in the Celtic revival and with **W.B. Yeats** founded the Irish Literary Theatre, forerunner of the Abbey Theatre. His own plays are not remembered. He had a gift for making enemies, and ill-considered remarks about his fellow writers earned him their contempt. He never read *A la recherche du temps perdu*, but nevertheless pronounced that Proust was 'trying to plough up fields with knitting needles'.

Moore Park was to be destroyed during the Irish civil war of 1922, but its landlord had returned to London in 1911. He settled in Pimlico, where he entertained a coterie of literary friends, recording these occasions in a collection of rambling essays, *Conversations in Ebury Street*. He also published several volumes of autobiography, including the frequently revised *Confessions of a Young Man*, which charmingly recalls his years in Paris. Yeats described Moore as 'a man carved out of a turnip'. Although unattractive, with a receding chin, walrus moustache and (according to his own description) 'champagne bottle shoulders', he attracted many women, notably the poet and editor Nancy Cunard, who had known him since the age of four, was incorrectly rumoured to be his daughter (her mother had been one of his lovers), and who wrote an affectionate memoir of him, *GM* (1956). Some doubt has been cast, however, on his volumes of memoirs recalling his conquests.

Fiction

A Modern Lover 3 vols 1883 [rev. as *Lewis Seymour and Some Women* 1917]; *A Mummer's Wife* 1885 [rev. 1918]; *A Drama in Muslin* 1886 [rev. as *Muslin* 1915]; *A Mere Accident* 1887 [rev. as *John Norton* 1895]; *Spring Days* 1888 [rev. 1912]; *Mike Fletcher* 1889; *Vain Fortune* 1891 [rev. 1895]; *Esther Waters* 1894 [rev. 1899, 1920]; *Celibates* 1895; *Evelyn Innes* 1898 [rev. 1898, 1901, 1908]; *Sister Teresa* 1901 [rev. 1909]; *The Untilled Field* (s) 1903 [rev. 1914, 1926, 1931]; *The Lake* 1905 [rev. 1906, 1921]; *The Brook Kerith* 1916 [rev. 1921, 1927]; *Héloïse and Abélard* 1921 [rev. 1925]; *Euphorian in Texas* (s) 1922; *In Single Strictness* (s) 1922; *Peronnik the Fool* (s) 1924; *Ulick and Soracha* 1926; *Aphrodite in Aulis* 1930; *A Flood* (s) 1930

Non-Fiction

Literature at Nurse 1885; *Parnell and His Island* 1887; *Confessions of a Young Man* (a) 1888 [rev. 1889, 1904, 1918, 1926]; *Impressions and Opinions* 1891; *Modern Painting* 1893 [rev. 1896, 1898]; *New Budget Extras No. 1* 1895; *Memoirs of My Dead Life* (a) 1906 [rev. 1915, 1928]; *Reminiscences of the Impressionist Painters* 1906; *Hail and Farewell* (a) 1915 [rev. 1925]: *Ave* 1911, *Salve* 1912, *Vale* 1914; *A Story-Teller's Holiday* 1918 [rev. 1928]; *Avowals* 1919; *Moore versus Harris* 1921; *In Single Strictness* 1922; *Conversations in Ebury Street* 1924 [rev. 1930]; *A Letter to The Times* 1925; *Celibate Lives* 1927; *Letters from George Moore to Edward Dujardin* [ed. J. Eglinton] 1929; *George Moore in Quest of Locale* 1931; *A Communication to My Friends* (a) 1931; *Letters of George Moore* [ed. J. Eglinton] 1942; *Letters to Lady Cunard* [ed. R. Hart-Davis] 1957; *George Moore in Translation: Letters to T. Fisher Unwin and Lena Milman 1894–1910* [ed. H.E. Gerber] 1968

Poetry

Flowers of Passion 1877; *Pagan Poems* 1881; *La Ballade de l'amant de coeur* 1886[?]; *The Talking Pine* 1931

Plays

Martin Luther [with B. Lopez] 1879; *The Strike at Arlingford* 1893; *The Bending of the Bough* 1900; *Diarmuid and Grania* [with W.B. Yeats] 1901 [pub. 1951]; *The Apostle* 1911; *Esther Waters* [from his novel] 1913; *Elizabeth Cooper* 1913 [rev. as *The Coming of Gabrielle* 1920]; *The Making of an Immortal* 1927 [pub. 1928]; *The Passing of the Essenes* 1930

Translations

Jacques Amyot *The Pastoral Loves of Daphnis and Chloë* [from Longus] 1924

Edited

Pure Poetry 1924

Biography

The Life of George Moore by Joseph Hone 1936; *A Portrait of George Moore* by John Freeman 1992

Lorrie (i.e. Marie Lorena) MOORE 1957–

Moore was born in Glens Falls, New York, the daughter of an insurance company executive, and grew up with two brothers and one sister in a household she describes as happy, but with 'much thwarted literary ambition' (her father had submitted short stories to the *New Yorker* in his youth). She graduated from St Lawrence University in 1978 and was, from 1982 to 1984, a lecturer in English at Cornell University, then taught creative writing at the University of Wisconsin.

Self-Help, her first collection of short stories, was published in 1985. Its use of narrative devices borrowed from psychological self-help manuals and witty, punning style – one character compulsively makes up lists in an attempt to combat feelings of 'listlessness' – made it an immediate success. Her first novel, *Anagrams*, was published the following year and was well received, although its structure, repeatedly shifting between a first-person heroine and the third-person *alter egos* she is constantly inventing, seemed to demonstrate the author's preference for the short-story form to which she has since returned. Her second collection, *Like Life*, is more pessimistic in tone. The comedy is still that of loss but conveyed with less exuberance than before. The title is lifted from one character's bemused attempts to come to terms with the diagnosis of pre-cancer: 'Isn't that ... like life?' she asks.

Her second novel, the quirkily titled *Who Will Run the Frog Hospital?*, is a bittersweet rites-of-passage story, in which a married woman on holiday with her husband in Paris recalls her friendship with another girl when they were teenagers and had taken a summer vacation job at a tacky American theme park called Storyland. Trapped by economic necessity in this fake world of childhood make-believe – which is based on scenes and characters from nursery rhymes and fairy-tales – the two friends are discovering the real world of adolescence, of delinquency, sex and unplanned pregnancy. The title refers to a time when the two girls tried to save frogs shot by local boys, but also to the world they are leaving behind them. While written with Moore's characteristic style and wit, this brief novel is deeply felt and very evocative of time and place, and has been judged her most impressive work to date.

Fiction

Self-Help (s) 1985; *Anagrams* 1986; *Like Life* (s) 1990; *Who Will Run the Frog Hospital?* 1994

Marianne (Craig) MOORE 1887–1972

The daughter of an engineer-inventor, Moore was born near St Louis, Missouri, but moved to Carlisle, Pennsylvania during her childhood after her father suffered a mental breakdown and was committed to a psychiatric hospital. She graduated from Bryn Mawr in 1909 and for the following four years taught typing, commercial law and sport at the Indian School in Carlisle. Her first two poems were published in the *Egoist* and **Harriet Monroe**'s *Poetry* in 1915, the year before she moved with her mother to Cheltenham, New Jersey. It was shortly afterwards that she began visiting nearby New York City, where she met a group of poets and critics associated with the magazine *Others*, among them **William Carlos Williams** and **Wallace Stevens**. She moved, again with her mother, to New York City in 1918 and lived there until her death, working initially as a secretary, private tutor and library assistant.

Her first volume, simply entitled *Poems* (1921), was published without her knowledge by two of her friends, **H.D.** (Hilda Doolittle) and Robert McAlmon, in London, where it was branded by **Edith Sitwell** as 'thick and uncouth'. *Observations*, published in America three years later, won the *Dial* Award of that year and established her reputation. Shortly afterwards, Moore became editor of the *Dial*, one of the leading literary periodicals of the time, a post which she retained until its closure in 1929. The Dial Press subsequently awarded her $2,000 in recognition of her 'distinguished service to American letters'.

Moore's poetry, like the best American poetry of her time produced by, for example, Stevens and Williams, is primarily concerned with the creative process and the relationship between poetry and the actual experience of living. 'Poetry', which went through several versions, remains one of her best-known pieces and faces the post-Imagist question: what is the basic relation between poetic emotion and real things? This was a small question around which to revolve her whole poetic career, and her obsession with it is reflected in the relatively small number of poems (some seventy in total) contained in her *Collected Poems* of 1951.

Perhaps the most famous image of Moore is a photograph taken by Esther Bubley at the Bronx Zoo in 1953, in which the poet, wearing a cartwheel straw hat, stands in front of two elephants. Many of her later poems are about animals, which she describes minutely and allusively, combining keen observation with bold imaginative leaps. Describing 'The Pangolin', for example, she refers to 'the fragile grace of the Thomas- / of Leighton Buzzard Westminster Abbey wrought- / iron vine'. Some of this allusiveness is explained in the notes she provided for her *Collected Poems*, where she writes: 'Since in *Observations*, and in anything I have written, there have been lines in which the chief interest is borrowed, and I have not yet been able to outgrow this hybrid method of composition, acknowledgements seem only honest.' Sources range from the works of Spenser to a leaflet about 'The World's Most Accurate Clocks' issued by the Bell Telephone Company. She never lost her zest for knowledge and learned to drive at sixty and to tango at seventy.

Despite her small output, Moore was the chief influence on the generation of poets who followed her, particularly **Elizabeth Bishop**, **Randall Jarell**, **Richard Wilbur** and **Robert Lowell**. **T.S. Eliot** wrote of her as 'one of the few who have done the language some service in my lifetime'. Moore, who never married, was said to have been intensely jealous of Eliot's second marriage in 1957.

Poetry

Poems 1921; *Marriage* 1923; *Observations* 1924; *Selected Poems* 1935; *The Pangolin and Other Verse* 1936; *What Are Years?* 1941; *Nevertheless*

1944; *A Face* 1949; *Collected Poems* 1951; *Like a Bulwark* 1956; *O to Be a Dragon* 1959; *Eight Poems* 1963; *Occasionem Cognosce* 1963; *The Arctic Ox* 1964; *The Complete Poems of Marianne Moore* 1967 [rev. 1981]; *Prevalent at One Time* 1970; *Unfinished Poems* 1972

Translations
Aldabert Stifter *Rock Crystal* [with E. Mayer] 1945; La Fontaine: *The Fables of La Fontaine* 1955, *Selected Fables of La Fontaine* 1995; Charles Perrault *Puss in Boots/The Sleeping Beauty/Cinderella* 1963

Non-Fiction
Predilections 1955; *Idiosyncrasy and Technique* 1958; *Letters from and to the Ford Motor Company* [with D. Wallace] 1958; *Dress and Kindred Subjects* 1965; *Poetry and Criticism* 1965

Plays
The Absentee [from M. Edgeworth's novel] 1962

Collections
A Marianne Moore Reader 1961; *Tell Me, Tell Me: Granite, Steel and Other Topics* 1968

Frank MOORHOUSE 1938–

Moorhouse, who was described by the *Times Literary Supplement* as the 'elder statesman of the Australian avant-garde', was born in Nowra, New South Wales, where his parents were in business. He left school in 1956 and became a cadet reporter on the Sydney *Daily Telegraph*, and served in the Australian Army and Reserves from 1957 to 1959. He studied at the University of Queensland, and was a reporter on the Wagga Wagga *Advertiser*, and later editor of the Lockhart *Review*. Active in the trade union movement, he was briefly editor of the *Australian Worker*, and was executive officer of the Workers' Educational Association from 1961 to 1965. Since then he has been a freelance journalist and writer.

His first short story had been published in *Southerly* in 1957, but the explicit treatment of sex in his fiction kept him out of the established literary magazines. In the 1960s he published in radical small magazines and soft-porn magazines, and his first collection, *Futility and Other Animals*, was produced by a publisher of the latter.

Moorhouse describes this and his subsequent collections as 'discontinuous narrative'. The stories, like those in his second collection, *The Americans, Baby*, describe the radical Bohemian fringe of Sydney in the 1960s, and are awash with sexual experimentation, drugs and trendiness. The so-called Balmain group of writers, artists and academics which flourished during the period undoubtedly provided Moorhouse with much of his material. *The Electrical Experience*, which won the National Award for Fiction in 1975, continues his examination of the American influence on Australian society, as does the film *The Coca-Cola Kid*, directed by Dusan Makavejev, which portrays an American marketing executive attempting to conquer the Australian soft-drinks market. Moorhouse adapted the screenplay from parts of the discontinuous narrative of his second and third collections.

In *The Everlasting Secret Family* he pushes his examination of contemporary mores and anxieties to a new peak with a story concerning a feminist at an academic conference who is raped by three Aborigines. His screenplay *The Disappearance of Azaria Chamberlain* is concerned with the notorious 'dingo baby' case, an Australian *cause célèbre* in which a mother accused of murdering her baby claimed that the child had been carried off by a wild dog. *Grand Days* is a big and carefully researched novel set in Geneva in the 1920s which documents the establishment of the League of Nations. The novel focuses on the relationship between a young Australian woman, Edith Campbell Berry, and an English soldier, Major Ambrose Westwood, and explores Moorhouse's joint concerns with political idealism and sexual experimentation.

Fiction
Futility and Other Animals (s) 1969; *The Americans, Baby* (s) 1972; *The Electrical Experience* (s) 1974; *Conference-ville* (s) 1976; *Tales of Mystery and Romance* (s) 1977; *The Everlasting Secret Family and Other Secrets* (s) 1980; *Selected Stories* (s) 1982 [as *The Coca Cola Kid* (s) 1985]; *Room Service* (s) 1985; *Forty-Seventeen* (s) 1988; *Lateshows* (s) 1990; *Grand Days* 1993; *Woman of High Direction* 1993

Edited
Coast to Coast 1973; *Days of Wine and Rage* 1980; *The State of the Art: the Mood of Contemporary Australia in Short Stories* 1983; *A Steele Rudd Selection: the Best Dad and Dave Stories with Other Rudd Classics* 1986; *Fictions 1988* 1988

Plays
Between Wars [with M. Thornhill] (f) 1974; *Conference-ville* [from his story] (tv) 1984; *The Disappearance of Azaria Chamberlain* (f) 1984; *The Coca-Cola Kid* (f) 1985

Edwin (George) MORGAN 1920–

Morgan was born in Glasgow and has spent most of his life in that city, celebrating it in a volume of

Glasgow Sonnets. He was educated at Glasgow High School and at Glasgow University, remaining at the latter, after graduating in 1947, as a lecturer in English and subsequently a professor until his retirement in 1980. His university education was interrupted by the Second World War, during which he served in the Middle East with the Royal Army Medical Corps.

A frequent translator of a wide variety of works from the Italian, Russian and French, Morgan has been influenced by European literature while remaining a distinctively Scottish poet. (One of his notable triumphs is *Wi' the Haill Voice*, a volume of Mayakovsky's poetry translated into Scots dialect.) He published his first volume, *The Vision of Cathkin Braes*, in 1952, and since then has been a prolific writer who has attempted many forms, from short lyrics such as 'Aberdeen Train' to the far reaches of concrete poetry, taking in along the way salutes to 1960s' icons ('Che' and 'The Death of Marilyn Monroe'), playful, semi-nonsensical rhymes about animals ('The white rhinoceros was eating phosphorus!') and poems about science and technology ('In Sobieski's Shield'). His poetry also shows the influence of Anglo-Saxon literature (his first translation was of *Beowulf* in 1952), particularly in his use of portmanteau words – though this is something he shares with many poets of the 1960s (notably the 'Liverpool poets'). He is a reminder of a time in Britain when poetry was considered fun and, although experimental, accessible to a wide audience, and his later work has proved him less ephemeral than some of his contemporaries.

Along with his fellow Scot, **Ian Hamilton Finlay**, he has been a leading British exponent of concrete poetry, whose experiments in this field include 'Archives', in which the phrase 'generation upon generation' is arranged to form a column consisting of the repeated and gradually disintegrating 'generation upon', and one of the most famous poems in this form, 'The Computer's First Christmas Card', a sort of Chinese whispers which starts with 'jollymerry / hollyberry' and ends with 'asMERRYCHR / YSANTHEMUM'. His concrete poems have appeared in exhibitons as well as anthologies both in Britain and abroad. 'Although a poem is / undoubtedly a "game" / it is not a game', he writes in 'Not Playing the Game', and his work is at once ludic and serious. One experiment involved writing poems inspired by newspaper headlines, in which he resembles **William Plomer**, though with very different results.

Morgan has also written opera libretti, several works of criticism, particularly on modern Scottish literature, and has edited numerous volumes of verse, notably the six-volume *Scottish Poetry*. He has received many awards from the Scottish Arts Council, a Cholmondeley Award in 1968, and was given the OBE in 1982.

Poetry

The Vision of Cathkin Braes 1952; *The Cape of Good Hope* 1955; *Scotch Mist* 1965; *Starryveldt* 1965; *Sealwear* 1966; *Emergent Poems* 1967; *Bestiary* 1968; *Gnomes* 1968; *The Second Life* 1968; *Penguin Modern Poets 15* [with A. Bold, E. Kamau Brathwaite] 1969; *Proverbfolder* 1969; *The Horseman's Word* 1970; *Twelve Songs* 1970; *The Dolphin's Song* 1971; *Glasgow Sonnets* 1972; *Instamatic Poems* 1972; *From Glasgow to Saturn* 1973; *Nuspeak 8* 1973; *The Whittrick* 1973; *The New Divan* 1977; *Colour Poems* 1978; *Star Gate* 1979; *Poems of Thirty Years* 1982; *4 Glasgow Subway Poems* 1983; *Grafts/Takes* 1983; *Sonnets from Scotland* 1984; *From the Video Box* 1986; *Newspoems* 1987; *Themes on a Variation* 1988; *Tales from Limerick Zoo* 1988; *Collected Poems* 1990; *Hold Hands Among the Atoms* 1991; *Three Scottish Poets* [with L. Lochhead, N. MacCaig] 1992; *Sweeping Out the Dark* 1994

Plays

The Charcoal-Burner (l) 1969 [np]; *Columba* (l) 1976 [np]; *Valentine* (l) 1976 [np]; *Spell* (l) 1979 [np]; *The Apple-Tree* 1982; *Master Peter Pathelin* 1983

Non-Fiction

Essays 1974; *East European Poets* 1976; *Hugh MacDiarmid* 1976; *Rites of Passage: Translations* 1976; *Edwin Morgan: an Interview* [with M. Walker] 1977; *Provenance and Problematics of 'Sublime and Alarming Images' in Poetry* 1977; *Twentieth-Century Scottish Classics* 1987; *Crossing the Border* 1990; *Nothing Not Giving Messages: Reflections on His Work and Life* [ed. H. Whyte] 1990

Edited

Collins Albatross Book of Longer Poems 1963; *Scottish Poetry* vols 1–6 [with G. Bruce, M. Lindsay] 1966–72; *New English Dramatists 14* 1970; *Scottish Satirical Verse: an Anthology* 1980; David Anderson and David MacLennan *Roadworks: Song Lyrics for Wildcat* 1987; *New Writing Scotland*: vols 5 and 6 [with C. Macdougall] 1987–8, vol. 7 [with H. Whyte] 1989

Translations

Beowulf 1952; *Poems from Eugenio Montale* 1959; *Sovpoems: Brecht, Neruda, Pasternak, Tsetayeva, Mayakovsky, Martynov, Yevtushenko* 1961; Sándor Weöres: *Sándor Weöres and Ferenc Juhász: Selected Poems* [with D. Wevill] 1970, *Eternal Moment: Selected Poems* [with others] 1988; *Wi' the Haill Voice: Poems by Mayakovsky* 1972; *Fifty Renascence Love-Poems* [ed. I. Fletcher] 1975; August Graf von Platen-Hallermünde: *Selected Poems* 1978; Edmond Rostand *Cyrano de Bergerac* 1992

Wright (Marion) MORRIS 1910–

Appropriately for a writer whose work has been centrally concerned with being an American, Morris was born in the middle of America, in Central City, Nebraska. He was educated at Crane College, Chicago, and Pomona College, Claremont, California, and subsequently lectured at a number of US universities. He was professor of English at California State University from 1962 to 1975.

My Uncle Dudley (1942) was the first document of his enduring interest in exploring the nature of Americanness and the American past. Set in the 1920s, the book is narrated by Uncle Dudley's young sidekick as the two of them go vagabonding, recapturing both the nineteenth-century pioneer life and the spirit of Huckleberry Finn.

In his dual career as a writer and photographer Morris is concerned with discovering (or rediscovering, since much of his work has a conservative and nostalgic tinge) the essence of Americanness as expressed in the frontier, Midwest, pragmatic spirit and its continuance into the twentieth century. Commenting on his themes and his own Midwest roots, Morris has said: 'I am not a regional writer, but the characteristics of this region have conditioned what I see, what I look for, and what I find in the world to write about.' He seems to have been influenced by the literary nationalism of the 1910s and 1920s, as expressed in the work of **Sherwood Anderson**, Waldo Frank and the photographer Alfred Stieglitz. The nostalgic element in Morris's work becomes less simplistic as his career progresses. His interest is not only in championing the American past but in a complex exploration of the nature of perceiving that past.

Although he had very little enthusiasm for the turbulent 1960s and the decade's values, he forced himself to examine them in books such as *One Day*, about the shooting of Kennedy and the assassin's character, and *In Orbit*, about motorcycle outlaws. Morris has never been fashionable but he has built up an impressive body of work, the formal complexity of which qualifies what might be simplistic themes in lesser hands. He is highly respected both as a writer and photographer and has received numerous awards, including membership of the American Academy in 1970. He remains something of a 'writer's writer'.

Fiction

My Uncle Dudley 1942; *The Man Who Was There* 1945; *The World in the Attic* 1949; *Man and Boy* 1951; *The Works of Love* 1952; *The Deep Sleep* 1953; *The Huge Screen* 1954; *The Field of Vision* 1956; *Love Among the Cannibals* 1957; *Ceremony in Lone Tree* 1960; *What a Way to Go* 1962; *Cause for Wonder* 1963; *One Day* 1965; *In Orbit* 1967; *Green Grass, Blue Sky, White House* (s) 1970; *Fire Sermon* 1971; *War Games* 1972; *A Life* 1973; *Here Is Einbaum* (s) 1973; *The Cat's Meow* (s) 1975; *Real Losses, Imaginary Gains* (s) 1976; *The Fork River Space Project* 1977; *Plains Song* 1980; *The Origin of Sadness* (s) 1984; *Collected Stories 1948–1986* (s) 1986

Non-Fiction

The Inhabitants 1946; *The Home Place* 1948; *The Territory Ahead* 1958; *A Bill of Rights, a Bill of Wrongs, a Bill of Goods* 1968; *God's Country and My People* 1968; *Love Affair: a Venetian Journal* 1972; *About Fiction: Reverent Reflections on the Nature of Fiction with Irreverent Observations on Writers, Readers and Other Abuses* 1975; *Structure and Artefacts: Photographs 1933–1954* 1975; *Conversations with Wright Morris: Critical Views and Responses* [ed. R.E. Knoll] 1977; *Earthly Delights, Unearthly Adornments: American Writers as Image Makers* 1978; *Will's Boy: a Memoir* 1981; *Wright Morris: Portfolio of Photographs* 1981; *Picture America* [with J. Alinder] 1982; *Wright Morris: Photographs and Words* [ed. J. Alinder] 1982; *The Writing of My Uncle Dudley* 1982; *Solo: an American Dreamer in Europe* (a) 1933–4 1983; *Time Pieces: the Photographs and Words of Wright Morris* 1983; *A Cloak of Light: Writing My Life* (a) 1985; *Time Pieces: Photography, Imagination and Writing* 1989; *Wright Morris: Origin of a Species* [with others] 1992

Edited

The Mississippi River Reader 1962

(Philip) Blake MORRISON 1950–

The son of two doctors, who shared a practice, Morrison was born and brought up near Skipton in Yorkshire. He was educated at Ermysteds Grammar School in Skipton and the University of Nottingham, where he took a degree in English literature. While at University College, London, writing a PhD on the Movement poets of the 1950s, he was offered the job of reviewing poetry for the *Times Literary Supplement*, thus inaugurating a distinguished career in literary journalism. He joined the staff of the *TLS* in 1978 as poetry and fiction editor, but within a month of his starting the job, a prolonged strike at *The Times* began, stopping publication of both the newspaper and its supplements. 'The only difference between being a student and my first job,' he remarked, 'was that I got a bigger salary.' In 1981 he joined the *Observer* as deputy literary editor to Terence Kilmartin, whom he succeeded in 1987, before moving at the beginning of 1990 to be first literary editor of the *Independent on Sunday*. He held this post until 1994, when he retired to concentrate upon his writing.

His first book, *The Movement: English Poetry and Fiction of the 1950s*, was based on his thesis and published in 1980, and two years later he edited with **Andrew Motion** an influential anthology, *The Penguin Book of Contemporary British Poetry*. His own first volume of poetry, *Dark Glasses*, was published in 1984 and immediately established his reputation as one of the leading poets of his generation, winning a Somerset Maugham Award. His long poem in nine sections, 'The Inquisitor', takes up half of *Dark Glasses*. Its central character is a spy and its theme secrecy and betrayal. His poetry was influenced by that of the Movement poets, in particular by his 'favourite', **Philip Larkin**. 'There is much in [Larkin] that is not explicable,' he has said, and this reflects his own desire not to be 'strictly *pin-downable*'. Reviewers noted that the volume's title was appropriate, since 'you can't see his eyes'.

Morrison's second volume of poetry, *The Ballad of the Yorkshire Ripper*, won the Dylan Thomas Memorial Award. It takes its title from another long poem, which first appeared in the *London Review of Books*, inspired by the notorious case of the serial killer Peter Sutcliffe, and unsurprisingly proved controversial. The 'Ballad' is written in Yorkshire dialect, spoken by an anonymous, enigmatic narrator. Other poems in the volume touch on similarly contentious topical issues such as the Sara Tisdale affair, the nuclear accident at Chernobyl and the US bombing of Libya.

He used another notorious crime – the abduction and murder in 1993 of the toddler James Bulger by two young boys – as the springboard for 'Little Angels, Little Devils', a 'documentary poem' broadcast by Channel 4 television in 1994. 'Did childhood die in 1993?' the poem asks in its opening line, then goes on to examine society's ambivalent attitude to children and innocence, illustrating this with extraordinary archive footage from films, television programmes, advertisements, paintings and photographs. Unlike some of his early poems, this one is absolutely direct.

Morrison has won numerous awards, including the Eric Gregory Award in 1980 and an E.M. Forster Award in 1988. His extremely frank, deeply moving and beautifully written meditation upon the final illness and death of his father, *And When Did You Last See Your Father?*, disturbed some critics and readers, but won both the *Esquire* and J.R. Ackerley Prizes. Its last chapter, from which it takes its title, poses the question: at what point during the process of dying and death does someone who is mortally ill stop being recognisably himself?

Based in London, married, with three children, Morrison is also the author of a critical study of **Seamus Heaney** and a book for children.

Poetry

Dark Glasses 1984 [rev. 1989]; *The Ballad of the Yorkshire Ripper* 1987; *Little Angels, Little Devils* (tv) 1994 [np]

Non-Fiction

The Movement: English Poetry and Fiction of the 1950s 1980; *Seamus Heaney* 1982; *And When Did You Last See Your Father?* (a) 1993

Edited

The Penguin Book of Contemporary British Poetry [with A. Motion] 1982

For Children

The Yellow House 1987

Toni (i.e. Chloe Anthony) MORRISON 1931–

Morrison was born Chloe Anthony Wofford in Lorain, a steel mill town in Ohio. Her family were migrants – sharecroppers on both sides who, like many black people in the South, lost their land and, at the turn of the century, went into the mines and mills of the industrialised North. 'We lived marginally,' she recalled; 'in 1933 we were paying four dollars a month rent in an apartment the landlord saw fit to burn us out of – while we were in it. He wanted to raise the rent.' She read voraciously as a child (from Jane Austen to Tolstoy), and was considered bright. In 1949 she went to Howard University, America's most distinguished black college. She looked forward to the thrill of intellectual stimulation but found that college was about 'getting married, buying clothes and going to parties … I giggled a lot, wise-cracked and tried to be with it – having no notion of what "it" was.'

After graduating she took a course in English at Cornell, where she 'read her head off', then returned to Howard to teach English. Here she met and married a Jamaican architect with whom she had a son in 1962. While she was pregnant with her second son, who was born in 1966, her marriage of six years broke up, perhaps providing the catalyst for her drift into a writers' group. Out of a story written for this group came Morrison's first novel, *The Bluest Eye* (1970), which is the story of Pecola Breedlove, a black girl who believes everything would be all right if only she had beautiful blue eyes.

In 1967 Morrison left teaching to work for a small textbook publishing house which moved to New York, where she subsequently became a senior editor at Random House. While holding down a full-time job and raising two children she continued to write. Through her work as a lecturer and publisher's editor, Morrison has done much to further black women's literature, by publishing new writers and by praising her contemporaries such as **Maya Angelou** and **Alice Walker**. Her own work caused **Allan Massie** to say that she is approaching greatness more surely than anyone else who has begun to publish in the USA in the last quarter of a century.

She is noted for entering in fiction worlds few others dare brave. *Beloved*, which deals with slavery and infanticide, was inspired by the true story of a woman who killed her baby after the infamous 1870s' Fugitive Slave Act in order to save the child from the slavery she had managed to escape. 'I thought at first it couldn't be written, but I was annoyed and worried that such a story was inaccessible to art,' she has said. 'In the end, I had to rely on the resilience and power of the characters – if they could live it all of their lives, I could write it.' Her work is informed by the world she grew up in – which accepted pain, storytelling and the supernatural as a way of life: in *Beloved*, the baby whose throat has been slit appears as an eighteen-year-old woman to haunt her mother. Her writing has a lyrical quality, characterised by one critic thus: 'That blues poetry is part of an extraordinary language, which "rememories" an oral culture without distorting it. Morrison's rhythmic, rhapsodic prose is like the songs of the men on the chain gang, making art out of pain.' Of her other novels, *Sula* won the National Book Critics Award, while *Jazz* is a fragmented narrative about the causes and consequences of a murder in Harlem in 1926.

Morrison herself is an imposing figure. She gives the impression that she would not suffer fools gladly but she laughs a lot and thinks the most important human quality is *joie de vivre*. She now holds the Albert Schweitzer chair at the State University of New York in Albany, nurturing several young writers through two-year fellowships that allow them 'to put their writing at the middle of their lives'.

Fiction

The Bluest Eye 1970; *Sula* 1974; *Song of Solomon* 1978; *Tar Baby* 1981; *Beloved* 1987; *Jazz* 1992

Non-Fiction

Playing in the Dark 1992; *Arguing Immigration* (c) [ed. N. Mills] 1994

Edited

Race-ing Jushce, En-gendering Power 1992

John (Clifford) MORTIMER 1923-

Born in Hampstead, north London, the only child of an eccentric barrister who continued to practise after he went blind (the subject of one of his son's best and best-known plays, *A Voyage around My Father*), Mortimer attended Harrow and Brasenose College, Oxford. Although he worked as a scriptwriter for the Crown Film Unit for a while after he graduated (an experience that inspired his acidic first novel, *Charade*, published in 1947), he was called to the Bar in 1948, and for many years pursued two successful careers: leading barrister and part-time writer. He became Queen's Counsel in 1966, and established a formidable reputation as a defending advocate in several celebrated cases touching on censorship (the *Oz* case, as well as the trials concerning **Hubert Selby Jr**'s novel *Last Exit to Brooklyn* and *Gay News*). He eventually retired from the Bar to concentrate on his writing, but has remained prominent as a supporter of the Labour Party (he is sometimes described as a 'champagne socialist') and a member of such liberal bodies as the Howard League for Penal Reform.

Between 1947 and 1956 he wrote six novels, mainly with legal backgrounds and traditional in manner (*Like Men Betrayed* is perhaps the most interesting of them), as well as a detective novel under the pseudonym Geoffrey Lincoln. He then turned to writing plays, finding success with a BBC radio play, *The Dock Brief*, and, recognising drama as his *métier*, following it with many short and full-length plays for radio, theatre and television. In 1978, he introduced his barrister character, *Rumpole of the Bailey*, in a volume of short stories, and has since followed him through several popular collections and television series. The novels he wrote in the 1980s were usually televised soon afterwards, and in some cases (such as *Paradise Postponed* and its sequel, *Titmuss Regained*) the idea was developed in tandem as prose fiction and television series. Such works are hardly novels at all in the traditional sense, but, if they can have few claims to be literature, possess great popular appeal.

Mortimer is very well known as a journalist – having been, for instance, at varying times the drama critic of the *New Statesman*, the *Evening Standard* and the *Observer* – and his interviews have been collected in book form. He has published two volumes of autobiography, and among his several television credits is the celebrated adaptation of **Evelyn Waugh**'s *Bri-*

deshead Revisited. His many translations of stage works include some of the farces of Feydeau and *Die Fledermaus* for the Royal Opera House, Covent Garden.

He married Penelope Fletcher (well known as the novelist **Penelope Mortimer**), in 1949 and was divorced from her in 1972, having had two children; he now lives with his second wife at his father's large house in Buckinghamshire, and they have two daughters.

Fiction

Charade 1947; *Rumming Park* 1948; *Answer Yes or No* [US *The Silver Hook*] 1950; *Like Men Betrayed* 1953; *The Narrowing Stream* 1954; *Three Winters* 1956; *No Moaning at the Bar* [as 'Geoffrey Lincoln'] 1975; *Will Shakespeare* 1977; *Rumpole of the Bailey* **(s)** 1978; *The Trials of Rumpole* **(s)** 1979; *Rumpole's Return* **(s)** 1980; *Regina v. Rumpole* **(s)** 1981; *Rumpole and the Golden Thread* **(s)** 1983; *Paradise Postponed* 1985; *Rumpole's Last Case* **(s)** 1987; *Rumpole and the Age of Miracles* **(s)** 1988; *Summer's Lease* 1988; *Rumpole à la Carte* **(s)** 1990; *Titmuss Regained* 1990; *The Rapstone Chronicles* 1991; *Dunster* 1992; *Rumpole on Trial* **(s)** 1992

Plays

Like Men Betrayed **(r)** 1955; *No Hero* **(r)** 1955; *Three Plays* [pub. 1958]: *The Dock Brief* **(r)**, **(tv)** 1957, **(st)** 1958, *I Spy* **(r)** 1957, **(tv)** 1958, **(st)** 1959, *What Shall We Tell Caroline?* 1958; *Three Winters* **(r)** 1958; *Ferry to Hong Kong* **(f)** [with L. Gilbert, V. Harris] 1959; *Lunch Hour and Other Plays* [pub. 1960]: *Call Me a Liar* **(r)**, **(tv)** 1958, **(st)** 1968, *David and Broccoli* **(tv)** 1960, *Lunch Hour* **(r)**, **(st)** 1960, **(f)** 1962, *Collect Your Hand Baggage* 1961, *Wrong Side* 1960; *The Encyclopedist* **(r)** 1961; *The Innocents* **(f)** [with W. Archibald, T. Capote] 1961; *Guns of Darkness* **(f)** 1962; *I Thank a Fool* **(f)** [with others] 1962; *Two Stars for Comfort* 1962; *The Running Man* **(f)** 1963; *A Voyage Round My Father* **(r)** 1963, **(st)** 1970 [pub. 1971]; *Bunny Lake is Missing* **(f)** [with P. Mortimer] 1964; *Education of an Englishman* **(r)** 1964; *A Rare Device* **(f)** 1965; *A Choice of Kings* **(tv)** 1966 [pub. in *Playbill 3* ed. A. Durband 1969]; *An Exploding Azalea* **(tv)** 1966; *A Flea in Her Ear* [from G. Feydeau's play] **(st)** 1966, **(f)** 1967 [pub. 1967]; *The Head Waiter* **(tv)** 1966; *The Judge* 1967; *The Other Side* **(tv)** 1967; *Desmond* **(tv)** 1968 [pub. in *The Best Short Plays 1971* ed. S. Richards 1971]; *Infidelity Took Place* **(tv)** 1968; *Cat Among the Pigeons* [from G. Feydeau's play] 1969 [pub. 1970]; *John and Mary* **(f)** 1969; *Come as You Are: Four Short Plays* [pub. 1971]: *Bermondsey* 1970, *Gloucester Road* 1970, *Marble Arch* 1970, *Mill Hill* 1970; *Married Alive* **(tv)** 1970; *The Captain of Köpenick* [from C. Zuckmayer's play] 1971; *Conflicts* [with others] 1971 [np]; *I, Claudius* [from R. Graves's novels] 1972 [np]; *Knightsbridge* **(tv)** 1972; *Swiss Cottage* **(tv)** 1972; *Collaborators* 1973; *A Little Place Off the Edgware Road, The Blue Film, The Destructors, The Case for the Defence, Chagrin in Three Parts, The Invisible Japanese Gentlemen, Special Duties, Mortmain* **(tv)** [from G. Greene's stories] 1975–6; *Heaven and Hell* 1976 [pub. 1978]: *The Fear of Heaven* **(r)** [as *Mr Luby's Fear of Heaven*], **(st)** 1976, *The Prince of Darkness* 1976 [rev. as *The Bells of Hell* 1977]; *The Lady from Maxim's* [from G. Feydeau's play] 1977; *Will Shakespeare* **(tv)** 1978; *Brideshead Revisited* **(tv)** [from E. Waugh's novel] 1981; *Unity* **(tv)** [from D. Pryce-Jones's book] 1981; *Edwin* **(r)** 1982 [pub. *Edwin and Other Plays* 1984]; *When That I Was* 1982 [np]; *The Ebony Tower* **(tv)** [from John Fowles's story] 1984; *A Little Hotel on the Side* [from M. Desvalliers and G. Feydeau's play] 1984 [pub. in *Three Boulevard Farces* 1985]; *Paradise Postponed* **(tv)** [from his novel] 1986; *Glasnost* **(r)** 1988; *Die Fledermaus or the Bat's Revenge* **(l)** [from L. Halévy and H. Meilhac's *Le Réveillon*; with T. Genée and C. Haffner; pub. 1989]; *John Mortimer Plays* [pub. 1989]; *Summer's Lease* **(tv)** [from his novel] 1989; *The Waiting Room* **(tv)** 1989; *Titmuss Regained* **(tv)** [from his novel] 1991

Non-Fiction

With Love and Lizards [with P. Mortimer] 1957; *Clinging to the Wreckage* **(a)** 1982; *In Character* 1983; *Character Parts* 1986

Edited

Famous Trials [with H. and J.H. Hodge] 1984; *The Penguin Book of Great Law and Order Stories* **(s)** 1990; *The Oxford Book of Villains* 1993

Penelope (Ruth) MORTIMER 1918–

Born Penelope Fletcher in Rhyl in North Wales, Mortimer was the daughter of an eccentric Church of England clergyman, who lost his faith in God, used the parish magazine to praise the Soviet persecution of the church, sent his daughter to seven different schools, culminating in a clergy daughters' school, and subjected her to prolonged 'clumsy petting'. Aged seventeen, she did a secretarial course in London, attended London University for a year, worked briefly for Butlin's holiday camps, and married her first husband, Charles Dimont, a journalist for Reuters, in 1937; she had four daughters – two with him, one with Kenneth Harrison, a scientist, and one with the poet Randall Swingler (although Dimont was registered as the father of all four children) – and they were divorced in 1947. (Mortimer describes all these experiences in her two fine autobiographical volumes, *About Time* and *About Time Too*.)

Her first novel, *Johanna*, was published the same year under the name of Penelope Dimont, and attracted little notice, but she began to develop a reputation as a freelance journalist, becoming a *New Yorker* writer and, among other things, being an agony aunt on the *Daily Mail* for two years under the name of Ann Temple. She really came to prominence as a novelist in the 1960s, and her best-known fictional work, *The Pumpkin Eater*, was published in 1962; her novels, specialising in detailed descriptions of the tensions threatening middle-class marriages, formed a notable component in the more up-to-date women's fiction being written in that decade.

In 1949, she had married the barrister and writer **John Mortimer**, with whom she had a son and a daughter, and from whom she was divorced in 1972. She collaborated with him on several works, including *With Love and Lizards*, an account of two months the family spent in Positano, Italy. In the late 1960s, she became the film critic of the *Observer*, succeeding **Penelope Gilliatt** in this position, and has continued to be prominent as a journalist, particularly for the *Sunday Times*.

Her most recent novel is *The Handyman*, which was published in 1983. Since then, she has preferred to concentrate on film scripts and television plays, and perhaps her most notable work in the latter field was her script for *Portrait of a Marriage*, the dramatisation of Nigel Nicolson's account of the partnership of his parents, Harold Nicolson and **V. Sackville-West**. Mortimer also published a sharp biography of the Queen Mother in 1986 and has become noted as one of the more discreet royal commentators. She has long divided her life between London and the country, and from 1978 lived mainly in the village of Chastleton in Oxfordshire, where she cultivated and wrote about the garden she had created. She has subsequently returned to live in London. One of her daughters is the actress Caroline Mortimer, who has performed in her mother's work.

Fiction

Johanna (as 'Penelope Dimont') 1947; *A Villa in Summer* 1954; *The Bright Prison* 1956; *Daddy's Gone A-Hunting* 1958; *Saturday Lunch with the Brownings* **(s)** 1960; *The Pumpkin Eater* 1962; *My Friend Says It's Bulletproof* 1967; *The Home* 1971; *Long Distance* 1974; *The Handyman* 1983

Non-Fiction

With Love and Lizards [with J. Mortimer] 1957; *About Time* **(a)** 1979; *Queen Elizabeth: a Life of the Queen Mother* [US *A Portrait of the Queen Mother*] 1986; *About Time Too* **(a)** 1993

Plays

The Renegade **(tv)** [from her story] 1961; *Bunny Lake Is Missing* **(f)** [with J. Mortimer] 1965; *Ain't Afraid to Dance* **(tv)** 1966; *Three's One* **(tv)** 1973; *A Summer Story* **(f)** 1988; *Portrait of a Marriage* [from N. Nicolson's biography] 1990

Nicholas MOSLEY 1923–

Mosley was born in London, the son of Oswald Mosley, future leader of the British Union of Fascists (about whom he wrote a remarkable two-volume biography), and his first wife Cynthia, who was the daughter of Lord Curzon and died in 1933. He succeeded an aunt as 3rd Baron Ravensdale in 1966 (the title had been created for Curzon in 1911 and allowed to pass through the female line for one generation) and his father as 7th baronet in 1980. He was educated at Eton, and then joined the army, serving as an infantry platoon commander in Italy from 1943 to 1945, being wounded and winning the Military Cross.

After the war, he read philosophy for a year at Oxford, but on his marriage to the painter Rosemary Salmond, in 1947, he retired to a hill-farm in North Wales to write. (He has always had sufficient private income to support himself.) His three early novels, of which *Spaces of the Dark* (1951) is the first, form the initial phase of his writing: fairly conventional in technique although elaborate stylistically. During this period, he was much influenced by meeting Raymond Raynes, the superior of an Anglican religious order, who converted him to his particular brand of Christianity. In the late 1950s Mosley was editor of *Prism*, a Christian magazine propagating Raynes's ideas, and he wrote Raynes's biography after his death.

The second phase of Mosley's novels begins with *Meeting Place* (1962); in these works, which include *Accident* (made into a celebrated Joseph Losey film from a screenplay by **Harold Pinter**), style is stripped down and content more experimental and abstract, in a manner reminiscent of the French *nouveau roman*. In 1970 Mosley was almost killed in a car accident, an event which seems ironical, since *Accident*, published in 1965, turns on a fatal crash and Mosley had actually appeared in the film in a cameo role.

For eight years during the 1970s he did not publish any novels but wrote film scripts (including one for Losey's *The Assassination of Trotsky*) and a life of the poet Julian Grenfell, to whom his first wife was related. *Catastrophe Practice* was the first of a projected seven-volume novel sequence in which each of the latter six novels was to examine one of the main characters introduced in the first. It was, however, brought to an end with the fifth volume, *Hopeful Monsters*, which was chosen as the Whitbread Book of the Year. The first in the series had been by far Mosley's most experimental work to date but the later volumes were in general more conservative in technique. Mosley dramatically defended his own style of novel, however, when, as one of the judges for the Booker Prize in 1991, he resigned from the panel on the grounds that the other judges were not interested in novels of ideas and had denied him a voice; he had particularly wished that *The Sins of the Father* by **Allan Massie** be shortlisted.

After a divorce from his first wife in 1974 (she died in 1991), he married a psychotherapist; he has four children by his first marriage and one by his second. His autobiography, *Efforts at Truth*, was published in 1994.

Fiction

Spaces of the Dark 1951; *The Rainbearers* 1955; *Corruption* 1957; *Meeting Place* 1962; *Accident* 1965; *Assassins* 1967; *Impossible Object* 1969; *Natalie Natalia* 1971; *Catastrophe Practice: Plays for Not Acting; and, Cypher: a Novel* 1979; *Imago Bird* 1980; *Serpent* 1981; *Judith* 1986; *Hopeful Monsters* 1990

Non-Fiction

Life Drawing 1954; *African Switchback* 1958; *The Life of Raymond Raynes* 1961; *Experience and Religion: a Lay Essay on Theology* 1965; *The Assassination of Trotsky* 1972; *Julian Grenfell: His Life and the Times of his Death 1888–1915* 1976; *Rules of the Game: Sir Oswald and Lady Cynthia Mosley 1896–1933* 1982; *Beyond the Pale: Sir Oswald Mosley and Family 1933–1980* 1983; *Efforts at Truth* (a) 1994

Plays

The Assassination of Trotsky (f) [from his book; with M. d'Amico] 1972; *Impossible Object* (f) [from his novel] 1975

Edited

Raymond Raynes *The Faith: Instructions on the Christian Faith* 1961

Andrew MOTION 1952–

From a brewing family, Motion was brought up in Essex and educated at Radley College and at University College, Oxford, where he studied English and was taught by **John Fuller**, who became his first publisher. While at Oxford, he met **Alan Hollinghurst**, who is the dedicatee of his long poem *Independence*, married in 1973 his first wife, and won the Newdigate Poetry Prize in 1976 for his poem *Inland*, which was published as a pamphlet by Fuller's Salamander Press. Between 1977 and 1981 he taught English at the University of Hull, where he befriended the librarian, **Philip Larkin**, publishing a study of his work in 1982. He then returned to Oxford to teach at St Anne's College.

Motion saw himself as a poet from the age of fifteen, though he admits this was largely a question of 'growing your hair long and offering your profile to the camera and being **Rupert Brooke**'. His first volume of poetry, *The Pleasure Steamers*, was published ten years later, in 1978, and immediately established him as a new poetic voice. His poetry is direct, clear, personal and frequently tells a story: indeed, his second volume was titled *Secret Narratives*; 'Skating', which was commissioned by *Poetry Review* and recalls his childhood, is a work of prose. In 1981, he was awarded the Arvon/*Observer* Poetry Prize for his poem 'The Letter'. A decade of his poetry, from 1974 to 1984, was collected in *Dangerous Play*, a volume dedicated to the memory of his mother, whose death in 1978 as the result of a riding accident ten years earlier Motion has often written about (notably in 'Skating' and the sequence of 'Anniversaries'). *Dangerous Play* was awarded the John Llewelyn Rhys Prize. Much of his poetry is based upon events in his life, notably friendships, love affairs, marriage and family life, a theme sounded in *Love in a Life*, which is principally about his second marriage (in 1985 to the literary editor Jan Dalley, with whom he has three children), but also includes poems about his first wife. Writing, he has said, is to '*burgle* life'. His work has been published in numerous newspapers and magazines, and he reviews regularly for the *Observer*, in which for a short period he wrote an opinion column.

Apart from Larkin, a principal influence on Motion's poetry has been **Edward Thomas**, about whose work he published a book in 1981, based on his thesis. *The Price of Everything*, with its references to the First World War, is in part an act of homage to Thomas. In 1982, he edited with **Blake Morrison** the contentious *Penguin Book of Contemporary British Poetry*; Motion was subsequently an influential poetry editor at Chatto & Windus (1983–9), and was for five years an editorial director of the company. His works of non-fiction include a joint biography of *The Lamberts* (the leading Australian painter George Lambert; his son, Constant Lambert, the composer; and his grandson, Kit Lambert, who was manager of the rock group the Who), which won a Somerset Maugham Award, and the authorised biography of Larkin, whose joint literary executor he became. His collection *Natural Causes*, which won a Dylan Thomas Award, had ended with a poem in memory of Larkin, entitled 'This is your subject speaking', but taking on the role of Jake Balokowsky (the imaginary American academic Larkin gleefully depicted working on this task in his 1968 poem 'Posterity') involved him in considerable controversy. Generally well received, the biography drew criticism from some of Larkin's friends – notably **Kingsley Amis**.

Motion has been at his least successful as a novelist. Having signed a contract to write a ten-volume *roman fleuve*, he produced a lamentable and poorly received first volume, *The Pale Companion*, which drew upon his own schooldays and dealt with homosexuality and the death of the protagonist's sister. A second volume, *Famous for the Creatures*, also received scant praise, and it was rumoured that the sequence had been abandoned in favour of a trilogy, but to date the third volume has yet to appear. This notwithstanding, in 1995 Motion was appointed to the Chair of Creative Writing at the University of East Anglia, as a successor to **Malcolm Bradbury**, whose 'messianic zeal for fiction', the *Daily Telegraph* observed, 'makes the appointment ... all the more surprising'.

Poetry

Inland 1976; *The Pleasure Steamers* 1978; *Independence* 1981; *Secret Narratives* 1983; *Dangerous Play* 1984; *Natural Causes* 1987; *Love in a Life* 1991; *The Price of Everything* 1994

Non-Fiction

The Poetry of Edward Thomas 1981; *Philip Larkin* 1982; *The Lamberts* 1986; *Philip Larkin: a Writer's Life* 1993

Fiction

The Pale Companion 1989; *Famous for the Creatures* 1991

Edited

The Penguin Book of Contemporary British Poetry [with B. Morrison] 1982

R(alph) H(ale) MOTTRAM 1883–1971

Mottram was born in Norwich, where he was to live for almost the whole of his long life, in the dwelling-house above Gurney's Bank, where his father, grandfather and great-grandfather had all been chief clerk. After education in a private school in Norwich and a year spent learning French in Lausanne, he followed his father into the bank (which had by then become part of the Barclays combine) in 1899. Mottram's father was a trustee of the marriage settlement of **John Galsworthy**'s wife. Galsworthy, whom Mottram met in 1904, encouraged his early, part-time, efforts to write, and two slim volumes of verse appeared under the name of J. Marjoram in 1907 and 1909.

Mottram enlisted at the outbreak of war in 1914, and was sent to Flanders in 1915, but was soon withdrawn from the trenches when it was discovered that he was one of the few soldiers who could make themselves understood in French. He spent the rest of the war collecting complaints of damage done by troops, experience which was to go into his only famous work of fiction, *The Spanish Farm Trilogy, 1914–1918*, published between 1924 and 1926, and revised for a single-volume edition in 1927. This was one of the first novels to take a detached, unsentimental (although moving) view of the war and remains one of the finest fictional works to emerge from the conflict. The first volume, *The Spanish Farm*, won the Hawthornden Prize and the trilogy was filmed in 1927. Sale of the film rights enabled Mottram to retire from the bank and become a full-time writer, varying this activity with much sitting on committees and public service: he was Lord Mayor of Norwich in 1953.

He wrote about sixty books in all, including many other novels, poetry, books on banking, industrial history, East Anglia, biographies of Norfolk worthies and a three-volume autobiography: very little of this seems destined to survive. Mottram had married, in 1918, Margaret ('Madge') Allen and they had three children; his wife died in 1970, and Mottram, by then very frail, moved from Norwich for the last year of his life, living with his widowed daughter at King's Lynn.

Fiction

The Spanish Farm Trilogy 1914–1918: *The Spanish Farm* 1924, *Sixty-Four, Ninety-Four!* 1925, *The Crime at the Vanderlynden's* 1926 [rev. in one volume 1927]; *The Dormer Trilogy*: *Our Mr Dormer* 1927, *The Boroughmonger* 1929, *Castle Island* 1931; *The Apple Disdained* 1928; *The English Miss* 1928; *Europa's Beast* 1930; *The New Providence* 1930; *The Headless Hound and Other Stories* (s) 1931; *Dazzle* 1932; *Home for the Holidays* 1932; *Through the Menin Gate* (s) 1932; *The Lame Dog* 1933; *Strawberry Time and The Banquet* (s) 1934; *Bumphrey's* 1934; *Early Morning* 1935; *Flower Pot End* 1935; *Time To Be Going* 1937; *There Was a Jolly Miller* 1938; *Miss Lavington* 1939; *You Can't Have It Back* 1939; *The Ghost and the Maiden* 1940; *The World Turns Slowly Round* 1942; *The Corbells at War* 1943; *Visit of the Princess* 1946; *The Gentleman of Leisure* 1948; *Come to the Bower* 1949; *One Hundred and Twenty-Eight Witnesses* 1951; *The Part That Is Missing* 1952; *Over the Wall* 1955; *Scenes That Are Brightest* 1956; *No One Will Ever Know* 1958; *Young Man's Fancies* 1959; *Musetta* 1960; *Time's Increase* 1961; *To Hell, with Crabb Robinson* 1962; *Happy Birds* 1964; *Maggie Mackenzie* 1965; *The Speaking Likeness* 1967; *Behind the Shutters* 1968

Non-Fiction

Ten Years Ago: Armistice and Other Memories 1928; *A History of Financial Speculation* 1929;

Three Personal Records of the War (c) [with J. Easton, E. Partridge] 1929; *Miniature Banking Histories* 1930; *John Crome of Norwich* 1931; *East Anglia* 1933; *Town Life* 1934; *Journey to the Western Front: Twenty Years After* 1936; *Portrait of an Unknown Victorian* 1936; *The Westminster Bank 1836–1936* 1936; *Noah* 1937; *Old England* 1937; *Success to the Mayor* 1937; *Autobiography with a Difference* (a) 1938; *The Window Seat; or, Life Observed by R.H. Mottram* 1954; *Another Window Seat; or, Life Observed by R.H. Mottram* 1957; *Trader's Dream: the Romance of the East India Company* 1939; *Bowler Hat: a Last Glance at the Old Country Banking* 1940; *Buxton the Liberator* 1946; *Hibbert Houses* 1947; *The Glories of Norwich Cathedral* 1948; *Norfolk* 1948; *Through Five Generations: a History of the Butterley Company* 1950; *The Broads* 1952; *If Stones Could Speak: an Introduction to an Almost Human Family* 1953; *John Galsworthy* 1953; *The City of Norwich Museums 1894–1954* 1954; *For Some We Loved: an Intimate Portrait of Ada and John Galsworthy* 1956; *Vanities and Verities* (a) 1958; *The Twentieth Century: a Personal Record* (a) 1969

Poetry

Repose and Other Verses [as 'J. Marjoram'] 1907; *New Poems* [as 'J. Marjoram'] 1909; *Poems New and Old* 1930; *Twelve Poems* 1968

Translations

Henry Daniel-Rops *The Misted Mirror* 1930

Edwin MUIR 1887–1959

Muir was born in the main island of Orkney, the youngest of six children of a crofting farmer. His family on both sides had been crofters for generations and the traditional life of Orkney remained a seminal source of Muir's imagination. In 1902, hit by the decline of crofting, the Muir family moved to a poor area of Glasgow, a traumatic transition to industrial society: within five years, both parents and two of their sons were dead. Muir himself was put to work as an office boy and clerk in various legal practices, an experience which later had to be expunged by psychoanalysis.

His early, hesitant, poems were contributed to A.R. Orage's influential magazine the *New Age*, and his life changed when in 1919 he married the energetic, cultivated and – some maintained – bullying Willa Anderson, moved with her to London, and became assistant to Orage on the magazine. From this time he began to write penetrating criticism and more mature poetry. From 1921 to 1924 the Muirs lived in various German-speaking areas, including Prague, and this led to their career as translators from the German: they were largely responsible for making the work of Kafka known in English. Outstanding among Muir's works of criticism was *The Structure of the Novel*.

His *First Poems* (1925) inaugurated his poetic career, and he wrote three novels in the late 1920s and early 1930s, which are noted for their poetic qualities. Muir's poetry, of which *The Labyrinth* is perhaps the outstanding volume, was plain in diction but marked by great seriousness and sincerity as well as a fruitful use of myth; much of the best of it was produced when he was over fifty. Muir was a well-known man-of-letters in the inter-war period, contributing a fortnightly signed notice of 'New Novels' to *The Listener*, worked for the British Council in Edinburgh during the Second World War, and later became director of its institutes in Prague and Rome. In the early 1950s he was warden of a working men's college near Edinburgh, and finally settled in Swaffham Prior in Cambridgeshire. He published an autobiography of great interest. An extremely modest, unassuming man, Muir was widely loved for his gentleness and integrity. He had one son.

Poetry

First Poems 1925; *Chorus of the Newly Dead* 1926; *Six Poems* 1932; *Variations on a Time Theme* 1934; *Journeys and Places* 1937; *The Narrow Place* 1943; *The Voyage and Other Poems* 1946; *The Labyrinth* 1949; *Prometheus* 1954; *One Foot in Eden* 1956

Fiction

The Marionette 1927; *The Three Brothers* 1931; *Poor Tom* 1932

Non-Fiction

We Moderns: Enigmas and Guesses 1918 [as 'Edward Moore']; *Latitudes* 1924; *Transition: Essays on Contemporary Literature* 1926; *The Structure of the Novel* 1928; *John Knox: Portrait of a Calvinist* 1929; *Scottish Journey* 1935; *Scott and Scotland: the Predicament of the Scottish Writer* 1936; *The Present Age from 1914* 1939; *The Story and the Fable* (a) 1940 [rev. as *An Autobiography* 1954]; *The Scots and Their Country* 1946; *The Estate of Poetry* 1962; *The Truth of Imagination: Some Uncollected Reviews and Essays* [ed. P.H. Butt] 1988

Translations

With Willa Muir: Gerhart Hauptmann: *The Dramatic Works* vols 8 and 9, 1925 and 1929, *The Island of the Great Mother* 1925; Lion Feuchtwanger: *Jew Süss* 1926, *The Ugly Duchess* 1927, *The Oil Islands and Warren Hastings* 1929, *Success* 1930, *Josephus* 1932, *The Jew of Rome* 1935, *The False Nero* 1937; Ernst Glaeser *Class of*

1902 1929; Ludwig Renn: *War* 1929, *After War* 1931; Franz Kafka: *The Castle* 1930, *The Great Wall of China and Other Pieces* 1933, *The Trial* 1937, *America* 1938, *Parables, in German and English* 1947, *In the Penal Settlement. Tales and Short Prose Works* (s) 1949; Emil Alphons Rheinhardt *The Life of Eleonara Duse* 1930; Kurt Henser *Inner Journey* 1932; Hermann Broch: *The Sleepwalkers* 1932, *The Unknown Quantity* 1935; Shalom Asch: *Three Cities* 1933, *Salvation* 1934, *Mottke the Thief* 1935, *Calf of Paper* 1936; Ernst Lothar: *Little Friend* 1933, *The Mills of God* 1935; Heinrich Mann *Hill of Lies* 1934; Erik Maria von Kühnelt-Leddihn *Night Over the East* 1936; Robert Neumann: *The Queen's Doctor* 1936, *The Woman Screamed* 1938; Georges Maurice Paléologue *The Enigmatic Czar* 1938; Carl Jakob Burchhardt *Richelieu* 1940; Zsolt Harsanyi: *Through the Eyes of a Woman* 1941, *Lover of Life* 1942

Bharati MUKHERJEE 1940–

Mukherjee was born in Calcutta, the daughter of a wealthy Brahmin industrialist and pharmaceutical chemist. She spent her earliest years as part of a traditional extended Bengali family in a large house in Calcutta, but aged eight went with her parents and two sisters to London, where they spent two and a half years and she first became fluent in English. After a brief period in Basle, the family returned to Calcutta in 1951, where she attended the English-speaking Loreto Convent School. She graduated in English from the University of Calcutta in 1959 and received her MA in English and ancient Indian culture at the University of Baroda, Gujarat, in 1961. Like her sisters, she then went to the USA to study, attending the University of Iowa's writers' workshop. Here she met Canadian writer (later professor) Clark Blaise, whom she married in 1963; they have two sons.

In 1964 and 1965 Mukherjee held her first lectureships at American universities, but in 1966 the couple moved to Montreal, and later Toronto, Canada, where she took up a post at McGill University, remaining there until 1980, having been appointed professor in 1978. Her first novel, *The Tiger's Daughter*, was published in 1972, and her second, the violent *Wife*, in 1975; both concern Indian immigrant women in the USA, although they are deliberately distanced from the author's own experience. Another early work of interest is *Days and Nights in Calcutta*, consisting of the journals that she and her husband kept on the first of two extended visits to India during the 1970s (during the second visit she was based in New Delhi, as director of the Shastri Institute in 1976–7). Meanwhile, Canada was becoming increasingly inhospitable to those of Asian origin in the wake of the admission of 5,000 Ugandan Asians in 1973; Mukherjee was attacked in a subway and thrown out of hotels. She objected to the official policy of 'multiculturalism', and decided in 1980 to return to the USA. Here she has taught at various colleges and universities, originally Skidmore College, Saratoga Springs, but later based in New York, where she taught creative writing at Columbia and was a professor at the City University. Since 1990, she has been a distinguished professor at the University of California at Berkeley.

Her celebrity as a writer began from the later 1980s, *The Middleman and Other Stories* gaining high praise and being followed by the acclaimed novel *Jasmine*, the odyssey of an Indian woman in various parts of the North American continent. These later fictions differ from her earlier books in that, while they still largely concern immigrants, they tell the story from the North American perspective. This tendency is even stronger in the next novel, *The Holder of the World*, where a modern American woman becomes involved with a seventeenth-century forebear who travels to the court of the Moghul emperor Aurangzebe, while her boyfriend builds up a database for a day in 1989: history meets the modern information explosion but the quest for identity remains. Mukherjee is a powerful occasional essayist, and has also published *The Sorrow and the Terror*, an account of an Air India plane crash off the Irish coast.

Fiction

The Tiger's Daughter 1972; *Wife* 1975; *Darkness* (s) 1985; *The Middleman and Other Stories* (s) 1988; *Jasmine* 1989; *The Holder of the World* 1993

Non-Fiction

Kautilya's Concept of Diplomacy: a New Interpretation 1976; *Days and Nights in Calcutta* [with C. Blaise] 1977, (f) 1991; *The Sorrow and the Terror* [with C. Blaise] 1987

Paul MULDOON 1951–

Muldoon is rare amongst younger Irish writers in that he found his own distinctive voice at once, without first producing imitations of **W.B. Yeats** or other Irish masters. Born in Portadown, County Armagh in Northern Ireland, he is the son of a market gardener and labourer and a schoolteacher. He spent his early life in a small village called The Moy and was educated at St Patrick's College, and then Queen's University, Belfast, where he graduated with a BA in English literature. He was taught there by **Seamus Heaney** and weekly discussion with him and other writers such as **Michael Longley** and **Derek Mahon** (known as the Belfast Group), encouraged Muldoon to complete a collection of poems, *Knowing My Place*, which was published when he was nineteen.

He had been writing poetry since he was fifteen, largely under the influence of **T.S. Eliot**, but realised that he was an impossible writer to imitate. The style of Muldoon's poetry cannot be compared to that of anyone else, for it is lyrical and traditional in form when it has to be, as in *New Weather*, and is otherwise tailored to the subject. Much of Muldoon's early work was concerned with his own family situation, where he tried to reconcile his mother's educated approach to life with his father's 'instinctive' rural wisdom and also his Republican outlook. The title-poem of *Why Brownlee Left*, which concerns a boy's search for his father, was taken from a medieval tale called *Immram Mael Duin*, from which the name Muldoon has evolved. Tensely structured with surreal images, the poem evokes a strong atmosphere of isolation and aloneness. His most ambitious work to date is *Madoc: a Mystery*. The title is also that of a poem by Southey about a twelfth-century legend, where Madoc, the youngest son of a Welsh king, sails to an unnamed country (presumably America), where he conquers a tribe of 'Aztecas'. Southey and Coleridge feature throughout the long sequence of linked fragments of varying lengths, complete in themselves. Rarely longer than a page, and often just a few lines, each poem is headed with the name of a philosopher or thinker, from ancient times to the present day. Muldoon often mixes allusions to myth with random references to anything that has caught his attention, literary or historical, but in *The Annals of Chile* his voice is more personal, as in sections of 'Yarrow' where he writes of his mother's harrowing death from cancer.

Muldoon worked from 1973 to 1986 as a radio producer. He married in 1987, the year he went to live in the USA, where he lectures in writing at Columbia and Princeton Universities.

Poetry

Knowing My Place 1971; *New Weather* 1973; *Spirit of Dawn* 1975; *Mules* 1977; *Names and Addresses* 1978; *Immran* 1980; *Why Brownlee Left* 1980; *Quoof* 1983; *The Wishbone* 1984; *Selected Poems 1968–1983* 1986; *Meeting the British* 1987; *Selected Poems 1968–1986* 1987; *Madoc: A Mystery* 1990; *Shining Brow* 1993; *The Annals of Chile* 1994; *The Prince of the Quotidian* 1994

Edited

The Scrake of Dawn 1979; *The Faber Book of Contemporary Irish Poetry* 1986; *The Essential Byron* 1989

Translations

Nuala Ni Dhoonhnaill *The Astrakhan Cloak* 1992

Alice MUNRO 1931–

Munro was born Alice Laidlaw in Wingham, Ontario, and lives in the rural Canada that is the background to her stories. Her father was an impoverished fox and mink farmer, and the family was of Scottish descent. Her mother suffered from Parkinson's disease, which meant that the young Munro took on a housekeeping role early in life. She has said that she was an unpopular child: 'my oddity just shone out of me', and she was glad of the time-consuming distractions of a domestic life: 'I would have been a lonely girl who didn't get asked to dance.'

At the age of eighteen she won a scholarship to university, where she met both the men she was to marry. At the time, Gerald Fremlin, a bleakly witty poet, ignored her and so in 1951 she married James Munro instead. They moved to the suburbs of Vancouver and she had three children. 'It was a good imitation of normal life,' she has said, and the same might be said for her short stories, which seem to be about everyday existence, but which have oddly disturbing undercurrents. She read a great deal – European and Russian literature as well as the work of older female American contemporaries such as **Carson McCullers**, **Eudora Welty** and **Flannery O'Connor**, whose books, with their rural settings, encouraged her to write about her own experience. She started to write in secret, but it took her many years to find her voice, and she did not publish her first book until 1964. *Dance of the Happy Shades* immediately established her, winning the Governor-General's Award.

Her marriage broke down in the 1970s, and she returned to Ontario, where Gerald Fremlin heard her on the radio and invited her out to lunch. They subsequently married, and she describes the circumstances as 'one of those incredibly romantic things you don't write about. It makes bad fiction.' What she does write about is small-town life in Canada, a subject which might seem parochial but which has not prevented her from becoming one of the most highly praised and respected writers of her generation, not only in Canada, but internationally. Furthermore, she has made her reputation in the short story, a form that publishers generally dislike.

Munro has written only one true novel, *Lives of Girls and Women*, although *The Beggar Maid* is sometimes considered one. Her high reputation has gone some way towards reclaiming a traditionally marginalised form – and small-scale, domestic subject-matter – for the literary mainstream. This has not been without obstacles, as is demonstrated by the troubled history of *The Beggar Maid*, originally published in Canada as *Who Do You Think You Are?* This volume collected together stories published in magazines such as the *New Yorker*, *Viva* and *Redbook*, about a number of different female protagonists. Munro's Canadian publishers wanted a volume containing stories about two characters, and wanted the stories equally distributed between them. This involved Munro in revising the stories, a process not as drastic as it may sound, for she acknowledged: 'I often write about the same heroine and give her a different name and a different occupation and a slightly different background because of something I want to do in the story. But her psychological makeup is no different.' At galley stage, however, Munro was unhappy with the result and so – at a cost to herself of over $2,000 – revised the entire book, adding two new stories, thus making it about a single protagonist, Rose (the beggar maid of the title). It now formed a discontinuous narrative which traced her life from an impoverished childhood through marriage, motherhood, divorce and several relationships. This new coherence allowed the book to be published in Britain as a novel, thus making it eligible for the Booker Prize, for which it was duly shortlisted. It is still the book by which she is best known, and contains some of her finest stories, notably 'Wild Swans', 'Mischief' and 'Simon's Luck'.

Amongst Munro's many skills are elision and compression: **Anne Tyler** has described her stories as 'so ample and fulfilling that they feel like novels. They present whole landscapes and cultures, whole families of characters.' **Cynthia Ozick** has called her simply 'our Chekhov'. She starts writing in longhand, but usually finds that of a first draft 'hardly a word is usable', and revises again and again until she is ready for 'the real writing which I do on the typewriter'. This process has resulted in some of the best short stories of the century, notably 'Fits' and 'Miles City, Montana' (in *The Progress of Love*), 'Accident' and 'Labor Day Dinner' (in *The Moons of Jupiter*), 'The Wilderness Station' (in *Open Secrets*) and the title-stories of *Dance of the Happy Shades* and *Friend of My Youth*. Among the prizes she has won have been a second Governor-General's Award for *The Beggar Maid* and the W.H. Smith Award for *Open Secrets*.

Fiction

Dance of the Happy Shades (s) 1968; *Lives of Girls and Women* 1971; *Something I've Been Meaning to Tell You* (s) 1974; *Who Do You Think You Are?* (s) 1978 [UK and US *The Beggar Maid* 1979]; *The Moons of Jupiter* (s) 1982; *The Progress of Love* (s) 1986; *Friend of My Youth* (s) 1990; *Open Secrets* (s) 1994

Iris (Jean) MURDOCH 1919–

Murdoch was born in Dublin, of Anglo-Irish stock on both sides; her father was a civil servant (the profession was thickly to populate his daughter's fiction) who had served as a cavalry officer in the First World War. When Murdoch was still an infant, the family moved to London, where she grew up largely in the western suburbs of Hammersmith and Chiswick, attending the Froebel Institute, Badminton School, Bristol, and, between 1938 and 1942, Somerville College, Oxford. She then herself entered the civil service, as an assistant principal, but left in 1944 to work with UNRRA, the United Nations relief organisation, in London, Belgium and Austria.

During this period she met such influential philosophers as Jean-Paul Sartre and Raymond Queneau, who fuelled her early interest in existentialism (her first published book, in 1953, was a study of Sartre). She left UNRRA in 1946 and, after a year without employment in London, took up a year's postgraduate studentship in philosophy at Newnham College, Cambridge, where she studied under another influential thinker, Ludwig Wittgenstein. In 1948, she became a fellow and tutor in philosophy at St Anne's College, Oxford, and remained there until 1963; since then, she has lived as a writer, although she also lectured at the Royal College of Art between 1963 and 1967.

Her first novel, *Under the Net*, was published in 1954, and she has been extremely prolific thereafter, publishing over twenty-five novels, although very little fiction in shorter forms. Certain themes – the pursuit of the good, for instance – and a common milieu of middle-class characters enduring complex emotional crises, usually in a darkly evoked London but sometimes in the country, dominate her work throughout. It is nevertheless possible to divide it into three phases: an early one, where the books are generally short and well-polished stylistically, as well as varying widely in genre, from the detailed realism of *The Sandcastle* to the Gothic fantasy of *The Unicorn*; a middle period, running roughly from *The Nice and the Good* to the Booker Prize-winning *The Sea, The Sea*, with longer and more complex novels, including several, such as *An Accidental Man* and *The Black Prince*, generally counted among her best; and a third stage, where the books are almost always over 500 pages in length, and are almost universally considered in need of a stringent editing that their author will not allow. Murdoch is generally counted among the leading post-war novelists, with an enormous readership, admirers who regard her as a literary saint, less numerous detractors who believe she is merely fashionable in her concerns.

Her other works consist of several philosophical treatises, among which *Metaphysics as a Guide to Morals* is an attempted *summa*; a number of plays, not always produced, of which one is in collaboration with **J.B. Priestley**; and a volume of poetry. In 1956 Murdoch married John Bayley, long a professor of English at Oxford, and more recently increasingly well known as a novelist and controversialist; the couple lived for many years at Steeple Ashton outside the city, and now reside in the academic suburb of North Oxford.

Fiction

Under the Net 1954; *The Flight from the Enchanter* 1956; *The Sandcastle* 1957; *The Bell* 1958; *A Severed Head* 1961; *An Unofficial Rose* 1962; *The Unicorn* 1963; *The Italian Girl* 1964; *The Red and the Green* 1965; *The Time of the Angels* 1966; *Under the Net* 1966; *The Nice and the Good* 1968; *Bruno's Dream* 1969; *A Fairly Honourable Defeat* 1970; *An Accidental Man* 1971; *The Black Prince* 1973; *The Sacred and Profane Love-Machine* 1974; *A Word Child* 1975; *Henry and Cato* 1976; *The Sea, The Sea* 1978; *Nuns and Soldiers* 1980; *The Philosopher's Pupil* 1983; *The Good Apprentice* 1985; *The Book and the Brotherhood* 1987; *The Message to the Planet* 1989; *The Green Knight* 1993

Non-Fiction

Sartre, Romantic Rationalist 1953 [as *Sartre, Romantic Realist* 1980]; *The Sovereignty of Good and Other Concepts: the Leslie Stephen Lecture* 1967; *The Sovereignty of Good* 1970; *The Fire and the Sun: Why Plato Banished the Artists* 1977; *Reynolds Stone* 1981; *Acastos: Two Platonic Dialogues* 1986; *The Existentialist Political Myth* 1989; *Metaphysics as a Guide to Morals* 1992

Plays

A Severed Head [from her novel; with J.B. Priestley] 1964; *The Servants and the Snow* [pub. 1973]; *Art and Eros* [from Acastos; pub. 1980]; *Above the Gods* [from Acastos] 1987; *The Black Prince* [from her novel] [pub. 1989]

Poetry

A Year of Birds 1978

Richard MURPHY 1927–

Born in County Mayo, Murphy was the son of a colonial administrator who became Mayor of Colombo and was later Governor of the Bahamas. He spent his early childhood in what was then Ceylon before being sent to boarding-school in England, first as a cathedral chorister at Canterbury Choir School, then at Wellington; from there he won a scholarship to Magdalen College, Oxford, and then went to the Sorbonne. He was aide-de-camp to his father as Governor of the Bahamas in 1948–9; other jobs included being a 'member of staff' at Lloyd's, night-watcher of a salmon river in Ireland, director of the English School in Canea, Crete, sheep farmer in the Wicklow mountains, and skipper of a Galway hooker. He has also taught evening classes at Morley College, was writer-in-residence at the University of Virginia, Charlottesville, in 1965, Compton Lecturer in Poetry at the University of Hull – where **Philip Larkin** was librarian – in 1969, and has held visiting professorships at various American universities.

Murphy started to emerge as one of the leading Anglo-Irish poets in the 1960s; fishermen and the sea are the inspiration of much of his early work, and he looks firmly at Ireland's past. Reviewing *New Selected Poems* in *Poetry Ireland*, Peter Denman appraised 'The Cleggan Disaster', from the collection *Sailing to an Island*, as 'one of the few successes among modern attempts at minor epic in poetry'. The narrative sequence *The Battle of Aughrim*, describing a sixteenth-century land battle between Catholics and Protestants, was equally successful. In *The Price of Stone* he confronts his own Irish upbringing. In later years, he has spent much time in Sri Lanka and that country's ancient history is the source of inspiration for *The Mirror Wall*. **Ted Hughes** has written that his verse is 'classical in a way that demonstrates what the classical strengths really are. It combines a high music with simplicity, force and directness in dealing with the world of action. He has the gift of epic objectivity; behind his poems we feel not the assertion of his personality, but the actuality of events, the facts and suffering of "history".'

Poetry

*The Archaeology of Love*1955; *Sailing to an Island* 1955; *The Woman of the House* 1959; *The Last Galway Hooker* 1961; *Three Irish Poets* [with T. Kinsella, J. Montague] 1961; *Six Irish Poets* [with others; ed. R. Skelton] 1962; *Sailing to an Island* 1964; *Penguin Modern Poets 7* [with J. Silkin, N. Tarn] 1966; *The Battle of Aughrim and The God Who Eats Corn* 1968; *High Island* 1974; *Selected Poems* 1979; *Care* 1982; *The Price of Stone* 1985; *The Mirror Wall* 1989; *New Selected Poems* 1989

Edited

The Mayo Anthology 1990

Thomas (Bernard) MURPHY 1935–

Murphy was born in Tuam, County Galway, educated at the Vocational School there and at the Vocational Teachers' Training College, Dublin. He did an apprenticeship as a fitter and welder and taught engineering at the Vocational School, Mountbellow, County Galway from 1957 to 1962. Between 1951, when he was only sixteen, and 1962, the year after *A Whistle in the Dark* was premiered at Stratford East after it had been turned down by the Abbey Theatre, he acted in and directed a number of plays. He was a member of the board of directors of the Abbey Theatre, Dublin, and since 1986 has been writer-in-association with them. He has also been Regents Lecturer at the University of California, Santa Barbara, writer-in-association with the Druid Theatre, Galway, and a founding member of Moli Productions, Dublin. The many awards he has received include the Irish Academy of Letters Award, Independent Newspapers Award, Harvey's Award (twice) and the *Sunday Tribune* Award. He became a member of the Irish Academy of Letters in 1982 and Aodána in 1984.

In his introduction to the Thomas Murphy issue of *Irish University Review* (Spring 1987) Christopher Murray discusses the violence, particularly in Murphy's early plays. His first play, *On the Outside* (1961), in which a couple of lads do not have the money for admittance to a dance, 'seethes with frustration'. In *A Whistle in the Dark* the 'pacifist' Michael Carney, taunted by his aggressive family for what they see as his cowardice, ends by killing his brother Des. In *Famine* John O'Connor kills his wife and child rather than witness their starvation. About this play Murphy has said: 'It's not about the history of the Irish Famine ... Famine to me meant twisted mentalities, poverty of love, tenderness and affection; the natural extravagance of youth wanting to bloom – to blossom – but being stalemated by a nineteenth-century mentality.' In *The Morning After Optimism*, which Murray calls 'the most experimental of his plays', once more brother kills brother. The setting of *The Sanctuary Lamp* is a darkened church which carries its own metaphor. *The Blue Macushla* is set in the night-club of that name, the theme, though applied to Ireland, resembling that of a Hollywood gangster movie. In the extraordinary *The Gigli Concert* the central figure, this time rich and successful, is obsessed with the voice of Gigli. In *Bailegangáire* the complaining grandmother is 'a female Job who longs to unburden herself of all her grievances before she dies' (Vivian Mercier). Colm Tóibín points out that in many of Murphy's plays there is a 'sense of dispossession, the sense of people being written out of history'. Murphy himself has said that 'people are increasingly adrift'. With **Brian Friel** in his different style, Murphy is without question Ireland's leading playwright.

Plays

On the Outside [with N. O'Donoghue] 1961 [pub. 1976]; *A Whistle in the Dark* 1961 [pub. 1971]; *The Fly Sham* **(tv)** 1963; *Veronica* **(tv)** 1963; *Famine* 1966 [pub. 1977]; *The Fooleen* **(tv)** [as *A Crucial Week in the Life of a Grocer's Assistant*] 1967, **(st)** 1969 [pub. 1978]; *The Orphans* 1968 [pub. 1974]; *Snakes and Reptiles* **(tv)** 1968; *Young Man in Trouble* **(tv)** 1970; *The Morning After Optimism* 1971 [pub. 1973]; *The White House* 1972 [np]; Trilogy: *The Moral Force* **(tv)** 1973, *The Policy* **(tv)** 1973, *Relief* **(tv)** 1973; *On the Inside* 1974 [pub. 1976]; *The Vicar of Wakefield* **(tv)** [from Goldsmith's novel] 1974; *The Sanctuary Lamp* 1975 [pub. 1976, rev. 1984]; *The J. Arthur Maginnis Story* 1976 [np]; *Conversations on a Homecoming* **(tv)** 1976, **[st)** 1985 [pub. 1986]; *Speeches of Farewell* **(tv)** 1976; *Epitaph under Ether* 1979 [np]; *The Blue Macushla* 1980 [np]; *Bridgit* **(tv)** 1981; *Fatalism* **(tv)** 1981; *The Informer* **(tv)** [from L. O'Flaherty's novel] 1981; *She Stoops to Conquer* **(tv)** [from Goldsmith's play] 1982; *The Gigli Concert* 1983 [pub. 1984]; *Bailegangáire* 1985 [pub. 1986]; *A Thief of a Christmas* 1985 [np]; *Too Late for Logic* 1989 [pub. 1990]; *A Whistle in the Dark and Other Plays* [pub. 1989]

Les(lie Allan) MURRAY 1938–

Murray was born in Nabiac, New South Wales where his parents ran a dairy farm in the Bunyah district. After Talee High School he studied English and German at the University of Sydney where he co-edited the magazines *Hermes* and *Arna* with the poet Geoffrey Lehmann. He left university in 1960 without receiving a degree (he eventually got one in 1969) and hitchhiked round Australia taking various casual jobs. He worked as a translator of technical material at the Australian National Library and lived in Europe from 1967 to 1968 before a brief stint as a civil servant in the prime minister's office in Canberra. From 1973 to 1980 he was an editor for *Poetry Australia* and he has been writer-in-residence at numerous universities.

Murray's first collection of poems, *The Ilex Tree*, which won the Grace Leven Prize in 1965, establishes many of the themes to which he has repeatedly returned. The volume takes its title from Virgil's *Eclogues* and focuses on the rural life of Murray's childhood, but despite the Arcadian allusion in the title, many of the poems emphasise the loneliness and drudgery of rural existence. In later collections such as *Lunch and*

Counter Lunch and *Ethnic Radio* he has concentrated on the increasing polarisation between modern urban values and Australia's rural – and Aboriginal – roots: 'When Sydney and the Bush meet now / there is no common ground.' This notion also informs *The Boys Who Stole the Funeral*, a narrative sequence of 140 sonnets which tells the story of two boys purloining a body from a Sydney funeral parlour.

David Malouf has described Murray as 'the most naturally gifted poet of his generation', while Clive James, reviewing *The Vernacular Republic* in 1982, asserted that he is concerned with defining a country 'not yet fully in possession of its own culture'. Murray has edited the *New Oxford Book of Australian Verse* in which he includes two of his own poems, 'Equanimity' and 'The Smell of Coal Smoke'. His numerous awards include the National Book Council Award in 1975 and 1985, and the Australian National Poetry Award in 1988 for *The Daylight Moon*. He was married in 1962, and has five children.

Poetry

The Ilex Tree [with G. Lehmann] 1965; *The Weatherboard Cathedral* 1969; *Poems against Economics* 1972; *Lunch and Counter Lunch* 1974; *Selected Poems: The Vernacular Republic* 1976; *Ethnic Radio* 1979; *The Boys Who Stole the Funeral* 1980; *Equanimities* 1982; *The Vernacular Republic: Poems 1961–1981* 1982; *The People's Otherworld* 1985; *The Daylight Moon and Other Poems* 1987; *The Idyll Wheel* 1989; *Dog Fox Field* 1990; *Collected Poems* 1991, *The Rabbiter's Bounty* 1991; *Translations from the Natural World* 1993

Non-Fiction

The Peasant Mandarin: Prose Pieces 1978; *Persistence in Folly: Selected Prose Writings* 1984; *The Australian Year: the Chronicles of Our Seasons and Celebrations* [with P. Solness] 1985; *The Gravy in Images* 1988; *Blocks and Tackles* 1990; *The Paperbark Tree* 1992

Edited

Anthology of Australian Religious Poetry 1986; *The New Oxford Book of Australian Verse* 1986 [rev. 1991]; *Fivefathers* 1994

Translations

Nikolai Sergeevich Trubetskoi *An Introduction to the Principles of Phonological Descriptions* [with H. Bluhme] 1968

L(eopold) H(amilton) MYERS 1881–1944

Myers was born in Cambridge, the son of F.W.H. Myers, the poet, man-of-letters and founder of the Society for Psychical Research; his mother was a famous society beauty. He was educated at Eton and Trinity College, Cambridge, which he left without taking a degree in 1901. As a young man he travelled in America, where he met, and eventually married in 1908, Elsie Palmer, the daughter of an American general and founder of the city of Colorado Springs; there were two daughters from this marriage. In 1906 he came into a substantial legacy and, as his friend **L.P. Hartley** put it, writing Myers' notice in the *Dictionary of National Biography*: 'his life in the main was leisured and uneventful'. He travelled extensively (visiting Ceylon in 1925), went out much in society (although a rebel against his class) and, during the First World War, having been rejected for military service, worked as a clerk in the trade department of the Foreign Office.

He had begun writing with a verse drama, *Arvat*, in 1908, and did not publish his first novel, *The Orissers*, until he was past forty. His major fictional achievement is the novel tetralogy eventually entitled *The Near and the Far* and published in collected form in 1943: this is set in an imaginary sixteenth-century India and, in dealing with a young man's search for spiritual satisfaction, explores by implication the need for spirituality in contemporary life. This work has been acclaimed by many critics and fellow writers as a masterpiece, and Myers was taken up strongly by the Leavisite school of criticism, but he has never been much generally read and is now in some danger of being forgotten.

Throughout his life Myers was subject to hypochondria and melancholia; from the late 1930s he became a communist, and the ruptures with former friends which this involved and his lack of practical political experience further damaged his mental health. During the Second World War he failed to complete an autobiographical work, despaired of writing and life, and killed himself, by veronal poisoning, at his Buckinghamshire home.

Fiction

The Orissers 1922; *The 'Clio'* 1925; *The Near and the Far* 1943 [US *The Root and the Flower* 1947]: *The Near and the Far* 1929, *Prince Jali* 1931, *Raj Amar* 1935, *The Pool of Vishnu* 1940; *Strange Glory* 1936

Plays

Arvat 1908

Edited

F.W.H. Myers *Human Personality and Its Survival of Bodily Death* 1907

N

Vladimir(ovich) NABOKOV 1899–1977

The son of a distinguished jurist and member of the Kerensky government, Nabokov was born in St Petersburg. His childhood was divided between that city and nearby Vyra, his mother's estate. Educated mainly by private tutors, he also attended Tenishev School from 1910 to 1919, when he and his family fled the Bolshevik revolution.

He spent the next three years in Cambridge, graduating from Trinity College with a degree in Romance and Slavic languages. From 1922 to 1937 he lived in Berlin, producing during this period – under the pseudonym Vladimir Sirin – his Russian work: nine novels, nearly fifty short stories and over 300 poems. His first novel, *Mashen'ka* (*Mary*, 1926), subtly merges physical deracination and emotional loss, and hints at the nostalgia which was to suffuse much of his subsequent work. However, published only by minor *émigré* outlets, Nabokov was unable to support through writing alone himself, his wife (whom he married in 1925) and his son. He therefore taught, translated, coached tennis and boxing and composed chess problems and crosswords. In 1937 he abandoned Nazi Germany for Paris, where he wrote the first of his eight English-language novels, *The Real Life of Sebastian Knight*. In 1940 he sailed for New York.

From 1941 to 1948 he held the posts of resident lecturer at Wellesley College, Massachusetts (teaching Russian), and research fellow in entomology at Harvard. He was an eminent lepidopterist, discovering and describing several new species. He was appointed, in 1948, associate professor of Slavic at Cornell, where his course, 'Masters of European Fiction', was 'enthusiastic, electric, evangelical'. (**Thomas Pynchon** may have been among his students.) Following the success of *Lolita*'s American publication in 1958, Nabokov emigrated again, this time to Switzerland. The Montreux Palace Hotel became his permanent home and he died there. He had been a US citizen since 1945.

Nabokov's transition to English was less a conversion than a reversion. He had been brought up in a trilingual household and claimed to have read English before he spoke Russian. (French was his third language.) Indeed, throughout his career, he preserved a remarkable artistic continuity and, when taxed with repeating himself, responded that 'originality has only its own self to copy'. Whether in the comic melodrama of *King, Queen, Knave* or the almost plotless *Look at the Harlequins!*, it is the precision of style and the cunning of construction that impress. Delicately, sensually, the physical properties of animate and inanimate are delineated and an aesthetic value is suggested. Where character is concerned, a moral value may be intimated as well – but the one cannot be assumed from the other. In fact, the typical Nabokovian anti-hero is a paranoid aesthete, a man who views people as *objets d'art* and treats them with accordant callousness. In this respect, the distance between creator and created is not great, for all of Nabokov's protagonists must submit to the cruel formalities of the novel in which they are trapped. Caught on the lepidopterist's pin, their role is to elicit neither sentiment nor sententiousness; they are elements in a composition.

Lolita remains Nabokov's best-known novel, largely because this elegant, witty and amoral story of a man's erotic obsession with a pubescent 'nymphet' – which is also a metaphor for Nabokov's love affair with the English language – was considered unfit for general publication in the USA and UK and was first issued in 1955 by the notorious Olympia Press in Paris. This is not, however, to underestimate the book itself, which is a masterpiece. Of his other books, *Bend Sinister* is a playful and allusive political fable about a totalitarian state, which is nevertheless concerned with the human heart, while *Pale Fire*, satirically inspired by his work as an academic, is a classic modernist text, which takes the form of a long poem embedded in an extensive, elaborate and (it transpires) spurious critical apparatus.

Although he was awarded the American National Medal for Literature in 1973, Nabokov received few official garlands in his lifetime. Death has not simplified his relationship with the critics. Adulation for his technical accomplishment is unstinting; but his subject-matter – and his apparent attitude toward it – has engendered in some a queasy repugnance. Little could be more disquieting than an absolute mastery of medium consorting with a sustained refusal to espouse the maxims of literary liberalism; and it is no surprise that Nabokov is accommodated best within a tradition which encompasses too Flaubert, Proust and Gide. Of his reputation

Nabokov said: '*Lolita* is famous, not I. I am an obscure, doubly obscure, novelist with an unpronounceable name.'

Fiction

As 'Vladimir Sirin': *Mary* 1926; *King, Queen, Knave* 1928; *The Defence* 1930; *The Eye* 1930; *Glory* 1932; *Laughter in the Dark* 1932; *Despair* 1936; *The Gift* 1938; *Invitation to a Beheading* 1938

As Vladimir Nabokov: *The Real Life of Sebastian Knight* 1941; *Bend Sinister* 1947; *Nine Stories* (s) 1947; *Lolita* 1955; *Pnin* 1957; *Nabokov's Dozen: Thirteen Stories* (s) 1958; *Pale Fire* 1962; *Quarter* (s) 1966; *The Waltz Invention* 1966; *Ada* 1969; *Transparent Things* 1972; *A Russian Beauty and Other Stories* (s) 1973; *Look at the Harlequins!* 1974; *Tyrants Destroyed and Other Stories* (s) 1975; *Details of a Sunset and Other Stories* (s) 1976

Translations

Three Russian Poets: Selections from Pushkin, Lermontov and Tyutchev 1944; Mikhail Lermontov *A Hero of Our Time* 1958; *The Story of Igor's Campaign. An Epic of the Twelfth Century* 1960; Pushkin *Eugene Onegin* 1964

Poetry

Poems 1959; *Poems and Problems* 1970; *Stikhi* 1979

Non-Fiction

Nikolai Gogol 1944; *Conclusive Evidence* (a) 1951 [rev. as *Speak, Memory* 1965]; *Strong Opinions* 1973; *The Nabokov–Wilson Letters: Correspondence Between Vladimir Nabokov and Edmund Wilson, 1940–1971* [ed. S. Karlinsky] 1979; *Lectures on Literature* [ed. F. Bowers] 1980; *Lectures on Russian Literature* [ed. F. Bowers] 1982; *Lectures on Don Quixote* [ed. F. Bowers] 1983

Collections

Nabokov. Criticism, Reminiscences, Translations and Tributes [ed. A. Appel] 1971

Biography

Vladimir Nabokov: the Russian Years by Brian Boyd 1990; *Vladimir Nabokov: the American Years* by Brian Boyd 1991

Shiva(dhar Srivinasa) NAIPAUL 1945–1985

Born in Port of Spain, Trinidad, Naipaul was the grandson of an Indian who had been brought to Trinidad as an indentured labourer to work on plantations. His father had become a journalist and married a daughter of Indian island landowners, and they had seven children, of whom another was the novelist **V.S. Naipaul**, thirteen years Shiva's senior. Naipaul's youth was dominated, as his brother's had been, by the struggle to win one of the four 'island scholarships' awarded each year to Oxford University, and he suffered a grim schooling. His aim achieved, he left for Oxford at eighteen, feeling 'haphazardly cobbled together from bits and pieces taken from everywhere and anywhere'. He graduated with a degree in Chinese, and having married a young English wife, with whom he was to have one son.

He embarked, as had his brother, on the career of writer and journalist (with the emphasis on the former) in England, and his first novel, *Fireflies* (1970), a story of the declining Indian elite on Trinidad, was published to considerable acclaim when he was twenty-five, winning him several prizes, including the John Llewellyn Rhys Prize. A second novel, *The Chip-Chip Gatherers*, another pessimistic vision of Trinidad, although less admired, won a Whitbread Award. He then seemed to desert the novel for travel writing and reportage, publishing many pieces in the *Spectator*, and books about travel in Africa and the mass-suicide at Jonestown, Guyana. In these writings he combined a dislike of many attitudes common in the Third World with an outsider's sceptical assessment of England. He made a return to the novel in 1983 with *A Hot Country*. Increasing success as a writer was cut short when he died of a heart-attack aged only forty. In his memory, the *Spectator* set up the Shiva Naipaul Prize, awarded each year for a travel essay by a writer under thirty-five; the first recipient was **Hilary Mantel**.

Fiction

Fireflies 1970; *The Chip-Chip Gatherers* 1973; *Black and White* 1980; *A Hot Country* 1983; *Beyond the Dragon's Mouth* (s) 1984; *Man of Mystery* (s) 1995

Non-Fiction

North and South: an African Journey 1978; *Unfinished Journey* 1986

V(idiadhar) S(urajprasad) NAIPAUL 1932–

Born in Trinidad, from a family of Indian Brahmin origin, and the son of a journalist, Naipaul was educated in Port of Spain and, like his brother **Shiva Naipaul** after him, won one of the four annual 'island scholarships' to Oxford, departing for England in 1950. On graduation, he immediately took up the career of freelance writer that he has followed since. From 1954 to 1956 he was a broadcaster for the BBC's *Caribbean Voices*, and from 1957 to 1961 he was a regular fiction reviewer for the *New Statesman*. He married in 1955.

His first three novels, beginning with *The Mystic Masseur* (1957), were comedies of life in Trinidad, but the high point of his early period as a novelist was *A House for Mr Biswas*, a pessimistic but rich study of a sensitive man's struggle with a lifetime in Trinidad. In 1961 Naipaul received a grant from the Trinidad government to travel in the Caribbean, and this inaugurated a long period in the 1960s and early 1970s in which he travelled widely in, among other countries, India, Uganda, Argentina, Iran, Pakistan, Malaysia and the USA, using the experience to fuel many non-fiction books – generally taking a pessimistic view of conditions in the developing world – as well as further impressive fictions, such as his horrifying portrait of emergent Africa, *A Bend in the River*. Naipaul is generally regarded as the leading novelist produced by the English-speaking Caribbean; he has received many literary prizes, including the Booker Prize for *In a Free State* in 1971, and is the subject of several critical books; he is, however, more admired than read.

In 1987, he returned to the novel form with *The Enigma of Arrival*, a clearly autobiographical novel about a writer of Caribbean origin now living in rural England and finding that a growing sense of homecoming is replacing the alienation and rootlessness of earlier life.

Fiction

The Mystic Masseur 1957; *The Suffrage of Elvira* 1958; *Miguel Street* 1959; *A House for Mr Biswas* 1961; *Mr Stone and the Knight's Companion* 1963; *A Flag on the Island* (s) 1967; *The Mimic Men* 1967; *The Loss of El Dorado* 1969; *In a Free State* (s) 1971; *Guerrillas* 1975; *A Bend in the River* 1979; *The Return of Eva Perón* (s) 1980; *Finding the Centre* (s) 1984; *The Enigma of Arrival* 1987; *A Way in the World* 1994

Non-Fiction

The Middle Passage. Impressions of Five Societies in the West Indies and South America 1962; *An Area of Darkness* 1964; *The Overcrowded Barracoon and Other Articles* 1972; *India: a Wounded Civilisation* 1977; *Among the Believers: an Islamic Journey* 1981; *A Turn in the South* 1989; *Homeless by Choice* [with R. Jhabvala, S. Rushdie] 1992

R(asipuram) K(rishnaswami) NARAYAN 1906–

Narayan was born in Madras, the fourth of eight children. His father, a middle-class Brahmin, worked as a headmaster for the government educational service and was often posted away from home. The family migrated to Mysore when Narayan was still a child and, to relieve the burden on his then pregnant mother, he was left to the care of the grandmother. His education at the local Christian Mission School was supplemented by her instruction in Tamil and readings from the Vedas. As the only child in the house, he was allowed to run free, as later evoked in *Swami and Friends*. Eventually he was called to Mysore by his father and enrolled in the Maharajah's College. He was awarded his BA, after failing several examinations, at the age of twenty-four, an exceptionally late age to graduate in India; the early story, 'Breach of Promise', was based on a farcical and half-hearted suicide attempted by the unwilling student. The period was also to yield him a second novel, *The Bachelor of Arts*.

Narayan worked, without success, as a teacher and editorial assistant before arriving at his vocation, a course of action aided by his father's indulgence and support. Sadly his father did not survive to see the publication of *Swami and Friends* (1935), his first novel. Personal tragedy followed with the death of Narayan's young wife of typhoid fever in 1939. They had one daughter, in regard for whom Narayan forced himself to come to terms with the loss, communicating with his dead wife through a medium and continuing the self-therapy in his fourth novel, *The English Teacher*. It is widely regarded as the first of his mature works, of which *The Guide* is the masterpiece. The tale of 'a confidence man turned saint' told in a complex series of flashbacks, it won the Sahitya Akademi Award, India's highest literary honour.

Narayan has long been considered the greatest living Indian writer although, as he writes in English, his books are a closed world to the vast majority of his countrymen. His characters, the fictional inhabitants of the fictional town of Malgudi (who come from a broad cross-section of Indian society), have been unanimously praised as 'true', and **Graham Greene**, upon whose recommendation Hamish Hamilton published his first novel, described him as 'the novelist I most admire in the English language ... one of the glories of English literature'. Narayan's evocation of Malgudi is often compared to **William Faulkner**'s Yoknapatawpha County although Narayan himself has described Faulkner as 'unreadable' and a 'bore'.

In 1974 he published *My Days: a Memoir*. He lives in Mysore but travels extensively in Europe and America. He is also the owner of Indian Thought Publications which publishes his work in India, both novels and short stories, many of which first appeared in the *New Yorker*. His brother, Laxman, is well known as a cartoonist.

Fiction

Swami and Friends 1935; *The Bachelor of Arts* 1937; *The Dark Room* 1939; *The English Teacher* 1945; *An Astrologer's Day and Other Stories* (s) 1947; *Mr Sampath* 1947; *The Financial Expert* 1952; *Waiting for the Mahatma* 1955; *Lawley Road and Other Stories* (s) 1956; *The Guide* 1958; *The Man-Eater of Malgudi* 1961; *Gods, Demons and Others* 1965; *The Sweet Vendor* 1967; *A Horse and Two Goats* (s) 1970; *The Ramayana* 1973; *The Painter of Signs* 1977; *The Mahabharata* 1978; *Malgudi Days* (s) 1982; *A Tiger for Malgudi* 1983; *Under the Banyan Tree and Other Stories* (s) 1985; *Talkative Man* 1986; *The World of Nagaraj* 1990; *The Grandmother's Tale* (s) 1992

Non-Fiction

Next Sunday 1955; *My Dateless Diary* 1960; *My Days* (a) 1974; *Reluctant Guru* 1974; *The Emerald Route* 1977; *A Writer's Nightmare* 1987; *A Story Teller's World* 1989

For Children

Salt and Sawdust: Stories and Table Talk 1993

(Frederic) Ogden NASH 1902–1971

Remembered by George Stevens as 'easy to imitate badly, impossible to imitate well', Nash was probably America's best-loved and most distinctive writer of light verse. He was born in Rye, New York, of an old Southern family; his great-great-great uncle was General Francis Nash, after whom Nashville is named. He attended Harvard for a year in 1920–1, but dropped out to earn a living. He had little success at teaching or selling bonds (he sold only one, and that was to his godmother), but he was better at copywriting, after which he moved across to publishing and then found a job on the editorial staff of the *New Yorker*.

His first book was a collaborative effort at a children's book, *The Cricket of Carador* (1925), with Joseph Alger. He published his first book of humorous verse, *Hard Lines*, in 1931, and never looked back; he was soon able to give up his *New Yorker* job and become a freelance writer. Nash's first published poem, 'Spring Comes to Murray Hill', is immediately distinctive, not least for its rhyming: 'I sit in an office at 244 Madison Avenue / And say to myself you have a responsible job, havenue?'

Inimitably mangled rhymes and aberrant metre are Nash's trademark. He was perfectly happy to rhyme 'vestibule' with 'indigestibule', or to observe that 'Spring is what winter / Always gazinta'. He claimed to be influenced by Julia Moore, the late nineteenth-century American versifier who is America's William McGonagall, and called himself a 'worsifier'. Nevertheless, his near-rhymes are, more often than not, neat and satisfying: 'If called by a panther / Don't anther.'

Nash's voice has a civilised silliness and an amiable cynicism, as in his famous observation that 'Candy / Is dandy / But liquor / Is quicker.' He has been placed in the tradition of so-called 'horse-sense comic writing', and he was a witty observer of American middle-class life in suburban homes and country clubs. He felt that 'the minor idiocies of humanity' were his field and characterised the body of his work as 'slightly goofy and cheerfully sour'. He always saw himself as a light entertainer but he could occasionally be serious, as in the moving final poem of *Hard Lines*, 'Old Men'. After considering how indifferent the world is to the death of the old, he suddenly changes the viewpoint: 'But the old men know when an old man dies.'

Nash became something of a national institution in his own lifetime, and was much in demand as a public figure. Throughout the 1940s and 1950s he featured in a number of panel games and radio and television shows, including those of Bing Crosby and Rudy Vallee, and he wrote with Kurt Weill and S.J. Perelman the 1943 Broadway hit musical *One Touch of Venus*. He married in 1931, and after becoming the father of two daughters he added mildly sentimental verse about children to his repertoire.

After a lifetime of mild hypochondria, which became a subject for his verse, Nash died of heart failure in 1971. Inevitably there were a number of tributes in the Nash manner, the best of which is perhaps that of Morris Bishop: 'Free from flashiness, free from trashiness / Is the essence of ogdenashiness. / Rich, original, rash and rational / Stands the monument ogdenational.'

Poetry

Free Wheeling 1931; *Hard Lines* 1931; *Happy Days* 1933; *Four Prominent So and Sos* 1934; *The Primrose Path* 1935; *The Bad Parent's Garden of Verse* 1936; *I'm a Stranger Here Myself* 1938; *The Face is Familiar* 1940; *Good Intentions* 1942; *Many Long Years Ago* 1945; *Versus* 1949; *Family Reunion* 1950; *The Private Dining Room* 1953; *You Can't Get There from Here* 1957; *Verses from 1929 On* 1959; *Everyone But Thee And Me* 1962; *Marriage Lines* 1964; *There's Always Another Windmill* 1968; *Bed Riddance* 1970; *The Old Dog Barks Backwards* 1972; *Ave Ogden!* [with J. Gleeson, B. Meyer] 1973; *I Wouldn't Have Missed It* 1975

For Children

The Cricket of Carador [with J. Alger] 1925; *Musical Zoo* 1947; *Parents Keep Out* 1951; *The Christmas That Almost Wasn't* 1957; *Custard the*

Dragon 1959; *A Boy Is a Boy* 1960; *Custard the Dragon and the Wicked Knight* 1961; *Girls Are Silly* 1962; *The New Nutcracker Suite and Other Innocent Verses* 1962; *Parents Keep Out* 1962; *The Adventures of Isabel* 1963; *A Boy and His Room* 1963; *The Untold Adventures of Santa Claus* 1964; *The Animal Garden* 1965; *The Cruise of the Aardvark* 1967; *The Mysterious Ouphe* 1967; *Santa Go Home* 1967; *The Scroobious Pip* [from E. Lear] 1968; *Custard the Dragon and the Wendingo* 1977

Edited

Nothing But Wodehouse 1932; *The Moon Is Shining Bright as Day* 1953; *I Couldn't Help Laughing* 1957; *Everybody Ought to Know* 1961

Plays

One Touch of Venus [with S.J. Perelman, K. Weill] 1944

Howard (Stanley) NEMEROV 1920–

Nemerov was born in New York City, educated at Fieldston, New York and Harvard, from which he graduated in 1941. On going down he joined the Royal Canadian Air Force, in which he served until the end of the war. He taught at various colleges and universities between 1946 and 1976 when he became Edward Mallinckrodt Distinguished University Professor at Washington University, St Louis. He was associate editor of *Furioso* from 1946 to 1951 and consultant in poetry to the Library of Congress in the mid-1960s. He has won an impressive number of awards and honours including Fellowship of the Academy of American Poets (1971) and the Levinson Prize from *Poetry (Chicago)* (1975).

Despite a critical notice in the *Hudson Review* for an early book, *The Image and the Law* – '[**W.H.**] **Auden** and [**Wallace**] **Stevens** are too persistently present for comfort' – he has gone on in his prolific way to become one of the leading American poets. His *Collected Poems* won the 1978 Pulitzer Prize for Poetry. He has irony, paradox and wit, combining humour with seriousness. In *Sentences* there are epigrams, narratives, riddles, meditations, many of them memorable: a standard Nemerov mix.

He has also published novels and short stories – winning a *Kenyon Review* Fellowship in Fiction in 1955 and the *Virginia Quarterly Review* Short Story Award in 1958 – and non-fiction, including collections of essays. He was married in 1944, and has three children.

Poetry

The Image and the Law 1947; *Guide to the Ruins* 1950; *The Salt Garden* 1955; *Small Moment* 1957; *Mirrors and Windows* 1958; *New and Selected Poems* 1960; *Five American Poets* [with others; ed. T. Gunn, T. Hughes] 1962; *Departure of the Ships* 1966; *The Blue Swallows* 1967; *A Sequence of Seven* 1967; *The Painter Dreaming in the Scholar's House* 1968; *The Winter Lightning* 1968; *Gnomes and Occasions* 1973; *The Western Approaches: Poems 1973–1975* 1975; *The Collected Poems of Howard Nemerov* 1977; *By A.L. Lebowitz's Pool* 1979; *Sentences* 1980; *A Spell Before Winter* 1981; *Gnomic Variations for Kenneth Burke* 1982; *Inside the Onion* 1984; *War Stories* 1987; *Trying Conclusions* 1991

Fiction

The Melodramatists 1949; *Federigo* 1954; *The Homecoming Game* 1957; *A Commodity of Dreams and Other Stories* (s) 1959; *Stories, Fables and Other Diversions* (s) 1971

Non-Fiction

Poetry and Fiction: Essays 1963; *Journal of the Fictive Life* 1965; *Reflexions on Poetry and Poetics* 1972; *Figures of Thought: Speculations on the Meaning of Poetry and Other Essays* 1978; *New and Selected Essays* 1985; *The Oak in the Acorn: On Remembrance of Things Past and Teaching Proust, Who Will Never Learn* 1987

Plays

Endor [pub. 1962]

Edited

Longfellow 1959; *Poets on Poetry* 1965

Collections

The Next Room of the Dream: Poems and Two Plays 1962; *A Howard Nemerov Reader* 1991

Biography

Howard Nemerov by Peter Meinke 1968

E(dith) NESBIT 1858–1924

Nesbit was born in Kennington, south London, the youngest daughter of the principal of a small local agricultural college. Her father died when she was three and her mother ran the college for some time before taking her family to Europe for the sake of the health of an elder daughter. When Nesbit was thirteen, the family returned to England and settled in Kent, and a love of that county remained with Nesbit all her life. She determined to be a poet and always felt that this was her true vocation, although most of the verse she wrote was far inferior to her prose. In 1877 she met Hubert Bland, who worked in a bank and was once described by **H.G. Wells** as 'a tawdry brain in the Fabian constellation'. They married in 1880, by which time Nesbit was seven months pregnant. Their married life, which survived financial and amatory difficulties, continued as unconventionally as it had started. Bland was swindled by a partner in the brush factory he had bought, and Nesbit was obliged to support the family by her writing (much of it hack work), an experience she drew upon in *The Railway Children*. Many of Nesbit's poems and stories appeared in magazines, anthologies and children's annuals. She was, for example, a regular contributor of fiction to the *Weekly Dispatch* (1887–94). She also edited a great many volumes of verse and prose for both adults and children. The Blands were founder members of the Fabian Society; Bland became a journalist, while Nesbit divided her time between her literary career, pamphleteering, and social work in the East End. Under the joint pseudonym 'Fabian Bland', they published short stories in radical journals, and a novel concerning Russian anarchists in London, *The Prophet's Mantle*.

When the children's governess became pregnant, Nesbit decided to adopt the child and rear it as her own. Some time later, she discovered that the child's father was Bland; the governess was to bear another of his children, also raised by Nesbit as her own. In 1893 the extended family removed to Eltham in Kent, where Nesbit's unconventional opinions, manner and dress caused much comment: she was as ardent a Baconian as she was a socialist; she linked arms with servants and smoked cigarettes in the streets; she bicycled wearing Liberty frocks. Here she wrote her famous children's novels, notably the stories concerning the Psammead (a sand fairy) and the Phoenix. The books sometimes contained Utopian themes, and the happy families she portrayed were in stark contrast to her own: her husband was an incorrigible philanderer, one baby was stillborn, and her third child, Fabian, died during minor surgery. During their marriage she made several bids to escape from Bland, and fell in love with **George Bernard Shaw** and Richard Le Gallienne, amongst others. Bland died in 1914 and Nesbit enjoyed a much happier relationship with 'Skipper' Tucker, a marine engineer whom she married in 1917. She also wrote a number of romantic novels for adults, few of which are read now, and was joint editor (with Spencer Pryse) of the *Neolith*, a short-lived magazine of literature and the arts (1907–8).

For Children

The Butler in Bohemia (s) [with O. Barron] 1894; *Doggy Tales* (s) 1895; *Pussy Tales* (s) 1895; *Romeo and Juliet and Other Stories* (s) 1897; *Royal Children of English History* (s) 1897; *A Book of Dogs* (s) 1898; *The Book of Dragons* (s) 1899; *The Story of the Treasure Seekers* 1899; *Nine Unlikely Tales for Children* (s) 1901; *The Wouldbegoods* 1901; *Five Children and It* 1902; *The Revolt of the Toys and What Comes of Quarrelling* (s) 1902; *Playtime Stories* (s) 1903; *The Rainbow Queen and Other Stories* (s) 1903; *Cat Tales* (s) [with R. Bland] 1904; *The New Treasure Seekers* 1904; *The Phoenix and the Carpet* 1904; *The Story of Five Rebellious Dolls* (s) 1904; *Oswald Bastable and Others* (s) 1905; *Pug Peter* (s) 1905; *The Railway Children* 1906; *The Story of the Amulet* 1906; *The Enchanted Castle* 1907; *Twenty Beautiful Stories from Shakespeare* (s) 1907; *The House of Arden* 1908; *The Old Nursery Stories* (s) 1908; *Harding's Luck* 1909; *Children's Stories from English History* (s) [with D. Ashley] 1910; *Children's Stories from Shakespeare* (s) 1910; *The Magic City* 1910; *The Wonderful Garden* 1911; *The Magic World* (s) 1912; *Wet Magic* 1913; *To the Adventurous* (s) 1923; *Five of Us and Madeleine* (s) [ed. R. Sharp] 1925

Fiction

The Prophet's Mantle [with H. Bland; as 'Fabian Bland'] 1885; *The Marden Mystery* 1894; *In Homespun* (s) 1896; *The Secret of Kyriels* 1899; *Thirteen Ways Home* (s) 1901; *The Red House* 1902; *The Literary Sense* (s) 1903; *The Incomplete Amorist* 1906; *Man and Maid* (s) 1906; *Daphne in Fitzroy Street* 1909; *Salone and the Head* 1909; *These Little Ones* (s) 1909; *Fear* (s) 1910; *Dormant* 1911; *The Incredible Honeymoon* 1916; *The Lark* 1922

Poetry

Lays and Legends 2 vols 1886, 1892; *The Better Part and Other Poems* 1888; *Leaves of Life* 1888; *A Pomander of Verse* 1895; *Songs of Love and Empire* 1898; *The Rainbow and the Rose* 1905; *Ballads and Lyrics of Socialism, 1883 to 1908* 1908; *Jesus in London* 1908; *Garden Poems* 1909; *Ballads and Verses of the Spiritual Life* 1911; *Many Voices* 1921

Non-Fiction

Wings and the Child, or the Building of Magic Cities 1913

Edited

Essays by Hubert Bland 1914

Biography

A Woman of Passion by Julia Briggs 1987

P(ercy) H(oward) NEWBY 1918–

Although Newby was born in Crowborough, Sussex, the son of a baker, his childhood was spent in various places in the Midlands and South Wales, and he was educated at Hanley Castle Grammar School, Worcester, and St Paul's College, Cheltenham. In 1939 he entered the army and served in the medical corps in France and Egypt. He was later seconded to teach English literature at Fuad I University in Cairo, and he remained there until 1946; a large number of his novels are set in the Middle East, an area with which he has retained contact since his wartime years there.

His first novel, *A Journey to the Interior*, was published in 1945 and in the same year he married; he has two children. In 1946 he returned to England, settled in Buckinghamshire, and worked as a freelance writer, bringing out a novel annually between 1947 and 1953 and making a reputation as one of the most promising writers of his generation. In 1949 he joined the BBC and had a long and very successful career as an administrator for almost thirty years, becoming controller of the Third Programme (later Radio 3) in 1958, and managing director of radio in 1975. As a novelist he belongs to a category of English writers, which perhaps includes **V.S. Naipaul** and **A.L. Barker**, whose critical reputations are high but who have not collected a large popular audience. He has won high praise for originality of characterisation, delicate farcical comedy (best seen in *The Picnic at Sakkara*) and the seriousness of theme evident in such novels as *Something to Answer For*, which won the first ever Booker Prize in 1969.

Since then his production has slowed down. *Coming in with the Tide*, published in 1991 and set in South Wales at the turn of the century, is the first historical novel he has written. His other books include a volume of short stories, early works on literature, two books about Egypt and a life of Saladin. He was chairman of the English Stage Company between 1978 and 1984 and was appointed CBE in 1972. He lives in Garsington, Oxfordshire.

Fiction

A Journey to the Interior 1945; *Agents and Witnesses* 1947; *Mariner Dances* 1948; *The Snow Pasture* 1949; *The Young May Moon* 1950; *A Season in England* 1951; *A Step to Silence* 1952; *The Retreat* 1953; *The Picnic at Sakkara* 1955; *Revolution and Roses* 1957; *Ten Miles from Anywhere and Other Stories* (s) 1958; *A Guest and His Going* 1960; *The Barbary Light* 1964; *One of the Founders* 1965; *Something to Answer For* 1969; *A Lot to Ask* 1973; *Kith* 1977; *Feelings Have Changed* 1981; *Leaning in the Wind* 1986; *Coming in with the Tide* 1991

Plays

The Reunion (r) 1970

For Children

The Spirit of Jem 1947; *The Loot Runners* 1949

Non-Fiction

Maria Edgeworth 1950; *The Novel 1945–1950* 1951; *The Third Programme* 1965; *The Egypt Story: Its Art, Its Monuments, Its People* [with F.J. Maroon] 1979; *Warrior Pharoahs: the Rise and Fall of the Egyptian Empire* 1980; *Saladin in His Time* 1983

Edited

Sir Richard Burton *A Plain and Literal Translation of the Arabian Nights' Entertainments* 1950

NGUGI wa Thiong'o (formerly James T. Ngugi) 1938–

That Ngugi's writing and the recent history of Kenya, his homeland, should be so inextricably linked is hardly surprising. His own education was interrupted for two years when the school he attended was closed down by the Mau Mau rebellion, and the struggle for Kenya's independence was to continue until 1963, the year of his graduation. His subsequent detention and self-exile underline the intensity of his commitment to political change in Africa. Yet his early fiction contains none of the overt Marxism which inspires, and to some critics mars, his mature writing.

He was born into a large, polygamous Kikuyu family and attended a Kikuyu school after a preparatory first year at the local mission. In 1959 he graduated from Alliance High School and in 1963 received his BA from Makere University in Uganda. A short spell on the Nairobi *Daily Nation* was followed by a further three years of study at Leeds University in England, and the beginning of the ideological reformation which is first apparent in *A Grain of Wheat*. His first published work, the play *The Black Hermit*, had been written in 1959 while he was still at school, but remained unperformed until 1962. Reviews were not encouraging, a setback overcome in 1964 with the publication of *Weep Not, Child*, Ngugi's first novel and the first English-language novel to have been published by an East African writer.

On his return to Kenya in 1967 he took up a post on the teaching staff of the English department of the University of Nairobi, a position he held until 1969 when the heavy-handedness of the administration's reaction to a student strike provoked his resignation. He taught in America briefly until offered a controlling post in the department from which he had resigned the previous year – a department that Ngugi was then instrumental in transforming to the Department of African Languages and Literature. He also renounced both his Christian name and the English language to write in Kikuyu and Swahili.

The publication in 1977 of *Petals of Blood*, which criticises not only the former rulers of Kenya, but also the new black elite which had displaced the colonial administration, proved to be another turning-point in his life. In the same year the authorities responded by banning *Nhaahika Ndeenda* (*I Will Marry When I Want*), a radio play written in collaboration with his wife, Ngugi wa Mirii, and dealing with the issues of wealth distribution and land ownership in the new Kenya. A house search and the confiscation of the greater part of his library was followed by detention without charge in January 1978. Ngugi was held incommunicado for two weeks before being moved to Kamitri maximum security prison where he was detained until December. He was finally released by presidential decree. The episode cost him his professorship. In 1982 his theatre group was banned from public performance and, fearing further reprisals, Ngugi began a prolonged self-exile in England.

Fiction

Weep Not, Child 1964; *The River Between* 1965; *A Grain of Wheat* 1967 [rev. 1993]; *Secret Lives* (s) 1974; *Petals of Blood* 1977; *Devil on the Cross* 1982; *Matigari* 1989

Plays

The Black Hermit 1962; *This Time Tomorrow* 1970; *This Time Tomorrow: Three Plays* [pub. 1978]; *I Will Marry When I Want* (r) [with Ngugi wa Mirii] 1980

Non-Fiction

Homecoming: Essays on African and Caribbean Literature, Culture and Politics 1972; *The Trial of Dedan Kimathi* [with Micere Githae Mugo] 1977; *Detained: a Prison Writer's Diary* 1981; *Writers in Politics: Essays* 1981; *Barrel of a Pen: Resistance to Repression in Neo-Classical Kenya* 1983; *Decolonising the Mind: the Politics of Language in African Literature* 1986; *Moving the Centre* 1993

Grace NICHOLS 1950–

Nichols was born and educated in Guyana, attending St Stephen's Scots School in Georgetown and the Progressive and Preparatory Institute, before graduating from the University of Guyana in 1967 with a diploma in communications. She stayed in Georgetown, working as a teacher for three years, then as a reporter on the national newspaper, and as an information assistant for the Guyanese government, becoming a freelance journalist before moving to England in 1977, where she has lived ever since.

Her first collection of poems, *I Is a Long-Memoried Woman*, was published in 1983 and was awarded the Commonwealth Poetry Prize. A carefully structured, chronological collection, it presents the reader with a history of the slave experience from the female viewpoint. The poems are deeply felt and intimately told in a first-person mixture of dialect and West Indian English, celebrating the individual strength and spirit underlying the shared experience of loss. Her second collection, *The Fat Black Woman's Poems*, published the following year, is a less planned and complete work, which adds the theme of the emigrant first arriving in England to a contemplation of childhood as seen through backward glances to the homeland. The first section of the book is written from the standpoint of a huge, loud-spoken black woman whose comments on the life around her add a knowing and pointed humour to the book.

Nichols was awarded a British Arts Council bursary in 1988, and published her first poetry collection for children, *Come on into My Tropical Garden*, in the same year. That was followed by *Lazy Thoughts of a Lazy Woman* in 1989. Although primarily known as a poet, Nichols has also written a novel, *Whole of a Morning Sky*, and a number of short-story collections for children. She lives with the poet and picture-book writer John Agard, with whom she has a daughter.

Poetry

I Is a Long-Memoried Woman 1983; *The Fat Black Woman's Poems* 1984; *Lazy Thoughts of a Lazy Woman and Other Poems* 1989

Fiction

Whole of a Morning Sky 1986

For Children

Trust You, Wriggly 1980; *Baby Fish and Other Stories* (s) 1983; *Leslyn in London* 1984; *The Discovery* 1986; *Come on into My Tropical Garden* 1988

Edited

Black Poetry 1988 [as *Poetry Jump Up* 1989]; *Can I Buy a Slice of Sky: Poems from Black, Asian and American Indians* 1991; *No Hickory, No Dickory, No Dock: a Collection of Caribbean Nursery Rhymes* [with J. Agard] 1991; *A Caribbean Dozen* 1994

Peter (Richard) NICHOLS 1927–

Born in Bristol (scene of many of his plays), the son of a salesman of grocery sundries, Nichols was educated at Bristol Grammar School. He entered the Royal Air Force towards the end of the war, seeing national service in India, Malaya, Singapore and Hong Kong and gravitating towards the entertainment corps, an experience which provided the background for his musical play *Privates on Parade*. He then studied at Bristol Old Vic Theatre School from 1948 to

1950, and subsequently 'struggled for five years to earn a living as an actor at a time when legit drama was run by a homosexual Mafia'. During this period he had a variety of other jobs such as park-keeper, English teacher in Italy and clerk. Changing course, he became a teacher in primary and secondary schools in London from 1957 to 1959.

He had been writing plays for some time, and in 1959 one of his plays won a BBC television playwriting competition: this launched him on a prolific career as a television playwright and freelance writer. He married Thelma Reed, a painter, in 1959, and they had four children: the eldest, Abigail, severely handicapped from birth, died in hospital at the age of ten. The family lived initially in Devon, moving to London in 1968, in various parts of which they have lived since. Great success as a stage playwright came to Nichols with *A Day in the Death of Joe Egg* (1967), a play about a couple coping with their severly handicapped daughter, which transferred to the West End after being put on by Nichols's friend Michael Blakemore at the Glasgow Citizens' Theatre: this, and his subsequent play, *The National Health*, a black comedy about the health service, won *Evening Standard* Best Play Awards.

Nichols has gone on to become one of Britain's most prolific and popular playwrights, taking the poignancies and comedies of ordinary middle-class life as his chief theme, and often incorporating elements of satire, vaudeville and musical comedy into his work. His autobiography, *Feeling You're Behind*, appeared in 1984.

Plays

After All **(tv)** 1959; *Walk on the Grass* **(tv)** 1959; *Promenade* **(tv)** 1959 [pub. in *Six Granada Plays* 1960]; *Ben Spray* **(tv)** 1961 [pub. in *New Granada Plays* 1961]; *The Big Boys* **(tv)** 1961; *The Reception* **(tv)** 1961; *The Heart of the Country* **(tv)** 1962; *Ben Again* **(tv)** 1963; *The Hooded Terror* **(tv)** 1963, **(st)** 1964; *The Continuity Man* **(tv)** 1963; *The Brick Umbrella* **(tv)** 1964; *Catch Us If You Can* **(f)** 1965; *When the Wind Blows* **(tv)** 1965; *Georgy Girl* **(f)** 1966; *A Day in the Death of Joe Egg*, **(st)**1967, **(f)** 1972; *Daddy Kiss It Better* **(tv)** 1968; *The Gorge* **(tv)** 1968 [pub. in *The Television Dramatist* ed. R. Muller 1973]; *Majesty* **(tv)** 1968; *Winner Takes All* **(tv)** 1968; *The National Health* **(st)** 1969, **(f)** 1972 [pub. 1970]; *Hearts and Flowers* **(tv)** 1970, *Neither Up Nor Down* **(st)** 1972 and *The Common* **(tv)** 1973 [pub. in *Plays: One* 1987]; *Forget-Me-Not-Lane* 1971; *Chez Nous* 1974; *The Freeway* [pub. 1975]; *Harding's Luck* [from E. Nesbit's novel] [np]; *Privates on Parade* **(f)** 1983; *Born in the Gardens* 1979 [pub. 1980]; *Passion Play* 1981; *Poppy* **(l)** [with M. Norman] 1982 [rev. 1991]; *Changing Places* **(f)** 1984; *A Piece of My Mind* 1987; *Plays: Two* [pub. 1991]

Non-Fiction

Feeling You're Behind **(a)** 1984

Robert (Malise Bowyer) NICHOLS
1893–1944

Born in Shanklin, on the Isle of Wight, the son of John Nichols, a minor poet, Nichols was educated at Winchester and Trinity College, Oxford, which he left to become a second lieutenant in the Royal Field Artillery. He was at the Front for only three weeks, when, according to his friend **Robert Graves**, he 'fell off a roof, went home as shell-shocked, slept with seventeen prostitutes in three weeks and got a bad dose'. In spite of his lack of front-line experience, Nichols published *Invocation*, a volume of 'War Poems and Others' in 1915. His second volume, *Ardours and Endurances*, made his reputation (comparatively short-lived) as one of the leading war poets. The war poems in the book (which also contains lighter works in the Georgian tradition, such as the long 'A Faun's Holiday' and numerous 'Phantasies') are arranged in sections to suggest the journey of the soldier-poet, from 'The Summons', through 'The Approach' to 'The Battle', 'The Dead' and 'The Aftermath'. Poems such as 'The Assault' attempt an impressionistic, onomatopoeic rendering of war experience with varying degrees of success ('On, on. Leăd. Leăd. Hail. / Spatter. Whirr! Whirr!'), while verses such as 'The Burial in Flanders', although undoubtedly sincere, are clumsy in their expressions of grief. Although Nichols was heterosexual, the startlingly homoerotic atmosphere of such poems attracted the attention of **Siegfried Sassoon**, who became a friend, depicting Nichols in his diaries 'alone and sex-ridden – the daemon of poesy leading him from gloom to gloom'. The popularity of *Ardours and Endurances* led to an appointment with the British Mission (Ministry of Information) in 1917, and he lectured on their behalf in New York.

Flamboyant and garrulous, Nichols was never troubled with self-doubts regarding his own abilities. For a time, his volcanic energy impressed Graves, Sassoon and **Edith Sitwell**. Less enamoured was **Aldous Huxley**, who recalled how the poet 'raved and screamed and hooted and moaned his filthy war poems' at a reading in 1917. Nichols's 'Keatsian' looks attracted the admiration of Edmund Gosse, but the older man's interest faded when he became the frequent recipient of long letters, the tone of which was 'as though Shakespeare were writing to an insect'. The archetypal failed romantic, he was constantly entangled with difficult women, including Nancy Cunard, who ended an affair by slapping his face in public. Hearing of Nichols's marriage in 1922 to the daughter of a bacon merchant, Graves was worried that the lunacy of Nichols's mother might descend to any offspring, but there were none.

Ill health and an unrevealed misdemeanour forced him to resign the Chair of English Literature at Tokyo Imperial University in 1924, a post he had held for three years, and he became instead script adviser to Douglas Fairbanks Sr in Hollywood. Nichols considered himself primarily a dramatist and, undiscouraged by the failure of two plays, produced the unmemorable *Wings over Europe* in 1930. Jealousy prompted his best-known work, *Fisbo*, a lengthy satire in the style of Alexander Pope, which shredded **Osbert Sitwell** but failed to antagonise him into a useful public row. Nichols never achieved his grandiose literary ambitions, although his *Anthology of War Poetry 1914–1918* was balanced and sensible. Deserted by his wife some years before his death, and generally forgotten as a writer, Nichols was remembered with amused affection by both Graves and Sassoon.

Poetry

Invocation 1915; *Ardours and Endurances* 1917; *Invocation and Peace Celebration Hymn* 1919; *Aurelia* 1920; *Fisbo* 1934; *A Spanish Triptych* 1936

Plays

Guilty Souls 1922; *Twenty Below* [with J. Tully] 1927; *Wings Over Europe* [with M. Browne] 1930

Fiction

Under the Yew 1928

Translations

Turgenev *Hamlet and Don Quixote* 1930

Edited

An Anthology of War Poetry 1914–1918 1943

Norman (Cornthwaite) NICHOLSON 1914–1987

Nicholson was born above his father's men's outfitting shop in Millom, Cumberland, and, except for two years, lived his entire life there, his roots in 'Thirty thousand feet of solid Cumberland'. His great-grandfather had farmed on the Furness peninsula and his grandfather came to Millom, when the town was prosperous, to run the horse and cart haulage for the new ironworks. He attended local schools. At fourteen he was 'Millom's champion reciter', a career that was cut short two years later when he contracted tuberculosis which left him with 'a chesty growl'. He came to poetry by a curious route. Sharing the country of Wordsworth (a little girl once addressed him as 'Mr Wordsworth'), he started aged twenty to absorb the modernists – in particular **T.S. Eliot, W.H. Auden** and **Stephen Spender** – and was ironically labelled a poet of the 1940s. The other decisive factor in his development was his conversion to Christianity at the age of twenty-two. Between the mid-1940s and the mid-1950s he published several collections of verse, but after *The Pot Geranium* wrote no poetry for ten years, confining himself to prose – a book on William Cowper (another strong influence), and books on the Lake District. In the mid-1940s he had also begun to write verse plays, the first of which, *The Old Man of the Mountains*, was produced at the Mercury Theatre, Notting Hill, in the same year as Ronald Duncan's *This Way to the Tomb.* He also wrote two novels, *The Fire of the Lord* and *The Green Shore.* After a decade of silence his verse became more colloquial and, following a period of neglect, his reputation grew steadily. In old age, among many awards, he won the Queen's Medal for Poetry (1977) and was appointed CBE.

He once said, 'To me Millom is the world. It is society writ small. It is the history of the last century or so, of the first phase of the industrial revolution – of boom and decline.' He saw himself, physically and spiritually, charting the same territory as Caedmon in the seventh century. In a poem celebrating the old poet he identifies his theme as 'the black thorns, the thickets of darkness / the ways and walls of a wild land / Where the spade grates on stone, on the grappling rose / And the Norse gods clamber on the Christian crosses'. For him continuity was everything. His father saw Halley's Comet in 1910; it was due to appear again in 1985, and in a moving poem written in 1970 he reviews his prospects of seeing it from the same house, perhaps from the same window. He married in 1956. When his wife died in 1982, ending a long and happy partnership, kindly neighbours would leave pies at the old man's door.

Poetry

Selected Poems [with K. Douglas, J. Hall] 1943; *Five Rivers* 1944; *Rock Face* 1948; *Prophecy to the Wind* 1950; *The Pot Geranium* 1954; *Selected Poems* 1966; *No Star on the Way Back* **(tv)** 1963 [pub. 1967]; *A Local Habitation* 1972; *Hard of Hearing* 1974; *Cloud on Black Combe* 1975; *Stitch and Stone* 1975; *The Shadow on Black Combe* 1978; *Sea to the West* 1981; *Selected Poems 1940–82* 1982

Plays
The Old Man of the Mountains 1945 [pub. 1946] [rev. 1950]; *Prophesy to the Wind* 1949 [pub. 1950]; *A Match for the Devil* 1953 [pub. 1955]; *Birth by Drowning* 1959 [pub. 1960]

Fiction
The Fire of the Lord 1944; *The Green Shore* 1947

Non-Fiction
Man and Literature 1943; *Cumberland and Westmorland* 1949; *H.G. Wells* 1950; *William Cowper* 1951; *The Lakers: the Adventures of the First Tourists* 1955; *Provincial Pleasures* 1959; *William Cowper* 1960; *Portrait of the Lakes* 1963 [rev. as *The Lakes* 1977]; *Enjoying It All* 1964; *Greater Lakeland* 1969; *Wednesday Early Closing* 1975

For Children
The Candy-Floss Tree [with F. Flynn, G. Mayer] 1984

Edited
An Anthology of Religious Verse Designed for the Times 1942; *Wordsworth: an Introduction and Selection* 1949; William Cowper: *Poems by William Cowper* 1951; *A Choice of Cowper's Verse* 1975; *The Lake District: an Anthology* 1977

Biography
Norman Nicholson by Philip Gardner 1973

Anaïs NIN 1903–1977

Nin was born in the Paris suburb of Neuilly, the first child of Joaquín Nin, a well-known composer and concert pianist of Spanish extraction, and Rosa, a French-Danish singer. She travelled widely during her childhood because of her father's work, but after her parents' separation when she was eleven, moved with her mother and two brothers to New York. It was on the boat to New York that she began the journal which she was to maintain almost obsessively for the rest of her life. She left school at sixteen and read avidly – especially anything she could find from the Catholic Church's Index Librorum Prohibitorum – and worked as a model for painters and fashion designers.

At the age of twenty she moved to Paris with her new husband, Hugh Guiler, a banker who later made underground films and illustrated some of her books under the name Ian Hugo. Bored and constrained by life as a banker's wife, Nin wrote her first book, *D.H. Lawrence: an Unprofessional Study* (1932), and was introduced to the then unpublished **Henry Miller** by the lawyer who drew up the contract for the book. Plunged into the Bohemian artistic life she had long craved, she became involved in a number of amorous liaisons, including an unusual triangular relationship with Miller and his wife, June, the precise details of which continue to fuel speculation. Despite her numerous affairs, Nin remained married to Guiler until her death though, at his request, all mentions of him were omitted from her published diaries.

Encouraged by Miller and other members of the Villa Seurat group of writers, Nin wrote *House of Incest*, a prose-poem, and her first novel, *Winter of Artifice*, both of which are elaborate and surreal explorations of the psyche influenced by the study of psychoanalysis she undertook with Otto Rank, with whom she also had an affair. According to her biographer Noel Riley-Fitch, Nin had been sexually abused by her father when young, and later in life seduced him, as is recorded in the unexpurgated diary finally published in 1993.

In the early 1940s she and Miller made money by supplying an elderly oil millionaire from Oklahoma with erotica at a rate of $200 per manuscript, half of which went to an agent. Their client was apparently insatiable, but could read a story once only, after which they were consigned to olive-drab steel cabinets in his office. He occasionally passed on instructions to Nin to 'Cut the Poetry!', but her elegant prose found a more appreciative audience when some of the stories were published as *Delta of Venus* and *Little Birds*. Nin recruited other writers to the task, including **George Barker** and **Robert Duncan**.

Nin returned to New York with Guiler, and thereafter lived in California and Mexico while working on the sequence of novels published collectively as *Cities of the Interior (Ladders to Fire, Children of the Albatross, The Four-Chambered Heart, A Spy in the House of Love* and *Solar Barque*). The sequence completed, she concentrated on preparing for publication the seven volumes of her diary which many commentators believe contain her finest work. Miller, always an enthusiastic promotor of his friends' work, compared the diary to the writings of St Augustine, Rousseau and Proust, to name but three, and it has been heralded as a major document of the women's movement. Nin died of cancer in Los Angeles.

Fiction
House of Incest 1936; *Winter of Artifice* 1939; *Under a Glass Bell* (s) 1944; *This Hunger* 1945; *Cities of the Interior* 1959 [rev. 1961]: *Ladders of Fire* 1946, *Children of the Albatross* 1947, *The Four-Chambered Heart* 1950, *A Spy in the House of Love* 1954, *Solar Barque* 1955 [rev. as *Seduction of the Minotaur* 1961]; *Collages* (s)

1964; *Delta of Venus* (s) 1977; *Waste of Timelessness and Other Early Stories* (s) 1977; *Little Birds* (s) 1979

Non-Fiction

D.H. Lawrence: an Unprofessional Study 1932; *Realism and Reality* 1946; *On Writing* 1947; *The Diary of Anaïs Nin* [ed. G. Stuhlmann] 7 vols 1966–81; *The Novel of the Future* 1968; *Unpublished Selections from the Diary* [ed. D. Schneider] 1968; *Nuances* 1970; *Paris Revisited* 1972; *Eidolons* [ed. R. Hart] 1973; *A Woman Speaks: the Lectures, Seminars, and Interviews of Anaïs Nin* [ed. E. Hinz] 1975; *Aphrodisiac* [with J. Boyce] 1976; *The Ineffable Frances Steloff* [with H. Miller] 1976; *In Favor of the Sensitive Man and Other Essays* 1976; *Portrait in Three Dimensions* 1979; *The Early Diary of Anaïs Nin* 4 vols 1980–85; *The White Blackbird* 1985; *Henry and June: from the Unexpurgated Diary of Anaïs Nin* 1986; *A Literate Passion: Letters of Anaïs Nin and Henry Miller 1932–1953* 1985

Biography

Anaïs Nin by Deidre Bair 1995

(Benjamin) Frank(lin) NORRIS 1870–1902

Norris was born in Chicago, the son of a wholesale jeweller and a former actress: his only brother to survive to adulthood was the novelist Charles G. Norris; his sister-in-law was the novelist Kathleen Thompson Norris. He grew up largely in San Francisco, and went to private school in Belmont, California. Believed to have some talent for drawing, he was sent to art school in London and then in Paris, where he immersed himself in the chivalric world of Froissart's Arthurian romances. His first published work was a long poem, *Yvernelle: A Tale of Feudal France* (1892), a work of little literary value.

Abruptly ordered home from Paris by his father, he attended the University of California and Harvard, a tall golden-haired college boy of the 1890s, now falling under the opposed literary influence of the naturalist Zola. In 1895 he sailed for South Africa to cover the political situation there for the *San Francisco Chronicle*: he took part in the Jameson expedition, suffered a severe bout of tropical fever, and was captured and expelled from the country by the Boers. He continued working as a newspaperman, suffering more fever when sent to cover the Cuban War in 1898, and later as a publisher's reader for Doubleday in New York. In 1900 he married; he had a daughter. Shortly before his death from peritonitis at the age of thirty-two, he had bought a ranch at Gilroy, California, where the family had hoped to settle.

Norris wrote much powerful if imperfect fiction: *McTeague*, a tragic story of a San Francisco dentist who becomes both thief and murderer, is justly admired, as are the two completed volumes of his projected trilogy, 'The Epic of the Wheat', the first of which, *The Octopus*, deals with the Californian wheat-growers and their stuggle with the railroad company, and the second, *The Pit*, with the commercial exploitation of wheat in Chicago. The third volume was to have covered the consumption of wheat in some famine-stricken area. But perhaps his finest novel is *Vandover and the Brute*, dealing with a decent man's descent into despair and madness. It was written at Harvard in 1894–5 and, for a while, believed lost in the San Francisco earthquake of 1906.

Fiction

Moran of the Lady Letty 1898; *Blix* 1899; *McTeague* 1899; *A Man's Woman* 1900; *The Octopus* 1901; *The Pit* 1903; *A Deal in Wheat and Other Stories* (s) 1903; *Complete Works* 1903 [rev. 1928]; *Shanghaied* 1904; *The Third Circle* (s) 1909; *Vandover and the Brute* 1914; *Frank Norris of 'The Wave'* (s) 1931

Non-Fiction

Responsibilities of the Novelist and Other Literary Essays 1903; *The Letters of Frank Norris* [ed. F. Wallace] 1956; *The Literary Criticism of Frank Norris* [ed. D. Pizer] 1964

Poetry

Yvernelle 1892

Harold NORSE (pseud. of Harold George Rosen) 1916–

Born illegitimate in the Bronx into a family of immigrant Lithuanian Jews and German Catholics, Norse grew up in Brooklyn. He began to write in childhood, then had a breakdown aged seventeen caused by his emerging homosexuality. From 1934 to 1938 he studied at Brooklyn college and edited the student newspaper; his assistant and fellow student, Chester Kallman, became his lover.

On 6 April 1939 Norse and Kallman went to a New York poetry reading given by **W.H. Auden**, **Christopher Isherwood** and **Louis MacNeice**. The youthful pair in the front row spent much of the reading winking and smiling at Auden and Isherwood and procured an invitation to visit the English writers' apartment for an interview. When Kallman turned up without Norse, Auden muttered something about his being 'the wrong blond', but speedily usurped Norse as Kallman's lover. Had the right blond visited the flat, one can speculate that Norse's career as a poet might have received the boost that distinguished connections can give. As it is, although he was a protégé of **William Carlos Williams** and knew the Beat poets as well as other notable writers, the audience for his own rangy free-verse poetry, of which he has published twelve volumes, has generally been small.

He published in magazines in the 1940s and was part of Greenwich Village Bohemia, meeting **Allen Ginsberg** in the subway, and living with **Tennessee Williams** while the latter was writing *The Glass Menagerie* (1944). Norse did many casual jobs at this time, but from 1949 to 1952 was an instructor in English at the Cooper Union Institute. In 1953 (the year his first collection, *The Undersea Mountain*, was published) he sold a Picasso a millionaire had given him, and used the proceeds to go to Europe, where he lived until 1968, teaching in Rome and later spending time at the famous 'Beat Hotel' in Paris. His one novel, *Beat Hotel*, derives from the 'cut-ups' of various texts he made at this time. From Norse's European period also dates what may be his most distinguished work, his translation of *The Roman Sonnets of G.G. Belli*, racy poetry written in the Roman dialect in the 1830s, which Norse translated, according to his own account, with the text in one hand and a Roman boy in the other. In 1968, suffering from hepatitis, he returned to the USA, was cured in Southern California, and in 1971 settled in San Francisco. There he founded the short-lived magazine *Bastard Angel* in 1972, taught briefly at San José State University, and benefited from the emerging homosexual publication movement (a national gay newspaper described him as 'The American Catullus' for his 1977 volume *Carnivorous Saint*). His raunchy autobiography, *Memoirs of a Bastard Angel*, has further increased his reputation.

Poetry

The Undersea Mountain 1953; *The Dancing Beasts* 1962; *Olé* 1966; *Karma Circuit: 20 Poems and a Preface* 1967; *Christmas on Earth* 1968; *Penguin Modern Poets 13* [with C. Bukowski, P. Lamantia] 1969; *Hotel Nirvana: Selected Poems 1953–73* 1974; *I See America Daily* 1974; *Carnivorous Saint: Gay Poems 1941–76* 1977; *Mysteries of Magritte* 1984; *The Love Poems 1940–85* 1986

Fiction

Beat Hotel 1975

Non-Fiction

Memoirs of a Bastard Angel (a) 1989; *The American Idiom: a Correspondence* 1991

Translations

The Roman Sonnets of G.G. Belli 1960

Alfred NOYES 1880–1958

Born in Wolverhampton, the son of a wealthy grocer, Noyes attended schools in Wales, before entering Exeter College, Oxford. Here he distinguished himself as an oarsman but failed to sit his final examinations because he was in London arranging the publication of his first book of poems, *The Loom of Years* (1902). This was praised by George Meredith, and Noyes soon became so popular that he could live by writing poetry. Each serial number of his epic poem, *Drake*, published in *Blackwood's Magazine* from 1906 to 1908, was awaited as eagerly as if it had been a novel and a copy accompanied Admiral Beattie through his wartime naval battles.

Noyes married Garnet Daniels, the daughter of an American colonel, in 1907 and he soon became popular as a lecturer in America; from 1914 to 1923 he was professor of modern English literature at Princeton. He was debarred from service in the First World War by defective eyesight (he was to go blind towards the end of his life), but returned to Britain during the war to serve in the Foreign Office and in France. While in England he was one of the few people shown Sir Roger Casement's diaries during his trial for treason, and believed these proved Casement to have been a homosexual; later, after having been attacked as an official traducer by **W.B. Yeats**, he changed his mind, and wrote a book aimed at clearing Casement of what he believed to be the slur of homosexuality.

Controversy was never far from Noyes; during the 1920s he converted to Roman Catholicism, and to a general hatred of modernism in literature was added a loathing of what he considered its immorality; he once succeeded in getting an auction of a copy of **James Joyce**'s *Ulysses* (1922) banned, and ordered fellow writer **Hugh Walpole** from his house for having recommended the book to a young girl. His own book on *Voltaire*, however, in which he attempted to prove that this writer was not anti-Christian, prompted the Holy Office to order a temporary suspension; this dispute was later resolved amicably, and Pope Pius XII apparently considered the charges against Noyes nonsensical.

Noyes's poetry is now largely forgotten, with the possible exception of 'The Highwayman'; another moving poem is 'Spring, and the Blind Children'. He also wrote plays, short stories and novels. With his second wife, whom he married in 1927, after his first wife's death, he had three children; they lived in North America during the Second World War, and in later years on the Isle of Wight.

Poetry
The Loom of Years 1902; *The Flower of Old Japan* 1903; *Poems* 1904; *The Forest of Wild Thyme* 1905 [rev. 1911]; *Drake* 2 vols 1906–8; *Poems* 1906; *Forty Singing Seamen and Other Poems* 1907; *The Golden Hynde and Other Poems* 1908; *The Enchanted Island and Other Poems* 1909; *In Memory of Swinburne* 1909; *Collected Poems* 4 vols 1910–27 [vols 1 and 2 rev. 1928–9]; *The Prayer for Peace* 1911; *Sherwood* 1911 [rev. as *Robin Hood* 1926]; *The Carol of the Fir Tree* 1913; *Tales of the Mermaid Tavern* 1913; *Two Christmas Poems* 1913; *The Wine-Press* 1913; *The Searchlights* 1914; *A Tale of Old Japan* 1914; *The Lord of Misrule and Other Poems* 1915; *A Salute from the Fleet and Other Poems* 1915; *Songs of the Trawlers* 1916; *The Avenue of the Allies and Victory* 1918; *The New Morning* 1918; *The Elfin Artist and Other Poems* 1920; *Selected Verse* 1921; *The Torch-Bearers* 3 vols 1922–30; *Songs of Shadow-of-a-Leaf and Other Poems* 1924; *Princeton, May 1917* 1925; *Dick Turpin's Ride and Other Poems* 1927; *Ballads and Poems* 1928; *The Strong City* 1928; *Twelve Poems* 1931; *The Cormorant* 1936; *Youth and Memory* 1937; *Wizards* 1938; *Orchard's Bay* 1939; *If Judgement Comes* 1941; *Poems of the New World* 1942; *Shadows on the Down and Other Poems* 1942; *Collected Poems* 1947; *The Assumption* 1950; *A Roehampton School Song* 1950; *A Letter to Lucian and Other Poems* 1956

Fiction
Walking Shadows (s) 1918; *Beyond the Desert* 1920; *The Hidden Player* (s) 1924; *The Return of the Scarecrow* 1929; *The Unknown God* 1934; *Youth and Memory* 1937; *The Last Man* 1940; *The Devil Takes a Holiday* 1955

Plays
Rada 1914

For Children
The Secret of Pooduck Island 1943; *Daddy Fell into the Pond and Other Poems for Children* 1952

Non-Fiction
Mystery Ships: Trapping 'U'-boats 1916; *What is England Doing?* 1916; *Open Boats* 1917; *Some Aspects of Modern Poetry* 1924; *New Essays and American Impressions* 1927; *The Opalescent Parrot* 1929; *Tennyson* 1932; *Voltaire* 1936 [rev. 1939]; *Pageant of Letters* 1940; *The Edge of the Abyss* 1942; *Portrait of Horace* 1947; *Two worlds for Memory* (a) 1953; *The Accusing Ghost: or Justice for Casement* 1957

Edited
The Magic Casement: an Anthology of Fairy Poetry 1908; Sir Walter Scott *The Minstrelsy of the Scottish Border* 1908; *The Temple of Beauty: an Anthology* 1910; *A Book of Princeton Verse* 1916; *The Helicon Poetry Series* 4 vols 1925; *The Golden Book of Catholic Poetry* 1946

Robert (Thomas) NYE 1939–

Nye was born in London. His father had risen from the working class to be a post office official in charge of a department for selling telephones in Essex; his mother was semi-illiterate. He attended Southend High School, but left at the age of sixteen, already determined to be a poet, and publishing his first poems in the *London Magazine* that year. Between the ages of sixteen and twenty-one, he did various jobs – newspaper reporter, milkman, market gardener, sanatorium orderly – but continued to devote himself to poetry. He married in 1959, and in 1961, the year his first volume of verse, *Juvenilia*, was published, he moved with his wife into a remote cottage in North Wales, where they lived with their three sons and he supported himself solely as a writer, a career never since abandoned. After a few years of poverty, his income was supplemented by regular reviewing. Since 1967 he has been poetry editor of the *Scotsman*, and from 1971 poetry critic of *The Times*; he also reviewed poetry for many years in the *Guardian*. He and his first wife were divorced in 1967, and in 1968 he married Aileen Campbell, a painter and poet who later trained as a Jungian psychologist, and they have three children, two of them Nye's stepchildren. They moved from the North Wales cottage, where they lived for ten years, to Edinburgh.

Although Nye continues to regard his primary vocation as that of poet, from the mid-1960s he began to branch out into other genres, including children's books, short stories and plays, and his eventual fame has come as a novelist. He published his first novel, *Doubtfire*, in 1967, but he broke through with his later novel *Falstaff*, which became a bestseller. It purports to be the autobiography of Sir John, told in old age to a series of secretaries, and with its rumbustious and experimental style it remains Nye's best-known work. It set the pattern, however, for future novels, all of which centre on historical and mythological figures, including Merlin, Sir Walter Raleigh, Gilles de Rais and Anne Hathaway, whose adventures are related with much exuberance, ribaldry and word-play.

This reliance on the common memory rather than the well of personal experience also marks Nye's collections of short stories, plays and books for children; the expression of feeling is wrung out of him through poetry, and during a long career he has published only five collections. From the early 1980s, he has concentrated on the novel; his plays, for instance, and a number of poetry anthologies he edited almost all date from the early 1970s. In 1977, after the success of *Falstaff*, he moved to Ireland, and continues to live there, in Cork, following a well-established career but one which may yet yield surprises.

Fiction

Doubtfire 1967; *Tales I Told My Mother* **(s)** 1969; *Penguin Modern Stories 6* **(s)** [with others] 1970; *The Same Old Story* 1971; *Falstaff* 1976; *Merlin* 1978; *Faust* 1980; *The Voyage of the Destiny* 1982; *The Facts of Life and Other Fictions* **(s)** 1983; *The Memoirs of Lord Byron* 1989; *The Life and Death of My Lord Gilles de Rais* 1990; *Mrs Shakespeare* 1993

Poetry

Juvenilia 1 1961; *Juvenilia 2* 1963; *Darker Ends* 1969; *Agnus Dei* 1973; *Divisions on a Ground* 1976; *A Collection of Poems 1955–1988* 1989

Plays

With Bill Watson: *Sawney Bean* 1969 [pub. 1970]; *Sisters* **(r)** 1969, **(st)** 1973 [pub. 1975]; *A Bloody Stupit Hole* **(r)** 1970 [np]; *Penthesilea* **(r)** 1971, **(st)** 1983 [from Kleist's play; pub. 1975]; *Reynolds, Reynolds* **(r)** 1971 [np]; *The Seven Deadly Sins* **(l)** [with J. Douglas] 1973 [pub. 1974]; *Mr Poe* 1974 [np]; *The Devil's Jig* **(r)** [from T. Mann; with H. Searle] 1980 [np]

For Children

March Has Horse's Ears 1966; *Taliesin* 1966; *Beowulf* 1968; *Three Tales* **(s)** 1970; *Wishing Gold* 1970; *Out of the World and Back Again* 1971; *Poor Pumpkin* 1971; *Cricket* 1975 [as *Once Upon Three Times* 1978]; *The Bird of the Golden Land* 1980; *Harry Pay the Pirate* 1981

Edited

A Choice of Sir Walter Raleigh's Verse 1972; *A Choice of Swinburne's Verse* 1973; *William Barnes: Selected Poems* 1973; *The English Sermon 1750–1850* 1976; *The Faber Book of Sonnets* 1976; *PEN New Poetry* 1986; Laura Riding: *First Awakenings* [with A.J. Clark, E. Friedmann] 1992, *A Selection of the Poems of Laura Riding* 1994

Joyce Carol OATES 1938–

Oates was born near the small city of Lockport in western New York State, the daughter of a tool-and-die designer who worked for more than forty years at the Harrison Radiator Company. She grew up at her grandparents' farm in Erie County, which was to become the Eden County of many of her novels and stories. She attended junior and senior high school in Lockport, and then Syracuse University from 1956 to 1960.

She had submitted her first novel to a publisher at the age of fifteen, and during her student years was turning out one a semester, and also seeing her early short stories published. She received an MA from the University of Wisconsin in 1961, and in the same year married Raymond Smith, a fellow English student, initially following him to Beaumont, Texas, where he held his first teaching post. In 1962, she herself was appointed to a teaching position in English at the University of Detroit, and the couple moved to this troubled Midwestern city, a second major geographical setting for her writing.

Her first collection of stories, *By the North Gate*, was published in 1963, and her first novel, *With Shuddering Fall*, announcing her familiar themes of violence, madness and sexual passion, in 1964. From then on a pattern of formidable productivity has been set, and she has now published more than forty-five novels and volumes of shorter fiction, as well as a great deal of work in other genres, such as poetry, journalism, criticism and drama. It is difficult to summarise the impact of such a vast output, but, although her fictional work embraces a wide and experimental variety of styles, it would be true to say that two factors draw it into a unity: a rootedness in American naturalism and American reality; and a preoccupation with violence which seems obsessional. Added to this is a Gothicism which perhaps derives from **William Faulkner**; in the 1980s, in fact, she experimented with a trilogy of novels, the first of them being *Bellefleur*, which attempted to update the Gothic genre. She rewrites little, often producing a story in an evening, and the result is sometimes that novels such as *them*, which deals with the Detroit race riots, can seem formless and undisciplined. On the other hand, *Unholy Loves*, an acute satire on academic life, suggests another of the many directions in which this always fertile talent could go.

She is less admired as a poet than as a critic, and such excursions into the world of sport as *On Boxing* seem marginal to her output. She taught at the University of Detroit until 1967, and then at the nearby University of Windsor, Ontario, until 1978, where she and her husband, who was also on the academic staff, founded the *Ontario Review* in 1974. She was a writer-in-residence at Princeton from 1979 to 1981, and since then has been a distinguished professor there, maintaining an undiminished flow of writing, as well as assisting her husband with the Ontario Review Press, which grew out of the magazine, and both of which they run from their home.

Fiction

By the North Gate (s) 1963; *With Shuddering Fall* 1964; *Upon the Sweeping Flood and Other Stories* (s) 1966; *A Garden of Earthly Delights* 1967; *Expensive People* 1968; *them* 1969; *Cupid and Psyche* (s) 1970; *The Wheel of Love and Other Stories* (s) 1970; *Wonderland* 1971; *Marriages and Infidelities* (s) 1972; *Do With Me What You Will* 1973; *A Posthumous Sketch* (s) 1973; *The Girl* (s) 1974; *The Goddess and Other Women* (s) 1974; *The Hungry Ghosts* (s) 1974; *Plagiarized Material* (s) [as 'Femandes/Oates'] 1974; *Where Are You Going, Where Have You Been?* (s) 1974; *The Assassins* 1975; *The Poisoned Kiss and Other Stories from the Portuguese* (s) [as 'Femandes/Oates'] 1975; *The Seduction and Other Stories* (s) 1975; *The Blessing* (s) 1976; *Childwold* 1976; *Crossing the Border* (s) 1976; *The Triumph of the Spider Monkey* 1976; *Daisy* (s) 1977; *Night-Side* (s) 1977; *A Sentimental Education* (s) 1978; *Son of the Morning* 1978; *The Step-Father* (s) 1978; *All the Good People I've Left Behind* (s) 1979; *Cybele* 1979; *The Lamb of Abyssalia* (s) 1979; *Unholy Loves* 1979; *Bellefleur* 1980; *A Middle-Class Education* (s) 1980; *Angel of Light* 1981; *A Bloodsmoor Romance* 1982; *Funland* (s) 1983; *Last Days* (s) 1984; *Mysteries of Winterthurn* 1984; *Wild Saturday and Other Stories* (s) 1984; *Solstice* 1985; *Wild Nights* (s) 1985; *Marya* 1986; *Raven's Wing* (s) 1986; *Lives of the Twins* [as 'Rosamond Smith'] 1987; *You Must Remember This* 1987; *The Assignation* (s) 1988; *American Appetites* 1989; *Soul-Mate* [as 'Rosamond Smith'] 1989; *Because It Is Bitter, and Because It Is My Heart* 1990; *I Lock My Door Upon Myself* 1990; *Heat and Other Stories* (s) 1991; *The*

Rise of Life on Earth 1991; *Black Water* 1992; *Where Is Here?* 1992; *Foxfire* 1993; *Haunted* 1994; *What I Lived For* 1994; *Will You Always Love Me?* (s) 1995; *Zombie* 1995

Poetry

Women in Love and Other Poems 1968; *Anonymous Sins and Other Poems* 1969; *Love and Its Derangements* 1970; *Woman Is the Death of the Soul* 1970; *In Case of Accidental Death* 1972; *Wooded Forms* 1972; *Angel Fire* 1973; *Dreaming America and Other Poems* 1973; *The Fabulous Beasts* 1975; *Public Outcry* 1976; *Abandoned Airfield 1977* 1977; *Season of Peril* 1977; *Snowfall* 1978; *Women Whose Lives Are Food, Men Whose Lives Are Money* 1978; *Celestial Timepiece* 1980; *The Stone Orchard* 1980; *Nightless Nights* 1981; *Invisible Woman* 1982; *Luxury of Sin* 1984; *The Time Traveller: Poems 1983–1989* 1989

Non-Fiction

The Edge of Impossibility: Tragic Forms in Literature 1972; *The Hostile Sun: the Poetry of D.H. Lawrence* 1974; *New Heaven, New Earth: the Visionary Experience in Literature* 1974; *Contraries* 1981; *The Profane Art* 1983; *On Boxing* 1987; *(Woman) Writer: Occasions and Opportunities* (a) 1988

Plays

The Sweet Enemy 1965 [np]; *Ontological Proof of My Existence* [with G. Prideaux] 1972 [pub. 1980]; *Sunday Dinner* 1970 [np]; *Miracle Play* 1973 [pub. 1974]; *Daisy* 1980 [np]; *Presque Isle* 1982 [np]; *The Triumph of the Spider Monkey* 1985 [pub. 1980]; *Lechery* 1985 [np]; *In Darkest America* 1990 [pub. 1991]; *I Stand Before You Naked* [pub. 1991]; *Twelve Plays* [pub. 1991]

Edited

Scenes from American Life: Contemporary Short Fiction (s) 1973; *The Best American Short Stories* (s) [with S. Ravenel] 1979; *Night Walks: a Bedside Companion* (s) 1982; *First Person Singular: Writers on Their Craft* 1983; *Story: Fictions Past and Present* [with B. Litzinger] 1985; *Reading the Fights* [with D. Halpern] 1988; *Best American Essays 1991* [with R. Abuan] 1992; *The Oxford Book of American Short Stories* (s) 1992; *The Sophisticated Cat* [with D. Halpern] 1992

(Josephine) Edna O'BRIEN 1930–

O'Brien was born in Tuamgraney, a small village in County Clare on the rural west coast of Ireland. She describes herself as a solitary child in a family of four, all of whom were taught at the National School in Scariff, and as a natural storyteller. She was sent to the Convent of Mercy in Loughreu from 1941 to 1946 when she left for Dublin, found work in a pharmacy and started writing small pieces for the Irish Press. She attended the Pharmaceutical College in Dublin, was awarded a licence in 1950, gave up pharmacy and then eloped to County Wicklow with the Czech/Irish writer Ernest Gébler, with whom she has two sons, one of them the writer **Carlo Gébler**. It was then that she began to read more widely to combat the loneliness of having been disowned by her family, discovering **James Joyce** and **W.B. Yeats** as well as a literary ambition of her own.

The couple moved to London where O'Brien began and finished *The Country Girls* (1960), her first novel, 'in a three week spasm'. The book recalls the author's own convent upbringing, following the fortunes of two contrasting young girls, the shy, sensitive Kate, and the brash, rebellious Baba, as they discover life and love, the candid treatment of which led to the banning of the book and its six successors by the Irish Censorship Board. *The Lonely Girl*, filmed and eventually republished in 1963 as *Girl with Green Eyes*, follows the country girls to town, and won O'Brien the Kingsley Amis Award in 1962. *Girls in Their Married Bliss* completes the trilogy in a mood of acid disillusionment (an epilogue added to the one-volume republication in 1986 concludes the story with Kate's suicide): O'Brien's own marriage had collapsed in the same year. Her next novel, *August Is a Wicked Month*, follows the heroine in her attempts to escape both herself and her unhappy marriage in a series of joyless holiday flings. 'The pain of loneliness, guilt, and loss', never far away in her writing, finally takes over when the wife returns to find that her son has been killed in a car crash. The near-impossibility of true love then became the theme of a series of novels about marriage, the realism of her early writing being replaced by monologue and reminiscence in *A Pagan Place* and *Night*. Comparisons with Colette, **J.D. Salinger** and Françoise Sagan gave way to sometimes huffy comparisons with the Joyce of Molly Bloom, particularly in America where the Irishness of her writing had previously put her in great favour. One notable dissident dismissed it as 'gift-wrapped porn'.

As a short-story writer, O'Brien has been published regularly in the *New Yorker*. Her first collection, *The Love Object*, was published in 1968; *Lantern Slides* won the Los Angeles *Times* Book Prize in 1990. She is also the author of plays, including one about **Virginia Woolf**; screenplays, including *James and Nora*, a study of Joyce's marriage; a book of poetry, *On the Bone*; and *Mother Ireland*, a tribute to her homeland.

Fiction

August Is a Wicked Month 1965; *Casualties of Peace* 1966; *The Country Girls Trilogy* 1987: *The Country Girls* 1960, *The Lonely Girl* [as *Girl with Green Eyes*] 1962, *Girls in Their Married Bliss* 1964, *A Pagan Place* 1970; *The Love Object* (s) 1968; *Zee & Co.* 1971; *Night* 1972; *A Scandalous Woman* (s) 1974; *Johnny I Hardly Knew You* 1977; *Mrs Reinhardt and Other Stories* (s) 1978; *Returning* (s) 1982; *Tales for the Telling* (s) 1986; *The High Road* 1988; *On the Bone* 1989; *Lantern Slides* 1990; *Time and Tide* 1992; *House of Splendid Isolation* 1994

Plays

Girl with Green Eyes (f) [from her novel *The Country Girls*] 1960; *Zee & Co.* (f) [from her novel] 1972; *A Pagan Place* [from her novel] 1973; *Virginia* 1981

Edited

Some Irish Loving 1979

Non-Fiction

Mother Ireland 1976

Flann O'BRIEN (pseud. of Brian O'Nolan) 1911–1966

O'Brien, whose original surname in its Gaelicised form was Ó Nualláin, adopted many other pseudonyms, of which the best known was that used in his *Irish Times* column, Myles na Gopaleen or gCopaleen ('Myles of the ponies'). He was born at Strabane in County Tyrone, one of twelve children of a customs and excise officer (the father had originally been known as Nolan, but his children preferred the other form). His father's profession took the family widely over Ireland, but most of O'Brien's childhood was spent in or near Dublin. He grew up speaking both Irish and English, and his literary and journalistic production uses both languages, with English predominating. He was educated by the Christian Brothers, and went up to University College, Dublin, in 1929. He studied there until 1935, but also spent a little documented year in Germany in 1933 and 1934, the first of several visits paid to the country during the Nazi years. He is said to have made an early marriage to Clara Ungerland, a nineteen-year-old German woman he met on a river cruise and who died one month after the wedding of galloping consumption, but whether these events took place cannot be confirmed.

In 1935 he entered the Irish civil service, being mostly concerned with town-planning matters, initially roads and drains. He published an early novel, *At Swim-Two-Birds* (1939), an experimental work in the tradition of **James Joyce** but differing from his work in its much greater geniality. At this period, O'Brien also wrote his other major novel, *The Third Policeman*, a much darker work playing experimental games with the detective genre, but it was turned down by his publisher, who suggested he write something more straightforward. Shocked by its rejection, he hid the manuscript away, and it was not found until after his death. He was in the process, however, of becoming much more well known as a journalist, particularly for the column he wrote in the *Irish Times* between 1940 and 1966, 'Cruiskeen Lawn' ('full little jug'), which is still regarded as one of the classic Irish satirical columns. His increasing alcoholism and Bohemian Dublin lifestyle meant that he wrote little more fiction for some years, except pulp detective stories published under the pseudonym of Stephen Blakesley. He married in 1948.

In 1953 he was forcibly retired from the civil service, and then began the darkest period of his life, surviving on his varied newspaper columns, until *At Swim-Two-Birds* was reissued by a London publisher in 1960 and unexpectedly made him famous as a novelist. Encouraged, he wrote two novels in the early 1960s (a fifth novel, in Irish, dates from the early 1940s). His life was, however, now plagued by frequent accidents and the effects of alcohol and smoking, and he eventually died of cancer of the pharynx in his mid-fifties. *The Third Policeman*, which many regard as his masterpiece, was published posthumously in 1967, as were several other books, mainly collections of his work, notably *The Best of Myles*.

Fiction

At Swim-Two-Birds 1939; *An Béal Bocht* [as 'Myles na gCopaleen'] 1941 [rev. as *The Poor Mouth* with P. Power 1973]; *The Hard Life* 1961; *The Dalkey Archive* 1964; *The Third Policeman* 1967

Plays

Faustus Kelly [as 'Myles na gCopaleen'] 1943

Translations

Brinsley MacNamara *Mairead Gillan* 1953

Non-Fiction

Cruiskeen Lawn [as 'Myles na gCopaleen'] 1943; *The Best of Myles* [ed. K. O'Nolan] 1968; *Myles from Dublin* [as 'George Knowall'; ed. M. Green] 1985

Collections

Stories and Plays 1973; *The Various Lives of Keats and Chapman and the Brother* [ed. B. Kiely] 1976; *Further Cuttings from the Cruiskeen Lawn* [ed. K. O'Nolan] 1976; *The Hair of the Dogma* [ed.

K. O'Nolan] 1976; *A Flann O'Brien Reader* [ed. S. Jones] 1978

Biography

Flann O'Brien by Peter Costello and Peter van de Kamp 1987

Kate O'BRIEN 1897–1974

O'Brien was born in Limerick (the Mellick of her novels), one of nine children of a prosperous horse dealer. Her mother died when she was five, and soon afterwards, O'Brien was sent to Laurel Hill Convent, where three elder sisters were already boarders. She spent twelve happy years there, which gave her the material for her delicate convent novel, *The Land of Spices*, which some critics consider her best work. In 1916 she went to University College, Dublin, and after graduating, spent a year as a governess in Bilbao, Spain, growing to love the country, which became, after Ireland, the chief setting for her fiction. She then worked for a while in the foreign languages department of the *Manchester Guardian*, and afterwards moved to London, where she taught in a convent school. In 1923 she married a Dutch journalist, Gustaaf Renier (who later became well known for the polemical book *The English, Are They Human?*), but the marriage lasted only a few months before their separation.

In 1926 O'Brien had a remarkable success with her first play, *Distinguished Villa*, and was to write several more plays, sometimes based on her novels and usually less successful than them. Articles and stories began to appear regularly, and her first novel, *Without My Cloak* (1931), like many of her books a chronicle of the Irish bourgeois society of her youth, received a warm reception, being soon followed by several others. During the 1930s O'Brien lived for several years with a woman friend in Spain until forced to leave in 1937 because of her Republican sympathies. A travel book, *Farewell, Spain*, came out of the experience. After the war, O'Brien returned to Ireland, living in the village of Roundstone, County Galway (in a house later occupied by the pop singer Sting). In 1946, she published her best-known novel, *That Lady*, which is set in the Spain of Philip II and which was called by Naomi Royde-Smith 'one of the finest historical novels in any European language'. Her last novel was *As Music and Splendour*; written largely in Rome, it is a restrained (although, for the time, courageous) treatment of lesbianism. She had written nine novels, and was also the author of travel books, a work on English diaries, a monograph on Teresa of Avila, a book of childhood autobiography, *Presentation Parlour*, and some distinguished journalism.

In 1965, feeling isolated in Ireland, she moved to Boughton in Kent, and eventually died in hospital in nearby Canterbury, two weeks after undergoing the amputation of a leg. Her work, very popular in the 1930s and 1940s, had suffered a serious eclipse by this time, and after her death was slow to develop a strictly literary reputation. This was undeserved, because, although she was no stylist, her work consistently and perceptively treats important themes, particularly the conflict between human feelings and the Catholic conscience, and evokes atmosphere with great skill. The republication of many of her novels by Virago in the 1980s was thus much to be welcomed.

Fiction

Without My Cloak 1931; *The Ante-Room* 1934; *Mary Lavelle* 1936; *Pray for the Wanderer* 1938; *The Land of Spices* 1941; *The Last of Summer* 1943; *That Lady* [US *For One Sweet Grape*] 1946; *The Flower of May* 1953; *As Music and Splendour* 1958

Non-Fiction

Farewell Spain 1937; *English Diaries and Journals* 1943; *Teresa of Avila* 1951; *My Ireland* 1962; *Presentation Parlour* (a) 1963

Plays

Distinguished Villa 1926; *The Bridge* 1927; *Set in Platinum* 1927; *The Silver Roan* 1927; *The Schoolroom Window* 1937; *The Last of Summer* 1945; *That Lady* 1949

Sean O'CASEY 1880–1964

The most famous dramatist of the Dublin tenements was born John Casey, a Protestant in mainly Catholic Dublin, and subsequently became a Catholic, an Irish nationalist, a socialist, and finally a communist. The youngest of eight children, he had little education, left school at fourteen, and taught himself to read and write. He saw many periods of unemployment in between jobs as a manual labourer, the longest of which lasted ten years before he was sacked and blackballed for trade union sympathies.

He was twenty-seven when his first work appeared in nationalist journals under the Gaelic name Sean O'Cathasaigh. He had earlier joined the Gaelic League and the Irish Republican Brotherhood. The latter organisation helped mastermind the Easter Rising of 1916, the subject of his powerful work *The Plough and the Stars*, which became the play most often revived by the Abbey Theatre. Its tragi-comic structure and the superb vitality of its poverty-stricken characters echoed the themes of his first popular play, *Juno and the Paycock* (1924). Rifts with **W.B. Yeats** and the Abbey Theatre, and a permanent abandonment of Irish nationalism caused O'Casey to settle in Devon, from where he continued his love–hate relationship with Ireland in subsequent plays, verse pamphlets, prose works and six much praised volumes of autobiography. His communism was 'personal' rather than 'political', and he enjoyed the friendships of many Conservatives, including Harold Macmillan. He married in 1928, and had two sons and a daughter.

Plays

Two Plays [pub. 1925]: *The Shadow of a Gunman* 1923, *Juno and the Paycock* 1924; *The Plough and the Stars* 1926; *The Silver Tassie* 1929 [pub. 1928]; *Within the Gates* 1934 [pub. 1933]; *Purple Dust* 1943 [pub. 1940]; *The Star Turns Red* 1940; *Red Roses for Me* 1943 [pub. 1942]; *Oak Leaves and Lavender* 1946; *Cock-a-Doodle Dandy* 1949; *The Bishop's Bonfire* 1955; *Five One Act Plays* [pub. 1958]; *The Drums of Father Ned* 1959 [pub. 1960]; *Behind the Green Curtains* 1962 [pub. 1961]; *The Harvest Festival* 1979

Non-Fiction

The Story of the Irish Citizen Army 1919; *The Flying Wasp* 1937; *I Knock at the Door* (a) 1939; *Pictures in the Hallway* (a) 1942; *Drums under the Windows* (a) 1945; *Inisfallen, Fare Thee Well* (a) 1949; *Rose and Crown* (a) 1952; *Sunset and Evening Star* (a) 1954; *Two Letters* 1963; *The Letters of Sean O'Casey* [ed. D. Krause] 4 vols 1975–92

Poetry

The grand Oul' Dame Britannia 1916; *Lament for Thomas Ashe* 1917; *Thomas Ashe* 1917; *The Story of Thomas Ashe* 1917; *England's Conscription Appeal to Ireland's Dead* 1918; *More Wren Songs* 1918; *Songs of the Wren* 2 vols 1918

Collections

Windfalls 1934; *The Green Crow* 1957; *Feathers from the Green Crow* [ed. R. Hogan] 1963; *Under a Coloured Cap* 1963; *Blasts and Benedictions* [ed. R. Ayling] 1967; *The Sean O'Casey Reader* [ed. B. Atkinson] 1968

Biography

Sean O'Casey by Garry O'Connor 1988

(Mary) Flannery O'CONNOR 1925–1964

O'Connor was born in Savannah, Georgia, into a Catholic family – a somewhat unusual thing in the 'Christ-haunted' Bible belt of the Southern States. In her birth in Georgia, Gothic literary output, and short, unhealthy life, she resembles another Southern Renaissance writer, **Carson McCullers**, and the two are perhaps sometimes confused, not least because of the similar forms their names take. O'Connor was probably the better writer, certainly the more consistently controlled one, and differs radically from McCullers in the non-sectarian religious dimension, with its deep symbolism, which pervades her work.

Her chief non-literary brush with fame happened when she was five, and trained a chicken to walk backwards, a triumph recorded by Pathé News. Otherwise her life was outwardly uneventful, darkened by the disseminated lupus, an intractable and painful skin disease, which she inherited from her father and from which both died early. The family moved to Millidgeville, Georgia, in 1938, where her father died three years later at the age of forty-five. She attended the Peabody High School, Georgia, the Georgia State College for Women, and the Writers' Workshop of the University of Iowa, graduating as master of fine arts in 1947, and being one of the earlier writers to benefit from this now famous course. Her first magazine story had appeared the previous year.

A good press greeted the first of her two novels, *Wise Blood*, in 1952. It deals with a young religious enthusiast who attempts to establish a church without Christ. The second novel, *The Violent Bear It Away*, has similar subject-matter – one boy's self-imposed mission to baptise another – and is generally accounted the better book. The finest work by O'Connor probably lies in her short stories, nineteen of which appeared in the two collections which were published during her life and shortly after her death. The *Complete Short Stories* of 1971 contains all thirty-one stories, the majority of them dealing with grotesque, even freakish, but often strangely inspired characters. There is also a collection of imaginative occasional prose and one of letters.

O'Connor suffered her first major attack of lupus in 1950, and from around 1955 was only able to get around on crutches. She lived with her mother in Milledgeville, but accepted as many invitations to lecture as she could and was in no way a recluse, rather a lively, slightly satirical woman. An abdominal operation in 1964 reactivated the lupus which quickly killed her.

Fiction

Wise Blood 1952; *A Good Man is Hard to Find* (s) 1955; *The Violent Bear It Away* 1960; *Everything That Rises Must Converge* (s) 1965; *The Complete Short Stories* (s) 1971

Non-Fiction
Mystery and Manners 1969; *Habit of Being* 1979

Frank O'CONNOR (pseud. of Michael Francis O'Donovan) 1903–1966

O'Connor was born in Cork, Ireland, the son of a poor soldier. He attended the Christian Brothers' School in Cork, and his rapt, solitary childhood there forms the material of many of his celebrated short stories. He could not afford a university education, but aged nineteen, an Irish nationalist, fought in the Irish civil war. He became a librarian, first in Cork, where he worked with Lennox Robinson, playwright and library adviser to the Carnegie Trust, and then in Dublin, where he entered literary society. George Russell (**AE**) printed many of his early short stories in the *Irish Statesman*.

His first volume of stories, *Guests of the Nation*, was published in 1931 and immediately won considerable acclaim; nine more collections were to follow at regular intervals throughout his life, and the qualities of wit, human warmth and close attention to Irish speech that these works show have led to his being regarded as a master of the short story. He also published two novels, but these have not won a reputation. In the 1930s, having become a friend of **W.B. Yeats**, he became a director of the Abbey Theatre, and during this period wrote five plays, some in collaboration; he was later to resign from the Abbey over questions of censorship.

During the Second World War, he came to London, where he worked for the Ministry of Information and delivered some notable radio broadcasts. After the war, he lived for a time in America, where he occupied a variety of lecturing posts but, pining for home, he returned to Dublin where he died. He was a noted translator of Irish language poetry (his own earliest writings had been in Irish), but his collection of such verse, *Kings, Lords and Commons*, was banned in Ireland because it contained an already-banned translation of the seventeenth-century satire 'The Midnight Court'. (This decision was eventually reversed after submissions to the Appeals Board.)

O'Connor wrote many general works including a life of Michael Collins, a study of Turgenev, studies of the short story and the novel, and a book based on his experiences of cycling around Ireland. He married twice, and had two sons and two daughters.

Fiction
Guests of the Nation (s) 1931; *The Saint and Mary Kate* 1932; *Bones of Contention and Other Stories* (s) 1936; *Dutch Interior* 1940; *Three Tales* (s) 1941; *Crab Apple Jelly* (s) 1944; *The Common Chord* (s) 1947; *Traveller's Samples* (s) 1951; *The Stories of Frank O'Connor* (s) 1953; *More Stories by Frank O'Connor* (s) 1954; *Domestic Relations* (s) 1957; *Collection Two* (s) 1964; *Collection Three* (s) 1969; *The Cornet Player Who Betrayed Ireland* (s) 1981

Non-Fiction
Death in Dublin: Michael Collins and the Irish Revolution [UK *The Big Fellow: a Life of Michael Collins*] 1937; *A Picture Book* 1943; *Towards an Appreciation of Literature* 1945; *The Art of the Theatre* 1947; *Irish Miles* 1947; *The Road to Stratford* 1948 [rev. as *Shakespeare's Progress* 1960]; *Leinster, Munster and Connaught* 1950; *The Mirror in the Roadway* 1957; *An Only Child* (a) 1961; *The Lonely Voice: a Study of the Short Story* 1962; *My Father's Son* (a) 1968; *The Backward Look: a Survey of Irish Literature* 1967; *W.B. Yeats: a Reminiscence* 1982

Translations
The Wild Bird's Nest: Poems from the Irish 1932; *Lords and Commons* 1938; *The Fountain of Magic: Translations from Irish Poetry mainly of the VII–XII Centuries* 1939; Eileen O'Connell *A Lament for Art O'Leary* 1940; Brian Merriman *The Midnight Court* 1945; *Kings, Lords and Commons* 1961; *The Little Monasteries* 1963; *A Golden Treasury of Irish Poetry* [with D. Greene] 1967

Plays
With Hugh Hunt *In the Train* 1937 [pub. in *The Genius of Irish Theatre* ed. S. Barnet, M. Berman, W. Burto 1960]; *The Invincibles* 1937 [pub. 1980]; *Moses' Rock* 1938 [np]

Time's Pocket 1938 [np]; *The Statue's Daughter* 1941 [np]

Poetry
Three Old Brothers and Other Poems 1936

Edited
Modern Irish Short Stories (s) 1957 [rev. as *Classic Irish Short Stories* 1985]; *A Book of Ireland* 1959

Biography
Voices by James Matthews 1983

Clifford ODETS 1906–1963

The son of Jewish immigrants from Lithuania, Odets was born in Philadelphia, but grew up in the Bronx in New York. His father became a successful printer, and Odets attended the Morris High School, but dropped out at the age of fifteen to pursue a career as an actor. He worked in radio and on the stage with the Theater Guild, practising the Stanislavsky method, and in 1931 he helped found New York's Group Theater, which specialised in putting on socially committed plays.

One of these was Odets' own first play, *Waiting for Lefty* (1935), which had won a competition organised by the New Theater League. It gained immediacy from its subject-matter, the involvement of the unions in the New York cab strike of 1934, and was an early example of agitprop theatre, directly addressing the audience as if it were attending a political meeting. The success of the play led to three more of Odets' works being mounted that year: his earlier play *Awake and Sing!*, *Till the Day I Die* (an anti-Nazi propaganda piece which formed the second half of the bill with the one-act *Waiting for Lefty*), and *Paradise Lost*. *Awake and Sing!* was a Depression drama set in the Jewish Bronx of Odets' own childhood; as its title suggests, it was an optimistic piece about the possibility of political change, though it acknowledged that this would involve struggle and sacrifice. Odets himself had joined the Communist Party the previous year, and he was seen as the new theatrical spokesman for the political left.

He never quite repeated the success of this *annus mirabilis* and was shortly lured to Hollywood as a screenwriter. His next play, *Golden Boy*, was very much in the Hollywood mode: the story of a young Italian-American who is a talented musician but fights his way out of poverty with his fists, becoming a professional boxer, then losing everything. It was made into film, directed by Rouben Mamoulian, in 1939. Like many writers before him, Odets spent much of his time in Hollywood doing rewrites and additional dialogue, becoming involved in the customary compromises of the film world. Something of his own feelings about this may be gauged from *The Big Knife*, a play in which a Broadway actor becomes a disillusioned film star. Although Odets wrote, and occasionally directed, screenplays (notably *None But the Lonely Heart*, 1944, adapted from the 1943 novel by **Richard Llewellyn**, and starring Ethel Barrymore), *The Big Knife* was filmed by others in 1955.

By this time Odets had alienated his former political allies by testifying before the House Un-American Activities Committee, and most people felt that he had sold out. Certainly his later work lacked much political punch and was characterised by cynicism rather than idealism, as in his screenplays for *Sweet Smell of Success* (1957) and *The Story on Page One* (1960), which he also directed. His penultimate play, *The Country Girl*, was about an actor with a drink problem attempting a comeback, and was successful both on stage and as a musical film (1954) with Bing Crosby and songs by Harold Arlen and Ira Gershwin.

Odets was married twice: first (and stormily) to the Austrian-born Hollywood actress Luise Rainer, then to Bette Grayson, with whom he had two children.

Plays

Three Plays [pub. 1935]: *Awake and Sing!* 1935, *Till the Day I Die* 1935, *Waiting for Lefty* 1935; *Paradise Lost* 1935 [pub. 1936]; *The General Died at Dawn* [from C.G. Booth's novel] 1936; *Golden Boy* 1937; *Rocket to the Moon* 1938 [pub. 1939]; *Six Plays* [pub. 1940]; *Clash by Night* 1941 [pub. 1942]; *The Russian People* [from K. Simonov's play] 1942 [np]; *Black Sea Fighters* **(f)** 1943; *None But the Lonely Heart* **(f)** [from R. Llewellyn's novel] 1944 [pub. in *Best Film Plays 1945* ed. J. Gassner, D. Nichols 1946]; *Deadline at Dawn* **(f)** [from W. Irish's novel] 1946; *Humoresque* **(f)** [from F. Hurst's novel; with Z. Gold] 1946; *The Big Knife* 1949; *The Country Girl* [pub. 1951; UK *Winter Journey* 1952, pub. 1953]; *The Flowering Peach* 1954; *Sweet Smell of Success* **(f)** [with E. Lehman] 1957; *The Story on Page One* **(f)** 1960; *Wild in the Country* **(f)** [from J.R. Salamanca's *The Lost Country*] 1961; *Big Mitch* **(tv)** 1963; *The Mafia Men* **(tv)** 1964; *The Silent Partner* 1972 [np]

Non-Fiction

Rifle Rule in Cuba [with C. Beals] 1935

Biography

Clifford Odets by Margaret Brenman-Gibson 1981

Eimar (Ultan) O'DUFFY 1893–1935

Born in Dublin, the son of a dentist appointed to the Viceroy of Ireland, O'Duffy was educated at Stonyhurst. Although he himself qualified as a dentist at University College, Dublin, he considered the profession distasteful, and never practised. Despite membership of the Irish Volunteers, he opposed the Easter Rising of 1916, and attempted to prevent Volunteers from forming in Belfast. His sprawling first novel, *The Wasted Island* (1920), which was clearly autobiographical, related his feelings against Padraig Pearse and the events leading to the uprising.

During the civil war of 1922, O'Duffy supported the Free State, and was given a post in the Department of External Affairs the following year. Finding the work untroublesome, he became one of the editors of the *Irish Review*, from 1922 to 1923. O'Duffy was disillusioned with the 'New Ireland', and abandoned Irish politics to continue writing in London where he lived from 1925. Unable to resist promotion of C.H. Douglas's theory of 'social credit', which also beguiled **Ezra Pound** and **Hugh MacDiarmid**, he became publicity agent for Lloyd George's Liberal supporters. His interest in economics was reflected in his satirical *Cuan-*

duine Trilogy. King Goshawk and the Birds, the first and most scathing volume, was stylistically a parody of the Irish sagas, and attacked capitalism. His most popular novel was *Life and Money*, which went into several editions.

A complex man, awkward to know, he died at New Malden in Surrey, leaving instructions for his incomplete autobiography to be destroyed. His work, although considered seriously in Ireland, has yet to be appreciated elsewhere.

Fiction

The Wasted Island 1920 [rev. 1929]; *The Lion and the Fox* 1922; *Printer's Errors* 1922; *Miss Rudd and Some Lovers* 1923; *Cuanduine Trilogy: King Goshawk and the Birds* 1926, *The Spacious Adventures of the Man in the Street* 1928, *Asses in Clover* 1933; *The Bird Cage* 1932; *Life and Money* 1932; *The Secret Enemy* 1932; *Head of a Girl* 1935

Poetry

A Lay of the Liffey and Other Verses 1919

Plays

Brierius Feast [pub. 1931]

Julia O'FAOLAIN 1932–

O'Faolain is the daughter of two Irish writers, **Sean O'Faolain** and his wife Eileen Gould, a children's author. She was born in London, towards the end of her father's four-year stay there, but her earliest years were spent in the Wicklow Mountains, where her father, newly successful, was living by his pen. The family moved to Dublin when she was five, and she grew up there in the society of many of Ireland's leading writers and other personalities. She was convent-educated by the nuns of the Sacred Heart, originally a French order, a link which led to her often spending her summers in France. She entered University College, Dublin, when aged seventeen, and the international connection was reinforced when scholarships financed two summers in Perugia, one in Venice, a postgraduate year at the University of Rome, and further postgraduate studies in Paris.

She returned to Dublin briefly, but then went to London in the mid-1950s, where at first she did various jobs but later became a supply teacher for the London County Council, simultaneously working as a translator at the Council of Europe in Strasbourg. Some early writing dates from this period, including a story in the *New Yorker*, but her career as an author really took off more than a decade later. In 1957, in Florence, she met the American Renaissance historian Lauro Martines, then doing his postgraduate thesis at Harvard, and married him the same year; they have one son. Martines obtained a post as assistant professor at Reed College, Oregon, and the couple lived at Portland from 1958 to 1962. They then spent four years in Florence, where Martines was a fellow at the Villa I Tatti and O'Faolain taught at the Scuola Interpreti. In 1966, Martines became a full professor at the University of California at Los Angeles, and initially the couple divided their time between there and France or Italy. In 1970, they bought a house in London, and have since lived between Britain and the USA, more recently centring themselves on the former, and frequently visiting Ireland.

O'Faolain's first collection of stories, almost equally divided between Irish and Italian settings, was published in 1968, and established her as a satirist with a gift for black comedy, and a sharply critical eye for the mores of her native country. Her first novel, *Godded and Codded* (1970), had to be withdrawn because of a libel threat in Britain, but appeared in America, under the title *Three Lovers*. It was perhaps a false start, because the action was almost exclusively sexual, but her subject-matter has broadened considerably since then, in a series of novels and short-story collections. Her most considerable novel is agreed to be her major treatment of Ireland, *No Country for Young Men*, but she has also written two historical novels, one set in sixth-century Gaul, and another, *The Judas Cloth*, about the Rome of Pio Nono. A feminist perspective dominates her work, which has consistently divided critical opinion. She writes surprisingly little from an American viewpoint. A well-known non-fictional work, written in collaboration with her husband, is *Not in God's Image*, an anthology about the historical position of women.

Fiction

Godded and Codded 1970 [US *Three Lovers* 1971]; *Man in the Cellar and Other Stories* (s) 1974; *Women in the Wall* 1975; *Melancholy Baby and Other Stories* (s) 1978; *No Country for Young Men* 1980; *Daughters of Passion* (s) 1982; *The Obedient Wife* 1982; *The Irish Signorina* 1984; *The Judas Cloth* 1992

Non-Fiction

Not in God's Image. Women in History [with L. Martines] 1973

Sean O'FAOLAIN 1900–1991

A generally accepted re-anglicisation of the Gaelic name adopted by John Francis Whelan; in its full Irish form it would be Seán ó Faoláin. He was born in Cork, the son of a constable in the

Royal Irish Constabulary, of a family which he later described as 'shabby genteels at the lowest possible social level', and educated at the Presentation Brothers College in Cork. His father's loyalty was to the British crown, and when the Easter Rising broke out in 1916 O'Faolain initially shared this, becoming converted to Republicanism soon after. He entered University College, Cork, in 1918 and at roughly the same time joined the Irish Volunteers, a forerunner of the IRA. He remained a student in Ireland on and off until 1926, and during the same period graduated from carrying out routine duties for the IRA to working in a bomb factory to acting as director of publicity during the civil war.

In 1926 he won a Commonwealth Fellowship to Harvard, and remained in America until 1929, teaching during the last year at Boston College. In 1928, he married Eileen Gould, a writer for children, who had been his companion in the IRA; their marriage endured sixty years, until her death in 1988, although disrupted by a number of affairs on his part, including two with the writers **Elizabeth Bowen** and Honor Tracy, and another with the American socialite Alene Erlanger. There were two children of the marriage, one of whom is the writer **Julia O'Faolain**.

From 1929 to 1933, O'Faolain taught at a Catholic training college in Twickenham, London, and during this period his friendship with the well-known editor Edward Garnett ripened. The result was the publication of his first book of stories, dealing with the Irish independence struggle, *Midsummer Night Madness* (1932), and a three-year subsidy from publishers Jonathan Cape to enable him to write. Thus armed, he returned to Ireland, and entered a prolific period lasting until 1942, during which he published another short-story collection, a play, three novels and three biographies of prominent Irish figures. Also, between 1940 and 1946, he edited *The Bell*, generally regarded as the best Irish journal of its time, in which he campaigned vigorously against censorship and the narrowness of Irish life.

From 1946 began a decade when he spent much time in Italy, and later he was frequently to be a visiting professor in the USA; his literary output changed too, in that he gave up the novel (a late fantasy appeared in 1979), and concentrated on producing collections of short stories as well as a wide range of non-fiction, including two travel books about Italy, critical works, a life of Cardinal Newman and an autobiography, *Vive Moi!* It is in the short story that his principal achievement is agreed to lie: his earlier stories dealt firmly with the Irish predicament, and later ones were Chekhovian studies of character, but all had economy and close observation. O'Faolain was the *doyen* of Irish men-of-letters of his time, although there is a hint of violence about him (an incident is recorded where he threw his sick mother down the stairs of her home and broke her leg). He was the director of the Arts Council of Ireland between 1957 and 1959.

Fiction

Midsummer Night Madness and Other Stories (s) 1932; *A Nest of Simple Folk* 1933; *Bird Alone* 1936; *A Purse of Coppers* (s) 1937; *Come Back to Erin* 1940; *Teresa and Other Stories* (s) 1947; *The Short Story* 1948; *The Stories of Sean O'Faolain* (s) 1958; *I Remember! I Remember!* (s) 1961; *The Heat of the Sun* (s) 1966; *The Talking Trees and Other Stories* (s) 1971; *Foreign Affairs and Other Stories* (s) 1976; *And Again?* 1979; *Collected Stories* (s) 3 vols 1980–82

Non-Fiction

The Life Story of Eamon de Valera 1933; *Constance Markievicz: a Biography* 1934; *King of the Beggars: a Biography of Daniel O'Connell* 1938; *An Irish Journey* 1940; *The Great O'Neill: a Biography of Hugh O'Neill, Earl of Tyrone, 1550–1616* 1942; *The Story of Ireland* 1943; *The Irish* 1947; *A Summer in Italy* 1949; *Newman's Way: a Biography of Cardinal Newman* 1952; *South to Sicily* 1953; *The Vanishing Hero: Studies of Novelists in the Twenties* 1956; *Vive Moi!* (a) 1965

Plays

She Had to Do Something 1938

Edited

The Autobiography of Theobald Wolfe Tone 1937; Samuel Lover *Adventures of Handy Andy* 1945

Biography

Sean O'Faolain by Maurice Harmon 1994

Liam O'FLAHERTY 1897–1984

O'Flaherty was born in the Aran Islands near Galway, Ireland, where his father was a farmer and fisherman. He grew up in extreme poverty, and this influenced him to become a socialist and rebel from an early age. Intended for the priesthood, he briefly attended a seminary, remaining at University College, Dublin, which he left to join the Irish Guards in 1914. Wounded in France in 1917, he was discharged from hospital suffering from acute melancholia, which he never completely dispelled. After a series of menial jobs in London, he sailed to South America in 1918, and lived as a beachcomber, later travelling through Canada and New York.

He returned to Ireland to fight half-heartedly in the 1922 civil war, on the Republican side, and the following year went to London to live and write. A poor first novel, *Thy Neighbour's Wife*, was accepted by Cape, largely to encourage him, but their reader, Edward Garnett, forced him to submit *The Black Soul* – which he wrote in an Oxfordshire farmhouse shared with **H.E. Bates** – page by page. The novel was acclaimed in Ireland, but attacked in London. O'Flaherty's fiction altered in style, and *The Informer* and other novels reflected his interest in violence, which was not always personally suppressed (**Samuel Beckett** recalled him constantly picking fights in Dublin pubs). In restaurants he would order all the wine for the evening at once, to ensure he would have it, a habit which arose from his deprived background.

Like his friend **Francis Stuart**, O'Flaherty never fitted into Dublin literary society. Always unsettled, he travelled to Russia, and wrote *I Went to Russia* on his return, a book remarkable for the little it says about that country. Despite his dislike of **W.B. Yeats**, he was elected a member of the Irish Academy of Letters, which the poet founded in 1932. A marriage which had failed ended with his wife's death in 1933, and O'Flaherty led a hectic social life, details of which he relates in a generally fictitious autobiography, *Shame the Devil*. Despite his querulous nature, he made friends, amongst them Oscar Wilde's son Vyvyan Holland, with whom he often stayed. After 1953, he wrote nothing, leaving a novel in Gaelic unfinished at his death. Handsome as a young man, O'Flaherty mellowed into gentle senility, and spent his last years cared for by a woman he had lived with intermittently for years. The claim that on his death-bed he returned to the religion he had forsaken as a student was generally disbelieved.

Fiction

Thy Neighbour's Wife 1923; *The Black Soul* 1924; *Spring Sowing* (s) 1924; *Civil War* (s) 1925; *The Informer* 1925; *The Child of God* (s) 1926; *Mr Gilhooey* 1926; *The Tent* (s) 1926; *The Fairy Goose and Two Other Stories* (s) 1927; *The Assassin* 1928; *Red Barbara and Other Stories* (s) 1928; *The House of Gold* 1929; *The Mountain Tavern* (s) 1929; *The Return of the Brute* 1929; *The Puritan* 1931; *Skerrett* 1932; *The Martyr* 1933; *Hollywood Cemetery* 1935; *Famine* 1937; *The Short Stories of Liam O'Flaherty* (s) 1937; *Two Lovely Beasts* (s) 1948; *Insurrection* 1950; *The Stories of Liam O'Flaherty* (s) 1956; *Liam O'Flaherty: Selected Stories* (s) [ed. D.A. Garrity] 1958; *The Short Stories of Liam O'Flaherty* (s) 1961; *The Pedlar's Revenge* (s) 1976

Non-Fiction

The Life of Tim Healy 1927; *Joseph Conrad: an Appreciation* 1929; *A Tourist's Guide to Ireland* 1929; *Two Years* (a) 1930; *A Cure for Unemployment* 1931; *The Ecstasy of Angus* 1931; *I Went to Russia* (a) 1931; *Shame the Devil* (a) 1934

Plays

Darkness 1926

Frank O'HARA 1926–1966

Many are those who might be the emblem of life in New York City. A strong contender is O'Hara, dead at forty after being knocked over by a jeep one night on Fire Island. It is a wonder that he lived as long as he did, for his was such a pell-mell existence that it was scarcely interrupted by sustaining an irretrievable bullet in the buttock from a passing thief in the lobby of his apartment building.

All this was a far cry from his Catholic upbringing in Baltimore. He later refused to speak of it, and never returned there after the death of his father and his mother's descent into alcoholism. His youthful penchant for Proust and music was not the rebellion it might have seemed, for there was an artistic side to the household. Certainly the navy cannot have known what it was getting when he enlisted towards the end of the war. For him, the most interesting aspect of one training documentary was its musical soundtrack, and, as for a posting to Key West: 'Its only excuse for being there is that **Wallace Stevens** wrote a poem about it.' Paradoxically enough, he was deployed on shore patrol in San Francisco to ensure that sailors did not stray into any gay bars. Out in the South Pacific, his time was apparently occupied in reading *Ulysses* and *The Magic Mountain* and composing music – the day of formal surrender was most notable for seeing 'a segment of landscape that would be perfect on a vase or parchment'.

He was better suited to Harvard, enrolling under the GI Bill of 1946 to study music and then English. He revelled in the company of his room-mate, the future author and artist **Edward Gorey**, much of whose time was spent in lolling on a chaise-longue and declaiming the novels of **I. Compton-Burnett**. Theirs was a life of continual culture, of absorbing movies, as it was to be in 1950s' Manhattan – that era when such Abstract Expressionists as Willem de Kooning came to the fore.

After attempts at playwrighting, O'Hara gravitated towards the Museum of Modern Art. Except for a brief stint on *Art News*, he remained there for the rest of his life, an increasingly significant figure in the art world and writing prolifically in odd moments (one collection was entitled *Lunch Poems*). Sedulous days of cataloguing and writing were in some contrast to nights of carousing in such haunts as the Cedar Tavern.

The period is suffused in cigarettes and alcohol. His poetry, invariably published in an *ad hoc* way, if at all, gained strength during the 1950s, influenced by **William Carlos Williams**, and might appear akin to the work of Jackson Pollock and de Kooning: broken-backed, it was avowedly occasional in its celebration of such friends and lovers as Larry Rivers and, especially, the ballet dancer Vincent Warren (they parted in 1961, to O'Hara's continued chagrin). In 1957 O'Hara had announced that he would no longer sleep with strangers, only friends, but such a manifesto did not preclude rivalry and bitchiness. Andy Warhol was shunned after the artist expressed an erotic urge to draw his feet; O'Hara's refusal led to his later being presented with an imaginary picture of his penis (it would have a certain value now, but the poet threw it away); a portrait of the real thing, erect, startled many a rich collector in 1954 at a show which included this Géricault-inspired work by Rivers.

Sometimes writing poems hundreds of lines long and at other times bagatelles, O'Hara is best taken *en masse*. His subtle work provides an unrivalled account of New York City and its denizens.

Poetry
A City Winter 1952; *Oranges* 1953; *Meditations in an Emergency* 1957; *Stones* [with L. Rivers] 1958; *Odes* [with M. Goldberg] 1960; *Second Avenue* 1960; *Featuring O'Hara* 1964; *Lunch Poems* 1964; *Love Poems* 1965; *Five Poems* 1967; *Odes* 1969; *Two Pieces* 1969; *Collected Poems* [ed. D. Allen] 1971; *Selected Poems* [ed. D. Allen] 1973; *Hymns of St Bridget* [with B. Berkson] 1974; *Poems Retrieved 1951–1966* [ed. D. Allen] 1975; *Early Poems 1946–1951* [ed. D. Allen] 1976

Plays
Try! Try! 1951 [rev. 1952; pub. in *Artists' Theatre* ed. H. Machiz 1960]; *Change Your Bedding* 1952 [np]; *Love's Labor* 1959 [pub. 1964]; *Awake in Spain* 1960; *The General Returns from One Place to Another* 1964 [np]

Non-Fiction
Jackson Pollock 1959; *Art Chronicles 1954–1966* 1974; *Standing Still and Walking in New York* [ed. D. Allen] 1975

Edited
Robert Motherwell: a Catalogue with Selections from the Artist's Writings 1966

Collections
A Frank O'Hara Miscellany 1974; *Early Writings* [ed. D. Allen] 1977

Biography
The Life of Frank O'Hara by Frank McShane 1980; *City Poet: the Life and Times of Frank O'Hara* by Brad Gooch 1993

John (Henry) O'HARA 1905–1970

Born in Pottsville, a small town in Pennsylvania (the Gibbsville of many of his novels and short stories), O'Hara was the first of eight children of a prosperous doctor of Irish descent. He passed a happy small-town childhood, not much disturbed by being expelled from two boarding-schools, but his father's early death meant that he did not go to Yale, as planned. He began work as a newspaperman in his native Pottsville, and then worked for another Pennsylvania paper before moving to New York, where he worked for various publications including *Time* magazine, as well as doing a large number of other jobs for brief periods.

By 1932 he had become a regular *New Yorker* contributor and was viewed as a rising writer. He had married an actress, Helen Petit, in 1931, who divorced him in 1933. In the same year he started writing his first novel, *Appointment in Samarra*, in a New York hotel bedroom, persuading a publisher to subsidise his work on it. When it was published in 1934 it met with great success and gained him immediate work in Hollywood as a scriptwriter. He was a professional writer thereafter and was to publish eighteen novels and 374 stories, his rate of work speeding up after 1953 when a near-fatal haemorrhage of his gastric ulcer caused him to abjure alcohol, of which he had been a great consumer. He spent the latter part of his life writing concentratedly at his large house, 'Linebrook', in the countryside near Princeton.

O'Hara wrote about the aspirations of middle-class materialistic Americans from the standpoint of one who shared their values; the case for this type of writing was put by Clarence A. Glasrud when he said: 'The academic world objects to O'Hara's view of life and literature, but if future generations seek an American Balzac to lay bare life in the United States from 1900 to 1970, they will find John O'Hara the most complete, the most accurate, and the most readable chronicler.' It is generally agreed that the finest of O'Hara's short stories are superior to his novels. He had a success with the volume *Pal Joey*, which he used as the basis for a musical play with songs by Rogers and Hart. It was subsequently made into a popular, though somewhat bowdlerised, film (1957), directed by George Sidney, with Frank Sinatra playing the part of the night-club singer. It is also agreed that his work deteriorated; his first two novels (the second was *BUtterfield 8*) are considered his best. Among English admirers of his plain but powerful prose was **John Braine**. He married three times, and had one daughter.

Fiction

Appointment in Samarra 1934; *BUtterfield 8* 1935; *The Doctor's Son and Other Stories* (s) 1935; *Hope of Heaven* 1938; *Files on Parade* (s) 1939; *Pal Joey* (s) 1940; *Pipe Night* (s) 1945; *Here's O'Hara* (s) 1946; *Hellbox* (s) 1947; *All the Girls He Wanted* (s) 1949; *A Rage to Live* 1949; *The Farmers Hotel* 1951; *Ten North Frederick* 1955; *A Family Party* 1956; *The Great Short Stories of O'Hara* (s) 1956; *Selected Short Stories* (s) 1956; *From the Terrace* 1958; *Ourselves to Know* 1960; *Sermons and Soda Water* 3 vols 1960; *Assembly* (s) 1961; *The Big Laugh* 1962; *The Cape Cod Lighter* (s) 1962; *Elizabeth Appleton* 1963; *49 Stories* (s) 1963; *The Hat on the Bed* (s) 1963; *The Horse Knows the Way* (s) 1964; *The Lockwood Concern* 1965; *Waiting for Winter* (s) 1966; *The Instrument* 1967; *And Other Stories* (s) 1968; *Lovey Childs* 1969; *The O'Hara Generation* (s) 1969; *The Ewings* 1972; *The Time Element and Other Stories* (s) [ed. A. Erskine] 2 vols 1972, 1974; *The Second Ewings* 1977

Plays

Five Plays [pub. 1961]

Non-Fiction

Sweet and Sour 1954; *My Turn* 1966; *A Cub Tells His Story* 1974; *An Artist Is His Own Fault: O'Hara on Writers and Writings* [ed. M.J. Bruccoli] 1977

Ben OKRI 1959–

Okri was born in Nigeria, but, shortly after his birth, his father came to England to study law. He spent his earliest years in Peckham, south London, and attended primary school there, but returned with his family to Nigeria aged seven. Back in Lagos, his father specialised in legal work with those who could not afford normal legal fees, but the family's own financial struggle meant that Okri was constantly being withdrawn from various schools. He educated himself largely at home, writing articles and stories from his teenage years, and completing his first novel at the age of nineteen.

This novel, *Flowers and Shadows*, was published in 1980, when Okri, having returned to England, was in the first year of a comparative literature course at Essex University. Lack of funds forced him to leave without taking a degree, and he moved to London, where, in the early 1980s, he often slept rough, but worked on more novels and stories. Two story collections and a further novel, *The Landscapes Within*, marked him out as a young writer of outstanding promise, writing a simple but resonant prose; but, although he won literary prizes and awards (an Arts Council bursary in 1984, the Aga Khan *Paris Review* Prize and the Commonwealth Writers' Prize for Africa in 1987), his work did not collect an immediate audience. His income was supplemented by work as poetry editor of *West Africa* and broadcasting for the BBC.

The turning-point in his fortunes came in 1991, when his third novel, *The Famished Road*, the moving story of Azaro, a Nigerian spirit child, won the Booker Prize (in a contest where, unusually, one of the most favoured candidates, **Timothy Mo**, had bet a substantial amount of money on Okri to win). Okri's next publication, in early 1992, was his first volume of poetry, *An African Elegy*, and he faced his middle-career in the 1990s firmly established as one of the leading younger writers of his time.

Fiction

Flowers and Shadows 1980; *The Landscapes Within* 1981; *Incidents at the Shrine* (s) 1986; *Stars of the New Curfew* (s) 1988; *The Famished Road* 1991; *Songs of Enchantment* 1993; *Astonishing the Gods* 1995

Poetry

An African Elegy 1992

Tillie OLSEN 1913–

Olsen was born Tillie Lerner in Omaha, Nebraska, the daughter of working-class parents who had fled Russia after the failed revolution of 1905. Her father, Samuel Lerner, was the secretary of the Nebraska Socialist Party. She left high school in the eleventh grade and worked at various times in domestic service, a warehouse, a food-processing factory and as a typist. At seventeen she joined the Young Communist League and was briefly imprisoned in Kansas City in 1932 for distributing political leaflets. *Yonnondio* was written during this period, but although a chapter was published in the *Partisan Review* in 1934, the complete novel remained unpublished until 1974. From 1936 she lived with Jack Olsen, a printer and politically active trade unionist whom she married in 1943. She brought up four daughters while continuing to work, and did not resume writing until the mid-1950s. *Silences*, which developed from a lecture given in 1962, describes her 'triple life' as mother, worker and aspirant writer.

The title-story of Olsen's collection, *Tell Me a Riddle*, describes the antagonism between an elderly couple and the wife's slow death from cancer, and won an O. Henry Award in 1961. Other stories anatomise social and racial divisions. Following critical acclaim for the collection she subsequently held a variety of academic posts at Amherst College, Stanford University, the Massachusetts Institute of Technology, the University of Massachusetts and the University of California, San Diego. She received doctorates from the University of Nebraska in 1979 and Knox College, Illinois in 1982. The manuscript of *Yonnondio* was unearthed some thirty years after its composition, revised and edited, and published in 1974. The title is taken from Whitman's poem of the same name, and is an Iroquois word meaning 'lament for the aborigines'. The novel is a despairing portrait of the lives of the impoverished Holbrook family, a lament for those who have left 'No picture, poem, statement, passing them to the future'. The essays collected in *Silences* further explore the silence of the dispossessed and creatively thwarted.

Fiction

Tell Me a Riddle (s) 1961; *Yonnondio* 1974

Non-Fiction

Silences 1978

Plays

I Stand Here Ironing 1981

Edited

Mother to Daughter, Daughter to Mother: Mothers on Mothering 1984

Charles OLSON 1910–1970

Increasingly recognised since his death as a major American poet in the line of **Ezra Pound** and **William Carlos Williams**, Olson was born in Worcester, Massachusetts, the son of a postman. Worcester was an ordinary industrial town, but the family summers spent on the coast at Gloucester were to be central to his life and work. He took a BA and MA at Wesleyan University, writing an important thesis on Herman Melville, and went on to study for a PhD at Harvard in the newly established American studies programme. He left without his doctorate after being awarded a Guggenheim Fellowship in 1939 for a book on Melville. This was abandoned during the war, in which Olson was assistant chief of the Foreign Language Division of the Office of War Information (he resigned in protest at the bureaucracy there). He also had a promising career with the Democratic Party's National Committee as a senior strategist (and was offered a post in the following administration by Roosevelt), but quit politics with the arrival of Truman, and instead concentrated on writing. In 1948 he took a temporary teaching post at Black Mountain College and stayed on to succeed Josef Albers as rector until the college closed in 1956. He then largely devoted himself to writing in Gloucester, accepting some more teaching at the State University of New York and at the University of Connecticut shortly before his death.

Olson's 1950 essay on 'projective verse' has had an immense impact on post-war American poetics, and forms a kind of manifesto for the poetics of the so-called Black Mountain school. Projective verse – now more usually called 'open' verse, although Olson's term better captures its dynamic aspect of moving energy – aims to be a poetry of process, in the tradition of romantic organicism, transcribing thoughts and being governed by the breath, the typographical possibilities of the typewriter, and the internal logic of sounds rather than imposed sense. It was a rallying cry against the dominant poetics of the time, which were governed by **T.S. Eliot** and the New Criticism and favoured careful, complex, ironic and well-finished writing. Olson's work on Melville had resurfaced in 1947 as *Call Me Ishmael*, with its celebrated opening: 'I take SPACE to be the central fact to man born in America ... I spell it large because it comes large here. Large, and without mercy.' His other major non-poetic book of the period is his *Mayan Letters*, written to **Robert Creeley** from Mexico, where Olson was studying hieroglyphics and the native New World civilisation of the Maya.

Olson's poems 'The Kingfishers' and 'In Cold Hell, in Thicket' (in the 1953 volume of the same name) are his purest demonstrations of projective verse and caused a sensation among sympathetically minded poets, if not among the public. His major poetic work is the long sequence *The Maximus Poems*. Addressed to the town of Gloucester, the sequence blends personal and social history, lamenting the contemporary world of corruption and big business, while looking forward to a recovered communal future. The Maximus of the title is at once a kind of man-at-large, an obscure second-century philosopher, Maximus of Tyre, and Olson himself, who was six-foot-seven. It has been compared to Pound's *Cantos* (1948–69) and to William Carlos Williams's *Paterson* (1946–58), and judged a failure as a whole, but none the less represents a highly important project within American verse.

More academic and Europhile critics predictably distrusted and dismissed Olson's free-form poetics, but he is the towering figure over post-war poets such as **Ed Dorn**, Creeley and **Robert Duncan**. He was one of the first to talk of a 'postmodern' verse, and to look to a wholly American field of possibility, choosing allegiance to the native civilisations such as the Maya rather than the corruption and imperialism of Europe. His definition and practice of projective verse (although it continues the innovations of Pound and Williams) is a milestone in post-war American poetry.

Poetry
Y & X 1948; *Letter for Melville* 1951; *This* 1952; *In Cold Hell, in Thicket* 1953; *The Maximus Poems 1–10* 1953; *The Maximus Poems 11–20* 1956; *O'Ryan 246810* 1958; *The Distances* 1960; *The Maximus Poems* 1960; *Maximus, from Dogtown* 1961; *Place; and Names* 1964; *O'Ryan 12345678910* 1965; *West* 1966; *Maximus Poems IV, V, VI* 1968; *Archaeologist of Morning* 1970; *The Maximus Poems* [ed. C. Boer, G.F. Butterick] 1975; *Spearmint and Rosemary* 1975; *The Horses of the Sea* 1976; *Some Early Poems* 1978; *Collected Poems* [ed. G.F. Butterick] 1988; *A Nation of Nothing but Poetry* [ed. G.F. Butterick] 1989

Non-Fiction
Call Me Ishmael: a Study of Melville 1947; *Mayan Letters* [ed. R. Creeley] 1953; *Anecdotes of the Great War* 1955; *Projective Verse* 1959; *A Bibliography on America for Ed Dorn* 1964; *Civil Liberties and the Arts* (c) [ed. W. Wasserstrom] 1964; *Signature to Petition* 1964; *Human Universe and Other Essays* [ed. D. Allen] 1965; *Proprioception* 1965; *Pleistocene Man: Letters from Charles Olson to John Clarke During October 1965* 1968; *Causal Mythology* 1969; *Letters for 'Origin'* [ed. A. Glover] 1969; *The Special View of History* [ed. A. Charters] 1970; *Additional Prose* [ed. G.F. Butterick] 1974; *Charles Olson and Ezra Pound: an Encounter at St Elizabeth's* [ed. C. Seelye] 1975; *In Adullam's Lair* 1975; *The Post Office* 1975; *Olson/Den Boer, a Letter* 1977; *Mythologies: the Collected Lectures and Interviews* [ed. G.F. Butterick] 1978–9; *Charles Olson and Robert Creeley: Complete Correspondence* [ed. G.F. Butterick] 1980; *Charles Olson and Cid Corman: Complete Correspondence* [ed. G. Evans] 1987–91; *In Love, in Sorrow: the Complete Correspondence of Charles Olson and Edward Dahlberg* [ed. P. Christensen] 1990

Plays
Apollonius of Tyana [pub. 1951]; *The Fiery Hunt and Other Plays* [ed. G.F. Butterick; pub. 1977]

Collections
Poetry and Truth: the Beliot Lectures and Poems [ed. G.F. Butterick] 1971

(Philip) Michael ONDAATJE 1943–

Ondaatje was born in Colombo, Sri Lanka, of mixed Dutch, Tamil and Sinhalese origins, writing Sinhalese and speaking Tamil until he was moved to London and Dulwich College, in 1953. Unhappy in England, he emigrated to Canada at the age of nineteen, enrolling at Bishop's University, Quebec, to complete his education, before moving to the University of Toronto. He graduated in 1965, by which time he had married his first wife, with whom he has two children. He taught at the University of Western Ontario until 1971, when he took a post at Glendon College, York University, Toronto.

His first book, *The Dainty Monsters*, a small press poetry collection, was published to local acclaim in 1967. Ondaatje was already associated with an emergent Canadian 'scene' – in 1966 his poems had been included in Raymond Souster's landmark anthology *New Wave Canada: the New Explosion in Canadian Poetry* – and the book also gave voice to the growing ecological awareness of that decade, ranging in scope from the lush landscape and wildlife of Sri Lanka to the hogs raised by Ondaatje on his farm in northern Toronto. A change of vision and a wider fame came in 1974 with the American publication of *The Collected Works of Billy the Kid: Left-Handed Poems* (1970), a fictionalised biography of the outlaw hero told in a mixture of poetry and prose. Ondaatje had also written and directed his first film, the independently produced *Sons of Captain Poetry*, in the same year. Influenced by the mythmaking of such films as Arthur Penn's *Bonnie and Clyde* (1967), he developed his interest in American lore and iconography in his first novel. In *Coming through Slaughter* (1976) the brief working life of famed New Orleans cornet-player Buddy Bolden, as retold through a mixture of documentary material and memoirs, is 'expanded ... polished to suit the truth of fiction', and made to represent a meditation on the 'artist's progression towards silence'. Bolden, normally regarded as the first genius of jazz, was committed at the age of thirty-one and spent the last twenty-four years of his life confined to an institution. The theme of violence, both psychic and actual, is continued in *There's a Trick with a Knife I'm Learning to Do: Poems 1963–1978.*

A return visit to Sri Lanka became the material of *Running in the Family* – part personal memoir, part travelogue and part family history – in which Ondaatje remembers and discovers the background to his father's eccentric and drink-overshadowed life. The honesty of the book and an equally personal poetry collection, *Secular Love*, describing the break-up of a marriage (he had recently separated from his second wife), increased his standing with those critics who had objected to the overly poetic nature of his early writing. *The English Patient*, his fourth novel, was joint winner of the Booker Prize in 1992.

Ondaatje is a keen farmer, who has registered his own breed of hogs. He has worked extensively in both theatre and film, and is the editor of the magazine *Mongrel Broadsides*. His critical writings include a book on fellow Canadian, the singer-poet Leonard Cohen.

Poetry
The Dainty Monsters 1967; *The Man with Seven Toes* 1969; *The Collected Works of Billy the Kid* 1970; *Rat Jelly* 1973; *Elimination Dance* 1978 [rev. 1980]; *There's a Trick with a Knife I'm Learning to Do: Poems 1963–1978* 1979 [UK *Rat Jelly and Other Poems 1963–1978* 1980]; *Secular Love* 1984; *Two Poems* 1986; *The Cinnamon Peeler* 1989

Fiction
Coming through Slaughter 1976; *In the Skin of a Lion* 1987; *The English Patient* 1992

Plays
Sons of Captain Poetry (f) 1970; *The Collected Works of Billy the Kid* 1973 [np]; *Coming through Slaughter* [from his novel] 1980 [np]

Non-Fiction
Leonard Cohen 1970; *Claude Glass* 1979; *Running in the Family* (a) 1982; *Tin Roof* 1982

Edited
The Broken Ark 1971 [rev. as *A Book of Beasts* 1979]; *Personal Fictions: Stories by Munro, Wiebe, Thomas and Blaise* (s) 1977; *The Long Poem Anthology* 1979; *The Brick Anthology* [with L. Spalding] 1989; *Brushes with Greatness: an Anthology of Chance Encounters with Greatness* [with R. Banks, D. Young] 1989; *The Faber Book of Contemporary Canadian Short Stories* (s) 1990; *From Ink Lake: an Anthology of Canadian Stories* (s) 1990

Eugene (Gladstone) O'NEILL 1888–1953

Born in New York, O'Neill was the son of one of the best-known melodramatic actors in America. His mother, also an actor, was a drug-addict, his brother an alcoholic, and he himself a man whose life was dogged by illness and depression. He attended Catholic boarding-schools, and then Princeton for a year, until he was suspended. He then spent several years knocking about the world: shipping as a seaman to Buenos Aires, prospecting for gold in Honduras, working as an actor and journalist. He married in 1907, but the marriage had been dissolved by 1912. (Their son, Eugene O'Neill Jr, was also a writer, and a Greek scholar, who committed suicide in 1950.)

In 1912, the year of his own suicide attempt, O'Neill entered a tuberculosis sanatorium and, while there, decided to become a playwright. He attended G.P. Baker's famous playwriting course at Harvard in 1914, and founded a new theatre group, the Provincetown Players, interested in producing the one-act plays he was now writing. Their production of his *Bound East for Cardiff*, one of a series of short early plays dealing with life at sea, in 1916, is generally regarded as the beginning of modern American theatre. His first full-length play was *Beyond the Horizon*, which won a Pulitzer Prize in 1920, and over the next fourteen years he became the most famous dramatist in the USA, alternating between realistic and symbolist dramas, and experimenting widely: he used, at various times, an all-black cast, a persistent drum-beat, mask-drama, juxtaposition of conventional dialogue with stylised interior monologue. Although *Anna Christie* and *Strange Interlude* both won Pulitzer Prizes, the greatest of his early plays is generally considered to be *Mourning Becomes Electra*, a trilogy in which he transferred the story of Aeschylus's *Oresteia* to the American Civil War.

From 1934, suffering from deep depression and affected by Parkinson's disease, O'Neill maintained a theatrical silence for twelve years, although he wrote many works which he destroyed, and in 1936 he became the first American dramatist to receive a Nobel Prize. He broke his silence with *The Iceman Cometh* in 1946, but most of his later plays were produced posthumously, including *Long Day's Journey into Night*, which is based on his own tortured family circumstances, and is often regarded as the greatest play to have come out of twentieth-century America. It was the fourth of his plays to win a Pulitzer Prize.

The argument about O'Neill's stature centres on his language; his *Times* obituary argued that 'his failure to form a style was to become his chief handicap as a dramatist', but others claim that his refusal to clothe tragic drama in conventional exalted language is in fact the key to his greatness. His second marriage was also dissolved (the couple had a son and also a daughter, Oonagh, who married Charlie Chaplin, and, like her brother, was cut out of O'Neill's will). His third marriage was to the well-known actress Carlotta Monterey, with whom he at last formed a happy partnership.

Plays
Thirst 1914, *Fog* 1917 and *The Web* 1924 [pub. in *Thirst and Other One Act Plays* 1914]; *Before Breakfast* [pub. in *The Provincetown Plays*] 1916; *Bound East for Cardiff* 1916 [pub. in *The Provincetown Plays*] 1916; *In the Zone* 1917, *The Long Voyage Home* 1917, *Ile* 1918, *The Moon of

the Caribbees 1918, *The Rope* 1918 and *Where the Cross Is Made* 1918 [pub. in *The Moon of the Caribbees and Six Other Plays of the Sea* 1919]; *The Sniper* 1917, *Abortion* 1959, *The Movie Man* 1959 and *Servitude* 1960 [pub. in *The Lost Plays of Eugene O'Neill* 1950]; *The Dreamy Kid* 1919 [pub. 1922]; *Beyond the Horizon* 1920; *Chris Christopherson* 1920 [np]; *Diff'rent* 1920 [pub. 1921]; *The Emperor Jones* 1920 [pub. 1921]; *Exorcism* 1920 [np]; *Anna Christie* 1921; *Gold* 1921; *The Straw* 1921; *The First Man* 1922; *The Hairy Ape* 1922; *All God's Chillun Got Wings* 1924; *The Ancient Mariner* [from Coleridge's poem] 1924 [pub. 1960]; *Desire under the Elms* 1924; *Welded* 1924; *The Fountain* 1925 [pub. 1926]; *The Great God Brown* 1926; *Lazarus Laughed* 1928 [pub. 1927]; *Marco Millions* 1928 [pub. 1927]; *Strange Interlude* 1928; *Dynamo* 1929; *Mourning Becomes Electra* 1931; *Ah, Wilderness!* 1933; *Days without End* 1933 [pub. 1934]; *The Iceman Cometh* 1946; *A Moon for the Misbegotten* 1947 [pub. 1952]; *Long Day's Journey into Night* 1956; *A Touch of the Poet* 1957; *Hughie* 1958 [pub. 1959]; *Inscriptions* [pub. 1960]; *More Stately Mansions* 1964 [pub. 1962]; *Children of the Sea and Three Other Unpublished Plays by Eugene O'Neill* [pub. 1972]; *The Personal Equation* [pub. 1988]

Non-Fiction

The Last Will and Testament of Silverdene Emblem O'Neill 1956; *'The Theatre We Worked For': the Letters of Eugene O'Neill to Kenneth MacGowan* [ed. R.M. Alvarez, J.R. Bryer] 1982; *'As Ever, Gene': the Letters of Eugene O'Neill to George Jean Nathan* [ed. A.W. and N.L. Roberts] 1987; *Selected Letters of Eugene O'Neill* [ed. T. Bogard, J.R. Bryer] 1988; *The Unknown O'Neill: Unpublished or Unfamiliar Writings of Eugene O'Neill* [ed. T. Bogard] 1988

Poetry

Poems 1912–1944 [ed. D. Gallup] 1980

Biography

O'Neill: Son and Playwright by Louis Shaeffer 1968; *O'Neill: Son and Artist* by Louis Shaeffer

---

(George) Oliver ONIONS 1873–1961

Onions was born in Bradford in Yorkshire, a county where his surname was common. He nevertheless decided that in combination with his forename it was faintly absurd and, to spare his two sons, he changed his name in 1918 to become (in private life) George Oliver. He had intended to be a painter, and left his working-class home in 1894 to study for three years at the Royal College of Art in London. Before publishing his first book, *The Compleat Bachelor*, in 1900, Onions worked as a draughtsman at the Harmondsworth Press (he always designed the wrappers for his own books), an experience which provided him with the material for a satire on popular journalism, *Little Devil Doubt*.

Despite his success, Onions was nervous and shy of his abilities, and there were often long gaps between his publications. *Whom God Hath Sundered*, a realistic novel about the rise of a businessman, is an acute psychological study, reworked from the trilogy *In Accordance with the Evidence* twelve years after its original publication. *Widdershins* is a memorable early collection of macabre short stories presented with a skill which Onions did not repeat in his later work. Despite the award of the James Tait Black Memorial Prize for *Poor Man's Tapestry* in 1946, Onions's reputation rests on *The Story of Ragged Robyn*, which is set in the late seventeenth century and focuses on injustice and cruelty.

A rapid but mumbling conversationalist, Onions was difficult to know well. Tall, heavy-set, with a square jaw, he was an amateur boxer in his younger days. A perfect foil for his reticence was Berta Ruck, whom he had met while she was an art student and married in 1909. She later became an enormously popular author of romances, and achieved a different sort of fame in 1922 when her name appeared on a tombstone in **Virginia Woolf**'s *Jacob's Room*. In reply to a solicitor's letter, Woolf replied that she had never heard of Miss Ruck, and the use of her name had been entirely inadvertent, an explanation the loyal Onions found incredible, although the quarrel was eventually settled amicably.

Fiction

The Compleat Bachelor 1900; *Tales from a Far Riding* 1902; *The Odd-Job Man* 1903; *Back o' the Moon* (s) 1906; *The Drakestone* 1906; *Admiral Eddy* (s) 1907; *Pedlar's Pack* 1908; *Draw in Your Stool* 1909; *Little Devil Doubt* 1909; *The Exception* 1910; *Good Boy Seldom* 1911; *Widdershins* (s) 1911; *In Accordance with the Evidence* 1912, *The Debit Account* 1913, *The Story of Louie* 1913 [trilogy rev. as *Whom God Hath Sundered* 1925]; *The Two Kisses* 1913; *A Crooked Mile* 1914; *Mushroom Town* 1915; *The New Moon* 1918; *A Case in Camera* 1920; *The Tower of Oblivion* 1921; *Peace in Our Time* 1923; *Ghosts in Daylight* (s) 1924; *The Spite of Heaven* 1925; *Cut Flowers* 1927; *The Painted Face* (s) 1929; *The Open Secret* 1930; *A Certain Man* 1931; *Catalan Circus* 1934; *Collected Ghost Stories* (s) 1935; *The Hand of Kornelius Voyt* 1939; *The Italian Chest* (s) 1939; *Cockrow* 1940; *The Story of Ragged Robyn* 1945; *Poor Man's Tapestry* 1946; *Arras of Youth* 1949; *A Penny for the Harp* 1952;

Bells Rung Backwards (s) 1953; *A Shilling to Spend* 1965

Patricia O'RANE

see Eleanor DARK

Joe (pseud. of John Kingsley) ORTON 1933–1967

'I'm from the gutter,' Orton once commented with characteristic defiance. 'And don't you ever forget it because I won't.' The son of a gardener employed by Leicester Council, he was born on the Saffron Lane Estate and educated locally. Eager to escape his background, he joined the Leicester Dramatic Society, making his debut in 1949 as a messenger in *Richard III*. He had originally been cast as Dorset – 'the son of some Queen (Elizabeth, I think)' – but looked too young, and since further parts were not forthcoming he joined other local societies. At the age of eighteen, he got a grant to go to the Royal Academy of Dramatic Art in London, where he met Kenneth Halliwell, a neurotic young man seven years his senior who became his mentor, lover and murderer. Under the heady influence of **Ronald Firbank**, they collaborated on a number of projects – including a novel called 'The Last Days of Sodom' and a 'satire in modern verse' entitled 'The Boy Hairdresser' – which they submitted to publishers without success. Their spare time was spent in defacing library books (pasting the photograph of a tattooed man on to a critical study of the works of **John Betjeman**, for example), and in 1962 they were prosecuted and sentenced to six months in jail, an experience which further fuelled Orton's inborn rage against authority.

His first play to be produced was *The Ruffian on the Stair*, based on 'The Boy Hairdresser', which was accepted by BBC radio in 1963. The title, taken from a poem by W.E. Henley, refers to death, and the plot revolves around a young man's attempt to exact revenge for the death of his brother, with whom he was having an incestuous relationship. It was broadcast the following year, and later revised for the stage. His first full-length play, *Entertaining Mr Sloane*, was also produced in 1964, at the New Arts Theatre in London, transferring to the West End largely due to the influence of **Terence Rattigan**, a champion of Orton's work. A bisexual comedy of manners in which a middle-aged brother and sister compete for the favours of a vicious young man, *Mr Sloane* established Orton's reputation for outrageous but highly polished comedy. Mannered and aphoristic in style, his plays used anarchic black comedy to satirise contemporary mores and sexual hypocrisy. Orton had learned from Firbank and Oscar Wilde the seditious uses of comedy and paradox, and presented a world in which accepted norms were gleefully turned upside-down. *Loot*, which won the *Evening Standard* and *Plays and Players* awards in 1966, concerns a couple of bank robbers who hide the money of the title in a coffin. (Orton returned from his mother's funeral with her dentures, which he presented to a member of the cast to use as a prop.) The homosexual relationship between the robbers is presented as completely natural, while the forces of law and order are embodied in a corrupt and brutal police inspector. *What the Butler Saw* is an elaborate and frantic farce set in a psychiatric unit and concludes with a character brandishing the (detached) penis of Winston Churchill. Unsurprisingly, these plays got him into trouble with the Lord Chamberlain's office, which frequently demanded excisions and alterations.

Although Halliwell's contribution to them was significant, the plays were performed and published under Orton's name. As Orton's star rose, Halliwell was seen as a mere hanger-on whose jealousy and insecurity did little to commend him to the theatrical world. They lived in a claustrophobic bedsitter in Islington, north London, the walls of which were covered in collage, and their relationship was further undermined by Orton's relentless promiscuity, which he recorded in detail in his unendearing diary.

After the success of *Loot*, Orton was approached to write a film script for the Beatles, but his proposed story of urban terrorism, which concluded with three of the male protagonists in bed with a woman, was not considered a suitable vehicle for the group. The script was returned without explanation in April 1967, but was bought by a producer for £10,000. By this time, Halliwell was being treated for severe depression, and four months later he beat Orton to death with a hammer, then committed suicide, leaving a note: 'If you read his diary all will be explained.'

Two months earlier, a double bill of *The Ruffian on the Stair* and *The Erpingham Camp* had been produced at the Royal Court Theatre under the grimly prophetic title *Crimes of Passion*, and when *What the Butler Saw* opened in a shambolic production two years after Orton's death, it was to disastrous notices, dismissed for example in the *Sunday Times* as 'a wholly unacceptable exploitation of sexual perversion'. Parallels were drawn between Orton's intentionally offensive work and his macabre end, with the obvious conclusion. He had died with his career in the ascendant, but his posthumous reputation took some time to recover. In 1975, Lindsay Anderson's new production of *What the Butler*

Saw led to a revaluation of that play, and a year later a complete edition of Orton's plays was edited by John Lahr, who went on to write a biography (the basis of Stephen Frears' 1987 film *Prick Up Your Ears*, scripted by **Alan Bennett**) and publish an edition of Orton's diaries. Orton once said that his ambition was 'to write a play as good as *The Importance of Being Ernest*', and if he never achieved this, he is none the less now regarded as Wilde's true heir, one of the British theatre's comic geniuses, whose work has not dated in the least and still retains its power to shock, subvert and entertain.

Plays
Entertaining Mr Sloane 1964; *Crimes of Passion* (st) 1967: *The Ruffian on the Stair* (r) 1964 [rev. 1966], *The Erpingham Camp* (tv) 1966; *Loot* 1966; *The Good and Faithful Servant* (tv) 1967 and *Funeral Games* (tv) 1968 [pub. 1970]; *What the Butler Saw* 1969; *The Complete Plays of Joe Orton* [ed. J. Lahr; pub. 1976]; *Up Against It* (f) 1985 [pub. 1979]

Fiction
Head to Toe 1971

Non-Fiction
The Orton Diaries [ed. J. Lahr] 1986

Biography
Prick Up Your Ears by John Lahr 1978

George ORWELL (pseud. of Eric Arthur Blair) 1903–1950

Orwell was born in Bengal, India; his father was a civil servant in the opium department, and his mother came from a family of old Burma hands. He and his sister Avril were brought to England to be educated, in Orwell's case at St Cyprian's, Eastbourne, and Eton College as a scholarship boy. He later attacked his preparatory school vitriolically (and libellously) in his essay 'Such, such were the joys' (1953, but not published in the UK until 1968), while at Eton he developed his antipathy towards the unfairness of the English class system and all its pretty nuances. He came to identify his own place in that system as among the 'lower-upper-middle class'.

After Eton, Orwell joined the Imperial Indian Police and spent five years in Burma as an assistant superintendent. The behaviour of the colonial officers there, and what he saw as the hypocrisies of imperialism, sickened him and he eventually resigned. Returning to England, he was determined to escape 'not merely from imperialism but from every form of man's dominion over man'.

Influenced by **Jack London**'s *People of the Abyss* (1903), he 'went native' in his own country by disguising himself as a tramp, sleeping in East End doss houses and spending time on the road. He also went to Paris and lived there for a while amongst the underclasses. His account of these experiences was published in 1933 as *Down and Out in Paris and London*. Despite regular review work for the *New Adelphi*, he was unable to make a living out of his writing and he took up a teaching post at a private school while completing his first novel, *Burmese Days* (1934). Both books were published under the pseudonym George Orwell, partly to protect his parents from possible ignominy over their itinerant, anti-imperialist son, and partly because the Orwell was a river he knew and liked.

In 1936 he married Eileen O'Shaugnessy, with whom he had a son. *Keep the Aspidistra Flying* (1936), written while he worked in a Hampstead bookshop, marks the end of his consciously literary phase: *The Road to Wigan Pier* – an account of industrial poverty in Wigan, Sheffield and Barnsley, published by the Left Book Club – signals the emergence of a mature, politicised, plain-speaking Orwell, who was undaunted by the inevitable accusations of posturing, or the graver charges of 'muck-raking' that were levelled at the book.

During the Spanish Civil War, he fought alongside the United Workers Marxist Party militia and was wounded by a shot through the throat. He never forgot the spirit of comradeship he experienced during this struggle, and his *Homage to Catalonia* demonstrates how his earlier political feelings had crystallised into a firm conviction in the moral rightness of democratic socialism, as well as the inhumanity of all forms of totalitarianism.

Poor health prevented him from joining the army during the Second World War, although he served in the Home Guard as a sergeant. He contributed 'London Letters' to *Partisan Review* and in 1941 joined the BBC, where he worked as a talks producer for the Indian section of the Corporation's Eastern Service. He continued to contribute essays, articles and reviews to numerous journals, and his peculiarly English brand of socialism, coupled with his plain, vigorous language, drew comparisons with Cobbett. He resigned from the BBC at the end of 1943 and became literary editor of *Tribune*. In November of that year, inspired by a small boy he had seen whipping a cart-horse, he started to write *Animal Farm*, his anti-Soviet farmyard fable. Publication was delayed until 1945, when the criticism of a British wartime ally became more acceptable.

After the war, Orwell lived mostly in Jura in the Western Isles of Scotland, where he struggled against recurrent outbreaks of tuberculosis to write the novel *Nineteen Eighty-Four*, his futurist vision of a post-war world governed by power-blocs. Despite its title, the novel was never meant to be prophetic; it was merely intended as a warning against the dangers of totalitarianism and allowing intellectuals to take over the state. It is his most misunderstood work, one in which some critics have even claimed to have seen evidence of a 'death-wish' and sado-masochistic tendencies. His wife had died in 1945, and during his final, long illness, in 1949, he married Sonia Brownell, who had worked at *Horizon*. Since his death, 'Orwellian' has become synonymous with the plain style of his journalism and essays, and for moral seriousness expressed with humour, simplicity and subtlety.

Fiction
Burmese Days 1934; *A Clergyman's Daughter* 1935; *Keep the Aspidistra Flying* 1936; *Coming Up for Air* 1939; *Animal Farm* 1945; *Nineteen Eighty-Four* 1949

Non-Fiction
Down and Out in Paris and London (a) 1933; *The Road to Wigan Pier* 1937; *Homage to Catalonia* (a) 1939; *Inside the Whale and other Essays* 1940; *The Lion and the Unicorn: Socialism and the English Genius* 1941; *Critical Essays* [US *Dickens, Dali and Others*] 1946; *The English People: an Essay* 1947; *Shooting an Elephant* 1950; *England, Your England and Other Essays* 1953; [rev. in US as *Such, Such Were the Joys* 1955]; *Decline of the English Murder and Other Essays* 1965; *The Collected Essays, Journalism and Letters* [ed. I. Angus, S. Orwell] 4 vols 1968; *The War Commentaries* 1985

Edited
E.M. Forster [with others] *Talking to India* 1943; *British Pamphleteers I: From the Sixteenth Century to the French Revolution* 1948

Biography
The Unknown Orwell by William Abrahams and Peter Stansky 1972; *Orwell* by Michael Shelden 1991

John (James) OSBORNE 1929–1994

Osborne was born in Fulham, where his father was known as a commercial artist. He was educated at Belmont College, Devon, and took up jobs in journalism and acting in repertory companies before he concentrated his talents on writing plays. He shot to fame in 1956 with *Look Back in Anger*, the first play of which he was the sole author, which opened at the Royal Court Theatre. The main character, Jimmy Porter, became known as the archetypal Angry Young Man. This play remains Osborne's best known, and is still hugely popular, although its artistic merits are now more questionable than they appeared at the time, and in later years it might be said to have been as much of a handicap to him as a source of strength: how does one grow old with grace when one is only recognised as being young and angry? This situation is made more complex by the fact that the label Angry Young Man is not universally accepted as being useful. Kenneth Allsop, in his lively book *The Angry Decade* (1963), has suggested that 'anger' is a misnomer: 'I think the more accurate word,' he writes, 'for this new spirit that has surged in during the fifties is dissentience.' He uses this word because it suggests the writers' disagreement with majority sentiment and opinion, rather than a revolutionary split from the establishment. Whatever the label, this mood, which Osborne to a certain degree helped mature into a serious artistic movement, is related to circumstances peculiar to the post-war era, above all to the vacuum left by the collapse of the writers of the 1930s, whose artistic and moral explorations had more or less dried up.

The Entertainer was Osborne's next major creation for the stage, and for many it remains his most accomplished play, both in terms of its dialogue and its technical aspects. Laurence Olivier played the lead in the first production in London, and this was taken as a symbolic marriage of the old theatre and the new type of play being produced by Osborne's generation. *The Entertainer*, an allegorical rather than a realistic play, concerns a song-and-dance man's last moments of sincere emotional intensity as he hears of the death of his son in Cyprus. *Luther* on the other hand, which explores the life and achievement of Martin Luther, has been criticised for being too realistic and blandly historical and has remained unpopular with theatre audiences and critics. Probably Osborne's most famous plays next to *Look Back in Anger* are *Inadmissible Evidence* and *A Patriot for Me*. The former presents the solitary world of Bill Maitland, a lawyer for whom life has lost its meaning because nobody considers his existence to be significant. His monologues dominate the play, and are considered brilliant. *A Patriot for Me*, meanwhile, ran into trouble with the Lord Chamberlain, who refused it a licence, mainly because of its homosexual theme: a spy story centring on the role of Alfred Redl, who spied for both the Austrians and the Russians, it is set just before the outbreak of the First World War.

Osborne has received a number of awards, including the Tony Award in 1963 for *Luther* and an Oscar in 1964 for the screenplay of Tony

Richardson's film of Henry Fielding's *Tom Jones*. Several of his plays have themselves been made into films: *Look Back in Anger* with Richard Burton and Claire Bloom in 1959, *The Entertainer* with Laurence Olivier in 1960 and *Inadmissible Evidence* with Nicol Williamson in 1968. But aside from his artistic life, he remained constantly in the public eye because of his marriages, of which there were five in total: he married Pamela Lane in 1951, the actress Mary Ure in 1957, the writer **Penelope Gilliatt** in 1963, Jill Bennett (who acted in several of his plays) in 1968 and Helen Dawson in 1978. His two savagely comic autobiographies, *A Better Class of Person* and *Almost a Gentleman*, were highly praised by the critics but Osborne was castigated for his disgraceful remarks about a number of these women – Jill Bennett in particular, whom he called 'the most evil woman I've ever come across'. He described her suicide as 'the coarse posturing of an overheated housemaid'. He also, on another level, insulted Anthony Creighton, with whom he collaborated on some plays, by calling him 'a cadging, homosexual drunk'. Creighton controversially made public his affair with Osborne shortly after the latter's death, and thus brought to light the playwright's bisexuality.

Plays

The Devil Inside Him [with S. Linden] 1950 [np]; *Personal Enemy* [with A. Creighton] 1955 [np]; *Look Back in Anger* **(st)** 1956, **(tv)** 1959; *The Entertainer* [with J. Addison] **(st)**, 1957, **(tv)** 1960; *Epitaph for George Dillon* [with A. Creighton] 1958; *The World of Paul Slickey* 1959; *A Matter of Scandal and Concern* **(tv)** 1960, **(st)** [as *A Subject of Scandal and Concern*] 1962; *Luther* 1961; *Plays for England* 1962 [pub. 1963]; *Tom Jones* **(f)** [from Fielding's novel] 1963; *Inadmissible Evidence* **(st)** 1964, **(f)** 1968 [pub. 1965]; *A Patriot for Me* 1965 [pub. 1966]; *A Bond Honoured* [from Lope de Vega's play] 1966; *The Charge of the Light Brigade* **(f)** [with C. Wood] 1968; *The Hotel in Amsterdam* 1968; *Time Present* 1968; *The Right Prospectus* **(tv)** 1970; *Very Like a Whale* **(tv)** 1971; *West of Suez* 1971; *A Gift of Friendship* **(tv)** 1972; *Hedda Gabler* [from Ibsen's play] 1972; *The Picture of Dorian Gray* [from O. Wilde's novel] 1973; *A Place Calling Itself Rome* [from Shakespeare's play *Coriolanus*] 1973; *A Sense of Detachment* 1973; *Jack and Jill* **(tv)** 1974 [pub. 1975]; *The End of Me Old Cigar* 1975; *Watch It Come Down* 1975; *Try a Little Tenderness* **(tv)** 1978; *You're Not Watching Me, Mummy* **(tv)** 1978; *God Rot Tunbridge Wells* **(tv)** 1985; *The Father* [from Strindberg's play] 1988; *Hedda Gabler* [from Ibsen's play] 1989; *Déjà Vu* 1991

Non-Fiction

A Better Class of Person **(a)** 1981; *The Meiningen Court Theatre 1866–1890* 1988; *Almost a Gentleman* **(a)** 1991

Wilfred (Edward Salter) OWEN 1893–1918

The oldest son of a railway clerk, Owen was born in Oswestry and educated at the Birkenhead Institute and the Shrewsbury Technical School. Encouraged to read widely by his ambitious and possessive mother, he absorbed the works of Shakespeare and the Romantic poets, particularly Keats, and began to write poetry of his own. He worked as a lay assistant in the parish of Dunsden near Reading, a post from which he resigned in mysterious circumstances. In 1913 he went to Bordeaux to teach English in the Berlitz School and remained in France as a private tutor when war was declared.

He returned to England in 1915, trained with the Artists' Rifles, and was gazetted a second lieutenant in the 2nd Battalion of the Manchester Regiment in the summer of 1916. His experience of trench warfare in France affected his poetry profoundly, replacing its Keatsian lushness with a stark realism about conditions there. He took as his subject 'the pity of War' and wrote of the passive suffering of the individual soldier, often with a homoerotic intensity. 'This book is not about heroes', he wrote of a projected volume of his poems. 'Nor is it about deeds, or lands, nor anything about glory, honour, might, majesty, dominion, or power, except War.'

While recovering from shell-shock at Craiglockhart Hospital near Edinburgh in 1917 he met a fellow patient, **Siegfried Sassoon**, who collaborated on what became his most famous poem, 'Anthem for Doomed Youth', encouraged him to further technical experiments (notably with half-rhyme), and introduced him to **Robert Graves** and literary London. He returned to France in September 1918, was awarded the Military Cross in October, and was killed on 4 November, a week before the end of the war.

Owen published only six poems during his life – three in the *Hydra*, the Craiglockhart magazine he edited, and three in the *Nation* – but his reputation grew swiftly after his death. The 1919 edition of **Edith Sitwell**'s anthology *Wheels* was dedicated to his memory and contained several of the poems by which he is best remembered: 'À Terre', 'The Show', 'The Sentry', 'Disabled', 'The Dead-Beat', 'The Chances' and 'Strange Meeting', which was acclaimed by John Middleton Murry in an article on 'The Condition of

Poetry' in the *Athenaeum* (5 December 1919) as 'the finest poem of the war'. This is a judgement which has since been frequently endorsed, and the poem, which depicts a post-mortem reconciliation between a British soldier and the German he has killed in battle, was placed at the beginning of the first edition of Owen's *Poems*, which was published in 1920, with a short introduction by Sassoon. This collected only twenty-three poems, including the savage 'Dulce et Decorum est', in which Owen graphically describes the effect of a gas attack; an expanded edition, edited with a memoir by **Edmund Blunden**, followed in 1931. Owen had a significant influence upon the writers of the 1930s, one of whom, **C. Day Lewis**, edited *The Collected Poems* in 1963. By this time, Owen was generally judged to be the finest poet of the First World War, not only because of his technical sophistication, but also because his unflinching but compassionate descriptions of soldiers 'who die as cattle' (as he put it in 'Anthem for Doomed Youth') had become the standard image of the war, one confirmed by numerous histories and memoirs of life in the trenches. Owen's emblematic status as the archetype of the War Poet was confirmed in 1961 when Benjamin Britten incorporated settings of several of the poems into the Latin text of his *War Requiem*.

In spite of destroying or mutilating a number of letters which contained 'the frightful implication of homosexuality', Owen's overly protective younger brother, Harold, produced a large edition of them in 1967, which added considerably not only to an understanding of Owen's complex personality, but also to the literature of the war.

Poetry

Poems [ed. S. Sassoon] 1920; *The Poems of Wilfred Owen* [ed. E. Blunden] 1931; *The Collected Poems of Wilfred Owen* [ed. C. Day Lewis] 1963; *War Poems and Others* [ed. D. Hibberd] 1973; *The Complete Poems and Fragments* [ed. J. Stallworthy] 2 vols 1983; *Wilfred Owen: the Poems* [ed. J. Silkin] 1985 [rev. as *The War Poems of Wilfred Owen* 1995]

Non-Fiction

Collected Letters [ed. J. Bell and H. Owen] 1967

Biography

Wilfred Owen by Jon Stallworthy 1974; *Wilfred Owen: the Last Year* by Dominic Hibberd 1992

Cynthia OZICK 1928–

Ozick was born and raised in Pelham Bay in the Bronx, the second child of a pharmacist and Hebrew scholar of Russian extraction, and niece of the Hebrew poet Abraham Regelson. After her local school she attended Hunter College, New York University, where she took her BA in English in 1949. The following year she gained her MA from Ohio State University having written her thesis on 'Parable in the Later Novels of **Henry James**'. She worked as an advertising copywriter for a Boston department store before getting married; she has one daughter and lives in New York.

Ozick began writing immediately after university and worked for seven years on a 'philosophical novel' influenced by her study of James before abandoning the project. Another seven years were devoted to her first novel, *Trust,* a long and complex work described by the author as 'about the failure of trust under every possible guise'. Published in 1966, the novel spans four decades of New York social and intellectual life and combines realism with symbolism, philosophy and literary allusion. After this she concentrated on short fiction, essays and book reviews, and became increasingly absorbed in Jewish cultural issues. 'Envy; or Yiddish in America', a comic story later collected in *The Pagan Rabbi*, was first published in *Commentary* in 1969, and is the story of a Yiddish poet consumed with envy for **Isaac Bashevis Singer**. It was intended as 'a great lamentation for the murder of Yiddish', but provoked charges of anti-Semitism in the Jewish press. Ozick taught English at New York University from 1964 to 1965, and has been the recipient of a Guggenheim Fellowship (1982) and a National Institute of Arts and Letters Living Award (1983).

Fiction

Trust 1966; *The Pagan Rabbi* (s) 1971; *Bloodshed and Three Novellas* (s) 1976; *Levitation* (s) 1982; *The Cannibal Galaxy* 1983; *The Messiah of Stockholm* 1987; *The Shawl* 1989

Non-Fiction

Art and Ardor: Essays 1983; *Seymour Adelman, 1906–1985* 1985; *Metaphor and Memory* 1988; *What Henry James Knew and Other Essays on Writers* 1993; *Portrait of the Artist as a Bad Character and Other Essays on Writing* 1994

P

Louise PAGE 1955–

Page was born in London, though she spent much of her childhood in Sheffield where she was educated at High Storrs Comprehensive School. She wanted to be an 'author' from a young age, and at ten was writing 'wonderfully long, extraordinary novels in which nobody did anything other than kiss passionately'. While at school she worked for an hour a day on a novel solemnly labelled 'Please Burn When I Die'. She studied drama and theatre arts at the University of Birmingham, graduating in 1976, and took a diploma in theatre studies at the University of Wales, Cardiff, in 1977. She has subsequently taught at Birmingham, and was the Yorkshire Television fellow in Creative Writing at Sheffield from 1979 to 1981, and resident playwright at the Royal Court Theatre from 1982 to 1983.

Her first play, *Want-Ad*, was produced in Birmingham in 1977, and in a revised version in London in 1979. During her early career she described herself as a radical, separatist feminist, though she has more recently described the need for 'a liberation of all people from preinformed ideology'. Through the subject of breast cancer, *Tissue* addresses the subject of how women see their bodies. Page's first substantial success came in 1982 when the Royal Court commissioned *Salonika*, a play about old age which she says she researched by talking to old ladies travelling on trains. An eighty-four-year-old mother and her daughter travel to Salonika to visit the grave of the father who died in the First World War, and the mother's seventy-four-year-old lover hitch-hikes to Greece to join her. The play celebrates life and deliberately challenges preconceptions about old age. Page has more recently written on the Falklands War and on nuclear war in *Goat*. She has also written for television and radio, and contributed scripts to *The Archers*, the long-running rural soap opera on BBC Radio 4.

Plays

Want-Ad 1977 [rev. 1979] [np]; *Glasshouse* 1977 [np]; *Saturday, Late September* (r) 1978; *Tissue* 1978 [pub. in *Plays by Women 1* ed. M. Wandor 1982]; *Hearing* 1979 [np]; *Lucy* 1979 [np]; *Agnus Dei* (r) 1980; *Flaws* 1980 [np]; *House Wives* 1981 [np]; *Salonika* 1982 [pub. 1983]; *Armistice* (r) 1983; *Falkland Sound/Voces de Malvinas* 1983 [np]; *Golden Girls* 1984 [pub. 1985]; *Real Estate* 1984 [pub. 1985]; *Beauty and the Beast* 1985 [pub. 1986]; *Goat* 1986 [np]; *Diplomatic Wives* 1988 [pub. 1989]; *They Said You Were Too Young* [pub. 1989]; *Plays: One* 1990; *Working Out* (r) 1991; *Another Nine Months* 1995 [np]

Grace PALEY 1922–

A New Yorker, born in what was then still the predominantly Jewish Bronx, Paley grew up in an environment remarkably helpful 'for the eventual making of literature'. There were 'lots of women in the kitchen talking, two strong languages, English and Russian, in my ear at home, and the language of my grandmother and the grownups in the street – Yiddish – to remind me of the person I really was.' Her Ukrainian-Jewish parents, Isaac and Manya Goodside, were already confirmed revolutionaries in their teens. Exiled to Siberia before they were twenty-one, they were eventually able to emigrate to the USA in 1905. Her mother died young; of her father, initially a doctor, later an artist, she has left a loving but pungent portrait in her autobiographical fragment 'A Conversation with My Father' in *Enormous Changes at the Last Minute*.

Breathing the air of politics from so early an age, it was, as Paley herself has said, 'natural that I should spend a lifetime trying to improve the world in legal and illegal ways'. Her early formal education was stormy, interrupted and eventually abandoned after unfinished courses at Hunter College and New York University. When only nineteen she married a cameraman, Jesse Paley, and, rather unexpectedly, given her later passionate and longstanding opposition to American militarism, spent the last few years of the Second World War having 'a lot of fun living in army camps'. After the war two children were born.

Since childhood she had written poems, but she found them increasingly unsatisfactory, feeling her work to be 'too literary'. She made her first tentative forays into prose, and the stories came swift and true, many of them making their way into her first published collection, *The Little Disturbances of Man* (1959). Although the book was warmly received, Paley had delayed publication for several years, fearing that her subject-matter – drawn from the wartime Bronx world of the old, the sick, the young and the female – was too 'trivial' for an age in which many authors were hell-bent on producing the Great American War Novel. In fact, politics permeate her writing, not as propaganda or polemic, but in its steady depiction of what she has called 'ordinary political people'. Her own political activism has ranged from local municipal campaigns to participation in national protest, against nuclear weapons, against the war in Vietnam and American interference in Nicaragua. With her second husband, Robert Nichols, a landscape gardener and author whom she married in 1972, she has lived in Chile; she has also visited China, Vietnam (in 1969) and Nicaragua. The once errant undergraduate has subsequently returned to teach at various universities, including Syracuse, Columbia, Sarah Lawrence College and City College, New York. Although she spends part of each year in her husband's Vermont house, she continues to inhabit the area of New York in which so many of her stories are set, and remains faithful to the form which brought her fame.

She dislikes the idea of the 'well-made story', 'not for literary reasons, but because it takes all hope away. Everyone, real or invented, deserves the open destiny of life.' (That last sentence is one approvingly quoted by Paley's contemporary **Jane Rule**.) 'I heard the voices' is how she has described her first attempts at writing prose fiction, and the phrase is entirely apt for one whose marvellous ear enables her to cast whole stories in the perfectly sustained idiom and accents of her richly idiosyncratic narrators. Certain characters recur throughout her work, including Virginia and Faith, two women who have raised their children alone, unlike Paley herself, although she acknowledges that Faith especially is in many ways an *alter ego* whose triumphs and failures sometimes mirror Paley's own. Her admitted twinship with Faith results in a fictional world whose characters are thereby free to argue with and reproach their creator, as in 'Listening', the last story in *Later the Same Day*. In its closing paragraphs Faith's friend, Cassie, rounds upon her for having 'told everybody's story but mine ... Where is Cassie? Where is *my* life?' Stunned by an onslaught which she recognises is justified, Faith apologises and asks forgiveness. The answer she receives, and which ends the story, is, one suspects, as much a message from Paley the woman to Paley the writer as it is an ultimatum from Cassie to Faith: 'You are my friend, I know that, Faith, but I promise you, I won't forgive you, she said. From now on, I'll watch you like a hawk. I do not forgive you.'

Fiction

The Little Disturbances of Man (s) 1959; *Enormous Changes at the Last Minute* (s) 1974; *Later the Same Day* (s) 1985; *The Collected Stories* (s) 1994

Poetry

Leaning Forward 1985; *Begin Again* 1992

Non-Fiction

365 Reasons Not to Have Another War 1989; *Long Walks and Intimate Talks* 1991

Mollie (Patricia) PANTER-DOWNES 1906–

The only child of a professional soldier, Panter-Downes was born in London, and spent two years of her early childhood with friends in Brighton, while her parents lived on the Gold Coast, where her father had been posted. They returned to England in 1914, and her father was killed at the Battle of Mons during the first month of the First World War. She and her mother subsequently lived in Surrey, where she was educated locally. She married in 1927, and subsequently had two daughters.

Her first novel, *The Shoreless Sea*, was published in 1923, when she was only seventeen, and was serialised in the *Daily Mirror*. 'It does not seem credible that an author so young could write with such assurance and with such a good sense of character,' the *Evening Standard* commented, and Panter-Downes found herself famous, her book advertised on buses and rushing through several editions. Her publishers capitalised on this success by issuing her second novel, *The Chase* (dedicated 'to my darling Daddie', and featuring a male protagonist), in an unusual dustjacket featuring a photograph of the youthful author, and emblazoned with the legend: 'A New Novel by Mollie Panter-Downes whose remarkable first book *The Shoreless Sea* is now in its 6th Edition'. This popularity was important, since she and her mother had been living on a small pension. It also led to the author having short stories published in magazines, although these have never been collected. Indeed, Panter-Downes has dismissed all five of her early novels, and insists that her reputation as a writer of fiction should be judged by her last novel, *One Fine Day*, which was serialised in the *Atlantic Monthly* and published in book form in 1947. The day of the title is a summer's one during the first year of peace after the Second World War. The novel details the uneventful life of a middle-class woman, who lives in a village near the Sussex coast. It is a novel in which nothing of note happens but a great deal of life is conveyed, particularly that life which the war had changed irrecoverably. Sharply observed, witty and beautifully written, it is an English classic.

In 1939 Panter-Downes started contributing a weekly 'Letter from London' to the *New Yorker*. A volume of her pieces appeared in 1940 as

Letter from England, and in 1971 William Shawn, the editor of the magazine, compiled a larger collection of her *London War Notes 1939–1945*, which forms a remarkable record of what it was like to live through the war for ordinary people. She continued as London correspondent for the magazine until 1987, and her two other works of non-fiction both appeared in the magazine in instalments. *Ooty Preserved* is an affectionate portrait of Ootacamund in southern India, which had been known in its Victorian heyday as 'the Queen of the Hill Stations'. *At the Pines* is a wonderfully evocative account of the latter years of Swinburne in the house he shared with Theodore Watts-Dunton on Putney Hill. It is an unusual and outstanding exercise in biography, less concerned with large events than with recreating the atmosphere of a quiet life of retirement. In this, and in its superb prose, it resembles *One Fine Day*.

Fiction
The Shoreless Sea 1923; *The Chase* 1925; *Storm Bird* 1930; *My Husband Simon* 1931; *Watling Green* 1943; *One Fine Day* 1947

Non-Fiction
Letter From England 1940; *Ooty Preserved* 1967; *At the Pines* 1971; *London War Notes 1939–1945* [ed. W. Shawn] 1971

For Children
Watling Green 1943

Sara PARETSKY 1947–

Born in Ames, Iowa and brought up in Kansas, Paretsky belonged to a family which presented the façade of the American Dream. In fact her parents' marriage was stormy. Her father, an abusive and violent man, was a first-generation American of Polish and Lithuanian descent; his wife was sixth-generation American and rooted in the WASP tradition, 'a good fifties stay-at-home mother, so scared by the outside world that she kept on having babies'. Paretsky, one of five children, was the only girl. She describes Kansas as the rockbed of American rural conservatism and her parents, both academics (her father was professor of microbiology at the University of Kansas), endorsed its traditions: their sons' education was paid for but their daughter had to earn her own way through college. As a result she attended the state university, which in the 1960s was 'a hotbed of radical feminism' – the exact opposite of what her parents wanted.

In 1966, at the end of her second year studying finance at the University of Kansas, Paretsky spent her vacation as a volunteer on a Presbyterian community programme looking after underprivileged children on the deprived south side of Chicago. She fell in love with the city, and returned there as soon as she had graduated, working for a company called Urban Research, running conferences on sexual equality in the workplace. When the company collapsed in 1974, she decided to do a PhD in history. She spent the month of her final examinations reading detective novels, but did not think at the time that she might want to write them herself. Instead she became a marketing manager for CNA, an insurance company, where she stayed for ten years.

The world of insurance provided the subject-matter for her first book, *Indemnity Only*, and also the stimulus to write it: 'I was thirty-one, sitting in a management meeting one day, working for this guy who was a total jack-off, sitting there with a balloon over my head saying "God, you are so stupid!", and my lips saying, "Gosh, Bill, what a wonderful idea!" And I thought, that's my detective, the person who's doing what I'm doing, only she gets to say what's in the balloon.' Paretsky wrote her first three novels while still at work, running a house, and singing in the local choir: 'Then one day I ran out of stamina and got hurt lifting too much paper at work. So I gave up work and took my chance on becoming a full-time writer.'

Each new novel tackles a different aspect of Chicago life: *Deadlock* is about the world of grain shipping in the Great Lakes; *Killing Orders* revolves around the Catholic church and the Marcincus scandal; *Bitter Medicine* concerns medical malpractice and the evils of religious fanaticism. *Toxic Shock*, which won the 1988 Crime Writers' Association Silver Dagger Award, was based on a scandal involving a Chicago company which had been monitoring its workers for asbestosis since the 1930s purely in order to predict how turnover would be affected by the resulting sickness and death, and which then disclaimed liability. *Burn Marks* is about corruption and collusion among Chicago politicians and the construction industry. Paretsky is a crime writer with a social conscience: she says, 'and V.I. Warshawski is the voice that I have for exploring those issues.' Her whisky-drinking fist-fighting heroine with self-doubts and an active sex-life has won wide acclaim for breaking new ground in a male-dominated genre. Paretsky was voted 1987 Woman of the Year by *Ms* magazine, and a body of feminist academics has grown up around her work.

Paretsky is married to a nuclear physicist twenty years her senior, whom she met when working briefly as a typist at the University of Chicago in the late 1970s.

Fiction
Indemnity Only 1982; *Deadlock* 1984; *Killing Orders* 1985; *Bitter Medicine* 1987; *Toxic Shock* 1988; *Burn Marks* 1990; *Guardian Angel* 1992; *V.I. Warshawski* 1993; *Tunnel Vision* 1994; *Windy City Blues* 1995

Edited
Beastly Tales (s) 1989; *A Woman's Eye* (s) 1991

Dorothy PARKER 1893–1967

The daughter of a prosperous garment manufacturer, Parker was born Dorothy Rothschild in West End, New Jersey, but was brought up in New York. Her mother, who was Scottish, died when Parker was still an infant; her father remarried shortly afterwards, and her childhood was not happy. Although half-Jewish, she was educated at the Blessed Sacrament convent school in New York and at Miss Dana's School in Morristown, New Jersey. After a brief spell playing the piano for dance classes, she submitted some poems to *Vogue*, with the result that she was recruited to the magazine and given a job writing captions for fashion photographs. Her gift for being economical and to the point may have developed here. She stayed for a year before being promoted in 1917 to become drama critic of *Vanity Fair*, as successor to **P.G. Wodehouse**. That same year she married Edwin Pond Parker II, who worked on Wall Street; the marriage was short-lived, since he immediately set off to fight in the First World War, and they separated shortly after he returned. They divorced in 1928, but she retained her married name, preferring it because it was not obviously Jewish.

It was with other writers on *Vanity Fair* that she became part of the Algonquin Round Table, which congregated at the famous New York hotel in order to lunch and exchange wisecracks. Amongst their number were Robert Benchley, Robert E. Sherwood, Alexander Woolcott, George S. Kaufman and Edna Ferber, but Parker was by far the most gifted of them. (Alan Rudolph's 1995 film *Mrs Parker and the Vicious Circle* is based on this period of her life.) Her sharpness, much appreciated at the Table, proved too much for *Vanity Fair* and she lost her job after writing a scathing review of *Caesar's Wife*, which starred Billie Burke, wife of the powerful impresario Florenz Ziegfeld. She subsequently worked on the *New Yorker*, becoming the epitome of its (in every sense) smart 1920s' incarnation. Between 1927 and 1933 she wrote the magazine's 'Constant Reader' book review column and rarely pulled her punches: 'The affair between Margot Asquith and Margot Asquith is one of the prettiest love stories in all literature.' During the last decade of her life she reviewed books for *Esquire*.

She published her first volume of poetry, characteristically titled *Enough Rope*, in 1926 and it became a bestseller, immediately establishing her reputation as a writer of short, witty lyrics, shot through with melancholy. It contained several of her most famous verses, including the cynical 'Unfortunate Coincidence' about protestations of mutual love, and 'Resumé', which catalogues the disadvantages of various forms of suicide, concluding mirthlessly: 'You might as well live.' Her short stories, originally published in the *New Yorker*, *Harper's*, *Cosmopolitan* and other such magazines, were in a similiar vein, as may be judged by the first collection, *Laments for the Living*, which appeared in 1930. The sparkle Parker gave off in conversation masked a depressive character and a life punctuated by abortion, suicide attempts and drinking bouts. Two of her best-known stories are 'Big Blonde', a tale about an alcoholic which won the O. Henry Award in 1929 and the admiration of **W. Somerset Maugham**, who said it had 'all the earmarks of a masterpiece'; and 'A Telephone Call', in which a desperate woman awaits a call from her lover, and which is one of several monologues she wrote.

Parker also had a career as a playwright and screenwriter, collaborating with **Elmer Rice** on the play *Close Harmony* in 1924, and with her second husband, Alan Campbell on the first version of *A Star Is Born* (1937). She had married Campbell in 1933, but the marriage was unstable, undermined by her drinking and his bisexuality. They divorced in 1947, but remarried three years later, separating again in 1953, but getting together once more for the last years of Campbell's life (he died in 1963). Her other principal Hollywood collaboration was with **Lillian Hellman** on the 1941 film version of the latter's play *The Little Foxes*. She also wrote a play about the life of Charles Lamb, *The Coast of Illyria*, on which she collaborated with Ross Evans. The dramatic work she was most proud of was *Ladies of the Corridor*, a play about a group of widows living in an Upper East Side hotel, which she wrote in the 1950s with Arnaud d'Usseau. Parker herself ended up in similar circumstances, an alcoholic recluse in a Manhattan hotel, with only her dog for company. She had twice attempted suicide during the 1920s, but surprised everyone by living to the age of seventy-three. A lifelong radical, who had travelled to Spain in 1937 to report on the Civil War, she left her estate to the National Association for the Advancement of Colored People.

Parker is chiefly remembered for her witticisms, many of which are familiar from dictionaries of quotations. At a Hallowe'en party she asked what game the guests were playing; on being told 'Ducking for apples', she replied: 'There, but for a typographical error, is the story of my life.' There was more substance to her than this, however, and *The Portable Dorothy Parker*, which she edited herself in 1944, revealed that she was someone of exceptional talent. If she was limited by her subject-matter – largely the unhappiness of human relationships, particularly for women – she made of it a substantial corpus of work. Hers is a poetry of disillusionment, which has echoes of **A.E. Housman** (whom she parodied in 'Cherry White'), though one poem outlined 'The Flaw in Paganism'. Her stories combine biting humour and plangent melancholy with genuine psychological insight, seen at its best, for example, in 'I Live on Your Visits' about a possessive mother and her teenage son.

Poetry
Enough Rope 1926; *Sunset Gun* 1928; *Death or Taxes* 1931; *Not So Deep as a Well* 1936; *Collected Poems* 1942

Fiction
Laments for the Living (s) 1930; *After Such Pleasures* (s) 1933; *Here Lies* (s) 1936; *Collected Stories* (s) 1942

Plays
Close Harmony [with E. Rice] 1924; *The Coast of Illyria* [with R. Evans] 1949; *Ladies of the Corridor* [with A. d'Usseau] 1953

Non-Fiction
The Collected Essays of Dorothy Parker 1970

Collections
The Portable Dorothy Parker 1944 [rev. as *The Collected Dorothy Parker* 1973]

Biography
You Might As Well Live by John Keats 1971

Frances PARKER

see Frances BELLERBY

Tim(othy Harold) PARKS 1954–

Parks was born in Manchester, the son of an evangelical clergyman, and attended various schools in Blackpool and London before going up to Cambridge. He graduated in 1977, and studied at Harvard, receiving his MA in 1979. He worked briefly as a writer for National Public Radio in Boston before returning to London where he worked in telephone sales. He had met Rita Baldassarre, also a translator, while at Harvard, and they married in 1979 and moved two years later to Verona where they have lived since and have two children. *Italian Neighbours* is a witty portrayal of Parks's life in Verona. Teaching English part-time, and later working as a translator, Parks continued to write novels as he had done since he was at Cambridge.

He had six manuscripts rejected before his first novel, *Tongues of Flame*, was runner-up in the Sinclair Prize competition for unpublished novels, after which it was accepted for publication in 1985, and went on to win a Somerset Maugham Award and a Betty Trask Award the next year. An account of fervent charismatic Christianity and its effects on the family of a suburban priest, the novel clearly draws on Parks's own upbringing, of which he has said: 'The choir, the church, morning service, evening service, prayer meetings, speaking in tongues and exorcisms. My father, you see, was very dedicated, and I was too, once.'

Parks's subsequent novels have continued to explore the tensions and complexities of family life. *Loving Roger*, which won the John Llewellyn Rhys Prize, concerns a woman whose passion drives her to murder her lover, and *Family Planning* examines the effects of a young man's mental illness on his parents and siblings. Parks draws on familiar, often autobiographical, material for his fiction, and has said: 'I have a radical determination to be accessible.' With *Cara Massimina* and its sequel *Mimi's Ghost* Parks mines a rich vein of black comedy. His numerous translations of Italian writers include Italo Calvino's posthumous fragmentary autobiography *The Road to San Giovanni*.

Fiction
Tongues of Flame 1985; *Loving Roger* 1986; *Home Thoughts* 1987; *Family Planning* 1989; *Cara Massimina* [as 'John MacDowell'] 1990 [US *Juggling the Stars* as Tim Parks] 1990; *Goodness* 1991; *Mimi's Ghost* 1995

Translations
Alberto Moravia: *Erotic Tales* 1985, *The Voyeur* 1986, *Journey to Rome* 1990; Antonio Tabucchi: *Indian Nocturne* 1988, *Vanishing Point, The Woman of Porto Pim* and *The Flying Creatures of Fra'Angelico* 1991; Fleur Jaeggy *Sweet Days of Discipline* 1991; Giuliana Tedeschi *A Place on Earth* 1991; Roberto Calasso *The Marriage of Cadmus and Harmony* 1992; Italo Calvino *The Road to San Giovanni* 1993

Non-Fiction
Italian Neighbours 1992

Alan (Stewart) PATON 1903–1988

Paton was born in Pietermaritzburg, the eldest of four children of English settlers. His father was a civil servant and his mother was a teacher; both were devout Anglicans, as was Paton. He was educated at Maritzburg College School until 1918, majored in science and received his BSc from the University of Natal in 1922 and a BEd in 1924. From 1925 to 1935 he taught, first at a Zulu school in Ixopo and then at Maritzburg College in Pietermaritzburg. From 1935 to 1948 he was able to put his political principles into practice as supervisor of the Diepkloof Reformatory, Johannesburg, the control of which passed from the Department of Prisons to the Department of Education under his guidance. When the Second World War interrupted his reforms, he signed up but was refused an army post because of his value to the civil service. From 1938 to 1947 he also acted as vice-chairman of Toc H.

His Christian convictions are evident in his first published work, a poem, *Meditation for a Young Boy Confirmed* (1944). In 1946 he undertook a tour of penal institutions in Europe and America but, upon reaching Sweden, a long postponed ambition to write creatively overtook him – 'the dam broke', as he put it – and he began *Cry, the Beloved Country*. He continued the novel in London, New York, Washington, Texas and Arizona, and completed it in 1947 in San Francisco. The book describes the loss and rediscovery of faith of a Zulu parson, as he searches for his lost sister through squalid townships and attempts to come to terms with his son's murder of a white man. It became a bestseller both in South Africa and worldwide and was adapted as a musical by **Maxwell Anderson** and Kurt Weill, *Lost in the Stars* (1949), and as a film under its original title (1951) with a screenplay written by Paton. He began his second novel, *Too Late the Phalarope*, whilst in London working on the film.

Paton continued to write essays and articles, many of which were first printed in Christian periodicals, calling for 'a spiritual solution' to his country's problems and in 1953, protesting distress over the political regime, he and his wife, Doris Farmer – they had married in 1928 and had two sons – withdrew for a year to live and work on a Toc H administered tuberculosis settlement for non-white patients in Natal. His political commitment came further to the fore when, in 1958, out of his Liberal Association, he founded the Liberal Party of South Africa, becoming its first president. The party disbanded ten years later when, because of new racial laws, it was to be made illegal, and Paton had his passport temporarily withdrawn. In 1963 he gave evidence in mitigation at the trial of Nelson Mandela and the following year published a biography of Jan Hofmeyr, the liberal vice-president in the Smuts administration.

The death of his wife in 1967 produced a moving tribute, *Kontakion for You Departed*, and in 1980 he published the first volume of his autobiography, *Towards the Mountain*. He lived in Natal to the end of his life with his second wife, and published a third and final novel, *Ah, But Your Land Is Beautiful*, at the age of seventy-eight. The second volume of his autobiography, *Journey Continued*, was published posthumously.

Fiction

Cry, the Beloved Country 1948; *Too Late the Phalarope* 1953; *Debbie Go Home* (s) [US *Tales from a Troubled Land*] 1961; *Ah, But Your Land Is Beautiful* 1981

Non-Fiction

Religious Faith and World Culture (c) [ed. A.W. Loos] 1951; *The Land and People of South Africa* 1955; *Hope for South Africa* 1958; *Thirteen for Christ* (c) [ed. M. Harcourt] 1963; *Hofmeyr* 1964; *The Long View* [ed. E. Callan] 1967; *Instrument of Thy Peace* 1968; *Kontakion for You Departed* [US *For You Departed*] 1969; *Apartheid and the Archbishop: the Life and Times of Geoffrey Clayton* 1973; *Knocking on the Door: Shorter Writings* 1975; *Towards the Mountain* (a) 1980; *Save the Beloved Country* 1987; *Journey Continued* (a) 1988

Plays

Cry, the Beloved Country (f) [from his novel] 1951; *Sponono* [with K. Shah] 1983

Poetry

Meditation for a Young Boy Confirmed 1944

Brian PATTEN 1946–

'Younger than the atom bomb', Patten was born in a working-class district of Liverpool, and educated at Sefton Park Secondary School. Leaving school at fourteen, he joined a local newspaper as a junior reporter and began publishing a magazine called *Underdog*, the first to give a voice to many of the then underground poets such as **Roger McGough** and **Adrian Henri**. It was the 'Liverpool Poets' movement that was largely responsible for popularising poetry in the 1960s, and it was mainly through Patten, the youngest of them, that the movement spread. He lived for a time in Paris with Guy Jequesson, a young French poet, at the offices of the communist paper *L'Humanité*, before moving on to Spain and Tangier. At eighteen he was in Dublin with **John Arden** and his wife **Margaretta D'Arcy**: later he returned to Liverpool and started a pop-music column.

His poems first came to public attention when he was fifteen – his first book, *Portraits*, was privately printed. **Edward Lucie-Smith** wrote of his next book, *Little Johnny's Confession*, containing poems mostly written in his late teens: 'There's a real freshness about these poems which is something quite apart from tricks of style ... The *Little Johnny* sequence has an amazing and very moving detachment, considering how young the author is.' It was reprinted within a few months of publication. It was the publication of his work along with that of Henri and McGough in *Penguin Modern Poets 10* that really caught the public imagination, however, and established Patten as a leading poet of the rising generation. The book was subtitled 'The Mersey Sound', a term previously applied to the pop music that had made Liverpool famous. The Beatles, of course, had come from Liverpool, and the book was published in the same year that the group issued their seminal album, *Sgt Pepper's Lonely Hearts Club Band*. Some of Patten's poems shared the surreal lyricism of Lennon and McCartney's songs.

Patten's next volume, *Notes to the Hurrying Man*, continued to develop this lyricism in a way that is far from gentle – a rocking-horse that comes to life, an astronaut who goes mad, a pornographer in his ruined garden, a greenfly asleep in the centre of its world – but the images remain simple and relate firmly to common experience. *Vanishing Trick* is a sequence of love poems that move through loss to celebration. *Grave Gossip* is once more distinguished by strong images: it contains poems about a man who gatecrashes Paradise, a mule that dreams of being articulate, a drunken tightrope-walker, a moth with 'no memory of flames', 'the animated telephones that menace'. *Love Poems* carries on from where *Vanishing Trick* left off.

Although Patten has continued to produce distinguished work, he is to some extent a victim of his own early celebrity, still best known for his work of the 1960s, when he was a leading figure on the poetry-reading circuit and once made a film with the Rolling Stones' Brian Jones for the BBC. He remains a prolific and fluent writer, who has also published a number of children's books – *Mr Moon's Last Case* won a special award from the Mystery Writers of America – and written plays.

Poetry

Portraits 1962; *Little Johnny's Confession* 1967; *Atomic Adam* 1967; *Penguin Modern Poets 10* [with A. Henri, R. McGough] 1967 [rev. 1974, 1983]; *Notes to the Hurrying Man: Poems Winter '66–Summer '68* 1969; *The Homecoming* 1970; *The Irrelevant Song* 1970; *Little Johnny's Foolish Invention* [with R. Sanesi] 1970; *Walking Out* 1970; *At Four O'Clock in the Morning* 1971; *The Irrelevant Song and Other Poems* 1971; *The Projectionist's Nightmare* 1971; *When You Wake Tomorrow* [with P. Benveniste] 1971; *And Sometimes It Happens* 1972; *Double Image* [with M. Baldwin, J. Fairfax] 1972; *The Eminent Professors and the Nature of Poetry* 1972; *The Unreliable Nightingale* 1973; *Vanishing Trick* 1976; *Grave Gossip* 1979; *Love Poems* 1981; *New Volume* [with A. Henri, R. McGough] 1983; *Storm Damage* 1988; *Grinning Jack* 1990; *The Magic Bicycle* 1993

Plays

The Hypnotic Island (**r**) 1977; *The Pig and the Junkie* 1977 [np]; *The Sly Cormorant* 1977 [np]; *The Man Who Hated Children* (**tv**) 1978; *The Ghosts of Riddle Me Heights* 1980 [np]; *Behind the Lines* [with R. McGough] 1982 [np]; *The Mouthtrap* [with R. McGough] 1982 [np]; *Blind Love* (**r**) 1983; *Mr Moon's Last Case* (**tv**) [from his novel] 1983; *Gargling with Jelly* [from his poems] 1988 [np]

For Children

The Elephant and the Flower (**s**) 1970; *Jumping Mouse* 1972; *Manchild* 1973; *Two Stories* (**s**) 1973; *Mr Moon's Last Case* 1975; *Emma's Doll* 1976; *The Sly Cormorant and the Fishes* 1977; *Gargling with Jelly* 1985; *Jimmy Tagalong* 1988; *Thawing Frozen Frogs* 1990; *Grizzelda Frizzle and Other Stories* (**s**) 1992; *Impossible Parents* 1994

Edited

The House That Jack Built: Poems for Shelter [with P. Krett] 1973; *Clare's Countryside: Natural History Poetry and Prose* 1981; *Gangsters, Ghosts, and Dragonflies: a Book of Story Poems* 1981; *The Puffin Book of Twentieth-Century Children's Verse* 1991

Tom (Neilson) PAULIN 1949–

Of Ulster Protestant stock (his maternal grandparents emigrated from Glasgow to Belfast in 1912), Paulin was born in Leeds, but grew up after 1953 in Belfast, attending the Annadale Grammar School. Although he has lived for much of his adult life in England, he has also spent considerable periods in Northern Ireland, and is usually regarded as one of the second generation of 'Ulster poets' who became prominent in English poetry from the 1960s and 1970s onwards. He studied English at Hull University, then did two years' research at Lincoln College,

Oxford, and from 1972 has been an academic at the University of Nottingham, first as lecturer and then as reader, later professor of poetry. He married Munjiet Kaur, who is of Indian origin, in 1973; they have two children.

His first volume of poetry was *Theoretical Locations* (1975), and in *A State of Justice*, published two years later, he established his insistence on political rather than personal emotion. He had developed strong left-wing convictions from early years and although originally, as an Ulster Protestant, a believer in the union with Great Britain, from the early 1980s he was converted to the idea of a 'sweet, equal republic', propagating this belief in his verse, much of which was designed to shock what he saw as complacent British attitudes. The same beliefs animated such polemical books as *Ireland and the English Crisis* and his play, *The Hillsborough Script*. Although the tone of his work was originally clipped and severe, making use, from *Liberty Tree* onwards, of dialect and frequent neologisms, by the time of *Walking a Line* (1994), a new playfulness and lyricism had become apparent.

Paulin is a well-known controversialist, and among the most celebrated of his disagreements was that with **Craig Raine** in the *London Review of Books* from early 1986, touched off by the poetry of **Geoffrey Hill** and developing into an argument about whether metrics can of their nature be political (Paulin contended that they can), and about *The Faber Book of Political Verse* which, like a similar anthology of vernacular verse, Paulin edited. In recent years Paulin's talent for controversy (and sometimes devastating criticism) has reached a wider audience with his appearances during the annual televising of the Booker Prize and on BBC 2's *Late Review*.

Besides the works mentioned, Paulin has published a study of **Thomas Hardy**, two updatings of Greek tragedy (*The Riot Act* is based on the *Antigone* of Sophocles, while *Seize the Fire* offers a treatment of the *Prometheus Bound* of Aeschylus), and the ambitious literary and political study, *Minotaur: Poetry and the Nation State*.

Poetry

Theoretical Locations 1975; *A State of Justice* 1977; *Personal Column* 1978; *The Strange Museum* 1980; *The Book of Juniper* 1981; *Liberty Tree* 1983; *The Argument at Great Tew* 1985; *Fivemiletown* 1987; *Seize the Fire* 1990; *Selected Poems 1972–1990* 1993; *Walking a Line* 1994

Non-Fiction

Thomas Hardy: the Poetry of Perception 1975; *A New Look at the Language Question* 1983; *Ireland and the English Crisis* 1984; *Minotaur: Poetry and the Nation State* 1992

Plays

The Riot Act [from Sophocles's play *Antigone*] 1985; *The Hillsborough Script* 1987

Edited

Henry James (s) [with P. Messent] 1982; *The Faber Book of Political Verse* 1986; *Hard Lines 3* [with F. Dubes, I. Dury] 1987; *The Faber Book of Vernacular Verse* 1990

Mervyn (Laurence) PEAKE 1911–1968

The son of a Congregational missionary doctor, Peake was born in Kuling, a hill station in central China, and was educated at the Tientsin Grammar School and Eltham College, Kent, a public school for the sons of missionaries. A spell at the Royal Academy Schools was followed by residence in an artists' colony on the Island of Sark, the setting of his allegorical, **T.F. Powys**-like novel, *Mr Pye*, in which the eponymous hero first sprouts wings, then horns. Peake excelled in portraiture and later returned to London to teach life-drawing at the Westminster School of Art. In 1937 he married one of his students, Maeve Gilmore, who was to become a successful painter; they had three children.

His first book was a fantastical pirate story for children, *Captain Slaughterboard Drops Anchor* (1939), which he illustrated himself. The entire first edition of this book was destroyed in a warehouse fire during the blitz (which is why the date of publication is sometimes given as that of the second edition, 1945). Peake himself was called up in 1940 and proved a hopelessly incompetent soldier. Left largely to his own devices, he did an enormous amount of writing and drawing and repeatedly applied to be an official war artist. In spite of the support of his commanding officer, he was always turned down, and his frustration contributed to the nervous breakdown he suffered. In 1942 he went absent without leave and, after a period in a military hospital and extended sick leave, he was invalided out of the army in May 1943. By this time he had been given a commission by the War Artists' Advisory Committee to make drawings and paintings in a glass factory which manufactured cathode ray tubes: this resulted in a volume of poems, *The Glassblowers*, which, with his novel *Gormenghast*, earned him the W.H. Heinemann Foundation Prize in 1950.

Gormenghast was the central volume of the richly detailed, Gothic trilogy for which Peake is best remembered. Set in a vast, sprawling castle, ossified by tradition and ritual, it follows the tribulations of Titus Groan, heir to the estate. Peake provided his own illustrations, and usually fixed his characters in his mind by drawing them before writing about them. The final volume of the trilogy, *Titus Alone*, is the darkest, influenced by Peake's visit in 1945 to Belsen concentration camp as an artist for the *Leader* magazine. By the time his drawings were published, Peake was already suffering from a form of Parkinson's disease, possibly the result of having unknowingly contracted encephalitis lethargica during the epidemic of 1917. He had already suffered a second nervous breakdown after the failure in 1957 of his play *The Wit to Woo* (the only one of the four he wrote to reach the stage), but the effects of his new illness were devastating, gradually robbing him of his faculties; he spent the last four years of his life in private nursing homes.

Like **Wyndham Lewis** and **Isaac Rosenberg**, Peake was distinguished both as an artist and as a writer. He provided illustrations for a number of books – amongst them such classics as *Alice in Wonderland*, *Treasure Island* and Grimms' *Household Tales* – and designed sets and costumes for the theatre. His poetry ranged from long **John Masefield**-like narratives, through appalled poems about Belsen, lyrical accounts of British industry, and a great deal of love poetry, to his splendid nonsense verse. He also adapted his own work for the radio, and published several short stories in assorted periodicals.

Fiction

The Gormenghast Trilogy: *Titus Groan* 1946, *Gormenghast* 1950, *Titus Alone* 1959 [rev. 1970]; *Mr Pye* 1953; *Sometime, Never* (s) [with W. Golding, J. Wyndham] 1956

Poetry

Shapes and Sounds 1941; *Rymes without Reason* 1944; *The Glassblowers* 1950; *The Rhyme of the Flying Bomb* 1962; *Poems and Drawings* 1965; *A Reverie of Bone* 1967; *A Book of Nonsense* 1972; *Selected Poems* 1972

For Children

Captain Slaughterboard Drops Anchor 1939; *Letters from a Lost Uncle* 1948

Plays

For Mr Pye – an Island (r) [from his novel *Mr Pye*] 1957, *Noah's Ark* and *The Wit to Woo* 1957 [pub. in *Peake's Progress* 1978]; *Titus Groan* (r) [from his novel] 1956

Non-Fiction

The Craft of the Lead Pencil 1946; *Drawings by Mervyn Peake* 1949; *Figures of Speech* (c) 1954; *Drawings of Mervyn Peake* [ed. H. Spurling] 1974

Collections

Mervyn Peake, Writings and Drawings [ed. M. Gilmore, S. Johnson] 1974; *Peake's Progress: Selected Writings and Drawings* [ed. M. Gilmore] 1978

Biography

Mervyn Peake by John Batchelor 1974

Walker PERCY 1916–1990

Percy was born in Birmingham, Alabama, to Leroy and Martha Pratt. His father, a lawyer, committed suicide when Percy was eleven and his mother was killed in a car crash two years later. With his two younger brothers he was then brought up by William Alexander Percy, his father's cousin, in Greenville, Mississippi. The son of a US senator, William was one of the great Southern personalities of his time, a lawyer, planter and poet whose 1941 autobiography, *Lanterns of the Levee*, is a powerful defence of the paternalism of the old Southern landowners. Visitors to his home included the black poet and activist **Langston Hughes**, and the young **William Faulkner**. Throughout his life, Percy paid tribute to his guardian's influence.

Percy was educated at Greenville High School, where he started writing, and at the University of North Carolina, Chapel Hill, where he graduated in chemistry in 1937. He studied medicine for four years at Columbia University's College of Physicians and Surgeons, and spent much of his time in New York cinemas, an experience upon which he was later to draw in *The Moviegoer*. In 1942, after a year as a pathologist carrying out autopsies at Bellevue Hospital in New York, he contracted pulmonary tuberculosis and had to spend two years in a sanatorium in the Adirondack Mountains. This period altered the course of his life: he was received into the Roman Catholic church, studied philosophy, and decided to become a writer rather than a doctor, having realised that what interested him was 'not the physiological and pathological processes with man's body but the problems of man himself, the nature and destiny of man'.

Left money by his adoptive father, he moved to New Orleans and subsequently to Covington, Louisiana, where he was married in 1946 to Mary Townsend, with whom he had two daughters. Influenced by Camus, Sartre and Dostoevsky, he began what was to be a long apprenticeship as a writer. He wrote two novels and a study of the philosophy of language, all of which were unpublished, before finding success with *The Moviegoer*, which won a National Book Award in 1962. The novel depicts a young man obsessed with the cinema, and treats the theme of alienation which recurs in Percy's work. His subsequent work is marked by a distinctively laconic voice which is both comic and deeply serious in its examination of moral and religious questions.

Although firmly rooted in the South, Percy rejected the tag of 'Southern writer' and said: 'Faulkner and all the rest of them were always going on about this tragic sense of history, and we're supposed to sit on our porches and talk about it all the time. I never did that. My South was always the New South. My first memories are of the Country Club, of people playing golf.' Like his adoptive father, Percy was politically active, involving himself in the civil rights movement and other liberal causes.

Fiction
The Moviegoer 1961; *The Last Gentleman* 1966; *Lancelot* 1977; *Love in the Ruins* 1971; *The Second Coming* 1980; *The Thanatos Syndrome* 1987

Non-Fiction
The Message in the Bottle 1975; *Lost in the Cosmos: the Last Self-Help Book* 1983; *Novel-Writing in an Apocalyptic Time* 1986

Biography
Pilgrim in the Ruins by Jay Tolson 1992

Caryl PHILLIPS 1958–

Phillips was born in St Kitts in the West Indies and his family emigrated in the same year to England where he was educated in Leeds and Birmingham. He read English at Queen's College, Oxford, graduating in 1979. As an undergraduate, he directed a number of plays and was a committee member of the Oxford University Experimental Theatre Club. In 1978 he founded the *Observer* Oxford Festival and the next year was the festival's artistic director. In the introduction to his travel book, *The European Tribe*, Phillips recounts how his theatrical activities so exhausted him that he had to have a period of rest during which he travelled to the USA. In Los Angeles he read **Richard Wright**'s *Native Son* (1940), which inspired him to want to write, suggesting 'a possibility of how I might be able to express the conundrum of my own existence'. Much of his subsequent work has explored the theme of black identity in a white-dominated society.

His first play, *Strange Fruit*, was produced at the Crucible Theatre, Sheffield, in 1980, and examines the predicament of a black schoolteacher struggling to raise two children and exist between two conflicting cultures. *The Wasted Years* won the BBC Giles Cooper Award as one of the best radio plays of 1984. His television documentary *Lost in Music* was filmed in St Kitts and New York and shown in 1985, and the following year he scripted a film about interracial cricket, *Playing Away*.

His first novel, *The Final Passage*, winner of the 1985 Malcolm X Prize, funded by the Greater London Council, is a bleak picture of the life of a young mulatto woman living on a Caribbean island in the 1950s. She suffers because of her mixed parentage, but on coming to London she continues to be defined by her colour, and is again the subject of abuse and racism. *Cambridge* is set in the eighteenth century, and again explores the theme of identity as the eponymous black hero temporarily finds a grudging acceptance in English society, only to be taken captive in Africa after the death of his white wife. *Crossing the River*, which was shortlisted for the 1993 Booker Prize, again tackles the subject of slavery and is an ambitious account of two brothers and a sister whose enslavement removes them to different continents in the nineteenth and twentieth centuries.

Phillips has been the resident dramatist at The Factory, London, writer-in-residence at Stockholm University and Amherst College, Massachusetts, and a member of the board of directors of the Bush Theatre, London.

Plays
Strange Fruit 1980 [pub. 1981]; *Where There Is Darkness* 1982; *The Shelter* 1983 [pub. 1984]; *The Hope and the Glory* (tv) 1984; *The Record* (tv) 1984; *The Wasted Years* (r) 1984 [pub. in *Best Radio Plays of 1984* 1985]; *Lost in Music* (tv) 1985; *Crossing the River* (r) 1986; *Playing Away* (f) 1986 [pub. 1987]; *The Prince of Africa* (r) 1987; *Writing Fiction* (r) 1991

Fiction
The Final Passage 1985; *Higher Ground* (s) 1986; *A State of Independence* 1986; *Cambridge* 1991; *Crossing the River* 1993

Non-Fiction
The European Tribe 1987

Jayne Anne PHILLIPS 1952–

Phillips was born in Buckhannon, West Virginia, a small town in a largely rural state which has its fictional counterpart in the Bellington of her writing. She was the middle child and only daughter of a father who worked in road construction and a mother who taught. She graduated from the University of West Virginia in 1974, having worked her way through college selling home improvements and domestic appliances door-to-door. She subsequently spent two unsettled years living briefly in Oakland and then Colorado, working in restaurants and motels, as a teacher at a remedial school, and in an amusement park, developing the insight into life as it is lived on the perimeter of the American Dream which was to surface in her early fiction.

She started writing poetry, on the strength of which she was accepted in 1976 on to a creative writing course run by the University of Iowa. Her first book, *Sweethearts*, a collection of twenty-four one-page prose pieces, was published the same year, beginning an association with and continuing support for the small presses; she has since published with Vehicle and Gunslinger in limited editions. She gave readings, taught creative writing and published a second volume of brief fictions before persuading Seymour Lawrence to take on *Black Tickets*. The book was published in 1979 and, whilst it was an immediate success, winning the Sue Kaufman Award for First Fiction and drawing tributes from **Raymond Carver**, **John Irving** and **Nadine Gordimer**, some critics were disturbed by the dense, mannered surface of much of the prose.

Her first novel, *Machine Dreams*, was five years in the writing and dispenses with the more overwrought technical effects of her earlier work. It interweaves letters, dreams and oral history to trace the slow disintegration of a small-town family from the Depression years to the Second World War and then Vietnam. She has held teaching posts at Boston University and at Brandeis University, and in 1985 she married a physician. She has one son and two stepsons.

Fiction

Sweethearts (s) 1976; *Counting* (s) 1978; *Black Tickets* (s) 1979; *How Mickey Made It* (s) 1981; *Fast Lanes* (s) 1984; *Machine Dreams* 1984; *Shelter* 1994

Harold PINTER 1930–

Pinter was born in the East End of London, the son of a Jewish tailor, and educated at Hackney Downs Grammar School where he acted in school productions. He went to the Royal Academy of Dramatic Art, but left after two unhappy terms and 'spent the next year roaming about a bit'. In 1949 he was fined by magistrates for having as a conscientious objector refused to do his national service. The following year he began to publish poems in *Poetry* (*London*) and got his first work as a professional actor on a BBC radio programme, *Focus on Football Pools*. After a brief return to training at the Central School of Speech and Drama, he toured Ireland from 1951 to 1952 with Anew McMaster's company performing Shakespeare, and in 1953 worked for Donald Wolfit's company in Hammersmith.

After four more years in provincial repertory theatre he wrote his first play, *The Room*, which was first produced by the Drama Department of Bristol University, and then by another student company at the Bristol Old Vic. The *Sunday Times* drama critic, Harold Hobson, praised this production, thereby attracting the attention of the impresario Michael Codron, who produced *The Birthday Party* at the Cambridge Arts Theatre and the Lyric, Hammersmith, in 1958. Most reviewers were extremely hostile (Milton Shulman began his review, 'Sorry, Mr Pinter, you're just not funny enough'), though the prescient Hobson described Pinter as possessing 'the most original, disturbing and arresting talent in theatrical London'. Another, perhaps unlikely, champion of his early plays was **Noël Coward**.

In rapid succession Pinter produced the body of work upon which his reputation rests. *The Dumb Waiter*, *The Caretaker* and *The Homecoming* explore a dark and enigmatic world of menace and uncertainty, and their precise 'meaning' has since kept an army of critics in employment. The word 'Pinteresque' has entered the language to describe the type of colloquial speech embodied in the faltering, pause-ridden dialogue of his characters. In the 1960s Pinter also wrote for radio and television, and wrote the screenplay for Joseph Losey's film *The Servant* based on **Robin Maugham**'s novel. The 1963 film of *The Caretaker*, directed by Clive Donner, was shot in a derelict house in Hackney near to Pinter's childhood home and funded by a consortium which included Peter Sellers, Richard Burton and Elizabeth Taylor. Pinter has since adapted numerous novels for the screen, notably **Penelope Mortimer**'s *The Pumpkin Eater*, **L.P. Hartley**'s *The Go-Between*, **F. Scott Fitzgerald**'s *The Last Tycoon*, **John Fowles**'s *The French Lieutenant's Woman*, **Margaret Atwood**'s *The Handmaid's Tale* and **Ian McEwan**'s *The Comfort of Stangers*. His *Proust Screenplay* is a remarkable exercise which translates the subtle nuances and motifs of Proust's novel into the language of cinema. The long-planned film, to be directed by Losey, was never made. Pinter has periodically directed for the stage, notably plays by **Simon Gray** and the London production of **David Mamet**'s controversial *Oleanna*. After *Betrayal* he wrote no new full-length plays until *Moonlight* in 1994, though short plays from this period include *A Kind of Alaska*, an account of a woman waking from a forty-year coma which was inspired by the case histories in Oliver Sacks's *Awakenings* (1973). His novel *The Dwarfs* was written in the early 1950s and formed the basis for the play of the same name. Published in 1990, it provides a fascinating insight into the formation of his work.

Pinter was married in 1956 to the actress Vivien Merchant with whom he has a son. She acted in many of his early plays. Following their divorce in 1980 he married the biographer and historian Lady Antonia Fraser. He has long been politically concerned, and expressed his outrage at the persecution of political prisoners after a journey to Turkey with **Arthur Miller** in 1985. The following year with his wife and other writers he established the so-called 'June 20th Society', a sometimes derided left-wing discussion group. He has won numerous awards, notably the Berlin Film Festival Silver Bear in 1963, BAFTA awards in 1965 and 1971, the Hamburg Shakespeare Prize in 1970, the Cannes Film Festival Palme d'Or in 1971 and the Commonwealth Award in 1981. He was made a CBE in 1966.

Plays

The Birthday Party and Other Plays [pub. 1960]: *The Room* 1957, *The Birthday Party* **(st)** 1958, **(f)** 1968, *The Dumb Waiter* 1960; *One to Another* [with J. Mortimer, N.F. Simpson] 1959 [pub. 1961]; *A Slight Ache and Other Plays* [pub. 1961]: *Pieces of Eight* **(c)** 1959, *A Slight Ache* **(r)** 1959, **(st)** 1961, *The Dwarfs* [from his novel] **(r)** 1960, **(st)** 1963, *A Night Out* **(r)**, **(tv)** 1960, **(st)** 1961; *The Caretaker* **(st)** 1960, **(f)** 1963; *Tea Party and Other Plays* [pub. 1967]: *Night School* **(tv)** 1960, **(r)** 1966, *Tea Party* **(tv)** 1965, **(st)** 1968, *The Basement* **(tv)** 1967, **(st)** 1968; *The Collection* **(tv)** 1961, **(st)** 1962 [pub. in *Three Plays* 1962; rev. 1978]; *The Compartment* **(f)** [pub. in *Project 1* with S. Beckett, E. Ionesco 1963]; *The Lover* **(st)**, **(tv)** 1963; *The Servant* **(f)** [from R. Maugham's novel] 1963; *The Pumpkin Eater* **(f)** [from P. Mortimer's novel] 1964; *The Quiller Memorandum* **(f)** [from A. Hall's *The Berlin Memorandum*] 1966; *Accident* **(f)** [from N. Mosley's novel] 1967; *The Go-Between* **(f)** [from L.P. Hartley's novel; pub. in *Five Screenplays*] 1971; *The Homecoming* **(st)** 1965, **(f)** 1976; *Landscape and Silence* [pub. 1969]: *Landscape* **(r)** 1968, **(st)** 1969, *Night* 1969, *Silence* 1969; *Old Times* 1971; *Monologue* **(tv)** 1973; *No Man's Land* 1975 [rev. 1991]; *The Proust Screenplay* [with B. Bray, J. Losey pub. 1977]; *Betrayal* **(st)** 1978, **(f)** 1983; *The Hothouse* 1980; *Family Voices* **(r)**, **(st)** 1981; *The French Lieutenant's Woman and Other Screenplays* [pub. 1982]: *The Last Tycoon* [from F.S. Fitzgerald's novel] **(f)** 1976, *Langrishe, Go Down* **(tv)** [from A. Higgins] 1978, *The French Lieutenant's Woman* **(f)** [from J. Fowles's novel] 1981; *A Kind of Alaska* 1982; *Other Places* 1982 [rev. 1984]; *Victoria Station* 1982; *The Big One* **(c)** 1983 [np]; *One for the Road* 1984; *Players* **(r)** 1985; *Mountain Language* 1988; *The Heat of the Day* **(tv)** [from E. Bowen's novel] 1989; *The Comfort of Strangers and Other Screenplays* **(f)** [pub. 1990]: *Turtle Diary* [from R. Hoban] 1985, *Reunion* [from F. Uhlman] 1989, *The Comfort of Strangers* [from I. McEwan's novel] 1990, *Victory* [from J. Conrad's novel] [nd]; *The Handmaid's Tale* **(f)** [from M. Atwood's novel] 1990; *Party Time* 1991; *Plays* [pub. 1991]; *The Trial* **(f)** [from Kafka's novel] 1992 [pub. 1993]; *Moonlight* 1993

Poetry

Poems [ed. A. Clodd] 1968; *I Know the Place* 1979; *Ten Early Poems* 1992

Fiction

The Dwarfs 1990

Edited

New Poems 1967: a PEN Anthology [with J. Fuller, P. Redgrave] 1967; *100 Poems by 100 Poets* [with A. Astbury, G. Godbert] 1986; *99 Poems in Translation* [with A. Astbury, G. Godbert] 1994

Non-Fiction

Mac 1968

Collections

Poems and Prose 1949–1977 1978 [rev. as *Collected Poems and Prose* 1986]; *Complete Works* 1990; *Pinter at Sixty* [ed. K.H. Burkman, J.L. Kundert-Gibbs] 1993

Ruth PITTER 1897–1992

Pitter was the daughter of two elementary-school teachers in the East End of London and was born at Ilford on the eastern fringes of the city. Her childhood there, and at the nearby suburb of Goodmayes, was influenced by two principal factors: the passion for literature inculcated by her parents; and their possession of a weekend cottage in Epping Forest, which gave her the love of nature which infused her poetry. She was educated at elementary school herself, and then at the Coburn School for Girls in Bow.

She began writing poetry in childhood and, when she was only thirteen her first work was published by A.R. Orage in the *New Age*. (When she died in 1992, a last link with that seminal journal was severed.) Orage and his flamboyant mistress, Beatrice Hastings, brought her into the society of writers, and **Hilaire Belloc** became enthusiastic for her work, paying for the printing of her first volumes in the 1920s and writing introductions to them.

Meanwhile, lacking university qualifications, she had spent much of the war working as a clerk in the War Office, and after it went to live in Suffolk where she helped an artistic couple called Jennings run the Walberswick Peasant Pottery Company, painting the pots and discovering a talent for carpentry. They later moved the business to London, and here she played an early role in the life of **George Orwell**, helping him find a room after he resigned from the Indian police in Burma and commenting on his jejune first manuscripts. In 1930, she and her lifelong companion Kathleen O'Hara started a craft business of their own in Chelsea where, in the manner of the Omega Workshops but on a smaller scale, they painted and decorated household objects such as trays. (Early in this occupation, Pitter sustained an eye injury which contributed to total blindness at the end of her life.) The Second World War put a stop to the business, and the two women found work at a crucible factory in Battersea, Pitter's despair there leading to a Christian conversion triggered by **C.S. Lewis**'s talks on the radio. After the war, Pitter attempted to reconstitute her craft business in London, but in 1952 moved to Long Crendon near Aylesbury in Buckinghamshire, where she lived for the rest of her life, initially enjoying some modest fame as a regular contributer to the *Brains Trust* in the 1950s, but becoming increasingly reclusive after the death of her partner in the 1970s.

Her status as a poet (her prose consists only of a few articles) is an interesting one: she published many collections until her muse began to falter from the mid-1960s; was admired by large numbers of writers; and received many honours, culminating in the Companionship of Literature in 1974; but her work, perhaps because it was resolutely attached to traditional forms, seldom attracted exegesis and is relatively little known even to the poetry-reading public. Religion and nature (particularly animals), treated generally with lyric seriousness but sometimes in a vein of rough humour, are the chief themes of a poetry varying greatly in quality but with great substance at its best. *Persephone in Hades* (a long poem), *A Trophy of Arms* and *The Ermine* were volumes of note, and a *Collected Poems* appeared in 1990. She is buried in the churchyard at Long Crendon; Bessie, the doll she had been given at the age of two and which had been her constant companion throughout her life, was buried with her.

Poetry

First Poems 1920; *First and Second Poems* 1927; *Persephone in Hades* 1931; *A Mad Lady's Garland* 1934; *A Trophy of Arms* 1936; *The Spirit Watches* 1939; *The Rude Potato* 1941; *The Bridge* 1945; *Ruth Pitter on Cats* 1947; *The Plain Facts, by a Plain but Amicable Cat* 1948; *The Ermine* 1953; *Still by Choice* 1966; *End of Drought* 1975; *A Heaven to Find* 1987; *Collected Poems* 1990

(Helen) Fiona PITT-KETHLEY 1954–

Pitt-Kethley was born in Edgware, Middlesex, daughter of a journalist and a civil servant, and a descendant of the eighteenth-century hymnist William Williams Pontycelyn. She was educated at Haberdashers' Aske's Girls' School, and at the Chelsea School of Art where she studied painting, graduating in 1976. She had a variety of jobs, teaching English, running an unsuccessful junk shop in Hastings, and as a film extra (she is a member of the Film Artistes Association).

She began publishing poems in magazines such as the *London Review of Books* and the *New Statesman*, and attracted considerable attention with her fifth collection – the first with a major publisher – *Sky Ray Lolly*. Written in a straightforward, earthy language, poems such as 'Bums', 'Wankers' and 'Men' take a sardonic look at contemporary sexual mores. Pitt-Kethley describes herself as a satirist of hypocrisy and has said: 'I castigate myself as much as other people.' Despite a fair amount of publicity and the vitality of her frequent public readings, her subsequent collections have sold only moderately, and she has developed a rather conspiratorial view of the supposed literary establishment. In 1986 she sued the Arts Council for racial discrimination – that year its poetry awards were reserved for writers from ethnic minorities – and in 1995 she wrote to the *London Review of Books* complaining that her publishers had axed their poetry list and railing against 'the Establishment which is so keen to shut me out again'. Other poets may feel that she is not unique in this.

She has edited an anthology of erotic prose and poetry, and *Journeys to the Underworld* is her entertaining account of a trip from Turin to Sicily initially undertaken to investigate the prophetic cult of the ancient Sybil, but rather more detailed in its frank account of her own sexual adventures. The eponymous protagonist of her first novel, *The Misfortunes of Nigel*, a novel about a sex-obsessed, misogynistic, wimpish writer married to a circus artiste, is apparently inspired by the poet **Hugh Williams**, who – so the author claims – failed to appreciate her sufficiently.

Poetry

London 1984; *Rome* 1985; *The Tower of Glass* 1985; *Gesta* 1986; *Sky Ray Lolly* 1986; *Private Parts* 1987; *The Perfect Man* 1989; *Dogs* 1993

Non-Fiction

Journeys to the Underworld 1988; *Too Hot to Handle* 1992; *The Pan Principle* 1995

Fiction

The Misfortunes of Nigel 1991; *The Maiden's Progress* 1992

Edited

The Literary Companion to Sex: an Anthology of Prose and Poetry 1992

David (Robert) PLANTE 1940–

As Plante himself has written, he was born 'in the same bed in which I was conceived' in a Franco-American parish of Providence, Rhode Island. He is the sixth son of Anclet, who worked in a factory which produced industrial tiles, and Albina Bison a housewife. On his father's side, Plante's great-grandmother was a Blackfoot Indian, but it was from his mother that he learned a little about Indian culture. He attended the local parochial Catholic grammar school, Notre Dame de Lourdes, where he was taught by French-Canadian nuns, and where the classes were held in French as well as English. The school was in favour of strict discipline, and Plante has said that he was terrified of the pastor who spoke no English. He was disturbed by repeated stories of how the French martyrs had been tortured and killed by the Indians, and his class gained the impression that because they were Catholics, they would be captured and tortured. These details were fictionalised in his novel *The Catholic*, which explores the emotional and physical response of a young homosexual to an unrequited affair, which is in part conditioned by his Catholic upbringing.

Plante's education continued at La Salle Academy, where he was taught by the Christian Brothers before going to the Jesuit-run Boston College. Later he went to the University of Louvain in Belgium to study philosophy and French literature for a year. When he returned to America, he worked as a researcher for *Hart's Guide to New York*, although he left before it was published to teach English to foreign students. He then taught French in Massachusetts, and while he was there he began to write fiction. When he was twenty-six he went to live in London and has been based there and in Italy ever since, with periods spent back in America at various universities, notably at the University of Tulsa, Oklahoma.

His first novel, *The Ghost of Henry James*, was published in 1970, and applied **Henry James**'s fictional methods to a contemporary story. Subsequent books were similarly experimental. His major achievement, however, is *The Francoeur Novels*, an intense trilogy of novels about a family of working-class Roman Catholic French Canadians, which clearly draws upon the author's memories of his own family and upbringing. The first two novels, *The Family* and *The Woods*, are narrated in the third person, but the last, *The Country*, is narrated in the first person by Daniel Francoeur, the son who leaves the family to go to college and then live abroad. The disintegration of the family is told in a spare, lucid prose, all the more moving for its restraint. 'No moralising, no philosophising, no dramatising of event or even of language,' **Philip Roth** commented of *The Country*, 'and yet Plante has written one of the most harrowing of contemporary novels.'

Of Plante's other novels, which are far better known in the USA (where he received an award from the American Academy of Arts and Letters in 1983) than in Britain, *The Catholic* has gained most attention, partly because of its extended and explicit scenes of homosexual love-making. It is also one of his best novels. He also caused controversy when he wrote a memoir of three *Difficult Women*: **Jean Rhys**, Sonia Orwell and Germaine Greer. Based upon his extensive diaries, the book presented frank portraits of three formidable figures with whom he had been involved. Objections were raised as to the ethics of such a book: in particular, the picture of Rhys in decrepit old age was thought exploitative (he had been helping her to write her memoirs at the time). Less squeamish readers recognised in the book the sort of ruthless honesty that is the hallmark of Plante's fiction. He has also written profiles for the *New Yorker*, and has had short stories published in anthologies, though they have yet to be collected in volume form.

Fiction

The Ghost of Henry James 1970; *Slides* 1971; *Relatives* 1972; *The Darkness of the Body* 1974; *Figures in Bright Air* 1976; *The Francoeur Novels* 1983: *The Family* 1978, *The Woods* 1981, *The Country* 1982; *The Foreigner* 1984; *The Catholic* 1985; *My Mother's Pearl Necklace* 1987; *The Native* 1987; *The Accident* 1991; *Annunciation* 1994

Non-Fiction

Difficult Women: a Memoir of Three 1983

Translations

Andreas Embiricos *Argo* [with N. Stangos] 1967

Alan (Frederick) PLATER 1935–

Plater was born in Jarrow where his father was an engineer, and studied architecture from 1953 to 1957 at King's College, a part of the University of Durham located in Newcastle. He worked in an architect's office in Hull in the late 1950s and qualified, becoming a member of the Royal Institute of British Architects in 1961, but had by this point begun to write for television and radio. His early plays concern northern working-class life.

In 1966 Plater met the songwriter Alex Glasgow with whom he has collaborated on various 'musical documentaries' influenced by Joan Littlewood's theatre. The most celebrated of these, *Close the Coalhouse Door*, depicts episodes from the life of a mining community, and enjoyed considerable success at the Playhouse Theatre, Newcastle, but less when it transferred to London in 1969. *Simon Says!* is a ferocious satire of the British establishment expressive of Plater's strongly left-wing political views. In 1970 he was one of the founders of the Hull Arts Centre (now the Humberside Theatre), and he has remained firmly committed to the regional theatre. His later stage plays include two adaptations from Bill Tidy's 'Fosdyke Saga' cartoon strip; *Rent Party*, a depiction of black life in Harlem in the 1930s first produced at the Theatre Royal, Stratford East; and *Sweet Sorrow*, a celebration of **Philip Larkin** written for the Hull Truck Theatre Company.

Plater is, however, best known as a writer for television, and his prolific output includes eighteen episodes of the police drama *Z-Cars* and thirty of *Softly, Softly*. His numerous adaptations for television include *The Barchester Chronicles* from the novels of Anthony Trollope, *Fortunes of War* from **Olivia Manning**'s novels, and *A Very British Coup* from Chris Mullin's novel, for which he won the 1988 BAFTA Writer's Award. He has also adapted his own trilogy of *Beiderbecke* novels which concern the comic misadventures of a pair of Leeds-based teachers turned amateur detectives.

Plater's novels sometimes feel as though they are television adaptations in the making. His numerous awards include the 1983 Sony Radio Award for *The Journal of Vasilije Bogdanovic*, and the 1987 Broadcasting Guild Award for *Fortunes of War*. He is the President of the Writers' Guild of Great Britain, and has been married twice, with three children from his first marriage.

Plays

Counting the Legs **(r)** 1961; *The Referees* **(tv)** 1961, **(st)** 1963 [np]; *The Smokeless Zone* **(r)** 1961; *The Mating Season* **(r)** 1962, **(st)** 1963 [pub. in *Worth a Hearing* ed. A. Bradley 1967]; *The Rainbow Machine* **(r)** 1962, **(st)** 1963 [np]; *A Smashing Day* **(tv)** 1962, **(st)** [with B. Kingsley, R. Powell] 1965 [np]; *So Long Charlie* **(tv)** 1963; *Ted's Cathedral* 1963 [np]; *A Quiet Night* **(tv)** 1963 [pub. in *Z Cars* ed. M. Marland 1968]; *See the Pretty Lights* **(tv)** 1963, **(st)** 1970 [pub. in *Theatre Choice* ed. M. Marland 1972]; *The Seventh Day of Arthur* **(r)** 1963; *Fred* **(tv)** 1964, **(r)** 1970; *The Nutter* **(tv)** 1965, **(st)** [as *Charlie Came to Our Town*] 1966 [np]; *Excursion* **(r)** 1966 [pub. in *You and Me* ed. A. Bradley 1973]; *The What on the Landing?* **(r)** 1967, **(st)** 1968 [np]; *Close the Coalhouse Door* 1968 [from S. Chaplin; with A. Glasgow] [pub. 1969]; *Hop Step and Jump* 1968 [np]; *On Christmas Day in the Morning* **(tv)** 1968 [pub. in *You and Me* 1973]; *Rest in Peace, Uncle Fred* **(tv)** 1968; *To See How Far It Is* **(tv)** 1968; *Don't Build a Bridge, Drain the River!* [with M. Chapman, M. Waterson] 1970 [np]; *And a Little Love Besides* 1970 [pub. in *You and Me* 1973]; *Simon Says!* [with A. Glasgow] 1970 [np]; *King Billy Vaudeville Show* [with others] 1971 [np]; *Seventeen Per Cent Said Push Off* **(tv)** 1972 [pub. in *You and Me* 1973]; *The Tigers Are Coming – OK?* 1972 [np]; *Tonight We Meet Arthur Pendlebury* **(tv)** 1972 [np]; *Brotherly Love* **(tv)** 1973; *The Slow Stain* **(r)** 1973; *Swallows on the Water* **(st)** 1973, **(r)** 1981 [np]; *When the Reds Go Marching In* 1973 [np]; *Annie Kenney* **(tv)** 1974 [pub. in *Act 3* ed. D. Self, R. Speakman 1979]; *Goldilocks and the Three Bears* **(tv)** 1974; *The Land of Green Ginger* **(tv)** 1974; *The Needle Match* **(tv)** 1974; *The Loner* **(tv)** 1975; *The Stars Look Down* **(tv)** [from A.J. Cronin's novel] 1975; *Tales of Humberside* [with J. Bywater] 1975 [np]; *Trinity Tales* **(st)**, **(tv)** [with A. Glasgow] 1975 [np]; *Willow Cabins* **(tv)** 1975; *5 Days in '55* **(r)** 1976; *Oh No – It's Selwyn Froggit* **(tv)** 1976; *Our Albert* 1976 [np]; *Practical Experience* **(tv)** 1976; *There Are Several Businesses Like Show Business* **(tv)** 1976; *We Are the Masters Now* **(tv)** 1976; *The Bike* **(tv)** 1977; *By Christian Judges Condemned* **(tv)** 1977; *Drums Along the Ginnel* 1977 [np]; *For the Love of Albert* **(tv)** 1977; *The Fosdyke Saga* [with B. Tidy] 1977 [pub. 1978]; *Fosdyke 2* [with B. Tidy] 1977 [np]; *Give us a Kiss, Christabel* **(tv)** 1977; *Middlemen* **(tv)** 1977; *Short Back and Sides* **(tv)** 1977 [pub. in *City Life* ed. D. Self 1980]; *Curriculee, Curricula* **(tv)** [with D. Greenslade] 1978; *Night People* **(tv)** 1978; *The Party of the First Part* **(tv)** 1978; *Well Good Night Then ...* **(st)** 1978, **(r)** [as *Who's Jimmy Dickenson?* 1986 [np]; *The Blacktoft Diaries* **(tv)** 1979; *Flambards* [from K.M. Peyton] **(tv)** 1979; *Reunion* **(tv)** 1979; *Tunes* **(r)** 1979; *The Good Companions* **(tv)** [from J.B. Priestley's novel] 1980; *The Barchester Chronicles* **(tv)** [from Trollope] 1981; *Get Lost* **(tv)** 1981; *On Your Way, Riley!* [with A. Glasgow] 1982 [np]; *Skyhooks* 1982 [np]; *Tolpuddle* **(r)** [with V. Hill] 1982; *Bewitched* **(tv)** [from E. Wharton's novel] 1983; *The Clarion Van* **(tv)** [from D.N. Chew] 1983; *The Consultant* **(tv)** (from J. McNeil's novel *Invitation to Tender*] 1983; *Feet Foremost* **(tv)** [from L.P. Hartley] 1983; *Pride of Our Alley* **(tv)** 1983; *A Foot on the Earth* 1984; *The Solitary Cyclist* **(tv)** [from A. Conan Doyle] 1984; *Thank You, Mrs Clinkscales* **(tv)** 1984; *The Beiderbecke Affair* **(tv)** [from his novel] 1985; *Coming Through* **(tv)** 1985; *A Murder Is Announced* **(tv)** [from A. Christie's story] 1985; *Prez* [with B. Cash] 1985; *Death Is Part of the Process* **(tv)** [from H. Bernstein's novel] 1986; *The Man with the Twisted Lip* **(tv)** [from A. Conan Doyle's story] 1986; *The Beiderbecke Tapes* **(tv)** [from his novel] 1987; *Fortunes of War* **(tv)** [from O. Manning] 1987; *The Beiderbecke Connection* **(tv)** 1988; *A*

Very British Coup **(tv)** [from C. Mullin's novel] 1988; *Campion* **(tv)** [from M. Allingham] 1989; *A Day in the Summer* **(tv)** [from J.L. Carr's novel] 1989; *Rent Party* 1989 [np]; *Going Home* 1990; *Sweet Sorrow* 1990; *I Thought I Heard a Rustling* 1991; *Misterioso* **(tv)** [from his novel] 1991; *Maigret and the Burglar's Wife* **(tv)** [from G. Simenon's novel] 1992; *The Patience of Maigret* **(tv)** [from G. Simenon's novel] 1992; *Selected Exits* [from G. Thomas's *A Few Selected Exits*] **(tv)** 1993; *Doggin' Around* **(tv)** 1994; *Frank Stubbs* **(tv)** 1994; *Oliver's Travels* **(tv)** [from his novel] 1995

Fiction

The Beiderbecke Affair 1985; *The Beiderbecke Tapes* 1986; *Misterioso* 1987; *The Beiderbecke Connection* 1992; *Oliver's Travels* 1994

For Children

The Trouble with Abracadabra 1975

Sylvia PLATH 1932–1963

Plath was born in Boston, Massachusetts, the daughter of German immigrant parents. Her father was a professor of biology at Boston University and a distinguished entomologist specialising in bees, and his protracted illness and death in 1940 exerted a profound influence on the personal mythology of Plath's subsequent poetry. She was educated at Gamaliel Bradford Senior High School (now Wellesley High School) and Smith College where she was befriended by the writer Olive Higgins Prouty. While still at college she published stories and poems in various magazines, and in the summer of 1953 worked on the college editorial board of *Mademoiselle* magazine in New York. Later that year she suffered a mental breakdown and attempted suicide, events which are described in her autobiographical novel, *The Bell Jar* (originally published under the pseudonym Victoria Lucas in 1963). In 1955 she won a Fulbright scholarship and attended Newnham College, Cambridge, and the following year met and married the poet **Ted Hughes**. The couple spent some time in America where Plath's temporary job as a secretary at a Massachusetts psychiatric clinic inspired her short story 'Johnny Panic and the Bible of Dreams'. The first of her two children was born in 1960, and after the publication of her first collection of poetry, *The Colossus*, she and Hughes moved to Devon where she wrote *The Bell Jar* and, in the last troubled year of her life, the poems of *Ariel* and *Winter Trees* which represent the flowering of her art. She committed suicide in London in 1963.

Few lives have been as meticulously documented as Plath's. Her journals are obsessively self-analytical (perhaps reflecting her years of psychiatric treatment) while her letters to her mother (collected in *Letters Home*) express an outward buoyancy at variance with the journals. Since her death, there have been several biographies and a good deal of vexed dispute about Hughes's behaviour in the marriage and his editing of her *Collected Poems* and *Journals*. Plath has become something of an icon for the feminist movement, but it is important not to confuse the biographical facts with the personal mythology of the poems. For example, the oppressive father-figure in a poem such as 'Daddy' ('Every woman adores a Fascist, / The boot in the face, the brute / Brute heart of a brute like you') yokes the child's rage at her father's loss with the suffering of the Holocaust, but has absolutely nothing to do with the man Otto Plath. Plath's *Three Women* was broadcast as a radio play by the BBC in 1962, and her *Collected Poems* won a posthumous Pulitzer Prize in 1982.

Poetry

The Colossus & Other Poems 1960; *Ariel* 1965; *Uncollected Poems* 1965; *Crossing the Water* 1971; *Crystal Gazer & Other Poems* 1971; *Fiesta Melons* 1971; *Lyonesse: Hitherto Uncollected Poems* 1971; *Winter Trees* 1971; *Pursuit* 1973; *Collected Poems* 1981

Fiction

The Bell Jar [as 'Victoria Lucas'] 1963

Plays

Three Women 1962

For Children

The Bed Book 1976

Non-Fiction

Letters Home: Correspondence 1950–1963 1975; *The Journals of Sylvia Plath* [ed. F. McCullough] 1982

Edited

American Poetry Now: a Selection of the Best Poems by Modern American Writers 1961

Collections

Johnny Panic and the Bible of Dreams, and Other Prose Writings 1977

Biography

Sylvia Plath by Lindsay W. Wagner-Martin 1987; *Bitter Fame* by Anne Stevenson 1989; *The Death and Life of Sylvia Plath* by Ronald Hayman 1991; *the Haunting of Sylvia Plath* by Jacqueline Rose 1991; *The Silent Woman* by Janet Macolm 1994

William (Charles Franklyn) PLOMER 1903–1973

The son of a minor civil servant with the Department of Native Affairs, who had previously had an erratic career as a soldier, policeman, journalist and sheep-farmer, Plomer was born in Pietersburg, South Africa. He spent his childhood shuttling between Africa and England, and went to preparatory schools in Johannesburg and Kent before being sent to Rugby, the war-time rigours of which he recalled in his contribution to **Graham Greene**'s *The Old School* (1934). On the boat to England in 1914 he was seduced by a young steward, an initiation for which he later said he was 'profoundly grateful'. He returned to South Africa in 1919 to complete his education at St John's College, Johannesburg, and during his last year at the school he started to write poetry.

His first job was as a veld sheep-farmer, and from this remote outpost he sent poems to **Harold Monro** back in England; Monro was encouraging but did not offer to publish any. By the summer of 1922 it was clear that Plomer did not have the makings of a sheep-farmer, although this experience furnished him with material for his first novel. He ran a trading station with his father, and began submitting articles about Zulu life to newspapers. His first novel, *Turbott Wolfe*, caused a scandal when it was published in 1926, since it dealt with miscegenation – 'A Nasty Book on a Nasty Subject', as the Natal *Advertiser* characterised it. Undeterred by this controversy, that same year he joined **Roy Campbell** as an editor on the satirical magazine *Voorslag* (or 'Whiplash'), to which they shortly recruited **Laurens van der Post**. The magazine proved too outspoken for the proprietors, and Plomer moved on to Japan, where he got a job as a tutor. His first volume of stories, *I Speak of Africa*, was published in 1927, along with his first volume of verse, *Notes for Poems*. His experiences in Japan provided material for a second volume of stories, *Paper Houses*, which was published in 1929, the year he arrived in London, 'a young man of English origins and Afro-Asian conditionings'. The Japanese years also formed the background to his discreetly homosexual novel *Sado*, originally intended as part of a much longer autobiographical novel.

In England he changed the pronunciation of his surname to 'Plumer'. After staying briefly with his parents in Pinner, Middlesex, he found lodgings in Bayswater, where his landlady was shortly afterwards murdered by her razor-wielding common-law husband. Plomer told his publisher **Virginia Woolf** that he had been obliged to clean up 'scraps of brain from the carpet', and this crime became the subject of his next novel, *The Case Is Altered*. By this time, he had become a well-known figure in literary circles and numbered **E.M. Forster**, **Christopher Isherwood**, **W.H. Auden**, **J.R. Ackerley** and **Stephen Spender** amongst his close friends. In 1937 he became a reader for the publisher Jonathan Cape, where during his thirty years with the firm he 'discovered' **Derek Walcott** ('the best poet in English of his generation'), but rejected **Barbara Pym**'s *An Unsuitable Attachment* (eventually published in 1982). His long report on the manuscript of **Malcolm Lowry**'s *Under the Volcano* (1947) prompted the author's famous 15,000-word defence of the novel. Other then unknown authors whose work he recommended for publication include **Alan Paton**, **John Fowles**, **Stevie Smith** and **Ian Fleming**. One major discovery was the diaries of the Victorian clergyman Francis Kilvert, which give a fascinating picture of rural life on the Welsh borders in the 1870s. Plomer undertook to edit the diaries himself, and their success gave rise to the Kilvert Society, which is still flourishing.

He continued to publish volumes of poetry and became well known for reviving the ballad form. He found his subjects – which were often of a macabre nature – in newspaper reports, advertisements and anecdotes. 'The Naiad of Ostend', for example, was suggested by an incident in the memoirs of Trollope's brother, while 'Headline History' was entirely made up of (probably invented) newspaper headlines. He also wrote fine lyric poetry and had a fruitful collaboration with Benjamin Britten, providing libretti for amongst others *Three Church Parables*, influenced by Noh drama, and *Gloriana*, an opera about Queen Elizabeth I.

During the war he became a regular contributor to **John Lehmann**'s *Penguin New Writing* under the pseudonym of Robert Pagan. He worked in naval intelligence and narrowly avoided prosecution for soliciting a sailor (it seems that Ian Fleming, who had recruited him to the job, intervened). In 1947 he met Charles Erdmann, with whom he was to spend the remainder of his life, much of it on the Sussex coast in 'Bungalow-land'. He wrote two volumes of discreet but very amusing memoirs, as well as further volumes of poetry and fiction and a biography of Cecil Rhodes. He was a natural wit and punster (translating 'a near-miss' in the war as 'une demi-vierge', for instance) and delighted friends by sending them Victorian and Edwardian postcards with absurd messages in his immaculate calligraphic handwriting.

Poetry

Notes for Poems 1927; *The Family Tree and Other Poems* 1929; *The Fivefold Screen* 1932; *Visiting the Caves* 1936; *Selected Poems* 1940; *In a Bombed House, 1941: Elegy in Memory of Anthony Butts* 1942; *The Dorking Thigh and Other Satires* 1945; *Borderline Ballads* 1955; *A Shot in the Park* 1955; *A Choice of Ballads* 1960;

Collected Poems 1960; *Conversations with My Younger Self* 1963; *Taste and Remember* 1966; *The Planes of Bedford Square* 1971; *Celebrations* 1972; *Collected Poems* 1973

Fiction

Turbott Wolfe 1926; *I Speak of Africa* **(s)** 1927; *Paper Houses* **(s)** 1929; *Sado* 1931 [US *They Never Came Back* 1932]; *The Case Is Altered* 1932; *The Child of Queen Victoria* **(s)** 1933; *The Invaders* 1934; *Four Countries* **(s)** 1949; *Museum Pieces* 1952; *A Brutal Sentimentalist* **(s)** [ed. E. Sano] 1969

Non-Fiction

Cecil Rhodes 1933; *Ali the Lion* 1936 [as *The Diamond of Jannina* 1970]; *Double Lives* **(a)** 1943; *At Home* **(a)** 1958

Plays

With Benjamin Britten: *Gloriana* **(l)** 1953; *Three Parables for Church Performance*: *Curlew River* **(l)** 1964, *The Burning Fiery Furnace* **(l)** 1966, *The Prodigal Son* **(l)** 1968

Edited

Haruko Ichikawa *A Japanese Lady In Europe* 1937; Francis Kilvert *Kilvert's Diary* 3 vols 1938–40 [rev. 1961]; Herman Melville: *Selected Poems* 1943, *Billy Budd* 1946; Anthony Butts *Curious Relations* [as 'William D'Arfey'] 1945; *New Poems 1961* [with H. Corke, A. Thwaite] 1961; *A Message in Code: the Diary of Richard Rumbold 1932–1960* 1964

Translations

Ingrid Jonker *Selected Poems* [with J. Cope] 1968

For Children

The Butterfly Ball and the Grasshopper's Feast [with A. Aldridge] 1973

Collections

Electric Delights [ed. R. Hart-Davies] 1978

Stephen POLIAKOFF 1952–

Poliakoff was born in London of Russian-Jewish stock, and despite a traditional English education at Westminster School and King's College, Cambridge, he has described himself as 'half Russian and wholly Jewish'. He began writing at an early age, having his first play, *Granny*, produced in 1969 while he was still at school, and has been a prolific playwright ever since, with eleven plays produced over the next six years.

His first big success came in 1975 with *Hitting Town*, in which a brother and sister spend a desolating night on the town, and this was thematically linked with his next play, *City Sugar*, produced the same year, in which an all-night radio phone-in programme acts as a linking and dominant image of urban isolation. *Strawberry Fields* continued to pursue this bleak vision of late twentieth-century life, and was the stage equivalent of a nihilist road movie. He once again dealt with the alienation of the young in his first film, *Runners* (directed by Charles Sturridge in 1983), in which parents attempted to trace children who had gone missing in London.

Travel was used as a metaphor in Poliakoff's first play for television, *Caught on a Train*, which also drew upon his Russian roots in its portrait of a domineering woman (played by Peggy Ashcroft) befriending and exploiting a young man as they travel across Europe. This is perhaps Poliakoff's best-known work and won a BAFTA Award. Ashcroft also took the lead role in *She's Been Away*, a moving play which addressed the problems of old age and failing faculties. Poliakoff's family background also informed his stage plays *Breaking the Silence* and *Playing with Trains*. His grandfather, Joseph Poliakoff, had been a distinguished inventor with the Marconi company, and the central character in *Breaking the Silence*, which is set in post-revolutionary Russia, was inspired by him. The later play explores the dual themes of why the British fail to properly exploit the skills of their inventors, allowing such enterprise to form part of the brain drain, and the effect of a man's genius upon his children.

Like his contemporary **Anthony Minghella**, Poliakoff became a director of his own screenplay when he made *Close My Eyes*, a stylish and sexy film which returns to the theme of sibling incest he first introduced in *Hitting Town*.

Plays

Granny 1969 [np]; *Bambi Ramm* 1970 [np]; *A Day with My Sister* 1971 [np]; *Lay-By* [with others] 1971 [pub. 1972]; *Pretty Boy* 1972 [np]; *Theatre Outside* 1973 [np]; *Berlin Days* 1973 [np]; *The Carnation Gang* 1974 [np]; *Clever Soldiers* 1974 [np]; *Heroes* 1975 [np]; *City Sugar* 1975 and *Hitting Town* 1975 [pub. 1976; rev. 1978]; *Strawberry Fields* 1977; *Stronger Than the Sun* **(tv)** 1977; *Shout Across the River* 1978 [pub. 1979]; *American Days* 1979; *Bloody Kids* **(tv)** 1980; *Caught on a Train* **(tv)** 1980 and *Favourite Nights* **(st)** 1981 [pub. 1982]; *The Summer Party* 1980; *Soft Targets* **(tv)** 1982 and *Runners* **(f)** 1983 [pub. 1984]; *Breaking the Silence* 1984; *Coming in to Land* 1987; *Hidden City* **(f)** 1987, **(tv)** 1989; *Playing with Trains* 1989; *She's Been Away* **(tv)** 1989; *Close My Eyes* **(f)** 1991; *Sienna Red* 1992

Katherine Anne (Maria Veronica Callista Russell) PORTER 1890–1980

Born at Indian Creek, Texas, Porter was the daughter of a farmer, and although her life was spent in many parts of the USA with periods in Europe, she remained distinctively a writer of the American South. She regarded writing as her vocation from an early age but, after education at convent school, she did many hack jobs in journalism and editing to support herself. She worked for a while as a reporter on the *Rocky Mountain News* in Denver, and then lived in Mexico.

From the late 1920s her subtle but powerful short stories began to attract attention in magazines, and in 1930 her first collection, *Flowering Judas*, was published: it was one of only a few slender volumes of short stories on which her great reputation largely rests, and which have given rise to frequent comparisons with **Katherine Mansfield**. Another outstanding volume was *Pale Horse, Pale Rider*, in which the novella *Noon Wine*, one of her finest works, appeared. During the 1930s she used a Guggenheim Fellowship to spend some years in Europe and married twice, being twice divorced.

The consensus of critical opinion is that her work declined in quality from the 1940s. In 1962, she published her only novel, *Ship of Fools*, a symbolic story of a sea voyage from Mexico to Germany during the rise of Nazism. Although it had great popular success, it has not been critically regarded as equal to her short stories. It had been promised as forthcoming from about 1940, and another much advertised book, a study of the Puritan Cotton Mather, on which she had allegedly been working since 1928, never appeared. In later life Porter shifted restlessly around the USA, but eventually died in Silver Spring, Maryland. Her final work was a study of the Sacco–Vanzetti case.

Fiction
Flowering Judas (s) 1930 [rev. as *Flowering Judas and Other Stories* 1935]; *Hacienda* 1934; *Noon Wine* 1937; *Pale Horse, Pale Rider* (s) 1939; *The Leaning Tower and Other Stories* (s) 1944; *Old Order* (s) 1955; *Ship of Fools* 1962; *A Christmas Story* (s) 1965

Plays
Noon Wine (tv) 1967

Non-Fiction
My Chinese Marriage 1921; *Outline of Mexican Popular Arts and Crafts* 1922; *The Never Ending Wrong* 1977

Edited
What Price Marriage? 1927; *Katherine Anne Porter's French Songbook* 1933

Translations
Fernandez de Lizardi *The Itching Parrot* 1942

Collections
The Days Before: Collected Essays and Occasional Writings 1952 [rev. 1970]

Biography
Katherine Anne Porter by Joan Givner 1982

Peter (Neville Frederick) PORTER 1929–

An Australian of mixed Scots and English ancestry, Porter has lived in England since 1951 – long enough to justify his own description of himself as 'an English poet with an Australian tang'. His memories of Australia are not sentimental: his childhood in Depression-era Brisbane and education in provincial Queensland could not have been more distant from the 'Australian dream' as lived in Sydney. At the large state school he attended half the pupils were forced by poverty to go barefoot. His mother died when he was nine, an experience which provided him with his own 'private iconography of hell'. The sense of a powerful, imposed isolation during this period of his life had the added effect of allowing him to develop, at least temporarily, a completely hermetic idea of himself as a poet, his first attempts at which were inspired by a self-confessed fantasy-identification with the great artists of the past.

The decision to leave Australia – he had left school aged seventeen to work on a provincial newspaper – was made for 'aesthetic reasons'. He felt determined to learn, despite the failure of his first compositions as pointed out by friends and acknowledged by himself, 'the hard lessons of technique and competition' and to prove his seriousness as a poet in an environment where poetry could be taken seriously. In London he worked as a bookseller and sometime clerk and joined the loosely assembled 'Group' meeting to discuss each other's work at the house of **Edward Lucie-Smith**. His first poem was published in 1957 and his first book, *Once Bitten, Twice Bitten*, in 1961, the year of his marriage. He has two daughters. Throughout the 1960s he worked in advertising.

His appearance in the second volume of the Penguin Modern Poets series in 1962 brought him to the attention of a wider audience both as a satirist of the 'thingyness' of modern life and as an elegist. Throughout the decade he was often to battle with modernism. 'There Are Too Many of Us' in *The Last of England* juxtaposes the 'world-eating evangelizers' of junk art and concrete poetry with 'the god of death who is immortal and cannot read'. Equally, his translations of Martial, collected in *After Martial*, demonstrate his principal preoccupation to be with the here-and-now even when looked at obliquely, through the distant past. The death of his wife in 1974 led him to write a series of reflections on the loss, *The Cost of Seriousness*.

Since leaving advertising, he has worked as editor, critic, journalist, poet, lecturer, radio dramatist and broadcaster. His home in Chelsea has long been a gathering place for London literati. A collected edition of his poetry was published in 1983 and reprinted in 1988, a success duplicated by the publication of a selected edition in 1989.

Poetry
Once Bitten, Twice Bitten 1961; *Penguin Modern Poets 2* [with K. Amis, D. Moraes] 1962; *Poems Ancient and Modern* 1964; *Solemn Adultery at Breakfast Creek* [with M. Jessett] 1968; *Words without Music* 1968; *A Porter Folio* 1969; *The Last of England* 1970; *Preaching to the Converted* 1972; *Jonah* [with A. Boyd] 1973; *A Share of the Market* 1973; *The Lady and the Unicorn* [with A. Boyd] 1975; *Living in a Calm Country* 1975; *Ariadne on Naxos* 1976; *The Cost of Seriousness* 1978; *Les Très Riches Heures* 1978; *English Subtitles* 1981; *The Animal Programme* 1982; *A Celebratory Ode* 1982; *Collected Poems* 1983; *Fast Forward* 1984; *Narcissus* [with A. Boyd] 1984; *The Run of Your Father's Library* 1984; *Machines* 1986; *The Automatic Oracle* 1987; *Mars* [with A. Boyd] 1988; *A Porter Selected* 1989; *Possible Worlds* 1989; *The Chair of Babel* 1992; *Millennial Fables* 1994

Edited
A Choice of Pope's Verse 1971; *New Poems 1971–72* 1972; *The English Poets: From Chaucer to Edward Thomas* [with A. Thwaite] 1974; *New Poetry 1* [with C. Osborne] 1975; *Poetry Supplement* 1980; *The Gregory Awards Anthology 1980* [with H. Sergeant] 1981; *Thomas Hardy* 1981; *The Faber Book of Modern Verse* 1982; *Christina Rossetti* 1986; *William Blake* 1986; *John Donne* 1988; *Martin Bell: Complete Poems* 1988; *William Shakespeare* 1988; *Lord Byron* [US *Byron*] 1989; *The Fate of Vultures: New Poetry of Africa* [K. Anyidoho, M. Zimunya] 1989; *William Butler Yeats* 1990; *Percy Bysshe Shelley* 1991; *Elizabeth Barrett Browning* 1992; *Robert Burns* 1992; *The Romantic Poets* [with G. Moore] 1992; *Robert Browning* 1993; *George Herbert* 1994; *Samuel Taylor Coleridge* 1994

Non-Fiction
Roloff Beny in Italy [with A. Thwaite] 1974; *The Shape of Poetry and the Shape of Music* 1980; *Sydney* 1980

Plays
The Siege Of Munster (r) 1971; *The Children's Crusade* (r) 1973; *All He Brought Back from the Dream* (r) 1978

For Children
Colour Me Cornwall (l) 1973; *Konrad of the Mountains* (l), (tv) [with C. Blyton] 1967 [pub. 1975]

Translations
Epigrams by Martial 1971; *After Martial* 1972; *Michelangelo: Life, Letters and Poetry* [with G. Bull] 1987

Chaim (Icyck) POTOK 1929–

Potok was born Hermann Harold Potok in New York City of Polish immigrant parents, was educated at Yeshiva University and the Jewish Seminary of America, being ordained rabbi in 1954 and gaining a PhD from the University of Pennsylvania in 1965. He was a chaplain with the US army from 1955 to 1957, more than fifteen months of his service being in Korea with a front-line medical battalion and an engineering combat battalion. He has been an instructor at the University of Judaism, Los Angeles, on the faculty of the Teachers' Institute, Jewish Theological Seminary, editor-in-chief of the Jewish Publication Society of America and chairman of its publications committee. In 1958 he married, and he has two daughters and a son.

The setting of his first novel, *The Chosen* (1967), is the tightly knit one of Hasidic Jews in Brooklyn whose uncompromising intensity divides them from the merely orthodox; even their *yeshiva* baseball team 'don't play to win. They play like it's the first of the Ten Commandments'. The friendship of two boys, both sons of rabbis, one liberal the other orthodox, is threatened by bitter family differences. In its sequel, *The Promise*, Reuven Malter from the liberal family is studying to be a rabbi, and is fiercely confronted by an orthodox teacher, while Danny Saunders has broken away from the destiny his conservative father had chosen for him. All Potok's novels have Jewish themes, but these two in particular clearly draw upon the author's own experiences as the liberal child of orthodox parents. Praised by many critics in Britain, *The Promise* was a Literary Guild Choice in America and achieved a first printing of 100,000 copies there. *My Name Is Asher Lev* is the story of the development of an artist, despite the fact that art is frowned on by his family ('Jews don't draw'); he produces two great crucifixions, for him the ultimate image of suffering, but they depict his own mother on the cross, and this blasphemy leads to his being sent into exile in Paris by his father's sect. Seventeen years later Potok wrote a sequel, *The Gift of Asher Lev*. He has also written non-fiction on the same theme, including *Wanderings: Chaim Potok's History of the Jews*.

Fiction
The Chosen 1967; *The Promise* 1970; *My Name Is Asher Lev* 1972; *In the Beginning* 1976; *The Book*

of Lights 1982; *Davita's Harp* 1985; *The Gift of Asher Lev* 1990; *I Am the Clay* 1992; *The Tree of Here* 1993; *The Sky of Now* 1994

Non-Fiction

Jewish Ethics 14 vols 1964–9; *The Jew Confronts Himself in American Literature* 1975; *Wanderings* 1978; *Ethical Living for a Modern World* 1985; *Tobiasse* 1986

(Helen) Beatrix POTTER 1866–1943

Born in South Kensington, London, the only daughter of Rupert Potter, a wealthy *rentier* whose money derived from the Lancashire cotton industry (as did his wife's), Potter passed a lonely and restricted childhood among governesses in her nursery. She found solace in the companionship of her brother Bertram, her love of animals – many of which she kept as pets and which she learnt to draw – and her attachment to the Lake District, where the family passed summer holidays. She never went to school, and as a young woman continued to live under the parental roof. In 1890 she published a small book of animal drawings called *A Happy Pair* accompanied by verses by Frederic Weatherley, who later wrote the song 'Roses of Picardy': she signed herself H.B.P.

In 1893 she wrote a letter to a child, Noël Moore, containing the first version of *The Tale of Peter Rabbit*, which was privately printed in an edition of 250 copies in 1901; the following year it was published by Frederick Warne and Co., who were to publish all her books for young children. She became attached to one of the publishers, Norman Warne, and in 1905 they became engaged to be married, but he died of leukaemia only a month later. In the same year she bought a farm in her beloved village of Sawrey in the Lake District, where she increasingly spent her time.

The years from 1906 to 1913 were her most productive period, when many of her masterpieces for children, such as *The Tale of Jemima Puddle-Duck* and *The Tale of Mr Tod*, were published. All her books concern animals, lovingly observed, recognisably feral despite their human dress and concern for social niceties, whose adventures are described in simple but witty and ironic prose. Potter oversaw the production and design of the books, insisting that they should be in a small format, for small hands to hold, with only a few (well-chosen) words on each page. She was never shy of introducing her readers to a sophisticated vocabulary: 'Children like a fine word occasionally,' she rightly asserted when Warne had queried 'soporific'. Her exquisite watercolour illustrations are an integral part of the book, often supplying additional information. **Graham Greene** once described her as the author of some of the century's best novels, and the stylish economy of her prose has earned her the admiration of many authors who themselves write for adults. An entire industry has sprung up around her small oeuvre: pottery, tea-towels, soft toys – all of which Potter herself, an astute businesswoman, would have endorsed, as she endorsed merchandising in her lifetime – as well as some lamentable cartoon films and new, large-format editions of her books, which she would deplore.

In 1913, while engaged in buying a larger farmhouse in Sawrey, she married a local solicitor, William Heelis. In 1923 she bought a substantial sheep farm, and she passed the rest of her life as a formidable and increasingly crusty countrywoman, reacting with extreme suspicion to English admirers of her work (her future biographer Margaret Lane was treated to 'the rudest letter I have ever received in my life'), although welcoming Americans. Her work deteriorated with her eyesight after 1918 and petered out around 1930. In 1966, the voluminous diary she had kept in code from the age of fifteen was published, after Leslie Linder had broken the code. At her death, she left several thousand acres of land, including Hill Top Farm, the setting of several of her books, to the National Trust.

For Children

A Happy Pair [with F. Weatherley] 1890; *The Tale of Peter Rabbit* 1901; *The Tailor of Gloucester* 1902; *The Tale of Squirrel Nutkin* 1903; *The Tale of Benjamin Bunny* 1904; *The Tale of Two Bad Mice* 1904; *The Pie and the Patty-Pan* 1905; *The Tale of Mrs Tiggy-Winkle* 1905; *The Story of a Fierce Bad Rabbit* 1906; *The Story of Miss Moppet* 1906; *The Tale of Mr Jeremy Fisher* 1906; *The Tale of Tom Kitten* 1907; *The Roly-Poly Pudding* 1908 [rev. as *The Tale of Samuel Whiskers* 1926]; *The Tale of Jemima Puddle-Duck* 1908; *The Tale of Ginger and Pickles* 1909; *The Tale of the Flopsy Bunnies* 1909; *The Tale of Mrs Tittlemouse* 1910; *Peter Rabbit's Painting Book* 1911; *The Tale of Timmy Tiptoes* 1911; *The Tale of Mr Tod* 1912; *The Tale of Pigling Bland* 1913; *Appley Dapply's Nursery Rhymes* 1917; *Tom Kitten's Painting Book* 1917; *The Tale of Johnny Town-Mouse* 1918; *Cecily Parsley's Nursery Rhymes* 1922; *Jemima Puddle-Duck's Painting Book* 1925; *Peter Rabbit's Almanac for 1929* 1929; *The Fairy Caravan* 1929; *The Tale of Little Pig Robinson* 1930; *Sister Anne* 1932; *Wag-by-Wall* 1944; *The Tale of the Faithful Dove* 1955; *The Sly Old Cat* 1971

Non-Fiction

The Journal of Beatrix Potter, 1881–1897 [ed. L. Linder] 1966 [rev. 1989]; *Beatrix Potter's*

Americans, Selected Letters 1982; *Beatrix Potter's Letters* [ed. J. Taylor] 1989

Collections

A History of the Writings of Beatrix Potter [ed. L. Linder] 1971

Biography

The Tale of Beatrix Potter by Margaret Lane 1946

Dennis (Christopher George) POTTER 1935–1994

Potter was born in the Forest of Dean, Gloucestershire, where his father was a miner, and brought up in a large extended family. He was educated at Bell's Grammar School, Coleford, and after doing national service read philosophy, politics and economics at New College, Oxford, graduating in 1959. While at Oxford he edited the university magazine *Isis* and appeared in a BBC documentary on class which gave him his first taste of notoriety when *Reynolds News* ran a story headlined 'MINER'S SON AT OXFORD ASHAMED OF HOME'. Undeterred, Potter joined the BBC as a graduate trainee and made a documentary, *Between Two Rivers*, which explored the same subject; in the same year he published *The Glittering Coffin*, a polemical attack on what he saw as the post-Suez national malaise.

In 1961, while working as a reporter on the *Daily Herald*, he suffered the first attack of psoriatic arthropathy which left him crippled and drug-dependent for much of the rest of his life. He worked as a television critic, bringing to the genre an astringent serious-mindedness, and during a period of remission from illness was an unsuccessful Labour candidate in the 1964 General Election, an experience he drew on for his fourth television play, *Vote Vote Vote for Nigel Barton*.

Potter was one of the first playwrights to write almost exclusively for television, and produced some of the medium's most innovative work. He repeatedly addressed the themes of sexuality, religion, politics and morality in a manner which frequently provoked controversy. *Son of Man* portrayed Christ as an earthy radical, while *Brimstone and Treacle* was banned by the BBC in 1976. Eventually broadcast in 1987, the play depicts a devil-like figure raping a brain-damaged girl and thereby effecting her recovery, and seems to suggest that good can paradoxically arise out of evil. *Blue Remembered Hills* is a powerful evocation of childhood in the Forest of Dean (a locale which recurs in Potter's work) which raises interesting psychological questions by casting adult actors as the children. Potter's finest work is perhaps the six-part 1978 series *Pennies from Heaven*, which depicts the dreams and romantic entanglements of sheet-music salesman Arthur Parker during the Depression years of the 1930s. Through the device of having the characters miming to the syrupy romantic songs of the period, Potter explores the relationship between personal and political deception and self-deception. He repeated the trick rather less successfully in his 1993 series on the post-Suez period, *Lipstick on Your Collar*.

His most explicitly autobiographical play, *The Singing Detective*, explores the past and fantasy life of Marlow, a writer of cheap detective fiction, as he lies in hospital paralysed with psoriasis. A scene in which the boy Marlow witnesses his mother's adultery in the Forest of Dean once again infuriated such guardians of public morality as Mary Whitehouse. His 1989 series *Blackeyes*, directed and narrated by Potter, is a study of sexual exploitation and pornography in which many critics detected a whiff of authorial lip-smacking in the camera's lingering appreciation of the model Jessica, and it led to Potter being dubbed 'Dirty Den' by the *Sun*.

Potter wrote four novels, and his adaptations for television included **Thomas Hardy**'s *The Mayor of Casterbridge*, **F. Scott Fitzgerald**'s *Tender Is the Night* and **Angus Wilson**'s *Late Call*. He also wrote the screenplays for a rather disappointing 1981 film version of *Pennies from Heaven* and for the 1983 film of Martin Cruz Smith's thriller *Gorky Park*. As a consequence of his drug treatment, Potter was diagnosed with incurable cancer in 1994 and, two months before his death, gave a long and moving television interview to his friend **Melvyn Bragg** in which he bemoaned the commercialisation and deterioration of television. This programme, which amounted to a final testament, won a BAFTA Award in 1995, the year that two last plays, *Cold Lazarus* and *Karaoke*, went into production. He was married in 1959, and had three children. He suffered the death of his wife only nine days before his own.

Plays

Alice **(tv)** 1965; *The Confidence Course* **(tv)** 1965; *The Nigel Barton Plays* [pub. 1968]: *Stand Up Nigel Barton* **(tv)** 1965, *Vote Vote Vote for Nigel Barton* **(tv)** 1965, **(st)** 1968; *Emergency-Ward 9* **(tv)** 1966; *Where the Buffalo Roam* **(tv)** 1966; *Message for Posterity* **(tv)** 1967; *A Beast with Two Backs* **(tv)** 1968; *The Bonegrinder* **(tv)** 1968; *Shaggy Dog* **(tv)** 1968; *Moonlight on the Highway* **(tv)** 1969; *Son of Man* **(st)**, **(tv)** 1969; *Angels Are So Few* **(tv)** 1970; *Lay Down Your Arms* **(tv)** 1970; *Casanova* **(tv)** 1971; *Paper Roses* **(tv)** 1971; *Traitor* **(tv)** 1971; *Follow the Yellow Brick Road* **(tv)** 1972 [pub. in *The Television Dramatist* ed. R. Muller 1973]; *Only Make Believe* **(tv)** 1973; *A Tragedy of Two Ambitions* **(tv)** [from

Hardy] 1973; *Joe's Ark* **(tv)** 1974, *Blue Remembered Hills* **(tv)** 1979 and *Cream in My Coffee* **(tv)** 1980 [pub. in *Waiting for the Boat* 1984]; *Schmoedipus* **(tv)** 1974; *Late Call* **(tv)** [from A. Wilson's novel] 1975; *Double Dare* **(tv)** 1976; *Where Adam Stood* **(tv)** [from E. Gosse's novel] 1976; *Brimstone and Treacle* **(st)** 1978, **(f)** 1982, **(tv)** 1987 [pub. 1978]; *The Mayor of Casterbridge* [from Hardy's novel] **(tv)** 1978; *Pennies from Heaven* **(tv)** 1978, **(f)** 1981 [pub. 1981]; *Blade on the Feather* **(tv)** 1980; *Rain on the Roof* **(tv)** 1980; *Sufficient Carbohydrate* **(st)** 1983, **(tv)** [as *Visitors*] 1987; *Gorky Park* [from M. Cruz Smith's novel] **(f)** 1984; *Dreamchild* **(f)** 1985; *Tender Is the Night* **(tv)** [from F.S. Fitzgerald's novel] 1985; *The Singing Detective* **(tv)** 1986; *Track 29* **(f)** 1988; *Christabel* **(tv)** [from C. Bielenberg's *The Past Is Myself*] 1988; *Blackeyes* **(tv)** [from his novel] 1989; *Secret Friends* **(tv)** 1991; *Lipstick on Your Collar* **(tv)** 1993; *Seeing the Blossom* 1994; *Cold Lazarus* 1995; *Karaoke* 1995

Fiction

Hide and Seek 1973; *Pennies from Heaven* [from his tv series] 1981; *Ticket to Ride* 1986; *Blackeyes* 1987

Non-Fiction

The Glittering Coffin 1960; *The Changing Forest* 1962; *Potter on Potter* [ed. G. Fuller] 1993

Ezra (Weston Loomis) POUND 1885–1972

Pound was born in Hailey, Idaho, and brought up from the age of four in Wyncote, Philadelphia, where his father became assistant assayer for the US Mint. He spent two years at the University of Pennsylvania where he befriended the young **William Carlos Williams**, and studied Anglo-Saxon and the Romance languages at Hamilton College, New York, from 1903 to 1906. His lifelong contempt for American academia seems to have originated from 1907 when his teaching career at Wabash College in Indiana was cut short after he had entertained an actress in his rooms.

He sailed for Europe in 1908 and in Venice published his first collection of poems, *A Lume Spento* (With Tapers Quenched). In London he lectured and wrote while vigorously cultivating the acquaintance of T.E. Hulme, **Ford Madox Ford**, **W.B. Yeats** and **Richard Aldington** (who described Pound as 'a small but persistent volcano in the dim levels of London literary society'). With Aldington and **H.D.** (Hilda Doolittle) he founded the Imagist school, and with **Wyndham Lewis** and the sculptor Henri Gaudier-Brzeska he founded Vorticism and published two issues of its manifesto, *Blast*, in 1914 and 1915. Pound was a selfless champion of Lewis, **T.S. Eliot** and **James Joyce**, and arranged the publication of their work in the *Egoist* and Harriet Monroe's *Poetry*. From this period date two of Pound's most widely read poems, 'Homage to Sextus Propertius' and 'Mauberly (1920)', and his versions from the Chinese, *Cathay*. In 1914 he married the artist Dorothy Shakespear with whom he had a son, and in 1922 began a lifelong relationship with the violinist Olga Rudge with whom he had a daughter. Shakespear seems to have tolerated this situation until the late 1960s when she was effectively estranged from Pound.

In Paris in the early 1920s Pound met **Ernest Hemingway**, **Gertrude Stein** and **e e cummings**, and undertook the editorial work which gave form to Eliot's *The Waste Land* (which is dedicated to him). In 1922, with characteristic aplomb, he announced that the Christian era was over and that the 'Pound Era' had begun. *The Pound Era* is the title of Hugh Kenner's groundbreaking 1972 study of the poet and his influence on modernism. In 1924 he moved to Rapallo in Italy where has interests turned increasingly to economics and C.H. Douglas's theory of social credit. His hatred for American capitalism and what he saw as usury contributed to his anti-Semitism and avowed admiration for Mussolini whom he met in 1933. From 1941 to 1943 he broadcast pro-fascist propaganda on Rome radio, and was arrested for treason by the partisans in 1945 and held in an American disciplinary centre in Pisa. He was subsequently returned to Washington where he was declared insane and incarcerated in St Elizabeth's Hospital. He continued to write, and the award of the 1949 Bollingen Prize for his *Pisan Cantos* created a great furore. Eliot, **Robert Frost** and others pressed for Pound's release, and in 1958 he returned to Italy where he lived until his death.

Pound's influence on the development of twentieth-century poetry, as both poet and critic, has been enormous, though his great life's work, the *Cantos*, is rarely read in its entirety. Pound dated the inception of his 'epic including history' to 1904 and published a final volume, *Late Cantos and Fragments*, in 1969. The whole is necessarily formless as it endeavours to fuse the personal with history, mythology, politics and economics in what Pound called, in an essay on Cavalcanti, 'a radiant world where one thought cuts through another with clean edge, a world of moving energies'.

Poetry

A Lume Spento 1908; *A Quinzaine for This Yule* 1908; *Exultations* 1909; *Personae* 1909; *Provenca* 1910; *Cazoni* 1911; *Ripostes* 1912; *Lustra* 1916; *The Fourth Canto* 1919; *Quia Pauper Amavi* 1919; *Hugh Selwyn Mauberley* 1920; *Umbra* 1920;

Poems 1918–1921 1921; *A Draft of XVI Cantos* 1925; *Personae – the Collected Poems of Ezra Pound* 1926; *A Draft of the Cantos 17–27* 1928; *Selected Poems* 1928; *A Draft of XXX Cantos* 1930; *Eleven New Cantos XXXI–XLI* 1934; *Homage to Sextus Propertius* 1934; *The Fifth Decade of Cantos* 1937; *Cantos LII–LXXI* 1940; *A Selection of Poems* 1940; *The Cantos* 1948; *The Pisan Cantos* 1948; *Selected Poems* 1949; *Seventy Cantos* 1950; *The Cantos of Ezra Pound* 1954; *Section: Rock Drill* 1955; *Diptych Rome–London* 1958; *Thrones* 1959; *Versi Prosaici* 1959; *Late Cantos and Fragments* 1969; *Selected Poems 1908–1965* 1975; *The Cantos* 1986

Non-Fiction
The Spirit of Romance 1910; *Gaudier-Brzeska* 1916; *Indiscretions* 1923; *Antheil and the Treatise on Harmony* 1924; *Ta Hio* 1928; *Imaginary Letters* 1930; *How to Read* 1931; *ABC of Economics* 1933; *ABC of Reading* 1934; *Make It New* 1934; *Jefferson and/or Mussolini* 1935; *Social Credit* 1935; *Polite Essays* 1937; *Guide to Kulchur* 1938; *What Is Money For?* 1939; *Carta da vista* 1942; *L'America, Roosevelt e le cause della guerra presente* 1944; *Oro e lavoro* 1944; *Introduzione alla natura economica degli S.U.A.* 1944; *Orientamenti* 1944; *'If This Be Treason ...'* 1948; *Patria Mia* 1950; *Confucian Analects* 1951; *The Letters of Ezra Pound 1907–1941* [ed. D.D. Paige 1950]; *Literary Essays* [ed. T.S. Eliot] 1954; *Impact* 1960; *Nuova economia editoriale* 1962; *EP to LU: Nine Letters Written to Louis Untermeyer* 1963; *Pound/Joyce: the letters of Ezra Pound to James Joyce* [ed. F. Read] 1967; *'Ezra Pound Speaking': Radio Speeches of World War II* [ed. L. Doob] 1978; *Letters to Ibbotson* [ed. M. Hurley, V.I. Mondolfo] 1979; *Ezra Pound and Dorothy Shakespear: Their Letters, 1909–1914* [ed. O. Pound, A. Walton Liz] 1984; *Pound/Lewis: the letters of Ezra Pound and Wyndham Lewis* [ed. T. Materer] 1985; *Pound/Zukofsky: the Selected Letters of Ezra Pound and Louis Zukofsky* [ed. B Ahearn] 1987; *Pound/The Little Review: the Letters of Ezra Pound to Margaret Anderson* [ed. M.J. Freidmann, T.L. Scott] 1988

Translations
The Spirit of Romance 1910; *Cathay* 1915; *Certain Noble Plays of Japan* 1916; *'Noh' or accomplishment* 1916; Fontenelle *Dialogues* 1917; Remy de Gourmont *The Natural Philosophy of Love* 1922; Édouard Estaunié *The Call of the Road* 1923; Confucius: *Digest of the Analects* 1937, *Testamento di Confucio* 1944, *Ciung lung l'asse che non vacilla: secondo dei libra Confuciani* 1945, *The Classic Anthology Defined by Confucius* 1954; Odon Por *Italy's Policy of Social Economics 1939/1940* 1941; Moscardino *Enrico Pea* 1956; Sophocles *Women of Trachis* 1956; *Love Poems of Ancient Egypt* 1962

Collections
Pavannes and Divisions 1918; *Instigations* 1920; *Alfred Venison's Poems* 1935; *Pavannes and Divagations* 1958

Biography
Ezra Pound by Charles Norman 1960 [rev. 1969]; *Ezra Pound: the Last Rower* by C. David Heyman 1976; *Ezra Pound: the Solitary Volcano* by John Tytell 1987

Anthony (Dymoke) POWELL 1905–

Powell was born in London, the son of a lieutenant-colonel and from a family of Welsh origin which had been providing officers to the army and navy for several generations. He was educated at Eton and Balliol College, Oxford, and at both institutions was the contemporary of many of the leading writers who established their reputations from the 1920s onwards; he was to share with them a subject-matter of social distinctions, proprieties and excitements, and with **Evelyn Waugh** and **Henry Green** is the outstanding writer of the group.

After leaving Oxford, he worked for nine years (1926–35) as a junior publisher at Duckworth, while establishing himself in literary and social London and publishing an accomplished first novel, *Afternoon Men*, in 1931, the first of five which appeared in the 1930s. In 1934 he married Lady Violet Pakenham, daughter of the 5th Earl of Longford and a member of one of Britain's most extensive aristocratic (and literary) clans. After working briefly as a film scriptwriter in England and Hollywood, he saw war service in the Welch Regiment and the Intelligence Corps, liaising with several foreign nations and attaining the rank of major.

Since the war, he has lived as a novelist and man-of-letters: he has been on the *Times Literary Supplement*, was literary editor of *Punch* (1952–8), and for nearly fifty years reviewed for the *Daily Telegraph*, until he resigned over a hostile notice the newspaper commissioned of one of his two volumes of collected journalism. Powell and his wife have two sons, one of whom is the television director Tristram Powell, and have lived since 1952 near Frome in Somerset.

Powell is considered to be one of the leading writers of his generation, and his work falls into a classical pattern of literary production, disturbed by only a few incongruities. There are first the five pre-war novels: elegant, comic, perhaps a little harsh. Then there is his post-war twelve-volume novel sequence, *A Dance to the Music of Time*, at once more relaxed and more complex in style than the earlier books, using the idea of life as a sort of formalised dance to produce a portrait of upper- and upper-middle-class life in his own time which is sometimes comic, sometimes melancholy, but never less than entertaining and unlikely to be surpassed. The two late novels which follow, *O, How the Wheel Becomes It!* and *The Fisher King*, are generally agreed to add little to his overall fictional achievement. His *oeuvre* has been rounded out in later years by four volumes of guarded but informative memoirs, later published under the collective title of *To Keep the Ball Rolling*, the two large volumes of collected journalism already mentioned, and (in 1995) *Journals 1982–6*, while his works also include a biography of the seventeenth-century writer John Aubrey, and two plays.

Fiction

Afternoon Men 1931; *Venusberg* 1932; *From a View to a Death* 1933; *Agents and Patients* 1936; *What's Become of Waring?* 1939; *A Dance to the Music of Time*: *A Question of Upbringing* 1951, *A Buyer's Market* 1952, *The Acceptance World* 1955, *At Lady Molly's* 1957, *Casanova's Chinese Restaurant* 1960, *The Kindly Ones* 1962, *The Valley of Bones* 1964, *The Soldier's Art* 1966, *The Military Philosophers* 1968, *Books Do Furnish a Room* 1971, *Temporary Kings* 1973, *Hearing Secret Harmonies* 1975; *O, How the Wheel Becomes It!* 1983; *The Fisher King* 1986

Plays

Two Plays [pub. 1971]

Non-Fiction

To Keep the Ball Rolling (a) 1983: *Infants of the Spring* 1976, *Messengers of Day* 1978, *Faces in My Time* 1980, *The Strangers Are All Gone* 1982; *Miscellaneous Verdicts* 1990; *Under Review* 1991; *A Reference for Mellors* 1994; *Journals 1982–6* 1995

Edited

Barnard Letters 1778–1824 1928; *Novels of High Society from the Victorian Age* 1947; *John Aubrey and His Friends* 1948; *Brief Lives and Other Selected Writings by John Aubrey* 1949

Dawn POWELL 1897–1965

Powell was born in Mount Gilead, Ohio; six years later, her mother died. Loathing the relatives with whom she was lodged (they had burned the stories which she had written and hidden under a mat), Powell repeatedly ran away, once armed with thirty cents from berry-picking. After graduating from Lake Erie College in Painesville, she arrived in New York City in 1918 and joined the naval reserve; not much was required of her, for the war soon ended. She adored the city, living the rest of her life in Greenwich Village, and it animates the novels which **Edmund Wilson** thought equal in their humour to the work of **Evelyn Waugh**, **Anthony Powell** and **Muriel Spark** (she herself favoured the *Satyricon*). Other novels, such as *Dance Night*, take place in Buckeye State, a version of Ohio; if the city novels are the more brittle in atmosphere, both locales feature desperation in love. As she once said: 'One must cling to whatever remnants of love, friendship, or hope above and beyond reason that one has, for the enemy is all around ready to snatch it.'

Hers was a pen that could be turned to anything in the way of journalism, all the more so after marrying an advertising man, Joseph Gousha, in 1920; funds were continually needed to support their disabled son. To further complicate matters, she conducted an open affair with Coburn 'Coby' Gilman, perhaps even forming a *ménage à trois* with Gousha. Frequently in the company of such people as **John Dos Passos**, Robert Benchley and **Ernest Hemingway**, she was a denizen of the Cedar Tavern rather than the Algonquin (she used to hit anybody who called her another **Dorothy Parker**). She often pops up in Wilson's diaries of the period, the hostess of wild parties. She and Wilson carried on a bizarre correspondence in which he became a seedy literary man, Ernest Wigmore, and she Mrs Humphrey Ward (later Raoul and Aurore).

Powell also attempted such plays as *Big Night* and *Jig Saw*, which had small productions in the 1930s, and wrote stories, but the novels endure best. To give the plots of any of these is to miss the point. They appear to have taken on a life of their own; as she put it: 'Fantastic designs made by real human beings earnestly labouring to maladjust themselves to fate. There are no philosophies for them to prove ... my characters are not slaves to an author's propaganda. I give them their heads. They furnish their own nooses.'

Although her novels did not stay in print, perhaps victims of an American failure to grasp irony, she kept writing. 'Somehow in the boom years, the boom was always lowered on me,' she said. She was undaunted by such misfortune as her husband's long struggle with cancer, but when she too was struck by it, in a less malign form, she allowed herself to succumb. (From the hospital, Wilson reported: 'she is, for her, terribly thin ... she was more cheerful and animated than I was.') In his last decade Wilson urged the merits of his old friend, but it was only the later enthusiasm of **Gore Vidal** and John Guare which ushered some of her fifteen novels back into print. Her voice is caught by Vidal, who recalls the opening night of his play *Visit to a Small Planet* (1957): 'Suddenly a voice boomed – tolled across the lobby. "Gore!" I stopped;

everyone stopped. From the cloakroom a small round figure, rather like a Civil War cannon ball, hurtled towards me and collided. As I looked down into that familiar round face with its snub nose and shining bloodshot eyes, I heard, the entire crowded lobby heard: "*How could you do this?* How could you *sell out* like this? To *Broadway*! To *Commercialism*! How could you give up *The Novel?* Give up the *security*? The security of knowing that every two years there will be – like clockwork – *that five-hundred-dollar advance*!" '

Fiction

Whither? 1925; *She Walks in Beauty* 1928; *The Bride's House* 1929; *Dance Night* 1930; *The Tenth Moon* 1932; *The Story of a Country Boy* 1934; *Turn, Magic Wheel* 1936; *The Happy Island* 1938; *Angels on Toast* 1940 [rev. as *A Man's Affair* 1956]; *A Time to Be Born* 1942; *My Home Is Far Away* 1944; *The Locusts Have No King* 1948; *Sunday, Monday and Always* (s) 1952; *The Wicked Pavilion* 1954; *A Cage for Lovers* 1957

Plays

Big Night 1933 [np]; *Jig Saw* 1934 [np]; *Lady Comes Across* 1941 [np]; *The Golden Spur* 1962 [np]

Collections

Dawn Powell at Her Best [ed. T. Page] 1995

John Cowper POWYS 1872–1963

Powys was the eldest of eleven children of a country parson, and from a long line of country parsons on both sides. His mother was related to the poets William Cowper and John Donne. Besides his two brothers, **T.F. Powys** and Llewelyn, who became well-known writers, another brother was a noted architect, and three sisters were respectively a novelist and poet, a portrait-painter and a world authority on old lace. He was born in Derbyshire, but grew up in Dorset and Somerset, a region which is important as the background to his major novels. After education at Sherborne and Corpus Christi, Cambridge, he became a university extension lecturer for a while, but then discovered the American lecture circuit, which he worked extensively every year between 1904 and 1934, at periods living in the USA during the summer also. His influence on American popular culture was an important one.

His parallel writing career developed slowly, and did not take off until he was in his fifties. He began with two volumes of poetry in 1896 and 1899; published his first novel, *Wood and Stone*, in 1915; but only achieved fame in 1929 with *Wolf Solent*, which was soon followed by the two other novels regarded as his best, *A Glastonbury Romance* and *Weymouth Sands*. His status as a writer is subject to much controversy: considered a major talent by some, he is thought unbearably prolix and pretentiously 'philosophical' by others, although almost all commentators agree that his work contains passages of great beauty.

From the mid-1930s he returned to England and worked as a writer, eventually settling in Wales, where he lived in his old age in a cottage at Blaenau Ffestiniog, receiving admirers from all over the world. He wrote until his death at the age of ninety, and his output was vast. His later novels are mainly historical romances and he published an *Autobiography* in 1934, which has been considered his masterpiece and compared in depth of self-analysis to Rousseau's *Confessions*. Other works include volumes of criticism and works of popular philosophy and guidance. He married in 1896 and had one son. His wife died in 1947, and in later years his companion was Phyllis Plainter, to whom many of his works are dedicated.

Fiction

Wood and Stone 1915; *Rodmoor* 1916; *Ducdame* 1925; *Wolf Solent* 1929; *The Owl, the Duck, and – Miss Rowe! Miss Rowe* 1930; *A Glastonbury Romance* 1932; *Weymouth Sands* 1934 [rev. as *Jobber Skald* 1935]; *Maiden Castle* 1936; *Morwyn* 1937; *Owen Glendower* 1940; *Porius* 1951; *The Inmates* 1952; *Atlantis* 1954; *The Brazen Head* 1956; *Up and Out* 1957; *All or Nothing* 1960

Non-Fiction

The War and Culture 1914; *Visions and Revisions* 1915; *Suspended Judgements* 1916; *The Complex Vision* 1920; *Psychoanalysis and Morality* 1923; *The Religion of a Sceptic* 1925; *The Meaning of Culture* 1929 [rev. 1939]; *In Defence of Sensuality* 1930; *A Philosophy of Solitude* 1933; *Autobiography* (a) 1934; *The Art of Happiness* 1935; *Enjoyment of Literature* 1938; *Moral Strife* 1942; *The Art of Growing Old* 1944; *Dostoevsky* 1946; *Obstinate Cymric: Essays 1935–1947* 1947; *Rabelais* 1948; *In Spite of: a Philosophy for Everyman* 1953

Poetry

Odes and Other Poems 1896; *Poems* 1899; *Wolf's-Bane Rhymes* 1916; *Mandragora* 1917; *Sapphire* 1922; *Lucifer* 1956

Translations

Homer and the Aether 1959

Biography

The Powys Brothers by Kenneth Hopkins 1962; *The Brothers Powys* by Richard Perceval Graves 1983

T(heodore) F(rancis) POWYS 1875–1953

Brother of the well-known writers **John Cowper Powys** and Llewelyn Powys, T.F. Powys is now generally thought to be the most substantial talent of the three, although even he does not have an indisputably major reputation. One of eleven children of a fiercely evangelical parson father, he inherited a heterodox version of Christianity, which informs all his best work. He was at various private schools and, in contrast to his two brothers, did not attend a university. In his twenties he attempted to farm for a while in East Anglia, but without much success. In 1905 he married and settled with his wife, using a small private income (supplemented in later life by small civil list pensions), in the Dorset village of East Chaldon, the model for the background to almost all his work. Here he cultivated a life of extreme simplicity, for many years never going further from the village than he could walk, eating the same meals every day. He had two children of his own and adopted a daughter.

For many years, there was very little market for his strange stories. His first volume, *Soliloquies of a Hermit* (1918), was a series of brief meditations, and later the influence of the sculptor Stephen Tomlin and **David Garnett** led to the publication of his short stories in *The Left Leg* (1923). He was eventually to publish eight novels and more than twenty collections of shorter pieces. His major novels are considered to be *Mr Weston's Good Wine*, an allegory in which God in disguise visits a Dorset village, and *Unclay*. All his work shares recurring themes of cruelty, personal simplicity, religion, myth and the supernatural. Powys was estimated highly by Leavisite critics, but others found his characterisation weak and his allegories unconvincing.

In the late 1930s he announced that he was giving up writing, but published a few final works in the 1940s, *Bottle's Path* and *God's Eyes a-Twinkle*. From around 1940, he was a semi-invalid and recluse, dying finally in the remote Dorset village of Mappowder.

Fiction

The Soliloquy of a Hermit 1916 [as *Soliloquies of a Hermit* (s) 1918]; *Black Bryony* 1923; *The Left Leg* (s) 1923; *Mark Only* 1924; *Mr Tasker's Gods* 1925; *Mockery Gap* 1925; *Innocent Birds* 1926; *A Strong Girl and The Bride* (s) 1926; *Mr Weston's Good Wine* 1927; *The House with the Echo* (s) 1928; *Fables* (s) 1929 [as *No Painted Plumage* (s) 1934]; *Kindness in a Corner* 1930; *Uncle Dottery* 1930; *Uriah on a Hill* 1930; *The White Paternoster and Other Stories* (s) 1930; *Unclay* 1931; *The Tithe Barn, and The Dove and the Eagle* (s) 1932; *The Two Thieves* (s) 1933; *Captain Patch* (s) 1935; *Make Thyself Many* 1935; *Bottle's Path and Other Stories* (s) 1946; *God's Eyes a-Twinkle* (s) 1947; *Rosie Plum and Other Stories* (s) [ed. F. Powys] 1966; *Come and Dine and Tadnol* (s) [ed. A.P. Riley] 1967

Biography

The Powys Brothers by Kenneth Hopkins 1962; *The Brothers Powys* by Richard Perceval Graves 1983

J(ohn) B(oynton) PRIESTLEY 1894–1984

Priestley was born and brought up in Bradford, where his father was a prosperous schoolmaster; his mother died when he was an infant. He attended Belle Vue Grammar School, and left at sixteen to work as a junior clerk at a Bradford wool firm. The job left him time to write, and already before the First World War his first articles had been accepted by London and local papers. He enlisted at the outbreak of war, and was lucky enough to survive two long spells in the front line in Flanders. He then studied with an ex-serviceman's grant at Trinity Hall, Cambridge, and, on leaving, took what may have seemed the audacious step of refusing several academic jobs for the life of a bookman in London.

He was immediately successful, producing the first of many books of essays, *Brief Diversions*, in 1922 (he was almost the last to serve the market of extemporising essays on very general themes) and following it with much criticism and journalism during the 1920s. In 1929 and 1930, his third and fourth novels, *The Good Companions* and *Angel Pavement*, the one a cheerful, picaresque story of a travelling concert party, the other a sombre survey of contemporary London, were immensely successful, and established him as one of the most popular writers of the age. He immediately embarked on a new career as a playwright, often managing his own plays, once writing a play in a week, just as he wrote a novel in nineteen days (in terms of production, the only comparisons with Priestley are nineteenth century: Dickens and Balzac). Such plays as *Time and the Conways*, *An Inspector Calls* and *I Have Been Here Before*, which accommodate ideas within an always effective framework, have been regularly revived from the 1970s to the 1990s, and perhaps constitute Priestley's most enduring legacy.

In wartime, Priestley's fame as a patriotic radio broadcaster was second only to Churchill's, although his pursuit of **P.G. Wodehouse** through the medium aroused criticism in dissenting quarters. His own brand of lifelong radicalism was later expressed in his early support for the Campaign for Nuclear Disarmament. He continued prolific into old age, as late as 1968 achieving success with the immensely long *The Image Men*, the story of two down-at-heel dons making a killing in the advertising world. Other highlights of his vast output include *English Journey*, a seminal work in arousing social conscience in the 1930s, *Literature and Western Man*, an ambitious survey of Western literature over the past 500 years, and the memoirs, *Margin Released*. Priestley had married for the first time, in 1919, Emily Tempest, who died young in 1925. In 1926 he married Mary ('Jane'), the former wife of the biographer and satirist D.B. Wyndham Lewis. This marriage ended in divorce in 1952, and the same year Priestley married the archaeologist Jacquetta Hawkes, with whom he lived in later years in a large house near Stratford-upon-Avon. His first two marriages brought him five children. Priestley has been dismissed as a middlebrow author, and his no-nonsense pipe-smoking image lent itself to caricature, but his creative power deserves respect, and his writing has survived longer than that of many comparable writers.

Plays

The Good Companions [from his novel; with E. Knoblock] 1931; *Dangerous Corner* 1932; *The Roundabout* 1932 [pub. 1933]; *Laburnum Grove* 1933 [pub. 1934]; *Eden End* 1934; *Cornelius* 1935; *Duet in Floodlight* 1935; *Bees on the Boat Deck* 1936; *Spring Tide* [as 'Peter Goldsmith'; with G. Billam] 1936; *I Have Been Here Before* 1937; *Mystery at Greenfingers* [pub. 1937]; *People at Sea* 1937; *Time and the Conways* 1937; *Music at Night* 1938; *When We Are Married* 1938; *Johnson over Jordan* 1939; *The Long Mirror* 1940; *Goodnight, Children* 1942; *Desert Highway* 1943 [pub. 1944]; *They Came to a City* 1943 [pub. 1944]; *The Golden Fleece* 1944 [pub. 1948]; *How Are They at Home?* 1944; *An Inspector Calls* 1946 [pub. 1947]; *Ever Since Paradise* 1947 [pub. 1950]; *Jenny Villiers* [pub. 1947]; *The Linden Tree* 1947 [pub. 1948]; *The Rose and Crown* [pub. 1947]; *The High Toby* 1948; *Home Is Tomorrow* 1948 [pub. 1949]; *The Olympians* (l) 1949; *Summer Day's Dream* 1949 [pub. 1950]; *Bright Shadow* 1950; *Dragon's Mouth* [with J. Hawkes] 1952; *Mother's Day* [pub. 1953]; *Private Rooms* [pub. 1953]; *Treasure on Pelican* [pub. 1953]; *Try It Again* [pub. 1953]; *A Glass of Bitter* [pub. 1954]; *Mr Kettle and Mrs Moon* 1955; *The Glass Cage* 1957; *A Severed Head* [from I. Murdoch's novel; with I. Murdoch] 1963

Fiction

Adam in Moonshine 1927; *Benighted* 1927; *The Good Companions* 1929; *Angel Pavement* 1930; *The Town Mayor of Miracourt* 1930; *Faraway* 1932; *Wonder Hero* 1933; *Albert Goes Through* 1933; *They Will Walk in the City* 1936; *The Doomsday Men* 1938; *Let the People Sing* 1939; *Black-Out in Gretley* (s) 1942; *Daylight on Saturday* 1943; *Three Men in New Suits* 1945; *Bright Day* 1946; *Going Up* 1950; *Lost Empires* 1950; *Festival at Farbridge* 1952; *The Other Place* 1953; *The Magicians* 1954; *Saturn Over the Water* 1961; *The Thirty-First of June* 1961; *The Shapes of Sleep* 1962; *Sir Michael and Sir George* 1964; *Lost Empires* 1965; *It's an Old Country* 1967; *The Image Men: Out of Town* 1968, *London End* 1969

Poetry

The Chapman of Rhymes 1918

Non-Fiction

Brief Diversions 1922; *Papers from Lilliput* 1922; *I for One* 1923; *Figures in Modern Art* 1924; *The English Comic Characters* 1925; *George Meredith* 1926; *Talking* 1926; *The English Novel* 1927; *Thomas Love Peacock* 1927; *Open House* 1927; *Apes and Angels* 1928; *The Balconinny* 1929; *English Humour* 1929 [rev. 1976]; *English Journey* 1934; *Midnight on the Desert* 1937; *Rain upon Godshill* 1939; *Britain Speaks* 1940; *Postscripts* 1940; *Out of the People* 1941; *British Women Go to War* 1943; *The Man-Power Story* 1943; *Letter to a Returning Serviceman* 1945; *The Sweet Dream* 1946; *Russian Journey* 1946; *The Arts under Socialism* 1947; *Theatre Outlook* 1947; *Delights* 1949; *Journey Down a Rainbow* [with J. Hawkes] 1955; *The Art of the Dramatist* 1957; *Thoughts in the Wilderness* 1957; *Topside, or, the Future of England* 1958; *Literature and Western Man* 1960; *William Hazlitt* 1960; *Charles Dickens: a Pictorial Biography* 1961; *Margin Released* 1962; *Man and Time* 1964; *The Moments and Other Pieces* 1966; *All England Listened* 1968; *Essays of Five Decades* 1969; *The Prince of Pleasure* 1969; *Anton Chekhov* 1970; *The Edwardians* 1970; *Over the Long High Wall* 1972; *Victoria's Heyday* 1972; *The English* 1973; *Outcries and Asides* 1974; *A Visit to New Zealand* 1974; *Particular Pleasures: Being a Personal Record of Some Varied Arts and Many Different Artists* 1975; *Found, Lost, Found: or, The English Way of Life* 1976; *Instead of the Trees* 1977; *Seeing Stratford* 1982

Edited

The Bodley Head Leacock 1952

Biography

J.B. Priestley by Vincent Brome 1988

F(rank) T(empleton) PRINCE 1912–

Born in Kimberley, South Africa, Prince was the son of a businessman and a teacher, both of whom were of English origin. After attending the Christian Brothers' College, he studied architecture for a year at the University of Witwatersrand, Johannesburg. His boyhood had been punctuated by several visits to England, and in 1931 he went there for good, initially to study at Balliol College, Oxford, where he took a first in English in 1934. He was a visiting fellow at the graduate college of Princeton University in 1935–6, and then returned to London to work in the Royal Institute of International Affairs.

His long poem 'An Epistle to a Patron' had been published by **T.S. Eliot** in the *Criterion* in 1935, and his first volume of verse, *Poems*, was published by Faber in 1938. In 1940 he was commissioned into the Intelligence Corps of the British army and served first at Bletchley Park, was later sent to Cairo, and ended the war as an interpreter in a camp for Italian prisoners-of-war. He married in 1943, and has two daughters. On demobilisation, he became a lecturer at Southampton University, and from 1957 to 1974 he was a professor of English there. Among the best known of his scholarly books is *The Italian Element in Milton's Verse* (1954). In the same year, his second volume of verse, *Soldiers Bathing and Other Poems*, was published, after a long interval of the sort that has characterised his poetic output. The title-poem remains his best-known single work; a poem of relaxed and sombre grandeur reflecting Prince's Roman Catholic faith, it is acknowledged as one of the finest works to come out of the Second World War. Nevertheless, it is only the highlight of an impressive poetic output which includes several long meditative poems often featuring the use of historic *personae*. His diction is uncompromisingly traditional, and he is not fashionable, but fellow poets recognise that as an artist dealing with love and the spiritual life he ranks high.

After retiring from Southampton, Prince entered a period of teaching at universities abroad from 1975 to 1983, during which he was in the USA, Jamaica and the Yemen; now retired, he lives in Southampton.

Poetry

Poems 1938; *Poems* 1941; *Soldiers Bathing and Other Poems* 1954; *The Stolen Heart* 1957; *The Doors of Stone: Poems 1938–1962* 1963; *Memoirs in Oxford* 1970; *Penguin Modern Poets 20* [with J. Heath-Stubbs, S. Spender] 1971; *Drypoints of the Hasidim* 1975; *Afterword on Rupert Brooke* 1976; *Collected Poems* 1979; *The Yüan Chên Variations* 1981; *Later On* 1983; *Fragment Poetry* 1986; *Walks in Rome* 1987; *Collected Poems 1935–1992* 1993

Non-Fiction

The Italian Element in Milton's Verse 1954; *In Defence of English* 1959; *William Shakespeare: the Poems* 1963

Edited

Milton: *Samson Agonistes* 1957, *Paradise Lost, Books I and II* 1962, *Comus and Other Poems* 1968; Shakespeare *The Poems* 1960

Translations

Sergio Baldi *Sir Thomas Wyatt* 1961

V(ictor) S(awdon) PRITCHETT 1900–

Pritchett was born in Ipswich, the son of a travelling salesman and Christian Scientist, and the vagaries of the lower-middle-class, the varieties of religious enthusiasm and the nature of English puritanism, are all constants of his work. He spent his childhood between the provinces and various London suburbs, attending Alleyn's School, Dulwich, and Dulwich College. He left at sixteen to work in the leather trade in Bermondsey. At twenty he went to Paris, where he worked as a shop assistant and salesman in the shellac, glue and photographic trades. He then became a correspondent for the *Christian Science Monitor*, first in Ireland, where the civil war was still being fought, and then in Spain. His first book was an account of his wanderings there, the now classic *Marching Spain* (1928). His first novel, *Clare Drummer*, appeared in the following year.

From 1926, when he became a critic for the *New Statesman*, thus inaugurating a long relationship, he has lived the life of a London man-of-letters, in later years travelling widely and holding several visiting professorships, primarily in the USA but also elsewhere. He married in 1936, and has two children. His son is the newspaper columnist Oliver Pritchett and his grandson the newspaper cartoonist Matthew ('Matt') Pritchett.

Pritchett was described in a 1994 *New Yorker* article as 'our language's presiding man of letters', and he has contributed with distinction to every literary genre except poetry, maintaining his creativity into extreme old age. He is, however, primarily associated with two forms, the short story and the essay. Complete collections of his work in these genres appeared in 1990 and 1991, the fruit of many separate volumes. The delicate evocation of unachieved relationships in such a story as 'Many Are Disappointed' make it an undoubted masterpiece, and other outstanding stories include 'The Camberwell Beauty', 'The Saint' and 'When My Girl Comes Home'. His essays range with both magisterial authority and sharp originality through the whole field of Western literature.

His novels, such as *Dead Man Leading* and *Mr Beluncle*, are perhaps less admired, but exemplify very well the idiosyncratic and slyly satirical tone that is characteristic of his work. *The Spanish Temper* is regarded as one of the classic English texts dealing with Hispanic themes, and is only one of many travel books. There are also two volumes of autobiography, *A Cab at the Door* and *Midnight Oil*, which provide a fascinating glimpse of bygone days, as well as authoritative accounts of such writers as Balzac, Turgenev and Chekhov, and some broadcast plays.

Pritchett was appointed CBE in 1968; knighted in 1975; has been the president of the Society of Authors since 1977; was appointed Companion of Literature in 1988; became a Companion of Honour in 1993; and received the PEN Golden Pen Award for long service to literature, also in 1993. Perhaps the only criticisms that might be levelled at his output are that he lacks passion and tends towards mass-production, but his status as a major writer is unquestioned.

Fiction

Clare Drummer 1929; *The Spanish Virgin and Other Stories* (s) 1930; *Shirley Sanz* 1932; *Nothing Like Leather* 1935; *Dead Man Leading* 1937; *You Make Your Own Life* (s) 1938; *It May Never Happen* (s) 1945; *Mr Beluncle* 1951; *Collected Stories* (s) 1956; *The Sailor, The Sense of Humour and Other Stories* (s) 1956; *When My Girl Comes Home* (s) 1961; *The Key to My Heart* (s) 1963; *The Saint and Other Stories* (s) 1966; *Blind Love* (s) 1969; *Penguin Modern Stories 9* (s) [with others] 1971; *The Camberwell Beauty* (s) 1974; *Selected Stories* (s) 1978; *On the Edge of the Cliff* (s) 1979; *Collected Stories* (s) 1982; *More Collected Stories* (s) 1983; *The Other Side of a Frontier* (s) 1984; *A Careless Widow and Other Stories* (s) 1989; *Complete Short Stories* (s) 1990 [US *Complete Collected Stories* 1991]

Non-Fiction

Marching Spain 1928; *In My Good Books* 1942; *Build the Ships: the Official Story of the Shipyards in War-Time* 1946; *The Living Novel* 1946; *Why Do I Write? An Exchange Between Graham Greene, Elizabeth Bowen and V.S. Pritchett* 1948; *Books in General* 1953; *The Spanish Temper* 1954; *London Perceived* 1962; *Foreign Faces* [US *The Offensive Traveller*] 1964; *New York Proclaimed* 1965; *The Working Novelist* 1965; *Dublin: a Portrait* 1967; *A Cab at the Door: Early Years* (a) 1968; *George Meredith and English Comedy* 1970; *Midnight Oil* (a) 1971; *The Gentle Barbarian: the Life and Work of Turgenev* 1977; *The Myth Makers: Essays on European, Russian and South African Novelists* 1979; *The Tale Bearers: Essays on English, American and Other Writers* 1980; *The Turn of the Years* [with R. Stone] 1982; *A Man of Letters: Selected Essays* 1985; *Chekhov: a Spirit Set Free* 1988; *At Home and Abroad* 1989; *Lasting Impressions: Essays 1961–1987* 1990; *The Complete Essays* 1991; *Balzac* 1993

Edited

This England 1938; Robert Louis Stevenson *Novels and Stories* 1945

Collections

The Other Side of the Frontier: a V.S. Pritchett Reader 1984

E(dna) Annie PROULX 1935–

Proulx is the eldest of five daughters born in Connecticut to a family of millworkers, inventors and artists. Her father was a French-Canadian immigrant, passionate to be American in all but surname, while her artist mother came from an Irish family devoted to the art of storytelling, and to whom Proulx attributes her talent. She went to college briefly in the 1950s, but left to get married. She has been married three times in all, but raised two of her three sons alone.

She learned to write through being a voracious reader. In 1963, aged twenty-eight, she went back to graduate college in Montreal to study history, which she says was 'invaluable training for novel writing'. She had no desire to teach, but, with three sons to feed, took up freelance journalism. Living in a remote shack near the Canadian border, she wrote for magazines about everything from the weather, cooking, canoeing, mice, cider, apples and lettuces to architecture. Whenever she could find time, she wrote short stories, set in the great American outdoors, but mirroring the lives of people in a constant state of flux. Most of them were published and by 1988 several had impressed an American publisher enough for him to offer her a contract for a collection of stories and a novel.

Heartsongs and Other Stories (1988) was her first book, followed by *Postcards*, a first novel of such dazzling accomplishment that it prompted the *New York Times* to say that she had come close to writing 'The Great American Novel'. With it she won the 1993 PEN/Faulkner Award. Set in the 1940s, *Postcards* is the story of Loyal Blood, a man who murders his girlfriend and spends the next forty years eking out an existence in the American West, haunted by guilt. The 'postcards' of the title are the book's *leitmotif*, handwritten epigraphs linking him to the past he has left behind.

Some years ago, on a canoeing trip, she discovered Newfoundland and fell in love with the place and its people. Out of this infatuation emerged her second novel, *The Shipping News*, which won her the *Irish Times* International

Fiction Prize, an American National Book Award and the 1994 Pulitzer Prize for Fiction. With its anti-hero Quoyle, one of life's losers until fate propels him to Newfoundland with two disturbed daughters and an eccentric aunt, it is as much about the island as it is about a man's search for himself. Each chapter is prefaced with a knot drawing and a piece of knot philosophy from *The Ashley Book of Knots*, another of Proulx's quirky, innovative devices for enhancing her narratives.

All her books are elemental. Place and weather are as integral to the plot as the characters. The historian in her demands copious and thorough research. She stays in the places she writes about, studies local history, maps, newspapers, reads telephone directories, studies dialect and talks to people. The result is a candour and grittiness more often associated with documentary than fiction. Yet it is this raw, vivid quality which make her books so singular and compelling.

She has lived in rural Vermont for most of her life, for some time in a house she built herself, but in 1995 she moved to Wyoming.

Fiction

Heart Songs and Other Stories (s) 1988 [rev. in UK as *Heart Songs* 1995]; *Postcards* 1992; *The Shipping News* 1993

J(eremy) H(alvard) PRYNNE 1936–

Famous for the sheer intractability of his work and for what has been termed 'an almost mythic quality of luminous opacity', Prynne was born in Kent. His father was an engineer; his mother, a descendant of Charles James Fox, was a woman of strong socialist principles who ran a private school. Prynne was educated at Beltane, St Dunstan's College and Jesus College, Cambridge. Apart from a short spell of national service, he has been a Cambridge don for most of his working life. After a year spent in the USA, he became a lecturer at Gonville and Caius College in 1962 and is married with two children.

Prynne's first volume of verse, his 1962 *Force of Circumstance and Other Poems*, was untypical both in being published by a major imprint, Routledge, and in its relative continuity with the English poetic tradition. It shows the influence of **Donald Davie** and **Charles Tomlinson**, although with an evident desire to move beyond both. Prynne's second volume, *Kitchen Poems*, showed his truer direction, and is more in line with the 'projective' free verse of **Charles Olson**. Its first line, 'The whole thing it is, the difficult', was the point of departure for several commentators and, like Prynne's subsequent work, it manifests an edgy play of associations that defy the audience to follow them closely, and instead force the reader to find meanings and significances which must often be largely of their own making. But however syntactically and semantically elusive it may often be, Prynne's work stands above the merely obscure by its austere and plangent beauty and the craft of its free verse.

Its difficulty has invited comparison with the work of **John Ashbery**, but Prynne's work is far less of a free-falling celebration of the disjointed postmodern experience and more sombre in its concern with notions of harm and damage. It is also altogether more austere (Prynne once condemned the music of Schoenberg as 'schmaltz') and far more politically engaged. Economics figures extensively, in such poems as 'Sketch for a Financial Theory of the Self', for example, and he has continually explored the grounds of subjectivity in its junctions with economics, the life sciences and collectivity: 'what *I am* is a special case of / what *we want* ... '. But to discuss his work, or to read of it, is empty without the distinctive experience of actually reading it. Here is a section from his acclaimed volume *The Oval Window*. 'Just a treat sod Heine you notice / the base going down, try to whistle / with a tooth broken. Safe in our hands / won't cut up rough, at all, pent up / and boil over. Fly my brother, he watches / at point of entry, only seeming to / have a heart for it. Thermal patchwork / will tell, sisal entreaty creams out.'

Critical responses to Prynne have inevitably focused on the inscrutability of his work: 'Either my critical intelligence is going to pot, or it's pretentious nonsense', wrote one detractor, while one of his greatest champions has written that: 'Hesitancy about confident interpretation of Prynne's work continues to seem necessary.' Prynne has pushed the innovations of modernism to their limits, in ways which Michael Grant considers make his work 'the most audacious of English postwar poetry'. **Peter Ackroyd**, associated with Prynne by Cambridge connections, has written that: 'By refusing to become a readily accessible and intelligible writer, he has ensured that poetry can no longer be treated as a deodorised museum of fine thoughts and fine feelings; he is creating, instead, a complete and coherent language.' Resistance to his work has taken the form of marginalisation rather than ridicule, and he has been published by the smallest of small presses and in samizdat editions. Outside of the mainstream, Prynne is regarded in some quarters as a major figure and a pervasive inspiration, the charismatic *éminence grise* of the so-called Cambridge School. He figures as 'Undark' in **Iain Sinclair**'s novel *Radon Daughters* (1994).

Poetry
Force of Circumstance and Other Poems 1962; *Kitchen Poems* 1968; *Aristeas* 1968; *Daylight Songs* 1968; *The White Stones* 1969; *Fire Lizard* 1970; *Brass* 1971; *Into the Day* 1972; *A Night Square* 1973; *Wound Response* 1974; *High Pink on Chrome* 1975; *News of Warring Clans* 1977; *Down Where Changed* 1979; *Poems* 1982; *The Oval Window* 1983; *Marzipan* 1986; *Bands Around the Throat* 1987; *Word Order* 1989

Non-Fiction
Stars, Tigers and the Shape of Words 1993

John (Sleigh) PUDNEY 1909–1977

Pudney was born in Langley, Buckinghamshire, the son of a farmer, and educated at Gresham's School, Holt, where he was a younger contemporary of **W.H. Auden**. On leaving school he worked for some years as a radio producer and scriptwriter for the BBC before joining the *News Chronicle* in 1937. He served as a war correspondent before joining the Royal Air Force in 1940 as one of the team of writers, which included **H.E. Bates**, instructed to put over the work done by air crews in a way that the public could understand. After the war he was a reviewer for the *Daily Express* and literary editor of *News Review* from 1948 to 1950, then entered publishing as a director of Putnam, where he built up a list of books on flying.

Before the war he had published two books of verse, *Spring Encounter* and *Open the Sky*, two collections of stories, and a novel, *Jacobson's Ladder*; but it was the war years that brought him fame. The twelve lines of 'For Johnny' were 'written on the back of an envelope in London during an air raid alert in 1941'. This little poem, more than any of the others Pudney wrote at the time, seemed to encapsulate the mood of the war in the air, doing what **Arthur Machen** had done in prose for the soldiers of the First World War: 'Do not despair / For Johnny-head-in-air; / He sleeps as sound / As Johnny under ground.' It first appeared in the *News Chronicle*, then in the collection *Dispersal Point*. It was read, with other of Pudney's poems, by Laurence Olivier on BBC radio and spoken by Michael Redgrave and John Mills in the film *The Way to the Stars* (1945), directed by Anthony Asquith, scripted by **Terence Rattigan**. (The film was given the title *Johnny in the Clouds* when released in America.) 'There never was a particular Johnny', Pudney wrote later. 'The twelve lines which forced themselves on me virtually ... in one go, were meant for them all. It is the same with the named individuals in other poems; the one stands for many.' His later work has been described as 'tougher and more ironic, closer to the vernacular, although perhaps in a self-conscious way'. Besides novels – *The Net* and *Thin Air* are probably the most substantial – and short stories, he published children's books, mainly in series, and non-fiction on a variety of subjects, including war, travel, aeroplanes, steam engines and a history of lavatories, *The Smallest Room*. He edited *Pick of Today's Short Stories* between 1949 and 1963.

Pudney was extraordinarily prolific, in spite of the fact that he suffered from alcoholism. (When writing his history of the travel agency Thomas Cook and Sons, he complained that he found the teetotalist lectures of the firm's founder 'thirsty reading'.) He was married twice, first in 1934 to Crystal, daughter of **A.P. Herbert**; the marriage was dissolved in 1955, when he married again. He died of throat cancer.

Poetry
Spring Encounter 1933; *Open the Sky* 1934; *Dispersal Point and Other Air Poems* 1942; *Beyond This Disregard* 1943; *South of Forty* 1943; *Almanack of Hope* 1944; *Flight above Clouds* 1944; *Ten Summers: Poems (1933–1943)* 1944; *Selected Poems* 1946; *Selected Poems* 1947; *Low Life* 1947; *Commemorations* 1948; *Sixpenny Songs* 1953; *Collected Poems* 1957; *The Trampoline* 1959; *Spill Out* 1967; *Spandrels* 1969; *Take This Orange* 1971; *Selected Poems 1967–1973* 1973; *Living in a One-Sided House* 1976

Plays
The Little Giant 1972; *Ted* **(tv)** 1972, **(st)** 1974

Fiction
And Lastly the Fireworks **(s)** 1935; *Jacobson's Ladder* 1938; *Uncle Arthur and Other Stories* **(s)** 1939; *It Breathed Down My Neck* **(s)** [US *Edna's Fruit Hat and Other Stories*] 1946; *Estuary* 1948; *The Europeans* **(s)** 1948; *Suffley Wanderers* 1948; *The Accomplice* 1950; *Hero of a Summer's Day* 1951; *The Net* 1952; *A Ring for Luck* 1953; *Trespass in the Sun* 1957; *Thin Air* 1961; *The Long Time Growing Up* 1971

For Children
Saturday Adventure 1950; *Sunday Adventure* 1951; *Monday Adventure* 1952; *Tuesday Adventure* 1953; *Wednesday Adventure* 1954; *Six Great Aviators* 1955; *Thursday Adventure* 1955; *Friday Adventure* 1956; *The Grandfather Clock* 1956; *Crossing the Road* 1958; *Spring Adventure* 1961; *The Hartwarp Dump* 1962; *The Hartwarp Light Railway* 1962; *Summer Adventure* 1962; *The Hartwarp Balloon* 1963; *Autumn Adventure* 1964; *The Hartwarp Bakehouse* 1964; *The Hartwarp Explosion* 1965; *Tunnel to the Sky* 1965; *Winter Adventure* 1965; *The Hartwarp Jets* 1967

Non-Fiction
The Green Grass Grew All Round 1942; *Who Only England Knows* 1943; *World Still There* 1945; *Music on the South Bank* 1951; *His Majesty King George VI* 1952; *The Queen's People* 1953; *The Thomas Cook Story* 1953; *The Smallest Room* 1954 [rev. as *The Smallest Room: With an Annexe* 1959]; *The Seven Skies* 1959; *Bristol Fashion* 1960; *Home and Away* (a) 1960; *A Pride of Unicorns* 1960; *The Camel Fighter* 1964; *The Golden Age of Steam* 1967; *Suez: De Lesseps' Canal* 1968; *A Draught of Contentment* 1971; *Crossing London's River* 1972; *Brunel and His World* 1973; *London's Docks* 1975; *Thank Goodness for Cake* 1978

Edited
Air Force Poetry [with H. Treece] 1944; *Laboratory of the Air: an Account of the Royal Aircraft Establishment of the Ministry of Supply, Farnborough* 1948; *Pick of Today's Short Stories* (s) 13 vols 1949–63; *Popular Poetry* 1953; *The Book of Leisure* 1957; *The Harp Book of Toasts* 1963; *The Batsford Colour Book of London* 1965; *Flight and Flying* 1968

James (Amos) PURDY 1923–

Purdy was born in the country near Ohio, the son of a businessman and lawyer of Huguenot farming stock. His mother was a descendant of James Otis, a signatory of the American Declaration of Independence. After studying at the University of Ohio, Purdy did postgraduate work at the University of Puebla, Mexico, and in Madrid. He subsequently worked as an interpreter in Latin America, France and Spain.

In 1956 he semi-privately published a short novel, *63: Dream Palace*, and a volume of short stories, *Don't Call Me by My Right Name*, copies of which he sent to established writers, including **Edith Sitwell**, who was impressed by their originality and declared him 'a *much* greater writer than [**William**] **Faulkner**'. She arranged for the two books to be published in England in one volume, even battling (unsuccessfully) to retain Purdy's use of the word 'motherfucker'. The book was subsequently published in an expanded, unexpurgated edition in 1961 as *Color of Darkness*, with an enthusiastic introductory essay by Sitwell. While her encomium was distinctly idiosyncratic and characteristically overstated, there is no doubt that her imprimatur helped to launch Purdy as a writer. The novella, which deals with homosexuality and violence, announced themes that were to occur throughout Purdy's work, and which have perhaps denied him the readership he deserves. He writes in the Gothic tradition of the American South, and early critics compared his work with that of **Carson McCullers** and **Truman Capote**. His vision is, however, bleaker than theirs, his world more brutal. In Purdy's novels, people are disembowelled, buried alive beneath piles of corpses, undergo crude abortions and are nailed to barn doors. A principal theme of his work is the exploitation and destruction of loveless innocents by adults. *Malcolm* is about someone whose sexual exploitation leads to his death, and has a cast in the best traditions of Southern Gothic: a midget, a black, bejewelled mortician, and so on. Of his later novels, perhaps the best is *In a Shallow Grave*, in which the redemption of suffering through love, another frequent theme, is shown in a bizarre, allegorical story about the courtship of a Virginian widow by a grotesquely injured Vietnam veteran.

His writing is dense, oblique and poetic, and he has an exact ear for Southern dialect. His reputation has always stood high amongst fellow writers such as **John Cowper Powys**, **Gore Vidal**, **Tennessee Williams** and **Dorothy Parker**, but he has at times been short of funds – he had to borrow money and clothes to travel to New York to meet Sitwell in 1957 – and has supplemented his income by teaching at New York University. He has also written poetry and plays (frequently publishing volumes containing a mixture of these and fiction), and *63: Dream Palace* has been used as the basis of an opera by Hans Werner Henze. He lives quietly in a small apartment in Brooklyn. Far from being austere, as the seriousness of his fiction might suggest, he is gently amusing, reflecting in conversation a comic element also present in his work.

Fiction
Don't Call Me by My Right Name and Other Stories (s) 1956; *63: Dream Palace* 1956; *63: Dream Palace: a Novella and Nine Stories* (s) 1957; *Malcolm* 1959; *Colour of Darkness* (s) 1961; *The Nephew* 1961; *Cabot Wright Begins* 1964; *Eustace Chisolm and the Works* 1967; *Jeremy's Version* 1970; *I am Elijah Thrush* 1972; *The House of the Solitary Maggot* 1974; *In a Shallow Grave* 1976; *Narrow Rooms* 1978; *Mourners Below* 1981; *On Glory's Course* 1984; *The Candles of Your Eyes* (s) 1986; *In the Hollow of His Hand* 1986; *Garments the Living Wear* 1989; *Out with the Stars* 1991

Plays
Mr Cough and the Phantom Sex 1960 [np]; *Cracks* 1963 [np]; *A Day After The Fair, True and Wedding Finger* [pub. 1977]; *Proud Flesh* [pub. 1981]; *Scrap of Paper* and *The Berrypicker* 1985 [pub. 1981]; *Ruthanna Elder* [pub. 1990]

Poetry
The Running Sun 1971; *Sunshine Is an Only Child* 1973; *Lessons and Complaints* 1978; *Sleep Tight*

1979; *The Brooklyn Branding Parlors* 1985; *Collected Poems* 1990

Collections

Children Is All 1962; *An Oyster Is a Wealthy Beast* 1967; *Mr Evening: a Story and Nine Poems* 1968; *On the Rebound: a Story and Nine Poems* 1970; *A Day After the Fair: a Collection of Plays and Stories* 1977

Barbara (Mary Crampton) PYM 1913–1980

The elder daughter of a solicitor, Pym was born in Oswestry, Shropshire, and educated at Huyton College, Liverpool, and St Hilda's College, Oxford, where she read English. By the time she was an undergraduate, the twin devotions of her life – to High Church Anglicanism and to English literature, particularly poetry – were already fully formed. While at Oxford, she wrote a story in which she and her sister were portrayed as 'spinsters of fifty', and this extraordinary projection laid the foundations of her first novel, *Some Tame Gazelle* (1950), and of the fictional world she made inimitably her own.

Her life after Oxford falls into two parts. During the first period, she lived at home with her parents or was involved in war work (she served as a rating in the Women's Royal Naval Service, then was promoted to third officer and worked as a naval censor, posted to Naples in 1944), began writing novels but was not published, and engaged in a long series of, generally unrequited, love affairs with unsuitable men. After the war, the romantic interest in her life faded – apart from one passion for a younger man in the 1960s, which led to one of her best novels, *The Sweet Dove Died*. She was employed from 1946 until her retirement in 1974 in editorial work at the International African Institute in London, and shared a succession of flats and houses with her sister in Pimlico, Barnes and Kilburn. She then went to live in a cottage in the Oxfordshire village of Finstock.

Between 1950 and 1961 she published six subtly comic novels detailing the quiet and apparently uneventful lives of English gentlewomen of a certain age. The archetypal Pym heroine has a rather dull job – one novel about indexers was originally titled 'A Thankless Task' – and involves herself in good works and the Anglican church. She is intelligent and well read, with a weakness for the lesser Victorian poets, and entertains mild and unrequited passions for men who by religious conviction or sexual persuasion tend not to be 'the marrying kind'. Pym's work for the African Institute is reflected in the almost anthropological studies she made of these 'excellent women', a phrase she used for the title of her second novel. The solidity of the world Pym describes is emphasised by characters who reappear from novel to novel.

In 1963 her seventh novel, *An Unsuitable Attachment*, was rejected by her publisher, with the result that no new novel appeared for sixteen years. 'It ought to be enough for anybody to be the Assistant Editor of *Africa*,' she confessed, 'but I find it isn't quite.' Characteristically, she was putting a brave face on circumstances which wounded her deeply. During the next fourteen years she submitted her work to twenty-one publishers, but her beautifully crafted novels of ordinary life amongst the dowdy middle classes were out of fashion and considered unlikely to be commercially viable. A reassessment of her work in 1977, when she was cited by David Cecil and **Philip Larkin** (with whom she had struck up a friendship) as one of the 'most underrated novelists of the century' in a survey organised by the *Times Literary Supplement*, coincided with the publication of what is perhaps her finest book, *Quartet in Autumn*. A grimly funny study of four old people, it was shortlisted for the Booker Prize, and led to a revival of interest in her work. She lived to enjoy the success of *The Sweet Dove Died*, but *A Few Green Leaves*, a moving meditation upon mortality set in an Oxfordshire village, was published after her death from cancer.

Pym's reputation has continued to grow, and a devoted cult has built up around her which has tended to obscure her genuine qualities. These include sharp observation, particularly when describing masculine complacency, a nice sense of irony, and an ability to convey the value and interest of all lives, however restricted. In his *TLS* contribution, Cecil described *Excellent Women* and *A Glass of Blessings* as 'the finest examples of high comedy to have appeared in England during the last seventy-five years'. A revealing and entertaining selection of her diaries, letters and notebooks was published in 1984 as *A Very Private Eye*, but this was followed by some inferior works. While *Crampton Hodnet* proved to be a welcome addition to the canon, *An Academic Question* and the fragments collected in *Civil to Strangers* should perhaps have remained in the bottom drawer.

Fiction

Some Tame Gazelle 1950; *Excellent Women* 1952; *Jane and Prudence* 1953; *Less Than Angels* 1955; *A Glass of Blessings* 1958; *No Fond Return of Love* 1961; *Quartet in Autumn* 1977; *The Sweet Dove Died* 1978; *A Few Green Leaves* 1980; *An Unsuitable Attachment* 1982; *Crampton Hodnet* 1985; *An Academic Question* 1986; *Civil to Strangers* 1987

Non-Fiction
A Very Private Eye [ed. H. Holt, H. Pym] 1984

Biography
A Lot to Ask by Hazel Holt 1990

Thomas (Ruggles) PYNCHON 1937–

Since the publication of his first novel, *V*, in 1963, Pynchon has effectively disappeared from public view, refusing to be interviewed or photographed, and relying on friends to protect his anonymity from a motley crowd of eccentric Pynchon enthusiasts who have made various attempts to track him down. His invisibility has inspired some curious rumours, notably the absurd suggestion in the 1970s that he is **J.D. Salinger**. There is in fact no doubt about Pynchon's identity, and his early life is amply documented (though questions still linger about the mysterious disappearance of his college records and photographs from Cornell ...).

He was born in Glen Cove, Long Island, into a long-established Massachusetts family. A great-grand-uncle of the same name wrote to Nathaniel Hawthorne in 1851 complaining about the use of the name 'Pyncheon' in *The House of the Seven Gables*. Pynchon's father was an industrial surveyor and later town supervisor of Oyster Bay where Pynchon attended the High School before entering Cornell in 1953 to study physics and engineering. His education was interrupted by a two-year spell in the US Navy, based at Norfolk, Virginia, and he majored in English upon returning to Cornell, attending a course run by **Vladimir Nabokov**, who later denied having any memory of him. He began publishing short stories while still at Cornell, and published work by his friend Richard Farina in the *Cornell Writer*. Five of his early stories are gathered in *Slow Learner*. After graduating in 1959, Pynchon moved to Greenwich Village where he worked on *V*, and in 1960 to Seattle where he was a technical writer for the Boeing Aircraft Corporation, and a member of the Minuteman Logistics Support Programme.

V is an elaborate comic confection, filled with parody and black humour, which spans the USA, Europe and Africa across the century. It contrasts the obsessive quest of a character called Stencil for the mysterious 'V' (what it or she may be is never clear) with the desultory life of a crowd known as 'The Whole Sick Crew'. Ideas of paranoia and conspiracy theory recur in *The Crying of Lot 49* in which the classic pattern of the detective story is reversed, the mystery surrounding a will growing to absurd and cosmic proportions by the end of the novel. Pynchon's undoubted masterpiece is *Gravity's Rainbow*, a black comedy set in Europe at the end of the Second World War, and containing over 400 characters. Pavlovian conditioning, probability theory, rocket technology, coprophilia and political conspiracy are just a few of the subjects touched on in a torrent of fabulously inventive language. The novel won a National Book Award in 1974 – Pynchon sent a stand-up comic to accept the award. It was also selected by the Pulitzer judges, but they were overruled by the Pulitzer advisory board, who deemed it both 'obscene' and 'unreadable'.

A lengthy silence preceded the much-hyped publication of *Vineland*, a gentler tale of radical politics in 1960s' California, which was generally judged a disappointment. However, Pynchon continues to inspire obsessive devotion among enthusiasts, and the journal *Pynchon Notes* contains essays on such esoteric matters as the treatment of international light-bulb cartels in *Gravity's Rainbow*.

Fiction
V 1963; *The Crying of Lot 49* 1966; *Gravity's Rainbow* 1973; *Slow Learner* (s) 1984; *Vineland* 1990

Non-Fiction
Deadly Sins [with others] 1993

Q

Peter (Courtney) QUENNELL 1905–1993

Quennell was the son of C.H.B. Quennell, an architect, but better known, with his wife Marjorie, as one of the pioneers in writing books for schools, particularly the famous series about 'Everyday Things' in various periods and countries. Quennell attended Berkhamsted School (where he was a contemporary of **Graham Greene**, whose father was headmaster), and where, in his sixth-form years, the inclusion of his work in *Public School Verse*, and the publication of an early volume of *Masques and Poems* (1922), seemed to herald a brilliant creative career. He attended Balliol College, Oxford, between 1923 and 1925, being sent down for a heterosexual affair – then less tolerated by the university than homosexual involvements – but immediately embarking on a grand literary and social career in the sparkling climate of the 1920s. In the next few years he published another volume of poems, a novel and a volume of short stories, but a more appropriate augury for his future was the frequency with which his name appeared in such journals as the *New Statesman*, *Life and Letters* and the *Criterion*; he was to become one of the last English writers to live primarily as a man-of-letters, someone devoted to the appreciation of the work of others.

The first of the many general books (primarily biographies) in which he carried out this task, *Baudelaire and the Symbolists*, was published in 1929. His course was set, but in the early years he had a few other posts, none of them pursued with much enthusiasm. From 1930 to 1931 (incredible as it seems now, since he held no degree) he was a professor at the Japanese government university, the Tokyo Bunrika Daigaku; in the later 1930s financial difficulties made him take up advertising; and during the war he served in the Ministry of Information and the Auxiliary Fire Service. In later life, he was a magazine editor – of the *Cornhill* and *History Today* – and continued his immense output of journalism, notably as book critic of the *Daily Mail* from 1943 to 1956, a post arranged for him by his friend Ann Rothermere, influential hostess and wife of the paper's proprietor.

Books continued to flow from him after his retirement from editorship in 1979, and he was knighted in the year before his death. He wrote and edited more than fifty volumes, and, although they are marked by no great originality, and consciously eschew personal feeling, their careful perfection of style, as well as delicious gift for irony, make them memorable. Perhaps his two-volume study of Byron is his finest work, but just as purely enjoyable are his two autobiographical books, *The Marble Foot* and *The Wanton Chase*, or such volumes of assorted reminiscences as *Customs and Characters* and *The Pursuit of Happiness*. Like his friend and rival critic **Cyril Connolly**, Quennell was known as a *coureur des femmes*, and, beside many mistresses, he had five wives. From the first four he eventually became divorced. He married, finally, Joan, Lady Peek, in 1967, and she survived him; he had a son from this marriage, as well as a daughter from the third.

Fiction

The Phoenix-Kind 1931; *Sympathy and Other Stories* (s) 1933

Poetry

Masques and Poems 1922; *Poems* 1926

Non-Fiction

Baudelaire and the Symbolists: Five Essays 1929 [rev. 1954]; *A Letter to Mrs Virginia Woolf* 1932; *A Superficial Journey through Tokyo and Peking* 1932; *Byron* 1934; *Byron: the Years of Fame* 1935 [rev. 1967]; *Victorian Panorama: a Survey of Life and Fashion from Contemporary Photographs* 1937; *'To Lord Byron': Feminine Profiles, Based Upon Unpublished Letters 1807–1824* [with G. Paston [pseudonym for E.M. Symonds] 1939; *Caroline of England: an Augustan Portrait* 1939; *Byron in Italy* 1941; *Four Portraits: Studies of the Eighteenth Century* 1945 [US *The Profane Virtues: Four Studies of the Eighteenth Century*] [rev. 1965]; *John Ruskin: the Portrait of a Prophet* 1949; *The Singular Preference: Portraits and Essays* 1952; *Spring in Sicily* 1952; *Hogarth's Progress* 1955; *The Sign of the Fish* 1960; *Shakespeare: the Poet and His Background* [US *Shakespeare: a Biography*] 1963; *Alexander Pope: the Education of a Genius 1688–1728* 1968; *Romantic England: Writing and Painting 1717–1851* 1970; *Casanova in London and Other Essays* 1971; *Samuel Johnson: His Friends and Enemies* 1972; *A History of English Literature* [with H. Johnson] 1973; *Who's Who in Shakespeare* [with H. Johnson] 1973; *The Marble Foot: an Autobiography 1905–1938* (a) 1976; *The Day Before Yesterday: a Photographic Album of Daily Life in Victorian and Edwardian England* 1978; *Vladimir Nabokov: His Life, His Work, His World:*

a Tribute 1979; *The Wanton Chase: an Autobiography from 1939* (**a**) 1980; *Customs and Characters: Contemporary Portraits* 1982; *The Pursuit of Happiness* 1988

Edited

Aspects of Seventeenth Century Verse 1933 [rev. 1947]; *The Private Letters of Princess Lieven to Prince Metternich 1820–1826* 1937; Henry Mayhew: *Mayhew's London* 1949, *London's Underworld* 1950, *Mayhew's Characters* 1951; Alexander Pope *The Pleasures of Pope* 1949; Byron: *Byron: Selections from Poetry, Letters and Journals* 1949, *Byron, a Self-Portrait: Letters and Diaries, 1798 to 1824* 1950, *Byronic Thoughts: Maxims, Reflections, and Portraits from the Prose and Verse of Lord Byron* 1960; *Selected Writings of John Ruskin* 1952; George Henry Borrow *The Bible in Spain* 1959; William Hickey *Memoirs of William Hickey* 1960 [as *The Prodigal Rake: Memoirs of William Hickey* 1962]; Henry de Montherlant *Selected Essays* 1960; *The Past We Share* [with A. Hodge] 1960; Thomas Moore *The Journal of Thomas Moore: 1818–1841* [rev. 1964]; *Marcel Proust, 1871–1922: a Centenary Volume* 1971; *Genius in the Drawing-Room: the Literary Salon in the Nineteenth and Twentieth Centuries* [US *Affairs of the Mind: the Salon in Europe and America from the Eighteenth to the Twentieth Century*] 1980; *A Lonely Business: a Self-Portrait of James Pope-Hennessy* 1981; *The Selected Essays of Cyril Connolly* [with C. Connolly] 1984; John Macdonald *Memoirs of an Eighteenth-Century Footman* 1985; Baudelaire *My Heart Is Laid Bare and Other Prose Writings* 1986

Translations

A. Buzurg Ibn Shahriyar *The Book of the Marvels of India* 1928; Anthony Hamilton *Memoirs of the Comte de Gramonk* 1930

R

Craig (Anthony) RAINE 1944–

Raine was born at Shildon, County Durham. His father, who is described in the autobiographical central prose section of the poetry collection *Rich*, had worked at a variety of trades including featherweight boxer and was disabled in the war. He attended Barnard's Castle, an Anglican public school, and went up to read English at Exeter College, Oxford, in 1963. He then did a BPhil, and subsequently read for a DPhil, but did not complete his thesis; simultaneously he had begun the first of a number of part-time lecturing appointments at various Oxford colleges. In 1972 he married Ann Pasternak Slater, a niece of the Russian poet Boris Pasternak; they have continued to make Oxford their home and have four children. In 1973 Raine turned from lecturing to freelance journalism, but returned to appointments at Lincoln, Exeter and Christ Church colleges between 1974 and 1979.

In the later 1970s he won a number of prizes in poetry competitions and served as book editor of the *New Review*, and his widespread public exposure as a poet begins from this time. His first collection was *The Onion, Memory* (1978), and his second, *A Martian Sends a Postcard Home*, established him as the leading member of a poetic school of 'Martians', to which poets such as **Christopher Reid** and **David Sweetman** are also considered to belong. In 1979 Raine again quit academe and, after spending time as co-editor of *Quarto* and poetry editor of the *New Statesman*, he became Faber and Faber's poetry editor in 1981, a post he used to become a considerable force, sometimes a controversial one, in the politics of the poetry world. Among his forays on public attention has been his poem adapted from Rimbaud, 'Arsehole', which caused protest when published in the *New Statesman* in 1983. As a poet Raine is marked above all by verbal cleverness and performance, although his third collection, *Rich*, introduces attempts to treat emotional themes more movingly. *History*, sections of which were serialised in the *New Yorker*, is a long poem which draws upon the lives of his own and his wife's families and forebears. Raine has also published an opera libretto based on Pasternak, *The Electrification of the Soviet Union*, a collection of literary essays, *Haydn and the Valve Trumpet*, and *'1953'*, an adaptation of Racine's tragedy *Andromaque*.

Poetry

The Onion, Memory 1978; *A Journey to Greece* 1979; *A Martian Sends a Postcard Home* 1979; *A Free Translation* 1981; *Rich* 1984; *History: the Home Movie* 1994

Non-Fiction

Hadyn and the Valve Trumpet 1990

Plays

The Electrification of the Soviet Union (l) [from B. Pasternak] 1986; *'1953'* [from Racine's play *Andromaque*] 1990

Edited

Kipling: *A Choice of Kipling's Prose* 1987, *Selected Poems* 1993; *1985 Observer/Arvon Anthology* [with A. Clampitt, A. Stevenson] 1987

Kathleen (Jessie) RAINE 1908–

Born in London, the daughter of an English master at a secondary school, Raine spent much of her childhood with an aunt in a remote village in Northumberland. The country of the Scottish borders has been an ideal landscape for her, inspiring much of her poetry, and is an area to which she has constantly returned, although financial necessity has compelled her to spend much of her life in London. She was educated at Girton College, Cambridge, where she specialised in biology and took her natural science tripos in 1929. Her first poems were printed at this time. Soon afterwards she married a young Cambridge don, Hugh Sykes Davies; later, she eloped with the poet **Charles Madge**; both marriages were dissolved. She went with her son and daughter to the border country, but returned to London during the war to work in various civil service departments.

During this period her first volume of poems, *Stone and Flower* (1943), with illustrations by Barbara Hepworth, appeared under the poetic entrepreneur Tambimuttu's imprint, Editions Poetry London. After the war, Raine lived by 'translation, book reviewing, evening classes at Morley College, all the usual improvisations'. In 1956, she became a fellow of Girton College and embarked on extensive research on Blake, a chief influence on her own poetry. In later years she had an intense relationship with the writer and naturalist Gavin Maxwell, frustrated by his homosexuality: this was the inspiration for the poems in *On a Deserted Shore*, published after his death.

Since 1945, Raine has been an uncompromising Platonist and symbolist, committed to the idea of man as a spiritual being. Her poetry has been praised for its exactness of natural observation, graceful lyric movement and clarity of diction, but her later work has often been considered too preoccupied with the ethereal at the expense of the earthy. Her *Collected Poems* appeared in 1981; other works include studies of Blake and **W.B. Yeats**, the essays in *Defending Ancient Springs* and four volumes of autobiography. Since 1981, she has been co-editor of the review *Temenos* (Classical Greek for 'sacred precinct').

Poetry

Stone and Flower: Poems 1935–43 1943; *Living in Time* 1946; *The Pythoness and Other Poems* 1952; *The Year One* 1952; *The Collected Poems of Kathleen Raine* 1956; *Christmas 1960* 1960; *The Hollow Hill and Other Poems 1960–1964* 1965; *Six Dreams and Other Poems* 1968; *Ninfa Revisited* 1968; *Pergamon Poets 4: Kathleen Raine and Vernon Watkins* [ed. E. Owen] 1968; *A Question of Poetry* 1969; *Penguin Modern Poets 17* [with D. Gascoyne, W.S. Graham] 1970; *The Lost Country* 1971; *On a Deserted Shore* 1973; *Three Poems Written in Ireland* 1973; *The Oval Portrait and Other Poems* 1977; *Fifteen Short Poems* 1978; *The Oracle in the Heart and Other Poems 1975–1978* 1980; *Collected Poems 1935–1980* 1981; *Precognition on the Isle of Eigg* 1986; *The Presence: Poems 1984–87* 1987; *Selected Poems* 1988; *To the Sun* 1988; *Golgonooza, City of Imagination* 1991; *Living with Mystery* 1992

Non-Fiction

William Blake 1951; *Coleridge* 1953; *Poetry in Relation to Traditional Wisdom* 1958; *Blake and England* 1960; *Defending Ancient Springs* 1967; *The Written Word* 1967; *Blake and Tradition* 1968; *William Blake* 1971; *Faces of Day and Night* (a) 1972; *Hopkins, Nature and Human Nature* 1972; *Years, the Tarot and the Golden Dawn* 1972; *Farewell Happy Fields* (a) 1973; *David Jones: Solitary Perfectionist* 1974; *Death-in-Life and Life-in-Death* 1974; *A Place, a State* [with J. Trevelyan] 1974; *The Land Unknown* (a) 1975; *The Inner Journey of the Poet* 1976; *Berkeley, Blake, and the New Age* 1977; *The Lion's Mouth* (a) 1977; *From Blake to 'A Vision'* 1978; *David Jones and the Actually Loved and Known* 1978; *Blake and the New Age* 1979; *Cecil Collins, Painter of Paradise* 1979; *'What is Man?'* 1980; *The Human Face of God: William Blake and the Book of Job* 1982; *The Inner Journey of the Poet and Other Papers* [ed. B. Keeble] 1982; *Yeats to Initiate: Essays on Certain Themes in the Writings of W.B. Yeats* 1984; *Poetry and the Frontiers of Consciousness* 1985; *The English Language and the Indian Spirit* 1986; *India Seen Afar* 1990; *Autobiographies* (a) 1991

Edited

Aspects de littérature anglaise 1918–1945 [with M.-P. Fouchet] 1947; Coleridge: *Letters of Samuel Taylor Coleridge* 1950, *Selected Poems and Prose of Coleridge* 1957; *Thomas Taylor the Platonist: Selected Writings* [with G. Mills Harper] 1969; *A Choice of Blake's Verse* 1974; *Shelley* 1974

Translations

Denis de Rougemont *Talk of the Devil* 1945; Paul Foulquie *Existentialism* 1948; Balzac: *Cousin Bette* 1948, *Lost Illusions* 1951; Calderón *Life's a Dream* [with R.M. Nadel] 1968; Lorca *Sun and Shadow* [with R.M. Nadel] 1972

John Crowe RANSOM 1888–1974

Born in Pulaski, Tennessee, the son of a Methodist minister, Ransom grew up in a number of small towns in that state. One great-uncle was a founder of the Ku Klux Klan, and the values and traditions of the old South remained a constant of Ransom's life and work. He attended Bowen School, Nashville, and Vanderbilt University in the same town, and then, as a Rhodes scholar, took a second undergraduate degree at Christ Church College, Oxford, graduating in *literae humaniores* in 1913. In 1914 he returned, as an instructor in English, to Vanderbilt, and remained there, after 1924 as a professor, until 1937, with a break from 1917 to 1919 when he served as an artillery officer in France.

While at Vanderbilt, he became one of the most influential figures in American letters, the acknowledged leader of the so-called Southern Renaissance in American poetry and literature, of the conservative political philosophy known as Southern agrarianism, and the related group of poets called the Fugitives, after their magazine the *Fugitive* (1922–5). Ransom's own achievement as a poet was concentrated very largely in the years covered by the magazine. His first poetic volume, *Poems about God* (1919), is immature, and almost all his mature work is concentrated in two volumes, *Chills and Fever* and *Two Gentlemen in Bonds*. He wrote few poems in the 1930s, and more appeared only intermittently thereafter.

As a poet Ransom is cerebral and rationalistic, but can communicate passion strongly while usually avoiding direct statement: he is undoubtedly among the major American poets of the century. From 1937 until his retirement in 1959 he was a professor of English at Kenyon College, Ohio, and during this period he became one of the most influential literary critics in America, the editor of the *Kenyon Review* and a pioneer of the New Criticism, which studied literary texts in terms of structure and texture rather than social and historical context. In later years he revised his earlier poems continuously, usually to their detriment. Ransom was a slight, courtly, pipe-smoking man; he married in 1920, and had a daughter and two sons.

Poetry

Poems about God 1919; *Chills and Fever* 1924; *Grace after Meat* 1924; *Two Gentlemen in Bonds* 1927; *Selected Poems* 1945

Non-Fiction

God without Thunder: an Unorthodox Defense of Orthodoxy 1930; *Topics for Freshman Writing* 1935; *The World's Body* 1938; *The Intent of the Critic* (c) [ed. D.A. Stauffer] 1941; *The New Criticism* 1941; *A College Primer of Writing* 1943; *Beating the Bushes: Selected Essays 1941–1970* 1972; *Selected Essays of John Crowe Ransom* [ed. J. Hindle, T.D. Young] 1984

Collections

Poems and Essays 1955

Edited

The Kenyon Critics 1951; *Selected Poems of Thomas Hardy* 1961

Arthur (Michell) RANSOME 1884–1967

The eldest child of Cyril Ransome, a professor of history at Leeds, Ransome went reluctantly to school at Rugby. He read science at the Yorkshire College (later Leeds University), which he left after two terms, to work as office boy for the publisher Grant Richards in 1901. For over a decade, he enjoyed a frugal Bohemian existence, writing stories and articles, the publication of which he invariably celebrated with friends over Australian burgundy and macaroni cheese. Holidays were spent in the Lake District, where he boated with a family who inspired the characters in his 'Swallows and Amazons' series of children's books.

Partly to escape a miserable marriage contracted in 1909, Ransome went to Russia in 1913, ostensibly to write a guide to St Petersburg, which he abandoned at the onset of the First World War. In his absence, he won a libel action brought by the famously litigious Lord Alfred Douglas against his biography of Oscar Wilde. He remained in Russia, as correspondent for the *Daily News*, and made friends with Lenin and other Bolsheviks. Because of his support of the Russian Revolution, he objected strongly to the British Foreign Office's interference in Russian affairs. He moved to Estonia in 1919, to live with Trotsky's former secretary, Evgenia Shelepin, whom he married after his divorce in 1924. Ransome built a ketch in which he sailed the Baltic in 1922, and about which he wrote in *Racundra's First Cruise*. After a short spell in Egypt as political correspondent to the Manchester *Guardian*, he reported from China, but he was bored and returned to England in 1926.

In 1929, aged forty-five he began *Swallows and Amazons*, published in 1930 by Cape, who preferred his *Rod and Line*, a volume of fishing essays. *Swallowdale* was a sequel to *Swallows and Amazons*, but the success of a third volume, *Peter Duck*, illustrated by himself, enabled him to continue the children's saga in nine other books. In 1936 *Pigeon Post* earned him the first Carnegie Medal for the best children's book. A sincere love of the country, small boats and plausible adventures, contributes to the series' popularity. Ransome's child characters were fully realised, as was the only prominent adult, Uncle Jim, whose idiosyncrasies resembled Ransome's own.

Ransome was large, bald and tweedy, with a Captain Pugwash moustache, but his humour belied this appearance. He must have been amused at his award of DLitt from Leeds University in 1952, but gratified to receive the CBE in 1953. His last book was *Mainly about Fishing*, although a posthumous autobiography was published in 1967.

For Children

The Child's Book of the Seasons 1906; *Highways and Byways in Fairyland* 1906; *Old Peter's Russian Tales* (s) 1916; *Aladdin and His Wonderful Lamp in Rhyme* 1919; *Swallows and Amazons* 1930; *Swallowdale* 1931; *Peter Duck* 1932; *Winter Holiday* 1933; *Coot Club* 1934; *Pigeon Post* 1936; *We Didn't Mean to Go to Sea* 1937; *Secret Water* 1939; *The Big Six* 1940; *Missee Lee* 1941; *The Pict and the Martyrs* 1943; *Great Northern?* 1947

Non-Fiction

The Souls of the Street and Other Little Papers 1904; *Pond and Stream* 1906; *Things in Our Garden* 1906; *Bohemia in London* 1907; *A History of Story-Telling* 1909; *Edgar Allan Poe* 1910; *The Hoofmarks of the Faun* 1911; *Oscar Wilde* 1912; *Portraits and Speculations* 1913; *The Elixir of Life* 1915; *Six Weeks in Russia in 1919* 1919; *The Soldier and Death* 1920; *The Crisis in Russia* 1921; *Racundra's First Cruise* 1923; *The Chinese Puzzle* 1927; *Rod and Line* 1929; *Fishing* 1955; *Mainly about Fishing* 1959; *The Autobiography of Arthur Ransome* (a) [ed. R. Hart-Davis) 1967

Collections
The Stone Lady, Ten Little Papers and Two Mad Stories 1905

Biography
The Life of Arthur Ransome by Hugh Brogan 1984

Raja RAO 1909–

With **Mulk Raj Anand** and **R.K. Narayan**, Rao is one of the three most distinguished English-language Indian writers of the mid-century. He was born into a well-known Brahmin family at Mysore, the son of a professor, and was educated at Madras University, taking his BA in 1929. Shortly afterwards, he left India for Europe, where he remained for a decade. He did postgraduate studies at the University of Montpellier and then at the Sorbonne, where he was a student of the well-known Professor Louis Cazamian.

His first short stories were published in English and French, and some of his early writing is also in an Indian language, Kannada. In 1931, he married a French academic, Camille Mouly; his second novel, *The Serpent and the Rope*, published in 1960, presents a remarkably objective analysis of the subsequent breakdown of their marriage. On his return to India, he spent much time travelling the sub-continent in a search for his spiritual heritage and also edited the literary magazine *Tomorrow*. His first novel, *Kanthapura* (1938), is an account of the impact of Ghandism on a south Indian village told in a racy, breathless style by an old woman, and was highly praised by **E.M. Forster** among others; a literary reputation was in the making which Rao has been able to sustain by the publication of scarcely half-a-dozen works of fiction throughout his life. He returned to the theme of Ghandism in south India during the 1930s in his highly praised volume of short stories, *The Cow of the Barricades*.

After the war, he spent much of his life in France and also travelled widely throughout the world. From 1965 he was a professor of philosophy at the University of Austin, Texas, teaching one semester a year and living in India when not teaching. He married an actress, Katherine Jones, in 1965, and has one son. He retired from full-time teaching in 1983, but maintains his links with the university. Rao's later work as a novelist has been more metaphysical and philosophical, as are his later short stories, *The Policeman and the Rose*. More lighthearted (for Rao) is *The Cat and Shakespeare*, in which the Hindu notion of *karma* is symbolised by a cat.

Fiction
Kanthapura 1938; *The Cow of the Barricades and Other Stories* (s) 1947; *The Serpent and the Rope* 1960; *The Cat and Shakespeare* 1965; *Comrade Kirillov* 1976; *The Policeman and the Rose* (s) 1977

Non-Fiction
Whither India [with I. Singh] 1948; *The Chess Master and His Moves* 1978

Edited
Changing India [with I. Singh] 1939; Jawaharlal Nehru *Soviet Russia: Some Random Sketches and Impressions* 1949

Frederic (Michael) RAPHAEL 1931–

Raphael was born in Chicago, to a British father and American mother, educated at Charterhouse from where, in 1950, he won a major classical scholarship to St John's College, Cambridge. Soon after going down in 1955 he married, and has two sons and a daughter (the painter Sarah Raphael). The following year he published his first novel, *Obbligato*, a satirical account of the popular music scene. Since then there has been a steady stream of novels on a variety of themes, television plays and screenplays from his own and other people's novels – he won an Oscar in 1966 for his screenplay of *Darling* and the British Screen Writers Award in three successive years (1965, 1966 and 1967) – stage plays and collections of stories.

Of the novels, *Orchestra and Beginners* is characteristic, opening with a country-house party in southern England seemingly untroubled by the declaration of war that morning. It proceeds to explore the middle-class values of the time through an examination of a marriage based on decidedly rocky principles. Very different is *Like Men Betrayed*, set in Greece, which is the study of a revolutionary, climaxing with the military coup of April 1967, while *Who Were You with Last Night?*, in which a husband has murderous thoughts about his wife, is a thriller which has been said to share 'the shocking ambiguity of *The Turn of the Screw*'. *Richard's Things*, in which a wife develops a passionate yet ambiguous relationship with the young woman who was with her husband when he died, is one of several of his novels he has adapted for the screen.

Of his other work for cinema and television, he is best known for his adaptations of **Thomas Hardy**'s *Far from the Madding Crowd* (1967), and his own *Darling*, the cynical story of the rise and fall of a model, and very much a story of the 1960s. Both films were directed by John Schlesinger and starred that icon of the period, Julie Christie. His most popular work for television was the series *The Glittering Prizes*, which followed a group of Cambridge undergraduates. He returned to this theme in *Oxbridge Blues*. Other television work includes distinguished adaptations of novels by **Roy Fuller** and **Geoffrey Household**. His Jewish origins, which have been a significant influence upon his work, became a central theme of his drama series *After the War*.

A novelist of much variety, he has been praised for his brilliant dialogue and the precision of his style. He has also written pungent, personal and controversial biographies of Byron and **W. Somerset Maugham** and been a regular reviewer of books, notably for the *Sunday Times*.

Fiction

Obbligato 1956; *The Earlsdon Way* 1958; *The Limits of Love* 1960; *A Wild Surmise* 1961; *The Graduate Wife* 1962; *The Trouble with England* 1962; *Lindmann* 1963; *Darling* 1965; *Orchestra and Beginners* 1967; *Like Men Betrayed* 1970; *Who Were You with Last Night?* 1971; *April, June and November* 1972; *Richard's Things* 1973; *California Time* 1975; *The Glittering Prizes* 1976; *Sleeps Six and Other Stories* **(s)** 1979; *Oxbridge Blues and Other Stories* **(s)** 1980; *Heaven and Earth* 1985; *Think of England* **(s)** 1986; *After the War* 1988; *A Double Life* 1993; *Latin Lover and Other Stories* **(s)** 1994

Plays

Bachelor of Hearts **(f)** [with L. Bricusse] 1958; *Lady at the Wheel* [with R. Beaumont, L. Bricusse, L. Hill] 1958 [np]; *Don't Bother to Knock* **(f)** [with D. Connan, F. Gotfurt] 1961; *The Executioners* **(tv)** 1961; *A Man on the Bridge* 1961 [np]; *Image of a Society* [from R. Fuller's novel] 1963; *Nothing But the Best* **(f)** 1963; *The Trouble with England* **(tv)** [from his novel] 1964; *Darling* **(f)** [from his novel] 1965; *The Island* [pub. in *Eight Plays 2* ed. M. Fellows 1965]; *Two for the Road* **(f)** [pub. 1967]; *Far from the Madding Crowd* **(f)** [from Hardy's novel] 1967; *A Severed Head* **(f)** [from I. Murdoch's novel] 1970; *Daisy Miller* **(f)** [from H. James's novel] 1974; *The Glittering Prizes* **(tv)** [from his novel] 1976; *Rogue Male* **(tv)** [from G. Household's novel] 1976; *Something's Wrong* **(tv)** 1978; An Early Life 1979 [np]; *From the Greek* 1979 [np]; *The Serpent Son* **(tv)** [with K. McLeish] 1979; *Of Mycene and Men* **(tv)** [with K. McLeish] 1979; *School Play* **(tv)** 1979; *The Best of Friends* **(tv)** 1980; *Richard's Things* [from his novel] 1981; *Oxbridge Blues* **(tv)** 1984; *After the War* **(tv)** [from his novel] 1989; *The King's Whore* **(f)** [from J. Toumier's novel; with A. Corti, D. Marlowe, D. Vigne] 1990; *The Man in the Brooks Brothers Shirt* **(f)** 1990

Non-Fiction

W. Somerset Maugham and His World 1977 [rev. 1989]; *Cracks in the Ice: Views and Reviews* 1979; *The List of Books: an Imaginary Library* [with K. McLeish] 1981; *Byron* 1982; *The Hidden I* 1990; *Of Gods and Men* 1994

Translations

The Poems of Catullus [with K. McLeish] 1978; Aeschylus *Plays One* and *Plays Two* [with K. McLeish, J.M. Walton] 1991

Edited

Bookmarks 1975

Terence (Mervyn) RATTIGAN 1911–1977

Born in Kensington, the son of a minor diplomat and grandson of a chief justice for the Punjab, Rattigan won a scholarship to Harrow, then in 1930 went up to Trinity College, Oxford, to read history. There he acted with the Oxford University Dramatic Society and developed a reputation for style and unconventionality: one fellow undergraduate, Michael Longson (later the author of the memoir *A Classical Youth*, 1985), felt compelled to visit him in his rooms to warn him about his behaviour. In 1933 *First Episode*, a play Rattigan had written in collaboration with Philip Heimann, opened at a theatre in Kew, and the following year enjoyed a brief run in London's West End. Rattigan persuaded his father to let him leave Oxford without a degree and give him an allowance with which to set himself up as a playwright in London.

He wrote a number of plays which were never produced, but in 1936 *French without Tears*, a comedy based on a long vacation from Oxford spent learning French at a crammer's in Boulogne, was a big West End success, running for over 1,000 performances. Rattigan was soon established as one of the leading playwrights of the day, earning at the age of twenty-five some £25,000 a year in royalties. His later reputation as an establishment figure is belied by his activities in the 1930s, when he was a contributor to the *New Statesman* and associated with left-wing causes. He had been a pacifist at Oxford, but now marched on Downing Street to demand assistance for the Republican cause in the Spanish Civil War (**T.C. Worsley** dedicated his 1971 novel about that war, *Fellow Travellers*, to Rattigan 'who also in his time chanted Arms for Spain'). *Follow My Leader*, a satire upon Hitler he had written with Anthony Maurice, was banned by the Lord Chamberlain in 1938, and

After the Dance, his much underrated play which criticises the cynicism and frivolity of the 1920s' generation, remained unrevived after its 1939 premiere until a fine television version in 1993.

During the Second World War, Rattigan served with the Royal Air Force, an experience recalled in *Flare Path.* He also wrote several screenplays for propagandist feature films, notably *The Day Will Dawn* (1942), the semi-documentary *Journey Together* (1944), and *The Way to the Stars* (1945), in which use was made of **John Pudney**'s famous poem 'For Johnny'. This last film was directed by Anthony Asquith, with whom Rattigan was to collaborate on screen adaptations of his best-known plays of the 1940s: *The Winslow Boy*, about a naval cadet falsely accused of stealing a postal order and based on the Archer-Shee case of 1908, and *The Browning Version*, about a desiccated schoolmaster (definitively played in the film by Michael Redgrave). Rattigan's particular territory was what **E.M. Forster** diagnosed as 'the undeveloped heart', a peculiarly British trait in which emotions are tamped down beneath a phlegmatic public exterior, and his skill lay in showing that this exterior was no more than a thin crust beneath which genuine passions seethed. What happens when such passions break the surface is demonstrated in *The Deep Blue Sea*, which opens with a woman attempting suicide. The estranged wife of a barrister, she has become involved with a caddish younger man, a former air ace with a drinking habit. The idea for the play arose when one of Rattigan's former lovers, a young actor called Kenneth Morgan, committed suicide, and a homosexual version of the play is said to have existed.

From the late 1950s onward, Rattigan's reputation went into decline, his 'well-made' plays dismissed as **John Osborne**'s generation of playwrights came to prominence. Rattigan reacted pugnaciously (and unwisely) by creating a fictional member of the audience, 'Aunt Edna', whom he believed playwrights should seek to please. Later plays were not, on the whole, up to the standard of his earlier work, although *Ross*, which deals with Lawrence of Arabia's service in the RAF, once again demonstrated Rattigan's compassionate understanding of complicated and repressed psychology, and his last play, *Cause Célèbre*, was a moving fictional version of the Rattenbury murder case of 1935. A younger playwright he did admire, and champion, was **Joe Orton**, whose *Entertaining Mr Sloane* (1964) he described as 'the most exciting and stimulating first play that I've seen in thirty (odd) years of playgoing'. He persuaded a West End theatre to take on the play, investing £3,000 in its transfer.

A courtly, urbane figure, with rooms in Albany and a Rolls-Royce with a personalised number-plate, Rattigan looked every inch the successful playwright, and he was knighted in 1971. He spent his last years in Bermuda, dying there of bone-marrow cancer. Undoubtedly, his homosexuality was a vital component of his work, not only in the subjects he chose (his penultimate project was a television film, as yet unproduced, about Nijinsky and Diaghilev, and in the original version of *Separate Tables*, the disgraced major had been exposed for homosexual activity rather than for advances to a young woman in a cinema), but also in the way he returned repeatedly to the theme of love thwarted by differences in age, class or the disapproval of society. Although he enjoyed a number of relationships – notably with the socialite and Member of Parliament Chips Channon – homosexuality was not legalised in Britain until the last decade of his life.

Since Rattigan's death, his reputation has gradually revived, and his plays have begun to take their place in the repertory once more. The theatrical fashion which at one point seemed destined to sweep his work into oblivion has itself passed, and his best plays are now becoming recognised as beautifully structured, deeply felt and psychologically complex investigations of the British character.

Plays

First Episode [with P. Heimann] 1933; *French without Tears* **(st)** 1936, **(f)** 1939 [pub. 1937]; *After the Dance* 1939; *Follow My Leader* [with A. Maurice] 1940; *Flare Path* 1942; *While the Sun Shines* 1942 [pub. 1944]; *Love in Idleness* 1944 [pub. 1945]; *The Winslow Boy* **(st)** 1946, **(f)** 1948 [pub. 1946]; *Playbill*: *The Browning Version* [**(f)** 1951] and *Harlequinade* 1948; *Adventure Story* 1949 [pub. 1950]; *Who Is Sylvia?* 1950 [pub. 1951]; *The Final Test* **(tv)** 1951; *The Deep Blue Sea* **(st)** 1952, **(f)** 1955; *The Sleeping Prince* 1953 [pub. 1954]; *Separate Tables* 1954 [pub. 1955]; *The Prince and the Showgirl* **(f)** 1957; *Variation on a Theme* 1958; *Ross* 1960; *Adventure Story* **(tv)** 1961; *Heart to Heart* **(tv)** 1962; *Man and Boy* 1963; *The VIPs* **(f)** 1963; *The Yellow Rolls-Royce* **(f)** 1965; *Nelson, a Portrait in Miniature* **(tv)** 1966; *All on Her Own* **(tv)** 1968; *A Bequest to the Nation* **(st)** 1970, **(f)** 1973; *In Praise of Love* 1973; *Cause Célèbre* **(r)** 1975, **(st)** 1977 [pub. 1978]

Biography

Terence Rattigan by Michael Darlow and Gillian Hodson 1979

Simon (Arthur Nöel) RAVEN 1927–

The son of a man of independent means who spent much of his life playing golf, Raven was born in Virginia Water, Surrey. By his own admission, his life has been something of a rake's progress, and a similarly gamey quality is the hallmark of his fiction. Indeed, the luckless, bisexual Fielding Gray, anti-hero of his novel sequence *Alms for Oblivion*, is an acknowledged self-portrait.

Raven won a top scholarship to Charterhouse, where he was educated alongside William Rees-Mogg (future editor of *The Times* and television's moral watchdog) and James Prior (future Tory cabinet minister), both of whom are alleged to be models for characters in *Alms for Oblivion*. Expelled from school, he went to do his national service, and, as Britain prepared to hand India back to the Indians, was sent to Bangalore as an officer cadet (a experience which furnished material for his novel *Sound the Retreat*). In 1948 he went to King's College, Cambridge, where he read classics and 'was taken up by some of the more glittering Fellows', including **E.M. Forster**, who provided (through his friend **J.R. Ackerley**, then literary editor of the *Listener*) an *entrée* to Grub Street. Raven graduated in 1951, the same year he married Susan Kilner (the journalist Susan Raven). They had one son and the marriage was brief: within two years Raven had entered the King's Shropshire Light Infantry, which he later described as 'a jolly, *louche*, discernibly amateur, above all *bachelor* sort of regiment'. He served as a captain in Germany (the background for *The Sabre Squadron*) and Kenya. He was obliged to resign his commission in 1957 (the year of his divorce) as the result of unpaid gambling debts, and embarked upon a career as a man-of-letters.

His first novel, *The Feathers of Death* (1959), concerned a homosexual scandal in the army which ends in murder and court martial, and announced themes which recur throughout his work. *The Rich Pay Late*, the first volume of *Alms for Oblivion*, was published in 1964, by which time he had also published a study of *The English Gentleman* (a fascinating background to the milieu and mores of his novels) and a miscellany of journalism, drama and autobiography, *Boys Will Be Boys*, the title-essay of which, originally published in *Encounter* in 1960, was a groundbreaking account of 'The Male Prostitute in London'. Nine further volumes of *Alms* appeared, providing a raffish picture of post-war life and centring on ten principal characters, taking them (unsequentially) from their far-from-innocent schooldays to an older-but-scarcely-wiser middle age. The sequence is uneven, but as a whole is brilliantly plotted, teeming with incident and excitement (much of it sexual), extremely funny and occasionally very moving. It provides a splendidly seedy and superbly vigorous counterpart to Anthony Powell's *A Dance to the Music of Time* (1951–75). Some of the characters recur in Raven's other novels, and the sequence is continued by a second *roman-fleuve*, *The First Born of Egypt*.

Raven has also written several volumes of scabrous memoirs, largely devoted to cricket (an enduring passion), the army and sexual and alcoholic excess. Perhaps surprisingly, *'Is There Anybody There?' said the Traveller* (subtitled 'Memoirs of a Private Nuisance', and the subject of a number of priggish reviews) is the only one of his books that has had to be withdrawn after a number of pending libel actions. Raven's classical education has paid dividends both in his style and in his subject-matter: his best books describe bad behaviour in fine prose. He has also written extensively for television, adapting the works of (amongst others) Anthony Trollope, **Nancy Mitford** and **Iris Murdoch**, and scripting *Edward and Mrs Simpson*, a hugely successful series about the abdication of Edward VIII.

Fiction

The Feathers of Death 1959; *Brother Cain* 1959; *Doctors Wear Scarlet* 1960; *Close of Play* 1962; *Alms for Oblivion*: *The Rich Pay Late* 1964, *Friends in Low Places* 1965, *The Sabre Squadron* 1966, *Fielding Gray* 1967, *The Judas Boy* 1968, *Places Where They Sing* 1970, *Sound the Retreat* 1971, *Come Like Shadows* 1972, *Bring Forth the Body* 1974, *The Survivors* 1976; *The Fortunes of Fingel* **(s)** 1976; *Sexton Blake and the Demon God* [with J. Garforth] 1978; *An Inch of Fortune* 1980; *The Roses of Picardie* 1980; *September Castle* 1983; *The First Born of Egypt*: *Morning Star* 1984, *The Face of the Waters* 1985, *Before the Cock Crow* 1986, *New Seed for Old* 1988, *Blood of My Bone* 1989, *In the Image of God* 1990, *The Troubadour* 1992; *Islands of Sorrow* 1994

Plays

Move up Country **(r)** 1960, **(tv)** 1961 [pub. 1965]; *Loser Plays All* **(r)** 1961; *A Present from Venice* **(r)** 1961; *Royal Foundation* **(tv)** 1961 [pub. 1965]; *A Friend in Need* **(r)** 1962; *The Gate of Learning* **(r)** 1962; *The Domesday School* **(r)** 1963 [pub. 1965]; *The High King's Tomb* **(r)** 1964 [pub. 1965]; *Panther Larkin* **(r)** 1964 [pub. 1965]; *The Scapegoat* **(tv)** 1964 [pub. 1965]; *The Scouncing Stoup* **(r)** 1964 [pub. in *New Radio Drama* 1966]; *Advise and Dissent* **(tv)** 1965; *The Gaming Book* **(tv)** 1965; *The Melos Affair* 1965; *Sir Jocelyn, the Minister Would Like a Word* **(tv)** 1965; *Triad* **(r)** [from Thucydides' *History of the Peloponnesian War*] 1965; *A Pyre for Private James* **(tv)** 1966; *Royal Foundation and Other Plays* [pub. 1966]; *A Soirée at Bossom's Hotel* **(tv)** 1966; *The Last Expedition* **(r)** 1967; *The Prisoners in the Cave* **(r)** 1967 [n.p.]; *The Tutor* **(r)** 1967; *The Case of Father Brendon* 1968 [np]; *Point Counter Point* **(tv)** [from A. Huxley's novel] 1968; *On Her Majesty's Secret Service* **(f)** [with R. Maibaum] 1969; *The*

Way We Live Now (tv) [from Trollope's novel] 1969; *The Human Element* (tv) [from Somerset Maugham's novel] 1970; *Unman Wittering and Zigo* (f) [from G. Cooper's play] 1971; *The Pallisers* (tv) [from Trollope] 1974; *Salvation* (r) 1974; *An Unofficial Rose* (tv) [from I. Murdoch's novel] 1974; *Edward and Mrs Simpson* (tv) 1978; *In Transit* (r) 1978; *Love in a Cold Climate* (tv) [from N. Mitford's novel] 1980; *The Search for Alexander the Great* (tv) [with G. Lefferts] 1981; *Ghost Stories* (tv) [from M.R. James] 1990

Non-Fiction

The English Gentleman: an Essay in Attitudes 1961 [US *The Decline of the Gentleman* 1962]; *Shadows on the Grass* (a) 1982; *The Old School* 1986; *The Old Gang* (a) 1988; *Bird of Ill Omen* (a) 1989; *'Is There Anybody There?' Said the Traveller* (a) 1990

Edited

The Best of Gerald Kersh 1960

Collections

Boys Will Be Boys 1963

Derek RAYMOND

see Robin COOK

Herbert (Edward) READ 1893–1968

Descended from a long line of yeoman farmers, Read was born at Muscoates Grange, Kirbymoorside, in the North Riding of Yorkshire. His first ten years were spent on this farm, but, after his father's death, he was sent, unhappily, to a school for orphans in Halifax and began employment as a clerk in the Leeds Savings Bank when aged sixteen. Through evening study, he was able to proceed to the University of Leeds, but his career there was cut short by the First World War, during which he served in the trenches, was promoted to captain, and received the Distinguished Service Order and the Military Cross. His first book of poems, much influenced by Imagism, was *Songs of Chaos* (1915).

After demobilisation, he worked for a time at the Treasury, but soon became an assistant curator at the Victoria and Albert Museum, a first move into the field of art criticism and history with which he was primarily associated. He entered London literary life, becoming a close friend of **T.S. Eliot**, and the most frequent contributor, after Eliot himself, to the *Criterion*. His first major work of literary criticism was *Reason and Romanticism* (1926), and in that year he also published his first *Collected Poems*. Read considered himself primarily a poet, but his spare, taut verses are rarely completely successful: the long poem 'The End of a War', reflecting his experience in the trenches, is perhaps one of his best. Read became a professor of fine art at Edinburgh University from 1931 to 1933, and then editor of the *Burlington Magazine* until 1939. He was for many years a regular and controversial art critic for the *Listener*, much valued by the combative literary editor, **J.R. Ackerley**, whom he coached in artistic matters over extended luncheons at the Café Royal (at the BBC's expense).

He published well over sixty books of very many types: many volumes of art criticism (in which he succeeded Roger Fry as the main English champion of modernism); studies of Wordsworth, Malory and Sterne; many volumes about education; two books about life in the trenches; verse drama; and the two books which are generally regarded as his finest achievements in imaginative terms: his autobiography of childhood, *The Innocent Eye*, and his allegorical, fantastic novel, *The Green Child*.

In later life, he held various academic posts, including professor of poetry at Harvard, and devoted much of his time to public affairs, committees and art administration: he was as well known for having been one of the founders of the Institute of Contemporary Arts as for sitting down on Ban the Bomb demonstrations in Trafalgar Square. He was knighted in 1953, and received the Dutch Erasmus Prize in 1966 for contributions to European culture. In his last years he left London and lived at Stonegrave in Yorkshire, near to where he had been born. He married twice, and had five children, one of whom is the novelist **Piers Paul Read**.

Poetry

Songs of Chaos 1915; *Eclogues* 1919; *Naked Warriors* 1919; *Mutations of the Phoenix* 1923; *Collected Poems 1913–1925* 1926; *The End of a War* 1931; *Poems 1914–1934* 1935; *Thirty-Five Poems* 1940; *A World within a War* 1944; *Moon's Face and Poems Mostly Elegiac* 1955

Fiction

Ambush (s) 1930; *The Green Child* 1935

Non-Fiction

English Pottery [with B. Rackham] 1924; *In Retreat* (a) 1925; *English Stained Glass* 1926; *Reason and Romanticism* 1926; *English Prose Style* 1928; *Phases of English Poetry* 1928; *The Sense of Glory* 1929; *Julien Benda and the New Humanism* 1930; *Wordsworth* 1930; *The Meaning of Art* 1931; *The Innocent Eye* (a) 1933; *Art Now: an Introduction to the Theory of Modern Painting and Sculpture* 1934; *Art and Industry* 1934; *Essential Communism* 1935; *Five on Revolutionary Art* (c) [ed. B. Rea] 1935; *In Defence of Shelley and Other Essays* 1936; *Art and Society* 1937; *Collected Essays in Literary Criticism* 1938; *Poetry and Anarchism* 1938; *Annals of Innocence and Experience* (a) 1940; *The Philosophy of Anarchism* 1940; *The Politics of the Unpolitical* 1943; *Education through Art* 1943; *The Education of Free Men* 1944; *A Coat of Many Colours* 1945; *The Psychopathology of Reaction in the Arts* 1948; *Coleridge as Critic* 1949; *Education for Peace* 1950; *Art and Evolution of Man* 1951; *Byron* 1951; *Contemporary British Art* 1951; *The Philosophy of Modern Art: Collected Essays* 1952; *The True Voice of Feeling* 1953; *Anarchy and Order* 1954; *The Art of Sculpture* 1956; *To Hell with Culture and Other Essays on Art and Society* 1953; *The Grass Roots of Art: Lectures on the Social Aspects of Art in an Industrial Age* 1955; *Icon and Idea: the Function of Art in the Development of Human Consciousness* 1955; *The Tenth Muse: Essays in Criticism* 1957; *Lynn Chadwick* 1958; *A Concise History of Modern Painting* 1959; *Creative Arts in American Education* [with T. Munro] 1960; *The Form of Things Unknown* 1960; *Truth is More Sacred* [with E. Dahlberg] 1961; *A Letter to a Young Painter* 1962; *Thc Contrary Experience* (a) 1963; *A Concise History of Modern Sculpture* 1964; *Form in Modern Poetry* 1964; *Henry Moore: a Study of His Life and Work* 1965; *The Origins of Form in Art* 1965; *Redemption of the Robot: My Encounter with Education through Art* 1966; *Art and Alienation: the Role of the Artist in Society* 1967; *Poetry and Experience* 1967; *Arp* 1968; *The Correspondence of Sir Herbert Read* [pub. in S. Berne's *The Unconscious Victorious and Other Stories*] 1969; *The Cult of Sincerity* (a) 1969; *Pursuits and Verities* 1983

Edited

Wilhelm Worringer *Form in Gothic* 1927; *The London Book of English Prose* [with B. Dobree] 1931; *The English Vision* 1933; *Unit One: the Modern Movement in English Architecture, Painting and Sculpture* 1934; A.R. Orage *Selected Essays and Critical Writings* [with D. Saurat] 1935; *Essays and Studies by Members of the English Association* vol 21 1936; T.E. Hulme *Speculations: Essays on Humanism and the Philosophy of Art* 1936; *Surrealism* 1936; *The Knapsack: a Pocket Book of Prose and Verse* 1939; *English Master Painters* 1940; *The Practice of Design: Essays by A. Morton and Others* 1946; *The London Book of English Verse* [with B. Dobree] 1949; *The Collected Works of C.G. Jung* [with M. Fordham] 20 vols 1953–79; *This Way Delight: a Book of Poetry for the Young* 1956; *The Acanthus History of Sculpture* [with H.D. Molesworth] 4 vols 1962 onwards; *Encyclopaedia of the Arts* 1966 [rev. as *The Thames and Hudson Dictionary of Art and Artists* 1985]; *Selected Works of Miguel de Unamuno* [with others] 1967 onwards

Plays

The Parliament for Women [pub. 1960]; *Aristotle's Mother* [pub. 1961]; *Lord Byron at the Opera* (r) [pub. 1963]

Translations

Rudolf Arnheim *Radio* [with M. Ludwig] 1936

Biography

The Last Modern by James King 1990

Piers Paul READ 1941-

Read was born at Beaconsfield, the third son of the poet and art critic **Herbert Read**, and lived in Buckinghamshire until he was eight when his family moved to North Yorkshire; his Roman Catholic upbringing and education have been crucial to his development as a writer. He attended Gilling Castle, York, Ampleforth and St John's College, Cambridge. On going down from Cambridge in 1962, he won a Ford Foundation Award as artist-in-residence in Berlin from 1963 to 1964. Returning to London, he became a sub-editor on the *Times Literary Supplement* in 1965. A Commonwealth Fund Award took him to New York in 1967, an experience that produced his 'American novel', *The Professor's Daughter*, with its background of student unrest. By then he had published his first book, *Game in Heaven with Tussy Marx*, Marx's youngest daughter's view of the world from heaven. His second, *The Junkers*, which won the Geoffrey Faber Memorial Prize, and *The Free Frenchman*, which won the Enid McLeod Literary Prize, both take a horrified look at Nazism. *Monk Dawson*, perhaps the most interesting of his early novels, concerns a priest who loses his vocation. Of his mature work the best novels are probably *A Married Man* and *Season in the West*, which won a James Tait Black Memorial Prize. They display Read's continuing concern with moral laxity. In *A Married Man* Clare sleeps with her husband's worthless friend without really knowing why; in *A Season in the West*, Laura Morton does the same thing with a dissident exile for no better reason than that 'everybody does it'. So far, Read's achievement has not received full recognition, at least in the UK, although his books have appeared round the world, not only in America – where several serious critical essays have been written on him – but translated into Dutch, the Scandinavian languages and Finnish, French, German, Spanish, Russian, Turkish, Polish, Japanese and Hebrew. He has also published two books of reportage: *Alive: the Story of the Andes Survivors*, which won the Thomas More Medal (USA) for 'the most distinguished contribution to Catholic Literature', and *The Train Robbers*. His several stories and articles, published in various places, have not as yet been collected. He married in 1967, and has four children.

Fiction
Game in Heaven with Tussy Marx 1966; *The Junkers* 1968; *Monk Dawson* 1969; *The Professor's Daughter* 1971; *The Upstart* 1973; *Polonaise* 1976; *A Married Man* 1979; *The Villa Golitsyn* 1981; *The Free Frenchman* 1986; *A Season in the West* 1988; *On the Third Day* 1990; *The Patriot* 1996

Non-Fiction
Alive: the Story of the Andes Survivors 1974; *The Train Robbers* 1978; *Quo Vadis?* 1991; *Ablaze* 1993

Hamish READE

see Simon GRAY

Peter READING 1946–

Born in Liverpool, the son of an electrical engineer, Reading attended Alsop High School in the city and then the Liverpool College of Art, where he trained as a painter and took a BA in 1967. He taught briefly at a large comprehensive school in Liverpool, and then returned to his former art college for two years as a lecturer in the department of art history. His first collection of poems, a pamphlet called *Water and Waste*, was published by a small press in 1970, and his first collection from a major publishing house was *For the Municipality's Elderly* in 1974. He married in 1968, and has one daughter. In 1970 the family moved to Shropshire, where Reading has worked as a labourer and in a variety of jobs in an animal-feed compounding mill. This life was broken between 1981 and 1983 when he had a fellowship at Sunderland Polytechnic; he then returned to the mill. He has also been a poetry reviewer for the *Sunday Times*.

He was briefly involved in controversy in 1984 when his poem 'Cub', published in the *Times Literary Supplement*, was thought by some to be anti-Semitic, and a correspondence, both for and against, ensued in the literary papers. With the publication of *Shitheads* in 1990 Reading had brought out fifteen volumes of verse inclusive of a selected poetry, *Essential Reading*. His literary reputation, initially slow to develop, has now become considerable: *Diplopic* won the first Dylan Thomas Prize in 1983, while *Stet* won the Whitbread Award for Poetry. Reading is a dark and ironic poet who dares to broach the unmentionable; he is formally inventive, and his often despairing tone is made more powerful by a strong vein of black humour. His reputation looks likely to advance still further.

Poetry
Water and Waste 1970; *For the Municipality's Elderly* 1974; *The Prison Cell and Barrel Mystery* 1976; *Nothing for Anyone* 1977; *Fiction* 1979; *Tom o'Bedlam's Beauties* 1981; *Diplopic* 1983; *C.* 1984; *Ukulele Music* 1985; *5x5x5x5x5* 1983; *Essential Reading* [ed. A. Jenkins] 1986; *Stet* 1986; *Final Demands* 1988; *Perduta Gente* 1989; *Shitheads* 1990; *Evagatory* 1992; *3 in 1* 1992; *Last Poems* 1994

John (Francisco) RECHY 1934–

One of America's most groundbreaking and controversial homosexual novelists of the 1960s, Rechy was born in El Paso, Texas, to Chicano parents. Spanish was his native language, and he learned English at school. He was educated at the University of Texas, El Paso, where he attended the New School for Social Research. He did military service in the US Army and was stationed in Germany. Since his success as a novelist, he has taught creative writing.

He shot to fame in 1963 with his first novel, *City of Night*, which became a notorious bestseller. The title is abridged from James Thomson's poem 'The City of Dreadful Night' (1874), and the book follows a seemingly autobiographical hustler from El Paso – the same figure recurs with different names in Rechy's books – across the bleak homosexual nightlife which is its author's terrain. Picaresque in structure, it drifts with its narrator through Los Angeles, New York, San Francisco, Hollywood, Chicago and New Orleans. *Numbers*, his second novel, follows a former hustler who returns to his old sexual arena to have thirty more sexual contacts and no more (thirty is printer's terminology for 'the end'). He seems to have achieved this aim – with the acts recounted in obsessive, quasi-pornographic detail – and to be leaving when he turns back for more. We leave him on number thirty-seven, with the clear implication that he is like a character trapped in a kind of Dantean hell. Other novels include *Rushes*, set in a leather bar of the same name, and *The Vampires*, a camp study in evil that features Blue, a male prostitute; Topaze, an exquisite dwarf; La Duquesa, a mourning drag queen; Tor, a muscleman; Savannah, the world's most beautiful woman; Malissa, 'queen of evil'; a couple of incestuous twins; a Catholic priest; and a voodoo priest and priestess. They are all brought together at a houseparty on an island, for what turns out to be a weekend of hell. Two other volumes of particular interest are *The Sexual Outlaw*, which is subtitled 'A Nonfiction Account, with Commentaries, of Three Days and Nights in the Sexual Underground', and which Rechy considered innovative in its autobiographical 'documentary' form; and *The Miraculous Day of Amalia Gomez*. Although Rechy rejects the Catholic faith in which he was raised, elements of it appear in some of his books and sometimes structure them – the Stations of the Cross, for example. His most ambitious novel to date is probably *Marilyn's Daughter*, which focuses on a woman who may be the daughter of Marilyn Monroe and incorporates historical 'faction'.

Inevitably critics have been divided over Rechy, and his critical fortunes have declined since his first book, which was highly praised by **James Baldwin** among others. One of his most incisive – if negative – readers has been **Terry Southern**, who placed him in 'the self-revelatory school of Romantic Agony'. Critics have noted a tortured Gothic extremity in his work, as well as an adolescent and histrionic quality. It is no surprise, perhaps, that his favourite writers include Dostoevsky, Poe, Hawthorne and **Tennessee Williams** – the last of these for 'the giant dragon emotions; the raging emotions; the screaming about life'.

Rechy is not shy about the merits of his work, insisting on his seriousness and literary status: he has said of *The Sexual Outlaw* that it is 'a daring novel in content and form; a grand and lasting artistic achievement'. Not least among his gratifications has been to teach a creative writing course for which he was himself turned down as a student, when it was run by **Pearl S. Buck**.

Fiction

City of Night 1963; *Numbers* 1967; *This Day's Death* 1970; *The Vampires* 1971; *The Fourth Angel* 1972; *The Sexual Outlaw* 1977; *Rushes* 1979; *Bodies and Souls* 1983; *Marilyn's Daughter* 1988; *The Miraculous Day of Amalia Gomez* 1991

Plays

Momma as She Became – Not as She Was 1978 [np]; *Tigers Wild* 1986 [np]

Peter (William) REDGROVE 1932–

Redgrove was born in Kingston, Surrey, and educated at Taunton School. Aged eighteen, and in the army for national service, he decided to fake loss of memory to be released; he was diagnosed an incipient schizophrenic and given insulin shock therapy fifty times, a treatment which drove him into convulsions and coma. He used this experience in his 1973 novel, *In the Country of the Skin*, which won the *Guardian* Fiction Prize, and it informs his later poetry, which explores the release from rationality and extreme states of consciousness.

He read natural science at Cambridge, where he also wrote poetry and became a friend of **Ted Hughes**. On coming down he worked in a variety of jobs connected with science: research chemist, teacher in Malaga, scientific journalist for *The Times*, reader for the BBC and writer of advertising brochures for scientific products. In 1955, he was a founder member of 'the Group', the informal association of poets that met under the direction of first **Philip Hobsbaum** and later **Edward Lucie-Smith**.

Redgrove's first collection of poetry, *The Collector and Other Poems*, was published in 1960, and was the first of around forty volumes and broadsheets of verse which, with several novels or 'metaphysical thrillers', numerous plays and some non-fiction works, make up a large output. He shares the concern with irrational nature of his friend Hughes, but is a more hermetic and surrealistic writer, and one who divides opinion more: he has often been convicted by critics of what a *Times Literary Supplement* reviewer called a 'purely verbal excitement', but has his enthusiastic defenders. Since the early 1960s he has lived mainly by writers' residencies: he spent a year as visiting poet at Buffalo University (1961–2), and has been Eric Gregory Poetry Fellow at Leeds (1966–83), resident author at Falmouth School of Art in Cornwall, Leverhulme Emeritus Fellow (1985–7), and writer-at-large in North Cornwall in 1988. He has also practised as a Jungian lay psychiatrist. For thirteen years he was married to the sculptor Barbara Redgrove and had three children with her. From 1969 he began to live with the poet and writer Penelope Shuttle; they had a child born in 1976 and married in 1980. Many of his later works are in collaboration with her, including *The Wise Wound*, a pioneering study of menstruation.

Poetry

The Collector and Other Poems 1960; *The Nature of Cold Weather and Other Poems* 1961; *At the White Monument and Other Poems* 1963; *The Force and Other Poems* 1966; *The God-Trap* 1966; *The Old White Man* 1968; *Penguin Modern Poets 11* [with D.M. Black, D.M. Thomas] 1968; *Work in Progress MDMLXVIII* 1969; *The Mother, the Daughter and the Sighing Bridge* 1970; *The Shirt, the Skull and the Grape* 1970; *The Bedside Clock* 1971; *Love's Journeys: a Selection* 1971; *Dr Faust's Sea-Spiral Spirit and Other Poems* 1972; *Two*

Poems 1972; *The Hermaphrodite Album* [with P. Shuttle] 1973; *Aesculapian Notes* 1975; *Sons of My Skin: Selected Poems 1954–1974* [ed. M. Peel] 1975; *The Fortifiers, the Vitrifiers and the Witches* 1977; *From Every New Chink of the Ark and Other New Poems* 1977; *Skull Event* 1977; *Ten Poems* 1977; *Happiness* 1978; *The White Night-Flying Moths Called 'Souls'* 1978; *The Weddings at Nether Powers and Other New Poems* 1979; *The First Earthquake* 1980; *The Apple-Broadcast and Other New Poems* 1981; *The Working of Water* 1984; *A Man Named East and Other New Poems* 1985; *The Mudlark Poems and Grand Buveur* 1986; *Explanation of Two Visions* 1986; *In the Hall of the Saurians* 1987; *The Moon Disposes: Poems 1954–1987* 1987 [rev. *Poems 1954–1987* 1989]; *The First Earthquake* 1989; *A Garland for the Black Goddess* 1989; *Dressed as for a Tarot Pack* 1990; *Under the Reservoir* 1992; *The Laborators* 1993; *My Father's Trapdoor* 1994; *Sex-Magic-Poetry-Cornwall* [ed. J. Robinson] 1994

Fiction

In the Country of the Skin 1973; *The Terrors of Dr Treviles* [with P. Shuttle] 1974; *The Glass Cottage* [with P. Shuttle] 1976; *The God of Glass* 1979; *The Sleep of the Great Hypnotist* 1979; *The Beekeepers* 1980; *The Facilitators* 1982; *One Who Set Out to Study Fear* (s) [from Grimm] 1989; *The Cyclopean Mistress* (s) 1993

Plays

The Nature of Cold Weather (r) 1961; *Mr Waterman* (r) 1961; *The White Monument* (r) 1963; *The Anniversary* (r) 1964; *The Sermon* (r) 1964 [pub. 1966]; *The Case* (r) 1965; *Double Bill* (r) 1965; *Three Pieces for Voices* [pub. 1972]; *In the Country of the Skin* (r) [from his novel] 1973; *Dance the Putrefact* (r) [with A. Smith-Masters] 1975; *The Holy Sinner* (r) [from T. Mann's novel] 1975; *Miss Carstairs Dressed for Blooding and Other Plays* [pub. 1976]; *The God of Glass* (r) [from his novel] 1977 [pub. 1979]; *The Hypnotist* 1978 [np]; *The Martyr of the Hives* (r) 1980 [pub. in *Best Radio Plays of 1980* 1981]; *Florent and the Tuxedo Millions* (r) 1982; *The Sin Doctor* (r) 1983; *Dracula in White* (r) 1984; *The Scientists of the Strange* (r) 1984; *Time for the Cat-Scene* (r) 1985; *The Valley of Trelamia* (r) 1986; *Stories from Grimm* (r) 1987

Non-Fiction

The Wise Wound: Menstruation and Everywoman [with P. Shuttle] 1978 [US *The Wise Wound: Eve's Curse and Everywoman* 1979]; *The Black Goddess and the Sixth Sense* 1987; *Alchemy for Women* [with P. Shuttle] 1995

Edited

Poet's Playground 1963; *Universities Poetry 7* 1965; *New Poems 1967* [with J. Fuller, H. Pinter] 1968; *Lamb and Thundercloud* 1975; *New Poetry 5* [with J. Silkin] 1979; *Cornwall in Verse* 1982

Henry REED 1914–1986

Reed was born in Birmingham, where his father was a master bricklayer and foreman in charge of forcing at Nocks' Brickworks; his mother was illiterate. He received much early education from his sister and was able to progress to King Edward VI Grammar School, Aston, and Birmingham University, taking his MA by thesis there in 1936. During the mid-1930s he began research for a life of **Thomas Hardy**, which he never completed but on which he worked until the mid-1950s. He was associated with the influential 'Birmingham Group' of writers, although he was principally a poet. After university, he became a freelance writer and journalist, eking out his income with a little teaching, and travelling frequently in Italy.

In 1941 he was called up into the army, and his period of training as an officer cadet, which partly consisted of a crash-course in Japanese, became the basis of a brilliant series of early poems which includes 'Lessons of War', of which the much anthologised 'Naming of Parts' is perhaps the most famous single English poem to come out of the Second World War. Some verses about sailors in Cape Town, which Reed submitted to the *Listener*, were returned by the disappointed literary editor, **J.R. Ackerley** – an early champion of Reed's work – because the BBC would not allow him to print a poem containing the word 'brothels'. Reed's period of army service lasted only a year, after which he was seconded to Naval Intelligence at Bletchley Park, where he spent the remainder of the war teaching Japanese to Wrens.

It was during the war that he met Michael Ramsbotham, with whom he was to have the major relationship of his life. After the war the two men lived at various addresses in the West Country, but they parted in 1950, at which date Reed moved back to London, living mainly in a flat in Upper Montagu Street. For fifteen years he was employed by the BBC, producing amongst other things a radio dramatisation of *Moby Dick*. He reviewed extensively, writing a study of the post-war novel, and was a noted translator, particularly of Balzac, Leopardi and Ugo Betti, whose reputation in Britain he was largely responsible for establishing. He also spent several periods as a visiting professor of poetry at Washington University, Seattle.

He continued to write poetry and verse plays, and was an exceptional parodist. 'Chard Whitlow' ('As we get older we do not get any younger ...') is an extremely funny parody of **T.S. Eliot**'s *Four Quartets*, while his radio plays about Hilda Tablet, an imaginary female composer, sent up biography, radio documentary and opera (notably the work of Benjamin Britten), and became a considerable cult. Reed was a self-critical, hypersensitive, depressive man who in later years succeeded in committing very little to paper. The last decade of his life was tragic, as he was by then heavily alcoholic, lived mostly on Complan, and was harried by the Inland Revenue. He published only two volumes of poetry during his lifetime, and has been largely remembered for a single poem, but an edition of his *Collected Poems* in 1991 seems likely to enhance his reputation.

Poetry

A Map of Verona 1946; *The Lessons of the War* 1970; *Collected Poems* [ed. J. Stallworthy] 1991

Plays

Moby Dick (r) [from H. Melville's novel] 1947; *Hilda Tablet and Others* (r) [pub. 1971]: *A Very Great Man Indeed* 1953, *The Private Life of Hilda Tablet* 1954, *A Hedge, Backwards* 1956, *The Primal Scream, as it were ...* 1958; *The Streets of Pompeii* (r) [pub. 1971]: *The Unblest* 1949, *The Monument* 1950, *The Streets of Pompeii* 1952, *Return to Naples* 1950, *The Great Desire I Had* 1952, *Vincenzo* 1955

Translations

Paride Rombi *Perdu and His Father* 1954; Ugo Betti: *The Burnt Flower Bed* 1957, *The Queen and the Rebels* 1957, *Summertime* 1957, *Crime on Goat Island* 1960; Dino Buzzati *Larger than Life* 1960; Balzac: *Père Goriot* 1962, *Eugénie Grandet* 1964; Natalia Ginzburg *The Advertisement* 1964

Non-Fiction

The Novel Since 1939 1946

Ishmael (Scott) REED 1938–

A leading satirist and experimental writer, Reed was born in Chattanooga, Tennessee. He graduated from the University of Buffalo in 1960 and was employed as a reporter on the Buffalo *Empire Star Weekly*. In 1965 he organised the first American Festival of Negro Art and co-founded two community newspapers: the *East Village Voice* in New York and *Advance* in New Jersey. He also founded the Yardbird Publishing Company in California in 1971 and, as editorial director, published and distributed the work of unknown ethnic writers.

He is best known as a novelist who examines both tangible and theoretical components of Western civilisation as repressive forces. His major concern is to establish an alternative black aesthetic called NeoHooDoo based on the cults of black people throughout the world. NeoHooDoo demands the reinstitution of primeval rites such as magic and voodoo, which Reed believes to be the true spiritual doctrine of the black people, and which will purge them of the taint and inhibiting grasp of Western civilisation. His most highly acclaimed writings use African and Afro-American folklore as a literary device to create fictions.

His first novel, *The Free-Lance Pallbearers* (1967), is a parody of the nineteenth-century Gothic genre; the story of a young man's search for self-knowledge in a stressful, power-mad Christian society. His first volume of poetry, *Catechism of a NeoAmerican HooDoo Church*, expounds the oral tradition of Africa. Subsequent works introduced the concept of NeoHooDoo and Reed's style of turning syncretism – the basis of NeoHooDoo – into a literary method, whereby he mixes standard English with dialect, slang, phonetic spelling, drawings and neology. The *New York Times Book Review* comments: 'His so-called nonsense words raise disturbing questions about the nature of language.' In 1973 Reed's novel *Mumbo Jumbo* and his volume of poetry *Conjure* were both nominated for National Book Awards. Four poems in *Conjure* were set to music, but an interested record company pulled out at the last minute because the poems dealt too explicitly with 'black magic'.

Reed has held lectureships at the Universities of California, Washington and Buffalo and he is an associate fellow of Yale University. In 1975 he won a Rosenthal Foundation Award, a Guggenheim Fellowship and an Academy Award. In 1980 he became president of the Before Columbus Foundation. He is married to the choreographer Carla Black and has two daughters. He lives in California.

Fiction

The Free-Lance Pallbearers 1967; *Yellow Back Radio Broke-Down* 1969; *Mumbo Jumbo* 1972; *The Last Days of Louisiana Red* 1974; *Flight to Canada* 1976; *The Terrible Twos* 1982; *Reckless Eyeballing* 1986; *The Terrible Threes* 1989; *Japanese by Spring* 1993

Poetry

Catechism of a NeoAmerican Hoodoo Church 1970; *Conjure* 1972; *Chattanooga* 1973; *A Secretary to the Spirits* 1978; *New and Collected Poems* 1988

Non-Fiction

The Rise and Fall and ...? of Adam Clayton Powell [as 'Emmett Coleman'] 1967; *Shrovetide in New Orleans* 1978; *God Made Alaska for the Indians* 1982; *Airing Dirty Laundry* 1993; *Oakland Rhapsody* 1995

Edited

19 Necromancers from Now 1970; *Yardbird Reader* (annual) 5 vols 1971–7; *Yardbird Lives!* [with A. Young] 1978; *Calafia: the California Poetry* 1979; *Quilt 2–3* [with A. Young] 2 vols 1981, 1982; *Writin' is Fightin': 37 Years of Boxing on Paper* 1988; *The Before Columbus Foundation Fiction Anthology* [with K. Trueblood, S. Wong] 1992

Jeremy REED 1954–

Born and raised in Jersey, Reed emerged from a childhood he described as 'solitary and dark-sided' to attend Essex University, where he gained a PhD in literature. Since the early 1970s, when his work started to appear in small-press publications and in periodicals, including the *New Statesman*, *Aquarius* and the *Literary Review*, he has been esteemed as one of the most imaginative, original and lyrical of contemporary British poets. He was a prize-winner in the National Poetry Competition in 1982 and received an Eric Gregory Award in the same year.

His early poetry staked out his principal territory, dealing with madness, psychic anguish, drugs and sexuality. In this, it is similar to his first novel, *The Lipstick Boys* (1984), which Reed has describd as 'a voyage to the other end of the night, a disturbing and harrowing study of sexual outsiders'. *By the Fisheries*, which won a Somerset Maugham Award in 1985, was a departure, in which Reed focuses on a single phenomenon made vivid with naturalistic metaphor, as the poem's titles suggest: 'Rain', 'Tulips', 'Snail'. **Seamus Heaney** wrote of this collection: 'It is full of rich and careful writing, dense with pleasure in words'. By 1987, Reed had a sufficiently high reputation to have a volume of *Selected Poems* published by Penguin. **Kathleen Raine** described him as 'the most imaginatively gifted poet since **Dylan Thomas**'. This was followed in 1990 by *Nineties*, another highly praised volume, which contained poems on the rock singer Lou Reed, about whom Reed (no relation) has written a biography. A later volume of poetry, *Black Sugar*, was subtitled characteristically 'Gay, Lesbian and Heterosexual Love Poems', and was described by **Peter Reading** as 'agreeably lascivious'.

Reed is also well known for his visually dynamic poetry readings: holding aloft one gloved hand, he recites his poems in a curious sing-song manner, occasionally through a human skull. There are those who think that his delivery does very little for his work, his monotonous chanting upsetting the natural rhythms of the verse and making it less easy to understand than it is on the page. Such decadent posturing has made it hard for some people to take his work seriously, although amongst his admirers are many poets of an earlier generation, including **David Gascoyne**, who wrote an introduction to Reed's translation of Novalis's *Hymnen an die Nacht* (1800).

His novels have also attracted serious critical attention from the likes of **Robert Nye** and **J.G. Ballard**. Amongst them are ones recreating the lives of the Marquis de Sade (*When the Whip Comes Down*) and Lautréamont (*Isidore*) – seminal figures in the decadent canon. Other novels, such as *Diamond Nebula*, are polymorphously perverse science fiction, while *Inhabiting Shadows*, with its impressionistic, druggy account of a group of adolescents in a seaside resort, seems to hark back to the world of the 1970s, where David Bowie (another of Reed's rock heroes) held glittering, bisexual sway.

Reed's major work of non-fiction is *Madness – the Price of Poetry*, a collection of essays which won widespread praise. Of his own relationship with the muse, he has commented: 'I've had a total commitment to poetry since the age of nine which has led me to sacrifice a career, security, relationships and personal possessions.'

Poetry

Target 1972; *Vicissitudes* 1974; *Diseased Near Deceased* 1975; *The Priapic Beatitudes* 1975; *Agate Paws* 1975; *Emerald Cat* 1975; *Ruby Onocentaur* 1975; *Blue Talaria* 1976; *Count Bluebeard* 1976; *The Isthmus of Samuel Greenberg* 1976; *Jack's in His Corset* 1978; *Saints and Psychotics* 1979; *Bleecker Street* 1980; *Walk on Through* 1980; *No Refuge Here* 1981; *A Long Shot to Heaven* 1982; *A Man Afraid* 1982; *The Secret Ones* 1983; *By the Fisheries* 1984; *Elegy for Senta* 1985; *Nero* 1985; *Skies* 1985; *Border Pass* 1986; *Selected Poems* 1987; *Engaging Form* 1988; *The Escaped Image* 1988; *Dicing for Pearls* 1990; *Nineties* 1990; *Anastasia in Purple Leopard Spots* 1992; *Around the Day, Alice* 1992; *Black Sugar* 1992; *Red-Haired Android* 1992; *Volcano Smoke at Diamond Beach* 1992; *Artaud* 1993; *Orange Pie* 1993; *Turkish Delight* 1993; *Kicks* 1994; *Pop Stars* 1994; *Segmenting the Black Orange* [ed. P. Scanlon] 1994; *Torch Lighters* [ed. P. Scanlon] 1994

Fiction

The Lipstick Boys 1984; *Blue Rock* 1987; *Red Eclipse* 1989; *Inhabiting Shadows* 1990; *Isidore* 1991; *When the Whip Comes Down* 1992; *Chasing Black Rainbows* 1994; *Diamond Nebula* 1994

Translations

Novalis *Hymns to the Night* 1989; Eugenio Montale *The Coastguard's House* 1990; Jean

Cocteau *Tempest of Stars* 1992; R. Desnos *Jellyfish* 1993

Non-Fiction

Madness – the Price of Poetry 1989; *Delirium: an Interpretation of Arthur Rimbaud* 1991; *Lipstick, Sex and Poetry* (a) 1991; *Waiting for the Man: a Biography of Lou Reed* 1994

James (i.e. John) REEVES 1909–1978

Born in north London, Reeves was the son of a company secretary and an ex-teacher, both devout and socially progressive Nonconformists. His early education was at a Montessori school in Hampstead Garden Suburb, and later he won open scholarships to Stowe and to Cambridge, where he graduated in English. For nearly twenty years, from 1933 to 1952, he taught English, first in schools and later in teacher training colleges. He pioneered new methods of teaching poetry which made him educationally influential and which he was later to expound in books such as *Teaching Poetry*. He was a disciple of **Robert Graves**, and his first volume of poetry, *The Natural Need* (1936), was published by the Seizin Press, run by Graves and **Laura Riding** in Majorca (Riding wrote the preface to the volume). Reeves married in 1936, and had a son and two daughters.

In 1952, influenced by failing eyesight and the desire to develop his creative work, Reeves became a full-time author, and from then on was very prolific, primarily as poet, but also as critic, author for children, anthologist and editor. His poetry had already been excluded by **W.B. Yeats** from *The Oxford Book of Modern Verse* on the grounds that he was 'too reasonable, too truthful', and that the Muses always prefer the embraces of 'gay, warty lads'. This notwithstanding, Reeves was an excellent poet, undoubtedly quiet and occasionally weakly neo-Georgian in manner, but with a dark facility for probing personal emotion which makes him considerable at his best. Instancing the guiltily homosexual love poem 'All Days But One', the critic Martin Seymour-Smith has written: 'Few have more memorably portrayed the pains, pleasures and sinister or unhappy nature of the conventional life than Reeves: few are more startling beneath a tranquil surface.' Perhaps because of these qualities, Reeves was the finest poet for children since **Walter de la Mare**. He was also a general author for children, published collections of folk-song, and, as editor of Heinemann's 'Poetry Bookshelf' series from 1951, edited many poets for it himself. He gave his recreations in *Who's Who* as music and Venice; in later years he lived at Lewes in Sussex.

Poetry

The Natural Need 1936; *The Imprisoned Sea* 1949; *XII Poems* 1950; *The Password and Other Poems* 1952; *The Talking Skull* 1958; *Collected Poems 1929–1959* 1960 [rev. 1974]; *The Questioning Tiger* 1964; *Selected Poems* 1967; *Subsong* 1969; *Poems and Paraphrases* 1972; *Arcadian Ballads* 1977; *The Closed Door* 1977

For Children

The Wandering Moon 1950; *Mulcaster Market: Three Plays for Young People* [pub. 1951]; *The Blackbird and the Lilac* 1952; *English Fables and Fairy Stories Retold* 1954; *The King Who Took Sunshine* 1954; *Pigeons and Princesses* 1956; *A Health of John Patch: a Ballad Operatta* 1957; *Prefabulous Animiles* 1957; *Mulbridge Manor* 1958; *Titus in Trouble* 1959; *Hurdy Gurdy: Selected Poems for Children* 1961; *The Ragged Robin* 1961; *Sailor Rumbelow and Britannia* 1961; *The Story of Jackie Thimble* 1964; *The Strange Light* 1964; *The Pillar Box Thieves* 1965; *The Road to a Kingdom* [from the Bible] 1965; *The Secret Shoemakers and Other Stories* (s) 1966; *The Cold Flame* [from Grimm] 1967; *Rhyming Will* 1967; *The Angel and the Donkey* 1969; *Heroes and Monsters: Legends of Ancient Greece Retold*: *Gods and Voyagers* 1969, *Islands and Palaces* 1971 [UK *Giants and Warriors* 1977]; *Mr Horrox and the Gratch* 1969; *The Trojan Horse* 1969; *How the Moon Began* 1971; *Maildun the Voyager* 1971; *The Path of Gold* 1972; *Complete Poems for Children* 1973; *The Voyage of Odysseus: Homer's Odyssey Retold* 1973; *The Lion That Flew* 1974; *Two Greedy Bears* 1974; *More Prefabulous Animilies* 1975; *A Clever Mouse* 1976; *Quest and Conquest: Pilgrim's Progress Retold* 1976; *Exploits of Don Quixote Retold* 1977; *Fables from Aesop* 1977; *Eggtime Stories* (s) 1978; *The Gnome Factory and Other Stories* (s) 1978; *The James Reeves Storybook* (s) 1978; *A Prince in Danger* 1979; *Snow-White and Rose-Red* [with J. Rodwell] 1979

Edited

J. Bronowski and others *The Quality of Education* [with D. Thompson] 1947; *The Writer's Way: an Anthology of English Prose* 1948; *Dialogue and Drama* [with N. Culpan] 1950; *Orpheus: a Junior Anthology of English Poetry* 1950; D.H. Lawrence *Selected Poems* 1951; *The Speaking Oak* 1951; John Donne *Selected Poems* 1952; Gerard Manley Hopkins *Selected Poems* 1953; John Clare *Selected Poems* 1954; *Green Broom* 1954; *Grey Goose and Gander* 1954; *The Holy Bible in Brief* 1954; *Strawberry Fair* 1954; *Yellow Wheels* 1954; *The Merry Go Round* 1955; Jonathan Swift: *Gulliver's Travels* 1955, *Selected Poems* 1966; Browning *Selected Poems* 1956; Jules Verne *Twenty Thousand Leagues Under the Sea* 1956; *The Modern Poet's World* 1957; *A Golden Land:*

Stories, Poems and Songs 1958; Cecil Sharp [ed.] *The Idiom of the People: English Traditional Verse* 1958; Emily Dickinson *Selected Poems* 1959; Coleridge *Selected Poems* 1959; *The Personal Vision* 1959; *Over the Ranges* [with W.V. Aughterson] 1959; *The Rhyming River* 4 vols 1959; *The Everlasting Circle: English Traditional Verse* 1960; Stephen Leacock *Unicorn Leacock* 1960; *The Taste of Courage: the War 1939–1945* [with D. Flower] 1960; Robert Graves *Selected Poetry and Prose* 1961; *Great English Essays* 1961; *The First Bible: an Abridgement for Young Readers* 1962; *Georgian Poetry* 1962; *Three Tall Tales* 1964; *The Cassell Book of English Poetry* 1965; *A New Canon of English Poetry* [with M. Seymour-Smith] 1967; *An Anthology of Free Verse* 1968; *The Christmas Book* 1968; *Homage to Turnbull Stickney* 1968; *One's None: Old Rhymes for New Tongues* 1968; *The Sayings of Dr Johnson* 1968; *The Poets and Their Critics: Arnold to Auden* 1969; *The Poems of Andrew Marvell* [with M. Seymour-Smith) 1969; *Chaucer: Lyric and Allegory* 1970; Thomas Gray *Selected Poems* 1973; *The Vein of Mockery: Twentieth Century Verse Satire* 1973; *Five Romantic Poets* 1974; Walt Whitman *Selected Poems* [with M. Seymour-Smith] 1976; *The Springtime Book: a Collection of Prose and Poetry* 1976; *The Autumn Book: a Collection of Prose and Poetry* 1977; Hardy *Selected Poems* [with R. Gittings] 1981

Non-Fiction

Man Friday: a Primer of English Composition and Grammar 1953; *The Critical Sense: Practical Criticism of Prose and Poetry* 1956; *Teaching Poetry* 1958; *A Short History of English Poetry 1340–1940* 1961; *Understanding Peotry* 1965; *Commitment to Poetry* 1969; *Inside Poetry* [with M. Seymour-Smith] 1970; *How to Write Poems for Children* 1971; *The Reputation and Writings of Alexander Pope* 1976; *The Writer's Approach to the Ballad* 1976

Translations

Frantisek Hrubin *Primrose and the Winter Witch* 1964; Pushkin *The Golden Cockerel and Other Stories* (s) 1969; Marie de France *The Shadow of the Hawk and Other Stories* (s) 1975

Plays

AD One 1974

Christopher (John) REID 1949–

Reid was born in Hong Kong, where his father worked for the Shell Oil Company. He was sent to preparatory school in England, aged seven, and then to Tonbridge. He went up to Exeter College, Oxford, in 1968. On graduating he was variously a part-time librarian at the Ashmolean Classics Library, an actor, a filing clerk, a flyman at the Victoria Palace Theatre, and what he describes as 'nanny/tutor'. None of these experiences has so far appeared directly in his verse; indeed he has protested: 'I am quite happy to be shrouded in mystery so far as my private life is concerned.'

In 1978 he won the Eric Gregory Award for his first book, *Arcadia*, which was offered in typescript. It also won both a Somerset Maugham Award and the Hawthornden Prize. Its subjects are largely taken from everyday life, but, as **Blake Morrison** noted in the *New Statesman*: 'After reading *Arcadia* you feel that the world is a stronger, richer, more various place than you'd supposed, a feeling which it takes commitment of a kind to produce.' The opening poems of his second collection, *Pea Soup*, also have domestic settings, but thereafter the poetry takes off into 'foreign and fictitious lands' as he considers the 'next-to-nothingness' of human existence. The volume ends with a handful of love poems. A more sombre note has crept in, but comedy still prevails.

For his third book, *Katerina Brac*, Reid goes further, having chosen to 'translate' the fictitious poet whose middle name, it is implied, is Brica. In 'A Box', he describes the ingredients of her shadowy life: 'a bed, a soup bowl, a landscape of mists and birches, / the words spoken by a pensive mother'. Reid has revealed that her 'homeland and language are never identified', adding playfully: 'I hope I have caught something of the original text.' He has been an influential poetry editor at Faber and Faber, and in 1990 edited the annual *Poetry Book Society Anthology*.

Poetry

Arcadia 1979; *Pea Soup* 1982; *Katerina Brac* 1985; *In the Echoey Tunnel* 1991

Edited

The Poetry Book Society Anthology 1989–90 1990

Forrest REID 1875–1947

Reid was the youngest of twelve children born into a comfortable middle-class Presbyterian family in Belfast, though only six of the children survived. His father had lost a great deal of money during the American Civil War. His mother was collaterally descended from Henry VIII's sixth wife, Catherine Parr. Throughout what he later described as his 'loveless' childhood in Mount Charles, a street near Queen's University, Reid was looked after by his beloved nurse Emma Holmes, and he describes her sudden and, to him, baffling departure soon after the death of his father in 1881 in his 'spiritual autobiography' *Apostate*.

After attending the Royal Academical Institution in Belfast and being briefly apprenticed to the tea trade, he took his BA at Christ's College, Cambridge, in 1908, where he met and dined with, amongst others, **Ronald Firbank**. He had already published two novels, *The Kingdom of Twilight* (1904) and *The Garden God* (1905), and knew **Henry James** well enough by this stage to dedicate the latter to him as a 'slight token of respect and admiration'. As soon as he read the book, however, James became eager to distance himself from its fairly obvious homosexual content: he demanded the dedication be removed before ceasing all correspondence with the author.

On graduation from Cambridge, Reid returned to write and, apart from some foreign travel, lived there quietly and uneventfully for the rest of his life, mainly in Ormiston Crescent. He reviewed for London papers such as the *Spectator* and cultivated friendships with numerous writers, including **E.M. Forster**, who would, along with **L.P. Hartley** and **Edwin Muir**, champion his work over the next few decades. The inspiration for most of his fiction seems to have come from the fact that he was a pederast, and the consequent desire to capture the beauty, spontaneity and grace of young boys and the friendships they develop both between one another and sometimes with older men. But this creative process was not so much one of sublimation as of celebration, for Reid openly befriended many boys during his lifetime.

Reid is now best remembered for his *Tom Barber Trilogy*, which traces the development of the eponymous hero between the ages of eleven and fifteen. The books were published in reverse chronological order: *Uncle Stephen* (1931), *The Retreat* (1936), and his masterpiece *Young Tom* (1944), which won a James Tait Black Memorial Prize for the best novel of that year. Reid has been accused of being sentimental and idealising about boys and boyhood in these novels, but countered such charges with the comment that he 'preferred the literature of escape, and what *I* should call the literature of imagination, for the escape is only from the impermanent to the permanent'.

As well as fiction, Reid published studies of his contemporaries **W.B. Yeats** and **Walter de la Mare**, a number of anthologies (on sleep and love, for instance), and a discussion of the illustrators of the 1860s (which is remarkably still in print, while most of his novels are not). He died in Warrenpoint, County Down, Northern Ireland. Despite a brief revival in the late 1980s, when a number of his books were reprinted for the first time in popular paperback editions, Reid remains a minor and largely unread novelist.

Fiction

The Kingdom of Twilight 1904; *The Garden God* 1905; *The Bracknels* 1911 [UK *Denis Bracknel* 1947]; *Following Darkness* 1912 [rev. as *Peter Waring* 1937]; *The Gentle Lover* 1913; *At the Door of the Gate* 1915; *The Spring Song* 1916; *A Garden by the Sea* (s) 1918; *Pirates of the Spring* 1919; *Pender Among the Residents* 1922; *Demphon* 1927; *Tom Barber Trilogy*: *Uncle Stephen* 1931, *The Retreat* 1936, *Young Tom* 1944; *Brian Westby* 1934; *Retrospective Adventures* 1941

Non-Fiction

W.B. Yeats. A critical study 1915; *Apostate* 1926; *Illustrators of the Sixties* 1928; *Walter de la Mare. A critical study* 1929; *Private Road* (a) 1940; *Notes and Impressions* 1942; *The Milk of Paradise. Some Thoughts on Poetry* 1946; *The Suppressed Dedication and Envoy of The Garden God* 1975

Translations

Greek Anthology 1943

Mary RENAULT (pseud. of Eileen Mary Challans) 1905–1983

The daughter of a moderately prosperous doctor of Huguenot stock, Renault was born and grew up in Forest Gate, east London. She attended Clifton Girls' School in Bristol and then St Hugh's College, Oxford, where she took a third in English in 1928. After university, wishing to be independent of her parents, she took various jobs as a clerk and in factories, but a severe bout of rheumatic fever forced her back to the parental home for convalescence. In 1933, however, she became a trainee nurse at the Radcliffe Infirmary in Oxford, where the harsh regime was alleviated by her meeting a fellow nurse, Julie Mullard, who was to be her partner until her death.

From 1936, Renault and Mullard spent a decade as nurses in various schools and hospitals, usually together but sometimes forced apart, especially during the war. At the same time, Renault was establishing herself as a writer, publishing her first novel, *Purposes of Love*, in 1939, and creating a minor scandal with what was for the time an exceptionally frank account of both heterosexuality and lesbianism. It was the first of six novels with contemporary settings, many of them drawing on her experience as a nurse, which she published up to 1953. In 1947, one of these won the MGM Award, a substantial cash prize offered to novelists whose work could possibly be turned into films, but the resulting problems of supertax were the motive-force impelling Renault and Mullard to emigrate in 1948 to South Africa, the racial problems of which were then little publicised. They lived initially in Durban, but after a decade moved to near Cape Town.

The last of the six contemporary novels, *The Charioteer*, is also the finest, and one of the first novels to deal openly and seriously with male homosexuality, a theme that is continued in the equally moving *The Last of the Wine*, the first of the eight remarkable fictional reconstructions of life in ancient Greece on which her reputation is largely based. A strong personal charge informs that book and, where this is lacking, the ancient Greek novels can sometimes seem a little wooden, but they cannot be faulted for historical accuracy and a rare empathy with Hellenic themes. Renault lived at the Cape until her death, a tough and intelligent woman who enjoyed her dogs and the company of homosexual men, fairly remote from political involvement but in sympathy with liberal movements in an increasingly repressive atmosphere. She was a member of the Progressive Party of South Africa and a president of the PEN Club of South Africa, although she eventually clashed with **Nadine Gordimer** over the latter's more radical stance within the country's politics.

Renault wrote little non-fiction, but her most important work within this field was *The Nature of Alexander* (Alexander the Great was also the subject of several of her later novels), while *The Lion in the Gateway* is a rather uninspired retelling for children of the story of the Persian Wars.

Fiction

Purposes of Love 1939; *Kind Are Her Answers* 1940; *The Friendly Young Ladies* 1944; *Return to Night* 1946; *North Face* 1948; *The Charioteer* 1953; *The Last of the Wine* 1956; *The King Must Die* 1958; *The Bull from the Sea* 1962; *The Mask of Apollo* 1966; *Fire from Heaven* 1970; *The Persian Boy* 1972; *The Praise Singer* 1979; *Funeral Games* 1981

Non-Fiction

The Nature of Alexander 1975

For Children

The Lion in the Gateway 1964

Biography

Mary Renault by David Sweetman 1993

Ruth (Kruse) RENDELL 1930–

Rendell's father, a man who 'looked for sorrow rather than happiness in life', had been born in poverty, and originally worked in Plymouth dockyard, but an eventual university education enabled him to qualify as a teacher of science and mathematics. Her maternal grandparents were a Swede and a Dane who came to England from Copenhagen in 1905; her mother, born in Sweden but brought up in Denmark, developed multiple sclerosis soon after she was born. With these two withdrawn parents, Rendell (née Grasemann) grew up, attending Loughton County High School, on the eastern fringes of London, in metropolitan Essex (the area features in several of her novels, including the terrifying *The Face of Trespass*).

After leaving school, she worked as a reporter and sub-editor on several local newspapers, but in 1950, when she was twenty, married fellow journalist Donald Rendell – he later became a parliamentary reporter for the *Daily Mail* – and stopped work two years later after their only son was born. For a decade she was a housewife and unpublished writer, trying several different genres and arriving at the detective story almost by chance. Her first crime novel, *From Doon with Death*, which introduced her detective Inspector Reginald Wexford of the small town of Kingsmarkham, was published in 1964. From that time, her reputation developed slowly but steadily, until, with the widespread televising of her books from the mid-1980s, she was firmly established, with **P.D. James**, as one of the leading crime novelists of her generation.

The texture of her books is sparser than those of James; she is more insistent in confronting the shocking and macabre aspects of life, and much more prolific. By the mid-1990s she had produced, in addition to a non-fiction book about Suffolk and much journalism, almost fifty crime novels and collections of short stories. They divide into three types: the books featuring Wexford (considerably under half the total, and becoming more infrequent with time), which are fairly straightforward police procedurals; the studies in abnormal psychology, which include such bizarre and unforgettable books as *The Lake of Darkness* and *A Judgement in Stone*; and the novels which she began to publish in the 1980s as Barbara Vine (she was often called Barbara when a child, and Vine was her great-grandmother's maiden name), which are the closest to mainstream fiction. The claim is sometimes made, as with James, that she straddles the gap between crime and literary fiction, and, while the quality of her prose may raise doubts about this, there is no denying the fertility of her imagination, its memorable strangeness and her chilling observation of city and suburban atmospheres.

She divorced her husband in 1975, remarried him in 1977, and has lived with him for the last decade on a remote estate near the Suffolk village of Polstead, lending out the estate cottages for writing purposes to such avant-garde younger novelists as **Martin Amis**, **Julian Barnes** and **Jeanette Winterson**. She has other homes in Regent's Park and Aldeburgh, but produces two books a year, her life 'all writing and publicity', apparently as driven and relentless as her work.

Fiction

As Ruth Rendell: *From Doon with Death* 1964; *To Fear a Painted Devil* 1965; *Vanity Dies Hard* 1965; *A New Lease of Death* 1967; *Wolf to the Slaughter* 1967; *The Secret House of Death* 1968; *The Best Man to Die* 1969; *A Guilty Thing Surprised* 1970; *No More Dying Then* 1971; *One Across, Two Down* 1971; *Murder Being Once Done* 1972; *Some Lie and Some Die* 1973; *The Face of Trespass* 1974; *Shake Hands for Ever* 1975; *A Demon in My View* 1976; *The Fallen Curtain* (s) 1976; *A Judgement in Stone* 1977; *A Sleeping Life* 1978; *Make Death Love Me* 1979; *Means of Evil* (s) 1979; *The Lake of Darkness* 1980; *Put on by Cunning* 1981; *The Fever Tree* (s) 1982; *Master of the Moor* 1982; *The Speaker of Mandarin* 1983; *The Killing Doll* 1984; *The Tree of Hands* 1984; *The New Girlfriend* (s) 1985; *An Unkindness of Ravens* 1985; *Live Flesh* 1986; *Collected Short Stories* (s) 1987; *Heartstones* 1987; *Talking to Strange Men* 1987; *The Veiled One* 1988; *The Bridesmaid* 1989; *The Best Man to Die* 1990; *Going Wrong* 1990; *Unguarded Hours* (s) [with H. Simpson] 1990; *The Copper Peacock and Other Stories* (s) 1991; *Kissing the Gunner's Daughter* 1992; *Mystery Cats* (s) [with others] 1992; *The Crocodile Bird* 1993; *Simisola* 1994

As 'Barbara Vine': *A Dark-Adapted Eye* 1986; *A Fatal Inversion* 1987; *The House of Stairs* 1988; *Gallowglass* 1990; *King Solomon's Carpet* 1991; *Asta's Book* 1993; *No Night Is Too Long* 1994

Non-Fiction

Ruth Rendell's Suffolk [with P. Bowden] 1989

Edited

A Warning to the Curious: the Ghost Stories of M.R. James 1987; *Undermining the Central Line* [with C. Ward] 1989

Kenneth REXROTH 1905–1982

Rexroth was born in South Bend, Indiana, but grew up in Chicago from the age of twelve. The son of a wholesale druggist, he had numerous jobs including factory worker, fruit picker, horse wrangler, forest patrolman, soda jerk and asylum attendant. He attended the Chicago Art Institute and the Art Students League in New York, but he was largely – and voraciously – self-taught, devouring the alternative syllabus of his generation: Japanese and Chinese poetry, Zen Buddhism, anthropology, mysticism and the Greeks. Politically he was an anarchist and pacifist, a sometime communist and sometime associate of the 'Wobblies' (the Industrial Workers of the World). He was married four times, and had two daughters by his third marriage.

Rexroth's first published book of poetry was *In What Hour* (1940), although *The Signature of All Things* (1950) is more representative of his mature style. Book publication came relatively late in a career that began when he was around seventeen and encompassed painting, translation, prose, jazz scholarship, playwriting and poetry. The breadth of his interests laid him open to criticism as a dilettante, but his various concerns all come together to create the Bohemian proto-Beat lifestyle for which he may be best remembered. He was the model for the anarchist lawyer Reinhold Cacoethes in **Jack Kerouac**'s *The Dharma Bums* (1958), and for Kenny in **James T. Farrell**'s *Studs Lonigan* (1932–5). He was an influence on the Beat movement and founded the San Francisco Poetry Center with younger poets **Lawrence Ferlinghetti** and **Allen Ginsberg**. Nevertheless he sought to distance himself from the Beat movement, criticising it for its lack of rigour and its sloppy artistic excesses.

Influenced himself by the poetry of **William Carlos Williams** and the second Chicago Renaissance, Rexroth remained a voice apart, refusing to subscribe to the then current academic orthodoxies of what he termed the 'corn belt metaphysical', the highly wrought poetry favoured by the New Criticism, full of paradox, irony and allusion. Instead his poetry has a directness and simplicity that aims for an intensity and clarity of communication based on the voice, and not the printed page. His detractors have found him to be a prime example of the 'chopped-up-prose' school of poetry. It is probably his unfashionably direct style and its striving for a kind of day-to-day mysticism or – as one critic termed it – a 'supercharged imagism' that accounted for his years in the critical wilderness: he was largely excluded from serious attention between the 1940s and the 1960s, and also failed to achieve the more popular audience of the Beats. Rexroth received numerous awards in the last two decades of his life, but there is still no consensus of opinion on the ultimate value of his work.

Poetry

In What Hour 1940; *The Phoenix and the Tortoise* 1944; *The Art of Worldly Wisdom* 1949; *The Signature of All Things* 1950; *The Dragon and the Unicorn* 1952; *A Bestiary for My Daughters Mary and Catherine* 1955; *In Defense of the Earth*

1956; *The Homestead Called Damascus* 1963; *Natural Numbers* 1963; *The Heart's Garden, The Garden's Heart* 1967; *Penguin Modern Poets 9* [with D. Levertov, W.C. Williams] 1967; *The Spark in the Tinder of Knowing* 1968; *Sky Sea Birds Trees Earth House Beasts Flowers* 1970; *New Poems* 1974; *On Flower Wreath Hill* 1976; *The Silver Swan: Poems Written in Kyoto 1974–1975* 1976; *The Morning Star* 1979

Translations

O.V. de Lubicz-Miłosz *Fourteen Poems* 1952; *100 Poems from the Japanese* 1955; *100 Poems from the Chinese* 1956; *30 Spanish Poems of Love and Exile* 1956; *100 Poems from the Greek and Latin* 1962; *Poems from the Greek Anthology* 1962; *Love and the Turning Earth: 100 More Classical Poems* 1970; *Love and the Turning Year: 100 More Chinese Poems* 1970; *100 Poems from the French* 1970; *The Orchid Boat: Women Poets of China* [with Ling O. Chung] 1972; Pierre Reverdy *Selected Poems* 1973; *100 More Poems from the Japanese* 1976; *Burning Heart: the Women Poets of Japan* [with I. Atsumi] 1977; Li Ch'ing-chao *Complete Poems* [with Ling O. Chung] 1979

Non-Fiction

Bird in the Bush: Obvious Essays 1959; *Assays* 1961; *An Autobiographical Novel* (a) 1966; *Classics Revisited* 1968; *The Alternative Society: Essays from the Other World* 1970; *With Eye and Ear* 1970; *American Poetry: In the Twentieth Century* 1971; *The Elastic Retort: Essays in Literature and Ideas* 1973; *Communalism: From Its Origins to the Twentieth Century* 1975

Edited

Selected Poems of D.H. Lawrence 1948; *The New British Poets* 1949; *Four Young Women: Poems* 1973; David Meltzer *Tens: Selected Poems 1961–1971* 1973; *The Selected Poems of Czesław Miłosz* 1973; Kazuko Shiraishi *Seasons of Sacred Lust* 1978

Plays

Beyond the Mountains 1951

Biography

A Life of Kenneth Rexroth by Linda Hamalian 1991

Jean RHYS (pseud. of Ella Gwendolen Rees Williams) 1890–1979

One of five children of a doctor of Welsh descent, Rhys was born at Roseau, Dominica, where her Creole mother's family had been planters. She came to England in 1907 and spent a term at the Perse School, Cambridge, then trained at the Academy of Dramatic Art. Her subsequent life was very similar to that of the protagonists of her books: 'I don't think I know what character is,' she once said. 'I just write about what happened.' Reduced to poverty after her father's death, she left drama school and worked as a chorus girl, a mannequin, and the ghost writer of a book about furniture; she also received a small allowance from a former lover. In 1919 she went to Holland and married Jean Lenglet, a Dutch songwriter and journalist, with whom she lived in Vienna, Budapest and Paris, often on the fringes of the criminal underworld. He was imprisoned in 1923 for currency offences and she divorced him in 1932. They had two children, a son who died in infancy, and a daughter.

'Down to her last three francs', she lived with **Ford Madox Ford** and his mistress, Stella Bowen, while her husband was in jail. Ford published her early stories in *Transatlantic Review*, wrote an introduction to her first book of 'sketches and studies of present-day Bohemian Paris', *The Left Bank* (1927), became her lover, and was portrayed by her as the philandering H.J. Heidler in *Postures* (which is now known by its American title, *Quartet*). She wrote three further novels which share the same spare, evocative style and a recognisable lonely, vulnerable central female character (later dubbed 'the Jean Rhys Woman'), drifting between London and Paris. She met her second husband, Leslie Tilden Smith, in 1927 and married him in 1932; he died in 1945.

After the publication of *Good Morning, Midnight* in 1939, Rhys was forgotten, and she lived for many years in the West Country, often in great poverty. An advertisement placed by a BBC producer in 1957 discovered her whereabouts and the writer **Francis Wyndham** befriended her and encouraged her to continue writing; her short stories were published in the *London Magazine* and *Art and Letters* and the novel she was working on, *Wide Sargasso Sea*, was published in 1966 to great critical acclaim, winning the W.H. Smith Award, the Royal Society of Literature Award and an Arts Council Bursary. The novel drew upon Rhys's memories of the West Indies to tell the story of Mrs Rochester, the mad wife portrayed briefly in Charlotte Brontë's *Jane Eyre* (1847). Her earlier novels were reissued, and her stories were collected in one volume. Jean Rhys became a cult, but the author herself – whose third husband, Max Hamer (whom she married in 1947 and who had also served a prison term), had died in 1964 – continued to live alone in her primitive Devon cottage at Cheriton FitzPaine, without telephone or television, drinking heavily but still writing. (**David Plante**'s notorious *Difficult Women*, 1983, contains a grim portrait of Rhys in old age.) A final volume of stories, *Sleep It Off, Lady*, appeared in 1976, and in 1978 she was made a CBE. Although she was always vague about the details of her life (she claimed not to know why her husbands were imprisoned and, when asked why her son had died, replied: '*Je n'sais pas.* I was never a good mother'), she nevertheless left an unfinished autobiography, which was published posthumously as *Smile Please*.

Fiction
The Left Bank (s) 1927; *Postures* 1928 [US *Quartet*]; *After Leaving Mr Mackenzie* 1930; *Voyage in the Dark* 1934; *Good Morning, Midnight* 1939; *Wide Sargasso Sea* 1966; *Tigers Are Better Looking* (s) 1968; *Sleep It Off, Lady* (s) 1976

Non-Fiction
Smile Please (a) [ed. D. Athill] 1979; *Letters 1931–66* [ed. D. Melly, F. Wyndham] 1984

Translations
Francis Carco *Perversity* 1928; Jean Lenglet *Barred* [as 'Edward de Nève'] 1932

Biography
Jean Rhys by Carloe Angier 1985

Anne RICE 1941–

Born Howard Ellen O'Brien in New Orleans in 1941, Rice was called Howard after her father, but hated the name and discarded it while at convent school. Her father was a postal worker with artistic leanings, sculpting and writing fiction, and her mother was a religious alcoholic who slept in the daytime, drank at night, and died when Anne was fourteen. Rice attended Texas Women's University and San Francisco State College, and did some graduate study at Berkeley in 1969–70. In 1961 she married her high-school sweetheart, the poet Stan Rice, and they have followed a stormy courtship with a stormy marriage. They have had two children, one of whom, Michelle, died of leukaemia.

Rice's distinctive brand of vampire fiction was launched with the publication of *Interview with the Vampire* in 1976. The writing was spurred by grief after the loss of five-year-old Michelle. Rice dreamed that there was something wrong with her child's blood before she became ill, and 'reincarnated' her in the book as the vampire child Claudia, who is granted eternal life at the age of six. The book was written in five weeks of night-long writing jags that Rice describes as 'white-hot, access-the-subconscious' sessions.

Rice has said of her first book: 'It had something to do with growing up in New Orleans, this strange, decadent city full of antebellum houses. It had something to do with my old-guard Catholic background. It had something to do with the tragic loss of my daughter and with the death of my mother when I was fourteen.' She adds that her characters are 'in some way … a perfect metaphor for me'. Rice's Catholic background has contributed a distinctively post-Catholic sense of the joys of sin to her work. It also contains a strong vein of sado-masochism, and Rice has written more conventional S/M erotica under the pseudonym of A.N. Roquelaure, along with romantic fiction as Anne Rampling.

Interview with the Vampire has sold over 4,000,000 copies, but critics are far from unanimous in praising her. Some laud the sensuality of her work and its innovations, which include her shift to the vampire's point of view, the ethical articulacy of her vampires, and her powerful remaking of the old connections between sexuality, fear and death. Others find her prose style turgid and her sensibility farcical and camp. The modernity of her work – the vampire Lestat, for example, in the 1985 novel of the same name, is a rock star – also divides readers. Inevitably *Interview with the Vampire* was made into a film (1994), and **Neil Jordan** was chosen as director. Rice was so appalled by the casting – in particular that of the anodyne Tom Cruise as Lestat – that she bought space in the press to denounce the film while it was in production. Cruise's performance, however, changed her mind and she subsequently gave the by now usefully controversial film her blessing. The critics were on the whole less enthusiastic. Subsequent works such as *The Mummy* have not achieved the success of her vampire novels, which seem to be underscored both by her sexuality and a strain of doominess in her psychic make-up: 'When I'm writing, the darkness is always there. I go where the pain is.'

Fiction
As Anne Rice: *The Vampire Chronicles*: *Interview with the Vampire* 1976, *The Vampire Lestat* 1985, *The Queen of the Damned* 1988, *The Tale of the Body Thief* 1992; *The Feast of All Saints* 1979; *Cry to Heaven* 1982; *The Mummy* 1989; *The Witching Hour* 1990; *Lasher* 1993

As 'A.N. Roquelaure': *The Claiming of Sleeping Beauty* 1983; *Beauty's Punishment* 1984; *Beauty's Release* 1985

As 'Anne Rampling': *Exit to Eden* 1985; *Belinda* 1989

Elmer RICE (pseud. of Elmer Leopold Reizenstein) 1892–1967

Born in New York, the son of German-Jewish immigrants, Rice left school at fourteen, worked as an office boy and then as a clerk in his cousin's law office, studying law in the evenings and graduating from the New York Law School *cum laude* in 1912. He did not, however, practise law but sent his first play, *On Trial* (1914), to a producer, who surprisingly accepted it immediately; the first play to use the film technique of flashback, it was hugely successful, making his name and enabling him to marry on the proceeds his first wife, with whom he had two children. After this first Broadway success, he spent several years working in amateur and experimental theatre, before returning to mainstream drama with *The Adding Machine*, an expressionistic play about the boredom of office life. *Street Scene*, about a New York tenement block, won him a Pulitzer Prize and was made into a successful film from his own screenplay (1931) and – in collaboration with Kurt Weill and **Langston Hughes** – a Broadway musical (1947). *We, the People* was a powerful protest against the hardship caused by the Depression, and Rice was a lifelong radical, both in his theatre and his conduct. He was a founder member of the American Civil Liberties Union and his most famous public stand came in 1951 when he resigned from the Playwright's Television Theatre when an actor was refused a part because of left-wing affiliations. Rice's last notable success was in 1945 with *Dream Girl*, although he continued to write plays into the 1960s and completed about fifty of them in total, always well crafted and thoughtful, although not touching the highest levels of inspiration.

He divorced his first wife in 1942 and married the actress Betty Field, who appeared in many of his plays; they had three children before this marriage too ended in divorce. Rice also wrote three novels, a book on the theatre, and an engaging autobiography called *Minority Report*. He lived for many years in New York before moving with his second wife to a farm in Connecticut.

Plays

On Trial **(st)** 1914, **(f)** 1928 [pub. 1919]; *The Iron Cross* [with F. Harris] 1917 [pub. 1965]; *The Home of the Free* 1918 [pub. 1934]; *For the Defense* **(st)** 1919, **(f)** 1922 [np]; *Wake Up, Jonathan* [with H. Hughes] 1920 [pub. 1928]; *Doubling for Romeo* 1921 [np]; *It Is the Law* **(st)** 1922, **(f)** 1924; *The Adding Machine* **(st)** 1923, **(f)** 1969 [pub. 1923]; *Close Harmony* [with D. Parker] 1924 [pub. 1929]; *The Mongrel* [from H. Bahr] 1924 [np]; *Is He Guilty?* [from R. Lothar] 1927 [np]; *Cock Robin* [with P. Barry] 1928 [pub. 1929]; *Wirim Amerika* 1928; *A Diadem of Snow* [pub. 1929]; *See Naples and Die* **(st)** 1929, **(f)** [as *Oh Sailor, Behave*] 1930; *Street Scene* **(st)** 1929, **(f)** 1931, **(l)** [with L. Hughes, K. Weill] 1947; *The Subway* 1929; *Counsellor-at-Law* **(st)** 1931, **(f)** 1933; *The Left Bank* 1931; *House in Blind Alley* [pub. 1932]; *Black Sheep* 1932 [pub. 1938]; *We, the People* 1933; *Between Two Worlds* 1934 [pub. 1935]; *Bus in Urbe* [pub. 1934]; *Exterior* [pub. 1934]; *The Gay White Way* [pub. 1934]; *Judgement Day* 1934; *Landscape with Figures* [pub. 1934]; *Not for Children* 1935; *The Passing of Chow-Chow* [pub. 1935]; *American Landscape* 1938 [pub. 1939]; *Flight to the West* 1940 [pub. 1941]; *Journey to Jerusalem* 1940 [np]; *Two on an Island* 1940; *A New Life* 1943 [pub. 1944]; *Dream Girl* **(st)** 1945, **(f)** 1948 [pub. 1946]; *The Grand Tour* **(st)** 1951, **(tv)** 1954 [pub. 1952]; *The Winner* 1954; *Cue for Passion* 1958 [pub. 1959]; *Love Among the Ruins* 1963

Fiction

A Voyage to Purila 1930; *Imperial City* 1937; *The Show Must Go On* 1949

Non-Fiction

Minority Report **(a)** 1954; *The Living Theatre* 1959

Biography

The Independence of Elmer Rice by Robert Hogan 1965; *Elmer Rice* by Frank Durham 1970

Adrienne (Cecile) RICH 1929–

Rich was born in Baltimore, in her Jewish father's workplace, 'a hospital in the Black ghetto, whose lobby contained an immense statue of Christ' – an appropriate start for a writer whose poetry, essays, speeches and journalism have consistently grappled with issues of race, faith and identity. Mixed Gentile and Jewish parentage, and the guilty silences and confusions surrounding it, helped create a child and woman 'split at the root'. Her 1982 essay of that title documents the long struggle to integrate her 'disconnected angles' – colour, class, faith, politics, sexuality, the increasing disability caused by rheumatoid arthritis – and 'bring them whole'.

Early education was provided at home, by her talented but frustrated musician mother, a white Protestant Southern Lady, 'obsessed with ancestry'. Already a (locally) performed playwright by the age of twelve, Rich was encouraged, monitored, goaded and praised by a relentlessly ambitious father whose 'investment in [her] intellect and talent was egotistical, tyrannical, opinionated, and terribly wearing'. Nevertheless he taught her 'to believe in hard work, to mistrust easy inspiration, to write and write … He made me feel, at a very young age, his power of language and that I could share it.' That power, and women's relation to it, has become one of the adult Rich's central preoccupations. By 1945 she 'was writing poetry seriously'.

Her departure from Baltimore, in 1947, for the North, and Radcliffe College, Cambridge, Massachusetts, marked her introduction to a world in which Jewish culture and identity could at last be openly celebrated, and Rich immersed herself in it joyfully. In 1951 her first collection of poems, *A Change of World*, published to tremendous critical acclaim, won the Yale Younger Poets Award. Her marriage in 1953 to Alfred Conrad, a Harvard economics professor who was, nevertheless, the 'wrong sort' of Brooklyn Jew, initiated a lengthy period of alienation from her bitterly disappointed father.

In the next six years Rich produced three sons; and, with other young faculty wives, struggled to fulfil the impossible American Dream version of the perfect wife and mother. Anger, despair, exhaustion, and the insights that accompanied them, later informed one of her strongest and most influential prose works, *Of Woman Born: Motherhood as Experience and Institution*. In 1955 her collection of verse, *The Diamond-Cutters and Other Poems*, brought heartache to Rich's desperately ambitious contemporary, **Sylvia Plath**, who believed she saw in Rich her only true rival. Plath was already dead when Rich's *Snapshots of a Daughter-in-Law* appeared in 1963, and ushered in that distinctive blend of passionate intellect, political ardour and consummate technical artistry which has consistently characterised her mature work. Throughout the 1960s and 1970s her political involvements intensified: the black civil rights movement, opposition to the war in Vietnam, and the emerging women's liberation movement all engaged her energies.

In 1966 she separated from her husband and moved to New York. (Their subsequent divorce was eventually followed by his suicide.) Rich's years of teaching literature across a range of American colleges and universities (including Brandeis, Rutgers, Cornell and Stanford) prompted characteristically witty and incisively elegant critical essays, on individual authors and on the uses and abuses of criticism itself. Constantly self-critical, her later work has addressed those difficult and painful intersections between race, class, age and gender which earlier feminist theorists, including the young Rich herself, often failed to recognise.

She has successfully combined political activism and critical enquiry with the creation of some of late twentieth-century America's finest poetry. The reaction of establishment critics to her work has often revealed more about their personal prejudices than about her poetic achievements. Swept to unusually early fame as 'a good girl', she endured a period in which her increasingly explicit lesbianism brought charges of 'stridency' from earlier champions. Rich has noted sardonically that an increasing acceptance of open homosexuality in literary and critical circles has mysteriously coincided with an apparent 'return' of her poetical powers.

In 1991, more than forty years after winning her first poetry award, she won the Commonwealth Poetry Prize.

Poetry

A Change of World 1951; *Poems* 1952; *The Diamond Cutters and Other Poems* 1955; *The Knight, After Rilke* 1957; *Snapshots of a Daughter-in-Law ... Poems 1954–1962* 1963; *Leaflets: Poems 1965–1968* 1968; *Focus* 1966; *Necessities of Life: Poems 1954–1962* 1966; *The Will to Change: Poems 1968–1970* 1971; *Diving into the Wreck: Poems 1971–1972* 1973; *Poems: Selected and New 1950–1974* 1974; *Twenty-One Love Poems* 1977; *The Dream of a Common Language: Poems 1974–1977* 1978; *A Wild Patience Has Taken Me This Far: Poems 1978–1981* 1981; *Sources* 1983; *The Fact of a Doorframe* 1984; *Your Native Land, Your Life* 1986; *Time's Power: Poems 1985–1988* 1989; *Atlas of the Difficult World: Poems 1988–1991* 1992; *Collected Early Poems, 1950–1971* 1993

Non-Fiction

Of Woman Born 1976; *Women and Honour: Some Notes on Lying* 1977; *Compulsory Heterosexuality and Lesbian Existence* 1980; *What Is Found There: Notebooks on Poetry and Politics* 1994

Fiction

On Lies, Secrets and Silence: Selected Prose 1966–1978 1979; *Blood, Bread and Poetry: Selected Prose 1979–1986* 1986

Plays

Ariadne 1939; *Not I, But Death* 1941

Translations

Poems by Ghalib [with A. Ahmad, W. Stafford; ed. A. Ahmad] 1969; Mark Insingel *Reflections* 1973; Francisco Alarcon *De amor oscuro* [with F. Aragon] 1990

Dorothy (Miller) RICHARDSON 1873–1957

Describing herself has having grown up 'in secluded suroundings in late Victorian England', Richardson was the daughter of a gentleman of leisure who dissipated his money by unwise investment. When she was seventeen, her parents' marriage broke up and she was forced to earn her living, first as a governess (briefly in

Germany), and later as a journalist and clerical worker and, for many years, a secretary in a firm of Harley Street dentists.

She became interested in women's suffrage and moved in avant-garde circles, becoming for a time one of the mistresses of **H.G. Wells**, the model for Hypo Wilson in her *Pilgrimage* sequence. (Mrs Wells was a childhood friend, and she met Wells through this connection.) Her literary friends encouraged her to write, and a chorus of praise attended the first volume of the sequence, *Pointed Roofs* (1915), which was regarded as officially launching the 'stream-of-consciousness' method in English fiction. The sequence, which takes place within the perceptions of one woman, Miriam Henderson, came to a temporary conclusion in 1938 with the twelfth volume, *Dimple Hill*, but a last instalment, *March Moonlight*, was published in 1967.

Pilgrimage was Richardson's life work, and she published only one other book, a history of the Quakers, and two essays. Highly praised by a minority of critics, *Pilgrimage* has attracted adverse criticism to the effect that the heroine is dull, that for a 'stream-of-consciousness' novelist Richardson is in fact extremely reticent, and that much of the writer's undoubted perceptiveness is wasted on minutiae.

Richardson lived for much of her life in London, and married the artist and illustrator Alan Odle in 1917 on the understanding that he had six months to live: he lived in fact for more than thirty years, during which time she nursed him devotedly. In later years, she lived in obscurity and eventually died in a nursing home in Beckenham, Kent. The growth of feminist criticism from the 1960s led to some revival of interest in her, and *Pilgrimage* was reissued by Virago in four volumes in 1969, but it is probably still true that this 'most abominably unknown contemporary writer', as **Ford Madox Ford** described her, is more often referred to than read.

Fiction

Pilgrimage: *Pointed Roofs* 1915, *Backwater* 1916, *Honeycomb* 1917, *Interim* 1919, *The Tunnel* 1919, *Deadlock* 1921, *Revolving Lights* 1923, *The Trap* 1925, *Oberland* 1927, *Dawn's Left Hand* 1931, *Clear Horizon* 1935, *Dimple Hill* 1938, *March Moonlight* 1967

Non-Fiction

Gleanings from the Works of George Fox 1914; *The Quakers Past and Present* 1914; *Jane Austen and the Inseparables: an Essay on Book Illustration* 1930

Biography

Dorothy Richardson by John Rosenberg 1973

Henry Handel (pseud. of Ethel Florence Lindesay) RICHARDSON 1870–1946

Richardson was born in Melbourne, the daughter of Walter, an Irish physician who had emigrated to Victoria during the gold rush of the 1850s, and English born Mary. The couple were prosperous by the time their first daughter was born after fifteen years of marriage, but while on a trip to England in 1873 they learned of Walter's financial collapse and returned to Australia. Walter died insane in 1879, 'a gentle broken creature who might have been a stranger', and Mary supported her family by working as postmistress in a small mining town, while sending Ethel to the Presbyterian Ladies' College in Melbourne. In 1887 Mary took her daughters to Leipzig for Ethel to study piano at the Conservatorium, but after three years Ethel felt unsuited to a career as a concert pianist, and became engaged to J.G. Robertson, a student of German, whom she married in Dublin in 1895. Living in Leipzig and later Strasbourg, she read widely in European literature, and began writing in the late 1890s at Robertson's prompting. Her account of 'a musician who failed to make good' in Leipzig eventually became her first novel, *Maurice Guest*, published in 1908, four years after she moved with her husband to London where he had been appointed the first professor of German at the University of London.

Richardson assumed her *nom de plume*, borrowed from an uncle, because 'there had been much talk in the press about the ease with which a woman's work could be distinguished from a man's; and I wanted to try out the truth of the assertion'. It may also be that the explicit treatment of loveless sex in *Maurice Guest* would not have assisted her husband's career. Her second novel, *The Getting of Wisdom*, deals with her schooldays in Melbourne, and reached a new audience in 1977 when it was made into a highly successful film. In 1912 she returned to Australia for the last time to research the trilogy, based on her father's life, which was to occupy her for two decades. *The Fortunes of Richard Mahony* describes the eponymous hero's life from his arrival in Australia in 1852 to his death in 1879, a study in rootlessness, dissatisfaction and the reversal of fortunes which is also a panoramic portrait of the young colony. Following completion of the trilogy, Richardson was a Nobel Prize nominee in 1932. She moved to Sussex after Robertson's death in 1933 and her last novel was a disappointing, semi-documentary study of the youth of Cosima Wagner. Her autobiography of her own youth was published posthumously in 1948.

Fiction

Maurice Guest 1908; *The Getting of Wisdom* 1910; *The Fortunes of Richard Mahony*: *Australia Felix* 1917, *The Way Home* 1925, *Ultima Thule* 1929; *The End of a Childhood and Other Stories* (s) 1934; *The Young Cosima* 1939

Non-Fiction

Two Studies 1931; *Myself When Young* (a) 1948

Mordecai RICHLER 1931–

Richler was brought up in the Jewish ghetto of Montreal centred on St Urbain Street, his parents having escaped from Russia and Poland during the pogroms in the early part of the century. He had a traditional Jewish upbringing, much influenced by his grandfather, who was a noted Hasidic scholar. In 1949 he went to Sir George Williams University in Montreal, but dropped out two years later and went to Europe, living for most of a year in Paris.

He was back in Montreal in 1952 and did some freelance work in the CBC newsroom, but in 1954, the year of publication of his first novel, *The Acrobats*, an immature work largely set in Spain, he emigrated to England, initially finding rooms with a Canadian film director in Swiss Cottage, London, and soon entering the worlds of film, television and freelance writing. He contributed to many magazines, and wrote several screenplays: perhaps the best-known film he worked on was *Life at the Top* (1965). He found his fictional form with his second novel, *Son of a Smaller Hero*, the first of a trio of novels – *The Apprenticeship of Duddy Kravitz* and *St Urbain's Horseman* are the others – which make intense but unsentimental use of Richler's youth, the escape from the ghetto and sexual confrontation with the Gentile. The sharply satirical and scatalogical *Cocksure*, which was banned by W.H. Smith, represents another side of his talent.

After a brief first marriage he married again in 1960, and subsequently had five children; in 1972 he took his whole family back to Canada, where he has lived in Westmount and then Montreal. He has been writer-in-residence at Carleton University, Ottawa, adviser to the Canadian Book-of-the-Month Club and other national bodies, and has turned increasingly to the theatre (*Duddy*, a musical based on his novel, appeared briefly in Edmonton in 1984). Since 1970, new novels have been published only at intervals of a decade. Among his other books are two for children and collections of essays: he has also edited the influential *Canadian Writing Today* (1970). In Canada he has several times been awarded the Governor-General's Literary Award, and both *St Urbain's Horseman* (1971) and *Solomon Gursky Was Here* (1990) were shortlisted for the Booker Prize.

Fiction

The Acrobats [US *Wicked We Love*] 1954; *Son of a Smaller Hero* 1955; *A Choice of Enemies* 1957; *The Apprenticeship of Duddy Kravitz* 1959; *The Incomparable Atuk* [US *Stick Your Neck Out*] 1963; *Cocksure* 1968; *The Street* (s) 1969; *St Urbain's Horseman* 1971; *Joshua Then and Now* 1980; *Solomon Gursky Was Here* 1990

Plays

Paid In Full (tv) 1958; *The Trouble with Benny* (tv) 1959 [np]; *The Apprenticeship of Duddy Kravitz* (tv) [from his novel] 1961, (f) 1974, (st) 1981; *It's Harder to be Anybody* (r) 1961; *No Love for Johnnie* (tv) [with N. Phipps] 1961; *Duddy* [from his novel] 1984 [np]; *Joshua Then and Now* (f) [from his novel] 1985

Non-Fiction

Hunting Tigers under Glass: Essays and Reports 1968; *Shovelling Trouble* 1972; *Notes on an Endangered Species and Others* 1974; *Creativity and the University* [with A. Fortier, R. May] 1975; *Images of Spain* [with P. Christopher] 1977; *The Great Comic Book of Heroes and Other Essays* [ed. R. Fulford] 1978; *Home Sweet Home: My Canadian Album* 1984

Edited

Canadian Writing Today 1970; *The Best of Modern Humour* 1983

For Children

Jacob Two-Two Meets the Hooded Fang 1975; *Jacob Two-Two and the Dinosaur* 1987

Conrad (Michael) RICHTER 1890–1968

Richter's family had emigrated to America from Germany in the mid-nineteenth century and settled in Pennsylvania, where Richter was born in Pine Grove. His father was a Lutheran minister who was ordained in 1904, when Richter was fourteen: the process of entry into the faith was thus observed at close quarters during the writer's adolescence. Leaving college in 1906, he worked as a farm labourer, bank clerk, salesman, lumberjack and journalist in and around Pittsburgh. He later acknowledged his journalism as a strong formative influence upon his concise literary style.

He published his first short story in 1913; his second, 'Brothers of No Kin', was published the following year, but Richter received only twenty-five dollars for it – and that a year later. Between 1913 and 1925 he published a large number of stories, but his first full-length work was the philosophical essay *Human Vibration* (1926), a bizarre theoretical mixture of physics, biomechanics and speculative argument. Richter explained it as the 'overture' which his fiction was to underpin, but this link has never convincingly been explained, either by the author or his critics.

Having married in 1915, he published his own magazine in order to gain a measure of financial stability. It failed after a year, but he continued to work as a publisher until 1928, when his wife became seriously ill. He sold his company and moved from Ohio to New Mexico where she was to convalesce. The move seemed to galvanise Richter, for in this new environment he began writing his best short stories and the novels for which he is most widely known. His work reflects his conservationist attitudes, contrasting the values of the cattle-plains with those of encroaching modernisation and urbanism. He celebrates the pioneer life, sometimes, as in *The Trees* (the first volume of a trilogy, *The Awakening Land*), in a historical setting (in this case, the eighteenth century). The trilogy's titles suggest the move from pioneering, through agriculture, towards urbanisation, with *The Trees* followed by *The Fields* and *The Town* (which won a Pulitzer Prize in 1951). Other fiction is set in farming communities and amongst Native Americans, and his short stories were collected under such titles as *Early Americana* and *The Rawhide Knot*.

Occasional stints as a screenwriter in Hollywood between 1937 and 1940 disappointed him, although two of his novels – *The Sea of Grass* and *Tacey Cromwell* – were made into films. In 1950 the Richters moved back to Pine Grove, where Richter continued to work until his death. *The Waters of Kronos*, which won a National Book Award, is a much admired late work which makes use of fantasy and symbolism.

Fiction

Brothers of No Kin and Other Stories (s) 1924; *Early Americana* (s) 1936; *The Sea of Grass* 1937; *The Awakening Land*: *The Trees* 1940, *The Fields* 1946, *The Town* 1950; *Tacey Cromwell* 1942; *The Free Man* 1943; *Always Young and Fair* 1947; *The Light in the Forest* 1953; *The Lady* 1957; *The Waters of Kronos* 1960; *A Simple Honourable Man* 1962; *The Grandfathers* 1964; *The Rawhide Knot and Other Stories* (s) 1978

Non-Fiction

Human Vibration 1926; *Principles in Bio-Physics* 1927; *The Mountain on the Desert* 1955

(John) Edgell RICKWORD 1898–1982

Born in Colchester, Essex, the son of the town's first borough librarian, Rickword attended a dame's school and Colchester Royal Grammar School before going straight into the army in autumn 1916 as a subaltern. He saw front-line action in France in the Royal Berkshire Regiment, losing the sight of one eye and winning the Military Cross. The poetry he wrote in the trenches was published in many periodicals at the time and includes the much anthologised 'Winter Warfare' ('Colonel Cold strode up the Line ...')

He went up to Pembroke College, Oxford, on a service scholarship but left in 1940 when he married Margaret McGrath, an Irishwoman, who had already borne his first child. He took up a life of literary freelancing in London, being the *Times Literary Supplement* reviewer of **T.S. Eliot**'s *The Waste Land* in 1922, publishing the first important English study of Rimbaud in 1924 and his own first collection of poems, *Behind the Eyes*, in 1921. In the mid-1920s his wife was confined to a mental asylum and their two daughters were brought up by foster-parents.

In 1925 a wealthy friend, Ernest Wishart, founded the *Calendar of Modern Letters*, one of the more influential literary magazines of the century, and Rickword became co-editor. The magazine (which lasted until 1927) published **Hart Crane**, **Robert Graves** and Eliot, and made a regular feature of its 'scrutinies', evaluations of established writers, often critical. F.R. Leavis was to acknowledge the influence of Rickword's periodical by naming his own literary journal *Scrutiny*. Wishart went on to found the publishing house which became Lawrence and Wishart, publishers to the Communist Party, and Rickword made a precarious living for some years working for them, and as editor from 1934 to 1938 of the Communist *Left Review* (he had long been a socialist and moved left during the 1930s).

He published three volumes of verse during his early career, but wrote little poetry after 1930, with the exception of the well-known political poem 'To the Wife of any non-intervention Statesman' in 1938. His long silence may explain why he has been neglected as a poet, but his dense and ironic work about his 'attitude to society, to love and to war' has been estimated as important by many eminent critics.

His first wife died, still confined, in 1944, and a month later he remarried; his second wife died in 1963. (His numerous affairs had included one with Nancy Cunard.) In the late 1940s Rickword edited a third journal, *Our Time*, but by then he was a neglected figure, being forced to buy back the entire edition of his *Collected Poems* in 1947 to republish it later. He took up bookselling, first on his own account at Deal, and later as manager of Collet's Bookshop in Hampstead (he had failed to get a job at Foyle's when Christina Foyle declined to pay him ten pounds a week). In 1966 he received one of three special Arts Council grants for writers who had long gone unrecognised, and two collections of his essays in the 1970s consolidated his reputation as a critic. In his last years, when he lived at Halstead, Essex, and later Islington, he went entirely blind.

Poetry
Behind the Eyes 1921; *Invocations to Angels and Happy New Year* 1928; *Twittingpan and Some Others* 1931; *Collected Poems* 1947; *Fifty Poems* 1970; *Behind the Eyes* 1976

Non-Fiction
Rimbaud: the Boy and the Poet 1924; *Milton: the Revolutionary Intellectual* 1940; *William Wordsworth 1779–1850* 1950; *Gillray and Cruikshank* [with M. Katanka] 1973; *Essays and Opinions 1921–1931* [ed. A. Young] 1974; *Literature in Society: Essays and Opinions II 1931–1978* [ed. A. Young] 1978

Edited
Scrutinies by Various Writers 2 vols 1928, 1931; *A Handbook for Freedom: a Record of English Democracy through Twelve Centuries* 1939; *Soviet Writers Reply to English Writers' Questions* 1948; Christopher Caudwell *Further Studies in a Dying Culture* 1949; *Radical Squibs and Loyal Ripostes: Satirical Pamphlets of the Regency 1819–1821* 1971

Fiction
Love One Another (s) 1929

Translations
François Porche *Charles Baudelaire: a Biography* 1928 [with D.M. Garman; jointly as 'John Mavin']; Marcel Coulon *Poet under Saturn: the Tragedy of Verlaine* 1932; Ronald Firbank *Harmonie* and *La Princesse aux soleils* 1974

Biography
Edgell Rickword: a Poet at War by Charles Hobday 1989

Laura RIDING (Jackson) 1901–1991

Riding was born Laura Reichenthal, but legally adopted the surname of Riding in 1927. Her father, a tailor and failed businessman in New York, was of Austrian Jewish origin; a lifelong socialist, he hoped that his daughter would become an American Rosa Luxemburg. She was educated at the Girls' High School, Brooklyn, and Cornell University. While a student, she began writing poems and married a history lecturer, Louis Gottschalk, from whom she was divorced in 1925. In the same year she left America for Europe, and was for more than a decade the companion and literary collaborator of fellow poet **Robert Graves**, living for a period with him and his first wife in a distinctly uncomfortable *ménage à trois*. Before the war, she published nine volumes of poetry and one of *Collected Poems*. Her poetry, elliptical and modernist in manner, is of undoubted difficulty, but has been highly esteemed by a number of influential critics.

In 1939, after breaking with Graves (whose popular success with *I, Claudius* and *Claudius the God* in 1934 enraged her), she returned to America where, in 1941, she married a farmer and unsuccessful writer, Schuyler Jackson, whose wife she had persecuted until the woman had to be removed to a mental institution. About this time, she rejected the writing of poetry as an inadequate guide to truth, and began work with her husband on a large-scale project, originally intended as a dictionary, which would attempt to define the true meaning of words. While working on this, the couple lived in Florida, where they supported themselves largely by the growing and shipping of citrus fruits. After her husband's death in 1968, Riding continued with the work alone: she completed it, in the form of an investigation of language entitled 'Rational Meaning: a New Foundation for the Definition of Words', but it failed to find a publisher. She broke a publishing silence of more than two decades in 1962, but her later works, which were few, made little public impact. They include *Progress of Stories*, an enlarged version, with prefaces, of her earlier story collection of 1935, and *The Telling*, in which she aimed to tell 'the story of human beings in the universe'. Earlier works include several books and pamphlets, as well as a pseudonymous novel, written with Graves, and *Lives of Wives*, a book about marriages in history. Having declared in the wake of Graves's success with the Claudius books that she 'didn't think she could sink so low' as to write historical fiction, she subsequently wrote *A Trojan Ending*, a novel about the siege of Troy.

It is for her relationship with and (frequently deleterious) influence upon Graves that Riding remains best known, a fact which would not have pleased her. Two biographies of Graves were published during her lifetime in which she appears in a less than flattering light, but she had no regard for what she called 'literary-critical-biographical spuriosities'. In any case, despite the scant regard that has been paid to her work, her own high opinion of her merits was wholly impregnable.

Poetry
The Close Chaplet 1926; *Voltaire, a Biographical Fantasy* 1927; *Love as Love, Death as Death* 1928; *Poems. A Joking Word* 1930; *Though Gently* 1930; *Twenty Poems Less* 1930; *Laura and Francisca* 1931; *The Life of the Dead* 1933; *Poet. A Lying Word* 1933; *Collected Poems* 1938; *The Poems of Laura Riding* 1986; *Progress of Stories* 1986; *First Awakenings* 1992

Fiction
Experts Are Baffled (s) 1930; *No Decency Left* [as 'Barbara Rich'] 1932; *Progress of Stories* (s) 1935 [rev. 1982]; *Convalescent Conversations* [with R. Graves; as 'Madeleine Vara'] 1936; *A Trojan Ending* 1937; *Description of a Life* 1980

Non-Fiction
A Survey of Modernist Poetry [with R. Graves] 1927; *Anarchism Is Not Enough* 1928; *Contemporaries and Snobs* 1928; *A Pamphlet against Anthologies* [with R. Graves] 1928; *Four Unposted Letters to Catherine* 1930; *Lives of Wives* 1939; *The Telling* 1972

Translations
Marcel LeGoff *Anatole France at Home* 1926; Georg Schwarz *Almost Forgotten Germany* [with R. Graves] 1936

Edited
Everybody's Letters 1933; *The World and Ourselves* 1938

Biography
Robert Graves: the Years with Laura by Richard Perceval Graves 1990; *In Extremis* by Deborah Baker 1993

Anne (Barbara) RIDLER 1912–

A highly accomplished poet, in whose work certain currently unfashionable traditions of English poetry remain alive, Ridler has led a life of considerable integrity close to the centre of the English literary establishment, in the quietest and least pejorative sense of that phrase. Born Anne Bradby, the daughter of a master at Rugby School, she was educated at King's College, London, and in 1938 married Vivian Ridler, the printer to Oxford University; their long marriage and four children have been the subjects of some of her finest poems. She worked for some years in the editorial department of Faber and Faber, during the reign of **T.S. Eliot**. As well as writing her own poetry and a large quantity of verse drama, she has been active as an editor, editing *The Faber Book of Modern Verse* and a number of books for Oxford University Press. She has edited editions of Thomas Traherne, George Darley and **Charles Williams**.

Ridler is a committed Anglican whose main poetic subject is love – domestic, mystical, but always with something of the larger signification that it has in Eliot and the religious **W.H. Auden**. Eliot and Auden are the most obvious influences on her work, along with the Metaphysical poets and the thought of Charles Williams. Her work is cerebral by English standards, and has a distinctive mix of Anglican sensuality and rigour in diction. The risks of love in the world are an attendant theme, with the fear of loss and harm and separation, resolved by a belief in the eternality of things or 'endurance of delight'. Her conception of time often resembles that of Traherne or Eliot. Other poems, such as 'For This Time', written in the blitz, are resolved in the manner of prayer.

Ridler's greatest single poem has been held to be 'A Matter of Life and Death', which follows the life of her son from conception to manhood. Equally characteristic is her late poem for her husband, 'Something Else', which is about love in old age, and the closing poem in the *Collected Poems*, 'The Halcyons', which takes its theme from Ovid's *Metamorphoses*: 'But something more is meant / By those myths of bird-changes. / That love continues blest / In different guises; / That immortality / Is not mere repetition: / It is a blue flash, / A kingfisher vision.'

Although her work is exactly the kind of thing that a good deal of post-war American poetry has been written in reaction against, Ridler's reputation seems more than secure. Her work is more than merely conservative in its distinctively twentieth-century awareness and edge; it shows the marks of an age of anxiety.

Poetry
Poems 1939; *A Dream Observed and Other Poems* 1941; *The Nine Bright Shiners* 1943; *The Golden Bird and Other Poems* 1951; *A Matter of Life and Death* 1959; *Some Time After and Other Poems* 1972; *Italian Prospect* 1976; *Dies Natalist* 1980; *Ten Poems* [with E.J. Scovell] 1984; *New and Selected Poems* 1988; *Collected Poems* 1994

Plays
Cain 1943; *The Shadow Factory* 1945 [pub. 1946]; *Henry Bly* 1947 and *The Mask, and the Missing Bridegroom* 1951 [pub. in *Henry Bly and Other Plays* 1950]; *The Trial of Thomas Cranmer* (l) [with B. Kelly] 1956; *The Departure* (l) [with E. Maconchy] 1961 [pub. in *Some Time After and Other Poems* 1972]; *Who Is My Neighbour?* 1961 [pub. 1963]; *The Jesse Tree* (l) [with E. Maconchy] 1970 [pub. 1972]; *The King of the Golden River* (l) [with E. Maconchy] 1975 [np]; *The Lambton Worm* (l) [with R.S. Johnson] 1978 [pub. 1979]

Translations
Cavalli: *Rosinda* 1973, *Eritrea* 1975, *La Calisto* 1984 [np]; Striggio *Orfeo* 1975; Badoaro *The*

Return of Ulysses 1978 [np]; Cicognini *Orontea* 1979 [np]; Grimani *Agrippina* 1982 [np]; Lorenzo da Ponte *Cosi Fan Tutte* **(st)** 1986, **(r)**, **(tv)** 1988 [pub. 1987]

Non-Fiction

Olive Willis and Downe House: an Adventure in Education 1967; *A Victorian Family Postbag* 1988; *Profitable Wonders: Aspects of Thomas Traherne* [with A.M. Allchin, J. Smith] 1989; *A Measure of English Poetry* 1991

Edited

Shakespeare Criticism 1919–1935 1936; *A Little Book of Modern Verse* 1941; Walter de la Mare *Time Passes and Other Poems* 1942; *Best Ghost Stories* **(s)** 1945; *The Faber Book of Modern Verse* 1941–51; Charles Williams: *The Image of the City and Other Essays* 1958, *Charles Williams: Selected Writings* 1961; *Shakespeare Criticism 1935–1960* 1963; *James Thomson: Poems and Some Letters* 1963; *Best Stories of Church and Clergy* **(s)** [with C. Bradby] 1966; *Thomas Traherne: Poems, Centuries and Three Thanksgivings* 1966; *Selected Poems of George Darley* 1979; *The Poems of William Austin* 1984

Elizabeth Madox ROBERTS 1881–1941

Roberts was born near Springfield, Kentucky; her parents were both descended from Kentucky settlers. Suffering from poor health all her life, she spent much of her childhood in the Colorado mountains, and lived for long periods in New York and California where she taught in various schools. Kentucky, however, remained dominant in her imagination, an area she made familiar through her poetry and fiction. To pass convalescent months, she wrote a volume of poetry, *In the Great Steep's Garden* (1915). It was not until she was well enough to attend the University of Chicago, at the age of thirty-six, however, that her poetic imagination developed in conjunction with her studies in philosophy and English. In 1921, she was awarded the Fiske Prize by the university for poems which were collected as *Under the Tree* and published the following year.

With her first novel, *The Time of Man* (1926), which was acclaimed at once, Roberts became better known as a writer of fiction than a poet. In lyrical, poetic prose, she incorporated the philosophical idealism of Bishop Berkeley with a powerful ability to graphically depict the essence of Kentucky. Based on *The Odyssey*, the novel chronicles the heroic twenty-year struggle of the female narrator, a migrant married to a Kentucky farm worker, against the poverty and the injustice of the human condition. **Glenway Wescott** declared that after it, 'no other author will ever have the right to call his place Kentucky'. He recalled Roberts as shy, but friendly, with a crown of yellow hair in her youth that turned to grey, giving her an overtly scholarly appearance.

The symbolic theme of wandering was further developed with *The Great Meadow*, where the female heroine takes part in the eighteenth-century trek from Virginia to Kentucky. *He Sent Forth a Raven*, a novel based on the story of Noah, was her only failure, characterising her worst tendency to over-poeticise the speech of her characters into obscurity and confusion. Fresher and simpler were short stories collected as *The Haunted Mirror*, and awarded the O. Henry Prize. Because of her poor health, which resulted in her early death from anaemia, Roberts's life was completely uneventful.

Fiction

The Time of Man 1926; *My Heart and My Flesh* 1927; *Jingling in the Wind* 1928; *The Great Meadow* 1930; *A Buried Treasure* 1931; *The Haunted Mirror* **(s)** 1932; *He Sent Forth a Raven* 1935; *Black Is My Truelove's Hair* 1938; *Not by Strange Gods* **(s)** 1941

Poetry

In the Great Steep's Garden 1915; *Under the Tree* 1922; *Song in the Meadow* 1940

Biography

Elizabeth Madox Roberts by Frederick P.W. McDowell 1963

Michael (i.e. William Edward) ROBERTS 1902–1948

Roberts was born in Bournemouth, where his parents had a shop, but spent much of his time on their farm in the New Forest. He was educated locally, then in 1920 went to King's College, London, on a scholarship to read chemistry. Two years later he won an exhibition to read mathematics at Trinity College, Cambridge, where he adopted the name Michael after his hero, the Russian scientist-poet Mikhail Lomonosov. He recorded that he 'read some mathematics and much humorous literature' (which later bore fruit when he edited the wide-ranging and inventive *Faber Book of Comic Verse*), and he was vice-president of the university's Socialist Society.

After graduating, he became a school teacher at the Royal Grammar School, Newcastle-upon-Tyne. Conditions on Tyneside during the late 1920s and early 1930s led to his joining the Communist Party, but he was expelled after one year. He also spent one summer working in Paris for the Fabian Research Bureau, investigating wages and production. During this period in the North, he also began writing poetry and published his first volume, *These Our Matins*, in 1930. The following year he came to London to be senior mathematics master at the Mercers' School in Holborn, and began to establish himself on the capital's literary scene. His two anthologies, *New Signatures* and *New Country*, which he published in collaboration with **John Lehmann**, were amongst the most important of the 1930s, and contained work by **W.H. Auden**, **C. Day Lewis**, **Stephen Spender**, **Christopher Isherwood** and other leading writers of the period. **Leonard Woolf**, who published *New Signatures* at the Hogarth Press, commented in his 1967 autobiography that it 'was and still is regarded as that generation's manifesto', and it carried an introductory essay by Roberts defining the state of contemporary poetry. Not everyone was impressed, and his rival editor, **Geoffrey Grigson**, complained of *New Country*: 'Roberts in a long preface "usses" and "ours" as though he were G.O.C. a new Salvation Army or a cardinal presiding over propaganda.'

A series of broadcasts he made for BBC radio in 1934 entitled 'Whither Britain?', in which he criticised armament manufacturers, industrialists, the media, and the 'uglification of landscape and wanton waste of our resources', led to his dismissal from the Mercers' School, after which he returned to the Royal Grammar School in Newcastle, where he remained until the outbreak of the Second World War. He had married Janet Adam-Smith, until then literary editor of the *Listener* and a champion of the Auden generation of poets, in 1935, and spent the next few years bringing up a family, doing literary journalism, and going on climbing holidays, since he was a keen mountaineer (he was bequeathed Snowdonia in Auden and **Louis MacNeice**'s 'Last Will and Testament'). His famous *Faber Book of Modern Verse* appeared in 1936, the same year he produced a second volume of *Poems*. In spite of the literary circles in which he moved, Roberts was less interested in politics than in aesthetics: 'Primarily poetry is an exploration of the possibilities of language,' he wrote; 'a too self-conscious concern with "contemporary" problems deflects the poet's efforts from his true objective'. While such well-known poems as 'In Our Time' address current issues, he rarely allows mere rhetoric to deflect him from a concern with language, and other familiar poems from this period, such as 'The Secret Springs', concentrate upon the natural world familiar to him as a hiker and climber. The clarity he demands from his contemporaries is apparent in his own highly lucid poetry. In the latter part of the decade he wrote two important books of criticism, *The Modern Mind*, which dealt with the use of English and its changes, and a study of the philosopher T.E. Hulme.

During the war he joined the BBC's European Service, broadcasting to the Nazi-occupied countries, then in 1945 became principal of the College of St Mark and St John in Chelsea, London. His last books include *The Recovery of the West*, which deals with the decay of Western society, and *The Estate Man*, which was unfinished at his death from leukaemia, and returns to one of the themes of his broadcasts, the wasting of the earth's resources.

Poetry

These Our Matins 1930; *Poems* 1936; *Orion Marches* 1939; *Collected Poems* 1958

Edited

New Signatures 1932; *Elizabethan Prose* 1933; *New Country* 1933; *The Faber Book of Modern Verse* 1936; *The Faber Book of Comic Verse* 1942

Non-Fiction

Critique of Poetry 1934; *Newton and the Origin of Colours* [with E.R. Thomas] 1934; *The Modern Mind* 1937; *T.E. Hulme* 1938; *The Recovery of the West* 1941; *The Estate Man* 1951

Collections

Selected Poems and Prose [ed. F. Grubb] 1980

Michèle (Brigitte) ROBERTS 1949–

Roberts was born in Bushey, Hertfordshire, the daughter of an English businessman and a French mother who was a teacher. She was educated at a convent and at Somerville College, Oxford, where she specialised in medieval literature, graduating in 1970. She held various jobs as a literary assistant, cook, teacher and pregnancy counsellor, and was poetry editor of the feminist magazine *Spare Rib* from 1975 to 1977; she had the same job at *City Limits* from 1981 to 1983. She was also in the early 1980s a writer-in-residence in the boroughs of Lambeth and Bromley.

She is a self-confessed lapsed Catholic, and her first novel, *A Piece of the Night*, which won the *Gay News* Literary Award in 1978, describes a woman's journey of self-discovery as she rebels against the teachings of the Catholic church and the expectations of orthodox family life to become a feminist and lesbian. Roberts has repeatedly addressed religious themes, and aroused some controversy with *The Wild Girl*, which takes the form of a 'fifth gospel', the life of Christ narrated by the prostitute Mary Magdalene. The novel was denounced by two Conservative Members of Parliament, but viewed more charitably by some critics for creating 'powerful new metaphors for a post-Christian society'. Roberts draws on her French-Catholic background in her most ambitious novel, *Daughters of the House*, which was shortlisted for the Booker Prize in 1992. This story of two sisters living in mid-century France is a dense, beautifully observed but sometimes rather impenetrable tapestry of ideas. Roberts's feminist concerns – 'I *need* to write in order to break through the silence imposed on women in this culture,' she has said – are also reflected in her poetry, which blends irreverence, erotic fantasy and social comment in often arresting language.

Fiction
A Piece of the Night 1978; *Tales I Tell My Mother* (s) [with others] 1978; *The Visitation* 1983; *The Wild Girl* 1984; *The Book of Mrs Noah* 1987; *More Tales I Tell My Mother* (s) [with others] 1988; *In the Red Kitchen* 1990; *Daughters of the House* 1992; *During Mother's Absence* (s) 1993; *Flesh and Blood* 1994

Poetry
Licking the Bed Clean: Five Feminist Poets (c) 1978; *Smile, Smile, Smile, Smile* (c) 1980; *The Mirror of the Mother: Selected Poems 1975–1985* 1986; *Psyche and the Hurricane* 1991; *All the Selves I Was* 1995

Plays
The Journeywoman 1988 [np]

Edited
Cutlasses and Earrings [with M. Wandor] 1977

E(ileen) Ar(buth)not ROBERTSON 1903–1961

Robertson was born and brought up in Surrey, the daughter of a country doctor, of an exuberant middle-class family ('a close little society that carried its bruises like banners', as she later described it). When she was fourteen, the family moved to Notting Hill, London; she was educated at the girls' public school, Sherborne (about which she wrote an amusing memoir for **Graham Greene**'s celebrated anthology, *The Old School*, 1934), and in Paris and Switzerland. She spent much of her childhood 'in small boats' and met her future husband while sailing. He was Henry Turner, General Secretary of the Empire (later Commonwealth) Press Union, and they married in 1927, had one son with whom they lived in Hampstead, London, and often cruised in their yacht to the Dutch and Belgian coasts.

She published nine novels, the first of which, *Cullum* (1928), caused a minor sensation with its sexual frankness, and the last of which, *The Strangers on My Roof*, came out posthumously in 1964. Her first real success came in 1931 with *Four Frightened People*, the story of four castaways in the Malayan jungle, which she wrote without ever having been to the Far East. Her work, although very popular in the 1930s and 1940s, has been slow to develop a strictly literary reputation – a contemporary critic described her as 'the finest yachtsman's novelist now writing in the English language' – but since her revival by the feminist publishers Virago in the 1980s, her novels have aroused some interest for their particular atmosphere, which combines outstanding descriptions of landscape, sailing and bird-life with a certain bright-eyed wit and cruelty in her young middle-class heroines.

During the Second World War she worked for the Ministry of Information in the Films Division and gave lectures on the cinema to the army. She was also a well-known broadcaster and, in her capacity as film critic for the BBC, became involved in a celebrated lawsuit in 1947 with the film company Metro-Goldwyn-Mayer, which tried to bar her from its films; she eventually lost her case in the House of Lords, but was considered to have won the moral victory. After thirty-four years of married life, her husband was killed in a sailing accident, and in consequence she committed suicide, the coroner bringing in a verdict of accidental death.

Fiction
Cullum 1928; *Three Came Unarmed* 1929; *Four Frightened People* 1931; *Ordinary Families* 1933; *Summer's Lease* 1940; *The Signpost* 1943; *Devices and Desires* 1954; *Justice of the Heart* 1958; *The Strangers on My Roof* 1964

Non-Fiction
Thames Portrait [with H.E. Turner] 1937

Edited
The Spanish Town Papers: Some Sidelights on the American War of Independence [with H.E. Turner] 1959

For Children
Mr Cobbett and the Indians [with H.H. Berry] 1942

Edwin Arlington ROBINSON 1869–1935

Robinson was born in Head Tide, Maine, and raised in nearby Gardiner, the small town which is the model for the Tilbury Town which recurs throughout his poetry. He graduated from Gardiner High School in 1888 and studied at Harvard from 1891 to 1893, but was forced to leave because of the decline in the family fortunes. His mother had died and his father expressed an increasingly eccentric interest in spiritualism. Robinson moved to New York where he worked at a variety of ill-paid jobs. His first volume of poetry, *The Torrent and the Night Before*, was printed at his own expense in 1896 and attracted little attention. The following year a revised and extended edition was published as *The Children of the Night*, a volume containing sketches of characters from Tilbury Town, and 'Credo', the poet's affirmation of his spiritual faith. Robinson was an inspector of subway construction when his next collection, *Captain Craig*, was published. It impressed Theodore Roosevelt, who secured him a less onerous job as a clerk in the New York Custom House, where he worked from 1904 to 1910. After *The Town Down the River*, which contains the frequently anthologised 'Miniver Cheevy', he was able to live by his writing.

Robinson was single-minded in his dedication to poetry. He never taught or married, and wrote little else, apart from two ill-fated forays into the theatre, *Van Zorn* and *The Porcupine*. He was insistent in his use of traditional forms, ranging from blank verse to Petrarchan sonnets and villanelles, and once described free verse, together with prohibition and motion pictures, as 'a triumvirate from hell, armed with the devil's instructions to abolish civilization'. His abiding concern with the individual's relationship to the transcendent places him firmly in the New England, Emersonian tradition. Published in 1917, *Merlin* is the first part of his Arthurian trilogy, completed by *Lancelot* and *Tristram*, lengthy narrative poems which describe the Arthurian figures as individuals driven by human impulses rather than supernatural forces.

Robinson won a Pulitzer Prize on three occasions: in 1922 for his *Collected Poems*, in 1925 for *The Man Who Died Twice*, and in 1928 for *Tristram*.

Poetry

The Torrent and the Night Before 1896 [rev. as *The Children of the Night* 1897]; *Captain Craig* 1902; *The Town Down the River* 1910; *The Man against the Sky* 1916; *Merlin* 1917; *Lancelot* 1920; *The Three Taverns* 1920; *Avon's Harvest* 1921; *Collected Poems* 1922; *Roman Bartholow* 1923; *The Man Who Died Twice* 1924; *Dionysus in Doubt* 1925; *Tristram* 1927; *Sonnets 1889–1927* 1928; *Cavender's House* 1929; *The Glory of the Nightingales* 1930; *Matthias at the Door* 1931; *Nicodemus* 1932; *Talifer* 1933; *Amaranth* 1934; *King Jasper* 1935; *Hannibal Brown* 1936

Non-Fiction

Letters to George Schmitt [ed. C.J. Weber] 1940; *Untriangulated Stars: Letters to Harry de Forest Smith 1890–1905* [ed. D. Sutcliffe] 1947; *Letters to Edith Brower* [ed. R. Cory] 1968

Plays

Van Zorn 1914; *The Porcupine* 1915

Edited

Selections from the Letters of Thomas Sergeant Perry 1929

Biography

Robinson: The Life of Poetry Louis O. Coxe 1968

Theodore (Huebner) ROETHKE 1908–1963

Born in the small town of Saginaw in Michigan, of German immigrant ancestry, Roethke was the son of the owner of a large greenhouse; imagery drawn from the cultivation of plants as well as the word 'green' recur obsessively through his poetic work. He was educated at high school in Saginaw, and then attended the nearby University of Michigan, where he received his BA in 1929, and, briefly, Harvard Law School. He made his career as a teacher in colleges and universities, working at Lafayette, Pennsylvania (where he also coached tennis), Pennsylvania State and Bennington and, from 1948 until his death, as a professor of English at the University of Washington in Seattle.

He began publishing his poetry in literary magazines from the early 1930s, and his first volume, *Open House*, was published in 1941, winning him high praise from **W.H. Auden**. By the publication of his fourth volume, *The Waking*, in 1953, he was firmly established as one of the leading American poets of his day. His early work was conventional in diction, and his style always remained imitative to an extent, but he evolved a complex, highly tensed manner for the expression of emotional disturbance which made him a great influence on the developing 'confessional' school of American poets, and **Sylvia Plath** in particular. *The Lost Son and Other Poems*, regarded as a spiritual autobiography, perhaps contains his best work.

Roethke received much encouragement during his career and was the recipient of many literary awards and prizes, but his life was a difficult one: he drank very heavily from early manhood and suffered frequent nervous breakdowns, leading to several spells in mental hospitals. He married in 1953, and his honeymoon was his first visit to Europe, although in the last decade of his life he paid several more visits. It is often stated incorrectly that Roethke died by suicide: he died following a heart-attack suffered in a swimming pool, after several years of ill health brought on by alcoholism. Two volumes of collected prose were published after his death.

Poetry

Open House 1941; *The Lost Son and Other Poems* 1948; *Praise to the End* 1951; *The Waking* 1953; *Words for the Wind* 1957; *The Far Fields* 1964; *The Collected Poems of Theodore Roethke* 1966; *Selected Poems* [ed. B. Roethke] 1969

Non-Fiction

I Am! Says the Lamb 1961; *On the Poet and his Craft: Selected prose of Theodore Roethke* [ed. R.J. Mills] 1965; *Selected Letters of Theodore Roethke* 1968; *Straw for the Fire: From the Notebooks of Theodore Roethke, 1943–1963* [ed. D. Wagoner] 1972

For Children

Party at the Zoo 1963; *Dirty Dinky and other Creatures* 1973

Fr(ederick William Serafino Austin Lewis Mary) ROLFE 1860–1913

The eldest son of a piano manufacturer and Dissenter, Rolfe was born in Cheapside. He usually styled himself Fr. Rolfe – variously taken to stand for Frederick or Father, and the name under which his best-known novel, *Hadrian the Seventh*, was published – but at other times he assumed the name and title of Baron Corvo. He left his school in Camden Town aged fifteen, in rebellion against his family. He became an unattached student at Oxford for a while, and was later appointed a master at a school for boy choristers at Oban by the Marquess of Bute.

In 1886 he converted to Roman Catholicism and began training as a priest, first at Oscott and then at the Scots College in Rome, but in 1890 he was rejected for the priesthood, an event that embittered the rest of his life, contributing to his paranoia and quarrelsomeness and helping to ensure his constant penury. He stayed on in Rome for a year, and then spent some time in Christchurch, Hampshire, and, as a tutor again, in Aberdeen. During the mid-1890s he was in Holywell in north Wales painting ecclesiastical banners, a job he held until he presented his employer with a demand for a £1,000 fee.

In 1898 he came to London determined to be a writer: his stories based on Italian folk-tales were published in the *Yellow Book* and then appeared in volume form as *Stories Toto Told Me* (1898). Rolfe was helped by a number of friends and patrons in London, including the writer Robert Hugh Benson (brother of **E.F. Benson**), but a pattern had become established by which he quarrelled with these patrons, wrote them vituperative letters, painted unpleasant portraits of them in his fiction and, on occasion, even blackmailed them. He became established enough as a writer to publish seven books – fiction, a history and a translation – during his lifetime: most notable among them is *Hadrian the Seventh*, in which his *alter ego*, George Arthur Rose, also a failed priest, becomes Pope. The book is written in the ornate style for which Rolfe became famous, and its wit and intensity have won it many admirers. It won a new audience in 1968 when Peter Luke adapted the story for a much admired play.

In 1908, after some years in Oxford, Rolfe moved to Venice, where he spent his last years: there, although sometimes helped by more patrons, he was often destitute, living in a boat or wandering the streets, and consoling himself with many homosexual encounters. He recounted his adventures and misfortunes in a series of scurrilous letters, which horrified his first biographer, who described them as 'an unwitting account, step by step, of the destruction of a soul'. They were later published as *The Venice Letters*, and his novel *The Desire and Pursuit of the Whole* (one of a number of posthumously published works and fragments) is loosely based upon his life in that city.

Indeed, much of Rolfe's fame was posthumous, and he is chiefly distinguished as the subject of a biographical masterpiece, *The Quest for Corvo*, by A.J.A. Symons, founder of the Wine and Food Society and brother of the writer **Julian Symons**. Described by its author as 'an experiment in biography', this book follows its title by detailing Symons's experiences in researching Rolfe's extraordinary life. Certain aspects of Rolfe's character were used by **Pamela Hansford Johnson** in depicting the eponymous anti-hero of her novel *The Unspeakable Skipton* (1959).

Fiction

Stories Toto Told Me (s) 1898; *In His Own Image* (s) 1901; *Hadrian the Seventh* 1904; *Don Tarquinio* 1905; *The Weird and the Wanderer* [with C.H.C. Pirie-Gordon] 1912; *The Desire and Pursuit of the Whole* 1934; *Hubert's Arthur* [with C.H.C. Pirie-Gordon] 1935; *Three Tales of Venice* (s) 1950; *Amico di Sandro* 1951; *The Cardinal Prefect of Propaganda and Other Stories* (s) 1957; *Nicholas Crabbe* 1958; *Don Renato* 1963; *The Armed Hands and Other Stories and Pieces* (s) 1972

Poetry

Tarcissus 1880; *Collected Poems* 1972

Translations

The Rubaiyat of Umar Khaiyam 1903

Non-Fiction

Letters to Grant Richards 1952; *Letters to C.H.C. Pirie-Gordon* [ed. C. Woolf] 1959; *Letters to Leonard Moore* [ed. B.W. Korn, C. Woolf] 1960;

The Letters of Baron Corvo to Kenneth Grahame 1962; *Letters to R.M. Dawkins* [ed. C. Woolf] 1962; *Without Prejudice: One Hundred Letters to John Lane* [ed. C. Woolf] 1963; *The Venice Letters* [ed. C. Woolf] 1974

Biography

The Quest for Corvo by A.J.A. Symons 1934; *Corvo* by Donald Weeks 1971; *Frederick Rolfe: Baron Corvo* by Miriam J. Benkovitz 1977

Isaac ROSENBERG 1890–1918

The eldest son of recent Lithuanian Jewish immigrants, Rosenberg was born (the survivor of twin boys) in Bristol, but moved to Stepney in London's East End in 1897. His father was cultured but poor and had a number of jobs, including that of a pedlar, while his mother took in washing and sewing; all seven of the family lived in a single room. Rosenberg left the local board school at the age of fourteen and helped supplement the family income by becoming an apprentice engraver with the famous firm of Hentschel's in Fleet Street, attending evening classes in painting at Birkbeck College. He disliked his job, but was rescued by patrons in 1911 and sent to the Slade School of Art, where he was a member of a particularly brilliant generation: a famous photograph of a school picnic in 1912 shows Rosenberg with Carrington, Stanley Spencer, C.R.W. Nevinson, Adrian Allison, Mark Gertler and David Bomberg. The last two were also from the East End and together with Rosenberg were popularly known as 'The Whitechapel Boys'. Rosenberg developed his twin talents of painting and poetry – fifty copies of a pamphlet entitled *Night and Day* were printed in 1912 – but left the school to face unemployment and persistent ill health.

He sailed to South Africa to join one of his sisters in June 1914, and did some paintings (most of which he lost overboard on the return journey), and wrote a draft of his verse play, *Moses*. He was in Cape Town when the First World War broke out. Isolated by race, class, education and geography from the prevailing mood of excited optimism, he greeted news of the war without enthusiasm, and eventually joined up for financial rather than patriotic reasons, since he was unable to find a job on his return to England in 1915. In April of that year he funded the printing of 100 copies of his volume *Youth* by selling three paintings to Edward Marsh, who was a generous but bemused patron, since Rosenberg's poetry was very different from his Georgian ideal. Despite his small size and his poor health, he was accepted as a private in the newly formed 'Bantam Battalion' of the Suffolk Regiment. (Not wishing to kill people, he had wanted to join a medical unit, but was too short.) Carrying a copy of Donne's poems and Sir Thomas Browne's *Religio Medici*, he reported for training at Bury St Edmunds. 'Falstaff's scarecrows were nothing to these,' he wrote of his fellow Bantams. 'Three out of every 4 have been scavengers, the fourth is a ticket-of-leave.' He loathed the army and found little of the mitigating camaraderie that sustained other rankers, since anti-Semitism was rife. He was sustained by contact with the civilian world of his patrons, such as Marsh and Sydney Schiff. He eventually sailed to France with the King's Own Royal Lancasters in June 1916, within days of *Moses* being printed in a pamphlet with a handful of poems. He spent some time in the front line, where he wrote one of his best-known poems, 'Break of Day in the Trenches', which was printed with 'Marching' in **Harriet Monroe**'s magazine *Poetry* in December. After several transfers, he was killed while on night patrol near Arras, as April Fool's Day dawned. His body was never found.

The date of Rosenberg's death would have struck him as a final irony, for he had endured a luckless life, and was by nature a melancholic jester, suffering from periodic bouts of depression. His 'Self Portrait in a Felt Hat, looking right, 1915' (now in London's National Portrait Gallery) perfectly captures his sardonic personality. His Jewish, working-class origins strongly influenced his paintings and set him apart from the mainstream soldier-poets, most of whom were Anglo-Saxon officers and gentlemen. Unlike them, he had no illusions about warfare in the first place, and his poetry, though highly sophisticated, is stark and unromantic. The lyricism of pre-war poetry occasionally surfaces, as in the beautiful 'Returning, we hear the larks', but the vivid 'Louse Hunting' and the brutal 'Dead Man's Dump' are perhaps more characteristic of his work.

His reputation as a writer was made gradually, but he is now generally regarded as one of the finest of the war poets. **Keith Douglas**, who is arguably the best poet of the Second World War (and was very much an officer and gentleman), paid tribute in 'Desert Flowers': 'Rosenberg I only repeat what you were saying'. Much of his poetry had been written in a barely legible scrawl on scraps of paper, and the first collection of his *Poems*, which appeared in 1922, was edited by Gordon Bottomley (a poet Rosenberg greatly admired), who took it upon himself to alter words he thought archaic or obscure and in other ways 'improve' the poems. A *Collected Works* appeared in 1937, with texts partly restored by Bottomley and his co-editor, and an introduction by **Siegfried Sassoon**. The definitive edition, which included poetry, prose, letters, paintings and drawings (but not *Moses*) was edited by Ian Parsons in 1979. *Moses* appeared in a scholarly facsimile edition, with reproductions of drafts, typescripts, proofs and published text, in 1990, with an introductory essay by **Jon Silkin**, a major champion of Rosenberg's work. An exhibition of paintings, manuscripts and memorabilia at Leeds University in 1959 confirmed Rosenberg's high standing as an artist.

Poetry

Night and Day 1912; *Youth* 1915; *Poems by Isaac Rosenberg* [ed. G. Bottomley] 1922

Plays

Moses [pub. 1916; rev. 1990]

Collections

The Collected Works of Isaac Rosenberg [ed. G. Bottomley, D. Harding] 1937; *The Collected Works of Isaac Rosenberg* [ed. I. Parsons] 1979

Biography

Isaac Rosenberg: the Half Used Life by Jean Liddiard 1975; *Isaac Rosenberg Poet and Painter* by Jean Moorcroft Wilson 1975; *Journey to the Trenches* by Joseph Cohen 1975

Alan (John) ROSS 1922–

Although born in Calcutta, where his father was manager of a tea company, Ross was sent to England at the age of seven to attend preparatory school. He went to Haileybury in 1936, and regarded the famous military school as 'essentially philistine', but was able to indulge there in what became a lifelong passion for cricket. In 1940 he entered St John's College, Oxford, to read modern languages. He played some cricket, but now found that poetry was competing for his attention. Poetry won: he had a single work published and left Oxford in 1942 'with the poems of **Louis MacNeice** in my pocket'. He joined the Royal Navy, serving on an Arctic convoy, and remained in the service for five years. He spent much of his spare time reading and writing poetry about his experience, which he sent to **John Lehmann**, who published them in *Penguin New Writing*. He later collected his poems about his naval career into one volume, *Open Sea*.

His first volume, *The Derelict Day*, was published in 1947 and won him an Atlantic Award of £500, given by the Rockefeller Foundation. He travelled to Corsica with the painter John Minton, and together they produced the illustrated book *Time Was Away*, one of several travel books Ross has written. He joined the British Council in 1947, serving in London, apart from a trip to Iraq as personal assistant to the Controller of Education, but resigned after four years to become a freelance writer. By this time he had become part of the London literary world, contributing to the *New Statesman*, *Horizon* and the *Times Literary Supplement*, and he met his wife at a party given by **Cyril Connolly**. He spent some time as a sports correspondent for the *Observer*, writing initially about football (he was a devotee of Tottenham Hotspur), and later (from 1953 to 1972) about cricket, travelling abroad to report test matches. He has produced numerous books on sport, including *The Cricketer's Companion*, a biography of the cricketing Prince Ranjitsinhji, and an anthology about horseracing, *The Turf*.

With his wide interests and contacts, he was the natural successor to John Lehmann as editor of the *London Magazine*, which he has seen through numerous crises, financial and legal, but has kept going ever since. He has attracted a cosmopolitan list of contributors, and has given space to both established writers and new talents, many of whose careers were launched in the magazine. **William Boyd** and **Graham Swift** are two notable alumni. An offshoot of the magazine has been London Magazine Editions, a small publishing house responsible for issuing such minor classics as **J. Maclaren-Ross**'s *Memoirs of the Forties* (1965) and **T.C. Worsley**'s *Flannelled Fool* (1967), as well as the poems of **Bernard Spencer**, the journals of the painter Keith Vaughan, and **Roy Fuller**'s autobiography.

Of his other volumes of poetry, *Death Valley* was a Poetry Book Society Choice. His work is distinguished by its clarity and craftsmanship, and for dealing with such subjects as sport in an intellectual manner. Amongst his other books are a study of the work of the war artists of the Second World War, and a book about the 1940s. He has published two fine volumes of autobiography, in which lively reminiscences of the many writers and artists he has known are interspersed with poems. Indeed, he has written that *Blindfold Games*, which describes his childhood, education and war service, 'represents the raw material out of which I began to write poetry'.

Poetry

Summer Thunder 1941; *The Derelict Day* 1947; *Poetry, 1945–1950* 1951; *Something of the Sea* 1954; *To Whom It May Concern* 1958; *African Negatives* 1962; *North from Sicily* 1965; *Poems, 1942–1967* 1967; *A Calcutta Grandmother* 1971; *Tropical Ice* 1972; *Taj Express* 1972; *Open Sea* 1975; *Death Valley and Other Poems* 1980; *After Pusan* 1995

Non-Fiction
Time Was Away: a Notebook in Corsica [with J. Minton] 1948; *The Forties: a Period Piece* 1950; *The Bandit on the Billiard Table* 1954 [rev. as *South to Sardinia* 1989]; *Australia '55: a Journal of the M.C.C. Tour* 1955; *Cape Summer and the Australians in England* 1957; *Through the Caribbean: the M.C.C. Tour of the West Indies 1959–1960* 1960; *The West Indies at Lord's* 1963; *Ranji: Prince of Cricket* 1983; *Blindfold Games* (a) 1986; *Cape Summer Cricket* 1986; *The Emissary* 1986; *The Kingswood Book of Cricket* 1986; *West Indies at Lord's* 1986; *Coastwise Lights* (a) 1988

With Patrick Eager: *A Summer to Remember* 1981; *Summer of the All Rounder* 1982; *Summer of Speed* 1983; *An Australian Summer* 1985; *West Indian Summer* 1988; *Tour of Tours* 1989

Edited
John Gay *Poems* 1950; F. Scott Fitzgerald *Borrowed Time* [with J. Ross] 1951; *The Gulf of Pleasure* 1951; *Abroad: Travel Stories* (s) 1957; *The Cricketer's Companion* 1960; *Stories from the London Magazine* (s) 1964; *Living in London* 1974; Lawrence Durrell *Selected Poems* 1977; *The Penguin Cricketer's Companion* 1981; *The Turf* 1982; *Colours of War* 1983; *Kiwis and Indians* [with P. Eager] 1983; *Living Out of London* 1984; *The London Magazine 1961–1985* 1986; *Signals* 1991

Translations
Phillipe Diolé: *The Under Sea Adventure* 1953, *The Seas of Sicily* 1955; P.D. Gaissea *The Sacred Forest* 1954; R. Merle *Death Is My Trade* 1954; Andrea Empeirikos *Amour, amour* [with N. Stangos] 1966

For Children
The Onion Man 1959; *Danger on Glass Island* 1960; *The Wreck of Moni* 1965; *A Castle in Sicily* 1965

Henry ROTH 1906–

Roth was born in Tysmenica, then part of the Austro-Hungarian Empire. His father was a waiter who took the family to New York a year or so later where they lived in Brooklyn, and then in Manhattan's predominantly Jewish Lower East Side. Roth was educated at the academically prestigious De Witt Clinton High School, and at the City College of New York where he found a mentor in Eda Lou Walton, a professor of English. He graduated in science in 1928 and stayed with Walton in her Greenwich Village house where he met writers and intellectuals such as **Hart Crane** and Margaret Mead.

With Walton's financial support and encouragement, he spent four years writing *Call It Sleep*, which received only moderate critical praise and few sales when it was published in 1934. The novel is a powerful description of Jewish immigrant experience seen from the viewpoint of a young boy. Roth joined the Communist Party in 1933, and was later to attribute his inability to finish another satisfactory novel to his efforts at writing about a proletarian hero in the correct spirit of socialist realism. For the next forty-odd years he published only occasional short stories in magazines such as the *New Yorker* and *Commentary*, and worked in a variety of jobs as a schoolteacher, precision tool grinder, hospital attendant, waterfowl farmer and, intermittently, as a private tutor.

Roth's standing altered drastically in the 1960s after the critics Alfred Kazin and Leslie A. Fiedler, in response to a request from *American Scholar* to name an unjustly neglected book, cited *Call It Sleep*. The 1964 paperback reissue of the novel became a bestseller, and critics praised the novel's stream-of-consciousness technique and the linguistic montage of Yiddish, Hebrew, English and colloquial speech, comparing it to Mark Twain's *Huckleberry Finn* (1884) and **James Joyce**'s *A Portrait of the Artist as a Young Man* (1916). Roth received a grant from the American Academy in 1965 and held the D.H. Lawrence Fellowship at the University of New Mexico in 1968, living during his tenure on the Frieda Lawrence ranch in Taos.

In 1979 he published a short autobiographical pamphlet, *Nature's First Green*, and in the same year began work on a projected six-volume sequence of autobiographical novels, *Mercy of a Rude Stream*. This is now reportedly virtually complete in some 3,000 manuscript pages, and two volumes have thus far been published. The first novel begins in 1914 in the same territory as *Call It Sleep*, and is interspersed with reflections addressed by the narrator to his word processor, which is known, significantly, as 'Ecclesias' (as in the biblical book of acceptance and release). Roth has also published *Shifting Landscape*, a collection of occasional pieces collated by his friend, the Italian translator of *Call It Sleep*, Mario Materassi. Roth was married in 1939, and has two sons.

Fiction
Call It Sleep 1934; *Mercy of a Rude Stream: A Star Shines Over Mt Morris Park* 1994, *A Diving Rock on the Hudson* 1995

Non-Fiction
Nature's First Green (a) 1979; *Shifting Landscape: a Composite 1925–1987* [ed. M. Materassi] 1987

Philip (Milton) ROTH 1933–

Roth was born in the Jewish neighbourhood of Newark, New Jersey, son of an insurance salesman of Austro-Hungarian stock. He attended Rutgers University for a year before transferring to Bucknell University, where he gained his BA in English. He went on to take an MA at the University of Chicago, and after military service he returned there as an English instructor from 1956 to 1958.

He was already publishing short stories in magazines and he left teaching to write full time after his first book of stories, *Goodbye, Columbus*, was accepted for publication in 1958. When it appeared in the following year it was immediately acclaimed and won him a Guggenheim Fellowship and a National Book Award. This book and the two novels that followed it deal in an unsparingly comic fashion with American Jewish identity and the various schisms in that community between orthodox Jews, tough-minded urban Jews like Roth himself, and philistine, *nouveau riche* suburban Jews like the family of the girl with whom the narrator of *Goodbye Columbus* falls in love. Roth was increasingly controversial in the Jewish community, being accused of a variety of offences, from shameless irreverence to encouraging anti-Semitism.

This controversy was dwarfed by the furore over Roth's 'masturbation novel', *Portnoy's Complaint*, which appeared in 1969 and alerted a whole generation of readers to the recreational uses of raw liver. Parts of it had appeared in magazines to such interest that Roth earned in the region of a million dollars from it before it was even published. It is the story of repressed Alexander Portnoy, who is infatuated with gentile girls and dominated by his overbearing Jewish Momma, as told from the couch to his psychoanalyst. Roth continued to mine this vein in *The Breast*, in which the hero is transformed into a giant breast.

Roth's popular career was established but critics were less impressed by the post-*Portnoy* books until Roth launched his fictional *alter ego* Nathan Zuckerman in *The Ghost Writer* (1979). Many critics consider this and the other Zuckerman books to be his best work. Later novels include *The Counterlife*, which won the National Critics Circle Award in 1987, and *Operation Shylock*, a satire in which an impostor called Philip Roth plans to lead the Jews out of Israel back to Europe. This last book mixes fact – the trial of John Demjanjuk for war crimes in Jerusalem in 1988 – and fantasy, and was shortlisted for the *Irish Times* International Prize. His memoir of his family, *Patrimony*, won the National Critics Circle Award in 1992.

Roth has attracted attention from heavyweight academic critics such as Harold Bloom, but not everyone admires him. Apart from those who find his work offensive, there are still those who find it unimpressive and declare that it does not survive rereading. Throughout his work Roth has evolved a worldly, 'wiseass' voice and comments: 'If the goal is to be innocent of all innocence, I'm getting there.' He is married to the actress Claire Bloom.

Fiction

Goodbye, Columbus (s) 1959; *Letting Go* 1962; *When She Was Good* 1967; *Portnoy's Complaint* 1969; *Our Gang* 1971; *The Breast* (s) 1972; *The Great American Novel* 1973; *My Life as a Man* 1974; *The Professor of Desire* 1977; *The Ghost Writer* 1979; *Novotny's Pain* (s) 1980; *Zuckerman Unbound* 1981; *The Anatomy Lesson* 1983; *The Prague Orgy* 1985; *The Counterlife* 1987; *Zuckerman Bound* 1989; *Deceptions* 1990; *Operation Shylock* 1993

Non-Fiction

The Facts: a Novelist's Autobiography (a) 1988; *Patrimony* (a) 1991

Plays

The Cherry Orchard [from Chekov's play] 1981; *The Ghost Writer* (tv) [from his novel; with T. Powell] 1983

Collections

Reading Myself and Others 1975

Bernice RUBENS 1923–

Born into a Jewish family in Cardiff, Rubens is the daughter of a credit draper – who, duped by a ticket salesman, had thought he was emigrating to New York – and a schoolteacher. Her family was made up of musical prodigies: all three of her siblings became professional musicians, and Rubens herself plays the cello, though she started comparatively late in life since the family could not afford this instrument. She once claimed that being a really good musician would mean more to her than her writing. She was educated at Cardiff High School for Girls, then took an honours degree in English at the University of South Wales in the same year that she married Rudi Nassauer, a writer. They have two children and separated in 1969.

Rubens taught English for two years in Birmingham before starting a career as a documentary film maker. *Stress*, which examined the problems of parents with mentally handicapped children, won the American Blue Ribbon Award in 1968. Other films in which she has been involved include United Nations projects.

Her first novel, *Set on Edge*, was published in 1960 and like many of those that followed drew upon her Jewish background. Of these early novels, the best known is *Madame Sousatzka*, about a music teacher and her young Jewish prodigy. It was later made into a film directed by John Schlesinger, with the prodigy turned into an Asian boy. Her fourth novel, *The Elected Member*, also concerns a child prodigy, but one who has grown up and failed, becoming a drug addict. The son of a rabbi, he becomes the family scapegoat – the elected member of the title – an Old Testament notion, but one also found in the theories of the psychiatrist R.D. Laing, whose work Rubens has acknowledged as an influence. The novel, which was published in the USA as *The Chosen People*, won the Booker Prize in 1970 and established Rubens as one of the leading writers of her generation.

The 'link between sanity and madness' in this novel is one that Rubens has continued to explore, as she investigates 'the ever changing meaning of those terms'. She frequently deals with people whose lives tread this fine boundary. Miss Hawkins in *A Five Year Sentence* (also shortlisted for the Booker Prize) abandons her plan to kill herself when presented with a five-year diary as a retirement-gift from her colleagues. Unable to function without being given orders, she starts writing the diary, then following the actions she has described. *Mr Wakefield's Crusade* is another black comedy in which an individual's life of failure is altered when a man drops dead in front of him in the post office. Rubens sets these bizarre events in a solidly realised world, thus making them perfectly plausible. Perhaps her most extraordinary novel, and one of her best, is *Spring Sonata*, which takes her obsession with child prodigies to the limit. Buster is not so much an infant as a foetal phenomenon, a gifted violinist before he is born. Listening from within the womb to his parents and grandparents, he decides to stay put, dodging the surgeon's hands during a caesarian section and taking the opportunity to snatch a violin and bow from the bedside. Described by the author as 'a fable', the novel is also an exploration of the tensions within families, specifically within Jewish ones, with the emphasis upon duty and guilt.

The family is explored more solemnly in *Brothers*, a long novel about four generations of a Russian Jewish family, taking in over 100 years of persecution and migration. *Kingdom Come* is another historical novel, based on the true story of a seventeenth-century Turkish Messiah. Although these more sober novels have had their admirers, Rubens is resigned to being most praised for what she modestly describes as her 'quirky little books'. In fact, beneath their zany surface, she is dealing with major themes: love, religion, the relationship between an individual and society and an individual and his or her family.

Fiction

Set on Edge 1960; *Madame Sousatzka* 1962; *Mate in Three* 1965; *The Elected Member* [US *The Chosen People*] 1969; *Sunday Best* 1971; *Go Tell the Lemming* 1973; *I Sent a Letter to My Love* 1975; *The Ponsonby Post* 1977; *A Five Year Sentence* 1978; *Spring Sonata* 1979; *Birds of Passage* 1981; *Brothers* 1983; *Mr Wakefield's Crusade* 1985; *Our Father* 1987; *Kingdom Come* 1990; *Mother Russia* 1992

(James) David RUDKIN 1936–

Rudkin was born in London, the son of a strict Evangelical Protestant pastor and his wife, a former millgirl and later teacher from rural Protestant Ulster. In his childhood, Rudkin and his siblings were forbidden the theatre, the cinema and all 'worldly' books as abodes of Satan – ironic for one who was to make his name as a playwright, albeit one much concerned with themes of guilt and redemption. Early in the Second World War, the family moved to Birmingham, where Rudkin was educated at the famous King Edward School from 1947 to 1955. He then did two years' national service, training and working as a cipher operator in the Rhineland, Germany, and rising to the rank of sergeant.

From 1957 to 1961 he attended St Catherine's College, Oxford, where he read classics, with music as a side-study. In his teenage years he had devoted great attention to symphonic music, and he also wrote some early fiction, but at Oxford, inspired by the work of **Harold Pinter**, he turned towards playwriting, and saw his first radio play, *No Accounting for Taste*, produced in 1960. From 1961, he taught classics and music at the County High School, Bromsgrove, Worcestershire.

In the meantime, however, a play written and performed at Oxford in 1960, *Afore Night Come*, was quickly taken up by Peter Hall in 1962 and produced by the Royal Shakespeare Company, an immense boost to a young writer. The play is inspired by Rudkin's own experience as a fruit picker, and tells how a gang of these workers set on one of their number and eventually kill him. The play moves from social realism to horror and an examination of original sin, and proved controversial during its brief run, being disrupted several times and helping to fuel the then current controversy over the Lord Chamberlain's censorship. Since then, Rudkin has maintained a steady flow of work in almost every dramatic medium – he has translated plays (ranging from Ibsen to Greek tragedy) and operatic libretti, written his own libretti, been a screen and television writer as well as a ballet scenarist – and from 1964 has lived entirely as a writer. Despite his early success, this was for many years not easy. For a decade after *Afore Night Come* he was forced to concentrate on radio and television work before the production of his next major stage play, *Ashes*, the story of an infertile couple's search for a child. Often long intervals have elapsed between the writing of a work and its production: this was true of the huge, fable-like *The Sons of Light*, which was mainly written in 1964 and 1965 but not produced until 1976. Rudkin's uncompromising stance and the controversial themes he tends to tackle – religion, violence, homosexuality, Northern Ireland – perhaps account for his difficulties in finding acceptance, but through his marked individuality as well as the sheer number of his works he has been recognised as a major playwright. He has written little prose, although in early years he functioned as a film critic. He married Alexandra Margaret Thompson, an actress, in 1967, and they have four children. Rudkin does not comment on his own work and gives no interviews; he is the dedicated and driven writer *par excellence*.

Plays

No Accounting for Taste **(r)** 1960; *Afore Night Come* 1960 [pub. in *New English Dramatists 7* 1963]; *The Stone Dance* **(tv)** 1963; *Moses and Aaron* **(l)** [with Schoenberg] 1965; *The Persians* **(r)** [from Aeschylus's play] 1965; *Children Playing* **(tv)** 1967; *Gear Change* **(r)** 1967; *Burglars* 1968 [np]; *House of Character* **(tv)** 1968, **(st)** [as *No Title*] 1974 [np]; *Blodwen, Home from Rachel's Marriage* **(tv)** 1969; *The Grace of Todd* **(l)** [with G. Crosse] 1969 [pub. 1970]; *Bypass* **(tv)** 1972; *The Filth Hunt* 1972 [np]; *Atrocity* **(tv)** 1973; *Cries from Casement as His Bones Are Brought to Dublin* **(r)**, **(st)** 1973 [pub. 1974]; *Ashes* 1973 [pub. 1977]; *Penda's Fen* **(tv)** 1974 [pub. 1975]; *The Ash Tree* **(tv)** [from M.R. James's story] 1975 [np]; *The Coming of the Cross* **(tv)** 1975; *Hecuba* **(r)** [from Euripides's play] 1975; *Pritan* **(tv)** 1975 [np]; *Cross* **(tv)** 1975; *The Sons of Light* 1976 [pub. 1981]; *Sovereignty under Elizabeth* 1977 [np]; *Hippolytus* [from Euripides's play] 1978 [pub. 1980]; *Hansel and Gretel* 1980 [np]; *Artemis 81* **(tv)** 1981; *The Living Grave* **(tv)** 1981; *The Triumph of Death* 1981; *Peer Gynt* [from Ibsen's play] 1982 [pub. 1983]; *Across the Water* **(tv)** 1983; *Space Invaders* 1984 [np]; *Will's Way* 1985 [np]; *The Saxon Shore* 1986; *Deathwatch* and *The Maids* [from J. Genet] 1987 [np]; *White Lady* **(tv)** 1987; *Broken Strings* **(l)** [with P. Vir] 1990 [np]; *The Green Knight* **(tv)** 1990; *Rosmersholm* **(r)** [from Ibsen's play] 1990; *When We Dead Awaken* [from Ibsen's play] 1990 [np]; *John Piper in the House of Death* 1991; *The Lovesong of Alfred J. Hitchcock* **(r)** 1993; *Symphonie Pathétique* 1993; *The Haunting of Mahler* **(r)** 1994

For Children

Burglars 1970 [pub. in *Prompt Two* ed. A. Durband 1976]

Jane RULE 1931–

Born in Plainfield, New Jersey, Rule spent her early life in a variety of American states, learning early that 'though all communities have rules, they are not the same ones in say, suburban California and rural Kentucky. So I grew up with some respect for the necessity of rules but also with some critical distance from any particular rule's usefulness or moral justification.' Critical distance was increased by a year in England (1952–3), with visits to European cities, when she 'discovered large numbers of people who did not feel simple gratitude for America's having won the war'. Returning to America in 1953 she encountered the rise of McCarthyism, and the reality of mass-hysteria enlisted in the fight against political and intellectual freedom. For a young, aspiring writer increasingly aware of her own lesbianism, America was not the place to be. In 1956 she moved to Vancouver, to live with Helen Sonthoff who has been her partner ever since (and who, as a friend of **W.H. Auden**, provided the occasion for a meeting between established poet and raw beginner which Rule has described vividly in her essay ' "Silly Like Us": a recollection').

For some twenty years Rule lectured in English at the University of British Columbia and served as the Director of International House. In Canada she found a landscape, people and tradition more congenial to her than her native America. In Vancouver itself she became part of a thriving artistic community and, despite being virtually unpublished for the first six years there, felt herself constantly supported by it. Her continuing gratitude to Canada, whose citizen she now is, permeates her work.

Rule's undiagnosed dyslexia kept the world of books closed until her thirteenth year when she finally found the key to 'the English language locked up in the matter on the page'. Making up for lost time she was, by sixteen, a voracious reader and an arrogant apprentice writer, 'with scorn for nearly every written word but my own'.

Already aware of her homosexuality, and of the almost universal hostility which its disclosure would arouse, she initially accepted the period's prevailing literary dictum of 'form before content', thereby enabling her to concentrate on style and technique whilst ignoring disquieting subject-matter. When, at the age of thirty-one, she read **Dorothy Baker**'s *Cassandra at the Wedding* (1962) and recognised a kindred spirit, her response was 'not one of joyful recognition but of alarm. I had been alone too long to know what to do with literary company.' By that time she had completed three novels (all unpublished) and some short stories. Macmillan of Canada expressed interest in the third novel but insisted that she find an American or English co-publisher. *Desert of the Heart* was accordingly published by the London house of Secker & Warburg (the first non-Canadian publisher to whom it was offered) in 1964. It was subsequently filmed in 1984 as *Desert Hearts*; directed by the feminist film-maker Donna Deitch, it proved to be one of a crop of commercially successful lesbian love stories released that year.

Engaging early with the re-emerging women's movement and with the advent of Gay Liberation, Rule began to question the received wisdom which placed form before content, and came to recognise the dangerous evasions it served. She carried her questioning further in *Lesbian Images*, a genuinely ground-breaking work in which she identified and explored strategies for survival employed by earlier generations of sexually ambiguous women writers, including **Gertrude Stein**, **Willa Cather**, **Elizabeth Bowen**, **I. Compton-Burnett** and Dorothy Baker. She has become a prominent and effective gay activist, an experienced broadcaster, mediator, platform speaker and polemicist: some of her best articles are collected in *A Hot-Eyed Moderate*. In 1983 her native country bestowed on her the Fund for Human Dignity Award of Merit 'for her contribution to the education of the American public about the lives of Lesbians and Gay Men'.

Rule has skilfully juggled the often conflicting demands of activism and writing. 'The political animal in each of us is competitive, egotistical and self-righteous', she has written, and its counterpart in fictional form is the novel centred on a single protagonist. Since *Against the Season* her novels have focused on seemingly very divergent groups and communities, and she has 'discovered [her] subject matter in the world we share in common'. A blind writer once said to her: 'You're the only writer I know who includes characters who happen to be physically handicapped. In most fiction, if they are there at all, it's *because* they're handicapped.' She herself has put it differently: 'When nobody looks funny to you any more, you are at home.'

Fiction

The Desert of the Heart 1964; *This Is Not For You* 1970; *Against the Season* 1971; *Theme For Diverse Instruments* (s) 1975; *The Young in One Another's Arms* 1977; *Contract with the World* 1980; *Outlander* (s) 1981; *Inland Passage and Other Stories* (s) 1985; *Memory Board* 1987; *After the Fire* 1989

Non-Fiction

Lesbian Images 1975; *A Hot-Eyed Moderate* 1985

Carol RUMENS 1944–

Rumens is the editor of *Poetry Review* and a prolific poet herself. She was born in London and educated at convent schools. Aged sixteen, she was reviewing concerts for the Croydon *Advertiser*. She entered London University in 1964 to study philosophy and 'left clutching marriage' in 1965. The poems in her first collection, *A Strange Girl in Bright Colours* (1973), deal with love, marriage and the contraceptive pill. She gave up her job as an advertising copywriter in 1981 on the publication of her third collection, *Unplayed Music*. The prevalent theme remained the feminine experience in love and family life, but the poetry was considered highly imaginative; the *Observer* remarked: 'She can make a poem blossom from carefully evocative detail to an ending of mystery and power.'

Following *Star Whisper*, which was a Poetry Book Society Choice, Rumens won a Cholmondeley Award in 1984. From 1983 to 1985 she held the post of writer-in-residence at the University of Kent at Canterbury. With the hope in mind that 'one day we will not feel obliged to think of writers in terms of gender at all', she edited *The Chatto Book of Post-Feminist Poetry 1964–1984*. The collection featured poetry by women all over the world and, apart from its title, was well received.

Her subsequent collections have expounded her deep attachment to Russia and Eastern Europe and her feelings about the paradoxes of exile and freedom. She is admired for her skill in transmitting thoughts, experiences and sensibilities not necessarily her own, but which are made real through observation and empathy. *From Berlin to Heaven* is an evocation of a journey where love, politics, faith and betrayal are entwined themes, sustained throughout a wide range of situations and moods.

Rumens' second husband is a Russian technical translator and she has two daughters. She comments: 'One of the things I admire about Soviet society is the attempt to create an atheist society. Until we can do that we are not really grown up.'

Poetry
A Strange Girl in Bright Colours 1973; *A Necklace of Mirrors* 1978; *Unplayed Music* 1981; *Scenes from the Gingerbread House* 1982; *Star Whisper* 1983; *Direct Dialling* 1985; *Icons, Waves* 1986; *Selected Poems* 1987; *The Greening of Snow Beach* 1988; *From Berlin to Heaven* 1990; *Thinking of Skins: New and Selected Poems* 1993

Fiction
Plato Park 1987

Plays
Nearly Siberia 1989

Non-Fiction
Jean Rhys: a Critical Study 1985

Edited
Making for the Open: the Chatto Book of Post-Feminist Poetry 1964–1984 1985; *Slipping Glances: Winter Poetry Supplement* 1985; *New Women Poets* 1990; E. Bartlett *Two Women Dancing: New Selected Poems* 1995

(Alfred) Damon RUNYON 1880–1946

Runyon was born Alfred Damon Runyan in Manhattan, the prairie town in Kansas. Newsprint and tall tales were in his blood: his father was an itinerant printer and publisher of small-town newspapers. The family moved to Pueblo, Colorado, where Runyon's mother died when he was seven. He was allowed to roam free within the town's juvenile street life, hanging out with young gangs and running messages in the red-light district. His father, meanwhile, spent his free time in bars, hustling drinks with embroidered stories about the Wild West and his own service under General Custer.

Runyan senior found his son a job on the Pueblo *Evening Press*. He was a news journalist at fifteen, and when a typographical slip rendered his name 'Runyon' he decided to keep it that way. After initial rejection as underage and undersized, Runyon managed to enlist in the US Army in 1898 and saw service in the Philippines, where he wrote for army papers. He left the service in 1899 and became an itinerant reporter, working on small dailies and sleeping in cheap hotels and hobo 'jungles'. His journalistic career shaped up with sports and political reporting on the Rocky Mountain *News*, and in 1908 he became a director of the Denver Press Club.

He began to publish verses and short stories in magazines, often humorous army tales featuring characters with picturesque nicknames, and moved to New York in 1911, living with a friend who offered him a room in exchange for help with his writing. Runyon was by now publishing his own work in magazines such as *Harper's*, and in 1911 he published a volume of potboiling verse called *The Tents of Trouble*, which earned him the sobriquet 'the [**Rudyard**] **Kipling** of Colorado'. He then found work on the Hearst daily the New York *American*, where the sports editor felt his full name was pompous and removed the Alfred from his byline.

Runyon's baseball coverage was enlivened with off-pitch stories, and even his field reportage was digressive and highly stylised: 'I always made covering a standard story like a big race or a ball game more or less of a stunt.' He wrote a boxing series and began to write editorials. By now he had a persona and a following, and in the early 1920s he settled into afternoons of typing and long nights among the sporting and journalistic crowd in the Broadway bars, including the all-night delicatessen Lindy's (which features as Mindy's in his fiction).

Runyon had married after his girlfriend persuaded him to give up drinking, but the marriage was not happy; he largely ignored his wife, who then died of drink herself, and he became estranged from his two children. His style was now well established, with pungent tales of the gambling, racing and criminal crowd, full of alliterative nicknames and humorous argot, and narrated in the 'historic present' (i.e. 'Yesterday I am walking down the street'), a stylistic peculiarity that was almost a trademark. He churned out an immense amount of material: columns; human-angle reportage; boxing and racing coverage; and shallow and stylised short stories, often with a certain tough-guy sentimentality. His underworld stories were particularly popular, and his 1932 collection *Guys and Dolls* was a bestseller, although its success was commercial rather than critical. In 1950 Frank Loesser and Abe Burrows's musical adaptation of one of the stories, 'The Idyll of Miss Sarah Brown', opened on Broadway under the title *Guys and Dolls*. Incorporating characters and episodes from some of Runyon's other stories, it was one of the great Broadway musicals, running for 1,200 performances and frequently revived thereafter. An inferior film, with Marlon Brando and Frank Sinatra, followed in 1955.

In 1938 Runyon developed throat cancer. His second marriage, to a young chorus-girl, was also on the rocks at this time. A 1944 operation left him unable to speak, a particularly cruel blow for a man whose life had revolved around all-night talk sessions in bars. His ashes were scattered out of a plane over Broadway, by the First World War air ace Eddie Rickenbacker.

Fiction
Guys and Dolls (s) 1932; *More Than Somewhat* (s) 1937; *Furthermore* (s) 1938; *Take It Easy* (s) 1938; *My Wife Ethel* (s) 1939; *My Old Man* (s) 1940; *The Turps* (s) 1951

Non-Fiction
Trials and Other Tribulations 1947

Poetry
The Tents of Trouble 1911

Plays
A Slight Case of Murder 1940

Collections
Damon Runyon from First to Last 1954

Biography
Damon Runyon by Jimmy Breslin 1991

(Ahmed) Salman RUSHDIE 1947–

Rushdie was born in Bombay two months before Indian independence. His father was a Cambridge-educated businessman, and his Muslim family spoke both English and Urdu. In 1961 Rushdie was sent to Rugby School in England and in 1964 the family reluctantly joined the Muslim exodus from India to Pakistan, settling in Karachi. He has spoken of the sense of cultural dislocation and divided loyalties he felt as a result of these experiences, and his work frequently reflects this sense of 'coming from too many places'. He read history at King's College, Cambridge, where he was involved with the Footlights theatre club, and after graduating in 1968, he spent a year acting with a fringe theatre group at Oval House in Kennington, before recognising the 'mediocrity' of his acting ability. From 1970 to 1981 he funded his writing by working intermittently as a freelance advertising copywriter for Ogilvy and Mather and Charles Barker, during which time he claims to have been responsible for the slogan 'Naughty but Nice', which was widely used to advertise cream.

Rushdie's first novel, *Grimus*, an exercise in fantastical science fiction which draws on the twelfth-century Sufi poem *The Conference of Birds*, received some critical recognition but sold few copies. *Midnight's Children* is a comic allegory of Indian history which revolves around the lives of narrator Saleem Sinai and the other 1,000 children born in the 'magic' hour immediately after the Declaration of Independence. The novel owes something to Latin American magic realism and to Günter Grass's *The Tin Drum* (1959), and won the Booker Prize in 1981 and the James Tait Black Memorial Prize in 1982, establishing Rushdie in the front rank of contemporary writers. Its savage satire of Indira Gandhi and her son Sanjay (who instigated a controversial sterilisation campaign) ensured that it was banned in India. In the shorter and more polemical *Shame* Rushdie turned his attention to the rival political dynasties of Pakistan with similar consequences.

With his fourth novel, *The Satanic Verses*, which won the Whitbread Award in 1988, Rushdie unwittingly achieved an international notoriety arguably unmatched by any other writer in history, though the reason for this had nothing to do with its literary merits. The novel itself is a dense patchwork of interlocking stories and dream sequences which explores ideas of identity, migration and religious revelation. The story concerns the lives of two Indian actors in England and there are distinct elements of autobiography in the background of one of these, Saladin Chamcha. The novel was immediately banned in India and South Africa on the grounds of blasphemy against Islam, and was burned on the streets of Bradford, Yorkshire. In February 1989 the Ayatollah Khomeini, the Iranian spiritual leader, issued a *fatwa* against Rushdie and 'those publishers who were aware of [the novel's] contents' which called on 'all zealous Muslims to execute them quickly'. Rushdie was forced into hiding under police protection, and various of his publishers and translators around the world have been the subject of attacks and even murder. His attempts to apologise for the distress caused to Muslims met with little success, and he has commented that the novel is an examination of 'the conflict between the secular and religious views of the world. Ironically, it is precisely this conflict which has now engulfed the book.' Malise Ruthven's *A Satanic Affair* (1990) provides a useful history of the events surrounding the *fatwa*.

Rushdie's other books include *The Jaguar Smile*, an account of a short trip to Nicaragua in 1986, and *Haroun and the Sea of Stories*, a children's book which is none the less an interesting meditation on the nature and function of storytelling. He has been married twice, in 1976 to Clarissa Luard, with whom he has a son, and in 1988 to the American writer Marianne Wiggins, a marriage which broke up under the pressure of the *fatwa* and their enforced hiding.

Fiction
Grimus 1975; *Midnight's Children* 1981; *Shame* 1983; *The Satanic Verses* 1988; *In Good Faith* 1990; *Imaginary Homelands* 1991; *East, West* 1994

Non-Fiction
The Jaguar Smile 1987

For Children
Haroun and the Sea of Stories 1990

Willy (i.e. William Martin) RUSSELL 1947–

Russell was born at Whiston, near Liverpool, and left school at the age of fifteen with one O-level in English. He was a hairdresser for six years, an experience, he says, that made him 'a good listener'. At the age of twenty, he decided to complete his education and went to college in order to improve his qualifications, after which he became a schoolteacher in Toxteth. He married in 1969, and started going to plays with his wife, who had always been a regular theatre-goer. His ambition to be a playwright was sparked when he saw a production of **John McGrath**'s *Unruly Elements* at Liverpool's Everyman Theatre in 1971. What he particularly noticed about this play was 'the poetry of common speech', and this has been a hallmark of his own work.

His first play, *Keep Your Eyes Down*, was produced that same year, but he made his name with *John, Paul, George, Ringo ... and Bert*, a musical about the Beatles. This had been commissioned by the Liverpool Everyman (of which he subsequently became an honorary director) and won the *Evening Standard* and London Theatre Critics awards for the best musical in 1974. Thereafter his plays have won widespread popular and critical acclaim. He has said that his work is concerned with the essential goodness of humanity, and although his characters are often depicted in bleak circumstances, there is an underlying optimism and warmth in his view of the world. This has inevitably led to accusations of sentimentality, but on the whole Russell manages to avoid this pitfall.

Two of his best-known plays have female protagonists. *Educating Rita*, which was inspired by his experiences at evening classes, is about a young working-class woman who decides to study English with the Open University. Much of the comedy arises from her fresh, unschooled reaction to the classics of English literature, but she is never patronised by the author, who recognises from his own experience that education is a means of escape from one's circumstances. *Shirley Valentine* is also about escape, and takes the form of a monologue by a housewife before and after a transforming holiday in Greece. Both plays were made into very successful films from Russell's own screenplays, featuring the actresses who originally created the roles on stage (Julie Walters, who won an Oscar, and Pauline Collins respectively).

Russell's other big success has been *Blood Brothers*, a Liverpudlian folk opera about a pair of twins separated at birth and brought up in completely different environments. (A singer-songwriter who performs on the folk circuit, Russell wrote the music and lyrics for this and several other of his plays.) It enjoyed a very long run in London's West End and played on Broadway. Russell has also written plays for television, the best of which was *Our Day Out*, the affecting story of a group of Liverpool schoolchildren on a coach outing with two teachers, one of whom is a disciplinarian, the other a liberal.

Plays
Keep Your Eyes Down 1971; *Playground* 1972; *Sam O'Shanker* 1972 [rev. 1973]; *King of the Castle* **(tv)** 1973; *Terraces* [pub. in *Second Playbill I* ed. A. Durband 1973; as *Terraces* 1979]; *When the Reds* [from A. Plater's play *The Tigers Are Coming – OK?*] 1973; *John, Paul, George, Ringo ... and Bert* 1974; *Break In* **(tv)** 1975 [pub. in *Scene Scripts 2* ed. M. Marland 1978]; *Breezeblock Park* 1975 [pub. 1978]; *The Cantril Tales* [with others] 1975; *The Death of a Young Man, Young Man* 1975; *I Read the News Today* **(r)** 1976 [pub. in *Home Truths* 1982]; *One for the Road* [as *Painted Veg and Parkinson* 1976; as *Dennis the Menace* 1978; as *Happy Returns* 1978] [pub. 1980; rev. 1986, pub. 1985]; *Our Day Out* **(tv)** 1977 [pub. in *Act I* ed. D. Self, R. Speakman 1979; rev. with B. Eaton, C. Mellors 1973, pub. 1974]; *Lies* **(tv)** 1978 [pub. in *City Life* ed. D. Self 1980]; *Politics and Terror* **(tv)** 1978 [pub. in *Wordplays* ed. A. Durband 1982]; *Stags and Hens* **(st)** 1978, **(f)** [as *Dancing through the Dark*] 1983 [pub. 1985]; *The Daughters of Albion* **(tv)** 1979; *The Boy with the Transistor Radio* **(tv)** [pub. in *Working* ed. D. Self] 1980; *Educating Rita* **(st)** 1980, **(f)** 1983 [pub. 1981]; *Blood Brothers* 1981 [pub. 1986]; *One Summer* **(tv)** 1983; *I Read the News Today* [pub. 1985]; *Shirley Valentine* **(st)** 1986, **(f)** 1990 [pub. 1988]

For Children
Tam Lin 1972

Poetry
Sam O'Shanker 1978

S

V(ictoria Mary) SACKVILLE-WEST 1892–1962

'Vita' Sackville-West's imagination was fired by her birthplace, the vast country house of Knole in Kent, seat of the Earls of Dorset, where she grew up, but which she was unable to inherit because, although the only child of Lord Sackville, she was female. Her lineage was aristocratic but heterodox, for her mother, the eccentric Lady Sackville, was the illegitimate daughter of a famous Spanish dancer (about whom Sackville-West was to write a book, *Pepita*) and the 2nd Lord Sackville: thus, Lady Sackville married her father's nephew.

Although predominantly (indeed voraciously) homosexual, Sackville-West married in 1913 the diplomat Harold Nicolson, who also preferred lovers of his own sex. The marriage had its crises, but in spite of extramarital affairs on both sides, it endured, producing two children, the art critic Benedict Nicolson and the publisher Nigel Nicolson, who in 1973 compiled *Portrait of a Marriage* from his mother's notes. The most serious threat to the marriage was posed by Violet Trefusis, the daughter of Mrs George Keppel, celebrated mistress of Edward VII. She and Sackville-West had met at school, and their relationship continued until after their respective marriages; at one point they 'eloped' to France and had to be brought back to England by their distraught husbands.

Enormously prolific, Sackville-West was a poet, novelist, critic, biographer (of Aphra Behn, Joan of Arc and others), hagiographer, anthologist and journalist. Several of her books, both fiction and non-fiction, centred upon Knole, as did the novel she inspired, *Orlando* (1928) by **Virginia Woolf**, one of a long catalogue of women who fell for Sackville-West's swashbuckling personal style. Other notable lovers included Hilda Matheson, head of the BBC Talks Department (who, assisted by **J.R. Ackerley** and Lionel Fielden – both also homosexual – produced a famous broadcast by the Nicolsons on the topic of marriage), and **Roy Campbell**'s wife, Mary. During the 1920s and 1930s she made frequent radio broadcasts for the BBC, and in 1927 won the Hawthornden Prize for her epic rural poem, *The Land*. In 1930 the Nicolsons bought Sissinghurst Castle, a near-derelict house in Kent which they restored and at which Sackville-West created a magnificent garden, her most enduring work. After the Second World War she wrote highly popular gardening articles for the *Observer*, and lectured for the British Council. A striking figure, habitually attired for gardening in jodphurs tucked into officer's lace-up boots, and a belted jacket over a blouse and pearls, she was once described as looking like Lady Chatterley and the gamekeeper rolled into one.

She was created a Companion of Honour in 1948, and continued to write up until her death from cancer. She enjoyed writing poetry most, though she mistrusted her own fluency – 'it's just like a pianola reeling off', she complained. Her verse is little read now, but her gardening books continue to be reprinted, and even her novels, which are not much regarded by the critics, have enjoyed a revival.

Poetry

Poems of West and East 1917; *Orchard and Vineyard* 1921; *The Land* 1926; *King's Daughter* 1929; *Invitation to Cast out Care* 1931; *Sissinghurst* 1931; *Twenty-Three Poems* 1931; *Collected Poems* 1933; *Solitude* 1938; *Selected Poems* 1941; *The Garden* 1946

Fiction

Heritage 1919; *The Dragon in Shallow Waters* 1921; *The Heir* (s) 1922; *The Challenge* 1923; *Grey Wethers* 1923; *Seductions in Ecuador* 1924; *The Edwardians* 1930; *All Passion Spent* 1931; *Family History* 1932; *Thirty Clocks Strike the Hour and Other Stories* (s) 1932; *The Dark Island* 1934; *Grand Canyon* 1942; *Devil at Westease* 1947; *The Easter Party* 1953; *No Signposts in the Sea* 1961

Non-Fiction

Knole and the Sackvilles 1922; *Passenger to Teheran* 1926; *Aphra Behn: the Incomparable Astrea* 1927; *Twelve Days: an Account of a Journey Across the Bakhiari Mountains* 1928; *Andrew Marvell* 1929; *The Women Poets of the Seventies* [ed. H. Granville-Barker] 1929; *Beginnings* (c) 1934; *Pepita. Biographies of Josefa Duran y Ortega and Victoria Sackville-West* 1937; *Some Flowers* 1937; *Saint Joan of Arc* 1937; *Country Notes* 1939; *Country Notes in Wartime* 1940; *English Country Houses* 1941; *The Eagle and the Dove* 1943; *The Woman's Land Army* 1944; *Nursery Rhymes* 1947; *Hidcote Manor Garden* 1951; *In Your Garden* 1951; *In Your Garden Again* 1953; *Walter de la Mare and The Traveller* 1953; *More for Your Garden* 1955; *Even More for Your Garden* 1958; *A Joy of Gardening* [ed. H.I. Popper] 1958; *Daughter of France: the Life of Anne Marie Louise d'Orléans* 1959; *Faces: Profiles of Dogs* 1961; *Portrait of a Marriage* (a) [ed. N. Nicolson] 1973; *Dearest Andrew: Letters from Vita Sackville-West to Andrew Reiber* 1980; *The Letters of Vita Sackville-West to Virginia Woolf* [ed. L. De Salvo, M.A. Lenska] 1984

Biography

Vita by Victoria Glendinning 1983

Lisa ST AUBIN DE TERÁN 1953–

Terán was born in Kensington and raised in Clapham and Wimbledon, where the family's neighbour was the travel writer Eric Newby. In her first volume of memoirs, *Off the Rails*, she paints a vivid portrait of her Guianese-born father, Jan Rynveld, a sometime poet, politician, diplomat and professor whose 'amorous imbroglios were so numerous as to resemble something out of the Keystone Cops'. She had a perfunctory education at James Allen's Girls' School in East Dulwich, discovering the 'recurring magic' of trains as she played truant and travelled to Brighton. At the age of sixteen she met and married Jaime Terán, a Venezuelan political exile who was a 'complete stranger' when they married. The couple travelled in Europe for two years before political changes allowed them to return to his estate in the Venezuelan Andes in 1972. Here she spent seven years managing a sugar plantation and absorbing the stories told by family and servants which would furnish her with the material for her first novel, *Keepers of the House* (1982), which won a Somerset Maugham Award.

After seven years in Venezuela, Terán returned to England where she married the poet **George MacBeth**, who encouraged her writing. *Keepers of the House* is a largely autobiographical story narrated by Lydia, who marries Diego Beltran when she is sixteen and returns to his estate. The weird and grotesque stories she hears from an old family servant give the novel a distinctly magical realist feel. Her second novel, *The Slow Train to Milan*, draws on her time travelling in Europe with her first husband and a band of his fellow political exiles. Her marriage to MacBeth broke down when she met the painter Robbie Duff-Scott whom they had commissioned to paint her portrait. MacBeth's 1991 novel *Another Love Story* is a thinly veiled account of these events. Terán's later novels include *Black Idol*, a fictionalised account of the poet Harry Crosby who killed his mistress Josephine Baker in New York, and *Joanna*, a grim family saga involving rape, child battering and insanity. Her second volume of memoirs, *A Valley in Italy*, describes her life with her third husband, Duff-Scott, and an extended family in a villa in Umbria. She has three children, one from each of her marriages.

Fiction

Keepers of the House 1982; *The Slow Train to Milan* 1983; *The Tiger* 1984; *The Bay of Silence* 1986; *Black Idol* 1987; *The Marble Mountain and Other Stories* (s) 1989; *Joanna* 1990; *Nocturne* 1992

Non-Fiction

Off the Rails: Memoirs of a Train Addict (a) 1989; *Venice: the Four Seasons* 1992; *A Valley in Italy: Confessions of a House Addict* (a) 1994

Poetry

The Streak 1980; *The High Place* 1985

Edited

Indiscreet Journeys: Stories of Women on the Road (s) 1989

SAKI (pseud. of Hector Hugh Munro) 1870–1916

Born in Burma, Saki lost his mother when he was three: she was killed by a cow – just the sort of bizarre accident which was to become a staple of his short stories. He was sent back to Devon with his brother, Charles, and his devoted sister, Ethel, to be brought up by a pair of maiden aunts, Augusta and 'Tom', who were manifestly unsuited for the job. Educated privately, he followed his father into the Burma military police, but rapidly became ill and returned to England, where he embarked on a career in political journalism. He wrote satirical parliamentary sketches for the *Westminster Gazette* (including those based upon Lewis Carroll's books, collected as *The Westminster Alice*), and became foreign correspondent in the Balkans for the *Morning Post*.

His first book was a bloodthirsty and irreligious account of *The Rise of the Russian Empire* (1900), modelled upon Gibbon. In 1901 the first of his 'Reginald' stories, featuring a Wildean young cynic-about-town, was published in the *Westminster Gazette*. Two volumes of 'Reginald' stories were published and these made his name. Further stories followed, in which languid aristocrats exchanged aphorisms and nature exploded terrifyingly into Edwardian drawing-rooms. Characterised by brevity, polish, wit and black humour, his stories expose the savagery which lay beneath the veneer of pre-war society, and include several classics of the genre. One volume was entitled *Beasts and Super-Beasts*, and Saki had an affinity with animals, the beastlier the better. In 'Tobermory', a cat gains the power of speech and disrupts a houseparty by repeating the guests' vicious comments about each other; in 'The Music on the Hill', a young woman who disbelieves in Pan is killed by a stag; in 'Gabriel-Ernst', a distinctly homoerotic tale of lycanthropy, a wolf-boy carries off a child. Many of his stories are undoubtedly (and refreshingly) cruel, and this may have something to do with his unhappy childhood: it is the sort of cruelty that a put-upon small boy, in wishful fantasies, might unleash upon horrible adults. Indeed, in 'Sredni Vashtar', perhaps his best story, he depicts with considerable relish a guardian (clearly based upon Aunt Augusta) being despatched by her ward.

Saki's homosexuality was apparently quite well known in pre-war literary circles, and frequently surfaces in his work. A taste for aphorisms is not the only characteristic that Reginald shares with Oscar Wilde. (One story, featuring naked choristers, bears the equivocal title 'Reginald's Choir Treat'.) Ethel Munro (who contributed a marvellous memoir to *The Square Egg*) was certainly shrewd enough to guess at the nature of her brother's relationships with a succession of young men, and after his death she burned all his private papers. Although in his forties, he enlisted as a ranker in the Royal Fusiliers, and was shot dead in the trenches by a sniper, his last words being: 'Put that bloody cigarette out.'

Apart from his stories, he wrote one full-length play, two dramatic sketches and two novels. Of these last, *When William Came* is a surprisingly sentimental 'invasion novel' about 'London under the Hohenzollerns', but *The Unbearable Bassington* is a far subtler book, a prophetic anthem for doomed youth which has affinities with **J.M. Barrie**'s *Peter Pan* (1904) and the poems of **A.E. Housman.**

Fiction
Reginald (s) 1904; *Reginald in Russia* (s) 1910; *The Chronicles of Clovis* (s) 1911; *The Unbearable Bassington* 1912; *When William Came* 1913; *Beasts and Super-Beasts* (s) 1914; *The Toys of Peace* (s) 1919

Non-Fiction
The Rise of the Russian Empire 1900; *The Westminster Alice* [with F. Carruthers Gould] 1902

Collections
The Square Egg and Other Sketches 1924

Biography
Saki by A.J. Langguth 1981

J(erome) D(avid) SALINGER 1919–

Salinger was born in New York, where he spent his early life, being educated at Manhattan public schools. He attended three universities without taking any degrees, as well as the Valley Forge Military Academy, and also spent a year in Europe in 1937 and 1938. He was in the US Army from 1942 to 1946, and served as an infantry staff sergeant in Europe from D-Day until the end of the war. He won five battle stars and is a member of the Légion d'Honneur, but it has been suggested that the disgust with modern life expressed in much of his work has its roots in the horrors of battle experience.

He began writing when he was fifteen, and his first short story was published in 1940 in *Story*. During the 1940s his stories began to appear frequently in many magazines including the *New Yorker*. His first novel, *The Catcher in the Rye*, was published in 1951, and has been perhaps the most popular 'serious' novel to be published in English-speaking countries since the war: its massive success somewhat embarrassed the publicity-shy Salinger. It is the story of Holden Caulfield, an adolescent boy adrift in New York and in rebellion against the adult world of 'phonies': the theme of sensitive youthful distaste for a corrupt world is Salinger's most constant subject. Almost all his other fiction, except for a number of the stories in *Nine Stories*, concerns the members of the hypersensitive and artistic Glass family, a central element being the suicide of the youthful Seymour Glass and the complex reactions of his siblings to it over a period of time.

Salinger's last published fiction was a story in the *New Yorker* in 1965: he claims to be still writing but simply not publishing. He has retired to his home in the New Hampshire hills, shunning all publicity and reacting violently to signs of public interest or criticism. The attempt by the British poet and critic **Ian Hamilton** to write a biography was consistently thwarted by Salinger, who issued writs and injunctions to prevent the book from being published. Salinger secured a Pyrrhic victory: Hamilton was obliged to rewrite the book (after it had already reached proof stage) as an account of his failure to write the biography he had planned, but the very public battle brought Salinger a great deal of unwelcome attention. He married in 1953 and divorced in 1976, having had one son and one daughter.

Fiction
The Catcher in the Rye 1951; *Nine Stories* (s) [UK *For Esmé – with Love and Squalor and Other Stories*] 1953; *Franny and Zooey* 1961; *Raise High the Roof Beam, Carpenters and Seymour: an*

Introduction (s) 1963; *The Complete Uncollected Short Stories of J.D. Salinger* (s) 1988

Biography

In Search of J.D. Salinger by Ian Hamilton 1988

John SALISBURY

see David CAUTE

Carl (August) SANDBURG 1878–1967

Sandburg was born Carl Johnson in Galesburg, Illinois, the son of poor Swedish immigrants. His father (who changed the family surname) was a blacksmith and later a railroad worker. Sandburg left school when he was thirteen, and for the next seven years did a wide variety of manual jobs in Illinois, Kansas, Nebraska and Colorado. He returned to Galesburg in 1898 with the trade of house-painter, but soon enlisted in the Spanish-American War. A fellow infantryman in Puerto Rico suggested that he should study, and while he was at Lombard College, Galesburg, a professor there, Philip Green Wright, encouraged his writing and was later to pay for the printing of his first volume of poems. After leaving college without graduating in 1902, he became a newspaperman and also became known as a socialist campaigner: by 1910 he was secretary to the mayor of Milwaukee. In the meantime he had married Lillian Steichen, sister of the famous photographer Edward Steichen, whose biography Sandburg was to write: they had three daughters.

In 1913 he moved to Chicago, and his poems began to appear in **Harriet Monroe**'s newly founded and influential magazine *Poetry*. The publication of *Chicago Poems* (1916) established him as a leading American poet writing in a free-verse Whitmanesque style and celebrating, with populist fervour, the men and ideals of the Midwestern area where he was born. After the First World War, he joined the *Chicago Daily News* as an editorial writer, and began to tour the country as a poetry-reciter and folk-singer; his book *The American Songbag* established him as a leading folklorist. Among many volumes of poetry perhaps outstanding was *The People, Yes*. His one novel, *Remembrance Rock*, was an epic saga of America; he wrote stories for children; and *Always the Young Strangers* was an engaging volume of early autobiography. His six-volume biography of his hero Abraham Lincoln was published between 1926 and 1939.

He retired from full-time journalism in 1932, but during the Second World War emerged once more to write a folksy, platitudinous, morale-boosting column for the Chicago *Times*. In later years he was a much-loved national bard, his only rival in this area being the far greater poet **Robert Frost.** After 1945 he lived for many years with his wife on a farm at Flat Rock, North Carolina, folk-singing and breeding goats; he died there, aged eighty-nine.

Poetry

Chicago Poems 1916; *Cornhuskers* 1918; *Smoke and Steel* 1920; *Slabs of the Sunburnt West* 1922; *Selected Poems* [ed. R. West] 1926; *Good Morning, America* 1928; *The People, Yes* 1936; *Bronze Wood* 1941; *Complete Poems* 1950; *Harvest Poems 1910–1960* 1960; *Six New Poems and a Parable* 1961; *Honey and Salt* 1963

Non-Fiction

Abraham Lincoln: the Prairie Years vols 1–2 1926; *Steichen, the Photographer* 1929; *Potato Face* 1930; *Mary Lincoln, Wife and Widow* 1932; *Abraham Lincoln: the War Years* vols 1–4 1939; *Lincoln Collector: the Story of Oliver R. Barrett's Great Private Collection* 1950; *Always the Young Strangers* (a) 1953; *Abraham Lincoln: the Prairie Years and the War Years* 1954; *The Letters of Carl Sandburg* [ed. H. Mitgang] 1968; *Ever the Winds of Change* (a) [ed. M. Sandburg] 1983

Edited

Songs of America 1926; *The American Songbag* 1927; *A Lincoln and Whitman Miscellany* 1938; *Carl Sandburg's New American Songbag* 1950

Fiction

Remembrance Rock 1948

For Children

Rootabaga Stories 1922; *Rootabaga Pigeons* 1923; *Abe Lincoln Grows Up* 1928; *Rootabaga Country* 1929; *Early Moon* 1930; *Prairie-Town-Boy* (a) 1955; *Wind Song* 1960

Collections

The Sandburg Range 1957; *Sweet Music: a Book of Family Reminiscence and Song* 1963; *Breathing Tokens: Carl Sandburg* [ed. M. Sandburg] 1978; *Fables, Foibles and Foobles* 1988

Biography

Carl Sandburg: a Study in Personality and Background by Karl W. Detzer 1941; *Carl Sandburg* by Harry L. Golden 1961; *Carl Sandburg* by Richard Crowder 1964; *Carl Sandburg* by North Callahan 1987

William (i.e. Norman Trevor) SANSOM 1912–1976

Sansom (who adopted the name William at a later date) was born in Camberwell, London, the son of a naval architect. He was educated at Uppingham, and also spent considerable periods of his youth travelling and studying in Europe. Later, he worked in a bank and an advertising agency, and played the piano in a night-club. He began writing while serving with the London Fire Service during the Second World War, and his short stories, vividly describing life during the blitz, began to appear in influential magazines such as *Penguin New Writing* and *Horizon*. His first collection, *Fireman Flower and Other Stories*, appeared in 1944.

After the war he worked for a while as a scriptwriter, but soon became a professional freelance author: he was helped initially by a literary bursary of £200 arranged between the publishers Hodder & Stoughton and the Society of Authors. He published a dozen collections of short stories and became one of the most highly regarded practitioners of this form in Britain. His subjects range enormously from the quirkily funny to the memorably macabre, but his work is always marked by a strong sense of physical detail and place (particularly London) and a sharp but sympathetic view of human relationships. He also wrote a number of novels, of which *The Body*, a tragi-comic tale of jealousy, is outstanding.

In 1954 he married the actress Evelyn Grundy, with whom he had one son; they lived at Hamilton Terrace, London, becoming known for their hospitality to close friends. Like his friend and fellow fireman **Henry Green**, Sansom was a heavy drinker; his wife also became one and the marriage was punctuated by violent altercations. His drunkenness often resulted in anti-social behaviour: on one occasion he had to be removed from the audience at a Miss World contest after being sick into his hat. His friend **Alan Ross** (who was with him at the time) judged that: 'Only **Philip Toynbee** threw up more often and in more places than Bill.'

Sansom's large output, which also encompassed travel books, children's stories, essays, songs and lyrics, was always professional, but in later years – when he was a crochety freelance author in poor health affecting Edwardian clothes and a bowler hat – it took on a distinct atmosphere of potboiling. At his best, however, Sansom was a remarkable and individual writer, whose work now seems to be unjustly neglected.

Fiction

Fireman Flower and Other Stories (s) 1944; *Three* (s) 1946; *The Equilibriad* 1948; *Something Terrible, Something Lovely* (s) 1948; *The Body* 1949; *The Passionate North* (s) 1950; *The Face of Innocence* 1951; *A Touch of the Sun* (s) 1952; *A Bed of Roses* 1954; *Lord Love Us* (s) 1954; *A Contest of Ladies* (s) 1956; *The Loving Eye* 1956; *Among the Dahlias* (s) 1957; *The Cautious Heart* 1958; *Selected Short Stories* (s) 1960; *The Last Hours of Sandra Lee* 1961 [as *The Wild Affair* 1964]; *The Stories of William Sansom* (s) 1963; *Goodbye* 1966; *The Ulcerated Milkman* (s) 1966; *Hans Feet in Love* 1971; *The Marmalade Bird* (s) 1973; *A Young Wife's Tale* 1974

Non-Fiction

Jim Braidy: the Story of Britain's Firemen [with J. Gordon, S. Spender] 1943; *Westminster in War* 1947; *South: Aspects and Images from Corsica, Italy and Southern France* 1948; *Pleasures Strange and Simple* 1953; *The Icicle and the Sun* 1958; *The Bay of Naples* 1960; *Blue Sky, Brown Studies* 1960; *Away to It All* 1964; *The Grand Tour* 1968; *Proust and His World* 1971; *The Birth of a Story* 1972

For Children

It Was Really Charlie's Castle 1953; *The Light That Went Out* 1953; *The Get-Well Quick Coloring Book* 1963; *Christmas* 1968

Edited

Choice: Some New Stories and Prose 1946; Edgar Allen Poe *The Tell-Tale Heart and Other Stories* (s) 1948

Frank SARGESON 1903–1982

New Zealand's first important writer not to become an expatriate, Sargeson was born in Hamilton, near Auckland, the son of a shopkeeper. He was educated at Hamilton High School and, from the age of seventeen, worked for five years in solicitors' offices, qualifying as a solicitor of the Supreme Court of New Zealand in 1926 but never practising. Instead, he took a ticket to Europe and tramped over hundreds of miles of England and the Continent, returning to New Zealand when his money ran out. Between 1928 and 1929 he worked as an estates clerk in a public department, but suffered a nervous breakdown, and went to live with his uncle on a sheep farm. From 1932 he lived at Takapuna, Auckland, at first in a beach hut, and worked as a farmhand, market-gardener, milkman and pantryman, and, for rather longer, as a journalist. In later years, with increasing fame as a writer, he moved into a small house in Auckland, never free of financial worries, but also helped to an extent by some generous private patronage in an increasingly freelance life.

His short stories began to appear from the mid-1930s, and he was greatly encouraged by **John Lehman**, editor of *Penguin New Writing*, who also published early works such as the novella *That Summer*. Sargeson's early work was rather orthodoxly left wing and satirical of the middle class; in later years his concerns became more broadly humane but with an abiding concern with the plight of the outsider and drifter and with the rituals and experiences which attend coming to manhood in society. He was also an undoubted master of the vernacular. In his first three decades as a writer he published largely novellas and short stories, but from the mid-1960s began a remarkable late flowering: six novels, several plays, three volumes of autobiography and many more stories. His richly humorous novel *Memoirs of a Peon*, detailing the progress of a latter-day Casanova, was perhaps the peak of his achievement. Sargeson was homosexual, bound by a number of long-term attachments. A legendarily modest and scrupulous man, he was famed for his generosity to younger New Zealand writers, notably **Janet Frame**.

Fiction

Conversation with My Uncle and Other Sketches (s) 1936; *A Man and His Wife* (s) 1940; *When the Wind Blows* 1945; *That Summer and Other Stories* (s) 1946; *I Saw in My Dream* 1949; *I for One* 1954; *Collected Stories 1935–1963* (s) [ed. B. Pearson] 1964 [rev. as *The Stories of Frank Sargeson* 1973]; *Memoirs of a Peon* 1965; *The Hangover* 1967; *Joy of the Worm* 1969; *Man of England Now* 1972; *Sunset Village* 1976; *En Route* 1979

Non-Fiction

Once Is Enough (a) 1973; *More Than Enough* (a) 1975; *Never Enough! Places and People Mostly* (a) 1977; *Conversation in a Train and Other Critical Writing* [ed. K. Cunningham] 1983

Edited

Speaking for Ourselves 1945

Plays

A Time for Sowing 1961 and *The Cradle and the Egg* 1962 [pub. in *Wrestling with the Angel* 1965]

William SAROYAN 1905–1981

The son of a Presbyterian minister and writer, who had emigrated from Armenia to New Jersey in 1905, Saroyan was born in Fresno, California. His father, who had been obliged to take farm-labouring work, died from peritonitis aged thirty-six in 1911 after drinking a forbidden glass of water given by his wife. She was forced to put Saroyan with his brother and two sisters in an orphanage, while she became a maid. Six years later, the family were reunited in Fresno. Saroyan, who had a completely unacademic record, left his Fresno public school at fifteen. His mother showed him some of his father's writing at the age of nine, and from that moment he determined to be a writer. To alleviate what he described as his family's 'most amazing and comical poverty' he became a street news-vendor at the age of seven, the first of many casual jobs, but at twenty-two he decided to 'loaf and write' and never work again.

In 1933, *Star* magazine paid him fifteen dollars for 'The Daring Young Man on the Flying Trapeze', an experimental short story about a young writer who starves to death. Without waiting for their approval, Saroyan told *Star* he would supply them with a new short story every day for a month and, after two weeks, was told to continue sending them in. A collection of short stories, titled after his first success, was published in 1934 and launched him on a career that would produce a torrent of stories, novels and plays. As much again as appeared in his lifetime remains unpublished.

Saroyan's first successful play was *My Heart's in the Highlands*, but in 1939 *The Time of Your Life* won a Pulitzer Prize, which the playwright declined on the grounds that commerce should not judge the arts. In the same year, he wrote a hit pop song with a cousin, 'C'mon-a My House', sung by Rosemary Clooney. He achieved financial security when MGM bought a film scenario, *The Human Comedy*, in 1941, and he joined the US Army and was posted to London in 1942 as part of a film unit attached to the Signal Corps. He narrowly avoided a court martial when *The Adventures of Wesley Jackson*, a novel he was given time off to write, turned out to be anti-war.

In 1943, Saroyan had married the seventeen-year-old Carol Marcus, but when one night in 1949 she unexpectedly revealed she was Jewish and illegitimate, he left at once for Paris and divorced her. They remarried in 1951, only to divorce again the following year. His wife (who subsequently married the actor Walter Matthau) was granted custody of their children, Aram who became a poet and wrote a book about his father, and Lucy who became an actress. Saroyan's private life was paralleled by his chaotic financial affairs, which were not improved by a declining interest in his writing, which some critics considered sentimental and unsuited to a post-war society. Working rapidly, and rarely revising anything, opening a sentence at random with the first thought that occurred, Saroyan's haphazard 'method' attracted as many detractors as admirers. The pattern he established following his final divorce, involved constant travel, drinking, gambling, and some writing. By the late 1960s, he managed to temper his restlessness, and by renewing his early challenge of writing a new piece every day, wrote himself out of debt and earned a vast income.

His last miscellaneous essays often reflected his search for a 'lost family', his symbolic yearning for universal brotherhood. This included twelve volumes of semi-fictionalised autobiography. When young, Saroyan dressed smartly to complement his dark good looks. In old age he tended to heaviness, favoured chunky, colourful home-knit jerseys, and grew a massive walrus moustache. Shortly before his death from cancer, he remarked: 'Everybody has got to die, but I have always believed an exception would be made in my case. Now what?' Half his ashes were buried in California, and the rest in Armenia.

Fiction
The Daring Young Man on the Flying Trapeze (s) 1934; *Inhale and Exhale* (s) 1936; *Three Times Three* (s) 1936; *A Gay and Melancholy Flux* (s) 1937; *Little Children* (s) 1937; *Love, Here Is My Hat* (s) 1938; *A Native American* (s) 1938; *The Trouble With Tigers* (s) 1938; *Peace, It's Wonderful* (s) 1939; *Three Fragments and a Story* (s) 1939; *My Name Is Aram* (s) 1940; *The Insurance Salesman and Other Stories* (s) 1941; *Saroyan's Fables* (s) 1941; *Razzle Dazzle* (s) 1942; *The Human Comedy* 1943; *Dear Baby* (s) 1944; *The Adventures of Wesley Jackson* 1946; *The Saroyan Special* (s) 1948; *The Fiscal Hoboes* (s) 1949; *The Assyrians* (s) 1950; *Rock Wagram* 1951; *Tracey's Tiger* 1951; *The Laughing Matter* 1953; *Mama I Love You* 1956; *The Whole Voyald* (s) 1956; *Papa You're Crazy* 1957; *Boys and Girls Together* 1963; *One Day in the Afternoon of the World* 1964; *The Ashtree Talkers* (s) 1977

Plays
The Man with the Heart in the Highlands 1938 [rev. as *My Heart's in the Highlands* 1939]; *The Time of Your Life* 1939 [pub. 1942]; *Hero of the World* 1940; *Love's Old Sweet Song* 1940; *The Ping Pong Circus* 1940; *Something about a Soldier* 1940; *A Special Announcement* 1940; *Subway Circus* 1940; *Sweeney in the Trees* 1940; *The Beautiful People* 1941; *Jim Dandy* 1941 [pub. 1947]; *The People with Light Coming Out of Them* 1941; *Hello Out There* 1941 [pub. 1942]; *Across the Board on Tomorrow Morning* 1942 [pub. 1941]; *The Agony of Little Nations* 1942; *Bad Men in the West* 1942; *Coming through the Rye* 1942; *Elmer and Lily* 1942; *The Poetic Situation in America* 1942; *Talking to You* 1942; *Get Away, Old Man* 1943 [pub. 1944]; *The Hungerers* 1945 [pub. 1939]; *A Decent Birth, a Happy Funeral* 1949; *Don't Go Away Mad* 1949; *Sam Ego's House* 1949; *The Son* 1950; *Opera, Opera* 1955; *The Cave Dwellers* 1957 [pub. 1958]; *Ever Been in Love with a Midget?* 1957; *The Slaughter of the Innocents* 1957 [pub. 1958]; *The Accident* 1958; *Cat, Mouse, Man, Woman* 1958; *Once Around the Block* 1959; *Settled Out of Court* [with H. Cecil] 1960; *Sam, the Highest Jumper of Them All* 1960 [pub. 1961]; *High Time along the Wabash* 1961; *Ah Man* [with P. Fricker] 1962; *The Doctor* 1963; *Patient, This I Believe* 1963; *The Playwright and the Public* 1963; *Dentist and Patient* 1968; *The Dogs* 1968; *Husband and Wife* 1968; *The New Play* 1970; *Armenians* 1974; *The Rebirth Celebrations of the Human Race at Artie Zabala's Off-Broadway Theatre* 1975

Non-Fiction
Those Who Write Them and Those Who Collect Them 1934; *Christmas* 1939; *Harlem as Seen by Hirschfield* 1939; *Hilltop Russians in San Francisco* 1941; *Why Abstract?* 1945; *The Twin Adventures* (a) 1950; *The Bicycle Rider in Beverley Hills* (a) 1953; *Here Comes, There Goes, You Know Who* (a) 1961; *A Note on Hilaire Hiler* 1962; *Not Dying* (a) 1963; *Short Drive, Sweet Chariot* (a) 1966; *Look at Us: Let's See: Here We Are: Speak Soft: I See, You See, We All See; Stop, Look, Listen; Beholder's Eye; Don't Look Now But Isn't That You? (us? US?)* 1967; *I Used to Believe I Had Forever: Now I'm Not so Sure* (a) 1968; *Letters from 74 rue Taitbout* (a) 1969; *Days of Life and Death and Escape to the Moon* (a) 1970; *Places Where I've Done Time* (a) 1972; *The Tooth and My Father* (a) 1974; *An Act or Two of Foolish Kindness* (a) 1976; *Famous Faces and Other Friends* (a) 1976; *Morris Hirschfield* 1976; *Sons Come and Go, Mothers Hang in Forever* (a) 1976; *Chance Meetings* (a) 1978; *Obituaries* (a) 1979; *Births* (a) 1981; *The Pheasant Hunter* 1986

For Children
Me 1963; *Horsey Gorsey and the Frog* 1968; *The Circus* 1986

Edited
Hairenik 1934–1939: an Anthology of Short Stories and Poems 1939; *Hayats'uts' Hovhannes* 1978

Collections
The Time of Your Life 1939; *The William Saroyan Reader* 1958; *After Thirty Years* 1964; *Madness in the Family* [ed. Leo Hamilton] 1988

Biography
Saroyan by Barry Gifford and Lawrence Lee 1984; *William Saroyan* ed. by Leo Hamalian 1987

(Eleanor) May SARTON 1912–95

Christened Eléanore Marie, Sarton had her names later Anglicised to Eleanor May. She was born in Wondelgem, Belgium, to the eminent Belgian historian of science (and founder of the magazine *Isis*) George Sarton and his wife, an English furniture and fabric designer. On the outbreak of the First World War, when Sarton was two, the family fled to England, and from there, in 1916, to Cambridge, Massachusetts, where George Sarton obtained a part-time teaching post at Harvard and where they became American citizens in 1924. Sarton was educated at Shady Hill, a progressive open-air school, and later at the Cambridge High and Latin School, also spending a year back in Belgium at the Institut Belge de Culture Française.

At the age of seventeen she first published her poems in the influential journal *Poetry*, and in the same year joined Eva Le Gallienne's Civic Repertory Theater in New York, training there for several years and founding her own Apprentice Theater, which disbanded in 1936. During the 1930s she also visited Europe frequently, spending a year in Paris in 1931–2 and forming a friendship with **Virginia Woolf**) during a visit to England in 1936. She also claimed to have had an affair with **Elizabeth Bowen**, and this became the subject of her first novel, *The Single Hound* (1938), although the character based upon herself was transformed into a man. From the late 1930s on, Sarton devoted herself to writing, and also taught at a wide variety of American universities and colleges, until retiring from this in 1972.

During her early career, after her first book of poetry was published in 1937, she tended to write volumes of poetry and novels (and, at first, short stories) in alternation; from the late 1950s she added the publication of memoirs and journals, and in all wrote more than forty books. As a poet she is direct and lyrical, and was long ignored by the critics as a sentimentalist, until the rise of the women's movement. Her early novels were set in Europe (the first with an American setting is *Faithful Are the Wounds*, based on the life and suicide of the Harvard writer and professor F.O. Matthiessen). Many of her novels spring directly from her own experience of youth, middle age and old age, and she is a writer of largely women's (and often, lesbian) experience, but one who does not ignore men. Among the best known of her volumes of journals is *After the Stroke*, dealing with her recovery from a minor stroke suffered in early 1986. Other works include some documentary film scripts produced during the Second World War and some unpublished plays.

In the 1970s and 1980s she became celebrated as a reader of her own work and coast-to-coast lecturer. She lived for most of her life on America's East Coast, and shared a house in Cambridge, Massachusetts for many years with Judith Matlock, a professor of English at Simmons College. In 1973 she moved into a clifftop house overlooking the sea in York, Maine.

Poetry

Encounter in April 1937; *Inner Landscape* 1939; *The Lion and the Rose* 1948; *The Land of Silence* 1953; *In Time Like Air* 1957; *Cloud, Stone, Sun, Vine* 1961; *A Private Mythology* 1966; *As Does New Hampshire* 1967; *Kinds of Love* 1970; *A Grain of Mustard Seed* 1971; *A Durable Fine* 1972; *Halfway to Silence* 1980; *Letters from Maine* 1984; *The Silence Now* 1988; *Endgame* 1992; *Collected Poems 1930–1993* 1993; *Coming into Eighty* 1995

Fiction

The Single Hound 1938; *The Bridge of Years* 1946; *Shadow of a Man* 1950; *A Shower of Summer Days* 1952; *Faithful Are the Wounds* 1955; *The Birth of a Grandfather* 1957; *The Fur Person* 1957; *The Small Room* 1961; *Joanna and Ulysses* 1963; *Mrs Stevens Hears the Mermaids Singing* 1965; *Miss Pickthorn and Mr Hare* 1966; *The Poet and the Donkey* 1969; *As We Are Now* 1973; *Punch's Secret* 1974; *Crucial Conversations* 1975; *A Walk through the Woods* 1976; *A Reckoning* 1978; *Anger* 1982; *The Magnificent Spinster* 1985; *The Education of Harriet Hatfield* 1989

Non-Fiction

I Knew a Phoenix: Sketches for an Autobiography (a) 1959; *The Movement of Poetry* (c) 1962; *Plant Dreaming Deep* (a) 1968; *The Leopard Land: Haniel and Alice Long's Santa Fe* 1972; *Journal of a Solitude* (a) 1973; *A World of Light: Portraits and Celebrations* 1976; *The House by the Sea: a Journal* 1977; *May Sarton: a Self-Portrait* (a) 1982; *At Seventy: a Journal* 1984; *After the Stroke* 1988; *Conversations with May Sarton* 1991; *A House of Gathering* 1993

Edited

Letters to May 1986

Plays

Toscanini (f) 1944; *Valley of the Tennessee* (f) 1944; *The Underground River* 1947

Collections

Sarton Selected 1991

Siegfried (Loraine) SASSOON 1886–1967

Although 'clothed in the gilded surname of Sassoon' (a leading Jewish family), Sassoon was brought up as an English country gentleman in the Weald of Kent. His parents separated when he was three, and he saw little of his father. Educated at Marlborough College and at Clare College, Cambridge, he was a dreamy youth, more interested in fox-hunting, cricket and poetry than in scholarship. After university, he lived at home and wrote poems which were privately printed in limited editions ('Melodious ramblings, published at a loss, / But gracefully reviewed by Edmund Gosse', as he later put it). Like many of his generation, he was greatly influenced by **John Masefield**, and his long poem *The Daffodil Murderer* (published under the pseudonym Saul Kain) is a skilful parody of the elder poet.

He enlisted as a trooper with the Sussex Yeomanry as the First World War loomed, then transfered to the Royal Welch Fusiliers, and served with reckless bravery as a captain on the Western Front, earning himself a Military Cross and the nickname 'Mad Jack'. He had gone to war with all the ideals of his class and upbringing, but disillusionment set in and his response (spurred by the deaths of friends and the suffering of his men) was a succession of bitter, satirical poems criticising the conduct of the war and the complacency of those at home. In 1917 he threw his MC into the Mersey and publicly declared his intention to refuse further orders; instead of being court-martialled, he was immediately sent to Craiglockhart, a neuraesthenics' hospital near Edinburgh, largely thanks to the intervention of his sometime friend and fellow Fusilier, **Robert Graves**. There he met, befriended and encouraged **Wilfred Owen**, before returning to his regiment, with whom he served in Palestine and France until July 1918, when he was accidentally shot in the head by one of his own regiment.

After the war, he helped to establish Owen's posthumous reputation, editing a volume of his poetry, and provided financial help for the indigent and ungrateful Graves, whose occasionally fanciful memoir, *Good-bye to All That* (1929), deeply upset him. He became involved in Labour pacifist politics and was literary editor of the socialist *Daily Herald*. He had a number of affairs: with the artist Gabriel Atkin, with Prince Philip of Hesse (grandson of Frederick III of Prussia), and with the self-styled aesthete, the Hon. Stephen Tennant, whose flamboyance was in stark contrast to Sassoon's military bearing – **Edith Sitwell** dubbed them 'Little Lord Fauntleroy and the Aged Earl'. Tennant spent much of the early 1930s convalescing from tuberculosis, and he eventually found Sassoon's bedside attentiveness oppressive. He broke off the relationship in 1933 and within months Sassoon married Hester Gatty. The marriage was unhappy, and the couple lived largely apart, but it produced a son upon whom Sassoon doted.

Sassoon had spent the 1920s attempting to find a new poetic voice for peacetime, and his poetry underwent a gradual transition from the strenuously satirical to the gently meditative, a shift confirmed by his conversion to Roman Catholicism in 1957. He lived reclusively in Wiltshire, where he wrote two subtle, sardonic trilogies about his childhood and the war. They traced the century's brutal loss of innocence by ironically juxtaposing the Western and rural Home Fronts, and reflected his 'craving to revisit the past and give the modern world the slip'. The first trilogy proved immensely popular and its first volume, *Memoirs of a Fox-Hunting Man*, was awarded the Hawthornden Prize. Although he was awarded the CBE in 1951 and the Queen's Medal for Poetry in 1957, Sassoon subsequently felt himself to be forgotten by the literary world, and his later years were spent in melancholy self-communion.

Poetry

Poems 1906; *Sonnets* 1909; *Sonnets and Verses* 1909; *Poems* 1911; *Twelve Sonnets* 1911; *Melodies* 1912; *An Ode for Music* 1912; *The Daffodil Murderer* [as 'Saul Kain'] 1913; *Discoveries* 1915; *Morning-Glory* 1916; *The Redeemer* 1916; *The Old Huntsman and Other Poems* 1917; *To Any Dead Officer* 1917; *Counter-Attack and Other Poems* 1918; *Four Poems* 1918; *A Literary Editor for the New London Daily Newspaper* 1919; *Picture-Show* 1919 [rev. 1920]; *The War Poems of Siegfried Sassoon* 1919; *Lines Written in the Reform Club* 1921; *Lingual Exercises for Advanced Vocabularians* 1923; *Recreations* [as 'anon.'] 1923; *Selected Poems* 1925; *Satirical Poems* 1926 [rev. 1933]; *The Heart's Journey* 1927 [rev. 1928]; *Nativity* 1927; *To My Mother* 1928; *A Suppressed Poem* 1929; *In Sicily* 1930; *On Chatterton* 1930; *Poems* [as 'Pinchbeck Lyre'] 1931; *To the Red Rose* 1931; *Prehistoric Burials* 1932; *The Road to Ruin* 1933; *Vigils* 1934; *Rhymed Ruminations* 1939; *Poems Newly Selected 1916–1935* 1940; *Collected Poems* 1947; *Common Chords* 1950; *Emblems of Experience* 1951; *The Tasking* 1954; *An Adjustment* 1955; *Sequences* 1956; *Poems by Siegfried Sassoon* [ed. D. Silk] 1958; *Lenten Illuminations; Sight Sufficient* 1958; *The Path to Peace* 1960; *Collected Poems 1908–1956* 1961; *An Octave: 8 September 1966* 1966; *The War Poems* [ed. R. Hart-Davis] 1983

Fiction

The Complete Memoirs of George Sherston [US *The Memoirs of George Sherston*] 1937: *Memoirs of a Fox-Hunting Man* [as 'anon.'] 1928, *Memoirs of an Infantry Officer* [as 'anon.'] 1930, *Sherston's Progress* 1936

Non-Fiction

The Old Century and Seven More Years (a) 1938; *On Poetry* 1939; *Early Morning Long Ago* (a) 1941; *The Weald of Youth* (a) 1942; *Siegfried's Journey 1916–1920* (a) 1945; *Meredith* 1948; *Something about Myself* (a) 1966; *Letters to a Critic* [ed. M. Thorpe] 1976; *Siegfried Sassoon Diaries* [ed. R. Hart-Davis] 3 vols 1981–5; *Letters to Max Beerbohm* [ed. R. Hart-Davis] 1986

Plays
Orpheus in Diloeryum [as 'anon.'] 1908; *Hyacinth* 1912; *Amyntas* 1912

Dorothy L(eigh) SAYERS 1893–1957

The only daughter of the Revd Henry Sayers, headmaster of Christ Church Choir School, Oxford, and later a clergyman in the Fen country, Sayers was one of the first women to take a degree at Oxford: a first in modern languages in 1915. She taught briefly at schools in England and France before becoming an advertising copywriter with S.H. Benson Ltd for nine years, until 1931. In this position, she influenced the history of advertising, and the background is used in one of her most successful detective novels, *Murder Must Advertise*. After publishing two volumes of poetry, she turned to detective fiction as the easiest method of becoming a professional writer. Her first detective novel, introducing her aristocratic detective Lord Peter Wimsey, was *Whose Body?* (1923). She followed this with eleven others which, if they have been much vilified, are still viewed as classics of the detective genre.

In 1924, after a casual affair, she gave birth in secret to an illegitimate son, with whose upbringing she refused to concern herself except financially and from whom, in his adult life, she was estranged. In 1926, she married Capt. Oswald Atherton Fleming, a former war hero turned wastrel, whom she supported through her writing: they had no children, and he died in 1950.

In later years, bored with detection, her writing turned to Christian apologia, and she became famous for her radio plays based on the life of Christ, *The Man Born to Be King*. In her final years she produced an intermittently beautiful translation of Dante's *Divina Commedia*. A notably eccentric woman, who gave her recreation as motorcycling, and as an undergraduate had become notorious for walking down Oxford High Street smoking a cigar, she died at the age of sixty-four of a coronary thrombosis brought on by drinking and chain-smoking.

Fiction
Whose Body? 1923; *Clouds of Witness* 1926; *Unnatural Death* 1927; *Lord Peter Views the Body* (s) 1928; *The Unpleasantness at the Bellona Club* 1928; *The Documents in the Case* 1930; *Strong Poison* 1930; *The Five Red Herrings* 1931; *Have His Carcase* 1932; *Hangman's Holiday* (s) 1933; *Murder Must Advertise* 1933; *The Nine Tailors* 1934; *Gaudy Night* 1935; *Busman's Honeymoon* 1937; *In the Teeth of the Evidence* (s) 1939; *Lord Peter* (s) 1972; *Talboys* 1972

Plays
Busman's Honeymoon [from her novel; with M. St C. Byrne] 1937; *The Zeal of Thy House* 1937; *He That Should Come* (r) 1938 [pub. 1939]; *The Devil to Pay* 1939; *Love All* 1940 [np]; *The Golden Cockerel* [from Pushkin's story] (r) 1941; *The Man Born to Be King* (r) 1941–2 [pub. 1943]; *The Just Vengeance* 1946; *Where Do We Go from Here?* [with others] (r) 1948; *The Emperor Constantine* 1951 [rev. as *Christ's Emperor* 1952]

Non-Fiction
The Greatest Drama Ever Staged 1938; *Strong Meat* 1939; *Begin Here: a Wartime Essay* 1940; *Creed or Chaos?* 1940; *The Mind of the Maker* 1941; *Why Work?* 1942; *The Other Six Deadly Sins* 1943; *Making Sense of the Universe* 1946; *Unpopular Opinions: Twenty-One Essays* 1946; *Creed or Chaos? and Other Essays on Popular Theology* 1947; *The Lost Tools of Learning* 1948; *Introductory Papers on Dante* 1954; *Further Papers on Dante* 1957; *The Days of Christ's Coming* 1960; *The Poetry of Search and the Poetry of Statement* 1963; *Christian Letters to a Post-Christian World* 1967; *Wilkie Collins: a Critical and Biographical Study* [ed. E.R. Gregory] 1977

Translations
Thomas the Troubadour's Romance of Tristan 1929; Dante: *The Heart of Stone, Being the Four Canzoni of the 'Pietra' Group* 1946, *Dante's Inferno* 1949, *Dante's Purgatorio* 1955, *Dante's Paradiso* [with B. Reynolds] 1962; *The Song of Roland* 1957

For Children
Even the Parrot: Exemplary Conversations for Enlightened Children 1944

Poetry
Op. I 1916; *Catholic Tales and Christmas Songs* 1918; *Lord, I Thank Thee* 1943; *The Story of Adam and Christ* 1955

Edited
Oxford Poetry [with others] 1918–19; *Tales of Detection* (s) 1936

Biography
Such a Strange Lady by Janet Hitchman 1975; *Dorothy L. Sayers* by James Brabazon 1981

William (Neil) SCAMMELL 1939–

Born in the village of Hythe in Hampshire, the son of a plumber and a chamber maid, Scammell left school aged fifteen without qualifications and spent ten years after that in a wide variety of jobs: copy boy on a newspaper, darkroom assistant for a photographic company and factory cleaner. For several years he travelled the world as a photographer aboard Cunard liners. In the early

1960s he went to London and worked first as a publicity officer for the trade body representing wholesalers of fresh fruit and vegetables and then as a male char, cleaning houses. From 1964 to 1967 he went to Bristol University to read English and philosophy, and since then his career has been made in adult education. From 1968 to 1975 he was tutor-organiser for North Gloucestershire for the Workers' Educational Association, and since then he has occupied his present post as lecturer in English for the Department of Continuing Education at Newcastle University, living at Cockermouth in Cumbria. He married the painter Jackie Scammell in 1964, and they have two sons.

After trying to write fiction for many years, Scammell turned to poetry in the mid-1970s and his first book, *Yes and No*, was published in 1979. He is a poet of private and literary themes, known for his attachment to strict verse metres (he is particularly skilled in the couplet), although he has also written free verse; his work is accessible and has a tone of affirmative physicality. Beside poetry, he has published an excellent study of **Keith Douglas**, and has become well known as a literary journalist, writing for, among others, the *Sunday Times*, *Observer*, *Spectator* and *London Magazine*. He has won numerous awards and sat on public bodies: he was the chairman of Northern Arts Literature Panel from 1982 to 1985, and is currently a commitee member of and selector for the Poetry Book Society.

Poetry

Yes and No 1979; *A Second Life* 1982; *Jouissance* 1985; *Eldorado* 1987; *Bleeding Heart Yard* 1992; *The Game: Tennis Poems* 1992; *Five Easy Pieces* 1993

Non-Fiction

Keith Douglas: a Study 1988

Edited

Adam's Dream: Poems from Cumbria and Lakeland [with R. Pybus] 1981; *Between Comets: For Norman Nicholson at 70* 1984; *The New Lake Poets* 1991; *The Poetry Book Society Anthology* 1992; *This Green Earth: a Celebration of Nature Poetry* 1992; Ted Hughes *Winter Pollen: Occasional Prose* 1994

Translations

Martin Sorescu *The Biggest Egg in the World* [with others] 1987

Vernon SCANNELL 1922–

Scannell was born in Spilsbury, Lincolnshire, and after a wandering early childhood, spent partly in Ireland, his family settled down at Aylesbury, Buckinghamshire, where he attended Queen's Park School, showing a talent for English composition and boxing. He left school aged fourteen, and during the Second World War served in the Gordon Highlanders, fighting in North Africa and later in Normandy, where in 1944, near Caen, he received a machine-gun burst through both legs, which led to his hospitalisation. While in a Scottish convalescent depot, after hostilities in Europe had ended, being unable to stand military life any longer, he deserted, and spent two years on the run in Leeds, earning his living by boxing in the professional ring and coaching children for examinations. He was eventually court-martialled, sent to an army mental hospital and discharged.

While in Leeds, he had managed to attend courses at the university. His first poems began to appear in magazines at this time, and his first collection, *Graves and Resurrections* in 1948, but he reached poetic maturity only with his fourth collection, *A Sense of Danger* (1962). His poetry is direct and truthfully detailed, inhabiting a world of urban violence and haunted by the memory of war. He has also written several novels, which are less admired than his poetry: *The Face of the Enemy*, dealing with ex-servicemen facing civilian life, is perhaps the best. Scannell freelanced for a time after his discharge, and then from 1955 to 1962 taught at Hazlewood School, Surrey: since then he has been a freelance writer and broadcaster. He married in 1954, and has five children. He has fulfilled several assignments as writer-in-residence, and *A Proper Gentleman* is an acerbic account of one of these, in the Oxfordshire village of Berinsfield. *The Tiger and the Rose* is a more general autobiography.

Poetry

Graves and Resurrections 1948; *A Mortal Pitch* 1957; *The Masks of Love* 1960; *A Sense of Danger* 1962; *Walking Wounded* 1965; *Epithets of War: Poems 1965–1969* 1969; *Pergamon Poets 8* [with J. Silkin; ed. D. Butts] 1970; *Company of Women* 1971; *Corgi Modern Poets in Focus 4* [with others; ed. J. Robson] 1971; *Selected Poems* 1971; *Incident at West Bay* 1972; *The Winter Man* 1973; *Meetings in Manchester* 1974; *The Loving Game* 1975; *A Mordern Tower Reading 1* [with A. Lykiard] 1976; *New and Collected Poems 1950–1980* 1980; *Winterlude* 1982; *Funeral Games and Other Poems* 1987; *Soldiering On* 1989; *Collected Poems 1950–1993* 1993

For Children

The Dangerous Ones 1970; *Mastering the Craft* 1970; *The Apple-Raid* 1974; *A Lonely Game* 1979; *Catch the Light* [with G. Harrison, L. Smith] 1982;

The Clever Potato 1988; *Love Shouts and Whispers* 1990; *Travelling Light* 1991; *On Your Cycle, Michael* [with A. Ross] 1992

Fiction

The Fight 1953; *The Wound and the Scar* 1953; *The Big Chance* 1960; *The Shadowed Place* 1961; *The Face of the Enemy* 1961; *The Dividing Night* 1962; *The Big Time* 1965; *Ring of Truth* 1983

Non-Fiction

Edward Thomas 1963; *The Tiger and the Rose* (a) 1971; *Three Poets, Two Children* [with D. Abse, L. Clark; ed. D. Badham-Thornhill] 1975; *Not without Glory: Poets of the Second World War* 1976; *A Proper Gentleman* (a) 1977; *How to Enjoy Poetry* 1983; *How to Enjoy Novels* 1984; *Argument of Kings* (a) 1987; *Drums of Morning: Growing Up in the Thirties* (a) 1992

Plays

A Man's Game (r) 1962; *A Door with One Eye* (r) 1963; *The Cancelling Dark* (r) [with C. Whelan] 1965

Edited

New Poems 1962 [with P. Beer, T. Hughes] 1962; *Your Attention Please: an Anthology from the Open University Poets* 1983; *Sporting Literature* 1987

Ann (Acheson) SCHLEE 1934–

Schlee was born in Greenwich, Connecticut, the daughter of Major-General Sir Duncan Cumming, a noted British colonial administrator in various parts of Africa, and his wife, a teacher from an American academic family, and of part-English part-Irish descent. Schlee spent her first eleven years with her mother in the USA, because her father's colonial duties and then the Second World War kept the family separated. She had, however, visited Africa as a child, and also lived for two years in Egypt before being sent to Downe House School, then in the village of Downe in Kent (and where **Elizabeth Bowen** was among former *alumnae*).

Schlee read English at Somerville College, Oxford, continuing to follow her parents to different parts of Africa during vacations, and then taught for two years at a school in America, but her wandering years were over on her marriage in 1957 to Nicholas Schlee, a sales conference organiser and painter. For thirty years the couple lived in Wandsworth, south-west London, where they raised a family of four.

Schlee began writing late, in her mid-thirties, and initially wrote novels for children, of which the first, *The Strangers*, was published in 1971, and the sixth and last, *The Vandal* (which won the *Guardian* Award for Children's Fiction), in 1978. She then made the transition to novels for adults with the finely crafted *Rhine Journey*, a story of hidden tensions on a cruise in Rhenish Prussia in the 1850s, which was shortlisted for the Booker Prize and established her reputation. It has remained her most popular book, but is eclipsed as a work of fiction by the two dense, complex novels which followed it, *The Proprietor* and *Laing*, both also set in the nineteenth century, and in the Scilly Isles and western Africa respectively. These books take as their theme the restrictions and challenges faced by humankind, set against the long perspective of historical events, material which Schlee handles with an impressive and tragic power.

In 1988 Schlee and her husband moved to near Reading in Berkshire, and for five years, from 1989, she taught at the public school Bradfield, retiring in 1994 to concentrate once more on her writing. She has also become well known as a teacher of creative writing and judge in literary competitions, being a member of the Booker Prize panel in 1991 and credited by one newspaper as being largely responsible for securing the award for **Ben Okri**.

For Children

The Strangers 1971; *The Consul's Daughter* 1972; *The Guns of Darkness* 1973; *Ask Me No Questions* 1976; *Desert Drum* 1977; *The Vandal* 1979

Fiction

Rhine Journey 1980; *The Proprietor* 1983; *Laing* 1987

Olive (Emilie Albertina) SCHREINER 1855–1920

Although Schreiner was born in Basutoland, southern Africa, her father was a Methodist missionary of German descent, who had married the daughter of a London Congregational minister in 1837. The Schreiners held various missionary posts, before settling at the Wittenberg mission station in 1854. The settlement was isolated and bleak, and the house the family occupied, part farm and part dwelling, was made from whitewashed clay. There was no school for Schreiner to attend, and she was educated by her rather strict mother. In 1861, the family moved to Healdtown, where Schreiner's father became head of the Wesleyan Native Industrial Training Institution, a position from which he was dismissed four years later, on the flimsy pretext that he had broken a rule by trading with the natives with whom he worked. Completely innocent, he moved to Balfour, where he set himself up as a general dealer, but the business failed when customers did not pay their bills. The effect of this, and the dismissal of her father who had spent his life helping the poor, led Schreiner to abandon her faith. After reading Herbert Spencer's *First Principles* (1862), she became a free-thinker.

From 1871 to 1881, during terms as a governess, Schreiner wrote a novel, *Undine* and several drafts of her most noted book, *The Story of an African Farm*, which reflected some of her experiences of early childhood and adolescence. She left South Africa for England in 1881 to find a publisher for her writing. The novel was finally accepted for Chapman & Hall by George Meredith, and was originally published under the pseudonym Ralph Iron. Dealing with the situation of an unmarried mother, and combining this with an attack on Christianity and a forceful plea for the emancipation of women, *The Story of an African Farm* became a sensation.

Schreiner was supported by Havelock Ellis, whom she had met in 1884, and who remembered her as having a 'short vigorous body – a beautiful head with large dark eyes, at once so expressive and so observant'. Later, they fell in love, but were never lovers. The book also brought the friendship of Rider Haggard and **Edward Carpenter**, but Schreiner was so plagued with visitors that she moved to the unfashionable East End of London to work without interruption.

In 1889, she returned to South Africa, where she stayed in Cape Town with her brother William, who introduced her to Cecil Rhodes. She was influenced by his conversation and ideas, but this did not prevent her from viciously attacking him later in her novel *Trooper Peter Halket of Mashonaland*, an exposé of Rhodes's abuse of power as prime minister, and an account of the resultant Jameson Raid. (Following Rhodes's resignation, he was succeeded as prime minister by William Schreiner). Meanwhile, in 1894, she had married Samuel Cronwright, who abandoned his career as a lawyer and Member of Parliament to help his wife. A daughter was born the following year, but lived only one day. Schreiner was strongly pro-Boer and her views were published in *An English South African Woman's View of the Situation*, just before the outbreak of the Boer War, but she published nothing more of substance until 1911, when *Woman and Labour*, regarded as a key feminist work, appeared.

Schreiner had suffered from asthma since the age of fifteen, but despite increasing ill health, she continued to write and involve herself in politics, occasionally painting for relaxation. During the First World War, she came alone to England; when in 1920 her husband arrived in London to escort her back, he failed to recognise the old woman who answered the door as his wife. She returned alone to Cape Town in September 1920, where she died in December. The bodies of her daughter and a favourite dog were exhumed, and buried with her in a hilltop cairn in the Karroo district, where she had once lived.

Fiction

The Story of an African Farm [as 'Ralph Iron'] 1883; *Dreams* (s) 1890; *Dream Life and Real Life* (s) 1893; *Trooper Peter Halket of Mashonaland* 1897; *Stories, Dreams, Allegories* (s) [ed. S.C. Cronwright-Schreiner] 1923; *From Man to Man* 1926; *Undine* 1928

Non-Fiction

The Political Situation in Cape Colony [with S.C. Cronwright-Schreiner] 1896; *An English South African Woman's View of the Situation* 1899; *A Letter on the Jew* 1906; *Closer Union: a Letter on South African Union and the Principles of Government* 1909; *Woman and Labour* 1911; *Thoughts on South Africa* 1923; *The Letters of Olive Schreiner 1876–1920* [ed. S.C. Cronwright-Schreiner] 1924; *Olive Schreiner: Letters* [ed. R. Rive] 1988 onwards; *My Other Self: the Letters of Olive Schreiner and Havelock Ellis* [ed. Y.C. Draznin] 1992

Biography

Olive Schreiner by Ruth First and Ann Scott 1980

Delmore SCHWARTZ 1913–1966

Schwartz was born in Brooklyn, New York, and after high school attended the University of Wisconsin in 1933, New York University where he took his BA in philosophy in 1935, and Harvard from 1935 to 1937. Thereafter he held a number of academic posts, as instructor in English composition and assistant professor at Harvard, and as visiting lecturer at New York University, Kenyon School of English, Indiana School of Letters, Princeton University, the University of Chicago, and Syracuse University. While still a student at Harvard he had edited a small literary magazine called *Mosaic*, and was later the editor of *Partisan Review* from 1943 to 1947 (and its associate editor from 1947 to 1953), poetry editor of *New Republic* from 1955 to 1957, and literary consultant to the New Directions publishing house. Schwartz was married twice.

His first book, *In Dreams Begin Responsibilities*, a collection including poetry, a story and a play, was published in 1938 by James Laughlin's newly established New Directions, and attracted immediate praise. A translation of Rimbaud's *A Season in Hell*, and the semi-autobiographical *Genesis*, an account in poetry and prose of a Jewish boy growing up in New York, further consolidated his reputation and established him as the leading figure in the generation of poets which included **Robert Lowell**, **Randall Jarrell** and **John Berryman**.

By the end of the 1940s Schwartz was increasingly afflicted with chronic depression and alcoholism and spent many spells in sanatoria. His 1950 collection *Vaudeville for a Princess* was badly received, and after 1953 he was said to have taken up the mantle of **Dylan Thomas** as the drunken 'house poet' at the White Horse Tavern in New York. His friend **Saul Bellow** closely modelled the character of Von Humboldt Fleisher in *Humboldt's Gift* (1975) on Schwartz. The novel's narrator, Charlie Citrine, describes his mentor thus: 'Poet, thinker, problem drinker, pill-taker, man of genius, manic depressive, intricate schemer, success story, he once wrote poems of great wit and beauty, but what had he done lately? Had he uttered the great words and songs he had in him? He had not. Unwritten poems were killing him.' Embittered and painfully aware of his own failure to fulfil his early promise, Schwartz died of a heart-attack in a Greenwich hotel.

Poetry
Vaudeville for a Princess and Other Poems 1950; *Summer Knowledge: New and Selected Poems, 1938–1958* 1959; *Last and Lost Poems* 1979

Fiction
The World Is a Wedding (s) 1948; *Successful Love and Other Stories* (s) 1961

Plays
Shenandoah 1969 [pub. 1941]

Non-Fiction
American Poetry at Mid-Century [with J. Crowe Ransom, J. Hall Wheelock] 1958; *Selected Essays of Delmore Schwartz* [ed. D.A. Dike, D.H. Zucker] 1970

Translations
Rimbaud *A Season in Hell* 1939

Edited
Syracuse Poems, 1964 1965

Collections
In Dreams Begin Responsibilities 1938; *Genesis, Book I* 1943

Biography
Delmore Schwartz by James Atlas 1977

Paul (Mark) SCOTT 1920–1978

Such was Scott's fictional identification with India that many people imagine he was a son of the Raj. In fact he was born in the north London suburb of Southgate, the younger son of a commercial artist. He was educated at Winchmore Hill Collegiate School, but was obliged to leave at the age of fourteen when the family's money ran out (a fate which was to befall Hari Kumar in *The Raj Quartet*). His father ignored Scott's early interest in poetry and art and directed him into a career in accountancy. It was through this job that he met an aesthetic estate agent called Gerald Armstrong who became his lover and introduced him to the writers of the *fin de siècle*, notably Oscar Wilde, whose novel *The Picture of Dorian Gray* (1891) was to have a profound and lasting influence upon him.

At the outbreak of war, Scott enlisted as a private. During training in England he became involved in some sort of homosexual scandal and determined to repress his sexual nature: within nine months he married Elizabeth (Penny) Avery, who also became a novelist; they had two daughters. In 1943 he went to India as a trainee officer and was bewitched by the country. After the war ended, he spent four months awaiting repatriation: 'I just sat there absorbing it all.' Back in England he worked as book-keeper for the Falcon and Grey Walls Press, a financially unsound publishing house later caricatured by **Muriel Spark** in *A Far Cry from Kensington* (1988). Aware of the company's impending collapse, Scott resigned and became a literary agent, working for the firm which eventually became David Higham Associates; his clients included Spark, **John Braine**, Morris West and M.M. Kaye.

His first novel was rejected by seventeen publishers and remains unpublished, but its successor, *Johnnie Sahib* (1952), won him an Eyre & Spottiswoode Literary Fellowship. His reputation was made the following year by *The Alien Sky*, which sold well and was adapted for radio and television; but subsequent books did not do as well, and Scott was debilitated by undiagnosed amoebiasis, which he attempted to combat by heavy drinking, a habit which persisted throughout his life. In 1960 he gave up his job to become a full-time writer. His dedication to this task was such that his family life began to disintegrate. 'It was as if he had exiled himself to the one room where there was nothing but the typewriter and the blank page,' one daughter recalled. 'It was the making of him as a writer, but the unmaking of him as a human being.'

The novel he knew he had it in him to write was the ambitious, multi-layered *Raj Quartet*, set in India during the period 1942–7. He visited India to refresh his memory and read innumerable books about the country before embarking upon this massive project. The *Quartet* was never as popular in his lifetime as it was to be in 1984, when it was adapted for the highly acclaimed television series, *The Jewel in the Crown*. Success came to Scott with *Staying On*, a pendant to the *Quartet*, which won the Booker Prize in 1977, but it came too late. His wife had been driven to a refuge for victims of domestic violence, and he was dying from cancer. He spent his last years between his Hampstead house and the University of Tulsa, Oklahoma, where he was a visiting fellow. Capable of immense charm, Scott was a tormented man who could only express himself through his books; the dark, despairing side of his character revealed by his biographer surprised many of his friends.

Fiction

Johnnie Sahib 1952; *The Alien Sky* [US *Six Days in Marapore*] 1953; *A Male Child* 1956; *The Mark of the Warrior* 1958; *The Chinese Love Pavilion* 1960; *The Birds of Paradise* 1962; *The Bender* 1963; *The Corrida at San Feliu* 1964; *The Raj Quartet*: *The Jewel in the Crown* 1966, *The Day of the Scorpion* 1968, *The Towers of Silence* 1971, *A Division of the Spoils* 1975; *Staying On* 1977; *After the Funeral* (s) 1979

Plays

Pillars of Salt [pub. in *Four Jewish Plays*] 1948; *Lines of Communication* (r) [from his novel *Johnnie Sahib*] 1951, (tv) 1952; *Sahibs and Memsahibs* (r) 1958

Poetry

I, Gerontius 1941

Non-Fiction

My Appointment with the Muse, Essays, 1961–75 [ed. S.C. Rees] 1986

Biography

Paul Scott by Hilary Spurling 1990

E(dith) J(oy) SCOVELL 1907–

Scovell was born in a suburb of Sheffield, where her father, a Londoner by birth, was the vicar. Scovell and her sisters were partly educated by governesses, then at Casterton School, Westmorland. From there, she went to Somerville College, Oxford, where she further developed her childhood interest in writing poetry and for a time edited *Fritillary*, the magazine of the women's colleges.

After taking her degree in 1930, she lived in London, working as a secretary, doing some reviewing, and occasionally having poems published in the *New Statesman* and other journals. Her first collection, *Shadows of Chrysanthemums*, appeared, printed on wartime austerity paper, in 1944, and was followed by two other volumes of verse until the mid-1950s. She had married the ecologist Charles Elton in 1937, and moved to Oxford, where he was a reader in animal ecology; their two children were born in the city during the war, and they lived there until Elton's death in 1991. Scovell took no further paid employment, but her life was diversified by much travel, because in the 1960s and 1970s she acted as field assistant and secretary for her husband during his research in Brazil and Panama. She has also often visited in later years the island of Montserrat in the West Indies, where her children live, and spent considerable periods in Edinburgh.

Scovell is a reflective and domestic poet whose work shares many characteristics with that of her near-contemporary (and friend) **Anne Ridler**. Between 1956 and the appearance of *The Space Between* in 1982, she did not have a volume published, and her work passed through a period of neglect, but especially since the appearance of her *Collected Poems* in 1988 and *Selected Poems* in 1991 she has been increasingly recognised as a writer of quiet but enduring worth, with fine observation of nature and an understated Christian ethic. A fair proportion of poems in the later volumes consist of translations of the Italian writer Giovanni Pascoli. Scovell continues to be based in Oxford, but often travels to London and elsewhere to give readings of her work.

Poetry

Shadows of Chrysanthemums and Other Poems 1944; *The Midsummer Meadow and Other Poems* 1946; *The River Steamer and Other Poems* 1956; *The Space Between* 1982; *Ten Poems* [with A. Ridler] 1984; *Listening to Collared Doves* 1986; *Collected Poems* 1988; *Selected Poems* 1991

(John) Peter SCUPHAM 1933–

Scupham was born in Liverpool. His father was head of the Educational Broadcasting Unit for the BBC. Educated at The Perse, Cambridge, and St George's in Harpenden, Scupham did his national service with the Royal Army Ordnance Corps, and graduated from Emmanuel College, Cambridge, in 1957. He then taught at Skegness Grammar School in Lincolnshire for four years, before becoming Chairman of the English Department at St Christopher's in Letchworth, Hertfordshire, a post he still holds. He was married in 1957, and has four children.

His first volume of poetry, *The Small Containers*, was published by the Phoenix Pamphlets

Poets Press in 1972. *Prehistories*, published in 1975, began a long-standing association with the Oxford University Press which published his *Selected Poems* in 1990. In 1974 he co-founded the Mandeville Press which specialises in hand-printed, short-run volumes of poetry. He is also an editor for the Cellar Press. Both houses are run from Hitchin in Hertfordshire, where Scupham now lives. In his own writing he has tended to shun 'the raw, the self-absorbed, the cosmic', to concentrate on exactness, formality and precision: 'Poetry,' he has said, 'is a game of knowledge, and I enjoy the complexity of rules that make the game worth playing'. The varied subject-matter of his poetry mirrors his own varied passions: for jazz, for archaeology and geology – 'The Nondescript' was written for the environmental pressure-group Friends of the Earth – and for his family, and the ties of family life. *The Hinterland*, *Summer Palaces* and *Winter Quarters* all contain poems on war and the military, to write which the poet drew on his father-in-law's First World War diaries and his own experience of national service. *Christmas Past*, published in 1981, was the first product of a long-running seasonal collaboration with fellow jazz fan and co-editor at Mandeville, the poet John Mole.

Poetry

The Small Containers 1972; *Children Dancing* 1972; *The Snowing Globe* 1972; *The Gift* 1973; *The Nondescript* 1973; *Prehistories* 1975; *The Hinterland* 1977; *A Mandeville Trokia* [with N. Powell, G. Szirtes] 1977; *Megaliths and Water* 1978; *Natura* 1978; *Summer Palaces* 1980; *Transformation Scenes* 1982; *Winter Quarters* 1983; *Under the Barrage* 1985; *Out Late* 1986; *The Air Show* 1988; *Selected Poems* 1990; *Watching the Perseids* 1990; *The Ark* 1994

With John Mole: *Christmas Past* 1981; *Christmas Games* 1983; *Christmas Visits* 1985; *Winter Emblems* 1986; *Christmas Fables* 1987; *Christmas Gifts* 1988; *Christmas Books* 1989; *Christmas Boxes* 1990

Plays

Duffy and the Devil (l) [with D. and R. Tutt] 1989

John SEDGES

see Pearl S. BUCK

Hubert SELBY Jr 1928–

The son of an engineer and apartment building manager in the New York borough of Brooklyn, Selby grew up there, receiving little formal education. He had attended the Peter Stuyvesant High School for one year, and was just fifteen, when he joined the US merchant marine in 1944 and, after some time on harbour duties, sailed to join the closing stages of the Second World War. In 1946, in Bremen, Germany, he was taken off ship suffering from tuberculosis, and spent the next four years hospitalised. He came close to death several times and eventually lost ten ribs and part of a lung, but it was while laid up that he began to write. (In these circumstances, his career curiously parallels that of the English writer **Alan Sillitoe** at the same period.)

He had begun drinking to excess and, on leaving hospital, returned to Brooklyn, drifting for a while into the sort of hopeless, vice-ridden life that he was later to memorialise in his famous *Last Exit to Brooklyn*. Two circumstances helped him: a friend, Gilbert Sorrentino, encouraged his literary ambitions; and his marrige to Inez Taylor, in 1955, with whom he had two children before their divorce in 1969, forced him to work to support his family. His first story appeared in a magazine in 1956, and during the period before *Last Exit to Brooklyn* was published in 1964 he held various jobs, including seaman and insurance clerk. The appearance of the book, which consists of six linked short stories, several of which had already appeared in magazines (the *Provincetown Review* was involved in an obscenity trial for publishing one in 1961), caused a sensation because of what was at the time an extremely frank treatment of homosexuality, rape and drug dependence. *Last Exit to Brooklyn* was banned in Italy and was the subject of a protracted legal action in Britain; although its reception in the USA was mixed, it made Selby's name, but the money he received landed him with further problems.

In 1967, he was arrested for possession of heroin and sentenced to a month in jail, where he freed himself from drug addiction by a period of 'cold turkey'. He then spent a further two years being cured of alcohol dependence before beginning to write again. His terrifying novel of 1971, *The Room*, draws on his knowledge of prison, and is regarded by some as his major work. Indeed, both these early books have been defended by many reputable critics against charges of obscenity on the grounds that Selby brings a compassionate and puritanical, indeed specifically Christian, perspective to his often extreme material. However, in the two novels that followed in the 1970s, there were signs that Selby was exhausting his vein, and since then his only prose fiction has been a collection of short stories, in 1986.

After a brief second marriage, he married again in 1969, and subsequently had two more children. The family moved to Los Angeles, where Selby has continued to live, writing screenplays and television scripts, teaching the occasional creative writing course, and emerging from a decade of obscurity in the 1980s to see his early books republished and younger writers become interested in the scarifying vision that informs his work.

Fiction

Last Exit to Brooklyn 1964; *The Room* 1971; *The Demon* 1976; *Requiem for a Dream* 1978; *Song of the Silent Snow* (s) 1986

Will SELF 1961–

Self was born in London. He was educated in Finchley and read philosophy, politics and economics at Exeter College, Oxford. On leaving university, he held various jobs ranging from roadsweeper to economic adviser for the Northern Territory Government of Australia. His cartoons first started appearing in 1982, and from then until 1988 he published them on a regular basis in a number of magazines and newspapers. His cartoon series 'Slump' ran in the *New Statesman* from 1984 to 1986.

Self's first work of fiction, *The Quantity Theory of Insanity* (1991), was shortlisted for the 1992 John Llewellyn Rhys Prize and won the 1993 Geoffrey Faber Memorial Prize. His developing reputation increased with his inclusion in the list of *Granta*/Book Trust 'Best of Young British Novelists' and his writing style was hailed for its clever use of language and persistently macabre tone. His twin novellas *Cock & Bull* in which a suburban housewife grows a penis and a rugby-player a vagina, explore notions of gender in a surreal style that is also darkly humorous. Some critics have felt that what they see as extravagant writing has been achieved at the expense of substance. *My Idea of Fun* was published in 1993 and received mixed reviews. Self's gymnastic use of language to create images that deliberately shock can seem both admirable and implausible. However, the character of the Fat Controller, a Mephistophelean figure who appears throughout the life of the book's protagonist, is drawn with inventive exuberance.

Grey Area and Other Stories presents a world in which clinical depression, the dank London sky and a tedious day at the office define the parameters of daily life. The book is grey in another sense, in that it occupies a limbo space between the short story, the novel and the essay. The characters and actions in the book exist in a parallel world that shifts uneasily in contour and geographical location.

Self's individual perspective on the world has not been achieved without a certain personal cost, and he has often frequented the cushioned parlours of therapists. He claims he owes his life to a particular psychoanalyst who persuaded his mother not to have an abortion while pregnant with the writer. He describes his approach to writing fiction being 'motivated by two conflicting aims. The first is to make people laugh at the absurdity of the world. The second is to get them to feel profoundly uncomfortable about it.' His use of language ranges from a high-flown literary style, rococo in its laterally associative twists, to a modern argot that captures a genuine linguistic pulse while underlining its distance from conventional mainstream English. His reaction to criticism of his work is characteristically defiant: 'I want to be misunderstood. And the other thing that amuses me is, I don't particularly want to be liked. Nobody goes into the business of writing satire to be liked.' Self is also a critic and journalist, writing for *The Times*, *Guardian*, *Independent*, *Sunday Times* and other newspapers and magazines. His essay on the essence of being English – 'Birth of the Cool' – published in the *Guardian* (6 August 1994) is a particularly fine and astute example of his journalistic writing. *Junk Mail* incorporates his collected writings, essays and cartoons. He is married and has two children.

Fiction

The Quantity Theory of Insanity 1991; *Cock & Bull* 1992; *My Idea of Fun* 1993; *Grey Area and Other Stories* (s) 1994

Collections

Junk Mail 1995

Robert (William) SERVICE 1874–1958

The eldest of ten children of a Scottish bank employee, Service grew up in Glasgow, attending Hillhead High School, and then worked for a few years in the Commercial Bank of Scotland. Thirsting for adventure, he travelled steerage to Canada in 1895, and spent ten years wandering up and down the Pacific coast, working at a great variety of outdoor jobs.

In 1905, having joined the Canadian Bank of Commerce, he went to the Yukon, where he lived for eight years, and witnessed the Yukon Gold Rush: out of this 'roughneck' experience came his first book of poetry, *Songs of a Sourdough* (1907), which includes his best-known poem, 'The Shooting of Dan McGrew'. Although almost forgotten now, this and subsequent volumes of verse, celebrations in ballad form of the life of the frontier, enjoyed enormous popularity in the earlier part of the century (*Songs of a Sourdough* alone had sold two million copies by 1940) and ensured Service financial independence for life. He moved to France in 1912, where he lived, except for a period back in Canada during the Second World War, for the rest of his life, owning villas in Monte Carlo and Brittany.

In later years he became an aesthetic pleasure-lover rather than, as previously, a [**Rudyard**] 'Canadian **Kipling**'. He published many more volumes of verse, with titles such as *Songs of a Sun-Lover*, *Lyrics of a Lowbrow*, *Carols of an Old Codger* and so on: he had always been more poetaster than poet. His many novels are mostly unmemorable, although perhaps *The Pretender*, with its charming scenes of the Latin Quarter, deserves revival. He married in 1913, and had one daughter.

Poetry

Songs of a Sourdough [US *The Spell of the Yukon*] 1907; *Ballads of a Cheechako* 1909; *Rhymes of a Rolling Stone* 1912; *Rhymes of a Red Cross Man* 1916; *Ballads of a Bohemian* 1921; *Twenty Bath-Tub Ballads* 1939; *Bar-Room Ballads* 1940; *Songs of a Sun-Lover* 1949; *Rhymes of a Roughneck* 1950; *Lyrics of a Lowbrow* 1951; *Rhymes of a Rebel* 1952; *Songs for My Supper* 1953; *Carols of an Old Codger* 1954; *Rhymes for My Rags* 1956; *Cosmic Carols* 1957; *Songs of the High North* 1958; *Later Collected Verse* 1960

Fiction

The Trail of Ninety-Eight 1910; *The Pretender* 1914; *The Poisoned Paradise* 1922; *The Roughneck* 1923; *The Master of the Microbe* 1926; *The House of Fear* 1927

Non-Fiction

Why not Grow Young? Living for Longevity 1928; *Ploughman of the Moon* (a) 1945; *Harper of Heaven* (a) 1948

Vikram SETH 1952–

Born in Calcutta, the son of a footwear consultant known in Delhi as 'Mister Shoe', and Ms Justice Seth, a high court judge, Seth (whose name is pronounced to rhyme with 'Gate') left India for Oxford, originally to study literature. Unwilling to plod through books that might bore him, he read instead philosophy, politics and economics. Tempted by Californian sunshine after rainy England, he continued with economics and classical Chinese at Stanford University, before taking a diploma at Nanjing University in China.

During the summer of 1981, Seth hitch-hiked through China to Tibet, a journey which resulted in idiosyncratic and exotic adventures, recorded in *From Heaven Lake*, which won the Thomas Cook Award for the best travel book of 1983. He returned to California, but while working on his doctoral thesis, he was overwhelmed by a chance encounter with Charles Johnson's 1977 translation of Pushkin's *Eugene Onegin* (1831). Inspired by this, Seth spent nine months writing *The Golden Gate*, a verse novel of nearly 600 sonnets of iambic tetrameter in thirteen chapters, a task which was generally rewarded with charges of lunacy. An unusual gift for language was already apparent in Seth's *Mappings*, and *The Humble Administrator's Garden*, where he revitalised traditional verse-form with his poetic personality. *All You Who Sleep Tonight* was memorable for the poem 'Soon' about a person dying of Aids, written entirely in monosyllables, reflecting as does much of Seth's work, a concern for society. A light-hearted collection of animal fables in verse, *Beastly Tales from Here and There*, was inspired by tales from several countries.

Seth has been influenced by the poet **Timothy Steele**, a leader of the New Formalist school of poetry, and the dedicatee of *The Golden Gate*. The strictures of the complex verse-form Seth had chosen for his novel paradoxically provided flexibility of scope for his reflections on aspects of Californian life. With perfect concealed artistry, the novel balances satire, fun and serious contemplation, revealing Seth's wide cultural awareness, and a generous love of humanity. Seth portrays himself anagrammatically in the novel as 'Kim Tarvesh'. *The Golden Gate* was hailed by **Gore Vidal** as 'The Great Californian Novel' and was awarded the Commonwealth Poetry Prize.

Seth abandoned 'the gray futility of his dank thesis' to write a novel about post-independence India, *A Suitable Boy*. Narrated in a traditional form, and embodying many themes and aspects of Indian life revealed through a wide selection of characters, the book was said at 1,349 pages to be the longest modern novel in the English language. It was attacked by Lord Gowrie, Chairman of the Booker Prize jury, and although heavily tipped to win the prize, did not make it on to the shortlist. Nevertheless, it won both the W.H. Smith Award and the Commonwealth Writer's Prize.

Seth has openly revealed his bisexuality in his verse and his somewhat puckish appearance has contributed to his popularity as a literary pop star. He has a love of music which has resulted not only in his 'untuneful singing of Schubert *lieder*' but also to his writing a libretto for *Arion and the Dolphin*, an opera by Alec Roth. He has also written 'Lynch & Boyle', an unpublished verse play about an English publishing house.

Poetry
Mappings 1981; *The Humble Administrator's Garden* 1985; *All You Who Sleep Tonight* 1990; *Beastly Tales from Here and There* 1993

Fiction
The Golden Gate 1986; *A Suitable Boy* 1993

Non-Fiction
From Heaven Lake 1983

Translations
Three Chinese Poets 1992

Plays
Arion and the Dolphin (l) [with A. Roth] 1994

For Children
Arion and the Dolphin [from his libretto] 1994

Anne (Harvey) SEXTON 1928–1974

Born in Newton, Massachusetts, in the area near Boston where she spent much of her life, Sexton was the daughter of a salesman: an ancestor, William Brewster, had sailed to America on the *Mayflower*. According to her own account, she was locked in a room until she was five: perhaps the disparity between this and her outwardly normal childhood laid the grounds for her mental instability. After attending Garland Junior College, she eloped, aged nineteen, with Alfred M. Sexton, an executive in a wool company, and had two daughters with him.

She worked as a salesgirl, fashion-model and librarian; then, in the mid-1950s, she had a nervous breakdown: 'as I came out of it (and if I ever really came out of it) I started to write poems'. She met **Sylvia Plath** at **Robert Lowell**'s poetry class and the two women spent much time discussing suicide. Sexton's poetry was always under the influence of Plath, and her first volume, *To Bedlam and Part Way Back* (1960), established her as a poet in a frankly confessional mode, describing, as it did, her descent into madness. Later volumes were to deal with abortion, surgery, tortured family relationships and the details of physical love, often with great lyrical intensity, though sometimes with crudity.

Her poetry brought her fame and contact with intellectuals. From 1961 to 1963 she was a scholar at Radcliffe, and grants paid for several sojourns in Europe. She was awarded a Pulitzer Prize for *Live or Die* in 1966, and lectured in creative writing at Boston University from 1969 to 1972, becoming a professor there in the latter year. All this time, her depressiveness and her fascination with death and religion continued, as demonstrated in her posthumous volumes, *The Death Notebooks* and *The Awful Rowing Towards God*, containing poetry whose power is somewhat vitiated by a too facile invocation of concentration-camp and other horrors. She was divorced from her husband in 1974, and it may have been renewed mental distress connected with this that led to her suicide in the same year.

Her posthumous career has also been controversial. In 1991 Diane West Middlebrook published a biography of Sexton which made use of tape-recordings of some 300 consultations between the poet and her therapist. Many felt that this violated medical as well as biographical ethics.

Poetry
To Bedlam and Part Way Back 1960; *All My Pretty Ones* 1962; *Selected Poems* 1964; *Live or Die* 1966; *Poems* [with T. Kinsella, D. Livingstone] 1968; *Love Poems* 1969; *Transformations* 1971; *The Book of Folly* 1972; *The Death Notebooks* 1974; *The Awful Rowing Towards God* 1975; *45 Mercy Street* [ed. L. Gray Sexton] 1976

For Children
With Maxine Kumin: *Eggs of Things* 1963; *More Eggs of Things* 1964; *Joey and the Birthday Present* 1971; *The Wizard's Tears* 1975

Non-Fiction
Anne Sexton: a Self-Portrait in Letters 1977

Biography
Anne Sexton by Diane West Middlebrook 1991

Peter (Levin) SHAFFER 1926–

Shaffer was born in Liverpool, the twin of Anthony Shaffer, author of the long-running play *Sleuth* (1970) and numerous screenplays. They were both educated at St Paul's and Trinity College, Cambridge, where they co-edited *Granta*. Before that they had served as 'Bevin Boys' at Chisley colliery, Kent. They published two novels together in the 1950s but thereafter their careers have taken different courses. Peter Shaffer spent the early 1950s in New York where he did various jobs. Returning to London, he was with Boosey & Hawkes, the music publisher, was literary critic of *Truth* from 1956 to 1957, and music critic of *Time and Tide* from 1961 to 1962.

His first stage play, *Five Finger Exercise*, a 'well-made' family drama which ran for a year at the Comedy Theatre, won the *Evening Standard* Award for best play. *The Royal Hunt of the Sun*, a spectacular play about the conquest of the Incas, was produced by the National Theatre, as was the farce *Black Comedy*. In the 1970s came *Equus* and *Amadeus*, which are generally regarded as his finest achievements. *Equus* is the psychological study of a youth who has savagely and inexplicably blinded some horses. It transpires that he has created his own distorted religion, and the play explores the roots of this, and its 'cure'. Presented in a stripped down but highly theatrical manner, with the horses represented by actors in wicker masks, the play was made into a more realistic, but less satisfactory film (1977), directed by Sidney Lumet from Shaffer's own screenplay. *Amadeus* explored the nature of genius by telling the story of Mozart from the point of view of his jealous rival, Salieri. It too was made into a film (1984), directed by Milosz Forman from Shaffer's screenplay. In spite of a misjudged central performance, the film won four Academy Awards for best picture, director and screenplay and for the actor playing Salieri. Both plays won Tony awards in New York, and *Amadeus* won the *Plays and Players* and London Critics awards.

More recently *Yonadab* used as one of its sources **Dan Jacobson**'s novel *The Rape of Tamar* (1970), while *Lettice and Lovage*, which starred Maggie Smith for a long run at the Globe Theatre, was a comedy concerning the rather too imaginative guide to the most boring stately home in England. Probably the most versatile of contemporary British playwrights (*Equus*, for example, could hardly be more different from *Lettice and Lovage*), Shaffer has enjoyed a phenomenal success founded on his mastery of stagecraft and the fecundity of his ideas. He was appointed CBE in 1987.

Plays

The Salt Land **(tv)** 1955; *Balance of Terror* **(tv)** 1957; *The Prodigal Father* **(r)** 1957; *Five Finger Exercise* 1958; *The Merry Roosters' Panto* 1962 [as *Cinderella* 1969]; *The Private Ear & The Public Eye* 1962; *Lord of the Flies* **(f)** [from W. Golding's novel; with P. Brook] 1963; *The Royal Hunt of the Sun* 1964; *Black Comedy* 1965; *White Lies* 1967 [rev. 1976]; *The Battle of Shrivings* 1970 [rev. as Shrivings 1975]; *The Public Eye* **(f)** [from his play] 1972; *Equus* **(st)** 1973, **(f)** 1977; *Amadeus* **(st)** 1979, **(f)** 1984; *The Collected Plays* [pub. 1982]; *Black Mischief* 1983; *Yonadab* [from D. Jacobson's novel *The Rape of Tamar*] 1985; *Lettice and Lovage* 1987; *Whom Do I Have the Honour of Addressing?* **(r)** 1989 [pub. 1990]; *Gift of the Gorgon* 1992 [pub. 1993]

Fiction

The Woman in the Wardrobe [as 'Peter Antony'] 1951; *How Doth the Little Crocodile?* [as 'Peter Antony'; with A. Shaffer] 1952; *Withered Murder* [with A. Shaffer] 1955

Karl (Jay) SHAPIRO 1913–

Shapiro was born in Baltimore, Maryland, where his father was in business, and spent a year at the University of Virginia in 1932–3 before holding various positions as a clerk, salesman and librarian. His first collection, *Poems*, was privately printed in 1935, and the next year he returned to academia, studying at Johns Hopkins University in Baltimore until 1939. His poetry was by now appearing widely in magazines such as the *New Yorker* and *Harper's*, and he attracted more attention when a selection appeared under the title 'Noun' in the New Directions anthology *Five Young American Poets*. Shapiro did a course in library training before being called up for the US Army in 1941.

His first full collection, *Person, Place and Thing*, was edited by his agent and fiancée, Evalyn Katz, and appeared in 1942 while he was serving in New Guinea. All bar one of the poems in his Pulitzer Prize-winning collection *V-Letter* were written while on active service with the Medical Corps in the South Pacific, though Shapiro was 'on guard against becoming a "war poet" '. After the war, he was a consultant to the Library of Congress for a year, and in 1948 took up the first of a succession of academic posts. The first, as associate professor of writing, was at Johns Hopkins, this despite his having described the university as 'the Oxford of all sickness' in *Person, Place and Thing*. His final academic post was as professor of English at the University of California from 1968 to 1984.

Shapiro was initially a traditionalist, writing in the manner of **W.H. Auden**, but he changed direction drastically with his 1964 collection, *The Bourgeois Poet*. Influenced by Whitman and the Beat poets, the collection is written in a loose free verse often indistinguishable from prose. In the 1970s he returned to traditional forms such as the sonnet and launched a movement known as 'The New Formalism'. Always inclined towards satire, he now satirised the pomp and self-importance of poetry itself. He was married to Evalyn Katz until their divorce in 1967, and had three children; he has been married twice since. Two out of a projected three volumes of autobiography have been published.

Poetry

Poems 1935; *Five Young American Poets* [with others] 1941; *Person, Place and Thing* 1942; *The Place of Love* 1942; *V-Letter and Other Poems* 1944; *Essay on Rime* 1945; *Trial of a Poet and Other Poems* 1947; *Poems 1940–1953* 1953; *Poems of a Jew* 1958; *The Bourgeois Poet* 1964; *There Was That Roman Poet Who Fell in Love at*

Fifty-Odd 1968; *White-Haired Lover* 1968; *Auden (1907–1973)* 1974; *Adult Bookstore* 1976; *Collected Poems 1940–1978* 1978; *Love and War, Art and God* 1984; *Adam and Eve* 1986; *New and Selected Poems 1940–1986* 1987

Non-Fiction
English Prosody and Modern Poetry 1947; *Bibliography of Modern Prosody* 1948; *Poets at Work* (c) [ed. C.D. Abbott] 1948; *Beyond Criticism* 1953 [as *A Primer for Poets* 1965]; *In Defence of Ignorance* 1960; *Start with the Sun: Studies in Cosmic Poetry* [with J.E. Miller Jr, B. Slote] 1960; *The Writer's Experience* [with R. Ellison] 1964; *A Prosody Handbook* [with R. Beum] 1965; *Randall Jarrell* 1967; *To Abolish Children and Other Essays* 1968; *The Poetry Wreck: Selected Essays 1950–1970* 1975; *Poet: an Autobiography in Three Parts* (a) 2 vols 1988–90; *The Younger Son* 1988

Plays
The Tenor [from F. Wedekind; with E. Lert, H. Weisgall] 1952 [pub. 1957]; *The Soldier's Tale* (l) [from C.F. Ramuz; with Stravinsky] 1968

Fiction
Edsel 1971

Edited
Modern American and Modern British Poetry [with L. Untermeyer, R. Wilbur] 1955; *American Poetry* 1960; *Prose Keys to Modern Poetry* 1962

Tom (i.e. Thomas Ridley) SHARPE 1928–

Born in London and educated at Lancing and Pembroke College, Cambridge, Sharpe did his national service in the Royal Marines. He went to South Africa in 1951 as a social worker for the Non-European Affairs Department, before becoming a teacher in Natal. He had a photographic studio in Pietermaritzburg from 1957 to 1961, when he was deported. Back in England he became a lecturer in history at the Cambridge College of Arts and Technology, a post he held from 1963 until 1971, when the success of his first novel, *Riotous Assembly*, allowed him to become a full-time writer. He married in 1969, and has three daughters.

Both *Riotous Assembly* and its sequel, *Indecent Exposure*, grew out of Sharpe's experiences in South Africa, and were savagely comic, furiously plotted and broadly drawn accounts of the apartheid regime. Their titles may be said to suggest his subsequent career as a writer of bawdy farces, the first of which, *Porterhouse Blue*, was set in a Cambridge college, and he returned to an academic setting with *The Great Pursuit*. The hero of *Wilt* and its sequels is, as Sharpe had been, a lecturer at a polytechnic, but one bent on murdering his wife.

While Sharpe is rarely less than intelligent, inventive and consistently funny – one of the hardest things for a writer to achieve – it is generally thought that later books lack the force of his first two novels. Part of the reason for this is that in South Africa Sharpe had a subject worthy of his satire, and these novels were fuelled by a genuine moral outrage. Subsequent books are equally energetic, but the targets have been softer: planners and stately-home owners in *Blott on the Landscape*, the upper classes in *Ancestral Vices* and the public schools in *Vintage Stuff*. Sharpe remains, however, a provider of popular entertainments, and both *Porterhouse Blue* and *Blott on the Landscape* have been successfully adapted for television by **Malcolm Bradbury**, a great admirer of his work.

Fiction
Riotous Assembly 1971; *Indecent Exposure* 1973; *Porterhouse Blue* 1974; *Blott on the Landscape* 1975; *Wilt* 1976; *The Great Pursuit* 1977; *The Throwback* 1978; *The Wilt Alternative* 1979; *Ancestral Vices* 1980; *Vintage Stuff* 1982; *Wilt on High* 1984

Plays
The South African 1961; *She Fell Among Thieves* [from D. Yates's novel] 1978

George Bernard SHAW 1856–1950

Shaw was born in Dublin, the son of a Protestant corn factor, a failure and drunkard. His parents were unhappy together and neglected him, a background which may have led, as a reaction, to his puritanism, and to a certain emotional stunting which makes his work more impressive for language, wit and ideas than for involvement with people. His mother deserted her husband and son in Shaw's teenage years to become a singer in London. After attending various genteel day schools, and being apprenticed at fifteen to a Dublin estate agent, he joined his mother and sisters in England in 1876.

For a decade he lived in great poverty, doing various jobs and writing five novels which no publisher would accept, but from the mid-1880s onwards he established himself as a considerable critic of art, music and literature for many papers, and a propagandist for Wagner and Ibsen. He was among the leaders in the foundation of the Fabian Society in 1884, becoming a well-known socialist public speaker. (In later life Shaw, who admired all dictatorship, was equally enthusiastic for Stalin and Mussolini.) His first play in the West End was *Widowers' Houses* in 1892, but for many years he was more successful in getting his plays printed than performed (he also suffered from censorship), and only towards 1910 was he successful as a playwright. With a total output of over fifty plays, he eventually became probably the wealthiest author of his time, leaving a fortune of £300,000 when he died. Early years of struggle, however, led to a breakdown in 1898, and he married the woman who nursed him through it, the wealthy Charlotte Payne Townshend. It was a *marriage blanc* at her request (his relationships with well-known actresses seem also to have been sexless), but a happy partnership until her death in 1943.

Shaw's plays are often revived, and delightful comedies such as *Androcles and the Lion* and *Pygmalion* are likely to retain their appeal, but it has been questioned whether he is as important a dramatist as he was a critic and social commentator: his attempts at 'great' plays, such as *Heartbreak House* and *Saint Joan*, have equivocal reputations.

He wrote much polemical non-fiction, advocating vegetarianism, spelling reform, the incompetence of Shakespeare, and the Bergsonian life-force: these, and his engaging personality, made him one of the most famous (and loved) public figures of his day. He died in his mid-nineties at the house in Ayot St Lawrence where he had lived since 1906. Among his best works is the allegorical novella of 1932, *The Adventures of the Black Girl in Her Search for God*, a genuinely moving work.

Plays

Widowers' Houses 1892 [pub. 1893]; *Plays Pleasant and Unpleasant* 2 vols [pub. 1898]: *Arms and the Man* 1894, *Candida* 1897, *The Man of Destiny* 1897, *Mrs Warren's Profession* 1898, *You Never Can Tell* 1898, *The Philanderer* 1905; *Three Plays for Puritans* [pub. 1901]: *The Devil's Disciple* 1897, *Caesar and Cleopatra* 1899, *Captain Brassbound's Conversion* 1899; *The Admirable Bashville* 1901 [pub. 1909]; *Man and Superman* 1903; *How He Lied to Her Husband* 1904 [pub. 1907]; *John Bull's Other Island* 1904 [pub. 1907]; *Major Barbara* 1905 [pub. 1907]; *Passion, Poison and Petrification* [pub. in *Harry Furniss's Christmas Annual*] 1905; *The Doctor's Dilemma* 1906 [pub. 1911]; *Getting Married* 1908 [pub. 1911]; *Translations and Tomfooleries* [pub. 1926]: *The Fascinating Foundling* 1909, *The Music-Cure* 1914, *Jitta's Attonement* [from S. Trebitsch] 1923, *The Glimpse of Reality* 1927; *Press Cuttings* 1909; *The Shewing-Up of Blanco Posnet* 1909 [pub. 1911]; *The Dark Lady of the Sonnets* 1910 [pub. 1914]; Misalliance 1910 [pub. 1914]; *Fanny's First Play* 1911 [pub. 1914]; *Overruled* 1912 [pub. 1916]; *Androcles and the Lion* 1913 [pub. 1916]; *Heartbreak House, Great Catherine and Playlets of the Great War* [pub. 1919]: *Great Catherine* 1913, *The Inca of Perusalem* 1916, *Augustus Does His Bit* 1917, *O'Flaherty, V.C.* 1917, *Annajanska, the Bolshevik Empress* 1918, *Heartbreak House* 1920; *Pygmalion* 1914 [pub. 1916]; *Back to Methuselah* 1922 [pub. 1921]; *Saint Joan* 1923 [pub. 1924]; *The Apple Cart* 1929 [pub. 1930]; *Too Good to Be True* 1932 [pub. 1934]; *On the Rocks* 1933 [pub. 1934]; *The Six of Calais* 1934 [pub. 1936]; *Village Wooing* 1934; *The Simpleton of the Unexpected Isles* 1935 [pub. 1936]; *The Millionairess* 1936; *Cymbeline Refinished* 1937 [pub. 1947]; *Geneva* 1938 [pub. 1939]; *'In Good King Charles's Golden Days'* 1939; *Buoyant Billions* 1949; *Shakes versus Shav* 1949 [pub. 1951]; *Farfetched Fables* 1950 [pub. 1951]

Non-Fiction

The Quintessence of Ibsenism 1891 [rev. 1913]; *The Perfect Wagnerite* 1898; *The Commonsense of Municipal Trading* 1904; *On Going to Church* 1905; *Dramatic Opinions and Essays* 2 vols 1907 [rev. as *Our Theatre in the Nineties* 1932]; *On Shakespeare* 1907: *The Impossibilities of Anarchism* 1908; *The Sanity of Art* 1908; *Socialism and Superior Brains* 1910; *Preface for Three Plays by Brieux* 1910; *Peace Conference Hints* 1919; *What I Really Wrote about the War* 1920; *The Intelligent Woman's Guide to Socialism and Capitalism* 1928 [rev. as *The Intelligent Woman's Guide to Socialism, Capitalism, Sovietism and Fascism* 1937]; *Bernard Shaw and Karl Marx: a Symposium 1884–1889* [ed. R.W. Ellis] 1930; *Ellen Terry and Bernard Shaw: a Correspondence* [ed. C. St John] 1931; *Pen Portraits and Reviews* 1931; *Music in London 1890–94* 3 vols 1932; *Advice to a Young Critic: Letters 1894–1928* [ed. E.J. West] 1933; *The Future of Political Science in America* 1933; *The Political Madhouse in America and Nearer Home* 1933; *London Music in 1888–89* 1937; *Shaw Gives Himself Away* (a) 1939; *Florence Farr, Bernard Shaw and W.B. Yeats* [ed. C. Baker] 1941; *Everybody's Political What's What* 1945; *Sixteen Self Sketches* (a) 1949; *Bernard Shaw's Rhyming Picture Guide to Ayot Saint Lawrence* 1951; *Bernard Shaw and Mrs Patrick Campbell: Their Correspondence* [ed. A. Dent] 1952; *George Bernard Shaw: Last Will and Testament* 1954; *Bernard Shaw's Letters to Granville Barker* [ed. C.B. Purdom] 1956; *The Illusions of Socialism* 1956; *My Dear Dorothea: a Practical System of Moral Education for Females Embodied in a Letter to a Young Person of that Sex* 1956; *Shaw on Theatre* [ed. E.J. West] 1958; *How to Become a Musical Critic* [ed. D.H. Laurence] 1960; *To a Young Actress: the Letters of Bernard Shaw to Molly Tompkins* [ed. P. Tompkins] 1960; *The Matter with Ireland* [ed. D.H. Greene, D.H. Laurence] 1962; *Platform and Pulpit* [ed. D.H. Laurence] 1962; *The Religious Speeches of Bernard Shaw* [ed. W.S. Smith] 1963; *The Rationalization of Russia* [ed. H.M. Geduld] 1964; *Complete Prefaces of George Bernard Shaw* 1965;

The Road to Equality: Ten Unpublished Lectures and Essays 1884–1918 [ed. H. Cavanaugh, L. Crompton] 1971

Fiction

Cashel Byron's Profession 1886; *An Unsocial Socialist* 1887; *Love Among the Artists* 1900; *The Irrational Knot* 1905; *Immaturity* 1930; *The Adventures of the Black Girl in Her Search for God* 1932; *An Unfinished Novel* [ed. S. Weintraub] 1958

Edited

Fabian Essays in Socialism 1889; *Fabianism and the Empire* 1900; L. Gronlund *The Cooperative Commonwealth* 1892

Collections

Short Stories, Scrapes and Shavings 1934

Biography

Bernard Shaw by Michael Holroyd, 5 vols 1988–92

Wilfred (John Joseph) SHEED 1930–

Sheed is the son of F.J. Sheed and Maisie Ward, who founded the Roman Catholic publishing house of Sheed & Ward a few months after they married in 1926. Born in London, Sheed spent his early life in America, where his parents opened a branch of the company in New York, but returned to England to be educated at Downside. Despite the fact that his family believed 'the business of life was finding God, the rest was negotiable', Sheed was not swamped with religion. An attack of polio when he was fourteen had forced him to read – something he had done previously only under protest. He graduated in English at Lincoln College, Oxford, in 1954, and married in 1957, the same year he gained his MA.

During a year in Australia he began his first novel, *A Middle Class Education*, which he completed on his return to America in 1959. This related the adventures of a 'Lucky Jim'-style protagonist, and although too long, was described by the critic Bernard Bergonzi as containing 'neat observations of the raddled face of contemporary Oxford'. He was an associate editor of the Catholic magazine *Jubilee* from 1959 to 1966, and drama critic and book editor for *Commonweal* magazine, and currently writes for the *New York Times Book Review*.

The mood of Sheed's fiction is comic, although he does not regard himself as a humorous writer. Nevertheless, he perfected his idiosyncratic form in *Office Politics*, a satire on American and English journalism which was nominated for a National Book Award. Since his childhood, Sheed had known Clare Boothe Luce, who told him when he was eighteen that he 'looked like a bright sort of chap who might have an idea now and then'. The friendship led to a biography of her, which received critical acclaim. His volume of essays, *The Morning After*, was praised for a seriousness which was 'concealed by a self de-bunking'. Lugubrious in appearance, Sheed, who lives in New Jersey with his second wife and three children, has lost interest in travelling, and is 'waiting for the world to come to its senses, or at least for my pool-game to improve'.

Fiction

A Middle Class Education 1960; *The Hack* 1963; *Square's Progress* 1965; *Office Politics* 1966; *The Blacking Factory and Pennsylvania Gothic* (s) 1968; *Max Jamison* [UK *The Critic*] 1970; *People Will Always Be Kind* 1973; *Transatlantic Blues* 1978; *The Boys of Winter* 1987

Non-Fiction

Joseph 1958; *The Morning After* 1971; *Three Mobs: Labor, Church and Mafia* 1974; *Muhammad Ali* 1975; *The Good Word and Other Words* 1978; *Clare Boothe Luce* 1982; *Frank and Maisie: a Memoir with Parents* (a) 1985; *The Kennedy Legacy: a Generation Later* [with J. Lowe] 1988; *Essays in Disguise* 1990; *Baseball and Lesser Sports* 1991; *My Life as a Fan* 1993

Edited

G.K. Chesterton: Essays and Poems 1958; *Sixteen Short Novels* 1985

Sam(uel) SHEPARD (Rogers VII) 1943–

Shepard was born in Fort Sheridan, Illinois, the son of a career serviceman. Overseas at the time of his son's birth, his father was to remain separated from his family for the next five years, during which time mother and son criss-crossed the country from base to base. His father's retirement from the airforce reunited them and they bought an avocado ranch in Duarte, California, where Shepard attended high school, graduated in 1960, and studied agricultural science for a year. The region provided him with an 'absolute cross-section of everything American' and whilst there he worked variously as a stable hand, orange picker, sheep shearer and as a 'hot-walker' at the Santa Anita race track, living in Duarte until his father's drinking made home life increasingly difficult. A violent argument finally persuaded him to leave and, after a short spell in the YMCA, he joined a touring theatre group, the Bishop's Company Repertory Players, with whom he travelled the country and arrived in New York.

Shepard's job as busboy at the Village Gate brought him into contact with Ralph Cook, then head waiter but about to found Theater Genesis, and it was at Cook's instigation that Shepard began to write, one-act plays at first, for production off-off-Broadway. *Cowboys* (1964), the first of these, has since been lost but *The Rock Garden* gained notoriety when a scene was incorporated into Kenneth Tynan's *Oh, Calcutta!* (1969). Shepard himself was later to attribute the originality of his early writing to 'ignorance' and it was not until 1970 with *Operation Sidewinder* that he staged a full-length play. Always diverse, he also played drums for a rock group, the Holy Modal Rounders, from 1968 to 1971.

The increased emphasis on character of his later plays began after Shepard, his wife and their infant son decamped to England to escape the self-destructiveness Shepard saw consuming his contemporaries in New York. In London he wrote and directed the first production of *Geography of a Horse Dreamer* and began a series of plays examining the demands success makes upon the artist. It is ironic then that Shepard's most critically acclaimed and successful plays – *Fool for Love*, which was later filmed with Shepard in the leading role, and *A Life of the Mind*, which won the New York Drama Critics' Circle Award in 1985 – have contributed less to his general fame than has a parallel if occasional career as a Hollywood actor. On stage his plays have remained confined to off-Broadway. As an actor his face has appeared on the cover of countless magazines. Film credits include the lead in *Country* (1974) and a supporting role in *The Right Stuff* (1983), for which he won an Oscar nomination. He also collaborated with director Wim Wenders on the film *Paris, Texas* which won the Palme d'Or at Cannes in 1984. He now lives with actress Jessica Lange with whom he has two children.

Plays

Cowboys 1964 [np]; *The Unseen Hand and Other Plays* [pub. 1971]: *The Rock Garden* 1964, *4-H Club* 1965, *Forensic and the Navigators* 1967, *The Holy Ghostly* 1969, *The Unseen Hand* 1969, *Operation Sidewinder* 1970, *Shaved Spirits* 1970, *Back Dog Beast Bait* 1971; *Five Plays* [pub. 1967]: *Chicago* 1965, *Icarus's Mother* 1965, *Fourteen Hundred Thousand* 1966, *Red Cross* 1966, *Melodrama Play* 1967; *Dog* 1965 [np]; *Rocking Chair* 1965 [np]; *Up to Thursday* 1965 [np]; *Mad Dog Blues and Other Plays* [pub. 1971]: *Cowboys #2* 1967, *Cowboy Mouth* [with P. Smith] 1971, *Mad Dog Blues* 1971; *La Turista* 1967; *Blue Bitch* **(tv)** 1972, **(st)** 1973 [np]; *The Tooth of Crime* 1972 [pub. 1974]; *Nightwalk* [with J.-C. van Itallie, M. Terry] 1973 [pub. in *Open Theatre* 1975]; *Action* 1974 [pub. 1975]; *Geography of a Horse Dreamer* 1974; *Little Ocean* 1974 [np]; *Angel City and Other Plays* [pub. 1976]: *Killer's Head* 1975, *Angel City* 1976, *Curse of the Starving Class* 1976; *Buried Child and Other Plays* [pub. 1979]: *Suicide in B Flat* 1976, *Buried Child* 1978, *Seduced* 1978; *Fool for Love* [pub. 1983]: *The Sad Lament of Pecos Bill on the Eve of Killing His Wife* 1976, *Fool for Love* **(st)** 1983, **(f)** 1985; *Inacoma* 1977; *Tongues* [with others] 1978 and *Savage/Love* [with others] 1979 [pub. in *Seven Plays* 1981]; *Jackson's Dance* [with J. Levy] 1980 [np]; *True West* 1980 [pub. 1981]; *Superstitions* [with C. Stone] 1983 [np]; *Paris, Texas* **(f)** [with W. Wenders] 1984; *A Lie of the Mind* 1985 [pub. 1987]; *The War in Heaven* **(r)** 1985, **(st)** 1987; *Far North* **(f)** 1988 and *States of Shock* 1991 [pub. 1993]; *Hawk Moon* 1989 [np]

Fiction

Rolling Thunder Logbook **(s)** 1977; *Motel Chronicles* **(s)** 1982

Collections

Hawk Moon: a Book of Short Stories, Poems and Monologues 1973; *Joseph Chaikin & Sam Shepard: Letters & Texts 1972–1984* [ed. B.V. Daniels] 1989

Martin SHERMAN 1938–

Sherman was born in Philadelphia, the son of an attorney of Russian descent, and educated at high school in Camden, New Jersey, and at Boston University where he studied theatre arts, graduating in 1960. He had various occasional jobs to support his writing and produced a number of plays which achieved little recognition before the 1974 New York production of *Passing By*, which is described by Simon Callow in his book *Being an Actor* (1984) as a play about 'two young men going to bed together, catching each other's hepatitis, recuperating together, and finally falling in love'.

Sherman again addressed gay themes in his best-known play, *Bent*, which was given a stage reading in Waterford, Connecticut, in 1978, and first staged properly in a 1979 production, directed by Roger Chetwyn and starring Ian McKellan, at the Royal Court Theatre, London. Set in Nazi Germany, the play focuses on the persecution and murder of homosexuals, and provoked considerable controversy in some quarters. The central character, Max, is a seedy petty criminal and drug user, whose humanity emerges in the degrading squalor of Dachau. It was one of the first plays to explore the fate of homosexual men under the Nazi regime and publicised a hitherto little-known aspect of the Holocaust. The play won the Elizabeth Hull–Kate Warriner Award from the Dramatists Guild in 1979 and has been produced in thirty-five countries. It was revived at the National Theatre in 1989, again starring McKellan in a Sean Mathias production.

Sherman's other plays include *Soaps*, a satirical parody of American soap operas; *Messiah*, in which a woman is struck mute having witnessed Cossack soldiers torturing her husband to death; and *When She Danced*, a comedy which depicts one day in the life of the dancer Isadora Duncan. *A Madhouse in Goa* comprises two subtly related plays, neither of which is set in Goa. The first play is a quirky comedy of manners, the second an account of an assortment of people on a Greek island devastated by nuclear accidents in which it emerges that one of the characters is the author of the first play. Sherman has been a playwright-in-residence at various American universities, and with the Ensemble Studio Theatre in New York, and has adapted **Alice Thomas Ellis**'s *The Clothes in the Wardrobe* for a much admired television film.

Plays

A Solitary Thing [with S. Silverman] 1963 [np]; *Fat Tuesday* 1966 [np]; *Next Year in Jerusalem* 1968 [np]; *The Night Before Paris* 1969 [pub. 1970]; *Things Went Badly in Westphalia* 1971 [pub. 1970]; *Passing By* 1974 [pub. in *Gay Plays 1* ed. M. Wilcox 1984]; *Cracks* 1975 [pub. in *Gay Plays 2* ed. M. Wilcox 1986]; *Soaps* 1975 [np]; *Rio Grande* 1976 [np]; *Bent* 1979; *Blackout* 1978 [np]; *Messiah* 1982; *When She Danced* 1985 [pub. 1988]; *A Madhouse in Goa* 1989; *The Clothes in the Wardrobe* **(tv)** [from A.T. Ellis's novel] 1993

R(obert) C(edric) SHERRIFF 1896–1975

The son of an insurance clerk for Sun Assurance, Sherriff gew up in Kingston-upon-Thames, where he attended the local grammar school and at the age of seventeen followed his father into the insurance company. A few months later he enlisted as a second lieutenant in the infantry at the outbreak of the First World War, was severely wounded at Ypres in 1917, spent six months in hospital, and was demobilised as a captain in 1918, returning to the insurance company.

As a schoolboy he had been interested mainly in sport and had had no literary ambitions, but from the early 1920s he became a sort of unofficial playwright-in-residence for the Kingston Rowing Club, of which he had been captain. He wrote a play based on his letters to his mother (to whom he was devoted) from the trenches, but this proved beyond the rowing club's resources. He subsequently offered it to several theatrical managements without success, but **George Bernard Shaw** eventually helped him get a production by the London Stage Society. A theatrical manager took up the play, *Journey's End*, and it opened at the Savoy Theatre, London, in early 1929, enjoying instantaneous success, running for 594 performances, and eventually being seen in more than twenty countries. It remains the most famous play to have been written anywhere about the First World War and is a theatrical piece of simple strength and great durability.

Sherriff remained a fairly persistent writer for most of his life, and produced nine more stage plays, several radio plays and half-a-dozen novels (written in collaboration with Vernon Bartlett, one of them an adaptation of *Journey's End*), but, while many of these pieces have merit, none has proved lasting.

Journey's End made Sherriff rich, and he used the money to go up to New College, Oxford, in 1931 at the age of thirty-five, moving on from there to Hollywood, where he became a scriptwriter. Among distinguished films in which he had a hand were *Good-bye Mr Chips* (1939) and *The Dam Busters* (1954). Moving back to England, he lived mainly in Esher, Surrey, with a farm in Devon. He never married and entitled his autobiography *No Leading Lady*. A gentle, kindly man, his hobbies meant much to him: when he became too old for sport, he was a keen archaeologist, helping to excavate a Roman villa and fort. He died in Kingston-upon-Thames hospital near to where he had been born.

Plays

Cornlow-in-the-Downs 1923 [np]; *Profit and Loss* 1923 [np]; *The Feudal System* 1925 [np]; *Mr Birdie's Finger* 1926 [np]; *Journey's End* 1928 [pub. 1929]; *Badger's Green* 1930; *Windfall* 1933 [np]; *Two Hearts Doubled* 1934 [np]; *St Helena* [with J. de Casalis] 1936 [pub. 1934]; *The Road Back* [with C. Kenyon] 1937 [np]; *That Hamilton Woman* 1941 [np]; *This Above All* 1943 [np]; *Odd Man Out* **(f)** [from F.L. Green's novel] 1947 [pub. in *Three British Screenplays* ed. R. Manvill 1950]; *Miss Mabel* 1948 [pub. 1949]; *Home at Seven* 1949 [pub. 1950]; *Quartet* **(f)** [from Somerset Maugham] 1949 [pub. 1948]; *Trio* **(f)** [from Somerset Maugham; with N. Langley, Somerset Maugham] **(f)** 1950; *No Highway in the Sky* 1951 [np]; *The White Carnation* 1953; *The Long Sunset* **(r)** 1955, **(st)** 1962 [pub. 1956]; *The Night My Number Came Up* 1955 [np]; *Storm Over the Nile* 1956 [np]; *The Telescope* 1960 [pub. 1961]; *Cards* *with Uncle Tom* 1958 [np]; *A Shred of Evidence* 1960 [pub. 1961]; *The Oyburn Story* (tv) 1963

Fiction
Journey's End [from his play; with V. Bartlett] 1930; *The Fortnight in September* 1931; *Greengates* 1936; *The Hopkin's Manuscript* 1939 [rev. as *The Cataclysm* 1958]; *Chedworth* 1944; *Another Year* 1948; *The Wells of St Mary's* 1962; *The Siege of Swayne Castle* 1973

Non-Fiction
No Leading Lady (a) 1968

For Children
King John's Treasure 1954

Carol SHIELDS 1935–

The third child of a sweet-factory manager and a schoolteacher, Shields was born Carol Warner in Chicago, and grew up in the suburb of Oak Park. Always a voracious reader, she was encouraged to write, but at Hanover College, Indiana, was more interested in 'falling in love and going to dances'. She nevertheless won a place on an exchange programme with Exeter University in England, where the atmosphere was more academic. It was here that she met Don Shields, a Canadian engineering graduate, whom she married in 1957, and with whom she has five children. Apart from three years in Manchester during the 1960s, where her husband was finishing his PhD, she has lived in Canada ever since.

At the age of twenty-nine, she submitted seven poems to a Canadian Broadcasting Corporation competition for young writers, and won. She has since published two volumes of poetry, and for her master's thesis at the University of Ottawa she wrote a study of the nineteenth-century Canadian writer Susanna Moodie, publishing it in 1976, the same year that she published her first novel, *Small Ceremonies*. Before this, she had written the occasional short story, which was almost always sold either to the CBC or the BBC. The protagonist of *Small Ceremonies*, published in the week of her fortieth birthday, is also a novelist who is writing a book about Moodie, and the novel announced themes to which Shields was to return in subsequent books – in particular family life, biography and plagiarism.

It was followed by *The Box Garden* and the twinned novels, *Happenstance* and *A Fairly Conventional Woman*, which tell the same story from the differing perspectives of a husband and a wife. They completed a quartet of clever, realistic tales of marriage and domesticity, and firmly established her reputation in Canada. The first of her books to be published in the UK was *Mary Swann*, originally published in Canada as *Swann: a Mystery*. The eponymous Swann is a Canadian poet whose work bears some resemblance to that of Emily Dickinson, but whose life ended when she was hacked to death by her husband. A 'Swann Symposium' is being arranged, and the book consists of the interlocking (and unreliable) narratives of four people with a vested interest in the poet and her work, concluding with an account of the conference in the form of a film script. Boldly experimental in form, witty in its observations, it instantly established Shields's reputation in the UK as a leading Canadian writer, alongside **Alice Munro** and **Margaret Atwood**. It was followed by a single volume containing *Happenstance* and *A Fairly Conventional Woman*. Subsequent novels have been published simultaneously in Canada and Britain, and include *The Republic of Love* – an engaging and shrewdly observed tale of love at first sight, which was shortlisted for the *Guardian* Fiction Prize – and *The Stone Diaries*.

Like *Small Ceremonies* and *Swann*, *The Stone Diaries* is concerned with biography, in particular with lives that go unrecorded. Similarly experimental in form, the book traces the life of Daisy Goodwill, an 'ordinary' Canadian woman, from her birth in rural Manitoba in 1905 to her death in Florida ninety years later, and includes family trees, recipes, lists, photographs of the characters, and other fragments which go to make up a life. 'I am very interested in the interstices,' Shields has said, 'the little nerve-ends that get us from one day to the next.' The novel was shortlisted for the Booker Prize and won the Governor-General's Award, the National Book Critics' Circle Award and the Pulitzer Prize for fiction.

Shields has also published two volumes of short stories, a selection from which was published in Britain under the title *Various Miracles*, and has had several plays produced and published. In 1989 the Canadian feminist literary magazine *Room of One's Own* ran a Carol Shields issue (Vol. 13, Nos 1 and 2). She lives in Winnipeg, where she teaches part-time at the University of Manitoba, but spends most summers in France.

Fiction
Small Ceremonies 1976; *The Box Garden* 1977; *Happenstance* 1980 and *A Fairly Conventional Woman* 1982 [UK *Happenstance* 1991]; *Various Miracles* (s) 1985 and *The Orange Fish* (s) 1989 [rev. in UK as *Various Miracles* 1994]; *Swann* 1987 [UK *Mary Swann* 1990]; *A Celibate Season* [with B. Howard] 1991; *The Republic of Love* 1992; *The Stone Diaries* 1993

Poetry
Others 1972; *Intersect* 1974

Plays
Departures and Arrivals 1984 [pub. 1990]; *Thirteen Hands* 1993; *Fashion Power Guilt and the Charity of Families* [with C. Shields] 1995 [np]

Non-Fiction
Susanna Moodie: Voice and Vision 1976

Jon SILKIN 1930–

Silkin was born in London, the son of a solicitor; his grandparents on both sides were Russian Jewish immigrants. He was educated at Wycliffe College and Dulwich College and worked briefly as a journalist before doing his national service from 1948 to 1950, working in the Education Corps. He then rejected his father's wish that he should take up the law, and spent six years as a manual labourer, working as a janitor in a cemetery and a bricklayer's mate while living in lodgings in various parts of London.

He founded the literary magazine *Stand*, in its original form, in 1952, and his first commercially published volume of poetry, *The Peaceable Kingdom*, appeared in 1954. From 1956 to 1958 he taught English to foreign students, and from 1958 to 1960 he became Eric Gregory Poetry Fellow at Leeds University, beginning to study for an undergraduate degree at the same time, and later carrying out postgraduate study and reviving *Stand*. In 1965 he received financial support for the magazine from Northern Arts, and moved with it to Newcastle-upon-Tyne, where he has edited it ever since. Over twenty years it has become one of the most influential literary magazines in the English-speaking world, to the left in its general political stance but offering a broad home to many distinguished writers.

Silkin is well-known in several countries, such as the USA, Australia and Israel, where he has given many poetry readings and taught creative writing. From 1991 he spent two and a half years in Japan as professor of English and American literature at the University of Tsutuba. While in Iowa in 1968 he met the American writer Lorna Tracy, whom he married in 1974 and who is a co-editor of *Stand*. He had four children by two of his previous relationships: his eldest son, Adam, died in a mental hospital aged one, an event which gave rise to one of his most moving poems, 'Death of a Son'. Silkin is an intense poet with strong messages to communicate, although his work often achieves great formal beauty also: among his most successful poems are the beautifully observed 'flower poems' in *Nature with Man*. Besides poetry, he has written a study of the poets of the First World War, a verse play, *Gurney*, and is a well-known editor of anthologies. In 1985 he edited a volume of **Wilfred Owen**'s poems for the Penguin Poetry Library, but after complaints from **Jon Stallworthy** that this edition infringed the copyright of his own edition of Owen's poems, the book was withdrawn. The *Times Literary Supplement* carried a lively correspondence between the editors on the nature of copyright in editing. The book eventually reappeared in a revised edition in 1994.

Poetry
The Portrait and Other Poems 1950; *The Peaceable Kingdom* 1954; *The Two Freedoms* 1958; *The Re-Ordering of the Stones* 1961; *Flower Poems* 1964; *Nature with Man* 1965; *Penguin Modern Poets 7* [with R. Murphy, N. Tarn] 1966; *Poems New and Selected* 1966; *Three Poems* 1969; *Amana Grass* 1971; *Killhope Wheel* 1971; *Air That Pricks Earth* 1973; *The Principle of Water* 1974; *A 'Jarapiri' Poem* 1975; *The Little Time-Keeper* 1976; *Two Images of Continuing Trouble* 1976; *Jerusalem* 1977; *Into Praising* 1978; *The Lapidary Poems* 1979; *The Psalms with Their Spoils* 1980; *Selected Poems* 1980; *Autobiographical Stanzas* 1984; *Footsteps on a Downcast Path* 1984; *The Ship's Pasture* 1986; *Selected Poems* 1988; *The Lens-Breakers* 1992

Plays
Gurney 1985

Non-Fiction
Isaac Rosenberg, 1890–1918 [with M. de Sausmarez] 1959; *Out of Battle: the Poetry of the Great War* 1972 [rev. 1987]

Edited
Living Voices: an Anthology of Contemporary Verse 1960; *New Poems 1960* [with A. Cronin, T. Tiller] 1960; *Poetry of the Committed Individual: a 'Stand' Anthology of Poetry* 1973; *New Poetry 5* [with P. Redgrove] 1979; *The Penguin Book of First World War Poetry* 1979 [rev. 1981]; *Stand One: an Anthology of Stand Magazine Short Stories* **(s)** [with M. Blackburn, L. Tracy] 1984; *Wilfred Owen: the Poems* 1985 [rev. as *The War Poems of Wilfred Owen* 1995]; *Best Short Stories from Stand Magazine* **(s)** [with L. Tracy, J. Wardle] 1988; *The Penguin Book of First World War Prose* [with J. Glover] 1989

Translations
Nathan Zach *Against Parting* 1968

Alan SILLITOE 1928–

Sillitoe was born in Nottingham, the son of an illiterate tannery labourer who became one of the long-term unemployed during the 1930s'

Depression. His childhood was thus lived against a background of poverty, family quarrels and frequent moves to escape rent, but redeemed by an early love of reading and ambition to become a writer. He failed the 11-plus examination, and left school at the age of fourteen, spending the four years until national service began in a number of manual jobs in Nottingham factories. His period in the Royal Air Force, where he was a wireless operator, took him to Malaya, but on his return he was discovered to have tuberculosis and spent a further sixteen months in an RAF hospital.

An intensive period of learning to write began about this time, and was continued during the following two years, when he lived largely at home. In 1951, staying at his aunt's cottage in Kent, he met a young American married woman, the poet **Ruth Fainlight**, and they decided to live abroad together. Between 1952 and 1958, they lived, largely on his air force pension, in the south of France, Majorca (where they became friends of **Robert Graves**) and mainland Spain, both developing their writing.

In 1958, they returned to England, and the latest of a series of novels that Sillitoe had been working on, *Saturday Night and Sunday Morning*, was published. This story of Arthur Seaton, a rebellious young Nottingham factory worker reluctantly developing towards some sort of maturity, won immediate notice, and, like the title-story of the succeeding collection, *The Loneliness of the Long-Distance Runner*, another account of working-class rebellion, was turned into a highly successful film by Tony Richardson. Sillitoe was accordingly labelled as one of the Angry Young Men of the 1950s, a title he perhaps deserved more than some, since a sense of the individual pitted against relentless authority and an atmosphere of social protest have continued to dominate his work. He has, however, been highly prolific, and, if it is true that his two earliest works, developing from the tradition of that other son of Nottingham, **D.H. Lawrence**, are still his best known, it is also evident that only **Kingsley Amis** among the Angry Young Men equally mastered the challenge of producing a significant and developing *oeuvre* in the changing world of the later twentieth century. Besides some twenty novels, Sillitoe's work comprises numerous collections of short stories and poetry as well as plays, screenplays, books for children and non-fictional works. A certain clumsiness, the product of a hard apprenticeship to writing, has perhaps never left his style, but novels such as *A Start in Life* and *The Widower's Son*, as well as his short stories, have been much admired.

He and Ruth Fainlight married in 1959, and had a son and adopted a daughter. They have been based largely in London, but have also spent significant periods abroad, in Tangier, Spain and Israel, in which Sillitoe also seems to echo Lawrence. After a period from the early 1970s spent between their London home and Kent, they now divide their time between London and France.

Fiction

Saturday Night and Sunday Morning 1958; *The Loneliness of the Long-Distance Runner* (s) 1959; *The General* 1960; *Key to the Door* 1961; *The Ragman's Daughter* (s) 1963; *The Death of William Posters* 1965; *A Tree on Fire* 1967; *Guzman, Go Home and Other Stories* (s) 1968; *A Start in Life* 1970; *Travels in Nihilon* 1971; *Men, Women and Children* (s) 1973; *The Flame of Life* 1974; *Raw Material* 1974; *The Widower's Son* 1976; *The Storyteller* 1979; *The Second Chance and Other Stories* (s) 1981; *Her Victory* 1982; *The Lost Flying Boat* 1983; *Down from the Hill* 1984; *Life Goes On* 1985; *Out of the Whirlpool* 1987; *The Far Side of the Street* 1988; *The Open Door* 1989; *Last Loves* 1990; *Leonard's War* 1991; *Snow Stop* 1993

Poetry

The Rats and Other Poems 1960; *A Falling Out of Love and Other Poems* 1964; *Love in the Environs of Voronezh and Other Poems* 1968; *Shaman and Other Poems* 1973; *Barbarians and Other Poems* 1974; *Storm and Other Poems* 1974; *Snow on the North Side of Lucifer* 1979; *Sun Before Departure. Poems 1974–82* 1982; *Tides and Stone Walls* 1986; *Three Poems* 1988; *Collected Poems* 1993

For Children

The City Adventures of Marmalade Jim 1967; *Big John and the Stars* 1977; *The Incredible Fencing Fleas* 1978; *Marmalade Jim at the Farm* 1980; *Marmalade Jim and the Fox* 1984

Non-Fiction

Road to Volgograd 1964; *Mountains and Caverns* 1975; *The Saxon Shore Way* [with F. Godwin] 1983; *Life Without Armour* (a) 1995

Plays

Saturday Night and Sunday Morning (f) [from his novel] 1960; *The Loneliness of the Long Distance Runner* (f) [from his story] 1962; *The Ragman's Daughter* (f) [from his story] 1972; *Pit Strike* 1977; *3 Plays* [pub. 1978]

Translations

Lope de Vega *All Citizens Are Soldiers* [with R. Fainlight] 1969

Collections

Every Day of the Week: an Alan Sillitoe Reader 1987

(Marvin) Neil SIMON 1927–

Simon was born in the Bronx, New York, the son of a Jewish garment salesman, and graduated unspectacularly from De Witt Clinton High School in 1943. Of his childhood he has said: 'I was constantly being dragged out of movies for laughing too loud. Now my idea of the ultimate achievement in comedy is to make the whole audience fall onto the floor, writhing and laughing so hard that some of them pass out.' He studied engineering at New York University under the auspices of the Army Air Force Reserve training programme, and served two years in Colorado from 1945 to 1946 before finding work as a mail-room clerk at Warner Bros. in New York, where his older brother Danny was writing publicity material.

The two brothers got their first break at CBS in 1947 writing for Robert Q. Lewis's radio show, and were soon turning out television scripts and gags for *Phil Silvers Arrow Show*, the *Tallulah Bankhead Show* and others. Their first glimpse of Broadway came in 1955 when some of their sketches were used in *Catch a Star!* The writing team separated in 1956, and Simon went on to write for Sid Caesar, Phil Silvers's *Sergeant Bilko*, *The Jerry Lewis Show*, *The Jacky Gleason Show* and numerous others. Of his long apprenticeship in television, he has said: 'It was very tedious, and I did it only to pay the rent.' He was married in 1953 to Joan Baim, a dancer who died in 1973, and has remarried twice. He has two children by his first marriage.

Simon's first original play, *Come Blow Your Horn*, was based on his and Danny's experience of living in a Manhattan bachelor apartment. It was produced in Pennsylvania in 1960, and the next year on Broadway where it was an immediate success, running for two seasons. There has followed a steady succession of lucrative comedies, many of which are rooted in autobiography. *Barefoot in the Park* draws on the early years of his marriage, *The Odd Couple* of his experience of sharing an apartment with a divorced friend, and *Chapter Two* on his recovery after the loss of his first wife. 'The theme is me, my outlook on life,' he has said. 'If you spread [my career] out like a map, you can chart my emotional life.' Simon has adapted many of his plays as films, as well as adapting the work of others. He has won countless awards, including a Pulitzer Prize for his 1991 play *Lost in Yonkers* (which was filmed in 1993) and four Tony awards (in 1965, 1970, 1985 and 1991). He owns the Eugene O'Neill Theater in New York, and is, according to the *Washington Post*, 'the richest playwright alive and arguably the richest ever in the history of the theater'.

Plays

Adventures of Marco Polo [with others] 1959; *Heidi* [from J. Spyri's novel; with W. Friedberg, C. Warnick] 1959; *Come Blow Your Horn* 1960 [pub. 1961]; *Barefoot in the Park* **(st)** 1962, **(f)** 1967 [pub. 1964]; *Little Me* [from P. Dennis's novel; with C. Coleman, C. Leigh] 1962 [pub. 1979]; *The Odd Couple* **(st)** 1965, **(f)** 1968 [pub. 1966]; *After the Fox* **(f)** 1966; *The Star-Spangled Girl* 1966 [pub. 1967]; *Sweet Charity* [from F. Fellini's screenplay *Nights of Cariria*; with C. Coleman, D. Fields] 1966; *Plaza Suite* **(st)** 1968, **(f)** 1971 [pub. 1969]; *Promises, Promises* [from I.A.L. Diamond and B. Wilder's screenplay *The Apartment*; with B. Bacharach, H. David] 1968 [pub. 1969]; *Last of the Red Hot Lovers* **(st)** 1969, **(f)** 1972 [pub. 1970]; *The Gingerbread Lady* 1970 [pub. 1971]; *The Out of Towners* **(f)** 1970; *The Prisoner of Second Avenue* **(st)** 1971, **(f)** 1975 [pub. 1972]; *The Comedy of Neil Simon* [pub. 1972]; *The Heartbreak Kid* **(f)** 1972; *The Sunshine Boys* **(st)** 1972, **(f)** 1975 [pub. 1973]; *The Good Doctor* [from Chekhov; with P. Link] 1973 [pub. 1974]; *God's Favorite* 1974 [pub. 1975]; *California Suite* **(st)** 1976, **(f)** 1978 [pub. 1977]; *Murder by Death* **(st)** 1976; *The Goodbye Girl* **(f)** 1977; *Chapter Two* **(st)** 1977, **(f)** 1979; *The Cheap Detective* **(f)** 1978; *They're Playing Our Song* [with C. Bayer Sager, M. Hamlisch] 1978 [pub. 1980]; *Collected Plays 2* [pub. 1979]; *I Ought to Be in Pictures* **(st)** 1980, **(f)** 1982 [pub. 1981]; *Seems Like Old Times* **(f)** 1980; *Fools* 1981 [pub. 1982]; *Only When I Laugh* **(f)** [with D. Simon] 1981; *Actors and Actresses* 1983; *Brighton Beach Memoirs* **(st)** 1983, **(f)** 1986 [pub. 1984]; *Max Dugan Returns* **(f)** 1983; *The Lonely Girl* **(f)** [with S. Daniels, E. Weinberger] 1984; *Biloxi Blues* **(st)** 1985, **(f)** 1988 [pub. 1986]; *The Slugger's Wife* **(f)** 1985; *Broadway Bound* **(st)** 1986, **(f)** 1992 [pub. 1987]; *Rumors* 1988; *Jake's Women* 1990 [pub. 1993]; *Collected Plays 3* [pub. 1991]; *Lost in Yonkers* 1991; *The Marrying Man* **(f)** 1991

N(orman) F(rederick) SIMPSON 1919–

Born in London, Simpson was the son of a glassblower at a lamp factory in Hammersmith. He was educated at Emanuel School in Battersea, and then worked for some years in a bank. During the Second World War, he was in the Royal Artillery and the Intelligence Corps, and saw service in Italy and the Middle East. He had done some teaching before entering the army in 1941, and from 1946 until 1963 he taught A-level English for the adult education programme of the City of Westminster College in London. He also took a part-time degree for adults at Birkbeck College.

He first came to notice as a playwright when his play *A Resounding Tinkle* won third prize in an *Observer* competition in 1957, and he soon became one of the playwrights associated with the theatrical revolution being carried out by the English Stage Company at the Royal Court Theatre. His *One Way Pendulum*, produced there in 1959, was one of the few productions on which they did not lose money, and has remained his most popular play. Further stage plays followed sporadically, being varied with much television script work, and in 1963 Simpson retired from his job to become a full-time writer.

He has been called Britain's leading dramatist of the Absurd, but both in terms of achievement and *réclame*, he must be considered a pale shadow of Ionesco: his plays typically feature fantastic and disconnected happenings in an English suburban setting, and attempts to vary his formula were not successful. He has published one novel, *Harry Bleachbaker*, which was based on his stage play *Was He Anyone?* From 1976 to 1978 he was literary manager of the Royal Court. He married in 1944, and has a daughter; they have lived on a canal boat roaming the inland waterways with their dog.

Plays

A Resounding Tinkle 1957 [pub. 1958]; *The Hole* 1958; *One Way Pendulum* **(st)** 1959, **(f)** 1964 [pub. 1960]; *The Form* 1961; *Oh* 1961 [pub. 1964]; *The Cresta Run* 1965 [pub. 1966]; *Make a Man* **(tv)** 1966; *Three Rousing Tinkles* **(tv)** 1966; *Four Tall Tinkles* **(tv)** 1967; *We're Due in Eastbourne in 10 Minutes* **(tv)** [in *Four Tall Tinkles*] 1967, **(st)** 1971; *Diamonds for Breakfast* [with R. Harwood, P. Rouve] **(f)** 1968; *World in Ferment* **(tv)** 1969; *Charley's Grants* **(tv)** 1970; *How Are Your Handles?* 1970 [np]; *Playback 625* 1970 [np]; *Thank You Very Much* **(tv)** 1971; *Was He Anyone?* 1972 [pub. 1973]; *Something Rather Effective* **(r)** 1972; *Elementary My Dear Watson* **(tv)** 1973; *Silver Wedding* **(tv)** 1974; *Six Sketches for Radio* **(r)** 1974; *Anyone's Gums Can Listen to Reason* 1977; *In Reasonable Shape* [pub. in *Play Ten* ed. R. Rook] 1977; *Wainwright's Law* **(tv)** 1980; *Inner Voices* [from E. da Filippo] 1983; *Napoli Milionara* [from E. da Filippo] 1991

Fiction

Harry Bleachbaker [from his play *Was He Anyone?*] 1976

Andrew (Annandale) SINCLAIR 1935–

Sinclair was born in Oxford, the son of a colonial officer, and educated at the Dragon School and Eton, where he was a King's Scholar. He won another scholarship to Trinity College, Cambridge, but was unable to take this up until he had done his national service. He joined the Coldstream Guards, and this experience provided material for his first novel, *The Breaking of Bumbo*, a black comedy set at the time of the Suez Crisis, which was an immediate success when published in 1959 – 'at the right time and as a left subject,' as he put it. He had finally gone to Cambridge (where he took a double first in history) the previous year, and his extracurricular activities are recalled in another comic novel, *My Friend Judas*, which he has subsequently described as overwritten, too much influenced by **Dylan Thomas**, whose *Adventures in the Skin Trade* (1955) he was to adapt for the stage in 1965. The novel was nevertheless a bestseller, and he adapted it for the stage during the year of publication.

After doing postgraduate work on the Prohibition era in America (subsequently publishing his doctorate as a book), Sinclair embarked on a career as an academic, becoming a founding fellow and director of historical studies at Churchill College, Cambridge, then a lecturer in American History at University College, London. He continued to publish both fiction and non-fiction (including *The Better Half*, a book on the emancipation of the American woman, which won a Somerset Maugham Award in 1967), worked on stage and film scripts, and in 1967 founded Lorrimer Publishing, which specialised in publishing classic screenplays, including **Graham Greene**'s *The Third Man*.

That same year he embarked on what he considers his major fictional achievement, *The Albion Triptych*. A satirical fantasy on English themes, concerned with history, myth and political power, it consists of three novels, widely spaced through three decades of Sinclair's writing life: *Gog* (1967), which is set in 1945 against the backdrop of the Labour election victory; *Magog* (1972), which traces the following twenty or so years of corruption and decline; and *King Ludd* (1988), which goes back into history to describe the Luddite rebellion. His friend **William Golding** described the novels as representing 'the land's history seen by flashes of lightning', and they have been much admired for their sweep, inventiveness and energy.

Sinclair has pursued a parallel career as a biographer and historian. Amongst his best-known works in this field are *War Like a Wasp*, a history of the arts in Britain in the 1940s; *The Red and the Blue*, an account of the breeding ground of the 'Cambridge spies'; and lives of Che Guevara, Dylan Thomas, **Jack London**, and J. Pierpont Morgan. His later biography of the painter Francis Bacon was less successful, showing signs of haste (it was one of several rival books written in the wake of Bacon's death), and apparently assembled from newspaper clippings. His interest in the cinema has led to books on the directors John Ford and Sam Spiegel, and to his writing and directing adaptations of his own *The Breaking of Bumbo* and Dylan Thomas's 1954 radio play *Under Milk Wood*, for which he won major awards at both the Venice and Cannes film festivals.

He has been married three times; the first, in 1960, was brief and resulted in a son; the second was to the biographer Miranda Seymour, with whom he also had a son; the third was to the writer and hostess Sonia Melchett.

Fiction

The Breaking of Bumbo 1959; *My Friend Judas* 1959; *The Project* 1960; *The Hallelujah Bum* [US *The Paradise Bum*] 1963; *The Raker* 1964; *The Albion Triptych*: *Gog* 1967, *Magog* 1972, *King Ludd* 1988; *The Surrey Cat* [also as *Cat*] 1976; *A Patriot for Hire* [also as *Sea of the Dead*] 1978; *The Facts in the Case of E.A. Poe* 1980; *Beau Bumbo* 1985; *The Far Corners of the Earth* 1991; *The Strength of the Hills* 1992

For Children

Inkydoo, the Wild Boy [also as *Carina and the Wild Boy*] 1976

Non-Fiction

Prohibition: the Era of Excess 1962; *The Available Man: the Life Behind the Masks of Warren Gamaliel Harding* 1965; *The Better Half: the Emanicipation of the American Woman* 1965; *A Concise History of the United States* 1967; *The Last of the Best: the Aristocracy of Europe in the Twentieth Century* 1969; *Guevara* [US *Che Guevara*] 1970; *Dylan Thomas: Poet of His People* [US *Dylan Thomas: No Man More Magical*] 1975; *Jack: a Biography of Jack London* 1977; *The Savage: a History of Misunderstanding* 1977; *John Ford* 1979; *Corsair: the Life of J. Pierpont Morgan* 1981; *The Other Victoria: the Princess Royal and the Great Game of Europe* 1981; *Sir Walter Raleigh and the Age of Discovery* 1984; *The Red and the Blue* 1985; *Spiegel: the Man Behind the Pictures* 1987; *War Like a Wasp* 1989; *The Need to Give: the Patron and the Arts* 1990; *The Naked Savage* 1991; *Francis Bacon: His Life and Violent Times* 1993; *Patron Is Not a Dirty Word* 1993; *The Sword and the Grail* 1993

Plays

My Friend Judas [from his novel] 1959; *The Chocolate Tree* **(tv)** 1963; *Old Soldiers* **(tv)** 1964; *Adventures in the Skin Trade* [from D. Thomas] 1965 [pub. 1967]; *The Breaking of Bumbo* **(f)** [from his novel] 1970; *Under Milk Wood* **(f)** [from D. Thomas's radio play] 1971 [pub. 1972]; *Martin Eden* **(tv)** [from J. London's novel] 1981; *The Blue Angel* [from J. von Sternberg's film] 1983

Translations

Selections from the Greek Anthology 1967; *Bolivian Diary: Ernesto Che Guevara* 1968; Jean Renoir *La Grande Illusion* [with M. Alexandre] 1968

Edited

Jack London *The Call of the Wild, White Fang and Other Stories* 1981, *The Sea-Wolf and Other Stories* **(s)** 1989, *Tales of the Pacific* **(s)** 1989; *The War Decade: an Anthology of the 1940s* 1989

Clive (John) SINCLAIR 1948–

Sinclair was born in London, the son of the director of a furniture company, and educated at his local grammar school. He went to the University of East Anglia in 1966 and read English and American literature under **Malcolm Bradbury**, graduating in 1969. After a year of postgraduate study at the University of California, Santa Cruz, he returned to England where he wrote his first novel. From 1973 to 1976 he worked as a copywriter for the advertising agency Young & Rubicam, before continuing postgraduate study at Exeter University and UEA. His thesis examined the life and work of **Isaac Bashevis Singer** and his brother Israel Joshua Singer and was published as *The Brothers Singer* in 1983, the same year he received his PhD. From 1980 to 1981 he held a British Council Bicentennial Arts Fellowship in California and lectured at Santa Cruz, and in 1988 was the British Council Guest Writer in Residence at Uppsala University in Sweden. From 1983 to 1987 he was the literary editor of the *Jewish Chronicle*. He married in 1979, has one child, and lives in St Albans.

Sinclair's first novel, *Bibliosexuality* (1973) is a grotesque comedy concerning a disorder which makes the sufferer desire an unnatural relationship with a book. His first collection of short stories, *Hearts of Gold*, which won a Somerset Maugham Award in 1980, demonstrates a similar sense of the grotesque, its narrators including a vampire and a Jewish giraffe. In both his fiction and non-fiction Sinclair demonstrates a consistent concern with Jewish issues, and has written a travel book about Israel, *Diaspora Blues*. Of his interest in Jewish history he has written: 'Being English, I look at the Holocaust and Israel as an inside-outsider.' His 1992 novel *Augustus Rex* is a comic phantasmagoria in which the Swedish playwright Strindberg signs a Faustian pact with Beelzebub and is raised from the grave in the 1960s to be subjected to a variety of temptations.

Fiction

Bibliosexuality 1973; *Hearts of Gold* (s) 1979; *Bedbugs* (s) 1982; *Blood Libels* 1985; *Cosmetic Effects* 1989; *For Good or Evil* 1991; *Augustus Rex* 1992

Non-Fiction

The Brothers Singer 1983; *Diaspora Blues: a View of Israel* 1987

Iain (Macgregor) SINCLAIR 1943–

Born in Cardiff, Sinclair is the son of a doctor practising in a Welsh mining town; his father was of Scottish, his mother of Welsh, ancestry. He attended Cheltenham College and, in the early 1960s, the London School of Film Technique in Brixton. He subsequently went to Trinity College, Dublin, for four years, where he read English and fine art, and where the performance of two plays, written with Christopher Bamford, attest his early interest in writing.

In 1967 he married, and subsequently had three children. He returned to London, where he supported himself for some years by a mixture of labouring jobs (dock work, packing, park gardening, brewery assistant) and some lecturing and documentary film work. He made a film for West German television about the poet **Allen Ginsberg**, and with the proceeds bought the house in Hackney, east London, where he still lives. (He also published *The Kodak Mantra Diaries*, a documentary account of Ginsberg in London.)

From the early 1970s he used largely his own Albion Village Press to publish small volumes, mostly of poetry, some of mixed poetry and prose. He is a member of the group of poets and writers loosely associated with the Cambridge don **J.H. Prynne**, and is included in the anthology of their works, *A Various Art* (1987). Another writer associated with the group is **Peter Ackroyd**, who acknowledges that the basic idea for his highly successful novel *Hawksmoor* (1985) was drawn from Sinclair's 1975 book of poetry and essays, *Lud Heat*.

From the late 1970s Sinclair moved into the world of book dealing, specialising in the work of the Beat poets and in *noir* popular fiction, and the sometimes seedy world of secondhand books provided much of the inspiration for his first, highly experimental, novel, *White Chappell, Scarlet Tracings*, which made his name in 1987. He has followed it with *Downriver*, a complex novel evoking the character of east London and the Thames estuary (it has twelve sections evoking areas ranging from Spitalfields to the Isle of Sheppey and including the River Thames itself). The book celebrates the traditional character of the area and deplores gentrification, but some reviewers, although conceding its richness, complained that *Downriver* did not flow. Among his other activities, Sinclair is poetry editor for the publisher Paladin.

Poetry

Back Garden Poems 1970; *Muscat's Wurm* 1972; *The Birth Rug* 1973; *Groucho Positive, Groucho Negative* 1973; *Brown Clouds* 1977; *The Penances* 1977; *Suicide Bridge* 1979; *Flesh Eggs and Scalp Metal* 1983; *Fluxions* 1983; *Autistic Poses* 1985; *Significant Wreckage* 1988

Fiction

White Chappell, Scarlet Tracings 1987; *Downriver* 1991; *Radon Daughters* 1994

Non-Fiction

The Kodak Mantra Diaries 1971; *The Shamanism of Intent: Some Flights of Redemption* 1991

Plays

An Explanation [with C. Bamford] 1963 [np]; *Cords* [with C. Bamford] 1964 [np]

Collections

Lud Heat 1975

May (i.e. Mary Amelia St Clair) SINCLAIR 1863–1946

Sinclair was born in Cheshire, the daughter of an alcoholic ship-owner whose business collapsed in the 1870s. Her education was confined to a single year at Cheltenham Ladies' College when she was eighteen. Encouraged by the famous headmistress Miss Beale, Sinclair read philosophy, psychology and Greek literature, and began to write poetry and fiction. Her first short story appeared in 1895 and her first novel, *Audrey Craven*, two years later, but she did not make much impact until the publication of *The Divine Fire* in 1904. Other memorable books of the early period were *The Three Sisters* and her biography of the Brontës.

She had moved to London in 1897, where she lived with her mother until the latter's death, supporting herself by her writing and by literary journalism. She was active in the Women's Suffrage Movement, and served with the Hoover Relief Commission and an ambulance unit during the First World War, later publishing a journal recording her experiences in Flanders. After the war, she was considered a leading novelist, and was an intimate of many of the great writers of the period, including **H.G. Wells**, **Ford Madox Ford**, **John Galsworthy**, **Ezra Pound**, **Richard Aldington**, **H.D.** and **Dorothy Richardson**. It was Sinclair who coined the term 'stream of consciousness' to describe Richardson's prose, and it was a style she too adopted in several novels. She was strongly influenced by her reading of Freud and Jung, the results of which may be seen in such novels as the partly autobiographical *Mary Oliver*, and in *Life and Death of Harriett Frean*, a brief, ironical case history of self-repression and pointless self-sacrifice which many regard as her masterpiece. She continued to write poetry, including *The Dark Night*, a long narrative poem, and published two philosophical books on idealism. *Uncanny Stories* is a volume of stories on supernatural themes, a subject which increasingly preoccupied her as she grew older. In the 1920s she wrote a number of satirical comedies, including *Mr Waddington of Wyck* and *A Cure of Souls*.

She never married, and led a quiet, outwardly uneventful life in London and in the country cottages she had in Yorkshire and Gloucestershire, but she produced a large volume of work. She made several trips to America, where her books were appreciated as much as, if not more than, they were in Britain. During the last fifteen years of her life she was gradually incapacitated by Parkinson's disease, and in 1932 she retired to Aylesbury in Buckinghamshire, where she died.

Fiction

Audrey Craven 1897; *Mr and Mrs Neill Tyson* 1898; *Two Sides of a Question* 1901; *The Divine Fire* 1904; *The Helpmate* 1907; *Kitty Tailleur* 1908; *The Creators* 1910; *The Combined Maze* 1913; *The Judgement of Eve and Other Stories* (s) 1914; *The Three Sisters* 1914; *Tasker Jevons* 1916; *The Tree of Heaven* 1917; *Mary Olivier* 1919; *The Romantic* 1920; *Mr Waddington of Wyck* 1921; *Ann Severn and the Fieldings* 1922; *Life and Death of Harriett Frean* 1922; *Uncanny Stories* (s) 1923; *Arnold Waterlow* 1924; *A Cure of Souls* 1924; *Far End* 1926; *The Rector of Wyck* 1925; *The Allinghams* 1927; *The History of Anthony Waring* 1927; *Fame* (s) 1929; *Tales Told by Simpson* (s) 1930; *The Intercessor and Other Stories* (s) 1931

Non-Fiction

The Three Brontës 1912; *A Journal of Impressions of Belgium* 1915; *A Defence of Idealism* 1917; *The New Idealism* 1922

Translations

Rudolf Sohm *Outlines of Church History* 1895; Theodor von Sosnosky *England's Danger* 1901

Poetry

Essays in Verse 1891; *The Dark Night* 1924

Upton (Beall) SINCLAIR 1878–1968

Sinclair was born in Baltimore, into a family of the ruined Southern aristocracy; his father was an inebriate liquor salesman. When Sinclair was ten, the family moved to New York, and there he supported himself through New York City College and Columbia University by writing vast numbers of stories for pulp magazines, often on naval themes. He later calculated that by the time he was twenty-one he had already put into print more words than are contained in Sir Walter Scott's *Waverley* novels, and this was to set the pattern for a life in which he was to publish almost 100 books.

At twenty, he became a socialist and a member of the Socialist Party, and this commitment was to remain with him, despite temporary allegiances to other parties. He published his first serious novel in 1901, and endured a period of poverty with his first wife, whom he had married in 1900 (they had one son and were divorced in 1911). In 1906, his fifth novel, *The Jungle*, a bitter exposé of conditions in the Chicago meat-packing industry, won him fame and fortune; it had the biggest social impact of any American book since Harriet Beecher Stowe's *Uncle Tom's Cabin* (1852), and it led to reform. Sinclair characteristically used the proceeds from the novel to found Helicon Hall, a Utopian community for left-wing writers, which burned down after a year.

He continued to write many more novels – including an eleven-volume series, published between 1940 and 1953, about the adventures of an international troubleshooter, Lanny Budd – but of his fiction only *The Jungle* is much read today. His was perhaps more a journalistic than a literary talent, and the long series of polemical books he published, particularly in the 1920s, had strong public impact, although it was a measure of the opposition to him that many of his works throughout his life were published at his own expense.

His many attempts to run for office, which included running for Governor of California on the Democratic ticket, were also unsuccessful. For almost forty years, from 1915, he lived in Pasadena, California; in 1953 he suddenly moved to the remote Arizona village of Buckeye. His second wife, whom he married in 1913, predeceased him in 1961, as did his third wife, in 1967.

Fiction

Springtime and Harvest 1901 [rev. as *King Midas*]; *The Journal of Arthur Stirling* 1903; *Prince Hagen* 1903; *Manassas* [rev. as *Theirs Be the Guilt* 1959] 1904; *The Jungle* 1906; *The Overman* 1907; *The*

Metropolis 1908; *The Moneychangers* 1908; *Samuel the Seeker* 1910; *Love's Pilgrimage* 1911; *Damaged Goods* [from E. Brieux's play *Les Avariés*] 1913; *Sylvia* 1913; *Sylvia's Marriage* 1914; *King Coal* 1917; *Jimmie Higgins* 1919; *100%: the Story of a Patriot* [UK *The Spy*] 1920; *They Call Me Carpenter* 1922; *The Millenium* 1924; *Oil!* 1927; *Boston* 1928; *Mountain City* 1930; *Roman Holiday* 1931; *The Wet Parade* 1931; *Co-Op* 1936; *The Gnomobile* 1936; *The Flivver King* 1937; *No Pasaran!* 1937; *Little Steel* 1938; *Our Lady* 1938; *World's End Series*: *World's End* 1940, *Between Two Worlds* 1941, *Dragon's Teeth* 1942, *Wide Is the Gate* 1943, *Presidential Agent* 1944, *Dragon Harvest* 1945, *A World to Win* 1946, *Presidential Mission* 1947, *One Clear Call* 1948, *O Shepherd, Speak!* 1949, *The Return of Lanny Budd* 1953; *Another Pamela* 1950; *What Didymus Did* 1954 [US *What Happened to Didymus* 1958]; *Affectionately, Eve* 1961

Plays
The Machine [pub. 1911]; *The Naturewoman* [pub. 1912]; *The Second-Story Man* [pub. 1912]; *Prince Hagen* [from his novel] 1921; *Hell* [pub. 1923]; *The Pot Boiler* [pub. 1924]; *Singing Jailbirds* 1924; *Bill Porter* [pub. 1925]; *Oil!* [from his novel] 1929; *The Wet Parade* (f) [from his novel] 1932; *Depression Island* [pub. 1935]; *Wally for Queen* [pub. 1936]; *Marie Antoinette* [pub. 1939]; *A Giant's Strength* [pub. 1948]; *The Enemy Had It Too* [pub. 1950]; *Cicero* [pub. 1960]

Non-Fiction
A Captain of Industry 1906; *The Industrial Republic* 1907; *Good Health and How We Won It* [with M. Williams] 1909; *The Fasting Cure* 1911; *The Sinclair–Astor Letters: Famous Correspondence between Socialist and Millionaire* 1914; *The Profits of Religion* 1918; *The Brass Check* 1920; *The Book of Life: Mind and Body* 1921; *The Book of Life: Love and Society* 1922; *The Goose-Step: a Study of American Education* 1923; *The Goslings: a Study of American Schools* 1924; *Mammonart* 1925; *Letters to Judd, an American Working Man* 1926; *The Spokesman's Secretary* 1926; *Money Writes!* 1927; *Mental Radio* 1930; *Candid Reminiscences* 1932 [rev. as *American Outpost* 1948]; *I, Governor of California, and How I Ended Poverty* 1933; *Upton Sinclair Presents William Fox* 1933; *EPIC Plan for California* 1934; *I, Candidate for Governor, and How I Got Licked* 1935; *What God Means to Me* 1936; *Terror in Russia? Two Views* [with E. Lyons] 1938; *Letters to a Millionaire* 1939; *Telling the World* 1939; *Limbo on the Loose* 1948; *A Personal Jesus* 1952 [rev. as *The Secret Life of Jesus* 1962]; *The Cup of Fury* 1956; *My Lifetime in Letters* 1960; *The Autobiography of Upton Sinclair* (a) 1962; *Sergei Eisenstein and Upton Sinclair: the Making and Unmaking of 'Que Viva Mexico!'* [ed. H.M. Geduld, R. Gottesman] 1970

Edited
The Cry for Justice 1915

Isaac Bashevis SINGER 1904–1991

Born in Radzymin (now part of Poland, then part of Tsarist Russia), Singer was the son and grandson of Hasidic rabbis and brother of the novelists Israel Joshua Singer and Esther Kreitman. When he was four, the family moved to Warsaw, where his father supervised a *beth din*, or rabbinical court, an institution described in Singer's volume of memoirs, *In My Father's Court*. Singer also spent several years of his youth living in Bilgoray, a traditional Jewish village, or *shtetl*, and attended the Warsaw Rabbinical Seminary. Under the influence of his elder brother, he decided to become a journalist for the Yiddish press in Poland: in 1932 he became co-editor to the magazine *Globus* after several years as a proof-reader.

He began creative work in Hebrew, but soon turned to Yiddish, beginning with short stories. His first novel in Yiddish, *Satan in Goray*, which appeared in an English version in 1955, was published in its original form in 1935. That same year, like his brother before him, he emigrated to the USA, parting from his first wife, Rachel, and son, Israel, who went to Moscow and later Palestine. In New York, he began to work for the Yiddish newspaper the *Jewish Daily Forward*: he continued to work, on a freelance basis, for the paper for the rest of his life, and many of his short stories were first published in it. In 1940 he married Alma Haimann, a refugee from Nazi Germany, who worked for many years in a New York department store.

In 1950 Singer's long chronicle novel, *The Family Moskat*, appeared in the form which was to bring him fame: joint publication in Yiddish and English. It is difficult to say whether he is essentially a Yiddish- or an English-language writer; he wrote originally in Yiddish, but collaborated actively with many distinguished translators, including **Saul Bellow**, but most frequently Cecil Hemley. Some of his works are only available in English and others only in Yiddish.

His chief subject was traditional Polish life, but he also wrote of modern America. His work deals much with magic and the supernatural; although his themes are often sombre, he is a richly comic and affirmative writer. His superb short stories are perhaps even better than his distinguished novels. After the mid-1960s he turned increasingly to writing for children. He was awarded the Nobel Prize for Literature in 1978 and was universally acknowledged as one of the world's leading Jewish writers. He lived on the Upper West Side, and gave as his recreation 'walking in the bad air of New York City'.

Fiction
Satan in Goray 1935 [tr. J. Sloan 1955]; *The Family Moskat* [tr. A.H. Gross] 1950; *Gimpel the Fool and Other Stories* (s) [tr. with others] 1957; *The Magician of Lublin* [tr. E. Gottlieb, J. Singer] 1960; *The Spinoza of Market Street and Other Stories* (s) [tr. with M. Glicklich] 1961; *The Slave* [tr. with C. Hemley] 1962; *The Manor* [tr. with E. Gottlieb, J. Singer; ed. C. Hemley with others] 1967; *Short Friday and Other Stories* (s) [tr. R. Witman with others] 1967; *The Seance and Other Stories* (s) [tr. M. Ginsburg with others] 1968; *The Estate* [tr. E. Gottlieb, J. Singer, E. Shub; ed. E. Shub] 1969; *The Fearsome Inn* [tr. with E. Shub] 1970; *A Friend of Kafka and Other Stories* (s) 1970; *Enemies: a Love Story* [tr. A. Shevrin, E. Shub] 1972; *A Crown of Feathers and Other Stories* (s) 1973; *The Mirror and Other Stories* (s) 1975; *Passions and Other Stories* (s) 1975; *Naftali the Storyteller and Other Stories* (s) [tr. with R.S. Finkel, J. Singer] 1976; *Shosha* [tr. J. Singer] 1978; *Old Love* 1979; *Reaches of Heaven* 1980; *The Golem* 1982; *The Penitent* 1983; *The Image and Other Stories* (s) 1985; *The Death of Methuselah and Other Stories* (s) 1988; *The King of the Fields* 1989; *Scum* [tr. R. Schwartz] 1991; *Certificate* [tr. L. Wolf] 1992

For Children
Translated with Elizabeth Shub: *Mazel and Schlimazel* 1966; *Zlateh the Goat and Other Stories* (s) 1966; *The Fearsome Inn* 1967; *When Schlemiel Went to Warsaw and Other Stories* (s) 1968; *Elijah the Slave* 1970; *Joseph and Koza* 1970; *Alone in the Wild Forest* 1971; *The Topsy-Turvy Emperor of China* 1971; *The Wicked City* 1972; *The Fools of Chelm and Their History* 1973; *Why Noah Chose the Dove* 1974; *A Tale of Three Wishes* 1975; *Naftali the Storyteller and His Horse, Sus and Other Stories* (s) 1976; *The Power of Light* (s) 1980

Non-Fiction
In My Father's Court (a) [tr. C. Kleinerman-Goldstein] 1966; *A Day of Pleasure: Stories of a Young Boy Growing Up in Warsaw* (a) [tr. with E. Shub] 1969; *A Little Boy in Search of God: Mysticism in a Personal Light* (a) 1976; *A Young Man in Search of Love* (a) 1978; *Tully Filmus; Selected Drawings* (c) [ed. Filmus] 1971; *A Certain Bridge; Isaac Bashevis Singer on Literature and Life* 1979; *Nobel Lecture* 1979

Plays
The Mirror 1973 [pub. 1975]; *Yentl* [with L. Napolin] 1974 [pub. 1978]; *Schlemiel the First* 1974; *Teibele and Her Demon* [with E. Friedman] 1978 [pub. 1984]

Translations
Knut Hamsen *Pan* 1928; Erich Maria Remarque: *All Quiet on the Western Front* 1930; *The Road Back* 1930; Thomas Mann *The Magic Mountain* 1930; Leon S. Glaser *From Moscow to Jerusalem* 1938

Edited
The Hasidim: Paintings, Drawings and Etchings [with I. Moscowitz] 1973

Khushwant SINGH 1915–

Singh was born in Hadali, India (now in Pakistan), the son of a Sikh builder, and educated at the Modern School, New Delhi, St Stephen's College, New Delhi, and Government College, Lahore, where he took his BA in 1934. He travelled to England where he read law at King's College, London, graduating in 1938, the same year he was called to the Bar at the Inner Temple. Returning to India, he was married the following year (the couple have two children), and was a practising lawyer at the High Court, Lahore, until 1947. He was press attaché for the Indian Government in Ontario and London from 1947 to 1952, on the staff of the Department of Mass Communications at UNESCO from 1954 to 1956, and editor of *Yejna*, a publication of the Indian Government's Planning Commission, from 1956 to 1958.

Since 1958 he has been a freelance writer, and an occasional visiting lecturer at Oxford, the University of Rochester, New York, Princeton, the University of Hawaii, Honolulu, and Swarthmore College, Pennsylvania. One of India's foremost historians and journalists, he has been editor of the *Illustrated Weekly of India*, the *National Herald, New Delhi* and the *Hindustan Times*. He was a member of the upper house of the Indian Parliament from 1980 to 1986. Awarded the Padma Bhushan in 1974 by the President of India, he returned the decoration in 1984 in protest at the siege of the Golden Temple in Amritsar.

Singh's first novel, *Train to Pakistan*, is an impassioned protest against the 1947 partition of India and Pakistan, unflinching in its description of the turmoil and violence unleashed by this act. Despite a similarly sardonic tone of disillusionment, his second novel and short stories rarely transcend the level of journalism. After a long break from fiction, Singh published *Delhi*, the result of twenty years' work, in 1990. A compendious, bawdy and cynically funny tale narrated by a contemporary Delhi journalist, the novel seeks to embody the history of the city from its earliest days. 'History provided me with the skeleton,' Singh wrote in a brief foreword. 'I covered it with flesh and injected blood and a lot of seminal fluid into it.'

Fiction
The Mark of Vishnu and Other Stories (s) 1950; *Mano Majra* [UK *Train to Pakistan*] 1956; *The Voice of God and Other Stories* (s) 1957; *I Shall Not Hear the Nightingale* 1959; *A Bride for the Sahib and Other Stories* (s) 1967; *Black Jasmine* (s) 1971; *Collected Short Stories* (s) 1989; *Delhi* 1990

Non-Fiction
The Sikhs 1953; *The Unending Trail* 1957; *The Sikhs Today: Their Religion, History, Culture, Customs, and Way of Life* 1959; *The Fall of the Kingdom of the Punjab* 1962; *Ranjit Singh: Maharajah of the Punjab 1780–1839* 1962; *A History of the Sikhs* 2 vols 1963, 1966; *Not Wanted in Pakistan* 1965; *Ghadar, 1915: India's First Armed Revolution* [with S. Singh] 1966; *Homage to Guru Gobind Singh* [with S.V. Singh] 1966; *Shri Ram: a Biography* [with A. Joshi] 1968; *Khushwant Singh's India: a Mirror for Its Monsters and Monstrosities* 1969; *Khushwant Singh's View of India: Lectures on India's People, Religions, History, and Contemporary Affairs* [ed. R. Singh] 1974; *Khushwant Singh on War and Peace in India, Pakistan, and Bangladesh* [ed. M. Singh] 1976; *Good People, Bad People* [ed. M. Singh] 1977; *Khushwant Singh's India without Humbug* [ed. R. Singh] 1977; *Around the World with Khushwant Singh* [ed. R. Singh] 1978; *Indira Gandhi Returns* 1979; *Editor's Page* [ed. R. Singh] 1981; *We Indians* 1982; *Delhi: a Portrait* 1983; *Punjab Tragedy* [with K. Nayar] 1984

Edited
A Note on G.V. Desani's 'All about H. Hatterr' and 'Hali' [with P. Russell] 1952; *Land of the Five Rivers: Stories of the Punjab* (s) [with J. Thadani] 1965; Sita Ram Kohli *Sunset of the Sikh Empire* 1967; *I Believe* 1971; *Love and Friendship* 1974; *Stories from India* (s) [with Q. Hyder] 1974; *Gurus, Godmen, and Good People* 1975

Translations
Jupji: the Sikh Morning Prayer 1959; Mohammed Ruswa *Umrao Jan Ada: Courtesan of Lucknow* [with M.A. Husain] 1961; Amrita Pritam *The Skeleton and Other Writings* 1964; Rajinder Singh Bedi *I Take This Woman* 1967; *Hymns of Guru Nanak* 1969; Satindra Singh *Dreams in Debris: a Collection of Punjabi Short Stories* (s) 1972; *Sacred Writings of the Sikhs* [with others] 1974; K.S. Duggal *Come Back, My Master and Other Stories* (s) [with others] 1978; *Shikwa and Jawab-i-Shikwa/Complaint and Answer: Iqbal's Dialogie with Allah* 1981

John SINJOHN

see John GALSWORTHY

C(harles) H(ubert) SISSON 1914–

Sisson comes of two established English traditions: small Westmorland manufacturers on his father's side; a West Country agricultural background on his mother's. But, his father's fortunes having declined to being an optician and watchmaker in Bristol, he was born and brought up in a working-class area of the city, proceeding to its university with a scholarship from state schools. He took a first in English and philosophy in 1934, and then spent two years in Germany and France on postgraduate scholarships. He completed no thesis, but in Paris came under the influence of Charies Maurras of *Action française*, which laid the foundation for the Johnsonian, Anglican and splenetic conservatism that has animated his work.

On his return to England, he entered the civil service, in which he rose inexorably within the Ministry of Labour, at first dealing with unemployment benefit at a routine level but retiring in 1973, slightly early, as director of occupational safety and health at the ministry's successor, the Department of Employment. In 1937, the year he entered the service, he married, and subsequently had two daughters. His career was interrupted by wartime service in the Intelligence Corps, at first briefly in Ireland, and later in Bengal and the North-West Province of India. Work took precedence for him over literary ambition, and in his early manhood he abjured poetry, confining his writing at first to social and political articles appearing in the *New English Weekly* under the editorship of Philip Mairet from 1937 to 1949.

He is, in fact, unusual as a writer in that, although his primary reputation is as a poet and critic, he did not publish his first volume of verse until he was forty-seven, and his first three books fall into neither genre: they were the novel *An Asiatic Romance* (1953); a volume of translations from Heine; and a comparative study of European civil service administrations. His first volume of poetry was published in 1959, but it was *In the Trojan Ditch* (1974), published shortly after his retirement to devote himself to writing, which consolidated his reputation as a major poet. One reviewer described his first collection as having 'the inviting yet menacing ripeness of a bruised pear', and certainly sombre themes such as death, ageing, lust and self-contempt play a major role in his work, where plain language does little to conceal deeply felt ambiguities and a sense of the irrelevance of the self. His hatred of the modern world finds more combative form in his three volumes of essays. His second novel, *Christopher Homm*, (published in 1965 but written much earlier, in 1952), follows the life of an ordinary failure from death backwards to birth, and is second only to the poetry in Sisson's creative work. He has made translations from several European languages (including Latin), while his criticism reached its maturity with *English Poetry 1900–1950: an Assessment*. In retirement, he has lived at Langport, a small town in Somerset, and a solidly maintained flow of publications has included his collected poems, a volume of autobiography, and a book of new verse, *What and Who*, in 1994.

Poetry
Poems 1959; *Twenty-One Poems* 1960; *The London Zoo* 1961; *Numbers* 1965; *The Disincarnation* 1967; *The Metamorphoses* 1968; *Roman Poems* 1968; *In the Trojan Ditch* 1974; *The Corridor* 1975; *Anchises* 1976; *Exactions* 1980; *Collected Poems 1943–1983* 1984; *God Bless Karl Marx* 1987; *Nine Sonnets* 1991; *Pattern* 1993; *What and Who* 1994

Fiction
An Asiatic Romance 1953; *Christopher Homm* 1965

Translations
Versions and Perversions of Heine 1955; *Catullus* 1966; Horace *The Poetic Art* 1975; Lucretius *De Rerum Natura: the Poem on Nature* 1976; *Some Tales of La Fontaine* 1979; Dante *The Divine Comedy* 1980; *The Song of Roland* 1983; Joachim Du Bellay *The Regrets* 1984; Virgil *The Aeneid* 1986; Racine: *Athaliah* 1987, *Britannicus* 1987, *Phaedra* 1987

Non-Fiction
The Spirit of British Administration and Some European Comparisons 1959; *Art and Action* 1965; *Essays* 1967; *English Poetry 1900–1950: an Assessment* 1971; *The Case of Walter Bagehot* 1972; *David Hume* 1976; *The Avoidance of Literature: Collected Essays* 1978; *Anglican Essays* 1983; *The Poet and the Translator* 1985; *On the Look-Out* (a) 1989; *In Two Minds: Guesses at Other Writers* 1990; *Is There a Church of England?* 1993

Edited
The English Sermon 1659–1750 1976; *Selected Poems of Jonathan Swift* 1976; David Wright *A South African Album* 1976; Hardy *Jude the Obscure* 1978; *Autobiographical and Other Papers of Philip Mairet* 1981; Ford Maddox Brown *The Rash Art* 1982; *Selected Poems of Christina Rossetti* 1984

Edith (Louisa) SITWELL 1887–1964

(Francis) Osbert (Sacheverell) SITWELL 1892–1969

Sacheverell SITWELL 1897–1988

The Sitwells made a lively contribution to the English artistic scene of the 1920s, where they first set out to shock the middle classes, and to rout the philistine. They were influenced by Sir George and Lady Ida, their mismatched parents, and their lonely Derbyshire country house, Renishaw Hall, helped to shape their imaginative world.

Edith left home for London in 1913, where she struggled to support herself by writing poetry. Her first collection, *The Mother* (1915), was critically ignored, but in 1916 she was noticed as editor of *Wheels*, a controversial poetry anthology. Published until 1921, it opposed Georgian poetry and, although generally unremarkable, it published a posthumous selection of poems by **Wilfred Owen**. Osbert's less sophisticated pacifist verse, informed by his experiences in the trenches, also appeared there.

Of the trio, the most precious was Sacheverell, whose first poems, *The People's Palace* (1918), attracted the praise of **Aldous Huxley**. Like his brother, Sacheverell was educated at Eton where he first wrote poetry, and served with the Grenadier Guards. After a few terms at Christ Church, Oxford, he returned with the fledgling composer William Walton, whose talents he promoted, to live in Osbert's house in Chelsea.

In 1924, after publishing several collections of verse, Osbert published a volume of short stories, *Triple Fugue*, which suggested that his gift for acerbic observation was better suited to prose. He instituted many pranks which annoyed the respectable, but also helped to liberate the arts from the stifling atmosphere of post-Edwardianism, and kept the Sitwells' name well publicised.

Meanwhile, Edith published experimental poetry, using complex rhythms, and juxtaposing senses, where the language of the visual defined the audile and vice versa. These culminated in the entertainment *Façade*, where abstract poems were declaimed through a megaphone against a background of music by Walton. First performed publicly in 1923, *Façade* was subsequently parodied by **Noël Coward**, who dismissed the Sitwells as 'two wiseacres and a cow'.

In 1924, Sacheverell's pioneering *Southern Baroque Art*, essays on art, music and architecture of seventeenth- and eighteenth-century Spain and Italy, was the first of many books which drew attention to then neglected subjects. Osbert followed with *Discursions on Travel, Art and Life*, but it was with a novel, *Before the Bombardment*, that he found his *métier*. Set in Scarborough, where Edith and Sacheverell were born, it vividly evokes the decay of upper and middle Edwardian society before the First World War.

In the early 1920s, the collective term 'the Sitwells' was useful publicity, but Edith disliked the image of them as an 'aggregate Indian god'. Attention-seeking literary fights in the public arena were anathema to the self-effacing Sacheverell, who preferred to promote the arts rather than himself. He escaped the circus in 1925 via a happy marriage to Georgia Doble, a Canadian beauty. This effectively broke up the trinity, forcing Edith and Osbert to confront their private lives. Osbert was a discreet homosexual who vented his bad temper and frustration by smashing cheap crockery on his doorstep. This ceased in 1930, when he installed David Horner in his life, unexpectedly letting slip, in Huxley's words, the 'heavy waxen mask' of his Hanoverian features. For Edith there was no solution. Taunted in the past by her parents about her looks, and their assertion that she would never attract a man, she deliberately emphasised her unusual appearance with heavy brocades, and rings which increased in size with her age. **Elizabeth Bowen** caught the impression Edith made by likening her to 'a high altar on the move'. What Lytton Strachey described as her 'anteater's nose' is apparent in her portrait by **Wyndham Lewis**. In 1924, the year of her nostalgic poem *The Sleeping Beauty*, Edith moved to Paris, where she fell hopelessly and painfully in love with the homosexual Russian artist Pavel Tchelitchew. Her 1929 volume of verse, *Gold Coast Customs*, satirised snobbish London society, after which she spent the next decade concentrating on prose works in order to make money. Characterised by her wit, these included studies of Alexander Pope (a book which helped to revive his reputation) and, appropriately, *The English Eccentrics*. Returning to England in 1939, she wrote poetry which expressed her anguish at the Second World War. *Street Songs*, containing the famous 'Still Falls the Rain', brought a new public, responsive to her affirmation of religious faith.

Osbert had steadily published poetry, short stories and essays, notably *Escape with Me!*, an account of travels in the East. Marooned in England during the war, he embarked upon a four-volume autobiography, *Left Hand, Right Hand!*, in which he disclosed events rather than emotion, but created a memorable comic character in his exaggerated portrait of his father, whose eccentric foibles he embroidered, much to his brother's displeasure and sister's delight. The memoirs were verbose, but Osbert's evocation of a vanished age proved popular in a drab post-war Britain, and consolidated his reputation.

Sacheverell had abandoned poetry since the publication of his *Collected Poems* in 1936; numerous prose studies of composers such as Liszt and Scarlatti followed, while *Poltergeists* disclosed a taste for the curious and the macabre. The horror of the Second World War, and especially a period in the Home Guard in Northamptonshire, influenced him fully to realise his imaginative and poetic gifts in a prose fantasy, *Splendours and Miseries*. This stressed his belief in mankind in a destructive age through the restorative power of the arts, rather than through any religious creed.

Edith became a Roman Catholic in 1955, a step which did little to dilute her *folie de grandeur*, which grew with her reputation. Already the *grande dame* of English letters, she was created a Dame Commander of the Order of the British Empire in 1954. Bolstered by belated literary and academic recognition, she maintained verbal attacks on old enemies, notably F.R. Leavis and **Geoffrey Grigson**, but was a tireless supporter of those she admired, such as **Dylan Thomas**.

A victim of Parkinson's disease from the early 1950s, Osbert wrote less as his health progressively deteriorated. A distraught Edith slowly succumbed to alcohol, but her last poems, *The Outcasts*, were marked by a new simplicity of language. She died in 1964, the same year that Osbert finally left David Horner, to live permanently in Montegufoni, the family home near Florence. Edith's posthumous autobiography, *Taken Care Of*, revealed little of her personal life, and nothing of the severed family loyalties.

Following Osbert's death from heart failure in 1969, a copy of *Before the Bombardment*, his favourite of his books, was placed in the urn containing his ashes. Sacheverell was emotionally paralysed for a time by his brother's betrayal in not leaving Montegufoni to him, and only later could he write of this, in *For Want of the Golden City*. Free of the overbearing power of his siblings, he was sustained by his two sons, and a happy marriage, which ended with the death of his wife in 1980. He spent his final years writing and publishing his verse, fulfilling his belief that he was, after all, born to be a poet.

Edith Sitwell

Poetry

The Mother 1915; *Twentieth Century Harlequinade* [with O. Sitwell] 1916; *Clowns'*

Houses 1918; *The Wooden Pegasus* 1920; *Façade* 1922; *Bucolic Comedies* 1923; *The Sleeping Beauty* 1924; *Poor Young People* [with O. and S. Sitwell] 1925; *Troy Park* 1925; *Elegy on Dead Fashion* 1926; *Poem for a Christmas Card* 1926; *Rustic Elegies* 1927; *Popular Song* 1928; *Five Poems* 1928; *Gold Coast Customs* 1929; *Collected Poems* 1930; *Epithalamium* 1931; *In Spring* 1931; *Jane Barston* 1931; *Five Variations on a Theme* 1933; *Selected Poems* 1936; *Poems Old and New* 1940; *Street Songs* 1942; *Green Song* 1944; *The Song of the Cold* 1945; *The Weeping Babe* 1945; *The Shadow of Cain* 1947; *The Canticle of the Rose* 1949; *Poor Men's Music* 1950; *Selected Poems* 1952; *Gardeners and Astronomers* 1953; *Collected Poems* 1957; *The Pocket Poets* 1960; *Music and Ceremonies* 1962; *The Outcasts* 1962; *Selected Poems of Edith Sitwell* 1965

Non-Fiction
Poetry and Criticism 1925; *Alexander Pope* 1930; *Bath* 1932; *The English Eccentrics* 1933; *Aspects of Modern Poetry* 1934; *Victoria of England* 1936; *Trio* [with O. and S. Sitwell] 1938; *English Women* 1942; *A Poet's Notebook* 1943; *Fanfare for Elizabeth* 1946; *A Notebook on William Shakespeare* 1948; *The Queens and the Hive* 1962; *Taken Care Of* (a) 1965; *Selected Letters* [ed. J. Lehmann, D. Parker] 1970

Fiction
I Live under a Black Sun 1937

Plays
The Last Party [pub. in *Twelve Modern Plays* 1938]

Edited
Wheels 1916–21; *Look! The Sun* 1941; *The Atlantic Book of British and American Poetry* 1958; *Swinburne: a Selection* 1960

For Children
Children's Tales from the Russian Ballet 1920

Collections
Fire of the Mind [ed. A. Harper, E. Salter] 1976

Biography
Façades by John Pearson 1978; *Edith Sitwell* by Geoffrey Elborn 1981; *Edith Sitwell: a Unicorn Among Lions* by Victoria Glendinning 1981

Osbert Sitwell

Fiction
Triple Fugue (s) 1924; *Before the Bombardment* 1926; *The Man Who Lost Himself* 1929; *Dumb Animal* (s) 1930; *Miracle on Sinai* 1933; *Those Were the Days* 1938; *Open the Door!* (s) 1941; *A Place of One's Own* (s) 1941; *The True Story of Dick Whittington* (s) 1946; *Alive – Alive oh!* (s) 1947; *Death of a God* (s) 1949; *Collected Stories* (s) 1953

Non-Fiction
Who Killed Cock-Robin? 1921; *Discursions on Travel, Art and Life* 1925; *The People's Album of London Statues* 1928; *Dickens* 1932; *Winters of Content* 1932; *Brighton* [with M. Barton] 1935; *Penny Foolish* 1935; *Trio* [with E. and S. Sitwell] 1938; *Escape with Me!* 1939; *Left Hand, Right Hand!* (a): *Left Hand, Right Hand!* 1944, *The Scarlet Tree* 1946, *Great Morning* 1947, *Laughter in the Next Room* 1948; *A Letter to My Son* 1944; *Sing High! Sing Low!* 1944; *The Novels of George Meredith* 1947; *Noble Essences or Courteous Revelations* (a) 1950; *Tales My Father Taught Me* (a) 1962; *Pound Wise* 1963; *Queen Mary and Others* 1973

Poetry
Twentieth Century Harlequinade [with E. Sitwell] 1916; *The Winstonburg Line* 1919; *Argonaut and Juggernaut* 1919; *At the House of Mrs Kinsfoot* 1921; *Out of the Flame* 1923; *Poor Young People* [with E. and S. Sitwell] 1925; *Winter the Huntsman* 1927; *England Reclaimed: a Book of Eclogues* 1927; *Miss Mew* 1929; *Collected Satires and Poems* 1931; *Three-Quarter Length Portrait of Michael Arlen* 1931; *Three-Quarter Length Portrait of the Viscountess Wimborne* 1931; *Mrs Kimber* 1937; *Selected Poems, Old and New* 1943; *Four Songs of the Italian Earth* 1948; *Demos the Emperor* 1949; *England Reclaimed and Other Poems* 3 vols 1949–58; *Poems about People* 1965

Plays
All at Sea [with S. Sitwell] 1927; *Gentle Caesar* [with R.J. Minney] 1943

For Children
Fee Fi Fo Fum! (s) 1959

Edited
Sober Truth [with M. Barton] 1930; *Victoriana* [with M. Barton] 1931; *Two Generations* 1940; *A Free House! The Writings of W.R. Sickert* 1947

Sacheverell Sitwell

Biography
Façades by John Pearson 1978

Poetry
The People's Palace 1918; *Dr Donne and Gargantua (First Canto)* 1921; *The Hundred and One Harlequins* 1922; *Dr Donne and Gargantua (Canto the Second)* 1923; *The Parrot* 1923; *The Thirteenth Caesar* 1924; *Poor Young People* [with E. and O. Sitwell] 1925; *Exalt the Eglantine* 1926; *The Cyder Feast* 1927; *Two Poems, Ten Songs* 1929; *Dr Donne & Gargantua (First Six Cantos)* 1930; *Canons of Giant Art* 1933; *Collected Poems* 1936; *Selected Poems* 1948; *The Archipelago of Daffodils* 1972; *Auricula Theatre* 1972; *A Charivari of Parrots* 1972; *Flowering Cactus* 1972; *The House of the Presbyter* 1972; *Lily Poems* 1972; *Lyra Varia* 1972; *Nigritian* 1972; *Rosario*

d'arabeschi 1972; *Ruralia* 1972; *The Strawberry Feast* 1972; *To Henry Woodward* 1972; *A Triptych of Poems* 1972; *Tropicalia* 1972; *Variations upon Old Names of Hyacinths* 1972; *Baraka and Dionysia* 1973; *Battles of the Centaurs* 1973; *Les Troyens* 1973; *Pastoral and Landscape with the Giant Orion* 1973; *To E.S.* 1973; *Twelve Summer Poems of 1962* 1973; *Two Poems* 1973; *An Indian Summer* 1974; *Badinerie* 1974; *A Look at Sowerby's English Mushrooms and Fungi* 1974; *L'Amour au théâtre italien* 1975; *Nymphis et Fontibus and Nymphaeum* 1975; *A Pair of Entractes for August Evenings* 1975; *Temple of Segesta* 1975; *Brother and Sister* 1976; *Credo* 1976; *Placebo* 1976; *A Second Triptych of Poems* 1976; *Serenade to a Sister* 1976; *Two Themes* 1976; *Diptycha Musica* 1977; *Dodecameron* 1977; *Little Italy in London* 1977; *Nine Ballads* 1977; *The Octogenarian* 1977; *Looking for the Gods of Light* 1978; *The Rose Pink Chapel* 1978; *Scherzo di capriccio* 1978; *Scherzo di fantasia* 1978; *Catalysts in Collusion* 1979; *Op. cit. et cetera* 1979; *Allotment or Assignment* 1980; *For a Floral Inebriation* 1980; *Nocturnae sivlani potenti* 1980; *Pomona lactodorum* 1980; *Prende la mia chitarra* 1981; *An Indian Summer: 100 Recent Poems* 1982

Non-Fiction

Southern Baroque Art 1942; *All Summer in a Day* (a) 1926; *German Baroque Art* 1927; *A Book of Towers* 1928; *The Gothick North* 3 vols 1929–30; *Beckford and Beckfordism* 1930; *Spanish Baroque Art* 1931; *Mozart* 1932; *Liszt* 1934; *Touching the Orient* 1934; *A Background for Domenico Scarlatti* 1935; *Conversation Pieces* 1936; *Dance of the Quick and the Dead* 1936; *Narrative Pictures* 1937; *La Vie Parisienne* 1937; *Edinburgh* 1938; *German Baroque Sculpture* 1938; *The Romantic Ballet in Lithographs of the Time* 1938; *Roumanian Journey* 1938; *Trio* [with E. and O. Sitwell] 1938; *Mauretania* 1940; *Poltergeists* 1940; *Sacred & Profane Love* 1940; *Valse des fleurs* 1940; *Primitive Scenes and Festivals* 1942; *Splendours and Miseries* 1943; *British Architects & Craftsmen* 1945; *The Hunters and the Hunted* 1947; *Morning, Noon and Night in London* 1948; *The Netherlands* 1948; *Theatrical Figures in Porcelain* 1949; *Spain* 1950; *Cupid and the Jacaranda* 1952; *Fine Bird Books* 1953; *Truffle Hunt* 1953; *Portugal and Madeira* 1954; *Old Garden Roses* 1955; *Denmark* 1956; *Great Flower Books* 1956; *Arabesque and Honeycomb* 1957; *Malta* 1958; *Bridge of the Brocade Sash* 1959; *Journey to the Ends of Time* 1959; *Golden Wall and Mirador* 1961; *The Red Chapels of Banteai Srei* 1962; *Monks, Nuns and Monasteries* 1965; *Southern Baroque Revisited* 1967; *Gothick Europe* 1969; *For Want of the Golden City* 1973; *A Notebook on My New Poems* 1974; *Notebook on Twenty Canons of Giant Art* 1975; *A Note for Bibliophiles* 1976; *A Retrospect of Poems (1972–1979)* 1979; *Sacheverell Sitwell's England* [ed. M. Raeburn] 1986

Fiction

Far from My Home (s) 1931

Plays

All at Sea [woth O. Sitwell] 1927

Collections

Selected Works 1955

Biography

Façades by John Pearson 1978; *Sacheverell Sitwell* by Sarah Bradford 1993

Elizabeth SMART 1913–1986

Born in Ottawa, Smart was the daughter of a wealthy barrister. Her parents entertained the leading members of Canadian society (they were also friendly with Sir Stafford Cripps), and Smart was educated at a series of private schools. She completed her education during the 1930s by making twenty-two trips between London and Canada, as well as visits to Germany, Sweden, Norway, the USA and Mexico, and a round-the-world voyage as a private companion. In London she studied at King's College for a year, and she spent some months in 1938–9 working on the Ottawa *Journal.*

During the 1930s she had an affair with the painter Jean Varda (often known as Yanko, and a friend of **Henry Miller** and **Anaïs Nin**) but, on reading the poems of the English poet **George Barker** in a bookshop, she conceived a passion for him and began telling everyone she wanted to marry him. During the early years of the war she was in North America and, having been given Barker's address by **Lawrence Durrell**, she wrote to him asking for a manuscript; she received in return a plea from Barker to rescue him and his wife from Japan, where they had been sent by the British Council. She paid for their passage to California in July 1940 and within a month had become Barker's lover. He did not, however, leave his wife and, pregnant with her first child by him in 1941, Smart turned their relationship into an ultimately tragic novel, published in 1945 as *By Grand Central Station I Sat Down and Wept*. Smart's parents (who had earlier blocked Barker's entry into Canada on grounds of moral turpitude) persuaded the Canadian prime minister to ban the book, and it did not appear there until 1975. In Britain it had a largely underground reputation until republished, much revised, in 1966. It has been described by **Brigid Brophy** as one of the 'half-dozen masterpieces of poetic prose in the world', and certainly, with its symphonic structure, web of allusions and lyrical outpourings, it must rank among the century's most innovative productions.

Wishing to leave Barker, Smart returned to England in 1942 and, although she saw him again, eventually having four children by him, they never married. Her later life was passed mainly in England where she worked for advertising agencies and as a journalist, being literary editor of *Queen* in the early 1960s before retiring to a cottage at Bungay in Suffolk in 1966. She broke a thirty-two-year publishing silence with her volume of poems, *A Bonus*, in 1977, and published her second novel, *The Assumption of the Rogues and Rascals*, in 1978. Her short list of books was completed by another of poems, one of mixed poetry and prose, and a cookery book. In her last years she spent some periods back in Canada as a writer-in-residence. She died suddenly, apparently of a heart-attack, while visiting London. Her early writings and two volumes of diaries have been published since her death. Her son, Sebastian Barker, is a writer and chairman of the Poetry Society in London.

Fiction

By Grand Central Station I Sat Down and Wept 1945; *The Assumption of the Rogues and Rascals* 1978; *Juvenilia: Early Writings of Elizabeth Smart* [ed. A. Van Wart] 1987

Poetry

A Bonus 1977; *The Collected Poems of Elizabeth Smart* 1992

Non-Fiction

Cooking the French Way [with A. Ryan] 1960 [rev. 1966]; *Necessary Secrets: the Journals of Elizabeth Smart 1933–1941* [ed. by A. Van Wart] 1986; *Autobiographies* (a) [ed. C. Burridge] 1987; *Elizabeth's Garden: Elizabeth Smart on the Art of Gardening* [ed. A. Van Wart] 1989; *On the Side of the Angels: the Second Volume of the Journals of Elizabeth Smart* [ed. A. Van Wart] 1994

Collections

In the Meantime 1985

Biography

By Heart by Rosemary Sullivan 1991

Jane SMILEY 1952–

Smiley was born in Los Angeles, but grew up in St Louis, Missouri. She read English at Vassar, graduating in 1971, after which she spent a year in England working on an archaeological site. She returned to America and lived in Iowa with her first husband, working in a factory that made soft toys. Her literary career began when she attended the writers' workshop at the University of Iowa and gained a PhD.

She published short stories in journals such as the *Atlantic* and *Mademoiselle* and won the O. Henry Award for 'Lily', a story about a prize-winning poet. This was collected, along with four other stories and the title novella in *The Age of Grief*, a volume that was nominated for the National Book Critics Circle Award. The novella is narrated by a dentist whose relationship with his wife (who is also his partner in the practice) undergoes a crisis. This hazardous passage in their marriage is mirrored by a virulent bout of flu which afflicts them and their three small daughters, and the story details the different kinds of demands made by love.

Her first novel, *Barn Blind*, was published in 1980 and, like much of her work, has a farm setting. It traces the disintegration of a family in Illinois, torn apart by the mother's obsessive passion for her horses. The novel was very well received and heralded a career in which she would become one of the leading chroniclers of the contemporary rural Midwest. *A Thousand Acres* was another novel about a dysfunctional family, a modern reworking of *King Lear*, set amongst the farming community in a richly realised Iowa. It won both a Pulitzer Prize and the National Book Critics Circle Award.

Her other novels include *The Greenlanders*, an epic set in the country of the title during the fourteenth century, and the unusual campus novel, *Moo*. Unlike previous novels, *Moo* is a broad comedy, a satire as much about agribusiness as academe, and it fields a large cast of characters. Faced with swingeing cuts, a Midwestern university becomes involved in several immoral schemes to raise funds, including the destruction of South American rain forests in order to mine for gold and an experiment to raise a vast pig, who becomes the novel's central symbol.

Smiley lives in Ames, Iowa, with her third husband and her three children, and teaches at Iowa State University.

Fiction

Barn Blind 1980; *At Paradise Gate* 1981; *Duplicate Keys* 1984; *The Age of Grief* (s) 1987; *The Greenlanders* 1988; *Ordinary Love and Goodwill* 1989; *A Thousand Acres* 1992; *Moo* 1995

Dodie (i.e. Dorothy Gladys) SMITH 1896–1990

Smith was born in Whitefield, Lancashire, but moved to Old Trafford, Manchester, after her mother had been widowed young. She was educated at Walley Range High School and St Paul's Girls' School in London before going to the Royal Academy of Dramatic Arts in 1914. She had a brief and unsuccessful career as an actress, and made a slow start as a writer. Using the pseudonym Charles Henry Percy, she sold a screenplay for £3. 10s., and this was made into the film *Schoolgirl Rebels* (1915), a mild comedy about school inspectors. In order to earn a living, she became a buyer at Heal's, the famous London furniture store, and it was here that she met her future husband, Alec Beesley, whom she married 'after seven years of all but legal marriage' in 1939.

Her early plays were written under the pseudonym C.L. Anthony, and the second of them, *Autumn Crocus*, made her reputation in 1931. In spite of a nerve-racking first night at the Lyric Theatre, interrupted by Bank Holiday rowdies in the audience, it received very good notices. A romance about an English schoolmistress and the proprietor of a Tyrolean *Gasthaus*, it was made into a film starring Ivor Novello and Fay Compton in 1934. By the outbreak of the Second World War, she had written five more plays and had 1,800 West End performances to her credit. *Dear Octopus* – which, like *Bonnet over the Windmill*, she co-directed – was described as 'the family play of the decade'. A well-observed comedy about an upper-class family celebrating a Golden Wedding, it remains her best-known play.

In January 1939, she and Beesley, who was now acting as her manager, travelled to America, finally settling in California, where she worked briefly as a screenwriter, while Beesley (the Second World War having broken out) worked as a counsellor with conscientious objectors. Through **John van Druten** they met **Christopher Isherwood**, with whom they had a close but troubled friendship. Isherwood was at that period living in a Vedanta monastery, but spent most weekends with the Beesleys, who were worried that he was dissipating his talents. Smith, a Christian Scientist, disapproved of both Isherwood's religion and his hypochondria, and she recorded this and other friendships in a voluminous diary, which shows her to be a sharp judge of character. For a brief period, Isherwood lived in the chauffeur's apartment of their house on Pacific Coast Highway, and it was the Beesleys who suggested to van Druten that he should adapt Isherwood's Berlin stories for the play *I Am a Camera* (1951). Although very dissimilar writers, they discussed each other's work in detail, and the Beesleys are the dedicatees of Isherwood's *The World in the Evening* (1954).

Smith had embarked on a new career as a novelist, and her first novel was *I Capture the Castle* (1948), the story of an eccentric and impoverished household, narrated in diary-form by one of the daughters. Described by van Druten (who had expected something more intellectual, since Smith was a great admirer of Proust and **Henry James**) as 'a lovely book to read over hot-buttered toast', it sold over one million copies, and remains the best known of her books for adults. Smith was desperately homesick for England, and relieved to return there to live in Finchingfield, Essex, in 1953. Apart from Beesley, the passion of her life was her succession of Dalmation dogs (whose lives were also recorded in minute and extensive detail in her diaries). These pets inspired her children's book *The Hundred and One Dalmations*, an exciting and rather terrifying story about puppies being kidnapped in order to make coats. In the fur-loving Cruella de Vil, Smith created one of the most alarming villainesses in children's fiction. The initial success of the novel was significantly enhanced when it was adapted for a feature-length cartoon film by the Walt Disney studio in 1961. It proved to be one of the company's most profitable films ever, generating a large income for Smith. *The Starlight Barking* is a sequel, and she also wrote one other children's book, *The Midnight Kittens*.

Although her plays had fallen out of fashion in the post-war years, Smith continued to write for the theatre, adapting Henry James's *The Reverberator* and her own *I Capture the Castle*. She had a late success with her four-volume autobiography, the fourth volume of which, *Look Back with Gratitude*, was published when she was eighty-nine. A fifth volume remained unfinished at her death.

Plays

Schoolgirl Rebels (f) [as 'Charles Henry Percy'] 1915; As Dodie Smith: *Call It a Day* 1935 [pub. 1936]; *Bonnet Over the Windmill* 1937; *Dear Octopus* 1938; *Lovers and Friends* 1943 [pub. 1947]; *The Uninvited* (f) [with F. Partos] 1944; *Darling, How Could You!* (f) [with L. Samuels] 1951; *Letter from Paris* [from H. James's *The Reverberator*] 1952 [pub. 1954]; *I Captured the Castle* [from her novel] 1954 [pub. 1955]; *These People, Those Books* 1958; *Amateur Means Lover* 1961 [pub. 1962]

As C.L. Anthony: *British Talent* 1923; *Autumn Crocus* 1931; *Service* 1932; *Touch Wood* 1934

Fiction

I Capture the Castle 1948; *The New Moon with the Old* 1963; *The Town in Bloom* 1965; *It Ends with Revelations* 1967; *A Tale of Two Families* 1970; *The Girl from the Candle-Lit Bath* 1978

For Children

The Hundred and One Dalmatians 1957; *The Starlight Barking* 1967; *The Midnight Kittens* 1978

Non-Fiction

Look Back with Love (a) 1974; *Look Back with Mixed Feelings* (a) 1978; *Look Back with Astonishment* (a) 1979; *Look Back with Gratitude* (a) 1985

Iain Crichton SMITH 1928–

Smith, whose Gaelic name is Iain Mac A'Gobhainn, was born and brought up in a township of 'thirty or forty houses' on the Isle of Lewis, one of three children; his father died when he was young and the family were very poor. He spoke only Gaelic until he was five, attended the village school and then the grammar school at Stornoway, which, with its equivalent at Portree on Skye, served all the Western Highlands and Islands. Before going to Aberdeen University in 1945, he had never left the island, and claims never to have travelled on a train until he was seventeen. He graduated in English in 1949, and did his national service with the Army Education Corps, reaching the rank of sergeant. In 1952 he began teaching at a secondary school at Clydebank, before moving in 1955 to the Oban High School, where he was to remain for the next twenty-two years. He now lives at Taynuilt in Argyll, and these phases are recalled in his verse autobiography, *A Life*, a sequence of deftly handled poems recalling scenes, moods and people.

Prolific, while maintaining a high standard of excellence, he writes both in English and Gaelic (which he translates himself into English). A recurring theme is isolation. In the title-poem of *Notebooks of Robinson Crusoe* – a long poem employing a variety of styles, including prose – Crusoe personifies loneliness, and his island resembles Lewis: 'Since I myself was brought up on an island (fairly isolated) I could use some of my own experience and material as part of the poem.' Exile and consequent loss of identity is another theme, as in *Exiles*, a Poetry Book Society Choice, translated from the Gaelic. The collection contains the moving poem 'Lolaire', about the tragedy on New Year's morning 1919, when a ship carrying 300 Lewismen returning from the war struck a rock a short distance from Stornoway: 'I imagine an elder of the church speaking as he is confronted with this mind-breaking event.' The elderly feature largely in his work, as in the prose poem 'To an Old Woman': 'You were not a scholar in your day. Many a morning did you gut herring, and your hands were sore with salt, and the keen wind on the edge of your knife, and your fingers frozen with fire.' His own favourite is the long poem *Deer on the High Hills*, inspired by three deer he saw by the roadside while driving from Glasgow to Oban: 'It was icy January and there they were / like debutantes on a smooth ballroom floor.' He has described how the poem seemed to pour out of him; he wrote 300 lines without a pause. Smith has also published a number of plays, novels, collections of stories, a study of the poetry of **Hugh MacDiarmid**, poems for children and translations from the Gaelic. He has doctorates from the universities of Glasgow, Dundee and Aberdeen, won the HM Silver Jubilee Medal in 1952 and 1977, was awarded the OBE in 1980, and many literary prizes.

Poetry

The Long River 1955; *New Poets* [with K. Gershon, C. Levenson] 1959; *Deer on the High Hills* 1960; *Thistles and Roses* 1961; *Boibuill is sanasan reice* 1965; *The Law and the Grace* 1965; *At Helensburgh* 1968; *Three Regional Voices* [with M. Longley, B. Tebb] 1968; *From Bourgeois Land* 1969; *Selected Poems* 1970; *Hamlet in Autumn* 1972; *Love Poems and Elegies* 1972; *Penguin Modern Poets 21* [with G.M. Brown, N. MacCaig] 1972; *Rabhdan is rudan* 1973; *Eadar fealla-dha is Glaschu* 1974; *Notebooks of Robinson Crusoe* 1974; *Orpheus and Other Poems* 1974; *Poems for Donalda* 1974; *The Permanent Island: Gaelic Poems* 1975; *Notebooks of Robinson Crusoe and Other Poems* 1975; *In the Middle* 1977; *Selected Poems 1955–1980* [ed. R. Fulton] 1982; *The Emigrants* 1983; *The Exiles* 1984; *Selected Poems* 1985; *A Life* 1986; *The Island and the Language* 1987; *A'bheinn oir* 1989; *Na speuclairean dubha* 1989; *The Village and Other Poems* 1989; *Na guthan* 1991; *Turas tro shaoghal falamh* 1991; *An dannsa mu dheireadh* 1992; *Collected Poems* 1992; *An Honourable Death* 1992; *Listen to the Voice* 1993

Plays

A 'chiurt 1966; *An coileach* 1966; *A Kind of Play* 1975; *Two by the Sea* 1975; *The Happily Married Couple* 1977; *Goodman and Death Mahoney* (r) 1980

Fiction

An dubh is an gorm (s) 1963; *Consider the Lilies* [US *The Alien Light*] 1968; *The Last Summer* 1969; *Maighstirean is ministearan* (s) 1970; *Survival without Error and Other Stories* (s) 1970; *My Last Duchess* 1971; *An t-adhar Ameireagenach* (s) 1973; *The Black and the Red* 1973; *Goodbye Mr Dixon* 1974; *An t-aonaran* 1976; *The Village* 1976; *The Hermit and Other Stories* (s) 1977; *An End to Autumn* 1978; *On the Island* 1979; *Murdo and Other Stories* 1981; *A Field Full of Folk* 1982; *The Search* 1983; *Mr Trill in Hades and Other Stories* (s) 1984; *The Tenement* 1985; *In the Middle of the Wood* 1987; *The Dream* 1990; *Selected Short Stories* (s) 1990; *Thoughts of Murdo* 1993

For Children

Iain am meag nan reultam 1970; *Norman and the Dolls and Mary and the Wooden Horse* 1976; *Little Red Riding Hood and the Wooden Door* 1977; *River, River* 1978; *Na h-aimmidhean* 1979; *The Dreamer* 1980

Non-Fiction

The Golden Lyric: an Essay on the Poetry of Hugh MacDiarmid 1967; *The House with the Green Shutters* 1987; *Towards the Human* 1987

Edited

Scottish Highland Tales (s) 1982; *Moments in the Glass House* 1987

Translations

Duncan Ben Macintyre *Ben Dorain* 1969; Sorley Maclean *Poems to Eimhir* 1971

Collections

Burn is aran 1960

Stevie (i.e. Florence Margaret) SMITH 1902–1971

Born in Hull, Smith moved at the age of three to a house in the north London suburb of Palmers Green, where she lived for the rest of her life. Her childhood was overshadowed by tubercular peritonitis, and by her mother's ill-health and unhappy marriage: 'Poor Daddy took one look at me and rushed away to sea.' She was brought up in straitened circumstances by her mother and the 'Lion Aunt' to whom she was utterly devoted and with whom she continued to live after her mother's death. She adopted the name Stevie when her riding was compared to that of the famous jockey Steve Donaghue. After school, she enrolled at Mrs Hoster's Secretarial Training College and found a lifetime's employment with the Newnes magazine publishing group, acting as private secretary to Sir Neville Pearson. After her poems were rejected by a publisher, she took his advice and wrote a novel, which she typed during office hours on company stationery. Despite the success of the autobiographical *Novel on Yellow Paper* (1936), she preferred her later novel *The Holiday*, 'so richly melancholy like those hot summer days when it is so full of that calm before autumn'. This melancholy is also a dominant feature of her poetry with its apparently childlike treatment of sombre themes: loss, loneliness and death (particularly suicide). Her poetry, 'based largely on Gregorian chant and hymn tunes', can also be very funny. Her first volume, *A Good Time Was Had by All*, was published in 1937, suitably accompanied (rather than illustrated) by her quirky drawings, which were to be an integral part of her books.

The success of her novel had provided her with an entrée to literary London: her poems and reviews appeared in a variety of magazines and she made friends with a number of writers, notably **Olivia Manning**, whose bridesmaid she was. She provided Manning with a model for the character of Nancy Claypole in *The Doves of Venus* (1955), but her review of the novel, which mildly criticised the hyper-sensitive Manning, brought the friendship to an end.

In the 1950s she found it difficult to place her poems in magazines and became clinically depressed. One result of this was her most famous poem, 'Not Waving but Drowning', written in April 1953; another was a nervous breakdown and her attempted suicide at her office the following July. She recovered, but did not return to Newnes, turning instead to reviewing once more, concentrating particularly on books with Christian themes. She was a troubled agnostic, and some of her pronouncements on Christianity caused controversy, notably her poem 'How do you see?', which prompted an enormous response when printed in the *Guardian* at Whitsun in 1964. By this time, her career had revived: her *Selected Poems* had appeared in 1962 to excellent reviews (notably one by **Philip Larkin**), she became a popular reciter of her own verse, and was honoured by both the Cholmondeley Award and the Queen's Gold Medal for Poetry.

Poetry

A Good Time Was Had by All 1937; *Tender Only to One* 1938; *Mother, What Is Man?* 1942; *Harold's Leap* 1950; *Not Waving but Drowning* 1957; *Selected Poems* 1962; *The Frog Prince* 1966; *Penguin Modern Poets 8* [with E. Brock, G. Hill] 1966; *The Best Beast* 1969; *Poems for Children* 1970; *Scorpion and Other Poems* 1972; *Collected Poems* 1975

Fiction

Novel on Yellow Paper 1936; *Over the Frontier* 1938; *The Holiday* 1949

Non-Fiction

Some Are More Human Than Others 1958; *Cats in Colour* 1959

Edited

The Batsford Book of Children's Verse 1970 [as *Favourite Verse* 1984]

Collections

Me Again: Uncollected Writings of Stevie Smith [ed. J. Barbera, W. McBrien] 1981

Biography

Stevie Smith by Frances Spalding 1988

C(harles) P(ercy) SNOW 1905–1980

Born in Leicester, the son of a clerk in a shoe factory and church organist, Snow attended Alderman Newton's School and Leicester University College, specialising in science. He did postgraduate work at Cambridge, becoming a fellow of Christ's College in 1930 and remaining a tutor until 1945. His work in infra-red spectroscopy failed, but he developed a strong identification with the ethos of Cambridge physicists, combined with a commitment to literature and a romantic leftism.

His first novel, *Death under Sail* (1932), was a detective story, as was his last, *A Coat of Varnish*. In 1940, he published *Strangers and Brothers* (later republished under the title *George Passant*), the first volume of his major work, the novel sequence of the same title, which he brought to a conclusion with its eleventh volume, *Last Things*, in 1970. The life of its narrator, Lewis Eliot, closely mirrors Snow's own, which turned from academe towards administration, government and public life.

During the war, Snow became director of technical and general staff at the Ministry of Labour, and he was subsequently in charge of recruiting scientists to government service. He received many honours from several countries, was created a life peer (Lord Snow of Leicester) in 1964 and joined Harold Wilson's first government as parliamentary secretary of the newly created Ministry of Technology. However, he left the government, conscious of failure, in 1966. His Rede Lecture, *The Two Cultures and the Scientific Revolution*, given at Cambridge in 1959, provoked a furious attack from the critic F.R. Leavis and led to a celebrated controversy.

A shambling, massive, heavy jowled figure, Snow was once unkindly described by the journalist Malcolm Muggeridge as 'wandering Grock-like through his own imaginary corridors of power', and he preserved throughout his life a certain naïve worldliness. A related attitude may weaken his novels, but, if they do not aspire to high art, the record they preserve of a leading professional man's vision of the twentieth century is a uniquely interesting one. Snow also published much charming and informative general writing on literature and science. After emotional turmoil in early life, he married the novelist **Pamela Hansford Johnson** in 1950, and they had one son.

Fiction

Death under Sail 1932; *New Lives for Old* [as 'anon.'] 1933; *The Search* 1934 [rev. 1958]; *Strangers and Brothers*: *Strangers and Brothers* 1940 [as *George Passant* 1970], *The Light and the Dark* 1947, *Time of Hope* 1950, *The Masters* 1951, *The New Men* 1954, *Homecomings* 1956, *The Conscience of the Rich* 1958, *The Affair* 1960, *Corridors of Power* 1964, *The Sleep of Reason* 1968, *Last Things* 1970; *The Malcontents* 1972; *In Their Wisdom* 1974; *A Coat of Varnish* 1978

Non-Fiction

The Two Cultures and the Scientific Revolution 1959; *Science and Government* 1961; *On Magnanimity* 1962; *A Postscript to Science and Government* 1962; *The Two Cultures: a Second Look* 1964; *Variety of Men* 1967; *The State of Siege* 1969; *Public Affairs* 1971; *Trollope: His Life and Art* 1975; *The Realists. Portraits of Eight Novelists* 1978; *The Physicists. A Generation That Changed the World* 1981

Plays

With Pamela Hansford Johnson: *Family Party* 1951; *Her Best Foot Forward* 1951; *The Pigeon with the Silver Foot* 1951; *Spare the Rod* 1951; *The Supper Dance* 1951; *To Murder Mrs Mortimer* 1951

Biography

Stranger and Brother by Philip Snow 1982; *C.P. Snow: an Oral Biography* by John Halperin 1983

Gary SNYDER 1930–

Formerly associated with the Beat poets, Snyder represents a distinctive strain in American culture, fusing the nature mysticism of Emerson with Zen Buddhism and a fascination with Asian cultures. He was born in San Francisco and grew up on his parents' poor farm north of Seattle. His father worked at a number of jobs and suffered from unemployment in the Depression. He was educated at Reed College, where he studied anthropology and literature, and the University of California at Berkeley, where he studied Oriental languages from 1952 to 1956 (he is fluent in Japanese and reads Chinese) and participated in the Bay Area Beat scene. He has been married three times, the first two relatively briefly, but the third to Masa Uehara has been more lasting and produced two children, Kai and Gen.

Snyder was the model for Japhy Ryder in **Jack Kerouac**'s 1960 novel *The Dharma Bums*, and he exemplifies the Beat tendency to hold numerous casual jobs and serve spells in the Merchant Marine or as a firewatcher. He has worked for Kodak and been a burglar-alarm installer; worked on an American oil tanker; and been a logger, a fire lookout, a manual labourer and a US forest trail crew worker at Yosemite. Most of these jobs accord with his belief in hard physical labour and simple living, and another strand in his radical American heritage is that of the Industrial Workers of the World organisation, or 'Wobblies'; his grandfather was a Wobbly organiser. He spent many years in Japan, returning to the USA to teach. He taught at Berkeley for a year (1964–5) and became a professor of English at the University of California, Berkeley, in 1985.

Snyder's 1951 baccalaureate thesis (published without revision in 1979 as *He Who Hunted Birds in His Father's Village: the Dimensions of a Haida Myth*) is a useful starting-point for understanding his values. It argues that the poet-shaman draws inspiration from the Mother Goddess, and he has said of it: 'I mapped out practically all my major interests and I've followed through on them ever since. Most of the things concerning my poetry are handled there in one way or another as well as my particular approach to history, psychological problems, nature of the mind, nature of mythology, function and forms of literature, and so forth.'

The first poem in his first book *Riprap* (1959), 'Mid August at Sourdough Mountain Lookout', is characteristic in its simplicity and immediacy: 'I cannot remember things I once read / A few friends, but they are in cities. / Drinking cold snow-water from a tin cup / Looking down for miles / Through still air.' As with **Ezra Pound**'s Imagism, there is an avoidance of the distancing abstraction of metaphor and symbol in favour of statement and the concrete thing itself. He has said of *Riprap* that he wrote the poems 'under the influence of the geology of the Sierra Nevada and the daily trail-crew work of picking up and placing granite stones in tight cobble patterns on hard slabs ... I tried writing poems of tough, simple short words, with the complexity far beneath the surface texture. In part the line was influenced by the five- and seven-character Chinese poems I'd been reading, which work like sharp blows on the mind.'

With **Allen Ginsberg**, Snyder has been a central figure in the American counter-culture, and the two of them presided over the Great Human Be-In held at Golden Gate Park in 1967. Snyder's development has been consistent. His *Songs for Gaia* contains hymns to the Earth Mother, and his belief in ecology and tribal life became fashionable once more in the Green 1990s. He has said that he thinks New York should be levelled and made into a buffalo pasture, and he has largely turned his back on Western values. His prosiness, excess simplicity and low valuation of the Western heritage have not won him admirers in all quarters, although *Turtle Island* won a Pulitzer Prize in 1975. He remains a solid member of America's alternative establishment and his values are clear: 'I pledge allegiance to the soil / of Turtle Island / one ecosystem / in diversity / under the sun / With joyful interpenetration for all.'

Poetry

Riprap 1959; *Myths and Texts* 1960; *Across Lamarack Col* 1964; *The Firing* 1964; *Hop, Skip and Jump* 1964; *Nanao Knows* 1964; *Dear Mr President* 1965; *Riprap and Cold Mountain Poems* 1965; *Six Sections from Mountains and Rivers without End* 1965; *A Range of Poems* 1966; *Three Worlds, Three Realms, Six Roads* 1966; *The Back Country* 1967; *The Blue Sky* 1969; *Regarding Wave* 1969; *Sours of the Hills* 1969; *Anasazi* 1971; *Manzanita* 1971; *The Fudo Trilogy* 1973; *Turtle Island* 1974; *All in the Family* 1975; *Songs for Gaia* 1979; *True Night* 1980; *Axe Handles* 1983; *Tree Zen* 1984; *The Fates of Rocks and Trees* 1986; *Left Out in the Rain* 1986; *The Practice of the Wild* 1990; *No Nature* 1992

Non-Fiction

Four Changes 1969; *Earth House Hold: Technical Notes and Queries to Fellow Dharma Revolutionaries* 1969; *The Old Ways* 1977; *On Bread and Poetry* 1977; *He Who Hunted Birds in His Father's Village* 1979; *Good Wild Sacred* 1984; *Passage through India* 1984

Edited

The Wooden Fish: Basic Sutras and Gathas of Rinzai Zen [with G. Kanetsuki] 1961

E(dith Anna) O(enone) SOMERVILLE 1858–1949

Martin ROSS (pseud. of Violet Florence Martin) 1862–1915

As 'Somerville and Ross', these two women, who were cousins, created the most celebrated and enduring collaboration in modern English-language literature. They were born into the heart of the Anglo-Irish gentry: Somerville's father was a lieutenant-colonel and her great-grandfather had been chief justice of Ireland, while Ross was the daughter of a deputy lieutenant of Ireland and sprang from the ancient Galway family of Ross, from which she derived her pseudonym. Somerville, one of a family of eight, had been born in Corfu where her father was commanding a battalion of the Buffs, and grew up largely at the family home at Drishane, County Cork. Ross, from a family of eleven, was brought up largely in Dublin, where her father had become a leader-writer for the *Morning Herald*.

Both women were in the main educated at home by governesses, but both attended Alexandra College Dublin, for brief periods, although they did not meet there. Somerville, who was a talented visual artist (she continued to exhibit in later life), studied art in Düsseldorf, Paris and London before meeting her cousin in 1886 and making her her model. The personal and literary partnership began soon after: they lived together, mainly at Somerville's home at Drishane, and their first book, *An Irish Cousin*, was published in 1889. They had published fifteen books before Ross's death, including novels and short stories, essays and travel; although they travelled widely, and often lived for months at a time in Paris, the scene of almost all their books is Ireland. Their undoubted masterpiece is *The Real Charlotte*, which provides an ironic and superbly written account of the vanished Anglo-Irish society, although *Some Experiences of an Irish R.M.*, lighthearted stories about a country magistrate, has proved more enduringly popular.

The women lived lives typical of their time and class, riding enthusiastically to hounds (Somerville was the first female Master of Foxhounds, of the West Carbery Hunt) and proving themselves stalwarts of the Munster Women's Franchise League. Somerville was the organist of Castlehaven parish church for seventy-five years. Ross suffered a severe fall from her horse in 1898 which made her an invalid and may have contributed to her early death. After Ross died, Somerville derived what comfort she could from Dame Ethel Smyth, with whom she toured Italy and the USA. She lived into her eighties at Drishane and continued to publish her books as being by 'Somerville and Ross', maintaining that she remained in contact with Ross through spiritualist means; but their quality was poorer. Somerville wrote and edited a few books under her own name alone and used the pseudonym 'Geilles Herring' for her earliest work; her cousin wrote one book alone; many of the books were illustrated by Somerville.

Fiction

An Irish Cousin [as 'Geilles Herring'] 1889 [rev. 1903]; *Some Experiences of an Irish R.M.* (s) 1899; *Naboth's Vineyard* 1891; *The Real Charlotte* 1894; *The Silver Fox* 1898 [rev. 1902]; *A Patrick's Day Hunt* [by Ross] 1902; *All on the Irish Shore* (s) 1903; *Further Experiences of an Irish R.M.* (s) 1908; *Dan Russel the Fox* 1911; *The Story of the Discontented Little Elephant* [by Somerville] 1912; *In Mr Knox's Country* (s) 1915; *Mount Music* 1919; *An Enthusiast* 1921; *The Big House of Inver* 1925; *French Leave* 1928; *Little Red Riding Hood in Kerry* 1934; *The Sweet Cry of Hounds* (s) 1936; *Sarah's Youth* 1938

Poetry

Slipper's ABC of Foxhunting [by Somerville] 1903

Non-Fiction

Through Connemara in a Governess Cart 1892; *In the Vine Country* 1893; *Beggars on Horseback: a Riding Tour of North Wales* 1895; *Some Irish Yesterdays* 1906; *Irish Memories* 1917; *Stray-Aways* 1920; *Wheel-Tracks* 1923; *The States through Irish Eyes* [as Somerville] 1930; *An Incorruptible Irishman* 1932; *The Smile and the Tear* 1933; *Records of the Somerville Family of Castlehaven and Drishane from 1174 to 1940* [by Somerville; with H. Boyle Townshend] 1940; *Notions in Garrison* 1941; *Happy Days! Essays of Sorts* 1946; *The Selected Letters of Somerville and Ross* [ed. G. Lewis] 1989

Edited

The Mark Twain Birthday Book [as Somerville] 1885; *Notes of the Horn: Hunting Verse, Old and New* [as Somerville] 1934

Collections

Maria and Some Other Dogs [ed. Somerville] 1949

Biography

Somerville and Ross by Maurice Collis 1968

Charles (Hamilton) SORLEY 1895–1915

The elder of twin sons of a Scottish professor of moral philosophy, Sorley was born in Aberdeen but moved to Cambridge in 1900, when his father took up a professorship at King's College there. He was educated at King's College Choir School before winning an open scholarship to the public school Marlborough College. He enjoyed his time there, but was far too intelligent to be taken in by the prevailing ethos, and felt free to mock it in the poems he contributed to the school magazine, the *Marlburian*. Although he was in the Officers' Training Corps and took part in team sports, his favourite occupation was the solitary one of running on the downs above the school, a landscape he recalls in the best known of his early poems, 'Marlborough' and 'The Song of the Ungirt Runners'.

He was on a walking tour in Germany, a country he loved, when the First World War broke out and was briefly imprisoned (for eight and a half hours) as a spy, before returning to England, where he enlisted in the 7th Battalion of the Suffolk Regiment. He served as a second lieutenant and was killed at the Battle of Loos at the age of twenty. His only volume of verse, *Marlborough and Other Poems* was published in 1916, and its title led to his being hailed as an archetype of the public-school subaltern. In fact, as his remarkable letters (published after the war) show, he was a rebellious, sceptical and astonishingly mature young man, who never for a moment thought warfare noble or glorious. 'I am full of mute and burning rage and annoyance and sulkiness about it,' he wrote of the outbreak of war. 'I could wager that out of twelve million eventual combatants there aren't twelve who really want it. And "serving one's country" is so unpicturesque and unheroic when it comes to the point.' He judged **Rupert Brooke** 'far too obsessed with his own sacrifice', and suggested that his own sonnet 'Whom We Therefore Ignorantly Worship', written in September 1914, 'should get a prize for being the first poem written since 4 August that isn't patriotic'.

Sorley's reputation as a poet rests on perhaps as few as five poems, but these are amongst the finest of the war. They include the much anthologised, and much misunderstood, 'All the Hills and Vales Along', an ironic marching song which concludes: 'Strew your gladness on earth's bed, / So be merry, so be dead.' Less ambiguous are two poems in which Sorley confronts extinction squarely and without conventional faith: the bleak sonnet 'Such, Such is Death' and what was probably the last poem he wrote, 'When You See Millions of the Mouthless Dead', which was found in his kit after his death.

Dead before **Wilfred Owen** even enlisted, Sorley knew none of his fellow soldier-poets, but his work was much admired by both **Siegfried Sassoon** and **Robert Graves**, while **John Masefield** (whose work Sorley discussed in a lecture delivered to his school's literary society) described his death as '*the* great loss of the war'.

Poetry

Marlborough and Other Poems 1916; *The Collected Poems of Charles Hamilton Sorley* [ed. J. Moorcroft Wilson] 1985

Non-Fiction

The Letters of Charles Sorley [ed. W.R. Sorley] 1919; *The Collected Letters of Charles Hamilton Sorley* [ed. J. Moorcroft Wilson] 1990

Collections

The Poems and Selected Letters of Charles Hamilton Sorley [ed. H.D. Spear] 1978

Biography

Charles Hamilton Sorley by Jean Moorcroft Wilson 1985

William SOUTAR 1898–1943

Born in Perth, Scotland, Soutar was first influenced to write poetry when he was an adolescent, but his early efforts were banal. Excelling on the sports field and in the classroom, he left Perth Academy to join the navy in 1916, serving for a time with the Grand Fleet in Scapa Flow. During this period, he wrote idealistic verse which did not reflect the distaste for militarism which he later expressed in prose. Demobilised in early 1919, he had spent the last few months of the war on sick leave, suffering from what was thought to be fallen arches.

After a false start at Edinburgh University studying medicine, he changed courses to study English literature under Herbert Grierson, the editor of Donne. Influenced by him, Soutar destroyed most of his juvenilia, but published undergraduate poems in 1923. These were written in English, and were unremarkable. Soutar met the young C.M. Grieve (**Hugh MacDiarmid**) in the same year and, partly through a stimulating correspondence, also began to write verse in the Scots dialect. Wide reading and a conscientious study of a Scots dictionary resulted in *Seeds in the Wind*, which in 1933 confirmed him as a leading figure of the Scottish Literary Renaissance.

Recurring back-trouble, which had frequently incapacitated him, was diagnosed as spondylitis and prevented him from qualifying as a teacher after his graduation. A series of operations failed, and for the last ten years of his life, he was unable to leave his bed. Otherwise extremely strong physically, Soutar underplayed his illness, and received his visitors wearing jacket, shirt and bow tie, set off by a large head of bushy hair. Forced introspection led him to keep diaries, which revealed his preoccupation with pacifism, and his concern for doomed humanity, themes upon which he elaborated in his poetry. As in the case of Burns, Soutar's weakest verse was in English, but Benjamin Britten's 1969 song-cycle, *Who Are These Children?*, successfully set poems by him in Scots and English. The day before his death from tuberculosis, Soutar wrote of his confused grip on reality, caused by his illness, but thirteen years previously he had noted in his diary that: 'There can be no entry for the day on which I die. Let me write it down now. "To accept Life is to give it beauty." '

Poetry

Gleanings by an Undergraduate [as 'W.S.'] 1923; *Conflict* 1931; *Seeds in the Wind* 1933 [rev. 1943]; *The Solitary Way* 1934; *Brief Words* 1935; *Poems in Scots* 1935 [rev. as *Poems in Scots in English* ed. W.R. Aitken 1961]; *A Handful of Earth* 1936; *Riddles in Scots* 1937; *In the Time of Tyrants* 1939; *But the Earth Abideth* 1943; *The Expectant Silence* 1944; *Collected Poems* [ed. H. MacDiarmid] 1948; *Poems of William Soutar* [ed. W.R. Aitken] 1988

Non-Fiction
Diaries of a Dying Man [ed. A. Scott] 1954

Biography
Still Life by Alexander Scott 1958

Terry SOUTHERN 1924–

Southern was born in Alvarado, Texas, and brought up in Dallas where his father was a pharmacist and his mother a dress designer. After high school he began to study medicine at the Southern Methodist University, but his studies were interrupted by two years' serving with the US Army in Europe. He attended the University of Chicago and Northwestern University in Illinois, where he graduated in 1948 before spending four years at the Sorbonne in Paris funded by the GI Bill. In Paris he began to publish stories in the *Paris Review*, and back in the USA endured considerable hardship while writing three unpublished novels. He has said that he finally found his voice and was able to write the satirical *Flash and Filigree* after discovering the novels of **Henry Green**, and it was Green who helped arrange the first publication of the novel in England in 1958. A black comic satire set in California, the novel is particularly vehement in its ridiculing of the medical profession and contains a television quiz show called 'What's My Disease?' in which celebrity panellists guess at the infirmities of guests.

Candy is a semi-pornographic reworking of Voltaire's *Candide* (1759) (heroine Candy's Christian charity is such that she just can't say 'No') written in collaboration with Mason Hoffenberg for Maurice Girodias's Olympia Press in Paris and published under the pseudonym Maxwell Kenton. It was promptly banned and reissued the following year as *Lollipop*, and was published in the USA in 1964. *The Magic Christian* is a satire on materialism in which a billionaire uses his wealth to perpetrate elaborate practical jokes such as inducing the residents of Chicago to scrabble for money in a vast mound of excrement. The novel came to the attention of the director Stanley Kubrick with whom Southern collaborated on the screenplay for *Dr Strangelove* (1963). Southern's satirical propensities are readily apparent in this adaptation of Peter George's novel, and he won the British Screen Writers Award in 1964 and was nominated for an Oscar in 1963. Southern went on to write other screenplays, often in collaboration with others, notably the disappointing 1965 film of **Evelyn Waugh**'s *The Loved One* (1948), which he wrote with **Christopher Isherwood**; Roger Vadim's camp science-fiction *Barbarella* (1967); *End of the Road*, which was adapted from **John Barth**'s 1958 novel; and the seminal film of 1960s' youth rebellion, Dennis Hopper's *Easy Rider* (1969), for which he was again nominated for an Oscar. Of his own novels, both *Candy* and *The Magic Christian* were adapted by other writers to make unmemorable films, the latter chiefly notable for casting the Beatle's drummer, Ringo Starr, alongside Peter Sellers.

Southern's latest novel to date is *Blue Movie*, a satire on pornography which many critics felt was pornography dressed up as satire, but which does contain some wonderfully comic portraits of Hollywood characters. *Red-Dirt Marijuana* is a collection of stories and essays. Southern was married in 1952, and has one son.

Fiction
Candy [as 'Maxwell Kenton'; with M. Hoffenberg] 1958 [rev. as *Lollipop* 1959; US *Candy* 1964]; *Flash and Filigree* 1958; *The Magic Christian* 1959; *Blue Movie* 1970

Plays
Easy Rider (f) [with P. Fonda, D. Hopper] 1969

Non-Fiction
The Journals of 'The Loved One' 1965; *The Rolling Stones on Tour* 1978

Collections
Red-Dirt Marijuana and Other Tastes 1967

Edited
Writers in Revolt [with R. Seaver, A. Trocchi] 1963

(Akinwande Olu)Wole SOYINKA 1934–

The double perspective of Soyinka's writing – the dramatic use of Yoruba legend, ritual and dance, and the European sense of stagecraft – is equally apparent in the author's life. Born in Abeokuta, Western Nigeria, the son of a school headmaster and a devout Christian mother, Soyinka was introduced to the older Yoruba spirit world as a teenager by his grandfather. Urged by his father to continue his studies at the University of Ibadan, he was initiated by his grandfather in a primitive scarification ceremony consecrated to Ogun, ancient saint of metals and metalwork and, more latterly, of the African roads upon which Soyinka has always feared death.

Whilst at Ibadan he published poems in the magazine *Black Orpheus*. He was given a scholarship to Leeds University in 1954, graduating with honours in 1957, before moving to the Royal Court Theatre as a play reader. During his eighteen months in London he wrote his early plays, *The Invention* and *The Swamp Dwellers*, successfully debuted in Nigeria upon Soyinka's return in 1959. He then became a research fellow in drama at the University of Ibadan where he stayed until 1961, studying folk drama, travelling the country and incorporating elements of ritual into his own work. *A Dance of the Forests* was premiered by Masks, an English-speaking amateur company of Soyinka's, as part of the Nigerian independence celebrations in 1960. Prophetically, in view of the war to come, the play warns against the dangers of a return to a mythical tribal past. It was the collective aim of the Mbari Writers and Artists Club Soyinka had formed to foster a modern Nigerian culture, capable of mediating between the old and the new. He wrote, acted and produced as well as working as a teacher in Ife between 1962 and 1963, when he resigned for political reasons. He went on to establish the Orisun Repertory in 1964.

His first novel, *The Interpreters*, was an enlargement upon his by now constant theme, pitting individual freedom against the sacrifices demanded by the false prophets of both progress and superstition. Published in 1965 it was awarded the *New Statesman* Literary Prize. Appointed senior lecturer in English at the University of Lagos that same year, he was arrested and tried for theft when he protested against the fraudulent election of a local chief. His growing international reputation was confirmed when the literary establishment petitioned for and secured his acquittal, prompting him to pay his first visit to the USA in 1967.

American productions of his work had been scheduled when the deteriorating situation in his homeland persuaded him to return. Arrested by government officials on 17 August 1967, Soyinka was imprisoned without trial, and moved to Kaduna where he was to spend the greater part of the next two years in solitary confinement. His offence seems to have been to attempt to mediate between the Biafran leader Ojukwu and the Nigerian state. The bloody conflict he had predicted duly took place, and with startling ferocity. Denied pen and paper, Soyinka was successful in smuggling fragments of his journal and scraps of poetry out of his prison cell, each one of which widened the international protest at his incarceration.

Freed in 1969, he became head of the drama department at the University of Ibadan, and in 1970 he took a troupe to America to perform *Madmen and Specialists*, written in prison. The specialist of the title is a doctor with a new training in torture which he returns to practise on his ageing father. *The Man Died: Prison Notes* and *A Shuttle in the Crypt*, a book of poems, both deal directly with his term in prison. His writing ever since – including *Season of Anomy*, his second novel – has been marked by an anger and a loathing of tyranny that won him the Amnesty Prisoner of Conscience Prize. He was awarded the Nobel Prize for Literature in 1986 and was named Commander of the Federal Republic of Nigeria in the same year. He has also edited *Poems of Black Africa*, a highly regarded anthology, and published *Aké*, an autobiography of his childhood.

Plays

The Invention 1955 [np]; *Three Plays* [pub. 1963]: *The Swamp Dwellers* 1958, *The Trials of Brother Jero* 1960, *The Strong Breed* 1964; *Kongi's Harvest* 1964; *The Lion and the Jewel* 1959 and *A Dance of the Forests* 1960 [pub. in *Five Plays* 1964]; *Camwood on the Leaves* (r) 1960 [pub. 1973]; *The Jero Plays* [pub. 1973]: *The Trials of Brother Jero* 1960, *Jero's Metamorphosis* 1975; *Before the Blackout* 1965 [pub. 1971]; *The Road* 1965; *The Detainee* (r) 1965; *Rites of Harmattan Solstice* 1966 [np]; *Madmen and Specialists* 1970 [pub. 1971]; *The Bacchae* [from Euripides] 1973; *Death and the King's Horseman* 1976 [pub. 1975]; *Opera Wonyosi* [from B. Brecht's play *Dreigroschenoper*] 1977 [pub. 1981]; *Golden Award* 1980 [np]; *Die Still, Dr Godspeak* (r) 1981; *A Play of Giants* 1984; *Requiem for a Futurologist* 1983 [pub. 1985]

Poetry

Idanre and Other Poems 1967; *Poems from Prison* 1969; *A Shuttle in the Crypt* 1972; *Ogun Abibmañ* 1976; *Mandela's Earth and Other Poems* 1989; *From Zia with Love* 1992

Fiction

The Interpreters 1965; *Season of Anomy* 1973

Non-Fiction

The Man Died: Prison Notes 1972; *Myth, Literature and the African World* 1976; *Aké: the Years of Childhood* (a) 1981; *Art, Dialogue and Outrage* 1988; *Isarà: a Voyage Around 'Essay'* 1989

Translations

D.O. Fagunwa *Forest of a Thousand Demons: a Hunter's Saga* 1968

Edited

Poems of Black Africa 1973

Muriel (Sarah) SPARK 1918–

The daughter of a Scottish-Jewish father, who was an engineer, and an English mother, Spark was born Muriel Camberg in the Morningside area of Edinburgh. She was educated at James Gillespie's High School for Girls in Edinburgh, where at the age of eleven she entered the class of Miss Christina Kay, 'that character in search of an author'. Miss Kay was later reborn as the protagonist of Spark's famous novel *The Prime of Miss Jean Brodie*, a book which also recreates the Edinburgh of the author's childhood. She left school at the age of seventeen with the intention of becoming a teacher, and continued her education at Heriot Watt College, after which she taught English, arithmetic and nature studies at the Hill School in exchange for shorthand and typing lessons. Her next job was as a secretary to the owner of the women's department store William Small & Sons, where her 'faculties of character-observation were somewhat sharpened'. In 1937, without quite knowing why, she became engaged and followed her fiancé to what was then Southern Rhodesia, where she married: 'I intended to stay for the pre-arranged three years and gain as much human experience as I could,' she later wrote, and her first volume of short stories, *The Go-Away Bird* (1958), is based upon her African life.

In 1938 she had a son, but already her marriage was in trouble, since (she later claimed) her husband was mentally unstable: 'He became a borderline case, and I didn't like what I found either side of the border.' With the outbreak of the Second World War, she took the opportunity to start divorce proceedings, and returned to Britain in 1944. She entrusted her son to the care of her parents and took a job in the Foreign Office, working with 'black propaganda', and living at the Helena Club in London's Lancaster Gate, which became the model for the May of Teck Club in her novel *The Girls of Slender Means*. After the war she got a job with the National Jewellers' Association's magazine, *Argentor*, then between 1947 and 1949 was editor of *Poetry Review*. She had written poetry while in Africa, and her first volume, *Fanfarlo*, was published in 1952.

During the 1950s, while she had part-time jobs in publishing and worked as a freelance literary journalist, she became involved with the poet and critic Derek Stanford, with whom she collaborated on several books, and who wrote an early study of her work in 1963. She has claimed that this book was highly inaccurate and took her revenge in later years by portraying Stanford in one of her novels and denouncing him in her autobiography. In 1951 her short story 'The Seraph and the Zambesi' won a competition organised by the *Observer* and this launched a career in which she has become one of Britain's leading writers of fiction. Her first novel, *The Comforters*, which was inspired by a brief period during which she suffered from hallucinations, was published in 1957 and praised by many, including **Evelyn Waugh**. She had become a Roman Catholic in 1954, and this and subsequent novels reflect her religious beliefs and have led to a preoccupation with good and evil. This is also related to her Scottish roots: 'Here in Scotland, people are more capable of perpetrating good and evil than anywhere else,' one of her characters says. 'I don't know why, but so it is.' As a child she read Scott's *Minstrelsy of the Scottish Borders* (1802–3). 'I still recall my first reading of "Fair Helen of Kirconnell",' she has said: 'its pure lyricism, its narrative economy and its stark savagery. It entered my bloodstream and has remained there ever since.'

Spark's short, sharp and very funny novels are evidence of this infection. At Heriot Watt College she had taken a course in précis-writing: 'I love economical prose, and would always try to find the briefest way to express a meaning.' In an age which favours sheer bulk as an indication of serious intent, Spark's svelte novels show just how much can be achieved in a small compass. A large part of the comedy in her books arises from this terseness, a ruthlessness of style and content. The longest of her novels is *The Mandelbaum Gate*, in which she deals with her Jewish heritage; although it won the James Tait Black Memorial Prize, it is not considered to be among her best books. These include *Memento Mori*, a black comedy about old age and the Four Last Things; *The Prime of Miss Jean Brodie*, popular stage, film and television versions of which have not altogether reflected its complexity as a moral fable; *The Girls of Slender Means*, a tragi-comedy about the operation of divine grace, which echoes Gerard Manley Hopkins's 'The Wreck of the Deutschland' (1875); *The Abbess of Crewe*, inspired by the Watergate hearings, but set in a convent; *Loitering with Intent* and *A Far Cry from Kensington*, both of which draw upon her recollections of London in the 1950s and have mordant female narrators; and *Symposium*, in which the influence of the Border Ballads is at its most pervasive. *Loitering with Intent* is of particular interest as a reflection of Spark's own views on the nature of fiction.

Spark has also written a large number of fine stories, many of which have been published in the *New Yorker*. They have been collected in four volumes, including a collected edition, and amongst them are several masterpieces of the genre, notably 'The Portobello Road' (narrated by the ghost of a murder victim) and 'Bang-Bang You're Dead' (based on a shocking incident in Africa). She has also written plays for the radio and interesting studies of Mary Shelley and **John Masefield**. In later years, Spark has lived in Italy, and now shares a house near Arezzo with the sculptor Penelope Jardine. *Curriculum Vitae* is a curiously oblique and only occasionally revealing volume of autobiography covering the first half of her life.

Fiction
The Comforters 1957; *The Go-Away Bird* (s) 1958; *Robinson* 1958; *Memento Mori* 1959; *The Bachelors* 1960; *The Ballad of Peckham Rye* 1960; *The Prime of Miss Jean Brodie* 1961; *The Girls of Slender Means* 1963; *The Mandelbaum Gate* 1965; *The Public Image* 1968; *The Driver's Seat* 1970; *Not to Disturb* 1971; *The Hothouse by the East River* 1973; *The Abbess of Crewe* 1974; *The Takeover* 1976; *Territorial Rights* 1979; *Loitering with Intent* 1981; *Bang-Bang You're Dead and Other Stories* (s) 1982; *The Only Problem* 1984; *The Stories of Muriel Spark* (s) 1985; *A Far Cry from Kensington* 1988; *Symposium* 1990; *Collected Stories of Muriel Spark* (s) 1994

Poetry
The Fanfarlo and other Verse 1952; *Collected Poems I* 1968; *Going Up to Sotheby's and Other Poems* 1982

Non-Fiction
Child of Light: a Reassessment of Mary Shelley 1951 [rev. as *Mary Shelley* 1988]; *John Masefield* 1953; *Emily Brontë: Her Life and Work* 1955; *Curriculum Vitae* (a) 1992; *The Essence of the Brontës: a Compilation with Essays by Muriel Spark* 1993

Edited
Tribute to Wordsworth [with D. Stanford] 1950; *Selected Poems of Emily Brontë* 1952; *My Best Mary: The Letters of Mary Shelley* [with D. Stanford] 1953; *The Brontë Letters* 1954; *Letters of John Henry Newman* [with D. Stanford] 1957

For Children
The Very Fine Clock 1968

Plays
Doctors of Philosophy 1962 [pub. 1963]

Collections
Voices at Play 1961

(Charles) Bernard SPENCER 1909–1963

Spencer was born in India, the son of Sir Charles Spencer, a high-court judge in Madras and collateral descendant of the Spencer-Churchills. He was sent to England as an infant, and brought up by various guardians and relatives. Educated at Marlborough College, he was a contemporary of Anthony Blunt, **Louis MacNeice** and **John Betjeman**, who was to describe him as a 'shy squirrel'. There he developed his talent for painting which was to have a strong influence on his essentially visual verse style. In 1928 he went up to read Greats at Corpus Christi, Oxford, and was part of a generation that included Isaiah Berlin, **Arthur Calder-Marshall** and **Stephen Spender**, with whom he edited *Oxford Poetry* in 1930. Spender recalls this friendship in his elegiac poem 'One More New Botched Beginning'. Between 1932 and 1940 he was variously a prep-school master, copywriter, scriptwriter and deputy to **Geoffrey Grigson**, then editor of *New Verse*.

To an extent at odds with the political commitment of his generation, he remained detached from it and found his true voice only after his exposure to Greece; as his friend **Lawrence Durrell** put it: 'Greece woke him up.' In 1940 he had been posted by the British Council to Salonika and, remaining with the organisation after the war, he was to live almost entirely abroad. After Greece was overrun, he became part of the intellectual group of exiles in Cairo which included Durrell and George Seferis. Isolation and the strains of the protracted struggle are the themes of the Cairo poems in his first collection, *Aegean Islands*. In 1936 he had married Nora Gibbs, an actress, who died in 1947. 'At Courmayer', published two years later, is his moving elegy to her. In his introduction to Spencer's *Collected Poems*, Roger Bowen points out that after her death there is in his work 'a pull away from the ecstatic towards the elegiac and, at its most extreme, an enclosing sense of alienation'.

In 1955 **John Lehmann** began publishing Spencer's poems in the *London Magazine*, and he was a leading contributor when **Alan Ross** took over the editorship. His second collection, *The Twist in the Plotting*, was privately printed by the University of Reading where the bulk of his literary papers were deposited after his death.

In 1961 he remarried, and had a son. As a poet he had embarked on one of his most productive periods; it ended when he became ill, the symptoms being blackouts and loss of memory. One night he wandered off on his own from a private clinic in Vienna where he was being treated; early the following morning he was found dead by the railway line, his skull fractured. He left many poems which were published eighteen years after his death as a second volume of *Collected Poems*.

Poetry
Aegean Islands and Other Poems 1946; *The Twist in the Plotting* 1960; *With Luck Lasting* 1963; *Collected Poems* [ed. A. Ross] 1965; *Corgi Modern Poets in Focus 5* [with others; ed. D. Abse] 1973; *Collected Poems* [ed. R. Bowen] 1981

Translations
George Seferis *The King of Asine and Other Poems* [with L. Durrell, N. Valaoritis] 1948

Edited
Oxford Poetry 1930 [with S. Spender] 1930; *Oxford Poetry 1931* [with R. Goodman] 1931

Stephen (Harold) SPENDER 1909–1995

The third son of a distinguished liberal journalist and his Jewish wife, who was of German-Danish descent and died young, Spender was born in London. He was educated at University College School, London, and University College, Oxford, where he initially read history before changing to take a degree in philosophy, politics and economics. His passion, however, was literature, and it was at Oxford that he met the writers with whom his name will always be associated: **C. Day Lewis**, **Louis MacNeice** (with whom he edited *Oxford Poetry 1929*) and pre-eminently **W.H. Auden**, who was to introduce him to **Christopher Isherwood**. In 1928 he printed on a hand press a small edition of Auden's *Poems*, as well as a pamphlet of his own *Nine Experiments*. His first proper volume, *Twenty Poems*, was published two years later.

During the 1920s he spent some time travelling in Austria and Germany and working on a semi-autobiographical novel, *The Temple*, which was considered unpublishable because of its homosexual content (it eventually appeared in a revised form in 1977). The rise of fascism and unemployment sharpened his political conscience and informed his poetry during the 1930s, and he was briefly a member of the Communist Party. He was a contributor to the *Daily Worker*, and his verse play *The Trial of a Judge* was performed by the Group Theatre in 1938. He worked for the Republican cause during the Spanish Civil War and with the National Fire Service and the Foreign Office in London during the Second World War. In 1939 he became the associate editor of *Horizon*, the enormously influential monthly 'Review of Literature and Art' that was being set up by **Cyril Connolly**, and was originally run from Spender's London flat. Although called up in 1941, he remained an important contributor to the magazine.

After the war, he worked for UNESCO and continued to publish poetry as well as fiction, critical and political works, translations and a volume of autobiography, *World within World*, which is not only a record of an age but also an extraordinarily frank record of what it is like to be young, idealistic and confused. From 1953 to 1967 he was co-editor of *Encounter*, but dissociated himself from the magazine when it was revealed that it received funding from the CIA. He took up a number of university appointments in both England and America, where he has held several chairs and visiting professorships. He was awarded the CBE in 1962, the Queen's Gold Medal for Poetry in 1971 and the PEN Golden Pen Award for long service to literature in 1995. He was knighted in 1983.

Spender's association with Auden has inevitably meant that his own poetry has suffered by comparison. In spite of their long friendship, however, their poetry is very different. His reputation stood high in the 1930s, with such poems as 'Airman', 'An Elementary School Classroom in a Slum', 'Port Bou', 'The Pylons' (a poem which became one of the best known of the decade and gave rise to the term 'Pylon Poets'), and the much anthologised 'The Truly Great'. Among many fine later works are his moving sequence of poems 'Elegy for Margaret', written in memory of his sister-in-law (the wife of the photographer Humphrey Spender), verses about his children, and poems about such friends as MacNeice, Connolly and Stravinsky, including one on 'Auden's Funeral'. Of his numerous translations, *The Duino Elegies* of Rilke and his version of Sophocles's Oedipus trilogy are considered outstanding. He continued to write poetry right up until his death and a late, reflective volume, *Dolphins*, was published on his eighty-fifth birthday.

He was also well known as a critic and commentator and a speaker at international literary conferences, and was a founder member of the Index on Censorship. His first marriage, to Inez Pearn in 1936, was brief, but in 1941 he married the concert pianist Natasha Litvin, with whom he had two children. His later years were unsettled by two controversies. In 1990 a writer called Hugh David announced that he was working on Spender's biography, with the poet's approval. Spender disputed this and took unsuccessful measures to prevent the book being published. Sloppily researched, poorly written and riddled with inaccuracies, the biography was published to derisory reviews in 1992. Almost immediately, Spender became embroiled in a legal battle with **David Leavitt**, whose novel *While England Sleeps* (1993) he claimed plagiarised passages describing his relationship with T.A.R. Hyndman (a former guardsman and the subject of several poems) in his own *World within World*. The book was eventually withdrawn, but published in a revised edition in America in 1995. Spender was always a public figure, and earlier books which contain portraits or caricatures of him include **Wyndham Lewis**'s *The Apes of God* (1930), Isherwood's *Lions and Shadows* (1939), Michael Nelson's *A Room in Chelsea Square* (1958) and **T.C. Worsley**'s *Fellow Travellers* (1971).

Poetry

Nine Experiments by S.H.S. 1928; *Twenty Poems* 1930; *Poems* 1933; *Vienna* 1934; *At Night* 1935; *The Still Centre* 1939; *Selected Poems* 1940; *Ruins and Visions* 1942; *Spiritual Exercises* 1943; *Poems of Dedication* 1946; *Returning to Vienna 1947*

1947; *The Edge of Being* 1949; *Sirmione Peninsula* 1954; *Collected Poems 1928–1953* 1955; *Inscriptions* 1958; *Selected Poems* 1965; *The Generous Days: Ten Poems* 1969 [rev. *The Generous Days* 1971]; *Art Student* 1970; *Descartes* 1970; *Recent Poems* 1978; *Collected Poems 1928–1985* 1985; *Dolphins* 1994

Plays

The Trial of a Judge 1938; *Danton's Death* [from G. Büchner's play; with G. Rees] 1939; *To the Island* 1951; *Mary Stuart* [from Schiller's play] 1957 [pub. 1959]; *Lulu* [from F. Wedekind] 1958; *Rasputin's End* 1963; *Oedipus Trilogy* [from Sophocles] 1983

Fiction

The Burning Cactus (s) 1936; *The Backward Son* 1940; *Engaged in Writing, and the Fool and the Princess* (s) 1958; *The Temple* 1977

Non-Fiction

The Destructive Element: a Study of Modern Writers and Beliefs 1936; *Forward from Liberalism* 1937; *The New Realism* 1939; *Life and Death of the Poet* 1942; *Jim Braidy: the Story of Britain's Firemen* [with J. Gordon, W. Sanson] 1943; *Botticelli* 1945; *Citizens in War – and After* 1945; *European Witness* 1946; *Poetry Since 1939* 1946; *The God That Failed* (c) [ed. R. Crossman] 1950; *World within World* (a) 1951; *Europe in Photographs* 1951; *Shelley* 1952; *Learning Laughter* 1952; *The Creative Element: a Study of Vision, Despair, and Orthodoxy among Some Modern Writers* 1953; *The Making of a Poem* 1955; *The Imagination in the Modern World: Three Lectures* 1962; *The Struggle of the Modern* 1963; *Ghika: Painting, Drawings, Sculpture* 1965; *Chaos and Control in Poetry* 1966; *The Year of the Young Rebels* 1969; *W.H. Auden: a Memorial Address* 1973; *Love-Hate Relations: a Study of Anglo-American Sensibilities* 1974; *Eliot* 1976; *Cyril Connolly: a Memoir* 1978; *Henry Moore: Sculptures in Landscape* 1978; *The Thirties and After: Poetry and People* (1933–1975) 1978; *Venice* 1979; *America Observed* 1979; *Letters to Christopher: Stephen Spender's Letters to Christopher Isherwood 1929–1939, with the Line of the Branch* [ed. L. Bartlett] 1980; *China Diary* 1982 [with D. Hockney] 1982; *Last Drawings of Christopher Isherwood by Don Bachardy* (c) 1990

Edited

Oxford Poetry 1929 [with L. MacNeice] 1929; *Oxford Poetry 1930* [with B. Spencer] 1930; *New Writing* [with C. Isherwood, J. Lehmann] 1938–9; *Poems for Spain* [with J. Lehmann] 1939; *A Choice of Romantic Poetry* 1947; Walt Whitman *Selected Poems* 1950; *New Poems 1956* [with D. Abse, E. Jenning] 1956; *Great Writings of Goethe* 1958; *Great German Short Stories* (s) 1960; *The Concise Encyclopedia of English and American Poets and Poetry* 1963 [rev. 1970]; *Encounters: an Anthology from the First Ten Years of 'Encounter' Magazine* [with I. Kristol, M.J. Laskey] 1963; Abba Kovner and Nelly Sachs *Selected Poems* 1971; Shelley: *A Choice of Shelley's Verse* 1971, *The Poems of Percy Bysshe Shelley* 1971; *D.H. Lawrence, Novelist, Poet, Prophet* 1973; *W.H. Auden: a Tribute* 1975; *Henry Moore* 1978; *Journals 1939–1983* [ed. J. Goldsmith] 1985; *In Irina's Garden* 1986

Translations

Lorca: *Poems* [with J.L. Gili] 1939, *Selected Poems* [with J.L. Gili] 1943; Rainer Marie Rilke *Duino Elegies* [with J.B. Leishman] 1939 [rev. 1948]; Ernst Toller *Pastor Hall* 1939; Paul Éluard *Le dur désir de durer* [with F. Cornford] 1950; *The Life of the Virgin Mary* 1951; Frank Wedekind *Five Tragedies of Sex* 1952; C.P. Cavafy *Fourteen Poems* [with N. Stangos] 1967

For Children

The Magic Flute: Retold 1966

Christopher St John SPRIGG

see Christopher CAUDWELL

Jean (Wilson) STAFFORD 1915–1979

Of Irish and Scottish ancestry, Stafford was born in Covina, California, and spent her earliest years in that state, the daughter of a former newspaper reporter who later wrote Westerns under the pseudonyms of Jack Wonder and Ben Delight. Between the ages ten and twenty-one she lived in the small town of Boulder in the Rocky Mountains of Colorado, where she attended the local high school. Her childhood, marked by poverty and family tension (she later recalled her 'grandfather's Sunday punishment room'), laid the foundations for the traumas which were to mark her adult life. She attended the University of Colorado, and then won a scholarship to study in Heidelberg for the academic year 1936–7, during which period she caught a venereal disease (it may have been syphilis or gonorrhea) from a chance encounter, the first of many serious illnesses which bedevilled her life and which were exacerbated by incurable alcoholism and heavy smoking.

On her return to America, she became an instructor at Stephens College, Columbia, Missouri, an experience that taught her she was no teacher, although she was to return to academe briefly several times. A little later she became a secretary on the *Southern Review*, and for the first time became part of the circle of East Coast writers which included **Randall Jarrell**, **Delmore Schwartz**, **Peter Taylor** and **Robert**

Lowell. She married Lowell in 1940 and, before their divorce in 1948, he had broken her nose in a fight, tried to strangle her, and crashed a car they were in, thus fracturing her skull. She spent the last year of their marriage in 'psychoalcoholic clinics'.

She won great success with her complex and highly assured first novel, *Boston Adventure* (1944), the story of a young female outsider breaking into Bostonian society (her own marriage to Lowell had been opposed by his patrician family). Her second novel, *The Mountain Lion*, showed great insight into the world of a lonely child, while her third, *The Catherine Wheel*, perhaps surrenders too much to the Gothic tendencies which always threatened her work. She never completed another novel and, although she continued to write finely crafted short stories, many of which appeared in the *New Yorker*, the flow of these dried up from 1957. An unhappy second marriage to the journalist Oliver Jensen in 1950 effectively ended within a year (they were divorced in 1955), and while her third marriage, to A.J. Liebling, the famous *New Yorker* writer on food and boxing, was happy, it was brought to an end in 1963 by his early death, which had been hastened by extreme obesity.

In her last two decades Stafford wrote mainly reviews, but also published two children's books and a book about Lee Harvey Oswald's mother. During this time she lived at East Hampton, increasingly becoming a recluse, locked in pathological hatred of her two sisters, enduring constant alcoholic falls and appalling ailments ('she was always having brain tumors and breast cancer,' said a friend). Early in 1979 she entered New York Hospital for the thirty-fourth time and died in March. She caused posthumous consternation by having left her estate to and appointed as literary executor her cleaning lady, Josephine Monsell, a woman of only rudimentary schooling.

Fiction

Boston Adventure 1944; *The Mountain Lion* 1947; *The Catherine Wheel* 1952; *Children Are Bored on Sunday* (s) 1953; *Bad Characters* (s) 1964; *The Collected Short Stories of Jean Stafford* (s) 1969

Non-Fiction

A Mother in History 1966

Edited

The Lion and the Carpenter and Other Tales (s) 1962

For Children

Arabian Nights 1962; *Elephi: the Cat with the High IQ* 1962

Biography

Jean Stafford by David Roberts 1988; *The Interior Castle* by Ann Hulbert 1992

Mary STAFFORD

see F.M. MAYOR

Jon (Howie) STALLWORTHY 1935–

Stallworthy is the son of two third-generation New Zealanders who came to England in 1934 for what was originally intended as a temporary stay: his father, Sir John Stallworthy, eventually became Nuffield Professor of Obstetrics at Oxford University. Stallworthy attended the Dragon School and Rugby, and began his army service in 1953, spending some time in Nigeria. He then attended Magdalen College, Oxford, where he won the Newdigate Prize for Poetry. He joined the Oxford University Press in 1959 and worked for them for almost twenty years in both London and Oxford, eventually becoming deputy academic publisher at Oxford from 1974 to 1977. He married in 1960, and has three children.

His early poetry was rather imitative and uncertain, but from the time of his third major collection, *Root and Branch*, he has revealed his talents as a fastidious poet of largely domestic and emotional themes, who occasionally branches out into public statement. Particularly interesting is *A Familiar Tree*, which takes his own family history as its theme and was inspired by the discovery that his great-great-grandfather had sailed from England as a missionary to the Marquesas in 1833. From 1977 to 1986 Stallworthy was John Wendell Anderson Professor of English at Cornell University in the USA, and from 1986 he has been reader, then professor, in English literature at Oxford. His later work has been more academic than poetic, although there are new poems in *The Anzac Sonata*. Among his chief academic interests is **Wilfred Owen**, of whose estate he is a trustee and whose complete poems he has edited, which has involved him in a noted literary controversy with fellow poet **Jon Silkin**. In February 1986 Stallworthy and his publishers Chatto & Windus were able to secure the withdrawal by Penguin Books of Silkin's edition of Owen's poems, claiming that Silkin had breached Stallworthy's copyright in his particular editing of the verse. A spirited correspondence ensued in the *Times Literary Supplement*, in which Stallworthy described himself as Silkin's friend, a relationship the latter deprecated. He has also written a highly praised biography of Owen and one of **Louis MacNeice**, and has edited the collected poems of **Henry Reed**.

Poetry
The Earthly Paradise 1958; *The Astronomy of Love* 1961; *Out of Bounds* 1963; *The Almond Tree* 1967; *A Day in the City* 1967; *Positives* 1969; *Root and Branch* 1969; *A Dinner of Herbs* 1970; *The Apple Barrel: Selected Poems 1956–1963* 1974; *Hand in Hand* 1974; *A Familiar Tree* 1978; *The Anzac Sonata: New and Selected Poems* 1986; *The Guest from the Future* 1989

Non-Fiction
Between the Lines: Yeats's Poetry in the Making 1963; *Vision and Revision in Yeats's 'Last Poems'* 1969; *Poets of the First World War* 1974; *Wilfred Owen* 1974; *Louis MacNeice* 1995

Edited
Yeats: Last Poems: a Casebook 1968; *New Poems 1970–1971* [with A. Brownjohn, S. Heaney] 1971; *The Penguin Book of Love Poetry* 1973 [US *A Book of Love Poetry* 1974]; Wilfred Owen: *The Complete Poems and Fragments* 1983, *The Poems of Wilfred Owen* 1985, *War Poems of Wilfred Owen* 1994; *The Oxford Book of War Poetry* 1984; *First Lines: Poems Written in Youth from Herbert to Heaney* 1987; Henry Reed *Collected Poems* 1991

Translations
Five Centuries of Polish Poetry, 1450–1970 [with J. Peterkiewicz, J.B. Singer] 1970; Alexander Blok *The Twelve and Other Poems* [with P. France] 1970 [as *Selected Poems* 1974]; Boris Pasternak *Selected Poems* [with P. France] 1983

C(hristian) K(arlson) STEAD 1932–

Stead was born in Auckland in the North Island of New Zealand, the city in which he has passed much of his life. His forebears, of English, Irish and Scots origin, were first recorded in New Zealand in 1832; his two forenames he derives from his maternal grandfather, a half-German half-Swedish sea captain, who was harbour master in a number of remote coral atolls in the South Seas. Stead's own father was an accountant, active in New Zealand Labour politics, his mother a music teacher. He was educated at Mount Albert Grammar School in Auckland, and at the university in the city, taking a first in English in 1953. He then went abroad for his doctoral studies, first to Bristol University for a year in 1954–5, followed by a year in London.

He began his teaching career in Australia, as a university lecturer at the University of New England, New South Wales, between 1956 and 1957. He then went back to London to complete his doctorate before returning to New Zealand in 1959. There his career was largely made at the University of Auckland. First he was a lecturer in English, between 1959 and 1961, then a senior lecturer from 1962 to 1964, an associate professor between 1964 and 1968, and professor of English until 1986, when he took early retirement. He had married in 1955, and subsequently had three children.

Stead has practised as a poet, critic, editor, short-story writer and novelist. His best-known single work probably remains *The New Poetic* (1964), a study of **W.B. Yeats**, **Ezra Pound**, **T.S. Eliot** and the Georgian poets which has been very widely used in university English courses. It is one of four critical works in which it is generally agreed he has proved himself an acute analyst of the modern movement. With creative writing, his earliest reputation was made as a poet who also contributed stories widely to magazines. His earliest poetic volume, *Whether the Will Is Free* (1964), collected his work from the mid-1950s and seemed accomplished but surprisingly conservative. However, in the nine volumes of verse he published up to 1990 he proved himself a poet who brought discipline to free forms and who could handle abstraction without resorting to myth.

His retirement from his academic post in 1986 was so that he could concentrate on fiction, and, after publishing a novel and story collection in 1971 and 1981, he produced further novels from the mid-1980s. He is more experimental in his novels than in his poetry, and can be said to function at the point where the autobiographical novel meets modernism, using personal and family experience to illuminate a wide variety of social problems in New Zealand, Britain and the USA. His campus novel, *The Death of the Body*, and *Sister Hollywood* are perhaps the best known. He has been intermittently active in Labour politics, and plays a senior role in literary life, without entirely escaping controversy. For instance, four authors of Maori or part-Maori origin, including Booker Prize-winner **Keri Hulme**, withdrew their stories at a late stage from *The Faber Book of Contemporary South Pacific Stories* (1994), Hulme delivering an attack on Stead's fitness to edit the volume.

Poetry
Whether the Will Is Free: Poems 1954–62 1964; *Crossing the Bar* 1972; *Quesada: Poems 1972–74* 1975; *Walking Westward* 1979; *Geographies* 1982; *Poems of a Decade* 1983; *Paris* 1984; *Between* 1988; *Voices* 1990

Fiction
Smith's Dream 1971; *Five for the Symbol* (s) 1981; *All Visitors Ashore* 1984; *The Death of the Body* 1986; *Sister Hollywood* 1989; *The End of the Century at the End of the World* 1992

Non-Fiction
The New Poetic: Yeats to Eliot 1964; *In the Glass Case: Essays on New Zealand Literature* 1981; *Pound, Yeats, Eliot and the Modernist Movement* 1986; *Answering to the Language: Essays on Modern Writers* 1989

Edited
New Zealand Short Stories (s) 1966; *Measure for Measure: a Casebook* 1971; *The Letters and Journals of Katherine Mansfield: a Selection* 1977; *Maurice Duggan: Collected Stories* (s) 1981; *The New Gramophone Room: Poetry and Fiction* [with E. Smither, K. Smithyman] 1985; *The Faber Book of Contemporary South Pacific Stories* (s) 1994

Christina (Ellen) STEAD 1902–1983

Now generally regarded as a major twentieth-century novelist, Stead lived a wandering life, which may have contributed to her comparative neglect in earlier decades. She was born in Sydney, the daughter of a naturalist Fabian socialist in the government fisheries department. Her mother died when she was young and she grew up as the responsible eldest child of a large family. She trained as a teacher, a job she found she did not enjoy, and later took a business course at night while doing various jobs, preparing for her move to London in 1928. Shortly afterwards she went to Paris, where she worked for some years in a bank and met her future husband, the broker and political economist William Blech, who also wrote romantic and historical novels under the pseudonym of William Blake.

Stead's first published fictional work was *The Salzburg Tales* (1934), a collection of stories told by different narrators in the manner of a modern *Decameron*. It was soon followed by her first novel, *Seven Poor Men of Sydney*, a story of working-class waterfront life and one of relatively few of her works to be set in Australia. In 1935 she and Blech went to Spain, but they left at the outbreak of the civil war and settled in America in 1937, where she became a senior scriptwriter at Metro-Goldwyn-Mayer in 1943 – she was to experience trouble in Hollywood during the McCarthy era because of her left-wing views – and taught a course on the novel at New York University. Her major work, *The Man Who Loved Children*, a savage story of conflict in a modern marriage and its consequences for the children involved, is set in America and has received increasing critical attention since its reprint there in 1965. Its fascinated understanding of obsessiveness and mastery of significant detail make it an important novel.

Stead went back to Europe in 1947 and settled in London in 1953; another fine novel, *Cotter's England*, carries the theme of family conflict into finely observed working-class Britain. The four novellas in *The Puzzleheaded Girl* are other major products of Stead's later years. After her husband's death in 1968, she returned to Australia, and gained increasing recognition from then on, largely owing to the growth of feminist criticism. Her final novel, *I'm Dying Laughing*, was published posthumously.

Fiction
The Salzburg Tales (s) 1934; *Seven Poor Men of Sydney* 1934; *The Beauties and Furies* 1936; *House of All Nations* 1938; *The Man Who Loved Children* 1940; *For Love Alone* 1944; *Letty Fox* 1946; *A Little Tea, a Little Chat* 1948; *The People with the Dogs* 1952; *Cotter's England* [US *Dark Places of the Heart* 1966] 1967; *The Puzzleheaded Girl* (s) 1967; *The Little Hotel* 1973; *Miss Herbert* 1979; *I'm Dying Laughing* 1987

Translations
Fernand Gignon *Colour of Asia* 1955; Jean Giltene *The Candid Killer* 1956; Auguste Piccard *In Balloon and Bathyscaphe* 1956

Edited
Modern Women in Love (s) [with 'William Blake'] 1945

Biography
Christina Stead by Hazel Rowley 1995

Timothy (Reid) STEELE 1948–

Steele was born in Burlington, Vermont. His father was a teacher and his mother a nurse. He spent his boyhood in New England, obtained his BA from Stanford University, California, and his PhD from Brandeis University, where his doctoral dissertation on the history of the conventions of detective fiction was directed by the poet J.V. Cunningham. He was married in 1979. He has taught at Stanford and the University of California at Los Angeles and now teaches at California State University, Los Angeles.

Steele's major work consists of three volumes of poetry and a book of criticism, *Missing Measures: Modern Poetry and the Revolt against Meter*. In this he examines the question of why metre, which was for so long the dominant force in English poetry, has in this century been so widely neglected. It is a book of great acuity and breadth of knowledge, and tackles a large subject from a number of different historical and literary angles. Steele believes that the modernist revolution led by **Ezra Pound** and **T.S. Eliot** created a long-lasting spate of free verse that resulted from poets 'merely following, by rote and habit, a procedure of writing and breaking up into lines, predictably mannered prose'.

Steele's own poetry is strongly metrical and very far from mannered. One of its great strengths is its colloquiality. This was apparent from his first book of poems, *Uncertainties and Rest*, indeed, from one of the first-written poems in that book, 'Coda in Wind'. This mysterious directness informs all Steele's work, not only in poems about love or death or nature – subjects that too many modern poets tackle (if at all) with unnecessary wariness and indirection – but also in his meditations on culture. The poems in particular bring the reader into an intimate conversational presence, full of a zest for the particularities of life in all their precariousness, of learning lightly worn and insight seasoned with humour. Steele's work is enjoyed in America by readers who appreciate the use of clarity, metre and rhyme, but he could not be said to have a wide audience. Like the poetry of **Robert Frost** and **Richard Wilbur**, lines of his come to mind long after we have read the poems, to delight or remind or suggest or console. One of the finest poets of this century, his work will in due course be widely recognised for its excellence; but it deserves to be better known now, especially in England, where he as yet has no publisher.

Poetry

Uncertainties and Rest 1979; *Sapphics Against Anger and Other Poems* 1986; *The Colour Wheel* 1994; *Sapphics and Uncertainties: Poems 1970–1986* 1995

Non-Fiction

Missing Measures 1990

Gertrude STEIN 1874–1946

Stein set out, almost singlehandedly she believed, to invent and define modernist literature. 'I was there to kill what was not dead,' she later wrote, 'the Nineteenth Century, which was so sure of evolution and prayers.' She was born in Allegheny, Pennsylvania, of educated German-Jewish immigrants. The family was wealthy enough to make numerous cultural trips, to Europe during Stein's childhood and then to send her for her education in 1893 to the Harvard Annex, now Radcliffe College, where she was in one of the first classes of women to earn an actual degree. Here she studied under William James, who would prove to be the most important philosopher of his age. The influence of James's pragmatism – his emphasis, that is, on a flowing consciousness and a consequently unstable and unreliable vision of reality – struck a chord with Stein, as it seemed to point a way of bringing fiction out of its naturalist nineteenth-century period. James felt very positively about Stein's work, and followed the unfolding of her career with great interest, even visiting her in Paris as late as 1908.

It was during 1903 that Stein had moved to Paris with her brother Leo, an art collector and very much a Harvard aesthete. They were both equally interested in the artistic life of the city, and soon acquired works by Cézanne, Matisse and Picasso. Stein became determined to do the same for literature as these artists had done for painting. She had brought with her a novel-in-progress, 'Quod Erat Demonstrandum' (posthumously published in 1951 as *Things as They Are*). Her first published fiction was *Three Lives* (1909), based on a reworking of a late Flaubert text called *Trois Contes*. Stein claimed that the book was 'the first definite step away from the nineteenth and into the twentieth century in literature', mainly because of its emphasis on what she called 'the continuous present'.

Stein developed her modernist literary style with greater success with *The Making of Americans*, written between 1906 and 1908 but not published until 1925, when her most famous literary pupil, **Ernest Hemingway**, helped transcribe it. In *Tender Buttons* – a series of short fragmented prose lyrics which attempt to reduce reality to a series of spatial structures – the influence of the artworks she had acquired is apparent.

When England declared war on Germany on 4 August 1914, Stein was visiting philosopher Alfred North Whitehead in England with her lover Alice B. Toklas, and they were unable to return to their home until October. After a brief trip to Majorca in 1915, the two returned to Paris because of their concern for the plight of their adopted country, and they joined the American Fund For French Wounded. Because Stein spoke excellent German, after the Armistice both she and Toklas were sent to start a civilian relief operation in Alsace. Stein wrote about her First World War experiences in *The Autobiography of Alice B. Toklas* and *Everybody's Autobiography* and both she and Toklas received the French government's Medaille de la Reconnaissance Française in 1922.

In 1934, Stein and Toklas travelled to New York, where her opera *Four Saints in Three Acts*, with music by Virgil Thomson and an all-black cast, had recently astonished audiences with what the *New York Times* called its 'spirit of inspired madness'. Stein had not seen her native country for thirty years and was treated as a celebrity. ('Gerty Gerty Stein Stein Is Back Home Home Back', ran one headline in an attempted approximation of the repetitive incantatory style of her opera libretto.) She toured the USA, giving a series of lectures on such subjects as 'The Gradual Making of the Making of America', 'Portraits and Repetition' and 'Poetry and Grammar', returning to Paris in 1935. The following year she lectured in England and stayed with **Lord Berners**, who wrote a ballet, *Wedding Bouquet* (1937), based on her play *They Must Be Wedded, To Their Wife*. She capitalised upon her public reputation in her book of memoirs, *Everybody's Autobiography*.

Although both women were Jewish, they remained in France during the Second World War, living at various country houses, under the protection of Pétain. Toklas gardened and cooked, while Stein wrote a children's book, *The World Is Round*, and a memoir of this period, *Wars I Have Seen*. In December 1944, they returned to Paris, where their apartment was unharmed, but within a year Stein was suffering from the cancer which killed her in 1946. Toklas lived on until 1967, assembling her recipes to produce *The Alice B. Toklas Cook Book* (1954), famous for its recipe for hash fudge ('which anyone could whip up on a rainy day'), and writing a lively volume of her own memoirs, *What Is Remembered* (1963).

Fiction

Three Lives **(s)** [from Flaubert's *Trois Contes*] 1909; *Concluding with As a Wife Has a Cow* **(s)** 1926; *Lucy Church, Amiably* 1930; *Matisse, Picasso and Gertrude Stein* **(s)** 1932; *Ida* 1941; *Blood on the Dining-Room Floor* 1948; *Things as They Are* 1950; *Mrs Reynolds and Five Earlier Novelettes* **(s)** 1953; *A Novel of Thank You* 1955

Plays

A Village 1928; *Four Saints in Three Acts* **(l)** [with V. Thomson] 1929 [pub. 1934]; *Old and Old* 1913, *A Movie* 1920, *Reread Another* 1921, *Objects Lie on a Table* 1922, *Saints and Singing* 1922, *Am I to Go or I'll Say So* 1923, *Capital Capitals* 1923, *A List* 1923, *Four Saints in Three Acts* 1927, *A Lyrical Opera Made by Two To Be Sung* 1928, *A Bouquet, Their Wills* 1928, *Deux soeurs qui sont pas soeurs* 1929, *At Present* 1930, *Louis XI and Madame Giraud* 1930, *Madame Recamier* 1930, *Parlor* 1930, *They Weighed Weighlayed* 1930, *Civilisation* 1931, *The Five Georges* 1931, *Lynn and the College de France* 1931, *Say It with Flowers* 1931, *They Must Be Wedded, To Their Wife* 1931; *A Wedding Bouquet* 1937; *Brewsie and Willie* 1946; *In Savoy* 1946; *Last Operas and Plays* [ed. C. Van Vechten] 1949; *Lucretia Borgia* 1968

Non-Fiction

The Making of Americans 1925; *Composition as Explanation* 1926; *Useful Knowledge* 1929; *Dix Portraits* 1930; *How to Write* 1931; *The Autobiography of Alice B. Toklas* **(a)** 1933; *Portraits and Prayers* 1934; *Lectures in America* 1935; *Narration* 1935; *The Geographical History of America* 1936; *Everybody's Autobiography* **(a)** 1937; *Picasso* 1938; *Paris France* 1940; *What Are Masterpieces?* 1940; *Wars I Have Seen* **(a)** 1945; *Four in America* 1947; *Two: Gertrude Stein and Her Brother and Other Early Portraits* 1951; *Gertrude Stein on Picasso* [ed. E. Burns] 1970; *Correspondence and Personal Essays* 1972; *The Letters of Gertrude Stein and Carl Van Vechten 1913–1946* 1986

Poetry

Tender Buttons 1914; *Stanzas in Meditation and Other Poems* 1956; *Bee Time Vine and Other Pieces* 1957

Translations

Georges Hugnet *Before the Flowers of Friendship Faded Friendship Faded* 1931

For Children

The World Is Round 1939

Collections

Geography and Plays 1922; *The First Reader, and Three Plays* 1946; *Selected Writings of Gertrude Stein* [ed. C. Van Vechten] 1946; *As Fine as Melanctha* 1954; *Painted Lace and Other Pieces* 1955; *Alphabets and Birthdays* 1957; *Look at Me Now and Here I am. Writings and Lectures 1909–1945* 1967; *Fernhurst, QED and Other Early Writings* [ed. L. Katz] 1972; *Previously Uncollected Writings of Gertrude Stein* 1974

Biography

Charmed Circle by James R. Mellow 1974; *Everybody Who Was Anybody* by Janet Hobhouse 1975; *Gertrude and Alice* by Diane Souhami 1991

John (Ernst) STEINBECK 1902–1968

Steinbeck was born at Salinas, centre of the Californian lettuce industry, and his native region of Monterey Bay is constantly present in his writings. He was of German–Irish ancestry; his father was a county treasurer and his mother a teacher. He attended the local high school and Stanford University, where he intermittently studied science without taking a degree. Afterwards, he went to New York to work as a reporter, a job from which he was fired; he then took many menial jobs including apprentice hod-carrier, apprentice painter, caretaker of an estate, surveyor and fruit-picker. He married in 1930.

His first three novels made him very little money, but success came in 1935 with *Tortilla Flat*, a lively story of the *paisanos* (peasants of mixed blood) of California. Steinbeck continued to write sympathetically of the rural dispossessed in such books as *Of Mice and Men*, and he reached his apogee with *The Grapes of Wrath*, the story of the Joad family who leave the 'dust bowl' of Oklahoma to work as fruit-pickers in California. The enormous success of this novel, and its awakening of America's social conscience, have led to its impact being compared to that of Harriet Beecher Stowe's *Uncle Tom's Cabin* (1852). During the war Steinbeck, now famous, did special writing assignments for the US forces and was a war correspondent. In later years, he did much special reporting abroad, divided his time between New York and California and continued to be a controversial figure because of his support for the underprivileged, but he did not receive the acclaim for later novels, such as *East of Eden*, that earlier ones had achieved. Sentimentality and occasional pretentiousness are perhaps the faults of his writing: a great flair for vivid and realistic description and a capacity to move the reader are among its virtues. Steinbeck was a large, reticent man, widely respected for his impressive human qualities; he was married three times, and had two children by his second marriage. He also wrote plays and short stories: superb examples of the latter are collected in *The Long Valley*. He was awarded a Nobel Prize in 1962.

Fiction

Cup of Gold 1929; *The Pastures of Heaven* 1932; *To a God Unknown* 1935; *Tortilla Flat* 1935; *In Dubious Battle* 1936; *Of Mice and Men* 1937; *Red Pony* 1937; *The Long Valley* (s) 1938; *The Grapes of Wrath* 1939; *The Moon Is Down* 1942; *Cannery Row* 1945; *The Pearl* 1947; *The Wayward Bus* 1947; *Burning Bright* 1950; *East of Eden* 1952; *Sweet Thursday* 1954; *The Short Reign of Pippin IV* 1957; *The Winter of Our Discontent* 1961

Non-Fiction

The Sea of Cortez [with E. Ricketts] 1941; *A Russian Journal* 1948; *The Log from The Sea of Cortez* 1951; *Once There Was a War* 1958; *Travels with Charley: In Search of America* 1962; *America and the Americans* 1966; *Journal of a Novel. The 'East of Eden' Letters* 1969; *Steinbeck: a Life in Letters* [ed. E. Steinbeck, R. Wallsten] 1975; *Letters to Elizabeth: a Selection of Letters from J. Steinbeck to Elizabeth Otis* [ed. F. Riggs, F. Shasky] 1978

Biography

John Steinbeck by Jay Parini 1994

(Francis) George STEINER 1929–

Steiner was born in Paris to Austrian Jewish parents but was brought by his family to the USA in 1940 just before the Nazi occupation. After receiving his BA from the University of Chicago in 1948 and an MA from Harvard in 1950, he gained a DPhil as a Rhodes scholar at Oxford in 1955. Apart from academic posts in America, Steiner has been mainly based at universities in England and Switzerland: from 1961 to 1969 he was a fellow of Churchill College, Cambridge, and later, he was professor of comparative literature at the University of Geneva. He was a staff member of the *Economist* from 1942 to 1956. He married in 1955 and has two children.

Steiner's writings are those of a *Kulturkritik* in the humanist tradition of Central European Jewry; they draw on an international range of learning and erudition that endows them with a distinctively authoritative gravitas. His first book, *Tolstoy or Dostoevsky: an Essay in the Old Criticism* (1959), was a broadside at the so-called New Criticism of the 1950s which sought to separate literature from the subjectivity of the author, from its historic background and from questions of morality, politics and ideology. This and all his subsequent work emphasises the inseparability of the text from the inwardness of the author and the particular spiritual and historical milieu in which that text was conceived. His sense of the complexity of the creative process has led him into a wide-ranging enquiry into the nature and forms of literary art and its connections with the history of ideas and culture.

Another of his characteristic concerns is with how the extermination of the Jews in the Second World War has profoundly affected the assumption that literature is a humanising influence. For Steiner the revelation of the Holocaust was that 'man can read Goethe or Rilke in the evening, that he can play Bach or Schubert, and go to his day's work at Auschwitz in the morning'. This contradiction imbues such works of criticism as *Language and Silence*, *In Bluebeard's Castle* and the novel *The Portage to San Cristobal of A.H.* One of the alternatives, says Steiner, is silence – the tacit admission that the civilising power of a culture based on 'the word' is an illusion. In *The Portage to San Cristobal of A.H.* (dramatised by **Christopher Hampton** in 1982), Steiner explores the disquieting implications for interpretation of the whole phenomenon of Adolf Hitler. Controversially, Steiner has Hitler defending his actions by using ideas and language from Steiner's own essays, including the contention that the Jews are responsible for pressing the 'blackmail of transcendence' on to a hitherto innocently pagan world. He also portrays Hitler as having a withered arm, which Steiner himself has.

His other principal work, *After Babel*, is a sophisticated investigation into the nature of language which asks the question: why is there such a diverse and 'destructive prodigality' of human language given the generally held evolutionary belief that language is essentially adaptive behaviour?

Steiner's cosmopolitan, polymathic virtuosity has contributed to the discourse of ideas in the fields of linguistics, cultural criticism, intellectual history and philosophical anthropology, and his reflections on what he sees as the religious tendency of the literary imagination remain a thorn in the side of the deconstructionists.

As well as literary journalism, Steiner has also written on chess, edited *The Penguin Book of Modern Verse in Translation*, produced two books of short stories, one of which won the Silver Pen Award in 1993, and a volume of collected verse.

Fiction

Anno Domini (s) 1964; *The Portage to San Cristobal of A.H.* 1981; *Proofs and Three Parables* (s) 1993

Non-Fiction

Tolstoy or Dostoevsky 1959; *The Death of Tragedy* 1961; *Language and Silence: Essays on Language, Literature, and the Inhuman* 1967; *Extraterritorial: Papers on Literature and the Language Revolution* 1971; *In Bluebeard's Castle: Some Notes towards the Redefinition of Culture* 1971; *Fields of Force: Fischer and Spassky in Reykjavik* [UK *The Sporting Scene: White Knights in Reykjavik*] 1973; *After Babel: Aspects of Language and Translation* 1975 [rev. 1992]; *On Difficulty and Other Essays* 1978; *Martin Heidegger* [UK as *Heidegger*] 1978; *Antigones: How the Antigone Legend Has Endured in Western Literature, Art, and Thought* 1984; *A Reading against Shakespeare: the W.P. Ker Lecture* 1986; *Real Presences: the Leslie Stephen Memorial Lecture* 1986

Edited

Homer: a Collection of Critical Essays [with R. Fagles] 1962; *The Penguin Book of Modern Verse in Translation* 1966 [as *Poem into Poem: World Poetry in Modern Verse Translation* 1970]

Poetry

Real Presences 1989

Collections

George Steiner: a Reader 1984

James STEPHENS 1880–1950

Stephens was born in Dublin, but the circumstances and date of his birth are vague. Stephens himself obscured much of his early life: too shamingly squalid or too embarrassingly respectable? Either seems possible. Most probably he was born in 1880, the son of an HMSO van driver. (**James Joyce** remained stubbornly convinced that he and Stephens shared a birthday of 2 February, 1882, and advanced the fact as a compelling reason why Stephens should complete *Finnegans Wake* if death prevented Joyce from doing so. Scholars still wrangle over whether he was serious or not.) His father died when Stephens was presumably just over two years old, and it seems almost certain that the next four years were lived in the precarious poverty of the Dublin slums. (Stephens's first novel, *The Charwoman's Daughter* would contain some of Irish fiction's earliest descriptions of that teeming world of hardship and disease.)

When he was six Stephens was awarded a place at the Meath Protestant Industrial School for Boys. To have committed the civil offence of begging in the streets was the usual passport to this institution but Dublin's poorest mothers prized places there because it gave their sons a sound commercial or technical education. There he acquired the habits of reading and writing – often on paper bags that had held sweets – and was an enthusiastic gymnast. This enthusiasm, which survived into adult life, stemmed in part from a delight in physical activity (he was always a passionate dancer), in part from a (vain) hope that exercise would aid his growth. Notwithstanding his tiny stature (below five feet in height) and somewhat odd appearance, he has been described as 'one of the three giants of Irish talk'.

On leaving school, Stephens went to live with the family of a schoolfellow, Tom Collins, taking various jobs as a solicitor's typist, reading widely and writing incessantly. At some point in 1901 Stephens left his job and the Collinses, who never heard from him again. What exactly he did for the next five years remains a mystery. He later claimed that, among other things, he had become a clown in a circus; wrestled, starving, with starving dogs as desperate as he for dusty loaves; and owed his survival to a Belfast prostitute who took him in, penniless and sick, and nursed him like a mother. Whatever the truth may be, in 1905 he re-emerged in Dublin with his first publication, a story called 'The Greatest Miracle', which appeared in the *United Irishman*, a magazine edited by Arthur Griffith who would eventually become the first president of the Irish Free State, and was Stephens's first truly intimate friend. When the *United Irishman* was suppressed in 1906 for its revoluntionary views, Griffith promptly founded Sinn Fèin instead. In its office the great **AE** (George Russell) found Stephens apparently when he was looking for a rising star to set against **W.B. Yeat**'s new protégé, **J.M. Synge**.

Under the partronage of AE, one of Ireland's foremost literary and political figures, Stephens was launched into a group of writers and thinkers who included **Oliver St John Gogarty**, Padraic Colum, Stephen MacKenna and Maud Gonne. He had also found the pleasures of family life, living now with a woman who had been abandoned by her husband; with her he had a son and a daughter. Later, when the husband's death released her, they at last married.

Stephens's first volume of poems, *Insurrections* (1909), had an enormous impact: readers were gripped by its mixture of stark subject-matter and startlingly fresh and direct diction. It had a strongly Blakean visionary quality, deeply suspicious of organised religion and man-made justice, ardently hymning the virtues of childlike innocence and kindliness. His reputation was further enhanced by the publication, in 1912, of *The Charwoman's Daughter*, a combination of social realism, political allegory, fairy-tale and love story. Encouraged, Stephens became a full-time writer, and continued with the work which was to be his most popular, *The Crock of Gold* (1912). He made sufficient money from his writing to move temporarily to Paris and to rent a flat there which remained a precious haven for the rest of his life. In 1913 *The Crock of Gold* was awarded the Edmund Polignac Prize, and elicited one of Yeats's rare public endorsements of a living Irish author.

Enemies and detractors would later claim that Stephens had become a 'professional Irishman'. They dubbed him 'The Leprechaun', a reference both to his diminutive stature and to a character in *The Crock of Gold*. Such dismissive attitudes ignored the long years spent studying Irish history, literature and myth, and failed to recognise that, deeply though Stephens loved Ireland, his primary allegiance was always to poetry. As early as 1907 his 'Essay on Poetry' had insisted that imagination is the basis of all great art and that poetry is the greatest art of all. He never shifted from this belief, although his own poetic practice would undergo great changes, influenced by his enquiries into Eastern philosophy and religion, and by his increasing desire to bring written poetry closer to the oral. His *Collected Poems* of 1926 contains many revisions made to further that aim. In 1916, the momentous year of the Easter Rising, Stephens wrote to a friend that his love of Ireland was separate from politics, although the same year he engaged directly with political concerns in *The Insurrection in Dublin*, his attempt at an objective account of the causes of the year's events which he hoped both sides might find useful.

For the next eight years he poured his hopes and learning into three works which grew from his enormous knowledge of Irish myth and legend: *Irish Fairy Tales*, *Deirdre*, and *In the Land of Youth*. They were to be his last substantial prose works. In 1925, for reasons never fully disclosed, he left Ireland and never lived there again. After a lecture tour in America, he bought a house in Wembley, near London and, apart from a sojourn in Gloucestershire during the Second World War, lived there for the rest of his life. Quickly he made a place for himself in literary London. He anthologised, lectured in America, made trips to Paris, deepened his readings of Indian and Buddhist writings, and made many broadcasts for the BBC. But he wrote comparatively little. Something in England, he told AE, had sucked his stories from him, 'like a jug that had been turned upside down'. The deep loneliness he felt increasingly was intensified by the death of his son, killed in an accident on the morning of Christmas Eve, 1937. Money worries, at least, were at last eased permanently by a civil list pension awarded in 1942. Honours were accorded him, including the degree of Doctor of Letters, bestowed by the University of Dublin in 1947.

Poetry

Insurrections 1909; *The Lonely Road and Other Poems* 1909; *The Hill of Vision* 1912; *Five New Poems* 1913; *Songs from the Clay* 1915; *The Adventures of Seumas Beg/The Rocky Road to Dublin* 1915; *Green Branches* 1916; *Hunger* [as 'James Esse'] 1918; *Reincarnations* 1918; *Little Things* 1924; *Christmas in Freelands* 1925; *A Poetry Recital* 1925; *Collected Poems* 1926; *The Optimist* 1929; *The Outcast* 1929; *Theme and Variations* 1930; *Strict Joy* 1931; *Stars Do Not Make a Noise* 1931; *Kings and the Moon* 1938; *A James Stephens Reader* [ed. L. Frankenberg] 1962

Fiction

The Charwoman's Daughter 1912; *The Crock of Gold* 1912; *Here Are Ladies* (s) 1913; *The Demi-Gods* 1914; *Irish Fairy Tales* (s) 1920; *Deirdre* 1923; *In the Land of Youth* (s) 1924; *Etched in Moonlight* (s) 1928; *Irish Stories and Tales* (s) [with others] [ed. D. Garrity] 1955

Non-Fiction

The Insurrection in Dublin 1916; *Arthur Griffith: Journalist and Statesman* 1922; *On Prose and Verse* 1928; *How St Patrick Saves the Irish* 1931; *James, Seumas and Jacques: Unpublished Writings by James Stephens* [ed. L. Frankenberg] 1964; *Letters of James Stephens* [ed. R.J. Finneran] 1974

Plays

Julia Elizabeth 1929

Edited

The Poetical Works of Thomas MacDonagh 1916; *English Romantic Poems* [with E.L. Beck,

R.H. Snow] 1933; *Victorian and Later Poets* [with E.L. Beck, R.H. Snow] 1934

G(ladys) B(ronwyn) STERN 1890–1973

Stern was given the second name Bertha by her parents but changed this to Bronwyn. She was born into a family of wealthy Jews who had emigrated to England from central Europe in the middle of the nineteenth century, and she grew up in some splendour in Holland Park until the Vaal River diamond smash, after which the Sterns lived more modestly in a series of private hotels. She attended Notting Hill High School until she was sixteen, and then travelled with her parents in Germany and Switzerland, attending a day school at Wiesbaden and being 'finished and given a lick and a high polish' at Montreux. She studied at the Academy of Dramatic Art in London, but did not realise her ambition of going on the stage. Instead, she published her first novel, *Pantomime*, in 1914, and was eventually to follow it with more than forty others, all dictated to a series of secretaries, who left their mistress with predictable regularity.

In 1919 she married a New Zealand journalist, Geoffrey Lisle Holdsworth, whom she had met through **Noël Coward**. He was handsome, but he suffered from clinical depression, and the marriage did not last long. Thereafter Stern concentrated her affections on a series of women friends, to whom she was known as 'Peter': what male intimates she had, such as **John van Druten** and his lover Jack Cohen, tended to be homosexual. During the 1920s and 1930s she travelled widely, living in Italy (for five years), Cornwall, New York, Hollywood and the south of France; in later years, she divided her life between London (where she had a flat in Albany) and a cottage in Berkshire.

She is best known today for her series of five 'Matriarch' novels, which appeared from 1924 to 1942, and which are sometimes referred to as *The Rakonitz Chronicles*. They are based on her own extended family: she uses the family names of Rakonitz and Czelovar for the fictional characters, while the figure of the Matriarch is a portrait of her great-aunt, Anastasia Schwabacher. Several other of her earlier novels have high spirits and verve, but the later ones are much weaker. She wrote a number of books of a general nature, such as two on Jane Austen (in collaboration with **Sheila Kaye-Smith**) and one on Robert Louis Stevenson, as well as plays (including *Raffle for a Bedspread*, 'a one-act play for women only') and many rather inconsequential autobiographies in which she proved, as Richard Altich said, 'the importance and the dignity of the individual whim'. Her voluminous literary production came to an end in the early 1960s.

Fiction

Pantomime 1914; *See-Saw* 1914; *Twos and Threes* 1916; *Grand Chain* 1917; *A Marrying Man* 1918; *Children of No Man's Land* 1919 [US *Debatable Ground* 1921]; *Larry Munro* 1920 [US *The China Shop* 1921]; *The Room* 1922; *The Back Seat* 1923; *Smoke Rings* (s) 1923; *The Rakonitz Chronicles*: *Tents of Israel* 1924 [US *The Matriarch* 1948], *A Deputy was King* 1926, *Petruchio* 1929 [US *Modesta* 1931], *Shining and Free* 1935, *The Young Matriarch* 1942; *Thunderstorm* 1925; *The Happy Meddler* [with G. Holdsworth] 1926; *The Dark Gentleman* 1927; *Jack a' Manory* (s) 1927; *Mosaic* 1930; *The Shortest Night* 1931; *Little Red Horses* 1932 [US *The Rueful Mating* 1933]; *Long Lost Father* 1932; *The Angs* 1933; *Pelican Walking* (s) 1934; *Oleander River* 1937; *The Ugly Dachshund* 1938; *The Woman in the Hall* 1939; *Long Story Short* (s) 1939; *A Lion in the Garden* 1940; *Dogs in an Omnibus* 1942; *These Reasonable Shores* 1946; *No Son of Mine* 1948; *A Duck to Water* 1949; *Ten Days of Christmas* 1950; *The Donkey Shoe* 1952; *A Name to Conjure With* 1953; *Johnny Forsaken* 1954; *For All We Know* 1955; *Seventy Times Seven* 1957; *The Patience of a Saint* 1958; *Unless I Marry* 1959; *Bernadette* 1960; *One Is Only Human* 1960; *Credit Title* 1961; *The Woman in the Hall* 1961; *Dolphin Cottage* 1962; *Promise Not to Tell* 1964

Non-Fiction

Bouquet (a): *Travels in the Wine Producing Regions of France* 1927, *Monogram* (a) 1936, *Another Part of the Forest* (a) 1941, *Trumpet Voluntary* 1944, *Benefits Forgot* (a) 1949, *All in Good Time* 1954, *The Way It Worked Out* 1956; *Talking of Jane Austen* [with S. Kaye-Smith] 1943; *More Talk of Jane Austen* [with S. Kaye-Smith] 1950; *He Wrote Treasure Island: the Story of Robert Louis Stevenson* 1954; *... And Did He Stop to Speak to You?* 1957

Plays

The Matriarch [from her novel] 1929 [with F. Vosper; pub. 1931]; *The Man Who Pays the Piper* 1931; *Gala Night at 'The Willows'* 1950 [with R. Croft-Cooke]; *Raffle for a Bedspread* [pub. 1953]

James (Andrew) STERN 1904–1993

Stern was born in County Meath in Ireland, on his father's side from the enormously rich German-Jewish banking firm of Stern Brothers, on his mother's from hard-riding Anglo-Irish aristocracy. His father was an officer in the British army; as he himself said, he had few early memories of ascendancy Ireland that did not include a horse. He was at Eton in the classic generation of **Harold Acton** and **Cyril Connolly**, but his background was not sufficiently central, and he did not write enough, to avoid marginalisation as a writer.

After Eton, he spent a year in army training at Sandhurst, but then, aged twenty, sailed for Rhodesia, assisting a Scottish farmer with a herd of Hereford cattle, and gaining the experience of social isolation and racial tension which informs the powerful stories of Africa in his first collection, *The Heartless Land* (1932). Returned from Africa, he spent some time in the family bank in England and Germany, but escaped from there to an assistant editorship at the *London Mercury* and later to the hotel room in Paris where Oscar Wilde had died, where he wrote his first stories in the early 1930s. In Paris, too, he met his wife of sixty years, Constanze ('Tania') Kurella, the refugee daughter of a Berlin doctor, whom he married in 1935. Together the Sterns translated many works by leading European writers, such as Brecht, Kafka, Thomas Mann and Stefan Zweig.

A fair proportion of Stern's work in the short story dates from the 1930s; the two later of his four collections contain little new work, and some of that is reportage. His second collection, *Something Wrong*, concentrates largely on the trials of upper-class youth, but the range of his stories in general is very wide, both socially and geographically, and they are all marked by a strong sense of physical and psychological reality; perhaps their only significant weakness is an occasional over-elaboration and uncertainty in linguistic register. Stern's only other major work is *The Hidden Damage*, a revealing account of a visit he made with an Allied team to immediately post-war Germany. This book, published in the USA in 1947, was lost by the publisher Roger Senhouse in his office, and did not appear in England until 1990.

The Sterns lived in the USA for much of the 1940s and 1950s, returning in 1957, and taking up residence at Hatch Manor in Wiltshire, where for many years they entertained their wide range of literary friends. Perhaps the most significant of these was **W.H. Auden**, whom they had known both in the 1930s and later in New York, where Tania Stern attempted to teach the poet callisthenics, the system of physical relaxation in which she was expert; despite his homosexuality, he believed himself a little in love with her. Auden appears in *The Hidden Damage* as Mervyn, and the poet spent his last Christmas, in 1972, at Hatch Manor, celebrating the experience in his poem 'Thank You, Fog'. Once again, Stern appears in this relationship as an adjunct to the celebrated rather than one whose own talent made him noteworthy. Perhaps the publication of a dozen fine uncollected stories would help to increase interest in this underrated writer.

Fiction

The Heartless Land (s) 1932; *Something Wrong* (s) 1938; *The Man Who Was Loved* (s) 1951; *The Stories of James Stern* (s) 1968

Non-Fiction

The Hidden Damage 1947

Translations

Stefan Zweig *Brazil: Land of the Future* 1942; Herman Ullstein *The Rise and Fall of the House of Ullstein* 1944; Hugo von Hofmannsthal *Selected Writings* vol. 1 [with M. Hottinger, T. Stern] 1952; Erich Maria Remarque *Spark of Life* 1952; *A Woman in Berlin* [anon.] 1954; Leo Lania *The Foreign Minister* 1957; *Thomas Mann: Last Essays* [with T. Stern, C. and R. Winston] 1959; Franz Kafka: *Letters to Milena* [with T. Stern; ed. W. Haas] 1953, *Description of a Struggle and The Great Wall of China* [with E. and W. Muir, T. Stern] 1960, *Letters to Felice* [with E. Duckworth; ed. J. Born, E. Heller] 1974; *Letters of Sigmund Freud 1873–1939* [with T. Stern; ed. E.L. Freud] 1961; Bertolt Brecht *The Caucasian Chalk Circle* [with W.H. Auden, T. Stern] 1963; Hermann Kesten *Casanova* [with R. Pick] 1963

Edited

Grimms' Fairy Tales [from M. Hunt's translation] 1944

Wallace STEVENS 1879–1955

Stevens's mother's family, the Zellers, was of Dutch origin and he was born in Reading, Pennsylvania. He was educated at Harvard, and spent a year as a journalist, then attended the New York Law School, and practised law in New York until 1916. During this period he began to write and publish poetry, some of it in **Harriet Monroe**'s magazine *Poetry*. His first play, *Three Travellers Watch a Sunrise*, won that magazine's prize for verse drama in 1916.

From 1916 until his death he worked for the Hartford Accident and Indemnity Company, starting in the legal department and rising to become a vice-president of the firm in 1934. A large, rather fat man, who liked good living and collecting pictures, and cultivated simplistic right-wing views, Stevens seems to have felt little tension between his two roles ('What! Wally a poet!', one of his colleagues is alleged to have remarked when being told). His poetry, pursued largely in private, is extremely difficult and modernistic, treating philosophical themes in a manner at once witty and profound, and his work is a major poetic corpus of the century.

He did not publish his first volume, *Harmonium*, until 1923, when he was forty-four; it attracted almost no notice, sold 100 copies and was remaindered, but it contains many of his best poems, including the famous 'Sunday Morning' (one of the finest modern poems about death), 'The Emperor of Ice Cream' and 'Thirteen Ways of Looking at a Blackbird'. He did not publish another volume for over a decade, but after that, they came steadily. Amongst subsequent volumes are *The Man with the Blue Guitar* and *Notes Towards a Supreme Fiction*, the title-poems of which are amongst his best-known work. The title of the latter volume suggests Stevens's cerebral qualities as a poet, and he is also a quintessentially American poet. When asked whether it was possible to speak of an American (rather than a British) poem, he replied characteristically: 'We live in two different physical worlds, and it is not nonsense to think that that matters.' In the year of his death he won a Pulitzer Prize for his *Collected Poems* and since then his reputation has continued to grow. His work besides poetry comprised one volume of essays, *The Necessary Angel: Essays on Reality and the Imagination*, which won a National Book Award. Stevens was married and had one daughter.

Poetry

Harmonium 1923; *Ideas of Order* 1935; *Owl's Clover* 1936; *The Man with the Blue Guitar and Other Poems* 1937; *Notes Towards a Supreme Fiction* 1942; *Parts of a World* 1942; *Esthétique du Mal* 1945; *Transport to Summer* 1947; *Three Academic Pieces* 1947; *A Primitive Like an Orb* 1948; *The Auroras of Autumn* 1950; *Selected Poems* 1953; *The Collected Poems of Wallace Stevens* 1954

Plays

Three Travellers Watch a Sunrise 1920 [pub. 1916]; *Carlos Among the Candles* 1917; *Bowl, Cat and Broomstick* [pub. in *The Palm at the End of the Mind* 1971]

Non-Fiction

The Necessary Angel: Essays on Reality and the Imagination 1951; *Letters of Wallace Stevens* [ed. H. Stevens] 1966

Collections

Opus Posthumous [ed. S.F. Morse] 1957; *The Palm at the End of the Mind* [ed. H. Stevens] 1971

Anne (Katherine) STEVENSON 1933–

Although her parents were both American, Stevenson was born in Cambridge, England, where her father was studying philosophy with Ludwig Wittgenstein. When she was six months old, the family returned to Cambridge, Massachusetts, where her father, 'a first generation intellectual', held a fellowship at Harvard. The family subsequently moved to New Haven when her father started teaching at Yale. Stevenson recalls her father, and his love of poetry and the piano, in her poem 'Elegy' (in *The Other House*). She was educated at the University High School at Ann Arbor and at the University of Michigan.

Her first volume of poetry, *Living in America*, was published in 1965 and was followed four years later by *Reversals*. Her third volume, *Correspondences*, made her reputation on both sides of the Atlantic. An epistolary poem, it traced the history of a New England family through 150 years and was written as 'an act of rebellion'. She says that she had decided to reject 'the whole scheme of academia ... Suddenly we were plunged into ... a world threatened by the atomic bomb, nuclear fission, a third world of oppression and suffering.' She describes the poem as 'grappling with this sense of American guilt'.

Her distaste for the America of the 1970s led to her coming to live in Britain, where she has had a distinguished career as an academic, holding a number of fellowships at universities in Scotland and England. In 1979 she founded the Poetry Bookshop in Hay-on-Wye with her third husband and a friend. She sees her work as constantly evolving: 'Each of my books, I think, represents a stage in a quest.' *The Fiction-Makers* was a Poetry Book Society Choice and arose from the idea that 'much of social behaviour is fiction-making'. The book contains 'Willow Song', a beautiful poem written in memory of her friend **Frances Horovitz**. Many of her poems evoke landscapes she has known, and the quality of her work is suggested in John Lucas's description of her in the *New Statesman* as 'scrupulously talented'.

Her biography of **Sylvia Plath,** *Bitter Fame*, has, like every other book written about the poet, aroused controversy. Extremely hostile to its subject, it was written with the co-operation of Plath's sister-in-law, Olwen Hughes, and is one of the books discussed in the American journalist Janet Malcolm's contentious book *The Silent Woman* (1994).

Poetry

Living in America 1965; *Reversals* 1969; *Correspondences* 1974; *Travelling Behind Glass: Selected Poems 1963–1973* 1974; *Cliff Walk* 1977; *Enough of Green* 1977; *A Mordern Tower Reading 3* [with others] 1977; *Sonnets for Five Seasons* 1979; *Green Mountain Black Mountain* 1982; *Minute by Glass Minute* 1982; *New Poems*

1982; *A Legacy* 1983; *Making Poetry* 1983; *Black Grate Poems* 1984; *The Fiction-Makers* 1985; *Winter Time* 1986; *Selected Poems: 1956–1986* 1987; *The Other House* 1990; *Four and a Half Dancing Men* 1993

Plays
Correspondences [from her poem] 1975; *Child of Adam* 1976

Non-Fiction
Elizabeth Bishop 1966; *Bitter Fame* 1989

Edited
Selected Poems by Frances Bellerby 1986; *Avron 1985 Anthology* [with A. Clampitt, C. Raine] 1987; *Netting the Sun: New and Collected Poems by Phoebe Hesketh* 1989

J(ohn) I(nnes) M(ackintosh) STEWART 1906–1995

Stewart wrote non-genre fiction under his own name and crime fiction under the partial pseudonym of Michael Innes. He was born just outside Edinburgh, the son of the director of education in the city, and from families of Scottish gentry on both sides. He attended Edinburgh Academy (where Robert Louis Stevenson had been a pupil) and Oriel College, Oxford, where he graduated in English in 1928. He then spent a year travelling in Germany and Austria with A.J.P. Taylor, who had been a university friend, before attracting the attention of Francis Meynell of the Nonesuch Press, who asked him to edit Florio's *Montaigne*, his first book, published in 1931. On the strength of this scholarly work, he obtained a lectureship at the University of Leeds, which he held from 1930 to 1935. While there, he married Margaret Harwick, a doctor, in 1932. She died in 1979, and they had five children; one son is the economist and academic Michael James Stewart, another is the novelist Angus Stewart, author of the remarkable *Sandel* (1968).

In 1935, he sailed to Adelaide in South Australia to take up a professorship there, and on the long sea voyage out the first of the Michael Innes detective novels was written. This was *Death at the President's Lodging*, published in 1936, which, in its college setting, panoply of learned allusions and donnish wit, set a sparkling tone for the many Innes novels which were to follow, at a rate of almost one a year, until the mid-1980s. Not all were straight detective stories – dramas of flight and pursuit (*The Man from the Sea*, *The Journeying Boy*) bulked large, as did novels which explored fantasy (*Appleby on Ararat*), and those which almost abandoned detection for a flow of witty talk (*Stop Press*) – but most shared the urbane figure of the aristocratic policeman John Appleby and a cultivated atmosphere which refused to tangle with the reality of modern crime. A similar world was explored in the novels he published as J.I.M. Stewart, which began with *Mark Lambert's Supper* in 1954, and again continued until the mid-1980s. Most famous were the novels describing the academic background of Duncan Pattullo, which formed a five-volume sequence entitled *A Staircase in Surrey*. Here, though, many critics felt that the tendency to descend into weariness and lack of cohesion that had threatened the style of Michael Innes had become more evident.

Stewart returned from Adelaide in 1945, was briefly a lecturer at Queen's University, Belfast, and from 1948 until 1973 a student (i.e. fellow) of Christ Church, Oxford, also holding a readership in English literature from 1969, and completing one academic tour of duty in the USA, at Seattle, in 1961. Perhaps his best-known academic books are his study of *Eight Modern Writers* and the well-received biography of **Thomas Hardy**, published in 1971. He has also published a radio play and short stories, and his *oeuvre* of more than 100 titles, linking the privileged world of the 1920s and 1930s with the culminating years of the century, was crowned by the elegant if not particularly revealing autobiography, *Myself and Michael Innes*, which appeared in 1987.

Fiction
As J.I.M. Stewart: *Mark Lambert's Supper* 1954; *The Guardians* 1957; *A Use of Riches* 1957; *The Man Who Wrote Detective Stories and Other Stories* (s) 1959; *The Man Who Won the Pools* 1961; *The Last Tresilians* 1963; *An Acre of Grass* 1966; *The Aylwins* 1967; *Vanderlyn's Kingdom* 1968; *Avery's Mission* 1971; *A Palace of Art* 1972; *Mungo's Dream* 1973; *A Staircase in Surrey*: *The Gaudy* 1975, *Young Pattullo* 1976, *A Memorial Service* 1976, *The Madonna of the Astrolabe* 1977, *Full Term* 1978; *Cucumber Sandwiches and Other Stories* (s) 1979; *Our England Is a Garden and Other Stories* (s) 1979; *Andrew and Tobias* 1980; *The Bridge at Arta and Other Stories* (s) 1982; *My Aunt Christina and Other Stories* (s) 1983; *A Villa in France* 1983; *An Open Prison* 1984; *The Naylors* 1985; *Parlour 4 and Other Stories* (s) 1986

As 'Michael Innes': *Death at the President's Lodging* [US *Seven Suspects*] 1936; *Hamlet, Revenge!* 1937; *Lament for a Maker* 1938; *Stop Press* [US *The Spider Strikes*] 1939; *The Secret Vanguard* 1940; *There Came Both Mist and Snow* [US *Comedy of Errors*] 1940; *Appleby on Ararat* 1941; *The Daffodil Affair* 1942; *The Weight of the Evidence* 1943; *Appleby's End* 1945; *From*

London Far [US *The Unsuspected Chasm*] 1946; *What Happened at Hazelwood?* 1946; *A Night of Errors* 1947; *The Journeying Boy* [US *The Case of the Journeying Boy*] 1949; *Three Tales of Hamlet* (s) [with R. Heppenstall] 1950; *Operation Pax* [US *The Paper Thunderbolt*] 1951; *A Private View* 1952 [US *One Man Show* 1952, as *Murder Is an Art* 1959]; *Christmas at Candleshoe* 1953 [as *Candleshoe* 1978]: *Appleby Talking* (s) [US *Dead Men's Shoes*] 1954; *The Man from the Sea* [US *Death by Moonlight*] 1955; *Old Hall, New Hall* [US *A Question of Queens*] 1956; *Appleby Plays Chicken* [US *Death on a Quiet Day*] 1957; *Appleby Talks Again* (s) 1957; *The Long Farewell* 1958; *Hare Sitting Up* 1959; *The New Sonia Wayward* [US *The Case of Sonia Wayward*] 1960; *Silence Observed* 1961; *A Connoisseur's Case* [US *The Crabtree Affair*] 1962; *Money from Holme* 1964; *The Bloody Wood* 1966; *A Change of Heir* 1966; *Appleby at Allington* [US *Death by Water*] 1968; *A Family Affair* [US *Picture of Guilt*] 1969; *Death at the Chase* 1970; *An Awkward Lie* 1971; *The Open House* 1972; *Appleby's Answer* 1973; *Appleby's Other Story* 1974; *The Appleby File* 1975; *The Mysterious Commission* 1975; *The 'Gay Phoenix'* 1977; *Honeybath's Haven* 1977; *The Ampersand Papers* 1978; *Going It Alone* 1980; *Lord Mullion's Secret* 1981; *Sheiks and Adders* 1982; *Appleby and Honeybath* 1983; *Carson's Conspiracy* 1984; *Appleby and the Ospreys* 1986; *Death at the President's Lodging* 1989; *The New Sonia Wayward* 1991; *Silence Observed* 1992

Plays

Strange Intelligence [as 'Michael Innes'] (r) 1947 [pub. 1948]

Non-Fiction

Educating the Emotions 1944; *Character and Motive in Shakespeare: Some Recent Appraisals Examined* 1949; *James Joyce* 1957 [rev. 1960]; *Eight Modern Writers* 1963; *Thomas Love Peacock* 1963; *Writers of the Early Twentieth Century* 1963; *Rudyard Kipling* 1966; *Joseph Conrad* 1968; *Shakespeare: Lofty Scene* 1971; *Thomas Hardy: a Critical Biography* 1971; *Myself and Michael Innes* (a) 1987

Edited

Montaigne's Essays: John Florio's Translation 1931; Wilkie Collins *The Moonstone* 1966; Thackeray *Vanity Fair* 1968

Tom STOPPARD 1937–

Stoppard was born Tomas Straussler in Zlin, Czechoslovakia, the son of a company doctor with the Bata shoe company. The family moved to Singapore in 1939, and he and his mother and brother were evacuated to Darjeeling in India following the Japanese invasion in 1942. After his father's death his mother married Kenneth Stoppard, a British Army major, and the family moved to Bristol. In England Stoppard completed his education at the Dolphin School, Nottinghamshire, and Pocklington School in Yorkshire. At the age of seventeen he became a reporter on the *Western Daily Press*, then a feature writer on the Bristol *Evening World* in 1958. He resigned in 1960 and supported himself with freelance work and, having moved to London, was the drama critic of Peter Cook's short-lived magazine *Scene* from 1962 to 1963, often using the *nom de plume* of William Boot (borrowed from **Evelyn Waugh**'s *Scoop*, 1938).

His first play, *A Walk on the Water*, was broadcast by the BBC in 1963; it concerns the unsuccessful and eccentric inventor George Riley, and was revised and produced in London in 1968 as *Enter a Free Man*. A number of short radio plays, including *The Dissolution of Dominic Boot*, were produced before the publication in 1966 of Stoppard's first and only novel, *Lord Malquist and Mr Moon*. A surreal, parodic, intellectual farce, the novel went almost entirely unnoticed. In the same year a group of Oxford students performed *Rosencrantz and Guildenstern Are Dead* at the Edinburgh Festival where its rapturous reception encouraged Kenneth Tynan and Laurence Olivier to stage the play at the Old Vic in April 1967. Of his feelings in 1966 Stoppard has said: 'There was no doubt in my mind whatsoever that the novel would make my reputation, and the play would be of little consequence.' The relative fortunes of the two ventures have undoubtedly determined the subsequent course of his career.

Further full-length plays have included *Jumpers*, an absurdist drama inspired by the idea of men landing on the moon; *Travesties*, which brings Lenin, **James Joyce** and Tristan Tzara together in Zurich in 1917; and *The Real Thing*, a comedy of marital infidelity in which a successful playwright falls for an actress, Annie. Stoppard has been married twice, to Jose Ingle and to Miriam Moore-Robinson (the doctor and television personality Miriam Stoppard), and has two children. In 1990 he left Miriam (they were divorced in 1992) for the actress Felicity Kendal, who had played the part of Annie in the 1982 production of *The Real Thing*. *Arcadia* is set in the nineteenth and twentieth centuries and explores the contrast between the Classical and Romantic sensibilities.

Stoppard has written a number of screenplays including *Brazil* (with Terry Gilliam and Charles McKeown), and has adapted several novels for the screen, notably **Graham Greene**'s *The Human Factor*, **J.G. Ballard**'s

Empire of the Sun, **John le Carré**'s *The Russia House* and **E.L. Doctorow**'s *Billy Bathgate*. A film version of *Rosencrantz and Guildenstern*, directed by the author, was released in 1991. His concern with east European politics is evident in *Squaring the Circle*, and in his adaptation of his friend Václav Havel's play *Largo Desolato*. His 1991 radio play *In the Native State*, set in the last days of the Raj, provided Peggy Ashcroft with her last role, and was subsequently reworked for the stage as *Indian Ink*.

Plays

A Walk on the Water **(tv)** 1963 [rev. as *The Preservation of George Riley* 1964], **(st)** [as *Enter a Free Man*] 1968; *The Dog It Was That Died and Other Plays* [pub. 1983]: *The Dissolution of Dominic Boot* **(r)** 1964, *'M' Is for Moon Among Other Things* **(r)** 1964, *Teeth* **(tv)** 1966, *Another Moon Called Earth* **(tv)** 1967; *The Gamblers* 1965 [np]; *If You're Glad, I'll Be Frank* **(r)** 1965 [pub. 1969]; *Rosencrantz and Guildenstern Are Dead* **(st)** 1966, **(f)** 1991 [pub. 1967]; *A Separate Peace* **(tv)** 1966 [pub. 1977]; *Tango* [from S. Mrozek's play] 1966 [pub. 1968]; *Albert's Bridge* **(r)** 1967, **(st)** 1969; *After Magritte* 1970 [pub. 1968]; *Neutral Ground* **(tv)** 1968; *The Real Inspector Hound* 1968; *The Engagement* **(tv)** 1970; *Where Are They Now?* **(r)** 1970 [pub. 1973]; *Dogg's Our Pet* 1971 [np]; *Jumpers* 1972; *The House of Bernarda Alba* [from Lorca's play] 1973; *Travesties* 1974 [pub. 1975]; *Boundaries* **(tv)** [with C. Exton] 1975 [pub. as *The Boundary* 1991]; *Eleventh House* **(tv)** [with C. Exton] 1975; *The Romantic Englishman* **(f)** [with T. Wiseman] 1975; *Dirty Linen* 1976; *The Fifteen-Minute Hamlet* 1976; *New-Found-Land* 1976; *Three Men in a Boat* **(tv)** [from J.K. Jerome's novel] 1976; *Albert's Bridge and Other Plays* [pub. 1977]; *Every Good Boy Deserves Favour* 1977 [pub. 1978]; *Professional Foul* **(tv)** 1977 [pub. 1978]; *Night and Day* 1978; *Despair* **(f)** [from V. Nabokov's novel; with T. Wiseman] 1979; *Dogg's Hamlet, Cahoot's Macbeth* 1979 [pub. 1980]; *The Human Factor* **(f)** [from G. Greene's novel] 1979; *Undiscovered Country* [from A. Schnitzler's *Das weite Land*] 1979 [pub. 1980]; *On the Razzle* [from J. Nestroy's *Einen Jux will er sich machen*] 1982 [pub. 1981]; *The Dog It Was That Died* **(r)** 1982 [pub. 1983]; *The Real Thing* 1982; *Four Plays for Radio* **(r)** [pub. 1984]; *Rough Crossing* [from F. Molnar's *Play at the Castle*] 1984 [pub. 1985]; *Squaring the Circle* **(tv)** 1984; *Brazil* **(f)** [with T. Gilliam, C. McKeown] 1985; *Dalliance* [from A. Schnitzler's *Liebelei*] 1986; *Empire of the Sun* [from J.G. Ballard's novel] 1987; *Hapgood* 1988; *The Russia House* **(f)** [from J. Le Carré's novel] 1990; *Stoppard: the Plays for Radio 1964–1983* **(r)** [pub. 1990]; *Billy Bathgate* **(f)** [from E.L. Doctorow's novel] 1991; *In the Native State* **(r)** 1991, **(st)** [rev. as *Indian Ink*] 1995; *Arcadia* 1993; *Television Plays 1965–1984* **(tv)** [pub. 1993]

Fiction

Introduction 2 **(s)** [with others] 1964; *Lord Malquist and Mr Moon* 1966

Translations

Vaclav Havel *Largo Desolato* 1985

David (Malcolm) STOREY 1933–

Storey was born in Wakefield, Yorkshire, the son of a coal-miner and one of four children. He left grammar school at the age of seventeen to study art in Wakefield and, whilst there, won a scholarship to the Slade School of Art where he studied from 1953 to 1956. At the same time he was obliged to return to Leeds each weekend to fulfil the terms of a fourteen-year contract he had signed in 1952 as player for Leeds Rugby League Club. Storey has written of his relief at escaping the West Riding, but the dual role forced upon him by the weekly trip back to the North proved impossible to sustain. The four-hour train journey each Friday evening came to represent the two irreconcilable sides to his temperament and 'it was in order', Storey wrote, 'to accommodate the two extremes of this northern, physical world and its southern, spiritual counterpart that I started making notes which two years later [1955], while I was still at the Slade, resulted in the writing of *This Sporting Life*'. He was unable to buy himself out of the contract until 1956. He married in the same year, supporting himself in London by teaching art classes and working, at one time or another, as a bus conductor, postman and tent erector.

In 1959, when *This Sporting Life* was accepted for publication, Storey and his wife were living in a single-room flat above a sweet shop in Kings Cross, in debt and with the first of their four children on the way. Storey had had six or seven novels rejected when he received a telegram informing him that the novel had received the Macmillan Award for Fiction although as yet still in manuscript. Storey used the £7,500 prize money to buy a white Jaguar. The novel was later adapted for the screen and won the International Film Critics' Prize at Cannes in 1963. *Flight into Camden* was published later in 1960 with equal success, winning a Somerset Maugham Award. *Radcliffe*, a dark, Lawrentian novel about a passionate and destructive relationship between two men which ends in murder, was published in 1963 but, disappointed with its reception and encouraged by Lindsay Anderson, Storey turned to drama. Anderson directed Storey's first five plays, all of which were first produced at the experimental Royal Court Theatre. *The Restoration of Arnold Middleton* which had originally been written in 1959, won the *Evening Standard* Drama Award in 1967. Storey returned to the world of Rugby League in *The Changing Room*, which won the New York Drama Critics' Circle Award for the best play of the 1972–3 season. In 1972, after a gap of nine years, he returned to fiction with *Pasmore*, a fragment salvaged from a long unfinished novel he had written in 1964. He won the Booker Prize in 1976 for *Saville*. In his play *The March on Russia* he returns to the setting of *In Celebration* to find the Pasmore family celebrating their sixtieth wedding anniversary. It recreates, for some critics, a world which had all but vanished. Storey's elder brother, Anthony, is also a writer.

Fiction

Flight into Camden 1960; *This Sporting Life* 1960; *Radcliffe* 1963; *Pasmore* 1972; *A Temporary Life* 1973; *Saville* 1976; *A Prodigal Child* 1982; *Present Times* 1984

Plays

This Sporting Life **(f)** [from his novel] 1963; *The Restoration of Arnold Middleton* 1967; *In Celebration* 1969; **(f)** 1976; *The Contractor* 1970; *The Changing Room* 1971; *Home* 1972; *Cromwell* 1973; *The Farm* 1973; *Grace* **(tv)** [from J. Joyce's story] 1974; *Life Class* 1974 [pub. 1975]; *Mother's Day* 1976; *Sisters* 1978; *Early Days* 1980; *Phoenix* 1984 [np]; *The March on Russia* 1989; *Stages* 1992; *Plays* [pub. 1992]; *Plays Two* [pub. 1994]

Poetry

Storey's Lives: Poems 1951–1991 1992

Jack Trevor STORY 1917–1991

Story's attitude to life, as two well-publicised bankruptcies, three wives, eight children and numerous affairs amply demonstrate, was summarised by the title of his most famous novel, *Live Now, Pay Later*. Permanently insolvent, he was the author of some forty published novels and 400 short stories, a successful newspaper columnist, screen and scriptwriter, a model with Ugly's agency and the presenter of a popular television show, *Jack on the Box*.

He was born in Hertford, but when his father died in the First World War, his mother moved to Cambridge and took up work as a cleaner to support her five children. Story educated himself, attending night school in radio and electronics, and then working for Pye Radio, Murphy's and Marconi, where his skill with words won him a job on the house magazine. A certain financial recklessness was evident from the beginning. The film rights to his first published novel, *The Trouble with Harry* (1949), were sold to Alfred Hitchcock for $500 and then resold by Hitchcock to Paramount for $20,000. With bailiffs in the house, repossession was delayed whilst Story, with the Inland Revenue in tow, dashed to Claridges where Hitchcock was staying. Eventually caught up with at Victoria Station, the director declined an offer to buy the rights to Story's second novel. To supplement his income, Story then wrote a series of pulp Westerns (under the pen-name Bret Harding), and Sexton Blake thrillers, producing twenty between 1951 and 1961. His love-life was also complicated. By 1955, with the Hitchcock film a success, Story had eight children and two wives to support. Even with the film rights to *Live Now, Pay Later* sold for the more realistic sum of £12,000, money quickly disappeared.

As a columnist for the *Guardian* in the early 1970s he became a national celebrity; his publicly unsuccessful pursuit of 'Maggie' forced her to flee the country to escape his attentions. The columns continued until Story was sacked and were later reprinted as *Letters to an Intimate Stranger*. His novels, influenced by **William Saroyan** and **George Orwell**, present a comic and, sometimes, bleak vision of the world. In *Morag's Flying Fortress*, a not unusually absurd example of its author's storytelling talent, a gang of drunken American pilots smuggle girls aboard a plane and stage flying orgies over Nazi Germany; they are shot down and stranded in what they believe to be occupied Europe but is, in fact, Suffolk; having wiped out a platoon of the Home Guard they steal a boat and head 'back' for England but are captured by the enemy. After the war the survivors of this mission 'go searching for their mistake'; Captain Alec Ranger, now a spy for NATO, has also inadvertently discovered thalidomide to be a by-product of Hitler's Final Solution but has been brainwashed into forgetting this as the novel begins.

Penniless again by the time of the novel's composition, Story was made writer-in-residence by Milton Keynes in the mid-1970s. Given a flat for the one-year duration of the post, he was to remain there for the rest of his life. In 1990 he was hospitalised and given a lithium cure. The recovery was temporary and Story was discovered dead at his typewriter the following year.

Fiction

The Trouble with Harry 1949; *Protection for a Lady* 1950; *Green to Pagan Street* 1952; *The Money Goes Round and Round* 1958; *Mix Me a Person* 1959; *Man Pinches Bottom* 1962; *Live Now, Pay Later* 1963; *Something For Nothing*

1963; *The Urban District Lover* 1964; *Company of Bandits* 1965; *I Sit in Hanger Lane* 1968; *Dishonourable Member* 1969; *The Blonde and the Boodle* 1970; *Hitler Needs You* 1970; *One Last Mad Embrace* 1970; *The Season of the Skylark* 1970; *Little Dog's Day* 1971; *Whistle While I'm Dead* 1971; *Crying Makes Your Nose Run* 1974; *Morag's Flying Fortress* 1976; *Up River* 1979; *The Screwrape Lettuce* 1980; *Albert Rides Again* 1990

Non-Fiction

Letters to an Intimate Stranger (a) 1972; *Story on Crime* 1975; *Jack on the Box* (a) 1979; *Dwarf Goes to Oxford* (a) 1987

(Julian) Randolph STOW 1935-

The son of a barrister and his wife, Stow was born in Geraldton, Western Australia, and educated at Guildford Grammar School, and the University of Western Australia. After working at a mission for aboriginals, he studied anthropological linguistics, and then went as a patrol officer to the Trobriand Islands, where he served as a government anthropologist. From these early experiences he developed a passionate interest in the primitive peoples and landscapes of Australia, and their effect on modern Australia. Later he taught English and Commonwealth literature at the Universities of Adelaide, Western Australia and Leeds.

Since 1960 he has been based in Sussex, the location of *The Girl Green as Elderflower*, begun in 1966 but not published until 1980. The long gestation of this novel suggests something of Stow's difficulties in living down his early success – three novels and a volume of poetry published before he was twenty-four. His semi-autobiographical novel, *The Merry-Go-Round in the Sea*, seems in retrospect to have been a too sudden valediction to the major influences and themes of his early work – the isolated community, the power of landscape, the sense of history, the force of myth, the figure of the 'bystander'. Only with the publication of *Visitants* have these preoccupations been both exorcised and transmuted. Winner of the 1979 Patrick White Award, Stow's habitual tale of the confrontation between cultures is dazzlingly told, Rashomon-like, by five voices, two white and three black. Set in Papua New Guinea, it draws on Stow's early experiences and is based on a *Marie Celeste* type of incident that occurred on a neighbouring island. It is at once his most lyrical and profound work.

All Stow's novels are characterised by an intense feeling for place and history, whether the location is Australia, the South Pacific, Sussex or, as in *The Suburbs of Hell*, the Fenlands. A chilling tale of murder, this book is nevertheless a disappointing return to his old fictional territory. **Joseph Conrad** – in particular *Heart of Darkness* (1902) – is his most obvious influence, and the ease with which Stow's ideas have taken imaginative root in the landscape of the Australian outback perhaps accounts for his early blossoming and later problems as a novelist. Conrad is a less productive influence in rural England, and the later novels flirt uneasily with self-parody. But *To the Islands* – winner in 1958 of the Miles Franklin Award for Australian Literature – and *Tourmaline*, are authentic Australian novels. As powerfully as **Patrick White**'s comparable *Voss* (1957), they grapple with the country's 'bitter heritage' of cruelty and pain, guilt and expiation; and – unlike Conrad – with hope and forgiveness.

Fiction

A Haunted Land 1956; *The Bystander* 1957; *To the Islands* 1958; *Tourmaline* 1963; *The Merry-Go-Round in the Sea* 1965; *Visitants* 1979; *The Girl Green as Elderflower* 1980; *The Suburbs of Hell* 1984

Poetry

Act One 1957; *Outsider: Poems 1956–62* 1962; *A Counterfeit Silence* 1969; *Poetry from Australia: Pergamon Poets 6* [with others; ed. H. Sergeant] 1969

Plays

Eight Songs for a Mad King 1969 [pub. 1971]; *Miss Donnithorne's Maggot* 1974 [pub. 1977]

For Children

Midnite: the Story of a Wild Colonial Boy 1967

Collections

Visitants, Episodes from Other Novels, Poems, Stories, Interviews and Essays [ed. A.J. Hassall] 1990

Julia (Frances) STRACHEY 1901–1979

Strachey was born in India, where her father (Lytton Strachey's youngest brother, Oliver) worked for the East India Railway. She was sent to England aged six to be brought up by a succession of aunts, most notably Alys ('Loo') Pearsall Smith, an eccentric Quaker and militant suffragist who had recently been deserted by her husband, Bertrand Russell. Strachey was educated at a grim boarding-school and Bedales, spending holidays at the villa I Tatti (another aunt had married Bernard Berenson). She went to Bedford College intending to study psychology, but was mistakenly enrolled to study physiology, so left after a term to study commercial art at the Slade School. Through her Bloomsbury connections she met the sculptor Stephen Tomlin, whom she married in 1927. Tomlin was a manic-depressive bisexual who drank heavily and was guiltily unfaithful.

Strachey had been writing and publishing short stories, and completed her first and most famous novel, *Cheerful Weather for the Wedding*, during this period. A short, sardonic account of a chaotic country wedding, it was published by the Hogarth Press in 1932 to considerable critical acclaim. Meanwhile her marriage collapsed and she had affairs with several men, notably **Rosamond Lehmann**'s husband, Wogan Philipps, and **Philip Toynbee**, who was fifteen years her junior. She subsequently married the painter Lawrence Gowing – 'I have only known two geniuses,' she said, 'and I was married to both of them.'

In the 1940s a number of her stories and sketches appeared in *Penguin New Writing*, the *New Statesman* and the *New Yorker*. She published one further novel, *The Man on the Pier*, (later reissued as *An Integrated Man*, her preferred title) and was felt by many never to have fulfilled her promise. **Virginia Woolf** described her as 'a gifted wastrel', but although Strachey claimed that her ideal life would be 'lying on a pink fur rug doing absolutely nothing', the truth was that her literary career was hampered by her own perfectionism. After her death, her friend Frances Partridge was entrusted with 'two *huge* suitcases' of papers which demonstrated considerable industry, and from which she pieced together a fragmentary autobiography, *Julia*.

Fiction

Cheerful Weather for the Wedding 1932; *The Man on the Pier* 1951 [as *An Integrated Man* 1978]

Non-Fiction

Julia: a Portrait of Julia Strachey by Herself and Frances Partridge (a) 1983

T(homas) S(igismund) STRIBLING 1881–1955

Born in Clifton, Tennessee, Stribling was looked after by an aunt for most of his childhood, because of difficulties in his parents' marriage. He was educated at Clifton Mason Academy and Normal College, Florence, Alabama, and subsequently taught at the latter for one term only. As he noted, he and the school 'parted with mutual relief'.

After graduating at the University of Alabama with a law degree in 1905, he spent a year in a lawyer's office from 1906, when he gratefully left the job. The experience provided invaluable material thirty years later, for his novel *The Sound Wagon*, which debunked businessmen and lawyers. He always had an urge to write, but the only way into 'literature' he knew was to become a rather old office boy at the age of twenty-three at the Nashville magazine-publisher Taylor-Trotwood. Predictably, he departed after a year, but he had also made a start on his writing career by producing moral adventure stories for Sunday-school magazines. This was extremely lucrative, and enabled him to travel to South America and through Europe, writing and earning as he went.

The first novel that Stribling acknowledged was *Birthright*, published in 1922. Several other unremarkable novels followed, but when he married in 1930, he abandoned trashy fiction to write more seriously. He won a Pulitzer Prize in 1932 for *The Store*, the first of a trilogy collectively called *The Forge* and named after the second novel published in the same year. The final part was *Unfinished Cathedral*. In this tale of the rise of a landowning family of the Mid-South, his tendency to melodrama and an over-simplistic plot are mitigated by his careful observation of landscape and those who inhabit it, and gift for reproducing authentic speech patterns. His fiction invariably took the side of the oppressed individual, subject to overwhelming forces of history and change. **William Faulkner**'s mythical histories, set in the same area, were at first overshadowed by Stribling's objective realism. Literary judgement has now seen a reversal of opinion.

Successful detective stories featured Professor Poggioli, psychologist and philosopher, whose sharp intelligence reflected Stribling's best qualities. He regarded himself to be 'urbane, satiric, lively, ironic, realistic, accurate, humorous and philosophical', and enjoyed his reputation for providing compulsive reading. Bald, tall and tanned, and with a rather lean Sherlock Holmesian face, he demonstrated great skill on the tennis court, and the chess board.

Fiction

The Cruise of the Dry Dock 1917; *Birthright* 1922; *Fombombo* 1923; *Red Sand* 1924; *Teeftallow* 1926; *Bright Metal* 1928; *East Is East* 1928; *Clues of the Caribbees* 1929; *The Backwater* 1930; *The Forge*: *The Store* 1931, *The Forge* 1931, *Unfinished Cathedral* 1934; *The Sound Wagon* 1935; *These Bars of Flesh* 1938; *Best of Poggioli Detective Stories* (s) 1975

Plays

Rope [with D. Wallace] 1928

Non-Fiction
Laughing Stock (a) [ed. R.K. Cross, J.T. McMillan] 1982

(Henry) Francis (Montgomery) STUART 1902-

Of Protestant Irish landed gentry, Stuart was brought from Australia to Northern Ireland when a few months old, following the suicide of his mentally ill father. He was educated at Rugby, and at the age of seventeen, alienated from his mother, he reluctantly married Iseult Gonne in Dublin. (A former mistress of **Ezra Pound**, Iseult had also been courted by **W.B. Yeats**, as indeed had her mother, the formidable Maude Gonne MacBride.)

Influenced by the poets of the Russian Revolution, Stuart published verse, and attempted a novel when he was interned for Republican activities in the 1922 civil war. After his release, a collection of poems, *We Have Kept the Faith*, was awarded a prize by the Royal Irish Academy when Stuart was twenty-one. Finding the rewards of poetry insufficient to maintain a family, however, Stuart turned to prose after selling his successful poultry farm. He was always a man of action, and his favoured pursuits of horse racing, women and religion frequently featured in his early novels. The first of these, *Women and God* (1931), was mediocre, but later fiction convinced Yeats that if 'luck' was with Stuart, he would be Ireland's 'genius'. The decline in the imaginative quality of the novels towards the late 1930s, paralleled the collapse of Stuart's marriage. Determined to place himself outside society, and in an effort to understand himself, he left his wife and two children to spend the war years teaching English and Irish literature at Berlin University. He was dubbed Ireland's 'Lord Haw-Haw' by **Cyril Connolly** for making broadcasts to Ireland from Germany which mainly stressed the importance of maintaining Southern Irish neutrality. In 1946, Stuart was imprisoned in Freiburg by the occupying French for reasons which are unclear, and was eventually released without charge. He remained in Freiburg, living in considerable poverty, and in 1948 and 1949 published *The Pillar of Cloud* and *Redemption*, novels which reflected his experiences of Germany and the aftermath of the war, a theme he developed intermittently in later books.

Exiled first in Paris and then London, Stuart married one of his former students, Madeleine Meissner, in 1954, after Iseult's death in the same year. He returned to live in Ireland in 1958, but after a productive silence, the autobiographical *Black List, Section H* re-established Stuart's reputation in 1971, and was followed by a series of experimental novels. Madeleine Stuart died in 1986 and, unconcerned about Irish opinion, the eighty-five-year-old Stuart married the artist Finola Graham the following year. He celebrated some of his lifelong obsessions with a dry humour in his late novel *A Compendium of Lovers*.

Fiction
Women and God 1931; *The Coloured Dome* 1932; *Pigeon Irish* 1932; *Glory* 1933; *Try the Sky* 1933; *Things to Live For* 1934; *The Angel of Pity* 1935; *In Search of Love* 1935; *The White Hare* 1936; *The Bridge* 1937; *Julie* 1938; *The Great Squire* 1939; *The Pillar of Cloud* 1948; *Redemption* 1949; *The Flowering Cross* 1950; *Good Friday's Daughter* 1952; *The Chariot* 1953; *The Pilgrimage* 1955; *Victors and Vanquished* 1958; *The Angels of Providence* 1959; *Black List, Section H* 1971; *Memorial* 1973; *A Hole in the Head* 1977; *The High Consistory* 1981; *Faillandia* 1985; *A Compendium of Lovers* 1990

Poetry
We Have Kept the Faith 1923; *We Have Kept the Faith: New and Selected Poems* 1982; *Night Pilot* 1988

Non-Fiction
Nationality and Culture 1924; *Mystics and Mysticism* 1929; *Racing for Pleasure and Profit* 1937; *Der Fall Casement* [with R. Weiland] 1940; *States of Mind* 1984; *The Abandoned Snail Shell* 1987

Biography
Francis Stuart by Geoffrey Elborn 1990

H(oward) O(vering) STURGIS 1855-1920

The son of Russell Sturgis, an American architect and banker, Sturgis was born into the high society of Boston, but because his father became a partner in Baring's Bank, he grew up a rich man in England. He went to Eton – where he was overshadowed by his handsome and athletic elder brother, Julian, who also became a novelist – and then to Trinity College, Cambridge. Having no need to work, he settled near Eton, at Windsor, at a house called Queen's Acre (always abbreviated to Qu'acre), where he lived with his lifelong companion, William Haynes Smith. Here he entertained a wide cross-section of late Victorian and Edwardian society: bachelor dons, American expatriates, Eton masters, and novelists, who included **Henry James** (a close friend) and **Edith Wharton**. 'Howdie' was often to be found knitting or stitching behind an embroidery frame while his guests talked. He and his bachelor guests cultivated a series of romantic friendships with Eton boys: among others, Sturgis fell for Percy Lubbock, later the expert on the novel, then a house captain. Sturgis's pleasant lifestyle was complemented by sojourns at Venice and Cape Cod.

Although known as a dilettante, he published three novels: the first, *Tim* (1891), was a sentimental story of friendship between boys at Eton, but Sturgis's reputation now rests on *Belchamber*, about a crippled English marquis. Sensitive and troubled by the corruption of society, Sainty is cuckolded by his wife but develops a strong relationship with the illegitimate baby with which she presents him. The novel was heavily criticised by Henry James before publication, so much so that Sturgis wanted to withdraw it. When it finally appeared in 1904, it made little stir, but since it was reprinted in 1935, it has won for its author, who never published again, at least a modest niche in twentieth-century literature.

Fiction

Tim 1891; *All That Was Possible* 1895; *Belchamber* 1904

William (Clark) STYRON 1925–

Stryon was born in Newport News, Virginia, the son of a marine engineer. His mother died when he was thirteen. He was educated at Christchurch School, Virginia, and Davidson College before joining the marines in 1943. Stationed in the Pacific, he reached Okinawa just as the war ended, and returned the same year to Duke University where he studied creative writing as an officer candidate.

Having graduated in 1947, he worked for six months as a manuscript reader for McGraw-Hill in New York (an experience which provided the setting for his 1979 novel *Sophie's Choice*), and then studied under Hiram Haydn at the New School for Social Research in New York. With Haydn's guidance he published two short stories in *American Vanguard* and worked on his first novel, *Lie Down in Darkness*. Published in 1951 to critical acclaim, the novel explores the disintegration of a genteel Southern family and is technically and thematically indebted to two other Southern writers, **Thomas Wolfe** and **William Faulkner**. The sounding rhetoric and the powerful sense of history in Styron's work mark him as firmly rooted in the Southern cultural tradition. In the same year, Styron was briefly reactivated by the Marines for reserve duty in the Korean War, after which he travelled in Italy and France. In 1952, in Paris, he joined George Plimpton in founding the *Paris Review*, and the following year he married. The couple settled in Connecticut and have four children.

Styron's work has consistently explored the themes of evil and the capacity for self-redemption, and what he has termed 'the catastrophic propensity of human beings to dominate one another'. His highly succesful novel *The Confessions of Nat Turner* (winner of a Pulitzer Prize in 1968) was a controversial examination of the famous 1831 Virginia slave uprising, in the form of the 'autobiography' of one of its leaders. **James Baldwin** had been Styron's houseguest during part of the writing and had warned him that he would 'catch it from black and white' for his attempt at portraying black psychology, as indeed he did. Equally controversial and successful was *Sophie's Choice*, an apparently autobiographical story of a young man's involvement with a Polish woman who has survived Auschwitz, but is clearly doomed. The novel was widely praised and won a National Book Award in 1980, but some questioned Styron's use of the death camps for an explicit tale of sexual obsession, and the notion that the eponymous Sophie somehow colludes in her own degradation and destruction. Perhaps unsurprisingly, Meryl Streep received an Oscar for her portrayal of this figure in Alan J. Pakula's 1982 film of the novel.

In the 1980s Styron suffered a lengthy period of mental illness, which is movingly described in *Darkness Visible*. He has since published *A Tidewater Morning*, three stories based on his childhood in 1930s' Virginia. Highly regarded in France, he has received two decorations: the Commandeur de l'ordre des arts et des lettres, and the Légion d'Honneur.

Fiction

Lie Down in Darkness 1951; *The Long March* 1955; *Set This House on Fire* 1960; *The Confessions of Nat Turner* 1967; *Sophie's Choice* 1979; *A Tidewater Morning* (s) 1993

Non-Fiction

This Quiet Dust 1982; *Darkness Visible: a Memoir of Madness* (a) 1990

Plays

In the Clap Shack 1973

David SWEETMAN 1943–

The poetry which has earned Sweetman his most serious attention is a relatively small part of a wider career in letters and media. Born in Northumberland and educated at the unversities of Durham and East Africa, Sweetman has worked for the British Council, in radio, as a television arts producer, and as an editor for a French publishing house. As well as poetry he has written acclaimed biographies and a number of educational books and books for children, specialising in African topics.

Sweetman's poetry is represented by a single 1981 volume of selected poems, *Looking into the Deep End*, which was a Poetry Book Society Choice. The poems in it are heavily reliant on metaphor and simile, and Sweetman is inseparable from that particular moment in British poetry when **Craig Raine** led the 'Metaphor Men', soon to be rechristened the 'Martians' after Raine's second volume, *A Martian Sends a Postcard Home* (1979). Poetry has always had striking comparisons as one of its resources, but during the Martian moment it seemed to consist of almost nothing else. Writing sceptically about this tendency in market terms, the academic critic David Trotter noted that: 'Faber and Faber, who missed out on Raine and [**Christopher**] **Reid**, have acknowledged as much by signing up another metaphorfiend, David Sweetman.'

Sweetman's poetry can read remarkably like second-rate Raine, but sometimes his similies are emotionally satisfying rather than just metaphysically ingenious, like an unhappy man 'turning his wedding ring round and round / as if to somehow lower the volume'. At best there is a distinctive quality to Sweetman's work, intellectually agile in its movement from one thought to another and distinctive in its emotional tendencies. Much of it is melancholy, concerned with moments of disillusion or death, and sometimes suffused with a subdued, guilty eroticism. In the title-poem of his collection the poet witnesses a drowned boy (to whom the swimming pool had been 'a fascinating parcel, crumpled paper / surface, reflections of ravelled string') and his eyes move to the boy's crotch. This provokes associations and memories that are 'hastily reinterred' when he dives 'cleanly' into the water, and his now blurred vision returns to 'that innocent game of green-tiled chess / that sun and water play / with boneless, shifting pieces'.

Critics have been divided over Sweetman, some praising his freshness and originality, others complaining about a certain preciousness in similes such as 'miners stare like Coptic saints', and in references to Telemann or Huysmans or Ottoman courtiers, along with a striving for significance in the invocation of Marx or Bakunin or Dresden or Hiroshima. The reception of his biographies has been less mixed. Within the understanding that it was to some extent 'popular', his 1990 biography of Vincent Van Gogh attracted intense praise. He has followed it with a 1993 biography of **Mary Renault**, which has been acclaimed for its original research and for its quietly sympathetic approach, and a controversial life of Gauguin. For the time being, at least, Sweetman's judicious and solid biographies seem to be winning a hare and tortoise race with his poetry.

Poetry

Looking into the Deep End 1981

Non-Fiction

Queen Nzinga: the Woman Who Saved Her People 1971; *Picasso* 1973; *Captain Scott* 1974; *The Borgias* 1976; *Patrice Lumumba* 1978; *Spies and Spying* 1978; *The How and Why Wonder Book of the Arab World* 1979; *Amulet* 1980; *Death in the Desert* 1980; *The Moyo Kids* 1980; *Skyjack Over Africa* 1980; *Bishop Crowther* 1981; *Women Leaders in African History* 1984; *The Love of Many Things: a Life of Vincent van Gogh* 1990; *Mary Renault: a Biography* 1993; *Paul Gauguin: a Complete Life* 1995

Graham SWIFT 1949–

Swift was born in London. He was educated at Dulwich College, Queens' College, Cambridge, and York University (to do a PhD on the works of Dickens), after which he spent ten years teaching English part-time at various colleges in London.

His first two novels, *The Sweet-Shop Owner* (1980) – which was later adapted, not entirely successfully, for the stage – and *Shuttlecock*, were small-scale works, remarkable for their recreation of the period around the Second World War, a conflict which had ended before Swift was even born. The former dealt with relations between the proprietor of the title and his daughter; the latter, which won the Geoffrey Faber Memorial Prize in 1983, is a paranoid thriller concerning a man's search for his father, a former war hero now in a mental hospital. Swift received considerable critical acclaim for these two novels and for *Learning to Swim*, a volume of stories published by **Alan Ross**, in whose *London Magazine* many of them had first appeared. In 1983 he was selected as one of the 'Best of Young British Novelists' in an influential Book Marketing Council promotion. This promise was fulfilled the same year when his ambitious novel *Waterland* was shortlisted for the Booker Prize and won the Winifred Holtby and the *Guardian* Fiction Prizes and the Italian Premio Grinzane Cavour (Swift's work has been widely translated). The Holtby award is made for a regional novel, and *Waterland* is amongst other things a remarkable evocation of the landscape and history of the Cambridgeshire fens. Narrated by a teacher – Swift's early career is apparent in the occasionally didactic tone of his writing – it was his longest and most ambitious work to date. Its intricate plot concerns incest, murder and abortion, and the effect of these events in his adolescence upon the narrator as a grown man. It combines melodrama and an almost documentary treatment of brewing, the life-cycle of the eel and other subjects to create an immensely impressive novel about the processes of storytelling and the nature of history. It was subsequently made into a much underrated film.

Swift's next novel, *Out of This World*, disappointed some critics, and there is no doubt that it is a flawed work, notably in its scenes cast in the form of a psychoanalysis session; it was also, however, a lucid and challenging book about the topical issue of terrorism. *Ever After* is more like *Waterland*, concerned with history in its double narrative of a contemporary academic and his great-great-grandfather, a Victorian torn between science and religion. It is also related to *Shuttlecock* in that the narrator is sifting through papers in order to make sense of a forebear's life. If subsequent novels have made less impact than *Waterland*, Swift nevertheless remains a novelist of scope and ambition, who tackles moral and historical themes to produce genuinely serious (though never solemn) works of fiction.

A keen fisherman, Swift was co-editor with his angling companion, the poet, critic and novelist David Profumo, of *The Magic Wheel: an Anthology of Fishing in Literature*. He has lived for many years with the literary journalist Candice Rodd, the dedicatee of several of his books.

Fiction

The Sweet-Shop Owner 1980; *Shuttlecock* 1981; *Learning to Swim and Other Stories* (s) 1982; *Waterland* 1983; *Out of This World* 1988; *Ever After* 1992

Edited

The Magic Wheel [with D. Profumo] 1985

Frank (Arthur) SWINNERTON 1884–1982

Born in Wood Green, London, Swinnerton was the youngest child of a copperplate engraver and his wife, a designer. Due to poor health and family poverty, his formal education was perfunctory, and he largely educated himself by reading **Henry James**, Ibsen and Louisa M. Alcott and by attending public lectures by such figures as **George Bernard Shaw**, **Hilaire Belloc** and **G.K. Chesterton**. His career in literature began in 1901 when he joined the publishing house of J.M. Dent as a receptionist, and in 1906 he was involved in the launch of Everyman's Library. In 1907 he moved to Chatto & Windus as a reader, and remained as an editor until 1926, when he left to concentrate on his own writing.

His first novel, *The Merry Heart*, was published in 1909 with the assistance of **Arnold Bennett**, who was to become his greatest friend and most important influence. Other close friends included **H.G. Wells**, **John Galsworthy**, **J.M. Barrie** and **Hugh Walpole**. Swinnerton was a prolific and varied author: forty-two novels, fifteen works of literary comment, and innumerable reviews, articles and interviews. Few of his novels are now read, although his contemporary tales about ordinary people are as fluent as those of his peers; and he occupies a fictional space somewhere between George Gissing and Wells. *The Casement*, *September*, *Young Felix* and *Harvest Comedy* in particular were highly regarded at the time for their naturalistic charm. The later novels are heavier and more dramatic, with the exception of *Death of a Highbrow*, a touching study of an elderly writer.

He was notoriously modest about his own talents: 'I was from the year 1902 writing novels,' he once said. 'They were not good novels, because I am not a genius. But they were very short.' This is not disingenuousness: of the critics of the time, Swinnerton was the shrewdest in recognising literary merit. *The Georgian Literary Scene*, as valuable today as it was upon publication in 1935, proves both his prescience and his modesty: his own name appears nowhere in the text. His most successful novel was *Nocturne*, the result of a challenge by the publisher Martin Secker to write a book about the events of a single night. Swinnerton later dismissed the novel as 'my albatross', and a 'stunt', but it remains his most innovative and enduring work – *Ulysses* as written by **Arnold Bennett**. The story of two sisters whose lives are inhibited by their invalid father, it is an honest yet poetic portrait of working-class life in London. His last work, appropriately *Arnold Bennett: a Last Word*, was published in 1978 when he was ninety-four. He was president of the Royal Literary Fund from 1962 to 1966 and remained a familiar literary figure until his death.

Fiction

The Merry Heart 1909; *The Young Idea* 1910; *The Casement* 1911; *The Happy Family* 1912; *On the Staircase* 1914; *The Chaste Wife* 1916; *Nocturne* 1917; *Shops and Houses* 1918; *September* 1919;

Coquette 1921; *The Three Lovers* 1922; *Young Felix* 1923; *The Elder Sister* 1925; *Summer Storm* 1926; *Tokefield Papers* 1927; *A Brood of Ducklings* 1928; *Sketch of a Sinner* 1930; *The Georgian House* 1933; *Elizabeth* 1934; *Harvest Comedy* 1937; *The Two Wives* 1940; *The Fortunate Lady* 1941; *Thankless Child* 1942; *A Woman in Sunshine* 1944; *English Maiden* 1946; *A Faithful Company* 1948; *The Cats and Rosemary* 1948; *The Doctor's Wife Comes to Stay* 1949; *A Flower for Catherine* 1950; *Master Jim Probity* 1952; *A Month in Gordon Square* 1953; *The Summer Intrigue* 1955; *The Woman from Sicily* 1957; *A Tigress in Prothero* 1959; *The Grace Divorce* 1960; *Death of a Highbrow* 1961; *Quadrille* 1965; *Sanctuary* 1966; *On the Shady Side* 1970; *Nor All Thy Tears* 1972; *Rosalind Passes* 1973; *Some Achieve Greatness* 1976

Non-Fiction

George Gissing: a Critical Study 1912; *R.L. Stevenson: a Critical Study* 1914; *A London Bookman* 1928; *Authors and the Book Trade* 1932; *The Georgian Literary Scene* 1935 [rev. 1938]; *Swinnerton: an Autobiography* (a) 1936; *The Reviewing and Criticism of Books* 1939; *Arnold Bennett* 1950 [rev. 1961]; *The Bookman's London* 1951 [rev. 1969]; *Londoner's Post: Letters from Gog and Magog* 1952; *The Adventures of a Manuscript, Being the Story of 'The Ragged Trousered Philanthropists'* 1956; *Figures in the Foreground* 1963; *A Galaxy of Fathers* 1966; *Reflections from a Village* (a) 1969; *Arnold Bennett: a Last Word* 1978

Edited

An Anthology of Modern Fiction 1937; Arnold Bennett: *Literary Taste* 1937, *The Journals of Arnold Bennett* 1954; William Hazlitt *Conversations of James Northcote* 1949

Julian (Gustave) SYMONS 1912–1995

Symons was born in London, the son of a Jewish father who had emigrated to England in the last years of the nineteenth century and who progressed from shopkeeper to auctioneer. He never knew his father's original name, and the family is not connected with the Decadent poet Arthur Symons, although the man-of-letters A.J.A. Symons, author of *The Quest for Corvo* (1934), was Julian's brother. He grew up in Battersea and Clapham, attending local schools, which he left aged fourteen. He found work as a shorthand-typist and later secretary in a small engineering company, where he stayed for eleven years, developing his literary interests in his spare time.

From 1937 he was the editor for three years of a small poetry magazine, *Twentieth Century Verse*, which was influential in introducing the work of new and American poets. Symons himself published two early volumes of verse, but wrote little poetry thereafter. In the later 1930s he was an orthodox Trotskyite for some years, and retained moderate left-wing sympathies. During the war he was initially a conscientious objector, but was eventually in the army from 1942 to 1944, completing almost half his service in hospital. He married in 1941, and the couple had two children, one of whom, a daughter, died young.

After the war, he worked for some years in an advertising agency, but became a freelance writer in 1947 with the help of **George Orwell**, who was vacating his column on the *Manchester Evening News* and recommended Symons for the post. His subsequent output, as journalist (reviewer for the *Sunday Times* among other papers), novelist and general writer, was very great. He is best known as an author of detective stories and, although he wrote no other fiction, he is noted, in novels such as *The Progress of a Crime* and *The End of Solomon Grundy*, for having brought much of the psychological depth and social criticism of the mainstream novel to the crime story. He wrote an influential history of the genre, *Bloody Murder*, and was much honoured for his work in fictional crime, succeeding **Agatha Christie** in 1976 as president of the Detection Club. His general works of history, biography and criticism include lives of Carlyle, Poe and Horatio Bottomley, accounts of true-life crime, studies of the General Strike and the expedition sent to relieve General Gordon, and *Makers of the New*, a lively history of modernism.

Fiction

The Immaterial Murder Case 1945; *A Man Called Jones* 1947; *Bland Beginning* 1949; *The Thirty-First of February* 1950; *The Broken Penny* 1953; *The Narrowing Circle* 1954; *The Paper Chase* 1956; *The Colour of Murder* 1957; *The Gigantic Shadow* 1958; *The Progress of a Crime* 1960; *Murder! Murder!* (s) 1961; *The Killing of Francie Lake* 1962; *The End of Solomon Grundy* 1964; *The Belting Inheritance* 1965; *Francis Quarles Investigates* (s) 1965; *The Man Who Killed Himself* 1967; *The Man Whose Dreams Came True* 1968; *The Man Who Lost His Wife* 1970; *The Players and the Game* 1972; *The Plot against Roger Rider* 1973; *A Three-Pipe Problem* 1975; *The Blackheath Poisonings* 1978; *Sweet Adelaide* 1980; *The Detling Murders* 1982; *The Tigers of Subtopia and Other Stories* (s) 1982; *The Name of Annabel Lee* 1983; *The Criminal Comedy of the Contented Couple* 1985; *The Kentish Manor Murders* 1988; *Death's Darkest Fate* 1990;

Portraits of the Missing 1991; *The Advertising Murders* 1992; *Something Like a Love Affair* 1992; *Murder under the Mistletoe* 1993

Poetry
Confusions about X 1939; *The Second Man* 1943; *The Object of an Affair and Other Poems* 1974; *Seven Poems for Sarah* 1986

Non-Fiction
A.J.A. Symons: His Life and Speculations 1950; *Charles Dickens* 1951; *Thomas Carlyle: the Life and Ideas of a Prophet* 1952; *Horatio Bottomley: a Biography* 1955; *The General Strike: a Historical Portrait* 1957; *A Reasonable Doubt: Some Criminal Cases Re-Examined* 1960; *The Thirties: a Dream Revolved* 1960; *The Detective Story in Britain* 1962; *Buller's Campaign* 1963; *England's Pride: the Story of the Gordon Relief Expedition* 1965; *Crime and Detection: an Illustrated History from 1840* 1966; *Critical Occasions* 1966; *Bloody Murder; From the Detective Story to the Crime Novel: a History* 1972; *Notes from Another Country* 1972; *The Tell-Tale Heart: the Life and Works of Edgar Allan Poe* 1978; *Conan Doyle: Portrait of an Artist* 1979; *The Modern Crime Story* 1980; *Critical Observations* 1981; *Great Detectives: Seven Original Investigations* 1981; *George Orwell* 1984; *Dashiell Hammett* 1985; *Two Brothers: Fragments of a Correspondence* 1985; *Makers of the New: the Revolution in Literature, 1912–1939* 1987

Edited
An Anthology of War Poetry 1942; *Selected Writings of Samuel Johnson* 1949; Thomas Carlyle *Selected Works, Reminiscences and Letters* 1955; A.J.A. Symons *Essays and Biographies* 1969; *Britain Between the Wars* 1973; Wilkie Collins *The Woman in White* 1974; *The Angry Thirties* 1976; Edgar Allan Poe *Selected Tales* (s) 1980; *Crime and Detection Quiz* 1983; *The Penguin Classic Crime Omnibus* 1984; Chekhov *The Shooting Party* 1986; *The Essential Wyndham Lewis* 1989

(Edmund) J(ohn) M(illington) SYNGE 1871–1909

Synge was born in a suburb of Dublin, the youngest of eight children of a barrister, a newly middle-class member of a family of Protestant County Wicklow landowners. His father died when Synge was an infant, but his mother, although somewhat impoverished, was able to put her son through private schools and Trinity College, Dublin. Afterwards he went to Germany with the idea of training as a violinist, and passed from there to Italy and then to Paris, one of many young Irish aesthetes gathered in the French capital, but living frugally. He was already suffering from Hodgkin's disease, which was to kill him before he was forty.

In 1898 he met **W.B. Yeats**, who advised him to return to Ireland and, in particular, to visit the remote Aran Islands off the Donegal coast, to express 'a life that has never found expression'. Synge followed the advice, also spending time wandering around County Wicklow, imbibing the speech of the peasantry. By the time Yeats, **Lady Gregory** and **George Moore** came to found the Abbey Theatre in 1903 (Synge was later to become a co-director of it), he had two one-act plays ready for them, *In the Shadow of the Glen* and *Riders to the Sea*. The latter is one of his most notable works, transferring the starkness of Greek tragedy to the Irish countryside. His work was controversial from the start, being considered by many to be a slur on traditional Irish life: the Abbey Theatre did not at first dare to put on *The Tinker's Wedding* because of its anti-theological implications, while *The Playboy of the Western World* provoked full-scale riots when it was performed in Ireland and America. The latter was, however, quickly recognised as a masterpiece of twentieth-century theatre, its spare, lyrical prose achieving a more essentially poetic effect than modern verse drama could accomplish.

Synge also wrote a descriptive prose work, *The Aran Islands*, and was a considerable poet. Despite the controversy that surrounded him, he was an extremely retiring invalid, and for much of his life wedded to celibacy. In his last months he was engaged to Maire O'Neill, a leading actress at the Abbey Theatre, but their courtship was conducted exclusively from his sick-bed. The play he left unfinished at his death, *Deirdre of the Sorrows*, is recognised as one of his greatest works.

Plays
In the Shadow of the Glen 1903 [pub. 1905]; *Riders to the Sea* 1904; *The Well of the Saints* 1904 [pub. 1905]; *The Playboy of the Western World* 1907; *The Tinker's Wedding* 1908; *Deirdre of the Sorrows* [unfinished] 1910

Fiction
The Aran Islands 1907; *In Wicklow, West Kerry, and Connemara* 1911; *When the Moon Has Set* 1968

Poetry
Poems and Translations 1909

Non-Fiction
Collected Letters [ed. A. Saddlemyer] 1983–4; *Autobiography of J.M. Synge Constructed from Manuscripts* (a) [ed. A. Price] 1985

T

(Newton) Booth TARKINGTON 1869–1946

Born in Indianapolis, Tarkington was the son of a lawyer. Despite dictating short stories to his sister when he was six, he had an early ambition to be an illustrator, but only ever managed to have one drawing published, in *Life* magazine. After preparatory school in Lafayette, he went to Princeton, New Jersey, in 1893, but failed to graduate because of his deficient Greek. He was later compensated for this by Princeton awarding him an Honorary MA and DLitt. Tarkington was unsettled for the early part of his life, travelling widely and dabbling in politics. He was elected to the Indiana Legislature as a Republican in 1902, the year he married. The death of a daughter, which he said was the greatest tragedy of his life, ruined the marriage, and the couple divorced in 1911. He remarried in the following year.

Extremely prolific, Tarkington had some success with plays, as well as fiction. Two novels won him Pulitzer Prizes: *The Magnificent Ambersons* in 1918, and *Alice Adams* – probably his finest book – in 1921. Much of his work was marred by excessive sentiment, however, a factor which has tarnished his current reputation. Both *Alice Adams* and *The Magnificent Ambersons* were made into distinguished films: the first in 1935 with Katherine Hepburn as the small-town young woman with absurd social ambitions; the latter in 1942 by Orson Welles, who also wrote the screenplay and narrated the story of the decline of a wealthy and influential family. One of his most popular books was *Penrod*, which recalls the world of Mark Twain's *Huckleberry Finn* (1884), though it is hardly in the same class. Set in Indiana before the First World War, it recounts the adventures of a twelve-year-old boy and his friends in a small town. He wrote two sequels and collected all three stories in an omnibus edition in 1931. An early victim of political correctness, *Penrod* was reissued in 1965 purged of its patronising but nevertheless affectionate and (for the period) authentic descriptions of the black characters.

Tarkington had a permanently diabolical facial expression which belied a charming and genial nature. Lanky and willowy as a youth, he started compulsive chain smoking when a boy. He remarked that his attaining old age would have surprised most of his contemporaries – had they outlived him. In his sixties he suffered cataracts which made him blind in one eye. Undaunted, he relentlessly dictated, and left a final novel, *Show Piece*, which was published posthumously. Perhaps a **John Galsworthy** of American fiction, examining Midwestern middle-class life and morals, Tarkington attempted satire without malice. He had tried, according to his second wife, 'to explore the truth and mystery of human nature'.

Fiction

The Gentlemen from Indiana 1899; *Monsieur Beaucaire* 1900; *The Two Vanrevels* 1902; *Cherry* 1903; *The Beautiful Lady* 1905; *The Conquest of Canaan* 1905; *In the Arena* (s) 1905; *His Own People* 1907; *The Guest of Quesnay* 1908; *Beasley's Christmas Party* 1909; *The Spring Concert* 1910; *The Flirt* 1913; *Growth*: *The Turmoil* 1914, *The Magnificent Ambersons* 1918, *The Midlander* 1923, 1927; *Harlequin and Columbine and Other Stories* (s) 1918; *Ramsey Milholland* 1919; *Alice Adams* 1921; *Gentle Julia* 1922; *The Fascinating Stranger and Other Stories* (s) 1923; *Women* 1925; *Selections from Booth Tarkington's Stories* (s) [ed. L. Holmes] 1926; *The Plutocrat* 1927; *Claire Ambler* 1928; *Young Mrs Greeley* 1929; *Mirthful Haven* 1932; *Wanton Mally* 1932; *Presenting Lily Mars* 1933; *Little Orvie* 1934; *Mr White, The Red Barn, Hell, and Bridewater* 1935; *Rumbin Galleries* 1937; *The Fighting Littles* 1941; *The Heritage of Hatcher Ide* 1941; *Kate Fennigate* 1943; *Image of Josephine* 1945; *The Show Piece* [unfinished] 1947; *Three Selected Short Novels* 1947

Plays

With Harry Leon Wilson: *The Man from Home* [pub. as *The Guardian* 1907; rev. 1934]; *Cameo Kirby* 1908; *If I Had Money* [rev. as *Getting a Polish*] 1909; *Your Humble Servant* 1910; *The Gibson Upright* 1919; *Up from Nowhere* 1919; *Tweedles* 1923 [pub. 1924]; *How's Your Health?* 1929 [pub. 1930]

As Booth Tarkington: *Beauty and the Jacobin* 1912; *The Country Cousin* [with J. Street] 1917 [pub. as *Ohio Lady* 1916]; *Mister Antonio* 1916 [pub. 1925]; *Poledin* 1920; *Clarence* 1921; *The Intimate Strangers* 1921; *The Wren* 1921 [pub. 1922]; *Rose Briar* 1922; *Magnolia* 1923; *The Trysting Place* 1923; *Bimbo, the Pirate* 1926 [pub. 1924]; *Station YYYY* 1927; *The Travelers* 1927; *Colonel Susan* 1931; *The Help Each Other Club* 1933 [pub. 1934]; *Lady Hamilton and Her Nelson* 1945

Non-Fiction
The Collector's Watnot [with H. Kahler, K. Roberts] 1923; *Looking Forward and Others* 1926; *The World Does Move* (a) 1928; *Some Old Portraits: a Book of Aboput Art and Human Beings* 1939; *As It Seems to Me* 1941; *Your Amiable Uncle: Letters to His Nephews* 1949; *On Plays, Playwrights, and Playgoers: Selections from the Letters from Booth Tarkington to George C. Tyler and John Peter Toohey 1918–1925* [ed. A.S. Downer] 1959

For Children
Penrod 1914; *Penrod and Sam* 1916; *Penrod Jasber* 1916; *Seventeen* 1916; *Penrod: His Complete Story* 1931

Biography
Booth Tarkington by James L. Woodress 1955

(John Orley) Allen TATE 1899–1979

Tate was born in Winchester, Kentucky, into a prosperous Southern family; three ancestors had been officers in the American Revolution. After a bookish early childhood spent largely at home, he attended Georgetown Preparatory School, and then Vanderbilt University, where he was a budding poet and brilliant student, graduating *magna cum laude* in 1922. He was the only undergraduate invited to join the Fugitives, an influential group of Southern poets associated with the literary magazine the *Fugitive*. Chief of the group was **John Crowe Ransom**, Tate's teacher at Vanderbilt, and it was associated with a conservative agrarian philosophy. After university, Tate worked for a while in his brother's coal office but, after losing the company $700 in a day by shipping some coal to Duluth that should have gone to Cleveland, he settled down to the more congenial career of freelance writer and university teacher. He had posts successively at Memphis, the University of North Carolina, Princeton and the University of Minnesota; in later life, he also held visiting professorships at Oxford, England, and in Rome.

His poetry was exceptionally assured from early volumes such as *Mr Pope and Other Poems*: he was a cerebral, involuted poet, having something in common with his master, Ransom, but perhaps lacking a little of the latter's individuality. His work, however, continued to develop throughout his life, particularly after his conversion to Roman Catholicism in 1950; *The Swimmers* is an excellent late volume of verse. His most famous single poem is 'Ode to the Confederate Dead'.

He also wrote a highly regarded novel about the old South, *The Fathers*, biographies of Stonewall Jackson and Jefferson Davis, and was an influential literary critic of the New Criticism, editing the *Sewanee Review*, becoming a pugnacious and sometimes disliked literary figure. He was married three times, and had two sons and one daughter; his first wife was the novelist Caroline Gordon, from whom he was divorced.

Poetry
The Golden Man and Other Poems [with R. Mills] 1923; *Mr Pope and Other Poems* 1928; *Three Poems* 1930; *Poems 1928–1931* 1932; *The Mediterranean and Other Poems* 1936; *Selected Poems* 1937; *Sonnets at Christmas* 1941; *The Winter Sea* 1944; *Poems: 1920–1945* 1948; *Poems: 1922–1947* 1948; *Two Concerts for the Eye to Sing, If Possible* 1950; *Poems* 1960; *The Swimmers* 1970; *Collected Poems of Allen Tate 1919–1976* 1977

Fiction
A Southern Harvest 1937; *The Fathers* 1938 [rev. 1977]

Non-Fiction
Stonewall Jackson: the Good Soldier 1928; *Jefferson Davis: His Rise and Fall* 1929; *Reactionary Essays on Poetry and Ideas* 1936; *America through the Essay* [with A.T. Johnson] 1938; *Reason in Madness: Critical Essays* 1941; *Invitation to Learning* [with H. Cairns] 1941; *On the Limits of Poetry: Selected Essays* 1949; *The Forlorn Demon: Didactic and Critical Essays* 1953; *The Man of Letters in the Modern World: Selected Essays 1928–1955* 1955; *Collected Essays* 1959 [rev. *Essays of Four Decades* 1969]; *The Literary Correspondence of Donald Davidson and Allen Tate* [ed. J.T. Fain, T.D. Young] 1974; *Memories and Essays: Old and New 1926–1974* 1976 [US *Memoirs and Opinions 1926–1974* 1975]; *Moonstruck: a Memoir of My Life in a Cult* [with J. Vitek] 1979; *The Republic of Letters in America: the Correspondence of John Peale Bishop and Allen Tate* [ed. J.J. Hindle,T.D. Young] 1981; *The Poetry Reviews of Allen Tate 1924–1944* [ed. A. Brown, F.N. Cheney] 1983; *The Lytle–Tate Letters* [ed. E. Sarcone, T.D. Young] 1987

Edited
The Complete Poetry and Selected Criticism of Edgar Allan Poe 1968; *Who Owns America* [with H. Agar] 1936; *American Harvest: Twenty Years of Creative Writing in the United States* [with J.P. Bishop] 1942; *The Language of Poetry* 1942; *Princetown Verse Between the Two Wars* 1942; *A Southern Vanguard* 1947; *The Collected Poems of John Peale Bishop 1892–1944* [with C. Gordon] 1948; *The House of Fiction* 1950; *Sixty American Poets 1896–1944* 1954; *Modern Verse in English 1900–1950* 1958; *The Arts of Reading* [with

J. Berryman, R. Ross] 1960; *T.S. Eliot: the Man and His Work* 1967

Elizabeth TAYLOR 1912–1975

The daughter of an insurance inspector, Taylor was born Elizabeth Coles in Reading, Berkshire, and attended the Abbey School (where Jane Austen was among the former *alumnae*). After leaving school, she worked as a governess – an experience she used in her second novel, *Palladian* – and in a library. In 1936 she married the director of a sweet factory, and settled in the Buckinghamshire village of Penn. **I. Compton-Burnett** once said of Taylor that she was 'a young woman who looks as if she never had to wash her gloves': elegant, fastidious and reserved (she asked regular correspondents such as **Robert Liddell** to destroy her letters), Taylor devoted herself to domestic life and to bringing up her two children, making time away from this to write her novels and short stories.

The first of her novels, *At Mrs Lippincote's*, was published in 1945; eleven others (as well as four volumes of short stories) followed at intervals, of which the last, *Blaming*, was published posthumously. All were distinguished by beautiful, astringent writing and great moral warmth. They are mostly set amongst the post-war middle classes, which partly explains why her work has been persistently underrated. However, Taylor was a member of the Communist Party in the 1930s, and she remained a supporter of the Labour Party and an atheist throughout her life. Her eye is unfailingly sharp as she mercilessly exposes the frailties of the human heart and investigates people's capacity for self-deception. The protagonist of *The Soul of Kindness*, for example, is someone who believes herself to be just that, but who leaves havoc in her wake, while *Angel* (an untypical novel which was nevertheless chosen as one of the best post-war novels in a 1984 Book Council promotion) is about a preposterous popular novelist who believes every word of the drivel she writes, but whose own life is ruined by snobbishness and arrogance. Unlike this woman, Taylor conveys romantic passion with subtlety and assurance, particularly when the partners are ill-matched, as they frequently are. *In a Summer Season* charts the marriage of Kate Heron and her feckless younger husband, while in *A Game of Hide and Seek* the childhood passion of Harriet for her effete cousin Vesey persists into adulthood and unbalances her life. Taylor deals with a broad range of characters, but is particularly good at depicting children and domestic staff (a succession of wily chars and the splendid sailor-turned-housekeeper in *Blaming*). *Mrs Palfrey at the Claremont*, a funny and unsentimental novel about the indignities and depredations of old age, was shortlisted for the Booker Prize in 1971.

Her short stories were published in numerous magazines – including the *New Yorker* and *Harper's Bazaar*, thus trouncing the notion that her work is too parochially 'English' – and include several classics of the genre: 'A Dedicated Man', 'The Ambush', 'Miss A. and Miss M.'. The title-novella of her first collection, *Hester Lilly*, is a miniature masterpiece: the story of a marriage gradually disintegrating with the arrival of the husband's much younger cousin, it seems almost transparent, yet like all Taylor's work has an extraordinary resonance, and lives on in the mind of the reader.

Her final novel, *Blaming*, was written during her last illness, when she knew she was dying of cancer, and deals with bereavement. Although unrevised, it shows no diminution of her powers, or of her sense of high comedy. In the years since her death, her reputation has continued to grow, with almost all her books brought back into print by the feminist press Virago.

Fiction

At Mrs Lippincote's 1945; *Palladian* 1946; *A View of the Harbour* 1947; *A Wreath of Roses* 1949; *A Game of Hide and Seek* 1951; *The Sleeping Beauty* 1953; *Hester Lilly and Other Stories* (s) 1954; *Angel* 1957; *The Blush and Other Stories* (s) 1958; *In a Summer Season* 1961; *The Soul of Kindness* 1964; *A Dedicated Man and Other Stories* (s) 1965; *The Wedding Group* 1968; *Mrs Palfrey at the Claremont* 1971; *The Devasting Boys and Other Stories* (s) 1972; *Blaming* 1976

For Children

Mossy Trotter 1967

Peter (Hillsman) TAYLOR 1917–1994

Taylor was born in Trenton, Tennessee, the son of a lawyer who for many years was president of an insurance company in St Louis; his maternal grandfather (whose name was also Taylor) had been a governor of Tennessee. Taylor left Trenton when he was seven, but later grew up in Nashville, St Louis and Memphis, and the slowly changing world of the traditional American South is the background to much of his later fiction. He went to Vanderbilt University in 1936 to study with **John Crowe Ransom** and, after an interlude at Southwestern College in Memphis, followed him to Kenyon; his other influential teachers were **Allen Tate** and **Robert Penn Warren**, both, like Ransom, *doyens* of Southern agrarianism and the New Criticism, while among his friends at Kenyon were **Randall Jarrell** and **Robert Lowell** (who was Taylor's room-mate).

Despite such lustrous connections, and early publication of poems and stories in the *Kenyon Review* and *Southern Review*, Taylor has been only moderately celebrated, no doubt because of his decision to confine his work largely to the modest world of the short story. After post-graduate work in Louisiana, he served with the US Army in England during the Second World War, attaining the rank of sergeant. His career after that was academic, like that of many American writers: he taught between 1946 and 1967, with intermissions for visiting lectureships, at the University of North Carolina, and then for twenty years at the University of Virginia in Charlottesville. He did not, however, produce any academic writing, preferring to reserve himself for creative work. He married Eleanor Lilly Ross, a poet, in 1940, and they had two children, continuing to live in Charlottesville, until his death.

Many of Taylor's stories first appeared in the *New Yorker*; the first collection of them was published in 1948, and he produced five more, as well as a volume of *Collected Stories* in 1969. He is widely regarded as among the most subtle and resonant of American short-story writers, gaining his effects from gentle irony and perception rather than from violence and bombast. Many of the pieces, in *In the Miro District and Other Stories* particularly, are verse narratives, and much of his prose in fact begins as verse, although he wrote little poetry as such. He also wrote a large number of one-act plays, an early novella, *A Woman of Means* (which is often considered one of his finest works), and a late novel, *A Summons to Memphis*, published in 1986. With Lowell and Penn Warren he edited a volume about Jarrell, who had died in an incident involving a car in 1965. Taylor was a distinguished writer who, if his work seems unexciting to a sensation-seeking age, may well be remembered when gaudier talents are forgotten.

Fiction

A Long Fourth and Other Stories (s) 1948; *A Woman of Means* 1950; *The Widows of Thornton* (s) 1954; *Happy Families Are All Alike* (s) 1959; *Miss Leonora When Last Seen and Fifteen Other Stories* (s) 1963; *The Collected Stories of Peter Taylor* (s) [ed. H.H. McAlexander] 1969; *In the Miro District and Other Stories* (s) 1977; *The Early Guest* 1982; *The Old Forest* (s) 1985; *A Summons to Memphis* 1986; *The Oracle at Stoneleigh Court* 1993

Plays

Tennessee Day in St Louis 1957; *A Stand in the Mountains* 1971 [pub. 1985]; *Presences: Seven Dramatic Pieces* [pub. 1973]

Non-Fiction

Conversations with Peter Taylor [ed. H.H. McAlexander] 1987

Edited

The Road and Other Modern Stories (s) 1979; *Randall Jarrell 1914–1965* [with R. Lowell, R. Penn Warren] 1967

Edith TEMPLETON 1916–

Templeton was born in Prague, into a family of rich landowners of the class immediately below the titled aristocracy. Her native language is German; her identity is Bohemian rather than Czech; English has been her literary language since her youth. Her first four years were spent in Vienna, but after her parents split up she rarely saw her father, and grew up with her dominating mother and grandmother between a grand Prague residence and a castle in the Bohemian countryside, the background of her first three novels (*Summer in the Country, Living on Yesterday, The Island of Desire*), and of many short stories published in the *New Yorker*.

Rupture with her privileged background came early: when she was fourteen, a poem she innocently published led to her being banned from all state-run schools in the Czech Republic; and four years later she came to England, which was to remain her base until 1955, the first of many places of exile. In 1938 she married an Englishman, William Stockwell Templeton, but the marriage was unhappy, lasting only a few years, and giving place to a long series of affairs. Templeton served with the American and British forces during the war and just after it, but never afterwards worked.

She published her first novel in 1950, and three more followed rapidly, as well as a masterly and highly popular travel book, *The Surprise of Cremona*. But her second marriage, to Dr Edmund Ronald, a Viennese, in 1955, put an end both to her English residence and the prolific phase of her career. He had been for twenty years personal physician to the King of Nepal, and she joined him in India and Nepal for five months, before beginning on a peripatetic life which took them to many fashionable spots of Europe. The pair, who had one son, had their longest residence in Estoril, from 1961 until the Portuguese Revolution drove them out in 1976. They then moved to Bordighera in Italy, where Dr Ronald died in 1984 and Templeton remains.

Two more novels were added in thirty years: *Gordon*, published in 1966 under the pseudonym Louise Walbrook, an extremely candid account of a sado-masochistic love affair banned in Germany for indecency; and the eccentric but compelling *Murder in Estoril*, published in 1992. Templeton's later novels and shorter fiction, with their heavy emphasis on sexual obsessiveness, are alien from the perfectly controlled and exquisitely formal depiction of traditional Bohemian society with which she began, but share with them a regard for truth which borders on cruelty, and an unsparing use of personal experience unexpected from one whose roots are in the European *beau monde*. Her work was largely forgotten for many years after her early success in the 1950s, but a series of reprints in the 1980s, as well as the powerful advocacy of **Anita Brookner**, have done much to increase knowledge of an unusual literary talent.

Fiction

Summer in the Country 1950 [UK *The Proper Bohemians* 1952]; *Living on Yesterday* 1951; *The Island of Desire* 1952; *This Charming Pastime* 1955; *Gordon* [as 'Louise Walbrook'] 1966 [US *The Demon's Feast* 1968]; *Three* [with A. Gould and Calvin Tentfield] 1971; *Murder in Estoril* 1992

Non-Fiction

The Surprise of Cremona 1954

Emma (Christina) TENNANT 1937–

Tennant was born in London into a socially prominent family. Her great-great-aunt was Margot Asquith and her father was Christopher Grey, the 2nd Lord Glenconner. She spent an isolated childhood in the family's 'fake baronial castle' in Scotland while her father was head of the Special Operations Executive in Cairo, and she attributes much of the fabulous imaginative world of her fiction to the experience of this 'lost Eden of my childhood'. After the war she was educated at St Paul's Girls' School, London, an eccentric finishing school in Oxford, and the Ecole du Louvre in Paris, where she studied the history of art. She was presented at court in 1956, and the following year married Sebastian Yorke, son of the novelist **Henry Green**.

Her first novel, *The Colour of Rain*, published pseudonymously, is a gentle satire of upper-class shallowness which is stylistically indebted to the work of her father-in-law. Her publisher submitted it for the 1964 Formentor Prize, and Tennant heard that one of the judges, Alberto Moravia, had hurled it into a wastepaper bin denouncing it as 'a symbol of the decadence of English writing today'. She has said that she was so crushed by this that she was unable to return to fiction for nearly a decade, and during this period worked for *Queen* magazine and as features editor of *Vogue*. The discovery of English science fiction, and the encouragement of **J.G. Ballard** and **Michael Moorcock**, led her to write *The Time of the Crack*, a tale of apocalyptic disaster in a futuristic London. Having found this vein of fantasy, Tennant has gone on to blend fantasy and realism in a succession of innovative novels. *Hotel de Dream* explores the imaginative world of the residents of a run-down Kensington hotel, while *The Bad Sister* is a twentieth-century feminist reworking of James Hogg's *Confessions of a Justified Sinner* (1824).

Much of her fiction draws its initial inspiration from literary sources: *Alice Fell* combines elements of Lewis Carroll and Greek myth in squalid 1950s' London, and *Queen of Stones* clearly mirrors **William Golding**'s *Lord of the Flies* (1954) in describing a group of schoolgirls reverting to savagery. *Tess* is a complex exploration of the world of **Thomas Hardy**'s heroine and of an actress who plays the part, while *Pemberley* and *An Unequal Marriage* are rather more straightforward sequels to Jane Austen's *Pride and Prejudice* (1813) and describe Darcy and Elizabeth's marriage.

Tennant was the founder and editor (from 1975 to 1978) of the literary magazine *Bananas* in which she published work by **Angela Carter**, Ballard and Moorcock, and she has been general editor of the Penguin Lives of Modern Women series since 1985. She has a son from her first marriage (the writer Matthew Yorke), and two daughters from two subsequent marriages, both of which ended in divorce.

Fiction

The Colour of Rain [as 'Catherine Aydy'] 1963; *The Time of the Crack* 1973; *The Last of the Country House Murders* 1974; *Hotel de Dream* 1976; *The Bad Sister* 1978; *Wild Nights* 1979; *Alice Fell* 1980; *Queen of Stones* 1982; *Woman Beware Woman* 1983; *Black Marina* 1985; *The Adventures of Robina, by Herself* 1986; *The House of Hospitalities* 1987; *A Wedding of Cousins* 1988; *The Magic Drum* 1989; *Two Women of London* 1989; *Sisters and Strangers* 1990; *Faustine* 1991; *Pemberley* 1993; *Tess* 1993; *An Unequal Marriage* 1994

For Children

The Boggart 1980; *The Search for Treasure Island* 1981; *The Ghost Child* 1984; *Dave's Secret Diary* 1991

Edited

Bananas 1977; *Saturday Night Reader* 1979

Plays

Frankenstein's Baby (tv) 1990

Paul (Edward) THEROUX 1941–

Theroux was born in Medford, Massachusetts, into what he has described in *Sunrise with Seamonsters* as 'a populous family of nine unexampled wits'. His father, a shoe salesman of French-Canadian ancestry, encouraged the children's interest in literature, and one of Theroux's brothers, Alexander, also became a writer. After Medford High School, Theroux studied medicine at the University of Maine and the University of Massachusetts where he became disillusioned with the American medical profession – 'they won't operate unless you've eaten money' – and switched to English.

He was arrested in 1962 for organising an anti-war demonstration and, having graduated in 1963, joined the Peace Corps to avoid being drafted, where he was posted to teach English in Malawi. His unwitting involvement with a group of guerillas accused of plotting the assassination of the country's president, Dr Hastings Banda, led to his expulsion in 1965, and he moved to Makerere University in Kampala, Uganda, where he met **V.S. Naipaul**, an important mentor and influence.

Theroux's first novel, *Waldo* (1967) is a comedy concerning the adventures of a precociously successful young American writer, while *Fong and the Indians* describes the misfortunes of a hapless Chinese grocer in a corrupt, newly independent East African state. Fearful of the growing political instability in Uganda, Theroux moved in 1968 to the University of Singapore. In 1972 he was writer-in-residence at the University of Virginia, after which he moved to London to live by his writing.

Despite favourable reviews, Theroux's early novels enjoyed only moderate success. He sprang to prominence with *The Great Railway Bazaar*, the fruit of a four-month railway journey across Asia which combines an acute observation of the grotesque and comic with a strongly personal tone. Theroux regards travel-writing as a form of autobiography, and has undertaken similar journeys in South America and China, and around the coast of the British Isles. His fiction draws on a similarly wide range of settings and frequently focuses on the experience of exile and cultural dislocation in a manner which has invited comparison with the work of **Joseph Conrad** and **Graham Greene**. *The Family Arsenal* is the violent story of an American former Vietnamese consul who becomes involved with IRA terrorists in a seedy corner of London, while *The Mosquito Coast* describes the efforts of Allie Fox, an obsessive American inventor, to found an ideal community in the jungle of Honduras. The novel won the James Tait Black Memorial Prize in 1982 and was filmed by Peter Weir (1986) with Harrison Ford playing Fox.

Theroux's later novels include *My Secret History*, the memoirs of an American writer which are prefixed with a disclaimer warning against autobiographical interpretation, and the darker *Chicago Loop* which explores the disastrous consequences of sexual fantasy and marital duplicity. Theroux won the 1978 Whitbread Award for *Picture Palace* and the 1989 Thomas Cook Award for *Riding the Iron Rooster*. He was married in 1967, and has two sons.

Fiction

Waldo 1967; *Fong and the Indians* 1968; *Girls at Play* 1969; *Murder in Mount Holly* 1969; *Jungle Lovers* 1971; *Sinning with Annie and Other Stories* (s) 1972; *Saint Jack* 1973; *The Black House* 1974; *The Family Arsenal* 1976; *The Consul's Wife* 1977; *Picture Palace* 1978; *World's End and Other Stories* (s) 1980; *The Mosquito Coast* 1981; *The London Embassy* 1982; *Doctor Slaughter* 1984; *O-Zone* 1986; *My Secret History* 1989; *Chicago Loop* 1991

Non-Fiction

V.S. Naipaul. An Introduction to His Work 1972; *The Great Railway Bazaar* 1975; *The Old Patagonian Express* 1979; *The Kingdom by the Sea: a Journey around the Coast of Great Britain* 1983; *Sailing through China* 1983; *Sunrise with Seamonsters: Travels and Discoveries 1964–1984* 1985; *The Imperial Way: Making Tracks from Peshawar to Chittagong* [with S. McCurry] 1985; *Riding the Iron Rooster* 1988; *The Happy Isles of Oceania* 1992. A new volume of travel writing, *The Pillars of Hercules*, about to come out

For Children

A Christmas Card 1978; *London Snow* 1979

D(onald) M(ichael) THOMAS 1935–

The son of a plasterer, Thomas was born into a working-class Methodist household in Cornwall. Although he was later to make a reputation as a novelist dealing explicitly with sexuality, his upbringing was sheltered. He was educated at Redruth Grammar School and the University High School, Melbourne, reaching puberty aboard a ship *en route* to Australia – 'a profound experience'. After two years' national service, during which period he learned Russian, he read English at New College, Oxford, where he also began his career as a poet.

After graduation in 1958, he married and became a schoolteacher at Teignmouth Grammar School. During his first marriage, he had a concurrent affair with the woman who eventually became his second wife. This unusual arrangement, which resulted in children with both women (two sons and a daughter), went on for some time; at first Thomas thought it was beneficial to all three parties, but when his wife eventually left him, he had a nervous breakdown and subsequently underwent analysis. He claims to have survived his depression by listening repeatedly to a record of Rodgers and Hammerstein's musical, *South Pacific*.

His first novel, *The Flute Player*, won the *Guardian*–Gollancz Fantasy Novel Prize in 1979, but he made his name with his second novel, *The White Hotel*, which he wrote when the Hereford College of Education, where he had been a lecturer for fifteen years, closed down in 1978. His experience of analysis, in which he had to remember and write down his dreams, was to be an important contribution to his writing, which examines the interplay between dreams and waking and the world of the imagination. More specifically, it inspired him to write an erotic poem using a female persona, a process feminist critics have seen as having more to do with male fantasy than female sexuality. This eventually grew to become *The White Hotel*, in which an opera singer undergoes psychoanalysis with Sigmund Freud, fantasises about sadomasochistic sex with Freud's son, discovers the death-wish and ends up being raped with a bayonet at Babi Yar. (Thomas was later accused of having plagiarised a book on the massacre by a Russian author; he frequently uses material from other sources and has challenged the whole concept of 'originality'.) The novel was published to mixed reviews (some of them very hostile) in Britain, but was very well received in the USA, where the paperback rights were sold for an astonishing $199,500. It then underwent a revaluation, was shortlisted for the Booker Prize, won the *Los Angeles Times* Prize for Fiction, and became an international bestseller, translated into fourteen languages. Sceptics have suggested that the book's success had less to do with a sudden surge of interest in the literary novel, than with its sensationalist subject-matter.

Certainly, none of Thomas's other novels have achieved a similar popular success, although they have had their enthusiastic admirers among the critics. His knowledge of Russia inspired the sequence *Russian Nights*, comprising five novels in which sex and psychoanalysis form the principal ingredients. The last volume, *Lying Together*, was described by the *Sunday Times* as an 'outpouring of priapic post-modernism'. As well as his fiction, Thomas has written several volumes of poetry and has translated the work of Pushkin, Yevtushenko and Anna Akhmatova. He has also written an autobiographical work, *Memories and Hallucinations*. He has taught English and creative writing in American universities.

Fiction
The Flute Player 1979; *The White Hotel* 1981; *Birthstone* 1982; *Russian Nights*: *Ararat* 1983, *Swallow* 1984, *Sphinx* 1986, *Summit* 1988, *Lying Together* 1990; *Flying into Love* 1992; *Pictures at an Exhibition* 1993

Poetry
Personal and Possessive 1964; *Penguin Modern Poets 11* [with D.M. Black, P. Redgrove] 1968; *Two Voices* 1968; *The Lover's Horoscope* 1970; *Logan Stone* 1971; *The Shaft* 1973; *Lillith Prints* 1974; *Symphony in Moscow* 1974; *Love and Other Deaths* 1975; *The Rock* 1975; *Orpheus in Hell* 1977; *The Honeymoon Voyage* 1978; *Dreaming in Bronze* 1981; *News from the Front* [with S. Kantanis] 1983; *Selected Poems* 1983; *The Puberty Tree* 1992

Translations
Anna Akhmatova: *Requiem, and Poem without a Hero* 1976, *Way of All the Earth* 1979, *You Will Hear Thunder* 1985; Pushkin *The Bronze Horseman* 1982; Yevgeny Yevtushenko *Invisible Threads* 1981, *A Dove in Santiago* 1982

Edited
The Granite Kingdom: Poems of Cornwall 1970; *Poetry in Crosslight* 1975; *Songs from the Earth: Selected Poems of John Harris, Cornish Miner, 1820–1884* 1977

Non-Fiction
Memories and Hallucinations (a) 1988

Dylan (Marlais) THOMAS 1914–1953

Thomas was born and grew up in Swansea, where his father was the senior English master at the grammar school. Childhood summers were spent at the Carmarthenshire dairy farm of his aunt, Ann Jones, an experience which inspired one of his most beautiful poems, 'Fern Hill'. After leaving school in 1931, he became a reporter on the South Wales *Evening Post* for fifteen months, but moved to London in 1934. In the same year, his first volume of verse, *18 Poems*, was published, making an immediate public impact and leading to his being taken up by the influential older poet **Edith Sitwell**.

In 1937, Thomas married Caitlin Macnamara, a young dancer who had been dismissed from the London Palladium, and they had three children. In London, Thomas became a formidable Bohemian, drinking and smoking heavily as well as indulging in frequent love-affairs. (Norman Cameron wrote a poem about him in which he complained that when Thomas sat on a sofa he left a stain on it.) However, the Thomases formed the habit of going for various periods to the Carmarthenshire village of Laugharne (where they settled permanently in 1949), and there Thomas, a dedicated artist as well as a drunk, could work while recovering from his London jags.

During the war, he was rejected as unfit for military service, and worked in a documentary film unit while also beginning his prolific career as a freelance BBC radio broadcaster. In 1946, he published perhaps his finest volume of poems, *Deaths and Entrances*, but in the last six years of his life wrote only six poems. He found it easier to write the radio pieces collected in *Quite Early One Morning*, the short stories of *Adventures in the Skin Trade*, and the famous radio play *Under Milk Wood*. His prose, which also includes the early fictionalised autobiography *Portrait of the Artist as a Young Dog*, while inferior to his poetry, is often highly entertaining.

In 1950 he made the first of four lecture tours of the USA, during which he drank continually. During the last of these he collapsed with alcoholic poisoning and died shortly afterwards. Thomas was one of the most famous and publicly loved poets of his time. Although attacks have been made on his poetry as obscure, immature and windily rhetorical, in perhaps a dozen of his finest lyrical poems – 'Fern Hill', 'The Hunchback in the Park', 'In Memory of Ann Jones' and 'Poems in October' among others – he justifies **Herbert Read**'s description of his work as 'the most absolute poetry that has been written in our time'.

Poetry

18 Poems 1934; *Twenty-Five Poems* 1936; *New Poems* 1943; *Deaths and Entrances* 1946; *Collected Poems 1934–1952* 1952; *In Country Sleep and Other Poems* 1952; *Galsworthy and Gawsworth* 1954; *Two Epigrams of Fealty* 1954; *Dylan Thomas: the Poems* [ed. D. Jones] 1971; *The Notebook Poems 1930–1934* [ed. R. Maud] 1989

Plays

The Doctor and the Devils **(f)** [pub. 1953]; *Under Milk Wood* **(r)** 1954, **(st)** 1956; *The Beach of Falesa* **(f)** [from R.L. Stevenson's *Island Nights' Entertainment*] [pub. 1963]; *Twenty Years A-Growing* [from M. O'Sullivan] **(f)** [pub. 1964]; *Me and My Bike* **(f)** [pub. 1965]; *Rebecca's Daughters* **(f)** 1994 [pub. 1965]; *A Dream of Winter* **(f)** and *The Londoner* **(f)** [pub. in *The Doctor and the Devils and Other Scripts* 1966]

Fiction

Portrait of the Artist as a Young Dog **(s)** 1940; *Selected Writings of Dylan Thomas* 1946; *Conversation about Christmas* 1954; *Quite Early One Morning* **(r)** 1954; *A Prospect of the Sea and Other Stories and Prose Writings* **(s)** [ed. D. Jones] 1955; *Adventures in the Skin Trade and Other Stories* **(s)** 1955; *Early Prose Writings* [ed. W. Davies] 1971; *The Death of the King's Canary* [with J. Davenport] 1976; *The Collected Stories* **(s)** 1983

Non-Fiction

Letters to Vernon Watkins [ed. V. Watkins] 1957; *Selected Letters of Dylan Thomas* [ed. C. Fitzgerald] 1966; *Poet in the Making: the Notebooks of Dylan Thomas* [ed. R. Maud] 1967; *The Collected Letters* [ed. P. Ferris] 1985

Collections

The Map of Love: Verse and Prose 1939

Biography

Dylan Thomas by Paul Ferris 1977

(Philip) Edward THOMAS 1878–1917

Born in Lambeth, London, the son of the staff clerk for light railways and tramways at the Board of Trade, Thomas was of Welsh descent on both sides. His father, a propagandist for positivism, wished to force him into the civil service, but the young Thomas, having conceived twin passions for literature and nature, determined on a career as a man-of-letters. After attending St Paul's School, he entered Oxford, at first only as a non-collegiate student, in 1897. He published his first book, *The Woodland Life*, in the same year.

His wife-to-be became pregnant with their first child while he was a student; after going down in 1900, he immediately began supporting his growing family (eventually to consist of a son and two daughters) as a hardworking literary hack, never publishing less than one book a year, reviewing assiduously for the *Manchester Guardian*, but finding that what he could earn often left himself and his family in harrowing poverty. They lived in rented cottages, first in Hampshire and later in Kent, a life made further miserable by Thomas's deep depression and fits of bad temper. His books were largely essayistic treatments of country and literary themes: although mostly forgotten now, some of them, such as the essays collected in *Rest and Unrest* and *Light and Twilight*, and his one novel, *The Happy-Go-Lucky Morgans*, have merit.

Thomas is remembered now as a poet, a vocation he came to late in his life, under the influence of his friend **Robert Frost**, whom he met in 1912. He wrote poetry under the pseudonym of Edward Easterway but only a few poems were printed (privately) during his lifetime, his *Poems* (1917) being in press at the time of his death. Sometimes wrongly classed as a Georgian poet, Thomas is much more: beneath the country idyll of such poems as 'Adlestrop', his melancholia led him to explore such themes as sexual incompatibility and the essential sadness at the heart of experience, and he became a major poet of the century, even now underrated. In 1915, he enlisted as a private in the Artists' Rifles, and was later promoted to lance-corporal and second lieutenant. He was sent to France in 1917, and killed the same year, at Arras. When he died, as fellow poet **Walter de la Mare** put it, 'there was shattered a mirror of England'.

Poetry

Six Poems [as 'Edward Eastaway'] 1916; *Poems* 1917; *Last Poems* 1918; *Collected Poems* 1920; *Two Poems* 1927

Fiction

Celtic Stories (s) 1911; *Norse Tales* (s) 1912; *The Happy-Go-Lucky Morgans* 1913; *Four-and-Twenty Blackbirds* (s) 1915

Non-Fiction

The Woodland Life 1897; *Horae Solitariae* 1902; *Oxford* [with J. Fulleylove] 1903; *Rose Acre Papers* 1904; *Beautiful Wales* [with R. Fowler] 1905; *The Heart of England* 1906; *Richard Jefferies: His Life and Work* 1909; *The South Country* 1909; *Feminine Influence on the Poets* 1910; *Rest and Unrest* 1910; *Windsor Castle* 1910; *The Isle of Wight* 1911; *Light and Twilight* 1911; *Maurice Maeterlinck* 1911; *The Tenth Muse* 1911; *Algernon Charles Swinburne* 1912; *George Borrow: The Man and His Books* 1912; *Keats* 1912; *Lafcadio Hearn* 1912; *The Country* 1913; *The Icknield Way* 1913; *Walter Pater* 1913; *In Pursuit of Spring* 1914; *The Life of the Duke of Marlborough* 1915; *A Literary Pilgrim in England* 1917; *Cloud Castle and Other Papers* 1922; *The Last Sheaf* 1928; *The Childhood of Edward Thomas* (a) 1938; *Edward Thomas: the Last Five Years* [ed. E. Farjeon] 1958; *Letters from Edward Thomas to Gordon Bottomley* [ed. R.G. Thomas] 1968

Edited

The Poems of John Dyer 1903; *The Pocket Book of Poems and Songs for the Open Air* 1907; *The Book of the Open Air* 2 vols 1908; *British Butterflies and Other Insects* 1908; *The Pocket George Borrow* 1912; *This England: an Anthology from Her Writers* 1915

Biography

Edward Thomas by Jan Marsh 1978

R(onald) S(tuart) THOMAS 1913–

Regarded by many as the most important Welsh poet now writing in English, Thomas was born in Cardiff, the son of a seaman. In early years he mainly lived in port towns with his mother while his father was at sea (they spent much of the First World War in Liverpool), but then his father found work with a ferry-boat company operating between Wales and Ireland, and he was raised at Holyhead (Caergybi). His formal education began late, but he attended the grammar school in Anglesey and then the University College of North Wales, where he took a BA in classics. On his mother's suggestion, he studied for the Anglican ministry, at St Michael's College, Llandaff; he was ordained deacon in 1936, and became a priest the following year. He was to serve as curate, vicar and rector in six different areas of Wales, mainly small towns set among remote rural communities, until he resigned, shortly before reaching the retiring age, in 1978. This background is an integral part of his poetry, and he writes as a priest, a Welshman and a countryman.

He had begun to write verse in the 1930s, but his work gathered momentum after he married Mildred Eldridge in 1940, because she already had a reputation as a painter, and he strove to complement her artistic activity. His poems began to appear in magazines in wartime, and his first collection, *The Stones of the Field*, was privately printed in 1946. He became well known a decade later, when his collection *Song at the Year's Turning* was recommended to a London publisher by the novelist **James Hanley**, and **John Betjeman** wrote a foreword to it. Thomas has now published numerous volumes of verse, and his *Collected Poems 1945–1990* appeared in 1993. He writes a poetry which has strong roots, albeit rather despairing ones. The contradictions are many: he has been an Anglican among Dissenters, an intellectual among peasants, a radical among the apolitical, theologically unorthodox but duty-bound not to communicate this to his parishioners, a Welsh nationalist who writes in English and had stammeringly to learn his preferred tongue. But he has made all this into an *oeuvre* of severe and simple power, limited perhaps but likely to be enduring. His best-known poetic character is Iago Prytherch, the archetypal Welsh countryman who appears in several poems, but later work moves from rural realism to religious and political statement.

In retirement, Thomas has lived at Pwllheli, Gwynedd. His wife, with whom he had one son, died after many years of illness in 1991. In the same year, he aroused controversy when, at an Eisteddfod public meeting, he urged those engaging in violence against English targets not to be timid; less than twenty-four hours later, two fire-bombs were defused in Bangor. It must be remembered, however, that seemingly extreme pronouncements from nationalist writers form part of a tradition going back at least to Saunders Lewis (1893–1985), the Welsh-language poet, novelist, dramatist and critic who spent time in prison. Thomas has published a certain amount of non-fiction, including a volume of *Selected Prose* and his autobiography, *Neb*, written in Welsh.

Poetry
The Stones of the Field 1946; *An Acre of Land* 1952; *The Minister* 1953; *Song at the Year's Turning* 1955; *Poetry for Supper* 1958; *Tares* 1961; *Penguin Modern Poets 1* [with L. Durrell, E. Jennings] 1962; *The Bread of Truth* 1963; *Pieta* 1966; *Not That He Brought Flowers* 1968; *H'm* 1972; *Laboratories of the Spirit* 1975; *The Way of It* 1977; *Frequencies* 1978; *Between Here and Now* 1981; *Later Poems, 1972–1982* 1983; *Ingrowing Thoughts* 1985; *Experimenting with an Amen* 1986; *Collected Poems 1945–1990* 1993

Edited
The Batsford Book of Country Verse 1961; *The Penguin Book of Religious Verse* 1962; *Selected Poetry of Edward Thomas* 1964; *George Herbert, a Choice of Verse* 1967

Non-Fiction
What is a Welshman 1974; *Selected Prose* 1983; *Neb* (a) 1985

Flora (Jane) THOMPSON 1876–1947

One of ten children of a country stonemason, Thompson was born Flora Timms and grew up in the village of Juniper Hill near Brackley on the Oxfordshire–Northamptonshire border (the village she calls Lark Rise in her trilogy, naming it after Juniper Hill's largest cornfield). After a rudimentary education at a village school, she began work, aged fourteen, as a post-office clerk at the nearby village of Fringford. She transferred in 1897 to the post-office in the Surrey village of Grayshott, where she lived in lodgings and began to read widely. While there, she met and married another postal worker, John Thompson, and they both went to work at the main post office in Bournemouth: their not particularly happy partnership lasted for the rest of her life, and they had three children. They lived in Bournemouth until 1916, and then in Liphook, where her husband was sub-postmaster, until 1928.

During her Bournemouth years, she began to earn some money from writing small essays, stories and poems, and was greatly encouraged by the minor poet (then considerably better known) Ronald Campbell Macfie: her volume of poems, *Bog Myrtle and Peat*, published in 1921, attracted little notice. In 1928 John became postmaster at Dartmouth, and while they lived there his wife spent much of her time organising the Peverel Society, a correspondence-course and support-group for aspirant writers. It was only when she was sixty, in 1937, that she received encouragement from Sir Humphrey Milford of the Oxford University Press to begin her life's work.

Lark Rise was the first book of the trilogy eventually collected as *Lark Rise to Candleford*, in 1945, at last winning her a modest literary fame. Not exactly fiction and not exactly autobiography, the trilogy is a most remarkable evocation of the village world of her childhood, being both realistic and poetic; the older, longer-lasting England of the open fields was never recorded from the inside, but the world of the late nineteenth-century agricultural labourers found its great chronicler in Thompson.

After her husband's retirement in 1940, the couple went to live at Brixham, where Thompson died, aged seventy. *Still Glides the Stream*, a novel based on her childhood, was published posthumously in 1948.

Fiction
Lark Rise to Candleford 1945: *Lark Rise* 1939, *Over to Candleford* 1941, *Candleford Green* 1943; *Still Glides the Stream* 1948

Poetry
Bog Myrtle and Peat 1921

For Children
Apple-Georgie Farm [with K. Thompson] 1936

Edited
The Peverel Book of Verse 2 vols 1928, 1929

Colin (Dryden) THUBRON 1939–

Thubron was born in London, the son of a professional soldier, and is descended from John Dryden on his mother's side; he has described his upbringing as 'very conventional and privileged'. His father became a military attaché in Canada and the young Thubron travelled extensively, enjoying an 'almost thoughtlessly happy childhood'. His parents presented him with 'an idea of marriage that meant I wasn't so insecure that I wanted to escape into some other relationship. They made me self-sufficient.' He is still unmarried, and the pain of unrequited love is the stuff of his recent fiction. After Eton he went into publishing, joining the editorial staff of Hutchinson in 1959, before becoming a freelance film-maker for the BBC in 1962, then going to New

York to join Macmillan there; after that he set out on his travels in the Middle East.

He travelled alone and on foot: 'To travel with just one other person means you create a bubble of Englishness from which you peer and say, "Isn't that funny, isn't that odd". If you are alone it is you who are odd and susceptible to whatever forces are about.' His early travel books lacked self-confidence, but with *Among the Russians* – of which Nikolai Tolstoy said: 'It is hard to think of a better travel book written in this century' – and *Behind the Wall*, about his travels in China, he became recognised as a leading writer in his field. It was much the same with the novels, which he began to publish in his late thirties. He has said of them: 'Basically my first novel, *The God in the Mountain*, was the least assured. I suppose I found some sort of "voice" in *Emperor*, and particularly in *A Cruel Madness* and *Falling*, which were reviewed rather more extensively.' The former, which won the Silver PEN Award, is set in a lunatic asylum. Reviewing the book, **Nina Bawden** commented: 'Obsessions lie on the edge of madness. In using madness to describe them Mr Thubron has hit that most disturbing target, the target of imaginative truth'. In *Falling*, whose central figure is in prison, he returns to the theme of obsessive love; **Anita Brookner** found his voice 'once heard … difficult to forget'. Thubron has described his method as a novelist thus: 'I've tended to scratch away at some distress in myself and discontent with the way things are.'

Fiction

The God in the Mountain 1977; *Emperor* 1978; A Cruel Madness 1984; *Falling* 1989; *Turning Back the Sun* 1991

Non-Fiction

Mirror to Damascus 1967; *The Hills of Adonis: a Quest in Lebanon* 1968; *Jerusalem* [with A. Duncan] 1969; *Journey into Cyprus* 1975; *The Royal Opera House* 1982; *Among the Russians* 1983; *Behind the Wall* 1987; *The Silk Road to China* 1989

James (Grover) THURBER 1894–1961

Born in Columbus, Ohio, Thurber was the son of Charles Leander (later surnamed Lincoln), a minor politician, and Mary Thurber, a formidable eccentric and practical joker whom her son chronicled in his comic autobiography, *My Life and Hard Times*. When Thurber was six his brother shot at him with an arrow, and he lost the sight of his left eye. He attended public schools in Columbus, and entered Ohio State University in 1913, graduating in 1919 after his war service. (The army had refused him because of his disability, and he was a code clerk in Washington and then at the American Embassy in Paris.)

He became a newspaperman from 1920, first in his native Columbus, later on the Paris edition of the *Chicago Tribune*, and then in New York. He met the well-known *New Yorker* writer E.B. White at a party, they became friends, and White took Thurber, hoping to get him a job, to *New Yorker* editor Harold Ross; the latter (surprisingly, given Thurber's lack of administrative ability or experience) made him managing editor of the magazine. After extricating himself from this position, Thurber remained on the staff until 1933, and was a frequent contributor until his death. It fell upon the young **Truman Capote**, in his capacity as general dogsbody at the *New Yorker*, to lead the purblind Thurber to and from meetings. (He was also obliged to escort Thurber to his trysts with one of the magazine's secretaries, waiting in the living-room while their passion was consummated, and afterwards helping Thurber back into his clothes.)

Thurber's first book, in collaboration with E.B. White, was the satirical *Is Sex Necessary?* (1929), and he was to publish more than twenty other volumes. His production consisted mainly of comic short stories and essays with complementary drawings of comic men, women and dogs. Most of these first appeared in the *New Yorker* and were then published in books with titles such as *My World and Welcome to It* or *The Beast in Me and Other Animals*. Thurber's theme in all his humour was the innocent lost amid the complexities of the modern world, and he exploited it well enough to be ranked among the most inspired humorists of the century. He also wrote children's books (notably the fantasy novel *The Thirteen Clocks*), a well-regarded play, *The Male Animal* (in collaboration with Elliot Nugent) and *The Years with Ross*, a memoir of the *New Yorker*. His 1947 story about a daydreamer, 'The Secret Life of Walter Mitty', was taken up by psychologists – a group he frequently satirised – and 'Walter Mitty Syndrome' was put forward in a British medical journal as a clinical condition which manifested itself in compulsive fantasising. (The syndrome reappeared in fictional form in 1959 in **Keith Waterhouse**'s novel *Billy Liar*.)

He married twice, and had one daughter. In later years, after several unsuccessful cataract operations on his remaining eye, he became blind, but the devoted nursing of his wife, Helen Wismer (whose own eyesight was saved only by an operation), enabled him to maintain his literary production and humour to the end. In later years, the couple lived at West Cornwall, Connecticut.

Fiction
The Owl in the Attic and Other Perplexities 1931; *The Middle-Aged Man on the Flying Trapeze* (s) 1935; *The Last Flower* 1939; *My World and Welcome to It* (s) 1942; *Thurber Country* (s) 1953

Non-Fiction
The Seal in the Bedroom and Other Predicaments 1931; *My Life and Hard Times* (a) 1933; *Let Your Mind Alone!* 1937; *The Beast in Me and Other Animals* 1948; *The Thurber Album* 1952; *Thurber's Dogs* 1955; *Alarms and Diversions* 1957; *The Years with Ross* (a) 1959; *Credos and Curios* 1962

Plays
The Male Animal [with E. Nugent] 1940; *Random House* 1940; *Many Moons* [from his children's book] 1947

For Children
Many Moons 1943; *The Great Quillow* 1944; *The White Deer* 1945; *The Thirteen Clocks* 1950; *The Wonderful O* 1957

Collections
Is Sex Necessary? [with E.B. White] 1929; *Cream of Thurber* 1939; *Fables for Our Time and Famous Poems Illustrated* 1940; *The Thurber Carnival* 1945; *Further Fables for Our Time* 1956; *Thurber and Company* 1966

Biography
Thurber by Burton Bernstein 1975

Anthony (Simon) THWAITE 1930–

Thwaite was born in Chester, but into a family which on both sides had lived for many generations in Yorkshire – on his father's side as North Riding tenant farmers. He had a wandering childhood because his father worked for Lloyds Bank, and he also spent much of the war in the USA as an evacuee. He then attended Kingswood School near Bath, did his national service in Libya (where he was able to indulge his lifelong interest in archaeology), and attended Christ Church, Oxford, from 1952 to 1955, where he formed part of an undergraduate group of poets.

His first job after leaving Oxford was as an English lecturer at Tokyo University until 1957, and his adult life has seen him varying a successful career as a man-of-letters in London with periods spent abroad, often as a lecturer. In 1985–6 he had a fellowship from the Japan Foundation, and he drew on this experience for his volume of poems *Letter from Tokyo*. He also spent two years in the mid-1960s as a university lecturer in Libya as well as making shorter stays in many other countries.

The other prong of his career made him literary editor of the *Listener* (1962–5) and of the *New Statesman* (1968–72). Since then he has been mainly a freelance writer, critic and editor, combining this with the co-editorship of *Encounter* (1973–85) and a directorship at the publishing house of André Deutsch from 1976, in which year he was also chairman of the Booker Prize judging panel. In the breadth of his interests – many appearances on television and radio, editorship of the letters and collected poems of **Philip Larkin**, frequent reviewing – Thwaite is the nearest thing to the leading man-of-letters that the late twentieth century can provide, although it lies in the nature of the age that he does not enjoy quite the prominence of a **Cyril Connolly** or **John Lehmann**.

His most purely creative work is as poet: his first volume of *Poems* was published in 1953, and the latest update of his *Collected Poems* in 1989. Persistent characteristics of his verse are the tendency to approach personal statement through personae, allusiveness, an interest in the past and an attachment to the iambic pentameter line. He married Ann Harrop in 1955 (she is well known as the biographer Ann Thwaite), and they have four daughters.

Poetry
Poems 1953; *Home Truths* 1957; *The Owl in the Tree* 1963; *The Stones of Emptiness: Poems 1963–66* 1967; *Penguin Modern Poets 18* [with A. Alvarez, Roy Fuller] 1970; *Points* 1972; *Inscriptions: Poems 1967–72* 1973; *Jack* 1973; *New Confessions* 1974; *A Portion for Foxes* 1977; *Victorian Voices* 1980; *Telling Tales* 1983; *Poems 1953–1983* 1984; *Letter from Tokyo* 1987; *Poems 1953–1988* 1989

Non-Fiction
Essays on Contemporary English Poetry: Hopkins to the Present Day 1957 [rev. as *Contemporary English Poetry: an Introduction* 1959]; *Japan in Colour* [with R. Beny] 1967; *The Deserts of Hesperides: an Experience of Libya* 1969; *Poetry Today 1960–1973* 1973 [rev. as *Poetry Today: a Critical Guide to English Poetry 1960–1984* 1985]; *Roloff Beny in Italy* [with P. Porter] 1974; *Beyond the Inhabited World: Roman Britain* 1976; *Twentieth-Century English Poetry* 1978; *Odyssey: Mirror of the Mediterranean* [with R. Beny] 1981

Edited
New Poems 1961 [with H. Corke, W. Plomer] 1961; *The Penguin Book of Japanese Verse* [with G. Bownas] 1964; *The English Poets: From Chaucer to Edward Thomas* [with P. Porter] 1974; *Poems for Shakespeare 3* 1974; *New Poetry 4* [with F. Adcock] 1978; *The Gregory Awards Anthology 1981–82* [with H. Sergeant] 1982; *Larkin at Sixty* 1982; *Poetry 1945 to 1980* [with

J. Mole] 1983; *Six Centuries of Verse* 1984; *Collected Poems of Philip Larkin* 1988

J(ohn) R(onald) R(euel) TOLKIEN 1892–1973

Although his father's family had migrated from Saxony to England several generations before, and both Tolkien's parents were natives of Birmingham, his father was working temporarily as a bank manager in South Africa when Tolkien was born at Bloemfontein. Shortly before his father's death in 1896, Tolkien's mother had returned to live near Birmingham with her son, then aged three. She died in 1904 and, having entered the Roman Catholic church some years previously, left Tolkien's upbringing in the hands of a priest, Father Francis Xavier Morgan. Tolkien's devout Catholicism throughout his life is an important background to his thought and work.

He attended the King Edward VI Grammar School in Birmingham (where a precocious talent for languages, especially early Germanic ones, was evident) and Exeter College, Oxford, from which he graduated with first class honours in English in 1915. He joined the army, fought at the Battle of the Somme, and was shortly afterwards invalided home to England with trench fever. He married a childhood sweetheart in 1916, and they had four children. From 1918 to 1920 he worked as an assistant on the *Oxford English Dictionary*; he was then a professor of English at the University of Leeds until 1925, in which year he returned to Oxford as a professor of Anglo-Saxon. From 1945 to 1959 he was Merton Professor of English.

He published a Middle English vocabulary and a study of *Beowulf*, but despite his profound knowledge of philology, his increasing attachment to fiction meant that few scholarly works flowed from him. From around 1912 he had begun to invent a new language called Elvish, and he increasingly felt the need to create a mythology based on the language, which eventually led to the world of Middle Earth, where most of his fiction is set. He was a member, with the writers **C.S. Lewis** and **Charles Williams**, of the Inklings, a group of Christian, conservative Oxford writers, and it was Lewis's prompting which led to the publication of *The Hobbit* in 1937. Tolkien followed this up with a much more extended and complex trilogy of Middle Earth, *The Lord of the Rings*. This was highly praised by critics of the stature of **W.H. Auden**, and from the mid-1960s onwards the book became the centre of an enormous cult, appealing greatly to the student population of the time. Many other critics, however, have believed its appeal to be sub-literary, and **Edmund Wilson** went so far as to call it 'juvenile trash'. Persecuted by his fans, Tolkien retreated to a house in Bournemouth whose address was kept secret; he returned to Merton to die. In 1977, *The Silmarillion*, the book of stories he worked on for fifty years and never finished, was published, having been completed by his son, Christopher.

Fiction
The Hobbit 1937; *Farmer Giles of Ham* 1949; *The Lords of the Rings* 1966: *The Fellowship of the Ring* 1954, *The Two Towers* 1954, *The Return of the King* 1955; *Smith of Wootton Major* 1967; *The Silmarillion* (s) [unfinished; with C. Tolkien] 1977

Poetry
Songs for the Philologists [with others] 1936; *The Adventures of Tom Bombadil and Other Verses from the Red Book* 1962; *The Road Goes Ever On* 1967; *Bilbo's Last Song* 1974

For Children
The Father Christmas Letters [ed. B. Tolkien] 1976

Non-Fiction
A Middle English Vocabulary 1922; *Chaucer as a Philologist* 1934; *Beowulf: the Monsters and the Critics* 1937

Edited
Sir Gawain and the Green Knight [with E.V. Gordon] 1925; *Ancrene Wisse* 1962

Translations
Sir Gawain and the Green Knight, Pearl and Sir Orfeo [ed. C. Tolkien] 1975

Collections
Tree and Leaf 1964; *The Tolkien Reader* 1966; *Tree and Leaf, Smith of Wootton Major, The Homecoming of Beorhtnoth Beorhthelm's Son* 1975

Biography
J.R.R. Tolkien by Humphrey Carpenter 1977

(Alfred) Charles TOMLINSON 1927–

Tomlinson was born at Stoke-on-Trent, the son of an estate agent's clerk. He was educated at the high school in Longton (one of the 'five towns' into which the district of the Potteries is traditionally divided), and from there went to Magdalene College, Cambridge, where he graduated in English in 1948. For a while he taught at an elementary school in Camden Town, north London, but from 1951 spent a year as secretary to Percy Lubbock, a wealthy and elderly aesthete, at his home on the Ligurian coast of Italy. He returned to postgraduate work and later a fellowship at the University of London, and in 1957 he was appointed lecturer at the University of Bristol, where his subsequent academic career was based. It took him through the posts of reader between 1968 and 1982, professor from 1982 to 1992, and finally emeritus professor after his retirement in the latter year, and was diversified by several spells as a visiting professor in North America. He married in 1948, and the couple had two daughters. He has lived for many years in Gloucestershire. He is also known for his drawing and painting, which have been widely exhibited.

Tomlinson published his first book of verse, *Relations and Contraries*, in 1951, the first of more than a dozen volumes which have interspersed his career at regular intervals. The first book was conventional mid-century verse, but from *The Necklace* (1955), under the influence of American and Continental poets (particularly **Wallace Stevens**), he was beginning to create a personal style which has been modified but not fundamentally altered over the years. One critic has described Tomlinson as 'a reliable and imaginative observer' in whose poetry there is 'very little personal reference or rage, passion, wild delight'. His is an assured and academic poetry, gaining its force from recurrent motifs such as water and stone, full of painterly effects, sometimes 'difficult' and allusive, strongest perhaps in its response to the varied landscapes the poet has known (Italy, the desert south-west of the USA, rural and industrial England). Its lack of human reference, and learning from foreign traditions, have denied it a high critical estimation in England, except from a few critics such as **Donald Davie**, but in America Tomlinson is widely regarded as an important poet. His own poetry is complemented by translations from several European languages (he is also the editor of *The Oxford Book of Verse in English Translation*), and by criticism. Among a small number of books which fall into other categories, *Some Americans: a Personal Record* is perhaps outstanding.

Poetry

Relations and Contraries 1951; *The Necklace* 1955; *Solo for a Glass Harmonica* 1957; *Seeing Is Believing* 1958; *A Peopled Landscape* 1963; *American Scenes and Other Poems* 1966; *To Be Engraved on the Skull of a Cormorant* 1968; *Penguin Modern Poets 14* [with A. Brownjohn, M. Hamburger] 1969; *The Way of a World* 1969; *Words and Images* 1972; *Written on Water* 1972; *The Way in and Other Poems* 1974; *The Shaft* 1978; *The Flood* 1981; *Notes from New York and Other Poems* 1984; *Collected Poems* 1985; *The Return* 1987; *Annunciations* 1989; *The Door in the Wall* 1992

Translations

Versions from Fydor Tyutchev 1903–1873 1960; *Castillian Ilexes: Versions from Machado* [with H. Gifford] 1963; *Renga: a Chain of Poems* [with others] 1972; *Ten Versions from Trilce* [with H. Gifford] 1974; *Translations* 1983

Non-Fiction

Some Americans: a Personal Record 1981; *Isaac Rosenberg of Bristol* 1982; *Poetry and Metamorphosis* 1983; *Sense of the Past: Three Twentieth Century British Poets* 1983

Edited

Selected Poems of William Carlos Williams 1976; *Selected Poems of Octavio Paz* 1979; *The Oxford Book of Verse in English Translation* 1980

H(enry) M(ajor) TOMLINSON 1873–1958

One of the more distinguished of the many English authors who have written mainly of ships and the sea, Tomlinson was born at Poplar in the East End of London, the son of a foreman at the West India Dock. When his father died, an uncle put the twelve-year-old boy to work in a city shipping office; with his mother's encouragement, however, he read widely and gained the rudiments of an education. He gradually began to contribute to the radical *Morning Leader*, and in 1904 was able to leave his clerkship to become a reporter on the paper. His new employers sent him on several voyages and expeditions, including one that took him (ostensibly as ship's purser) 2,000 miles up the Amazon and Madeira rivers in Brazil, an experience that inspired his first book, *The Sea and the Jungle*, published in 1912.

During the First World War, Tomlinson was official correspondent for the *Daily News* (with which the *Leader* had amalgamated) and *The Times* at British General Headquarters in France from 1914 to 1917. In the latter year he became literary editor, under H.W. Massingham, of the *Nation*, and remained there until he followed his editor in resigning over a policy dispute in 1923. From then onwards, he was a freelance writer and journalist.

His best-known novel, *Gallion's Reach*, was published with great success in 1927: vivid, intense, poetic, it is a work that is never likely to be entirely forgotten. His later novels, perhaps, did not match it, but his many volumes of finely ruminative travel essays, written in an elaborate and beautiful prose, are now most unjustly neglected: *London River*, dealing with the landscape he knew best, is among the finest of these books.

In the 1930s he also became well known as an anti-war polemicist. He married in 1899, and the couple had three children.

Fiction

Old Junk 1918 [rev. 1933]; *Under the Red Ensign* 1926; *Gallion's Reach* 1927; *Illusion:1915* 1928; *Côte d'Or* 1929; *All Our Yesterdays* 1930; *Out of Soundings* 1931; *The Snows of Helicon* 1933; *South to Cadiz* 1934; *Below London Bridge* 1934; *Mars His Idiot* 1935; *All Hands!* 1937; *The Day Before* 1939; *Ports of Call* 1942; *Morning Light* 1946; *The Haunted Forest* 1951; *The Trumpet Shall Sound* 1957

Non-Fiction

The Sea and the Jungle 1912; *London River* 1921; *Waiting for Daylight: Essays* 1922; *Tidemarks* 1924; *Gifts of Fortune, and Hints for Those about to Travel* 1926; *Thomas Hardy* 1929; *Between the Lines. An Address on Modern Tendencies in Literature* 1930; *Norman Douglas* 1931 [rev. 1952]; *The Wind Is Rising* 1941; *The Turn of the Tide* 1945; *The Face of the Earth* 1950; *Malay Waters* 1950; *Trinity Congregational Church* 1952; *A Mingled Yarn* (a) 1953

Edited

Great Sea Stories of All Nations (s) 1930; *An Anthology of Modern Travel Writing* 1936

John Kennedy TOOLE 1937–1969

Born in New Orleans, Toole was the only son of a car salesman and a teacher. Beloved of his mother, Thelma, Toole was a precocious child who, at the age of ten, prefaced a school composition with the words: 'I find myself very highly learned and smart. I have been praised so many times in my ten years of life that this epic should speak for itself.' He was educated at Tulane University where he gained an MA in 1959, after which he taught English at colleges in New York and Louisiana.

While serving in the army in Puerto Rico (he taught English to Spanish-speaking recruits), Toole wrote *A Confederacy of Dunces*. In 1962 he submitted the manuscript to Simon & Schuster and was encouraged to revise it by the editor Bob Gottlieb, who had recently discovered **Joseph Heller**'s *Catch-22*. Toole did so, and there ensued another three years of rewriting before the publisher finally rejected the novel. Disheartened, Toole made only one half-hearted effort at offering the work to another publisher, and in 1969 at Biloxi, Mississippi, he committed suicide by connecting a length of hose to his car exhaust.

The story of Toole's literary success is entirely posthumous. In 1974 his now widowed mother unearthed the manuscript and dispatched it to more than a dozen publishers, all of whom rejected it. In 1979 the noted Southern novelist **Walker Percy** was teaching a summer course at Loyola University in New Orleans, and Thelma foisted the now tattered manuscript upon him. Reluctantly, Percy looked at it – 'first with the sinking feeling that it was not bad enough to quit, then with a prickle of interest, then a growing excitement, and finally an incredulity: surely it was not possible it was so good'. Percy arranged for its publication by Louisiana University Press, the novel won a Pulitzer Prize in 1981, and Toole was granted full 'neglected genius' status. The novel takes its title from Swift's observation that a genius is known 'by this sign, that the dunces are all in confederacy against him', and describes the larger-than-life comic figure of Ignatius J. Reilly as he battles with the dunces of New Orleans.

There was one surviving piece of juvenilia, the short novel *The Neon Bible* written when Toole was sixteen. For legal reasons Thelma withheld it from publication, but after her death in 1984, and protracted legal proceedings between members of the family, her appointed executor Kenneth Holditch finally arranged publication. There are reportedly two biographies of Toole in progress, 'Momma's Boy' by Barbara McIntosh and 'Genius Among the Dunces' by Holditch, but it seems likely that these two have fallen prey to the litigation which has surrounded Toole's estate.

Fiction

A Confederacy of Dunces 1980; *The Neon Bible* 1989

Jeff(rey) TORRINGTON 1935–

Torrington was born in the Gorbals, Glasgow, and raised by his mother after her separation from his father, an army cook. At the age of nine he discovered the McNeil Street library, and describes himself as having been 'mainlining on books ever since'. At thirteen he spent nine months in a sanatorium recovering from tuberculosis and here conceived the ambition to become a writer. He did not return to school, and has since had a variety of jobs including cinema projectionist, banana packer and steam-engine fireman. He worked for more than a decade at the Linwood car plant outside Glasgow, before accepting redundancy in 1981 after he was diagnosed as suffering from Parkinson's disease.

In his twenties Torrington started submitting stories to newspapers and magazines. The first of his published stories was 'Scared to Death' –

concerning a spider in a banana-packing factory – which was accepted by the Glasgow *Evening News* in 1960. Shortly after this he joined a writers' circle led by Edward Scouller who advised him to give up murder mysteries and ghost stories with surprise endings and write seriously about what he knew. The eventual outcome, the fruit of nearly thirty years' work, was his remarkable first novel, *Swing Hammer Swing!*, a richly comic portrait of the last days of the old Gorbals tenements. After leaving Linwood, Torrington studied at Glasgow University with **James Kelman** and **Philip Hobsbaum** (a mentor to other writers such as **Alasdair Gray**). Kelman read *Swing Hammer Swing!* after its seventh complete rewrite and sent it to his own publisher who immediately agreed to publish it. It received widespread critical praise and won the Whitbread Book of the Year Award in 1992. Torrington's second book is a collection of interconnected stories set in a car factory. He married in 1960, and has three children.

Fiction

Swing Hammer Swing! 1992; *The Devil's Carousel* (s) 1995

(Theodore) Philip TOYNBEE 1916–1981

Toynbee was born in Oxford, the second son of the great historian Arnold Toynbee, author of the twelve-volume *A Study of History* (1934–61), and Rosalind, daughter of Gilbert Murray, the leading classical scholar of his day. A rebellious youth, he ran away from Rugby aged seventeen, incited by Esmond Romilly who had already been sacked from Wellington and was editing the anti-public-school magazine *Out of Bounds*, supported by **George Bernard Shaw**, from a left-wing bookshop in Bloomsbury. Toynbee himself was hauled back to Rugby to be formally expelled. He later wrote a memoir of Romilly and another friend, Jasper Ridley, entitled *Friends Apart*. At Oxford he continued his rebellion against authority and joined the Communist Party. A visit to Nazi Germany, however, left him with an abiding detestation of fascism and he became genuinely politically committed. He supported the Republic in the Spanish Civil War, visiting Spain in the winter of 1936–7. Immediately before the Second World War he was editor of the Birmingham *Town Crier*. He joined the Welsh Guards on the outbreak of war and was later commissioned in the Intelligence Corps. In 1942–4 he was seconded to the Ministry of Economic Warfare.

His first three novels are conventional, the best of them being *A School in Private*, set in a preparatory school. Its first chapter was originally published as a short story in *Horizon*. It was not until after the war, in *Tea with Mrs Goodman* and *The Garden to the Sea*, that he found his true voice; in them the influence of Jung is clearly detectable. In 1950 he joined the *Observer*, first as a foreign correspondent, then chief reviewer, setting a new standard of weekly criticism which lasted to the end of his life. He was a convert to Anglicanism in his late fifties. 'The melancholy deepened', **Peter Vansittart** recalled. 'He was eventually rescued by religion, baffling to old friends, and perhaps owing something to his intellectual but for long stormy relations with his father, with whom he was eventually and happily reconciled.' He describes his conversion in *Towards the Holy Spirit*, *Part of a Journey* and *End of a Journey* which was published posthumously. In 1961 he had started writing *Pantaloon: or the Valediction*, the first volume of an ambitious autobiographical poem; *Two Brothers*, *A Learned City* and *Views from a Lake* were to follow. The *Times Literary Supplement* commented of *A Learned City*: 'Up to date, this work seems to resemble an elaborate stained-glass window blown into fragments by the great diapason of a too powerful organ, but which devotion and persistence can reassemble into brilliant sections of the main design.'

He married twice, and from his first marriage had two daughters – Polly Toynbee becoming a well-known journalist – and a son and two daughters from his second. He died at Lydney, Gloucestershire, just short of his sixty-fifth birthday.

Fiction

The Savage Days 1937; *A School in Private* 1941; *The Barricades* 1943; *Tea with Mrs Goodman* [US *Protolamium*] 1947; *The Garden to the Sea* 1953

Poetry

Pantaloon 1961; *Two Brothers* 1964; *A Learned City* 1966; *Views from a Lake* 1968

Non-Fiction

Friends Apart: a Memoir of Esmond Romilly & Jasper Ridley 1954; *The Fearful Choice: a Debate on Nuclear Policy* [with others] 1958; *Comparing Notes: a Dialogue across a Generation* [with A. Toynbee] 1958; *Thanatos: a Modern Symposium* [with M. Richardson] 1963; *Towards the Holy Spirit* 1973; *Christians Then and Now* 1979; *Part of a Journey* 1979; *End of a Journey* 1981

Translations

Raymond Cartier and Henri de Kérillis *Kérillis on the Causes of the War* 1939

Edited

Underdogs: Eighteen Victims of Society 1961; *The Distant Drum: Reflections of the Spanish Civil War* 1976

Biography
Faces of Philip by Jessica Mitford 1984

Barbara TRAPIDO 1941–

Trapido was born in Cape Town, South Africa, where her father is a mathematician. Her mother, who was a painter and dress designer, abandoned her career in favour of maternal duties. Trapido's Cape Town school was an all-white, all-female establishment, where the pupils dressed in navy serge gymslips and panama hats. She graduated in English and education in Cape Town, and her dislike of the South African political system encouraged her to leave her native country for the UK in 1963. She had married the previous year, and her husband is a lecturer in South African politics at Lincoln College, Oxford. After teaching in schools in London and Durham, and in a remand centre, she ran a pre-school playgroup.

Trapido began her first novel, *Brother of the More Famous Jack* when she was thirty, but put it aside for eight years when both her children were at school. Completed when she was forty, the novel won a Whitbread Award for fiction in 1982. Described as 'one of our most beguiling literary comediennes', she has been influenced by Shakespeare's comedies because of their balance of satire and romance, and by Jane Austen for the way in which she points dialogue. She does not write autobiographically, despite the shrewdly observed classroom experiences in *Temples of Delight*. For her, the 'magic is to look *through* a mirror when I write, not merely to look in the mirror. Imaginary characters glow more for me, far more than "real-life" ones.' Particularly fond of music, which she finds the most powerful of the arts, she is a thwarted singer who would like her dialogue to 'sound more like recitative'.

Fiction
Brother of the More Famous Jack 1982; *Noah's Ark* 1984; *Temples of Delight* 1990; *Juggling* 1994

Ben(jamin) TRAVERS 1886–1980

Travers was the greatest British master of theatrical farce of the century. He was the scion of a family owning one of the oldest wholesale grocery businesses in the City; a great-grandfather had been surgeon to Queen Victoria. On leaving Charterhouse in 1904, he was apprenticed to the family business, and spent some time in Malacca, where, according to his own account, he spent much of his time reading the plays of Arthur Wing Pinero, which he found in the local library. From 1911 onwards he found more congenial occupation than 'tea-clearing', working for the avant-garde publisher John Lane (who had brought out *The Yellow Book* in the 1890s), and during the war he worked as a flight instructor in the Royal Naval Air Service. In 1916 he married Violet Mouncey, with whom he was to have three children.

When the war ended, his wife's private income enabled the family to settle in Somerset and Travers to turn his hand to writing. He began as a novelist, and several of his farces were based on his own novels. His first theatrical farce to have a great success was *A Cuckoo in the Nest* (1925), and this was followed, until 1932, by eight more immensely successful 'Aldwych farces' (so called because they were all produced at the Aldwych Theatre, and with a resident company which included such actors as Robertson Hare and Ralph Lynn). These farces, with their elaborate plotting, firmly delineated characters and command of language, have made for Travers a lasting reputation. Perhaps *Rookery Nook* and *Plunder* are among the best of them. During the 1930s and 1940s, Travers continued to write farces, but with less success, while a serious play, *Chastity, My Brother*, based on the life of St Paul, did not meet with much favour. The death of his wife in 1951 began a period of withdrawal for Travers, when he spent much time abroad; but unlike some other playwrights of his generation, he was to have a splendid swansong. The ending of the Lord Chamberlain's censorship of the theatre in 1968 encouraged him to write an outspoken comedy about a middle-aged woman's awakening to sex, *The Bed Before Yesterday*, which was produced in 1975 and is considered by many to be his finest work.

Travers's other works include two volumes of autobiography, a volume of cricket reminiscences, and film treatments of most of his farces. A genial clubman and great character, Travers was revealed by a *Sunday Times* feature published months before his death at the age of ninety-four to combine firm Christianity with outspoken support of permissiveness, a taste for the novels of **Vladimir Nabokov** with the practice of doing backward somersaults every morning before breakfast. He expressed an (unfulfilled) desire to have the words 'This is where the real fun starts' inscribed on his tombstone.

Plays
The Dippers [from his novel] 1922; *The Three Graces* [from C. Lombardi and A.M. Willner's play] 1924 [np]; *A Cuckoo in the Nest* **(st)** 1925, **(f)** [with A.R. Rawlinson] 1933; *Rookery Nook* [from

his novel] (**st**) 1926, (**f**) 1930; *Thark* (**st**) 1927, (**f**) 1932; *A Little Bit of Fluff (Skirts)* (**f**) [with W. Dryden, R. Spence] 1928; *Mischief* [from his novel] 1928; *Plunder* 1928 [pub. 1931]; *A Cup of Kindness* (**st**) 1929, (**f**) 1934; *A Night Like This* (**st**) 1930, (**f**) 1932; *Turkey Time* (**st**) 1931, (**f**) 1933 [pub. 1934]; *Dirty Work* (**st**) 1932, (**f**) 1934; *A Bit of a Test* 1933; *Just My Luck* (**f**) 1933; *Up to the Neck* (**f**) 1933; *Lady in Danger* (**f**) 1934; *Fighting Stock* (**f**) 1935; *Foreign Affairs* (**f**) 1935; *Stormy Weather* (**f**) 1935; *Chastity, My Brother* 1936; *Dishonour Bright* (**f**) 1936; *O Mistress Mine* 1936 [rev. *Nun's Veiling* 1953]; *Pot Luck* (**f**) 1936; *For Valour* (**f**) 1937; *Banana Ridge* (**st**) 1938, (**f**) [with W.C. Mycroft, L. Storm] 1941 [pub. 1939]; *Old Iron* (**f**) 1938; *Second Best Bed* (**f**) 1938; *So This Is London* (**f**) [with others] 1939; *Spotted Dick* 1939; *She Follows Me About* 1943 [pub. 1945]; *Outrageous Fortune* 1947 [pub. 1948]; *Uncle Silas* (**f**) 1947; *Potter* (**tv**) 1948; *Picture Page* (**tv**) 1949; *Runaway Victory* 1949; *Wild Horses* 1952 [pub. 1953]; *Fast and Loose* (**f**) [with A.R. Rawlinson] 1954; *Corker's End* 1968; *The Bed Before Yesterday* 1975 [pub. in *Five Plays* 1977]; *After You with the Milk* 1985

Fiction

The Dippers 1920; *A Cuckoo in the Nest* 1922; *Rookery Nook* 1923; *Mischief* 1925; *The Collection Today* (**s**) 1929; *Hyde Side Up* 1933

Non-Fiction

Vale of Laughter (**a**) 1957; *A-Sitting on the Gate* (**a**) 1978; *94 Declared; Cricket Reminiscences* 1981; *The Book of Crouch End* 1990

Edited

The Leacock Book 1930; *Pretty Pictures, Being a Selection of the Best American Pictorial Humour* 1932

Rose TREMAIN 1943–

Tremain was born Rose Thomson in London and educated at Crofton Grange School. She spent a year at the Sorbonne in Paris before going to the University of East Anglia, a choice she made so as to be able to study English literature with **Angus Wilson**. She graduated in 1967 and taught for two years at Lyndhurst House School in London before becoming an assistant editor at the British Printing Corporation. From 1972 she did part-time research work and wrote two books – a history of the Women's Suffrage Movement, and an illustrated biography of Stalin – for a New York publisher.

Her first novel, *Sadler's Birthday* (1976), is a powerful and evocative portrayal of an elderly butler shuffling through the rooms of the deserted mansion he has inherited from his erstwhile employer and recalling the events of his life in service. She returned to the theme of old age in *The Cupboard*, in which an elderly and now-neglected novelist recounts her life and the history of the twentieth century to a young journalist, but in common with her former teacher, Angus Wilson, Tremain is chary of repeating herself, and has consistently worked in different forms.

Restoration, which was shortlisted for the Booker Prize and won the *Sunday Express* Award in 1989, is an ambitious and complex novel which defies easy categorisation. Set during the early reign of Charles II, it follows the fluctuating career of a surgeon who is appointed physician to the King's spaniels, and includes powerful accounts of the plague and the Great Fire of London. Tremain was a creative writing fellow at the University of Essex from 1979 to 1980, and has been an occasional lecturer in creative writing at UEA since then. In 1983 she was chosen as one of the 'Best of Young British Novelists' in the Book Marketing Council promotion. She won the Dylan Thomas Award in 1984 for her story 'The Colonel's Daughter', and the Giles Cooper Award in 1985 for her radio play *Temporary Shelter*, and has twice received the Angel Literary Award: for *The Swimming Pool Season* in 1985 and *Restoration* in 1989. She has been married twice, and now lives with the biographer Richard Holmes; she has a daughter from her first marriage.

Fiction

Sadler's Birthday 1976; *Letter to Sister Benedicta* 1978; *The Cupboard* 1981; *The Colonel's Garden and Other Stories* (**s**) 1984; *The Swimming Pool Season* 1985; *The Garden of the Villa Mollini and Other Stories* (**s**) 1987; *Restoration* 1989; *Sacred Country* 1992; *Evangelista's Fan and Other Stories* (**s**) 1994

Plays

The Wisest Fool (**r**) 1976; *Blossom* (**r**) 1977; *Dark Green* (**r**) 1977; *Don't Be Cruel* (**r**) 1978; *Leavings* (**r**) 1978; *Halleluiah, Mary Plum* (**tv**) 1978; *Down the Hill* (**r**) 1979; *Findings on a Late Afternoon* (**tv**) 1980; *Half Time* (**r**) 1980; *Mother's Day* 1980; *A Room for the Winter* (**tv**) 1981; *Yoga Class* 1981; *Hell and McLafferty* (**r**) 1982; *Moving on the Edge* (**tv**) 1983; *The Birdcage* (**r**) 1984; *Temporary Shelter* (**r**) 1984 [pub. in *Best Radio Plays of 1984* 1985]; *Daylight Robbery* (**tv**) 1986; *Will and Lou's Boy* (**r**) 1986; *The Kite Flier* (**r**) 1987

Non-Fiction

The Fight for Freedom for Women 1973; *Stalin: an Illustrated Biography* 1975

For Children

Journey to the Volcano 1985

Robert TRESSELL (pseud. of Robert P. Noonan) ?1870–1911

Tressell, whose middle name may have been Philippe, was a house- and sign-painter of Irish extraction. For at least the greater part of his life, he was himself part of the submerged classes of which he wrote, and he also seems to have been a secretive man: as a result there is almost no definite information about his life before about 1890. It is not known for certain where he was born – London, Dublin and Liverpool have been suggested – nor at what date, nor who his parents were: it has been suggested, but cannot be proved, that he came of a gentleman's family, had been well-educated, and left home never to return, aged sixteen (one report dating from his later life claims that he knew seven languages). It is certain that in the early 1890s he emigrated to South Africa, where he worked as a decorator, and in 1891 married a certain Elizabeth Hartel: who she was is not known, nor in what circumstances the marriage ended, but a daughter, Kathleen Noonan, was born in 1892.

In 1901, Tressell and his daughter went (or returned) to England, where they lived at Hastings (the 'Mugsborough' of *The Ragged Trousered Philanthropists*), and he worked as a house-painter for various local firms, meanwhile becoming involved in the nascent Labour Party. He wrote his 400,000-word novel between about 1906 and 1910, and when it was finished, he offered it to three publishers, who all refused it. In 1910 he decided to emigrate to Canada, leaving Kathleen with relations. He travelled as far as Liverpool, where the tubercular illness from which he had suffered for many years worsened; he was admitted to the Walton Street workhouse, where he died. His daughter took work as a maid in London after his death, keeping her father's manuscript in her trunk. It was eventually read by a neighbour of her employers, and published, in truncated and very mutilated form, in 1914 (with the author's pseudonym misspelt as Tressall). The original manuscript was discovered in 1946 and a full version, restored by F.C. Ball, published in 1955. *The Ragged Trousered Philanthropists* is a bitter and vivid attack on the condition of life in the building trades at the beginning of the century, and since the 1920s socialists have considered it a classic of working-class literature. Although its general reputation is perhaps more limited, few serious twentieth-century English novels have had a wider audience.

Fiction

The Ragged Trousered Philanthropists 1914 [rev. 1955]

William TREVOR (Cox) 1928–

One of the most highly regarded of currently practising novelists, Trevor was born in Mitchelstown in County Cork, the son of a Protestant bank official not of the ascendancy class, and grew up in various Irish provincial towns, attending a number of schools as well as Trinity College, Dublin. In his fiction he has been equally at home describing provincial Ireland and the life of the London suburbs and English provinces. Trevor's early artistic practice was as a sculptor and he did not begin to write seriously until in his early thirties. After university, he taught for some years, then came to England in 1953, and worked for some time, unhappily, as a copywriter in an advertising agency. He married in 1952, and the couple have two sons.

He published an early novel in the style of **Anthony Powell**, *A Standard of Behaviour*, in 1956, but then did not write again for several years: it was not until 1964, with the publication of *The Old Boys*, the story of eight octogenarian members of the committee of a public school's old boys' association, that his reputation was made. It received both the praises of **Evelyn Waugh** and the Hawthornden Prize; soon afterwards, Trevor gave up advertising and has lived ever since as a writer, well known for both novels and short stories and for television and radio plays, often adapted from his prose fiction.

Trevor's novels tend to concern a shabby–genteel world evoked with wit and pathos and in a spare, ceremonious prose: among his most highly regarded novels are *The Love Department* and *The Children of Dynmouth*, which won a Whitbread Award. Trevor has said: 'I am not much interested in myself either as a person or as a writer. I tend rather to write and to leave it at that.' In accordance with this dictum, he lives in Devon, spends much time gardening, plays little role in literary life and claims never to discuss writing.

Over the years, Trevor has returned to particular themes, one of which is the world of schools; *Old School Ties* collects together extracts from novels and stories on this theme along with autobiographical fragments. A selection of Trevor's Irish stories bore the ironic title, *The Distant Past*. As the word 'Ireland' has come to be associated with the Troubles, so the interaction between Ireland and England, the countries of Trevor's birth and adoption, is a theme which has gradually darkened. This is to be seen particularly in the collection of stories *Beyond the Pale* (the title refers to 'the English Pale', a historic term for that part of Ireland under English rule), and in the novel *Fools of Fortune*, which won him a second Whitbread Award.

Trevor is a member of the Irish Academy of Letters, and has received the Royal Society of Literature Award (for *Angels at the Ritz*), the Allied Irish Banks Prize for Literature, a third Whitbread Award for his novel *Felicia's Journey*, and an honorary CBE. He has thrice been shortlisted for the Booker Prize, once for *Reading Turgenev*, one of the two novellas published in the single volume *Two Lives*, and subsequently published separately. *Excursions in the Real World* is a volume of biographical, autobiographical and travel pieces, originally published in assorted journals between 1970 and 1992.

Fiction

A Standard of Behaviour 1958; *The Old Boys* 1964; *The Boarding House* 1965; *The Love Department* 1967; *The Day We Got Drunk on Cake* **(s)** 1967; *Mrs Eckdorff in O'Neill's Hotel* 1969; *Miss Gomez and the Bretheren* 1971; *The Ballroom of Romance* **(s)** 1972; *Elizabeth Alone* 1973; *Angels at the Ritz* **(s)** 1975; *The Children of Dynmouth* 1976; *Old School Ties* **(s)** 1976; *Lovers of Their Time* **(s)** 1978; *The Distant Past* **(s)** 1979; *Other People's Worlds* 1980; *Beyond the Pale* **(s)** 1981; *Fools of Fortune* 1983; *The Stories of William Trevor* **(s)** 1983; *The News from Ireland* **(s)** 1986; *The Silence in the Garden* 1988; *Nights at the Alexandra* 1987; *Family Sins* **(s)** 1990; *Reading Turgenev* 1991; *Two Lives* 1991; *Felicia's Journey* 1994

Non-Fiction

A Writer's Ireland 1984; *Excursions in the Real World* 1993

Plays

The Baby-Sitter **(tv)** 1965; *The Elephant's Foot* 1965; *Walk's End* **(tv)** 1966; *The Girl* **(tv)** 1967 **(st)** 1968; *A Night with Mrs Da Tanka* **(tv)** 1968 **(st)** 1972; *The Penthouse Apartment* **(r)** 1968; *The Mark-2 Wife* **(tv)** 1969; *Going Home* **(r)** 1970 **(st)** 1972; *The Italian Table* **(tv)** 1970; *The Boarding House* **(r)** [from his novel] 1971; *The Grass Widows* **(tv)** 1971; *The Old Boys* [from his novel] 1971; *O Fat White Woman* **(tv)** 1972; *The Schoolroom* **(tv)** 1972; *Access to the Children* **(tv)** 1973; *The 57th Saturday* 1973; *The General's Day* **(tv)** 1973; *Marriages* 1973; *Miss Fanshawe's Story* **(tv)** 1973; *An Imaginative Woman* **(tv)** [from T. Hardy's story] 1973; *A Perfect Relationship* **(r)** 1973; *Love Affair* **(tv)** 1974; *Eleanor* **(tv)** 1974; *Mrs Acland's Ghosts* **(tv)** 1975; *Scenes from an Album* **(r)** 1975; *The Statue and the Rose* **(tv)** 1975; *Two Gentle People* **(tv)** [from G. Greene's story] 1975; *Afternoon Dancing* **(tv)** 1976; *The Love of a Good Woman* **(tv)** [from his story] 1976; *The Girl Who Saw a Tiger* **(tv)** 1976; *Newcomers* **(tv)** 1976; *The Nicest Man in the World* **(tv)** 1976; *Voices from the Past* **(tv)** 1976; *Attracta* **(r)** 1977; *Another Weekend* **(tv)** 1978; *Last Wishes* **(tv)** 1978; *Memories* **(tv)** 1978; *Matilda's England* **(tv)** 1979; *The Old Curiosity Shop* **(tv)** [from Dickens' novel] 1979; *Beyond the Pale* **(r)** 1980; *The Happy Autumn Fields* **(tv)** [from E. Bowen's story] 1980; *Secret Orchards* **(tv)** [from J.R. Ackerley and D. Petre] 1980; *The Blue Dress* **(r)** 1981; *Autumn Sunshine* **(tv)** 1981 **(r)** 1982 [pub. in *Best Radio Plays of 1982*] 1983; *Elizabeth Alone* **(tv)** [from his novel] 1981; *The Ballroom of Romance* **(tv)** [from his story] 1982; *Travellers* **(r)** 1982; *Mrs Silly* **(tv)** 1983; *One of Ourselves* **(tv)** 1983; *Aunt Suzanne* **(tv)** 1984; *Broken Homes* **(tv)** [from his story] 1985; *The News from Ireland* **(r)** [from his story] 1986; *The Children of Dynmouth* **(tv)** [from his novel] 1987; *Events at Drimalghleen* **(tv)** [from his story] 1991 [np]

Lionel TRILLING 1905–1975

Trilling was born in New York City, where he was to live almost the whole of his life. He was the son of immigrant parents of Eastern European Jewish origin, although his mother had been born in the East End of London. His father was a tailor, who later, rather unsuccessfully, entered the wholesale fur business. Trilling attended New York public schools, and then Columbia University, where he took his MA in English in 1926, and where almost his whole academic career was to be spent. After graduating, he taught for a year at the University of Wisconsin, and then at Hunter College, New York. While at the latter, he married, in 1929, Diana Rubin, who, as Diana Trilling, was, like her husband, to attain fame as a writer and critic. She outlives him, and in 1993 published a memoir of her marriage, *The Beginning of the Journey*. They had one son.

Trilling returned to Columbia as an instructor in 1932, and from the 1930s was well known as a critic, being particularly associated with the re-emergence of the *Partisan Review* from 1937, and its leftist criticism of American culture. His early years teaching at Columbia were stormy – at one point, he risked dismissal because of his radicalism, and through anti-Semitic prejudice – but he was eventually to progress through the normal ranks of assistant professor and associate professor, attaining a full professorship in 1948. He was George Edward Woodberry Professor of literary criticism between 1965 and 1970, university professor between 1970 and 1974, and professor emeritus in the last year of his life. He also in later years held visiting appointments at Oxford, including a fellowship of All Souls. He was profoundly shocked by the student protests at Columbia in 1968, although he always remained broadly liberal in his criticism of movements more extreme than those he had once espoused.

Trilling was recognised as one of the three or four most significant literary critics of his time. He stresses the centrality of literature to human life, and was thus opposed to the New Criticism, with its concentration on the autonomous attributes of writing, and to later trends, with their emphasis on language. In contrast, in his three most important books of essays – *The Liberal Imagination*, *The Opposing Self* and *Beyond Culture* – he addressed the broadest questions of Freudianism, morality, politics and aesthetics, while his studies of **E.M. Forster** and Matthew Arnold revealed his rootedness in English literature of the nineteenth and early twentieth centuries.

He resembles **Edmund Wilson** and **Cyril Connolly** in that, although he was primarily a critic, his work also had an important creative dimension. His first published work had been a story in the *Menorah Journal* in 1925. His one novel, *The Middle of the Journey* (1947), is perhaps too directly concerned with the splits in the American left in the 1930s to have much continuing significance, but the title-story of his posthumously published volume of stories, 'Of This Time, of That Place', is recognised as a fictional masterpiece, and serves to remind us that a writer can attain mastery just as well in a glancing moment as in an impressive *oeuvre*.

Fiction

The Middle of the Journey 1947; *Of This Time, of That Place and Other Stories* (s) [ed. D. Trilling] 1979

Non-Fiction

Matthew Arnold 1939; *E.M. Forster* 1943; *The Liberal Imagination: Essays on Literature and Society* 1950; *Freud and the Crisis of Our Culture* 1955; *The Opposing Self: Nine Essays in Criticism* 1955; *A Gathering of Fugitives* 1956; *Beyond Culture* 1965; *Prefaces to the Experience of Literature* 1979

Edited

Mark Twain *The Adventures of Huckleberry Finn* 1948; *The Portable Matthew Arnold* 1949; *The Letters of John Keats* 1950 [US *Selected Letters* 1951]; John O'Hara *Selected Short Stories* (s) 1956; Ernest Jones *The Life and Works of Sigmund Freud* [with S. Marcus] 1961 [rev. 1964]; *Oxford Anthology of English Literature*: *Romantic Poetry and Prose* [with H. Bloom] 1973, *Victorian Prose and Poetry* [with H. Bloom] 1973; *The Last Decade* 1982; *Speaking of Literature and Society* 1982

Dalton TRUMBO 1905–1976

Trumbo was born in Montrose, Colorado, and after high school attended the University of Colorado from 1924 to 1925. When his father died in 1925 he moved to Los Angeles to support his family and worked for nine years on the night shift at a bakery. During this period he attended the University of California at Los Angeles and the University of Southern California, and wrote short stories and six novels, all of which were rejected.

The publication of his essays and stories in magazines such as *Vanity Fair*, and the post of managing editor of the *Hollywood Spectator*, enabled him to leave the bakery, and in 1935 he published his first novel, *Eclipse*, a satire concerning a self-made businessman's confrontation with provincial culture. He became a reader at Warner Bros., began a prolific career as a scriptwriter, and in 1939 won a National Book Award for his anti-war novel *Johnny Got His Gun*. In the same year he married (he subsequently had three children), and in 1945 was a war correspondent with the US Army Air Forces.

During the 1940s Trumbo was a highly successful scriptwriter earning $200,000 a year from his contract with Metro-Goldwyn-Mayer; but he was also a member of the Communist Party, a campaigner for union rights and the National Chairman of Writers for Roosevelt. In 1947 he was subpoenaed by the House Un-American Activities Committee and was one of the 'Hollywood Ten' group of writers and actors, which included Ring Lardner Jr, who refused to state whether they were, or ever had been, members of the Communist Party. Trumbo left the stand declaring: 'This is the beginning of the American concentration camp!' The Ten were charged with contempt and, despite the support of Humphrey Bogart, John Huston and others, later convicted. Trumbo was fired by Metro-Goldwyn-Mayer and in 1950, after the Supreme Court's rejection of their appeal, was imprisoned for a year. *The Time of the Toad* is his account of the 'American Inquisition'.

In exile in Mexico during the 1950s he produced screenplays under a variety of pseudonyms; *The Brave One*, written by 'Robert Rich' won an Oscar in 1957, which caused some consternation and confusion. His public rehabilitation continued when he was credited by Kirk Douglas for his script for *Spartacus* (1960) and publicly hired by Otto Preminger for *Exodus* in 1960. He wrote scripts for further films including *The Sandpiper* (1965), *The Fixer* (1968) and *Papillon* (1973). In 1971 he wrote and directed a film version of *Johnny Got His*

Gun. He died of a heart-attack in 1976, having failed to complete a last novel, *Night of the Aurochs*, which was subsequently prepared for publication by Robert Kirsch.

Fiction

Eclipse 1935; *Washington Jitters* 1936; *Johnny Got His Gun* 1939; *The Remarkable Andrew* 1941; *Night of the Aurochs* [unfinished; ed. R. Kirsch] 1979

Plays

The Biggest Thief in Town 1949; *The Brave One* **(f)** [as 'Robert Rich'] 1956; *Exodus* **(f)** 1960; *Spartacus* **(f)** 1960; *The Sandpiper* **(f)** 1965; *The Fixer* **(f)** [from B. Malamud's novel] 1968; *Johnny Got His Gun* **(f)** [from his novel] 1971; *Papillon* **(f)** [from H. Charrière's autobiography] 1973

Non-Fiction

Harry Bridges 1941; *The Time of the Toad* 1949; *The Devil in the Book* 1956; *Additional Dialogue: Letters of Dalton Trumbo, 1942–1962* [ed. H. Manfull] 1970

Frank (i.e. John Francis) TUOHY 1925–

Tuohy was born in Uckfield, Sussex, and educated at Stowe School (where he was a contemporary of Alan Maclean, who was to become his publisher) and King's College, Cambridge. After graduating in 1946 he became a lecturer at Turku University in Finland, the first in a series of academic posts. Until his retirement in 1989 he taught English at universities in Brazil, Poland and Japan, and was writer-in-residence at Purdue University, Lafayette, Indiana, during the 1970s.

Tuohy has consistently drawn on his experience of foreign countries in his work, which he has characterised as arising out of 'the interaction between two cultures'. His first novel, *The Animal Game* (1957), describes a young Englishman saved from a potentially ruinous relationship with an alluring yet corrupt Brazilian woman. His most highly regarded novel is *The Ice Saints* which won both the James Tait Black and the Geoffrey Faber Memorial Prizes in 1965. The novel presents a vivid portrait of the grim condition of Poland in the years of the post-Stalin thaw, and explores subtle moral questions through the story of an English woman trying unsuccessfully to rescue her sister's Polish-born son and take him back to the material prosperity of England.

In three collections of short stories Tuohy has explored such different milieux as upper-class life in Kensington and the world of Latin American drug traffickers. He won the Katherine Mansfield-Menton award in 1960 for the title-story of *The Admiral and the Nuns*, the E.M. Forster Memorial Award in 1972 for *Fingers in the Door*, and the Heinemann Award in 1979 for *Live Bait*. He has also written a highly regarded biography of **W.B. Yeats**.

Fiction

The Animal Game 1957; *The Warm Nights of January* 1960; *The Admiral and the Nuns with Other Stories* **(s)** 1962; *The Ice Saints* 1964; *Fingers in the Door* **(s)** 1970; *Live Bait and Other Stories* **(s)** 1978; *The Collected Stories* **(s)** 1984

Non-Fiction

Portugal 1970; *Yeats* 1976

Plays

The Japanese Student **(tv)** 1973

Amos TUTUOLA 1920–

Tutuola was born in Abeokuta, one of the biggest towns in Western Nigeria. His father was a farmer, and Tutuola was educated at a school run by the Salvation Army in Abeokuta, Lagos High School and the Anglican Central School, Abeokuta. On his father's death in 1939 his education was interrupted and he returned to the farm. In 1942 he joined the Royal Air Force as a blacksmith and on being demobilised in 1946, after a period of unemployment, he got a job as messenger in the Department of Labour in Lagos. 'I was still in this hardship and poverty when one night it came to my mind to write my first book – *The Palm-Wine Drinkard*,' he said. 'I wrote it within a few days successfully because I was a story-teller when I was in the school. So since then I have become a writer.' He joined the Nigerian Broadcasting Corporation in 1956. He is a founder member of the Mbari Club, a writers' and publishers' organisation in Ibadan.

The Palm-Wine Drinkard has a memorable opening: 'I was a palm-wine drinkard since I was a boy of ten years of age. I had no other work more than to drink palm wine in my life.' When a beautiful girl is seduced by a man who takes her into the woods and changes himself into a skull to haunt her, it is the 'drinkard', searching for his tapster, who rescues her. Tutuola believes in ghosts. 'Ghosts are still in our mind. When I was a child of about ten ... you would see ghosts and sheep killed for sacrifices ... There were more spirits and ghosts in those days than in the present time. Because the smell of the railway smoke, gas transport – like the big forests have been cut down – that's all; they [the spirits] keep away and away. They don't like to smell it.' Tutuola's novels and stories are pervaded by them, as his descriptive titles show, and are full of journeys. For instance, in *The Witch-Herbalist of the Remote Town* a young hunter goes in search of a distant witch-doctor to acquire a potion to make his wife pregnant.

Tutuola's first language is Yoruba. Though he writes in English, he speaks and reads it with difficulty. It is this isolation, it has been suggested, that makes him unique. His ungrammatical prose, his eccentric spelling and punctuation combine with his unfettered imagination, based on old African storytelling traditions, to give his work charm and immediacy.

His reputation has always been high in Europe and America, but in his native Nigeria it was felt that his style and subject-matter were backward-looking and that he was prized abroad chiefly as a 'primitive'. This changed when Nigeria achieved independence, and he is now well regarded at home as well as abroad.

In 1947 he married, and has four sons and four daughters. His books have been translated into several languages.

Fiction

The Palm-Wine Drinkard and His Dead Palm-Wine Tapster in the Deads' Town 1953; *My Life in the Bush of Ghosts* 1954 [rev. as *The Wild Hunter in the Bush of Ghosts* ed. B. Lindfors 1983]; *Simbi and the Satyr of the Dark Jungle* 1955; *The Brave African Huntress* 1958; *Feather Woman of the Jungle* 1962; *Ajaiyi and His Inherited Poverty* 1967; *The Witch Herbalist of the Remote Town* 1981; *Pauper, Brawler and Slanderer* 1987; *The Village Witch Doctor and Other Stories* (s) 1990

Anne TYLER 1941-

Tyler was born in Minneapolis, Minnesota, the daughter of an industrial chemist and a social worker. For much of her childhood the family lived among various Quaker communities in the rural south before settling in Raleigh, North Carolina. At Duke University she majored in Russian, and was taught a composition course by Reynolds Price, who was to prove a significant influence on her work. She did graduate studies at Columbia University, was the Russian bibliographer at Duke, and worked in the law library of McGill University, Montreal, before stopping work in 1967 to write full-time.

She began publishing short stories in magazines such as *Seventeen*, *Harper's* and the *New Yorker*, and has continued to do so, though her stories remain as yet uncollected. In her first novel, *If Morning Ever Comes* (1964), Tyler adumbrates many of the themes and concerns which dominate her work, examining family life and the minute detail of quotidian existence in the story of a young man returning from Columbia to North Carolina and attempting to deal with his family's expectations of him.

Critics have often commented on the absence of character development in this and other of Tyler's novels, but she rejects the notion of characters learning through their experience and has expressed an 'utter lack of faith in change'. Her exploration of the notion of the family is most explicit in *Dinner at the Homesick Restaurant*, where the story of the family is viewed from the perspective of each character in turn. Inexplicably closed off from each other, they are still strongly bound together by memories, and a major tension of the novel is in the two opposing forces of disintegration and cohesion. Absent-minded, meditative Ezra, the youngest son, runs the restaurant of the title where he creates a sort of surrogate family atmosphere for various characters seeking emotional fulfilment.

The Accidental Tourist, which won the National Book Critics Circle Award in 1986 and was made into a successful film starring William Hurt and Kathleen Turner, is one of her best and funniest novels. It concerns a man who writes travel guides for those who want to 'take trips without a jolt', a metaphor for emotional detachment, which he eventually overcomes through his relationship with a dog trainer and her vulnerable young son. *Ladder of Years* takes its title from an old people's home where the residents ascend up the floors of the building as they grow more infirm, and demonstrates her increasing concern with the transience of life.

She has won an award from the American Academy for *Earthly Possessions* in 1977, and a Pulitzer Prize in 1989 for *Breathing Lessons*. She was married in 1963 to Taghi Modarressi, a child psychologist, and has two daughters.

Fiction

If Morning Ever Comes 1964; *The Tin Can Tree* 1965; *A Slipping-Down Life* 1970; *The Clock Winder* 1972; *Celestial Navigation* 1974; *Searching for Caleb* 1976; *Earthly Possessions* 1977; *Morgan's Passing* 1980; *Dinner at the Homesick Restaurant* 1982; *The Accidental Tourist* 1985; *Breathing Lessons* 1988; *Saint Maybe* 1991; *Ladder of Years* 1995

Edited

The Best American Short Stories 1983 (s) [with S. Ravenel] 1983

U

Barry (Forster) UNSWORTH 1930–

Unsworth was born into a family of miners in a village in County Durham, and was educated at Stockton-on-Tees Grammar School and Manchester University, where he read English. During most of the 1960s he taught English in Greece and Turkey. Out of this experience came his second novel, *The Greeks Have a Word for It*, which satirises various members of the British Council. Married, with three daughters, he returned to England in order to educate his children. He settled in Cambridge, where he taught English as a foreign language.

His first novel, *The Partnership*, was published in 1966, but it was not until 1973, when his fourth novel, *Mooncranker's Gift* won the Heinemann Fiction Award, that he achieved any real success. Although critically acclaimed, his first three novels earned barely £500. 'One way or another, either in advance royalties, commissions or fellowships or Arts Council grants, I've managed to work and pay my way with occasional teaching thrown in through the years,' he said in 1982. This has included a two-year period as visiting literary fellow at the University of Newcastle, enabling him to return to his native north-east. His next novel, *The Stick Insect*, was made into a television play, but it was with *Pascali's Island* in 1980 that he found popular as well as critical success. Set on an Aegean Island during the last days of the Ottoman Empire, it was a sophisticated and atmospheric spy story. It was shortlisted for the Booker Prize and made into a well-regarded film in 1988, written and directed by James Dearden. *The Rage of the Vulture* was also set in the ruins of the Ottoman Empire in the days of the last sultan. Events of the 1970s mirror the past in *Stone Virgin*, one of Unsworth's best novels, which is set in Venice (where the author has lived), and concerns a conservationist and the statue of the title, a fifteenth-century madonna modelled from a prostitute.

In 1985 Unsworth was writer-in-residence at the University of Liverpool, a period during which his first marriage broke down and he suffered writer's block. His experiences of a great city in decay affected him deeply: 'It was very politicised and I fell into the trap of becoming politicised as a writer,' he later acknowledged. 'The whole situation brought out my guilt that as someone from a Northern working-class background, I have been writing about remote, exotic subjects.' He was attempting to write a novel about the slave trade, but ended up writing a novel about a writer unable to write a novel about the slave trade. *Sugar and Rum* had mixed reviews, but it cleared the way for *Sacred Hunger*, the novel he had intended to write. A vast book, it is set in the eighteenth century and tells the story of one family's involvement in the slave trade, as ship-owners and ship's doctor. It was joint winner of the 1992 Booker Prize, and Unsworth subsequently worked with Peter Hall on a television adaptation.

Unsworth is a major historical novelist, but although his books are highly atmospheric, his principal interest is in character and in moral dilemmas. He has yet to achieve the popularity of a writer such as **J.G. Farrell**, whose concerns were similar, but his reputation continues to grow. He now lives in Umbria, Italy.

Fiction

The Partnership 1966; *The Greeks Have a Word for It* 1967; *The Hide* 1970; *Mooncranker's Gift* 1973; *The Big Day* 1976; *The Stick Insect* 1976; *Pascali's Island* [US *Idol Hunter]* 1980; *The Rage of the Vulture* 1982; *Stone Virgin* 1985; *Sugar and Rum* 1988; *Sacred Hunger* 1992

John (Hoyer) UPDIKE 1932–

Updike was born in Shillington, a small town in Pennsylvania, the model of his fictional towns of Olinger and Brewer. His father taught at the high school there and kept a farm, while his mother, who had Dutch blood, was an aspirant writer, whose novel, *Enchantment*, was published under her maiden name of Linda Grace Hoyer in 1971. Updike's childhood was tormented by stammering and psoriasis, but his mother encouraged him in his literary ambition, and he chose Harvard as his university in 1950 because it was the location of the Harvard *Lampoon*. He married his first wife, Mary Pennington, while at college.

In the summer he graduated, he had his first story accepted by the *New Yorker*, inaugurating his long connection with the magazine. He spent a year in England, at the Ruskin School of Drawing and Fine Art in Oxford, and then was taken on by the *New Yorker* as a 'Talk of the Town' reporter. In 1957, wishing to devote himself to writing, Updike left full-time employment and New York itself for the small town of Ipswich, Massachusetts, where he lived for seventeen years and which is the model for Tarbox in his novel *Couples*. This was the classic period of his writing life. He was selling stories regularly to the *New Yorker*, many of them based on the town around him, and publishing many books after his volume of poetry, *The Carpentered Hen*, appeared in 1958, and his first novel, *The Poorhouse Fair*, set in an old people's home in New Jersey, in 1959. During this period he became one of the most successful American writers, making half a million dollars from the film rights to *Couples* alone.

Updike has an extremely large *oeuvre*, consisting of well over a dozen novels, eleven volumes of short stories, several collections of poetry, much criticism, and an extremely frank autobiography (really a series of six sketches of themes in his life), *Self Consciousness*. He writes in a dense, brilliant and poetic style, making a special feature of the exact anatomisation of human situations and the physical world, owing something here to the work of **Henry Green**, whose novels, he has said, 'made more of a stylistic impact upon me than those of any other writer in English, living or dead'. Four of his most celebrated novels, *The Rabbit Tetralogy*, published at intervals of roughly a decade from 1960, follow Harry 'Rabbit' Angstrom, a typical middle-American, from his days as a teenage basketball star to his final decline. Also highly admired is *The Coup*, unusually for Updike set outside America and about corruption in an African state. If criticisms are levelled at him, they are usually that he writes coldly and obsessively about sex, and that, while his aspirations are those of a great writer, his actual achievement is middlebrow. Later novels, such as *Memories of the Ford Administration* and *Brazil*, are generally considered to mark a falling-off, but demonstrate his continuing fertility. Much of his finest work is in his stories, of which *The Afterlife and Other Stories* deals with the onset of old age. He is also a magisterial essayist, and his range in this field is now equalled only by that of **V.S. Pritchett**.

In 1977, Updike was divorced from his first wife, with whom he had four children, one of whom is the writer David Updike. He then remarried, also in 1977. He lives in a large seaside house about twenty-five miles from Boston, Massachusetts. He is the subject of **Nicholson Baker**'s book *U and I* (1991).

Fiction

The Poorhouse Fair 1959; *The Same Door* (s) 1959; *Rabbit Angstrom* 1995: *Rabbit, Run* 1961, *Rabbit Redux* 1971, *Rabbit Is Rich* 1981, *Rabbit at Rest* 1990; *Pigeon Feathers* (s) 1962; *The Centaur* 1963; *Of the Farm* 1966; *The Music School* (s) 1967; *Couples* 1968; *Bech* (s) 1970; *Museums and Women* (s) 1972; *A Month of Sundays* 1975; *Marry Me* 1976; *The Coup* 1978; *Your Lover Just Called* (s) 1979; *Problems and Other Stories* (s) 1980; *Bech Is Back* (s) 1982; *The Witches of Eastwick* 1984; *Roger's Version* 1986; *Forty Stories* 1987; *Odd Jobs* (s) 1992; *Memories of the Ford Administration* 1993; *Brazil* 1994; *The Afterlife and Other Stories* (s) 1995

Poetry

The Carpentered Hen and Other Tame Creatures 1958; *Hoping for a Hoopoe* 1958; *Telephone Poles and Other Poems* 1964; *Midpoint and Other Poems* 1969; *Six Poems* 1973; *Tossing and Turning* 1977; *An Oddly Lovely Day Alone* 1979; *Sixteen Sonnets* 1979; *Five Poems* 1980; *Jester's Dozen* 1984; *Facing Nature* 1986

Plays

Three Texts from Early Ipswich 1968; *Buchanan Dying* 1974

Non-Fiction

Assorted Prose 1965; *Picked-Up Pieces* 1975; *Hugging the Shore* 1983; *Self-Consciousness* (a) 1989; *Just Looking* 1990

Edited

The Year's Best American Short Stories (s) 1984

Collections

Other John Updike: Poems, Short Stories, Prose, Plays 1982

Edward (Falaise) UPWARD 1903–

The son of a doctor, Upward was born in Romford, Essex. He received a conventional middle-class education at Repton and Corpus Christi College, Cambridge, but began early to revolt against his background and developed the Marxist convictions he has retained throughout his life. It was at Repton that he met **Christopher Isherwood**, who was to remain a close friend until the latter's death in 1986. (Isherwood always sent Upward drafts of his books and described him as 'the judge before whom all my work must stand trial and from whose verdict, much as I sometimes hate to admit it, there is no appeal'.) At Cambridge the two young men created Mortmere, 'a fantastic village' peopled with grotesques, about which they wrote a number of stories. Isherwood described it as 'a sort of anarchist paradise in which all accepted social and moral values were turned upside down and inside out and every kind of extravagant behaviour was possible and usual'. Although he declared that 'Mortmere was the mad nursery in which [Upward] grew up as a writer, and no future evaluation of his work will be able to ignore it,' most of the saga remained unpublished until 1994. Upward had allowed 'The Railway Accident' to appear in a slightly bowdlerised form under the pseudonym of Allen Chalmers (the name he had been given in Isherwood's 1939 volume of autobiography, *Lions and Shadows*) in the American anthology *New Directions* in 1949, but he destroyed a large number of the stories. His early ambition was to be a poet, and his first book was a volume of verse, *Buddha* (1924).

He became a schoolmaster in 1928 and remained in this profession until 1962: he taught English in various schools, including (for twenty-nine years) Alleyn's School, Dulwich, and from 1936 to 1939 he was on the editorial board of *Ploughshare*, the journal of the Teachers' Anti-War Movement. He enjoyed a legendary reputation as a novelist of promise in the 1920s without actually publishing anything and was regarded as a literary and political mentor by those writers associated with **W.H. Auden**, whose early poetry was influenced by Upward's imaginative landscape. Two stories appeared in **Michael Roberts**'s *New Country* in 1933, but his first novel, *Journey to the Border*, did not appear until 1938, by which time he had been a member of the Communist Party for four years. He rejected his early fantasies and concentrated on explicitly political fiction, outlining this change of direction in his 'Sketch for a Marxist Interpretation of Literature', a severely doctrinaire essay he contributed to **C. Day Lewis**'s *The Mind in Chains* (1937). Many commentators felt like **John Lehmann** that Upward's 'imaginative gift [was being] slowly killed in the Iron Maiden of Marxist dogma'. In fact, a survey of the stories he has written over the years (many of them published in **Alan Ross**'s *London Magazine*) would show the coherence of Upward's vision, and the connections between his early and late work, in which dreams and hallucinations figure largely.

Journey to the Border, which draws upon his work as a private tutor in the 1920s, is a transitional work that mixes the surreal elements of Mortmere with a new commitment to political change. Something of the struggle between his lyrical gifts and his political aspirations can be found in his major work, *The Spiral Ascent*, an autobiographical trilogy of novels in which Alan Sebrill, a Marxist poet and schoolmaster, attempts to write verse that will serve the Party cause. Like Sebrill, Upward left the Communist Party shortly after the Second World War, since it appeared to be abandoning its Marxist–Leninist principals and introducing unacceptable reforms. In this, as in much else, Upward was influenced by his wife, Hilda Percival, a schoolteacher whom he had married in 1936. (They had two children, and she died in 1995.) The trauma of leaving the Party resulted in Upward suffering a breakdown, and he published nothing between 1949 and 1962, when the first volume of *The Spiral Ascent* appeared. The trilogy cost Upward a great deal to write, but it stands as his testimony and as an invaluable record of the involvement of the British intelligentsia in the century's grassroots left-wing politics.

Upward's political commitment has undoubtedly prevented his books from achieving the readership they deserve (and resulted in difficulties in finding a publisher during the Cold War era), but it is also what has activated him as a writer. His determination to pursue this career without deviating from his principals, despite scant recognition from critics or readers, is in itself a defiant act of faith. He has always had his champions, however, and a late flowering in 1994 saw the publication of *The Mortmere Stories*, a revised edition of *Journey to the Border* (introduced by **Stephen Spender**) and a new volume of stories (introduced by Frank Kermode).

Fiction

Journey to the Border 1938 [rev. 1994]; *The Spiral Ascent* 1977: *In the Thirties* 1962, *The Rotten Elements* 1969, *No Home But the Struggle* 1977; *The Railway Accident and Other Stories* (s) 1969; *The Night Walk and Other Stories* (s) 1987; *The Mortmere Stories* [with C. Isherwood] (s) 1994; *An Unmentionable Man* (s) 1994

Poetry

Buddha 1924

Fred(erick Burrows) URQUHART 1912–

The son of a chauffeur, Urquhart was born in Edinburgh and spent his childhood on various estates in Fife, Perthshire and Wigtownshire, where his father was employed. He won a bursary to Stranraer High School, and had this transferred to Broughton Secondary School when his family moved back to Edinburgh. He left school at fifteen and between 1927 and 1935 worked in an Edinburgh bookshop. During this period he wrote three novels, none of which were published, but his fourth, written while he was unemployed, was published in 1938 as *Time Will Knit*.

Meanwhile, he had been trying to sell his short stories, some of which were unlikely to be published because of their subject-matter: 'No Fields of Amaranth', for example, was 'about a young homo and the man he'd picked up on holiday'. He was involved in a protracted correspondence with **J.R. Ackerley**, the literary editor of the *Listener*, who gave him a great deal of advice without accepting anything for publication. He had better luck with **John Lehmann** and other editors, and his stories began to appear in *Penguin New Writing*, the *London Mercury*, *Left Review* and other journals. His first volume of short stories, *I Fell for a Sailor*, appeared in 1940. This volume was noted for the skill with which Urquhart portrays the lives of working-class women. This has been a constant theme of his work, and he has pursued it with sympathy but without sentimentality, winning the praise of both **Compton Mackenzie** and **Stevie Smith**.

During the war he worked on a farm in Kincardineshire, and the harsh realities of rural life became another of his themes in such fine stories as 'The Dying Stallion'. Kincardineshire became his equivalent of **Thomas Hardy**'s Wessex, and many of his best stories are set in and around the fictional town of Auchencairn. One of these stories, 'The Ploughing Match', won the Tom Gallon Award and became the title-story of his second volume of collected stories. In 1944 he became farm secretary on the Duke of Bedford's estate at Woburn Abbey, then returned to London at the end of the war to work as a literary agent and become part of the Bohemian group of artists and writers who congregated in Fitzrovia (at one time he shared a flat with **Rhys Davies**). He enjoyed a number of affairs, including one with the son of **Harold Monro**, and in 1947 met Peter Wyndham Allen, who had been a dancer with the Sadler's Wells and other leading ballet companies. They were to live together until the latter's death in 1990. In 1973 he was responsible for extensively revising William Freeman's ten-year-old *Everyman's Dictionary of Fictional Characters*, 'ruthlessly eliminating' characters from the novels of such writers as Walter Besant, Mary Cholmondeley and F. Marion Crawford.

Although Urquhart has written several novels, his principal achievement has been in the short story. Alexander Reid described him as 'Scotland's leading short story writer of the century', and his work has been praised by many writers, including **V.S. Pritchett**, **Naomi Mitchison**, **George Orwell**, and **Edwin Muir**, who described Urquhart's use of dialect as 'exuberant, natural, coarse, witty ... the very idiom of Scottish working-class speech'. His work is far better known in Scotland than elsewhere and a selection of his stories, *Full Score*, was published by the Aberdeen University Press in 1989. After many years living in Sussex, he returned to his native Edinburgh, where he continues to write and work on his memoirs, a section of which was published in *New Writing Scotland* in 1993.

Fiction

Time Will Knit 1938; *I Fell for a Sailor* (s) 1940; *The Clouds Are Big with Mercy* (s) 1946; *Selected Stories* (s) 1946; *The Last GI Bride Wore Tartan* (s) 1948; *The Ferret Was Abraham's Daughter* 1949; *The Year of the Short Corn* (s) 1949; *The Last Sister* (s) 1950; *Jezebel's Dust* 1951; *The Laundry Girl and the Pole* (s) 1955; *Collected Stories* (s) 2 vols: *The Dying Stallion* 1967, *The Ploughing Match* 1968; *Palace of Green Days* 1979; *A Third Book of Modern Scottish Stories* (s) [with G.M. Brown, C. Smith] 1979; *A Diver in China Seas* (s) 1980; *Proud Lady in a Cage* (s) 1980; *Seven Ghosts in Search* (s) 1983; *Full Score* (s) 1989

Edited

No Scottish Twilight (s) [with M. Lindsay] 1947; *WSC: a Cartoon Biography of Winston Churchill* 1955; *Scottish Short Stories* (s) 1957; *The Cassell Miscellany (1848–1958)* 1958; *Scotland in Colour* [with K. Scowen] 1961; *Everyman's Dictionary of Fictorial Characters* [with W. Freeman, E.N. Pennell] 1973; *Modern Scottish Short Stories* (s) [with G. Gordon] 1978; *The Book of Horses* 1981

V

Laurens VAN DER POST 1906-

Van der Post, whose family were some of the first Dutch settlers in South Africa, was born in Philippolis, and educated at Grey College, Bloemfontein. After leaving school he became a reporter on the Natal Advertiser, then joined **Roy Campbell** and **William Plomer** as Afrikaans editor on their satirical magazine *Voorslag* ('Whiplash'). He subsequently travelled with Plomer to Japan in 1926, but unlike the older man returned to South Africa and journalism. In 1927 he went to live in England, working as a journalist, and marrying in 1929. He returned to South Africa to become leader writer for the *Cape Times* in 1930, and published his first novel, *In a Province*, about the involvement of two South African youths with communists, in 1934.

During the Second World War he served with the British Army in the Middle and Far East and was taken prisoner by the Japanese in 1942. This experience resulted in his novel *A Bar of Shadow* and a book of three stories, *The Seed and the Sower*, one of which, depicting the relationship between a captive and the Japanese camp commandant, was the basis of a popular film, *Merry Christmas Mr Lawrence* (1982), principally famous for casting two rock stars (David Bowie and Ryuichi Sakamoto) in the principal roles. He stayed on in Java after the war to work with Lord Mountbatten, acting as military attaché in Batavia from 1945 to 1947, and earning himself a CBE.

He led a peripatetic life after the war, farming in the Orange Free State, undertaking several exploratory missions in Africa for the Colonial Development Corporation and the British government, one to the Kalahari in 1952, which he describes in *The Lost World of the Kalahari*. He has also been a keen conservationist, winning many awards and becoming an adviser to the Prince of Wales on environmental issues, a role that has proved highly controversial. His experiences as a soldier, farmer and explorer inform much of his fiction, in which old-fashioned adventure stories are combined with philosophical speculation and a vivid evocation of landscape, particularly the landscape of South Africa. He has also written several volumes of non-fiction, including a book about Jung, who has been an important influence upon him. His first marriage, from which there were two children, ended in divorce in 1947 and two years later he married the writer Ingaret Giffard.

Fiction

In a Province 1934; *The Face Beside the Fire* 1953; *A Bar of Shadow* 1954; *Flamingo Feather* 1955; *The Seed and the Sower* (s) 1963; *The Hunter and the Whale* 1967; *A Story Like the Wind* 1972; *A Far-Off Place* 1974; *A Mantis Carol* 1975

Non-Fiction

Venture to the Interior 1952; *The Dark Eye in Africa* 1955; *Race Prejudice as Self-Rejection* 1957; *The Lost World of the Kalahari* 1958; *The Heart of the Hunter* 1961; *Patterns of Renewal* 1962; *Intuition, Intellect, and the Racial Question* 1964; *Journey into Russia* [US *A View of All the Russias*] 1964; *A Portrait of All the Russias* 1967; *A Portrait of Japan* 1968; *African Cooking* [with others] 1970; *The Night of the New Moon, August 6, 1945 ... Hiroshima* [US *The Prisoner and the Bomb*] 1970; *Man and the Shadow* 1971; *Jung and the Story of Our Time* 1975; *First Catch Your Eland* 1977; *Yet Being Someone Other* 1982; *Testament to the Bushman* [with J. Taylor] 1984; *A Walk with a White Bushman* 1986; *About Blady: a Pattern Out of Time* 1991

Biography

Laurens van der Post by Frederic I. Carpenter 1969

John (William) VAN DRUTEN 1901–1957

Van Druten was born in London. His father was a Dutch banker, and his mother, although English-born, was herself of Dutch ancestry. He attended University College School and, although his inclinations were literary, he was obliged by his father to study the law. He served articles of clerkship with a City solicitor and was himself admitted solicitor in 1923, meanwhile having had early articles accepted by *Punch* and other magazines. His distaste for the practical details of the law led to his taking up a post from 1923 to 1926 as Special Lecturer in English law and legal history at the University College of Wales, Aberystwyth, and he used the time to turn his hand to playwriting.

His first play, *The Return Half*, was staged in 1924 with the then unknown John Gielgud in the title role, and in the following year *Young Woodley* had an immense success in New York. (It was initially banned in Britain on what now seems the incredible grounds that it attacked the public-school system.) A period began, which lasted until the early 1950s, when van Druten turned out at least one well-made play almost every year, achieving great popular success as a dramatist. For some years he lived an unsettled life between Britain and the USA, often spending summers in the south of France in the 1930s with his lover, Jack Cohen, the novelist **G.B. Stern**, and other members, both male and female, of the homosexual society of the time. In 1939 he finally decided on America, buying a ranch in Thermal, California, in 1940, living there (with considerable periods in New York) for the rest of his life, and becoming a naturalised American citizen in 1944.

His most famous play came in 1951, when he wrote *I Am a Camera*, his adaptation of the story of Sally Bowles from **Christopher Isherwood**'s 1939 book *Goodbye to Berlin* (two friends of Isherwood, **Dodie Smith** and Alec Beesley, suggested the idea of an adaptation to van Druten in order to make Isherwood some money, although van Druten was unaware that this was their motive). Several of van Druten's plays, in fact, had been adaptations, of authors ranging from **Rebecca West** to **E.F. Benson**, but *I Am a Camera* – which was filmed (1955, from a screenplay by **John Collier**) and subsequently turned into the musical (1966) and the film (1972) *Cabaret* – was the most phenomenally successful. Van Druten's playwriting career was then almost at its end, as he retired to his ranch in his last years, having become obsessed with the religious ideas of the mystic Gerald Heard, who also strongly influenced Isherwood. His last work, *The Widening Circle*, one of three autobiographical books, was an exposition of his religious beliefs.

Van Druten also wrote four novels, the first an adaptation of his play *Young Woodley*, and the last, *The Vicarious Years*, which appeared in 1955, many years after the others, semi-autobiographical. From 1931 onwards he wrote many Hollywood screenplays, including that of the well-known *Gaslight* (1944), adapted from the play by **Patrick Hamilton**, and starring Ingrid Bergman, and he also directed many of his own later plays on Broadway as well as the Rodgers and Hammerstein musical *The King and I* (1951). It was as a consummate man of the stage that van Druten was celebrated in his own time and is half-forgotten today.

Plays

The Return Half 1924 [np]; *Young Woodley* 1925 [pub. 1926]; *Chance Acquaintance* 1927 [np]; *Diversion* 1927 [pub. 1928]; *The Return of the Soldier* [from R. West's novel] 1928; *After All* 1929; *Hollywood Holiday* [with B. Levy] 1931; *London Wall* 1931; *Sea Fever* [from M. Pagnol's play *Marius*; with A. Lee] 1931 [np]; *There's Always Juliet* 1931; *Somebody Knows* 1932; *Behold, We Live* 1932; *The Distaff Side* 1933; *Flowers of the Forest* 1934; *Most of the Game* 1935 [pub. 1936]; *Gertie Maude* 1937; *Leave Her to Heaven* 1940 [pub. 1941]; *Old Acquaintance* 1940 [pub. 1941]; *The Damask Cheek* 1942 [pub. 1943]; *Solitaire* [from E. Corle's novel] 1942; *The Voice of the Turtle* 1943 [pub. 1944]; *I Remember Mama* [from K. Forbes's *Mama's Bank Account*] 1944 [pub. 1945]; *The Mermaids Singing* 1945 [pub. 1946]; *The Druid Circle* 1947 [pub. 1948]; *Make Way for Lucia* [from E.F. Benson] 1948 [pub. 1949]; *Bell, Book and Candle* 1950 [pub. 1951]; *I Am a Camera* [from C. Isherwood] 1951 [pub. 1952]; *I've Got Sixpence* 1952 [pub. 1953]; *Dancing in the Chequered Shade* 1955 [np]

Fiction

Young Woodley [from his play] 1929; *A Woman on Her Way* 1930; *And Then You Wish* 1936; *The Vicarious Years* 1955

Non-Fiction

The Way to the Present (a) 1938; *Playwright at Work* (a) 1953; *The Widening Circle* (a) 1957

Peter VANSITTART 1920–

Vansittart was born in Bedford, and educated at Haileybury and Worcester College, Oxford, to which he won a major scholarship in modern history. He was for twenty years a teacher, and between 1947 and 1959 was director of the Burgess Hill School in Hampstead, north London, a mixed independent school for seven- to seventeen-year-olds run on liberal lines; it no longer exists. Since that time, a small private income has allowed him to live the life of a writer. He was married in his youth, and has one daughter. He has continued to live in Hampstead, with a second home at Ipswich in East Anglia.

Vansittart published his first novel, *I Am the World*, about a dictator, when he was twenty-two, and since then has produced over twenty novels (he does not write short stories). He began with modern settings, as in *Broken Canes*, a book about a progressive school, but soon moved on to the historical and legendary subjects which he has continued to exploit. It would not be too much to say that he has almost singlehandedly redeemed the historical novel from its middle-brow reputation, for his writing is very cryptic and dense, his style mainly spare but with passages of great lyrical beauty. A persistent sense of symbolic and mythical dimensions beneath the surface to his prose makes him fascinating but not always easy to read. Certain historical periods preoccupy him: the ancient world, as in such later novels as *The Wall*, about the decaying Roman empire of the reign of Aurelian, or *A Choice of Murder*, centring on the ambiguous figure of Timolean of Syracuse; and pre-Reformation and Reformation Germany, as in *The Friends of God* and *A Safe Conduct*. The legendary themes he has handled include *Lancelot*, *The Death of Robin Hood* and *Parsifal*. His novels have not been issued in paperback, but he continues to enjoy the esteem of a small but discriminating band of readers who like demanding, rewarding fiction.

His autobiography, *Paths from a White Horse*, is another dense read, more revealing of his tastes and interests than his personal life. He has compiled heavily annotated anthologies on such varied subjects as the First World War, London, the French Revolution and royalty, edited **John Masefield**'s war letters and has published non-fiction for children on historical subjects. *The Ancient Mariner and the Old Sailor* is about the use of language, while *In the Fifties* is a fascinating anecdotal portrait of that decade, mainly focusing on literary London, and generously frank about the many *faux pas* he made as a young writer dealing with older ones.

Fiction

I Am the World 1942; *Enemies* 1947; *The Overseer* 1949; *Broken Canes* 1950; *A Verdict of Treason* 1952; *A Little Madness* 1953; *The Game and the Ground* 1956; *Orders of Chivalry* 1958; *The Tournament* 1959; *A Sort of Forgetting* 1960; *Carolina* 1961; *The Friends of God* [US *The Siege*] 1963; *The Lost Lands* 1964; *The Story Teller* 1968; *Pastimes of a Red Summer* 1969; *Landlord* 1970; *Quintet* 1976; *Lancelot* 1978; *The Death of Robin Hood* 1981; *Harry* 1981; *Three Six Seven* 1983; *Aspects of Feeling* 1986; *Parsifal* 1988; *The Wall* 1990; *A Choice of Murder* 1992; *A Safe Conduct* 1995

For Children

The Dark Tower: Tales from the Past 1965; *The Shadow Land: More Tales from the Past* 1967; *Green Knights, Black Angels: the Mosaic of History* 1969

Non-Fiction

Vladivostok 1972; *Dictators* 1973; *Worlds and Underworlds: Anglo-European History through the Centuries* 1974; *Flakes of History* 1978; *The Ancient Mariner and the Old Sailor* 1985; *Paths for a White Horse* (a) 1985; *In the Fifties* 1995

Edited

Voices from the Great War 1981; *Voices: 1870–1914* 1985; *John Masefield's Letters from the Front 1915–17* 1985; *Voices from the Revolution* 1989

Carl VAN VECHTEN 1880–1964

Van Vechten, novelist, photographer and critic, was born in Cedar Rapids, Iowa. His father was a banker, who went broke but started afresh in life insurance and made a small fortune. Van Vechten escaped from Cedar Rapids to read English at the University of Chicago, where **Robert Herrick** was one of his tutors, but he soon lost interest in academic studies, developing instead his passion for music, and landing a job as a reporter with the *Chicago American* newspaper. After a certain youthful flippancy led to his being fired for 'lowering the tone of the Hearst newspapers', he went to New York where he persuaded **Theodore Dreiser** to allow him to write on Richard Strauss's *Salome* for *Broadway Magazine*; he also became the assistant music critic on the *New York Times* and was responsible for 'discovering' Stravinsky, Satie, Gershwin and others, as well as introducing jazz and ragtime to a wider audience. Sent to Europe by his father, in 1907 Van Vechten married Anna Elizabeth Snyder, who divorced him in 1912. He was the Paris correspondent of the *New York Times*, reporting chiefly on the arts; his popular first novel, *Peter Whiffle* (1922), describes a young man's experiences in Paris before the First World War. He also met Mabel Dodge (Luhan), who introduced him to 'hundreds of extraordinary people', including **Gertrude Stein**, whose American lecture tour of 1934–5 he facilitated and whose literary executor he eventually became.

Back in America in 1914, he married the Russian-born Jewish actress Fania Marinoff, an event which prompted his first wife to press for alimony payments which, by mutual consent, had never been paid. Unable and unwilling to make the back-payments, Van Vechten was jailed for four months, spending his term in a cell with a piano and a constant stream of visitors. He continued to have money problems until his elder brother's death in 1928 when he inherited \$1,000,000. By this time he had published several volumes of criticism, six of his seven stylish novels, in which fantasy and satire are combined and which had a distinct vogue in the 1920s, and had edited an anthology of cat stories, *Lords of the Housetops*. He was spending most of his time in clubs and at parties where the Harlem Renaissance was in full swing. A passionate admirer of black culture, he promoted the careers of many artists, most notably the 'poet laureate of Harlem', **Langston Hughes**, who provided songs for Van Vechten's (still) controversial novel *Nigger Heaven*. Set in Harlem, this book divided the black artistic community, many people objecting to its title and dismissing it as exploitative. He also introduced **Ronald Firbank** to American readers and revived interest in Herman Melville.

In the 1930s Van Vechten gave up writing to become one of the twentieth century's best portrait photographers. His images of writers, artists and performers are well known and often reproduced, while his collection of 'Photographs of Celebrated Negroes' at the University of New Mexico is an unrivalled archive. He also founded numerous other public collections connected with black arts and letters, the George Gershwin Memorial Collection of Music and Musical Literature, and the Anna Mabel Pollock Memorial Library of Books about Cats at Yale. His collection of pornographic scrapbooks was left to the Beinecke Library at Yale University.

Fiction
Peter Whiffle 1922; *The Blind Bow-Boy* 1923; *The Tattoed Countess* 1924; *Firecrackers* 1925; *Nigger Heaven* 1926; *Spider Boy* 1928; *Feathers* (s) 1930; *Parties* 1930

Non-Fiction
Music After the Great War 1915; *Music and Bad Manners* 1916; *Interpreters and Interpretations* 1917; *The Merry-Go-Round* 1918; *The Music of Spain* 1918; *In the Garret* 1920; *The Tiger in the House* 1920; *Red* 1925; *Excavations* 1926; *Sacred and Profane Memories* (a) 1932; *Fragments* (a) 1955

Edited
Lords of the Housetops (s) 1921; Rimsky-Korsakov *My Musical Life* 1924; Gertrude Stein: *Selected Writings of Gertrude Stein* 1946, *Last Operas and Plays* 1949

Biography
Carl Van Vechten and the Irreverent Decades by Bruce Kellner 1968

Gore (Eugene Luther) VIDAL 1925–

Vidal was born in West Point, New York, but was raised near Washington, DC, in the house of his grandfather, Thomas P. Gore, a popularist US senator from Oklahoma. From his early years he was immersed in the milieu of popular politics, and his grandfather's extensive library made possible a serious study of literature and history. He has always been characterised as a dissenting insider, a reputation he appropriately achieved in his own family: they were conventional Democrats, but he began writing satirical and scathing attacks on the Kennedy administration, and consequently became known as an 'unconventional' Democrat.

During the summer after his 1943 graduation from Phillips Exeter Academy, Vidal joined the Enlisted Reserve Corps of the US Army, which sent him to study engineering at Virginia Military Institute. However, most of his time in the Corps was spent writing, and he later drew upon his experiences as first mate on Freight Ship 35 in the Alaskan Harbor Craft Detachment in his first novel, *Williwaw* (1946).

On his discharge, he wrote and took an assortment of jobs before formulating 'a five-year plan: an all-out raid on television which could make me enough money to live the rest of my life'. His third novel, *The City and the Pillar*, written when he was still only twenty-one, made him famous. It dealt with the life of a typical, all-American lower-middle-class boy who happened to be homosexual, and it gained instant notoriety. He has recently claimed, with absolute justification, that his novel 'broke the mould' of gay American fiction, which had up until then dealt with 'shrieking queens or lonely bookish boys who married unhappily and pined for Marines'. The book was reissued in 1965 with a different ending, which was supposed to distance it still further from stereotype, though whether it was successful in its attempt is debatable: in the first edition the protagonist, Jim, kills his boyhood lover Bob, while in the second he merely rapes him. Vidal's 1968 novel *Myra Breckinridge* (dedicated to his friend **Christopher Isherwood**) was also an attack on sexual norms and the myth of machismo American culture. A transsexual comedy which also parodies the cult of the Hollywood film star, it was described by **Brigid Brophy** as a masterpiece of 'the high baroque comedy of bad taste', and was followed in 1974 by a sequel, *Myron*. His concern with such matters has inevitably brought him into conflict with more macho American fiction writers, the most obvious being **Norman Mailer**, with whom Vidal publicly fell out in 1971 following an article in the *New York Review of Books*. In it, Vidal had said that Mailer regarded women as objects to be poked, humiliated and killed; Mailer was not the only one to see this as being a rather unsubtle allusion to a recent stabbing incident involving Mailer and his wife. The public animosity had still not abated by 1978, when the two writers were due to appear on an American chat show, only to end up being recorded separately after refusing to share the same stage.

In 1960 Vidal was nominated as the Democratic candidate for the congressional seat in New York State's 29th District, which had always been firmly Republican. He gained 20,000 votes more than the presidential candidate John F. Kennedy in that District, but lost the election by some 25,000 votes. This defeat may well have done him more good than anything else in his career, as his success as a novelist and political commentator is arguably the result of this detour in his political career. He briefly returned to his origins in 1982, when he launched a mildly amusing (and maybe self-mocking) campaign in California for the US Senate.

One of his most consuming interests has been the political *mise-en-scène*, and in a series of novels he has been retelling the history of the USA as experienced by one family and its connections. They are, in chronological order of subject, *Burr*, *Lincoln*, *1876*, *Empire*, *Hollywood* and *Washington, D.C.* His fat volume of collected essays, *United States*, published in 1993, won a National Book Award, and is essential reading for those interested in contemporary American politics and literature. Vidal is still a controversial critic of the US establishment, though his fame is now such that he has co-starred with Tim Robbins in the film *Bob Roberts* (1994). Vidal is also the author of three detective novels, published in the 1950s under the pseudonym of Edgar Box.

Fiction

Williwaw 1946; *In a Yellow Wood* 1947; *The City and the Pillar* 1948 [rev. 1965]; *The Season of Comfort* 1949; *Dark Green, Bright Red* 1950; *A Search for the King* 1950; *The Judgement of Paris* 1953; *Messiah* 1955; *A Thirsty Evil* (s) 1958; *Julian* 1964; *Washington D.C.* 1967; *Myra Breckinridge* 1968; *Two Sisters* 1970; *Burr* 1974; *Myron* 1974; *1876* 1976; *Kalki* 1978; *Creation* 1981; *Duluth* 1983; *Lincoln* 1984; *Empire* 1987; *Hollywood* 1989; *Live from Golgotha* 1992

As 'Edgar Box': *Death in the Fifth Position* 1954; *Death Likes It Hot* 1955; *Death Before Bedtime* 1965

Non-Fiction

Rocking the Boat 1963; *Reflections Upon a Sinking Ship* 1969; *Homage to Daniel Shays* 1972 [UK *Collected Essays 1952–1972* 1974; rev. as *On Our Own Now* 1976]; *Great American Families* [with others] 1975; *Matters of Fact and Fiction* 1977; *Sex Is Politics and Vice Versa* 1979; *Pink Triangle and Yellow Star and Other Essays* 1982; *The Second American Revolution* 1982; *Vidal in Venice* 1985; *Armageddon?* 1987; *Screening History* 1992; *United States: Essays 1952–1992* 1993; *Palimpsest* (a) 1995

Plays

Visit to a Small Planet 1957; *The Best Man* 1960; *On the March to the Sea* 1962; *Romulus* 1963; *Weekend* 1968; *An Evening with Richard Nixon* 1972

Barbara VINE

see Ruth RENDELL

Kurt VONNEGUT Jr 1922–

A fourth generation German-American, the son and grandson of Indianapolis architects, Vonnegut was educated in the public schools of the city. In 1940 he went to Cornell University, New York, for three years, where he studied biology and chemistry but did not take a degree. He enlisted in the US infantry and, as a battalion scout in Europe, was captured by the Germans at the Battle of the Bulge in December 1944, and sent to a prisoner-of-war camp in Dresden. During the fire-bombing of Dresden in February 1945 he was a prisoner in a meat-locker under a slaughterhouse, and was later employed by the Germans as a miner of corpses in the town; these experiences were to be used by him twenty-five years later in what most critics still regard as his major novel, *Slaughterhouse-Five*.

In 1945 he married a childhood friend: they had two daughters and a son, and also adopted the three children of Vonnegut's sister, who died in her forties of cancer. On demobilisation, Vonnegut did graduate studies in anthropology at the University of Chicago, while also working as a police reporter at the City News Bureau there; from 1947 to 1950 he was employed in the public relations department of the General Electric Company.

Vonnegut's first novel, *Player Piano*, satirising the impersonal values of vast corporations, was published in 1952. About this time, he became a freelance writer, and supported himself for several years by writing science-fiction stories for popular magazines. His second novel, *The Sirens of Titan*, was also standard science fiction, and for many years the 'science-fiction/fantasy' tag applied to Vonnegut's work was one that he was clearly unhappy with. He supplemented his income for many years with jobs such as teacher in a home for disabled children and dealer in space-age technology, but the novels that he wrote from *Mother Night* – hip, breezy, impressionistic and pessimistic in style – caught the public mood of the 1960s and early 1970s, and his highly distinctive style and appearance only added to his popularity. From the 1970s onwards Vonnegut has continued to publish fiction, and his thirteenth novel, *Hocus Pocus*, appeared in 1990, but the critical consensus is that his work has declined. He has also published essays, autobiography and drama.

In 1979 he divorced his first wife and married the photographer Jill Krementz, with whom he has one child. In 1984 he made a suicide attempt (his mother had committed suicide in 1944), and recurrent bouts of depression, both personal and about the fate of the planet, of which his fiction takes an increasingly dystopic view, have been surmounted by him with difficulty in later years.

Fiction
Player Piano 1952; *The Sirens of Titan* 1959; *Canary in a Cathouse* (s) 1961; *Mother Night* 1962; *Cat's Cradle* 1963; *God Bless You, Mr Rosewater* 1965; *Welcome to the Monkey House* (s) 1968; *Slaughterhouse-Five* 1969; *Breakfast of Champions* 1973; *Slapstick* 1976; *Jailbird* 1979; *Deadeye Dick* 1982; *Galápagos* 1985; *Bluebeard* 1987; *Hocus Pocus* 1990; *The Face* 1992

Non-Fiction
Wampeters, Foma, and Granfalloons 1974; *Palm Sunday* (a) 1981; *Fates Worse than Death* (a) 1992

Plays
Happy Birthday, Wanda June 1970; *Between Time and Timbuctu* (tv) 1972

For Children
Sun Moon Star [with I. Chermayeff] 1980

W

Helen (Jane) WADDELL 1889–1965

Born in Tokyo, the youngest of twelve children of a missionary from Banbridge, County Down, Waddell came from a long tradition of Presbyterianism on both sides of the family. She grew up largely in Ulster, and obtained a first in English from Queen's University, Belfast, in 1911. Her first book, *Lyrics from the Chinese*, was published in 1913. She looked after her ageing stepmother until the latter's death, then in 1920 went up to Somerville College, Oxford, to read for a research degree. At Oxford, she first came into contact with the *Carmina Burana*, reading of which was to direct her life-work into the study of medieval Latin lyrics and humanism.

She taught at Bedford College, London, in 1922–3, then studied in Paris for two years. From 1927 to 1936 she published the series of books that made her name and took the literary world by storm: among them, *The Wandering Scholars*, a seminal study of the Goliardic Latin lyric poets; *Mediæval Latin Lyrics*, sensitive although sometimes free translations from medieval Latin; and *Peter Abelard*, a novel of twelfth-century Paris and Brittany and perhaps her greatest work.

As a result of these books, she received countless invitations to lecture, came to number **George Bernard Shaw, Max Beerbohm** and **Siegfried Sassoon** among her friends, to breakfast with Stanley Baldwin, lunch with Queen Mary and dine with General de Gaulle. Her last substantial work was *Poetry in the Dark Ages* in 1948; her later years were darkened by mental paralysis. She never married, and lived for many years in Primrose Hill, London.

Fiction

Peter Abelard 1933

Non-Fiction

The Wandering Scholars 1927; *Poetry in the Dark Ages* 1948

Plays

The Spoiled Buddha 1919; *The Abbé Prévost* 1935 [pub. 1931]

Translations

Lyrics from the Chinese 1913; *Mediæval Latin Lyrics* 1929; Prévost *The History of Chevalier des Grieux and Manon Lescaut* 1931; Marcel Aymé *The Hollow Field* 1933; *Beasts and Saints* 1934; *The Desert Fathers* 1936; Jacques [as 'Guy Robin'] *A French Soldier Speaks* 1941; Milton *Lament for Damon* 1943; *Plucking the Rushes: an Anthology of Chinese Poetry in Translations* [with E. Pound, A. Waley; ed. D. Holbrook] 1968; *Aphrodite: a Mythical Journey in Eight Episodes* 1970; *More Latin Lyrics from Virgil to Milton* [ed. F. Corrigan] 1976

Edited

A Book of Medieval Latin for Schools 1931

For Children

The Fairy Ring 1921; *Stories from Holy Writ* 1949 [rev. as *The Story of Saul* ed. E. Moss 1966]; *The Princess Splendour and Other Stories* (s) 1969

John (Barrington) WAIN 1925–1994

Born in Stoke-on-Trent, the son of a successful dentist who was the first of his family to emerge from the working class, Wain passed a restrictive but bookish childhood, was educated at Newcastle-under-Lyme Grammar School and, after being rejected for army service, went to St John's College, Oxford. From 1947 to 1955 he was a lecturer in English at Reading University, but eventually gave this up to follow the career of freelance author and journalist, having already some broadcasting experience and a contract from the *Observer* to review books.

He published his first novel, *Hurry on Down* (US title *Born in Captivity*), in 1953, and this story of provincial Charles Lumley who opts to join the working class had a large success at the time and led to Wain being labelled an Angry Young Man (although his publisher issued a statement that he was not one). At the same time, he was also becoming known as one of the leading younger poets of the Movement, a loose grouping of writers whose stance was ironic and anti-romantic, and he continued to publish novels and verse, although more sparsely from the 1970s onwards. Of his poetry, the long poem *Wildtrack* is perhaps especially notable; in the novel, it is generally considered that he failed to fulfil his promise, although *A Winter in the Hills* was admired and *Young Shoulders* won a Whitbread Award; perhaps some of his best work was in volumes of short stories such as *The Life Guard*.

Wain also produced a great many critical and edited works, wrote a prize-winning life of

Samuel Johnson, received numerous honours, and held a number of visiting professorships, including the professorship of poetry at Oxford from 1973 to 1978. Like many of those who were dubbed Angry Young Men, he adopted a bluff right-wing tone in his later work. He married twice, and had three sons.

Fiction

Hurry on Down [US *Born in Captivity*] 1953; *Living in the Present* 1955; *The Contenders* 1958; *A Travelling Woman* 1959; *Nuncle and Other Stories* (s) 1960; *Strike the Father Dead* 1962; *The Young Visitors* 1965; *Death of the Hind Legs and Other Stories* (s) 1966; *The Smaller Sky* 1967; *A Winter in the Hills* 1970; *The Life Guard and Other Stories* (s) 1971; *King Caliban and Other Stories* (s) 1978; *The Pardoner's Tale* 1978; *Young Shoulders* 1982; *Comedies* 1990

Poetry

Mixed Feelings 1951; *A Word Carved on a Sill* 1956; *Weep Before God* 1961; *Wildtrack* 1965; *Letters to Five Artists* 1969; *The Shape of Feng* 1975; *Poems 1949–1979* 1981

Plays

Harry in the Night 1975; *Frank* 1984

Non-Fiction

Interpretations: Essays on 12 English Poems 1955; *Preliminary Essays* 1957; *Gerard Manley Hopkins: an Idiom of Desperation* 1959; *Sprightly Running: Part of an Autobiography* (a) 1962; *Essays on Literature and Ideas* 1963; *The Living World of Shakespeare: a Playgoer's Guide* 1964; *A House for the Truth* 1972; *Samuel Johnson* 1974; *Professing Poetry* 1977

Edited

Contemporary Reviews of Romantic Poetry 1953; *Fanny Burney's Diary* 1960; *Anthology of Modern Poetry* 1963; *Macbeth: a Casebook* 1968; *Othello: a Casebook* 1971; Samuel Johnson: *Johnson as Critic* 1973, *Johnson on Johnson* 1976; Hardy: *The Dynasts* 1966, *Selected Shorter Poems of Thomas Hardy* 1975, *The New Wessex Selection of Thomas Hardy's Poetry* 1978; *Personal Choice: a Poetry Anthology* 1978; *An Edmund Wilson Celebration* 1978; *Anthology of Contemporary Poetry* 1979; *Everyman's Book of English Verse* 1981; *The Private Memoirs and Confessions of a Justified Sinner: James Hogg* 1983

Tom WAKEFIELD 1935–

Born into a mining family in Cannock, Staffordshire, Wakefield was educated at Rugeley Grammar School and London University. After graduating he became a teacher, rising to the position of headmaster of Downsview School, an appointment he held until the success of his Isobel Quirk trilogy – *Trixie Trash, Star Ascending, Isobel Quirk in Orbit* and *The Love Siege* – enabled him to become a full-time writer.

Wakefield's first book, *He's Much Better, He Can Smile Now* (1974), a non-fiction work which grew out of his educational experiences, anticipated his fictional territory: the lives of marginal individuals; and his second non-fiction work, *Some Mothers I Know,* featured the principal characteristic of that territory: strong, independent women. His novels are peopled by individuals apparently surplus to the requirements of society, or at least outside the predominant social group: the old, the single, the disadvantaged – outsiders owing to age, sexuality, marital status or disability. His major strength as a novelist is his ability to uncover emotional depths and inner resources in seemingly unpromising characters, a capacity which makes him part of a fictional tradition which has helped to humanise and preserve aspects of British culture and life.

The critical success of the Isobel Quirk trilogy and his childhood memoir, *Forties' Child,* led to the award of the North West Arts Council Literary Fellowship in 1980–2, based in Lancaster. During this time he wrote *Mates* and *Drifters,* which constitute his most overt statement on the homosexual condition: that it entails either mating or drifting. He is a sensitive writer on homosexuality, although his belief that we need many identities makes it only one of several themes in his fiction. For his work to date he received an Oppenheim Award for Literature in 1983.

All Wakefield's novels express his feeling for the extraordinariness of ordinary life, and he reveals an all-encompassing sympathy for the world and a belief in its capacity for love. At his weakest he tends towards whimsy; but his optimism is based on a spiritual and political faith in the innate goodness of people. His novels have progressively demonstrated his mastery of the form, developing increasingly complex structures to express his imaginative vision. His own favourite is *Lot's Wife,* but *War Paint* is his most ambitious and impressive to date: a condition-of-England novel which tenderly mixes comedy and tragedy.

Fiction

Trixie Trash, Star Ascending 1977; *Isobel Quirk in Orbit* 1978; *The Love Siege* 1979; *Mates* 1983; *Drifters* 1984; *The Discus Throwers* 1985; *The Variety Artistes* 1987; *Lot's Wife* 1989; *Secret Lives: Three Novellas* [with P. Gale, F. King] 1991; *War Paint* 1993

Non-Fiction

He's Much Better, He Can Smile Now 1974 [rev. as *Special School* 1977]; *Some Mothers I Know, Living with Handicapped Children* 1978; *Forties' Child* (a) 1980

Edited

The Ten Commandments 1992

Derek (Alton) WALCOTT 1930–

Walcott was born at Castries, St Lucia, in the West Indies, and has said that 'the advantage of being brought up in a place that hasn't much money is that the libraries have only the best'. His father died when Walcott was very young; his mother, who was a teacher, exposed him to poetry from the beginning since she declaimed Shakespeare as she went round the house. He was inspired by the hymns of the Methodist church to become a poet at the age of eight. 'The only valid breathing verse is the quatrain,' he has said; 'the quatrain of my childhood, my church.'

He was educated at St Mary's College, Castries, and the University College of the West Indies in Kingston, Jamaica. From 1953 to 1957 he taught at schools on several Caribbean islands, then turned to journalism. He was a feature writer for *Public Opinion* in Kingston, then a feature writer and drama critic for the Trinidad *Guardian*. In 1950, he had founded the St Lucia Arts Guild. He has married three times, has a son and two daughters, and divides his time between the Caribbean and the USA, where he has been a professor of poetry at the University of Boston and has taught creative writing.

He published a first volume of *25 Poems* when he was eighteen, and a volume of selected poems, *In a Green Night*, appeared when he was thirty-two. In an early poem, Walcott asks himself how he can choose 'Between this Africa and the English tongue I love?' He has written in both standard English and in West Indian dialect, but his abiding concern is with his African roots. The tension is not only between language, but also between differing loyalties: 'And I, whose ancestors were slave and Roman have seen both sides of the imperial foam.' One of his best-known poems, 'A Far Cry from Africa', describes his struggle to find allegiance with Kenya during the Mau Mau revolt against British rule. He was awarded the Queen's Medal for Poetry in 1988.

His most ambitious work to date is the epic poem *Omeros*, which takes its title from the Greek word for 'Homer'. It consists of sixty-four chapters divided into seven books, and its principal characters are two poor Caribbean fishermen with the classical names of Achille and Philocrete. In this immensely complex poem, recorded history is mingled with an account of the sufferings of exile.

Walcott was the founding director of the Little Carib Theatre (which later became the Trinidad Theatre Workshop) from 1959 to 1971, and he has written a large number of plays for stage and radio. Of these *Dream on Monkey Mountain* was commissioned by the Royal Shakespeare Company in the late 1960s, but never performed because of a shortage of black actors. It was subsequently produced in the USA, and is considered to be his most impressive play. He has collaborated on several musicals with Galt MacDermott (best known for the hippy musical *Hair*). Of these, *O Babylon!*, which is about Rastafarian squatters pressed to make way for a hotel for an impending visit of Haile Selassie, was produced at London's Riverside Studios in 1976. His plays include elements of pantomime and vaudeville, drawing on a wide range of theatrical traditions, and combine realism with fable and fantasy. His dramatic work is very much of a piece with his poetry and, taken together as poet and playwright, Walcott is not only the leading writer to emerge from the Caribbean, but stands among the most important of all contemporary writers.

Poetry

25 Poems 1948; *Epitaph for the Young* 1949; *Poems* 1951; *In a Green Night: Poems 1948–60* 1962; *Selected Poems* 1964; *The Castaway and Other Poems* 1965; *The Gulf and Other Poems* 1969; *Another Life* 1973; *Sea Grapes* 1976; *The Star-Apple Kingdom* 1980; *Selected Poetry* [ed. W. Brown] 1981; *The Fortunate Traveller* 1982; *Midsummer* 1984; *Collected Poems 1948–84* 1986; *The Arkansas Testament* 1987; *Omeros* 1990; *Poems 1965–1980* 1993

Plays

Cry for a Leader 1950; *Henri Christophe* 1950; *Robin and Andrea* [pub. 1950]; *Senza Alcun Sospetto* (r) 1950, (st) [as *Paolo and Francesca*] 1951; *The Price of Mercy* 1951; *Three Assassins* 1951; *Harry Oernier* 1952; *The Charlatan* 1954; *Crossroads* 1954; *The Sea at Dauphin* 1954; *The Golden Lions* 1956; *The Wine of the Country* 1956; *Ti-Jean and His Brothers* 1957; *Drums and Colours* 1958; *Jourmard* 1959; *Malcochan* 1959; *Batai* 1965; *Dream on Monkey Mountain* 1967 [pub. in *Dream on Monkey Mountain and Other Plays* 1972]; *Franklin* 1969 [rev. 1973]; *In a Fine Castle* 1970; *The Joker of Seville* [with G. MacDermott] 1974 [pub. 1978]; *O Babylon!* [with G. MacDermott] 1976 [pub. 1978]; *Remembrance* 1977; *The Snow Queen* (tv) 1977; *Pantomime* 1978; *Marie Laveau* [with

G. MacDermott] 1979; *The Isle is Full of Noises* 1982; *Beef, No Chicken* 1985 [pub. in *Three Plays* 1986]; *The Odyssey* 1993

Biography
Derek Walcott: Poet of the Islands by Ned Thomas 1980

Arthur (David) WALEY 1889–1966

One of the most distinguished Orientalists of the century (and one who never filled a permanent academic post), whose translations of Chinese classical lyric rank as fine English poems in their own right, Waley was born Arthur David Schloss, the son of an economist and Fabian socialist. His mother's maiden name was Waley, and the family assumed that name in 1914. He had a conventional upper-class education at Rugby and King's College, Cambridge, where he read classics and formed part of the brilliant intellectual circle collected there in pre-1914 days.

He worked briefly in Spain with a South American export business, but in 1913, partly through the influence of the painter Walter Sickert, was appointed assistant keeper of the Print Room in the British Museum. He taught himself Chinese and Japanese in his spare time, publishing a first volume of *Chinese Poems* in 1916. By 1918 he was ready to publish *A Hundred and Seventy Chinese Poems*, which had an enormous success and was followed by several other similar collections. Waley's translations show immense subtlety of rhythm and concreteness of imagery; his influence on modern English poetry has been diffuse but pervasive. He resigned from the museum in 1930 for health reasons and, apart from being a supplementary lecturer in the School of Oriental Studies and working in the Far Eastern Section of the Ministry of Information during the Second World War, did not hold a permanent post again.

He published only a few original poems and stories, but among other distinguished translations from Chinese and Japanese were his versions of *The Tale of Genji* and *Monkey*. He never visited the Far East, and was for many years a resident of Gordon Square, Bloomsbury, known to the Bloomsbury Group, a reserved and courtly figure riding to the museum on his bicycle. His companion for many years was the elegant Beryl de Zoete, an expert on Far Eastern dance forms, but he also carried on a long relationship with Alison Grant Robinson; after de Zoete's death he moved to Hampstead where he was looked after by Grant Robinson, whom he married a month before his death.

Translations
Chinese Poems 1916; *A Hundred and Seventy Chinese Poems* 1918; *Japanese Poetry: the 'Uta'* 1919; *More Translations from the Chinese* 1919; *The Nō Plays of Japan* 1921; *The Temple and Other Poems* 1923; Lady Murasaki *The Tale of Genji* 6 vols 1925–33; *The Pillow-Book of Sei Shōnagon* 1928; R. Wilhelm *The Soul of China* 1928; *The Lady Who Loved Insects* 1929; *The Travels of an Alchemist: the Journey of the Taoist Ch'Ang Ch'un from China to the Hindukush* 1931; *The Book of Songs* 1937; *The Analects of Confucius* 1938; Wu Ch'Êng-Ên *Monkey* 1942; *Folk Songs from China* [with I. Gass, Tzu-Jen Ku] 1943; *Chinese Poems* 1946 [rev. 1961]; *The Real Tripitaka* 1952; *The Nine Songs: a Study of Shamanism in Ancient China* 1955; *Seventy Seven Poems by Alberto de Lacerda* [with A. de Lacerda] 1955; *Ballads and Stories from Tun-huang* 1960

Non-Fiction
An Index of Chinese Artists Represented in the Sub-Department of Oriental Prints and Drawings in the British Museum 1922; *Zen Buddhism and Its Relation to Art* 1922; *An Introduction to the Study of Chinese Painting* 1923; *The Originality of Japanese Civilisation* 1929; *A Catalogue of Paintings Recovered from Tun-Huang by Sir Aurel Stein* 1931; *The Way and Its Power: a Study of the Tao Tê Ching and Its Place in Chinese Thought* 1934; *Three Ways of Thought in Ancient China* 1939; *The Life and Times of Po Chü-I, 772–846 A.D.* 1949; *The Poetry and Career of Li-Po, 701–762 A.D.* 1951; *Ynan Mei* 1956; *The Opium War through Chinese Eyes* 1958

Collections
The Secret History of the Mongols and Other Pieces 1964

Alice (Malsenior) WALKER 1944–

Walker was born the youngest of eight children, in Eatonton, Georgia. Her father was active in the civil rights movement and defied death threats to cast the first black vote in the county. His attempts at farming were less successful and Walker's mother had to work as a maid as well as on the farm to help out. A shooting accident in 1952 deprived Walker of sight in one eye, which made her eligible for a scholarship offered to disabled students at Spelman College. She applied and won, studying there from 1961 to 1963, when she was offered a scholarship at Sarah Lawrence in New York. She now sits on the college board of trustees.

In the summer of 1964 she travelled around Africa, became pregnant and, on her return to

America and Sarah Lawrence, had an abortion. The poems in *Once* (1968), her first published volume, were nearly all written in the week immediately after the operation. The book opens with these lines from Camus which might be taken as descriptive of Walker's own childhood: 'Poverty was not a calamity for me. It was always balanced by the richness of light ... circumstances helped me. To correct a natural indifference I was placed halfway between misery and the sun.' She received her BA in 1965 and found a publisher shortly after graduating. In 1966 she moved to Mississippi to work on the voter registration programme and met her husband, the civil rights lawyer Melvyn Leventhal. They married the following year in New York and returned to Mississippi to become the first legally married mixed-race couple to have lived in Jackson. Her first novel, *The Third Life of Grange Copeland* (1970), was written there and completed three days before the birth of her daughter. Her second, *Meridian*, is the 'spiritual and political biography' of a black, female political activist in the civil rights movement which also manages to describe the 'progress of a saint'. The title recalls the birthplace of James Chaney, one of three political activists murdered by the Ku Klux Klan in 1964, and the book has been taught, Gloria Steinem points out, 'as part of some American history as well as literature courses'.

Walker has held numerous academic posts, teaching at Jackson, Boston, Yale, Berkeley and Brandeis, and is highly respected as a critic and essayist. In 1976 she was divorced from her husband. Recognition on a national scale came in 1982 with the publication of her third novel, *The Color Purple*, which won both a Pulitzer Prize and a National Book Award. The 1985 Steven Spielberg film of the novel, for which Walker acted as consultant, was an Academy Award nominee. The passionate interest in folklore which she makes use of in her fiction – the tale of Louvinie in *Meridian* is an example – is not limited to that of her own culture, and in 1984 she co-founded and ran Wild Trees Press, publishing Balinese writers amongst others. She now lives in the Japantown area of San Francisco.

Fiction

The Third Life of Grange Copeland 1970; *In Love and Trouble* (s) 1973; *Meridian* 1976; *You Can't Keep a Good Woman Down* (s) 1981; *The Color Purple* 1982; *The Temple of My Familiar* 1989; *Finding the Green Stone* 1991; *Possessing the Secret of Joy* 1992

Poetry

Once 1968; *Revolutionary Petunias and Other Poems* 1973; *Good Night Willie Lee, I'll See You in the Morning* 1979; *Horses Make a Landscape Look Beautiful* 1984; *Her Blue Body Everything We Know* 1991

Non-Fiction

In Search of Our Mothers' Gardens: Womanist Prose 1983; *Living by the Word: Selected Writings 1973–1987* 1988; *To Hell with Dying* 1988

Hugh (Seymour) WALPOLE 1884–1941

Born in Auckland, New Zealand, Walpole was the eldest of three children of the then incumbent of St Mary's Cathedral, Auckland – an Englishman by birth, who later became Bishop of Edinburgh. Walpole came to England when five years old, and was educated at King's School, Canterbury (where, like **W. Somerset Maugham**, he was miserably unhappy), and Emmanuel College, Cambridge. His parents had intended him to take Holy Orders, but six months in a mission for seamen persuaded him not to become a priest. He became briefly an assistant master at Epsom College, an experience on which he drew for his fine early novel *Mr Perrin and Mr Traill*.

In 1909 he published his first novel, *The Wooden Horse*, went to London, and met many writers, of whom **Henry James** became a particular patron and friend. (It is said that Walpole, by then a practising homosexual, offered his young body to the Master, only to be fastidiously refused.) In the First World War Walpole served with the Russian Red Cross in Galicia and was later in charge of the Anglo-Russian propaganda bureau in Petrograd, witnessing the March Revolution.

From 1919 he entered fully on the career of all-purpose avuncular man-of-letters, making many lecture tours in the USA, chairing many committees, publishing a great number of novels and volumes of short stories, receiving the CBE and a knighthood. Although he sacrificed what might have been a real talent for serious fiction to middlebrow success, he had a deep need for approval, and Maugham's unkind caricature of him as the novelist Alroy Kear in *Cakes and Ale* (1930) did much to poison his last years. These were, however, also marked by emotional happiness with the policeman Harold Cheevers, who remained his faithful companion until his death at his home near Keswick.

Fiction

The Wooden Horse 1909; *Maradick at Forty* 1910; *Mr Perrin and Mr Traill* 1911; *The Prelude to Adventure* 1912; *Fortitude* 1913; *The Duchess of Wrexe* 1914; *The Golden Scarecrow* (s) 1915; *The Dark Forest* 1916; *The Green Mirror* 1918; *Jeremy*

1919; *The Secret City* 1919; *The Captives* 1920; *The Thirteen Travellers* (s) 1921; *The Young Enchanted* 1921; *The Cathedral* 1922; *Jeremy and Hamlet* 1923; *The Old Ladies* 1924; *Portrait of a Man with Red Hair* 1925; *Harmer John* 1926; *Jeremy at Crale* 1927; *Wintersmoon* 1928; *The Silver Thorn* (s) 1928; *Farthing Hall* [with J.B. Priestley] 1929; *Hans Frost* 1929; *The Herries Chronicle*: *Rogue Herries* 1930, *Judith Paris* 1931, *The Fortress* 1932, Vanessa 1933, *The Bright Pavilions* 1940; *Above the Dark Circus* 1931; *Four Fantastic Tales* (s) 1932; *All Souls Night* (s) 1933; *Captain Nicholas* 1934; *Cathedral Carol Service* (s) 1934; *The Inquisitor* 1935; *A Prayer for My Son* 1936; *John Cornelius* 1937; *Head in Green Bronze* (s) 1938; *Joyful Delaneys* 1938; *The Sea Tower* 1939; *The Blind Man's House* 1941; *The Killer and the Slain* 1942; *The Old Ladies* 1942; *Katherine Christian* 1944; *Mr Huffam* (s) 1948

Non-Fiction

Joseph Conrad 1916; *The Art of James Branch Cabell* 1920; *The Crystal Box* 1924; *The English Novel: Some Notes on Its Evolution* 1925; *Reading: an Essay* 1926; *My Religious Experience* 1928; *A Letter to a Modern Novelist* 1932; *The Apple Trees: Four Reminiscences* (a) 1932; *Extracts from a Diary* 1934; *Claude Houghton: Appreciations* [with C. Dane] 1935; *A Note on the Origins of the Herries Chronicle* 1940; *The Freedom of Books* 1940; *Roman Foundation* 1940; *Some Women Are Motherly* 1943

Plays

Robin's Father 1918 [np]; *The Young Huntress* 1933 [np]; *Vanessa* (f) 1934; *David Copperfield* (f) [from Dickens's novel; with H. Estabrook] 1934; *Little Lord Fauntleroy* [from F.H. Burnett's novel; with others] 1936; *The Cathedral* [from his novel] 1937

Edited

The Waverley Pageant: Best Passages from the Novels of Sir Walter Scott 1932; *Famous Stories of Five Centuries* (s) [with W. Partington] 1934; *A Second Century of Creepy Stories* (s) 1937; *The Nonesuch Dickens* [with others] 23 vols 1937–8

For Children

A Stranger 1926

Biography

Hugh Walpole by Rupert Hart-Davis 1952

Marina WARNER 1946–

Half British, half Italian, Warner was born in London. As a convent schoolgirl, she was so devoted to the Virgin Mary that, dissatisfied with the school's beribboned Grotto to the Virgin, she constructed her own; at seventeen she abandoned religion altogether. Her education continued at Lady Margaret Hall, Oxford, where she studied French and Italian.

Warner is a novelist, a critic and a historian. She is author of three erudite studies of mythology – all considered powerfully original feminist polemics – the reviews of which reached the extremes of eulogy and vituperation. The first of them, *Alone of All Her Sex* (1976), examined the texts, art and fables that have shaped Mariolatry over the centuries and their effect on the treatment and self-regard of women. *Monuments and Maidens* won the Fawcett Prize in 1986. An exploration of why the female form has been hijacked to personify a huge range of concepts since before the time of Hesiod, the book was praised for its studies of the Statue of Liberty and Fleet Street's depiction of Margaret Thatcher during the Falklands War.

In 1987 Warner was invited to California to research fairy-tales at the John Paul Getty Center for the History of Art and Humanities. Here, she completed her third novel, *The Lost Father*, which was shortlisted for the Booker Prize in 1988 and was regional winner of the Commonwealth Writers Prize in 1989. Set in Mussolini's Italy, it reverses her previous stance and is the story of an ordinary man whose life-story is mythologised by the women in his family.

As a contributor to Chatto & Windus's series of *CounterBlast* pamphlets in 1989, whose goal was to provide 'a forum for voices of dissent', she chose to write on the modern attitude towards children and entitled her treatise 'Into the Dangerous World'. Warner has reviewed for the *Times Literary Supplement* and the *London Review of Books*, and been a regular contributor to the *Independent*. She lives in London with her husband, the painter John Dewe Matthews and has one son.

Fiction

In a Dark Wood 1977; *The Skating Party* 1982; *The Lost Father* 1988; *Indigo* 1992; *L'Atalante* 1993; *Mermaids in the Basement* (s) 1993

Non-Fiction

Alone of All Her Sex: the Myth and Cult of the Virgin Mary 1976; *Queen Victoria's Sketch Book* 1979; *Joan of Arc* 1981; *Monuments and Maidens: the Allegory of the Female Form* 1985; *Imagining a Democratic Culture* 1991; *The Dragon Empress: Life and Times of Tz'u-hsi* 1993; *Richard Wentworth* 1993; *Modern Myths: 1994 Reith Lectures* 1994

For Children

The Crack in the Tea-Cup 1979; *The Impossible Day* 1982; *The Wobbly Tooth* 1984

Rex (Ernest) WARNER 1905–1986

The son of a clergyman of the Church of England and a schoolteacher, Warner grew up in Gloucestershire, attended St George's, Harpenden, and Wadham College, Oxford, where he read classics and then English. At Oxford he was a member of the circle including **W.H. Auden** and **C. Day Lewis** (Auden commemorated his friendship with Warner in the fourth of the 'Six Odes' in *The Orators*, 1932). Until 1945, Warner taught in various schools in England and, briefly, Egypt, while gaining some reputation as poet and novelist. His early *Poems* of 1937 combine sensuality with leftist polemic; his three early novels *The Wild Goose Chase, The Professor* and *The Aerodrome* were hailed as attempting something new in English fiction, namely to adapt to it some of the techniques and political insights characteristic of Kafka: *The Aerodrome*, about a neo-fascist air force invading an English village, is his most substantial novel.

After the war, Warner worked briefly with the Allied Control Commission in Berlin, was director of the British Institute in Athens from 1945 to 1947, and later made his career mainly as a professor in various American universities. During this period, he became increasingly known for his translations of the Greek and Roman classics and books on ancient subjects. His novels, too, were now based on classical themes and were perhaps more noted for exposition than imagination. In his last years, Warner returned to England and lived in Oxfordshire. He married three times: Frances Chamier Grove, in 1929, with whom he had two sons and a daughter and from whom he was divorced; Barbara, Lady Rothschild, in 1949, with whom he had a daughter; and Frances Chamier Grove, for the second time, in 1966.

Fiction

The Kite 1936; *The Wild Goose Chase* 1937; *The Professor* 1938; *The Aerodrome* 1941; *Why Was I Killed?* 1943; *Men of Stones* 1949; *Escapade* 1953; *Pericles the Athenian* 1963; *The Converts* 1967

Translations

The Medea of Euripides 1944; Aeschylus *Prometheus Bound* 1947; *Xenophon: The Persian Expedition* 1949, *History of My Times* 1966; Euripides: *Hippolytus* 1949, *Helen* 1951, *The Vengeance of the Gods* 1954; *Men and Gods. Stories from Ovid* 1950; Thucydides: *History of the Peloponnesian War* 1954, *Athens at War* 1970; Plutarch *The Fall of the Roman Republic* 1958, *Moral Essays* 1971; *Caesar's War Commentaries* 1960; Sepheres: *Poems* 1960, *On the Greek Style* 1966; *Confessions of St Augustine* 1963

Non-Fiction

English Public Schools 1945; *The Cult of Power* 1946; *John Milton* 1949; *E.M. Forster* 1950; *Views of Attica and Its Surroundings* 1950; *Greek and Trojans* 1951; *Ashes to Ashes. A Post-Mortem on the 1940–1951 Tests* 1951; *Eternal Greece* [with M. Hurlimenn] 1953; *The Greek Philosophers* 1958; *The Young Caesar* 1958; *Imperial Caesar* 1960; *Men of Athens* 1972

Poetry

Poems 1937

Edited

John Bunyan *The Pilgrim's Progress* 1951; *Look up at the Skies. The Poems and Prose of Gerard Manley Hopkins* 1972

Sylvia Townsend WARNER 1893–1978

Tha daughter of a housemaster at Harrow School, Warner was educated privately and haphazardly, and then spent a period working in a munitions factory during the First World War. From 1918 to 1928, she was on the editorial board of a scholarly work, the ten-volume *Tudor Church Music* (she was a noted scholar of fifteenth-and sixteenth-century music).

Her first volume of verse, *The Espalier*, appeared in 1925, and her first novel, *Lolly Willowes*, a story of witchcraft, in 1926. The latter was the first book to be chosen as Book of the Month in the USA and gained her a large public there; from 1935 onwards, the *New Yorker* would publish more than 140 of her short stories, later collected in many volumes. She was a noted hostess in inter-war London, and it was **T.F. Powys** (whose biography she long contemplated but never completed) who introduced her to her lifelong lover, Valentine Ackland, with whom she lived, in London and later in Dorset, until Ackland's death in 1969. Both women joined the Communist Party in 1935 and went to Spain several times in opposition to General Franco: *After the Death of Don Juan* translates the famous libertine's story to a contemporary political setting.

Warner published several novels, mainly on historical and folklore themes, the last of which appeared in 1954: perhaps *The Corner That Held Them*, an intense study of life in a fourteenth-century nunnery, is the most remarkable of these. She was also a distinguished (and still somewhat neglected) poet, whose *Opus 7*, a novel in verse, is only one of many memorable volumes. She wrote poetry until the end; later works include a translation of Proust and a biography of **T.H. White**. Her last stories, *Kingdoms of Elfin*, are set firmly in her favourite fictional territory of the numinous and faraway.

Fiction

Lolly Willowes 1926; *Mr Fortune's Maggot* 1927; *Some World Far from Ours* (s) 1929; *The True Heart* 1929; *Elinor Barley* 1930; *This Our Brother* 1930; *A Moral Ending and Other Stories* (s) 1931; *The Rainbow* 1932; *The Salutation* (s) 1932; *More Joy in Heaven and Other Stories* (s) 1935; *Summer Will Show* 1936; *After the Death of Don Juan* 1938; *24 Short Stories* (s) 1939; *The Cat's Cradle-Book* (s) 1940; *The People Have No Generals* 1941; *A Garland of Straw and Other Stories* (s) 1943; *The Museum of Cheats* (s) 1947; *The Corner That Held Them* 1948; *The Flint Anchor* 1954; *Winter in the Air and Other Stories* (s) 1955; *A Spirit Rises* (s) 1962; *Sketches from Nature* 1963; *A Stranger With a Bag and Other Stories* (s) 1966 [US *Swans on an Autumn River*]; *The Innocent and the Guilty* (s) 1971; *Kingdoms of Elfin* (s) 1977; *Scenes of Childhood* (s) 1981; *One Thing Leading to Another and Other Stories* (s) 1984; *Selected Stories* 1988

Poetry

The Espalier 1925; *Time Importuned* 1928; *Opus 7* 1931; *Whether a Dove or a Seagull* [with V. Ackland] 1933; *Two Poems* 1945; *Boxwood* 1957 [rev. 1960]; *King Duffus and Other Poems* 1968; *Azreal and Other Poems* 1978 [UK *Twelve Poems* 1980]; *Collected Poems* [ed. C. Harman] 1983; *Selected Poems* 1985

Non-Fiction

The Oxford History of Music Vol. 1 1929; *Somerset* 1949; *Jane Austen 1775–1817* 1951; *T.H. White: a biography* 1967; *Letters* [ed. W. Maxwell] 1982; *The Diaries of Sylvia Townsend Warner* [ed. C. Harman] 1994

Edited

The Weekend Dickens 1932; *The Portrait of a Tortoise: Extracted from the Journals of Gilbert White* 1946

Translations

Proust *By Way of Saint-Beuve* 1958; Jean René Huguenin *A Place of Shipwreck* 1963

Biography

Sylvia Townsend Warner by Claire Harman 1989

Robert Penn WARREN 1905–1989

Warren was once described by the *New York Times* as 'a man of letters on the old-fashioned, outsize scale'. He received seventeen honorary degrees and almost every major literary award available to writers in the USA. He is the only American writer to have won a Pulitzer Prize for both poetry and fiction – for the two volumes *Promises* (1957) and *Now and Then* (1978) and for the novel *All the King's Men* (1946). Born in Guthrie, Kentucky, the son of a banker with unfulfilled literary intentions, Warren had an early ambition to be a naval officer, but the summer before his entry to Vanderbilt University he suffered an eye injury which almost resulted in complete blindness: throughout his life he wore a glass eye.

As an undergraduate, he was recruited as the youngest member of the Fugitives (later the Agrarians), an esteemed literary group associated with Vanderbilt University and Nashville in the 1920s and 1930s, whose most important members, apart from Warren himself, were **John Crowe Ransom** and **Allen Tate**. In 1930 he contributed to their famous manifesto assessing the Southern economy, 'I'll Take My Stand', and co-founded *The Fugitive* poetry magazine. Warren's academic career continued with a Rhodes Scholarship to Oxford University, followed by teaching posts at Yale, Vanderbilt University, Southwestern College, Memphis, Louisiana State University and Minnesota University. He was professor of English at Yale from 1950, and professor emeritus from 1973. Between 1935 and 1942 he edited the distinguished literary quarterly the *Southern Review*.

Warren's first novel was *Night Rider* (1940); its subject was the tobacco wars of Kentucky in the early 1900s. He achieved fame in 1946 with *All the King's Men*, a tale about good intentions declining into megalomania and corruption; it became both a play and a film and has been translated into twenty languages. As one of the most eminent critics in the USA and a leading exponent of the New Criticism, he collaborated with Cleanth Brooks to write the immensely influential *Understanding Poetry* and *Understanding Fiction*. He edited numerous historical and sociological treatises, and in such books as *Who Speaks for the Negro?* examined democracy and expressed concern for the issues of racism and capital punishment.

With his second marriage in 1952, to the novelist Eleanor Clark, and the birth of his daughter, Warren emerged from what he felt was a ten-year period of poet's block, and wrote poetry prolifically until the end of his life. He regarded himself as a poet first and a novelist and critic second, and explained: 'A poem for me and a novel are not so different. They start much the same way, on the same emotional journey, and can go either way.' The tone of his later poetry seems personal, often autobiographical, though he claimed always to write of ideals and the outside world. The themes most consistently pursued in both his poetry and fiction have been the struggle for knowledge and identity and the mystery of time. He is admired for his rambling, conversational rhythm and majestic, ballad-like narrative.

In 1986 Warren became America's first Poet Laureate, and three years later died at his summer home in the Vermont woods aged eighty-four. He will be remembered as a teacher and as a major modern writer whose range of achievements will be hard to match.

Poetry
Thirty-Six Poems 1936; *Eleven Poems on the Same Theme* 1942; *Selected Poems 1923–1943* 1944; *Brother to Dragons* 1953; *To a Little Girl, One Year Old, in a Ruined Fortress* 1956; *Promises: Poems 1954–1956* 1957; *You, Emperors, and Others: Poems 1957–1960* 1960; *Selected Poems: New and Old 1923–1966* 1966; *Incarnations: Poems 1966–1968* 1968; *Audubon* 1969; *Or Else: Poem/Poems 1968–1974* 1974; *Selected Poems 1923–1975* 1977; *Now and Then: Poems 1976–1978* 1978; *Two Poems* 1979; *Being Here: Poetry 1977–1980* 1980; *Love* 1981; *Rumour Verified: Poems 1979–1980* 1981; *Chief Joseph of the Nez Percé* 1983; *New and Selected Poems 1923–1985* 1985

Fiction
Night Rider 1940; *At Heaven's Gate* 1943; *All the King's Men* 1946; *Blackberry Winter* (s) 1946; *The Circus in the Attic and Other Stories* (s) 1948; *World Enough and Time* 1950; *Band of Angels* 1955; *The Cave* 1959; *Wilderness* 1961; *Flood* 1964; *Meet Me in the Green Glen* 1971; *A Place To Come To* 1977

Plays
Proud Flesh 1947; *All the King's Men* [from his novel] 1959 [pub. 1960]

Non-Fiction
John Brown: the Making of a Martyr 1929; *I'll Take My Stand: the South and the Agrarian Tradition* 1930; *Segregation: the Inner Conflict in the South* 1956; *Remember the Alamo!* 1958; *Selected Essays* 1958; *The Gods of Mount Olympus* 1959; *The Legacy of the Civil War: Meditations on the Centennial* 1961; *Who Speaks for the Negro?* 1965; *A Plea in Mitigation: Modern Poetry and the End of an Era* 1966; *Homage to Theodore Dreiser* 1971; *John Greenleaf Whittier's Poetry: an Appraisal and a Selection* 1971; *Democracy and Poetry* 1975; *Jefferson Davis Gets His Citizenship Back* 1980

With Cleanth Brooks: *Understanding Poetry* 1938; *Understanding Fiction* 1943; *Modern Rhetoric: With Readings* 1949; *Fundamentals of Good Writing: A Handbook of Modern Rhetoric* 1950

Keith (Spencer) WATERHOUSE 1929–

Waterhouse was born in Leeds, described in his memoir of youth, *City Lights*, as 'the rhubarb capital of the universe'. His father was a costermonger who died when Waterhouse was three. Having failed his 11-plus examination he left school at fourteen and had various jobs, including one as an undertaker's clerk which later provided the background for his best-known novel, *Billy Liar*. After national service with the Royal Air force, he joined the *Yorkshire Evening Post* in 1951 as a trainee reporter, before moving to London, and in 1956 during a newspaper strike wrote his first novel, *There Is a Happy Land*, a comic account of a young boy's fantasies and mimicry of adult behaviour which anticipates many of the themes of his second novel, *Billy Liar*, and its 1975 sequel, *Billy Liar on the Moon*.

In 1958 he gave up full-time journalism and throughout the 1960s, in collaboration with his long-time friend Willis Hall, produced a steady stream of plays, musical revues and screenplays. Of this period Waterhouse recalls: 'When we had our office in Mayfair we'd work from nine until 12.30 in the morning, and then go out to places like the Treetrunk Club, Tommy's or the George and Pussy, and get blind roaring drunk until 2am.' Films they wrote include *Whistle Down the Wind* (1961), *A Kind of Loving* (1962) and John Schlesinger's classic *Billy Liar* (1963). During a spell in Hollywood they worked (uncredited) on Alfred Hitchcock's *Torn Curtain* (1966), and Waterhouse recalls: 'Hitchcock treated his writers much better than his actors and we had a bigger trailer than Paul Newman.'

Waterhouse was a columnist on the *Daily Mirror* from 1970 to 1986, but, after Robert Maxwell's acquisition of the paper, moved to the *Daily Mail*. His later novels include *Maggie Muggins*, concerning a day in the life of a hapless alcoholic woman, and *Unsweet Charity*, a caustic view of provincial life and the charity business. His highly successful 1989 play *Jeffrey Bernard Is Unwell* was adapted from the *Spectator* 'Low Life' column written by the notorious Soho drinker and *roué*, its title deriving from the magazine's frequently printed ironic explanation of the column's absence. Waterhouse's concern with declining standards of literacy was reflected in his membership of the Kingman Committee on the Teaching of the English Language from 1987 to 1988. He has been married twice, and has three children from his first marriage. He has won numerous awards for his journalism and was awarded the CBE in 1991.

Plays
The Town That Wouldn't Vote (r) 1951; *There is a Happy Land* (r) [from his novel] 1962; *How Many Angels* (tv) 1964; *The Woolen Bank Forgeries* (r) 1964; *Inside George Webley* (tv) 1968; *Queenie Castle* (tv) 1970; *The Warmonger* (tv) 1970; *Budgie* (tv) 1971–2; *By Endeavour Alone* (tv)

1973; *The Last Phone-In* (r) 1976; *Briefer Encounter* (tv) 1977; *The Upchat Line* (tv) 1977; *The Upchat Connection* (tv) 1978; *The Big Broadcast of 1922* (r) 1979; *Charlie Muffin* (tv) [from B. Freemantle's novels] 1979; *Public Lives* (tv) 1979; *West End Tales* (tv) 1981; *Steafel Variations* [with P. Tinniswood, D. Vosburgh] 1982; *The Happy Apple* (tv) [from J. Pulman's play] 1983; *This Office Life* (tv) [from his novel] 1984; *Charters and Caldicott* (tv) 1985; *Mr and Mrs Nobody* 1986 [pub. 1992]; *Andy Capp* (tv) 1988; *The Great Paper Chase* (tv) [from A. Delaro's *Slip Up*] 1988; *Jeffrey Bernard Is Unwell* 1989 [pub. 1991]; *Bookends* [from C. Brown's *The Marsh Marlowe Letters*] 1990 [pub. 1992]

With Willis Hall: *Billy Liar* [from his novel] (st) 1960, (f) 1963, (tv) 1973–4; *Celebration* 1961; *Whistle Down the Wind* (f) 1961; *All Things Bright and Beautiful* 1962 [pub. 1963]; *England, Our England* [with D. Moore] 1962 [pub. 1964]; *A Kind of Loving* (f) [from S. Barstow's novel] 1962; *The Sponge Room* 1962 [pub. 1969]; *Squat Betty* 1962; *The Valiant* (f) 1962; *Happy Moorings* (tv) 1963; *Man in the Middle* (f) 1963; *West Eleven* (f) 1963; *Come Laughing Home* 1964 [pub. 1965]; *Say Who You Are* 1965 [pub. 1966]; *Pretty Polly* (f) 1967; *Joey, Joey* 1968 [pub. 1978]; *Children's Day* 1969 [pub. 1975]; *Lock Up Your Daughters* (f) 1969; *Who's Who* 1971 [pub. 1974]; *The Card* [from A. Bennett's novel; with T. Hatch, J. Trent] 1973; *Lost Empires* [from J.B. Priestley's novel] 1985; *Budgie* [from his tv series; with D. Bluck, M. Shuman] 1988 [np]

Fiction

There Is a Happy Land 1957; *Billy Liar* 1959; *Jubb* 1963; *The Bucket Shop* 1968; *Billy Liar on the Moon* 1975; *Office Life* 1978; *Maggie Muggins* 1981; *In The Mood* 1983; *Mrs Pooter's Diary* 1983; *Thinks* 1984; *The Collected Letters of a Nobody* 1986; *Our Song* 1988; *Wishful Thinking* 1989; *Bimbo* 1990; *Unsweet Charity* 1992

Non-Fiction

With Guy Deghy: *The Café Royal: Ninety Years of Bohemia* 1955; *How to Avoid Matrimony* 1957; *The Joneses: How to Keep up with Them* 1959; *The Higher Joneses* 1961

Britain's Voice Abroad [with P. Cave] 1957; *The Future of Television* 1958; *The Passing of the Third Floor Buck* 1974; *Mondays, Thursdays* 1976; *Rhubarb Rhubarb and Other Noises* 1979; *Daily Mirror Style* 1981 [rev. as *Waterhouse on Newspaper Style* 1989]; *Fanny Peculiar* 1983; *Waterhouse at Large* 1985; *The Theory and Practice of Lunch* 1986; *The Theory and Practice of Travel* 1989; *English Our English* 1991; *Sharon and Tracy and the Rest: the Best of Keith Waterhouse in the Daily Mail* 1992; *City Lights: a Street Life* (a) 1994

For Children

With Willis Hall: *Worzel Gummidge* (tv) [from B.E. Todd; pub. as *Worzel Gummidge's Television Adventures*] 1979; *The Trials of Worzel Gummidge* 1980; *Worzel Gummidge Goes to the Seaside* 1980; *The Television Adventures of Worzel Gummidge and Aunt Sally* 1981; *Worzel Gummidge's Birthday* 1981; *The Irish Adventures of Worzel Gummidge* 1984; *Worzel Gummidge, a Musical* [with D. King] 1984; *Worzel Gummidge at the Fair* 1985; *Worzel Gummidge Down Under* 1987

Translations

With Willis Hall: Eduardo de Filippo: *Saturday, Sunday, Monday* 1973 [pub. 1974], *Filumena* 1977 [pub. 1978]

Edited

Writer's Theatre [with W. Hall] 1967

Vernon (Phillips) WATKINS 1906–1967

Watkins, who spent almost all his life in a junior position as a bank cashier, is the epitome of the many twentieth-century writers of all nationalities (Kafka, Cavafy, Pessoa, **Wallace Stevens**) who have worked at commercial jobs without ever losing their devotion to art. He was born in Maesteg, South Wales, where his father was in a rapidly rising position in Lloyd's Bank; his promotions soon took the family to Bridgend, Llanelli and, in 1913, Swansea, more precisely the nearby peninsula of Gower, where Watkins spent most of his life, writing poetry from childhood. His family sent him to preparatory school and Repton, at which (and also at Cambridge) he was a contemporary of **Christopher Isherwood** and **Edward Upward**, who invented 'Reynard Moxon' in their 'Mortmere' stories specifically to 'terrorise' him. Nevertheless, he was happy at Repton, but left Magdalene College suddenly after a stormy scene with the Master, A.C. Benson (brother of the novelist **E.F. Benson**), and returned to South Wales, where his father insisted he enter at Lloyd's Bank at the Cardiff branch.

After two years, he had a nervous breakdown and returned to Repton, where he attempted to assault his old headmaster, Dr Geoffrey Fisher. Once in a nursing home, he experienced a Christian conversion, and returned to the bank, this time the Swansea branch, where he worked thenceforth with unwavering devotion and often considerable incompetence. In the 1930s, although he was not yet published, his life was enriched by his meeting with **W.B. Yeats** and his important friendship with **Dylan Thomas** during the latter's early Swansea period. Watkins used to meet Thomas and several others during his lunch-breaks to read and discuss poetry at the city's Kardomah café. Watkins's first volume, *Ballad of the Mari Lwyd and Other Poems*, was published in 1941 and quickly established a reputation which further collections consolidated. His parents had been Welsh-speaking, but he had not learnt the language from them, and, despite his use of such figures as the sixth-century bard Taliesin, whom he shares with **Charles Williams**, Watkins's work is nurtured more by English and Continental poetry. It is carefully crafted, exploiting an unaffected lyricism which became unfashionable even as he wrote, but has often been found valuable by those returning to it. He also published translations during his lifetime and little else; some essays have appeared posthumously.

During the war, he served as a policeman in the Royal Air Force, but after such exploits as hosing a rank of airmen during a fire drill was transferred to the Intelligence Service. Here he met his wife and, despite an initial setback when his best man, Dylan Thomas, failed to turn up at the wedding, the couple were to be happy and have five children. In 1964, two years before his retirement from the bank, he held a visiting professorship at the University of Seattle, which seemed to open up a rewarding new career for him. However, after a period of teaching at the University of Swansea, and a return to Seattle, he died suddenly. It was revealed after his death that this unassuming and much-loved man had been among several poets under consideration for the Poet Laureateship of England.

Poetry

Ballad of the Mari Lwyd and Other Poems 1941; *The Lamp and the Veil* 1945; *The Lady with the Unicorn* 1948; *The Death Bell* 1954; *Cypress and Acacia* 1959; *Affinities* 1962; *Selected Poems 1930–1960* 1967; *Fidelities* 1968; *Uncollected Poems* 1969; *The Influences* 1976; *Elegy for the Latest Dead* 1977; *The Ballad of the Outer Dark and Other Poems* 1979; *The Breaking of the Waves* 1979; *The Collected Poems of Vernon Watkins* 1986

Translations

Heinrich Heine *The North Sea* 1951; *Selected Verse Translations* 1977

Non-Fiction

Yeats and Owen: Two Essays 1981

Alec (i.e. Alexander Raban) WAUGH 1898–1961

Waugh was born in West Hampstead, London, the elder son of the publisher and man-of-letters Arthur Waugh. He was educated at Sherborne, which he adored, but from which he was expelled for homosexuality. He immediately set to and wrote, at the age of seventeen, his first and most famous novel, *The Loom of Youth*, which (having been rejected by six publishers) was published in 1917 on the day he sailed to France as an officer in the Machine Gun Corps. The novel's unsensational account of homosexual romances, its criticism of the ascendancy of games, and its laconic depiction of cheating and other vices caused an enormous furore, provoking reams of journalism and even whole books intended to refute the author's libel upon the public-school system. In contrast to the accepted view of public schools as presented in innumerable anodyne school stories of the period, Waugh's novel suggested that such establishments were directed more by a code of expediency than one of honour, a criticism that was seen as particularly offensive, since at the time the British army was being led by young officers fresh from the schools. The book remained continuously in print until 1962 and has been reissued several times since.

Waugh published some undistinguished war poetry and was taken prisoner in 1918. After the armistice he returned to England and married Barbara, daughter of the short story writer W.W. Jacobs; the marriage was never consummated – Waugh blamed his own (hetero)sexual inexperience – and they were divorced in 1923. He became a 'literary adviser' at his father's firm, Chapman & Hall, and consolidated his literary reputation with occasional journalism and further volumes of fiction and non-fiction, including a factual study of *Public School Life*. In 1926 he resigned his job in order to devote his time to travel and writing, and in 1932 married Joan Chirnside, an Australian, with whom he had three children. During the Second World War he re-enlisted and served with the British Expeditionary Force in France and in Intelligence in the Middle East.

After the war he spent most of his time abroad, in New York, France and Tangier, writing popular and bestselling novels and several volumes of memoirs and travel writing. His career was constantly overshadowed by that of his infinitely more gifted younger brother, **Evelyn Waugh**, although they wrote very different books and professed no rivalry. Indeed, when the journalist Nancy Spain suggested in the *Daily Express* that Evelyn was jealous of Alec's (considerable) financial success, Evelyn sued: 'I had an exhilarating expedition to the Law Courts,' he reported, 'and came out two thousand pounds (tax free) to the good.' Alec gave evidence on his brother's behalf; relations between them were cordial rather than warm, however.

Having divorced his second wife in 1969, Waugh married Virginia Sorensen, an American writer of children's books and novels of Mormon life.

Fiction

The Loom of Youth 1917; *The Prisoners of Mainz* 1919; *Pleasure* (s) 1921; *The Lonely Unicorn* 1922; *Roland Whateley* 1922; *Card Castle* 1924; *Kept* 1925; *Love in These Days* 1926; *The Last Chukka* (s) 1928; *Nor Many Waters* 1928; *Portrait of a Celibate* 1929; *Three Score and Ten* 1929; *'Sir' She Said* 1930 [rev. as *Love in Conflict* 1977]; *So Lovers Dream* 1931; *Leap Before You Look* 1932; *No Quarter* 1932; *That American Woman* 1932; *Tropic Seed* 1932; *Playing with Fire* 1933; *Wheels within Wheels* 1933; *The Balliols* 1934; *Pages in Woman's Life* (s) 1934; *Jill Somerset* 1936; *Eight Short Stories* (s) 1937; *Getting Their Own Way* 1938; *No Truce with Time* 1941; *His Second War* 1944; *Unclouded Summer* 1948; *Guy Renton* 1952; *Island in the Sun* 1956; *Fuel for the Flame* 1960; *My Place in the Bazaar* (s) 1961; *The Mule on the Minaret* 1965; *A Spy in the Family* 1970; *The Fatal Gift* 1973; *Brief Encounter* [from N. Coward's play] 1975; *Married to a Spy* 1976

Non-Fiction

Public School Life: Boys, Parents, Masters 1922; *Myself when Young: Confessions* (a) 1923; *On Doing What One Likes* 1926; *The Coloured Countries* 1930; *'Most Women ...' 1931; Thirteen Such Years* 1932; *The Sunlit Caribbean* 1948; *The Lipton Story: a Centennial Biography* 1950; *Where the Clocks Strike Twice* 1951; *Merchants of Wine* 1957; *The Sugar Islands* 1958; *In Praise of Wine* 1959; *The Early Years of Alec Waugh* (a) 1962; *A Family of Islands: a History of the West Indies from 1492–1898* 1964; *My Brother Evelyn and Other Profiles* 1967; *Wines and Spirits* 1968; *Bangkok: the Story of a City* 1970; *A Year to Remember: a Reminiscence of 1931* (a) 1975; *The Best Wine Last. The Years 1932–1969* (a) 1978

Poetry

Resentment 1918

Evelyn (Arthur) (St John) WAUGH 1903–1966

Born into a publishing family in West Hampstead, Waugh was educated at Lancing College and at Hertford College, Oxford, where he mixed with the aesthetic set led by **Harold Acton**, and devoted the majority of his time to homosexual romance and alcoholic indulgence. After leaving Oxford, he enrolled briefly at Heatherley's Art School (he was an accomplished draughtsman and illustrated several of his own books), before becoming, disastrously, a preparatory-school master at Arnold House in North Wales. He was unhappy enough in this post to attempt suicide (his plans to drown himself were, he claimed, foiled by a jellyfish), and he eventually resigned, but the experience provided a wealth of material for his first novel, *Decline and Fall* (1928).

A family connection with Holman Hunt had fuelled Waugh's admiration for the Pre-Raphaelites and his first book was an admirable (though not well received) biography of D.G. Rossetti. He made his name, however, as a member and principal chronicler in his early novels of the Bright Young Things of the 1920s. His early satires of society life were characterised by black comedy (the publishers of *Rossetti* turned down *Decline and Fall* on the grounds of indecency) and were strongly influenced by **Ronald Firbank**. After the collapse of his short-lived first marriage to the Hon. Evelyn Gardiner, however, his vision darkened and he wrote the embittered tragi-comedy of adultery, *A Handful of Dust.* Therapeutic trips to Africa and South America led to a number of travel books and his employment as a foreign correspondent, notably in Abyssinia, *Scoop*, his cynical 'Novel about Journalists', and the splendidly tasteless *Black Mischief*, inspired by the coronation of the Emperor Haile Selassie, made use of his experiences. One of his principal themes as a writer was the battle between civilisation and barbarism, not only in Africa, but equally in Mayfair and in California, the setting of *The Loved One*, a comic novella about American funeral customs originally published in *Horizon*.

He became an ardent convert to Roman Catholicism in 1930, married Laura Herbert in 1937 (they had seven children), and served as an officer in the Marines during the Second World War. Disenchantment with the war and his unpopularity as a soldier led to his taking leave in order to write *Brideshead Revisited*, a great popular success but one which dismayed the critics because of its lush nostalgia and the author's genuflections before the aristocracy and the Catholic faith. (In 1980 it became a lavish and acclaimed television series, scripted by **John Mortimer**.) His *Sword of Honour* trilogy charts the increasing disillusionment of a romantic, middle-aged, Catholic gentleman during the war.

Waugh spent the remainder of his life in Somerset, where he adopted the role of a cantankerous Tory squire, clad in loud tweeds and affecting an ear trumpet. He played the part so well – he was beastly to his children, rude to Americans, and fulminated against the modern world – that it was difficult to tell how much of his bad behaviour was real and how much of it was a carefully maintained comic persona. Amongst his recreations was suing for libel in order to fund little luxuries: a successful action against the *Daily Express* for suggesting he was professionally jealous of his brother, **Alec Waugh**, paid for a Wilton rug specially commissioned to a prize-winning design of 1851 (he developed a taste for the uglier artefacts of the Victorian era).

He wrote several works of non-fiction, including biographies of Ronald Knox and the Jesuit martyr Edmund Campion and an autobiography of his youth, *A Little Learning*. An acute sense of *accidie* led to a breakdown in 1954, which he recorded in hilarious and pitiless detail as *The Ordeal of Gilbert Pinfold*, the novel that contains his most penetrating self-portrait. He was an indefatigable and brilliant writer of letters and kept an intermittent and extremely scurrilous diary: editions of both have been published. He founded something of a literary dynasty: among his children are the editor, columnist and novelist Auberon Waugh and the biographer Margaret Fitzherbert, and several of his grandchildren have become writers. He died, before his time, of a heart-attack on Easter Day, having attended Mass with his family.

Fiction

Decline and Fall 1928; *Vile Bodies* 1930; *Black Mischief* 1932; *A Handful of Dust* 1934; *Mr Loveday's Little Outing and Other Sad Stories* (s) 1936; *Scoop* 1938; *Put Out More Flags* 1942; *Work Suspended* 1942 [rev. as *Work Suspended and Other Stories* (s) 1948, 1951, 1967, 1983]; *Brideshead Revisited* 1945 [rev. 1960]; *Scott-King's Modern Europe* 1947; *The Loved One* 1948; *Helena* 1950; *Sword of Honour 1965*: *Men at Arms* 1952, *Officers and Gentlemen* 1955, *Unconditional Surrender* 1961; *Love Among the Ruins* 1953; *The Ordeal of Gilbert Pinfold* 1957; *Basil Seal Rides Again* 1963

Non-Fiction

Rossetti: His Life and Works 1928; *Labels: a Mediterranean Journey* 1930; *Remote People* 1931; *Ninety-Two Days* 1934; *Edward Campion: Jesuit and Martyr* 1935; *Waugh in Abyssinia* 1936; *Robbery under Law: the Mexican Object-Lesson* 1939; *When the Going Was Good* 1946; *Wine in Peace and War* 1947; *The Holy Places* 1952; *The Life of the Right Reverend Ronald Knox* 1959; *A Tourist in Africa* 1960; *A Little Learning* (a) 1964; *The Diaries of Evelyn Waugh* [ed. M. Davie] 1976; *A Little Order: a Selection from Evelyn Waugh's Journalism* [ed. D. Gallagher] 1977; *The Letters of Evelyn Waugh* [ed. M. Amory] 1980; *The Essays, Articles and Reviews of Evelyn Waugh* [ed. D. Gallagher] 1983; *Mr Wu and Mrs Stitch: the Letters of Evelyn Waugh and Lady Diana Cooper* [ed. A. Cooper] 1991

Edited

A Selection from the Occasional Sermons of Monsignor Ronald Knox 1949

Biography

Evelyn Waugh by Christopher Sykes 1975 [rev. 1977]; *Evelyn Waugh* by Martin Stannard, 2 vols 1986, 1992; *Evelyn Waugh* by Selina Hastings 1994

Mary (Gladys) WEBB 1881–1927

Webb was born Mary Meredith in the small village of Leighton, under the Wrekin, in Shropshire, the county in which she lived for most of her life and which all her novels celebrate. She was the daughter of a schoolmaster of Welsh descent and his Scots wife, the eldest of their six children. She spent her girlhood mainly at The Grange, a small country house near Much Wenlock, receiving much of her schooling at home: here she began to write poetry and prose at an early age, and her mystical attachment to her native countryside began to form.

Her adult life was largely a struggle against recurrent illness and near-poverty. In 1912, she married a schoolmaster and fellow native of Shropshire, Henry Webb, and, after two years in Weston-super-Mare, they returned to Shropshire, where they worked as market gardeners and had their own stall at Shrewsbury market. During this period, she began to write novels seriously, and *The Golden Arrow* was the first to be published, in 1916. In 1921 she and her husband moved to Hampstead in London, where they lived until her early death six years later.

Her writing continued bound to its native soil, and her best novel, the exquisitely poetical *Precious Bane*, was published in 1924, doing little to relieve her financial circumstances during her lifetime. Six months after her death it was read and publicly praised by the then prime minister, Stanley Baldwin: an extraordinary popular success for the novel began, but also the process by which it was to be dismissed as a middlebrow 'saga of the soil' and its literary merit to be undervalued. Her reputation was further undermined with the publication in 1932 of *Cold Comfort Farm*, an enormously successful novel in which **Stella Gibbons** parodied rural tales such as Webb's. For a time, it was largely thanks to Gibbons that Webb's work was remembered at all, but there has recently been a revival of interest in the genuine (rather than parodied) article.

Fiction
The Golden Arrow 1916; *Gone to Earth* 1917; *The House in Dromer Forest* 1920; *Seven for a Secret* 1922; *Precious Bane* 1924; *Armour Wherein He Trusted* **(s)** [unfinished] 1929; *The Chinese Lion* 1937

Poetry
Poems and The Spring of Joy 1928; *Fifty-One Poems* 1946

Collections
The Spring of Joy 1937; *Mary Webb: Collected Prose and Poems* 1977

Jerome WEIDMAN 1913–

Born in New York City to Jewish parents, Weidman was educated at City College from 1931 to 1933, after which he attended Washington Square College for one year. He graduated in law in 1937, after a three-year course at New York University Law School, and published his first novel, *I Can Get It for You Wholesale*, in the same year. The novel traced the sudden rise of a Jewish clerk to become a garment manufacturer: regarded by some critics as unpleasant in its detailing of the sordid aspects of life in the clothing industry, it was unusually realistic for its time. Two other novels and a book of short stories were published before 1942, when Weidman married and began work for the Office of War Information. He thinly disguised his experiences in a satirical novel, *Too Early to Tell*.

Weidman's early fiction depends for its effect on brutal realism, including graphic descriptions of sex before this was at all common. Extremely prolific, Weidman has often written of the problem of the American Jew in a Gentile society. Although he has in recent years reduced the shock value of his fiction, his admission that he writes to make money has perhaps stood in the way of a carefully planned novel which he has yet to write. His *Fourth Street East*, a fictional 'memoir' relating the childhood of Benny Kramer, a Manhattan boy from the Lower East Side, was affectionate and portrayed a time that has passed. Two other novels featuring Kramer followed, but reverted to a slick, ready-formulated pattern that has often typified Weidman's work. It has been suggested that Weidman is 'clever about people without being really wise about them'. Nevertheless, his literary ambitions are genuine, for he has said that in a lifetime of constant reading he has been trying to 'pick up a few pointers for what [**Ernest**] **Hemingway** has called the serious trade of novel writing', and that the only writers who have taught him anything useful have been **Joseph Conrad** and **James Gould Cozzens**. He has also written plays and film scripts and published a collection of essays and a travel book.

Fiction
I Can Get It for You Wholesale 1937; *What's in It for Me?* 1938; *The House That Could Whistle 'Dixie'* **(s)** 1939; *I'll Never Go There Any More* 1941; *The Lights Around the Shore* 1943; *Too Early to Tell* 1946; *The Captain's Tiger* **(s)** 1947; *The Price Is Right* 1959; *The Hand of the Hunter* 1951; *Give Me Your Love* 1952; *The Third Angel* 1953; *Your Daughter Iris* 1955; *A Dime a Throw* **(s)** 1957; *The Enemy Camp* 1958; *Before You Go* 1960; *Nine Stories* **(s)** 1960; *My Father Sits in the Dark* **(s)** 1961; *The Sound of Bow Bells* 1962; *Where the Sun Never Sets* **(s)** 1964; *Word of Mouth* 1964; *The Death of Dickie Draper* **(s)** 1965; *Other People's Money* 1967; *The Center of the Action* 1969; *Fourth Street East* 1970; *I, And I Alone* **(s)** 1972; *Last Respects* 1972; *Tiffany Street* 1974; *The Temple* 1975; *A Family Fortune* 1978; *Counselors-at-Law* 1980

Plays
The Damned Don't Cry **(f)** [with H. Medford] 1950; *The Eddie Cantor Story* **(f)** [with T. Sherdeman, S. Skolsky) 1953; *Slander* **(f)** 1957; *Fiorello!* [with G. Abbott] 1959 [pub. 1960]; *Tenderloin* [with G. Abbott] 1960 [pub. 1961]; *I Can Get It for You Wholesale* [from his novel] 1962; *Cool Off!* [with H. Blackman] 1964 [np]; *Reporter* **(tv)** 1964; *Pousse-Café* [with Duke Ellington] 1966 [np]; *Ivory Tower* [with J. Yaffe] 1968 [pub. 1969]; *Asterisk!* 1969; *The Mother Lover* 1969

Non-Fiction
Letter of Credit 1940; *Back Talk* 1963; *Praying for Rain* **(a)** 1986

Edited
The W. Somerset Maugham Sampler 1943; *Traveler's Cheque* 1954; *The First College Bowl Question Book* [with others] 1961

(Maurice) Denton WELCH 1915–1948

The youngest of three brothers, Welch was born in Shanghai, where his father was a businessman. His relationship with his mother, who died when he was eleven, was an important influence on his emotional development. Welch would later draw obsessively upon his memories of her domineering presence during his early years, when she encouraged him to love her with abandon and possession, treating him as a surrogate for her largely indifferent husband. She was a Christian Scientist, a fact which determined Welch's entry into St Michael's, Uckfield, a preparatory school which catered almost exclusively for the sons of those adhering to this religious system.

'All my childhood was spent in travelling backwards and forwards from China to England and England to China,' Welch later recalled, but he was finally sent for good to Repton, where, as he suggests in his autobiographical volume *Maiden Voyage* (1943), he felt uncomfortable, isolated and unliked. In his final term, when he was sixteen, he absconded. With the support of both his parents he afterwards attended the Goldsmiths School of Art.

Welch was by most accounts a highly promising art student, but on 7 June 1935, when he was twenty-three, he was knocked from his bicycle by a motorist, an accident which was the consequence of carelessness rather than drama and which is described with calm, controlled despair in Welch's posthumously published *A Voice through a Cloud*, the manuscript of which he left incomplete at his death. Despite the crippling spinal injury and the kidney complaints which resulted from the accident, Welch struggled to return to some sort of normal lifestyle. He was encouraged to write by, amongst others, **Edith Sitwell**, and in certain ways it is possible to argue, as **Jocelyn Brooke** immediately did, that the consequences of the accident inspired Welch more than might otherwise have been the case, and so helped him to discover his true vocation as a writer. Having said that, he continued to paint and many of his works appeared in *Vogue* over the years. He also provided decorations for his own books.

Welch was obsessed with minutiae and his work is characterised by a precisely detailed observation of himself and the minor events he had experienced when fully able, such as a visit to an antique shop or a walk by a river, which took on enormous significance after his accident. His work, with the exception of *In Youth Is Pleasure*, is unashamedly autobiographical, and is therefore not surprisingly pervaded by an acute sense of transience and decay. The major, recurring theme is adolescence, both his own and that of the virile and active boys to whom he frequently refers in his sixty or so short stories and in his *Journals* (originally published in 1952, but re-edited in 1984 in an unexpurgated edition), which cover the period 1942–8 and which run to some 200,000 words. The most celebrated of the short stories, all of which are collected in *Fragments of a Life Story*, is 'When I Was Thirteen' (first published in *Horizon* in 1944), which ends with the protagonist's elder brother berating him for having spent the night with a school friend: 'Bastard, Devil, Harlot, Sod.' The first publishers of Welch's work were disturbed by this frank admission of the author's homosexual propensity, an aspect of his novels which, however, has helped to ensure his cult status.

Welch despised his own looks, thought himself unworthy of others' love and affection, and caricatured his admittedly unusual facial characteristics in his numerous self-portraits. His love for younger members of his own sex took on an increasingly hopeless and doomed air, but in 1945 he befriended Eric Oliver and the two remained lovers until Welch's death. Oliver afterwards eagerly adopted the role of Welch's literary executor, and the writer's reputation has remained as high as it has largely because of the initial enthusiasm of his friend.

Fiction

In Youth Is Pleasure 1945; *Brave and Cruel* (s) 1949; *A Lost Sheaf* (s) 1951; *Fragments of a Life Story* (s) [ed. M. De-la-Noy] 1987

Non-Fiction

Maiden Voyage (a) 1943; *A Voice through a Cloud* (a) [unfinished] 1950; *The Denton Welch Journals* [ed. J. Brooke] 1952; *I Left My Grandmother's House* 1958; *The Journals of Denton Welch* [ed. M. De-la-Noy] 1984

Poetry

Dumb Instrument [ed. J.-L. Chevalier] 1976

Fay (i.e. Franklin) WELDON 1931–

Conceived and brought up in New Zealand, where her family had emigrated, but born in Worcestershire, Weldon is the daughter of Margaret Jepson, a successful romantic novelist of the 1930s, and Dr Frank Birkinshaw. They divorced when Weldon was five, and she eventually returned to England with her mother and sister to live with their grandmother in Belsize Park, London. She attended South Hampstead High School for Girls, and went on to St Andrews University, where she gained an MA in economics and psychology. After a stint in the civil service, during which she had her first child, she moved to Saffron Walden to run a cake shop with her sister and a friend – all three of them young unmarried mothers. She had a spell answering problem-page letters for the *Daily Mirror*, and then became an advertising copywriter for Ogilvy, Benson and Mather, part of the team that dreamed up the famous slogan 'Go to work on an egg'.

She was married briefly in her early twenties to a much older Cambridge teacher. In 1961 she married the antique-dealer, musician and painter Ron Weldon and had three more sons. While still a copywriter, she began writing plays for television and radio, including the award-winning first episode of *Upstairs, Downstairs*. In 1966, her television play *The Fat Woman's Tale* was broadcast, and in 1967, she turned it into a novel, *The Fat Woman's Joke*, which became the first of her caustic diatribes about the sex war, from a woman's point of view. Published three years before Germaine Greer's *The Female Eunuch*, it had the courage to talk about sexual and social inequality with wit and intelligence, in a way that had not been attempted before. Unwittingly, Weldon had tapped into a collective female consciousness, and found herself a figurehead for the new radical feminism.

The Fat Woman's Joke was followed by many further novels, among which perhaps the best known are: *Praxis* (which was nominated for the Booker Prize in 1978); *Puffball*; *The Life and Loves of a She-Devil*, which she successfully adapted for television in 1986, but which became something of a travesty as a Hollywood movie (*She-Devil*, 1989) starring Meryl Streep and Roseanne Barr; and *The Cloning of Joanna May*, also adapted for television. Her novels *Affliction* and *Splitting* relate directly to the breakdown of her own marriage after thirty years. She has written three collections of short stories, over fifty original plays and adaptations for television, radio and the stage, and one ecological libretto called *A Small Green Space*, which was performed by the English National Opera. Her non-fiction work includes *Letters to Alice on First Reading Jane Austen*, a study of **Rebecca West** and a pamphlet attacking censorship and the state of postmodern Britain called *Sacred Cows*. She describes herself as 'a leftish, humanist feminist'.

She is an ardent campaigner on behalf of writers and their treatment at the hands of exploitative publishers, and reserved her most vituperative rant on the subject for her speech as chairman of judges at the 1983 Booker Prize dinner. But she is a vociferous protester on many subjects, from the *fatwa* imposed on **Salman Rushdie** to psychotherapy, which she blames for the ultimate break-up of her marriage.

As her global perceptions have deepened, the direct passions of the early feminist fables have opened out into complex cautionary tales, occasionally with a supernatural or mystical slant, that have become increasingly polemical. 'What is fiction without polemic?' she says. 'Why do you want fiction if you don't want the world explained to you?' What does not change is the nature of Weldon's heroines, who are survivors all.

In 1974, despite her protests, her husband insisted that they moved to Somerset, and for the next twenty years, she divided her life between a smallholding near Glastonbury and two terraced houses in Kentish Town which served as office and London base. After he left her, in 1992, there were two years of separation, and then he died, eight hours before the divorce papers were stamped, in 1994. She remarried later that year.

Fiction

The Fat Woman's Joke 1967; *Down Among the Women* 1971; *Female Friends* 1975; *Remember Me* 1976; *Little Sisters* 1978; *Praxis* 1978; *Puffball* 1980; *Watching Me Watching You* **(s)** 1981; *The President's Child* 1982; *The Life and Loves of a She-Devil* 1983; *Polaris and Other Stories* **(s)** 1985; *The Shrapnel Academy* 1986; *Heart of the Country* 1987; *The Hearts and Lives of Men* 1987; *The Rules of Life* 1987; *Leader of the Band* 1988; *The Cloning of Joanna May* 1989; *Darcy's Utopia* 1990; *Moon over Minneapolis* 1991; *Growing Rich* 1992; *Life Force* 1992; *A Question of Timing* 1992; *Affliction* 1994; *A House Divided* 1995; *Splitting* 1995

Plays

A Catching Complaint **(tv)** 1966; *The Fat Woman's Play* **(tv)** [from her novel] 1966; *Dr De Waldon's Therapy* **(tv)** 1967; *Fall of the Goat* **(tv)** 1967; *The 45th Unmarried Mother* **(tv)** 1967; *Goodnight Mrs Dill* **(tv)** 1967; *What About Me* **(tv)** 1967; *Hippy Hippy Who Cares* **(tv)** 1968; *Ruined Houses* **(tv)** 1968; *£13083* **(tv)** 1968; *The Three Wives of Felix Hull* **(tv)** 1968; *Venus Rising* **(tv)** 1968; *The Loophole* **(tv)** 1969; *Permanence in Mixed Doubles* 1969 [pub. 1970]; *Office Party* **(tv)** 1970; *Poor Mother* **(tv)** 1970; *Hands* **(tv)** 1972; *The Lament of an Unmarried Father* **(tv)** 1972; *A Nice Rest* **(tv)** 1972; *The Spider* **(r)** 1972; *A Splinter of Ice* **(tv)** 1972; *Times hurries On* [pub. in *Scene Scripts* ed. M. Marland 1972]; *Comfortable Words* **(tv)** 1973; *Desirous of Change* **(tv)** 1973; *Housebreaker* **(r)** 1974; *In Memoriam* **(tv)** 1974; *Mr Fox and Mr First* **(r)** 1974; *Words of Advice* 1974; *Aunt Tarry* **(tv)** [from E. Bowen's story] 1975; *The Doctor's Wife* **(r)** 1975; *Friends* 1975; *Poor Baby* **(tv)** 1975; *The Terrible Tale of Timothy Bagshott* **(tv)** 1975; *Moving House* 1976; *Act of Rape* **(tv)** 1977; *Mr Director* 1978; *Action Replay* 1978 [pub. 1980; as *Love among the women* 1984]; *Polaris* **(r)** 1978 [pub. in *Best Radio Plays* 1979]; *All the Bells of Paradise* **(r)** 1979; *Weekend* **(r)** 1979; *Pride and Prejudice* **(tv)** [from J. Austen's novel] 1980; *Honey Ann* **(tv)** 1980; *Life for Christine* **(tv)** 1980; *After the Prize* 1981 [as *Wood Worm* 1984]; *I Love My Love* **(r)** 1981 **(st)** 1982 [pub. 1984]; *Little Mrs Perkins* **(tv)** [from P. Mortimer's story] 1982; *Redundant! or, The Wife's Revenge* **(tv)** 1983; *Out of the Undertow* **(tv)** 1984; *Jane Eyre* [from C. Brontë's novel] 1986; *The Life and Loves of a She-Devil* **(tv)** [from her novel] 1986 **(f)** [as *She-Devil*] 1990; *Heart of the*

Country **(tv)** [from her novel] 1987; *The Hole in the Top of the World* 1987; *The Cloning of Joanna May* **(tv)** [from her novel] 1992

For Children

Wolf the Mechanical Dog 1988; *Party Puddle* 1989

Non-Fiction

Simple Steps to Public Life [with P. Anderson, M. Stoff] 1980; *Letters to Alice on First Reading Jane Austen* 1985; *Rebecca West* 1985; *Sacred Cows* 1989

H(erbert) G(eorge) WELLS 1866–1946

Wells was born in Bromley, Kent, where his parents were unsuccessful shopkeepers. Their marriage was unhappy and his father effectively abandoned the family around 1880 when Wells's mother became the housekeeper at Uppark in Sussex, a location he draws on in the early pages of his novel *Tono-Bungay*. He had a chequered education and was briefly a chemist's assistant and a draper's apprentice, an experience he was vividly to recall in *Kipps* and *The History of Mr Polly*. He became a pupil teacher at Midhurst Grammar School, and by studying at night earned himself a scholarship to London University's Normal School of Science where the lectures of T.H. Huxley exerted a considerable influence on him. He had two brief spells as a schoolteacher and was a tutor for the University Correspondence College in the late 1880s before beginning to publish articles in the *Pall Mall Gazette* and elsewhere. He married his cousin Isabel in 1891, but the marriage collapsed after three years when he eloped with Amy Catherine Robbins (who was known at Wells's insistence as Jane). Throughout his life Wells was a dedicated womaniser and an advocate of sexual freedom, but despite notorious affairs with, among others, **Dorothy Richardson**, Odette Keun, **Elizabeth von Arnim** and **Rebecca West** (with whom he had a son, Anthony West), he remained married to Jane until her death in 1927. They had two sons.

Wells sprang to prominence with highly inventive science-fiction romances such as *The Time Machine*, *The Invisible Man* and *The War of the Worlds*, futuristic fantasies which none the less explore his social and political concerns. Perhaps the best of them is *The Island of Dr Moreau*, in which a shipwrecked man witnesses a disgraced vivisectionist's frightening creation of human–animal hybrids. In *Anticipations* and *A Modern Utopia* Wells outlines his notion of an ideal society. The books were treated seriously at the time, but now read as a curious mixture of social engineering, intellectual elitism, racism and his own quirky version of socialism. He was a member of the Fabian Society in the early 1900s but fell out with other members, notably **George Bernard Shaw**. Literary feuds were to become something of a speciality: his friendship with **Henry James** ended in 1915 after he launched a splenetic attack on James in *Boon*, and he had a lengthy, very public dispute with **Hilaire Belloc** after Belloc had attacked his *Outline of History*.

Wells was a prolific and often slapdash writer, but his best work is of enduring value, particularly the almost Dickensian novels of lower-middle-class life such as *Tono-Bungay*, *The History of Mr Polly* and *Love and Mr Lewisham*, in which he evokes his own background. *Ann Veronica* draws on his elopement with Jane in describing a young student's affair with her teacher and is a plea for sexual liberation. It was denounced as immoral and removed from public libraries, and sold very well in consequence. Despite the frequent scandals, Wells's stature was such that he attended meetings of the League of Nations Union after the First World War and, in the 1930s, visited the USSR and the USA to conduct interviews with both Stalin and F.D. Roosevelt. His increasingly pessimistic outlook is evident in *The Shape of Things to Come* (the basis of Alexander Korda's 1936 film *Things to Come*) and in his last doom-laden book, *Mind at the End of Its Tether*. His science-fiction novels have inspired a variety of film and stage versions, and his tale of Martian invaders, *The War of the Worlds*, was made by Orson Welles into a now notorious radio play broadcast in 1947.

Fiction

Select Conversations with an Uncle, Now Extinct and Two Other Reminiscences **(s)** 1895; *The Stolen Bacillus* **(s)** 1895; *The Time Machine* 1895; *The Wonderful Visit* 1895; *The Island of Dr Moreau* 1896; *The Wheels of Chance* 1896; *The Invisible Man* 1897; *The Plattner Story and Others* **(s)** 1897; *Thirty Strange Stories* **(s)** 1897; *The War of the Worlds* 1898; *When the Sleeper Wakes* 1898; *Tales of Space and Time* **(s)** 1899; *Love and Mr Lewisham* 1900; *The First Men in the Moon* 1901; *The Sea Lady* 1902; *Twelve Stories and a Dream* **(s)** 1903; *The Food of the Gods* 1904; *Kipps* 1905; *A Modern Utopia* 1905; *In the Days of the Comet* 1906; *The War in the Air* 1908; *Ann Veronica* 1909; *Tono-Bungay* 1909; *The History of Mr Polly* 1910; *The Country of the Blind* **(s)** 1911; *The Door in the Wall* **(s)** 1911; *The New Machiavelli* 1911; *Marriage* 1912; *The Passionate Friends* 1913; *The Wife of Sir Isaac Harman* 1914; *The World Set Free* 1914; *Bealby* 1915; *The Research Magnificent* 1915; *Mr Britling Sees It Through* 1916; *The Soul of a Bishop* 1917; *Joan and Peter* 1918; *The Undying Fire* 1919; *The Secret Places of the Heart* 1922; *Men Like Gods*

1923; *The Dream* 1924; *Christina Alberta's Father* 1925; *The World of William Clissold* 1926; *The Complete Short Stories of H.G. Wells* (s) 1927; *Meanwhile: the Picture of a Lady* 1927; *The Autocracy of Mr Parnham* 1930; *The Bulpington of Blup* 1932; *Brynhild* 1937; *The Camford Visitation* 1937; *Star Begotten* 1937; *Apropos of Dolores* 1938; *The Holy Terror* 1939; *Babes in the Darkling Wood* 1940; *All Aboard for Arafat* 1940; *You Can't Be Too Careful* 1941

Non-Fiction

Honours Physiology [with R.A. Gregory] 1893; *Textbook of Biology* 1893; *Certain Personal Matters* 1897; *Anticipations of the Reaction of Mechanical and Scientific Progress upon Human Life and Thought* 1901; *The Discovery of the Future* 1902; *Mankind in the Making* 1903; *The Future in America* 1906; *Socialism and the Family* 1906; *Will Socialism Destroy the Home?* 1907; *This Misery of Boots* 1907; *New Worlds for Old* 1908; *First and Last Things* 1908; *Floor Games* 1911; *Liberalism and Its Party* 1913; *Little Wars* 1913; *An Englishman Looks at the World* 1914; *The War That Will End War* 1914; *Boon* 1915; *What Is Coming* 1916; *The Elements of Reconstruction* 1916; *War and the Future* 1917; *God the Invisible King* 1917; *In the Fourth Year* 1918; *The Idea of a League of Nations* 1919; *The Way to a League of Nations* 1919; *History Is One* 1919; *The Outline of History* 1920; *Russian in the Shadows* 1920; *The New Teaching of History* 1921; *The Salvaging of Civilisation* 1921; *Washington and the Hope of Peace* 1922; *A Short History of the World* 1922; *Socialism and the Scientific Motive* 1923; *The Story of a Great Schoolmaster* 1924; *A Year of Prophesying* 1924; *Mr Belloc Objects to the Outline of History* 1926; *Democracy under Revision* 1927; *Playing at Peace* 1927; *The Book of Catherine Wells* 1928; *Mr Blettsworthy on Rampole Island* 1928; *The Open Conspiracy* 1928; *The Way the World Is Going* 1928; *The Adventures of Tommy* 1929; *The Common Sense of World Peace* 1929; *Imperialism and the Open Conspiracy* 1929; *The Science of Life* [with J. Huxley, G.P. Wells] 1930; *The Way to World Peace* 1930; *What Are We Going to Do with Our Lives?* 1931; *The Work, Wealth and Happiness of Mankind* 1931; *After Democracy* 1932; *The Shape of Things to Come 1933; Experiment in Autobiography* (a) 1934; *Stalin–Wells Talk* 1934; *The New America: the New World* 1935; *The Anatomy of Frustration* (a) 1936; *The Croquet Player* 1936; *The Brothers* 1938; *World Brain* 1938; *The Fate of Homo Sapiens* 1939; *The New World Order* 1939; *In Search of Hot Water* 1939; *The Rights of Man* 1940; *The Common Sense of War and Peace* 1940; *The Conquest of Time* 1942; *The New Rights of Man* 1942; *Phoenix* 1942; *Science and the World Mind* 1942; *Crux Ansat* 1943; *'42 to '44* 1944; *The Happy Turning* 1945; *Mind at the End of Its Tether* (a) 1945; *Henry James and H.G. Wells* [ed. L. Edel, G.N. Ray] 1958; *Arnold Bennett and H.G. Wells* [ed. H. Wilson] 1960; *George Gissing and H.G. Wells* [ed. R.A. Gettmann] 1961

Biography

The Time Traveller by Jeanne and Norman Mackenzie 1973; *H.G. Wells and Rebecca West* by Gordon N. Ray 1974; *H.G. Wells* by Anthony West 1984; *H.G. Wells and Rebecca West* by J.R. Hammond 1991

Eudora (Alice) WELTY 1909–

The most distinguished surviving writer of the Southern Renaissance, Welty was born in Jackson, Mississippi, where she has lived almost all her life in the house built in mock-Tudor style for her father, the president of a life insurance company, in the 1920s. He was of German-Swiss, her mother (an active Virginian) of British and French Huguenot ancestry. Welty's childhood was happy, and after attending public school in Jackson she studied at the Mississippi State College for Women between 1925 and 1927, then the University of Wisconsin, and finally the Columbia University School of Business from 1930 to 1931, where she took an advertising course. She then worked briefly on the staff of the *New York Times Book Review* before returning to Mississippi. In 1933 she took a job as publicity agent for the State Office of the Works Progress Administration (part of the New Deal), under whose auspices she travelled widely around the South, recording what she saw in a series of photographs which were published in 1971 under the title *One Time, One Place.*

She had a little initial difficulty in getting her literary work accepted, but her stories began to appear from the later 1930s in magazines, and a collection of them was published as *A Curtain of Green* in 1941. It was followed in 1942 by *The Robber Bridegroom*, a mixture of fairy-tale and ballad, using a traditional Mississippi theme of the wooing of the daughter of a plantation owner. In 1975 the book was turned into a musical. Welty's work consists largely of short stories, and of her novels the majority are really novellas. Perhaps *Delta Wedding* and *The Optimist's Daughter*, which won a Pulitzer Prize in 1973, are outstanding. She is a writer rooted in her native region, but also fascinated by the larger theme of the individual's struggle to retain his or her own values, sometimes against the pervasive and intrusive Southern family. Despite accusations in the radical 1960s that she was an apologist for an outdated culture, Welty does not in fact glorify either the ante-bellum or the post-bellum

South. Her vision is essentially a comic and subtle one, and by the time her *Collected Stories* appeared her work was well recognised and understood.

In later life she has taught at various colleges, and received many honours, but has continued to live an essentially private life in Jackson. Her other work is varied, and includes an autobiography of childhood, a collection of interviews, a volume of collected book reviews, and her co-editorship of *The Norton Book of Friendship*. Most of these have appeared since her active fiction-writing ceased in the early 1970s.

Fiction

A Curtain of Green and Other Stories (s) 1941; *The Robber Bridegroom* 1942; *The Wide Net and Other Stories* (s) 1943; *Delta Wedding* 1946; *The Golden Apples* (s) 1949; *The Ponder Heart* 1954; *The Bride of Innisfallen and Other Stories* (s) 1955; *The Shoe Bird* 1964; *Thirteen Stories* (s) 1965; *Losing Battles* 1970; *The Optimist's Daughter* 1972; *The Eye of the Storm* 1978; *Collected Stories of Eudora Welty* (s) 1982

Non-Fiction

Place in Fiction 1957; *Three Papers on Fiction* 1962; *One Time, One Place: Mississippi in the Depression* 1971; *Conversations with Eudora Welty* 1984; *Eudora* (a) 1984; *One Writer's Beginnings* 1984; *Eye of the Story: Selected Essays & Reviews* 1990; *A Writer's Eye: Collected Book Reviews* 1994

Edited

The Norton Book of Friendship [with R.A. Sharp] 1991

Glenway (Gordon) WESCOTT 1901–1987

Wescott was born, the son of a poor farmer, near Kewaskum, Wisconsin, the state whose reluctant laureate he became in his earlier fiction. His childhood was marked by conflict with his father and attachment to his mother, and while attending high school he lived with an uncle and others to escape his home. In 1917 he entered the University of Chicago and, although he left after a year, he had become president of the Poetry Society and mixed with his literary peers.

In 1919 he met Monroe Wheeler, who was to publish much of his work, and who was also his lover until both men died in their late eighties. (Theirs was not an exclusive relationship: for some years they lived in a *ménage à trois* with a lover of Wheeler, the photographer George Platt Lynes, while Wescott had attachments of his own.) After a brief spell as a department-store clerk, Wescott, accompanied by Wheeler, embarked on a six-year period, between 1919 and 1925, when they lived restlessly between New Mexico, the Berkshires, New York, England and Germany; then, until 1933, they lived mainly in France. The travelling lifestyle was financed by private patrons and secretary-companion jobs (to the critic and poet Yvor Winters and the family of the singer Elena Gerhardt, for example), while, when they were in New York, **Marianne Moore** brought home-made soup to the pair's basement.

Wescott had begun by publishing Imagist poetry, but during the 1920s he made a reputation not much inferior to **F. Scott Fitzgerald**'s with two novels and a volume of short stories set in a traditional Wisconsin, viewed with both love and detachment, and in which the habitual *alter ego* of his fiction, the writer Alwyn Tower, begins to appear. **Ernest Hemingway** draws an unsympathetic picture of Wescott at this period as Roger Prentiss in *The Sun Also Rises* (1926). About the time of Wescott's return to America, the California millionairess Barbara Harrison (who had helped Wheeler in his publishing ventures) married Wescott's brother, and for many years the entire clan, sometimes including George Platt Lynes, lived between New York apartments and two New Jersey farmhouses. In the 1930s Wescott published no fiction and rather desultory non-fiction, but the early 1940s saw him publish his two remaining novels: *The Pilgrim Hawk*, an elegant and subtle meditation on the relation between the artist and his subject-matter which is generally considered his major work, and *Apartment in Athens*, a slighter story about the Nazi occupation of Greece.

His subsequent forty-year silence was alleviated by many writerly public works and much social life. (Wheeler was publications director for the Museum of Modern Art, a position which placed the two men near the centre of New York society.) In his last years Wescott retired entirely to New Jersey, although Wheeler continued energetically to travel. Wescott's only major publication in later years was the eccentric volume of essays, *Images of Truth*, and his reputation fell into abeyance. It may, however, be revived by the publication of his journals, *Continual Lessons*.

Fiction

The Apple of the Eye 1924; *... Like a Lover* (s) 1926; *The Grandmothers* 1927; *Good-Bye, Wisconsin* (s) 1928; *The Babe's Bed* 1930; *The Pilgrim Hawk* 1940; *Apartment in Athens* 1945

Non-Fiction

Fear and Trembling 1932; *A Calendar of Saints for Unbelievers* 1932; *Images of Truth: Remembrances and Criticism* 1962; *The Best of All*

Possible Worlds: Journals, Letters and Remembrances 1914–1937 1975; *Continual Lessons: The Journals of Glenway Wescott, 1937–1955* [ed. R. Phelps] 1990

Plays
The Best One-Act Plays of 1940 [with others; M. Mayorga] 1941

Poetry
The Bitterns 1920; *Natives of Rock: XX Poems 1921–1922* 1925

Edited
The Maugham Reader 1950; *Short Novels of Colette* 1951

Arnold WESKER 1932–

Wesker was born in East London, the son of a Russian Jewish immigrant tailor who married the daughter of a Hungarian Jewish family. During the Second World War he was evacuated to five different places in England and Wales. He returned to London in 1943 to learn furniture-making at Upton House School, Hackney, where he became interested in acting. Accepted by the Royal Academy of Dramatic Art in 1948, but unable to take his place due to the lack of a grant, he worked as a bookseller, a carpenter's mate and a seed-sorter, until called up to do his national service in the Royal Air Force in 1950. After demobilisation in 1952, he worked again at odd jobs in Norfolk, where he met his future wife.

After training as a pastrycook in London, Wesker worked in a Paris kitchen for nine months to earn the fee to attend the London School of Film Technique. His ambition to make films altered when he entered a playwriting competition in the *Observer* with *The Kitchen*. Further encouraged by a performance of **John Osborne**'s *Look Back in Anger* in 1956, he immediately afterwards wrote *Chicken Soup with Barley*, which won him an Arts Council Award and enabled him to marry in 1958. The first of the *Wesker Trilogy* (which was completed by *Roots* and *I'm Talking about Jerusalem*), it earned the praise of Kenneth Tynan, who declared that its author was 'potentially a very important playwright'. The term 'kitchen-sink drama' – originally derogatory, but subsequently merely descriptive – was coined to describe Wesker's work, and to differentiate it and its kind from the customary plays of West End theatre, which had middle-class settings and themes.

In 1961, Wesker, in the company of other writers, 'incited civil disobedience' against the use of nuclear weapons, and was imprisoned for one month. On his release, he accepted the directorship of Centre 42, a movement based in The Roundhouse, London, which aimed to make the arts more widely accessible through trade union support. Wesker temporarily gave up playwriting to concentrate on this, but finally dissolved the Centre in 1970.

The success of his early plays prejudiced the reception of his later work. British audiences were ambivalent about *Their Very Own and Golden City*, a visionary play set in the future, although it won the Italian Premio Marzotto Prize. Wesker has also written short stories, including *Love Letters on Blue Paper*, which he adapted as a television play. *Caritas*, a chamber opera with music by Robert Saxton and a libretto by Wesker adapted from his grim 1981 play based upon the life of a thirteenth-century anchoress, was commissioned to open the 1991 Huddersfield Contemporary Music Festival.

Plays
The Wesker Trilogy 1960: *Chicken Soup with Barley* 1958 [pub. 1959], *Roots* 1959, *I'm Talking about Jerusalem* 1960; *The Kitchen* **(st)** 1959, **(f)** 1961 [pub. 1960]; *Chips with Everything* 1962; *Their Very Own and Golden City* 1965 [pub. 1966]; *The Four Seasons* 1965 [pub. 1966]; *The Friends* 1970; *The Old Ones* 1972 [pub. 1973]; *The Wedding Feast* 1974 and *The Merchant* 1976 [pub. in *Wesker, 4* 1981]; *The Journalists* 1981 [pub. 1975]; *Love Letters on Blue Paper* [from his story] **(tv)** 1975, **(st)** 1977 [pub. 1978]; *Caritas* **(st)** 1981, **(l)** [with R. Saxton] 1991 [pub. 1981]

Fiction
Six Sundays in January **(s)** 1971; *Love Letters on Blue Paper* 1974; *Said the Old Man to the Young Man* **(s)** 1978; *Love Letters on Blue Paper and Other Stories* **(s)** 1980

Non-Fiction
Fears of Fragmentation 1970; *As Much as I Dare: an Autobiography 1932–1959* **(a)** 1994

For Children
Fatlips 1978

Collections
The Journalists: a Triptych 1979

Mary WESLEY (pseud. of Mary Wellesley) 1912–

Descended from the Duke of Wellington's eldest brother, Wesley comes from an old army family, the daughter of a colonel and granddaughter of a general. She grew up 'relatively parentless', and was largely brought up by governesses, although she was close to her grandmother, Lady Dalby, who died when Wesley was thirteen. She married in 1937 in order to escape her home life. Her first husband was a barrister ten years her senior, with whom she settled in London. At the outbreak of the Second World War she was, as she put it, 'roped into intelligence' and worked at Bletchley Park. In 1941 she moved to Cornwall with her two children, leaving her husband in London; they never lived together again. In the autumn of 1944 she met the journalist Eric Siepman. He was married, but had lost contact with his estranged wife and could not find her in order to divorce her. He and Wesley finally married in 1952, had a son, and lived together in straitened circumstances (he was a great 'walker out of jobs') until Siepman's death from Parkinson's disease in 1970. By this time, Wesley was living on social security, and survived on a widow's pension through the next decade.

In the late 1960s she had published two books for children. Her first book for adults, *Jumping the Queue*, was published in 1983 when she was seventy, and inaugurated an extraordinary late flowering which led to her being one of Britain's most popular novelists. *Jumping the Queue* is a black comedy about a widow who is foiled in a suicide attempt and embarks on a new relationship with a man wanted for murder. It was followed a year later by *The Camomile Lawn*, which made her reputation, and remains perhaps her best book. Based on a family of five young cousins, the novel follows their tangled relationships during the Second World War, moving forward to view the survivors in old age as they congregate for a funeral. Wesley captures the atmosphere of the home front, where the dislocations forced by the war and the sense of imminent death sharpen the senses and loosen inhibitions. It is a light, sexy, witty and moving story with distinctly dark undertones, and was made into a classy television series (directed by Peter Hall), which caused considerable controversy because of the language and sex scenes. Part of Wesley's appeal is that this rather grand-sounding elderly woman should be writing in so uninhibited a fashion. She replies that she has 'lived for a hell of a long time and I've noticed and heard things that I've been able to put down on paper'.

If subsequent novels have not been quite up to the standard of *The Camomile Lawn*, they have nevertheless attracted their admirers and a growing readership. Several of the books have been adapted for television, not always successfully: the 1995 adaptation of *The Vacillations of Poppy Carew*, although decorative and well acted, could make little of the story of a young woman attempting to choose between several men. Wesley, who converted to Roman Catholicism in the 1960s and gets some of her best ideas while at Mass, is aware that time is against her: 'I'm much nearer to death than anything else,' she has said. 'Each time I start a book, I pray I'll live to finish it.'

Fiction

Jumping the Queue 1983; *The Camomile Lawn* 1984; *Harnessing Peacocks* 1985; *The Vacillations of Poppy Carew* 1986; *Not That Sort of Girl* 1987; *Second Fiddle* 1988; *A Sensible Life* 1990; *A Dubious Legacy* 1992; *Three Novels* 1992; *Imaginative Experience* 1994

For Children

The Sixth Seal 1969; *Speaking Terms* 1969; *Haphazard House* 1983; *Magic Landscapes* 1991

Nathanael WEST (pseud. of Nathan Wallenstein Weinstein) 1903–1940

West was born in New York, the eldest child of Russian Jewish immigrants – his father worked in construction – and was educated at the De Witt Clinton High School, where he showed a keen talent for drawing but none for writing. He left without graduating, but forged his school record in order to get into Tuft University, Massachusetts, which he was advised to leave after two months. Fortunately, there was another, more industrious, student at Tufts called Nathan Weinstein and somehow the records of these namesakes were 'confused', leaving West with sufficient credits to enter Brown University. (It is unclear whether West himself falsified the records, or whether a besotted secretary did so on his behalf.) At Brown he was a conspicuous success, overcoming the University's casual anti-Semitism, and charming his fellow students with his drawing, playing the banjo and introducing new dances. He did not mix with his fellow Jews, and was always uneasy about his racial background.

He graduated in 1924 and went to Paris where, influenced by Surrealism and Dada, he wrote his first book, *The Dream Life of Balso Snell*. He returned to New York in 1926 to become assistant manager of a hotel (a career he later gave to Homer Simpson in *The Day of the Locust*). *Balso Snell* attracted little attention when published in 1931, but by this time West was working on *Miss Lonelyhearts*, a dark fable about an agony columnist. This was published in 1933 to great acclaim, but the publisher went bankrupt and only a couple of hundred copies of the book reached the shops. Twentieth Century Fox bought the film rights, however, and brought West to Hollywood as a scriptwriter. The previous year West had edited, with **William Carlos Williams**, three issues of *Contact* magazine, and in 1933 he was assistant editor of *Americana*.

His characteristically macabre novella *A Cool Million* failed to achieve the critical success of *Miss Lonelyhearts*. He was unable to sell his short stories, and in 1935 he joined Republic Studios, working mostly on unmemorable films made from scripts to which several hack writers had contributed. Hollywood did, however, provide him with the setting of his last great novel, the apocalyptic *The Day of the Locust*, which demonstrated West's command of pictorial and cinematic techniques. The novel enjoyed mixed reviews and disappointing sales. His forays into the theatre were still less successful: *Even Stephen*, a play he wrote with S.J. Perelman (who had married West's adored sister, Laura), failed to reach the stage, while *Good Hunting*, written with Joseph Shrank, was produced on Broadway but taken off after only two days.

In April 1940 he married Eileen McKenny, who had featured in the famous *New Yorker* stories of her sister Ruth McKenny (collected in 1938 as *My Sister Eileen*, and the basis of Leonard Bernstein's musical *Wonderful Town*). Eight months later, on 22 December, the Wests were returning from a hunting trip in Mexico. Pulling onto a highway, West failed to notice a stop sign and crashed into another car. His pointer bitch survived, but both he and Eileen died of head injuries.

In spite of West's personal distance from his roots, emphasised by his change of name, his work often has Jewish themes and characters: 'Miss Lonelyhearts' is a scapegoat figure, while Pitkin is the ultimate *schlemiel*. He was an insatiable reader of European literature, and his sensibility – in particular his taste for the grotesque – led to his being undervalued by his contemporary American audience.

Fiction

The Dream Life of Balso Snell 1931; *Miss Lonelyhearts* 1933; *A Cool Million or, The Dismantling of Lemuel Pitkin* 1934; *The Day of the Locust* 1939

Plays

Good Hunting [with J. Shrank] 1932

Biography

Nathanael West by Jay Martin 1970

Rebecca WEST (pseud. of Cicily Isabel Fairfield) 1892–1983

West was the daughter of Charles Fairfield, an Irish journalist, unsuccessful entrepreneur and soldier-of-fortune, as well as a well-known defender of extreme individualism, whose abandonment of his family in 1901 was quickly followed by his death, leaving them in straitened circumstances. West's mother took her to her native Edinburgh, where she was educated, and which she left at eighteen to train for the stage at the Royal Academy of Dramatic Art in London. After a year, she became a journalist instead, joining the staff of *The Freewoman* and contributing to many other radical and feminist papers. At about this time she adopted her pseudonym, the name of the strange and dominant heroine of Ibsen's *Rosmersholm* (1886).

Initially influenced by the Pankhursts, she soon moved further to the left, and was one of those who insisted that suffrage on its own would not solve women's problems. In 1912 she wrote an outspoken criticism of **H.G. Wells**'s *Marriage*, which led to a meeting. She was the very type of the liberated young woman he had dealt with in his novel *Ann Veronica* (1909), and she became, for ten years, the most celebrated of his many mistresses (her own liaisons were almost equally numerous). The relationship between Wells and West was often stormy, but she bore him a son, the novelist Anthony West, with whom she did not get on in later years, and who, after his parents' deaths, wrote bitter memoirs of both of them.

West's first book was a study of **Henry James**, published in 1916, and her first novel, *The Return of the Soldier*, appeared in 1918. Her novels abound in strong female characters and weaker, self-indulgent male ones. Her vision is tragic, sometimes, as in *The Judge*, melodramatic. For many years she was more highly regarded as a travel writer, particularly for her account of what was then Yugoslavia, *Black Lamb and Grey Falcon*. She had travelled extensively in that country in the 1930s with her banker husband, Henry Andrews, whom she had rather surprisingly married in 1930 (he died in 1968). She was equally well known as a journalist, and once again achieved renown with *The Meaning of Treason* in 1949, based on her articles in the *New Yorker* about the trial of William Joyce and other British traitors of the Second World War. (In this interest in trial reporting and travel, combined with a novelistic career, she resembles another redoubtable writer, **Sybille Bedford**.)

In later life West published what are usually regarded as her two best novels, *The Fountain Overflows* and *The Birds Fall Down*, both rich and detailed accounts of growing up before the First World War. For twenty years she struggled with another novel, which was eventually published posthumously as the title-piece of *The Only Poet and Other Stories*. West retained her formidable energy into all but the very last of her ninety-one years, and since her death a revival of interest in women writers of the early part of the century has maintained her reputation, bringing several unpublished works to the attention of the public.

Fiction

The Return of the Soldier 1918; *The Judge* 1922; *Harriet Hume* 1929; *The Harsh Voice* (s) 1935; *The Thinking Reed* 1936; *Cousin Rosamund: The Fountain Overflows* 1956, *This Real Night* 1984, *Cousin Rosamund* [unfinished] 1985; *The Birds Fall Down* 1966; *Sunflower* 1986; *This Real Night* 1987; *The Only Poet and Other Stories* (s) 1992

Non-Fiction

Henry James 1916; *1900* 1928; *The Strange Necessity* 1928; *Lions and Lambs* [as 'Lynx'] 1929; *Elegy: an In Memoriam Tribute to D.H. Lawrence* 1930; *Ending in Earnest: a Literary Log* 1931; *A Letter to a Grandfather* 1933; *St Augustine* 1933; *The Modern 'Rake's Progress'* 1934; *Black Lamb and Grey Falcon* 1942; *The Meaning of Treason* 1949 [rev. as *The New Meaning of Treason* 1964]; *A Train of Powder* 1955; *The Court and the Castle: Some Treatments of a Recurrent Theme* 1957; *The Vassall Affair* 1963; *The Young Rebecca: Writings of Rebecca West 1911–1917* 1982; *Family Memories* 1987

Biography

H.G. Wells and Rebecca West by Gordon N. Ray 1974; *Rebecca West* by Victoria Glendinning 1987; *H.G. Wells and Rebecca West* by J.R. Hammond 1991

Edith (Newbold) WHARTON 1862–1937

Wharton was born Edith Newbold Jones in New York into a family of the American aristocracy (her great-grandfather was a general in the American War of Independence), and grew up, largely tended by governesses and tutors, in New York, Newport and Paris. She married Edward Robbins Wharton, who came from an old Virginia family, in 1885, and the couple subsequently spent a great deal of time travelling for amusement in Europe, settling in France in 1910.

The marriage was unhappy: he suffered from mental illness and she from nervous complaints, and to gain relief she turned to writing. Her first book was a volume of *Verses* in 1878, but her poetry is now forgotten. In 1897 she published a book on interior decoration (and she would continue to write the occasional volume of non-fiction), but she made her literary reputation with short stories published in *Scribners' Magazine*. Her first collection, *The Greater Inclination*, was published in 1899, and her first novel, *The Touchstone* in 1900. Her first popular success came five years later with *The House of Mirth*, which sold 100,000 copies within ten days of publication. Part of the reason for this was that Wharton's frank portrait of the sexual hypocrisies of high New York society was considered scandalous. It announced her principal themes and inaugurated a distinguished career as a witty chronicler of what was known as 'the old New York'. Amongst its admired successors were *The Custom of the Country*, in which a spoilt young woman from a *nouveau riche* background achieves her social ambitions but is never satisfied, and the uncharacteristic *Ethan Frome*, a stark New England tragedy. *The Age of Innocence*, which is set in the past of 1872, and describes a man torn between his young fiancée and her disgraced but fascinating cousin, is arguably her finest novel and won her a Pulitzer Prize in 1921. Her work is admired for its satirical humour and realism, its sense of the individual's struggle against the collective ethos, and its observant account of social change.

Wharton divorced her husband in 1913, and had another long-term, fundamentally unsatisfactory relationship with Walter Berny. Her one brief, fully sexual relationship was with Morgan Fullerton, and this, presumably, was the inspiration for 'Beatrice Palmato', an explicitly erotic short story about the incestuous relationship between a father and his daughter, one fragment of which was discovered amongst Wharton's papers after her death. During the First World War, her work with Belgian refugees fleeing the German advance earned her the Légion d'Honneur from the French government: she took charge of an entire orphanage of 600 children, organised a workshop for unemployed women, and ran charitable restaurants, feeding the dispossessed at below cost.

She spent her later years as a *doyenne* of European cosmopolitan society, passing the summers at her Villa Colombe at St Brice, near Paris, and wintering at Hyères on the Riviera. Described as having 'a finished manner and air', Wharton could be formidable – her great friend and mentor **Henry James** sometimes referred to her as 'the Angel of Desolation' – although she was fundamentally kind. Her literary reputation slumped after her death, but has since revived in tandem with that of James, with whose work her own bears some affinity of tone and theme. Her work has begun to attract film-makers, and Martin Scorsese's intelligent and sumptuous adaptation of *The Age of Innocence* (1993) won her many new admirers. Her final, incomplete novel, *The Buccaneers*, which describes the experiences of a group of young American women let loose amongst the British aristocracy, was controversially and crudely adapted for a BBC television series in 1995.

Fiction

The Greater Inclination (s) 1899; *The Touchstone* 1900; *Crucial Instances* (s) 1901; *The Valley of Decision* 2 vols 1902; *Sanctuary* 1903; *The Descent of Man and Other Stories* (s) 1904; *The House of Mirth* 1905; *The Fruit of the Tree* 1907; *Madame de Treymes* 1907; *The Hermit and the Wild Woman and Other Stories* (s) 1908; *Tales of Men and Ghosts* (s) 1910; *Ethan Frome* 1911; *The Reef* 1912; *The Custom of the Country* 1913; *Xingu and Other Stories* (s) 1916; *Summer* 1917; *The Marne* 1918; *The Age of Innocence* 1920; *The Glimpses of the Moon* 1922; *A Son at the Front* 1923; *Old New York* 4 vols 1924; *The Mother's Recompense* 1925; *Here and Beyond* (s) 1926; *Twilight Sleep* 1927; *The Children* 1928; *Hudson River Bracketed* 1929; *Certain People* (s) 1930; *The Gods Arrive* 1932; *Human Nature* (s) 1933; *The World Over* (s) 1936; *Ghosts* (s) 1937 [as *The Ghost Stories of Edith Wharton* 1973]; *The Buccaneers* [unfinished] 1938 [rev. by M. Mainwaring 1993]; *The Stories of Edith Wharton* (s) [ed. A. Brookner] 2 vols 1988–9

Poetry

Verses 1878; *Artemis to Actaeon and Other Verse* 1909; *Twelve Poems* 1926

Non-Fiction

The Decoration of Houses [with O. Codman Jr] 1897; *Italian Villas and Their Gardens* 1904; *Italian Backgrounds* 1905; *A Motor-Flight through France* 1908; *Fighting France, from Dunkerque to Belfort* 1915; *French Ways and Their Meaning* 1919; *In Morocco* 1920; *The Writing of Fiction* 1925; *A Backward Glance* (a) 1934

Translations

Hermann Sudermann *The Joy of Living* 1902

Edited

The Book of the Homeless 1916

Biography

Edith Wharton by Shari Benstock 1994

William WHARTON (pseud.) 1925–

'William Wharton' is the pseudonym of a writer and painter who lives in Paris. Unlike certain other publicity-shy American writers, he seems to enjoy the fuss surrounding his well-guarded anonymity. He willingly gives interviews and is happy for the back of his head to be photographed. One interviewer has remarked that his skull may yet become more famous than that of the Piltdown Man. Wharton believes that 'notoriety destroys most twentieth-century creative people'.

Born to working-class parents in Philadelphia, Wharton graduated from Upper Darby High School in 1943 and was immediately drafted into the army. Serving in Germany he rose to the rank of captain and was seriously injured by shrapnel in 1945. In Paris he underwent plastic surgery for facial injuries, and has worn a beard ever since. On his return to the USA, he studied painting at the University of California, Los Angeles, and later received a doctorate in psychology. He married in 1949, and worked as a teacher throughout the 1950s to support his family of four children. With the proceeds from renovating and letting two properties in Los Angeles, Wharton moved in 1961 to Paris to pursue a career as a painter. He had successful exhibitions in New York in 1963 and in London, New York and Munich in 1968, but now only sells his work privately.

Wharton had written since his discharge from the army, partly as a therapeutic activity, but had never submitted a manuscript to a publisher. A friend showed the manuscript of *Birdy* to Knopf in 1977 and it was published two years later. The novel was hugely successful, won a National Book Award in 1979, and (updated to the Vietnam War) was made into a distinguished film by Alan Parker in 1984. Birdy is a war veteran traumatised by his experience of combat who is now in an asylum where he comes to believe that he is a bird. Wharton has bred canaries since childhood, as does the character in the novel.

His subsequent novels contain considerable elements of autobiography. *A Midnight Clear* exploits further his experience of the war in France, and *Scumbler* describes the life of an American painter struggling to support his family in Paris. *Dad*, dedicated 'To the women in my life: Mother, sister, daughters, wife', investigates the relationships between fathers and sons through three generations of one family. Wharton's daughter Kate and her family were killed in a car crash in 1988 evidently caused by stubble-burning, and *Franky Furbo* contains a moving dedication to the family. *Last Lovers*, which some reviewers found both sentimental and misogynistic, depicts an affair between an American businessman who abandons his job and his family to become a painter, and a blind, seventy-one-year-old French virgin.

Fiction

Birdy 1979; *Dad* 1981; *A Midnight Clear* 1982; *Scumbler* 1984; *Pride* 1985; *Tidings* 1988; *Franky Furbo* 1989; *Last Lovers* 1991; *Wrongful Deaths* 1994

Malachi (pseud. of Marjorie Olive) WHITAKER 1895–1975

The eighth of eleven children, Whitaker was born Marjorie Taylor in Bradford. Her father was a bookbinder, and she was already a voracious and wide-ranging reader by the age of four. The fruits of her father's craft proved invaluable; the man himself was less cherished and her autobiography contains chilling passages which reveal the coldly corrosive dislike long felt by the daughter for a father who 'was the sort of man who occasionally took us to a theatre for a birthday treat when we were young, and made us come out before the end'. From her mother came the habit of language which startles and delights with its mixture of precision and vivid idiosyncrasy ('she ... has called me a variegated flat fish more than once') a love of walking and wandering. Like her daughter in later life, she engaged for years in a slightly sheepish quest for the God she knew was somewhere, though somehow she never quite found him.

Hating school, though loving the three-mile walk to it, full of the sounds, sights and smells of an uniquely Yorkshire landscape which later provided both the physical and emotional setting for much of her work, the young Whitaker 'was forced to make a world of [her] own to get along at all'. At home she was conscientiously naughty, thereby securing the 'punishment' of early bed, which guaranteed reading time uninterrupted by household chores and babyminding, and also the opportunity to draft the remarkable stream of words she produced throughout childhood and adolescence. By the time she was twenty she had already written a 50,000-word autobiography and a novel whose pages were later blown overboard into the Channel. With the outbreak of war in 1914 she supplemented her meagre wages by writing 'windy martial verses' for a Christmas card manufacturer. In 1917 she married Leonard Whitaker, then serving in the army. After demobilisation they lived for a spell in Rouen, where he busied himself with various entrepreneurial activities. They experienced early in their marriage dramatic swings between poverty and affluence, a pattern which continued, and which she, at least, rather enjoyed.

The lost novel proved to be the last thing Whitaker wrote for some years. Then, to her utter astonishment, one day in 1926 or 1927 ('I am not sure of the year') she wrote a story at one sitting. Unlike her adolescent productions, which imitated the work of other writers, this story, 'Sultan Jekker', imitated no one. She sent it to the *Adelphi* magazine, mainly because she admired its chief contributor, **D.H. Lawrence**. The editor, John Middleton Murry, gave the story a fittingly enthusiastic reception (Whitaker's work was later often compared to that of Murry's wife, **Katherine Mansfield**, who had died almost four years earlier). In the eighteen months following September 1927 the *Adelphi* published five of her stories. Moreover, Murry helped to secure publication in 1929 of her first collection, *Frost in April*. Her first reviewer, Humbert Wolfe, was cautious, finding her work 'like a piece of fog cut out and preserved', a judgement which pleased her. The book was warmly received, and praise from **Arnold Bennett**, the period's most powerful reviewer, clinched the matter.

In the decade following *Frost in April* Whitaker produced three more volumes of short stories and an autobiography, all of which won plaudits. She was constantly compared to Mansfield, and to Chekhov, which gratified but embarrassed her. The silent years, from 1920 to 1927, were in part spent waiting for children, who never came: 'The only physical result of our union was that I was left with the itch, which complaint was very rife in the army, though never mentioned in the popular novels of the period.' Childlessness, the 'itch', and the possible connection between the two forms yet another link with Mansfield, although the Whitakers, unlike the Murrys, were able to adopt children, a boy and a girl. They run like a gleaming thread of poignant hope through Whitaker's remarkable and perfect 1939 autobiography, *And So Did I*, its title taken from Coleridge's 'The Ancient Mariner': 'And a thousand thousand slimy things / Lived on; and so did I'. Written in the months preceding the outbreak of the Second World War, this deceptively 'simple' work, with its often *faux-naif* style, conversational tone and apparently rambling form, is in fact a skilfully ordered and meditative enquiry into those things which give human lives meaning, hope and purpose: a celebration and a statement of faith to set against the dark days surely coming.

It was to be virtually the last thing she wrote. She had, she told family and friends, written herself out, and would therefore stop. For her readers, those thirty-six silent years, between 1939 and her death at the age of eighty, raised a host of tantalising and ultimately frustrating questions.

Fiction

Frost in April (s) 1929; *No Luggage?* (s) 1930; *Five for Silver* (s) 1932; *Honeymoon and Other Stories* (s) 1934; *The Autobiography of Ethel Firebrace* [as 'anon.'; with G. Taylor] 1937; *The Crystal Fountain and Other Stories* (s) [ed. J. Hart] 1984

Non-Fiction

And So Did I (a) 1939

Antonia WHITE (pseud. of Eirene Botting) 1899–1980

White, who wrote under her mother's maiden name, was born Eirene Botting. Her father was Cecil Botting, a senior classics master at St Paul's School (and one half of Hillard and Botting, authors of the well-known Greek and Latin elementary textbooks). He dominated her early life, and converted to Catholicism when she was seven. She was sent to school at the Convent of the Sacred Heart, Roehampton, whence she was expelled, aged fifteen, for writing a novel in the style of Ouida. Her convent upbringing is described in *Frost in May* (1933), the first and most famous of her tetralogy of autobiographical novels.

She left St Paul's Girls' School early, against her father's wishes, and took many jobs, including that of actress in provincial repertory. Her first husband, whom she married in 1921, was alcoholic and impotent (their marriage is described in the third of the novels, *The Sugar House*) and the relationship led to the first of her attacks of severe mental illness, and her hospitalisation at the behest of her father. After her marriage was annulled, she became pregnant by a protégé of her father and had an abortion. She subsequently married a kindly homosexual, and in the later 1920s she worked as an advertising coywriter for Crawfords until sacked. She married her third husband, Tom Hopkinson, subsequently the editor of *Picture Post*, in 1930 and became a well-known journalist herself, being fashion editor of the *Daily Mirror* for a period. The late 1930s saw further serious mental illness and, in 1938, the annulment of her marriage, but during the war she worked at the BBC and in the Foreign Office. From 1945 onwards she supported herself as a freelance writer and journalist.

The last three novels of her tetralogy (in which Nanda Grey of the earlier novel becomes Clara Batchelor), are noted for unflinching self-analysis, and were written and published in quick succession in the early 1950s after a long period of writer's block. She also published a volume of short stories, *Strangers*, and an account in letters of her reconversion to Catholicism, *The Hound and the Falcon*. In addition she translated many novels from the French, notably works by Colette. She lived for many years in Kensington, London, but eventually died in a Sussex convent nursing home, leaving a note begging her two daughters' forgiveness. Her death was followed, in 1985, by a sustained literary controversy between the daughters, Susan Chitty and Lyndall Hopkinson, about their conflicting memoirs of their mother.

Fiction
Frost in May 1933; *The Lost Traveller* 1950; *The Sugar House* 1952; *Beyond the Glass* 1954; *Strangers* (s) 1954 [rev. 1981]

Translations
Maupassant *A Woman's Life* 1949; Colette: *The Shackle* 1951, *The Cat* 1953, *Claudine at School* 1956, *Claudine in Paris* 1958, *Claudine Married* 1960, *Claudine and Annie* 1962; *The Stories of Colette* (s) 1958, *The Innocent Libertine* 1968; H.C. Bordeaux *A Pathway to Heaven* 1952; A. Carrel *Reflections on Life* 1952; M. Duras *A Sea of Troubles* 1953; S. Groussard *A German Officer* 1955; P.A. Lesort: *The Wind Bloweth Where It Listeth* 1955, *The Branding Iron* 1958; Christine Arnothy *I Am Fifteen and I Don't Want to Die* 1956, *The Charlatan* 1959, *The Serpent's Bite* 1961, *The Captive Cardinal* 1964, *Those Who Wait* 1957, *It Is Not So Easy to Live* 1958; Fanny Rouget *The Swing* 1958; J.-M. Langlois-Berthelot [as 'J.M. Montguerre'] *Thou Shalt Love* 1958; L. Masson *The Tortoises* 1959, *The Whale's Tooth* 1963; E. Mahyère *I Will Not Serve* 1959; J. Storm *Till the Shadow Passes* 1960; A.E.A. Fabre-Luce *The Trial of Charles de Gaulle* 1963; P. Leulliette *St Michael and the Dragon* 1964; Thérèse de Saint Phalle *The Candle* 1968; *The Memoirs of the Chevalier d'Éon* 1970; G. Simenon *The Glass Cage* 1973; Paul-Gabriel Bouce *The Novels of Tobias Smollett* [with P.-G. Bouce] 1976; Voltaire *The History of Charles XII, King of Sweden* 1976

Plays
Three in a Room 1947

Non-Fiction
The Hound and the Falcon 1966; *As Once in May: the Early Autobiography of Antonia White* [ed. S. Chitty] 1983; *Diaries* [ed. S. Chitty] 2 vols 1991, 1992

For Children
Minka and Curdy 1957; *Living with Minka and Curdy* 1970

Biography
Now to My Mother by Susan Chitty 1985; *Nothing to Forgive: a Daughter's Life of Antonia White* by Lyndall Hopkinson 1988

Edmund (Valentine) WHITE 1940–

White was born in Cincinnati, though both his parents were Texans. His mother was a child psychologist and his father – whom White has described as 'a real misanthrope' – an engineer. Following his parents' divorce in 1947 he was raised by his mother in Texas and Chicago, and spent the summers with his father. He graduated from the University of Michigan in 1962 and moved in the same year to Greenwich Village, New York, where he worked as a staff writer in the book division of Time Inc. He was aware of being homosexual from a young age and his parents, regarding this as an illness, sent him to a number of 'insane psychiatrists', one of whom pronounced him 'unsalvageable'. In New York he explored the gay scene and was present at the Stonewall Tavern in 1969 on the occasion of the infamous police raid and subsequent riot which gave birth to the gay liberation movement.

White left Time in 1970 (though his introduction to paleolithic man, *The First Men*, was still to be published) and was an editor on the *Saturday Review* from 1972 to 1973. His first novel, *Forgetting Elena*, was rejected by twenty-two publishers before being published in 1973. The narrator is a victim of amnesia who wakes up in a cottage on an East Coast island resort and observes the snares and niceties of social etiquette as he tries to ascertain his own identity. The novel was highly praised by reviewers and by **Vladimir Nabokov**, a writer who has clearly influenced White. He continued to write while teaching creative writing at Johns Hopkins University, where he was associate professor from 1977 to 1979, and at Columbia, Yale and New York Universities. The successful publication of *A Boy's Own Story* in 1982, and the consequent award of a Guggenheim Fellowship, enabled him to move the following year to Paris, where he continues to live. The novel is a partially autobiographical *Bildungsroman*, though White has described himself as being considerably more sexually precocious than the rather diffident hero: 'I seduced a pianist in a bar in Acapulco – the Club de Pesca – when I was twelve.' *Caracole* is set in an imaginary city where a young man is caught up in a world of deceit, greed and vanity, while *The Beautiful Room Is Empty* is a sequel to *A Boy's Own Story* which takes the hero through his college years and ends with the Stonewall riot.

Initially White was resistant to the political function expected of a 'gay writer', fearful of becoming 'some sort of common denominator of public relations man to all gay people'. The onset of Aids and the resultant political backlash against the gay community has, however, made his work more consciously political. He is HIV-positive and has written about the effects of Aids in the short-story collection *The Darker Proof* a collaboration with **Adam Mars-Jones**. He has edited *The Faber Book of Gay Short Fiction*, and written an acclaimed biography of Jean Genet.

Fiction

Forgetting Elena 1973; *Nocturnes for the King of Naples* 1978; *A Boy's Own Story* 1982; *Caracole* 1985; *The Darker Proof* (s) [with A. Mars-Jones] 1987 [rev. 1988]; *The Beautiful Room Is Empty* 1988; *Skinned Alive* 1995

Non-Fiction

When Zeppelins Flew [with P. Wood] 1969; *The First Men* [with D. Browne] 1973; *The Joy of Gay Sex* [with C. Silverstein] 1977; *States of Desire: Travels in Gay America* 1980; *Genet: a Biography* 1993; *The Burning Library: Essays* [ed. D. Bergman] 1994

Edited

The Faber Book of Gay Short Fiction (s) 1991; *The Selected Writings of Jean Genet* 1993

Patrick (Victor Martindale) WHITE 1912–1990

The son of a wealthy Australian sheep-farmer, White was born in London during one of his family's periodic trips to Europe, and his earlier life was spent between England and Australia. He was sent from Australia to an English public school, Cheltenham College, and then returned to work for some years on two sheep stations in New South Wales, where he began to write novels. From 1932 to 1935 he read modern languages at King's College, Cambridge, and then spent some years in London, writing poetry, plays and novels: his first published volume was of verse, *The Ploughman* (1935). His first novel, *Happy Valley* (1939), established him as a young writer in a difficult modernist manner, although his work was to become stylistically more conservative from the 1950s onwards.

At the end of the 1930s he spent some time in America, and then enlisted as an intelligence officer in the Royal Air Force, serving during the war in North Africa, the Middle East and Greece, where he met his lifelong companion, Manoly Lascaris. After the war he returned with Lascaris to Australia, to a small property at Castle Hill near Sydney. White's first major novel was *The Aunt's Story*, an account of an independent Australian spinster, one of many lonely, solipsistic people in his fiction, searching for significance in their lives. Another major novel was *Voss*, the powerful and symbolic story of a German visionary leading an expedition into the Australian desert, while *Riders in the Chariot*, again a story of misfits and eccentrics, is the first of his novels to be set in his imaginary Sydney suburb, Sarsaparilla. White himself moved into Sydney in 1964, where he lived in the upmarket area of Centennial Park: there, after a long aloofness from public life, he became a strong supporter of Gough Whitlam's Labour government of 1972 to 1975.

Later works include *The Vivisector*, *The Twyborn Affair* and an autobiography, *Flaws in the Glass*: the latter two works confront his homosexuality, a theme not dealt with before. Although his works were often bitterly received in Australia, he was universally regarded as the major novelist of that country, and received a Nobel Prize in 1973. While his novels have been criticised for descent into rhetoric, their poetic and visionary qualities have won them an undisputed status.

The author of plays and short stories, as well as the novels for which he is best known, White described writing as a gift, a privilege and a burden. 'It's hell,' he once remarked. 'In another life I wouldn't like to be an artist.' His cantankerous nature was given full rein in *Patrick White Speaks*, an unrepentant final book, published in the year of his death, principally composed of invective aimed at late twentieth-century life. On his own instructions, news of his death was not made public until his funeral had taken place.

Fiction

Happy Valley 1939; *The Living and the Dead* 1941; *The Aunt's Story* 1948; *The Tree of Man* 1955; *Voss* 1957; *Riders in the Chariot* 1961; *The Burnt Ones* (s) 1964; *The Solid Mandala* 1966; *The Vivisector* 1970; *The Eye of the Storm* 1973; *The Cockatoos* (s) 1974; *A Fringe of Leaves* 1976; *The Twyborn Affair* 1979; *Three Uneasy Pieces* 1988

Plays

The Ham Funeral 1961; *The Season at Sarsaparilla* 1961; *A Cheery Soul* 1962; *Night On Bald Mountain* 1964; *Big Toys* 1978; *Netherwood* 1983; *Signal Driver* 1983

Poetry

The Ploughman and Other Poems 1935

Non-Fiction

Flaws in the Glass (a) 1981; *Patrick White Speaks* (a) 1990; *The Letters of Patrick White* [ed. D. Marr] 1995

Edited

Alex Xenophon Demirjian Gray *Memoirs of Many in One* 1986

Biography

Patrick White by David Marr 1991

T(erence) H(anbury) WHITE 1906–1964

Born in Bombay, the son of a district superintendant in the Indian police, White was the product of an unhappy marriage which broke up when he was a teenager – a background which may have contributed to later trauma. After early years in India, he was sent home to England for a conventional education in the army class of a public school, Cheltenham College. He then read English at Queen's College, Cambridge, and became a schoolmaster at another public school, Stowe.

He began to write in the school holidays, and published several volumes of fiction and verse; his first book was a volume of poems, *Loved Helen* (1929). Aged thirty, he resigned from Stowe on a capital of £100 and settled down in a cottage in a wood. *The Sword in the Stone*, the first volume of his Arthurian epic *The Once and Future King*, was published with great success, being chosen as Book of the Month in the USA. The tetralogy, which mixes fantasy, burlesque and feeling, is a great and still underestimated work of the imagination. (A fifth volume, discovered posthumously, was published as *The Book of Merlyn* in 1977.) Thereafter, White lived as a freelance writer, in Ireland during the war years and later in the Channel Islands. Other notable books were *Mistress Masham's Repose*, a classic book for children; *The Goshawk*, an account of how he trained a hawk; and *The Book of Beasts*, a translation from a Latin bestiary.

White was a bearded six-footer, devoted to blood sports and a succession of Irish red setter bitches, frequently drunk and quarrelsome, and of fierce right-wing prejudices; he was also a man of warm friendships and much generosity. He was unmarried, attempted heterosexual relationships, but was by temperament a lover of boys ('so puny in fact, so overwhelming in feeling,' as his biographer, **Sylvia Townsend Warner**, put it). He spent his last years in extended tours of Europe and America; he died, on his return to Europe, on board ship at the Piraeus, and is buried in the Protestant cemetery at Athens.

For Children

The Sword in the Stone 1939, *The Witch in the Wood* 1940 and *The Ill-Made Knight* 1941 [rev. in 1 vol. with *The Candle in the Wind* as *The Once and Future King* 1958]; *Mistress Masham's Repose* 1946; *The Master* 1957; *The Book of Merlyn* 1977

Fiction

Dead Mr Nixon [with R. McNair Scott] 1931; *They Winter Abroad* [as 'James Aston'] 1932; *Darkness at Pemberley* 1933; *Farewell Victoria* 1933; *First Less* [as 'James Aston'] 1933; *Earth Stopped* 1934; *Gone to Ground* 1935; *The Elephant and the Kangaroo* 1948; *The Maharajah and Other Stories* (s) [ed. K. Sprague] 1981

Non-Fiction

England Have My Bones 1936; *Burke's Steerage, or, The Amateur Gentleman's Introduction to Noble Sports and Pastimes* 1938; *The Age of*

Scandal: an Excursion through a Minor Period 1950; *The Goshawk* 1951; *The Scandalmonger* 1952; *The Godstone and the Blackymor* 1959; *America at Last: the American Journal of T.H. White* 1965; *The White/Garnett Letters* [ed. D. Garnett] 1968; *Letters to a Friend: the Correspondence Between T.H. White and L.J. Potts* [ed. F. Gallix] 1982

Poetry
Loved Helen and Other Poems 1929; *The Green Bay Tree* 1929; *Verses* 1962; *A Joy Proposed* 1983

Translations
The Book of Beasts 1954

Biography
T.H. White by Sylvia Townsend Warner 1967

John (Robert) WHITING 1917–1963

The son of an army officer, Whiting was born in Salisbury, but at the age of five moved to Northampton, when his father became a solicitor there. According to his own account, his childhood was a happy one, and he did moderately well at 'a minor public school in Taunton'. On leaving school in 1935 – he claimed never to have passed an examination in his life – he went to the Royal Academy of Dramatic Art, where he discovered that he was very nervous on stage. When the Second World War was declared, he registered as a conscientious objector, then changed his mind and enlisted as an anti-aircraft gunner. He was sent to a camp on Merseyside in 1940, and married the same year. He was commissioned in 1942 and discharged in late 1944, and his experience of the war was to have a profound effect on him. It was during the war that he began to write stories and poems, and this decided him to become a writer.

In the immediate post-war years he completed four stage plays – which represent half his total output – as well as four plays for radio and some prose fiction. Two of the stage plays, *No More A-Roving* and *The Conditions of Agreement*, were never produced during his lifetime. The first of his plays to be produced was *A Penny for a Song* in 1951, a light historical drama set against a military background on the Dorset coast in 1804. His first (mixed) success was *Saint's Day*, which won a competition run by the Arts Theatre: it found favour with the theatrical community, but received hostile notices from the critics, with the exception of J.C. Trewin. A far darker piece than its predecessor (though in fact written before it), it is concerned, as Whiting's later plays were to be, with spiritual struggles, and is set on 25 January, traditionally the date on which St Paul was converted to Christianity. '*Saint's Day* was written immediately after the dropping of the first atom bomb,' noted **Christopher Fry** (one of the judges of the competition and, with **T.S. Eliot**, an important influence upon Whiting). 'The whole play shudders with the fact.' Military themes were once more to the fore in *Marching Song*, about a general being tried for cowardice.

Although this last play received more respectful notices than his earlier ones, Whiting abandoned the theatre for several years to work on film scripts, translating the plays of Jean Anouilh and André Obey, and writing dramatic criticism for the *London Magazine*. His pieces for the magazine were published posthumously as *John Whiting on Theatre*. In 1960 Peter Hall commissioned him to write a play based on **Aldous Huxley**'s *The Devils of Loudon* (1952). *The Devils*, which showed the influence of Brecht, proved to be his last completed play and his first critical and commercial success; it remains the work by which he is best known, and was the basis of Ken Russell's highly controversial film (1970) of the same name.

Whiting died of a brain tumour at the age of forty-five, but despite his short life and small output, he proved to be highly influential. A collected volume of his plays was published after his death, as well as a further volume of his writings, *The Art of the Dramatist*, which contains stories, criticism and lectures. He was, says his biographer, 'the first British dramatist to catch ... the profound unease of the mid-twentieth century ... No other dramatist since [**George Bernard**] **Shaw** can compare with Whiting in stature except **John Arden**.' While this may be overstating the case, a number of leading playwrights, including **Harold Pinter**, have been influenced by him.

Although married, Whiting had a number of affairs. An actress friend who had collapsed at his funeral and had to be supported by his wife subsequently committed suicide, leaving her estate to Whiting's children. In 1965 an annual award for playwrights was founded in his memory.

Plays
Paul Southman (r) 1946; *Eye Witness* (r) 1947; *The Stairway* (r) 1949; *Love's Sweet Song* (r) 1950; *A Penny for a Song* 1951 [pub. 1963]; *Saint's Day* 1951 [rev. 1962; pub. 1963]; *Single Plays* 1951 [pub. 1964]; *Marching Song* 1954; *The Gates of Heaven* 1956; *The Devils* 1961; *No Why* 1964 [pub. 1961]; *The Conditions of Agreement* 1965; *No More A-Roving* (r) 1979, (st) 1987 [pub. 1975]; *The Collected Plays of John Whiting* [pub. 1969]

Non-Fiction
John Whiting on Theatre 1966

Translations
André Obey *Sacrifice to the Wind* 1959; Jean Anouilh *Madame de –* 1959, *Traveller without Luggage* 1959

Collections
The Art of the Dramatist 1970

Biography
The Dark Journey by Eric Salmon 1979

Anna WICKHAM (pseud. of Edith Alice Mary Harper) 1884–1947

The daughter of a piano repairer and his wife, a hypnotist and character reader, Wickham was born in Wimbledon, but went with her parents as a child to Australia and grew up there, attending Sydney High School. She told her father when aged ten that she would be a poet and took her pseudonym from the Brisbane street, Wickham Terrace, where the vow was made. She also wrote under the name of Edith Harper and published her first novel of verse as John Oland.

As a young woman, she gave elocution lessons in Sydney, but aged twenty-one she 'came home from Australia', as she put it, 'with the idea of writing verse; wrote some.' She spent a year studying drama at the Tree Academy of Acting, and another at the Paris Conservatoire, having singing lessons with de Reszke, but gave this up to marry the solicitor and astronomer Patrick Hepburn, with whom she had four sons (one of whom died young). Her husband was passionately opposed to her writing poetry, and when *Songs of John Oland* was brought out by a feminist vanity press in 1911 he had her committed to a lunatic asylum for six weeks.

From the teens of the century she moved in literary circles, and numbered **D.H. Lawrence**, **Malcolm Lowry** and **Dylan Thomas** as her friends, but recognition for her own work came only slowly. Her first commercial collection, *The Contemplative Quarry* (1915), was followed by few others, and a collection of her work had to wait for the feminist era she anticipated, and eventually appeared thirty-seven years after her death. Her reputation in America had always been high, and nowadays there is a general consciousness of the importance of her poetry which if sometimes crude, is always powerful, individual and copious in imagery. Although very self-critical, she believed in spontaneity and was gifted at producing extempore light verse, dashing off poems on postcards to friends or on the kitchen wall, which was covered in her writing and that of family and friends.

Relations with her husband were always difficult and characterised by violent quarrels. In 1926 they separated, and for a while Wickham and her children lived in the house of **Harold Monro**, who had published her volume *The Little Old House*. The family subsequently got back together again, but on Christmas Day 1929 Hepburn fell off a mountainside during a solitary walking tour in the Lake District and died of exposure, an end Wickham had 'foreseen' in her 1921 poem 'The Homecoming'. She was left with very little money and so let rooms in her house in Hampstead, advertising 'Stabling for Poets Painters and their Executives. Saddle your Pegasus here. Creative Moods respected. Meals at all Hours'. She carried on a passionate relationship, largely by correspondence, with the lesbian writer Natalie Barney, whom she had met during a visit to Paris in 1922. In the 1930s her two elder sons became a well-known tap-dancing act, billed initially as the Two Madisons, but appearing later, and internationally, as the Hepburn Brothers.

She lost several manuscripts and almost all her letters when her house was hit by a fire-bomb in 1943. During the war she ministered to her three fighting sons and retained a high sense of her poetic mission, writing in 1940: 'As soon as I have time, I hope to put up a record in English poetry, equal to what will be the allied victory in arms.' This hope was unfulfilled; suffering from depression, she hanged herself in 1947 when the Chancellor of the Exchequer increased the duty on cigarettes, of which she was an insatiable consumer.

Poetry
Songs of John Oland 1911 [as 'John Oland']; *The Contemplative Quarry* 1915; *The Man with a Hammer* 1916; *The Little Old House* 1921; *Selected Poems* 1936

Collections
The Writings of Anna Wickham: Free Woman and Poet [ed. R.D. Smith] 1984

Richard (Purdy) WILBUR 1921–

Wilbur, whose father was the artist Lawrence Wilbur, grew up in the countryside of New Jersey. 'If my poems are unfashionably towards nature,' he has written, 'I must blame this warp on a rural, pleasant and somewhat solitary childhood.' He was educated at Amherst before serving in the US Army in Italy and Germany, during which time he began writing poetry. He graduated from Harvard in 1947 and stayed on as a teacher until 1954. He has held posts in the English departments of Wellesley, Wesleyan

University and was writer-in-residence at Smith College during the 1970s.

His first volume of poetry, *The Beautiful Changes*, was published in 1947. Wilbur was writing at the same time as **Robert Lowell** and **James Dickey**, two poets who threw themselves energetically into the fashionable realm of personal, confessional poetry. Wilbur remained on the outside. He preferred to find an emotional outlet by writing of the spiritual essence of tangible things and their significant effect on a distanced narrator. He is praised for his use of language and his detailed images. Contemporary critics, though they admired his work, were used to poetry that abounded with spiritual crisis, and declared his voice overly academic and lacking in passion. **Randall Jarrell** commented: 'Mr. Wilbur never goes too far but he never goes far enough.' Lowell, in *Notebook 1967–68*, wrote: 'Until *The Mind-Reader*, one might not know that Wilbur had a family', and went on to define two categories into which American poets could be placed – the 'cooked' and the 'raw'. Wilbur he found to be undeniably 'cooked'.

In 1957 Wilbur won a National Book Award and a Pulitzer Prize for *Things of This World*. He is a distinguished translator of Molière and is considered the most accessible and sensitive American translator of poetry and drama in verse. He is most celebrated for his translation of Voltaire's *Candide*. 'It is the province of poets to make some order in the world,' he has said. 'But people can't afford to forget that there is a reality in things which survive all orders, great and small.'

Poetry

The Beautiful Changes and Other Poems 1947; *Ceremony and Other Poems* 1950; *Things of This World* 1956; *Poems 1943–1956* 1957; *Advice to a Prophet and Other Poems* 1961; *Loudmouse* 1963; *The Pelican from a Bestiary of 1120* 1963; *The Poems of Richard Wilbur* 1963; *Prince Souvanna Phouma: an Exchange between Richard Wilbur and William Jay Smith* 1963; *Complaint* 1968; *Walking to Sleep* 1969; *Opposites* 1973; *Seed Leaves* 1974; *The Mind-Reader* 1976; *Verses on the Times* [with W.J. Smith] 1978; *Seven Poems* 1981; *New and Collected Poems* 1988; *Runaway Opposites* 1995

Plays

From Molière: *The Misanthrope* 1955; *Tartuffe* 1965 [pub. 1963]; *School for Wives* 1971; *The Learned Ladies* 1977 [pub. 1978]; *The School for Husbands* 1991

Andromache [from Racine's play] 1982; *Phaedra* [from Racine's play] 1986

Non-Fiction

Emily Dickinson: Three Views [with L. Bogan, A. MacLeish] 1960; *Responses: Prose Pieces 1953–1976* 1976; *On My Own Work* 1983

Translations

Candide [from Voltaire's story] 1956 [pub. 1957]; *The Whale and Other Uncollected Translations* 1982

James WILCOX 1949–

Wilcox was born in Hammond, Louisiana, and is a writer rooted in the American South of his childhood. His father, a horn player, was head of the music department at the University of Southeastern Louisiana, and Wilcox was educated at Yale, where he took fiction seminars with **Robert Penn Warren** and wrote a novel as his thesis. His early career was in publishing, working first at Random House in New York, where he was the editor of **Cormac McCarthy** amongst others and rose to become an associate editor before leaving in 1977 to join Doubleday. He had been encouraged as a writer by **Toni Morrison**, who also worked at Random House, and in 1978 gave up publishing to devote himself to writing full time.

He wrote numerous short stories, which he submitted unsuccessfully to magazines, living off the earnings from an unproduced screenplay. Eventually, in 1981, the *New Yorker* accepted one of his stories, and two years later his first novel, *Modern Baptists* was published. Set in the imaginary Louisiana town of Tula Springs, it was a droll comedy about two half-brothers, one a glamorous convicted drug-dealer, the other a gentle, middle-aged bachelor. Praised by **Anne Tyler**, the novel established his reputation as a highly original comic writer. Tula Springs was also the setting of his next three novels, including *Sort of Rich*, in which the female middle-aged protagonist, tired of New York, comes to the town to marry one of its inhabitants. *Polite Sex* reverses this interchange, and traces the experiences of two young Tula Springs women who come to New York in search of their destiny. *Guest of a Sinner* is set entirely in New York and, like several of Wilcox's novels, has a not altogether likeable protagonist, a handsome forty-two-year-old pianist unable to commit himself to a relationship.

Wilcox has created a rich and fully realised fictional world, and characters (such as the Pickens brothers from *Modern Baptists*) sometimes reappear from book to book. Much of his comedy arises from misunderstandings or from the characters' capacity for self-deception, as well as from elaborate plotting, as in *Miss*

Undine's Living Room (the third of the Tula Springs novels) in which Olive Mackie blunders through life with sublime self-assurance – and gets away with it. Wilcox's novels have attracted excellent reviews and a devoted following, but have never gained him the reputation or financial rewards he deserves. Indeed his plight was the subject of a disheartening feature in a special double issue of the *New Yorker*, which was otherwise a celebration of fiction. He lives in New York.

Fiction

Modern Baptists 1983; *North Gladiola* 1985; *Miss Undine's Living Room* 1987; *Sort of Rich* 1989; *Polite Sex* 1991; *Guest of a Sinner* 1993

Michael (Denys) WILCOX 1943–

Wilcox was born in Totnes, Devon, to which town Alleyn Court School, founded by his grandfather and of which his father was then headmaster, had moved during wartime. The school moved back to Westcliffe-on-Sea, near Southend, where Wilcox grew up. His father, Denys Wilcox, who had been captain of the Essex cricket team and who came from a well-known cricketing family, died when Michael was nine. Wilcox attended Malvern College, and trained as a teacher at Isleworth, teaching in schools from 1966 to 1974, first in Newbury, and then in the north-east, where he became the head of the English department in a large Newcastle comprehensive school.

He had become acquainted with the playwright C.P. Taylor, who encouraged him in his long-held ambition to be a playwright himself. In 1974, still not having written a play, he gave up his job and lived in a small rented house in Jesmond in Newcastle, where he wrote his first play, *Rents*, the account of two young rentboys servicing homosexuals, which was his first major success when eventually produced in 1979. Wilcox's earliest plays were seen in small theatres and arts centres in the north of England, and he developed connections with well-known groups such as the Traverse Theatre Company; with Taylor, he was a founding member of the Northern Playwrights' Society, set up in an attempt to stimulate theatre in this region. As his success as a playwright developed, he was able to move to a series of rented cottages in the Northumberland countryside.

Wilcox has developed into a prolific playwright, and is also well known as an author of television scripts; his plays often deal with homosexual themes, but these tend to be incidental to the main story. *Accounts*, for example, details the hard life of two adolescent brothers (one homosexual, the other not), who have been left to help their mother run a farm in the north of England. *Lent*, which is perhaps his finest play, is about a small boy left at a preparatory school (founded by his family) during the holidays, and clearly draws upon his own childhood. Both plays have been adapted for television. *Massage* is a refreshingly unhysterical study of a man attracted to underage boys.

Among Wilcox's work for television are scripts for the popular *Inspector Morse* series, based on the novels of Colin Dexter, and his other writings have included scripts for British Medical Association videos on subjects such as Aids and drug abuse; the libretto for *Tornrak*, an opera by John Metcalf; and articles for such magazines as *Opera Now*. Since 1984, he has been editor of the Methuen Gay Plays series which has reissued forgotten plays alongside new work.

In 1991 he published *Outlaw in the Hills*, a diary of a year in his life, 1989, which also contains autobiographical material relating to earlier periods. He lives in Northumberland, where he continues a family interest in cricket, which began when George Wilcox founded Whitehaven Cricket Club in 1838.

Plays

The Boy Who Cried Stop 1974 [np]; *Dekka & Dava* 1974 [pub. 1987]; *The Atom Bomb Project* 1975 [np]; *Grimm Tales* 1975 [np]; *Roar Like Spears* 1975 [np]; *The Blacketts of Bright Street* 1977 [np]; *The Phantom of the Fells* 1977 [np]; *Pioneers* 1977 [np]; *Mowgli* 1978 [np]; *Standard Procedure* **(tv)** 1978; *Rents* 1979 [pub. 1983]; *Accounts* **(st)** 1981, **(tv)** 1983 [pub. in *Gay Plays* 1984]; *Cricket* **(tv)** 1982; *Midnight Feast* **(tv)** 1982 [pub. 1987]; *Burnt Futures* **(tv)** 1983; *In Disgrace* **(tv)** 1983; *Lent* **(st)** 1983, **(tv)** 1985 [pub. 1983]; *78 Revolutions* 1984 [np]; *Massage* 1986 [pub. 1987]; *Green Fingers* 1990 [pub. 1991]; *Tornrak* **(l)** [with J. Metcalf] 1990 [np]

Non-Fiction

Outlaw in the Hills: a Writer's Year **(a)** 1991

Edited

Gay Plays 4 vols 1984–90 1984

Thornton (Niven) WILDER 1897–1975

Born in Madison, Wisconsin, the son of a newspaper editor of Calvinist leanings, Wilder grew up partly in China and partly in the USA. He attended Berkeley, Yale and Princeton, served briefly in the First World War, and from 1921 to 1928 was a housemaster in a school. His first novel, *The Cabala*, appeared in 1926, but it was *The Bridge of San Luis Rey*, a philosophical novel built from a series of accidental relationships surrounding the collapse of a bridge in Peru in 1704, which really made his reputation, both nationally and internationally, in 1927.

From 1930 to 1936, he was a lecturer on literature at the University of Chicago, and his later life varied several prestigious academic appointments with periods spent as a cultural ambassador for the USA abroad. During the Second World War he was in the Army Air Corps Intelligence Division as a major and saw service in Italy. His best novel is probably *Heaven's My Destination*, about a book salesman who is also a modern saint. Wilder developed an equally successful career as a playwright, with hits such as *Our Town* and *The Skin of Our Teeth*, the latter of which caused him some trouble over allegations that it was plagiarised from **James Joyce**'s *Finnegans Wake* (1939). Both were awarded Pulitzer Prizes. A later play, *The Matchmaker*, was turned into the successful musical *Hello, Dolly!* (1963). In 1948 Wilder published *The Ides of March*, a novel about Julius Caesar, but did not return to the form until the disappointing *The Eighth Day* (1967). In the early 1960s he had spent several years in seclusion in a small Arizona town working on dating the plays of Lope de Vega.

A writer who persisted in addressing the eternal verities in his work, Wilder has been variously considered a significant minor talent or a middlebrow charlatan. That said, he won a Pulitzer Prize three times, a National Book Award for *The Eighth Day*, and in 1965 was the first recipient of the National Medal for Literature. He was homosexual, and in his last years shared his Connecticut home with his sister, the novelist Isabel Wilder.

Fiction

The Cabala 1926; *The Bridge of San Luis Rey* 1927; *The Woman of Andros* 1930; *Heaven's My Destination* 1935; *The Ides of March* 1948; *The Eighth Day* 1967; *Theophilus North* 1973

Plays

The Angel That Troubled the Waters and Other Plays [pub. 1928]; *The Long Christmas Dinner and Other Plays in One Act* [pub. 1931]; *Our Town* 1938; *The Merchant of Yonkers* 1939 [rev. as *The Matchmaker* 1954, as *Hello Dolly!* 1963]; *The Skin of Our Teeth* 1942; *Our Century* 1967; *The Alcestiad* [with H.E. Herlitschka] 1977

Non-Fiction

James Joyce, 1882–1941 1944; *American Characteristics and Other Essays* 1979; *The Journals of Thornton Wilder 1939–1961* [ed. D. Gallup] 1985

Charles (Walter Stansby) WILLIAMS 1886–1945

Williams was the son of a clerk living in Holloway and working in the City, a man of some culture, whose educational influence on his son was strong. When Williams was eight, the family moved to St Albans, and he attended the school attached to the abbey there. In 1901 he won an intermediate scholarship to University College, London, but was forced to withdraw after two years because of financial difficulties, getting work at a Methodist bookshop in Holborn. In 1908, he entered the London office of the Oxford University Press, in whose service he was to remain until his death, first as a proof reader but later as a trusted editor and literary adviser. Meanwhile, he graduated from attending classes at a working-men's institute to becoming a lecturer in English literature at the City Literary Institute and other adult education establishments. He became well known for his inspirational address, so much so that his disciples even formed themselves into an order, called the 'Companions of the Co-inherence', to propagate his ideas on Christianity and magic.

In 1912, the writers Wilfrid and Alice Meynell were instrumental in arranging for the publication of his first book, *The Silver Stair*, a sonnet sequence addressed to his future wife, Florence ('Michal') Conway, whom he married in 1917, and with whom he had one son. He was judged unfit for service in the First World War. His highly productive period as a writer began from around 1927, and he produced over thirty books of poetry, plays, literary criticism, fiction, historical biography and theology. His novels, which can be loosely described as supernatural thrillers, include such titles as *Descent into Hell*, *War in Heaven* and *All Hallows' Eve*. Perhaps his best book of criticism is *The Figure of Beatrice*, an original and influential interpretation of Dante.

The best-known period of Williams's life began with the outbreak of the Second World War, when the Press and its employees moved to Oxford. There he was quickly taken into the circle of Christian writers, of whom the best-known members were **C.S. Lewis** and **J.R.R. Tolkien**, called the Inklings, and meeting regularly at the Eagle and Child (or 'Bird and Baby') public house. Lewis had already met and greatly admired Williams and he arranged for him to give a course of university lectures in English, although he had no formal qualifications. Williams's hopes of remaining at the university were cut short by his death shortly after VE-Day. These last years had also seen the publication of his best-known work, the two volumes of Arthurian poetry, written in a stressed prosody, called

Taliessin through Logres and *The Region of the Summer Stars*. These poems, like all of Williams's work, have attracted controversy. During his lifetime, his influential admirers included **T.S. Eliot**, **Dorothy L. Sayers** and **W.H. Auden**, but his work has gone severely out of fashion since then, as the brand of Christian mysticism he represented, the denial of any distinction between the natural and supernatural, recedes further and further from the taste of the time.

Poetry
The Silver Stair 1912; *Poems of Conformity* 1917; *Divorce* 1920; *Windows of Night* 1925; *Heroes and Kings* 1930; *Taliessin through Logres* 1938; *The Region of the Summer Stars* 1944

Fiction
War in Heaven 1930; *Many Dimensions* 1931; *The Place of the Lion* 1931; *The Greater Trumps* 1932; *Shadows of Ecstasy* 1933; *Descent into Hell* 1937; *All Hallows' Eve* 1945

Non-Fiction
Poetry at Present 1930; *The English Poetic Mind* 1932; *Bacon* 1933; *Reason and Beauty in the Poetic Mind* 1933; *A Short Life of Shakespeare with the Sources* 1933; *James I* 1934; *Rochester* 1935; *Queen Elizabeth* 1936; *Henry VII* 1937; *Stories of Great Names* 1937; *He Came Down from Heaven* 1938; *The Descent of the Dove* 1939; *Witchcraft* 1941; *The Forgiveness of Sins* 1942; *The Figure of Beatrice: a Study of Dante* 1943; *Flecker of Dean Close* 1946

Plays
A Myth of Shakespeare 1929; *Three Plays* [pub. 1931]; *Thomas Cranmer of Canterbury* 1936; *Judgement at Chelmsford* 1939; *The House of the Octopus* 1945; *Seed of Adam and Other Plays* [ed. A. Ridler] 1948

Edited
Poems of Home and Overseas [with V.H. Collins] 1921; *A Book of Longer Modern Verse* 1926; *A Book of Victorian Narrative Verse* 1926; *The Oxford Book of Regency Verse* [with H.S. Milford, F. Page] 1928; *The New Book of English Verse* 1935; *The New Christian Year* 1941; *The Passion of Christ* 1941; *The Letters of Evelyn Underhill* 1943; *Solway Ford and Other Poems by Wilfred Gibson* 1945

Collections
Arthurian Torso 1948

Heathcote WILLIAMS 1941–

Williams was born in Helsby, Cheshire. The son of a lawyer, he read law at Christ Church, Oxford, after attending Eton, and failed his degree in 1962. In London in the early 1960s he wrote *The Speakers*, a documentary novel describing the lives of four Hyde Park orators, most notably the self-styled 'King of the Gypsies', Billy MacGuinness, with whom Williams lived. In his first play, *The Local Stigmatic*, produced at the Traverse Theatre in Edinburgh, and subsequently at the Royal Court, Williams first addressed what he has called the 'psychic capitalism' of the mass media. Al Pacino has made a low-budget film version of the play. His best-known play, *AC/DC*, winner of the *Evening Standard* Award among others in 1970, portrays a wild, visceral confrontation between hippies and schizophrenics. The play's climactic trepanning scene draws upon his acquaintance with Bart Huges, the Dutch advocate of trepanation as a means of mind-expansion. As an associate editor of the *Transatlantic Review* (1963–72) Williams published a dialogue between Huges and Joey Mellen, a Chelsea antiques dealer who had trepanned himself.

In the 1970s Williams was involved in a number of political and counter-culture movements. Under the auspices of the Albion Free State, he ran a squatting agency for the homeless in Notting Hill ('We Will Squat the Building of Your Choice'). He co-founded two underground magazines, *Suck* (with Germaine Greer and others) and the *Fanatic*. In 1974 he sued the police for breaking up the Windsor Free Festival.

His 1977 play *Hancock's Last Half-Hour* is a poignant portrayal of the comedian's final moments before his suicide in a hotel room. Williams is a proficient conjuror (despite having once suffered serious burns while fire-eating) and a former member of the Magic Circle, and his television play *What the Dickens!* depicts Charles Dickens putting on a conjuring show for Thackeray, Carlyle and others. In the 1980s his attention turned to environmental issues with the long polemical poems *Whale Nation*, *Sacred Elephant* and *Autogeddon*, the last of which was performed on television by Jeremy Irons in 1991. These poems were published in lavishly illustrated volumes which, for poetry, sold in remarkable quantities. The more lyrical *Falling for a Dolphin* describes his own experience of swimming with a hermit dolphin off the Irish coast. He has performed stage, radio and television readings of the poems, as well as acting in films such as Derek Jarman's *The Tempest* (1980), in which he plays Prospero, and David Leland's *Wish You Were Here* (1987), which is based on the early life of Cynthia Payne, the celebrated Streatham procuress.

Plays
The Local Stigmatic 1966 [pub. in *Traverse Plays* 1965]; *Malatesta* (tv) 1969; *AC/DC* 1970 [pub.

1972]; *Remember the Truth Dentist* [with B. Flagg] 1974 [np]; *The Speakers* [from his novel; with W. Gaskill, M. Stafford-Clark] 1974 [pub. 1980]; *Very Tasty* [with T. Allen] 1975 [np]; *Anatomy of a Space Rat* 1976 [np]; *Hancock's Last Half-Hour* 1977; *The Immortalist* 1977 [pub. 1978]; *Playpen* 1977 [np]; *At It* 1982 [np]; *What the Dickens!* **(tv)** 1983

Poetry

Falling for a Dolphin 1988 [rev. 1990]; *Whale Nation* 1988; *Sacred Elephant* 1989; *Autogeddon* 1991

Fiction

The Speakers 1964

Non-Fiction

Manifestoes, Manifesten 1975

Hugo (Mordaunt) WILLIAMS 1942–

Williams, son of actor and playwright Hugh Williams and brother of the actor Simon Williams, was born in Windsor and educated at Eton College. He was seventeen when the *London Magazine* first published his poetry, and a year later he was given the job of assistant editor. He has also worked for *News Review* and from 1980 to 1988 was poetry editor and television critic for the *New Statesman*. His first collection of poetry, *Symptoms of Love*, was published in 1965 and won the Eric Gregory Award. He has since won a Cholmondeley Award (1970) and the Geoffrey Faber Memorial Prize (1979).

Writing Home was a Poetry Book Society Choice in 1985. The collection is an evocation of his father, of boarding-school, and of a theatrical family in the 1950s. It was praised for its almost childlike directness and sustained combination of humour and nostalgia, whether describing the haunting Mr Ray overseeing letter-writing at Eton, or his father playing Scrabble with the family, throwing cigarettes from a train to a couple of soldiers, or applying his stage make-up. Williams takes letters as his theme, and in the poem 'An Actor's War' uses material from his father's sad, funny letters to his wife from North Africa.

'I always wanted to be an actor, but never liked performing in public,' Williams has said. 'Poems in the end are the lines you give yourself ... I think of my poems as humorous dramatic monologues, one-man shows, "pieces of action", self-satires, above all as entertainments'. He describes *Self-Portrait with a Slide* as a preparation for old age. The character of Sonny Jim, his *alter ego* in the collection, is 'a knockabout character left over from the sixties, a slightly retarded, randy failed artist and party-goer not unlike myself'.

Williams has lectured on the creative writing course at the University of East Anglia and was theatre critic for the short-lived *Sunday Correspondent*. His wife, Hermine Demoraine, is a tightrope walker whose acclaimed autobiography was published in 1989. They have one daughter.

Poetry

Symptoms of Love 1965; *Sugar Daddy* 1970; *Some Sweet Day* 1975; *Love Life* 1979; *Writing Home* 1985; *Selected Poems* 1989; *Self-Portrait with a Slide* 1990

Fiction

All the Time in the World 1966; *No Particular Place to Go* 1980

Nigel WILLIAMS 1948–

Williams was born in Cheadle, Cheshire, the son of a headmaster and the youngest of three boys, all of whom became writers. He was educated at Highgate School and Oriel College, Oxford, where he read history. He joined the BBC and worked on **Melvyn Bragg**'s book programme *Read All about It* and on *Arena*. He subsequently became executive editor of *Bookmark* and has directed various programmes, including one on **George Orwell** (1983) and **James Fenton**'s documentary *Cambodian Witness* (1987).

His early plays have a strongly political edge, often treating themes of alienation and destructiveness. The best-known of them are *Class Enemy*, an exploration of urban tensions set in a classroom which won the *Plays and Players* Award for most promising playwright in 1978; *Line'Em* which portrays a confrontation between pickets and the army; and *W.C.P.C.*, in which a policeman acts as an *agent provocateur* in men's public lavatories, only gradually to discover that virtually the whole police force is involved in a gay conspiracy. The comic note struck by this play is also evident in *The Adventures of Jasper Ridley*, in which the hero is sponsored by Prince Charles as Unemployed Person of the Year.

Williams was a joint winner of a Somerset Maugham Award in 1978 for his first novel, *My Life Closed Twice*, and his subsequent comic novels include *Star Turn*, about an incorrigible liar who invents a fantasy world populated with such figures as Freud, Churchill, Proust and Goebbels, and *Witchcraft*, in which a hack screenwriter is haunted by the seventeenth-century witch-finder he is researching for a film.

His greatest success has been with *The Wimbledon Poisoner*, which concerns Henry Farr, a solicitor and amateur historian whose fascination with a historical poisoner leads him to various farcical and ill-fated attempts at murdering his shrewish wife. 'What makes murder funny,' Williams has said, 'is that almost everyone would like to have a go at it, and only afterwards do they get the dark thoughts.' Williams adapted the novel for a 1995 television series starring Robert Lindsay and Alison Steadman, and he has written further novels about the hapless Farr. His talent for comic parody is evident in *2½ Men in a Boat*, a modern reworking of Jerome K. Jerome's masterpiece, which features a BBC executive in an Armani suit babbling into his mobile phone throughout the boat trip. Williams is married, and has three children.

Fiction

My Life Closed Twice 1977; *Jack Be Nimble* 1980; *Charlie* [from his tv play] 1984; *Star Turn* 1985; *Black Magic* 1986; *Witchcraft* 1987; *The Wimbledon Poisoner* 1990; *They Came from SW19* 1992; *East of Wimbledon* 1993; *2½ Men in a Boat* 1993; *Scenes from a Poisoner's Life* 1994

Plays

Double Talk 1976 [np]; *Snowwhite Washes Whiter, and Deadwood* 1977 [np]; *Talkin' Blues* **(tv)** 1977; *Class Enemy* 1978; *Real Live Audience* **(tv)** 1978; *Easy Street* 1979 [np]; *Line 'Em* 1980; *Sugar and Spice* 1980; *Trial Run* 1980; *Baby Talk* **(tv)** 1981; *Let 'Em Know We're Here* **(tv)** 1981; *The Adventures of Jasper Ridley* 1982 [pub. 1983]; *W.C.P.C.* 1982 [pub. 1983]; *Johnny Jarvis* **(tv)** [from his novel for children] 1983; *Charlie* **(tv)** 1984; *Deathwatch* [from J. Genet] 1985 [np]; *My Brother's Keeper* 1985; *Breaking Up* **(tv)** 1986; *Country Dancing* 1986 [pub. 1987]; *As It Was* [from H. Thomas] 1987 [np]; *Nativity* 1989 [np]; *Centrepoint* **(tv)** 1990; *Kremlin Farewell* **(tv)** 1990; *Skallagrigg* **(tv)** [from W. Horwood's novel] 1994; *The Wimbledon Poisoner* **(tv)** [from his novel] 1995

For Children

Johnny Jarvis 1983

Tennessee (i.e. Thomas Lanier) WILLIAMS 1911–1983

The son of a travelling salesman, Williams was born in rural Mississippi at the home of his grandfather, an episcopal clergyman. When he was twelve, his father moved the family to St Louis, but this was not a success. Williams, his unstable sister Rose (to whom he was devoted) and his mother, an ageing Southern belle, could not adjust to what they saw as the family's worsened circumstances, and Williams began at an early age to exorcise these tensions through writing. He studied briefly at the University of Missouri, but was forced to leave in order to work as a clerk for the shoe company which employed his father. This proved to be a grinding experience which, combined with the atmosphere in the family home and anxiety about his homosexuality, led to his suffering a partial breakdown. He recovered rapidly, however, and when he began to sell some of his writing, he was able to complete his education at the University of Iowa.

After graduation he returned to a variety of menial jobs while he tried to establish himself as a playwright. Early plays were performed by amateur dramatic companies, while *The Battle of Angels* (1940) was put on by the Theatre Guild in Boston without much success. During this period he collaborated with **Donald Windham** on an adaptation of **D.H. Lawrence**'s short story 'You Touched Me'. In 1943 he was rescued from a job as a cinema usher by his agent, who managed to negotiate a six-month contract as a screenwriter at Metro-Goldwyn-Mayer. In his spare time from the studio he worked on *The Glass Menagerie*, a haunting play about a young woman trapped in a childlike world, awaiting the 'gentleman callers' that had characterised her mother's youth. The play is clearly autobiographical (narrated by the protagonist's brother, Tom) and was first suggested to Williams by letters he received from his sister, who had undergone a leucotomy in 1937. This had left her in a permanent twilight world, and the trauma of this was to have a profound effect on Williams's life and work. *The Glass Menagerie* opened in 1944 and was an immedate popular and critical success. In subsequent plays, he was often to draw on his own family background in his depiction of women driven or destroyed by frustrated sexuality or mental instability. His work is the most significant theatrical manifestation of what became known as Southern Gothic, and he shares many themes with novelists such as **Carson McCullers** and **Truman Capote**.

A Streetcar Named Desire, in which the neurotic Southern belle Blanche DuBois is raped by the brutal Stanley Kowalski, established Williams as the leading playwright of his generation, and also launched the career of Marlon Brando, who reprised the role of Stanley in the 1951 film version, for which Williams wrote the screenplay. His subsequent plays did not have quite the same impact – in particular the mystical *Camino Real* is a muddled allegory about life and death which is set in Mexico and introduces figures from history and literature such as Casanova,

Byron and Don Quixote – but he returned to form in 1955 with *Cat on a Hot Tin Roof*, in which the eponymous 'cat' is a sexually frustrated woman whose husband is obsessed by a former room-mate who has committed suicide. This was one of several plays in which Williams's own homosexuality formed a submerged theme (submerged even more in film adaptations, which have tended to tone down the sexual aspects of his work). *Suddenly Last Summer* is a particularly heady brew of mental instability, homosexuality, mother-fixation and cannibalism, and once again draws upon the life of Rose.

Williams also wrote several volumes of short stories; three novels, including *The Roman Spring of Mrs Stone*, about an ageing beauty and a gigolo (a theme he returned to in one of the best of his later plays, *Sweet Bird of Youth*); numerous screenplays; and a volume of highly unreliable memoirs. In the 1960s he entered a period when his plays were less successful and he became increasingly dependent on drink and drugs. He was also a hypochondriac who took numerous medications, and the most constant of his lovers, Frank Merlo, often acted at his nurse. He continued to write, however, and among later plays the elegiac *Vieux Carré*, set in New Orleans, has great charm. He eventually died in a New York hotel room, where he accidentally choked to death on the cap of a bottle of barbiturates. Since his death there have been assorted wrangles over his estate and various projected biographies. In spite of the occasional crudity of his work, he remains one of the century's most individual and admired playwrights, and his high reputation seems unlikely to fade.

Plays

Battle of Angels **(st)** 1940, **(f)** [as *The Fugitive Kind*] 1960 [pub. 1945]; *The Glass Menagerie* **(st)** 1944, **(f)** 1964; *Wagons Full of Cotton and Other One-Act Plays* [pub. 1945]; *You Touched Me!* [with D. Windham] 1945 [pub. 1947]; *A Streetcar Named Desire* **(st)** 1947, **(f)** 1951; *American Blues: Five Short Plays* [pub. 1948]; *Summer and Smoke* **(st)** 1948, **(f)** 1961; *The Rose Tattoo* **(st)** 1951, **(f)** 1955; *I Rise in Flame, Cried the Phoenix* 1951; *Camino Real* 1953; *Cat on a Hot Tin Roof* **(st)** 1955, **(f)** 1958; *Lord Byron's Love Letter* **(l)** 1955; *Baby Doll* **(f)** 1956; *Suddenly Last Summer* **(st)** 1958, **(f)** 1959; *The Night of the Iguana* **(st)** 1959, **(f)** 1964 [pub. 1962]; *Sweet Bird of Youth* **(st)** 1959, **(f)** 1962; *Period of Adjustment* **(st)** 1961, **(f)** 1962 [pub. 1960]; *The Milk Train Doesn't Stop Here Anymore* 1962 [pub. 1964]; *The Eccentricities of a Nightingale* 1964; *The Gnädiges Fräulein* 1967; *The Mutilated* 1967; *Kingdom of Earth* 1968; *In the Bar of a Tokyo Hotel* 1969; *The Two-Character Play* 1969; *Dragon Country: a Book of Plays* [pub. 1969]; *Small Craft Warnings* 1972; *Out Cry* 1973; *Vieux Carré* 1979; *Crève Coeur* 1980

Fiction

One Arm and Other Stories **(s)** 1948; *The Roman Spring of Mrs Stone* 1950; *Hard Candy* **(s)** 1954; *Three Players of a Summer Game and Other Stories* **(s)** 1960; *The Knightly Quest* **(s)** 1966; *Eight Mortal Ladies Possessed* **(s)** 1974; *Moise and the World of Reason* 1975; *The Bag People* 1982; *It Happened the Day the Sun Rose and Other Stories* **(s)** 1982

Poetry

Blue Mountain Ballads 1946; *In the Winter of Cities* 1956; *Androgyne, Mon Amour* 1977

Non-Fiction

Where I Live: Selected Essays [ed. C.R. Day, B. Woods] 1973; *Memoirs of Tennessee Williams* **(a)** 1975; *Tennessee Williams: Letters to Donald Windham, 1940–1965* [ed. D. Windham] 1976

Biography

The Kindness of Strangers by Donald Spoto 1985

William Carlos WILLIAMS 1883–1963

Ezra Pound's often cited injunction to 'make it new' is combined in Williams's work with the need to 'make it local', to make use of the materials he himself knew. For Williams those materials were a lifetime spent almost entirely in Rutherford, New Jersey, his birthplace. Two lifetimes even, in that for most of his life he was a doctor. His mother, a Puerto Rican, was a painter and his father a businessman who never gave up British nationality. Despite the literary interests of his parents, his own early enthusiasm at school was for mathematics and science. It was later, at medical school at the University of Pennyslvania, from which he graduated as an MD in 1906, that the urge to write began to assert itself.

His first models were Keats and Whitman, whose influence is apparent in his first published volume, *Poems*, privately printed in 1909. After that the influence of Pound, whom he met in his first year at Pennsylvania, begins to exert its weight. Pound introduced him to Hilda Doolittle (**H.D.**), and Williams's second book, *The Tempers*, is of the Imagist school. His own experience of the world had also grown. In 1909 he began work as an intern in the notorious 'Hell's Kitchen' area of New York, but after post-graduate study in Leipzig he was able to establish a private practice in Rutherford which he was to run for the next forty years. He claimed that he gleaned material for his fiction from his patients. In 1912 he married Florence Hermann, herself an immigrant, whose family story was to be the subject-matter of Williams's *White Mule* trilogy. They had two sons.

The publication in 1922 of **T.S. Eliot**'s *The Waste Land* and its immediate success was the next big influence on his development. He later described it as a 'catastrophe of the magnitude of an atom bomb', for to him it represented a turning away from the American towards European influences (he was a committed anglophobe). Eliot's continuing fame was to remain a source of irritation especially in that Williams's own work received little attention until the publication of *Paterson*. After *Spring and All* (1923), containing what many consider to be his finest poetry, he published no more verse for eleven years. The five *Paterson* books, a *Prelude*-like account of the growth of a poet's mind, published between 1946 and 1958, made him a father-figure to a new generation of American poets such as **Allen Ginsberg** and **Robert Lowell**. In them he introduced the 'variable foot'; this was intended to approximate the rhythms of American speech, and was his major stylistic innovation.

In the late 1940s he suffered the first of several strokes and he was forced to retire from medicine in 1951. The enforced lay-off made his poetry no less affirmative than before, only more reflective. His continuing ill health was also the reason given for the withdrawal of a promised post at the Library of Congress, although a newspaper campaign alleging communist affiliations may well have contributed to the official decision. The offer was eventually reinstated but only for the final few months of the original term and it was never renewed. A partially paralysing stroke in 1958 and an operation for cancer the following year confined him to bed, for which he wrote *Pictures from Brueghel*, his last book. It won a Pulitzer Prize in 1962. He died in his sleep the following year.

Poetry

Poems 1909; *The Tempers* 1913; *Al Que Quiere* 1917; *Sour Grapes* 1921; *Go Go* 1923; *Collected Poems 1921–1931* 1934; *An Early Martyr and Other Poems* 1935; *Adam and Eve and the City* 1936; *The Complete Collected Poems* 1938; *The Broken Span* 1941; *The Wedge* 1944; *Paterson* 5 vols 1946–58 [rev. in 1 vol. 1963]; *The Clouds* 1948; *The Pink Church* 1949; *Collected Later Poems* 1950 [rev. 1963]; *The Desert Music and Other Poems* 1954; *Journey to Love* 1955; *Pictures from Brueghel and Other Poems* 1962

Fiction

The Great American Novel 1923; *A Voyage to Pagany* 1928; *The Knife of the Times and Other Stories* (s) 1932; *A Novelette and Other Prose* (s) 1932; *White Mule Trilogy*: *White Mule* 1937, *In the Money* 1940, *The Build-Up* 1952; *Life along the Passaic River* 1938; *The Farmers' Daughters: the Collected Stories* (s) 1961

Non-Fiction

In the American Grain 1925; *A Beginning on the Short Story* 1950; *The Autobiography* (a) 1951; *Selected Essays* 1954; *I Wanted to Write a Poem: the Autobiography of the Works of a Poet* [ed. E. Heal] 1958; *Yes, Mrs Williams: a Personal Record of My Mother* 1959; *The Embodiment of Knowledge* [ed. R. Loewinsohn] 1974; *A Recognizable Image: William Carlos Williams on Art and Artists* 1978

Plays

A Dream of Love 1948; *Many Loves and Other Plays: the Collected Plays* [pub. 1961]

Translations

P. Soupault *Last Nights in Paris* 1929; Y. Goll *Landless John* 1944; P. Espinosa *The Dog and the Fever* [with R.H. Williams] 1954; *Sappho. A Translation* 1957

Collections

Kora in Hell 1920; *Spring and All* 1923

Henry WILLIAMSON 1897–1977

The son of a bank clerk, Williamson was educated at Colfe's Grammar School, London, although he came from a Dorset background. He enlisted as a private at the beginning of the First World War, was commissioned, and returned at the age of twenty-three, grey-haired and mentally scarred. After a brief period as a reporter on the *Weekly Dispatch*, he tried to live for a while on his war pension and the proceeds of a weekly article on nature, sleeping sometimes on the Thames Embankment or on a haystack in Kent. Later, inspired by Richard Jefferies' country classic *The Story of my Heart* (1883), he retired to a cottage on Exmoor to write.

His first novel, *The Beautiful Years*, which begins *The Flax of Dreams* tetralogy, was published in 1921, but phenomenal success came his way with the publication in 1927 of *Tarka the Otter*, an animal fable which won the Hawthornden Prize and is still his most famous book. He was now able to set up as a farmer in Norfolk with his wife and family, writing equally about nature and human society. Less well known than the animal books, but amongst his finest works, is *The Patriot's Progress*, the story of a City clerk who enlists as a private in the First World War and becomes emblematic of the hapless ranker. Vividly expressionist in style, it was written in collaboration with the Tasmanian artist William Kermode, whose lino-cut illustrations act as 'shuttering to my verbal concrete', as Williamson put it. This is one of the most extraordinary books to emerge from the First World War, a savage indictment of what war does to people, conveyed in mimetic shellbursts of language.

In the 1930s Williamson's detestation of war led to his becoming an admirer of Hitler and Sir Oswald Mosley, a connection which resulted in him being briefly interned at the beginning of the Second World War. In later years, he lived largely in Devon, and in 1951 published the first of fifteen semi-autobiographical novels in a sequence called *A Chronicle of Ancient Sunlight*, which he brought to a conclusion in 1969. The story of one man, Phillip Maddison (the cousin of Willie Maddison in *The Flax of Dreams*), living through the twentieth century, it has been less read than the contemporaneous *romans fleuves* by **Anthony Powell** and **C.P. Snow**, although Williamson's admirers claim that, while uneven, it is not inferior. Williamson was twice married and twice divorced; he had six children by his first marriage and one by his second.

Fiction

The Flax of Dreams: *The Beautiful Years* 1921, *Dandelion Days* 1922, *The Dream of Fair Women* 1924, *The Pathway* 1928; *The Peregrine's Saga* (s) [US *Sun Brothers*] 1923; *The Old Stag Stories* (s) 1926; *Tarka the Otter* 1927; *The Linhay on the Downs* (s) 1929; *The Patriot's Progress* [with W. Kermode] 1930; *The Village Book* (s) 1930; *The Labouring Life* (s) 1932; *The Gold Falcon* 1933; *Star-Born* 1933; *Salar the Salmon* 1935; *The Phasian Bird* 1948; *A Chronicle of Ancient Sunlight*: *The Dark Lantern* 1951, *The Donkey Boy* 1952, *How Dear Is Life* 1954, *Young Phillip Maddison* 1953, *A Fox under My Cloak* 1955, *The Golden Virgin* 1957, *Love and the Loveless* 1958, *A Test to Destruction* 1960, *The Innocent Moon* 1961, *It Was the Nightingale* 1962, *The Power of the Dead* 1963, *The Phoenix Generation* 1965, *A Solitary War* 1966, *Lucifer Before Sunrise* 1967, *The Gale of the World* 1969; *Tales of Moorland and Estuary* (s) 1953

Non-Fiction

The Lone Swallows 1922; *A Soldier's Diary of the Great War* 1929; *The Wet Flanders Plain* 1929; *The Wild Red Deer of Exmoor* 1931; *On Foot in Devon* 1933; *Devon Holiday* 1935; *Goodbye West Country* (a) 1937; *The Children of Shallowford* 1939; *Genius of Friendship: T.E. Lawrence* 1941; *The Story of a Norfolk Farm* 1941; *Norfolk Life* [with L. Rider Haggard] 1943; *The Sun in the Sands* (a) 1945; *Clear Water Stream* (a) 1958; *In the Woods* (a) 1960; *The Weekly Dispatch* 1983; *Days of Wonder* 1987; *From a Country Hilltop* 1988; *Some Notes on the Flax of Dream and Other Essays* 1988

Edited

An Anthology of Modern Nature Writing 1936; *Nature in Britain* 1936; Richard Jefferies: *Hodge and His Masters* 1937, *Richard Jefferies: Selections of His Work with Details of His Life and Circumstances, His Death and Immortality* 1937; James Farrar *The Unreturning Spring* 1950; *My Favourite Country Stories* (s) 1966

For Children

Scribbling Lark 1949

Biography

Henry: an Appreciation by Dan Farson 1982

Calder (Baynard) WILLINGHAM Jr 1922–

Willingham was born in Atlanta, Georgia, where his father was a hotel manager, and educated at a military academy, the Citadel in Charleston, South Carolina, from 1940 to 1941, when he went to the University of Virginia, Charlottesville. He moved to New York in 1943 and four years later established a reputation as an *enfant terrible* of the literary scene with the publication of his first novel, *End as a Man*. This naturalistic and violent account of life in a Southern military academy placed Willingham, in the view of some critics, in the forefront of the post-war generation of writers which included **William Styron** and **Norman Mailer**. One reviewer, however, commented that it read 'with the imbecile monotony of scrawls on lavatory walls'. The controversy was further fuelled when the New York Society for the Suppression of Vice unsuccessfully brought obscenity charges against the novel's publisher in 1947. Willingham subsequently adapted the novel for the stage and it opened off-Broadway in September 1953, moving a month later to Broadway.

In the first half of the 1950s he produced a steady stream of novels of which *Natural Child*, about a group of Young New York Bohemians, is probably the most notable. His big novels of the 1960s, *Eternal Fire* and *Providence Island*, attracted little attention, and Willingham, always dismissive of the critics, has commented: 'Doubtful as I am of the literary world, it has nonetheless been a profound astonishment to me that these novels have not received far greater recognition.'

Since the late 1950s he has enjoyed a successful career as a scriptwriter, collaborating with Stanley Kubrick and Jim Thompson on *Paths of Glory* (1957) and subsequently writing screenplays for *The Graduate* (1967), *Little Big Man* (1970), *Thieves Like Us* (1974) and many other films. In 1991 a film version of his romantic comedy *Rambling Rose* was released, scripted by Willingham, directed by Martha Coolidge, and starring Robert Duvall and Laura Dern. He believes that his 'serious work has been done in the novel' and writes scripts to support his family of six children.

Fiction

End as a Man 1947; *Geraldine Bradshaw* 1950; *The Gates of Hell* **(s)** 1951; *Reach to the Stars* 1951; *Natural Child* 1952; *To Eat a Peach* 1955; *Eternal Fire* 1963; *Providence Island* 1969; *Rambling Rose* 1972; *The Big Nickel* 1975; *The Building of Venus Four* 1977

Plays

End as a Man [from his novel] 1953; *Thou Shalt Not Commit Adultery* [with D. Reisman] **(tv)** 1978; *Rambling Rose* **(f)** [from his novel] 1992

A(ndrew) N(orman) WILSON 1950–

Wilson was born in Stone, Staffordshire; his forebears had been pottery owners and his father was for a time a managing director at the Wedgwood pottery. He attended Rugby School and, from 1969 to 1972, New College, Oxford, While there, in 1971, he married Katherine Duncan-Jones, a fellow in English at Somerville College and his tutor; they had two daughters. After graduation, he did freelance teaching for the university and at a crammer's; spent a year at St Stephen's House, Oxford, training for the Anglican priesthood; and from January 1975 taught for five terms at Merchant Taylors' School. From 1976 to 1981, Wilson was a junior lecturer first at St Hugh's and then at New College, Oxford.

His first novel, *The Sweets of Pimlico*, was published in 1977, winning the John Llewelyn Rhys Prize, and setting him off on a prolific fictional career. At around the same time, he became active as a journalist and reviewer, and by 1986 was to have written around half a million words in these genres. On losing his Oxford lectureship in 1981, he moved straight into the literary editorship of the *Spectator*, but was fired two years later for alleged misconduct in the handling of a review. This was only one of several controversies in which this celebrated 'young fogey' became involved, and on which his reputation thrived. In the mid-1980s he announced a (much publicised) retreat from the limelight to concentrate on his more serious writing, but he was soon back as literary editor of the *Evening Standard* and the author of many other columns. In 1985 he published a spirited defence of the Christian religion, *How Can We Know?*, but six years later appeared before the public with the pamphlet *Against Religion*, a blast of the trumpet soon followed up with his lavish life of Jesus, one of six full-length contributions to the biographical genre, the most notable of which are his lives of Tolstoy and **C.S. Lewis**.

Besides various general works, he had also published fourteen novels by the mid-1990s, of which the three beginning with *Incline Our Hearts* (1988) form a trilogy. They show a ready talent for fiction, uniting absurdist comedy with Tory satire, rarely failing to involve and sometimes move the reader. Amongst his other books are *Stray*, the remarkable fictional 'autobiography' of an alley-cat, which **Brigid Brophy** declared a classic to rank alongside Anna Sewell's *Black Beauty* (1877), and *Lilibet*, a verse narrative about the early life of Queen Elizabeth II, which was published anonymously as 'by a Loyal Subject of Her Majesty'.

He and his first wife split up in the later 1980s, and he moved from Oxford to London; in 1991 he married Ruth Guilding, an art historian with whom he lives in Camden Town.

Fiction

The Sweets of Pimlico 1977; *Unguarded Hours* 1978; *Kindly Light* 1979; *The Healing Art* 1980; *Who Was Oswald Fish?* 1981; *Wise Virgin* 1982; *Scandal* 1983; *Gentlemen in England* 1985; *Love Unknown* 1986; *Stray* 1987; *Incline Our Hearts* 1988; *A Bottle of Smoke* 1990; *Daughter of Albion* 1991; *The Vicar of Sorrows* 1993

Non-Fiction

The Laird of Abbotsford: a View of Sir Walter Scott 1980; *The Life of John Milton* 1983; *Hilaire Belloc* 1984; *How Can We Know?* 1985; *Penfriends from Porlock* 1988; *Tolstoy* 1988; *C.S. Lewis: a Biography* 1990; *Against Religion* 1991; *The Faber Book of Church and Clergy* 1992; *Jesus* 1992; *The Rise and Fall of the House of Windsor* 1993

Poetry

Lilibet 1984

Edited

Sir Walter Scott *Ivanhoe* 1982

Angus (Frank Johnston) WILSON 1913–1991

The youngest child (by thirteen years) of financially insecure parents, Wilson was born in Bexhill-on-Sea and – until he entered Westminster School – had a peripatetic childhood, the family moving from hotel to hotel depending upon the fluctuating fortunes of his father, who was an inveterate gambler. Something of his parents' circumstances are apparent in the elderly Calverts in *Late Call*, while the father in *No Laughing Matter* is given Wilson *père*'s own nickname of 'Billy Pop'. He learned all about genteel poverty at an early age, and several of his short stories are set in dowdy hotels and amongst people 'intent upon the cultivation of a Knightsbridge exterior with a Kensington purse'.

At Merton College, Oxford, Wilson studied medieval history and developed his acting talents; he was a brilliant mimic, as became apparent when he began to write. After graduation, he pursued a number of jobs before joining the British Library in 1937. He eventually became deputy to the superintendent of the Reading Room, but during the war he worked for the Foreign Office at Bletchley Park and suffered a nervous breakdown. He began writing short stories as therapy and the first of them were published in *Horizon*. By this time he had returned to the Library, where he was responsible for replacing books lost in the blitz, and he wrote at the weekends. Although he made his reputation with two volumes of satirical short stories which perfectly suited the brave new post-war world, he remained at the Library until 1955, by which time he had published his first novel, *Hemlock and After*, which drew upon his homosexual experiences and displayed his unrivalled gift for depicting social unease. But this time he had met Tony Garrett, another employee of the Library, with whom he was to spend the remainder of his life.

A voracious reader, Wilson became a regular reviewer, notably for the *Observer*, and a visiting professor to a number of American universities. From 1963 to 1978 he was a lecturer at the University of East Anglia, and with **Malcolm Bradbury** set up a notably successful creative writing course: **Kazuo Ishiguro** and **Ian McEwan** were amongst the alumni. He was a fellow and subsequently president of the Royal Society of Literature, chairman of the National Book League and a member of the Arts Council, while his honours included a CBE (1968), a CLit (1972) and a knighthood (1980). His novels share the liberal concerns of **E.M. Forster** and the elaborate plotting and rich characterisation of Dickens. Truly Victorian in scope and ambition, they are nevertheless very much novels of the twentieth century: anyone who wants to know what England was like between the 1920s and the 1950s will find no sharper portrait. As well as three volumes of short stories and eight novels, Wilson wrote plays, travel books, criticism (of himself – *The Wild Garden* – and others), social history and biographical studies of **Rudyard Kipling**, Zola and Dickens.

In the late 1970s his reputation appeared to be in a decline which his disappointing final novel, *Setting the World on Fire*, did nothing to halt. Furthermore, the policies of the Thatcher government seemed to him to be wholly alien to the England he loved, and in 1985 he left his Suffolk home and passionately tended garden to settle in France, where he was properly honoured as a writer. The inexplicable popular and critical neglect of his work was such that when he developed encephalitis and returned to England, he had to be supported by a pension from the Royal Literary Fund and money raised by friends. His faculties severely impaired, he spent his last years in a nursing home where he eventually died of a stroke.

None of Wilson's novels was in print at his death, although a revival of interest seemed imminent. His talent was perhaps too diverse for a country where 'cleverness' is distrusted – although highly readable and immensely funny, the novels are allusive and make use of parody and alienating techniques – but his work has a solidity which makes much post-war fiction seem flimsy. His early novels examine the strengths and failings of liberal humanism, while later ones are determinedly experimental: *The Old Men at the Zoo* uses animals symbolically and contains a nuclear holocaust; *No Laughing Matter* is a panorama of English life in the twentieth century in the form of a parodic family saga; while *As If By Magic* was described by Malcolm Bradbury as 'our best British attempt at the inter-cultural global novel'. Wilson's assessment of the similarly experimental, and similarly neglected, **Henry Green** applies equally to himself: 'He remains one of the few really considerable English novelists of our time'.

Fiction

The Wrong Set and Other Stories (s) 1949; *Such Darling Dodos and Other Stories* (s) 1950; *Hemlock and After* 1952; *For Whom the Cloche Tolls* [with P. Jullian] 1953; *Anglo-Saxon Attitudes* 1956; *A Bit off the Map and Other Stories* (s) 1957; *The Middle Age of Mrs Eliot* 1958; *The Old Men at the Zoo* 1961; *Late Call* 1964; *No Laughing Matter* 1967; *As If By Magic* 1973; *Setting the World on Fire* 1980

Non-Fiction

Emile Zola: a Study of His Novels 1952; *Tempo: the Impact of TV on the Arts* 1964; *The Wild Garden; or, Speaking of Writing* 1964; *The World of Charles Dickens* 1970; *The Naughty Nineties* 1976; *The Strange Ride of Rudyard Kipling* 1977; *Diversity and Depth in Fiction* 1983; *Reflections in a Writer's Eye* 1986

Edited

Writers of East Anglia 1977; *East Anglia in Verse* [with T. Garrett] 1982; *The Viking Portable Dickens* 1983

Plays

The Mulberry Bush 1956

Biography

Angus Wilson by Margaret Drabble 1995

Colin (Henry) WILSON 1931–

Wilson was born in Leicester where his father worked in a shoe factory. Despite leaving Gateway Secondary Technical School at the age of sixteen with a mediocre academic record, he had been a voracious reader from a precociously young age, possibly in consequence of his mother's influence, and had written since the age of nine. At thirteen the autodidactic child, known to his friends as 'the Professor', wrote a paper on Einstein, and a year later attempted a book-length summary of the world's entire scientific knowledge: 'My ambition was to develop the atomic bomb, and when this was done in 1945, I lost interest in science.' From 1947 he worked in a variety of jobs, including spells as a laboratory assistant at his old school and as a civil servant in the tax office. Following two years as an aircraftman in the Royal Air Force, he spent the winter of 1950–1 in Paris and Strasbourg before moving to London. He continued to read and write voraciously and in 1954 started the novel subsequently published as *Ritual in the Dark*.

Sleeping on Hampstead Heath during the summer and writing in the British Museum, Wilson started *The Outsider* the next year. The book attracted immediate attention when it was published in May 1956, and the author of a study of the existential 'outsider' which draws on the lives and work of Camus, Nietzsche and Kafka among others, found himself splashed across the pages of the popular press. A talent for self-publicity, and the occasion when his future father-in-law publicly threatened to horsewhip him for supposed depravity, doubtless contributed to the public interest. In contrast to the critical success of *The Outsider*, *Religion and the Rebel* was almost universally panned, but the indefatigable Wilson remained steadfast in his intent 'to finish as the greatest writer European civilization has produced', and has since produced a formidable body of philosophical work in an attempt to define an 'optimistic existentialism'. The extremes of human consciousness and behaviour are central to Wilson's thinking, and he has undertaken studies of crime, violence, sexual oddity and the occult. His fiction, beginning in 1960 with *Ritual in the Dark*, in which a young writer is drawn into the world of a mass murderer, explores many of the same themes.

After a short-lived marriage in the early 1950s, Wilson remarried in 1960, and the couple have three children. He has held a number of visiting professorships at American universities, and lives in Cornwall.

Non-Fiction

The Outsider 1956; *Religion and the Rebel* 1957; *The Age of Defeat* [US *The Stature of Man*] 1959; *Encyclopaedia of Murder* [with P. Pitman] 1961 [rev. 1984]; *The Strength to Dream* 1962; *Origins of the Sexual Impulse* 1963; *Rasputin and the Fall of the Romanovs* 1964; *Brandy of the Damned* 1964 [US *Chords and Discords* 1966; rev. as *Colin Wilson on Music* 1967]; *Beyond the Outsider* 1965; *Eagle and Earwig* 1965; *Introduction to the New Existentialism* 1966; *Sex and the Intelligent Teenager* 1966; *Voyage to a Beginning* **(a)** 1966; *Bernard Shaw: a Reassessment* 1969; *A Casebook of Murder* 1969; *Poetry and Mysticism* 1969; *The Strange Genius of David Lindsay* [with J.B. Pick, E.H. Visiak] 1970 [US *The Haunted Man* 1979]; *The Occult* 1971; *L'Amour: the Ways of Love* [with P. Rimaldi] 1972; *New Pathways in Psychology: Maslow and the Post-Freudian Revolution* 1972; *Order of Assassins: the Psychology of Murder* 1972; *Strange Powers* 1973; *Tree by Tolkien* 1973; *A Book of Booze* 1974; *Hermann Hesse* 1974; *Jorge Luis Borges* 1974; *Wilhelm Reich* 1974; *The Craft of the Novel* 1975; *Mysterious Powers* 1975 [rev. as *Mysteries of the Mind* 1978]; *The Unexplained* 1975; *Enigmas and Mysteries* 1976; *The Geller Phenomenon* 1976; *Colin Wilson's Men of Mystery* 1977; *Mysteries: an Investigation into the Occult, the Paranormal and the Supernatural* 1978; *Science Fiction as Existentialism* 1978; *The Book of Time* [with J. Grant] 1980; *Frankenstein's Castle* 1980; *Starseekers* 1980; *The War against Sleep: the Philosophy of Gurdjieff* 1980 [rev. 1986]; *Anti-Sartre, with an Essay on Camus* 1981; *Poltergeist! A Study in Destructive Haunting* 1981; *The Quest for Wilhelm Reich* 1981; *Witches* 1981; *Access to Inner Worlds: the Story of Brad Absetz* 1983; *Encyclopaedia of Modern Murder 1962–1982* [with D. Seaman] 1983 [rev. 1989]; *A Criminal History of Mankind* 1984; *Lord of the Underworld: Jung and the Twentieth Century* 1984; *Psychic Detectives: the Story of Psychometry and the Paranormal in Crime Detection* 1984; *Afterlife* 1985; *The Bicameral Critic* **(c)** [ed. H.F. Dossar] 1985; *Rudolf Steiner: the Man and His Vision* 1985; *An Essay on the 'New' Existentialism* 1986; *The Goblin Universe* [with T. Holiday] 1986; *The Laurel and Hardy Theory of Consciousness* 1986; *Scandal! An Encyclopaedia* [with D. Seaman] 1986; *The Encyclopaedia of Unsolved Mysteries* 1987; *Jack the Ripper: Summing Up and Verdict* [with R. Odell] 1987; *The Musician as Outsider* 1987; *Aleister Crowley: the Nature of the Beast* 1988; *Autobiographical Reflections* **(a)** 1988; *The*

Magician from Siberia 1988; *The Mammoth Book of True Crime* 1988; *The Mammoth Book of Time* [ed. J. Grant] 1988; *Beyond the Occult: a Twenty Year Investigation into the Paranormal* 1989; *The Decline and Fall of Leftism* 1989; *The Misfits: A Study of Sexual Outsiders* 1989; *Written in Blood: a History of Forensic Detection* 1989; *The Serial Killers: a Study in the Psychology of Violence* [with D. Seaman] 1990; *The Mammoth Book of the Supernatural* 1991

Fiction

Ritual in the Dark 1960; *Adrift in Soho* 1961; *Man without a Shadow* [US *The Sex Diary of Gerard Sorme*] 1963; *The World of Violence* [US *The Violent World of Hugh Greene*] 1963; *Necessary Doubt* 1964; *The Glass Cage* 1966; *The Mind Parasites* 1967; *The Philospher's Stone* 1969; *The God of the Labyrinth* 1970 [US *The Hedonists* 1971]; *The Killer* [US *Lingard*] 1970; *The Black Room* 1971; *The Return of the Lloigor* (s) 1974; *The Schoolgirl Murder Case* 1974; *The Space Vampires* 1976; *The Janus Murder Case* 1984; *The Personality Surgeon* 1986; *Spider World*: *The Delta* 1987, *The Tower* 1987, *The Desert* 1988, *Fortress* 1989; *The Sex Diary of a Metaphysician* 1988

Plays

The Metal Flower Blossom 1958; *Viennese Interlude* 1960; *Strindberg* 1970 [as *Pictures in a Bath of Acid* 1971, as *Strindberg: a Fool's Decision* 1975]; *Mysteries* 1979

Edited

The Directory of Possibilities [with J. Grant] 1981

Edmund WILSON 1895–1972

Widely regarded at the most eminent American man-of-letters of the century, Wilson was born in New Jersey, of an American Puritan family that had filled the learned professions for generations: his father was a lawyer and politician. He attended boarding-school and Princeton University, where he was a close friend of **F. Scott Fitzgerald**, whose posthumously published works he edited. He was briefly a reporter on the New York *Evening Sun* after leaving Princeton, and then served in the US Army during the First World War in the USA, England and France. He was managing editor of *Vanity Fair* in 1920–1 and associate editor of the *New Republic* from 1926 to 1931. After that, he was largely freelance, contributing widely to the most prestigious American magazines, although from 1944 to 1948 he was the regular literary critic on the *New Yorker*.

His first volume (written with John Peale Bishop) was *The Undertaker's Garland* (1922), a compilation in prose and verse about death, but it was *Axel's Castle*, an immensely influential study of Symbolism, which established Wilson as a leading critic. During the 1930s he was considerably to the left, writing in approving fashion about a visit to the Soviet Union; *To the Finland Station*, a historical study of Marxism, was another influential book. Wilson also wrote a novel set in Greenwich Village, *I Thought of Daisy*; a praised volume of short stories, *Memoirs of Hecate Country*; plays and verse: like **Cyril Connolly** in England, a figure he closely resembles, he straddled the gap between criticism and creation. Wilson also wrote many other works of criticism and essays as well as books about the literature of the American Civil War, the American Indian, the Dead Sea Scrolls, Canada, and an irascible account of his own dealings with the income tax authorities. Wilson was four times married and with three of his wives had one child each; his most celebrated wife was his third, the novelist **Mary McCarthy**, with whom he had a son.

Non-Fiction

Axel's Castle: a Study in the Imaginative Literature of 1870–1930 1931; *The American Jitters: a Year of the Slump* [as *Devil Take the Hindmost*] 1932; *The Triple Thinkers: Ten Essays in Literature* 1938 [rev. as *The Triple Thinkers: Twelve Essays on Literary Subjects* 1948]; *To the Finland Station* 1940; *The Boys in the Back Room: Notes on California Novelists* 1941; *The Wound and the Bow: Seven Studies in Literature* 1941; *Europe without Baedeker: Sketches among the Ruins of Italy, Greece, and England* 1947 [rev. with *Notes from a European Diary: 1963–64* 1966]; *Classics and Commercials: a Literary Chronicle of the Forties* 1950; *The Shores of Light: a Literary Chronicle of the Twenties and Thirties* 1952; *The Scrolls from the Dead Sea* 1955; *Red, Black, Blond and Olive: Studies in Four Civilizations: Zuni, Haiti, Soviet Russia, Israel* 1956; *A Piece of My Mind: Reflections at Sixty* 1957; *The American Earthquake: a Documentary of the Twenties and Thirties* 1958; *Apologies to the Iroquois* 1960; *Patriotic Gore: Studies in the Literature of the American Civil War* 1962; *The Cold War and Income Tax: a Protest* 1963; *The Bit Between My Teeth: a Literary Chronicle of 1950–1965* 1965; *O Canada: an American's Notes on Canadian Culture* 1965; *A Prelude: Landscapes, Characters and Conversations from Earlier Years of My Life* 1967: *The Fruits of the MLA* 1968; *Upstate: Records and Recollections of Northern New York* 1971; *A Window on Russia for the Use of Foreign Readers* 1972; *The Devils and Canon Barham: Ten Essays on Poets, Novelists, and Monsters* 1973; *The Twenties: From Notebooks and Diaries of the Period* [ed. L. Edel] 1975; *Letters on Literature and Politics 1912–1972* [ed. D. Aaron] 1977

Poetry

Three Reliques of Ancient Western Poetry Collected by Edmund Wilson from the Ruins 1951; *Wilson's Christmas Stocking* 1953; *A Christmas Delirium* 1955; *Night Thoughts* 1961; *Holiday Greetings* 1966

Plays

The Crime in the Whistler Room 1924 [pub. in *This Room and This Gin and These Sandwiches* 1937]; *Discordant Encounters* 1926; *The Little Blue Light* 1950; *Five Plays* [pub. 1954]; *The Duke of Palermo and Other Plays* [pub. 1969]

Fiction

I Thought of Daisy 1929 [rev. with *Galahad* 1957]; *Memoirs of Hecate County* (s) 1946 [rev. 1958]

Edited

F. Scott Fitzgerald: *The Last Tycoon: an Unfinished Novel by F. Scott Fitzgerald, Together with The Great Gatsby and Selected Stories* 1941, *The Crack-Up: With Other Uncollected Pieces, Note-Books and Unpublished Letters* 1945; *The Last Tycoon: an Unfinished Novel* 1949; *The Shock of Recognition: the Development of Literature in the United States Recorded by the Men Who Made It* 1943 [rev. 1955]; *The Collected Essays of John Peale Bishop* 1948; Chekhov *Peasants and Other Stories* (s) 1956

Collections

The Undertaker's Garland [with J. Peale Bishop] 1922; *Poets, Farewell!* 1929; *Travels in Two Democracies* 1936; *Note-Books of Night* 1942

Donald WINDHAM 1920–

Born in Atlanta, Georgia, Windham was the second son of parents who divorced when he was young. He was educated at the local high school. After graduating, he worked for a year in a Coca-Cola Company barrel factory and then left permanently for New York, where in 1942 he met the actor Sandy Campbell and embarked on an emotional and creative partnership which lasted until Campbell's death in 1988.

Windham's other significant early acquaintance was **Tennessee Williams**, with whom he collaborated on a dramatisation of **D.H. Lawrence**'s story 'You Touched Me' (produced in 1943 but not published until 1947). The Broadway success of the play in 1945 allowed him to retire as editor of *Dance Index* (1943–5) and travel to Italy, where he finished his first novel, *The Dog Star* (1950). Inspired by the Actaeon myth, the book is an austere portrait of small-town life. His next novel, *The Hero Continues*, loosely based on the career of Williams, is a disturbing study of an artist's decline from integrity to disintegration. In his later memoirs of Williams, *As if ...*, and **Truman Capote**, *Footnote to a Friendship*, Windham again addressed, with perception and sympathy, the dangers of public success.

The award of a Guggenheim Fellowship in 1960 allowed Windham to return to Italy, where he worked on the autobiographical pieces that eventually became his childhood memoir *Emblems of Conduct*. His Joycean, epiphanic collection of short stories, *The Warm Country* (published in 1960 with an introduction by **E.M. Forster**), also features pieces based on his childhood in Atlanta (most notably 'The Starless Air'), as well as stories inspired by Italy; the latter influence bearing fullest expression in his third and best novel, *Two People*. A cool revision of Thomas Mann's *Death in Venice* (1912), it describes a casual but significant affair between an American man and an Italian youth in Rome. Like all of Windham's work, the novel is about individuals locked in suffocating circumstances, while uncharacteristically reaching some sort of positive resolution.

Neither of Windham's subsequent novels have achieved the success of *Two People*, and he appears to have turned from fiction to memoirs, becoming the Boswell of his generation. As limited as his output has been, he remains nevertheless one of America's subtlest and most subversive prose stylists.

Fiction

The Dog Star 1950; *The Hero Continues* 1960; *The Warm Country* (s) 1960; *Two People* 1965; *Tanaquil* 1972; *Stone in the Hourglass* 1981

Plays

You Touched Me! [from D.H. Lawrence's story; with T. Williams] 1943 [pub. 1947]

Non-Fiction

Emblems of Conduct (a) 1964; *Footnote to a Friendship* 1983; *As if ...* 1985; *Lost Friendships* 1987; *The Roman Spring of Alice Toklas* 1987

Edited

Tennessee Williams' Letters to Donald Windham 1940–1965 1976

Jeanette WINTERSON 1959–

Winterson was a foundling, adopted by Pentecostal Evangelists who raised her in Accrington, Lancashire, to be a missionary after her adoptive mother saw a 'sign' indicating that the child could save the world. Her first novel, *Oranges Are Not the Only Fruit*, is semi-autobiographical. In it

Mr Winterson, a shy man who is employed in a television factory, is a shadowy figure, unlike his wife, who looms large. She is a domineering woman who teaches her daughter to read and write and sends her to school (which she calls 'The Breeding Ground') only when prosecution is threatened. On the novel's publication, her mother sent Winterson a note that read 'You are the child of the devil. Love Mother', and severed all ties.

At twelve, Winterson was preaching from the hustings and saving souls. At sixteen she was having her first love affair, with an older woman, a fish-filleter, and she was thrown out of the Elim Pentecostal Church and her home. She spent two years driving an ice-cream van and doing make-up in a funeral parlour that belonged to a family friend; a year working in Calderstone's Mental Hospital; and from there she went to St Catherine's Oxford. She actually failed the University's entrance exams, but travelled down personally to upbraid the authorities and secured a place to read English.

After leaving university, she came to London and had an affair with a wealthy stockbroker, although she had a job at the Gateways, a famous lesbian club. His subsequent proposal, with the promise of a country cottage, was turned down, and Winterson did a number of short-term jobs (including a stint as dogsbody at the Roundhouse theatre 'writing programme notes, clearing up') before deciding upon a career in publishing. She approached the feminist press Pandora and, although she did not get the job, her descriptions of her upbringing so intrigued the editor who had interviewed her that it was suggested she went away and wrote a book about it.

Oranges Are Not the Only Fruit won the Whitbread Award for a first novel, but was followed by *Boating for Beginners*, which retells the story of Noah and was published against the advice of friends. It got the bad reviews is deserved and it is said that Winterson subsequently attempted to suppress its reissue, though it is now back in print. A year later she published what many people regard as her best book, *The Passion*, an inventive historical novel about Napoleon's chef and a beautiful courtesan. It was turned down by her former publisher, but eventually won the John Llewelyn Rhys Prize and prompted **Gore Vidal** to describe Winterson as 'the most interesting young writer' to have appeared 'in twenty years'. Many people agreed, and her next novel, *Sexing the Cherry*, a fantastical trip into the reign of Charles II, won an E.M. Forster Award.

Winterson's subsequent career has been, to say the least of it, controversial. This is partly due to particular episodes in her life and career (*Written on the Body* supposedly contains an unflattering portrait of a novelist with whose wife Winterson had enjoyed an affair), but the principal debate is about her literary standing. She has attracted a hard core of admirers, and is still considered in some quarters 'the leader of her literary generation ... a master of the English language', as the 'Notes on the Authors' put it in *The Penguin Book of Lesbian Short Stories* (1993), edited by her partner, the academic and critic Margaret Reynolds. This is an opinion that Winterson herself shares, frequently stating as much in public. She is perhaps unique in choosing one of her own novels as her Book of the Year in a newspaper round-up, and of her first novel she has written: 'The most important thing about *Oranges* is not its wit nor its warmth, but its new way with words.' Her way with words (particularly in the novel *Art and Lies*) has been questioned by some critics, and she has visited the houses of reviewers who have ventured criticism in order to castigate them in person. This overweening self-regard has not endeared her to everyone, and has tended to obscure her genuine gifts.

Even Winterson's detractors, however, praised her three-part television adaptation of *Oranges Are Not the Only Fruit*, which (unlike most adaptations) actually improved on the original novel. It outraged the tabloid newspapers because of its explicit, but highly decorous, scenes of lesbian lovemaking, but it delighted the critics and won several awards. Her volume of non-fiction, *Art Objects*, which includes essays on writers she admires such as **Virginia Woolf** and **Gertrude Stein**, as well as some observations about the importance of art and artists, neatly sums up her position in its subtitle: *Essays on Ecstasy and Effrontery*.

Fiction

Oranges Are Not the Only Fruit 1985; *Boating for Beginners* 1986; *The Passion* 1987; *Sexing the Cherry* 1989; *Written on the Body* 1992; *Art and Lies* 1994

Non-Fiction

Art Objects: Essays on Ecstasy and Effrontery 1995

Plays

Oranges Are Not the Only Fruit (**tv**) [from her novel] 1989

P(elham) G(renville) WODEHOUSE 1881–1975

Born in Guildford, Surrey, Wodehouse was of aristocratic ancestry on both sides, and his father was a judge in Hong Kong. He was taken to

Hong Kong when a baby, but sent home, with his elder brothers, aged two. He attended private schools and lived with a series of aunts, who were often married to country clergymen, an atmosphere which was to form a vital component in his 100 or so works of fiction. His public school was Dulwich College, to which he remained devoted all his life. On leaving school, he worked for two years, very miserably, in a London bank, but left in 1903, already having achieved some success as a freelance writer, to write the 'By the Way' column on the London *Globe*.

His first novel, a school story entitled *The Pothunters*, was published in 1902, and during the Edwardian age many of his school stories were serialised in *The Captain* and other boys' papers. The first of Wodehouse's great comic characters, Psmith (the 'P' is silent), emerged from these school stories. A languid, dandiacal Old Etonian, who affects left-wing views and addresses everyone as 'Comrade', he turns up at Sedleigh halfway through *Mike* (1909) and runs away with the story. His later career is charted in several novels (*Psmith in the City*, *Psmith Journalist* and so on), and he turns up in another Wodehouse locale, Blandings Castle, in *Leave It to Psmith*. Blandings is the Shropshire seat of the pig-loving Lord Emsworth, one of Wodehouse's most endearing creations. His most famous characters, however, are the ineffable Bertie Wooster and his imperturbable manservant Jeeves, who first appeared in a collection of short stories in 1919. Wooster is the archetypal 'silly ass', a true innocent in a world of scheming aunts and lovesick young women, and this lands him in all manner of scrapes, from which the supremely intelligent and unflappable Jeeves has to extricate him. Wodehouse's world seems 'period', but it never really existed – it is 'a world as timeless as that of *A Midsummer Night's Dream* and *Alice in Wonderland*', as his great admirer **Evelyn Waugh** put it. Similarly, his language, which seems archaic, though related to Edwardian schoolboy patois, is largely his own invention and provides much of the comedy. He is also a master of the inventive simile and metaphor.

In 1904, Wodehouse had visited the USA for the first time, and he soon began spending part of every year there: he was also there, being unfit for service, during the years of the First World War. (A great many of his novels and stories have American settings.) In 1914, he married a widow, Ethel Rowley, and adopted her daughter (there were no children of the Wodehouse marriage). His wife became his indefatigable business manager and outlived him, dying in 1984.

From the 1920s his writing was immensely popular in England (*'Petroniumne dicam an Terentium nostrum?'* – 'Shall I call him our Petronius or our Terence?' – said the Public Orator at Oxford when admitting him for a DLitt.), and he was also extremely well known as a writer of lyrics and books for innumerable musical comedies. He had a particularly fruitful collaboration with the Gershwins on Jerome Kern and Guy Bolton, resulting in such shows as *Sitting Pretty*, and *O, Kay!*; these prove him to be a lyricist in the same class as Cole Porter. His most famous song (with music by Kern) is 'Bill', originally written in 1918, but incorporated in Kern and Oscar Hammerstein's *Show Boat* in 1927.

From 1934 the Wodehouses lived in the south of France, and were still there when the Nazis invaded in 1940. Wodehouse was taken to Berlin and persuaded to make broadcasts on German radio; these were totally innocuous in content, but aroused an immense reaction in England, where there were calls in Parliament that he be tried for treason. After the war, he could not be guaranteed immunity from prosecution in Britain and he settled in the USA, becoming an American citizen in 1955: he continued writing to the end. As his *Times* obituary put it, Wodehouse remained a late Victorian teenager into his eighties, dressed in a blue blazer, shy, immensely fond of sports. His comic world has made him as much loved by intellectuals as by an immense public: **Sinclair Lewis** said that he was 'not an author but a whole department of rather delicate art'. He was knighted just weeks before his death in Long Island when he was ninety-three.

Fiction

The Pothunters 1902; *A Prefect's Uncle* 1903; *Tales of St Austin's* (s) 1903; *The Gold Bat* 1904; *The Head of Kay's* 1905; *Love Among the Chickens* 1906; *Not George Washington* [with H. Westbrook] 1907; *The White Feather* 1907; *Mike* 1909 [rev. as *Mike at Wrykin* and *Mike and Psmith* 1953]; *The Swoop* 1909; *The Intrusion of Jimmy* [UK *A Gentleman of Leisure*] 1910; *Psmith in the City* 1910; *The Prince and Betty* 1912 [rev. as *Psmith Journalist* 1915]; *The Little Nugget* 1913; *The Man Upstairs* (s) 1914; *Something Fresh* [US *Something New*] 1915; *Uneasy Money* 1916; *The Man with Two Left Feet* (s) 1917; *Piccadilly Jim* 1917; *A Damsel in Distress* 1919; *My Man Jeeves* (s) 1919; *Their Mutual Child* 1919 [UK *The Coming of Bill* 1920]; *The Little Warrior* 1920 [UK *Jim the Reckless* 1921]; *Indiscretions of Archie* (s) 1921; *The Adventures of Sally* 1922 [US *Mostly Sally* 1923]; *The Clicking of Cuthbert* (s) 1922 [US *Golf without Tears* 1924]; *Three Men and a Maid* 1922; *The Inimitable Jeeves* (s) [US *Jeeves*] 1923; *Leave It to Psmith* 1923; *Bill the Conqueror* 1924; *Ukridge* (s) 1924 [US *He Rather Enjoyed It* 1926]; *Carry On, Jeeves* (s) 1925; *Sam the Sudden* [US

Sam in the Suburbs] 1925; *The Heart of a Goof* (s) 1926 [US *Divots* 1927]; *Meet Mr Mulliner* (s) 1927; *The Small Bachelor* 1927; *Money for Nothing* 1928; *Fish Preferred* [UK *Summer Lightning*] 1929; *Mr Mulliner Speaking* (s) 1929; *Very Good, Jeeves* (s) 1930; *Big Money* 1931; *If I Were You* 1931; *Doctor Sally* 1932; *Hot Water* 1932; *Heavy Weather* 1933; *Blandings Castle* (s) 1935; *Right Ho, Jeeves* [US *Brinkley Manor*] 1934; *Thank You, Jeeves* 1934; *Enter Psmith* 1935 [rev. as *Mike and Psmith* 1953]; *The Luck of the Bodkins* 1935 [US rev. 1936]; *Laughing Gas* 1936; *Young Men in Spats* (s) 1936; *Lord Emsworth and Others* (s) [US *Crime Wave at Blandings*] 1937; *Summer Moonshine* 1937; *The Code of the Woosters* 1938; *Uncle Fred in the Springtime* 1939; *Eggs, Beans and Crumpets* (s) 1940; *Quick Service* 1940; *Money in the Bank* 1942; *Joy in the Morning* 1946; *Full Moon* 1947; *Spring Fever* 1948 [rev. as *The Old Reliable* 1951]; *Uncle Dynamite* 1948; *The Mating Season* 1949; *Nothing Serious* (s) 1950; *Barmy in Wonderland* [US *Angel Cake*] 1952; *Pigs Have Wings* 1952; *Ring for Jeeves* [US *The Return of Jeeves*] 1953; *Jeeves and the Feudal Spirit* 1954 [US *Bertie Wooster Sees It Through* 1955]; *French Leave* 1956; *Something Fishy* [US *The Butler Did It*] 1957; *Cocktail Time* 1958; *A Few Quick Ones* (s) 1959; *How Right You Are, Jeeves* [UK *Jeeves in the Offing*] 1960; *The Ice in the Bedroom* 1961; *Service with a Smile* 1961; *Stiff Upper Lip, Jeeves* 1963; *Biffen's Millions* [UK *Frozen Assets*] 1964; *The Brinkmanship of Galahad Threepwood* [UK *Galahad at Blandings*] 1965; *Plum Pie* (s) 1966; *The Purloined Paperweight* [UK *Company for Henry*] 1967; *Do Butlers Burgle Banks?* 1968; *A Pelican at Blandings* 1969 [US *No Nudes Is Good Nudes* 1970]; *The Girl in Blue* 1970; *Much Obliged, Jeeves* [US *Jeeves and the Tie That Binds*] 1971; *Pearls, Girls and Monty Bodkin* 1972 [US *The Plot That Thickened* 1973]; *Bachelors Anonymous* 1973; *Aunts Aren't Gentlemen* 1974 [US *The Cat Nappers* 1975]; *Sunset at Blandings* [unfinished; ed. R. Usborne] 1977

Plays

After the Show [with H. Westbrook] 1911; *A Gentleman of Leisure* [with J. Stapleton] 1911 [as *A Thief in the Night* 1913]; *Brother Alfred* [with H. Westbrook] 1913; *Pom, Pom* [with A. Caldwell] 1916; *The Golden Moth* [with F. Thompson] 1921; *The Cabaret Girl* [with P. Grossmith] 1922; *The Beauty Prize* [with P. Grossmith] 1923; *Hearts and Diamonds* [with G. John] 1926; *The Play's the Thing* [from F. Molnar] 1926; *Her Cardboard Lover* [with V. Wyngate] 1927; *Good Morning, Bill* [with L. Fodor] 1927; *A Damsel in Distress* [with I. Hay] 1928; *Baa, Baa, Black Sheep* [with I. Hay] 1929; *Candle-Light* [from S. Geyer] 1929; *Leave It to Psmith* [with I. Hay] 1930; *The Inside Stand* 1935; *Arthur* [from F. Molnar] 1947; *Game of Hearts* [from F. Molnar] 1947; *Nothing Serious* 1950

With Guy Bolton: *Have a Heart* 1917; *Leave It to Jane* 1917; *Oh, Boy!* 1917; *The Girl Behind the Gun* 1918; *Oh, Lady! Lady!* 1918; *O, My Dear!* 1918; *See You Later* 1918; *The Blue Mazurka* 1921; *Pat* 1922; *Sitting Pretty* 1924; *O, Kay!* 1926; *The Nightingale* 1927; *Who's Who* 1934; *Anything Goes* 1936; *Don't Listen, Ladies* [from S. Guitry] 1948; *Phipps* 1951; *Come On, Jeeves* 1954; *Leave It to Jeeves* 1971

Non-Fiction

The Globe by the Way Book [with H. Westbrook] 1908; *Louder and Funnier* 1932; *Bring on the Girls: the Improbable Story of Our Life in Musical Comedy with Pictures to Prove It* (a) [with G. Bolton] 1953; *Performing Flea: a Self-Portrait in Letters* (a) 1953; *America: I Like You* 1956; *Over Seventy* (a) 1957

Biography

P.G. Wodehouse by David A. Jasen 1974 [rev. 1981]; *P.G. Wodehouse* by Frances Donaldson 1982

Thomas (Clayton) WOLFE 1900–1938

Born in Asheville, North Carolina, Wolfe was the son of a sepulchral mason and his wife, who lived apart from him and ran a boarding-house. His early background and his savagely recriminative family are described in *Look Homeward, Angel* (1929), the first of his novels (all of which are autobiographical). He entered the state university at fifteen and emerged in 1920 'a hulking, shaggy slow-moving young colossus' (he was almost six foot six tall and had a frame to match). He went to Harvard to try George Pierce Baker's well-known course in playwriting, but found little success in writing plays, so became an instructor in English at the Washington Square college of New York University and spent some time in Europe while writing his first novel. This was massively undisciplined, like all Wolfe's work, but the famous editor Maxwell Perkins of Scribners took him on and helped him shape the novel into publishable form. (Wolfe's works always owed a lot to wholesale editing, first by Perkins and later by Edward C. Aswell of Harpers.)

By February 1930 the royalties from *Look Homeward, Angel* were such as to allow Wolfe to give up his academic job; he sailed for Paris and, returning to New York, worked on *Of Time and the River*, in which he takes Eugene Gant, the hero of his first novel, into the toils of the literary life. (In his last two novels, the hero, still based on himself, is renamed George Webber.)

Of major importance in Wolfe's life was his relationship, from 1925 to 1930, with a woman almost twenty years his senior, the stage-designer Aline Bernstein: she is the Esther Jack of his last two posthumous novels, *The Web and the Rock* and *You Can't Go Home Again*. Wolfe died before reaching forty of a cerebral infection following pneumonia and surgery. He had led an unhealthy life in which he poured out his immense novels at night subsisting on coffee, baked beans and cigarettes. Despite the many longueurs in his work, Wolfe is rich in all the gifts of the novelist, and must be accounted a major writer. Although he is rigidly autobiographical, one of his strongest suits is the understanding of others. The short stories in *From Death to Morning* contain some of his best work.

Fiction
Look Homeward, Angel 1929; *Of Time and the River* 1935; *From Death to Morning* (s) 1935; *The Face of a Nation* [ed. J.H. Wheelock] 1939; *The Web and the Rock* 1939; *You Can't Go Home Again* 1940; *The Hills Beyond* (s) 1941; *A Stone, a Leaf, a Door* 1945; *A Western Journal* 1951; *The Short Novels of Thomas Wolfe* [ed. C.H. Holman] 1961; *The Mountains* [ed. P.M. Ryan] 1970; *The Complete Short Stories of Thomas Wolfe* (s) [ed. F.E. Skipp] 1987

Non-Fiction
The Story of a Novel 1936; *The Autobiography of an American Novelist* (a) [ed. L. Field] 1983

Plays
Mannerhouse 1948; *Welcome to Our City* [ed. R.S. Kennedy] 1983

Biography
Look Homeward: a Life of Thomas Wolfe by David Herbert Donald 1987

Tom (i.e. Thomas Kennerly) WOLFE 1931–

Wolfe was born in Richmond, Virginia, into a long-established patrician Southern family. His father was a farmer and professor of agronomy, and editor of the magazine *Southern Planter*. After attending St Christopher's School, Richmond, and Washington and Lee University, Lexington, Wolfe went in 1952 to Yale where he gained his PhD in American studies four years later. A college baseball player, he tried out for the New York Giants as a pitcher, was unsuccessful, and began work the same year as a reporter on the *Springfield Union*. After three years there and a further three at the *Washington Post* where he grew disillusioned with the 'pale beige tone' of the journalism, he moved in 1962 to Manhattan and worked on the *Herald Tribune*'s *New York* magazine whilst freelancing for other magazines including *Esquire*. In 1963, unable to complete an article on Californian customised cars for *Esquire*, he wrote up his notes in a long, detailed stream-of-consciousness intended to help someone else to write the piece. The editor, Byron Dobell, ran the notes precisely as they were, and so was born the famous high-octane Wolfe prose style, characterised by hyperbole and the wild use of punctuation.

The Kandy-Kolored Tangerine-Flake Streamline Baby is a collection of twenty-two articles on the latest trends and heroes of pop culture. Further books established Wolfe as a leading practitioner of the 'New Journalism', which he defined as 'the use by people writing nonfiction of techniques which heretofore had been thought of as confined to the novel or the short story'. Perhaps the most notorious example of this is *Radical Chic*, his account of the party given by Leonard and Felicia Bernstein for the Black Panthers, which contains the innermost private thoughts of its protagonists. *The Electric Kool-Aid Acid Test* describes the antics of **Ken Kesey** and the Merry Pranksters, and *From Bauhaus to Our House* is a powerful diatribe against modern architecture.

In 1978 he married Sheila Berger, art director of *Harper's*, with whom he has three children, and the next year published *The Right Stuff*, an examination of the early NASA space programme which pushed his work further in the direction of fiction. The book won the 1980 National Book Award for non-fiction, and was filmed by Philip Kaufman in 1983 with **Sam Shepard** in one of the leading roles. Wolfe believes that the American novel has lost sight of its essential duty to address social reality (an argument spelled out in his 1990 essay 'Stalking the Billion-Footed Beast') and had long contemplated a novel on the nineteenth-century model which would describe New York City. His hugely successful first novel, *The Bonfire of the Vanities*, was a rigorously researched description of Wall Street greed, political corruption and racial tension which endeavoured to return the attentions of the novel to 'the dirt of everyday life'. Brian de Palma's 1990 film version bears little relationship to the novel and was critically panned and accused of racism.

Non-Fiction
The Kandy-Kolored Tangerine-Flake Streamline Baby 1965; *The Electric Kool-Aid Acid Test* 1968; *The Pumphouse Gang* 1968; *Radical Chic* and *Mau-Mauing the Flak Catchers* 1970; *The Painted Word* 1975; *Mauve Gloves & Madmen, Clutter & Vine* 1976; *The Right Stuff* 1979; *In Our Time* 1980; *From Bauhaus to Our House* 1981; *The Purple Decades* 1983; *The New America* 1989

Fiction

The Bonfire of the Vanities 1987

Plays

The Right Stuff (f) [from his novel] 1983; *The Bonfire of the Vanities* (f) [from his novel] 1990

Edited

The New Journalism 1973

Tobias (Jonathan Ansell) WOLFF 1945–

Wolff was born in Birmingham, Alabama; his father was of German Jewish and his mother of Irish ancestry. His brother Geoffrey Wolff is a biographer and novelist, and the two siblings have won fame partly because both have written memoirs of their childhoods, which differed widely because, when Tobias was five, their mother left their father and each brother went with a different parent. Geoffrey Wolff has told the story of his childhood with his father, an aeronautical engineer and alcoholic conman who served a prison sentence for car theft, in *The Duke of Deception* (1979).

Rosemary Wolff was a secretary, and the odyssey across America on which she led her younger son is chronicled in *This Boy's Life*, a memoir which uses some of the techniques of fiction and which brought Wolff to wide public notice on its publication in 1989. It is a story of how the mother and son went West to mine uranium, how Wolff grew up delinquent, partly in a town called Concrete in Washington State, how his childhood was disfigured by his mother's marriage to a hated stepfather, and how he used forged letters of recommendation to get into an exclusive Eastern preparatory school, the Hill School in Pennsylvania.

He was expelled, however, and soon joined the army, serving four years, which included a year's tour of duty in Vietnam, chronicled in his second volume of memoirs, *In Pharoah's Army*. After returning to America, he was eventually successful in entering Hertford College, Oxford, where he studied between 1969 and 1972. Back in the USA he worked briefly and unsuccessfully as a reporter on the *Washington Post*, before going to California to take jobs as a waiter, a night-watchman and a high-school teacher. In 1975, his thirtieth year, his life changed in three important ways: he published an early novel, *Ugly Rumours*; he married a teacher of art history, later to become a social worker, with whom he was to have three children; and he received a fellowship from Stanford University, where he subsequently taught. In 1980 he became writer-in-residence at Syracuse University in upstate New York.

His writing career proper is generally considered to have begun with the twelve short stories of *In the Garden of the North American Martyrs*, published in 1981 (this appeared in Britain as *Hunters in the Snow*), and it was for his short stories that he was largely known, and with which he achieved a considerable reputation, before the publication of *This Boy's Life*. His stories are often classified, along with the writing of **Raymond Carver** and **Richard Ford**, as part of a 1980s' school of 'dirty realism', with their unsparing concentration on the underside of American life, although this classification perhaps fails to do justice to the taut lyricism of Wolff's work. Besides the books mentioned, he has also published *The Barracks Thief*, a novella making effective use of his knowledge of army life, which won the 1985 PEN/Faulkner Award.

Fiction

Ugly Rumors 1975; *In the Garden of the North American Martyrs* 1981 [UK *Hunters in the Snow* 1982]; *The Barracks Thief* 1984; *Back in the World* (s) 1985

Non-Fiction

This Boy's Life (a) 1989; *In Pharoah's Army* (a) 1994

Edited

Matters of Life and Death: New American Stories (s) 1982

Nicholas WOLLASTON 1926–

Wollaston was born in the Cotswolds, the second of three children. His mother (who died when he was sixteen) was a niece of Beatrice Webb, while his father had been an explorer, with an African mountain, two plants and a Tibetan rabbit named after him; he subsequently became a tutor of King's College, Cambridge, where in 1930 he was shot dead by a deranged undergraduate. Wollaston was educated at Winchester, which he loathed, before enlisting in the navy; he saw no action in the war, but served as a navigator on a minesweeper in 1945–6. The following year he continued his education at King's, but left after four terms to climb a Saharan mountain. He then worked for an uncle's leather business in Kenya. In 1952 he returned to Cambridge to complete his degree.

He has spent much of his life travelling (the Pacific Islands, China, Israel, India, Central and South America, South-East Asia, the West Indies), often on journalistic assignments for a variety of newspapers and magazines, and is the author of several travel books – including *Red Rumba*, described by **Graham Greene** as 'perhaps the best travel book since Patrick Leigh Fermor's *The Traveller's Tree*', and *Winter in England* – and *The Man on the Ice Cap*, a biography of the explorer Augustus Courtauld. He married in 1961, and worked briefly as a writer in the African service of the BBC. His first novel, *Jupiter Laughs*, was published in 1967, and he has subsequently published six others, including *Mr Thistlewood*, which concerns the Cato Street conspiracy to blow up the Cabinet in 1820, and *Café de Paris*, an evocative and superbly structured book based upon the bombing of the celebrated London night-club in 1940 (an event also used by **Anthony Powell** in *A Dance to the Music of Time*, 1951–75).

Of the poet protagonist of this last novel, Wollaston wrote: 'He saw that life was a monster but on balance it was worthwhile, and his duty – his delight – was to celebrate the news. He must offer something in return for life, for the people who shared its vast conspiracy. He would do it in words chosen for their tone and rhythm and the patterns in which he could make them dance.' This is what Wollaston does in his own work, for he is a careful, lyrical writer. Equally remarkable is his autobiographical book, *Tilting at Don Quixote*, in which he uses Cervantes' knight as the archetype of the writer, creating romance out of the mundane; a journey through La Mancha in Quixote's footsteps is counterpointed with a vividly impressionistic account of Wollaston's own life.

Fiction

Jupiter Laughs 1967; *Pharaoh's Chicken* 1969; *The Tale Bearer* 1972; *Eclipse* 1974; *Mr Thistlewood* 1985; *The Stones of Bau* 1987; *Café de Paris* 1988

Non-Fiction

Handles of Chance 1956; *China in the Morning* 1960; *Red Rumba* 1962; *Winter in England: a Traveller in His Own Country* 1965; *The Man on the Ice Cap* 1980; *Tilting at Don Quixote* (a) 1990

Charles (Gerald) WOOD 1932–

Wood was born in St Peter Port, Guernsey. From a young age he was involved in his parents' theatre company and he has described his reluctant debut on the stage as 'like a circumcision rite'. The working-class theatre he knew in his youth is reflected in the affectionate portrait of the Harris family in his play *Fill the Stage with Happy Hours*. He was educated at Chesterfield Grammar School and King Charles I School, Kidderminister, and attended Birmingham College of Art before losing his grant in 1950. Craving an order he felt lacking in his family's chaotic life, he enlisted with the 17th/21st Lancers with whom he served for five years. He married in 1954 – the couple have two children – and from 1955 worked in a variety of jobs as factory worker, scenic artist and stage manager, and as a layout artist on the Bristol *Evening Post* from 1959 to 1962.

Wood started writing for the fledgling medium of television, and drew on his experience of the army for *Prisoner and Escort*, a half-hour play broadcast in 1961. The play was subsequently included in the triple bill of stage plays, *Cockade*. He has returned repeatedly to the subjects of army life and war, notably in *Dingo*, an ironic comment on the portrayal of Second World War heroes in films, and *H*, a portrait of General Henry Havelock. *Dingo* was considered subversive when it was submitted to the Lord Chamberlain's Office, but was eventually produced in Bristol and at the Royal Court. Wood has had a long and fruitful collaboration with the film director Richard Lester for whom he scripted *The Knack* (adapted from the play by **Ann Jellicoe**), the Beatles' film *Help*! and others. He also collaborated with **John Osborne** on the script of Tony Richardson's *The Charge of the Light Brigade*.

Wood's writing career went into a decline in the 1970s, something reflected in his BBC television series *Don't Forget to Write*, which starred George Cole as a scriptwriter. Having met the Falklands veteran Robert Lawrence, Wood wrote *Tumbledown* for the BBC. The play created a predictable media furore – 'PLAY STABS NATION IN BACK', screamed one tabloid newspaper – and went on to win a BAFTA Award and a Royal Television Society Award in 1989. He has subsequently adapted **Alan Judd**'s *A Breed of Heroes* (1981) for television and **Beryl Bainbridge**'s *An Awfully Big Adventure* (1989) for the cinema.

Plays

Prisoner and Escort (tv) 1961; *Traitor in a Steel Helmet* (tv) 1961; *Cowheel Jelly* (r) 1962; *Not at All* (tv) 1962; *Cockade* 1963 [pub. in *New English Dramatists 8* 1965]; *Drill Pig* (tv) 1964; *Tie Up the Ballcock* (st) 1964, (f) 1967 [pub. in *Second Playbill 3* 1973]; *Don't Make Me Laugh* 1965 [np]; *Help!* [with M. Behm] (f) 1965; *The Knack ... and How to Get It* [from Ann Jellicoe's play; with R. Lester] (f) 1965; *Meals on Wheels* 1965 [np]; *Fill the Stage with Happy Hours* 1966 [pub. in *New English Dramatists 11* 1967]; *Dingo* 1967 [pub. 1969]; *Drums along the Avon* (tv) 1967; *How I Won the War* (f) 1967; *The Charge of the Light Brigade* [with J. Osborne] (f) 1968; *Labour* 1968 [np]; *The Long Day's Dying* (f) 1968; *A Bit of a Holiday* (tv) 1969; *The Bed-Sitting Room* [with J. Antrobus] (f) 1969; *H* 1969 [pub. 1970]; *Colliers Wood* 1970 [np]; *The Emergence of Anthony*

Purdy, Esq. (tv) 1970; *A Bit of Family Feeling* (tv) 1971; *Welfare* 1971 [np]; *A Bit of Vision* (tv) 1972; *Next to Being a Knight* (r) 1972; *Veterans* 1972; *A Bit of an Adventure* (tv) 1974; *The Can Opener* [from V. Lanoux's play] 1974 [np]; *Death or Glory Boy* (tv) 1974; *Mützen ab* (tv) 1974; *Jingo* 1975 [np]; *Love Lies Bleeding* (tv) 1976; *The Script* 1976 [np]; *Do As I Say* (tv) 1977; *Don't Forget to Write* (tv) 1977–9; *Has 'Washington' Legs?* 1978; *Cuba* (f) 1980; *The Garden* 1982 [np]; *Red Monarch* [from Y. Krotkov] (tv) 1983; *Puccini* (tv) 1984; *Red Star* 1984 [np]; *Wagner* (tv) 1984; *Dust to Dust* (tv) 1985; *Across from the Garden of Allah* 1986 [np]; *My Family and Other Animals* [from G. Durrell's book] (tv) 1987; *Tumbledown* (tv) 1988 [pub. 1987; *The Plantagenets* [from Shakespeare] 1989; *Man, Beast and Virtue* [from Pirandello] 1990; *A Breed of Heroes* (tv) [from A. Judd's novel] 1994; *An Awfully Big Adventure* (f) [from B. Bainbridge's novel] 1995

Leonard (Sidney) WOOLF 1880–1969

Woolf was born in London into a Jewish family. His father was a barrister who died in 1892 leaving the family in straitened circumstances. Woolf was educated at St Paul's School before winning a scholarship to read classics at Trinity College, Cambridge. In Cambridge in 1902 he was elected to the intellectual society the Apostles, other members of which at the time included **E.M. Forster**, Roger Fry, John Maynard Keynes, Lytton Strachey and the philosopher G.E. Moore, who was to exert a great influence on the members of the Bloomsbury Group. Despite his intellectual brilliance, Woolf did badly in his civil service entrance examinations, which compelled him reluctantly to accept work with the Colonial Civil Service. He served in Ceylon from 1904 to 1911, an experience which was to provide him with the material for his first novel, *The Village in the Jungle* (1913). The novel was widely praised for its penetrating understanding of Asian life. Woolf returned to London in 1911, and the following year married Virginia Stephen, whom he had first met in 1904. His only other novel, *The Wise Virgins*, is a portrait of their courtship, as in a very different way is **Virginia Woolf**'s novel *Night and Day* (1919). Woolf might well have gone on to write more fiction but for the failure of this book, the frequent mental breakdowns of his wife, during which he was a conscientious nurse, and other circumstances which led him into a life of publishing and political writing.

In 1913 he joined the Fabian Society, and at the instigation of Sidney Webb began writing for the *Nation*. He was an unsuccessful Labour Party parliamentary candidate in 1922. In 1917 he and Virginia established their own publishing house, the Hogarth Press, whose first book was *Two Stories*, 'Three Jews' by Leonard and 'The Mark on the Wall' by Virginia. They went on to publish some of the most important writers of the age, including **T.S. Eliot, Katherine Mansfield** and **Gertrude Stein**, and were at the centre of the Bloomsbury Group of writers, intellectuals and artists. Woolf's most important books include three volumes of political philosophy: *After the Deluge*; *Quack, Quack!*, an impassioned attack on the irrationalism in fascist politics and the work of philosophers such as Spengler and Bergson; and *Barbarians at the Gate*, which deals with the imminent threat of war.

After Virginia's suicide in 1941, Woolf fell in love with Trekkie Parsons, the wife of the publisher Ian Parsons, and she was to be a close companion for the rest of his life. His five-volume autobiography and the 1990 collection of his letters provide a fascinating insight into his life and times.

Fiction

The Village in the Jungle 1913; *The Wise Virgins* 1914; *Two Stories* (s) [with V. Woolf] 1917; *Stories of the East* (s) 1921

Non-Fiction

The Control of Industry by Co-Operators and Trade Unionists 1914; *Education and the Co-Operative Movement* 1914; *The Control of Industry by the People through the Co-Operative Movement* 1915; *Co-Operation and the War* 2 vols 1915; *International Government* 1916; *The Future of Constantinople* 1917; *After the War* 1918; *Co-Operation and the Future of Industry* 1919; *International Economic Policy* 1919; *Economic Imperialism* 1920; *Empire and Commerce in Africa* 1920; *Mandates and Empire* 1920; *Scope of the Mandates under the League of Nations* 1921; *Socialism and Co-Operation* 1921; *International Co-Operative Trade* 1922; *Fear and Politics: a Debate at the Zoo* 1925; *Essays on Literature, History, Politics, Etc.* 1927; *Hunting the Highbrow* 1927; *Imperialism and Civilization* 1928; *The Way of Peace* 1928; *After the Deluge* 3 vols 1931–53; *The Modern State* 1933; *Labour's Foreign Policy* 1934; *Quack, Quack!* 1935; *The League and Abyssinia* 1936; *Barbarians at the Gate* 1939; *The War for Peace* 1940; *The International Post-War Settlement* 1944; *Foreign Policy: the Labour Party's Dilemma* 1947; *What is Politics?* 1950; *Sowing: 1880–1904* (a) 1960; *Growing: 1904–1911* (a) 1961; *Diaries in Ceylon* 1962; *Beginning Again: 1911–1918* (a) 1964; *A Calendar of Consolation* 1967; *Downhill All the Way: 1919–1939* (a) 1967; *The Journey Not the Arrival Matters: 1939–1969* (a) 1969; *Letters of Leonard Woolf* [ed. F. Spotts] 1990

Edited
The Framework of a Lasting Peace 1917; *Fabian Essays on Co-Operation* 1923; *The Intelligent Man's Way to Prevent War* 1933; Virginia Woolf: *The Death of a Moth and Other Essays* 1942, *A Haunted House and Other Stories* (s) 1943, *The Moment and Other Essays* 1947, *The Captain's Death Bed and Other Essays* 1950, *A Writer's Diary* 1953, *Virginia Woolf and Lytton Strachey: Letters* [with J. Strachey] 1956, *Granite and Rainbow: Essays* 1958

Plays
The Hotel 1939

(Adeline) Virginia WOOLF 1882–1941

Woolf was born into the late Victorian intellectual aristocracy: her father was Sir Leslie Stephen, the founder of the *Dictionary of National Biography*, while her mother was a member of the Duckworth publishing family. She grew up at the family home at Hyde Park Gate, London, and, although, as a woman, she was denied much formal education, she benefited from constant intellectual society. Her happiest days while a child were spent on family holidays at St Ives, Cornwall (the radiant atmosphere of which is evoked by her finest novel, *To the Lighthouse*), but her youth was marred by a series of emotional shocks. Two family members went mad, her half-brother Gerald Duckworth sexually abused her (an experience which left her with a lifelong unresponsiveness to sex), and her mother died while Woolf was in her early teens. Growing to adulthood, she herself showed signs of serious mental instability, and her father's death in 1904 was followed by one of the most severe of her frequent mental breakdowns: she thought she heard the birds talking in classical Greek and Edward VII swearing in the garden; she tried to throw herself out of a window.

More positively, she moved, with her sister Vanessa and two brothers, to Gordon Square, the first of her homes in Bloomsbury, where (by the standards of the day) they lived unconventionally and entertained her brothers' Cambridge friends: this was the nucleus of the famous Bloomsbury Group. From 1905 she began to write for the *Times Literary Supplement*, and she was always as well known for her criticism and essays as for her novels. In 1912, she married the administrator and publisher **Leonard Woolf**: their marriage (although largely sexless) was devoted, and he shielded her from recurrent madness. In 1917 they founded the Hogarth Press – which was to publish her later novels and other noted modernist works – largely for therapeutic reasons. She also had a number of lesbian relationships, including one with fellow writer **V. Sackville-West**, who inspired her novel *Orlando*, subtitled by Woolf 'A Biography'.

Her first novel, *The Voyage Out*, (1915), was traditional in form, but the series of modernist and impressionistic novels on which her fame rests began in 1922 with *Jacob's Room*. Further monuments of her technique were *Mrs Dalloway* and *The Waves*. During the inter-war period she was at the centre of literary society both in London and at her home in Rodmell, Sussex. It was at Rodmell that, fearing further madness, she loaded her pockets with stones and drowned herself in the River Ouse.

For many years Woolf was usually regarded as the least of the great modernists, adept at catching moments of experience, but limited in subject-matter, character-creation and social range, but her reputation has been augmented by feminist criticism (her *A Room of One's Own* is considered a seminal feminist text). Her literary legacy is completed by the many published volumes of her sharp, gossipy and amusing *Letters* and *Diary*: the latter, with its poetical record of the creative process, is much liked by her admirers.

Fiction
The Voyage Out 1915; *Two Stories* (s) [with L. Woolf] 1917; *Night and Day* 1919; *Monday or Tuesday* (s) 1921; *Jacob's Room* 1922; *Mrs Dalloway* 1925; *To the Lighthouse* 1927; *Orlando* 1928; *The Waves* 1931; *The Years* 1937; *Between the Acts* 1941; *A Haunted House and Other Stories* (s) [ed. L. Woolf] 1943; *Mrs Dalloway's Party* (s) 1973

Non-Fiction
The Common Reader 2 vols 1925, 1932; *A Room of One's Own* 1929; *Flush* 1933; *Three Guineas: On the Part that Women Can Play in the Prevention of War* 1938; *Roger Fry: a Biography* 1940; *Contemporary Writers* 1965; *The Letters of Virginia Woolf* 5 vols [ed. N. Nicolson] 1975–9; *Moments of Being: Unpublished Autobiographical Writings* (a) [ed. J. Schulkind] 1976; *The Diary of Virginia Woolf* [ed. A.O. Bell] 5 vols 1977–84; *The Letters of Virginia Woolf to Vita Sackville-West* [ed. L. DeSalvo, M.A. Leaska] 1984

Edited by Leonard Woolf: *The Death of the Moth and Other Essays* 1942; *The Moment and Other Essays* 1947; *The Captain's Death Bed and Other Essays* 1950; *A Writer's Diary* 1953; *Virginia Woolf and Lytton Strachey: Letters* [with J. Strachey] 1956; *Granite and Rainbow: Essays* 1958

Plays
Freshwater 1935 [pub. 1976]

For Children
Nurse Lugton's Golden Thimble (s) 1966

Biography
Virginia Woolf by Quentin Bell, 2 vols 1972; *Virginia Woolf: a Winter's Life* by Lyndall Gordon 1984

T(erence) C(uthbert) WORSLEY 1907–1977

Worsley was born in Durham, the son of a highly eccentric, cricket-loving vicar who was appointed Dean of Llandaff Cathedral in 1926. He suffered an appalling childhood trauma when his youngest brother, with whom he was bathing in the sea, was drowned, a death for which he inevitably felt responsible. He was educated at Marlborough College, where he distinguished himself on the cricket field, and St John's College, Cambridge, where he read classics. He became an assistant master at Wellington College, the famous military public school, in 1929, the same year in which his father, who was inclined to preach against sex and jazz, mysteriously absconded from the Deanery at Llandaff, abandoning his family for a short succession of mistresses.

Worsley spent five years battling with 'the Old Guard' of traditionalists at Wellington, and was suspected of being behind the anti-school magazine *Out of Bounds*, edited by Giles and Esmond Romilly, the rebellious nephews of Winston Churchill. He left in order to write a novel, but was shortly recruited to teach at Gordonstoun, which had recently been founded by the eccentric Kurt Hahn. After a political argument with Hahn, Worsley left and embarked upon a career in literary journalism. During the Spanish Civil War he worked as a driver with a moblile blood-transfusion unit, and his first book, *Behind the Battle* (1939), was a work of non-fiction recalling this experience. That same year he collaborated with **W.H. Auden** on a pamphlet, *Education Today and Tomorrow*, and he was to write further books about the public-school system. He also joined the staff of the *New Statesman*, but left when war broke out to become an education officer with the Royal Air Force, a job he was obliged to abandon after a nervous breakdown. Although homosexual, he attempted marriage, having undergone psychoanalysis. Predictably, the marriage was a failure, and he subsequently had three long-term relationships with men.

He returned to the *New Statesman* in 1946, remaining there as drama critic until 1958, when he joined the *Financial Times*. Amongst the first to recognise the cultural importance of television, he moved in 1964 from writing theatre reviews to becoming the newspaper's television critic. His writings on the theatre were collected in *The Fugitive Art*, while those on television were published as *Television: the Ephemeral Art*. He also was frequently heard on the radio, and was a regular guest on *The Critics*, a weekly forum on the arts.

He recalled his early life, and that of his father, in a frank and funny memoir, *Flannelled Fool*, which was published in 1967, the year in which homosexuality was partly decriminalised. Subtitled 'A Slice of a Life in the Thirties', it was published by **Alan Ross**, and extremely well received, regarded as something of a pioneer work in the field of homosexual confessional literature. His first novel, *Five Minutes, Sir Matthew*, was published in 1969, and is a study of two actors, father and son, with deeply antipathetic views on the theatre. His only other novel, *Fellow Travellers*, is a fictionalised memoir of the Spanish Civil War in which several of his friends and contemporaries are easily recognisable. The central character, Martin Murray, has been acknowledged by **Stephen Spender** as a portrait of himself, while Harry Watson is clearly based on his guardsman friend, T.A.R. Hyndman; Gavin Blair Summers is based on Giles Romilly, although the background of Pugh Griffiths also matches Romilly's. The book is assembled like a dossier, with letters, notes, extracts from diaries and novels, and it entertainingly dissects the political and sexual tangles of the period. It was dedicated to **Terence Rattigan**, whose plays Worsley had championed when they fell out of fashion.

Worsley retained the appearance of a sporting schoolmaster: tall, lanky, with short hair and round spectacles. In 1972 he was forced to retire because of his worsening emphysema. He invested all his capital in a trawler in which he hoped to sail the Mediterranean and to berth permanently in some harbour close to a casino, since he was a keen and sometimes reckless gambler. This enterprise proved disastrous, however, and he was obliged to sell the boat at a loss. He ended up in Brighton, where, increasingly incapacitated by his illness, he took his own life with a massive overdose of sleeping pills.

Fiction
Five Minutes, Sir Matthew 1969; *Fellow Travellers* 1971

Non-Fiction
Behind the Battle (a) 1939; *Education Today and Tomorrow* [with W.H. Auden] 1939; *Barbarians and Philistines: Democracy and the Public Schools* 1940; *The End of the 'Old School Tie'* 1941; *Shakespeare's Histories at Stratford* [with J. Dover Wilson] 1951; *The Fugitive Art: Dramatic*

Commentaries 1947–1951 1952; *Flannelled Fool* (a) 1967; *Television: the Ephemeral Art* 1970

David (John Murray) WRIGHT 1920–

Wright was born in Johannesburg, South Africa, to English-speaking parents; his father was a stockbroker in the family firm. At the age of seven Wright contracted scarlet fever which left him totally deaf. In his autobiography and history of deaf education, *Deafness: a Personal Account*, he writes: 'My becoming deaf when I did – if deafness had to be my destiny – was remarkably lucky. By the age of seven a child will have grasped the essentials of language, as I had.' He was taught by private tutors, and at thirteen his mother accompanied him to England in order that he might attend the Northampton School for the Deaf. He left the school in 1939 and went up to Oriel College, Oxford, where he read English. Having graduated in 1942, he moved to London were he held 'a sort of non-job' on the staff of the *Sunday Times* for five years while spending his time in the pubs of Fitzrovia where he met **Dylan Thomas, Roy Campbell, George Barker** and other poets of his generation.

Wright had written poetry since shortly after going deaf, and feels that the reading and writing of poetry may have become for him a substitute for music. In 1949 he published his 'first and let us hope worst' collection, *Poems*, and travelled the following year in Italy on the proceeds of an Atlantic Award for Literature. He was married in 1951 to Phillipa Reid, an actress with the mobile Century Theatre; following her death, Wright remarried in 1987. In the 1950s he supported himself with translation and editorial work (he edited *The Faber Book of Twentieth Century Verse* with his old Oxford friend **John Heath-Stubbs**) and was co-editor of two literary magazines, *Nimbus* (1955–6) and *X* (1959–62). In the 1960s he jointly wrote three travel books with Patrick Swift, and was the Gregory Fellow in Poetry at Leeds University from 1965 to 1967.

Wright's verse has repeatedly returned to the two themes of his deafness – the autobiographical *Monologue of a Deaf Man* is his most explicit and emotional treatment of the subject – and his South African origins. He has returned to the country on many occasions, primarily for his mother's sake, but has been a fierce and ironic critic of its racial policies.

Poetry
Poems 1949; *Moral Stories* 1954; *Monologue of a Deaf Man* 1958; *Adam at Evening* 1965; *Poems* 1966; *Nerve Ends* 1969; *Corgi Modern Poets in Focus 1* [with others; ed. D. Abse] 1971; *A South African Album* [with others; ed. C.H. Sisson] 1976; *To the Gods the Shades* 1976; *A View of the North* 1976; *Metrical Observations* 1980; *Selected Poems* 1988; *Elegies* 1990

Non-Fiction
Roy Campbell 1961; *Deafness* (a) 1969 [rev. 1990] With Patrick Swift: *Algarve* 1965 [rev. 1971]; *Minho and North Portugal* 1968; *Lisbon* 1971

Edited
The Forsaken Garden: an Anthology of Poetry 1824–1909 [with J. Heath-Stubbs] 1950; *The Faber Book of Twentieth Century Verse* [with J. Heath-Stubbs] 1953 [rev. 1965, 1975]; *South African Stories* 1960; *Seven Victorian Poets* 1964; *The Mid-Century: English Poetry 1940–60* 1965; *Longer Contemporary Poems* 1966; *The Penguin Book of English Romantic Verse* 1968; Thomas De Quincey *Recollections of the Lakes and the Lake Poets* 1970; Edward Trelawny *Records of Shelley, Byron, and the Author* 1973; *The Penguin Book of Everyday Verse: Social and Documentary Poetry 1250–1916* 1976; Hardy: *Selected Poems* 1978, *Under the Greenwood Tree* 1978; Edward Thomas *Selected Poems and Prose* 1981; *An Anthology from X: a Quarterly Review of Literature and the Arts 1959–1962* [with P. Swift] 1988

Translations
Beowulf 1957; Chaucer: *The Canterbury Tales* 1964, *The Canterbury Tales: a Verse Translation with an Introduction and Notes* 1985

Judith (Arundell) WRIGHT 1915–

Wright, a poet, critic, historian, writer of short stories and children's fiction, conservationist and passionate activist for Aboriginal land rights, is one of Australia's most eminent poets. Born near Armidale, northern New South Wales, she inherited the myths and ethos of an old pioneering family. After her mother's death in 1927 she was educated under her grandmother's supervision until attending New England Girls' School, Armidale, from 1929 to 1933. After informal studies in English and philosophy at Sydney University, she toured Britain and Europe in 1937–8, and, on returning to Sydney, worked as a secretary-stenographer. From 1938 onwards she published poems in a range of literary periodicals, and published her first collection, *The Moving Image*, in 1946. She had moved to Queensland as clerk at the Universities Commission, then in 1945 became University Statistician at the University of Queensland. In Queensland she met her lifelong partner, the writer and philosopher Jack McKinney, whom she later married. Their daughter was born in 1950.

Wright's early poems received wide acclaim and were justly celebrated for their visionary naming and imaging of the Australian landscape. But the poetic conviction they embodied went further and deeper than the term 'nature poet' usually suggests. Poetry was, for her, an urgent task, a response to the need to bridge the human and the natural world, since for her, the source of both life and language was 'the living earth from which we have separated ourselves, but of which we are part and in which we cannot help participating'. Her 1952 essay 'The Writer and the Crisis' did not address what other writers saw as the crisis – the Cold War, and other external threats – but a much greater menace: the failure of the language to meet the demands made on it. This intensely political perception of how the land and the language interact, mediated by human values – or their corruption – flowered through the 1970s and 1980s in a richly textured programme both of writing and of activism. The writing included poetry, histories, essays and commentary; the activism focused on conservation issues and Aboriginal land rights. This commitment was not easy, nor without cost: vested interest groups did their best to marginalise her voice and to deride her work. Despite the power of the forces ranged against her, she remained resolute in her conviction that the continued destruction of the natural environment, and the concomitant destruction of the Aboriginal people, must prove equally catastrophic in the end for the perpetrators.

Unlike other poets of her generation, Wright did not seek refuge within university or professional life. She remained a countrywoman, receiving occasional subsidies for particular projects, and otherwise undertaking a miscellany of literary tasks, remarking once: 'I have always been a person who worked for my living in one way or another. I've always had to do a lot of hack work: writing school plays for the A[ustralian] B[roadcasting] C[omission], and doing children's books, generally doing housewife jobs in literature, you might say.'

Wright's honours and awards are numerous. In addition to honorary doctorates from many Australian universities, she has received the Grace Leven Prize for Poetry (1950 and 1972); the Encyclopaedia Britannica award (1964); the Robert Frost Award (1975); the ASAN World Prize for Poetry (1984); and the Queen's Medal for Poetry (1992).

Poetry

The Moving Image 1946; *Woman to Man* 1949; *The Gateway* 1953; *The Two Fires* 1955; *Birds* 1962; *Five Senses* 1963; *City Sunrise* 1964; *The Other Half* 1966; *Collected Poems 1942–1970* 1971; *Alive: Poems 1971–1972* 1973; *Fourth Quarter and Other Poems* 1976; *The Double Tree: Selected Poems 1942–1976* 1978; *Phantom Dwelling* 1985; *A Human Pattern* 1990; *Through Broken Glass* 1992; *Collected Poems 1942–1985* 1994

Fiction

The Nature of Love 1966

Non-Fiction

The Generations of Men 1959; *Charles Harpur* 1963; *Preoccupations in Australian Poetry* 1965; *Henry Lawson* 1967; *Because I Was Invited* 1975; *The Coral Battleground* 1977; *The Cry for the Dead* 1981; *Born of the Conquerors* 1991; *Going on Talking* 1992

Edited

Australian Poetry 1948 1948; *A Book of Australian Verse* 1956; *New Land, New Language* 1957

Kit (i.e. Christopher) WRIGHT 1944–

Famed as the tallest poet in Britain, Wright was born in Kent, the son of a preparatory school master who loved literature and had a gift for communication. He was educated at Berkhamsted School, where he was fortunate enough to have excellent English teachers. At the age of seventeen he met **Vernon Scannell**: 'He was very encouraging,' Wright recalled. 'He didn't like what I'd written much, but he could see something in it. Until then I thought poetry was written by people like Wordsworth.' Under Scannell's influence he began to read contemporary poets such as **Ted Hughes**, and wrote continuously. After taking a degree at New College, Oxford, he went to Ontario to lecture in English at Brock University, and he has also held the post of fellow-commoner in creative arts at Trinity College, Cambridge. Between 1970 and 1975 he was the education secretary of the Poetry Society of London.

His poems first appeared in a joint publication, *Treble Poets I* (1974), and two years later he was one of six poets chosen to tour the USA in the Bicentennial Poetry Exchange. His first volume, *The Bear Looked Over the Mountain*, was published in 1977 and won the Geoffrey Faber Memorial Prize and the Poetry Society's Alice Hunt Bartlett Prize. His work is witty, compassionate and highly accessible, and he has read his poems in theatres, schools, colleges and public houses all over the country, as well as on radio and television. He is clearly influenced by popular culture and the media, particularly the cinema. 'How the South East Was Lost', for example, recalls his childhood in the language of a Western film. In spite of a superficial flippancy, his poetry has a firm moral base – one of his best-known poems is 'Frankie and Johnny in 1955', a sympathetic account of Ruth Ellis, the last woman to be hanged in Britain – and is always inventive. Part of its jokiness is related to his height: self-conscious about it as an adolescent, he became 'the joker who could make everyone laugh'. He has written several volumes of poetry for children, though claims that he has to 'put on a different pair of shoes' to do so. He has been described as 'the first light verse master since **Gavin Ewart**'.

Poetry
Treble Poets I [with E. Maslen, S. Miller] 1974; *The Bear Looked Over the Mountain* 1977; *Bump-Starting the Hearse* 1983; *From the Day Room* 1983; *Poems 1974–1983* 1988; *Short Afternoons* 1989

For Children
Arthur's Father 4 vols 1978; *Rabbiting On and Other Poems* 1978; *Hot Dog and Other Poems* 1981; *Professor Potts Meets the Animals of Africa* 1981; *Tigerella* 1993; *One of Your Legs Is Both the Same!* 1994

Edited
Soundings: a Selection of Poems for Speaking Aloud 1975; *Poems for 9-Year-Olds and Under* 1984; *Poems for Over 10-Year-Olds* 1984

Richard (Nathaniel) WRIGHT 1908–1960

Often regarded as one of the most influential black writers of the twentieth century, Wright was born on a sharecroppers' farm near Natchez, Mississippi, the grandson of a slave. His mother was a teacher, but became paralysed when Wright was five years old, and so he was passed around impoverished relatives and grew into an unruly boy who left school aged fifteen and proceeded to educate himself. He took a job as a post office clerk in Memphis, Tennessee, where he circumvented the 'whites only' rule of the library by forging letters, purportedly from a white borrower, requesting that 'this nigger boy' be allowed to take out books. On his arrival in Chicago in 1932 he joined the Communist Party, and in 1934 joined the Federal Writers' Project there. He later moved to New York, writing for magazines such as the *Daily Worker* and the *New Masses*.

His first book, a collection of four novellas entitled *Uncle Tom's Children*, depicted the oppression of the black people of the South, and was published to great critical acclaim in 1938. The following year Wright was awarded a Guggenheim Fellowship, and in 1940 his novel *Native Son* jolted the literary scene. He wrote it while living in Brooklyn and it immediately became a bestseller. It tells the story of Bigger Thomas, a furious, isolated, ignorant and completely unsympathetic young black who takes a job as a chauffeur, is treated as an equal by the white communist friends of his employer's daughter and in his confusion ends up murdering both her and his own girlfriend, with which action he feels he has finally defined himself. The novel illustrated Wright's most pervading ideology: that the environment is usually the principal villain, and that what is bad for the American Negro is bad for America. **James Baldwin** described *Native Son* as 'the most powerful and celebrated statement we have yet of what it means to be a Negro in America', and the *Nation* commented that 'it lays bare with a ruthlessness that spares neither race, the lower depths of the human and social relationships of blacks and whites'. This was exactly what Wright set out to do throughout his career as a writer. He adapted the novel for the stage (it was produced on Broadway by Orson Welles), and appeared in the title-role in an Argentinian film version of the book in 1950.

With the money he made from *Native Son*, Wright bought his mother a house in Chicago and travelled to Mexico. In 1946 he was invited to France by the French government and was fêted there by literary Paris as the greatest living black writer. The following year he settled permanently in Paris with his second wife, Ellen (who had been head of his CP cell in New York), and their daughter. Existentialist writers such as Sartre and de Beauvoir were in tune with his ideas about alienation, and his later fiction combined racial themes with existential metaphysics. His range as a writer extended from works such as *White Man, Listen!*, a collection of essays dealing with the emotional reactions and relations of different races, to surrealistic novellas like *The Man Who Lived Underground*, the story of an excluded black who spies on and robs society from his vantage point in the sewers. In his later years, Wright devoted much of his energy to supporting the national independence movements in Africa; *Black Power* describes his experiences in the Gold Coast (now Ghana), whose independence he helped secure.

Although he had left the CP in 1944 (for reasons explained in his contribution to Richard Crossman's celebrated 1950 symposium, *The God That Failed*), Wright became a victim of the McCarthy witch-hunts and came under the surveillance of the CIA and the FBI. Frequent interrogations at the US embassy in Paris added to his already serious difficulties: his popularity as a writer had been on the wane for some time (which brought him financial problems), and his health was further undermined by visits to Africa. His attempts to move to England were thwarted by the British and US governments, who conspired to keep him in France, where he died in poverty.

Fiction

Uncle Tom's Children: Four Novellas 1938 [rev. 1940]; *Native Son* 1940; *The Outsider* 1953; *Savage Holiday* 1954; *The Long Dream* 1958; *Lawd Today* 1963

Non-Fiction

Twelve Million Black Voices 1941; *Black Boy: a Record of Childhood and Youth* (a) 1945; *The God That Failed* (c) [ed. R.H.S. Crossman] 1950; *Black Power* 1954; *The Color Curtain: a Report on the Bandung Conference* 1956; *Pagan Spain* 1957; *White Man, Listen!* 1957; *American Hunger* (a) 1977

Plays

Native Son [from his novel] 1941

Collections

Eight Men 1961; *A Richard Wright Reader* [ed. M. Fabre, E. Wright] 1978

Biography

The Unfinished Quest of Richard Wright by Michel Fabre 1973

Francis (Guy Percy) WYNDHAM 1924–

The son of a professional soldier who died in 1941, Wyndham was born in London. His mother was the writer Violet Wyndham, the daughter of **Ada Leverson**, and he is also related to **Henry Green**. His family moved to Wiltshire when he was four, and he was educated at Eton and (briefly) at Christ Church, Oxford, which he left to serve in the army in 1943. Although he was invalided out with tuberculosis after a few months, he did not return to Oxford to complete his education. Instead, he started writing short stories, although they did not appear in volume form until 1975, having lain 'forgotten in the back of a drawer for nearly thirty years'.

Meanwhile, he embarked on a career in literary journalism, reviewing fiction for the *New Statesman*, the *Observer* and the *London Magazine*. Between 1953 and 1958 he worked in publishing, then became literary editor of *Queen* magazine in 1959. In 1964 he joined the staff of the *Sunday Times* as a feature writer, remaining there until 1980. His mislaid short stories, all of which were vignettes of life on the home front during the Second World War, were eventually published in 1975, prompting **Nina Bawden** to comment: 'They are perfect stories, crisply turned, very resonant, and it is lucky for us that he found them.' Ten years later he produced a second volume, *Mrs Henderson and Other Stories*, which **Alan Hollinghurst** in the *Times Literary Supplement* described as 'a vindication of the dictum that less is more' – a phrase that may be said to sum up Wyndham's career. At the age of sixty-three he won the Whitbread Award for a first novel with *The Other Garden*, which once again had a background of the home front during the Second World War. Set in the English countryside, it is narrated by an adolescent boy, and described the effects of the war on his family. Just over 100 pages long, it was widely and enthusiastically reviewed. It is his best-known book and his only novel to date.

Wyndham is also the co-author of a biography of Trotsky, and the co-editor (with Diana Melly) of a volume of **Jean Rhys**'s letters. He had befriended Rhys in the late 1950s, when she was living as a recluse and had been long forgotten, and even thought dead. He encouraged her to start writing again, and is well known as a mentor of other writers, notably **Bruce Chatwin**. The high quality of his journalism is on show in *The Theatre of Embarrassment*, which gathers together book reviews, essays, profiles, interviews and reminiscences. The title-piece is an outstanding essay about the role of embarrassment in the performing arts, while elsewhere he writes about a wide range of people from Honoré de Balzac to Hylda Baker, taking in **John Updike, Elizabeth Bowen**, P.J. Proby, Colette, Ada Leverson and the Kray twins along the way.

His 'secret aim', he has written, when recreating an interview was to invest it 'with the intensity and clear-cut shape of a good short story'. His small output has made him something of a legend, but if he has not achieved much public recognition for his fiction and remains little known to the reading public, his reputation amongst his fellow writers is very high indeed.

Fiction

Out of the War (s) 1975; *Mrs Henderson and Other Stories* (s) 1985; *The Other Garden* 1987

Non-Fiction

The Theatre of Embarrassment 1991

Edited

Jean Rhys Letters, 1931–1966 [with D. Melly] 1984; Bruce Chatwin *Photographs* and *Notebooks* [with D. King] 1993

John WYNDHAM (Parkes Lucas Beynon Harris) 1903–1969

John Beynon Harris wrote under his own name and under the pseudonyms John Beynon, Lucas

Parkes, Johnson Harris and, most famously, John Wyndham. Born in Knowle in Warwickshire, the son of a barrister, he lived a peripatetic life with his mother and younger brother (the writer Vivian Beynon Harris) after his parents' separation when he was eight. He attended a number of prep schools before going to Bedales where he remained until 1921. After school he attempted a number of careers, briefly studying agriculture and the law, and doing short stints in advertising and commercial art. Throughout these years a small allowance from his parents enabled him to write science-fiction short stories, which were influenced by his early reading of **H.G. Wells**, though he had little success with them.

In 1929 he read a copy of *Amazing Stories* and immediately turned his hand to writing for the American science-fiction pulp magazines. The story 'Worlds to Barter' was a success when it was published by *Wonder Stories* in 1931 and was followed by a succession of inventive and popular stories which demonstrate a degree of literary virtue and moral vision rare in the genre. In 1935 he made his first foray into the English market when the magazine *Passing Show* serialised 'The Secret People', an improbable tale concerning a couple kidnapped by pygmies under the Sahara. It was published as a book the next year under the name of John Beynon, as was a second serial, 'Stowaway to Mars', which became *Planet Plane* in volume form.

Wyndham worked as a censor in the civil service from 1940 to 1943 before joining the Royal Signal Corps with whom he served during the invasion of Normandy. After a few years of failure, his reputation was quickly established with the publication of *The Day of the Triffids* in 1951. The novel describes the chaos which ensues after a stellar explosion has blinded most people and a disastrous agricultural experiment has produced the giant carnivorous plants which threaten the survival of mankind. An undistinguished film version of the novel was released in 1963, though *The Midwich Cuckoos* was more successfully filmed as *Village of the Damned* in 1960.

Wyndham's subsequent novels, *The Chrysalids* most notable among them, deal with similar kinds of catastrophe. His very English style led one critic to describe him as 'the Trollope of science fiction', but his influence on the renaissance of English science fiction in the 1950s was considerable. He married a teacher in 1963, and lived a quiet, country life in Petersfield, Hampshire.

Fiction

As 'John Wyndham': *The Day of the Triffids* 1951; *The Kraken Wakes* 1953; *Jizzle* (s) 1954; *The Chrysalids* 1955; *Re-Birth* 1955; *Sometime, Never* (s) [with W. Golding, M. Peake] 1956; *The Midwich Cuckoos* 1957; *The Outward Urge* [with 'Lucas Parkes'] 1959; *The Seeds of Time* (s) 1959; *Trouble with Lichen* 1960; *Consider Her Ways* (s) 1961; *Chocky* 1968

As 'John Beynon': *Foul Play Suspected* 1935; *The Secret People* 1935; *Planet Plane* 1936 [as *Stowaway to Mars* 1972]

Dornford YATES (pseud. of Cecil William Mercer) 1885–1960

The son of a solicitor of King's Bench Walk, and a cousin of H.H. Munro (**Saki**), Yates was educated at Harrow and University College, Oxford. He was called to the Bar in 1909, and the following year he was a junior barrister working on the case of Dr Crippen, but he did not practise the law for long. In the First World War he saw service in Egypt and Salonika and rose from lieutenant to captain.

He had begun writing before the war, and his first success was *The Courts of Idleness* (1920). He was to become one of the most popular middlebrow writers of the 1920s and 1930s, specialising in light, sometimes farcical, adventure stories featuring well-born heroes and romantic settings. *Berry and Co.* and *Jonah and Co.* introduced two popular characters who were to feature in a long series of novels.

In the 1920s, troubled by chronic rheumatism, Yates went to live at Pau in southern France. He married twice, and had one son. In 1940 he had to flee France before the German advance; he went to South Africa, where he volunteered for the army, attained the rank of major, and eventually saw service with the Southern Rhodesian forces. After the war, he settled at Umtali in Rhodesia, where he died. His autobiography, *As Berry and I were Saying*, which was followed by *B-Berry and I Look Back*, took the unusual form of a dialogue between himself and four of his own fictional characters.

Fiction

The Brother of Daphne (s) 1914; *The Courts of Idleness* 1920; *Anthony Lyveden* 1921; *Berry and Co* (s) 1921; *Jonah and Co.* (s) 1922; *Valerie French* 1923; *And Five Were Foolish* (s) 1924; *As Other Men Are* (s) 1925; *The Stolen March* 1926; *Blind Corner* 1927; *Perishable Goods* 1928; *Blood Royal* 1929; *Maiden Stakes* (s) 1929; *Fire Below* 1930; *Adele and Co.* (s) 1932; *Safe Custody* 1932; *Storm Music* 1934; *She Fell Amongst Thieves* 1935; *And Berry Came Too* (s) 1936; *She Painted Her Face* 1937; *This Publican* 1938; *Gale Warning* 1939; *Shoal Water* 1940; *Period Stuff* (s) 1942; *An Eye for a Tooth* 1943; *The House That Berry Built* 1945; *Red in the Morning* 1946; *The Berry Scene* (s) 1947; *Cost Price* 1949; *Lower Than Vermin* 1950; *Ne'er-Do-Well* 1954; *Wife Apparent* 1956

Non-Fiction

As Berry and I Were Saying (a) 1952; *B-Berry and I Look Back* (a) 1958

Plays

Eastward Ho! [with O. Asche] 1919

Biography

Dornford Yates by A.J. Smithers 1982

W(illiam) B(utler) YEATS 1865–1939

Yeats was born in Dublin and developed an intense relationship with his father, the eloquent and much respected (though commercially unsuccessful) portrait painter John Butler Yeats. Though he initially followed in his father's footsteps and wished to be a painter, Yeats eventually abandoned this calling in order to devote himself to serious writing. His relationship with his father would nevertheless influence profoundly the nature of the poetry, as J.B. Yeats never tired of instructing his son that a successful poet must recognise that life and art are inextricably linked. It is important to understand that the stated intention of the poet was that he should live to produce a body of work (which eventually included volumes of autobiography, letters, essays, philosophical speculation and of course poetry) which would, by drawing on his obsessive observation of the philosophical, mystical, political and spiritual aspects of the world, objectively reflect the whole universal system in microcosm.

While a student at the Dublin School of Art, Yeats had met George Russell (**A.E.**). Both were interested in mysticism, and in 1885 they founded the Dublin Lodge of the Hermetic Society, the first meeting of which Yeats chaired. His interest in the occult never wavered, though it is most obviously reflected in his earliest poems, collected in *The Wanderings of Oisin and Other Poems* (1889), which also draw on his interest in Irish politics and Irish folk-tales. He was one of the founding members of the Irish Literary Society (1891–2) and at this time first met **Lady Gregory**, who was then a leading figure in the Irish literary revival and who was consequently involved in the encouragement of Irish national drama. They became close friends, and her estate, Coole Park, was the setting for several of his poems. Yeats's play *The Countess Kathleen* was performed in Dublin in 1892 and

Cathleen ni Houlihan (1902) was one of the first plays of the Abbey Theatre Company, of which Yeats, Lady Gregory and **J.M. Synge** were joint directors.

Yeats met the committed revolutionary Maud Gonne in 1889, and was immediately captivated by her charm, power and beauty. He courted her unsuccessfully for thirteen years and the fact that his poetry during this period became increasingly concerned with Irish nationalism is in no small way attributable to his desire to impress Gonne with his own political convictions. How sincere those convictions were can be gauged by a comment he later made about the effect Gonne had upon him: 'If she said the world was flat or the moon an old caubeen tossed up into the sky,' he wrote, 'I would be proud to be of her party.' Her eventual marriage to Major John MacBride in 1903 was a shattering blow for Yeats, and his poetry tellingly moves away from politics during the next few years. He later commented that he thought her to have been 'destroyed' by revolutionary fervour. His ambivalent attitude towards the power and passion associated with political insurrection is reflected in one of his best-known collections *Michael Robartes and the Dancer*, which contains 'Easter, 1916', a poem which, like Marvell's 'An Horatian Ode' (1650), famously manages not to come down in support of any one side involved in the nationalist rising, despite displaying superficial admiration for the revolutionaries.

In 1917 Yeats married George Hyde-Lees, and in the following years developed his system of symbolism and masks in volumes such as *A Vision*, *Seven Poems and a Fragment* and *Words for Music Perhaps and Other Poems*. He was also, from 1922 to 1928, a senator in the Irish Free State, and in 1923 received the Nobel Prize for Literature. Though he died and was buried in France, his body was disinterred and brought back for burial to Drumcliff in Sligo in 1948. His major achievements are perhaps his two collections *The Tower* and *The Winding Stair*, which were written after he had moved away from his romantic origins towards the modernism of **T.S. Eliot** and **Ezra Pound** (both of whom he had earlier met in London). Yeats is now widely considered to be one of the most important writers of poetry in English in the twentieth century.

Poetry

Mosada 1886; *The Wanderings of Oisin and Other Poems* 1889; *Poems* 1895; *The Wind Among the Reeds* 1899; *The Shadowy Waters* 1900; *In the Seven Woods* 1903; *Poems, 1899–1905* 1906; *The Poetical Works of William B. Yeats* 2 vols 1906, 1907; *Poems Lyrical and Narrative* 1908; *Poems* 1909; *The Green Helmet and Other Poems* 1910 [rev. 1912]; *Poems Written in Discouragement* 1913; *Responsibilities* 1914; *The Wild Swans at Coole* 1917; *Michael Robartes and the Dancer* 1921; *Later Poems* 1922; *Seven Poems and a Fragment* 1922; *The Cat and the Moon and Certain Poems* 1924; *October Blast* 1927; *The Tower* 1928; *A Packet for Ezra Pound* 1929; *The Winding Stair* 1929; *Words for Music Perhaps and Other Poems* 1932; *The Winding Stair and Other Poems* 1933; *The King of the Great Clock Tower* 1934; *Wheels and Butterflies* 1934; *A Full Moon in March* 1935; *New Poems* 1938; *On the Boiler* 1939

Plays

The Countess Kathleen 1892; *The Land of Heart's Desire* 1894; *Cathleen ni Hoolihan* [with Lady Gregory] 1902; *Where There Is Nothing* 1902; *The Hour Glass and Other Plays* [pub. 1904]; *The King's Threshold* 1904; *Deirdre* 1907; *The Golden Helmet* 1908; *The Unicorn from the Stars and Other Plays* [with Lady Gregory; pub. 1908]; *Two Plays for Dancers* [pub. 1919]; *Four Plays for Dancers* [pub. 1921]; *The Player Queen* 1922; *Plays in Prose and Verse* [pub. 1922]; *Sophocles' King Oedipus* 1928; *Stories of Michael Robartes and His Friends* [pub. 1929]; *Collected Plays of W.B. Yeats* [pub. 1934]; *The Words upon the Window Pane* 1934; *The Herne's Egg* 1938; *Diarmuid and Grania* [with G. Moore] 1951

Non-Fiction

Ideas of Good and Evil 1903; *Discoveries* 1908; *Synge and the Ireland of His Time* 1911; *The Cutting of an Agate* 1912; *Reveries over Childhood and Youth* (a) 1915; *Per Amica Silentia Lunae* 1918; *Four Years, 1887–1891* (a) 1921; *The Trembling of the Veil* (a) 1922; *Essays* 1924; *The Bounty of Sweden* (a) 1925; *Estrangement* (a) 1926; *The Death of Synge and Other Passages from an Old Diary* 1928; *Letters to the New Island* 1934; *Dramatis Personae* (a) 1935; *Modern Poetry* 1936; *A Speech and Two Poems* 1937; *The Autobiography of W.B. Yeats* (a) 1938; *Essays by W.B. Yeats 1931–1936* 1937; *If I Were Four-and-Twenty* 1940; *Letters on Poetry from W.B. Yeats to Dorothy Wellesley* 1940; *Florence Farr, Bernard Shaw and W.B. Yeats* 1941; *W.B. Yeats and T. Sturge Moore: Their Correspondence, 1901–1937* 1953; *W.B. Yeats: Letters to Katharine Tynan* 1953; *The Letters of W.B. Yeats* 1954; *The Senate Speeches of W.B. Yeats* 1961; *The Correspondence of Robert Bridges to W.B. Yeats* [ed. R.J. Finneran] 1977; *Theatre Business: the Correspondence of the First Abbey Theatre Directors: W.B. Yeats, Lady Gregory and J.M. Synge* 1982

Fiction

John Sherman and Dhoya (s) 1891; *The Celtic Twilight* (s) 1893; *The Secret Rose* (s) 1897; *The*

Tables of the Law and The Adoration of the Magi (s) 1897; *Stories of Red Hanrahan* (s) 1904

Edited
Fairy and Folk Tales of the Irish Peasantry 1888; *Stories from Carleton* 1889; *Representative Irish Tales* 1891; *Irish Fairy Tales* 1892; Blake: *The Poems of William Blake* 1893, *The Works of William Blake* 1893; *A Book of Irish Verse* 1895; *Sixteen Poems by William Allingham* 1905; *Some Essays and Passages by John Eglinton* 1905; *Twenty-One Poems by Lionel Johnson* 1905; *Poems of Spenser* 1906; *Twenty-One Poems by Katharine Tynan* 1907; *Poetry and Ireland* 1908; J.M. Synge: *Poems and Translations by John M. Synge* 1909, *Deirdre of the Sorrows* 1910; *Selections from the Writings of Lora Dunsany* 1912; *The Oxford Book of Modern Verse, 1892–1935* 1936; *The Ten Principal Upanishads* [with S.P. Swami] 1937

Collections
A Vision 1925; *Last Poems and Two Plays* 1939

Biography
Yeats by Frank Tuohy 1976

Andrew (John) YOUNG 1885–1971

Born in Elgin, Scotland, Young was educated at school and university in Edinburgh, and entered the Presbyterian ministry. His father paid for the publication of his first book of poems, *Songs of Night* (1910), and this was followed by many other slim volumes of verse which, after the publication in 1933 of *Winter Harvest*, began to gain Young a substantial reputation as a meditative poet of nature and religion. In 1920, he came south to England and was minister of the Presbyterian church in Hove. In 1939 he was received into the Church of England, and became vicar of Stonegate in Sussex in 1941, retiring in 1959. He was also a canon of Chichester Cathedral, and lived in retirement at Yapton. He was long married to Janet Green, a lecturer in English literature, and they had two children.

Most of his poems are short, but in 1952 he published a long eschatological poem, *Into Hades*, which was later combined with the visionary *A Traveller in Time*, to produce *Out of the World and Back*, a major poetic testament. A shy, austere man, Young was an English parson-poet in the tradition of Herrick and Herbert, with the erudite hobbies typical of the breed; in his case, he was a noted expert on botany, and his book *A Prospect of Flowers* is as well known as his poetry. Although Young is a minor rather than a major poet, his reputation stands high with several critics and is likely to increase.

Poetry
Songs of Night 1910; *Boaz and Ruth and Other Poems* 1920; *Death of Eli and Other Poems* 1921; *Thirty-One Poems* 1922; *The Bird Cage* 1926; *The Cuckoo Clock* 1928; *The New Shepherd* 1931; *Winter Harvest* 1933; *The White Blackbird* 1935; *Speak to the Earth* 1939; *The Green Man* 1947; *Collected Poems* 1950 [rev. 1960]; *Into Hades* 1952; *Quiet as Moss* 1956; *Out of the World and Back* 1958; *Burning as Light* 1967; *The New Poly-Olbion* 1967; *Poetic Jesus* 1967

Plays
The Adversary [pub. 1923]; *Nicodemus* 1937

Non-Fiction
A Prospect of Flowers 1945; *A Retrospect of Flowers* 1950; *A Prospect of Britain* 1956; *The Poet and the Landscape* 1962

E(mily) H(ilda) YOUNG 1880–1949

The daughter of a shipbroker, Young was born and brought up in Northumberland. She was educated at Gateshead Grammar School and Penrhos College in Wales. In 1902 she married a solicitor who practised in Bristol, and that city became the Radstowe of her novels, the first of which, *A Corn of Wheat*, was published in 1910. Although over-age, her husband enlisted in the army in the First World War and was killed at Ypres in 1917. Young spent the war working in a munitions factory and as a groom in a stables. In 1918 she moved to the South London suburb of Sydenham to become part of an extraordinary and secret *ménage à trois*. Her lover, Ralph Henderson, was headmaster of Alleyn's School, and his wife had been a friend of Young's husband since their schooldays together in Bristol. Henderson's position made it imperative that these domestic arrangements were disguised, and Young officially lived in a self-contained flat in the Hendersons' house. Henderson and Young were both passionate climbers and spent their holidays together in Snowdonia, the Alps and the Dolomites, 'chaperoned' by trusted friends. When Henderson retired in 1940, he and Young moved to Bradford-on-Avon, Wiltshire, where they remained until Young's death from lung cancer.

Unsurprisingly, secrecy is the theme of Young's best-known novel, *Miss Mole*, which won the James Tait Black Memorial Prize in 1930, and was televised by the BBC as *Hannah* in 1980. Hannah Mole is a domestic servant, her blood relationship to a prominent Radstowe figure kept secret when she is installed as housekeeper to a widowed Nonconformist minister. Like the creature after which she is named,

Hannah works underground, gradually undermining the harsh dogma of her master's creed and replacing it with a more humane vision of God. Questions of morality and propriety, the restrictions of family life and the difficulties of those who attempt to break them continue to be the themes of Young's subsequent novels. In spite of this subject-matter, however, she is a far from dour novelist, and her books are leavened with wit. It is appropriate that this subversive writer should at first have been mistaken by reviewers for a man. Her work was rediscovered in the 1980s by the feminist press Virago, which has reissued several of her best-known books, notably *The Misses Mallett* (originally published as *The Bridge Dividing*) and the semi-autobiographical *William*.

Fiction

A Corn of Wheat 1910; *Yonder* 1912; *Moor Fires* 1916; *The Bridge Dividing* 1922 [as *The Misses Mallett* 1927]; *William* 1925; *The Vicar's Daughter* 1928; *Miss Mole* 1930; *Jenny Wren* 1932; *The Curate's Life* 1934; *Celia* 1938; *Chatterton Square* 1948

For Children

Caravan Island 1940; *River Holiday* 1947

Francis Brett YOUNG 1884–1954

Born at Halesowen (the Halesby of his novels), Young came from Midlands medical families on both sides and, after leaving Epsom College, he took a medical degree at Birmingham University. He began to practise at Brixham in 1907 and remained there until 1914, except for a period of two years when he was a ship's surgeon on a voyage to Japan.

His first novel, *Undergrowth*, written in collaboration with a young brother, was published in 1913, and in these early years he was considered to be among the most promising of the younger novelists. In 1915 he joined the Royal Army Medical Corps, and served in the campaign in German East Africa, catching malaria and being invalided out: he later wrote an excellent book, *Marching on Tanga*, describing his wartime experiences. He had married Jessie Hankinson, who was for some time an opera singer, in 1908, and after the war he and his wife lived at Anacapri, from which he turned out a novel more or less every year, winning wide popular success with *Portrait of Clare* and *My Brother Jonathan*. Young's novels depend greatly on his Midlands and medical background; his work was always more or less autobiographical. He also wrote poetry, short stories, a book of criticism and two plays, and he was a minor composer, setting some of **Robert Bridges**' poems to music.

On their return to England in the 1930s the Youngs lived in a large house at Pershore, Worcestershire, and, as his novels grew lengthier, he increasingly lost what critical reputation he had had. During the Second World War he exhausted himself writing *The Island*, a vast poem celebrating English history, and, broken in health, he retired to spend his last years living in South Africa (whose history he had also tried to celebrate, in an immense, never-completed, trilogy). In his last years he published only *In South Africa*, a guide for prospective settlers. He died in a Cape Town nursing home and his last novel was published posthumously.

Fiction

Undergrowth [with E. Brett Young] 1913; *Deep Sea* 1914; *The Dark Tower* 1915; *The Iron Age* 1916; *The Crescent Moon* 1918; *The Francis Brett Young Omnibus* 1932: *The Young Physician* 1919, *The Black Diamond* 1921, *The Red Knight* 1921; *The Tragic Bride* 1920; *Pilgrim's Rest* 1922; Woodsmoke 1924; *Sea Horses* 1925; *Portrait of Clare* [US *Love Is Enough*] 1927; *The Key of Life* 1928; *My Brother Jonathan* 1928; *Black Roses* 1929; *Jim Redlake* [US *The Redlakes*] 1930; *Mr & Mrs Pennington* 1931; *Blood Oranges* 1932; *The House under the Water* 1932; *The Cage Bird and Other Stories* (s) 1933; *This Little World* 1934; *White Ladies* 1935; *Far Forest* 1936; *Portrait of a Village* 1937; *They Seek a Country* 1937; *The Christmas Box* (s) 1938; *Dr Bradley Remembers* 1938; *The City of Gold* 1939; *Cotswold Honey and Other Stories* (s) [US *The Ship's Surgeon and Other Stories*] 1940; *Mr Lucton's Freedom* 1940; *Man about the House* 1942; *Wistanlow* 1956

Poetry

Five Degrees South 1917; *Poems 1916–1918* 1919; *The Island* 1944

Plays

Captain Swing [with W.E. Stirling] 1919; *The Furnace* [with W. Armstrong] 1928

Non-Fiction

Robert Bridges: a Critical Study 1913; *Marching on Tanga* (a) 1917 [rev. 1919]; *In South Africa* 1952

Translations

Edwin Cerio *That Capri Air* [with others] 1929

Edited

A Century of Boys' Stories 1935

Z

Adam ZAMEENZAD 19??–

Zameenzad was born in Nairobi, where his Pakistani parents were schoolteachers. Skittish about his age, he changes his date of birth according to his mood. When last asked, he hit upon 1954, but those who know him suggest he was probably born in the 1940s. When the family returned to Pakistan he spent much of his time in what seems to have been a politically formative experience: the Zameenzads lived like small feudal gentry and he would climb over the walls of their relatively comfortable farm to fraternise with their landless tenantry, among whom he found an open-hearted warmth and companionship otherwise missing from his life. After his parents separated he lived with his mother in a state of increasing financial hardship, which he looks back on as first-hand experience of poverty and its degrading effects.

Zameenzad was privately educated in Pakistan on the English public-school model and travelled in India, Europe, the USA and Canada. He received further education in Pakistan, England and America, taking a BA in English, economics, philosophy and Persian, followed by an MA in English and American literature, as well as courses in education, TEFL, social welfare and law. He lectured in English at Punjab University, was a social welfare officer in Lahore and currently teaches English in Romford.

His first novel, *The Thirteenth House*, won the David Higham Award for the best first novel in 1987 and was praised by **Doris Lessing** as 'unusually talented and vigorous'. It is the unhappy story of Zahid, an underdog of a clerk in Karachi. Zahid and his family suffer from oppression and poverty and grasp at the false hopes offered by a guru with disastrous results.

Zameenzad's own life has been something of a spiritual quest. He was born a Muslim and has been through Buddhism and Catholicism before finding satisfaction as a Quaker, and he now describes himself as a Catholic with Quaker leanings. He combines this with faith in communism, believing that communism is not anti-Christian ('Marx's biggest blunder') and that the two are ultimately inseparable: 'Christ's Kingdom on earth and a communist Utopia are one and the same thing.'

All Zameenzad's novels are cries against injustice, whether the subject is African starvation (*My Friend Matt and Hena the Whore*) or the Rabelaisian picaresque of an illegal immigrant in Britain (*Cyrus Cyrus*). He gave the royalties from *My Friend Matt and Hena the Whore* to famine relief, and states his main ambition is 'just to change the world'.

Fiction

The Thirteenth House 1987; *My Friend Matt and Hena the Whore* 1988; *Love Bones and Water* 1989; *Cyrus Cyrus* 1990; *Gorgeous White Female* 1995

Benjamin ZEPHANIAH 1958–

Zephaniah, one of the most popular of the UK's performing poets, was born in Birmingham but spent much of his childhood in Jamaica. Back in Birmingham, after several brushes with the law, one with the National Front, and a short spell in prison, he began chanting his poetry to local audiences and soon developed a following. At twenty-two, he arrived in London, and his first book, *Pen Rhythm*, was published by a community press in 1981.

Since *The Dread Affair*, a passionate, vivid and often humorous condemnation of all aggression, Zephaniah has toured the UK, Europe and the Caribbean, and was shortlisted (amidst much discourse) for the post of artist-in-residence at Cambridge.

Playing the Tune was his first work written for the theatre and was performed at the Theatre Royal, Stratford East. He has also written *Job Rocking*, which was performed at the Riverside Studios. In 1988–9, as writer-in-residence for the City of Liverpool and African Arts Collective, he taught at schools and community centres. As a recording artist, Zephanian has worked with several Jamaican reggae bands including The Wailers. His album, *Us and Dem*, was released in 1990 and, like the previous ones, *Big Boys Don't Make Girls Cry* and *Free South Africa*, features his poetry.

Poetry

Pen Rhythm 1981; *The Dread Affair* 1985; *A Rasta Time in Palestine* 1987; *City Psalms* 1992; *Inna Liverpool* 1992

Plays

Playing the Tune 1985; *Job Rocking* 1987; *Streetwise* 1990

Louis ZUKOFSKY 1904–1978

Personally and poetically obscure, Zukofsky is America's great unsung modernist. He was born in Manhattan's Jewish ghetto into a Russian Jewish immigrant family, and was initially raised speaking Yiddish. As a child he saw Shakespeare, Ibsen, Strindberg and Tolstoy, all in Yiddish ('Even Longfellow's Hiawatha was to begin with read by me in Yiddish') and his Jewishness always remained an important part of his intensely intellectual creative identity.

He went to Columbia University at the age of sixteen, gaining his MA at twenty. At this period he was committed to the modernist avant-garde as an elite revolutionary force, and he wrote his first major poem, 'Poem Beginning "The" ', in 1926. This came complete with annotations and was written in the wake of **T.S. Eliot**'s *The Waste Land* (1922). He soon became a correspondent and friend of **Ezra Pound**, who persuaded **Harriet Monroe** to appoint Zukofsky as guest editor for an issue of her magazine *Poetry*. This was the 'objectivist' issue, which provoked such negative reader responses that Monroe dissociated herself from it in the following issue. He also edited *An Objectivist's Anthology*. Objectivism was in some respects comparable to Pound's Imagism, but it placed greater formalist emphasis on the poem as an opaque verbal object in its own right, and not a glimpse of reality. It was also more historically and politically oriented, since Zukofsky was an enthusiastic scholar of Marxism. In 1928 he began his greatest work, which was to occupy him for the rest of his life: his poem *'A'*. (In 1946 he was to write that 'a case can be made for the poet giving some of his life to the use of the words *the* and *a*: both of which are weighted with as much epos and historical destiny as one man can perhaps resolve.') *'A'* is a vast work in sections, like Pound's *Cantos*, but considerably more all-embracing and difficult. His other major poem is 'Mantis', written in two parts, the second of them an explanation of the first. It is on one level a political allegory about a hapless praying mantis in the New York subway, which is both a symbol of the proletariat and a creature that the tall, thin, ascetic Zukofsky had some affinities with.

He was held in high regard by other poets. Pound once described him as 'the only intelligent man in America', and in 1931 Eliot joined Pound and **Marianne Moore** in sponsoring him for a Guggenheim Fellowship (unsuccessfully, as it turned out.) Pound also dedicated his *Guide to Kulchur* (1938) to Zukofsky, although the two men later quarrelled over politics. In 1947 he took a job at the New York Polytechnic Institute in Brooklyn, teaching creative writing to engineering students, and he remained there until 1966. Characteristically he said he preferred this to teaching at a liberal arts college, since engineering or science was closer to his concept of poetry.

Zukofsky's disregard for a wide readership was partly a pragmatic response to its absence, and partly a hermeticism of spirit, rooted in the Jewish Cabalistical thinking of his background, which divides the world into a small minority of initiates and the great exoteric mass of the profane and ignorant. His family was especially important to him: his wife Celia was a composer and his son Paul a highly talented violinist, for whom Zukofsky wrote his novel *Little*, about a violin prodigy. One of his more curious works from the second half of his career is his translation of Catullus, based on close sound equivalence rather than sense.

Zukofsky's work still awaits a wider readership, despite the esteem in which he has been held by other poets such as **William Carlos Williams** and **Robert Creeley**. Hugh Kenner, a Pound specialist and no stranger to difficult poetry, has called *'A'* 'the most hermetic poem in the language, which they will still be elucidating in the 22nd century'.

Poetry

55 Poems 1941; *Anew* 1946; *Some Time* 1956; *Barely and Widely* 1958; *Five Statements for Poetry* 1958; *'A' 1–12* 1959; *I's* 1963; *After I's* 1964; *Found Objects* 1964; *An Unearthing* 1965; *Finally a Valentine* 1965; *I Sent Thee Late* 1965; *Iyyob* 1965; *'A' 14* 1967; *'A' 13–21* 1969; *Autobiography* 1970; *All the Collected Short Poems* 2 vols 1965, 1966; *'A' 24* 1972; *'A' 22 & 23* 1975; *'A'* 1978; *80 Flowers* 1978; *Gamut* 1978; *Complete Short Poetry* 1991

Fiction

It Was 1959 [rev. as *Ferdinand* 1968]; *Little: a Fragment for Careenagers* 1967; *Little* 1970; *Collected Fiction* 1990

Non-Fiction

A Test of Poetry 1948; *Bottom: On Shakespeare* [with C. Zukofsky] 1963; *Prepositions* 1967; *The Gas Age* 1969

Edited

An Objectivist's Anthology 1932; *Le Style Apollinaire* 1934

Translations

Catullus: *Catullus Fragmenta* [with C. Zukofsky] 1968, *Catullus* [with C. Zukofsky] 1969

Plays

Arise, Arise 1965 [pub. 1973]